THE
DICTIONARY
OF
NATIONAL BIOGRAPHY

The Concise Dictionary

PART II
1901–1970

Oxford New York Toronto Melbourne

OXFORD UNIVERSITY PRESS

1982

Oxford University Press, Walton Street, Oxford OX2 6DP

London Glasgow New York Toronto
Delhi Bombay Calcutta Madras Karachi
Kuala Lumpur Singapore Hong Kong Tokyo
Nairobi Dar es Salaam Cape Town
Melbourne Auckland

and associate companies in
Beirut Berlin Ibadan Mexico City

© *Oxford University Press 1982*

British Library Cataloguing in Publication Data
The Concise Dictionary of National Biography.
Part 2 : 1901–1970
1. Great Britain—Biography—Dictionaries
920'.0092   CT773
ISBN 0-19-865303-4

Printed in Great Britain by
Richard Clay (The Chaucer Press) Ltd,
Bungay, Suffolk

NOTE

THIS volume of the *Concise DNB* contains epitomes of the lives of all those persons appearing in the *Dictionary of National Biography* who died between 1901 and 1970. It supersedes the previous twentieth-century *Concise DNB*, edited by Helen Palmer, which went up to 1950. The epitomes for 1951–1970 are entirely new, and opportunity has also been taken to make some corrections in those portions of the text which have appeared in print before.

The publishers wish to record the debt which this volume owes to H. F. Oxbury, CMG, who compiled the final twenty years of epitomes, together with those of Queen Victoria and some other entrants from the *Concise DNB to 1900*, and saw the work through the press.

CONCISE DICTIONARY

OF

NATIONAL BIOGRAPHY

1901-1970

ABBEY, EDWIN AUSTIN (1852-1911), artist; born and educated at Philadelphia; studied art at academy there; worked for Harper's publishing firm, New York, 1871; came to England, 1878; won earliest fame as pen-and-ink illustrator; exhibited at Royal Academy 'A Milkmaid', in black and white, 1885; exhibited also at Royal Institute of Painters in Water Colours, 1883-7, and elsewhere; brilliant in pastel work; exhibited his first oil-painting at Royal Academy, 'A Mayday Morning', 1890; ARA, 1894; RA, 1898; best-known works include 'Fiammetta's Song', 1894, 'O Mistress Mine', 1899, 'Columbus in the New World', 1906; painted official picture of 'The Coronation of King Edward VII', 1903-4; executed notable mural decorations for Boston public library, the state capitol of Pennsylvania, and Royal Exchange; hon. LLD, Pennsylvania; memorial exhibition at Royal Academy, 1912.

ABBEY, JOHN ROLAND (1894-1969), book collector; educated privately; served on western front in the Rifle brigade, 1915-16; director, Kemp Town Brewery, Brighton, 1919; rejoined Rifle brigade, 1939-43; succeeded father as chairman of the brewery, 1943; started book collection, 1929, formed complete collections of books from Kelmscott, Ashendene, and Gregynog Presses; interested in modern bindings; ordered from Paul Bonet, 1937; also collected English colour-plate books; bought collection of illuminated manuscripts formed by C. H. St. John Hornby [q.v.]; commissioned series of handsomely produced catalogues: G. D. Hobson [q.v.], *English Bindings, 1490-1940, in the Library of J. R. Abbey* (1940), A. R. A. Hobson, *French and Italian Collectors and their Bindings, illustrated from examples in the Library of J. R. Abbey* (1953), and J. J. G. Alexander and A. C. de la Mare, *The Italian Manuscripts in the Library of Major J. R. Abbey* (1969); his own publications include *Scenery of Great Britain and Ireland in Aquatint and Lithography, 1770-1860* (1952), *Life in England, in Aquatint and Lithography, 1770-1860* (edited by E. Jutro, 1953), and *Travel in Aquatint and Lithography*

(1956-7, 2 vols); largest English book collector of his time.

ABBOTT, EDWIN ABBOTT (1838-1926), teacher and scholar; educated at City of London School and St. John's College, Cambridge (senior classic, 1861); ordained, 1862; headmaster of City of London School, which under him provided an intellectual training unsurpassed in any other English school and produced many distinguished pupils, 1865-89; introduced study of comparative philology into sixth-form curriculum, of chemistry into entire upper school, and of English literature throughout the school; broad churchman; Hulsean lecturer at Cambridge, 1876; works include *Shakespearian Grammar* (1870), *Bacon and Essex* (1877), *Johannine Vocabulary* (1905), *Johannine Grammar* (1906), *Philochristus* (1878), *Onesimus* (1882), and *Silanus the Christian* (1906).

ABBOTT, EVELYN (1843-1901), classical scholar; educated at Lincoln grammar school, Somerset College, Bath, and Balliol College, Oxford; spinal accident paralysed his lower limbs for life, 1866; BA and MA, 1873; master at Clifton, 1870-3; fellow and tutor of Balliol, 1874-1901; edited *Hellenica*, 1880, and 'Heroes of the Nations' series, contributing *Pericles*, 1891; collaborated with Lewis Campbell [q.v.] in *Life of Jowett*, 1897; wrote *History of Greece*, 3 vols., 1888-1900; hon. LLD, St. Andrews, 1879.

À BECKETT, ARTHUR WILLIAM (1844-1909), humorist; son of Gilbert Abbott à Beckett [q.v.]; educated at Honiton and Felsted; entered Civil Service, 1862; left to engage in journalism, 1865; editor of *Sunday Times*, 1891-5; contributed to *Punch*, 1874-1902; works include *The À Becketts of Punch* (1903) and *Recollections of a Humourist* (1907); joined Church of Rome, 1874.

ABEL, SIR FREDERICK AUGUSTUS, baronet (1827-1902), chemist; educated at

Royal College of Chemistry; demonstrator of chemistry at St. Bartholomew's Hospital, 1851; lecturer at Royal Military Academy, Woolwich, 1852; collaborated in *Handbook of Chemistry*, 1854; chemist to war department, Woolwich, 1856; chief official authority on explosives; invented cordite, 1889; FRS, 1860; received royal medal, 1887; knighted, 1883; president of Chemical and other societies; president, British Association, Leeds, 1890; organizing secretary and director, Imperial Institute, and baronet, 1893; an accomplished musician.

ABELL, SIR WESTCOTT STILE (1877-1961), naval architect and surveyor; educated at Royal Naval Engineering College, Keyham, and Royal Naval College, Greenwich, 1897-1900; lost right hand in accident, 1897; entered Royal Corps of Naval Constructors; posted to Devonport dockyard, 1900; professional private secretary to Sir Philip Watts [q.v.], director of naval construction, 1904-7; professor of naval architecture, Liverpool, 1909; member of Institution of Naval Architects, 1909; member of Board of Trade committee to examine application of Merchant Shipping Act to internationalization of load lines, 1913; chief ship surveyor, Lloyds, 1914-28; served on Admiralty committees concerned with shipping during 1914-18 war; KBE, 1920; professor of naval architecture, Armstrong College, Durham University, 1928; designed Channel train ferry; president, Institute of Marine Engineers, 1924-5; master, Worshipful Company of Shipwrights, 1931-2; president, Smeatonian Society of Civil Engineers, 1941; publications include *The Ship and her Work* (1923), *The Safe Sea* (1932), and *The Shipwright's Trade* (1948).

ABERCONWAY, first BARON (1850-1934), barrister and business man. [See McLAREN, CHARLES BENJAMIN BRIGHT.]

ABERCONWAY, second BARON (1879-1953), industrialist. [See McLAREN, HENRY DUNCAN.]

ABERCORN, second DUKE OF (1838-1913). [See HAMILTON, JAMES.]

ABERCROMBIE, LASCELLES (1881-1938), poet and critic; educated at Malvern College; lecturer in poetry, Liverpool, 1919-22; professor of English literature, Leeds, 1922-9, London, 1929-35; Goldsmiths' reader, Oxford, 1935-8; FBA, 1937; a distinguished 'metaphysical' poet; works include *Mary and the Bramble* (1910), *The Sale of St. Thomas* (1931), and many critical studies.

ABERCROMBIE, SIR (LESLIE) PATRICK (1879-1957), architect and professor of town planning; brother of Lascelles Abercrombie

[q.v.]; educated at Uppingham; articled architect in Manchester and Liverpool; practised in Chester from 1920; assistant lecturer, Liverpool School of Architecture, 1907-9; research fellow, town planning and civil design, and lecturer, building construction and Gothic architecture, 1909-15; professor of civic design, 1915-35; of town planning, University College, London, 1935-46; first prize for replanning Dublin, 1916; produced regional scheme, Doncaster coalfield, 1922; reported on Sheffield (1924, 1931), Thames Valley (1929) Cumberland (1932), Wye Valley and East Suffolk (1935); influential in founding Council for the Preservation of Rural England, 1926; campaigned for London's Green Belt; member of royal commission on distribution of industrial population, 1937-9; produced *County of London Plan*, 1943, and *Greater London Plan*, 1944; post-war planning included Plymouth, Hull, Edinburgh, Clyde region, West Midlands, Cyprus, Hong Kong, Addis Ababa; FRIBA, 1925, royal gold medal, 1946; knighted, 1945.

ABERDARE, third BARON (1885-1957), athlete. [See BRUCE, CLARENCE NAPIER.]

ABERDEEN AND TEMAIR, first MARQUESS OF (1847-1934), statesman, and **ABERDEEN AND TEMAIR, MARCHIONESS OF** (1857-1939). [See GORDON, JOHN CAMPBELL.]

ABERHART, WILLIAM (1878-1943), Canadian provincial politician and evangelist; BA, Queen's University, Kingston, 1906; principal, Crescent Heights high school, Calgary, 1915-35; founded Calgary Prophetic Bible Institute, 1918; from 1932 advocated Social Credit; premier and minister of education, Alberta government, with Social Credit majority, 1935-43; failed to apply Social Credit principles, from which he increasingly turned towards a more orthodox administration.

ABNEY, SIR WILLIAM DE WIVE-LESLIE (1843-1920), photographic chemist and education official; educated at Rossall; joined Royal Engineers, 1861; entered science and art department, South Kensington, 1877; assistant director for science, 1884; director, 1893; principal assistant secretary, Board of Education, 1899-1903; pioneer in advancement of practical photography, photographic emulsion-making, spectro-photography, colour analysis, and colour vision; FRS, 1876; Rumford medallist, 1882; KCB, 1900; works include *Instruction in Photography* (1870) and *Treatise on Photography* (1875).

ABRAHAM, CHARLES JOHN (1814-1903), first bishop of Wellington, New Zealand;

educated at Eton and King's College, Cambridge; fellow of King's, 1836-49; BA, MA, 1840; DD, 1859; ordained, 1838; master at Eton, 1839-49; joined Bishop Selwyn [q.v.] in New Zealand, 1850; principal of St. John's College, Auckland, 1850; archdeacon of Waitemata, 1853; bishop of Wellington, 1858-70; coadjutor bishop of Lichfield, 1870; prebendary (1872) and canon (1876) of Lichfield; helped to organize Selwyn College, Cambridge, 1882; author of devotional works.

ABRAHAM, WILLIAM (1842-1922), labour politician and trade-union leader; popularly known by his eisteddfod pen-name, 'Mabon'; began work as pit-boy; pioneer of trade-unionism among Welsh miners and first miners' representative to enter parliament from Welsh coal-field; MP, Rhondda division, 1885-1918, Rhondda West, 1918-20; first president of South Wales Miners' Federation, 1898; advocate of sliding-scale system; exercised moderating influence; PC, 1911.

ABU BAKAR TAFAWA BALEWA, ALHAJI SIR (1912-1966), prime minister of the Federation of Nigeria. [See TAFAWA BALEWA.]

ABUL KALAM AZAD, MAULANA (1888-1958), Indian Minister of education. [See AZAD.]

ACLAND, SIR ARTHUR HERBERT DYKE, thirteenth baronet, of Columb John, Devon (1847-1926), politician and educational reformer; third son of Sir J. D. Acland, eleventh baronet [q.v.]; educated at Rugby and Christ Church, Oxford; liberal MP, Rotherham division, 1885-99; authority on educational questions; entered cabinet as vice-president of the Committee of Council on Education, 1892; secured raising of age of compulsory attendance at school from ten to eleven; retired from active politics, 1895; opposed education bill, 1902; works include *A Handbook of the Political History of England* with C. Ransome (1882).

ACTON, SIR EDWARD (1865-1945), judge; educated at Uppingham and Wadham College, Oxford; called to bar (Inner Temple), 1891; practised in Manchester and Liverpool; a county court judge, 1918-20; his promotion to the King's Bench division (1920-34) created a precedent; knighted, 1920.

ACTON, JOHN ADAMS- (1830-1910), sculptor. [See ADAMS-ACTON.]

ACTON, SIR JOHN EMERICH EDWARD DALBERG, eighth baronet, and first BARON ACTON (1834-1902), historian and moralist; born at Naples of Shropshire Roman Catholic family; educated at Paris, Oscott, and privately

at Edinburgh; studied history and criticism at Munich under Döllinger, 1848-54; visited America in 1855, Russia in 1856, Italy, with Döllinger, in 1857; settled at the family seat, Aldenham, 1858; whig MP for Carlow, 1859-65; formed friendship with Gladstone; became joint proprietor of the monthly *Rambler*, which was converted under Acton's editorship in 1862 to a quarterly, *The Home and Foreign Review*, representing liberal catholic opinions; in it advocated on liberal grounds Döllinger's reunion of Christendom, 1864; stopped the *Review* on threat of papal veto; contributed to the weekly *Chronicle*, 1867-8; wrote much for the revived quarterly, *North British Review*; at Rome with Gladstone, 1866; baron, 1869; strenuous in opposition to adoption by Catholic Church of dogma of papal infallibility; published his views in *Letters from Rome on the [Œcumenical] Council*, 1870; criticized in letters to *The Times* Gladstone's pamphlet on *The Vatican Decrees*, 1874; FSA, 1876; from 1879 spent winter at Cannes, autumn in Tegernsee, Bavaria, and parts of spring and summer in London; wrote for reviews; helped to found *English Historical Review*, 1886; hon. LLD, Cambridge, 1888; hon. DCL, Oxford, 1887; hon. fellow of All Souls, 1891; lord-in-waiting to Queen Victoria, in Gladstone's fourth administration, 1892 (KCVO, 1897); regius professor of modern history, Cambridge, from 1895 to death; hon. fellow of Trinity College; lectured at Cambridge on French Revolution; planned and edited preliminary draft of *Cambridge Modern History* (1899-1912); died at Tegernsee; library of 59,000 volumes purchased from family and presented to Cambridge University, 1903; independent works (posthumously published) include *Lectures on Modern History* (1906), *The History of Freedom* and *Historical Essays and Studies* (1907), and *Lectures on the French Revolution* (1910); all display vast erudition, epigrammatic style, and passion for political righteousness and liberty of conscience.

ACWORTH, SIR WILLIAM MITCHELL (1850-1925), expert on railway economics; educated at Uppingham and Christ Church, Oxford; specialized in railway transport economics, becoming greatest expert in the world on the relationship between railways and governments; served on railway commissions in England, Ireland, Canada, Southern Rhodesia, India, Austria, and Germany; chief work, *The Elements of Railway Economics* (1905); knighted, 1921; KCSI, 1922.

ADAM, JAMES (1860-1907), Platonist; educated at Aberdeen University and Caius College, Cambridge; first class, classical tripos,

1884; first chancellor's medallist, 1884; fellow and classical lecturer of Emmanuel College, Cambridge, 1884; tutor, 1890; senior tutor, 1900; visited Greece, 1890; supporter of degrees for women; edited Plato's *Apology* (1887), *Crito* (1888), *Euthyphro* (1890), *Protagoras* (1893), and *Republic* (1902), the last a standard work; Gifford lectures (*The Religious Teachers of Greece*), delivered at Aberdeen 1902, 1904, and 1905, published posthumously, 1908; *The Vitality of Platonism*, collected essays, 1911.

ADAM SMITH, SIR GEORGE (1856–1942), Old Testament scholar and theologian [See SMITH]

ADAMI, JOHN GEORGE (1862–1926), pathologist; BA, Christ's College, Cambridge (first class in both parts of the natural sciences tripos, 1882–4); professor of pathology and bacteriology, McGill University, Montreal, 1892; assistant director of medical services to Canadian expeditionary force, 1914–18; vice-chancellor of Liverpool University, 1919; raised by appeal £360,000 for university expansion; chief work, *Principles of Pathology* (2 vols., 1908–9); FRS, 1905.

ADAMS, JAMES WILLIAMS (1839–1903), army chaplain in India; educated at Hamlin and Porter's school, Cork, and Trinity College, Dublin; BA, 1861; ordained, 1863; chaplain at Calcutta, Peshawar, and Kashmir, 1866–75; saw much active service at Kabul, 1878, and elsewhere; risked life at Villa Kazi, 1879; VC, 1881; at battle of Kandahar, 1880; settled in England, 1886; hon. chaplain to Queen Victoria, 1900; hon. MA, Dublin, 1903.

ADAMS, SIR JOHN (1857–1934), educationist; educated at Glasgow University; principal of Free Church training colleges in Aberdeen (1890), Glasgow (1898); of London Day Training College and first professor of education, London, 1902–22; knighted, 1925.

ADAMS, WILLIAM BRIDGES- (1889–1965), theatrical producer. [See BRIDGES-ADAMS.]

ADAMS, WILLIAM DAVENPORT (1851–1904), journalist and compiler; son of W. H. D. Adams [q.v.]; educated at Merchant Taylors' School and Edinburgh University; editor of provincial papers, and dramatic critic, 1878–1904; compiled an unfinished *Dictionary of the Drama* (1904) and other works.

ADAMS, WILLIAM GEORGE STEWART (1874–1966), public servant; educated at St. John's grammar school, Hamilton, Glasgow

University, and Balliol College, Oxford; first class, *lit. hum.*, and modern history, 1900–1; tutor, Borough Road Training College, Isleworth, 1901–2; lecturer in economics, Chicago University, 1902; lecturer in economics and secretary, university extension, Manchester University, 1903; superintendent, statistics and intelligence, Irish Department of Agriculture and Technical Instruction, Dublin, under Sir Horace Plunkett [q.v.], 1905; reader in political theory and institutions, Oxford, 1910; fellow, All Souls; Gladstone professor, 1912; warden of All Souls, 1933–45, in succession to Lord Chelmsford [q.v.]; founded and edited the *Political Quarterly*, 1914–16; joined staff of Ministry of Munitions, 1915; private secretary to prime minister, Lloyd George, 1917–18; member, committee on examinations for the Civil Service, 1918; member, royal commission on universities of Oxford and Cambridge, 1919–22, pro-vice-chancellor, Oxford, 1939–45; farmed on Boars Hill, Oxford; took leading part in establishing National Federation of Young Farmers' Clubs, 1923; member, Development Commission, 1923–49; chairman, National Council of Social Service, 1919–49; delegate to China from universities' China committee, 1931–2; CH, 1936; hon. DCL Oxford; hon. LLD Glasgow, and Manchester; hon. fellow, All Souls, 1945.

ADAMS-ACTON, JOHN (1830–1910), sculptor; educated at Lady Byron's school, Ealing; studied art under Matthew Noble [q.v.] and at Royal Academy Schools (1853–8); won Academy's travelling studentship, 1858; at Rome till 1865; executed many notable London memorials; regularly exhibited at Royal Academy till 1892.

ADAMSON, SIR JOHN ERNEST (1867–1950), educationist; trained as a teacher, St. Mark's College, Chelsea; BA, London, 1894; principal, Normal College, Pretoria, 1902–5; director of education, Transvaal, 1905–24; with J. C. Smuts [q.v.] introduced an educational system synthesizing Boer-Republican and British elements; master, and professor of education, Rhodes University College, Grahamstown, 1925–30; knighted, 1924.

ADAMSON, ROBERT (1852–1902), philosopher; educated at Daniel Stewart's Hospital, Edinburgh, and Edinburgh University; graduated in philosophy, 1871; assistant professor, 1871–4; joined editorial staff of *Encyclopaedia Britannica* (9th ed.), contributing many philosophical articles; appointed professor of philosophy and political economy at Owens College, Manchester, 1876; supported admission of women students at Manchester; hon.

LLD, Glasgow, 1883; professor of logic at Aberdeen (1893–5) and Glasgow (1895); published various works on Greek and modern philosophy and logic; earlier work idealistic, but later work naturalistic and realistic; library presented to Manchester University.

ADCOCK, SIR FRANK EZRA (1886–1968), historian of Greece and Rome; educated at Wyggeston grammar school, Leicester, and King's College, Cambridge; first class, parts i and ii, classical tripos, 1908–9; Craven scholar, 1908; fellow, King's College, 1911; lecturer in classics; lay dean of King's, 1913–19; served in intelligence division, Admiralty, 1914–18; OBE, 1917; joined J. B. Bury and S. A. Cook [qq.v.], in editing the *Cambridge Ancient History*; succeeded Bury as chief editor, 1927; professor of ancient history, Cambridge, 1925–51; president, Roman Society, 1929–31; FBA 1936; worked in Foreign Office, 1939–43; president, Classical Association, 1947–8; vice-provost of King's, 1951–5; knighted, 1954; number of hon. degrees; publications include, *The Roman Art of War under the Republic* (1940), *The Greek and Macedonian Art of War* (1957), *Roman Political Ideas and Practice* (1959), *Caesar as Man of Letters* (1956), *Thucydides and his History* (1963), and *Marcus Crassus, Millionaire* (1966).

ADDERLEY, CHARLES BOWYER, first BARON NORTON (1814–1905), statesman; educated at Christ Church, Oxford; BA, 1835; pioneer of town planning at Saltley, near Birmingham, 1937; tory MP for Northern division of Staffordshire, 1841–78; opposed Peel's free trade policy, 1846; interested in colonial development; helped to found Church of England colony of Canterbury in New Zealand, and the Colonial Reform Society, 1849; persistently advocated colonial self-government; introduced reformatory schools bill, 1852; responsible for Young Offenders Act, 1854; admitted to Privy Council as vice-president of the education committee, 1858; passed a first Local Government Act, 1858; under-secretary for the Colonies, 1866; defended action of Governor Eyre [q.v.] in Jamaica; carried through British North America Act (1867) creating the Dominion of Canada; KCMG, 1869; chairman of sanitary commission, 1871; president of Board of Trade, 1874–8; passed merchant shipping bill, legalizing 'loadline', 1876; baron, 1878; advocated free education and opposed payment by results, 1882; a strong churchman and writer on religious topics; a competent musician and art critic; memorial hall at Saltley.

ADDISON, CHRISTOPHER, first VISCOUNT ADDISON (1869–1951), statesman; educated at Trinity College, Harrogate, Sheffield medical school, and St. Bartholomew's; qualified, 1892; professor of anatomy, Sheffield, 1897–1901; appointed lecturer in anatomy, Charing Cross Hospital, 1901; liberal MP, Hoxton, January 1910; advised Lloyd George on national health insurance bill and other welfare measures; parliamentary secretary, Board of Education, 1914, Ministry of Munitions, 1915; introduced new techniques of 'war socialism'; PC 1916; canvassed liberal support for Lloyd George as prime minister, December 1916, and himself became minister of munitions; minister of reconstruction, July 1917; president, Local Government Board, January 1919; first minister of health, June 1919; his Housing, Town Planning Act of 1919 provided state assistance for local authority housing but proved unacceptably costly; minister without portfolio, April–July 1921; lost his seat, 1922; joined labour party; MP, Swindon, 1929–31, 1934–5; parliamentary secretary, Ministry of Agriculture, 1929; minister, June 1930; declined to support 'national' government, 1931; baron, 1937; viscount, 1945; KG, 1946; leader of labour peers, 1940; as leader of House of Lords, 1945–51, obtained passage of far-reaching legislative programme; secretary of state for dominions, 1945–7; lord privy seal, 1947–51; paymaster general, 1948–9; lord president of the Council, 1951; publications include *Politics from Within* (2 vols., 1924), *Practical Socialism* (2 vols., 1926), and *Four and a Half Years* (2 vols., 1934).

ADLER, HERMANN (1839–1911), chief rabbi; born at Hanover; son of Nathan Marcus Adler [q.v.]; brought to London, 1845; educated at University College School and University College; BA, London, 1859; consecrated, 1859; studied theology at Prague, 1860; Ph.D, Leipzig, 1862; principal of Jews' Theological College, London, 1863, being subsequently tutor, chairman of council, and president; first minister of Bayswater synagogue, 1864–91; delegate chief rabbi for his father, 1879; chief rabbi, 1891–1911; active in social reform; president of Jewish Historical Society of England and vice-president of Anglo-Jewish and other associations; hon. LLD, St. Andrews, 1899; hon. DCL, Oxford, and CVO, 1909; published works on Jewish subjects.

ADSHEAD, STANLEY DAVENPORT (1868–1946), architect and professor of town planning; served articles in Manchester; practised in London; FRIBA, 1905; first lecturer (1909), professor (1912–14), civic design, Liverpool; first professor, town planning, London,

1914–35; influential authority in a new field of aesthetics; work (in association) included reconstruction of duchy of Cornwall estate, Kennington, building of Dormanstown, and many housing schemes; selected site for Lusaka.

AE (pseudonym) (1867–1935), Irish writer. [See RUSSELL, GEORGE WILLIAM.]

AGA KHAN, AGA SULTAN SIR **MOHAMMED SHAH** (1877–1957), third holder of the title; born in Karachi; succeeded father, 1885, and assumed responsibility as 48th head of the Ismaili sect, 1893; urged his followers (11–12 million) to integrate with countries they lived in and exercised a moderating influence; member, viceroy's legislative council, 1902–4; led Moslem deputation to Lord Minto [q.v.], 1906; first president, All-India Moslem League, 1906–12; supported Allies in both world wars; undertook mission of reassurance to Egypt, 1915; published *India in Transition* (1918) advocating South Asian Federation with self-governing India at centre; chairman All-Indian Moslem conference, 1928, which formulated attitude to India's future; chairman, British-India delegation to Round Table conferences, 1930–2; helped secure unanimous report from Linlithgow joint select committee (1933–4); leader, Indian delegation, League of Nations Assembly, 1932, 1934–7; president of Assembly, 1937; retained influence over his own community but gradually yielded Moslem political leadership to M. A. Jinnah [q.v.]; published *Memoirs* (1954); successful breeder of racehorses which five times won the Derby; KCIE, 1898; GCIE, 1902; GCSI, 1911; GCVO, 1923; GCMG, 1955; PC, 1934; died at Versoix near Geneva and buried at Aswan.

AGATE, JAMES EVERSHED (1877–1947), dramatic critic; educated at Giggleswick and Manchester grammar school; entered father's business selling cotton cloth; contributed dramatic criticism to *Daily Dispatch* (1906), *Manchester Guardian* (1907–14); dramatic critic, *Saturday Review* (1921–3), *Sunday Times* (1923–47), BBC (1925–32); his diaries, *Ego* (9 vols., 1935–48), recorded chiefly the books, plays, personalities, club talk, and Bohemian life of his time.

AGNEW, SIR **JAMES WILLSON** (1815–1901), prime minister of Tasmania; born in co. Antrim; educated at University College, London, at Paris, and Glasgow; MRCS, 1838; MD, Glasgow, 1839; went to Sydney, NSW (1840), and practised there; colonial assistant surgeon at Hobart, 1845; helped to found Tasmanian Royal Society; member of legislative council, 1877; minister without portfolio, 1877–80;

visited England, 1880; premier of Tasmania, 1886–7; KCMG, 1894.

AGNEW, SIR **WILLIAM,** first baronet (1825–1910), art dealer; son of Thomas Agnew, printseller and mayor of Salford, 1850–1; educated at Swedenborgian school, Salford; partner in father's firm, 1850; helped to form many private art collections; purchased Gainsborough's 'Duchess of Devonshire' for 10,000 guineas, 1876; benefactor to National Gallery; joined firm of Bradbury & Evans, proprietors of *Punch*, 1870; liberal MP for SE. Lancashire and Stretford, 1880–86; helped to found National Liberal Club; baronet, 1895.

AGNEW, SIR **WILLIAM GLADSTONE** (1898–1960), vice-admiral; grandson of Sir William Agnew [q.v.]; entered navy, 1911; specialized in gunnery; commander, 1932; captain 1937; commanded cruiser *Aurora*, 1940–3; as senior officer, Force K, Mediterranean (1941–3), took full part in invasion of N. Africa, Italy, and Sicily; CB, 1941; DSO, 1943, bar, 1944; rear-admiral and KCVO, 1947; director of personal services, 1947–9; retired and vice-admiral, 1950.

AIDÉ, CHARLES HAMILTON (1826–1906), author and musician; born in Paris; son of Armenian merchant; educated privately and at Bonn University; joined British army; devoted to music, art, and literature; entertained largely in London; published *Eleanore* (1856) and other verse; a prolific musical composer and accomplished amateur artist; published many society novels showing French influence, and occasionally wrote for the stage.

AIKMAN, GEORGE (1830–1905), painter and engraver; educated at Edinburgh high school; after working in Manchester and London joined his father, an Edinburgh engraver, as partner; studied at Royal Scottish Academy life class; first exhibited at Scottish Academy, 1850; ARSA, 1880; exhibited at Royal Academy, 1874–1904; mainly confined himself to landscape; practised etching and mezzotint.

AINGER, ALFRED (1837–1904), writer, humorist, and divine; educated at University College School, at Joseph King's boarding school, London, 1849, and at King's College (under F. D. Maurice, q.v.); entered Trinity Hall, Cambridge, 1856; contributed to university magazine, *The Lion* (1857–8); BA (law), 1860; MA, 1865; ordained, 1860; assistant master at collegiate school, Sheffield, 1864–6; reader at Temple, 1865–93; friend of Tennyson; contributor to *Macmillan's Magazine*, 1859–96;

admiring student of Charles Lamb's writings; wrote life of Lamb, 1882; edited Lamb's essays, 1883, poems, etc., 1884, and letters, 1888; contributor to this Dictionary; a popular lecturer and preacher; canon of Bristol, 1887-1903; select preacher at Oxford, 1893; master of the Temple, 1894 till death; chaplain to Queen Victoria and King Edward VII; other works include life of Crabbe (1903) and *Lectures and Essays* (posthumous, 2 vols., 1905).

AINLEY, HENRY HINCHLIFFE (1879-1945), actor; trained under (Sir) Frank Benson [q.v.]; became famous (1902) as Paolo in *Paolo and Francesca;* later especially memorable as Malvolio in *Twelfth Night* (1912), Joseph Quinney in *Quinneys'* (1915), Hassan (1923), and James Fraser in *The First Mrs. Fraser* (1929); his early romantic and later masterful performances were equally distinguished and his diction notably fine.

AIRD, SIR JOHN, first baronet (1833-1911), contractor; privately educated; joined father's business; constructed several gas and water reservoirs at home and abroad; chief partner, 1870; carried out much railway and dock work; best known by the construction of dams at Aswan and Assiut, 1898-1902; conservative MP for N. Paddington, 1887-1905; first mayor of Paddington, 1900; baronet, 1901; enthusiastic art collector and freemason.

AIREDALE, first BARON (1835-1911), ironmaster.[See KITSON, James.]

AITCHISON, CRAIGIE MASON, LORD AITCHISON (1882-1941), lord justice-clerk of Scotland; MA (1903), LLB (1906), Edinburgh; called to Scottish bar, 1907; KC, 1923; foremost criminal advocate especially in defence; lord advocate, PC, 1929; MP (labour, later national labour), Kilmarnock division, 1929-33; lord justice-clerk, 1933-41.

AITCHISON, GEORGE (1825-1910), architect; educated at Merchant Taylors' School; articled to father, 1841; entered Royal Academy Schools, 1847; BA, London, 1851; visited Rome, 1853; succeeded father in his business and as architect to London and St. Katharine Docks Co., 1861; friend of Lord Leighton [q.v.]; ARA, 1881; RA, 1898; professor of architecture at Academy; president of Royal Institute of British Architects, 1896-9; contributor to this Dictionary.

AITKEN, ALEXANDER CRAIG (1895-1967), mathematician; born at Dunedin, New Zealand; educated at Otago Boys' high school,

and Otago University; served in New Zealand Expeditionary Force at Gallipoli and in France, 1915-17; post graduate scholar, Edinburgh University, under (Sir) E. T. Whittaker [q.v.], 1923; D.Sc., 1926; lecturer in actuarial mathematics, Edinburgh, 1925; reader in statistics, 1936; succeeded Whittaker as professor of mathematics, 1946-65; FRS, 1936; vice-president, Royal Society of Edinburgh; hon. degrees Glasgow and New Zealand; FRSL, 1964; hon. FRSNZ; hon. fellow, Faculty of Actuaries; published *Gallipoli to the Somme. Recollections of a New Zealand Infantryman* (1963, Hawthornden prize).

AITKEN, WILLIAM MAXWELL, first BARON BEAVERBROOK (1879-1964), newspaper proprietor; born in Canada; educated at local school; entered law firm and sold insurance and bonds; formed Canada Cement Company, 1909; came to England and, helped by Bonar Law, became conservative MP, Ashton-under-Lyne, 1910; knighted, 1911; Canadian Government representative, British GHQ, France, 1914; baronet, 1916; assisted Lloyd George to supersede Asquith as prime minister, 1916; baron, 1917; minister of information, PC, 1918; purchased *Daily Express*, 1916; launched *Sunday Express*, 1918; acquired *Evening Standard* 1923; supported Edward VIII in abdication crisis, 1936; advocated 'Empire Free Trade'; supported Neville Chamberlain over Munich, 1938; minister of aircraft production, 1940; minister of supply, 1941; lord privy seal, 1943-5; opposed entry to European Common Market, 1961; publications include *Politicians and the War* (2 vols., 1928-32), *Man and Power 1917-18* (1956), *The Decline and Fall of Lloyd George* (1963), and *The Divine Propagandist* (1962).

AKERS, SIR WALLACE ALAN (1888-1954), chemist; educated at Aldenham School and Christ Church, Oxford; first class, chemistry, 1909; worked for Brunner Mond, 1911-24, Borneo Company, 1924-8, ICI, 1928-53; in charge at Billingham, 1931-7; a director of ICI, 1941, research director, 1944-53; directed British work on atomic energy, 1941-6; knighted, 1946; FRS, 1952.

AKERS-DOUGLAS, ARETAS, first VISCOUNT CHILSTON (1851-1926), statesman; educated at Eton and University College, Oxford; conservative MP, East Kent, 1880-5, St. Augustine's division, 1885-1911; opposition whip, 1883; parliamentary secretary to Treasury, 1885-6, 1886-92; chief opposition whip, 1892-5; first commissioner of works, 1895-1902; secretary of state, home department, 1902-5; viscount, 1911; PC, 1891; GBE, 1920.

AKERS-DOUGLAS, ARETAS, second Viscount Chilston (1876–1947), diplomatist; son of first Viscount Chilston [q.v.]; educated at Eton; entered diplomatic service, 1898; second secretary, 1905; served in Athens, Rome, Vienna, and several Balkan posts; first secretary, 1912; diplomatic secretary to Lord Curzon [q.v.], 1919–21; minister, Vienna (1921–8), Budapest (1928–33); ambassador, Moscow, 1933–8; gained the personal regard of Litvinoff and established a *modus vivendi* with the Soviet Union; succeeded father, 1926; GCMG, 1935; PC, 1939.

ALANBROOKE, first Viscount (1883–1963), field-marshal. [See Brooke, Alan Francis.]

ALBANI, Dame MARIE LOUISE CÉCILIE EMMA (1852–1930), singer; born Lajeunesse, at Chambly, near Montreal; brought to Europe by her father, *c.* 1867; studied singing in Paris and Milan; made her début in Bellini's *La Sonnambula* at Messina, adopting name of 'Albani', 1870; went to London, 1871; made her début in Royal Italian opera at Covent Garden, 1872; sang almost every season there until 1896; also sang frequently on the continent and toured the United States and Canada; equally successful in oratorio; married Ernest Gye, 1878; retired, 1911; DBE, 1925.

ALCOCK, Sir JOHN WILLIAM (1892–1919), airman; obtained aviator's certificate, 1912; joined Royal Naval Air Service, 1914; instructor at Eastchurch flying school; stationed in Eastern Mediterranean, 1916; awarded DSC, and taken prisoner by Turks, 1917; with (Sir) Arthur Whitten Brown [q.v.], first to make non-stop flight of Atlantic (16 hours 27 minutes), 1919; KBE, 1919.

ALDENHAM, first Baron (1819–1907), merchant and scholar. [See Gibbs, Henry Hucks.]

ALDERSON, Sir EDWIN ALFRED HERVEY (1859–1927), lieutenant-general; joined army, 1876; took part in Egyptian campaign, 1882; in command of Mounted Infantry helped to quell Matabele revolt in South Africa, 1896; commanded Mounted Infantry in South African war, 1901–2; major-general, 1907; lieutenant-general, 1914; commanded Canadian Army Corps in France, 1915–16; inspector-general of Canadian forces, 1916–18; KCB, 1916.

ALDERSON, HENRY JAMES (1834–1909), major-general; born at Quebec; entered Royal Military Academy, Woolwich, 1848; served in Crimea; present on special mission at bombardment of Charleston, USA, 1864; major-general,

1892; president of ordnance committee, War Office, 1891–6; director of Armstrong, Whitworth & Co., 1897 till death; KCB, 1891.

ALDINGTON, EDWARD GODFREE, ('RICHARD') (1892–1962), writer; educated at Dover College and University College, London; became avant-garde poet, and, with Ezra Pound and Hilda Doolittle, originated the 'Imagists', 1912; appointed literary editor, the *New Freewoman*, renamed *The Egoist*, 1914; published *Images 1910–1915*, 1915; served in army in France and Flanders, 1916; reviewed French books for the *Times Literary Supplement*, 1919–20; assistant editor, *Criterion*, 1921; left England for France, 1928; published first novel, *Death of a Hero*, 1929; published seven novels, short stories, poetry, and works of criticism, 1929–39; made the United States his headquarters, 1935–47; published autobiography, *Life for Life's Sake*, 1941; other publications include *Wellington* (1946), *Pinorman*, *Personal Recollections of Norman Douglas*, *Orioli*, *and Prentice* (1954), and *Lawrence of Arabia* (1955).

ALDRICH-BLAKE, Dame LOUISA BRANDRETH (1865–1925), surgeon; entered London School of Medicine for Women, 1887; first woman master in surgery, 1895; surgeon to Elizabeth Garrett Anderson Hospital and Royal Free Hospital; DBE, 1925.

ALEXANDER, Mrs (pseudonym) (1825–1902), novelist. [See Hector, Annie French.]

ALEXANDER, ALBERT VICTOR, Earl Alexander of Hillsborough (1885–1965), politician; son of a blacksmith; left school at thirteen; became boy clerk in office of Bristol school board; transferred to school management department of Somerset County Council, 1903; chief clerk, 1919; lay preacher; secretary local branch of National Association of Local Government Officers; vice-president, local Co-operative Society; served in Artists' Rifles, 1914–18; secretary to parliamentary committee of Co-operative Congress, 1920; labour and co-operative MP, Hillsborough division of Sheffield, 1922–31, and 1935–50; parliamentary secretary, Board of Trade, under Sidney Webb [q.v.], 1924; first lord of Admiralty, 1929; negotiated London Naval Treaty, 1930; did not join 'national' government; lost seat, 1931; returned to Commons, 1935; first lord again under Churchill, 1940–5; continued as first lord under Attlee, 1945; member of Cabinet delegation to India, with Sir Stafford Cripps and Lord Pethick-Lawrence [qq.v.], 1946; first minister of Defence, 1946–50; raised to peerage, 1950; chancellor, Duchy of Lancaster, 1950–1; succeeded Lord Jowitt [q.v.] as labour leader in House of Lords, 1955-64; vice-president, Chelsea football club; master, Bakers' Company; PC, 1929; CH, 1941; elder brother, Trinity House, 1941; viscount, 1950; earl, 1963; KG, 1964; hon. LLD Bristol and Sheffield.

ALEXANDER, BOYD (1873-1910), African traveller and ornithologist; educated at Radley College, 1887-91; joined army, 1893; at Kumasi, 1900; studied bird life in West Africa; explored Lake Chad, 1904-5; made detailed survey of West African continent, 1905-6; Royal Geographical Society's medallist, 1908; continued exploration of West Africa, 1908-10; murdered by natives, Kenya; published *From the Niger to the Nile*, 2 vols., 1907.

ALEXANDER, SIR GEORGE (1858-1918), actor-manager, whose original name was GEORGE SAMSON; made first professional appearance at Nottingham, 1879; engaged by (Sir) Henry Irving [q.v.] for Lyceum Theatre, 1881; re-engaged, and accompanied Irving to America, 1884-5; leading man of Lyceum company, 1885-9; manager of St. James's Theatre, London, 1891-1918; produced, among other notable plays, (Sir) A. W. Pinero's *The Second Mrs. Tanqueray*, 1893; knighted, 1911; possessed a fine stage presence and acted with distinction.

ALEXANDER, HAROLD RUPERT LEOFRIC GEORGE, first EARL ALEXANDER OF TUNIS (1891-1969), field-marshal; educated at Harrow and Royal Military College, Sandhurst; commissioned in Irish Guards, 1911; served in France, 1914-19; MC, 1915; DSO, 1916; rapid promotion; acting brigadier-general in command of 4th Guards brigade, 1918; member, Allied Relief Commission in Poland under (Sir) Stephen Tallents [q.v.], 1919-20; fought against Bolsheviks in Latvia; commanded his regiment in Constantinople and Gibraltar, 1922-3; posted to Staff College, 1926-7; attended Imperial Defence College, 1930; held staff appointments, 1931-4; commanded Nowshera brigade on North-West Frontier, 1934; CSI, 1936; major-general, 1937, youngest general in British Army; commanded 1st division in France under Sir John Dill [q.v.], 1939; supervised evacuation of Dunkirk, 1940; lieutenant-general, 1940; commanded army in Burma, 1942; almost captured by Japanese on evacuation of Rangoon; on good terms with Chiang Kai Shek and General J. W. Stilwell; withdrew British army to India; commander-in-chief, Middle East, 1942; prepared for decisive battle at El Alamein; worked well with General Montgomery (later Viscount Montgomery of Alamein); victory at Alamein,

followed by Casablanca Conference, 1943; appointed deputy c.-in-c. to General Eisenhower; set up 18th Army Group; fought Tunisian campaign; c.-in-c., 15th Army Group in invasion of Sicily, and Italy; succeeded Sir Maitland (later Lord) Wilson [q.v.] as c.-in-c., Mediterranean; field-marshal; all Italy overrun, and over a million Germans surrender, 1945; governor-general, Canada, 1946-52; last British governor-general; produced official dispatches on his campaigns, published in *London Gazette*; minister of defence, 1952-4, director, Alcan, and other companies; allowed his memoirs to be ghosted in *Sunday Times*; published, 1962; devoted much time to painting in retirement; CB, 1938; KCB, and GCB, 1942; viscount, 1946; GCMG, and KG, 1946; PC, 1952; OM, 1959; colonel, Irish Guards, 1946-69; constable of Tower of London, 1960-5; grand master of Order of St. Michael and St. George; elder brother, Trinity House; president, MCC, 1955; numerous foreign decorations.

ALEXANDER, SAMUEL (1859-1938), philosopher; born at Sydney, New South Wales; educated at Melbourne University and Balliol College, Oxford; first class, *lit. hum.*, 1881; fellow of Lincoln, 1882-93, the first professing Jew to receive a fellowship at Oxford or Cambridge; studied experimental psychology in Germany; published *Moral Order and Progress* (1889); professor of philosophy (1893-1924) at Manchester where he lived until his death and was greatly loved; bent upon a comprehensive system of ontological metaphysics, completed his task in his Gifford lectures published as *Space, Time, and Deity* (2 vols., 1920), a work of sweeping design which marked the end of an epoch; later turned to aesthetic theory and wrote *Beauty and Other Forms of Value* (1933); FBA, 1913; OM, 1930.

ALEXANDER, WILLIAM (1824-1911), archbishop of Armagh; educated at Tonbridge School and Brasenose College, Oxford; influenced by Newman; BA, 1847; ordained 1847; gained a Denyer prize with essay on 'Divinity of Christ', 1850; won university prize for sacred poem, 1860; bishop of Derry, 1867-93; DD, Oxford, 1867; archbishop of Armagh and primate of all Ireland, 1893 till death; published *St. Augustine's Holiday and other Poems*, 1886; an eloquent preacher and lecturer; Bampton lectures (1876) on *Witness of the Psalms to Christ and Christianity*; published commentaries on the Johannine Epistles (1881 and 1889) and other theological works; hon. DCL, Oxford, 1876, LLD, Dublin, 1892, D.Litt., Oxford, 1907; GCVO, 1911; married in 1850 Cecil Frances Alexander, born Humphreys, hymn writer [q.v.].

ALEXANDER-SINCLAIR, Sir EDWYN SINCLAIR (1865-1945), admiral; entered navy, 1879; captain, Royal Naval College, Osborne (1905-8), *Temeraire* (1913-15); commodore, first light cruiser squadron, 1915; gave 'enemy in sight' signal leading to battle of Jutland, 1916; rear-admiral, sixth light cruiser squadron, 1917; led surrendered German fleet into Rosyth, 1918; commanded first battle squadron, Atlantic fleet (1922-5), and China station (1925-7), the Nore (1927-30); admiral, 1926; KCB, 1919; GCB, 1930.

ALEXANDRA CAROLINE MARY CHARLOTTE LOUISE JULIA (1844-1925), of Denmark, queen-consort of King Edward VII; eldest daughter of Prince Christian of Schleswig-Holstein-Sonderburg-Glücksburg, who became king of Denmark as Christian IX in 1863; born in Copenhagen; formally betrothed to Albert Edward, Prince of Wales, at the palace of Laeken, near Brussels, 9 Sept. 1862; married to him in St. George's chapel, Windsor, 10 Mar. 1863; owing to the seclusion of the widowed Queen Victoria, became with her husband the leader of English society; quickly secured the affection of the British people; sympathized strongly with Denmark in her struggle with Prussia for the possession of the duchies of Schleswig-Holstein, 1863-4; gave birth to her eldest son, afterwards the Duke of Clarence, 1864; visited Denmark and Sweden with the prince, 1864; gave birth to her second son, afterwards King George V, 1865; paid first of her four visits with her husband to Ireland, 1868; accompanied the prince on a foreign tour which included a visit to Egypt, 1868-9; kept aloof from foreign politics except for showing sympathy with her relations in Denmark, Greece, and Russia; frequently visited Russia; travelled to St. Petersburg at some personal risk to be with her sister the Empress Marie after the assassination of Alexander II, 1881; received the Order of the Garter after the accession of Edward VII, 1901; both as princess and as queen notable for her charities, which gave great stimulus to beneficent work on part of wealthy and influential people; withdrew into comparative retirement on the death of Edward VII, 1910; continued her interest in the London Hospital and in many schemes to alleviate suffering; 'Alexander Day' instituted for benefit of hospitals, 1913; lay in state in Westminster Abbey prior to her burial at Windsor; renowned for her beauty. For the names and dates of birth of her younger children see Edward VII.

ALEXANDRA VICTORIA ALBERTA EDWINA LOUISE DUFF, Princess Arthur of Connaught, Duchess of Fife (1891-1959),

elder daughter of first Duke of Fife and of Princess Louise [q.v.]; succeeded father, 1912; married Prince Arthur [q.v.], 1913; nursed at St. Mary's Hospital, Paddington, from 1915; SRN, 1919; with husband in South Africa, 1920-3; subsequently nursed at University College and Charing Cross hospitals; RRC, 1925; ran Fife Nursing Home in London, 1939-49.

ALGER, JOHN GOLDWORTH (1836-1907), journalist and author; first wrote for *Norfolk News*; joined *The Times* parliamentary reporting staff, 1866; assistant to Baron de Blowitz, *The Times* correspondent in Paris, 1874-1902; wrote on by-ways of French Revolution.

ALGERANOFF, HARCOURT (1903-1967), dancer and ballet master; changed name from Harcourt Algernon Leighton on joining Anna Pavlova's company, 1921; founder-member, Markova-Dolin Company, 1935; joined De Basil Ballet Russe, 1936; joined International Ballet, 1943; appeared in character roles including the astrologer in *Le Coq d'Or*, Pierrot in *Carnaval*, Dr Coppélius in *Coppélia* and Carabosse in *The Sleeping Beauty*; worked in Australian Children's Theatre, 1954; ballet master, Borovansky ballet, 1959; guest artist, Australian ballet, 1962-3; ballet master, North Western Ballet Society, Australia; published *My Years with Pavlova* (1957).

ALINGTON, first Baron (1825-1904), sportsman. [See Sturt, Henry Gerard.]

ALINGTON, CYRIL ARGENTINE (1872-1955), headmaster and dean; educated at Marlborough and Trinity College, Oxford; first class, *lit. hum.*, 1895; fellow of All Souls, 1896; deacon, 1899; priest, 1901; DD, 1917; assistant master, Marlborough, 1896-9, Eton, 1899-1908; headmaster, Shrewsbury School, 1908-17, Eton, 1917-33; chaplain to king, 1921-33; dean of Durham, 1933-51; works include *Twenty Years, 1815-35* (1921), the Shrewsbury and Eton *Fables*, and some detective and other novels.

ALISON, Sir ARCHIBALD, second baronet (1826-1907), general; son of Sir Archibald Alison [q.v.]; educated at Glasgow University; joined army, 1846; served at Sebastopol, 1855; wounded in second relief of Lucknow, 1857; CB, 1861; prominent in Ashanti war, 1873-4; received thanks of parliament and KCB, 1874; major-general, 1877; commanded Highland brigade at Tel-el-Kebir, 1882; lieutenant-general, 1882; commanded force in Egypt, 1883; in command of Aldershot division, 1883-8;

GCB, 1887; hon. LLD, Cambridge, Edinburgh, and Glasgow; member of India Council, 1889-99; wrote on military topics for *Blackwood's Magazine*.

ALLAN, SIR WILLIAM (1837-1903), engineer and politician; joined navy as engineer, 1857; taken prisoner at capture of Charleston, USA, 1861; manager of North-Eastern Engineering Company, 1868; founded Scotia engine works, Sunderland, 1886; director of Albyn shipping line there; radical MP for Gateshead, 1893 till death; knighted, 1902; published many volumes of Scottish verse, including *Hame-spun Lilts* (1874), *Lays of Leisure* (1883), and *Songs of Love and Labour* (1903).

ALLBUTT, SIR THOMAS CLIFFORD (1836-1925), physician; BA Gonville and Caius College, Cambridge, 1860; consulting physician in Leeds, 1861-89; commissioner in lunacy in London, 1889-92; regius professor of physic at Cambridge, 1892-1925; chief work, *System of Medicine* (1896-9); FRS, 1880; KCB, 1907; PC, 1920.

ALLEN, SIR CARLETON (KEMP) (1887-1966), legal scholar; born in Australia; educated at Newington College, Sydney, Sydney University, and New College, Oxford; first class, jurisprudence, 1912; Eldon law scholar, 1913; produced and acted in OUDS; served in army in France, 1915-18, MC; appointed lecturer in law, University College, Oxford, 1919; called to bar (Lincoln's Inn), 1919; Stowell Civil Law fellow, University College, 1920; dean, 1922-6; junior proctor, 1924-5; delivered Tagore lectures in Calcutta, 1926; published *Law in the making*, 1927; professor of jurisprudence, Oxford, 1929-31, fellow, University College, 1929; Oxford secretary, Rhodes trustees, warden, Rhodes House, 1931-52; publications include *Bureaucracy Triumphant* (1931), *Law and Orders* (1945), *The Queen's Peace* (1953), and *Aspects of Justice* (1958); JP Oxford, 1941; KC, 1945; FBA, 1944; DCL Oxford, 1932; knighted, 1952.

ALLEN, GEORGE (1832-1907), engraver and publisher; started in life as a joiner; Ruskin's pupil and assistant at Working Men's College, 1854; studied mezzotint; illustrated *Modern Painters*; undertook publication of Ruskin's works at his residence at Orpington, 1871; removed offices to London, 1890; published library edition of Ruskin's works, 1903-11; skilled geologist, mineralogist, and botanist.

ALLEN, (HERBERT) WARNER (1881-1968), journalist and author; educated at Charterhouse and University College, Oxford; Taylorian Spanish scholar, 1903; Paris correspondent of *Morning Post*, 1908; published edition of translation by James Mabbe [q.v.] of Spanish *Celestina*, 1908; served as official representative of British press in France and Italy, 1914-19; published *The Unbroken Line* (1916), and *Our Italian Front* (with paintings of Martin Hardie [q.v.], 1920); CBE, 1920; chevalier of Legion of Honour; foreign editor, *Morning Post*, 1925-8; London editor, *Yorkshire Post*, 1928-30; contributed to *Saturday Review*; served in Ministry of Information, 1940-1; published number of books on wine, including *The Wines of France* (1924), *Sherry* (1933), and *A History of Wine: Great Vintage Wines from the Homeric Age to the Present Day* (1961); collaborated with E. C. Bentley [q.v.] on *Trent's Own Case* (1936); also published mystical writings, including *The Uncounted Hour* (1936), and *The Timeless Moment* (1946).

ALLEN, SIR HUGH PERCY (1869-1946), musician and musical statesman; influenced by Dr F. J. Read; his assistant at Chichester Cathedral, 1887; B.Mus. (1892), D.Mus. (1898), Oxford; organ scholar, Christ's College, Cambridge, 1892; organist, St. Asaph Cathedral (1897), Ely (1898-1901), New College, Oxford (1901-18, fellow, 1908); professor of music (1918-46), Oxford, where he obtained creation of music faculty (1944); director, Royal College of Music, 1918-37; director of music, University College, Reading (1908-18), and Cheltenham Ladies' College (1910-18); conducted Oxford and London Bach choirs; an inspiring influence particularly among young people; sought to spread the love of music and teach people to make music—often better than they knew how; his pre-eminent musico-political position perhaps obscured his eminence as a musician; knighted, 1920; KCVO, 1928; GCVO, 1935.

ALLEN, SIR JAMES (1855-1942), New Zealand statesman; born in South Australia; educated at Clifton and St. John's College, Cambridge; MHR, Dunedin East (1887-90) and Bruce, South Otago (1892-1920); minister of finance and education, 1912-15, of defence, 1912-20; created a New Zealand division of Royal Navy, 1913; dispatched expeditionary force, October 1914; introduced war pensions (1915) and conscription (1916); high commissioner, London, and Dominion representative, League of Nations, 1920-7; member, NZ legislative council, 1927-41; KCB, 1917; GCMG, 1926.

ALLEN, JOHN ROMILLY (1847-1907), archaeologist; made life study of pre-Norman art in Great Britain; edited *Cambrian Archaeological Journal*, 1889 till death; FSA Scotland, 1883; Rhind lecturer in archaeology at Edinburgh, 1885; chief work, *Celtic Art in Pagan and Christian Times*, 1904.

ALLEN, PERCY STAFFORD (1869-1933), president of Corpus Christi College, Oxford, and Erasmian scholar; educated at Clifton and Corpus Christi College; professor of history, Government College, Lahore, 1897-1901; fellow of Merton College, Oxford, 1908-24; president of Corpus Christi College, 1924-33; published (with his wife) *Opus Epistolarum Des. Erasmi Roterodami* (11 vols., 1906-47), a masterly critical edition, and other Erasmian studies; FBA, 1923.

ALLEN, REGINALD CLIFFORD Baron Allen of Hurtwood (1889-1939), labour politician; educated at Berkhamsted School, Bristol, and Peterhouse, Cambridge; general manager, *Daily Citizen*, 1911-15; thrice imprisoned as conscientious objector; chairman, ILP, 1922-6; director, *Daily Herald*, 1925-30; supported 'national' labour group, 1931-6; baron, 1932.

ALLEN, ROBERT CALDER (1812-1903), captain RN; as master of *Dido* suppressed Malay pirates of Borneo, 1842-4; harbour master at Malta (1866) and Devonport (1867); employed at Deptford dockyard, 1867-70; CB, 1877.

ALLENBY, EDMUND HENRY HYNMAN, first Viscount Allenby of Megiddo (1861-1936), field-marshal; educated at Haileybury and Sandhurst; commissioned in Inniskillings (6th Dragoons), 1882; served in Bechuanaland and Zululand expeditions; captain, 1888; adjutant, 1889; entered Staff College, 1896; major, 1897; adjutant to 3rd Cavalry brigade, 1898; established a sound reputation in the South African war, commanding his regiment at Bloemfontein and a column during the final rounding-up operations; CB, 1902; major-general, 1909; inspector-general of cavalry, 1910-14; commanded cavalry in the retreat from Mons and at first battle of Ypres; the V Corps at second battle of Ypres (1915); the Third Army (1915-17) at Arras; KCB, 1915; taking command of the Egyptian expeditionary force (June 1917) he invigorated the troops and in Oct. attacked the Turks, breaking through at Gaza and driving them northward beyond Jaffa; captured Jerusalem, 9 Dec. 1917; the German offensive in France delayed a further attack until 19 Sept. 1918 when with a huge right-wheel movement he drove the Turks into the hills whilst the cavalry swept down into the plain of

Megiddo and passed through the valley of Jezreel down to the Jordan near Beisan; in Oct. Damascus was captured on the 1st, Aleppo on the 26th, and an armistice signed with the Turks on the 30th; in one of the most notable campaigns of cavalry employed in strategic mass his inspiration, thrustfulness, and the confidence he inspired were priceless assets; of strong character and violent temper, he was always animated by the highest sense of duty, simple and sincere, thorough in everything; field-marshal and viscount, 1919; special high commissioner for Egypt, 1919-25; obtained recognition of Egypt as a sovereign state, 1922.

ALLERTON, first Baron (1840-1917), politician. [See Jackson, William Lawies.]

ALLIES, THOMAS WILLIAM (1813-1903), theologian; educated at Eton and Wadham College, Oxford; BA, 1832; MA, 1837; fellow, 1833-41; came under tractarian influence; ordained, 1838; joined Church of Rome, 1850; secretary of Catholic poor school committee, 1853-90; actively promoted Catholic primary education; first professor of modern history, Catholic University of Ireland, 1855; works include *A Life's Decision* (1880) and *The Formation of Christendom* (8 vols., 1865-9), showing St. Peter's predominance in history.

ALLINGHAM, MARGERY LOUISE (1904-1966), crime novelist; educated at Perse high school, Cambridge; left school at fifteen and wrote fiction for *Sexton Blake* and *Girls' Cinema*; published *The Crime at Black Dudley*, introducing Albert Campion, 1929; followed by *Mystery Mile*, introducing Campion's manservant, Lugg, 1930; wrote for the *Strand* magazine, and produced number of other mystery stories, including *Flowers for the Judge* (1936); continued writing during 1939-45 war; post-war novels include *The Tiger in the Smoke* (1952), *The Beckoning Lady* (1955), and *Cargo of Eagles* (completed by her husband, Philip Youngman Carter, 1968).

ALLMAN, GEORGE JOHNSTON (1824-1904), mathematician; BA, Trinity College, Dublin, 1844; LLB, 1853; LLD, 1854; professor of mathematics, Queen's College, Galway, 1853-93; FRS, 1884; hon. D.Sc., Dublin, 1882; wrote *History of Greek Geometry from Thales to Euclid*, 1889; a friend of Comte and a positivist.

ALMA-TADEMA, Sir LAWRENCE (1836-1912), painter; born in Holland; entered Antwerp Academy, 1852; exhibitor at Paris Salon and gold medallist, 1864; settled in London, 1870; received letters of denization,

1873; ARA, 1876; RA, 1879; knighted, 1899; OM, 1907; works (numbering 408) chiefly depict subjects from the Merovingian period, ancient Egypt, Greece, and Rome; they show profound archaeological knowledge.

ALMOND, HELY HUTCHINSON (1832-1903), headmaster of Loretto School; educated at Glasgow University and Balliol College, Oxford; BA, 1855; MA, 1862; tutor at Loretto (preparatory) school, 1857; master at Merchiston School, Edinburgh, 1858; became proprietor of Loretto, 1862; raised school to public school standard, revolutionized Scottish school methods, attached great importance to physical exercise, diet, and clothing, and improved stamina of his pupils; published educational writings and sermons.

ALTHAM, HARRY SURTEES (1888-1965), schoolmaster, cricket historian and administrator; educated at Repton and Trinity College, Oxford; cricket blue, 1911-12; assistant master, Winchester, 1913; served in army during 1914-18 war, DSO, MC; returned to Winchester 1919; housemaster, 1927-47; played cricket for Hampshire; president, MCC, 1959-60; treasurer, 1950-63; chairman, test selection committee, 1954; chairman, MCC Youth Cricket Association, 1952-65; president, English Schools Cricket Association, 1951-7; president, Hampshire county cricket club, 1947-65; CBE, 1957; publications include *History of Cricket* (1926), *The World of Cricket* (1966), and contributions to *Wisden* and the *Cricketer*.

ALTRINCHAM, first BARON (1879-1955), administrator and politician. [See GRIGG, EDWARD WILLIAM MACLEAY.]

ALVERSTONE, VISCOUNT (1842-1915), judge. [See WEBSTER, RICHARD EVERARD.]

AMBEDKAR, BHIMRAO RAMJI (1891-1956), Indian statesman; an Untouchable; educated at Elphinstone College, Bombay, Columbia University, New York, and London School of Economics; D,Sc., 1923; called to bar (Gray's Inn), 1922; practised in Bombay; founded Society for Welfare of Outcastes, 1924, and People's Education Society, 1945; nominated to Bombay legislature, 1927; professor, Government Law College, Bombay, 1928; represented Depressed Classes at Round Table conferences, 1930-2; negotiated Poona Pact favourable to Untouchables with M. K. Gandhi [q.v.], 1932; joined executive council in charge of labour, 1942; member of Constituent Assembly, 1946; minister for law, 1947-51; principal architect of India's independent constitution, 1948; became Buddhist, 1956.

AMEER ALI, SYED (1849-1928), Indian jurist and Islamic leader; born at Cuttack in Orissa; educated at Calcutta University; first Moslem to graduate MA; came to England, 1870; called to bar (Inner Temple), 1873; practised law in Calcutta; founded first Moslem political organization in India, 1877; member of Bengal legislative council, 1878; chief presidency magistrate, 1879; one of three Indian additional members of governor-general's council, 1883; judge of high court of Calcutta, 1890; retired to England, 1904; PC, 1909; achieved international position as protagonist of Islam.

AMERY, LEOPOLD CHARLES MAURICE STENNETT (1873-1955), statesman and writer; born in India; educated at Harrow and Balliol College, Oxford; first class, *lit. hum.*, 1896; fellow of All Souls, 1897-1955; on staff of *The Times*, 1899-1909; in South Africa, 1899-1900; became passionate advocate of British imperialism; edited and largely wrote *Times* history of South African war (7 vols., 1900-9); called to bar, (Inner Temple), 1902; conservative MP, South Birmingham (Sparkbrook), 1911-45; joined cabinet secretariat, 1916; parliamentary under-secretary, Colonial Office, 1919-21; parliamentary and financial secretary, Admiralty, 1921; PC, 1922; first lord of Admiralty, 1922-4; colonial secretary, 1924-9, and dominions secretary, 1925-9; unsuccessfully advocated imperial preference with zeal amounting to bigotry; established Empire Marketing Board; dominion status defined at imperial conference of 1926; secretary of state for India, 1940-5; worked to bring India to independence within the Commonwealth; CH, 1945; published *My Political Life* (3 vols., 1953-5); a founder of the Empire Parliamentary Association; a Rhodes trustee, 1919-55; his elder son executed for treason, 1945.

AMHERST, WILLIAM AMHURST TYSSEN-, first BARON AMHERST OF HACKNEY (1835-1909), bibliophile and Norfolk landowner; educated at Eton and Christ Church, Oxford; conservative MP for West Norfolk, 1880-5, and South-West Norfolk, 1885-92; baron, 1892; enthusiastic collector of books, works of art, and Egyptian papyri; well known as cattle breeder, shot, and yachtsman; lost much of his fortune through the fraud of his trustee; art collection and library, including seventeen 'Caxtons' and illuminated MSS, sold, 1906-9.

AMOS, SIR (PERCY) MAURICE (MACLARDIE) SHELDON (1872-1940), jurist and judge in Egypt; son of Sheldon Amos [q.v.]; first class, moral sciences tripos (1893-5),

Trinity College, Cambridge; called to bar (Inner Temple), 1897; judge of Cairo native court (1903), of native court of appeal, 1906–12; director, Khedivial School of Law, 1913–15; judicial adviser, Egyptian government, 1917–25; Quain professor of comparative law, London, 1932–7; KBE, 1922; KC, 1932.

AMPTHILL, second BARON (1869–1935). [See RUSSELL, ARTHUR OLIVER VILLIERS.]

AMULREE, first BARON (1860–1942), lawyer and industrial arbitrator. [See MACKENZIE, WILLIAM WARRENDER.]

ANDERSON, SIR ALAN GARRETT (1877–1952), shipowner and public servant; son of Elizabeth Garrett Anderson [q.v.]; scholar of Eton and Trinity College, Oxford; joined family shipping firm, joint founders of Orient Line; director of the P & O Company; controller of shipping supply, Royal and Merchant navies, 1917–18; deputy governor, Bank of England, 1925–6; director, Suez Canal Company, 1927–52; director, Midland Railway, from 1911; chairman of Railway Executive and controller of railways, 1941–5; chairman, Wheat Executive, 1916–17, and Cereals Control Board, 1939–41; MP, City of London, 1935–40; KBE, 1917; GBE, 1934.

ANDERSON, ALEXANDER (1845–1909), labour poet under pseudonym of 'Surfaceman'; railway platelayer in native village of Kirkconnel, 1862; self-taught; sent to *People's Friend* (1870) verses collected in *A Song of Labour and other Poems* (1873); visited Italy; attracted the favourable notice of Carlyle and Lord Houghton; assistant librarian at Edinburgh University, 1880–3, 1886–1909; works, including *Songs of the Rail* (1878), show lyric power and vivid vision in dealing with railway and humble Scottish child life.

ANDERSON, ELIZABETH (1836–1917), better known as MRS ELIZABETH GARRETT ANDERSON; physician; born GARRETT; after struggles to study medicine obtained licence to practise of Society of Apothecaries, 1865; opened dispensary for women and children which developed into New Hospital for Women, Euston Road (Elizabeth Garrett Anderson Hospital), 1866; senior physician there, 1866–92; MD, Paris, 1870; married J. G. S. Anderson, 1871; member of British Medical Association, 1873; on staff of London School of Medicine for Women, 1875–1903; helped to improve status of women.

ANDERSON, GEORGE (1826–1902), Yorkshire cricketer; member of All England XI, 1857–64; visited Australia, 1863; captain of Yorkshire county club; had good defence and hitting power as batsman; actuary of Bedale Savings Bank, 1873–94.

ANDERSON, SIR HUGH KERR (1865–1928), physiologist and administrator; educated at Harrow and Gonville and Caius College, Cambridge; BA, 1887; MB, 1891; university lecturer in physiology; fellow of his college, 1897; master of Gonville and Caius College, 1912–28; member of royal commission on universities of Oxford and Cambridge, 1919; member of Cambridge commission, 1923; FRS, 1907; knighted, 1922; author jointly and alone of many important scientific papers.

ANDERSON, JOHN, first VISCOUNT WAVERLEY (1882–1958), administrator and statesman; educated at George Watson's College and the university, Edinburgh; graduated in mathematics, chemistry, and natural philosophy; studied uranium at Leipzig; first in civil service examination, 1905; joined Colonial Office; moved to National Health Insurance Commission, 1912, Ministry of Shipping, 1917, and Local Government Board, 1919; chairman, Board of Inland Revenue, 1919–22; joint under-secretary, Dublin, 1920–22; permanent under-secretary, Home Office, 1922–32; governor of Bengal, 1932–7; independent nationalist MP for Scottish Universities, 1938–50; lord privy seal, 1938; home secretary and minister of home security, 1939; lord president of the Council with over-all responsibility for organizing civilian and economic resources, including atomic energy, 1940–3; chancellor of the Exchequer, 1943–5; chairman, atomic energy advisory committee, 1945–8; chairman, Port of London Authority and Covent Garden Opera Trust, 1946–58; FRS, 1945; Romanes lecturer, Oxford, 1946; CB, 1918; KCB, 1919; GCB, 1923; GCIE, 1932; GCSI, 1937; PC, 1938; viscount, 1952; OM, 1957.

ANDERSON, SIR KENNETH ARTHUR NOEL (1891–1959), general; born in India; educated at Charterhouse and Sandhurst; commissioned in Seaforth Highlanders, 1911; served on western front (MC) and in Palestine, 1914–18; commanded 11th Infantry brigade in withdrawal to Dunkirk, 1940; commanded Eastern Task Force for successful landings in French North Africa, and subsequently First British Army, November 1942; unable to take Tunis until May 1943; held home commands (1943–5) and East African command (1945–6); governor, Gibraltar, 1947–52; CB, 1940; KCB, 1943; general, 1949.

ANDERSON, MARY REID (1880–1921), women's labour organizer; born MACARTHUR;

general secretary, Women's Trade Union League, 1903; formed National Federation of Women Workers and helped to create National Anti-Sweating League, 1906; married W. C. Anderson, 1911; member of reconstruction and other committees, 1914-18; British representative at labour conference in America, 1920.

ANDERSON, STANLEY ARTHUR CHARLES (1884-1966), engraver, etcher, and water-colour painter; educated at Merchant Venturers' Technical College, Bristol; apprenticed at fifteen to his father, engraver; studied at Bristol School of Art; won open etching scholarship, 1909; studied at Royal College of Art, London, under (Sir) Francis (Frank) Short [q.v.]; exhibited at Royal Academy, 1909; associate, and later fellow, Royal Society of Painter-Etchers (and Engravers); engaged on munitions work, 1914-18; produced drypoints, including 'Durer's House, Nürnberg,' 'St. Nicholas, Prague', and 'The Reading Room'; appointed visiting instructor in etching, Goldsmiths' College School of Art, 1925-40; painted much sought-after water-colour landscapes; representative of British line-engraving and drypoint, Venice Biennial International Art Exhibition, 1938; taught at British School, Rome, 1930-52; ARA, 1934; RA, 1941; CBE, 1951; work widely exhibited; represented in British, Victoria and Albert, Fitzwilliam, and Ashmolean museums.

ANDERSON, STELLA (1892-1933), better known as STELLA BENSON, novelist; worked and travelled widely; married (1921) J. C. O'G. Anderson of customs service in China where she lived thereafter; published short stories, travel sketches, poems, and novels, including the popular *Tobit Transplanted* (1931), combining fantasy with realism, satire with pity.

ANDERSON, SIR THOMAS McCALL (1836-1908), physician; MD, Glasgow University, 1858; after study abroad, appointed lecturer in medicine in Andersonian Institute, 1860; first physician of hospital for diseases of skin, Glasgow, 1861 till death; professor of clinical medicine (1874-1900) and of practice of medicine (1900-8) at Glasgow University; had large consulting practice; knighted 1905; wrote many medical treatises, including *Lectures on Clinical Medicine* (1877) and *On Diseases of the Skin* (1887).

ANDERSON, SIR WARREN HASTINGS (1872-1930), lieutenant-general; educated at Royal Military College, Sandhurst; joined army, 1890; served in South African war, 1899-1902; employed in various directorates at War Office, 1906-10; appointed to general staff

of Southern command, Salisbury, 1911; served in European war, 1914-18; with First Army as major-general, general staff, 1917-18; commandant of Staff College, 1919-22; chief general staff officer with Allied Army of Black Sea, 1922-3; commanded Baluchistan district, Quetta, 1924-7; quartermaster-general of the forces and lieutenant-general, 1927; KCB, 1922.

ANDREWES, SIR FREDERICK WILLIAM (1859-1932), pathologist and bacteriologist; educated at Christ Church, Oxford; qualified at St. Bartholomew's Hospital; FRCP, 1895; lecturer (1897), professor (London University, 1912-27) in pathology, St. Bartholomew's; an influential teacher; played an important part in developing the relationship between bacteriology and public health and clinical medicine; researches included lymphadenoma, arteriosclerosis, and classification of variant strains of dysentery bacilli; FRS, 1915; knighted, 1920.

ANDREWS, SIR JAMES, baronet (1877-1951), lord chief justice of Northern Ireland; graduated from Trinity College, Dublin, 1899; called to Irish bar, 1900; KC, 1918; lord justice of appeal, Supreme Court, 1921-37; lord chief justice of Northern Ireland, 1937-51; competent, businesslike, and manifestly fair; bencher, King's Inns, 1920, Inn of Court of Northern Ireland, 1926; baronet, 1942.

ANDREWS, THOMAS (1847-1907), metallurgical chemist and ironmaster; succeeded father as head of Wortley ironworks, 1871; made valuable metallurgical researches; FRS, 1888; received gold medal, Society of Engineers, 1902; advocate of technical education.

ANGELL, SIR (RALPH) NORMAN (1872-1967), publicist; educated at lycée in St. Omer and Geneva University; took variety of jobs in America, 1889-97; returned to Paris, 1898; edited the *Daily Messenger*; came to notice of Lord Northcliffe [q.v.]; editor, *Continental Daily Mail*, 1904; published first book, *Patriotism under Three Flags*, 1903; followed by *Europe's Optical Illusion*, 1909, republished as *The Great Illusion*, 1910; its theme that armed aggression did not pay; the first practical discussion of the possibility of preventing war; gave up editorship and propagated his ideas in lectures and writing, especially in the *New Republic*; one of pioneers of the idea of a League of Nations; labour MP, North Bradford, 1929-31; knighted, 1931; awarded Nobel peace prize, 1933; published *After All*, 1951.

ANGUS, JOSEPH (1816-1902), Baptist divine and biblical scholar; MA, Edinburgh, 1837; entered Baptist ministry, 1838; secretary of Baptist Missionary Society, 1841; president of Baptist College at Stepney, 1849-57, and at Regent's Park, 1859-93; hon. DD, Brown University, USA, 1852; member of first London school board, 1870; published useful handbooks to the Bible (1853), English language (1864), English literature (1866); *Baptist Authors and History, 1527-1800* (1896), and other works.

ANGWIN, Sir (ARTHUR) STANLEY (1883-1959), engineer; B.Sc. (Eng.), East London College, 1907; entered Post Office engineering department; commanded 52nd Divisional Signal Company, 1914-18; DSO, MC; assisted in construction of large PO radio stations; in charge of radio branch, 1928; assistant (1932), deputy (1935), and PO engineer-in-chief (1939-46), chairman, Cable & Wireless, 1947-51, of Commonwealth Telecommunications Board, 1951-6, and of Radio Research Board, 1947-52; president, 1943-4, and Faraday medallist, 1953, Institution of Electrical Engineers; knighted, 1941; KBE, 1945; KCMG, 1957.

ANNANDALE, THOMAS (1838-1907), surgeon; MD, Edinburgh, 1860, winning gold medal; house-surgeon, Edinburgh Royal Infirmary, 1860; assistant surgeon, 1865; surgeon, 1871; FRCS Edinburgh, 1863, and England, 1888; regius professor of clinical surgery, Edinburgh University, 1877; DCL, Durham, 1902; Annandale gold medal in clinical surgery founded at Edinburgh University; published works on surgery.

ANSON, Sir WILLIAM REYNELL, third baronet (1843-1914), warden of All Souls College, Oxford; educated at Eton and Balliol College, Oxford; fellow of All Souls, 1867; read for bar and practised on Home circuit; succeeded father, 1873; Vinerian reader in English law at Oxford, 1874; as (first lay) warden of All Souls, 1881-1914, preserved its historical continuity while loyally accepting reforms; vice-chancellor, 1898; unionist MP, Oxford University, 1899 till death; parliamentary secretary, Board of Education, 1902-5; strenuously opposed to finance bill, 1909, and Parliament Act, 1911; PC, 1911; combined real learning with wide knowledge of affairs; chief works, *The Principles of the English Law of Contract* (1879) and *The Law and Custom of the Constitution* (part I, 1886, part II, 1892).

ANSTEY, F. (pseudonym), humorous writer. [See GUTHRIE, THOMAS ANSTEY.]

ANSTEY, FRANK (1865-1940), Australian journalist and politician; emigrated from London; labour member, Victorian assembly

(1902-10), Commonwealth parliament (1910-34); strongly opposing conscription, attacked money power in the *Kingdom of Shylock*, 1917; minister for health and repatriation, 1929-31; gifted orator and forceful writer.

ANTAL, FREDERICK (1887-1954), art historian; born in Budapest; studied art history in Vienna; worked at Museum of Fine Arts, Budapest, 1914-19; moved to Berlin, 1922, England, 1933; naturalized, 1946; studied sixteenth-century Italian art and influence on it of earlier periods; works include *Florentine Painting and its Social Background. The Bourgeois Republic before Cosimo de' Medici's Advent to Power*, (1948), studies on Fuseli and his contemporaries, (1956) and Hogarth, (1962).

APPLETON, Sir EDWARD VICTOR (1892-1965), physicist; educated at Hanson School, Bradford, and St. John's College, Cambridge; first class, parts i and ii, natural sciences tripos, 1913-14; did research in mineralogy; served in army, 1914-18; fellow, St. John's College, Cambridge, 1919; assistant demonstrator in physics, Cavendish laboratory, 1920; research on radio waves and the ionosphere; Wheatstone professor of physics, King's College, London, 1924-36; Jacksonian professor of natural philosophy, Cambridge, 1936-9; secretary, Department of Scientific and Industrial Research, 1939-49; closely concerned with development of radar and secret work on atomic bomb; principal and vice-chancellor, Edinburgh University, 1949; founder and editor-in-chief of *Journal of Atmospheric and Terrestrial Physics*, 1950; president, International Union of Scientific Radio, 1934-52; FRS 1927; awarded Nobel prize, 1947; KCB, 1941; GBE, 1946; hon. degrees and decorations from many countries; published *Thermionic Vacuum Tubes, and their Applications* (1932); gave Reith lectures on 'Science and the Nation', 1956.

ARBER, AGNES (1879-1960), botanist; born ROBERTSON; B.Sc., London, 1899; first class, parts i and ii, natural sciences tripos; Newnham College, Cambridge, 1901-2; Quain student in biology, 1903-8, lecturer in botany, 1908-9, University College, London; married E. A. N. Arber, 1909; researched on plant anatomy in Balfour laboratory, Newnham, until 1927, thereafter at home; published works on *Herbals* (1912), *Water-plants* (1920), *Monocotyledons* (1925), *Gramineae* (1934), *The Natural Philosophy of Plant Form* (1950), and *The Mind and the Eye* (1954); FRS, 1946.

ARBER, EDWARD (1836-1912), man of letters; Admiralty clerk, 1854-78; professor of

English, Mason College, Birmingham, 1881–94; produced *English Reprints* (1868–71), *Transcript of the Registers of the Company of Stationers of London, 1554–1640* (1875–94), and *Term Catalogues, 1668–1709* (1903–6).

ARBERRY, ARTHUR JOHN (1905–1969), orientalist; educated at Portsmouth grammar school and Pembroke College, Cambridge; first class, parts i and ii, classical tripos, 1926–7; first class, parts i and ii, oriental languages tripos, 1929; research fellow, Pembroke College, 1931; head, classics department, Cairo University, 1932–4; assistant librarian, India Office, London,1934–9; served in Ministry of Information, 1939–44; professor of Persian, London University, 1944; professor of Arabic, and head, Near and Middle East Department, 1946; Sir Thomas Adams's professor of Arabic, Cambridge, 1947–69; published over sixty works including the *Mawāqif and Mukhātabāt* of Niffari (1935), and a translation of the Koran (1935); Litt.D. Cambridge, 1936; FBA, 1949; hon. D.Litt., Royal University of Malta, 1963.

ARBUTHNOT, Sir ALEXANDER JOHN (1822–1907), Anglo-Indian official and author; elder brother of General Sir Charles George Arbuthnot [q.v.]; educated at Rugby under Dr Arnold, 1832–9, and at Haileybury, 1840–41; writer for East India Company, 1840; compiled papers relating to public instruction in Madras province, 1854; first director of public instruction, Madras, 1855; vice-chancellor of Madras University, 1871–2, and of Calcutta University, 1878–80; chief secretary to Madras government, 1862; acting governor, Feb.–May 1872; KCSI, 1873; appointed member of governor-general's council, 1875; CIE, 1878; opposed to reduction of salt and cotton duties in India, 1877–9; demurred to Lord Lytton's aggressive Afghan policy, 1879; settled in Hampshire, 1880; member of India Council, 1887–97; contributor to this Dictionary, 1885–1901; author of *Life of Clive* (1898) and *Memories of Rugby and India* (posthumous, 1910).

ARBUTHNOT, FORSTER FITZGERALD (1833–1901), orientalist; born at Belgaum, Bombay; educated at Haileybury; in Bombay civil service, 1853–78; friend of Sir Richard Burton [q.v.]; compiler of books on oriental themes, the chief being *Persian Portraits* (1887) and *Arabic Authors* (1890); inaugurated (1891) new series of 'Oriental Translation Fund'.

ARBUTHNOT, Sir ROBERT KEITH, fourth baronet (1864–1916), rear-admiral; entered navy, 1877; succeeded father, 1889; commander, 1897; flag captain at Portsmouth to (Lord) Fisher [q.v.], 1903–4; removed from his ship in consequence of a speech offensive to Germany, 1910; commodore of third destroyer flotilla, 1910; rear-admiral, 1912; second-in-command of second battle squadron, 1913; in command of first cruiser squadron, 1915; killed at battle of Jutland; posthumous KCB.

ARCH, JOSEPH (1826–1919), politician; itinerant agricultural labourer, 1835–72; instrumental in forming Warwickshire Agricultural Labourers' Union, 1872; organizing secretary (afterwards president) of newly founded National Agricultural Labourers' Union, 1872; liberal MP, North-West Norfolk, 1885–6, 1892–1902; did more than any other man of his time to improve conditions of agricultural workers.

ARCHER, JAMES (1823–1904), painter; studied art at Trustees' Academy, Edinburgh; ARSA, 1850; RSA, 1858; exhibited many works at Scottish Academy, including 'The Child John in the Wilderness', 1842, and 'Rosalind and Celia' (diploma work), 1854; executed oil and chalk portraits; removed to London, 1863; exhibited regularly at Royal Academy, 1850–1900; most successful in costume pictures and portraits of children; chief portraits include Sir George Trevelyan (1872), Irving (1892), Sir Daniel Macnee (1877); work of refined quality and akin to that of the pre-Raphaelites.

ARCHER, WILLIAM (1856–1924), critic and journalist; educated at Edinburgh University; settled in London, 1878; devoted himself to study of the theatre and became dramatic critic to several newspapers; largely contributed to raising of standard of English stage; worked for the abolition of theatrical censorship and formation of a national theatre; works include edition of the collected works of Ibsen (1906–7), *Masks or Faces?* (1888), *Play-making* (1912), and *The Green Goddess* (1921), his one successful play.

ARCHER-HIND (formerly HODGSON), RICHARD DACRE (1849–1910), Greek scholar and Platonist; educated at Shrewsbury and Trinity College, Cambridge; Craven scholar, 1871; chancellor's medallist and BA, 1872; fellow, 1873; tutor, 1878; published editions of *Phaedo* (1883) and *Timaeus* (1888) and *Translations into Greek Verse and Prose* (1905); ardent gardener and musician.

ARDAGH, Sir JOHN CHARLES (1840–1907), major-general RE; educated at Trinity College, Dublin; entered Royal Military Academy, Woolwich, 1858; superintended construction of Fort Popton at Milford Haven,

1860; deputy assistant quartermaster-general for intelligence at War Office, 1876; sent to Constantinople to report on its defence, 1876; reported on Montenegrin and Bulgarian defences; at congress of Berlin, 1878; CB (civil), 1878; British commissioner for delimitation of Turco-Greek frontier, 1881; sent to Egypt in charge of intelligence department, 1882; restored Alexandria after its bombardment; present at battles of Tel-el-Kebir (1882) and El Teb (1884); CB (military), 1884; commandant at Cairo during Khartoum relief expedition, 1884; assistant adjutant-general for defence at War Office, 1887; private secretary to viceroy of India, 1888-94; CIE, 1892; KCIE, 1894; director of military intelligence, 1896-1901; compiled statement of military resources of Boers in South African war, 1899; KCMG, 1902; British (army) delegate at conference for revision of Geneva convention, 1906; hon. LLD, Dublin, 1897; a skilful water-colour artist.

ARDEN-CLARKE, Sir CHARLES NOBLE (1898-1962), colonial governor; educated at Rossall School; served in Machine Gun Corps, 1917; set aside classical scholarship to Emmanuel College, Cambridge and joined Colonial Service, 1920; served in Nigeria, 1920-36; assistant resident commissioner, Bechuanaland, 1936; resident commissioner, 1937; resident commissioner, Basutoland, 1942; governor, Sarawak, 1946; governor, Gold Coast, 1949-57; guided Gold Coast to independent Ghana; chairman, UN Good Offices Committee on Namibia, 1958; chairman, Royal African Society, and Royal Commonwealth Society for the Blind, 1959; chairman, National Council for Supply of Teachers Overseas, 1960; member, advisory commission on Central Africa, 1960; CMG, 1941; knighted, 1946; KCMG, 1948; GCMG, 1952.

ARDEN-CLOSE, Sir CHARLES FREDERICK (1865-1952), geographer; commissioned in Royal Engineers, 1884; with Survey of India, 1889-93; on German East Africa frontier commission, 1898; chief instructor, surveying, Chatham, 1902-5; wrote *Text Book of Topographical and Geographical Surveying*, (1905); head of War Office geographical section, 1905-11; obtained colonial surveys and progress with Carte Internationale du Monde au Millionième; director-general, Ordnance Survey, 1911-22; undertook second geodetic levelling of UK; redesigned UK 1-inch map; published map of Roman Britain; supplied armies with 32 million maps during war; colonel, 1912; CB, 1916; KBE, 1918; FRS, 1919; hon. Sc.D. Cambridge.

ARDILAUN, first BARON (1840-1915), philanthropist. [See GUINNESS, SIR ARTHUR EDWARD.]

ARDITI, LUIGI (1822-1903), musical conductor and composer; born at Crescentino, Piedmont; produced and conducted his *La Spia* at New York, 1856; conductor of opera at Her Majesty's Theatre, London, 1858-67, and at Covent Garden, 1869; conducted first performance of Gounod's *Faust*, 1863; toured with Adelina Patti, 1882-7; composed songs 'Il Bacio', 1860, and 'L'Ardita', 1862; published *My Reminiscences*, 1896.

ARDWALL, LORD (1845-1911), Scottish judge. [See JAMESON, ANDREW.]

ARGYLL, ninth DUKE OF (1845-1914), governor-general of Canada. [See CAMPBELL, JOHN DOUGLAS SUTHERLAND.]

ARKELL, WILLIAM JOSCELYN (1904-1958), geologist; educated at Wellington College and New College, Oxford; first class, geology, 1925; college lecturer in geology, 1929-33; research fellow, New College, 1933-40, Trinity College, Cambridge, 1947-58; published monographs on *British Corallian Lamellibranchia*, 1929-37, *Ammonites of the English Corallian Beds*, 1935-48, and *English Bathonian Ammonites*, 1951-8; also *Jurassic System in Great Britain*, (1933), *Jurassic Geology of the World*, (1956), *Geology of Oxford* and *Oxford Stone*, (1947); FRS 1947.

ARKWRIGHT, Sir JOSEPH ARTHUR (1864-1944), bacteriologist; educated at Wellington College and Trinity College, Cambridge; qualified at St. Bartholomew's Hospital, 1889; on staff of Lister Institute of Preventive Medicine, 1906-27; investigations included diphtheria, dysentery, trench fever, typhus, and animal diseases; made fundamental researches on bacterial variation; FRCP, 1916; FRS, 1926; knighted, 1937.

ARLEN, MICHAEL (1895-1956), novelist; an Armenian born in Bulgaria; educated at Malvern College; naturalized, 1922; best known for *The Green Hat* (1924), a cynical, sophisticated yet sentimental glittering romance of London café society which proved an influential best-seller; other novels include *Lily Christine* (1929) and *Men Dislike Women* (1931); settled in France, then New York.

ARLISS, GEORGE (1868-1946), actor; real name AUGUSTUS GEORGE ANDREWS; acted in America, 1901-23; returned to London (1923) in *The Green Goddess*, later (1928) the first of his famous talking films; these included *Disraeli*, *Voltaire*, and *Cardinal Richelieu*; his performances dignified and thoughtful, but miniature, studies imperfectly concealing his own distinctive features and charm of voice and personality.

ARMES, PHILIP (1836-1908), organist and composer of sacred music; chorister at Rochester Cathedral, 1848-50; articled to J. L. Hopkins [q.v.], organist there, 1850; assistant, 1856; organist of Durham Cathedral, 1862 till death; Mus.Doc., Oxford, 1864, and Durham, 1874; first professor of music, Durham, 1897.

ARMOUR, JOHN DOUGLAS (1830-1903), Canadian judge and jurist; born near Peterborough, Ontario; BA, Toronto University, 1850; called to bar, 1853; QC, 1867; puisne judge (1877) and chief justice (1887) of court of queen's bench; chief justice of Ontario, 1900; judge of supreme court of Canada, 1902; hon. LLD, Toronto University, 1902; represented Canada on Alaska boundary tribunal.

ARMSTEAD, HENRY HUGH (1828-1905), sculptor; studied design at Somerset House and Royal Academy Schools; designer and worker in metal for Messrs. Hunt & Roskell till 1863; studied sculpture; visited Rome, 1863-4; employed on Albert Memorial (1864) and many public buildings; also executed effigies and imaginative works, including 'Remorse' (in Tate Gallery); ARA, 1875; RA, 1879; taught in Academy Schools from 1875.

ARMSTRONG, EDWARD (1846-1928), historian and teacher; son of John Armstrong, the younger [q.v.]; educated at Bradfield and Exeter College, Oxford; BA and fellow of Queen's College, Oxford, 1869; senior bursar, 1878-1911; pro-provost, 1911-22; warden of Bradfield, 1910-25; undertook modern history teaching for Queen's College, 1883; a stimulating lecturer and teacher; became leading English authority on the Italian renaissance; Dante scholar; works include *Elizabeth Farnese* (1892), *The French Wars of Religion* (1892), *Lorenzo de' Medici* (1896), and *The Emperor Charles V* (1902); FBA, 1905.

ARMSTRONG, Sir GEORGE CARLYON HUGHES, first baronet (1836-1907), newspaper proprietor; born at Lucknow; military cadet in East India Company's army, 1855; served in Indian Mutiny; retired with rank of captain, 1861; editor from 1871, and proprietor from 1875, of *Globe*, conservative newspaper, in which he prematurely disclosed terms of Salisbury-Schouvaloff treaty, 1878; acquired newspaper *The People*, 1882; baronet, 1892.

ARMSTRONG, HENRY EDWARD (1848-1937), chemist and educationist; studied at Royal College of Chemistry under (Sir) Edward Frankland [q.v.]; taught at St. Bartholomew's Hospital, 1870-82; professor of chemistry, London Institution, 1871-84, Central Institution (later Technical College) at South Kensington, 1884-1913; with W. E. Ayrton and W. C. Unwin [qq.v.] introduced higher technical education for engineers and chemists; researches included the chemistry of naphthalene, theory of aqueous solution, the crystallography of organic compounds; brought about great reforms in teaching both elementary and advanced science; FRS, 1876.

ARMSTRONG, THOMAS (1832-1911), artist; studied art in Manchester and Paris (under Ary Scheffer), and with the Barbizon school; exhibited landscape and figure pieces at Royal Academy, 1865-77, and Grosvenor Gallery, 1877-81; art director, South Kensington Museum, 1881-98; acquired for museum many important foreign works of art; revived English enamelling, 1886; CB, 1898.

ARMSTRONG, WILLIAM (1882-1952), actor and producer; educated at Heriot's School, and studied music at the university, Edinburgh; founded its dramatic club; acted in Stratford and Glasgow; joined Liverpool Repertory Theatre, 1914; as producer (1922) and director (1923) until 1944 made it a great school of acting, developing natural talents in a flexible, sensitive style; with Sir Barry Jackson [q.v.] in Birmingham, 1945-7; CBE, 1951.

ARMSTRONG-JONES, Sir ROBERT (1857-1943), alienist; qualified at St. Bartholomew's Hospital, 1880; FRCS, 1885; FRCP, 1907; junior medical officer, Royal Earlswood Institution (1880-2), Colney Hatch (1882-8); medical superintendent, Earlswood (1888-93), Claybury (1893-1916); made Claybury renowned for modern methods of treatment of mental diseases; knighted, 1917.

ARNOLD, Sir ARTHUR (1833-1902), radical politician and writer; at first surveyor and land agent; early published sensational novels; government inspector of public works under Local Government Board, 1864; issued *History of the Cotton Famine*, 1864; editor of *Echo*, evening paper, 1868-75; described travels through East (1875) in *Through Persia by Caravan*, 1877; wrote for reviews radical essays, which he collected in *Social Politics*, 1878; MP for Salford, 1880-5; of strong philhellenic sympathies; alderman of London County Council, 1889-1904, and chairman, 1895-6; knighted, 1895; hon. LLD, Cambridge, 1897.

ARNOLD, Sir EDWIN (1832–1904), poet and journalist; educated at King's College, London, and University College, Oxford; BA, 1854; MA, 1856; obtained Newdigate prize, 1852; published *Poems Narrative and Lyrical*, 1853; principal of Deccan College, Bombay, 1856–61; wrote *History of the Marquis of Dalhousie's Administration*, 1862–5; leader-writer on *Daily Telegraph* from 1861; a chief editor, 1873; supported Turkey in Russo-Turkish war and Lord Lytton's Indian policy; CSI, 1877; KCIE, 1888; published *The Light of Asia*, a poem, 1879; travelled along Pacific coast and in Japan, 1889; published picturesque accounts of tour; visited America, 1891; a voluminous poet and oriental translator; his collected poetical works appeared in 1888.

ARNOLD, GEORGE BENJAMIN (1832–1902), organist and composer, mainly of church music; pupil of George William Chard and S. S. Wesley [qq.v.]; Mus.Bac., Oxford, 1853; Mus. Doc., 1860; organist of Winchester Cathedral, 1865–1902.

ARNOLD, Sir THOMAS WALKER (1864–1930), orientalist; BA, Magdalene College, Cambridge; teacher in philosophy, Mohammedan Anglo-Oriental College, Aligarh, 1888–98; worked for reform of Islam; professor of philosophy, Government College, Lahore, 1898–1904; assistant librarian, India Office, 1904–9; educational adviser for Indian students in England, 1909–20; professor of Arabic and Islamic studies, London University, 1921–30; knighted, 1921; FBA, 1926; works include *The Preaching of Islam* (1896), *The Caliphate* (1924), *Painting in Islam* (1928), and *The Islamic Book* (1929).

ARNOLD, WILLIAM THOMAS (1852–1904), author and journalist; born at Hobart, Tasmania; educated at Rugby and University College, Oxford; BA, 1876; won Arnold prize for *The Roman System of Provincial Administration*, 1879; writer on politics and occasionally on drama for *Manchester Guardian*, 1879–98; helped to found Manchester School of Art; edited Keats, 1884; wrote *Studies in Roman Imperialism* (posthumous, 1906).

ARNOLD-FORSTER, HUGH OAKELEY (1855–1909), author and politician; grandson of Arnold of Rugby [q.v.]; educated at Rugby and University College, Oxford; BA, 1877; MA, 1900; called to bar, 1879; private secretary to his adoptive father, W. E. Forster [q.v.], chief secretary for Ireland, 1880; joined Cassell & Co., 1885; prepared educational manuals, including 'Citizen Reader' series (1886); secretary of Imperial Federation League and advocate of naval efficiency, 1884; unionist MP for West Belfast, 1892–1906, and for Croydon, 1906–9; wrote much on army questions, 1892–1900; chairman of commission of land inquiry in South Africa, 1900; secretary of the Admiralty, 1901; supported tariff reform; secretary of state for war in A. J. Balfour's administration, 1903–5; reorganized War Office; a fluent speaker and writer; author of *English Socialism of To-day*, 1908.

ARROL, Sir WILLIAM (1839–1913), engineering contractor; began construction of Dalmarnock ironworks, near Glasgow (eventually largest British structural steelworks), 1872; new Tay bridge (1882–7), Forth bridge (1883–90), and steel work for Tower bridge, London (1886–94), constructed by his firm; knighted, 1890; liberal unionist MP, South Ayrshire, 1892–6.

ARTHUR OF CONNAUGHT, Princess (1891–1959). [See Alexandra Victoria Alberta Edwina Louise Duff.]

ARTHUR FREDERICK PATRICK ALBERT (1883–1938), prince of Great Britain and Ireland; son of first Duke of Connaught [q.v.] and grandson of Queen Victoria; educated at Eton and Sandhurst; commissioned in 7th Hussars, 1901; saw active service in South African war and in 1914–18; governor-general of South Africa, 1920–23; raised nearly £2 million as chairman of Middlesex Hospital; KG, 1902; PC, 1910.

ARTHUR WILLIAM PATRICK ALBERT, Duke of Connaught and Strathearn (1850–1942), third son of Queen Victoria; entered Royal Military Academy, Woolwich, 1866; commissioned in Royal Engineers, 1868; transferred to Royal Artillery; in Canada with 1st battalion, Rifle Brigade, 1869–70; captain, 1871; created Duke of Connaught and Strathearn and Earl of Sussex, and transferred to 7th Hussars, 1874; assistant adjutant-general, Gibraltar, 1875–6; lieutenant-colonel, 1st battalion, Rifle Brigade, 1876; married Princess Louise of Prussia, 1879; major-general, 1880; commanded 1st Guards brigade with distinction in Egyptian war, 1882; commanded in Bombay (1886–90), Portsmouth (1890–3), Aldershot (1893–8), Ireland (1900–4), and Mediterranean (1907–9); inspector-general of the forces, 1904–7; lieutenant-general, 1889; general, 1893; field-marshal, 1902; with son, Prince Arthur [q.v.], renounced succession to duchy of Saxe-Coburg-Gotha, 1899; attended Delhi durbar, 1903; opened Union parliament, South Africa, 1910; governor-general, Canada, 1911–16; opened the new chambers of the Indian legislature, 1921; interested in horticulture and freemasonry but pre-eminently in the army.

ARTHUR, WILLIAM (1819-1901), Wesleyan divine; missionary in India, 1839-41; in France, 1846-8; secretary of Wesleyan Missionary Society, 1851-68; honorary secretary, 1888-91; principal of Methodist College, Belfast, 1868-71; president of Wesleyan Conference, 1866; his voluminous writings include *The Tongue of Fire* (1856) and works on the papacy.

ASCHE, (THOMAS STANGE HEISS) OSCAR (1871-1936), actor, manager, author, and producer; born at Geelong, Australia; studied acting in Norway, his father's country; joined Benson repertory company, 1893; in management at the Adelphi (1904) and His Majesty's (1907); produced memorable Shakespearian revivals, playing a notable Othello; wrote libretto of *Chu Chin Chow* (His Majesty's, 1916-21) and produced *The Maid of the Mountains* at Daly's, 1917.

ASHBEE, CHARLES ROBERT (1863-1942), architect, craftsman, and town planner; BA, Cambridge, 1886; articled to G. F. Bodley [q.v.]; founded the Guild of Handicraft (in London, 1888-1902, Chipping Campden, 1902-7), notable for its printing; as civic adviser and secretary to the Pro-Jerusalem Society (1918-22) responsible for much cleaning, restoration, and preservation, the revival of Arab crafts, and many ideas incorporated in plans for development of the modern city of Jerusalem.

ASHBOURNE, first BARON (1837-1913), lord chancellor of Ireland. [See GIBSON, EDWARD.]

ASHBY, ARTHUR WILFRED (1886-1953), agricultural economist; educated at Tysoe village school, Ruskin College (diploma in economics and political science) and Institute for Research in Agricultural Economics, Oxford, and Wisconsin University; wrote history of allotments and smallholdings, 1917; with Board of Agriculture, 1917-19; at the Institute at Oxford, 1919-24; head of department of agricultural economics, Aberystwyth, 1924-45; professor, 1929; director, the Institute at Oxford, 1946-52; member of Agricultural Wages Board from 1924; influential in foundation of Agricultural Economics Society and Milk Marketing Board; CBE, 1946.

ASHBY, HENRY (1846-1908), physician; studied medicine at Guy's Hospital; MRCS, 1873; MB, 1874; MD, 1878; FRCP, 1890; at Liverpool, 1875-8, and at Manchester from 1878 to death; lecturer on children's diseases in Owens College, Manchester, 1880-1908; president of various medical societies; enjoyed world-wide repute as expert on children's diseases; works include *Diseases of Children* (1899), *Notes on Physiology* (1878), and *Health in the Nursery* (1898).

ASHBY, THOMAS (1874-1931), archaeologist; educated at Winchester and Christ Church, Oxford; first class, *lit. hum.*, 1897; first student (1901), assistant director (1903), director (1906-25), British School at Rome, which after 1915 embraced art and architecture as well as archaeology; revised Platner's *Topographical Dictionary of Ancient Rome*, 1929; wrote *Aqueducts of Ancient Rome*, 1935; FBA, 1927.

ASHER, ALEXANDER (1835-1905), solicitor-general for Scotland; educated at Edinburgh University; called to Scottish bar, 1861; appointed advocate deputy, 1870; liberal MP for Elginshire, 1881-1905; QC, 1881; solicitor-general for Scotland, 1881-5, 1886, 1892-4; hon. LLD, Aberdeen, 1883, Edinburgh, 1891.

ASHFIELD, BARON (1874-1948), chairman of the London Passenger Transport Board. [See STANLEY, ALBERT HENRY.]

ASHLEY, EVELYN (1836-1907), author; son of seventh Earl of Shaftesbury [q.v.]; educated at Harrow and Trinity College, Cambridge; MA, 1858; private secretary to Palmerston, 1858-65; visited Garibaldi in Italy, 1860; helped to produce *The Owl*, society journal, 1864; called to bar, 1863; treasurer of county courts, 1863-74; completed Lord Dalling's unfinished *Life of Palmerston*, 5 vols., 1870-6; liberal MP for Poole, Dorset, 1874-80, and Isle of Wight, 1880-5; under-secretary to Board of Trade in Gladstone's ministry, 1880-2, and to Colonial Office, 1882-5; ecclesiastical commissioner, 1880-5; made five unsuccessful attempts to enter parliament as liberal unionist, 1886-95; PC, 1891; five times mayor of Romsey, 1898-1902.

ASHLEY, WILFRID WILLIAM, BARON MOUNT TEMPLE (1867-1938), politician; son of Evelyn Ashley [q.v.]; educated at Harrow and Magdalen College, Oxford; conservative MP, Blackpool, 1906-18, Fylde division, 1918-22, New Forest division, 1922-32; minister of transport, 1924-9; reorganized the ministry and planned 'roundabouts'; PC, 1924; baron, 1932.

ASHLEY, SIR WILLIAM JAMES (1860-1927), economic historian; BA, Balliol College, Oxford, 1881; his interest in economic history stimulated by lectures of Arnold Toynbee [q.v.]; also influenced by German school of economic historians; fellow of Lincoln College, 1885; professor of political economy and constitutional history, Toronto University, 1888-92; professor of economic history, Harvard, 1892; first professor of commerce, Birmingham University, 1901-25; vice-principal, 1918-25; served on royal commission on agriculture, 1919; member of agricultural tribunal of investigation, 1922-4; member of committee on industry and trade, 1924; knighted, 1917; works include *An Introduction to English Economic History and Theory* (2 parts, 1888-93), *The Tariff Problem* (1903), *The Economic Organisation of England* (1914), and *The Bread of our Forefathers* (1928).

ASHMEAD BARTLETT, SIR ELLIS (1849-1902), politician. [See BARTLETT.]

ASHTON, THOMAS GAIR, first BARON ASHTON OF HYDE (1855-1933), industrialist, philanthropist, and politician; educated at Rugby and University College, Oxford; carried on family cotton business in Manchester and Hyde where he was a generous supporter of technical and secondary education; liberal MP, Hyde division, 1885-6, Luton division, 1895-1911; noted for knowledge of finance; baron, 1911.

ASHTON, THOMAS SOUTHCLIFFE (1889-1968), economic historian; educated at Ashton-under-Lyne secondary school and Manchester University; MA, 1910; lecturer to trade unionists; lecturer in economics, Sheffield University, 1912; and Birmingham University, 1919; senior lecturer in economics, Manchester, 1921; reader in public finance and currency, 1927; publications include *Iron and Steel in the Industrial Revolution* (1924), *The Coal Industry of the Eighteenth Century* (with Joseph Sykes, 1929), and *An Eighteenth Century Industrialist: Peter Stubs of Warrington* (1939); professor of economic history, London School of Economics, 1944-54; published *The Industrial Revolution* (1948), *An Economic History of England: the Eighteenth Century* (1955), and *Economic Fluctuations in England 1700-1800* (1959); FBA, 1951; Ford's lecturer, Oxford, 1953; hon. vice-president, Royal Historical Society, 1961; honorary degrees from Nottingham, Manchester, and Stockholm.

ASHTON, WINIFRED (1888-1965), playwright and novelist under name of CLEMENCE DANE; educated in England, Germany, and Switzerland; studied at the Slade School; went on stage, 1913, but soon ceased to act, and took up writing; adopted pen name from St. Clement Danes in the Strand; produced three novels, including *Regiment of Women* (1917); her successful play *A Bill of Divorcement*, produced in West End, 1921; followed by *Will Shakespeare* in blank verse, 1921; wrote other novels and plays and film-scripts, including *Adam's Opera* (with Richard Addinsell, 1928), *Wild Decembers* (1932), *The Happy Hypocrite* (1936), and *Eighty in the Shade* (1959); best of her novels, *Broome Stages* (1931); CBE, 1953.

ASHWELL, LENA MARGARET (1872-1957), actress; daughter of C. Ashwell B. Pocock; educated in Toronto; studied singing at Lausanne Conservatoire and Royal Academy of Music; acted for (Sir) George Alexander, Sir Henry Irving, (Sir) Charles Wyndham, and (Sir) H. Beerbohm Tree [qq.v.]; became leading actress, especially in portrayal of suffering; opened Kingsway Theatre, 1907; provided concerts and plays for troops at front, 1914-18; OBE, 1917; ran Lena Ashwell Players (1919-29), working in suburbs, and after 1924, at Century Theatre to provide good plays at low prices with opportunities for young actors; twice married.

ASKWITH, GEORGE RANKEN, BARON ASKWITH (1861-1942), arbitrator; educated at Marlborough and Brasenose College, Oxford; first class, history, 1884; called to bar (Inner and Middle Temples), 1886; KC, 1908; controller-general, commercial, labour, and statistical department, Board of Trade, 1909-11; chief industrial commissioner, 1911-19; notable conciliator; KCB, 1911; baron, 1919.

ASLIN, CHARLES HERBERT (1893-1959), architect; educated at Sheffield central school and university; associate (1920), fellow (1932), bronze medallist (1951), and president (1954-6) of Royal Institute of British Architects; borough architect, Rotherham, 1922; deputy county architect, Hampshire, 1926; borough architect, Derby, 1929; redeveloped town's central areas; county architect, Hertfordshire, 1945-58; developed prefabricated construction of schools using space, light, and colour to create new kind of environment; CBE, 1951; hundredth school opened, 1955.

ASQUITH, ANTHONY (1902-1968), film director; youngest son of Herbert Henry Asquith (later Earl of Oxford and Asquith) and his second wife Margaret (Margot) [q.v.]; educated at Summer Fields, Oxford, Winchester, and Balliol College, Oxford; joined British International Film Company, 1926; president, the Association of Cinematographic Technicians, 1937-68; directed films, including

A Cottage on Dartmoor (1929), *Tell England* (1930–1), *Pygmalion* (1938), *French without Tears* (1939), *Quiet Wedding* (1940), *Fanny by Gaslight* (1944), *The Winslow Boy* (1948), *The Browning Version* (1950), *The Importance of Being Ernest* (1951), *The Doctor's Dilemma* (1958), and *The Yellow Rolls-Royce* (1964).

ASQUITH, LADY CYNTHIA MARY EVELYN (1887–1960), writer; daughter of eleventh Earl of Wemyss; married Herbert Asquith, 1910; private secretary (1918–37) to Sir J. M. Barrie [q.v.] most of whose fortune she inherited; cultivated friendships with writers and artists; works include three volumes of reminiscences, anthologies of ghost and children's stories, novels, and biographies; her *Diaries (1915–18)* published 1968.

ASQUITH, CYRIL BARON ASQUITH OF BISHOPSTONE (1890–1954), judge, fourth son of H. H. Asquith (Earl of Oxford and Asquith) [q.v.]; scholar of Winchester and Balliol College, Oxford; first class, *lit. hum.*, 1913; Hertford, Craven, Ireland, and Eldon scholar; fellow of Magdalen, 1913; called to bar (Inner Temple), 1920; appointed to Law Revision Committee, 1934; KC, 1936; recorder of Salisbury, 1937; judge of King's Bench division (knighted), 1938; lord justice of appeal (PC), 1946; lord of appeal in ordinary and life peer, 1951; refused lord chancellorship, 1951; chairman, commission on higher education in colonies, 1943–4, and royal commission on equal pay, 1944–6; with J. A. Spender [q.v.] wrote life of Asquith (2 vols., 1932).

ASQUITH, EMMA ALICE MARGARET (generally known as MARGOT), COUNTESS OF OXFORD AND ASQUITH (1864–1945), sixth daughter of Sir Charles Tennant [q.v.]; married Herbert Henry Asquith [q.v.] as his second wife, 1894; 'unteachable and splendid', her magnetic personality made her a legend in her lifetime but politically always a potential liability; greatly influenced taste and fashion; publications include *Autobiography* (2 vols., 1920–2), *More Memories* (1933), and *Off the Record* (1943).

ASQUITH, HERBERT HENRY, first EARL OF OXFORD AND ASQUITH (1852–1928), statesman; educated at City of London School; BA, Balliol College, Oxford; first classes in classical moderations and *lit. hum.*; president of Union, 1874; called to bar (Lincoln's Inn), 1876; devoted leisure to politics as ardent Gladstonian liberal; MP, East Fife, 1886–1918; as junior counsel for C. S. Parnell [q.v.] before Parnell commission, his reputation made at bar and enhanced in House of Commons, 1888; QC, 1890; member of cabinet as home secretary,

1892–5; returned to bar in years of opposition, 1895–1905; joined 'Liberal Imperial' group during South African war, 1899–1902; chancellor of the Exchequer, 1905–8; an orthodox, thrifty, progressive financier; made first provision for old age pensions, 1908; prime minister, 1908–16; succeeded at moment of ebb in liberal tide; appointed Lloyd George chancellor of the Exchequer; budget of 1909, designed to pay for increased navy and social programme, rejected by House of Lords; passed by both Houses after general election, 1910; turned his attention to question of House of Lords veto; held second election on House of Lords question, after obtaining from King George V 'hypothetical understanding' that if government obtained 'sufficient majority' he would create peers to overcome any resistance which Lords might offer; parliament bill passed, after great opposition, by majority of 17, 1911; repudiated charge of 'coercing the king'; introduced home rule bill for Ireland, 1912; this produced great agitation in Ulster under leadership of Sir Edward Carson [q.v.], one of most serious incidents of which was intimation of group of officers at Curragh camp, Ireland, that they would rather be dismissed service than help to coerce Ulster, March 1914; assumed secretaryship of war in order to reassure public; bill passed into law, Sept. 1914, but suspended for duration of 1914–18 war; Welsh disestablishment bill similarly dealt with; opposed enfranchisement of women until 1918, and became special target of militant suffragettes; in foreign policy worked in complete accord with foreign secretary, Sir Edward Grey [q.v.]; while seeking to conciliate Germany, considered two essentials to be loyalty to French *entente* and security against challenge of increasing German fleet; concentrated on navy, while maintaining expeditionary force and territorial army; in face of crisis of July–Aug. 1914 convinced that honour and policy alike required British intervention; accomplished feat of bringing cabinet and country to almost unanimous conclusion that this was so; his government handled first stages of 1914–18 war skilfully and successfully, but both it and public unprepared for prolonged and devastating struggle; failure of Dardanelles expedition (Feb. 1915) and agitation over munitions brought purely liberal government to end, May 1915; formed coalition cabinet, in which principal unionist leaders and one labour member included; new government torn by incessant struggles, especially that between Lord Kitchener and Lloyd George over latter's desire (also opposed by British and especially French commanding officers) to transfer chief part of army from Western to Eastern front; evacuation of Dardanelles, necessitated by Salonika expedition (Oct. 1915), combined with ill success of allied arms in France, did much to sap credit of coalition government; met troubles with fortitude and quietly prepared for Somme campaign and institution of compulsory service; visited Ireland after Easter rebellion, 1916; vainly urged adoption of home rule; violently and unscrupulously attacked by newspapers towards end of 1916; series of manœuvres in favour of his displacement by Lloyd George produced his resignation and that of his principal liberal colleagues, Dec. 1916; as leader of opposition, gave general support to government, but dissented from prime minister over issue involved by (Sir) F. Maurice's letter to newspapers, 1918; as result, with other non-coalition liberals, not returned to parliament at election of 1918; returned at Paisley by-election, 1920; devoted himself to Irish question, advocating dominion home rule; gave liberal support to labour to enable it to take office after election of Dec. 1923; defeated at Paisley, 1924; created Earl of Oxford and Asquith, 1925; KG, 1925; resigned liberal leadership, 1926.

ASQUITH OF YARNBURY, BARONESS (1887-1969), political figure. [See BONHAM CARTER, (HELEN) VIOLET.]

ASTBURY, SIR JOHN MEIR (1860-1939), judge; educated at Manchester grammar school and Trinity College, Oxford; BCL, 1883; called to bar (Middle Temple), 1884; bencher, 1903; practised on Chancery side in Manchester; QC and migrated to London, 1895; liberal MP, Southport division, 1906-10; Chancery judge, 1913-29; PC, 1929; his judgment that the general strike (1926) was illegal was influential in causing its collapse.

ASTBURY, WILLIAM THOMAS (1898-1961), physicist, crystallographer, and molecular biologist; educated at Longton high school and Jesus College, Cambridge; first class, parts i and ii, natural sciences tripos, 1920-1; worked at University College, London, 1921, and Royal Institution, 1923; lecturer, Leeds University, set up textile physics laboratory, 1928; reader in textile physics, 1937; professor, head of Department of Biomolecular Structure, 1945; demonstrated that a protein could exhibit different properties and activities through differences in folding pattern; and carried out research on structure of nucleic acids; FRS, 1940; member of Council, 1946-7; hon. founder member, British Biophysical Society, 1961; Sc.D Cambridge; doctor *honoris causa*, Strasburg University, 1946.

ASTON, FRANCIS WILLIAM (1877-1945), experimental physicist; educated at Malvern College and Mason College, Birmingham; undertook research with P. F. Frankland [q.v.] and later J. H. Poynting [q.v.]; research assistant to Sir J. J. Thomson [q.v.] (1910-13) working on mass-analysis of positive rays by the parabola method; Clerk Maxwell student, Cambridge, 1913; technical assistant, Royal Aircraft Establishment, 1914-18; returning to Cambridge built his first mass-spectrograph and achieved velocity focusing of positive rays; established the phenomenon of isotopy, the whole number rule, and the rule of odd and even; research fellow, Trinity College, Cambridge, 1920; FRS, 1921; Nobel prize for chemistry, 1922; next turned to analysis of the metallic elements and clearly established distinction between even- and odd-numbered elements; built a second mass-spectrograph of greater precision with which, between 1927 and 1935, a complete resurvey of the elements was made.

ASTON, SIR GEORGE GREY (1861-1938), major-general; born in Cape Colony; educated at Westminster and Royal Naval College, Greenwich; joined Royal Marine Artillery, 1879; completed course (1891) and served as instructor (1904-7), Staff College, Camberley; brigadier-general on staff of Lord Methuen [q.v.] in South Africa, 1908-13; colonel-commandant, Royal Marine Artillery, 1914-17; major-general, 1917; KCB, 1913.

ASTON, WILLIAM GEORGE (1841-1911), Japanese scholar; BA, Queen's University, Ireland, 1862; MA, 1863; hon. D.Litt., 1890; appointed student interpreter in British consular service at Yedo, Japan, 1864; consul at Hiogo, 1880-83; British consul-general in Korea, 1884-6; Japanese secretary to British legation at Tokyo, 1886-9; CMG, 1889; published Japanese grammars, 1869-72; translated *Ancient Chronicles of Japan*, 1896; wrote works on *Japanese Literature*, 1899, and *Shinto*, 1905; formed unique collection of native Japanese books.

ASTOR, NANCY WITCHER, VISCOUNTESS ASTOR (1879-1964), politician and hostess; born in United States; married Robert Gould Shaw, 1897; (divorced, 1903), came to England, 1904; married Waldorf (later second Viscount Astor, q.v.), 1906; converted to Christian Science, 1914; conservative MP Sutton division, Plymouth, 1919-45; first woman to sit in Commons; hostess to politicians and other influential people at Cliveden; CB 1937; hon. freeman, Plymouth, 1959.

ASTOR, WALDORF, second VISCOUNT ASTOR (1879-1952), public servant; born in New York; educated at Eton and New College,

Oxford; married Mrs Nancy Witcher Shaw, 1906 [q.v.]; settled at Cliveden; became Christian Scientist; unionist MP (1910-19) for and benefactor of Plymouth; became virtual proprietor of *Observer*, 1911; succeeded father, 1919; parliamentary secretary, Ministry of Food (1918) of Health (1919-21); supporter of Round Table group, Chatham House (chairman of council, 1935-49), League of Nations, etc.; during thirties 'Cliveden set' became unjustly associated with appeasement; collaborated in several works on agriculture; successful breeder of racehorses; established *Observer* trust, 1945.

ATCHERLEY, Sir RICHARD LLEWEL-LYN ROGER (1904-1970), air marshal; educated at Oundle School; cadet, Royal Air Force College; instructor, Central Flying School, 1926; member, RAF High Speed Flight for Schneider Trophy contest, 1929; exhibition flying in United States and test pilot; commanded night fighter squadron, 1939-40; group captain; commanded night fighter training airfield; sector commander, 1942; commanded group in North Africa, 1943; commanded RAF Cadet College, 1946-8; commander-in-chief Pakistan air force, 1949; commanded No. 12 Fighter Group, 1951; commander-in-chief, Flying Training Command, 1955-8; AFC, (1940), bar (1942), air vice-marshal, 1951; air-marshal, 1956; OBE, 1941; CBE, 1945; KBE, 1956; CB, 1950.

ATHLONE, EARL OF (1874-1957). [See CAM-BRIDGE, ALEXANDER AUGUSTUS FREDERICK WILLIAM ALFRED GEORGE.]

ATHOLL, DUCHESS OF (1874-1960), public servant. [See STEWART-MURRAY, KATHARINE MARJORY.]

ATHOLSTAN, BARON (1848-1938), newspaper proprietor. [See GRAHAM, HUGH.]

ATKIN, JAMES RICHARD, BARON ATKIN (1867-1944), judge; born in Brisbane; educated at Christ College, Brecon, and Magdalen College, Oxford; called to bar (Gray's Inn), 1891; KC, 1906; built up a wide practice in common law and commercial actions; judge of King's Bench division (knighted), 1913; lord justice of appeal (PC), 1919; lord of appeal in ordinary (life peerage), 1928-44; made important contributions in many fields of English law; chairman, lord chancellor's committee on crime and insanity, 1922-3; of the Council of Legal Education, 1919-34; FBA, 1938.

ATKINS, Sir IVOR ALGERNON (1869-1953), organist and choirmaster; B.Mus., Oxford, 1892; organist, Worcester Cathedral,

1897-1950; librarian, 1933-53; enlarged repertory of Three Choirs Festival, revived it, 1920, and conducted for last time, 1948; introduced works of Sir Edward Elgar and Sir Walford Davies [qq.v.]; established Bach's *St. Matthew Passion* as regular feature; edited the *St. John Passion* (1929), Brahms's *Requiem*, and the *Worcester Psalter* (1948); knighted, 1921.

ATKINSON, SIR EDWARD HALE TINDAL (1878-1957), lawyer; educated at Harrow and Trinity College, Oxford; called to bar (Middle Temple), 1902, bencher 1929, treasurer, 1948; recorder, Southend-on-Sea, 1929; director of public prosecutions, 1930-44; advised on wartime defence regulations and was involved in many important cases including spy cases; chairman, Central Price Regulation Committee, 1945-53; KCB, 1932.

ATKINSON, JOHN, BARON ATKINSON (1844-1932), judge; born at Drogheda; graduated, Queen's College, Galway, 1861; called to bar (Ireland), 1865, (Inner Temple), 1890; QC, 1880; solicitor-general for Ireland, 1889-92; attorney-general for Ireland, 1892 and 1895-1905; conservative MP, North Londonderry, 1895-1905; lord of appeal, 1905-28; PC and life peer, 1905.

ATKINSON, ROBERT (1839-1908), philologist; studied romance languages on continent, 1857-8; BA, Trinity College, Dublin, 1863; MA, 1866; LLD, 1869; hon. D.Litt., 1891; professor of romance languages from 1869, and of Sanskrit from 1871; skilled in romance, Sanskrit, Tamil, Telugu, Hebrew, Persian, Arabic, Chinese, Celtic, and Coptic languages; edited *Vie de Seint Auban*, Norman-French poem (1876); member (1875), secretary (1878-1901), president (1901-8) of Royal Irish Academy; Todd professor of Celtic languages there, 1884; did much pioneer work in Celtic grammar and language; edited with translation and glossary middle Irish work, *The Passions and Homilies from the Leabhar Breac*, 1897; co-editor of *The Irish Liber Hymnorum* (2 vols., 1898).

ATTHILL, LOMBE (1827-1910), obstetrician and gynaecologist; BA and MB, Trinity College, Dublin, 1849; MD, 1865; assistant physician to Rotunda Hospital, Dublin, 1851; first doctor in Ireland to perform ovariotomy successfully; master of Rotunda Hospital, 1875; introduced Listerian principles there; president of Dublin Obstetrical Society (1874-6) and of Royal Academy of Medicine in Ireland (1900-3; author of *Clinical Lectures on Diseases Peculiar to Women* (1871) and *Recollections of an Irish Doctor* (posthumous, 1911).

ATTLEE, CLEMENT RICHARD, first EARL ATTLEE (1883-1967), statesman; educated at Haileybury and University College, Oxford 1901-4; called to bar (Inner Temple), 1905; manager, Haileybury House, boys' club, Stepney, 1907; joined independent labour party, 1908; lecturer in social administration, London School of Economics, 1913; served in army at Gallipoli, and in Mesopotamia and France, 1914-18; mayor, Stepney, 1919; labour MP Limehouse, 1922-50; parliamentary private secretary to Ramsay MacDonald, 1922; under-secretary of state for war, 1923; served on Simon Commission, 1927-9; chancellor, Duchy of Lancaster, 1930; postmaster-general, 1931; deputy leader, Parliamentary labour party, 1931-5; leader, 1935; published *The Will and the Way to Socialism*, 1935; and *The Labour Party in Perspective*, 1937; lord privy seal, 1940-2; deputy prime minister, 1942; lord president of the Council, 1943; attended foundation conference of UN at San Francisco, 1945; prime minister, 1945-51; carried legislation for nationalization of key industries and establishment of welfare state; granted independence to India and Pakistan; decided Britain should manufacture atomic bomb; MP, West Walthamstow, 1950; labour government defeated in 1951 election; retired as leader of the labour party, 1955; PC 1935; CH, 1945; OM, 1951; earl, 1955; KG, 1956; hon. bencher, 1946; FRS, 1947; hon. fellow, University College, Oxford (1942), Queen Mary College, London (1948), and LSE (1958).

ATTWELL, MABEL LUCIE (1879-1964), illustrator; educated at the Coopers' Company School, London; attended classes at Regent Street and Heatherley's art schools; contributed drawings to *The Tatler* and *Bystander*; illustrated works in the 'Raphael House Library of Gift Books' including *Mother Goose* (*c*.1909), and *Alice in Wonderland* (1910); followed by *Hans Andersen's Fairy Tales* (1913), *The Water Babies* (1915) and *Grimms' Fairy Tales* (1925); illustrated stories written by Queen Mary of Romania, *Peeping Pansy* (1919) and *The Lost Princess* (1924); illustrated Barrie's *Peter Pan and Wendy* (1921); designed picture postcards for Messrs. Valentine & Sons; Attwell pictures of children figured on all kinds of nursery equipment; *Lucie Attwell Annual* published, 1922-74, ten years after her death.

AUBREY, MELBOURN EVANS (1885-1957), Baptist minister; educated at Cardiff Baptist College and Mansfield College, Oxford; ordained, 1911; pastor, St. Andrew's Street church, Cambridge, 1913-25; secretary, Baptist Union of Great Britain and Ireland, 1925-50, president, 1950-1; moderator, Free Church

Federal Council, 1936; CH, 1937; chairman, Churches' Committee for Christian Reconstruction in Europe; vice-president, British Council of Churches, 1948-50; member, central committee, World Council of Churches, 1948-54.

AUMONIER, JAMES (1832-1911), landscape painter; at first a designer of calicoes; practised landscape painting in Kensington Gardens and Epping Forest; first exhibited at the Royal Academy, 1871; subjects mainly English countryside; visited Venice, 1891; original member of Institute of Oil Painters; works include 'Sheep-Washing' (Tate Gallery) and 'Sunday Evening'.

AUSTEN, HENRY HAVERSHAM GODWIN- (1834-1923), explorer and geologist. [See GODWIN-AUSTEN.]

AUSTEN, SIR WILLIAM CHANDLER ROBERTS- (1843-1902), metallurgist. [See ROBERTS-AUSTEN.]

AUSTEN LEIGH, AUGUSTUS (1840-1905), provost of King's College, Cambridge; educated at Eton and King's College, Cambridge; BA (fourth classic), 1863 MA, 1866; fellow, 1862; ordained, 1865; tutor, 1868-81; dean, 1871-3, 1882-5; vice-provost, 1877-89; provost, 1889-1905; vice-chancellor of the university, 1893-5; successful administrator of the college; published *History of King's College*, 1899.

AUSTIN, ALFRED (1835-1913), poet laureate; educated at Stonyhurst and Oscott; called to bar (Inner Temple), 1857; leader-writer (chiefly on foreign affairs) to *Standard*, 1866-96; joint-editor with W. J. Courthope [q.v.] of *National Review*, 1883-7; sole editor, 1887-95; poet laureate, 1896; published twenty volumes of verse—satires, lyrics, narrative, and dramatic poems, 1871-1908; prose works include three novels, political, critical, personal, and miscellaneous writings; at his best in prose garden-diaries and poetry of the countryside; failed in attempts to treat philosophic themes in epic and dramatic form.

AUSTIN, HERBERT, BARON AUSTIN (1866-1941), motor manufacturer; educated at Rotherham grammar school and Brampton College; went to Australia, 1884; apprenticed at Langlands foundry, Melbourne; joined Wolseley Sheep Shearing Machine Co.; returned to Birmingham, 1893; produced first Wolseley three-wheel car, 1895; founded Austin Motor Co., Ltd., 1905; skilled engineer of foresight and determination; popularized motoring with the Austin Seven (1922) and Twelve (1921); conservative MP, King's Norton, 1918-24; chairman of 'shadow' aero engine committee, 1937-40; generous benefactor, especially to hospitals; KBE, 1917; baron, 1936.

AUSTIN, JOHN LANGSHAW (1911-1960), philosopher; educated at Shrewsbury and Balliol College, Oxford, first class, *lit. hum.*, and fellow of All Souls, 1933; fellow and tutor, Magdalen, 1935; White's professor of moral philosophy and fellow of Corpus Christi, 1952-60; believed in co-operative discussion, a powerful influence on development of philosophy; his *Philosophical Papers* published, 1961, and *How to do Things with Words* and his lectures on perception (*Sense and Sensibilia*), 1962; FBA 1958.

AVEBURY, first BARON (1834-1913), banker, man of science, and author. [See LUBBOCK, SIR JOHN.]

AVORY, SIR HORACE EDMUND (1851-1935), judge; educated at King's College, London, and Corpus Christi College, Cambridge; third class, law, 1873; called to bar (Inner Temple), 1875; bencher, 1908; joined South-Eastern circuit; an official prosecuting counsel at Central Criminal Court, 1889; senior counsel, 1899; KC, 1901; acquired a leading practice in 'Crown paper' matters; commissioner of assize for South-Eastern (1909) and Northern (1910) circuits; judge (1910), senior judge (1923-35) of King's Bench division; a silent judge with a cold manner; his summings up were usually models of lucidity and accuracy; knighted, 1910; PC, 1932.

AYERST, WILLIAM (1830-1904), divine; born at Danzig; scholar of Caius College, Cambridge; BA, 1853; MA, 1856; Hulsean (1855) and Norrisian (1858) prizeman; went to India, 1859; chaplain to Khyber field force, 1879-81; founded Ayerst Hall, Cambridge, 1884; published theological works.

AYLWARD, GLADYS MAY (1902-1970), missionary; daughter of a postman; educated at elementary school; worked as shopgirl, nanny, and parlourmaid; accepted for training as missionary, 1929; rejected as unsuitable; heard that a small mission in North China needed helper who must pay passage out; worked as parlourmaid to save necessary money; joined Mrs Lawson at the mission in Yangsheng, Shansi province, 1932; became Chinese citizen, 1936; during Japanese war led a hundred children to safety across the Yellow River, 1940; worked as Bible-woman among lepers; returned to England, 1949; Alan Burgess wrote her biography, *The Small Woman* (1957), later made into film, 'The Inn of the Sixth Happiness' (1959); settled in Taiwan, 1958-70; died in Taipei.

AYRTON, WILLIAM EDWARD (1847-1908), electrical engineer and physicist; educated at University College, London; BA, 1867; studied electricity at Glasgow under Lord Kelvin [q.v.]; in Indian telegraph service, 1863-73; professor of physics and telegraphy at Tokyo, 1873-8; did much experimental work there with Prof. John Perry; professor at City and Guilds of London Institute, 1879, and at Central Technical College, South Kensington, 1884-1908; with Perry invented surface contact system for electric railways (1881) and a series of portable electrical measuring instruments (1882-91); made researches with Thomas Mather, FRS, from 1891 onwards; pioneer of electricity as a motive power, 1879; widely consulted as electrical engineer; FRS, 1881; royal medallist, 1901; president of Institution of Electrical Engineers, 1892, and of Physical Society, 1890-2; adviser to Admiralty and Board of Trade; pioneer and organizer of technical education; supporter of women's rights; author of *Practical Electricity* (1887) and 151 scientific papers.

AZAD, MAULANA ABUL KALAM (1888-1958), Indian minister of education; born in Mecca; received orthodox Moslem education in India; joined revolutionary movement; started *Al Hilal*, Urdu weekly, 1912, *Al Balagh*, 1915; interned 1916-20; thereafter leading figure in non-co-operation movement and frequently in prison; president of Congress, 1923 and 1940-6; chief spokesman of Congress during constitutional discussions; as education minister, 1947-58, chief architect of educational policy of new India; founder-president Indian Council for Cultural Relations; prepared translation, with commentary, of Koran.

AZARIAH, SAMUEL VEDANAYAKAM (1874-1945), first Indian bishop of an Anglican diocese; of Tamil peasant stock; worked for YMCA from 1896; secretary, National Missionary Society of India, 1906-9; ordained, 1909; head of Dornakal mission, 1909-12; bishop of Dornakal, 1912-45; strong advocate of an indigenous ministry and Church union.

BABINGTON SMITH, Sir HENRY (1863-1923), civil servant. [See Smith.]

BACHARACH, ALFRED LOUIS (1891-1966), food scientist, publicist, and musician; educated at St. Paul's School and Clare College, Cambridge; worked in Wellcome Chemical Research Laboratories, 1915-18; Wellcome Chemical Works, 1919; worked with Joseph Nathan & Co. Ltd. (later, Glaxo Laboratories), 1920-50; carried out research on artificial baby foods; head, Publicity Services Group, 1950-5; publications include *Science and Nutrition* (1938), and (as editor) *The Nation's Food* (1946); accomplished pianist, edited and published books, including *The Musical Companion* (1934), *Lives of the Great Composers* (1935), *British Music of Our Time* (1946), and *The Music Masters* (1948-54); fellow and vice-president, Royal Institute of Chemistry; president, Nutrition Society.

BACKHOUSE, Sir EDMUND TRELAWNY, second baronet (1873-1944), historian, and authority on China; educated at Winchester and Merton College, Oxford; student interpreter, British legation, Peking, 1897; professor, Peking University, 1903-13; succeeded father, 1918; collaborated in works with J. O. P. Bland and Sir Sidney Barton [qq.v.]; lived until his death as a recluse in Peking; his lifework, including an English-Chinese dictionary and historical writings, destroyed by Japanese, 1937.

BACKHOUSE, Sir ROGER ROLAND CHARLES (1878-1939), admiral of the fleet; entered navy, 1892; specialized in gunnery; commander, 1909; captain, 1914; served with (Lord) Jellicoe [q.v.], 1914-15; commanded light cruiser in Harwich force, 1915-16; flag captain to commander of battle cruiser force, 1916-18; director of naval ordnance, 1920-3; commanded battleship *Malaya* (1923-4) and third battle squadron (1926-7), Atlantic fleet; rear-admiral, 1925; as third sea lord and controller of the navy (1928-32) sought to prevent economy from going too far; second-in-command, Mediterranean fleet, 1932-4; admiral, 1934; commander-in-chief, home fleet, 1935-8; first sea lord, 1938; retired through ill health (June), promoted admiral of the fleet (July), 1939; KCB, 1933; GCB, 1938; GCVO, 1937; brother of Sir E. T. Backhouse [q.v.].

BACON, JOHN MACKENZIE (1846-1904), scientific lecturer and aeronaut; BA, Trinity College, Cambridge, 1870; ordained, 1870;

fellow of the Royal Astronomical Society, 1888; witnessed solar eclipses at Vadsö (9 Aug. 1896) and at Buxar, India (Jan. 1898); made his first balloon ascent, Aug. 1888; successfully experimented from balloon in acoustics and wireless telegraphy; crossed Irish Channel in balloon, Nov. 1902; popular lecturer and scientific writer; author of *By Land and Sky* (1900) and *The Dominion of the Air* (1902).

BACON, Sir REGINALD HUGH SPENCER (1863-1947), admiral; entered navy, 1877; specialized in torpedo; DSO, Benin expedition, 1897; captain, 1900; superintended construction, trial, and crew training of submarines, 1900-4; naval assistant to (Lord) Fisher [q.v.] (1904-5) mainly concerned with designs committee for *Dreadnought* and battle cruisers; first captain of *Dreadnought*, 1906; director of naval ordnance, 1907-9; rear-admiral, 1909; managing director, Coventry Ordnance Works, 1909-15; designed and took to France 15-inch howitzer guns, 1915; commanded Dover Patrol, 1915-17; vice-admiral, 1915; protected shipping and supported French sea flank; many actions included defeat of German destroyers by the *Swift* and *Broke*, 1917; dismissed with (Lord) Jellicoe [q.v.], Dec. 1917; served in Ministry of Munitions, 1918-19; admiral, 1918; publications include autobiography, books on Dover Patrol, *The Jutland Scandal* (1924), and biographies of Fisher (1929) and Jellicoe (1936); KCB and KCVO, 1916.

BADCOCK, Sir ALEXANDER ROBERT (1844-1907), general, Indian staff corps; educated at Harrow and Addiscombe; in commissariat department of Indian army, 1864-95; superintended supplies for Kabul-Kandahar field force, 1880; CB, 1880; collected transport for Sudan, 1885; commissary general-in-chief, 1890; CSI for services in Chitral, 1895; quartermaster-general, India, 1895; lieutenant-general, 1900; member of council of India, 1902-7; KCB, 1902.

BADDELEY, MOUNTFORD JOHN BYRDE (1843-1906), compiler of 'Thorough Guide' books for pedestrians (1884-1906); educated at King Edward's grammar school, Birmingham, and Clare College, Cambridge; BA, 1868; assistant master at Sheffield grammar school, 1880-4; settled in lake district, which he popularized as pleasure resort.

BADELEY, HENRY JOHN FANSHAWE, Baron Badeley (1874-1951), clerk of the

Parliaments and engraver; educated at Radley and Trinity College, Oxford; entered Parliament Office, 1897; principal clerk, Judicial Office, 1919, and clerk assistant of the Parliaments, 1930; clerk of the Parliaments, 1934-49; KCB, 1935; baron, 1949; fellow of and regular exhibitor with Royal Society of Painter-Etchers and Engravers.

BADEN-POWELL, ROBERT STEPHEN-SON SMYTH, first BARON BADEN-POWELL (1857-1941), lieutenant-general, and founder of the Boy Scouts and Girl Guides; son of Baden Powell and brother of Sir G. S. Baden-Powell [qq.v.]; educated at Charterhouse; gazetted to 13th Hussars in India, 1876; specialized in reconnaissance and scouting; took part in Zululand (1888), Ashanti (1895-6), and Matabeleland (1896) campaigns; intelligence officer for Mediterranean, 1891-3; commanded 5th Dragoon Guards in India, 1897-9; with emphasis on observation, deduction, and initiative, described his course of instruction in *Aids to Scouting* (1899); sent to South Africa to raise two regiments, 1899; held Mafeking during siege of 217 days; relieved, 17 May 1900; major-general, 1900; raised South African Constabulary, 1900-3; inspector-general of cavalry, 1903-7; established Cavalry School at Netheravon (1904), *Cavalry Journal* (1905); lieutenant-general, 1907; commanded Northumbrian division of Territorials, 1908-10; retired (1910) in order to devote himself to the Boy Scout movement; held trial camp, 1907; his *Scouting for Boys* (1908) led to countrywide formation of troops; later established Sea Scouts, Girl Guides, Wolf Cubs, and Rover Scouts; acclaimed chief scout of world at first jamboree, 1920; latterly devoted himself to international aspects of movement; wrote many books illustrated by himself; KCVO and KCB, 1909; baronet, 1922; baron, 1929; OM, 1937; died in Kenya.

BAILEY, SIR ABE, first baronet (1864-1940), South African financier and statesman; born in Cape Colony; educated in England; built up the Abe Bailey and London and Rhodesian mining group; member Cape (1902-5), Transvaal (1907-10), and Union (1915-24) assemblies; an active worker for South African Union; his houses in Cape Town and London were centres for exchange of views; famous breeder and owner of race-horses; KCMG, 1911; baronet, 1919.

BAILEY, CYRIL (1871-1957), classical scholar; educated at St. Paul's School and Balliol College, Oxford; Hertford and Craven scholar; first class, *lit. hum.*, 1894; classical fellow and tutor, Exeter College, 1894; fellow of Balliol, 1902; a superb classical teacher and notable Oxford personality; public orator, 1932-9; delegate of University Press, on council of Lady Margaret Hall (chairman, 1921-39), and president of Bach Choir; works include *Epicurus* (1926), text of Lucretius, with commentary and translation (1947), and works on Roman religion; FBA, 1933; CBE, 1939.

BAILEY, SIR EDWARD BATTERSBY (1881-1965), geologist; educated at Kendal grammar school and Clare College, Cambridge; first class, parts i and ii, natural sciences tripos, 1901-2; field geologist, Scottish branch, Geological Survey of Great Britain, 1902-29; professor, geology, Glasgow University, 1929-37; director, Geological Survey of Great Britain and Museum of Practical Geology, 1937-45; served in France with Royal Garrison Artillery, 1914-18; MC, croix de guerre with palms, chevalier de la Légion d'Honneur; publications include *Introduction to Geology* (with J. Weir and W. J. McCallien, 1939), *Tectonic Essays, mainly Alpine* (1935), *Geological Survey of Great Britain* (1952) and biographies of Sir Charles Lyell (1962) and James Hutton (1967) [qq.v.]; FRS 1930; hon. doctorates from Belfast, Birmingham, Cambridge, Edinburgh, Glasgow, and Harvard; knighted, 1945.

BAILEY, FREDERICK MARSHMAN (1882-1967), explorer and naturalist; educated at Edinburgh Academy, Wellington, and Royal Military College, Sandhurst; posted to Middlesex Regiment in India, 1900; member of mission to Lhasa led by (Sir) Francis Younghusband [q.v.], 1903-4; explored trade routes in Tibet, 1904; studied natural history of Tibet, 1905; expedition exploring China-Tibet border, 1911; explored, with Captain Henry Morshead, the Tsangpo gorges, 1913; discovered the blue poppy (meconopsis baileyi); wounded in France and Gallipoli 1915; served in India as political officer, 1916-38; publications include *China-Tibet-Assam* (1945), *Mission to Tashkent* (1946), and *No Passport to Tibet* (1957).

BAILEY, SIR GEORGE EDWIN (1879-1965), mechanical engineer and industrialist; educated at Loughborough grammar school and University College, Nottingham; joined Brush Electrical Engineering Co. as apprentice; joined British Westinghouse Co., as draughtsman, 1907; superintendent, 1913; works manager, Metropolitan-Vickers Electrical Co., 1919; director and general manager, manufacture, 1927; co-ordinated production, Associated Electrical Industries Ltd., 1929; chairman, Metropolitan Vickers Electrical Co.,

1944; chairman, Associated Electrical Industries, 1951-4; director, 1954-7; involved in production of radar and other equipment for navy and airforce, 1938-45; member, Engineering and Industrial Panel, Ministry of Labour and National Service; CBE, 1941; knighted, 1944.

BAILEY, JOHN CANN (1864-1931), critic and essayist; educated at Haileybury and New College, Oxford; works include *Dr. Johnson and his Circle* (1913), *Milton* (1915), *Shakespeare* (1929); contributed regularly to *The Times Literary Supplement*, the *Quarterly*, *Edinburgh*, *Fortnightly*, and other reviews, investing traditional views with freshness and interest.

BAILEY, KENNETH (1909-1963), biochemist; educated at Orme boys' school, Newcastle under Lyme, and Birmingham University, first class honours; B.Sc., 1931; Ph.D., 1933; Beit fellow, Royal College of Science, Imperial College, London, 1933; studied structure of proteins; worked at Porton and in Low Temperature Research Station, Cambridge, 1939; joined biochemical laboratory, Cambridge; Cambridge Ph.D., 1944; assistant director, research, and fellow, Trinity College, 1948; Cambridge Sc.D.; FRS, 1953; edited (with Hans Neurath) *The Proteins* (1954); discovered tropomyosin (1946).

BAILEY, MARY, LADY BAILEY (1890-1960), airwoman; daughter of fifth Lord Rossmore; married Sir Abe Bailey [q.v.], 1911; obtained pilot's licence, 1927; flew solo Croydon-Cape Town and back, Mar. 1928-Jan. 1929; awarded Britannia Trophy and DBE, 1930.

BAILEY, PHILIP JAMES (1816-1902), author of *Festus*; educated at Nottingham and Glasgow University; called to bar (Lincoln's Inn), 1840; published *Festus*, 1839 (2nd edn., revised, 1845, eleventh or jubilee edn. 1889), the work admired by Tennyson, Browning, and the pre-Raphaelites: published *The Angel World, and other Poems* (1850), *The Mystic, and other Poems* (1859), and *The Universal Hymn* (1867); erroneously regarded as father of 'spasmodic' school of poetry which Aytoun satirized in *Firmilian*, 1854; awarded civil list pension of £100, 1856; settled in Jersey, 1864; returned to England, 1876; hon. LLD, Glasgow, 1901.

BAILHACHE, SIR CLEMENT MEACHER (1856-1924), judge; educated at City of London School and London University; LLB, 1877; called to bar (Middle Temple), 1889; joined South Wales circuit; specialized in commercial law; KC, 1908; judge and knighted, 1912; dealt with many problems in Commercial Court

during European war, 1914-18; less successful as criminal judge than in civil cases.

BAILLIE, CHARLES WALLACE ALEXANDER NAPIER ROSS COCHRANE-, second BARON LAMINGTON (1860-1940); educated at Eton and Christ Church, Oxford; assistant private secretary to Lord Salisbury, 1885; conservative MP, North St. Pancras, 1886-90; succeeded his father, 1890; governor of Queensland, 1895-1901, of Bombay, 1903-7; his main interest was the welfare of both the Empire and Eastern peoples with whom he urged a better understanding; GCMG, 1900; GCIE, 1903.

BAILLIE, SIR JAMES BLACK (1872-1940), vice-chancellor of Leeds University; educated at Edinburgh and Trinity College, Cambridge; professor of moral philosophy, Aberdeen, 1902-24; vice-chancellor of Leeds, 1924-38; served on a number of industrial arbitration committees; knighted, 1931.

BAIN, ALEXANDER (1818-1903), psychologist, logician, and writer on education; graduated at Marischal College, Aberdeen, 1840; assistant to professor of moral philosophy there, 1841; visited London and made the acquaintance of John Stuart Mill, Grote, Lewes, and Carlyle; wrote *The Study of Character, including an Estimate of Phrenology*, 1861; assistant secretary, metropolitan sanitary commission, 1848-50; lecturer at Bedford College for Women; published *The Senses and the Intellect* (1855) and *The Emotions and the Will* (1859); professor of logic and English in Aberdeen University, 1860-80; published *Mental and Moral Science* (1868), *Logic* (1870), and *Mind and Body* (1872); LLD, Edinburgh, 1869; assisted in editing *Grote's Aristotle*, 1872; edited Grote's minor works, 1873; founded the periodical *Mind*, 1876; contributed *Education as a Science* to 'International Scientific' series, 1879; elected lord rector of Aberdeen University, 1890; received civil list pension of £100, 1895; one of first to apply to psychology the results of physiological investigations; lucid exponent of a *posteriori* school of psychology; his system of philosophy materialistic; a utilitarian in ethics; *Autobiography* published, 1904.

BAIN, FRANCIS WILLIAM (1863-1940), scholar and writer; educated at Westminster and Christ Church, Oxford; first class, *lit. hum.* (1886), and association football blue; fellow of All Souls, 1889-96; professor of history and political economy, Deccan College, Poona, 1893-1919; wrote successful Hindu love-stories supposedly translated from Sanskrit, including *A Digit of the Moon* (1899) and *The Substance of a Dream* (1919); CIE, 1918.

BAIN, Sir FREDERICK WILLIAM (1889-1950), chemical industrialist; educated at Banff Academy; served in Ministry of Munitions, 1916-18; became director of United Alkali Company; entered Imperial Chemical Industries combine, 1926; vice-chairman (1931), later chairman, General Chemicals group; an executive manager, ICI (1939), director (1940), deputy chairman (1945-50); chairman, chemical control board, Ministry of Supply, 1941-4; knighted, 1945.

BAIN, ROBERT NISBET (1854-1909), historical writer and linguist; shorthand writer in solicitor's office; acquired unaided twenty foreign European tongues; assistant at British Museum from 1883 to death; wrote *Gustavus III and his Contemporaries, 1746-92* (1894), *The First Romanovs, 1613-1725* (1905), *Scandinavia, 1513-1900* (1905), *Slavonic Europe* (1908), *Charles XII* (1895, for 'Heroes of the Nations' series); translated fairy tales from Russian, Finnish, and Ruthenian, 1893-4.

BAINBRIDGE, FRANCIS ARTHUR (1874-1921), physiologist; BA and MD, Trinity College, Cambridge; professor of physiology, Durham University, 1911, and at St. Bartholomew's Hospital, London, 1915; FRS, 1919; author of *The Physiology of Muscular Exercise* (1919).

BAINES, FREDERICK EBENEZER (1832-1911), promoter of the post-office telegraph system; at fourteen constructed telegraphic apparatus; in service of Electric Telegraph Company, 1848-55; clerk in General Post Office, 1855; proposed cable to connect England with Australia; advocated (1856) government acquisition of existing telegraph systems, which was carried out in 1870; surveyor-general for telegraph business, 1875; proposed telegraphic communication around British sea coast (1878), which was carried out in 1892; inspector-general of mails, 1882-93; organized parcel post service, 1883; CB, 1885; enthusiastic Volunteer; edited *Records of Hampstead*, 1890; wrote *Forty Years at the Post-Office*, 2 vols., 1895.

BAIRD, ANDREW WILSON (1842-1908), colonel RE; educated at Addiscombe (1860) and Royal Military Academy, Woolwich (1861); special assistant engineer of Bombay harbour defence works, 1861-5; served in Abyssinian expedition under Lord Napier [q.v.], 1868; assistant superintendent of Indian trigonometrical survey, 1869; made pioneer study of tidal observations by harmonic analysis in England (1870) and on gulf of Cutch (1872); organized at Poona new department of survey along coast lines from Aden to Rangoon, 1877;

FRS, 1884; mint master at Calcutta, 1889-97; colonel, 1896; CSI, 1897; published *Notes on the Harmonic Analysis of Tidal Observations* (1872) and *Spirit-Levelling Operations of the Great Trigonometical Survey of India* (1885).

BAIRD, JOHN LOGIE (1888-1946), television pioneer; studied electrical engineering in Glasgow; began television research with make-shift equipment, 1922; first demonstrated true television, 26 Jan. 1926; made first transatlantic transmission and demonstration of colour and stereoscopic television, 1928; his system used by British Broadcasting Corporation, 1929-37; showed big-screen television, 1930; gave first ultra-short-wave transmission, 1932.

BAIRNSFATHER, CHARLES BRUCE (1888-1959), cartoonist; born in India; educated at United Services College; abandoned army and art successively for electrical engineering; with Royal Warwickshire Regiment in France, 1914; captain, 1915; contributed to *Bystander* front-line cartoons of Old Bill and Bert; his play *The Better 'Ole* produced, 1917; officer-cartoonist, 1916-18; official cartoonist US Army in Europe, 1942-4; less successful with peacetime humour.

BAIRSTOW, Sir EDWARD CUTHBERT (1874-1946), musician; studied under John Farmer and Sir Frederick Bridge [qq.v.]; B.Mus. (1894), D.Mus. (1900), Durham; organist of parish church, Wigan (1899), Leeds (1906), York Minster (1913-46); professor of music, Durham, 1929-46; conducted choral societies and judged at competition festivals; mainly influential through his pupils; knighted, 1932.

BAIRSTOW, Sir LEONARD (1880-1963), professor of aviation; educated at elementary and secondary schools, Halifax, and Royal College of Science, London; joined Engineering Department, National Physical Laboratory, 1904; principal in charge, aerodynamics section, 1909; appointed to Air Board, 1917; professor, aerodynamics, Imperial College, London, 1920; Zaharoff professor of aviation and head of Department of Aeronautics, 1923-45; chairman, Aeronautical Research Council, 1949-52; FRS, CBE, 1917; hon. fellow, Royal Aeronautical Society, 1945; considerable contributor to knowledge of stability of aircraft; served on committees investigating the R. 38 and R. 101 airship accidents; knighted, 1952.

BAJPAI, Sir GIRJA SHANKAR (1891-1954), Indian statesman; educated at Merton College, Oxford; entered Indian civil service, 1914; held secretariat and advisory posts;

member of council for education department, 1940; agent-general for India in United States, 1941–7; secretary-general for external affairs, 1947–52; governor of Bombay, 1952–4; influential champion of Commonwealth bond; KBE, 1935; KCSI, 1943.

BAKER, SIR BENJAMIN (1840–1907), civil engineer; joined staff of Sir John Fowler [q.v.]; partner, 1875; won high repute as consulting engineer; with Fowler engaged on underground communications of London; helped to construct metropolitan railway (1861) and district railway (1869); consulting engineers for first 'tube' railways, City and South London (1890), Central London (opened 1900), and Bakerloo railways; studied theory of resistance of materials, such as brickwork and metals; constructed Forth bridge on cantilever principles (1883–90); KCMG, 1890; advised on engineering projects in Egypt; designed dams at Aswan, 1898–1902, and at Assiut; KCB, 1902; designed Avonmouth dock and other works in England, Canada, and United States; president of Institution of Civil Engineers, 1895; fellow (1890) and vice-president (1896–1907) of Royal Society; hon. D.Sc., Cambridge, 1900, LLD, Edinburgh, 1890, M.Eng., Dublin, 1892; memorial window in Westminster Abbey.

BAKER, HENRY FREDERICK (1866–1956), mathematician; educated at the Perse School and St. John's College, Cambridge; bracketed senior wrangler, 1887; Smith's prize, 1889; fellow, 1888–1956; college lecturer, 1890; university lecturer, 1895–1914 (Cayley lecturer, 1903–14); Lowndean professor of astronomy and geometry, 1914–36; his early interest was in theory of algebraic functions but his real love became geometry; achieved no major breakthrough but was an inspiring teacher; works include *Abel's Theorem and the Allied Theory, including Theta Functions* (1897), *Multiply Periodic Functions* (1907), and *Principles of Geometry* (6 vols., 1922–33); FRS, 1898, Sylvester medallist, 1910; president (1910, 1911) and De Morgan medallist (1905), London Mathematical Society.

BAKER, SIR HERBERT (1862–1946), architect; educated at Tonbridge School; worked under (Sir) Ernest George [q.v.]; went to South Africa, 1892; restored and subsequently built new Groote Schuur for Cecil Rhodes [q.v.]; modelled his work on old Dutch houses; built chancel and part of nave of Anglican cathedral in Cape Town and the memorial to Rhodes above Groote Schuur; joined Milner's 'kindergarten' (1902) and brought new spirit in building to Transvaal; built many beautiful houses in Johannesburg and Government

House, the Union Buildings, and first portion of Anglican cathedral at Pretoria; also built nave of cathedral at Salisbury and Government House, Nairobi; collaborated with Sir Edwin Lutyens [q.v.] from 1912 as architect for New Delhi; other works include India House, South Africa House, and Church House, London, and Rhodes House, Oxford; RA, 1932; knighted, 1926; KCIE, 1930.

BAKER, HERBERT BRERETON (1862–1935), chemist; educated at Blackburn and Manchester grammar schools and Balliol College, Oxford; first class, natural science, 1883; chemistry master, Dulwich College, 1886–1902; headmaster, Alleyn's School, 1902–4; Lee's reader in chemistry, Christ Church, Oxford, 1904–12; professor, Imperial College of Science, 1912–32; an experimentalist exceptionally skilled in manipulating intensively ('Baker') dry substances; FRS, 1902.

BAKER, JAMES FRANKLIN BETHUNE- (1861–1951), professor of divinity. [See BETHUNE-BAKER.]

BAKER, SHIRLEY WALDEMAR (1835–1903), Wesleyan missionary and premier of Tonga; emigrated to Australia, 1853; sent as missionary to Tonga, South Pacific, 1860; negotiated treaty between Tonga and Germany recognizing Tonga as an independent kingdom, 1874; appointed premier, 1881; revised constitution; set up Wesleyan Free Church there, 1885; all-powerful with king, but unpopular with chiefs; removed by British high commissioner, 1890; died at Haapai.

BALDWIN, STANLEY, first EARL BALDWIN OF BEWDLEY (1867–1947), statesman and three times prime minister; son of Worcestershire ironmaster and cousin of Rudyard Kipling [q.v.]; educated at Harrow and Trinity College, Cambridge; entered family business; conservative MP, Bewdley division of Worcestershire, 1908–37; parliamentary private secretary to Bonar Law, 1916–17; joint financial secretary to Treasury, 1917–21; PC, 1920; realized 20 per cent. of his estate to purchase War Loan for cancellation, 1919; president of Board of Trade, 1921–2; his speech at Carlton Club meeting (19 Oct. 1922) revealed intense distrust of Lloyd George and brought coalition to an end; as chancellor of the Exchequer (1922–3) criticized for his settlement of the American debt; succeeded Bonar Law as prime minister, 22 May 1923; devoted to the parliamentary way, he set himself to diminish class hatred, retain national unity, and take the bitternesss out of political life; gained confidence of the nation in speeches which conveyed a plain, provincial, intensely patriotic, and completely English personality; lost majority in seeking mandate for protection (Nov. 1923) and gave way to labour government (Jan. 1924); became prime minister with a reunited party, 4 Nov. 1924; his fairminded leadership and notably his broadcasts brought the nation successfully through the general strike (1926) and gained him a unique moral ascendancy, but he did little, before or after, to pacify the coal industry; criticized for indolence in office and despite solid achievements of his government defeated by labour at general election, 1929; attacked by Rothermere and Beaverbrook press for not fully supporting their empire free trade campaign and by (Sir) Winston Churchill and the 'diehards' for supporting dominion status for India; nevertheless retained leadership of his party which recognized in him its greatest electoral asset; lord president of Council in Ramsay MacDonald's 'national' government, 1931-5; welcomed return to protection and presided over imperial economic conference, Ottawa, 1932; successfully supported Government of India bill, 1935; was slow to admit and reluctant to face German menace and recognized that the country, being profoundly pacifist, would not, without education, agree to rearmament; became prime minister for third time, June 1935; at general election (Nov.) supported a new defence programme while adhering to League covenant; failed to impose real sanctions on Italy but compelled by public opinion to disavow Hoare-Laval pact ceding Ethiopian territory to Italy; introduced measure of rearmament and appointed a minister for co-ordination of defence, 1936; handled the abdication crisis (1936) with dignity and sure judgement; resigned, 28 May 1937, and created KG and earl; a religious man with a serious view of life he possessed a genius for waiting upon events, a profound understanding of the middle classes whose dislike of foreigners and intellectuals he shared, and an acute sense of the House of Commons.

BALDWIN BROWN, GERARD (1849-1932), historian of art. [See BROWN.]

BALEWA, ALHAJI SIR ABU BAKAR TAFAWA (1912-1966), prime minister of the Federation of Nigeria. [See TAFAWA BALEWA.]

BALFOUR, SIR ANDREW (1873-1931), expert in tropical medicine and public health, and novelist; MD, Edinburgh, and DPH and Rugby blue, Cambridge; director (1902-13) of Wellcome Tropical Research Laboratories in Khartoum whence he banished malaria; founded Wellcome Bureau of Scientific Re-search, London, 1913; organized sanitary reforms, Mediterranean and East African theatres of war; director, London School of Hygiene and Tropical Medicine, 1923-31; wrote novels of historical adventure; KCMG, 1930.

BALFOUR, ARTHUR, first BARON RIVER-DALE (1873-1957), industrialist; educated at Ashville College, Harrogate; joined Seebohm & Dieckstall, Sheffield, which sold crucible steel; managing director, 1899; developed high speed steel; firm became Arthur Balfour & Co., 1915; served on innumerable war and post-war committees and on departmental advisory councils; KBE, 1923; chairman, Industry and Trade Commission, 1924-9; and Advisory Council for Scientific and Industrial Research, 1937-46; member, Economic Advisory Council; led RAF training mission to Canada, 1939; tireless promoter of overseas trade; baronet, 1929; baron, 1935; GBE, 1942.

BALFOUR, ARTHUR JAMES, first EARL OF BALFOUR (1848-1930), philosopher and statesman; profoundly influenced by his mother's personality and distinguished home circle; educated at Eton and Trinity College, Cambridge; much interested in scientific development, with which he kept abreast; engaged in metaphysical studies; his philosophical works include *A Defence of Philosophic Doubt* (1879) and *Foundations of Belief* (1895); his energies diverted from philosophy to politics by influence of his uncle, the third Marquess of Salisbury [q.v.]; conservative MP, Hertford, as supporter of Disraeli's last administration, 1874; held seat until 1885; his career favourably influenced by fact that public life was still strongly coloured by aristocratic influences; *arbiter elegantiarum* of coterie known as the 'Souls'; enjoyed enormous social prestige; extended wide hospitality at Whittingehame, his East Lothian home; associated in parliament with 'Fourth Party'; occupied mediatorial position between Lord Randolph Churchill [q.v.], leader of 'Fourth Party', and his uncle, thus increasing his own consequence in conservative counsels; president of Local Government Board, 1885-6; MP, East Manchester, 1885-1906; secretary for Scotland, 1886; entered cabinet, 1886; chief secretary for Ireland, 1887; took power to suppress National League in any district he thought desirable; suppression of riots, consequent on prosecution, under Crimes Act, of William O'Brien [q.v.], won for him title of 'Bloody Balfour'; 'resolute government' achieved its policy; features of his administration include Land Purchase Act (1891), Light Railways Act (1889), and establishment of Congested

Districts Board (1890); conservative leader in Commons and first lord of Treasury, 1891-2, 1895-1902; prime minister, 1902-5; his position at once rendered difficult by outbreak of struggle between tariff reform and free trade sections of unionist party; his own attitude governed by two considerations: to keep party together and to secure 'liberty of fiscal negotiation'; disruption of cabinet and departure of extremists became inevitable, but produced rapid development of party friction; meantime his administration made its mark in domestic and foreign policy; undertook military reconstruction; set up Committee of Imperial Defence, a consultative non-party council of experts and statesmen, assisted by secretarial staff; his most important pieces of domestic legislation, education bill (1902) and Licensing Act (1904); in foreign policy inaugurated Franco-British *entente*, a striking departure from nineteenth-century tradition; also renewed Anglo-Japanese agreement, 1905; resigned premiership, for which he had provided constitutional recognition and official precedence previously unknown, Dec. 1905; defeated at East Manchester and following in House reduced to small remnant, Jan. 1906; MP, City of London, (Mar.) 1906-22; as leader of opposition, put up good fight against education bill (1906), licensing bill (1908), and budget (1909); leading member of abortive conference of party leaders, June-Nov. 1910; refused to help to form coalition government; conservative party again split over attitude to be adopted to parliament bill, 1911; resigned leadership owing to rejection of his advice to surrender by 'diehard' opponents of bill, Nov. 1911; turned his attention to Irish question, recommending division of North and South Ireland; gave assurance of support to bolder section of cabinet on outbreak of European war, Aug. 1914; resumed membership of Committee of Imperial Defence; attended meetings of 'war council' or 'inner cabinet' convened by prime minister from Nov. 1914; first lord of the Admiralty in coalition government, May 1915; quickly acquired knowledge of important questions of naval administration and brought serenity to previously distracted department; withdrawal from Gallipoli and battle of Jutland (May 1916) both fell within his term of office; created special department to deal with submarine menace, Nov. 1916; took no part in organizing cabal which ousted Asquith from power, but also took no exception to it; his assistance essential to formation of new coalition, in which he became foreign secretary, 1916-19; owing to influence of Lloyd George, dyarchy created in foreign affairs; went to USA as head of diplomatic mission, thus seizing important opportunity, Apr. 1917; influence of Lloyd George on his

general foreign policy drove him towards courses such as humiliation of Germany which he wished to avoid; issued Balfour Declaration in favour of Jewish national home in Palestine, 1917; this project took shape at peace of Versailles; dominated situation at peace conference, 16 Feb.-8 Mar. 1919; altered it vastly for the better; later largely abrogated his office and no longer participated in principal discussions; his direct responsibility for peace treaty thus reduced; treaty of St. Germain with Austria (Sept. 1919) his particular contribution to settlement; resigned, retaining place in cabinet as lord president of the Council (Oct.) 1919-22; leading British delegate at Washington naval conference, 1921-2; recommended general cancellation of war debts, 1922; largely influenced financial rehabilitation of Austria under auspices of League of Nations, 1922; KG, 1922; earl, 1922; lord president of the Council, 1925-9; responsible for foundation of Committee of Civil Research; so-called Balfour Definition prepared way for Statute of Westminster (1931), 1926; president of British Association, 1904; of British Academy, 1921-30; Gifford lecturer, Glasgow University, 1913-14, 1922-3; OM, 1916.

BALFOUR, LADY FRANCES (1858-1931), churchwoman, suffragist, and author; daughter of eighth Duke of Argyll [q.v.]; married (1879) E. J. A. Balfour, brother of A. J. Balfour [q.v.]; ardently supported Church of Scotland and organized rebuilding of Crown Court church, London; a mistress of invective in the cause of women's suffrage; published several memoirs and her reminiscences.

BALFOUR, GEORGE WILLIAM (1823-1903), physician; uncle of R. L. Stevenson [q.v.]; studied medicine at Edinburgh and at Vienna; MD and LRCS Edinburgh, 1845; FRCP, 1861; physician to Royal Infirmary, 1867-82; published *Introduction to Study of Medicine* (1865), *Diseases of the Heart and Aorta* (1876), and *The Senile Heart* (1894); librarian and president (1882-4) of College of Physicians of Edinburgh; hon. LLD, Edinburgh, 1884, St. Andrews, 1896.

BALFOUR, GERALD WILLIAM, second EARL OF BALFOUR (1853-1945), politician and psychical researcher; brother of A. J. Balfour [q.v.] whom he succeeded in earldom by special remainder, 1930; educated at Eton and Trinity College, Cambridge; fellow, 1878; conservative MP, Central Leeds, 1885-1906; chief secretary for Ireland, 1895-1900; carried Local Government (Ireland) Act, 1898; PC, 1900; president, Board of Trade, 1900-5; with his sister, Eleanor Mildred Sidgwick [q.v.], devoted himself to psychical research, especially the trance utterances of mediums.

BALFOUR, HENRY (1863-1939), anthropologist; educated at Charterhouse and Trinity College, Oxford; second class, natural science (biology), 1885; curator of Pitt-Rivers Museum, Oxford, 1891-1939; fellow of Exeter College from 1904; taught technology and prehistoric archaeology for Oxford diploma in anthropology from 1907; accorded personal title of professor, 1935; travelled very widely; outstanding as an observer, collector, classifier, and craftsman; works include *The Evolution of Decorative Art* (1893) and many contributions to learned journals; FRS, 1924.

BALFOUR, SIR ISAAC BAYLEY (1853-1922), botanist; B.Sc., Edinburgh University, 1873; professor of botany, Glasgow University, 1879; Sherardian professor of botany, Oxford University, 1884; professor of botany, Edinburgh University, king's botanist in Scotland, and keeper of royal botanic garden, 1888-1922; FRS, 1884; KBE, 1920.

BALFOUR, JOHN BLAIR, first BARON KINROSS (1873-1905), Scottish judge; educated at Edinburgh University; passed to Scottish bar, 1861; became a foremost advocate; liberal MP for Clackmannan, 1880-99; solicitor-general for Scotland, 1880; lord advocate, 1881-5, 1886, and 1892-5; PC, 1883; hon. LLD, Edinburgh, 1882; took prominent part in carrying Local Government Act for Scotland, 1894; lord president of the Court of Session, 1899; baron, 1902.

BALFOUR, SIR THOMAS GRAHAM (1858-1929), author and educationist; son of Thomas Graham Balfour [q.v.]; BA, Worcester College, Oxford, 1882; called to bar (Inner Temple), 1885; director of technical education for Staffordshire, 1902; general director of education for Staffordshire, 1903; knighted, 1917; works include *Educational Systems of Great Britain and Ireland* (1898) and life of his cousin, Robert Louis Stevenson [q.v.] (1901).

BALFOUR OF BURLEIGH, sixth BARON (1849-1921), statesman. [See BRUCE, ALEXANDER HUGH.]

BALFOUR-BROWNE, WILLIAM ALEXANDER FRANCIS (1874-1967), entomologist; educated at St. Paul's School and Magdalen College, Oxford; hockey blue, 1894-5; director, Sutton Broad Laboratory, 1902-6; taught biology, Queen's College, Belfast, 1906-13; taught entomology, Cambridge University, 1913-15; served with RAMC in France, 1915-

16; professor, entomology, Imperial College, London University, 1925-30; wrote three Ray Society monographs on British water beetles (1940, 1950, 1958); published *Water Beetles and Other Things* (1964); fellow, Royal Society of Edinburgh.

BALL, ALBERT (1896-1917), airman; joined army, 1914; seconded to Royal Flying Corps, 1916; largely contributed to ascendancy of British over German air service established at battle of the Somme (July); killed in air fight; MC and DSO, 1916; VC (posthumous), 1917; destroyed forty-three aeroplanes and one balloon; greatest fighter pilot of air service of his time.

BALL, FRANCIS ELRINGTON (1863-1928), historian and antiquary; son of John Thomas Ball [q.v.]; educated privately; worked for unionist cause in Ireland; works include *A History of the County of Dublin* (1902-20), *The Correspondence of Jonathan Swift, D.D.* (1910-14), *The Judges in Ireland, 1221-1921* (1926), *Swift's Verse; an Essay* (1929); the greatest authority of the time on Swift.

BALL, SIR (GEORGE) JOSEPH (1885-1961), intelligence officer, party administrator, and businessman; educated at King's College School and King's College, London; called to bar (Gray's Inn), 1913; worked as civilian official, Scotland Yard; served in MI5, 1914-27; persuaded by J. C. C. (later first Viscount) Davidson [q.v.] to join conservative party organization as director of publicity, 1927-30; inserted agents in labour party headquarters; director, Conservative Research Department, 1930-9; deputy chairman, Security Executive, 1940-2; chairman, Henderson's Transvaal Estates and other companies; director, Consolidated Goldfields of South Africa and Beaumont Property Trust; chairman, Hampshire Rivers Catchment Board, 1947-53; OBE, 1919; KBE, 1936.

BALL, JOHN (1861-1940), golfer; his father owned Royal Hotel, Hoylake, headquarters of Royal Liverpool Golf Club; won the amateur championship, 1888, 1890, 1892, 1894, 1899, 1907, 1910, 1912; first amateur to win the open championship, 1890; a magnificent iron player of great accuracy, but less happy on the green.

BALL, SIR ROBERT STAWELL (1840-1913), astronomer and mathematician; son of Robert Ball [q.v.]; BA, Trinity College, Dublin; professor of applied mathematics and mechanism, Royal College of Science, Dublin, 1867-74; Andrews professor of astronomy, Dublin University, and royal astronomer of Ireland, 1874-92; Lowndean professor of astronomy, Cambridge, 1892-1913; FRS, 1873; knighted, 1886; popularizer of astronomy; most distinguished as mathematician; published researches on the theory of screw motions and their relations.

BALLANCE, Sir CHARLES ALFRED (1856-1936), surgeon; qualified at St. Thomas's Hospital where he became aural surgeon (1885), assistant surgeon (1891), surgeon (1900-19); chief surgeon, Metropolitan Police, 1912-26; chiefly interested in nerve repair; KCMG, 1918.

BANBURY, FREDERICK GEORGE, first Baron Banbury of Southam (1850-1936), politician; educated at Winchester; head of Frederick Banbury & Sons, stockbrokers, 1879-1906; conservative MP for Peckham division, 1892-1906, City of London, 1906-24; baronet, 1903; PC, 1916; baron, 1924; authority on finance and an uncompromising champion of the old order.

BANCROFT, MARIE EFFIE, Lady Bancroft (1839-1921) (DNB 1922-30, under Sir Squire Bancroft), actress and theatrical manager; born Wilton; wife of Sir Squire Bancroft [q.v.]; began career in burlesque in London, 1856; opened Prince of Wales's Theatre, London, with H. J. Byron [q.v.], 1865; comedy actress of genius; joined Roman Church, 1885.

BANCROFT, Sir SQUIRE BANCROFT (1841-1926), actor and theatrical manager, whose original name was Squire Butterfield; went on the stage, 1861; engaged by Marie Wilton (see above) for Prince of Wales's Theatre, London, 1865; took surname of Bancroft, 1867; joint manager with his wife of Prince of Wales's, 1867; rebuilt and opened Haymarket Theatre, 1880; retired, 1885; effected great reforms in theatrical art and business; a good actor whose special part was the 'swell'; knighted 1897.

BANDARANAIKE, SOLOMON WEST RIDGEWAY DIAS (1899-1959), fourth prime minister of Sri Lanka; educated at St. Thomas's College, Colombo, and Christ Church, Oxford; secretary and junior treasurer of Union; called to bar (Inner Temple); elected to Colombo municipal council and secretary of Ceylon National Congress, 1927; elected to legislature, 1931; minister of local administration, 1936; became Buddhist; formed Sinhala Mahasabha to represent Buddhist interests, 1937; joined with Ceylon National Congress to form United National Party, 1947; minister of health and local government, 1947; resigned and formed Sri Lanka Freedom Party,

1951; leader of opposition, 1952; established the People's United Front; prime minister, 1956; unable to control extremists; assassinated by Buddhist monk.

BANKES, Sir JOHN ELDON (1854-1946), judge; educated at Eton and University College, Oxford; called to bar (Inner Temple), 1878; KC, 1901; judge of King's Bench division, 1910-15; lord justice of appeal, 1915-27; knighted, 1910; PC, 1915; GCB, 1928.

BANKS, Sir JOHN THOMAS (1815?-1908), physician; BA and MB, Trinity College, Dublin, 1837; MD, 1843; physician to House of Industry Hospital, Dublin, 1843-1908; censor (1847-8) and president (1869-71) of College of Physicians, Ireland; king's professor of medicine at Trinity College, 1849-68; regius professor of physic, Dublin University, 1880-98; first president of Royal Academy of Medicine, Ireland, 1882; an expert in mental disease; advocated psychological study for medical students; popular in Dublin society; KCB, 1889; hon. D.Sc., Royal University, 1882; hon. LLD, Glasgow, 1888; high sheriff, co. Monaghan, 1891.

BANKS, LESLIE JAMES (1890-1952), actor; scholar of Glenalmond and Keble College, Oxford; went on stage, 1911; received facial wound in war; reached west end by 1921 and became much in demand as stage and film actor with wide range, i.e. Claudius in *Hamlet*, Bottom in *A Midsummer Night's Dream*, the lead in *Goodbye Mr. Chips* (1938) and *Life with Father* (1947).

BANKS, Sir WILLIAM MITCHELL (1842-1904), surgeon; MD, Edinburgh University, 1864; FRCS, 1869; surgeon to Paraguay government; demonstrator, lecturer, and professor of anatomy, Liverpool Infirmary school of medicine, 1868-94; emeritus professor, 1894; surgeon, Royal Infirmary, Liverpool, 1877-1902; knighted, 1899; hon. LLD, Edinburgh; advocated operation for cancer of breast; raised Liverpool medical school to well-organized faculty of the university; collector of early medical literature; contributor to scientific journals; Banks memorial lectureship founded in Liverpool University.

BANNERMAN, Sir HENRY CAMPBELL- (1836-1908), prime minister. [See Campbell-Bannerman.]

BANTING, Sir FREDERICK GRANT (1891-1941), Canadian surgeon and physiologist; qualified at Toronto University, 1916; MC, Cambrai, 1918; became highly qualified surgeon; suggested that the enzymes of the external secretion of the pancreas might, in the course of preparing the extract, destroy the internal secretion which researchers could not find; provided with facilities for research at Toronto University by J. J. R. Macleod [q.v.], 1921; with C. H. Best obtained the antidiabetic hormone insulin; successfully treated first human being, Jan. 1922; head of Banting and Best Department of Medical Research created by Ontario legislature, 1923-41; shared Nobel prize for medicine with Macleod, 1923; KBE, 1934; FRS, 1935; placed in charge of Canadian medical research (1939) and lost life on flight to England for consultation.

BANTOCK, SIR GRANVILLE RANSOME (1868-1946), composer; studied at Royal Academy of Music; director of music, New Brighton, 1897-1900; introduced Sibelius to British public; principal, music school, Midland Institute, Birmingham, 1900-8; professor of music, Birmingham University, 1908-34; his works, frequently on oriental themes, include 'Hebridean' symphony, choral cantata 'Omar Khayyám', choral symphony 'Atalanta in Calydon', and an opera *The Seal-Woman*; his versatility prejudiced the development of a personal style; chiefly distinguished for his instrumental mastery; knighted, 1930.

BARBELLION, W. N. P. (pseudonym) (1889-1919), diarist and biologist. [See CUMMINGS, BRUCE FREDERICK.]

BARBIROLLI, SIR JOHN (GIOVANNI BATTISTA) (1899-1970), musical conductor and cellist; educated at St. Clement Dane's school, London, and Trinity College of Music; public début as cellist, 1911; studied at Royal Academy of Music, 1912-16; cellist, 1916-24; début as conductor, 1926; conductor, Covent Garden Opera touring company, 1929; conductor, Scottish Orchestra, 1933; conductor, Philharmonic Symphony Orchestra, New York, 1936-42; conductor, Hallé Orchestra, 1943-68; conductor laureate for life, 1968; conductor-in-chief, Houston Symphony Orchestra, 1961-7; knighted, 1949; CH, 1969; fellow, Royal Academy of Music (1928); hon. degrees, Manchester (1950), Dublin (1952), Sheffield (1957), London (1961), Leicester (1964), and Keele (1969).

BARBOUR, SIR DAVID MILLER (1841-1928), Indian civil servant and economist; BA, Queen's University, Ireland, 1862; entered Indian civil service, 1862; under-secretary to government of India in finance department, 1872; successively accountant-general in Punjab, Madras, and Bengal; secretary to

government of India in department of finance and commerce, 1882; argued vainly in favour of reintroduction of bimetallism by international agreement, 1885-8; finance member of governor-general's council, 1888; advocated as only alternative to bimetallism adoption of gold standard by government of India; left India, 1893; member of Sir Henry Fowler's committee on Indian currency, 1898; served on many commissions and committees of imperial importance; works include *The Theory of Bimetallism* (1885) and *The Standard of Value* (1912); KCSI, 1889; KCMG, 1899.

BARCROFT, SIR JOSEPH (1872-1947), physiologist; educated at Friends' School, Bootham, Leys School, and King's College, Cambridge; first classes, natural sciences tripos, 1896-7; prize fellow, 1899; studied haemoglobin and physiology of life at high altitudes; invented differential blood gas manometer, 1908; chief physiologist, gas-warfare centre, Porton, 1917-18, 1939-41; reader (1919), professor (1925-37) of physiology, Cambridge; extended researches to the spleen and physiology of the developing foetus; publications include *The Respiratory Function of the Blood* (1914) and *Researches on Pre-Natal Life* (1946); FRS, 1910; knighted, 1935.

BARDSLEY, JOHN WAREING (1835-1904), bishop of Carlisle; educated at Trinity College, Dublin; BA, 1859; MA, 1865; DD, Lambeth, 1887; archdeacon of Liverpool, 1886; bishop of Carlisle, 1892-1904; capable organizer and fluent preacher of evangelical doctrine.

BARGER, GEORGE (1878-1939), chemist; son of Dutch engineer; educated at Utrecht and London and Cambridge universities; head of chemistry department, Goldsmiths' College, New Cross, 1909-13; on staff of Medical Research Committee, 1914-19; professor of medical chemistry, Edinburgh, 1919-37; of chemistry, Glasgow, 1937-9; his lifelong interest in ergot led him to investigate alkaloids and simpler nitrogenous compounds of biological importance; FRS, 1919.

BARING, EVELYN first EARL OF CROMER (1841-1917), statesman, diplomatist, and administrator; entered Woolwich, 1855; commissioned, and accompanied battery to Ionian Islands, 1858; aide-de-camp to high commissioner, Sir Henry Knight Storks [q.v.]; followed chief to Malta and thence on special mission to Jamaica, 1864; returned to England, 1867; entered Staff College; private secretary to Lord Northbrook [q.v.], viceroy of India, 1872; CIE, 1876; went out to Cairo as first British commissioner of 'Caisse de la Dette' created to deal with liabilities of Khedive Ismail, 1877; resigned, 1879; on deposition of Ismail and succession of Tewfik appointed British controller in Egypt, 1879; financial member of viceroy's council in India, 1880–3; KCSI, 1883; British agent and consul-general in Egypt, 1883; as representative of the one power occupying country in force, had *de facto* to impose British will; 'advised' Egyptian government to withdraw temporarily from Sudanese provinces; reluctantly consented to General Gordon's mission to Sudan, 1884; urged raising of loan for Egyptian irrigation and expedition to relieve Gordon [q.v.]; after Gordon's death insisted on effectual evacuation of all Sudan except Suakin, 1885; turned his attention to interior and obtained loan of a million to spend on irrigation, 1885; so successful that surplus appeared in Treasury accounts for 1889; improved railways, administration of justice, and education; CB, 1885; KCB, 1887; GCMG, 1888; co-operated harmoniously with Khedive Tewfik; on his death supported Tewfik's son. Abbas, against Sultan's candidate, 1892; created Baron Cromer, 1892; experienced difficulties with Abbas, but nationalist movement checked by appearance of British battalion in Cairo, 1894; surveys for Nile dam put in hand, 1895; acted as minister of war, and supported (Lord) Kitchener [q.v.], sirdar of Egyptian army, against Egyptian cabinet and British government during reconquest of Sudan, 1896–8; enforced Anglo–Egyptian arrangement which he had devised for excluding internationalism from reconquered Sudan, 1898; brought about revision of land assessment, completion of land survey, lowering of land tax, creation of land bank, and subjection of interior and education to advisers; created viscount, 1899; earl, 1901; resigned position in Egypt, 1907; made maiden speech in House of Lords, 1908; took lead of free traders; president of Dardanelles commission, 1916; works include *Modern Egypt* (1908), *Ancient and Modern Imperialism* (1910), *Abbas II* (1915), and three volumes of *Political and Literary Essays* (reviews of new books written for *Spectator* from 1912 onwards); remained strong whig through life; while not a genius, possessed powerful and versatile talents, whose full exercise was ensured by a strong character and vigorous constitution.

BARING, MAURICE (1874–1945), poet and man of letters; educated at Eton and Trinity College, Cambridge; in diplomatic service, 1898–1904; special correspondent in Manchuria, Russia, and the Balkans, 1904–12; served with Lord Trenchard [q.v.] in Royal Air Force, 1915–19; works include volumes of verse, novels, autobiography *The Puppet Show*

of Memory (1922), and an anthology especially revealing of his character and culture *Have You Anything to Declare?* (1936).

BARING, ROWLAND THOMAS, second EARL OF CROMER (1877–1953), lord chamberlain to the household; son of first Earl of Cromer [q.v.]; born in Cairo; educated at Eton; entered diplomatic service, 1900; at Foreign Office, 1907–11; managing director, Baring Brothers, 1913; succeeded father, 1917; assistant private secretary to King George V, 1916–20; lord chamberlain, 1922–38; a tactful censor of plays; PC, 1922.

BARING, THOMAS GEORGE, first EARL OF NORTHBROOK (1826–1904), statesman; son of first Baron Northbrook [q.v.] and nephew of Sir George Grey [q.v.]; gentleman commoner of Christ Church, Oxford (BA, 1846); whig MP for Penryn and Falmouth, 1857; civil lord of Admiralty, 1857; under-secretary in India Office (1859–64), at War Office (1861 and 1868), at Home Office (1864); secretary to the Admiralty (1866); governor-general of India, 1872–6; rigidly controlled finance and modified local taxation; successfully met the Bengal famine by well-designed measures of relief; opposed to Lord Salisbury's policy of external aggression in Afghanistan; supported Indian cotton tariff in opposition to Lord Salisbury; entertained Prince of Wales on his visit to India (winter of 1875–6); left India, Apr. 1876; showed as viceroy genuine sympathy with Indians and strict impartiality; created earl, 1876; appointed first lord of the Admiralty under Gladstone, 1880; one of the ministers responsible for dispatch of General Gordon [q.v.] to the Sudan; special commissioner to Egypt, 1884, and advocate of single British control there; attacked by W. T. Stead [q.v.] in articles on navy, 1885; opposed home rule, 1886, and tariff reform, 1903; active in local administration of Hampshire on retirement from political office in 1886; high steward of Winchester, 1889.

BARING-GOULD, SABINE (1834–1924), divine and author; BA, Clare College, Cambridge, 1856; rector of Lew-Trenchard, Devonshire, 1881; voluminous writer on folklore, travel, and on the West of England; chief works, *The Origin and Development of Religious Belief* (1869–70) and *The Lives of the Saints* (1872–7); his best-known novel, *The Broom Squire* (1896); wrote 'Onward Christian Soldiers'.

BARKER, SIR ERNEST (1874–1960), scholar; educated at Manchester grammar school and Balliol College, Oxford; first class, *lit. hum.*, 1897, and history, 1898; classical fellow, Merton, 1898–1905; lecturer, modern history,

Wadham, 1899-1909; fellow and lecturer, St. John's, 1909-13; fellow and tutor, New College, 1913-20; principal, King's College, London, 1920-8; professor of political science and fellow of Peterhouse, Cambridge, 1928-39; translated Otto von Gierke and Ernst Troeltsch; works include *Political Thought of Plato and Aristotle* (1906), *Political Thought in England from Herbert Spencer to Today* (1915), *Traditions of Civility* (1948), *From Alexander to Constantine* (1956), *Social and Political Thought in Byzantium* (1957); a philosopher-historian he sought to reconcile English individualism with his strong 'sense for the community'; FBA, 1947; knighted, 1944.

BARKER, HARLEY GRANVILLE GRAN-VILLE- (1877-1946), actor, producer, dramatist, and critic. [See GRANVILLE-BARKER.]

BARKER, SIR HERBERT ATKINSON (1869-1950), manipulative surgeon; apprenticed to bone-setter; established himself in London with remarkable success; had great self-confidence and real gift of healing; in his lifelong fight for recognition of status of manipulative surgeon aroused hostility of medical profession towards an unqualified practitioner and became subject of prolonged controversy; knighted 1922.

BARKER, DAME LILIAN CHARLOTTE (1874-1955), first woman assistant prison commissioner; trained at Whitelands College, Chelsea; became LCC, elementary school-teacher; lady superintendent, Woolwich Arsenal, 1915-19; CBE, 1917; executive officer, Central Committee on Women's Training and Employment, 1920-3; governor, girls' Borstal, Aylesbury, 1923-35; improved conditions and exerted strong personal influence; first woman assistant prison commissioner, responsible for all women's prisons, 1935-43; DBE, 1944.

BARKER, THOMAS (1838-1907), mathematician; scholar of Trinity College, Cambridge, senior wrangler (1862) and first Smith's prizeman; professor of pure mathematics at Owens College, Manchester, 1865-85; successful teacher; laid stress on logical basis of mathematics; endowed professorship of cryptogamic botany at Manchester.

BARKLA, CHARLES GLOVER (1877-1944), physicist; educated at Liverpool Institute and University College, and at Trinity and King's colleges, Cambridge; BA by research, 1903; worked at Liverpool University, 1902-9; professor of physics, King's College, London, 1909-13; of natural philosophy, Edinburgh, 1913-44; worked on the polarization of X-rays; examined connection between atomic weight of the radiator and the absorptiveness of the radiation; distinguished between harder K-radiation and softer L-radiation; strong supporter of the wave theory of X-rays; FRS, 1912; Nobel prize, 1917.

BARLING, SIR (HARRY) GILBERT, baronet (1855-1940), surgeon and academic administrator; qualified as a chemist; and at St. Bartholomew's Hospital in medicine, 1879; FRCS, 1881; pathologist (1879), assistant surgeon (1885), surgeon (1891-1915) at the General Hospital, Birmingham; professor of surgery (1893-1912), dean of medical faculty (1905-12), vice- (later pro-) chancellor (1913-33), Birmingham University; a leading negotiator in the establishment of the United Hospital and chairman of the Hospitals Centre Scheme; baronet, 1919.

BARLOW, SIR (JAMES) ALAN (NOEL), second baronet (1881-1968), public servant; son of (Sir) Thomas Barlow [q.v.]; educated at Marlborough and Corpus Christi College, Oxford; first class, *lit. hum.*, 1904; clerkship, House of Commons, 1906; junior examiner, Board of Education, 1907; Ministry of Munitions, 1915-18; Ministry of Labour, in charge of Training Department, 1919-33; principal private secretary to prime minister, Ramsay Macdonald, 1933-4; under secretary, Treasury, 1934-8; joint second secretary, 1938-48; trustee, National Gallery, 1948-55; JP Bucking-hamshire, hon. fellow, Corpus Christi College, Oxford.

BARLOW, SIR THOMAS, first baronet (1845-1945), physician; studied at Owens College, Manchester, and University College, London; qualified, 1871; assistant physician, Hospital for Sick Children, Great Ormond Street, 1875; physician, 1885-99; assistant physician, University College Hospital, 1880; physician, 1885-1910; Holme professor of clinical medicine, 1895-1907; best known for researches on infantile scurvy, which he identified as a separate disease from rickets; also distinguished between tuberculous and simple meningitis and made other important discoveries in children's diseases; held court appointments under Queen Victoria, King Edward VII, and King George V; baronet, 1900; KCVO, 1901; FRS, 1909.

BARLOW, SIR THOMAS DALMAHOY (1883-1964), industrialist and public servant; son of (Sir) Thomas Barlow and younger brother of (Sir) J. Alan Barlow [qq.v.]; educated at Marlborough and Trinity College, Cambridge; joined family textile business, Barlow &

Jones Ltd., of Bolton and Manchester, 1904; on board, District Bank, 1922; chairman, 1947-60; chairman, Lancashire Industrial Development Council; president, Manchester Chamber of Commerce, 1931-3; director-general, civilian clothing, 1941-5; chairman, Council of Industrial Design, 1944-7; member of council, Royal College of Art; KBE, 1934; GBE, 1946.

BARLOW, WILLIAM HAGGER (1833-1908), dean of Peterborough; scholar of St. John's College, Cambridge; took honours in four triposes; MA, 1860; BD, 1875; DD, Christ Church, Oxford, 1895; principal of CMS training college, Islington, 1875-82; vicar of Islington, 1887-1901; chairman of Islington vestry, 1887-1900; prebendary of St. Paul's, 1898; dean of Peterborough, 1901; select preacher at Oxford and Cambridge.

BARLOW, WILLIAM HENRY (1812-1902), civil engineer; son of Peter Barlow [q.v.]; apprenticed at Woolwich dockyard; in Constantinople erected machinery and buildings for Turkish ordnance, 1832-8; engineer to several English railways, 1838-44; principal engineer, Midland Railway, 1844; consulting engineer, 1857; invented Barlow saddleback rail, 1849; laid out southern portion of London and Bedford line, including St. Pancras station, 1862-9; made many researches in sound and electricity, in the theory of structures, and in the strength of beams; FRS, 1850; member of court of inquiry into Tay bridge disaster, 1879; designed new Tay bridge, 1882; member (1845), Telford medallist (1849), and president (1879-80) of Institution of Civil Engineers; one of first civil members of ordnance committee, 1881.

BARNABY, SIR NATHANIEL (1829-1915), naval architect; draughtsman in royal dockyard, Woolwich, 1852; entered naval construction department of Admiralty, 1854; head of the staff of Sir E. J. Reed [q.v.], chief constructor of the navy, 1863; chief naval architect (styled director of naval construction, 1875), 1872-85; designed battleships, including the *Inflexible* type and *Admiral* class, and cruisers of the *Mersey* 'protected' and *Orlando* 'belted' class; KCB, 1885; author of *Naval Development in the Nineteenth Century* (1902).

BARNARDO, THOMAS JOHN (1845-1905), philanthropist; clerk in Dublin, 1859; was 'converted', 1862; did evangelizing work in Dublin slums; entered London Hospital as missionary medical student, 1866; FRCS Edinburgh, 1879; founded East End juvenile mission for destitute children, 1867; opened (1870) boys' home in Stepney, which developed into

'Dr. Barnardo's Homes'; founded Girls' Village Home, Barkingside, Essex, 1876; sent first party of boys to Canada, 1882; at his death had assisted 250,000 children.

BARNES, ERNEST WILLIAM (1874-1953), bishop of Birmingham; educated at King Edward's School, Birmingham, and Trinity College, Cambridge; bracketed second wrangler, 1896; president of Union and first class, part ii, mathematical tripos, 1897; Smith's prize and fellow of Trinity, 1898; assistant lecturer, 1902, tutor, 1908-15; FRS, 1909; deacon, 1902; priest 1903; master of Temple, 1915-19; canon of Westminster, 1918; bishop of Birmingham, 1924-53; known for 'gorilla' sermons supporting evolutionary theory; attacked doctrine of Real Presence and had no understanding of sacramentalism; involved in series of controversies; successfully sued for slander by Cement Makers Federation, 1941; his *Rise of Christianity* (1947), precluding recognition of miracles, aroused fierce opposition and strongly criticized in Convocation by Archbishops of Canterbury and York; of high character and evident, if naïve, intellectual honesty; also a mathematician of some note.

BARNES, GEORGE NICOLL (1859-1940), statesman; worked as an engineer in London; assistant secretary (1892), general secretary (1896-1908), Amalgamated Society of Engineers; labour MP, Blackfriars (later Gorbals) division, Glasgow, 1906-22; minister of pensions, 1916-17; entered war cabinet, 1917; resigned from labour party (1918) and remained in government in order to influence peace conference as minister plenipotentiary; chiefly responsible for institution of International Labour Organization; represented Great Britain at its first conference, Washington, 1919; delegate at first League of Nations Assembly, 1920; resigned from government, 1920; PC, 1916; CH, 1920.

BARNES, SIR GEORGE REGINALD (1904-1960), broadcasting director and college principal; nephew of Sir Kenneth Barnes [q.v.]; educated at Osborne, Dartmouth, and King's College, Cambridge; first class, historical tripos, 1925; appointed assistant master, Dartmouth, 1927; assistant secretary, Cambridge University Press, 1930; joined BBC, 1935; director of talks, 1941; assistant controller of talks, 1945; inaugurated Third Programme, 1946; director of spoken word, 1948, of television at time of great expansion, 1950; knighted at Lime Grove studios, 1953; principal, University College of North Staffordshire (Keele), 1956-60.

BARNES, JOHN GORELL, first BARON GORELL (1848-1913), judge; BA, Peterhouse,

Cambridge; called to bar (Inner Temple), 1876; succeeded to great junior practice of Sir J. C. Mathew [q.v.], 1881; QC, 1888; judge of Probate, Divorce, and Admiralty division, 1892; president, 1905-9; PC, 1905; baron, 1909; chairman of county courts committee, 1909, of copyright committee leading to Act of 1911, and of royal commission on divorce, 1909-12; delivered many important judgments in court of first instance, Court of Appeal, Privy Council, and House of Lords.

BARNES, SIR KENNETH RALPH (1878-1957), principal of the Royal Academy of Dramatic Art; brother of Violet and Irene Vanbrugh [qq.v.]; educated at Westminster and Christ Church, Oxford; became freelance journalist; appointed principal, Academy of Dramatic Art, 1909; obtained royal charter, 1920, recognition as charity (income-tax exemption), 1926, of acting as fine art (rate exemption), 1930; a firm believer in social and spiritual importance of the theatre; Academy's Gower Street building completed, 1931; Vanbrugh Theatre opened, 1954; retired, 1955; knighted, 1938.

BARNES, ROBERT (1817-1907), obstetric physician; pupil of George Borrow [q.v.]; studied at University College, London, and St. George's Hospital; MD, 1848; FRCP, 1859; obstetric physician at London (1863), St. Thomas's (1865), and St. George's (1875) hospitals; president of Obstetrical Society, 1865-6; established British Gynaecological Society, 1884; pioneer of operative gynaecology; Lumleian lecturer at College of Physicians, 1873; hon. FRCS, 1883; author of *Obstetrical Observations* (1870) and other works.

BARNES, SYDNEY FRANCIS (1873-1967), cricketer; professional in Lancashire League cricket; played in three test matches, 1901-2; played for Lancashire, 1902-4; Staffordshire, 1904-35; test side in Australia, 1907-8; and in England, 1909; and Australia, 1911-12; took four wickets for one run in five overs at Melbourne; first-class cricket bowling average, 719 wickets at 17.09 and batting average, 173 innings, 1,563 runs at 12.70; remembered as a remarkable bowler.

BARNES, SIR THOMAS JAMES (1888-1964), Treasury solicitor; educated at Mercers School, London; articled to firm of solicitors; admitted solicitor, 1911; clerk, lord chancellor's department, 1911; served in Royal Naval Volunteer Reserve, 1914-17; legal adviser, Ministry of Shipping, 1919; solicitor to Board of Trade, 1920; procurator-general and Treasury solicitor, 1934-53; served on Ever-

shed committee on Supreme Court practice and procedure, 1947-53; member, Monopolies Restrictive Practices Commission, 1954-9; knighted, 1927; KCB, 1938; GCB, 1948.

BARNES, WILLIAM EMERY (1859-1939), divine; educated at Peterhouse, Cambridge; first class, theology, 1881; ordained, 1883; fellow of Peterhouse from 1889; Hulsean professor of divinity, Cambridge, 1901-34; a painstaking and simple teacher and a fine Hebrew, Rabbinic, and, in particular, Syriac scholar; of strong conservative and devotional tendencies, he was cautious in his use of the new approach to Old Testament studies.

BARNETT, DAME HENRIETTA OCTAVIA WESTON (1851-1936), social reformer; born ROWLAND; married (1873) S. A. Barnett [q.v.]; first nominated woman guardian, 1875; co-founder of Children's Country Holidays Fund and founder of London Pupil Teachers' Association, 1884; active advocate of settlement ideal in United States; formed Hampstead Garden Suburb Trust, 1903; founded Henrietta Barnett School, Hampstead, 1901; DBE, 1924.

BARNETT, LIONEL DAVID (1871-1960), orientalist; educated at high school, Institute, and University College, Liverpool, and Trinity College, Cambridge; first class, parts i and ii, classical tripos, 1894-6; Craven scholar (1894) and student (1897); assistant keeper, oriental printed books, British Museum, 1899-1908 and 1948-60; keeper, 1908-36; compiled ten descriptive catalogues; professor of Sanskrit, University College, London, 1906-17; lecturer in Sanskrit, 1917-48, in ancient Indian history and epigraphy, 1922-48, School of Oriental Studies; wrote mostly on Indological subjects for both specialist and general reader; FBA, 1936; CB, 1937.

BARNETT, SAMUEL AUGUSTUS (1844-1913), divine and social reformer; BA, Wadham College, Oxford; rector of St. Jude's, Whitechapel, 1873-94; canon of Bristol, 1894-1906; canon, and finally subdean, of Westminster, 1906-13; first warden of Toynbee Hall, which he had largely helped to found, 1884-96; instrumental in building Whitechapel Art Gallery; founded Education Reform League, 1884; attacked housing problem; wrote on religious and social questions.

BARODA, SIR SAYAJI RAO, MAHARAJA GAEKWAR OF (1863-1939), born at Kavlana, Bombay; chosen to replace unsatisfactory ruling maharaja, 1875; invested with governing powers, 1881; despite over-centralization, excessive caution, prolonged absenteeism, and some friction with the British, he raised Baroda from chaos to a model state; influential at first and second Indian Round Table conferences; GCSI, 1887; GCIE, 1919.

BARON, BERNHARD (1850-1929), tobacco manufacturer and philanthropist; born in South Russia; as a boy migrated to the United States; entered tobacco industry; settled in England and formed cigarette machine company, 1895; developed business of Carreras Limited, 1903-29; a munificent benefactor to charities.

BARR, ARCHIBALD (1855-1931), inventor of range-finders; graduated in engineering, Glasgow, 1876; professor and founder of engineering laboratories at Yorkshire College 1884-9, Glasgow, 1889-1913; with William Stroud founded the firm which designed naval range-finders, height-finders, fire-control, and other precision instruments; FRS, 1923.

BARRETT, WILSON (1846-1904), actor and dramatist, whose original name was WILLIAM HENRY BARRETT; began life as printer; first appeared on stage, 1864; played Tom Robinson in *It's Never too Late to Mend* and in *East Lynne*, 1867; first produced *Jane Shore* at Leeds, 1875; opened Court Theatre, London, 1879; then introduced Madame Modjeska to London public, 1880; began notable management of Princess's Theatre, June 1881; very successful in melodramas, *The Lights o' London*, *The Romany Rye*, *The Silver King*, and *Claudian*, 1881-3; his interpretation of Hamlet provoked controversy, 1884; wrote *Hoodman Blind* (in collaboration with H. A. Jones, q.v.), 1885, and other plays; toured America, 1886-7, paying five subsequent visits; returned to Princess's Theatre (1888), appearing in *Ben-my-Chree*; opened new Olympic Theatre, London, 1890; played Othello at Liverpool, 1891; brought out new pieces at Grand Theatre, Leeds, including his version of Hall Caine's *Manxman* (1894) and *The Sign of the Cross* (1895), which ran prosperously at the Lyric Theatre, London, next year, and was followed by his *The Daughters of Babylon*, 1897; paid two visits to Australia, 1898 and 1902; produced at Adelphi his *The Christian King*, 1902; a popular impersonator of classical and melodramatic roles despite his stilted delivery and gesture.

BARRIE, SIR JAMES MATTHEW, baronet (1860-1937), playwright and novelist; son of a hand-loom weaver of Kirriemuir, Forfarshire; educated at Glasgow and Dumfries academies and Edinburgh University; MA, 1882; leader-writer and sub-editor, *Nottingham Journal*, 1883; wrote sketches for *St. James's Gazette* in some of which Kirriemuir appeared as

'Thrums'; moved to London, 1885; published *When a Man's Single* (in *British Weekly* under signature 'Gavin Ogilvy', 1887-8); *Auld Licht Idylls* (1888); *A Window in Thrums* (1889); *My Lady Nicotine* (1890); his first novel *The Little Minister* (1891); *Margaret Ogilvy* (1896, a tribute to his mother); *Sentimental Tommy* (1896) and *Tommy and Grizel* (1900), stories of a small boy's fantasies; his plays include *Ibsen's Ghost* (1891); *Walker, London* (1892); *The Professor's Love Story* (1892, a marked success in the United States); a successful dramatization of *The Little Minister* (1897); *The Wedding Guest* (1900); *Quality Street* (1902, a sentimental comedy); *The Admirable Crichton* (1902); *Little Mary* (1903); *Peter Pan* (1904, a children's play, annually revived; the statue of Peter Pan in Kensington Gardens was privately erected by Barrie in 1912); *Alice-Sit-by-the-Fire* (1905); *What Every Woman Knows*, one of Barrie's greatest successes (1908); *Rosalind* (1912); *A Kiss for Cinderella* (1916); *Dear Brutus* (1917); *Mary Rose* (1920); *The Boy David* (1936); a most individual writer who rarely drew his inspiration from the circumstances of the day; had very few failures; some of his one-act plays, such as *The Twelve-Pound Look* (1910) and *Shall We Join the Ladies?* (1921), are masterpieces of construction; lord rector of St. Andrews (1919) and chancellor of Edinburgh (1930); baronet, 1913; OM, 1922.

BARRINGTON, RUTLAND (1853-1922), actor and vocalist, whose original name was GEORGE RUTLAND BARRINGTON FLEET; went on the stage, 1874; acted with Richard D'Oyly Carte [q.v.] at Opera Comique and Savoy Theatre, 1877-88; played several original parts in Gilbert and Sullivan operas; played again at the Savoy, 1889-94; acted with George Edwardes's company at Daly's Theatre, 1894 and 1896-1904.

BARRINGTON-WARD, SIR LANCELOT EDWARD (1884-1953), surgeon; brother of R. M. Barrington-Ward [q.v.]; educated at Westminster, Bromsgrove, and Edinburgh University; qualified in medicine, 1908; Ch.M., 1913; at Great Ormond Street Hospital for Sick Children from 1910; consultant, 1914; also at Wood Green and Royal Northern hospitals; surgeon to royal household, 1937; extra surgeon to Queen, 1952; KCVO, 1935; wrote *Abdominal Surgery of Children* (1928); edited *Royal Northern Operative Surgery* (1939).

BARRINGTON-WARD, ROBERT McGOWAN (1891-1948), journalist; educated at Westminster and Balliol College, Oxford; worked on *The Times*, 1913-14; served in France (MC and DSO) and realized ineffectiveness of war, 1914-18; assistant editor,

Observer, 1919–27; assistant editor (1927), deputy editor (1934), editor (1941–8) of *The Times*; wrote extensively on Anglo–German relations, for long seeking solution other than war; more successful on domestic affairs, notably unemployment.

BARRY, ALFRED (1826–1910), primate of Australia and canon of Windsor; second son of Sir Charles Barry [q.v.]; educated at King's College, London, and Trinity College, Cambridge; fourth wrangler, second Smith's prizeman, and seventh classic, 1848; MA, 1851; DD, 1866; principal of Cheltenham College, 1862–8, and of King's College, London, 1868–83; canon of Worcester, 1871, of Westminster, 1881, and of Windsor, 1891–1910; archbishop of Sydney and primate of Australia, 1884–9; Boyle lecturer, 1876–8; Bampton lecturer, Oxford, 1892; Hulsean lecturer, Cambridge, 1894; rector of St. James, Piccadilly, 1895–1900; effective preacher of broad church doctrine; hon. DCL, Oxford, 1870; publications include *What is Natural Theology?* (1877) and *The Teacher's Prayer Book* (1884).

BARRY, ERNEST JAMES (1882–1968), professional sculler; won Doggett's Coat and Badge, 1903; beat George Towns (British champion since 1899), 1908; successfully defended *Sportsman* Challenge Cup for British championship, 1911–12; defeated Dick Arnst for world championship, 1912; defeated Harry Pearce, of Australia, 1913; lost to J. Fulton of Australia, 1919; defeated Fulton, and won fifth world championship, 1920; after retirement, worked as professional coach in Denmark and Germany; appointed King's Bargemaster; retired because of ill health, 1952.

BARRY, Sir GERALD REID (1898–1968), journalist and administrator; educated at Marlborough and Corpus Christi College, Cambridge; joined *Saturday Review*, 1921; editor, 1925–30; editor, *Week-end Review*, 1930–4; features editor, *News Chronicle*, 1934; managing editor, 1936–47; director-general, Festival of Britain, 1948–52; publications include *This England* (1933), and *Report on Greece* (1945); knighted, 1951; hon. ARIBA (1940); FRSA.

BARRY, Sir JOHN WOLFE WOLFE- (1836–1918), civil engineer. [See WOLFE-BARRY.)

BARSTOW, Sir GEORGE LEWIS (1874–1966), civil servant and chairman of the Prudential Assurance Company; educated at Clifton College and Emmanuel College, Cambridge; first class, parts i and ii, classical tripos,

1895–6; entered Local Government Board, 1896; Treasury, 1898; assistant secretary, supply services, 1919–27; government director, Anglo-Persian (later Anglo-Iranian) Oil Company, 1927–46; director, Prudential Assurance Company, 1928; deputy chairman, 1935; chairman, 1941–53; president, Swansea University College, 1929–55; deputy chairman, Court of London University; CB, 1913; KCB, 1920.

BARTHOLOMEW, JOHN GEORGE (1860–1920), cartographer; manager of father's firm, 'Edinburgh Geographical Institute', 1889; published great atlases of Scotland (1895) and England and Wales (1903); greatest achievement '*The Times' Survey Atlas of the World*, completed 1921; improved system of layer colouring for marking contours.

BARTLET, JAMES VERNON (1863–1940), ecclesiastical historian; educated at Highgate School and Exeter College, Oxford; first class, theology, 1887; tutor (1889), senior tutor (1890–1900), professor of church history (1900–28), Mansfield College, Oxford; secretary, Oxford Society of Historical Theology, 1894–1936; although never ordained he was a recognized Congregational leader.

BARTLETT, Sir ELLIS ASHMEAD (1849–1902), politician; born at Brooklyn, USA; came to England in boyhood; BA, Christ Church, Oxford, 1872; MA, 1874; president of Union, 1873; called to bar, 1877; visited Serbia and Bulgaria, 1877–8; conservative MP for Eye, 1880–4, and for Ecclesall from 1885 to death; strenuous and grandiloquent advocate of British imperialism; conducted *England*, first conservative penny weekly newspaper, 1880–98; civil lord of the Admiralty, 1885, 1886–92; knighted, 1892; with Turkish army in war with Greeks and taken prisoner, 1897; in South Africa, 1899; published *Shall England keep India?* (1886), *The Battlefields of Thessaly* (1897), and other works.

BARTLETT, Sir FREDERIC CHARLES (1886–1969), psychologist; educated at home due to ill health, and at London University; first class, philosophy, 1910; MA, 1912; and St. John's College, Cambridge; first class, part i, moral sciences tripos, 1914; assistant director, labora-tory of experimental psychology, 1914; director, 1922; first professor, experimental psychology, Cambridge, 1931–52; member, flying personnel research committee, Air Ministry; member, Medical Research Council; CBE, 1941; knighted, 1948; publications include *Remembering* (1932); editor *The British Journal of Psychology*, 1924–48; FRS, 1932; president (1950) and hon. fellow (1954), the British

Psychological Society; hon. degrees, Athens, Princeton, Louvain, London, Edinburgh, Oxford, and Padua.

BARTLEY, Sir GEORGE CHRISTOPHER TROUT (1842-1910), founder of the National Penny Bank; educated at University College School; science examiner (1860) and assistant director of science division (1866-80) in education department, South Kensington; wrote pamphlets on social questions from 1870, including poor law reform, 1873-6; advocated old age pensions, 1872; founded Middlesex Penny Bank (1872) and the National Penny Bank (1875), which met with rapid success; conservative MP for North Islington, 1885-1906; KCB, 1902.

BARTON, Sir EDMUND (1849-1920), Australian statesman; born in Sydney; called to bar, 1871; member of New South Wales legislative assembly five times between 1879 and 1900; speaker, 1883-7; attorney-general, 1889 and 1891-3; leader of federal convention, 1897; largely responsible for new constitution bill; first prime minister of Australian Commonwealth, 1901-3; GCMG, 1902; judge of Australian high court of justice, 1903.

BARTON, JOHN (1836-1908), missionary; nephew of Bernard Barton [q.v.]; BA, Christ's College, Cambridge, 1859; MA, 1863; founded Cambridge University Church Missionary Union; principal of new cathedral missionary college, Calcutta, 1865; vicar of Holy Trinity, Cambridge, 1877-93; chief secretary, Church Pastoral Aid Society, 1893-8; wrote geographical works.

BARTON, Sir SIDNEY (1876-1946), diplomatist; educated at St. Paul's School; posted to Peking in consular service, 1895; Chinese secretary, 1911; consul-general, Shanghai, 1922-9; energetic in protecting British interests; minister in Addis Ababa, 1929-36; won confidence of Emperor; organized security of British community and remained until after Italians captured the city; KBE, 1926; KCVO, 1931; GBE, 1936.

BASHFORTH, FRANCIS (1819-1912), ballistician; BA, St. John's College, Cambridge (second wrangler); fellow, 1843; ordained, 1850; rector of Minting, near Horncastle, 1857-1908; professor of applied mathematics, Woolwich, 1864-72; carried out and reported on series of important ballistic experiments, 1864-80; invented chronograph for determining air-resistance to projectiles, 1865.

BASS, MICHAEL ARTHUR, first BARON BURTON (1837-1909), brewer; educated at Harrow and Trinity College, Cambridge; BA, 1860; MA, 1864; extended father's brewing business; liberal MP for Stafford (1865-8), East Staffordshire (1868-85), and Burton division of Staffordshire (1885-6); personal friend of Gladstone; baronet, 1882; baron, 1886; became liberal unionist on home rule question; supported tariff reform, 1903; frequent host of King Edward VII; KCVO, 1904; generous benefactor to Burton-on-Trent; built ferrybridge and many churches there; an art connoisseur.

BASSETT-LOWKE, WENMAN JOSEPH (1877-1953), model maker; joined father's Northampton engineering firm; founded Bassett-Lowke Ltd., producing scale model toys and exhibition models; opened shops in London, Manchester, and Edinburgh; provided models for Normandy invasion, 1944; edited *Model Railway Handbook* from 1906; publications include Penguin *Book of Trains* (1941) and Puffin book *Locomotives* (1947).

BATEMAN, HENRY MAYO (1887-1970), cartoonist; educated at Forest Hill House, London; studied drawing and painting at Westminster School of Art and the Goldsmith's Institute; worked in London studio of Charles van Havenmaet; first published drawings in *Scraps*, 1903; began long association with *Tatler*, 1904; first one-man exhibition *Satires and Caricatures*, Brook St. Gallery, 1911; drew theatrical caricatures for *Tatler*, 1912; began 'The Man Who' series, illustrating social gaffs and mishaps; published in *Punch*, 1915; second one man show, Leicester Galleries, 1919; published collections of drawings, and had number of exhibitions arranged, 1921-78.

BATES, CADWALLADER JOHN (1853-1902), antiquary; of old Northumbrian descent; educated at Eton and Jesus College, Cambridge; BA, 1872; MA, 1875; travelled much in Poland and the Carpathians; practical farmer, revived famous Kirklevington shorthorns; developed taste for hagiography and joined Roman Church, 1893; recognized authority on medieval history of Northumbria; wrote *Border Holds* (1891) and *History of Northumberland* (1895).

BATES, Sir PERCY ELLY, fourth baronet (1879-1946), merchant and ship-owner; educated at Winchester; partner in Edward Bates & Sons, merchants and shipowners, 1900-46; succeeded brother, 1903; joined Cunard Steam-Ship Company, 1910; deputy chairman, 1922; chairman, 1930-46; chairman (1934-46) of Cunard White Star limited company formed to complete the *Queen Mary* (1934) and build *Queen Elizabeth* (1938) for Atlantic crossing; directed commercial shipping services, 1914-

18, and served on advisory council, Ministry of War Transport, 1939-45; director of *Morning Post*, 1924-37; GBE, 1920.

BATESON, SIR ALEXANDER DING-WALL (1866-1935), judge; educated at Rugby and Trinity College, Oxford; called to bar (Inner Temple), 1891; bencher, 1920; junior counsel to Admiralty, 1909; KC, 1910; judge of Probate, Divorce, and Admiralty division, 1925-35; possessed of sound common sense and the capacity for decision.

BATESON, MARY (1865-1906), historian; daughter of William Henry Bateson [q.v.]; student of Newnham College, Cambridge; placed second in first class, historical tripos, 1887; lecturer at Newnham from 1888 till death; under the influence of Mandell Creighton [q.v.] she made special study of monastic history; published *The Register of Crabhouse Nunnery* (1889) and *Origin and Early History of Double Monasteries* (1899); turned to municipal history, editing *Records of the Borough of Leicester* (3 vols., 1899-1905), *The Charters of the Borough of Cambridge* (1901), *The Cambridge Gild Records* (1903), and *Grace Book B* (2 vols., 1903-5); also edited works for Early English Text, Camden, and other antiquarian societies; her *Laws of Breteuil* (1900-1) and *Borough Customs* (2 vols., 1904-6) show masterly scholarship; wrote *Mediaeval England* (in 'Story of the Nations' series, 1903); contributed to *Social England, Cambridge Modern History*, and to this Dictionary; Warburton lecturer, Manchester University, 1905; ardent advocate of women's emancipation.

BATESON, WILLIAM (1861-1926), biologist; son of William Henry Bateson and brother of Mary Bateson [qq.v.]; educated at Rugby and St. John's College, Cambridge; took up study of embryology; visited United States; fellow of St. John's College, Cambridge, 1885-1910; champion of Mendelism, 1900-4; professor of biology at Cambridge, 1908-9; director of John Innes Horticultural Institution, Merton, 1910-26; FRS., 1894; founder of experimental study of heredity and variation which he termed 'genetics'.

BATHURST, CHARLES, first VISCOUNT BLEDISLOE (1867-1958), agriculturist and public servant; educated at Sherborne, Eton, and University College, Oxford; called to bar (Inner Temple), 1892; studied at Royal Agricultural College, 1893-6; introduced concept of landownership as a profession; a founder of Country Landowners' Association; conservative MP, South Wiltshire, 1910-18; parliamentary secretary, Ministry of Food,

1916-17; chairman, royal commission sugar supplies, and director of sugar distribution, 1917-19; KBE, 1917; baron, 1918; parliamentary secretary, Ministry of Agriculture, 1924-8; PC, 1926; governor-general of New Zealand, 1930-5; viscount, 1935; chairman, royal commission on closer union of Rhodesias and Nyasaland, 1938.

BATSFORD, HARRY (1880-1951), publisher, bookseller, and author; educated at Henley House and City of London schools; entered family business, 1897; chairman and managing director, 1917-51; published (and also wrote) books on British architecture and topography such as the 'British Heritage' series; honorary ARIBA, 1926.

BATTENBERG, PRINCE LOUIS ALEXANDER OF (1854-1921), admiral of the fleet. [See MOUNTBATTEN, LOUIS ALEXANDER.]

BAUERMAN, HILARY (1835-1909), metallurgist; original student at School of Mines, 1851; assistant to Geological Survey, 1855; geologist to North American boundary commission, 1858-63; surveyor for Indian and Egyptian governments, 1867-9; lecturer in metallurgy at Firth College, Sheffield, 1883; professor at Ordnance College, Woolwich, 1888-1906; wrote much for technical journals; published *Metallurgy of Iron* (1868) and *Systematic Mineralogy* (1881).

BAX, SIR ARNOLD EDWARD TREVOR (1883-1953), composer; studied at Royal Academy of Music; adopted Celtic, later Nordic, idiom; tone poems include *In the Faëry Hills* (1910), *The Garden of Fand* (1916), *Tintagel* (1917), *The Tale the Pine Trees Knew* (1931); wrote seven symphonies (1922-39); two works for piano and orchestra for Harriet Cohen [q.v.]; Cello Concerto (1932); Violin Concerto (1937-8); and much chamber and choral music; a 'brazen Romantic' his work was robust yet wistful; master of King's Musick, 1942-52, of the Queen's, 1952-3; knighted, 1937; KCVO, 1953.

BAXTER, LUCY (1837-1902), writer on art under pseudonym of LEADER SCOTT; daughter of William Barnes [q.v.], the Dorsetshire poet; married (1867) Samuel Thomas Baxter at Florence, where she lived till death; collaborated in literary research with John Temple Leader [q.v.]; works on Italian art include *The Cathedral Builders*, 1899 and 1900.

BAYLEY, SIR STEUART COLVIN (1836-1925), administrator in India; son of William Butterworth Bayley [q.v.]; posted to Lower

Bengal, 1856; chief commissioner of Assam, 1878; lieutenant-governor of Bengal, 1887; political secretary to India Office, 1890; member of India Council, 1895–1905; KCSI, 1878; GCSI, 1911.

BAYLIS, LILIAN MARY (1874–1937), theatrical manager; studied violin under J. T. Carrodus [q.v.]; taught music in Johannesburg; returned to England (1898) to assist her aunt, Emma Cons, in management of Royal Victoria Hall (the 'Old Vic') then a philanthropic temperance music-hall; developed its musical activities; becoming sole manager (1912) made it 'the home of Shakespeare' and a leading theatre; acquired Sadler's Wells Theatre for opera and ballet, 1931; CH, 1929.

BAYLIS, THOMAS HENRY (1817–1908), lawyer and author; educated at Harrow and Brasenose College, Oxford; BA, 1835; MA, 1841; called to bar, 1856; QC, 1875; judge of court of passage, Liverpool, 1876–1903; works include *The Temple Church* (1893) and legal treatises.

BAYLISS, SIR WILLIAM MADDOCK (1860–1924), physiologist; educated at University College, London, and Wadham College, Oxford; returned to University College, 1888; carried out physiological research in collaboration with Ernest Henry Starling [q.v.] with whom he discovered secretin, 1902; professor of general physiology, University College, 1912; connected with Physiological Society, 1885–1924; FRS, 1903; Croonian lecturer, 1904; knighted, 1922; chief work *Principles of General Physiology* (1915).

BAYLISS, SIR WYKE (1835–1906), painter and writer; took special interest in architecture; exhibited at Royal Academy 'La Sainte Chapelle', 1865; president (1888–1906) of Royal Society of British Artists, where he exhibited 'St. Peter's, Rome', 1888, 'The Cathedral, Amiens', 1900; writings include *Rex Regum* (1898) and *Olives, the Reminiscences of a President* (1906); FSA, 1870; knighted, 1897.

BAYLY, ADA ELLEN (1857–1903), novelist under the pseudonym of EDNA LYALL; of deeply religious temperament; ardent supporter of women's emancipation and of all political liberal movements; supported Charles Bradlaugh's fight for religious liberty (1880–5); published her first story, *Won by Waiting* (1879); best works were *Donovan* (1882), admired by Gladstone, its sequel, *We Two* (1884), and *In the Golden Days* (1885); her *Autobiography of a Slander* (1887) had wide vogue; *Doreen* (1894), *Autobiography of a Truth* (1896), and *The Hinderers* (1902),

touched on political questions; her clear style, constructive faculty, and skilful characterization of young girls were qualified by her earnest political purpose; fond of music and travel.

BAYLY, SIR LEWIS (1857–1938), admiral; entered navy; specialized in torpedo; commanded destroyer flotilla, 1907; rear-admiral, 1908; president, War College, 1908–11; commanded battle cruiser squadron (1911–13), third battle squadron (1913–14), first battle squadron and Channel fleet (1914), Western Approaches, 1915–19; conducted a vigorous anti-submarine campaign, from 1917 with mixed Anglo-American force; admiral, 1917; KCB, 1914; KCMG, 1918.

BAYNES, NORMAN HEPBURN (1877–1961), historian; educated at Eastbourne College and New College, Oxford; called to bar (Lincoln's Inn), 1903; tutor under Law Society, 1903–16; assistant, history department, University College, London, 1913; reader, history of the Roman Empire, 1919; professor, Byzantine history, 1931; fellow, University College, 1936; hon. professor, Byzantine history, 1937–42; emeritus professor; Foreign Research and Press Service, Oxford, 1939–45; returned to Byzantine studies in London, 1945; publications include *Byzantine Empire* (1925), *Historia Augusta* (1926), and chapters on Byzantine emperors in *Cambridge Medieval History* (vols i and ii); FBA, 1930; hon. fellow, Westfield College, London (1937), and New College, Oxford (1947); doctor *honoris causa*, St. Andrews (1934), Oxford (1942), Durham (1946), Cambridge (1949), and London (1951).

BEACH, SIR MICHAEL EDWARD HICKS, first EARL ST. ALDWYN (1837–1916), statesman. [See HICKS BEACH.]

BEALE, DOROTHEA (1831–1906), principal of Cheltenham Ladies' College; educated in Paris (1847–8) and at Queen's College Harley Street (1848); mathematical tutor there, 1849; head teacher of clergy daughters' school, Casterton, Westmorland (the Lowood of Charlotte Brontë's *Jane Eyre*), 1857–8; successful principal of Cheltenham Ladies' College, 1858–1906; evidence before endowed schools inquiry commission (1865) gave immense impetus to girls' education; founded St. Hilda's College, Cheltenham, first English training college for secondary women teachers, 1885, and St. Hilda's Hall, opened at Oxford in 1893 with a view to giving teachers in training the benefit of a year at Oxford; president of Headmistresses' Association, 1895–7; ardent supporter of women's suffrage; received freedom of borough of Cheltenham, 1901; hon.

LLD, Edinburgh, 1902; collaborated in *Work and Play in Girls' Schools*, 1898.

BEALE, LIONEL SMITH (1828-1906), physician and microscopist; medical student at King's College, London; assistant to Sir Henry Acland [q.v.] at Oxford, 1847; MB, 1851; resident physician at King's College Hospital, London, 1850-51; professor of physiology and anatomy at King's College, 1853-69, of pathological anatomy, 1869-76, and of medicine, 1876-96; FRCP, 1859; Lumleian lecturer, 1875; FRS, 1865; delivered Croonian lectures, 1865; his many works include *The Microscope and its Application to Clinical Medicine* (1854), and *The Structure and Growth of the Tissues* (1865), *Disease Germs* (1872), *Bioplasm* (1872), and other works, foreshadowing by microscopic investigation modern conceptions of bacterial disease; first to practise method of fixing tissues by injections; discovered 'Beale's cells'; skilful draughtsman and illustrator of his own works; president of Microscopical Society, 1879-80; long speculated on philosophy, publishing *Life Theories* (1870-1) and *Our Morality* (1887).

BEARDMORE, WILLIAM, BARON INVERNAIRN (1856-1936), shipbuilder; served apprenticeship at Parkhead Forge, controlled by his father; founded William Beardmore & Co., world-famous for building men-of-war, merchant ships, and the R. 34, first airship to make double crossing of Atlantic; baronet, 1914; baron, 1921.

BEARSTED, first VISCOUNT (1853-1927), joint-founder of the Shell Transport and Trading Company. [See SAMUEL, MARCUS.]

BEATRICE MARY VICTORIA FEODORE (1857-1944), princess of Great Britain and Ireland; fifth daughter and constant companion of Queen Victoria by whose instructions she transcribed and edited the Queen's diaries after her death; married (1885) Prince Henry Maurice of Battenberg [q.v.]; governor of Isle of Wight, 1896-1944.

BEATTIE-BROWN, WILLIAM (1831-1909), Scottish landscape painter; studied at Trustees' Art Academy, Edinburgh; exhibited at the Royal Scottish Academy from 1848 till death, at Royal Academy and elsewhere, Scottish highland landscapes; showed much technical skill and accuracy; RSA, 1884; diploma work 'Coire-na-Faireamh', 1883.

BEATTY, SIR (ALFRED) CHESTER (1875-1968), mining engineer, art collector, and philanthropist; born in New York; educated at Westminster School, Dobbs Ferry, New York, Columbia University (School of Mines), and Princeton; graduated as mining engineer; appointed consulting engineer and assistant general manager, Guggenheim Exploration Company, 1903; settled in England and founded Selection Trust Limited, 1913; developed zinc and lead mines in Siberia and Serbia, and diamond mines in West Africa; aided development of Copperbelt in Northern Rhodesia (later Zambia); became British subject, 1933; retired and settled in Dublin, 1950; built up unique collection of Indian and Persian miniatures and manuscripts and early Bibles;, also collected rare books and pictures; presented Impressionist pictures to National Gallery in Dublin, 1950; handed over to Irish nation thirteenth-century *Book of Hours*, 1955; financed cancer research, setting up Chester Beatty Research Institute (later the Institute of Cancer Research: Royal Cancer Hospital); knighted, 1954; first hon. citizen of the Irish Republic.

BEATTY, DAVID, first EARL BEATTY (1871-1936), admiral of the fleet; born in Cheshire of old Irish stock; entered *Britannia*, 1884; DSO, 1896, for daring leadership of gunboat force on the Nile; specially promoted commander (1898) for further services on Nile; captain (1900) for services during Boxer rising; rear-admiral (1910, the youngest flag officer for over a hundred years); naval secretary to (Sir) Winston Churchill, 1912-13; commanded battle cruiser squadron in Grand Fleet, 1913-16; urged equipment of Cromarty and Rosyth as operational bases; his prompt intervention at Heligoland Bight (28 Aug. 1914) turned disaster to success; intercepted Admiral Hipper's force at Dogger Bank (24 Jan. 1915) severely damaging the *Seydlitz* and destroying the *Blücher*; owing to damage inflicted on his flagship he was unable to retain his leadership, his signals were misinterpreted, and the enemy escaped; vice-admiral 1915; on 30 May 1916 left Rosyth with six battle cruisers and the fifth battle squadron to join (Lord) Jellicoe [q.v.] and the Grand Fleet in an attempt to draw the German forces into the Skagerrak; on the afternoon of 31 May he sighted the German battle cruisers and a fierce battle ensued in which the *Indefatigable* and the *Queen Mary* blew up; nevertheless, the British fire was beginning to tell when the High Sea Fleet was sighted and he reversed his course northwards in order to lead Admiral Scheer into Jellicoe's clutches; after an hour of dogged fighting he turned sharply eastward to prevent Hipper from sighting the British fleet which was thus enabled to complete its deployment unobserved; he then took up his position in the van and by 6.30 p.m. the main battle fleets were in action (battle of Jutland); he thus carried out the duties assigned to him with conspicuous success; commander-in-chief, Grand Fleet, 1916-19; encouraged senior officers to use their initiative; introduced a system of plotting positions on a synchronized basis; a firm believer in the convoy system, he ran his own to and from Norway; accepted surrender of German High Sea Fleet, Nov. 1918; admiral (Jan.), admiral of the fleet (Apr.), 1919; first sea lord, 1919-27; withstood attempts to reduce naval strength; advocated a strongly defended base at Singapore and Admiralty control of the Fleet Air Arm; confirmed the creation of the department of scientific research and established the Admiralty experimental laboratory at Teddington; KCB, 1914; KCVO and GCB, 1916; GCVO, 1917; OM and earl, 1919; PC, 1927; lord rector of Edinburgh University, 1917-36; of dauntless moral and physical courage; in moments of crisis his brain worked with absolute clarity; his judgement was sound and his decisions were the result of careful reflection.

BEATTY, SIR EDWARD WENTWORTH (1877-1943), Canadian man of business; educated at Upper Canada College and Toronto University; called to the bar (Ontario), 1901; Dominion KC, 1915; vice-president (1914), president (1918-42), Canadian Pacific Railway; chancellor, McGill University, 1921-43; GBE, 1935.

BEAUCHAMP, seventh EARL (1872-1938), politician. [See LYGON, WILLIAM.]

BEAVER, SIR HUGH EYRE CAMPBELL (1890-1967), industrialist and engineer; educated at Wellington College; served in Indian police; personal assistant to Sir Alexander Gibb [q.v.], consulting engineer, 1922; led team surveying ports of Canada, 1931; secretary and partner in Gibb's firm, 1932; director-general, Ministry of Works, 1940-5; managing director, Guinness's, 1946-60; chairman, British Institute of Management, 1951; chairman, Advisory Council of the Department of Scientific and Industrial Research, 1954-6; president, Federation of British Industries, 1957-9; knighted, 1943; KBE, 1956; hon. degrees, Cambridge, Trinity College, Dublin, and National University of Ireland.

BEAVERBROOK, first BARON (1879-1964), newspaper proprietor. [See AITKEN, WILLIAM MAXWELL.]

BEAZLEY, SIR JOHN DAVIDSON (1885-1970), classical archaeologist; educated at King Edward VI School, Southampton, Christ's Hospital, and Balliol College, Oxford; first class, *lit. hum.*, 1907; student of Christ Church and tutor in classics, 1908-25; Lincoln professor of classical archaeology, 1925-56; expert in Attic-verse painting; knighted, 1949; CH, 1959; FBA, 1927; publications include *Attic Red-figure Vases in American Museums* (1918), *Greek Vases in Poland* (1928), *Potter and Painter in Ancient Athens* (1945), *The Berlin Painter* (1964), and *Greek Sculpture and Painting* (with B. Ashmole, 1932); hon. degrees from Oxford, Cambridge, Glasgow, Durham, Reading, Paris, Lyon, Marburg, and Thessalonika; hon. fellow, Balliol and Lincoln Colleges.

BECKETT, SIR EDMUND, fifth baronet, and first BARON GRIMTHORPE (1816-1905), lawyer, mechanician, and controversialist; educated at Eton and Trinity College, Cambridge; BA, 1838; MA, 1841; LLD, 1863; called to bar, 1841; QC, 1854; recognized as leader of parliamentary bar, 1860; chancellor of province of York, 1877-1900; baron, 1886; engaged in theological controversy; opposed to New Testament revision and to ritualism; first president of Protestant Churchmen's Alliance, 1889; interested in ecclesiastical architecture, designing several Yorkshire churches; wrote *A Book on Building, Civil and Ecclesiastical*, 1876; prominent and generous, from 1877 onwards, in restoration of St. Albans Cathedral, the cause of much controversy; fond of mechanical invention; published *A Rudimentary Treatise on Clock and Watch Making* (1850, often reprinted); designed clocks for 1851 exhibition, for the clocktower in Houses of Parliament (1859), and for St. Paul's Cathedral (finished 1893); president of Horological Institute, 1868; author of *Astronomy without Mathematics* (1865), *Life of John Lonsdale, Bishop of Lichfield* (his father-in-law, 1868), and controversial works.

BEDDOE, JOHN (1826-1911), physician and anthropologist; studied medicine at University College, London (BA, 1851), and at Edinburgh (MD, 1853); on medical staff in Crimea; physician to Bristol Royal Infirmary, 1862-73; FRCP, 1873; made in youth pioneer ethnological researches all over Europe; observed hair and eye colours in west of England (1846) and Orkney (1853); FRS, 1873; hon. LLD, Edinburgh, 1891; Rhind lecturer there on 'The Anthropological History of Europe', 1891; wrote *Contributions to Scottish Ethnology* (1853), *The Races of Britain* (1885), and *Memories of Eighty Years* (1910).

BEDFORD, eleventh DUKE OF (1858-1940), and **BEDFORD**, DUCHESS OF (1865-1937). [See RUSSELL, HERBRAND ARTHUR.]

48

BEDFORD, WILLIAM KIRKPATRICK RILAND (1826-1905), antiquary and genealogist; BA, Brasenose College, Oxford, 1848; MA, 1852; secretary of Union, 1848; rector of Sutton Coldfield, 1850-92; authority on antiquities of Sutton Coldfield; official genealogist and historian of order of St. John of Jerusalem; wrote on cricket, archery, and other topics.

BEDSON, Sir **SAMUEL PHILLIPS** (1886-1969), experimental pathologist and virologist; educated at Abbotsholme School and Durham University; B.Sc., 1907; graduated in medicine, 1912; studied bacteriology at Pasteur Institute, Paris; research work at Lister Institute, London; MD, Durham, 1914; wounded in 1914-18 war; RAMC pathologist, 1916-19; lecturer in bacteriology, Durham Medical School, 1919-21; research at Lister Institute, on foot and mouth disease, 1921-6; Freedom research fellow, London Hospital, 1926-34; important work on the nature of viruses; professor of bacteriology, London Hospital Medical College, 1934-52; pathologist in Emergency Medical Service, 1939-44; director, pathology division, London Hospital, 1946; consultant adviser in pathology, Ministry of Health, 1949-60; in charge of British Empire Campaign Virus Unit, Middlesex Hospital, 1952-62; FRS, 1935; FRCP, 1945; knighted, 1956; hon. D.Sc., Belfast (1937), and Durham (1946).

BEECHAM, THOMAS (1820-1907), patent medicine vendor; patented in 1847 a new formula for pills; employed 'Worth a guinea a box' as his advertising motto; built large factories at St. Helens and New York (1885) and on continent; benefactor to South Lancashire.

BEECHAM, Sir **THOMAS** (1879-1961), conductor; grandson of Thomas Beecham [q.v.]; educated at Rossall School and Wadham College, Oxford, 1897-8; founded St. Helens Orchestral Society; studied music in London, 1900, conducted first public orchestral concert, London, 1905; founded New Symphony Orchestra, 1906; first Covent Garden opera season, 1910; presented Diaghilev's Russian Ballet, 1911; introduced Chaliapin, 1913; knighted and succeeded to baronetcy, 1916; first Delius Festival in London, 1929; founded London Philharmonic Orchestra, 1932; formed Royal Philharmonic Orchestra, 1946; conducted extensively in London, Paris, and United States, 1950-7; CH, 1957; publications include *A Mingled Chime* (1944).

BEECHING, HENRY CHARLES (1859-1919), dean of Norwich, man of letters; BA, Balliol College, Oxford; rector of Yattendon,

Berkshire, 1885-1900; professor of pastoral theology, King's College, London, 1900; canon of Westminster, 1902; dean of Norwich, 1911 till death; select preacher at Oxford, Cambridge, and Dublin; writer of verse and essayist.

BEERBOHM, Sir HENRY MAXIMILIAN (MAX) (1872-1956), author and cartoonist; half-brother of Sir Herbert Beerbohm Tree [q.v.]; educated at Charterhouse and Merton College, Oxford; popular in London's artistic and fashionable life; dramatic critic, *Saturday Review*, 1898-1910; married an American actress and settled in Rapallo, 1910; the best essayist, parodist, and cartoonist of his age, his satire was ruthless and urbane, product of a highly cultivated intelligence; works include *The Happy Hypocrite* (1897), *Zuleika Dobson* (1911), and *Seven Men* (1919), volumes of essays and of cartoons; knighted, 1939; buried in St. Paul's.

BEEVOR, CHARLES EDWARD (1854-1908), neurologist; studied medicine at University College Hospital, London, and abroad; MD, 1881; FRCP, 1888; physician to National Hospital for the Epileptic, London, and Great Northern Hospital; made valuable researches on the localization of cerebral functions and cerebral arterial circulation; president of Neurological Society, 1907-8; published standard treatises on anatomy and nervous diseases.

BÉGIN, LOUIS NAZAIRE (1840-1925), cardinal and archbishop of Quebec; born in French Canada; educated at Quebec and Rome; professor of theology, Laval University, Quebec, 1868-84; bishop of Chicoutimi, 1888-91; archbishop of Quebec, 1898-1925; cardinal, 1914.

BEILBY, Sir GEORGE THOMAS (1850-1924), industrial chemist; educated at Edinburgh University; joined Oakbank Oil Company as chemist: turned his attention to production of cyanides, 1890; made detailed study of the flow of solids; interested in economical use of fuel; first chairman and director of Fuel Research Board, 1917; FRS, 1906; knighted, 1916; published *Aggregation and Flow of Solids* (1921).

BEIT, ALFRED (1853-1906), financier and benefactor; born at Hamburg of Jewish family; joined firm of diamond merchants, 1870; went to Kimberley, South Africa, 1875; formed independent business there, 1878; joined firm of J. Porges & (Sir) Julius Wernher [q.v.], 1882; settled in London, 1888, and formed firm of Wernher, Beit & Co., 1890; intimate friend and adviser of Cecil Rhodes [q.v.], with whom he amalgamated chief Kimberley diamond mines as the De Beers Consolidated Mines, 1888; developed the Transvaal gold mines, and evolved the Great Deep Level scheme, 1891; an original director of British South Africa Company, 1889; implicated with Rhodes in Jameson raid, 1895; generous contributor to South African war funds, 1899–1902; an ardent imperialist; formed fine collection of pictures and bronzes; founded Beit professorship of colonial history at Oxford, 1905; benefactor to Imperial College of Technology, London, to Rhodesia, to London, and Hamburg charities, and to National Gallery; thirty fellowships for medical research founded in his memory, 1909.

BEIT, SIR OTTO JOHN, first baronet (1865–1930), financier and philanthropist; born at Altona, Germany; came to England, and was naturalized, 1888; director of British South Africa Company, 1910; KCMG, 1920; baronet, 1924; as residuary legatee and trustee (1906) of his brother Alfred Beit [q.v.], administered vast wealth philanthropically.

BEITH, JOHN HAY (1876–1952), writer as IAN HAY; grandson of Alexander Beith [q.v.]; educated at Fettes and St. John's College, Cambridge; taught at Durham School (1902–6) and Fettes (1906–12); publications include *Pip* (1907), *A Knight on Wheels* and *Lighter Side of School Life* (1914), *The First Hundred Thousand* (1915), and *Housemaster* (1936); adapted his novels for the theatre, notably *Tilly of Bloomsbury*, 1919, and proved effective collaborator for stage; CBE, 1918; director, war office public relations, 1938–41.

BELCHER, JOHN (1841–1913), architect; partner in father's City practice, 1865–75; abandoned Gothic for Italian renaissance style; designs include offices of Institute of Chartered Accountants in the City, 1890; Victoria and Albert Museum, 1891; Whiteley's stores, Bayswater, 1912; offices of London Zoological Society, 1913; ARA, 1900; RA, 1900; president of Royal Institute of British Architects, 1904.

BELISHA, (ISAAC) LESLIE HORE-, BARON HORE-BELISHA (1893–1957), politician. [See HORE-BELISHA.]

BELL, ALEXANDER GRAHAM (1847–1922), inventor of the telephone and educator of the deaf; educated at Edinburgh and London universities; emigrated to Canada, 1870; carried on work for deaf in United States; professor of vocal physiology, school of oratory, Boston University, 1873; constructed first rough telephone in Boston, 1875; introduced telephone into England and France, 1877; financed Volta Bureau, Washington, for increasing knowledge relating to deaf; interested in eugenics; naturalized American citizen, 1874.

BELL, (ARTHUR) CLIVE (HEWARD) (1881–1964), art critic; educated at Marlborough and Trinity College, Cambridge; in Paris studying art, 1903–4; married Vanessa, daughter of Sir Leslie Stephen [q.v.], 1907; formed nucleus of 'Bloomsbury group'; with Roger Fry [q.v.]; selected exhibits for first Post-Impressionist exhibition in London, 1910; publications include *Art* (1914), *Proust* (1928), and *Enjoying Pictures* (1934).

BELL, SIR CHARLES ALFRED (1870–1945), Indian administrator; educated at Winchester and New College, Oxford; posted to Bengal in Indian civil service, 1891; political officer, Sikkim, 1908–18, 1920–1; negotiated treaty with Bhutan, 1910; intimate friend of thirteenth Dalai Lama; adviser on Tibetan affairs during negotiations between Tibet, China, and Great Britain, 1913–14; visited Lhasa, 1920–1; publications include *Manual of Colloquial Tibetan* (1905), *Tibet Past and Present* (1924), and *Portrait of the Dalai Lama* (1946); KCIE, 1922.

BELL, CHARLES FREDERIC MOBERLY (1847–1911), manager of *The Times*; born at Alexandria; joined Egyptian mercantile firm in Alexandria, 1865; partner, 1873; correspondent to *The Times* on Egyptian questions from 1865 onwards; entered on journalism, 1875; founded *Egyptian Gazette*, 1880; published *Khedives and Pashas* (1884) and *Egyptian Finance* (1887); appointed manager of *The Times* 1890; improved its business organization; started in connexion with the paper a literary organ, *Literature* (1897–1901), which was replaced by the *Literary Supplement*; subsequently added other supplements; pioneer of wireless press messages across Atlantic; published *The Times Atlas* (1895), *Encyclopaedia Britannica* (9th edn, 1898), and *History of South African War* (7 vols., 1900–9); established The Times Book Club, 1905, which came into conflict with publishers and booksellers; managing director of *The Times* publishing company, formed 1908.

BELL, SIR FRANCIS HENRY DILLON (1851–1936), New Zealand lawyer and statesman; educated at Auckland, Dunedin, and St. John's College, Cambridge; senior optime, 1873; called to bar (Middle Temple), 1874; crown solicitor, Wellington, 1879–1911, except in 1893–6 when MHR for Wellington; KC, 1907; leader of legislative council, 1912–28; minister of internal affairs and immigration, 1912–15; attorney-general 1918–26; introduced state forestry and extended land transfer system; minister for external affairs, 1923-6; acting prime minister, 1921 and 1923; prime minister, 14-30 May 1925; a consummate authority on legal and constitutional questions and devoted supporter of imperial unity; KCMG, 1915; GCMG, 1923; PC, 1926.

BELL, GEORGE KENNEDY ALLEN (1883-1958), bishop of Chichester; educated at Westminster and Christ Church, Oxford; deacon, 1907; priest, 1908; curate, Leeds parish church, 1907; tutor (1910), student (1911), Christ Church; domestic chaplain to Archbishop Davidson [q.v.], 1914-24; dean of Canterbury, 1924-9; instituted dramatic performances in cathedral; bishop of Chichester, 1929-58; introduced new methods of evangelism; encouraged the arts; supported Establishment but sought freedom in matters of worship and doctrine; a leader of the Life and Work Movement from 1925; condemned indiscriminate bombing of Germany in war of 1939-45; first chairman, central committee, World Council of Churches, 1948-54; hon. president, 1954-8; chairman, Church of England council on foreign relations, 1945-58; joint chairman, Anglican-Methodist conversations, 1956; published works in cause of unity, biography of Randall Davidson (2 vols., 1935), and a paperback *Christianity and World Order* (1940).

BELL, GERTRUDE MARGARET LOW-THIAN (1868-1926), traveller, archaeologist, and government servant; granddaughter of Sir Isaac Lowthian Bell [q.v.]; educated at Lady Margaret Hall, Oxford; first woman to obtain first class in modern history, 1888; became notable alpinist; journeyed from Jerusalem to Konia in Asia Minor, 1905; with Sir W. M. Ramsay [q.v.] explored Bin-bir-Kilisse, near Isaura, 1907; explored Ukhaidfr near Kerbela, 1909 and 1911; made unsuccessful attempt to penetrate central Arabia, 1913-14; appointed to Arab intelligence bureau, Cairo, 1915; assistant political officer, Basra, 1916; oriental secretary to civil commissioner, Baghdad, 1917, and to high commissioner of Iraq, 1920; helped to inaugurate national museum at Baghdad, 1923; died at Baghdad; published works on travel and archaeology; *Letters*, 2 vols., 1927.

BELL, SIR (HAROLD) IDRIS (1879-1967), scholar; educated at Nottingham high school and Oriel College, Oxford; assistant, department of manuscripts, British Museum, 1903; deputy keeper, 1927; keeper, 1929-44; hon. reader in papyrology, Oxford, 1935-50; hon. fellow, Oriel College, 1936; OBE, 1920; CB, 1936; knighted, 1946; president, British Acad-

emy, 1946-50; publications include *Jew and Christians in Egypt* (1924), chapters on Roman Egypt in *Cambridge Ancient History* (vols x and xi), and *The Development of Welsh Poetry* (1936).

BELL, SIR HENRY HESKETH JOUDOU (1864-1952), colonial administrator; born in West Indies; became colonial civil servant; served in Gold Coast and Bahamas; appointed administrator, Dominica, 1900; governor, Uganda Protectorate, 1906, Northern Nigeria, 1909, Leeward Islands, 1912, Mauritius, 1916-24; his main interest always in development; recognized authority on witchcraft; KCMG, 1908; GCMG, 1925;

BELL, HORACE (1839-1903), civil engineer; as engineer of Indian public works department did much work on Indian state railways, 1862-94; published *Railway Policy in India* (1894) and for Indian natives school primer on India (3rd edn. 1893) and the *Laws of Wealth* (1883).

BELL, SIR ISAAC LOWTHIAN, first baronet (1816-1904), metallurgical chemist and pioneer in industrial enterprise; educated at Bruce's Academy, Newcastle upon Tyne; trained in physics and chemistry in Germany, Denmark, Edinburgh, and Paris; joined the ironworks of his father's firm, Messrs. Losh, Wilson & Bell, at Walker, Tyneside, 1836; married Margaret, daughter of Hugh Lee Pattinson [q.v.], 1842; started with father-in-law chemical works at Washington, near Gateshead, 1852; meanwhile in 1844 with two brothers, Thomas and John, formed firm of Bell Brothers, leasing a blast-furnace at Wylam-on-Tyne; in 1845 on death of his father assumed direction of Walker works; in 1854 Bell Brothers started Clarence works on Tees opposite Middlesbrough with three blast-furnaces to smelt Cleveland ore; helped to construct Cleveland railway to bring ironstone to the works; limestone quarries and collieries were also acquired in Weardale; when the Cleveland railway was purchased by the North-Eastern railway in 1865 Bell became a director for life; steel was made at Clarence from 1889, and important steelworks were built there; a salt bed was found near Clarence in 1874, and thenceforth proved profitable; Bell Brothers ultimately employed in their mines, collieries, and ironworks 6,000 workpeople; Bell studied iron manufacture abroad and was an efficient student of applied science; president of Iron and Steel Institute (Bessemer gold medallist, 1874), 1873-5, of Institution of Mechanical Engineers, 1884, and of Society of Chemical Industry, 1889; received Albert medal of Society of Arts,

1895; published results of experimental researches in *Chemical Phenomena of Iron Smelting* (1872) and *Principles of the Manufacture of Iron and Steel* (1884); mayor of Newcastle upon Tyne, 1854-5 and 1862-3; liberal MP for the Hartlepools, 1875-80; FRS, 1875; baronet, 1885.

BELL, JAMES (1824-1908), chemist; educated at University College, London; entered Inland Revenue chemical laboratory, Somerset House, 1846; principal there, 1874-94; FRS, 1884; CB, 1889; president of Institute of Chemistry, 1888-91; wrote scientific works.

BELL, SIR THOMAS (1865-1952), shipbuilder; born in India; educated at King's College School and Royal Naval Engineering College, Devonport; engineering manager, Clydebank yard, John Brown & Co., 1899; managing director, 1909-35, at time of development of turbines in ships such as the *Carmania*, *Lusitania*, *Aquitania*, *Queen Mary*, and *Hood*; KBE, 1917; deputy controller, dockyards and war shipbuilding, 1917-18.

BELL, VALENTINE GRAEME (1839-1908), civil engineer; pupil of Sir James Brunlees [q.v.], 1859; constructed railways in Jamaica, 1880-3; director of public works there, 1887-1908; CMG, 1903.

BELL, VANESSA (1879-1961), painter; daughter of Sir Leslie Stephen [q.v.], and sister of Virginia (Woolf, q.v.); educated at home; studied at Royal Academy Schools, 1901-4; married (Arthur) Clive (Howard) Bell [q.v.], 1907; four paintings exhibited at the 'Second Post-Impressionist Exhibition' in London, 1912; collaborated with Roger Fry [q.v.] in the Omega workshops, 1913-19; partnership with Duncan Grant in mural decoration, 1913; contributed regularly to group exhibitions, 1919; first London exhibition of her own work, 1922 followed by others in 1930-56.

BELLAMY, JAMES (1819-1909), president of St. John's College, Oxford; educated at Merchant Taylors' School and St. John's College, Oxford; BA, 1841; MA, 1845; ordained, 1843; BD, 1850; DD, 1872; president of St. John's College, 1871-1909; member of university commission, 1877-9; vice-chancellor, 1886-90; a strong conservative and accomplished musician.

BELLEW, HAROLD KYRLE (1855-1911), actor; son of John C. M. Bellew [q.v.]; joined the merchant service; made London début as actor, 1875; took Shakespearian and other parts with Adelaide Neilson and (Sir) Henry Irving

[qq.v.], 1876-86; chief roles were Jack Absolute and Orlando; was associated with Mrs Brown-Potter at Gaiety Theatre, 1887-97; from 1902 to death acted in America; died at Salt Lake City.

BELLMAN, SIR (CHARLES) HAROLD (1886-1963), a pioneer of the building society movement; left school at thirteen; worked in Railway Clearing House, London, 1900-14; principal assistant, Ministry of Munitions, 1915-18; joined board of Abbey Road Building Society, 1918; assistant secretary, 1920; secretary, 1921; general manager, 1927; managing director, 1930; chairman, 1937; (when Abbey Road merged with National), chairman and joint managing director, Abbey National, 1944; chairman, National Association of Building Societies, 1933-6; chairman, Building Societies Association, 1937; president, International Union of Building Societies, 1934-8; MBE, 1920; knighted, 1932; hon LLD, Washington, 1939; publications include *The Building Society Movement* (1927), *The Thrifty Three Millions* (1935), and *Bricks and Mortals* (1949).

BELLO, SIR AHMADU, SARDAUNA OF SOKOTO (1910-1966), prime minister of Northern Nigeria; educated at provincial school, Sokoto, and Moslem Katsina Training College; teacher in Sokoto, 1931; district head, Rabah, 1934; member of Sultan's Council, Sardauna, 1938; visited England to study local government, 1948; member, Northern House of Assembly, 1949; president, Northern Peoples Congress; first minister of works for the North, 1951; first prime minister, Northern Region, 1954-66; CBE, 1953; KBE, 1959; assassinated in military coup, 1966; founder and first chancellor, Ahmadu Bello University.

BELLOC, JOSEPH HILAIRE PIERRE RENÉ (1870-1953), poet and author; born at St. Cloud; after death of father brought to Sussex by English mother; educated at Oratory School, Birmingham, and Balliol College, Oxford; first class, history, and president of Union, 1895; wrote *The Bad Child's Book of Beasts* (1898) and others in same vein; made high-spirited contribution to liberal journalism opposing South African war; published *Paris* (1900), *Robespierre* (1901), and *The Path to Rome* (1902); showed particular talent for describing, and sketching, places and people; naturalized, 1902; liberal MP, South Salford, 1906-10, but considered party system a corrupt collusion; maintained steady output of biographies, topographical works, and political novels, including *Pongo and the Bull* (1910); literary editor, *Morning Post*, 1906-10; founded *Eye-Witness*, 1911; predicted exchange of freedom for security in *The Servile State* (1912);

completed Lingard's [q.v.] *History of England* from 1689 to 1910 (1915); contributed military commentary to *Land and Water* throughout 1914-18 war; his later work of uneven quality and written under financial pressure, but included *Europe and the Faith* (1920) and much writing in the Roman Catholic cause; in series of biographies elaborated idea of Reformation as movement of rich against poor; engaged in controversy with H. G. Wells, Dean Inge, and Dr. Coulton [qq.v.]; regarded writing as unsatisfactory trade and most wished to be remembered for his poetry, in which he had gift for expression of deeply felt personal emotion; *The Cruise of the 'Nona'* (1925) otherwise his most personal memorial.

BELLOWS, JOHN (1831-1902), printer and lexicographer; published *French-English Dictionary*, 1872 (which had a wide circulation), and other dictionaries; a keen archaeologist, he discovered traces of Roman city wall at Gloucester, 1873; visited the Dukhobortsi and Count Tolstoy in Russia, 1892; visited America, 1901; hon. MA, Harvard; a Quaker, teetotaller, and vegetarian; wrote on travel, religion, and politics.

BEMROSE, WILLIAM (1831-1908), writer on wood-carving; succeeded to father's printing business at Derby, 1857;' wrote *Manual of Woodcarving*, 1862—which reached standard rank—and kindred works; author of authoritative works on porcelain; skilled amateur artist; chairman of Derby school board, 1886-1902; pioneer of Volunteer movement.

BENDALL, CECIL (1856-1906), professor of Sanskrit at Cambridge; Sanskrit exhibitioner, Trinity College, Cambridge; migrated to Caius, 1877; BA, 1879; fellow of Caius, 1879-86; in oriental department, British Museum, 1882-98; professor of Sanskrit at University College, London, 1885-1903, and at Cambridge, 1903-6; acquired several oriental MSS while travelling in India, 1884-5, 1898-9; edited much Sanskrit literature; left oriental MSS and books to Cambridge University; an expert Indian palaeographer and keen musician.

BENHAM, WILLIAM (1831-1910), theologian; studied under F. D. Maurice [q.v.] at King's College, London; ordained, 1857; divinity tutor, St. Mark's College, Chelsea, 1857-65; professor of modern history, Queen's College, Harley Street, 1866-71; rector of St. Edmund the King, Lombard Street, 1882-1910; hon. canon of Canterbury, 1888; his voluminous writings include *Life of Archbishop Tait*, with Randall Davidson [q.v.], (1891) and *Old London Churches* (1908); edited 'Ancient

and Modern Library of Theological Literature'; wrote for *Church Times* under name of 'Peter Lombard', 1890-1910.

BENN, SIR ERNEST JOHN PICKSTONE, second baronet, of Old Knoll, Lewisham (1875-1954), publisher, economist, and individualist; brother of W. W. Benn, Viscount Stansgate [q.v.]; educated at lycée Condorcet, Paris, and Central Foundation School, London; joined Benn Brothers, publishing trade papers, 1891; succeeded father, 1922; founded book publishing company, 1923; introduced Benn's Sixpenny Library; as co-founder of Individualist Bookshop (1926) and lifelong freetrader, launched campaign culminating in foundation of Society of Individualists, 1942; wrote voluminously in libertarian cause; CBE, 1918.

BENN, WILLIAM WEDGWOOD, first VISCOUNT STANSGATE (1877-1960), parliamentarian; brother of Sir Ernest Benn [q.v.]; educated at lycée Condorcet, Paris, and University College, London; first class, French, 1898; liberal MP, St. George's 1906-18, Leith, 1918-27; junior lord of Treasury, 1910-14; served in yeomanry and RNAS, 1914-18; DSO, DFC; labour MP, North Aberdeen, 1928-31; secretary of state for India, 1929-31; PC, 1929; authorized statement on dominion status, 1929; ordered arrest of M. K. Gandhi [q.v.] for civil disobedience, 1930; declined to join 'national' government, 1931; labour MP, Gorton division of Manchester, 1937-42; viscount 1942; secretary of state for air, 1945-6; president, Inter-Parliamentary Union, 1947-57.

BENNETT, ALFRED WILLIAM (1833-1902), botanist; BA, London, 1853; MA, 1855; B.Sc., 1868; bookseller and publisher, 1858-68; lecturer on botany, Bedford College, 1868; translated and edited Julius Sachs's *Lehrbuch der Botanik*, 1875; translated and published works on alpine plants; made researches into cryptogamic plants; collaborated in *Handbook of Cryptogamic Botany* (1889) and other works.

BENNETT, EDWARD HALLARAN (1837-1907), surgeon; BA, MB, and M.Ch., Trinity College, Dublin, 1859; FRCS Ireland, 1863; president, 1884-6; MD, 1864; university anatomist, Dublin University, 1864; professor of surgery, Trinity College, Dublin, 1873; an authority on bone fracture; discovered 'Bennett's fracture', 1881; formed collection of fractures at pathological museum, Trinity College, Dublin.

BENNETT, (ENOCH) ARNOLD (1867-1931), novelist, playwright, and man of letters;

son of a Wesleyan Methodist solicitor, of Hanley, Staffordshire; artistic, musical, and bookish family; educated at Middle School, Newcastle under Lyme; solicitors' clerk in London, 1888-93; published 'A Letter Home' in *Yellow Book* (1895) and a novel, *A Man from the North*, 1898; assistant editor (1893) and editor (1896-1900) of *Woman*; published *Anna of the Five Towns* (1902), the first of the serious novels about life in the Potteries upon which his fame rests; lived in Paris from 1902, writing plays, romances, articles, and novels, including his best work *The Old Wives' Tale* (1908), and *Clayhanger* (1910), *Hilda Lessways* (1911), *The Card* (1911); his plays, *What the Public Wants* (1909), *Milestones* (with Edward Knoblock, 1912), and *The Great Adventure* (1913), brought him huge popularity; returned to England, 1912; *These Twain* (1916) completed the 'Clayhanger' trilogy; a powerful journalist in aid of the war effort; henceforward mixed with the rich and powerful whom he satirized in *The Pretty Lady* (1918) and *Lord Raingo* (1926); partner in management of Lyric Theatre, Hammersmith, and much occupied in patronage of the arts and social engagements; contributed weekly literary causerie to *Evening Standard*; his versatility and habit of writing at different levels of seriousness made him an easy target for detractors, but his best work is full of wisdom, truth, and humour.

BENNETT, GEORGE MACDONALD (1892-1959), chemist; first class, chemistry, London, 1912, and part i, natural sciences tripos, Cambridge, 1915; senior demonstrator, Guy's Hospital medical school, 1921; lecturer in organic chemistry, Sheffield, 1924, Firth professor of chemistry, 1931; professor of organic chemistry, King's College, London, 1938; government chemist, 1945-59; made wide-ranging contributions to chemical knowledge; FRS, 1947; CB, 1948.

BENNETT, PETER FREDERICK BLAKER, BARON BENNETT OF EDGBASTON (1880-1957), industrialist; educated at King Edward's School, Five Ways, Birmingham; co-founder of Thomson Bennett Ltd., 1907; amalgamated with Joseph Lucas Ltd., 1914, to manufacture combined ignition and lighting systems; chairman and managing director, 1948; director-general, tanks and transport (Ministry of Supply), 1939-40, of emergency services organization (Ministry of Aircraft Production), 1940-1; chairman, Automatic Gun Board, 1941-4; conservative MP, Edgbaston, 1940-53; parliamentary secretary, Ministry of Labour, 1951-2; knighted, 1941; baron, 1953.

BENNETT, RICHARD BEDFORD, VISCOUNT BENNETT (1870-1947), Canadian prime minister; graduated in law, Dalhousie University, and called to New Brunswick bar, 1893; KC, 1905; conservative member North-West Territories legislature, 1898-1905, Alberta, 1909-11; of Canadian House of Commons, for Calgary, 1911-17, 1925-38; director-general of national service, 1916-17; led conservative party in opposition, 1927-30; prime minister, 1930-5; finance minister 1930-2; strongly supported imperial preference; at imperial conference (1930) pressed for economic conference held at Ottawa, 1932; discussed trade agreement with United States, 1933; announced New Deal of far-reaching reforms (1935) and lost election; retired to England, 1938; PC, 1930; viscount, 1941.

BENSON, ARTHUR CHRISTOPHER (1862-1925), man of letters and master of Magdalene College, Cambridge; son of Edward White Benson and brother of Robert Hugh Benson [qq.v.]; scholar of Eton and King's College, Cambridge; assistant master at Eton, 1885; very successful housemaster, 1892-1903; went to live at Cambridge, 1903; fellow of Magdalene College, Cambridge, 1904; master, 1915-25; voluminous writer; works include life of his father (1899), *Fasti Etonenses* (1899), *Selections from the Correspondence of Queen Victoria*, with Viscount Esher (q.v., 1907), essays, literary criticisms, short biographies, and novels.

BENSON, EDWARD FREDERIC (1867-1940), author; son of E. W. Benson [q.v.] and brother of A. C. and R. H. Benson [qq.v.]; educated at Marlborough and King's College, Cambridge; first class in classics; an uncontrollably prolific writer whose talent consequently never fully matured; novels of social satire include the 'Dodo' and 'Lucia' series and *Secret Lives* (1932); wrote also plays, biographies, novels of school and university life, and highly efficient stories of the supernatural; his reminiscences have value as sources for social history.

BENSON, SIR FRANCIS ROBERT (FRANK) (1858-1939), actor-manager; educated at Winchester and New College, Oxford; first professional appearance in Irving's *Romeo and Juliet*, 1882; opened under own management, 1883; provided Shakespearian festival at Stratford for thirty-years from 1886, producing all but two of Shakespeare's plays; gave eight London seasons; knighted at Drury Lane Theatre following a performance of *Julius Caesar*, 1916; although not a great actor or teacher, by the opportunities which he afforded young actors he made his company a nursery for the English Stage.

BENSON, GODFREY RATHBONE, first BARON CHARNWOOD (1864-1945), liberal politician and man of letters; brother of Sir Frank Benson [q.v.]; educated at Winchester and Balliol College, Oxford; first class, *lit. hum.*, 1887; liberal MP, Woodstock division, 1892-5; active in church, charitable, and municipal work; publications include a widely known biography of Abraham Lincoln, 1916; baron, 1911.

BENSON, SIR REGINALD LINDSAY (REX) (1889-1968), merchant banker; educated at Eton and Balliol College, Oxford; served with 9th Lancers, 1910; aide-de-camp to viceroy of India, 1913-14; in France, 1914-16; MC, DSO, croix de guerre and Legion of Honour; military secretary to governor of Bombay, 1921; on special mission to Batum, 1922; joined Robert Benson & Co., 1924; chairman, 1935; rejoined army, 1939; military attaché, Washington, 1941-4; rejoined Robert Benson & Co. Ltd., 1944; on merger with Lonsdale Investment Trust, chairman, 1947-59; enthusiast for Anglo-American understanding; knighted, 1958.

BENSON, RICHARD MEUX (1824-1915), divine and founder of a religious order; life student and BA, Christ Church, Oxford; vicar of Cowley, near Oxford, 1850; formed society of mission priests of St. John the Evangelist, 1866; superior, 1866-90; vicar of Cowley St. John, 1870-86; established branch house of Cowley Fathers in Boston, USA, 1870-1; revisited Boston, 1892-9; returned to Cowley, 1899, and remained there till death; consistently loyal to Anglican position; numerous works include *Manual of Intercessory Prayer* (1863).

BENSON, ROBERT HUGH (1871-1914), Catholic writer and apologist; son of Edward White Benson [q.v.]; educated at Eton and Trinity College, Cambridge; took Anglican orders, 1894; member of the Community of the Resurrection, Mirfield, 1898-1903; received into Roman Church, 1903; popular as preacher and writer of press articles; author of numerous works of historical fiction and novels of modern life.

BENSON, STELLA (1892-1933), novelist. [See ANDERSON, STELLA.]

BENT, SIR THOMAS (1838-1909), prime minister of Victoria; born in New South Wales; market gardener; member for Brighton in Victoria parliament, 1871-94; entered James Service [q.v.] ministry as vice-president of the board of public works, 1880; commissioner of railways, 1881-3; chairman of first railways standing committee, 1887-9; speaker, 1892-4; ruined by 'land boom', 1893; engaged in dairy farming, 1894-1900; re-elected member for Brighton, 1900; minister for railways and works, 1902-3; prime minister, 1904-8; KCMG, 1908; opposed to labour party.

BENTLEY, EDMUND CLERIHEW (1875-1956), writer; educated at St. Paul's School and Merton College, Oxford; president of Union, 1898; called to bar (Inner Temple), 1902; joined *Daily News*, 1901; on *Daily Telegraph*, 1912-34; invented 'Clerihew' verse form published in *Biography for Beginners* (with illustrations by G. K. Chesterton, q.v.) and subsequent volumes; his distinctive detective story, *Trent's Last Case* (1913), became a classic.

BENTLEY, JOHN FRANCIS (1839-1902), architect; entered firm of London architects, 1855; joined Church of Rome, 1862; architectural work showed desire for perfection in detail and soundness of construction; did much ornamental work for churches and private residences—Convent of the Sacred Heart, Hammersmith, 1866, Carlton Towers, Selby, 1874, and St. Mary of the Angels, Bayswater; built Church of the Holy Rood, Watford, 1887-92, and convent of the Immaculate Conception, Bocking Bridge, Braintree, 1897; decorated St. Botolph, Aldgate and Bishopsgate, and other City churches; designed and built in Byzantine style Roman Catholic cathedral at Westminster, 1894, the materials being brickwork, masonry, and concrete, free from iron spans.

BENTON, SIR JOHN (1850-1927), civil engineer; assistant engineer in Indian public works department, 1873; effected great reforms in irrigation while working for Burmese government, 1897-1902; chief engineer and secretary to Punjab government, 1902; inspector-general of irrigation for all India, 1906-12; carried out Triple Canals scheme and constructed Upper Swat Valley canal; KCIE, 1911.

BERESFORD, LORD CHARLES WILLIAM DE LA POER, BARON BERESFORD (1846-1919), admiral; entered navy, 1859; conservative MP, Waterford, 1874-80; took part in bombardment of Alexandria (captain and mentioned in dispatches), 1882, and Nile expedition, 1884-5; CB for rescue of Sir C. W. Wilson [q.v.], 1885; MP, East Marylebone, 1885-6-9; fourth naval lord of Admiralty,

1886-8; rear-admiral, 1897; MP, York (1897-1900), Woolwich (1902-3); vice-admiral, 1902; chief in command of Channel squadron and KCB, 1903; commander-in-chief, Mediterranean, 1905; admiral, 1906; commander-in-chief of Channel fleet, 1907-9; opposed Admiralty policy; MP, Portsmouth, 1910-16; retired and received GCB, 1911; baron, 1916.

BERGNE, Sir JOHN HENRY GIBBS (1842-1908), diplomatist; son of J. B. Bergne [q.v.]; educated at London University; entered Foreign Office, 1861; superintendent of treaty department, 1881-94; superintendent of commercial department and examiner of treaties, 1894-1902; British delegate at international copyright conference, Berne (1886) and Berlin (1908), as well as at conferences for abolition of bounties on sugar at Brussels (1899 and 1901), signing convention, 1902; KCMG, 1888; KCB, 1903; rendered important service to Authors' Society from 1890 to death; member of Alpine Club.

BERKELEY, Sir GEORGE (1819-1905), colonial governor; born in island of Barbados; BA, Trinity College, Dublin, 1842; colonial secretary and controller of customs of British Honduras, 1845; lieutenant-governor of St. Vincent, 1864; governor-in-chief of West Africa settlements, 1873, and of Leeward Islands, 1874-81; KCMG, 1881.

BERKELEY, RANDAL MOWBRAY THOMAS (RAWDON), eighth Earl of Berkeley (1865-1942), scientist; established right to peerage, 1891; educated in France; served as an officer in navy, 1878-87; researched on crystal structure in Oxford laboratories; built own laboratory at Foxcombe and carried out extensive research on osmotic pressure, 1898-1928; established validity of indirect vapour-pressure method of measurement; inherited Berkeley estates, 1916; FRS, 1908.

BERNARD, Sir CHARLES EDWARD (1837-1901), Anglo-Indian administrator; educated at Rugby and Haileybury; entered Indian civil service, 1857; secretary to Sir Richard Temple [q.v.] and Sir George Campbell [q.v.], 1871-8; chief commissioner of British Burma, 1880-7; secretary of revenue department, India Office, 1887-1901; KCSI, 1886; prominent athlete; edited memoirs of Sir George Campbell, 1893.

BERNARD, JOHN HENRY (1860-1927), archbishop of Dublin and provost of Trinity College, Dublin; born at Sooree, Bengal; BA, Trinity College, Dublin, 1880; fellow, 1884-1902; ordained, 1886; Archbishop King's

lecturer (afterwards professor) in divinity, Trinity College, 1888-1911; his teaching left a lasting impress on Church of Ireland; treasurer of St. Patrick's Cathedral, 1897; dean of St. Patrick's, 1902; promoted friendly relations between Churches of England and Ireland; bishop of Ossory, 1911; archbishop of Dublin, 1915; took prominent part in Irish convention summoned by British government, 1917; provost of Trinity College, 1919; raised efficiency of college and improved its amenities; loyally accepted Free State government, 1922; works include commentary on Kant's *Dialect of the Pure Reason* (1889) and translation of his *Kritik of the Judgment* (1892), edition of the *Liber Hymnorum* (1898), and commentary on St. John's Gospel (1928).

BERNARD, THOMAS DEHANY (1815-1904), divine; brother of Mountague Bernard [q.v.]; BA, Exeter College, Oxford, 1838; chancellor's prize essayist, 1839; ordained, 1840; Bampton lecturer, 1864; rector of Walcot, Bath, 1864-86; canon of Wells, 1868-1901; chancellor of Wells Cathedral from 1879 to death; of evangelical sympathies; published theological works.

BERNERS, fourteenth Baron (1883-1950), musician, artist, and author. [See Tyrwhitt-Wilson, Sir Gerald Hugh.]

BERRY, Sir GRAHAM (1822-1904), prime minister of Victoria; draper in Chelsea, 1848; emigrated to Victoria, 1852; purchased *Collingwood Observer*, 1860; member for Collingwood as advanced liberal protectionist in legislative assembly, 1861-5; purchased *Geelong Register*, 1866; member for Geelong, West, 1868; treasurer in J. A. Macpherson's ministry, 1870, and in Sir C. Gavan Duffy's ministry, 1871-2; thrice prime minister, 1875-6, 1877-80, and 1880-1; formed with James Service (q.v., 1883) coalition government which passed useful measures; agent-general to colony in London, 1886-91; returned to Melbourne, 1891; treasurer in William Shiel's ministry, 1892; speaker of legislative assembly, 1894-7; a fervent advocate of democratic principles.

BERRY, (JAMES) GOMER, first Viscount Kemsley (1883-1968), newspaper proprietor; brother of (Henry) Seymour Berry (later Lord Buckland, q.v.) and of William Ewart Berry (later Lord Camrose, q.v.); educated at Merthyr Tydfil; apprenticed to *Merthyr Tydfil Times*; joined William in newspaper partnership, 1902-37; divided business, 1937; chairman, Allied Newspapers (renamed Kemsley Newspapers in 1943); largest newspaper owner in Britain including *Daily Sketch* (later *Daily*

Graphic), *Sunday Graphic*, and *Sunday Times*, and over twenty provincial newspapers; energies concentrated on *Sunday Times*; supported Neville Chamberlain's pacific approach to Hitler (1938); and Eden's policy on Suez, 1956; fought number of libel actions; chief target of labour MPs in debate which led to royal commission on the press, 1947-9; sold *Daily Graphic* to Lord Rothermere, 1952; sold control of *Sunday Times* to Roy H. Thomson (later Lord Thomson of Fleet), 1959; last survivor of old style self-made newspaper barons; baronet, 1928; baron, 1936; viscount, 1945; GBE, 1959.

BERRY, SIDNEY MALCOLM (1881-1961), Congregational Church leader; son of Congregational minister; educated at Tettenhall College, Wolverhampton, and Clare College, Cambridge; went on to Mansfield College, Oxford, 1903; ordained, 1906; minister at Chorlton-cum-Hardy, Manchester, 1909; Carr's Lane Chapel, Birmingham, 1912-23; secretary, Congregational Union, 1923-48; delivered Warrack lectures, *Vital Preaching*, 1936; contributed religious meditations to *Sunday Times*; hon. DD, Glasgow, 1936; moderator, Free Church Federal Council, 1934-7; chairman, Congregational Union, 1947; supported moves for British Council of Churches, 1942; minister-secretary, International Congregational Council, 1949.

BERRY, WILLIAM EWERT, first VISCOUNT CAMROSE (1879-1954), newspaper proprietor; employed by W. W. Hadley [q.v.] on *Merthyr Times*; moved to London, 1898; launched *Advertising World*, 1901; thereafter worked in association with his brother, future Viscount Kemsley [q.v.]; founded *Boxing* (1909) and other periodicals; purchased *Sunday Times*, 1915; editor-in-chief until 1937; acquired *Financial Times* with St. Clement's Press, 1919, Weldon's group, Kelly's Directories, and *Graphic* publications; baronet, 1921; founded Allied (later Kemsley) Newspapers, 1924, with future Lord Iliffe [q.v.] to acquire *Daily Dispatch* and a number of provincial newspapers; other acquisitions included *Daily Sketch*, Amalgamated Press (1926), Edward Lloyd, Ltd. (1927), and *Daily Telegraph* (1928); baron, 1929; on dissolution of association with Iliffe and Kemsley (1937) retained *Daily Telegraph*, Amalgamated Press, and *Financial Times;* viscount, 1941.

BERTIE, FRANCIS LEVESON, first VISCOUNT BERTIE OF THAME (1844-1919), diplomatist; educated at Eton; entered Foreign Office, 1863; private secretary to Robert Bourke (later Lord Connemara, q.v.), 1874-80; assistant under-secretary of state for foreign affairs, 1894-1903; ambassador at Rome, 1903-4, and Paris, 1905-18; largely instrumental in preserving the Anglo-French *entente*; KCB, 1902; GCB, 1908; baron, 1915; viscount, 1918.

BESANT, ANNIE (1847-1933), theosophist, educationist, and Indian politician; born WOOD; married (1867) Frank Besant, a clergyman, whom she left in 1873; joined National Secular Society, 1874, Fabian Society, 1885; organized matchmakers' strike, 1888, and formed their union; converted to theosophy; president, Theosophical Society, 1907-33; from 1895 active in India where she attempted to reconcile theosophy with Hinduism; founded Central Hindu College (1899) and girls' school (1904) which became nucleus of a Hindu university, 1916; claimed her adopted son as messiah, 1909; initiated Home Rule for India League, 1916; president, Indian National Congress, 1918; lost her political influence by protesting against violence, 1919.

BESANT, SIR WALTER (1836-1901), novelist; educated at King's College, London, and at Christ's College, Cambridge; 18th wrangler, 1859; MA, 1863; master at Leamington College, 1860; senior professor at Royal College, Mauritius, 1861-7; settled in London, 1867; published *Early French Poetry* (1868), *The French Humourists* (1873), and other works; secretary of Palestine Exploration Fund, 1868-86; wrote with E. H. Palmer [q.v.] *Jerusalem*, 1871; edited *Survey of Western Palestine*, 1881; as contributor to *Once a Week*, 1869, he became acquainted with the editor, James Rice [q.v.], with whom he collaborated in several novels (1871-81), including *The Golden Butterfly* (1876), *By Celia's Arbour* (1878), and *The Chaplain of the Fleet* (1879); from 1882 continued fiction without collaboration, mainly based on historical incident, e.g. *Dorothy Forster* (1884) and *For Faith and Freedom* (1888); his *All Sorts and Conditions of Men* (1882) and *Children of Gibeon* (1886) called attention to social evils in east London, and stimulated the foundation of 'The People's Palace', Mile End, for intellectual improvement and rational amusement, 1887; helped to found Society of Authors, 1884; started and edited *The Author*, 1890; defined authors' financial position in *The Pen and the Book*, 1899; knighted, 1895; commenced survey of London (1894) which appeared in ten elaborate volumes (1902-12); edited 'Fascination of London' series from 1897; a keen freemason; interested in antiquities of Hampstead, where he long resided; other works include *The Eulogy of Richard Jefferies* (1888), *Captain Cook* (1889), and (with W. J. Brodribb q.v.) *Constantinople* (1879); his *Autobiography* appeared in 1902.

BESICOVITCH, ABRAM SAMOILO-VITCH (1891–1970), mathematician; born in Russia; educated at university of St. Petersburg; professor of mathematics, Perm University, 1917; professor, Pedagogical Institute, Leningrad, 1920–4; escaped to Copenhagen, 1924; lecturer, Liverpool University, 1926; lecturer, Cambridge University, 1927; Cayley lecturer, 1928; fellow, Trinity College, 1930–70; Rouse Ball professor of mathematics, 1950–8; FRS, 1934; published *Almost Periodic Functions* (1955).

BESSBOROUGH, ninth EARL OF (1880–1956), governor-general of Canada. [See PONSONBY, VERE BRABAZON.]

BETHAM-EDWARDS, MATILDA BARBARA (1836–1919), novelist and writer on French life. [See EDWARDS.]

BETHUNE-BAKER, JAMES FRANKLIN (1861–1951), professor of divinity; educated at King Edward's School, Birmingham and Pembroke College, Cambridge; graduated in classics; first class, part ii, theology, 1886; deacon, 1888; priest, 1889; fellow of Pembroke, 1891–1951, dean, 1891–1906; BD, 1901; DD, 1912; Lady Margaret's professor of divinity, 1911–35; edited *Journal of Theological Studies*, 1903–35; sought advancement of liberal Christianity; works include *Introduction to the Early History of Christian Doctrine* (1903) and *The Way of Modernism* (1927); FBA, 1924.

BETTERTON, HENRY BUCKNALL, BARON RUSHCLIFFE (1872–1949), politician; educated at Rugby and Christ Church, Oxford; called to bar (Inner Temple), 1896; conservative MP, Rushcliffe division, 1918–34; parliamentary secretary, Ministry of Labour, 1923, 1924–9; minister of labour, 1931–4; first chairman, Unemployment Assistance Board, 1934–41; chairman, nurses' salaries committee, 1943; baronet, 1929; PC, 1931; baron, 1935.

BEVAN, ANEURIN (1897–1960), politician; received elementary education in Tredegar; became miner and energetic union worker; at Central Labour College, London, 1919–21; after considerable unemployment became lodge disputes agent, 1926; labour MP, Ebbw Vale, 1929–60; supporter of and influenced by Sir Stafford Cripps [q.v.]; a founder, contributor to, and editor (1942–5) of *Tribune*; opposed government throughout war; elected to national executive, 1944; minister of health and housing, 1945–51; obtained passage of National Health Service Act (1946) and co-operation of medical and dental professions for introduction of Health Service, 1948; minister of labour, Jan.

1951; resigned Apr. 1951 over proposal to introduce health charges; in shadow cabinet, 1952–4; visited Russia and China with Attlee, 1954; stood unsuccessfully for parliamentary party leadership, 1955; party treasurer, 1956; shadow minister for colonial, then foreign affairs, from 1956; opposed unilateral nuclear disarmament, 1957; deputy leader, parliamentary party, 1959; frequently in conflict with his own party from which he was once (1939) expelled and several times nearly so, he was nevertheless a vitalizing influence; after Churchill probably the best speaker in the House; PC, 1945.

BEVAN, ANTHONY ASHLEY (1859–1933), orientalist and biblical scholar; brother of E. R. Bevan [q.v.]; educated at Lausanne, Strasbourg, and Trinity College, Cambridge; first class, Semitic languages, 1887; fellow and lecturer in oriental languages, 1890; lord almoner's professor of Arabic, 1893–1933; mainly interested in classical Arabic and Hebrew, and in Manichaeism; edited the *Naķā'id of Jarir and al-Farazdak* (3 vols., 1905–12) and wrote a masterly *Commentary on the Book of Daniel* (1892); FBA, 1916–28.

BEVAN, EDWYN ROBERT (1870–1943), scholar, historian, philosopher, and publicist; brother of A. A. Bevan [q.v.]; educated at Monkton Combe and New College, Oxford; first class, *lit. hum.*, 1892; influenced by Friedrich von Hügel [q.v.]; active member of Society of Jews and Christians and Student Christian Movement; through devotion to Christian religion came to study the Hellenistic age, upon which he became a recognized authority; lecturer in Hellenistic history and literature, King's College, London, 1922–33; publications include *The House of Seleucus* (2 vols., 1902), *Jerusalem under the High Priests* (1904), *Stoics and Sceptics* (1913), *Indian Nationalism* (1913), *Hellenism and Christianity* (1921), *Christianity* (1932), *Symbolism and Belief* (1938); FBA, 1942.

BEVAN, WILLIAM LATHAM (1821–1908), archdeacon of Brecon; educated at Rugby and Balliol College, Oxford; BA, 1842; MA, 1845; ordained, 1844; held living of Hay, Breconshire, 1845–1901; canon of St. David's, 1879–93; archdeacon of Brecon, 1895–1907; an accomplished linguist and Welsh Church historian; wrote *History of St. David's* (1888), works on ancient geography, and pamphlets in defence of the Welsh Church.

BEVERIDGE, WILLIAM HENRY, first BARON BEVERIDGE (1879–1963), social reformer and economist; educated at Charterhouse and Balliol College, Oxford; first class, *lit. hum.*, 1901; studied law; BCL, 1903; sub-warden, Toynbee Hall, London, 1903; appointed leader-writer on 'social problems', *Morning Post*, 1905; advocated labour exchanges and state insurance; personal assistant to (Sir) Winston Churchill, Board of Trade, 1908; assistant secretary, 1913; transferred to Ministry of Munitions, 1915; returned to Board of Trade, 1916; drafted new Unemployment Insurance Act; British representative, Inter-Allied Food Mission, 1918; permanent secretary, Ministry of Food, 1919; resigned from civil service; appointed director, London School of Economics, 1919-37; master, University College, Oxford, 1937-44; FBA; under-secretary, Ministry of Labour, 1940; chairman, social services inquiry, 1941; report on *Social Insurance and Allied Services*, 1942; second report, *Full Employment in a Free Society*, published, 1944; 'Beveridge plan' became blueprint for welfare state legislation of 1944-8; liberal MP Berwick on Tweed, 1944-5; CB, 1916; KCB, 1919; baron, 1946; publications include *Unemployment: a Problem of Industry* (1909), *Power and Influence* (1953), and *Prices and Wages in England from the Twelfth to the Nineteenth Century* (first volume published 1939); hon. LLD, London, Aberdeen, Birmingham, Chicago, Columbia, Melbourne, Paris, and Oslo, and other hon. degrees; hon. fellow, Balliol, Nuffield, and University Colleges, Oxford.

BEVIN, ERNEST (1881-1951), trade-union leader and statesman; of illegitimate birth; became farm worker at eleven; casual worker in Bristol at thirteen; van driver for mineral-water firm, 1901; active in Bristol Socialist Society and Right to Work movement; full-time official of Dockers' Union, 1911, assistant national organizer, 1913, national organizer, 1914, assistant general secretary, 1920; brought to trade-union movement massive self-confidence, great negotiating ability, and conviction of need for centralized authority; delegate to Trades Union Congress, 1915: on executive of Transport Workers' Federation, 1916; became known as 'Dockers' KC', 1920, when he persuaded dockers instead of striking to take claim to court of inquiry where he won by brilliant advocacy; led Council of Action in boycotting dispatch of arms for use against Russian revolutionaries; ruthlessly negotiated merger of fourteen unions into Transport and General Workers Union of which he became general secretary, 1922; saw that advance for workers lay in negotiation from strength and by 1926 had absorbed twenty-seven unions into his own; added 100,000 members from Workers' Union, 1929; member of general council of TUC from 1925; helped change climate of industrial relations with shift

towards conciliation; served on Macmillan committee on finance and industry and Economic Advisory Council; turned *Daily Herald* into popular national newspaper; chairman, TUC, 1936-7; a powerful opponent of pacificism in labour party and consistently urged rearmament; MP, Central Wandsworth (1940-50), and East Woolwich (1950-1); minister of labour and national service (1940-5) with responsibility for all manpower and labour questions and power to conscript and direct labour; carried out vast redeployment of manpower with minimum of industrial trouble; member of War Cabinet; chairman, Production Executive; served on many committees and was one of the most helpful members of the government with informed interest in economic problems, post-war reconstruction, and international relationships; foreign secretary, 1945-51; involved in protracted negotiations over peace treaties; failed to obtain extension of wartime treaty with Russia but contained her expansion of influence in Europe until in 1947 he made it clear to the United States that Britain could no longer do so alone and obtained American intervention; initiated Organization for European Economic Co-operation, 1947, to take advantage of Marshall Aid; obtained treaty of Dunkirk with France (1947), treaty of mutual assistance with France, Belgium, Holland, and Luxembourg (1948), and North Atlantic Treaty to include USA and Canada (1949); unable to solve Palestine problem, handed it over to United Nations and considering their decision wrong withdrew British administration; lord privy seal, Mar. 1951; PC, 1940; 'a turn-up in a million', his impact was massive and creative, in the trade-union movement, the wartime administration, and the postwar labour government in which he was Attlee's closest and most loyal assistant with a powerful voice in all major decisions.

BEWLEY, Sir EDMUND THOMAS (1837-1908), Irish lawyer and genealogist; BA, Trinity College, Dublin, 1857; MA, 1863; LLD, 1885; called to Irish bar, 1862; QC, 1882; regius professor of feudal and English law in Dublin University, 1884-90; judge of supreme court of judicature of Ireland, 1890-8; knighted, 1898; made important genealogical researches and published legal treatises.

BHOPAL, HAMIDULLAH, Nawab of (1894-1960), educated at Mahommedan Anglo-Oriental College; BA, Allahabad, 1915; chief secretary, Bhopal, 1916-22; minister of law and justice, 1922-6; assumed ruling powers, 1926; chancellor, Chamber of Princes, 1931-2 and 1944-7; a delegate to Round Table conference, 1930; negotiated to safeguard princes' rights;

subsequently advocated division of India into Hindustan, Pakistan, and Rajasthan (confederation of princely states); surrendered state to Indian Union, 1947; an all-round sportsman and internationally known polo player; GCIE, 1929; GCSI, 1932.

BHOWNAGGREE, Sir MANCHERJEE MERWANJEE (1851-1933), Indian lawyer and politician; educated Bombay University; called to bar (Lincoln's Inn), 1885; as judicial counsellor in Bhavanagar introduced reforms in law administration; conservative MP, North-East Bethnal Green, 1895-1906; for long the leading Indian permanently resident in Great Britain; KCIE, 1897.

BICESTER, first Baron (1867-1956), banker. [See Smith, Vivian Hugh.]

BICKERSTETH, EDWARD HENRY (1825-1906), bishop of Exeter; son of Edward Bickersteth (1786-1850, q.v.); BA, Trinity College, Cambridge, 1847; MA, 1850; hon. DD, 1885; thrice chancellor's medallist for English verse, 1844-5-6; Seatonian prizeman, 1854; ordained, 1848; vicar of Christ Church Hampstead, 1855-85; active supporter of Church Missionary Society; dean of Gloucester, Jan. 1885; bishop of Exeter, Apr. 1885-1900; composer of 'O Brothers, lift your voices', 'Peace, Perfect Peace' (in *From Year to Year*, 1883), and other hymns; editor of *The Hymnal Companion to the Book of Common Prayer*, 1870; author of *Yesterday, To-day, and For Ever* (1866, 17th edn. 1885) and other poetical works; published commentary on the New Testament (1864) and many devotional works.

BIDDER, GEORGE PARKER (1863-1953), marine biologist; grandson of G. P. Bidder [q.v.]; educated at Harrow and Trinity College, Cambridge; Sc.D., 1916; worked at and gave financial support to marine biological laboratories at Plymouth and Naples; a leading authority on sponges; managing director (1897-1908), chairman (1915-19), Cannock Chase Colliery Company.

BIDDULPH, Sir MICHAEL ANTHONY SHRAPNEL (1823-1904), general; educated at Woolwich; awarded Royal Humane Society's medal for life saving, 1842; with Royal Artillery in Crimea; in trenches at siege of Sevastopol, 1854; on special telegraph construction service in Asia Minor, 1855-9; CB, 1873; in command of Rohilkhand district, 1875-8, and of Quetta field force in Afghan war, 1878-9; present at occupation of Kandahar; accomplished successful march with the Thal Chotiali field force, 1879; KCB, 1879; commanded Rawalpindi

district in India, 1880; promoted colonel commandant, 1885, and general, 1886; president of ordnance committee, 1887-90; keeper of regalia at Tower of London, 1891-6; GCB, 1895; gentleman usher of the black rod from 1896 to death.

BIDDULPH, Sir ROBERT (1835-1918), general; served in Crimea, Indian Mutiny, and China; private secretary to (Viscount) Cardwell [q.v.], 1871; high commissioner and commander-in-chief, Cyprus, 1879-86; director-general of military education, 1888-93; general, 1892; governor and commander-in-chief, Gibraltar, 1893-1900; army purchase commissioner, 1904; KCMG, 1880; GCB, 1899.

BIDWELL, SHELFORD (1848-1909), pioneer of telephotography; BA, Caius College, Cambridge, 1870; LLB and MA, 1873; called to bar, 1873; began researches into photoelectric properties of selenium, 1880; described instrument for electrically transmitting pictures of natural objects to a distance along a wire, 1881; published many papers on telegraphic photography, physics, and kindred subjects; FRS, 1886; Sc.D., Cambridge, 1900.

BIFFEN, Sir ROWLAND HARRY (1874-1949), geneticist, plant breeder, and professor of agricultural botany; educated at Cheltenham grammar school and Emmanuel College, Cambridge; first class, parts i and ii, natural sciences tripos, 1895-6; professor of agricultural botany, 1908-31; director, Plant Breeding Institute, 1912-36; proved (1903) that inheritance of a 'constitutional' character in wheat conformed with Mendelian principles; transformed plant breeding from empiricism to orderly scientific method; FRS, 1914; knighted, 1925.

BIGG, CHARLES (1840-1908), classical scholar and theologian; educated at Manchester grammar school and Corpus Christi College, Oxford; BA, 1862; MA, 1864; DD, 1876; Hertford scholar, 1860; won Gaisford Greek prose prize, 1861, and Ellerton theological essay, 1864; ordained, 1863; classical tutor of Christ Church, Oxford, 1863; headmaster of Brighton College, 1871-81; chaplain of Corpus Christi College, 1881-7; Bampton lecturer on 'The Christian Platonists of Alexandria', 1886; hon. canon of Worcester, 1889-1901; regius professor of ecclesiastical history at Oxford and canon of Christ Church, 1901-8; published *Neoplatonism* (1895), *The Origins of Christianity* (1909), and other works.

BIGGE, ARTHUR JOHN, Baron Stamfordham (1849-1931), private secretary to King George V; educated at Rossall School and Royal

Military Academy; commissioned in Royal Artillery, 1869; served in Kaffir and Zulu wars, 1878–9; through death of his friend, the Prince Imperial, became known to Queen Victoria; her assistant private secretary, 1880–95, private secretary, 1895–1901; KCB, 1895; private secretary to Prince (later King) George, 1901–31; PC, 1910; baron, 1911; showed persistent industry, tact, wisdom, sure grasp of affairs, and unswerving rectitude and impartiality, bearing delicate responsibilities with sagacity and resourcefulness and giving confidence to the new King; an intimate friend of Randall Davidson [q.v.].

BIGHAM, JOHN CHARLES, first VISCOUNT MERSEY (1840–1929), judge, called to bar (Middle Temple), 1870; joined Northern circuit and specialized in commercial business; QC, 1883; powerful advocate and skilful cross-examiner; MP, Exchange division of Liverpool, 1895–7; judge of Queen's Bench division, 1897; member of royal commission for revision of martial law sentences in South Africa, 1902; president of railway and canal commission, 1904; president of Probate, Divorce, and Admiralty division, 1909–10; PC, 1909; baron, 1910; viscount, 1916; heard appeals in House of Lords and attended judicial committee of Privy Council; heard appeals from prize court during European war.

BIKANER, MAHARAJA SHRI SIR GANGA SINGH BAHADUR, MAHARAJA OF (1880–1943); educated at Mayo College, Ajmer; assumed ruling powers, 1898; obtained water for his territory, extended railways, and provided hospitals and schools; introduced representative assembly (1913) and elected majority (1937); first chancellor, Chamber of Princes, 1921–6; attended Round Table conferences, 1930–1; supported all-India federation until alarmed by Congress activities; served in China (1900) and France and Egypt (1914–15); signatory to treaty of Versailles; honorary major-general, 1917; general, 1937; GCIE, 1907; GCSI, 1911; GCVO, 1919.

BILES, SIR JOHN HARVARD (1854–1933), naval architect; apprenticed at Portsmouth dockyard; studied at Royal School of Naval Architecture and Royal Naval College; joined Admiralty, 1877; chief designer to J. & G. Thomson, Clydebank, 1880–90; the *City of Paris* and *City of New York* (1887) revolutionary in design and construction; professor of naval architecture, Glasgow, 1891–1921; a talented and sympathetic teacher; travelled widely from 1907 as consultant, particularly in America and India; served on many government committees; knighted, 1913; KCIE, 1922.

BING, GERTRUD (1892–1964), scholar; born in Hamburg; educated in a Hamburg school and at Munich and Hamburg Universities; Ph.D., 1921; joined Kulturwissenschaftliche Bibliothek Warburg, 1922; librarian and personal assistant to Aby Warburg; assistant to Fritz Saxl, 1929; assisted in removal of Bibliothek Warburg to England, 1933; assistant director, Warburg Institute, London University, 1944; director, and professor of history of the classical tradition, 1955–9; hon. fellow, Warburg Institute; hon. D.Litt., Reading, 1959; naturalized, 1946; edited Warburg's *Gesammelte Schriften* (1932), Saxl's *Lectures* (1957), and contributed a study of Warburg to *La Rinascita del paganesimo antico*, Florence (1966).

BINNIE, SIR ALEXANDER RICHARDSON (1839–1917), civil engineer; executive engineer in public works department of India, 1867; chief engineer for Bradford waterworks, 1875–90, and to London County Council, 1890–1901; knighted, 1897; reported on water supplies of Malta, Petrograd, and Ottawa.

BINNIE, WILLIAM JAMES EAMES (1867–1949), civil engineer; son of Sir A. R. Binnie [q.v.]; educated at Rugby, Trinity College, Cambridge, and Karlsruhe Polytechnic; joined father's practice, 1902; became recognized authority on flooding and water conservation; his most notable work the Gorge dam, Hong Kong; commissioner for Aswan dam (1928) and Nile electricity scheme (1937).

BINYON, (ROBERT) LAURENCE (1869–1943), poet, art-historian, and critic; scholar of St. Paul's School and Trinity College, Oxford; second class, *lit hum.*, 1892; entered department of printed books, British Museum, 1893; transferred to prints and drawings, 1895; assistant keeper, 1909; in charge of sub-department, oriental prints and drawings, 1913; keeper, 1932–3; fostered appreciation of oriental art and the English water-colourists; works include *Painting in the Far East* (1908); editions of Blake's *Woodcuts* (1902), *Drawings and Engravings* (1922), and *Engraved Designs* (1926); the poem 'For the Fallen' (Sept. 1914); two odes: 'The Sirens' (1924) and 'The Idols' (1928); plays: *Attila* (1907) and *Arthur: A Tragedy* (1923); *Collected Poems* (2 vols., 1931); deeply interested in the speaking of verse and in versification; CH, 1932.

BIRCH, GEORGE HENRY (1842–1904), architect and archaeologist; designed decoration for St. Nicholas Cole Abbey, London, 1882; president of Architectural Association, 1874–5; FSA, 1885; curator of Soane's Museum,

Lincoln's Inn Fields, from 1894 to death; wrote *London Churches of the Seventeenth and Eighteenth Centuries*, 1896.

BIRCH, Sir (JAMES FREDERICK) NOEL (1865-1939), general; educated at Marlborough and Royal Military Academy; commissioned in Royal Artillery, 1885; served in South Africa; major, 1900; took 7th brigade, RHA, to France, 1914; CRA, 7th division and I Corps, 1915; artillery adviser, GHQ, 1916-19; major-general, 1917; director-general, Territorial Army, 1921-3; master-general of the ordnance, 1923-7; general, 1926; KCMG, 1918.

BIRD, (CYRIL) KENNETH 'Fougasse' (1887-1965), cartoonist; educated at Farnborough Park School, Cheltenham College and King's College, London; B.Sc., 1908; worked in naval dockyard, Rosyth, 1909; joined Royal Engineers, 1914; severely wounded, 1915, and unable to walk for five years; started to draw; first drawing accepted by *Punch* signed 'Fougasse', 1916; regular contributor to *Punch*, 1917; published *A Gallery of Games*, 1921; and *Drawn at a Venture* (with preface by A. A. Milne, q.v.), 1922; art editor, *Punch*, 1937; editor, 1949; large number of drawings and posters for various ministries, including 'Careless Talk Costs Lives', 1940-5; published *You Have Been Warned* (with W. D. H. McCullough), 1935; fellow, King's College, London, 1936; CBE, 1946.

BIRD, HENRY EDWARD (1830-1908), chess player; by profession an accountant; wrote pamphlets on railway finance; a leading chess player in tournaments, 1851-99; won first prize in British Chess Association tournament (1889), winning 9½ out of 10 games; wrote many books on chess of mediocre value, including *Chess Practice* (1882), *Modern Chess* (1887 and 1889), and *Chess Novelties* (1895).

BIRD, ISABELLA LUCY (1831-1904), traveller and authoress. [See BISHOP.]

BIRD, Sir JAMES (1883-1946), naval and aircraft constructor; a naval architect; director (1919), owner (1923-8), Supermarine Aviation Works, Southampton; produced seaplanes (winning Schneider Trophy, 1927) and flying-boats designed by R. J. Mitchell [q.v.]; sold out to Vickers; supervised production of Spitfires, 1939-45; knighted, 1945.

BIRDWOOD, Sir GEORGE CHRISTOPHER MOLESWORTH (1832-1917), Anglo-Indian official and author; born at Belgaum, Western India: MD, Edinburgh University; medical practitioner in Bombay, 1858-68;

entered statistics and commerce department of India Office, 1878; prolific writer on Indian art, etc.; knighted, 1881; KCIE, 1887.

BIRDWOOD, HERBERT MILLS (1837-1907), Anglo-Indian judge; brother of Sir G. C. M. Birdwood [q.v.]; born at Belgaum, Western India; educated at Edinburgh University and Peterhouse, Cambridge; 23rd wrangler and fellow, 1858; MA, 1863; LLM, 1879; LLD, 1890; called to bar, 1889; hon. fellow of Peterhouse, 1901; went to Bombay, 1859; as judge of Ratnagiri district (1871-81) gained reputation for independence; judicial commissioner at Karachi, 1881-5; judge of Bombay high court, 1885-92; member of Bombay council, 1892-7; CSI, 1893; vice-chancellor of Bombay University, 1891-2; wrote on Indian flora; returned to England (1897), and practised before Privy Council on Indian appeals.

BIRDWOOD, WILLIAM RIDDELL, first BARON BIRDWOOD (1865-1951), field-marshal; born in India; son of H. M. Birdwood [q.v.]; educated at Clifton; commissioned in Royal Scots Fusiliers, 1883; served in India, 1885-99, South Africa, 1899-1902; assistant military secretary, 1902-4, military secretary, 1905-9, to Lord Kitchener [q.v.] in India; chief staff officer Mohmand Field Force, 1908; DSO; major-general, 1911; commanded Kohat independent brigade, 1909-12; secretary to army department, 1912-14; corps commander, Australian and New Zealand troops in Egypt, Dec. 1914; sent to Dardanelles, Feb. 1915; landed Anzac Corps on western shore of Gallipoli peninsula (Anzac Cove), 25 Apr.; held on for six months; evacuated Dardanelles Army, Dec. 1915-Jan. 1916; commanded I Anzac Corps in France, Mar. 1916-Nov. 1917; combined Australian Corps, Nov. 1917-May 1918; Fifth Army from May 1918; general 1917; baronet, 1919; commanded Northern Army, India, 1920-4; commander-in-chief, India, 1925-30; field-marshall, 1925; master of Peterhouse, Cambridge, 1930-8; CB, 1911; KCB, 1917; GCB, 1923; KCSI, 1915; GCSI, 1930; KCMG, 1914; GCMG, 1919; GCVO, 1937; baron, 1938.

BIRKENHEAD, first EARL OF (1872-1930), lord chancellor. [See SMITH, FREDERICK EDWIN.]

BIRKETT, WILLIAM NORMAN, first BARON BIRKETT (1883-1962), barrister and judge; educated at day school, Ulverston, and higher grade school, Barrow; worked as apprentice in father's drapers' business; studied at

Emmanuel College, Cambridge, 1907-11; president of the Union, 1910; called to bar (Inner Temple), 1913; practised in Birmingham, 1913-20; entered London chambers of Sir Edward Marshall Hall [q.v.], 1920; KC, 1924; briefed in many famous cases, 1925-39; knighted, 1941; judge, King's Bench, 1941; alternate British judge, Nuremberg war criminals trial, 1945; Appeal judge, 1950-6; baron, 1958; hon. LLD Cambridge, 1958; hon. fellow, Emmanuel College, 1946; hon. degrees, London, Birmingham, and Hull; president, Pilgrims, succeeding Lord Halifax [q.v.], 1958.

BIRLEY, Sir OSWALD HORNBY JOSEPH (1880-1952), painter; educated at Harrow and Trinity College, Cambridge; studied in Paris at Julian's and in Madrid copying from Velazquez; settled in London, 1906, establishing sound reputation as portrait painter producing high-quality faithful likenesses with insight and sympathy; painted most of royal family, wartime statesmen, and military leaders; knighted, 1949.

BIRMINGHAM, GEORGE A. (pseudonym), (1865-1950), novelist. [See HANNAY, JAMES OWEN.]

BIRRELL, AUGUSTINE (1850-1933), author and statesman; son of Baptist minister and grandson of Henry Grey [q.v.]; educated at Amersham Hall School, Caversham, and Trinity Hall, Cambridge; second class, law, 1872; called to bar (Inner Temple), 1875; developed steady Chancery practice; QC, 1895; Quain professor of law, London, 1896-9; became known as capable and versatile reviewer; *Obiter Dicta* (1884), included essays on Browning, Cellini, Carlyle, and actors; published a second series and *Charlotte Brontë*, 1887; liberal MP, West Fife (1889-1900) and North Bristol (1906-18); president, Board of Education, 1905-7; framed unsuccessful education bill of 1906; chief secretary for Ireland, 1907-16; established National University of Ireland, 1908; good-natured but indolent; confined himself to day-to-day administration and took no part in high policy; mainly advised by John Redmond [q.v.]; failed to realize danger of the Sinn Feiners and resigned after Easter rebellion, 1916; later publications include a sketch of his father-in-law, Frederick Locker-Lampson [q.v.], 1920; an entertaining writer of wit and humour.

BIRRELL, JOHN (1836-1901), orientalist; MA, St. Andrews University, 1855; entered Church of Scotland ministry, 1861; professor of Hebrew and oriental languages, St. Mary's College, St. Andrews, 1871; DD, Edinburgh,

1878; as examiner of Scottish secondary schools (1876-88) he originated university local examinations at St. Andrews.

BISCOE, CECIL EARLE TYNDALE-(1863-1949), missionary and educationist in Kashmir. [See TYNDALE-BISCOE.]

BISHOP, EDMUND (1846-1917), liturgiologist and historian; served in Education Office, 1864-85; joined Roman Church, 1867; discovered, transcribed, and annotated 'Collectio Britannica' of 300 papal letters preserved at British Museum and published by editors of *Monumenta Germaniae Historica*, 1880; his works include *Liturgica Historica* (1918).

BISHOP, ISABELLA LUCY (1831-1904), traveller and authoress; born BIRD; for reasons of health lived much in open air; visited Canada and America, 1854; published *The English-woman in America* (1856); made her home in Edinburgh, 1858, encouraging crofter emigration to Canada, 1862-6, and attacking Edinburgh poverty in *Notes on Old Edinburgh* (1869); visited Australia and New Zealand, 1872, Sandwich Islands, 1873, Rocky Mountains, 1873; in the result published two notable works, *The Hawaian Archipelago* (1875) and *A Lady's Life in the Rocky Mountains* (1879); visited Japan, 1878, and wrote *Unbeaten Tracks in Japan* (1880); married in 1881 Dr Bishop; on husband's death (1886) she studied medicine; went to India, 1889, visiting Kashmir, Tibet, Persia, Armenia, 1889-90; founded hospitals in Kashmir and Punjab and elsewhere, and wrote in favour of missions; first lady fellow of Royal Geographical Society, 1892; made extensive tour in Japan, Korea, and China, 1894-7; published *Korea and her Neighbours* (1898); a good photographer, a fearless traveller, and keen observer.

BLACKBURN, HELEN (1842-1903), pioneer of woman's suffrage; secretary to National Society for Women's Suffrage, 1874-95; editor of *Englishwoman's Review*, 1881-1903; wrote standard work on *Women's Suffrage* (1902), and other books on women's political and economic position.

BLACKBURNE, JOSEPH HENRY (1841-1924), chess player; employed in hosiery business, Manchester; turned attention to chess, 1857-9; won first prize in Manchester Chess Club's tournament, 1861-2; competed in London international tournament, 1862; won first prize in Berlin tournament, 1881; specialized in simultaneous displays with and without sight of board; most popular figure in English chess world.

BLACKER, (LATHAM VALENTINE) STEWART (1887–1964), soldier, inventor, and explorer; educated at Cheltenham College, Bedford School and Sandhurst; entered Indian Army and served in Afghanistan, Turkistan, and Russia; associated with development of 3.5-inch infantry mortar, 1905; obtained flying certificate, 1911; transferred to Royal Flying Corps, 1914; seriously wounded three times, 1915–17; OBE, 1920; member, Imperial General Staff, 1924–8; member, first expedition to fly over Everest, 1933; published *First Over Everest* (1933); invented anti-tank weapon upon which PIAT was based.

BLACKETT, Sir BASIL PHILLOTT (1882–1935), financial administrator; born in Calcutta of missionary parents; educated at Marlborough and University College, Oxford; first class, *lit. hum.*, and entered Treasury, 1904; secretary, royal commission on Indian finance, 1913–14; Treasury representative, Washington, 1917–19; controller of finance, Treasury, 1919–22; finance member, viceroy's council, India, introducing major reforms, 1922–8; a director of Bank of England, 1929; chairman, Imperial and International Communications Company, 1929–32; after 1931 a convert to 'planned' money and a prophet of the 'sterling area'; KCB, 1921; KCSI, 1926.

BLACKLEY, WILLIAM LEWERY (1830–1902), divine and social reformer; educated at Brussels and Trinity College, Dublin; BA, 1850; MA, 1854; ordained, 1854; an energetic parish priest; wrote much on national insurance, thrift, and pauperism; his compulsory state insurance scheme (1878) examined by House of Commons committee (1885–7) and reported upon adversely; his persistent agitation stimulated movements for old age pensions in England and colonies; vicar of St. James-the-Less, Westminster, 1889–1902; an eloquent speaker and accomplished linguist; his *Collected Essays* on thrift and national insurance, 1880, were re-edited with memoir, 1906.

BLACKMAN, FREDERICK FROST (1866–1947), plant physiologist; educated at Mill Hill, St. Bartholomew's Hospital, and St. John's College, Cambridge; demonstrator (1891), lecturer (1897), reader (1904–36), Cambridge Botany School; developed vigorous school of research mainly in plant respiration; showed bulk of exchange of carbon dioxide between foliage and air to be via stomata; pioneer in application of physico-chemical ideas to biological problems; FRS, 1906.

BLACKMAN, VERNON HERBERT (1872–1967), plant physiologist; educated at

City of London and King's College Schools, St. Bartholomew's Hospital Medical School, and St. John's College, Cambridge; first class, parts i and ii, natural sciences tripos, 1894–5; fellow, St. John's College, 1898–1904; Sc.D. (Cambridge), 1906; assistant, Department of Botany, British Museum (Natural History), 1896; professor, botany, Leeds University, 1907; professor, plant physiology and pathology, Imperial College, London, 1911–37; FRS 1913; concerned with development of Fruit Research Station, East Malling; dominant figure in development of experimental botany in Britain; edited *Annals of Botany* (1922–47); hon. doctorates, Allahabad and Benares, 1937.

BLACKWELL, ELIZABETH (1821–1910), the first woman doctor of medicine; emigrated from Bristol to New York, 1832; opened school in Cincinnati, 1838; studied medicine at Geneva, NY, 1847; MD, 1849; studied in Paris and at St. Bartholomew's Hospital, London, 1849; returned to America, 1850; opened dispensary in New York (1853), afterwards the New York Infirmary and College for Women; with sister, Emily, and Marie Zackrzewska, despite opposition, she opened hospital in New York conducted entirely by women, 1857; her name placed on British medical register, 1859; trained nurses for American civil war, 1864; her hospital granted charter, 1865; first professor of hygiene there, 1868; settled in England and founded National Health Society, 1871; accepted chair of gynaecology at newly founded London School of Medicine for Women, 1875; her writings include *Laws of Life* (1872), *Moral Education of the Young . . . under Medical and Social Aspects* (2nd edn. 1879), *Pioneer Work: Autobiographical Sketches* (1895), *Essays in Medical Sociology* (2 vols., 1902).

BLACKWOOD, ALGERNON HENRY (1869–1951), author; educated at Wellington College and Edinburgh University; spent ten precarious years in North America, described in *Episodes before Thirty* (1923), before entering journalism; progressed from tales of supernatural to novels of fancy and fantasy such as *The Centaur* (1911), *Pan's Garden* (1912), and *A Prisoner in Fairyland* (1913); CBE, 1949.

BLACKWOOD, FREDERICK TEMPLE HAMILTON-TEMPLE, first Marquess of Dufferin and Ava (1826–1902), diplomatist and administrator; born at Florence; son of Helen Selina Sheridan [q.v.]; educated at Eton and Christ Church, Oxford, 1844–6; created Baron Clandeboye in English peerage, 1850; supported liberal policy; lord-in-waiting to Queen Victoria, 1849–52 and 1854–8; attaché to Lord John Russell's mission at abortive conference at Vienna for ending Crimean war, 1855; after yachting voyage to Iceland and Spitsbergen, published *Letters from High Latitudes*, 1856; appointed British commissioner to assist Lord Dalling [q.v.] in inquiry into massacres in Levant, 1860; his proposal for independent governor adopted; KCB, 1861; under-secretary for India (1864–6) and to War Office (1866–8) under Palmerston; chancellor of duchy of Lancaster under Gladstone, 1868; created earl, 1871; as governor-general of Canada (1872–8) pacified agitators and strengthened spirit of federal union among Canadians; GCMG, 1876; hon. DCL, Oxford, 1879; appointed by Beaconsfield British ambassador at St. Petersburg, 1879; transferred to Constantinople, 1881; at Cairo reconstructed Egyptian administration after Arabi Bey's defeat at Tel-el-Kebir, 1882; advocated in his report representative institutions and municipal self-government; in Egypt created legislative council and assembly; GCB, 1883; succeeded Lord Ripon [q.v.] as governor-general of India, 1884; dealt tactfully with Indian land question; conciliated Amir of Afghanistan, 1885; strengthened British rule in India by improving railway communications with Afghan border, increasing the army, and constituting a new force of Burma military police; annexed Upper Burma, 1886; retired from India, 1888; promoted to marquess, 1888; made freeman of City of London, 1889; ambassador at Rome, 1889–91; lord rector of St. Andrews University, 1891; ambassador at Paris, 1891–6; improved relations between England and France in regard to Siam and the Congo; warden of cinque ports, 1891–5; hon. LLD, Cambridge, 1891; lord rector of Edinburgh University, 1901; suffered loss of money and reputation as chairman of London and Globe Finance Corporation, 1901.

BLAIR, ERIC ARTHUR (1903–1950), author under pseudonym of GEORGE ORWELL; scholar of Eton; studied poverty by experience; works include literary essays and political satire; developed highly individual socialism through *The Road to Wigan Pier* (1937) and *Animal Farm* (1945) to *Nineteen Eighty-Four* (1949).

BLAKE, EDWARD (1833–1912), Canadian lawyer and politician; born in Upper Canada; called to bar, 1856; prime minister of Ontario, 1871–2; minister of justice, 1875–7; leader of liberal opposition, 1880–8; nationalist member for South Longford, Ireland, 1892–1907; died in Toronto.

BLAKE, DAME LOUISA BRANDRETH ALDRICH- (1865–1925), surgeon. [See ALDRICH-BLAKE.]

BLAKISTON, HERBERT EDWARD DOUGLAS (1862–1942), president of Trinity College, and vice-chancellor of Oxford; educated at Tonbridge and Trinity College; first class, *lit. hum.*, 1885; ordained; fellow, chaplain, and lecturer (1887), tutor (1892), senior tutor and junior bursar (1898), president (1907–38), estates bursar (1915–38); vice-chancellor, 1917–20.

BLAMEY, SIR THOMAS ALBERT (1884–1951), field-marshal; born in New South Wales; commissioned in Australian Army, 1906; at Staff College, Quetta, 1911–13; major, 1914; served in Gallipoli and France; DSO, 1917; chief commissioner of police, Victoria, 1925–36; lieutenant-general, 1939; commanded Australian Army Corps in Libya, Anzac Corps in Greece; rapidly expanded and organized home forces as commander-in-chief, Australian Army, 1941–5; commanded Allied Land Forces in New Guinea, 1942–3; field-marshal, 1950; knighted, 1935.

BLAND, EDITH, MRS HUBERT BLAND (1858–1924), writer of children's books, poet, and novelist, under the name E. NESBIT; daughter of John Collis Nesbit [q.v.]; married Hubert Bland, 1880, T. T. Tucker, 1917; best known for her children's books, such as *The Story of the Treasure Seekers* (1899).

BLAND, JOHN OTWAY PERCY (1863–1954), writer on Chinese affairs; worked in China from 1883; agent, British and Chinese Corporation, Ltd., 1906–16; *The Times* correspondent successively in Shanghai and Peking, 1897–1910; many works on China include *China under the Empress Dowager* (1910, with (Sir) Edmund Backhouse, q.v.).

BLAND-SUTTON, SIR JOHN, baronet (1855–1936), surgeon. [See SUTTON.]

BLANDFORD, GEORGE FIELDING (1829–1911), physician; educated at Tonbridge, Rugby, and Wadham College, Oxford; BA, 1852, MA and MB, 1857, MD, 1867; studied medicine at St. George's Hospital, London; lecturer there, 1865–1902; FRCP, 1869; Lumleian lecturer, 1895; acquired large lunacy practice in London from 1863; president of Medico-Psychological Association of Great Britain, 1877, and of psychological section, British Medical Association, 1894; chief work was *Insanity and its Treatment* (1871); interested in art, literature, and music.

BLANESBURGH, BARON (1861–1946), judge. [See YOUNGER, ROBERT.]

BLANEY, THOMAS (1823-1903), physician and philanthropist; apprenticed to medical department, East India Company, 1836; studied at Grant Medical College, Bombay, 1851-5; made special study of fevers and plague; coroner of Bombay, 1876-93; generous benefactor to poor of Bombay; founded 'Blaney' school for poor white children; original member of Bombay corporation, 1872; obtained municipal water supply, sheriff of Bombay, 1875 and 1888; CIE, 1894.

BLANFORD, WILLIAM THOMAS (1832-1905), geologist and zoologist; studied at Royal School of Mines (1852-4) and Freiberg in Saxony; obtained post in Indian Geological Survey, 1854; investigated coalfield near Talchir (1854-7) and geology of Burma (1860); appointed deputy superintendent, he surveyed Bombay presidency, 1862-6; attached to Abyssinian expedition, 1867; published works on the geology of Abyssinia, 1870, and of India, 1879; settled in London, 1881; edited for government works on Indian fauna, contributing two vols. on mammals (1888, 1891) and two on birds (1895, 1898); president of Geological Society, 1888-90; FRS, 1874; CIE, 1904.

BLATCHFORD, ROBERT PEEL GLANVILLE (1851-1943), journalist, author, and socialist; sergeant in 103rd regiment and later a clerk in Northwich; wrote for *Sunday Chronicle*, 1885-91; lost job on becoming socialist and with three others founded the *Clarion* (1891) as socialist weekly; his articles reprinted as *Merrie England* (1893) popularized both socialism and his paper; later lost favour through supporting South African war, attacking orthodox religion, and giving warnings against Germany; publications include *Britain for the British* (1902) and *The Sorcery Shop* (1907).

BLAYDES, FREDERICK HENRY MARVELL (1818-1908), classical scholar; commoner of Christ Church, Oxford, 1836; Hertford scholar, 1838; BA, 1840; MA, 1843; toured in France and Italy, 1840-1; ordained, 1842; vicar of Harringworth, Northamptonshire, 1843-84; edited Aristophanes, Aeschylus, and Sophocles; chief work was *Aristophanis comoediae*, 1882-93, devoted mainly to verbal and textual criticism and emendation; hon. LLD, Dublin, 1892; Ph.D., Budapest; presented his classical library to St. Paul's School, 1901; interested in homoeopathy and music.

BLEDISLOE, first VISCOUNT (1867-1958), agriculturist and public servant. [See BATHURST, CHARLES.]

BLENNERHASSETT, SIR ROWLAND, fourth baronet (1839-1909), political writer; of Roman Catholic parentage; educated at Stonyhurst, Christ Church, Oxford, and Louvain University; formed friendship with Döllinger at Munich (1864) and with Lord Acton [q.v.] (1862); helped to start *Chronicle*, a literary organ of liberal catholicism, 1867; admirer of Bismarck; liberal MP for Galway city, 1865-74, and for Kerry county, 1874-85; at first supported but later opposed home rule; interested in Irish education and land question; advocated peasant proprietorship, 1884; president of Queen's College, Cork, 1897-1904; PC Ireland, 1905; wrote frequently in leading periodicals on political subjects, and published speeches and addresses.

BLIND, KARL (1826-1907), political refugee and author; born at Mannheim, Germany; entered Heidelberg University, 1845; engaged there in revolutionary political agitation; imprisoned at Bonn for circulating pamphlet *Deutscher Hunger und Deutsche Fürsten*, 1847; proscribed by Baden government for share in democratic risings, 1848; took refuge in Alsace; imprisoned at Strasbourg; prominent in rising at Staufen (24 Sept.); released by revolutionaries from prison at Bruchsal, 1849; helped to set up provisional government for Baden, 1 June 1849; its representative in Paris; exiled in turn from France (1849) and Belgium (1852); settled in England, 1852; entertained leading European political exiles; introduced Mazzini to Swinburne; championed nationalist and democratic causes in all countries from 1878 till death; wrote much on Indian and German mythology.

BLOGG, HENRY GEORGE (1876-1954), coxswain of Cromer lifeboat; joined crew, 1894; second coxswain, 1902; coxswain, 1909-47; a superb seaman inspiring crew's complete confidence; received Royal National Life-boat Institution's gold medal thrice, silver medal four times, Empire Gallantry medal, 1924, George Cross, 1941.

BLOMFIELD, SIR REGINALD THEODORE (1856-1942), architect; educated at Haileybury and Exeter College, Oxford; joined his uncle (Sir) A. W. Blomfield [q.v.], 1881; set up own practice, 1884; restored many houses including Chequers; designed 'Cross of Sacrifice' and Menin Gate at Ypres for Imperial War Graves Commission; other works include Lady Margaret Hall, Oxford, Lambeth bridge, and the Regent Street quadrant; publications include *History of Renaissance Architecture in England 1500-1800* (2 vols., 1897) and *History of French Architecture, 1494-1774* (4 vols.,

1911-21); enjoyed many controversies, notably over his design for an office building in Carlton Gardens (1932); ARA, 1905; RA, 1914; FRIBA, 1906; knighted, 1919.

BLOOD, Sir BINDON (1842-1940), general; educated at Queen's College, Galway, and Indian Military Seminary; commissioned in Royal Engineers, 1860; served mainly in India; saw active service in Jowaki expedition (1877-8), Zulu war (1879), second Afghan war (1880), at Tel-el-Kebir, 1882; chief staff officer, Chitral relief force (KCB, 1896); commanded Malakand and Buner field forces (1896-8); major-general, 1898; commanded in Eastern Transvaal, 1901, and Punjab, 1901-7; general, 1906; chief royal engineer, 1936-40; successes due to brilliant staff work and strategy.

BLOOD, Sir HILARY RUDOLPH ROBERT (1893-1967), colonial administrator; educated at Irvine Royal Academy and Glasgow University; MA, 1914; served in Ceylon civil service, 1920-30; colonial secretary, Grenada, 1930-4; Sierra Leone, 1934-42; governor, Gambia, 1942-7; Barbados, 1947-9; and Mauritius, 1949-54; constitutional commissioner, British Honduras, 1959; Zanzibar, 1960; chairman, constitutional commission, Malta, 1960; chairman, Royal Commonwealth Society for the Blind, 1962-5; chairman, Royal Society of Arts, 1963-5; CMG, 1934; KCMG, 1944; GBE, 1953; hon. LLD, Glasgow, 1944.

BLOOMFIELD, GEORGIANA, Lady (1822-1905), maid of honour to Queen Victoria, 1841-5; married second Baron Bloomfield [q.v.], 1845, whom she accompanied on his diplomatic missions; published *Reminiscences of Court and Diplomatic Life*, 2 vols., 1883.

BLOUET, LÉON PAUL, 'Max O'Rell' (1848-1903), humorous writer; born in Brittany; cavalry officer in Franco-German war; French master at St. Paul's School, 1876-87; wrote *John Bull et son île* (1887) and *Jonathan and his Continent* (1889), vivacious pictures of English and American life; died in Paris.

BLOUNT, Sir EDWARD CHARLES (1809-1905), Paris banker and promoter of French railways; of Roman Catholic family; attaché to Paris embassy, 1829; in Rome, where he met (Cardinal) Wiseman [q.v.] and the future Napoleon III, 1830; started banking business in Paris, which failed at the revolution, 1848; promoted railway enterprise in France; formed company (Chemin de fer de l'Ouest), 1838; chairman till 1894; constructed railways from Paris to Rouen (1843), and from Amiens to Boulogne (1845); supported French royal

family, 1848; in 1852 started in Paris a new banking business which proved successful and was merged in Société Générale of Paris, 1870; remained in Paris during siege of 1870; British consul there, Jan.-Mar. 1871; CB, 1871; KCB, 1878; commander of Legion of Honour; long president of British Chamber of Commerce in Paris; devoted to horse-racing; published *Recollections*, 1902.

BLUMENFELD, RALPH DAVID (1864-1948), editor; born in Wisconsin, the son of a newspaper owner; joined *New York Herald*, 1887; head of London office, 1890; New York superintendent, 1893-4; sold linotype machines in England, 1894-1900; news editor, *Daily Mail*, 1900-2; foreign editor, *Daily Express*, 1902-4; editor, 1904-32 (editor-in-chief from 1924); chairman, London Express Newspapers Company, 1915-48; expert in front-page presentation of news and influential in shaping popular journalism; naturalized, 1907.

BLUMENTHAL, JACQUES [JACOB] (1829-1908), composer of songs; born in Hamburg; settled in London, 1848; pianist to Queen Victoria; composed songs which won lasting popularity.

BLUNT, Lady ANNE ISABELLA (1837-1917) (DNB 1922-30, under W. S. Blunt whom she married in 1869), daughter of William Noel, first Earl of Lovelace; Arabic scholar, horse-woman, musician, traveller, authoress; Baroness Wentworth, 1917; died in Cairo.

BLUNT, WILFRID SCAWEN (1840-1922), traveller, politician, and poet; in diplomatic service, 1858-69; inherited Crabbet Park, Sussex, 1872; travelled extensively; with his wife, penetrated mysterious territory of Nedj, 1878; visited India, 1878, 1883-4; as result became anti-imperialist; published *Ideas about India*, 1885; criticized British occupation of Egypt; took up cause of Ireland; principally memorable for his poetry, notably for his sonnets and lyrics; a brilliant talker and agreeable host; kept stud farm.

BLYTHSWOOD, first BARON (1835-1908), scientist. [See CAMPBELL, ARCHIBALD CAMPBELL.]

BLYTON, ENID MARY (1897-1968), writer for children; educated at St. Christopher's school for girls, 1907-15; training as teacher, Ipswich, 1916-18; teaching in Kent, 1919; nursery governess, Surbiton, 1920; poem published at age of fourteen; short stories and poems published in 1921-2; first book *Child Whispers*, 1922; prolific output for forty years;

six titles published, 1935; eleven titles published, 1940; by 1965 over four hundred titles published; first major children's author to appear in paperback editions; publications include *Famous Five* and *Adventure* and *Noddy* series; wrote and edited the *Enid Blyton Magazine*.

BODDA PYNE, LOUISA FANNY (1832–1904), soprano vocalist; made début in London, 1842, and Paris, 1847; toured in America (1854–7) with William Harrison (1813–68, q.v.); produced at Lyceum and Covent Garden new English operas, 1857–62; married Frank Bodda, 1868; received civil list pension, 1896.

BODINGTON, Sir NATHAN (1848–1911), vice-chancellor of Leeds University; BA, Wadham College, Oxford, 1872; MA, 1874; fellow and tutor of Lincoln College, Oxford, 1875–85; hon. fellow, 1898; first professor of Greek, Mason College, Birmingham, 1881; appointed (1882) professor of Greek and principal of Yorkshire College, Leeds; vice-chancellor of Victoria University, Manchester, 1896–1900; promoted independent university for Leeds, which was founded with Bodington as vice-chancellor, 1904; extended the university's activities in scientific and technical directions; hon. Litt.D., Manchester, 1895, LLD, Aberdeen, 1906; knighted, 1908.

BODKIN, Sir ARCHIBALD HENRY (1862–1957), lawyer; grandson of Sir W. H. Bodkin [q.v.]; educated at Highgate School; called to bar (Inner Temple), 1885; junior Treasury counsel at Old Bailey, 1892, senior counsel, 1908; a leading criminal advocate; in chambers with uncle Sir Harry Poland [q.v.]; prosecuted G. J. Smith, 1915; prepared case against Sir Roger Casement [q.v.]; recorder of Dover, 1901–20, 1931–47; director of public prosecutions, 1920–30; cases included prosecution of Horatio Bottomley and of Clarence Hatry [qq.v.]; knighted 1917; KCB, 1924.

BODKIN, THOMAS PATRICK (1887–1961), museum director and art critic; born in Dublin; educated at Belvidere College, Clongowes Wood College, and Royal University of Ireland; called to bar (King's Inn), 1911; as art collector, influenced by Sir Hugh Lane [q.v.]; member of commission on Irish coinage design, 1926; director, National Gallery of Ireland, 1927–35; director, Barber Institute and professor of fine arts, Birmingham University, 1935; publications include *The Approach to Painting* (1927), *Hugh Lane and his Pictures* (Verona, 1932), *Four Irish Landscape Painters* (1920), and *Dismembered Masterpieces* (1945); hon. professor of history of fine arts,

Dublin, hon. ARIBA, hon. degrees, National University of Ireland and Dublin University.

BODLEY, GEORGE FREDERICK (1827–1907), architect; descended from Sir Thomas Bodley [q.v.]; pupil of George Gilbert Scott [q.v.], 1845–50; first exhibited at Royal Academy, 1854; designed (1860–70) many private houses and churches, including St. Michael's, Brighton; partner with Thomas Garner [q.v.], 1896–98, with whom he designed the churches of Holy Angels, Hoar Cross (in simple style), St. Augustine, Pendlebury, 1874 (an elaborate structure), and All Saints, Cambridge; built independently churches at Clumber and Eccleston, and Community Church, Cowley, Oxford; did work at Oxford (at Magdalen, and the tower of 'Tom Quad' at Christ Church) and at Cambridge (the 'Bodley' group of buildings at King's College and Queens' College chapel); other works were cathedral at Hobart, Tasmania, and school board offices on Thames embankment; prepared plans for episcopal cathedral of SS Peter and Paul, Washington, 1906; ARA, 1882; RA, 1902; his work is a later nineteenth-century counterpart of that of the architects of the Oxford Movement, Pugin, Scott, and Street; combined ecclesiological knowledge with sound taste; friend of Morris, Burne-Jones, Madox Brown, and Rossetti; published small volume of verse, 1899; FSA, 1885; hon. DCL, Oxford, 1907.

BODY, GEORGE (1840–1911), canon of Durham; BA, St. John's College, Cambridge, 1862; MA, 1876; hon. DD, Durham, 1885; rector of Kirby Misperton, Yorkshire, 1870; 'canon missioner' of Durham, 1883–1911; lecturer in pastoral theology at King's College, London, 1909; combined evangelical fervour with tractarian principles; published many devotional works.

BOLDERO, Sir HAROLD ESMOND ARNISON (1889–1960), physician and medical administrator; educated at Charterhouse and Trinity College, Oxford; qualified at Middlesex Hospital, 1915; DM, 1925; on staff of Evelina Hospital for Children, 1921–34; assistant physician, Middlesex Hospital, 1922, dean of medical school, 1934–54; fellow, 1933, registrar, 1942–60, Royal College of Physicians; its representative in National Health Service negotiations and on Joint Consultants' Committee; knighted, 1950.

BOLS, Sir LOUIS JEAN (1867–1930), lieutenant-general; entered army, 1887; served in South African war, 1899–1902; appointed to Staff College, 1910; served in European war, 1914–18; took prominent part in battle of

Messines ridge, June 1917; as General Allenby's chief of staff captured Jerusalem, Dec. 1917; chief administrator, Palestine, 1919; governor and commander-in-chief, Bermuda, 1928; CB, 1915; KCMG, 1918; KCB, 1919.

BOMBERG, DAVID GARSHEN (1890–1957), painter; studied at City and Guilds Institute, and Westminster and Slade art schools; founder-member, London Group, 1913; works commissioned by Canadian war memorial and Zionist Organization; travelled widely; after 1945 taught at Borough Polytechnic; painted mostly landscapes, still-life, and human head, with magnificent draughtsmanship and unique sonority of colour; died in obscurity but subsequently recognized.

BOMPAS, HENRY MASON (1836–1909), county court judge at Bradford from 1896; brother of William Carpenter Bompas [q.v.]; BA, St. John's College Cambridge (fifth wrangler), 1858; barrister (Inner Temple), 1863.

BOMPAS, WILLIAM CARPENTER (1834–1906), bishop of Selkirk; ordained, 1859; went as missionary to Mackenzie River, 1865; arrived at Fort Yukon, 1869; bishop of Athabasca, 1874, of Mackenzie River, 1884, and of Selkirk, 1890; spent last years at Caribou Crossing, where he built a church, 1904; resigned bishopric, 1905; author of *The Diocese of Mackenzie River*, 1888; translated hymns, prayers, and portions of Bible for Canadian Indians.

BONAR, JAMES (1852–1941), political economist; son of A. A. Bonar [q.v.]; educated at Glasgow Academy and University, Leipzig, Tübingen, and Balliol College, Oxford; first class, *lit. hum.*, 1877; university extension lecturer in east London, 1877–80; examiner in Civil Service Commission, 1881–1907; deputy master, Royal Mint, Ottawa, 1907–19; publications include *Malthus and His Work* (1885), *Philosophy and Political Economy* (1893), *Catalogue of the Library of Adam Smith* (1894) on whom he was a leading authority, and many papers in economic journals; FBA, 1930.

BONAR LAW, ANDREW (1858–1923), statesman. [See LAW.]

BOND, SIR (CHARLES) HUBERT (1870–1945), psychiatrist and administrator; MB (1892), B.Sc., public health (1893), Edinburgh; first medical superintendent, Ewell Colony for Epileptics, 1903–7, LCC mental hospital, Long Grove, 1907–12; a commissioner in lunacy (1912) and later, until 1945, senior com-

missioner, Board of Control; raised level of medical work in mental hospitals; obtained voluntary admission, 1930; KBE, 1929.

BOND, SIR ROBERT (1857–1927), premier of Newfoundland; born at St. John's, Newfoundland; MA, Edinburgh University; entered Newfoundland house of assembly, 1882; speaker, 1884; colonial secretary in administration of Sir W. V. Whiteway [q.v.], 1889–94, 1895–7; immediately encountered two difficult international problems with which he was to be closely concerned for next twenty years: fishery disputes between Newfoundland and France, and negotiations between Newfoundland, Canada, and USA about fishery rights and reciprocal trade; chairman of deputation sent by Newfoundland to abortive Ottawa conference to discuss question of Newfoundland entering Canadian federation, 1895; his courage in succeeding months saved Newfoundland from financial collapse; premier and colonial secretary, 1900–9; modified 'Reid contract', 1901; fishery disputes with France settled, 1904; with USA, 1910; leader of opposition, 1909–14; KCMG, 1901.

BOND, WILLIAM BENNETT (1815–1906), primate of all Canada from 1904; a native of Truro; emigrated to Newfoundland; ordained, 1840; curate (1848–60) and rector (1860–78) of St. George's, Montreal; became canon of Christ Church Cathedral, 1866; dean of Montreal, 1872; bishop, 1878; archbishop and metropolitan, 1901; took foremost part in expansion of Canadian Church; president of Bishop's College, Lennoxville (hon. MA, 1854); hon. LLD, McGill University, 1870; strong evangelical and temperance reformer.

BONDFIELD, MARGARET GRACE (1873–1953), trade-union leader and first British woman cabinet minister; apprenticed to drapery trade; assistant secretary, National Union of Shop Assistants, 1898–1908; helped Mary Macarthur [q.v.] found National Federation of Women Workers, 1906, assistant secretary, 1915; on amalgamation, chief woman officer with National Union of General and Municipal Workers, 1921–38; elected to Parliamentary Committee, Trades Union Congress, 1918; first woman chairman of TUC, 1923; labour MP Northampton, 1923–4, Wallsend, 1926–31; parliamentary secretary Ministry of Labour, 1924; minister of labour, 1929–31; declined to join 'national' government; chairman, Women's Group on Public Welfare, 1939–49; PC, 1929; CH, 1948.

BONE, JAMES (1872–1962), journalist; left school at fourteen to work in Laird Line office,

Glasgow, 1886; joined *North British Daily Mail*; wrote with (Professor) A. M. Charteris [q.v.] *Glasgow in 1901* (1901) illustrated by his brother (Sir) Muirhead Bone [q.v.]; joined London office of *Manchester Guardian*, 1902; London editor, 1912-45; important contribution, the London letter of the *Guardian*; published also *The London Perambulator* (1925); survived sinking of *Western Prince* by torpedo, 1940; CH, 1947; hon. ARIBA, 1927.

BONE, SIR MUIRHEAD (1876-1953), water-colour painter and etcher; apprenticed architect; studied at Glasgow School of Art; settled in London, 1901; his drypoints soon gained recognition; produced landscapes, portraits, and was especially effective with buildings; first official war artist; published *The Western Front* (2 vols., 1917), with text by C. E. Montague [q.v.]; influential in foundation of Imperial War Museum; knighted, 1937.

BONE, STEPHEN (1904-1958), painter and art critic; son of Sir Muirhead Bone [q.v.]; educated at Bedales and Slade School; most successful with oil landscapes; art critic, *Manchester Guardian*.

BONE, WILLIAM ARTHUR (1871-1938), chemist and fuel technologist; graduated in chemistry from Owens College, Manchester; head of chemistry department, Battersea Polytechnic, 1896-8; lecturer in chemistry and metallurgy, Owens College, 1898-1906; professor of fuel and metallurgy, Leeds, 1906-12; of chemical technology, London, 1912-36; established a department of fuel technology at Royal College of Science, 1912; formulated the hydroxylation theory of hydrocarbon combustion; with C. D. McCourt developed the incandescent surface combustion (Bonecourt) process; researches included the constitution of coal and blast-furnance reactions; FRS, 1905.

BONHAM-CARTER, SIR EDGAR (1870-1956), jurist and administrator; grandson of G. W. Norman [q.v.]; educated at Clifton and New College, Oxford; played rugby for Oxford and England; called to bar (Lincoln's Inn), 1895; legal secretary to Sudan government, 1899-1917; devised country's civil and criminal law; senior judicial officer, Baghdad, 1917-19; judicial adviser, Mesopotamia, 1919-21; on executive committee, National Trust, 1927-50; chairman, First Garden City, Ltd., Letchworth, 1929-39; founder and first chairman, British School of Archaeology in Iraq, 1932-50; KCMG, 1920.

BONHAM CARTER, (HELEN) VIOLET, BARONESS ASQUITH OF YARNBURY (1887-1969),

political figure; only daughter of Herbert Henry Asquith (later Earl Oxford and Asquith); educated privately; supported her father on political platforms, 1905-18; close friend of (Sir) Winston Churchill; bitter opponent of Lloyd George; married (Sir) Maurice Bonham Carter, 1915; president, Women's Liberal Federation (1923-5 and 1939-45); president, Liberal Party Organisation, 1945-7); defeated in elections at Wells (1945) and Colne Valley (1951); vice-chairman, United Europe Movement, 1947; president, Royal Institute of International Affairs, 1964-9; governor, BBC, 1940-6; member, royal commission on the Press, 1947-9; governor, Old Vic, 1945; trustee, Glyndebourne Arts Trust, 1955; first woman to give Romanes lecture at Oxford, 1963; published *Winston Churchill as I Knew Him* (1965); DBE, 1953; baroness, 1964; hon. LLD Sussex, 1963.

BONNEY, THOMAS GEORGE (1833-1923), geologist; BA, St. John's College, Cambridge; ordained, 1857; fellow of his college, to which in 1861 he returned, 1859; tutor, 1868; lecturer in geology, 1869; Yates-Goldsmid professor of geology, University College, London, 1877-1901; assistant general secretary, British Association, and settled in Hampstead, 1881; returned to Cambridge, 1905; president, British Association, 1910; FRS; devoted himself especially to study of petrology, alpine geology in particular, and later to ice-work, on which he wrote copiously.

BONNEY, (WILLIAM FRANCIS) VICTOR (1872-1953), gynaecologist; qualified at Middlesex Hospital, 1896; obstetric tutor, 1903; assistant (1908), senior (1930-7) gynaecological surgeon; raised gynaecology to major branch of surgery and devised superb operative technique; with Sir Comyns Berkeley wrote *A Textbook of Gynaecological Surgery* (1911) and other works.

BONWICK, JAMES (1817—1906), Australian author and archivist; left England, 1840; conducted model school in Hobart, Van Diemen's Land (now Tasmania); joined in rush to Victorian goldfields, 1852; appointed archivist to government of New South Wales, 1884; issued officially *Historical Records of New South Wales* (7 vols., 1893-1901); wrote on early Australian history, and *An Octogenarian's Reminiscences*, 1902.

BOOT, JESSIE, first BARON TRENT (1850-1931), man of business and philanthropist; succeeded father as a herbalist in Nottingham; opened first chemist's shop, 1877; built up world's largest retail chemists' undertaking;

added libraries and other departments; began manufacture of own drugs, 1892; many benefactions to Nottingham included the new University College; knighted, 1909; baronet, 1917; baron, 1929.

BOOTH, CHARLES (1840-1916), shipowner and writer on social questions; partner in Alfred Booth & Co., shipowners, Liverpool, 1862; author of *Life and Labour of the People in London* (1891-1903), which gives a comprehensive and illuminating picture of London in the last decade of the nineteenth century; largely contributed to passing of Old Age Pensions Act, 1908; FRS, 1899; PC, 1904.

BOOTH, HUBERT CECIL (1871-1955), engineer; educated at College and County schools, Gloucester; studied civil and mechanical engineering at City and Guilds Central Institution; designed 'Great Wheels' at Blackpool, Vienna, and Paris; set up as consulting engineer, 1901; invented vacuum cleaner, 1901, and developed its industrial possibilities; chairman, British Vacuum Cleaner Company, 1901-52.

BOOTH, WILLIAM (1829-1912), popularly known as 'General' Booth, founder of the Salvation Army; apprenticed to Nottingham pawnbroker; underwent experience of conversion, 1844; removed to London, 1849; itinerant preacher of Methodist New Connexion, 1852; married Catherine Mumford (Mrs Catherine Booth, q.v.), 1855; began work as independent revivalist, 1861; started Christian Mission in Whitechapel, 1865; adopted for it the title 'Salvation Army', 1878; championed degraded poor in great cities and attracted multitudes; interest in social reform led to publication of *In Darkest England and the Way Out* (1890); more interesting as a man than as founder of anything new in religion or politics; entirely ignorant of theology, obscurantist in intellectual sphere, narrow and uncompromising in religion; his son Bramwell (see below) real organizer of Army; yet Booth probably changed more lives for better than any other religious emotionalist for centuries.

BOOTH, WILLIAM BRAMWELL (1856-1929), Salvation Army leader; son of William Booth [q.v.]; chief of staff, Salvation Army, 1880-1912; 'general', 1912-29; associated with W. T. Stead [q.v.] in campaign resulting in Criminal Law Amendment Act, 1885; secured Army's right to hold open-air meetings; visited European capitals, India, Australia, United States, etc. on behalf of Salvation Army.

BOOTHBY, GUY NEWELL (1867-1905), novelist; born at Adelaide, South Australia;

educated in England; produced melodrama in Adelaide, 1888-91; settled in England, 1894; wrote fifty sensational novels, the best dealing with Australian life; the most popular was *A Bid for Fortune, or Dr. Nikola's Vendetta*, 1895.

BOOTHMAN, SIR JOHN NELSON (1901-1957), air chief marshal; commissioned in Royal Air Force, 1921; won Schneider Trophy, 1931; assistant chief of air staff, technical requirements, 1945-8; AOC, Iraq, 1948-50; controller of supplies (air), Ministry of Supply, 1950-3; held Coastal Command, 1953-6; air chief marshal, 1954; KBE, 1951.

BORDEN, SIR ROBERT LAIRD (1854-1937), Canadian statesman; born in Nova Scotia; called to bar, 1878; QC, 1891; conservative MP, Halifax, 1896; leader of conservative party, 1901; a reformer at heart; his manifesto (1907) called for controls unattractive to conservatives; urged expropriation of Grand Trunk Pacific Railway (1909) and the building of a Canadian navy; Canadian prime minister, 1911-20; frequently accused of splitting Canada by forming a two-party government to enforce conscription, 1917; but the split was already there and despite failure to bridge the rift he played a large part in making Canada a nation; attended imperial conference (1917), peace conference (1919), and Washington conference (1921-2); represented Canada on League of Nations council and chief Canadian delegate to Assembly in 1930; PC, 1912; GCMG, 1914.

BORTHWICK, ALGERNON, BARON GLENESK (1830-1908), proprietor of the *Morning Post*; son of Peter Borthwick [q.v.]; educated in Paris and at King's College School, London; foreign correspondent of *Morning Post* in Paris, 1850; succeeded father as editor, 1852; helped to produce with Evelyn Ashley [q.v.] the society journal *The Owl*, 1864-70; bought *Morning Post*, 1876; reduced price from 3*d.* to 1*d.*, 1881; knighted, 1880; suggested 'Primrose League' in *Morning Post*, 1883; conservative MP for South Kensington, 1885-95; carried measure amending law of libel, 1888; advocated tariff reform, and supported Lord Randolph Churchill [q.v.]; baronet, 1887; baron, 1895; married Alice, daughter of Thomas Henry Lister [q.v.], 1870; handed over control of paper (1895) to son Oliver (*d.* 1905); keenly interested in theatre; restored to solvency Chelsea Hospital for Women, of which he was president in 1905.

BOSANQUET, BERNARD (1848-1923), philosopher; cousin of Sir F. A. Bosanquet [q.v.]; BA, Balliol College, Oxford; fellow of University College, Oxford, 1870-81; settled in

London in order to engage in philosophical writing and social work, 1881; member of London Ethical Society and London School of Ethics and Social Philosophy, for which he lectured; also worked for Charity Organisation Society; professor of moral philosophy, St. Andrews University, 1903-8; his works include *Knowledge and Reality* (1885), *Logic* (1888), *A History of Aesthetic* (1892), and *The Philosophical Theory of the State* (1899); Gifford lecturer, Edinburgh, 1911, 1912; with F. H. Bradley [q.v.], last great representative of nineteenth-century school of British Hegelianism; value of his philosophy lies in its combination of universality and concreteness.

BOSANQUET, Sir FREDERICK ALBERT (1837-1923), common serjeant; son of S. R. Bosanquet [q.v.]; educated at Eton and King's College, Cambridge; called to bar (Inner Temple) and joined Oxford circuit, 1863; QC, 1882; common serjeant, 1900-17; additional judge, Central Criminal Court, 1917; knighted, 1907; joint-author of *A Practical Treatise on the Statutes of Limitations . . .* (1867).

BOSANQUET, ROBERT CARR (1871-1935), archaeologist; scholar of Eton and Trinity College (first class, classics), Cambridge; director (1900-5), British School at Athens; professor of classical archaeology (1906-20) at Liverpool where an Institute of Archaeology was established; excavated in Egypt, Greece, and Crete, and Roman sites in Wales.

BOSWELL, JOHN JAMES (1835-1908), major-general; entered Bengal army, 1852; commanded detachment of 3rd and 6th Punjab infantry in Indian Mutiny, 1857, and 2nd Sikh infantry in Afghan war, 1878-80; at battle of Kandahar, 1880; CB, 1881; major-general, 1885.

BOSWELL, PERCY GEORGE HAMNALL (1886-1960), geologist; B.Sc., London, 1911; studied geology at Imperial College; D.Sc., 1916; wartime geological adviser, Ministry of Munitions; OBE, 1918; George Herdman professor of geology, Liverpool, 1917-30; professor of geology, Imperial College, 1930-8; FRS, 1931; contributed to economic and engineering applications of geology, advising on road and water undertakings; publications include *On the Mineralogy of the Sedimentary Rocks* (1933) and papers on East Anglian stratigraphy and the stratigraphy and tectonics of the Silurian rocks of Denbighshire.

BOSWORTH SMITH, REGINALD (1839-1908), schoolmaster and author. [See SMITH.]

BOTHA, LOUIS (1862-1919), South African soldier and statesman; born in Natal; brought up on large farm near Vrede in Orange Free State from 1869; served as volunteer under Lukas Meyer, landrost of Utrecht, in successful campaign for restoration of Dinizulu, son of Zulu chief, Cetywayo, 1884; commissioner to delimit farms on Zulu territory assigned to volunteers after campaign; obtained Waterval farm in 'New Republic' which (1888) joined Transvaal; field-cornet and native commissioner at Vryheid; mobilized burgher force during Jameson raid, 1895; member for district in first volksraad, 1897; supporter of P. J. Joubert's liberal views on uitlander franchise; on declaration of war with England mustered commando at Vryheid, 1899; distinguished himself at battle of Dundee and promoted assistant-general (Oct.); forced Sir Redvers Buller [q.v.] to abandon attempt to relieve Ladysmith by pushing across Tugela at Colenso and making direct attack on Joubert's force (Dec.); urged immediate advance, but was overruled; saved situation at Tabanyama and Spion Kop (Jan. 1900); forced Buller to retire at Vaalkrantz (Feb.); promoted commandant-general of Transvaal on Joubert's death (March); reorganized commandos; made last stand for Johannesburg and Pretoria at Doornkop (May); defeated at closely contested battle of Diamond Hill (June); carried on guerrilla warfare against British, 1900-2; submitted, after obtaining reasonable terms, at Vereeniging, May 1902; visited England, 1902; lived thenceforth chiefly at Pretoria; with other leading Boers founded nationalist organization 'Het Volk', 1905; formed ministry under British crown, 1907; headed Transvaal delegation at Union convention, 1908-9; formed strong political combination with J. C. Smuts [q.v.]; first prime minister of Union of South Africa, 1910-19; dealt satisfactorily with problems of Indian immigrants and unrest on Rand; wholeheartedly supported Great Britain during European war, 1914-18; put down serious revolt of Dutch against intervention in war, 1914-15; commanded successful campaign against German South-West Africa, resulting in unconditional surrender of German forces and colony, 1915; with Smuts attended Versailles peace conference as South African delegate, 1919; died at Pretoria.

BOTTOMLEY, GORDON (1874-1948), poet and dramatist; educated at Keighley grammar school; obliged by ill health to live in seclusion; published representative choice of verse in *Poems of Thirty Years* (1925); verse plays include *King Lear's Wife* (1915) and *Gruach* (1921).

BOTTOMLEY, HORATIO WILLIAM (1860–1933), journalist and financier; nephew of G. J. Holyoake [q.v.]; ran away from an orphanage; shorthand writer, Supreme Court of Judicature; founded *Hackney Hansard* (1884) and similar publications; bankrupt, 1891; conducted own defence and acquitted of fraud, 1893; promoted many companies few of which paid their shareholders; founded *John Bull*, 1906; liberal MP, South Hackney, 1906–12; acquitted of fraud, 1909; squandered enormous sums; bankrupt, 1911; re-established himself by 'patriotic' use of gifts as journalist and speaker; independent MP, South Hackney, 1918–22; new enterprises ended in imprisonment, 1922–7; died in want and obscurity.

BOUCHERETT, EMILIA JESSIE (1825–1905), advocate of women's progress; of old Lincolnshire family; went to London, 1859; helped to found Society for the Promotion of Employment of Women, 1860; organized first petition to parliament for women's franchise, 1866; founded and edited *Englishwoman's Review*, 1866–71; contributed to leading reviews on women's questions.

BOUCICAULT, DION, the younger (1859–1929), actor-manager; son of Dion Boucicault [q.v.]; born in New York; first appeared on stage there, 1879; in partnership in Australia, 1886–96; with Charles Frohman at Duke of York's Theatre, 1901–15; original producer of Barrie's *Peter Pan*, 1904; manager, New Theatre, 1915; brilliant stage-director; chief parts, Sir William Gower (Pinero's *Trelawny of the 'Wells'*) and Mr Pim (Milne's *Mr. Pim Passes By*).

BOUGHTON, GEORGE HENRY (1833–1905), painter and illustrator; travelled and studied art in British Isles, 1856; exhibited in America, 1858; worked in Paris, 1860–2; exhibited at Royal Academy from 1863 till death; ARA, 1879; RA, 1896; treated mainly peasant life of Brittany and Holland, and New England history; works include 'Weeding the Pavement' (1882), 'A Dutch Ferry' (1883), 'The Road to Camelot' (1898); wrote and illustrated *Sketching Rambles in Holland*, 1885.

BOUGHTON, RUTLAND (1878–1960), composer; educated at Aylesbury Endowed School; worked with concert agency, 1892–8; studied at Royal College of Music, 1898–1901; held post at Midland Institute, Birmingham, 1905–10; with Reginald Buckley organized Glastonbury Festival, 1914–16, 1919–26; his opera *The Immortal Hour*, based on the Celtic drama by Fiona Macleod [q.v.], performed at Glastonbury, 1914, Birmingham, 1921, London, 1922; other works included *Bethlehem*

(1915) and *Queen of Cornwall* (1924); granted civil list pension, 1938.

BOURCHIER, ARTHUR (1863–1927), actor-manager; BA, Christ Church, Oxford; prime mover in foundation of Oxford University Dramatic Society, 1884; went on the stage, 1889; manager of Royalty Theatre, London, 1895–6; of Garrick Theatre, 1900–6; his productions included *Merchant of Venice* (1905) and *Macbeth* (1906); his best parts those requiring broad and hearty treatment, such as Henry VIII, Bottom, and Old Bill (in *The Better 'Ole*); died in Johannesburg.

BOURCHIER, JAMES DAVID (1850–1920), *The Times* correspondent in Balkan peninsula with headquarters first at Athens, then at Sofia, 1892–1918; entrusted with many secret negotiations preceding Balkan alliance, 1911–12; held unique position in Balkans.

BOURDILLON, Sir BERNARD HENRY (1883–1948), colonial governor; scholar of Tonbridge and St. John's College, Oxford; entered Indian civil service, 1908; seconded to Iraq, 1919; counsellor, 1924–9; negotiated Anglo-Iraq treaty of 1926; colonial secretary, Ceylon, 1929–32; governor of Uganda, 1932–5; of Nigeria, where he encouraged education and government anti-leprosy work, 1935–43; KBE, 1931; KCMG, 1934; GCMG, 1937.

BOURINOT, Sir JOHN GEORGE (1837–1902), writer on Canadian constitutional history; born at Sydney, Cape Breton, of Huguenot extraction; BA, Trinity College, Toronto; hon. LLD, 1889; founded and edited *Halifax Herald*, 1860; chief reporter of Nova Scotia assembly from 1861; chief clerk of Canadian House of Commons from 1880 till death; his *Manual of the Constitutional History of Canada* (1888) was standard textbook; president of Royal Society of Canada, 1892; KCMG, 1898.

BOURKE, ROBERT, BARON CONNEMARA (1827–1902), governor of Madras; younger brother of sixth Earl of Mayo [q.v.]; educated at Trinity College, Dublin; called to bar (Inner Temple), 1852; acquired large parliamentary practice; conservative MP for King's Lynn, 1868–86; under-secretary for foreign affairs under Disraeli (1874–80) and Salisbury (1885–6); PC, 1880; governor of Madras, 1886; baron and GCIE, 1887; by tact and industry improved relations with the people; improved sanitary conditions and railway communications in the presidency; resignation (1890) followed by divorce from wife, the daughter of James Andrew, Marquess of Dalhousie [q.v.], 1891; remarried, 1894.

BOURNE, FRANCIS ALPHONSUS (1861-1935), cardinal; educated at St. Edmund's College, Ware, St. Sulpice, Paris, and Louvain University; deacon, 1883; priest, 1884; first rector of diocesan seminary, Wonersh, 1891-7; a domestic prelate, 1895; coadjutor (1896) and bishop (1897) of Southwark; archbishop of Westminster, 1903-35; cardinal, 1911; much occupied with saving the voluntary schools; a man of prayer and a great pastor.

BOURNE, GILBERT CHARLES (1861-1933), zoologist and oarsman; educated at Eton and New College, Oxford; rowed in winning crews, 1881-2; first class, natural science, 1885; first director, Marine Biological Laboratory, Plymouth; Linacre professor of zoology and comparative anatomy, Oxford, 1906-21; researches mainly in morphology; FRS, 1910; outstanding rowing coach; his theory on structure of racing boats adopted by several designers.

BOURNE, HENRY RICHARD FOX (1837-1909), social reformer and author; born in Jamaica; pupil of Henry Morley [q.v.] at University College, London; wrote popular accounts of England's colonial expansion; owner of *Examiner*, 1870-3; editor of *Weekly Dispatch*, a radical organ, 1876-87; secretary of Aborigines Protection Society from 1889 till death; fervent champion of all native races; chief works were lives of Sir Philip Sidney (1862) and John Locke (1876), and *English Newspapers*, 2 vols., 1887.

BOURNE, ROBERT CROFT (1888-1938), politician and oarsman; son of G. C. Bourne [q.v.]; educated at Eton and New College, Oxford; second class, history, 1911; stroked four successive winning university crews, 1909-12; conservative MP, Oxford City, 1924-38; deputy chairman of ways and means, 1931-8; PC, 1935.

BOUSFIELD, HENRY BROUGHAM (1832-1902), first bishop of Pretoria; BA, Caius College, Cambridge, 1855; MA, 1858; vicar of Andover, 1870-8; bishop of Pretoria, 1878-1902; wrote *Six Years in the Transvaal*, 1886.

BOWATER, SIR ERIC VANSITTART (1895-1962), industrialist; educated at Charterhouse; served in Royal Artillery, 1913-17; joined family business, W. V. Bowater & Sons, paper merchants; planned first Bowater paper mill; managing director, 1926; chairman, Bowater Organization, 1927-62; director-general, Ministry of Aircraft Production, 1940-5; controller, 1945; knighted, 1944; chairman, Bowater-Scott Corporation, manufacturers of soft tissues, 1956; set up newsprint mills in United States, 1953-9; officer of the Legion of Honour; FRSA, 1948.

BOWDEN, FRANK PHILIP (1903-1968), experimental physicist; born in Tasmania; educated at Hutchins School, Hobart, and the university of Tasmania; B.Sc., 1924; M.Sc., first class honours, 1925; research student, Gonville and Caius College, Cambridge, 1927; fellow, 1929; D.Sc. (Tasmania), 1933; Sc.D. (Cambridge), 1938; research on surface science related to friction and lubrication; further research in Australasia, 1939-45; returned to Cambridge, 1945; reader in physical chemistry, 1946; reader in physics, 1957; professor (*ad hominem*) in surface physics, 1966; adviser to Tube Investments Ltd., 1953; director, English Electric Company, 1958; chairman, Executive Committee, National Physical Laboratory, 1955-62; FRS, 1948; CBE, 1956.

BOWEN, EDWARD ERNEST (1836-1901), schoolmaster and song writer; brother of Lord Bowen [q.v.]; educated at Trinity College, Cambridge; Bell university scholar, 1855; fourth classic, 1858; fellow, 1859; master at Harrow, 1859-1901; advocated closer relations between master and pupils; recommended creation of modern side, which he directed, 1869-93; wrote for reviews on military and theological topics; collected his *Harrow Songs and other Verses*, 1886, which included 'Forty Years On', 1872; efficient cricketer, footballer, skater, and mountaineer; liberal in politics; bequeathed property to Harrow School.

BOWER, FREDERICK ORPEN (1855-1948), botanist; educated at Repton, Trinity College, Cambridge, Würzburg, and Strasbourg; lecturer in botany, South Kensington, 1882-5; regius professor, Glasgow, 1885-1925; established well-equipped department with world-wide reputation in morphological botany housed in first botanical institute in Great Britain; publications include *The Origin of a Land Flora* (1908), *The Ferns* (3 vols., 1923-8), and *Primitive Land Plants* (1935), FRS, 1891; president, British Association, 1930.

BOWES, ROBERT (1835-1919), bookseller, publisher, and bibliographer; head of Macmillan and Bowes' (Bowes and Bowes) bookshop, Cambridge, 1863 till death; published researches on Cambridge University printers and Cambridge books; prominent in Cambridge social and civic life.

BOWES-LYON, CLAUDE GEORGE, fourteenth (Scotland) and first (UK) EARL OF STRATHMORE AND KINGHORNE (1855-1944); succeeded father, 1904; married (1881)

Nina Cecilia Cavendish-Bentinck; his youngest daughter, Elizabeth Angela Marguerite, married the Duke of York (afterwards King George VI), 1923; created earl in UK peerage, 1937.

BOWHILL, Sir FREDERICK WILLIAM (1880–1960), air chief marshal; born in India; served sixteen years in Merchant Service; qualified as pilot and joined Royal Flying Corps, Naval Wing, 1913; in Mesopotamia for relief of Kut; DSO, 1918; air marshal and KCB, 1936; held Coastal Command, 1937–41; secured sinking of *Bismarck*, 1941; air chief marshal, 1939; in Canada organizing RAF Ferry Command, 1941–3; held Transport Command, 1943–5; British member, International Civil Aviation Organization, 1945–6; chief aeronautical adviser, Ministry of Civil Aviation, 1946–57.

BOWLBY, Sir ANTHONY ALFRED, first baronet (1855–1929), surgeon; born at Namur; son of T. W. Bowlby [q.v.]; educated at St. Bartholomew's Hospital, where he filled offices, 1881–1919; FRCS, 1881; president, 1920–3; served as surgeon in South African war, 1899; advisory consulting surgeon to British forces in France in 1914–18 war; knighted, 1911; CMG, 1901; KCMG, 1915; KCVO, 1916; KCB, 1919; baronet, 1923; fine organizer and administrator.

BOWLER, HENRY ALEXANDER (1824–1903), painter; exhibited landscapes at Royal Academy, 1847–71; assistant director for art at South Kensington, 1876–91.

BOWLES, THOMAS GIBSON (1842–1922), politician; started society paper, *Vanity Fair*, 1868; conservative MP, King's Lynn, 1892–1906; joined liberal party, 1906; liberal MP, King's Lynn, 1910; brought action against Bank of England (1913) which entailed passing of Act making legal the provisional collection of taxes; an authority on international and maritime law.

BOWLEY, Sir ARTHUR LYON (1869–1957), statistician; educated at Christ's Hospital and Trinity College, Cambridge; tenth wrangler, 1891; Sc.D., 1913; taught statistics at London School of Economics from 1895; reader, 1908; professor, 1919–36; at University College, Reading, lecturer in mathematics (1900–7), in economics (1913–19), professor in both (1907–13); director, Oxford University Institute of Statistics, 1940–4; a practitioner in applied statistics in field of social studies; developed sampling techniques in their application to social studies exemplified in *Livelihood and Poverty*, with A. R. Burnett-Hurst, 1915; studied definition and measurement of national income; edited London and Cambridge Economic Service, 1923–45; contributed to *New Survey of London Life and Labour*, 1930–5; a founder of the Econometric Society, 1933; FBA, 1922; knighted, 1950.

BOYCE, Sir RUBERT WILLIAM (1863–1911), pathologist and hygienist; studied medicine at University College, London; MB London, 1889; published *Textbook of Morbid Histology*, 1892; first professor of pathology, University College, Liverpool, 1894; promoted establishment of Liverpool University, and creation of medical chairs there; helped to found Liverpool School of Tropical Medicine, 1898; examined epidemics of yellow fever at New Orleans and British Honduras, 1905; published books on subject, and formed at Liverpool a bureau of yellow fever; FRS, 1902; knighted, 1906.

BOYCOTT, ARTHUR EDWIN (1877–1938), pathologist and naturalist; educated at Hereford Cathedral School, Oriel College, Oxford, and St. Thomas's Hospital; BM, 1902; lecturer in pathology, Guy's Hospital, 1907–12; professor of pathology, Manchester (1912–15) and London (University College Hospital, 1915–35); investigated the physiology and pathology of blood; edited *Journal of Pathology and Bacteriology* for eleven years; chief recreation the ecology of British snails; FRS, 1914.

BOYD, HENRY (1831–1922), principal of Hertford College, Oxford; BA, Exeter College, Oxford; ordained, 1854; perpetual curate of St. Mark's, Victoria Docks, 1862–74; fellow of Hertford College, Oxford, 1875; principal, 1877–1922; vice-chancellor of Oxford University, 1890; master of the Drapers' Company, 1896–7; commissioned (Sir) T. G. Jackson [q.v.] to enlarge and partly rebuild Hertford College, the most notable improvements of which were a new hall and chapel completed 1907; a high churchman and conservative; an accomplished painter in water-colours and a keen sportsman.

BOYD, Sir THOMAS JAMIESON (1818–1902), lord provost of Edinburgh; head of uncle's publishing house of Oliver and Boyd, 1873–98; master of Merchant Company of Edinburgh, 1869; reformed educational foundations of the corporation, 1870; promoted New Royal Infirmary, Edinburgh, 1879; lord provost of Edinburgh, 1877–82; knighted, 1881; FRS Edinburgh.

BOYD CARPENTER, WILLIAM (1841–1918), bishop of Ripon. [See CARPENTER.]

BOYLE, SIR COURTENAY EDMUND (1845-1901), permanent secretary of the Board of Trade, 1893-1901; born in Jamaica; educated at Charterhouse and Christ Church, Oxford; played for Oxford at cricket and tennis, 1865-7; keen angler; private secretary to Lord Spencer, Irish viceroy, 1868; CB, 1885; assistant secretary to Local Government Board, 1885, and to Board of Trade, 1886; KCB, 1892; published *Method and Organization in Business* (1901) and edited *Mary Boyle, her Book*, 1901.

BOYLE, SIR EDWARD, first baronet (1848-1909), legal writer; surveyor for twenty years; called to bar, 1887; QC, 1898; baronet, 1904; conservative MP for Taunton, 1906-9; joint-author of important legal treatises on rating and compensation.

BOYLE, GEORGE DAVID (1828-1901), dean of Salisbury; son of Lord Boyle [q.v.]; educated at Charterhouse and Exeter College, Oxford; BA, 1851; MA, 1853; vicar of Kidderminster, 1867-80; dean of Salisbury, 1880-1901; published *Recollections* (1895) and *Salisbury Cathedral* (1897).

BOYLE, RICHARD VICARS (1822-1908), civil engineer; pupil of Charles Vignoles [q.v.]; chief engineer of Longford and Sligo railway, 1846-7; district engineer on East Indian railway, 1853-68; distinguished himself in Indian Mutiny at Arrah, 1857; appointed field officer to (Sir) Vincent Eyre [q.v.]; CSI, 1869; engineer in chief for imperial Japanese railways, 1872-7; travelled much after retirement.

BOYLE, WILLIAM HENRY DUDLEY, twelfth EARL OF CORK AND ORRERY (1873-1967), admiral of the fleet; entered *Britannia* as naval cadet, 1887; midshipman, 1889; commander, Naval Intelligence Department, Admiralty, 1909-11; captain, naval attaché, Rome, 1913-15; commanded *Fox* in Red Sea and Indian Ocean, 1915-17; commanded *Repulse* in Grand Fleet, 1917, and *Lion*, 1918; CB, 1918; rear-admiral, 1923; vice-admiral, 1928, president, Royal Naval College, Greenwich; KCB, 1931; admiral, 1932; commander-in-chief, Home Fleet, 1933; succeeded to title, 1934; commander-in-chief, Portsmouth, 1937-9; admiral of the fleet, 1938; flag officer, Narvik, 1940; president, Shaftesbury Homes and Arethusa Training Ship, 1942-53; published *My Naval Life*, 1942.

BOYS, SIR CHARLES VERNON (1855-1944), physicist; educated at Marlborough; graduated at Royal School of Mines; metropolitan gas referee, 1897-1939; invented fused quartz fibre suspension for galvanometers and

other instruments; made precise determination of Newtonian constant of gravitation; worked for many years on an automatic recording calorimeter to measure calorific value of gas; invented rotating lens camera for study of formation of lightning discharges, 1900; a prolific inventor and expert witness in patent cases; FRS, 1888; knighted, 1935.

BRABAZON, HERCULES BRABAZON (1821-1906), painter; educated at Harrow and Trinity College, Cambridge; BA, 1844; MA, 1848; inherited Brabazon estates in Ireland and property in Sussex; studied art at Rome, 1844-7; made frequent foreign tours; painted with Ruskin in France; influenced by Turner; held exhibition of work at Goupil Gallery, 1892; showed gift of colour and sensitiveness to nature; ardent pianist and model landlord; Brabazon gallery at Sedlescombe, Sussex, opened to public.

BRABAZON, JOHN THEODORE CUTH-BERT MOORE-, first BARON BRABAZON OF TARA (1884-1964), aviator and politician; educated at Harrow and Trinity College, Cambridge; apprenticed in Darracq works, Paris; won *Circuit des Ardennes*, 1907; first Englishman to pilot heavier-than-air machine under power in England, 1909; first pilot's certificate issued by Royal Aero Club, 1910; served with Royal Flying Corps in 1914-18 war; MC, commander of Legion of Honour; conservative MP, Chatham division of Rochester, 1918-1929; Wallasey, 1931-42; parliamentary private secretary to (Sir) Winston Churchill, 1919; parliamentary secretary, Ministry of Transport, 1923-4 and 1924-7; minister of transport, PC, 1940; minister of aircraft production, 1941-2; president, Royal Aero Club; president, Royal Institution; chairman, Air Registration Board; baron, 1942; GBE 1953; won Curzon Cup three times; member, Royal Yacht Squadron; captain, Royal and Ancient Golf Club, St. Andrews.

BRABAZON, REGINALD, twelfth EARL OF MEATH (1841-1929); clerk in Foreign Office, 1863; served in diplomatic service, 1868-77; devoted himself to philanthropic work, 1873; founded Hospital Saturday Fund Committee; annual amount collected rose from £6,463 (1874) to over £100,000 (1929); his other social works include the preservation of many of London's open spaces; succeeded father, 1887; originated idea of celebrating 'Empire Day'.

BRACKEN, BRENDAN RENDALL, VIS-COUNT BRACKEN (1901-1958), politician and publisher; born at Templemore, Tipperary; absconded from Jesuit College, Mungret; sent to Australia, 1916; settled in England, 1919; spent two terms at Sedbergh; became preparatory school teacher; through J. L. Garvin [q.v.], obtained useful introductions and became lifelong faithful supporter of (Sir) Winston Churchill; a director of Eyre & Spottiswoode, 1925; founded *Banker*; acquired *Financial News*, *Investors Chronicle*, and *Practitioner* and joint ownership (1929) with Sir Henry Strakosch (q.v.) of *The Economist*; conservative MP, North Paddington, 1929-45, Bournemouth, 1945-51; PPS to Churchill, 1939-41; PC, 1940; popular minister of information, 1941-5; first lord of Admiralty, 1945; founded *History Today*; chairman, *Financial Times* amalgamated with *Financial News*, and from 1945 of Union Corporation; viscount, 1952; generous benefactor of Churchill College, Cambridge.

BRACKENBURY, SIR HENRY (1837-1914), general, and writer on military subjects; entered Woolwich, 1854; professor of military history at Woolwich, 1868; accompanied (Viscount) Wolseley [q.v.] to Ashanti, 1873, Cyprus, 1878, Zululand, 1879, and Egypt, 1884; deputy assistant quartermaster-general and head of intelligence branch at headquarters, 1886-91; military member of council of viceroy of India, 1891-6; director-general of ordnance during South African war, 1899-1902; GCB, 1900; PC, 1904; published works on military subjects.

BRACKLEY, HERBERT GEORGE (1894-1948), air transport pioneer; DSO, commanding bomber squadron in France, 1917; made first Newfoundland-New York flight, 1919; with British air mission in Japan, 1921-4; superintendent, Imperial Airways, 1924-39; introduced new aircraft and surveyed new routes including England-Australia flying-boat route, 1935; served in Coastal and Transport commands, 1939-45; air-commodore, 1943; assistant to chairman, BOAC, 1945-8; chief executive, British South American Airways Corporation, 1948; drowned at Rio de Janeiro.

BRADBURY, JOHN SWANWICK, first BARON BRADBURY (1872-1950), civil servant; educated at Manchester grammar school and Brasenose College, Oxford; entered Treasury; worked on 1909 budget; planned financial fabric of health-insurance scheme; insurance commissioner, 1911-13; joint permanent secretary, Treasury, 1913-19; provided new issue of currency notes, 1914; mainly responsible for introducing war savings certificates; delegate, Reparation Commission, 1919-25; chairman, Food Council, 1925-9; KCB, 1913; baron, 1925.

BRADDON, SIR EDWARD NICHOLAS COVENTRY (1829-1904), premier of Tasmania; assistant in government railways, Calcutta, 1854; superintendent of excise and stamps in Oudh, 1862; and of statistics, 1868; inspector-general of registration, 1871-7; retired to Tasmania, 1878; entered house of assembly, 1879; minister of lands and works and of education, 1887; agent-general for colony in London, 1888-93; KCMG, 1891; premier of Tasmania, 1894-9; PC and hon. LLD, Cambridge, 1897; senior member for Tasmania in first Australian Commonwealth parliament, 1901; wrote *Thiry Years of Shikar*, 1895.

BRADDON, MARY ELIZABETH (1837-1915), novelist. [See MAXWELL.]

BRADFORD, SIR EDWARD RIDLEY COLBORNE, first baronet (1836-1911), Anglo-Indian administrator and commissioner of the Metropolitan Police, London; joined East India Company, 1853; served with 6th Madras cavalry in Indian Mutiny, 1857, and with Mayne's Horse, 1858-9; mauled by tiger and lost left arm, 1863; served in various districts as political agent; superintendent of thagi and dakaiti, dealing with sedition cases, 1874; chief commissioner of Ajmir, 1878; introduced municipal government there; KCSI, 1885; secretary in political and secret departments, India Office, 1887; commissioner of police in London, 1890-1903; allayed police disaffection, June 1890; improved police organization and welfare; GCB, 1897; GCVO and baronet, 1902.

BRADFORD, SIR JOHN ROSE, baronet, of Mawddwy (1863-1935), physician and physiologist; educated at University College School and University College, London; B.Sc., 1883; MD, 1889; FRS, 1894; FRCP, 1897; authority on kidney diseases; assistant physician (1889), physician (1900-23), and one of the greatest clinical teachers at University College Hospital; physician, Seamen's Hospital, 1905-19; promoted study of tropical diseases, creation of Sleeping Sickness Bureau and of Beit fellowships in medical research; senior medical adviser, Colonial Office, 1912-24; consulting physician in France, 1914-19; president, Royal College of Physicians, 1926-31; KCMG, 1911; baronet, 1931.

BRADLEY, ANDREW CECIL (1851-1935), literary critic; son of Charles Bradley [q.v.], brother of F. H. and half-brother of G. G. Bradley [qq.v.]; educated at Cheltenham and Balliol College, Oxford; first class, *lit. hum.*, 1873; fellow (1874), honorary fellow (1912) of Balliol; first professor of literature and history,

University College, Liverpool, 1882-90; professor of English language and literature, Glasgow, 1890-1900; of poetry, Oxford, 1901-6; FBA, 1907; Gifford lecturer, Glasgow, 1907-8; works include *Shakespearean Tragedy* (1904) and *Oxford Lectures on Poetry* (1909); prominent in creation of English Association (1906), president, 1911; among the greatest English critics of Shakespeare; founded a research fellowship at Balliol.

BRADLEY, FRANCIS HERBERT (1846-1924), philosopher; son of Charles Bradley [q.v.], and half-brother of G. G. Bradley [q.v.]; BA, University College, Oxford, 1869; fellow of Merton College, 1870; lived retired life at Merton until just before his death; LLD, Glasgow University, 1883; hon. FBA., 1923; published *Ethical Studies* (1876), *The Principles of Logic* (1883), *Appearance and Reality, a Metaphysical Essay* (1893), and *Essays on Truth and Reality* (1914); also many essays, published as *Collected Essays* (2 vols., 1935); chief inspiration of his thought was Hegel, although he disclaimed the name of Hegelian because there is nothing in his work corresponding to Hegel's 'objective logic'.

BRADLEY, GEORGE GRANVILLE (1821-1903), dean of Westminster and schoolmaster; son of Charles Bradley [q.v.]; educated at Rugby and University College, Oxford; first class classic and fellow, 1844; master at Rugby, 1845-58; headmaster of Marlborough 1858-70; raised number of pupils and enlarged school buildings; successful Latin prose teacher; close friend of Tennyson and of Dean Stanley [qq.v]; master of University College, Oxford, 1870-81; raised standard of scholarship there; member of university commission and canon of Worcester, 1880; dean of Westminster, 1881-1902; restored masonry and finances of the abbey, and revised the system of interments; buried there; his *Latin Prose Composition* (1881) much used.

BRADLEY, HENRY (1845-1923), philologist and lexicographer; corresponding clerk to Sheffield cutlery firm, 1863-83; mastered knowledge of modern European languages and of Greek and Latin, and acquired acquaintance with Hebrew; took up literary work in London, 1884; appointed an editor of the *New English Dictionary*, 1889; senior editor, 1915; moved to Oxford, 1896; hon. D.Litt., Oxford, 1914; works include *Making of English* (1904); an authority on Old and Middle English and on British place-names.

BRAGG, SIR WILLIAM HENRY (1862-1942), physicist; educated at King William's College and Trinity College, Cambridge; third

wrangler, 1884; Elder professor of mathematics and physics, Adelaide, 1886-1909; Cavendish professor of physics, Leeds, 1909-15; Quain professor, University College, London, 1915-23; studied radioactivity and ionization of X-rays; with son, (Sir) W. L. Bragg, established for first time how atoms are arranged in crystals such as diamond, founded modern science of crystallography, shared Nobel prize for physics (1915), and published *X-Rays and Crystal Structure* (1915); worked on underwater acoustics for Admiralty, 1914-18; Fullerian professor of chemistry and director of Davy-Faraday laboratory, Royal Institution, 1923-42; fellow (1907), president (1935-40) of Royal Society; president, British Association, 1928; KBE, 1920; OM, 1931.

BRAID, JAMES (1870-1950), golfer; born at Earlsferry where he caddied and won boys' competitions; continued to play while working as joiner in St. Andrews and Edinburgh; a clubmaker for Army & Navy Stores, London, 1893-6; professional Romford (1896-1904), at Walton Heath (1904-50); won open championship, 1901, 1905, 1906, 1908, 1910; *News of World* tournament, 1903, 1905, 1907, 1911; French championship, 1910.

BRAILSFORD, HENRY NOEL (1873-1958), journalist; first class, logic and moral philosophy, Glasgow University, 1894; wrote successively for *Manchester Guardian*, *Daily News*, *Nation* (1907-23), and *New Statesman* (from thirties until 1946); joined independent labour party, 1907; edited *New Leader*, 1922-6; supported women's suffrage, colonial independence, and early days of Soviet Russia; attacked 'secret diplomacy', Versailles Treaty, and the fascist and later communist regimes; considered war result of economic rivalry of great Powers; publications include *The War of Steel and Gold* (1914), *The Russian Workers' Republic* (1921), *Socialism for To-day* (1925), *Rebel India* (1931), *Shelley, Godwin, and their Circle* (1913), and *Voltaire* (1935).

BRAIN, DENNIS (1921-1957), virtuoso horn-player; educated at St. Paul's School and Royal Academy of Music; third generation player; principal horn in Royal Air Force Central Band and Orchestra, 1939-46; thereafter a brilliantly assured and popular soloist; killed in car crash.

BRAIN, WALTER RUSSELL, first BARON BRAIN (1895-1966), physician and medical statesman; educated at Mill Hill School and New College, Oxford; BA, 1919; London Hospital, 1920; BM, B.Ch. (Oxon), 1922; DM, 1925; FRCP, 1931; studied neurology, London

Hospital; physician, Maida Vale Hospital, 1925; assistant physician, London Hospital, 1927; physician, Moorfields Hospital, 1930-7; director, London Hospital unit for investigation of carcinomatous neuropathies; president, Royal College of Physicians, 1950-7; knighted, 1952, baronet, 1954; baron, 1962; FRS, 1964; hon. fellow, New College, Oxford, 1952; hon. degrees, Oxford, Manchester, Southampton, Wales, Belfast, and Durham; president, Association of Physicians, 1956; president, Association of British Neurologists, 1958; editor, *Brain*, 1954-66; publications include *Mind, Perception and Science* (1951), *Speech Disorders* (1961), *Diseases of the Nervous System* (1933, 6th edn., 1962), and *Science and Man* (1966).

BRAITHWAITE, DAME (FLORENCE) LILIAN (1873-1948), actress; educated at London high schools; appeared regularly and successfully in west end, 1900-48; at first played beautiful suffering heroines in drawing-room dramas; established herself as dramatic actress with Florence Lancaster in *The Vortex* (1924) and as light comedienne in *The Truth Game* (1928); her greatest success *Arsenic and Old Lace* (1942-6); DBE, 1943.

BRAITHWAITE, SIR WALTER PIPON (1865-1945), general; commissioned in Somerset Light Infantry from Sandhurst, 1886; staff officer in South Africa, 1899-1902; commandant, Staff College, Quetta, 1911-14; chief of general staff to Sir Ian Hamilton [q.v.] at Gallipoli, 1915; commanded 62nd division (in England, 1915-17, in France, 1917-18) and IX Corps (1918-19); general, 1926; adjutant-general to forces, 1927-31; governor, Chelsea Hospital, 1831-8; KCB, 1918.

BRAMBELL, FRANCIS WILLIAM ROGERS (1901-1970), zoologist; born in Dublin; educated at Aravon School and Trinity College, Dublin; BA, 1922; Ph.D. (Dublin), 1924; research scholar, University College, London, 1924-6; D.Sc. (London), 1927; lecturer in zoology, King's College, London, 1927-30; Lloyd Roberts professor of zoology, University College of North Wales, 1930-68; course of marine biology led to establishment of Marine Sciences Laboratory, Menai Bridge; initiated postgraduate M.Sc. course in animal reproduction; dean of science faculty, 1939-43; vice-principal, 1948-50 and 1956-8; member, Agricultural Research Council; member, editorial board, *Journal of Embryology and Experimental Morphology*; FRS, 1949; council member, 1954-6; CBE, 1966; member, University Grants Committee, 1960-8; published *The Development of Sex in Vertebrates* (1930),

and *The Transmission of Passive Immunity from Mother to Young* (1970).

BRAMPTON, BARON (1817-1907), judge. [See HAWKINS, HENRY.]

BRAMWELL, SIR BYROM (1847-1931), physician; educated at Cheltenham; qualified in medicine, Edinburgh, 1896; physician, Newcastle Royal Infirmary, 1874-9; set up as consultant in Edinburgh, 1879; pathologist (1882), assistant physician (1885), physician (1897-1912), Edinburgh Royal Infirmary; a born diagnostician and brilliant clinical teacher; publications include *Diseases of the Spinal Cord* (1881), *Atlas of Clinical Medicine* (3 vols., 1892-6), and *Clincil Studies* (8 vols., 1903-10); FRSE, 1886; FRCP (London), 1923; knighted, 1924.

BRAMWELL, SIR FREDERICK JOSEPH, baronet (1818-1903), engineer; brother of Baron Bramwell [q.v.]; apprenticed to John Hague, London engineer and inventor of system for propelling trains by atmospheric pressure, 1834; studied methods of steam propulsion; as manager to Hague, constructed locomotive for Stockton and Darlington railway, 1843; started for himself, 1853; developed legal and consultative side of profession; first to practise as technical advocate; a recognized authority on municipal and waterworks engineering; adviser to all London water companies; constructed sewage disposal scheme for Portsmouth; member (1854) and president (1874) of Institution of Mechanical Engineers; president of British Association at Bath, 1888, and of Society of Arts, 1901; hon. secretary of Royal Institution, 1885-1900; chairman of City and Guilds Institute, 1878-1903; knighted, 1881; FRS, 1873; hon. DCL, Oxford, 1886, LLD, Cambridge, 1892; baronet, 1889; predicted supersession of steam engine by internal combustion engine, 1881.

BRANCKER, SIR WILLIAM SEFTON (1877-1930), major-general and air vice-marshal; commissioned, Royal Artillery, 1896; served South African war, 1900-2, and India, 1903-12; learned to fly and appointed to staff of director-general of military aeronautics, 1913; deputy-director, 1914; director of air organization, War Office, 1916-17; deputy director-general, military aeronautics, 1917; on Air Council, 1918; KCB, 1918; promoted commercial flying from 1919; director of civil aviation, 1922; air vice-marshal, 1925; killed in airship R. 101 disaster near Beauvais, France.

BRAND, HENRY ROBERT, second VISCOUNT HAMPDEN and twenty-fourth BARON DACRE (1841-1906), governor of New South

Wales, 1859-9; served in Coldstream Guards; liberal MP for Hertfordshire, 1868-74; for Stroud, 1880-6; surveyor-general of ordnance, 1883-5; opposed to home rule; succeeded father, 1892; GCMG, 1899.

BRAND, HERBERT CHARLES ALEXANDER (1839-1901), commander RN; entered navy, 1851; at capture of Canton and Taku forts, 1858; helped Edward John Eyre [q.v.] to suppress negro revolt in Morant Bay, 1865; president of court martial on ringleaders, who were put to death; charge of murder against him in London dismissed, 1867; retired with rank of commander, 1883.

BRAND, ROBERT HENRY, first BARON BRAND (1878-1963), banker and public servant; educated at Marlborough and New College, Oxford; first class, modern history, 1901; fellow, All Souls, 1901; joined Lord Milner's staff (Milner's kindergarten), South Africa, 1902; permanent secretary to Intercolonial Council of the Transvaal and Orange River Colony, 1904; joined Lazard Brothers, bankers, and remained with them for fifty years, 1909-60; member, editorial board, *Round Table*; set up Imperial Munitions Board, Ottawa, 1915; attended Paris peace conference; co-operated with (Lord) J. M. Keynes [q.v.]; head of British Food Mission, Washington, 1941; senior British member, British-American Combined Food Board, 1941-4; chief Treasury representative, Washington, 1944-6; CMG (1910), hon. DCL Oxford (1937), baron, 1946; published *The Union of South Africa* (1909), *War and National Finance* (1921), and *Why I am not a Socialist* (1923).

BRANDIS, SIR DIETRICH (1824-1907), forest administrator and botanist; born at Bonn of good Hanoverian family; studied botany at Athens, Bonn, and Copenhagen; Ph.D., 1848; put in charge of and saved threatened teak forests in Burma, 1856-62; inspector-general of Indian forests, 1864; founded at Dehra Dun school for native foresters, 1878; published *The Forest Flora of North-West and Central India*, 1874; FRS, 1875; CIE, 1878; KCIE, 1887; hon. LLD, Edinburgh, 1889; leaving India, he advised English and American forestry students, 1888-96; divided his time between Bonn and Kew; chief work *Indian Trees* (1906); died at Bonn.

BRANGWYN, SIR FRANK (FRANÇOIS GUILLAUME) (1867-1956), artist; born in Bruges of Welsh Roman Catholic parentage; worked for William Morris [q.v.], c.1882-4; travelled on commission in Europe, the Near East, and South Africa; settled in Hammer-

smith, 1896, and Ditchling, 1924; at centre of art nouveau movement; a bold and vigorous painter, best known for large murals; his 'British Empire' panels rejected by House of Lords went ultimately to Swansea guildhall; ARA, 1904; RA, 1919; knighted, 1941; Royal Academy retrospective exhibition, 1952; honorary citizen of Bruges where there is a Brangwyn Museum.

BRASSEY, THOMAS, first EARL BRASSEY (1836-1918); son of Thomas Brassey [q.v.]; educated at Rugby and University College, Oxford; liberal MP, Hastings, 1868-86; interested in wages question and naval matters; toured round world in his yacht *Sunbeam*, 1876-7; civil lord of Admiralty, 1880-4; KCB, 1881; parliamentary secretary to Admiralty, 1884-5; first produced *Brassey's Naval Annual*, 1886; baron, 1886; governor of Victoria, 1895-1900; lord warden of cinque ports, 1908-13; earl, 1911.

BRAY, CAROLINE (1814-1905), authoress; sister of Charles Christian [q.v.], Mary [q.v.], and Sara Hennell [see below]; married Charles Bray of Coventry [q.v.]; intimate friend of George Eliot [q.v.] from 1842; entertained Emerson (1848) and Herbert Spencer [q.v.] at Coventry; wrote *Physiology and the Laws of Health* (1860) and *Our Duty to Animals* (1871); initiated Coventry Society for Prevention of Cruelty to Animals, 1874; her sister SARA HENNELL (1812-99) was author of *Christianity and Infidelity* (1857), of *Essay on the Sceptical Tendency of Butler's 'Analogy'* (1859, of classical rank, commended by Gladstone), and *Present Religion as a Faith owning Fellowship with Thought* (3 vols., 1865-87).

BRAY, SIR REGINALD MORE (1842-1923), judge; BA, Trinity College, Cambridge, 1865; called to bar (Inner Temple), 1868; recorder of Guildford, 1891-1904; QC, 1897; as a leader employed in many types of litigation; judge of King's Bench division, 1904-23; knighted, 1904; considered to be a very able puisne judge.

BRAZIL, ANGELA (1868-1947), writer; founded a genre with her girls' school stories which were immediately successful and praised for their realism; wrote over fifty, including *A Fourth Form Friendship* (1911), *The School by the Sea* (1914), and *Monitress Merle* (1922).

BRENNAN, LOUIS (1852-1932), mechanical engineer; his dirigible torpedo for coast defence purchased by British government for over £100,000; supervised its manufacture, 1887-96; CB, 1892; worked on helicopters for Air Ministry, 1919-26.

BRENTFORD, first VISCOUNT (1865-1932), statesman. [See HICKS, WILLIAM JOYNSON-.]

BRERETON, JOSEPH LLOYD (1822-1901), educational reformer; educated at Rugby and University College, Oxford; BA, 1846; MA, 1857; Newdigate prize, 1844; rector of West Buckland, North Devon, 1852-67, and of Little Massingham, King's Lynn, 1867-1901; permanently injured in railway accident, 1882; established 'county' schools in Devon (1858) and in Norfolk (1871), and advocated national education on county basis; founded (1873) Cavendish 'County' College at Cambridge to connect county schools with the universities; the college, a financial failure, was dissolved, 1892; published religious verse.

BRESSEY, Sir CHARLES HERBERT (1874-1951), civil engineer; educated at Forest School, Walthamstow and on continent; practised as an architect; served in Royal Engineers, 1914-18; joined Ministry of Transport as a road engineer, 1919; chief engineer, 1921-35; engineer in charge, Highway Development Survey of Greater London, 1935-8; knighted, 1935.

BRETT, JOHN (1831-1902), landscape painter; entered Royal Academy Schools, 1854; his pre-Raphaelite pictures, 'The Stone Breaker' (1858) and 'Val d'Aosta' (1859), praised by Ruskin; painted mainly Cornish seascapes, including 'Mounts Bay' (1877), 'Cornish Lions' (1878), 'Britannia's Realm' (1880); treated subjects in minute scientific detail; a keen astronomer; ARA, 1881.

BRETT, REGINALD BALIOL, second VISCOUNT ESHER (1852-1930), government official; son of first viscount [q.v.]; educated at Eton and Trinity College, Cambridge; private secretary to Marquess of Hartington [q.v.], 1878-85; liberal MP, Penryn and Falmouth, 1880-5; withdrew to Orchard Lea, near Windsor, and admitted to Queen Victoria's private circle; secretary, Office of Works, 1895-1902; superintended diamond jubilee, 1897; succeeded father, 1899; superintended funeral of Queen Victoria, 1901, and coronation of Edward VII, 1902; turned his attention to army reform; member of royal commission which inquired into military preparations for and conduct of South African war, 1902; chairman of War Office reconstruction committee, 1903; report, recommending, among other points, creation of Army Council and general staff for army, accepted, 1904; creation of post of chief of general staff introduced system into War Office; joined Committee of Imperial Defence, 1904; supported army reforms of Lord Haldane [q.v.] and acted as liaison between King and ministers; on confidential mission in France from Sept. 1914 onwards; KCB, 1902; GCB, 1908; PC, 1922.

BREWER, Sir ALFRED HERBERT (1865-1928), organist and composer; chorister at Gloucester Cathedral, 1877-80; organist, St. Michael's church, Coventry, 1886-92; director of music, Tonbridge School, 1892-6; organist of Gloucester Cathedral, 1896-1928; a sound choir-trainer and superlative performer; conducted with distinction eight Three Choirs' Festivals; founded Gloucestershire Orchestral Society, 1905; an industrious composer; FRCO, 1895; knighted, 1926.

BREWTNALL, EDWARD FREDERICK (1846-1902), painter; studied at Lambeth School of Art; exhibited at Royal Water Colour Society (1875-1900) and (mostly oils) at Royal Academy (1872-1900).

BRIDGE, Sir CYPRIAN ARTHUR GEORGE (1839-1924), admiral; born at St. John's, Newfoundland; entered navy, 1853; present at operations in White Sea under Sir E. Ommanney [q.v.], 1854; subsequently served in East Indies, Mediterranean (under Sir W. F. Martin, q.v.), Irish station, and West Indies; commander, 1869; captain, 1877; during period on half-pay (1877-81) studied beginnings of German navy; commander of the *Espiègle* on Australia station and deputy commissioner for Western Pacific, 1881-5; director of intelligence department, Admiralty, 1889-94; commander-in-chief, Australia station, 1894-8; vice-admiral, 1898; commander-in-chief in China, 1901; contributed to successful issue of Anglo-Japanese treaty, 1902; strongly opposed plan of establishing permanent naval base at Wei-hai-wei; admiral, 1903; retired, 1904; assessor on international commission on Dogger Bank 'incident', 1904; member of Mesopotamia commission of inquiry, 1916; KCMG, 1899; GCB, 1903; studied war throughout his life; published several naval works.

BRIDGE, FRANK (1879-1941), musician; studied at Royal College of Music; played violin and viola in several string quartets and frequently conducted orchestras; compositions include: (chamber music) string Quartets in E minor and G minor; 'Phantasie' string Quartet, Trio, and Quartet for piano and strings; a Quintet and a Sextet; (for orchestra) symphonic poem 'Isabella', a suite 'The Sea', and 'Phantasm' for piano and orchestra.

BRIDGE, Sir JOHN FREDERICK (1844-1924), organist, composer, and musical antiquary; chorister at Rochester Cathedral, 1851-9; organist, Holy Trinity church, Windsor, 1865-9; organist, Manchester Cathedral, 1869-75; permanent deputy organist, Westminster Abbey, 1875; organist, 1882; a gifted choir-trainer; directed music at Queen Victoria's jubilee, 1887, Edward VII's coronation, 1902, George V's coronation, 1911, etc.; professor of harmony and counterpoint, Royal College of Music, 1883-1924; Gresham professor of music, Gresham College, London, 1890; composed much church and choral music.

BRIDGE, THOMAS WILLIAM (1848-1909), zoologist; assistant at museum of zoology, Cambridge University, 1870; BA, Trinity College, 1876; MA, 1880; professor of biology, Mason College, Birmingham, 1880; successful teacher and organizer; FRS, 1903; made researches into anatomy of fishes.

BRIDGEMAN, SIR FRANCIS CHARLES BRIDGEMAN (1848-1929), admiral of the fleet; entered navy, 1862; flag captain to Sir Michael Culme-Seymour in Channel squadron, Mediterranean, and at Portsmouth; rear-admiral, 1903; commander-in-chief, home fleet, 1907-9, 1911; second sea lord at Admiralty, 1910-11; first sea lord, 1911-12; loyally supported Lord Fisher [q.v.] and (Sir) Winston Churchill; vice-admiral of United Kingdom, 1920; KCB, 1908; died in Bahamas.

BRIDGES, EDWARD ETTINGDENE, first BARON BRIDGES, (1892-1969), public servant; only son of Robert Seymour Bridges [q.v.], poet laureate; educated at Eton and Magdalen College, Oxford; first class, *lit. hum.*, 1914; served in army in France, 1914-17; MC; fellow, All Souls, 1920-7; assistant principal, Treasury, 1919; assistant secretary, 1934; head, Treasury division controlling expenditure on supply of the armed forces, 1935; secretary to the cabinet, the Committee of Imperial Defence, the Economic Advisory Council and Ministry for Co-ordination of Defence, 1938; responsibilities divided with Sir H. L. (later Lord) Ismay [q.v.], 1940; secretary to the Treasury, 1945-56; chairman, National Institute for Research into Nuclear Energy, 1957, Fine Arts Commission, 1957-68, British Council, 1959-67, and Pilgrim Trust, 1965-8; chancellor, Reading University, 1959; chairman, Oxford Historic Buildings Fund, 1957; GCB (1944), GCVO (1946), PC (1953), baron (1957), KG (1965); FRS, 1952; hon. degrees, Oxford, Cambridge, London, Bristol, Leicester, Liverpool, Reading, and Hong Kong; hon. fellow, All Souls and Magdalen Colleges, Oxford, and London School of Economics and Political Science.

BRIDGES, SIR (GEORGE) TOM (MOLESWORTH) (1871-1939), lieutenant-general; nephew of Robert Bridges [q.v.]; commissioned in Royal Artillery, 1892; served in South

African war; captain, 1900; raised and commanded Tribal Horse, Somaliland, 1902-4; DSO, 1905; at St. Quentin (Aug. 1914) rallied two British battalions with toy whistle and drum; commanded 19th division, 1915-17; major-general, 1917; headed military missions in United States (1918) and with army of Orient, 1919; governor of South Australia, 1922-7; KCMG, 1919; KCB, 1925.

BRIDGES, JOHN HENRY (1832-1906), positivist philosopher; son of Charles Bridges [q.v.]; educated at Rugby and Wadham College, Oxford; BA, 1855; fellow of Oriel, 1855; MB, 1859; emigrated to Australia on marriage, 1860; returned to England, 1861; medical inspector to Local Government Board, 1870-98; under influence of Richard Congreve [q.v.] he studied Comtist philosophy; lectured to London Positivist Society, 1870-1900; works include translation of Comte's *Politique Positive I.* (1865 and 1875) and *Five Discourses on Positive Religion* (1882); wrote much on social reform and on health questions; delivered Harveian oration, 1892; edited Roger Bacon's *Opus Majus*, 1897 (reissued with many corrections and emendations, 1900); *Essays and Addresses* published posthumously, 1907.

BRIDGES, ROBERT SEYMOUR (1844-1930), poet laureate; educated at Eton; BA, Corpus Christi College, Oxford, 1867; friends of his youth included V. S. S. Coles [q.v.], D. M. Dolben, and G. M. Hopkins; entered as a medical student at St. Bartholomew's Hospital, London, 1869; MB, 1874; held various medical appointments, including that of casualty physician at St. Bartholomew's (1877-9), until he retired on account of his health, 1881; travelled in Egypt, Syria, and in various European countries during these years; issued his first volume of poems, which received a long and appreciative review by Andrew Lang [q.v.] in the *Academy*, 1873; published anonymously *The Growth of Love* (1876); lived at Yattendon, Berkshire, 1882-1904; produced, in collaboration with H. E. Wooldridge [q.v.] and H. C. Beeching [q.v.], *The Yattendon Hymnal*, 1895-9; writings of the Yattendon period include his eight dramas and quasi-dramas and one long narrative poem, *Eros and Psyche* (1885); wrote *On the Prosody of Paradise Regained and Samson Agonistes* (1889); republished it, with additions, 1893; this, together with his practice as poet, inaugurated new development of English verse; settled at Chilswell House, which he built on Boar's Hill overlooking Oxford, 1907; issued one-volume edition of *Poems* (excluding plays), 1912; poet laureate, 1913; with others, founded Society for Pure English (which did not become active until 1919), 1913; compiled anthology,

The Spirit of Man (1916); published *New Verse* (1925); his *magnum opus*, *The Testament of Beauty*, appeared on his eighty-fifth birthday, 1929; OM, 1929.

BRIDGES, SIR WILLIAM THROSBY (1861-1915), general; joined Australian head-quarters staff, 1902; chiefly memorable as commandant of Duntroon Royal Military College, 1910-14; commanded first Australian contingent in European war; mortally wounded at Gallipoli; KCB, 1915.

BRIDGES-ADAMS, WILLIAM (1889-1965), theatrical producer; educated at Bedales and Worcester College, Oxford; played with OUDS, and was associated with ADC production at Cambridge, 1907; assisted in stage management and directed in repertory, 1908-1917; director, annual Shakespeare Festival, Stratford-upon-Avon, 1919-34; directed at Stratford twenty-nine out of thirty-six plays in the First Folio, *The Merry Wives of Windsor* at the Lyric Hammersmith (1923), and *Much Ado about Nothing* at the New Theatre (1926); directed *Oedipus Rex* at Covent Garden (1936); worked for British Council, 1939-44; published *The Irresistible Theatre* (1957); CBE, 1960.

BRIDIE, JAMES (pseudonym), (1888-1951), playwright. [See MAVOR, OSBORNE HENRY.]

BRIERLY, JAMES LESLIE (1881-1955), international lawyer; educated at Charterhouse and Brasenose College, Oxford; first class, jurisprudence, 1905; fellow of All Souls, 1906; called to bar (Lincoln's Inn), 1907; professor of law, Manchester, 1920-2; Chichele professor of international law and diplomacy, Oxford, 1922-47; Montague Burton professor of international relations, Edinburgh, 1948-51; elected member of United Nations International Law Commission, 1948, chairman, 1951; works include *The Law of Nations* (1928), *Outlook for International Law* (1944); edited *British Year Book of International Law*, 1929-36; CBE, 1946.

BRIGGS, JOHN (1862-1902), Lancashire cricketer; obtained fame as slow left-hand bowler; paid six visits to Australia, 1884-97; played for England; mainstay of Lancashire county team, 1883-94; destructive on slow wickets.

BRIGHT, JAMES FRANCK (1832-1920), master of University College, Oxford; son of Richard Bright, MD [q.v.]; educated at Rugby and University College, Oxford; first class in law and modern history, 1854; ordained, 1856; master of modern school, Marlborough College,

1855-72; lecturer in modern history at University College, Oxford, 1873; fellow, 1874; helped to raise status of college and to establish intercollegiate lecture system; master, 1881-1906; took active part in university and municipal affairs; liberal in theology and politics; author of *History of England* (5 vols., 1875-1904).

BRIGHT, WILLIAM (1824-1901), church historian; educated at Rugby and University College, Oxford; first class, classics (BA, 1846, MA, 1849, DD, 1869); fellow, 1847-68; theological tutor, Trinity College, Glenalmond, and Bell lecturer in ecclesiastical history, 1851-8; regius professor of ecclesiastical history and canon of Christ Church, Oxford, 1868-1901; forcible and humorous lecturer; voluminous writings include *A History of the Church*, A.D. *313-451* (1860), *Waymarks of Church History* (1894), *The Age of the Fathers* (2 vols., 1903), sermons, and some first-rate hymns.

BRIGHTMAN, FRANK EDWARD (1856-1932), liturgiologist; educated at Bristol grammar school and University College, Oxford; second class, *lit. hum.* (1879) and theology (1880); deacon, 1884; priest, 1885; an original librarian of Pusey House, 1884-1903; fellow of Magdalen, 1902-32; joint-editor, *Journal of Theological Studies*, 1904-32; specialist on oriental rites; publications include *The English Rite* (2 vols., 1915); prebendary of Lincoln, 1902; FBA, 1926.

BRIGHTWEN, ELIZA (1830-1906), naturalist; born ELDER; married George Brightwen, banker, 1855; studied natural history at her home at Stanmore; works include *Wild Nature won by Kindness* (1890) and *Inmates of my House and Garden* (1895); most popular naturalist of day; her autobiography edited by (Sir) Edmund Gosse [q.v.], 1909.

BRIND, SIR (ERIC JAMES) PATRICK (1892-1963), admiral; educated at Osborne and Dartmouth; midshipman, 1909; at battle of Jutland, 1916; commander, 1927; captain, 1933; appointed to Admiralty Tactical Division, 1934; chief of staff to commander-in-chief Home Fleet, 1940; involved in destruction of *Bismarck*, 1941; CBE; rear-admiral, 1942; assistant chief of naval staff, Admiralty, 1942-4; CB, 1944; served in Pacific Fleet, 1944-6; vice-admiral, 1945; KCB, 1946; president, Royal Naval College, Greenwich, 1946-9; commander-in-chief Far East Station, 1949-51; admiral, 1949; organized escape of *Amethyst* from the Yangtse; GBE, 1951; commander-in-chief Allied Forces Northern Europe, new NATO command, 1951-3.

BRISE, Sir EVELYN JOHN RUGGLES-
(1857-1935), prison reformer. [See RUGGLES-
BRISE.]

BRITTAIN, VERA MARY (1893-1970),
writer, pacifist, and feminist; educated at St.
Monica's, Kingswood, and Somerville College,
Oxford; served as VAD in France and Malta
during 1914-18 war; returned to Oxford and
formed close friendship with Winifred Holtby,
author of *South Riding*; latter's death in 1935
commemorated in *Testament of Friendship*,
1940; married (Sir) George Catlin, 1925;
published *Testament of Youth*, 1933; joined
Peace Pledge Union, 1936; actively pacifist
during 1939-45 war; published *Testament of
Experience*, 1957; hon. D.Litt. Mills College,
California; mother of Shirley Williams, labour
cabinet minister, and leading social democrat.

BROADBENT, Sir WILLIAM HENRY,
first baronet (1835-1907), physician; studied
medicine at Owens College, Manchester, and
Paris; MB, London, 1858; MD, 1860; physician
to London Fever Hospital, 1860-79, and
elsewhere; at St. Mary's Hospital, pathologist,
1860, lecturer in comparative anatomy, 1861,
physician from 1865, and lecturer in medicine,
1871-88; made researches into cancer, paralysis,
and aphasia; advanced theory known as 'Broadbent's hypothesis' of hemiplegia; fellow, 1869,
senior censor, 1895, Croonian lecturer, 1887,
and Lumleian lecturer, 1891, at Royal College
of Physicians; Lettsomian lecturer of Medical
Society of London, 1874; member of royal
commission on fever hospitals, 1881; wrote
valuable treatise on heart disease, 1897;
physician to members of royal family from 1891;
baronet, 1893; KCVO, 1901; prominent in
public movements for prevention and cure of
tuberculosis; hon. member of many foreign
societies; FRS, 1897; received numerous
honorary degrees; *Collected Papers* edited by
son, 1908.

BROADHURST, HENRY (1840-1911),
labour leader; son of a stonemason; worked as
stonemason, 1853-72; settled in London, 1865;
chairman of masons' committee agitating for
increased pay, 1872; led his trade union to fix its
headquarters permanently in London, giving
central committee power to negotiate for whole
membership, thus establishing representative
democracy in trade unions; elected secretary of
Labour Representation League, 1873; elected
secretary of parliamentary committee of Trades
Union Congress, 1875; worked actively for
amendments to Factory Acts, for employers'
liability, and the like; liberal MP for Stoke-on-
Trent, 1880-85; member of royal commissions

on housing of working classes (1884) and on the
aged poor (1892); ardent agitator for extension
of franchise; MP for Bordesley, 1885-6; under-
secretary in home department, 1886; MP for
West Nottingham, 1886-92; in later years out of
sympathy with trade-union developments; MP
for Leicester, 1894-1906; published autobiography, 1901.

BROCK, Sir OSMOND DE BEAUVOIR
(1869-1947), admiral of the fleet; entered navy,
1882; served under (Lord) Beatty [q.v.] at
battles of Heligoland, Dogger Bank, and
Jutland; his chief of staff, 1916-19; rear-
admiral, 1915; deputy chief of naval staff,
1919-22; held Mediterranean (1922-5) and
Portsmouth (1926-9) commands; admiral,
1924; admiral of the fleet, 1929; KCVO, 1917;
GCB, 1929.

BROCK, Sir THOMAS (1847-1922), sculptor; pupil of J. H. Foley [q.v.], whom he
succeeded as supplier of monuments and official
statuary; Academy gold medallist in sculpture,
1969; executed many London statues, including
memorial to Queen Victoria in front of Buckingham Palace in collaboration with Sir Aston
Webb [q.v.]; his equestrian statue of Black
Prince at Leeds also notable; executed many
portrait busts; RA, 1891; hon. DCL, Oxford,
1909; KCB, 1911.

BRODETSKY, SELIG (1888-1954), mathe-
matician and Zionist leader; born in Ukraine;
settled in London, 1893; senior wrangler,
Trinity College, Cambridge, 1908; Ph.D.,
Leipzig; lecturer in Bristol, 1914, Leeds, 1919;
professor, 1924-48; joined World Zionist
Executive, 1928; president, Board of Deputies
of British Jews, 1940-9, of British Zionist
Federation, 1948, of Hebrew University,
Jerusalem, 1949-52.

BRODRIBB, CHARLES WILLIAM (1878-
1945), journalist, scholar, and poet; scholar of
St. Paul's School and Trinity College, Oxford;
joined *The Times*, 1904; an assistant editor and
special writer, 1914-45; chiefly interested in
classical scholarship, English literature, and
London antiquities.

BRODRIBB, WILLIAM JACKSON (1829-
1905), translator; educated at St. John's College, Cambridge; sixth classic, 1852; fellow,
1856; vicar of Wootton Rivers, Wilts., 1860-
1905; translator and editor of works of Tacitus.

BRODRICK, GEORGE CHARLES (1831-
1903), warden of Merton College, Oxford;
educated at Eton and Balliol; first class in
classics (1853) and in law and history (1854);

president of Union, 1854-5; BA, 1854; MA, 1856; DCL, 1886; elected fellow of Merton, 1855; called to bar, 1859; joined staff of *The Times*, 1860; made several unsuccessful attempts to enter parliament as liberal; lucid writer on politics; published *Political Studies* (1879), *English Land and English Landlords* (1881), and *Memories and Impressions* (1900); member of London school board, 1877-9; warden of Merton, 1881-1903; active in university reform; wrote popular histories of Merton (1885) and of the university (1886); opposed Gladstone's Irish policy.

BRODRICK, (WILLIAM) ST. JOHN (FREMANTLE), ninth VISCOUNT MIDLETON and first EARL OF MIDLETON (1856-1942), statesman; educated at Eton and Balliol College, Oxford; president of the Union, 1878; conservative MP, West Surrey (1880-5), Guildford division (1885-1906); financial secretary to the War Office, 1886-92; under-secretary for war, 1895-8; for foreign affairs, 1898-1900; PC, 1897; secretary of state for war, 1900-3; for India, 1903-5; succeeded father, 1907; leader of southern unionists in Ireland; negotiated truce, 1921; KP, 1916; earl, 1920.

BROMBY, CHARLES HAMILTON (1843-1904), son of Charles Henry Bromby [q.v.]; called to bar, 1867; attorney-general of Tasmania, 1876-8; efficient Italian student.

BROMBY, CHARLES HENRY (1814-1907), second bishop of Tasmania; educated at Uppingham and St. John's College, Cambridge; BA, 1837; MA, 1840; DD, 1864; first principal of Cheltenham training college, 1843-64; bishop of Tasmania, 1864-82; reorganized finances of Tasmanian Church on its disestablishment; assistant bishop of Lichfield, 1882-91, and of Bath and Wells, 1891-1900; wrote pamphlets on education.

BROODBANK, SIR JOSEPH GUINNESS (1857-1944), public servant; entered East and West Dock Company's service, 1872; secretary, 1889; of London and India Docks Company, 1901; Board of Trade member of Port of London Authority, 1909-20; member of many committees on transport; published *History of the Port of London* (2 vols., 1921); knighted, 1917.

BROOK, NORMAN CRAVEN, BARON NORMANBROOK (1902-1967), secretary of the cabinet; educated at Wolverhampton grammar school and Wadham College, Oxford; hon. fellow, Wadham, 1949; entered home civil service, 1925; principal private secretary to Sir John Anderson (later Viscount Waverley, q.v.),

1938-42; deputy secretary, cabinet; permanent secretary, Ministry of Reconstruction, 1943-5; secretary of the cabinet, 1947-62; also joint permanent secretary, Treasury, and head of home civil service, 1956-62; chairman of governors, BBC, 1964-7; CB, 1942; KCB, 1946; GCB, 1951; PC, 1953; baron, 1963.

BROOKE, ALAN ENGLAND (1863-1939), biblical scholar and provost of King's College, Cambridge; scholar of Eton and King's; fellow, 1889; dean, 1894-1918; provost, 1926-33; deacon, 1891; priest, 1904; canon, 1916; editor with Norman McLean [q.v.] of larger Cambridge edition of Septuagint; Ely professor of divinity, 1916-26; FBA, 1934.

BROOKE, ALAN FRANCIS, first VISCOUNT ALANBROOKE (1883-1963), field-marshal; educated in France and at the Royal Military Academy, Woolwich; served with Royal Field Artillery in Ireland, 1902-6; and in India, 1906-9; transferred to Royal Horse Artillery, 1909; served on western front, 1914-18; DSO, and bar; promoted from lieutenant to lieutenant-colonel; Staff College, Camberley; instructor, 1923; Imperial Defence College, 1927; brigadier, commanded School of Artillery, Larkhill, 1929-32; instructor, IDC, 1932-4; major-general, 1935; lieutenant-general in command Anti-Aircraft Corps, 1938; commander-in-chief, Southern Command, 1939; fought as corps commander in France; organized evacuation of troops; commander-in-chief, Home Forces, 1940; chief of Imperial General Staff, 1941; chairman, Chiefs of Staff Committee, 1941-5; field-marshal, 1944; organized British contribution to military war effort from time of defeats of 1941 to victories of 1944-5; master gunner of St. James's Park, 1946-56; director, Midland Bank, 1949-63; chancellor, Queen's University, Belfast, 1949-63; lord lieutenant, county of London and constable of the Tower, 1950; KCB, 1940; GCB, 1942; baron, 1945; viscount, 1946; KG, 1946; OM, 1946; president, Zoological Society, 1950-4.

BROOKE, SIR CHARLES ANTHONY JOHNSON (1829-1917), second raja of Sarawak; born JOHNSON; joined uncle, Sir James Brooke [q.v.], first raja, 1852; succeeded to title and assumed name Brooke, 1868; worked for progress and development of Sarawak.

BROOKE, RUPERT CHAWNER (1887-1915), poet; educated at Rugby and King's College, Cambridge; became conspicuous figure in university life; published volume of *Poems*, 1911; fellow of King's, 1912; travelled in United States, Canada, New Zealand, and

South Sea islands, 1913-14; joined Royal Naval Division, 1914; died at Scyros; *1914 and other Poems* published posthumously; fellowship dissertation on John Webster and *Letters from America* have also been published.

BROOKE, STOPFORD AUGUSTUS (1832-1916), divine and man of letters; BA, Trinity College, Dublin; ordained, 1857; minister of proprietary chapel of St. James, York Street, London, 1866-76, and of Bedford chapel, Bloomsbury, 1876-95; seceded from English Church, 1880; celebrated preacher; works include *Life and Letters of Frederick W. Robertson* (1865) and *Primer of English Literature* (1876).

BROOKE, ZACHARY NUGENT (1883-1946), historian; educated at Bradfield and St. John's College, Cambridge; fellow of Gonville and Caius College, 1908-46; professor of medieval history, 1944-6; joint-editor, *Cambridge Medieval History*; published *The English Church and the Papacy* (1931) and *A History of Europe, 911-1198* (1938); FBA, 1940.

BROOKE-POPHAM, SIR (HENRY) ROBERT (MOORE) (1878-1953), air chief marshal; educated at Haileybury and Sandhurst; captain 1904; qualified as pilot, 1911; commanded Royal Flying Corps squadron, 1912-14; staff officer with BEF in France, 1914-18, providing administrative and technical support for RFC; DSO and AFC; controller, aircraft production, 1918; air commodore, Royal Air Force, 1919; director of research, 1920-1; commanded RAF Staff College, 1921-6; commanded Fighting Area, 1926-8, Iraq, 1928-30; air vice-marshal, 1924; air marshal, 1931; commandant, Imperial Defence College, 1931-3; commander-in-chief Air Defence of Great Britain, 1933-5; air chief marshal and inspector-general, 1935; commanded in Middle East during Abyssinian crisis 1935-6; retired, 1937; governor of Kenya, 1937-9; headed training missions to Canada and South Africa, 1939-40; commander-in-chief Far East, 1940-1; CMG, 1918; CB, 1919; KCB, 1927; GCVO, 1935.

BROOKING ROWE, JOSHUA (1837-1908), Devonshire antiquary. [See ROWE.]

BROOM, ROBERT (1866-1951), palaeontologist; educated at Hutcheson's grammar school and the university, Glasgow; MB, CM, 1889; practised medicine intermittently in South Africa after 1897; professor of zoology and geology, Victoria College, Stellenbosch, 1903-10; curator of fossil vertebrates, Transvaal Museum, 1934-6; studied Permian reptiles of North America and South Africa and the Australopithecines; gave remarkably intelligible and accurate accounts of his studies in over four hundred papers; published *Finding the Missing Link* (1950); FRS, 1920.

BROTHERHOOD, PETER (1838-1902), civil engineer; partner in Compton Street engine works, Goswell Road, 1867; introduced Brotherhood engine, for driving torpedoes by means of compressed air, 1872; designed air compressor, and servo-motor for torpedoes, 1876.

BROUGH, BENNETT HOOPER (1860-1908), mining expert; nephew of Lionel Brough [q.v.]; successively student and teacher at Royal School of Mines, 1878-82; published *Treatise on Mine Surveying*, 1888; secretary of Iron and Steel Institute, 1893-1908.

BROUGH, LIONEL (1836-1909), actor; brother of William Brough and Robert Barnabas Brough [qq.v.]; errand boy in *Illustrated London News* office, 1848; made début on stage at Lyceum, 1854; assistant publisher of *Daily Telegraph*, 1855; originated street selling of newspapers; gave monologues at Regent Street Polytechnic, 1862; introduced 'Pepper's Ghost', 1863; resumed acting as profession, 1864; ability recognized in his representation of Ben Garner in *Dearer than Life*, 1868; successful as Tony Lumpkin and other comedy characterizations from 1869 onwards; joint lessee with Willie Edouin [q.v.] of Toole's Theatre, 1884; visited America (1886) and South Africa (1888); acted with company of Sir H. Beerbohm Tree [q.v.] from 1894 to death; pre-eminent in burlesque; noted for simple drollery and affectation of blank stolidity.

BROUGH, ROBERT (1872-1905), painter; studied at Royal Scottish Academy life school, Edinburgh, and in Paris; exhibited at Royal Scottish and Royal academies; ARSA, 1904; portrait painter of virile powers, uniting simplicity and breadth; chief works were 'Master Philip Fleming', 1900, and 'Sir Charles Tennant's family', 1905; killed in railway accident.

BROUGHTON, RHODA (1840-1920), novelist; novels include *Cometh Up as a Flower* and *Not Wisely but too Well* (1867), *Goodbye, Sweetheart* (1872), *Nancy* (1873), *Joan* (1876), *Belinda* (1883), *Doctor Cupid* (1886), *Foes-in-law* (1900), and *A Waif's Progress* (1905).

BROWN, (ALFRED) ERNEST (1881-1962), politician; educated at local school in Torquay and employed as clerk; Baptist lay preacher;

joined Sportsman's battalion, 1914; commissioned in Somerset Light Infantry, 1916; MM and MC; liberal MP Rugby, 1923-4; MP Leith, 1927-45; became liberal national, 1931; parliamentary secretary, Ministry of Health, 1931-2; secretary, Department of Mines, 1932-5; minister of labour, 1935-40; secretary of state for Scotland, 1940; minister of health, 1941-3; chancellor of Duchy of Lancaster, 1943; minister of aircraft production, 1945; chairman of European committee of UNRRA; 1944-5; president, Baptist Union of Great Britain and Ireland, 1948-9; PC, 1935; CH, 1945.

BROWN, Sir ARTHUR WHITTEN (1886-1948), air navigator and engineer; trained as electrical engineer; observer in Royal Flying Corps, 1915; obtained pilot's licence, 1918; navigator to (Sir) John Alcock [q.v.] in first nonstop Atlantic flight, 16 hours 27 minutes, 14-15 June 1919; representative of Metropolitan-Vickers, 1919-39; trained Royal Air Force pilots in navigation and engineering, 1939-45; KBE, 1919.

BROWN, DOUGLAS CLIFTON, Viscount RUFFSIDE (1879-1958), speaker of the House of Commons; educated at Eton and Trinity College, Cambridge; joined militia, 1900; in France and Belgium, 1914-18; major, 1919; coalition unionist MP, Hexham division, 1918-23, 1924-51; deputy chairman, 1938, chairman, Jan. 1943, of ways and means; speaker, Mar. 1943-1951; his equable temperament invaluable to the House during post-war return to party political warfare; PC, 1941; viscount, 1951.

BROWN, ERNEST WILLIAM (1866-1938), mathematician and astronomer; scholar and fellow (1889) of Christ's College, Cambridge; sixth wrangler, 1887; professor of mathematics, Haverford College, Pennsylvania, 1891-1907, and Yale, 1907-32; mainly interested in developing his complete and accurate theory of the motion of the moon; published his *Tables* in three volumes, 1919; FRS, 1897.

BROWN, FREDERICK (1851-1941), professor of fine art; studied at South Kensington; head of Westminster School of Art (1877-93), of Slade School of Fine Art (1893-1917); reformed teaching of art and attracted many distinguished pupils whose individuality he encouraged; leader in foundation of New English Art Club, 1886.

BROWN, GEORGE DOUGLAS (1869-1902), novelist; son of farmer; educated at Glasgow University (MA, 1890) and Balliol

College, Oxford, 1891-5; published a novel of Scottish life, *The House with the Green Shutters*, under pseudonym of George Douglas, 1901.

BROWN, Sir GEORGE THOMAS (1827-1906), veterinary surgeon; professor of veterinary science at Royal Agricultural College, Cirencester, 1850-63; connected with veterinary department of Privy Council, 1865-89; at Board of Agriculture, 1889-93; knighted, 1898; professor (from 1881) and principal (1888-94) of Royal Veterinary College; published *Animal Life in the Farm* (1885) and departmental reports.

BROWN, GERARD BALDWIN (1849-1932), historian of art; nephew of H. S. Leifchild [q.v.]; educated at Uppingham and Oriel College, Oxford; first class, *lit. hum.*, 1873; fellow of Brasenose, 1874-7; Watson Gordon professor of fine art, Edinburgh, 1880-1930; works include *From Schola to Cathedral* (1886), *The Arts in Early England* (6 vols., 1903-37), *The Care of Ancient Monuments* (1905), and *The Glasgow School of Painters* (1908); FBA, 1924.

BROWN, HORATIO ROBERT FORBES (1854-1926), historian of Venice; born at Nice; while a boy at Clifton College made acquaintance of J. A. Symonds [q.v.] who exercised influence over his intellectual tastes; settled with his mother at Venice, 1879; appointed to succeed R. L. Brown [q.v.] in work of calendaring Venetian state papers concerning English history preserved at the Frari; compiled calendars covering years 1581-1613 (1894-1905); his other works include *Life on the Lagoons* (1884); *Venetian Studies* (1887); *The Venetian Printing Press* (1891); *Venice, an Historical Sketch* (1893); *John Addington Symonds, a Biography* (1895); *Studies in the History of Venice* (2 vols., 1907); *Letters and Papers of John Addington Symonds* (1923); *Dalmatia* (1925); visited England every summer; left Venice during 1914-18 war, but returned in 1919; died at Belluno; hon. LLD, Edinburgh, 1900.

BROWN, Sir JOHN (1880-1958), lieutenant-general, Territorial Army; educated at Magdalen College School, Brackley; ARIBA, 1921; FRIBA, 1930; practised as architect in Northampton and London; joined Volunteers, 1901; served in Dardanelles and Palestine during 1914-18 war; DSO; a dynamic commander of 162nd (East Midland) Infantry Brigade, 1924-8; chairman, British Legion, 1930-4; appointed deputy director-general, Territorial Army, and major-general, 1937; lieutenant-general and deputy adjutant-general (T), 1939; director-general, Territorial Army and inspector-general welfare and education, 1940-1; CBE, 1923; CB, 1926; KCB, 1934.

BROWN, JOSEPH (1809-1902), barrister; BA, Queens' College, Cambridge, 1830; MA, 1833; called to bar, 1845; treasurer of Middle Temple, 1878-9; had large commercial practice; initiated publication of *Law Reports*, 1865; CB, 1892.

BROWN, OLIVER FRANK GUSTAVE (1885-1966), fine art dealer; educated at St. Paul's School and in France; joined the Leicester Galleries, 1903; partner in firm of Ernest Brown and Phillips, 1914; exhibited foreign artists, Matisse (1919), Camille Pissarro (1920), Picasso (1921), Degas (1923), Van Gogh (1923), Gauguin (1924), Cézanne (1925), Renoir (1926), and Chagall (1935); arranged displays of sculpture by Epstein, Rodin, and Henry Moore; published *Exhibition* (autobiography), 1968; served on arts panel of Arts Council, 1949-54; OBE, 1960.

BROWN, PETER HUME (1849-1918), historian; MA, Edinburgh University; Sir William Fraser's professor of ancient history at Edinburgh, 1901-18; works include biographies of George Buchanan (1890), John Knox (1895), and Goethe (1920), and *History of Scotland* (1899-1909); editor of *Register of the Privy Council of Scotland*, 1898.

BROWN, THOMAS GRAHAM (1882-1965), neurophysiologist and mountaineer; educated at Edinburgh Academy and Edinburgh University; B.Sc., 1903; graduated with honours in medicine, 1906; Muirhead demonstrator in physiology, Glasgow, 1907-10; Carnegie fellow, Liverpool, 1910; MD, 1912; D.Sc., 1914; lecturer in experimental physiology, Man-chester University, 1913-15; served with RAMC, 1915-19; professor, physiology, university of Wales, Cardiff, 1919-47; MRCP (Edinburgh), 1921; FRS, 1927; climbed Brenva face of Mont Blanc with F. S. Smythe [q.v.], 1927; climbed in Alaska, the Himalayas, and the Karakoram; climbed Mont Blanc by six separate routes; climbed the Matterhorn, 1952, in his seventieth year; editor, *Alpine Journal*, 1949-53; published *Brenva* (1944), and *The First Ascent of Mont Blanc* (with Sir Gavin de Beer, 1957).

BROWN, SIR WALTER LANGDON LANGDON- (1870-1946), physician and regius professor of physic in the university of Cambridge. [See LANGDON-BROWN.]

BROWN, WILLIAM FRANCIS (1862-1951), Roman Catholic bishop; educated at high school, Dundee, Trinity College, Glenalmond, and University College School, London; became Roman Catholic, 1880; priest 1886; parish priest, St. Anne's, Vauxhall, 1892-1951; vicar-general, 1904, provost, 1916, auxiliary bishop, Southwark, 1924-51; appointed apostolic visitor to Scotland, 1917; an expert on educational problems; largely responsible for Education (Scotland) Act of 1918.

BROWN, WILLIAM HAIG (1823-1907), master of Charterhouse. [See HAIG BROWN.]

BROWN, WILLIAM JOHN (1894-1960), union leader and member of Parliament; educated at Sir Roger Manwood's grammar school, Sandwich; civil service clerk, 1910-19; general secretary, Civil Service Clerical Association, 1919-42; parliamentary secretary, 1942-50; labour MP, West Wolverhampton, 1929-31; strong defender of liberty, he became critical of socialist discipline and trade-union domination; independent MP for Rugby, 1942-50; a fluent speaker and writer and edifying advocate of fundamental moral values.

BROWN, WILLIAM MICHAEL COURT (1918-1968), medical research worker; educated at Fettes College, Edinburgh, and St. Andrews University; BSc., 1939; qualified, 1942; joined department of radiotherapy, Edinburgh Royal Infirmary, 1943; fellow, faculty of Radiologists, deputy director, radiotherapy department, 1950; director, clinical effects of radiation research unit, Medical Research Council, at Western General Hospital Edinburgh, 1956-68; studied leukaemogenic effects of radiation and human population cytogenetics; published findings in *Chromosome Studies on Adults* (1966), *Human Population Cytogenetics* (North-Holland, 1967), and in a review in the *British Medical Bulletin* (25, No. 1, 1969); OBE, 1957; FRCPE, 1965; hon. professor, Edinburgh, 1967.

BROWNE, EDWARD GRANVILLE (1862-1926), Persian scholar and orientalist; BA, Pembroke College, Cambridge; MB, 1887; fellow of Pembroke, 1887; visited Persia, Oct. 1887 to Oct. 1888; Persia thenceforth the central object of his studies; university lecturer in Persian, 1888; Sir Thomas Adams's professor of Arabic, 1902; his works include *A Traveller's Narrative, written to illustrate the Episode of the Bab* (1891), *The New History of Mirza Ali Muhammad the Bab* (1893), *A Year amongst the Persians* (1893), *Literary History of Persia until the time of Firdausi* (1902-24), *History of the Persian Revolution, 1905-1909* (1910), *Arabian Medicine* (1921), edited *History of Ottoman Poetry* by E. J. W. Gibb [q.v.]; FBA, 1903.

BROWNE, GEORGE FORREST (1833–1930), bishop of Stepney and later of Bristol; BA, Catharine Hall, Cambridge, 1856; ordained, 1858; fellow of St. Catharine's College, 1863; secretary to university commission, 1877–81; Disney professor of archaeology, 1887; canon of St. Paul's Cathedral, 1892; suffragan bishop of Stepney, 1895; bishop of Bristol, 1897–1914; DD, Cambridge, 1896; hon. DD, Oxford, 1908; FBA, 1903; published many works.

BROWNE, SIR JAMES CRICHTON- (1840–1938), physician and psychologist; qualified at Edinburgh, 1861; medical director, West Riding Asylum, 1866; lord chancellor's visitor in lunacy, 1875–1922; co-editor, *Brain*, 1878–85; emphasized importance of prodromal symptoms; FRS, 1883; knighted, 1886; treasurer, Royal Institution, 1889.

BROWNE, SIR JAMES FRANKFORT MANNERS (1823–1910), general; joined Royal Engineers, 1842; served in Canada, 1845–51, and in Ireland, 1851–3; in command of 1st company of royal sappers and miners, Chatham, 1854; in trenches at Sevastopol, 1855; conspicuous at the Redan; led right attack, severely wounded and invalided home, Nov. 1855; CB, 1855; commanded engineers in Bombay presidency at Poona, 1859; governor of Royal Military Academy, Woolwich, 1880–7; general, 1888; colonel commandant of Royal Engineers, 1890; KCB, 1894.

BROWNE, SIR SAMUEL JAMES (1824–1901), general; born in India; joined 46th Bengal native infantry, 1840; served in second Sikh war, 1848–9; commanding officer of 2nd Punjab cavalry, mainly on Peshawar frontier, 1851–63; at siege of Lucknow, 1858; awarded VC for gallantry against rebels at Sirpura, Aug. 1858; KCSI, 1876; lieutenant-general, 1877; as military member of governor-general's council, he made preparations for Afghan war, 1878–9; commanded 1st division of Peshawar field force, capturing fortress of Ali Masjid (Nov.) and Jellalabad (Dec.) 1878; prepared scheme for advance on Kabul, 1879; occupied Gandamak, 1879; KCB, 1879; GCB, 1891; general, 1888; inventor of sword belt adopted in the army.

BROWNE, SIR STEWART GORE- (1883–1967), soldier, and settler and politician in Northern Rhodesia (Zambia). [See GORE-BROWNE.]

BROWNE, THOMAS (1870–1910), painter and black-and-white artist; apprenticed to Nottingham firm of lithographic printers, 1884–91; went to London, 1895; contributed black-and-white sketches to London and American periodicals; created comic types of American journalism, Weary Willie and Tired Tim; founded colour-printing firm of Tom Browne & Co. at Nottingham, 1897; exhibited water-colours at Royal Academy, 1898–1901; published *Tom Browne's Comic Annual*, 1904–5.

BROWNE, WILLIAM ALEXANDER FRANCIS BALFOUR- (1874–1967), entomologist. [See BALFOUR-BROWNE.]

BROWNING, SIR FREDERICK ARTHUR MONTAGUE (1896–1965), soldier and courtier; educated at Eton and Royal Military College, Sandhurst; served in France with Grenadier Guards, 1915–18; DSO, croix de guerre; acting captain, 1918; captain, 1920; major, 1928; adjutant, Sandhurst, 1924–8; commanded 2nd battalion, Grenadier Guards, 1935–9; commanded experimental airborne formation, 1940; 1st Airborne division, 1941; lieutenant-general, commanded 1st Airborne corps, 1944; concerned with planning Arnhem operation, 1944; chief of staff, Supreme Allied Commander, South East Asia, 1944–6; military secretary to secretary of state for war, 1946–8; comptroller of the household of Princess Elizabeth and the Duke of Edinburgh, 1946–52; treasurer and controller to Duke of Edinburgh, 1952–9; CB, 1943; KBE, 1946; KCVO, 1953; GCVO, 1959; married novelist (Dame) Daphne du Maurier, 1932.

BROWNING, SIR MONTAGUE EDWARD (1863–1947), admiral; entered navy, 1876; specialized in gunnery; inspector of target practice, 1910–13; rear-admiral, 1911; rear-admiral, third battle squadron, Grand Fleet, 1913–15; commanded third cruiser squadron, 1915–16; commander-in-chief, North America and West Indies, 1916–18; vice-admiral, 1917; commanded fourth battle squadron, 1918; president, allied naval armistice commission, 1918–19; admiral, 1919; second sea lord, 1919–20; commander-in-chief, Devonport, 1920–3; KCB, 1917; GCMG, 1919; GCB, 1924; GCVO, 1933.

BROWNING, OSCAR (1837–1923), schoolmaster, fellow of King's College, Cambridge, and historian; educated at Eton and King's; fellow of King's, 1859; assistant master at Eton, 1860–75; a popular housemaster and an educational reformer; dismissed from Eton on unsubstantiated charges of misconduct; returned to King's; lecturer in history, 1880; university lecturer, 1883; threw himself into task of reforming and reorganizing the college; made his influence felt informally among undergraduates in many directions; principal of

Cambridge University day training college for teachers, which he largely helped to found, 1891-1909; left Cambridge, 1908; spent latter years in Rome, where he died; a keen radical; wrote historical manuals.

BRUCE, ALEXANDER HUGH, sixth BARON BALFOUR OF BURLEIGH (1849-1921), statesman; representative peer for Scotland, 1876-1921; chairman of numerous commissions, 1882-1917; secretary for Scotland, 1895-1903; worked to promote union of Church of Scotland and United Free Church; PC, 1892; KT, 1901; GCMG, 1911.

BRUCE, CHARLES GRANVILLE (1866-1939), soldier, mountaineer, and traveller; son of first Baron Aberdare [q.v.]; educated at Harrow and Repton; commissioned in Oxfordshire and Buckinghamshire Light Infantry, 1887; transferred to 5th Gurkha Rifles, 1889; evolved system of training for mountain warfare; originated Frontier Scouts; saw service on North-West Frontier, 1891-8; travelled extensively in Himalayas; in Nanga Parbat expedition, 1895; lieutenant-colonel, 1913; wounded at Gallipoli; commanded Independent Frontier brigade, Bannu (1916-19), North Waziristan field force (1917), and served in Afghan war, 1919; invalided out as brigadier-general, 1920; CB, 1918; organized and led Mount Everest expeditions of 1922 and 1924.

BRUCE, CLARENCE NAPIER, third BARON ABERDARE (1885-1957), athlete; succeeded his father, 1929; educated at Winchester and New College, Oxford; called to bar (Inner Temple), 1911; played cricket, golf, rackets, and tennis for Oxford; notable cricketer for Middlesex, 1919-29; amateur rackets champion, 1922 and 1931; open champion, 1932; won doubles ten times; USA tennis champion, 1930, England, 1932 and 1938; chairman, National Association of Boys' Clubs, 1943-57; on executive committee of International Olympics, 1931-57; died in car accident.

BRUCE, SIR DAVID (1855-1931), discoverer of the causes of Malta fever and sleeping sickness; born in Melbourne; educated at Stirling high school and graduated in medicine at Edinburgh, 1881; commissioned in Army Medical Service; found organism later known as *Brucella melitensis* and proved it the cause of Malta fever (later established goat's milk as the source of infection); assistant professor of pathology, Army Medical School, Netley, 1889-94; in Zululand (1894) established *Trypanosoma brucei* as cause of tsetse-fly disease and nagana; supervised investigation of sleeping sickness (Uganda, 1903) proving it a trypanosome disease carried by tsetse; surgeon-general, 1912; commandant, Royal Army Medical College, Millbank, 1914-19; directed research on trench fever and tetanus; knighted, 1908; KCB, 1918; president, British Association, 1924.

BRUCE, SIR GEORGE BARCLAY (1821-1908), civil engineer; apprentice to Robert Stephenson & Co., 1836-41; resident engineer of Royal Border bridge, 1845; chief engineer of Madras railway, 1853-6; constructed many English and foreign railway lines, including railway and pier at Huelva, Spain, 1873-6; Telford medallist (1851) and president (1887-8) of Institution of Civil Engineers; knighted, 1888; interested in Presbyterian church work and education; member of London school board, 1882-5.

BRUCE, SIR HENRY HARVEY (1862-1948), admiral; entered navy, 1875; King's harbour master, Bermuda, and captain, 1905; commanded successively the *Blenheim*, *Arrogant*, *Prince George*, *Defence*, and (1913-15) the battleship *Hercules*; first superintendent (1915-20) of new Rosyth dockyard where his efficient organization was invaluable to the fleet; vice-admiral, 1922; admiral, 1926; KCB, 1920.

BRUCE, STANLEY MELBOURNE, first VISCOUNT BRUCE OF MELBOURNE (1883-1967), Australian prime minister and diplomatist; educated at Melbourne grammar school and Trinity Hall, Cambridge; rowing blue, 1904; called to bar (Middle Temple), 1907; returned to Australia, 1911; served in 1914-18 war with Royal Fusiliers at Gallipoli; MC; croix de guerre avec palme; elected to Australian Parliament, 1918; Australian delegate to League of Nations, 1921; Treasury minister, 1922; prime minister, 1923-9; minister without portfolio, 1931; led Australian delegation to Ottawa Conference, 1932; resident minister, London, 1932; high commissioner, 1932-45; viscount, 1947; director, P & O; chairman, Finance Corporation for Industry, 1947-57; chairman, World Food Council; chancellor, Australian National University, Canberra; president, Leander Club, 1948-52; CH, 1927; FRS, 1944; hon. degrees, Cambridge, Oxford, St. Andrews, Glasgow, Edinburgh, Leeds, Toronto, and all Australian universities; published *The Imperial Economic Situation* (1931), and *Rowing, Notes on Coaching* (1936).

BRUCE, VICTOR ALEXANDER, ninth EARL OF ELGIN and thirteenth EARL OF KINCARDINE (1849-1917), statesman and viceroy of India; son of James Bruce, eighth earl [q.v.]; born in Canada; educated at Glenalmond, Eton, and Balliol College, Oxford; succeeded father, 1863; held office in liberal government of 1886; viceroy of India, 1893-8; his period of office marked by political unrest, frontier disturbances, plague, and famine; proved too weak for position which he had reluctantly accepted; KG, 1899; chairman of royal commission which inquired into military preparations for South African war, 1902, and of commission on Free Church case, 1905; colonial secretary in Campbell-Bannerman's government, 1905-8.

BRUCE, WILLIAM SPEIRS (1867-1921), polar explorer and oceanographer; studied medicine at Edinburgh University; went on various polar expeditions, 1892-9; planned Scottish national Antarctic expedition to explore Weddell Sea, 1902-4; published its *Report*; explored Spitsbergen between 1906 and 1920; received medals of various geographical and scientific societies.

BRUCE LOCKHART, SIR ROBERT HAMILTON (1887-1970), diplomatist and writer. [See LOCKHART.]

BRUNT, SIR DAVID (1886-1965), meteorologist; educated at Abertillery Intermediate (later County) school, University College of Wales, Aberystwyth, and Trinity College, Cambridge; first class, parts i and ii, mathematical tripos, 1909-10; Isaac Newton student, 1911-13; lecturer in mathematics, Teachers' Training College, Caerleon, 1914-16; served in Royal Engineers, 1916-19; superintendent, Army Services Division, Meteorological Office (part of Air Ministry from 1921); chairman, Meteorological Sub-Committee of the Chemical Warfare (later Defence) Committee, 1921-42; professor, meteorology, Imperial College, 1934-52; FRS, 1939; secretary, Royal Society, 1948-57; knighted, 1949; KBE, 1959; Sc.D., Cambridge, 1940; hon. doctorates, Wales (1951) and London (1960); president, Royal Meteorological Society, 1942-4; president, Physical Society, 1945-7; chairman, Electricity Supply Research Council of the Central Electricity Authority, 1952; publications include *Combination of Observations* (1917) and *Physical and Dynamical Meteorology* (1934).

BRUNTON, SIR THOMAS LAUDER, first baronet (1844-1916), physician; MD, Edinburgh University; lecturer in materia medica and pharmacology, St. Bartholomew's Hospital, London; discovered use of amyl nitrite in treatment of *angina pectoris*; Goulstonian lecturer, 1877; Lettsomian lecturer, 1886; Croonian lecturer, 1889; Harveian orator, 1894;

baronet, 1908; publications include *Text-book of Pharmacology and Therapeutics* (1885).

BRUSHFIELD, THOMAS NADAULD (1828-1910), lunacy specialist and antiquary; studied medicine at London Hospital, 1845-9; MRCS, 1850; MD, St. Andrews, 1862; medical superintendent at Brookwood Asylum, 1865-82; pioneer of 'non-restraint' treatment of insane; settled at Budleigh Salterton, 1882; published *Raleghana* (1896-1907), *Ralegh Miscellanea* (1909-10), and a Ralegh bibliography (1886); FSA, 1899; a founder of Devon and Cornwall Record Society.

BRYCE, JAMES, VISCOUNT BRYCE (1838-1922), jurist, historian, and statesman; son of James Bryce the younger (1806-77, q.v.); grandson of James Bryce the elder (1767-1857, q.v.); educated at Glasgow high school, Belfast Academy, Glasgow University, and Trinity College, Oxford; obtained first classes in classical moderations, *lit. hum.*, and law and modern history; distinctions included Craven scholarship, Vinerian law scholarship, and presidency of Union; fellow of Oriel College, 1862-89; called to bar (Lincoln's Inn, which he entered, 1862), 1867; joined Northern circuit; practised until 1882; won Arnold prize at Oxford with essay on Holy Roman Empire, 1863 (published, 1864, securing for him European reputation); as assistant commissioner reported on schools of Lancashire, Shropshire, Worcestershire, Monmouthshire, and eight Welsh counties for schools inquiry commission, 1865-6; assisted development of university education at Manchester, where he lectured on law, 1868-74; regius professor of civil law at Oxford, 1870-93; began revival of study of Roman law; visited St. Petersburg and Moscow, southern Russia, Caucasus, and Armenia, 1876; published *Transcaucasia and Ararat*, 1877; acquired deep and lasting interest in affairs of Near East; principal advocate of Armenian nation in England; founder and first president of Anglo-Armenian Society; liberal MP, Tower Hamlets division, 1880-5; for South Aberdeen, 1885-1906; established himself in parliament as authority on Eastern question; associated with foundation of *English Historical Review*, 1886; published *The American Commonwealth* (the outcome of three visits to USA, 1870, 1881, 1883), 1888, aiming to portray 'the whole political system of the country in its practice as well as its theory'; book acquired high reputation in USA; visited India, 1888-9; member of Gladstone's last cabinet as chancellor of duchy of Lancaster; member of cabinet committee which prepared Irish home rule bill; president, Board of Trade, 1894; chairman of royal commission on secondary education, 1894-5, which laid foundations of new administrative structure of secondary schools; spent autumn in South Africa, 1895; published *Impressions of South Africa*, 1897; his attitude towards Boers gained him unpopularity during South African war, 1899-1902; chief secretary for Ireland, 1905-6; relied too much on views of under-secretary (Lord) MacDonnell [q.v.]; his most notable achievement there setting up of commission on Irish university education; ambassador at Washington, 1907-13; extremely popular with American people; exerted conciliatory influence in Canada; viscount, 1914; member of Hague Tribunal, 1914; presided over commission to consider alleged German excesses, Sept. 1914; devoted remainder of his life to forwarding project of League of Nations; chairman of conference on reform of House of Lords, 1918; published *Modern Democracies*, 1921; received degrees from thirty-one universities; FRS, 1893; FBA, 1902; OM, 1907.

BRYDON, JOHN McKEAN (1840-1901), architect; studied architecture in Liverpool and Italy; settled in London, 1868; built at Bath municipal buildings (1891-5), technical schools (1895-6), Victoria Art Gallery (1901), and pump-room extensions; also London School of Medicine for Women (1897-9) and other public and private buildings; designed local government board and education offices in Whitehall, 1898.

BUCHAN, ALEXANDER (1829-1907), meteorologist; MA, Edinburgh, 1848; at first schoolmaster; secretary of Scottish Meteorological Society, 1860; organized compilation of meteorological statistics in Scotland; inaugurated observatory on Ben Nevis, 1883; appointed member of Meteorological Council, 1887; librarian of Royal Society of Edinburgh, 1878-1906; published *The Handy Book of Meteorology* (1867), a recognized textbook; made valuable contributions to *Journal of Scottish Meteorological Society* and to Royal Society; FRS, 1898; hon. LLD, Glasgow, 1887.

BUCHAN, CHARLES MURRAY (1891-1960), footballer and journalist; played for Sunderland, 1911-25, as captain from 1920; received six England caps; captain of Arsenal, 1925-8, where he developed defence in depth and laid foundation of future success; leading sports writer, *Daily News*, later *News Chronicle*; founded *Charles Buchan's Football Monthly*, 1951.

BUCHAN, JOHN, first BARON TWEEDSMUIR (1875-1940), author, and governor-general of

Canada; son of Scottish Free Church minister; educated at Hutcheson's grammar school and the university, Glasgow, and Brasenose College Oxford; first class, *lit. hum.*, and president of the Union, 1899; called to bar (Middle Temple), 1901; bencher, 1935; earned living by journalism; assistant private secretary to Lord Milner [q.v.] in South Africa, 1901-3; literary adviser to T. A. Nelson, the publisher, 1907; major in Intelligence Corps, 1916; subordinate director, Ministry of Information, 1917-18; conservative MP, Scottish Universities, 1927-35; lord high commissioner to the General Assembly, Church of Scotland, 1933-4; governor-general of Canada, 1935-40; wrote throughout his life; tales of adventure include *Prester John* (1910), *The Thirty-Nine Steps* (1915), *Greenmantle* (1916), and *Mr. Standfast* (1919); *Huntingtower* (1922) opened a new series based on Glasgow memories; historical novels include *Midwinter* (1923), *The Blanket of the Dark* (1931); biographical studies include *Lord Minto* (1924), *Montrose* (1928), *Sir Walter Scott* (1932), *Oliver Cromwell* (1934), and *Augustus* (1937); wrote also *History of the Great War* (revised edn. 1921-2) and *History of the Royal Scots Fusiliers* (1925); CH, 1932; baron and GCMG, 1935; PC, 1937; GCVO, 1939; chancellor, Edinburgh University, 1937-40.

BUCHANAN, GEORGE (1827-1905), surgeon; MA, Glasgow, 1846; studied medicine at Andersonian University, Glasgow; MD, St. Andrews, 1849; surgeon in Crimea, 1856; professor of anatomy, Andersonian University, and surgeon to Glasgow Royal Infirmary from 1860; professor of clinical surgery, Western Infirmary, 1874-1900.

BUCHANAN, GEORGE (1890-1955), politician; born in Gorbals, Glasgow, of radical parentage; became engineers' pattern-maker; a member of independent labour party and active trade-unionist and street orator; labour MP, Gorbals, 1922-48; president, Patternmakers' Union, 1932-48; as joint undersecretary for Scotland, 1945-7, initiated slum clearances; minister of pensions, 1947-8; PC, 1948; first chairman, National Assistance Board, 1948-53, his life devoted to championship of the poorest.

BUCHANAN, SIR GEORGE CUNNINGHAM (1865-1940), civil engineer; chief engineer, Dundee Harbour Trust, 1896-1901, Rangoon Port Trust, 1901-15; administered port of Basra, 1915-17; on Indian Munitions Board, 1917-19; reported on transport problems, South Africa (1922), Australia (1925); knighted, 1915; KCIE, 1917.

BUCHANAN, SIR GEORGE SEATON (1869-1936), expert in public health; son of Sir George Buchanan [q.v.]; qualified from St. Bartholomew's, 1891; medical inspector, Local Government Board, 1895; chief inspector of foods, 1906-11; chief assistant medical officer, 1911-19; a senior medical officer, Ministry of Health, 1919-34; promoted co-operation in international health; knighted, 1922.

BUCHANAN, SIR GEORGE WILLIAM (1854-1924), diplomatist; born at Copenhagen; son of Sir Andrew Buchanan [q.v.]; entered diplomatic service, 1876; third secretary, Rome, 1878; second secretary, Tokyo, 1879; chargé d'affaires, Darmstadt, 1893; British agent, Venezuela boundary arbitration tribunal, Paris, 1898; agent and consul-general, Sofia, 1903; ambassador at St. Petersburg, 1910; advocated Anglo-Russian alliance, 1914; after outbreak of 1914-18 war directed his energies to obtaining maximum effort from Russia and later to combating pro-German influences and demands for separate peace; left Russia, Jan. 1918; advocated armed intervention in Russia; ambassador at Rome, 1919-21; KCMG, 1909; GCMG, 1913; GCB, 1915; PC, 1910.

BUCHANAN, JAMES, BARON WOOLAVINGTON (1849-1935), philanthropist and racehorse owner; made a fortune as a distiller of whisky; won St. Leger (1916) with Hurry On, Derby (1922) with Captain Cuttle, and both races and Eclipse Stakes (1926) with Coronach; benefactor of British Museum, Edinburgh University, Middlesex Hospital, etc.; baronet, 1920; baron, 1922; GCVO, 1931.

BUCHANAN, SIR JOHN SCOULAR (1883-1966), aeronautical engineer; educated at Allan Glen's School; marine apprentice, G. & J. Weir of Glasgow; exhibitioner, Royal Technical College, 1906; inspector of factories, Newcastle, 1908-14; technical officer attached to Royal Naval Air Service, 1914-18; posted to Royal Aircraft Establishment, Farnborough, 1918; joined Air Ministry Research and Development Department, 1919; presented papers to Royal Aeronautical Society on light aeroplane trials and on Schneider trophy seaplane race, 1924-6; pointed to need for RAF high speed flight; subsequent races in 1927-31 won by Britain; assistant director, research and development (aircraft), Air Ministry, 1930-40; transferred to Ministry of Aircraft Production, 1940; assistant chief executive, 1942-5; CBE, 1934; knighted, 1944; president, Royal Aeronautical Society, 1949-50; technical director, Short Bros. and Harland, 1945-8; chairman, London and South-Eastern Regional Board for Industry, 1949-60.

BUCHANAN, ROBERT WILLIAMS (1841-1901), poet and novelist; son of a socialist and secularist tailor who owned and edited several socialist journals in Glasgow from 1850; finished education at Glasgow University, where David Gray [q.v.] was a fellow student; went to London, 1860; wrote for *Athenæum* and other periodicals; under T. L. Peacock's influence produced his 'pseudo-classic poems' *Undertones* (1863); *London Poems* (1866) established his reputation; settled near Oban (1866-74), writing much narrative poetry and prose; chief prose work of that period was *Master Spirits* (1874); granted civil list pension, 1870; satirized Swinburne and others in 'The Session of the Poets' in *Spectator* (1866), and attacked the pre-Raphaelites—especially D. G. Rossetti—in article in *Contemporary Review*, 'The Fleshly School of Poetry' (1871), the prelude to a long and bitter controversy; won libel action against Swinburne (1875), and finally made amends to Rossetti (1881-2); settled in London, 1877; published long series of novels, the chief being *The Shadow of the Sword* (1876) and *God and the Man* (1881), both of which he dramatized; from 1880 to 1897 wrote and produced a long series of plays; published *Ballads of Life, Love, and Humour* (1882); visited America, 1884, producing his melodrama *Alone in London*; his *A Man's Shadow* produced at Haymarket, 1889-90; later dramatic successes were *The Charlatan* (1894) and *The Strange Adventures of Miss Brown* (1895); lost fortune in disastrous speculation; of combative temperament; had strong lyric gift, and abundant but ill-regulated force in fiction and drama; his collected *Poems* (3 vols.) appeared in 1874, and *Poetical Works* in 1884 and 1901.

BUCHANAN, WALTER JOHN (JACK) (1890-1957), actor and theatre manager: educated at Glasgow Academy; a singer and dancer of nonchalant grace he produced and played the lead in musical comedies, 1922-43, including *Sunny* (1926) and *Stand up and Sing* (1931); equally popular in London and New York; first appeared as straight comedy actor, 1944, in *The Last of Mrs. Cheyney* by F. Lonsdale [q.v.]; his films were often versions of his musical comedy successes; his extensive theatre interests included financing the building of Leicester Square Theatre.

BUCK, SIR PETER HENRY (1880-1951), ethnologist and politician; born in New Zealand of Irish father and Maori mother; received tribal upbringing; educated at Te Aute College and Otago Medical School; qualified, 1904; medical officer of health to Maoris, 1905-8; MP, 1909-14; with New Zealand expeditionary force, 1914-18; DSO; director of Maori hygiene, 1919-27; lecturer, Bishop Museum, Hawaii, 1927, director, 1936-51; appointed visiting professor of anthropology, Yale, 1932; made profound study of and published works on Polynesian culture of highest scientific value; KCMG, 1946; first Maori FRS, NZ.

BUCKLAND, WILLIAM WARWICK (1859-1946), legal scholar; first class, law (1884), Gonville and Caius College, Cambridge; fellow and called to bar (Inner Temple), 1889; lecturer, 1895; tutor, 1903; senior tutor, 1912-14; president, 1923-46; regius professor of civil law, 1914-45; publications include *The Roman Law of Slavery* (1908) and *Text-book of Roman Law from Augustus to Justinian* (1921); FBA, 1920.

BUCKLE, GEORGE EARLE (1854-1935), editor of *The Times* and man of letters; scholar of Winchester and New College, Oxford; first class, *lit. hum.*, 1876, history, 1877; fellow of All Souls, 1877-85; called to bar (Lincoln's Inn), 1880; assistant to editor of *The Times*, 1880; editor, 1884; continued traditional support of government of the day; administrative changes following change of proprietorship (1908) made his position difficult; resignation solicited, 1912; completed (in four volumes, 1914-20) *Life of Benjamin Disraeli* begun by W. F. Monypenny [q.v.] and the selection and editing (from 1862, in two series each of three volumes, 1926-32) of the *Letters of Queen Victoria*.

BUCKLEY, HENRY BURTON, first BARON WRENBURY (1845-1935), judge; educated at Merchant Taylors' School and Christ's College, Cambridge; ninth wrangler, 1868; fellow of Christ's, 1868-82; called to bar (Lincoln's Inn), 1869; bencher, 1891; QC, 1886; judge of Chancery division, 1900; lord justice of appeal, 1906-15; PC, 1906; baron, 1915; active in judicial and legislative work of House of Lords; author of classic *The Law and Practice under the Companies Acts* (1873; 11th edn. 1930); founded scholarship in political economy, Cambridge, 1904.

BUCKMASTER, STANLEY OWEN, first VISCOUNT BUCKMASTER (1861-1934), lord chancellor and statesman; educated at Aldenham School and Christ Church, Oxford; called to bar (Inner Temple), 1884; member of Lincoln's Inn, 1902; bencher, 1910; developed large Chancery practice; KC, 1902; counsel to Oxford University, 1910-13; liberal MP, Cambridge borough (1906-10), Keighley division (1911-15); solicitor-general (knighted), 1913-15; director, Press Bureau, 1914-15; lord chancellor (PC and baron), 1915-16; thereafter an appellate judge in House of Lords and judicial committee of Privy Council; a 'consummately equipped judge' and powerful orator on many subjects; chairman of governors, Imperial College of Science and Technology, 1923-34; GCVO, 1930; viscount, 1933.

BUCKTON, GEORGE BOWDLER (1818-1905), entomologist; crippled for life at age of five; student at Royal College of Chemistry, 1848-55; from 1865 made important researches in natural history; published *Monograph on British Aphides* (4 vols., 1876-83) and other entomological works; FRS, 1857.

BUDGE, SIR (ERNEST ALFRED THOMPSON) WALLIS (1857-1934), Assyriologist and Egyptologist; studied Hebrew and Syriac with Charles Seager and Samuel Birch [qq.v.]; sent to Cambridge under William Wright [q.v.] by Gladstone, 1878; scholar of Christ's College, 1881; assistant at British Museum, 1883; keeper, department of Egyptian and Syrian antiquities, 1894-1924; created efficient department; made many purchases and arranged speedy publication of texts with translations; deciphered the hieratic papyri in the museum including *Teaching of Amenemapt*, 1924; edited standard text of *Book of the Dead*, 1898; other works include *Egyptian Hieroglyphic Dictionary* (1920) and *Rise and Progress of Assyriology* (1925); knighted, 1920.

BULFIN, SIR EDWARD STANISLAUS (1862-1939), general; educated at Stonyhurst and Trinity College, Dublin; commissioned in Yorkshire regiment, 1884; captain, 1895; served on staff, South Africa, 1898-1901; commanded mobile column, 1901-2; major-general, 1914; commanded 'Bulfin's force' at first Ypres (1914), 28th division at second Ypres (1915), 60th division in France, Salonika, and Egypt (1915-17), and XXI Army Corps (1917-19) in Allenby's Palestine campaign; KCB, 1918; general, 1925.

BULLEN, ARTHUR HENRY (1857-1920), English scholar; son of George Bullen [q.v.]; educated at City of London School and Worcester College, Oxford; edited works of sixteenth- and seventeenth-century dramatists and song writers; carried on Shakespeare Head Press at Stratford-upon-Avon, 1904-20.

BULLER, ARTHUR HENRY REGINALD (1874-1944), botanist and mycologist; B.Sc. (London), Mason College, Birmingham, 1896; studied in Leipzig and Munich; assistant lecturer in botany, Birmingham, 1901-4; professor, university of Manitoba, 1904-36; published *Researches on Fungi* (7 vols., 1909-50) and *Essays on Wheat* (1919); FRS, 1929.

BULLER, Sir REDVERS HENRY (1839-1908), general; born at Downes, Crediton; educated at Eton; received commission in army, 1858; after service in Benares and China joined fourth battalion of the 60th (the King's Royal Rifle Corps) at Quebec, 1862; attracted notice of (Viscount) Wolseley [q.v.] in Red River expedition, 1870; chief intelligence officer to Wolseley in Ashanti, 1873; CB, 1874; commanded with success the frontier light horse in sixth Kaffir war in South Africa, 1878-9; received VC for gallant rescues, June 1879; aide-de camp to Queen Victoria, colonel, and CMG, 1879; chief of staff to Sir H. Evelyn Wood [q.v.] in South Africa, 1881; chief of intelligence staff to Wolseley in Egypt, 1882; present at Tel-el-Kebir; KCMG, 1882; commanded first infantry brigade at El Teb and Tamai under Sir Gerald Graham [q.v.], 1884; chief of staff in relief expedition of Khartoum, 1884; made masterly retreat from Gabat to Korti; KCB, 1885; was sent to Ireland (Aug. 1886) under Salisbury administration to restore order in Kerry; made under-secretary for Ireland and Irish privy counsellor; adjutant-general, 1890-7; a successful and economical administrator at War Office; reorganized and combined the supply and transport services; GCB, 1894, and general, 1896; in command at Aldershot, 1898; sent to South Africa in chief command, 1899; arrived in Cape Town (Oct.); moved to relief of Ladysmith; on arrival of Lord Roberts [q.v.], moved Natal army on Colenso (Dec.), where he was defeated; was defeated again at Spion Kop (Jan. 1900) and at Vaalkrantz; after much fighting at Hlangwane heights, he relieved Ladysmith (28 Feb.); his leadership was severely criticized; he entered Dundee (15 May); reached Volksrust (11 June); came in touch with main army (4 July); decisively defeated Boers at Bergendal (27 Aug.) and finally broke down the Boer resistance; marched north to Lydenburg and to Pretoria (10 Oct.); was thanked by Lord Roberts for his services; reached England (9 Nov.); received freedom of Southampton, Exeter, and Plymouth; GCMG, 1901; resumed command (Jan. 1901) of Aldershot division; the appointment was sharply criticized in press; Buller made indiscreet defence, and was removed from command, Oct. 1901; equestrian statue at Exeter with inscription 'He saved Natal'.

BULLER, Sir WALTER LAWRY (1838-1906), ornithologist; born in New Zealand; editor-in-chief of *Maori Messenger*, 1861; came to England as secretary to agent-general for New Zealand, 1871; called to bar, 1874; practised in New Zealand supreme court till 1886; wrote *History* (1873) and *Manual* (1882) of the birds of New Zealand; FRS, 1876; KCMG, 1886.

BULLOCH, WILLIAM (1868-1941), bacteriologist, pathologist, and medical historian; qualified at Aberdeen, 1890; studied pathology on the continent; bacteriologist to Lister Institute (1895), London Hospital (1897); first Goldsmiths' Company's professor of bacteriology, London University, 1919-34; FRS, 1913; researched in serology, immunology, and medical history and biography; interested in haemophilia; invented 'Bulloch jar' for cultivating anaerobic bacteria, 1900; published a *History of Bacteriology*, 1938.

BULWER, Sir EDWARD EARLE GASCOYNE (1829-1910), general; nephew of Lord Dalling [q.v.] and of Edward Bulwer-Lytton the novelist [q.v.]; joined 23rd Royal Welsh Fusiliers, 1849; went with regiment to Scutari, 1854; took part in crossing of the Alma and in attack on the Redan, 1854-5; served in relief of Lucknow, 1857-8; occupied fort of Gosainganj, and helped to restore tranquillity, Sept. 1858; victorious at Jabrowli and Purwa (Oct.) over superior forces; CB, 1859; thenceforth held only staff employment; as adjutant-general at headquarters for auxiliary forces (1873-9) helped to weld the regular and auxiliary forces under Lord Cardwell's short-service system; inspector-general of recruiting, 1880-6; KCB, 1886; general 1893; lieutenant-governor and commander of troops in Guernsey, 1889-94; colonel of the Royal Welsh Fusiliers, 1898; GCB, 1905.

BULWER-LYTTON, VICTOR ALEXANDER GEORGE ROBERT, second EARL OF LYTTON (1876-1947); son of first Earl of Lytton [q.v.]; succeeded father, 1891; educated at Eton and Trinity College, Cambridge; a passionate and practical idealist; his belief in free trade and devotion to cause of women's suffrage debarred him from orthodox success as a conservative; held junior appointments, 1916-22; governor of Bengal, 1922-7; chairman of League of Nations mission to Manchuria which condemned Japanese aggression, 1932; chairman, Council of Aliens, 1939-41; gave much time to the arts and social services; PC, 1919; GCIE, 1922; GCSI, 1925; KG, 1933.

BUNSEN, ERNEST DE (1819-1903), theologian; son of Christian, Baron von Bunsen, Prussian diplomatist; born in Rome; educated in Berlin; served in German army and in Prussian legation, 1837-49; settled in London, 1850; keen literary and biblical student; chief work was *Biblical Chronology*, 1874; an accomplished musician.

BUNSEN, Sir MAURICE WILLIAM ERNEST DE, baronet (1852-1932), diplomatist. [See DE BUNSEN.]

BUNTING, Sir PERCY WILLIAM (1836–1911), social reformer and editor of *Contemporary Review*; grandson of Jabez Bunting [q.v.]; BA, Pembroke College, Cambridge, 1859; called to bar, 1862; promoted forward movement in Methodism; helped to found Leys School, Cambridge, 1873, the West London Mission, 1887, and National Free Church Council, 1892; founded National Vigilance Association, 1883; edited on liberal lines *Contemporary Review* (1882–1911), giving social reform a prominent place; promoted moral purity and international amity.

BURBIDGE, EDWARD (1839–1903), liturgiologist; BA, Emmanuel College, Cambridge, 1862; MA, 1865; vicar of Backwell, Somerset, 1882–1902; became prebendary of Wells, 1887; published *Liturgies and Offices of the Church* (1885), a standard authority, besides devotional works.

BURBIDGE, FREDERICK WILLIAM (1847–1905), botanist; on staff of *Garden*, 1870–7; chief published work *Cultivated Plants* (1877) praised by Gladstone and Hooker; went to Borneo, 1877; brought back new plants; published chronicles of travel, 1880; the 'Burbidgea nitida' named after him; curator of botanical gardens, Trinity College, Dublin, 1880–1905; hon. MA, Dublin, 1889.

BURBURY, SAMUEL HAWKSLEY (1831–1911), mathematician; educated at Shrewsbury and St. John's College, Cambridge; won many university prizes for classics; fifteenth classic and second wrangler, 1854; fellow, 1854; wrote with Henry William Watson [q.v.] important works on electricity and magnetism; FRS, 1890.

BURDETT-COUTTS, ANGELA GEORGINA, Baroness Burdett-Coutts (1814–1906), philanthropist; daughter of Sir Francis Burdett (1770–1844, q.v.), and grand-daughter through her mother of Thomas Coutts [q.v.], banker; at her father's town house in St. James's Place she met leading politicians, scientists, and literary men; inherited from the Duchess of St. Albans, second wife of Thomas Coutts, his property and share in bank, 1837; assumed additional surname of Coutts; removed to 1 Stratton Street, 1837; 'the richest heiress in all England'; entertained English and foreign celebrities during sixty years; intimate with royal family, Duke of Wellington, Sir Robert Peel, Disraeli, Gladstone, Duke of Cambridge, Napoleon III, Empress Eugénie, Tom Moore, Samuel Rogers, Dickens, Sir William and Joseph Hooker, Sir Henry Irving, and many others; took active part in banking business; personally administered her private charities; interested in schemes of social reform; munificent benefactor to Church of England; built and endowed St. Stephen's, Westminster (1847), adding schools in 1849, and other London churches; established Westminster Technical Institute, 1893; endowed (1847) bishoprics of Cape Town and Adelaide; founded (1857) bishopric of British Columbia, with large endowments; introduced sewing and cookery into elementary schools; endowed two geological scholarships at Oxford, 1861; presented Schimper's herbarium of mosses to Kew; subsidized Ragged School Union; supported shoeblack brigades from 1851, and aided in subsidizing training ships for poor boys from 1874; helped to found National Society for the Prevention of Cruelty to Children, 1884; active in reform of east London women's industries; started (1860) 'sewing school' for women in Spitalfields, providing food, housing, and medical attendance; helped Spitalfields weavers sending many to the colonies, 1863–9; instituted Flower Girls' Brigade, 1879; founded Burdett-Coutts working youths' club in Shoreditch, 1875; adding gymnasium, 1891; befriended costermongers at Bethnal Green, providing stables for their donkeys; intense lover of animals; leader of Royal Society for Prevention of Cruelty to Animals; instituted scheme of prize essays; encouraged goat breeding; pioneer of model tenements in Bethnal Green (1862), and of garden city aims in 'Holly Village' on her Holly Lodge estate at Highgate; started fish and vegetable market scheme for east London, 1864; but owing to opposition of existing vested interests the Columbia market (opened 1869) was a failure, although it encouraged improvement in methods of food distribution; sought to improve conditions and industry among Irish poor; sent Irish emigrants to Canada from 1863 onwards; revived fishing industry at Baltimore, co. Cork, where she inaugurated a fishery training school, 1887; aided colonial and missionary effort in Borneo and Africa; friend and supporter of Sir H. M. Stanley [q.v.]; stimulated cotton industry of South Nigeria; raised to peerage, 1871; received freedom of cities of London (1872) and Edinburgh (1874), and of several City companies; liberally helped Turkish peasantry in Russo-Turkish war, 1877; received order of Medjidie, 1878; sent out hospital equipment for wounded in Zulu war, 1879; married William Lehman Ashmead-Bartlett, 1881; entertained General Gordon [q.v.] before he went to Sudan, 1884; compiled 'Woman's Work in England' for Chicago exhibition, 1893; buried in Westminster Abbey.

BURDON, JOHN SHAW (1826–1907), missionary bishop of Victoria, Hong Kong (1874–97), and Chinese scholar; joined Church

Missionary Society, 1850; at Shanghai (1853), Hangchow (1859), and Shaohsing (1860-1); in Ningpo when city was captured by Chinese rebels, December 1861; went as missionary pioneer to Peking, 1862; chaplain to British legation there, 1865-72; one of translators of New Testament and Prayer Book into Chinese, 1872.

BURDON-SANDERSON, SIR JOHN SCOTT, baronet (1828-1905), regius professor of medicine at Oxford; studied medicine at Edinburgh University (MD., 1851) and in Paris; medical officer of health for Paddington, 1856-67; FRS, 1867; Croonian lecturer (1867, 1877, and 1889); appointed Jodrell professor of physiology at University College, London, 1874; fellow (1871), Harveian orator (1878), Baly medallist (1880), and Croonian lecturer (1891) of the Royal College of Physicians; first Waynflete professor of physiology (and fellow of Magdalen College, Oxford), 1882-95; regius professor of medicine at Oxford, 1895-1903; made important experimental researches into physiology, and contributed to current progress of pathology; president of British Association, 1893; member of royal commissions on hospitals for infectious diseases (1883), on consumption of tuberculous meat and milk (1891), and on university of London (1892); baronet, 1899; hon. LLD, Edinburgh, and D.Sc., Dublin; wrote much in scientific journals; memoir with selected papers and addresses edited by his widow, 1911.

BURGE, HUBERT MURRAY (1862-1925), headmaster of Winchester College, bishop of Southwark, and afterwards of Oxford; born at Kingston, Jamaica; BA, University College, Oxford, 1886; fellow and tutor, 1890-1900; headmaster, Repton College, 1900-1; headmaster, Winchester College, where he brought the curriculum more into line with modern views, 1901-11; bishop of Southwark, 1911-19; bishop of Oxford, 1919-25; honorary fellow, University College, 1907; select preacher at Oxford, 1899-1902 and 1920-1; a trusted counsellor in church affairs.

BURGH CANNING, HUBERT GEORGE DE, second MARQUESS and fifteenth EARL OF CLANRICARDE (1832-1916), Irish landed proprietor; liberal MP, co. Galway, 1867-71; succeeded father, 1874; life spent resisting movement to limit Irish landlord's power; vast estates compulsorily transferred to Congested Districts Board, 1915.

BURKITT, FRANCIS CRAWFORD (1864-1935), professor of divinity; educated at Harrow and Trinity College, Cambridge; a wrangler, 1886; first class, theology, 1888; university lecturer in palaeography, 1903; Norrisian professor of divinity, 1905-35; publications include 'Text and Versions' in *Encyclopaedia Biblica* (1903), indispensable edition of the old Syriac Gospels (1904), *The Gospel History and its Transmission* (1906), *Eucharist and Sacrifice* (1921), liturgical and Franciscan studies, and works on Manichaeism and Gnosticism; champion of eschatological interpretation of aims and teaching of Jesus; practising member of Church of England of modernist school; FBA, 1905.

BURN, ROBERT (1829-1904), scholar and archaeologist; educated at Shrewsbury and Trinity College, Cambridge; senior classic, 1852; fellow of Trinity, 1854-1904; author of *Rome and the Campagna,* 1871.

BURN-MURDOCH, JOHN (1852-1909), lieutenant-colonel; entered Royal Engineers, 1872; served in Afghan war, 1878-80; wounded at storming of Asmai Heights, 1879; field engineer in Egyptian war, 1882; at battle of Tel-el-Kebir; prominent in seizure of Zagazig, Sept. 1882; officer commanding engineer of Indian state railways, 1893; lieutenant-colonel, 1900.

BURNAND, SIR FRANCIS COWLEY (1836-1917), playwright, author, and editor of *Punch*; educated at Eton and Trinity College, Cambridge; founded Cambridge Amateur Dramatic Club; became Roman Catholic, 1858; joined staff of *Punch*, 1862; editor, 1880-1906; increased the paper's reputation; humorist; author of burlesques and adaptations of popular French farces; knighted, 1902.

BURNE, SIR OWEN TUDOR (1837-1909), major-general; joined 20th East Devonshire regiment, 1855; ordered to India (1857), studying Hindustani on voyage; active under (Sir) T. H. Franks [q.v.] in events leading to relief of Lucknow, 1858; military secretary to Sir Hugh Rose (later Lord Strathnairn, q.v.), commander-in-chief in India, 1860; returned to England, 1865; aided in suppression of Fenian conspiracy, 1867; private secretary to Lord Mayo [q.v.], viceroy of India, 1868-72; CSI, 1872; political aide-de-camp to the Duke of Argyll, secretary of state for India, 1872; head of political and secret department of India Office in London, 1874; private secretary to Lord Lytton [q.v.], Indian viceroy, 1876-8; CIE, 1878; KCSI, 1879; member of council of India, 1886-96; major-general, 1889; GCIE, 1896; wrote in periodicals on Eastern questions; published *Memories,* 1907.

BURNELL, CHARLES DESBOROUGH (1876-1969), oarsman; educated at Eton and

Magdalen College, Oxford; member of four consecutive crews in university boat races, 1895-8; won Grand Challenge Cup each year, 1898-1901; member, Leander crew which won gold medal in Olympic Regatta, 1908; served with London Rifle Brigade on western front, 1914-19; DSO, 1919; steward, Royal Regatta, 1919; umpire at Henley for over forty years; umpired university boat race, 1927-30; president, Leander, 1954-7, and Henley Rowing Club up to 1969; JP, 1934; deputy lieutenant, Berkshire, 1936; OBE, 1954; senior partner in Wise and Burnell, stockbrokers.

BURNET, JOHN (1863-1928), classical scholar; educated at Edinburgh University and Balliol College, Oxford; first classes in classical moderations (1884), and *lit. hum.* (1887); prize fellow of Merton College, Oxford, 1889; professor of Greek, St. Andrews, 1891-1926; an inspiring and successful teacher, able administrator, and enthusiastic educationist; his greatest contribution to scholarship his critical edition in 'Oxford Classical Texts' series of whole text of Plato's works, which superseded all previous editions, 1900-13; other works include an edition of *Nichomachean Ethics* of Aristotle (1899), commentaries on the *Phaedo* (1911), *Euthyphro, Apology,* and *Crito* (1924), *Early Greek Philosophy* (1892), and *Greek Philosophy, Part I, Thales to Plato* (1914).

BURNET, Sir JOHN JAMES (1857-1938), architect; educated at Western Academy, Glasgow and École des Beaux-Arts, Paris; practised with father in Glasgow; moved to London, 1905; works include (Glasgow): Athenaeum, Barony church, and cenotaph; (London): Kodak building, Unilever House, King Edward VII galleries for British Museum; knighted, 1914; FRIBA, 1897; ARA, 1921; RA, 1925.

BURNETT, Sir CHARLES STUART (1882-1945), air chief marshal; in Royal Flying Corps (1914-19) commanded successively No. 12 Squadron, the Fifth Wing, and the Palestine Brigade; wing commander, Royal Air Force, 1919; air officer commanding, Iraq, 1932-5; air marshal, 1936; commander-in-chief, Training Command, 1936-9; inspector-general of Royal Air Force, 1939-40; chief of air staff, Royal Australian Air Force, 1940-2; air chief marshal, 1945; KCB, 1936.

BURNETT, Dame IVY COMPTON- (1884-1969), novelist. [See COMPTON-BURNETT.]

BURNETT, Sir ROBERT LINDSAY (1887-1959), admiral; brother of Sir Charles Burnett [q.v.]; educated at Bedford School; entered *Britannia*, 1903; qualified as physical train-

ing instructor; captain, 1930; commodore, Chatham barracks, 1939-40; rear-admiral, 1940; flag officer, Home Fleet minelaying squadron (1940-2), destroyer flotillas (1942-3), tenth cruiser squadron (1943-4); vice-admiral and DSO, 1943; a determined escorter of convoys to Russia; commander-in-chief, South Atlantic, 1944-6, Plymouth, 1947-50; admiral, 1946; chairman, White Fish Authority, 1950-4; OBE, 1925; CB, 1942; KBE, 1944; KCB, 1945; GBE, 1950; hon. LLD Aberdeen.

BURNETT-STUART, Sir JOHN THEODOSIUS (1875-1958), general; educated at Repton and Sandhurst; joined Rifle Brigade, 1895; served in India and South African war; DSO; in France, 1914-18; major-general and deputy adjutant-general, 1917-18; commanded Madras district, 1920-2; director, military operations and intelligence, War Office, 1922-6; commanded 3rd division, Southern command, 1926-30, with higher direction of Experimental Mechanized Force; GOC troops in Egypt, 1931-4; general, 1934; GOC-in-C Southern command, 1934-8; CB, 1917; KCB, 1932; GCB, 1937; CMG, 1916; KBE, 1923.

BURNEY, Sir (CHARLES) DENNISTOUN, second baronet (1888-1968), naval inventor; only son of (Admiral Sir) Cecil Burney [q.v.]; entered *Britannia*, 1903; joined *Exmouth* as midshipman, 1905; joined destroyer *Afridi*, 1909, and *Crusader*, used for experimental anti-submarine work; interested in anti-submarine work including the use of seaplanes; carried out research at Bristol aviation works, 1911-14; commanded destroyer *Velox*, 1914; joined the *Vernon*, Portsmouth Torpedo School; responsible for development of explosive paravane; took out patents for use by merchant ships; CMG, 1917; retired from navy as lieutenant-commander, 1920; succeeded to baronetcy, 1929; joined Vickers Ltd. as consultant; published *The World, the Air, and the Future* (1929); conservative MP Uxbridge, 1922-9; formed Airship Guarantee Company; appointed (Sir) Barnes Wallis as chief designer, 1923; airship development halted by R. 101 disaster, 1930; employed during 1939-45 war on secret experimental work; patents taken out between 1915 and 1962 included aerial gliding bombs, gun-fired rockets, and sonar apparatus for detecting fish shoals.

BURNEY, Sir CECIL, first baronet (1858-1926), admiral of the fleet; entered navy, 1871; flag captain, home fleet, 1902-4; inspecting captain of all boys' training ships, 1905-9; rear-admiral, 1909; commanded fifth cruiser squadron, 1911; Atlantic squadron, 1911-12; commanded international naval force at

Antivari on Montenegrin coast and international force occupying Scutari, 1913; commanded second and third fleets, 1913; went to first battle squadron of Grand Fleet, being second-in-command under Jellicoe [q.v.], 1914; present at battle of Jutland, 1916; admiral, 1916; second sea lord at Admiralty, 1916; commander-in-chief, coast of Scotland, 1917; at Portsmouth, 1919-20; admiral of the fleet, 1920; KCB, 1913; baronet, 1921.

BURNHAM, first BARON (1833-1916), newspaper proprietor. [See LEVY-LAWSON, EDWARD.]

BURNHAM, fourth BARON (1890-1963), newspaper proprietor and soldier. [See LAWSON, EDWARD FREDERICK.]

BURNHAM, first VISCOUNT (1862-1933), newspaper proprietor. [See LAWSON, HARRY LAWSON WEBSTER LEVY-.]

BURNS, DAWSON (1828-1909), temperance reformer; became secretary of National Temperance Society, 1846; Baptist pastor at Salford, 1851; helped to found United Kingdom Alliance, 1853; wrote for *Alliance News* from 1856; actively promoted temperance legislation; contributed annual letters to *The Times* on 'national drink bill' (1886-1909); a director of Liberator Building Society, which failed in 1892; wrote books on temperance subjects.

BURNS, JOHN ELLIOT (1858-1943), labour leader and politician; apprenticed as an engineer in London; keen cricketer and precocious politician and trade-unionist; acquired socialist and radical ideas; tramped continent studying labour conditions; joined Social Democratic Federation, 1884; represented it at industrial remuneration conference, 1885; left it (1889) and founded Battersea Labour League henceforth the main basis of his political work; became well known as a speaker in London and a notable orator for Metropolitian Radical Federation; twice arrested and in 1887 sentenced to six weeks' imprisonment; prominent leader in dock strike, 1889; for some years a delegate to Trades Union Congress but dropped out of trade-union and socialist movements; represented Battersea on London County Council, 1889-1907, and in parliament, 1892-1918; refusing to join Independent Labour Party became increasingly associated with the liberals; president of Local Government Board, 1905-14; first artisan to reach cabinet rank; aroused socialist hostility by action over the Poplar board of guardians, 1906, and by opposition to poor law reforms recommended by the royal commission, 1909; president of Board of Trade, 1914; believed that war could have been averted, resigned, and took no further part in public life; his collection of books on old London presented to the County Council by Lord Southwood [q.v.], 1943; PC, 1905.

BURNSIDE, WILLIAM (1852-1927), mathematician; educated at St. John's and Pembroke colleges, Cambridge; bracketed second wrangler, 1875; first Smith's prizeman, fellow, and lecturer, Pembroke, 1875; professor of mathematics, Royal Naval College, Greenwich, 1885-1919; concerned with ballistics for gunnery and torpedo officers, mechanics and heat for engineer officers, and dynamics for naval constructors; FRS, 1893; president, London Mathematical Society, 1906-8; honorary fellow, Pembroke, 1900; his publications include *Theory of Groups* (1897) and over 150 papers.

BURRELL, SIR WILLIAM (1861-1958), art collector; sold family shipping firm, 1917; began collecting as a boy; acquired (1916) and lavishly furnished Hutton Castle; between 1911 and 1944 spent average of £20,000 annually on acquiring works of art; presented collection of some 8,000 objects to city of Glasgow, 1944, adding £450,000 to provide new museum; most interested in later Middle Ages and Renaissance art, including tapestries, stained and painted glass; also Chinese pottery and porcelain; a trustee of the Tate and the National Gallery of Scotland; knighted, 1927.

BURROUGHS (afterwards TRAILL-BURROUGHS), SIR FREDERICK WILLIAM (1831-1905), lieutenant-general; joined 93rd Highlanders, 1848; served under Lord Clyde [q.v.] in Crimea, 1854-5; at battles of the Alma and Balaclava, and at siege of Sevastopol; recommended for VC for services at the Secunderabagh during second relief of Lucknow in Indian Mutiny; brevet major, 1858; took part in North-West frontier expedition, 1863; commanded 93rd Highlanders at Ambela; lieutenant-general, 1881; CB, 1873; KCB, 1904.

BURROWS, CHRISTINE MARY ELIZABETH (1872-1959), principal successively of St. Hilda's and the Oxford Home-Students (later St. Anne's); educated at Cheltenham Ladies' College and Lady Margaret Hall and St. Hilda's Hall, Oxford; tutor, St. Hilda's, 1894; vice-principal, 1896; succeeded mother as principal, 1910-19; principal, Oxford Home-Students, 1921-9; a pioneer at St. Hilda's and an experienced administrator of the future St. Anne's; set a high standard for her students.

BURROWS, MONTAGU (1819-1905), Chichele professor of modern history at Oxford; entered Royal Naval College, 1832; at bombardment of Acre, 1840; commander, RN, 1852; went to Magdalen Hall, Oxford, 1853; took first classes in classics and modern history, 1856-7; retired from navy, 1862; joined party of moderate churchmen; original member of English Church Union; helped to found Keble College, 1870; Chichele professor of modern history, 1862-1905; fellow of All Souls, 1865; chief works were *The Worthies of All Souls* (1874) and *The Cinque Ports* (1888); edited vols. ii and iii of *Collectanea* for Oxford Historical Society; contributed to this Dictionary.

BURT, THOMAS (1837-1922), trade-unionist and liberal politician; began pitwork, 1847; early took part in trade-unionism; general secretary, Northumberland Miners' Mutual Confidence Association, 1865-1913; liberal MP, Morpeth, 1874-1918; his name associated with reforms such as Employers' Liability Act (1880), factory and workshop legislation, and improved Mines Acts; secretary to Board of Trade, 1892-5; PC, 1906.

BURTON, first BARON (1837-1909), brewer. [See BASS, MICHAEL ARTHUR.]

BURTON, SIR MONTAGUE MAURICE (1885-1952), multiple tailor; born in Lithuania of Jewish parentage; came alone to England, 1900; set up as general outfitter in Chesterfield, 1903; had five men's tailor shops with headquarters in Sheffield and manufacturing in Leeds by 1913; had four hundred shops, and factories and mills, by 1929, when the company went public; made a quarter of the British uniforms in 1939-45 war and a third of demobilization clothing; vastly improved working conditions in the trade; favoured organized labour and collective bargaining; endowed chairs in industrial relations, Leeds and Cardiff (1929), Cambridge (1930); of international relations, Jerusalem (1929), Oxford (1930), Edinburgh (1948), and London (1936); knighted, 1931.

BURY, JOHN BAGNELL (1861-1927), classical scholar and historian; double first class (classics and philosophy), Trinity College, Dublin, 1882; fellow of Trinity, 1885; Erasmus Smith professor of modern history, 1893; regius professor of Greek, 1898; of modern history at Cambridge, 1902; fellow of King's College, Cambridge, 1903; honorary fellow of Oriel College, Oxford; FBA; recipient of many honorary degrees; between 1881 and 1892 turned from philology to history, particularly that of later Roman Empire; works include *The Nemean Odes of Pindar* (1890), *The Isthmian Odes of Pindar* (1892), *History of the Later Roman Empire from Arcadius to Irene* (1889), *History of the Roman Empire . . . to the death of Marcus Aurelius* (1893), a seven-volume edition of Gibbon's *Decline and Fall* (1896-1900), *History of Greece to the death of Alexander the Great* (1900), *Life of St. Patrick* (1905), a series of papers on public law and administration in later Roman Empire (1906-11), *History of the Eastern Roman Empire from the fall of Irene to the accession of Basil I* (1912), *History of Freedom of Thought* (1914), *Idea of Progress* (1920), *History of the Later Roman Empire from the death of Theodosius I to the death of Justinian* (1923); an editor of *Cambridge Ancient History*; an objective historian, who emphasized unity and continuity of European history.

BUSHELL, STEPHEN WOOTTON (1844-1908), physician and Chinese archaeologist; studied at Guy's Hospital; house-surgeon, 1866; MB, 1866; physician to British legation at Peking, 1867-1900; CMG, 1897; authority on and collector of Chinese porcelain and pottery; chief works were *Oriental Ceramic Art* (10 vols., 1897) and *Chinese Art* (2 vols., 1904); translated native works on Chinese pottery and porcelain.

BUSK, RACHEL HARRIETTE (1831-1907), writer on folklore; daughter of Hans Busk the elder [q.v.]; joined Church of Rome, 1858; lived in Rome from 1862; wrote much on Roman politics; published folk-tales of Spain (1870), Austria (1874), and Italy (1887).

BUTCHER, SAMUEL HENRY (1850-1910), scholar and man of letters; son of Samuel Butcher [q.v.]; educated at Marlborough and Trinity College, Cambridge; senior classic and chancellor's medallist, 1873; member of 'The Apostles'; fellow of Trinity, 1874-5; migrated on marriage (1876) to University College, Oxford, where he was a successful tutor; published with Andrew Lang [q.v.] translation of *Odyssey* (1879); appointed professor of Greek in Edinburgh University, 1882; member of Scottish universities commission, 1889-1900; published *Some Aspects of the Greek Genius* (1891) and *Aristotle's Theory of Poetry and Fine Art* (1895); helped to organize unionist party in Edinburgh, 1886; resigned professorship after wife's death, and removed to London, 1903; lectured at Harvard University, 1904; member of two royal commissions on university education in Ireland, 1901-3 and 1906-7; succeeded Sir Richard Jebb [q.v.] as unionist MP for Cambridge University, 1906; spoke frequently and effectively on educational and Irish questions; part founder (1903) and president (1907) of Classical Association of England; president of British Academy, 1909; appointed trustee of British Museum, 1908; edited Demosthenes' speeches, 2 vols. 1903-7.

BUTLER, ARTHUR GRAY (1831-1909), headmaster of Haileybury; son of Dr George Butler [q.v.]; educated at Rugby and University College, Oxford; president of Union, 1853; Ireland scholar and first class, classics, 1853; fellow of Oriel, 1856; master at Rugby, 1858-62; first headmaster of reconstituted Haileybury College, 1862-7; showed great organizing capacity; dean and tutor of Oriel, 1875-97; honorary fellow, 1907; wrote dramas and verse.

BUTLER, ARTHUR JOHN (1844-1910), Italian scholar; son of William John Butler [q.v.]; educated at Eton and Trinity College, Cambridge; Bell university scholar (1864) and eighth classic (1867); fellow of Trinity, 1869; examiner under Council on Education, 1870-87; became partner in Rivington's publishing firm, 1887; edited *Calendars of Foreign State Papers*, 1899-1910; professor of Italian at University College, London, from 1898 till death; translated Dante's *Purgatory* (1880), *Paradise* (1885), and *Hell* (1892); translated many French and German works; ardent alpinist from boyhood; had unparalleled knowledge of Oetzthal Alps; member of Alpine Club, 1886; edited *Alpine Journal*, 1890-3.

BUTLER, EDWARD JOSEPH ALOYSIUS (in religion DOM CUTHBERT) (1858-1934), Benedictine abbot and scholar; educated at Downside; novice, 1876; priest, 1884; headmaster, Downside, 1888-92; superior, Benet House, Cambridge, 1896-1904; abbot of Down-side, 1906-22; judicious scholar; works include *Benedictine Monachism* (1919) and *Western Mysticism* (1922).

BUTLER, ELIZABETH SOUTHERDEN, LADY (1846-1933), painter; born THOMPSON; sister of Alice Meynell [q.v.]; married (1877) (Sir) W. F. Butler [q.v.]; 'The Roll Call' (Academy, 1874) instantly popular and purchased by Queen Victoria; thereafter painted mainly military subjects with precision and masterly draughtsmanship.

BUTLER, FRANK HEDGES (1855-1928), balloonist and pioneer of flying; made his first balloon ascent, 1901; suggested formation of Aero Club, which was registered 1901; completed one hundred balloon ascents, including some records, by 1907; inspired Aero Club to foster early development of flying in England; a keen traveller; FRGS, 1877.

BUTLER, SIR (GEORGE) GEOFFREY (GILBERT) (1887-1929), historian; grandson

of George Butler (1774-1853, q.v.); MA, Trinity College, Cambridge; fellow of Corpus Christi College, Cambridge, 1910; entered news department of Foreign Office, 1915; director, British bureau of information in United States at New York, 1917-19; returned to Corpus; active as librarian and teacher of diplomatic history; promoted conservative principles among Cambridge undergraduates; burgess, Cambridge University, 1923-9; on royal commission on government of Ceylon, 1927; KBE, 1919.

BUTLER, SIR HAROLD BERESFORD (1883-1951), public servant; scholar of Eton and Balliol College, Oxford; first class, *lit. hum.* and fellow of All Souls, 1905; joined Local Government Board, 1907, Home Office, 1908; secretary, Foreign Trade Department, 1916; at Ministry of Labour, 1917; secretary-general, first International Labour Conference, 1919; deputy director International Labour Office, 1920-32; director, 1932-8; first warden, Nuffield College, Oxford, 1939-43; southern regional commissioner for civil defence, 1939-42; minister in charge of British Information Service, Washington, 1942-6; CB, 1919; KCMG, 1946.

BUTLER, HENRY MONTAGU (1833-1918), headmaster of Harrow School, dean of Gloucester, master of Trinity College, Cambridge; son of Dr George Butler [q.v.]; educated at Harrow and Trinity College, Cambridge; senior classic and fellow of Trinity, 1855; headmaster of Harrow, 1860-85; reconciled inspirations of past to aspirations of present; dean of Gloucester, 1885-6; master of Trinity, 1886 till death; publications include volumes of school and university sermons.

BUTLER, JOSEPHINE ELIZABETH (1828-1906), social reformer; daughter of John Grey of Dilston [q.v.]; married George Butler [q.v.]; supported early stages of movement for women's higher education; settling in Liverpool (1866), established homes for work-girls and fallen women, whose cause she fought as secretary to Ladies' National Association for Repeal of Contagious Diseases Act, 1869-85; contributed to repeal of Act, 1886; caused reform of law affecting 'white slave traffic' in continental countries; wrote numerous pamphlets and volumes, including memoirs of her father (1896), her husband (1892), and sister, Madame Meuricoffre (1901), *Personal Reminiscences of a Great Crusade* (1896), and *Life of St. Catherine of Siena* (1898).

BUTLER, SIR MONTAGU SHERARD DAWES (1873-1952), Indian administrator and master of Pembroke College, Cambridge; brother of Sir S. H. and Sir G. G. Butler [qq.v.]; educated at Haileybury and Pembroke College, Cambridge; first class, parts i and ii, classical tripos, 1894-5; president of Union, and fellow, 1895; honorary fellow, 1925; entered Indian civil service, 1896, and posted to Punjab; president, legislative council, 1921; secretary, government of India department for education, health, and lands, 1922; president, Council of State, 1924; governor, Central Provinces, 1925-33; lieutenant-governor, Isle of Man, 1933-7; master of Pembroke, 1937-48; CIE, 1909; CVO, 1911; CB, 1916; CBE, 1919; knighted, 1924; KCSI, 1924.

BUTLER, Sir RICHARD HARTE KEATINGE (1870-1935), lieutenant-general; educated at Harrow and Sandhurst; commissioned in Dorsetshire regiment, 1890; captain, 1894; fought with distinction in South African war and in France (1914-15); deputy chief of staff to (Lord) Haig [q.v.], 1915-18; commanded III Corps in final offensive, 1918; KCMG, 1918; KCB, 1919; lieutenant-general, 1923; GOC-in-C, Western command, 1924-8.

BUTLER, SAMUEL (1835-1902), philosophical writer; educated at Shrewsbury and St. John's College, Cambridge; twelfth classic, 1858; abandoned intention of taking holy orders; emigrated to New Zealand, 1859; his success as sheepbreeder detailed in *A First Year in Canterbury Settlement* (1863); returned to England, 1864-5; studied painting at Heatherley's school; exhibited at Royal Academy; published anonymously *Erewhon*, a *jeu d'esprit* (1872), and *The Fair Haven* (1873), an ironic defence of Christian evidences; lost money in unsound investments; wrote *Life and Habit* (1877), in which he contested Darwin's law of natural selection; in *Evolution Old and New* (1879), in *Unconscious Memory* (1880), and in *Luck or Cunning* (1886), he pursued his attack on the Darwinian banishment of mind from the universe; an original topographer of Italian Switzerland and Italian art critic; published *Alps and Sanctuaries of Piedmont and the Canton Ticino* (1881) and kindred works; studied and composed music in London with H. Festing Jones; a keen Homeric student; published *The Authoress of the Odyssey* (1897) and prose translations of *Iliad* (1898) and *Odyssey* (1900); other works were *Life of Samuel Butler, Bishop of Lichfield* (2 vols., 1896), *Shakespeare's Sonnets Reconsidered* (1899), and *Erewhon Revisited* (1901); an autobiographical novel, *The Way of All Flesh* (1903), and *Essays on Life, Art and Science* appeared posthumously; a terse and lucid iconoclast.

BUTLER, Sir (SPENCER) HARCOURT (1869-1938), Indian administrator; brother of Sir G. G. G. Butler [q.v.]; educated at Harrow under uncle, H. M. Butler [q.v.], and as ICS probationer at Balliol College, Oxford; posted to North-West Provinces, 1890; secretary, famine commission, 1901; secretary, foreign department, 1907-10; in charge of education, 1910-15; lieutenant-governor, Burma (1915-18), United Provinces (1918-21); as governor introduced dyarchical system in United Provinces, 1921-3, and Burma, 1923-7; chairman, Indian States committee, 1927-9; KCSI, 1911; GCSI, 1928.

BUTLER, Sir WILLIAM FRANCIS (1838-1910), lieutenant-general and author; descended from tenth Earl of Ormonde [q.v.]; of Roman Catholic family; obtained commission in 69th Foot, 1858; in India, 1860; in Channel Islands, 1866, where he met Victor Hugo; look-out officer on frontier in Canada, 1868; sent on mission to Red River settlement and to Saskatchewan, 1870; published history of 69th Foot (1870), *The Great Lone Land* (1872), and *The Wild North Land* a vivid description of his Canadian experiences (1873); joined Sir Garnet Wolseley's expedition to Ashanti, 1873; failed in attempt to reach Kumasi, 1873-4; major and CB, 1874; went on special service in Natal, 1875; engaged on duty in England, 1875-9; married Elizabeth Thompson, the artist, 1877; served in Zulu war, 1879, in Egyptian war, including Tel-el-Kebir, 1882; was charged with provision of boats on Nile for relief of General Gordon [q.v.], 1884; prominent in victory at Kirbekan, Feb. 1885; brigadier-general under Sir F. C. A. Stephenson [q.v.] at Giniss, 1885; KCB, 1886; commanded garrison of Alexandria, 1890; in command of a brigade at Aldershot, 1893; transferred to SE district, 1896; commanded troops in South Africa, Oct. 1898; a strong pro-Boer, he was sceptical of grievances of uitlanders; resigned, Aug. 1899; in England assumed command of Western district (1899-1905); lieutenant-general, 1900; GCB, 1906; Irish privy counsellor, 1909; wrote lives of Gordon (1889), Sir Charles Napier (1890), and Sir George Colley (1899); his *Autobiography* published posthumously, 1911.

BUTLIN, Sir HENRY TRENTHAM, first baronet (1845-1912), surgeon; educated at St. Bartholomew's Hospital, London; full surgeon, 1892, and lecturer in surgery at the Hospital school, 1897; president of British Medical Association, 1910-11, of Royal College of Surgeons, 1909-11; baronet, 1911; wrote *Diseases of the Tongue* (1885).

BUTT, DAME CLARA ELLEN (1872-1936), singer; educated at South Bristol high school and scholar of Royal College of Music; in 1892 began a career of unexampled popularity on concert platform; of splendid appearance, she possessed a magnificent contralto voice remarkable for its broad effect; 'Sea Pictures' and other music was composed for her by Sir Edward Elgar [q.v.]; married (1900) Robert Kennerley Rumford, baritone; DBE, 1920.

BUTTERWORTH, GEORGE SAINTON KAYE (1885-1916), composer; BA, Trinity College, Oxford; researched in English folk-music; set parts of A. E. Housman's *A Shropshire Lad* to music; his masterpiece, a rhapsody, produced 1913; killed in action in France.

BUXTON, NOEL EDWARD NOEL-, first BARON NOEL-BUXTON (1869-1948), politician and philanthropist. [See NOEL-BUXTON.]

BUXTON, PATRICK ALFRED (1892-1955), medical entomologist; educated at Rugby and Trinity College, Cambridge; first class, parts i and ii, natural sciences tripos, 1914-15; qualified at St. George's Hospital, 1917; entomologist to government of Palestine, 1921-3; led research expedition on filariasis to Samoa, 1923-6; head of entomology department, London School of Hygiene and Tropical Medicine, 1926-55; played influential role in insecticide research; works include *Animal Life in Deserts* (1923) and *Natural History of Tsetse Flies* (1955); FRS, 1943; CMG, 1947.

BUXTON, SYDNEY CHARLES, EARL BUXTON (1853-1934), statesman; son of Charles Buxton [q.v.]; educated at Clifton and Trinity College, Cambridge; liberal MP, Peter-borough (1883-5), Poplar (1886-1914); under-secretary for colonies, 1892-5; postmaster-general, 1905-10; president of Board of Trade, 1910-14; governor-general of South Africa, 1914-20; worked harmoniously with Louis Botha [q.v.] and inspired great affection; PC, 1905; GCMG and viscount, 1914; earl, 1920.

BUXTON, SIR THOMAS FOWELL, third baronet (1837-1915), governor of South Australia, 1895-8; educated at Harrow and Trinity College, Cambridge; succeeded father, 1858; liberal MP, King's Lynn, 1865-8; GCMG, 1899.

BUZZARD, SIR (EDWARD) FARQUHAR, first baronet (1871-1945), physician; educated at Charterhouse and Magdalen College, Oxford; qualified at St. Thomas's Hospital; on staff of National Hospital for Nervous Diseases,

Queen Square, Royal Free Hospital, St. Thomas's Hospital, Belgrave Hospital for Children; regius professor of medicine, Oxford, 1928-43; played leading part in foundation of Nuffield Institute for Medical Research (1935), chair and institute of social medicine (1943), and in extending scope of Oxford medical school; physician-in-ordinary to the King, 1932-6; KCVO, 1927; baronet, 1929.

BYNG, JULIAN HEDWORTH GEORGE, VISCOUNT BYNG OF VIMY (1862-1935), field-marshal; youngest son of second Earl of Strafford and grandson of first earl [q.v.]; educated at Eton; gazetted from militia to 10th Hussars, 1883; captain, 1889; passed Staff College, 1894; major, 1898; raised and commanded South African Light Horse (1900-1) proving a leader of the highest quality; commanded 10th Hussars (1902), Cavalry School, Netheravon (1904), 2nd Cavalry brigade, Eastern command (1905), 1st Cavalry brigade, Aldershot (1907), Territorial East Anglian division (1910), in Egypt (1912); major-general, 1909; commanded 3rd Cavalry division at first Ypres (1914), Cavalry Corps (1915), and IX Corps at Suvla (1915) where he drew up successful plan for evacuation; commanded Canadian Corps (1916-17) winning its confidence and affection; captured Vimy ridge, Apr. 1917; commander, Third Army (1917-19); prepared the disappointing Cambrai offensive (1917); put up strong resistance to German offensive in Mar. 1918 and played great part in final allied offensive; KCMG, 1915; general, 1917; baron, 1919; governor-general of Canada, 1921-6; proved extremely popular but criticized for handling of constitutional crisis of 1926 in which he refused a dissolution to W. L. Mackenzie King [q.v.]; chief commissioner of Metropolitan Police (1928-31) despite much initial opposition from labour party; tightened up discipline, introduced reforms, and established strong hold over the force; viscount, 1928; field-marshal, 1932.

BYRNE, SIR EDMUND WIDDRINGTON (1844-1904), judge; called to bar, 1867; QC 1888; a successful leader in Chancery cases; conservative MP for Walthamstow, 1892-7; appointed judge in Chancery division and knighted, 1897; an accurate and painstaking, but slow, judge.

BYRON, ROBERT (1905-1941), traveller, art critic, and historian; educated at Eton and Merton College, Oxford; published fruits of travels in *The Byzantine Achievement* (1929), *An Essay on India* (1931), *First Russia, then Tibet* (1933), and *The Road to Oxiana* (1937) an inquiry into origins of Islamic art; drowned on way to Egypt.

BYWATER, INGRAM (1840-1914), Greek scholar; educated at University and King's College schools, London; BA, Queen's College, Oxford; fellow of Exeter College, 1863-84; travelled abroad with the Mark Pattisons; reader in Greek at Oxford, 1884; regius professor of Greek, 1893-1908; lived in Lon-don during vacations; president of Oxford Aristotelian Society, c.1885-1908; delegate of Oxford University Press, 1879-1914; contributor to *Oxford English Dictionary*; bibliophile; bequeathed collection of books to Bodleian Library; works include edition of Fragments of Heraclitus (1877), of works of Priscianus Lydus (1886), and monumental edition of *Poetics* of Aristotle (1909); much of his work anonymous.

C

CABLE, (ALICE) MILDRED (1878-1952), missionary; educated at Guildford high school; for China Inland Mission studied medicine; joined Evangeline French [q.v.] in Hwochow, 1902; with Francesca French [q.v.] they undertook educational work for girls; in 1923-39 the trio travelled the Gobi Desert preaching the Gospel; subsequently worked in England for the Bible Society.

CADBURY, GEORGE (1839-1922), cocoa and chocolate manufacturer and social reformer; joined father's cocoa factory in Birmingham, 1856; with his elder brother took entire control of business; 1861; moved works to Bournville, 1879; built houses, primarily for his employees, on adjoining land, 1893-1900; founded Bournville Village Trust, 1900; associated welfare and education with housing; an enthusiastic worker for adult school movement; a practising Quaker and keen liberal.

CADMAN, JOHN, first BARON CADMAN (1877-1941), scientist and public servant; B.Sc., Durham, 1899; trained as mining engineer; government inspector, 1902-8; professor of mining, Birmingham, 1908-20; organized first school of petroleum technology; investigated Persian oilfields, 1913; technical adviser (1921), director (1923) and later chairman, Anglo-Persian Oil Company; served on Advisory Council, Department of Scientific and Industrial Research (1920-8, 1934-9) and on Fuel Research Board (1923-41); chairman, civil aviation inquiry, 1937-8; principal government adviser on oil, 1939-41; FRS, 1940; KCMG, 1918; baron, 1937.

CADOGAN, SIR ALEXANDER GEORGE MONTAGU (1884-1968), diplomatist; educated at Eton and Balliol College, Oxford; entered diplomatic service, 1908; attaché, Constantinople, 1909; in Foreign Office, 1912; Vienna, 1913-14; Foreign Office, 1914-18; private secretary to parliamentary secretary of state, 1919-20; head of League of Nations section; 1921-33; minister, Peking, 1934-5; deputy under-secretary of state; 1936; permanent under-secretary, 1938-46, British representative at United Nations headquarters, 1946-50; chairman, British Broadcasting Corporation, 1952-7; CB, 1932; KCB, 1941; CMG, 1926; KCMG, 1934; GCMG, 1939; PC, 1946; OM, 1951; hon. fellow, Balliol, 1950; *The Diaries of Sir Alexander Cadogan, 1938-45*, (edited by David Dilks), published, 1971.

CADOGAN, GEORGE HENRY, fifth EARL CADOGAN (1840-1915), conservative statesman; succeeded father, 1873; under-secretary of state for war, 1875; for colonies, 1878; lord privy seal, 1886-92; KG, 1891; lord-lieutenant of Ireland, 1895-1902.

CAILLARD, SIR VINCENT HENRY PENALVER (1856-1930), administrator; commissioned in Royal Engineers, 1876; president, council of administration of Ottoman Public Debt and financial representative of England, Holland, and Belgium in Constantinople, 1883-98; chiefly responsible for subsequent success of administration; served on board of Messrs. Vickers, 1898-1927; financial director, 1906; an ardent tariff reformer; knighted, 1896; died in Paris.

CAINE, SIR (THOMAS HENRY) HALL (1853-1931), novelist; educated at Liverpool elementary school; studied architecture; friend of Dante Gabriel Rossetti [q.v.] who suggested Isle of Man as unusual and apt setting for successful romantic stories including *The Deemster* (1887) and *The Manxman* (1894); *The Christian* (1897) and *The Eternal City* (1901) astonishingly popular; member of House of Keys, 1901-8; KBE, 1918; CH, 1922.

CAINE, WILLIAM SPROSTON (1842–1903), politician and temperance advocate; joined father's metal business at Egremont, 1861–78; early devoted himself to temperance movement at Liverpool; radical MP for Scarborough, 1880–5; supported local option, 1880; civil lord of Admiralty, 1884; MP for Barrow in Furness, 1886–92; helped to organize 'liberal unionists' against Gladstone's home rule policy, 1886; left unionist party on licensing question, 1890; liberal MP for East Bradford, 1892–5, and for Camborne, 1900–3; an able advocate of self-government in India; served on royal commissions on administration of Indian expenditure (1895–6) and liquor licensing laws (1896–9).

CAIRD, EDWARD (1835–1908), master of Balliol College, Oxford, and philosopher; brother of John Caird [q.v.]; educated at Glasgow and St. Andrews universities, 1850–7; influenced by Goethe through Carlyle's work; Snell exhibitioner at Balliol College, Oxford, 1860–3; BA, 1863; member of the 'Old Mortality Club' with T. H. Green, Swinburne, and others; a 'radical' in politics, religion, and philosophy; fellow and tutor of Merton, 1864–6; professor of moral philosophy at Glasgow, 1866–93; an early advocate of higher education of women; published *A Critical Account of the Philosophy of Kant* (2 vols., 1877–89), monograph on *Hegel* (1883), and *The Social Philosophy and Religion of Comte* (1885); also *The Evolution of Religion* (1893) and *The Evolution of Theology in the Greek Philosophy* (2 vols., 1904), both based on Gifford lectures, of 1891–2 and 1900 respectively; succeeded Benjamin Jowett [q.v.] as master of Balliol, 1893; supported grant of degrees to women and education of working men at Oxford; resisted bestowal of honorary degree on Cecil Rhodes, 1899; resigned mastership, 1907; received honorary degrees from St. Andrews (1883), Oxford (1891), Glasgow (1894), and Cambridge (1898); an original fellow of British Academy, 1902.

CAIRD, SIR JAMES, baronet, of Glenfarquhar, Kincardine (1864–1954), shipowner and a founder of the National Maritime Museum; educated at Glasgow Academy; as owner of Scottish Shire Line, 1903–17, co-operated in opening up west coast trade with Antipodes; financed restoration of *Victory* and preservation of *Implacable*; obtained valuable collections for, and in all gave over a million and a quarter pounds to, National Maritime Museum, opened in 1937; baronet, 1928.

CAIRNES, WILLIAM ELLIOT (1862–1902), captain and military writer; son of John Elliot Cairnes [q.v.]; joined army, 1882; promoted captain, 1890; military critic of *West-minster Gazette* during South African war, 1899–1901; published anonymously *An Absent-Minded War* (1900), full of pungent and epigrammatic criticism, and other books on military topics.

CAIRNS, DAVID SMITH (1862–1946), Scottish theologian; nephew of John Cairns [q.v.]; studied at United Presbyterian College, Edinburgh; ordained, 1895; professor of dogmatics and apologetics, United Free Church (later Christ's) College, Aberdeen, 1907–37; principal, 1923–37.

CAIRNS, SIR HUGH WILLIAM BELL (1896–1952), neuro-surgeon; born in Australia; MB, BS, Adelaide University, 1917; Rhodes scholar, Balliol College, Oxford; rowing blue, 1920; FRCS, 1921; at London Hospital by 1932 had a neuro-surgical unit introducing technique studied with Harvey Cushing in Boston; professor of surgery, Oxford, and fellow of Balliol, 1937–52; organized special hospital for head injuries, 1940–5; KBE, 1946.

CALDECOTE, first VISCOUNT (1876–1947), lawyer and statesman. See INSKIP, THOMAS WALKER HOBART.]

CALDECOTT, SIR ANDREW (1884–1951), colonial governor; educated at Uppingham and Exeter College, Oxford; joined Malayan civil service, 1907; chief secretary, Federated Malay States, 1931–3; colonial secretary, Straits Settlements, 1933–5; governor of Hong Kong, 1935–7, of Ceylon, 1937–44; reported on and recommended positive approach to Ceylon's constitutional reform; CBE, 1926; CMG, 1932; knighted, 1935; KCMG, 1937; GCMG, 1941.

CALDERON, GEORGE (1868–1915), dramatist; son of P. H. Calderon, RA [q.v.]; plays include *The Fountain* (1909), *The Little Stone House* (1911), *Revolt* (1912); *Tahiti*, impressions of the South Seas, published posthumously, 1921; good linguist and Slavonic student; missing at Dardanelles.

CALKIN, JOHN BAPTISTE (1827–1905), organist and composer; after holding several posts as organist, became professor at Guildhall School of Music, 1883; composed much sacred music, church services, and hymn tunes.

CALLAGHAN, SIR GEORGE ASTLEY (1852–1920), admiral; entered navy, 1866; captain, 1894; commanded British naval brigade which relieved legations in Peking during Boxer rising, 1900; CB, 1900; captain of Portsmouth dockyard, 1903–4; rear-admiral, 1905; commanded new fifth cruiser squadron, 1907–8;

second-in-command, Mediterranean station, 1908-10; KCVO, 1909; vice-admiral, 1910; commander-in-chief, home fleets, 1911-14; GCVO, 1912; removed from command on outbreak of war on account of age, 1914; commander-in-chief at the Nore, 1915-18; GCB, 1916; admiral of the fleet, 1917-18; Bath King of Arms, 1919.

CALLENDAR, HUGH LONGBOURNE (1863-1930), physicist; BA, Trinity College, Cambridge, 1885; fellow, 1886; professor of physics, Royal Holloway College, 1888; at McGill University, Montreal, 1893; Quain professor of physics, University College, London, 1898; professor of physics, Royal College of Science, London, 1902; FRS, 1894; Rumford medallist, 1906; president, Physical Society of London, 1910; CBE, 1920; carried out elaborate investigations into steam and thermometry, the present scale of temperature being based on his work; works include *The Properties of Steam* (1920) and several editions of *Steam Tables* (1915, 1922, 1927).

CALLENDER, Sir GEOFFREY ARTHUR ROMAINE (1875-1946), naval historian and first director of National Maritime Museum; educated at St. Edward's School and Merton College, Oxford; taught history at Osborne (1905-21) and Dartmouth (1921-2); professor of history, Greenwich, 1922-34; organizer and first director, National Maritime Museum, 1934-46; knighted, 1938.

CALLOW, WILLIAM (1812-1908), watercolour painter; brother of John Callow (1822-78, q.v.); studied in London (1823-9) and in Paris (1829-41); his 'View of Richmond', exhibited at Paris Salon (1831), attracted attention; taught in family of King Louis Philippe; exhibited over 1,400 drawings at Society of Painters in Water Colours; exhibited also in oils at British Institution (1848-67) and Royal Academy (1850-76); taught in London from 1855 to 1882; pupils included Empress Frederick and Lord Dufferin; wrote *Autobiography*, 1908.

CALLWELL, Sir CHARLES EDWARD (1859-1928), major-general; entered Royal Field Artillery, 1878; in intelligence branch of War Office, 1887-92; appointed to staff of Sir Redvers Buller [q.v.] on outbreak of South African war, 1899; in command of mobile column, Western Transvaal and Cape Colony, 1900-2; retired, 1909; director of military operations and intelligence, War Office, 1914-16; major-general and KCB, 1917; acquired reputation as military writer; works include *Small Wars* (1896), *The Dardanelles* (1919), and *Field-Marshal Sir Henry Wilson, his life and diaries* (2 vols., 1927).

CALMAN, WILLIAM THOMAS (1871-1952), zoologist; educated at the high school and university, Dundee; B.Sc., 1895; D.Sc., 1900; joined British Museum (Natural History), 1903; deputy keeper, 1921, keeper, 1927-36, zoology department; became leading carcinologist; published (1909) Crustacea volume of *Treatise on Zoology*, edited by Sir Ray Lankester [q.v.] and *The Life of Crustacea* (1911); secretary, Ray Society, 1919-46; president, Linnean Society, 1934-7; FRS, 1921; CB, 1935; hon. LLD, St. Andrews.

CALTHORPE, sixth BARON (1829-1910), agriculturist. [See GOUGH-CALTHORPE, AUGUSTUS CHOLMONDELEY.]

CALTHORPE, Sir SOMERSET ARTHUR GOUGH- (1864-1937), admiral of the fleet; entered navy, 1878; commander, 1896; captain, 1902; naval attaché, Russia, Norway, Sweden, 1902-5; rear-admiral, 1911; commanded second cruiser squadron, 1914-16; second sea lord, 1916-17; British commander-in-chief, Mediterranean, 1917-19; with skilful and rapid diplomacy concluded armistice with Turkey, 1918; high commissioner there, 1918-19; admiral, 1919; commander-in-chief, Portsmouth, 1920-3; admiral of the fleet, 1925; KCB, 1916; GCMG, 1919.

CAM, HELEN MAUD (1885-1968), historian, educated at home and Royal Holloway College, London University; first class, history, 1907; studied at Bryn Mawr College, Philadelphia, 1908; MA, London, 1909; teacher at Cheltenham Ladies' College, 1909-12; assistant lecturer, Holloway College, 1912-19; staff lecturer, 1919-21; Pfeiffer research fellow, Girton College, Cambridge, 1921-6; lecturer in history for college, 1926; and university, 1929; vice-mistress, Girton, 1944-8; Zemurray Radcliffe professor, Harvard, 1948-54; president, International Commission for the History of Assemblies of Estates, 1949-60; elected to British Academy, 1945; hon. doctorates, Smith College, Mount Holyoke College, North Carolina University, and Oxford; hon. fellow, Somerville College, 1964; vice-president, Selden Society, 1962-5; vice-president, Royal Historical Society, 1958; CBE, 1957; publications include *Local Government in Francia and England, 768-1034*, (1912), *Liberties and Communities in Medieval England: collected studies in administration and topography* (1944), and *Law Finders and Law Makers in Medieval England*, (1962).

CAMBRIDGE, second DUKE OF (1819-1904), field-marshal. [See GEORGE WILLIAM FREDERICK CHARLES.]

CAMBRIDGE, ALEXANDER AUGUSTUS FREDERICK WILLIAM ALFRED GEORGE, EARL OF ATHLONE (1874-1957), brother of Queen Mary [q.v.]; took family name of Cambridge and title of Athlone, 1917; educated at Eton and Sandhurst; served in army, 1894-1919; DSO, South African war; chairman of committee on needs of medical practitioners which recommended foundation of postgraduate medical school associated with London University, 1921; chancellor, London University, 1932-55; a tactful and successful governor of South Africa (1923-31) and of Canada (1940-6); governor and constable of Windsor Castle, 1931-57; married Princess Alice, daughter of Duke of Albany [q.v.], 1904; GCVO, 1904; GCB, 1911; GCMG, 1923; KG, 1928; PC, 1931; appointed grand master of order of St. Michael and St. George, 1936.

CAMERON, SIR DAVID YOUNG (1865-1945), painter and etcher; educated at Glasgow Academy; studied at Glasgow School of Art and Mound School, Edinburgh; executed over 500 etchings or drypoints; travelled and exhibited on continent; associate engraver (1911), painter (1916), Royal Academy; RA, 1920; war artist for Canadian government, 1917; joined British School at Rome, 1919; chairman, faculty of painting, 1925-38; master painter for decoration of St. Stephen's Hall, 1925; King's painter and limner in Scotland, 1933-45; member, Royal Fine Art Commission and National Art-Collections Fund; knighted, 1924.

CAMERON, SIR DONALD CHARLES (1872-1948), colonial governor; born in British Guiana; educated at Rathmines School, Dublin; served in colonial secretariat, British Guiana, 1891-1904; assistant colonial secretary, Mauritius, 1904-7; criticized section of its public service; transferred to Southern Nigeria, 1908; central secretary after amalgamation of Northern and Southern Nigeria, 1914-21; chief secretary to the government, 1921-4; with Sir Hugh Clifford [q.v.] brought administration to high level of efficiency; governor of Tanganyika, 1925-31; believed in indirect rule, established legislative council, reorganized African civil service, and encouraged education; strongly opposed union with Kenya and Uganda; governor of Nigeria, 1931-5; KBE, 1923; GCMG, 1932.

CAMERON, SIR (GORDON) ROY (1899-1966), pathologist; born in Australia, educated in village schools and Queen's College, Melbourne, 1916-22; resident medical officer, Melbourne Hospital, 1923; lecturer in pathology, Melbourne University, 1924; deputy director, Walter and Eliza Hall Institute,

Melbourne, 1925-7; worked in University College medical school, London, 1927; Graham scholar in pathology, 1928; Beit fellow, 1930-3; reader in morbid anatomy, University College medical school, 1934; professor, 1937; and director, Graham department, 1946-64; seconded to Chemical Defence Experimental Station, Porton, 1939-45; assistant editor, *Journal of Pathology and Bacteriology*, 1935-55; hon. FRCP, 1941; FRS, 1946; hon. LLD, Edinburgh (1956), and Melbourne (1962); member, Agricultural Research Council, 1947-56; member, Medical Research Council, 1952-6; knighted, 1957; founder president, College of Pathologists, 1962; publications include *The Pathology of the Cell* (1952).

CAMM, SIR SYDNEY (1893-1966), aircraft designer, educated at Royal Free School, Windsor; apprentice woodworker; secretary, Windsor Model Aeroplane Club; joined Martinsyde aeroplane company, Brooklands, 1914-23; senior draughtsman, Hawker Engineering Company, 1923; chief designer, 1925-66; elected to board, Hawker Siddeley Aviation, 1935-66; supervised design of light biplane, Cygnet, 1925; concerned with development of military planes, 1925; designed Hawker Hart day bomber, and the Fury; studied fighter monoplanes, 1933, and designed the Hurricane, 1934, the Typhoon, Tempest, and Sea Fury; with advent of jet engine, 1942, designed the Sea Hawk, and Hunter, which gained world speed record, 1953; concerned with design of vertical take off and landing fighter, 1958-66; fellow, Royal Aeronautical Society, 1932; president, 1954-5; chairman, technical board, Society of British Aircraft Constructors, 1951-3; CBE, 1941; knighted, 1953.

CAMPBELL, ARCHIBALD CAMPBELL, first BARON BLYTHSWOOD (1835-1908), scientist; born at Florence; served in Crimea, 1854; conservative MP for Renfrewshire (1873-4) and for West Renfrewshire (1885-92); created baronet (1890) and baron (1892); made valuable researches in astronomical and physical science; hon. LLD, Glasgow, and FRS, 1907.

CAMPBELL, BEATRICE STELLA (1865-1940), better known as MRS PATRICK CAMPBELL, actress; born TANNER; married (1884) Patrick Campbell, (killed in South Africa, 1900); made professional début in Liverpool (1888) and London (1890); an actress in the grand manner; endowed with dark beauty (her mother was Italian), a rich and expressive voice, and a gift for portraying passionate, complex women; remarkably successful as Paula in *The Second Mrs. Tanqueray*, 1893; other parts included Fédora (1895), Magda (1896), the Rat Wife in

Little Eyolf (1896), Lady Macbeth (1898), Mrs Daventry (1900), Mélisande in French to the Pelléas of Sarah Bernhardt (1904), Hedda Gabler (1907), Lady Patricia (1911), and Eliza Doolittle (1914) in *Pygmalion* by her firm friend, G. B. Shaw [q.v.]; her later performances less successful; married (1914) G. F. M. Cornwallis-West.

CAMPBELL, FREDERICK ARCHIBALD VAUGHAN, third EARL CAWDOR (1847-1911), first lord of the Admiralty; educated at Eton and Christ Church, Oxford; conservative MP for Carmarthenshire, 1874-85; succeeded to earldom, 1898; took active part in Pembrokeshire local affairs; as chairman of Great Western Railway, greatly improved the service, 1895-1905; first lord of the Admiralty in A. J. Balfour's government, 1905; took leading part in conservative opposition to Lloyd George's budget, 1909, and in drafting resolutions for reform of the House of Lords, 1910; president of Institution of Naval Architects, 1905; an able debater and administrator.

CAMPBELL, GORDON (1886-1953), vice-admiral; educated at Dulwich College; entered *Britannia*, 1900; brilliant commander of Q-(decoy tramp) ships, 1915-17; sank three submarines; VC, DSO, two bars; captain, 1917; flag captain to commander-in-chief, Coast of Ireland Patrol, 1917-18; commanded *Tiger*, 1925-7; retired as rear-admiral, 1928; vice-admiral, 1932; recounted exploits in *My Mystery Ships* (1928); national MP, Burnley, 1931-5.

CAMPBELL, (IGNATIUS) ROYSTON DUNNACHIE (1901-1957), poet and translator, known as ROY CAMPBELL; born in Durban; educated at high school; came to Europe, 1919; lived in maritime Provence, 1928-33, in Spain, 1933-6, in Portugal, 1952-7; became Roman Catholic, 1935; published *The Flaming Terrapin* (1924), *The Wayzgoose* (1928), *Adamastor* (1930), *The Georgiad* (1931), *Taurine Provence* (1932), *Flowering Reeds* (1933), *Mithraic Emblems* (1936), his pro-Franco *Flowering Rifle* (1939) which aroused liberal criticism, *Talking Bronco* (1946), *Light on a Dark Horse* (1951); translations from Spanish, French, and Portuguese included poems of St. John of the Cross, *Les Fleurs du Mal* (1952), and two novels by Eça de Queiroz; killed in car crash.

CAMPBELL, JAMES HENRY MUSSEN, first BARON GLENAVY (1851-1931), Irish lawyer and politician; senior moderator, classics and history, Trinity College, Dublin; called to bar (Ireland), 1878; QC, 1892; unionist MP, St.

Stephen's Green (1898-1900), Dublin University (1903-16); solicitor-general, Ireland, 1901-5; attorney-general, 1905, 1916; lord chief justice of Ireland, 1916-18; baronet, 1917; lord chancellor of Ireland, 1918-21; baron, 1921; over-dominant chairman of Irish Free State senate, 1922-8.

CAMPBELL, SIR JAMES MACNABB (1846-1903), Indian official and compiler of the *Bombay Gazetteer*; son of John McLeod Campbell [q.v.]; MA, Glasgow, 1866; joined Indian civil service, 1869; compiler of *Bombay Gazetteer* (to which he contributed much on ethnology), 34 vols., 1873-1901; CIE, and hon. LLD, Glasgow, 1885; held customs appointments at Bombay from 1891; did much service in combating plague at Bombay, 1897-8; KCIE, 1897; triennial memorial medal founded by Royal Asiatic Society; collected much material on Indian history and folklore.

CAMPBELL, DAME JANET MARY (1877-1954), medical officer; educated at Brighton high school; MB, London School of Medicine for Women, 1901; MD, 1904; MS, 1905; assistant school medical officer, London School Medical Service, 1904-7; medical officer, Board of Education, 1907-19; senior medical officer, maternity and child welfare, Ministry of Health and chief woman adviser, Board of Education, 1919-34; DBE, 1924.

CAMPBELL, JOHN CHARLES (1894-1942), major-general; educated at Sedbergh; joined Royal Artillery, 1914; commanded 4th Regiment, Royal Horse Artillery, in North Africa, 1940-1; organized mobile ('Jock') columns; DSO; commanded 7th support group, 1941-2; VC for leadership at battle of Sidi Rezegh, 21-2 Nov. 1941; major-general, 1942; killed in motor accident.

CAMPBELL, JOHN DOUGLAS SUTHERLAND, ninth DUKE OF ARGYLL (1845-1914), governor-general of Canada, 1878-83; liberal MP, Argyllshire, 1868; married Princess Louise [q.v.], daughter of Queen Victoria, 1871; unionist MP, South Manchester, 1895; succeeded father, 1900.

CAMPBELL, LEWIS (1830-1908), classical scholar; educated at Glasgow University and Balliol College, Oxford; first class, *lit. hum.*, 1853; fellow of Queen's College, 1855-8; honorary fellow of Balliol, 1894; vicar of Milford, Hampshire, 1858-63; professor of Greek at St. Andrews, 1863-92; Gifford lecturer on *Religion in Greek Literature*, 1894-5 (published 1898); edited Plato's *Theætetus*, 1861, *Sophistes* and *Politicus*, 1867; made special study of chronology of Plato's dialogues; published complete edition of Sophocles (2 vols., 1875–81), and translated Sophocles (1883) and Aeschylus (1890) into English verse; completed Jowett's edition of Plato's *Republic*, 3 vols., 1894; collaborated in *Life of Jowett*, 1897; died at Brissago, Lake Maggiore.

CAMPBELL, SIR MALCOLM (1885–1948), racing motorist; educated at Uppingham; established profitable business insuring newspapers against libel actions; served in Royal Flying Corps, 1914–18; won over 400 motor-racing trophies including 200-mile race at Brooklands, 1927, 1928; after 1925 set up series of land-speed records reaching over 300 m.p.h. in 1935; made water-speed record of 141.74 m.p.h., 1939; knighted, 1931.

CAMPBELL, (RENTON) STUART (1908–1966), journalist; educated at Lavender Hill School and Wandsworth Technical Institute secondary school; reporter on *Hendon and Finchley Times* and other provincial newspapers; reporter, Manchester office, *News Chronicle*, 1933; joined *Daily Mirror*, London, 1935; assistant editor, *Sunday Pictorial*, 1937; editor, 1940–6; managing editor, the *People*, 1946–57; editor, 1957–66; member, Press Council, 1961–4; flair for investigating and exposing abuses.

CAMPBELL, SIR RONALD HUGH (1883–1953), diplomatist; educated at Haileybury; entered Foreign Office, 1907; minister at Paris, 1929–35, at Belgrade, 1935–9; ambassador to France, 1939–40; unable to persuade French Government to remove overseas; to Portugal, 1940–5; averted crisis over allied occupation of Timor; obtained allied facilities in Azores and vital supplies of wolfram; CMG, 1917; KCMG, 1936; GCMG, 1940; PC, 1939.

CAMPBELL, WILLIAM HOWARD (1859–1910), missionary and entomologist; MA, Edinburgh, 1880; BD, 1882; served London Missionary Society in India, 1884–1909; published theological works in the Telugu language, on which he was a leading authority; made valuable collection of Indian moths; died at Bordighera.

CAMPBELL-BANNERMAN, SIR HENRY (1836–1908), prime minister; born in Glasgow; son of Sir James Campbell, wholesale draper at Glasgow; assumed name of Bannerman under maternal uncle's will, 1872; educated at Glasgow University (1851–3, hon. LLD, 1883) and Trinity College, Cambridge (BA, 1858, MA, 1861); partner in father's business till 1868; liberal MP for the Stirling Burghs from 1868 till death; financial secretary to War Office in

Gladstone's first (1871–4) and second (1880–2) administrations; secretary to the Admiralty, 1882–4; as chief secretary for Ireland, enhanced his reputation, 1884–5; became secretary for war and entered cabinet in Gladstone's third administration, Feb.—June 1886; actively supported Gladstone's home rule proposals in office and opposition; served on royal commission of inquiry into administration of naval and military departments, 1888–90; again secretary for war in Gladstone and Rosebery administrations, 1892–5; established forty-eight-hour week at Woolwich Arsenal, 1894; the allegation that the War Office under his control was inadequately supplied with cordite caused defeat of Rosebery's government, 1895; created GCB; a member of committee of inquiry into Jameson raid, which exonerated imperial and South African governments of complicity, 1896–7; succeeded as leader of the liberal party in House of Commons on Harcourt's resignation, 1899; opposed South African policy of Joseph Chamberlain [q.v.]; advocated conciliatory measures and 'rights of self-government' to conquered Boer states; denounced policy of enforcing unconditional surrender on the Boers; divided the liberal party by describing English warfare in South Africa as 'methods of barbarism' (June 1901), and by his support of home rule; opposed A. J. Balfour's education bill, 1902; reunited liberal party in opposition to Chamberlain's fiscal policy, 1903; attacked importation of Chinese indentured labour into South Africa; developed liberal programme, advocating small holdings, payment of members, retrenchment in public expenditure, curtailment of House of Lords veto; postponed consideration of home rule question to prevent party schism, 1904–5; became prime minister on Balfour's resignation, Dec. 1905; was for first time formally admitted in that capacity to fourth place of precedence among the King's subjects; the liberals routed unionists in general election, Jan. 1906; Campbell-Bannerman's government ended Chinese labour in South Africa, and established full responsible government in Transvaal and Orange Free State; introduced education bill for public control of public expenses of education and for abolition of religious tests for teachers (withdrawn owing to opposition of House of Lords), trades disputes bill (passed), and plural voting bill (rejected by House of Lords); Campbell-Bannerman emphatic in warning that will of people must prevail against House of Lords; in foreign affairs advocated arbitration for settling international disputes and limitation of armaments; welcomed members of Russian duma in London, 1906; favoured two-power naval standard, but advocated alliances with greatest naval powers; his article in the *Nation* on the eve of Hague peace conference (May 1907), urging the limitation of armaments, excited German mistrust which prevented discussion of the question at the conference; his government carried through parliament (1907) Haldane's army scheme, the Criminal Appeal Act, Deceased Wife's Sister's Marriage Act, Small Holdings Act for England and Wales, and motion for restricting power of House of Lords (June); became 'father of the House of Commons', May 1907; hon. DCL, Oxford, and LLD, Cambridge, 1907; went owing to ill health to Biarritz, Nov. 1907-Jan. 1908; resigned on grounds of continued ill health and died Apr. 1908; a strenuous, uncompromising fighter and strong party man, he was a fearless, sagacious, and optimistic leader; a good linguist and raconteur, he composed his set speeches with fastidious care, was ready in debate, and abounded in shrewd wit and humour; he supported women's suffrage; monument placed by parliament in Westminster Abbey, 1912.

CAMPION, GILBERT FRANCIS MONTRIOU, BARON CAMPION (1882-1958), clerk of the House of Commons; educated at Bedford School and Hertford College, Oxford; first class, *lit. hum.*, 1905; clerk in House of Commons, 1906; clerk assistant, 1930; clerk of the House, 1937-48; published *Introduction to the Procedure of the House of Commons* (1929) and, with D. W. S. Lidderdale, *European Parliamentary Procedure* (1953); edited fourteenth (1946) and fifteenth (1950) editions of Sir T. Erskine May [q.v.] on *Parliamentary Practice*; CB, 1932; KCB, 1938; GCB, 1948; baron, 1950; hon. DCL Oxford.

CAMROSE, first VISCOUNT (1879-1954), newspaper proprietor. [See BERRY, WILLIAM EWERT.]

CANNAN, CHARLES (1858-1919), scholar and university publisher; educated at Clifton and Corpus Christi College, Oxford; fellow, classical tutor, and dean, Trinity College, Oxford, 1884; junior bursar, 1887; delegate of University Press, 1895; secretary to delegates, 1898-1919; profoundly influenced development of the Press and made London office real department of the university; student of Aristotle and anonymous editor of *Selecta ex Organo Aristoteleo Capitula* (1897).

CANNAN, EDWIN (1861-1935), economist; brother of Charles Cannan [q.v.]; educated at Clifton and Balliol College, Oxford; an outstanding teacher of economics (professor from 1907) at London School of Economics and Political Science, 1895-1926; played essential part in developing its main tradition; a severe critic of classical economists yet deeply imbued with their spirit; entered vigorously into the monetary controversies of the day; works include *Elementary Political Economy* (1888), *History of the Theories of Production and Distribution in English Political Economy, 1776-1848* (1893), *History of Local Rates in England* (1896, 2nd edn. 1912), *The Economic Outlook* (1912), *Wealth: A Brief Explanation of the Causes of Economic Welfare* (1914), *A Review of Economic Theory* (1929), and the standard edition of Adam Smith's *Wealth of Nations* (2 vols., 1904).

CANNING, SIR SAMUEL (1823-1908), a pioneer of submarine telegraphy; helped Charles Bright [q.v.] to construct and lay first Atlantic cable, 1857-8; manufactured and laid Atlantic cables of 1865 and 1866, by means of *Great Eastern* steamship; knighted, 1866.

CANNON, HERBERT GRAHAM (1897-1963), zoologist; educated at Wilson's grammar school, Camberwell, and Christ's College, Cambridge; first class, part i, natural sciences tripos, 1918; left Cambridge for post as naturalist, Board of Fisheries laboratory, Conway; returned to take BA, 1919; demonstrator, Department of Zoology, Imperial College, London, 1920; professor, zoology, Sheffield University, 1926; research into embryology, feeding mechanisms and general anatomy of crustacea; FRS, Edinburgh, 1927; FRS, 1935; publications include *The Evolution of Living Things* (1958), and *Lamarck and Modern Genetics* (1959).

CANTON, WILLIAM (1845-1926), poet and journalist; born in Chusan; journalist in London, Glasgow, and in London again; works include *A Lost Epic and Other Poems* (1887), *The Invisible Playmate* (1894), *W. V. Her Book* (1896), *A Child's Book of Saints* (1898), *In Memory of W. V.* (1901), and *History of the British and Foreign Bible Society* (5 vols., 1904-10).

CAPE, HERBERT JONATHAN (1879-1960), publisher; began as errand-boy for Hatchard's; traveller, then manager, with George Duckworth, 1904-20; manager, Medici Society, 1920; with George Wren Howard founded Jonathan Cape, 1921; to hard work, shrewdness, and integrity added a flair for publishing; began with successful reissue of *Travels in Arabia Deserta* by C. M. Doughty [q.v.]; his authors included T. E. Lawrence, Duff Cooper, Mary Webb [qq.v.], H. E. Bates, Ian and Peter Fleming, (Dame) C. V. Wedgwood, Hugh Lofting, and Arthur Ransome [q.v.]; and the Americans, Ernest Hemingway, Sinclair Lewis, and Eugene O'Neill.

CAPEL, THOMAS JOHN (1836-1911), Roman Catholic prelate; chaplain of English-speaking Catholics at Pau and successful proselytizer, 1860-8; prominent figure in London society from 1868; the Monsignor Catesby of Disraeli's *Lothair*, 1870; carried on unsuccessfully Catholic public schools at Kensington, 1873-82; migrated to America, 1883; subsequently prelate in charge of Roman Catholic Church of Northern California; died at Sacramento; wrote controversial tracts.

CAPES, WILLIAM WOLFE (1834-1914), historical scholar; BA, Queen's College, Oxford; fellow and tutor of Queen's, 1856; rector of Bramshott, Hampshire, 1869-1901; university reader in ancient history, 1870-87; fellow and tutor of Hertford College, 1876-86; wrote books on ancient history; canon of Hereford, 1903-14; edited *Charters and Records of Hereford Cathedral* (1908).

CAPPER, SIR THOMPSON (1863-1915), major-general; born at Lucknow; joined army, 1882; DSO for services in South Africa 1899-1902; first commandant of Indian Staff College, Quetta; CB, 1910; major-general and inspector of infantry, 1914; commanded 7th division in Belgium, 1914-15; KCMG, 1915; died of wounds.

CARDEN, SIR SACKVILLE HAMILTON (1857-1930), admiral; entered navy, 1870; captain, 1899; rear-admiral, 1908; admiral superintendent, Malta dockyard, 1912; as vice-admiral, commanded British battle squadron associated with French forces in Mediterranean, Sept. 1914; at request of Admiralty, drew up detailed plan for forcing Dardanelles, involving systematic demolition of fortifications and subsequent invasion of peninsula, 1915; commanded operations against Dardanelles, Feb. 1915; relinquished command, owing to ill health, Mar. 1915; KCMG, 1916; retired with rank of admiral, 1917.

CARDEW, PHILIP (1851-1910), major RE; joined Royal Engineers, 1871; applied electricity to military purposes; instructor in electricity at Chatham, 1882; designed galvanometer for measuring large currents of electricity, 1882; invented the voltmeter, 1885, the vibratory transmitter for telegraphy (his most important discovery), 'separators', and other electrical devices; first electrical adviser to Board of Trade and promoted major, 1882; partner with Sir William Preece [q.v.] & Sons, 1898; contributed many papers to scientific societies.

CAREY, ROSA NOUCHETTE (1840-1909), novelist; published forty novels, including *Nellie's Memories* (1868), *Wee Wifie* (1869), *Uncle Max* (1887), *Only the Governess* (1888), all with wide vogue; an orthodox upholder of high-church principles.

CARLILE, WILSON (1847-1942), founder of the Church Army; inherited family silk business but ruined by a slump; deacon, 1880; priest, 1881; curate at St. Mary Abbots, Kensington, 1880-2; sought contact with working classes; founded Church Army (1882) as strictly non-party lay society within Church of England; established training colleges for working men (1884) and women (1887) for whom Convocation established the office of evangelist (men, 1897; women, 1921); social work included Church Army homes for ex-prisoners, tramps, etc.; his evangelical and social work a stimulus to other independent efforts; rector of St. Mary-at-Hill, City of London, 1891-1926; prebendary, St. Paul's Cathedral, 1906-42; CH, 1926.

CARLING, SIR ERNEST ROCK (1877-1960), surgeon and pioneer in radiotherapy; educated at the Royal grammar school, Guildford, and King's College, London; qualified at Westminster Hospital, 1901; FRCS, 1904; assistant surgeon, 1906, surgeon, 1919, honorary consulting surgeon, 1942; largely planned new Westminster Hospital, medical school, nurses home, and radium annex; member of National Radium and Atomic Energy commissions and Medical Research Council; chairman, International Commission of Radiological Protection; adviser to Home Office, Ministries of Labour and Health, and World Health Organization; knighted, 1944.

CARLISLE, ninth EARL OF (1843-1911), amateur artist. [See HOWARD, GEORGE JAMES.]

CARLISLE, COUNTESS OF (1845-1921), promoter of women's political rights and of temperance reform. [See HOWARD, ROSALIND FRANCES.]

CARLYLE, ALEXANDER JAMES (1861-1943), political philosopher, ecclesiastical historian, and social reformer; educated at Exeter College, Oxford; ordained, 1888; fellow, University College, 1893-5; rector, St. Martin and All Saints, Oxford, 1895-1919; mainstay of the Christian Social Union; advocated planned social reform and the unity of the Churches; works include the lucid and influential *History of Mediaeval Political Theory in the West* (6 vols., 1903-36), partly in collaboration with his brother, Sir R. W. Carlyle [q.v.], and *Political Liberty* (1941); canon of Worcester, 1930-4; FBA, 1936.

CARLYLE, BENJAMIN FEARNLEY, DOM AELRED (1874-1955), founder of the Benedictine community of Prinknash Abbey; educated at Blundell's and Newton Abbot College; medical student, St. Bartholomew's Hospital, 1892-6; Anglican Benedictine oblate, 1893; ordained, 1904; founder (1902) and abbot (1902-21) of Benedictine community which settled at Caldey island (1906), was received into the Roman Catholic Church (1913), and moved to Prinknash (1928); solemnly professed and ordained, 1914; resigned to work in Canadian mission fields, 1921; returned to community, 1951.

CARLYLE, SIR ROBERT WARRAND (1859-1934), Indian civil servant and scholar; educated at Glasgow University and Balliol College, Oxford; posted to Bengal, 1880; chief secretary, Bengal government, 1904-7; secretary, revenue and agriculture, government of India, 1907-10; member, governor-general's council, 1910-15; KCSI, 1911; collaborated with brother, A. J. Carlyle [q.v.], in *History of Mediaeval Political Theory in the West*.

CARMAN, WILLIAM BLISS (1861-1929), poet; born in New Brunswick; his most noteworthy poetry, *Songs from Vagabondia* series (1894, 1896, 1900), published in collaboration with Richard Hovey; essentially a nature poet; died in Connecticut.

CARMICHAEL, SIR THOMAS DAVID GIBSON-, eleventh baronet, BARON CARMICHAEL (1859-1926), overseas administrator and art connoisseur; BA, St. John's College, Cambridge; succeeded father, 1891; liberal MP, Midlothian, 1895-1900; governor of Victoria, Australia, 1908-11; of Madras, 1911-12; of Bengal, 1912-17; guided Bengal safely through troublous times; baron, 1912; trustee of National Gallery, etc.; sales of his collections (1902, 1926) events in art world.

CARNARVON, fifth EARL OF (1866-1923), Egyptologist, [See HERBERT, GEORGE EDWARD STANHOPE MOLYNEUX.]

CARNEGIE, ANDREW (1835-1919), manufacturer and philanthropist; born at Dunfermline; emigrated with parents to Pennsylvania, 1848; rose, by means of successful enterprises in manufacture of railway lines, bridges, coaches, and locomotives, from position of telegraph boy in Pittsburg (1850) to that of foremost ironmaster in America (1881); Carnegie Steel Company (with profits of $40 million) formed, 1899; sold his steel concern and retired from business in order to devote himself to distribution of his wealth, 1901; benefactions include endowment of libraries in United States, British Isles, Canada, etc. (1882-1919), 'hero' funds in United States (1904), Great Britain (1908), and European countries (1909-11), and Palace of Peace at The Hague (1903); works include *The Gospel of Wealth* (1900).

CARNEGIE, JAMES, sixth *de facto* and ninth *de jure* EARL OF SOUTHESK (1827-1905), poet and antiquary; collected gems, pictures, and Asiatic cylinders; obtained the title (forfeited in 1715) of Earl of Southesk, 1855; created KT and peer of United Kingdom, 1869; his first poetical work, *Jonas Fisher* (anonymous, 1875), was erroneously assigned by hostile reviewer in *Examiner* to Robert Buchanan [q.v.]; devoted later years to antiquarian research; hon. LLD, St. Andrews, 1872, and Aberdeen, 1875.

CARNOCK, first BARON (1849-1928), diplomatist. [See NICOLSON, SIR ARTHUR.]

CARÖE, WILLIAM DOUGLAS (1857-1938), architect; son of Danish consul, Liverpool; educated at Ruabon and Trinity College, Cambridge; articled to J. L. Pearson [q.v.]; architect to ecclesiastical commissioners and Charity Commission (1895-1938) and to Canterbury, Durham, Southwell, St. David's, Brecon, and Jerusalem cathedrals; designed internal fittings for numerous medieval churches and restored many medieval buildings; representative of closing phase of Gothic revival; FRIBA, 1890.

CARPENTER, ALFRED FRANCIS BLAKENEY (1881-1955), vice-admiral; nephew of Edward Carpenter [q.v.]; entered navy, 1897; specialized in navigation; commander, 1915; joined Roger (later Lord) Keyes [q.v.] at Admiralty, 1917; assisted in planning attack on Zeebrugge and Ostend; commanded *Vindictive*, 1918, and successfully brought her alongside the Mole for attack on Zeebrugge; VC and captain; wrote *The Blocking of Zeebrugge* (1921); commanded *Carysfort*, 1921-3; Chatham dockyard, 1924-6, *Benbow*, 1927-8, *Marlborough*, 1928-9; rear-admiral and retired, 1929; vice-admiral, 1934.

CARPENTER, EDWARD (1844-1929), writer on social subjects; BA, Trinity Hall, Cambridge; fellow, 1868; ordained, 1869; relinquished fellowship and orders, 1874; joined staff of university extension movement, 1874; visited USA, where he made particular friends with Walt Whitman, 1877; settled in Sheffield, 1878; lived at Millthorpe, a Derbyshire hamlet near Chesterfield, 1883-1922; engaged in literary work, market gardening, and sandal-making; moved to Guildford, 1922; his life a reaction against Victorian convention and respectability; abjured his social class; neither philosopher nor economist; his works include *Towards Democracy* (1883), *England's Ideal* (1885), and *Civilization, its Cause and Cure* (1889).

CARPENTER, GEORGE ALFRED (1859-1910), physician; MRCS, 1885; MB, London, 1886; MD, 1890; made special study of, and published works on, children's diseases; physician to Evelina Hospital, Southwark; founded (1904) and edited *British Journal of Children's Diseases*; helped to found (1900) Society for Study of Disease in Children, and edited its *Reports*, 1900-8.

CARPENTER, Sir (HENRY CORT) HAROLD (1875-1940), metallurgist; nephew of J. Estlin Carpenter [q.v.]; educated at St. Paul's School and Merton College, Oxford; first class, natural science, 1896; in charge of departments of chemistry and metallurgy, National Physical Laboratory, 1901-6; professor of metallurgy, Manchester, 1906-13, Royal School of Mines, 1913-40; FRS, 1918; knighted 1929.

CARPENTER, JOSEPH ESTLIN (1844-1927), Unitarian divine; son of W. B. Carpenter, the biologist [q.v.]; MA, University College, London; trained for Unitarian ministry; minister at Clifton, Bristol, 1866-9, and in Leeds, 1869-75; professor of ecclesiastical history, comparative religion, and Hebrew, Manchester New College, Gordon Square, London, 1875; vice-principal, 1885-99; migrated with college to Oxford, 1889; principal, 1906-15; Wilde lecturer in comparative religion, Oxford University, 1914-24; his numerous theological works include *The Composition of the Hexateuch, an Introduction* (1902); an able exponent of higher criticism.

CARPENTER, ROBERT (1830-1901), cricketer; had various professional engagements, 1854-60; played for Players *v.* Gentlemen, 1859-73; with Tom Hayward and George Tarrant raised standard of Cambridgeshire cricket; visited America (1859) and Australia (1863-4); elegant batsman, unsurpassed for back play.

CARPENTER, WILLIAM BOYD (1841-1918), bishop of Ripon; BA, St. Catharine's College, Cambridge; royal chaplain, 1879; canon of Windsor, 1882; bishop of Ripon, 1884-1911; clerk of the closet, 1903-10, 1911-18; canon of Westminster, 1911-18; KCVO, 1912; prolific writer and notable preacher.

CARR, Sir CECIL THOMAS (1878-1966), public lawyer; educated at Bath College and

Trinity College, Cambridge; first class, part i, classical tripos, 1899; third class, part ii, law tripos, 1901; called to bar (Inner Temple), 1902; practised on Western Circuit, 1902-14; served in army in India, 1914-18; LLD Cambridge, 1920; assistant editor, Revised Statutes and Statutory Rules and Orders (later, Statutory Instruments), 1919; editor, 1923; lectured in United States, 1935-40; counsel to the Speaker, 1943-55; member, Statute Law Committee, 1943-65; knighted, 1939; KCB, 1947; bencher, Inner Temple, 1944; KC, 1945; hon. degrees, Columbia (1940), London (1952), and Queen's University, Belfast (1954); FBA, 1952; hon. fellow, Trinity College, Cambridge, 1963; election secretary, and then chairman, the Athenaeum, 1943-56; president, Selden Society, 1958-61; publications include *Delegated Legislation* (1921) and *Concerning English Administrative Law* (1941).

CARRINGTON, Sir FREDERICK (1844-1913), general; joined army, 1864; CMG for services in Transvaal, 1880; commanded Carrington's Horse in Bechuanaland expedition, 1885; KCMG, 1887; major-general 1895; KCB, 1897; served in South African war, 1899-1902.

CARR-SAUNDERS, Sir ALEXANDER MORRIS (1886-1966), biologist, sociologist, and academic administrator; educated at Eton and Magdalen College, Oxford; first class, zoology, 1908; demonstrator, comparative anatomy; studied biometrics in London; secretary, research committee, Eugenics Education Society; sub-warden, Toynbee Hall, 1910-14; called to bar (Inner Temple), 1914; served with Army Service Corps, 1914-18; first Charles Booth professor, social science, Liverpool, 1923-37; director, London School of Economics, 1937-56; member Asquith commission on higher education in the colonies, 1943; chairman of committees and commissions aiding development of universities and colleges in dependent territories, 1947-62; knighted, 1946; FBA, 1946; KBE, 1957; hon. doctorates, Glasgow, Columbia, Natal, Dublin, Liverpool, Cambridge, Grenoble, Malaya, and London; hon. fellow, Peterhouse, Cambridge, University College of East Africa and London School of Economics; publications include *The Population Problem* (1922), *The Professions* (with P. A. Wilson, 1933), and *New Universities Overseas* (1961).

CARRUTHERS, ALEXANDER DOUGLAS MITCHELL (1882-1962), explorer and naturalist, educated at Haileybury and Trinity College, Cambridge; trained in land survey and taxidermy; with British Museum expedition to

Ruwenzori, 1905–6; travelled in Russian Turkestan and borders of Afghanistan, 1907–8; research on wild sheep; explored, with John H. Miller and Morgan Philips Price, desert of Outer Mongolia and upper Yenesei river, 1910; hon. secretary, Royal Geographical Society, 1916–21; publications include *Unknown Mongolia* (1913), *Arabian Adventure* (1935), and *Beyond the Caspian, a Naturalist in Central Asia* (1949).

CARSON, EDWARD HENRY, BARON CARSON (1854–1935), Ulster leader and lord of appeal in ordinary; born in Dublin; educated at Portarlington School and Trinity College, Dublin; called to Irish bar, 1877; junior counsel to attorney-general, 1887; QC, 1889; solicitor-general for Ireland, 1892; MP, Dublin University, 1892–1918; called to English bar (Middle Temple), 1893; QC, 1894; acknowledged a leading advocate after Oscar Wilde's libel action against Queensberry, 1895; solicitor-general for England, 1900–5; knighted, 1900; PC, 1905; leader of Irish unionists in House of Commons (1910) and of movement for a provisional government in Ulster, 1911; Ulster Volunteer Force raised and covenant drafted, 1912; unsuccessfully moved exclusion of whole province of Ulster from home rule bill (Jan. 1913); preparations pushed forward in Ulster; plans adopted for provisional government; accepted chairmanship of central authority; corresponded with Asquith who offered 'county option' with time limit of six years (Mar. 1914); (Sir) Winston Churchill threatened force if offer refused; Curragh incident and Larne gun-running revealed danger of civil war; 'county option' bill passed Commons but Lords substituted permanent exclusion of province of Ulster; with James Craig [q.v.] represented Ulster at subsequent party conference; when this broke down 'county option' was adhered to but owing to the imminence of war the bill was not proceeded with; a party truce was proclaimed; the home rule bill, however, passed into law although its operation was suspended until the end of the war; nevertheless, Carson offered the services of Ulster Volunteers; attorney-general, May 1915–Oct. 1916; first lord of Admiralty, Dec. 1916 to July 1917 when he entered the war cabinet; agreed to setting up of Irish convention, 1917; resigned Jan. 1918 on learning that Lloyd George intended to introduce home rule bill for whole of Ireland; MP, Duncairn division of Belfast, 1918–21; repeated his opposition to policy of home rule but did not oppose the Government of Ireland bill, 1920, since its defeat would have brought into operation the Act of 1914; resigned leadership of Ulster unionists, 1921, but always continued to watch their interests; lord of appeal in ordinary,

1921–9; a powerful, conscientious, and fearless advocate whose disregard for money compared with duty was exemplified in the Archer-Shee case (1910); at his best in cross-examination and in his appeal to the jury; a great orator of passionate sincerity; distinguished by a moral grandeur of character and a charming personality.

CARTE, RICHARD D'OYLY (1844–1901), promoter of English opera; joined father's musical instrument business, 1861; composed operettas; founded successful concert agency, 1870; engaged in theatrical management; produced Gilbert and Sullivan's *Trial by Jury*, 1875, *Sorcerer*, 1877, *H.M.S. Pinafore*, 1878, *Pirates of Penzance*, 1880, and *Patience*, 1881; built Savoy Theatre, the first public building in England lighted by electricity, 1881; produced *Iolanthe*, 1882, *Mikado*, 1885, *Yeomen of the Guard*, 1888, *Gondoliers*, 1889, and operas by other composers; had five touring companies in America; erected magnificent but unremunerative Royal English Opera House (1891) at Cambridge Circus, London, which became (1892) the prosperous Palace Theatre of Varieties; raised standard of musical taste in the English theatre.

CARTER, SIR EDGAR BONHAM- (1870–1956), jurist and administrator. [See BONHAM-CARTER.]

CARTER, (HELEN) VIOLET BONHAM-, BARONESS ASQUITH OF YARNBURY (1887–1969), political figure. [See BONHAM-CARTER.]

CARTER, HOWARD (1874–1939), painter and archaeologist; draughtsman for Egypt Exploration Fund, 1891–9; inspector-in-chief of monuments, Upper Egypt and Nubia, for Egyptian government, 1899–1903; worked with fifth Earl of Carnarvon [q.v.], 1908–23; discovered tomb of Amenophis I and (1922) of Tutankhamūn.

CARTER, HUGH (1837–1903), painter; exhibited at Royal Academy, 1859–1902, and Royal Institute of Painters in Water Colours, mainly subject paintings; work shows delicate colouring and subtle delineation of character.

CARTER, THOMAS THELLUSSON (1808–1901), tractarian divine; educated at Eton and Christ Church, Oxford; first class, *lit. hum.*, 1831; ordained, 1832; rector of Clewer, 1844–80; founded House of Mercy, Clewer, for fallen women, 1849; and sisterhood, 1852; hon. canon of Christ Church, 1870; influenced by *Tracts for the Times*; with Liddon and Pusey drew up declaration in defence of confession, 1873; a founder of English Church Union and Confraternity of the Blessed Sacrament; thrice prosecuted for ritual excesses; had much influence among high churchmen as organizer and devotional author.

CARTON, RICHARD CLAUDE (1856-1928), dramatist, whose original name was CRITCHETT; son of George Critchett [q.v.]; his plays include *Lord and Lady Algy* (1898); his comedies, tending towards farce, were full of witty dialogue, and poked fun at aristocracy.

CARTON DE WIART, SIR ADRIAN (1880-1963), lieutenant-general; educated at the Oratory School, Edgbaston, and Balliol College, Oxford; fought in South African war; commissioned in 4th Dragoon Guards; ADC to commander-in-chief South Africa, 1905; joined Camel Corps, 1914; lost an eye; DSO; during 1914-18 war wounded eight times, lost left hand; VC, 1916; commanded series of infantry brigades; second-in-command to General Louis Botha [q.v.], leader of British Military Mission to Poland, 1918; commanded mission, 1919; resigned commission and settled in Poland, 1924-39; returned to England and commanded 61st division with headquarters at Oxford, 1939; sent on military mission to Yugoslavia; his plane shot down, Italian prisoner, 1941; sent to Lisbon by Italians as intermediary, 1943; (Sir) Winston Churchill's personal representative to Chiang Kai-shek in China, 1943-6; hon. fellow, Balliol College, 1947; hon. doctorate, Aberdeen; CMG, 1918; CB, 1919; KBE, 1945; published *Happy Odyssey*, 1950.

CARVER, ALFRED JAMES (1826-1909), master of Dulwich College; BA, Trinity College, Cambridge, 1849; Bell university scholar, 1846; fellow of Queens' College, 1850-3; master of Alleyn's College, Dulwich, 1858-83, which under him was subdivided into two schools, Dulwich College (public) and Alleyn's School (secondary); rebuilt Dulwich College, 1870; broadminded and energetic in educational matters; founded Carver memorial prize for modern languages; hon. DD, Lambeth, 1861; hon. canon of Rochester, 1882.

CARY, ARTHUR JOYCE LUNEL (1888-1957), author; born in Londonderry; educated at Clifton and Trinity College, Oxford; studied art in Edinburgh and Paris, 1907-9; served in Nigerian political service, 1913-20; settled in Oxford, 1920; after 1932 produced sixteen novels, many short stories, two long poems, etc.; works include four African novels, (1932-9); an autobiographical novel *A House of Children* (1941); and two trilogies, one concerned with art (*Herself Surprised, To Be a Pilgrim, The Horse's*

Mouth, 1941-4), the other with politics (*Prisoner of Grace, Except the Lord, Not Honour More*, 1952-5); and his aesthetic credo *Art and Reality* (1956).

CASE, THOMAS (1844-1925), Waynflete professor of philosophy and president of Corpus Christi College, Oxford; BA, Balliol College, Oxford; tutor, Corpus Christi College, 1876; Waynflete professor of moral and metaphysical philosophy and fellow of Magdalen College, 1889; president of Corpus Christi College, 1904-24; resigned professorship, 1910; opposed changes in church, state, and university; notable writer of letters to *The Times*.

CASEMENT, ROGER DAVID (1864-1916), British consular official and Irish rebel; entered British consular service, 1892; his official report on administration of Congo Free State (1903) largely contributed towards its abolition (1908); his report on atrocities committed by agents of Peruvian Amazon Company (1910) published as blue book (1912); knighted and retired, 1911; returned to Ireland, 1913; on committee of Irish National Volunteers, 1913; visitied Berlin as propagandist for Irish nationalism, Nov. 1914; landed from German submarine at Tralee, 1916; arrested and hanged as traitor.

CASEY, WILLIAM FRANCIS (1884-1957), editor of *The Times*; born in Cape Town; educated at Castleknock College and Trinity College, Dublin; called to Irish bar, 1909; had two plays produced by Abbey Theatre, 1908; on staff of *The Times*, 1914-52; posted to Washington, 1919, Paris, 1920; chief foreign sub-editor, 1923; foreign leader-writer, 1928; deputy editor, 1941; editor, 1948-52.

CASH, JOHN THEODORE (1854-1936), physician; MB, Edinburgh, 1876; researches with (Sir) T. L. Brunton [q.v.] represented new approach to elucidation of action of drugs and helped synthetic chemists in discovering new remedies; regius professor of materia medica and therapeutics, Aberdeen, 1886-1919; FRS, 1887.

CASSEL, SIR **ERNEST JOSEPH** (1852-1921), financier and philanthropist; born at Cologne; associated with financial house of Bischoffsheim and Goldschmidt, in London, with whom he laid the foundations of his immense fortune, 1870-84; naturalized, 1878; undertook independent business after 1884; enterprises in which he was concerned include establishment of Swedish Central Railway on sound financial basis; reorganization of Louisville and Nashville Railway in America; arrangement of finances of Mexican Central Railway;

issuing of Mexican, Chinese, and Uruguay government loans; financing of construction of Nile dams at Aswan and Assiut and formation of National Bank of Egypt; creation of State Bank of Morocco and National Bank of Turkey; formed valuable collection of old masters and other works of art; racing activities led to formation of friendship with Prince of Wales, afterwards King Edward VII; gave away about £2 million to charities during lifetime; distinctions include KCMG (1899) and PC (1902).

CASSELS, SIR ROBERT ARCHIBALD (1876-1959), general; born in India; educated at Sedbergh and Sandhurst; commissioned, 1896; served in India from 1897; in Mesopotamia, 1915-18; commanded 11th Cavalry brigade, 1917-20; a distinguished cavalry leader at Sharqat in final offensive against Turks, 1918, DSO; major-general, 1919; cavalry adviser, India, 1920-3; commandant, Peshawar district, 1923-7; adjutant-general, India, 1928-30; general, 1929; GOC-in-C, Northern command, India, 1930-4; commander-in-chief, India, 1935-41; CB, 1918; KCB, 1927; GCB, 1933; GCSI, 1940.

CASSELS, WALTER RICHARD (1826-1907), theological critic; partner in firm of Peel, Cassels & Co. at Bombay till 1865; member of legislative council of Bombay, 1863-5; published anonymously *Supernatural Religion* (2 vols., 1874, vol. 3, 1877), impugning credibility of miracles and authenticity of the New Testament; engaged in controversy with Bishop Lightfoot [q.v.], 1874-89; an enthusiastic art collector; also published poems.

CATES, ARTHUR (1829-1901), architect; pupil of Sydney Smirke, RA [q.v.]; architect to land revenues of the crown, 1870; vice-president (1888-92) and chairman of board of examiners (1882-96) of Royal Institute of British Architects, where he initiated progressive examinations.

CATHCART, EDWARD PROVAN (1877-1954), physiologist; educated at Ayr Academy; MB, Glasgow, 1900; MD, 1904; D.Sc., 1908; at Lister Institute, 1902-5; Grieve lecturer in physiological chemistry, Glasgow, 1906-15; professor of physiology, London Hospital medical school, 1915-19; Gardiner professor of physiological chemistry, Glasgow, 1919-28, regius professor of physiology, 1928-47; published *Physiology of Protein Metabolism* (1912) and *The Human Factor in Industry* (1928); studied people's dietary habits and energy expenditure; chairman, Scottish health services committee, and Industrial Fatigue Research Board; on General Medical Council, 1933-45; FRS, 1920; FRSE, 1932; CBE, 1924.

CATTO, THOMAS SIVEWRIGHT, first BARON CATTO (1879-1959), governor of the Bank of England; educated at Peterhead Academy and Heaton School, Newcastle upon Tyne; employed with merchants in Batoum and Baku; joined MacAndrews & Forbes as London manager, 1904; second in command in Smyrna, 1906; a vice-president in New York, 1909-19; Admiralty representative on Russian commission to United States, 1915-17; on British Food Mission, 1917; head of British Ministry of Food in North America, 1918; chairman of Andrew Yule & Co., 1919-40; in Calcutta, 1919-28; became partner in Morgan Grenfell & Co., 1928, and returned to London; a director of Bank of England, 1940; director-general, equipment and stores, Ministry of Supply, 1940; financial adviser to Treasury, 1940-4; as governor of the Bank of England, 1944-9, ensured its nationalization with maximum retention of operational independence; CBE, 1919; baronet, 1921; baron, 1936; PC, 1947.

CAVAN, tenth EARL OF (1865-1946), field-marshal. [See LAMBART, FREDERICK RUDOLPH.]

CAVE, GEORGE, VISCOUNT CAVE (1856-1928), lawyer and statesman; BA, St. John's College, Oxford; called to bar (Inner Temple), 1880; began practice on Chancery side; KC and recorder of Guildford, 1904; standing counsel to Oxford University, 1913; concerned himself with local administration in Surrey; unionist MP, Kingston division of Surrey, 1906-18; won respect of all parties in parliament; PC, solicitor-general, and knighted, 1915; home secretary, 1916-19; lord of appeal in ordinary and created viscount, 1918; lord chancellor, 1922-4-8; most important case on which he adjudicated reference to Privy Council of rival claims of Canada and Newfoundland in regard to boundary of respective territories in Labrador, 1927; chancellor of Oxford University, 1925; chairman of numerous commissions and committees.

CAVELL, EDITH (1865-1915), nurse; entered London Hospital as probationer, 1895; first matron of Dr Depage's clinic—the Berkendael medical institute at Brussels, 1907; in charge of institute when it became a Red Cross hospital, 1914; assisted allied soldiers to escape from Belgium, 1914-15; arrested by Germans, 5 Aug. 1915, and placed in solitary confinement at St. Gilles; brought to trial, 7 Oct.; confessed that her efforts had been successful and was therefore condemned to death by court martial; shot, 12 Oct.

CAVENDISH, SPENCER COMPTON, MARQUESS OF HARTINGTON and eighth DUKE OF DEVONSHIRE (1833-1908), statesman; eldest son of seventh earl [q.v.] and brother of Lord Frederick Cavendish [q.v.]; MA, Trinity College, Cambridge, 1854; hon. LLD, 1862; liberal MP for North Lancashire, 1857; became Marquess of Hartington, 1858; moved successful motion of want of confidence in Derby ministry, June 1859; met President Lincoln in United States, 1862; under-secretary at War Office, 1863; secretary of state for war and member of cabinet in Lord Russell's government, 1866; supported Gladstone on Irish Church disestablishment, and lost his seat, 1868; elected for Radnor boroughs, 1869; postmaster-general in Gladstone's first administration; introduced bill for voting by ballot, 1870; chief secretary for Ireland, 1870-4; passed 'coercion bill'; led liberal party on Gladstone's resignation, 1875; favoured maintenance of Turkish dominion in Russo-Turkish war, 1877; severely criticized government's Afghan policy, 1878; received freedom of city of Glasgow, 1877; lord rector of Edinburgh University, 1879; MP for North-East, Lancashire, 1880-5; refused premiership; secretary of state for India under Gladstone, 1880-2; withdrew British forces from Afghanistan; secretary of state for war, 1882-5; partly responsible for sending General Gordon [q.v.] to Sudan, and for failure to rescue him from Khartoum; his whig principles brought him into collision with radical leaders, Joseph Chamberlain and Sir Charles Dilke [qq.v.]; mediator between government and House of Lords on question of franchise and redistribution bills, 1884; consistently opposed Gladstone's home rule policy, and favoured coercive measures in Ireland; MP for Rossendale, Lancashire, 1885-91; refused office under Gladstone, 1886; opposed the first home rule bill, and with Chamberlain founded new party of liberal unionists, who combined with conservatives under Lord Salisbury to drive Gladstone from power, 1886; declined premiership, 1886 and 1887; independently supported Salisbury's government, 1887-92; chairman of royal commissions on administration of the naval and military departments, 1888-90, and on relations between employers and employed, 1891; succeeded father as Duke of Devonshire, Dec. 1891; and as chancellor of Cambridge University; moved rejection of Gladstone's second home rule bill in House of Lords, 1893; joined Salisbury's coalition government as president of the Council, 1895-1902; held same office under A. J. Balfour, 1902-3; strongly opposed to Chamberlain's fiscal schemes, 1902; resigned office owing to Balfour's pronouncements on fiscal policy, 1903; opposed tariff reform in House of Lords, 1904; KG, 1902; frequently entertained King Edward VII; generous landlord and public-spirited benefactor; keen sportsman; speeches well constructed and logical; of transparent honesty, simplicity of purpose, and disinterestedness; died at Cannes.

CAVENDISH, VICTOR CHRISTIAN WILLIAM, ninth DUKE OF DEVONSHIRE (1868-1938); educated at Eton and Trinity College, Cambridge; unionist MP, West Derbyshire, 1891-1908; whip, 1901; financial secretary to Treasury, 1903-5; succeeded uncle (see above), 1908; governor-general of Canada, 1916-21; colonial secretary, 1922-4; active in preparations for British Empire Exhibition of 1924; PC, 1905; KG, 1916.

CAWDOR, third EARL (1847-1911), first lord of the Admiralty. [See CAMPBELL, FREDERICK ARCHIBALD VAUGHAN.]

CAWOOD, SIR WALTER (1907-1967), scientist and civil servant; educated at Archbishop Holgate's grammar school, York and Leeds University; B.Sc. (London), 1929; Ph.D. (Leeds) in chemistry, 1932; research worker; joined Air Ministry, 1938; engaged on chemical warfare research; improvised pre-set range finder for use in aircraft attacking flying bombs; deputy director, Royal Aircraft Establishment, Farnborough, 1947-53; responsible for aeronautical research programme, Ministry of Aviation, 1953-9; chief scientist, War Office, 1959-64; chief scientist, Ministry of Aviation, 1964-7; fellow, Royal Aeronautical Society; CBE, 1953; CB, 1956; KBE, 1965.

CAWTHORNE, SIR TERENCE EDWARD (1902-1970), otologist; educated at Denstone College and King's College Hospital medical school; MRCS, LRCP, 1924; FRCS, 1930; held junior appointments in ear and throat departments of King's College Hospital; assistant surgeon, ear, nose, and throat department, 1932; full surgeon, 1939; aural surgeon, National Hospital for Nervous Diseases, 1936; consulting adviser in otolaryngology, Ministry of Health, 1948-67; specialized in neurological aspects of deafness, Ménière's disease, and paralysis of facial nerve; clinical director, Wernher research unit on deafness, 1958-64; president, Royal Society of Medicine, 1962-4; hon. MD Upsala, (1963), hon. LLD Syracuse, New York, (1964), and hon. FRCS Ireland, (1966); knighted, 1964.

CECIL, EDGAR ALGERNON ROBERT GASCOYNE-, VISCOUNT CECIL OF CHELWOOD (1864-1958), a creator of the League of Nations; third son of third Marquess of Salisbury; educated at Eton and University College, Oxford; president of Union, 1885; called to bar (Inner Temple), 1887; practised at parliamentary bar; QC, 1899; conservative MP, East

Marylebone, 1906–10; independent MP, Hitchin, 1911–23; parliamentary under-secretary, foreign affairs, 1915–18; minister of blockade, 1916–18; assistant secretary of state, foreign affairs, 1918–19; his memorandum, 1916, was a basis of the League Covenant; dominated debates at Paris conference commission on League, 1919; delegate for South Africa at League Assemblies, 1920–2; in charge of League affairs, as lord privy seal, 1923, as chancellor of duchy of Lancaster, 1924–7; delegate to preparatory disarmament commission, 1926, to Geneva conference on naval disarmament, 1927; resigned over cabinet's refusal to accept naval parity with America; chairman, departmental committee on League affairs and British representative on preparatory disarmament commission, 1929–31; president, League of Nations Union, 1923–45; organized peace ballot, 1934–5; Nobel peace prize, 1937; viscount, 1923; PC, 1915; CH, 1956; chancellor, Birmingham University; rector, Aberdeen.

CECIL, LORD EDWARD HERBERT GASCOYNE- (1867–1918), soldier and civil servant; son of third Marquess of Salisbury; entered army, 1887; joined Egyptian government as under-secretary of state in ministry of finance, 1905; financial adviser, 1912; acted as high commissioner, 1914–15; his *Leisure of an Egyptian Official* published in 1921.

CECIL, HUGH RICHARD HEATHCOTE GASCOYNE-, BARON QUICKSWOOD (1869–1956), politician and provost of Eton; youngest son of third Marquess of Salisbury; educated at Eton and University College, Oxford; first class, modern history, and fellow of Hertford, 1891; conservative MP, Greenwich, 1895–1906; burgess for Oxford University, 1910–37; a creator of and active in Church Assembly; failed to obtain Commons acceptance of Revised Prayer Book, 1927–8; provost of Eton, 1936–44; a strong conservative and Anglican and accomplished orator; PC, 1918; baron, 1941.

CECIL, JAMES EDWARD HUBERT GASCOYNE-, fourth MARQUESS OF SALISBURY (1861–1947), eldest son of third Marquess of Salisbury; educated at Eton and University College, Oxford; conservative MP, Darwen division (1885–92), Rochester (1893–1903); succeeded father, 1903; chairman, church parliamentary committee, 1893–1900; commanded 4th Bedfordshire battalion in South Africa, 1899–1900; under-secretary for foreign affairs, 1900–3; lord privy seal, 1903–5; president of Board of Trade, 1905; became recognized as leader of the old conservatives; opposed the finance bill (1909) and parliament bill (1911)

maintaining neither had mandate from the electorate; never ceased thereafter to agitate for reform of House of Lords; unsuccessfully introduced bill to that effect, 1934; chairman of supreme appeal tribunal for conscientious objectors, 1916–18; demanded firm government for Ireland and opposed treaty of 1921; a leader in seeking to free conservatives from the Lloyd George coalition; lord president of the Council (1922–4) and chancellor of the duchy of Lancaster (1922–3); chairman of sub-committee of Committee of Imperial Defence which recommended development of chiefs of staff committee and appointment of a chairman to deputize for prime minister on the Committee of Imperial Defence; lord privy seal, 1924–9; leader of House of Lords, 1925–9; of opposition in the Lords, 1929–31; strongly opposed Government of India bill, 1935; became concerned over lack of defence preparations; advocated conscription (1938) and a national government (1939); president, National Union of Conservative and Unionist Associations, 1942–5; a most regular attendant in parliament and authoritative speaker on many issues, especially ecclesiastical, moral, and industrial problems, military, local government, imperial, and foreign affairs; frequently in conflict with his own leaders and still more so with the liberal and labour parties; yet enjoyed immense popularity and exercised unique influence; of deepest personal religion; chairman of Canterbury House of Laymen, 1906–11; vitally interested in his three estates and gave much time to local work; GCVO, 1909; KG, 1917.

CECIL, ROBERT ARTHUR TALBOT GASCOYNE-, third MARQUESS OF SALISBURY (1830–1903), prime minister; lineal descendant of Robert Cecil, first Earl of Salisbury [q.v.]; born at Hatfield; educated at Eton and Christ Church, Oxford; fellow of All Souls, 1853; in Australia, 1851–3; conservative MP for Stamford, 1853–68; made speeches on property, religious education, and foreign affairs; married (1857) with father's disapproval; wrote much for reviews, beginning (1860) a long series of articles in *Quarterly Review*; contributed 'Theories of Parliamentary Reform' to *Oxford Essays* (1858) which showed distrust of democracy and aversion from change; vigorous in opposition to liberal government, 1859–66; influenced public opinion by his pungent articles in *Quarterly Review*; succeeded brother as Viscount Cranborne, 1865; became secretary for India and privy counsellor in Derby administration, 1866–7; resigned post on Disraeli's introduction of household suffrage reform bill, 1867; opposed Gladstone's motion for Irish Church disestablishment, 1868; succeeded father, Apr. 1868; criticized Irish Land and

Education Acts of 1870, and Gladstone's abolition of army purchase; unsuccessfully resisted Universities Tests Abolition Act, 1871; succeeded Derby as chancellor of Oxford University, 1869; urged appointment of universities commissions, 1877; chairman of Great Eastern Railway, 1868–72; secretary for India in Disraeli's government, 1874–8; criticized Disraeli's public worship regulation bill, 1874; favoured 'forward policy' in Afghanistan, inaugurated by Lord Lytton [q.v.], 1876; opposed to Russian designs in Asia and powers at Constantinople, 1876; his proposal to reorganize Bulgaria, under control of governors nominated by Sultan, refused by the Porte; became secretary for foreign affairs, 1 Apr. 1878; issued 'Salisbury Circular' (2 Apr.), requiring submission of the treaty of San Stefano to a European conference, and emphatically declaring against creation of a 'big Bulgaria' as a menace to Europe; with Lord Beaconsfield he represented England at congress of Berlin (June–July 1878), ratifying treaty, by which the idea of a 'big Bulgaria' was abandoned, Austria was entrusted with administrative control of Bosnia and Herzegovina, Russia obtained Batum as free port, and England secured protectorate of Cyprus; created KG, June 1878; leader of opposition in Lords on death of Beaconsfield, 1881; negotiated privately with liberal government the provisions of a redistribution bill supplementary to franchise bill, 1884; a trenchant critic of liberal foreign policy in Sudan, Egypt, and Afghanistan; prime minister and foreign secretary, June 1885; successfully negotiated with Russia on Afghan frontier question; secured eastern frontier of India by annexation of Burma; raised Egyptian loan and supported unity of Eastern Rumelia and Bulgaria; passed bill for housing of working classes; his party in a minority at the general election in Nov. 1885; resigned on passing of vote of censure in House of Commons, Feb. 1886; again prime minister and first lord of the Treasury, July 1886; foreign secretary, Dec. 1886; inaugurated first colonial conference, 1887; Local Government Act and Closure Act passed, 1888; by granting royal charter to British East Africa Company he recovered English hold over upper sources of Nile, 1888; granted charter to British South Africa Company, 1889; acquired English protectorate over Zanzibar, giving up Heligoland to Germany and recognizing French protectorate in Madagascar, 1890; passed Free Education Act, 1891; supported appointment of Parnell commission, 1888; defeated at general election, 1892; formed coalition ministry as premier and foreign secretary with Duke of Devonshire [q.v.] and Joseph Chamberlain [q.v.], 1895; lord warden of cinque ports, 1895; his government's measures included Workmen's Compensation Act (1897), Criminal Evidence Act (1898), and Inebriates Act (1898); pacified America in dispute over the boundary between British Guiana and Venezuela, 1895–6; prevented European intervention in Spanish-American war, 1898; facilitated American control of Panama canal by Hay-Pauncefote treaty, 1901; secured Wei-hai-wei for England and policy of open door in China, developing British enterprise in China, 1897; caused French to relinquish claims in the Sudan, 1898–9; concluded secret treaty with Germany regarding Portuguese Africa; refused (1897) to avenge Armenian massacres without approval of 'Concert of Europe', which under his leadership established autonomy in Crete, 1899; supporter of international arbitration and Hague conference, 1899; refused idea of foreign mediation in South African war, 1900; resigned foreign secretaryship and became lord privy seal, Nov. 1900; resigned premiership to nephew, A. J. Balfour, July 1902; died at Hatfield; a close student of science and theology, he was president of the British Association in 1894, and was interested in electricity in later life; contemning the impracticable and scorning sentiment and cant, he was a master of satire; cautious in diplomacy, he regarded democracy as inimical to individual freedom, had strong faith in the historical continuity of government, and was a severe critic of the radical idea of progress; his foreign policy was one of pacific imperialism; GCVO, 1902; DCL, Oxford, 1869; LLD, Cambridge, 1888; hon. student of Christ Church, 1894; *Essays: Foreign Politics* were posthumously republished (1905) from the *Quarterly Review*, to which he contributed thirty-three articles, 1860–83.

CENTLIVRES, ALBERT VAN DE SANDT (1887–1966) chief justice of South Africa and chancellor of Cape Town University; educated at South African College School, South African College and New College, Oxford; Rhodes scholar; BA, 1909; BCL, 1910; called to bar (Middle Temple) 1910; advocate, Cape Provincial division, South Africa, 1911; parliamentary draftsman, 1920; KC, 1927; High Court judge, South-West Africa, 1922; Cape Provincial division, 1932–4; puisne judge, 1935; Apellate division, 1939; chairman, public service inquiry commission, 1944; chief justice, 1950–7; made important judgments in constitutional law; chancellor, Cape Town University, 1951; hon. DCL, Oxford; hon. LLD, Melbourne, Cape Town, Witwatersrand, and Rhodes University; hon. fellow, New College, Oxford; hon. bencher, Middle Temple.

CHADS, SIR HENRY (1819–1906), admiral; entered navy, 1834; served in East Indies and on west coast of Africa; commander-in-chief at the Nore, 1876-7; admiral, 1877; KCB, 1887.

CHADWICK, HECTOR MUNRO (1870-1947), Anglo-Saxon scholar and historian of early literature; first class, classical tripos, parts i and ii, 1892-3, Clare College, Cambridge; fellow, 1893; university lecturer in Scandinavian, 1910; professor of Anglo-Saxon, 1912-45; reformed English studies at Cambridge; publications include *The Origin of the English Nation* (1907), *The Heroic Age* (1912), and *Growth of Literature* (3 vols., 1932-40, with his wife); FBA, 1925.

CHADWICK, ROY (1893-1947), aeronautical engineer; joined A. V. Roe & Co., 1911; designs include the Avro Baby and Avro Avian; training planes Tutor and Anson; Manchester twin-engined and Lancaster four-engined heavy bombers; the York, Lincoln, and Lancastrian; the Tudor, first pressurized civil aircraft; the Shackleton and Avro Delta plan form aircraft; killed on test flight.

CHALMERS, JAMES (1841-1901), missionary and explorer; joined London Missionary Society, 1861; ordained, 1865; worked at Rarotonga in South Seas, 1866-76; first white man to visit villages of New Guinea, 1876; explored Gulf of Papua; discovered mouths of Purari river, 1879; planted line of mission posts from Papuan gulf to Louisiade Archipelago, 1884; of great service in foundation of protectorate at Port Moresby, 1884; visited R. L. Stevenson [q.v.] at Samoa, 1890; worked in Fly river district at Saguane, 1892-4; sailed for Goaribari Island, Apr. 1901; killed and eaten by savages at Dopima; a simple, zealous, courageous missionary; wrote accounts of work in New Guinea; autobiography in Lovett's *Life*, 1902.

CHALMERS, SIR MACKENZIE DAL-ZELL (1847-1927), judge, parliamentary draftsman, and civil servant; BA, Trinity College, Oxford; called to bar (Inner Temple), 1869; joined Home circuit, 1872; published *Digest of the Law of Bills of Exchange*, 1878; drafted bill which issued in Bills of Exchange Act, 1882; standing counsel to Board of Trade, 1882; helped to reform law of bankruptcy; county court judge, Birmingham, 1884-96; drafted bill which issued in Sale of Goods Act, 1894; legal member of viceroy of India's council, 1896-9; assistant parliamentary counsel, 1899; first parliamentary counsel, 1902; permanent under-secretary of state, home department, 1903-8; third work of codification issued in Marine Insurance Act, 1906; KCB, 1906.

CHALMERS, ROBERT, BARON CHALMERS (1858-1938), civil servant and master of Peterhouse, Cambridge; educated at City of London School and Oriel College, Oxford; second class, natural science, 1881; passed first into Civil Service, 1882; appointed to Treasury; chairman, Board of Inland Revenue, 1907-11; KCB, 1908; permanent secretary, Treasury, 1911-13, and jointly, 1916-19; governor of Ceylon, 1913-16; under-secretary for Ireland, 1916; baron, 1919; master of Peterhouse, 1924-31; FBA, 1927; his pomposity and cynicism concealed sensitivity and many benefactions.

CHAMBERLAIN, (ARTHUR) NEVILLE (1869-1940), statesman; son of Joseph Chamberlain and half-brother of Sir Austen Chamberlain [qq.v.]; educated at Rugby and Mason College, Birmingham; unsuccessfully attempted to grow sisal on father's estate in Bahamas, 1890-7; became outstanding figure in Birmingham industrial life; elected to city council, 1911; established only municipal savings bank, 1916; appointed director-general of national service by Lloyd George (1916) but given neither authority nor equipment; resigned 1917; conservative MP for a Birmingham division, 1918-40; postmaster-general, 1922; chancellor of the Exchequer, 1923-4; minister of health, 1923, 1924-9; successfully tackled immediate housing problem; his Local Government Act (1929) reformed the poor law; while in opposition (1929-31) reorganized conservative central office; chancellor of the Exchequer, 1931-7; secured adoption of general tariff, 1932; took chief political initiative in increasing air estimates, 1934; active in cabinet discussion of foreign affairs; supported League sanctions against Italy, 1935; succeeded Baldwin as prime minister, May 1937; tried to bring Italy and Germany back into the comity of nations in order to avert a war for which Britain was unprepared; negotiations with Italy led to agreement but less successful with Germany; after German invasion of Austria (Mar. 1938) stated that if France was called upon by reason of German aggression to go to aid of her ally Czechoslovakia, Britain's intervention would be 'well within the bounds of probability'; with German army ready to strike Czechoslovakia France proved unready to fulfil her obligations and Daladier suggested to Britain some 'exceptional procedure'; Chamberlain therefore flew to Berchtesgaden to discuss a peaceful settlement with Hitler (15 Sept. 1938) who demanded self-determination for Sudeten Germans; flew to Godesberg, 22 Sept., to announce acceptance by Britain, France, and Czechoslovakia; but Hitler then demanded immediate evacuation by Czechs and conference ended in deadlock; sent warning to Hitler that if France entered into hostilities with Germany Britain would support her; when war seemed inevitable was invited, with Mussolini and Daladier, to meet Hitler at Munich; Godesberg terms so modified that Czech government accepted them; agreement signed, 30 Sept.; acclaimed as saviour of peace on return to England; after invasion of Czechoslovakia (Mar. 1939) announced full support to Poland and after Italian invasion of Albania (Apr. 1939) gave same assurance to Greece and Romania and made long-term agreement with Turkey; denied reports that Britain would abandon Poland after Russo-German pact (21 Aug.); declared war (3 Sept. 1939) after German invasion of Poland (1 Sept.); labour leaders refused to serve under him; resigned 10 May 1940 after withdrawal from Norway; lord president of the Council in (Sir) Winston Churchill's government; resigned owing to illness, 1 Oct., declining titular honours; exceptionally efficient on the business side of the premiership and a lucid and masterful leader of his cabinet.

CHAMBERLAIN, Sir CRAWFORD TROTTER (1821-1902), general; brother of Sir N. B. Chamberlain [q.v.]; cadet in Bengal army, 1837; served in Afghan war (1839), in Sikh war (1845-9), and Momund expedition (1854); promoted lieutenant-colonel for services in Indian Mutiny; CSI, 1866; commanded Gwalior district, 1866-9, and Oudh division, 1974-9; general, 1880; GCIE, 1897.

CHAMBERLAIN, HOUSTON STEWART (1855-1927), political writer; nephew of Sir N. B. Chamberlain and Sir C. T. Chamberlain [qq.v.]; lived successively at Dresden, Vienna, and Bayreuth; student of Wagner; cut himself adrift from England and became naturalized as a German, 1916; his reputation as a thinker mainly rests on *Die Grundlagen des neunzehnten Jahrhunderts* (1899), published in English as *The Foundations of the Nineteenth Century* (1911); this is a broad survey of European culture, not simply a glorification of everything German.

CHAMBERLAIN, JOSEPH (1836-1914), statesman; educated at University College School; member of screw-manufacturing firm in Birmingham, 1854-74; early became interested in question of social reform, which made him adopt radical views; chairman of National Education League of Birmingham, 1868, and of National Education League, 1870; mayor of Birmingham, 1873-5; helped to improve housing and sanitation of Birmingham and to develop municipal social reform generally; colleague of John Bright [q.v.] in parliamentary representation of Birmingham, 1876; formed close offensive and defensive alliance with (Sir)

Charles Dilke [q.v.]; carried out reorganization of liberal party, 1877; became president of Board of Trade and entered second Gladstone cabinet, 1880; out of sympathy with majority of colleagues in ministry of 1880-5; held strongly patriotic and national opinions on foreign affairs combined with extreme radical views on internal matters; opposed to policy of coercion in Ireland and entered into negotiations with Charles Stewart Parnell [q.v.] resulting in Kilmainham treaty, 1882; strongly opposed introduction of Irish land purchase bill without accompanying Irish local government bill, 1885; opposed Egyptian policy of government and advocated relief of General Gordon [q.v.], 1884; deplored loss of Cameroons to England, 1884; proved his capacity at Board of Trade by responsibility for two measures relating to merchant shipping (1880), electric lighting bill (1881), Act for reforming law of bankruptcy (1883), Patent Act (1883), and merchant shipping bill (1884); gave great offence to Queen Victoria by violence of his language at this time and incurred remonstrance from Gladstone for advocacy of unauthorized liberal programme; had friendly discussion with Gladstone on subject of liberal programme, Oct. 1885; MP, West Birmingham, 1885; president of Local Government Board in third Gladstone cabinet, Feb.-Mar. 1886; resigned on introduction of home rule bill which he had consistently opposed; voted against bill, June 1886; sacrificed future leadership of liberal party and even risked political extinction rather than comply with demands of Parnell; resignation of Lord Randolph Churchill [q.v.] (Dec. 1886) dispelled his hopes of organizing new national party; went on mission to United States to negotiate treaty regarding North American fisheries, Nov. 1887-Mar. 1888; visited Egypt and was impressed by benefits accruing to that country from British occupation, 1889; joined third Salisbury cabinet as secretary of state for colonies, 1895; as secretary kept two objects in view; first, the tightening of the bond between Great Britain and the self-governing colonies, and secondly, the development of the resources of the crown colonies together with an increase of trade between them and Great Britain; gave much attention to improving position of West Indies; waged successful campaign on question of health in tropical countries and was instrumental in setting on foot two special schools of tropical medicine; during his secretaryship British possessions in West Africa extended by effective occupation of territory behind Gold Coast and Lagos and by placing of Royal Niger Company's territories under control of the Colonial Office (1900); most memorable work performed in connexion with dominions, especially South Africa; soon after taking up office was confronted with difficulties created by the Jameson raid; accused of complicity in raid; his position with regard to South Africa made easier by appointment of high commissioner, Sir Alfred (afterwards Viscount) Milner [q.v.], in whom he could place entire confidence, 1897; proposed meeting between Milner and President Kruger to discuss grievances of Transvaal uitlanders, chief of these being franchise question; Bloemfontein conference (May–June 1899), honest attempt on part of British authorities to reach *modus vivendi*, broke down through Kruger's intractability; in spite of conciliatory attitude adopted by Chamberlain, war was rendered inevitable; visited South Africa on conclusion of war, 1902; first secretary of state to visit British colony in connection with political questions; helped forward reconciliation between rival races and parties in Cape Colony and dealt successfully with Boers of Transvaal and capitalists of Rand; vetoed suspension of Cape Colony constitution, 1902; piloted Commonwealth of Australia bill through House of Commons, 1900; although mainly preoccupied with colonial questions while in unionist government, largely responsible for Workmen's Compensation Act, 1897; weakest side of his programme shown in his excursions into foreign policy; resigned office on return from South Africa, 1903; induced to take this step by government's refusal to include in its programme granting of preference to imperial wheat; his conviction of need for closer imperial union, which he believed would most readily be achieved through commercial union, caused him to change his economic policy from free trade to tariff reform; proclaimed himself convinced imperialist at Liverpool, Oct. 1903; devoted himself to explaining and popularizing his views during three years' campaign, 1903–6; obliged to withdraw from public life owing to ill health for remaining years, 1906–14; Chamberlain's most outstanding qualities were loyalty to his friends and freedom from jealousy; a collected edition of his speeches published 1914; contributed numerous articles to the *Fortnightly Review* and *Nineteenth Century*.

CHAMBERLAIN, Sir (JOSEPH) AUSTEN (1863–1937), statesman; son of Joseph Chamberlain and half-brother of Neville Chamberlain [qq.v.]; educated at Rugby and Trinity College, Cambridge; visited France and Germany and conceived deep love for former; liberal unionist MP, East Worcestershire (1892–1914), West Birmingham (1914–37); civil lord of Admiralty, 1895–1900; financial secretary to Treasury, 1900–2; postmaster-general, 1902–3; PC, 1902; chancellor of the Exchequer, 1903–5; in opposition attempted to leaven conservative mass with

protectionist doctrine; on resignation of A. J. Balfour (1911) he and Walter (later Viscount) Long [q.v.] as rivals for leadership of conservative party stood down in favour of Bonar Law who at first relied greatly upon him; secretary of state for India, 1915; resigned after revelation of mismanagement of Mesopotamian campaign, 1917; member of war cabinet, Apr. 1918; chancellor of the Exchequer, 1919–21; produced three sound budgets; conservative leader, 1921; dissatisfaction with his support of Lloyd George culminated in Carlton Club meeting (19 Oct. 1922) which brought coalition and his leadership to an end; foreign secretary, 1924–9; denounced protocol of Geneva; secured signature of Locarno pact and KG, 1925; regularly attended council of League of Nations; tried to improve Anglo-Egyptian relations; his policy of solidarity with France unpopular in many quarters; first lord of Admiralty in 'national' government, Aug.–Oct. 1931; declined further office to make room for younger men and exercised greatest influence as elder statesman; Nobel peace prize, 1925.

CHAMBERLAIN, Sir NEVILLE BOWLES (1820–1902), field-marshal; brother of Sir C. T. Chamberlain [q.v.]; obtained commission in East India Company's army, 1837; joined 16th Bengal native infantry, 1838; at storming of Ghazni, 1839; at Kandahar, 1841; several times wounded; in march from Kandahar to Kabul, 1842; served in Gwalior campaign, 1843; military secretary to governor of Bombay, 1846–8; conspicuous in passage of the Chenab under Lord Gough [q.v.], 1849; in charge of organization of military police for Punjab, 1852; went to South Africa for health, 1852–4; in command of Punjab irregular force, 1854; adjutant-general of Bengal army in Indian Mutiny, 1857; repulsed mutineers before Delhi (July); CB, 1857; broke up Sikh conspiracy at Dera Ismail Khan, 1858; forced his way against Mahsuds to Kaniguram, 1860; KCB, 1863; led force against Wahabi fanatics on Ambela pass, 1863; severely wounded; accompanied Duke of Edinburgh to India, 1869; lieutenant-general, 1872; GCSI, 1873; GCB, 1875; general, 1877; commanded Madras army, 1876–81; returned to England, 1881; field-marshal, 1900.

CHAMBERS, DOROTHEA KATHARINE (1878–1960), lawn tennis champion; born DOUGLASS; married Robert Lambert Chambers, 1907; Olympic gold medallist, 1908; won Wimbledon singles seven times in 1903–14; narrowly defeated by Suzanne Lenglen in memorable match in 1919; captained British winning Wightman Cup team, 1925, winning her singles and doubles matches at age of forty-six.

CHAMBERS, Sir EDMUND KERCHEVER (1866-1954), historian of the English stage and civil servant; educated at Marlborough and Corpus Christi College, Oxford; first class, *lit. hum.*, 1889; entered Education Department, 1892; second secretary, 1921-6; first president of Malone Society, 1906-39; produced many editions of English classics; edited *Early English Lyrics* (1907) and *Oxford Book of Sixteenth Century Verse* (1932); works include *The Medieval Stage* (2 vols., 1903), *The Elizabethan Stage* (4 vols., 1923), *Shakespeare* (2 vols., 1930), and biographies of Coleridge (1938) and Matthew Arnold (1947); CB, 1912; FBA, 1924; KBE, 1925.

CHAMBERS, RAYMOND WILSON (1874-1942), scholar and writer on English language and literature; graduated in English, University College, London, 1894; Quain student, 1899; fellow, 1900; librarian, 1901; assistant professor of English, 1904; Quain professor, 1922-41; work ranged from early Germanic legend to Shakespeare and Milton, from textual study to biography; publications include *Beowulf* (1914) and *Thomas More* (1935); FBA, 1927.

CHAMIER, STEPHEN HENRY EDWARD (1834-1910), lieutenant-general, royal Madras artillery; born in Madras; nephew of Frederick Chamier [q.v.]; joined Madras artillery, 1853; served in Indian Mutiny at Cawnpore and Lucknow, 1857-8; commanded first battery artillery, Hyderabad, 1858; inspector-general of ordnance, Madras, 1881-6; lieutenant-general and CB, 1886; a good musician (Mus. Bac., Dublin).

CHAMPNEYS, BASIL (1842-1935), architect; son of W. W. and brother of Sir F. H. Champneys [qq.v.]; educated at Charterhouse and Trinity College, Cambridge; set up as architect, 1867; works include: (Oxford) Indian Institute, Mansfield College; (Cambridge) Archaeological Museum, Newnham College; John Rylands memorial library, Manchester; happiest in stone and late Gothic style; published *Memoirs and Correspondence* (1900) of his friend Coventry Patmore [q.v.].

CHAMPNEYS, Sir FRANCIS HENRY, first baronet (1848-1930), obstetrician; son of W. W. Champneys [q.v.]; BA, Brasenose College, Oxford; studied medicine at St. Bartholomew's Hospital; assistant obstetric physician, St. George's Hospital, and obstetric physician, General Lying-in Hospital, 1880; obstetric physician, St. George's, 1885; physician accoucheur, St. Bartholomew's, 1891-1913; did much to promote Midwives Act (1902); chairman, Central Midwives Board, 1903-30; baronet, 1910.

CHANCE, Sir JAMES TIMMINS, first baronet (1814-1902), manufacturer and lighthouse engineer; scholar of Trinity College, Cambridge; seventh wrangler, 1838; joined father's glass business at Birmingham; made improvements in dioptric apparatus for lighthouses from 1850 onwards; joined Michael Faraday [q.v.] in experiments with apparatus at Whitby southern lighthouse, 1860; read classic papers on the subject before Institution of Civil Engineers, 1867, for which he was awarded the Telford medal and premium; retired from business, 1872; benefactor to Birmingham institutions; presented West Smethwick Park to town, 1895; endowed Chance school of engineering in Birmingham University, 1900; baronet, 1900.

CHANCELLOR, Sir JOHN ROBERT (1870-1952), soldier and administrator; educated at Blair Lodge, Polmont and Royal Military Academy; commissioned in Royal Engineers, 1890; served in Dongola and Tirah expeditions; DSO, 1898; assistant military secretary, Committee of Imperial Defence, 1904; secretary, Colonial Defence Committee, 1906-11; major, 1910; lieutenant-colonel, 1918; governor of Mauritius, 1911-16, of Trinidad and Tobago, 1916-21, of Southern Rhodesia, 1923-8; high commissioner for Palestine and Trans-Jordan, 1928-31; CMG, 1909; KCMG, 1913; GCMG, 1922; GCVO, 1925; GBE, 1947.

CHANNELL, Sir ARTHUR MOSELEY (1838-1928), judge; son of Sir W. F. Channell [q.v.]; BA, Trinity College, Cambridge; called to bar (Inner Temple), 1863; joined South-Eastern circuit; specialized in local government; QC, 1885; recorder of Rochester, 1888-97; judge of Queen's Bench division and knighted, 1897; retired, 1914; PC, 1914; member of Judicial Committee in prize court appeals, 1916-21.

CHANNER, GEORGE NICHOLAS (1842-1905), general, Indian staff corps; born at Allahabad; joined Indian army, 1859; served in Ambela campaign, 1863; won VC for bravery in Burkit Putus Pass, 1875; served in Afghan war, 1878-80; commanded Kuram field force at capture of Peiwar Kotal, 1879; commanded 1st brigade of Hazara field force in Black Mountain expedition, 1888; CB, 1899; brigadier-general in command of Assam district, 1892-6; general, 1899.

CHAPLIN, HENRY, first Viscount Chaplin (1840-1923), politician and sportsman; gentleman commoner, Christ Church, Oxford, 1859-60; formed friendship with Prince of Wales; inherited Blankney Hall, Lincolnshire, from uncle, 1859; his primary interests, hunting and racing; master of Burton hunt, 1865; won the Derby with Hermit, 1867; follower and admirer of Disraeli; keen protectionist; conservative MP, mid-Lincolnshire (known after 1885 as Sleaford division), 1868-1906; for Wimbledon, 1907-16; in opposition, 1880-5; disagreed in party matters with Lord Randolph Churchill [q.v.]; chancellor of duchy of Lancaster, 1885-6; entered cabinet as president of Board of Agriculture, 1889; in charge of Small Holdings Act; president of Local Government Board, 1895-1900; introduced Agricultural Rating Act (1896), Vaccination Act (1898), and Housing Act (1900); representative of agriculture on tariff reform commission, 1903; leader of opposition to coalition government, 1915; created viscount, 1916; his fortune impaired by hospitality and cost of stables and kennels.

CHAPMAN, DAVID LEONARD (1869-1958), chemist; brother of Sir Sydney Chapman [q.v.]; educated at Manchester grammar school and Christ Church, Oxford; first class, chemistry, 1893; on chemistry staff of Manchester University, 1897-1907; fellow and tutor in charge of science teaching, Jesus College, Oxford, 1907-44; FRS, 1913; his theoretical method basic to detonation studies.

CHAPMAN, EDWARD JOHN (1821-1904), mineralogist; professor of mineralogy in University College, London (1849-53), and Toronto University (1853-95); published researches into minerals and fossils of Canada, 1864.

CHAPMAN, ROBERT WILLIAM (1881-1960), scholar and university publisher; educated at Dundee high school, St. Andrews (first class, classics), and Oriel College, Oxford; first class, *lit. hum.*, 1906; appointed assistant secretary to Charles Cannon [q.v.], at the Clarendon Press, 1906; secretary to the delegates, 1920-42; edited Jane Austen's novels (5 vols., 1923, with K. M. Metcalfe) and letters (2 vols., 1932); published *Jane Austen—A Critical Bibliography* (1953); edited Boswell's *Tour to the Hebrides* and Johnson's *Journey to the Western Islands* (1924), Boswell's *Note Book 1776-77* (1925), *Johnson and Boswell Revised by Themselves and Others* (1928), *Papers Written by Dr. Johnson and Dr. Dodd, 1777* (1926), and *The Letters of Samuel Johnson with Mrs. Thrale's Genuine Letters to Him* (3 vols., 1952); wrote *Cancels* (1930) and many articles on bibliography and kindred subjects; FBA, 1949; CBE, 1955; fellow, Magdalen College, Oxford, 1931; hon. LLD St. Andrews, 1933.

CHAPMAN, SYDNEY (1888-1970), mathematician and geophysicist; educated at elementary school, Royal Technical Institute, Salford,

Manchester University and Trinity College, Cambridge; first class, part ii, mathematics tripos, 1910; senior assistant, Greenwich Observatory, 1910-14; college lecturer, Trinity College, Cambridge, 1914; returned to Greenwich, 1916-18; research into gas viscosity, heat conduction and diffusion, 1912-19; FRS, 1919; professor, natural philosophy, Manchester, 1919-24; chief professor, mathematics, Imperial College, London, 1924; worked for Army Council, 1943-5; Sedleian professor, natural philosophy, Oxford; fellow, Queen's College, 1946-53; president, organizing committee, International Geophysical Year, 1957-8; publications include *Solar Plasma, Geomagnetism and Aurora* (1964), *Atmospheric Tides* (with R. S. Lindzen, 1970), and *Solar-Terrestrial Physics* (with S. I. Akasofer, 1972).

CHAPMAN, Sir SYDNEY JOHN (1871-1951), economist and civil servant; brother of D. L. Chapman [q.v.]; educated at Manchester grammar school and Owens College; BA, London, 1891; first class, parts i and ii, moral sciences tripos, 1897-8, Trinity College, Cambridge; Stanley Jevons professor of political economy, Manchester, 1901-18; joined Board of Trade, 1915; permanent secretary, 1919-27; chief economic adviser to Government, 1927-32; on Import Duties Advisory Committee, 1932-9; chairman, arc lamp carbon pool, controller of matches, and vice-chairman, Central Price Regulation Committee in war of 1939-45; published *Lancashire Cotton Industry* (1904), *Work and Wages* (3 vols., 1904-14), and *Outlines of Political Economy* (1911); CBE, 1917; CB, 1919; KCB, 1920.

CHARLES, JAMES (1851-1906), portrait and landscape painter; entered Royal Academy Schools, 1872; exhibited regularly at Academy (1875-1904) and Grosvenor Gallery; paintings include 'Christening Sunday', 1887; 'Will it Rain?', 'Milking Time', 'In Spring Time', 1896; in Paris, 1889-90; in Venice, 1891; skilful in sunlight effects; works shown at winter exhibition of Royal Academy, 1907; represented in many public galleries.

CHARLES, ROBERT HENRY (1855-1931), archdeacon of Westminster and biblical scholar; elder brother of Sir R. H. Charles; educated at Queen's College, Belfast, and Trinity College, Dublin; deacon, 1883; priest, 1884; after curacies in London settled in Oxford, 1891; despite lack of sympathetic understanding acquired vast and accurate knowledge of Apocalyptic literature; edited Oxford edition of *The Apocrypha and Pseudepigrapha of the Old Testament in English* (1913); published editions of *Book of Jubilees* (1895), *Enoch* (1906), *Testaments of the*

Twelve Patriarchs (1908), the *Apocalypse of St. John* (1920), and a *Critical and Exegetical Commentary on the Book of Daniel* (1929); FBA, 1906; canon of Westminster, 1913; archdeacon, 1919.

CHARLESWORTH, MARTIN PERCIVAL (1895-1950), classical scholar; first class, parts i and ii, classics tripos, Jesus College, Cambridge, 1920-1; fellow and lecturer, St. John's College, 1923; tutor, 1925-31; president, 1937-50; ordained, 1940; university lecturer (1926), Laurence reader (1931-50) in classics; joint-editor, *Cambridge Ancient History*; FBA, 1940.

CHARLEY, SIR WILLIAM THOMAS (1833-1904), lawyer; BA, St. John's College, Oxford, 1856; DCL, 1868; called to bar, 1865; helped to reorganize conservative party in London and Lancashire, 1867; conservative MP for Salford, 1868-80; strong supporter of Disraeli; common serjeant, 1878-92; knighted and QC, 1880; vigorous defender of Church of England; enthusiastic Volunteer; author of legal and other works.

CHARLOT, ANDRÉ EUGENE MAURICE (1882-1956), showman; born in Paris; educated at Lycée Condorcet and Conservatoire; joint-manager, Alhambra, London, 1912; naturalized, 1922; produced 36 west-end revues in 1915-35; also light comedy, farce, and musical plays, including *Wonder Bar* (1930); introduced restaurant floor shows, 1922, and BBC programmes; moved to Hollywood, 1937.

CHARNWOOD, first BARON (1864-1945), liberal politician and man of letters. [See BENSON, GODFREY RATHBONE.]

CHAROUX, SIEGFRIED JOSEPH (1896-1967), sculptor; born and educated in Vienna; served in Austro-Hungarian army, 1915-17; studied at Vienna School of Fine Arts, 1919-24; and at Vienna Academy, 1924-8; worked as political cartoonist; first exhibition, 1925; settled in England, 1935; became naturalized, 1946; bronze bust of Sir Stafford Cripps [q.v.] shown at Royal Academy, 1946; terracota figure, 'Youth: Standing Boy' acquired through Chantrey Bequest for Tate Gallery, 1948; other works included 'The Islanders' for South Bank Festival of Britain exhibition, 1951, 'Civilisation: The Judge' for the Law Courts, and 'The Cellist', outside the Royal Festival Hall, London; ARA, 1949; RA, 1956; hon. professor, Republic of Austria, 1958.

CHARRINGTON, FREDERICK NICHOLAS (1850-1936), philanthropist and temperance reformer; left father's brewery to take up religious and temperance work in its vicinity with extreme, sometimes eccentric, enthusiasm; founded Tower Hamlets mission; original member, London County Council, 1889-95.

CHARTERIS, ARCHIBALD HAMILTON (1835-1908), biblical critic; MA, Edinburgh, 1853; hon. DD, 1868, LLD, 1898; joined Church of Scotland ministry, 1858; professor of biblical criticism, Edinburgh University, 1868-98; published conservative theological works on canonicity of New Testament, 1880-7; organized practical Christian effort in Scotland; revived order of deaconesses; started and edited a periodical, *Life and Work*, 1879; supporter of foreign missions; moderator of General Assembly, 1892.

CHASE, DRUMMOND PERCY (1820-1902), last principal of St. Mary Hall, Oxford, 1857-1902; BA, first class, *lit. hum.*, Oriel College, Oxford, 1841; fellow, 1842-1902, and tutor; DD, 1880; vicar of St. Mary's, Oxford, 1855-63, 1876-8; wrote pamplets on academic questions; edited with translation Aristotle's *Nicomachean Ethics*, 1847.

CHASE, FREDERIC HENRY (1853-1925), bishop of Ely; BA, Christ's College, Cambridge; ordained, 1876; lecturer in theology, Pembroke College, 1881-90; at Christ's College, 1893-1901; principal, Cambridge Clergy Training School, 1887-1901; president, Queens' College, 1901-6; Norrisian professor of divinity, Cambridge, 1901-5; bishop of Ely, 1905-24; bestowed much labour, both in Convocation and on committees, on revision of Book of Common Prayer; published many learned theological works.

CHASE, MARIAN EMMA (1844-1905), water-colour painter; studied under Margaret Gillies [q.v.]; exhibited at Royal Academy and elsewhere, 1866-1905; member of Institute of Painters in Water Colours, 1876; truthful and delicate painter of flowers, fruit, and still life.

CHASE, WILLIAM ST. LUCIAN (1856-1908), lieutenant-colonel; born at St. Lucia, West Indies; joined Bombay army, 1875; served in Afghan war, 1879-80; won VC in sortie from Kandahar, 1880; took part in Zhob Valley (1884) and Lushai (1889-90) expeditions, and Tirah campaign, 1897-8; lieutenant-colonel, 1899; CB, 1903.

CHATFIELD, ALFRED ERNLE MONTACUTE, first BARON CHATFIELD (1873-1967), admiral of the fleet; educated at St. Andrew's School, Tenby; entered *Britannia* at Dartmouth,

1886; joined *Iron Duke*, 1888; specialized in gunnery; lieutenant in *Caesar*, 1899; captain, commanded *Albermarle* in Atlantic fleet, 1909; flag captain in *Lion* with David (later Admiral of the Fleet first Earl) Beatty [q.v.], 1913; fought in battles of Heligoland Bight (1914), Dogger Bank (1915), and Jutland (1916); flag captain and chief of staff to Beatty in *Iron Duke*, 1916; fourth sea lord, Admiralty, 1919; assistant chief of staff, 1920; rear-admiral, 1920; senior naval delegate at Washington, 1921; concerned with negotiations for naval armaments limitation, 1921–2; commanded third cruiser squadron, Mediterranean, 1923–5; third sea lord and controller of the navy, 1925–9; vice-admiral, 1926; commander-in-chief, Atlantic fleet in *Nelson*, 1929; admiral, Mediterranean fleet, in *Queen Elizabeth*, 1930–2; first sea lord, 1933–8; secured naval control of Fleet Air Arm, 1937; admiral of the fleet, 1935; baron, 1937; minister for co-ordination of defence, PC, OM, 1939; post abolished, 1940; hon. degrees, Oxford and Cambridge; published *The Navy and Defence* (1942) and *It Might Happen Again* (1947).

CHATTERJEE, SIR ATUL CHANDRA (1874–1955), Indian civil servant; BA, Calcutta, 1892; studied at King's College, Cambridge, 1895; posted to United Provinces, 1897; revenue secretary, 1917, chief secretary, 1919; secretary (1921), council member (1923), central department of industries; high commissioner, London, 1925–31; on India Council, 1931–6; adviser to secretary of state, 1942–7; represented India at International Labour Conferences; president, 1927; vice-president, 1932, president, 1933 (leading its delegation to world economic conference), International Labour Office; represented India at League Assembly, 1925, London naval conference, 1930, Ottawa conference, 1932; wrote *A Short History of India* (with W. H. Moreland, 1936) and *The New India* (1948); KCIE, 1925; KCSI, 1930; GCIE, 1933; hon. LLD Edinburgh.

CHAUVEL, SIR HENRY GEORGE (1865–1945), general; born in New South Wales; captain, Queensland military staff, 1896; served in South African war; trained mounted troops, 1902–11; adjutant-general, Australian forces, 1911–14; commanded 1st Light Horse brigade in Egypt and Gallipoli, 1914–15; 1st Australian Infantry division, Gallipoli and Egypt, 1915–16; Anzac Mounted division in Egypt, 1916–17; Desert Column, 1917; Desert Mounted Corps, in Allenby's advance in Palestine, 1917–19; inspector-general, Australian forces, 1919–30; chief of general staff, 1923–30; KCMG, 1917; GCMG, 1919.

CHAVASSE, CHRISTOPHER MAUDE (1884–1962), bishop of Rochester; eldest (and twin) son of the Revd. Francis James Chavasse [q.v.]; educated at Magdalen College School, Oxford, Liverpool College, and Trinity College, Oxford; blue for lacrosse and athletics; he and twin brother represented Britain at the Olympic Games, 1908; ordained to curacy, St. Helens, Lancashire; chaplain in the Forces, 1914–18; MC, croix de guerre; deputy assistant chaplain-general in IXth Corps, 1918; vicar, St. George's, Barrow in Furness, 1919; rector, St. Aldates, Oxford, 1922–8; opposed revision of the Prayer Book; first master, St. Peter's Hall, 1929–39; bishop of Rochester, 1940–60; chairman of the commission which produced *Towards the Conversion of England* (1945); OBE, 1936; TD, 1940; DL, 1959; DD (Lambeth); hon. fellow, Trinity College, Oxford (1955) and St. Peter's (1949).

CHAVASSE, FRANCIS JAMES (1846–1928), bishop of Liverpool; BA, Corpus Christi College, Oxford; ordained, 1870; rector of St. Peter-le-Bailey, Oxford, 1877–89; principal of Wycliffe Hall, Oxford, 1889–1900; bishop of Liverpool, 1900–23; the effective founder of Liverpool Cathedral; projector of St. Peter's Hall, Oxford; a hard-working evangelical churchman and effective preacher.

CHEADLE, WALTER BUTLER (1835–1910), physician; BA, Caius College, Cambridge, 1859; MB, 1861; explored western Canada, 1862–4; published *The North-west Passage by Land*, 1865; FRCP, 1870; senior censor, 1898; Lumleian lecturer, 1900; on active staff of St. Mary's Hospital, 1867–1904; physician to Children's Hospital, Great Ormond Street, 1869–92; first (1877) defined nature of infantile scurvy; wrote on feeding of infants, 1889, and on infantile rheumatism, 1899; supported claims of medical women.

CHEATLE, ARTHUR HENRY (1866–1929), otologist; assistant surgeon in aural surgery, King's College Hospital medical school, 1899; aural surgeon, lecturer in aural surgery, King's College Hospital, 1910; Hunterian professor, Royal College of Surgeons, 1906; transformed otology into scientific entity; his chief original work related to anatomy of temporal bone; collected specimens of temporal bones and presented 700 to Hunterian Museum; CBE, 1919.

CHEESMAN, ROBERT ERNEST (1878–1962), explorer and naturalist; educated at Merchant Taylors' School and Wye Agricultural College; keen ornithologist, elected to British Ornithologists' Union, 1908; enlisted in the Buffs, 1914; assistant to deputy director of agriculture, Mesopotamia, 1917; private secretary to high commissioner, Iraq, 1920–3; fellow,

Royal Geographical Society, 1920; did mapping work on Arabian coast, 1921, and in Arabian desert, 1923-4; joined Sudan political service; consul, North-West Ethiopia, 1925; explored and mapped the Blue Nile and Lake Tana, 1926-33; retired, 1934; OBE, 1923; CBE, 1935; head of Ethiopian intelligence section, Sudan Defence Force, 1940; oriental counsellor, Addis Ababa, 1942-4; publications include *Notes on Vegetable Growing in Mesopotamia in 1917* (1918), *In Unknown Arabia* (1926), and *Lake Tana and the Blue Nile* (1936).

CHEETHAM, SAMUEL (1827-1908), arch-deacon of Rochester; fellow of Christ's College, Cambridge, 1850; tutor, 1853-8; MA, 1853; DD, 1880; professor of pastoral theology, King's College, London, 1863-82; honorary canon of Rochester, 1878; archdeacon of South-wark, 1879, and of Rochester, 1882; FSA, 1890; Hulsean lecturer, Cambridge, 1896-7; edited with Sir William Smith [q.v.] *Dictionary of Christian Antiquities*, 1875-80; completed history of church by Charles Hardwick [q.v.].

CHELMSFORD, second BARON (1827-1905), general. [See THESIGER, FREDERIC AUGUSTUS.]

CHELMSFORD, first VISCOUNT (1868-1933), viceroy of India. [See THESIGER, FREDERIC JOHN NAPIER.]

CHERMSIDE, SIR HERBERT CHARLES (1850-1929), lieutenant-general; grandson of Sir R. A. Chermside [q.v.]; passed first into and first out of Royal Military Academy, Woolwich; commissioned to Royal Engineers, 1870; employed continuously on foreign service, 1876-99; served in Turkey, Anatolia, and Egypt; governor-general, Red Sea littoral, 1884-6; military attaché, Constantinople, 1889-96; British military commissioner and commander of British troops, Crete, 1897-9; major-general, 1898; commanded Curragh district, 1899, 1901; commanded third division in South Africa, 1900; governor of Queensland, 1902-4; retired with rank of lieutenant-general, 1907; KCMG, 1897.

CHERRY-GARRARD, APSLEY GEORGE BENET (1886-1959), polar explorer; educated at Winchester and Christ Church, Oxford; assistant zoologist with Captain R. F. Scott [q.v.] on second Antarctic expedition; made Winter Journey, 1911, with E. A. Wilson [q.v.] and Bowers to Emperor Penguin rookery at Cape Crozier; went alone to One Ton Depôt, Mar. 1912, hoping to meet Scott's returning party; member of subsequent search party; wrote *The Worst Journey in the World* (1922).

CHERWELL, VISCOUNT (1886-1957), scientist and politician. [See LINDEMANN, FREDERICK ALEXANDER.]

CHESTERTON, GILBERT KEITH (1874-1936), poet, novelist, and critic; educated at St. Paul's School; became journalist and chose to remain so in style and manner; edited *New Witness*, 1916-23, and *G.K.'s Weekly*, 1925-36; publications include: *The Wild Knight* (1900), *The Ballad of the White Horse* (1911), and *The Queen of Seven Swords* (1926); fantastic stories: *The Napoleon of Notting Hill* (1904) and *Manalive* (1912); studies of *Robert Browning* (1903), *Charles Dickens* (1906), *The Victorian Age in Literature* (1913), and *R. L. Stevenson* (1927); and detective stories beginning with *The Innocence of Father Brown* (1911); social and religious thought developed in controversy with G. B. Shaw, H. G. Wells, and Robert Blatchford [qq.v.], and influenced by friendship with Hilaire Belloc and marriage (1901) with Frances Blogg, an Anglo-Catholic; defined social credo in *What's Wrong with the World?* (1910); followed *Heretics* (1905) with *Orthodoxy* (1908) in which he saw Christianity as answer to riddle of universe; his fascinating but inaccurate *Short History of England* (1917), and *Irish Impressions* (1919) and *The New Jerusalem* (1920) marked stages in progress towards his reception into Roman Catholic Church, 1922; published *St. Francis of Assisi* (1923), *The Everlasting Man* (1925), and *St. Thomas Aquinas* (1933); after 1908 lived at Beaconsfield where, childless, he surrounded himself with children and yearly grew a fatter and more legendary figure, with flapping hat, cloak, and sword-stick; absent-minded, high-spirited, and good-natured almost to weakness; but rock-like in maintaining his ideas.

CHETWODE, SIR PHILIP WALHOUSE, seventh baronet, and first BARON CHETWODE (1869-1950), field-marshal; educated at Eton; commissioned in 19th Hussars, 1889; saw active service in Burma (1892-3) and South Africa (1899-1902); captain, 1897; succeeded father, 1905; commanded his regiment, 1908-12; the 5th Cavalry brigade (1914-15) and 2nd Cavalry divison (1915-16) in France and Flanders; Desert Column in Egypt, 1916-17; XX Corps in advance on Jerusalem, 1917-18; conceived and directed stoke which turned Turks' resistance; military secretary, War Office, 1919-20; deputy chief of imperial general staff, 1920-2; held Aldershot command, 1923-7; chief of general staff, India, 1928-30; commander-in-chief, 1930-5; field-marshal, 1933; KCB, 1918; GCSI, 1934; OM, 1936; baron, 1945.

CHEVALIER, ALBERT (1861-1923), comedian; went on stage, 1877; made first appearance in London music-hall at London Pavilion, 1891; an immediate success; engaged in all principal London and provincial music-halls; his songs, such as 'My Old Dutch', immensely popular; toured successfully in America; gave series of recitals annually in London; later appeared again on regular stage; sang over a hundred songs, eighty of which he composed.

CHEYLESMORE, second BARON (1843-1902), mezzotint collector. [See EATON, WILLIAM MERITON.]

CHEYLESMORE, third BARON (1848-1925), major-general. [See EATON, HERBERT FRANCIS.]

CHEYNE, THOMAS KELLY (1841-1915), Old Testament scholar; educated at Worcester College, Oxford, and at Göttingen; vice-principal of St. Edmund Hall, Oxford, 1864-8; fellow of Balliol, 1868-82; joined Old Testament revision company, 1884; Oriel professor of interpretation of scripture at Oxford and canon of Rochester, 1885-1908; disciple in biblical criticism of Heinrich von Ewald; initiated scholarly critical movement in England; author of many books on the Old Testament; his later work spoiled by extravagant views.

CHEYNE, SIR (WILLIAM) WATSON, first baronet (1852-1932), bacteriologist and surgeon; qualified at Edinburgh, 1875; house-surgeon to Lister in Edinburgh and at King's College Hospital where he became assistant-surgeon (1880), surgeon, and professor, 1891-1917; pioneer in antiseptic surgery and emphasized value of preventive medicine; FRS, 1894; baronet, 1908; MP, Scottish universities, 1917-22.

CHIFLEY, JOSEPH BENEDICT (1885-1951), Australian prime minister; born in New South Wales; mainly self-educated; joined railways, 1903; qualified as driver, 1914; dismissed as leading trade-unionist but reinstated, 1917; a founder of Australian Federated Union of Locomotive Enginemen, 1920; MHR for Macquarie, 1928-31, 1940-51; minister for defence, 1931; on royal commission on banking, 1936-7; closest confidant of John Curtin [q.v.]; member of war cabinet and its production executive, 1941-5; treasurer, 1941-9; established federal monopoly of direct taxation, 1942; expanded functions of Commonwealth Bank; minister for post-war reconstruction, 1942-5; successfully introduced measures for demobilization, resettlement, welfare, and development, and Australian participation in international planning; prime minister, 1945-9; leader of opposition, 1950-1; attended prime ministers'

conferences, 1946 and 1949; created Australian National Works Council (1943), Australian National University, Trans-Australia Airlines, Snowy Mountains Hydro-Electricity scheme and other national enterprises; reformed institutions for industrial arbitration and conciliation; PC, 1946.

CHILD, HAROLD HANNYNGTON (1869-1945), author and critic; scholar of Winchester and Brasenose College, Oxford; from 1902 wrote for *The Times Literary Supplement* and literary and dramatic criticism and special articles for *The Times*; dramatic critic, *Observer*, 1912-20; wrote stage-histories for the 'New Shakespeare'; contributed widely to *Cambridge History of English Literature*.

CHILD, THOMAS (1839-1906), minister of the 'new church'; became Swedenborgian preacher, 1872; published much in support of 'new church' principles; chief work *Root Principles in Rational and Spiritual Things*, 1905, a reasoned reply to Haeckel.

CHILD-VILLIERS, MARGARET ELIZABETH, COUNTESS OF JERSEY (1849-1945). [See VILLIERS.]

CHILD-VILLIERS, VICTOR ALBERT GEORGE, seventh EARL OF JERSEY, and tenth VISCOUNT GRANDISON (1845-1915), colonial governor. [See VILLIERS.]

CHILDE, VERE GORDON (1892-1957), prehistorian; born in Sydney; educated at Church of England grammar school, Sydney University, and Queen's College, Oxford; B.Litt., 1916; first class, *lit. hum.*, 1917; librarian, Royal Anthropological Institute, 1925-7; Abercromby professor of prehistoric archaeology, Edinburgh, 1927-46; professor of prehistoric European archaeology and director, Institute of Archaeology, London, 1946-56; works include *The Dawn of European Civilization* (1925), *Prehistoric Migrations in Europe* (1950), *New Light on the Most Ancient East* (1952), and *Prehistory of European Society* (1958); FBA, 1940.

CHILDERS, ROBERT ERSKINE (1870-1922), author and politician; son of R. C. Childers [q.v.]; BA, Trinity College, Cambridge; clerk in House of Commons, 1895-1910; served with city imperial volunteer battery of Honourable Artillery Company in South African war; published *The Riddle of the Sands*, outcome of yachting expeditions to German coast, 1903; increasingly concerned with Irish affairs; convert to home rule, 1908; employed in Royal Naval Air Service, 1914-19; subsequently devoted himself to working for complete independence of Ireland as republic; accompanied Irish republican envoys to Paris, 1919; member for county Wicklow of self-constituted Dáil Éireann, 1921; principal secretary to Irish delegation which negotiated treaty with British government, 1921; irreconcilable to treaty; joined Republican army; arrested, court-martialled, and shot.

CHILDS, WILLIAM MACBRIDE (1869-1939), educationist; educated at Portsmouth grammar school and Keble College, Oxford; second class, history, 1891; lecturer (1893) in history at future (1902) University College, Reading; principal, 1903; vice-chancellor (1926), after his devoted efforts had secured its conversion into a university; retired, 1929.

CHILSTON, first VISCOUNT (1851-1926), statesman. [See AKERS-DOUGLAS, ARETAS.]

CHILSTON, second VISCOUNT (1876-1947), diplomatist. [See AKERS-DOUGLAS, ARETAS.]

CHIROL, SIR (IGNATIUS) VALENTINE (1852-1929), traveller, journalist, and author; born and educated abroad; clerk in Foreign Office, 1872-6; travelled widely, especially in Near East, 1876-92; *The Times* correspondent in Berlin, 1892; observed with anxiety ill will towards England of German officials and press; in charge of *The Times* foreign department, London, 1896; favoured Anglo-Japanese alliance, French *entente*, understanding with Russia, and good relations with USA; retired, 1912; increasingly concerned with India (which he visited seventeen times); member of royal commission on Indian public services, 1912-14; knighted, 1912; his works include three books on India.

CHISHOLM, HUGH (1866-1924), journalist and editor of the *Encyclopaedia Britannica*; called to bar (Middle Temple), 1892; on staff of *St. James's Gazette*, 1892-9; joint-editor of eleven new volumes of *Encyclopaedia Britannica* (1902-3), 1900; editor-in-chief, 1903; eleventh edition published, 1910-11; city (financial) editor of *The Times*, 1913-20; again editor of *Encyclopaedia Britannica* (12th edn., 1922), 1920.

CHOLMONDELEY, HUGH, third BARON DELAMERE (1870-1931), pioneer settler in Kenya; educated at Eton; succeeded father, 1887; took up 100,000 acres in Njoro district of East Africa Protectorate, 1903; inaugurated research on wheat breeding; first president, Colonists' Association, 1903; unofficial member of legislative council, 1907; member for Rift Valley, 1920-31; believed in white settlement and envisaged an East African dominion; KCMG, 1929.

CHRISTIANSEN, ARTHUR (1904-1963), editor; educated at Wallasey grammar school; reporter, *Wallasey and Wirral Chronicle*, 1920; London editor, *Liverpool Evening Express*, 1925; news editor, *Sunday Express*, 1926; assistant editor, 1928; editor, *Daily Express*, 1933-58; published *Headlines all my Life*, 1961.

CHRISTIE, JOHN (1882-1962), founder of Glyndebourne Opera; educated at Eton and Trinity College, Cambridge; BA, 1905; master at Eton, 1906-1922; served with Kings' Royal Rifle Corps, 1914-16, MC; came into ownership of Glyndebourne estate, 1920; married (Grace) Audrey Laura St. John-Mildmay, opera singer, 1931; constructed opera house and opened season with *Le nozze di Figaro*, with active support from his wife, 1934; successful seasons annually to 1939; productions discontinued during 1939-45 war; productions at Edinburgh Festival, 1947-9; opera revived at Glyndebourne with financial support from John Lewis Partnership, 1950; 'The Glyndebourne Festival Society' formed, 1952; set up charitable trust, 1954; CH, 1954.

CHRISTIE, RICHARD COPLEY (1830-1901), scholar and bibliophile; educated at Lincoln College, Oxford; first class, law and history; BA, 1853; MA, 1855; called to bar (Lincoln's Inn), 1857; professor of ancient and modern history, 1854-66, political economy and commercial science, 1855-66, and jurisprudence and law, 1855-69, holding chairs in plurality at Owens College, Manchester; governor and member of council, 1870; member of council and university court, Victoria University, 1880-96; chancellor of diocese, Manchester, 1872-94; joint-legatee of Sir Joseph Whitworth [q.v.], 1887; benefactor of Owens College, to which he bequeathed his library; contributed to *Dictionary of National Biography*, and *Encyclopaedia Britannica*; publications include *Etienne Dolet, the Martyr of the Renaissance* (1880); succeeded James Crossley [q.v.] as chairman, Chetham Society, 1883-1901; hon. LLD, Manchester, 1895.

CHRISTIE, SIR WILLIAM HENRY MAHONEY (1845-1922), astronomer; son of S. H. Christie [q.v.]; BA, Trinity College, Cambridge, 1868; fellow of Trinity, 1869; chief assistant, Royal Observatory, Greenwich, 1870; astronomer royal, 1881; improved equipment and enlarged buildings of observatory; made several expeditions to observe solar eclipses; FRS, 1881; president, Royal Astronomical Society, 1890-2; KCB, 1904; died at sea.

CHRYSTAL, GEORGE (1851–1911), mathematician; educated at Aberdeen and Peterhouse, Cambridge; second wrangler and Smith's prizeman, and fellow of Corpus, 1875; made researches for verifying 'Ohm's law', 1876; professor of mathematics at Edinburgh University, 1879–1911; FRS Edinburgh, 1880, and general secretary, 1901–11; in later life worked out theories on oscillations in lakes, for which he received royal medal at Royal Society of London, 1911; hon. LLD, Aberdeen (1887) and Glasgow (1911); published standard handbook on *Algebra*, 1886–9.

CHUBB, SIR LAWRENCE WENSLEY (1873–1948), protagonist of open space preservation and other amenities; secretary of National Trust, 1895; of Commons Preservation Society, 1896–1948; also of Scapa Society, 1916–48, National Playing Fields Association, 1928–47, and Coal Smoke Abatement Society, 1902–29; knighted, 1930.

CHURCH, SIR WILLIAM SELBY, first baronet (1837–1928), physician; BA, University College, Oxford, 1860; Lee's reader in anatomy and senior student, Christ Church, Oxford, 1860–9; entered St. Bartholomew's Hospital, 1862; full physician, St. Bartholomew's, 1875–1902; DM, 1868; president, College of Physicians, 1899–1905; president, Royal Society of Medicine, 1908–10; baronet, 1901; KCB, 1902; served on several royal commissions.

CHURCHILL, SIR WINSTON LEONARD SPENCER (1874–1965), statesman and indomitable war leader; elder son of Lord Randolph Spencer Churchill [q.v.], and Jennie, daughter of Leonard Jerome of New York; grandson of seventh Duke of Marlborough [q.v.]; educated at Harrow and Sandhurst; commissioned in 4th Queen's Own Hussars, 1895; saw action in Cuba with Spanish forces, and made report for *Daily Graphic*; joined his regiment in India and saw service on North-West Frontier; published *The Story of the Malakand Field Force*, 1898; took part in battle of Omdurman, 1898; published *The River War*, 1899; having resigned his commission, went to South Africa, in arrangement with the *Morning Post*; captured by the Boers and escaped, 1899; published *London to Ladysmith, via Pretoria* and *Ian Hamilton's March* (1900); unionist MP for Oldham, 1900; unwilling to support tariff reform proposals of Joseph Chamberlain [q.v.], joined liberal party, 1904; published *Lord Randolph Churchill*, 1906; liberal MP, North-West Manchester, 1906; parliamentary under-secretary for the colonies; PC, 1907; published *My African Journey*, 1908; president of the Board of Trade, and MP Dundee, 1908; married Clementine Hozier,

1908; in co-operation with Lloyd George, concerned to alleviate distress of working people; introduced bill to insure against unemployment; home secretary, 1910; introduced Mines Act, 1911; superintended 'the battle of Sidney Street', 1911; first lord of the Admiralty, 1911; reorganized the Navy Board and set out to modernize and strengthen the Royal Navy; ensured that the fleet was ready for war, 1914; advocated naval attack on the Dardanelles, 1915; forced to resign from Admiralty, as condition made by Bonar Law for coalition government with Asquith, 1915; blamed for failure of Gallipoli campaign; chancellor of Duchy of Lancaster; resigned and rejoined army; in command of 6th battalion, Royal Scots Fusiliers in France, 1916; returned to political life; minister of munitions, 1917; coalition liberal MP Dundee, 1918; secretary for war (and air), 1918–21; colonial secretary, 1921–2; supported Lloyd George when coalition government ended, 1922; defeated at Dundee; bought Chartwell manor; CH, 1922; published *World Crisis* (1923); conservative MP Epping, 1924–45; chancellor of the Exchequer, 1924–9; carried through return of sterling to the gold standard, 1925; edited the *British Gazette* during general strike, 1926; published *My Early Life* (1930); resigned from conservative shadow cabinet and opposed the India bill, 1931; supported Edward VIII in abdication crisis, 1936; published *Great Contemporaries*, 1937; challenged prime minister Neville Chamberlain's foreign policy and relations with Adolf Hitler; on eve of Second World War returned to Admiralty, 1939; coalition government necessitated by British failure in Norway; on refusal of labour leaders to serve under Chamberlain and resignation of latter, prime minister and minister of defence, 1940; amid military disasters of 1940 inspired British people to believe in ultimate victory; leader of conservative party on death of Chamberlain, 1940; paid tribute to Battle of Britain fighter pilots, 'Never has so much been owed by so many to so few'; appointed Sir Alan Brooke (later Viscount Alanbrooke) [q.v.] as commander-in-chief, Home Forces; secured aid of United States through close association with president Roosevelt; leased bases in Caribbean to United States in return for fifty destroyers to assist in battle of the Atlantic; warned Joseph Stalin of impending German invasion of Russia, 1941; formulated Atlantic Charter in agreement with Roosevelt; when Japanese attacked Pearl Harbor, decided that entry of United States into the war ensured final victory, 1941; agreed with Roosevelt's proposal for united allied command in south-west Pacific, 1941; gravely ill on visit to forces in North Africa but recovered, 1943; success of British and American forces in North Africa and Sicily followed by disagreement with Roosevelt on future campaign in Europe, 1943; distrust of Russian postwar intentions; second front agreed, to take place in 1944; welcomed appointment of General Eisenhower as supreme allied commander in Europe; dissuaded by King George VI from viewing landings in Normandy on D-Day, 1944; influence on war policy declined as allied forces advanced into Europe; met Roosevelt and Stalin at Yalta; agreed on establishment of United Nations, 1945; on German surrender and end of war in Europe, proposed to Clement Attlee, leader of the labour party that wartime coalition should continue until defeat of Japan; on refusal of Attlee, resigned and called election; defeated, 1945; MP Woodford, 1945-64; argued need for European unity and western co-operation in 'cold war' (speech at Fulton, Missouri) 1946; prime minister again, 1951; KG, 1953; published *The Second World War* (6 vols., 1948-54) and *A History of the English-Speaking Peoples* (1956-8); resigned as prime minister, 1955; given state funeral in London, 1965; a remarkable Englishman and notable orator; held honorary degrees from over twenty universities; lord warden of Cinque Ports, 1941-65; OM, 1946; Nobel prize for literature, 1953; achievements as a painter recognized; honorary Royal Academician Extraordinary, 1948.

CHUTER-EDE, JAMES CHUTER, BARON CHUTER-EDE, (1882-1965), parliamentarian; educated at Epsom National Schools, Dorking high school, Battersea Pupil Teachers' Centre, and Christ's College, Cambridge; assistant master, Surrey elementary schools, member, Epsom Urban District Council, 1908; member, Surrey County Council, 1914; served in army during 1914-18 war; labour MP, Mitcham, 1923; MP, South Shields, 1929-31, and 1935-64; parliamentary secretary, Ministry of Education, 1940-5; home secretary, 1945-51; involved in controversy regarding the abolition of capital punishment; PC, 1944; CH, 1953; DL (Surrey); hon. MA, Cambridge (1943), hon. doctorates from Bristol (1951), Sheffield (1960), and Durham (1954); chairman, BBC advisory council; deputy leader, House of Commons, 1947; leader, 1951; became life peer, 1964.

CILCENNIN, VISCOUNT (1903-1960), politician. [See THOMAS, JAMES PURDON LEWES.]

CLANRICARDE, second MARQUESS OF (1832-1916), Irish landed proprietor. [See BURGH CANNING, HUBERT GEORGE DE.]

CLANWILLIAM, fourth EARL OF (1832-1907), admiral of the fleet. [See MEADE, RICHARD JAMES.]

CLAPHAM, SIR ALFRED WILLIAM (1883-1950), archaeologist; educated at Dulwich College; articled to architect; joined Royal Commission on Historical Monuments, 1912; technical editor, 1913-33; secretary to commissioners, 1933-48; combined wide historical learning with sound architectural knowledge; publications include *English Romanesque Architecture* (2 vols., 1930-4) and *Romanesque Architecture in Western Europe* (1936); FBA, 1935; knighted, 1944.

CLAPHAM, SIR JOHN HAROLD (1873-1946), historian; educated at Leys School and King's College, Cambridge; first class, history, 1895; fellow, 1898-1904, 1908-46; professor of economics, Leeds, 1902-8; returned to King's College, 1908; tutor, 1913-28; vice-provost, 1933-43; professor of economic history, 1928-38; publications include *An Economic History of Modern Britain* (3 vols., 1926-38) and *History of Bank of England* (3 vols., 1944-58); joint-editor, first volume of *Cambridge Economic History of Europe* (1941); fellow (1928), president (1940-5), British Academy; knighted, 1943.

CLARENDON, sixth EARL OF (1877-1955), public servant. [See VILLIERS, GEORGE HERBERT HYDE.]

CLARK, ALBERT CURTIS (1859-1937), classical scholar; educated at Haileybury and Balliol College, Oxford; first class, *lit. hum.*, 1881; classical fellow, Queen's (lecturer, 1882-7, tutor, 1887-1913); university reader in Latin, 1909-13; Corpus Christi professor, 1913-34; works include full-scale edition of Cicero's *Pro Milone* (1895) followed by four volumes of his orations, an edition of Asconius (1907), *The Vetus Cluniacensis of Poggio* (1905), and *The Descent of Manuscripts* (1918); overstressed importance of stichometry; FBA, 1916.

CLARK, SIR ALLEN GEORGE (1898-1962), entrepreneur and industrialist; born in United States; came to Britain, 1905; educated at Felsted School, 1913-15; served in British army and Royal Flying Corps, 1915-18; purchased share in 'Plessey' engineering company; company entered electronics business, 1922; designed and built first portable radio sets in Britain; and later, first commercially produced television set in the world; manufactured aero-engine equipment, electronics, and munitions during 1939-45 war; supervised expansion of 'Plessey' in post-war period; involved in silicon technology, 1952; became naturalized, 1927; knighted, 1961; council member, Society of British Aircraft Constructors, and Telecommunication Engineering and Manufacturing Association.

CLARK, JAMES (JIM) (1936–68), racing motorist; educated at Lovetts School, Edinburgh; left at sixteen to work on family sheep farm; took part in local motor club events; first post-war sports car driver to lap British circuit at over 100 m.p.h., 1958; joined Colin Chapman's Lotus team, 1960; collided with von Tripp at Monza (von Tripp and fourteen spectators killed), 1961; achieved seven grand prix victories in a season and world championship, 1963; won second world championship, 1965; killed in race at Hockenheim, 1968; OBE, 1964; first hon. burgess of Duns, 1965; Jim Clark memorial Room at Duns contains many of his trophies.

CLARK, JOHN WILLIS (1833–1910), man of science and archaeologist; son of William Clark [q.v.]; educated at Eton and Trinity College, Cambridge; first class, classics, 1856; fellow, 1858; superintendent of museum of zoology, Cambridge, 1866–91; registrar of Cambridge University, 1891–1910; did much to forward endowment of university library; supporter of Cambridge Amateur Dramatic Club; wrote many works on Cambridge history, including *Architectural History of the University and Colleges of Cambridge* (with Robert Willis [q.v.], 4 vols., 1886), *Concise Guide to Cambridge* (1898), and on Barnwell Priory (1897, 1907); collaborated in *Life of Professor Sedgwick*, 2 vols., 1890; also wrote *The Care of Books*, 1901; bequeathed collections of books on Cambridge to university library.

CLARK, Sir WILLIAM HENRY (1876–1952), civil servant; son of John Willis Clark [q.v.]; educated at Eton and Trinity College, Cambridge; first class, classical tripos, 1897; appointed to Board of Trade, 1899; private secretary to Lloyd George, 1906–10; member for commerce and industry, Indian executive council, 1910–16; head of commercial intelligence, Board of Trade, 1916; comptroller-general, Department of Overseas Trade, 1917–28; high commissioner in Canada, 1928–34; in South Africa, 1934–9; chairman, Imperial Shipping Committee, 1939–40; CSI, 1911; KCSI, 1915; KCMG, 1930; GCMG, 1937.

CLARK KERR, ARCHIBALD JOHN KERR, Baron Inverchapel (1882–1951), diplomatist; born near Sydney; educated at Bath College and Heidelberg University; entered diplomatic service, 1905; minister to Guatemala, 1925, Chile, 1928, Sweden, 1931–5; ambassador to Iraq, 1935, to China, 1938, to Russia, 1942–6; present at conferences in Moscow and those at Tehran (1943), Yalta, and Potsdam (1945); established good relationship

with Stalin whose esteem he retained to the end; special ambassador to Netherlands East Indies, 1945; ambassador to United States, 1946–8; KCMG, 1935; GCMG, 1942; PC, 1944; baron, 1946.

CLARKE, Sir ANDREW (1824–1902), lieutenant-general and colonial official; joined Royal Engineers, 1844; dispatched to Van Diemen's Land, 1846; transferred to New Zealand, 1848; surveyor-general of Melbourne, 1853; served on legislative council; drafted bill for new constitution for colony, 1854; surveyor-general and commissioner of lands, 1856; inaugurated railways there, 1857; appointed to command of Royal Engineers at Colchester (1859) and Birmingham (1862); served in Ashanti, 1863; director of engineering works at Admiralty, 1864–73; CB, 1869; extended docks at Chatham, Portsmouth, and elsewhere; advocated English purchase of Suez Canal, 1870; KCMG, 1873; successful governor of the Straits Settlements, 1873–5; suppressed piracy; head of the public works department in India, 1877–80; CIE, 1877; returned to England, 1880; inspector-general of fortifications, 1882; paid close attention to defences of coaling stations and commercial harbours; advocated widening of Suez Canal, 1882; vice-president of international committee on subject, 1884; urged construction of railway from Suakin to Berber, 1884; GCMG, 1885; lieutenant-general, 1886; subsequently director of British North Borneo Company, which named Clarke province after him; agent-general for Victoria, 1891–4, 1897–1902.

CLARKE, Sir CASPAR PURDON (1846–1911), architect, archaeologist, and museum director; trained as architect, he held posts at South Kensington Museum from 1867; made for the museum purchases in Near East (1876), and in Spain and Italy (1879); keeper of India Museum (1883–92), and art collections (1892) there; director of South Kensington Museum, 1896–1905; director of Metropolitan Museum, New York, 1905–10; designed many London buildings, including National School of Cookery, 1887; lectured and wrote much on architecture and Eastern crafts; FSA, 1893; CIE, 1883; knighted, 1902.

CLARKE, CHARLES BARON (1832–1906), botanist; from Trinity College, Cambridge, migrated to Queens'; third wrangler, 1856; fellow (1857) and lecturer in mathematics (1858–65); MA, 1859; called to bar, 1860; joined staff of Presidency College, Calcutta, 1865; inspector of schools in eastern Bengal; obtained 7,000 specimens of Indian plants; superintendent of Calcutta botanical gardens, 1869–71; helped Sir Joseph Hooker [q.v.], at Kew in his

Flora of British India, 1879-83; inspector of schools in India, 1883-7; settled on retirement at Kew, 1887, continuing work on Indian botany; president of Linnean Society, 1894-6; FRS, 1882; wrote on Bengal botany, political economy, geography, and ethnology.

CLARKE, SIR CHARLES NOBLE ARDEN- (1898-1962), colonial governor. [See ARDEN-CLARKE.]

CLARKE, SIR EDWARD GEORGE (1841-1931), lawyer and politician; educated at College House, Edmonton, and City Commercial School; helped father in silversmith's shop, 1854-8; clerk, India Office, 1859; resigned (1860) to read for bar, supporting himself by journalism; called (Lincoln's Inn), 1864; QC, 1880; a most eminent common law leader; practised until 1914; conservative MP, Southwark (1880), Plymouth (1880-1900); solicitor-general (knighted), 1886-92; declined office and freely criticized government in 1895-1900 and was asked to resign seat; MP, City of London, 1906; PC, 1908; sent own obituary to *The Times*.

CLARKE, SIR FRED (1880-1952), educationist; educated at elementary school and Technical College, Oxford; first class, modern history, Oxford, 1903; appointed master at Diocesan Training College, York, 1903; professor of education at Hartley University College, Southampton, 1906; at South African College, Cape Town, 1911; at McGill University, Montreal, 1929; adviser at Institute of Education, London, 1935, director, 1936-45; works include *Foundations of History-Teaching* (1929), *Education and Social Change* (1940), and *Freedom in the Educative Society* (1948); knighted, 1943.

CLARKE, GEORGE SYDENHAM, BARON SYDENHAM OF COMBE (1848-1933), administrator; passed first into and out of Royal Military Academy; gazetted to Royal Engineers, 1868; lecturer at Royal Indian Engineering College, 1871-80; secretary to colonial defence committee, 1885-92, to royal commission on navy and army administration, 1888-90; superintendent, royal carriage department, Woolwich Arsenal, 1894-1901; member of War Office reconstruction committee (1904) and Committee of Imperial Defence; governor of Victoria (1901-3) and Bombay (1907-13); chairman of royal commission on venereal diseases (1913) and of central appeal tribunal, 1915-16; KCMG, 1893; FRS, 1896; baron, 1913.

CLARKE, HENRY BUTLER (1863-1904), historian of Spain; educated at St. Jean-de-Luz and Wadham College, Oxford; BA, 1888;

studied Spanish history and literature; Taylorian teacher of Spanish at Oxford, 1890-2; Fereday fellow of St. John's, Oxford, 1894; lived at St. Jean-de-Luz from 1891; wrote *Spanish Grammar* (1892), *History of Spanish Literature* (1893), *Modern Spain, 1815-1898* (1906); an Arabic scholar.

CLARKE, LOUIS COLVILLE GRAY (1881-1960), connoisseur, collector, and museum director; studied history, Trinity Hall, Cambridge (fellow, 1929), and anthropology, Exeter College, Oxford; travelled widely; curator, University Museum of Archaeology and Ethnology, Cambridge, 1922-37; director, Fitzwilliam Museum, Cambridge, 1937-46; gave museum over 2,700 works of art in his lifetime and bequeathed his own collections mainly to it; had intuitive understanding and extensive knowledge of every kind of art.

CLARKE, SIR MARSHAL JAMES (1841-1909), South African administrator; joined Royal Artillery, 1863; lieutenant-colonel, 1883; resident commissioner of Basutoland, 1884-93, of Zululand, 1893-8, of Rhodesia, 1898-1905; won confidence of Africans; KCMG, 1886.

CLARKE, MAUDE VIOLET (1892-1935), historian; obtained first classes in history at Belfast and Lady Margaret Hall, Oxford; became history tutor (1919), fellow (1922), and vice-principal (1933), Somerville College, Oxford, and university lecturer, 1930; works include *Medieval Representation and Consent*, 1936.

CLARKE, THOMAS (1884-1957), journalist, author, and broadcaster; educated at higher grade school, Bolton and Ruskin Hall, Oxford; special writer, *Daily Dispatch*, 1907; news editor, *Daily Sketch*, 1909; foreign staff, *Daily Mail*, 1911; night news editor, 1914-16, news editor, 1919-22; assistant editor, *Melbourne Herald*, 1923-6; managing editor, *Daily News*, 1926, then editor and director, *News Chronicle*, until 1933; publications include *My Northcliffe Diary* (1931); broadcast to Latin America, 1942-8.

CLASPER, JOHN HAWKS (1836-1908), boat-builder and oarsman; son of Henry Clasper, Newcastle oarsman and boat-builder, and inventor of outrigger; won many sculling races at regattas from 1854; with father won pair-oar championship of the Tyne, 1858; expert trainer; improved sliding seat and keelless boat; invented countervail; built successful Cambridge eight-oared boats of 1870-3; became leading builder of racing boats.

CLAUSEN, SIR GEORGE (1852-1944), painter; studied at South Kensington and in

Paris; influenced by Bastien-Lepage; painted mainly agricultural life; also pure landscape, and some portraits and interiors; preoccupied with effects of light, especially figures seen against the sun, but retained solidity of form; ARA, 1895; RA, 1908; knighted, 1927.

CLAUSON, ALBERT CHARLES, BARON CLAUSON (1870-1946), judge; nephew of Lord Wrenbury [q.v.]; educated at Merchant Taylors' School and St. John's College, Oxford; first class, *lit. hum.*, and called to bar (Lincoln's Inn), 1891; KC, 1910; a leading equity practitioner; judge of the Chancery division, 1926-38; lord justice of appeal, 1938-42; knighted, 1926; PC, 1938; baron, 1942.

CLAXTON, BROOKE (1898-1960), Canadian politician; born in Montreal; educated at Lower Canada College and McGill University; BCL and called to Quebec bar, 1921; specialized in insurance law; associate professor, commercial law, McGill, 1930-44; MP, 1940-54; minister of national health and welfare, 1944-6; of national defence, 1946-54; effective organizer of the liberal party and its policies; chairman, Canada Council, 1957-60; honorary degrees from many universities.

CLAY, SIR HENRY (1883-1954), economist; educated at Bradford grammar school and University College, Oxford; lecturer for Workers' Educational Association, 1909-17; at Manchester, Stanley Jevons professor of political economy, 1922-7; professor of social economics, 1927-30, and regular contributor to *Guardian*; joined Bank of England, 1930; economic adviser to the governor, 1933-44; to Board of Trade and Ministry of War Transport, 1941-4; warden of Nuffield College, Oxford, 1944-9; works include *Economics: an Introduction for the General Reader* (1916), *The Post-War Unemployment Problem* (1929), and *The Problem of Industrial Relations* (1929); a founder (1938) of National Institute of Economic and Social Research; favoured private enterprise, reduced government expenditure, balanced budgets, and retention of gold standard; knighted, 1946.

CLAYDEN, PETER WILLIAM (1827-1902), journalist and author; active Unitarian minister, 1855-68; secretary of newly founded Free Church Union, 1868; as leader-writer and assistant editor of *Daily News* (1868-96) he increased its influence as liberal nonconformist organ; supported Gladstone's anti-Turkish views, 1876-80, and advocated support of Armenians, 1896-7; published memoirs of Samuel Rogers (1877-9) and religious and political works.

CLAYTON, SIR GILBERT FALKINGHAM (1875-1929), soldier and administrator; joined Egyptian army, 1900; retired from army and transferred to Sudan government service, 1910; director of military intelligence, Cairo, 1914; brigadier-general, 1917; adviser to ministry of interior, Egypt, 1919-22; chief secretary, Palestine, 1922-5; negotiated treaty of Jeddah, disposing of differences between Great Britain and Hejaz, 1927; KBE, 1919; KCMG, 1926; died at Baghdad.

CLEMENTI, SIR CECIL (1875-1947), colonial administrator and traveller; educated at St. Pauls' School and Magdalen College, Oxford; posted to Hong Kong, 1899; land officer and police magistrate, 1903-6; travelled widely; journeyed from Andijan to Kowloon, 1907-8; colonial secretary, British Guiana, 1913-22; mapped route from Kaieteur Falls to summit of Mount Roraima; colonial secretary, Ceylon, 1922-5; governor of Hong Kong, 1925-30; of Straits Settlements and high commissioner for Malay States, 1930-4; KCMG, 1926; GCMG, 1931.

CLERK, SIR DUGALD (1854-1932), mechanical engineer; studied at Anderson's College, Glasgow, and Yorkshire College of Science, Leeds; worked with Glasgow and Birmingham firms on theory and design of gas engine; founded (1888) with G. C. (later Lord) Marks the firm of Marks and Clerk, consulting engineers and patent agents; patented an engine working on the Clerk (two-stoke) cycle, 1881; researches embodied in *The Gas, Petrol and Oil Engine*, 1909; FRS, 1908; KBE, 1917.

CLERK, SIR GEORGE RUSSELL (1874-1951), diplomatist; born in India; educated at Eton and New College, Oxford; joined Foreign Office, 1898; head of war department, 1914; attended Rome conference and went with Lord Milner [q.v.] to Russia, 1917; conducted peace negotiations in Bucharest and Budapest, 1919; appointed minister to Czechoslovakia, 1919; ambassador to Turkey, 1926, to Belgium, 1933, to France, 1934-7; CMG, 1908; KCMG, 1917; GCMG, 1929; CB, 1914; PC, 1926.

CLERKE, AGNES MARY (1842-1907), historian of astronomy; resided in Italy, 1867-77, writing on astronomical and literary subjects, mainly for *Edinburgh Review*; published *Popular History of Astronomy during the Nineteenth Century* (1885), *System of the Stars* (1890), and *Problems in Astrophysics* (1903); contributed lives of astronomers to this Dictionary; elected honorary member of Royal Astronomical Society, 1903; an accomplished musician.

CLERKE, ELLEN MARY (1840-1906), translator of Italian verse, poet, and novelist; sister of A. M. Clerke [q.v.]; published *Fable and Song in Italy* (1899) and astronomical monographs.

CLERY, Sir CORNELIUS FRANCIS (1838-1926), major-general; joined army, 1858; professor of tactics, Royal Military College, Sandhurst, 1872-5; served in Zulu war, 1879; against Arabi Pasha, 1882; on Suakin and Gordon relief expeditions, 1884-5; commandant, Staff College, Camberley, 1888-93; served in South African war, 1899-1900; retired, 1901; KCB, 1899; KCMG, 1900; published *Minor Tactics* (1875).

CLEWORTH, THOMAS EBENEZER (1854-1909), educational controversialist; BA, St. John's College, Cambridge, 1883; rector of Middleton, Lancashire, 1888; honorary canon of Manchester, 1902; founded Church Schools Emergency League, 1903, for maintenance of church schools and of church teaching in elementary schools; organized demonstrations against liberal government's education bill, 1906; wrote many educational leaflets.

CLIFFORD, Sir BEDE EDMUND HUGH (1890-1969), colonial governor; born in New Zealand; educated in Melbourne, Australia; worked with firm of surveyors and as officer on a tramp steamer; commissioned in Royal Engineers, 1914; aide-de-camp to governor-general, Australia, 1917; private secretary, 1918-20; secretary to Prince Arthur of Connaught [q.v.], governor-general of South Africa, 1921; secretary to Earl of Athlone [q.v.], 1924; imperial secretary to South African high commission, 1924; first white man to cross Kalahari desert, 1928; first British representative to Union of South Africa, 1928; governor, Bahamas, 1931; governor, Mauritius, 1937; and Trinidad, 1942-5; CMG, 1923; CB, 1931; KCMG, 1933; GCMG, 1945; published autobiography, *Proconsul*, 1964.

CLIFFORD, FREDERICK (1828-1904), journalist and legal writer; joined parliamentary staff of *The Times*, 1852; joint proprietor of *Sheffield Daily Telegraph*, 1863; helped to found Press Association, 1868; assistant editor of *The Times*, 1877-83; called to bar, 1859; published standard textbook on private bill practice (1870) and *The History of Private Bill Legislation* (2 vols., 1885-7); student of and writer on agricultural questions.

CLIFFORD, Sir HUGH CHARLES (1866-1941), colonial administrator; son of Sir H. H. Clifford [q.v.]; joined Malay civil service, 1883;

British resident, Pahang, 1896-9, 1901-3; colonial secretary; Trinidad and Tobago, 1903-7; Ceylon, 1907-12; governor, Gold Coast (1912-19), Nigeria (1919-25), Ceylon (1925-7), Straits Settlements (1927-9); KCMG, 1909; GCMG, 1921.

CLIFFORD, JOHN (1836-1923), Baptist leader; converted 1850; baptised, 1851; student at Midland Baptist College, Leicester, 1855-8; pastor, Praed Street Baptist church, Paddington, 1858; growth of congregation eventually necessitated building of Westbourne Park chapel, one of the most notable places of worship in London; an ardent evangelical who believed that religion was concerned with the whole of life; showed great sympathy with the masses; exercised liberalizing influence in theology; withstood demand of C. H. Spurgeon [q.v.] that Baptist Union of Great Britain and Ireland should cease to tolerate 'down-grade' developments in theological outlook and of modern biblical criticism and adopt a definite creed, 1887; opposed state aid for denominational schools and led 'passive resistance' to Education Act of 1902; president, Baptist World Alliance, 1905-11; a voluminous writer; CH, 1921.

CLIVE, Sir ROBERT HENRY (1877-1948), diplomatist; educated at Haileybury and Magdalen College, Oxford; entered diplomatic service, 1902; first secretary, 1915; consul-general, Munich (1923), Tangier (1924-6); held series of difficult and frustrating posts as minister in Tehran (1926-31) and ambassador in Tokyo (1934-7) and Brussels (1937-9); KCMG, 1927; PC, 1934; GCMG, 1936.

CLODD, EDWARD (1840-1930), banker and author; clerk, London Joint Stock Bank, 1862; secretary, 1872-1915; attended lectures and read industriously; joined Royal Astronomical Society, 1869; resigned and joined Folk-Lore Society, 1878; president, 1895 and 1896; helped to found Johnson Club, 1884, and Omar Khayyám Club, 1892; chairman, Rationalist Press Association, 1906; friend of many distinguished scientific and literary men; entertained nearly all the eminent later Victorians at his Aldeburgh home; works include *The Childhood of the World* (1873), *Jesus of Nazareth* (1880), *The Story of Creation* (1888), *Memories* (1916).

CLOSE, Sir CHARLES FREDERICK ARDEN- (1865-1952), geographer. [See ARDEN-CLOSE.]

CLOSE, MAXWELL HENRY (1822-1903), geologist; BA, Trinity College, Dublin, 1846; MA, 1867; minister in Church of Ireland, 1848-61; made valuable researches in Irish glacial geology; president of Irish Royal Geological Society, 1878–9; treasurer of Royal Irish Academy, 1879–1903; promoted study of Irish language; wrote anonymously on physics and astronomy.

CLOWES, Sir WILLIAM LAIRD (1856–1905), naval writer; educated at King's College, London; fellow, 1895; published poem *Meroe*, 1876; made reputation as naval correspondent to *The Times*, 1890–5; influenced naval estimates by anonymous articles in *Daily Graphic* on 'The Needs of the Navy', 1893; compiled *The Royal Navy*, 7 vols., 1897–1903; knighted, 1902; granted civil list pension, 1904; excellent linguist.

CLUNES, ALEC (ALEXANDER DE MORO SHERRIFF) (1912–1970), actor, stage director, and theatre manager; educated at Cliftonville, Margate; after experience with amateur groups, joined Old Vic Company, 1934; leading actor at Shakespeare Memorial Theatre, Stratford-upon-Avon, 1939; toured with Old Vic Company, 1941; founded Arts Theatre Group of Actors, London, 1942, manager, director and actor, 1942–53; assisted new dramatists; staged Christopher Fry's *The Lady's Not for Burning*, 1948; and John Whitings' *Saint's Day*, 1951; succeeded Rex Harrison as Professor Higgins in *My Fair Lady*; published *The British Theatre*, 1964.

CLUNIES-ROSS, GEORGE (1842–1910), owner of Cocos and Keeling Islands; born in Cocos Islands; grandson of John Clunies-Ross, first owner of the islands; studied engineering at Glasgow, 1862; returned to Cocos Islands, 1862; introduced modern machinery and scientific methods in coconut industry, and restored dwindling family fortunes; control of islands transferred to governor of Ceylon, 1878, and to governor of Straits Settlements, 1886; islands incorporated as part of Singapore, 1903.

CLUNIES ROSS, Sir IAN (1899–1959), veterinary scientist and scientific administrator. [See Ross.]

CLUTTON, HENRY HUGH (1850–1909), surgeon; BA, Clare College, Cambridge, 1873; MB, 1879; MC, 1897; entered St. Thomas's Hospital, 1872; full surgeon, 1891; FRCS, 1876; president of Clinical Society, 1905; specialist on diseases of bones and joints; described 'Clutton's joints', a knee affection in children, 1886.

CLUTTON-BROCK, ARTHUR (1868–1924), essayist, critic, and journalist; BA, New College, Oxford; called to bar (Inner Temple), 1895; art critic to *The Times*, 1908; frequently contributed to *The Times Literary Supplement*; in time his outlook became less aesthetic and more moralistic; his works include *Shelley, the Man and the Poet* (1909) and *Thoughts on the War* (1914–15).

CLYDE, JAMES AVON, Lord Clyde (1863–1944), lord justice-general of Scotland; educated at Edinburgh Academy and University; first class, classics, 1884; passed advocate, 1887; KC, 1901; solicitor-general for Scotland, 1905; lord advocate, 1916–20; PC, 1916; led brilliantly for British South Africa Company before Privy Council regarding ownership of unalienated lands of Southern Rhodesia, 1918; unionist MP, West (later North) Edinburgh, 1909–20; lord justice-general of Scotland and lord president of the Court of Session, 1920–35; chairman, royal commission on the Court of Session and office of sheriff principal 1926.

CLYDESMUIR, first Baron (1894–1954), public servant. [See Colville, David John.]

CLYNES, JOHN ROBERT (1869–1949), labour leader; worked in Oldham textile mill; district organizer for National Union of Gasworkers and General Labourers (later of General and Municipal Workers), 1891; secretary, Lancashire district, 1896; president of union, 1912–37; represented union at inaugural conference, labour representation committee (1900); on labour party national executive, 1909–39; labour MP, North-Eastern (later Platting) division of Manchester, 1906–31, 1935–45; parliamentary secretary to food controller, 1917; food controller and PC, 1918; chairman, parliamentary labour party, 1921; defeated by Ramsay MacDonald, 1922; lord privy seal and deputy leader of House of Commons, 1924; home secretary, 1929–31; refused party leadership, 1931; of sound judgement and unobtrusively influential.

COADE, THOROLD FRANCIS (1896–1963), headmaster of Bryanston School; educated at Glebe House School, Hunstanton, Harrow, Royal Military College, Sandhurst, and Christ Church, Oxford; assistant master Harrow, 1922–32; headmaster, Bryanston, 1932–59; one of the great headmasters of his time; edited *Harrow Lectures in Education* (1931), and *Manhood in the Making* (1939).

COATALEN, LOUIS HERVÉ (1879–1962), automobile engineer and aero-engine designer; born in France; educated at École des Arts et Métiers, Cluny; worked as draughtsman with De Dion-Bouton et Cie and other French motor manufacturers; joined Crowden Motor Car Co., Leamington Spa, 1900; chief engineer, Humber

Ltd., Coventry; partnered William Hillman of Coventry, 1907; chief engineer, Sunbeam Motor Car Co., Wolverhampton, 1909; produced succession of Sunbeam, Talbot, and Darracq cars and aero engines; won number of grand prix races, 1922-7; Sunbeam-Coatalin engines fitted to R. 34 airship; left Sunbeam Co., and returned to France, 1930; chairman and managing director, Freins Hydrauliques Lockhead, Paris, and chairman, S. A. Bougies K. L. G.; worked on diesel aero-engines; president, Société des Ingenieurs de l'Automobile, 1953; commandant, Légion d'Honneur.

COATES, ERIC (1886-1957), composer; studied composition and viola at Royal Academy of Music; played viola with Queen's Hall Orchestra, 1910-19; composer of songs and light music including *Miniature Suite*, *London Suite*, 'Three Bear' fantasy, 'Stonecracker John', 'Calling all Workers', and 'By the Sleepy Lagoon'; his work is fresh, melodious, and light-hearted; published autobiography *Suite in Four Movements* (1953).

COATES, JOSEPH GORDON (1878-1943), prime minister of New Zealand; educated at Matakohe public school; became farmer; MHR, Kaipara, 1911-43; postmaster-general, 1919-25; minister of justice (1919-20), public works (1920-6), railways (1923-8), and native affairs (1921-8); prime minister, 1925-8; PC, 1926; served in coalition government (1931-5) as minister for public works (1931-3), transport (1931-5), and finance (1933-5); bold measures to deal with depression included Agricultural Emergency Powers Act, depreciation of exchange rate, and establishment of Reserve Bank; member of dominion war cabinet, 1940-3.

COBB, GERARD FRANCIS (1838-1904), musician; BA, Trinity College, Cambridge (first class, classics tripos and moral sciences tripos), 1861; fellow, 1863; junior bursar, 1869-94; strongly advocated Roman and Anglican reunion; prolific composer of songs and church music; chief work 'A Song of Trafalgar' (1900); president of National Cyclists' Union, 1878.

COBB, JOHN RHODES (1899-1952), racing motorist; educated at Eton and Trinity Hall, Cambridge; achieved lap records at Brooklands from 1929; and world land-speed record, Bonneville salt flats, 1938, 1939, 1947 (394.2 m.p.h.); died on Loch Ness attempting water-speed record.

COBBE, SIR ALEXANDER STANHOPE (1870-1931), general; born in India; educated at Wellington and Sandhurst; commissioned in South Wales Borderers, 1889; transferred to Indian staff corps, 1892; DSO (Ashanti), 1900; VC (Somaliland), 1903; commanded III Indian Corps, Mesopotamia, 1916-19; KCB, 1917; military secretary, India Office, 1920-6 and 1930-1; general, 1924; commander-in-chief, Northern command, India, 1926-30.

COBBE, FRANCES POWER (1822-1904), philanthropist and religious writer; studied history, astronomy, and philosophy; influenced by Theodore Parker, whose works she edited (14 vols., 1863-71), and by Kant's ethics; published anonymously *The Theory of Intuitive Morals*, 1855; travelled much abroad, especially in Italy, where she met Mazzini; associated with Mary Carpenter [q.v.] in her ragged school and reformatory work, 1858; engaged in workhouse philanthropy at Bristol, 1859; advocated woman's suffrage and admission of women to university degrees, 1862; on staff of *Echo*, 1868-75, she investigated cases of destitution; helped to promote Matrimonial Causes Act, 1878; joint-secretary of National Anti-Vivisection Society, 1875-84; a frank and lucid writer; an occasional preacher in Unitarian chapels; voluminous writings include *Broken Lights* (1864), *Darwinism in Morals* (1872), *The Duties of Women* (1881); Autobiography, 2 vols., 1904.

COBDEN-SANDERSON, THOMAS JAMES (1840-1922), bookbinder and printer; called to bar (Inner Temple), 1871; in touch with William Morris and Burne-Jones families; studied bookbinding under Roger de Coverley, 1883; opened his own workshop, 1884; carried on Doves bindery, Hammersmith, 1893-1921, and, with (Sir) Emery Walker [q.v.], Doves Press, 1900-16; employed a revived fifteenth-century French type; an important figure in 'arts and crafts' movement.

COCHRAN, SIR CHARLES BLAKE (1872-1951), showman; educated at Brighton grammar school; in United States, 1891-7, obtained theatrical experience; set up as theatrical agent in London, promoting boxing and wrestling matches, music-hall acts (including Houdini), roller-skating, and circuses; produced Reinhardt's *The Miracle* (1911), *The Better 'Ole* by Bruce Bairnsfather [q.v.] (1917); reopened London Pavilion, 1918, for successful run of revues, including *Dover Street to Dixie* with Florence Mills (1923); introduced Dolly Sisters in *The League of Notions* at redecorated New Oxford Theatre (1921); other promotions included appearances by Sarah Bernhardt, Eleanora Duse, Sacha Guitry, and Chaliapin, the Chauve Souris company, Diaghilev ballet, Wells–Beckett and Beckett–Carpentier fights, and Suzanne Lenglen tennis exhibitions; his association with (Sir) Noël Coward included

On with the Dance (1925), *Bitter Sweet* (1929), *Private Lives* (1930), *Cavalcade* (1931), and *Conversation Piece* (1934); other productions included *Evergreen* with Jessie Matthews (1930), *Escape Me Never* with Elisabeth Bergner (1933), and by Sir A. P. Herbert and Vivian Ellis *Big Ben* (1946) and *Bless the Bride* (1947); knighted, 1948.

COCHRANE, DOUGLAS, MACKINNON BAILLIE HAMILTON, twelfth EARL OF DUNDONALD (1852-1935), lieutenant-general; educated at Eton; entered 2nd Life Guards, 1870; his rides with dispatches to announce seizure of Gakdul Wells and death of General Gordon [q.v.] and fall of Khartoum made him famous; succeeded father, 1885; commanded 2nd Life Guards, 1895-9; fought in South Africa, 1899-1900; commanded and reorganized Canadian Militia, 1902-4; lieutenant-general, 1906; KCVO, 1907.

COCHRANE-BAILLIE, CHARLES WALLACE ALEXANDER NAPIER ROSS, second BARON LAMINGTON (1860-1940). [See BAILLIE.]

COCKCROFT, SIR JOHN DOUGLAS (1897-1967), physicist; educated at Todmorden secondary school and Victoria University, Manchester; M.Sc. Tech., 1922; St. John's College, Cambridge, first class, part ii, mathematics tripos, 1924; research student, Cavendish Laboratory, under Sir Ernest (later Lord) Rutherford [q.v.], 1924-39; supervised Mond Laboratory, 1935; worked on radar with Sir H. T. Tizard [q.v.], 1938-40; chief superintendent, Air Defence Research and Development Establishment, Christchurch, 1940-3; in charge of Montreal laboratory and building of NRX heavy water reactor at Chalk River, Canada, 1943-6; director, atomic energy research station, Harwell, 1946-59; first member for research, Atomic Energy Authority, 1954-9; first master, Churchill College, Cambridge, 1959-67; CBE, 1944; knighted, 1948; KCB, 1953; OM, 1957; Nobel prize for physics (with E. T. S. Walton), 1951; Atoms for Peace award, 1961.

COCKERELL, DOUGLAS BENNETT (1870-1945), bookbinder; educated at St. Paul's School; bookbinder under T. J. Cobden-Sanderson [q.v.], 1893; established own workshop, 1898; merged with W. H. Smith & Son's bindery as controller, 1905-14; founded own bindery at Letchworth, 1924; valuable works entrusted to him included *Codex Sinaiticus*; taught at Central School of Arts and Crafts, 1896-1935; published *Bookbinding and the Care of Books* (1901).

COCKERELL, SIR SYDNEY CARLYLE (1867-1962), museum director and bibliophile; educated at St. Paul's School; joined family firm, George J. Cockerell & Co. as clerk, 1884; partner, 1889-91; secretary to Kelmscott Press, 1891-6; secretary to Wilfrid Scawen Blunt [q.v.], 1896; partner of process-engraver, (Sir) Emery Walker [q.v.], 1900; director, Fitzwilliam Museum, Cambridge, 1908-37; European advisor to Felton Trustees of National Gallery of Victoria, Melbourne, 1936; fellow, Jesus College, Cambridge, 1910-16; fellow, Downing College, 1932-7; hon. Litt.D., Cambridge, 1930; knighted, 1934.

COCKS, ARTHUR HERBERT TENNYSON SOMERS-, sixth BARON SOMERS (1887-1944), chief scout for Great Britain and the British Commonwealth and Empire, and governor of Victoria. [See SOMERS-COCKS.]

CODNER, MAURICE FREDERICK (1888-1958), painter; educated at Stationers' Company School and Colchester Art School; portrait painter of distinguished personages preferably in ceremonial attire; sitters included King George VI (1951) and Queen Elizabeth the Queen Mother (1952), and many businessmen for board-room portraits; in landscape favoured trees, moving water, or snow scenes.

COGHLAN, SIR CHARLES PATRICK JOHN (1863-1927), first premier of Southern Rhodesia; born in Cape Colony; solicitor; elected to legislative council as representative for Bulawayo, 1908; a representative of Southern Rhodesia at national convention which had for its object formation of South African Union, 1908-9; acquiesced in decision that Southern Rhodesia should for the present remain outside Union; opposed amalgamation with Northern Rhodesia, 1917; on grant of responsible government, head of first ministry of Southern Rhodesia, 1923; knighted, 1910.

COHEN, SIR ANDREW BENJAMIN (1909-1968), colonial administrator; educated at Malvern College and Trinity College, Cambridge; first class, parts i and ii, classical tripos, 1930-1; assistant principal, Board of Inland Revenue, 1931; transferred to Colonial Office, 1932; organised supplies, Malta, 1940-3; assistant secretary, Colonial Office, 1943; assistant under-secretary of state, African division, 1947-52; assisted colonial secretary, Arthur Creech Jones [q.v.], to prepare for self-government in Africa; began *The Journal of African Administration*; governor, Uganda, 1952-7; permanent representative, Trusteeship Council at United Nations, 1957-60; published *British Policy in Changing Africa*, 1959; set up Department of

Technical Co-operation, 1961; permanent secretary, Ministry of Overseas Development, 1964-8; OBE, 1942; CMG, 1948; KCMG, 1952; KCVO, 1954; hon. LLD Queen's University, Belfast, 1960.

COHEN, ARTHUR (1829-1914), lawyer; BA, Magdalene College, Cambridge; first professing Jew to graduate at Cambridge; called to bar (Inner Temple), 1857; specialized in commercial law; junior counsel for Great Britain in *Alabama* arbitration at Geneva, 1872; QC, 1874; liberal MP, Southwark, 1880-7; to avoid by-election refused judgeship, 1881; standing counsel to India Office, 1893; counsel for Great Britain in Venezuela arbitration at The Hague, FBA, and member of royal commission on trade unions, 1903; PC, 1905; chairman of royal commission on shipping combinations, 1906.

COHEN, HARRIET (1896-1967), pianist; educated at Royal Academy of Music, 1909-15; won numerous prizes as a student; studied piano under Tobias Matthay; joined staff of Matthay School as professor, 1922; carried out concert tours in Britain, Europe, and the United States; career interrupted by tuberculosis, 1925; lost use of right hand, 1948; performed *Concertino for Left Hand*, composed for her by Sir Arnold Bax, 1950; CBE, 1938; hon. doctorate, National University of Ireland, 1960; autobiography, *A Bundle of Time*, published posthumously, 1969.

COHEN, Sir ROBERT WALEY (1877-1952), industrialist; grandson of Jacob Waley [q.v.]; educated at Clifton and Emmanuel College, Cambridge; joined Shell company, 1901; negotiated merger with Royal Dutch oil company, 1906; a director and chief assistant to managing director; encouraged university men into industry; petroleum adviser to Army Council in 1914-18 war; KBE, 1920; retired from Shell, 1928; chairman, African and Eastern company, 1929; negotiated merger with Niger Company into United Africa Company, 1929; resigned 1931; leading figure in Anglo-Jewry, opposed to Zionism, yet main creator of Palestine Corporation.

COILLARD, FRANÇOIS (1834-1904), protestant missionary under the Paris Missionary Society in the Zambezi region; born at Asnières-les-Bourges, Cher, France, of Huguenot family; sent to Basutoland, 1857; worked at Leribé, 1859-79; interpreter between Sir Theophilus Shepstone [q.v.] and Basuto chief Makotoko at Witzie's Hoek, 1865; baptized Makotoko, 1868; completed church at Leribé, 1871; went on evangelizing expedition to Banyai territory, 1877; taken prisoner at Bulawayo by Lobengula; established strong mission stations in Barotse territory, 1882; promoted native confidence in later British administration; published *Sur le Haut Zambèse* (1889), translated into English (1897).

COKAYNE, GEORGE EDWARD (1825-1911), genealogist; BA, Exeter College, Oxford, 1848; MA, 1852; called to bar, 1853; Norroy (1882) and Clarenceux (1894) King of Arms; FSA, 1866; published *G.E.C.'s Complete Peerage* (8 vols., 1887-98) and *Baronetage* (5 vols., 1900-6).

COKE, THOMAS WILLIAM, second EARL OF LEICESTER, of Holkham (1822-1909), agriculturist; son of first earl [q.v.]; an ardent agriculturist and forester; greatly improved his Holkham estate; lord-lieutenant of Norfolk, 1846-1906; keeper of privy seal of the duchy of Cornwall, 1870-1901; KG, 1873; a whig in politics.

COKER, ERNEST GEORGE (1869-1946), engineer; graduated at Edinburgh and Cambridge; assistant (later associate) professor, civil engineering, McGill University, 1898-1905; professor, Finsbury Technical College, 1905-14; University College, London, 1914-34; experimented in use of photo-elastic methods of determining stress distribution; published *A Treatise on Photo-Elasticity* (with L. N. G. Filon, q.v., 1931); FRS, 1915.

COLE, GEORGE DOUGLAS HOWARD (1889-1959), university teacher, writer, and socialist; educated at St. Paul's School and Balliol College, Oxford; first class, *lit. hum.*, and fellow of Magdalen, 1912; joined Fabian Society and independent labour party and became persuasive advocate of guild socialism; joined Fabian (later Labour) Research Department, 1913; honorary secretary, 1916-24; research adviser to Amalgamated Society of Engineers, 1915; tutor, London University, 1921-5; reader in economics, Oxford, 1925-44; Chichele professor of social and political theory, 1944-57; faculty fellow, Nuffield College, 1939-44, research fellow, 1957-9, sub-warden, 1942-3; chairman, Nuffield Social Reconstruction Survey, 1941-4; secretary, New Fabian Research Bureau, 1931-5, chairman, 1937-9; chairman, Fabian Society, 1939-46, 1948-50, president, 1952-9; influential in Workers' Educational Association; founded Tutors' Association; member, Economic Advisory Council, 1930; writer from 1918, a director, 1947, chairman, 1956, *New Statesman*; works include *The World of Labour* (1913), *Self-Government in Industry* (1917), *Guild Socialism Re-stated* (1920), *The Next Ten Years in British Social and Economic Policy* (1929), *The Intelligent Man's Guide*

Through World Chaos (1932), *Practical Economics* (1937), *The Condition of Britain* (with Margaret Cole, 1937), *The Common People* (with R. W. Postgate, 1938), *British Working Class Politics, 1832-1914* (1941), *Chartist Portraits* (1941), *A Century of Co-operation* (1946), *The Intelligent Man's Guide to the Post-War World* (1947), *Post-War Condition of Britain* (1956), *Socialist Thought* (4 vols., 1953-8); with Margaret Cole wrote twenty-nine detective stories in 1923-42; passionate in feeling but lucid in exposition he had exceptional qualities as a teacher.

COLEBROOK, LEONARD (1883-1967), bacteriologist; educated at Guildford grammar school, Westbourne high school, Bournemouth, and Christ's College, Blackheath; studied medicine at the London Hospital Medical College, and St. Mary's Hospital; MB, BS, (London), 1906; assistant, inoculation department, St. Mary's Hospital Medical School, 1907; worked on vaccine therapy and salvarsan; worked on wound infections while in RAMC, 1914-18; joined scientific staff, Medical Research Council, 1919; worked on dental caries; seconded to work with Sir Almroth E. Wright [q.v.] at St. Mary's Hospital, 1929; studied puerperal fever; honorary director, research laboratories, Queen Charlotte's Maternity Hospital, 1930; established value of sulphonamides; colonel, RAMC, bacteriological consultant to BEF, 1939; worked on infection and treatment of burns, 1940; director, burns investigation unit of the Medical Research Council, 1942-8; hon. fellow, Royal College of Obstetricians and Gynaecologists, 1944; FRS, 1945; hon. D.Sc., Birmingham, 1950; hon. fellow, Royal College, of Surgeons, 1950; published biography of Almroth Wright, 1954.

COLEMAN, WILLIAM STEPHEN (1829-1904), book-illustrator and painter; published *Our Woodlands, Heaths, and Hedges* (1859) and *British Butterflies* (1860); illustrated many books on natural history from 1858 onwards; executed numerous water-colour landscapes, etchings, and pastel work; exhibited at Dudley Gallery, 1865-79; established Minton Art Pottery Studio, Kensington Gore.

COLERIDGE, BERNARD JOHN SEYMOUR, second BARON COLERIDGE (1851-1927), judge; son of first Baron Coleridge [q.v.]; BA, Trinity College, Oxford, 1875; called to bar (Middle Temple), 1877; joined Western circuit; liberal MP, Attercliffe division of Sheffield, 1885-94; QC, 1892; succeeded father, 1894; judge, King's Bench division, 1907-23; tried the murderer, J. A. Dickman, 1910; a zealous humanitarian and talented musician.

COLERIDGE, MARY ELIZABETH (1861-1907), poet, novelist, and essayist; as a child she wrote verse and romance; first novel, *The Seven Sleepers of Ephesus* (1893), praised by R. L. Stevenson [q.v.]; her *Poems Old and New* (1907) and *Gathered Leaves* (1910), stories and essays, appeared posthumously.

COLERIDGE, STEPHEN WILLIAM BUCHANAN (1854-1936), author and anti-vivisectionist; son of first Baron Coleridge [q.v.]; educated at Trinity College, Cambridge; called to bar (Middle Temple), 1886; clerk of assize, South Wales circuit, 1890-1936; lectured and wrote on English literature; a founder of the NSPCC; his denunciation of vivisection was extreme.

COLERIDGE-TAYLOR, SAMUEL (1875-1912), musical composer; of negro birth; composition pupil of (Sir) C. V. Stanford [q.v.] at Royal College of Music; his *Hiawatha's Wedding Feast* produced 1898, followed by *Death of Minnehaha* (1899) and final section (1900); whole produced at Albert Hall, 1900; compositions include incidental music for plays by Stephen Phillips [q.v.] and 'A Tale of Old Japan' (poem by Alfred Noyes), 1911; student and apostle of African negro music.

COLES, CHARLES EDWARD [Pasha] (1853-1926), reformer of Egyptian prisons; born in India; entered Indian police department, 1873; director-general, Egyptian prisons, 1897-1913; transformed prisons by securing new buildings and improving their maintenance; had prisoners taught trades; CMG, 1900; died at Biarritz.

COLES, VINCENT STUCKEY STRATTON (1845-1929), divine and hymn writer; educated at Eton and Balliol College, Oxford; deacon, 1869; priest, 1870; succeeded father as rector of Shepton Beauchamp, Somerset, 1872; a librarian of Pusey House, Oxford, 1884; principal, 1897-1909; warden, community of the Epiphany, Truro, 1910-20; widely known as preacher, missioner, and spiritual guide; greatly influenced undergraduates; published meditations, sermons, and hymns.

COLLEN, SIR EDWIN HENRY HAYTER (1843-1911), lieutenant-general; joined Royal Artillery, 1863; served in Abyssinian war, 1867-8; assistant controller-general in second Afghan war, 1880; served in Eastern Sudan expedition at Tamai and Thankul, 1885; military secretary to Indian government, 1887-96; military member of governor-general's council, 1896; improved military equipment and mobilization of Indian army; lieutenant-general, 1905; CB,

1897; GCIE, 1901; member of War Office regulations committee, 1901-4; wrote history of *The Indian Army*, 1907.

COLLES, HENRY COPE (1879-1943), musical historian and critic; studied at Royal College of Music, Worcester College, Oxford (B.Mus., 1904), and under (Sir) Walford Davies [q.v.]; assistant to J. A. Fuller-Maitland [q.v.] on *The Times*, 1905-11; musical editor, 1911-43; edited third edition of *Grove's Dictionary of Music and Musicians* (1927), fourth edition and supplementary volume (1940); publications include *The Growth of Music* (1912-16) and *Symphony and Drama, 1850-1900* (1934).

COLLETT, SIR HENRY (1836-1901), colonel, Indian staff corps; joined Bengal army, 1855; served in Oudh during Indian Mutiny, 1858-9, in Assam, 1862-3, and Abyssinian campaign, 1868; quartermaster-general in Afghan war, 1878-80; commanded 23rd pioneers at Kandahar, 1880; CB, 1881; colonel, 1884; held command in Chin Lushai expedition, 1889-90; prominent in expedition to Manipur; KCB, 1891; commanded Peshawar district, 1892-3; on retirement (1893) prepared at Kew handbook on the flora of Simla (published 1902).

COLLIE, JOHN NORMAN (1859-1942), chemist and mountaineer; educated at Charterhouse, Clifton College, University College, Bristol, and Würzburg (Ph.D., 1884); assistant to (Sir) William Ramsay [q.v.], University College, London, 1887-96; professor of chemistry, Pharmaceutical Society College, London, 1896-1902; of organic chemistry, London University, 1902-28; worked mainly in field of organic chemistry; FRS, 1896; climbed in the Himalaya with A. F. Mummery [q.v.] and notably in the Canadian Rockies.

COLLIER, JOHN (1850-1934), painter and writer on art; son of first Baron Monkswell [q.v.]; educated at Eton and Slade School of Art; at his best in portraiture, in which he achieved a sober veracity; concerned with accuracy rather than interpretation; publications include *A Primer of Art*, 1882; a rationalist and son-in-law of T. H. Huxley [q.v.].

COLLINGS, JESSE (1831-1920), politician; in business in Birmingham, 1850-79; prominently associated with Joseph Chamberlain's programme of municipal reform; associated with Joseph Arch [q.v.] and programme of land reform; employed phrase 'three acres and a cow'; liberal MP, Ipswich (1880-6), Bordesley, Birmingham (1886-1918); PC, 1892; undersecretary to home department, 1895-1902.

COLLINGWOOD, CUTHBERT (1826-1908), naturalist; BA, Christ Church, Oxford, 1849; MB, 1854; surgeon and naturalist in HMS *Rifleman* and *Serpent* in China seas, 1866-7; published researches in marine zoology, 1868; a prominent Swedenborgian; died in Paris; wrote *The Travelling Birds* (1872) and theological works in verse and prose.

COLLINGWOOD, SIR EDWARD FOYLE (1900-1970), mathematician, medical administrator, and university leader; educated at Osborne, Dartmouth, and Trinity College, Cambridge, after being invalided from navy, 1913-21; studied mathematics under G. H. Hardy [q.v.]; MA, 1925; Ph.D., 1929; elected to council, Trinity College, 1930; JP 1935; high sheriff, Northumberland, 1937; served in RNVR, in 1939-45 war; chief scientist, Admiralty mine design department, 1943-5; returned to mathematics, 1948; Sc.D. Cambridge, 1959; FRS, 1965; hon. LLD, Glasgow, 1965; treasurer, London Mathematical Society, 1960-9; president, 1969; founder member, Newcastle Regional Hospital Board; chairman, 1953-68; member, Medical Research Council, 1960-8; hon. D.Sc. Durham; 1950; chairman, council of Durham University, 1963-70; CBE, 1946; knighted, 1962.

COLLINGWOOD, ROBIN GEORGE, (1889-1943), philosopher and historian; educated at Rugby and University College, Oxford; first class, *lit. hum.*, and philosophy; fellow of Pembroke, 1912; university lecturer in philosophy and Roman history, 1927-35; Waynflete professor of metaphysical philosophy, 1935-41; made drawings of all important Roman inscriptions in Britain and published *Roman Britain and the English Settlements* (with J. N. L. Myres, 1936, for *Oxford History of England*); his *Essay on Philosophical Method* (1933) modified and developed doctrine of English idealists through meeting contemporary criticism; other publications include *Essays on Metaphysics* (1940) and *The New Leviathan* (1942); FBA, 1934.

COLLINS, JOHN CHURTON (1848-1908), author and professor of English; educated at King Edward's School, Birmingham and Balliol College, Oxford; BA, 1872; greatly interested in literature in his youth; coached London candidates for Civil Service examinations from 1873, and wrote for the press and magazines; made friends with Swinburne; edited Cyril Tourneur's works (1878), and Lord Herbert of Cherbury's poems (1881); contributor to the *Quarterly Review* from Oct. 1878, and many of his articles there were republished independently; successful lecturer for Oxford and London university extension from 1880; long agitated with good ultimate effect for academic recognition of English literature at Oxford; urged his views in *The Study of English Literature* (1891) and in periodicals; an outspoken critic of current literature in *Saturday Review*, 1894–1906; collected essays in *Ephemera Critica* (1901), *Studies in Shakespeare* (1904), *Studies in Poetry and Criticism* (1905), and *Voltaire, Montesquieu, and Rousseau in England*, (1905); professor of English at university of Birmingham, 1904 to death; hon. Litt.D., Durham, 1905; zealous amateur student of criminology; brilliant conversationalist; drowned at Oulton Broad near Lowestoft.

COLLINS, JOSEPHINE (JOSÉ) (1887–1958), actress and singer; illegitimate daughter of music-hall artist Lottie Collins; appeared with (Sir) Harry Lauder [q.v.] in music-hall and pantomime; worked in New York, 1911–16; appeared at Daly's 1916–21, and the Gaiety, 1922–4; subsequently less successful; wartime favourite as Teresa in *The Maid of the Mountains* by Frederick Lonsdale [q.v.], 1917–20; a soprano of clear true warmth and tempestuous personality; thrice married.

COLLINS, MICHAEL (1890–1922), Irish revolutionary leader and chairman of provisional government of Irish Free State in 1922; born county Cork; held clerkships in London, 1906–16; active member of Irish Republican Brotherhood there; returned to Ireland, 1916; took part in Easter rebellion; imprisoned for short time in England; organizing genius of Volunteer and Sinn Fein movement; minister for home affairs on declaration of independence and institution of provisional constitution by Sinn Fein MPs, 1919; director of organization and subsequently of intelligence for Irish Volunteers; on supreme council of Irish Republican Brotherhood; all Irish revolutionary organizations declared illegal by British government, Sept. 1919; responsible for raising loans for movement; one of five Irish delegates who negotiated treaty with British government, 1921; chairman and minister of finance of provisional government, Jan. 1922; faced with organized opposition to treaty; on outbreak of civil war (June), commanded Free State army; reduced opposition in Dublin; killed by irregulars.

COLLINS, RICHARD HENN, BARON COLLINS (1842–1911), judge; scholar of Trinity College, Dublin, 1860; migrated to Downing College, Cambridge, 1863; fourth classic; 1865; honorary fellow, 1885; called to bar, 1867; QC, 1883; expert in litigation between rival municipalities and on railway companies; a lucid and precise advocate; judge of Queen's Bench, 1891; judicial member and chairman of railway and

canal commission, 1894; appointed to Court of Appeal and PC, 1897; master of the rolls, 1901; lord of appeal and life peer, 1907; represented Great Britain on tribunal dealing with Venezuelan boundary, 1899; chairman (1904) of commission of inquiry which led to Criminal Appeal Act of 1907; first president of Classical Association, 1903.

COLLINS, WILLIAM EDWARD (1867–1911), bishop of Gibraltar; BA, Selwyn College, Cambridge, 1887; DD, 1903; ordained, 1890; professor of ecclesiastical history, King's College, London, 1893–1904; helped to found Church Historical Society (1894–1904), contributing several historical studies; bishop of Gibraltar, 1904–11; visited Persia and Asiatic Turkey, 1907; died at sea off Smyrna.

COLNAGHI, MARTIN HENRY (1821–1908), picture dealer and collector; son of printseller; at first organizer of railway advertising; expert and buyer of pictures, helping to form several private collections; took Flatou's Gallery in Haymarket (1877–88) and Marlborough Gallery (1888), where he exhibited ancient and modern works of art; authority on Dutch and Flemish schools; discovered Van Goyen; privately purchased in 1896 Colonna or Ripaldi Raphael, which was subsequently sold to J. P. Morgan for £80,000; bequeathed several pictures and his fortune to National Gallery to form Martin Colnaghi bequest.

COLOMB, SIR JOHN CHARLES READY (1838–1909), writer on imperial defence; entered Royal Marines, 1854; retired, 1869; devoted rest of life to advocacy of strong military and naval defence; conservative MP for Bow and Bromley, 1886–92, and for Great Yarmouth, 1895–1906; KCMG, 1888; PC, 1903; chief works are *The Defence of Great and Greater Britain* (1879) and *Imperial Federation* (1886).

COLQUHOUN, ROBERT (1914–1962), painter of figure subjects, and MACBRYDE, ROBERT (1913–1966), painter of still life and figure subjects; the former educated at the Kilmarnock Academy and the Glasgow School of Art; the latter also educated at the Glasgow School of Art where they met, became inseparable companions, known as the Roberts; first joint exhibition at Kilmarnock, 1938; moved into studio in London which became centre for painters and writers, 1941–7; exhibitions in London, 1942–50; moved to studio at Lewes, 1947–9; designed costumes and scenery for Massine's ballet *Donald of the Burthens* (1951); Colquhoun alone made designs for *King Lear* (Stratford, 1953); last joint exhibition at Kaplan

Gallery, 1960; after Colquhoun's death in 1962, MacBryde moved to Dublin where he was killed by a car, 1966.

COLTON, Sir JOHN (1823-1902), premier of South Australia; went with father from Devonshire to Australia, 1839; founded firm of merchants at Adelaide; retired, 1883; mayor of Adelaide, 1874-5; in house of assembly, 1865-78; liberal commissioner of public works, 1868-70; treasurer, 1875-6; premier, 1876-7, and again, 1884-5; carried land and income tax bill; temperance advocate; KCMG, 1891.

COLVILE, Sir HENRY EDWARD (1852-1907), lieutenant-general; joined Grenadier Guards, 1870; served in Sudan, 1884-5 (CB, 1885), and in Egypt, 1885-8; acting commissioner in Uganda protectorate, 1893; commanded expedition against Kabarega, king of Unyoro, 1894; KCMG, 1895; commanded Guards brigade at Belmont, Modder River, and Magersfontein, Nov.-Dec. 1899; in command of new ninth division, he hemmed in Gen. Cronje at Paardeberg, Feb. 1900; at occupation of Bloemfontein (Mar.); ruined his career by failure to relieve General Broadwood's column at Sanna's Post and Colonel Spragge's force at Lindley (Mar.-May); ordered to command a brigade at Gibraltar, but recalled to England (Nov.); retired as lieutenant-general, Jan. 1901; defended himself in *The Work of the Ninth Division* (1901); wrote *The History of the Soudan Campaign* (1889).

COLVILLE, DAVID JOHN, first BARON CLYDESMUIR (1894-1954), public servant; educated at Charterhouse and Trinity College, Cambridge; conservative MP, North Midlothian, 1929-43; secretary, Department of Overseas Trade, 1931-5; parliamentary undersecretary of state for Scotland, 1935-6; financial secretary to Treasury, 1936-8; Scottish secretary, 1938-40; governor of Bombay, 1943-8; PC, 1936; GCIE, 1943; baron, 1948; governor of BBC, 1950-4.

COLVILLE, Sir STANLEY CECIL JAMES (1861-1939), admiral; entered *Britannia*, 1874; served in Zulu war (1879), and Egyptian (1882), Nile (1884-5), and Dongola (1896) campaigns; commander, 1892; captain, 1896; rear-admiral, 1906; vice-admiral, 1911; admiral, 1914; commanded first battle squadron, home fleet, 1912-14; responsible for defences of Scapa Flow, 1914-16; commander-in-chief, Portsmouth, 1916-19; vice-admiral of United Kingdom and lieutenant of Admiralty, 1929-39; KCB, 1912; GCVO, 1915; GCMG, 1919; GCB, 1921.

COLVIN, Sir AUCKLAND (1838-1908), Indian and Egyptian administrator; born at Calcutta; son of John Russell Colvin [q.v.]; in Indian service, 1858-78; English controller of Egyptian finance, 1880; as acting consul-general he quelled insurrection in Egypt, Sept. 1881; by articles in *Pall Mall Gazette* influenced English recognition of responsibility in Egypt; financial adviser to khedive; KCMG, 1881; financial member of viceroy's council in India, 1883; advocated international recognition of bimetallism; increased salt duty and imposed export duty on petroleum, 1887-8; lieutenant-governor of North-West Provinces, Nov. 1887; improved water supply and drainage stystem; uncompromising opponent of Indian National Congress; CIE, 1883; KCSI, on retirement, 1892; wrote life of his father (1895) and *Making of Modern Egypt* (1906).

COLVIN, IAN DUNCAN (1877-1938), journalist, biographer, and poet; educated at Inverness College and Edinburgh University; on staff of Allahabad *Pioneer* (1900-3) and *Cape Times* (1903-7); as leader-writer of deadly satiric touch on *Morning Post* (1909-37) earned title of 'keeper of the tory conscience'; severely critical of compromise and concession; works include biographies of Jameson, Dyer, and Carson, and satirical verse.

COLVIN, Sir SIDNEY (1845-1927), critic of art and literature; BA, Trinity College, Cambridge, 1867; fellow, 1868; Slade professor of fine art, Cambridge, 1873-85; director, Fitzwilliam Museum, 1876-83; keeper, department of prints and drawings, British Museum, 1883-1912; greatly improved arrangement and housing of collection; many important gifts and bequests secured through his influence; his most notable acquisition, the Malcolm collection of drawings and engravings, purchased 1895; knighted, 1911; works include *John Keats, His Life and Poetry* (1917); *Vailima Letters* (1895) of R. L. Stevenson [q.v.] to whom he was much attached; and editions of Stevenson's correspondence and works.

COLVIN, Sir WALTER MYTTON (1847-1908), brother of Sir A. Colvin [q.v.]; born in Burma; educated at Rugby and Trinity College, Cambridge; called to bar, 1871; acquired vast criminal practice in Allahabad; knighted, 1904.

COMMERELL, Sir JOHN EDMUND (1829-1901), admiral of the fleet; entered navy, 1842; gained the VC for gallantry in the Baltic, 1854; CB, 1866; commander-in-chief on west coast of Africa, 1871, and on North America station, 1882-5; KCB, 1874; conservative MP for Southampton, 1885-8; admiral, 1886; GCB,

1887; commander-in-chief at Portsmouth, 1888; admiral of the fleet, 1892.

COMMON, ANDREW AINSLIE (1841–1903), astronomer; early devoted himself to astronomy; with a silver-on-glass mirror of 3-foot diameter he made first successful photograph of the comet of June 1881; photographed great nebula in Orion, 1882; awarded gold medal of Royal Astronomical Society, 1884; completed 5-foot equatorial reflecting telescope, 1891; made small mirrors for observing eclipses for Royal Society, South Kensington, and Greenwich; invented telescopic gun sight for use in army and navy; FRS, 1885; president of Royal Astronomical Society, 1895–7; hon. LLD, St. Andrews, 1891.

COMPER, SIR (JOHN) NINIAN (1864–1960), church architect; educated at Glenalmond and Ruskin's Art School, Oxford; articled to G. F. Bodley and T. Garner [qq.v.]; a fervent Anglo-Catholic; his understanding of the purpose of a church as a roof over a free-standing altar was in advance of his time and made his sensitive workmanship unmistakable; built fifteen and restored and redecorated many more churches, both Anglican and Roman Catholic; favoured white interiors with strong clear colours; sought 'unity by inclusion' of styles and preferred Christ in Glory to Christ crucified; St. Mary's, Wellingborough, his *chef d'œuvre*; knighted, 1950.

COMPTON, LORD ALWYNE FREDERICK (1825–1906), bishop of Ely; son of second Marquess of Northampton [q.v.]; educated at Eton and Trinity College, Cambridge; fourteenth wrangler, 1848; rector of Castle Ashby, 1852–78; dean of Worcester, 1879–86; lord high almoner, 1882; bishop of Ely, 1886–1905; simple, direct preacher of high-church views; a keen archaeologist, he caused diocesan documents to be arranged and catalogued.

COMPTON-BURNETT, DAME IVY (1884–1969), novelist; educated at Addiscombe College, Hove, Howard College, Bedford, and Royal Holloway College; wrote first novel, *Dolores*, 1911; *Pastors and Masters* published, 1925; other novels include *Brothers and Sisters* (1929), *Daughters and Sons* (1937), *Parents and Children* (1941), and *Mother and Son* (1955); awarded James Tait Black memorial prize; CBE, 1951; DBE, 1967; C.Lit., 1968; hon. D.Litt., Leeds, 1960.

COMRIE, LESLIE JOHN (1893–1950), astronomer and computer; born in New Zealand; trained in chemistry, Auckland University College; Ph.D., St. John's College, Cambridge,

1924; entered Nautical Almanac Office, 1925; deputy superintendent, 1926; superintendent, 1930–6; completely revised *Nautical Almanac* for 1931; founded Scientific Computing Service, 1936, for large-scale numerical computation; produced *Hughes' Tables for Sea and Air Navigation* (1938), *Chambers's Six-Figure Mathematical Tables* (2 vols., 1948–9), and many others; FRS, 1950.

CONDER, CHARLES (1868–1909), artist; was sent to Sydney, New South Wales, 1884; studied and exhibited at Melbourne, where his 'The Hot Wind' (1890) attracted notice; returned to England, 1890; studied under Cormon in Paris; exhibited his works there (1891 and 1896), especially designs for fans; married and settled in Chelsea, 1901; designed lithographs; painted landscapes in oils; most characteristic work was in water-colours, delicately toned, on panels of white silk; style akin to that of Watteau.

CONDER, CLAUDE REIGNIER (1848–1910), colonel Royal Engineers, Altaic scholar, and Palestine explorer; grandson of Josiah Conder and cousin of Charles Conder [qq.v.]; joined Royal Engineers, 1870; continued scientific survey of Western Palestine, for Palestine Exploration Fund, 1872–3, surveying country west of Jordan; completed with (Lord) Kitchener [q.v.] *Memoirs of the Survey*, 7 vols., 1880; published *Tent Work in Palestine* (1878) and *Hand-book to the Bible* (1879); discovered ancient city of Kadesh; completed survey of 500-square miles of country across Jordan, 1881; joined in Egyptian campaign, 1882, and in Bechuanaland expedition, 1884; at Plymouth on ordnance survey, 1887; in charge of engraving department at Southampton, 1888–95; directed relief of distress in Ireland, 1895; commanding royal engineer at Weymouth, 1895; brevet-colonel, 1899; hon. LLD, Edinburgh, 1891; authority on Hittite and Altaic languages; voluminous writings include *Altaic Hieroglyphs and Hittite Inscriptions* (1887), *Palestine* (1891), *The Tell Amarna Tablets*, with translation and description (1893), and works for Palestine Pilgrims Text Society.

CONGREVE, SIR WALTER NORRIS (1862–1927), general; entered army, 1883; gazetted into Rifle Brigade, 1885; press censor on headquarter staff of Sir Redvers Buller [q.v.], 1899; VC for gallantry at Colenso, Dec.; participated in Lord Roberts's march to Pretoria; present at Poplar Grove, Dreifontein, surrender of Bloemfontein, and subsequent operations in Transvaal; assistant military secretary to Lord Kitchener [q.v.], 1901; commanded School of

Musketry, Hythe, 1909; commanded 18th Infantry brigade, with rank of brigadier-general, 1911; took part in battle of Aisne, Sept. 1914; commanded 6th division, May 1915; commanded XIII Army Corps, which reached Montauban ridge, outstanding feature of Somme attack, 1 July 1916; participated in battle of Arras, 1917; commanded VII Corps, 1918; transferred to X Corps; commanded British forces in Egypt, 1919-23; full general, 1922; general commanding-in-chief, Southern command, Salisbury, 1923; governor of Malta, 1925; KCB, 1917; died at Valetta.

CONINGHAM, SIR ARTHUR (1895-1948), air marshal; born in Brisbane; educated at Wellington College, New Zealand; served in Samoa and Gallipoli in Canterbury Mounted Rifles, whence his nickname 'Maori'; transferred to Royal Flying Corps, 1916-18; DSO, 1917; flight-lieutenant, Royal Air Force, 1919; air commodore, 1939; commanded No. 4 Group (night bombers), 1939-41; Desert Air Force supporting Eighth Army in North Africa (including El Alamein battle), 1941-3; Allied Tactical Air Forces in capture of Tunisia, Pantelleria, Sicily, and southern Italy, 1943-4; 2nd Tactical Air Force in England and Normandy, 1944-5; Flying Training Command, 1945-7; the original architect of tactical air forces; KCB, 1942; KBE, 1946.

CONNARD, PHILIP (1875-1958), painter; studied textile design, South Kensington; taught at Lambeth Art School; member of new English Art Club; became known as a decorative artist, his commissions including panels for ballroom at Delhi and the *Queen Mary*; his water-colours notable for subtle apprehension of atmosphere. ARA, 1918; RA, 1925; CVO, 1950.

CONNAUGHT AND STRATHEARN, DUKE OF (1850-1942), [See ARTHUR WILLIAM PATRICK ALBERT.]

CONNEMARA, BARON (1827-1902), governor of Madras. [See BOURKE, ROBERT.]

CONNOR, RALPH (1860-1937), (pseudonym), divine and author. [See GORDON, CHARLES WILLIAM.]

CONNOR, SIR WILLIAM NEIL (1909-1967), journalist; educated at Glendale grammar school; left school at sixteen; book-keeper, Arks Publicity; copy writer, advertising agency, J. Walter Thompson, 1922; joined staff of *Daily Mirror* and started 'Cassandra' column, 1935; hard hitting journalist, critical of Churchill government; joined army and worked with (Sir) Hugh Cudlipp, producing the *Union Jack*,

1942-5; returned to *Mirror*, 1946; lifelong professional critic of the Establishment; knighted, 1966.

CONQUEST, GEORGE AUGUSTUS (1837-1901), actor-manager, whose original name was OLIVER; manager of Grecian Theatre, 1872-85; sole lessee of Surrey Theatre from 1885; retired from stage, 1894; acted with great melodramatic power; best known as acrobatic pantomimist and animal impersonator; reinvented 'flying' by 'invisible' wires; co-author of over a hundred original or adapted melodramas; chief were *Sentenced to Death* (1875), *Mankind* (1881), and *The Crimes of Paris* (1883).

CONRAD, JOSEPH (1857-1924), master mariner and novelist; of Polish parentage; born near Mohilow in Poland; his original name, TEODOR JOSEF KONRAD KORZENIOWSKI; his boyhood greatly influenced by his father (died 1869), a cultivated Polish patriot; went to sea as registered seaman in French merchant marine, 1874; first landed in England, 1878; qualified as third mate, 1880; second mate, 1881; first mate, 1883; naturalized as British subject under name of Joseph Conrad, 1886; ship's master, 1886; sailed Malay Archipelago, 1887 and 1888; later explored the Congo; visited Ukraine, 1890; left merchant service, 1894; had begun to write a story (1889) which was published as *Almayer's Folly*, 1895; this was welcomed by small but discriminating section of public; followed it with *An Outcast of the Islands* (1896) and *The Nigger of the 'Narcissus'* (1897); had many years to wait for general recognition; ill health and straitened means rendered him peculiarly liable to periods of despairing gloom; published *Lord Jim* (1900), *Nostromo* (1904), *The Mirror of the Sea* (1906), *The Secret Agent* (1907); *Under Western Eyes* (1911); his emergence as one of the great novelists of the day due to publication of *Chance* in New York, 1914; thenceforth uncritically and therefore immoderately praised; his other chief works, in addition to volumes of stories, include *The Arrow of Gold* (1919), *The Rescue* (1920), and *The Rover* (1922); his remarkable life-story enabled him to use many strange and picturesque experiences as material for fiction; his most notable achievement, creation of much fine prose in language not his own.

CONSTANT, HAYNE (1904-1968), mechanical and aeronautical engineer; educated at King's College Choir School, Cambridge, King's School Canterbury, Technical Institute in Folkestone, Sir Roger Manwoods' School, Sandwich, and Queens' College, Cambridge; first class, mechanical sciences tripos, 1927; joined engine department, Royal Aircraft

Establishment, Farnborough, 1928-34; lecturer, Imperial College, 1934-6; in charge Supercharger Section, RAE, 1936; designed gas turbine engine for driving propellers, 1937; head of Engine Department, RAE, 1941; proposed development which, built by Metropolitan Vickers, became prototype of modern jet engine, 1943; deputy director, National Gas Turbine Establishment, 1946; director, 1948-60; FRS, 1948; CBE, 1951; CB, 1958; scientific advisor to Air Ministry, 1960-2; chief scientist (RAF), Ministry of Defence , 1964; publications include *Gas Turbines and their Problems* (1948).

CONWAY, ROBERT SEYMOUR (1864-1933), classical scholar and comparative philologist; educated at City of London School and Gonville and Caius College, Cambridge; first class, classics, 1885-7; professor of Latin at Cardiff (1893-1903) and Manchester (1903-29); publications include *The Italic Dialects* (2 vols., 1897), an edition of Livy (in collaboration, 1914-35), and works on Virgil; FBA, 1918.

CONWAY, WILLIAM MARTIN, Baron Conway of Allington (1856-1937), art critic and collector and mountaineer; educated at Repton and Trinity College, Cambridge; professor of art at Liverpool (1885-8) and Cambridge (1901-4); presented his 100,000 photographs of art and architecture to Courtauld Institute; a mountaineer who preferred passes to peaks, he missed few alpine seasons between 1872 and 1901; knighted (1895) for explorations in Karakorams; made first crossing of Spitsbergen, 1896; unionist MP, combined English universities, 1918-31; baron, 1931; publications include *Zermatt Pocket Book* (1881), *Woodcutters of the Netherlands in the Fifteenth Century* (1884), and *The Art Treasures in Soviet Russia* (1925).

CONYBEARE, FREDERICK CORNWALLIS (1856-1924), Armenian scholar; BA, University College, Oxford, 1879; fellow, 1880-7; searched collections of Armenian manuscripts for material for textual criticism of Greek classics from ancient Armenian versions; studied church history and textual criticism of Septuagint and New Testament; made numerous discoveries; catalogued Armenian manuscripts in British Museum (1913) and Bodleian Library (1918); member of Rationalist Press Association, 1904-15.

CONYNGHAM, Sir GERALD PONSONBY LENOX- (1866-1956), geodesist. [See LENOX-CONYNGHAM.]

COOK, ARTHUR BERNARD (1868-1952), classical scholar and archaeologist; scholar of St. Paul's School and Trinity College, Cambridge;

first class, parts i and ii, classical tripos, 1889-91; fellow of Trinity, 1893-9; lecturer in classics, 1900, fellow, 1903, vice-president, 1935, Queens' College; professor of Greek, Bedford College, London, 1893-1907; reader in Greek (1907-31), Laurence professor (1931-4), classical archaeology, Cambridge; works include *The Metaphysical Basis of Plato's Ethics* (1895) and *Zeus, a Study in Ancient Religion* (3 vols., 1914, 1925, 1940); FBA, 1941.

COOK, ARTHUR JAMES (1883-1931), miners' leader; a miner in South Wales for twenty-one years; studied at Central Labour College, 1911-12; district agent, 1919; secretary, Miners' Federation of Great Britain, 1924-31; leading figure in general strike, 1926; warm-hearted and impulsive he was an agitator rather than a negotiator.

COOK, Sir BASIL ALFRED KEMBALL- (1876-1949), civil servant. [See KEMBALL-COOK.]

COOK, Sir EDWARD TYAS (1857-1919), journalist; educated at Winchester and New College, Oxford; joined staff of *Pall Mall Gazette*, 1883; editor, 1890-2; editor of newly founded *Westminster Gazette*, 1893-6, and of *Daily News*, 1896-1901; edited library edition of Ruskin's *Works* (38 vols., 1903-11); biographer of Ruskin (1911), Florence Nightingale (1913), and Delane (1915); joint manager of Press Bureau, 1915-19; knighted, 1912; KCB, 1917.

COOK, Sir FRANCIS, first baronet (1817-1901), merchant and art collector; entered father's drapery firm in City of London, 1833; partner, 1843; head, 1869; first visited Portugal, 1841; bought and restored Monserrate palace, Cintra, 1856; renewed prosperity of district; created Viscount Monserrate, 1864; at Doughty House, Richmond Hill (1860), he formed fine art collection of foreign schools; FSA, 1873; established Alexandra House, South Kensington, for women music and art students, 1885; baronet, 1886; left estate of £1,500,000.

COOK, Sir JOSEPH (1860-1947), Australian politician; Staffordshire coalminer; emigrated to Australia, 1885; held office in New South Wales parliament under (Sir) George Reid [q.v.], 1894-9; entered federal parliament, 1901; leader, free trade party, 1908; defence minister under Alfred Deakin [q.v.] in fusion of anti-labour forces, 1909-10; laid foundations of compulsory military training; prime minister, 1913-14; led opposition, supporting war effort, 1914-17; entered coalition as navy minister, 1917; member of imperial war cabinet, 1918;

delegate to peace conference, 1919; Commonwealth treasurer, 1920; high commissioner, London, 1921-7; PC, 1914; GCMG, 1918.

COOK, STANLEY ARTHUR (1873-1949), Semitist, biblical scholar, archaeologist, and student of religion; first class, Semitic languages, Gonville and Caius College, Cambridge, 1894; fellow, 1900; lecturer in Hebrew (1904-32) and comparative religion (1912-20); university lecturer in Aramaic, 1931; regius professor of Hebrew, 1932-8; joint-editor, *Cambridge Ancient History*; editor, *Quarterly Statement*, Palestine Exploration Fund, 1902-32; FBA, 1933.

COOKE, GEORGE ALBERT (1865-1939), regius professor of Hebrew and canon of Christ Church, Oxford; educated at Merchant Taylors' School and Wadham College, Oxford; second class, theology, 1888; ordained, 1889; Oriel professor of interpretation of Holy Scripture and canon of Rochester, 1908-14; regius professor of Hebrew and canon of Christ Church, 1914-36; published *Text-book of North-Semitic Inscriptions*, 1903.

COOLIDGE, WILLIAM AUGUSTUS BREVOORT (1850-1926), mountaineer and historian; born near New York; BA, Exeter College, Oxford; fellow of Magdalen College, 1875-1926; ordained, 1882; from 1885 devoted himself to Swiss mountaineering, geography, and history, and editing of books on climbing and surveying; died at Grindelwald.

COOPER, SIR ALFRED (1838-1908), surgeon; studied medicine at St. Bartholomew's Hospital and in Paris; MRCS, 1861; FRCS, 1870; surgeon to West London and other hospitals; knighted, 1902; a keen Volunteer; helped to found Rahere lodge of freemasons; author of *Syphilis and Pseudo-Syphilis*, 1884.

COOPER, ALFRED DUFF, first VISCOUNT NORWICH (1890-1954), politician, diplomatist, and author; son of Sir Alfred Cooper [q.v.]; educated at Eton and New College, Oxford; entered Foreign Office, 1913; served with Grenadier Guards, 1917-18, DSO; married, 1919, Lady Diana Manners; resigned from Foreign Office, 1924; conservative MP, Oldham, 1924-9, St. George's, Westminster, 1931-45; financial secretary, War Office, 1928-9, 1931-4; to Treasury, 1934-5; secretary for war, 1935-7; first lord of Admiralty, 1937; resigned over Munich agreement, 1938; minister of information, 1940-1; chancellor of Duchy of Lancaster, 1941-3; resident cabinet minister, Singapore, 1941-2; chairman, cabinet committee on security, 1942-4; British representa-

tive, French Liberation Committee, Algiers, 1944; ambassador to France, 1944-7; treaty of alliance obtained, 1947; publications include *Talleyrand* (1932), *Haig* (2 vols., 1935-6), and autobiography *Old Men Forget* (1953); PC, 1935; GCMG, 1948; viscount, 1952.

COOPER, CHARLOTTE (1870-1966), lawn tennis champion. [See STERRY.]

COOPER, SIR DANIEL, first baronet (1821-1902), Australian merchant; a native of Lancashire; visited Sydney in boyhood; after education in London, joined mercantile firm in Sydney, 1843; elected to legislative council of New South Wales, 1849; raised funds for Crimean war sufferers, 1854; speaker of New South Wales, 1856-9; knighted, 1857; settled in London, 1859; did much service in Lancashire cotton crisis; baronet, 1863; agent-general for New South Wales, 1897-9; KCMG, 1880; GCMG, 1890.

COOPER, EDWARD HERBERT (1867-1910), novelist; BA, University College, Oxford, 1890; Paris correspondent of *New York World*; visited and wrote on Finland, 1901; gained distinction as author of sporting novels and imaginative stories for children after manner of Lewis Carroll; chief works were *Mr. Blake of Newmarket* (1897), *Wyemark and the Sea Fairies* (1897), and *Wyemark and the Mountain Fairies* (1900).

COOPER, SIR (FRANCIS) D'ARCY, baronet (1882-1941), industrialist; educated at Wellington College; partner in family firm of chartered accountants, 1910; retrieved Lord Leverhulme [q.v.] from financial difficulties, 1921; succeeded him as chairman of Lever Brothers, 1925; by merging with Margarine Union (1929) created Unilever, one of world's largest businesses; chairman, executive committee, export council, Board of Trade, 1940-1; baronet, 1941.

COOPER, JAMES (1846-1922), Scottish divine; MA, Aberdeen, 1867; parochial minister, 1873-98; regius professor of ecclesiastical history, Glasgow, 1898; moderator, General Assembly of Church of Scotland, 1917; a keen ecclesiologist and 'Scoto-Catholic'.

COOPER, JAMES DAVIS (1823-1904), wood-engraver; apprenticed to Josiah Whymper [q.v.]; a successful wood-engraver in the 1860s; worked with Randolph Caldecott [q.v.] on the Macmillan editions of Washington Irving's *Old Christmas*, 1876, and *Bracebridge Hall*, 1877; engraved illustrations for books by Darwin and Huxley, and for Stanley's *In Darkest Africa*,

1890; skilful interpreter of landscape and groups of natives and animals.

COOPER, SIR (THOMAS) EDWIN (1874–1942), architect; practised in partnership until 1910, thereafter independently; works include Marylebone town hall (1911), Port of London Authority building (1912–22), Lloyd's (1928), National Provincial Bank (1931–2), Devonport School of Pathology and Nurses' Home, Greenwich, Star and Garter Home, Richmond, Cowdray Club, Cavendish Square, the South London Hospital for Women, and the School of Biochemistry, Cambridge; FRIBA, 1903; ARA, 1930; RA, 1937; knighted, 1923.

COOPER, THOMAS MACKAY, BARON COOPER OF CULROSS (1892–1955), lord justice-general of Scotland; educated at George Watson's College and Edinburgh University; first class, classics, 1912; LLB with distinction, 1914; advocate, 1915; KC, 1927; conservative MP, West Edinburgh, 1935–41; solicitor-general for Scotland, 1935; lord advocate, 1935–41; lord justice-clerk (as Lord Cooper), 1941–6; lord justice-general and lord president of Court of Session, 1947–54; chairman, hydro-electric development in Scotland (1941–2) and other committees; a founder (1934) of and writer for Stair Society; PC, 1935; baron, 1954.

COOPER, THOMAS SIDNEY (1803–1902), animal painter; born at Canterbury; at first a coach-painter, then a scene-painter; went to London, 1823, copying at British Museum; admitted to Royal Academy Schools; drawing master (1823–7) at Canterbury and elsewhere; visited Brussels, 1827, where he produced lithographs and gained friendship of Verboeckhoven, Belgian animal painter; found chief models in seventeenth-century Dutch school; settled in England, 1831; exhibited 48 pictures at British Institution, 1833–63, and 266 at Royal Academy, 1833–1902; ARA, 1845; RA, 1867; best pictures include 'Drovers crossing New-bigging Muir in a Snowdrift, East Cumberland' (1860), 'The Shepherd's Sabbath' (1866), 'Milking Time in the Meadows' (diploma picture, 1869), and studies of bulls: 'The Monarch of the Meadows' (1873) and 'Separated but not Divorced' (1874); settled at 'Vernon Holme', Canterbury, 1848; published *My Life*, 1890; CVO, 1901; converted birthplace into 'Sidney Cooper Gallery of Art', 1865, and presented it to town of Canterbury, 1882.

COOPER, THOMPSON (1837–1904), biographer and journalist; son of Charles Henry Cooper [q.v.]; compiled *Athenae Cantabrigienses*, vol. 1, 1858, vol. 2, 1861; FSA, 1860; sub-editor (1861) and parliamentary reporter (1862)

of *Daily Telegraph*; connected with *The Times* as parliamentary reporter, 1866–86; summary writer of debates in House of Commons, 1886–98, and in House of Lords, 1898–1904; contributed 1,422 articles to this Dictionary, 1885–1901, mainly on Roman Catholic divines and early Cambridge graduates; compiled *Biographical Dictionary*, 1873, and Supplement, 1883; edited *Men of the Time*, 1872–84; joined Church of Rome at early age.

COPE, SIR ALFRED WILLIAM (1877–1954), civil servant; entered as boy clerk; preventive inspector, Customs and Excise, 1908; second secretary, Ministry of Pensions, 1919; assistant under-secretary, Dublin, and clerk of Irish Privy Council, 1920–2; undertook secret negotiations leading to truce of 1921; CB, 1920; KCB, 1922.

COPELAND, RALPH (1837–1905), astronomer; went to Australia, 1853–8; studied astronomy at Göttingen, 1865–9; published *First Göttingen Catalogue of Stars*, 1869; Ph.D., 1869; went on German Arctic expedition to explore east coast of Greenland, reaching latitude 75° 11.5′ N., 1870; FRAS, 1874; observed transit of Venus at Mauritius, Dec. 1874; discovered great tree fern (*Cyathea copelandi*) on South Atlantic island of Trinidad; in charge of Lord Lindsay's observatory at Dunecht, Aberdeen, 1876; there made many cometary discoveries; observed transit of Venus at Jamaica, 1882; catalogued Lord Crawford's astronomical literature, 1890; edited *Copernicus, a Journal of Astronomy*, 3 vols., 1891–4; astronomer royal for Scotland and professor of astronomy at Edinburgh University, 1889.

COPINGER, WALTER ARTHUR (1847–1910), professor of law, antiquary, and bibliographer; called to bar, 1869; published *Law of Copyright in Literature and Art*, 1870; settled in Manchester as equity draftsman and conveyancer; published works on conveyancing, 1872, 1875; professor of law in Owens College, Manchester, 1892; LLD, Lambeth, 1889; founder and first president of Bibliographical Society, London, 1892; published *Supplement to Hain's Repertorium Bibliographicum* (1895–8) and *Incunabula Biblica* (1892); set up private press at Manchester, 1893; also published works on genealogy (*History of the Copingers*, 1882), heraldry, and manorial history, including *History of Suffolk* (5 vols., 1904–7) and *Manors of Suffolk* (7 vols., 1905–11); member of Catholic Apostolic Church; wrote on *Predestination*, 1889; an accomplished musical composer.

COPISAROW, MAURICE (1889–1959), chemist; born in Russia; naturalized, 1915;

M.Sc., Manchester, 1914; D.Sc., 1925; trained in research by Chaim Weizmann [q.v.]; with Ministry of Munitions, 1915-18, undertaking extensive research on TNT; on research staff, British Dyestuffs Corporation, 1919-22; thereafter unemployed through ill health and loss of sight; nevertheless investigated wide variety of subjects with great originality, including rock formation, biochemical causes of malignant growth, action of enzymes, and agricultural topics.

COPPARD, ALFRED EDGAR (1878-1957), story-writer and poet; finished school at nine; confidential clerk, Eagle Ironworks, Oxford, 1907-19; first of many collections of short stories, *Adam and Eve and Pinch Me*, published by Golden Cockerel Press, 1921; first of five volumes of lyrics, *Hips and Haws*, 1922; his *Selected Tales* (1946) published by Book of the Month Club of America; wrote with wholly individual quality of imagination and wisdom.

COPPIN, GEORGE SELTH (1819-1906), actor and Australian politician; obtained some fame in English provinces as capable comedian, 1841; eloped to Sydney, 1843; built theatre at Adelaide, 1846; made fortune there, but lost it in speculation, 1851; made another theatrical fortune in Melbourne, 1852-4; toured in England, 1854; induced G. V. Brooke [q.v.] to join him in Australia; opened Theatre Royal, Melbourne, 1856 (rebuilt 1872), of which he was managing director till death; organized first grand opera season in Australia; toured with the Keans [qq.v.] in America, 1864-5; member of legislative council of Victoria, 1858-64, 1874-89; advocated federation of colonies, intercolonial free trade, and acclimatization; first to import camels and English thrushes into Australia; founded Sorrento-on-the-Sea, where Mount Coppin is named after him.

COPPINGER, RICHARD WILLIAM (1847-1910), naval surgeon and naturalist; MD, Dublin, 1870; surgeon of HMS *Alert* on voyage of polar exploration with (Sir) George S. Nares [q.v.], 1875, and on exploring cruise in Patagonian and Polynesian waters, 1878-82; inspector-general of hospitals and fleets, 1901-4; author of *The Cruise of the Alert, 1878-82* (1883).

CORBET, MATTHEW RIDLEY (1850-1902), landscape painter; exhibited at Royal Academy, 1875-1902; ARA, 1902; painted mainly Italian scenery, e.g. 'Morning Glory' (1894) and 'Val d'Arno—Evening' (1901) in Tate Gallery.

CORBETT, EDWARD JAMES (JIM) (1875-1955), destroyer of man-eating tigers, naturalist, and author; born at Naini Tal; educated at St. Joseph's College; served (1895-1914) with Bengal and North Western Railway; took Kumaon force to France, 1917-18; became expert photographer of big game; killed ten man-eaters in 1907-38; granted freedom of forests of India; moved to Nyeri, Kenya, 1947; wrote *The Man-Eaters of Kumaon* (1946), *The Man-Eating Leopard of Rudraprayag* (1948), and *The Temple Tiger* (1954).

CORBETT, JOHN (1817-1901), promoter of the salt industry in Worcestershire and benefactor; studied engineering; joined father's firm of carrier of merchandise by canal boats; bought Stoke Prior salt works near Droitwich, 1852; transformed enterprise from a failure to success; in twenty-five years he raised annual output from 26,000 tons of salt to 200,000; generous to his work-people; liberal MP for Droitwich, 1874-85; opposed to home rule; liberal unionist MP for Mid-Worcestershire, 1886-92; advocated women's suffrage; a generous benefactor to Stourbridge, Droitwich, Birmingham University, and other Midland institutions.

CORBETT, Sir JULIAN STAFFORD (1854-1922), naval historian; BA, Trinity College, Cambridge, 1875; called to bar (Middle Temple), 1879; took up fiction writing, 1886, but later turned to serious historical writing; history lecturer, Royal Naval War College, Greenwich, 1902; Ford's lecturer, Oxford, 1903; knighted, 1917; works include *Drake and the Tudor Navy* (1898), *The Successors of Drake* (1900), *England in the Mediterranean, 1603-1714* (1904), *Fighting Instructions, 1530-1816* (1905), *England in the Seven Years War* (1907), *The Campaign of Trafalgar* (1910), *The Spencer Papers* (1913-14), *Naval Operations* (official history of the European war at sea, first three volumes, 1920-3).

CORBOULD, EDWARD HENRY (1815-1905), water-colour painter; son of Henry Corbould [q.v.]; first exhibited at Royal Academy, 1835; mainly worked in water-colours; exhibited 250 drawings at New Water Colour Society from 1837, his works including 'The Canterbury Pilgrims assembled at the Old Tabard Inn', 1840; instructor of historical painting to the royal family, 1851-76; designed book illustrations for Waverley novels and other works; many works were popular in engravings; his 'Lady Godiva' is in National Gallery of New South Wales.

CORELLI, MARIE (pseudonym), (1855-1924) novelist. [See MACKAY, MARY.]

CORFIELD, WILLIAM HENRY (1843–1903), professor of hygiene and public health; BA, Magdalen College, Oxford (first class in mathematics and physics), 1864; medical fellow of Pembroke College, 1865–75; Radcliffe travelling fellow, 1867; studied hygiene in Paris and Lyons; MB, Oxford, 1868; MD, 1872; FRCP London, 1875; professor of hygiene and public health at University College, London, 1869; medical officer for St. George's, Hanover Square, 1872–1900; pioneer of house sanitation; first consulting sanitary adviser to Office of Works, 1899; president of Epidemiological Society, 1902–3; died at Marstrand, Sweden; wrote *Dwelling Houses: their Sanitary Construction and Arrangements* (1880), *Laws of Health* (1880), and kindred works.

CORK AND ORRERY, twelfth EARL of. (1873–1967) [See BOYLE, WILLIAM HENRY DUDLEY.]

CORNFORD, FRANCES CROFTS (1886–1960), poet; daughter of Sir Francis Darwin [q.v.]; married F. M. Cornford [q.v.], 1909; their home a meeting-place for artists and men of letters; her first *Poems* published, 1910; *Collected Poems*, 1954; Queen's medal, 1959; a minor poet of the 'Georgian' period and after; also published verse translations from Russian and French.

CORNFORD, FRANCIS MACDONALD (1874–1943), classical scholar; educated at St. Paul's School and Trinity College, Cambridge; first class, parts i, and ii, classical tripos, 1895–7; fellow, 1899; lecturer in classics, 1904; Brereton-Laurence reader in classics, 1927; Laurence professor of ancient philosophy, 1931–9; in forefront of modern Platonic scholars; works include *Thucydides Mythistoricus* (1907), *From Religion to Philosophy* (1912), *The Origin of Attic Comedy* (1914), *Plato's Theory of Knowledge* (1935), *Plato's Cosmology* (1937), *Plato and Parmenides* (1939), and *Microcosmographia Academica* (1908), a skit on university politics; FBA, 1937.

CORNISH, CHARLES JOHN (1858–1906), naturalist; BA, Hertford College, Oxford, 1885; classical master at St. Paul's School from 1885 till death; contributed natural history articles to *Spectator* and *Country Life*; author of *Life at the Zoo* (1895); his *Animal Artisans and other Studies of Birds and Beasts* (1907) has memoir by his widow.

CORNISH, FRANCIS WARRE WARRE- (1839–1916), teacher, author, and bibliophile [See WARRE-CORNISH.]

CORNISH, VAUGHAN (1862–1948), geographer; brother of C. J. Cornish [q.v.]; educated at St. Paul's School and Owens College, Manchester; B.Sc., 1888; D.Sc., 1901; travelled widely studying surface waves; published results in *Ocean Waves and Kindred Geophysical Phenomena* (1934); lectured on strategical geography to naval and military officers, 1914–18; published historical study *The Great Capitals* (1923) emphasizing the forward position of capital cities; worked for preservation of English countryside.

CORNWALLIS, SIR KINAHAN (1883–1959), administrator and diplomatist; born in New York; educated at Haileybury and University College, Oxford; joined Sudan civil service, 1906, Egyptian civil service, 1912, GHQ Cairo, intelligence, 1914, Arab Bureau, 1915; director, Arab Bureau, 1916–20; personal adviser to King Feisal of Iraq, 1921–33; negotiated treaty giving Iraq independence, 1930; retired, 1935; ambassador to Iraq, 1941–5; held out in embassy against rebel pro-axis government until troops arrived from India, 1941; thereafter obtained co-operation from new government; in Foreign Office, 1945–6; CMG, 1926; knighted, 1929; KCMG, 1933; GCMG, 1943.

CORNWELL, JAMES (1812–1902), writer of school books; organizer of country schools for British and Foreign School Society, 1835; principal of society's Borough Road training college, 1846–85; from 1841 published simple and useful school books on grammar, composition, and arithmetic; his *School Geography* (1847) passed through ninety editions.

CORRY, MONTAGU WILLIAM LOWRY, BARON ROWTON (1838–1903), politician and philanthropist; educated at Harrow and Trinity College, Cambridge; BA, 1861; called to bar, 1863; private secretary to Disraeli, 1866–81; attended him as secretary of embassy at congress of Berlin, 1878; CB, 1878; created Baron Rowton, 1880; succeeded to Rowton Castle estate, 1889; acquired by will Lord Beaconsfield's correspondence and papers, 1881; trustee of Guinness Trust [artisan dwellings] Fund, 1889; resolving to provide a poor man's hotel, he built 'Rowton House', Vauxhall, 1892; other Rowton houses erected at King's Cross (1896), Newington (1897), Hammersmith (1899), Whitechapel (1902), and Camden Town (1905); KCVO, 1897; PC, 1900.

CORY, JOHN (1828–1910), philanthropist, coal-owner, and shipowner; son of Richard Cory, coal exporter of Cardiff and advocate of teetotalism; joined father's firm, 1844; established foreign coal depots, which numbered eighty at his death; acquired Pentre, Rhondda colliery, 1868; vice-chairman of Barry Dock and Railway Company; strong advocate of teetotalism, and supporter of Salvation Army, Band of Hope Union, and Dr Barnardo's Homes; benefactor to Cardiff institutions to amount of £50,000 a year; left by will £250,000 for charitable purposes.

CORYNDON, Sir ROBERT THORNE (1870–1925), South African administrator; born in Cape Colony; sent by Cecil Rhodes [q.v.] to Mafeking and attached to Bechuana border police; private secretary to Rhodes; first British resident and British South Africa Company's representative with Lewanika, paramount chief of the Barotse, 1897; administrator of North-Western Rhodesia, 1900–7; resident commissioner, Swaziland, 1907–16; of Basutoland, 1916–17; governor and commander-in-chief; Uganda, 1917–22; governor and commander-in-chief, Kenya Colony, and high commissioner, Zanzibar, 1922; KCMG, 1919; died at Nairobi.

COSGRAVE, WILLIAM THOMAS (1880–1965), first president of the Executive Council of the Irish Free State; born in Dublin; educated by Christian Brothers; entered father's business (licensed vintner); joined with Arthur Griffith [q.v.] in founding Sinn Fein, 1905; Sinn Fein representative, Dublin corporation, 1909–22; fought with Irish Volunteers, captured and condemned to death, sentence commuted, 1916; released, 1917; won Kilkenny City by-election as Sinn Fein abstentionist, 1917; returned unopposed at Kilkenny North, 1918; minister for local government, 1919–21; represented Carlow-Kilkenny, 1921–7; Cork Borough, 1927–43; supporter of Anglo-Irish treaty, 1921; chairman, Provisional Government, 1922; president, Dail Eireann, 1922; first elected president, Executive Council, 1922–32; also minister of Finance, 1922–3; minister of Defence, 1924, and minister of External Affairs, 1927; successfully founded new Irish State; leader of opposition party, Fine Gail, 1933–44; retired from Dail Eireann, 1944; chairman, Irish Racing Board, 1946–56; KGC, 1st class, Pian Order, 1925; hon. doctorates, Catholic University of Washington, Columbia, Cambridge, Dublin, and the National University of Ireland.

COSTAIN, Sir RICHARD RYLANDES (1902–1966), industrialist; educated at Merchant Taylor's School, Crosby, and Rydal School, Colwyn Bay; joined family building firm, Richard Costain & Sons, 1920; joint managing director, 1927; managing director, 1929; floated public company, 1933; built Dolphin Square flats, London, 1935; director of works, Ministry of Works, during 1939–45 war; built the Festival of Britain, established Costain group, comprising eighty autonomous companies, 1950; president, London Master Builders' Association, 1950; chairman, Harlow Development Corporation, 1950; knighted, 1954; president, Export Group for the Construction Industries, 1955–7.

COTTON, JACK (1903–1964), property developer; educated at King Edward VI grammar school, Birmingham, and Cheltenham College; articled clerk in firm of estate agents and surveyors, 1921; set up his own firm, 1924; financed development of residential and commercial property in Birmingham, 1932–9; built shadow-factories during 1939–45 war; took over Chesham House (Regent Street) Ltd. in London and set up City Centre Properties, 1955; merged with 'City and Central' and 'Murrayfield' and created biggest property company in the world, 1960; commemorated in Cotton Terraces, Zoological Gardens, Regents Park, London; founded chair of architecture and fine arts, Hebrew University, Jerusalem, and chairs of biochemistry at Royal College of Surgeons and the Weizmann Institute, Israel.

COUCH, Sir ARTHUR THOMAS QUILLER-, ('Q') (1863–1944), Cornishman, man of letters, and professor of English literature. [See QUILLER-COUCH.]

COUCH, Sir RICHARD (1817–1905), judge; called to bar, 1841; recorder of Bedford, 1858–62; puisne judge of high court of Bombay, 1862; chief justice, 1866; knighted, 1866; chief justice at Calcutta, 1870–75; tried Gaekwar of Baroda on charge of poisoning Colonel Robert Phayre [q.v.], 1875; PC, 1875; member of judicial committee of Privy Council, 1881–1901.

COULTON, GEORGE GORDON (1858–1947), historian and controversialist; educated at Felsted and St. Catharine's College, Cambridge; ordained (1883–4) but feeling unable to continue turned to teaching; Birkbeck lecturer in ecclesiastical history, Trinity College, Cambridge, 1910; university lecturer in English and fellow of St. John's, 1919; possessed remarkable knowledge of Middle Ages; his strongly expressed religious views involved him in prolonged controversy with Roman Catholics; publications include Five Centuries of Religion (4 vols., 1923–50) and Medieval Panorama (1938); FBA, 1929.

COUPER, Sir GEORGE EBENEZER WILSON, second baronet (1824–1908), Indian administrator; born in Nova Scotia; joined Bengal civil service, 1846; secretary at Lucknow to chief commissioners of Oudh, 1856–7; served with distinction in siege of Lucknow, 1857; CB, 1860; chief secretary of North-West Provinces government, 1859; succeeded to baronetcy, 1861; judicial commissioner of Oudh, 1863; as chief commissioner (1871-6) he revised land assessments; first lieutenant-governor of North-Western Province and chief commissioner of Oudh, 1877; efficiently met famine of 1877-8; encouraged Indian industrial enterprises; KCSI, 1877; CIE, 1878; retired, 1882.

COUPLAND, Sir REGINALD (1884-1952), historian of the British Empire and Commonwealth; educated at Winchester and New College, Oxford; first class, *lit. hum.*, 1907; fellow and lecturer, Trinity College, 1907-14; Beit lecturer in colonial history, 1913-18 professor, 1920-48; member of royal commission on superior civil services in India, 1923-4, on Palestine, 1936-7; with Cripps mission to India, 1942; published works on the American revolution, the history of East Africa, and Indian problems and politics; CIE, 1928; KCMG, 1944; FBA, 1948; hon. D.Litt. Durham.

COURT BROWN, WILLIAM MICHAEL (1918-1968), medical research worker. [See BROWN.]

COURTAULD, AUGUSTINE (1904-1959), Arctic explorer; cousin of Samuel Courtauld [q.v.]; educated at Charterhouse and Trinity College, Cambridge; BA, 1926; explored in Greenland with (Sir) James Wordie, 1926 and 1929; with 'Gino' Watkins [q.v.], spending five months alone in ice-cap station, 1930-1; Polar medal, 1932; organized expedition to map and climb Watkins Range, 1935; an experienced yachtsman; active in local government and community service.

COURTAULD, SAMUEL (1876-1947), industrialist and art patron; educated at Rugby; director of family silk firm, 1915; chairman, 1921-40; his emphasis on individuality of worker and importance of spiritual values influential on industrial relationships; gave £50,000 for purchase of French paintings by Tate Gallery, 1923; presented many paintings from his own collection to London University and endowed Courtauld Institute of Art; largely responsible for bringing Warburg Institute to London; trustee of Tate and National galleries.

COURTHOPE, WILLIAM JOHN (1842-1917), civil servant, poet, and literary critic; BA, New College, Oxford; chancellor's English essay prizeman, 1868; entered Education Office, 1869; civil service commissioner, 1887-1907; professor of poetry at Oxford, 1895-1900; CB, 1895; completed (1881-9) standard edition of

Pope's *Works* begun by Whitwell Elwin [q.v.]; wrote *History of English Poetry*, 1895-1910; writer of verse.

COURTNEY, LEONARD HENRY first BARON COURTNEY OF PENWITH (1832-1918), journalist and statesman; BA, St. John's College, Cambridge (second wrangler); leader-writer to *The Times*, 1865-81; professor of political economy, University College, London, 1872-5; liberal MP, Liskeard, 1875; under-secretary for Home Office, 1880; secretary of Treasury, 1882-4; deputy speaker, 1886-92; baron, 1906; anti-imperialist and zealot for proportional representation.

COURTNEY, WILLIAM LEONARD (1850-1928), philosopher and journalist; born at Poona; BA, University College, Oxford, 1872; fellow of Merton, 1872; headmaster, Somersetshire College, 1873; fellow and philosophy tutor, New College, Oxford, 1876; joined staff of *Daily Telegraph*, 1890; chief dramatic critic and literary editor, mid-nineties until 1925; editor of *Fortnightly Review*, 1894-1928.

COUSIN, ANNE ROSS (1824-1906), hymn writer; born CUNDELL; married William Cousin, Presbyterian minister, 1847; best known by 'The sands of time are sinking', composed in 1854, and published in her *Immanuel's Land and other Pieces* (1876), and by 'King Eternal! King Immortal'.

COWAN, Sir WALTER HENRY, baronet, of the Baltic and of Bilton (1871-1956), admiral; entered navy, 1884; served in Benin (1897) and other expeditions; at Omdurman and Fashoda, 1898; DSO; aide-de-camp to Sir Reginald Wingate [q.v.] in pursuit of Khalifa, 1899; to Kitchener [q.v.] in South Africa, 1900; captain, 1906; flag captain to (Sir) Osmond Brock [q.v.] in *Princess Royal*, 1915; at Jutland, 1916; commodore, first light cruiser squadron, 1917-19, serving in Baltic, 1919; rear-admiral, 1918; vice-admiral, 1923; commanded battle cruiser squadron, 1921-2, Scotland, 1925-6, America and West Indies, 1926-8; admiral, 1927; retired, 1931; with Commandos and later with Indian Regiment in North Africa, 1940-2; taken prisoner at Bir Hacheim, 1942; repatriated, 1943; with Commandos in Italy, 1943-4; bar to DSO, 1944; retired, 1945; MVO, 1904; CB 1916; KCB, 1919; baronet, 1921.

COWANS, Sir JOHN STEVEN (1862-1921), general; educated at Sandhurst; joined Rifle Brigade, 1881; passed Staff College with distinction, 1891; deputy assistant quartermaster-general in movements branch of War Office to supervise transport of troops to Egypt; major,

1898; colonel, 1903; served at Aldershot, 1903-6; in India, 1906-10; director-general of Territorial Force, 1910; quartermaster-general, 1912-19; KCB, 1913; lieutenant-general, 1915; proved to be administrative genius, and carried out with scarcely a hitch during four years of war (1914-18) enormous expansion in necessary army services—barrack and hospital accommodation, food supplies, motor transport, horses, clothing, general stores, and personnel; GCMG, 1918; general and GCB, 1919; joined Roman Church shortly before his death.

COWARD, SIR HENRY (1849-1944), musician; elementary schoolmaster, 1870-87; founded and directed (1876-1933) Sheffield Tonic Sol-fa Association (later Sheffield Musical Union); B.Mus., Oxford, 1889; chorus master (1896-1908) of Sheffield musical festival which achieved world-wide fame for choral excellence; knighted, 1926.

COWDRAY, first VISCOUNT (1856-1927), contractor. [See PEARSON, WEETMAN DICKINSON.]

COWELL, EDWARD BYLES (1826-1903), scholar and man of letters; early devoted himself to oriental literature; studied Sanskrit, Persian, and Arabic; joined father's business of maltster at Ipswich, 1842; contributed articles to *Westminster Review* on oriental and Spanish literature; friend of Edward FitzGerald [q.v.]; went to Magdalen College, Oxford; BA, 1854; met Tennyson, Thackeray, Jowett, Max Müller, and Horace H. Wilson [qq.v.]; catalogued oriental MSS for Bodleian; as undergraduate he translated oriental works, and wrote on Persian poetry; his edition of Vararuci's *Prākrta-Prakāśa* (1854) established his reputation as Sanskrit scholar; professor of English history in Presidency College, Calcutta, 1856-64; instituted MA course at Calcutta University; sent copy of MS of Omar Khayyám to Edward FitzGerald; principal of Sanskrit College, Calcutta, 1858; edited many native works for Asiatic Society of Bengal in *Bibliotheca India*; edited and translated the *Kusumāñjali*, 1864, and other works; returned to England, 1864; first professor of Sanskrit in university of Cambridge, 1867-1903; fellow of Corpus Christi College, 1874; a founder of Cambridge Philological Society, 1868; with pupils he issued series of important Sanskrit texts and translations; continued to study Persian and Spanish; also took up archaeology, architecture, Welsh poetry, botany, and geology; awarded first gold medal of Royal Asiatic Society, 1898; hon. LLD, Edinburgh, 1875; hon. DCL, Oxford, 1896; original member of British Academy, 1902.

COWEN, SIR FREDERIC HYMEN (1852-1935), composer and conductor; studied in London, Leipzig, and Berlin; made name as pianist, 1868; conductor for Philharmonic Society (1888-92, 1900-7), Hallé Orchestra (1896-9), of Handel festivals, 1903-12, 1920, 1923; compositions include 'Scandinavian' symphony (1880), operas, cantatas, and many songs; knighted, 1911.

COWIE, WILLIAM GARDEN (1831-1902), bishop of Auckland; law scholar of Trinity Hall, Cambridge; BA, 1855; DD, 1869; DD, Oxford, 1897; ordained, 1854; chaplain to forces in India, 1857; present at Lucknow, Aliganj, and other battles; in Afghan campaign, 1863-4; bishop of Auckland, New Zealand, 1869-1902; conciliated Maoris and encouraged native ministry; primate of New Zealand, 1895; published *Our Last Year in New Zealand* (1888).

COWLEY, SIR ARTHUR ERNEST (1861-1931), orientalist and Bodley's librarian; educated at St. Paul's School and Trinity College, Oxford; graduated, 1883; sub-librarian at Bodleian Library in charge of oriental department, 1899; fellow of Magdalen and tutor in rabbinic Hebrew literature, 1902; Bodley's librarian, 1919-31; edited *Concise Catalogue of Hebrew Printed Books in the Bodleian Library*, 1929; publications include *The Samaritan Liturgy* (1909), *Aramaic Papyri of the 5th Century B.C.* (1923), and *The Hittites* (1920); FBA, 1914; nominated for knighthood, 1931.

COWPER, FRANCIS THOMAS DE GREY, seventh EARL COWPER (1834-1905), lord-lieutenant of Ireland; nephew of William Francis Cowper, Baron Mount Temple [q.v.]; educated at Eton and Christ Church, Oxford; first class in law and history, 1855; succeeded father as Earl Cowper, 1856; pioneer of Volunteer movement; inherited baronies of Butler and Dingwall, 1871, and of Lucas, 1880, as well as vast property in Bedfordshire, Hertfordshire, Nottinghamshire, and Lancashire; did useful county work in Bedfordshire and Hertfordshire; liberal politician; KG, 1865; represented Board of Trade in House of Lords, 1871-3; appointed lord-lieutenant of Ireland under Gladstone, 1880; favoured renewal of Coercion Act, arrested Parnell, and suppressed Land League, Oct. 1881; resigned office, Apr. 1882; left Dublin two days before Phoenix Park murders; drafted new coercion bill, which was passed by his successor, 1882; lord-lieutenant of Bedfordshire; active in opposition to Gladstone's home rule bill of 1886; chairman of Manchester Ship Canal commission, 1885, of royal commission on working of Irish Land Acts of 1881 and 1885,

1886-7, and of London University commission, 1892.

COX, ALFRED (1868-1954), general practitioner and medical secretary of British Medical Association; entered medical profession as dispenser-assistant; MB, BS, Durham, 1891; practised in Gateshead, 1891-1908; worked for new constitution for British Medical Association obtained in 1903; deputy secretary, BMA, 1908; secretary, 1912-32; secretary to Central Medical War Committee, 1914-18; OBE; conducted negotiations with Ministry of Health; worked for affiliation of Canadian and South African medical associations; a founder of the Association Professionnelle Internationale des Médecins, 1925.

COX, GEORGE (called SIR GEORGE) WILLIAM (1827-1902), historical writer; born at Benares; scholar of Rugby and Trinity College, Oxford; BA, and MA, 1859; accompanied Bishop Colenso [q.v.] to South Africa, 1853-4; defended theological views of Colenso, whose biography he published, 1888; wrote much on Greek, Indian, and English history; chief work, *History of Greece* (2 vols., 1874), later superseded; won wide popularity with *Tales from Greek Mythology* (1861); claimed baronetcy of Cox of Dunmanway, 1877 (disallowed, 1911); rector of Scrayingham, Yorkshire, 1881-97; chosen bishop of Natal by Colenso's adherents, was refused consecration by Archbishop Benson [q.v.], 1886; received civil list pension, 1896.

COX, HAROLD (1859-1936), economist and journalist; educated at Tonbridge School and Jesus College, Cambridge; a senior optime, 1882; secretary of Cobden Club, 1899-1904; liberal MP, Preston, 1906-9; editor, *Edinburgh Review*, 1912-29; collaborated with Sidney Webb [q.v.] in *Eight Hours Day* (1891) but later opposed socialism and wrote *Economic Liberty* (1920).

COX, LESLIE REGINALD (1897-1965), palaeontologist; educated at Owen's School, Islington, and Queens' College, Cambridge; first class, parts i and ii, natural sciences tripos, 1920-1; Sc.D. Cambridge, 1937; assistant keeper, department of geology (later palaeontology), British Museum (Natural History), South Kensington, 1922; contributed to *Treatise on Invertebrate Palaeontology*, published by Geological Society of America, and published 'New light on William Smith [q.v.] and his work' (*Proc. Yorks. Geol. Soc.*, vol. 25 (1942), pp. 1-99); vice-president, Geological Society of London, 1952-4; vice-president, Palaeontographical Society, 1957 onwards; president, the Geologists' Association, 1954-6; president,

Palaeontological Association, 1964-5; president, Malacological Society, 1957-60; FRS, 1950; OBE, 1958.

COX, SIR PERCY ZACHARIAH (1864-1937), soldier, administrator, and diplomatist; educated at Harrow and Sandhurst; commissioned in 2nd Cameronians (in India), 1884; joined Indian staff corps, 1889; assistant political resident, Zeila, British Somaliland protectorate, 1893; transferred to Berbera, 1894; by defeat of Rer Hared clan (1895) established reputation for decision and ability; political agent and consul at Muscat, 1899-1904; restored good relations with Sultan Feisal; acting political resident in Persian Gulf and consul-general, 1904-9; resident, 1909-14; secured lease of frontage on Euphrates to Anglo-Persian Oil Company, 1909; won confidence of Abdulaziz ibn Saud; KCIE, 1911; chief political officer, Indian expeditionary force 'D' in Mesopotamia, 1914-18; negotiated treaty with ibn Saud, 1915; KCSI, 1915; GCIE, 1917; acting minister, Tehran, 1918-20; high commissioner in Iraq, 1920-3; took over government during illness of King Feisal; signed treaty of alliance and appointed GCMG, 1922.

COZENS-HARDY, HERBERT HARDY, first BARON COZENS-HARDY (1838-1920), judge; BA, London University; called to bar (Lincoln's Inn), 1862; practised as Chancery junior; QC, 1882; MP, North Norfolk, 1885-99; raised to bench and knighted, 1899; lord justice of appeal and PC, 1901; master of the rolls, 1907; baron, 1914.

CRADDOCK, SIR REGINALD HENRY (1864-1937), Indian civil servant; educated at Wellington and Keble College, Oxford; posted to Central Provinces, 1884; chief commissioner, 1907-12; KCSI, 1911; home member, viceroy's council, 1912-17; lieutenant-governor, Burma, 1917-22; initiated Rangoon University scheme; conservative MP for combined English universities, 1931-7.

CRADOCK, SIR CHRISTOPHER GEORGE FRANCIS MAURICE (1862-1914), admiral; entered navy, 1875; commanded naval brigade which led allied forces at storming of Taku forts, and directed relief of Tientsin settlement, 1900; published his idealistic conception of the naval career in *Whispers from the Fleet*, 1907; rear-admiral, 1910; KCVO, 1912; appointed to command of North America and West Indies station, 1913; on outbreak of 1914-18 war faced with difficult task of keeping North and South Atlantic free for British trade; received ambiguous orders from Admiralty; defeated in engagement with German squadron under

Admiral von Spee off Coronel and went down on flagship, 1 Nov.

CRAIG, EDWARD HENRY GORDON (1872-1966), artist and stage designer; only son of Edward William Godwin, architect, and (Dame) (Alice) Ellen Terry [q.q.v.], actress; educated at Bradfield College and at Heidelberg; played Arthur in *The Dead Heart* with (Sir) Henry Irving [q.v.]; adopted Craig as stage name; ceased to act and started *The Page* magazine, 1897; designed and directed Henry Purcell's [q.v.] *Dido and Aeneas*, 1900; and *Acis and Galatea*, 1902; settled in Florence, 1907; issued theatre magazine, *The Mask*; designed *Hamlet* at the Art Theatre, Moscow, 1911; staged Ibsen's *The Pretenders* in Copenhagen, 1926; publications include *On the Art of the Theatre* (1911), *The Theatre Advancing* (1921), etchings, *Scene* (1923), *Woodcuts and Some Words* (1924); biographies include *Henry Irving* (1930), *Ellen Terry and Her Secret Self* (1931), and his memoirs, *Index to the Story of My Days* (*1872-1907*), (1957); RDI, of the Royal College of Art, 1938; CH, 1958; president, Mermaid Theatre, London, 1964.

CRAIG, ISA (1831-1903), poetical writer. [See KNOX.]

CRAIG, JAMES, first VISCOUNT CRAIGAVON (1871-1940), statesman; born in Belfast; educated at Merchiston Castle, Edinburgh; became a stockbroker; MP, East Down (1906-18), Mid-Down (1918-21); baronet, 1918; parliamentary secretary to Ministry of Pensions (1919-20) and Admiralty (1920-1); chief lieutenant to Sir Edward (later Lord) Carson [q.v.] in opposing home rule; concerned mainly with organizing means of resistance in Ulster; first prime minister of Northern Ireland, 1921-40; restored order and introduced measures providing a new educational system, social services, improved housing, drainage, and agricultural position, and encouraging new industries; viscount, 1927.

CRAIG, SIR JOHN (1874-1957), steelmaster; educated at Dalziel public school, Motherwell; entered David Colville's Dalzell ironworks as office boy, 1888; director, 1910; chairman, 1916-56; pursued policy of expansion in both wars; founded Colvilles Ltd., 1934; by 1937 obtained capacity of 1,100,000 tons of ingots and decided to reconstruct Clyde Ironworks; a founder of National Federation of Iron and Steel Manufacturers, 1918; president, Iron and Steel Institute, 1940-2; knighted, 1943.

CRAIG, WILLIAM JAMES (1843-1906), editor of Shakespeare; educated at Trinity College, Dublin; BA, 1865; MA, 1870; a private

coach in London for army and civil service candidates, 1874-6; professor of English at University College, Aberystwyth, 1876-9; resumed teaching in London, 1879-98; a devoted student of Shakespeare; edited 'The Oxford Shakespeare' (1 vol., 1894), the 'Little Quarto Shakespeare' (40 vols., 1901-4), and the 'Arden Shakespeare' (40 vols. by different hands), to which he contributed the volume on *King Lear*, 1901.

CRAIGAVON, first VISCOUNT (1871-1940), statesman. [See CRAIG, JAMES.]

CRAIGIE, PEARL MARY TERESA (1867-1906), novelist and dramatist under the pseudonym of JOHN OLIVER HOBBES; born at Chelsea, Mass., USA; eldest child of American parents who settled in London in her infancy; married (Feb. 1887) Reginald Walpole Craigie, whom she divorced (July 1895); joined Church of Rome, 1892; began serious education in early married life; published in 1891 *Some Emotions and a Moral*, a first novel of cynical flavour which was well received; maintained her success in *The Sinner's Comedy* (1892), *The School for Saints* (1897), *The Serious Wooing* (1901), *Robert Orange* (1902), and many other works of fiction; attempted drama with less acceptance in *The Ambassador* (1898), *A Repentance* (1899), *The Wisdom of the Wise* (1900), and some other plays which failed to attract; described in *Imperial India*, 1903, the coronation durbar at Delhi in 1903; wrote many miscellaneous essays and sketches; her command of epigram—humorous, caustic, and cynical—gives her work its value.

CRAIGIE, SIR ROBERT LESLIE (1883-1959), diplomatist; educated at Heidelberg; entered Foreign Office, 1907; assistant under-secretary, 1935; ambassador to Japan, 1937; interned, 1941; repatriated, 1942; British representative on United Nations War Crimes Commission, 1945-8; on Geneva conference for protection of war victims, 1949; CMG, 1929; CB, 1930; KCMG, 1936; PC, 1937; GCMG, 1941.

CRAIGIE, SIR WILLIAM ALEXANDER (1867-1957), lexicographer and philologist; educated at West End Academy, Dundee, St. Andrews (honours in classics and philosophy, 1888), and Oriel College, Oxford; first class, *lit. hum.*, 1892; assistant to Latin professor, St. Andrews, 1893-7; worked on *New English Dictionary*, 1897-1933; joint editor, 1901-33; Taylorian lecturer in Scandinavian languages, Oxford, 1904-16; Rawlinson and Bosworth professor of Anglo-Saxon, 1916-25; professor of English, Chicago, 1925-36, editing *Dictionary* of American English completed in 1944; worked on *Dictionary of Older Scottish Tongue* to letter I; works include *Primer of Burns* (1896), *Scandinavian Folk-Lore* (1896), *Specimens of Icelandic Rímur* (3 vols., 1952); a founder of Anglo-Norman Text Society, 1938, of Icelandic Rímur Society, 1947; FBA, 1931; knighted, 1928.

CRAIGMYLE, first BARON (1850-1937), lawyer and politician. [See SHAW, THOMAS.]

CRAIK, SIR HENRY, first baronet (1846-1927), civil servant, politician, and man of letters; BA, Balliol College, Oxford; entered Education Department, 1870; secretary, Scottish Education Department, 1885-1904; conservative MP, Glasgow and Aberdeen universities, 1906, and for Scottish universities, 1918; KCB, 1897; PC, 1918; baronet, 1926; works include *Life of Swift* (1882) and *The State and Education* (1883).

CRANBROOK, first EARL OF (1814-1906), statesman. [See GATHORNE-HARDY, GATHORNE.]

CRANE, WALTER (1845-1915), artist; apprenticed to W. J. Linton [q.v.], wood-engraver, 1859; illustrated several series of picture books, chiefly for children, 1863-96; master of Art-Workers' Guild, 1888, 1889; twice president of Arts and Crafts Exhibition Society; ARWS, 1888; associated with William Morris [q.v.] in Socialist League; principal of Royal College of Art, South Kensington, 1898; painted water-colour landscapes and works in oil.

CRAVEN, HAWES (1837-1910), scene-painter, whose full name was Henry Hawes Craven Green; apprenticed to John Gray, scene-painter; worked at Olympic, Drury Lane, and Covent Garden theatres and Theatre Royal, Dublin, 1857-64; painted scenes at Lyceum for C. A. Fechter [q.v.], 1863-4; at Prince of Wales's for *Play* and *School*, 1868-9; worked for Sir Henry Irving [q.v.] at Lyceum from 1878 to 1902; also for Savoy, Her Majesty's, and Garrick theatres, 1885-1905; developed scenic realism and stage illusion to fullest legitimate limits; excelled in landscape.

CRAVEN, HENRY THORNTON (1818-1905), dramatist and actor, whose real name was HENRY THORNTON; made début at Fanny Kelly's theatre, Soho, 1841; played Malcolm to Macready's Macbeth at Macready's farewell performance, Drury Lane, 20 Feb. 1851; married Eliza Nelson, 1852; acted with wife at Sydney and Melbourne, 1854-7; his dramas *Milky White* (1864) and *Meg's Diversion* (1866) were successful in London and provinces; made final appearance in his last play *Too True*, 1876.

CRAWFORD, twenty-seventh EARL OF (1871-1940), politician and art connoisseur. [See LINDSAY, DAVID ALEXANDER EDWARD.]

CRAWFORD, twenty-sixth EARL OF (1847-1913), astronomer, collector, and bibliophile. [See LINDSAY, JAMES LUDOVIC.]

CRAWFORD, OSBERT GUY STANHOPE (1886-1957), archaeologist; born in Bombay; educated at Marlborough and Keble College, Oxford; graduated in geography, 1910; observer with Royal Flying Corps, 1917-18; archaeology officer, Ordnance Survey, 1920-46, revising and compiling maps with archaeological information; began survey of megalithic monuments and series of period maps; developed use of air photography; founded, 1927, and edited, 1927-57, *Antiquity*; works include *Field Archaeology* (1932), *Air Survey and Archaeology* (1924), *Air Photography for Archaeologists* (1929), and *Castle and Churches in the Middle Nile Region* (1953); FBA, 1947; CBE, 1950.

CRAWFURD, OSWALD JOHN FREDERICK (1834-1909), author; son of John Crawfurd [q.v.]; educated at Eton and Merton College, Oxford; entered Foreign Office, 1857; consul at Oporto, 1867-91; CMG, 1890; published novels and sketches of Portuguese life, essays, and dramas.

CRAWFURD, SIR RAYMOND HENRY PAYNE (1865-1938), physician and scholar; educated at Winchester and New College, Oxford; qualified at King's College Hospital where he became assistant physician (1898) and physician (1905-30); wrote on history of medicine; fellow (1901), registrar (1925-38), Royal College of Physicians; knighted, 1933.

CREAGH, SIR GARRETT O'MOORE (1848-1923), general; joined army, 1866; awarded VC during Afghan war, 1879-80; political resident and general officer in command at Aden; general officer commanding British expeditionary force in China, 1901; subsequently held various commands in India; general, 1907; commander-in-chief, India, 1909-14; KCB, 1902.

CREAGH, WILLIAM (1828-1901), major-general and administrator; first of old Roman Catholic military family to become a protestant; joined 19th Bombay infantry, 1845; served in Punjab campaign, 1848-9; in Mutiny, at defeat of Tantia near Jhansi, 1858, and at his capture, 1859; administered native state of Dhar, 1861-2; commanded his regiment in second Afghan war, 1878-9; made military road from Jacobabad to Dhadar (109 miles), and from Dhadar over Bolan pass to Darwaza (63 miles); major-general, 1870.

CREECH JONES, ARTHUR (1891-1964), politician. [See JONES, A. C.]

CREED, JOHN MARTIN (1889-1940), divine; educated at Wyggeston School, Leicester, and Gonville and Caius College, Cambridge; ordained, 1913; dean and lecturer in theology, St. John's College, Cambridge, 1919-26; Ely professor of divinity and canon of Ely, 1926-40; FBA, 1939; publications include critical commentary on St. Luke's Gospel (1930) and *The Divinity of Jesus Christ. A Study in the History of Christian Doctrine since Kant* (1938).

CREED, SIR THOMAS PERCIVAL (1897-1969), lawyer and educationist; educated at Wyggeston School, Leicester, and Pembroke College, Oxford; served with Leicestershire Regiment in France, 1915-19; MC; entered Sudan political service, 1922; called to bar (Lincoln's Inn), 1925; seconded to legal department; Sudan Government, 1926-31; seconded to Iraqi Government, 1931-5; high court judge, Khartoum, 1935; chief justice, Sudan, 1936; legal secretary, member, governor-general's council, 1941-8; CBE, 1943; KBE, 1946; KC, 1948; secretary, King's College, London, 1948; principal, Queen Mary College, 1952-1967; vice-chancellor, London University, 1964-7; hon. fellow, Pembroke College, Oxford, 1950; hon. LLD, Leicester, 1965; hon. bencher, Lincoln's Inn.

CREIGHTON, MANDELL (1843-1901), scholar, historian, and bishop of London; educated at Durham grammar school and Merton College, Oxford; first class, *lit. hum.*; fellow, 1866; tutor, 1867; ordained, 1870; accepted college living of Embleton, 1875; rural dean, Alnwick, 1879; published *History of the Papacy* (vols. i and ii, 1882; vols. iii and iv, 1887; vol. v, 1894); first Dixie professor of ecclesiastical history, and fellow, Emmanuel College, Cambridge, 1884; first editor, *English Historical Review*, 1886-91; canon of Worcester, 1885; of Windsor, 1890; bishop of Peterborough, 1891; of London, 1897-1901; first president, Church Historical Society, 1894-1901; Hulsean lecturer, 1893-4, and Rede lecturer, 1895, Cambridge; Romanes lecturer, Oxford, 1896; publications include *The Age of Elizabeth* (1876), *Cardinal Wolsey* (1888), and *Queen Elizabeth* (1896), with sermons, lectures, and contributions to the *Dictionary of National Biography*; DD, Oxford and Cambridge, and number of hon. degrees.

CREMER, ROBERT WYNDHAM KETTON- (1908-1969), biographer and historian. [See KETTON-CREMER.]

CREMER, SIR WILLIAM RANDAL (1838-1908), peace advocate; carpenter's apprentice; mixed in trade-unionism in London, 1852; promoter of Amalgamated Society of Carpenters and Joiners, 1860; secretary of British section of International Working Men's Association, 1865; friend of Mazzini and Garibaldi; as secretary of newly formed Workmen's Peace Association from 1871 to death, he persistently advocated in America and on continent international arbitration; awarded Nobel peace prize, 1903; secretary (1889-1908) of Interparliamentary Union; radical MP for Haggerston, 1885-95 and 1900-8; opposed new independent labour movement; edited *Arbitrator*, monthly peace journal, from 1889; knighted, 1907.

CREWE-MILNES, ROBERT OFFLEY ASHBURTON, second BARON HOUGHTON, and MARQUESS OF CREWE (1858-1945), statesman; son of Baron Houghton [q.v.]; educated at Harrow and Trinity College, Cambridge; succeeded father, 1885; viceroy of Ireland, 1892-5; PC, 1892; earl, 1895; KG, 1908; lord president of the Council, 1905-8, 1915-16; lord privy seal, 1908-11, 1912-15; as leader of House of Lords from 1908 secured considerate hearing for minority view by presenting government case with moderation and courage; colonial secretary, 1908-10; saw realization of South African union; attended constitutional conference, 1910; secretary of state for India, 1910-15; marquess, 1911; president, Board of Education, 1916; resigned with Asquith; chairman, London County Council, 1917; ambassador in Paris, 1922-8; secretary of state for war, 1931; leader of independent liberals in House of Lords, 1936-44; wrote biography of his father-in-law, Lord Rosebery (1931); chancellor of Sheffield University, 1918-44.

CRICHTON-BROWNE, SIR JAMES (1840-1938), physician and psychologist. [See BROWNE.]

CRIPPS, CHARLES ALFRED, first BARON PARMOOR (1852-1941), lawyer and politician; scholar of Winchester and New College, Oxford; first classes, history, jurisprudence, civil law; called to bar (Middle Temple), 1877; QC, 1890; attorney-general to successive Princes of Wales, 1895-1914; KCVO, 1908; conservative MP, Stroud (1895-1900), Stretford (1901-6), and Wycombe (1910-14) divisions; baron and PC, 1914; opposed war and studied plans for future world order; lord president of Council in labour governments, 1924 (specially responsible for League of Nations affairs) and 1929-31; vicar-general, Canterbury (1902-24) and York (1900-14); first chairman of House of Laity, 1920-4.

CRIPPS, Sir (RICHARD) STAFFORD (1889–1952), statesman and lawyer; son of C. A. Cripps (Lord Parmoor, q.v.); his mother a sister of Beatrice Webb [q.v.]; educated at Winchester and University College, London; called to bar (Middle Temple), 1913; assistant superintendent, Queensferry explosives factory, 1915–18; for six years treasurer of World Alliance to promote international friendship through the Churches; became vegetarian and teetotaller due to lifelong ill health; nevertheless by immense energy and mental ability achieved brilliant success at bar, especially in patent and compensation cases; KC, 1927; joined labour party, 1929; solicitor-general (and knighted), 1930; labour MP, East Bristol, 1931–50; declined to join 'national' government, 1931; became extreme socialist and leading member of Socialist League; proposed United Front, 1936, supported by foundation of *Tribune*, 1937; proposed wider Popular Front, 1938; expelled from labour party, 1939–45; ambassador to Russia, 1940–2; organized pact of mutual assistance after German invasion of Russia, 1941; PC, 1941; leader of House of Commons and lord privy seal, 1942; undertook unsuccessful mission to India, 1942; minister of aircraft production, 1942–5; president, Board of Trade, 1945–7; member of cabinet mission to India, 1946; minister for economic affairs, 1947; chancellor of Exchequer, 1947–50; obtained voluntary wages and dividend restraint through sheer moral superiority, 1948–9; trusted for the evident reality of his Christian principles and his high ideals in personal and public life; devalued the pound, Sept. 1949; resigned through ill health, 1950; rector of Aberdeen University, 1942–5; FRS, 1948; CH, 1951.

CRIPPS, WILFRED JOSEPH (1841–1903), writer on plate; of ancient Cirencester family; BA, Trinity College, Oxford, 1863; MA, 1866; called to bar, 1865; published *Old English Plate* (1878, 9th edn. 1906), *Old French Plate* (1880), and kindred works; made valuable archaeological researches in Cirencester; FSA, 1890; CB, 1889.

CROCKER, HENRY RADCLIFFE- (1845–1909), dermatologist. [See RADCLIFFE-CROCKER.]

CROCKETT, SAMUEL RUTHERFORD (1860–1914), novelist; MA, Edinburgh University; studied at New College, Edinburgh, 1882–6; minister of Free Church, Penicuik, Midlothian, 1886; wrote *The Stickit Minister* (1893), *The Raiders* and *The Lilac Sunbonnet* (1894); resigned ministry, 1895.

CROFT, HENRY PAGE, first BARON CROFT (1881–1947), politician; educated at Eton,

Shrewsbury, and Trinity Hall, Cambridge; strong advocate of tariff reform; conservative MP, Christchurch division (later Bournemouth), 1910–40; formed strongly imperial 'national party', 1917–22; returning to conservative fold agitated for tariffs and opposed Indian self-government; joint parliamentary under-secretary for war, 1940–5; baronet, 1924; baron, 1940; PC, 1945.

CROFT, JOHN (1833–1905), surgeon; entered St. Thomas's Hospital, 1850; MRCS, 1854; FRCS, 1859; on staff at St. Thomas's from 1860; surgeon, 1871–91; early adopted Listerian methods; introduced 'Croft's splints' for leg fractures; contributed papers to medical societies' *Transactions*.

CROFTS, ERNEST (1847–1911), historical painter; exhibited at Royal Academy, 1874–1910; ARA, 1878; RA, 1896; keeper and trustee, 1898; treated historical subjects, mainly of Napoleonic and Cavalier periods; chief works include 'Marlborough after Ramillies' (1880), 'Charles I on the way to Execution' (1883), 'Napoleon and the Old Guard at Waterloo' (1895), and 'The Funeral of Queen Victoria' (1903); FSA, 1900.

CROKE, THOMAS WILLIAM (1824–1902), Roman Catholic archbishop of Cashel; studied for priesthood in Paris and Belgium and at Irish College of Rome; DD, 1847; professor of ecclesiastical history at Catholic University, Dublin, under Cardinal Newman [q.v.]; president of St. Colman's College, Fermoy, 1858–68; friend of Cardinal Manning [q.v.] from 1870; Catholic bishop of Auckland, New Zealand, 1870–5; archbishop of Cashel, 1875–1902; encouraged athletics and temperance; a strong nationalist and supporter of land agitation.

CROMER, first EARL OF (1841–1917), statesman, diplomatist, and administrator. [See BARING, EVELYN.]

CROMER, second EARL OF (1877–1953), lord chamberlain to the household. [See BARING, ROWLAND THOMAS.]

CROMPTON, HENRY (1836–1904), positivist and advocate of trade unions; son of Sir Charles Crompton [q.v.]; BA, Trinity College, Cambridge, 1858; clerk of assize on Chester and North Wales circuit, 1858–1901; called to bar, 1863; became ardent positivist, 1859; advocate of social reform; strong supporter of trade unions; author of *Industrial Conciliation* (1876) and *Our Criminal Justice* (1905).

CROMPTON, RICHMAL (1890–1969), author. [See LAMBURN, RICHMAL CROMPTON.]

CROMPTON, ROOKES EVELYN BELL (1845–1940), engineer; naval cadet in Crimean war, 1856; educated at Harrow; served in Rifle Brigade, 1864–75; introduced steam road-haulage in India; bought partnership in Chelmsford engineering firm, 1875; began making electric-light plant; installations included the Mansion House and Law Courts; formed Kensington Court electricity supply company (1886) advocating direct-current system; advised Indian government on electrical projects, 1890–9; took volunteer corps of electrical engineers to South Africa, 1900; CB, 1901; consultant to War Office on mechanical transport; engineer-member of Road Board, 1910; employed on tank development, 1914–18; FRS, 1933.

CROOKES, SIR WILLIAM (1832–1919), man of science; educated at Royal College of Chemistry, Hanover Square, London; assistant in the college, 1850–4; superintendent of meteorological department, Radcliffe Observatory, Oxford, 1854; lecturer in chemistry at Chester training college, 1855; lived in London from 1856 onwards; founded (1859) and for many years edited *Chemical News*; discovered thallium, 1861; FRS, 1863; carried out experiments on properties of highly rarefied gases and on elements of 'rare earths'; separated uranium-X from uranium, 1900; knighted, 1897; OM, 1910; president of various scientific societies, including the Royal Society (1913–15).

CROOKS, WILLIAM (1852–1921), labour politician; a cooper by trade; supported by 'Will Crooks's wages fund' from 1892; lectured and taught in Poplar; mayor of Poplar (first labour mayor in London), 1901; labour MP, Woolwich, 1903–6, 1906–10, 1910–18; PC, 1916.

CROOKSHANK, HARRY FREDERICK COMFORT, VISCOUNT CROOKSHANK (1893–1961), politician; educated at Eton and Magdalen College, Oxford; served with Grenadier Guards in 1914–18 war; severely wounded; served in diplomatic service, 1919–24; conservative MP Gainsborough, 1924–56; parliamentary under-secretary, Home Office, 1934–5; secretary for mines, 1935–9; financial secretary to Treasury, 1939–43; PC, 1939; postmaster-general, 1943–5; minister of health and leader of House of Commons, 1951–2; lord privy seal, 1952–5; CH, 1955; viscount, 1956; chairman, Historic Churches Preservation Trust, 1956; chairman, Political Honours Scrutiny Committee, 1958; high steward of Westminster, 1960; hon. DCL Oxford, 1960.

CROSS, CHARLES FREDERICK (1855–1935), analytical chemist; educated at King's College, London (BSc, 1878), Zürich, and Manchester; with E. J. Bevan set up as analytical and consulting chemists, 1885; his discovery (1892) of viscose made possible artificial silk (rayon); pioneer in production of viscous films and of cellulose acetate (Celanese), 1894; publications include *Cellulose* (with Bevan, 1895); FRS, 1917.

CROSS, KENNETH MERVYN BASKERVILLE (1890–1968), architect; son of Alfred William Stephen Cross, architect; educated at Felsted School and Gonville and Caius College, Cambridge, 1911–12; studied at university school of architecture; partnered his father in private practice, 1922; architect to number of local authorities and commercial firms; specialist in design and construction of swimming baths; published (with his father) *Modern Public Baths and Wash-Houses* (1930); fellow, Royal Institute of British Architects, 1931; president, 1956–8; served on number of committees and boards; hon. DCL, Durham.

CROSS, RICHARD ASSHETON, first VISCOUNT CROSS (1823–1914), statesman; educated at Rugby and Trinity College, Cambridge; called to bar (Inner Temple), 1849; joined Northern circuit; conservative MP, Preston, 1857–62; partner in Parr's bank, Warrington; chairman, 1870; MP, South-West Lancashire, 1868; home secretary, 1874–80; introduced among others the following Acts: Licensing Act (1874), Artisans' Dwellings, Factory, Employers and Workmen, and Conspiracy and Protection of Property Acts (1875), and Factories and Workshops Act (1878); in opposition, 1880–5; home secretary, 1885–6; secretary for India, 1886–92; viscount, 1886; privy seal, 1895–1900; retired, 1902; FRS; GCB, 1880; GCSI, 1892; close personal friend of Queen Victoria.

CROSSMAN, SIR WILLIAM (1830–1901), major-general RE; joined Royal Engineers, 1848; sent to Western Australia (1852) to superintend construction of public works by convicts; recalled to England, 1856; secretary to royal commission on defences of Canada, 1862; in Japan and China, 1866–70; joined special commission to inquire into resources of Griqualand West, 1875; CMG, 1877; first inspector of submarine mining defences, 1876–81; reported on Hong Kong, Singapore, and Australian defences, 1881–2, and on Jamaica and West India Islands finances, 1884; KCMG, 1884; major-general, 1886; MP for Portsmouth, 1885–6 as liberal, and 1886–92 as liberal unionist.

CROSTHWAITE, SIR CHARLES HAUKES TODD (1835–1915), Indian civil servant; entered Indian civil service, 1857; CSI and chief commissioner of Burma, 1887; KCSI, 1888; lieutenant-governor of North-Western Provinces and Oudh, 1892-5; on council of India, 1895-1905; wrote books on India.

CROWDY, DAME RACHEL ELEANOR (1884-1964), social reformer; educated at Hyde Park New College, London; trained as nurse at Guy's Hospital, 1908; joined Voluntary Aid Detachments, 1911; lecturer and demonstrator at National Health Society, 1912-14; with (Dame) Katherine Furse [q.v.] established VAD during 1914-18 war; in charge of VADs on the continent, 1914; Royal Red Cross medals, 1916-17; DBE, 1919; chief of Social Questions and Opium Traffic Section at League of Nations, 1919; went with International Typhus Commission to Poland, 1920-1; carried out social work in many countries abroad, 1931-9; Regions' adviser to Ministry of Information, 1939-46.

CROWE, SIR EDWARD THOMAS FREDERICK (1877-1960), public servant; born in Ionian Islands; educated at Bedford grammar school; joined consular service in Japan, 1897; commercial attaché, 1906, counsellor, Tokyo, 1918-24; director of foreign division, 1924-8, comptroller-general, 1928-37, Department of Overseas Trade; vice-president, 1937-60 (president, 1942-3), Royal Society of Arts; on Fleming committee on public schools, 1942-4; CMG, 1911; knighted, 1922; KCMG, 1930.

CROWE, EYRE (1824-1910), artist; son of Eyre Evans Crowe and brother of Sir Joseph Archer Crowe [qq.v.]; studied art in Paris under Paul Delaroche, 1839-44, and at Royal Academy Schools; became secretary to W. M. Thackeray [q.v.], 1851; accompanied him to America, 1852-3; exhibited at Royal Academy, 1848-1904; ARA, 1875; chief works include 'Brick Court, Middle Temple, April 1774' (1863) and 'The Queen of the May' (1879); later inspector under science and art department, South Kensington; published *With Thackeray in America*, 1893.

CROWE, SIR EYRE ALEXANDER BARBY WICHART (1864-1925), diplomatist; son of Sir J. A. Crowe [q.v.]; born at Leipzig; educated in Germany; entered Foreign Office, 1885; submitted to Sir Edward Grey (later Viscount Grey of Fallodon, q.v.) 'Memorandum on present state of British relations with France and Germany', 1907; secretary to British delegation at second peace conference at The Hague, 1907; British delegate at international maritime conference, London, which drew up abortive

Declaration of London, 1908-9; British agent in arbitration at The Hague between British and French governments in Savarkar case, 1911; assistant under-secretary of state for foreign affairs, 1912; upheld view that *entente* bound Great Britain to support France, 1914; a British plenipotentiary at peace conference, Versailles, 1919; permanent under-secretary of state for foreign affairs, 1920-5; KCMG, 1911.

CROZIER, WILLIAM PERCIVAL (1879-1944), journalist; educated at Manchester grammar school and Trinity College, Oxford; first class, *lit. hum.*, 1902; joined *Manchester Guardian* under C. P. Scott [q.v.], 1903; military critic, 1918; editor, 1932-44; modernized paper and maintained its liberal policies.

CRUIKSHANK, ROBERT JAMES (1898-1956), journalist; had little formal education; a foreign correspondent on *Daily News*, 1919; diplomatic correspondent, 1924-8; New York correspondent, 1928-36; managing editor, *Star*, 1936-41; director, American division, Ministry of Information, 1941-5; CMG, 1945; a director, *Daily News* and *News Chronicle*, 1945; editor, *News Chronicle*, 1948-54; publications include *The Double Quest* (1936), *Roaring Century* (for centenary of *News Chronicle*, 1946), and *The Moods of London* (1951).

CRUM, WALTER EWING (1865-1944), Coptic scholar; educated at Eton and Balliol College, Oxford; studied Egyptology in Paris and Berlin; edited many Coptic texts; catalogued Coptic manuscripts in British Museum (1905) and John Rylands Library, Manchester (1909); produced definitive Coptic dictionary (6 parts, 1929-39); FBA, 1931.

CRUMP, CHARLES GEORGE (1862-1935), archivist; junior clerk, Public Record Office, 1888; senior assistant keeper, 1916-23; insisted on need for study of records in their setting as products of an administrative machine; publications include *History and Historical Research* (1928).

CRUTTWELL, CHARLES ROBERT MOWBRAY FRASER (1887-1941), historian; son of C. T. Cruttwell [q.v.]; educated at Rugby and Queen's College, Oxford; first class, *lit. hum.*, 1910, modern history, 1911; fellow of All Souls, 1911; of Hertford College, 1919; principal of Hertford, 1930-9; publications include *History of the Great War, 1914-1918* (1934).

CRUTTWELL, CHARLES THOMAS (1847-1911), historian of Roman literature; scholar of St. John's College, Oxford; first class in classics, 1870; fellow of Merton, 1870; tutor,

1874-7; headmaster of Bradfield College, 1877-80, and of Malvern, 1880-5; canon of Peterborough, 1903; chief work was *A History of Roman Literature*, 1877.

CUBITT, WILLIAM GEORGE (1835-1903), colonel Indian staff corps; born in Calcutta; joined 13th regiment Bengal native infantry, 1853; won VC for bravery at Chinhut near Lucknow, 1857; served in Duffla expedition 1874-5; in Afghan war, 1880, with Akha expedition, 1883-4, and Burmese expedition, 1886-7; colonel, 1883; retired, 1892.

CUDLIPP, PERCIVAL THOMAS JAMES (1905-1962), journalist; educated at Howard Gardens secondary school, Cardiff; left at thirteen to work on *South Wales Echo*; precocious talent for witty verse; reporter on *Evening Chronicle*, Manchester, 1924; dramatic critic, *Sunday News*, London, 1925; wrote topical lyrics for Co-optimists revue; special writer, *Evening Standard*, 1929; assistant editor, 1931; editor, 1933; editorial manager, *Daily Herald*, 1938; succeeded Francis Williams (later Lord Francis-Williams, q.v.) as editor, 1940-53; columnist, *News Chronicle*; founder-editor, *New Scientist*, 1956.

CULLEN, WILLIAM (1867-1948), chemist and metallurgist; studied chemistry at Andersonian College, Glasgow, and metallurgy and mining in Freiberg; joined Nobel's Explosives, Glasgow, 1890; general manager, British South African Explosives Company, Modderfontein, 1901-15; worked mainly on smokeless powders; improved mining conditions; active in education organizations developing into university of Witwatersrand which he represented in England; served in Ministry of Munitions, 1915-19, and remained in England as consultant.

CULLINGWORTH, CHARLES JAMES (1841-1908), gynaecologist and obstetrician; studied at Leeds School of Medicine; MRCS, 1865; lecturer on medical jurisprudence, Owens College, Manchester, 1879; professor of obstetrics and gynaecology, 1883-8; obstetric physician at St. Thomas's Hospital, 1888-1904; FRCP, 1887; helped to found *Journal of Obstetrics and Gynaecology*; hon. DCL, Durham, 1893, LLD, Aberdeen, 1904; investigated causation of pelvic peritonitis; published *Clinical Illustrations of the Diseases of the Fallopian Tubes and of Tubal Gestation* (1895) and other books and papers.

CULLIS, WINIFRED CLARA (1875-1956), physiologist; educated at King Edward VI high school for girls, Birmingham, and Newnham College, Cambridge; demonstrator in physi-

ology, Royal Free Hospital medical school, 1901; lecturer and head of physiology department, 1908-41; D.Sc., 1908; in London University: reader, 1912, professor, 1920, first holder of Jex-Blake chair of physiology, 1926-41; head, women's section, British Information Services, New York, 1941-3; lecturer in Middle East, 1944-5; co-founder British and International Federations of University Women; deputy chairman, English-Speaking Union; first woman member of Physiology Society; CBE, 1929.

CUMMINGS, ARTHUR JOHN (1882-1957), journalist and author; brother of B. F. Cummings [q.v.]; educated at Rock Park School, Barnstaple; worked successively on *Devon and Exeter Gazette*, *Rochdale Observer*, *Sheffield Telegraph*, and *Yorkshire Post*; successively assistant editor, deputy editor, and political editor, *Daily News* (later *News Chronicle*), 1920-55; a forceful crusader for radicalism; publications include *The Moscow Trial* (1933), *The Press and a Changing Civilisation* (1936), and *This England* (1945); president, Institute of Journalists, 1952-3.

CUMMINGS, BRUCE FREDERICK (1889-1919), diarist and biologist; generally known by pseudonym W. N. P. BARBELLION; in Natural History Museum, South Kensington, 1911-17; published extracts from his diaries, *The Journal of a Disappointed Man*, 1919.

CUNINGHAM, JAMES McNABB (1829-1905), surgeon-general; born at Cape of Good Hope; MD, Edinburgh University, 1851; hon. LLD, 1892; joined Bengal medical service, 1851; sanitary commissioner for Bengal presidency, 1869-75, and for Indian empire, 1875-85; surgeon-general, 1880; CSI, 1885; published *A Sanitary Primer for Indian Schools*, 1879.

CUNNINGHAM, ANDREW BROWNE, VISCOUNT CUNNINGHAM OF HYNDHOPE (1883-1963), admiral of the fleet; son of Daniel John Cunningham [q.v.]; educated at Edinburgh Academy and Stubbington House, Fareham; entered *Britannia* as cadet, 1897; midshipman in *Doris* at outbreak of South African war, 1899; joined Naval Brigade, 1900; lieutenant, 1904; commanded destroyer *Scorpion*, 1911-18; served at Dardanelles; commander, 1915; DSO; served in Dover Patrol under Roger (later Lord) Keyes [q.v.], 1918; bar to DSO; captain, 1919; second bar to DSO, 1920; captain (D), 6th destroyer flotilla, 1922; 1st flotilla, 1923; flag captain and chief staff officer to (Sir) Walter Cowan [q.v.]; commanded battleship, *Rodney*, 1929; commodore, Royal Naval Barracks, Chatham, 1931; naval aide-de-camp to the king, 1932; rear-admiral (destroyers) in Mediterranean, 1933;

CB, 1934; vice-admiral, 1936; commanded Battle Cruiser Squadron, 1937–8; deputy chief of naval staff, Admiralty, 1938; admiral, commander-in-chief, Mediterranean, 1939; KCB; attacked Taranto harbour, 1940, decreasing threat to British convoys to Greece and Crete; defeated Italian fleet off Cape Matapan, 1941; head of British Admiralty delegation to Washington, 1942; 'Allied naval commander Expeditionary Force' under General Eisenhower; covered 'Torch' landings in North Africa, 1942; admiral of the fleet, commander-in-chief, Mediterranean, 1943; first sea lord, 1943–5; GCB, 1941; baronet, 1942; baron, 1945; viscount, and OM, 1946; published *A Sailor's Odyssey* (1951); lord high commissioner, General Assembly, Church in Scotland, 1950 and 1952; lord rector Edinburgh University, 1945–8; president, Institution of Naval Architects, 1948–51.

CUNNINGHAM, DANIEL JOHN (1850–1909), professor of anatomy; son of John Cunningham (1819–93, q.v.); graduated in medicine at Edinburgh, 1874; professor of anatomy at Trinity College, Dublin, 1883–1903, and Edinburgh University, 1903–9; published *Manual of Practical Anatomy* (2 vols., 1893–4) and other scientific works; made many anthropological researches, published by Royal Irish Academy; received hon. degrees from Dublin, St. Andrews, Glasgow, and Oxford; FRS, 1891.

CUNNINGHAM, Sir GEORGE (1888–1963), Indian civil servant; educated at Fettes and Magdalen College, Oxford; captained Oxford and Scotland rugby teams; entered Indian civil service, 1910; posted to Punjab, 1910–13; joined Indian political service; political agent, North Waziristan, 1922–3; counsellor at Kabul, 1925–6; private secretary to Lord Irwin (later the Earl of Halifax, q.v.), 1926–31; governor, North-West Frontier province, 1937–46; retired, but recalled by M. A. Jinnah [q.v.] when province became part of Pakistan, 1947–8; rector, St. Andrews University, 1946; CIE, 1925; CSI, 1931; KCIE, 1935; KCSI, 1937; GCIE, 1946; hon. LLD St. Andrews and Edinburgh; hon. fellow, Magdalen College, 1948.

CUNNINGHAM, Sir JOHN HENRY DACRES (1885–1962), admiral of the fleet; entered *Britannia*, 1900; midshipman in *Gibraltar*; lieutenant, 1905; specialized in navigation; served as navigator in *Berwick*, *Russell*, and *Renown* during 1914–18 war; commander, 1917; navigator, *Hood*, 1920; commander, Navigation School, 1922; captain, 1924; served in variety of posts, including director of plans, Admiralty;

rear-admiral, 1936; assistant chief of naval staff (air), 1937; fifth sea lord and chief of naval air services, 1938; vice-admiral, 1939; commanded first cruiser squadron in Mediterranean; fourth sea lord, 1941–3; CB, 1937; KCB, 1941; admiral, commander-in-chief, Levant, 1942; commander-in-chief, Mediterranean, 1943; succeeded Lord Cunningham of Hyndhope [q.v.], (no relation), as first sea lord, 1946; admiral of the fleet, 1948; chairman, Iraq Petroleum Company, 1948–58.

CUNNINGHAM, WILLIAM (1849–1919), pioneer economic historian; BA, Trinity College, Cambridge (bracketed senior in moral sciences tripos), 1872; vicar of Great St. Mary's, Cambridge, 1887–1908; professor of economics, King's College, London, 1891–7; fellow of Trinity, 1891; archdeacon of Ely, 1907–19; FBA, 1903; best-known work *The Growth of English Industry and Commerce* (7 editions, 1882–1910).

CUNNINGHAME GRAHAM, ROBERT BONTINE (1852–1936), traveller, scholar, etc. [See GRAHAM.]

CURRIE, Sir ARTHUR WILLIAM (1875–1933), general; born in Canada; educated at Strathroy, Ontario; abandoned teaching for brokerage and insurance; joined militia as a gunner, 1897; in 1914–18 war commanded 2nd Canadian Infantry brigade (1914), 1st Canadian division (1915), Canadian Corps (1917–19); planned and carried through battle of Hill 70 (Aug. 1917) and fought at Passchendaele and Amiens; KCMG, 1917; general, 1919; principal and vice-chancellor, McGill University, 1920–33.

CURRIE, Sir DONALD (1825–1909), founder of the Castle Steamship Company; joined Cunard Steamship Company, Liverpool, 1844; established branches at Havre, Paris, Bremen, and Antwerp, 1849–54; formed 'Castle' Shipping Company, Liverpool, with sailing ships between Liverpool and Calcutta, 1862; made London port of departure, 1865; formed new line of communication between England and Cape Town, 1872; his company amalgamated with Union Steamship Company, under name of Union-Castle Mail Steamship Company, 1900; instrumental in hoisting British flag at St. Lucia Bay, 1883; conveyed troops and stores to South Africa, 1899; liberal MP for Perthshire, 1880–5; represented West Perthshire as liberal unionist, 1886–1900; enlightened landowner in Perthshire and western islands; benefactor to University College Hospital, London, Edinburgh University, United Free Church of Scotland, and Belfast institutions; a hall founded in his memory at Cape Town University, 1910; CMG, 1881; GCMG, 1897; hon. LLD, Edinburgh, and freeman of Belfast, 1907; owned fine collection of Turner's works.

CURRIE, Sir JAMES (1868–1937), educationist; educated at Fettes, Edinburgh University, and Lincoln College, Oxford; joined Egyptian education service, 1899; first director of education in Sudan and first principal of Gordon Memorial College, 1900–14; director of training in ministries of Munitions (1916–18) and Labour (1918–21); director, Empire Cotton Growing Corporation, 1922–37; chairman of governing body of Imperial College of Tropical Agriculture, Trinidad, 1927–37; KBE, 1920; KCMG, 1933.

CURRIE, MARY MONTGOMERIE, Lady Currie (1843–1905), authoress under the pseudonym of Violet Fane; married first, Henry Sydenham Singleton, an Irish landowner, 1864; secondly, Sir Philip Henry Wodehouse (afterwards Baron) Currie [q.v.], 1894; figures as 'Mrs. Sinclair' in *New Republic* by W. H. Mallock [q.v.], 1877; published *Denzil Place; a Story in Verse* (1875), *Collected Verses* (1880), and other poetical works; *Poems* collected in 2 vols., 1892; wrote in prose *Edwin and Angelina Papers* (1878), *Collected Essays* (1902), and novels; accompanied second husband to Constantinople, 1894, and to Rome, 1898–1903.

CURRIE, PHILIP HENRY WODEHOUSE, Baron Currie (1834–1906), diplomatist; joined Foreign Office, 1854; accompanied Lord Salisbury to Constantinople (1876) and Berlin (1878); CB, 1878; KCB, 1885; permanent under-secretary of state for foreign affairs, 1889–93; appointed British ambassador at Constantinople, 1893; took leading part in securing protection and redress of Armenians from Sultan of Turkey, 1895; gave at embassy refuge to grand vizier Said Pasha from molestation of Sultan, 1895; helped to secure favourable terms for Greece and autonomy for Crete after Turco-Greek war, 1897; ambassador at Rome, 1898–1903; baron, 1899.

CURRIE, Sir WILLIAM CRAWFORD (1884–1961), shipowner and director of the Peninsular and Oriental Steam Navigation Company and of the British India Steam Navigation Company; educated at Glasgow Academy, Fettes, and Trinity College, Cambridge; rugby blue; joined David Strathie, Glasgow chartered accountants, 1906; qualified as chartered accountant, 1910; assistant, Mackinnon, Mackenzie & Co., Calcutta; partner, 1918; sheriff of Calcutta, 1921–2; member, Bengal Legislative Council, 1921–5; president,

Bengal Chamber of Commerce, 1924; member, Council of State for India; knighted, 1925; partner in Inchcape family firm, Gray, Dawes & Co., 1925; member, Imperial Shipping Committee, 1926–30; president, Chamber of Shipping of the United Kingdom, 1929–30; deputy chairman and managing director, P & O and BISN Companies, 1932; chairman, 1938; member, Advisory Council, Ministry of War Transport, and director, Liner Division, 1942–5; GBE, 1947; director, Inchcape & Co. Ltd., 1958; president, Institute of Marine Engineers, 1945–6; high sheriff, Buckinghamshire, 1947; prime warden, Worshipful Company of Shipwrights; director of number of other firms in banking and insurance and member of committees concerned with shipping.

CURTIN, JOHN (1885–1945), prime minister of Australia; educated in Victoria state schools; secretary, Victorian branch, Timber Workers' Union, 1911–15; of Anti-Conscription League, 1915–17; editor, *Westralian Worker*, 1917–28; labour MHR, Fremantle, 1928–31, 1934–45; leader of labour party in opposition, 1935–41; pledged full support to war effort, 1939; prime minister, 1941–5; a great leader; obtained majority in both houses, 1943; introduced measure of conscription; asserted Australia's right to be consulted on conduct of war in Pacific; acknowledged Australia's dependence on United States but claimed treatment as partner; urged closer commonwealth consultation; nominated PC, 1942.

CURTIS, EDMUND (1881–1943), historian; first class, modern history, Keble College, Oxford, 1904; history lecturer, Sheffield, 1905–14; professor of modern history, Trinity College, Dublin, 1914–39; Lecky professor, 1939–43; works include *History of Medieval Ireland* (1923).

CURTIS, LIONEL GEORGE (1872–1955), public servant; educated at Haileybury and New College, Oxford; called to bar (Inner Temple), 1902; in South Africa, 1899–1909; town clerk of Johannesburg, 1901–3; assistant colonial secretary, Transvaal, 1903–7; as head of Milner's kindergarten prepared Selborne memorandum, 1907; created 'closer union' societies, 1907–9; a founder of the *Round Table*, 1910; Beit lecturer on colonial history, Oxford, 1912; studied Commonwealth problems, 1911–16; joined constitutional discussions in India, 1916–17; published *Dyarchy*, 1920; brought about foundation of Royal Institute of International Affairs, 1920–1; research fellow, All Souls, 1921; adviser to Colonial Office on Irish affairs, 1921–4; published *Civitas Dei* (3 vols., 1934–7); CH, 1949.

CURTIS, WILLIAM EDWARD (1889–1969), experimental physicist; educated at Owen's School, Islington, and Imperial College, London, 1907–10; first class, physics and astronomy; demonstrator and research student under (Professor) Alfred Fowler [q.v.]; sapper, Royal Naval Division, 1914; served in Gallipoli; resumed studies in spectroscopy, Imperial College, 1918; lecturer, Sheffield University; reader, King's College, London; professor, Armstrong College, Newcastle upon Tyne, 1926–55; FRS, 1934; during 1939–45 war worked on camouflage and applied explosives; sub-rector, King's College, Newcastle, 1947–50; president, Institute of Physics, 1950–2; CBE, 1967; published many scientific papers; a skilful educational administrator and lecturer.

CURZON, GEORGE NATHANIEL, MARQUESS CURZON OF KEDLESTON (1859–1925), statesman; educated at Eton; early became absorbed by passion for the East; assailed by first symptoms of spinal curvature, which throughout deprived his character of elasticity, 1878; proceeded to Balliol College, Oxford, 1878; president of Union and secretary of Canning Club; BA, 1882; won Lothian and Arnold historical essay prizes, 1883; fellow of All Souls, 1883; MP, Southport division of Lancashire, 1886–92; began to manifest intense interest in foreign and colonial policy; threw himself into society of Crabbet Club and 'Souls'; entered upon period of his great journeyings, 1887; these journeys were (1) Canada, USA, Japan, China, Singapore, Ceylon; India, 1887–8; (2) Russia, Caucasia, Turkestan, 1888–9; (3) Persia, 1889–90; (4) USA, Japan, China, Cochin China, Siam, 1892; (5) the Pamirs, Afghanistan, Kabul, course of Oxus, 1894; result of these travels embodied in *Russia in Central Asia* (1889), *Persia and the Persian Question* (1892), and *Problems of the Far East* (1894); they earned for him reputation as a leading authority on Asiatic affairs, who was at once a xenophobe and a nationalist; rule and defence of India became for him ambition and cause; under-secretary, India Office, 1891; parliamentary under-secretary for foreign affairs, 1895–8; disapproved of Lord Salisbury's policy of what he regarded as undue passivity in matters of Armenian atrocities, Venezuela, and Cretan question; felt keenly about French encroachments in Siam and German occupation of Kiao-chow; established his reputation with parliament and public; chosen viceroy of India and created Baron Curzon of Kedleston, 1898; entered upon term of office, Jan. 1899; his seven years' viceroyalty falls into two periods, divided by durbar of 1903; during first, admired in India and supported at home; during second, his popularity in India waned and his relations with Whitehall

were increasingly embittered; held scales even between different nationalities committed to his rule; adopted cautious policy on North-West Frontier, but adventurous one in Persian Gulf; studied intensively land assessment and educational problem; concluded treaty with Nizam of Hyderabad which settled Berar question, 1902; durbar of Jan. 1903 marked summit of his viceregal splendour; set his stamp upon art and archaeology of India; secured partition of Bengal, thereby losing popularity with Indian opinion, 1905; came into bitter conflict with Lord Kitchener [q.v.], commander-in-chief in India, who during viceroy's absence in England (1904) raised question of unsatisfactoriness of system of dual control of Indian army; reached compromise with Kitchener, which was approved by home government, July 1905; resigned, owing to failure of home government to appoint new supply member of viceroy's council nominated by himself, Aug.; his return from India (Nov.) followed by eleven years' disappointment and domestic sorrow; chancellor of Oxford University, 1907; immersed himself in question of university reform and staved off government inquiry for many years; entered House of Lords as Irish representative peer, 1908; his advice largely secured bare majority of peers in support of liberal government's proposals, Aug. 1911; his main activities during these years concerned with Royal Geographical Society, National Gallery, and preservation of ancient monuments; created Earl Curzon of Kedleston, 1911; lord privy seal in coalition cabinet, 1915; opposed to evacuation of Gallipoli; pressed for compulsory service; in charge of shipping control committee, 1916; president of Air Board, 1916; active as member of inner war cabinet (Dec. 1916), leader in House of Lords, and lord president of Council; as leading conservative found difficulty in acting as spokesman of government pledged to liberal concessions; subservient to Lloyd George in Irish matters; his surrender over question of women's suffrage lost him prestige with conservatives; in absence of A. J. Balfour at peace conference in Paris, foreign minister 'in interim', Jan. 1919; succeeded Balfour as foreign secretary, Oct.; unable to secure free hand in foreign policy; his Persian policy a failure; more successful in dealing with Egyptian affairs; involved in intense difference of opinion with Lloyd George over Graeco-Turkish question; owing to fact of repeated clandestine negotiations being carried on by prime minister's secretariat in favour of Lloyd George's anti-Turkish policy, convinced that he could no longer support Lloyd George's coalition; foreign secretary in Bonar Law's ministry, Nov. 1922; dominated abortive conference of Lausanne and restored British prestige in Turkey, 1922–3; this was most striking of his diplomatic triumphs; less sure in handling European diplomacy, but foreshadowed Dawes and Young schemes for reparations payments; bitterly disappointed at not being designated prime minister, 1923; at Foreign Office magnanimously supported Stanley Baldwin; obtained satisfaction from Soviet government over trade agreement with Great Britain; failed to mediate between France and Germany; lord president of the Council, 1924; KG, 1916; marquess, 1921.

CURZON-HOWE, SIR ASSHETON GORE (1850-1911), admiral; great-grandson of Richard, first Earl Howe [q.v.]; entered navy, 1863; accompanied and instructed royal princes on cruise in *Bacchante*, 1879-82; commander, 1882; served in Vitu expedition in East Indies, 1890; CB, 1890; as commodore in North America station, he averted civil war in Nicaragua, 1894; CMG, 1896; second-in-command on China station, 1903; KCB and vice-admiral, 1905; commander-in-chief of Mediterranean fleet, 1908; admiral and GCVO, 1909; commander-in-chief at Portsmouth, 1910.

CUSHENDUN, BARON (1861-1934), Irish politician. [See McNEILL, RONALD JOHN.]

CUSHNY, ARTHUR ROBERTSON (1866-1926), pharmacologist; MB, CM, Aberdeen, 1889; professor of pharmacology, Michigan University, at Ann Arbor, 1893-1905; at University College, London, 1905-18; professor of materia medica and pharmacology, Edinburgh, 1918-26; carried out researches on action of *digitalis glucosides*, the functions of the kidneys, and pharmacological actions of optical isomers; works include *Textbook of Pharmacology and Therapeutics* (1899).

CUST, HENRY JOHN COCKAYNE (1861-1917), politician and journalist; great-grandson of first Baron Brownlow; BA, Trinity College, Cambridge; unionist MP, Stamford division of Lincolnshire, 1890-5, Bermondsey, 1900-6; editor of *Pall Mall Gazette*, 1892-6; chairman of central committee for national patriotic organizations, 1914.

CUST, SIR LIONEL HENRY (1859-1929), art historian; assistant, department of prints and drawings, British Museum, 1884; director, National Portrait Gallery, 1895-1909; superintended installation of collection at St. Martin's Place; surveyor of king's pictures, 1901-27; joint-editor, *Burlington Magazine*, 1909-19; made special studies of Dürer and Van Dyck; works include *Van Dyck* (1900), *Authentic Portraits of Mary, Queen of Scots* (1903), and *The Royal Collection of Paintings* (1905-6).

CUST, ROBERT NEEDHAM (1821-1909), orientalist; grandson of first Baron Brownlow; educated at Eton; joined Indian civil service; served in Sikh war, 1845; administered Hoshiarpur and Ambala districts; reported on Punjab district after second Sikh war, 1849; called to bar, 1857; judicial commissioner of Amritsar, 1861; joined legislative council, 1864; retired from Indian service, 1867; studied oriental languages; published over sixty volumes on oriental philology or religion; chief were *Modern Languages of the East Indies* (1878), *Linguistic and Oriental Essays* (7 series, 1880-1904); joined Royal Asiatic Society, 1851; secretary, 1878-99; interested in missionary work; hon. LLD, Edinburgh, 1885; attended coronations of William IV, Queen Victoria, and Edward VII; *Memoir* published, 1899.

CUSTANCE, HENRY (1842-1908), jockey; won the Cesarewitch, 1858 and 1861; the Derby, 1860, 1866, 1874; One Thousand Guineas, 1867; St. Leger, 1866; published *Riding Recollections and Turf Stories* (1894).

CUSTANCE, SIR REGINALD NEVILLE (1847-1935), admiral; assistant director (1886-90), director (1899-1902), naval intelligence; rear-admiral, Mediterranean fleet, 1902-4; vice-admiral, 1906; second-in-command, Channel fleet, 1907-8; admiral, 1908; being outspokenly critical of Admiralty policy was retired in 1912; advocated study of naval warfare and building of many ships of moderate tonnage, powerful armament, and slow speed in preference to large armoured cruisers; KCMG, 1904.

CUTTS, EDWARD LEWES (1824-1901), antiquary; BA, Queens' College, Cambridge, 1848; after holding curacies, he visited East to report on position of Syrian and Chaldean Churches; published many works on archaeology and ecclesiastical history, including *Turning-Points of English Church History* (1874) and *History of Early Christian Art* (1893).

D

D'ABERNON, Viscount (1857–1941), financier and diplomatist. [See VINCENT, SIR EDGAR.]

DADABHOY, Sir MANECKJI BYRAMJI (1865–1953), Indian lawyer, industrialist, and parliamentarian; a Parsi; graduated from St. Xavier's College, Bombay; called to bar (Middle Temple), 1887; member, Nagpur municipal corporation, 1890–1930; government advocate, 1896; acquired many industrial interests; a governor, Imperial Bank of India, 1920–32; member of legislative council from 1908; of Council of State from 1921; president, 1932–46; delegate, Round Table conference, 1931; a nationalist who valued the Commonwealth tie; CIE, 1911; knighted, 1921; KCIE, 1925; KCSI, 1936.

DAFOE, JOHN WESLEY (1866–1944), Canadian journalist; editor, Winnipeg *Free Press*, 1901–44; made it leading exponent of liberal ideas; secured liberal support for coalition enforcing conscription, 1917; his support of W. L. Mackenzie King [q.v.] tempered by criticism of foreign policy; crusaded for Canadian nationhood and collective security through League of Nations; enthusiastic upholder of war effort, 1939–44.

DAIN, Sir (HARRY) GUY (1870–1966), general medical practitioner; educated at King Edward's grammar school and Mason College, Birmingham; MRCS, LRCP, 1893; MB, BS, London, 1894; hospital appointments in Birmingham; general practice in Selly Oak; interested in medico politics; member first Insurance Committee, and chairman, first Panel Committee; member, Insurance Acts Committee, London, 1917; chairman, 1924–36; member, General Medical Council, 1934–61; chairman, British Medical Association, 1943–9; leader of medical profession at inception of National Health Service; hon. LLD Aberdeen, 1939; hon. MD Birmingham, 1944; FRCS, 1945; knighted, 1961.

DAKIN, HENRY DRYSDALE (1880–1952), biochemist; educated at Merchant Taylors', Leeds modern school, and Yorkshire College; B.Sc., Manchester, 1901; after further research spent remainder of his life in charge of C. A. Herter's private biochemical laboratory, New York, and issued regular reports on his work; made wartime contribution to antiseptic treatment of wounds: married Herter's widow, 1916; FRS, 1917.

DALE, Sir DAVID, first baronet (1829–1906), ironmaster; born in Bengal; in service of Stock-

ton and Darlington Railway Company; embarked on extensive shipbuilding enterprises at Hartlepool, 1866; director of North-Eastern Railway Company, 1881; chairman of Sunderland Iron Ore Company, 1902; pioneer of arbitration in industrial disputes; served on several industrial commissions; part founder (1869) and president (1895) of Iron and Steel Institute; baronet, 1895; hon. DCL, Durham, 1895; memorial chair of economics founded at Armstrong College, Newcastle, 1909.

DALE, Sir HENRY HALLETT (1875–1968), physiologist-pharmacologist; educated at the Leys School and Trinity College, Cambridge; first class, parts i and ii, natural sciences tripos, 1896–8; scholar, St. Bartholomew's Hospital, London, B.Ch. (Cambridge), 1902; held studentships, University College, London, 1902–4; appointed to research post in physiology, Wellcome Research Laboratories, 1904; director, 1906–14; director, Biochemistry and Pharmacology, the projected Institute for Medical Researches, 1914; chairman, committee of departmental directors, 1923; director, Institute of Medical Research, 1928–42; director, Royal Institution and Fullerian professor of chemistry, 1942–6; trustee, Wellcome Trust, 1936; chairman, 1938–60; scientific adviser, 1960–8; period of research from 1906 to 1936 covered fundamental discoveries in pharmacology, including effects of histamine and acetylcholine, and acceptance of international standards for hormones, vitamins, and drugs; awarded Nobel prize for physiology and medicine (with Otto Loewi), 1936; secretary, Royal Society, 1925–35; president, 1940–5; chairman, scientific advisory committee to the cabinet, 1942–7; member, Medical Research Council, 1942–6; chairman of numerous committees concerned with scientific progress, such as medical and biological application of nuclear physics, 1945–9; president, Royal Society of Medicine, 1948–50, and of the British Council, 1950–5; FRS, 1914; CBE, 1919; knighted, 1932; GBE, 1943; OM, 1944; hon. fellow, Trinity College, Cambridge, University College, London, and the Chemical Society; numerous hon. degrees, and other academic honours.

DALEY, Sir (WILLIAM) ALLEN (1887–1969), medical officer of health; educated at Merchant Taylors' School, Crosby, and Liverpool University; B.Sc. London, in chemistry, 1906; MB, Ch.B., Liverpool, first class, 1909; MB, BS, London, 1910; DPH, Cambridge, 1911; MD, London, 1912; resident medical officer, London Fever Hospital, 1911; medical officer of health, Bootle, 1911–20; Blackburn,

1920–5; Hull, 1925–9; principal medical officer, London County Council, 1929; deputy to (Sir) Frederick Menzies [q.v.], 1938; county medical officer, 1939–48; FRCP; knighted, 1944; hon. physician to the King; member of council, Royal College of Physicians; president, Central Council for Health Education; trustee and governor of hospitals and other bodies; vice-chairman, academic board, Royal Postgraduate Hospital, Hammersmith; lectured during retirement in England and abroad.

DALLINGER, WILLIAM HENRY (1842–1909), Wesleyan minister and biologist; Wesleyan minister, 1861–80; governor and president of Wesley College, Sheffield, 1880–8; made classical investigations into life-history of 'flagellates' or 'monads' and into abiogenesis; improved microscopical technique; president of Royal Microscopical Society (1884–7); FRS, 1880; hon. D.Sc., Dublin, 1892, DCL, Durham, 1896; edited and rewrote Carpenter's *The Microscope and its Revelations*, 1890.

DALRYMPLE-HAY, SIR HARLEY HUGH (1861–1940), civil engineer. [See HAY].

DALTON, (EDWARD) HUGH (JOHN NEALE), BARON DALTON (1887–1962), labour politician; son of Canon John Neale Dalton, chaplain to Prince George (later George V); educated at Summer Fields, Oxford, Eton, and King's College, Cambridge; close friend of Rupert Brooke [q.v.]; called to bar (Middle Temple), 1914; served in France and Italy during 1914–18 war; published *With British Guns in Italy* (1919); lecturer, London School of Economics; D.Sc., 1921; reader, London University, 1920–36; published *Principles of Public Finance* (1923), *Towards the Peace of Nations* (1928), and *Practical Socialism for Britain* (1935); labour MP, Peckham, 1924–9, Bishop Auckland, 1929–31; parliamentary under-secretary to Arthur Henderson [q.v.], at Foreign Office; refused to serve in 'national' Government of 1931; MP, Bishop Auckland again, 1935–59; PC; minister of economic warfare, 1940; president, Board of Trade, 1942; recommended establishment of Ministry of Fuel and Power and National Coal Board; chancellor of the Exchequer, 1945; appointed Lord Keynes [q.v.] as personal adviser; nationalized Bank of England, 1946; faced sterling crisis, 1947; resigned after giving budget information prematurely to *Star*, 1947; chancellor, Duchy of Lancaster, 1948; minister of town and country planning, 1950; life peerage, 1960; published memoirs, *Call Back Yesterday* (1953), *The Fateful Years (1931–1945)* (1957), and *High Tide and After* (1962); hon. fellow LSE; hon. bencher, Middle

Temple, 1946; hon. degrees at Manchester, Durham, and Sydney.

DALTON, ORMONDE MADDOCK (1866–1945), classical scholar and medieval archaeologist; educated at Harrow and New College, Oxford; first class, *lit. hum.*, 1888; entered department of British and medieval antiquities, British Museum, 1895; assistant keeper, 1909; keeper, 1921–8; publications include official guides and catalogues and *Byzantine Art and Archaeology* (1911); FBA, 1922.

DALZIEL, DAVISON ALEXANDER, BARON DALZIEL (1854–1928), newspaper proprietor and financier; nephew of Brothers Dalziel [qq.v.]; a founder of Dalziel's News Agency, 1893; director of public companies, especially of overland transport undertakings; conservative MP, Brixton, 1910; baronet, 1919; baron, 1927.

DALZIEL, EDWARD (1817–1905), draughtsman and wood-engraver; second of Brothers Dalziel; partner with brother George as engraver, publisher, and printer in London, 1839–93; painter in oils and water-colours; designed woodcut illustrations in Dalziel's *Arabian Nights* (1864) and *Bible Gallery* (1880); joint-author of *The Brothers Dalziel* (1901).

DALZIEL, GEORGE (1815–1902), draughtsman and wood-engraver; eldest of Brothers Dalziel; founded firm of the Brothers Dalziel in London, 1839; worked much (1840–50) with Ebenezer Landells [q.v.], engraving blocks for *Punch* and *Illustrated London News*; issued long series of books with woodcut illustrations which superseded steel engravings; engraved works by Millais, Birket Foster, du Maurier, Sir John Tenniel, and Harrison Weir; cut illustrations to Edward Lear's *Book of Nonsense* (1862), and Lewis Carroll's nursery classics; engraved all illustrations in *Cornhill Magazine*, 1859, and in *Good Words* from 1862; illustrated Staunton's *Shakespeare* (1858–61) and Goldsmith's works (1865); when wood-engraving was superseded by cheaper photo-mechanical processes, Dalziel chiefly engaged in illustrating papers, viz. *Fun*, 1870–93, *Hood's Comic Annual*, from 1871, and *Judy*, 1872–88; wrote stories and verse; joint-author of *The Brothers Dalziel* (1901).

DALZIEL, JAMES HENRY, BARON DALZIEL OF KIRKCALDY (1868–1935), politician and newspaper proprietor; liberal MP, Kirkcaldy, 1892–1921; proprietor of *Reynolds' Weekly*, *Pall Mall Gazette*, etc.; knighted, 1908; PC, 1912; a supporter of Lloyd George; not successful as political director of *Daily Chronicle*, 1918–21; baronet, 1918; baron, 1921.

DALZIEL, THOMAS BOLTON GIL-CHRIST SEPTIMUS (1823-1906), draughtsman; seventh and youngest of the Brothers Dalziel; joined brothers' firm, 1860; devoted himself to drawing on wood; water-colour and charcoal artist of landscape; illustrated entirely *Pilgrim's Progress* (1865), and partly Dalziel's *Arabian Nights* (1864) and *Bible Gallery* (1880).

DAMPIER, SIR WILLIAM CECIL DAMPIER (1867-1952), formerly WHETHAM, scientist and agriculturist; first class, parts i and ii, natural sciences tripos, 1888-9, Trinity College, Cambridge; fellow, 1891-1952; lecturer, 1895-1922; FRS, 1901; farmed estate in Dorset, 1918-26; member, Agricultural Wages Board, 1925-42, and various agricultural committees; a development commissioner, 1933-51; first secretary, Agricultural Research Council, 1931-5; knighted and changed name to Dampier, 1931.

DANE, CLEMENCE (pseudonym) (1888-1965), playwright and novelist. [See ASHTON, WINIFRED.]

DANE, SIR LOUIS WILLIAM (1856-1946), Indian civil servant; educated at Kingstown School, Dublin; posted to Punjab, 1876; chief secretary to Punjab government, 1898-1900; resident of Kashmir, 1901-3; foreign secretary, government of India, 1903-8; headed mission to Kabul and negotiated treaty with Afghanistan, 1904-5; lieutenant-governor, Punjab, 1908-13; KCIE, 1905; GCIE, 1911.

DANIEL, CHARLES HENRY OLIVE (1836-1919), scholar and printer; BA, Worcester College, Oxford; fellow, 1863; provost, 1903-19; his private press produced books from about 1845 to 1903; these include reprints and works of contemporaries, notably of Robert Bridges [q.v.]; two of the most interesting productions, *The Garland of Rachel* (1881) and *Our Memories* (1893); revived Fell type.

DANIEL, EVAN (1837-1904), writer on the Prayer Book; vice-principal (1863) and principal (1866-94) of St. John's Training College, Battersea; BA, Dublin, 1870; MA, 1874; member of London school board, 1873-9; vicar of Horsham, 1894-1904; honorary canon of Rochester, 1879; wrote *The Prayer Book, its History and Contents*, 1877 (20th edn. 1901).

DANQUAH, JOSEPH BOAKYE (1895-1965), lawyer and politician in the Gold Coast (Ghana); worked as clerk in solicitor's office and (later) in Supreme Court; secretary to his half-brother (Sir) Nana Ofori Atta, a paramount chief; entered University College, London,

1921; BA, LLB, London, 1926; Ph.D, 1927; called to bar (Inner Temple); set up legal practice in Gold Coast; secretary, 'Youth Conference', 1930; established *Times of West Africa*, first Gold Coast daily, 1931; made proposals for constitutional reform, 1943; formed United Gold Coast Convention (UGCC), 1947; struggled for power in UGCC with Dr Kwame Nkrumah who left UGCC and formed Convention People's Party (CPP); in opposition to CPP government, 1951; lost seat in legislature, 1954; formed United Party, but defeated by Nkrumah for presidency of independent republic of Ghana, 1960; imprisoned, 1961; released, 1962; president, Ghana Bar Association, 1963; imprisoned again for opposing Nkrumah's one-party state, 1964; died in prison, 1965.

DANVERS, FREDERIC CHARLES (1833-1906), writer on engineering; writer in old East India house, 1853; joined newly formed India Office, 1858; senior clerk of public works department, 1867; registrar and superintendent of records, 1884-98; wrote *History of the Portuguese in India* (1894) and lists of factory and marine records of the East India Company (1897).

D'ARANYI, JELLY (1893-1966), violinist. [See under FACHIRI, ADILA (1886-1962).]

DARBISHIRE, HELEN (1881-1961), scholar, critic, and principal of Somerville College, Oxford; educated at Oxford high school and Somerville; Pfeiffer scholar; first class, English, 1903; lecturer, Royal Holloway College, London; English tutor, Somerville, 1908; fellow, 1922; university lecturer, 1927-31; principal of Somerville, 1931-45; hon. fellow, 1946; visiting professor, Wellesley College, Massachusetts, 1925-6; collaborated with Ernest de Selincourt [q.v.] on Wordsworth; published *Wordsworth's Poems published in 1807* (1914), *The Manuscript of Paradise Lost, Book I* (1931), and *The Early Lives of Milton* (1932); completed the Clarendon Wordsworth, offered a new text of *Paradise Lost* (1952), and published the complete poems of Milton in Oxford Standard Authors series (1958); other publications include *The Poet Wordsworth* (1950), and a new edition of the *Journals of Dorothy Wordsworth* (1958); chairman of Dove Cottage, 1943-61; FBA, 1947; hon. D.Litt., Durham and London; CBE, 1955.

DARBYSHIRE, ALFRED (1839-1908), architect; apprenticed at Manchester; built Comedy Theatre, Manchester, and altered Lyceum, London, for (Sir) Henry Irving [q.v.], 1878; an amateur actor; played with Irving, 1865, and Helen Faucit [q.v.], 1879; FSA, 1894;

expert in heraldry; wrote *A Chronicle of the Brasenose Club, Manchester* (2 vols., 1892-1900) and *Memoirs* (1897).

D'ARCY, CHARLES FREDERICK (1859-1938), archbishop of Armagh and primate of all Ireland; educated at high school and Trinity College, Dublin; ordained, 1884; vicar and dean of Belfast, 1900-3; bishop of Clogher (1903), Ossory, Ferns, and Leighlin (1907), Down, Connor, and Dromore (1911); openly supported Ulster unionists; archbishop of Dublin (1919), of Armagh, and primate of all Ireland, 1920-38; FBA, 1927; publications include *Idealism and Theology* (1899) and *Providence and the World Order* (1932).

DARLING, CHARLES JOHN, first BARON DARLING (1849-1936), judge; called to bar (Inner Temple), 1874; bencher, 1892; joined Oxford circuit; wrote for *St. James's Gazette* and other periodicals; QC, 1885; conservative MP, Deptford, 1888-97; appointed to Queen's Bench division, 1897; deputized for lord chief justice, 1914-18; PC, 1917; retired, 1923; baron, 1924; excellent in summing-up but frequently lost jury's respect through levity; judgments characterized by close reasoning; presided over Steinie Morrison (1911) and Armstrong (1922) cases and Crippen (1910) and Casement (1916) appeals.

DARWIN, BERNARD RICHARD MEIRION (1876-1961), essayist and sports writer; grandson of Charles Darwin [q.v.]; educated at Summerfields, Oxford, Eton and Trinity College, Cambridge; golf blue, 1895-7, captain, 1897; called to bar (Inner Temple), 1903; decided to write full-time, mostly about golf, 1908; contributed articles to *The Times*, 1907-53; and to *Country Life*, 1907-61; served in Royal Army Ordnance Corps in 1914-18 war; captained British golf team in first Walker Cup match in United States, 1922; won the President's Putter, 1924; captain, Royal and Ancient Golf Club, 1934; made golf reporting a branch of literary journalism; wrote foreword to *Oxford Dictionary of Quotations*, 1941; other publications include *Green Memories* (1928), a book on W. G. Grace (1934), *Pack Clouds Away* (1941), and *The World that Fred Made* (1955); CBE, 1937.

DARWIN, SIR CHARLES GALTON (1887-1962), physicist; son of (Sir) George Darwin, FRS [q.v.], and grandson of Charles Robert Darwin [q.v.], author of *The Origin of Species* (1859); educated at Marlborough College and Trinity College, Cambridge; fourth wrangler, part i, and first class, part ii, mathematical tripos, 1909-10; joined Ernest (later Lord)

Rutherford [q.v.], and Niels Bohr as Schuster lecturer in mathematical physics at Manchester, 1910-12; worked with H. G. J. Moseley [q.v.] on diffraction of X-rays, 1912-14; during 1914-18 war served with Royal Engineers and Royal Flying Corps; MC; fellow and lecturer, Christ's College, Cambridge, 1919-22; FRS, 1922; first Tait professor of natural philosophy, Edinburgh University, 1924; produced papers on atomic structure; master, Christ's College, Cambridge, 1936; director, National Physical Laboratory, 1938-49; first director of British office for improving Anglo-American scientific war co-operation, Washington, 1941; involved in liaison over atomic bomb; scientific adviser to the War Office; on return to NPL promoted production of Pilot ACE, the first digital computer available to British industry; published *The Next Million Years* (1952); member, University Grants Committee, 1943-53; president, Physical Society, 1941-4; president, Eugenics Society, 1953-9; KBE, 1942; hon. degrees from Bristol, Manchester, St. Andrews, Trinity College (Dublin), Delhi, Edinburgh, Chicago, and California; hon. fellow, Christ's College (1939), and Trinity College, Cambridge (1953); vice-president, Royal Society, 1939.

DARWIN, SIR FRANCIS (1848-1925), botanist; son of Charles Darwin [q.v.]; BA, Trinity College, Cambridge, 1870; his father's secretary and research assistant, 1875-82; university lecturer in botany, Cambridge, 1884; fellow of Christ's College, 1886; honorary fellow, 1908; reader in botany, 1888-1904; rendered twofold service to science: first, published *Life and Letters of Charles Darwin* (1887) and *More Letters of Charles Darwin* (1903); secondly, increased knowledge of vegetable physiology; FRS, 1882; knighted, 1913; received many honorary degrees.

DARWIN, SIR GEORGE HOWARD (1845-1912), mathematician and astronomer; son of Charles Darwin [q.v.]; educated at Clapham grammar school and Trinity College, Cambridge; BA (second wrangler), 1868; fellow of Trinity, 1868-78; FRS, 1879; Plumian professor of astronomy and experimental philosophy at Cambridge, 1883-1912; earliest scientific papers, originating with memoir (1876) *On the Influence of Geological Changes on the Earth's Axis of Rotation*, deal solely with the earth; next series concerned with earth-moon system and the part played by 'tidal friction' in its development; later series surveys solar system; authority on many scientific subjects, including tidal theory, geodesy, and dynamical meteorology; president of British Association when it visited South Africa, and KCB, 1905; collected works published, 1907-11.

DARWIN, SIR HORACE (1851-1928), civil engineer; son of Charles Darwin [q.v.]. BA, Trinity College, Cambridge, 1874; devoted his life to designing and manufacturing scientific apparatus for Cambridge natural science school; FRS, 1903; KBE, 1918.

DASHWOOD, EDMÉE ELIZABETH MONICA (1890-1943), authoress under penname, E. M. DELAFIELD; daughter of novelist Mrs Henry de la Pasture; married (1919) Major A. P. Dashwood; wrote *Thank Heaven Fasting* (1932) and other novels, three plays, and notably *The Diary of a Provincial Lady* (1930) and its sequels.

DAUBENEY, SIR HENRY CHARLES BARNSTON (1810-1903), general; joined 55th Foot, 1829; served in Coorg campaign, 1832-4; in Chinese war, 1841-2; CB, 1842; recommended for VC for conspicuous bravery at Inkerman, 1854; inspector of army clothing, 1858-69; KCB, 1871; GCB, 1884; general, 1880.

DAVENPORT, HAROLD (1907-1969), mathematician; educated at Accrington grammar school, Manchester University, and Trinity College, Cambridge; wrangler, part ii, mathematical tripos, 1929; research student in number theory, Cambridge; fellow, Trinity College, 1932; collaborated with (Professor) H. Heilbroun; assistant lecturer, Manchester, 1937; studied geometry of numbers; Sc.D., Cambridge, 1938; FRS, 1940; professor of mathematics, University College of North Wales, Bangor, 1941; Astor professor, University College, London, 1945; succeeded A. S. Besicovitch [q.v.] as Rouse Ball professor of mathematics, Cambridge, 1958; carried out research on quadratic forms; published *The Higher Arithmetic*, 1952; president, London Mathematical Society, 1957-9; hon. D.Sc., Nottingham University.

DAVENPORT-HILL, ROSAMOND (1825-1902), educational administrator [See HILL.]

DAVEY, HORACE, BARON DAVEY (1833-1907), judge; educated at Rugby and University College, Oxford; BA (double first in classics and mathematics), 1856; fellow, 1854; honorary fellow, 1884; MA, 1859; hon. DCL, 1894; called to bar, 1861; acquired large practice in Chancery courts; QC, 1875; practised in Rolls courts; an unrivalled 'case' lawyer; liberal MP for Christchurch, 1880-5, and Stockton on Tees, 1888-92; solicitor-general under Gladstone, Feb.-July 1886; knighted, 1886; appointed lord justice of appeal and PC, 1893;

raised to peerage as lord of appeal in ordinary, 1894; pronounced in favour of the men in trade-union appeals; largely responsible for Street Betting Act, 1906; his judgments invariably concise and lucid; interested in literature, especially modern French; chairman of commission to make statutes for university of London, 1898; FBA, 1905; collaborated in work upon costs in Chancery, 1865.

DAVID, ALBERT AUGUSTUS (1867-1950), headmaster and bishop; educated at Exeter School and Queen's College, Oxford; first class, *lit. hum.*, 1889; deacon, 1894; priest, 1895; taught at Bradfield, Rugby, and his college; headmaster, Clifton College, 1905-9; of Rugby, 1909-21; bishop of St. Edmundsbury and Ipswich, 1921-3; of Liverpool, 1923-44.

DAVID, SIR (TANNATT WILLIAM) EDGEWORTH (1858-1934), geologist; educated at Magdalen School and New College, Oxford, and Normal School of Science and Technology; assistant geological surveyor to government of New South Wales, 1882-91; professor of geology, Sydney, 1891-1924; led expedition to Ellice Islands, 1897; FRS, 1900; with (Sir) Douglas Mawson in Shackleton expedition reached and recorded south magnetic pole, 1909; in 1914-18 war raised mining battalion of Australian Tunnellers and responsible for mining under Messines ridge; DSO, 1918; geologist to British armies, 1918-19; KBE, 1920; thereafter engaged in compiling his *Geology of the Commonwealth of Australia*.

DAVIDS, THOMAS WILLIAM RHYS (1843-1922), oriental scholar; Ceylon civil servant, 1866-76; studied Pali and early Buddhism; founded Pali Text Society, 1881; honorary professor of Pali and Buddhist literature, University College, London, 1882-1912; professor of comparative religion, Manchester, 1904-15; works include *Buddhism* (1878) and important Dīgha Nikāya volumes (1886-1921); responsible for a Pali dictionary, issued 1921-5.

DAVIDSON, ANDREW BRUCE (1831-1902), Hebraist and theologian; MA, Marischal University, Aberdeen, 1849; assistant to John Duncan (1796-1870, q.v.), 1858, whom he succeeded as professor of Hebrew and oriental languages, New College, Edinburgh, 1863-1902; stimulating teacher; member of Old Testament revision company, 1870-84; received hon. degrees from Aberdeen, Edinburgh, Glasgow, and Cambridge; made lifelong research into language, historical exegesis, and theology of Old Testament; published works on Hebrew grammar (1874) and syntax (1894); commentaries on 'Job' (1884), 'Ezekiel' (1892), and 'Nahum, Habakkuk, Zephaniah' (1896), in

Cambridge Bible; *Biblical and Literary Essays* (1902); and sermons.

DAVIDSON, CHARLES (1824–1902), water-colour painter; member of New (1849–53) and Old (1858) Water-Colour Societies, where he exhibited over 800 works, mainly typical English landscapes.

DAVIDSON, JAMES LEIGH STRACHAN- (1843–1916), classical scholar. [See STRACHAN-DAVIDSON.]

DAVIDSON, JOHN (1857–1909), poet; schoolmaster in Scotland from 1872 to 1889; published *Scaramouch in Naxos* (1889) and other plays before settling in London, 1889; published *Perfervid*, a novel (1890), *Fleet Street Eclogues* (1893), which proved his genuine poetic gift, *Ballads and Songs* (1894), his most popular work, *Fleet Street Eclogues* (2nd ser. 1896), and *New Ballads* (1897); abandoned lyric for drama, writing original plays and translating foreign ones; finally (1901–8), wrote series of 'Testaments' expounding a materialistic and aristocratic philosophy; awarded civil list pension, 1906; committed suicide by drowning at Penzance.

DAVIDSON, JOHN COLIN CAMPBELL, first VISCOUNT DAVIDSON (1889–1970), politician; educated at Westminster School and Pembroke College, Cambridge; called to bar (Middle Temple), 1913; private secretary to colonial secretaries, Lord Crewe, Lewis (later Viscount) Harcourt [q.q.v.], and Bonar Law, 1910–16; went with Bonar Law to Treasury, 1916; friend of Stanley (later Earl) Baldwin; drafted 'coupon' for 1918 general election; conservative MP, Hemel Hempstead, 1920–3; parliamentary private secretary to Bonar Law, and later, to Stanley Baldwin; used influence to end coalition government and ensure return of Bonar Law, 1922 supported Baldwin for succession to Bonar Law; chancellor of Duchy of Lancaster; re-elected for Hemel Hempstead, 1924–37; parliamentary and financial secretary to the Admiralty; chairman, conservative party, 1926–30; PC, 1928; chancellor of Duchy of Lancaster again, 1931; chairman, Indian States Inquiry Committee, 1932; member, Joint Select Committee, 1933, leading to Government of India Act, 1935; viscount, 1937; served in Ministry of Information, 1940–1; CB, 1919; CH, 1923; GCVO, 1935.

DAVIDSON, SIR JOHN HUMPHREY (1876–1954), major-general; born in Mauritius; educated at Harrow and Sandhurst; joined King's Royal Rifle Corps, 1896; in South Africa, 1899–1902; DSO; instructor, training and tactics, Staff College, 1912–14; with III Corps in France, 1914; operations officer to Sir Douglas (later Earl) Haig [q.v.], 1915; director of military operations, Dec. 1915–18; major-general, 1918; CB, 1917; KCMG, 1919; conservative MP, Fareham, 1918–31; published *Haig: Master of the Field* (1953).

DAVIDSON, JOHN THAIN (1833–1904), Presbyterian minister; minister at Salford, 1859–62, and Islington, 1862–91; inaugurated in 1868, and continued till 1891, Sunday afternoon services for non-church-going people at Agricultural Hall; moderator of synod of Presbyterian Church of England, 1872; a powerful preacher; published several volumes of addresses.

DAVIDSON, RANDALL THOMAS, BARON DAVIDSON OF LAMBETH (1848–1930), archbishop of Canterbury; born of Presbyterian parentage; educated at Harrow and Trinity College, Oxford; deacon, 1874; priest, 1875; resident chaplain to Archbishop Tait [q.v.], 1877–82; dean of Windsor and Queen Victoria's domestic chaplain, 1883; on confidential terms with both Queen and Archbishop Benson [q.v.]; bishop of Rochester, 1891–5; published life of Archbishop Tait, 1891; bishop of Winchester, 1895–1903; actively engaged in ritual crisis in the Church, 1898–1901; spoke frequently in House of Lords on social questions; archbishop of Canterbury, 1903; his task during twenty-five years' primacy was to maintain unity and vindicate comprehensiveness of Church of England and to strengthen her witness in life of nation; member of royal commission on ecclesiastical discipline, 1904; first step towards Prayer Book revision taken when, as result, letters of business issued to Convocations, 1906; opposed education bill (1906) and Deceased Wife's Sister Marriage Act (1907); presided over fifth Lambeth Conference of bishops of Anglican communion, 1908; gave evidence of wide knowledge of missionary problems and deep interest in development of Anglican Church abroad; attempted to promote Anglo-German friendship, 1909, 1912; crowned King George V and played full part in parliament bill controversy, 1911; resisted legislation disestablishing and disendowing Welsh Church; as result of demands from Bishop Gore [q.v.] and others for new declarations of Church's faith for purpose of denouncing modernism, persuaded Convocation to pass (1914) important resolution, to which, in 1918, Gore appealed; as result of appeal of Frank Weston [q.v.], bishop of Zanzibar, called upon to deal with Kikuyu controversy, 1914; his answer, although attacked by both sides, upheld comprehensiveness of Anglican Church, 1915; exercised steadying influence throughout 1914–18 war, firmly supporting allied cause, but refraining from anti-German rhetoric; protested against use by British troops of poison gas and air reprisals; authorized National Mission of Repentance and Hope, 1916; gave public support to inauguration of League of Nations; warmly supported Church of England Assembly (Powers) Act, 1919; presided over sixth Lambeth Conference, which appealed for reunion of Christendom on wider basis than had previously been conceived, 1920; trusted and admired by Free Churchmen; drew much closer relations between Anglican Church and Orthodox Churches; protested against religious persecution by Soviet government, 1922-3; gave friendly cognizance to Malines conversations, 1921-6; gave great lead to forces of peace by his appeal during general strike, 1926; deeply disappointed by failure of Prayer Book revision, 1927-8; resigned archbishopric and created baron, 1928; an able administrator, who in general policy pursued middle course.

DAVIE, THOMAS BENJAMIN (1895-1955), pathologist, teacher, vice-chancellor; born in Cape Colony; educated at Stellenbosch University; honours in science, 1914; teachers' diploma, 1916; first class in medicine, Liverpool, 1928; MRCP and MD, 1931; junior lecturer (1929-31), senior lecturer (1933-5) in pathology, Liverpool; appointed professor of pathology, Bristol, 1935, Liverpool, 1938; of applied pathology and medical dean, 1945; organized blood-transfusion services, 1939-45; principal and vice-chancellor, Cape Town University, 1948-55; inspiring protagonist of academic freedom; published *Textbook of Pathology* (with J. H. Dible, 1939); FRCP, 1940.

DAVIES, CHARLES MAURICE (1828-1910), author; MA, Durham University, 1852; DD, 1864, fellow, 1849; published novels attacking high-church practices; articles in *Daily Telegraph* republished as *Unorthodox London* (1873), *Orthodox London* (2 vols., 1874-5), and *Mystic London* (1875).

DAVIES, CLEMENT EDWARD (1884-1962), lawyer and politician; educated at Llanfyllin county school and Trinity Hall, Cambridge; first class, parts i and ii, law tripos, 1906-7; called to bar (Lincoln's Inn), 1909; successful commercial practice in London, 1910-14; adviser to procurator-general on enemy activities in neutral countries and on the high seas, 1914; seconded to Board of Trade; secretary to president, Probate, Divorce, and Admiralty division, 1918-19, and to the master of the Rolls, 1919-23; junior Treasury counsel, 1919-25; KC, 1926; chairman, Montgomery-

shire quarter-sessions, 1935-62; on board of Unilever, 1930-41; liberal MP, Montgomeryshire, 1929-62; reported on provision of public health services and housing, 1939; supported Churchill for prime minister, 1940; leader of liberal party, 1945-56; PC, 1947; hon. fellow, Trinity Hall, 1950; bencher, Lincoln's Inn, 1953; hon. LLD, university of Wales, 1955.

DAVIES, DAVID, first BARON DAVIES (1880-1944), public benefactor; inherited Welsh industrial properties and gave much public service to Wales; founded King Edward VII Welsh National Memorial Association (1911) to combat tuberculosis and provided administrative centre for it and Welsh League of Nations Union in Temple of Peace and Health, Cardiff (1938); liberal MP, Montgomeryshire, 1906-29; parliamentary private secretary to Lloyd George, 1916-17; founded New Commonwealth movement advocating international police force, 1932; baron, 1932.

DAVIES, SIR (HENRY) WALFORD (1869-1941), musician; scholar of Royal College of Music; organist, Temple church, 1898-1923; professor of music, Aberystwyth (1919-26) and chairman, National Council of Music for university of Wales (1919-41): Gresham professor of music, London, 1924-41; organist, St. George's chapel, Windsor, 1927-32; master of the king's musick, 1934-41; broadcast very successful courses of lectures; compositions include 'Everyman' (cantata, 1904), 'Solemn Melody' (organ and strings) and 'God be in my head' (1908), and 'Song of St. Francis' (1912); knighted 1922; KCVO, 1937.

DAVIES, JOHN LLEWELYN (1826-1916), theologian; BA, Trinity College, Cambridge (bracketed fifth in classical tripos), 1848; fellow of Trinity, 1850; ordained, 1851; influenced by F. D. Maurice [q.v.]; rector of Christ Church, Marylebone, 1856-89; vicar of Kirkby Lonsdale, Westmorland, 1889-1908; promoter of movements for education of women and working men; liberal in politics; broad churchman; helped to found National Church Reform Union, 1870; noted preacher; author of numerous theological works.

DAVIES, ROBERT (1816-1905), philanthropist; was put by father in charge of foundry at Carnarvon; with brothers founded successful shipowning firm at Menai Bridge; made large anonymous benefactions to Calvinistic Methodist chapels, to Welsh Methodist Mission in India, and other institutions; his younger brother RICHARD (1818-96) was liberal MP, Anglesey, 1868-86.

DAVIES, (SARAH) EMILY (1830-1921), promoter of women's education; sister of John

Llewelyn Davies [q.v.]; with others organized college for women opened at Hitchin, 1869, and transferred to Cambridge (Girton College), 1873; honorary secretary, 1867-1904; mistress, 1873-5; pioneer in woman's suffrage movement.

DAVIES, WILLIAM HENRY (1871-1940), poet and author; tramped in United States for several years and later peddled in England; *The Soul's Destroyer, and other poems* (1905) brought influential help from critics including G. B. Shaw [q.v.] who wrote preface for *Autobiography of a Super-Tramp* (1908); further autobiography appeared in two novels and *Later Days* (1925); his *Collected Poems* (1943) contains 636 pieces of effortless ease, delicacy, and perfection which put him with the happiest of the Elizabethans; awarded civil list pension, 1919.

DAVIES, WILLIAM JOHN ABBOTT (1890-1967), rugby footballer; educated at Pembroke Dock grammar school, the Royal Naval College, Keyham, and the Royal Naval College, Greenwich; played rugby with United Services Club, Portsmouth, 1910; capped for England at fly-half, 1913; served in the *Iron Duke* and *Queen Elizabeth* during 1914-18 war; assistant constructor, Portsmouth; OBE, 1919; played in partnership at half-back with C. A. Kershaw for United Services, Portsmouth and England, 1919-23; never appeared in a losing England side in the international championship; won 22 caps; captained Royal Navy and Hampshire; England selector, 1923-6; attached to staff of commander-in-chief, Mediterranean Fleet, 1935-8; chief constructor, 1939; assistant director, warship production, 1942; superintendent, warship production on the Clyde, 1946; director, merchant shipbuilding and repairs, 1949-50; president, Civil Service Football Club, 1937-66; wrote *Rugby Football* (1923) and *How to Play Rugby Football* (1933).

DAVIES, SIR WILLIAM (LLEWELYN) (1887-1952), librarian; educated at Portmadoc county school and Aberystwyth; BA, 1909 (honours in Welsh); MA, 1912; assistant librarian, National Library of Wales, 1919; librarian, 1930-52; acquired for preservation over 3 million documents relating to Wales; used every opportunity to bring library into contact with the Welsh people and was leading spirit in organizations promoting Welsh intellectual life; knighted, 1944; hon. LLD university of Wales.

DAVIS, CHARLES EDWARD (1827-1902), architect and antiquary; son of a Bath architect; Bath city architect and surveyor, 1863-1902;

discovered thermal baths (1869), well (1877-8), Great bath (1880-1), and Circular bath (1884-6), all Roman; his reconstruction of the Queen's bath (1886-9) evoked much criticism; had extensive private practice; FSA, 1850; published *Mineral Baths of Bath*, 1883.

DAVIS, HENRY WILLIAM CARLESS (1874-1928), historian, and editor of the *Dictionary of National Biography* from 1919 to 1928; BA, Balliol College, Oxford, 1895; fellow of All Souls College, 1895-1902; history lecturer at New College, 1897; at Balliol College, 1899; fellow of Balliol, 1902; helped to organize 'trade clearing house', which expanded into 'war trade intelligence department', 1915; vice-chairman of department, 1915-19; returned to Oxford and became editor of *Dictionary of National Biography*, 1919; professor of modern history, Manchester, 1921; Ford's lecturer at Oxford, 1924-5; regius professor of modern history at Oxford and fellow of Oriel College, 1925; CBE, 1919; FBA, 1925; an influential teacher; works include *Charlemagne* (1900), *England under the Normans and Angevins* (1905), *Medieval Europe* (1911), *Regesta Regum Anglo-Normannorum* (vol. i, 1913); *The Age of Grey and Peel* (Ford's lectures, 1930); responsible for 1912-1921 volume of this Dictionary (1927).

DAVITT, MICHAEL (1846-1906), Irish revolutionary and labour agitator; of Roman Catholic peasant stock; joined Fenians, 1865; organizing secretary of Irish Republican Brotherhood, 1868; sentenced to fifteen years' penal servitude for treason-felony, 1870; released, 1877; met Henry George in America, 1878; introduced land agitation into Irish movement, and founded Land League of Ireland, 1879; arrested, tried, convicted, and released, 1880; organized American Land League and Ladies' Land League, 1880; imprisoned at Portland, 1881-2; elected MP for county Meath while a prisoner, Feb. 1882; induced C. S. Parnell [q.v.] to found National League after suppression of Land League, June 1882; imprisoned for sedition, Jan.-May 1883; with Henry George advocated land nationalization, 1882-5; prominent as a respondent in Parnell commission which examined charges, brought by *The Times*, that he intended to bring about by violence complete independence of Ireland; made five days' speech (24-31 Oct. 1889), published as *The Defence of the Land League*, 1891; edited the *Labour World*, organ of British labour movement (1890-1); anti-Parnellite MP for North Meath, 1892, and for South Mayo, 1895-9; helped to found United Irish League, 1898; attacked Land Purchase Act, 1903; visited Russia to show sympathy with revolutionary party, 1903-5; stood for reconciliation of extreme with constitutional nationalism, and of democracy with nationality; a collectivist and secularist; 'Davitt memorial church' erected at Straide; author of *Leaves from a Prison Diary* (1884) and *The Fall of Feudalism in Ireland* (1904).

DAWBER, SIR (EDWARD) GUY (1861-1938), architect; worked under Sir Thomas Deane and Sir Ernest George [qq.v.]; practised in Cotswolds and from 1891 in London designing mainly country houses; works include Eyford Park, Gloucestershire, Ashley Chase, Dorset, the Foord Almshouses, Rochester; prominent in establishing (1926) Council for Preservation of Rural England; ARA, 1927; RA, 1935; knighted, 1936.

DAWKINS, RICHARD McGILLIVRAY (1871-1955), scholar; educated at Totnes grammar school, Marlborough College, King's College, London, and Emmanuel College, Cambridge; first class, parts i and ii, classical tripos, 1901-2; fellow, 1904; honorary fellow, 1922; director, British School of Archaeology, Athens, 1906-14; Bywater and Sotheby professor of Byzantine and modern Greek, Oxford, 1920-39; fellow of Exeter, 1922; honorary fellow, 1939; FBA, 1933; works include *Modern Greek in Asia Minor* (1916) and translation with commentary of *Chronicle of Makhairas* (2 vols., 1932).

DAWKINS, SIR WILLIAM BOYD (1837-1929), geologist, palaeontologist, and antiquary; BA, Jesus College, Oxford, 1860; member, Geological Survey of Great Britain, 1861-9; curator, Manchester museum, 1869; first professor of geology, Owens College (afterwards Manchester University), 1874-1908; excavated caves in order to carry out researches into contemporaneity of man with extinct mammals; FRS, 1867; knighted, 1919; works include *Cave Hunting* (1874) and *Early Man in Britain and his Place in the Tertiary Period* (1880).

DAWSON, BERTRAND EDWARD, VISCOUNT DAWSON OF PENN (1864-1945), physician; educated at St. Paul's School and University College, London; B.Sc., 1888; qualified at London Hospital, 1890; assistant physician, 1896; physician, 1906-45; consulting physician in France, 1914-19; considered it duty of medical profession to promote national health; chairman, Ministry of Health consultative council on medical services (1919-20) whose report foreshadowed national health service; member of British Medical Association planning commission on whose work the white paper of 1944 was to some extent based; spokesman of medical profession in House of Lords after 1920; chairman, Army Medical Advisory Board, 1936-45; president, Royal Society of Medicine (1928-30), British Medical Association (1932 and 1943), Royal College of Physicians (1931-8); physician-in-ordinary to King George V, 1914-36; instrumental in saving his life, 1928; attended him in his last illness; KCVO, 1911; GCVO, 1917; baron, 1920; PC, 1929; viscount, 1936.

DAWSON, (GEORGE) GEOFFREY (1874-1944), twice editor of *The Times*; born ROBINSON; assumed name of Dawson, 1917; educated at Eton and Magdalen College, Oxford; first class, *lit. hum.*, 1897; fellow of All Souls, 1898-1944; entered Colonial Office, 1899; served under Lord Milner [q.v.] in South Africa, 1901-5; editor, Johannesburg *Star*, 1905-10; joined *The Times*, 1911; editor, 1912-19, 1923-41; especially intimate with Baldwin, whose Indian policy he supported, as also Neville Chamberlain's policy of appeasing Germany.

DAWSON, GEORGE MERCER (1849-1901), geologist; son of Sir John William Dawson [q.v.]; born in Nova Scotia; as geologist and botanist to Canadian-American boundary commission formed large natural history collection, 1873-5; appointed to Canadian Geological Survey; made scientific researches in North-West and British Columbia; director of Survey, 1895; CMG and FRS, 1891; president of Royal Society of Canada, 1894; collaborated in *Comparative Vocabularies of the Indian Tribes of British Columbia*, 1884.

DAWSON, JOHN (1827-1903), trainer of racehorses; brother of Matthew Dawson [q.v.]; settled at Newmarket, 1861; trainer to Prince Batthyany, Sir Robert Jardine [q.v.], and others; trained winners of Cesarewitch (1878), Derby (1875), St. Leger (1876), Two Thousand Guineas (1876 and 1898); trained for a time King Edward VII's Perdita II.

DAWTRY, FRANK DALMENY (1902-1968), secretary of the National Association of Probation Officers; educated at Sheffield central secondary school; worked in Sheffield steelworks, 1917; unemployed, 1921-3; appointed clerk/book-keeper, Sheffield Council of Social Service, 1923; secretary, Personal Services Committee, 1927; secretary, Wakefield Discharged Prisoners' Aid Society, 1927; resident DPA secretary and welfare officer, HM Prison, Wakefield, 1937; organizer, Maidstone Prison, 1944-6; secretary, National Council for the Abolition of the Death Penalty, 1946; general secretary, National Association of Probation Officers, 1948-67; active member of number of committees concerned with penal reform, welfare of offenders, and pacifism; fellow, Royal

Society of Arts, MA *honoris causa* (Leeds), 1963; MBE 1957; OBE, 1967.

DAY, SIR JOHN CHARLES FREDERIC SIGISMUND (1826-1908), judge; of Roman Catholic parentage; called to bar, 1849; QC, 1872; treasurer of Middle Temple, 1896; as editor of Roscoe's *Evidence at Nisi Prius* and of annotated edition of the Common Law Procedure Act of 1852, became authority on new methods of pleading and practice; successful in breach of promise and libel cases; judge of Queen's Bench division, 1882-1901; knighted, 1882; stern criminal judge; chairman of royal commission to inquire into Belfast riots, 1886, and member of Parnell commission, 1888; PC, 1901; discriminating art collector.

DAY, LEWIS FOREMAN (1845-1910), decorative artist; considerably influenced contemporary ornament; lecturer at Royal Society of Arts and Royal College of Art, South Kensington; art examiner and adviser to Board of Education from 1890; publications include *Windows* (1897), *Lettering in Ornament* (1902), *Nature and Ornament* (2 vols., 1908-9).

DAY, WILLIAM HENRY (1823-1908), trainer and breeder of racehorses; first successful as jockey; trained winners of Two Thousand Guineas (1855 and 1859), Brigantine (winner of Oaks and Ascot Cup, 1869), and Foxhall (winner of Grand Prix, Cesarewitch, and Cambridgeshire, 1881, and Ascot Cup, 1882); formed breeding stud at Alvediston near Salisbury, 1873; lost fortune by land speculation; published *The Racehorse in Training* (1880) and *Reminiscences* (1886).

DEACON, GEORGE FREDERICK (1843-1909), civil engineer; assisted Cromwell Varley [q.v.] in laying second Atlantic cable, 1865; consulting engineer at Liverpool (1865-71); as borough engineer (1871-80) laid inner circle tramway rails (1877) and introduced wood paving there; as water engineer (1871-90) invented and introduced Deacon waste water meter, 1873; designed masonry dam in Vyrnwy valley, to supply water for Liverpool by means of aqueduct 76 miles long; waterworks opened, 1892; work carried out on scientific and aesthetic lines; constructed waterworks for Merthyr Tydfil and other towns from 1890; reported on London water-supply, 1897; hon. LLD, Glasgow, 1902.

DEAKIN, ALFRED (1856-1919), Australian politician; born at Melbourne; admitted to Victorian bar, 1877; liberal MP, West Bourke, 1879, but resigned immediately; re-elected, 1880; minister of water-supply and commis-

sioner of public works in Berry-Service coalition, 1883; also solicitor-general, 1883; chief secretary and minister of water-supply in coalition with Duncan Gillies [q.v.], 1886-90; visited England as representative of Victoria at colonial conference, 1887; promoted federation movement in Victoria, 1891-8; played important part in discussions of constitution bill in London, 1900; attorney-general in first Commonwealth ministry, 1901; prime minister, 1903-4, 1905-8, and 1909-10.

DEAKIN, ARTHUR (1890-1955), trade-union leader; of illegitimate birth; became a steel worker in Wales; appointed official of Dock, Wharf, Riverside, and General Workers' Union, 1919; assistant district secretary, North Wales area, Transport and General Workers' Union, 1922; national secretary for General Workers' group of the union, 1932-5; assistant general secretary under and loyal supporter of Ernest Bevin [q.v.], 1935; acting general secretary, 1940-6; general secretary, 1946-55; emerged from Bevin's shadow to become the dominant trade-union figure; supported government policies on productivity, profits, and wage restraint; became strongly anti-communist; helped to form International Confederation of Free Trade Unions and persuaded his union to ban communists from office, 1949; twice refused knighthood; CH, 1949; PC, 1954.

DEANE, SIR JAMES PARKER (1812-1902), judge; educated at Winchester and St. John's College, Oxford; BCL, 1834; DCL, 1839; called to bar (Inner Temple), 1841; QC, 1858; treasurer, 1878; obtained large practice in probate and divorce and ecclesiastical courts; vicar-general of province of Canterbury, 1872; legal adviser to Foreign Office, 1872-86; PC and knighted, 1885.

DEARMER, PERCY (1867-1936), divine; educated at Christ Church, Oxford; deacon, 1891; priest, 1892; secretary, London Christian Social Union, 1891-1912; vicar of St. Mary's, Primrose Hill, 1901-15; first professor of ecclesiastical art, King's College, London, 1919-36; canon of Westminster, 1931; sought to improve standards of art and music in public worship; largely responsible for editing of *English Hymnal* (1906) and *Songs of Praise* (1925); chairman, League of Arts, 1921-36; hon. ARIBA.

DEBENHAM, FRANK (1883-1965), geographer; born in New South Wales; educated at King's School, Parramatta, and Sydney University; joined Antarctic expedition led by Captain Robert Scott [q.v.], as geologist and cartographer, 1910; served in Oxfordshire and Buckinghamshire Light Infantry during

1914-18 war; Royal Geographical Society lecturer, Cambridge, 1919; fellow, Gonville and Caius College, 1920; tutor, 1923-8; director, Scott Polar Research Institute, 1926-46; reader, geography department, Cambridge, 1927; first professor of geography, 1931-49; published numerous papers and books, including *The Polar Regions* (1930), *In the Antarctic* (1952), and *Antarctica; the Story of a Continent* (1959); edited the *Polar Record*; OBE, 1919; hon. fellow, Royal Geographical Society, 1965; hon. degrees from Sydney, Perth, and Durham.

DE BUNSEN, Sir MAURICE WILLIAM ERNEST, baronet (1852-1932), diplomatist; educated at Rugby and Christ Church, Oxford; entered diplomatic service, 1877; first secretary, Paris, 1902-5; minister, Lisbon, 1905; ambassador, Madrid, 1906-13; unofficial mediator in Franco-Spanish dispute over Morocco, 1911-12; ambassador, Vienna, 1913-14; headed mission to South America, 1918; KCVO, 1905; PC, 1906; baronet, 1919.

DE BURGH, WILLIAM GEORGE (1866-1943), philosopher; educated at Winchester and Merton College, Oxford; first class, *lit. hum.*, 1889; lecturer in classics and philosophy, Reading, 1896; professor of philosophy, 1907-34; guided development of college into a university; publications include *The Legacy of the Ancient World* (1924); FBA, 1938.

DE BURGH CANNING, HUBERT GEORGE, second MARQUESS and fifteenth EARL OF CLANRICARDE (1832-1916), Irish landed proprietor. [See BURGH CANNING.]

DE CHAIR, Sir DUDLEY RAWSON STRATFORD (1864-1958), admiral; nephew of Sir Harry H. Rawson [q.v.]; born in Canada; entered *Britannia*, 1878; captain, 1902; assistant controller to Sir John (later Earl) Jellicoe [q.v.], 1908-11; rear-admiral, 1912; naval secretary to first lord, 1913-14; commanded tenth cruiser squadron, 1914-16, patrolling North Sea; KCB, 1916; naval adviser, Ministry of Blockade, 1916-17; commanded third battle squadron in Channel, 1917-18; vice-admiral, 1917; protested at dismissal of Jellicoe; commanded Coastguard and Reserves, 1918-21; admiral, 1920; president, inter-allied commission on enemy warships, 1921-3; governor, New South Wales, 1923-30; KCB, 1916; KCMG, 1933; hon. LLD McGill University.

DEEDES, Sir WYNDHAM HENRY (1883-1956), soldier and social worker; educated at Eton; commissioned in King's Royal Rifle Corps, 1901; served in South African war; in Malta, 1908-10; seconded to Turkish *gendarmerie*, 1910-14; served in intelligence, London, Gallipoli, and Cairo, 1914-18; major and DSO, 1916; brigadier-general, 1918; civil secretary in Palestine, 1920-3; CMG, 1919; knighted, 1921; settled in Bethnal Green; supporter of Zionist movement, Turkish Centre, and National Council of Social Service.

DE FERRANTI, SEBASTIAN ZIANI (1864-1930), electrical engineer and inventor. [See FERRANTI.]

DE HAVILLAND, Sir GEOFFREY (1882-1965), aircraft designer and manufacturer; educated at St. Edward's School, Oxford; trained in mechanical engineering at Crystal Palace Engineering School, 1900-3; apprenticed with Williams and Robinson, Rugby; draughtsman with Wolseley Tool and Motor Car Co., Birmingham, 1905; designed his own aeroplane and taught himself to fly, 1908-10; designed important range of aeroplanes, 1912-14; designed and flew military aircraft during 1914-18 war; founded de Havilland Aircraft Co., 1920; pioneered the Moth (flown by Amy Johnson [q.v.] and other long-distance flyers); de Havilland Comet won MacRobertson International England to Australia race, 1934; produced the Mosquito during 1939-45 war; designed the jet Vampire fighter; built the Comet jet airliner, 1949; OBE, 1918; Air Force Cross, 1919; CBE, 1934; KB, 1944; OM, 1962; council member, Fauna Preservation Society.

DE HAVILLAND, GEOFFREY RAOUL (1910-1946), test pilot; educated at Stowe School; entered father's aircraft company, 1928; taught flying, 1932-5; became test pilot, 1935; chief test pilot, 1937; tested the Flamingo (1939), Mosquito (1940), Vampire (1943), and Hornet (1944); killed testing a jet aircraft.

DELAFIELD, E. M. (pseudonym) (1890-1943), authoress. [See DASHWOOD, EDMÉE ELIZABETH MONICA.]

DE LA MARE, WALTER JOHN (1873-1956), poet, novelist, and anthologist; educated at St. Paul's Cathedral Choristers' School; worked until 1908 for Anglo-American Oil Co.; his *Songs of Childhood* (1902) followed by a large output of poems, stories, novels, books for children, and anthologies; among the best known are *The Return* (1910), *The Listeners* (1912), *Peacock Pie* (1913), *Memoirs of a Midget* (1921), *Behold, this Dreamer* (1939), and *The Traveller* (1946); the poetry which irradiated all his work was built upon a questing disruptive approach to the certainties of life, oblique, tentative, and sometimes bizarre; CH, 1948; OM, 1953; buried in St. Paul's.

DELAMERE, third BARON (1870-1931), pioneer settler in Kenya. [See CHOLMONDELEY, HUGH.]

DE LA RAMÉE, MARIE LOUISE (1839-1908), novelist under the pseudonym of OUIDA; introduced to W. Harrison Ainsworth [q.v.], 1859, who published in *Bentley's Miscellany* seventeen short tales by her, 1859-60; her forty-five novels, which chiefly deal with military and fashionable life, include *Under Two Flags* (1867), *Puck* (1870), *Two Little Wooden Shoes* (1874), *Moths* (1880), and *Bimbi, Stories for Children* (1882); many were translated into foreign languages and successfully dramatized; settled permanently in Florence, 1874, where she entertained expensively; lived from 1894 in poverty at Sant' Alessio; was awarded civil list pension, 1904; died at Viareggio; cynical, artificial in manner, and quick at repartee, she always wrote sympathetically of Italian peasants and dogs; in her later works, *Views and Opinions* (1895) and *Critical Studies* (1905), she opposed militarism, women's suffrage, and vivisection.

DE LA RUE, SIR THOMAS ANDROS, first baronet (1849-1911), printer; son of Warren De la Rue [q.v.]; joined father's firm, c.1871; as head (1889-96) increased firm's reputation for artistic production of English and foreign postage stamps; baronet, 1898.

DE LÁSZLÓ, PHILIP ALEXIUS (1869-1937), painter. [See LÁSZLÓ DE LOMBOS.]

DELEVINGNE, SIR MALCOLM (1868-1950), civil servant and reformer; first class, *lit. hum.*, Trinity College, Oxford, 1891; entered Home Office; deputy under-secretary of state, 1922-32; shaped much industrial legislation and authority on control of dangerous drugs; British representative, labour commission, peace conference, 1919; took large part in establishing International Labour Office; chairman, Safety in Mines Research Board, 1939-47; KCB, 1919; KCVO, 1932.

DELIUS, FREDERICK (1862-1934), musician; son of wool merchant of German origin; educated at Bradford grammar school and International College, Isleworth; his parents opposing a musical career, he went to plant oranges in Florida where he learnt and then taught pianoforte and violin; studied at Leipzig Conservatorium and met Grieg who dispelled parental opposition; devoted himself to composing, living in Paris, later at Grez-sur-Loing; established as a composer in Germany; his later popular success in England mainly due to Sir Thomas Beecham who directed recording

of main works; these include operas: *Koanga* and *A Village Romeo and Juliet*; choral works: 'Sea Drift', 'A Mass of Life', 'Song of the High Hills', 'Songs of Farewell'; tone poem: 'Over the Hills and Far Away'; and rhapsody 'Brigg Fair'; work inspired by beauties of nature and human emotions but criticized for lack of 'form' and monotony of texture; lost sight and use of limbs, 1924; CH, 1929.

DELL, ETHEL MARY (1881-1939), novelist. [See SAVAGE.]

DELLER, SIR EDWIN (1883-1936), principal of London University; became clerk at fourteen; in service of Kent education committee; LLB, London, 1911; a secretary in academic department (1912), academic registrar (1921), principal (1929-36) of London University; knighted, 1935.

DE MONTMORENCY, JAMES EDWARD GEOFFREY (1866-1934), legal scholar; LLB, Peterhouse, Cambridge, 1890; called to bar (Middle Temple), 1892; Quain professor of comparative law, London (1920-32), and dean of faculty of laws (1930-2); a prolific writer.

DE MONTMORENCY, RAYMOND HARVEY, third VISCOUNT FRANKFORT DE MONTMORENCY (1835-1902), major-general; joined army, 1854; served in Crimea; recommended for VC; in Indian Mutiny, 1857-8; in Abyssinian expedition, 1867; commanded frontier force in Sudan, 1886-7; directed British field column during operations on Nile, 1887; major-general, 1889; succeeded to peerage, 1889.

DE MORGAN, WILLIAM FREND (1839-1917), artist, inventor, and author; son of Augustus de Morgan [q.v.]; educated at University College School and University College, London; entered Academy Schools, 1859; early made acquaintance of pre-Raphaelite circle; experimented in manufacture of stained glass and tiles; established pottery industry in Chelsea, 1871; rediscovered process of making coloured lustres; joined William Morris [q.v.] at Merton Abbey, 1882-8; erected factory at Fulham, 1888; retired, 1905, and firm dissolved, 1907; wintered in Florence, 1890-1914; ware employed for decorative panels in steamships, etc.; published novels in later life: *Joseph Vance* (1906), *Alice-for-Short* (1907), *Somehow Good* (1908), *It Never Can Happen Again* (1909), *An Affair of Dishonour* (1910), *A Likely Story* (1911), *When Ghost Meets Ghost* (1914); two unfinished stories published posthumously.

DEMPSEY, SIR MILES CHRISTOPHER (1896-1969), general; educated at Shrewsbury and Sandhurst; commissioned into Royal Berkshire Regiment, 1915; served in France, 1916-18; MC; in Iraq, 1919-20; served with his regiment and as staff officer, 1920-39; commanded 13th Infantry brigade in France, 1940; DSO; lieutenant-general, in command of XIII Corps, Eighth Army in North Africa and Italy, 1943; commanded Second Army in invasion of Europe, 1944; KCB, 1944; received surrender of Hamburg, 1945; commanded Fourteenth Army in succession to Sir William (later Viscount) Slim [q.v.]; commander-in-chief, Allied Land Forces, South East Asia; KBE, 1945; general, 1946; commander-in-chief, Middle East, 1946-7; director of number of companies; chairman, Racehorse Betting Control Board, 1947-51; deputy lieutenant, Berkshire, 1950; commander-in-chief (designate) UK Land Forces, 1951-6; GBE, 1956.

DENMAN, GERTRUDE MARY, LADY DENMAN (1884-1954), public servant; daughter of W. D. Pearson, first Viscount Cowdray [q.v.]; privately educated; married third Baron Denman, 1903; one son, one daughter; chairman, National Federation of Women's Institutes and largely responsible for the movement's success, 1917-46; chairman, Family Planning Association, 1930-54; of Cowdray Club, 1932-53; president, Ladies' Golf Union, 1932-8; director, Women's Land Army, 1939-45; used her talent for organizing with good sense, impartiality, and humour; CBE, 1920; DBE, 1933; GBE, 1951.

DENNEY, JAMES (1856-1917), theologian; MA, Glasgow; studied theology at Glasgow Free Church College; professor of systematic and pastoral theology at the college, 1897; professor of New Testament language, literature, and theology, 1899; principal, 1915-17; noted expository preacher; worked for reunion of Free Church with established Church of Scotland; best known for writings on doctrines of the person and work of Christ.

DENNISTON, ALEXANDER GUTHRIE (ALASTAIR) (1881-1961), public servant (intelligence); educated at Bowdon College, Cheshire, and at the universities of Bonn and La Sorbonne; taught at Merchiston Castle School, 1906-9; taught languages at Osborne; played hockey for Scotland; played leading part in setting up Room 40 OB, the establishment responsible for decoding enemy communications during 1914-18 war; head of Government Code and Cypher School, 1919-42; the School (at Bletchley Park) mastered the German Enigma cypher machine, 1941; when organization divided into military and civil wing, 1942,

appointed head of civil wing, 1942-5; returned to teaching (at a Leatherhead preparatory school), 1945; CBE, 1933; CMG, 1941.

DENNISTON, JOHN DEWAR (1887-1949), classical scholar; scholar of Winchester and New College, Oxford; second class, *lit. hum.*, 1910; fellow of Hertford, 1913-49; publications include *Greek Literary Criticism* (1924) and *The Greek Particles* (1934), and editions of Cicero's *Philippics I and II* (1926) and Euripides' *Electra* (1939); joint-editor, *Oxford Classical Dictionary* (1949); FBA, 1937.

DENNY, SIR ARCHIBALD, first baronet (1860-1936), shipbuilder and engineer; educated at Dumbarton Academy, Lausanne, and Royal Naval College, Greenwich; partner in family shipbuilding and engineering firm, 1883; designer and builder of high-speed passenger (notably cross-Channel) vessels; baronet, 1913.

DENNY, SIR MAURICE EDWARD, second baronet, of Dumbarton (1886-1955), engineer and shipbuilder; son of Sir Archibald Denny [q.v.]; succeeded him, 1936; educated at Tonbridge School, Lausanne and Heidelberg universities, and Massachusetts Institute of Technology; first class, naval architecture; became a partner in family firm, 1911, a director, 1918, vice-chairman, 1920, chairman, 1922-52; promoted research and collaborated in Denny-Brown stabilizer and two types of torsion meters; CBE, 1918; KBE, 1946; hon. LLD Glasgow.

DENT, EDWARD JOSEPH (1876-1957), musical scholar; scholar of Eton and King's College, Cambridge; Mus.B., 1899; fellow of King's, 1902-8; professor of music, Cambridge, 1926-41; a founder of British Music Society and International Society for Contemporary Music; translated operas by Mozart, Verdi, etc.; edited and produced works of Purcell; made arrangement of *Beggar's Opera*; publications include *Foundations of English Opera* (1928) and biography of Busoni (1933); FBA, 1953; hon. D.Mus. Oxford.

DENT, JOSEPH MALABY (1849-1926), publisher; apprenticed to bookbinders in Darlington and London; opened his own business in Hoxton, 1872; moved to Great Eastern Street, c.1881; set up as publisher, 1888; productions include 'Temple Library'; 'Temple' (pocket) Shakespeare; 'Temple Classics'; 'Medieval Towns' series; 'Haddon Hall Library'; 'Everyman's Library' (1906), a working library of the world's literature, edited by Ernest Rhys [q.v.]; moved to Bedford Street,

Strand, 1898; opened model printing and binding factory at Garden City, Letchworth, 1907; erected Aldine House, Bedford Street, 1911.

DERBY, seventeenth EARL OF (1865-1948). [See STANLEY, EDWARD GEORGE VILLIERS.]

DERBY, sixteenth EARL OF (1841-1908), governor-general of Canada. [See STANLEY, FREDERICK ARTHUR.]

D'ERLANGER, SIR GERARD JOHN REGIS LEO (1906-1962), investment banker, company director, and airman; educated at Eton and in Paris; qualified as chartered accountant, 1933; obtained pilot's 'A licence' at Airwork Flying School, Heston, 1931; joined Myers & Company, 1934; member, London Stock Exchange, 1935; joined board of Hillman's Airways (later British Airways Ltd.), 1934; commandant, Air Transport Auxiliary, 1939; director, British Overseas Airways Corporation, 1940; managing director, British European Airways 1946; chairman, 1947-9; chairman, BOAC, 1956-60; knighted, 1958; chairman, City and International Trust and director of other trust companies, 1960-2.

DE ROBECK, SIR JOHN MICHAEL, baronet (1862-1928), admiral of the fleet; entered navy, 1875; rear-admiral, 1911; admiral of patrols, 1912-14; commanded ninth cruiser squadron, Aug. 1914; patrolled one of mid-Atlantic areas, with base at Finisterre; second-in-command to Sir Sackville Carden [q.v.] in naval expedition to Dardanelles, 1915; took over command, Mar.; won approval of all associated with disastrous attempt to force Dardanelles; in supreme command of all naval forces which carried out withdrawal of army from Dardanelles, 1916; vice-admiral commanding second battle squadron, 1916; received grant of £10,000 and baronetcy, 1919; commander-in-chief, Mediterranean fleet, 1919-22; of Atlantic fleet, 1922-4; KCB, 1916; admiral, 1920; admiral of the fleet, 1925.

DE SAULLES, GEORGE WILLIAM (1862-1903), medallist; worked for Joseph Moore [q.v.], medallist; engraver to the Royal Mint, 1893; designed Edward VII coronation (1902), South Africa (1899-1902), and Ashanti (1900) medals; executed dies for new Queen Victoria coins, 1893; designed English coins after King Edward VII's accession, 1902; exhibited at Royal Academy, 1898-1903.

DESBOROUGH, BARON (1855-1945), athlete, sportsman, and public servant. [See GRENFELL, WILLIAM HENRY.]

DE SELINCOURT, ERNEST (1870-1943), scholar and literary critic. [See SELINCOURT.]

DE SOISSONS, LOUIS EMMANUEL JEAN GUY DE SAVOIE-CARIGNAN, VISCOUNT D'OSTEL, BARON LONGROY (1890-1962), architect and town planner; born in Canada; educated in England and France; articled to J. H. Eastwood; Tite prizeman, 1912; student, British School, Rome, 1913; served with RASC during 1914-18 war; OBE, 1918; appointed by (Sir) Ebenezer Howard [q.v.] to plan Welwyn Garden City, 1920; member, Central Housing Advisory Committee; senior partner, Louis de Soissons, Peacock, Hodges and Robertson, responsible for restoration of Nash terraces in Regent's Park and a great variety of architectural work; architect to Duchy of Cornwall; FRIBA, MTPI, and SADG, 1923; ARA, 1942; RA, 1953; treasurer, 1959; member, Royal Fine Arts Commission, 1949-61; CVO, 1956.

DE STEIN, SIR EDWARD SINAUER (1887-1965), merchant banker; educated at Eton and Magdalen College, Oxford; called to the bar (Inner Temple), 1912; served in King's Royal Rifle Corps, 1914-18; entered City of London, and formed merchant bank, Edward de Stein & Co., in partnership with the Hon. John Mulholland, 1926; director of finance (raw materials), Ministry of Supply, 1941; knighted, 1946; a philanthropist, interested in Toc H and boys' clubs; chairman, British Red Cross finance committee, 1949; contributed to *Punch* and published *Poets in Picardy* (1919); exhibited at Royal Society of Painters in Water Colours; gave Lindisfarne Castle to National Trust.

DE SYLLAS, STELIOS MESSINESOS (LEO) (1917-1964), architect; educated at Haberdashers' Aske's and Christ's College, Finchley; studied architecture at Bartlett School, University College, London, 1933, and the Architectural Association School; as student editor launched *Focus* quarterly, 1938; helped to start Co-operative Partnership (later Architects Co-Partnership), 1939; joined Research and Experiments Department, Ministry of Home Security; architect and planning officer to Government of Barbados, 1945; returned to England, 1947; established branch office for his partnership in Nigeria, 1954; responsible for winning design for St. Paul's Cathedral Choir School, 1962; member of council, Architectural Association, 1956; chairman, RIBA Commonwealth Architects' Conference, 1963.

DES VŒUX, SIR (GEORGE) WILLIAM (1834-1909), colonial governor; born at Baden of Huguenot descent; educated at Charterhouse and Balliol College, Oxford; went to Toronto, 1856; practised at bar; became stipendiary magistrate in British Guiana where he championed natives, 1863; as administrator of St. Lucia (1869-78) reorganized and codified old French system of law; governor of Fiji, 1880-5; Newfoundland, 1886, and Hong Kong, 1887-91; GCMG, 1893; published *My Colonial Service*, 1903.

DETMOLD, CHARLES MAURICE (1883-1908), animal painter, and etcher; influenced by Japanese art; with twin brother Edward Julius produced portfolio of etchings of birds and animals of remarkable technical ability, 1898; illustrated *Jungle Book* for Rudyard Kipling [q.v.], 1903; committed suicide.

DE VERE, AUBREY THOMAS (1814-1902), poet and author; son of Sir Aubrey de Vere, second baronet [q.v.]; educated at Trinity College, Dublin; came early under Wordsworth's influence; was intimate with Sir Henry Taylor, Tennyson, Robert Browning, and R. H. Hutton [qq.v.]; visited Rome, 1839; travelled in Italy, 1843-4; published *The Waldenses and other Poems* (1842), and *English Misrule and Irish Misdeeds* (1848) in which he showed Irish sympathies and criticized English methods; joined Roman Church, 1851; nominal professor of political and social science in new Dublin Catholic University, 1854; interested in Irish legend and history; voluminous works include *The Legends of St. Patrick* (1872), *Critical Essays* (3 vols., 1887-9), *Recollections* (1897), and dramas.

DE VERE, Sir STEPHEN EDWARD, fourth baronet (1812-1904), translator of Horace; brother of Aubrey Thomas de Vere [q.v.]; called to Irish bar, 1836; entered Roman communion, 1848; liberal MP for Limerick, 1854-9; published *Translations from Horace*, 1886; succeeded to baronetcy, 1880.

DEVERELL, Sir CYRIL JOHN (1874-1947), field-marshal; educated at Bedford School; gazetted to West Yorkshire regiment, 1895; in France commanded 4th battalion, East Yorkshire regiment (1915), 20th brigade, 7th division (1915-16), 3rd division (1916-19), at Arras, Ypres, Cambrai and in final offensive; quartermaster-general (1927-30) and chief of general staff (1930-1), India; held Western (1931-3), Eastern (1933-6) commands, England; general, 1933; field-marshal, 1936; chief of imperial general staff, 1936-7; KBE, 1926; GCB, 1935.

DE VILLIERS, JOHN HENRY, first BARON DE VILLIERS (1842-1914), South African judge;

born in Cape Colony; called to bar (Inner Temple), 1865; began practice at Cape bar, 1866; member of house of assembly, Cape Colony, 1867; attorney-general, 1872; chief justice, 1873; worked for South African federation; knighted, 1877; member of royal commission which drew up Pretoria convention, 1881; KCMG, 1882; PC, 1897; first colonial judge on judicial committee of Privy Council; president of national convention, 1908-9; baron, 1910; first chief justice of South African Union.

DEVINE, GEORGE ALEXANDER CASSADY (1910-1966), actor and theatre director; educated at Clayesmore School and Wadham College, Oxford; president, Oxford University Dramatic Society; played Mercutio in 1932 OUDS production of *Romeo and Juliet*, directed by (Sir) John Gielgud; business manager to Motley, stage designers; met Michel Saint-Denis, director of the *Compagnie des Quinze*; with Saint-Denis set up London Theatre Studio, 1936-9; fought in Royal Artillery in Burma during 1939-45 war; rejoined Saint-Denis at Old Vic Centre, 1945-52; directed opera, and Shakespeare repertory at Stratford-upon-Avon, 1954-5; artistic director, English Stage Company, 1956; CBE, 1958.

DEVLIN, JOSEPH (1871-1934), Irish politician; Roman Catholic born in Belfast; entered journalism and Irish nationalist movements; a leading organizer of National Volunteers (1913); member of Irish convention (1917-18); MP at Westminster (1902-22, 1929-34) and in Northern Ireland parliament after 1921; witty debater and powerful speaker.

DEVONPORT, first VISCOUNT (1856-1934), business man. [See KEARLEY, HUDSON EWBANKE.]

DEVONS, ELY (1913-1967), statistician and economist; educated at a number of schools, including North Manchester municipal high school and Manchester University; persuaded by Harold Laski [q.v.] to study economics; first class honours, 1934; awarded research fellowship; published article in *Economic Journal*, 1935; economic assistant on staff of Cotton Trades Organizations, Manchester, 1935; joined Central Economic Information Service, 1940; reorganized presentation of official statistics; worked in Ministry of Aircraft Production; head of planning department, 1944; assisted in first issue of OEEC statistical bulletins; reader (and later) professor of applied economics, Manchester, 1945; professor of commerce, London School of Economics,

1959-65; member, Local Government Commission, 1959-65; publications include *Planning in Practice* (1950), *Introduction to British Economic Statistics* (1956), and *Essays in Economics* (1961).

DEVONSHIRE, eighth DUKE OF (1833-1908), statesman. [See CAVENDISH, SPENCER COMPTON.]

DEVONSHIRE, ninth DUKE OF (1868-1938). [See CAVENDISH, VICTOR CHRISTIAN WILLIAM.]

DEWAR, SIR JAMES (1842-1923), natural philosopher; educated at Edinburgh University; Jacksonian professor of natural experimental philosophy, Cambridge, 1875-1923; Fullerian professor of chemistry, Royal Institution, 1877-1923; his early subjects of investigation included the specific heat of Graham's hydrogenium, for which he employed vacuum jacket, indispensable for liquefaction of gases, and precursor of 'Thermos' flask; made several biological researches; soon after 1875 joined G. D. Liveing [q.v.] in series of spectroscopic investigations; at Royal Institution carried out important work on liquefaction of gases; his use of charcoal for production of very high vacua greatly contributed to subsequent advances made by atomic physics; always anxious to employ liquid gases in opening up new fields of research; with Sir F. A. Abel [q.v.] invented cordite; FRS, 1877; knighted, 1904.

DE WET, CHRISTIAAN RUDOLPH (1854-1922), Boer general and politician; born in Orange Free State republic; served in Transvaal war, 1880-1; member of Free State volksraad; led roving life; made name by capture of British force at Nicholson's Nek, Oct. 1899; promoted general; captured Lord Roberts's convoy at Watervaal, Feb. 1900; commander-in-chief, Free State forces, 1900; restored morale of Boer army; kept field for two years by employing guerrilla tactics, of which he became greatest modern exponent; delegate to national convention, 1908-9; supported J. B. M. Hertzog [q.v.] against Louis Botha [q.v.]; in favour of secession from British Empire; with General Beyers, planned armed revolt against Botha's government, Oct. 1914; defeated (Nov.) and captured (Dec.); found guilty of high treason, but released after short imprisonment.

DE WIART, SIR ADRIAN CARTON (1880-1963), lieutenant-general. [See CARTON DE WIART.]

DE WINTON, SIR FRANCIS WALTER (1835-1901), major-general and South African administrator; served in Crimea, 1854; military attaché at Constantinople, 1878-83; KCMG, 1884; GCMG, 1893; administrator-general of Congo, 1885-6; suppressed rebellion of Yonnies on West African coast, 1887; CB, 1888; as commissioner to Swaziland deemed British protectorate impracticable, 1889; major-general, 1890; controller of household of George, Duke of York, 1892; hon. LLD, Cambridge, 1892.

DE WORMS, HENRY, BARON PIRBRIGHT (1840-1903), politician; close friend of Count von Beust, Austrian statesman, who introduced him to Disraeli, 1867; conservative MP for Greenwich, 1880-5, and for Toxteth, 1885-95; parliamentary secretary to Board of Trade, 1885-6, 1886-8; under-secretary for colonies, 1888-92; first Jewish PC, 1888; president of international conference in London on sugar bounties, 1887; baron, 1895; president of Anglo-Jewish Association, 1872-86; FRS; wrote scientific and political works.

DEWRANCE, SIR JOHN (1858-1937), mechanical engineer; educated at Charterhouse and King's College, London; took control of family engineering business, 1879; started research laboratory; later encouraged research by team-work; inventions related mainly to steam fittings and boiler mountings; chairman, Babcock & Wilcox, Ltd. (1899-1937) and from 1914 of Kent Coal Concessions, Ltd.; KBE, 1920.

D'EYNCOURT, SIR EUSTACE HENRY WILLIAM TENNYSON-, first baronet (1868-1951), naval architect. [See TENNYSON-D'EYNCOURT.]

DIBBS, SIR GEORGE RICHARD (1834-1904), premier of New South Wales; born in Sydney; joined father-in-law's sugar refining business there, 1857; formed shipping business, 1859; member of legislative assembly of New South Wales, 1874; treasurer and colonial secretary, 1883-5; premier, Oct.-Dec. 1885; colonial secretary, 1886-7; premier and colonial secretary, Jan.-Mar. 1889 and 1891-4; became protectionist; KCMG, 1892; managing director of New South Wales Savings Banks, 1897-1904.

DIBDIN, SIR LEWIS TONNA (1852-1938), ecclesiastical lawyer, judge, and administrator; BA, St. John's College, Cambridge, 1874; called to bar (Lincoln's Inn), 1876; official counsel to attorney-general in charity matters, 1895-1901; chancellor of Rochester, Exeter, and Durham dioceses; KC, 1901; dean of the Arches, 1903-34; gave judgment in Deceased Wife's Sister case, 1908; vicar-general of province of Canterbury, 1925-34; first church estates commissioner, 1905-30; initiated pension and other beneficial measures; with A. L. Smith [q.v.] prepared historical section of report on which Enabling Act (1919) was based; influential in Church Assembly; knighted, 1903.

DICEY, ALBERT VENN (1835–1922), jurist; brother of E. J. S. Dicey [q.v.]; BA, Balliol College, Oxford; president of Union; fellow of Trinity College, Oxford, 1860–72; won Arnold historical essay prize, 1860; called to bar (Inner Temple), 1863; junior counsel to commissioners of Inland Revenue, 1876–90; Vinerian professor of English law and fellow of All Souls, 1882–1909; QC, 1890; an ardent unionist; his principal works *Introduction to the Study of the Law of the Constitution* (1885), *Digest of the Law of England* (1896), and *Law and Public Opinion in England* (1905).

DICEY, EDWARD JAMES STEPHEN (1832–1911), author and journalist; BA, Trinity College, Cambridge, 1854; president of Union; leader-writer for *Daily Telegraph*, 1861; wrote accounts of visits to America (1862), Russia, Holy Land, and Egypt (1867–70); editor of *Observer*, 1870–89; keenly interested in affairs of Eastern Europe, Egypt, and South Africa; called to bar (Gray's Inn), 1875; CB, 1886; influenced public opinion by his knowledge, humour, judgement, and vivid style; author of works on Egypt and Bulgaria.

DICK, Sir WILLIAM REID (1878–1961), sculptor; served five-year apprenticeship as stonemason's assistant; studied at School of Art, Glasgow, 1906–7; exhibited at Royal Academy, 1908; served in army in France and Palestine, 1916–18; ARA, 1921; RA, 1928; president, Royal Society of British Sculptors, 1933–8; member, Royal Fine Art Commission, 1928, and Royal Mint Advisory Committee, 1934–5; KCVO, 1935; King's (later Queen's) sculptor in ordinary for Scotland, 1938; Royal Scottish academician, 1939; important works include the 'Lion' above the Menin Gate, Ypres (1927), and the 'Eagle' surmounting the RAF memorial at Westminster; other memorials include those to Lord Irwin (later first Earl of Halifax, q.v.), New Delhi (1932), David Livingstone [q.v.] at Victoria Falls (1931–3) and President Roosevelt, Grosvenor Square, London (1948); also the recumbent effigy of George V for the King's tomb designed by Sir Edwin Lutyens [q.v.] in St. George's Chapel, Windsor, a similar figure of Queen Mary, and the stone statue of George V, with Sir Giles Gilbert Scott [q.v.] as architect, Westminster Abbey; other works include a large number of portrait busts.

DICK-READ, GRANTLY (1890–1959); obstetrician and advocate of natural childbirth;

educated at Bishop's Stortford College and St. John's College, Cambridge; qualified at London Hospital, 1914; practised in Woking and Harley Street, and in Johannesburg (1949–53); publications include *Natural Childbirth* (1933) and *Revelation of Childbirth* (1942); focused attention on 'fear-tension-pain' syndrome and its relief by relaxation.

DICKINSON, GOLDSWORTHY LOWES (1862–1932), humanist, historian, and philosophical writer; son of L. C. Dickinson [q.v.]; educated at Charterhouse and King's College, Cambridge; first class, classics, 1884; fellow, 1887; lecturer in political science, 1896–1920; publications include *The Development of Parliament during the Nineteenth Century* (1895), *The Greek View of Life* (1896), *Justice and Liberty* (1908) and other dialogues in the Socratic tradition, *The International Anarchy, 1904–1914* (1926), *After Two Thousand Years* (1930); drafted schemes for and worked actively for foundation of a 'League of Nations' (a term he may have invented).

DICKINSON, HENRY WINRAM (1870–1952), historian of engineering and technology; educated at Manchester grammar school; studied engineering at Owens College and with Beardmore's, Glasgow; entered Science Department (later Science Museum), South Kensington, 1895; assistant keeper, machinery division, 1900; keeper of mechanical engineering, 1924–30; a founder of Newcomen Society (1920); its secretary until 1951; editor of its *Transactions* until 1950; published biographies of James Watt (1936) and others, and *A Short History of the Steam Engine* (1939).

DICKINSON, HERCULES HENRY (1827–1905), dean of the Chapel Royal, Dublin, from 1868; MA, Trinity College, Dublin, 1849; DD, 1866; vicar of St. Ann's, Dublin, 1855–1902; professor of pastoral theology, Dublin University, 1894; member of royal commission for licensing reform, 1896–9; helped to found Alexandra College for Women, Dublin, 1866; warden, 1866–1902; wrote theological works.

DICKINSON, LOWES (CATO) (1819–1908), portrait painter; visited Italy and Sicily, 1850–3; friend of pre-Raphaelites; met F. D. Maurice [q.v.], and with others formed Christian socialist movement; helped to found in 1854 Working Men's College, where he taught drawing; exhibited at Royal Academy (1848–91) portraits of Queen Victoria, the Prince of Wales, Lord Kelvin, Cobden, Grote, Gladstone, Bright, and General Gordon; memorial art studentship founded at Working Men's College, 1909.

DICKSEE, SIR FRANCIS BERNARD (FRANK) (1853-1928), painter; came of artistic family; studied at Royal Academy Schools; his first great popular success, 'Harmony' (Royal Academy exhibition, 1877); thenceforth exploited quasi-historical, romantic, and sentimental vein or else vein of melodrama in modern setting; his numerous pictures include 'Too Late' (1883), 'The Crisis' (1891), and 'The Light Incarnate' (1922); for a time also fashionable portrait painter, especially popular with women sitters; RA, 1891; president of Royal Academy, 1924; knighted, 1925.

DICKSON, SIR COLLINGWOOD (1817-1904), general; born at Valenciennes; son of Sir Alexander Dickson [q.v.]; joined Royal Artillery, 1835; served in Spain, 1837-40; instructed Turkish artillery at Constantinople, 1841-5; received VC (1855) for bravery at Sebastopol, 1854; distinguished in battle of Inkerman, 1855; in Ireland, 1856-62; CB, 1865; served on fortifications committee, 1868-9; inspector-general of artillery, 1870-5; KCB, 1871; general, 1877; president of ordnance committee, 1881-5; GCB, 1885; accomplished linguist.

DICKSON, SIR JAMES ROBERT (1832-1901), Australian statesman; educated at Glasgow high school; worked in City of Glasgow Bank; emigrated to Australia, 1854; entered Bank of Australasia; became auctioneer in Queensland, 1862; promoted Royal Bank of Queensland; member of Queensland House of Assembly for Enoggera, 1872-87; minister of works, 1876-87; resigned and resided in Europe, 1887-92; re-elected for Bulimba, 1892-3, and 1896; minister for home affairs, Queensland, and later, premier, 1898; advocated formation of Australian commonwealth; represented Queensland at London conference, 1900; minister of defence, first Australian government, 1900; CMG, 1897; KCMG, 1901; hon. DCL, Oxford, 1900.

DICKSON, WILLIAM PURDIE (1823-1901), professor of divinity and translator; studied at St. Andrews (1837-44) for Presbyterian ministry; DD, 1864; professor of biblical criticism, Glasgow University, 1863-73; professor of divinity, 1873-95; hon. LLD, Edinburgh, 1885; translated Mommsen's *History of Rome* (4 vols., 1862-7) and *Roman Provinces* (1887).

DICKSON-POYNDER, SIR JOHN POYNDER, sixth baronet, and BARON ISLINGTON (1866-1936). [See POYNDER.]

DIGBY, WILLIAM (1849-1904), Anglo-Indian publicist; prepared six volumes of Ceylon *Hansard*, 1871-6; edited *Madras Times*, 1877-9; urged lord mayor of London's Southern Indian famine relief fund, 1877; CIE, 1878; edited *Western Daily Mercury*, 1880-2; secretary of National Liberal Club, 1882-7; advocate of self-government in India; works include *Prosperous British India*, 1901.

DILKE, SIR CHARLES WENTWORTH, second baronet (1843-1911), politician and author; son of first baronet [q.v.]; scholar of Trinity Hall, Cambridge; senior legalist, LLB, 1866; LLM, 1869; twice president of Union; enthusiastic oarsman; called to bar, 1866; toured round world with William Hepworth Dixon [q.v.], 1866-7, publishing his experience in *Greater Britain* (1868), the title being his own invention; radical MP for Chelsea, 1868-86; opposed W. E. Forster's education bill; expressed strong republican views in the country and in parliament; succeeded to baronetcy and proprietorship of *Athenæum* and of *Notes and Queries*, 1869; married his first wife, 1872; frequently visited Paris; friend of Gambetta and republican leaders; published anonymously a satirical brochure, *The Fall of Prince Florestan of Monaco*, 1874; made second tour round world, 1875; prominent in parliament, 1874-80; seconded resolution for extension of county franchise to agricultural labourers, 1879; attacked conservative government's South African policy; leader of radical section of Gladstone's government, 1880; under-secretary to the Foreign Office, 1880-2; chairman of royal commission for negotiating commercial treaty with France, 1881-2; became friend of Prince of Wales; entered cabinet as president of Local Government Board, 1882-5; chairman of royal commission on housing of working classes, 1884; conducted redistribution bill through House of Commons; jointly responsible for sending General Gordon [q.v.] to the Sudan, 1884; opposed to Gladstone's Irish policy, 1885; co-respondent in divorce suit, *Crawford* v. *Crawford and Dilke*, 1885-6; was rejected by electors of Chelsea, 1886; was largely ostracized from public life owing to divorce court proceedings; married his second wife, Emilia Frances (see below), widow of Mark Pattison [q.v.], 1885; pursued close study of English and imperial problems, publishing *Problems of Greater Britain*, 2 vols., 1890; visited Greece and Constantinople, 1887-8, and India, 1888-9; MP for Forest of Dean, 1892-1911; spoke in parliament mainly on industrial, foreign, and imperial affairs; made art and literary bequests to the National Portrait Gallery and other institutions.

DILKE, EMILIA FRANCES, LADY DILKE (1840-1904), historian of French art; born

STRONG; after private education at Oxford, studied art at South Kensington, 1859-61; married her first husband, Mark Pattison [q.v.], rector of Lincoln College, Oxford, Sept. 1861; published *Renaissance of Art in France* (1879), and embodied subsequent researches in *Art in the Modern State* (1888), and in volumes on French painters (1889), architects and sculptors (1900), and engravers and draughtsmen (1902), all of the eighteenth century; at the same time she wrote stories of mystical temper and actively promoted improvement in the social and industrial condition of working women, joining the Women's Trades Union League and attending the Trades Union Congresses, 1884-1904; after Pattison's death in 1884 she married (1885) Sir Charles Wentworth Dilke (see above), an early friend, and thenceforth identified herself with his fortunes.

DILL, SIR JOHN GREER (1881-1944), field-marshal; educated at Cheltenham College and Sandhurst; joined 1st battalion, Leinster regiment, in South Africa, 1901; captain, 1911; brigade-major in France, 1914-16; GSO 2, Canadian Corps, 1916-17; GSO 1, 37th division, 1917; chief of operations branch, GHQ, 1918; DSO, 1915; colonel, 1920; army instructor, Imperial Defence College, 1926-9; chief general staff officer, Western command, India, 1929-30; major-general, 1930; commandant, Staff College, 1931-4; director, military operations and intelligence, 1934-6; lieutenant-general, 1936; commanded forces in Palestine, 1936-7; GOC-in-C, Aldershot, 1937-9; commanded I Corps in France, 1939-40; general, 1939; field-marshal, 1941; vice-chief of imperial general staff, Apr. 1940; chief of imperial general staff, May 1940; felt compelled to discourage early offensive operations and considered by (Sir) Winston Churchill to be over-cautious and obstructive; visited Greece, Turkey, and Yugoslavia, Feb.-Apr. 1941; at first opposed sending British forces to Greece; changed his mind but was disappointed in efficacy of aid from allies on the spot; under prolonged strain his health deteriorated; relinquished appointment, Dec. 1941; governor-designate of Bombay, 1941-2; accompanied Churchill to United States, Dec. 1941; remained there as senior British representative on combined chiefs of staff committee in Washington; established himself in confidence of American official and military circles; died in Washington and buried in Arlington cemetery; KCB, 1937; GCB, 1942.

DILL, SIR SAMUEL (1844-1924), classical scholar, historian, and educationist; BA, Lincoln College, Oxford; fellow and tutor, Corpus Christi College, Oxford, 1869; high master, Manchester grammar school, 1877-88; effective as teacher and administrator; professor of Greek, Queen's College, Belfast (later Queen's University), 1890-1924; knighted, 1909; wrote three books on Roman society.

DILLON, EMILE JOSEPH (1854-1933), philologist, author, and journalist; studied oriental languages under Renan and at several European universities, acquiring also various modern languages; correspondent of *Daily Telegraph* (1887-1914) in Russia and on special missions elsewhere including Armenia (1894), Crete (1897), second Dreyfus trial (1899) and Boxer rising (1900); leader-writer in five languages and adventurous chronicler of events; adviser to Count Witte at Portsmouth peace conference (1905) and to many other statesmen and diplomatists, but left little permanent mark.

DILLON, FRANK (1823-1909), landscape painter; exhibited at Royal Academy, 1850-1907; joined Royal Institute of Painters in Water Colours, 1866; frequently visited Egypt from 1854, painting many Egyptian scenes; also studied Japanese art; friend of Mazzini and Hungarian revolutionaries.

DILLON, HAROLD ARTHUR LEE-, seventeenth VISCOUNT DILLON (1844-1932), antiquary; educated at Bonn University; served in Rifle Brigade, 1862-74; succeeded father, 1892; first curator of the armouries of the Tower of London, 1892-1913; encyclopaedic knowledge of medieval and later periods; chairman of trustees of National Portrait Gallery (1894-1928) to which he presented portraits from the Ditchley collection; FBA, 1902; CH, 1921.

DILLON, JOHN (1851-1927), Irish nationalist politician; son of J. B. Dillon [q.v.]; qualified in surgery, Dublin; supported C. S. Parnell [q.v.] against Isaac Butt [q.v.], 1879; accompanied Parnell to USA on campaign on behalf of Land League, 1879; elected MP, Tipperary, in absence; strongly advocated agrarian agitation; twice imprisoned in Kilmainham jail, 1881; MP, East Mayo, 1885-1918; close personal ally of William O'Brien [q.v.]; with O'Brien promulgated 'plan of campaign', 1886; again imprisoned, 1888; rearrested, but with O'Brien escaped via France to USA, 1890; absent from Ireland during Parnell divorce proceedings; reimprisoned, 1891; on release, declared unconditionally for anti-Parnellites; chairman of anti-Parnellite group, 1896; bent on extirpating landlord power in Ireland; supported John Redmond [q.v.] as chairman of united party, 1900; split with O'Brien, 1904; chief member of Irish leader's 'cabinet'; instrumental in procuring establishment of National University, 1908;

with Redmond represented Irish nationalists at Buckingham Palace conference on Ulster question, July 1914; joined in abortive attempt to bring in home rule by agreement, July 1916; chairman of Irish party, 1918; with Sinn Fein leaders resisted Irish conscription, 1918; defeated at East Mayo and withdrew from public life, 1918; an inspiring orator whose knowledge was enormous and labour endless.

DIMBLEBY, RICHARD FREDERICK (1913-1965), journalist and broadcaster; educated at Mill Hill School; worked on family paper, the *Richmond and Twickenham Times*, 1931; BBC news reporter, 1936; experience in Spanish civil war; BBC war correspondent with BEF, 1939; covered campaigns in Middle East, East Africa, Greece, and the Western Desert; flew with Bomber Command over Germany, 1943; first reporter into Belsen; commented on television on state occasions, and in 'Panorama'; chairman, 'Twenty Questions' on radio; carried out television commentary on Winston Churchill's funeral, 1965; public subscription funded, in his memory, cancer research fellowship at St. Thomas's Hospital; OBE, 1945; CBE, 1959; hon. LLD Sheffield, 1965.

DIMOCK, NATHANIEL (1825-1909), theologian; BA, St. John's College, Oxford, 1847; MA, 1850; held livings in Kent, 1848-87; wrote much on church doctrine with profound erudition from evangelical point of view; influential member of Bishop Creighton's 'Round Table Conference' on holy communion doctrine and ritual, 1900; voluminous works include *The Doctrine of the Sacraments* (1871), *Curiosities of Patristic and Medieval Literature* (3 pts., 1891-5); a memorial edition of his works was published, 1910-11.

DINES, WILLIAM HENRY (1855-1927), meteorologist; BA, Corpus Christi College, Cambridge, 1881; private correspondence tutor in mathematics; devoted himself increasingly to dynamical and physical sides of meteorology; leading exponent of experimental meteorology in England; took up subject of wind-force and developed pressure-tube anemometer; directed work on upper air for British Meteorological Office, 1905; undertook searching examination of problem of solar and terrestrial radiation; president, Royal Meteorological Society, 1901-2; FRS, 1905.

DIX, GEORGE EGLINGTON ALSTON, Dom GREGORY (1901-1952), monk of Nashdom Abbey; educated at Westminster and Merton College, Oxford; deacon, 1924; priest, 1925; entered Anglican Benedictine community, 1926; solemnly professed at Nashdom,

1940; prior, 1948-52; an historian and liturgical scholar; BD and DD Oxford; publications include *The Shape of the Liturgy* (1945).

DIXIE, LADY FLORENCE CAROLINE (1857-1905), authoress and traveller; born DOUGLAS; married Sir Alexander Beaumont Churchill Dixie, eleventh baronet, 1875; hunted big game in Africa, Arabia, and Rocky Mountains; explored Patagonia, 1878-9; correspondent for *Morning Post* in Zulu war, 1879; denounced Irish Land League agitation, 1880-3; alleged without proof that she was a victim of Fenian outrage near Windsor, 1883; advocated complete sex equality; chief works were *Across Patagonia* (1880) and *Songs of a Child* (1902-3).

DIXON, SIR ARTHUR LEWIS (1881-1969), civil servant; educated at Kingswood School, Bath, and Sidney Sussex College, Cambridge; ninth wrangler, 1902; entered home Civil Service and served in Home Office, 1903-46; in charge of 'war-measures' division, 1914-18; secretary to committee under Lord Desborough [q.v.] to review police service, 1919-20; in charge of police division; assistant under-secretary, 1932; in charge of fire-brigades division; created wartime National Fire Service; modernized both police and fire services; knighted, 1941; promoted to principal assistant under-secretary of state; published *Atomic Energy for the Layman* (1950).

DIXON, HENRY HORATIO (1869-1953), professor of botany; educated at Rathmines School and Trinity College, Dublin; senior moderator, natural science, 1891; university professor of botany, 1904-49; appointed director of botanic garden, 1906, keeper of the herbarium, 1910; created School of Botany, Trinity College, Dublin, opened in 1907; appointed professor of plant biology in Trinity College, 1922; worked on transpiration and water relations of plants; publications include *Transpiration and the Ascent of Sap in Plants* (1914) and *Practical Plant Biology* (1922); FRS, 1908; hon. fellow, Trinity College, Dublin.

DIXON, SIR PIERSON JOHN (1904-1965), diplomatist; educated at Bedford School and Pembroke College, Cambridge; first class, parts i and ii, classical tripos, 1925-7; Porson prize; Craven scholarship, 1926; junior research fellow, 1927; entered foreign service, 1929; served in Madrid (1932), Ankara (1936), and Rome (1938); returned to Foreign Office, 1940; principal private secretary to Anthony Eden (later Earl of Avon), 1943; and to Ernest Bevin [q.v.], 1945; ambassador, Prague, 1948; deputy under-secretary, 1950-4; permanent United Kingdom representative at United Nations,

1954-60; ambassador, Paris, 1960-5; CMG, 1945; CB, 1948; KCMG, 1950; GCMG, 1957; published *The Iberians of Spain* (1940), *Farewell, Catullus* (1953), *The Glittering Horn* (1958), and *Pauline* (1964).

DIXON, SIR ROBERT BLAND (1867-1939), engineer vice-admiral; elder brother of W. E. Dixon [q.v.]; educated at Royal Naval Engineering College, Keyham, Royal Naval College, Greenwich; engineer commander, 1904; captain, 1915; engineer manager, Portsmouth, 1912-17; assistant director of dockyards at Admiralty (1917), assistant engineer (1919), deputy engineer (1920), engineer-in-chief of fleet, 1922-8; engineer rear-admiral (1919), vice-admiral, 1922; KCB, 1924; maintained Admiralty lead in engineering; retained use of steam turbine.

DIXON, WALTER ERNEST (1870-1931), pharmacologist; younger brother of Sir R. B. Dixon [q.v.]; entered St. Thomas's Hospital; MD, 1898; lecturer (1909), first reader (1919-31) in pharmacology, Cambridge; influential in obtaining recognition of importance of pharmacology; author of *A Manual of Pharmacology* (1905) and *Practical Pharmacology* (1920); FRS, 1911.

DOBBS, SIR HENRY ROBERT CONWAY (1871-1934), Indian civil servant and administrator of Iraq; scholar of Winchester and Brasenose College, Oxford; chief commissioner, Baluchistan, 1917; foreign secretary to government of India, 1919-23; negotiated treaty with Afghanistan, 1920-1; KCIE, KCSI, 1921; high commissioner (1923-9) in Iraq where he gave wise and sympathetic guidance.

DOBELL, BERTRAM (1842-1914), bookseller and man of letters; befriended James Thomson [q.v.], and arranged independent publication of *City of Dreadful Night*, 1880; his great achievement recovery of poetical works of Thomas Traherne (1903), followed by prose *Centuries of Meditations* (1908).

DOBSON, FRANK OWEN (1886-1963), sculptor; educated at Harrow Green School, Leyton Technical School, and Hospitalfield Art Institute, Arbroath; continued artistic training at City and Guilds School, Kennington; first one-man exhibition of paintings and drawings, 1914; on active service with Artists' Rifles, 1914-18; exhibited (as the only sculptor) in Group X exhibition in London, 1920; first one-man sculpture show, 1921; designed backdrops for (Dame) Edith Sitwell's [q.v.] *Façade*, 1922; president, London Group, 1923-7; designed and printed fabrics, 1930; first exhibited at Royal Academy, 1933; professor of sculpture, Royal College of Art, 1946-53; made large plaster group, 'London Pride' for 1951 Festival of Britain exhibition; other sculptures included 'Susanna' (1923, Tate Gallery), 'Lopokova' (1924, Arts Council), 'Head of a Girl' (1925), 'Truth' (1930, outside Tate Gallery); ARA, 1942; RA, 1953; associate, Royal Society of British Sculptors, 1938; CBE, 1947.

DOBSON, (HENRY) AUSTIN (1840-1921), poet and man of letters; entered Board of Trade, 1856; principal clerk in marine department, 1884-1901; verse includes *Vignettes in Rhyme* (1873), *Proverbs in Porcelain* (1877), and *At the Sign of the Lyre* (1885); prose works include lives of Hogarth (1879), Fielding (1883), Steele (1886), Goldsmith (1888), Horace Walpole (1890), Samuel Richardson (1902), Fanny Burney (1903), and *Thomas Bewick and his Pupils* (1884).

DOBSON, SIR ROY HARDY (1891-1968), aircraft engineer; apprenticed to T. & R. Lees, Hollinwood, near Manchester; joined (Sir) E. A. V. Roe [q.v.] & Co. Ltd., 1914; works manager, 1919; general manager, 1934; director, 1936; managing director, 1941; managing director, Hawker Siddeley Group Ltd., 1958; chairman, A. V. Roe Canada Ltd. and Orenda Engines Ltd. in Toronto, 1951; director, Canadian Imperial Bank of Commerce, 1955-66; chairman, Hawker Siddeley Group, 1963-7; responsible for production of Lancaster bomber, and for design of Vulcan bomber; CBE, 1942; knighted, 1945; hon. fellow, Royal Aeronautical Society, 1956.

DODD, FRANCIS (1874-1949), painter and etcher; studied at Glasgow School of Art, in Paris and Italy; his portraits of naval and military commanders (1914-18) in Imperial War Museum; active member of Manchester Academy and New English Art Club; exhibited at Royal Academy from 1923; ARA, 1927; RA, 1935.

DODGSON, CAMPBELL (1867-1948), critic and historian of art; scholar of Winchester and New College, Oxford; first class, *lit. hum.*, 1890; entered department of prints and drawings, British Museum, 1893; keeper, 1912-32; leading authority on early German prints; edited *Print Collector's Quarterly*, 1921-36; bequeathed collection to British Museum; FBA, 1939.

DODGSON, FRANCES CATHARINE (1883-1954), artist; daughter of W. A. Spooner [q.v.]; studied at Ruskin School, Oxford, and Royal Academy and Slade schools; married

Campbell Dodgson [q.v.], 1913; her portrait-drawings much in demand for her ability to capture a likeness but she was too modest to work professionally.

DODS, MARCUS (1834–1909), Presbyterian divine and biblical scholar; son of Marcus Dods [q.v.]; MA, Edinburgh University, 1854; hon. DD, 1891; minister of Renfield Free church, Glasgow, 1864–89; a published sermon (1877) questioning verbal inspiration caused much discussion; professor of New Testament criticism in New College, Edinburgh, 1889; libelled for his views on inspiration at General Assembly, 1890; principal of New College, 1907–9; works include commentaries on Genesis (1888) and 1 Corinthians (1889), *The Bible, its Nature and Origin* (Bross lectures, 1905); *Letters*, edited by son (2 vols., 1910–11).

DOHERTY, HUGH LAWRENCE (1875–1919), lawn tennis player; made name at lawn tennis as Cambridge undergraduate; winner of All England singles championship at Wimbledon, 1902–6; with his brother, R. F. Doherty, doubles champion eight times between 1897 and 1905; American national champion, 1903.

DOLLING, ROBERT WILLIAM RAD-CLYFFE [FATHER DOLLING] (1851–1902), divine and social reformer; educated at Trinity College, Cambridge; did social work in Dublin, 1870–8; intimate with 'Father' Stanton [q.v.] and Alexander Mackonochie [q.v.]; warden of St. Martin's Postman's League in south London, 1879–82; vicar of St. Agatha's, Landport, 1885–95; resigned owing to disputes with bishop on questions of ritual; wrote *Ten Years in a Portsmouth Slum* (1869); as vicar of St. Saviour's, Poplar (1898–1901), he influenced social and municipal affairs; unconventional preacher.

DOLMETSCH, (EUGENE) ARNOLD (1858–1940), musician and musical craftsman; born in France; naturalized, 1931; studied in Brussels and at Royal College of Music, London; revived early English instrumental music; rediscovered John Jenkins and William Lawes [qq.v.]; made new lutes, harpsichords, etc.; re-established recorder as instrument of popular music.

DONALD, Sir JOHN STEWART (1861–1948), Indian civil servant; educated at Bishop Cotton School, Simla; entered Punjab provincial service, 1882; transferred to central government, 1890; British representative, Indo-Afghan frontier commission, 1894; resident in Waziristan, 1908–20; acting chief commis-

sioner, North-West Frontier Province, 1913–15; KCIE, 1915.

DONALD, Sir ROBERT (1860–1933), journalist; founded *Municipal Journal* (1893) and *Municipal Year Book* (1897); edited *Daily Chronicle*, 1902–18; resigned on sale of paper to Lloyd George group; warned that limitation of editorial freedom would be a national danger; chairman, publicity committee, British Empire Exhibition (1924), and Empire Press Union, 1915–26; GBE, 1924.

DONALDSON, Sir JAMES (1831–1915), educationist, classical and patristic scholar; MA, Aberdeen; rector of Edinburgh high school, 1866–81; professor of humanity, Aberdeen University, 1881; principal and vice-chancellor of St. Andrews University, 1889; knighted, 1907; author of patristic works; helped to establish compulsory primary education in Scotland.

DONALDSON, St. CLAIR GEORGE ALFRED (1863–1935), successively archbishop of Brisbane and bishop of Salisbury; son of Sir S. A. Donaldson [q.v.]; educated at Eton and Trinity College, Cambridge (first classes in classics and theology); deacon, 1888; priest, 1889; head of Eton mission, Hackney Wick, 1891–1900; rector of Hornsey, 1901–4; bishop (1904), archbishop (1905–21) of Brisbane; built cathedral; secured right of entry for denominational teachers into state schools; increased number of clergy born and trained in Australia; bishop of Salisbury, 1921–35; first chairman, Missionary Council, 1921–33; chairman, Canterbury convocation committee on church and marriage, 1931–5; KCMG, 1933.

DONAT, (FRIEDERICH) ROBERT (1905–1958), actor; educated at Manchester central school; joined Sir Frank Benson [q.v.], 1924; at Liverpool Playhouse, 1928, Festival Theatre, Cambridge, 1929; created Gideon Sarn in *Precious Bane* by Mary Webb [q.v.], 1931; two Camerons in *A Sleeping Clergyman* by James Bridie [q.v.], 1933; film work for Sir Alexander Korda [q.v.] included *Mr. Chips* by James Hilton [q.v.]; played Becket in *Murder in the Cathedral*, 1953; despite asthma had great purity of diction and beautiful voice.

DONKIN, BRYAN (1835–1902), civil engineer; grandson of Bryan Donkin [q.v.]; studied engineering in Paris; apprentice (1856) and partner (1868) in grandfather's engineering works; pursued valuable researches into design and construction of heat engines and steam boilers; perfected glass 'revealer' for showing condensation effects in the cylinder; made inquiry into motive power from blast-furnace gases; wrote *A Text-Book on Gas, Oil and Air Engines* (1894).

DONNAN, FREDERICK GEORGE (1870-1956), physical chemist; born in Colombo; educated at Belfast Royal Academy and Queen's College; Ph.D., Leipzig, 1896; at University College, London, worked under Sir William Ramsay [q.v.], 1898-1903; Sir John Brunner professor of physical chemistry, Liverpool, 1904-13; succeeded Ramsay at University College, 1913-37; research consultant, Brunner Mond, 1920-6; on research council, ICI, 1926-39; internationally known as colloid chemist; FRS, 1911; CBE, 1920.

DONNELLY, Sir JOHN FRETCHEVILLE DYKES (1834-1902), major-general Royal Engineers; born at Bombay; joined Royal Engineers, 1853; served in battle of Inkerman and in trenches before Sevastopol, 1854; conspicuous in assault on the Redan, June 1855; recommended for VC; retired with honorary rank of major-general, 1887; assisted Sir Henry Cole [q.v.] in reorganizing science and art department at South Kensington, 1858; inspector for science; 1859; arranged for payment of teachers by results of examinations of pupils, 1859; became 'director of science', 1874; supervised science schools and institutions throughout country; secretary and permanent head of science and art department, 1884; retired, 1889; CB, 1886; KCB, 1893; exhibited water-colour sketches and etchings at Royal Academy (1888-1901), published military pamphlets.

DONNET, Sir JAMES JOHN LOUIS (1816-1905), inspector-general of hospitals and fleets from 1875; born at Gibraltar; MD, St. Andrews, 1857; entered navy, 1840; at capture of Acre; surgeon in Arctic expedition with Sir Erasmus Ommanney [q.v.], 1850-1; in Pacific, 1854; KCB, 1897.

DONOGHUE, STEPHEN (1884-1945), jockey and trainer; won first race at Hyères, 1905; champion jockey, 1914-23; won Derby on Pommern (1915), Gay Crusader (1917), Humorist (1921), Captain Cuttle (1922), Papyrus (1923), and Manna (1925); Queen Alexandra Stakes, Ascot, on Brown Jack, 1929-34; won the Oaks and retired, 1937.

DONOUGHMORE, sixth EARL OF (1875-1948), chairman of committees of the House of Lords. [See HELY-HUTCHINSON, RICHARD WALTER JOHN.]

DOODSON, ARTHUR THOMAS (1890-1968), mathematician and oceanographer; educated at Rochdale secondary school and Liverpool University; first class, B.Sc. (chemistry and mathematics), 1911; first class, mathematics, 1912; handicapped by deafness; worked with Messrs. Ferranti, and Corporation of Manchester, 1912-16; appointed to statistics post, University College, London, 1916; worked on ballistics for War Office; M.Sc., Liverpool, 1914; D.Sc., 1919; secretary, Tidal Institute, Liverpool; associate director, 1929-45; made intensive studies of coastal flooding; president, finance committee, International Union of Geodesy and Geophysics, 1954; hon. lecturer, Liverpool University; FRS, 1933; hon. fellow, Royal Society of Edinburgh, 1953; CBE, 1956.

DORRIEN, Sir HORACE LOCKWOOD SMITH- (1858-1930), general. [See SMITH-DORRIEN.]

DOUBLEDAY, HERBERT ARTHUR (1867-1941), publisher and genealogist; educated at Dulwich College and London University; launched *Victoria History of the Counties of England*; chief editor, 1901-3; for Vicary Gibbs [q.v.] published *The Complete Peerage*; became assistant editor, 1916; editor, 1920-41.

DOUGHTY, CHARLES MONTAGU (1834-1926), poet and traveller; uncle of C. H. M. Doughty-Wylie [q.v.]; BA, Caius College, Cambridge; began travels as poor student, 1870; after settling at Damascus for a year to learn Arabic, joined Meccan pilgrimage caravan, 1876; travelled to Medain Salih, Hail, Kheybar, and the Kasim in Central Asia, finishing at Jiddah, 1878; unique value of journey began when, leaving caravan and proclaiming himself Englishman and Christian, he gathered new information about geography and geology of north-western Arabia and gained understanding of Arab character and nomad life; published *Travels in Arabia Deserta*, his first experiment in Elizabethan English, 1888; devoted remainder of his life to poetry; this includes *The Dawn in Britain* (1906-7) and *Adam Cast Forth* (1908).

DOUGHTY-WYLIE, CHARLES HOTHAM MONTAGU (1868-1915), soldier and consul; joined army, 1889; served in India, Egypt, South Africa, and China; military consul for Konia province of Asia Minor, and later Cilicia, 1906; saved Christian communities at Adana, 1909; CMG and consul-general at Addis Ababa, Abyssinia, 1909; CB, 1912; on staff of Gallipoli expedition, 1915; killed while leading brilliant charge on 'Hill 141'; posthumous VC.

DOUGLAS, Sir ADYE (1815-1906), premier of Tasmania; emigrated to Tasmania, and admitted to bar, 1839; five times mayor of Launceston; elected to legislative council, 1855; opposed transportation; urged claims of Tasmania to responsible government; premier and chief secretary, 1884–6; first agent-general for colony, 1886–7; chief secretary, 1892–4; president of legislative council, 1894–1904; knighted, 1902.

DOUGLAS, LORD ALFRED BRUCE (1870–1945), poet; educated at Winchester and Magdalen College, Oxford; his friendship with Oscar Wilde [q.v.] provoked his father, Lord Queensberry [q.v.], to action which led to Wilde's conviction; edited the *Academy*, 1907–10; thereafter involved in many legal actions; sentenced for libelling (Sir) Winston Churchill, 1923; published verse, books on Wilde, and *Autobiography*, 1929.

DOUGLAS, SIR CHARLES WHITTINGHAM HORSLEY (1850–1914), general; joined army, 1869; served in South African war, 1899–1901; adjutant-general, War Office, 1904–9; KCB, 1907; general, 1910; GCB, 1911; inspector-general, home forces, 1912; chief of imperial general staff, 1914.

DOUGLAS, CLAUDE GORDON (1882–1963), physiologist; educated at Wellington College, Wyggeston grammar school, Leicester, and Magdalen College, Oxford; first class, natural science, 1904; scholar, London University, 1905; BM, B.Ch., 1907; fellow and lecturer in natural science, St. John's College, Oxford, 1907–50; DM, 1913; collaborated with J. S. Haldane [q.v.] on breathing in man, 1913–14; worked in RAMC during 1914–18 war on physiological aspects of gas warfare; MC, 1916; CMG, 1919; university demonstrator, 1927; reader, 1937; titular professor, 1942–53; involved in government committee work on wide variety of problems such as chemical warfare, conditions in mines, pneumoconiosis, and diet and energy requirements; vice-president, St. John's College, dean and keeper of the Groves; FRS, 1922; member, council of Royal Society, 1928–30; *ad hominem* professor, Oxford, hon. fellow, St. John's College, 1950.

DOUGLAS, CLIFFORD (HUGH) (1879–1952), originator of the theory of Social Credit; educated at Stockport grammar school; became an engineer; served in Royal Flying Corps, 1914–18; reorganized production and cost accounting of Royal Aircraft Factory, Farnborough; in *Economic Democracy* (1920) advocated subsidies to liberate prices from cost of production; adviser to government of Alberta (1935–6) where Social Credit League obtained office, 1935; his theories gradually abandoned.

DOUGLAS, GEORGE (pseudonym) (1869–1902), novelist. [See BROWN, GEORGE DOUGLAS.]

DOUGLAS, GEORGE CUNNINGHAME MONTEATH (1826–1904), Hebraist; BA, Glasgow University, 1843; DD, 1867; joined Free Church at disruption; professor of Hebrew at Glasgow theological college, 1857–1902; member of Glasgow school board; member of Old Testament revision company, 1870–84; wrote on conservative lines *The Old Testament and its Critics* (1902).

DOUGLAS, (GEORGE) NORMAN (1868–1952), writer; born Douglass, of mainly Scottish parentage, at Thüringen; educated at Uppingham and Karlsruhe Gymnasium; in the diplomatic service, 1893–6; assistant editor, *English Review*, 1912–15; first gained recognition with *Siren Land* (1911), *Fountains in the Sand* (1912), and *Old Calabria* (1915); published *South Wind* (1917) followed by *They Went* (1920), *Alone* (1921), *Together* (1923), *Looking Back* (1933), and *Late Harvest* (1946); an ardent lover of both sexes he lived mainly in Italy, settling for long periods in Florence and in Capri where he died.

DOUGLAS, SIR (HENRY) PERCY (1876–1939), hydrographer; entered *Britannia*, 1890; chose surveying branch; superintendant of charts, 1910–15; did invaluable work as surveying officer, Dardanelles, 1915–16; captain, 1915; director, naval meteorological service, 1917–18; prepared for Zeebrugge and Ostend exploits, 1918; assistant hydrographer, 1919–21; hydrographer of the navy, 1924–32; rear-admiral, 1927; vice-admiral, 1931; KCB, 1933; developed Douglas–Schafer sounding gear; advocated echo-sounding; gave much attention to preparation of the *Discovery*.

DOUGLAS, WILLIAM SHOLTO, BARON DOUGLAS OF KIRTLESIDE (1893–1969), marshal of the Royal Air Force; educated at Tonbridge and Lincoln College, Oxford; served in France with Royal Field Artillery, 1914; transferred to Royal Flying Corps, 1915; commanded No. 84 Squadron; MC, 1916; DFC, 1919; pilot for Handley-Page Aircraft Co., 1919; rejoined air force as squadron-leader, 1920; served in flying schools, staff appointments, and Imperial Defence College, 1920–36; air vice-marshal, 1938, on Air Ministry staff; assistant (in 1940, deputy) chief of air staff; succeeded Sir Hugh (later Lord) Dowding [q.v.] as commander-in-chief, Fighter Command, 1940; air marshal; KCB, 1941; commander-in-chief, Middle East; air chief marshal, 1942; commander-in-chief, Coastal Command, 1944; commander-in-chief, British Air Forces of Occupation, 1945; GCB; marshal of the Royal Air Force; succeeded

General Montgomery (later Viscount Montgomery of Alamein) as commander-in-chief, British Forces in Germany, and military governor, British zone, 1946–7; baron, 1948; director, BOAC, 1948; chairman, BEA, 1949–64; hon. fellow, Lincoln College, 1941; president, International Air Transport Association, 1956.

DOUGLAS, Sir WILLIAM SCOTT (1890–1953), civil servant; grandson of William Scott Douglas (1815–83, q.v.); educated at George Heriot's School and Edinburgh University; served in Customs and Excise, 1914–20, 1926–9; with Reparation Commission, Paris, 1920–6; in Ministry of Labour, 1929–37; secretary, Department of Health, Scotland, 1937–9; in charge of establishment division, Treasury, 1939–42; permanent secretary, Ministry of Supply, 1942–5; and Ministry of Health, with Aneurin Bevan [q.v.] setting up National Health Service, 1945–51; CB, 1938; KBE, 1941; KCB, 1943; GCB, 1950.

DOUGLAS-PENNANT, GEORGE SHOLTO GORDON, second Baron Penrhyn (1836–1907), landowner; conservative MP for Carnarvonshire, 1866–8, 1874–80; succeeded to peerage and Penrhyn estate and Bethesda slate quarries, 1886; championed free labour and refused to recognize trade-union officials in strikes at his quarries, which he closed down, 1897 and 1900; succeeded in libel action against *Clarion*, the socialist newspaper, 1903; ardent sportsman; won Goodwood Cup (1898) and Ascot Gold Vase (1894); a strong tory and churchman; founded North Wales Property Defence Association.

DOUGLAS-SCOTT-MONTAGU, JOHN WALTER EDWARD, second Baron Montagu of Beaulieu (1866–1929), pioneer of motoring; trained as competent mechanic and engine-driver; conservative MP, New Forest division of Hampshire, 1892–1905; succeeded father, 1905; his energetic advocacy helped extension of motoring and development of modern motor-car; member of Road Board, 1910–19; adviser on mechanical transport services to government of India, 1915–19; interested in aviation.

DOVE, Dame (JANE) FRANCES (1847–1942), founder of Wycombe Abbey School; educated at Queen's College, Harley Street, at a boarding-school, and Girton; MA, Dublin, 1905; assistant mistress, Cheltenham, 1874–7; joined staff of St. Leonards School, 1877; headmistress, 1882–96; founded Wycombe Abbey School, 1896; retired, 1910; DBE, 1928.

DOVE, JOHN (1872–1934), journalist; educated at Rugby and New College, Oxford; called to bar, 1898; joined Milner's 'kindergarten' as assistant town clerk (later town clerk), Johannesburg, 1903; chairman, Transvaal land settlement board, 1907–11; editor, *Round Table*, 1920–34.

DOVER WILSON, JOHN (1881–1969), Shakespearian scholar. [See Wilson.]

DOWDEN, EDWARD (1843–1913), critic; brother of John Dowden [q.v.]; BA, Trinity College, Dublin; professor of English literature, Trinity College, 1867; works include *Shakspere, His Mind and Art* (1875), *Shakspere Primer* (1877), and *Life of Shelley* (1886).

DOWDEN, JOHN (1840–1910), bishop of Edinburgh; son of staunch Presbyterian; BA, Trinity College, Dublin, 1861; DD, 1876; Pantonian professor of theology at Glenalmond, 1874–80; principal of theological hall of Scottish Episcopal Church, 1880–6, and canon of St. Mary's Cathedral, Edinburgh, 1880; bishop of Edinburgh, 1886–1910; hon. LLD, Edinburgh, 1904; founded Scottish History Society (1886), editing several of its publications; Rhind lecturer before Society of Antiquaries of Scotland, 1901; published *The Celtic Church in Scotland* (1894), *The Workmanship of the Prayer Book* (1899), *The Medieval Church in Scotland* (1910), sermons and pamphlets.

DOWDING, HUGH CASWALL TREMENHEERE, first Baron Dowding (1882–1970), air chief marshal; educated at Winchester and the Royal Military Academy, Woolwich; joined Royal Garrison Artillery; served in Gibraltar, Ceylon, Hong Kong, and India; entered Staff College, Camberley, 1912–13; obtained Royal Aero Club's pilot certificate, 1913; RFC Reserve officer; commandant RFC, Dover Camp, 1914; commanded No. 16 squadron in France, 1915; brigadier-general, 1917; permanent officer in RAF, 1919; chief staff officer, Air HQ, Iraq, 1924; director of training, Air Ministry, 1926; member, Air Council, 1930; AOC-in-C, Fighter Command, 1936–40; commanded during the Battle of Britain; appointed to post in Air Ministry, 1941–2; CMG, 1919; CB, 1928; KCB, 1933; GCVO, 1937; GCB, 1940; baron, 1943; in retirement, wrote books and articles on spiritualism and theosophy.

DOWIE, JOHN ALEXANDER (1847–1907), religious fanatic; born in Edinburgh; emigrated to Adelaide, 1860; Congregational minister near Adelaide, 1871; prominent in Sydney religious, social, and political life; built tabernacle for 'divine healing' at Melbourne, Victoria, 1882; removed to Chicago, 1890; opened Zion's tabernacle there, 1893; proprietor and overseer of Zion City on Lake Michigan, 1900; announced himself as 'Elijah the Restorer'; enjoyed for a time wide notoriety and published organ *Leaves of Healing*; visited England (1903 and 1904) with little success; deposed by officers of his church, 1906; unsuccessful in lawsuit for restitution of church funds and property; died at Illinois.

DOWNEY, RICHARD JOSEPH (1881–1953), Roman Catholic archbishop; born at Kilkenny, Ireland; entered Liverpool diocesan junior seminary, 1894, St. Joseph's College, Upholland, 1901; ordained, 1907; DD, Gregorian University, Rome, 1911; worked for Catholic Missionary Society preaching and lecturing, 1911–26; professor of dogmatic theology, Upholland, 1926; vice-rector, 1927; archbishop of Liverpool, 1928–53; purchased site for cathedral and appointed Sir Edwin Lutyens [q.v.] as architect, 1930; champion of voluntary schools and the bishops' spokesman on education; a genial personality and accomplished controversialist.

DOYLE, Sir ARTHUR CONAN (1859–1930), author; nephew of Richard Doyle [q.v.]; MB, Edinburgh University; his early novels include *Micah Clarke* (1887), *The White Company* (1890), and *The Sign of Four* (1890); attracted attention chiefly by series of short stories entitled *The Adventures of Sherlock Holmes*, which began in *Strand Magazine*, 1891; used this creation of his fancy in many stories short and long, so that Sherlock Holmes became familiar figure to millions of readers; served as physician in South African war, 1899–1902; published *The War in South Africa. Its Cause and Conduct*, a vindication of England (1902); in later years absorbed by subject of spiritualism; knighted, 1902.

DOYLE, JOHN ANDREW (1844–1907), historian; inherited property from grandfather, Sir John Easthope [q.v.]; educated at Eton and Balliol College, Oxford; BA, 1867; fellow of All Souls, 1867–1907; took active part in local affairs in Breconshire; advocated rifle shooting at the universities; authority on dog and racehorse breeding; closely studied American history; published *Summary History of America* (1875), *The English in America* (1882), *The Puritan Colonies* (2 vols., 1887), *The Middle Colonies* (1907), *The Colonies under the House of Hanover* (1907), and *Essays on Various Subjects* (posthumous, 1911).

DRAX, Sir REGINALD AYLMER RANFURLY　　　PLUNKETT-ERNLE-ERLE- (1880–1967), admiral. [See PLUNKETT.]

DRAYTON, HAROLD CHARLES GILBERT (always known as HARLEY) (1901–1966), financier; son of an LCC gardener; left school at thirteen for job as office boy in St. David's Group of companies; befriended by J. S. Austen, solicitor responsible for Lord St. David's [q.v.] investment trust companies; flair for figures; manager, Securities Agency, 1928; *de facto* head, St. David's Group, 1938; responsible for management and liquidation of Lloyd George Political Fund; launched Industrial and Commercial Finance Corporation, 1945; chairman, larger trusts in Group; chairman, British Electric Traction, 1947; backed independent television; chairman, Mitchell Cotts, and United Newspapers; director, Midland Bank, and Eagle Star Insurance; chairman, Association of Investment Trusts; treasurer, Institute of Directors, 1946; Plumton estate, near Bishop's Stortford, left to him by J. S. Austen; high sheriff of Suffolk, 1957.

DREDGE, JAMES (1840–1906), civil engineer and journalist; joint-editor and proprietor of *Engineering* from 1879 till death; British commissioner at many international exhibitions, including Chicago (1893) and Brussels (1897); CMG, 1898; published *Modern Examples of Road and Railway Bridges* (1872) and *Electric Illumination* (2 vols., 1882).

DRESCHFELD, JULIUS (1846–1907), physician and pathologist; born in Bavaria of Jewish parents; educated at Owens College, Manchester, and Manchester Royal School of Medicine; MD, Würzburg, 1864; surgeon in Bavarian army, 1864; settled in Manchester, 1869; on staff of Manchester Royal Infirmary, 1873–1907; FRCP, 1883; Bradshawe lecturer, 1887; professor of pathology (1881–91) and of medicine (1891–1907) at Victoria University, Manchester; almost forestalled Pasteur in researches on hydrophobia, 1882–3; an expert neurologist; published numerous scientific papers in English and German journals; memorial volume with biography and bibliography published, 1908.

DREW, Sir THOMAS (1838–1910), architect; pupil of Sir Charles Lanyon [q.v.], 1854; FRIBA, 1889; president of Royal Hibernian Academy, 1900; knighted, 1900; hon. LLD, Dublin, 1905; president of Royal Society of Antiquaries of Ireland, 1895–7; designed Rathmines town hall, 1889, and Belfast Cathedral, 1899.

DREYER, Sir FREDERIC CHARLES (1878–1956), admiral; son of J. L. E. Dreyer [q.v.]; educated at Royal School, Armagh; entered *Britannia*, 1891; became a gunnery specialist and awarded £5,000 for inventions; flag commander (1910–12); CB, 1914; captain (1915–16) to Sir John (later Earl) Jellicoe [q.v.]; present at Jutland; director of naval ordnance, 1917–18; of naval artillery and torpedoes, 1918–19; of gunnery division, 1920–2; rear-admiral, 1923; assistant chief of naval staff, 1924–7; commanded battle cruiser squadron, 1927–9; vice-admiral, 1929; deputy chief of naval staff, 1930–3; admiral and KCB, 1932; commander-in-chief, China station, 1933–6; retired, 1939, but returned to service, 1939–43; GBE, 1937; published *How to get a First Class in Seamanship* (1900), and *The Sea Heritage* (1955).

DREYER, GEORGES (1873–1934), pathologist; born at Shanghai of Danish parentage; MD (1900) and demonstrator of pathology, Copenhagen; professor of pathology, Oxford, 1907–34; revolutionized teaching and extended department in building of Sir William Dunn School of Pathology; instituted (1915) Standards Laboratory providing standardized reagents for serological diagnosis; carried out research with passionate precision of technique; naturalized, 1912; CBE, 1919; FRS, 1921.

DREYER, JOHN LOUIS EMIL (1852–1926), astronomer; born at Copenhagen; studied mathematics and astronomy at Copenhagen University; assistant, Earl of Rosse's observatory, Birr Castle, Parsonstown, Ireland, 1874; assistant, Dublin University observatory, Dunsink, 1878; director, Armagh observatory, 1882–1916; works include *New General Catalogue of Nebulae and Clusters of Stars, being the Catalogue of Sir John Herschel, revised, corrected, and enlarged* (1888), supplements (1895 and 1908), *Tycho Brahe* (1890), *History of the Planetary Systems from Thales to Kepler* (1906), *Scientific Papers of Sir William Herschel* (1912), *Tychonis Brahe Opera Omnia* (1913–26).

DRINKWATER, JOHN (1882–1937), playwright, poet, and actor; educated at Oxford high school; entered assurance company; with (Sir) Barry Jackson founded Pilgrim Players, later (1909) the Birmingham Repertory Theatre, for which he produced, acted, and wrote; works include poems; verse plays; historical plays including *Abraham Lincoln* (1918) and *Oliver Cromwell* (1921); and comedy *Bird in Hand* (1927).

DRIVER, SAMUEL ROLLES (1846–1914), regius professor of Hebrew and canon of Christ Church, Oxford, 1883–1914; BA, New College, Oxford; fellow, 1870; member of Old Testament revision company, 1875–84; works include *A Treatise on the Use of the Tenses in Hebrew*

(1874), *Introduction to the Literature of the Old Testament* (1891), and Old Testament commentaries.

DRUCE, GEORGE CLARIDGE (1850–1932), botanist; qualified as pharmaceutical chemist, Northampton, 1873; opened chemist shop in Oxford, 1879; helped to found Ashmolean Natural History Society of Oxfordshire; Fielding curator in department of botany, 1895–1932; secretary, British Botanical Exchange Club, 1903–32; travelled and collected widely; his classifications unorthodox; published floras of Oxfordshire, Berkshire, Buckinghamshire, and Northamptonshire, and *Comital Flora of the British Isles* (1932); FRS, 1927.

DRUMMOND, Sir GEORGE ALEXANDER (1829–1910), senator in the parliament of Canada (1880–1910) and president of the Bank of Montreal from 1905; emigrated from Edinburgh to Montreal, 1854; founded Canada Sugar Refining Company, 1879; president of Montreal Board of Trade, 1886–8; KCMG, 1904; CVO, 1908; philanthropist and art collector.

DRUMMOND, Sir JACK CECIL (1891–1952), nutritional biochemist; educated at Roan School, Greenwich, King's College School, and East London College; first class, chemistry, 1912; at Cancer Hospital Research Institute, 1914–19; at University College, London, 1919; reader, 1920; professor of biochemistry, 1922–45; scientific adviser, Ministry of Food, 1940–6; adviser on nutrition to SHAEF (1944), to Control Commission, 1945; director of research, Boots Pure Drug Company, 1945–52; published *The Englishman's Food* (with Anne Wilbraham, 1939); knighted and FRS, 1944; murdered with wife and daughter while camping in French Alps.

DRUMMOND, JAMES (1835–1918), Unitarian divine; son of Revd W. H. Drummond [q.v.]; BA, Trinity College, Dublin; studied theology at Manchester New College, London; pastor at Cross Street chapel, Manchester, 1860–9; lecturer at Manchester New College, 1869–85; principal (London), 1885–9, (Oxford), 1889–1906; an eloquent preacher and independent theological thinker and writer.

DRUMMOND, JAMES ERIC, sixteenth EARL OF PERTH (1876–1951), first secretary-general of the League of Nations; succeeded his half-brother, 1937; educated at Eton; entered Foreign Office, 1900; private secretary to prime minister Asquith (1912–15), to foreign secretaries Grey and Balfour [qq.v.] (1915–18); with British delegation to peace conference, 1918–19;

first secretary-general of the League of Nations, 1919-33; recruited excellent talent and made Geneva the political centre for negotiating European problems; ambassador to Italy, 1933-9; awarded Wateler peace prize, 1931; became a representative peer of Scotland (1941) and deputy leader of liberal party in the Lords (1946): CB, 1914; KCMG, 1916; GCMG, 1934; PC, 1933; hon. DCL Oxford, and LLD Liverpool.

DRUMMOND, SIR PETER ROY MAXWELL (1894-1945), air marshal; born and educated in Perth, Western Australia; served in Gallipoli, 1915; with Royal Flying Corps in Palestine, 1916-18; MC, 1917; DSO and bar, 1918; flight-lieutenant, Royal Air Force, 1919; deputy chief, Royal Australian Air Force, 1925-9; senior air staff officer (1937-41), deputy AOC-in-C to Sir A. W. (later Lord) Tedder (1941-3), Middle East; air member for training, 1943-5; lost flying to Canada; KCB, 1943.

DRUMMOND, WILLIAM HENRY (1854-1907), Canadian physician and poet; emigrated from Ireland to Canada, 1865; as telegraph operator (1869) at Bord-à-Plouffe he first met with French-speaking backwoodsmen, whom he faithfully represents in *The Habitant* (1897), *Johnny Courteau* (1901), and *The Voyageur* (1905); graduated in medicine at Bishop's College, Montreal, 1884; professor of medical jurisprudence there from 1895; hon. LLD, Toronto, 1902.

DRURY, (EDWARD) ALFRED (BRISCOE) (1856-1944), sculptor; assistant to Aimé Jules Dalou in Paris, 1881-5; executed sculptures for main entrance, Victoria and Albert Museum (1908), for gate pillars of Victoria Memorial (1911); portrait-sculpture includes Sir Joshua Reynolds at Burlington House (1931); ARA, 1900; RA, 1913.

DRURY-LOWE, SIR DRURY CURZON (1830-1908), lieutenant-general; BA, Corpus Christi College, Oxford, 1853; joined 17th Lancers, 1854; served in Crimea and Indian Mutiny; commanded regiment in Zulu war, 1879-80; CB, 1879; distinguished as cavalry commander in Egyptian war, 1882; helped in victories of Kassasin and Tel-el-Kebir; occupied Cairo, receiving surrender of Arabi Pasha; KCB, 1882; inspector-general of cavalry at Aldershot, 1885-90; lieutenant-general, 1890; GCB, 1895.

DRYLAND, ALFRED (1865-1946), chartered civil engineer; borough surveyor, Deal, 1883; assistant county surveyor, Kent, 1890; county surveyor, Herefordshire (1898),

Wiltshire (1906), Surrey (1908); county engineer, Middlesex, 1920-32; responsible for Great West, New Cambridge, North Circular roads, Barnet and Watford by-passes, and Western Avenue.

DRYSDALE, CHARLES VICKERY (1874-1961), electrical engineer, physicist, and social philosopher; educated at Finsbury Technical College and Central Technical College, South Kensington; associate head, Applied Physics and Electrical Engineering Department, Northampton Institute, London, 1896-1910; D.Sc. (London), 1901; designed electrical measuring instruments; joined Admiralty Experimental Station, 1918; scientific director, 1920; superintendent, Admiralty Research Laboratory, 1921-9; director, scientific research, Admiralty, 1929-34; co-secretary, (with his wife), Malthusian League, 1907; editor, the *Malthusian*; president, 1921-52; member, National Birth Control Council, 1930; fellow, Physical Society, 1898; member of council, 1936-9; fellow, Institute of Physics; vice-president, 1932-6; president, Optical Society, 1904; fellow, Royal Society of Edinburgh, 1921; fellow, Royal Statistical Society; member, Board of Managers, Royal Institution, 1934-6; OBE, 1920; CB, 1932.

DRYSDALE, LEARMONT (1866-1909), musical composer; while student at Royal Academy of Music, London (1888-92), he wrote notable orchestral compositions; his 'Tam o' Shanter', 1891, *The Plague* (musical play), 1896, and *The Red Spider* (light opera), 1898, met with great success; composed original settings of Scots lyrics and arranged folk-songs.

DU CANE, SIR EDMUND FREDERICK (1830-1903), major-general RE, and prison reformer; joined Royal Engineers, 1848; organizer of convict labour in Western Australia and magistrate, 1851-6; designed new land works at Dover and Plymouth, 1858-63; inspector-general of military prisons and chairman of board of directors of convict prisons, 1869; reorganized county and borough prisons; his scheme (1873) transferring control and cost of local prisons to government legalized by Prison Act of 1877; CB, 1873; KCB, 1877; became chairman of prison commissioners to administer the Act; inaugurated registration of criminals; suggested composite portraiture to Sir Francis Galton [q.v.]; major-general, 1877; clever water-colour painter; wrote *The Punishment and Prevention of Crime* (1885).

DUCKETT, SIR GEORGE FLOYD, third baronet (1811-1902), archaeologist and lexicographer; educated at Harrow and Christ

Church, Oxford; joined army, 1832; compiled *Technological Military Dictionary* (1848) in German, English, and French; recognized abroad but ignored at home; succeeded to baronetcy, 1856; FSA, 1869; published genealogical history of Duckett family, and several volumes on the charters and records of Cluniac foundations, 1877-93; wrote *Anecdotal Reminiscences of an Octo-nonagenarian* (1895).

DUCKWORTH, SIR DYCE, first baronet (1840-1928), physician; MD, Edinburgh, 1863; medical tutor, St. Bartholomew's Hospital, 1865; full physician, 1883-1905; joint-lecturer on medicine, 1890-1901; FRCP, 1870; retired from treasurership of College after nearly forty years' service, 1923; Lumleian lecturer, 1896; Harveian orator, 1898; knighted, 1886; baronet, 1909; chief work *Treatise on Gout* (1889); essentially exponent of art of medicine.

DUCKWORTH, WYNFRID LAURENCE HENRY (1870-1956), anatomist; nephew of Sir Dyce Duckworth [q.v.]; educated at Birkenhead school and Jesus College, Cambridge; first class, parts i and ii, natural sciences tripos, 1892-3; qualified at St. Bartholomew's Hospital; MD, 1905; Sc.D., 1906; fellow of Jesus, 1893-1956; master, 1940-5; university lecturer in physical anthropology, 1898-1920; senior demonstrator of anatomy, 1907-20; reader in human anatomy, 1920-40; publications include *Morphology and Anthropology* (1904) and *Prehistoric Man* (1912).

DU CROS, SIR ARTHUR PHILIP, first baronet (1871-1955), pioneer of pneumatic tyre industry; born in Dublin, son of Harvey du Cros who became chairman of the Pneumatic Tyre company, 1889; joined his father, 1892; joint managing director, 1896; founded industry in Coventry; founded Dunlop Rubber Company, 1901, and devoted twenty-five years to its development; conservative MP, Hastings, 1908-18; coalition unionist MP, Clapham, 1918-22; urged military use of motor transport and of aviation; generous with his personal fortune but lost most of it when his company failed; his house near Bognor used for convalescence of King George V, 1929; baronet, 1916.

DUDGEON, LEONARD STANLEY (1876-1938), pathologist; qualified (1899) at St. Thomas's Hospital where he became director of pathology and bacteriology (1905-38) and dean of medical school (1928-38); professor of pathology, London, 1919-38; publications include *Bacterial Vaccines and their Position in Therapeutics* (1927); FRCP, 1908; CMG, 1918.

DUDGEON, ROBERT ELLIS (1820-1904), homoeopath; MD, Edinburgh, 1841; practised homoeopathy in London, 1845; edited *British Journal of Homoeopathy*, 1846-84; translated works by Hahnemann, 1849-50; helped to found Hahnemann Hospital and school of homoeopathy, Bloomsbury Square, 1850; secretary (1848) and president (1878 and 1890) of British Homoeopathic Society; invented Dudgeon's sphygmograph, 1878; wrote on *Homoeopathy*, 1854, and optics.

DUDLEY, second EARL OF (1867-1932), lord-lieutenant of Ireland and governor-general of Australia. [See WARD, WILLIAM HUMBLE.]

DUFF, SIR ALEXANDER (LUDOVIC) (1862-1933), admiral; entered navy, 1875; captain, 1902; director, naval mobilization, 1911-14; rear-admiral, fourth battle squadron, Grand Fleet, 1914-16; director, anti-submarine division, 1916-17; assistant chief of naval staff, 1917-19; vice-admiral and KCB, 1918; commander-in-chief, China station, 1919-22; admiral, 1921.

DUFF, SIR BEAUCHAMP (1855-1918), general; entered Royal Artillery from Woolwich, 1874; served in India; passed out of Staff College with distinction, 1889; military secretary to commander-in-chief in India, Sir George Stuart White [q.v.], 1895-9; military secretary to White, and later on Lord Roberts's staff, during South African war, 1899-1900; CB and returned to India, 1901; assisted Lord Kitchener [q.v.] to reorganize Indian army, 1903-9; chief of staff and KCVO, 1906; KCB, 1907; secretary of military department at India Office, 1909; KCSI, 1910; GCB, 1911; commander-in-chief and military member of council in India, 1914-16; GCSI and recalled to England to give evidence before Mesopotamia commission, 1916; assigned large share of blame for failure of Mesopotamian operations, although partially exonerated on account of over-great responsibility.

DUFF, SIR JAMES FITZJAMES (1898-1970), professor of education and vice-chancellor; educated at Winchester and Trinity College, Cambridge; first class, part i, classical tripos; second class, part ii, economics tripos, 1920-1; lecturer in education, Armstrong College, Newcastle upon Tyne, 1922; educational superintendent, Northumberland County Council, 1925; senior lecturer in education, Manchester, 1927; professor, 1932-7; warden, Durham colleges, 1937-60, alternating as vice-chancellor with Lord Eustace Percy [q.v.] at Newcastle; member of commissions, including higher education in the colonies, 1943-5; chairman, academic advisory committee leading to foundation of Sussex University, 1961; governor, BBC, 1959-65; knighted, 1949; hon. degrees from Aberdeen (1942), Durham (1950), and Sussex (1964); mayor of Durham, 1960; lord lieutenant, Durham, 1964.

DUFF, SIR LYMAN POORE (1865-1955), chief justice of Canada; BA, Toronto, 1887; LLB, 1889; called to bar, Ontario (1893), British Columbia (1895); QC, 1900; judge, Supreme Court of British Columbia (1904), of Canada (1906); chief justice, 1933-44; combined great intellectual force with rare capacity for legal analysis; his contribution mainly in the constitutional field; member, Judicial Committee of Privy Council, 1919-46; bencher of Gray's Inn, 1924; an active liberal but declined public office; GCMG, 1934; hon. degrees from nine universities.

DUFF, SIR MOUNTSTUART ELPHINSTONE GRANT (1829-1906), statesman and author. [See GRANT DUFF.]

DUFFERIN AND AVA, first MARQUESS OF (1826-1902), diplomatist and administrator. [See BLACKWOOD, FREDERICK TEMPLE HAMILTON-TEMPLE.]

DUFFY, SIR CHARLES GAVAN (1816-1903), Irish nationalist and colonial politician; engaged in journalism in Dublin, 1836; started the *Nation* as proprietor and editor, 1842; gathered a brilliant staff of 'Young Irelanders' who developed a strong nationalist sentiment; produced 'The Library of Ireland', a shilling series of Irish biography, poetry, and criticism; was accused of sedition with Daniel O'Connell [q.v.], 1844; opposed O'Connell's Irish federal plan, 1844; called to Irish bar, 1845; became intimate with Thomas Carlyle [q.v.], 1845; formed Irish Confederation, 1847; suggested formation of independent Irish party in House of Commons, 1848; was arrested, the *Nation* being suppressed for advocating rebellion, 1848-9; joined Irish Tenant League for fixity of tenure, fair rents, and free sale; independent MP for New Ross, 1852-5; emigrated to Australia, 1855; barrister at Melbourne; member of the house of assembly, Victoria, 1856; minister of land and works, 1857-9 and 1862-5; carried Duffy's Land Act to facilitate acquisition of land by immigrants; prime minister, 1871-2; KCMG, 1873; speaker of the house of assembly, 1876-80; spent remainder of life in south of Europe in literary work; wrote *Young Ireland, 1840-50* (2 vols., 1880-3), *Life of Thomas Davis* (1890), *Conversations with Thomas Carlyle* (1892), and *My Life in Two Hemispheres* (1898); died at Nice.

DUFFY, SIR FRANK GAVAN (1852-1936), Australian judge; son of Sir C. G. Duffy [q.v.];

educated at Stonyhurst and Melbourne University; called to Victorian bar, 1874; KC, 1901; justice of High Court of Australia, 1913; chief justice, 1931-5; KCMG, 1929; PC, 1932; unfavourable to extended interpretation of Commonwealth powers.

DUFFY, PATRICK VINCENT (1836-1909), landscape painter; keeper of the Royal Hibernian Academy, 1871-1909.

DUKE, SIR FREDERICK WILLIAM (1863-1924), Indian civil servant; educated at Arbroath high school; entered Indian civil service, 1882; assigned to Bengal, where he served in districts, 1884-1908; magistrate of Howrah, 1897-1902; commissioner of Orissa, 1905-8; chief secretary to government of Bengal, 1908; last lieutenant-governor of Bengal, 1911; senior member of council of governor of Bengal, 1912-14; member of council of India, 1914-20; engaged in discussions on framework of Indian government; accompanied E. S. Montagu [q.v.] to India, 1917-18; permanent under-secretary of state, India Office, 1920-4; KCIE, 1911.

DUKE, HENRY EDWARD, first BARON MERRIVALE (1855-1939), judge and politician; born in Devon and educated locally; entered press gallery of House of Commons, 1880; called to bar (Gray's Inn), 1885; joined Western circuit; QC, 1899; recorder of Devonport (1897-1914) and Plymouth (1897-1900); unionist MP, Plymouth (1900-6), Exeter (1910-18); PC, 1915; chief secretary for Ireland after Easter rebellion, 1916-18; his methods in an impossible task criticized as too conciliatory; knight and lord justice of appeal, 1918; president of the Probate, Divorce, and Admiralty division, 1919-33; baron, 1925; serious, imperturbable counsel, formidable in cross-examination; dignified and urbane judge.

DUKES, ASHLEY (1885-1959), dramatist, critic, and theatre manager; educated at Silcoates School and Manchester University; B.Sc., 1905; taught science in London, Munich, and Zürich; studied German theatre; drama critic on *New Age*, 1909-12, *Vanity Fair*, 1912-14, *Star*, 1913-14, *Illustrated Sporting and Dramatic News*, 1920-4; married (Dame) Marie Rambert, 1918; made many adaptations, translations, and dramatizations, including *Jew Süss* (1929); his own best-known play *The Man With a Load of Mischief*, 1924; opened Mercury Theatre in Ladbroke Road, 1933, for poets' drama and his wife's ballet.

DULAC, EDMUND (1882-1953), artist; born in Toulouse; studied at university and art school, Toulouse; settled in London; naturalized, 1912; illustrated books; designed costumes and scenery for theatre, medal for King's poetry prize, and from 1926 covers for *American Weekly*; designed stamps for George VI's reign and coronations of 1937 and 1953; also Free French colonial stamps and first French liberation stamp; a decorator of delicate and patient craftsmanship.

DU MAURIER, SIR GERALD HUBERT EDWARD BUSSON (1873-1934), actormanager; son of G. L. P. B. du Maurier [q.v.]; educated at Harrow; perfected a nonchalant style of acting of remarkable technique; played in *The Admirable Crichton, Peter Pan, Dear Brutus*, and other Barrie plays; in management at Wyndham's (1910-25), later at St. James's; knighted, 1922; skilful producer; his theatrical career singularly successful.

DUNCAN, SIR ANDREW RAE (1884-1952), public servant; educated at Irvine Royal Academy and Glasgow University; MA, 1902; LLB, 1911; practised as solicitor; secretary to government shipbuilding committees, 1916-18; coal controller, 1919-21; knighted, 1921; chairman, Central Electricity Board, 1927-35; of British Iron and Steel Federation, 1935-40, 1945-52; iron and steel controller, 1939; president, Board of Trade, 1940, 1941-2; minister of supply, 1940-1, 1942-5; MP, City of London, 1940-50; a governor of Bank of England, 1929-40; pioneer of large-scale industrial management; sympathetic to nationalized services but opposed to nationalized industries; GBE, 1938; PC, 1940; hon. LLD Glasgow.

DUNCAN, GEORGE SIMPSON (1884-1965), New Testament scholar; educated at Forfar Academy, Edinburgh University (first class, classics, 1906), and Trinity College, Cambridge; first class, part i, classical tripos, 1909; ordained in Church of Scotland; army chaplain, 1915; appointed to GHQ of Sir Douglas (later Earl) Haig [q.v.]; OBE; professor, biblical criticism, St. Andrews, 1919; principal, St. Mary's College, St. Andrews, 1940; vicechancellor, 1952-3; moderator, General Assembly of Church of Scotland, 1949; concerned with production of New English Bible, 1947; president, Society for New Testament Studies, 1948; hon. doctorates from Edinburgh, Glasgow, and St. Andrews; publications include *Saint Paul's Ephesian Ministry* (1929), *Galatians* (in the series of Moffatt Commentaries, 1934), *Jesus Son of Man* (1948), and *Douglas Haig as I Knew Him* (1966).

DUNCAN, SIR PATRICK (1870-1943), South African statesman and governor-general;

born in Scotland; educated at George Watson's College and the university, Edinburgh, and Balliol College, Oxford; first class, *lit. hum.*, 1893; retained lifelong interest in Greek philosophy and published translation of the *Phaedo* (1928); entered Inland Revenue Department under (Lord) Milner [q.v.], 1894; colonial treasurer, Transvaal, 1901-3; colonial secretary, 1903-7; acting lieutenant-governor, 1906-7; called to bar (Inner Temple), 1908; practised in Johannesburg; KC, 1924; judicial commissioner for high commission territories, 1929-33; member of South African parliament, 1910-20, 1921-36; by his fair-mindedness and freedom from racial prejudice exercised a moderating influence; published *Suggestions for a Native Policy* (1912) criticizing racial segregation and emphasizing duty of Europeans to enable natives to share in benefits of European culture; deeply interested in social welfare and education; minister of interior, health, and education under J. C. Smuts [q.v.], 1921-4; his chief lieutenant in South African Party in opposition, 1924-33; minister of mines in coalition under J. B. M. Hertzog [q.v.], 1933-6; governorgeneral of South Africa, 1937-43; on defeat of Hertzog over Union neutrality, 4 Sept. 1939, refused his request for a dissolution as undesirable in the circumstances and invited Smuts to form a government; GCMG and nominated PC, 1937.

DUNDAS, LAWRENCE JOHN LUMLEY, second MARQUESS OF ZETLAND (1876-1961), public servant and author; son of first Marquess, viceroy of Ireland (1889-92); educated at Harrow and Trinity College, Cambridge; aide-de-camp to Lord Curzon [q.v.], viceroy of India, 1900; conservative MP, Hornsey division of Middlesex, 1907-16; member, royal commission on public services in India, 1912-14; governor of Bengal, 1917-22; GCIE, 1917; GCSI, 1922; PC; member, Indian Round Table conferences, 1930-2; succeeded Sir Samuel Hoare (later Viscount Templewood, q.v.) as secretary of state for India, 1935-40; wounded when Sir Michael O'Dwyer [q.v.] was assassinated in London; active supporter of Territorial movement; steward of Jockey Club, 1928-31; active in county business in Yorkshire; president, Royal Geographical Society, 1922-5; chairman, National Trust, 1931-45; publications include *Lands of the Thunderbolt: Sikhim, Chumbi and Bhutan* (1923), *India: a Bird's-eye View* (1924), and *The Heart of Âryâvarta; A Study of the Psychology of Indian Unrest* (1925); also the authorized biography of Earl Curzon (3 vols., 1928) and of the first Earl of Cromer (1932); hon. LLD, Cambridge and Glasgow; hon. Litt.D., Leeds; FBA, 1929; KG, 1942.

DUNDONALD, twelfth EARL OF (1852-1935), lieutenant-general. [See COCHRANE, DOUGLAS MACKINNON BAILLIE HAMILTON.]

DUNEDIN, VISCOUNT (1849-1942), judge. [SEE MURRAY, ANDREW GRAHAM.]

DUNHILL, THOMAS FREDERICK (1877-1946), composer; studied at Royal College of Music; joined staff, 1905; piano professor, Eton, 1899-1908; compositions include chamber music and numerous songs; 'Elegiac Variations' (1922); Symphony in A minor (1923); one-act opera *The Enchanted Garden* and pageant play *The Town of the Ford* (1925); light opera *Tantivy Towers* (1931); children's opera *Happy Families* (1933); and overture 'Maytime' (1945).

DUNHILL, SIR THOMAS PEEL (1876-1957), surgeon; born in Australia; MB, Melbourne, 1903; MD, 1906; on surgical staff, St. Vincent's Hospital, Melbourne, 1908-14; improved operation for goitre; served as surgeon in France, 1914-18; joined G. E. Gask [q.v.] in St. Bartholomew's Hospital surgical professorial unit, 1920; established large private practice as general surgeon; surgeon to King, 1930; sergeant-surgeon, 1939; extra-surgeon, 1952; CMG, 1919; KCVO, 1933; GCVO, 1949.

DUNLOP, JOHN BOYD (1840-1921), inventor and pioneer of the pneumatic rubber tyre; veterinary surgeon in Belfast, 1867; fitted tricycle with pneumatic instead of solid rubber tyres, 1887; formed business which ultimately developed into Dunlop Rubber Company Ltd.; invention revolutionized cycling and made possible motor road vehicle.

DUNMORE, seventh EARL OF (1841-1907), explorer. [See MURRAY, CHARLES ADOLPHUS.]

DUNNE, SIR LAURENCE RIVERS (1893-1970), metropolitan magistrate; educated at Eton and Magdalen College, Oxford; served in 60th Rifles in 1914-18 war; MC, and croix de guerre with palms; called to bar (Inner Temple), 1922; metropolitan magistrate, 1936; chief metropolitan magistrate, 1948-60; bencher, Inner Temple; knighted, 1948; deputy chairman (1948-64) and chairman (1964-6), Berkshire quarter-sessions.

DUNPHIE, CHARLES JAMES (1820-1908), art critic and essayist; educated at Trinity College, Dublin; founded *Patriotic Fund Journal*, 1854-5; art and dramatic critic to *Morning Post*, 1856-95; published *Wildfire* (1876), *Sweet Sleep* (1879), and *The Chameleon* (1888), semi-cynical and fluent essays.

DUNRAVEN and MOUNT-EARL, fourth EARL OF (1841-1926), Irish politician. [See QUIN, WINDHAM THOMAS WYNDHAM-.]

DUNROSSIL, first VISCOUNT (1893-1961), Speaker of the House of Commons. [See MORRISON, WILLIAM SHEPHERD.]

DUNSANY, eighteenth BARON OF (1878-1957), writer. [See PLUNKETT, EDWARD JOHN MORETON DRAX.]

DUNSTAN, SIR WYNDHAM ROWLAND (1861-1949), chemist and director of the Imperial Institute; educated at Bedford School and abroad; professor of chemistry, Pharmaceutical Society's school of pharmacy, 1886-96; director, scientific and technical department, Imperial Institute, 1896; director of the Institute, 1903-24; greatly developed its work of investigating resources of the Empire; a vice-president of Aristotelian and Chemical societies; FRS, 1893; KCMG, 1924.

DU PARCQ, HERBERT, BARON DU PARCQ (1880-1949), judge; educated at Victoria College, Jersey, and Exeter College, Oxford; called to bar (Middle Temple), 1906; KC, 1926; recorder of Portsmouth (1928-9); Bristol (1929-32); inquired into disturbances at Dartmoor prison, 1932; judge of King's Bench division (knighted), 1932; lord justice of appeal (PC), 1938; lord of appeal in ordinary (life peer), 1946-9; chairman, royal commission on justices of peace, 1946-8, Council of Legal Education, 1947-9.

DUPRÉ, AUGUST (1835-1907), chemist; born at Mainz; studied chemistry at Giessen and Heidelberg (Ph.D., 1855); came to London, 1855; assistant demonstrator at Guy's Hospital; discovered presence of copper in vegetable and animal tissues; lecturer in toxicology at Westminster Hospital medical school, 1863-97; made special inquiries into Thames purification and sewage treatment; officially engaged from 1873 in researches on explosives; examined Fenian 'infernal machines', 1882-3; a member of ordnance research board, 1906; evolved original methods of analysis and testing for safety; published many scientific papers; FRS, 1875.

DURAND, SIR HENRY MORTIMER (1850-1924), Indian civil servant and diplomatist; born in India; son of Sir H. M. Durand [q.v.]; entered Indian civil service, 1870; posted to Bengal, 1873; attaché in foreign department where, owing to proficiency in oriental languages, he rose rapidly in political or diplomatic department; foreign secretary in India, 1884-94; conducted mission to Amir of Afghanistan, whereby 'Durand line' between British and Afghan areas of influence was demarcated, 1893; inspired confidence in subordinates and Asiatic rulers with whom he came in contact; minister plenipotentiary at Tehran, 1894-1900; ambassador at Madrid, 1900-3; at Washington, 1903-6; KCIE, 1889; his premature abandonment of his Indian career a mistake; successful neither in Persia nor USA.

DURNFORD, SIR WALTER (1847-1926), provost of King's College, Cambridge; son of Richard Durnford [q.v.], bishop of Chichester; BA and fellow, King's College, Cambridge, 1869; master at Eton, 1870-99; returned to Cambridge, 1899; vice-provost of King's, 1909; provost, 1918-26; GBE, 1919.

DUTT, ROMESH CHUNDER (1848-1909), Indian official, author, and politician; born in Calcutta; educated at Presidency College, Calcutta; called to English bar, 1871; joined Bengal civil service, 1871; CIE, 1892; commissioner of Burdwan, 1894, and of Orissa, 1895-7; member of Bengal legislative council, 1895; settled in London, 1897; as president of National Congress at Lucknow (1899) condemned government's land revenue policy; lecturer on Indian history, University College, London, 1898-1904; revenue minister of state of Baroda, 1904-7; prime minister, 1909; Indian member of royal commission on Indian decentralization, 1907-8; works include *History of Bengali Literature* (1877), *History of Civilization in Ancient India* (3 vols., 1888-90), *Economic History of British India, 1757-1837* (1902), and translations from the Sanskrit.

DUTTON, JOSEPH EVERETT (1874-1905), biologist; MB and CM at Victoria University, Liverpool, and Holt fellow in pathology, 1897; went with Liverpool school of tropical medicine to Nigeria to study the mosquito, 1900; discovered in Gambia first trypanosome in man which caused sleeping-sickness, 1901; at Stanley Falls (1904) discovered cause of tick fever in man; succumbed to fever at Kosongo.

DUVEEN, JOSEPH, BARON DUVEEN (1869-1939), art dealer, patron, and trustee; son of Sir J. J. Duveen [q.v.]; entering father's business became world's foremost art dealer, selling great masters in America; trustee, Wallace Collection (1925-39), National (1929-36) and National Portrait (1933-9) galleries; a princely benefactor to these and to Tate Gallery and British Museum; knighted, 1919; baronet, 1927; baron, 1933.

DUVEEN, SIR JOSEPH JOEL (1843-1908), art dealer and benefactor; born in Holland; settled at Hull as general dealer, 1866; helped to form many private collections of Nankin and oriental porcelain; extended business to tapestry, pictures, and objets d'art, 1879; benefactor to National and Tate galleries; added 'Turner wing' to Tate Gallery, 1908; knighted, 1908.

DYER, REGINALD EDWARD HARRY (1864-1927), brigadier-general; born in India; educated at Royal Military College, Sandhurst; commissioned to Queen's Royal regiment, 1885; transferred to Indian army; saw active service, 1886-1908; achieved notable success when commanding operations in south-eastern Persia, 1916; commanded training brigade at Jullundur, India, 1916; restored order after outbreak of serious and unprovoked civil disturbance at Amritsar, Apr. 1919; three of his actions gave rise to bitter controversy and led in Oct. to appointment by government of India of committee to inquire into proceedings; as result, his resignation demanded by commander-in-chief in India, 1920.

DYER, SIR WILLIAM TURNER THISELTON- (1843-1928), botanist. [See THISELTON-DYER.]

DYKE, SIR WILLIAM HART, seventh baronet (1837-1931), politician; educated at Harrow and Christ Church, Oxford; famous rackets player; an originator of lawn tennis; conservative MP, West Kent (1865-8), Mid-Kent (1868-85), Dartford (1885-1906); chief whip, 1874; succeeded father, 1875; PC, 1880; vice-president of Committee of Council on Education, 1887-92.

DYSON, SIR FRANK (WATSON) (1868-1939), astronomer; educated at Bradford grammar school and Trinity College, Cambridge; second wrangler, 1889; fellow, 1891; chief assistant at Royal Observatory, Greenwich, 1894; astronomer royal for Scotland, 1906; astronomer royal, 1910-33; gave particular attention to reduction and discussion of meridian observations; improved time service; keenly interested in total eclipses of sun; FRS, 1901; knighted, 1915; KBE, 1926; president, International Astronomical Union, 1928-32.

DYSON, SIR GEORGE (1883-1964), musician; began to play music at age of five and to compose at seven; studied at Royal College of Music with (Sir) C. V. Stanford [q.v.], 1900; awarded Mendelssohn travelling scholarship, 1904; first major composition, *Siena*, played at Queen's Hall, 1907; teacher at Royal Naval College, Osborne, 1908-11, Marlborough,

1911-14, and Rugby, 1914; served with Royal Fusiliers in 1914-18 war; D.Mus., Oxford, 1917; major, RAF to organize military bands, 1919; with (Sir) H. Walford Davies [q.v.] composed *The Royal Air Force March*; music master, Wellington, 1921, and master of music, Winchester, 1924; director, Royal College of Music, 1937-52; published books include *The Progress of Music* (1932), and *Fiddling While Rome Burns* (autobiography, 1954); compositions include *The Canterbury Pilgrims* (1931), *Nebuchadnezzar* (1935), and *Music for Coronation* (1953); trustee, Carnegie United Kingdom Trust, 1942;

chairman, 1955-9; knighted, 1941; KCVO, 1953; FRCM, 1929; hon. RAM, 1937; FICS, 1950; FRSCM, 1963; hon. LLD, Aberdeen, 1942, and Leeds, 1956; first president, National Federation of Music Societies, 1935.

DYSON, WILLIAM HENRY (WILL) (1880-1938), cartoonist and etcher; born in Australia; came to London, 1909; cartoonist on *Daily Herald*, 1913-25, 1931-8; of extreme radical outlook and ardent exponent of social credit; exhibited etchings in London and New York.

E

EADY, CHARLES SWINFEN, first BARON SWINFEN (1851-1919), judge; called to bar (Inner Temple), 1879; QC, 1893; Chancery judge and knighted, 1901; lord justice of appeal and PC, 1913; master of the rolls, 1918; baron, 1919.

EADY, SIR (CRAWFURD) WILFRID (GRIFFIN) (1890-1962), public servant; educated at Clifton and Jesus College, Cambridge; first class, classical tripos, 1912; entered home Civil Service, 1913; served in India Office, Home Office, and Department of Foreign Trade, 1913-17; transferred to Ministry of Labour, 1917-38; deputy under-secretary of state, Home Office, 1938-40; deputy chairman, Board of Customs and Excise, 1940, chairman, 1940-2; second secretary, Treasury, 1942-52; worked with Lord Keynes [q.v.] on post-war loan agreements; led British delegation to Washington for talks regarding suspension of sterling convertibility requirement, 1947; director, Richard Thomas and Baldwins, and Steel Company of Wales, 1952; CMG, 1932; CB, 1934; KBE, 1939; KCB, 1942; GCMG, 1948; hon. fellow, Jesus College, Cambridge, 1945; director, Old Vic Trust, Glyndebourne Arts Trust, and National Film Institute; principal, Working Men's College.

EARDLEY-WILMOT, SIR SAINTHILL (1852-1929), forester. [See WILMOT.]

EARLE, JOHN (1824-1903), philologist; BA, Magdalen College, Oxford, 1845; MA, 1849; fellow of Oriel, 1848; professor of Anglo-Saxon, 1849-54 and 1876-1903; rector of Swanswick, 1857-1903; wrote much on Anglo-Saxon; chief works were *Two of the Saxon Chronicles Parallel*, with introduction, notes, and glossary (1865), *A Book for the Beginner in Anglo-Saxon*

(1866), and *The Philology of the English Tongue* (1866) his most popular work, *Anglo-Saxon Literature* (1884), and *The Deeds of Beowulf* (1892); an accomplished Dante scholar.

EARLE, SIR LIONEL (1866-1948), civil servant; educated at Marlborough and Merton College, Oxford; assistant secretary, royal commission on Paris Exhibition, 1898; private secretary to Lord Dudley [q.v.], 1902-3, Lord Crewe [q.v.], 1907-10, (Lord) Harcourt [q.v.], 1910-12; permanent secretary, Office of Works, 1912-33; KCB, 1916; GCVO, 1933.

EAST, SIR ALFRED (1849-1913), painter and etcher; studied art at Glasgow and in Paris; RA exhibitor, 1883; ARA, 1899; RA, 1913; president of Royal Society of British Artists, 1906; knighted, 1910; primarily an interpreter of landscape.

EAST, SIR CECIL JAMES (1837-1908), general; joined 82nd regiment, 1854; served in Crimea (1855) and Indian Mutiny (1857), in Lushai expedition (1871-2) and Zulu war (1879); commanded first brigade in Burmese expedition, 1886-7; CB, 1887; governor of Royal Military College, Sandhurst, 1893-8; lieutenant-general, 1896; general, 1902; KCB, 1897.

EAST, SIR (WILLIAM) NORWOOD (1872-1953), criminal psychologist; educated at King's College School; qualified at Guy's Hospital, 1897; MD, 1901; joined Prison Medical Service, 1899; medical inspector of prisons, 1924; a commissioner of prisons, 1930-8; investigated psychological treatment of prisoners; publications include *An Introduction to Forensic Psychiatry in the Criminal Courts* (1927) and *Medical Aspects of Crime* (1936); knighted, 1947.

EASTLAKE, CHARLES LOCKE (1836-1906), keeper of the National Gallery, London; secretary to Royal Institute of British Architects, 1866-77; as keeper and secretary to National Gallery, 1878-98, rearranged the paintings there; wrote *A History of the Gothic Revival* (1871), and *Notes on the Principal Pictures* at Milan (1883), Paris (1883), Munich (1884), and Venice (1888).

EASTON, HUGH RAY (1906-1965), stained-glass artist; educated at Wellington College and university of Tours; influenced in designing stained glass by (Sir) J. Ninian Comper [q.v.]; trained with Blacking Co. Guildford; set up studio at Cambridge; commander, RNVR, in Ministry of Information, 1939-45; set up workshop at Harpenden, 1945; designed over 250 windows, including work in cathedrals at Canterbury, Durham, Ely, Exeter, Winchester, and many others; also work in barracks, hospitals, town halls, and colleges; greatest achievement, 'Battle of Britain' window, east end of Westminster Abbey.

EATON, HERBERT FRANCIS, third BARON CHEYLESMORE (1848-1925), major-general; joined Grenadier Guards, 1867; retired as major-general without having seen active service, 1899; succeeded brother, 1902; chairman, National Rifle Association, when Bisley became best shooting centre in Empire, 1903-25; engaged in multifarious philanthropic works; collected military medals; KCMG, 1919.

EATON, WILLIAM MERITON, second BARON CHEYLESMORE (1843-1902), mezzotint collector; educated at Eton; succeeded to peerage, 1891; formed largest private mezzotint collection; bequeathed some 10,000 mezzotint portraits to British Museum, which were exhibited, 1905-10.

EBBUTT, NORMAN (1894-1968), journalist; educated at Willaston School, Nantwich; worked in Germany, Spain, France, and Russia, 1909-14; joined *The Times*, 1914; served in RNVR, 1914-18; rejoined *The Times*, 1918; chief correspondent in Germany, 1926-37; expelled for anti-Nazi reporting, 1937; his messages did not influence Neville Chamberlain's policy, but helped people in Britain to realize the nature of Hitler's regime; paralysed in consequence of stroke, 1937-68.

EBSWORTH, JOSEPH WOODFALL (1824-1908), editor of ballads; younger son of Joseph Ebsworth [q.v.]; spent his youth in Edinburgh, where his father kept a bookshop; studied art; employed in Manchester by Faulkner Bros., lithographers, 1848; exhibited four water-colour views of Edinburgh at Scottish Academy, 1849; wrote verse and prose for Scottish press, 1850-60; BA, St. John's College, Cambridge, 1864; MA, 1867; took Anglican orders, 1864; after serving cures, chiefly at Bradford, was vicar of Molash, Kent, 1871-94; devoted himself to editing old collections of 'drolleries' and ballads for the Ballad Society; produced *Bagford Ballads* (from the British Museum), 1876-80, and completed a reprint of *The Roxburgh Ballads* (vols. v-ix), 1883-99; FSA, 1881.

ECCLES, WILLIAM HENRY (1875-1966), physicist and engineer; educated at Royal College of Science, South Kensington; B.Sc., first class, physics, 1898; demonstrator in physics department; assistant to Marconi, 1899-1901; D.Sc., London; head of mathematics and physics department, South Western Polytechnic, Chelsea, 1901; reader in graphic statics, University College, London, 1910; adviser on use of wireless to War Office and other government departments, 1914-18; succeeded Silvanus Thompson [q.v.] as professor, applied physics and electrical engineering, City and Guilds College, Finsbury, 1916-26; private consulting engineer; leading physicist in radio science; president, Institution of Electrical Engineers, 1926-7; president, Physical Society, 1928-30; president, Institute of Physics, 1929; FRS, 1921; fellow, Imperial College.

ECKERSLEY, THOMAS LYDWELL (1886-1959), theoretical physicist and engineer; grandson of T. H. Huxley [q.v.]; educated at Bedales School and University College, London; studied engineering and mathematics at Trinity College, Cambridge, 1911; served in Royal Engineers, 1914-18 on problems of wireless telegraphy; with Marconi company, 1919-46, as theoretical research engineer; worked especially on direction-finding and long-distance short-wave communication links; FRS, 1938.

EDDINGTON, SIR ARTHUR STANLEY (1882-1944), mathematician and astrophysicist; educated at Owens College, Manchester (B.Sc., 1902), and Trinity College, Cambridge; senior wrangler, 1904; fellow, 1907; chief assistant to astronomer royal, 1906-13; devised mathematical method for analysing stellar motions; Plumian professor of astronomy, Cambridge, 1913-44; director of the observatory from 1914; in *Stellar Movements and the Structure of the Universe* (1914) created the new subject of stellar dynamics; on Einstein's theory of 'general relativity' prepared for Physical Society a report (1918) expanded as *The Mathematical Theory of Relativity* (1923); made important original contributions to the theory, his interpretation of which led him to conviction that the true foundation of natural philosophy must be in epistemology; his researches on internal constitution of stars led to discovery of the mass-luminosity relation and made possible the modern theory of stellar evolution; solved the nature of 'white dwarfs'; collected his investigations in *The Internal Constitution of the Stars* (1926); devoted his later years to constructing a comprehensive doctrine combining and transcending the theories of quantum-mechanics and relativity; his final systematic presentation published posthumously as *Fundamental Theory* (1946); wrote also many semi-popular works; FRS, 1914; knighted, 1930; OM, 1938.

EDDIS, EDEN UPTON (1812-1901), portrait painter; exhibited at Royal Academy, 1834-81; painted subjects of rustic genre and children, and portraits including Macaulay (1850), Lord Overstone (1851), Lord Coleridge (1878), Sydney Smith, and Theodore Hook.

EDE, JAMES CHUTER CHUTER-, BARON CHUTER-EDE (1882-1965), parliamentarian. [See CHUTER-EDE.]

EDGE, SIR JOHN (1841-1926), Indian judge; educated at Trinity College, Dublin; called to Irish bar (King's Inns, Dublin), 1864; to English bar (Middle Temple), 1866; QC, 1886; chief justice, high court of judicature, North-Western Provinces of India, at Allahabad, and knighted, 1886; judicial member of council of India, 1899-1908; PC and member of Judicial Committee, 1909-26.

EDGE, SELWYN FRANCIS (1868-1940), pioneer motorist; born in Australia; educated in London; left Dunlop Tyre Co. to go into motor business, 1896; founded S. F. Edge, Ltd., for sale of Napier cars; sold out to Napiers, 1912; successful in bicycle, motor, and motor-boat racing; pioneer in use of farm tractor.

EDGEWORTH, FRANCIS YSIDRO [originally YSIDRO FRANCIS] (1845-1926), economist and statistician; half-nephew of Maria Edgeworth [q.v.]; BA, Balliol College, Oxford; professor of political economy, King's College, London, 1888; Tooke professor of economic science and statistics, 1890; Drummond professor of political economy, Oxford, 1891-1922; fellow of All Souls College, 1891; successively editor, chairman of editorial board, and joint-editor, *Economic Journal*, 1891-1926; FBA, 1903; works include *Mathematical Psychics* (1881) and *Metretike* (1887); contributed regularly to learned journals, especially on

theory of probability; approached moral sciences with mathematical bias.

EDMONDS, SIR JAMES EDWARD (1861-1956), military historian; educated at King's College School and Royal Military Academy, Woolwich; gazetted to Royal Engineers, 1881; instructor in fortification, Woolwich, 1890-5; joined intelligence division, 1899; in South Africa with Lords Kitchener and Milner [qq.v.], 1901-4; in charge of MO5 (later MI5), 1907-11; colonel, 1909; GSO 1, 4th division, 1911-14; with engineer-in-chief, GHQ France, 1914-18; retired as brigadier-general, 1919; director, historical section, military branch, Committee of Imperial Defence, 1919-49; himself wrote the history of the western front with compression and lucidity using British, allied, and enemy material with equal care; CB, 1911; CMG, 1916; knighted, 1928; hon. D.Litt. Oxford; wrote *A Short History of World War I* (1951).

EDOUIN, WILLIE (1846-1908), comedian, whose real name was WILLIAM FREDERICK BRYER; toured through Australia and the Far East, 1857; joined Lydia Thompson [q.v.] in New York, 1870; first appeared in London, 1874; opened Toole's Theatre successfully with *The Babes, or Whines from the Wood*, 1884; scored successes in *Our Flat* (1889-90), *La Poupée* (1897), and *The Girl from Kay's* (1902); admirable for grotesquerie and whimsicality.

EDRIDGE-GREEN, FREDERICK WILLIAM (1863-1953), authority on colour perception; studied at Durham and St. Bartholomew's Hospital; qualified, 1887; MD, 1889; FRCS, 1892; published *Memory* (1888), *Colour-Blindness and Colour Perception*, (1891), and *Physiology of Vision* (1920); his tests for colour-blindness officially adopted, 1915; appointed ophthalmic adviser to Board of Trade, 1920, and subsequently to London Pensions Board and Ministry of Transport; CBE, 1920.

EDWARD VII (1841-1910), KING OF GREAT BRITAIN AND IRELAND AND OF THE BRITISH DOMINIONS BEYOND THE SEAS, EMPEROR OF INDIA, born at Buckingham Palace on 9 Nov. 1841, was eldest son and second child of Queen Victoria and Prince Albert. Baptized at St. George's chapel, Windsor, on 25 Jan. 1842, he was named Albert after his father and Edward after his mother's father, Edward, Duke of Kent. From childhood he spoke English and German, and early learned French, of which in adult years he had an exceptional mastery. In 1846 he paid a first visit to Cornwall and Wales, in 1848 to Scotland, and in 1849 to Ireland. His parents, prompted by Baron Stockmar, bestowed great care on his education, and watched his development closely. He was subjected to a strict discipline under his parents' eyes, at first by private tutors; but he was always impatient of serious study. His permitted recreations included the theatre and music, and he practised elocution and drawing with some success. Practically isolated from boys of his own age, he had small opportunity of playing games, but he rode well and acquired a life-long love of horses and dogs. With his parents he attended the opening of the Great Exhibition (1 May 1851) and of the Crystal Palace (June 1854), and he went to Paris for the first time on a visit to Napoleon III and the Empress Eugénie, Aug. 1855. Walking tours through Dorset (Aug. 1856) and through the English lakes (1857) were followed by a longer foreign tour for study and sightseeing down the Rhine and into Switzerland (July–Oct. 1857). Confirmed at Windsor on 1 Apr. 1858, he was provided with a semi-independent household at White Lodge in Richmond Park (May–Nov. 1858), and on 9 Nov. 1858, his seventeenth birthday, when he received a very solemn admonition from his parents, he was gazetted colonel in the army unattached and nominated KG. A governor, Col. Robert Bruce, was appointed next day. With Bruce he visited at Potsdam his eldest sister, who had lately married Prince Frederick of Prussia, and afterwards made a four months' sojourn in Rome (Jan.–Apr. 1859). He visited Pope Pius IX and met Robert Browning and Frederic Leighton [qq.v.]. On his way home he stayed with King Pedro at Lisbon. Next summer his father sent him to Edinburgh to study under Lyon (later Baron) Playfair [q.v.]. In spite of publicly expressed doubt as to the wisdom of pursuing too academic a training, he matriculated at Oxford on 17 Nov. 1859 as a member of Christ Church, and remained in residence till the end of summer term, 1860. On 8 July 1860 he set out for Canada on the invitation of the Canadian government, accompanied by the Duke of Newcastle, secretary of state for the colonies, and a distinguished suite. It was the first visit of a royal prince to a British colony. At Montreal (4 Sept.) he opened the great railway bridge across the St. Lawrence river, and at Ottawa he laid the foundation stone of the parliament building. Passing to the United States, he was cordially welcomed at Washington by President Buchanan, and planted a chestnut tree by Washington's tomb at Mount Vernon. He returned to Plymouth, 15 Nov. The expedition had the effect of strengthening the loyalty of Canada to the mother country and of increasing the good feeling between England and the United States. After spending the rest of the year at Oxford, the Prince continued his education at Trinity College, Cambridge, through 1861. In Aug. he joined the 2nd battalion of Grenadier Guards in camp at the Curragh, and next month made a third tour in Germany, where a first meeting was arranged for him with his future wife, Princess Alexandra, eldest daughter of Prince Christian of Schleswig-Holstein-Sonderburg-Glücksburg. Her father was next heir to the throne of Denmark, which he ascended as Christian IX on 15 Nov. 1863. The first meeting of the Prince and Princess Alexandra was in the cathedral at Speier, 24 Sept. 1861; each made a good impression on the other. Resuming residence at Cambridge for the Michaelmas term, he was elected a bencher of the Middle Temple (31 Oct.), and opened the new library at the Inn. On 13 Dec. he was summoned from Cambridge to Windsor owing to the illness of his father, who died next day. Thereupon his widowed mother claimed that full control which his father had hitherto exercised; Queen Victoria never ceased to think of the Prince of Wales save as a boy. But his views of life broadened on reaching man's estate, and he chafed against the perpetual tutelage to which his mother sought to subject him. In accordance with plans formed by his father he made a tour in the Holy Land, Feb.–May 1862, accompanied among others by Arthur Penrhyn Stanley [q.v.]; on his return journey he was entertained at Constantinople by the Sultan, at Athens by the King of Greece, and at Fontainebleau by the Emperor Napoleon III. On 9 Sept. 1862 he was formally betrothed to Princess Alexandra at King Leopold's palace of Laeken; on 1 Nov. 1862 the Queen gave formal assent to the union, the announcement of which evoked enthusiasm in England and Denmark, but was received coolly in Germany. At the end of the year his sister, the crown princess of Prussia, with her husband, accompanied the Prince of Wales on a tour through the Mediterranean. Meanwhile a separate establishment was formed for him at home. The estate of Sandringham in Norfolk was purchased for £220,000 out of the income accumulated during his minority from the duchy of Cornwall, which was the heir-apparent's appanage. Marlborough House in London was provided at the public expense. On 25 Feb. 1863 he held a levée for his mother at St. James's Palace, and thus first performed a ceremonial function. On 5 Feb. 1863 he took his seat in the House of Lords at the opening of parliament. Parliament granted him an annuity of £40,000 and one of £10,000 to his bride, Princess Alexandra, with a prospective annuity of £30,000 in case of widowhood. The marriage took place in St. George's chapel, Windsor, on 10 Mar. 1863, and many festivities followed. On 2 May he first attended the Royal Academy banquet. On 7 June he received the freedom of the City of London. On 16 June he was made hon. DCL at Oxford, and next year hon. LLD at Cambridge. While spending much time as a country gentleman and sportsman and visiting annually Scotland, the Riviera, and Homburg, he assumed the role of leader of fashionable life in London, encouraging the lighter social amusements, and forming a wide and cosmopolitan circle of acquaintances. His mother excluded him from all political responsibilities, but he carried on and extended his father's work of charity and public utility. He became president of the Society of Arts (22 Oct. 1863) and president of St. Bartholomew's Hospital (18 Mar. 1867), and was thenceforth indefatigable in inaugurating public buildings, and in presiding at charity festivals. He thrice visited Ireland at this period—in 1865, when he opened the International Exhibition at Dublin (8 May); in 1868 (with the Princess), when he was made a Knight of St. Patrick and hon. LLD, Trinity College, Dublin; and in 1871, when he opened the Royal Agricultural Society's Exhibition. Despite exclusion from political business, the Prince interested himself in foreign affairs. His request for access to Foreign Office papers was practically refused by the Queen on account of his alleged want of discretion. Frank in expression of sympathy with Denmark during the Schleswig-Holstein crisis (1864), he sought first-hand intelligence from men in public life at home and abroad. In 1866 he first went to Russia for the wedding of his wife's sister Dagmar to Tsarevitch Alexander. A long tour (Nov. 1868–May 1869) embraced Copenhagen, Stockholm, Berlin, Vienna, Egypt (where he inspected the newly completed Suez Canal), Constantinople, and Paris. He won golden opinions at foreign courts. During the Franco-German war (1870) he manifested sympathy with France and showed much kindness to Napoleon III and his family in their English exile. Persistent rumours of the Prince's addiction to frivolous amusements grew in 1870, when he appeared in the witness-box to deny imputations in Sir Charles Mordaunt's action for divorce against his wife (Feb.). The sensational press abounded in scandalous insinuations which were reproduced in 'The coming K——', a clever parody of Tennyson's *Idylls*, and many like satires in verse. Meanwhile the Prince opened for the Queen the Thames Embankment (13 July 1871) and presided over a series of international exhibitions at South Kensington which were continued for four years without success (1871–4). In Nov. 1871 he was attacked by typhoid and was gravely ill for a month. After attending a national thanksgiving for his recovery at St. Paul's Cathedral (27 Feb. 1872), he completed his convalescence in a

Mediterranean yachting tour, revisiting Pope Pius IX at Rome. Anxiety over the Prince's illness revived enthusiasm for the monarchy. Gladstone, with whom the Prince's relations were always friendly, vainly urged the Queen to provide the Prince with regular employment either in India or Ireland. He was made field-marshal on 10 June 1875. He was at Vienna for the opening of the International Exhibition, May 1873, and at St. Petersburg for the marriage of his brother, the Duke of Edinburgh, to the Tsar's daughter, 1874. He was tactfully entertained at Birmingham by Joseph Chamberlain [q.v.], then mayor, on 3 Nov. 1874. A tour in India followed, Oct. 1875 to May 1876; he was officially entertained by the Indian government at Bombay, Madras, and Calcutta, and enjoyed much sport as guest of native princes. The personal tie between the princes of India and English royalty was greatly strengthened by his visit. In Europe his interest in France steadily grew; there he was friendly with all political parties and ranks. As president of the British section of the Paris Exhibition, 1878, he advocated at opening ceremonies a good understanding between the two countries. He formed the acquaintance of Gambetta. Though Lord Beaconsfield shared the Queen's doubts of his discretion, the Prince favoured the conservative leader's anti-Russian policy, 1876–8. On 6 May 1879 he voted in the House of Lords for the deceased wife's sister bill. Becoming intimate with Sir Charles Dilke [q.v.], under-secretary for foreign affairs (1880–2) in Gladstone's second ministry, he sought to aid in Paris Dilke's negotiations with the French government for an Anglo-French commercial treaty. He attended the funeral at St. Petersburg, Mar. 1881, of Tsar Alexander II, who had been assassinated. His offer to serve in the Egyptian campaign was refused, July 1882. Openly disapproving the recall of Sir Bartle Frere [q.v.] from the Cape, 1880, he condemned the pusillanimity of liberal policy in Egypt and the Sudan, 1884, and publicly deplored the sacrifice of General Gordon [q.v.]. Appointed member of the royal commission on housing, Feb. 1884, he was friendly with all his fellow commissioners, who included two representatives of labour, Henry Broadhurst and Joseph Arch [qq.v.]. Revisiting Ireland, Apr. 1885, he was coolly received by nationalists, but generally met with a cordial welcome.

In middle life the Prince was active in freemasonry, becoming grand master of the order, Sept. 1875. Appointed trustee of the British Museum, May 1881, he performed his duties regularly. He helped to establish the Royal College of Music, 1882–3. His public activities took him to all parts of the country. He inaugurated the Mersey Tunnel, 28 Apr.

1886, Truro Cathedral, 1880-7, and the Tower Bridge, 1886-94; was four times president of the Royal Agricultural Society; helped to organize at South Kensington international Fisheries Exhibition (1883), Health Exhibition (1884), Inventions and Music Exhibition (1885), and India and Colonies Exhibition (1886), and interested himself in founding the Imperial Institute to celebrate Queen Victoria's jubilee, 1887.

On 10 Mar. 1888 the Prince celebrated his silver wedding. He showed much sympathy with his widowed sister, the Empress Frederick, when her husband, Emperor Frederick III of Germany, died of cancer (15 June 1888) two months after his accession. Though there was mutual affection between the Prince and his nephew, the new emperor William II, they often caused one another passing irritation. The Prince warmly welcomed Emperor William on his first visit to England after his accession, Aug. 1889. On his eldest daughter's marriage, 1889, the Prince invited additional pecuniary provision for his family. Parliament granted an extra annuity of £36,000 to terminate six months after Queen Victoria's death.

The Prince remained a constant patron of the theatre in London, maintaining friendly relations with Sir Henry Irving [q.v.] and leading actors. But his chief amusement from middle life onwards was horseracing, training horses at Newmarket and often occupying rooms at the Jockey Club there; he thrice won the Derby— with Persimmon (1896), Diamond Jubilee (1900), and with Minoru (1909); he won the Grand National at Liverpool with Ambush II (1900). For a time he was prominent in yacht-racing, and won many prizes with the *Britannia*, a vessel designed for him in 1892. His indulgence in sport was deemed excessive by the austere, and on 5 June 1891 he somewhat alienated public opinion by his evidence in the Tranby Croft case [see under WILSON, ARTHUR brother of the first BARON NUNBURNHOLME], when it was admitted that he played baccarat for high stakes; writing privately to Archbishop Benson of Canterbury [q.v.] (13 Aug. 1891), he denied sympathy with gambling. During Lord Salisbury's ministry (1886-92), while taking no part in home politics, he frequently expressed his desire for a good understanding between England and France when visiting the latter country. He engaged in sporting tours in Romania, in Hungary and in Austria (1888), and was the guest of Baron Hirsch, the Jewish millionaire, in Hungary (1894); in the latter year he twice visited Russia, for the marriage of his wife's niece, Xenia, to the Grand Duke Alexander Michálovitch (July), and for the funeral of his wife's brother-in-law, Tsar Alexander III (Oct.). During Gladstone's last ministry (1892-4) much official intelligence was communicated to him with the Queen's reluctant consent; he became a member of the old age pensions commission, 1893; his friendship with Gladstone continued till the statesman's death; he was a pall-bearer at Westminster Abbey at Gladstone's funeral on 25 May 1898, and was president of the national memorial committee (July). In 1895, when Salisbury formed a new administration, the Prince was accorded the cabinet members' right of receiving the foreign dispatches. He devoted some energy to encouraging medical research, helping to found the National Society for the Prevention of Consumption (21 Dec. 1888), and the national leprosy fund which commemorated the tragic heroism of Father Damien (June 1889). At Queen Victoria's diamond jubilee (June 1897) he received the new dignity of grand master and principal grand cross of the Order of the Bath, and he inaugurated in honour of the jubilee the Prince of Wales's Hospital Fund for London, which was renamed on his accession King Edward VII Fund. The difficulties with the Transvaal Republic which led in 1899 to the South African war alienated the goodwill of Europe towards England; on 4 Apr. 1900 the Prince while travelling with the Princess to Denmark was shot at without injury by a youth, Sipido, at the Gare du Nord, Brussels; though president of the British section of the Paris Exhibition of 1899 he, contrary to his custom, did not attend the inauguration in Paris.

The Prince took the oaths of sovereignty under the style of Edward VII the day after Queen Victoria's death at Osborne on 22 Jan. 1901. His speech to the Privy Council, which was his own composition and was delivered without notes, pronounced his full determination to be a constitutional sovereign in the strictest sense of the word. During the new session of parliament, which he opened on 14 Feb. 1901, the additional title was bestowed on him of 'King of the British Dominions beyond the Seas'; he was allowed an annual grant of £470,000 apart from the income of the duchy of Lancaster (£60,000 a year), and the expense of maintaining the royal palaces and royal yachts. Friends' anticipations that he would prove unequal to his new station were belied. He held no well-defined views of domestic legislation, but was deeply interested in foreign policy, and was punctual in formal business. His old circle of friends remained unchanged, and he indulged in all his former amusements, but he gave new splendour to royal ceremonials and exercised a brilliant hospitality in London, which became the headquarters of the court after an interval of forty years. He spent little time at Windsor. He had inherited Balmoral and

Osborne from his mother; but he was not often at Balmoral, and Osborne he abandoned, converting it into a convalescent home for army and naval officers, 9 Aug. 1902. He welcomed the close of the South African war (31 May 1902), and attended a thanksgiving service in St. Paul's (8 June); the coronation was appointed for 26 June, on a scale of exceptional magnificence; but two days before, the King was compelled to submit to an operation for perityphlitis, from which he made a good recovery, and the postponed ceremony took place on 9 Aug. After a yachting cruise to Scotland he made a royal progress through south London, and lunched with the lord mayor at the Guildhall, 24 Oct. Meanwhile Lord Salisbury resigned the premiership (11 July 1902), and was succeeded by A. J. Balfour.

King Edward travelled much abroad during the reign, in accordance with early practice; he attended the funeral of his sister, Empress Frederick, at Friedrichshof (5 Aug. 1901), and after visiting Homburg went to Copenhagen, where he met Tsar Nicholas. His already large kinship with foreign sovereigns was extended by the election of his son-in-law Prince Charles of Denmark as King of Norway (Oct. 1905) and by the marriage of his niece Princess Ena of Battenberg to Alphonso XIII, King of Spain (31 May 1906); wits of Paris thenceforth called him 'l'oncle de l'Europe'. His friendly relations with foreign countries remained much as before his accession. There was no foundation for the current belief of a personal hostility to Germany. The German Emperor visited Sandringham for celebration of the King's sixty-first birthday, 9 Nov. 1902. In spring of 1903 a long foreign tour brought the King to Lisbon, Rome, and Paris, with Charles Hardinge (later Baron Hardinge of Penshurst, q.v.), then assistant under-secretary of the Foreign Office, in attendance. In Paris, which he had not visited for three years, he regained his former popularity, and conspicuously helped to improve the relations between the two countries. The *entente cordiale* was concluded on 8 Apr. 1904. He was entertained by the German Emperor at Kiel on 29 June 1904. Thenceforth he spent three or four months each year abroad. Several weeks each spring were passed at Biarritz, and a like period of the autumn at Marienbad. He cruised in the Mediterranean during springs of 1905, '6, '7, and '9, meeting the Kings of Greece, Spain, and Italy; and he visited the northern courts of Sweden, Denmark, and Norway, Apr. 1908. Frequently passing to and from Paris, he enjoyed varied intercourse with French society. Only during 1905 did he fail to visit Germany, where the press imputed to him anti-German tendencies. In Aug. 1906, 1907, and 1908 he met the

German Emperor in Germany on good terms. The Emperor paid a state visit to Windsor, 11-18 Nov. 1907, and King Edward returned the compliment at Berlin, Feb. 1909. He met the Emperor of Austria thrice—once at Gmunden (Aug. 1905) and twice at Ischl (Aug. 1907-8). He visited the Tsar of Russia at Reval (9 June 1908) and the Tsar was his guest at Cowes (Aug. 1909); some resentment was shown in England at the King's friendly relations with the Tsar, owing to English sympathy with Russian revolutionary movements.

Meanwhile at home he continued to identify himself with philanthropic work and public improvements. He laid the foundation stone of Liverpool Cathedral, 19 June 1904; opened university buildings at Sheffield (1905), Leeds (1908), and Birmingham (1909), and he thrice visited Ireland—in 1903, 1904, and 1907. He intervened little in domestic politics, though he was interested in appointments. With his third prime minister, Sir Henry Campbell-Bannerman, he developed cordial relations. Of Campbell-Bannerman's colleagues Lord Carrington was an old friend, and with (Lord) Haldane [q.v.] he was soon very intimate. On Campbell-Bannerman's retirement, Apr. 1908, the King invited Asquith to fill the vacant office. When the House of Lords threatened the rejection of Lloyd George's budget of 1909, the King vainly urged on the conservative leaders the impolicy of that action. On the defeat of the budget in the Lords (30 Nov.) and the continuance of the liberals in office after the general election, the King raised no remonstrance to his ministers' proposal to curtail for the future the veto of the Lords. The King disliked the situation but abstained from interference; the controversy was still in progress at his death. During his usual spring visit to Biarritz in 1910 he suffered severely from bronchial trouble, but he returned to London (27 Apr.) apparently in good health. Taken ill in London on 2 May, he died at Buckingham Palace on 6 May, and his only surviving son was proclaimed George V. The dead King lay in state in Westminster Hall (16-20 May), and was buried in the vault below St. George's chapel, Windsor.

King Edward eminently satisfied contemporary conditions of kingship; of cosmopolitan temperament, he spoke with equal ease English, French, and German; he revived the ceremonial splendour of the crown, and proved himself an admirable representative of the nation abroad. The austere deemed his addiction to pleasure excessive; but his support of philanthropic causes silenced criticism. He had an expert faculty for business, and distributed his energies over a wide field; he gathered orally very varied stores of knowledge, and remembered personal details with great accuracy; he never seems to have forgotten a face. Personally courageous, he admired every manifestation of heroism. He was greatly attached to dumb animals. Memorials were erected in all parts of the Empire.

By his wife, Queen Alexandra, the King had two sons, Albert Edward (*b.* 8 Jan. 1864, *d.* 14 Jan. 1892) and his successor, George (*b.* 3 June 1865); a third son, John (*b.* 6 Apr. 1871) lived only a day; the two surviving sons were both educated in youth as naval cadets. Of the King's three daughters, Princess Louise (*b.* 20 Feb. 1867), afterwards Princess Royal, married the Duke of Fife (27 July 1889); the second, Princess Victoria, was born on 6 July 1868; and the third, Maud (*b.* 26 Nov. 1869), married in 1896 Prince Charles of Denmark, afterwards Haakon VII, King of Norway.

EDWARD OF SAXE-WEIMAR, PRINCE (1823-1902), field-marshal; brought up by his aunt, Queen Adelaide; playfellow of Queen Victoria and George, Duke of Cambridge; joined army, 1841; served with Grenadier Guards in Crimea with distinction; ADC to Lord Raglan [q.v.] (1855) and to Queen Victoria (1855-9); commanded forces in Ireland, 1885-90, and 1st Life Guards from 1888 to death; KCB, 1881; GCB, 1887; KP, 1890; GCVO, 1901; hon. LLD, Dublin, 1891; field-marshal, 1897.

EDWARDS, ALFRED GEORGE (1848-1937), successively bishop of St. Asaph and first archbishop of Wales; educated at Llandovery College and Jesus College, Oxford; deacon, 1874; priest, 1875; headmaster, Llandovery College, 1875-85; chaplain and secretary to bishop of St. David's, 1885-9; bishop of St. Asaph, 1889; leading negotiator for the Church over Welsh disestablishment; secured terms better than those originally proposed; thereafter occupied with reorganization and financial consolidation as first archbishop of Wales, 1920-34.

EDWARDS, EBENEZER (1884-1961), miners' leader; worked in collieries from age of twelve; went to Ruskin College, Oxford, on miners' scholarship, 1908; president, Ashington Miners' lodge, 1912; financial secretary, Northumberland miners, 1920; member, executive, Miners' Federation, 1926; labour MP, Morpeth, 1929-31; vice-president, Miners' Federation, 1930; president, 1931; general secretary, 1932-1944; secretary, National Union of Mineworkers, 1944-6; chief labour relations officer, National Coal Board, 1946-53; secretary, Miners' International Federation,

1934-46; president, Trades Union Congress, 1944-5.

EDWARDS, SIR **FLEETWOOD ISHAM** (1842-1910), lieutenant-colonel RE; joined Royal Engineers, 1863; accompanied Sir John Simmons [q.v.] to Berlin congress, 1878; keeper of the privy purse to royal household, 1895; KCB, 1887; lieutenant-colonel, 1890; PC, 1895; GCVO, 1901; intimate adviser of Queen Victoria.

EDWARDS, HENRY SUTHERLAND (1828-1906), author and journalist; joined *Punch* staff, 1848; wrote light drama with R. B. Brough and A. S. Mayhew [qq.v.]; served as *The Times* correspondent in Poland (1862-3) and in Franco-Prussian war (1870-1); first editor of *Graphic*, 1869; published *History of Opera* (2 vols., 1862), lives of Rossini (1869) and Sims Reeves (1881), *Personal Recollections* (1900), and translations.

EDWARDS, JOHN PASSMORE (1823-1911), editor and philanthropist; at first a lawyer's clerk at Truro; represented at Manchester (1845) the London *Sentinel*, founded in interests of Anti-Corn Law League; in London advocated early closing, chartism, and international peace; after unsuccessful publishing ventures he purchased *Building News*, 1862, and *Echo*, first half-penny newspaper, 1876, which he successfully edited and controlled till 1896; supported all progressive movements; president of London Reform Union, 1894, and of Anti-Gambling League; an enthusiastic member of Peace Society; denounced Crimean and South African wars; liberal MP for Salisbury, 1880-5; founded some seventy free libraries, hospitals, and convalescent homes in United Kingdom, as well as an art gallery at Newlyn, a settlement in London, and University Hall in Clare Market; endowed scholarship in English literature, Oxford, 1902; declined knighthood; published *A Few Footprints*, an autobiography (1905).

EDWARDS, LIONEL DALHOUSIE ROBERTSON (1878-1966), sporting artist; trained at W. Frank Calderon's School of Animal Painting, London; member, London Sketch Club; drawings accepted by *Country Life*, 1898; first exhibition at Porlock, 1904; served in Army Remount Service during 1914-18 war; member, Royal Institute of Painters in Water Colours, 1927; exhibited at Royal Academy, 1931; publications include *My Hunting Sketch Book*, vols. i and ii (1928 and 1930), *Famous Foxhunters* (1932), *My Irish Sketch Book* (1938), *Scarlet and Corduroy* (1941), *Reminiscences of a Sporting Artist* (1947), and *Thy Servant, the Horse* (1952); held exhibitions at Tryon Gallery, London, the last one, 1964.

EDWARDS, MATILDA BARBARA BETHAM- (1836-1919), novelist and writer on French life; mainly self-educated; travelled widely in France among republican and anti-clerical circles; wrote novels for sixty years; edited writings of Arthur Young [1741-1820, q.v.].

EDWARDS, SIR OWEN MORGAN (1858-1920), man of letters; began life as itinerant preacher in ministry of Welsh Calvinistic Methodists; first class in modern history, Balliol College, Oxford; tutorial fellow of Lincoln College, Oxford, 1889-1907; chief inspector of Welsh education, 1907-20; edited Welsh magazines and wrote books on Wales; knighted, 1916.

EGERTON, SIR ALFRED CHARLES GLYN (1886-1959), scientist; educated at Eton and University College, London; first class, chemistry, 1908; instructor at Royal Military Academy, 1909-13; worked on explosives, 1914-18; at Clarendon Laboratory, Oxford, 1919-36; reader in thermodynamics, 1921-36; professor of chemical technology, Imperial College of Science, 1936-52; researched problems of combustion; war services included reorganization of British Central Scientific Office, Washington, 1942; FRS, 1926; knighted, 1943; honorary degrees from universities at home and abroad.

EGERTON, SIR CHARLES COMYN (1848-1921), field-marshal; joined army, 1867, and Indian army, 1871; DSO, 1891; CB, 1895; KCB, 1903; led successful expedition against 'Mad Mullah' of Somaliland, 1903-4; full general, 1904; member of council of India, 1907-17; field-marshal, 1917.

EGERTON, HUGH EDWARD (1855-1927), historian; BA, Corpus Christi College, Oxford; first Beit professor of colonial history, Oxford, 1905-20; fellow of All Souls College, 1905; works include *Short History of British Colonial Policy* (1897), *Federations and Unions within the British Empire* (1911), *Causes and Character of the American Revolution* (1923).

ELGAR, SIR EDWARD WILLIAM, baronet (1857-1934), composer; son of an organist and music seller in Worcester; studied violin but gradually turned to composition in which he was encouraged by his wife (married, 1889); works include 'Froissart' overture (1890), 'The Light of Life' and 'Scenes from the Saga of King Olaf' (1896); 'Enigma Variations' (1899)

and 'The Dream of Gerontius' (1900) set the seal upon his genius; there followed 'The Apostles' (1903), 'The Kingdom' (1906), 'Pomp and Circumstance' marches (1901-7), overture 'Cockaigne' (1901), 'Introduction and Allegro' for strings (1905), Symphony in A Flat (1908), in E Flat (1911), violin concerto (1910), 'Falstaff' (1913), violoncello concerto (1919); composed little after wife's death in 1920; master of king's musick, 1924-34; knighted, 1904; OM, 1911; baronet, 1931.

ELGAR, FRANCIS (1845-1909), naval architect; student at Royal School of Naval Architecture, South Kensington, 1864-7; chief assistant to Sir Edward Reed [q.v.], 1871; adviser on naval construction to Japanese government, 1879-81; served on departmental committee of Board of Trade, whose report led to regulations fixing maximum loadline for merchant ships, 1883; first professor of naval architecture, Glasgow University, 1883-6; director of dockyards, 1886-92; consulting naval architect of Fairfield Shipbuilding Company, 1892-1907; hon. LLD, Glasgow, 1885; FRS, 1895; FSA, 1896; member of tariff reform commission, 1904; published *The Ships of the Royal Navy*, 1875.

ELGIN, ninth EARL OF (1849-1917), statesman and sometime viceroy of India. [See BRUCE, VICTOR ALEXANDER.]

ELIAS, JULIUS SALTER, VISCOUNT SOUTH-WOOD (1873-1946), newspaper proprietor; entered Odhams Brothers, 1894; director, 1898; managing director, 1920; chairman, 1934; printed first issue of *John Bull* launched by Horatio Bottomley [q.v.], 1906; bought paper from him, 1920; acquired the *People* (1925), 51 per cent holding of *Daily Herald* (1929), *Illustrated*, *News Review*, *Sporting Life*, *Woman*, and others; opened works at Watford (1936), printing for Odhams and other concerns, such as Illustrated Newspapers, Ltd., of which he was chairman; baron, 1937; viscount, 1946.

ELIOT, SIR CHARLES NORTON EDGE-CUMBE (1862-1931), diplomatist and orientalist; educated at Cheltenham and Balliol College, Oxford; first class, *lit. hum.*, and fellow of Trinity, 1884; entered diplomatic service, 1887; serving in Near East; British representative, Samoan commission, 1899; KCMG, 1900; commissioner for British East African Protectorate, 1900; resigned on question of policy, 1904; vice-chancellor, Sheffield (1905-12) and Hong Kong (1912-18) universities; PC, 1919; ambassador to Japan, 1920-6; works include *Turkey in Europe* (1900) and *Hinduism and Buddhism* (1921).

ELIOT, Sir JOHN (1839-1908), meteorologist; educated at St. John's College, Cambridge; second wrangler and first Smith's prizeman, 1869; held mathematical professorships in India until he became meteorological reporter to the Indian government, 1886; director-general of Indian observatories, 1899-1903; FRS, 1895; CIE, 1897; KCIE, 1903; advocated organization of meteorological work on imperial basis; published departmental reports and *Climatological Atlas of India* (1906); an accomplished musician.

ELIOT, THOMAS STEARNS (1888-1963), poet, playwright, critic, editor, and publisher; born in St. Louis, Missouri; educated at Smith Academy, St. Louis, 1898-1903, Milton Academy, Massachusetts, 1905, and Harvard University; BA, 1909; MA, 1910; composed first mature poem, 1910; awarded travelling scholarship; studied the *Appearance and Reality* of F. H. Bradley [q.v.] at Merton College, Oxford, under supervision of Harold Joachim [q.v.], 1914; met Ezra Pound and decided to settle in England; 'Prufrock' published in *Poetry* (Chicago), 1915; teacher at High Wycombe grammar school, 1915, and Highgate junior school, 1916; worked in Lloyds Bank, London, 1917-25; published *Prufrock and Other Observations* (1917), *Poems* (1919), *Ara Vos Prec* (1920), *The Sacred Wood* (1920), and *The Waste Land* (1922); became British subject, 1927; founded and edited the *Criterion*, quarterly review, 1922; director, Faber & Gwyer (later Faber & Faber), 1925; revisited United States to lecture at Harvard, 1932-3; received Nobel prize for literature, 1948; OM, 1948; hon. fellow, Merton College, Oxford, and Magdalene College, Cambridge; eighteen honorary degrees; Legion of Honour; published poems include *Sweeney Agonistes* (1932) and *Four Quartets* (1944), four poems which had been published separately as *Burnt Norton* (1935), *East Coker* (1940), *The Dry Salvages* (1941), and *Little Gidding* (1942); poetic drama includes *Murder in the Cathedral* (1935), *The Family Reunion* (1939), *The Cocktail Party* (1949), *The Confidential Clerk* (1953), and *The Elder Statesman* (1958); other publications include *After Strange Gods* (1934), *On Poetry and Poets* (1957), and *To Criticize the Critic* (1965); a man of letters *par excellence* of the English-speaking world.

ELKAN, BENNO (1877-1960), sculptor; born at Dortmund of Jewish parentage; studied in Munich, Karlsruhe, Paris, and Rome; influenced by Rodin and Bartholomé; his 'Germany Mourns Her Heroes' (1913-14) became Frankfurt's war memorial; settled in England, 1933; naturalized, 1946; exhibited portrait heads and medals at Royal Academy; executed bronze candelabra for Westminster Abbey, Buckfast Abbey, Israel's Parliament House, etc.; OBE, 1957.

ELLERMAN, Sir JOHN REEVES, first baronet (1862-1933), financier and shipowner; trained as accountant; joined (1892) board of Leyland line and became chairman; founded Ellerman line, 1901; absorbed one concern after another including (1916) Wilson's of Hull; joined Lord Northcliffe [q.v.] as shareholder in *The Times* and Associated Newspapers; held bulk of shares of *Sphere* and *Tatler* group; dealt in London property, etc.; had passion for secrecy and obscured transactions under guise of trust companies; estate worth over £36½ million; baronet, 1905; CH, 1921.

ELLERY, ROBERT LEWIS JOHN (1827-1908), astronomer; went to Melbourne, 1851; superintendent of newly founded government observatory at Williamstown, 1853; director of geodetic survey of Victoria, 1858-74; government astronomer at Melbourne, 1863-95; prepared catalogues of star places; helped in photographic chart of the whole sky; a founder and president (1856-84) of Royal Society of Victoria; FRAS, 1859; FRS, 1873; CMG, 1889; work recorded in *Astronomical Results of the Melbourne Observatory*, 1869-88, and in the Melbourne *General Catalogues*, 1874 and 1890.

ELLES, Sir HUGH JAMIESON (1880-1945), general; educated at Clifton and Woolwich; commissioned in Royal Engineers, 1899; captain, 1908; commanded Tank Corps in France, 1916-18; improved tactical training and first employed tanks successfully at Cambrai, 1917; DSO, 1916; commanded Tank Corps training centre, Wool, 1919-23; major-general, 1928; director of military training, 1930-3; lieutenant-general, 1934; master-general of the ordnance, 1934-8; general, 1938; civil defence commissioner for south-west England, 1939-45; KCMG, 1919; KCVO, 1929; KCB, 1935.

ELLICOTT, CHARLES JOHN (1819-1905), bishop of Gloucester; educated at St. John's College, Cambridge; Bell university scholar, 1838; BA, 1841; MA, 1844; Platt fellow, 1845-8; Hulsean lecturer, 1859, and professor, 1860; professor of New Testament exegesis at King's College, London, 1858; dean of Exeter, 1861-3; bishop of Gloucester and Bristol, 1863, and of Gloucester on division of sees, 1897; restored cathedral, promoted church extension, formed Church Aid Society; member of royal commission on ritual (1867-70) and other committees; chairman of New Testament revision company, 1870-81; published commentaries on Pauline epistles (1856) and on the New Testament (3 vols., 1878-9, abridged edition for schools, 14 vols., 1878-83), the Old Testament (5 vols., 1882-4), and the complete Bible (7 vols., 1897); member of Alpine Club, 1871-1904.

ELLIOT, ARTHUR RALPH DOUGLAS (1846-1923), politician; brother of fourth Earl of Minto [q.v.]; BA, Trinity College, Cambridge; called to bar (Inner Temple), 1870; liberal MP Roxburghshire, 1880; liberal unionist MP, 1886-92; MP, Durham, 1898-1906; financial secretary to Treasury, Apr. to Sept. 1903; edited *Edinburgh Review*, 1895-1912.

ELLIOT, SIR GEORGE AUGUSTUS (1813-1901), admiral; born at Calcutta; entered navy, 1827; in command of ships on North America station, in Channel, and in Baltic, 1843-55; rear-admiral, 1858; superintendent of Portsmouth dockyard, 1863-5; admiral, 1874; conservative MP for Chatham, 1874-5; commander-in-chief at Portsmouth, 1875; KCB, 1877.

ELLIOT, GILBERT JOHN MURRAY KYNYNMOND, fourth EARL OF MINTO (1845-1914), governor-general of Canada and viceroy of India; educated at Eton and Trinity College, Cambridge; entered army, 1867; served in many wars in different parts of the world, 1870-82; military secretary to governor-general of Canada, Lord Lansdowne [q.v.], 1883-5; succeeded father, 1891; governor-general of Canada, 1898-1904; maintained cordial relations with colonial secretary, Joseph Chamberlain [q.v.], and Sir Wilfrid Laurier [q.v.], liberal premier, and acquired great popularity throughout the Dominion; his period of office era of great prosperity for Canada; directly responsible for Canadian troops taking part in South African war; viceroy of India, 1905-10; worked in complete harmony with secretary of state, John (afterwards Viscount) Morley [q.v.]; claimed to have initiated Morley-Minto reforms of 1909; established friendly relations with Indian princes; KG, 1910.

ELLIOT, SIR HENRY GEORGE (1817-1907), diplomatist; born at Geneva; son of second Earl of Minto [q.v.], and brother-in-law of Lord John Russell [q.v.]; in Tasmania, 1836-9; entered diplomatic service, 1841; British envoy at Copenhagen, 1858; sent on special mission to Naples to congratulate Francis II on accession to throne of Two Sicilies, and to obtain constitutional reforms, 1859; neglect of his representations led to seizure of Sicily by Garibaldi, 1860; left Naples

on union of Italy under King Victor Emanuel, 1860; sent on special mission to Greece, Apr. 1862; assisted in arranging new constitution under new King George I; succeeded Sir James Hudson [q.v.] as British envoy at Turin, 1863; appointment roused much criticism and political controversy; transferred to Constantinople and made PC, 1867; British representative at opening of Suez Canal; GCB, 1869; often in conflict with Russian ambassador; criticized for delay in reporting 'Bulgarian atrocities' of 1876; British plenipotentiary at conference at Constantinople, 1876; ambassador at Vienna, 1877-84; engaged in critical negotiations between conclusion of San Stefano treaty and the meeting of the congress at Berlin, 1878; caused Gladstone to disavow his attack on Austrian government, 1880; retired to England, 1884.

ELLIOT, WALTER ELLIOT (1888-1958), politician; educated at Glasgow Academy and University; first class, science, 1910, medicine, 1913; conservative MP, Lanark, 1918-23, Kelvingrove, 1924-45, 1950-8, Scottish Universities, 1946-50; under-secretary for health for Scotland, 1923, 1924-6; under-secretary for Scotland, 1926-9; financial secretary to Treasury, 1931-2; PC, 1932; minister of agriculture, 1932-6; secretary of state for Scotland, 1936-8; minister of health, 1938-40; became freelance journalist and broadcaster; chairman, commission on higher education in West Africa, 1943-4; rector, Aberdeen University, 1933-6, Glasgow University, 1947-50; FRS, 1935; FRCP, 1940; CH, 1952; honorary degrees from four Scottish universities.

ELLIOTT, SIR CHARLES ALFRED (1835-1911), lieutenant-governor of Bengal; son of Henry Venn Elliott [q.v.]; joined East India Company, 1856; served in Mutiny; assistant commissioner in Oudh till 1863, where he collected information about its history and folklore; secretary to government of North-West Provinces, 1870-7; concerned chiefly with settlement and revenue questions; directed famine relief operations in Mysore; issued famous report on Indian famines of 7 July 1878; chief commissioner of Assam, 1881; chairman of committee of inquiry into Indian public expenditure, 1886; CSI, 1878; KCSI, 1887; lieutenant-governor of Bengal, 1890-5; prosecuted survey and compiled record of rights in Bihar, despite opposition from zemindars; strong advocate of economy; adopted firm attitude to sedition in native press; retired, 1895; served on London school board, 1897-1900, and on education committee of London County Council, 1904-6; published *Chronicles of Oonao* (1862) and reports on Indian administration.

ELLIOTT, EDWIN BAILEY (1851-1937), mathematician; educated at Magdalen School and College, Oxford; first class, mathematics, 1873; fellow and tutor of Queen's College, 1874-92; lecturer, Corpus Christi College, 1884-93; first Waynflete professor of pure mathematics, 1892-1921; FRS, 1891; his *Introduction to the Algebra of Quantics* (1895) for long the recognized textbook.

ELLIOTT, THOMAS RENTON (1877-1961), physician and physiologist; educated at Durham School and Trinity College, Cambridge; first class, parts i and ii, natural sciences tripos, 1900-1; research on autonomic nervous system; medical student, University College Hospital, London, 1906; MD, 1908; assistant physician, and Beit fellow 1910; FRS, 1913; FRCP, 1915; served in France, 1914-18; DSO, 1918; CBE, 1919; professor of medicine, University College, 1919-39; served three terms on Medical Research Council; Beit and Welcome trustee; member, Goodenough committee on medical education, 1942-4; fellow, Clare College, Cambridge, 1908.

ELLIS, Sir ARTHUR WILLIAM MICKLE (1883-1966), physician and regius professor of medicine in the university of Oxford; born in Toronto; educated at Upper Canada College and Toronto University; BA, 1906; MB, 1908; MD; resident in pathology, Lakeside Hospital, Cleveland, Ohio; demonstrator in pathology, Western Reserve University medical school, 1909-10; assistant resident physician, Hospital of the Rockefeller Institute, New York, 1911-14; served in Canadian Army Medical Corps, 1914-18; OBE, 1917; friend of Sir William Osler [q.v.]; assistant director, medical unit, London Hospital, 1920; first professor of medicine, London Hospital, 1924-43; MRCP, 1920; FRCP, 1929; advisor in medicine, Ministry of Health, 1939; regius professor of medicine, Oxford, 1943-8; knighted, 1953; made distinguished contributions to knowledge of kidney diseases.

ELLIS, FREDERICK STARTRIDGE (1830-1901), bookseller and author; started bookselling business in Covent Garden, 1860; removed to New Bond Street, 1872; long official buyer for British Museum; edited Henry Huth's catalogue of books, 5 vols., 1880; friend and publisher of William Morris, Rossetti, and Ruskin [qq.v.]; friend also of Swinburne and Burne-Jones [qq.v.]; compiled concordance to Shelley, 1892; edited Morris's Kelmscott editions of Caxton's *Golden Legend* (1892) and Cavendish's *Life of Wolsey* (1893).

ELLIS, HENRY HAVELOCK (1859-1939), pioneer in the scientific study of sex, thinker,

critic, essayist, and editor; educated at private schools in Surrey; taught in New South Wales, 1875-9; influenced by writings of James Hinton [q.v.]; saw search for scientific truth as source of artistic satisfaction and chose study of sex as his life-work; medical student at St. Thomas's Hospital, 1881-9; began 'Mermaid' series of dramatists contemporary with Shakespeare; edited 'Contemporary Science' series, 1889-1915; developed relations of passionate attachment to Olive Schreiner [q.v.] which lasted until her death in 1920, and to Edith Mary Oldham Lees, authoress, whom he married in 1891; publications include *Studies in the Psychology of Sex* (7 vols., 1897-1928), *Man and Woman* (1894), *Little Essays of Love and Virtue* (1922), *The Dance of Life* (1923), *Impressions and Comments* (3 vols., 1914-24), and *My Life* (1940).

ELLIS, JOHN DEVONSHIRE (1824-1906), civil engineer and metallurgist; son of Birmingham brass manufacturer; purchased with Sir John Brown [q.v.] small Atlas engineering works at Sheffield, 1854; increased capital to 3 millions sterling; managing director (on their conversion into limited liability company), 1864-1905; produced iron plates for British ironclads by new and cheaper welding process; perfected a compound armour of steel and iron; received Bessemer gold medal from Iron and Steel Institute, 1889, of which he became vice-president in 1901.

ELLIS, ROBINSON (1834-1913), classical scholar; educated at Rugby and Balliol College, Oxford; first class, classical mods and *lit. hum.*; Ireland scholar, 1855; fellow of Trinity College, Oxford, 1858; professor of Latin, University College, London, 1870-6; reader in Latin at Oxford, 1883; Corpus professor of Latin, 1893; vice-president of Trinity, 1879-93; FBA, 1902; works include *Commentary on Catullus* (1876), edition of Ovid's *Ibis* (1881), and editions of many minor Latin authors.

ELLIS, THOMAS EVELYN SCOTT-, eighth BARON HOWARD DE WALDEN and fourth BARON SEAFORD (1880-1946), writer, sportsman, and patron and lover of the arts. [See SCOTT-ELLIS.]

ELLIS, Sir WILLIAM HENRY (1860-1945), civil engineer; son of J. D. Ellis [q.v.]; educated at Uppingham; worked for Tannett, Walker & Co., Leeds, 1878-87; entered John Brown & Co., 1887; director, 1906; managing director, 1919; master cutler of Sheffield, 1914-18; intrepid mountaineer; endowed trust for Swiss guides, 1938; GBE, 1918.

ELPHINSTONE, Sir (GEORGE) KEITH (BULLER) (1865-1941), engineer; educated at

Charterhouse; partner (1893), later until 1941 chairman, Elliott Brothers; designed first continuous-roll chart recorder; installed original electric speed-recording apparatus, Brooklands; invented speedometers for motor-cars; collaborated in developing fire-control tables for navy; KBE, 1920.

ELSIE, LILY (1886-1962), musical-comedy actress; at age of eight showed faculty for mimicking vaudeville celebrities such as (Sir) Harry Lauder and Vesta Tilley [qq.v.]; played title-role in *Little Red Riding Hood* at Queen's Theatre, Manchester, 1896; engaged by George Edwardes to play many small parts; had great success in *The Merry Widow* at Daly's Theatre, 1907, *The Dollar Princess*, 1909, and *The Count of Luxembourg*, 1911; noted for her beauty, charm, and elegance; married and retired from stage, 1911; returned for charity performances during 1914-18 war, including *The Admirable Crichton* by (Sir) James Barrie [q.v.]; played in *The Truth Game* by Ivor Novello [q.v.], 1928.

ELSMIE, GEORGE ROBERT (1838-1909), Anglo-Indian civilian and author; joined Indian service, 1858; called to bar, 1871; commissioner at Peshawar, 1872-8 and 1885; commissioner of Lahore, 1880-2; judge of Punjab chief court, 1882-5; vice-chancellor of Punjab University, 1885-7; member of legislative council, 1888; first financial commissioner, 1889; CSI, 1894; returned to England, 1894; hon. LLD, Aberdeen, 1904; published *Lumsden of the Guides* (1899), *Thirty-five Years in the Punjab* (1908), and other works.

ELTON, OLIVER (1861-1945), scholar and critic; educated at Marlborough and Corpus Christi College, Oxford; first class, *lit. hum.*, 1884; independent lecturer in English literature, Owens College, Manchester, 1890-1900; King Alfred professor, Liverpool, 1901-25; works include translation of first nine books of *Danish History* of Saxo Grammaticus (1894); *Introduction to Michael Drayton* (1895); *The Augustan Ages* (1899); life of F. York Powell, with selection from his letters, etc. (2 vols., 1906); *A Survey of English Literature: 1780-1830* (2 vols., 1912), *1830-1880* (2 vols., 1920), *1730-1780* (2 vols., 1928); *C. E. Montague, a Memoir* (1929); *The English Muse, a Sketch* (1933); edited *Prefaces and Essays* of George Saintsbury [q.v.], 1933; published verse translations in *Verse from Pushkin and Others* (1935) and *Evgeny Onegin* (1937); FBA, 1924.

ELVIN, SIR (JAMES) ARTHUR (1899-1957), founder of Wembley Stadium; received elementary education; became observer in Royal Flying Corps and taken prisoner; obtained demolition contract after British Empire Exhibition at Wembley; purchased stadium and inaugurated greyhound racing, 1927; built Empire Pool with removable floor for staging spectacles; secured Olympic Games for Wembley, 1948; knighted, 1946.

ELWES, GERVASE HENRY (CARY-) (1866-1921), singer; educated at Christ Church, Oxford; in diplomatic service, 1891-5; adopted singing as career, 1902; sang in Paris, London, Holland, Belgium, Germany, and America; accidentally killed at Boston, USA.

ELWES, HENRY JOHN (1846-1922), traveller, botanist, and entomologist; collected birds, butterflies, and moths, but chiefly plants; travelled in Turkey, Asia Minor, India, Tibet, Mexico, North America, Chile, Russia, Siberia, Formosa, China, Japan; hunted big game; FRS, 1897; works include 'The Geographical Distribution of Asiatic Birds' (*Proceedings* of the Zoological Society, June 1873), *Monograph of the Genus Lilium* (1880), *Trees of Great Britain and Ireland* (with Augustine Henry, 1906-13).

ELWORTHY, FREDERICK THOMAS (1830-1907), philologist and antiquary; travelled much in Europe, collecting charms and amulets; FSA, 1900; good linguist and draughtsman; editorial secretary of Somerset Archaeological Society, 1891-6; published works on West Somerset dialect, on the evil eye (1895), and on other superstitions.

EMERY, WILLIAM (1825-1910), archdeacon of Ely; first boy to enter new City of London School, 1837; entered Corpus Christi College, Cambridge; fifth wrangler, 1847; MA, 1850; BD, 1858; fellow, 1847-65; keen advocate of Volunteer movement; organized first church congress at Cambridge, 1861; permanent secretary of congress from 1869; archdeacon of Ely, 1864-1907; canon, 1870; organized first diocesan conference.

EMMOTT, ALFRED, first BARON EMMOTT (1858-1926), politician and cotton spinner; entered father's cotton-spinning firm, Oldham, 1879; joined Oldham town council, 1881; mayor, 1891; maintained lifelong association with Oldham; liberal MP, Oldham, 1899-1911; chairman of committees, House of Commons, 1906-11; baron, 1911; under-secretary of state for colonies, 1911; first commissioner of works, 1914; left office (1915) and began creation of War Trade Department, which he directed until 1919; this department designed to maintain British export trade without weakening blockade; president, world cotton congress, Manchester and Liverpool, 1921; interested in education.

ENSOR, SIR ROBERT CHARLES KIRK-WOOD (1877-1958), journalist and historian; scholar of Winchester and Balliol College, Oxford; first class, *lit. hum.* and president of Union, 1900; joined *Manchester Guardian*, 1901; called to bar (Inner Temple), 1905; leader-writer, *Daily News*, 1909-11, *Daily Chronicle*, 1912-30; on executive of Fabian Society, 1907-11, 1912-19; appointed (1930) to write 1870-1914 volume of Oxford History of England, published 1936; the work authoritative, just, and immensely readable; fellow, Corpus Christi College, Oxford, 1937-46; wrote as 'Scrutator' for *Sunday Times* from 1941; served on royal commissions on population (1944-9) and the press (1947-9); knighted, 1955; hon. fellow, Balliol and Corpus Christi Colleges, Oxford.

ENTWISTLE, WILLIAM JAMES (1895-1952), scholar; born in China; educated at mission school, Chefoo, and Robert Gordon's College and the university, Aberdeen; first class, classics, 1916, Fullerton scholar, 1918; spent 1920 in Madrid learning Spanish, Catalan, and Portuguese; lecturer in Spanish, Manchester University, 1921; Stevenson professor of Spanish, Glasgow, 1925; King Alfonso XIII professor, Oxford, 1932-52; obtained honour school of Portuguese and inclusion of Catalan and Spanish-American literature in syllabus; joint-editor, *Modern Language Review*, 1934-48; general editor, Great Languages Series, 1940-52, and linguistics contributions to *Chambers's Encyclopaedia*; publications include *The Arthurian Legend in the Literatures of the Spanish Peninsula* (1925), *The Spanish Language* (1936), *European Balladry* (1939), and *Aspects of Language* (1953); FBA, 1950; hon. LLD Aberdeen, and Glasgow.

EPSTEIN, SIR JACOB (1880-1959), sculptor; born in New York City of Jewish parentage; studied sculpture in America and at Julian's Academy, Paris; settled in London, 1905; naturalized, 1911; an original member of London Group; his nude figures for British Medical Association Strand building by Charles Holden [q.v.] caused outcry (1908) as did his memorial to W. H. Hudson [q.v.] (Hyde Park, 1925) and other stone carvings rooted in early or primitive sculpture until recognition came late in life; commissions included Christ in Majesty (Llandaff Cathedral), St. Michael and the Devil (Coventry Cathedral); his Lazarus bought by New College, Oxford; his more acceptable portrait bronzes included wartime Service chiefs and (Sir) Winston Churchill; KBE, 1954; hon. DCL, Oxford.

ERNLE, BARON (1851-1937), administrator, author, and minister of agriculture. [See PROTHERO, ROWLAND EDMUND.]

ESDAILE, KATHARINE ADA (1881-1950), art historian; born McDOWALL; niece of E. W. Benson [q.v.]; educated at Notting Hill high school and Lady Margaret Hall, Oxford; married (1907) A. J. K. Esdaile; made intensive study of post-medieval sculpture in England; publications include *English Monumental Sculpture since the Renaissance* (1927), *The Life and Works of Louis François Roubiliac* (1928), *Temple Church Monuments* (1933), and *English Church Monuments* (1946).

ESHER, second VISCOUNT (1852-1930), government official. [See BRETT, REGINALD BALIOL.]

ESMOND, HENRY VERNON (1869-1922), actor and dramatist, whose original name was HENRY JACK; joined the stage, 1885; acted for Sir G. Alexander [q.v.] at St. James's Theatre, 1893-1900; wrote about thirty plays, including *One Summer's Day* (1897) and *When We were Twenty-one* (1901).

ETHERIDGE, ROBERT (1819-1903), palaeontologist; cousin of John Beddoe [q.v.]; curator of museum of British Philosophical Institution, 1850-7; palaeontologist to Geological Survey, 1863; assisted T. H. Huxley [q.v.] as lecturer at Royal School of Mines; made list of 18,000 species of fossils (catalogued, 1888); assistant keeper in geology at Natural History Museum, 1881-91; FRS, 1871; president of Geological Society, 1880-2; assistant editor of *Geological Magazine*, 1865-1903; works include *Stratigraphical Geology and Palaeontology* (1887).

EUAN-SMITH, SIR CHARLES BEAN (1842-1910), soldier and diplomatist; born at George Town, British Guiana; joined Madras infantry, 1859; served in Abyssinian expedition, 1867, in Persia, 1870-1, and in Zanzibar, 1872; CSI, 1872; consul at Muscat, 1879; in Afghan war in Lord Roberts's expedition to Kandahar; consul-general at Zanzibar, 1887; persuaded sultan to place himself under protection of Great Britain, 1890; CB, 1889; KCB, 1890; British envoy in Morocco, 1891; retired, 1893; hon. DCL, Oxford, 1893.

EUMORFOPOULOS, GEORGE (1863-1939), collector of Chinese and other works of art; born in Liverpool of Greek parentage;

entered Ralli Brothers, merchants, of London; became vice-president; retired, 1934; collected ceramics of Sung and earlier periods, Chinese archaic bronzes and jades, sculptures and paintings, also modern European art; Chinese portion acquired by national museums in 1934.

EVA, (pseudonym) (1826–1910), Irish poetess. [See O'DOHERTY, MARY ANNE.]

EVAN-THOMAS, SIR HUGH (1862–1928), admiral; joined navy, 1876; accompanied Princes Albert Victor and George on *Bacchante* cruise, 1877–80; served in Mediterranean, 1894–7, 1900–2; temporary naval secretary to first lord of Admiralty, 1905; held post (confirmed, 1906) until 1908; promoted to flag rank, 1912; second-in-command of first battle squadron, 1913–15; transferred to command of fifth battle squadron, 1915; in this post took notable part in battle of Jutland, 1916; his fine force brilliantly supported (Lord) Beatty [q.v.] and saw some of heaviest fighting of day; vice-admiral, 1917; admiral, 1920; commander-in-chief at the Nore, 1921–4; KCB, 1916.

EVANS, SIR ARTHUR JOHN (1851–1941), archaeologist; son of Sir John Evans [q.v.]; educated at Harrow and Brasenose College, Oxford; first class, modern history, 1874; travelled in Balkans; studied language, antiquities, and customs at Ragusa, 1876–82; collected vases and coins in Italy and Sicily and prehistoric seal-stones in Crete; keeper, Ashmolean Museum, Oxford, 1884–1908; acquired (1894) a share of estate at Kephála, near Candia (Knossos); excavated there, 1899–1907; found an elaborate palace of late Bronze Age (c.1700–1400), with many clay tablets inscribed in 'linear' script, superimposed upon earlier buildings with 'Kamárais' pottery (c.2000), brilliant frescoes, and imported Egyptian and Babylonian objects; published preliminary scheme of classification of this new prehistoric culture into 'early', 'middle', and 'late' Minoan, each subdivided into periods I, II, and III (1904), *Scripta Minoa* (2 vols., 1909–52), and *The Palace of Minos at Knossos* (4 vols., 1921–35); carried out skilful and extensive reconstruction and restoration; conveyed property in trust to British School of Archaeology at Athens, 1926; FRS, 1901; a founder of British Academy, 1902; president, British Association, 1916–19; knighted, 1911.

EVANS, SIR CHARLES ARTHUR LOVATT (1884–1968), physiologist; educated at Birmingham secondary school and Municipal Technical School; at sixteen worked in Mason Science College, Birmingham; lecturer in physiology, Handsworth Technical School,

1902–8; demonstrator, Birmingham Midland Institute, 1904–7; interim lecturer, Birmingham University; B.Sc. London, 1910; awarded Sharpey scholarship in Physiology Department, University College, London, 1910; studied medicine; MRCS, LRCP, 1916; D.Sc.; joined RAMC; worked on gas warfare at RAMC College, Millbank; professor, experimental physiology, Leeds, 1918; joined staff of National Institute for Medical Research, Hampstead, 1919–22; professor of physiology, St. Bartholomew's Hospital, 1922–6; Jodrell professor, University College, London, 1926–39, and 1945–9; hon. LLD, Birmingham, 1934; worked at Chemical Defence Experimental Establishment, Porton Down, during 1939–45 and in 1949–67; produced fourteen editions of Starling's *Principles of Human Physiology* (1930–58); published *Recent Advances in Physiology*; chairman, Royal Veterinary College, 1949–63; member, Medical Research Council, 1947–50; chairman, Military Personnel Research Committee, 1948–53; FRS, 1925; fellow, University College, London, and Royal Veterinary College; LLD London, 1957; knighted, 1951.

EVANS, DANIEL SILVAN (1818–1903), Welsh scholar and lexicographer; rector of Llanymawddwy, 1862–76, and Llanwrin, 1876–1903; hon. BD, Lampeter, 1863; lecturer in Welsh at University College of Wales, 1875–83; hon. LLD, 1901; research fellow of Jesus College, Oxford, 1897; honorary canon of Bangor, 1888; chancellor of the cathedral, 1895; chief work *Dictionary of the Welsh Language* (5 parts, A–E, 1887–1906); other works include *English–Welsh Dictionary* (2 vols., 1852–8) and volumes of poems.

EVANS, EDMUND (1826–1905), wood-engraver and colour printer; apprenticed to Ebenezer Landells [q.v.], 1840–7; worked for *Illustrated London News*; made colour prints of illustrations by Birket Foster for Pfeiffer's *Travels in the Holy Land*, 1852; prepared first 'yellow-back' illustrated cover, 1853; printed in oil colour from wood blocks; gained chief fame by colour printing of works by Walter Crane, Randolph Caldecott, and Kate Greenaway [qq.v.] including Crane's *The Baby's Opera* (1877) and Greenaway's *Under the Window* (1879).

EVANS, EDWARD RATCLIFFE GARTH RUSSELL, first BARON MOUNTEVANS (1880–1957), admiral; entered navy from mercantile marine training ship *Worcester*; second officer of relief ship *Morning* to first Antarctic expedition of (Sir) Robert Scott [q.v.], 1902–4; second-in-command Scott's second expedition and captain of *Terra Nova*, 1910-13; accompanied Scott (Jan. 1912) to within 150 miles of pole; served in Dover Patrol, 1914-18; commanded *Broke* which rammed enemy destroyer off Dover, 1917; DSO and captain; commanded *Repulse*, 1926-8; rear-admiral, 1928; commanded Australian squadron, 1929-31; Africa station, 1933-5; as acting high commissioner suspended Tshekedi [q.v.], 1933; commanded at Nore, 1935-9; vice-admiral, 1932; admiral, 1936; regional civil defence commissioner, London, 1939-45; rector of Aberdeen University, 1936-42; received freedom of Dover, 1938; CB, 1913; KCB, 1935; baron, 1945.

EVANS, SIR (EVAN) VINCENT (1851-1934), journalist; managing director, Chancery Lane Safe Deposit and Offices Company; secretary, National Eisteddfod Association (1881-1934), and Honourable Society of Cymmrodorion (1887-1934); governor, museum, library, and university of Wales; knighted, 1909; CH, 1922.

EVANS, GEORGE ESSEX (1863-1909), Australian poet; emigrated from London to Australia, 1881; while sheep-farming contributed poems to *Queenslander* from 1882; appointed district registrar at Toowoomba, Queensland, 1888; chief works were *Loraine, and other Verses* (1898) and *The Secret Key* (1906); poems distinctively inspired by Australian life and environment.

EVANS, SIR GUILDHAUME MYRDDIN-(1894-1964), civil servant. [See MYRDDIN-EVANS.]

EVANS, HORACE, BARON EVANS (1903-1963), physician; educated at Liverpool College, City of London School and London Hospital Medical College; qualified, 1925; MB, BS, 1928; MD, MRCP, 1930; FRCP, 1938; trained under (Sir) Arthur Ellis [q.v.]; assistant director, medical unit, 1933; assistant physician, London Hospital, 1936; physician, 1947; consulting physician to the Royal Navy, Royal Masonic Hospital, and other hospitals; succeeded Lord Dawson of Penn [q.v.] as physician to Queen Mary, 1946, to King George VI, 1949 and to Queen Elizabeth, 1952-63; KCVO, 1949; GCVO, 1955; baron, 1957; hon. D.Sc., university of Wales; hon. FRCS; made intensive study of Bright's disease; president, Medical Society of London; vice-president, Royal College of Physicians; Croonian lecturer, 1955.

EVANS, SIR JOHN (1823-1908), archaeologist and numismatist; son of Arthur Benoni Evans and brother of Sebastian Evans [qq.v.]; entered uncle's paper-making business at Nash Mills, Hemel Hempstead, 1840; made scientific researches, especially into water-supply, exploring water-bearing strata in his own district; made fine collection of stone and bronze implements, fossil remains, and medieval antiquities; formed collections of ancient British money, of Anglo-Saxon and English coins; first to place study of ancient British coinage on scientific basis; FSA, 1852; FRS, 1864; president of Geological Society, 1874-6, and of Numismatic Society, 1874-1908; joint-editor of *Numismatic Chronicle* from 1861; received hon. degrees from Oxford, Cambridge, Dublin, and Toronto; KCB, 1892; interested in Hertfordshire local affairs; high sheriff of county, 1881; portions of collections of implements are in Ashmolean Museum, Oxford; works include *Flint Implements in the Drift* (1860), *The Coins of the Ancient Britons* (1864) with *Supplement* (1890), *The Ancient Stone Implements of Great Britain* (1872), *The Ancient Bronze Implements . . . of Great Britain* (1881).

EVANS, JOHN GWENOGVRYN (1852-1930), Welsh palaeographer; Unitarian pastor, 1876-80; gave up ministry owing to ill health and devoted himself to editing Welsh texts; works include *Red Book Mabinogion* (with Sir J. Rhŷs, q.v., 1887), *Facsimile of the Black Book of Carmarthen* (1888), *Red Book Bruts* (with Rhŷs, 1890), *Book of Llan Dâv* (with Rhŷs, 1893), *White Book Mabinogion* (1907), *Book of Aneurin* (1907), *Book of Taliesin* (1910), *Report on Manuscripts in the Welsh Language* (1898-1910).

EVANS, MEREDITH GWYNNE (1904-1952), physical chemist; educated at Leigh grammar school and Manchester University; first class, chemistry, 1926; remained at Manchester, latterly as full lecturer, until 1939; professor of physical chemistry, Leeds, 1939-49, Manchester, 1949-52; worked in field of the mechanisms of chemical reactions; FRS, 1947.

EVANS, SIR SAMUEL THOMAS (1859-1918), politician and judge; admitted solicitor, 1883; liberal MP, mid-Glamorganshire, 1890-1910; called to bar (Middle Temple), 1891; QC, 1901; recorder of Swansea, 1906-8; solicitor-general, 1908; president of Probate, Divorce, and Admiralty division, 1910; GCB, 1916; reputation as judge rests on series of judgments in prize delivered during European war, 1914-18; most notable cases: the *Kim* in which he applied doctrine of 'continuous voyage' to carriage of contraband goods; the *Leonora*, in which he held that so-called 'reprisals' Order in Council of 16 Feb. 1917 was not inconsistent with established principles of international law; and the *Möwe, Roumanian, Hamborn*, and *Zamora*.

EVANS, SEBASTIAN (1830-1909), journalist; brother of Sir John Evans [q.v.]; scholar of Emmanuel College, Cambridge; BA, 1853; MA, 1856; LLD, 1868; window designer at Messrs. Chance's glass works near Birmingham, 1857-67; an ardent conservative; edited *Birmingham Daily Gazette*, 1867-70; called to bar, 1873; part founder and editor of *People*, 1878-81; exhibited at Royal Academy in oil, water-colour, and black-and-white; translated *The High History of the Holy Graal*, 1898; published verse showing feeling for medieval beauty.

EVANS, SIR (WORTHINGTON) LAMING WORTHINGTON-, first baronet (1868-1931); politician; solicitor, 1890-1910; conservative MP, Colchester (1910-29), St. George's, Hanover Square (1929-31); secretary of state for war, 1921-2, 1924-9; postmaster-general, 1923-4; baronet, 1916; PC, 1918.

EVATT, HERBERT VERE (1894-1965), Australian judge and statesman; born in New South Wales; educated at Fort Street high school, Sydney and Sydney University; BA, 1915; MA, 1917; LLB, 1918; LLD, 1924; admitted to bar of New South Wales; joined Australian labour party; elected to State legislative assembly, 1925; became independent member, 1927; took silk, 1929; appointed justice of the High Court of Australia, 1930; published *The British Dominions as Mandatories* (1934), *The King and His Dominion Governors* (1936), *Injustice Within the Law* (1937), and *Australian Labor Leader* (1940); resigned, and elected ALP member, House of Representatives, 1940; attorney-general and minister for external affairs, 1941-9; attended San Francisco Conference, 1945; president, UN General Assembly, 1948-9; deputy leader, federal ALP, 1949, leader, 1951-60; involved in Petrov affair, 1954-5; chief justice, Supreme Court, New South Wales, 1960-2; PC, 1942; received many honorary degrees; speeches published as *Foreign Policy of Australia* (1945), *Australia in World Affairs* (1946), and *The United Nations* (lectures at Harvard, 1948).

EVE, SIR HARRY TRELAWNEY (1856-1940), judge; educated at Exeter College, Oxford; called to bar (Lincoln's Inn), 1881; QC, 1895; bencher, 1899; liberal MP, Ashburton division, Devon, 1904-7; judge in Chancery division, 1907-37; knighted, 1907; PC, 1937; a thoroughly sound judge, notably on charitable trusts, with very comprehensive grasp of equity.

EVERARD, HARRY STIRLING CRAW-FURD (1848-1909), writer on golf; settled at St. Andrews, winning golf prizes; works include

History of the Royal and Ancient Club of St. Andrews (1907).

EVERETT, JOSEPH DAVID (1831-1904), man of science; entered Glasgow College, 1854; BA, 1856; MA, 1857; professor of mathematics in King's College, Windsor, Nova Scotia, 1859-64; returned to Glasgow, 1864; professor of natural philosophy, Queen's College, Belfast, 1867-97; FRS, 1879; fellow of Royal University of Ireland; settled in London, 1898; works include *Units and Physical Constants* (1875), *Outlines of Natural Philosophy* (1887); translated Deschanel's *Physics* (1870).

EVERETT, SIR WILLIAM (1844-1908), colonel; joined army, 1864; vice-consul at Erzerum, 1878; consul at Kurdistan, 1882-7; CMG, 1886; professor of military topography at Sandhurst, 1888-92; colonel, 1893; technical adviser for delimitation of Sierra Leone frontier, 1895; commissioner for Niger frontier, 1896-8, and Togoland frontier, 1900; KCMG, 1898.

EVERSHED, (FRANCIS) RAYMOND, first BARON EVERSHED (1899-1966), judge; educated at Clifton College and Balliol College, Oxford; served in Royal Engineers in France, 1918-19; called to bar (Lincoln's Inn), 1923; KC, 1933; bencher, 1938; chairman, central price regulation committee, 1939-42; regional controller, Nottinghamshire, Derbyshire, and Leicestershire coal-producing region, 1942-4; justice, Chancery Division, 1944; knighted; lord justice, Court of Appeal, 1947; chairman, committee on practice and procedure in the Supreme Court, 1947-53; master of the rolls, 1949-62; baron, 1956; lord of appeal in ordinary, 1962-5; British member, Permanent Court of Arbitration, The Hague, 1950; president, Bar Musical Society, 1952-66; chairman, Pilgrim Trust, 1959-65; chairman, British Council Law Committee, 1956-66; hon. fellow, Balliol College, 1947; FSA, 1950; president, Clifton College, 1951; treasurer, Lincoln's Inn, 1958; many hon. degrees.

EVERSHED, JOHN (1864-1956), astronomer; educated privately; observed solar eclipses, 1896, 1898, 1900, 1905, 1927, 1936; appointed assistant director, Kodaikanal Observatory, India, 1906; director, 1911-23; thereafter until 1953 maintained a solar observatory at Ewhurst; his discoveries included 'Evershed effect'—a radial circulation of gases in sunspots; devised method for detecting and measuring Doppler shifts; FRS, 1915; CIE, 1923.

EVERSLEY, BARON (1831-1928), statesman. [See SHAW-LEFEVRE, GEORGE JOHN.]

EVES, REGINALD GRENVILLE (1876–1941), painter; educated at University College School and Slade School; gifted in portraying a characteristic expression; prominent sitters included Thomas Hardy, Sir Frank Benson, Stanley Baldwin, Lord Jellicoe, and Leslie Howard [qq.v.]; ARA, 1933; RA, 1939; official war artist, 1940–1.

EWART, ALFRED JAMES (1872–1937), botanist; educated at Liverpool; Institute and University College, Liverpool; B.Sc., London, 1893; Ph.D., Leipzig, 1896; government botanist (1905–21) and professor of botany and plant physiology (1905–37), Melbourne; FRS, 1922; publications include *Physics and Physiology of Protoplasmic Streaming in Plants* (1903) and *Flora of Victoria* (1930).

EWART, CHARLES BRISBANE (1827–1903), lieutenant-general; brother of Sir John Alexander Ewart [q.v.]; joined Royal Engineers, 1845; served in Crimea, 1854–6; commanding royal engineer of London district, 1866–71, at Dover, 1877–9, and Gibraltar, 1879–82; CB, 1869; member of ordnance committee, 1884; brigadier-general in Sudan, 1885; lieutenant-governor of Jersey, 1887–92; lieutenant-general, 1888; colonel commandant, RE, 1902.

EWART, Sir JOHN ALEXANDER (1821–1904), general; born in Bombay; joined army, 1838; served in Crimea; recommended for VC for bravery at assault of the Secunderabagh in Indian Mutiny, 1857; CB, 1858; commanded 78th Ross-shire Buffs, 1859–64; general, 1884; colonel, 92nd Gordon Highlanders, 1884–95; KCB, 1887; GCB, 1904; published *The Story of a Soldier's Life*, 2 vols., 1881.

EWART, Sir JOHN SPENCER (1861–1930), lieutenant-general; son of Sir J. A. Ewart [q.v.]; joined Queen's Own Cameron Highlanders, 1881; served in Egyptian war (1882), Nile expedition (1884–5), in Sudan (1885–6), and in Lord Kitchener's Sudan campaign (1898); served with distinction, South African war, 1899–1902; entered War Office, 1902; director, military operations at headquarters, 1906–10; adjutant-general to forces and second military member of Army Council, 1910–14; lieutenant-general, 1911; GOC-in-C, Scottish command, 1914–18; KCB, 1911.

EWING, Sir (JAMES) ALFRED (1855–1935), engineer; educated at Dundee high school and Edinburgh University; professor of mechanical engineering and physics, Tokyo, 1878–83; devised instruments for measuring and recording earthquakes; professor of engineering, Dundee (1883), of mechanism and applied mechanics, Cambridge, 1890–1903; tripos instituted, 1892; laboratory founded, 1894; director of naval education, 1903–16; organized the system of scientific and engineering training; supervised deciphering of intercepted German wireless messages, 1914–17; principal and vice-chancellor, Edinburgh (1916–29), at time of unexampled development and expansion; publications include *Magnetic Induction in Iron and other Métals* (1891) and *Thermodynamics for Engineers* (1920); FRS, 1887; KCB, 1911; president, British Association, 1932.

EWINS, ARTHUR JAMES (1882–1957), chemist; educated at Alleyn's School; joined Wellcome Physiological Research Laboratories, 1899; B.Sc., London, 1906; assistant successively to John Mellanby and George Barger [qq.v.]; in charge of chemical division, 1909; studied distribution and action of acetylcholine; D.Sc., London, 1914; with National Institute for Medical Research, 1914–17; director of research, May and Baker, 1917–52; his researches on chemotherapy of infections included production of Sulphapyridine (M & B 693); FRS, 1943.

EYRE, EDWARD JOHN (1815–1901), governor of Jamaica; emigrated to Australia and engaged in sheep-farming, 1833; began journeys into unknown sand deserts of interior as magistrate and protector of aborigines, 1836; starting from Adelaide (1840), passed with a single native round head of Great Australian Bight, and reached King George's Sound, 1841; published experiences in 2 vols., 1845; revisited England, 1845; lieutenant-governor of New Zealand, 1846–53; governor of St. Vincent, 1854–60; appointed governor of Jamaica, 1864; forcibly suppressed Morant Bay native rebellion, proclaiming martial law, Oct. 1865; confirmed sentence of death for high treason passed by Lieutenant H. C. A. Brand [q.v.] on George William Gordon, a coloured member of legislature, and over 600 other persons; denounced in England for cruelty; temporarily suspended and tried by royal commission of inquiry at Kingston, Jan.–Mar. 1866; commended for promptitude, but was blamed for unnecessary rigour, and recalled, 1866; retired to Shropshire; supported by Kingsley and Carlyle, Ruskin and Tennyson, but condemned by 'Jamaica committee', including J. S. Mill, Huxley, Tom Hughes, Herbert Spencer, and Goldwin Smith; was made defendant in abortive legal proceedings which aimed at bringing him to trial for murder (Mar. 1867 and June 1869); legal expenses paid from public funds, 1872; received pension as retired colonial governor, 1874.

F

FABER, Sir GEOFFREY CUST (1889–1961), publisher; educated at Rugby and Christ Church, Oxford; first class, classical mods, 1910, and *lit. hum.*, 1912; joined Oxford University Press, 1913; served in France and Belgium during 1914–18 war; fellow, All Souls; called to bar (Inner Temple), 1921; negotiated with (Sir) Maurice Gwyer [q.v.], and became chairman of Faber & Gwyer Ltd., 1923; transformed firm to Faber & Faber Ltd., 1927; member of council, Publishers' Association, 1934; president, 1939–41; first chairman, National Book League, 1945; succeeded Geoffrey Dawson [q.v.] as college estates bursar, 1923–51; published poems include *Interflow* (1915), *In the Valley of Vision* (1918), and *The Buried Stream* (1941); other works include *Oxford Apostles* (1933), *Jowett* (1957), and an edition of John Gay [q.v.] in the Oxford Poets series (1926); knighted, 1954; resigned as chairman, Faber & Faber Ltd., 1960; president, 1960–1.

FABER, OSCAR (1886–1956), consulting engineer; educated at St. Dunstan's College, Catford; graduated from Central Technical College, 1907; fellow, 1929; D.Sc., London, 1915; chief engineer, Trollope & Colls, 1912; practised as consulting engineer from 1921; concerned with structural design of new Bank of England, underpinning of Durham Castle, Mulberry harbour project, rebuilding of House of Commons; early realized potentialities of reinforced concrete and importance of mechanical and electrical services in large buildings; CBE, 1951.

FACHIRI, ADILA ADRIENNE ADALBERTINA MARIA (1886–1962), and her sister, **D'ARANYI,** JELLY (1893–1966), violinists; born in Budapest into a musical family; taught by Béla Bartók and Jenö Hubay; as children, entered Budapest Academy of Music; Adila won Artists diploma, 1906; went to Berlin as pupil of Joseph Joachim, her uncle; made début in Berlin, 1907; introduced Jelly as partner in Bach's D minor Concerto for Two Violins; distantly related to Bertrand (later Lord) Russell [q.v.]; welcomed in England; settled in England at outbreak of 1914–18 war; Adila married Alexander P. Fachiri, 1915; his death restored her to concert platform, 1939; in demand as teacher; died in Florence, 1962; Jelly's career began at fourteen; during 1920–40, established reputation in Europe and America; formed friendship and partnership with Myra Hess [q.v.]; carried out charity recitals in English cathedrals, 1933; CBE, 1946; died in Florence, 1966.

FAED, JOHN (1819–1902), artist; brother of Thomas Faed [q.v.]; for forty years painted miniatures; ARSA, 1847; RSA, 1851; later painted figure subjects; chief were 'Burd Helen', 'The Cottar's Saturday Night', 1854, 'The Poet's Dream', 1883, and 'The Wappinschaw'; exhibited at Royal Academy, 1862–80.

FAGAN, JAMES BERNARD (1873–1933), actor-manager, producer, and playwright; educated at Trinity College, Oxford; acted with (Sir) F. R. Benson and (Sir) H. B. Tree [qq.v.]; plays include *The Rebels* (1899), *And So To Bed* (1926), and *The Improper Duchess* (1931); took over management of Court Theatre, London, 1918; opened Oxford Playhouse, 1923.

FAGAN, LOUIS ALEXANDER (1845–1903), etcher and writer on art; grandson of Robert Fagan [q.v.]; born at Naples; served in British legation at Caracas under father, 1866–7; friend of Sir Anthony Panizzi [q.v.], whose biography he wrote, 2 vols., 1880; in prints department at British Museum, 1869–94; wrote *Handbook* to department (1876) and other works on engraving and engravers; exhibited etchings at Royal Academy, 1872–81; a popular lecturer; historian of the Reform Club, 1886; died at Florence.

FAIRBAIRN, ANDREW MARTIN (1838–1912), Congregational divine; early joined Evangelical Union founded by Dr James Morison [q.v.]; studied at theological college of Union in Edinburgh; minister at Bathgate, 1860; in Aberdeen, 1872; principal of Airedale theological college, Bradford, 1877–86; first principal of Mansfield College, Oxford, 1886–1909; chief works *Christ in Modern Theology* (1893) and *Philosophy of the Christian Religion* (1902).

FAIRBAIRN, STEPHEN (1862–1938), oarsman; born in Australia; educated at Geelong grammar school and Jesus College, Cambridge; rowed four times for Cambridge; worked on Australian sheep station, 1884–1904; successfully coached Jesus and other rowing clubs, 1905–38; sought looseness and ease with attention on the oar and moving the boat rather than on body position; instituted 'Head of the River' race, Putney–Mortlake, 1925.

FAIRBRIDGE, KINGSLEY OGILVIE (1885–1924), founder of farm schools overseas; born at Grahamstown, Cape Colony; studied forestry at Oxford, 1908–11; conceived idea of emigrating children to learn farming; with his wife opened farm school near Perth, Western Australia, 1913, now model for other farm schools.

FAIREY, Sir (CHARLES) RICHARD (1887-1956), aircraft manufacturer; educated at Merchant Taylors' School and Finsbury Technical College; qualified as electrical engineer; joined Short Brothers, 1913; founded own aircraft company, 1915, and became leading and farseeing aircraft designer; produced over one hundred different types of aircraft including fast bomber Fox (1925), Long Range Monoplane (1928, gained long-distance non-stop record, 1933), wartime Swordfish, Gyrodyne helicopter (1948), and Fairey Delta (1956, first plane to exceed 1,000 m.p.h.); member, Aeronautical Research Committee, 1923-6; awarded Wakefield medal for design of variable camber wing, 1936; deputy director, British air mission in USA, 1940, director, 1942-5; knighted, 1942.

FAIRFIELD, Baron (1863-1945), judge. [See GREER, (FREDERICK) ARTHUR.]

FAIRLEY, Sir NEIL HAMILTON (1891-1966), physician; born in Inglewood, Victoria, Australia; educated at Scotch College, Melbourne, and Melbourne University; qualified in medicine, first class, 1915; commissioned in Australian Army Medical Service, 1916; worked in Lister Institute, London, 1919; MRCP (London); DPH (Cambridge); returned to Australia, 1920; medical research officer, Bombay Bacteriological Laboratory, 1922; settled in London at Hospital for Tropical Diseases, 1929; carried out research on hepatitis; in Cairo, discovered sulphaguanidine to be a cure for bacillary dysentery, 1941; consultant to Australian troops in Middle East Force, 1940-1; concerned with prevention of malaria; brigadier, director of medicine, Australian Medical Service, South Pacific Area, 1942; proved effectiveness of mepacrine as antimalarial drug; Wellcome professor of tropical medicine, London University, 1946-9; FRS, 1942; OBE, 1918; CBE, 1941; KBE, 1950; hon. secretary, Royal Society of Tropical Medicine and Hygiene, 1930-51; president, 1951-3.

FALCKE, ISAAC (1819-1909), art collector and benefactor to the British Museum; made collections of majolica and lustre ware (now in Wallace Collection), of fifteenth- and sixteenth-century bronzes (now at Berlin), and of Wedgwood china (presented to British Museum), as well as of Chinese and other porcelain.

FALCONER, LANOE (pseudonym) (1848-1908), novelist. [See HAWKER, MARY ELIZABETH.]

FALCONER, Sir ROBERT ALEXANDER (1867-1943), Canadian educationist; born in Canada; educated at Queen's Royal College School, Trinidad, London and Edinburgh universities, and in Germany; lecturer (1892), professor of New Testament Greek (1895), Pine Hill Presbyterian theological college, Halifax, Nova Scotia; principal, 1904-7; president, Toronto University, 1907-32; fellow (1916), president (1931-2), Royal Society of Canada; KCMG, 1917.

FALKINER, CÆSAR LITTON (1863-1908), Irish historian; son of Sir Frederick Richard Falkiner [q.v.]; BA, Dublin, 1886; MA, 1890; called to Irish bar, 1887; assistant land commissioner, 1898-1908; wrote *Illustrations of Irish History and Topography*, 1904; edited Ormonde Papers for Historical MSS Commission, 5 vols., 1902-8, and Swift's letters, 1908; killed while mountaineering at Chamonix.

FALKINER, Sir FREDERICK RICHARD (1831-1908), recorder of Dublin; BA, Trinity College, Dublin, 1852; called to Irish bar, 1852; QC, 1867; recorder of Dublin, 1876-1905; knighted, 1896; PC, Ireland, 1905; prominent in Church of Ireland questions; published *Literary Miscellanies* (1909), and wrote on Swift's portraits; died in Madeira.

FALKNER, JOHN MEADE (1858-1932), author and antiquary; educated at Marlborough and Hertford College, Oxford; tutor to sons of (Sir) Andrew Noble [q.v.]; later secretary, and finally chairman of Noble's company, Sir W. G. Armstrong, Whitworth & Co.; honorary librarian to dean and chapter of Durham; honorary reader in palaeography, Durham; wrote a *History of Oxfordshire* (1899), poetry, and three romances: *The Lost Stradivarius* (1895), *Moonfleet* (1898), *The Nebuly Coat* (1903).

FANE, VIOLET (pseudonym) (1843-1905), authoress. [See CURRIE, MARY MONTGOMERIE, LADY CURRIE.]

FANSHAWE, Sir EDWARD GENNYS (1814-1906), admiral; entered navy, 1828; commander in East Indies, 1844; suppressed piracy at Borneo, 1845; a lord of the Admiralty, 1865; vice-admiral, 1870; CB, 1871; commander-in-chief on North America station, 1870-3, and at Portsmouth, 1878-9; KCB, 1881; GCB, 1887.

FARJEON, BENJAMIN LEOPOLD (1838-1903), novelist; emigrated from London to Australia, 1855; thence went to New Zealand, settling at Dunedin; joint-editor and proprietor of *Otago Daily Times*, 1861; returned to England, 1868; published melodramatic novels; chief were *Grif* (1866, dramatized 1891), *London's Heart* (1873), *The Mystery of M. Felix* (1890); best was *Devlin the Barber* (1888).

FARJEON, ELEANOR (1881-1965), writer; daughter of Benjamin Leopold Farjeon [q.v.], novelist; established as a writer by *Martin Pippin in the Apple-Orchard*, 1921; other books include *A Nursery in the Nineties* (1935), *Ladybrook* (1931), *Humming Bird* (1936), *Miss Granby's Secret* (1940), *Ariadne and the Bull* (1945), and *Edward Thomas: The Last Four Years* (1958); her poems include *Pan-Worship* (1908), and *Nursery Rhymes of London Town* (1916); plays written in collaboration with her brother, Herbert, include *The Glass Slipper* (St. James's, 1944), and *The Silver Curlew* (Arts, 1949); most notable work, her children's books, which include *Silver-Sand and Snow* (1951), *The Children's Bells* (1957), and *The Little Bookroom* (1955).

FARMER, EMILY (1826-1905), water-colour painter; best known for her groups of children and genre subjects; exhibited at Royal Academy and elsewhere; chief work was 'Deceiving Granny'.

FARMER, JOHN (1835-1901), musician; studied music at Leipzig and Coburg; worked in father's lace business, 1853-7; ran away to Zürich, and taught music there; on staff of Harrow School, 1864-85; composed numerous Harrow school songs; organist of Balliol College, Oxford, 1885-1901; early champion in England of Bach and Brahms; published works include oratorios and fairy opera; edited *Gaudeamus* (1890) and other song collections.

FARMER, Sir JOHN BRETLAND (1865-1944), botanist; first class, natural science, Magdalen College, Oxford, and university demonstrator in botany, 1887; fellow, 1889-97; assistant professor, Royal College of Science, 1892; professor, 1895-1929; director, biological laboratories, 1913-29; trained workers in applied botany and gave important service to colonial agriculture; with J. E. Salvin-Moore responsible for cytological terms 'meiosis' and 'meiotic'; FRS, 1900; knighted, 1926.

FARNELL, LEWIS RICHARD (1856-1934), rector of Exeter College, Oxford, and classical scholar; educated at City of London School and Exeter College; first class, *lit. hum.*, 1878; fellow, 1880; classical lecturer, 1883; sub-rector, 1883-93; senior tutor, 1893-1913; rector, 1913-28; university lecturer, classical archaeology, 1903-14; Wilde lecturer, 1908-11; vice-chancellor,

1920-3; FBA, 1916; works include *The Cults of the Greek States* (5 vols., 1896-1906).

FARNINGHAM, MARIANNE (pseudonym) (1834-1909), hymn writer and author. [See HEARN, MARY ANNE.]

FARNOL, (JOHN) JEFFERY (1878-1952), novelist; privately educated; studied at Westminster School of Art; in USA, 1902-10; with *The Broad Highway* (1910), *The Money Moon* (1911), and *The Amateur Gentleman* (1913) set tone of large output of popular romances.

FARQUHAR, JOHN NICOL (1861-1929), missionary and oriental scholar; BA, Christ Church, Oxford; professor, London Missionary Society's college, Calcutta, 1891; professor of comparative religion, Manchester, 1923-9; works include *Primer of Hinduism* (1912) and *Outline of the Religious Literature of India* (1920).

FARQUHARSON, DAVID (1840-1907), landscape painter; treated chiefly the Perthshire and western highlands; lived at Edinburgh, 1872-82, London, 1882-97, and Cornwall from 1897; exhibited at Royal Scottish and Royal academies; ARSA, 1882; ARA, 1904; chief works were 'The Links of Forth' (1883), 'In a Fog' (1897), 'Full Moon and Spring Tide' (1904), and 'Birnam Wood' (1906).

FARRAR, ADAM STOREY (1826-1905), professor of divinity and ecclesiastical history at Durham from 1864; BA, St. Mary Hall, Oxford, 1850; DD, 1864; tutor at Wadham College, 1855; won Denyer prizes for theology, 1853-4; published Bampton lectures, *A Critical History of Free Thought*, 1862; canon of Durham, 1878.

FARRAR, FREDERIC WILLIAM (1831-1903), dean of Canterbury; born at Bombay; son of India missionary; educated at King's College, London, under F. D. Maurice [q.v.]; scholar of Trinity College, Cambridge, 1852; member of 'Apostles'; won chancellor's medal for English verse, 1852; fourth classic, 1854; fellow of Trinity, 1856; MA, 1857; DD, 1874; master at Marlborough, 1853-5, and at Harrow, 1855-70; while at Harrow published school stories: *Eric, or Little by Little* (1858), *Julian Home* (1859), and *St. Winifred's* (1862), besides *An Essay on the Origin of Language* (1860), which attracted Darwin's attention, *Chapters on Language* (1865), and *Families of Speech* (1870); FRS, 1866; urged serious teaching of science; edited *Essays on a Liberal Education*, 1867; published *Seekers after God* (1868) and Hulsean lecture at Cambridge, *The Witness of History to Christ* (1871); headmaster of Marlborough, 1871-6;

visited Palestine, 1870; published his *Life of Christ* (12 editions, 1874), *Life of St. Paul* (1879), *The Early Days of Christianity* (1882), and *Lives of the Fathers* (1889); rector of St. Margaret's, and canon of Westminster, 1876-95; a successful preacher; chaplain to House of Commons, 1890-5; archdeacon of Westminster, 1883; evoked criticism by challenging doctrine of eternal punishment in sermons at the Abbey, 1877, which were published in *Eternal Hope* (1878); in answer to reply by E. B. Pusey [q.v.], modified his views in *Mercy and Judgement* (1881); lectured in America, 1885; Bampton lecturer at Oxford, 1886; dean of Canterbury, 1895-1903; restored and repaired the cathedral.

FARREN, ELLEN, known as NELLIE FARREN (1848-1904), actress; daughter of Henry Farren [q.v.]; played leading parts at Olympic Theatre (1864-8), in burlesque and in comedy characters, in which she rivalled Mrs Keeley [q.v.]; with company of John Hollingshead [q.v.] at Gaiety Theatre from 1868; won great popularity as principal boy; her chief successes included Sam Weller (1871) and Thaddeus in Byron's *The Bohemian G' Yurl* (1877); in old comedy she shone as Pert in *London Assurance* (1866), Lydia Languish in *The Rivals* (1874), and Maria in *Twelfth Night* (1876); combined pathos with humour as Clemency Newcome in Dickens's *Battle of Life* (1873); visited America and Australia, 1888-91; subsequently appeared only in benefit performances till 1903; unbounded spirits, good-humour, drollery, and sympathy made her a universal favourite.

FARREN, WILLIAM (1825-1908), actor; natural son of William Farren (1786-1891, q.v.); acted at Haymarket in juvenile tragedy or light comedy, 1853-67; roles included Captain Absolute, and Guibert in Browning's *Colombe's Birthday*, 1853; at Vaudeville from 1871, where he was the original Sir Geoffrey Champneys in Byron's *Our Boys*, 1875-8; other parts were Sir Peter Teazle, 'a masterpiece of sheer virtuosity', and Sir Anthony Absolute; started Conway-Farren old comedy company at the Strand, 1887; retired to Rome, 1898.

FARREN, SIR WILLIAM SCOTT (1892-1970), aeronautical engineer; educated at the Perse School and Trinity College, Cambridge; first class, part i, mathematical tripos, and mechanical sciences tripos, 1912-14; head of aerodynamics department, Royal Aircraft Factory, Farnborough; assisted in design of SE5a and CE1 flying boat; joined Armstrong Whitworth Aircraft, 1918; lecturer in aeronautics and engineering, Cambridge, 1920; fellow, Trinity College, 1933; lecturer on strength of aircraft structure, Royal College of

Science, 1922-31; deputy director of scientific research, Air Ministry, 1937; director, technical development, Ministry of Aircraft Production, 1940; director, Royal Aircraft Establishment, Farnborough, 1941-6; technical director, Blackburn Aircraft Co., 1946; with A. V. Roe, Manchester, 1947-61; director Hawker Siddeley Aviation, 1959-61; major projects, design of Vulcan 'V Bomber' and 'Blue Steel' missile; MBE, 1918; CB, 1943; knighted, 1952; FRS, 1945; hon. D.Sc., Manchester; president, Royal Aeronautical Society, 1953-4.

FARRER, AUSTIN MARSDEN (1904-1968), philosopher, theologian, and biblical scholar; educated at St. Paul's and Balliol College, Oxford; first class, classical hon. mods., *lit. hum.*, and theology, 1925-8; Craven scholar; deacon, 1928; ordained, 1929; chaplain and tutor, St. Edmund Hall, Oxford, 1931-5; fellow and chaplain, Trinity College, 1935-60; warden, Keble College, 1960-8; hon. fellow, Trinity College, 1963; FBA, 1968; publications include *Finite and Infinite* (1943), *The Freedom of the Will* (1958), *Saving Belief* (1964), and *Love Almighty and Ills Unlimited* (1962); a great preacher.

FARRER, WILLIAM (1861-1924), historian of Lancashire and feudal genealogist; acquired collections for new history of Lancashire made by J. P. Earwaker [q.v.]; works include co-editing of *Victoria History of the County of Lancaster* (1906-14), *Early Yorkshire Charters* (1914-16), *Feudal Cambridgeshire* (1920), *Honors and Knights Fees* (1923-4; two other volumes posthumously published, 1925, 1927); died in Norway.

FARWELL, SIR GEORGE (1845-1915), judge; BA, Balliol College, Oxford; called to bar (Lincoln's Inn), 1871; QC, 1891; judge of Chancery division and knighted, 1899; lord justice of appeal and PC, 1906-13; gave decision in Taff Vale case, 1900.

FAUSSET, ANDREW ROBERT (1821-1910), divine; scholar of Trinity College, Dublin, where he obtained many classical prizes; BA, 1843; DD, 1886; vicar of St. Cuthbert's, York, 1859-1910; eloquent evangelical preacher and sound scholar; edited *Comedies of Terence*, 1844, and works by Homer, Livy, and Euripides; published *Guide to the Study of the Book of Common Prayer* (1894) and many theological works and commentaries.

FAWCETT, DAME MILLICENT (1847-1929), better known as MRS HENRY FAWCETT, leader of the women's suffrage movement; born

Garrett; sister of Elizabeth Anderson [q.v.]; married Henry Fawcett [q.v.], 1867; promoted women's education at Cambridge; joined first women's suffrage committee, 1867; made first public speech on subject, 1868; worked to secure to married women legal right to their own property and for protection of girls; visited Ireland repeatedly to speak against home rule, 1887-95; leader of ladies' commission of inquiry into Boer concentration camps in South Africa, 1901; as president (1897) of influential constitutional National Union of Women's Suffrage Societies opposed militant suffragettes, 1905-14; suspended propaganda and encouraged women's war work, 1914-18; retired from presidency of National Union after first women's suffrage victory, 1918; DBE, 1925.

FAY, Sir SAM (1856-1953), railway general manager; educated at Blenheim House School, Fareham; joined London and South Western Railway, 1872; secretary and general manager, Midland and South Western Junction Railway, 1892; superintendent of the line, Waterloo, 1899; general manager, Great Central Railway, 1902-22; developed Immingham dock and knighted there, 1912; director-general of movements and railways and member of Army Council, 1918-19; director of number of companies; published *The War Office at War* (1937).

FAY, WILLIAM GEORGE (1872-1947), actor and producer; educated at Belvedere College, Dublin; possessed genius for character and comedy acting; creations at Abbey Theatre (1904-8) included Christy Mahon in *The Playboy of the Western World*; acted and produced in England and latterly appeared in films.

FAYRER, Sir JOSEPH, first baronet (1824-1907), surgeon-general and author; son of naval commander; lived in Westmorland, where he met Wordsworth; MRCS, 1847; FRCP, 1872; FRCS, 1878; MD, Rome, 1849; joined Indian medical service, 1850; assistant surgeon in Pegu war, 1852; civil surgeon of Lucknow, 1856; prominent in Mutiny; MD, Edinburgh, 1859; professor of surgery at Medical College, Calcutta, 1859-74; president of Asiatic Society of Bengal, 1867; CSI, 1868; accompanied Duke of Edinburgh (1870) and Prince of Wales (1875) on Indian tours, of which he published *Notes*, 1876; president of medical board of India Office, 1873-95; KCSI, 1876; FRS, 1877; hon. LLD, Edinburgh and St. Andrews; baronet, 1896; in chief work, *The Thanatophidia of India* (1872), first advocated permanganate treatment of venomous snake bites; wrote *Recollections of my Life* (1900).

FEARNSIDES, WILLIAM GEORGE (1879-1968), geologist; educated at Wheelwright grammar school, Dewsbury, and Sidney Sussex College, Cambridge; first class, parts i and ii, natural sciences tripos, 1900-1; Harkness scholar in geology; fellow, Sidney Sussex, 1904; Taylor and college lecturer, 1908; university demonstrator in petrology, 1909; authority on Lower Palaeozoic of Wales; first Sorby professor of geology, Sheffield University, 1913-45; during 1914-18 war became authority on moulding sands; vice-president, Midland Institute of Mining Engineers, 1928-32; advised Attock Oil Co. in India and Burma, 1930, and 1938; geological adviser to Anglo-Iranian Oil Co. on oil exploration in Britain; president, Geological Society, 1943-5; geological advisor to National Coal Board West Midlands Division, 1947; FRS, 1932; dean of faculty of pure science, Sheffield, 1931-4; hon. fellow, Sidney Sussex, 1946.

FEETHAM, RICHARD (1874-1965), judge, and last survivor of 'Milner's kindergarten'; educated at Marlborough and New College, Oxford, 1895-7; called to bar (Inner Temple), 1899; joined legal staff of London County Council; a friend of Lionel G. Curtis [q.v.], who invited him to South Africa to join Sir Alfred (later Viscount) Milner [q.v.] in the Transvaal, 1901; town clerk, Johannesburg, 1903-5; legal advisor to Lord Selborne [q.v.]; member of Transvaal Legislative Council, 1907-10; one of the intellectual architects of South African unification; member (for Parktown) of South African Legislative Assembly, 1915-23; member, Southborough committee on Indian reforms, 1918; judge, Transvaal division, South African Supreme Court, 1923; chairman, Ulster boundary commission, 1924; chairman of number of other commissions; judge-president, Natal Provincial division, Supreme Court, 1931; judge, Appellate division, 1939-44; vice-chancellor, university of Witwatersrand, 1938-48; chancellor in succession to J. H. Hofmeyr [q.v.], 1949-61; opposed apartheid in higher education; CMG, 1924; hon. LLD, Witwatersrand, 1949, and Natal, 1958.

FELKIN, ELLEN THORNEYCROFT (1860-1929), better known as ELLEN THORNEYCROFT FOWLER, novelist; daughter of first Viscount Wolverhampton [q.v.]; married A. L. Felkin, 1903; attracted attention with novel *Concerning Isabel Carnaby* (1898); wrote many others.

FELLOWES, EDMUND HORACE (1870-1951), clergyman and musical scholar; educated at Winchester and Oriel College, Oxford; B.Mus., 1896; deacon, 1894; priest, 1895; minor canon, St. George's chapel, Windsor, 1900-51; edited English music of *c*.1545-1645: 36 vols. of madrigals, 32 of lute songs, 20 of Byrd's music; published *The English Madrigal Composers* (1921) and *William Byrd* (1936); honorary librarian, St. Michael's College, Tenbury Wells, 1918-48; MVO, 1931; CH, 1944; hon. D.Mus. Dublin, Oxford, and Cambridge; hon. fellow, Oriel College.

FELLOWES, SIR EDWARD ABDY (1895-1970), clerk of the House of Commons; educated at Marlborough, 1909-14; commissioned in Queen's Royal West Surrey Regiment, 1914; MC; appointed to clerkship in House of Commons, 1919; clerk to Public Accounts Committee; second clerk assistant, 1937; commanded Palace of Westminster company of Home Guard, 1942-4; CB, 1945; clerk assistant, 1948; succeeded Sir Frederic Metcalfe as clerk of the House, 1954-61; KCB, 1955; joint-editor (with (Sir) T. G. B. Cocks) of *Treatise on the Law, Privileges, Proceedings, and Usage of Parliament* (1957); published *Selection from the Volumes of Decisions from the Chair* (1960); interested in commonwealth legislatures; CMG, 1953; president, Association of Secretaries General of Parliaments, 1956-60; chairman, General Advisory Council of BBC, 1962-7, and Hansard Society; chairman and then president, Study of Parliament Group, 1964-70; member, Royal Society of Arts.

FENN, GEORGE MANVILLE (1831-1909), novelist; a short sketch, 'In Jeopardy', was accepted by Dickens for *All the Year Round*, 1864; soon wrote stories for boys, embodying natural history studies, producing more than 170 volumes; editor of *Cassell's Magazine*, 1870, and proprietor of *Once a Week*, 1873; wrote dramatic criticism and farces.

FENWICK, ETHEL GORDON (1857-1947), pioneer of nursing reform; born MANSON; matron, St. Bartholomew's Hospital, 1881-7; married (1887) Bedford Fenwick; founded (Royal) British Nurses' Association (1887), the National and International Councils of Nurses, and (1926) British College of Nurses; led movement which obtained (1919) state registration of nurses.

FERGUSON, FREDERIC SUTHERLAND (1878-1967), bibliograper; educated at the Grocers' Company School, Hackney Downs, London, and King's College, London; entered firm of Bernard Quaritch, antiquarian booksellers, 1897; served with Cameronians, 1916-18; rejoined Quaritch, 1919; managing director, 1928-43; retired from Quaritch, 1947; worked in British Museum on catalogue of early British books, 1947-67; made outstanding contribution to *A Short-Title Catalogue of Books Printed in England, Scotland & Ireland, and of English Books Printed Abroad, 1475-1640* (1926); published (with R. B. McKerrow [q.v.]) *Title-Page Borders used in England and Scotland, 1485-1640* (1932); hon. MA, Oxford, 1955; hon. LLD Edinburgh, 1955; president, Bibliographical Society, 1948-50; president, Edinburgh Bibliographical Society, 1935-6; president, International Antiquarian Booksellers' Association, 1934.

FERGUSON, HARRY GEORGE (1884-1960), engineer and inventor; son of an Irish farmer; founded garage business in Belfast; first man to fly in Ireland (1909), using monoplane designed and built by himself; advocated mechanization of farm work; produced revolutionary design for light manœuvrable farm tractor, 1935; his partnership with Henry Ford, 1939-47, repudiated after Ford's death; founded his own companies in Detroit and Great Britain; sued Fords and obtained 9.25 million dollars compensation; merged with Canadian Massey-Harris farm machinery concern, 1953; chairman, 1953-4; worked thereafter on producing safer road vehicles.

FERGUSON, MARY CATHERINE, LADY FERGUSON (1823-1905), biographer; born GUINNESS; married Sir Samuel Ferguson [q.v.], 1848, and shared his social and literary activities; wrote *The Story of the Irish before the Conquest* (1868) and a life of her husband (2 vols., 1896).

FERGUSON, RONALD CRAUFORD MUNRO-, VISCOUNT NOVAR (1860-1934), politician; served in Grenadier Guards, 1879-84; liberal MP, Ross and Cromarty (1884-5), Leith Burghs (1886-1914); PC, 1910; governor-general of Australia, 1914-20; GCMG, 1914; viscount, 1920; secretary for Scotland, 1922-4; KT, 1926.

FERGUSSON, SIR CHARLES, seventh baronet of Kilkerran (1865-1951), soldier and administrator; son of Sir James Fergusson [q.v.]; succeeded him, 1907; educated at Eton and Sandhurst; commissioned in Grenadier Guards, 1883; with Egyptian Army, 1896-1903; major-general, 1908; inspector of infantry, 1909-12; commanded 5th division, Ireland, 1913; held officers to their duty during Curragh incident, 1914; took division to France, 1914; commanded II Corps, 1914-16; XVII Corps, 1916-18; fought at Arras, 1917; broke Hindenburg Line and took Cambrai, 1918; military governor, Cologne, Dec. 1918-Aug. 1919; general, 1921; governor-general, New Zealand,

1924-30; chairman, West Indies closer union commission, 1932; CB, 1911; KCB, 1915; GCB, 1932; KCMG, 1918; GCMG, 1924.

FERGUSSON, SIR JAMES, sixth baronet of Kilkerran (1832-1907), governor of Bombay; educated at Rugby and University College, Oxford; succeeded to baronetcy, 1849; entered Grenadier Guards; served in Crimea; conservative MP for Ayrshire, 1855-7, 1859-68; under-secretary for India, 1866, and to Home Office, 1867; PC and governor of South Australia, 1868-73, and of New Zealand, 1873-5; KCMG, 1875; governor of Bombay, 1880-5; sought welfare of Deccan peasantry by modifying assessment of land revenue and granting remissions in times of scarcity; created first agricultural department; greatly developed the port of Bombay, and promoted rural and urban self-government; GCSI, 1885; MP for North-East Manchester, 1885-1906; under-secretary at Foreign Office, 1886-91; postmaster-general, 1891-2; killed in earthquake at Kingston, Jamaica.

FERGUSSON, SIR (JOHN) DONALD (BALFOUR) (1891-1963), civil servant; educated at Berkhamsted School and Magdalen College, Oxford; first class, modern history, 1914; served with army in France, 1914-18; entered Treasury, 1919; private secretary to six chancellors of the Exchequer; assistant secretary, 1934; permanent secretary, Ministry of Agriculture and Fisheries, 1936-45; organized department to implement policy of reviving farming industry and increasing production in war; permanent secretary, Ministry of Fuel and Power, 1945-52; director, Prudential Assurance Company, and Agricultural Mortgage Corporation, 1952; CB, 1935; KCB, 1937; GCB, 1946.

FERMOR, SIR LEWIS LEIGH (1880-1954), geologist; educated at Wilson's grammar school, Camberwell and Royal School of Mines; B.Sc., London, 1907; D.Sc., 1909; joined Geological Survey of India, 1902; superintendent, 1910; director, 1930-5; did important work on manganese and on coal in Korea State and the Kargali coal-seam; wrote memoirs on Indian Archaean rocks; a founder and first president, National Institute of Sciences in India, 1933; reported on mining industry in Malaya, 1939; FRS, 1934; knighted, 1935.

FERRANTI, SEBASTIAN ZIANI DE (1864-1930), electrical engineer and inventor; obtained work with Siemens Brothers at Charlton, near Woolwich, 1881; patented Ferranti alternator, which won him recognition among electrical engineers, 1882; set up independent business in London for manufacture

of electrical apparatus, 1883; engineer to Grosvenor Gallery Electric Supply Corporation, 1886; chief electrician to London Electric Supply Corporation, 1887; planned station which, on his initiative, was erected at Deptford, to supply all London north of Thames; his scheme crippled by Electric Lighting Act (1888), which reduced area intended to be supplied and permitted competition of low-power-stations within this area; devoted himself to private business as manufacturing engineer, 1892; established Ferranti Limited near Oldham, 1896; took out 176 patents, 1882-1927; pioneer in high-voltage systems and originator of long-distance transmission of high-power electical current; FRS, 1927; died at Zürich.

FERRERS, NORMAN MACLEOD (1829-1903), master of Caius College, Cambridge, and mathematician; educated at Eton and Caius College; senior wrangler and first Smith's prizeman, 1851; fellow, 1854; tutor, 1865; elected master, 1880; DD, 1881; hon. LLD, Glasgow, 1883; FRS, 1877; took prominent but conservative part in university affairs; published *Trilinear Co-ordinates* (1861) *Spherical Harmonics* (1877), and wrote much for mathematical journals; edited *Quarterly Journal of Mathematics*, 1855-91.

FERRIER, SIR DAVID (1843-1928), physician; MA, Aberdeen University, 1863; MB, Edinburgh University, 1868; began long service with King's College, London, as demonstrator of physiology, 1871; successively assistant physician, physician, and consulting physician; professor of neuropathology, 1889-1908; began researches on electrical excitation of brain, 1873; published *The Functions of the Brain*, 1876; carried out experiments with monkeys; originator, by his advocacy, no less than surgeons themselves, of modern cerebral surgery; FRS, 1876; knighted, 1911.

FERRIER, KATHLEEN MARY (1912-1953), singer; educated at Blackburn high school; became Post Office telephonist at fourteen; after marriage (1935) taught piano; after 1937 progressed from local to national and world fame as a singer of glorious voice and warm personality; Benjamin Britten composed *Rape of Lucretia* for her (Glyndebourne, 1946); also notable in *Orpheo* and in works by Mahler; CBE, 1953; died prematurely of cancer.

FESTING, JOHN WOGAN (1837-1902), bishop of St. Albans; descended from Michael C. Festing [q.v.]; BA, Trinity College, Cambridge, 1860; DD, 1890; vicar of Christ Church, Albany Street, 1878-90; bishop of St. Albans, 1890; businesslike administrator;

president of the Universities' Mission to Central Africa, 1892-1902.

FETHERSTON HAUGH, Sir HERBERT MEADE- (1875-1964), admiral. [See MEADE-FETHERSTON HAUGH.]

FFOULKES, CHARLES JOHN (1868-1947), first curator of the Imperial War Museum and master of the armouries of the Tower of London; educated at Radley, Shrewsbury and St. John's College, Oxford; B.Litt., 1911; master of the armouries, Tower of London, 1912-38; published *Inventory and Survey* (2 vols., 1916); suggested and made collection for Imperial War Museum (opened 1920); curator, 1917-33; publications include *The Armourer and his Craft* (1912); CB, 1934.

FIELD, (AGNES) MARY (1896-1968), producer of children's films; educated at Surbiton high school and Bedford College, London; worked as teacher of English and history; realized potentialities of films for educational purposes; joined British Instructional Films, 1926; transferred to Gaumont British Instructional Films, 1934; partnered Percy Smith, in making *The Secrets of Nature* and *The Secrets of Life*; awarded diplôme d'honneur at International Cinema Festival, Brussels, with *The Tortoiseshell Butterfly* (1935); executive producer for J. Arthur Rank's Children's Entertainment Films Department, 1944; produced *Bush Christmas* (1947); executive officer, Children's Film Foundation, 1951; fellow, British Film Academy, and Royal Photographic Society; chairman, Brussels International Centre of Films for Children; OBE, 1951; published *Good Company* (1952); hon. FRPS, FBKS, and FBFA.

FIELD, Sir (ARTHUR) MOSTYN (1855-1950), admiral; entered navy, 1869; served in surveying ships, 1876-84; commanded *Dart* (surveying coasts of Australia, 1884-9), *Egeria* (China station, 1890-3), *Penguin* (South Pacific islands, 1896-9), and *Research* (Scotland and Ireland, 1900-4); hydrographer of the navy, 1904-9; completed many surveys including Scapa Flow; rear-admiral, 1906; vice-admiral, 1910; admiral, 1913; Admiralty representative, Port of London Authority, 1909-25; FRS, 1905; KCB, 1911.

FIELD, Sir FREDERICK LAURENCE (1871-1945), admiral of the fleet; entered navy, 1884; superintendent of signal schools, 1912-14; commanded *King George V* at Jutland, 1916; chief of staff to Sir Charles Madden [q.v.], 1916-18; director of torpedoes and mines, 1918-20; rear-admiral, 1919; third sea lord,

1920-3; KCB, 1923; commanded special service squadron on world cruise, 1923-4; KCMG, 1924; deputy chief of naval staff, 1925-8; commander-in-chief, Mediterranean, 1928-30; admiral, 1928; first sea lord, 1930-3; prior to trouble at Invergordon (1931) had declined responsibility if cuts in naval pay were proportionately greater than those in other services; admiral of the fleet and GCB, 1923.

FIELD, WALTER (1837-1901), painter; son of Edward W. Field [q.v.]; worked in oils and water-colours, mainly on landscapes and Thames scenery; exhibited at Old Water-Colour Society and Royal Academy; helped to found Hampstead Heath Protection Society; most popular works were 'The Milkmaid singing to Isaak Walton' and 'Henley Regatta'.

FIELD, WILLIAM VENTRIS, BARON FIELD (1813-1907), judge; called to bar, 1850; QC, 1864; obtained large commercial practice on Midland circuit; made judge of court of Queen's Bench, 1875; showed great learning but quick temper on bench; gave judgment which was upheld on appeal in many important cases; created PC and raised to peerage on retirement, 1890.

FIFE, DUCHESS OF (1891-1959). [See ALEXANDRA VICTORIA ALBERTA EDWINA LOUISE DUFF, PRINCESS ARTHUR OF CONNAUGHT.]

FIFE, DUCHESS OF (1867-1931), princess royal of Great Britain and Ireland. [See LOUISE VICTORIA ALEXANDRA DAGMAR.]

FIGGIS, JOHN NEVILLE (1866-1919), historian and divine; BA, St. Catharine's College, Cambridge; Mirfield father, 1907; works include *The Divine Right of Kings* (1896), *From Gerson to Grotius* (1907), and *Churches in the Modern State* (1913).

FILDES, Sir (SAMUEL) LUKE (1844-1927), painter; adopted popular anecdotal, melodramatic vein of painting; chief pictures in this style, 'Applicants for admission to a casual ward' (1874) and 'The Doctor' (1891); gradually devoted himself almost exclusively to portraiture; painted state portraits of King Edward VII, Queen Alexandra, King George V; RA, 1887; knighted, 1906; KCVO, 1918.

FILON, LOUIS NAPOLEON GEORGE (1875-1937), mathematician; born in France, son of tutor to Prince Imperial; naturalized, 1898; BA, University College, London, 1896; lecturer in pure mathematics, 1903-12; Goldsmid professor of applied mathematics and mechanics, 1912-37; director, university observatory, 1929-37; vice-chancellor, 1933-5; his outstanding contributions were in mechanics of continuous media; greatest achievement was theory of 'generalized plane stress'; FRS, 1910; CBE, 1933.

FINBERG, ALEXANDER JOSEPH (1866-1939), writer on the history of English art; educated at King's College, London; art critic to several papers; completed arrangement and compiled inventory of Turner bequest; published *Life of J. M. W. Turner* (1939); founded Walpole Society (1911) to encourage study of British art.

FINCH, GEORGE INGLE (1888-1970), physicist; born in New South Wales; educated at Wolaroi College, New South Wales, and Federal Technical high school, Zürich; research chemist, Royal Arsenal; demonstrator, Imperial College, London; served in army during 1914-18 war; MBE, 1917; demonstrator, Imperial College, 1919; lecturer in electrochemistry, 1921; professor, applied physical chemistry, 1936-52; director, National Chemical Laboratory, India, 1952-7; research associated with wear in internal combustion engines, action of lubricants, and electrodeposition; scientific advisor, fire division, Ministry of Home Security, 1941; FRS, 1938; president, Physical Society, 1947-9; chevalier de la légion d'honneur, 1952; skilled mountaineer; pioneered use of oxygen, Everest expedition, 1922; president, Alpine Club, 1959; publications include *The Making of a Mountaineer* (1924), and *Climbing Mount Everest* (1930).

FINCH-HATTON, HAROLD HENEAGE (1856-1904), imperialist politician; son of tenth Earl of Winchelsea [q.v.]; went to Queensland, 1876; engaged in cattle-farming and gold prospecting till 1883, when he returned to England; published *Advance, Australia* (1885); a founder of Imperial Federation League; conservative MP for Newark, 1895-8; an ardent sportsman.

FINLAY, ROBERT BANNATYNE, first VISCOUNT FINLAY (1842-1929), lord chancellor; called to bar (Middle Temple), 1867; QC, 1882; liberal MP, Inverness Burghs, 1885-92, 1895-1906; solicitor-general and knighted, 1895; attorney-general, 1900-6; assisted in preparing British case in Venezuelan boundary arbitration, 1899; led for Great Britain and Canada in Alaska boundary arbitration, 1903; for Great Britain in Venezuelan claims arbitration, 1903; for Canada in Newfoundland fisheries arbitration, 1910; MP, Edinburgh and St. Andrews universities, 1910; lord chancellor, 1916-18;

member of Permanent Court of Arbitration (1920) and judge of Permanent Court of International Justice (1921-8) at The Hague; baron, 1916; viscount, 1919.

FINLAY, WILLIAM, second VISCOUNT FINLAY (1875-1945), judge; son of first Viscount Finlay [q.v.]; educated at Eton and Trinity College, Cambridge; called to bar (Middle Temple), 1901; junior counsel, Board of Inland Revenue, 1905-14; KC, 1914; judge of King's Bench division, 1924; succeeded father, 1929; lord justice of appeal and PC, 1938; British representative, United Nations war crimes commission, 1945; KBE, 1920; GBE, 1945.

FINLAYSON, JAMES (1840-1906), Scottish physician; MB, Glasgow, 1867; MD, 1869; hon. LLD, 1899; physician to Western Infirmary, Glasgow, from 1875 to death, and to Royal Children's Hospital, 1883-98; voluminous works include *Clinical Manual* (1878), *Life of Peter Lowe* (1889), and *Life of Robert Watt* (1897).

FINNIE, JOHN (1829-1907), landscape painter and engraver; pupil of William Bell Scott [q.v.]; master of School of Art, Liverpool, 1855-96; president of Liverpool Academy, 1887-8; early experimented in etching and engraving; practised mezzotint from 1886, exhibiting at Royal Academy and elsewhere.

FINZI, GERALD RAPHAEL (1901-1956), composer; privately educated; studied music with (Sir) Edward Bairstow [q.v.], 1918-22; professor of composition, Royal Academy of Music, 1930-3; rooted in English music, English poetry, and the Wessex countryside, his music is restrained and contemplative; orchestral works include a Clarinet (1949) and Violoncello (1956) concerto; large choral works include *Intimations of Immortality* (1950) and *In Terra Pax* (1954); most notable for his vocal music especially for solo voice; verse settings include Hardy, Shakespeare, and Robert Bridges.

FIRTH, SIR CHARLES HARDING (1857-1936), historian; educated at Clifton and Balliol College, Oxford; first class, modern history, 1878; made magnificent collection of books mainly on English seventeenth century; excelled in handling historical evidence; edited texts including *Memoirs of the Life of Colonel Hutchinson* (1885) and the *Clarke Papers* (4 vols., 1891-1901); contributed seventeenth-century lives to this Dictionary; wrote *Oliver Cromwell* (1900), *Cromwell's Army* (1902), *The House of Lords during the Civil War* (1910); completed history by S. R. Gardiner [q.v.] with

The Last Years of the Protectorate (2 vols., 1909); history lecturer, Pembroke College, Oxford, 1887-93; fellow of All Souls, 1902; regius professor of modern history, 1904-25; his criticisms of the Oxford system of teaching history were without effect; trustee of National Portrait Gallery, 1908-29; member of Historical Manuscripts Commission and of royal commission on public records (1910-19); a founder and first president (1906-10) of Historical Association; FBA, 1903; knighted, 1922.

FIRTH, JOHN RUPERT (1890-1960), professor of general linguistics; educated at Keighley grammar school and Leeds University; first class, history, 1911; joined Indian education service, 1915; professor of English, university of Punjab, Lahore, 1919-28; senior lecturer in phonetics, University College, London, 1928-38; senior lecturer, School of Oriental and African Studies, 1938-56; head of phonetics and linguistics department, 1941-56; reader in linguistics and Indian phonetics, London University, 1940; professor of general linguistics, 1944-56; established general linguistics as recognized university subject.

FIRTH, SIR WILLIAM JOHN (1881-1957), industrialist; began as office boy; entered tinplate trade, 1901; associated with Henry Folland in development of Grovesend Steel & Tinplate Co.; amalgamated with Richard Thomas and established selling agency, 1923; chairman of Richard Thomas, 1931-40; established continuous wide strip mill at Ebbw Vale; settled in South Africa, 1947, and died there; knighted, 1932.

FISCHER WILLIAMS, SIR JOHN (1870-1947), international lawyer. [See WILLIAMS.]

FISHER, ANDREW (1862-1928), Australian statesman; migrated from Scotland to Queensland, 1885; labour member for Gympie, 1893-6; regained seat, 1899; joined seven-days' ministry of Anderson Dawson (first labour ministry in world) as secretary for railways and public works, 1899; member for Wide Bay in first Commonwealth parliament, 1901; leader of federal labour party, 1907; prime minister, 1908-9; returned to power, 1910-13; measures passed included establishment of Commonwealth bank, strong defence policy, and inauguration of new capital at Canberra; again prime minister, 1914-15; wholeheartedly supported Great Britain in European war; high commissioner of Australia in London, 1915-21; devoted to British Empire.

FISHER, HERBERT ALBERT LAURENS (1856-1940), historian, statesman, and warden of New College, Oxford; brother of Sir W. W. Fisher [q.v.]; educated at Winchester and New College, Oxford; first class. *lit. hum.*, and fellow, 1888; taught modern history which he studied at Paris and Göttingen; member of royal commission on public services in India, 1912-17; vice-chancellor, Sheffield University, 1912-16; encouraged the arts and the application of science to industry; president of Board of Education, 1916-22; introduced Education Act of 1918; established state scholarships to universities, percentage grants for teachers' salaries, and the School Certificate; proposed part-time continuation schools up to age of eighteen; MP, Hallam division of Sheffield (1916-18), combined English universities (1918-26); warden of New College, 1925-40; publications include *Bonapartism* (1908), biography (1910) and *Collected Papers* (1911) of F. W. Maitland [q.v.], *James Bryce* (2 vols., 1927), *History of Europe* (3 vols., 1935); FBA, 1907, president, 1928-32; FRS, 1920; OM, 1937.

FISHER, JAMES MAXWELL McCONNELL (1912-1970), ornithologist; educated at Eton and Magdalen College, Oxford; ornithologist, Oxford University Expedition to Spitsbergen, 1933; taught at Bishop's Stortford College; assistant curator, Zoological Society, London, 1936-9; honorary secretary, British Trust for Ornithology, 1938-44; worked in Oxford, 1940-6; primary interest in sea-birds; landed from helicopter on Rockall, 1955; natural history editor, Collins, publishers, 1946-54; deputy chairman, National Parks (later Countryside) Commission, 1968-70; fellow, Royal Geographical Society, Linnaen Society, Geological Society, and Zoological Society; publications include *Birds as Animals* (1939, reissued 1954), *Watching Birds* (1940), *The Birds of Britain* (1942), *The Fulmar* (1952), and *Rockall* (1956).

FISHER, JOHN ARBUTHNOT, first BARON FISHER (1841-1920), admiral of the fleet; born in Ceylon; entered navy, 1854; qualified in gunnery school, *Excellent*, and joined *Warrior*, first 'ironclad', 1863; on staff of *Excellent*, 1846-9, 1872-6; devoted himself to development of torpedo; captain, 1874; went to Admiralty for first time, 1876; at sea, 1876-82; CB for services in Egypt against Arabi Pasha, 1882; captain of gunnery school, Portsmouth, 1883-6; director of ordnance and torpedoes at Admiralty, 1886-90; rear-admiral, 1890; third sea lord and controller of navy, 1892-7; KCB, 1894; vice-admiral, 1896; commander-in-chief, North America and West Indies station, 1897; commanded Mediterranean fleet, 1899-1902; ensured increased efficiency in every department; admiral, 1901; GCB, 1902; second sea lord, with charge of personnel of fleet, 1902-3; introduced important administrative reforms, notably common entry scheme in training of naval officers at Osborne, 1903; commander-in-chief, Portsmouth, 1903; member, War Office reconstruction committee, 1903-4; first sea lord, 1904-10; OM, 1904; organized redistribution of fleet to meet growing German menace; advocated designing of *Dreadnought* type of battleship and battle cruiser, 1905; Cawdor memorandum, statement of Admiralty policy, published before fall of conservative government, 1905; obliged to diminish programme under liberal government; estranged from Lord Charles Beresford [q.v.], commander-in-chief of Channel fleet, 1907; issued programme of eight battleships, 1909-10; baron, 1909; returned to Admiralty as first sea lord after outbreak of European war, 1914; responsible for battle of Falkland Islands, 1914; disapproved of naval attempt to force passage of Dardanelles and resigned, 1915; one of the greatest administrators in history of royal navy.

FISHER, SIR (NORMAN FENWICK) WARREN (1879-1948), civil servant; scholar of Winchester and Hertford College, Oxford; second class, *lit. hum.*, 1902; entered Board of Inland Revenue, 1903; commissioner, 1913; deputy chairman, 1914; chairman, 1918; member, committee on staffs, 1918-19; permanent secretary to Treasury and head of Civil Service, 1919-39; increased Treasury establishment and recruited from other departments; fostered team-work between Treasury and departments and secured acceptance (1925) for his proposal that permanent secretaries should always be accounting officers, thus combining responsibility for policy and finance; strove for conception of the service as a unity and for highest standard of conduct in the service; recommended independence of thought and speech to senior civil servants in their relations with ministers; much occupied with organization for war and preoccupied with rearmament; defence commissioner, north-western region, 1939-40; special commissioner organizing services and clearance in London, 1940-2; KCB, 1919; GCVO, 1928.

FISHER, ROBERT HOWIE (1861-1934), Scottish divine; educated at George Watson's College and Edinburgh University; ordained, 1885; minister at Morningside (1900) and St. Cuthbert's (1914-25), Edinburgh; university lecturer, 1911-16; editor of Church of Scotland's official magazine *Life and Work*, 1902-25; did important work on Church union committee.

FISHER, SIR RONALD AYLMER (1890-1962), mathematical statistician and geneticist; educated at Harrow and Gonville and Caius College, Cambridge; wrangler with distinction, 1912; taught mathematics and physics, 1915-19; statistician, Rothamsted Experimental Station, 1919-33; fellow, Caius, 1920; Sc.D. Cambridge, 1926; FRS, 1929; published *Statistical Methods for Research Workers* (1925), and *The Genetical Theory of Natural Selection* (1930); succeeded Karl Pearson [q.v.] as Galton professor of eugenics, University College, London, 1933-43; Arthur Balfour professor of genetics, Cambridge, 1943-57; president, Caius College, 1956; research fellow, Division of Mathematical Statistics of the Commonwealth Scientific and Industrial Research Organization, Adelaide, 1959-62; knighted, 1952; awarded many academic honours; hon. D.Sc., London.

FISHER, SIR WILLIAM WORDSWORTH (1875-1937), admiral; brother of H. A. L. Fisher [q.v.]; entered *Britannia*, 1888; qualified as gunnery lieutenant, 1900; commander, 1906; captain, 1912; commanded battleship *St. Vincent*, 1912-17; present at Jutland; CB, 1918; director, anti-submarine division, 1917-18; chief of staff to Sir John De Robeck [q.v.] in Mediterranean (1919-22) and Atlantic (1922-4); rear-admiral, 1922; director of naval intelligence, 1926-7; fourth sea lord, 1927-8; vice-admiral, 1928; deputy chief of naval staff, 1928-30; commanded first battle squadron, Mediterranean, 1930-2; admiral and commander-in-chief, Mediterranean, 1932-6; exercised fleet in night-fighting; his forces greatly increased owing to Italo-Abyssinian war, 1935; commander-in-chief, Portsmouth, 1936-7; KCB, 1929; GCB and GCVO, 1935; his wide interests gave him an understanding of men which kept morale at a very high standard.

FISON, LORIMER (1832-1907), Wesleyan missionary and anthropologist; left Caius College, Cambridge, for Australia, 1855; Wesleyan missionary at Fiji, 1863-71; studied native kinship; joined Alfred William Howitt [q.v.] in anthropological research in New South Wales and Victoria, 1871-5; again in Fiji, 1875-84, writing on Fijian land tenure and antiquities; settled at Melbourne, 1888-1905; published *Tales from Old Fiji* (1904); awarded civil list pension, 1905; died at Essenden, Victoria.

FITCH, SIR JOSHUA GIRLING (1824-1903), inspector of schools and educational writer; BA, London University, 1850; MA, 1852; joined staff of Borough Road training college, Southwark, 1852; principal, 1856, proving a stimulating teacher of method; assistant commissioner of schools in Yorkshire, 1865–7, of elementary schools in the large northern towns, 1869, and of endowed schools, 1870–7; chief inspector of schools for eastern division, 1883–5; inspector of elementary training colleges for women in England and Wales, 1885–94; visited America and reported on American education, 1888; helped in foundation of Girton and Girls' Public Day School Company; published *Lectures on Teaching* (1881) and *Educational Aims and Methods* (1900); his account of *The Chautauqua Reading Circles* (1888) suggested the National Home Reading Union; hon. LLD, St. Andrews, 1888; knighted, 1896.

FITZALAN OF DERWENT, first VISCOUNT (1855–1947). [See HOWARD, EDMUND BERNARD FITZALAN-.]

FITZALAN-HOWARD, HENRY, fifteenth DUKE OF NORFOLK (1847–1917). [See HOWARD.]

FITZCLARENCE, CHARLES (1865–1914), brigadier-general; joined army (Royal Fusiliers), 1885; VC for gallantry during South African war, 1900; transferred to Irish Guards, 1900; served with distinction as brigadier-general commanding 1st (Guards) brigade in 1st division, 1914; killed at head of Irish Guards in Flanders.

FITZGERALD, GEORGE FRANCIS (1851–1901), natural philosopher; son of William FitzGerald [q.v.]; BA (in mathematics) at Trinity College, Dublin, 1871; fellow, 1877; Erasmus Smith professor of natural and experimental philosophy, 1881–1901; developed electro-magnetic theory of radiation; made researches into electric waves and electrolysis; FRS, 1883; president of Physical Society, 1892–3; prominent in Irish educational affairs; *Scientific Writings* collected by Sir Joseph Larmor [q.v.], 1902.

FITZGERALD, SIR THOMAS NAGHTEN (1838–1908), surgeon; LRCS Ireland, 1857; FRCS, 1884; surgeon at Melbourne Hospital from 1858 till death; an able operator; knighted, 1897; surgeon in South African war, and CB, 1900.

FITZGIBBON, GERALD (1837–1909), lord justice of appeal in Ireland; scholar of Trinity College, Dublin; hon. LLD, 1895; called to Irish bar, 1860, and to English bar, 1861; leader of Munster circuit; successful advocate in *O'Keefe* v. *Cullen* case, 1873, and *Bagot* v. *Bagot* will case, 1878; solicitor-general for Ireland, 1877–8; promoted lord justice of appeal, 1878;

PC, 1900; served on various Irish educational commissions; chairman of commission on Irish educational endowments, 1885–97; an active freemason, prominent in affairs of Church of Ireland; friend of Lord Randolph Churchill [q.v.] and other politicians.

FITZMAURICE, BARON (1846–1935). [See PETTY-FITZMAURICE, EDMOND GEORGE.]

FITZMAURICE, SIR MAURICE (1861–1924), civil engineer; articled to Sir Benjamin Baker [q.v.]; engaged under Baker and Sir John Fowler [q.v.] in construction of Forth bridge; entered service of London County Council under Sir A. R. Binnie [q.v.], 1892; chief resident engineer to Egyptian government, 1898; engaged on construction of Aswan dam, 1898–1901; chief engineer to London County Council, 1901–12; knighted, 1912; FRS, 1919.

FITZMAURICE-KELLY, JAMES (1857–1923), historian of Spanish literature; began to make name for himself as authority on Spain, 1886; published *Life of Miguel de Cervantes Saavedra* (1892) and *History of Spanish Literature* (1898); first Gilmour professor of Spanish language and literature, Liverpool University, 1909–16; Cervantes professor of Spanish language and literature, King's College, London, 1916–20; FBA, 1906.

FITZPATRICK, SIR DENNIS (1837–1920), Indian civil servant; entered Indian civil service, 1858; established reputation over case of Begum Samru of Sardhana, 1866–72; in legislative department of Indian government, 1874–85; KCSI, 1890; lieutenant-governor of Punjab, 1892–7; on council of India, 1897; vice-president, 1901; GCSI, 1911.

FITZPATRICK, SIR (JAMES) PERCY (1862–1931), South African statesman and author; born in Cape Colony; educated at Downside; moved from Cape to Transvaal; joined Hermann Eckstein & Co., 1892; partner, 1898–1907; member of 'reform committee' and imprisoned after Jameson raid; president, Witwatersrand chamber of mines (1902); leading supporter of employment of Chinese labour; Transvaal delegate to national convention, 1908–9; MP, Pretoria East, 1910–20; presented Delville Wood to nation as a memorial and initiated Armistice day two-minute silence; knighted, 1902; KCMG, 1911; publications include *The Transvaal from Within* (1899) and *Jock of the Bushveld* (1907).

FITZROY, EDWARD ALGERNON (1869–1943), speaker of the House of Commons; son of third Baron Southampton; educated at Eton and Sandhurst; served in 1st Life Guards,

1889-91, 1914-18; country gentleman and county councillor in Northamptonshire; conservative MP, South Northamptonshire (later Daventry division), 1900-6, 1910-43; deputy chairman of ways and means, 1922-4, 1924-8; speaker, 1928-43; retained confidence of all parties; PC, 1924; his widow created Viscountess Daventry in her own right, 1943.

FLANAGAN, BUD (WEINTROP, CHAIM REEVEN) (1896-1968), comedian; son of Polish immigrants; went to school in Petticoat Lane; call-boy at age of ten; employed in various jobs in United States, 1910-15; served with Royal Artillery in France, 1915-18; met Chesney Allen, his future partner; formed unsuccessful double act with Roy Henderson, 1919-22; met Chesney Allen again, 1926; made London début together at Holborn Empire, 1929; success; first royal command performance, 1930; under Val Parnell, joined the 'Crazy Gang'; appeared in numerous west-end shows, 1931-9; performed for the troops, 1941-5; Chesney Allen retired owing to ill health, 1945; 'Crazy Gang' continued to be successful, 1945-59; among Flanagan and Allen films were, *Underneath the Arches* (1937), *We'll Smile Again* (1942), and *Here comes the Sun* (1945); OBE, 1959; appeared in fifteen variety command performances.

FLEAY, FREDERICK GARD (1831-1909), Shakespearian scholar; BA, Trinity College, Cambridge (13th wrangler), 1853; obtained places in four triposes; MA, and ordained, 1856; relinquished orders, 1884; engaged as schoolmaster, 1856-76; published *Hints on Teaching* (1874); interested in phonetics; edited *The Spelling Reformer*, 1880-1; compiled *Life of Shakespeare* (1886), *A Chronicle History of the London Stage, 1559-1642* (1890), and a *Biographical Chronicle of the English Drama, 1559-1642* (2 vols., 1891); in later life turned to Egyptology and Assyriology, published *Egyptian Chronology* (1899).

FLECK, ALEXANDER, first BARON FLECK (1889-1968), industrial chemist; educated at Saltcoats public school and Hillhead high school, Glasgow; entered Glasgow University as laboratory boy; joined teaching staff, 1911; joined staff of Glasgow Radium Committee, 1913; D.Sc., 1916; chief chemist, Castner Kellner Alkali Co., 1917; works manager, 1919; planned new works for Imperial Chemical Industries at Billingham, 1926; managing director, General Chemicals Division, 1931; chairman, Billingham Division, 1937; appointed to ICI Board, 1944; chairman, Scottish Agricultural Industries, 1947-51; deputy chairman, ICI, 1951; chairman, 1953-60; chairman, Coal

Board Organization Committee, 1953-5; chairman, Scientific Advisory Council, 1958-65; president, British Association for the Advancement of Science, 1958; FRS, 1955; hon. degrees from many universities; president, Society of Chemical Industry, 1960-2; director, Midland Bank, 1955; chairman, Nuclear Safety Advisory Committee, 1960-5; chairman, International Research and Development Company, president of the Royal Institution, 1963; KBE, 1955; baron, 1961.

FLECKER (HERMAN) JAMES ELROY (1884-1915), poet and dramatist; educated at Dean Close School (Cheltenham), Uppingham, and Trinity College, Oxford; entered consular service, 1908; sent to Constantinople, 1910; vice-consul at Beirut, 1911-13; died at Davos, Switzerland; his finest poem, 'The Golden Journey to Samarkand'; collected poems published, 1916; two plays published posthumously: *Hassan* (1922) and *Don Juan* (1925).

FLEMING, SIR ALEXANDER (1881-1955), bacteriologist; educated at Darvel, Kilmarnock Academy, and the Polytechnic Institute, Regent Street, London; became shipping clerk; entered St. Mary's Hospital medical school, 1901; obtained senior entrance scholarship and virtually every prize and scholarship available during his studentship; qualified, 1906; MB, BS, London, 1908; FRCS, 1909; appointed assistant bacteriologist to (Sir) Almroth Wright [q.v.] in St. Mary's inoculation department, 1906; lecturer in bacteriology, St. Mary's medical school, 1920; professor in bacteriology, London University, 1928-48; principal, Wright-Fleming Institute of Microbiology, 1946-54; served with Royal Army Medical Corps, 1914-18, studying treatment of war wounds; with Wright advocated physiological principles rather than use of antiseptics for treatment of war wounds; observed dangers of hospital cross-infection; discovered antimicrobial substance lysozyme, the body's natural antibiotic, 1922; discovered pencillin by observing action of mould on a culture of staphylococcus, 1928; realized its potentialities as non-toxic, non-irritant anti-septic but lacked resources to develop it; showed sulphonamides to be bacteriostatic not bactericidal; combined intuitive and original observation with great technical skill and inventiveness; FRS, 1943; FRCP and knighted, 1944; received Nobel prize (jointly), 1945; doctorates from many universities at home and abroad; rector of Edinburgh University, 1951-4; buried in St. Paul's Cathedral.

FLEMING, SIR ARTHUR PERCY MORRIS (1881-1960), engineer; educated at Portland

House Academy, Newport; studied electrical engineering at Finsbury Technical College; joined British Westinghouse (later Metropolitan-Vickers), 1900; became chief engineer, transformer department, Manchester; superintendent, 1913; manager of company's education department, 1917; founded, 1920, research department which by 1929 had a very large high-voltage laboratory; developed demountable high-power thermionic valves used in radar stations; director of research and education, 1931-54; knighted, 1945; joint author of number of books on engineering subjects.

FLEMING, DAVID HAY (1849-1931), historian, antiquary, and critic; educated at Madras College, St. Andrews; studied local antiquities and ecclesiastical history; joined the Original Seceders, 1899; contributed to *British Weekly* and *The Bookman* as champion of Scottish reformation and the covenanters; works include *Mary Queen of Scots* (1897) and *The Reformation in Scotland* (1910).

FLEMING, DAVID PINKERTON, LORD FLEMING (1877-1944), Scottish judge; educated at Glasgow high school and Glasgow and Edinburgh universities; admitted advocate, 1902; advocate depute, 1919; KC, 1921; conservative MP, Dumbartonshire, 1924-6; solicitor-general for Scotland, 1922-3, 1924-6; senator, College of Justice in Scotland, 1926-44; chairman, London appellate tribunal for conscientious objectors, 1940-2; of committee to consider means of developing association between public schools and the general educational system, 1942-4.

FLEMING, GEORGE (1833-1901), veterinary surgeon; served as veterinary surgeon in Crimea and China, 1855-60, in Syria and Egypt, 1867; principal veterinary surgeon to army, 1883-90; CB, 1887; president of Royal College of Veterinary Surgeons, 1880-4, 1886-7; secured passage of Veterinary Surgeons Act, 1881; hon. LLD, Glasgow, 1883; wrote *Horse Shoes and Horse Shoeing* (1869) and *Veterinary Sanitary Science* (2 vols., 1875).

FLEMING, IAN LANCASTER (1908-1964), writer; brother of Peter Fleming, traveller and writer; educated at Eton, Sandhurst, and privately in Austria, Germany, and Switzerland; joined Reuters, 1931; worked as banker and stockbroker in London, 1933-9; personal assistant to director of naval intelligence, 1939-45; foreign manager, Kemsley group of newspapers, 1945-59; creator of secret-agent, James Bond ('007'); published *Casino Royale* (1953), *Goldfinger* (1959), *The Spy Who Loved Me* (1962), *You Only Live Twice* (1964), and

The Man with the Golden Gun (1965); also wrote children's stories, *Chitty-Chitty-Bang-Bang*, which was filmed, and *Dr No* (1958), *Moonraker* (1958), and *On Her Majesty's Secret Service* (1963).

FLEMING, JAMES (1830-1908), canon of York; educated at Shrewsbury and Magdalene College, Cambridge; BA, 1853; BD, 1865; vicar of St. Michael, Chester Square, 1873-1908; canon of York, 1876; chaplain to royal family from 1876; Whitehead professor of preaching and elocution at London College of Divinity from 1880; a strong protestant, he supported Kensit's agitation against ritualism; charged with plagiarism of Dr Talmage's sermons, 1887; a popular preacher and graceful speaker; published *The Art of Reading and Speaking* (1896) and sermons.

FLEMING, SIR (JOHN) AMBROSE (1849-1945), electrical engineer and inventor of the wireless valve; B.Sc., University College, London, 1870; studied at South Kensington under (Sir) Edward Frankland and F. Guthrie [qq.v.] and at St. John's College, Cambridge; first class, natural sciences tripos, 1880; fellow, 1883; professor of electrical engineering, University College, London, 1885-1926; obtained new laboratories, 1893; studied alternating current transformers and high-voltage transmission and published *The Alternate Current Transformer* (2 vols., 1889-92); a leading expositor of photometry and the apostle of the potentiometer; closely associated with Marconi's wireless transmissions; invented the thermionic valve, 1904; published *The Principles of Electric Wave Telegraphy* (1906) and *The Propagation of Electric Currents in Telephone and Telegraph Conductors* (1911); strong supporter of J. L. Baird [q.v.]; a leading teacher and popular scientific lecturer; FRS, 1892; knighted, 1929.

FLEMING, SIR SANDFORD (1827-1915), Canadian engineer; born in Scotland; went to Canada, 1845; chief railway engineer to government of Nova Scotia, 1864; chief engineer of Inter-Colonial Railway, 1867-76; engineer-in-chief of Canadian Pacific Railway, 1871-80; headed 'Ocean to Ocean' expedition, 1872; procured laying of Pacific cable, 1902; KCMG, 1897; died at Halifax, Nova Scotia.

FLETCHER, SIR BANISTER FLIGHT (1866-1953), architect and architectural historian; son of Banister Fletcher [q.v.]; educated at Norfolk county school, King's and University colleges, London; joined father, 1884; partner and ARIBA 1889; FRIBA, 1904; president, 1929-31; London University extension lecturer, 1901-38; called to bar (Inner

Temple), 1908; with father published *A History of Architecture on the Comparative Method* (1896; 6th and revised edn., 1921; 17th edn., 1961); knighted, 1919.

FLETCHER, CHARLES ROBERT LESLIE (1857-1934), historian; educated at Eton and Magdalen College, Oxford; first class, modern history, 1880; fellow of All Souls, 1881; fellow and tutor of Magdalen, 1889-1906; an unconventional, stimulating teacher; delegate of Clarendon Press, 1905-27; works include *An Introductory History of England* (5 vols., 1904-23).

FLETCHER, Sir FRANK (1870-1954), headmaster; scholar of Rossall School and Balliol College, Oxford; first class, *lit. hum.*, 1893; won Craven, Ireland, and Derby scholarships; master at Rugby, 1894; headmaster of Marlborough, 1903-11; of Charterhouse, 1911-35; knighted 1937; hon. fellow of Balliol, 1924; president, Classical Association, 1946; published an edition of the sixth book of the *Aeneid* (1941).

FLETCHER, JAMES (1852-1908), naturalist; entered London bank, 1871; transferred to Ottawa, 1875; employed in parliamentary library there, 1876-87; botanist to dominion experimental farms, 1887; founded Ottawa Field Naturalists' Club, to whose *Transactions* he made valuable contributions on botany and entomology.

FLETCHER, REGINALD THOMAS HERBERT, Baron Winster (1885-1961), politician; entered *Britannia*, 1899; served during 1914-18 war in destroyers; lieutenant-commander, 1922; retired from navy, 1924; liberal MP, Basingstoke, 1923-4; joined labour party, 1929; labour MP, Nuneaton, 1935; rejoined navy, 1939, commander; parliamentary private secretary to A. V. Alexander (later Earl Alexander of Hillsborough, q.v.), first lord of the Admiralty, 1940-1; baron, 1942; minister of civil aviation; PC, 1945-6; policy embodied in Civil Aviation Act, 1946; governor and commander-in-chief, Cyprus, 1946-9; KCMG, 1948.

FLETCHER, Sir WALTER MORLEY (1873-1933), physiologist and administrator; educated at University College School, London, and Trinity College, Cambridge; first class, parts i and ii, natural sciences tripos, 1894-5; fellow, 1897; tutor, 1905-14; his investigation of 'respiration' of frog's muscle laid foundation of modern ideas of cellular activity; FRS, 1915; qualified in medicine, St. Bartholomew's Hospital, 1900; secretary, Medical Research Committee (later Council),

1914-33; established its authority by his enthusiasm and wise use of scientific talent; advised Sir William Dunn trustees and Rockefeller Foundation on endowments; KBE, 1918; CB, 1929.

FLETT, Sir JOHN SMITH (1869-1947), geologist; educated at George Watson's College and the university, Edinburgh; graduated in arts, science, and medicine; lecturer in petrology; petrographer to Geological Survey of Great Britain, 1901-11; in charge of Scottish branch, 1911-20; director of Survey, 1920-35; reorganized and enlarged it and transferred it to Geological Museum, South Kensington; FRS, 1913; KBE, 1925.

FLEURE, HERBERT JOHN (1877-1969), professor of geography and anthropology; educated at home in Guernsey and Aberystwyth University College of Wales; first class, zoology, 1901; awarded college fellowship and studied at Zoological Institute, Zurich, 1902-4; D.Sc., Wales, 1904; assistant lecturer (later lecturer) in zoology, geology, and botany, Aberystwyth, 1904-8; head of department of zoology, 1908-10; professor of zoology and lecturer in geography, 1910-17; professor of anthropology and geography, Aberystwyth, 1917-30; first professor of geography, Manchester, 1930-44; visiting professor to overseas universities, 1944-50: hon. lecturer, University College, London, 1946; publications include *Human Geography in Western Europe* (1918), *The Corridors of Time* (10 vols. in collaboration with H. J. E. Peake, q.v., 1927-56), and *Natural History of Man in Britain* (1951); president, Geographical Association, 1948; FRS, 1936; hon. fellow, Royal Geographical Society and member of many other societies and institutes; hon. degrees from Edinburgh, Wales, and Bowdoin.

FLINT, ROBERT (1838-1910), philosopher and theologian; professor of moral philosophy, St. Andrews University, 1864-76; professor of divinity, Edinburgh University, 1876; LLD, Glasgow; DD, Edinburgh; Baird lecturer, 1876-7; Stone lecturer at Princeton, USA, 1880; delivered Croall, 1887-8, and Gifford, 1908-9, lectures; voluminous writings include his Baird lectures, *Theism* (1877) and *Antitheistic Theories* (1879), *Socialism* (1894), *Hindu Pantheism* (1897), and *Agnosticism* (1903, Croall lecture).

FLINT, Sir WILLIAM RUSSELL (1880-1969), artist; educated at Daniel Stewart's College, Edinburgh, and Royal Institution School of Art; apprenticed as lithographic artist; moved to London, 1900; illustrator to medical publications, worked with paper makers, and later, as staff artist to *Illustrated London News*, 1903-7; freelance illustrator of books, including the *Imitation of Christ* and *Morte d'Arthur*; exhibited water-colours at Royal Academy, 1905; also painted in oils; ARA, 1924; RA, 1933; president, Royal Society of Painters in Water-Colours, 1936-56; knighted, 1947; retrospective exhibition at Royal Academy, 1962.

FLOREY, HOWARD WALTER, Baron Florey (1898-1968), experimental pathologist and main creator of penicillin therapy; born in Adelaide; educated at Kyre College and St. Peter's Collegiate School, and Adelaide University medical school; first class honours; MB, B.Sc., 1921; Rhodes scholar, 1921; enrolled in Department of Physiology under Sir Charles Sherrington [q.v.], and at Magdalen College, Oxford; first class, physiology, 1923; studied blood flow in brain capillaries; continued this study at Pathology Department, Cambridge, 1924; B.Sc. Oxford, 1925; studied in United States, 1925; Freedom Research fellow, London Hospital, 1926; Huddersfield lecturer in pathology, Cambridge, 1927; fellow, Gonville and Caius College, and director of medical studies; Ph.D. Cambridge, 1927; studied 'lysozyme' in mucus discovered by (Sir) Alexander Fleming [q.v.] in 1922; Joseph Hunter professor of pathology, Sheffield, 1932; succeeded Georges Dreyer [q.v.] as professor of pathology, Oxford, 1935-62; fellow, Lincoln College; made Oxford School of Pathology among best in the world; carried out research with (Sir) Ernest B. Chain on penicillin; effectiveness proved on patients at Radcliffe Infirmary, Oxford, 1941-2; tried out penicillin on war wounds in North Africa with (Sir) Hugh Cairns [q.v.], 1943; published *Antibiotics* (2 vols., with six of his collaborators, 1949); FRS, 1941; knighted, 1944; shared Nobel prize for medicine with Chain and Fleming, 1945; chancellor, Australian National University, 1965; president, Royal Society, 1960; provost, Queen's College, Oxford, 1962; baron, OM, 1965; many hon. degrees and other academic honours.

FLOWER, Sir CYRIL THOMAS (1879-1961), deputy keeper of the public records; educated at St. Edward's School, Oxford, and Worcester College, 1899-1901; entered Public Record Office, 1903; called to bar (Inner Temple), 1906; began fifty years' work on Curia Regis Rolls, 1910; private secretary to director of contracts, War Office, 1914; served in France with Royal Garrison Artillery, 1916: croix de guerre; returned to Public Record Office, 1919; deputy keeper, 1938-47; CB, 1939; responsible for evacuating records for safety during 1939-45 war; and for their return post-war; chairman,

council of British Records Association; hon. director, Institute of Historical Research, 1939-44; vice-president, Society of Antiquaries, 1939-43; president, Selden Society, 1949-52; knighted, 1946; FBA, 1947; edited *Public Works in Mediaeval Law* (vol. xxxii, 1915, vol. xl, 1923), and prepared an *Introduction to the Curia Regis Rolls, 1199-1230* (vol lxii, 1943); also edited *Curia Regis Rolls, Vol I, Richard I-2 John*, published in 1922, to the fourteenth volume (1230-2), published in 1961.

FLOWER, ROBIN ERNEST WILLIAM (1881-1946), scholar and poet; educated at Leeds grammar school and Pembroke College, Oxford; first class, *lit hum.*, 1904; assistant, department of manuscripts, British Museum, 1906; deputy keeper, 1929-44; publications include vol. ii, *Catalogue of Irish Manuscripts in British Museum* (1926), *Poems and Translations* (1931), and *The Western Island* (1944); FBA, 1934.

FLOYER, ERNEST AYSCOGHE (1852-1903), explorer; served in Indian telegraph station, 1869-76; explored interior of Baluchistan, 1876-7; published *Unexplored Baluchistan*, 1882; inspector-general of Egyptian telegraphs, 1878-1903; superintended experiments of tree and plant cultivation in Egypt; commanded expedition to desert between the Nile and the Red Sea, 1891; rediscovered emerald mines in Sikait and Zabbara; described his explorations in various scientific journals.

FLUX, Sir ALFRED WILLIAM (1867-1942), economist, statistician, and civil servant; bracketed senior wrangler, St. John's College, Cambridge, 1887; fellow, 1889; lecturer in political economy, Owens College, Manchester, 1893; professor, 1898; professor, McGill University, 1901-8; statistical adviser, Board of Trade, 1908; in charge of statistical department, 1918-32; responsible for first census of production, index of prices, indexes on industrial activity, and estimate of national income based on product; knighted 1934.

FOAKES JACKSON, FREDERICK JOHN (1855-1941), divine. [See Jackson.]

FOGERTY, ELSIE (1865-1945), founder and principal of Central School of Speech Training and Dramatic Art; established her school at Royal Albert Hall, 1898; obtained London University recognition for diploma in dramatic art, 1923; CBE, 1934.

FOLLEY, (SYDNEY) JOHN (1906-1970), biochemist; educated at Swindon and North

Wilts secondary school and Technical Institution, and Manchester University; first class, chemistry, 1927; researched on colloids; M.Sc.; joined biochemical laboratory; Ph.D. (Manchester), 1931; assistant lecturer, biochemistry, Liverpool University; research assistant, physiology department, National Institute for Research in Dairying, Shinfield, Reading, 1932–45; head of department, 1945–70; worked on physiology of lactation; FRS, 1951; hon. doctorate, Ghent, 1964; member of many learned societies and advisory bodies; wrote numerous scientific papers and reviews.

FOOT, ISAAC (1888–1960), politician; educated at Plymouth public school and Hoe grammar school; qualified as solicitor, 1902; sustained seven electoral defeats between 1910 and 1945; liberal MP, Bodmin, 1922–4, 1929–35; parliamentary secretary for mines, 1931; resigned over Ottawa agreements, 1932; member of India round table conference and joint select committee; PC, 1937; president, liberal party, 1947; a lifelong local preacher and vice-president of Methodist Church, 1937–8; lord mayor of Plymouth, 1945–6; for many years president of Cromwell Association; a voracious reader and natural orator; exercised a formative influence on his sons, four of whom entered public life with distinction.

FORBES, SIR CHARLES MORTON (1880–1960), admiral of the fleet; educated at Dollar Academy; entered *Britannia*, 1894; specialized in gunnery; flag commander of *Iron Duke* at Jutland, 1916; DSO; captain, 1917; deputy director, Naval Staff College, 1921–3; director of naval ordnance, 1925–8; rear-admiral, 1928; third sea lord and controller, 1932–4; vice-admiral, 1933; second-in-command, Mediterranean Fleet, 1934–6; KCB, 1935; admiral, 1936; commander-in-chief, Home Fleet, 1938–40; inflicted heavy losses on Germans during battle for Norway; admiral of the fleet and GCB, 1940; commander-in-chief, Plymouth, 1941–3; maintained vigorous operations against the enemy; mounted attack on St. Nazaire; prepared for invasion of France.

FORBES, GEORGE WILLIAM (1869–1947), prime minister of New Zealand; progressive farmer; liberal member for Hurunui, 1908–43; whip, 1912; party leader, 1925–8; minister of agriculture and deputy prime minister, 1928–30; prime minister with labour support and minister of finance and foreign affairs, 1930–1; prime minister in coalition with J. G. Coates [q.v.], 1931–5; minister for railways and foreign affairs (1931–5), for native affairs (1934–5); attorney-general, 1933–5; leader of opposition, 1935–6; PC, 1930.

FORBES, JAMES STAATS (1823–1904), railway manager and art connoisseur; studied engineering under I. K. Brunel [q.v.]; joined Great Western railway, twice subsequently refusing post of general manager; general manager of London, Chatham and Dover railway, 1861; restored its fortunes; chairman of directors, 1873–99; secured good terms for the railway on its amalgamation with South Eastern railway, 1899; as director of London District railway, 1870–1901, overcame difficulties of competition; collector of pictures, especially by nineteenth-century artists.

FORBES, (JOAN) ROSITA (1890–1967), traveller and writer; published first book *Unconducted Wanderers* (1919), a story of a journey round the world, ending in North Africa; disguised as Moslem, travelled across Libyan desert to oasis of Kufara, 1920–1; published *The Secrets of the Sahara: Kufara* (1921); other accounts of the journeys include *El Raisini, the Sultan of the Mountains* (1924), *From Red Sea to Blue Nile* (1925), *Conflict: Angora to Afghanistan* (1931); and her autobiographies *Gypsy in the Sun* (1944) and *Appointment in the Sun* (1946), abridged and reissued as *Appointment in the Sun* (1949).

FORBES, STANHOPE ALEXANDER (1857–1947), painter; educated at Dulwich College; studied at Lambeth School of Art, Royal Academy Schools, and in France; a leading member of the 'Newlyn School'; founded school of art in Newlyn, 1899; ARA, 1892; RA, 1910.

FORBES-ROBERTSON, SIR JOHNSTON (1853–1937), actor. [See ROBERTSON.]

FORBES-SEMPILL, WILLIAM FRANCIS, nineteenth BARON SEMPILL (1893–1965), representative peer for Scotland, engineer, author, and airman; educated at Eton; apprenticed to Rolls-Royce Ltd., Derby, 1910–13; served during 1914–18 war in Royal Flying Corps and Royal Naval Air Service; transferred to Royal Air Force; colonel, 1918; personal assistant to Sir (William) Sefton Brancker [q.v.], 1918; headed many missions overseas promoting sale of British aircraft, 1919–25; extensive flying experience in light aircraft; president, Royal Aeronautical Society, 1927–30; active in many other aeronautical directions; chairman, London Chamber of Commerce, 1931–5; succeeded as Lord Sempill, 1934; rejoined Naval Air Service, 1939–41; publications include *The Air and the Plain Man* (1931).

FORD, EDWARD ONSLOW (1852–1901), sculptor; studied painting at Antwerp and modelling at Munich; ARA, 1888; RA, 1895; important works are Rowland Hill (1881), Irving as Hamlet (1883), Gordon (1890), Shelley memorial at Oxford (1892) and Queen Victoria memorial at Manchester (1901); executed also bronze statuettes, such as 'Folly', 'Peace', and 'Echo'.

FORD, FORD MADOX (1873-1939), author and critic; son of Francis Hueffer [q.v.]; changed name, 1919; educated at University College School, London; works include *Rossetti* (1902), *The Fifth Queen* historical trilogy (1906-8), the Tietjens series of war novels (*Some Do Not*, 1924, etc.); founded *English Review* (1908), *Transatlantic Review* (1924); after 1922 lived in France and America.

FORD, PATRICK (1837-1913), Irish-American journalist and politician; born in Galway; taken to Boston, USA, 1841; founded and conducted *Irish World*, chief organ of Irish-Americans and supporter of successive Irish movements, 1870-1913; died at Brooklyn, USA.

FORD, WILLIAM JUSTICE (1853-1904), cricketer and writer on cricket; educated at Repton and St. John's College, Cambridge; master at Marlborough, 1877-86; one of hardest hitters known; historian of Middlesex and Cambridge University cricket clubs.

FORDHAM, Sir HERBERT GEORGE (1854-1929), writer on cartography; lived at Odsey, Hertfordshire, devoting himself to local administration, from 1891; chairman, Cambridgeshire county council, 1904-19; knighted, 1908; formed map collection and became a leading European authority on cartography; published bibliographical studies of early maps of France and England, articles on road-making and map-making in the sixteenth, seventeenth, and eighteenth centuries, etc.

FORESTER, CECIL SCOTT (1899-1966), novelist; born Cecil Lewis Troughton Smith; took name Forester for professional purposes, 1923; educated at Alleyn's School and Dulwich College; first book, *Payment Deferred* (1926) followed by *Brown on Resolution* (1929), *Death to the French* (1932), *The African Queen* (1935), and *The General* (1936); also published biography of *Nelson* (1929); correspondent in Spain during civil war, 1936-7; created Horatio Hornblower in *The Happy Return* (1937), followed by many other Hornblower stories, including *a Ship of the Line*, awarded James Tait Black memorial prize, 1938; settled in United States and contributed to the *Saturday Evening Post*, 1945; further publications include

The Hornblower Companion (1964), *The Ship* (1943), *The Commodore* (1945), and *The Man in the Yellow Raft* (1969).

FORESTIER-WALKER, Sir FREDERICK WILLIAM EDWARD FORESTIER (1844-1910), general; entered army, 1862; served in Kaffir war, 1877-8; CB, 1878; in Zulu war, 1879; in Bechuanaland expedition, 1884-5; CMG, 1886; commanded troops in Egypt, 1890-5; KCB, 1894; lieutenant-general commanding Western district of England, 1895-9; in command of lines of communication in South African war, 1899-1901; general 1902; governor of Gibraltar, 1905-10; GCMG, 1900.

FORMAN, ALFRED WILLIAM (1840-1925), man of letters; brother of H. B. Forman [q.v.]; translated Wagner's operas, Victor Hugo's plays, etc.

FORMAN, HENRY BUXTON (1842-1917), man of letters; entered Post Office, 1860; CB, 1897; retired, 1907; principal works include editions of *Poetical Works* (1876) and *Prose Works* (1880) of Shelley, of *Letters of John Keats to Fanny Brawne* (1878), and *Poetical Works and Other Writings of John Keats* (1883).

FORMBY, GEORGE (1904-1961), comedian; born George Hoy Booth, his father, a music hall comedian, having adopted name of Formby; blind until age of two; became stable-boy at age of seven; after his father's death, decided to do his father's act, 1921; adopted his own style with ukelele, 1925; made first comedy talkie film, *Boots! Boots!*, 1934; made many other successful films; entertainer of the troops; OBE, 1946; successful stage appearance in *Zip Goes a Million*, 1952; made over 200 gramophone records and twenty-two films; successful with Russian audiences.

FORREST, Sir GEORGE WILLIAM DAVID STARCK (1845-1926), historian of India; born in India; BA, St. John's College, Cambridge, 1870; entered Bombay educational service, 1872; director of records, Bombay, 1888; of Imperial Record Office, Calcutta, 1891-1900; CIE, 1899; knighted, 1913; edited Indian state papers; wrote *History of the Indian Mutiny* (1904, 1912) and lives of Lords Roberts (1914) and Clive (1918).

FORREST, JOHN first BARON FORREST (1847-1918), Australian explorer and conservative politician; born in Western Australia; entered survey department, 1865; surveyor-general, 1883; first premier of Western Australia, 1890-1901; minister of defence in first Commonwealth ministry, 1901-3; KCMG,

1901; held office intermittently, 1903-18; first Australian politician raised to peerage, 1918; died at sea.

FORSTER, EDWARD MORGAN (1879-1970), novelist and man of letters; great-grandson of Henry Thornton [q.v.]; educated at Tonbridge and King's College, Cambridge; an Apostle, friend of G. Lowes Dickinson [q.v.], and one of the Bloomsbury set; published first novel, *Where Angels Fear to Tread* (1905); visited India with Lowes Dickinson, 1912; revisited India; private secretary to Maharaja of Dewas State Senior, 1921; fellow, King's College, 1927-33; hon. fellow, 1946; lived in King's, 1946-70; CH, 1953; OM, 1969; eight honorary degrees; publications include *The Longest Journey* (1907) *A Room with a View* (1908), *Howard's End* (1910), *A Passage to India* (awarded the Femina Vie-Heureuse and James Tait Black memorial prizes, 1924), and *Abinger Harvest* (1936); assisted in writing libretto of Benjamin Britten's opera, *Billy Budd*; homosexual novel *Maurice*, written in 1914, published only in 1971 after his death; works translated into twenty-one languages.

FORSTER, HUGH OAKELEY ARNOLD- (1855-1900), author and politician. [See ARNOLD-FORSTER.]

FORSTER, SIR MARTIN ONSLOW (1872-1945), chemist; Ph.D., Würzburg, 1892; D.Sc., London, 1899; demonstrator in chemistry, Royal College of Science, 1895; assistant professor, 1902-13; a director, British Dyes, Ltd., 1916; director, Salters' Institute of Industrial Chemistry, 1918-22; of Indian Institute of Science, Bangalore, 1922-33; investigated mainly reactions of camphor and its derivatives; FRS, 1905; knighted, 1933.

FORSYTH, ANDREW RUSSELL (1858-1942), mathematician; educated at Liverpool Collegiate Institution and Trinity College, Cambridge; senior wrangler and fellow, 1881; college lecturer, 1884; professor of mathematics, University College, Liverpool, 1882-3; Sadleirian professor of pure mathematics, Cambridge, 1895-1910; professor of mathematics, Imperial College of Science, 1913-23; published *Treatise on Differential Equations* (1885) and *Theory of Functions* (1893); FRS, 1886.

FORTESCUE, GEORGE KNOTTES-FORD (1847-1912), librarian; entered department of printed books, British Museum, 1870; keeper, 1899-1912; main achievement, *Subject Index of Modern Works added to the Library of the British Museum* (1880-1910).

FORTESCUE, HUGH, third EARL FORTESCUE (1818-1905), whig MP for Plymouth, 1841-52, Barnstaple 1852-4, and Marylebone, 1854-9; raised to peerage in his father's barony, 1859; succeeded father in earldom, 1861; secretary to Poor Law Board, 1847-51; visited barracks and hospitals in Crimea, 1856; became liberal unionist, 1886; wrote many pamphlets on local government, health of towns, and middle-class education; purchased reversion to great part of Exmoor, 1897, and encouraged stag hunting.

FORTESCUE, SIR JOHN WILLIAM (1859-1933), military historian; educated at Harrow and Trinity College, Cambridge; private secretary to governors of Windward Islands (1880-2) and New Zealand (1886-90); librarian at Windsor Castle, 1905-26; publications include *History of the British Army* (13 vols., 1899-1930), *County Lieutenancies and the Army, 1803-1814* (1909), *British Statesmen of the Great War, 1793-1814* (1911), *Wellington* (1925); vigorous, lucid, and graphic writer; KCVO, 1926.

FOSS, HUBERT JAMES (1899-1953), musician and writer; grandson of Edward Foss [q.v.]; scholar of Bradfield College; joined Oxford University Press, 1921; founded (1924) and until 1941 headed musical department publishing works of young as well as established composers; these included Constant Lambert, Peter Warlock, E. J. Moeran, Holst, Ralph Vaughan Williams [qq.v.] and Walton, van Dieren, Rubbra, Rawsthorne, and Britten; also published books on music such as *Oxford Companion to Music* (1938) by Percy Scholes [q.v.]; after 1942 became freelance musician, author, and broadcaster.

FOSTER, SIR CLEMENT LE NEVE (1841-1904), inspector of mines; son of Peter Le Neve Foster [q.v.]; entered School of Mines, 1857; D.Sc., London, 1865; examined mineral resources of Sinaitic peninsula, 1868; employed in Venezuela and North Italy, 1868-72; inspector of mines in Cornwall, 1872-80, and North Wales, 1880-1901; professor at Royal School of Mines, 1890-1904; sustained serious cardiac injury in Snaefell lead mine, 1897; FRS, 1892; knighted, 1903; works include *Ore and Stone Mining* (1894) and *Mining and Quarrying* (1903).

FOSTER, SIR GEORGE EULAS (1847-1931), Canadian statesman; born in New Brunswick; studied at New Brunswick, Edinburgh, and Heidelberg universities; classical professor, New Brunswick, 1871-9; conservative MP, King's County (1882-96), York

County (1896-1900), North Toronto (1904-21); senator, 1921-31; minister of marine and fisheries (1885-8), finance (1888-96), trade and commerce (1911-21); an imperialist and advocate of imperial preference, he became an ardent internationalist; in domestic politics moved to a position of advanced liberalism; KCMG, 1914; PC (Canada), 1885, (United Kingdom), 1916.

FOSTER, Sir HARRY BRAUSTYN HYLTON HYLTON- (1905-1965), speaker of the House of Commons. [See HYLTON-FOSTER.]

FOSTER, JOSEPH (1844-1905), genealogist; nephew of Myles Birket Foster [q.v.]; early devoted to genealogical research; compiled pedigrees of Quaker families (1862-72), of county families of Lancashire (1873), and of Yorkshire (1874); published his *Peerage, Baronetage, and Knightage* (1879); compiled *Men at the Bar* (1888), *Admissions to Gray's Inn* (1899); edited and supplemented J. L. Chester's *London Marriage Licences* (1887), and his *Oxford Matriculation Register* in *Alumni Oxonienses* (8 vols., 1887-91); subsequent writings on heraldry were severely criticized.

FOSTER, Sir MICHAEL (1836-1907), physiologist; BA, London, 1854; studied science at University College; MB, 1858; MD, 1859; professor of practical physiology, University College, 1869; praelector of physiology, Trinity College, Cambridge, 1870; professor of physiology in the university from 1883; FRS, 1872; with T. H. Huxley [q.v.] developed method of practical laboratory work; introduced practical classes at Cambridge, for which he cooperated in *Textbook for the Physiological Laboratory* (1873), *The Elements of Embryology* (1874), and *A Course of Elementary Practical Physiology* (1876); stimulated original research; keenly interested in gardening, hybridized several plants; wrote a Primer (1890) and History (1901) of Physiology; joint-editor of Huxley's *Scientific Memoirs,* 1898-1902; founded *Journal of Physiology,* 1878; served on several royal commissions; president of British Association and KCB, 1899; liberal unionist MP for London University, 1900-6; opposed his party's education bill of 1902; defeated as liberal candidate, 1906.

FOSTER, Sir (THOMAS) GREGORY, first baronet, of Bloomsbury (1866-1931), provost of University College, London; educated at University College School and University College, London; taught in English department (1894-1904); secretary (1900), provost (1904-29), and virtually second founder of University College;

vice-chancellor, London University, 1928-30; knighted, 1917; baronet, 1930.

FOTHERINGHAM, JOHN KNIGHT (1874-1936), historian and authority on ancient astronomy; educated at City of London School and Merton College, Oxford; first classes, *lit. hum.* and modern history; fellow of Magdalen, 1909-16; reader in ancient history, London, 1912-20; in ancient astronomy and chronology, Oxford, 1925-36; FBA, 1933; edited Jerome's version of Eusebius (1923); established chronology of the Babylonian dynasties.

FOUGASSE [pseudonym] (1887-1965), cartoonist. [See BIRD, (CYRIL) KENNETH.]

FOULKES, ISAAC (1836-1904), Welsh author and editor; issued at Liverpool cheap reprints of Welsh classics, 1877-88; edited *Cymro,* weekly Welsh newspaper; edited biographical dictionary of eminent Welshmen, 1870.

FOWLE, THOMAS WELBANK (1835-1903), theologian and writer on poor law; BA, Oriel College, Oxford, 1858; MA, 1861; president of Union, 1858; friend of T. H. Green and A. V. Dicey [qq.v.]; vicar of Islip, 1875-1901; as poor law guardian advocated abolition of outdoor relief; published *The Poor Law* in 'English Citizen' series, 1881, which took standard rank; advocated creation of parish and district councils and old age pensions, 1892.

FOWLER, ALFRED (1868-1940), astrophysicist; studied mechanics at Normal School of Science, South Kensington; demonstrator in astronomical physics, 1888-1901; assistant professor of physics, 1901-15; professor of astrophysics, 1915-23; Yarrow research professor, Royal Society, 1923-34; identified previously unknown celestial spectra with spectra obtained in laboratory; first secretary of International Astronomical Union, 1919; FRS, 1910.

FOWLER, ELLEN THORNEYCROFT (1860-1929), novelist. [See FELKIN.]

FOWLER, HENRY HARTLEY, first VISCOUNT WOLVERHAMPTON (1830-1911), statesman; born in Sunderland; settled in Wolverhampton as solicitor, 1855; engaged in municipal affairs; mayor of Wolverhampton, 1863; first freeman of borough, 1892; liberal MP for Wolverhampton, 1880-5, for eastern division of city, 1885-1908; lucid and moderate speaker in House of Commons; became under-secretary for home affairs, 1884; financial secretary to Treasury and PC, 1886; keen critic of financial policy of conservative government, 1886-92; entered cabinet as president, Local Government Board, 1892; carried parish councils bill; became secretary for India, 1894; powerfully urged the reimposition of duties on cotton goods imported into India, 1895; GCSI, 1895; director, 1897, and president, 1901, of the National Telephone Company; supported South African war and opposed tariff reform; became chancellor of the duchy of Lancaster, 1905; viscount, Apr. 1908; made lord president of the Council, Oct. 1908-June 1910; hon. LLD, Birmingham, 1909.

FOWLER, HENRY WATSON (1858-1933), lexicographer; educated at Rugby and Balliol College, Oxford; second class, *lit. hum.*, 1881; assistant master at Sedbergh, 1882-99; with his brother F. G. Fowler (1870-1918) produced for Oxford University Press a translation of Lucian (4 vols., 1905), *The King's English* (1906), and *Concise Oxford Dictionary of Current English* (1911); subsequent publications include a pocket edition (1924) and *A Dictionary of Modern English Usage* (1926), his most original work; his literary essays collected in *More Popular Fallacies* (1904), *Si Mihi-!* (1907), *Between Boy and Man* (1908), and *Some Comparative Values* (1929).

FOWLER, SIR JAMES KINGSTON (1852-1934), physician; qualified at King's College Hospital, 1874; assistant physician (1880) and physician (1891), Brompton and Middlesex hospitals; served on Colonial Office committees; edited *A Dictionary of Practical Medicine* (1890); warden of Beaulieu Abbey; KCVO, 1910; KCMG, 1932.

FOWLER, SIR RALPH HOWARD (1889-1944), mathematician and mathematical physicist; scholar of Winchester and Trinity College, Cambridge; wrangler, 1911; fellow, 1914; college lecturer in mathematics, 1920; Plummer professor of mathematical physics, 1932-44; after 1939 co-ordinated North American with British war research; knighted, 1942; published notable paper 'Dense Matter' (*Monthly Notices*, Royal Astronomical Society, 1927) and *Statistical Mechanics* (1929); OBE, 1925; knighted, 1942; FRS, 1925.

FOWLER, THOMAS (1832-1904), president of Corpus Christi College, Oxford; schoolfellow at King William's College, Isle of Man, of T. E. Brown [q.v.], a life-long friend; BA, Merton College, Oxford, 1854; first classes in classics and mathematics; DD, 1886; tutor (1855) and sub-rector (1857-81) of Lincoln College; honorary fellow, 1900; advocated teaching of natural science and abolition of tests at Oxford,

1877; professor of logic, 1873-89; president of Corpus Christi College, 1881; vice-chancellor, 1901; hon. LLD, Edinburgh, 1882; published *Deductive Logic* (1867), *Inductive Logic* (1870), monographs on Locke (1880), Bacon (1881), and Shaftesbury and Hutcheson (1882), *Progressive Morality* (1884), and *The Principles of Morals* (2 parts, 1886); edited Bacon's *Novum Organum* (1878), and Locke's *Conduct of the Understanding* (1881); wrote histories of Corpus, 1893 and 1898.

FOWLER, WILLIAM WARDE (1847-1921), historian and ornithologist; educated at Marlborough and Lincoln College, Oxford; fellow of Lincoln, 1872; tutor, 1873; sub-rector, 1881-1906; retired to Kingham, his country home, 1910; works include writings on social and religious life of ancient Rome; knowledge of birds shown in *Kingham Old and New* (1913).

FOX, SIR CYRIL FRED (1882-1967), archaeologist and museum director; educated at Christ's Hospital, London, 1895-8; employed as clerk at Bovine Tuberculosis Research Station, Stansted; served in Essex Yeomanry during 1914-18 war; read archaeology at Magdalene College, Cambridge, 1919; published Ph.D thesis, *The Archaeology of the Cambridge Region* (1923); fellow, Society of Antiquaries of London; assistant, Museum of Archaeology and Ethnology, Cambridge, 1923; keeper of archaeology, National Museum of Wales, 1924; director, 1926-48; member, Ancient Monuments Board, Ministry of Works; publications include *The Personality of Britain* (1932), *Life and Death in the Bronze Age* (1959), and *Monmouthshire Houses: a study of building techniques and smaller house plans from the 15th to the 17th centuries* (3 vols, with Lord Raglan, 1951-4); president, Museums Association, 1934; FBA, 1940; knighted, 1935; hon. D.Litt., Wales, 1947; hon. fellow, Magdalene College, 1952.

FOX, DAME EVELYN EMILY MARIAN (1874-1955), pioneer worker in mental health; born and educated in Switzerland; studied at Somerville College, Oxford, 1898; worked at Women's University Settlement, Southwark; honorary secretary, Central Association for the Mentally Defective (from 1946 National Association for Mental Health), 1913-51; first honorary secretary, Child Guidance Council, 1927; served on LCC mental hospitals committee, 1914-24; CBE, 1937; DBE, 1947.

FOX, SIR FRANCIS (1844-1927), civil engineer; son of Sir Charles Fox [q.v.]; partner with father and brother in firm of Sir Charles Fox & Sons, civil and consulting engineers,

1861; assisted brother and (Sir) James Brunlees [q.v.] in constructing Mersey tunnel, 1880-6; responsible for much railway work in England and South Africa; much employed in superintending treatment and preservation of ancient buildings, including cathedrals of Winchester (1905-12), Lincoln, Peterborough, Exeter, and St. Pauls, London; knighted, 1912.

FOX, HAROLD MUNRO (1889-1967), zoologist; son of Georg Gotthilf Fuchs; educated at Brighton College and Gonville and Caius College, Cambridge; BA, 1911; lecturer in zoology, Royal College of Science, London; joined Army Service Corps and changed his name to Fox, 1914; lecturer, Cairo, 1919; fellow, Caius College, 1920; organized expedition to study fauna of the Suez Canal, 1924; research on red-green blood of fan-worm *Spirographis*; professor, zoology, Birmingham, 1927-41; published *Selene, or Sex and the Moon* (1928); transferred to Bedford College, London, 1941-54; FRS, 1937; Fullerian professor of physiology, Royal Institution, 1953-6; hon. D.Sc., Bordeaux University, 1965; edited for forty years *Biological Reviews of the Cambridge Philosophical Society*; published papers on ostracod crustaceans, 1962-7; an international authority on respiratory pigments.

FOX, SIR LIONEL WRAY (1895-1961), chairman of the Prison Commission; educated at Heath grammar school, Halifax, and Hertford College, Oxford; served with Duke of Wellington's Regiment during 1914-18 war; MC; joined Home Office, 1919; secretary to Prison Commission, 1925-34; deputy receiver to Metropolitan Police district, 1934; acting receiver, 1941-2; chairman, Prison Commission, 1942-60; published *The Modern English Prison* (1934), and *The English Prison and Borstal Systems* (1952); also wrote for *Encyclopaedia Britannica* and the *British Journal of Delinquency*; president, United Nations European Consultative Group on prevention of crime and treatment of offenders, 1951-60; CB, 1948; knighted, 1953.

FOX, SAMSON (1838-1903), inventor and benefactor; founded Leeds Forge Company, 1874; patented Fox corrugated boiler furnaces, 1877, and pressed steel under-frames for railway wagons, 1886; first employed in England water gas on large scale for metallurgical and lighting purposes; pioneer of acetylene industry in Europe; generous benefactor to Royal College of Music, 1889; mayor of Harrogate, 1889-91.

FOX, TERENCE ROBERT CORELLI (1912-1962), chemical engineer; educated at Regent Street Polytechnic secondary school

and Jesus College, Cambridge; first class, mechanical sciences tripos, 1933; mechanical engineer, ICI, 1933-7; Cambridge university demonstrator, 1937; fellow, King's College, 1941; university lecturer, 1945; Shell professor of chemical engineering, Cambridge, 1946-59; a great teacher and promoter of research.

FOX BOURNE, HENRY RICHARD (1837-1909), social reformer and author. [See BOURNE.]

FOX STRANGWAYS, ARTHUR HENRY (1859-1948), schoolmaster, music critic, and founder-editor of *Music and Letters*. [See STRANGWAYS.]

FOX-STRANGWAYS, GILES STEPHEN HOLLAND, sixth EARL OF ILCHESTER (1874-1959), landowner and historian; succeeded father, 1905; educated at Eton and Christ Church, Oxford; devoted himself to managing his estates, breeding racehorses, and publications including *The Home of the Hollands, 1605-1820* and *Chronicles of Holland House, 1820-1900* (1937) and other works based on Holland House papers; worked on notebooks of George Vertue [q.v.] for Walpole Society; a trustee of British Museum from 1931 and of National Portrait Gallery from 1922; chairman, Royal Commission on Historical Monuments, 1934-57; GBE, 1950.

FOXWELL, ARTHUR (1853-1909), physician; BA, St. John's College, Cambridge, 1877; MB, 1883; MD, 1891; MRCS, 1881; FRCP, 1892; Bradshawe lecturer, 1889; resident pathologist at general hospital, Birmingham, from 1884; professor of therapeutics at Birmingham University and M.Sc., 1906; published *Essays on Heart Disease* (1896).

FOXWELL, HERBERT SOMERTON (1849-1936), economist and bibliographer; BA, London, 1867; senior, moral sciences tripos, St. John's College, Cambridge, 1870; fellow, 1874; director of economic studies, 1912; professor of political economy, University College, London (1881-1927); his historical bent was in opposition to 'orthodox' English economics; lectured at London School of Economics (1895-1922) on his special subject of banking and currency; his first library on economic subjects (1740-1848) presented to London University by Goldsmiths' Company, 1903; a second collection purchased by Harvard University, 1929; FBA, 1905.

FOYLE, WILLIAM ALFRED (1885-1963), bookseller; educated at Dame Alice Owen's School, Islington; appointed clerk to (Sir)

Edward Marshall Hall, KC, [q.v.], 1902; opened second-hand bookshop in Islington, 1903; energetic searcher for rare books with increasing reputation, opened shop in Charing Cross Road, 1907; ambitious to create greatest bookshop in the world; began trade in new books, 1912; introduced Foyle's literary lectures; and founded the Book Club, 1937; collected many rare books and illuminated manuscripts, and formed a great library at Beeleigh Abbey, Essex; his work continued by his daughter, Christina.

FRAENKEL, EDUARD DAVID MORTIER (1888-1970), classical scholar; born in Berlin; educated at Askanisches Gymnasium, Berlin University, 1906-9, and Göttingen, 1909-12; privat-dozent, Berlin, 1917; professor extraordinarius, 1920; published *Plautinisches in Plautus* (1922); professor, Kiel, 1923; Göttingen, 1928; Freiburg im Breisgau, 1931; helped to found periodical *Gnomon*; forbidden to teach by Nazis, 1933; moved to Oxford; Bevan fellow, Trinity College, Cambridge, 1934; with assistance from A. E. Housman [q.v.], Corpus professor of Latin, Oxford, 1934-53; naturalized, 1939; published *Agamemnon* (3 vols. 1950); hon. fellow, Corpus Christi College, Oxford; published other books on Horace, Aristophanes, and on Latin word-order and prose rhythm, 1957-64; one of most learned classical scholars of his time; FBA, 1941-64; hon. degrees from West Berlin, Urbino, St. Andrews, Florence, Fribourg, and Oxford.

FRAMPTON, SIR GEORGE JAMES (1860-1928), sculptor and craftsman; studied at Royal Academy Schools and in Paris; powerfully attracted towards 'arts and crafts' movement; associated with *Studio* magazine, through which he wielded great influence in Germany; employed combination of different materials in one work; his works include series of statues of Queen Victoria, 'Peter Pan' in Kensington Gardens (1912), and Edith Cavell, St. Martin's Place (1920); RA, 1902; knighted, 1908.

FRANCIS-WILLIAMS, BARON (1903-1970), author, journalist, and publicist. [See WILLIAMS, EDWARD FRANCIS.]

FRANKAU, GILBERT (1884-1952), novelist; educated at Eton; entered family business as cigar merchant; after service in 1914-18 war became novelist with business-like efficiency; *Peter Jackson, Cigar Merchant* (1920) received popular acclaim and subsequent novels appearing with clockwork regularity had wide readership; they include *Masterson* (1926), *Three Englishmen* (1935), and *Unborn Tomorrow*

(1953); his characters larger than life, heroes dashing and tough, his sympathies with the extreme Right, but his narrative style compelling and his research meticulous.

FRANKFORT DE MONTMORENCY, third VISCOUNT (1835-1902), major-general, [See DE MONTMORENCY, RAYMOND HARVEY.]

FRANKLAND, PERCY FARADAY (1858-1946), chemist; son of Sir Edward Frankland [q.v.]; educated at University College School, Royal School of Mines, and Würzburg (Ph.D.); B.Sc., London, 1881; demonstrator under his father, South Kensington, 1880-8; professor of chemistry, University College, Dundee, 1888-94; at Birmingham, 1894-1919; built up very strong school of chemistry; dean of faculty of science, 1913-19; worked on water analysis, bacteriology, chemistry of optically active compounds, and stereochemistry; FRS, 1891.

FRANKLIN, CHARLES SAMUEL (1879-1964), radio telecommunications engineer; educated at Finsbury Technical College; joined Wireless Telegraph and Signal Company (later Marconi's Wireless and Telegraph Co.), 1899; acted as wireless operator for Marconi, 1902; worked closely with Marconi, 1903-24; with Marconi, developed short-wave beam system, 1916-20; installed short-wave station at Poldhu, 1923-4; assisted in design and installation of transmitter and aerial system of 2LO, London's first broadcasting station, 1922; designed transmitters and aerial system, BBC station, Alexandra Palace, world's first regular television service, 1936; retired from Marconi Co., 1939; received many awards; CBE, 1949.

FRANKS, ROBERT SLEIGHTHOLME (1871-1964), theologian; educated at Sir William Turners grammar school, Redcar, and St. John's College, Cambridge; studied theology at Mansfield College, Oxford, under A. M. Fairbairn [q.v.]; ordained, 1900; lecturer in theology, Woodbrooke, Selly Oak, 1904; principal, Western College, Bristol, 1910-39; publications include *The New Testament Doctrines of Man, Sin and Salvation* (1908), *A History of the Doctrine of the Work of Christ* (2 vols. 1918), *The Atonement* (1934), and *The Doctrine of the Trinity* (1953); contributed to the *Dictionary of Christ and the Gospel*, edited by James Hastings [q.v.]; D.Litt., Oxford, 1919; hon. LLD, Bristol, 1928.

FRASER, ALEXANDER CAMPBELL (1819-1914), philosopher; educated at Edinburgh University; ordained to Free Church ministry, 1844; professor of logic and metaphysics in Edinburgh Free Church theological college, 1846-56, and at Edinburgh University, 1856-91; Gifford lecturer, 1894-6; FBA, 1903; works include edition of Berkeley's *Works* and *Life and Letters* (1871), study of Locke (1890), and edition of his *Essay concerning Human Understanding* (1894); a stimulating teacher whose philosophical standpoint was theism based on moral faith.

FRASER, Sir ANDREW HENDERSON LEITH (1848-1919), Indian civil servant; entered Indian civil service, 1869; served in Central Provinces, 1871-98; secretary in home department and later chief commissioner of Central Provinces, 1898; president of Indian police commission, 1901; lieutenant-governor of Bengal and KCSI, 1903; had to face storm roused by partition, 1905; retired, 1908.

FRASER, CLAUD LOVAT (1890-1921), artist and designer; designed settings and costumes for *As You Like It* and *The Beggar's Opera*, 1920; decorated books; produced designs for theatre, rhyme sheets, broadsides, etc.

FRASER, DONALD (1870-1933), missionary; educated at Glasgow high school, university, and Free Church Hall; promoter of Student Christian Movement; ordained and went to Nyasaland, 1896; trekked south with the Ngoni (1902) and founded Loudon which became a Christian community and the centre of a new civilization; moderator, United Free Church of Scotland, 1922-3; foreign mission secretary, 1925-33.

FRASER, Sir FRANCIS RICHARD (1885-1964), professor of medicine and first director of the British Postgraduate Medical Federation; son of Thomas Richard Fraser [q.v.]; educated at Edinburgh Academy and Christ's College, Cambridge; first class, part i, natural sciences tripos, 1907; MB (Edinburgh), 1910; MD, 1922; assistant, Rockefeller Institute, New York, 1912; instructor and assistant physician, 1914; served with Royal Army Medical Corps, 1915-18; consulting physician to Army of the Rhine, 1919-20; professor, St. Bartholomew's Hospital medical school, 1920, succeeding Sir Archibald Garrod [q.v.]; first professor of medicine, British Postgraduate Medical School, Hammersmith, 1934; seconded to Ministry of Health, 1939; director-general, Emergency Medical Services, 1941; knighted, 1944; director, British Postgraduate Medical Federation, 1946-60; deputy vice-chancellor, London University, 1947-9; hon. LLD, Edinburgh and London; carried out research in many fields including poliomyelitis, heart disease, thyroid disease, and disorders of neuro-humoral transmission.

FRASER, HUGH, first Baron Fraser of Allander (1903-1966), draper, company chairman, and philanthropist; educated at Glasgow Academy and Warriston School, Moffat; entered his father's drapery business, 1919; managing director, 1924; chairman, 1927; acquired other businesses and floated House of Fraser Ltd., 1948; acquired John Barker group, London, 1957, and Harrods, 1959; took over the *Glasgow Herald*, 1964; undertook expansion of tourist industry in Scottish Highlands, including facilities at Aviemore, 1959; created Hugh Fraser Foundation, 1960; baronet, 1961; baron, 1964; hon. LLD St. Andrews, 1962.

FRASER, PETER (1884-1950), prime minister of New Zealand; born in Scotland; emigrated, 1910; settled in Wellington as watersider; played large part in organization of New Zealand labour party, 1916; represented Wellington Central, 1918-50; secretary, parliamentary labour party, 1919-35; minister of education and health, 1935-40; prime minister, 1940-49; exerted influence beyond New Zealand, making reputation for sound sense and statesmanship; negotiated agreement with Australia, 1944; assumed international stature as a leader of 'small nations' at meetings of United Nations general assembly; PC, 1940; CH, 1946.

FRASER, SIMON JOSEPH, fourteenth (sometimes reckoned sixteenth) Baron Lovat and forty-first MacShimi (1871-1933); succeeded father, 1887; raised and commanded Lovat Scouts in South African war; originated cotton-planting in Sudan; first chairman, Forestry Commission, 1919-27; KT, 1915.

FRASER, Sir THOMAS RICHARD (1841-1920), pharmacologist; born at Calcutta; MD, Edinburgh University; professor of materia medica and clinical medicine at Edinburgh, 1877-1917; FRS, 1877; president of Royal College of Physicians, Edinburgh, 1900-2; knighted, 1902; his researches on poisons important.

FRASER, WILLIAM, first Baron Strathalmond (1888-1970), industrialist; educated at Glasgow Academy; joined father's firm, Pumpherston Oil Co., 1909; director, 1913; joint managing director, 1915; during 1914-18 war created Scottish Oil Agency; CBE, 1918; chairman, Inter-Allied Petroleum Specifications Commission, 1918; when Scottish Oils Ltd. taken over by Anglo-Persian Oil Co. (later British Petroleum), director in charge of production, 1923; deputy chairman, under Sir John (later Lord) Cadman [q.v.], 1928; chairman, 1941-56; pre-eminent in the oil industry on Middle East affairs; petroleum adviser to War Office, 1935-45; chairman, oil supply advisory committee, 1951-2; knighted, 1939; baron, 1955; hon. LLD, Birmingham, 1951; director, Burmah Oil Co., Great Western Railway, and National Provincial Bank.

FRAZER, ALASTAIR CAMPBELL (1909-1969), pharmacologist and food scientist; educated at Lancing College and St. Mary's Hospital Medical School, London University; MB, B.Ch., 1932; taught physiology and pharmacology, 1932-42; reader in pharmacology, Birmingham University, 1942; professor, 1943-67; Sir Halley Stewart research fellow, 1937-45; Ph.D (London), 1941; MD (Birmingham), 1943; D.Sc., (London), 1945; started honours degree course in medical biochemistry, 1953; altered name of department from pharmacology to medical biochemistry, 1956; adviser to government on overseas developments; member (and later, chairman), committee on safety of drugs, 1963; member, Agricultural Research Council; first director-general, British Nutrition Foundation, 1967; CBE, 1962; publications included *Malabsorption Syndromes* (1968).

FRAZER, SIR JAMES GEORGE (1854-1941), social anthropologist; educated at Glasgow University and Trinity College, Cambridge; first class, classical tripos, 1878; fellow, 1879-1941; influenced towards anthropology by Sir E. B. Tylor and William Robertson Smith [qq.v.]; wrote (1888) on 'Taboo' and 'Totemism' for ninth edition of *Encyclopaedia Britannica*; published *The Golden Bough* (1st edn. 1890; 2nd edn. 1900; 3rd edn. completed 1915); collected and classified similar phenomena from all parts of the world and worked out by induction a continuous development in human institutions based upon the concept of survival as a link between the various stages; began *The Golden Bough* as an attempt to find an answer to the riddle of the King of the Wood in the Arician grove of Diana on Lake Nemi; developed it into separate dissertations on *The Magic Art* (1911), *Taboo and the Perils of the Soul* (1911), *The Dying God* (1911), *Adonis, Attis, Osiris* (1914), *Spirits of the Corn and of the Wild* (1912), *The Scapegoat* (1913), *Balder the Beautiful* (1913), and *Aftermath* (1936); claimed that the cycle depicted 'the long evolution by which the thoughts and efforts of man have passed through the successive stages of Magic, Religion and Science'; placed on record an array of facts of even greater value than his theories; although capable of brilliant and far-reaching

hypotheses he failed really to understand the implications of his data; he ignored the findings of the psycho-analytical school and was out of touch with current sociological theory; other books include *Psyche's Task* (1909), re-issued as *The Devil's Advocate* (1928); *Totemism and Exogamy* (4 vols., 1910); *The Belief in Immortality and the Worship of the Dead* (1913); *Folk-lore in the Old Testament* (1918); *The Worship of Nature* (1926); *The Gorgon's Head* (1927); *The Fear of the Dead in Primitive Religion* (3 vols., 1933-6); and *Anthologia Anthropologica* (unclassified notebooks; 4 vols., 1938-9); an edition of Pausanias's *Description of Greece* (1898), of the *Fasti* of Ovid (1929), of the letters of William Cowper (1912); professor of social anthropology, Liverpool, 1907-22; FBA, 1902; FRS, 1920; knighted, 1914; OM, 1925.

FREAM, WILLIAM (1854-1906), writer on agriculture; student at Royal College of Science, Dublin, 1872-5; B.Sc., London, 1877; professor of natural history, Royal Agricultural College, Cirencester, 1877-9; visited Canada 1884, 1888, and 1891, writing accounts of Canadian agriculture; first Steven lecturer on agricultural entomology, Edinburgh, 1890-1906; edited Royal Society's *Journal*, 1890-1900; agricultural correspondent of *The Times*, 1894-1906; published *The Elements of Agriculture* (1891).

FRÉCHETTE, LOUIS HONORÉ (1839-1908), Canadian poet and journalist; born near Quebec; at Quebec edited *Journal de Québec*; clerk of legislative council, 1889-1908; published French verse, including *Mes Loisirs* (1863), *Les Oiseaux de neige* (1880), *Les Fleurs Boréales* (1881), and *La Légende d'un peuple* (1887); CMG, 1897; president of Royal Society of Canada; wrote also *La Noël au Canada*, a prose collection of tales, and dramas; a poet of French-Canadian patriotism; influenced by Victor Hugo.

FREEDMAN, BARNETT (1901-1958), artist; son of east-end Russian Jewish immigrants; self-educated during five years in hospital; at Royal College of Art, 1922-5; subsequently taught there and at the Ruskin, Oxford; official war artist, 1941-6; especially successful as commercial designer and book illustrator; a distinguished letterer and typographer; his knowledge of and skill in lithography stimulated interest in the craft; CBE, 1946; RDI, 1949.

FREEMAN, GAGE EARLE (1820-1903), writer on falconry; vicar of Macclesfield Forest, 1856-89, and Askham, 1889-1903; devoted leisure to hawking; wrote on the subject in *Field* newspaper; collaborated with F. H. Salvin [q.v.] in *Falconry; Its Claims, History and Practice*, 1859; published *Practical Falconry* (1869) and volumes of verse.

FREEMAN, JOHN (1880-1929), poet and critic; junior clerk in Liverpool Victoria Friendly Society, 1893; secretary and director, 1927; his poetical works include *Stone Trees* (1916), *Presage of Victory* (1916), *Memories of Childhood* (1919), *Poems New and Old* (1920), *Music* (1921), *The Grove* (1924), *Collected Poems* (1928), *Last Poems* (posthumous, 1930); in prose he wrote *The Moderns* (1916), *Portrait of George Moore* (1922), *English Portraits* (1924), *Herman Melville* (1926).

FREEMAN, JOHN PEERE WILLIAMS-(1858-1943), archaeologist. [See WILLIAMS-FREEMAN.]

FREEMAN, SIR RALPH (1880-1950), civil engineer; educated at Haberdashers' School and Central Technical College; joined engineering firm of Sir Douglas Fox & Partners (after 1938 Freeman, Fox & Partners), 1901; partner, 1912; senior partner, 1921-50; especially talented in design of steel bridges, the most notable being Sydney Harbour bridge; knighted, 1947.

FREEMAN, SIR WILFRID RHODES, first baronet (1888-1953), air chief marshal; educated at Rugby and Sandhurst; gazetted to Manchester Regiment, 1908; transferred to Royal Flying Corps, 1914; served in France and Middle East; MC, 1915; DSO, 1916; commanded Central Flying School, 1925-7; AOC, Trans-Jordan and Palestine, 1930-3; commandant, RAF Staff College, 1934-6; Air Council member for research and development, 1936-40, and for production, 1938-40; air chief marshal, 1940; vice-chief of air staff, 1940-2; chief executive, Ministry of Aircraft Production, 1942-5; CB, 1932; KCB, 1937; GCB, 1942; baronet, 1945.

FREEMAN-MITFORD, ALGERNON BERTRAM, first BARON REDESDALE, of the second creation (1837-1916), diplomatist and author. [See MITFORD.]

FREEMAN-THOMAS, FREEMAN, first MARQUESS OF WILLINGDON (1866-1941), governor-general of Canada and viceroy of India; educated at Eton and Trinity College, Cambridge; captain of cricket, 1889; liberal MP, Hastings (1900-6), Bodmin division (1906-10); baron, 1910; governor of Bombay, 1913-18; of Madras, 1919-24; broke down tradition of British social exclusiveness; viscount, 1924; chairman, Anglo-Chinese Boxer indemnity

mission, 1926; governor-general of Canada, 1926-30; viceroy of India, 1931-6; met civil disobedience with firmness; earl and PC, 1931; marquess, 1936; represented government at New Zealand centennial celebrations and headed trade mission to South America, 1940.

FREETH, FRANCIS ARTHUR (1884-1970), industrial chemist; educated at Yardleys, Birkenhead, Audlem grammar school and Liverpool University; first class, chemistry, 1905; M.Sc., 1906; joined Brunner Mond & Co., alkali manufacturers, 1907; chief chemist, 1909; recalled from army service for work on production of ammonium nitrate for explosives, 1915; OBE; D.Sc., Liverpool, 1924; FRS, 1925; worked for ICI when Brunner Mond amalgamated, 1926; retired, 1937; re-engaged by ICI, 1944-52; introduced ICI Fellowship Scheme; encouraged promising young chemists; early ICI executives mostly 'Freeth men'.

FREMANTLE, SIR EDMUND ROBERT (1836-1929), admiral; son of first Baron Cottesloe [q.v.]; entered navy, 1849; served in *Queen* on Mediterranean station and in *Spartan* on China station; commanded *Eclipse* during Maori war in New Zealand, 1864-7; captain, 1867; as commander of paddle-steamer *Barracouta* took part in Gold Coast operations against Ashantis, 1873-4; severely wounded; commanded *Lord Warden*, 1877; *Invincible*, 1879; twice saved life while in command of these ships; senior naval officer, Gibraltar, 1881-4; rear-admiral, 1885; second-in-command of Channel squadron, 1886-7; commander-in-chief, East Indies, 1888-91; commanded successful punitive expedition against Sultan of Vitu in British East Africa, 1890; vice-admiral, 1890; commander-in-chief, China, 1892-5; at Devonport, 1896-9; admiral, 1896; retired, 1901; KCB, 1889; writer on naval subjects, being particularly interested in problems of shipbuilding policy, and consistently opposing contention that days of great ships were over.

FRENCH, EVANGELINE FRANCES (1869-1960) and FRANCESCA LAW FRENCH (1871-1960), missionaries in China; worked with Mildred Cable [q.v.].

FRENCH, SIR HENRY LEON (1833-1966), civil servant; educated privately and at King's College, London; joined Civil Service and posted to Board of Agriculture, 1901; promoted to second secretary, 1934; seconded to Board of Trade as director of food (defence plans), 1936; permanent secretary, Ministry of Food, 1939-45, under Lord Woolton [q.v.]; reputation rests on remarkable success of the Ministry during the 1939-45 war; on retirement from civil service filled posts in film industry; director-general, British Film Producers Association, 1946-57; president, 1957-8; first chairman of Festival Gardens Ltd.; chairman, Unesco Co-operating Body on Mass Communications (1946-53); OBE, (1918); CB, (1920); KBE, (1938); KCB, (1942); and GBE, (1946).

FRENCH, JOHN DENTON PINKSTONE, first EARL OF YPRES (1852-1925), field-marshal; joined Suffolk artillery militia, 1870; gazetted to 8th (shortly transferred to 19th) Hussars, 1874; served in Lord Wolseley's expedition for relief of General Gordon [q.v.], 1884-5; commanded 19th Hussars, 1888; colonel and assistant-adjutant-general, War Office, 1895; on outbreak of South African war dispatched to Natal to command mounted troops under Sir G. S. White [q.v.], 1899; then sent to Cape Colony, which he virtually cleared of invaders before arrival of Lord Roberts [q.v.] in South Africa, Jan. 1900; relieved Kimberley, 15 Feb.; secured surrender of Boers at Paardeberg (27 Feb.) and turned Boer front at Poplar Grove (7 Mar.) and Driefontein (10 Mar.); commanded Johannesburg district, Nov.; lieutenant-general, 1902; commander-in-chief, Aldershot, 1902-7; general, 1907; inspector-general of forces, 1907; responsible for total reform in conduct of military manœuvres; chief of imperial general staff, 1912; field-marshal, 1913; resigned appointment owing to Curragh incident, 1914; commander-in-chief, British expeditionary force in France, Aug. 1914; lamentably failed to conform to French plan of action; British expeditionary force transferred to Flanders, Oct.; made fruitless attempt to break German line at Neuve Chapelle, Mar. 1915; made abortive attempt to seize Aubers ridge, hoping thereby to facilitate capture of Lille, May; renewed attempts made at Festubert and Givenchy (May and June), equally disastrous owing to inadequate forces; battle of Loos (25-28 Sept.) no more successful; owing to dissatisfaction felt with his conduct of operations, resigned position as commander-in-chief, Dec.; created viscount and appointed commander-in-chief of home forces, 1916; achieved satisfactory results in training troops for overseas and protecting Great Britain from air attacks; lord-lieutenant of Ireland, where position of affairs grew increasingly unsatisfactory, 1918-21; earl, 1922.

FRERE, MARY ELIZA ISABELLA (1845-1911), author; daughter of Sir Bartle Frere, first baronet [q.v.]; went with father to Bombay, 1863; collected Indian folk-lore tales in *Old Deccan Days* (1868); wrote poems; accompanied father to South Africa, 1877, and later travelled in Egypt and Palestine.

FRERE, WALTER HOWARD (1863-1938), bishop of Truro, historian, and liturgiologist; son of P. H. Frere [q.v.]; educated at Charterhouse and Trinity College, Cambridge; deacon, 1887; priest, 1889; joined community of Resurrection, 1892; superior 1902-13, 1916-22; bishop of Truro, 1923-35.

FRESHFIELD, DOUGLAS WILLIAM (1845-1934), mountain explorer and geographer; educated at Eton and University College, Oxford; recorded some of his explorations in *Exploration of the Caucasus* (1896) and *Round Kangchenjunga* (1903); promoted advancement of geography; president, Royal Geographical Society, 1914-17; edited *Alpine Journal*, 1872-80.

FREYBERG, BERNARD CYRIL, first BARON FREYBERG (1889-1963), lieutenant-general; educated at Wellington College, New Zealand; qualified in dentistry, Otago, 1911; commissioned in Royal Naval Volunteer Reserve, 1914; friend of (Sir) Winston Churchill; served at Antwerp and in Gallipoli; helped carry coffin at burial of Rupert Brooke [q.v.], 1915; DSO; fought in France with rank of lieutenant-colonel, 1916; awarded VC at battle of the Somme; brigadier-general, commanded 88th brigade of 29th division near Passchendaele; bar to DSO; second bar to DSO, 1918; wounded six times during 1914-18 war; croix de guerre with palms; commanded 1st battalion, Manchester Regiment, 1929; major-general, 1934; retired from service, 1937; director, Birmingham Small Arms Co., 1937-9; recalled to army to command 2nd New Zealand Expeditionary Force, 1939; served in Egypt and Crete; fought at Alamein under Bernard Montgomery (later Viscount Montgomery of Alamein), 1942; and on the Mareth Line; lieutenant-general; took New Zealand division to Italy; attempted to break through at Casino; took part in advance of Eighth Army north of Rome; third bar to DSO, 1945; governor-general, New Zealand, 1945-52; lieutenant-governor and deputy constable, Windsor Castle, 1952; moulded the New Zealand division into superb fighting machine; played important part in relations between British and New Zealand Governments; CB, 1936; KCB, and KBE, 1942; CMG, 1919; GCMG, 1946; baron, 1951; hon. LLD, St. Andrews (1922), and Oxford (1945).

FREYER, SIR PETER JOHNSTON (1851-1921), surgeon; MD and MS, Queen's University of Ireland; served in Indian medical service, 1875-96; surgeon to St. Peter's Hospital for Stone, London, 1897-1914; CB and KCB, 1917.

FRIESE-GREENE, WILLIAM (1855-1921), pioneer of cinematography. [See GREENE.]

FRITH, WILLIAM POWELL (1819-1909), artist; studied art under Sass and at Royal Academy Schools; exhibited at Royal Academy from 1840 subject pictures, such as 'English Merrymaking a Hundred Years ago', 1847; ARA, 1845; RA, 1853; chief works were 'Ramsgate Sands' (1853), 'Derby Day' (1858, now in Tate Gallery), 'The Railway Station' (1862), 'Charles II's last Whitehall Sunday' (1867); unsuccessfully attempted to rival Hogarth's morality pictures, 1878-80; other works were 'Charles Dickens' (1859) and 'Uncle Toby and the Widow Wadman' (1866, Tate Gallery); visited Holland, 1850 and 1880, and Italy, 1875; CVO, 1908; published *John Leech, his Life and Work* (1891), *Reminiscences* (1887), and *Further Reminiscences* (1888).

FRITSCH, FELIX EUGEN (1879-1954), algologist; educated at Warwick House School, Maida Vale; B.Sc., London, 1898: D.Sc., 1905; D.Phil., Munich, 1899; in charge botany department, East London College, 1907-48; professor in London University, 1924-48; campaigned for foundation of Freshwater Biological Association (1929) and biological station at Wray Castle; publications include *The Structure and Reproduction of the Algae* (2 vols., 1935-45) and five textbooks with Sir Edward Salisbury; FRS, 1932; president Linnean Society, 1949-52.

FROWDE, HENRY (1841-1927), publisher; manager of London office of Oxford University Press, 1874-1913; styled 'Publisher to the Universty of Oxford', 1883.

FRY, CHARLES BURGESS (1872-1956), sportsman; exhibitioner of Repton and scholar of Wadham College, Oxford; blue for cricket, athletics, and association football; played football for England; tied world's record long jump, 1893; first played cricket for Sussex, 1894; for Hampshire, 1909; in test cricket, 1899; captained England team which defeated Australia and South Africa, 1912; considered greatest batsman of his time; made 30,886 runs, averaging over 50, with 94 centuries; edited and directed *Fry's Magazine;* director of training ship *Mercury*, 1908-50; published books on cricket, and autobiography, *Life Worth Living* (1939).

FRY, DANBY PALMER (1818-1903), legal writer; clerk in Poor Law Board, 1836-78; reported chartist proceedings at Kennington common, 1848; called to bar, 1851; legal adviser (1878-82) to Local Government Board; original member of Philological Society, 1842, and of Early English Text Society, 1864; published standard legal handbooks.

FRY, SIR EDWARD (1827-1918), judge; great-grandson of Joseph Fry [q.v.]; in business in Bristol, 1843-8; BA, University College, London, 1851; called to bar (Lincoln's Inn), 1854; published *A Treatise on the Specific Performance of Contracts* (1858); acquired large practice in Chancery and company work and at parliamentary bar; QC, 1869; practised in turn before Vice-Chancellors James and Bacon, afterwards migrating to Rolls court; appointed additional judge in Chancery division and knighted, 1877; principal legal work performed on rule committee of judges; sat in Court of Appeal, 1883-92; in later life played important part in international affairs; judge on Hague tribunal, 1900; arbitrator at Hague between United States and Mexico in pious funds of California dispute, 1902-3; British legal assessor on commission appointed to deal with Dogger Bank incident, 1904; played active part at second Hague conference, 1907; arbitrator between France and Germany over Casablanca incident, 1908-9; FRS, 1883; GCB, 1907; a strong Quaker.

FRY, JOSEPH STORRS (1826-1913), cocoa manufacturer and Quaker philanthropist: brother of Sir Edward Fry [q.v.]; partner in family business in Bristol, 1855; president of 'London Yearly Meeting' of Friends for fifteen years.

FRY, ROGER ELIOT (1866-1934), art critic and artist; son of Sir Edward Fry [q.v.]; of Quaker stock; educated at Clifton and King's College, Cambridge; first class, parts i and ii, natural sciences tripos, 1887-8; turned from science to painting and, after visits to Italy, to connoisseurship; became an enthralling lecturer, communicating his own excitement; art critic for *Athenaeum*, 1901; edited Reynolds's *Discourses*, 1905; director, Metropolitan Museum of Art, New York, 1905-10; interest in modern painting aroused (1906) by work of Cézanne; organized exhibitions of French Post-Impressionists at Grafton Galleries, 1910 and 1912; became champion of modern art in face of virulent attack; founded Omega workshops for manufacture of well-designed articles of daily use, 1913; established a position as a critic which made him greatest influence on taste since John Ruskin [q.v.]; his writings collected in *Vision and Design* (1920) and *Transformations* (1926); maintained that response to art is a pure, disinterested activity, 'a reaction to a relation and not to sensations or objects or persons or events'; other publications include *Cézanne*

(1927) and *Henri Matisse* (1930); rejected for Slade professorship at Oxford (1910, 1927), at Cambridge (1904); elected at Cambridge, 1933; his own painting matured as a careful, learned naturalism; at his best as a painter of architecture; a founder and supporter of the London Group.

FRY, SARA MARGERY (1874-1958), reformer; sixth daughter of Sir Edward Fry [q.v.]; educated privately, at the future Roedean and Somerville College, Oxford; read mathematics; librarian of Somerville, 1899-1904; warden, women's hostel, Birmingham University, 1904-14; worked in France for Friends' War Victims Relief Committee, 1914-18; member of labour party, 1918-39; set up London house with brother Roger Fry [q.v.], 1919; served on University Grants Committee, 1919-48; secretary, Penal Reform League, amalgamated (1921) as Howard League for Penal reform, 1919-26; subsequently chairman or vice-chairman; became deeply involved in campaign for abolition of capital punishment; appointed magistrate, 1921; education adviser to Holloway Prison, 1922; principal of Somerville, 1926-31; increasingly occupied with international aspects of penal reform; member, Colonial Office advisory committee on penal reform, 1936; lectured in China, 1933, and in United States, 1942-3; served as governor of BBC and proved a good broadcaster, taking early part in 'Any Questions?' and the Brains Trust; publications include *The Future Treatment of the Adult Offender* (1944) and *Arms of the Law* (1951).

FRY, THOMAS CHARLES (1846-1930), schoolmaster and dean of Lincoln; BA, Pembroke College, Cambridge; ordained, 1871; headmaster of Berkhamsted School, 1887-1910; improved buildings and equipment; raised educational standard and increased numbers; dean of Lincoln, 1910; raised nearly £100,000 for cathedral restoration fund, 1921-30; paid three visits overseas to collect funds.

FRYATT, CHARLES ALGERNON (1872-1916), merchant seaman; entered Great Eastern Railway Company's service as able seaman on Harwich to Antwerp steamship route, 1892; made 143 trips in command of vessels during European war, 1914-16; ship *Brussels* attacked by German submarine, Mar. 1916; commended by Admiralty for plucky escape; captured by German destroyers, June; court-martialled and shot as *franc-tireur* at Bruges, July.

FULLER, SIR CYRIL THOMAS MOULDEN (1874-1942), admiral; entered navy, 1887; captain, 1910; senior naval officer, Cameroons, 1914-16; DSO, 1916; assistant

director, plans division, 1917; director, 1918-20; headed British naval section, peace conference, 1919; rear-admiral, 1921; third sea lord, 1923-5; vice-admiral, 1926; KCB, 1928; commander-in-chief, North America and West Indies station, 1928-30; admiral, 1930; second sea lord and chief of naval personnel, 1930-2.

FULLER, JOHN FREDERICK CHARLES (1878-1966), major-general; educated at Malvern College and Sandhurst, 1897-8; commissioned into 43rd, the 1st battalion, Oxfordshire Light Infantry; fought in South African war; served in India; entered Staff College, 1913; wrote for the *Army Review*; fought in France, 1915-16; posted to Tank Corps, 1916; wrote articles advocating new model army based on tanks and other modern equipment; senior instructor, Staff College, 1923; published *The Foundations of the Science of War* (1925); major-general 1930; retired, 1933; published *Lectures on F.S.R.*, vol. iii (1932), on use of tanks; published ten books, including *Memoirs of an Unconventional Soldier* (1936) and *The Decisive Battles of the Western World, and their Influence upon History* (3 vols., 1954-6); DSO, 1917; CBE, 1926; CB, 1930.

FULLER, SIR (JOSEPH) BAMPFYLDE (1854-1935), Indian administrator and author; educated at Marlborough; served in North-West and Central provinces; chief commissioner, Assam, 1902-5; lieutenant-governor (1905), Eastern Bengal and Assam, where considered lacking in patience in handling hostility to partition and his 'resignation' accepted (1906); publications include *The Empire of India* (1913) and *The Tyranny of the Mind* (1935); KCSI, 1906.

FULLER, SIR THOMAS EKINS (1831-1910), agent-general for Cape Colony, 1902-7; editor of *Cape Argus*, South Africa, 1864; advocated university and responsible government in Cape Colony; general manager of Union Steamship Company at Cape Town, 1875-98; represented Cape Town in house of assembly, 1878-1902; supported policy of Cecil Rhodes [q.v.]; KCMG, 1904; wrote monograph on Rhodes, 1910.

FULLER-MAITLAND, JOHN ALEXANDER (1856-1936), musical critic and connoisseur. [See MAITLAND.]

FULLEYLOVE, JOHN (1845-1908), landscape painter; studied drawing and architecture; travelled much in England and abroad; member of Royal Institute of Painters in Water Colours, 1879; exhibited in London water-colour drawings of 'Petrarch's Country' (1886), Oxford

(1888), Cambridge (1890), Versailles (1894), Greece (1896), and Palestine (1902); also painted in oils; many drawings were reproduced as illustrations to books; an able architectural draughtsman and accomplished water-colour artist.

FURNEAUX, WILLIAM MORDAUNT (1848-1928), schoolmaster and dean of Winchester; BA, Corpus Christi College, Oxford, 1872; ordained, 1874; assistant master, Marlborough College, 1874—82; headmaster of Repton School, 1882-1900; set school on firm financial basis and improved the buildings; dean of Winchester, 1903-19; successfully raised funds to save cathedral fabric from collapse, 1905-12.

FURNESS, CHRISTOPHER, first BARON FURNESS (1852-1912), shipowner and industrialist; early joined brother's firm of wholesale provision merchants; set up as shipowner, 1877; amalgamated with other firms; liberal MP, Hartlepools, 1891-5, 1900-10; knighted, 1895; baron, 1910.

FURNISS, HARRY (1854-1925), caricaturist and illustrator; educated in Dublin; came to London, 1873; joined staff of *Illustrated London News*, 1876; chosen by (Sir) Henry Lucy [q.v.] to illustrate his 'Essence of Parliament' in *Punch*, 1881; 'called to the Table' at *Punch* office, 1884; resigned from staff, 1894; embarked on unsuccessful journalistic ventures, 1894-5; illustrated complete editions of Dickens (1910) and Thackeray (1911); took up new art of the cinematograph, 1912.

FURNISS, HENRY SANDERSON, BARON SANDERSON (1868-1939), principal of Ruskin College; almost blind from birth; second class, modern history, Hertford College, Oxford, 1893; tutor and lecturer in economics, Ruskin College, 1907; principal, 1916-25; baron, 1930.

FURNIVALL, FREDERICK JAMES (1825-1910), scholar and editor; educated at private schools, University College, London, and Trinity Hall, Cambridge; BA, 1847; MA, 1850; enthusiastic oarsman from youth; improved design of sculling boats, 1845; barrister of Gray's Inn, 1849; interested in social reform; joined Christian socialists; made acquaintance of John Ruskin [q.v.], 1849; helped to found Working Men's College, London, 1854; taught English grammar and literature there; developed college's social life; became an outspoken agnostic; his relations strained with F. D. Maurice [q.v.], principal of the college; applied principles of co-operation to literary study; member of Philological Society from

1847, and sole secretary from 1862 to death; became (1861) editor of the Philological Society's suggested English Dictionary, which developed into the *New English Dictionary* of Oxford; contributed to that work through life; founded Early English Text Society, 1864, and Chaucer Society, 1868; edited Chaucer's works from MS, and thereby furnished new material from textual study; edited *Percy Ballads* with J. W. Hales, 1868, and founded Ballad Society; established New Shakespere Society, 1873; resented Swinburne's sarcastic attack on society's methods; engaged in undignified controversy with Halliwell-Phillipps and with Swinburne [qq.v.], 1881; wrote elaborate preface for 'Leopold' Shakespeare, 1876; supervised issue of photographic facsimiles of Shakespeare quartos (43 vols., 1880-9); founded Wiclif and Browning societies, 1881; the Browning Society greatly extended the poet's popularity; founded Shelley Society, 1886; lost substantial fortune inherited from his father on failure of Overend and Gurney's Bank, 1867; received civil list pension, 1884; hon. Ph.D., Berlin, 1884; hon. fellow of Trinity Hall, 1902; hon. D.Litt., Oxford, 1901; original FBA, 1902; devoted to rowing till death, he founded National Amateur Rowing Association, to include all ranks of society, 1891, and Hammersmith (renamed Furnivall) Sculling Club for girls and men, 1896.

FURSE, CHARLES WELLINGTON (1868-1904), painter; pupil of Alphonse Legros [q.v.] at Slade School, London, and of Julian in Paris; first exhibited at Royal Academy, 1888; gained early distinction as portrait painter; developed original and spontaneous manner; skilful painter of horses; went in search of health to South Africa, 1895; made decorative paintings for Liverpool town hall in manner of Tintoretto, 1898-1901; married daughter of John Addington Symonds [q.v.] (Dame Katharine Furse), 1900, and settled next year at Camberley, producing his 'Return from the Ride' and 'Diana of the Uplands' (both in Tate Gallery), 'Lord Charles Beresford', 'Lord Roberts' (unfinished); ARA, 1903; died of tuberculosis; a keen art critic and energetic worker.

FURSE, DAME KATHARINE (1875-1952), pioneer Service woman; fourth daughter of J. A. Symonds [q.v.]; spent youth in Davos; educated privately; married C. W. Furse [q.v.], 1900; enrolled as VAD, 1909; took first unit to France, 1914; organized VAD Department in London, 1914-17; appointed commandant-in-chief, 1916; GBE, 1917; director, Women's Royal Naval Service, 1917-19; subsequently worked in Switzerland and took up skiing; also worked for Girl Guides.

FUST, HERBERT JENNER- (1806-1904), cricketer. [See JENNER-FUST.]

FYFE, DAVID PATRICK MAXWELL, EARL OF KILMUIR (1900-1967), lord chancellor; educated at Watson's College, Edinburgh, and Balliol College, Oxford; called to bar (Gray's Inn), 1922; in chambers at Liverpool with (Sir) George Lynskey [q.v.]; KC, 1934; recorder of Oldham, 1936; conservative MP, West Derby division of Liverpool, 1935-45; solicitor-general, knighted, 1942; attorney-general, PC, 1945; deputy chief prosecutor at Nuremberg; home secretary, 1951; GCVO, 1953; lord chancellor and Viscount Kilmuir, 1954-62; dismissed from cabinet by Macmillan; accepted an earldom but refused to take any further part in legal or political affairs; chairman, Plessey Company; hon. fellow, Balliol, and other academic distinctions; rector, St. Andrews (1955-8); visitor, St. Anthony's College, Oxford, 1953.

FYFE, HENRY HAMILTON (1869-1951), writer; educated at Fettes; joined *The Times*; moved to *Morning Advertiser*, 1902; edited *Daily Mirror*, 1903-7; special correspondent, *Daily Mail*, 1907-18; reported retreat from Mons, 1914; in Russia, 1915-16; in Spain, Portugal, Italy, USA, 1917; at Crewe House, 1918; editor, *Daily Herald*, 1922-6; wrote for *Daily Chronicle*, 1926-30, and subsequently for *Reynolds' News*; versatile miscellaneous writer of novels, plays, biographies, etc.

FYFE, SIR WILLIAM HAMILTON (1878-1965), headmaster and university vice-chancellor; educated at Fettes and Merton College, Oxford; first class, *lit. hum.*, 1901; taught at Radley, 1902-4; fellow and principal of the postmasters, Merton College, 1904-19; tutor in classical honour mods.; commissioned in territorials in 1914-18 war; headmaster, Christ's Hospital, 1919-30; vice-chancellor, Queen's University, Kingston, Ontario, 1930-6; principal, Aberdeen University, 1936-48; chairman, advisory council on education in Scotland; member, inter-university council for higher education in the colonies; knighted, 1942; hon. doctorates from Canadian, British, and American universities; fellow, Royal Society of Canada, 1932; hon. fellow, Merton College, 1948; chairman, governors of Gordonstoun, 1945-8; publications include translations of works of Tacitus and Aristotle.

FYLEMAN, ROSE AMY (1877-1957), writer for children; educated privately; studied and taught singing; became regular contributor of verse to *Punch* after publication of 'There are fairies at the bottom of our garden!', 1917; *Fairies and Chimneys* (1918) and other collections of verse and of tales became nursery favourites; wrote children's plays; and, with Thomas Dunhill [q.v.], a children's opera, 1933.

G

GADDUM, SIR JOHN HENRY (1900-1965), pharmacologist; educated at Rugby and Trinity College, Cambridge; first class, part i, mathematical tripos, 1920; entered University College Hospital, London, 1922; qualified, 1924; failed Cambridge MB, but later awarded Sc.D.; began pharmacological research at Wellcome Research Laboratories, 1925; assistant to (Sir) Henry Dale [q.v.] at National Institute for Medical Research, 1928-34; professor of pharmacology, Cairo University, 1934-5; professor of pharmacology, University College, London, 1935; professor, College of the Pharmaceutical Society, London, 1938; during 1939-45 war, worked at Chemical Defence Research Station, Porton; professor of materia medica, Edinburgh University, 1942-58; director, Institute of Animal Physiology, Cambridge, 1958; FRS, 1945; fellow, Royal Society of Edinburgh; knighted, 1964; hon. LLD Edinburgh; did important research on the development of specific and sensitive methods of biological assay; publications include textbook on pharmacology (1940), translated into German, Spanish, and Japanese; Gaddum Memorial Trust created by British Pharmacological Society, 1966.

GADSBY, HENRY ROBERT (1842-1907), musician; professor of harmony at Guildhall School of Music from 1880; published choral and orchestral cantatas, part songs, and anthems, and textbooks on harmony (1883) and sight singing (1897).

GAINFORD, first BARON (1860-1943), politician and man of business. [See PEASE, JOSEPH ALBERT.]

GAIRDNER, JAMES (1828-1912), historian; son of John Gairdner [q.v.]; clerk in Public

Record Office, 1846-93; joint-editor of *Calendar of Letters and Papers of the Reign of Henry VIII*; edited *Paston Letters*, etc.

GAIRDNER, SIR WILLIAM TENNANT (1824-1907), professor of medicine at Glasgow; son of John Gairdner [q.v.]; MD, Edinburgh, 1845; hon. LLD, 1883; pathologist and physician to Royal Infirmary there; professor of medicine at Glasgow University, 1862-90; medical officer of health, Glasgow, 1863-72; attractive lecturer; made many original researches in connection with heart and lung diseases; FRS, 1892; KCB, 1898; works include *Clinical Medicine* (1862) and *The Physician as Naturalist* (1889).

GAITSKELL, HUGH TODD NAYLOR (1906-1963), politician; educated at Winchester and New College, Oxford; first class, philosophy, politics, and economics, 1927; tutored by G. D. H. Cole [q.v.]; joined labour party as an undergraduate; lecturer in political economy, University College, London, 1928-38; head of department and university reader, 1938; temporary civil servant, during 1939-45 war, in Ministry of Economic Warfare, and Board of Trade; labour MP South Leeds, 1945-63; parliamentary secretary, Ministry of Fuel and Power, 1946; minister and PC, 1947; minister of state for economic affairs, 1950; chancellor of the Exchequer, 1950-1; opposed to policies of Aneurin Bevan [q.v.]; defeated Bevan and Herbert Morrison (later Lord Morrison of Lambeth, q.v.) for party leadership, 1955; defeated at general election, 1959; opposed party resolution in favour of unilateral disarmament; opposed British entry into EEC, 1962; CBE, 1945; hon. DCL, Oxford, 1958.

GALE, FREDERICK (1823-1904), cricketer and writer on cricket under the pseudonym of 'Old Buffer'; works include *Public School Matches and those we meet there* (1853) and *The Game of Cricket* (1887); brother-in-law of Walter Severn and a friend of Ruskin [qq.v.].

GALLACHER, WILLIAM (1881-1965), working-class agitator and politician; left school at twelve and worked as grocer's delivery boy; apprenticed as brass-finisher, 1895; worked as ship's steward, 1909-10; joined Social Democratic Federation, 1906; chairman, Clyde Workers' Committee; sentenced to imprisonment for sedition, 1916, and for rioting, 1919; supported Russian revolution and Soviet Union, 1917; published (with J. R. Campbell, q.v.) *Direct Action* (1919); after personal discussion with Lenin, accepted view that workers should take part in parliamentary politics; helped to found British communist party, 1920-1; served

further periods of imprisonment for sedition, 1921 and 1925; communist MP West Fife, 1935-50; publicly opposed the Munich Agreement, 1938; chairman, communist party, 1943-56; president, 1956-63; publications include *The Chosen Few: a Sketch of Men and Events in Parliament* (1940), *The Case for Communism* (1949), and *The Tyrants' Might is Passing* (1954).

GALLOWAY, SIR WILLIAM (1840-1927), mining engineer; researched into causes of mine explosions, which he came to believe largely depended on floating coal dust; this view, at first considered unorthodox, was finally confirmed by other investigators; professor of mining, University College, Cardiff, 1891-1902; knighted, 1924.

GALLWEY, PETER (1820-1906), Jesuit preacher and writer; studied at Stonyhurst; entered Society of Jesus, 1836; in charge of Farm Street church, London, 1857-69 and 1877-1906; provincial of Jesuits in England, 1873-6; published *Lectures on Ritualism* (2 vols., 1879), pamphlets, and sermons.

GALSWORTHY, JOHN (1867-1933), playwright and novelist; educated at Harrow and New College, Oxford; called to bar (Lincoln's Inn), 1890; published early novels under pseudonym John Sinjohn; *The Man of Property* (1906) began the *Forsyte Saga* sequence (published collectively, 1922) in which he wrote with an exact and not unsympathetic observation of Victorian upper-class commercial society, continuing the story of its supersession by a more easy-going generation in the trilogy *A Modern Comedy* (1929); his remarkably successful plays (including *The Silver Box*, 1906, *Justice*, 1910, and *The Skin Game*, 1920) state a theme with great simplicity of construction and dialogue; OM, 1929; Nobel prize, 1932.

GALTON, SIR FRANCIS (1822-1911), founder of the science of 'eugenics'; born at Birmingham; entered Trinity College, Cambridge, 1840; travelled in Syria and Egypt, 1844; published account of exploration, into interior of Damaraland, in *Tropical South Africa* (1853); FRS, 1856; general secretary of British Association, 1863-7; in *Meteorographica* (1863) he pointed out importance of 'anticyclones', a word coined by himself; began researches into laws of heredity, 1865; initiated anthropometric laboratory at Health Exhibition, 1884-5, collecting impressions of fingers; proved permanence of fingerprints; published *Finger Prints* (1893) and *Finger Print Directory* (1895); influenced by his cousin Charles Darwin's *Origin of Species* he investigated the heritability of genius; published his results in *Hereditary Genius* (1869), *English Men of Science* (1874), *Human Faculty* (1883), *Natural Inheritance* (1889), and *Noteworthy Families* (1906); founded eugenics laboratory, 1904, and research fellowship and scholarship, 1907, at University College, London; initiated quarterly journal, *Biometrika*, 1901; knighted, 1909; by will left £45,000 for foundation of chair of eugenics in London University; wrote *Memories of My Life* (1908).

GAME, SIR PHILIP WOOLCOTT (1876-1961), air vice-marshal, governor of New South Wales, and commissioner of the metropolitan police; educated at Charterhouse and Royal Military Academy, Woolwich; joined Royal Artillery, 1895; served in South Africa, 1901-2; attended Staff College, 1910; served in France, 1914-16; DSO, 1915; served on Royal Flying Corps staff of Brigadier-General (later Viscount) Trenchard [q.v.], 1916; joined Royal Air Force, 1918; director of Training and Organization (air commodore), 1919; air vice-marshal, 1922; air member for personnel, Air Ministry, 1923-9; governor, New South Wales, 1930-5, dismissed labour party premier, J. T. Lang, a controversial decision; succeeded Lord Trenchard as commissioner, metropolitan police, 1935-45; CB, 1919; KCB, 1924; GBE, 1929; KCMG, 1935; GCVO, 1937; GCB, 1945.

GAMGEE, ARTHUR (1841-1909), physiologist; born at Florence; brother of Joseph Gamgee (1828-86, q.v.); MD, Edinburgh University, 1862; lecturer on physiology at Royal College of Surgeons, Edinburgh, 1863-9; fellow, 1872; studied under Kühne at Heidelberg and Ludwig at Leipzig, 1871; FRS, 1872; FRCP, London, 1896; first Brackenbury professor of physiology in Owens College, Manchester, 1873-85; practised in Switzerland from 1889; visited America, 1902 and 1903; extended knowledge of physical and chemical properties of haemoglobin; made elaborate research on diurnal variations of temperature of human body; works include *Textbook of Physiological Chemistry* (2 vols., 1880-93); died in Paris.

GANDHI, MOHANDAS KARAMCHAND (1869-1948), Indian political leader and social reformer; son of chief minister in Porbandar; of Vaisya sub-caste of merchants; called to bar (Inner Temple), 1891; studied many religious works; influenced especially by the *Bhagavad Gita* and New Testament; experimented in asceticism and in 1906 took vow of complete sexual abstinence; in South Africa (1893-1914) a prosperous lawyer and leader of Indian opposition to racial discrimination; evolved technique of passive resistance or 'Satyagraha' (truth-force); secured compromise agreement

with Smuts, 1914; raised Indian ambulance units in South African war, Zulu expedition, and in 1914; returned to India, 1915; conducted recruiting campaign, 1918; established *ashram* at Sabarmati; moved to Wardha, near Nagpur, 1933; in opposition to Rowlatt Act, 1919, introduced the *hartal* (strike with prayer and fasting); extended activities of Indian National Congress to the villages; secured its acceptance of non-co-operation policy, 1920; discontinued civil disobedience for fear of violence, 1922; imprisoned, 1922-4; undertook first major fast, for Hindu-Moslem unity, 1924; withdrawing from politics travelled country preaching handspinning, Hindu-Moslem unity, and abolition of untouchability; adopted loin-cloth, shawl, and sandals and became widely known as the Mahatma and regarded as a saint; marched to sea to break salt law and imprisoned, 1930; released for talks with Lord Irwin, 1931; attended second Round Table conference in London, 1931; attracted much publicity but proved unsuccessful in conference; threatened further civil disobedience and imprisoned, 1932; fasted to prevent separate electorates for depressed classes and obtained alternative suggestions, 1932; released at beginning of 21-day self-purification fast, 1933; resigned from Congress (1934), but remained its mentor; persuaded it to accept office, 1937; resumed leadership, 1940-1; opposed force even in defence of India; imprisoned after 'Quit India' motion, 1942-4; failed to reach agreement on Hindu-Moslem problems with M. A. Jinnah [q.v.]; totally opposed to but unable to prevent partition (1947); fasted in Calcutta (1947) and Delhi (1948) to obtain communal harmony; shot dead by militant Hindu.

GANN, THOMAS WILLIAM FRANCIS (1867-1938), archaeologist; educated at King's School, Canterbury, and Middlesex Hospital; medical officer, British Honduras, 1894-1923; excavated remains of ancient Maya civilization notably at Lubaantun (1904), Coba and Ichpaatun (1926), Tzibanche (1927), and Noh Mul (1936, 1938); lecturer in Central American archaeology, Liverpool, 1919-38.

GARBETT, CYRIL FORSTER (1875-1955), archbishop of York; son of a clergyman and nephew of Edward and James Garbett [qq.v.]; educated at Farnham and Portsmouth grammar schools; Gomm scholar, Keble College, Oxford; president of Union, 1898; at Cuddesdon acquired fidelity to a strict rule of life; deacon, 1899; priest, 1901; curate (1899-1909) and vicar (1909-19), St. Mary's, Portsea; as bishop of Southwark, 1919-32, was an indefatigable visitor, became expert on problems of bad housing and malnutrition, and provided diocese with twenty-five new churches; chairman, BBC, religious advisory committee, 1923-45; until outbreak of war had more relaxed life as bishop of Winchester, 1932-42; archbishop of York, 1942-55; assiduously attended House of Lords, used broadcasting and the press to interpret the Church to the nation; publications include *The Claims of the Church of England* (1947), *Church and State in England* (1950), and *In an Age of Revolution* (1952); in 1943-55 visited Russia, North America, Malaya, Australia, many European countries, and the Holy Land; a stern disciplinarian and inhibited by shyness, he was formidable to his clergy but mellowed in later life; not an original thinker but very widely read and a talented synthesist whose purpose was to build a bridge between the secular and sacred views of life; much trusted by the laity whose problems he understood and who respected his common sense and sanctity; PC, 1942.

GARCIA, MANUEL (PATRICIO RODRIGUEZ] (1805-1906), singer and teacher of singing; born at Zafra, Spain; studied harmony in Paris; as professor at Paris conservatoire he taught Jenny Lind [q.v.]; published famous *Traité complet de l'art du chant* (1847); professor of singing at Royal Academy of Music, London, 1848-95; invented laryngoscope (1854), which became universally used in medicine and surgery; made CVO on hundredth birthday, 1905.

GARDINER, SIR ALAN HENDERSON (1879-1963), Egyptologist and linguist; brother of H. B. Gardiner [q.v.], the composer; educated at Charterhouse, the Sorbonne, and Queen's College, Oxford; first class, Hebrew and Arabic, 1901; co-operated in preparation of Egyptian dictionary in Berlin, 1902-12; reader in Egyptology, Manchester, 1912-14; publications include *The Admonitions of an Egyptian Sage* (1909), *Notes on the Story of Sinuhe* (1916), *Egyptian Letters to the Dead* (1928), *Late-Egyptian Stories* (1932), *Ancient Egyptian Onomastica* (3 vols., 1947), and *Egyptian Grammar* (1927); edited the *Journal of Egyptian Archaeology*, 1916-21, 1934, and 1941-6; discovered pictographic ancestor of Phoenician alphabet, 1915; also published *The Theory of Speech and Language* (1932), *The Theory of Proper Names* (1940), and *Egypt of the Pharaohs* (1961); D.Litt., Oxford, 1910; FBA, 1929; hon. degrees, Durham and Cambridge; hon. fellow, Queen's College, Oxford, 1935; knighted, 1948.

GARDINER, ALFRED GEORGE (1865-1946), author and journalist; joined *Northern Daily Telegraph*, 1886; editor, *Daily News*, 1902-19; supporter of liberal political and social reforms and of Asquith's conduct of war; in sharp disagreement with Lloyd George and

severe critic of Versailles treaty; publications include life of Sir William Harcourt (2 vols., 1923) and volumes of portrait sketches.

GARDINER, HENRY BALFOUR (1877-1950), composer; educated at Charterhouse and New College, Oxford; studied under Ivan Knorr at Frankfurt; compositions include 'Shepherd Fennel's Dance' and (for chorus and orchestra) 'News from Whydah' and 'April' (1912) and 'Philomena' (1923); finding post-war musical atmosphere uncongenial ceased composition, 1924; generous benefactor of contemporary musicians.

GARDINER, SAMUEL RAWSON (1829-1902), historian; BA, Christ Church, Oxford, 1851; married daughter of Edward Irving [q.v.], 1856; settled in London to study history of Puritan revolution, supporting himself meanwhile by teaching; lecturer (1872-7) and professor of modern history (1877-85) at King's College, London; published first instalment of his *History of England* (1603-42) in 1863, and last in 1882; collective edition in 10 vols., 1883-4; there followed *The Great Civil War* (3 vols., 1886-91) and *The History of the Commonwealth and Protectorate* (3 vols., 1895-1901); his unfinished *Last Years of the Protectorate* was completed by (Sir) Charles Firth [q.v.], 2 vols., 1909; his historical work shows scientific arrangement of material, minute accuracy, impartiality, but lacks the picturesque style of Froude or Macaulay; wrote also *The Thirty Years War* (1874) and *The Puritan Revolution* (1876) for 'Epochs of English History' series; *Student's History of England* (3 vols., 1890); edited *Constitutional Documents of the Puritan Revolution* (1889); was editor of *English Historical Review*, 1891-1901; director of Camden Society, 1869-97, he edited twelve of its volumes, and also contributed to publications of Navy Records Society, Scottish History Society, and to this Dictionary; awarded civil list pension, 1882; research fellow of All Souls, Oxford, 1884-92, and of Merton, 1892-1902; received many honorary distinctions at home and abroad and was Ford's lecturer, Oxford, 1896.

GARDINER, SIR THOMAS ROBERT (1883-1964), civil servant; educated at Lurgan College, county Armagh, Royal High School, Edinburgh and Edinburgh University; MA; first class, history and economic science, 1905; entered Post Office, 1906; private secretary to the secretary of the Post Office, (Sir) Q. E. C. Murray [q.v.], 1914-17; controller, London Postal Service, 1926-34; deputy director-general, Post Office, 1934-6; director-general, 1936-9, and 1940-5; permanent secretary, Ministry of Home Security, 1939-40; chairman,

Stevenage New Town Development Corporation, 1947-8; chairman, National Dock Labour Board; government director, Anglo-Iranian Oil Co., 1950-3; member of a number of public committees and commissions; KBE, 1936; KCB, 1937; GBE, 1941; GCB, 1954; hon. LLD, Edinburgh, 1949.

GARDNER, ERNEST ARTHUR (1862-1939), classical scholar and archaeologist; brother of Percy Gardner [q.v.]; educated at City of London School and Gonville and Caius College, Cambridge; first class, classical tripos, parts i and ii, 1882-4; director, British School of Archaeology, Athens, 1887-95; professor of archaeology, London, 1896-1929; vice-chancellor, 1924-6; works include *Handbook of Greek Sculpture* (1896-7).

GARDNER, PERCY (1846-1937), classical archaeologist and numismatist; brother of E. A. Gardner [q.v.]; educated at City of London School and Christ's College, Cambridge; first class, classical and moral sciences triposes, 1869; assistant in department of coins and medals, British Museum, 1871; Disney professor of archaeology, Cambridge, 1880; editor, *Journal of Hellenic Studies*, 1880-96; professor of classical archaeology, Oxford, 1887-1925; works include *New Chapters in Greek History* (1892), *History of Ancient Coinage 700-300 B.C.* (1918), and *New Chapters in Greek Art* (1926); FBA, 1903; exponent of 'evolutional' Christianity in *Exploratio Evangelica* (1899), etc.

GARGAN, DENIS (1819-1903), president of Maynooth College; ordained, 1843; professor of humanity, 1845, and of ecclesiastical history, 1859; vice-president, 1885, and president, 1894, of Maynooth; received King Edward VII there, 1903; published *The Ancient Church of Ireland* (1864).

GARNER, THOMAS (1839-1906), architect; fellow pupil of Sir George Gilbert Scott [q.v.] with G. F. Bodley [q.v.], whose partner he became, 1869-97; did much ecclesiastical, domestic, and collegiate architecture in Oxford, Cambridge, and London; in later life joined Church of Rome; joint-author of *The Domestic Architecture of England during the Tudor Period* (1908).

GARNER, WILLIAM EDWARD (1889-1960), chemist; educated at Market Bosworth grammar school and Birmingham (honours in chemistry, 1912) and Göttingen universities; appointed lecturer, University College, London, 1919; reader in physical chemistry, 1924; Leverhulme professor of physical and inorganic chemistry, Bristol, 1927-54; carried out systema-tic study of the kinetics of solid reactions and of heterogeneous catalysis; in 1939-45 associated with notable developments in armaments and munitions; CBE, 1946; FRS, 1937.

GARNETT, CONSTANCE CLARA (1861-1946), translator of Russian classics; born BLACK; educated at Brighton high school and Newnham College, Cambridge; married (1889) Edward Garnett; translations include whole of Turgenev, Dostoevski, Gogol, and (virtually) Chekhov, and Tolstoy's two novels.

GARNETT, JAMES CLERK MAXWELL (1880-1958), educationist and secretary of League of Nations Union; scholar of St. Paul's and Trinity College, Cambridge; sixteenth wrangler, 1902; called to bar (Inner Temple), 1908; examiner, Board of Education, 1904-12; principal, Manchester College of Technology, 1912-20; secretary, League of Nations Union, 1920-38; impelled by strong sense of Christian service; publications include *World Loyalty* (1928); CBE, 1919.

GARNETT, RICHARD (1835-1906), man of letters and keeper of printed books in the British Museum; elder son of Richard Garnett [q.v.]; privately educated; early developed linguistic and literary aptitudes; entered British Museum library, 1857, and won favour of Sir Anthony Panizzi [q.v.]; assistant keeper of printed books and superintendent of the reading room, 1875-90; superintended printing of catalogue; keeper of printed books, 1890-9; his chief acquisitions noticed in *A Description of 300 Notable Books*, privately printed, 1899; president of Library Association, 1892-3; engaged through life in literary work, poetic, critical, and biographical; his most important publications were hitherto unpublished *Relics of Shelley* (1862) and *The Twilight of the Gods* (1888, new edn. 1903), apologues of pleasantly cynical flavour in Lucian's vein; other writings include *Io in Egypt and other Poems* (1859, new edn. 1893); brief biographies of Milton and Carlyle (1877), Emerson (1888), Edward Gibbon Wakefield (1898), and a *History of Italian Literature* (1897); a contributor to this Dictionary; hon. LLD, Edinburgh, 1883; CB, 1895.

GARRAN (formerly GAMMAN), **ANDREW** (1825-1901), Australian journalist and politician; BA, London, 1845; MA, 1848; migrated to Adelaide, 1851; assistant editor (1856) and editor (1873-85) of *Sydney Morning Herald*; member of legislative council of New South Wales, 1887; president of royal commission on strikes, 1890, whose report led to Trades Disputes Conciliation Act, 1892; president of arbitration council, 1892; vice-president of executive council, 1895-8; edited *Picturesque Atlas of Australasia* (1886).

GARRARD, APSLEY GEORGE BENET CHERRY- (1886-1959), polar explorer. [See CHERRY-GARRARD.]

GARRETT, FYDELL EDMUND (1865-1907), publicist; BA, Trinity College, Cambridge, and president of Union, 1887; joined staff of *Pall Mall Gazette*; sent for phthisis cure to South Africa, 1889; intimate with Cecil Rhodes [q.v.] and President Kruger; described experiences in *In Afrikanderland*, 1891; returned to London, writing for *Pall Mall* and *Westminster* gazettes, 1891-5; translated Ibsen's *Brand*, 1894; as editor of *Cape Times*, 1895-1900, he influenced public affairs in South Africa; member of Cape parliament, 1898-1902; advocated united autonomous South Africa; settled in Devonshire, 1904; wrote also *The Story of an African Crisis* (1897).

GARRETT ANDERSON, ELIZABETH (1836-1917), physician. [See ANDERSON.]

GARROD, SIR ALFRED BARING (1819-1907), physician; studied medicine at University College Hospital; MD, 1843; physician and professor of materia medica and therapeutics and a professor of clinical medicine at University College Hospital, 1849-63, and at King's College Hospital, 1863-74; FRCP, 1856; Gulstonian lecturer, 1857; Lumleian lecturer, 1883; FRS, 1858; knighted, 1887; made valuable researches into gout; wrote *Treatise on Gout and Rheumatic Gout* (1859).

GARROD, SIR (ALFRED) GUY (ROLAND) (1891-1965), air chief marshal; educated at Bradfield College and University College, Oxford; leading member of OUDS; commissioned in Leicestershire Regiment, 1914; seconded to Royal Flying Corps, 1915; DFC; MC; permanently commissioned in Royal Air Force, 1919; filled number of Staff College appointments, 1919-39; air commodore, 1936; air vice-marshal, 1939; director of equipment, Air Ministry; air member for training on Air Council, 1940-3; deputy Allied air commander-in-chief, South-East Asia, 1943; commander-in-chief, RAF, Mediterranean and Middle East, 1945; head of RAF delegation, Washington; air chief marshal; retired, 1948; held several business posts; warden, Bradfield, 1959; hon. fellow, University College, Oxford (1917); hon. LLD Aberdeen; OBE, 1932; GBE, 1948; CB, 1941; KCB, 1943.

GARROD, SIR ARCHIBALD EDWARD (1857-1936), physician and biochemist; son of Sir A. B. Garrod [q.v.]; educated at Marlborough, Christ Church, Oxford, and St. Bartholomew's Hospital; assistant physician, 1903; full physician, 1912; consulting physician to Mediterranean forces, 1915-19; KCMG, 1918; regius professor of medicine, Oxford, 1920-7; works include *Inborn Errors of Metabolism* (1909) and *The Inborn Factors in Disease* (1931); FRS, 1910.

GARROD, HEATHCOTE WILLIAM (1878-1960), scholar; educated at Bath College and Balliol College, Oxford; first class, *lit. hum.*, and Newdigate prize, 1901; tutor at Merton College, 1904-25; fellow of Merton from 1901, his rich uncommon personality made him a presiding genius there; professor of poetry, 1923-8; publications include edition of Statius (1906); *Oxford Book of Latin Verse* (1912); *Wordsworth: Lectures and Essays* (1923); edited Keats for Oxford English Texts, 1939; with Mrs Allen completed edition by P. S. Allen [q.v.] of *Letters of Erasmus*; original works include *Oxford Poems* (1912) and *Epigrams* (1946); CBE, 1918; FBA, 1931; hon. fellow, Merton College; hon. D.Litt. Durham, and LLD Edinburgh.

GARSTANG, JOHN (1876-1956), archaeologist; educated at Blackburn grammar school and Jesus College, Oxford; reader in Egyptian archaeology, Liverpool, 1902; professor of methods and practice of archaeology, 1907-41; organized Institute of Archaeology; director, Jerusalem School of Archaeology, 1919-26, and Palestine Department of Antiquities, 1920-6; founded British Institute of Archaeology at Ankara, 1948; excavated in Egypt, Asia Minor, and Sudan; in Jericho, 1930-6; thereafter in Turkey; publications include *The Land of the Hittites* (1910), *Joshua Judges* (1931), *The Heritage of Solomon* (1934), and *Prehistoric Mersin* (1953); CBE, 1949; hon. LLD Aberdeen.

GARSTIN, SIR WILLIAM EDMUND (1849-1925), engineer; born in India; entered Indian public works department, 1872; joined group of Indian engineers appointed to reorganize irrigation system of Egypt, 1885; inspector-general of irrigation, Egypt, 1892; under-secretary of state in Ministry of Public Works, 1892; his notable system of water storage and control comprised construction of great dam at Aswan and barrages at Assiut, Esna, and Zifta; after freeing of Sudan from Dervish rule (1898) cleared Bahr el Jebel and Bahr el Ghazal from 'sudd'; responsible for formation of Sudan irrigation service; supervised care of buildings and antiquities of Egypt; adviser to Public Works department, 1905-8; KCMG, 1897; earned profound gratitude of Egyptian people for his irrigation work.

GARTH, Sir RICHARD (1820–1903), chief justice of Bengal, 1875–86; educated at Eton and Christ Church, Oxford; in university cricket XI, 1839–42; called to bar, 1847; QC, 1866; conservative MP for Guildford, 1866–8; knighted, 1875; able judge but partisan controversialist; opposed to Bengal tenancy bill; promoted Legal Practitioners Act, 1879; PC, 1888; supported Indian National Congress in *A Few Plain Truths about India* (1888).

GARVIE, ALFRED ERNEST (1861–1945), theologian and Church leader; first class, theology, Mansfield College, Oxford, 1892; professor at Hackney and New colleges, London, 1903; principal, New College, 1907, Hackney, 1922, of combined colleges, 1924–33; publications include *The Ritschlian Theology* (1899), *The Christian Doctrine of the Godhead* (1925), *The Christian Ideal for Human Society* (1930), and *The Christian Belief in God* (1932).

GARVIN, JAMES LOUIS (1868–1947), editor of the *Observer*; joined *Newcastle Chronicle*, 1891; *Daily Telegraph*, 1899; wrote as 'Calchas' in *Fortnightly Review*; advocated military preparedness and tariff reform; edited *Outlook*, 1905–6; *Pall Mall Gazette*, 1912–15; *Observer*, 1908–42; established remarkable personal influence; made *Observer* financial success, strong political force, and new pattern of Sunday paper, combining news with full treatment of arts; supported Lloyd George's war policy; criticized Versailles treaty and in *The Economic Foundations of Peace* (1919) sought system Germany could enter as an equal; favoured negotiation with Hitler from strength but insisted Munich must be last concession; strong supporter of 'national' government (1931) and lifelong advocate of coalitions in emergency; wrote for *Sunday Express*, 1942–5; *Daily Telegraph*, 1945–7; edited 13th and 14th editions of *Encyclopaedia Britannica*; published *Life of Joseph Chamberlain* (3 vols., 1932–4); CH, 1941.

GASELEE, Sir ALFRED (1844–1918), general; joined army, 1863; served chiefly in Indian wars and expeditions, 1863–1900; CB, 1891; KCB, 1898; commanded British expeditionary force for relief of legations in Peking, 1900; full general, 1906; commanded Northern army in India, 1907–8; GCB, 1909.

GASELEE, Sir STEPHEN (1882–1943), librarian, scholar, and connoisseur; scholar of Eton and King's College, Cambridge; Pepysian librarian, Magdalene College, 1908–19; fellow, 1909–43; served in Foreign Office, 1916–19; librarian and keeper of the papers there, 1920–43; honorary librarian, Athenaeum, 1928–43; publications include Coptic texts; text of Petronius' *Satyricon* (1910); editions of Apuleius (1915), Parthenius (1916), and Achilles Tatius (1917) for Loeb Classical Library; and *The Oxford Book of Medieval Latin Verse* (1928); presented valuable book collections to Cambridge university library; FBA, 1939; KCMG, 1935.

GASK, GEORGE ERNEST (1875–1951), surgeon; educated at Dulwich College; qualified from St. Bartholomew's Hospital, 1898; FRCS, 1901; council member, 1923–39; assistant surgeon to (Sir) D'Arcy Power [q.v.], 1907; consulting surgeon, Fourth Army; DSO, 1917; CMG, 1919; formed surgical professorial unit, Bart's, after 1918; retired 1935; with J. Paterson Ross wrote pioneer study of *The Surgery of the Sympathetic Nervous System* (1934); temporary surgeon, Radcliffe Infirmary, Oxford, 1939–45; succeeded Lord Moynihan [q.v.] as chairman, editorial committee, *British Journal of Surgery*.

GASKELL, WALTER HOLBROOK (1847–1914), physiologist; born at Naples; BA, Trinity College, Cambridge (26th wrangler), 1869; studied physiology at Leipzig, 1874–5; MD, Cambridge, 1878; FRS, 1882; university lecturer in physiology, 1883; fellow of Trinity Hall, 1889; his physiological researches revolutionized current ideas of action of the heart, and of cardiac disease; aroused great controversy by his theory of the origin of vertebrates.

GASQUET, FRANCIS NEIL (in religion Dom Aidan) (1846–1929), cardinal and historian; entered Benedictine novitiate at Downside priory, 1866; master at Downside school, 1870; priest, 1874; prior, 1878–85; his priorship turning-point in history of Downside; resigned owing to ill health; engaged in historical research; his publications include *Henry VIII and the English Monasteries* (1888–9); chairman of papal commission for reorganization of English Benedictine Congregation, 1899; president of papal commission for revising text of vulgate, 1907; cardinal deacon, 1914; successfully countered propaganda of central powers; prefect of archives of Holy See, 1917; librarian of Holy Roman Church, 1919; cardinal priest, 1924; died in Rome.

GASTER, MOSES (1856–1939), scholar and rabbi; born and educated in Bucharest; rabbinical diploma, Breslau, 1881; expelled from Romania for activities on behalf of Jews; naturalized British subject, 1893; chief rabbi of Sephardi Jews in England, 1887–1918; a founder and president of English Zionist Federation; publications include *The Sword of Moses* (1896), *Rumanian Bird and Beast Stories* (1915), and *Samaritan Eschatology* (1932).

GATACRE, SIR WILLIAM FORBES (1843-1906), major-general; joined 77th Foot in Bengal, 1862; instructor in surveying at Sandhurst, 1875-9; commanded regiment at Secunderabad, 1884; in command of Bombay district, 1894-7, and 3rd brigade of relief force in Chitral expedition, 1895; CB; received Kaisar-i-Hind gold medal for services in Bombay plague, 1897; commanded brigade in advance up Nile for recovery of Khartoum; major-general and KCB; known as 'General Backacher', 1898; in South African war he defended railway from East London to Bethulie, 1899; was forced to retreat at Stormberg (Dec.); joined main army at Bloemfontein, Mar. 1900; occupied Dewetsdorp and sent detachment on to Reddersburg; detachment surrounded and surrendered owing to Gatacre's failure to relieve it (Apr.); removed from command; commanded Eastern district at Colchester, 1900-3; explored rubber forests in Abyssinia, 1905; died of fever at Iddeni.

GATENBY, JAMES BRONTË (1892-1960), zoologist; born in New Zealand; educated at Wanganui Collegiate School, St. Patrick's College, Wellington, and Jesus College, Oxford; first class, zoology, 1916; professor of zoology and comparative anatomy, Trinity College, Dublin, 1921-59, of cytology, 1959-60; an inspiring teacher; studied cytoplasmic structures in many animals from protozoans to man.

GATER, SIR GEORGE HENRY (1886-1963), administrator; educated at Winchester and New College, Oxford; diploma in education, Oxford, 1909; assistant director of education, Nottingham, 1912; served in army in Gallipoli, Egypt, and France, 1914-18; DSO and bar; croix de guerre; commander of the Legion of Honour; CMG, 1918; director of education, Lancashire, 1919-24; education officer, London County Council, 1924-33; clerk to LCC, 1933-9; succeeded (Sir A. C.) Cosmo Parkinson [q.v.] as permanent under-secretary of state for the colonies, 1939, and 1942-7; short periods in Ministry of Home Security and Ministry of Supply, 1940-2; chairman, School Broadcasting Council, and member, BBC General Advisory Council; hon. fellow, New College and the Royal College of Music; warden of Winchester College, 1951-9; knighted, 1936; KCB, 1941; GCMG, 1944.

GATES, REGINALD RUGGLES (1882-1962), botanist, geneticist, and anthropologist; born near Middleton, Nova Scotia; ancestors included Sir John Gates, and Major-General Horatio Gates [qq.v.]; educated at Middleton high school and Mount Allison University, Sackville, Nova Scotia; first class, science, 1903; demonstrator in botany, McGill University;

senior fellow, Chicago University, Ph.D., 1908; wrote many books on genetics, including *The Mutation Factor in Evolution, with particular reference to Oenothera* (1915); paid first visit to Europe, 1910; worked at Imperial College of Science, London, 1911; lecturer in biology, St. Thomas's Hospital, 1912-14; associate professor of zoology, California University, 1915-16; reader, botany department, King's College, London, 1919; professor, 1921-42; FRS, 1931; research fellow in botany and anthropology, Harvard University; made outstanding contributions to anthropology and human genetics as well as to botany and cytology; married Dr Marie Charlotte Carmichael Stopes [q.v.]; marriage annulled; vice-president, Linnean Society; vice-president, Royal Anthropological Institute; fellow, King's College, London; LLD, Mount Allison; D.Sc., London.

GATHORNE-HARDY, GATHORNE, first EARL OF CRANBROOK (1814-1906), statesman; educated at Eton and Oriel College, Oxford; BA, 1836; MA, 1861; honorary fellow, 1894; called to bar, 1840; obtained lead on sessions and at parliamentary bar; conservative MP for Leominster, 1856-65; under-secretary for Home Department, 1858-9; active champion of Church of England; defeated Gladstone in parliamentary election for Oxford University, 1865; president of the Poor Law Board and PC, 1866; introduced poor law amendment bill, 1867; home secretary after Hyde Park riots, May 1867-8; dealt firmly with Fenian conspirators; in opposition warmly attacked Irish Church disestablishment bill, 1869; secretary of state for war under Disraeli, 1874-8; opposed public worship regulation bill, 1874; introduced regimental exchanges bill, 1875; supported Disraeli's pro-Turkish policy; succeeded Lord Salisbury as secretary for India, 1878; raised to peerage as Viscount Cranbrook, 1878; sanctioned Vernacular Press Act of 1878; supported Lord Lytton's forward policy on North-West frontier; justified coercion of Amir Shere Ali; approved separation of Kandahar from Kabul, 1880; GCSI, 1880; sat on royal commission on cathedral churches, 1879-85; lord president of the Council, 1885-92; created earl, 1892; denounced Gladstone's home rule bill, 1893; hon. DCL, Oxford, 1865; hon. LLD, Cambridge, 1892; good debater and platform speaker; ardent sportsman and broad churchman.

GATTY, ALFRED (1813-1903), vicar of Ecclesfield and author; BA, Exeter College, Oxford, 1836; MA, 1839; DD, 1860; vicar of Ecclesfield from 1839 till death; sub-dean of York Minster, 1862; published *The Bell: its Origin and History* (1847), *Sheffield Past and Present* (1873), verse, biographies, and sermons.

GAUVAIN, SIR HENRY JOHN (1878-1945), surgeon and specialist in tuberculosis; first class, natural sciences, St. John's College, Cambridge, 1902; qualified at St. Bartholomew's Hospital, 1906; medical superintendent, Lord Mayor Treloar Cripples' Home (later Orthopaedic Hospital), Alton, 1908-45; established first hospital school (1912), marine branch (1919), private clinics (1925); knighted, 1920.

GEDDES, AUCKLAND CAMPBELL, first BARON GEDDES (1879-1954), public servant; brother of Sir Eric Geddes [q.v.]; educated at George Watson's College and the university, Edinburgh; qualified in medicine, 1903; MD, 1908; professor of anatomy, Royal College of Surgeons, Dublin, 1909; McGill University, Montreal, 1913-14; director of recruiting, War Office, 1916; CB, KCB, and PC, 1917; MP, Basingstoke, 1917-20; minister of national service, 1917-19; president, Board of Trade, 1919-20; ambassador in Washington, 1920-4; GCMG, 1922; chairman, royal commission on food prices (1924), of Rio Tinto Company and founding chairman, Rhokana Corporation; baron, 1942; although blind, he wrote *The Forging of a Family* (1952).

GEDDES, SIR ERIC CAMPBELL (1875-1937), politician, administrator, and man of business; educated at Merchiston Castle School, Edinburgh; deputy general manager, North Eastern Railway, 1914; inspector-general, transportation, all theatres of war, 1916-17; controller of navy, May-July 1917; first lord of Admiralty, 1917-18; unionist MP, Cambridge borough, 1917-22; minister of transport, 1919-21; carried through railway amalgamation; chairman of 'Geddes Axe' committee on national economy, 1921-2; chairman of Dunlop Rubber Company and Imperial Airways; knighted, 1916; PC, 1917; GCB, 1919.

GEDDES, SIR PATRICK (1854-1932), biologist, sociologist, educationist, and town planner; educated at Perth Academy; studied under T. H. Huxley [q.v.] and in Paris; microscopy made impossible by attack of blindness; demonstrator in botany, Edinburgh; professor of botany, Dundee, 1889-1914; established 'world's first sociological laboratory' (1892) in Outlook Tower, Castlehill, Edinburgh, where he arranged stories of regional interpretation; interests centred increasingly on civics and town planning; his Cities and Town Planning Exhibition shown in Europe and India; professor of civics and sociology, Bombay, 1920-3; settled (1924) at Montpellier and built an unofficial 'Scots College'; an evolutionist; his concept of the relationship environment-function-organism led to place-work-folk; collaborated

with Sir J. A. Thomson in *The Evolution of Sex* (1889), etc.; knighted, 1932.

GEDYE, (GEORGE) ERIC (ROWE) (1890-1970), journalist; educated at Clarence School, Weston-super-Mare, and Queen's College, Taunton; matriculated, London University; served in Gloucestershire Regiment in France, 1914-18; served on Inter-Allied Rhineland High Commission, 1919-22; appointed local correspondent (Cologne) of *The Times* and *Daily Mail*, 1922; 'special correspondent' of *The Times*, 1923-5; Central European correspondent, *The Times*, Vienna, 1925; moved to *Daily Express*, 1926; and to *Daily Telegraph*, 1929; expelled from Austria by Gestapo, 1938; worked for *Daily Telegraph* in Prague; Moscow correspondent of the *New York Times*, 1939; correspondent in Turkey, 1940-1; worked again in Vienna for the *Daily Herald*, the *Observer*, and the *Manchester Guardian*, 1945; MBE, 1946; head of evaluation for Radio Free Europe, 1954-61; publications include *A Wayfarer in Austria* (1928), *The Revolver Republic* (1930), *Heirs to the Hapsburgs* (1932). and *Fallen Bastions* (1939).

GEE, SAMUEL JONES (1839-1911), physician; studied medicine at University College, London; MB, 1861; MD, 1865; FRCP, 1870; physician and lecturer at St. Bartholomew's Hospital from 1868 to death; Gulstonian (1871), Bradshaw (1892), and Lumleian (1899) lecturer at Royal College of Physicians; had wide knowledge of early medical literature; published *Auscultation and Percussion* (1870) and *Medical Lectures and Aphorisms* (1902).

GEIKIE, SIR ARCHIBALD (1835-1924), geologist; educated at Edinburgh high school and university; joined Scottish branch of Geological Survey, 1855; director in Scotland, 1867; director-general for Great Britain, 1882; retired, 1901; Murchison professor of geology and mineralogy, Edinburgh, 1871-81; FRS, 1865; knighted, 1891; KCB, 1907; OM, 1913; his greatest contributions to geological science concerned with past volcanic history of Great Britain; furthered study of Scottish glacial deposits; works include *History of Volcanic Action during the Tertiary Period in Britain* (1888) and *Ancient Volcanoes of Great Britain* (1897).

GEIKIE, JOHN CUNNINGHAM (1824-1906), religious writer; Presbyterian minister in Canada and in England, 1848-73; hon. LLD, Edinburgh, 1891; awarded civil list pension, 1898; works include *Life and Words of Christ* (2 vols., 1877) and *Hours with the Bible* (10 vols., 1881-4).

GELL, SIR JAMES (1823-1905), Manx lawyer and judge; admitted to Manx bar, 1845; edited for government Manx statute laws (1836-48); attorney-general, 1866-98; first deemster, 1898; clerk of the rolls, 1900-5; deputy-governor, 1897-1902; knighted, 1877; CVO, 1902.

GELLIBRAND, SIR JOHN (1872-1945), major-general; born in Tasmania; educated at King's School, Canterbury, and Sandhurst; retired (captain) from Manchester regiment to Tasmania, 1912; served with Australian forces at Gallipoli; brilliant brigade commander in France; commanded 3rd Australian division, 1918; DSO, 1916; KCB, 1919; commissioner of police, Victoria, 1920-2; nationlist member (Denison, Tasmania), federal parliament, 1925-8.

GENÉE, DAME ADELINE (1878-1970), ballet dancer; born in Jutland of parents named Jensen; came under guardianship of an uncle with stage name, Genée, and trained as dancer; first public appearance in Christiania (Oslo), 1888; danced in Berlin and Munich, where she danced Swanilda in *Coppélia*, 1896; settled in London, appearing at the Empire, Leicester Square, 1897; undertook tours of America, Australia, and New Zealand and visited Copenhagen and Paris; danced before Edward VII and Queen Alexandra, 1905; danced in London at the Coliseum, Alhambra, and the Albert Hall, 1914-17; first president, the Association of Operatic Dancing (later the Royal Academy of Dancing), formed to improve standards of teaching, 1920-54; Genée gold medal instituted, 1931; sponsored and led the English Ballet Company on visit to Copenhagen, 1932 (first group of British ballet dancers to go abroad); DBE, 1950; commander of the Order of Dannebrog; hon. D.Mus., London.

GEORGE V (1865-1936), KING OF GREAT BRITAIN, IRELAND, AND THE BRITISH DOMINIONS BEYOND THE SEAS, EMPEROR OF INDIA; born at Marlborough House 3 June 1865, second son of the Prince and Princess of Wales, later King Edward VII and Queen Alexandra; at Windsor Castle (7 July) baptized George Frederick Ernest Albert; passed his childhood in an atmosphere of sustained happiness and affection and remained devoted to his parents throughout life; with his elder brother, the Duke of Clarence [q.v.], joined the *Britannia* as naval cadet, 1877; set his heart upon a naval career; in *Bacchante* made a cruise to the West Indies (1879-80) and to South America, South Africa, Australia, Japan, and China (1880-2); separating for first time from his brother joined corvette *Canada* for service on North America station, 1883; promoted sub-lieutenant, 3 June 1884; KG,

1884; secured first class in seamanship, gunnery, and torpedo work at Royal Naval College, Greenwich, 1885; served on Mediterranean station until 1888; received freedom of City of London, June 1889; commanded gunboat *Thrush*, 1890-1; promoted commander, 1891; his naval career ended by death of Duke of Clarence, 14 Jan. 1892; created Duke of York (1892) and provided with apartments in St. James's Palace ('York House') and York Cottage at Sandringham; in Chapel Royal, St. James's Palace, 6 July 1893, married Princess Victoria Mary (May) of Teck, who had previously been engaged to his elder brother; their close companionship henceforward exemplified a lofty standard of family life; the following children were born to them: Prince Edward (1894), Prince Albert (1895), Princess Mary (1897), Prince Henry (1900), Prince George (1902), and Prince John (1905); on death of Queen Victoria (22 Jan. 1901) became Duke of Cornwall and his public duties increased; obtained the services of Sir Arthur Bigge (later Lord Stanfordham, q.v.) whose knowledge and devotion proved inestimable asset; with the Duchess toured Australia and New Zealand (1901), opening the new Commonwealth parliament in Melbourne (9 May); on return journey visited South Africa and crossed and recrossed Canada; created Prince of Wales and Earl of Chester, 9 Nov. 1901; maintained uniformly harmonious relations with his father who put state papers at his disposal; formed habit of attending parliamentary debates; played golf and lawn tennis; perfected his shooting, sailed his famous yacht *Britannia*, and collected postage stamps of British Empire; in the last three pastimes he was an expert in his own right; paid several visits to the courts of Europe; with the Princess made an extensive tour of India (1905-6); visited Canada (1908).

On death of his father was proclaimed King, 6 May 1910; felt himself ill equipped in face of a complex political situation and little known to his subjects; profoundly impressed by his coronation in Westminster Abbey, 22 June 1911; celebrations in London followed by a review at Spithead of largest naval fleet ever assembled and by visits to Ireland and Wales; left with the Queen in Nov. for a state visit to India; at the magnificent coronation durbar (12 Dec.) at Delhi announced that the seat of government was to be transferred there from Calcutta. On 16 Nov. 1910 Asquith and Crewe obtained from him a 'hypothetical understanding' that should need arise in the next parliament he would agree to create a sufficient number of peers to secure the passage of the parliament bill designed to restrict the powers of the House of Lords; when this was revealed the Lords accepted the bill (10 Aug. 1911) in an atmosphere of intense political excitement; the bill had the effect of increasing the responsibility of the King when a bill, thrice rejected by the Lords, came up for royal assent, since he alone must decide whether an appeal to the country would be justified; this situation became imminent in 1913 when the Lords twice rejected the home rule bill; as the Irish situation worsened, the King urged restraint on all in an endeavour to prevent civil strife; invited representatives of all parties to a conference at Buckingham Palace at which the speaker presided, July 1914; conference failed but the controversy was laid aside upon the outbreak of the 1914-18 war.

In 1912 the King had told Prince Henry of Prussia, the Emperor's brother, that in the event of Austria and Germany going to war with Russia and France, England would come to the assistance of the latter; on 26 July 1914 when war seemed imminent he informed Prince Henry that England still hoped not to be drawn in; this was misinterpreted as an assurance of neutrality; resided for most of war at Buckingham Palace where he ordered that no alcohol should be consumed and maintained strict adherence to rationing regulations; paid numerous visits to hospitals, naval and military formations, factories, etc.; distributed 58,000 decorations; paid five visits to Grand Fleet and seven to armies in France and Belgium; fractured his pelvis when thrown from his horse during a visit in 1915; adopted the name of Windsor for royal house and family, 1917; at the end of the war he and the Queen were greeted with unparalleled demonstrations of affection.

With Queen visited Belfast (June 1921) to open Ulster parliament and earnestly appealed to all Irishmen for forbearance; urged Lloyd George to make fresh overtures, which resulted in a truce; at a critical stage in subsequent negotiations (Sept.) secured a more conciliatory tone in note from Lloyd George to Mr de Valera and an invitation to Sinn Fein representatives to meet Lloyd George; agreement signed, 6 Dec. On resignation (1923) of Bonar Law, who intimated that he would prefer not to tender advice on his successor, summoned Stanley Baldwin in preference to Lord Curzon [q.v.] who was in the House of Lords; on sending for Ramsay MacDonald (Jan. 1924), first labour prime minister, assured him that he might count on his assistance in every way; was immediately at home with the labour ministers; granted dissolution to MacDonald in Oct.; asked his people to forget all bitterness after general strike, 1926; by the Statute of Westminster (1931) parliament ceased to control the overseas dominions and the King alone constituted the bond between them and the home country; in broadcasts each Christmas Day from 1932 he

sent a personal message of kindliness to British homes throughout the world and asserted a simple faith in the continued guidance of a divine Providence; established singular hold upon the affections of his people; contracted a streptococcal infection (Nov. 1928) which necessitated an operation for drainage of chest (12 Dec.) and for some weeks the issue remained in doubt; although not completely recovered attended a service of thanksgiving in St. Paul's Cathedral, 7 July 1929; when MacDonald tendered the resignation of his government owing to internal dissension arising out of the financial crisis (Aug. 1931), urged him to consider an all-party government and called a conference of the party leaders which resulted in the 'national' government; had by now a great store of political experience so that ministers were apt not only to render but to seek advice; the extent to which he had become the father of his people was demonstrated when he celebrated his silver jubilee, 1935; died at Sandringham, 20 Jan. 1936, after an illness which was short and peaceful in its close; the funeral took place at Windsor, 28 Jan.

In person he was neatly made and slightly below middle height; his voice strong and resonant, his eyes arrestingly blue; his naval training had implanted habits of discipline, and his mode of life was extremely regular; he was quick to check infractions of traditional observances and duties; a sound churchman, with the habit of daily Bible reading, he always attended Sunday morning service; keenly interested in all three fighting Services and conscientious in his perusal of state documents; in private life his pursuits were those of the English country gentleman; in his mistrust of cleverness, his homespun common sense, dislike of pretension, ready sense of ludicrous and devotion to sport, he possessed qualities which appealed to Englishmen of all classes.

GEORGE VI (1895-1952), KING OF GREAT BRITAIN, IRELAND, AND THE BRITISH DOMINIONS BEYOND THE SEAS; born at York Cottage, Sandringham, 14 Dec. 1895; second son of Duke and Duchess of York, afterwards King George V and Queen Mary; baptized Albert Frederick Arthur George; served in navy, 1909-17; qualified as pilot in RAF, 1919; at Trinity College, Cambridge, 1919-20; became president of Industrial Welfare Society and made industrial areas his special interest; inaugurated Duke of York camps for boys from public schools and industry, 1921; KG, 1916; created Duke of York, 1920; KT, 1923; PC, 1925; married in Westminster Abbey, 26 Apr. 1923, Lady Elizabeth Angela Marguerite Bowes-Lyon, daughter of fourteenth Earl of Strathmore and Kinghorne [q.v.]; together they established a pattern of family life and devotion to duty; two children were born to them; Princess Elizabeth (1926) and Princess Margaret (1930); with Duchess toured New Zealand and Australia, 1927; proclaimed King, 12 Dec. 1936, upon abdication of his elder brother, King Edward VIII; crowned in Westminster Abbey, 12 May 1937; with Queen paid state visit to France, 1938, and visited Canada and United States, 1939; despite speech impediment resumed father's tradition of Christmas Day broadcast to Empire, 1939; entirely confident of outcome of the war; remained in London throughout and was indefatigable in sympathetic visiting of bombed areas; created George Cross and Medal mainly for civilian gallantry, 1940; popularity of monarchy manifest during celebrations at end of European war; with Queen and Princesses visited Southern Africa, 1947; underwent right lumbar sympathectomy operation, 1949; opened Festival of Britain, 3 May 1951; his left lung removed, 23 Sept. 1951; died in his sleep at Sandringham, 6 Feb. 1952; buried in St. George's chapel, Windsor; earned respect and affection of nation for courage in assuming monarchy for which he had not been trained; had the simple religious faith and many of the characteristics of his father.

GEORGE EDWARD ALEXANDER EDMUND, DUKE OF KENT (1902-1942), fourth son of King George V; served in navy until 1929; visited Canada (1927), South America (1931), and South Africa (1934); created Duke of Kent, 1934; staff officer, Training Command, Royal Air Force, 1940; killed in air crash.

GEORGE WILLIAM FREDERICK CHARLES, second DUKE OF CAMBRIDGE, EARL OF TIPPERARY AND BARON CULLODEN (1819-1904), field-marshal and commander-in-chief of the army; only son of Adolphus Frederick, first duke [q.v.]; born at Hanover; sent to England, 1830; GCH, 1825; KG, 1835; served in Hanoverian army, 1836; settled in England on Queen Victoria's accession, 1837; contracted a morganatic marriage, 1840; made brevet-colonel; commanded 17th Lancers at disturbances at Leeds, 1842; commanded troops at Corfu, 1843-5; GCMG and major-general, 1845; commanded Dublin district, 1847-52; succeeded to dukedom, 1850; KP, 1851; commanded a division in Crimea, 1854; present at Alma and Inkerman; GCB, 1855; succeeded Lord Hardinge [q.v.] as general commanding in chief, 1856; general and PC, 1856; colonel of artillery and engineers, 1861; president of National Rifle Association, 1859; helped to found Staff College; field-marshal, 1862; as general commanding in chief he was subordinated to war minister by War Office Act of

1870; opposed such innovations as short service, formation of army reserve, and linking of battalions; commander-in-chief with sole control of supply, 1887; difficulties with secretary for war led to inquiry into naval and military administration (1888-90), and to his enforced resignation, 1895; as chief personal ADC to Queen Victoria he undertook for her many social duties; opened London international exhibition, 1862; president of Christ's Hospital and London Hospital for over fifty years; ranger of Hyde Park, 1852, and of Richmond Park, 1857; KT, 1881; elder brother of Trinity House, 1885; received freedom of City of London, 1857; paid last visit to Germany, 1903.

GEORGE, DAVID LLOYD, first EARL LLOYD-GEORGE OF DWYFOR (1863-1945), statesman. [See LLOYD GEORGE.]

GEORGE, SIR ERNEST (1839-1922), architect; articled to London architect; joined Royal Academy Schools and won gold medal for architecture, 1859; in partnership with Harold Peto (until 1893), obtained countless commissions for elaborate domestic architecture which he executed in Tudor and Jacobean styles; office became fashionable training-ground for young architects; 'discovered' Netherlands for nineteenth century; skilfully adapted details of Flemish and Dutch work of early Renaissance for use in London; his practice of large-scale domestic architecture ruined by liberal legislation from 1906 onwards; retired, 1920; knighted, 1911; RA, 1917.

GEORGE, GWILYM LLOYD-, first VISCOUNT TENBY (1894-1967). [See LLOYD-GEORGE.]

GEORGE, HEREFORD BROOKE (1838-1910), historical writer; BA, New College, Oxford, 1860; MA, 1862; tutor, 1867-91; pioneer of military history at Oxford; published *Battles of English History* (1895), *Napoleon's Invasion of Russia* (1899), *New College, 1856-1906* (1906); made first ascent of Gross Viescherhorn, 1862; edited *Alpine Journal*, 1863-7; published *The Oberland and its Glaciers* (1866); lost fortune by failure of West of England and South Wales Bank, at Bristol, 1880.

GEORGE, LADY MEGAN LLOYD (1902-1966), politician. [See LLOYD-GEORGE.]

GERARD, (JANE) EMILY (1849-1905), novelist; sister of Sir Montagu Gilbert Gerard [q.v.]; married Chevalier Miecislas de Laszowski, 1869; lived in Galicia; with sister Dorothea collaborated in four novels; wrote independently six novels, including *The Voice of a Flower* (1893).

GERARD, SIR MONTAGU GILBERT (1842-1905), general; of Catholic parentage; joined army, 1864; served in second Afghan war, 1878-80; present at Kassassin and Tel-el-Kebir and CB, 1882; sent on secret mission to Persia, 1881 and 1885; British military attaché at St. Petersburg, 1892; negotiated in Pamirs boundary dispute, 1895; commanded first-class district, Bengal, 1899; KCSI, 1897; KCB, 1902; general, 1904; chief British attaché in Manchuria in Russo-Japanese war, 1904; died at Irkutsk on way home; published *Diaries of a Soldier and a Sportsman* (1903).

GERE, CHARLES MARCH (1869-1957), artist; studied at Gloucester and Birmingham art schools; illustrated books for William Morris and St. John Hornby [qq.v.]; settled in Painswick, 1904; painted landscapes with figures in oil, tempera, and water-colour; most successful with small landscapes of Cotswold countryside and water-colour portraits of children; ARA, 1934; RA, 1939; paintings to be seen in Tate and galleries at Birmingham and Liverpool.

GERMAN, SIR EDWARD (1862-1936), composer (originally EDWARD GERMAN JONES); studied at Royal Academy of Music; works include incidental music for Shakespearian productions, *Merrie England* (1902), *A Princess of Kensington* (1903), and *Tom Jones* (1907); wrote march and hymn for coronation, 1911; knighted, 1928; combined artistic achievement with popular appeal.

GERTLER, MARK (1891-1939), painter; born in Spitalfields of Austro-Jewish parents; studied at Slade School; by his talents, vivacity, and exotic beauty gained early entry into artistic and intellectual circles; work profoundly original, of masterly craftsmanship, large and firm design, in rich and harmonious colour; committed suicide.

GIBB, SIR ALEXANDER (1872-1958), engineer; educated at Rugby; worked with (Sir) John Wolfe-Barry [q.v.]; then joined family firm; constructed Rosyth naval base; chief engineer, ports construction, France and Belgium, 1916-18; director-general, civil engineering, Ministry of Transport, 1919-21; established consulting firm, 1922; designs included London Zoo aquarium, Singapore naval base, Cook graving dock, Sydney; collaboration in Galloway hydro-electric scheme and Mulberry harbour; CB and KBE, 1918; GBE, 1920; FRS, 1936; hon. LLD, Edinburgh.

GIBB, SIR CLAUDE DIXON (1898-1959), engineer; born in South Australia; studied at South Australian School of Mines and Adelaide University; graduated in engineering, 1923; joined firm of Sir C. A. Parsons [q.v.] in England, 1924; chief engineer, 1929; general manager, 1937; joint managing director, 1943; with Ministry of Supply, 1940-5; as chairman and managing director from 1945 re-equipped and expanded Parsons; collaborated in first designs for gas-cooled nuclear power plants; formed Nuclear Power Plant Company; FRS, 1946; knighted, 1945; KBE, 1956; honorary degrees, London and Durham.

GIBB, ELIAS JOHN WILKINSON (1857-1901), orientalist; educated at Glasgow University; early studied Arabic, Persian, and Turkish languages and literatures; translated Ottoman prose and verse, 1879-84; published a detailed *History of Ottoman Poetry*, vol. i, 1900 (vols. ii-vi edited by E. G. Browne [q.v.] after Gibb's death, 1902-9); fine oriental library divided among British Museum, Cambridge University, and British Embassy at Constantinople.

GIBBINGS, ROBERT JOHN (1889-1958), wood-engraver, author, and book designer; studied at Slade School and Central School of Arts and Crafts; first honorary secretary of and exhibited with Society of Wood Engravers, 1920; owned Golden Cockerel Press, 1924-33; employed engravers such as Eric Ravilious and Eric Gill [qq.v.]; himself illustrated nineteen out of the seventy-two books produced; lecturer in book production, Reading University, 1936-42; wrote and ilustrated *Sweet Thames Run Softly* (1940) and other 'river books'.

GIBBINS, HENRY DE BELTGENS (1865-1907), writer on economic history; born in Cape Colony; Cobden prizeman at Oxford, 1890; D.Litt., Dublin, 1896; wrote works on industrial and commercial history; edited Methuen's 'Social Questions of the Day' series, 1891.

GIBBON, SIR (IOAN) GWILYM (1874-1948), civil servant; served in Local Government Board, 1903-19; assistant secretary, Ministry of Health, 1919; director, local government division, 1934-5; leader in framing Rating and Valuation (1925) and Local Government (1929) Acts; fostered expansion of social services and town planning; knighted, 1936; left about £50,000 to Nuffield College, Oxford.

GIBBS, HENRY HUCKS, first BARON ALDENHAM (1819-1907), merchant and scholar; MA, Exeter College, Oxford, 1844; joined family firm of bankers, 1843; head of firm, 1875;

director of Bank of England, 1853-1901; governor, 1875-7; wrote many pamphlets advocating bimetallism; published *A Colloquy on Currency*, 1893; helped to found *St. James's Gazette*, 1880; conservative MP for City of London, 1891-2; served on royal commission on stock exchange, 1877-8; baron, 1896; benefactor to Keble College, Oxford; leading member of English Church Union from 1862; bought advowson of and restored church at Aldenham; helped to restore St. Albans Abbey; fond of shooting, he lost right hand in gun accident, 1864; helped in preparation of *New English Dictionary*; edited works for Early English Text Society and Roxburgh Club; Spanish scholar and bibliophile; trustee of National Portrait Gallery; FSA, 1885.

GIBBS, Sir PHILIP ARMAND HAMILTON (1877-1962), writer; educated at home; worked for publishing house of Cassell; editor of Tillotson's literary syndicate; literary editor, *Daily Mail*, 1902; moved to *Daily Express*, and then to *Daily Chronicle*, 1908; also worked with *Daily Graphic*; war correspondent during 1914-18 war; KBE, 1920; chevalier of the Legion of Honour; toured United States lecturing, 1919; resigned from *Daily Chronicle*, 1920; edited *Review of Reviews*, 1921-2; war correspondent with *Daily Sketch*, 1939; publications include *Founders of the Empire* (1899), *The Street of Adventure* (1909), *The Soul of War* (1915), *Realities of War* (1920), *The Middle of the Road* (1922), *Adventures in Journalism* (1923), *Since Then* (1930), *Across the Frontiers* (1938), *America Speaks* (1942), *The Pageant of the Years* (1946), and *Life's Adventure* (1957).

GIBBS, VICARY (1853-1932), genealogist and gardener; son of first Lord Aldenham [q.v.]; educated at Eton and Christ Church, Oxford; conservative MP, St. Albans, 1892-1904; developed famous gardens at Aldenham House; edited (1910-20) and financed first five volumes of new and scholarly edition of *Complete Peerage*.

GIBSON, EDWARD, first BARON ASHBOURNE (1837-1913), lord chancellor of Ireland; BA, Trinity College, Dublin; called to Irish bar, 1860; QC, 1872; conservative MP, Dublin University, 1875; Irish attorney-general, 1877-80; baron, 1885; lord chancellor of Ireland with seat in cabinet, 1885, 1886-92, 1895-1905.

GIBSON, GUY PENROSE (1918-1944), airman; educated at St. Edward's School, Oxford; commissioned in Royal Air Force, 1936; took part in first attack (on Kiel canal) of war; served in Fighter Command, 1940-2; wing commander, 1942; superb commander of No. 106 (bomber) squadron, 1942-3; DSO and bar; commanded special squadron for, and led successful, attack on Möhne dam, May 1943; VC; operations officer, No. 55 base, 1944; crashed in Holland.

GIBSON, Sir JOHN WATSON (1885-1947), contracting engineer; worked for Lord Cowdray [q.v.] in charge of construction of Sennar dam and irrigation works in Sudan; founded own firm working for Sudan and Egyptian governments; directed construction of Mulberry harbours for invasion of Normandy; knighted, 1945.

GIBSON, WILFRID WILSON (1878-1962), poet; educated at private schools; published *The Stonefolds* (1907); moved to London, 1912; contributed to the five volumes of *Georgian Poetry* (1912-22) of (Sir) Edward Marsh [q.v.]; helped, with Rupert Brooke [q.v.] and others, to produce *New Numbers*; legacy from Rupert Brooke enabled him to live as a poet; served in Army Service Corps, 1917-19; published *Collected Poems* (1926), and fourteen other books including *The Alert* (1941) and *Within Four Walls* (1950).

GIBSON, WILLIAM PETTIGREW (1902-1960), keeper of the National Gallery; educated at Westminster and Christ Church, Oxford; abandoned medicine for history of art; lecturer and assistant keeper, Wallace Collection, 1927-36; reader, history of art, London University, and deputy director, Courtauld Institute, 1936-9; keeper of National Gallery, 1939-60.

GIFFARD, Sir GEORGE JAMES (1886-1964), general; educated at Rugby and the Royal Military College, Sandhurst; commissioned into the Queen's Royal Regiment, 1906; seconded to King's African Rifles, 1911; served in East Africa, 1913-18; DSO; croix de guerre; selected for Staff College, Camberley, 1919; served in number of staff posts, 1920-33; aide-de-camp to the King, 1935-6; major-general; inspector-general, African Colonial Forces, 1936; CB, 1938; military secretary to secretary of state for war, Leslie (later Lord) Hore-Belisha [q.v.], 1939; commander-in-chief, West Africa, 1940; general, KCB, 1941; commanded Eastern Army, India, 1943; commander-in-chief, 11th Army Group, under Lord Louis (later Earl) Mountbatten, 1943; gave firm support to William (later Viscount) Slim [q.v.] and the 14th Army; disagreement with Mountbatten led to dismissal, 1944; GCB; aide-de-camp general to the King, 1945-6; president, Army Benevolent Fund; colonel, the Queen's Royal Regiment, 1945-54.

GIFFARD, HARDINGE STANLEY, first EARL OF HALSBURY (1823-1921), lord chancellor;

son of Stanley Lees Giffard [q.v.]; BA, Merton College, Oxford; called to bar (Inner Temple), 1850; joined South Wales circuit, 1851; practised at Old Bailey and at Middlesex sessions at Clerkenwell; junior prosecuting counsel at Central Criminal Court, 1859; QC, 1865; leading counsel for Governor E. J. Eyre [q.v.], 1867-8; second counsel for Tichborne claimant, Arthur Orton [q.v.], 1871-2; solicitor-general and knighted, 1875; led for crown in *Franconia* case, 1876; conservative MP, Launceston, 1877; took active part in parliament and law courts over case of Charles Bradlaugh [q.v.]; greatest forensic triumph, case of *Belt* v. *Lawes*, 1882; lord chancellor, 1885-6, 1886-92, 1895-1905; baron, 1885; earl, 1898; largely responsible for Land Transfer Act (1897) and Criminal Evidence Act (1898); presided over production of complete digest of *Laws of England* (1905-16); led 'diehards' among peers against parliament bill, 1911.

GIFFEN, SIR ROBERT (1837-1910), economist and statistician; apprenticed to lawyer in Glasgow; took up journalism, 1860; sub-editor of *Globe*, 1862-6; assistant editor of *Economist*, 1868-76; chief of statistical department to Board of Trade, 1876-97; the commercial (1882) and the labour (1892) departments were subsequently included in his control; served on many royal commissions; edited *Journal of Royal Statistical Society*, 1876-91; helped to found Economic Society, 1890; criticized Gladstone's home rule finance, 1893; liberal unionist from 1886, and finally unionist free trader; KCB, 1895; strong individualist; advocated 'free banking'; voluminous writings include *Essays in Finance* (2 series, 1880-6), *The Case against Bimetallism* (1892), and *Economic Enquiries and Studies* (2 vols., 1904).

GIFFORD, EDWIN HAMILTON (1820-1905), schoolmaster and theologian; BA, St. John's College, Cambridge (senior classic and fifteenth wrangler), 1843; fellow, 1843-4; honorary fellow, 1903; headmaster of King Edward's School, Birmingham, 1848-62; archdeacon of London and canon of St. Paul's, 1884; wrote *Voices of the Past* (1874) and edited with translation Eusebius's *Praeparatio Evangelica* (5 vols., 1903).

GIGLIUCCI, COUNTESS (1818-1908), oratorio and operatic prima donna. [See NOVELLO, CLARA ANASTASIA.]

GILBERT, SIR ALFRED (1854-1934), sculptor; studied at Heatherley's, Royal Academy Schools, under (Sir) Edgar Boehm [q.v.], at École des Beaux-Arts, and in Italy; founded reputation with works such as 'Icarus'; settled in England, 1884; ARA, 1887; RA, 1892; lived in Bruges, 1901-26; works include memorials to Shaftesbury (the fountain and 'Eros' in Piccadilly Circus), the Duke of Clarence (at St. George's, Windsor), and Queen Alexandra (at Marlborough Gate, St. James's); knighted, 1932.

GILBERT, SIR JOSEPH HENRY (1817-1901), agricultural chemist; son of Joseph and Ann Gilbert [qq.v.]; studied agricultural chemistry at Glasgow, London, and at Giessen under Liebig; co-worker with John Bennet Lawes [q.v.] in Rothamsted agricultural experiments; president of Chemical Society, 1882-3; FRS, 1860; professor of rural economy at Oxford, 1884-90; knighted, 1893.

GILBERT, SIR WILLIAM SCHWENCK (1836-1911), dramatist; son of William Gilbert (1804-90, q.v.); used his own pet name 'Bab' as pseudonym in later life; BA, London, 1857; joined militia, 1857; retired as major, 1883; clerk in Privy Council office (education department), 1857-61; called to bar, 1863; regular contributor to *Fun* from 1861; illustrated books by father, 1863 and 1869; his 'Yarn of the Nancy Bell' refused by *Punch* as 'too cannibalistic', 1866; this and other 'Bab' ballads appeared in *Fun* from 1866 to 1871 and were published in volume form as *Bab Ballads* (1869) and *More Bab Ballads* (1873); became playwright with *Dulcamara*, a successful burlesque, 1866; wrote many other extravaganzas; collaborated with Frederick Clay [q.v.] in musical sketches for the German Reeds [qq.v.], 1869-72; introduced by Clay to (Sir) Arthur Sullivan [q.v.], 1871; first collaborated with Sullivan in a burlesque, *Thespis*, 1871; wrote blank verse fairy comedy, *The Palace of Truth* (1870), *Pygmalion and Galatea* (1871), *The Wicked World* (1873); wrote series of comedies (some under pseudonym of F. L. Tomline) for Marie Litton [q.v.] at Court Theatre; *The Happy Land* (1873) roused much enthusiasm and public excitement; produced also serious plays, including *Charity* (1874), *Broken Hearts* (1875), and *Dan'l Druce* (1876); collaborated with Sullivan for D'Oyly Carte's opera company in long series of comic operas, viz. *Trial by Jury* (1875), *The Sorcerer* (1877), *H.M.S. Pinafore* (1878), *The Pirates of Penzance* (produced in New York, 1879), *Patience* (1881); operas transferred to Savoy Theatre, 1881; later 'Savoy' operas were *Iolanthe* (1882), *Princess Ida* (1884), *The Mikado* (1885)—the most popular work—*Ruddigore* (1887), *The Yeomen of the Guard* (1888), *The Gondoliers* (1889); separated from Sullivan and Carte owing to financial disagreement, 1890-3; collaborated with Alfred Cellier [q.v.] in *The Mountebanks*, 1892; resumed collaboration with Sullivan in *Utopia, Limited* (1893) and *The*

Grand Duke (1896); produced *Fallen Fairies* with (Sir) Edward German [q.v.], 1909, and *The Hooligan*, a serious sketch, 1911; built and owned Garrick Theatre, 1889; interested in astronomy, beekeeping, and horticulture; knighted, 1907; plays show literary grace, whimsical humour (known as 'Gilbertian' humour), urbane satire, good taste, and lyric excellence; master of stage management; fond of epigram and repartee; a kindly cynic; dramatic works collected in *Original Plays* (4 series, 1876-1911) and in *Original Comic Operas* (8 parts, 1890).

GILES, HERBERT ALLEN (1845-1935), Chinese scholar and author; son of J. A. Giles [q.v.]; educated at Charterhouse; served in China consular service, 1867-93; professor of Chinese, Cambridge, 1897-1932; fostered intelligent understanding of Chinese culture by a stream of books ranging from technical to popular.

GILES, PETER (1860-1935), philologist and master of Emmanuel College, Cambridge; educated at Aberdeen University and Gonville and Caius College, Cambridge; first class, classical tripos, parts i and ii, 1884-7; fellow of Emmanuel, 1890; master, 1911-35; vice-chancellor, Cambridge, 1919-21; university reader in comparative philology, 1891-1935; published *Short Manual of Comparative Philology* (1895).

GILL, (ARTHUR) ERIC (ROWTON) (1882-1940), stone-carver, engraver, typographer, and author; attended Chichester Art School; in London studied lettering under Edward Johnston [q.v.], architecture, and masonry; became a letter-cutter; carved first figure direct from stone, 1909; became Roman Catholic (1913) and carved Stations of the Cross, Westminster Cathedral (1913-18); sculptures include 'Mankind' (Tate Gallery), work at Broadcasting House, and 'Creation of Adam' at Geneva; designed ten printing types including 'Perpetua' and 'Gill Sans-serif'; illustrations include the 'Canterbury Tales' and the 'Four Gospels'; books include *The Necessity of Belief* (1936) and *Autobiography* (1940); ARA, 1937.

GILL, SIR DAVID (1843-1914), astronomer; educated at Dollar Academy, Marischal College and the university, Aberdeen; established in Aberdeen time service similar to that installed in Edinburgh by Charles Piazzi Smyth [q.v.], 1863; in charge of private observatory erected at Dunecht by Lord Lindsay (afterwards twenty-sixth Earl of Crawford, q.v.), 1872-6; took part in observations of transit of Venus at Mauritius, 1874; chief result of expedition revelation of

possibilities of heliometer for astronomical measurements; made successful expedition to island of Ascension in order to measure with heliometer distance of Mars from Earth and thus derive sun's distance, 1877; H.M. astronomer at Cape of Good Hope, 1879-1907; obtained larger heliometer with which he redetermined sun's distance and determined mass of Jupiter; pioneer in application of photography to astronomy, and carried out photographic survey of Southern heavens, 1885-98; organized geodetic survey of South Africa; KCB, 1900.

GILLIATT, SIR WILLIAM (1884-1956), obstetrician; educated at Wellingborough College; qualified from Middlesex Hospital, 1908; MD, 1910; FRCS, 1912; appointed assistant obstetric and gynaecological surgeon, King's College Hospital, 1916; senior surgeon, 1925-46; on honorary staff, Samaritan Hospital, 1926-46; attended Princess Marina, Duchess of Kent, for birth of her children and Princess Elizabeth for birth of Prince Charles (1948) and Princess Anne (1950); surgeon-gynaecologist to the Queen, 1952-6; CVO, 1936; knighted, 1948; KCVO, 1949; FRCP, 1947; president, Royal Society of Medicine, 1954-6.

GILLIES, DUNCAN (1834-1903), premier of Victoria; emigrated from Glasgow to Ballarat goldfields, 1852; led miners in resistance to government, 1853-4; member of Ballarat mining boards, 1858; member of legislative assembly, 1859; president of board of land and works, 1868; commissioner of railways and roads, 1872-5; minister of agriculture, 1875-7; commissioner of railways in Service government, 1880, and in Service-Berry coalition, 1883-6; premier and treasurer, 1886-90; extended revenue, expenditure, and railways; passed Irrigation Act, 1886; supported Australian federation; agent-general for Victoria in London, 1894-7; speaker of Victoria house of assembly, 1902-3.

GILLIES, SIR HAROLD DELF (1882-1960), plastic surgeon; born in New Zealand; educated at Wanganui College and Gonville and Caius College, Cambridge; qualified from St. Bartholomew's Hospital, 1908; FRCS, 1910; in charge of pioneer centres for plastic surgery in both wars; plastic surgeon to St. Bartholomew's, LCC, and RAF; first president of British Association of Plastic Surgeons, 1946; published *Plastic Surgery of the Face* (1920) and *The Principles and Art of Plastic Surgery* (with D. R. Millard, 2 vols., 1975); knighted, 1930; played golf for England, and was a proficient painter in oils.

GILMOUR, SIR JOHN, second baronet, of Lundin and Montrave (1876-1940), politician;

unionist MP, East Renfrewshire (1910–18), Pollok division of Glasgow, 1918–40; PC, 1922; secretary for Scotland, 1924–9; minister of agriculture, 1931–2; home secretary, 1932–5; minister of shipping, 1939–40; succeeded father, 1920; GCVO, 1935.

GILSON, JULIUS PARNELL (1868–1929), palaeographer and scholar; BA, Trinity College, Cambridge; assistant, department of manuscripts, British Museum, 1894; assistant keeper, 1909; keeper and Egerton librarian, 1911; largest work *A Catalogue of Western Manuscripts in the Old Royal and King's Collection* (with Sir G. F. Warner, q.v., 1921).

GINNER, ISAAC CHARLES (1878–1952), artist; born in Cannes of British parentage; inspired by Van Gogh whose ideals and the continental movements in art he introduced to British painters after settling in London, 1910; exhibited with Camden Town and London groups; founder-member, Cumberland Market Group, and member of New English Art Club; sought complete transposition of nature, working *en plein air*; ARA, 1942; CBE, 1950.

GINSBERG, MORRIS (1889–1970), sociologist and moral philosopher; born in Lithuania; migrated to England, 1910; read philosophy at University College, London; BA, first class, 1913; MA, 1915; part-time assistant to L. T. Hobhouse [q.v.] at London School of Economics, 1914; reader, 1924; professor of sociology, 1929; emeritus professor, 1954; collaborated with L. T. Hobhouse and G. C. Wheeler on *The Material Culture and Social Institutions of the Simpler Peoples* (1915); other publications include *Moral Progress* (1944), *The Idea of Progress: A Revaluation* (1953), *Reason and Unreason in Society* (1947), and *On Justice in Society* (1965); FBA, 1953; hon. degrees, London, Glasgow, and Nottingham; hon. fellow LSE; president, Aristotelian Society, 1942–3; Huxley medal, 1953; Herbert Spencer lecturer, 1958.

GINSBURG, CHRISTIAN DAVID (1831–1914), Old Testament scholar; born at Warsaw of Jewish parentage; educated at rabbinic school at Warsaw; became Christian, *c.*1847; came to England; naturalized, 1858; original member of Old Testament revision company, 1870; published first volume of his principal work, edition of *The Massorah* (1880).

GIRDLESTONE, GATHORNE ROBERT (1881–1950), orthopaedic surgeon; qualified at St. Thomas's Hospital, 1908; worked with Sir Robert Jones [q.v.]; with him formed Central Council for Care of Cripples, 1920; from 1915 in charge of military orthopaedic centre, Oxford,

which became (1922) Wingfield (rebuilt 1933 as Wingfield-Morris) Orthopaedic Hospital; professor of orthopaedic surgery, Oxford, 1937–9; also responsible for establishment of Churchill Hospital, Oxford.

GIROUARD, DÉSIRÉ (1836–1911), Canadian judge; born in Quebec province; DCL, McGill University, 1874; called to bar of Lower Canada, 1860; QC, 1880; published treatise on bills of exchange, 1860; conservative MP for Jacques Cartier constituency, 1878–95; opposed execution of Louis Riel [q.v.], 1885; judge of the supreme court of Canada, 1895–1911.

GIROUARD, Sir (EDOUARD) PERCY (CRANWILL) (1867–1932), railway engineer and colonial administrator; born at Montreal; son of Désiré Girouard [q.v.]; commissioned in Royal Engineers, 1888; director of railways in Sudan (1896–8) and South Africa (1899–1902); KCMG, 1900; high commissioner (1907), governor (1908–9), Northern Nigeria; constructed first railway; governor, British East Africa Protectorate, 1909–12; director-general, munitions supply, 1915–17.

GISSING, GEORGE ROBERT (1857–1903), novelist; left Owens College, Manchester, in disgrace for America, where he wandered penniless until 1877; studied literature and philosophy at Jena; returned to England, 1878; published *Workers in the Dawn*, 1880; found an appreciative reader in Frederic Harrison [q.v.], to whose sons he became tutor, 1882; gained precarious livelihood by occasional journalism; published *The Unclassed* (1884), *Demos* (1886), and other novels illustrating degrading effects of poverty on character; visited Naples, Rome, and Athens; published *A Life's Morning* (1888), *The Nether World* (1889), *The Emancipated* (1890), *New Grub Street* (1891), *Born in Exile* (1892), and *The Odd Women* (1893); revisited Italy with H. G. Wells [q.v.], 1897, recording some experiences and impressions in *By the Ionian Sea* (1901); in Rome he found material for historical romance *Veranilda* (published posthumously, 1907); on return to England wrote *The Town Traveller* (1898) and *Our Friend the Charlatan* (1901); died of pneumonia at St. Jean-de-Luz; other works include critical study of *Charles Dickens* (1898), *The Private Papers of Henry Ryecroft* (1903), and *The House of Cobwebs* (1906).

GLADSTONE, HERBERT JOHN, Viscount Gladstone (1854–1930), statesman; youngest son of W. E. Gladstone, prime minister; educated at Eton; BA, University College, Oxford, 1876; liberal MP, Leeds,

1880-5, West Leeds, 1885-1910; a liberal whip and junior lord of Treasury, 1881-5; financial secretary at War Office, 1886; under-secretary, Home Office, 1892-4; first commissioner of works, 1894-5; chief liberal whip, 1899; preserved neutrality within party during South African war, 1899-1902; secretary of state for home affairs, 1905-10; carried through parliament twenty-two bills, which showed growing tendency towards bureaucracy; Court of Criminal Appeal established, 1907; deeply interested in problem of young offender; instituted Borstal system and children's courts; first governor-general and high commissioner of Union of South Africa, 1910-14; worked in full harmony with Louis Botha [q.v.]; viscount, 1910; manifested pious pugnacity on behalf of father's reputation; published *After Thirty Years* (1928).

GLADSTONE, JOHN HALL (1827-1902), chemist; studied chemistry in London and Giessen; FRS, 1853; Fullerian professor of chemistry at Royal Institution, 1874-7; president of Physical (1874) and Chemical (1877-9) societies; hon. D.Sc., Trinity College, Dublin, 1892; made pioneer researches into chemistry in relation to optics; early student of spectroscopy; discovered copper-zinc union for decomposition of water; member of London school board, 1873-94; advocate of technical education, manual instruction, and spelling reform; works include *Theology and Natural Science* (1867), *Michael Faraday* (1872), *Miracles* (1880), and hymns.

GLAISHER, JAMES (1809-1903), astronomer and meteorologist; assistant at Cambridge University observatory, 1833-5, and at Greenwich, 1835-8; chief of magnetic and meteorological department there, 1838-74; improved instruments; published *Hydrometrical Tables* (1847); organized voluntary system of precise meteorological observation throughout England, 1847; prepared meteorological reports for registrar-general, 1847-1902; helped to establish daily weather report for *Daily News*, 1849; FRS, 1849; secretary of Royal Meteorological Society, 1850-72; defined relations between weather and cholera epidemics and water-supply; made balloon ascents for meteorological observations with Henry Coxwell [q.v.], 1862-6; published observations in *British Association Reports*, 1862-6, and account of his ascents in *Voyages Aériens* (1869, translated into English as *Travels in the Air*, 1871); was also interested in astronomy and mathematical science; completed and published *Factor Tables* (3 vols., 1879-83); on committee of Palestine Exploration Fund; translated Flammarion's *Atmosphere* and Guillemin's *World of Comets* (1876).

GLAISHER, JAMES WHITBREAD LEE (1848-1928), mathematician, astronomer, and collector; son of James Glaisher [q.v.]; BA, Trinity College, Cambridge (second wrangler), 1871; fellow, 1871-1928; lecturer, 1871-1901; member of council of Royal Astronomical Society, 1874-1928; president, 1886-8, 1901-3; FRS, 1875; maintained lifelong connection with London Mathematical Society; member of council, 1872-1906; president, 1884-6; worked hard on behalf of British Association; published nearly four hundred papers on pure mathematics and astronomy; in middle life took up study of pottery, on which he became recognized authority; bequeathed his collection to Fitzwilliam Museum, Cambridge.

GLAZEBROOK, MICHAEL GEORGE (1853-1926), schoolmaster; BA, Balliol College, Oxford, 1877; high master of Manchester grammar school, 1888; ordained, 1890; headmaster of Clifton College, where he encouraged music, 1891-1905; canon of Ely, 1905; leader of 'Modern Churchmen' movement; works include *Lessons from the Old Testament* (1890), *The End of the Law* (1911), and *Faith of a Modern Churchman* (1918).

GLAZEBROOK, Sir RICHARD TETLEY (1854-1935), physicist; fifth wrangler (1876), fellow (1877), senior bursar (1895), Trinity College, Cambridge; assistant director, Cavendish Laboratory, 1891; principal, University College, Liverpool, 1898-9; first director, National Physical Laboratory, 1899-1919; gave priority to construction of units and standards and after 1909 carried out important research in aeronautics, notably on conditions of stability; Zaharoff professor of aviation, Imperial College of Science, 1920-3; fellow (1882), foreign secretary (1926-9), Royal Society; knighted, 1917; KCB, 1920; KCVO, 1934.

GLEICHEN, Lady FEODORA GEORGINA MAUD (1861-1922), sculptor; daughter of Prince Victor of Hohenlohe-Langenburg [q.v.]; pupil of Alphonse Legros [q.v.]; exhibited regularly at Royal Academy; works include public memorials and portrait busts; first woman member (posthumous) of Royal Society of British Sculptors.

GLENAVY, first Baron (1851-1931), Irish lawyer and politican. [See CAMPBELL, JAMES HENRY MUSSEN.]

GLENESK, Baron (1830-1908), proprietor of the *Morning Post*. [See BORTHWICK, ALGERNON.]

GLENNY, ALEXANDER THOMAS (1882-1965), immunologist; educated at

Alleyn's School and Chelsea Polytechnic; B.Sc., London; joined Wellcome Physiological Research Laboratories, 1899; head of Immunology Department, 1906; responsible for production of antitoxins; made important contributions to schedules of immunization of man and animals, including immunization against diphtheria and tetanus; FRS, 1944; retired from Wellcome Laboratories, 1947.

GLENVIL HALL, WILLIAM GEORGE (1887-1962), politician. [See HALL.]

GLOAG, PATON JAMES (1823-1906), theological writer; student of Edinburgh and St. Andrews universities; thrice visited Germany (1857-67) and studied German theological literature; Church of Scotland minister of Galashiels, 1871-92; Baird lecturer, 1879; moderator of General Assembly, 1889; professor of biblical criticism in Aberdeen University, 1896-9; hon. DD, St. Andrews, 1867, and LLD, Aberdeen, 1899; wrote much on New Testament exegesis.

GLOAG, WILLIAM ELLIS, LORD KINCAIRNEY (1828-1909), Scottish judge; called to Scottish bar, 1853; advocate depute, 1874; sheriff of Perthshire, 1885; raised to bench as Lord Kincairney, 1889.

GLOVER, TERROT REAVELEY (1869-1943), classical scholar and historian; educated at Bristol grammar school and St. John's College, Cambridge; first class, classical tripos, parts i and ii, 1891-2; fellow, 1892; teaching fellow, 1901; professor of Latin, Queen's University, Kingston, Ontario, 1896-1901; university lecturer in ancient history, Cambridge, 1911-39; orator, 1920-39; president, Baptist Union, 1924, Classical Association, 1938; works include *Studies in Virgil* (1904), *The Conflict of Religions in the Early Roman Empire* (1909), *The Jesus of History* (1917), and *Democracy in the Ancient World* (1927).

GLYN, ELINOR (1864-1943), novelist; born SUTHERLAND; educated privately and read widely; made successful entry into society; married (1892) Clayton Glyn, landowner; wrote *The Visits of Elizabeth* (1900) and other 'society' novels; passionate romances, notably *Three Weeks* (1907) and *His Hour* (1910); and more serious character studies, including *Halcyone* (1912) containing recognizable portraits of her friends Lord Curzon and F. H. Bradley [qq.v.]; *It* (1927) made word synonymous with personal magnetism; script-writer in Hollywood, 1920-9.

GODFREY, DANIEL (1831-1903), bandmaster of Grenadier Guards, 1856-96; composed famous 'Guards' waltz, 1863, and much popular military music; toured with band in America, 1876; promoted second lieutenant, 1887.

GODFREY, WALTER HINDES (1881-1961), architect and antiquary; educated at Whitgift grammar school, Croydon, and Central School of Arts and Crafts; joined architectural section of London County Council, 1900; member of Committee for the Survey of London, 1901; prepared four volumes of Surrey, 1909-27; left LCC for private practice, 1903; formed partnership of Wratten and Godfrey, 1905; prepared illustrations and architectural studies in Survey of London Series; employed in Accounts Division, Ministry of Munitions, 1915-19; resumed private practice; FRIBA, 1926; retired from practice but undertook restoration of Herstmonceaux Castle and other buildings, 1932-9; director, National Buildings Record (later the National Monuments Record), 1941-60; published many books, including *A History of Architecture in London* (1911), *The Story of Architecture in England* (2 vols., 1928 and 1931), and *The English Almshouse* (1955); member of royal commission on historical monuments, 1944, and other committees; CBE, 1950.

GODFREY, WILLIAM (1889-1963), cardinal, seventh archbishop of Westminster; educated at St. John's, Kirkdale, Ushaw College, Durham and the Venerable English College, Rome; doctorates in philosophy and theology at the Gregorian University, 1913 and 1917; ordained priest, 1916; curate, St. Michael's, Liverpool, 1917; classics master, Ushaw College, 1919; professor of philosophy, and, in 1928, of dogmatic theology; published *The Young Apostle* (1924), and *God and Ourselves* (1927); contributed to *Ushaw Magazine*, 1921-30; succeeded Archbishop Hinsley [q.v.] as rector, English College, Rome, 1930; domestic prelate to Pope Pius XI; member of the Supreme Council for the Propagation of the Faith; counsellor in the papal mission for George VI's coronation, 1937; apostolic visitor of all seminaries in Great Britain, 1938; archbishop of Liverpool, 1953; succeeded Cardinal Griffin [q.v.], as seventh archbishop of Westminster, 1956; created cardinal, 1958; member of Central Preparatory Commission to examine and recast the *schemata* for discussion in the second Vatican Council, 1961.

GODKIN, EDWIN LAWRENCE (1831-1902), editor and author; son of James Godkin [q.v.]; BA, London, 1851; left law for authorship; published *The History of Hungary* (1853); *Daily News* correspondent in Crimea; settled in United States, 1856; called to New York bar,

1858; supported North in civil war, 1862; edited New York *Nation*, which by its independence and literary power influenced American public opinion, 1865-99; the paper was recognized organ of independent 'Mugwumps', 1884-94; Godkin denounced system of Tammany Hall and caused defeat of Tammany, 1894; opposed American annexation of Hawaii and Philippines, high tariffs, and bimetallism; revisited England, 1889; his philosophical radical views expounded in *Reflections and Comments* (1895), *Problems of Modern Democracy* (1896), and *Unforeseen Tendencies of Democracy* (1897); hon. DCL, Oxford, 1897; Godkin memorial lectures on citizenship founded at Harvard University.

GODLEE, Sir RICKMAN JOHN, baronet (1849-1925), surgeon; nephew of Lord Lister [q.v.]; studied medicine at London University; MB, 1872; MS, 1873; assistant surgeon, University College Hospital, 1877; full surgeon, 1885; professor of clinical surgery, 1892; Holme professor of clinical surgery, 1900; emeritus professor, 1914; performed first operation for removal of tumour from the brain, 1884; stimulated development of thoracic surgery; FRCS, 1876; president, Royal Society of Medicine, 1916-18; baronet, 1912; KCVO, 1914; works include *Atlas of Human Anatomy* (1880) and life of Lister (of whom he had intimate personal knowledge, 1917).

GODLEY, Sir ALEXANDER JOHN (1867-1957), general; cousin of A. D. Godley and Lord Kilbracken [qq.v.]; educated at Haileybury, United Services College, and Sandhurst; gazetted to Royal Dublin Fusiliers, 1886; served in South Africa, 1896-7 and 1899-1901; appointed to command New Zealand military forces and major-general, 1910; commanded New Zealand expeditionary force, 1914-18; commanded New Zealand and Australian division at Gallipoli, II Anzac Corps in France, 1916-18, XXII Corps, 1918-19; military secretary to Churchill, 1920-2; commander-in-chief, British Army of Rhine, 1922-4; GOC, Southern Command, 1924-8; governor of Gibraltar, 1928-33; general, 1923; KCMG, 1914; KCB, 1916; GCB, 1928.

GODLEY, ALFRED DENIS (1856-1925), classical scholar and man of letters; BA, Balliol College, Oxford, 1878; tutor and fellow of Magdalen College, 1883-1912; public orator, 1910-25; writer of light humorous or satiric verse and prose; commentator on and translator of Herodotus, Tacitus, and Horace; 'an almost perfect writer of elegant Latin' in his Creweian orations.

GODLEY, (JOHN) ARTHUR, first BARON KILBRACKEN (1847-1932), civil servant; son of

J. R. Godley [q.v.]; educated at Rugby and Balliol College, Oxford; assistant (1872-4) and principal (1880-2) private secretary to Gladstone; fellow of Hertford College, 1874-81; permanent under-secretary of state for India, 1883-1909; established efficiency of India Office on firm basis; KCB, 1893; baron, 1909.

GODWIN, GEORGE NELSON (1846-1907), Hampshire antiquary; chaplain of the forces, 1877-90; antiquary and historian of Hampshire and neighbouring counties; published *Civil War in Hampshire, 1642-5* (1882), *Bibliotheca Hantoniensis* (1891), and other works.

GODWIN-AUSTEN, HENRY HAVERSHAM (1834-1923), explorer and geologist; son of R. A. C. Godwin-Austen [q.v.]; educated at Royal Military Academy, Sandhurst; joined 24th Foot, 1851; served in second Burmese war, 1852; attached to Great Trigonometrical Survey of India to assist with first survey of Kashmir, 1856; permanent topographical assistant to survey, 1860; discovered and surveyed great Karakoram glaciers, etc., 1861; served on political mission of (Sir) Ashley Eden [q.v.] to Bhutan in eastern Himalaya, 1863; explored Naga hills, 1873-4; accompanied first Dafla expedition, 1875; carried out important geological investigations; retired, owing to ill health, 1877; FRS, 1880.

GOGARTY, OLIVER JOSEPH ST. JOHN (1878-1957), surgeon, man of letters, and wit; born in Dublin; educated at Stonyhurst and Clongoweswood; qualified in Dublin, 1907; practised as nose and throat surgeon; moved into literary and political circles, gaining reputation as great Irish wit; member of first Senate of Irish Free State; moved to London, 1937, to United States, 1939; published *As I was Going Down Sackville Street* (1937) and other volumes of reminiscences, novels, and *Collected Poems* (1951).

GOLD, Sir HARCOURT GILBEY (1876-1952), oarsman; educated at Eton and Magdalen College, Oxford; stroked Eton to victory in Ladies' Plate (1893-5), Oxford (1896-8), and Leander thrice in Grand Challenge Cup; coached two winning Olympic eights (1908, 1912); appointed steward of Henley Regatta, 1909; committee member, 1919; chairman, 1945; president, 1952; OBE, 1918; knighted, 1949.

GOLDIE, Sir GEORGE DASHWOOD TAUBMAN (1846-1925), founder of Nigeria; commissioned in Royal Engineers, 1865-7; first visited West Africa, 1877; discovered some British firms in Niger delta engaged in competition and not attempting to open up trade with rich interior; formed United African Company, 1879; launched National African Company, 1881; French companies, which shortly appeared on scene, amalgamated with it, 1884; situation further complicated by summoning of West African conference in Berlin, 1884-5; race between England, France, and Germany for Africa inaugurated; secured grant of royal charter to company under title of Royal Niger Company Chartered and Limited, which remained trading company while charged with administrative duties, 1886; appointed political administrator and deputy-governor; governor, 1895; concluded numerous treaties with native rulers of interior; secured end of struggle with Germany (1893) and France (1898) and final definition of respective spheres; although British sphere smaller than originally projected, secured half a million square miles of most fertile, highly mineralized and thickly populated portion of West Africa; territory known as Nigeria, 1897; carried out successful campaigns against slave-raiding Mohammedan states of Nupé and Ilorin, 1897; legal status of slavery thereupon abolished in company's territory; company's charter surrendered to British crown, 1900; KCMG, 1887; FRS, 1902.

GOLDSCHMIDT, OTTO (1829-1907), pianist and composer; born of Jewish parents at Hamburg; studied pianoforte at Leipzig under Mendelssohn, who greatly influenced him; in London, 1848; toured in America with Jenny Lind [q.v.], 1851; married her, 1852; settled in England, 1856; pianoforte professor at Royal Academy of Music, 1863; vice-principal, 1866-8; produced *Ruth* at Hereford musical festival, 1867; became conductor of newly founded 'Bach choir', 1876; composed 'Music, an Ode', 1898, and works for pianoforte.

GOLDSMID, Sir FREDERIC JOHN (1818-1908), major-general; born at Milan; joined 37th Madras native infantry, 1839; served in China (1840), in Crimea, and in Indian Mutiny; arranged for telegraph construction along coast of Gwadar, 1861; superintended carrying of wires from Europe across Persia to India, 1864; director-general of Indo-European telegraph; negotiated Anglo-Persian telegraph treaty, 1865; CB, 1866; constructed telegraph line across whole of Persia; described the experience in *Travel and Telegraph*, 1874; commissioner for delimitation of Persian and Baluchistan boundary, 1871; investigated Persian and Afghan claims to Seistan; recorded proceedings in *Eastern Persia*, 1870-2 (2 vols., 1876); KCSI, 1871; controller of crown lands in Egypt, 1880-3; organized intelligence department in campaign of 1882; established adminis-

trative system in Congo, 1883; published *Life of Sir James Outram* (2 vols., 1880); accomplished oriental linguist; vice-president of Royal Asiatic Society, 1890-1905.

GOLDSMID-MONTEFIORE, CLAUDE JOSEPH (1858-1938), Jewish biblical scholar and philanthropist. [See MONTEFIORE.]

GOLLANCZ, Sir HERMANN (1852-1930), rabbi, Semitic scholar; born at Bremen; brother of Sir Israel Gollancz [q.v.]; BA, University College, London, 1873; preacher at various synagogues in London, Manchester, and Dalston, 1876-92; first minister at Bayswater synagogue, Harrow Road, 1892-1923; obtained rabbinic degree in Galicia, which caused great controversy, 1897; his degree finally recognized and the requirements for obtaining rabbinical diploma in Great Britain defined; Goldsmid professor of Hebrew, University College, London, 1902-23; knighted, 1923; undertook much public work outside interests of Jewish community.

GOLLANCZ, Sir ISRAEL (1863-1930), scholar and man of letters; brother of Sir Hermann Gollancz [q.v.]; BA, Christ's College, Cambridge, 1887; lecturer in English at Cambridge, 1896-1906; professor of English language and literature, King's College, London, 1903-30; a founder, original fellow, and secretary of British Academy, 1902-30; knighted, 1919; in the first rank as English and Shakespearian scholar.

GOLLANCZ, Sir VICTOR (1893-1967), publisher; brother of Sir Hermann and Sir Israel Gollancz [qq.v.]; educated at St. Paul's School and New College, Oxford; commissioned in Northumberland Fusiliers, 1914; seconded to Repton School as master, 1916-18; began publishing with Benn Brothers, 1920; joined by Douglas Jerrold [q.v.], 1923; founded own firm, with Stanley Morison [q.v.] as typographer, 1928; first success, R. C. Sherriff's *Journey's End*; list included Daphne du Maurier, Elizabeth Bowen, and Dorothy Sayers [q.v.]; founded Left Book Club, 1936; good public speaker in support of nuclear disarmament and other causes; started Save Europe Now movement, 1945; published *A Year of Grace* (1950), *From Darkness to Light* (1956), and *Reminiscences of Affection* (1968); awarded Peace prize of German Book Trade, 1960; knighted, 1965.

GOOCH, GEORGE PEABODY (1873-1968), historian; educated at Eton, King's College, London, and Trinity College, Cambridge; first class, historical tripos, 1894; Lightfoot scholar,

1895; studied in Berlin and Paris, 1895-6; taught at Mansfield House, the Working Men's College, and Toynbee Hall; distressed by South African war; liberal MP, Bath, 1906-10; joint editor with J. Scott Lidgett [q.v.] of *Contemporary Review*, 1911-60; assisted by Lord Acton [q.v.] in preparation of *English Democratic Ideas in the Seventeenth Century* (1898); published *History and Historians in the Nineteenth Century* (1913) and *Germany and the French Revolution* (1920); criticized policy of Sir Edward Grey (later, Viscount Grey of Fallodon, q.v.) leading to 1914-18 war; joined Sir Adolphus Ward [q.v.] in editing the *Cambridge History of British Foreign Policy* (3 vols., 1922-3), and with H. W. V. Temperley [q.v.], edited *British Documents on the Origins of the War* (13 vols., 1926-38); other publications include *Before the War: Studies in Diplomacy* (1936-8), and *Frederick the Great* (1947); hon. doctorates, Durham and Oxford; FBA (1926); hon. fellow, Trinity College, Cambridge (1935); CH, 1939; OM, 1963.

GOODALL, FREDERICK (1822-1904), artist; taught by father, Edward Goodall [q.v.]; exhibited at Royal Academy, 1839-1902; ARA, 1852; RA, 1862; early works show influence of Sir David Wilkie [q.v.], e.g. 'The Tired Soldier' (1842) and 'The Village Holiday' (1847); visits to Egypt (1858-9 and 1870) determined subject of later pictures, as 'The Nubian Slave' (1864), 'The Flight into Egypt' (1884), 'Sheep Shearing in Egypt' (1892); also painted English landscape and portraits; showed technical ability but little inspiration; published gossiping *Reminiscences*, (1902).

GOODE, SIR WILLIAM ATHELSTANE MEREDITH (1875-1944), journalist and financial adviser; represented Associated Press with Admiral Sampson throughout Spanish–American war, and in London, 1898-1904; managing editor, *Standard*, 1904-10; joint news editor, *Daily Mail*, 1911; director, cables department, Ministry of Food, 1917-19; British director, relief in Europe, 1919-20; reported on economic conditions in Central Europe, 1920; president, Austrian section, Reparation Commission, 1920-1; financial agent in London to Hungary, 1923-41; chief security officer, Ministry of Food, 1939-42; KBE, 1918.

GOODEN, STEPHEN FREDERICK (1892-1955), engraver; educated at Rugby and Slade School; illustrated books in line-engraving, notably the Nonesuch *Bible* (5 vols., 1925-7); ARA, 1937; RA, 1946; CBE, 1942.

GOODENOUGH, FREDERICK CRAUFURD (1866-1934), banker; born in Calcutta; educated at Charterhouse and Zürich University; admitted solicitor; secretary of Barclay & Co. Ltd. (1896), general manager (1903), director (1913), chairman (1917); a convinced imperialist; formed Barclays Bank (Dominion, Colonial, and Overseas); founded London House as hall of residence in London chiefly for dominion students, 1930; member of council of India, 1918-30; hon. DCL, Oxford, 1933.

GOODENOUGH, SIR WILLIAM EDMUND (1867-1945), admiral; son of J. G. Goodenough [q.v.]; entered navy, 1880; first captain of Dartmouth, 1905-7; appointed to *Southampton* as commodore, first (later second) light cruiser squadron, 1913-16; trained his captains to know his mind enabling them to act without instructions; fought at Heligoland Bight (1914), Dogger Bank (1915), and Jutland (1916); rear-admiral, second battle squadron, 1916-18; superintendent, Chatham dockyard, 1919-20; KCB, 1919; vice-admiral, 1920; commander-in-chief, Africa station, 1920-2; the Nore, 1924-7; admiral, 1925; criticized Admiralty administration but declined post of second sea lord, 1925; GCB, 1930; president, Royal Geographical Society, 1930-3; chairman, British Sailors' Society.

GOODENOUGH, SIR WILLIAM MACNAMARA, first baronet (1899-1951); banker; son of F. C. Goodenough [q.v.]; educated at Wellington College and Christ Church, Oxford; joined Barclays Bank; local director, Oxford, 1923; director of Bank, 1929; vice-chairman, 1934; deputy chairman, 1936; chairman, 1947-51; director, Barclays DCO, 1933; deputy chairman, 1937; chairman, 1943-7; chairman of Nuffield trusts for developing Oxford medical school and founding (1937) Nuffield College; of Nuffield Provincial Hospitals Trust and (1943) of Nuffield Foundation; chairman, inter-departmental committee on medical schools, 1942-4; baronet, 1943; a founder of the Oxford Society and curator of University Chest; hon. LLD, Manchester.

GOODEY, TOM (1885-1953), nematologist; educated at Northampton grammar school and Birmingham University; B.Sc., 1908; worked at Rothamsted Experimental Station and zoology department, Birmingham; at London School of Tropical Medicine, 1921-6; at Institute of Agricultural Parasitology, 1926-47; head of nematology department, Rothamsted, 1947-52; published *Plant Parasitic Nematodes and the Diseases they Cause* (1933) and *Soil and Fresh-water Nematodes* (1951); sang professionally under name of Roger Clayson; FRS, 1947; OBE, 1950.

GOODHART-RENDEL,HARRY STUART (1887-1959), architect; grandson of Lord Rendel who left him the life interest of his fortune, 1913; educated at Eton and Mulgrave Castle, Yorkshire; Mus.B., Trinity College, Cambridge, 1909; began architectural practice, 1909; had comprehensive knowledge of Victorian architecture from which his own work was a vigorous and original development; became Roman Catholic in middle life; later work mainly concerned with churches; his buildings less important than his devoted services to the profession as a scholarly personality of eloquence and wit; president, Architectural Association, 1924-5; of Royal Institute of British Architects, 1937-9; of Design and Industries Association, 1948-50; Slade professor of fine art, Oxford, 1933-6; director, Architectural Association school of architecture, 1936-8; CBE, 1955; publications include *Nicholas Hawksmoor* (1924); governor, Sadler's Wells.

GOODMAN, JULIA (1812-1906), portrait painter; born SALAMAN; exhibited at Royal Academy, 1838-1901; painted over 1,000 portraits in oils or pastels.

GOODRICH, EDWIN STEPHEN (1868-1946), zoologist; educated in Pau; entered Slade School, 1888; assistant to (Sir) E. R. Lankester [q.v.] and commoner of Merton College, Oxford, 1892; first class, natural science, 1895; fellow, 1900; Aldrichian demonstrator in comparative anatomy, 1898; Linacre professor, 1921-45; researched and travelled widely; distinguished between the nephridium and coelomoduct; established true nature of differences between various types of fish-scales; publications include *Cyclostomes and Fishes* (1909), *Living Organisms* (1924), and *Studies on the Structure and Development of Vertebrates* (1930); FRS, 1905.

GOOSSENS, SIR EUGENE (1893-1962), conductor and composer of music; educated at the Muzick-Conservatorium, Bruges, the Liverpool College of Music, and the Royal College of Music, London, 1907; began professional career as violinist in Queen's Hall Orchestra under Sir Henry J. Wood [q.v.], 1912; began conducting career with various opera companies, 1916; conducted for Diaghilev Ballet; choral work, *Silence*, performed at Gloucester Three Choirs' Festival, 1922; conductor, Rochester Philharmonic Orchestra, New York State, 1923-31; conductor, Cincinnati Symphony Orchestra, 1931-47; chevalier of Legion of Honour, 1934; conducted first performance of his own opera *Don Juan de Mañara* at Covent Garden, 1937; conductor, Sydney Symphony Orchestra, 1947-56; knighted, 1955; other compositions include *The Apocalypse*, a choral work, and

Judith, another opera; published autobiography *Overture and Beginners* (1951).

GORDON, ARTHUR CHARLES HAMILTON-, first BARON STANMORE (1829-1912), colonial governor; son of fourth Earl of Aberdeen [q.v.]; MA, Trinity College, Cambridge; lieutenant-governor of New Brunswick, 1861; governor of Trinidad, 1866-70; Mauritius, 1871-4; Fiji, 1875-80; New Zealand, 1880-3; Ceylon, 1883-90; KCMG, 1871; GCMG, 1878; baron, 1893.

GORDON, CHARLES WILLIAM (1860-1937), divine, and author under the pseudonym of RALPH CONNOR; born and educated in Ontario; ordained Presbyterian minister, 1890; minister of St. Stephen's church, Winnipeg, 1894-1936; CMG, 1935; his romantic novels with religious motif such as *The Sky Pilot* (1899) and *The Man from Glangarry* (1901) were Canadian best-sellers.

GORDON, GEORGE STUART (1881-1942), president of Magdalen College, Oxford, and professor of poetry; educated at Glasgow University and Oriel College, Oxford; first class, *lit. hum.*, 1906; prize fellow in English literature, Magdalen College, 1907; professor at Leeds, 1913-22; Merton professor of English literature, Oxford, 1922-8; president of Magdalen, 1928-42; professor of poetry, 1933-8; vice-chancellor, 1938-41; publications include notable articles in *The Times Literary Supplement*, school editions of Shakespeare, and posthumously *Anglo-American Literary Relations* (1942), *Shakespearian Comedy* (1944), and *Lives of Authors* (1950).

GORDON, ISHBEL MARIA, (MARCHIONESS OF ABERDEEN AND TEMAIR (1857-1939); born MARJORIBANKS; married future first marquess (see below), 1877; devoted to religious and humanitarian pursuits and to liberalism; initiated Onward and Upward Association; in Canada founded Victorian Order of Nurses, 1898; president for many years of International Council of Women; GBE, 1931.

GORDON, JAMES FREDERICK SKINNER (1821-1904), Scottish antiquary; MA, St. Andrews, 1842; in charge of St. Andrew's Episcopal church, Glasgow, 1844-90; pioneer in abolition of ruinous tenements in Glasgow; published *The Ecclesiastical Chronicle for Scotland* (4 vols., 1867), *A History of Glasgow* (1872), and topographical works; enthusiastic freemason.

GORDON, JOHN CAMPBELL, seventh EARL OF ABERDEEN and first MARQUESS OF

ABERDEEN AND TEMAIR (1847-1934), statesman; third son of fifth earl; educated at St. Andrews University and University College, Oxford; succeeded to title, 1872; became a constant liberal; lord-lieutenant of Ireland, 1886, 1906-15; governor-general of Canada, 1893-8; PC, 1886; KT, 1906; marquess, 1916; lord rector of St. Andrews, 1913-16; much occupied in social welfare.

GORDON, SIR JOHN JAMES HOOD (1832-1908), general; entered army, 1849; served in Indian Mutiny and in Afridi expedition, 1877-8; prominent in Afghan war, 1878-9; CB, 1879; commanded troops in expeditions to Karmana and against Malikshahi Waziris, 1880, and in Mahsud Waziris expedition, 1881; commanded brigade in Burmese expedition, 1886-7; assistant military secretary at headquarters, 1890-6; general, 1894; member of India Council, 1897-1907; KCB, 1898; GCB, 1908; published history of Sikhs, 1904.

GORDON, MERVYN HENRY (1872-1953), medical bacteriologist; grandson of William Buckland [q.v.]; educated at Marlborough, Keble College, Oxford, and St. Bartholomew's Hospital; BM, 1898; DM, 1903; on staff of pathology department, St. Bartholomew's, 1898-1923; thereafter remained in department as staff member of Medical Research Council; studied streptococci, transmission of bacteria through the air, cerebro-spinal fever, filtrable viruses; directed a team studying Hodgkin's disease; member, Army Pathological Advisory Committee from 1909; served in RAMC, 1914-18; CMG, 1917; CBE, 1919; FRS, 1924; hon. LLD, Edinburgh.

GORDON, SIR THOMAS EDWARD (1832-1914), general; joined army, 1849; served in Indian Mutiny, 1857-9, and Afghan war, 1879-80; CB, 1881; attached to legation at Tehran, 1889-93; full general, 1894; KCB, 1900.

GORDON-LENNOX, CHARLES HENRY, sixth DUKE OF RICHMOND and first DUKE OF GORDON (1818-1903), lord president of the Council; son of fifth Duke of Richmond [q.v.]; ADC to Duke of Wellington, 1842-52; conservative MP for West Sussex, 1841-60; president of Poor Law Board and PC, 1859; succeeded father, 1860; KG, 1867; president of Board of Trade, 1867-9; leader of conservative party in House of Lords, 1868; lord president of Council, 1874-80; introduced the agricultural holdings (1875) and elementary schools (1876) bills; created first Duke of Gordon, 1876; carried contagious diseases (animals) bill, 1877; reorganized veterinary department of Privy

Council; chairman of royal commission on agriculture (1879-82), whose report led to Agricultural Holdings Act, 1883, and creation of Board of Agriculture; a mediator between liberal government and House of Lords in franchise bill crisis of 1884; secretary for Scotland, 1885-6; chancellor of Aberdeen University, 1861; hon. LLD, 1895; DCL, Oxford, 1870; LLD, Cambridge, 1894; member (1838) and president (1868 and 1883) of Royal Agricultural Society; improved famous Southdown sheep at Goodwood, and shorthorns at Gordon castle.

GORDON-TAYLOR, SIR GORDON (1878-1960), surgeon; educated at Robert Gordon's College and the university, Aberdeen; qualified from Middlesex Hospital, 1903; B.Sc., London, 1904; FRCS, 1906, council member, 1932-48; appointed assistant surgeon, Middlesex Hospital, 1907; surgeon, 1920; surgeon to Fourth Army in France in 1914-18 war and surgeon rear-admiral, 1939-45; published *The Dramatic in Surgery* (1930), and *The Abdominal Injuries of Warfare* (1939); CB, 1942; KBE, 1946.

GORE, ALBERT AUGUSTUS (1840-1901), surgeon-general; MD, Queen's University, Ireland, 1858; LRCS Ireland, 1860; served with army medical staff in West Africa (1861), Ashanti war (1873), and Egypt (1882); principal medical officer to forces in India; CB, 1899; wrote account of his campaigns.

GORE, CHARLES (1853-1932), bishop successively of Worcester, Birmingham, and Oxford; educated at Harrow and Balliol College, Oxford; first class, *lit. hum.*, and fellow of Trinity, 1875; deacon, 1876; priest, 1878; a lifelong Anglo-Catholic; vice-principal of Cuddesdon College, 1880-3; 'principal librarian', Pusey House, 1884-93; mainly through personal relations exercised very strong influence on the religious life of the university; active on behalf of Christian Social Union; formed (1892), and until 1901 was superior of, the Community of the Resurrection, a brotherhood of celibate priests 'having all things in common'; established it at Mirfield, 1898; in his essay 'The Holy Spirit and Inspiration' in *Lux Mundi* (1889) to the distress of his friends concluded that the humanity of Christ entailed certain limitations of consciousness; canon of Westminster, 1894; bishop of Worcester, 1902; of the new diocese of Birmingham to the creation of which he contributed almost all his private fortune, 1905; established excellent relations with civic authorities, Free Churchmen, and evangelicals; unfailingly interested in Workers' Educational Association; supported the budget of 1909; a strong and generally successful disciplinarian; translated (1911) to Oxford which was less responsive to his masterful personal influence; his decisive temper drew him into constant controversy; led protest against consecration of Hensley Henson [q.v.], 1917-18; resigned (1919) and moved to London; dean of theological faculty, King's College, 1924-8; attended Malines conversations; supported revision of the Prayer Book, 1927-8; works include *The Ministry of the Christian Church* (1888), *The Body of Christ* (1901), *The Basis of Anglican Fellowship* (1914), *The Reconstruction of Belief* (1926), *Christ and Society* (1928); through a mind and character of singular force exercised upon his Church an influence unequalled in his generation.

GORE, GEORGE (1826-1908), electro-chemist; head of Institute of Scientific Research, Birmingham, 1880; discovered amorphous antimony and electrolytic sounds; FRS, 1865; improved methods of electro-plating; wrote *The Art of Electrometallurgy* (1877) and philosophic works; hon. LLD, Edinburgh, 1877; awarded civil list pension, 1891; left estate of £5,000 to Royal Society and Royal Institution.

GORE, JOHN ELLARD (1845-1910), astronomical writer; engineer in Indian works department, 1868-79; published *Catalogue of Known Variable Stars* (1884), *The Worlds of Space* (1894), and *The Stellar Heavens* (1903); FRAS, 1878; member of Royal Irish Academy.

GORE, WILLIAM GEORGE ARTHUR ORMSBY-, fourth Baron Harlech (1885-1964), statesman and banker. [See Ormsby-Gore.]

GORE-BROWNE, Sir STEWART (1883-1967), soldier, and settler and politician in Northern Rhodesia (Zambia); educated at Harrow and Royal Military Academy, Woolwich; commissioned to Royal Field Artillery; served in South Africa; appointed to Anglo-Belgian boundary commission to determine boundary between Northern Rhodesia and Belgian Congo, 1911-14; served in France during 1914-18 war; DSO, 1917; returned with rank of lieutenant-colonel, 1921; settled in Northern Rhodesia; elected to Legislative Council, 1935; leader of unofficial members; nominated to represent African interests, 1938; knighted, 1945; lost support of Europeans and resigned unofficial leadership, 1946; supported (Sir) Roy Welensky in opposing rule of Colonial civil service; resigned from Legislative Council, 1951; failed to make come-back with United National Independence Party, founded by Kenneth Kaunda, 1962.

GORELL, first Baron (1848-1913), judge. [See Barnes, John Gorell.]

GORER, PETER ALFRED ISAAC (1907-1961), biologist and geneticist; educated at Charterhouse and Guy's Hospital, London; B.Sc., 1929; MRCS, LRCP, 1932; studied genetics with J. B. S. Haldane [q.v.] at University College, London; joined Lister Institute, 1934; colleague of D. W. W. Henderson [q.v.]; studied genetics of individuality by investigating marker substances (antigens) on the surface of red cells and tissues; returned to Guy's Hospital, 1940-6; collaborated on immunogenetics with George D. Snell and Sally Lyman Allen at Bar Harbour, Maine, 1946-7; reader in experimental pathology, Guy's Hospital, 1947; published many papers; one of the most important contributors to the study of organ and tissue graft rejection, tumour immunity, and the genetics of immune responsiveness; FRS, 1960.

GORST, Sir JOHN ELDON (1835-1916), lawyer and politician; BA, St. John's College, Cambridge (third wrangler); went to New Zealand, 1860; returned to England and called to bar, 1865; conservative MP, borough of Cambridge, 1866; MP, Chatham, and QC, 1875; member of 'fourth party', 1880-4; solicitor-general and knighted, 1885; under-secretary of state for India, 1886; MP, Cambridge University, 1892-1906; last vice-president of committee of Privy Council on education, 1895-1902; left conservative party over Chamberlain's fiscal campaign.

GORST, Sir (JOHN) ELDON (1861-1911), consul-general in Egypt; son of Sir J. E. Gorst [q.v.]; born in New Zealand; called to bar, 1885; attaché to British agency at Cairo, 1886; controller of direct revenues, 1890-2; adviser to ministry of the interior, 1894; financial adviser, 1898-1904; CB, 1900; KCB, 1902; assistant under-secretary of state to Foreign Office, 1904-7; consul-general in Egypt, 1907-11; promoted municipal and local self-government there; passed law for enlarging powers of provincial councils, 1910; broad-minded administrator, with financial and linguistic ability; GCMG, 1911.

GORT, sixth Viscount (1886-1946), field-marshal. [See Vereker, John Standish Surtees Prendergast.]

GOSCHEN, GEORGE JOACHIM, first Viscount Goschen (1831-1907), statesman; grandson of George Joachim Göschen, a Leipzig publisher, and son of a London banker; educated in London and Saxe Meiningen, at Rugby and Oriel College, Oxford; first class, *lit. hum.* and president of the Union, 1853; founded 'Essay Club' at Oxford, 1852; entered father's banking firm; in South America on business, 1854-6; director of Bank of England, 1858; published *Theory of the Foreign Exchanges* (1861) which attracted wide attention; liberal MP for City of London, 1863-80; joined Lord Russell's ministry as vice-president of Board of Trade, 1865, and entered the cabinet as chancellor of the duchy of Lancaster, 1866; president of Poor Law Board in Gladstone's first administration, 1868-71; reformed local government system; first lord of the Admiralty, 1871-4; his refusal to reduce estimates largely responsible for dissolution of government in 1874; investigated financial position of Egypt at viceroy's invitation, 1876; opposed to his party's plan of equalization of borough and county franchise, 1877; MP for Ripon, 1880-5; declined viceroyalty of India, 1880; went on special embassy to Sultan, to compel Turks to carry out treaty of Berlin as regards Greece, Montenegro, and Bulgaria, 1880-1; refused secretaryship for war, 1882, and speakership of Commons, 1883; opposed to radicalism of Joseph Chamberlain and Sir Charles Dilke [qq.v.]; out of sympathy with Gladstone's foreign policy, and Parnell's policy of home rule; elected for East Edinburgh, defeating extreme radical candidate, 1885; joined Lord Hartington [q.v.] in forming liberal unionist party, and helped to defeat Gladstone's home rule bill, 1886; chancellor of Exchequer in succession to Lord Randolph Churchill [q.v.] in Lord Salisbury's government, 1886-92; MP for St. George's, Hanover Square, 1887-1900; converted national debt from 3 per cent. to $2\frac{1}{2}$ per cent. stock, 1888; his firmness prevented financial panic in Baring crisis, 1890; strenuous in opposition to Gladstone's home rule bill of 1893; first lord of the Admiralty in Salisbury's third administration, 1895-1900; made vast increases in naval strength and naval estimates; created viscount, 1900; published life of grandfather (1903) and *Essays on Economic Questions* (1905); opposed Chamberlain's fiscal policy, 1903; showed remarkable consistency of character as statesman; powerful speaker; busy in non-political affairs; ecclesiastical commissioner, 1882; strong supporter of extension of university teaching in London from 1879; hon. DCL, Oxford, 1881; hon. LLD, Aberdeen and Cambridge, 1888, and Edinburgh, 1890; chancellor of Oxford University, 1903.

GOSLING, HARRY (1861-1930), trade-union leader; apprenticed as Thames waterman, 1875; took leading part in struggle to improve working conditions of dock and river labour throughout country; ardent trade-unionist; alderman of London County Council, 1898-1925; labour MP, Whitechapel and St. George's, 1923-30; minister of transport and paymaster-general, 1924.

GOSSAGE, SIR (ERNEST) LESLIE (1891-1949), air marshal; educated at Rugby and Trinity College, Cambridge; served in Royal Flying Corps, 1915-18; DSO, 1919; commanded No. 11 (fighter) group, 1936-40; Balloon Command, 1940-4; Air Training Corps, 1944-6; KCB, 1941.

GOSSE, SIR EDMUND WILLIAM (1849-1928), poet and man of letters; son of P. H. Gosse [q.v.]; worked in catalogue section of British Museum, 1865-75; began to study literature, from which he had been debarred by rigid upbringing; first writer to introduce Ibsen to English public; as writer for reviews earned reputation of sound critic; translator to Board of Trade, attached to commercial department, 1875; Clark lecturer in English literature, Trinity College, Cambridge, 1885-90; attacked for inaccuracy by J. Churton Collins [q.v.], 1886; librarian to House of Lords, 1904-14; contributed weekly literary articles to *Sunday Times*, 1918-28; knighted, 1925; his numerous volumes of poetry, criticism, and biography include *Collected Poems* (1911), *Seventeenth Century Studies* (1883), and *Father and Son* (1907).

GOSSELIN, SIR MARTIN LE MARCHANT HADSLEY (1847-1905), diplomatist; secretary of embassy at Brussels, Madrid, Berlin, and Paris, 1885-98; joint British delegate in customs tariffs conferences at Brussels, 1889-90; secretary of international conference for suppression of African slave-trade, 1889-90; CB, 1890; member of commission for delimitation of French and English possessions about river Niger, 1898; KCMG, 1898; British envoy at Lisbon, 1902-5; GCVO, 1904; joined Roman Church, 1878.

GOSSET, WILLIAM SEALY (known as 'STUDENT') (1876-1937), statistician and industrial research scientist; scholar of Winchester and New College, Oxford; first class, natural science, 1899; joined Arthur Guinness, Son & Co. (1899) and became head brewer (1937); applied statistical methods to problems of the chemist and biologist and did pioneer work in agricultural experimentation.

GOTCH, JOHN ALFRED (1852-1942), architect and author; educated at Kettering grammar school, Zürich, and King's College, London; in partnership with Charles Saunders in Kettering; built houses, schools, banks, war memorials, etc.; admirer of the old building crafts; no sympathy with modern trends; PRIBA, 1923-5; authoritative publications include *Architecture of the Renaissance in England*

(2 vols., 1891), *The Growth of the English House* (1909), *Inigo Jones* (1928).

GOTT, JOHN (1830-1906), bishop of Truro; MA, Brasenose College, Oxford, 1854; DD, 1873; vicar of Leeds, 1873-85; founded Leeds clergy school, 1875; dean of Worcester, 1886; bishop of Truro, 1891-1906; inherited valuable library, which included the four Shakespeare folios; published *The Parish Priest of the Town* (1887).

GOTT, WILLIAM HENRY EWART (1897-1942), lieutenant-general; educated at Harrow and Sandhurst; commissioned in King's Royal Rifle Corps, 1915; nicknamed 'Strafer'; taken prisoner, 1917; commanded 1st battalion in Egypt, 1938-9; GSO 1, 7th Armoured division, 1939; commanded its support group in successful campaign against Italians, 1940-1; 7th Armoured division in autumn offensive, 1941; XIII Corps, 1942; outstanding desert leader; selected by (Sir) Winston Churchill to command Eighth Army but shot down returning to Cairo, Aug. 1942; DSO and bar, 1941; CB, 1942.

GOUGH, SIR CHARLES JOHN STANLEY (1832-1912), general; born in India; joined army, 1848; VC for gallantry during Indian Mutiny; KCB for services in Afghan war, 1881; general, 1891; retired, with GCB, 1895.

GOUGH, HERBERT JOHN (1890-1965), engineer and expert on metal fatigue; educated at Regent Street Polytechnic Technical School, London and University College School; apprenticed to Messrs. Vickers, Sons & Maxim, 1909-13; designer draughtsman; B.Sc. London (and later, D.Sc. and Ph.D.); joined staff of National Physical Laboratory, 1914-38; superintendent, engineering department, 1930; served with Royal Engineers, 1914-19; MBE (military), 1919; at NPL established causes of metal fatigue and developed new methods of design; published *The Fatigue of Metals* (1924); first director of scientific research, War Office, 1938; director-general of scientific research and development, Ministry of Supply, 1942-5; engineer-in-chief, Lever Brothers and Unilever Ltd., 1945-55; CB, 1942; FRS, 1933; council member, 1939-40.

GOUGH, SIR HUBERT DE LA POER (1870-1963), general; elder son of (General Sir) Charles John Stanley Gough, GCB, VC [q.v.]; educated at Eton and Sandhurst; commissioned in 16th Lancers, 1889; served in India and in the South African war; instructor, staff College, 1903-6; commanded 16th Lancers; brigadier-general commanding 3rd Cavalry Brigade at the Curragh, 1911; prepared to resign rather than

initiate military operations against Ulster, 1914; major-general; lieutenant-general; commanded 7th division and 1st Corps in France, 1915; commanded Fifth Army in third Ypres campaign, 1917; replaced by Sir Henry Rawlinson [q.v.], 1918; retired with rank of full general, 1922; KCB, 1916; KCVO, 1917; GCMG, 1919; published his account of 1918 campaign in *Fifth Army* (1931); amends made by Lloyd George in his *War Memoirs* (1936); GCB, 1937; commanded a London zone Home Guard, 1940-2; published memoirs, *Soldiering On* (1954).

GOUGH, SIR HUGH HENRY (1833-1909), general; brother of Sir C. J. S. Gough [q.v.]; born at Calcutta; joined Bengal army, 1853; served in Indian Mutiny; at siege of Delhi and relief of Cawnpore; won VC for gallantry at Alambagh, 1857; conspicuous for bravery at Lucknow, 1858; commanded 12th Bengal cavalry in Abyssinia campaign, 1868; CB, 1868; commanded cavalry of Kuram field force, 1878-9; brigadier-general of communications with Kabul field force; commanded cavalry brigade in march to Kandahar, 1880; KCB, 1881; general, 1894; GCB, 1896; keeper of crown jewels in London, 1898; published *Old Memories* (1897).

GOUGH, JOHN EDMUND (1871-1915), brigadier-general; son of Sir C. J. S. Gough [q.v.]; born in India; joined army, 1892; VC, 1903; served on staff of (Lord) Haig [q.v.], 1914-15; killed in France; posthumous KCB, 1915.

GOUGH-CALTHORPE, AUGUSTUS CHOLMONDELEY, sixth BARON CALTHORPE (1829-1910), agriculturist; educated at Eton and Merton College, Oxford; MA, 1855; succeeded brother in peerage, 1893; started at Elvetham famous herd of shorthorn cattle, Southdown sheep, and Berkshire pigs; generous donor of land to Birmingham city (1894) and university (1900).

GOUGH-CALTHORPE, SIR SOMERSET ARTHUR (1864-1937), admiral of the fleet. [See CALTHORPE.]

GOULD, SIR FRANCIS CARRUTHERS (1844-1925), cartoonist; drew caricatures for amusement while broker and jobber on stock exchange; member of *Pall Mall Gazette* staff, 1890; of *Westminster Gazette*, 1893-1914; assistant editor, 1896; a keen radical; knighted, 1906; edited his own paper, *Picture Politics*, 1894-1914; his political cartoons show wit, wealth of ideas, faculty for seizing a likeness, and urbanity; his favourite subject Joseph Chamberlain [q.v.].

GOULD, NATHANIEL (1857-1919), known as NAT GOULD, novelist; journalist in Australia, 1884-95; first book, *The Double Event* (1891), great success; wrote about 132 books all concerned with horse-racing.

GOULDING, FREDERICK (1842-1909), master printer of copper plates; 'devil' to J. A. M. Whistler [q.v.] in printing some of his etchings, 1859; friend of Sir F. S. Haden [q.v.] from 1862; printed works by Whistler, Haden, Alphonse Legros [q.v.], Rodin, and others; assistant to Legros in etching class at National Art Training School, 1876-82; succeeded Legros, 1882-91; first master printer to Royal Society of Painter-Etchers, 1890.

GOWER, (EDWARD) FREDERICK LEVESON- (1819-1907), politician. [See LEVESON-GOWER.]

GOWER, SIR HENRY DUDLEY GRESHAM LEVESON (1873-1954), cricketer; grandson of first Baron Leigh [q.v.]; captain of cricket at Winchester (1892) and Oxford (1896); of MCC team in South Africa, 1909-10; of Surrey, 1908-10; president, Surrey County Cricket Club, 1929-40; test match selector; associated with Scarborough Cricket Festival for fifty years; knighted, 1953.

GOWERS, SIR ERNEST ARTHUR (1880-1966), public servant; son of (Sir) William Richard Gowers [q.v.]; educated at Rugby and Clare College, Cambridge; first class, classical tripos, 1902; entered Civil Service; posted to Inland Revenue Dept and transferred to India Office; called to the bar (Inner Temple), 1906; private secretary to several ministers; principal private secretary to David Lloyd George, chancellor of the Exchequer, 1911; chief inspector, National Health Insurance Commission (England), 1912-17; secretary to Conciliation and Arbitration Board for Government Employees, 1917; director of production, Mines Department (Board of Trade), 1919; permanent under-secretary for mines, 1920; chairman, Board of Inland Revenue, 1927-30; chairman, coal commission, 1938; chairman, manpower sub-committee of Committee of Imperial Defence; regional commissioner for civil defence, London region, 1939; senior regional commissioner, 1941; chairman of number of other commissions and committees; chairman, National Hospitals for Nervous Diseases, London, 1948-57; one of greatest public servants of his day; published *Plain Words: a Guide to the Use of English* (1948), written at invitation of Sir Edward (later Lord) Bridges [q.v.]; other publications include *ABC of Plain Words* (1951), and a revision of *Modern English Usage* (1965),

by H. W. Fowler [q.v.]; hon. D.Litt., Manchester; hon. associate, Royal Institute of British Architects; CB, 1917; KBE, 1926; KCB, 1928; GBE, 1945; GCB, 1953.

GOWERS, SIR WILLIAM RICHARD (1845-1915), physician; educated at University College, London; MRCS, 1867; on staff of hospital for paralysed and epileptic, Queen Square, London, 1870; and on that of University College Hospital, 1872; FRS, 1887; specialized in neurology; chief work, *A Manual of Diseases of the Nervous System* (1886).

GOWRIE, first EARL OF (1872-1955), soldier and governor-general of Australia. [See HORE-RUTHVEN, ALEXANDER GORE ARKWRIGHT.]

GRACE, EDWARD MILLS (1841-1911), cricketer; studied medicine at Bristol; MRCS, 1865; practised at Thornbury from 1869 till death; coroner for West Gloucestershire, 1875-1909; played cricket for Gentlemen *v.* Players between 1862 and 1869; visited Australia with George Parr's team, 1863; with brothers raised Gloucestershire to first-class cricketing county and played for England *v.* Australia, 1880; unorthodox and forcible batsman; pioneer of 'pull' stroke; unrivalled as fielder at point; played till age of seventy.

GRACE, WILLIAM GILBERT (1848-1915), cricketer; brother of Edward Mills Grace [q.v.]; studied medicine in Bristol and London; MRCS, 1879; surgeon in Bristol, 1879-99; played cricket for Gentlemen. *v.* Players between 1865 and 1906; supreme as batsman in England, 1866-76; long series of extraordinary scores reached its zenith in 1876 with 400, not out; with brothers started, and made first-class, Gloucestershire county eleven, 1870; visited Australia, 1873 and 1891; played for England *v.* Australia, 1880 and 1882; presented with £5,000 by *Daily Telegraph* fund, 1895; in 43 years' career made 126 centuries, scored 54,896 runs, and took 2,876 wickets; first-rate as bowler and fielder as well as batsman; his powerful physique capable of great endurance; known to public as 'W. G.' and widely recognized by his thick black beard.

GRAHAM, HENRY GREY (1842-1906), writer on Scottish history; educated at Edinburgh University; Church of Scotland minister in Glasgow, 1884-1906; published life of *Rousseau* (1882), *Social Life of Scotland in the 18th Century* (2 vols., 1899), and *Scottish Men of Letters of the 18th Century* (1901).

GRAHAM, HUGH, BARON ATHOLSTAN (1848-1938), newspaper proprietor; born in

Quebec; founded *Montreal Star* (1869), *Family Herald, Weekly Star*, and *Montreal Standard*; acquired *Montreal Herald*; uncannily skilful in assessing news values and public taste; strong protectionist and imperialist; gave generously to hospitals and medical research, etc.; knighted, 1908; baron, 1917.

GRAHAM, JOHN ANDERSON (1861-1942), missionary; educated at Glasgow high school and Edinburgh University; ordained and appointed missionary of Church of Scotland Young Men's Guild in Kalimpong, 1889; made it centre of educational, welfare, and religious advance; founded children's (Dr Graham's) homes, 1900; first missionary moderator, General Assembly, Church of Scotland, 1931; died in Kalimpong.

GRAHAM, ROBERT BONTINE CUN-NINGHAME (1852-1936), traveller, scholar, Scottish nationalist, socialist, etc.; educated at Harrow and in Brussels; rode with 'gauchos' in Spanish America; became friend of 'Buffalo Bill' in Mexico; equally at home in Spain, Morocco, or Scotland; liberal MP, North-West Lanarkshire, 1886-92; became ardent socialist, devoted to William Morris [q.v.]; imprisoned (1887) after Trafalgar Square riot; first president, national party of Scotland, 1928; works include studies of old Spanish life and the *conquistadores*, and volumes of stories, essays, and sketches; died at Buenos Aires; new city of Don Roberto (Argentina) named after him.

GRAHAM, SIR RONALD WILLIAM (1870-1949), diplomatist; educated at Eton and abroad; entered diplomatic service, 1892; first secretary, 1904; counsellor of embassy, Cairo, 1907; adviser, Egyptian ministry of interior, 1910-16; represented British commanding officer with Egyptian administration, 1914-16; assistant under-secretary, Foreign Office, 1916-19; minister to Holland, 1919-21; ambassador to Italy in an increasingly difficult period, 1921-33; signed four-power pact on behalf of Britain, 1933; KCMG, 1915; PC, 1921; GCVO, 1923.

GRAHAM, THOMAS ALEXANDER FERGUSON (1840-1906), artist; studied at Edinburgh under Scott Lauder [q.v.]; in Paris (1860), Brittany (1862), Venice (1864), Morocco (1885); painted much in Fifeshire fishing villages; in 'The Passing Salute' and 'The Siren' showed command of colour and sense of atmosphere; hon. RSA, 1883; friend of Orchardson and Pettie [qq.v.].

GRAHAM, WILLIAM (1839-1911), philosopher and political economist; BA, Trinity College, Dublin 1867; MA, 1870; vindicated Berkeley's philosophy in *Idealism*, 1872; lecturer in mathematics at St. Bartholomew's Hospital, 1877; his *The Creed of Science* (1881) praised by Darwin and Gladstone; professor of jurisprudence and political economy in Queen's College, Belfast, 1882-1909; hon. Litt.D., Dublin, 1905; other works include *Socialism New and Old* (1890) and *English Political Philosophy from Hobbes to Maine* (1899).

GRAHAM, WILLIAM (1887-1932), labour leader; educated at George Heriot's School, Edinburgh; became journalist; joined independent labour party, 1906; elected to Edinburgh town council, 1906; MA, Edinburgh, 1915; labour MP, Central Edinburgh, 1918-31; served on royal commissions on income-tax and Oxford and Cambridge, and many committees; financial secretary to Treasury, 1924; president of Board of Trade, 1929-31; responsible for coal-mines bill, several overseas missions, and industrial inquiries; PC, 1924.

GRAHAM BROWN, THOMAS (1882-1965), neurophysiologist and mountaineer. [See BROWN.]

GRAHAM-HARRISON, SIR WILLIAM MONTAGU (1871-1949), parliamentary draftsman; born HARRISON; changed name on marriage, 1900; educated at Wellington and Magdalen College, Oxford; first class, jurisprudence, 1894; fellow of All Souls, 1895; called to bar (Lincoln's Inn), 1897; member, London school board, 1900-3; entered parliamentary counsel office, 1903; solicitor for HM Customs and Excise, 1913; second parliamentary counsel, 1917; first, 1928-33; KC, 1930; KCB, 1926.

GRAHAM-LITTLE, SIR ERNEST GORDON GRAHAM (1867-1950), physician, and member of parliament for London University; born in India; educated in South Africa; BA, Cape University, 1887; graduated in medicine, London, 1893; physician, skin department, and lecturer in dermatology, St. Mary's Hospital, 1902; consulting physician, 1934; member, London University senate, 1906-50; independent MP, London University, 1924-50; first elected to oppose Haldane's proposed reforms; sturdy progressive individualist; implacable opponent of national health service; knighted, 1931.

GRAHAME, KENNETH (1859-1932), author; educated at St. Edward's School, Oxford; clerk in Bank of England, 1879; secretary of Bank of England, 1898-1908; publications include *The Golden Age* (1895), *Dream Days* (1898), and *The Wind in the Willows* (1908).

GRAHAME-WHITE, CLAUDE (1879–1959), pioneer aviator and aircraft manufacturer; educated at Bedford grammar school; trained as engineer; obtained pilot's certificate from French Aero Club and made notable flights in England and America, 1910; set up London Aerodrome and aviation company at Hendon, 1911; trained pilots and made advances in design and manufacture; his factories and aerodrome purchased by Government after 1914–18 war; wrote many books on flying.

GRANET, Sir (WILLIAM) GUY (1867–1943), barrister, railway administrator and chairman; educated at Rugby and Balliol College, Oxford; called to bar (Lincoln's Inn), 1893; secretary, Railway Companies' Association, 1900; general manager, Midland Railway Company, 1906; director, 1918; chairman, 1922; chairman, London Midland and Scottish Railway, 1924–7; on royal commission on Civil Service, 1912–14; national economy committee, 1921–2; director-general, movements and railways, 1917; chairman, British and Allied provisions commission, 1918; knighted, 1911; GBE, 1923.

GRANT, Sir (ALFRED) HAMILTON, twelfth baronet, of Dalvey (1872–1937), Indian civil servant; son of Sir Alexander Grant [q.v.]; educated at Fettes and Balliol College, Oxford; served in foreign and political department, North-West Frontier Province; deputy secretary, Indian foreign department, 1912; foreign secretary, 1914–19; successfully conducted preliminary negotiations (1919) for treaty with Afghanistan (1921); chief commissioner, North-West Frontier Province, 1919–22; KCIE, 1918; KCSI, 1922; succeeded brother, 1936.

GRANT, Sir CHARLES (1836–1903), Anglo-Indian administrator; brother of Sir Robert Grant [q.v.]; entered Bengal civil service, 1858; foreign secretary to Indian government, 1881; KCSI, 1885; edited *Central Provinces Gazetteer* (1870).

GRANT, GEORGE MONRO (1835–1902), principal of Queen's University, Kingston, Canada, from 1877; born in Nova Scotia; educated at Glasgow University; hon. DD, 1877; Presbyterian missionary in Nova Scotia, 1860–3; united the Presbyterian Church throughout Canada, and inaugurated first General Assembly, 1875; moderator, 1889; secured state endowment of School of Mines in Queen's University, 1893; twice (1872 and 1883) travelled through Canada, describing his experiences in *Ocean to Ocean* (1873) and *Picturesque Canada* (1884); strong imperialist;

wrote *Religions of the World* (1894); president of Royal Society of Canada, 1891; CMG, 1901.

GRANT, Sir ROBERT (1837–1904), lieutenant-general; born at Bombay; son of Sir Robert Grant [q.v.]; joined Royal Engineers, 1854; served in West Indies and North America, 1857–65; in command of Royal Engineers at Aldershot (1877–80), Plymouth (1880), Woolwich (1881), and in Scotland (1884); and with Nile expeditionary force, 1885; CB, 1889; inspector-general of fortifications and major-general, 1891; lieutenant-general, 1897; carried out works of defence and barrack construction; KCB, 1896; GCB, 1902.

GRANT DUFF, Sir MOUNTSTUART ELPHINSTONE (1829–1906), statesman and author; son of James Grant Duff [q.v.]; MA, Balliol College, Oxford, 1853; LLB, London, and called to bar, 1854; liberal MP for Elgin Boroughs, 1857–81; became authority on questions of foreign policy; under-secretary of state for India, 1868–74, and for colonies, 1880; PC, 1880; governor of Madras, 1881–6; CIE, 1881; GCSI, 1887; on return to England (1887) devoted himself to literature; published *Notes from a Diary* (14 vols., 1897–1905), a valuable contribution to social history; travelled much in Europe and Asia; lord rector of Aberdeen University, 1866–72; president of Royal Geographical and Historical societies; FRS, 1901; trustee of British Museum, 1903; wrote also *Studies of European Politics* (1866) and *A Victorian Anthology* (1902).

GRANTHAM, Sir WILLIAM (1835–1911), judge; called to bar, 1863; QC, 1877; treasurer of Inner Temple, 1904; successful in circuit work; conservative MP for East Surrey, 1874–85, and for Croydon, 1885–6; made judge of Queen's Bench division and knighted, 1886; industrious, energetic, but garrulous judge; his decisions as judge in election petition cases (1906) were severely criticized; rebuked by prime minister for indiscreet speech from bench, 1911; model country gentleman, and good judge of horses.

GRANVILLE-BARKER, HARLEY GRANVILLE (1877–1946), actor, producer, dramatist, and critic; first London stage appearance, 1892; acted and produced for Stage Society; great friend of G. B. Shaw [q.v.]; made Shaw's name as dramatist and his own as director, in partnership with J. E. Vedrenne at Court Theatre, 1904–7; produced repertory for Charles Frohman including Galsworthy's *Justice* (1910); his productions of *The Winter's Tale* and *Twelfth Night* (1912) and *Midsummer Night's Dream* (1914) set completely new standard and proved profoundly influential; made great success with Shaw's *Fanny's First Play* (1911) and Arnold Bennett's *The Great Adventure* (1913); opened season at St. James's with *Androcles and the Lion* (1913); war interrupted his career, which he virtually abandoned after his second marriage, 1918; chairman, British Drama League, 1919-32; wrote valuable prefaces for edition of Shakespeare, 1923-46; director, British Institute, Paris, 1937-9; his own plays included *Weather Hen*, *The Marrying of Ann Leete*, *Waste*, *The Voysey Inheritance*, and *The Madras House*.

GRAVES, ALFRED PERCEVAL (1846-1931), author and educationist; son of Charles Graves [q.v.]; educated at Trinity College, Dublin; Home Office clerk, 1869-75; inspector of schools (1875-1910), for many years in Taunton, from 1895 in Southwark; with (Sir) C. V. Stanford [q.v.] published *Songs of Old Ireland* (1882) and *Songs of Erin* (1892); poet, essayist, and anthologist on Irish subjects.

GRAVES, GEORGE WINDSOR (1873?-1949), comedian; gave unique presentation of comic elderly men; memorable performances in *The Merry Widow* (1907) as Baron Popoff and in *Me and My Girl* (1937); played comedy lead in Drury Lane pantomimes, 1909-15; successful also on music-halls and films.

GRAY, Sir ALEXANDER (1882-1968), economist and poet; educated at Dundee high school and Edinburgh University; first class, mathematics, 1902, and economic science, 1905; placed second in civil service examination to John Anderson (later Viscount Waverley, q.v.), a lifelong friend; posted to Local Government Board, 1905-9; transferred to Colonial Office, 1909-12; National Health Insurance Commission, 1912-19; Ministry of Health, 1919-21; Jeffrey professor of political economy, Aberdeen, 1921-35; Edinburgh, 1935-56; publications include *The Scottish Staple at Veere* (1909), *Development of Economic Doctrine* (1931), and *Socialist Tradition, Moses to Lenin* (1946); produced volumes of translations of songs and ballads from German, Dutch, and Danish, including *Songs and Ballads Chiefly from Heine* (1920), *Arrows* (1932), and *Four and Forty* (1954); also published his own verse in English or in Scottish dialect, including *Any Man's Life* (1924) and *Gossip* (1928); served on number of government committees; chairman, Scottish Schools Broadcasting Council; member, Fulbright Commission; president, Scottish Economic Society, 1960-3; CBE, 1939; knighted, 1947; hon. degrees from four universities, including Aberdeen and Edinburgh.

GRAY, Sir ARCHIBALD MONTAGUE HENRY (1880-1967), dermatologist; educated at Cheltenham College and University College and Hospital, London; MRCS, LRCP, MB London, 1903; BS, 1904; resident and junior appointments at University College Hospital and Hospital for Women, Soho Square, 1904-9; MD, 1905; MRCP, 1907; FRCS, 1908; fellow, University College; first obstetric registrar, University College Hospital; physician for diseases of the skin, University College Hospital, 1909-46; served at War Office in RAMC, 1914-18; consulting dermatologist with army in France, 1918-19; CBE; consulting dermatologist to RAF; in charge of skin department, Hospital for Sick Children, Great Ormond Street, 1920-34; made important observations on rare disease, *sclerema neonatorum*, 1926; editor, the *British Journal of Dermatology*, 1916-29; president, British Association of Dermatology, 1938-9; represented London University on governing body of Postgraduate Medical School, Hammersmith, 1935-60; contributed to sections on skin diseases in official publications; president, Royal Society of Medicine, 1940-2; advisor in dermatology to Ministry of Health, 1948-62; chairman, medical committee, University College Hospital, 1926-35; member, London University senate, 1929-50; dean of faculty of medicine, 1932-6; member, General Medical Council, 1950-2; hon. LLD London, 1958; knighted, 1946; KCVO, 1959.

GRAY, BENJAMIN KIRKMAN (1862-1907), economist; son of Congregational minister; Unitarian minister 1894-7; engaged in social work in London, 1898-1902; developed strong socialistic views; published *History of English Philanthropy* (1905) and *Philanthropy and the State* (1910).

GRAY, GEORGE BUCHANAN (1865-1922), Congregational minister; BA, University College, London, 1886; first class, school of oriental studies, Oxford, 1891; tutor, Mansfield College, Oxford, 1891; ordained Congregational minister, 1893; professor of Hebrew and exegesis of Old Testament, Mansfield, 1900-22; a stimulating teacher and original investigator; interested in problems of social welfare; works include *Hebrew Proper Names* (1896), *Commentary on Numbers* (1903), *Commentary on Isaiah I-XXVII* (1912), *Forms of Hebrew Poetry* (1915), *Commentary on Job* (completing work of S. R. Driver, q.v., 1921), and *Sacrifice in the Old Testament* (1925).

GRAY, GEORGE EDWARD KRUGER (1880-1943), designer; born KRUGER; added Gray, 1918; diploma in design, Royal College of

Art; work includes King George V and VI silver coinage; great seals of King George VI and Canada; stained-glass windows, Eltham Palace; and heraldic designs.

GRAY, HERBERT BRANSTON (1851–1929), schoolmaster; BA, Queen's College, Oxford; assistant master, Westminster, 1875; ordained, 1877; headmaster of Louth grammar school, 1879; of St. Andrew's College, Bradfield, 1880–1910; saved college from extinction, supplying new consitution and buildings and raising numbers from 50 to 300; founded Bradfield Greek Play; on retirement travelled, wrote, and held two livings.

GRAY, LOUIS HAROLD (1905–1965), physicist and radiobiologist; educated at Latimer School, Christ's Hospital and Trinity College, Cambridge; first class, parts i and ii, natural sciences tripos, 1926–7; admitted to Cavendish Laboratory under Sir Ernest (later Lord) Rutherford [q.v.], 1928; Ph.D., fellow of Trinity, 1930; senior physicist and Prophit scholar of Royal College of Surgeons at Mount Vernon Hospital, Middlesex, 1933–46; set up physics laboratory to measure radiation in treatment of cancer; planned and built 400 kV neutron generator; recruited by Sir Edward Mellanby [q.v.] to head laboratory side of the radiotherapeutic unit, Hammersmith Hospital, 1946; research in basic science of radiobiology; director, British Empire Cancer Research Campaign research unit, Mount Vernon Hospital, 1953–65; FRS, 1961; hon. D.Sc., Leeds, 1962; president British Institute of Radiology, 1950; founder and first chairman, Association for Radiation Research.

GREAVES, WALTER (1846–1930), painter; son of a Chelsea waterman; pupil of J. A. M. Whistler [q.v.]; painted portraits and landscape subjects, largely of Chelsea district, in oil and water-colour; later works show influence of Whistler; work not 'discovered' and exhibited until 1911.

GREEN, ALICE SOPHIA AMELIA (1847–1929), better known as MRS STOPFORD GREEN, historian; born STOPFORD; came from Ireland to England, 1874; married J. R. Green, the historian [q.v.], 1877; became ardent radical and home ruler; took up study of early Irish history; settled in Dublin, c.1917; works include *Henry the Second* (1888), *Town Life in the Fifteenth Century* (1894), *The Making of Ireland and its Undoing* (1908), and *A History of the Irish State to 1014* (1925).

GREEN, CHARLES ALFRED HOWELL (1864–1944), archbishop; scholar of Charter-

house and Keble College, Oxford; deacon, 1888; priest, 1889; vicar of Aberdare, 1893–1914; archdeacon of Monmouth and canon of Llandaff, 1914–21; bishop of Monmouth, 1921; of Bangor, 1928–44; archbishop of Wales, 1934–44; published *The Setting of the Constitution of the Church in Wales* (1937).

GREEN, FREDERICK WILLIAM EDRIDGE- (1863–1953), authority on colour perception. [See EDRIDGE-GREEN.]

GREEN, GUSTAVUS (1865–1964), aero-engine designer; established cycle business at Bexhill, 1897; took out first patent for internal combustion engine, 1900; founded Green Motor Patents Syndicate, manufacturing small stationary engines and motor-cycles, 1906–14; Green water-cooled V.8 engine used in airship *Gamma*, 1910; produced most successful British aero-engines, 1909–14; international prizes won by aeroplanes and motor boats equipped with Green engines; technical director, Green Engine Co. Ltd., 1912–25; a gifted mechanic rather than a trained engineer; hon. companion, Royal Aeronautical Society, 1958.

GREEN, SAMUEL GOSNELL (1822–1905), Baptist minister and bibliophile; joined Baptist ministry, 1844; classical tutor (from 1851) and president (1863–76) of Horton College, Bradford; editor (1876), editorial secretary (1881), and historian (1899) of the Religious Tract Society; published *Handbook to Grammar of Greek Testament* (1870), *Christian Ministry to the Young* (1883), and other theological works; president of Baptist Union, 1895; chairman of editorial committee of *Baptist Hymnal*; assisted Mrs Rylands in forming John Rylands library, Manchester, 1899; hon. DD, St. Andrews, 1900.

GREEN, WILLIAM CURTIS (1875–1960), architect; educated at Newton College; articled to John Belcher [q.v.] and trained at Royal Academy Schools under Phené Spiers [q.v.]; commenced practice, 1898; FRIBA, 1909, and Royal medallist, 1942; works include Wolseley House (later Barclays Bank), Westminster Bank, and Stratton House, Piccadilly; London Life Association, King William Street; Scotland Yard, Embankment; Cambridge University Press, Euston Road; Dorchester Hotel, Park Lane; Queen's Hotel, Leeds; Stockgrove Park near Leighton Buzzard; a scholarly architect who produced work of lasting English quality; ARA, 1923; RA, 1933; president, Architectural Association.

GREENAWAY, CATHERINE (KATE) (1846–1901), artist; studied art at South Kensington, at Heatherley's, and under Alphonse

Legros [q.v.] in Slade School, London; exhibited at Royal Academy, 1877; won fame as exponent of child life and inventor of original children's books; published *Under the Window* (1879), *Kate Greenaway's Birthday Book* (1880), *Mother Goose* (1881), *Little Ann and other Poems* (1883); also *Language of Flowers* (1884), *Marigold Garden* (1885), and an annual *Almanack* from 1883 to 1895; work much admired by John Ruskin [q.v.]; created a gallery of children with quaint costumes and unhackneyed accessories; designed beautiful book-plates.

GREENE, HARRY PLUNKET (1865–1936), singer; born in Ireland; educated at Clifton; professor of singing, Royal Academy (1911–19) and Royal College (1912–19) of Music; first appeared as bass-baritone, 1888; established reputation in *Lieder* and Brahms songs, and as interpreter of (his father-in-law) Parry's cantatas and oratorios and of Stanford's Irish song-cycles.

GREENE, WILFRID ARTHUR, BARON GREENE (1883–1952), judge; educated at Westminster and Christ Church, Oxford; first class, *lit. hum.*, 1906; fellow, All Souls, 1907; Vinerian scholar and called to bar (Inner Temple), 1908; bencher, 1925; KC, 1922; lord justice of appeal, knighted, and PC, 1935; master of Rolls, 1937–49; baron, 1941; lord of appeal in ordinary, 1949–50; chairman of committees on company law (1925), trade practices (1930), international communications (1931), and beet-sugar industry (1935); principal, Working Men's College, 1936–44; chairman, National Buildings Record Office, 1941–5; DCL, Oxford, and number of honorary degrees.

GREENE, WILLIAM FRIESE- (1855–1921), pioneer of cinematography; began life as travelling photographer; experimented with J. A. R. Rudge, of Bath, on reproduction, by camera and projecting lantern, of synthesis of motion, 1882–4; established photographic business in London; after many experiments produced sensitized celluloid ribbon-film, 1889; film patented, and projected scene exhibited, 1890.

GREENE, SIR (WILLIAM) GRAHAM (1857–1950), civil servant; educated at Cheltenham College; entered Admiralty, 1881; assistant private secretary and head of private office to successive first lords, 1887–1902; principal clerk, in charge of personnel branch and carrying educational reforms into effect, 1902; assistant secretary, Admiralty, 1907; permanent secretary, 1911; retired by Lloyd George (1917); appointed by (Sir) Winston Churchill as secretary, Ministry of Munitions, 1917–20; KCB, 1911.

GREENBRIDGE, ABEL HENDY JONES (1865–1906), writer on ancient history and law; born at Barbados; BA, Balliol College, Oxford, 1888; MA, 1891; D.Litt., 1904; fellow (1889) and tutor (1902) of Hertford College; fellow of St. John's, 1905; published *Infamia* (1894), *Handbook on Greek Constitutional History* (1896); edited William Smith's and Gibbon's histories of Rome; commenced a new *History of Rome* (vol. i, 1904); accurate and critical historian.

GREENWELL, WILLIAM (1820–1918), archaeologist; brother of Dora Greenwell [q.v.]; BA, Durham; minor canon of Durham, 1854–1907; librarian to dean and chapter, 1862–1907; rector of St. Mary-the-Less, Durham, 1865–1918; documents edited include *Feodarium Prioratus Dunelmensis* (1872); FRS, 1878.

GREENWOOD, ARTHUR (1880–1954), politician; educated at higher grade school and Yorkshire College, B.Sc., Leeds, 1905; lecturer in economics, 1913; moved to London, 1914; civil servant, 1916–20; secretary, labour party research department, 1920–43; responsible for much of the party's constructive thought after 1931 and united intellectual and trade-union wings of party; MP, Nelson and Colne, 1922–31, Wakefield, 1932–54; parliamentary secretary, Ministry of Health, 1924; minister of health, 1929–31; responsible for Widows', Orphans', and Old Age Contributory Pensions Act, 1929; elected deputy leader of labour party, 1935; treasurer, 1943–54; member, war cabinet, 1940–2, without portfolio; responsible for economic affairs (1940) and reconstruction (1941–2); appointed Beveridge committee; lord privy seal, 1945–7; paymaster-general, 1946–7; minister without portfolio, 1947; PC, 1929; CH, 1945; hon. LLD Leeds.

GREENWOOD, FREDERICK (1830–1909), journalist; at first a printer's reader; published novels, 1854–60; first editor of *Queen* illustrated paper, 1861–3; contributed his novel *Margaret Denzil's History* serially to *Cornhill Magazine*, 1863; succeeded W. M. Thackeray [q.v.] as editor of *Cornhill*, 1862–8; founded with George Smith [q.v.], and edited, *Pall Mall Gazette*, 1865; secured its triumph in 1866 by three papers by his brother James, 'A Night in a Casual Ward'; an independent conservative and vigilant student of foreign affairs; suggested to Lord Beaconsfield purchase of Suez Canal shares, 1875; attacked Gladstone's anti-Turkish attitude, 1876–8; left *Pall Mall Gazette*, on its conversion by a new owner into a radical organ, for newly founded conservative *St. James's Gazette*, 1880; advocated occupation of Egypt, 1882; retired from *St. James's*, 1888; founded and edited weekly *Anti-Jacobin*, 1891-2; opposed South African war, 1899; friend of George Meredith [q.v.]; published *The Lover's Lexicon* (1893), *Imagination in Dreams* (1894).

GREENWOOD, HAMAR, first VISCOUNT GREENWOOD (1870-1948), politician; born in Canada; officer in militia; BA, Toronto, and moved to England, 1895; called to bar (Gray's Inn), 1906; KC, 1919; liberal MP, York (1906-10), Sunderland (1910-22); 'anti-socialist', East Walthamstow (1924-9); served in France, 1914-16; secretary, Overseas Trade Department, 1919-20; chief secretary for Ireland, 1920-2; reinforced Royal Irish Constabulary with undisciplined recruits ('Black and Tans') whose violence he defended; took little part in discussions preceding Irish treaty; honorary treasurer, conservative party, 1933-8; baronet, 1915; baron, 1929; viscount, 1937.

GREENWOOD, THOMAS (1851-1908), promoter of public libraries; library assistant at Sheffield; founded in London (1871) firm of printers of trade journals, which he edited; advocated rate-supported libraries; published *Public Libraries* (1866), *Museums and Art Galleries* (1888), and *Greenwood's Library Year Book* (1897); presented library of Edward Edwards (1812-86, q.v.), as well as his own 'Library for Librarians' (1906), to Manchester public library; wrote life of Edwards, 1902.

GREER, (FREDERICK) ARTHUR, BARON FAIRFIELD (1863-1945), judge; educated at Old Aberdeen grammar school and the university; called to bar (Gray's Inn), 1886; practised in Liverpool and London; KC, 1910; judge of King's bench division and knighted, 1919; presided over trial for murder of F. R. Holt, 1920; lord justice of appeal, 1927-38; his dissenting judgments several times approved in the Lords; member (1917), chairman (1934-6), Council of Legal Education; PC, 1927; baron, 1939.

GREET, SIR PHILLIP BARLING BEN (1857-1936), actor-manager; first London appearance, 1883; entered (1886) upon lifework with series of open-air ('pastoral') performances of Shakespeare; revived *Everyman*, 1902; with Lilian Baylis [q.v.] at Old Vic, 1914-18; gave Shakespearian performances for London County Council schools, etc., 1918-22; with Regent's Park open-air theatre, 1933-5; knighted, 1929.

GREG, SIR WALTER WILSON (1875-1959), scholar and bibliographer; son of William Rathbone Greg and grandson of James Wilson [qq.v.]; educated at Harrow and Trinity College, Cambridge; graduated, 1897; librarian, 1907-13; honorary fellow, 1941; joined Bibliographical Society, 1898; spent sixty years working on descriptive bibliography of English drama in four volumes (1939-59); general editor (1906-39), president (1939-59), Malone Society; close friend of R. B. McKerrow and A. W. Pollard [qq.v.]; edited the Henslowe *Diary* and *Papers* (1904-8) and wrote *Pastoral Poetry and Pastoral Drama* (1906) for A. H. Bullen [q.v.]; his editions of manuscript plays included *Sir Thomas More* (1911); other publications include *Dramatic Documents from The Elizabethan Playhouses* (1931), *English Literary Autographs, 1550-1650* (1925-32), *Marlowe's 'Doctor Faustus' 1604-1616* (1950), and *The Editorial Problem in Shakespeare* (3rd edn., 1954); FBA, 1928; knighted, 1950; hon. D.Litt. Oxford and LLD Edinburgh; hon. fellow, Trinity College, Cambridge.

GREGO, JOSEPH (1843-1908), writer on art; collected works by Rowlandson, Morland, and Cruikshank [qq.v.]; published *Rowlandson the Caricaturist* (2 vols., 1880), *Cruikshank's Water Colours* (1904), *Thackerayana* (1875), *History of Parliamentary Elections* (1886), and *Pictorial Pickwickiana* (2 vols., 1899); organized picture exhibitions; invented a system of reproducing eighteenth-century colour prints.

GREGORY, SIR AUGUSTUS CHARLES (1819-1905), Australian explorer and politician; born at Farnsfield, Nottinghamshire; joined survey department of West Australia, 1841; explored interior of continent; discovered pastoral and mineral wealth of Murchison district; undertook unsuccessful expeditions in search of F. W. L. Leichhardt [q.v.], 1855, 1858; explored rivers of east coast of Australia, 1855-6; surveyor for Queensland, 1859-75; member of legislative council, 1882; first mayor of Toowong, 1902; sat on commission of inquiry into condition of aborigines, 1876-83; CMG, 1875; KCMG, 1903; joint-author with brother Francis of *Journals of Australian Exploration* (1884).

GREGORY, EDWARD JOHN (1850-1909), painter; studied at Royal Academy, 1869; contributed sketches to *Graphic* till 1875; best work exhibited at Royal Institute of Painters in Water Colours, of which he became president in 1898; exhibited mainly portraits at Royal Academy; ARA, 1883; RA, 1898; best-known pictures in oil are 'Marooning' (1887), 'Boulter's Lock—Sunday Afternoon' (1898); in water-colours, 'A Look at the Model' and 'Souvenir of the Institute'.

GREGORY, FREDERICK GUGENHEIM (1893–1961), plant physiologist; born Fritz Gugenheim, son of Jewish expatriate from Germany; educated at Owen's School, Islington and Imperial College, London; first class, botany; ARCS, 1914; B.Sc., 1915; DIC, 1917; M.Sc., 1920; D.Sc., 1921; worked on physiology on greenhouse crops, Cheshunt Experimental Station, 1914–17; victim of abuse for people with German names; changed name by deed poll, 1916; appointed to post in Research Institute of Plant Physiology, Imperial College, 1917; interested in analysis of plant growth; studied growth of barley crop, 1919; experimented with effects of major plant nutrients in controlling growth; collaborated with F. J. Richards [q.v.], 1926–58; visited Sudan to advise on cotton growing by irrigation, 1928; assistant professor of plant physiology, Imperial College, under V. H. Blackman [q.v.], 1929–37; succeeded Blackman as professor, 1937; director, Research Institute, 1947–58; FRS, 1940; served on Council, 1949–51; left legacy to create Frederick Gregory Fund for benefit of Botany Department, Imperial College.

GREGORY, ISABELLA AUGUSTA, Lady (1852–1932), playwright and poet; born Persse; married Sir W. H. Gregory [q.v.], 1880; a founder with W. B. Yeats and G. A. Moore [qq.v.] (1899) of Irish Literary Theatre; and with Yeats and J. M. Synge [q.v.] (1904) of Abbey Theatre, Dublin; director, 1904–32; wrote twenty-seven plays including *The Image* (1910), *The Golden Apple* (1916), *The Dragon* (1920), and *Sancho's Master* (1928); translated Gaelic sagas.

GREGORY, JOHN WALTER (1864–1932), geologist and explorer; graduate of London University; assistant in geological department, British Museum, 1887–1900; professor of geology, Melbourne, 1900–4, Glasgow, 1904–29; made world-wide expeditions; drowned in Peru; publications include *The Great Rift Valley* (1896), *Geography, Structural, Physical, and Comparative* (1909), and works on bryozoa; FRS, 1901.

GREGORY, Sir RICHARD ARMAN, baronet (1864–1952), author, scientific journalist, and editor of *Nature*; received elementary school education; obtained scholarship to Normal School of Science; fellow student and lifelong friend of H. G. Wells [q.v.]; first classes, astronomy and physics, 1887; computer, Solar Physics Committee, and assistant to (Sir) Norman Lockyer [q.v.], 1889–92; his assistant editor on *Nature*, 1893; editor, 1919–38; scientific editor, Macmillans, 1905–39; joint-founder and editor, *School World* (incorporated in *Journal of Education*, 1918), 1899–1939; joined British Association, 1896; president, 1939–46; a founder and first secretary (1901) of Association's educational science section; active in British Science Guild and obtained its merger with British Association, 1936; publications include *Honours Physiography* (with H. G. Wells, 1893) and *Discovery or the Spirit and Service of Science* (1916; revised as Penguin, 1949); knighted, 1919; baronet, 1931; FRS, 1933.

GREGORY, ROBERT (1819–1911), dean of St. Paul's; MA, Corpus Christi College, Oxford, 1846; DD, 1891; vicar of St. Mary-the-Less, Lambeth, 1853–73; served on royal commission on ritual, 1867; canon of St. Paul's, 1868; helped to improve cathedral fabric and finances; a zealous member of Lower House of Canterbury Convocation from 1868; defended Athanasian Creed and ritual; member of education commission, 1886; dean of St. Paul's, 1890; able administrator; published *Elementary Education* (1895) and sermons; autobiography edited by W. H. Hutton [q.v.], 1912.

GREIFFENHAGEN, MAURICE WILLIAM (1862–1931), painter; studied at Royal Academy Schools and exhibited at Academy from 1884; ARA, 1916; RA, 1922; executed striking portraits of men and large-scale decorative work; headmaster, life department, Glasgow School of Art, 1906–29.

GRENFELL, BERNARD PYNE (1869–1926), papyrologist; BA, Queen's College, Oxford, 1892; researched in Egypt, 1893-4-5; carried out excavations at the Fayum and Behneseh, 1895-6-7, 1898–1902, 1903–6; made important discoveries for Graeco-Roman period; found quantities of papyri; professor of papyrology at Oxford, 1908; FBA, 1905; works, mostly in collaboration with his friend A. S. Hunt [q.v.], include *Revenue Laws of Ptolemy Philadelphus* (1896), *An Alexandrian Erotic Fragment* (1896), *The Oxyrhynchus Papyri* (17 vols., 1898–1927), *The Amherst Papyri* (1900–1), *The Tebtunis Papyri* (1902–7); an exemplary editor.

GRENFELL, EDWARD CHARLES, first Baron St. Just (1870–1941), banker and politician; cousin of W. H. Grenfell, Lord Desborough [q.v.]; educated at Harrow and Trinity College, Cambridge; manager, J. S. Morgan & Co. (from 1909 Morgan, Grenfell & Co.), London, 1900; partner, 1904–41; a director of Bank of England, 1905–40; conservative MP, City of London, 1922–35; baron, 1935.

GRENFELL, FRANCIS WALLACE, first Baron Grenfell (1841–1925), field-marshal;

joined army, 1859; went on service to South Africa, 1874; staff officer during last Kaffir war, 1878; took part in final defeat of Zulus at Ulundi, 1879; assistant-adjutant-general to Sir Garnet Wolseley [q.v.] in Egyptian expedition, 1882; sirdar of Egyptian army, 1885-92; consolidated Egyptian hold on Suakin, 1891; reorganized Egyptian forces; inspector-general, reserve forces, War Office, 1894; commanded British garrison in Egypt, 1897; governor and commander-in-chief, Malta, 1899; created baron, 1902; full general, 1904; commander-in-chief, Ireland, 1904-8; field-marshal, 1908.

GRENFELL, GEORGE (1849-1906), Baptist missionary and explorer of the Congo; did pioneering work up Lower Congo river; reached Stanley Pool, 1881; surveyed Congo as far as equator, 1884; discovered Ruki river; navigated the Ikelemba; met cannibals in Bangala region; reached 'Grenfell' falls, the most northerly point yet achieved in Congo basin, 1884; explored affluents of Congo from east and south, and discovered the Batwa dwarf tribes, 1885; went up main stream of the Kasai, and Kwa and Mfini to Lake Leopold II, and the Kwango to the Kingunji rapids, 1886; received by King Leopold at Brussels, 1887; made chevalier and Belgian plenipotentiary for settlement with Portugal of the Luanda frontier, 1891-3; at Bolobo, 1893-1900; explored Aruwimi river (1900-2) and established missionary station at Yalemba, 1903; died of blackwater fever at Basoko.

GRENFELL, HUBERT HENRY (1845-1906), expert in naval gunnery; joined navy, 1859; made first designs of hydraulic mountings for naval ordnance, 1869; naval adviser at Berlin congress, 1878; invented self-illuminating night sights for naval ordnance, 1891; helped to found Navy League.

GRENFELL, JULIAN HENRY FRANCIS (1888-1915), soldier and poet; son of W. H. Grenfell, Lord Desborough [q.v.]; educated at Eton and Balliol College, Oxford; entered army (1st Dragoons), 1910; accompanied regiment to France and DSO, 1914; died of wounds at Boulogne; best-known poem 'Into Battle' (1915).

GRENFELL, SIR WILFRED THOMASON (1865-1940), medical missionary and author; educated at Marlborough and London Hospital; qualified, 1886; joined Royal National Mission to Deep-Sea Fishermen, 1887; founded and developed (1893-1935) Labrador Medical Mission; raised most of the funds himself, after 1912 by means of the International Grenfell Association for which he lectured and wrote;

first honorary DM, Oxford, 1907; KCMG, 1927.

GRENFELL, WILLIAM HENRY, BARON DESBOROUGH (1855-1945), athlete, sportsman, and public servant; educated at Harrow and Balliol College, Oxford; president, University Boat and Athletic clubs; master, draghounds; liberal MP, Salisbury (1880-2, 1885-6), Hereford City (1892-3, resigned); conservative MP, Wycombe division, 1900-5; baron, 1905; active in local government; for thirty-two years chairman, Thames Conservancy Board; president, London Chamber of Commerce; won Thames punting championship, 1888-90; twice swam Niagara; climbed in Alps; shot in Rockies, India, and Africa; president, Olympic Games, 1908; of Bath Club (1894-1942), Marylebone Cricket Club, Lawn Tennis and Amateur Fencing associations, Coaching and Four-in-Hand clubs; chairman, Pilgrims of Great Britain, 1919-29; GCVO, 1925; KG, 1928.

GRENVILLE, FRANCES EVELYN, COUNTESS OF WARWICK (1861-1938), born MAYNARD; married (1881) Lord Brooke (fifth Earl of Warwick, 1893); celebrated beauty and member of the Prince of Wales's circle; converted to socialism by Robert Blatchford [q.v.]; thereafter used her position, fortune, and energies in charitable and social interests.

GREY, ALBERT HENRY GEORGE, fourth EARL GREY (1851-1917), statesman; son of Charles Grey [q.v.]; educated at Harrow and Trinity College, Cambridge; liberal MP, South Northumberland, 1880-5, and for Tyneside division of county, 1885-6; concentrated his energies on promotion of imperial unity; joined board of directors of British South Africa Company, 1889; administrator of Rhodesia, 1896-7; very successful governor-general of Canada, 1904-11; succeeded uncle, 1894.

GREY, CHARLES GREY (1875-1953), writer on aviation; grandson of John Grey, nephew of Josephine Butler [qq.v.]; educated at Erasmus Smith School, Dublin, and Crystal Palace School of Engineering; joined *Autocar*, 1905; joint-editor, *Aero*, 1909; founded (1911) and edited (1911-39) *Aeroplane*, the most widely read aviation newspaper; campaigned for preservation of Royal Air Force, for British aircraft industry, and for Imperial Airways; edited *Jane's All the World's Aircraft*, 1916-41.

GREY, SIR EDWARD, third baronet, and VISCOUNT GREY OF FALLODON (1862-1933), statesman and bird-lover; brought up at Fallodon, Northumberland, where from boyhood he watched birds and fished; inherited estate and title from grandfather, Sir George Grey [q.v.], 1882; educated at Winchester and Balliol College, Oxford; sent down for incorrigible idleness, 1884; from Mandell Creighton [q.v.] learnt moderate democratic liberalism and a sense of public duty which was to keep him in politics despite strong preference for country life; at Fallodon enlarged the famous ponds and tamed many varieties of duck; private secretary to Sir Evelyn Baring (later Earl of Cromer) and H. C. E. Childers [qq.v.], 1884; liberal MP, Berwick-on-Tweed, 1885-1916; parliamentary under-secretary, Foreign Office, 1892-5; made declaration on French encroachment on Nile, 1895; member, West Indian royal commission, 1897; his support of South African war sowed seeds of distrust in him as foreign secretary felt by more radical members of his party; director (1898), chairman (1904), North Eastern Railway; foreign secretary, 1905-16; established cordial relations with Campbell-Bannerman who supported his policy of friendship with France; let France and Germany know that in his opinion England would fight in event of German aggression; continued 'military conversations' with France initiated by Lord Lansdowne [q.v.]; Algeciras conference (1906) on Morocco resulted in Germany accepting compromise because England had rallied to France; determined to secure friendship of United States and began intimate correspondence with Theodore Roosevelt; principal author of Anglo-Russian agreement (1907) as only means of avoiding clashes in Tibet, Afghanistan, or Persia and of preventing Russia entering German orbit; his *ententes* with France and Russia disliked by many liberals although later he was criticized for not having turned them into alliances; this he would not have done, even had public opinion permitted, lest France or Russia should thereby feel encouraged to provoke unjustified war; renewed alliance with Japan, 1911; by his firmness and tact over Congo atrocities secured reform and retained Belgian friendship; published (1912) Putumayo blue book based on reports of Roger Casement [q.v.]; during Agadir crisis (1911) strongly supported by Lloyd George whose speech made Germany realize that Britain would fight if she attacked France; agreed to unsuccessful attempt (1912) by Lord Haldane [q.v.] to persuade Germany to limit growth of navy; concessions to remove German complaint of encirclement included Baghdad railway agreement (1913) and willingness to allow her a large share of Portuguese colonies in Africa should they be sold; refused to make a secret agreement on this or any other matter except as wartime measure; supported by France and Germany (1913) in averting European war after Serbian defeat of Turkey; unsuccessful after murder of Archduke Francis Ferdinand

(1914) because Germany refused a conference and backed Austria; his handling of crisis kept country united and gained American support; despite difficulties arising from blockade maintained excellent relations with America; concluded secret treaty of London (Apr. 1915) with Italy whose extortionate terms he accepted, since the Allies might have lost the war had Italy joined the central powers; deliberately sacrificed chance of preserving eyesight to continuance of duty; created viscount, July 1916; resigned Dec. 1916; president of League of Nations Union from 1918; became increasingly blind; chancellor of Oxford University, 1928-33; wrote *Twenty-five Years, 1892-1916* (2 vols., 1925); his *Fly Fishing* (1899), *Fallodon Papers* (1926), and *The Charm of Birds* (1927) place him in front rank of English nature writers; PC, 1902; KG, 1912; FRS, 1914.

GREY, MARIA GEORGINA (1816-1906), promoter of women's education; born SHIRREFF; sister of Emily Shirreff [q.v.]; married William Thomas Grey, 1841; initiated Girls' Public Day School Company, 1872, which in 1929 had twenty-five schools; founded Maria Grey Training College for women teachers, 1878; published a novel, and works on women's enfranchisement and education.

GRIERSON, SIR GEORGE ABRAHAM (1851-1941), Indian civil servant and philologist; educated at St. Bees, Shrewsbury, and Trinity College, Dublin; served in Bengal presidency, 1873-98; completed *Linguistic Survey of India* (1898-1928) describing 179 separate languages and 544 dialects belonging to five distinct families; retired from India, 1903; publications include *Seven Grammars of the Dialects and Subdialects of the Bihārī Language* (8 parts, 1883-7) and *Bihār Peasant Life* (1885), *The Modern Vernacular Literature of Hindustan* (1889); series of articles and books on Kashmīrī culminating in *A Dictionary of the Kashmīrī Language* (1916-32), *The Piśāca Languages of North-Western India* (1906) and *Torwālī* (1929); and studies of Prakrit, Ahom, and the Kuki-Chin languages; KCIE, 1912; FBA, 1917; OM, 1928.

GRIERSON, SIR HERBERT JOHN CLIFFORD (1866-1960), scholar; educated at the Gymnasium and university, Aberdeen (graduated, 1887), and Christ Church, Oxford; first class, *lit. hum.*, 1893; first professor of English, Aberdeen, 1894; regius professor of rhetoric and English literature, Edinburgh, 1915-35; rector, 1936-9; edited the poems of Donne (2 vols., 1912), Byron (1923), Milton (1925), Scott's letters (in collaboration, 12 vols., 1932-7), *The Oxford Book of Seventeenth Century Verse* (with

G. Bullough, 1934); other works include *Cross Currents in English Literature of the XVII Century* (1929) and biography of Scott (1938); FBA, 1923; knighted, 1936; hon. degrees from many universities.

GRIERSON, Sir JAMES MONCRIEFF (1859-1914), lieutenant-general; joined Royal Artillery from Woolwich, 1878; employed on intelligence work in India; passed first into Staff College, 1883; head of Russian section of intelligence division, 1889; military attaché at Berlin, 1896; served in South African war and China, 1900; CB, 1901; director of military operations at War Office and major-general, 1904; largely helped to lay foundations of military co-operation between Great Britain and France; held home commands, 1906-14; KCB, 1911; author of military works; died in France on way to front.

GRIFFIN, BERNARD WILLIAM (1899-1956), cardinal; studied at Cotton and Oscott colleges and English College, Rome; ordained, 1924; obtained doctorates in theology and canon law; secretary to successive archbishops of Birmingham, 1927-37; appointed auxiliary bishop, Birmingham, 1938, and vicar-general of diocese; archbishop of Westminster, 1943; cardinal, 1946; travelled widely in cause of international understanding; deplored financial burden to Catholics of 1944 Education Act and kept Catholic hospitals outside National Health Service.

GRIFFIN, Sir LEPEL HENRY (1838-1908), Indian civil servant; joined Indian civil service in Punjab, 1860; compiled standard accounts of Punjab families in *Punjab Chiefs* (1865), *The Law of Inheritance to Sikh Chiefships* (1869), and *The Rajas of the Punjab* (1870); superintendent of Kapurthala state from 1875; helped to establish Abdur Rahman on Afghan throne and reconciled him to English policy, 1880; CSI, 1878; KCSI, 1881; agent-general to governor-general in central India, 1881-9; advocated use of Indian vernaculars in teaching in India; joint-founder of *Asiatic Quarterly Review*, 1885; settled in England, 1889; chairman of Imperial Bank of Persia; works include *Ranjit Singh* (in 'Rulers of India' series, 1892).

GRIFFITH, ALAN ARNOLD (1893-1963), aero-engineer; educated at Douglas secondary school and Liverpool University; B.Eng., first class, mechanical engineering, 1914; M.Eng., 1917; D.Eng., 1921; joined Royal Aircraft Factory (later Royal Aircraft Establishment), Farnborough, 1915; published important paper 'The Theory of Rupture' (*Phil. Trans. A*, 1920) on behaviour of materials; investigated propel-ler problems and turbine blading; demonstrated that gas turbine was a feasible aircraft engine, 1926; principal scientific officer, Air Ministry Laboratory, South Kensington, 1928-31; head of Engine Department, Farnborough, 1931; conducted axial compressor experiments with Hayne Constant [q.v.]; research engineer, Rolls-Royce, Derby, 1939-60, directly responsible to E. W. (later Lord) Hives [q.v.] for aero-engine research; research led directly to production of Rolls-Royce RB 108 engines, designed specifically for jet lift and vertical take-off; FRS, 1941; CBE, 1948.

GRIFFITH, ARTHUR (1872-1922), Irish political leader; at an early age associated himself with Irish nationalist movement; employed in Transvaal, 1896-9; returned to Ireland to take part in founding weekly newspaper which was to be organ of "98 clubs' separatist movement; edited *The United Irishman*, 1899-1906; preached doctrine that Irish self-government could never be attained through parliamentary action at Westminster, a doctrine which did not then appeal to most Irishmen; at convention of 'National Council' advocated adaptation to Irish conditions of 'Hungarian' method of passive resistance under title of 'Sinn Fein' policy, 1905; Sinn Fein ('Ourselves Alone') enunciated doctrine of national self-help and self-reliance; aimed at independent self-government for Ireland; believer in use of physical force; president of Sinn Fein party, 1910; associated himself with Irish National Volunteers, 1913; opposed any Irishman joining British army during European war; his papers suppressed, 1914-15; took no part in Easter rebellion, 1916; support for Sinn Fein increased after 1916; several times imprisoned, 1916-21; resigned presidency of party in favour of Mr de Valera, 1917; MP, East Cavan, 1918; minister for home affairs on formation of Dail Eireann, 1919; acting-president, 1919-20; carried out 'Hungarian' policy against British government; with Michael Collins [q.v.] and three other plenipotentiaries negotiated with British government and finally secured treaty, 1921; president of Dail Eireann, 1922; the real creator of autonomous Irish state.

GRIFFITH, FRANCIS LLEWELLYN (1862-1934), Egyptologist; educated at Sedbergh, Highgate, and Queen's College, Oxford; official student of Egypt Exploration Fund, 1884; assistant in department of British and medieval antiquities and ethnography, British Museum, 1888-96; assistant professor of Egyptology, University College, London, 1892-1901; honorary lecturer, Manchester, 1896-1908; reader (1901), honorary professor (1924-32), Oxford; created Griffith Egyptological Fund at Oxford, 1907; in charge of Oxford excavations in Nubia; prolific and accurate scholar with genius for decipherment; his *Stories of the High Priests of Memphis* (1900) first established demotic on firm scientific basis; bequeathed his papers, estate, and Egyptological library to found Griffith Institute at Oxford; FBA, 1924.

GRIFFITH, RALPH THOMAS HOTCH-KIN (1826-1906), Sanskrit scholar; BA, Queen's College, Oxford, 1846; Boden Sanskrit scholar, 1849; professor of English literature (1853) and headmaster (1854) of Benares Government College; inspector of schools, 1856; principal of Benares College, 1861-78; founded *Pandit*, college monthly journal, 1866; director of public instruction in North-West provinces, 1878-85; CIE, 1885; retired to Kotagiri; published verse translations, embodying spirit of originals, of Sanskrit classics, including Kālidāsa's *Kumárasambhava* (1853), the *Rámáyan of Válmíki* (5 vols., 1870-3), hymns of the *Rigveda* (4 vols., 1889-92), *Sámaveda* (1893), and *Atharvaveda* (2 vols., 1895-6).

GRIFFITHS, ARTHUR GEORGE FREDERICK (1838-1908), inspector of prisons and author; born at Poona; joined 63rd regiment, 1855; served in Crimea; brigade major at Gibraltar, 1864-70; inspector of prisons in England, 1878-96; historian of Millbank jail, 1875, and of Newgate, 1884; editor of *Army and Navy Gazette*, 1901-4; best known as writer of some thirty novels of prison life and detective stories; wrote also *Fifty Years of Public Service* (1904).

GRIFFITHS, ERNEST HOWARD (1851-1932), physicist; educated at Owens College, Manchester, and Sidney Sussex College, Cambridge; fellow, 1897; a university coach; principal and professor of experimental philosophy, University College of South Wales, Cardiff, 1901-18; deeply interested in completion (1909) of Viriamu Jones memorial research laboratory; his researches included the boiling-point of sulphur, the latent heat of evaporation of water, freezing-points of dilute aqueous solutions, and other accurate measurements of heat; FRS, 1895.

GRIFFITHS, EZER (1888-1962), physicist; educated at Aberdare intermediate school and University College, Cardiff; first class, physics; fellow, university of Wales; D.Sc.; joined heat section of National Physical Laboratory, Teddington, 1915-53; one of leading world authorities on heat insulation, heat transfer, evaporation, and related matters, such as refrig-

eration; visited Australia to examine problems of transportation of apples, 1923; worked on such problems as vapour trails made by modern aircraft, 1939-45; carried out experiments for Medical Research Council and the Admiralty on effects of radiation on man; vice-president, Physical Society; chairman, Research Committee, Institute of Refrigeration, 1936-8; chairman, governing body of Twickenham Technical College; member of many scientific committees; FRS, 1926; OBE, 1950; published *Methods of Measuring Temperature* (1918), *Pyrometers* (1926), and *Refrigeration, Principles and Practice* (1951); also responsible for articles in Sir R. T. Glazebrook's [q.v.] *Dictionary of Applied Physics* and in Sir T. E. Thorpe's [q.v.] *Dictionary of Applied Chemistry*.

GRIFFITHS, Sir JOHN NORTON-, first baronet (1871-1930), engineer. [See NORTON-GRIFFITHS.]

GRIGG, EDWARD WILLIAM MACLEAY, first BARON ALTRINCHAM (1879-1955), administrator and politician; born in Madras; grandson of Sir Edward Deas Thomson [q.v.]; scholar of Winchester and New College, Oxford; on staff of *Times*, 1903-5 and (as head of colonial department) 1908-13; joint-editor, *Round Table*, 1913; in France with Guards division, 1914-18; MC; DSO; military secretary to Prince of Wales, 1919-20; CMG; CVO, 1919; KCVO, 1920; a private secretary to Lloyd George, 1920-2; liberal MP, Oldham, 1922-5; secretary, Rhodes Trust, 1923-5; governor of Kenya, 1925-30; KCMG, 1928; conservative MP, Altrincham, 1933-45; parliamentary secretary, Ministry of Information, 1939-40; joint parliamentary under-secretary, War Office, 1940-2; minister resident, Middle East, 1944-5; PC, 1944; baron, 1945; editor, *National Review*, 1948-55; publications include *The Faith of an Englishman* (1936), and *Kenya's Opportunity* (1955).

GRIGG, Sir (PERCY) JAMES (1890-1964), public servant; educated at Bournemouth (secondary) school and St. John's College, Cambridge; first class, parts i and ii, mathematical tripos, 1910-12; headed list of entrants to administrative class of Civil Service, and appointed to Treasury, 1913; served in Eastern Europe and Office of External Ballistics, 1915-18; principal private secretary to five chancellors of the Exchequer, 1921-30; chairman, Board of Customs and Excise; chairman, Board of Inland Revenue; finance member, viceroy's executive council, India, 1933-9; permanent secretary, War Office, 1939; collaborated closely with Sir Alan Brooke (later Viscount Alanbrooke, q.v.), chief of imperial general staff; appointed by

Churchill as secretary of state for war, 1942–5; nationalist MP, East Cardiff, 1942–5; first British executive director, International Bank for Reconstruction and Development, 1946; director of companies; published autobiography, *Prejudice and Judgment* (1948); KCB, 1932; KCSI, 1936; PC, 1942; hon. fellow, St. John's College, Cambridge, 1943; hon. LLD, Bristol, 1946; hon. bencher, Middle Temple, 1954.

GRIGGS, WILLIAM (1832–1911), inventor of photo-chromo-lithography; technical assistant to director of Indian Museum, 1855; photolithographer in India Office till 1885; invented photo-chromo-lithographic process; set up works at Peckham, 1868; published plates for Forbes Watson's *Textile Manufactures . . . of India* (1866); reproduced facsimile editions of *Mahābhāsya* (1871), Shakespeare quartos (43 vols., 1881–91), and Ashbee reprints; issued from 1881 some 200 *Portfolios of Industrial Art* of all countries, and from 1884 *Journal of Indian Art and Industry*.

GRIMBLE, SIR ARTHUR FRANCIS (1888–1956), colonial administrator, broadcaster, and writer; born in Hong Kong; educated at Chigwell School and Magdalene College, Cambridge; entered colonial service, 1914; posted to Gilbert and Ellice Islands; resident commissioner, 1926; administrator and colonial secretary, St. Vincent, 1933–6; governor, Seychelles, 1936–42, Windward Isles, 1942–8; obtained popular success recounting his Pacific experiences for BBC, subsequently published as *A Pattern of Islands* (1952) and *Return to the Islands* (1957); CMG, 1930; KCMG, 1938.

GRIMTHORPE, first BARON (1816–1905), lawyer, mechanician, and controversialist. [See BECKETT, SIR EDMUND.]

GROOME, FRANCIS HINDES (1851–1902), Romany scholar and miscellaneous writer; son of Robert Hindes Groome [q.v.]; interested in gipsy life from boyhood; lived with gipsies at home and abroad; settled down to literary work in Edinburgh, 1876; joined staff of Messrs. Chambers, 1885; sub-editor of *Chambers's Encyclopaedia* (10 vols. 1888–92); published *In Gipsy Tents* (1880), *Kriegspiel* (a Romany novel, 1896), and *Gypsy Folk Tales* (1899); had wide knowledge of Jacobite literature.

GROSE, THOMAS HODGE (1845–1906), registrar of Oxford University, (1877–1906); scholar of Balliol College, Oxford; BA, first class, classics and mathematics, 1868; president of the Union, 1871; fellow and tutor of Queen's College, Oxford, from 1871; helped

T. H. Green [q.v.] in editing Hume's works, 1874–5: ardent alpinist.

GROSSMITH, GEORGE (1847–1912), entertainer and singer in light opera; gave 'humorous and musical recitals'; employed in Gilbert and Sullivan's series of comic operas, first at Opera Comique and then at Savoy Theatre, 1877–89; 'created' many of the chief parts; with brother wrote *Diary of a Nobody* (1894).

GROSSMITH, GEORGE, the younger (1874–1935), actor-manager and playwright; son of George Grossmith [q.v.]; originated the 'dude' in musical comedy; at Gaiety Theatre, 1901–13; later notable performances included *Kissing Time* (1919) and *No, No, Nanette* (1925); in management after 1914; wrote musical plays including *The Spring Chicken*, *Rogues and Vagabonds* (first modern type revue in London, 1905), and *The Bing Boys are Here*; introduced cabaret entertainment, 1922.

GROSSMITH, WALTER WEEDON (1854–1919), comedian; brother of George Grossmith, the younger [q.v.]; excelled in part of 'dudes' and small, underbred, unhappy men.

GROSVENOR, RICHARD DE AQUILA, first BARON STALBRIDGE (1837–1912), railway administrator and politician; son of second Marquess of Westminster [q.v.]; liberal MP, Flintshire, 1861–86; PC, 1872; created baron and became liberal unionist, 1886; director of North Western Railway Company, 1870; chairman, 1891–1911.

GUBBINS, JOHN (1838–1906), breeder and owner of racehorses; settled at Bruree, county Limerick, 1886, and bred horses and hounds; bred Galtee More (winner of Two Thousand Guineas, St. Leger, and Derby, 1897), and Ard Patrick (Eclipse Stakes, 1903); headed list of winning owners, 1897.

GUEDALLA, PHILIP (1889–1944), historian and essayist; educated at Rugby and Balliol College, Oxford; president of Union; first class, modern history, 1912; called to bar (Inner Temple), 1913; five times defeated as liberal candidate, 1922–31; publications include *Palmerston* (1926), *Gladstone and Palmerston* (1928), *The Duke* (1931), *The Queen and Mr. Gladstone* (2 vols., 1933), *The Hundred Days* (1934), *The Hundred Years* (1936), *Mr. Churchill* (1941), and *The Two Marshals* (1943).

GUEST, FREDERICK EDWARD (1875–1937), politician and promoter of aviation; brother of first Viscount Wimborne [q.v.]; educated at Winchester; private secretary to his cousin (Sir) Winston Churchill; liberal MP, 1910-22, 1923-9; conservative MP, 1931-7; PC, 1920; secretary of state for air, 1921-2; master, Guild of Air Pilots and Air Navigators, 1932.

GUEST, Sir IVOR CHURCHILL, third baronet, and first Viscount Wimborne (1873-1939), politician; son of first Lord Wimborne; brother of F. E. Guest [q.v.]; educated at Eton and Trinity College, Cambridge; conservative MP, Plymouth, 1900-6; liberal MP, Cardiff Boroughs, 1906-10; PC and baron (as Ashby St. Ledgers), 1910; paymaster-general, 1910-12; succeeded father, 1914; lord-lieutenant of Ireland, 1915-18; viscount, 1918.

GUGGENHEIM, EDWARD ARMAND (1901-1970), authority on thermodynamics; educated at Charterhouse and Gonville and Caius College, Cambridge; first class, part i, mathematical tripos, and part ii, natural sciences (chemistry) tripos, 1921-3; research under (Professor Sir) Ralph H. Fowler [q.v.], 1923-5; continued scientific studies in Denmark, 1925-31; worked under Professor F. G. Donnan [q.v.] at University College, London; wrote first book *Modern Thermodynamics by the methods of W. Gibbs* (1933); held number of lecturing posts in the United States and England, 1932-9; Sc.D., Cambridge; worked for Royal Navy and Montreal Laboratory of Atomic Energy, 1939-46; suggested successful means of neutralizing German magnetic mines; FRS, 1946; professor of chemistry, Reading University, 1946-66; hon. life member, Faraday Society, 1967; member of number of scientific committees; published many scientific papers and books.

GUGGISBERG, Sir FREDERICK GORDON (1869-1930), soldier and administrator; born at Toronto; commissioned in Royal Engineers, 1889; employed by Colonial Office on special survey of Gold Coast Colony and Ashanti, 1902; director of surveys there, 1905; in Southern Nigeria, 1910; served in France, 1914-18; governor of Gold Coast, 1919; his policy directed to government by and for Africans; governor of British Guiana, 1928-9; KCMG, 1922; his writings include *Handbook of the Southern Nigeria Survey* (1911).

GUINNESS, Sir ARTHUR EDWARD second baronet, and first Baron Ardilaun (1840-1915), philanthropist; son of Sir Benjamin Guinness, first baronet [q.v.]; BA, Trinity College, Dublin; succeeded father, 1868; head of Guinness brewery, Dublin, 1868-77; conservative MP, Dublin City, 1868-9, 1874-80; baron, 1880; munificent benefactor to Dublin and to Irish Church.

GUINNESS, EDWARD CECIL, first Earl of Iveagh (1847-1927), philanthropist; son of Sir B. L. Guinness [q.v.]; shared in management of Guinness's brewery; chairman of company, 1886; baronet, 1885; baron, 1891; viscount, 1905; earl, 1919; FRS, 1906; benefactions include: £250,000 to erect workmen's dwellings in London and Dublin; £250,000 for destroying Dublin slum property; £250,000 to Lister Institute of Preventive Medicine, London; bequeathed to the nation portion of Ken Wood estate and art collection for Ken Wood house.

GUINNESS, HENRY GRATTAN (1835-1910), divine and author; ordained as evangelist, 1857; preached in England, Ireland, America, and on the continent, 1857-72; at Dublin helped in 'conversion' of Dr Barnardo [q.v.], 1866; founded East London Institute for training missionaries, 1873, 'Livingstone Inland Mission' in Congo, 1878, and other missions in South America and India, all of which were amalgamated into 'Regions Beyond Missionary Union', 1899; made missionary tour of the world, 1903; collaborated with wife in *The Approaching End of the Age* (1878) and *The Divine Programme of the World's History* (1888); he also published grammars of the Congo language.

GUINNESS, Sir RUPERT EDWARD CECIL LEE, second Earl of Iveagh (1874-1967), philanthropist; chairman, Guinness Company, brewers, 1927-62; educated at Eton and Trinity College, Cambridge; won diamond sculls, 1895-6; won Kings Cup at Cowes, 1903; director, Arthur Guinness, Son. & Co. Ltd., 1899; worked with Irish hospital in South Africa, 1900; CMG, 1901; member, London County Council; conservative MP Haggerston, Shoreditch, 1908-10; set up Woking Park Farm, training establishment for emigrants to Canada; MP SE Essex, (later Southend), 1912-27; governor, Lister Institute of Preventive Medicine; chairman, Wright-Fleming Institute of Microbiology; financed research into production of clean milk at Rothamsted Institute; commanded London division, RNVR, 1914; ADC to the King, 1916; first chairman, Tuberculin Tested Milk Producers Association, 1920; assisted in development of Research Institute in Dairying, University College, Reading; hon. D.Sc. Reading; chancellor; chairman, Chadacre Agricultural Institute; became Earl of Iveagh and inherited Elveden estate, 1927; converted Elveden into efficient, economic farming unit; lieutenant of the City of London; deputy lieutenant, Surrey and Essex, chairman, Guinness Trust and Iveagh Trust; presented Iveagh House, Dublin, to Republic of Ireland, 1939; CB, 1911; KG, 1955; FRS, 1964; chancellor,

Trinity College, Dublin; married (1903) ONSLOW, GWENDOLEN FLORENCE MARY (1881-1966), philanthropist; elder daughter of fourth Earl of Onslow [q.v.]; conservative MP Southend (succeeding husband), 1927-35; member, National Prisoners of War Fund; CBE, 1920; chairman, National Union of Conservative and Unionist Associations.

GUINNESS, WALTER EDWARD, first BARON MOYNE (1880-1944), statesman; son of E. C. Guinness, first Earl of Iveagh [q.v.]; educated at Eton; served in South African and 1914-18 wars; DSO and bar, 1917-18; conservative MP, Bury St. Edmunds, 1907-31; undersecretary for war, 1922-3; financial secretary to Treasury, 1923-4, 1924-5; PC, 1924; minister of agriculture, 1925-9; baron, 1932; chairman, royal commissions on Durham University (1934) and West Indies (1938-9); colonial secretary, 1941-2; deputy minister of state, Cairo, 1942; minister, Jan. 1944; assassinated, Nov. 1944.

GULLY, WILLIAM COURT, first VISCOUNT SELBY (1835-1909), speaker of the House of Commons; son of James Manby Gully [q.v.]; BA, Trinity College, Cambridge, 1856; president of Union; called to bar, 1860; QC, 1877; established good practice in commercial cases; liberal MP for Carlisle, 1892; recorder of Wigan, 1892; elected speaker of the House of Commons, 1895; distinguished for dignity, courtesy, and impartiality; ordered forcible removal of Irish members from the House, Mar. 1901; resigned speakership and raised to peerage, 1905; hon. LLD, Cambridge, 1900, DCL, Oxford, 1904.

GUNN, BATTISCOMBE GEORGE (1883-1950), Egyptologist; educated at Bedales, Westminster and Allhallows; published The Instruction of Ptah-hotep (1906); worked at Harageh (1913-14), El-Amarna (1921-2), and Saqqara (1924-7); published Studies in Egyptian Syntax (1924); assistant keeper, Egyptian Museum, Cairo, 1928-31; curator, Egyptian section, University Museum, Philadelphia, 1931-4; professor of Egyptology, Oxford, 1934-50; edited Journal of Egyptian Archaeology, 1935-9; FBA, 1943.

GUNN, SIR JAMES (1893-1964), painter; educated at Glasgow high school, Glasgow School of Art, Edinburgh College of Art and the Académie Julien, Paris; commissioned with 10th Scottish Rifles, 1914-18; painted portraits of prime ministers, field-marshals, judges, dons, bankers and writers and artists, including Belloc, Chesterton, and Baring (1932) and Delius (1933); painted George VI and family at Windsor, and state portrait of Queen Elizabeth II (1956); followed Augustus John [q.v.] as president, Royal Portrait Society, 1953; ARA, 1953; RA, 1961; knighted, 1963; hon. degrees, Manchester and Glasgow.

GÜNTHER, ALBERT CHARLES LEWIS GOTTHILF (1830-1914), zoologist; born in Württemberg; took medical degree at Tübingen, 1858; appointed to staff of British Museum and naturalized, 1862; keeper of zoological department, 1875-95; vice-president of Royal Society, 1875-6; gold medallist of Royal and Linnean societies; works include Catalogue of Fishes in the British Museum (1859-70) and Introduction to the Study of Fishes (1880).

GUNTHER, ROBERT WILLIAM THEODORE (1869-1940), zoologist and antiquary; son of A. C. L. G. Günther [q.v.]; educated at Magdalen College, Oxford; lecturer (1894), tutor (1896), in natural science, fellow (1897-1928); university reader in history of science, 1934-9; created Museum of the History of Science, Oxford, 1935; publications include 'Early Science in Oxford' series (1920-45).

GURNEY, SIR HENRY LOVELL GOLDSWORTHY (1898-1951), colonial civil servant; educated at Winchester and University College, Oxford; played golf for university; served in Kenya, 1921-35, 1936-44; secretary, East African Governors Conference co-ordinating defence and supply problems of the territories, 1938-44; chief secretary, 1941; colonial secretary, Gold Coast, 1944-6; as chief secretary, Palestine (1946-8), and high commissioner, Malaya (1948-51), worked closely with military authorities at time of great terrorist activity; CMG, 1942; knighted, 1947; KCMG, 1948; ambushed and killed, 1951.

GURNEY, HENRY PALIN (1847-1904), scientist; scholar of Clare College, Cambridge; fourteenth wrangler, 1870; fellow, 1870-83; partner of firm of Wren & Gurney, examination coaches, 1877; published book on crystallography, 1875; from 1894 to death principal of Durham College of Science (renamed Armstrong College, 1901), Newcastle upon Tyne; founded department of mineralogy and crystallography there, 1895; hon. DCL, Durham, 1896; killed in alpine accident near Arolla.

GUTHRIE, SIR JAMES (1859-1930), portrait painter and president of the Royal Scottish Academy; almost entirely self-trained; associated with 'Glasgow School'; portrait painter in Glasgow, 1885; president of Royal Scottish Academy, 1902-19; removed to Edinburgh;

RSA, 1892; knighted, 1903; painted 'War Statesmen' group for National Portrait Gallery, London.

GUTHRIE, THOMAS ANSTEY (1856-1934), humorous writer under pseudonym of F. ANSTEY; educated at King's College School and Trinity Hall, Cambridge; *Vice-Versa* (1882), *The Brass Bottle* (1900), and *Only Toys!* (1903) the most successful of his fantastic novels based on magic; in contributions to *Punch* developed talent for burlesque and parody.

GUTHRIE, WILLIAM (1835-1908), legal writer; educated at Glasgow and Edinburgh universities; called to Scottish bar, 1861; registrar of friendly societies for Scotland, 1872; hon. LLD, Edinburgh, 1881; sheriff-principal at Glasgow, 1903; edited many legal works.

GUTTERIDGE, HAROLD COOKE (1876-1953), barrister and professor; born in Naples; educated there, at Leys School and King's College, Cambridge; first class, history (1898), law (1899); called to bar (Middle Temple), 1900; KC, 1930; Sir Ernest Cassel professor of industrial and commercial law, London, 1919-30; reader, later professor, in comparative law, Cambridge, 1930-41; publications include *Smith's Mercantile Law*, 13th edn. (1931), *Bankers' Commercial Credits* (1932), and *Comparative Law* (1946).

GUY, SIR HENRY LEWIS (1887-1956), chartered mechanical engineer; educated at Penarth county school; diploma in mechanical and electrical engineering, University College, South Wales, 1909; joined British Westinghouse; chief engineer, mechanical department, Metropolitan-Vickers Electrical Company, 1918-41; secretary, Institution of Mechanical Engineers, 1941-51; member, advisory councils of Ministry of Supply and Department of Scientific and Industrial Research, committee of National Physical Laboratory, and many other technical bodies; FRS, 1936; CBE, 1943; knighted, 1949.

GWATKIN, HENRY MELVILL (1844-1916), historian, theologian, and conchologist; BA, St. John's College, Cambridge; Dixie professor of ecclesiastical history, Cambridge, 1891-1916; Gifford lecturer, Edinburgh, 1903; works include *Studies of Arianism* (1882), *The Knowledge of God* (1906), and *Early Church History* (1909); a distinguished teacher.

GWYER, SIR MAURICE LINFORD (1878-1952), lawyer and civil servant; educated at Highgate, Westminster, and Christ Church, Oxford; fellow, All Souls, 1902; called to bar (Inner Temple), 1903; KC, 1930; on legal staff, National Health Insurance Commission, 1912, Ministry of Shipping, 1917; legal adviser, Ministry of Health, 1919-26; Treasury solicitor and King's proctor, 1926-34; first parliamentary counsel to Treasury, 1934-7; drafted Government of India Bill; chief justice of India, 1937-43; vice-chancellor, Delhi University, 1938-50 CB, 1921; KCB, 1928; KCSI, 1935; GCIE, 1948; hon. degrees from Oxford and Indian universities.

GWYNN, JOHN (1827-1917), scholar and divine; BA and fellow, Trinity College, Dublin; country parson in county Donegal, 1864-82; regius professor of divinity, Trinity College, Dublin, 1888-1917; published numerous Syriac studies and edition of *Book of Armagh* (1913).

GWYNN, STEPHEN LUCIUS (1864-1950), author and Irish nationalist; son of John Gwynn and grandson of W. S. O'Brien [qq.v.]; educated at St. Columba's College, Dublin, and Brasenose College, Oxford; first class, *lit. hum.*, 1886; nationalist MP, Galway City, 1906-18; on Irish convention, 1917; published *The Queen's Chronicler and Other Poems* (1901), lives of Scott (1930), Swift (1933), Goldsmith (1935), and R. L. Stevenson (1939), and many books on Ireland.

GWYNNE, HOWELL ARTHUR (1865-1950), journalist; educated at Swansea grammar school; special and war correspondent, Reuter's, 1893-1904; organized its South African war service; 1899-1902; editor, *Standard*, 1904-11; *Morning Post*, 1911-37; supported tariff reform and by his intimacy with all those of note who shared his convictions maintained prestige (but not circulation) of the paper; CH, 1938.

GWYNNE-VAUGHAN, DAME HELEN CHARLOTTE ISABELLA (1879-1967), botanist, and leader of women's services in both world wars; educated at Cheltenham Ladies' College and King's College, London; B.Sc. in botany, 1904; D.Sc., for thesis on fungi, 1907; demonstrator for V. H. Blackman [q.v.], University College, London, 1904; assistant lecturer, Royal Holloway College, 1905; lecturer, University College of Nottingham, 1907; head of department of botany, Birkbeck College, London, 1909; fellow of Linnean Society; founded, with Dr Louisa Garrett Anderson, University of London Suffrage Society, 1907; joint chief controller, Women's Army Auxiliary Corps in France, 1917; CBE, 1918; commandant, Women's Royal Air Force, 1918-19; DBE, 1919; professor of botany, Birkbeck College, 1921; member, royal commission on food prices, 1924; president, Mycological

Society, 1928; GBE, 1929; first director, Auxiliary Territorial Service, 1939-41; disagreements with senior officers brought about enforced resignation; returned to Birkbeck College,

1941-4; hon. LLD Glasgow, 1920; published many scientific studies and two textbooks on fungi; also her autobiographical *Service with the Army* (1942).

H

HACKER, ARTHUR (1858-1919), painter; studied at Royal Academy Schools and in Paris; painted subject pictures, chief being 'The Annunciation' (1892), London street scenes, and portraits; ARA, 1894; RA, 1910.

HACKING, SIR JOHN (1888-1969), chartered electrical engineer; educated at Burnley grammar school and Leeds Technical Institute; engineering experience with Newcastle upon Tyne Electric Supply Co., 1908-13; joined Merz & McLellan, electrical consultants, and engaged on electrification of railways in Argentina and India and other projects in South Africa, 1915-33; deputy chief engineer, Central Electricity Board, 1933-44; chief engineer, 1944-7; maintained electricity supply in difficult wartime conditions; deputy chairman (operations), British Electricity Authority, 1947-53; returned to Merz & McLellan as consultant, 1953-66; knighted, 1949; president, Electrical Research Association, British Electrical Power Convention, and Institution of Electrical Engineers, 1951-2.

HADDON, ALFRED CORT (1855-1940), anthropologist; educated at Christ's College, Cambridge (fellow, 1901); first class, natural sciences tripos (comparative anatomy), 1878; professor of zoology, Royal College of Science, Dublin, 1880-1901; studied marine biology in Torres Strait (1888-9) and determined to save vanishing ethnological data; lecturer in physical anthropology, Cambridge, 1894-8; in ethnology, 1900-9; reader, 1909-25; planned Cambridge anthropological expedition (1898-9) to Torres Strait, New Guinea, and Sarawak (6-vol. report published, 1901-35); helped to make University Museum of Archaeology and Ethnology a primary centre for research; FRS, 1899.

HADEN, SIR FRANCIS SEYMOUR (1818-1910), etcher and surgeon; studied medicine at the Sorbonne and at Grenoble; FRCS, 1857; formed large private practice in London and did much public work; published pamphlets strongly opposed to cremation; devoted leisure to etching, mainly of landscape, from 1843 onwards; influenced by J. A. M. Whistler [q.v.], whose half-sister he married; chief works include 'Thames Fishermen', 'Shere Mill Pond', 'Breaking up of the *Agamemnon*'; his drypoints executed in 1877, 'Windmill Hill', 'Sawley Abbey', 'Essex Farm', 'Boat House', show vigorous style; founder and president of the Society of Painter-Etchers, 1880; knighted, 1894; member of several French artistic societies; exhibited at Royal Academy, 1860-85; worked also in water-colour and black chalk; chief collections of work are in British Museum and New York Public Library; pioneer of scientific criticism of Rembrandt's etchings; published *The Etched Work of Rembrandt* (1879), *About Etching* (1879), and kindred works.

HADFIELD, SIR ROBERT ABBOTT, baronet (1858-1940), metallurgist and industrialist; educated at Collegiate School, Sheffield; entered father's steel works; chairman and managing director, 1888; an enlightened employer; established laboratory and built up strong experimental organization; discovered manganese steel, 1882; other investigations included silicon and alloy steels, production of sound steel ingots, deformation of steel at high velocities, and corrosion; studied history of metallurgy; publications include *Metallurgy and its Influence on Modern Progress* (1925) and *Faraday and his Metallurgical Researches . . .* (1931); knighted, 1908; baronet, 1917; FRS, 1909.

HADLEY, WILLIAM WAITE (1866-1960), editor of the *Sunday Times*; educated at village and night schools; apprenticed to *Northampton Mercury*; joined *Rochdale Observer*, 1887; editor, 1893-1908; managing editor, *Northampton Mercury* group, 1908-23; parliamentary correspondent, *Daily Chronicle*, 1924-30; assistant editor, 1931, editor, 1932-50, *Sunday Times*; maintained notable team of regular contributors including Ernest Newman, James Agate, (Sir) Desmond MacCarthy, and (Sir) R. C. K. Ensor [qq.v.]; published *Munich: Before and After* (1944).

HADOW, GRACE ELEANOR (1875-1940), principal of the Society of Oxford Home-Students (now St. Anne's College) and pioneer in social work; sister of Sir W. H. Hadow [q.v.]; first class, English, Somerville College, Oxford, 1903; tutor in English, Lady Margaret Hall, 1906, lecturer, 1909-17; secretary, Barnett House, Oxford, 1920-9; principal, Society of Home-Students, 1929-40; vice-chairman, National Federation of Women's Institutes, 1916-40.

HADOW, Sir (WILLIAM) HENRY (1859-1937), scholar, educationist, and critic and historian of music; educated at Malvern and Worcester College, Oxford; first class, *lit. hum.*, 1882; lecturer (1884), fellow, tutor, and dean (1889); B.Mus., 1890; principal, Armstrong College, Newcastle, 1909-19; vice-chancellor, Sheffield University, 1919-30; chairman, consultative committee, Board of Education, 1920-34; established music as part of a liberal education; works include *Sonata Form* (1896) and *English Music* (1931); edited *Oxford History of Music* and contributed *The Viennese Period* (vol. v, 1904); knighted, 1918.

HAGGARD, Sir HENRY RIDER (1856-1925), novelist; went to South Africa as secretary to Sir H. E. G. Bulwer, governor of Natal, 1875; master and registrar of Transvaal high court, 1878; returned to England and settled in Norfolk, 1879; travelled round world as member of dominions royal commission, 1912-17; knighted, 1912; KBE, 1919; had two interests, agriculture and romantic writing; works include *Rural England* (1902), *King Solomon's Mines* (1885), *She* (1887), and *Ayesha* (1905).

HAIG, DOUGLAS, first Earl Haig (1861-1928), field-marshal; educated at Clifton, Oxford, and Sandhurst; gazetted to 7th Hussars, 1885; entered Staff College, 1896; special service officer employed in Sudan campaign, 1898; accompanied Major-General French (afterwards Earl of Ypres, q.v.) to South Africa as staff officer, 1899; conspicuous for ingenuity, enterprise, and brilliant staff work; column commander, 1900-2; inspector-general of cavalry to Lord Kitchener [q.v.] in India, 1903-6; major-general, 1904; a director on general staff at War Office, 1906; responsible for scheme of imperial general staff and drafting of first British field service regulations; chief of staff to Sir O'Moore Creagh [q.v.] in India, 1909-11; commanded at Aldershot, 1911; aide-de-camp general, 1914; took to France I Army Corps on outbreak of European war, Aug. 1914; urged on Lord Kitchener necessity of creating great national army; conducted orderly retreat of I Corps after battle of Mons. Aug.-Sept.; won confidence of his men and of commander-in-chief; at first battle of Ypres (19 Oct.-22 Nov.) his magnificent defence, imperturb-

able calm, and tactical skill made him national figure; commanded First Army, 1915; planned successful attack on Neuve Chapelle, which marked beginning of new epoch in war, Mar.; entrusted with attack on Loos, Sept.; succeeded French as commander-in-chief, Dec.; at Joffre's request prepared counter-offensive to relieve Verdun which resulted in battle of Somme, July-Nov. 1916; directed to conform to instructions of General Nivelle, new French commander-in-chief, at allied conference at Calais, 1917; Vimy ridge captured by British, but failure of French attacks produced breakdown of French morale, to cover which British continued attacks, capturing Messines ridge (June) and attacking Ypres front (July); after exhausting effort Passchendaele captured, Nov.; situation critical for Allies in winter of 1917-18; expected German blow fell upon British Third and Fifth Armies, Mar. 1918; secured appointment of Foch to control operations on western front, Mar.; his courage and resolution inspired his men to resist second great German attack in Flanders, Apr.; loyally supported Foch; his conviction that time had come for supreme effort, issuing in decision to make general advance against enemy, secured victory in 1918; led armies to Rhine; commander-in-chief, home forces, 1919-21; received thanks of parliament, £100,000, and created earl, 1919; devoted himself to cause of ex-servicemen; president of British Legion and chairman of United Services Fund.

HAIG BROWN, WILLIAM (1823-1907), master of Charterhouse; educated at Christ's Hospital and Pembroke College, Cambridge; second classic, 1846; MA, 1849; LLD, 1864; fellow, 1848; honorary fellow, 1899; appointed headmaster of Charterhouse School, 1863; active in advocating removal of school from London to Godalming, 1864; school opened at Godalming, 1872; numbers rose from 150 in 1872 to 500 in 1876; made many additions and improvements; known as 'our second founder'; appointed master of the Charterhouse, 1897; honorary canon of Winchester, 1891; published *Charterhouse, Past and Present* (1879), *Carthusian Memories* (verse, 1905), and other works.

HAIGH, ARTHUR ELAM (1855-1905), classical scholar; BA, Corpus Christi College, Oxford (first class, classics), 1878; fellow of Hertford College, 1878-86; fellow and tutor of Corpus, 1901; author of *The Attic Theatre* (1889) and *The Tragic Drama of the Greeks* (1896).

HAILEY, (WILLIAM) MALCOLM, Baron HAILEY (1872-1969), public servant; educated at Merchant Taylors' School and Corpus

Christi College, Oxford; first class, classical hon. mods. and *lit. hum.*, 1892-4; entered Indian civil service; posted to Punjab, 1896; colonization officer, Jhelum Canal Colony, 1901; served in secretariat of Punjab and Government of India, 1907-12; first chief commissioner, Delhi, 1912-18; finance member, viceroy's Executive Council, 1919-22; home member and leader of government bloc in Legislative Assembly, 1922; attached great importance to principle of dyarchy; governor, Punjab, 1924-8; governor, United Provinces, 1928-34; attended Round Table Conference, 1930-1; baron, 1936; director, survey of Africa, proposed by J. C. Smuts [q.v.]; published *An African Survey* (1938); employed on missions to African colonies and Belgian Congo, 1939-40; chairman of committees concerned with research in Africa; wrote books on Africa, last of which *The Republic of South Africa and the High Commission Territories* was published in 1963; KCSI, 1922; GCSI, 1932; GCIE, 1928; GCMG, 1939; PC, 1949; OM, 1956; hon. fellow, Corpus Christi College, Oxford, and many other academic honours; Rhodes trustee, 1946-66; few men contributed so much to the transition from bureaucratic rule to democracy in India; few so much to the peaceful transfer of power in Africa.

HAILSHAM, first VISCOUNT (1872-1950), statesman and lord chancellor. [See HOGG, DOUGLAS McGAREL.]

HAINES, SIR FREDERICK PAUL (1819-1909), field-marshal; joined 4th (King's Own) regiment, 1839; served in 1st Sikh war; military secretary to Lord Gough [q.v.], 1846-9; served at Alma and Balaclava; prominent at Inkerman; military secretary to Sir Patrick Grant [q.v.] at Madras, 1856-60; commanded Mysore division, 1865-70; commander-in-chief at Madras, 1871-5; KCB, 1871; lieutenant-general, 1873; commander-in-chief in India, 1876-81; superintended Afghan war, 1878-9; general and GCB, 1877; CIE, 1878; GCSI, 1879; differed from the viceroy, Lord Lytton [q.v.], in regard to the plans of attack on Afghanistan, 1878-9; suggested the relief of Kandahar by Roberts's force from Kabul, 1880; declined baronetcy; field-marshal, 1890.

HAKING, SIR RICHARD CYRIL BYRNE (1862-1945), general; commanded 5th brigade, 1911-14; 1st division, 1914-15; XI Corps, 1915-18, in France and Italy; chief British representative, allied armistice commission, 1918-19; headed British military mission to Russia and Baltic, 1919; commanded allied troops, East Prussia plebiscite area, 1920; League of Nations high commissioner, Danzig,

1921-3; commanded British troops in Egypt, 1923-7; general, 1925; KCB, 1916.

HALCROW, SIR WILLIAM THOMSON (1883-1958), civil engineer; educated at George Watson's College and university of Edinburgh; trained with Thomas Meik & Sons; chief engineer, contracting firm, Topham, Jones, and Railton, 1910; consultant in partnership with C. S. Meik from 1921; designed and constructed ordnance factories, deep-level shelters, Phoenix units of Mulberry harbours; chairman, engineers' panels reporting on Severn Barrage (1944) and hydro-electric projects in Rhodesia (1951); knighted, 1944; president, Institute of Civil Engineers, 1946-7.

HALDANE, ELIZABETH SANDERSON (1862-1937), sister of J. S. and R. B. (Viscount) Haldane [qq.v.]; a lifelong liberal; worked under Octavia Hill [q.v.]; helped to establish similar organization in Edinburgh, 1884; first woman member, Scottish Savings Committee, 1916; first woman JP in Scotland, 1920; member of Scottish universities' committee (1909), royal commission on Civil Service (1912), General Nursing Council (1928), vice-chairman, Territorial Forces Nursing Service; a governor, Birkbeck College; Carnegie trustee, 1914-37; publications include translations of Descartes and Hegel; CH, 1918.

HALDANE, JOHN BURDON SANDERSON (1892-1964), geneticist; son of John Scott Haldane, and brother of Richard Burdon Haldane (later Viscount Haldane) and Elizabeth Sanderson Haldane (Naomi Mitchison) [qq.v.]; educated at Dragon School, Oxford, Eton, and New College, Oxford; first class, mathematical mods. and *lit. hum.*, 1912-14; served in Black Watch in France, Mesopotamia, and India, 1914-18; fellow, New College, 1919; reader in biochemistry, Cambridge, under (Sir) F. Gowland Hopkins [q.v.]; dismissed because cited in divorce case, 1925; appealed and reinstated, 1926; also 'officer in charge of genetical investigations', John Innes Horticultural Research Station, 1927-36; FRS, 1932; Fullerian professor of physiology, Royal Institution, 1930-2; professor of genetics, and then, of biometry, University College, London, 1933-57; member, Biometry Research Unit, Indian Statistical Institute, Calcutta, 1957; became Indian citizen, 1961; head of Laboratory of Genetics and Biometry, Bhubaneswar, 1962; major contribution to science, the re-establishment of Darwinian natural selection as accepted mechanism of evolutionary change; investigated disaster in submarine *Thetis* and did other work for Royal Navy and Royal Air Force, 1939-45; publications include *Daedalus* (1924), *Possible Worlds*

1927), *The Inequality of Man* (1932), and *My Friend Mr. Leakey* (1937), a collection of children's stories; chairman, editorial board, the *Daily Worker*, 1940-50; hon. D.Sc. (Oxford), 1961; hon. fellow, New College, 1961; doctorate of Paris University, and many other academic honours.

HALDANE, JOHN SCOTT (1860-1936), physiologist and philosopher; brother of R. B. (Viscount) Haldane and Elizabeth Haldane [qq.v.]; educated at Edinburgh Academy and University and at Jena; graduated in medicine (Edinburgh), 1884; demonstrator in physiology, Oxford, 1887; reader, 1907-13; investigated suffocative gases in coal-mines and wells; reported (1896) to home secretary on causes of death in colliery explosions and fires; introduced new methods for gas analysis and for investigating physiology of the respiration and blood; investigated physiology of lung respiration, cardiac output, and function of kidney in relation to body's physiological requirements; developed (1907) 'stage decompression' for bringing deep-divers to surface; worked (1914-18) on identification of war-gases and pathology and treatment of their effects; director (1912-36), mining research laboratory, supported by colliery owners; investigated ventilation, rescue apparatus, underground fires, and pulmonary disease; publications include *Respiration* (1922) and *The Sciences and Philosophy* (1929); FRS, 1897; CH, 1928.

HALDANE, RICHARD BURDON, Viscount Haldane (1856-1928), statesman, lawyer, and philosopher; grandson of James Alexander Haldane [q.v.]; educated at Edinburgh Academy and universities of Göttingen and Edinburgh (MA, with first class honours in philosophy); called to bar (Lincoln's Inn), 1879; most successful in type of legal work involving consideration of legal principles rather than mere interpretation of facts; junior to Horace (afterwards Lord) Davey [q.v.], 1882; liberal MP, East Lothian, 1885-1911; QC, 1890; 'went special', 1897; employed in Canadian and Indian cases and other appeals before Privy Council; also dealt with considerable number of appeals before House of Lords, including important United Free Church of Scotland case (1904); lord chancellor, 1912-15; secured increase in number of lords of appeal and raised judicial committee of Privy Council to position in which it commanded increasing confidence at home and in dominions; secretary of state for war, 1905-12; took office in difficult conditions, chief of which was indifference of liberal party to army; made extensive study of whole problem of army organization; imperial general staff created, 1909; carried through House of Commons legislation necessary to give effect to schemes of army reform; these included formation of militia into special reserve, creation of Officers' Training Corps, and improvement of medical and nursing services under Territorial system; created viscount, 1911; sent by cabinet on abortive mission to Germany, 1912; owing to this later groundlessly accused of pro-German sympathies; interested in higher education and administration, one of his principal efforts on behalf of university organization being directed towards establishment of provincial or 'civic' universities; chairman (1904) of small committee whose recommendations resulted in creation of University Grants Committee; chairman of royal commission on university education in London, 1909; worked towards increasing efficiency of public administration; after leaving office (1915) became progressively estranged from official liberal party; lord chancellor in labour administration, 1924; led small number of labour peers who formed official opposition in House of Lords, 1925-8; FRS, 1906; FBA, 1914; his works include *The Pathway to Reality* (1903), *The Reign of Relativity* (1921), *The Philosophy of Humanism* (1922), and *Human Experience* (1926).

HALE-WHITE, Sir WILLIAM (1857-1949), physician; son of William Hale-White (Mark Rutherford) [q.v.]; qualified at Guy's Hospital, 1879; demonstrator of anatomy, 1881; assistant physician, 1885; full physician, 1890-1918; developed large consulting practice; retired, 1927; an editor of *Guy's Hospital Reports*, 1886-93, and of *Quarterly Journal of Medicine*; wrote on many branches of medicine and on medical history; publications include *Materia Medica, Pharmacology and Therapeutics* (1892); medical biographies published in *Guy's Hospital Reports*; *Great Doctors of the Nineteenth Century* (1935); and *Keats as Doctor and Patient* (1938); KBE, 1919.

HALFORD, FRANK BERNARD (1894-1955), aircraft engine designer; educated at Felsted and Nottingham University College; joined Royal Flying Corps, 1914; redesigned Austro-Daimler engine for DH4; worked with Sir Harry Ricardo, 1919-23, for Aircraft Disposal Company, 1924-7; technical director, Napier engine company, 1935-44, de Havilland Engine Company, 1944-55; designed Cirrus (1925) and Gipsy (1928) engines; Rapier and Sabre for Napier; and for jet propulsion the Goblin, Ghost, Gyron, Sprite, and Spectre engines; CBE, 1948; president, Royal Aeronautical Society, 1951-2.

HALIBURTON, ARTHUR LAWRENCE, first Baron Haliburton (1832-1907), civil servant; son of Thomas Chandler Haliburton [q.v.]; born and educated at Windsor, Nova Scotia; called to bar there, 1855; joined commissariat department of British army, 1855; served in Crimea and in Canada; civilian assistant director of supplies and transports, 1869; director, 1879; supervised victualling of eight campaigns; KCB, 1885; permanent undersecretary for war, 1895–7; GCB, 1897; baron, 1898; vigorously defended short military service in *The Times*, 1897; in later years advocated universal military training.

HALIFAX, second VISCOUNT (1839–1934). [See WOOD, SIR CHARLES LINDLEY.]

HALIFAX, first EARL OF (1881–1959), statesman. [See WOOD, EDWARD FREDERICK LINDLEY.]

HALL, SIR (ALFRED) DANIEL (1864–1942), educationist, administrator, and scientific research worker; scholar of Manchester grammar school and Balliol College, Oxford; first class, natural science, 1884; first principal, South Eastern Agricultural College, Wye, Kent, 1894–1902; director, Rothamsted Experimental Station, 1902–12; member of Development Commission, 1910–17; secretary, Board of Agriculture, 1917–20; chief scientific adviser, Ministry of Agriculture, 1920–7; director, John Innes Horticultural Institution, 1927–39; publications include *The Soil* (1903), *The Book of the Rothamsted Experiments* (1905), *A Pilgrimage of British Farming* (1913), *The Apple* (1933, with M. B. Crane), *The Genus Tulipa* (1940), and *Reconstruction and the Land* (1941); FRS, 1909; KCB, 1918.

HALL, ARTHUR HENRY (1876–1949), engineer; educated at Clifton and Trinity Hall, Cambridge; first class, mechanical sciences, 1898; served engineering apprenticeship; assistant mechanical engineer, Woolwich Arsenal, 1905; assistant superintendent, 1914–17; director, torpedoes and mines production, Admiralty, 1917–19; superintendent, airship construction, Cardington, 1926–8; chief superintendent, Royal Aircraft Establishment, Farnborough, 1928–41; CB, 1937.

HALL, SIR ARTHUR JOHN (1866–1951), physician; educated at Rugby, Caius College, Cambridge, and St. Bartholomew's Hospital; qualified, 1889; FRCP, 1904; on staff of Sheffield Royal Hospital, 1890–1931; taught physiology and pathology in and built up Sheffield Medical School into faculty of medicine, Sheffield University (1905); dean, 1911–16; professor of medicine, 1915–31; authority on encephalitis lethargica; knighted, 1935; hon.

D.Sc. Sheffield, 1928; president, Association of Physicians of Great Britain and Ireland, 1931.

HALL, CHRISTOPHER NEWMAN (1816–1902), Congregationalist divine; son of John Vine Hall [q.v.]; ordained, 1842; accomplished preacher of evangelical fervour; minister of Surrey chapel, 1854; LLB, London, 1856; chairman of Congregational Union, 1866; built Christ Church, Westminster Bridge Road, 1876; resigned pastorate, 1892; hon. DD, Edinburgh, 1902; wrote many devotional works and hymns; autobiography published, 1898.

HALL, SIR EDWARD MARSHALL (1858–1927), lawyer; called to bar (Inner Temple), 1883; first real opportunity, murder case (*Rex v. Hermann*), when he obtained verdict of manslaughter, 1894; QC, 1898; conservative MP, Southport division, Lancashire, 1900–6; East Toxteth division, Liverpool, 1910–16; practice temporarily affected by Court of Appeal's criticism of his conduct in libel case against *Daily Mail*, 1901; greatest triumph, Russell divorce case, 1923; recorder of Guildford, 1916; knighted, 1917; at his best, a powerful advocate.

HALL, FITZEDWARD (1825–1901), philologist; born at Troy, New York; went to India, 1846; tutor (1850) and professor of Sanskrit (1853) at Benares government college; served in Sepoy mutiny; settled in London, 1862; professor of Sanskrit at King's College, and librarian to India Office; while in India discovered many Sanskrit MSS, and edited many Sanskrit and Hindi literary works and treatises of Hindu philosophy; edited works for Early English Text Society, 1864–9, published many philological researches, and from 1878 helped in preparation of *New English Dictionary* and Joseph Wright's *Dialect Dictionary*; hon. DCL, Oxford, 1860; hon. LLD, Harvard, 1895.

HALL, HARRY REGINALD HOLLAND (1873–1930), archaeologist; BA, St. John's College, Oxford, 1895; assistant, department of Egyptian and Assyrian antiquities, British Museum, 1896; took part in Egypt Exploration Fund excavations in winters of 1903–4–5–6; deputy keeper, department of Egyptian and Assyrian antiquities, 1919; keeper, 1924; FBA, 1926; works include *The Oldest Civilisation of Greece* (1901) and *The Ancient History of the Near East* (1913).

HALL, HUBERT (1857–1944), archivist; educated at Shrewsbury; entered Public Record Office, 1879; senior clerk, 1892; assistant keeper, 1912–21; secretary, royal commission on public records, 1910–18; reader in palaeography and economic history, London, 1896-1926; publications include *Studies in English Official Historical Documents* (1908).

HALL, SIR JOHN (1824-1907), premier of New Zealand; born at Hull; employed at London General Post Office, 1843-52; emigrated to Lyttelton, New Zealand, for sheep-farming, 1852; first mayor of Christchurch, 1862-3; elected for Christchurch to first house of representatives, 1855; colonial secretary, 1856; postmaster-general and electric telegraph commissioner, 1866-9; called to legislative council, 1872; colonial secretary, 1872-3; opposed (secular) Education Act of 1877; premier, 1879-82; KCMG, 1882; his ministry repealed Sir George Grey's land-tax, passed triennial parliaments bill and universal suffrage bill; Hall introduced woman's suffrage amendment into electoral bill, 1893.

HALL, WILLIAM GEORGE GLENVIL (1887-1962), politician; educated at Friends' School, Saffron Walden; clerk in Barclay's Bank, London; did social work in East End; joined independent labour party, 1905; although a Quaker, enlisted and served in army, 1914-18; financial officer, labour party, 1919-39; unsuccessful candidate in 1922-4 elections; labour MP, Portsmouth Central, 1929-31; MP, Colne Valley, Yorkshire, 1939-62; called to bar (Gray's Inn), 1933; financial secretary to Treasury, 1945-50; PC, 1947; British representative, United Nations Assembly, 1945, 6, and 8; chairman, Parliamentary labour party, 1950-2; attended Consultative Assembly, Strasbourg, 1950-2; member, BBC advisory council, 1952; president, United Kingdom Alliance, 1959.

HALL, SIR (WILLIAM) REGINALD (1870-1943), admiral; entered navy, 1884; commanded battle cruiser *Queen Mary*, 1913-14; introduced many reforms including three-watch system; director, intelligence division, Admiralty, 1914-18; recruited large staff whose deciphering of German naval signals became a principal factor in naval warfare; intercepted German messages to other countries including Zimmermann telegram proposing German-Mexican offensive alliance; its publication influential on American entry into war; instrumental in capturing Sir Roger Casement [q.v.]; employed counter-agents and devised many ruses to deceive Germans; rear-admiral, 1917; vice-admiral, 1922; admiral, 1926; conservative MP, West Derby division of Liverpool (1918-23), Eastbourne (1925-9); KCMG, 1918.

HALL, (WILLIAM) STEPHEN (RICHARD) KING-, BARON KING-HALL

(1893-1966), writer, and broadcaster on politics and international affairs. [See KING-HALL.]

HALLÉ (formerly NORMAN-NERUDA), WILMA MARIA FRANCISCA, LADY (1839-1911), violinist; daughter of Josef Neruda; born in Moravia; appeared in London, 1849; married Ludwig Norman, 1864, and secondly Sir Charles Hallé [q.v.], 1888; teacher in Berlin, 1898; violinist to Queen Alexandra, 1901.

HALLIBURTON, WILLIAM DOBINSON (1860-1931), physiologist and biochemist; educated at University College School and University College, London; B.Sc., 1879; MD, 1884; professor of physiology, King's College, 1890-1923; FRS, 1891; pioneer in biochemistry; publications include *Chemical Physiology and Pathology* (1891), *Essentials of Chemical Physiology* (1893), 'Halliburton's' *Physiology* (1896; 19 editions); edited first seven volumes of *Physiological Abstracts*.

HALLIDAY, SIR FREDERICK JAMES (1806-1901), first lieutenant-governor of Bengal; joined Bengal civil service, 1825; judicial and revenue secretary in Bengal 1838; secretary in the home department, 1849-53; first lieutenant-governor of Bengal, 1854-9; reorganized police, improved roads, and advanced education; initiated Calcutta University, 1856; suppressed rising of Santal tribes, 1855; helped to check Sepoy mutiny in Bengal, 1858; KCB, 1860; involved in long controversy—from 1857 to 1888—with a subordinate, William Tayler [q.v.], commissioner of Patna, whom he removed from his office, 1857; member of council of India, 1868-86; an accomplished musician, he frequently performed in concerts in Bengal, London, and elsewhere.

HALSBURY, first EARL OF (1823-1921), lord chancellor. [See GIFFARD, HARDINGE STANLEY.]

HALSEY, SIR LIONEL (1872-1949), admiral; entered navy, 1885; commanded *New Zealand*, 1912-15; captain of fleet to (Lord) Jellicoe [q.v.], 1915-16; fourth sea lord, 1916; third sea lord, 1917-18; rear-admiral, 1917; commanded second battle cruiser squadron in *Australia*, 1918-19; commanded *Renown* as chief of staff to Prince of Wales visiting Canada and United States (1919), Australia, New Zealand, West Indies, etc. (1920); comptroller and treasurer to Prince of Wales, 1920-36; vice-admiral, 1921; admiral, 1926; KCMG, 1918.

HAMBLEDEN, second VISCOUNT (1868-1928), philanthropist. [See SMITH, WILLIAM FREDERICK DANVERS.]

HAMBLIN SMITH, JAMES (1829–1901), mathematician. [See SMITH.]

HAMBOURG, MARK (1879–1960), pianist; born in Southern Russia; brought to London, 1889, and played as infant prodigy; studied in Vienna and made début at Vienna Philharmonic Symphony Concert as adult pianist, 1895; naturalized, 1896; began recording for HMV with the 'Moonlight Sonata', 1909; made worldwide tours playing chamber music and concertos but especially remembered as recitalist in grand virtuoso manner; last public performance, 1955.

HAMBRO, SIR CHARLES JOCELYN (1897–1963), merchant banker; educated at Eton and Sandhurst; served in Coldstream Guards in France, 1916–18; MC; secretary, C. J. Hambro & Son, family firm; played important part in merger with British Bank of Northern Commerce and establishment of Hambros Bank, 1921; director, Bank of England, 1928; established, under (Lord) Montagu C. Norman's [q.v.] direction, the exchange control division, 1932–3; chairman, Great Western Railway, 1940–5; joined Ministry of Economic Warfare, 1939; worked in Special Operations Executive; visited Sweden, 1940; KBE, 1941; deputy head, SOE, 1941–2; executive chief, succeeding (Sir) Frank Nelson [q.v.], 1942–3; head of British raw materials mission, Washington, 1943–5; chairman, Hambros Bank, 1961–3.

HAMIDULLAH, NAWAB OF BHOPAL (1894–1960). [See BHOPAL.]

HAMILTON, CHARLES HAROLD ST. JOHN (1876–1961), writer, better known as FRANK RICHARDS; educated at schools in West London; received first cheque for a short story at age of seventeen; wrote (under pseudonyms) for the *Gem* and the *Magnet*, boys' papers, in particular, about Greyfriars School and its 'Famous Five', including Billy Bunter, 1907–40; wrote Bunter scripts for television, 1952; created other schools, including Rookwood in the *Boys' Friend* and Cliff House School, with Bessie Bunter in the *School Friend*.

HAMILTON, DAVID JAMES (1849–1909), pathologist; studied medicine at Edinburgh; worked at pathology at Vienna and Paris, 1874–6; professor of pathology at Aberdeen, 1882–1908; pioneer of bacteriological diagnosis of diphtheria and typhoid; investigated 'braxy' and 'louping ill' in sheep; published standard textbook on pathology (2 vols., 1889–94); FRS, 1908; hon. LLD, Edinburgh, 1907.

HAMILTON, SIR EDWARD WALTER (1847–1908), Treasury official; son of Walter Kerr Hamilton [q.v.]; educated at Eton and Christ Church, Oxford; private secretary successively to Robert Lowe and Gladstone, 1872–85; Treasury official, 1885–1907; permanent secretary to the Treasury (with Sir George Murray, q.v.), 1902–7; KCB, 1894; GCB, 1906; PC, 1908; specially connected with Goschen's financial measures, of which he wrote an account, 1889; published monograph on Gladstone, 1898.

HAMILTON, EUGENE JACOB LEE- (1845–1907), poet and novelist. [See LEE-HAMILTON.]

HAMILTON, LORD GEORGE FRANCIS (1845–1927), statesman; son of first Duke of Abercorn [q.v.]; conservative MP, Middlesex, 1868–84; for Ealing division of county, 1885–1906; under-secretary for India, 1874–80; introduced Indian loans bill, 1874; first lord of Admiralty, 1885–6, 1886–92; his administration period of extensive naval reform and construction; secretary of state for India, 1895–1903; decided not to evacuate Chitral, 1895; faced with revolt of Waziris and Afridis, 1897; had to deal with problems of famine and plague; his suggestion of dispatch of troops to Cape in South African war accepted, 1899; resigned as free trader, 1903; chairman of royal commissions on poor law and unemployment, 1905–9; and on Mesopotamian campaign, 1916–17.

HAMILTON, SIR IAN STANDISH MONTEITH (1853–1947), general; educated at Wellington College and Sandhurst; posted to 92nd Highlanders, India, 1873; wounded at Majuba Hill, Natal, 1881; aide-de-camp to (Lord) Roberts [q.v.], 1882–90; assistant adjutant-general, musketry, Bengal, 1890–3; military secretary to Sir George White [q.v.], 1893–5; deputy quartermaster-general, Simla, 1895–7; commandant, Musketry School, Hythe, 1898; in South Africa (1899–1902) showed conspicuous gallantry at Elandslaagte; present at siege of Ladysmith; commanded mounted infantry division in advance on Pretoria; KCB, 1900; chief of staff to Lord Kitchener [q.v.]; commanded four columns in final drive in Western Transvaal; military secretary, War Office, 1900–3; quarter-master-general, 1903–4; headed military mission with Japanese, 1904–5; published *A Staff Officer's Scrap Book* (2 vols., 1905–7); held Southern command, 1905–9; general, 1907; adjutant-general, 1909–10; GCB, 1910; his memorandum favouring voluntary enlistment published as *Compulsory Service* (1910); GOC-in-C, Mediterranean command, and inspector-general of oversea forces, 1910-14; commanded Central Force, 1914-15; appointed to command Anglo-French army to assist navy in forcing Dardanelles, Mar. 1915; first attack (25 Apr.) within ace of success; three further attacks failed for lack of men or ammunition; made final attempt with landing at Suvla, 6 Aug.; recalled Oct. 1915; GCMG, 1919; published *Gallipoli Diary* (2 vols., 1920); lord rector, Edinburgh University, 1932-5.

HAMILTON, JAMES, second DUKE OF ABER-CORN (1838-1913); son of James Hamilton, first duke [q.v.]: conservative MP, county Donegal, 1860-80; succeeded father, 1885; KG, 1892; official figurehead of Irish landlords in land war; resisted home rule.

HAMILTON, JOHN ANDREW, VISCOUNT SUMNER (1859-1934), judge; educated at Manchester grammar school and Balliol College, Oxford; first class, *lit. hum.*, 1881; president of Union, 1882; prize fellow of Magdalen, 1882-9; called to bar (Inner Temple), 1883; bencher, 1909; joined Northern circuit; voluminous contributor to this Dictionary; developed large Commercial Court practice; KC, 1901; judge of King's Bench division, 1909; never reserved a judgment; promoted to Court of Appeal and PC, 1912; lord of appeal in ordinary, 1913-30; life peerage, 1913; viscount 1927; took leading part in appeals from Prize Court; judgments remarkable for style and wit as well as legal merits; a British delegate to Reparation Commission, 1919; GCB, 1920; chairman, royal commission on compensation for war damage, 1921.

HAMILTON, SIR RICHARD VESEY (1829-1912), admiral; entered navy, 1843; served on Arctic voyages, 1850-4; in Crimean and second Chinese wars, 1855-7; attained flag rank, 1877; vice-admiral, 1885; commander-in-chief, China station, 1885-8; admiral and KCB, 1887; first sea lord, 1889-91; GCB, 1895.

HAMILTON FAIRLEY, SIR NEIL (1891-1966), physician. [See FAIRLEY.]

HAMILTON FYFE, SIR WILLIAM (1878-1965), headmaster and university vice-chancellor. [See FYFE.]

HAMMOND, SIR JOHN (1889-1964), animal scientist; educated at Gresham's School, Holt, Edward VI middle school, Norwich, and Downing College, Cambridge; diploma in agriculture, 1910; served in France in Royal Norfolk Regiment, 1914-16; physiologist, Animal Nutrition Institute, Cambridge, 1920; superintendent, Animal Research Station, 1931-54;

interested in artificial insemination; supervised Downing College Estate farms, 1939-45; first president, British Society of Animal Production; CBE, 1949; knighted, 1960; FRS, 1933; fellow, Downing College, 1936; many hon. doctorates and other distinctions; publications include *The Physiology of Reproduction in the Cow* (1927) and *Growth and Development of Mutton Qualities in the Sheep* (1932); on editorial boards of *Journal of Agricultural Science*, *Empire Journal of Experimental Agriculture*, and *Journal of Dairy Research.*

HAMMOND, JOHN LAWRENCE LE BRETON (1872-1949), journalist and historian; educated at Bradford grammar school and St. John's College, Oxford; second class, *lit. hum.*, 1895; edited liberal *Speaker*, 1899-1907; secretary, Civil Service Commission, 1907-13; special correspondent, *Manchester Guardian*, 1919-49; married (1901) Lucy Barbara Bradby; with her wrote *The Village Labourer, 1760-1832* (1911), *The Town Labourer, 1760-1832* (1917), *The Skilled Labourer, 1760-1832* (1919), *Lord Shaftesbury* (1923), *The Rise of Modern Industry* (1925), *The Age of the Chartists* (1930), *James Stansfeld* (1932), and *The Bleak Age* (1934); wrote also *Charles James Fox* (1903), *C. P. Scott* (1934), and *Gladstone and the Irish Nation* (1938); FBA, 1942.

HAMMOND, WALTER REGINALD (1903-1965), cricketer; educated at Cirencester grammar school; played for Gloucestershire at age of seventeen; established as fine all-rounder, 1925; in MCC side in West Indies; made 238 not out at Bridgetown; first player since W. G. Grace [q.v.] to score 1,000 runs in May, (1927); played in all five tests in South Africa, 1927-8; first encounter with Australians, 1928-9; made 905 runs in victorious series in Australia; headed English averages with 3,323 runs, 1933, and in every season but one for the rest of his career; played finest test match innings at Lords, 240 in 6 hours, 1938; captained Gloucestershire and England; served in Royal Air Force, 1939-45; retired at end of 1946-7 tour to Australia; made over 50,000 runs in his career, took over 700 wickets, and over 800 catches; in 140 test innings, scored 7,249 runs, with 22 centuries, took 83 wickets, and made 110 catches; published *Cricket My Destiny* (1946).

HAMPDEN, second VISCOUNT (1841-1906), governor of New South Wales. [See BRAND, HENRY ROBERT.]

HAMSHAW THOMAS, HUGH (1885-1962), palaeobotanist. [See THOMAS, HUGH HAMSHAW.]

HANBURY, ELIZABETH (1793–1901), centenarian and philanthropist; of Quaker parentage; visited prisons with Elizabeth Fry [q.v.]; married Cornelius Hanbury, 1826; active in anti-slavery movement and prison reform; aided in her work by her daughter CHARLOTTE (1830–1900), who established mission at Tangier for ameliorating lot of Moorish prisoners; autobiography published, 1901.

HANBURY, SIR JAMES ARTHUR (1832–1908), surgeon-general; MB, Trinity College, Dublin, 1853; entered army medical service; FRCS England, 1887; served in Afghan war, 1878–9; at battle of Tel-el-Kebir, 1882; KCB, 1882; surgeon-general of forces in Madras, 1888–92; surgeon-general, 1892.

HANBURY, ROBERT WILLIAM (1845–1903), politician; educated at Rugby and Corpus Christi College, Oxford; conservative MP for Tamworth, 1872–8, for North Staffordshire, 1878–80, and for Preston, 1885–1903; ceaselessly attacked policy of Gladstone's government, 1892–5; PC and financial secretary of the Treasury, 1895–1900; joined cabinet as president of Board of Agriculture, 1900.

HANBURY-WILLIAMS, SIR JOHN COLDBROOK (1892–1965), industrialist; educated at Wellington College; served in France with 10th Royal Hussars, 1914–18; travelled in China and Japan on family business; joined Courtaulds Ltd., 1926; director, Snia Viscosa, Italian associate company, 1928–65; initiated developments leading to manufacture of 'cellophane' and (with Imperial Chemical Industries) nylon; managing director, Courtaulds, 1935; deputy chairman, 1943; succeeded Samuel Courtauld [q.v.] as chairman, 1946–62; made the firm one of largest industrial concerns in Britain; gentleman usher, 1931–46; knighted, 1950; CVO, 1956; director, Bank of England, 1936–63; in charge of foreign-exchange control, 1940–1; refused to succeed Lord Catto [q.v.] as governor, 1949; served at Ministry of Economic Warfare, 1942; high sheriff, county of London, 1943 and 1958; actively associated with charitable causes and member of number of committees and missions.

HANCOCK, ANTHONY JOHN (TONY) (1924–1968), comedian; educated at Durlston Court, Swanage, and Bradfield College; enlisted in Royal Air Force, 1942; toured with ENSA and Ralph Reader 'Gang Shows'; appeared at Windmill Theatre, London, 1948; met immediate success with 'Hancock's Half-Hour' on BBC radio, 1954; owed much to Bill Kerr, Kenneth Williams, and particularly, Sid James, as well as his scriptwriters, Alan Simpson and Ray Galton; transferred to television, 1956; success ended when Sid James and the scriptwriters were abandoned; last of cherished line of English comedians, but success brought boredom and disaster; died by his own hand.

HANCOCK, SIR HENRY DRUMMOND (1895–1965), civil servant; educated at Haileybury and Exeter College, Oxford; served in Sherwood Foresters and Intelligence Corps, 1914–18; entered administrative class, civil service, 1920; appointed to Ministry of Labour; private secretary to Sir Horace Wilson, permanent secretary; deeply involved in unemployment problems, 1928–38; private secretary to J. H. Thomas [q.v.], lord privy seal; member of staff of National Assistance Board; transferred to Home Office, 1938; organized financing of Civil Defence; secretary-general, British Purchasing Commission, Washington, 1941; CMG, 1942; deputy secretary, Ministry of Supply, 1942–5; deputy secretary Ministry of National Insurance, 1945; KBE, 1947; succeeded Sir Thomas Phillips [q.v.], as permanent secretary, 1949; KCB, 1950; permanent secretary, Ministry of Food, 1951; chairman, Board of Inland Revenue, 1955–8; chairman, Local Government Commission for England; member of boards of Booker Bros., McConnell & Co., and Yorkshire Insurance Co.; GCB, 1962.

HANDLEY, THOMAS REGINALD (TOMMY) (1892–1949), radio comedian; produced own radio revues from 1925; presented brilliantly absurd weekly cartoon of daily life in ITMA ('It's That Man Again!'), 1939–49; introduced 'Office of Twerps', seaside resort 'Foaming-at-the-Mouth', post-war 'Island of Tomtopia', and notable characters including Mrs Mopp and Funf; first radio show royal command performance, 1942.

HANDLEY PAGE, SIR FREDERICK (1885–1962), aircraft designer. [See PAGE.]

HANKEY, MAURICE PASCAL ALERS, first BARON HANKEY (1877–1963), secretary to the cabinet; educated at Rugby School; joined Royal Marine Artillery, 1895; sword of honour, Royal Naval College; served in *Ramillies*, 1899–1902; transferred to Naval Intelligence, Admiralty 1905–7; returned to sea as intelligence officer, 1907; assistant secretary, Committee of Imperial Defence, 1908; secretary, 1912; chief of war cabinet secretariat, 1916; secretary to the cabinet, secretary, Committee of Imperial Defence, and clerk to the Privy Council, 1923–38; secretary on British side, Paris peace conference, 1919, and conferences at Washington, 1921, Genoa, 1922, Lausaune,

1932, and The Hague, 1929 and 1930; secretary of Imperial conferences, 1921, 1923, 1926, 1930, and 1937, and the London naval conference, 1930; created the cabinet secretariat, 1916, and laid down principles which have since guided its performance; baron, 1939; joined Neville Chamberlain's war cabinet, as minister without portfolio; chancellor, Duchy of Lancaster and paymaster-general in Churchill's government, 1940-2; chairman, cabinet's Scientific Advisory Committee; FRS, 1942; publications include *The Supreme Command 1914-18* (2 vols., 1961), *Government Control in War* (1945), *Diplomacy by Conference* (1946), and *The Science and Art of Government* (1951); CB, 1912; KCB, 1916; GCB, 1919; GCMG, 1929; GCVO, 1934; PC, 1939; hon. degrees from Oxford, Cambridge, Edinburgh, and Birmingham.

HANKIN, ST. JOHN EMILE CLAVERING (1869-1909), playwright; educated at Malvern and Merton College, Oxford; wrote plays of realistic frankness; his *The Return of the Prodigal* (1905), *The Charity that began at Home* (1906), *The Cassilis Engagement* (the most popular of his plays, 1907), were published in 1907 as *Three Plays with Happy Endings*; pushed realism further in *The Last of the De Mullins* (1908); cynically satirized middle-class convention and sentiment.

HANLAN (properly HANLON), EDWARD (1855-1908), Canadian oarsman; born at Toronto; rowing champion of Canada, 1877, of America, 1878, of England, 1879, and of the world, 1880; retained the last title, 1881-4.

HANNAY, JAMES OWEN (1865-1950), novelist under pseudonym GEORGE A. BIRMINGHAM; educated at Haileybury and Trinity College, Dublin; deacon, 1888; priest, 1889; rector of Westport, county Mayo, 1892-1913; Mells, Somerset, 1924-34; canon of St. Patrick's, 1912-22; vicar, Holy Trinity, Kensington, 1934-50; many humorous novels include *General John Regan* (1913) and *Send for Dr. O'Grady* (1923).

HANNAY, ROBERT KERR (1867-1940), Scottish historian; educated at Glasgow and Oxford; curator, historical department, Register House, Edinburgh, 1911-19; professor of ancient (Scottish) history and palaeography, Edinburgh, 1919-40; historiographer-royal for Scotland, 1930-40; works include *The Archbishops of St. Andrews* (with Sir John Herkless; 5 vols., 1907-15).

HANWORTH, first VISCOUNT (1861-1936), judge. [See POLLOCK, ERNEST MURRAY.]

HARARI, MANYA (1905-1969), publisher and translator; born at Baku, daughter of Grigori Benenson, a Jewish financier; family emigrated to London, 1914; educated at Malvern Girls' College and Bedford College, London; BA, 1924; visited Palestine and married Ralph Andrew Harari [q.v.], 1925; became Roman Catholic, 1932; worked on *Dublin Review*; edited her own periodical, the *Changing World*, 1940-2; joined Political Warfare Department as translator; founded Harvill Press with Marjorie Villiers, 1946; continued as director when the Press became subsidiary of Collins, 1954; publisher and joint translator, with Max Hayward, of Boris Pasternak's *Dr. Zhivago*, 1958; also published other Russian authors such as Alexander Solzhenitsyn and Ilya Ehrenburg; visited Palestine and Russia, 1948-61; working on her memoirs when she died; *Memoirs* published in 1972.

HARARI, RALPH ANDREW (1892-1969), merchant banker, art scholar and collector; born in Cairo; son of (Sir) Victor Harari Pasha; leading member of Egypt's Anglo-Jewish community; educated at Lausaune and Pembroke College, Cambridge; first class, parts i and ii, economics tripos, 1912-13; boxing blue; served under Sir Edmund (later Viscount), Allenby [q.v.] in Palestine, 1914; finance officer to (Sir) Ronald Storrs [q.v.], military governor of Jerusalem; director, trade and commerce, under Sir Herbert (later Viscount) Samuel [q.v.], 1920-5; economic adviser, GHQ Middle East, 1939; worked in Department of Political Warfare; OBE; managing director, S. Japhet & Co., London merchant bankers, 1945; collector and authority on Islamic metalwork, and collector of Beardsley drawings; also collected an important album of Hokusai (1760-1949) sketches, and other Japanese art; with his wife Manya [q.v.] noted for hospitality in their London home.

HARBEN, SIR HENRY (1823-1911), pioneer of industrial life assurance; accountant of Prudential Mutual Assurance Association, 1852; started scheme of life assurance for working classes and proved its practicability, 1854; secretary, 1856; actuary, 1870; resident director, 1873; chairman, 1905; president, 1907; knighted, 1897; master of Carpenter's Company, 1893; founded working men's convalescent home at Rustington, 1895; represented Hampstead on London County Council, 1889-94; first mayor of Hampstead, 1900; generously supported local charities; published *Mortality Experience of the Prudential Assurance Company* (1871).

HARCOURT, AUGUSTUS GEORGE VERNON (1834-1919), chemist; BA, Balliol

College, Oxford; assistant to (Sir) B. C. Brodie [q.v.]; Lee's reader in chemistry and a senior student of Christ Church, Oxford, 1859; tutor, 1864-1902; FRS, 1863; researched on rate of chemical change, on coal gas, and chloroform as anaesthetic.

HARCOURT, LEVESON FRANCIS VERNON- (1839-1907), civil engineer. [See VERNON-HARCOURT.]

HARCOURT, LEWIS, first VISCOUNT HARCOURT (1863-1922), politician; son of Sir William Harcourt [q.v.]; private secretary to father, 1881-6, 1892-5, 1895-1904; liberal MP, Rossendale division of Lancashire, 1904-16; helped to found Free Trade Union; first commissioner of works, 1905-10; with Viscount Esher [q.v.] founded London Museum, 1911; secretary of state for colonies, 1910-15; his main interest economic and scientific development; returned to Office of Works, 1915-16; viscount, 1917.

HARCOURT, SIR WILLIAM GEORGE GRANVILLE VENABLES VERNON (1827-1904), statesman; born at York; son of William Vernon Harcourt and grandson of Edward Harcourt [qq.v.]; of Plantagenet descent; educated privately and at Trinity College, Cambridge; member of 'Society of Apostles', and president of Union Debating Society, 1849; contributed to *Morning Chronicle* while an undergraduate; called to bar, 1854; acquired large practice at the parliamentary bar; wrote regularly for newly founded *Saturday Review*, 1855-9; contributed to *The Times* under signature of 'Historicus' many letters on international law in regard to American war from 1861 onwards; letters were published separately as *Letters by Historicus on . . . International Law* (1863) and *American Neutrality* (1865); QC, 1866; member of neutrality laws commission, 1869; served also on royal commissions on naturalization laws, 1870, and on extradition, 1878; Whewell professor of international law at Cambridge, 1869-87; contributed further letters to *The Times* on parliamentary reform, redistribution of seats, and Irish Church disestablishment, 1866-9; engaged in party politics, 1867; liberal MP, Oxford, 1868; declined post of judge-advocate-general; active in discussion on Irish Church bill; candid critic of liberal government; chairman of committee whose deliberations resulted in registration of parliamentary voters bill, 1871; championed religious equality in debates on elementary education bill and on university tests bill, 1870; advocated abolition of purchase of commissions in army, 1871; opposed payment of election expenses by constituencies:

urged law reform, and helped to pass the Judicature Act of 1873; member of select committee on Civil Service expenditure, 1873; succeeded Sir Henry James [q.v.] as solicitor-general and knighted, 1873; while in opposition, he supported public worship regulation bill, 1874; opposed royal titles bill and merchant shipping bill, 1876; vigorously denounced Turkey, 1876-8; severely criticized conservative government's policy in Afghanistan and South Africa, 1878-9; by speeches and letters to *The Times* greatly influenced political opinion; made home secretary and PC in Gladstone's administration, Apr. 1880; on defeat at Oxford, was returned MP for Derby, May 1880; introduced Ground Game Act, giving occupier equal right with landlord to kill ground game; recommended central control of London water supply, 1880; advocated birch instead of detention for juvenile offenders, and proposed commission of inquiry into industrial schools, 1881; during troubles in Ireland he carried peace preservation (Ireland) bill (1881), prevention of crimes (Ireland) bill (1882), and explosive substances bill (1883); by his firmness stamped out the dynamite conspiracy in London, 1883; improved labour conditions in coal-mining, introduced local government board (Scotland) bill, which was rejected by the Lords, 1883; introduced but abandoned London government bill, 1884; active in franchise agitation, 1884; replaced clause in registration bill abolishing electoral disqualification on receipt of medical relief, 1885; dissociated himself from Joseph Chamberlain's radicalism, 1885; joined Gladstone's cabinet as chancellor of the Exchequer on Gladstone's acceptance of home rule, Feb.-July 1886; criticized new conservative government, and attempted to reunite liberal party; attacked Irish coercion policy of Lord Salisbury's government; censured government's treatment of *The Times*' attacks on Parnell; supported Irish land bill and Allotments Act, 1887; opposed cession of Heligoland to Germany, 1890; persuaded Gladstone to repudiate Parnell's leadership of Irish party after divorce proceedings, Nov. 1890; in his speeches which won him popularity through the country he urged much domestic reform; opposed A. J. Balfour's Irish local government bill, 1892; again chancellor of the Exchequer in Gladstone's fourth administration, 1892; had charge of local veto bill, which was abandoned; passed home rule bill through Commons; carried parish councils bill, which was greatly amended by the Lords, 1894; bitterly denounced upper house; served under new prime minister, Lord Rosebery, as leader of the House of Commons, 1894; introduced death duties budget which imposed a single graduated tax on real and personal property and revolutionized existing system of taxation; raised income-tax and duties on beer and spirits; this budget established his financial reputation; he passed local government bill for Scotland; introduced local liquor control bill, 1895; passed fourth budget (May 1895); resigned on defeat of government on motion dealing with cordite supply (June); was defeated at Derby (July) and elected for West Monmouth; denounced advance of Anglo-Egyptian army in Sudan, and urged inquiry into Jameson raid; as member of committee he made searching examination of Cecil Rhodes [q.v.], but defended committee's report from radical attack, 1897; opposed unionist education bill (1896), and agricultural rating bill; supported Gladstone's censure of Armenian massacres in opposition to Lord Rosebery, 1896; urged annexation of Crete by Greece; championed protestantism and attacked ritualism in his letters to *The Times* on 'Lawlessness in the Church', which led to certain reforms, 1898; resigned leadership of liberal party, Dec. 1898; was opposed to extreme imperialist policy; censured English conduct of South African war, Jan. 1900; denounced the war as 'unjust and engineered', 1901; protested against introduction of forced Chinese labour into South Africa, 1903; resisted proposed tax on imported corn and Balfour's education bill, 1902; reiterated faith in free trade in opposition to Chamberlain's fiscal reform proposals, 1903; declined peerage, 1902; honorary fellow, Trinity College, Cambridge, 1904; succeeded to family estates at Nuneham, Oxfordshire, 1904; last of the old school of parliamentarians; speeches abound in argument and irony; an aristocrat by instinct; fond of gardening and dairy-farming.

HARCOURT-SMITH, Sir CECIL (1859-1944), archaeologist and director of the Victoria and Albert Museum; scholar of Winchester; entered department of Greek and Roman antiquities, British Museum, 1879; assistant keeper, 1896; keeper, 1904-9; chairman, re-organization commission, South Kensington exhibits, 1908; director and secretary, Victoria and Albert Museum, 1909-24; adviser, royal art collections, 1925; surveyor, royal works of art, 1928-36; knighted, 1909; KCVO, 1934.

HARDEN, Sir ARTHUR (1865-1940), chemist; first class, chemistry, Owens College, Manchester, 1885; lecturer in chemistry, Manchester University, 1888; chemist to Lister Institute (1897), head of department of biochemistry, 1907-30; investigated fermentation of sugar by various bacteria; discovered essential part played by phosphorylation and dephosphorylation in breakdown of sugar by yeast and in fermentation by other micro-organisms; joint-editor, *Biochemical Journal*,

1913-37; FRS, 1909; shared Nobel prize, 1929; knighted, 1936.

HARDIE, JAMES KEIR (1856-1915), socialist and labour leader; miner in Lanarkshire, 1866; dismissed as agitator, 1878; took up journalism and began to work for organization of miners; successively miners' county agent for Lanarkshire and secretary for Ayrshire; secretary of Scottish miners' federation, 1886; left liberals and became chairman of newly formed Scottish labour party, 1888; founded *Labour Leader*, 1889; independent labour MP, South West Ham, 1892-5; chairman of newly formed independent labour party, 1893-1900, and 1913-15; MP, Merthyr Burghs, 1900-15; first leader of labour party in parliament, 1906-7; excellent speaker; did more than any man to create British political labour movement.

HARDIE, MARTIN (1875-1952), artist and museum official; nephew of John Pettie [q.v.]; educated at St. Paul's School and Trinity College, Cambridge; entered Victoria and Albert Museum, 1898; keeper, departments of painting and of engraving, illustration, and design, 1921-35; talented etcher and water-colourist; publications include *The British School of Etching* (1921), book on Samuel Palmer [q.v.] (1928) whose work he rediscovered, and history of British water-colour painting (3 vols., 1966-8); CBE, 1935.

HARDIE, WILLIAM ROSS (1862-1916), classical scholar; MA, Edinburgh; scholar, BA, and fellow, Balliol College, Oxford; tutor, 1885-95; professor of humanity, Edinburgh, 1895-1916; brilliant composer and teacher; wrote *Lectures on Classical Subjects* (1903), *Latin Prose Composition* (1908), *Silvulae Academicae* (1911), *Res Metrica* (1920).

HARDIMAN, ALFRED FRANK (1891-1949), sculptor; scholar, British School, Rome, 1920-2; style strongly decorative tending to hardness, based on Roman, early fifth-century Greek, and Etruscan work; executed statue of Haig, Whitehall, and statues on eastern half of County Hall; ARA, 1936; RA, 1944.

HARDING, Sir EDWARD JOHN (1880-1954), civil servant; educated at Dulwich College and Hertford College, Oxford; joined Colonial Office, 1904; called to bar (Lincoln's Inn), 1912; secretary, dominions royal commission, 1912-17; deputy secretary, Imperial Conferences, 1923, 1926; assistant under-secretary, Dominions Office, 1925; permanent under-secretary, 1930-9; high commissioner in South Africa, 1940-1; CMG, 1917; CB, 1926; KCMG, 1928; KCB, 1935; GCMG, 1939.

HARDING, GILBERT CHARLES (1907-1960), broadcasting and television star; educated at Royal Orphanage, Wolverhampton and Queens' College, Cambridge; became Roman Catholic, 1929, and schoolmaster; joined BBC, 1939; served in Canada, 1944-7; thereafter conducted popular programmes (Brains Trust, 'What's My Line?', etc.) with refreshing but stormy spontaneity; a humane and learned individual at odds with the Establishment and himself; published autobiography, *Along My Line* (1953).

HARDINGE, ALEXANDER HENRY LOUIS, second BARON HARDINGE OF PENSHURST (1894-1960), private secretary to King Edward VIII and King George VI; succeeded father, first Lord Hardinge of Penshurst [q.v.], 1944; educated at Harrow and Trinity College, Cambridge; appointed assistant private secretary to King George V, 1920; principal private secretary to King Edward VIII, 1936, to King George VI, 1936-43; PC, 1936; MVO, 1925; CVO, 1931; CB, 1934; GCVO and KCB, 1937; GCB, 1943.

HARDINGE, CHARLES, BARON HARDINGE OF PENSHURST (1858-1944), statesman; grandson of first Viscount Hardinge [q.v.]; educated at Harrow and Trinity College, Cambridge; entered Foreign Office, 1880; served successively in Constantinople, Berlin, Washington, Sofia, Bucharest, Paris, Tehran, and St. Petersburg; assistant under-secretary of state, 1903-4; accompanied King Edward VII on tour of western European capitals; PC, KCMG, KCVO, 1904; ambassador to Russia, 1904-6; permanent under-secretary of state, 1906-10, 1916-20; attended King on visits to Europe and (1908) Russia; viceroy of India, 1910-16; work for social betterment included establishment of Moslem (Aligarh) and Hindu (Benares) universities; seriously wounded by bomb thrown during state entry of Delhi, 1912; censured by commission of inquiry into Mesopotamia expedition, 1917; ambassador in Paris, 1920-2; baron, 1910; KG, 1916.

HARDWICKE, SIR CEDRIC WEBSTER (1893-1964), actor; educated at King Edward VI grammar school, Stourbridge, and Bridgnorth School; studied at the (Royal) Academy of Dramatic Art, 1912; joined the Shakespeare Company of (Sir) Frank Benson [q.v.], 1913; with Old Vic Company, 1914; served in army, 1914-21; joined company of (Sir) Barry Jackson [q.v.] at Birmingham Repertory Theatre, 1922; went with Jackson to Court Theatre, London, 1924; made success in long running *The Farmer's Wife* (1925), followed by *Yellow Sands* (1926), the *Show Boat* (1928), *The Apple Cart*

(1929), and *The Barretts of Wimpole Street* (1930-1); further successes included Abel Drugger in *The Alchemist* (1932), and the doctor in *The Late Christopher Bean* (1933); knighted, 1934; made first appearance on New York stage in *Shadow and Substance* (1938); played in films in Hollywood, 1939-45; returned to London, 1944; joined Old Vic Company, 1948, but later returned to New York and died there; published *Let's Pretend* (1932), and *A Victorian in Orbit* (1961).

HARDWICKE, sixth EARL OF (1867-1904), politician. [See YORKE, ALBERT EDWARD PHILIP HENRY.]

HARDY, FREDERIC DANIEL (1827-1911), painter of domestic subjects and portraits; exhibited at Royal Academy, 1851-98; pictures fetched high prices; represented in public galleries in London, Leicester, Wolverhampton, and Leeds.

HARDY, GATHORNE GATHORNE-, first EARL OF CRANBROOK (1814-1906), statesman. [See GATHORNE-HARDY.]

HARDY, GODFREY HAROLD (1877-1947), mathematician; educated at Cranleigh, Winchester, and Trinity College, Cambridge; fourth wrangler, 1898; fellow, 1900; Cayley lecturer in mathematics, 1914; Savilian professor of geometry, Oxford, 1920; Sadleirian professor of pure mathematics, Cambridge, 1931-42; published *A Course of Pure Mathematics* (1908), four volumes in 'Cambridge Mathematical Tracts', *An Introduction to the Theory of Numbers* (1938, with E. M. Wright), *Divergent Series* (1949), and over 350 original papers; much of best work done in collaboration, notably with Professor J. E. Littlewood; with him and Professor George Pólya published *Inequalities* (1934); worked also (1914-20) with Srinivasa Ramanujan; edited his collected works and published *Ramanujan* (1940); contributed to genetics 'Hardy's law' on transmission of dominant and recessive Mendelian characters; FRS, 1910.

HARDY, HERBERT HARDY COZENS-, first BARON COZENS-HARDY (1838-1920), judge. [See COZENS-HARDY.]

HARDY, SAM (1882-1966), footballer; attended Newbold church school, Chesterfield; joined Chesterfield football club, as goalkeeper, 1903; transferred to Liverpool, 1905; began international career, 1907; capped twenty-one times for England, 1907-20; in 1908-9 England defeated Wales, Ireland, and Scotland without conceding a goal; moved to Aston Villa, 1912;

won two F.A. cup medals; moved to Notting-ham Forest, 1921; retired after 552 League appearances, 1925; a model professional; licensee of the 'Gardener's Arms', Chesterfield.

HARDY, THOMAS (1840–1928), poet and novelist; came of native Dorset stock on both sides; inherited tradition of rural music, sacred and profane; his experiences of rural life sup-plied many years later rich material for his art; educated at private school at Dorchester; pupil of ecclesiastical architect at Dorchester, 1856; continued to study Latin and Greek; became acquainted with William Barnes [q.v.], the Dorset poet; carried out architectural work in London, 1862–7; employed by (Sir) A. W. Blomfield [q.v.]; his interest at this time centred almost entirely on poetry; his first novel, *The Poor Man and the Lady*, accepted for publica-tion (1869), but he destroyed the manuscript on being advised by George Meredith [q.v.] not to publish; gradually abandoned architecture for fiction; published *Desperate Remedies* (1871), *Under the Greenwood Tree* (1872), *A Pair of Blue Eyes* (1873), *Far from the Madding Crowd* (1874), *The Hand of Ethelberta* (1875), and *The Return of the Native* (1878); thenceforth for nearly twenty years his fiction was not only his profession (a fact which he somewhat regretted as he wished to devote himself to poetry), but an art of noble form, amazing wealth of substance, and profound significance; his later novels include *The Trumpet Major* (1880), *Two on a Tower* (1882), *The Mayor of Casterbridge* (1886), *The Woodlanders* (1887), *Tess of the D'Urber-villes* (1891), and *Jude the Obscure* (1895); abandoned fiction, which had now served his turn, practically and artistically, for poetry; published two collections of lyrics, *Wessex Poems* (1898) and *Poems of Past and Present* (1901); published his greatest single achieve-ment, dramatic epic on the Napoleonic theme, *The Dynasts*, in three parts, 1903, 1906, 1908, a grand exhibition of absolute determinism; pub-lished *Time's Laughingstocks* (a collection of lyrics) (1909); this inaugurated period of wholly lyrical activity and one of daring development, only exception being *The Famous Tragedy of the Queen of Cornwall* (1923).

HARDY, Sir WILLIAM BATE (1864–1934), biologist; educated at Framlingham and Gon-ville and Caius College, Cambridge; first class, natural sciences tripos (zoology), 1888; fellow, 1892; tutor, 1900–18; university lecturer in physiology, 1913; first chairman of Food Investigation Board, 1917–28; director of food investigation, 1917–34; superintendent, Low Temperature Research Station, Cambridge, 1922–34; FRS, 1902; knighted, 1925; pioneer in colloid chemistry, the molecular physics of films, surfaces, and boundary conditions, static friction, and lubricants.

HARE, AUGUSTUS JOHN CUTHBERT (1834–1903), author; born in Rome; nephew of Augustus and Julius Hare [qq.v.]; BA, Univer-sity College, Oxford, 1857; lived mostly in Italy and on Riviera, 1863–70; published *Memorials of a Quiet Life* (i.e. of Mrs Augustus Hare, his aunt) (3 vols., 1872–6); accomplished water-colour artist; book and art collector; compiled numerous guidebooks—to Rome (2 vols., 1871), London (1878), Italy (5 vols., 1883–4), and France (4 vols., 1890–5); also published *The Story of My Life* (6 vols., 1896–1900), and *Life of Baroness Bunsen* (2 vols., 1878).

HARE, Sir JOHN (FAIRS) (1844–1921), actor, whose original name was JOHN FAIRS; made first professional appearance at Liverpool, 1864; acted chiefly in plays by T. W. Robertson [q.v.], with Prince of Wales's company, Lon-don, 1865–74; actor-manager, with W. H. Kendal [q.v.], of Court Theatre, 1874–9, and St. James's Theatre, 1879–88; manager of Garrick Theatre, 1889–95; toured in America and provinces; knighted, 1907; helped con-siderably to mould and develop modern English acting tradition which avoids both formality and rhetoric.

HAREWOOD, sixth EARL OF (1882–1947). [See LASCELLES, HENRY GEORGE CHARLES.]

HARINGTON, Sir CHARLES (1872–1940), general; known as TIM; educated at Cheltenham and Sandhurst; posted to 2nd battalion, Liver-pool (later King's) regiment, 1892; major-general, general staff, under General Plumer [q.v.], 1916–18; deputy chief, imperial general staff, 1918–20; GOC-in-C, Army of Black Sea (1920–1), allied forces of occupation in Turkey (1921–3); handled Chanak crisis (1922) with tact and skill; held Northern command (1923–7), Western command, India (1927–31), Aldershot (1931–3); general, 1927; governor of Gibraltar, 1933–7; KCB, 1919; GCB, 1933.

HARKER, ALFRED (1859–1939), petro-logist; educated at St. John's College, Cam-bridge; fellow, 1885; demonstrator in geology, 1884; university lecturer, 1904; reader in petro-logy, 1918–31; surveyed Western Isles for Geo-logical Survey of Scotland, 1895–1905; works include *Natural History of Igneous Rocks* (1909) and *Metamorphism* (1932); FRS, 1902.

HARLAND, HENRY (1861–1905), novelist; born at St. Petersburg of American parents; commenced literary career under pseudonym Sidney Luska; showed mastery of short story in

Grey Roses (1895) and *Comedies and Errors* (1898); literary editor of *Yellow Book*, 1894-7; chief works were *The Cardinal's Snuff Box* (1900) and *My Friend Prospero* (1904).

HARLECH, fourth BARON (1885-1964), statesman and banker. [See ORMSBY-GORE, WILLIAM GEORGE ARTHUR.]

HARLEY, ROBERT (1828-1910), mathematician; Congregational minister at Brighouse and Leicester, 1854-72, and subsequently at Oxford and in Australia; vice-principal of Mill Hill School, 1872-81; principal of Huddersfield College, 1882-5; made researches in higher algebra, especially in the theory of quintics; FRS, 1863.

HARMAN, SIR CHARLES EUSTACE (1894-1970), judge; educated at Eton and King's College, Cambridge; commissioned in Middlesex Regiment; prisoner of war, 1915-18; returned to Cambridge; first class, parts i and ii, classical tripos; BA, 1920; called to bar (Lincoln's Inn) 1921; KC, 1935; judge, Chancery division, 1947; knighted; Appeal Court judge and PC, 1959; bencher, Lincoln's Inn, 1939; treasurer, 1959; retired, 1970.

HARMSWORTH, ALFRED CHARLES WILLIAM, VISCOUNT NORTHCLIFFE (1865-1922), journalist and newspaper proprietor; largely self-educated; helped to give his younger brothers start in life; took up freelance journalism, 1882; gained practical experience with publishing firm in Coventry, 1885-7; with his brother Harold (afterwards Viscount Rothermere, q.v.) formed general publishing business in London (Amalgamated Press), 1887; issued from it growing number of periodicals, including *Answers*, 1888; acquired *Evening News*, 1894; founded *Daily Mail*, an elaborately planned halfpenny morning newspaper, which was his greatest achievement in creative journalism and opened new epoch in Fleet Street, 1896; enlisted services of number of skilled writers; advertised and stimulated inventions of day, financed exploration schemes, and offered prizes for various kinds of skill, including aviation feats later on; founded *Daily Mirror*, 1903; first decade of twentieth century zenith of his career; won remarkable position in English life and amassed huge fortune; leased famous Tudor mansion, Sutton Place, Surrey, where he entertained stream of British and foreign visitors, 1899-1917; created baronet, 1903; baron, 1905; chief proprietor of *The Times* on formation of company, 1908; never had more difficult task than in struggle to put *The Times* on its feet; question of price, which he changed seven times in little more than seven years, constant anxiety;

at critical moment wholly responsible for saving it from extinction, but when he died it needed steadiness added to vitality; the 1914-18 war made him a public figure; placed himself at head of all popular movements of moment; initiated *The Times* fund for sick and wounded which reached nearly £17 million; paid several visits to armies in France and Italy; on invitation of Lloyd George undertook British war mission in USA, 1917; created viscount, 1918; after the armistice became estranged from and bitterly attacked Lloyd George; influence of *The Times* potent in bringing about Irish settlement, 1921; celebrated twenty-fifth birthday of *Daily Mail* at Olympia with great magnificence, 1921; made world tour, 1921-2; a consummate journalist, who changed whole course of British journalism by making it both lively and prosperous.

HARMSWORTH, HAROLD SIDNEY, first VISCOUNT ROTHERMERE (1868-1940), newspaper proprietor; younger brother of Alfred Harmsworth, Viscount Northcliffe [q.v.], whose firm (later the Amalgamated Press) he entered, 1888; developed rare financial ability; with brother acquired *Evening News*, 1894; founded *Daily Mail*, 1896; baronet, 1910; baron, 1914; took over *Daily Mirror*, 1914; founded *Sunday Pictorial*, 1915, and Glasgow *Record*; air minister, 1917-18; viscount, 1919; controlled Associated Newspapers (*Daily Mail*, *Evening News*, *Sunday Dispatch*), 1922-32; founded a professorship at Oxford, two at Cambridge; other benefactions included gifts to Middle Temple, Foundling Hospital, municipal art galleries.

HARPER, SIR GEORGE MONTAGUE (1865-1922), lieutenant-general; entered Royal Engineers, 1884; served in South African war, 1899-1900; deputy-director, military operations, War Office, 1911; went to France on outbreak of European war, 1914; commanded 51st (Highland) division, Territorial Force, 1915-18; took part in battles of Somme (1916), Arras, Ypres, and Cambrai (1917); commanded IV Army Corps, 1918; lieutenant-general, 1919; commander-in-chief, Southern command, 1919; KCB, 1918.

HARRADEN, BEATRICE (1864-1936), novelist; educated at Cheltenham, Queen's and Bedford colleges; BA, London; worked for female suffrage and emancipation; wrote *Ships That Pass in the Night* (1893) and less successful novels on similar theme.

HARREL, SIR DAVID (1841-1939), Irish administrator and public servant; entered Royal Irish Constabulary, 1859; resident magistrate, 1879; chief commissioner, Dublin metropolitan police, 1883-93; under-secretary for Ireland, 1893-1902; later served on trade disputes boards in England; Irish PC, 1900; knighted, 1893; GCB, 1920.

HARRINGTON, TIMOTHY CHARLES (1851-1910), Irish politician; founded *Kerry Sentinel*, 1877; imprisoned under Coercion Acts, 1881 and 1883; as secretary of National League, devised 'Plan of Campaign' for land war, 1886; nationalist MP for county Westmeath, 1883-5, and for Dublin (harbour) division, 1885-1910; called to Irish bar, 1887; counsel for C. S. Parnell [q.v.] in Parnell commission, 1888-9; remained faithful to Parnell after Parnell's divorce action; helped to reunite Irish party, 1900; lord mayor of Dublin, 1901-4.

HARRIS,   FREDERICK   LEVERTON (1864-1926), politician and art collector; shipowner; conservative MP, Tynemouth, 1900-6; unionist MP, Stepney, 1907-10, East Worcestershire, 1914-18; largely contributed to defeat of bill embodying Declaration of London, 1911; joined trade division of Admiralty, 1914; largely devised and directed blockade of Germany, 1915-19; PC, 1916; bequests include 'grangerized' life of Fanny Burney to National Portrait Gallery and majolica collection to Fitzwilliam Museum, Cambridge.

HARRIS, GEORGE ROBERT CANNING, fourth BARON HARRIS (1851-1932), cricketer and administrator; born in Trinidad; son of third Baron Harris [q.v.]; succeeded father, 1872; played in Eton (1868-70) and Oxford (1871, 1872, 1874) elevens; captained Kent, 1875-89; prominent in early matches with Australia; English captain, 1880, 1884; under-secretary for India, 1885-6, for war, 1886-9; governor of Bombay, 1890-5; popularized cricket among Indians; president, MCC, 1895; GCIE, 1890; GCSI, 1895; CB, 1918.

HARRIS, (HENRY) WILSON (1883-1955), journalist and author; nephew of James Rendel Harris [q.v.]; a Quaker; educated at Plymouth College and St. John's College, Cambridge; president of Union, 1905; joined *Daily News*, 1908; League of Nations Union, 1923, editing *Headway*; editor, *Spectator*, 1932-53; independent MP, Cambridge University, 1945-50; publications include autobiography, *Life So Far* (1954); hon. LLD St. Andrews.

HARRIS, JAMES RENDEL (1852-1941), biblical scholar, archaeologist, and orientalist; educated at Plymouth grammar school and Clare College, Cambridge; third wrangler, 1874; fellow, 1875; professor, Johns Hopkins

University, 1882-5, Haverford College, 1885-92; lecturer in palaeography, Cambridge, 1893-1903; director of studies, Woodbrooke Quaker settlement, 1903-18; curator, eastern manuscripts, John Rylands library, Manchester, 1918-25; found Syriac versions of *Apology of Aristides* (1889) and *Odes of Solomon* (1909); published specialist studies and devotional writings; FBA, 1927.

HARRIS, JAMES THOMAS ('FRANK') (1856-1931), author, editor, and adventurer; spent early years in Ireland and America; became lifelong systematic amorist; used editorship of *Fortnightly Review*, 1886-94, for own social advancement; aimed unsuccessfully at socialist premiership; made *Saturday Review*, 1894-8, most brilliant weekly of the time; ruined reputation with later journalistic ventures; his biographies of Wilde (1916) and Shaw (1931), and *My Life* (1925-30) reveal delusions of greatness and reckless unreliability but contain valuable sidelights.

HARRIS, JOHN WYNDHAM PARKES LUCAS BEYNON (1903-1969), writer, best known as JOHN WYNDHAM; educated at Bedales, 1918-21; wrote science fiction for American magazines, 1930; published as serial in *The Passing Show* 'The Secret People' under name John Beynon, 1935; this and *Stowaway to Mars* published in book form, 1936; worked as civil servant and served in Royal Corps of Signals, 1939-45; published *The Day of the Triffids*, under name John Wyndham (1951); made into film (1963); other published novels include *The Kraken Wakes* (1953), *The Chrysalids* (1955), *The Midwich Cuckoos* (1957), filmed as 'Village of the Damned' (1960), and *The Outward Urge* (1959); also published collections of short stories including *The Seeds of Time* (1956) and *Consider her Ways* (1961); a notable writer of science fiction.

HARRIS, SIR PERCY ALFRED, first baronet (1876-1952), politician; educated at Harrow and Trinity Hall, Cambridge; called to bar (Middle Temple), 1899; joined family firm, Bing, Harris; liberal member for South-West Bethnal Green of LCC (1907-34, 1946-52) and of Parliament, 1922-45; MP, Market Harborough, 1916-18; chief liberal whip, 1935-45; publications include *London and its Government* (1913, rewritten, 1931); baronet, 1932; PC, 1940.

HARRIS, THOMAS LAKE (1823-1906), mystic; emigrated with parents from Buckinghamshire to America, 1828; organized 'independent   Christian   congregation'   on Swedenborgian principles in New York, 1848;

claimed to be a medium; published, from 1850 onwards, lengthy poems which were (he claimed) revealed to him in trances; edited *Herald of Light*, 1857-61; paid visits to England, 1859-61, 1865-6; set up a community near Wassaic, 1861; joined in America by Laurence Oliphant [q.v.], with whose money Harris purchased farms at Brocton, Lake Erie, 1886, and engaged in vine-growing; exerted complete sway over Oliphant until legally compelled to restore Oliphant's property, 1881; removed to Santa Rosa, 1875; advocated theory of celestial marriages, 1876; proclaimed his attainment of immortality, 1891; depicted by Oliphant as David Masollam in *Masollam* (1886); published much mystical prose and verse.

HARRIS, TOMÁS (1908-1964), artist, art dealer, and intelligence officer; educated at University College School, London; won scholarship to Slade School of Art, 1923-6; studied at British Academy, Rome; decided to become art dealer, 1928; joined his father at Spanish Art Gallery, Bruton Street, London; continued to own gallery after father's death, 1943; brought to England great variety of valuable Spanish *objets d'art*; joined security service, 1940; one of principal organizers of 'Operation Garbo', which seriously misled Germans about allied invasion plans, 1944; OBE, 1945; after 1939-45, spent much time in Spain, painting and designing tapestries; made magnificent collection of etchings and lithographs by Goya (now in British Museum); wrote standard work on etchings of Goya (2 vols., 1964).

HARRISON, FREDERIC (1831-1923), author and positivist; BA, Wadham College, Oxford; fellow, 1854-6; called to bar (Lincoln's Inn), 1858; at Oxford adopted liberal tenets and positivist doctrines, the latter owing to influence of Richard Congreve [q.v.]; president of English Positivist Committee, 1880-1905; influenced life and thought of his time by vigour as man of practical experience; served on royal commission on trade unions, 1867-9; secretary, royal commission for digesting the law, 1869-70; professor of jurisprudence, constitutional and international law for Council of Legal Education, 1877-89; travelled widely; a prolific writer on historical and literary subjects; his works include *Cromwell* (1888) and *Chatham* (1905), both in 'Twelve English Statesmen' series; *William the Silent* (1897), *Ruskin* (1902), and *Theophano: the Crusade of the Tenth Century*, a romance (1904).

HARRISON, HENRY (1867-1954), Irish nationalist and writer; educated at Westminster School and Balliol College, Oxford; MP for mid-Tipperary, 1890-2; strong supporter of C. S. Parnell [q.v.] and his widow; commissioned, Royal Irish Regiment, 1915; MC, OBE; Irish correspondent, *The Economist*, 1922-7; edited *Irish Truth*, 1924-7; supplied corrections to *History of 'The Times'* (vol. iv, 1952) over Richard Pigott [q.v.] forgeries; publications include *Parnell Vindicated* (1931) and *Parnell, Joseph Chamberlain and Mr. Garvin* (1938); hon. LLD Dublin.

HARRISON, JANE ELLEN (1850-1928), classical scholar; educated at Cheltenham Ladies' College and Newnham College, Cambridge, 1874-9; second class, classical tripos, 1879; studied archaeology in London and paid three visits to Greece, 1880-98; lecturer in classical archaeology at Newnham College (her home until 1922), 1898; took up study of Russian during 1914-18 war; works include *The Mythology and Monuments of Ancient Greece* (1890), *The Prolegomena to the Study of Greek Religion* (1903), and *Themis, a Study of the Social Origins of Greek Religion* (1912).

HARRISON, MARY ST. LEGER (1852-1931), novelist under pseudonym of LUCAS MALET; daughter of Charles Kingsley [q.v.]; married Revd William Harrison, 1876; successful novelist in romantic tradition although considered extremely outspoken; publications include *Colonel Enderby's Wife* (1885) and *The Wages of Sin* (1891).

HARRISON, REGINALD (1837-1908), surgeon; MRCS, 1859; FRCS, 1866; surgeon at Royal Infirmary, Liverpool (1867-89), and lecturer on anatomy (1865) and on surgery (1872) at medical school there; helped to convert school into medical faculty of Liverpool University; vice-president (1894-5) of Royal College of Surgeons; Hunterian professor of surgery, 1890-1, and Bradshaw lecturer, 1896; president of Medical Society of London, 1890; established system of street ambulances in Liverpool.

HARRISON, SIR WILLIAM MONTAGU GRAHAM- (1871-1949), parliamentary draftsman. [See GRAHAM-HARRISON.]

HART, SIR BASIL HENRY LIDDELL (1895-1970), military historian and strategist; educated at St. Paul's School and Corpus Christi College, Cambridge; commissioned in King's Own Yorkshire Light Infantry, 1914-16; as adjutant, training units, Volunteer Force, evolved new methods of instruction and battle drill, 1917-21; helped to compile post-war *Infantry Training* manual; transferred to Army Education Corps, 1921-7; published views on new approach to warfare in *Strategy—The*

Indirect Approach (1929); military correspondent, the *Daily Telegraph*, 1925-35; correspondent and defence adviser, *The Times*, 1935-9; published over thirty books on military leaders and *The Real War* (1930), and *The Ghost of Napoleon* (1933); unofficial adviser to minister of war, Leslie (later Lord) Hore-Belisha [q.v.], 1937-8; suggested reforms distrusted by military establishment; consistently opposed to 'total war', 1939-45; published *The Other Side of the Hill* (1948), *The Rommel Papers* (1953), *The Tanks* (2 vols., 1959), *Memoirs* (2 vols., 1965), and *History of the Second World War* (1970); denied that nuclear weapons would preclude non-nuclear warfare; hon. D.Litt., Oxford, 1964; hon. fellow, Corpus Christi College, Cambridge, 1965; knighted, 1966.

HART, SIR RAYMUND GEORGE (1899-1960), air marshal; educated at Simon Langton School, Canterbury; commissioned in Royal Flying Corps, 1917; ARCS, Imperial College of Science, 1921; joined Royal Air Force, 1924; worked on radar from 1936; deputy director of radar, 1941-3; chief signals officer, Allied Expeditionary Air Force, 1943-5, British Air Force, 1945-6; at Air Ministry: in charge of technical service plans (1947-9), director-general of engineering (1951-5), controller, engineering and equipment, 1956-9; OBE, 1940; CBE, 1944; CB, 1946; air marshal and KBE, 1957.

HART, SIR ROBERT, first baronet (1835-1911), inspector-general of customs in China; BA, Queen's College, Belfast, 1853; MA, 1871; hon. LLD, 1882; entered Chinese consular service, 1854; assistant at Ningpo, 1855-8, and at Canton, 1858; secretary to allied Anglo-French commissioners at Canton; joined Chinese imperial maritime customs service as deputy commissioner at Canton, 1859; organized customs service at Foochow and other ports, 1861-3; commissioner of customs at Shanghai, 1863; inspector-general, 1863-1906; met C. G. Gordon [q.v.] after Taiping rebellion, 1864; reconciled Gordon and Li Hung Chang; remained in Peking from 1864 till 1908; revisited Europe only twice, in 1866 and 1878; practical creator of Chinese imperial customs; controlled also imperial posts from 1896; helped to settle difficulties between Great Britain and China, 1875, which resulted in Chefoo convention, 1876, and those in Formosa between China and France, 1885; advocated necessary reforms in China, 1894-5; besieged in British legation at Peking during Boxer outbreak, 1900; organized native customs service, 1901; helped in re-establishment of Manchu dynasty; published experiences in *These from the Land of Sinim* (1901); authority terminated by Chinese

government, which subordinated his service to a board of Chinese officials, 1906; KCMG, 1882; GCMG, 1889; baronet, 1893.

HARTINGTON, MARQUESS OF (1833-1908), statesman. [See CAVENDISH, SPENCER COMPTON.]

HARTLEY, ARTHUR CLIFFORD (1889-1960), engineer and inventor; educated at Hymers College, Hull, and City and Guilds College, London; B.Sc. (Eng.), London, 1910; qualified as pilot and worked on armaments with Royal Flying Corps, 1914-18; joined engineering division, Anglo-Persian Oil Company, 1924; chief engineer, 1934-51; at Ministry of Aircraft Production (1939-45) with others produced FIDO (fog dispersal) and HAIS and HAMEL pipelines for PLUTO; in private practice invented hoister for loading tankers at sea; CBE, 1944; president, Institution of Civil Engineers, 1959.

HARTLEY, SIR CHARLES AUGUSTUS (1825-1915), civil engineer; served in Crimean war; chief engineer to European commission of Danube, 1856-1907; earned sobriquet 'father of the Danube' for work in clearing and making navigable Sulina and St. George estuaries; knighted, 1862; left Romania, 1872; KCMG, 1884; member of international technical commission of Suez Canal, 1884-1906; advice sought by Indian, Russian, and American governments.

HARTOG, SIR PHILIP(PE) JOSEPH (1864-1947), educationist; brother of N. E. Hartog [q.v.]; educated at University College School and Owens College, Manchester; B.Sc., 1882; assistant chemistry lecturer, 1891; academic registrar, London University, 1903-20; enthusiastic promoter of School of Oriental Studies; on Calcutta university commission, 1917; vice-chancellor, Dacca, 1920-5; on Indian public service commission, 1926-30; chairman, Indian education committee, 1928-9; published *The Writing of English* (1907), *Words in Action* (1947); and with E. C. Rhodes *An Examination of Examinations* (1935) and *The Marks of Examiners* (1936); knighted 1926; KBE, 1930.

HARTREE, DOUGLAS RAYNER (1897-1958), scientist; educated at Bedales and St. John's College Cambridge; Ph.D., 1926; professor of applied mathematics (1929-37), of theoretical physics (1937-45), Manchester; Plummer professor of mathematical physics, Cambridge, 1946-58; improved calculation of trajectories and became a leader in science of computation; invented 'self-consistent field' method for solution of atomic problems; developed methods of automatic control; published *Numerical Analysis* (1952); FRS, 1932.

HARTSHORN, VERNON (1872-1931), miners' leader and politician; miners' agent for Maesteg, 1905; elected to national executive, Miners' Federation, 1911; labour MP, Ogmore division of Glamorganshire, 1918-31; postmaster-general and PC, 1924; member of Indian statutory commission, 1927-30; lord privy seal, 1930-1.

HARTSHORNE, ALBERT (1839-1910), archaeologist; son of Charles Henry Hartshorne [q.v.]; secretary of Archaeological Institute of Great Britain, 1876-83, 1886-94; edited *Archaeological Journal*, 1878-92; FSA, 1882; published *Old English Glasses* (1897), *Oxford, 1691-1712* (1910); and works on monumental effigies and English churches.

HARTY, SIR (HERBERT) HAMILTON (1879-1941), musician; held organ appointments, Belfast and Dublin; established himself in London as accompanist, 1900; conducted Hallé Orchestra, Manchester, 1920-33; composer in romantic tradition; works include songs, violin Concerto (1909), tone-poem 'With the Wild Geese' (1910), cantata 'The Mystic Trumpeter' (1913), 'Irish' Symphony (1924), arrangements of Handel's 'Water Music' and 'Music for Royal Fireworks', and suite from works of John Field; knighted, 1925; commemorated by chair of music, Queen's University, Belfast.

HARVEY, HILDEBRAND WOLFE (1887-1970), marine biologist; educated at Gresham's School, Holt, and Downing College, Cambridge, 1906-10; served in RNVR, 1914-18; on staff of Marine Biological Association, Plymouth, 1921-58; worked on physical oceanography of western English channel; produced number of papers on biological chemistry and physics of sea water, plankton production, and the rate of diatom growth, 1928-33; FRS, 1945; CBE, 1958; one of the founders of systematic research on the biological productivity of the sea.

HARVEY, SIR JOHN MARTIN MARTIN- (1863-1944), actor-manager. [See MARTIN-HARVEY.]

HARVEY, OLIVER CHARLES, fourth baronet, and first BARON HARVEY OF TASBURGH (1893-1968), diplomatist; educated at Malvern College and Trinity College, Cambridge; first class, part i, historical tripos, 1914; served in France, Egypt, and Palestine, 1914-18; entered diplomatic service, 1919; head of Chancery, Paris, 1931; private secretary to Anthony Eden (later Earl of Avon), 1936; and to Lord Halifax [q.v.], 1938; minister in Paris, 1939; reappointed as Eden's private secretary, 1941-3; assistant under-secretary, Foreign Office, 1943; succeeded Sir Orme Sargent [q.v.] as deputy under-secretary, 1946; worked closely with Ernest Bevin [q.v.] as he had done with Eden; ambassador, Paris, following Duff Cooper (later Viscount Norwich, q.v.), 1948-54; baron, 1954; succeeded as fourth baronet, 1954; trustee, Wallace Collection; chairman, Franco-British Society; CMG, 1937; CB, 1944; KCMG, 1946; GCMG, 1948; GCVO, 1950; published *Diplomatic Diaries, 1937-1940* (ed. John Harvey, 1970) and *War Diaries, 1941-1945* (ed. John Harvey, 1978).

HARWOOD, BASIL (1859-1949), musician and composer; educated at Charterhouse and Trinity College, Oxford; organist, St. Barnabas, Pimlico, 1883; Ely Cathedral, 1887; Christ Church, Oxford, 1892-1909; church and organ compositions include Services in A flat and E minor and Sonata in C sharp minor; edited *Oxford Hymn Book*.

HARWOOD, SIR HENRY HARWOOD (1888-1950), admiral; entered navy, 1903; specialized in torpedo; captain, 1928; commanded South American division, 1936-40; fought battle of River Plate, resulting in scuttling of *Admiral Graf Spee*, Dec. 1939; rear-admiral and KCB, 1939; assistant chief of naval staff, Admiralty, 1940-2; commander-in-chief, Mediterranean (later Levant), 1942-3; engaged in flank support and seaborne supply of Eighth Army; commander-in-chief, Orkneys and Shetlands, 1944-5; admiral and invalided out, 1945.

HASLETT, DAME CAROLINE HARRIET (1895-1957), electrical engineer; educated at Haywards Heath high school; qualified in general and electrical engineering; first secretary, Women's Engineering Society, 1919, and editor of its journal *Woman Engineer*; founded (1924) and directed until 1956 Electrical Association for Women and edited its organ *Electrical Age*; encouraged domestic use of electricity; served on British Electricity Authority and its successors, 1947-57; CBE, 1931; DBE, 1947.

HASSALL, CHRISTOPHER VERNON (1912-1963), poet, biographer, playwright, and librettist; son of John Hassall [q.v.], painter; educated at Brighton College and Wadham College, Oxford, spent some years as actor; won Hawthornden prize with *Penthesperon* (poems),

1938; commissioned, 1941; joined Army Education Corps, 1942; published *The Slow Night* (1949), *The Red Leaf* (1957), and *Bell Harry, and Other Poems* (posthumously, 1963); plays include *Christ's Comet*, *The Player King*, and *Out of the Whirlwind*; wrote libretti for many composers, including Walton's *Troilus and Cressida* (1954); also published biographies, including *Edward Marsh: Patron of the Arts* (1959, awarded the James Tate Black memorial prize), and *Rupert Brooke* (1964); collaborated with Ivor Novello [q.v.] on lyrics for *Glamorous Night* (1935); fellow, Royal Society of Literature; governor, London Academy of Music and Dramatic Art.

HASSALL, JOHN (1868–1948), poster artist; educated at Newton Abbot College and Neuenheim College, Heidelberg; studied farming in Canada and art in Antwerp and Paris; began poster designing, 1895; work humorous, robust, and simple, with direct advertising message and high decorative standard; also illustrated children's books.

HASTIE, WILLIAM (1842–1903), professor of divinity at Glasgow; MA, Edinburgh, 1867; BD, 1869; Croall lecturer, 1892; hon. DD, Edinburgh, 1894; principal of Church of Scotland college, Calcutta, 1878; relieved of post, 1883; professor of divinity at Glasgow, 1895; translated many German theological works; wrote *Theology as Science* (1899), *The Theology of the Reformed Church* (1904), as well as *La Vita Mia*, a sonnet sequence (1896), *The Vision of God* (1898), and other verse.

HASTINGS, JAMES (1852–1922), editor and divine; engaged in pastoral work of Scottish Free Church, 1884–1911; founder and editor of *Expository Times*, 1889–1911; works include *Dictionary of the Bible* (1898–1904), and *Encyclopaedia of Religion and Ethics* (1908–21).

HASTINGS, SIR PATRICK GARDINER (1880–1952), lawyer; educated at Charterhouse; called to bar (Middle Temple), 1904; succeeded to chambers of Sir Horace Avory [q.v.], 1910; KC, 1919; labour MP, Wallsend, 1922–6; attorney-general and knighted, 1924; his withdrawal of prosecution of Campbell of *Workers' Weekly* for sedition precipitated government's downfall; became eminent leader of common law bar until retirement in 1948; appeared in spectacular cases such as the actions between Dr Stopes and Dr Sutherland [qq.v.], for libel by H. J. Laski [q.v.], and the Savidge tribunal; his gift for simplification expedited hearing of commercial cases; brilliant in cross-examination and addressed jury with wit and incisive appeal to intelligence; publications include *Auto-*

biography (1948), and *Famous and Infamous Cases* (1950); also wrote plays including *The Blind Goddess* (1947).

HATRY, CLARENCE CHARLES (1888–1965), company promoter and financier; educated at St. Paul's School, London; took over bankrupt business on death of his father, 1906; established as insurance broker, 1911; purchased and resold Leyland Motors with large profit; registered his Commercial Bank of London, 1920; a shrewd gambler who made large gains and large losses; registered the Austin Friars Trust group, 1927; firms taken over and reorganized as the Drapery Trust and Allied Ironfounders; with another of his companies, Corporation and General Securities, attempted to take over United Steel, 1929; failure of this project led to heavy losses by investors, prosecution and admission of forgery; sentenced by Mr Justice Avory [q.v.] to fourteen years penal servitude, 1930; released by home secretary, 1939; published *Light out of Darkness* (1939); continued his career as financier, with fluctuating fortunes, 1940–65.

HATTON, HAROLD HENEAGE FINCH- (1856–1904), imperialist politician. [See FINCH-HATTON.]

HATTON, JOSEPH (1841–1907), novelist and journalist; editor successively of *Bristol Mirror* (1863–8), *Gentleman's Magazine* (1868), *Sunday Times*, and *People* (1892); published American experiences in *Today in America* (2 vols., 1881); accompanied (Sir) Henry Irving [q.v.] to America, 1883; chronicled Irving's *Impressions of America* (2 vols., 1884), and *Reminiscences* of J. L. Toole [q.v.] (2 vols., 1889); published novels *Clytie* (1872—which he dramatized) and *By Order of the Czar* (1890), *The New Ceylon* (1882), *Journalistic London* (1882), and other works.

HATTON, SIR RONALD GEORGE (1886–1965), horticultural scientist; educated at Brighton College, Exeter School, and Balliol College, Oxford; worked as farm labourer; published *Folk of the Furrow* under pen-name, Christopher Holdenby (1913); studied agriculture at Wye College, Kent, 1912; acting director East Malling Research Station, 1914; director, 1918–49; developed East Malling into leading fruit research institute in the world; classified and standardized rootstocks; collaborated effectively with John Innes Horticultural Institute and Institute of Plant Physiology, Imperial College of Science, London; instrumental in starting the *Journal of Pomology*, 1919, which became the *Journal of Horticultural Science*, 1948; joint editor, 1924–47; first director,

Imperial (later Commonwealth) Bureau of Horticulture and Plantation Crops, 1929; vice-president, Royal Horticultural Society, 1952; CBE, 1934; knighted, 1949; FRS, 1944; fellow, Wye College, 1949.

HAVELOCK, SIR ARTHUR ELIBANK (1844-1908), colonial governor; son of William Havelock [q.v.]; joined army, 1862, retired, 1877; held administrative posts in West Indies, 1874-81; CMG, 1880; governor of West African settlements, 1881; forcibly settled frontier dispute with Liberia, and occupied territories between rivers Sherbro and Mano, 1882-3; KCMG, 1884; governor of Natal (1886-9), of Ceylon (1890-5), where he extended railways and abolished levy on rice cultivation, of Madras (1895-1901), and of Tasmania (1901-4); GCMG, 1895; GCIE, 1896; GCSI, 1901.

HAVELOCK, SIR THOMAS HENRY (1877-1968), naval mathematician; educated at Durham College of Physical Science, Newcastle upon Tyne, and St. John's College, Cambridge; B.Sc., 1895; first class, part ii, mathematics tripos, 1901; suffered serious injury in an accident, 1898; Gregson fellow, St. John's College, 1903; lecturer in applied mathematics, Armstrong College (formerly Durham College of Science), 1906; D.Sc., 1907; FRS, 1914; professor of applied mathematics, 1915; his injury precluded active service, 1914-18; carried out research on optics and naval hydrodynamics; professor of pure and applied mathematics, 1928-45; hon. member, Royal Institution of Naval Architects, 1943; hon. DCL, Durham, 1958; hon. D.Sc., Hamburg, 1960; hon. fellow, St. John's College, Cambridge, 1945; knighted, 1951; succeeded Sir Westcott Abell [q.v.] as honorary acting head of Department of Naval Architecture, 1941-4; vice-principal, Armstrong College, 1933-7; sub-rector, King's College, Newcastle upon Tyne, 1937-42.

HAVERFIELD, FRANCIS JOHN (1860-1919), Roman historian and archaeologist; scholar of Winchester and New College, Oxford; a senior student of Christ Church, 1892-1907; Camden professor of ancient history and fellow of Brasenose, 1907-19; created scientific study of Roman Britain; works include *Romanization of Roman Britain* (1905) and *Roman Occupation of Britain* (Ford's lectures, published 1924).

HAVILLAND, SIR GEOFFREY DE (1882-1965), aircraft designer and manufacturer. [See DE HAVILLAND.]

HAWEIS, HUGH REGINALD (1838-1901), author and preacher; grandson of Thomas Haweis [q.v.]; showed musical ability, especially as violinist, from boyhood; lame through hip disease; BA, Trinity College, Cambridge, 1859; started university magazine *Lion*; travelled for health in Italy, 1859-60; incumbent of St. James's, Westmoreland Street, Marylebone, from 1866 till death; by means of somewhat sensational methods filled the church; organized 'Sunday evenings for the people'; pioneer of Sunday opening of museums; published *Music and Morals* (1871), *My Musical Life* (1884), *Old Violins* (1898); theological writings include *Thoughts for the Times* (1872), *Winged Words* (1885), and *Christ and Christianity* (5 vols., 1886-7); successful lecturer on music; Lowell lecturer at Boston, 1885; toured the colonies, 1893; described experiences in *Travel and Talk* (2 vols., 1896); visited Rome, 1897; married in 1867 MARY (d. 1898), daughter of Thomas Musgrave Joy [q.v.]; she published *Chaucer for Children* (1877), *Chaucer for Schools* (1880), and other works; a capable artist, she exhibited at Royal Academy.

HAWEIS, MARY (d. 1898). [See under HAWEIS, HUGH REGINALD.]

HAWKE, SIR (EDWARD) ANTHONY (1895-1964), judge; son of (Sir) (John) Anthony Hawke [q.v.], judge; educated at Charterhouse and Magdalen College, Oxford; called to bar (Middle Temple) 1920; joined Western circuit and Devon sessions; junior prosecuting counsel, Central Criminal Court, 1932; third senior prosecuting counsel, 1937; second senior prosecuting counsel, 1942; senior prosecuting counsel, 1945-50; recorder, Bath, 1939-50; deputy chairman, Hertfordshire quarter-sessions, 1940-50; bencher, 1942; chairman, county of London quarter-sessions, 1950; knighted; common sergeant, City of London, 1954; recorder of London, 1959; treasurer, Middle Temple, 1962; edited fifteenth edition of Roscoe's *Criminal Evidence*.

HAWKE, SIR (JOHN) ANTHONY (1869-1941), judge; first class, jurisprudence, St. John's College, Oxford, 1891; called to bar (Inner Temple), 1892; KC, 1913; recorder of Plymouth, 1912-28; attorney-general to Prince of Wales, 1923-8; judge, King's Bench division, 1928-41; conservative MP, St. Ives, 1922-3, 1924-8; knighted, 1928.

HAWKE, MARTIN BLADEN, seventh BARON HAWKE OF TOWTON (1860-1938), cricketer; played in Eton (1878, 1879) and Cambridge (1882, 1883, 1885) elevens; captained Yorkshire, 1883-1910; took teams to India, West Indies, South Africa, etc.; president, MCC, 1914-18; succeeded father, 1887.

HAWKER, MARY ELIZABETH (1848–1908), novelist writing under the pseudonym of LANOE FALCONER; granddaughter of Peter Hawker [q.v.]; gained success as novelist by *Mademoiselle Ixe* (1890), translated into many foreign languages, *Cecilia de Noel* and *The Hôtel d'Angleterre*, both 1891.

HAWKINS, SIR ANTHONY HOPE (1863–1933), novelist under pseudonym of ANTHONY HOPE; educated at Marlborough and Balliol College, Oxford; first class, *lit. hum.*, 1885; president of Union, 1886; abandoned practice at bar after success (1894) of *The Prisoner of Zenda*, a modern romance of adventure, and the delicately witty *Dolly Dialogues*; later works included Ruritanian stories, analytical novels of character, and plays; knighted, 1918.

HAWKINS, HENRY, BARON BRAMPTON (1817–1907), judge; son of solicitor; called to bar, 1843; QC, 1858; obtained foremost place among leaders of bar; defended Simon Bernard (1852), Sir John Dean Paul (q.v., 1855), and Miss Sugden in Lord St. Leonard's will case (1875–6); appeared for defence against Arthur Orton [q.v.], claimant in Tichborne ejectment case; led for crown in criminal action against claimant for perjury, 1873–4; a master in cross-examination; largely employed in compensation and election petition cases; appointed judge of Queen's Bench division and knighted, 1876; transferred to Exchequer division; tried the Stauntons for murder in the 'Penge case', 1877; tried other murder cases, and unjustly obtained the nickname of 'Hanging Hawkins'; an admirable criminal judge, patient and thorough; favoured leniency for first offences; as civil judge less successful, being verbose, tautological, and contradictory in judgments; resigned judgeship and sworn PC, 1898; baron, 1899; fond of horse-racing; joined Roman Catholic communion after retirement from bench; *Reminiscences* (2 vols., published, 1904).

HAWKINS, HERBERT LEADER (1887–1968), geologist and palaeontologist; educated at Reading School, Kendal grammar school, and Manchester University; first class, geology; M.Sc., 1910; lecturer, University College of Reading, 1909; professor of geology, 1920–52; D.Sc., Manchester; FRS, 1937; specialist on fossil Echinoides; published 102 scientific papers between 1909 and 1965; also *Humanity in Geological Perspective* (1939); president, geological section, British Association for the Advancement of Science, 1936; president, Geological Society, 1941–2; president, Palaeontographical Society, 1943–66; consulting geologist, Thames Valley Water Board, 1961.

HAWORTH, SIR (WALTER) NORMAN (1883–1950), chemist; first class, chemistry, Manchester, 1906; researched on chemistry of terpenes; studied at Göttingen; lecturer, St. Andrews, 1912–20; worked on carbohydrates; professor of chemistry, Newcastle, 1920–5; Birmingham, 1925–48; built up important school of carbohydrate chemistry; produced first chemical synthesis of a vitamin (vitamin C), 1932; FRS, 1928; shared Nobel prize, 1937; knighted, 1947.

HAWTHORN, JOHN MICHAEL (1929–1959), racing motorist; only son of motor engineer and racing motor-cyclist; educated at Ardingly College; raced at Goodwood, 1952; won French grand prix for Ferrari, 1953 and 1958; Spanish grand prix, 1954; Le Mans for Jaguar, 1955; world champion and British Automobile Racing Club gold medal, 1958; died in road accident as had his father in 1954.

HAWTREY, SIR CHARLES HENRY (1858–1923), actor; joined the stage, 1881; rewrote a German farce which he entitled *The Private Secretary* and which ran 1884–6; altogether managed eighteen London theatres, including the Globe and the Comedy, and produced about one hundred plays; knighted, 1922; excelled in staging farce and light comedy; made capital out of his facial immobility.

HAY, SIR HARLEY HUGH DALRYMPLE- (1861–1940), civil engineer; joined London and South Western Railway; resident engineer, Waterloo and City Railway, 1894; his system of constructing tube railways became standard practice; consulting engineer, London Underground, 1902–40; constructed Bakerloo, Northern, and Piccadilly lines; other works include widening of Richmond bridge, tunnel under river Hugli, Calcutta (1931), London Post Office railway (1928), and secret intercommunication tunnels between government offices; knighted, 1933.

HAY, IAN (pseudonym) (1876–1952), writer. [See BEITH, JOHN HAY.]

HAYES, EDWIN (1819–1904), marine painter; studied art in Dublin; exhibited at Royal Academy, 1845–1904; member of Royal Hibernian Academy, 1870; chief works were 'Off Dover' (1891), 'Crossing the Bar' (1895); represented in Tate and other public galleries.

HAYMAN, HENRY (1823–1904), honorary canon of Carlisle and headmaster of Rugby; educated at Merchant Taylors' School and St. John's College, Oxford; BA, 1845; MA, 1849;

DD, 1870; fellow, 1844-55; treasurer of the Union; headmaster of St. Olave's, Southwark, 1855-9, of Cheltenham, 1859-68, and Bradfield, 1868-9; elected headmaster of Rugby, under protest from masters, 1869; instituted unsuccessful proceedings against governors for dismissal, 1874; nominated to crown living of Aldingham, Lancashire, 1874; honorary canon of Carlisle, 1884; published classical translations, an edition of Homer's *Odyssey* (3 vols., 1881-6), sermons, and essays.

HAYNE, CHARLES HAYNE SEALE- (1833-1903), liberal politician. [See SEALE-HAYNE.]

HAYWARD, JOHN DAVY (1905-1965), anthologist and bibliophile; educated at Gresham's School, Holt, and King's College, Cambridge, 1923-7; afflicted with muscular dystrophy; confined to invalid chair; formed salon where T. S. Eliot, (Sir) Geoffrey Faber [qq.v.], and F. V. Morley met together; publications include *Collected Works* of Rochester (1926), *Complete Poetry and Selected Prose* of Donne (1929), Swift's *Gulliver's Travels* (1934), and Swift's *Selected Prose Works* (1949); compiled anthologies, including *Nineteenth Century Poetry* (1932), *T. S. Eliot: Points of View* (1941), *Seventeenth Century Poetry* (1948), *Dr. Johnson* (1948), *The Penguin Book of English Verse* (1956), and *The Oxford Book of Nineteenth Century English Verse* (1964); CBE, 1953; editorial adviser to Cresset Press; editorial director, the *Book Collector*; vice-president, Bibliographical Society; he and T. S. Eliot lived together, 1946-57; left books and letters from authors, including T. S. Eliot, to the library of King's College, Cambridge, where there is now a Hayward Room.

HAYWARD, ROBERT BALDWIN (1829-1903), mathematician; BA, St. John's College, Cambridge (fourth wrangler), 1850; fellow, 1852-60; mathematical master at Harrow, 1859-93; published original researches; FRS, 1876; original member of Alpine Club, 1858.

HAZLITT, WILLIAM CAREW (1834-1913), bibliographer and man of letters; grandson of William Hazlitt [q.v.], the essayist; works include *History of . . . Republic of Venice* (1858, etc.) and *Handbook to Popular, Poetical, and Dramatic Literature of Great Britain* (1867).

HEAD, BARCLAY VINCENT (1844-1914), Greek numismatist; entered coins department, British Museum, 1864; keeper, 1893-1906; produced important series of Greek coin catalogues; chief work, *Historia Numorum* (1887).

HEAD, SIR HENRY (1861-1940), neurologist; educated at Charterhouse, and Halle, Cambridge, and Prague universities, and University College Hospital; MB (Camb.), 1890; successively registrar, assistant physician (1896), full physician, consulting physician, London Hospital; established 'Head's Areas' by his thesis 'On disturbances of sensation, with special reference to the pain of visceral disease' (*Brain*, 1893-6); FRS, 1899; FRCP, 1900; editor of *Brain*, 1905-21; underwent experimental operation in nerve division performed by James Sherren (described in *Brain*, 1908); publications include *Aphasia and Kindred Disorders of Speech* (2 vols., 1926); knighted, 1927.

HEADLAM, ARTHUR CAYLEY (1862-1947), bishop of Gloucester; brother of Sir J. W. Headlam-Morley [q.v.]; scholar of Winchester and New College, Oxford; first class, *lit. hum.*, and fellow of All Souls, 1885; deacon, 1888; priest, 1889; rector of Welwyn, 1896-1903; principal, King's College, London, 1903-12; divided theological department from secular faculties; regius professor of divinity, Oxford, 1918-23; bishop of Gloucester, 1923-45; closely interested in Christian reunion; chairman, Church of England council on foreign relations, 1933-45; editor, *Church Quarterly Review*, 1901-21; prolific, vigorous theological writer of enlightened conservatism; CH, 1921.

HEADLAM, WALTER GEORGE (1866-1908), scholar and poet; educated at Harrow and King's College, Cambridge; gained many university classical prizes, 1885-7; BA, 1887; MA, 1891; Litt.D., 1903; fellow, 1890; rarely surpassed in Greek versions of English poetry; translated poems by Meleager, 1890, and Aeschylus, 1900-8; his *Letters and Poems* (1900) edited with memoir by his brother.

HEADLAM-MORLEY, SIR JAMES WYCLIFFE (1863-1929), political historian; BA, King's College, Cambridge, 1887; fellow, 1890-6; professor of Greek and ancient history, Queen's College, London, 1894-1900; staff inspector of secondary schools, Board of Education, 1902-14; historical adviser to propaganda department, Wellington House, 1914-17; to Foreign Office, 1920-8; knighted, 1929; works include *Bismarck and the German Empire* (1899) and *The History of Twelve Days* (1915), material for third edition of which formed vol. xi of G. P. Gooch and H. W. V. Temperley's *British Documents on the Origins of the War* (1926).

HEAL, SIR AMBROSE (1872-1959), furniture designer and dealer; educated at Marlborough; joined family firm, 1893; chairman, 1913-53; introduced furniture combining functional utility with simplicity of line; extended scope of business to include antiques, carpets, curtains, etc.; co-founder, Design and Industries Association, 1915; knighted, 1933; RDI, 1939; Albert medallist of Royal Society of Arts for services to industrial design, 1954; wrote number of books including *The English Writing-Masters and their Copy-Books, 1570-1800* (1931).

HEALY, JOHN EDWARD (1872-1934), journalist; educated at Trinity College, Dublin; joined Dublin *Daily Express*; became editor; Dublin correspondent of *The Times*, 1899-1907; editor, *Irish Times*, 1907-34; actively defended the union during home rule controversy; opposed partition and any broadening of gap between Dublin and London.

HEALY, TIMOTHY MICHAEL (1855-1931), Irish political leader and first governor-general of Irish Free State; educated by the Christian Brothers; clerk in Newcastle (1871) and London (1878); contributed weekly parliamentary letter to the *Nation*; organized Canadian tour for C. S. Parnell [q.v.], 1880; arrested but acquitted during agrarian agitation, 1880; MP, Wexford, 1880-3; expert at obstruction; showed constructive ability in debate on Land Act, 1881; imprisoned for six months (1883) during further agrarian agitation; called to Irish bar, 1884; QC, 1899; Parnellite MP, county Monaghan (1883-5), South Londonderry (1885-6), North Longford (1887-92); became increasingly distrustful of Parnell; recommended his temporary retirement from Irish leadership when Gladstone refused to co-operate further with him after the O'Shea divorce case, 1890; anti-Parnellite MP, North Louth, 1892-1910; accused John Dillon and T. P. O'Connor [qq.v.] of subservience to English liberalism; expelled from National League, 1895; supported choice of John Redmond [q.v.] as chairman of nationalists (1900) but expelled from party (1902); friendly with chief secretary, George Wyndham [q.v.]; founded 'All for Ireland' League (1910) to promote Irish cause by conciliation; MP, North-East Cork, 1910-18; in general sympathy with Sinn Fein; fulfilled duties as governor-general (1922-8) with social tact.

HEARN, MARY ANNE, 'MARIANNE FARNINGHAM' (1834-1909), hymn writer and author; edited *Sunday School Times* from 1885; published forty volumes of essays and hymns, which included 'Watching and Waiting for me'; autobiography published, 1907.

HEATH, CHRISTOPHER (1835-1905), surgeon; son of Christopher Heath [q.v.];

studied medicine at King's College, London; MRCS, 1856; FRCS, 1860; became surgeon (1866) and Holme professor of clinical surgery (1875) at University College Hospital; at Royal College of Surgeons, Hunterian professor of surgery (1886-7), Bradshaw lecturer (1892), Hunterian orator (1897), and president (1895); visited America, 1897; great teacher of anatomy and surgery, but backward in new bacteriological science; published many works on anatomy and surgery.

HEATH, SIR (HENRY) FRANK (1863-1946), academic and scientific administrator; educated at Westminster, University College, London, and Strasbourg; professor of English, Bedford College, 1890-6; assistant registrar and librarian, London University, 1895; academic registrar, 1901-3; director of special inquiries and reports, Board of Education, 1903-16; in charge of universities branch, 1910-16; joint-secretary, royal commission on university education in London, 1909-13; member, Treasury advisory committee on university college grants, 1909-11; proposed formation and first secretary of Advisory Council (1915-27) and executive Department of Scientific and Industrial Research (1916-27); KCB, 1917.

HEATH, SIR LEOPOLD GEORGE (1817-1907), admiral; entered Royal Naval College, Portsmouth, 1830; served in Mediterranean and East Indies; commander, 1847; employed at destruction of Lagos, 1850, and in Crimean war, 1853-4; principal agent of transports and CB, 1855; commodore in command in East Indies, 1867-70; KCB, 1870; vice-president of ordnance select committee, 1870-1; rear-admiral, 1871; admiral, 1884; published *Letters from the Black Sea* (1897).

HEATH, SIR THOMAS LITTLE (1861-1940), civil servant and authority on ancient mathematics; educated at Clifton and Trinity College, Cambridge; first class, parts i and ii, classical tripos (1881-3); twelfth wrangler, 1882; passed first into civil service, 1884; entered Treasury; assistant secretary, 1907; joint permanent secretary, controlling administrative side, 1913; comptroller-general, National Debt office, 1919-26; KCB, 1909; KCVO, 1916; a leading authority on Greek mathematics; made accessible in modern notation works of Diophantus (1885), Apollonius of Perga (1896), Archimedes (1897), and Euclid (1908); other publications include *A History of Greek Mathematics* (2 vols., 1921) and *Greek Astronomy* (1932); FRS, 1912; FBA, 1932.

HEATH ROBINSON, WILLIAM (1872-1944), cartoonist and book-illustrator. [See ROBINSON.]

HEATHCOTE, JOHN MOYER (1834-1912), tennis player; BA, Trinity College, Cambridge; pupil of Edmund Tompkins, professional tennis champion; amateur tennis champion, 1859-81, 1883, 1886.

HEATON, Sir JOHN HENNIKER, first baronet (1848-1914), postal reformer; went to Australia, 1864; journalist in Sydney; settled in London, 1884; conservative MP, Canterbury, 1885-1910; by his postal reform campaign (opened 1886) won imperial penny postage (1898), Anglo-American (1908), Anglo-Australian (1905-11); baronet, 1911.

HEAVISIDE, OLIVER (1850-1925), mathematical physicist and electrician; by self-training acquired remarkable skill in application of mathematics to electrodynamics; pursued his studies in Camden Town, 1876-89; subsequently worked at Paignton, Newton Abbot, and Torquay, in his latter days living in great seclusion; propounded theory of 'surface conduction'; enriched and clarified language of electrodynamics; introduced 'Expansion Theorem', or operational calculus; his 'distortionless' case of wide application in general dynamics; FRS, 1891.

HECTOR, ANNIE FRENCH (1825-1902), novelist writing as Mrs Alexander; daughter of Robert French, Dublin solicitor; settled with parents in London; friend of W. H. Wills [q.v.]; married (1858) Alexander Hector (d. 1875), explorer, merchant, and archaeologist; lived in Germany and France, 1876-82, and St. Andrews, 1882-5; published her best-known novel, *The Wooing o't* (1873), and over forty others.

HECTOR, Sir JAMES (1834-1907), Canadian geologist; MD, Edinburgh, 1856; surgeon and geologist on exploring expedition of John Palliser [q.v.] to western North America, 1857-60; discovered Hector's Pass; director of geological survey of New Zealand, 1865; reported on fossiliferous formations of New Zealand; observed volcanic and glacial phenomena; CMG, 1875; KCMG, 1887; FRS, 1866; Lyell medallist of Geological Society, 1876; died at Wellington, NZ; published *Outlines of New Zealand Geology* (1886).

HEILBRON, Sir IAN MORRIS (1886-1959), chemist; educated at Glasgow high school and Royal Technical College, and Leipzig; lecturer; Royal Technical College, 1909-14; served in army, 1914-18; DSO; professor of organic chemistry, Royal Technical College, 1919-20; Liverpool, 1920-33; Manchester, 1933-5, Sir Samuel Hall professor of chemistry,

1935-8; Imperial College of Science, 1938-49; director, Brewing Industry Research Foundation, 1949-58; pioneer of organic chemical research developed for therapeutic and industrial use (i.e. vitamins, steroids, acetylenic derivatives, penicillin, DDT); FRS, 1931; knighted, 1946.

HEINEMANN, WILLIAM (1863-1920), publisher; set up business in London, 1890; published novels by R. L. Stevenson, Kipling, Sarah Grand, Flora Annie Steel, John Galsworthy, Joseph Conrad, H. G. Wells, etc.; plays by Sir Arthur Pinero, W. Somerset Maugham, Israel Zangwill, etc.; produced International Library of translations of fiction; commissioned translations of Dostoevsky, Turgenev, Tolstoy, Ibsen, Björnson, Rolland; chief literary enterprise, Loeb Classical Library of translations.

HELE-SHAW, HENRY SELBY (1854-1941), engineer; apprenticed in Bristol; graduated from University College, Bristol, 1880; first professor of engineering, Bristol, 1881-5; Liverpool, 1885-1904; Transvaal Technical Institute, 1904-5; principal, 1905-6; returned to England as consulting engineer; many inventions included instruments for recording wind velocities; integrating machines; friction clutch widely used on motor vehicles; hydraulic transmission gear and steering gear for ships; automatic variable-pitch airscrew; demonstrated nature of streamline flow; organized Liverpool trials on commercial motor vehicles, 1897; influential in introduction (1920) of national certificate scheme of training engineers; FRS, 1899.

HELENA VICTORIA, Princess; Victoria Louise Sophia Augusta Amelia Helena (1870-1948); elder daughter of Prince and Princess Christian; her interests philanthropic, social, and musical.

HELLMUTH, ISAAC (1817-1901), bishop of Huron; born near Warsaw of Hebrew parents; joined Church of England in Liverpool, 1841; emigrated to Canada, 1844; ordained and made professor of Hebrew at Bishop's College, Lennoxville, 1846; DD, Lambeth, 1853; DCL, Trinity College, Toronto, 1854; visited England (1861) and collected funds for new Huron evangelical theological college at London, Ontario; first principal, 1863; bishop of Huron, 1871-3; founded Western University, 1878; coadjutor to Robert Bickersteth, bishop of Ripon [q.v.], 1883-4.

HELY-HUTCHINSON, RICHARD WALTER JOHN, sixth Earl of Donoughmore (1875-1948), chairman of committees of

House of Lords; educated at Eton and New College, Oxford; succeeded father, 1900; under-secretary for war, 1903-5; chairman of committees, 1911-31; of Ceylon constitutional commission, 1927-8; of committee on ministers' powers, 1929-31; PC, 1918.

HEMMING, GEORGE WIRGMAN (1821-1905), mathematician and law reporter; BA, St. John's College, Cambridge (senior wrangler and first Smith's prizeman), and fellow, 1844; barrister of Lincoln's Inn, 1850; treasurer, 1897; QC, 1875; junior counsel to Treasury, 1871-5; published works on calculus, 1848, and plane trigonometry, 1851; edited *Equity Cases* and *Chancery Appeals*, subsequently merged in *Law Reports* of Chancery division.

HEMPHILL, CHARLES HARE, first BARON HEMPHILL (1822-1908), lawyer and politician; son of Mrs Barbara Hemphill [q.v.]; BA, Trinity College, Dublin, 1843; called to Irish bar, 1845; acquired large practice in Leinster circuit; QC, 1860; Irish county court judge from 1863 till passing of County Courts (Ireland) Act, 1877; serjeant-at-law, 1882; supported Gladstone's home rule policy (1886); Irish solicitor-general under Gladstone, 1892-5; Irish PC, 1895; liberal MP for North Tyrone, 1895-1906; baron, 1906.

HENDERSON, ARTHUR (1863-1935), labour leader and statesman; apprenticed as moulder; district delegate, Ironfounders Union, and member of Newcastle city council, 1892; labour MP, Barnard Castle (1903-18), Widnes (1919-22), East Newcastle (1923), Burnley (1924-31), and Clay Cross division (1933-5); secretary, labour party, 1911-34; PC, 1915; adviser on labour in cabinet, 1915-17 (president, Board of Education, 1915, paymaster-general, 1916, minister without portfolio, 1916); visited Russia, 1917; resigned (1917) after recommending sending delegates to Stockholm international socialist conference; strengthened and broadened labour party organization; chief whip, 1914, 1921-3, 1925-7; responsible for statement of aims 'Labour and the Nation', 1928; home secretary, 1924; foreign secretary, 1929-31; forced resignation of Lord Lloyd [q.v.] in effort to reach Egyptian agreement; sent ambassador to Russia; established British leadership in seeking secure foundations for international peace; led labour party opposition to MacDonald's 'national' government and lost his seat, 1931; presided over world disarmament conference, 1932-4; Nobel peace prize, 1934.

HENDERSON, SIR DAVID (1862-1921), lieutenant-general; entered army, 1883; served in South African war, 1899-1902; learned to fly,

1911; director of military training, War Office, 1912-13; director-general of military aeronautics, 1913; general officer commanding Royal Flying Corps, 1914-17; KCB, 1914; lieutenant-general, 1917; KCVO, 1919.

HENDERSON, DAVID WILLIS WILSON (1903-1968), microbiologist; educated at Hamilton Academy, West of Scotland Agricultural College and Glasgow University; B.Sc. (Agriculture), 1926; adviser in dairy bacteriology, Ministry of Agriculture and Fisheries; on staff of Armstrong College (then part of Durham University), 1927; M.Sc., 1930; Carnegie research fellow, Lister Institute of Preventive Medicine (London), 1931; Beit memorial research fellow, 1932-5; on bacteriological staff, Lister Institute, 1935-46; Ph.D. (London), 1934; D.Sc., 1941; dealt mainly with immunology of salmonella and clostridium species; member of biology group, Chemical Defence Experimental Establishment, Porton, 1940; played key role in Anglo-American liaison on microbiological defence; US medal of freedom, bronze palm, 1946; chief superintendent (later director), Microbiological Research Establishment, 1946-64; continued interest in viral aerosols up to 1967; president, Society for General Microbiology, 1963-5; CB, 1957; FRS, 1959.

HENDERSON, GEORGE FRANCIS ROBERT (1854-1903), colonel and military writer; son of William George Henderson [q.v.]; scholar of St. John's College, Oxford; entered Sandhurst, 1876; commanded a company at Tel-el-Kebir, 1882; served in ordnance store department, 1885-90; published *The Campaign of Fredericksburg* (1886); instructor at Sandhurst in military topography and tactics, 1890; exercised great influence as professor of military art and history at Staff College, 1892-9; published *Stonewall Jackson and the American Civil War* (2 vols., 1898); accompanied Lord Roberts [q.v.] to the Cape, 1899; director of military intelligence and CB, 1900; died at Aswan.

HENDERSON, GEORGE GERALD (1862-1942), chemist; graduated in chemistry and arts, Glasgow; head of chemistry department, Queen Margaret College, 1889-92; Freeland professor, Royal Technical College, 1892-1919; built up close relations with manufacturing interests; obtained new laboratories; regius professor, Glasgow University, 1919-37; created new chemistry institute; FRS, 1916.

HENDERSON, SIR HUBERT DOUGLAS (1890-1952), economist; educated at Aberdeen grammar school, Rugby, and Emmanuel College, Cambridge; first class, part ii, economics tripos, and president of Union, 1912; acquired liberal and reforming views; secretary, Cotton Control Board, 1917-19; lecturer in economics, Cambridge, 1919-23; editor, *Nation and Athenaeum*, 1923-30; assistant, later joint, secretary, Economic Advisory Council, 1930-4; fellow, All Souls, 1934-52; active in establishment of Institute of Statistics; member, West Indies royal commission, 1938-9; adviser at Treasury, 1940-4; Drummond professor of political economy, Oxford, 1945-51; member, royal commission on population, 1944, chairman, 1946; chairman, statutory committee on unemployment insurance, 1945-8; warden, All Souls, 1951-2; his acute critical powers sought practical solutions to economic problems within political possibilities; FBA, 1948; knighted, 1942; president, Royal Economic Society, 1950.

HENDERSON, JOSEPH (1832-1908), portrait and marine painter; studied art at Trustees' Academy, Edinburgh; treated genre subjects till 1871, when he devoted himself to the sea; painted in oil and water-colour; exhibited at Royal Academy, 1871-86; best pictures shown at the Royal Glasgow Institute; 'The Flowing Tide' is in the Glasgow Gallery.

HENDERSON, SIR NEVILE MEYRICK (1882-1942), diplomatist; educated at Eton and abroad; entered diplomatic service, 1905; minister plenipotentiary, 1924; served at Cairo, 1924-8; Paris, 1928-9; minister plenipotentiary at Belgrade, 1929-35; formed close friendship with King Alexander; ambassador at Buenos Aires, 1935-7; Berlin, 1937-9; had mystical conception of role as mediator and no preconceived dislike of authoritarian government; sincerely believed in and encouraged British government's policy of appeasement until after occupation of Czechoslovakia, Mar. 1939; published *Failure of a Mission* (1940); KCMG, 1932; PC, 1937; GCMG, 1939.

HENDERSON, SIR REGINALD GUY HANNAM (1881-1939), admiral; entered navy through *Britannia*; served in *Erin* at Jutland, 1916; on Admiralty war staff 1916-19; his trial convoy led to adoption of system, 1917; captain, 1917; rear-admiral, 1929; rear-admiral, aircraft-carriers, 1931-3; vice-admiral, 1933; third sea lord and controller, 1934-9; responsible for great expansion of navy; KCB, 1936; GCB and admiral, 1939.

HENDERSON, WILLIAM GEORGE (1819-1905), dean of Carlisle; BA, Magdalen College, Oxford (first class classic), 1840; MA, 1843; DCL, 1853; DD, 1882; fellow, 1847-53; headmaster of Victoria College, Jersey, 1852-62, and of Leeds grammar school, 1862-84;

dean of Carlisle, 1884; edited for Surtees Society the Latin missals of York and Hereford.

HENLEY, WILLIAM ERNEST (1849-1903), poet, critic, and dramatist; born at Gloucester; pupil there of T. E. Brown [q.v.]; a cripple from boyhood; in Edinburgh infirmary, 1873-5; his *Hospital Verses*, some of which were published in *Cornhill Magazine* (July 1875), led the editor, (Sir) Leslie Stephen [q.v.], to visit him and introduce him to R. L. Stevenson [q.v.]; worked in Edinburgh on staff of *Encyclopaedia Britannica*, 1875; settled in London (1877-8) as editor of weekly journal, *London*; while editor of *Magazine of Art* (1882-6) championed Rodin and J. A. M. Whistler [q.v.]; became (1889) editor of *Scots Observer*, renamed in 1891 the *National Observer*, an imperialist weekly paper; encouraged young authors; edited *New Review*, 1894-8; obtained poetic fame by *Book of Verses*, 1888; wrote *London Voluntaries* (1893), *For England's Sake* (patriotic songs, 1900), and *Hawthorn and Lavender* (lyrics, 1901); published literary *Views and Reviews* (1890) and a companion volume on art (1902); collaborated with R. L. Stevenson in the plays *Deacon Brodie, Beau Austin, Admiral Guinea*, and *Macaire*; initiated series of reprints of 'Tudor Translations', 1892; joint-compiler of *Slang Dictionary*, 1894-1904; contributed essay on Burns's life to, and helped in, centenary edition of poetry of Robert Burns, 4 vols., 1896-7; compiled *Lyra Heroica*, 1891; hon. LLD, St. Andrews, 1893; granted civil list pension, 1898; portrayed as 'Burly' in R. L. Stevenson's essay, 'Talk and Talkers'; works collected in 6 vols., 1908.

HENNELL, SARA (1814-1905). [See under BRAY, CAROLINE.]

HENNESSEY, JOHN BOBANAU NICKERLIEU (1829-1910), deputy surveyor-general of India, 1883-4; born at Fatehpur; worked on trigonometrical survey in jungle tracts of Bengal and in Punjab from 1844; studied mathematics at Jesus College, Cambridge, 1863-5; organized the reproduction of survey sheets in India by means of photo-zincographic process, 1865; superintendent of survey, 1874; FRS, 1875; hon. MA, Cambridge, 1876; CIE, 1885.

HENNESSY, HENRY (1826-1901), physicist; brother of Sir John Pope-Hennessy [q.v.]; professor of physics at Roman Catholic University, Dublin, 1855-74; at Royal College of Science, Dublin, 1874-90; vice-president of Royal Irish Academy, 1870-3; FRS, 1858; made valuable researches in meteorology and climatology.

HENRIQUES, SIR BASIL LUCAS QUIXANO (1890-1961), club leader and magistrate; born into old-established Jewish family; educated at Harrow and University College, Oxford, 1909-13; under influence of Claude J. G. Montefiore and (Sir) Alexander H. Paterson [qq.v.], became active in liberal Jewish movement and in boys' clubs; opened the Oxford and St. George's in the East Jewish Boys Club, 1914; served in the army and commanded first tank to be used in war, 1914-18; returned to Oxford and St. George's Club (later, Bernhard Baron Settlement), 1918-47; magistrate, 1924-55; chairman, East London Juvenile Court, 1936-55; CBE, 1948; knighted, 1955; publications include *Club Leadership* (1933), *The Indiscretions of a Magistrate* (1950), and *Club Leadership Today* (1951).

HENRY, SIR EDWARD RICHARD, baronet, of Campden House Court (1850-1931), commissioner of Metropolitan Police; entered Indian civil service; inspector-general of police, Bengal, 1891; evolved system of classifying fingerprints which has been generally adopted; assistant commissioner, Metropolitan Police, in charge of CID, 1901-3; commissioner (1903) at time of increasing public disorder due to unemployment, suffragettes, etc.; his determined efforts to improve police conditions could not prevent a strike, 1918; resigned and created baronet; KCVO, 1906; GCVO, 1911.

HENRY, MITCHELL (1826-1910), Irish politician; studied medicine at Manchester; FRCS, 1854; MP for county Galway, 1871-85; supported Isaac Butt [q.v.]; opposed Gladstone's Irish university bill on question of sectarian education; persistently denounced over-taxation of Ireland, 1874-7; bought large estate of Kylemore, county Galway, and reclaimed bog land; disapproved of Land League operations; elected MP for Blackfriars division of Glasgow, 1885; voted against home rule, 1886; retired on defeat, 1886; chairman of firm of A. & S. Henry, merchants, of Manchester, 1889-93.

HENSCHEL, SIR GEORGE (1850-1934), musician; born at Breslau of Polish-Jewish parentage; studied in Leipzig and Berlin; a baritone singer of singular vitality and authority; his own accompanist; first conductor, Boston Symphony Orchestra (1881-4), Scottish Orchestra, Glasgow (1893-5); founder, organizer, and first conductor, London symphony concerts, 1886-97; gave remarkable duet recitals with first wife, 1881-1901; continued as conductor, concert singer, and composer, 1909-14; last broadcast, 1934; naturalized, 1890; knighted, 1914.

HENSON, HERBERT HENSLEY (1863-1947), bishop successively of Hereford and Durham; educated at private school, Broadstairs; non-collegiate student, Oxford; first class, modern history, and fellow of All Souls, 1884; deacon, 1887; priest, 1888; vicar of Barking, 1888-95; chaplain, Ilford Hospital, 1895-1900; canon of Westminster and rector of St. Margaret's, 1900-12; proctor in convocation, 1903-12; dean of Durham, 1912-18; his orthodoxy suspect and consecration as bishop of Hereford (1918) strongly opposed by Anglo-Catholics; bishop of Durham, 1920-39; of increasingly liberal churchmanship, great eloquence, and independence of mind; took strenuous part in ecclesiastical conflicts; published sermons and addresses, *Ad Clerum* (1937), *Bishoprick Papers* (1946), and *Retrospect of an Unimportant Life* (3 vols., 1942-50).

HENSON, LESLIE LINCOLN (1891-1957), actor-manager; educated at Cliftonville College and Emanuel School, Wandsworth; joined concert party, 1910; gained immediate success as Henry in *To-Night's the Night* (Gaiety, 1915); organized entertainment for troops, 1916-18; appeared in series of musical comedies at Winter Garden and from 1935 at Gaiety; in management with Tom Walls [q.v.] produced *Tons of Money* (1922) and series of Aldwych farces; worked for ENSA 1939-45; thereafter his brand of humour began to stale.

HENTY, GEORGE ALFRED (1832-1902), writer for boys; educated at Westminster and Caius College, Cambridge; served in Crimea with hospital commissariat department; adopted journalism, 1865; correspondent to *Standard* during Austro-Italian war (1866), in Abyssinia (1867-8), Franco-German war (1870-1), in Ashanti (1873-4), and on Prince of Wales's Indian tour (1875), and elsewhere; published his first boys' book, *Out in the Pampas* (1868), and from 1876 brought out some three volumes a year, dealing mainly with military history; edited *Union Jack*, 1880-3; keen yachtsman; also published some twelve orthodox novels.

HERBERT, AUBERON EDWARD WILLIAM MOLYNEUX (1838-1906), political philosopher and author; son of third Earl of Carnarvon [q.v.]; educated at Eton and St. John's College, Oxford; fellow, 1855-69; joined army, 1858; resigned, 1862; president of Union at Oxford, 1862; DCL, 1865; present during Prusso-Danish war, 1864; in America during civil war, and at Sedan in Franco-German war; private secretary to Sir Stafford Northcote (later Earl of Iddesleigh, q.v.), 1866-8; abandoned conservative views; liberal MP for Nottingham, 1870-4; declared himself a republican, 1872;

supported Joseph Arch [q.v.] and Agricultural Labourers' Union, 1872; championed Charles Bradlaugh [q.v.], 1880; became ardent disciple of Herbert Spencer [q.v.] and an agnostic; published *A Politician in Trouble about his Soul*, 1884; issued a monthly *Organ of Voluntary Taxation*, 1890-1901, and *The Voluntaryist Creed*, 1908; engaged in farming in the New Forest from 1874 till death; relinquished sport on becoming vegetarian; interested in prehistoric remains and in psychic research; also wrote verse.

HERBERT, AUBERON THOMAS, eighth BARON LUCAS (1876-1916), politician and airman; son of A. E. W. M. Herbert [q.v.]; BA, Balliol College, Oxford; succeeded maternal uncle, 1905; held office in liberal government, 1908-15; PC, 1912; joined Royal Flying Corps, 1915; missing in France.

HERBERT, GEORGE EDWARD STANHOPE MOLYNEUX, fifth EARL OF CARNARVON (1866-1923), Egyptologist; son of fourth earl [q.v.]; educated at Eton and Trinity College, Cambridge; succeeded father, 1890; wintered in Egypt from 1903; began sixteen years' collaboration in excavation with Howard Carter [q.v.], 1907; after 1914-18 war obtained concession to excavate in valley of the Tombs of the Kings near Thebes; tomb of King Tutankhamūn, of Dynasty XVIII, discovered, Nov. 1922; actual burial chamber and store-chamber unsealed, Feb. 1923; a wealth of treasures revealed; died at Cairo.

HERBERT, SIR ROBERT GEORGE WYNDHAM (1831-1905), colonial official; scholar of Eton and Balliol College, Oxford; fellow of All Souls from 1854 till death; DCL, 1862; private secretary to Gladstone, 1855; called to bar, 1858; went to Queensland as colonial secretary, 1859; member of legislative council and first premier, 1860-5; permanent under-secretary for the colonies in London, 1871-92; agent-general for Tasmania, 1893-6; KCB, 1882; GCB, 1902; hon. LLD, Cambridge, 1886.

HERDMAN, SIR WILLIAM ABBOTT (1858-1924), marine naturalist; MA, Edinburgh University; assistant to Sir C. W. Thomson [q.v.] in working on *Challenger* collections; first Derby professor of natural history, Liverpool University, 1881-1919; devoted attention to co-ordination of fishing industry with scientific research; his encouragement of plankton survey in Irish Sea especially noteworthy; FRS, 1892; CBE, 1920; knighted, 1922.

HERFORD, BROOKE (1830-1903), Unitarian divine; brother of William Henry Her-

ford [q.v.]; Unitarian minister from 1851; founded and edited *Unitarian Herald*, 1861; minister at Strangeways, Manchester (1864-76), in Chicago (1876-82), and Boston (1882-92); preacher to Harvard University and DD, 1891; minister of Rosslyn Hill chapel, Hampstead, 1892-1901; president of British and Foreign Unitarian Association, 1898-9; published *Brief Account of Unitarianism* (1903), sermons, tracts, and hymns.

HERFORD, CHARLES HAROLD (1853-1931), scholar and critic; nephew of B. and W. H. Herford [qq.v.]; eighth classic, Trinity College, Cambridge, 1879; professor of English, University College, Aberystwyth (1887-1901), Manchester (1901-21); the most accomplished English literary scholar of his age; publications embrace also the literature of Germany, Italy, Norway, and Russia; concentrated on ideas and characters of authors; FBA, 1926.

HERFORD, WILLIAM HENRY (1820-1908), writer on education; studied for Unitarian ministry at York and Manchester; spent three years in Bonn and Berlin; imbibed Pestalozzi's and Froebel's educational ideas on visit to Pestalozzian school at Hofwyl, near Berne, 1847; managed Pestalozzian school at Lancaster, 1850-61; inspiring preacher, teacher, and lecturer in Manchester from 1863; directed co-educational school for younger children at Fallowfield, Manchester, 1873-86; published *The School: an Essay towards Humane Education* (1889), *The Student's Froebel* (1893), and translated works on education.

HERKOMER, SIR HUBERT VON (1849-1914), painter; born in Bavaria; brought to England, 1857; naturalized; studied at South Kensington art schools; achieved great success with picture, 'The Last Muster—Sunday at the Royal Hospital, Chelsea', 1875; ARA, 1879; portraits include those of Wagner, Ruskin, Lord Kelvin, and Marquess of Salisbury; subject-pictures include 'Found' (1885) and 'On Strike' (1890); RA, 1890; founded and directed school of art at Bushey, 1883-1904; Slade professor of fine art at Oxford, 1885-94; knighted, 1907; wrote on etching, etc.

HERRING, GEORGE (1832-1906), philanthropist; at first a turf commission agent and owner of racehorses; subsequently made fortune as financial commission agent in City of London; chairman of City of London Electric Lighting Company; devoted wealth to London Sunday Hospital Fund from 1899, to Salvation Army land scheme at Boxted, Essex, and to North-West London Hospital, Camden Town; bequests to charities under will totalled £900,000.

HERRINGHAM, Sir WILMOT PARKER (1855-1936), physician; educated at Winchester, Keble College, Oxford, and St. Bartholomew's Hospital; assistant physician, 1895; full physician, 1904-19; especially interested in kidney diseases; honorary general secretary to Association of Physicians on its foundation (1907) and to international medical congress (1913); knighted, 1914; vice-chancellor, London University, 1912-15; consulting physician in France, 1914-18; KCMG, 1919; chairman of General Nursing Council (1922-6), governors of Old Vic (1921-9), and council of Bedford College.

HERSCHEL, ALEXANDER STEWART (1836-1907), astronomer; son of Sir John Herschel [q.v.]; born in South Africa; BA, Trinity College, Cambridge (twentieth wrangler), 1859; studied meteorology at Royal School of Mines, London, 1861; professor of physics at Glasgow, 1866-71, and at Durham College, Newcastle, 1871-86; reported on observations of meteors to British Association, 1862-81; experimented in photography; FRAS, 1867; FRS, 1884; DCL, Durham, 1886; observed solar eclipse in Spain, 1905.

HERTSLET, Sir EDWARD (1824-1902), librarian of the Foreign Office; assistant to father, Lewis Hertslet [q.v.], at Foreign Office library, 1840; sub-librarian, 1855; librarian, 1857; attached to mission to Berlin congress and knighted, 1878; CB, 1874; KCB, 1892; compiled *Hertslet's Commercial Treaties* (vols. xii-xix, 1871-95), *Recollections of the Foreign Office* (1901), and other works.

HERTZ, JOSEPH HERMAN (1872-1946), chief rabbi of the United Hebrew Congregations of the British Empire; migrated to America from Slovakia, 1884; Ph.D., Columbia University, graduated as rabbi, and appointed to Syracuse, 1894; Johannesburg, 1898; Orach Chayim congregation, New York, 1911; chief rabbi, United Hebrew Congregations of British Empire, 1913-46; combative conservative; strong Zionist; CH, 1943.

HERTZOG, JAMES BARRY MUNNIK (1866-1942), prime minister of South Africa; graduated from Victoria College, Stellenbosch (1888), and Amsterdam University (1892); judge, Orange Free State, 1895-9; commanded Boer southern forces, 1899-1902; attorney-general and minister of education, Orange River Colony, 1907-10; represented Smithfield in Union parliament, 1910-40; minister of justice, 1910-12; formed national party, 1913; prime minister, 1924-39; of native affairs, 1924-9; external affairs, 1929-39; gave Union its own flag, citizenship, and ministers plenipotentiary; substituted Afrikaans for Dutch; formed coalition with Smuts, 1932; united party, 1933; carried through segregation policy, notably disfranchisement of Cape Bantu, 1936; defeated on proposal of neutrality, 5 Sept. 1939.

HESELTINE, PHILIP ARNOLD (1894-1930), writer on music, and musical composer under the pseudonym of PETER WARLOCK; influenced by Frederick Delius, D. H. Lawrence [qq.v.], and Bernard van Dieren; compositions include songs and part songs, instrumental and choral works; his music that of a belated Elizabethan.

HESS, Dame (JULIA) MYRA (1890-1965), pianist; studied at Guildhall School of Music and the Royal Academy of Music, 1903-8; gave first concert, conducted by (Sir) Thomas Beecham [q.v.], at Queen's Hall, London, 1907; first great success in Holland with Concertgebouw Orchestra, 1912; made American début at Aeolian Hall, New York, 1922; played regularly with London String Quartet, and partnered Hungarian violinist, Jelly d'Aranyi [q.v.]; organized daily chamber-music concerts at National Gallery, London, 1939-45; last concert appearance, Royal Festival Hall, London, 1961; increasingly troubled by arthritis of the hands and severe circulatory problems; disliked recording, but best recording probably Beethoven's Sonata in E major, Op. 109 (1954); piano transcription (1926) of chorale-setting from Bach's Cantata No. 147 ('Jesu, Joy of Man's Desiring') achieved immense popularity; CBE, 1936; DBE, 1941; hon. degrees from number of universities.

HETHERINGTON, Sir HECTOR JAMES WRIGHT (1888-1965), university vice-chancellor; educated at Dollar Academy and Glasgow University, 1905-10; lecturer in moral philosophy under Sir Henry Jones [q.v.], 1910-14; warden, Glasgow University Settlement; professor of logic and philosophy, University College, Cardiff, 1915-20; principal and professor of philosophy, University College, Exeter, 1920-4; professor, moral philosophy, Glasgow, 1924-7; vice-chancellor, Liverpool University, 1927-36; principal, Glasgow University, 1936-61; hon. ARIBA; chairman, Committee of Vice-Chancellors and Principals, 1943-7 and 1949-52; associated with many charitable trusts; chairman of number of committees, including the Colonial Universities Grants Committee, 1942-8; recognized as doyen of vice-chancellors throughout the commonwealth; chairman or president of over fifty educational and charitable bodies; publications include *Social Purpose* (with J. H. Muirhead, q.v., 1918), *International Labour Legislation* (1920), and *Life and Letters of Sir Henry Jones* (1924); knighted, 1936; KBE, 1948; GBE, 1962; hon. degrees from many universities and professional bodies.

HEWART, GORDON, first VISCOUNT HEWART (1870-1943), lord chief justice of England; educated at Bury and Manchester grammar schools and University College, Oxford; second class, *lit. hum.*, 1891; called to bar (Inner Temple), 1902; KC, 1912; bencher, 1917; liberal MP, Leicester, 1913-22; solicitor-general, 1916-19; attorney-general, 1919-22; lord chief justice, 1922-40; brilliant advocate; less successful as judge through tendency to forget he was no longer an advocate; president, Classical Association, 1926, English Association, 1929; knighted, 1916; PC, 1918; baron, 1922; viscount, 1940.

HEWETT, SIR JOHN PRESCOTT (1854-1941), Indian civil servant; educated at Winchester and Balliol College, Oxford; posted to North-Western Provinces and Oudh, 1877; under-secretary, Indian home department, 1886; deputy secretary, 1890; secretary, 1895-1902; secretary, royal commission on opium, 1893; chief commissioner, Central Provinces, 1902-4; in charge of Indian commerce and industry department, 1904-7; encouraged economic activities of Geological Survey and establishment of steel industry at Jamshedpur; lieutenant-governor, United Provinces, 1907-12; organized Naini Tal industrial conference and Allahabad exhibition; presided over durbar committee, 1911; conservative MP, Luton, 1922-3; KCSI, 1907; GCSI, 1911; KBE, 1917.

HEWINS, WILLIAM ALBERT SAMUEL (1865-1931), political economist, historian, and politician; second class, mathematics, Pembroke College, Oxford; organizer and first director, London School of Economics, 1895-1903; secretary, Tariff Commission (1903-17), chairman (1920-2); conservative MP, Hereford City, 1912-18; fought six unsuccessful elections; prolific writer on economic and imperial matters.

HEWLETT, MAURICE HENRY (1861-1923), novelist, poet, and essayist; made his name by romantic fiction, but regarded himself as a poet; works include *The Forest Lovers* (1898), *The Life of Richard Yea-and-Nay* (1900), *The Queen's Quair* (1904), *Rest Harrow* (1910), and *The Song of the Plow* (1916).

HIBBERT, SIR JOHN TOMLINSON (1824-1908), politician; BA, St. John's College, Cambridge, 1847; MA, 1851; called to bar, 1849;

liberal MP for Oldham, 1862-74, 1877-85, 1892-5; held subordinate offices in Gladstone's four administrations; served on various commissions, and was interested in poor law reform; PC, 1886; KCB, 1893; constable of Lancaster castle, 1907-8.

HICHENS, ROBERT SMYTHE (1864-1950), novelist; educated at Clifton; studied music and journalism; published *The Green Carnation* (1894, a satire on Oscar Wilde, q.v.), many romances frequently in desert settings, notably *The Garden of Allah* (1904), macabre stories, and several plays.

HICHENS, (WILLIAM) LIONEL (1874-1940), man of business; educated at Winchester and New College, Oxford; a dispatch rider in South Africa, 1899-1900; joined Milner's 'kindergarten' as treasurer, Johannesburg (1901), Transvaal (1902-7), and inter-colonial council; rebuilt and rehabilitated Cammell Laird & Co. as chairman, 1910-40; showed remarkable understanding of labour; pioneer in seeing industrial problem as a moral question.

HICKS, EDWARD LEE (1843-1919), bishop of Lincoln; BA, Brasenose College, Oxford; fellow of Corpus Christi College, Oxford, 1866; rector of Fenny Compton, Warwickshire, 1873; principal of Hulme Hall, Manchester, 1886; canon of Manchester and rector of St. Philip's, Salford, 1892; bishop of Lincoln, 1910; writer on Greek epigraphy; a strong teetotaller.

HICKS, SIR (EDWARD) SEYMOUR (GEORGE) (1871-1949), actor-manager and author; educated at Victoria College, Jersey; spent two years with the Kendals [qq.v.]; wrote and acted in *Bluebell in Fairyland* (1901); *The Cherry Girl* (1903); *The Catch of the Season* (1904); *The Gay Gordons* (1907); *The Dashing Little Duke* (1909); built Aldwych Theatre, 1905; opened the Hicks, 1906; had memorable season at the Garrick, 1922; a versatile comedian; knighted, 1935.

HICKS, GEORGE DAWES (1862-1941), philosopher; educated at Royal grammar school, Guildford, Owens College, Manchester, Manchester College, Oxford, and Leipzig; Ph.D., 1896; minister, Unity Church, Islington, 1897-1903; professor of moral philosophy, University College, London, 1904-28; leading authority on Kant and Berkeley; set out his theory of knowledge in *Critical Realism* (1938); FBA, 1927.

HICKS, GEORGE ERNEST (1879-1954), trade-unionist; educated at village school; national organizer, 1912, general secretary,

1919, Operative Bricklayers' Society; first general secretary, Amalgamated Union of Building Trade Workers, 1921-40; chairman, Trades Union Congress, 1926-7; labour MP, East Woolwich, 1931-50; parliamentary secretary, Ministry of Works, 1940-5; CBE, 1946.

HICKS, ROBERT DREW (1850-1929), classical scholar; BA, Trinity College, Cambridge; fellow, 1876-1929; became blind, 1900; among the most learned students of Greek philosophy in his generation; chief work, edition of Aristotle's *De Anima* (1907).

HICKS, WILLIAM JOYNSON-, first VIS-COUNT BRENTFORD (1865-1932), statesman; educated at Merchant Taylors' School; admitted solicitor, 1887; conservative MP, North-West Manchester (1908-10), Brentford (1911-18), Twickenham (1918-29); especially interested in road traffic and aviation; baronet, 1919; in 1923 successively postmaster- and paymaster-general, financial secretary to Treasury with cabinet seat (PC), and minister of health; home secretary, 1924-9; dealt successfully with general strike, 1926; passed Shops Act, 1928; president, National Church League, 1921; prominent in defeat of Prayer Book revision, 1927-8; viscount, 1929.

HICKS BEACH, SIR MICHAEL EDWARD, ninth baronet, and first EARL ST. ALDWYN (1837-1916), statesman; educated at Eton and Christ Church, Oxford; succeeded to baronetcy, 1854; conservative MP, East Gloucestershire, 1864-85, West Bristol, 1885-1906; parliamentary secretary of Poor Law Board, 1868; under-secretary for home department, 1868; in opposition, 1869-74; chief secretary for Ireland, and showed sympathy with reform, 1874-8; entered cabinet, 1876; colonial secretary, 1878-80; chiefly preoccupied with South African affairs; followed general lines of policy of predecessor, the fourth Earl of Carnarvon [q.v.]; supported Sir Bartle Frere [q.v.] until Oct. 1878; sympathized with Frere throughout, but after that date British policy in South Africa largely controlled by Disraeli; in opposition, 1880-5; attacked government for treatment of General Gordon [q.v.], 1884; chancellor of Exchequer and leader of House of Commons, 1885-6; as leader of opposition, conducted victorious anti-home rule campaign, 1886; made way for Lord Randolph Churchill [q.v.] as chancellor of Exchequer and leader of Commons, 1886; Irish secretary, 1886-7; president of Board of Trade, 1888-92; chancellor of Exchequer, 1895-1902; opponent of tariff reform; chairman, royal commission on ecclesiastical discipline, 1904-6; viscount, 1906; earl, 1915.

HIGGINS, EDWARD JOHN (1864-1947), third general of Salvation Army; helped to establish it in United States; subsequently dealt with its foreign affairs; British commissioner, 1911-18; chief of staff, 1918-29; played leading part in deposition of Bramwell Booth [q.v.]; succeeded Booth, 1929; retired, 1934.

HIGGINS, SIR JOHN FREDERICK ANDREWS (1875-1948), air marshal; educated at Charterhouse and Woolwich; commissioned in Royal Artillery, 1895; seconded to Royal Flying Corps, 1912; commanded successively squadron, wing, and brigade in France, 1914-18; AOC, Iraq, 1924-6; air member, supply and research, Air Council, 1926-30; KBE, 1925; KCB, 1928; air marshal, 1929; AOC-in-C, India, 1939-40.

HILBERY, SIR (GEORGE) MALCOLM (1883-1965), judge; educated at University College School, London; called to bar (Gray's Inn), 1907; served in RNVR during 1914-18 war; bencher, Gray's Inn, 1927; recorder of Margate, 1927; KC, 1928; commissioner of assize, South-Eastern circuit, 1935; judge, King's Bench division; knighted, 1935; chairman, Berkshire quarter-sessions, 1946-63; senior puisne judge, Queen's Bench division; PC, 1959-62; treasurer, Gray's Inn, 1941; collector of Dutch masters; published *Duty and Art in Advocacy* (1946); chairman, board of governors, Royal Masonic Hospital; prominent member of Thames Yacht Club.

HILDITCH, THOMAS PERCY (1886-1965), chemist; educated at Owen's School, North Islington and University College, London; first class, B.Sc., 1907; associate of Institute of Chemistry, 1908; fellow, 1911; undertook research work in London, Jena, and Geneva, 1907-11; D.Sc. (London), 1911; technical research chemist, Joseph Crosfield & Sons, soap manufacturers, 1911-26; first Campbell Brown professor of industrial chemistry, Liverpool, 1926-51; carried out important research on natural fats; published *The Chemical Constitution of Natural Fats* (four editions, 1940-64); FRS, 1942; CBE, 1952.

HILES, HENRY (1828-1904), musical composer; held many posts as organist; Mus.Bac., Oxford, 1862; Mus.Doc., 1867; lecturer on harmony and composition at Owens College, Manchester, 1876; professor at Royal Manchester College of Music; composed organ music, glees, cantatas, oratorios, anthems; published educational works on music; edited *Wesley Tune Book*.

HILL, ALEXANDER STAVELEY (1825-1905), barrister and politician; BA, Exeter

College, Oxford, 1852; DCL, 1855; fellow, 1854-64; called to bar, 1851; QC, 1868; treasurer of Inner Temple, 1886; acquired large parliamentary and probate practice; recorder of Banbury, 1866-1903; conservative MP for Coventry and Staffordshire, 1868-1900; PC, 1892; early advocate of tariff reform; formed large cattle ranch in Canada, 1871; hon. LLD, Toronto, 1892; wrote *Practice of the Court of Probate* (1859).

HILL, ALSAGER HAY (1839-1906), social reformer; LLB, Trinity Hall, Cambridge, 1862; called to bar, 1864; urged more scientific classification of paupers and national system of labour registration for unemployed, 1867-8; established labour exchange in London (1871) and edited *Labour News*, organ of communication between masters and men seeking work; vice-president of National Sunday League, 1876-90; published *Rhymes with Good Reason* (1870-1) and other verse.

HILL, SIR ARTHUR WILLIAM (1875-1941), botanist; educated at Marlborough and King's College, Cambridge; first class, natural sciences tripos parts i and ii, 1897-8; fellow, 1901; university demonstrator in botany, 1899; lecturer, 1904; assistant director, Royal Botanic Gardens, Kew, 1907; director, 1922-41; edited publications and extended activities and amenities of the Gardens; FRS, 1920; KCMG, 1931.

HILL, SIR (EDWARD) MAURICE (1862-1934), judge; son of G. B. N. Hill [q.v.]; educated at Haileybury and Balliol College, Oxford; first class, *lit. hum.*, 1884; called to bar (Inner Temple), 1888; bencher, 1917; KC, 1910; judge of Probate, Divorce, and Admiralty division (1917-30) trying mainly the many Admiralty cases resulting from war; his memorandum on freedom of seas the basis of article 297 of treaty of Versailles and embodied in Maritime Ports Convention, 1923.

HILL, FRANK HARRISON (1830-1910), journalist; studied for Unitarian ministry at Manchester under Dr James Martineau [q.v.], 1846-51; adopted profession of journalism; editor of *Northern Whig*, Belfast, 1861-5; alone of journalists in Ireland supported North in American war; friend of Harriet Martineau, Browning, and Crabb Robinson [qq.v.]; assistant editor of *Daily News*, 1865; editor, 1869-86; made paper an influential party organ; wrote with keen insight and caustic pen; retired owing to opposition to Gladstone's home rule policy, 1886; leader-writer of the *World*, 1886-1906; published *Political Portraits* (1873) and *Life of George Canning* (1881).

HILL, GEORGE BIRKBECK NORMAN (1835-1903), editor of Boswell's *Life of Johnson*; grandson of Thomas Wright Hill [q.v.]; BA, Pembroke College, Oxford, 1858; while at the university joined Old Mortality Club, and came to know Burne-Jones and Rossetti [qq.v.]; DCL, 1871; private schoolmaster, 1858-75; elaborately edited Boswell's *Life of Johnson* (6 vols., 1887), *Johnson's Letters* (2 vols., 1892), *Johnsonian Miscellanies* (2 vols., 1897), and *Johnson's Lives of the English Poets* (3 vols., 1905); described a tour in Scotland as *Footsteps of Samuel Johnson* (1889); published also life of his uncle, Sir Rowland Hill (2 vols., 1880), and *Letters of Rossetti to William Allingham* (1897); his *Letters* published, 1903 and 1906.

HILL, SIR GEORGE FRANCIS (1867-1948), numismatist; educated at University College School, University College, London, and Merton College, Oxford; first class, *lit. hum.*, 1891; entered department of coins and medals, British Museum, 1893; keeper, 1912; director and principal librarian of museum, 1931-6; launched successful campaign for purchase of *Codex Sinaiticus*; publications include *Corpus of Italian Medals before Cellini* (2 vols., 1930); FBA, 1917; KCB, 1933.

HILL, SIR LEONARD ERSKINE (1866-1952), physiologist; son of G. B. N. Hill and brother of Sir Maurice Hill [qq.v.]; educated at Haileybury; qualified in medicine from University College Hospital, 1889; lecturer in physiology, London Hospital, 1895; professor, 1912-14; head, department of applied physiology, National Institute for Medical Research, 1914-30; researches covered the physiology of the circulatory and respiratory systems, including decompression of divers, measurement of efficiency of ventilation, significance of solar radiation for health; published *The Physiology and Pathology of the Cerebral Circulation* (1896); FRS, 1900; knighted, 1930; hon. LLD Aberdeen; fellow, University College, London.

HILL, LEONARD RAVEN- (1867-1942), artist, illustrator, and cartoonist. [See RAVEN-HILL.]

HILL, OCTAVIA (1838-1912), philanthropist; granddaughter of Dr Thomas Southwood Smith [q.v.]; early influenced by Christian socialists, especially F. D. Maurice [q.v.]; helped in artistic training by John Ruskin [q.v.]; opened school with sisters in Marylebone; impressed by urgency of housing problem in which she interested Ruskin, 1864; first houses for improvement purchased through him, 1865; undertaking placed on business footing; through fund raised by friends, enabled to devote herself to housing reform, 1874; appointed manager of Southwark property by ecclesiastical commissioners, 1884; owed much to co-operation of sisters and workers trained by her; supporter of Charity Organisation, Kyrle, and Commons Preservation societies; co-founder of National Trust; held aloof from political measures for social reform.

HILL, SIR RODERIC MAXWELL (1894-1954), air chief marshal; nephew of Sir George Francis Hill [q.v.]; educated at Bradfield College and University College, London; joined Royal Flying Corps, 1916; became highly skilled pilot; MC; in charge of experimental flying, Farnborough, 1917-23; served at Hinaidi (1924-6), Cairo (1926-7), Oxford (1930-2); deputy director, repair and maintenance, Air Ministry, 1932-6; commanded in Palestine and Trans-Jordan, 1936-8; director, technical development, Air Ministry, 1938-40; air vice-marshal, 1939; air marshal, 1940; director-general, research and development, Ministry of Aircraft Production, 1940-1; controller, technical services, British Air Commission, United States, 1941-2; commandant, RAF Staff College, 1942-3; air marshal commanding Air Defence of Great Britain, 1943-4, Fighter Command, 1944-5; planned defence against flying bombs and made courageous decision to redeploy forces more effectively after attack had begun, 1944; KCB, 1944; Air Council member for training, 1945-6, for technical services, 1947-8; air chief marshal, 1947; rector, Imperial College of Science and Technology, 1948-54.

HILL, ROSAMOND DAVENPORT- (1825-1902), educational administrator; daughter of Matthew Davenport Hill [q.v.]; settled with parents at Bristol, 1851; assisted her father and Mary Carpenter [q.v.]; visited Mettray reformatory, 1855; founded on Mettray principles girls' industrial school at Bristol, 1866; visited prisons in Australia, 1872-4; published *What We Saw in Australia* (1874); member of London school board, 1879-97; visited schools in United States and Canada, 1888; resisted provision of free meals in London schools; wrote *Elementary Education in England* (1893).

HILLS, ARNOLD FRANK (1857-1927), shipbuilder and philanthropist; BA, University College, Oxford; joined directorate of Thames Ironworks and Shipbuilding Company, Blackwall; lived in Canning Town, giving much time and thought to improvement and recreation of firm's workpeople, 1880-5; considerable prosperity enjoyed by firm, 1890-1900; after launching of the *Thunderer*, Thames Ironworks, unable to compete with northern shipyards, compelled to close down, 1911.

HILLS, SIR JOHN (1834-1902), major-general; born in Bengal; joined Bombay engineers, 1854; in Persian expedition, 1857; field engineer in Abyssinian expedition, 1867; commandant of Bombay sappers at Kirkee, 1871-83; commanded division of Kandahar field force, 1879-80; at defence of Kandahar; CB, 1881; commanding royal engineer of Burma expeditionary force, 1886-7; major-general, 1887; KCB, 1900; published *The Bombay Field Force* (1880) and *Points of a Racehorse* (1903).

HILTON, JAMES (1900-1954), novelist; educated at Sir George Monoux grammar school, Walthamstow, Leys School, and Christ's College, Cambridge; first class, English tripos, 1921; obtained fame with *Lost Horizon* (1933) and *Good-Bye Mr. Chips* (1934) a tender portrait of his schoolmaster father; became Hollywood script-writer; his later novels equally successful but less likely to endure; died at Long Beach, California.

HIND, ARTHUR MAYGER (1880-1957), historian of engraving; educated at City of London School and Emmanuel College, Cambridge; first class, classics, 1902; entered prints and drawings department, British Museum, 1903; keeper, 1933-45; Slade professor of fine art, Oxford, 1921-7; catalogued fifteenth-century Italian engravings (7 vols., 1938 and 1948), Rembrandt's etchings (1912), and drawings by Dutch and Flemish artists in the British Museum (4 vols., 1915-31); published *History of Engraving and Etching* (1908), of *Woodcut* (1935), and *Engraving in England in the Sixteenth and Seventeenth Centuries* (3 vols., 1952-64); served in Army Service Corps, 1915-18; OBE; hon. LLD Glasgow.

HIND, HENRY YOULE (1823-1908), geologist and explorer; went to Canada, 1846; professor of chemistry and geology in Trinity University, Toronto, 1853-64; conducted government explorations and geological surveys; explored Nova Scotia goldfields, 1869-71; discovered extensive cod banks off shore above Belle Isle, 1876; published accounts of his explorations; arranged scientific evidence in fisheries controversy between America and Canada, 1877; DCL, King's College, Windsor, Nova Scotia.

HIND, RICHARD DACRE ARCHER- (1849-1910), Greek scholar and Platonist. [See ARCHER-HIND.]

HINDLEY, SIR CLEMENT DANIEL MAGGS (1874-1944), railway engineer and public servant; educated at Dulwich and Trinity College, Cambridge; assistant engineer, East

Indian Railway, 1897; secretary, 1914; deputy general manager, 1918; general manager, 1920-1; chief commissioner, Indian railways, 1922-8; chairman, Racecourse Betting Control Board, 1928-44; knighted, 1925; KCIE, 1929.

HINDLEY, JOHN SCOTT, VISCOUNT HYNDLEY (1883-1963), business man and administrator; educated at Weymouth College; engineering apprentice, in Durham Colliery; changed over to commercial side of industry; made reputation in large-scale distribution of coal; member of coal controller's export advisory committee, 1917; chairman, Stephenson, Clarke Ltd., coal exporters, 1938; on boards of several other companies; director, Bank of England, 1931; commercial adviser to Mines Department; controller-general, Ministry of Fuel and Power, 1942-3; first chairman, National Coal Board, 1946-51; knighted, 1921; baronet, 1927; baron, 1931; GBE, 1939; viscount, 1948.

HINGESTON-RANDOLPH (formerly HINGSTON), **FRANCIS CHARLES** (1833-1910), antiquary; BA, Exeter College, Oxford, 1855; MA, 1859; rector of Ringmore, Devonshire, 1860-1910; early developed antiquarian tastes; edited Capgrave's chronicle and other works in Rolls series, 1858-60; prebendary of Exeter Cathedral, 1885; edited *Episcopal Registers* of diocese, 11 parts, 1886-1909; wrote also on church architecture.

HINGLEY, SIR BENJAMIN, first baronet (1830-1905), ironmaster; head of father's chainmaking firm at Netherton, 1865; manufactured anchors; influential in Midlands in preservation of industrial peace; mayor of Dudley, 1890; liberal MP for North Worcestershire, 1885; joined liberal unionists, 1886; returned to liberalism, 1892; baronet, 1893.

HINGSTON, SIR WILLIAM HALES (1829-1907), Canadian surgeon; born in province of Quebec; studied medicine at McGill University and Edinburgh; LRCS, 1852; hon. FRCS, 1900; practised in Montreal from 1854; joined staff of Hôtel Dieu, 1860; founded Women's Hospital; professor of clinical surgery at Laval University, 1878; first to remove the tongue and lower jaw in one operation, 1872; mayor of Montreal, 1875-6; knighted, 1895; senator, 1896; wrote on climate of Canada and on vaccination.

HINKS, ARTHUR ROBERT (1873-1945), astronomer and geographer; BA, Trinity College, and second assistant, Cambridge Observatory, 1895; chief assistant, 1903-13; worked on determination of solar parallax; lecturer in surveying and cartography, 1908-13; assistant secretary, Royal Geographical Society, 1913; secretary, 1915-45; published *Map Projections* (1912) and *Maps and Survey* (1913); FRS, 1913.

HINKSON, KATHARINE (1861-1931), better known as KATHARINE TYNAN, poet and novelist; lived in Ireland until her marriage (1883) with Henry Albert Hinkson, barrister and novelist; published over 100 novels, collections of poems, and five volumes of autobiography valuable for personal accounts of such figures as Parnell.

HINSHELWOOD, SIR CYRIL NORMAN (1897-1967), physical chemist and biochemist; educated at Westminster city school and Balliol College, Oxford; worked as chemist in Explosives Supply Factory, Queensferry, 1916-19; assistant chief laboratory chemist; research fellow, Balliol College, 1920; tutorial fellow, Trinity College, 1921-37; first book, *Kinetics of Chemical Change in Gaseous Systems* (1926), a milestone in chemical literature; succeeded Frederick Soddy [q.v.] as Dr Lee's professor, teaching physical and inorganic chemistry, 1937-64; moved to Exeter College; worked in new Physical Chemistry Laboratory, provided by Lord Nuffield [q.v.], 1941; took up study of chemical kinetics of living cells, 1936-67; senior research fellow, Imperial College, London, 1964; published *The Chemical Kinetics of the Bacterial Cell* (1946), *The Structure of Physical Chemistry* (1951), and *Growth, Function and Regulation in Bacterial Cells* (with A. C. R. Dean, 1966); FRS, 1929; president, Chemical Society, 1946-8; president, Royal Society, 1955-60; knighted, 1948; shared Nobel prize with Semenov, 1956; OM, 1960; president, British Association, 1964-5; hon. fellow, Trinity, Balliol, Exeter, and St. Catharine's Colleges, Oxford; delegate, Oxford University Press, 1934; a clever linguist and a competent painter in oils; trustee, British Museum; chairman of number of advisory committees, including education committee of Goldsmiths' Company.

HINSLEY, ARTHUR (1865-1943), cardinal; studied at Ushaw and English College, Rome; DD, Gregorian University; ordained, 1893; founder and headmaster (1899-1904), St. Bede's grammar school, Bradford; taught at Wonersh and pastor of Sutton Park (1904-11) and Sydenham (1911-17); rector, English College, Rome, 1917-30; domestic prelate, 1917; titular bishop of Sebastopolis, 1926; visitor apostolic to Africa, 1927; delegate apostolic and titular archbishop of Sardis, 1930-4; archbishop of Westminster, 1935–43; cardinal, 1937; launched 'Sword of the Spirit' movement, 1940.

HIPKINS, ALFRED JAMES (1826–1903), musical antiquary; accomplished interpreter of Chopin on piano; studied history of keyboard instruments; published *Musical Instruments, Historic, Rare and Unique* (1881); FSA, 1886; left collection of musical instruments to Royal College of Music.

HIRST, FRANCIS WRIGLEY (1873–1953), economist and liberal writer; educated at Clifton, Wadham College, Oxford (first class, *lit. hum.*, and president of Union, 1896), and London School of Economics; joint-editor, *Essays in Liberalism* (1897); called to bar (Inner Temple), 1899; assisted Lord Morley [q.v.] with biography of Gladstone; active in forming League against Imperialism and Militarism; wrote regularly for *Speaker* and *Nation*; editor, *The Economist*, 1907–16, supporting principles of peace, economy, and individual liberty; recruited competent staff and himself wrote the policy leaders; publications include biography of Adam Smith (1904), *Early Life and Letters of John Morley* (2 vols., 1927), *Economic Freedom and Private Property* (1935); unsuccessful liberal candidate for Parliament, 1910 and 1929.

HIRST, GEORGE HERBERT (1871–1954), cricketer; left village school at age of ten; obtained Yorkshire county trial, 1889; established himself with 99 wickets for 14.39, 1893; in 1906 made 2,385 runs (average 45.86) and took 208 wickets for 16.50; altogether took 2,727 wickets for 18.77 and made 36,203 runs averaging 34.05; played without distinction in nine test matches; a dangerous left-arm bowler and right-handed batsman favouring the hook and the pull; chief coach at Eton, 1919–37; honorary member, MCC, 1949.

HIRST, HUGO, BARON HIRST (1863–1943), chairman of General Electric Company; born and educated in Munich; naturalized, 1883; managing director, General Electric Company, 1900–43; chairman, 1910–43; made it a leading firm manufacturing and supplying electric equipment; recognized expert on international trading; baronet, 1925; baron, 1934.

HITCHCOCK, SIR ELDRED FREDERICK (1887–1959), man of business; educated at Burford grammar school; obtained Oxford diploma in economics, 1910; secretary, 1910, warden, 1917–19, Toynbee Hall; obtained interest in sisal company, Bird & Co. (Africa) Ltd., 1926; chairman, 1950–9; managing director of its Tanganyika estates, 1939–59; chairman or vice-chairman, Sisal Growers Association,

1946–59; established Sisal Marketing Association, 1949; obtained establishment of department of antiquities in Tanganyika; knighted, 1955; died in Tanga.

HIVES, ERNEST WALTER, first BARON HIVES (1886–1965); educated at Redlands School, Reading; worked in Reading garage and other garage jobs, 1903–8; joined Rolls-Royce Company, 1908; test driver; directed development of notable aero-engines, including the Kestrel and the R engine, precursor of the Merlin, which powered the Hurricanes and Spitfires of World War II; general works manager, 1936; managing director, 1946; chairman, 1950–7; under his direction Rolls-Royce led the world in design, development, and manufacture of gas turbine aero-engines; MBE, 1920; CH, 1943; baron, 1950; hon. D.Sc., Nottingham, 1949; hon. LLD, Cambridge, 1951; and hon. D.Sc., London, 1958.

HOARE, JOSEPH CHARLES (1851–1906), bishop of Victoria, Hong Kong, from 1898; BA, Trinity College, Cambridge, 1874; DD, 1898; joined CMS Mid-China mission at Ningpo, 1875; founded there training college for native evangelists; drowned in Castle Peak Bay while on preaching tour; contributed much to vernacular literature.

HOARE, SIR REGINALD HERVEY (1882–1954), diplomatist; grandson of Lord Arthur Hervey [q.v.]; educated at Eton; entered diplomatic service, 1905; counsellor in Turkey, 1924–8; in Egypt, 1928–31; minister to Persia, 1931, to Romania, 1935; protested against atrocities of Nazi-style regime; mission withdrawn on imposition of economic warfare on Romania, 1941; managing partner, family bank, 1944–54; CMG, 1926; KCMG, 1933.

HOARE, SIR SAMUEL JOHN GURNEY, second baronet and VISCOUNT TEMPLEWOOD (1880–1959), statesman; educated at Harrow and New College, Oxford; first class, history, 1903; played rackets and lawn tennis for Oxford; conservative MP, Chelsea, 1910–44; succeeded father, 1915; lieutenant-colonel with military mission to Russia, 1916–17, to Italy, 1917–18; secretary of state for air, 1922–4, 1924–9; PC, 1922; went on first civil air flight to India, 1926–7; GBE, 1927; secretary of state for India, 1931–5; a principal witness before joint select committee on Indian constitution, 1933–4, and piloted government of India bill through Commons; GCSI, 1934; foreign secretary, 1935; made notable speech to League Assembly on collective security, Sept. 1935; resigned when his plan with Laval for solution of Italy's claims on Abyssinia seemingly belied his speech and was repudiated by the cabinet, Dec. 1935; first lord of Admiralty, 1936-7; home secretary, 1937-9; defended Munich agreement; recruited for ARP and WVS; lord privy seal and member of war cabinet, 1939-40; secretary of state for air, 1940; ambassador to Spain, 1940-4; viscount, 1944; president, Lawn Tennis Association, 1932-56; of Howard League for Penal Reform, 1947-59; chancellor, Reading University, 1937-59; publications include *Ambassador on Special Mission* (1946), *Nine Troubled Years* (1954), and *Empire of the Air* (1957).

HOBART, Sir PERCY CLEGHORN STANLEY (1885-1957), major-general; born in India; educated at Clifton and Woolwich; joined Royal Engineers, 1904; served in India, 1906-15, 1921-7; in France, Mesopotamia, and Egypt, 1915-18; DSO; MC; OBE, 1919; joined Royal Tank Corps, 1923, and appointed its head, 1933; raised and commanded 1st Tank brigade, 1934-7; major-general and director of military training, 1937-8; raised 7th Armoured division, 1938-9; his advanced views unacceptable; re-employed at Churchill's behest; commanded 11th Armoured division, 1941-2; raised 79th Armoured division, 1942; trained it for and took part in invasion of Europe, 1944-5; CB, 1939; KBE, 1943.

HOBBES, JOHN OLIVER (pseudonym) (1867-1906), novelist and dramatist. [See CRAIGIE, PEARL MARY TERESA.]

HOBBS, Sir JOHN BERRY (JACK) (1882-1963), cricketer; son of groundsman at Jesus College, Cambridge; played cricket for Cambridgeshire, 1902; given trial for Surrey, 1903; awarded Surrey county cap, 1905; went with MCC side on Australian tour, 1907-8; played for England against South Africa, 1909-10; and against Australia, 1911-12; with W. Rhodes set up record of 323 for the first wicket at Melbourne; toured South Africa, and made 1,489 runs, 1913-14; made 11 centuries, and Surrey became champions, 1914; at retirement in 1934 had made 61,237 runs including 197 centuries; made hundredth century, 1923; scored century in final test at the Oval when England recovered the Ashes, 1926; played his last test in Australia, 1928-9; final century in first-class cricket made at Old Trafford, 1934; played altogether in 61 test matches and made 5,410 runs; life member, Surrey Club, 1935; life member, MCC, 1949; knighthood, 1953, the first conferred on a professional cricketer; published *My Cricket Memories* (1924) and *My Life Story* (1935).

HOBDAY, Sir FREDERICK THOMAS GEORGE (1869-1939), veterinary surgeon; MRCVS, 1892; FRCVS, 1897; began practice in Kensington, 1900; improved surgical methods; so successful in operating to relieve 'roaring' that horses so treated were termed 'hobdayed'; editor, *Veterinary Journal*, 1906-39; commanded a veterinary hospital in France and Italy, 1915-19; principal and professor of surgery, Royal Veterinary College, 1927; by 1936 had raised £285,000 for reconstruction, but alleged defects in college administration caused enforced retirement; knighted, 1933.

HOBHOUSE, ARTHUR, BARON HOBHOUSE (1819-1904), judge; son of Henry Hobhouse [q.v.]; educated at Eton and Balliol College, Oxford; BA (first class classic), 1840; called to bar, 1845; acquired large Chancery practice; QC, 1862; treasurer of Lincoln's Inn, 1880-1; charity commissioner, 1866; one of three commissioners for reorganizing endowed schools, 1869-72; law member of council of governor-general of India, 1872-7; responsible for Specific Relief Act, 1877; opposed Lord Lytton's Afghan policy; KCSI, 1877; member of judicial committee of Privy Council, 1881-1901; raised to peerage to try appeal cases in House of Lords, 1885; member of London school board, 1882-4; alderman of London County Council, 1889; life by L. T. Hobhouse and J. L. Hammond [qq.v.], 1905.

HOBHOUSE, EDMUND (1817-1904), bishop of Nelson, New Zealand, and antiquary; brother of Baron Hobhouse [q.v.]; educated at Eton and Balliol College, Oxford; BA, 1838; DD, 1858; fellow of Merton College, Oxford, 1841-58; vicar of St. Peter's-in-the-East, Oxford, 1843-58; bishop of Nelson, 1858-65; assistant to bishops Selwyn and Maclagan at Lichfield, 1869-81; edited records for Somerset Record Society (1887-94), which he helped to found.

HOBHOUSE, HENRY (1854-1937), pioneer in local government; nephew of Baron Hobhouse [q.v.]; educated at Eton and Balliol College, Oxford; first class, *lit. hum.*, 1875; called to bar (Lincoln's Inn) and county magistrate, 1880; with Sir Robert Wright [q.v.] wrote *Outline of Local Government and Local Taxation* . . . (1884); liberal (1885), liberal unionist (1886-1906) MP, East Somerset; chairman, Somerset county council, 1904-24; a country squire expert in county administration, especially interested in education and agriculture; PC, 1902.

HOBHOUSE, Sir JOHN RICHARD (1893-1961), shipowner; son of Henry Hobhouse [q.v.]; nephew of Beatrice Webb [q.v.]; educated at Eton and New College, Oxford; served in Royal Garrison Artillery, 1914-18; MC, 1917; director, Ocean Steam Ship Company, Liverpool, 1920; chairman, 1953-7; director, Royal Insurance Company, 1933-61; chairman, 1954-7; chairman, Liverpool Steam Ship Owners' Association, 1941-2; chairman, National Association of Port Employers, 1948-50; magistrate, 1929-57; knighted, 1946; treasurer, council of Liverpool University, 1942-8; president, 1948-54; pro-chancellor, 1948-57; hon. LLD, 1958; chairman, council of Liverpool School of Tropical Medicine, 1949-55.

HOBHOUSE, LEONARD TRELAWNY (1864-1929), philosopher and journalist; nephew of Baron Hobhouse [q.v.]; BA, Corpus Christi College, Oxford, 1887; tutor of Corpus, 1890; fellow, 1894; on staff of *Manchester Guardian*, 1897-1902; first professor of sociology, London University, 1907-29; works include *The Labour Movement* (1893), *The Theory of Knowledge* (1896), *Mind in Evolution* (1901), *Morals in Evolution* (1906), *Development and Purpose* (1913), *The Metaphysical Theory of the State* (1918), *The Rational Good* (1921), *The Elements of Social Justice* (1922), *Social Development* (1924); died at Alençon, Normandy.

HOBSON, ERNEST WILLIAM (1856-1933), mathematician; brother of J. A. Hobson [q.v.]; educated at Derby School, Imperial College of Science and Christ's College, Cambridge; senior wrangler and fellow, 1878; private coach until 1903; university lecturer (1883), Stokes lecturer (1903), Sadleirian professor (1910-31); leader in reforming the tripos; his *Theory of Functions of a Real Variable* (1907, final form, 2 vols., 1926-7) of first importance to English mathematics; other works include *Treatise on Plane Trigonometry* (1891), *The Domain of Natural Science* (1923), and *Spherical and Ellipsoidal Harmonics* (1931); FRS, 1893.

HOBSON, GEOFFREY DUDLEY (1882-1949), historian of bookbindings; educated at Harrow and University College, Oxford; first class, modern history, 1903; with others acquired auctioneering firm of Sotheby, 1909; leading authority on bookbindings; published *Maioli, Canevari and Others* (1926), *English Binding before 1500* (1929), and *Bindings in Cambridge Libraries* (1929).

HOBSON, JOHN ATKINSON (1858-1940), economist and publicist; brother of E. W. Hobson [q.v.]; educated at Derby School and Lincoln College, Oxford; an original thinker on economics with approach of sociologist; developed theory of under-consumption in *The Physiology of Industry* (1889, with A. F. Mummery, q.v.) and later works; prescribed such remedies for inequality as steeply graduated

taxation, extension of social services, and nationalization of monopolies in *The Industrial System: an Inquiry into Earned and Unearned Income* (1909) and subsequently; *The Evolution of Modern Capitalism: a Study of Machine Production* (1894) indispensable introduction to nineteenth-century economic history; studied search for new markets and opportunities for investment in *Imperialism* (1902); wrote frequently (1906-20) in the *Nation*.

HOBSON, SIR JOHN GARDINER SUMNER (1912-1967), attorney-general; educated at Harrow and Brasenose College, Oxford; called to bar (Inner Temple), 1938; served in Northamptonshire Yeomanry in France and North Africa, 1940-5; OBE, 1945; chairman, Rutland and Bedfordshire quarter-sessions; recorder, Northampton, 1958-62; KC, 1957; conservative MP, Warwick and Leamington, 1957; solicitor-general, 1962; attorney-general, knighted, 1962; PC, 1963; prosecuted the spy, Vassall, 1962; involved in Enahoro case, 1963; returned to private practice, 1964-7.

HOCKING, JOSEPH (1860-1937), novelist and preacher; ordained in United Methodist Free Church, 1884; minister, Woodford Green Union church, 1887-1910; used fiction to convey religious ideas to popular public; his fifty-three books include *The Woman of Babylon* (1906) and *The God That Answers by Fire* (1930).

HOCKING, SILAS KITTO (1850-1935), novelist and preacher; brother of J. Hocking [q.v.]; ordained in United Methodist Free Church, 1870; minister, Duke Street, Southport, 1883-96; thereafter devoted himself to writing, lecturing, and liberal politics; a best-selling author of healthy fiction.

HODGE, JOHN (1855-1937), labour leader; steel smelter at Motherwell; secretary to British Steel Smelters' Association on its foundation (1886); believed in conciliation and arbitration; president, Iron and Steel Trades Confederation, until 1931; served on national committee of labour party, 1900-15; MP, Gorton, 1906-23; first minister of labour, 1916-17; PC, 1917; minister of pensions, 1917-19; opposed to Association's participation in general strike, 1926; a constitutionalist relying on education and legislation to achieve his socialist ideals.

HODGETTS, JAMES FREDERICK (1828-1906), commander and archaeologist; in East India Company's fleet in Burmese war, 1851; commander in Indian navy; professor of seamanship in Berlin, St. Petersburg, and Moscow till 1881; published stories for boys and archaeological works.

HODGKIN, THOMAS (1831-1913), historian; BA, University College, London; partner in banking firm at Newcastle, 1859-1902; active Quaker; chief works, *Italy and her Invaders* (1879-99) and *History of England from the Earliest Times to the Norman Conquest* (1906).

HODGKINS, FRANCES MARY (1869-1947), painter; born, educated, and studied and practised art in Dunedin, New Zealand; visited Europe, 1901-3; returned and settled, 1906; in Paris, 1908-14; thereafter in England; maintained herself with difficulty; unsuccessfully attempted commercial designing; made contract with Alex. Reid and Lefevre, 1932; awarded civil list pension, 1942; a leading exponent of modern English painting, notable colourist, and one of the most gifted and original of women painters.

HODGSON, RALPH EDWIN (1871-1962), poet; in his teens travelled to America and was employed in Thalia Theatre, New York; worked in London as black-and-white artist; edited *Fry's Magazine of Outdoor Life* for C. B. Fry [q.v.], 1912; founded 'The Sign of the Flying Fame', publishers, with Claud Lovat Fraser [q.v.] and Holbrook Jackson, 1913; first published poem 'The Storm Thrush' in *Saturday Review*, 1904; first collection of poems, *The Last Blackbird and Other Lines* (published, 1907); *Poems* followed (1917); served in forces, 1914-18; keen love of countryside; inspired campaign to end trade in birds' feathers for women's apparel, 1920; this led to the Plumage Act, 1921; lecturer at Sendai University, Japan, 1924-38; published *The Skylark and Other Poems* (1958); and *Collected Poems* (1961).

HODGSON, RICHARD DACRE (1849-1910), Greek scholar and Platonist. [See Archer-Hind.]

HODGSON, Sir ROBERT MacLEOD (1874-1956), diplomatist; educated at Radley and Trinity College, Oxford; hockey blue; BA, 1897; vice-consul, Marseilles, 1904-6; commercial agent, Vladivostock, 1906-19; commercial counsellor in Russia, 1919-21; official agent, British commercial mission to Russia, 1921-4; chargé d'affaires, 1924-7; minister to Albania, 1928-36; British agent to Franco's Burgos administration, 1937-9; chargé d'affaires, Spain, Feb.-Apr. 1939; CMG, 1920; KBE, 1925; KCMG, 1939.

HODGSON, SHADWORTH HOLLWAY (1832-1912), philosopher; BA, Corpus Christi College, Oxford; chief works, *Time and Space* (1865), *Theory of Practice* (1870), *Philosophy of Reflection* (1878), and *Metaphysic of Experience* (1898).

HODSON, HENRIETTA (1841-1910), actress; went on stage, 1858; met (Sir) Henry Irving [q.v.] (1860), and with him went to Manchester; popular burlesque actress; first appeared in London, 1866; joined Queen's Theatre company, 1867; married as second husband (1868) Henry Labouchere [q.v.], who acquired sole control of Queen's Theatre, 1870; appeared there as Imogen in *Cymbeline*, 1871; assumed management of Royalty Theatre, Oct. 1871; inaugurated system of the unseen orchestra; revived *Wild Oats* (Dec. 1871), and won applause as Peg Woffington, 1875; introduced Mrs Langtry to stage, 1881; retired to Florence, 1903.

HOEY, FRANCES SARAH (Mrs Cashel Hoey) (1830-1908), novelist; born Johnston; married John Cashel Hoey, CMG, 1858; contributed short stories to *Chambers's Journal*, 1865-94; wrote eleven novels dealing with fashionable society; helped Edmund Yates [q.v.] in his novels; often visited Paris; translated many French and Italian works; received civil list pension, 1892.

HOFMEYR, JAN HENDRIK (1845-1909), South African politician; born and educated at Cape Town; edited *Ons Land* from 1871; formed Farmers' Association at Cape Town, 1878, amalgamated with the Afrikander Bond, 1883; chairman till 1895; member for Stellenbosch of Cape parliament, 1879-95; effective leader of constitutional Afrikanderdom; made the Dutch a political force; refused premiership, 1884; member of executive council of Cape Colony; as delegate in London (1887) proposed closer imperial union by means of imperial tariff of customs; negotiated for the British government with President Kruger the Swaziland convention, 1890; supporter of Cecil Rhodes [q.v.] till Jameson raid of 1895; initiated Bloemfontein conference between (Lord) Milner [q.v.] and President Kruger, 1899; advocated conciliation after war, 1903; favoured federation in South Africa.

HOFMEYR, JAN HENDRIK (1894-1948), South African statesman; nephew of J. H. Hofmeyr [q.v.]; educated at South African College School; honours in classics (1909), mathematics (1910), Cape University; awarded Rhodes scholarship, 1910; first class, *lit. hum.*, Balliol College, Oxford, 1916; professor of classics (1917-24), principal (1919-24), vice-chancellor (1926-30), and chancellor (1938-48), university of the Witwatersrand; administrator of Transvaal, 1924-9; MP, Johannesburg

North, 1929-48; minister of interior, public health, and education, 1933-6; mines, education, and labour and social welfare, 1936-8; finance and education, 1939-48; mines and education, and deputy prime minister, 1948; upheld Cape liberal tradition on native question; resigned from cabinet (1938) over appointment of unsuitable representative of native affairs; from united party caucus (1939) over Asiatics bill; PC, 1945.

HOGARTH, DAVID GEORGE (1862-1927), scholar and traveller; BA, Magdalen College, Oxford, 1885; Craven travelling fellow, 1886; travelled and excavated in Asia Minor, Cyprus, and Egypt, 1887-96; fellow of Magdalen, 1886; director of British School of Archaeology, Athens, 1897-1900; joined (Sir) Arthur Evans [q.v.] in beginning of excavations at Knossos, 1900; keeper of Ashmolean Museum, Oxford, 1908-27; developed museum, especially in departments of Cretan and Hittite archaeology; commander, Royal Naval Volunteer Reserve, 1915-19; director of Arab Bureau, Cairo, 1916; FBA, 1905; CMG, 1918; works include, *A Wandering Scholar in the Levant* (1896), *The Nearer East* (1902), *The Penetration of Arabia* (1904).

HOGG, DOUGLAS McGAREL, first VISCOUNT HAILSHAM (1872-1950), statesman and lord chancellor; son of Quintin Hogg [q.v.]; educated at Eton; entered family sugar firm; interested himself in Polytechnic; called to bar (Lincoln's Inn), 1902; KC, 1917; bencher, 1920; attorney-general to Prince of Wales, 1920-2; conservative MP, St. Marylebone, 1922-8; attorney-general 1922-4, 1924-8; lord chancellor, 1928-9, 1935-8; leader of opposition in House of Lords, 1930-1; secretary of state for war and leader of House of Lords, 1931-5; a British delegate, Ottawa conference (1932) and world economic conference (1933); lord president of the Council, 1938; knighted and PC, 1922; baron, 1928; viscount, 1929.

HOGG, QUINTIN (1845-1903), philanthropist; son of Sir James Weir Hogg [q.v.]; educated at Eton; partner in London firm of Hogg, Curtis, and Campbell, sugar merchants; started ragged school for boys at Charing Cross, 1864-5; purchased Royal Polytechnic Institution, Regent Street, and opened it for athletic, intellectual, spiritual, and social recreation, 1882; opened day school there and organized holiday tours (1886) and a labour bureau (1891); his success led to spread of polytechnic movement in London; alderman of London County Council, 1889-94; published *The Story of Peter* (religious addresses) (1900).

HOLDEN, CHARLES HENRY (1875-1960), architect; apprenticed in Manchester and studied at art school and technical college; joined C. R. Ashbee [q.v.] in London, 1897; assistant to Percy Adams, 1899, partner, 1907; works include buildings for Law Society, Chancery Lane; British Medical Association (Rhodesia House), Strand; Royal Northern Hospital; London University; Bristol public library and royal infirmary; for Frank Pick [q.v.] designed 55 Broadway and many tube stations; his abstract architectural composition, austere and simple, provided for sculptures by Eric Gill, Sir Jacob Epstein [qq.v.], Henry Moore, etc.; prepared plan for Canterbury and, with (Lord) Holford for City of London; ARIBA, 1906; FRIBA, 1921; royal medallist, 1936.

HOLDEN, HENRY SMITH (1887-1963), academic botanist and forensic scientist; educated at Castleton school and Victoria University, Manchester; assistant lecturer, University College, Nottingham; M.Sc. Manchester, 1911; bacteriologist, Royal Naval Hospital, Plymouth, 1916-19; professor of botany, Nottingham, 1932-6; D.Sc., 1921; fellow, Linnean Society, 1910; FRS Edinburgh, 1927; director, forensic science laboratory under Home Office, 1936-46; director, Metropolitan Police Laboratory, 1946-51; provided scientific evidence in number of notorious criminal cases; forensic science adviser to Home Office, 1951-8; CBE, 1958; published many papers on botanical subjects, mainly in *Annals of Botany* and the *Journal of the Linnean Society*.

HOLDEN, LUTHER (1815-1905), surgeon; studied at St. Bartholomew's Hospital, Berlin, and Paris; MRCS, 1838; FRCS, 1844; president, 1878; Hunterian orator, 1881; demonstrator (1846), lecturer on surgical anatomy (1859-71), surgeon (1865) at St. Bartholomew's Hospital; on retirement (1881) spent much time in foreign travel; good linguist, classic, and sportsman; primarily interested in anatomical study of surgery; published *Human Osteology* (2 vols., 1855) and other medical works.

HOLDER, SIR FREDERICK WILLIAM (1850-1909), first speaker of the Australian Commonwealth house of representatives, 1901-9; born in South Australia; editor and proprietor of *Burra Record*; mayor of Burra, 1886-90; member of legislative assembly of South Australia from 1887; treasurer of colony, 1889-90, 1894-9; premier, 1892 and 1899; commissioner of public works, 1893; KCMG, 1902.

HOLDERNESS, SIR THOMAS WILLIAM, first baronet (1849-1924), Indian civil servant;

born in New Brunswick; passed into Indian civil service, 1870; proceeded to India, 1872; employed as secretary in board of revenue, etc., at Allahabad, 1876; under-secretary to government of India, revenue department, 1881-5; director of land records and agriculture, 1888; assisted in famine relief operations, 1896-7; secretary to government of India, revenue and agricultural department, 1898-1901; secretary, revenue, statistics, and commerce department, India Office, 1901-12; permanent under-secretary of state, India Office, 1912-19; baronet, 1920.

HOLDICH, SIR THOMAS HUNGERFORD (1843-1929), Anglo-Indian frontier surveyor; entered Royal Engineers, 1862; sent to India, 1865; temporary assistant surveyor, Bhutan expedition, 1865-6; subsequently appointed permanently to survey department; began long connection with North-West Frontier as survey officer with Southern Afghanistan field force, 1878; served on Russo-Afghan boundary commission, 1884-6; superintendent of frontier surveys, 1892-8; member of Pamirs commission, 1895; commissioner for Perso-Baluch boundary, 1896; for Argentine-Chile boundary, 1902-3; KCMG, 1902.

HOLDSWORTH, SIR WILLIAM SEARLE (1871-1944), lawyer; educated at Dulwich and New College, Oxford; first class, history, 1893, jurisprudence, 1894; called to bar (Lincoln's Inn), 1896; bencher, 1924; fellow, St. John's College, Oxford, 1897; All Souls reader in English law, 1910; Vinerian professor, 1922-44; member, Indian States inquiry committee (1928) and committee on ministers' powers (1929-32); publications include *Sources and Literature of English Law* (1925), *Some Lessons from our Legal History* (1928), and *History of English Law* (13 vols., 1903-52); FBA, 1922; knighted, 1929; OM, 1943.

HOLE, SAMUEL REYNOLDS (1819-1904), dean of Rochester and author; BA, Brasenose College, Oxford, 1844; MA, 1878; published *Hints to Freshmen* (1847); vicar of Caunton, 1850-87; prebendary of Lincoln, 1875; enthusiastic huntsman, sportsman, and gardener; close friend of John Leech [q.v.] from 1858, who introduced him to W. M. Thackeray [q.v.]; contributed to *Punch*; a successful rose-grower and organizer of national show; his *Book about Roses* (1869) popularized rose growing; a popular preacher and platform orator; dean of Rochester and DD, Lambeth, 1887; lectured in United States, 1894; a humorous and charming letter-writer; *Letters* edited with memoir by G. A. B. Dewar, 1907; published *Memories* (1892), *More Memories* (1894), and *Then and Now* (1901); wrote hymns, sermons, and addresses.

HOLIDAY, HENRY (1839-1927), painter and worker in stained glass; his best-known painting, 'Dante and Beatrice', exhibited 1883; worked for Powell Glass Works, Whitefriars, as designer of cartoons for stained glass, from 1863; examples of his work to be seen all over England, in Europe and in the United States.

HOLLAMS, SIR JOHN (1820-1910), solicitor; admitted solicitor, 1844; president of Law Society, 1878-9; knighted, 1902; published *Jottings of an Old Solicitor* (1906).

HOLLAND, SIR EARDLEY LANCELOT (1879-1967), obstetrician; educated at Merchiston Castle, Edinburgh, and King's College Hospital, London; qualified, 1903; MB, BS, London, FRCS, 1905; held number of appointments in London hospitals and worked in Berlin; obstetric registrar and tutor, King's College Hospital, 1907; MD; MRCP, 1908; appointed obstetric and gynaecological surgeon, London Hospital, 1916-46, but only took up work in 1919; meanwhile, served in France with RAMC; researched into causes of stillbirth, 1922; held number of public appointments concerned with gynaecology; adviser to Ministry of Health, 1937-40; president, Royal College of Obstetricians and Gynaecologists, 1943-6; author of a *Manual of Obstetrics*; edited (with Aleck Bourne) two volumes of *British Obstetric and Gynaecological Practice* (1955); edited the *Journal of Obstetrics and Gynaecology of the British Empire*; made special study of pregnancy of Princess Charlotte [q.v.] who died in childbirth in 1817; hon. degrees and other academic honours; knighted, 1947.

HOLLAND, SIR (EDWARD) MILNER (1902-1969), lawyer; educated at Charterhouse and Hertford College, Oxford; BCL, 1927; called to bar (Inner Temple), 1927; substantial practice, mainly in Chancery division; assistant reader in equity, Council of Legal Education, 1931; reader, 1935; served in RASC, 1939-45; deputy director, Personal Services, War Office, 1943; CBE, 1945; KC, 1948; bencher, Lincoln's Inn, 1953; attorney-general, Duchy of Lancaster, 1951; chairman, General Council of the bar, 1957-8, and 1962-3; member of Council on Tribunals, 1958-62; vice-chairman, Inns of Court Executive Council, 1962; chairman, London Rented Housing Survey, 1963; knighted, 1959; KCVO, 1965.

HOLLAND, HENRY SCOTT (1847-1918), theologian and preacher; BA, Balliol College, Oxford; a senior student of Christ Church,

1870-84; canon of St. Paul's, 1884-1911; position became identified with social and economic questions; helped to found Christian Social Union, and edited *Commonwealth*, 1895-1912; contributed to *Lux Mundi*, 1889; regius professor of divinity at Oxford, 1911-18; raised standard required for divinity degrees; published little except sermons and articles, largely owing to indifferent health; liberal in politics and theology, but high churchman.

HOLLAND, SIR HENRY THURSTAN, second baronet, and first VISCOUNT KNUTSFORD (1825-1914); son of Sir Henry Holland [q.v.], first baronet; BA, Trinity College, Cambridge; called to bar (Inner Temple), 1849; assistant under-secretary for colonies, 1870-4; succeeded father, 1873; conservative MP, Midhurst, 1874-85; PC, 1885; secretary of state for colonies, 1888-92; baron and GCMG, 1888; viscount, 1895.

HOLLAND, SIR HENRY TRISTRAM (1875-1965), eye surgeon, missionary, and philanthropist; grandson of Canon Henry Baker Tristram [q.v.]; educated at Loretto School and Edinburgh University Medical School; MB; Ch.B., 1899; joined Punjab Mission of Church Missionary Society, 1900-48; worked in CMS Hospital, Quetta; established reputation for cataract surgery; FRCSE, 1907; performed over 60,000 operations for cataract; Quetta Hospital destroyed in earthquake, 1935; helped to raise funds for rebuilding; founder-member, Royal Commonwealth Society for the Blind; secretary, CMS Punjab Medical Executive Committee; Kaiser-i-Hind silver medal, 1910; gold medal, 1925; bar, 1931; CIE, 1929; knighted, 1936; hon. member, Ophthalmology Section, Royal Society of Medicine; vice-president, Pakistan Society; published *Frontier Doctor* (1958).

HOLLAND, SIR SIDNEY GEORGE (1893-1961), prime minister of New Zealand; educated at West Christchurch district high school; employed in hardware firm at age of fifteen; enlisted, 1915-17; formed Midland Engineering Co., with his brother; succeeded his father as conservative MP, Christchurch North, 1935; leader of national party, 1940; prime minister, 1949-57; and minister of finance, 1949-54; middle-of-the-road conservative government; joined Australia and United States in ANZUS treaty, 1951; joined Manila Pact, 1954; devoted to British Empire and the United States; PC, 1950; CH, 1951; GCB, 1957.

HOLLAND, SIR SYDNEY GEORGE, third baronet, and second VISCOUNT KNUTSFORD (1855-1931), hospital administrator and re-

former; son of first Viscount Knutsford [q.v.]; succeeded father, 1914; educated at Wellington and Trinity Hall, Cambridge; called to bar (Inner Temple), 1879; as a dock company director and chairman of Poplar Hospital (1891-6) put both concerns on their feet; chairman, London Hospital, 1896-1931; raised £5 million and transformed standard of nursing and administration.

HOLLAND, SIR THOMAS ERSKINE (1835-1926), jurist; son of Revd T. A. Holland [q.v.]; BA, Magdalen College, Oxford, 1858; called to bar (Lincoln's Inn), 1863; Chichele professor of international law and diplomacy, Oxford, 1874-1910; fellow of All Souls, 1875-1926; KC, 1901; FBA, 1902; knighted, 1917; works include edition of text of *De Jure Belli* of Alberico Gentili (q.v., 1917), *Letters to 'The Times' on War and Neutrality* (1909, 1921), *The Elements of Jurisprudence* (1880, 13th edn., 1924); this, his most important book, belongs to school of English analytical jurisprudence founded by John Austin [q.v.].

HOLLAND, SIR THOMAS HENRY (1868-1947), geologist and educational administrator; studied at South Kensington and Owens College, Manchester; assistant superintendent, Geological Survey of India, 1890; director, 1903-10; reorganized and extended its work; professor of geology, Manchester, 1910-18; president, Indian industrial commission, 1916; munitions board, 1917; member, viceroy's executive council, 1920-1; rector, Imperial College of Science and Technology, 1922-9; principal and vice-chancellor, Edinburgh University, 1929-44; negotiated fusion of divinity faculty with New College of Church of Scotland; president, British Association, 1929; FRS, 1904; KCIE, 1908; KCSI, 1918.

HOLLINGSHEAD, JOHN (1827-1904), journalist and theatrical manager; after some commercial travelling, became joint-editor with W. Moy Thomas [q.v.] of *Weekly Mail*, 1856; joined staff of *Household Words*, 1857; wrote *The Birthplace of Podgers*, 1858; his contributions to press were republished in several volumes, 1859-62; one of first contributors to *Cornhill Magazine*, 1859; dramatic critic to *Daily News*, 1863-8; occasional contributor to *Punch*; took part in such public movements as reform of entertainment licensing regulations and of copyright law; stage director of Alhambra, 1865-8; first manager of Gaiety Theatre, 1868-88; inaugurated matinées; produced mainly burlesque, but also operas and serious drama; introduced Ibsen to English public in *Pillars of Society*, 1880; brought complete Comédie Française company, including Sarah Bernhardt and the Coquelins, to London, 1879; also managed for short time Opera Comique, London; lost fortune in theatrical speculation; published *My Lifetime* (2 vols., 1895) and *Gaiety Chronicles* (1898).

HOLLINGWORTH, SYDNEY EWART (1899-1966), geologist; educated at Northampton School and Clare College, Cambridge; first class, parts i and ii, natural sciences tripos, 1920-1; at Cambridge came under influence of J. E. Marr and Alfred Hacker [qq.v.]; joined staff of Geological Survey of Great Britain, 1921; assigned to unit in Cumberland; London D.Sc., 1931; transferred to West Midlands, 1935; with J. H. Taylor [q.v.], worked on Jurassic ironstones, chief domestic source of iron ore for expansion of steel industry, 1939-45; succeeded W. B. R. King [q.v.] as Yates-Goldsmid professor of geology, University College, London, 1945-66; became leading expert on 'porphyry'-type copper deposits in the Andes; fellow, Geological Society of London, 1922; secretary, 1949-56; president, 1960-2; hon. member, Geologist's Association, 1964; consultant to Metropolitan Water Board.

HOLLIS, SIR LESLIE CHASEMORE (1897-1963), general; educated at St. Lawrence College, Ramsgate; commissioned in Royal Marine Light Infantry, 1915; posted to *Duke of Edinburgh* in first cruiser squadron, commanded by Sir Robert Arbuthnot [q.v.]; at battle of Jutland, 1916; served in various ships, 1921-7; passed naval staff course, 1928; intelligence officer on staff of commander-in-chief, Africa Station; attached to plans division, Admiralty, 1932-6; secretary, Joint Planning Sub-committee under Sir Maurice (later Lord) Hankey, and Hastings (later Lord) Ismay [qq.v.], 1936-46; secretary, Chiefs of Staffs Committee, 1938; deputy secretary, Committee of Imperial Defence, 1940-6; acting major-general, 1943; succeeded Ismay as deputy military secretary to the cabinet, 1946-9; commandant general, Royal Marine Corps, 1949-52; CBE; CB; KBE, 1946; full general, and KCB, 1951; published biography of Prince Philip, Duke of Edinburgh, *The Captain General* (1961), *One Marine's Tale* (1956), and *War at the Top* (1959).

HOLLOWELL, JAMES HIRST (1851-1909), advocate of unsectarian education; early became temperance agent and lecturer; joined Congregational ministry, 1875, serving in London, Nottingham, and Rochdale; founded Northern Counties Education League for unsectarian education; organized 'passive resistance movement', 1903; an untiring pamphleteer.

HOLMAN HUNT, WILLIAM (1827-1910), painter. [See HUNT.]

HOLME, CHARLES (1848-1923), founder and editor of *The Studio* magazine; in woollen business, 1871-92; became interested in art; founded *The Studio*, a magazine of fine and applied art, 1893; primary aim not commercial, but illustration and furtherance of good design.

HOLMES, ARTHUR (1890-1965), geologist; educated at Gateshead high school and Imperial College, London; B.Sc., 1909; ARCS, 1910; DIC; contracted malaria and blackwater fever in East Africa; demonstrator in geology, Imperial College, 1912-20; research into radioactivity, petrogenesis and physical geology; chief geologist, oil prospecting company, Burma, 1921-4; reader in geology, Durham; professor, 1924; regius professor, Edinburgh, 1943-56; professor emeritus, 1956; made important contributions to geological implications of radio-activity and the origin of alkali-rich igneous rocks; FRS, 1942; hon. LLD Edinburgh, 1960; many other academic honours; publications include *The Age of the Earth* (1913), *The Nomenclature of Petrology* (1920), *Petrographic Methods and Calculations* (1921), and *Principles of Physical Geology* (1944).

HOLMES, AUGUSTA (MARY ANNE) (1847-1903), composer; born in Paris; pupil of César Franck; published compositions from 1862 onwards; chief were 'In Exitu Israel' (1874), 'Irlande' (1882, a symphonic poem), 'Ode Triomphale' (1889), *La Montagne noire* (opera); became Roman Catholic, 1902; died at Versailles.

HOLMES, SIR CHARLES JOHN (1868-1936), landscape painter and art critic; educated at Eton and Brasenose College, Oxford; entered publishing; co-editor, *Burlington Magazine*, 1903-9; Slade professor, Oxford, 1904-10; director, National Portrait Gallery (1909-16), National Gallery (1916-28); organized photograph and publications departments; knighted, 1921; KCVO, 1928; a self-taught painter of mountain scenes and industrial subjects belonging to no school but his own; publications include *Constable and his Influence on Landscape Painting* (1902).

HOLMES, SIR GORDON MORGAN (1876-1965), neurologist; born in Dublin; educated at Dundalk Educational Institute and Trinity College, Dublin; BA, 1897; qualified in medicine, 1898; studied neuro-anatomy at Frankfurt; house-physician, National Hospital for Nervous Diseases, Queen Square, London,

1901; MD, 1903; FRCP, 1914; pathologist and director of research, 1904; honorary physician, 1909; worked with (Sir) Henry Head [q.v.] on clinical neurophysiology at London Hospital, 1911; consulting neurologist to British Army, 1914; CMG, 1917; CBE, 1919; succeeded Head as editor of *Brain*, 1922-37; FRS, 1933; made notable contributions to the physiology of the cerebral cortex, thalamus, and cerebellum; retired as consulting physician from National Hospital, 1941; knighted, 1951; received many academic honours; publications include *Introduction to Clinical Neurology* (1946).

HOLMES, SIR RICHARD RIVINGTON (1835-1911), librarian of Windsor Castle, 1870-1906; son of John Holmes [q.v.]; assistant at British Museum, 1854; archaeologist to Abyssinian expedition, 1868; purchased Abyssinian MSS for British Museum; rearranged and augmented the royal collections; KCVO, 1905; a delicate pen draughtsman, water-colour artist, stained-glass designer, and designer of bookbindings; FSA, 1860; compiled lives of Queen Victoria (1897), King Edward VII (1910), and works on Windsor.

HOLMES, THOMAS (1846-1918), policecourt missionary and philanthropist; ironmoulder, 1858-79; police-court missionary, Lambeth (1885-9), North London (1889-1905); founded Home Workers' Aid Association to redress conditions of sweated female labour, 1904; secretary to Howard Association for prison and criminal law reform, 1905-15; wrote on problems connected with crime.

HOLMES, THOMAS RICE EDWARD (1855-1933), historian and classical scholar; educated at Merchant Taylors' School and Christ Church, Oxford; master, St. Paul's School, 1886-1909; wrote *History of the Indian Mutiny* (1883) and many works on Julius Caesar.

HOLMES, TIMOTHY (1825-1907), surgeon; BA, Pembroke College, Cambridge (forty-second wrangler and twelfth classic), 1847; hon. fellow, 1900; M.Ch., 1900; FRCS, 1853; housesurgeon at St. George's Hospital; surgeon, 1867-87; Hunterian professor of surgery at Royal College of Surgeons, 1872; president of Royal Medical Society, 1900; edited *A System of Surgery* (4 vols., 1860-4); wrote life of Sir Benjamin Brodie (1898).

HOLMES, SIR VALENTINE (1888-1956), lawyer; educated at Charterhouse and Trinity College, Dublin; senior moderator, classics, 1911; called to bar (Inner Temple), 1913; bencher, 1935; joined (Sir) Leslie Scott [q.v.]; moved to own chambers, 1929, specializing in libel suits; junior counsel to Treasury in common-law matters, 1935-45; KC, 1945; retired, 1949; legal consultant, Shell Oil group, 1950-6; knighted, 1946.

HOLMYARD, ERIC JOHN (1891-1959), teacher, historian, and interpreter of science; educated at Sexey's School, Bruton, and Sidney Sussex College, Cambridge; first class, parts i and ii, natural sciences tripos (1910-12); head of science department, Clifton College, 1919-40; for ICI first editor, *Endeavour*, 1940-54, and editor (with C. Singer, q.v.), *History of Technology* (5 vols., 1954-8); contributed to study of alchemy.

HOLROYD, SIR CHARLES (1861-1917), painter-etcher and director of National Gallery, London; studied art under Alphonse Legros [q.v.] at Slade School, London; fellow of Society of Painter-Etchers, 1885; travelled in Italy, 1889-91 and 1894-7; exhibited at Royal Academy, but chiefly excelled as etcher of figure subjects, landscapes (English and Italian), and portraits; first keeper of Tate Gallery, 1897-1906; knighted, 1903; director of National Gallery, 1906-16; wrote *Michael Angelo Buonarrotti* (1903).

HOLROYD, HENRY NORTH, third EARL OF SHEFFIELD (1832-1909), patron of cricket; grandson of first earl [q.v.]; in diplomatic service, 1852-6; conservative MP for East Sussex, 1857-65; president of Sussex county cricket club from 1879; arranged first-class cricket matches at Sheffield Park; took English team to Australia, 1891-2; sold his Gibbon MSS to British Museum, 1895; died at Beaulieu.

HOLST, GUSTAV THEODORE (1874-1934), composer, whose original name was GUSTAVUS THEODORE VON HOLST; born in Cheltenham of Swedish ancestry; educated at the grammar school; studied composition at Royal College of Music; influenced by Bach and Wagner, later by English folk-song and the Tudor composers; trombone player with Carl Rosa Company and Scottish Orchestra; director of music, St. Paul's Girls' School (1905-34) and Morley College; professor of composition, Royal College of Music (1919-24), University College, Reading (1919-23); compositions of incomparable sureness of touch and clarity of texture; works include 'St. Paul's' suite for strings (1913), 'The Planets' (1914-17), 'The Hymn of Jesus' (1917), 'Ode to Death' (1919), operas *The Perfect Fool* (1922) and *At the Boar's Head* (1925), the 'Keats' choral symphony (1925), and 'Egdon Heath' (1927).

HOLYOAKE, GEORGE JACOB (1817-1906), co-operator and secularist; at first a tinsmith and whitesmith of Birmingham; joined Birmingham reform league, 1831; became a Chartist, 1832; joined Owenites, 1838; present at Birmingham Chartist riots, 1839; minister to Owenites at Worcester, 1840; became rationalist; edited *Oracle of Reason*, 1841; sentenced at Gloucester to six months' imprisonment for blasphemy, 1842; published account of trial, 1851; went to London, 1843, advocating freedom of thought and co-operation and co-partnership; edited *Reasoner*, 1846, and *Leader*, 1850; invented the term 'secularism', which he explained in pamphlet, 1854; advocated abolition of paper duties, extension of the franchise, and electoral reform; twice visited Canada and United States to study economic conditions; started *The Secular Review*, 1876; chief publications include *A History of Co-operation* (1875-7), *Self-help by the People* (1855), lives of Tom Paine (1851), Robert Owen (1859), and John Stuart Mill (1873), besides his autobiographical books: *Sixty Years of an Agitator's Life* (2 vols., 1892) and *Bygones worth Remembering* (1905).

HONE, EVIE (1894-1955), artist; born and died in Dublin; descended from brother of Nathaniel Hone [q.v.]; studied in London and France; worked with Albert Gleizes, 1921-31; exhibited abstract painting in Paris, London, and Dublin; turned to stained glass through interest in Rouault; became known for unique richly colourful work in churches and public places, including large window for Eton College chapel.

HOOD, ARTHUR WILLIAM ACLAND, first BARON HOOD (1824-1901), admiral; entered navy, 1836; commander, 1854; served in China, 1857-8, and in North America station, 1862-6; in charge of Royal Naval College, Portsmouth, 1866; director of naval ordnance, 1869-74; CB, 1871; commanded Channel fleet, 1878-82; first sea lord of Admiralty, 1885; admiral and KCB, 1885; GCB, 1889; baron, 1892.

HOOD, SIR HORACE LAMBERT ALEXANDER (1870-1916), rear-admiral; entered navy, 1882; served in Nile campaign, 1897-8; DSO for attack on dervishes at Illig, Somaliland, 1904; commanded Naval College, Osborne, 1910-13; rear-admiral, 1913; naval secretary to first lord of Admiralty, 1914; commanded third battle squadron of Grand Fleet, with flag on board *Invincible*, 1915; blown up at battle of Jutland, 1916; posthumous KCB.

HOOK, JAMES CLARKE (1819-1907), painter; studied at Royal Academy Schools;

exhibited at Academy from 1839; ARA, 1850; RA, 1860; subjects mainly old-fashioned genre of historical anecdote until 1854, when he visited Clovelly; thenceforth treated English coast scenery; praised by John Ruskin [q.v.]; best-known works were 'Pamphilus Relating his Story' (1844), 'Luff, Boy!' (1859), 'The Samphire Gatherer' (1875), 'The Stream' (1885, now in Tate Gallery); also painted portraits.

HOOKE, SAMUEL HENRY (1874-1968), biblical scholar and oriental linguist; educated at Wirksworth and St. Mark's School, Windsor, and Jesus College, Oxford; first class, theology, 1910; second class, oriental languages, 1912; Flavelle associate professor of oriental languages and literature, Victoria College, Toronto, 1913-25; Rockefeller fellow in anthropology studying oriental languages in London, 1926; Samuel Davidson professor of Old Testament Studies, London, 1930-42; master at Blundell's School; examining chaplain to bishop of Coventry; visiting professor, Ghana, 1958; president Folk Lore Society, 1936-7; fellow, Society of Antiquaries, 1937; president, Society for Old Testament Study, 1951; hon. DD Glasgow, 1950; hon. D.Th. Uppsala, 1957; hon. fellow, Jesus College, 1964; editor *Palestine Exploration Quarterly* for twenty-three years; edited *Myth and Ritual* (1933) and *The Labyrinth* (1935); wrote numerous books on Christian faith and the Old Testament.

HOOKER, SIR JOSEPH DALTON (1817-1911), botanist and traveller; son of Sir William Jackson Hooker [q.v.]; graduated MD, Glasgow, 1839; inherited from father a passion for botanical research; naturalist in Sir J. C. Ross's Antarctic expedition, 1839-43; published botanical results in six volumes, 1844-60; became intimate with Charles Darwin [q.v.], and collaborated with him in researches into the origin of species from 1843; greatly advanced the knowledge of geographical distribution of species; rejected theory of multiple origins in *Flora Antarctica*, 1844-7; appointed botanist to Geological Survey, 1845; FRS, 1847; explored Sikkim, part of Eastern Nepal, and passes into Tibet, 1848-9, making observations in geology and meteorology; travelled in Eastern Bengal, 1850-1, collecting plants representing some 700 species; published *Himalayan Journals* (1854) and *Illustrations of Sikkim-Himalayan Plants* (1855); appointed assistant director of Kew, 1855; published in 1860 his *Introductory Essay on the Flora of Tasmania*, in which he adopted the newly propounded Darwin-Wallace theory that species are derivative and mutable; went on scientific expedition to Syria and examined cedar grove on Mt. Lebanon, 1860; began with George Bentham [q.v.] the *Genera Plantarum*,

1862 (last part issued, 1883); succeeded father as director at Kew, 1865; published *Handbook of the New Zealand Flora* (1867) and *Student's Flora of the British Islands* (1870); explored with John Ball (1818-89, q.v.) the Great Atlas in Morocco; reached Tagherot Pass; president of Royal Society, 1873-8; visited the rocky mountains of Colorado and Utah, 1877; published report of botanical researches, 1881; published *Flora of British India* (7 vols., 1883-97) in which he described nearly 17,000 species; retired from Kew, 1885; edited *Journal* of Sir Joseph Banks [q.v.], 1896, and wrote on Indian Flora in *Imperial Gazetteer of India*, 1904; CB, 1869; KCSI, 1877; GCSI, 1897; OM, 1907; received many honours from English and foreign scientific societies and universities; buried at Kew.

HOOPER, SIR FREDERIC COLLINS, first baronet (1892-1963), industrialist; educated at Sexey's School, Bruton, and University College, London; commissioned in Dorset Regiment; served on western front, 1914-18; recruited to Lewis's of Liverpool by F. J. Marquis (later the Earl of Woolton, q.v.), 1922; joint managing director, 1940-2; set up Political Research Centre for conservative party, 1942-3; director of business training, Ministry of Labour, 1945-6; managing director, Schweppes mineral water company, 1948-63; built up Schweppes international reputation; published *Management Survey* (1948); member of number of committees and advisory boards; adviser on recruiting to minister of defence, 1960; knighted, 1956; baronet, 1962.

HOPE, ANTHONY (pseudonym) (1863-1933), novelist. [See HAWKINS, SIR ANTHONY HOPE.]

HOPE, JAMES FITZALAN, first BARON RANKEILLOUR (1870-1949), parliamentarian; son of J. R. Hope-Scott and grandson of fourteenth Duke of Norfolk [qq.v.]; educated at Oratory School and Christ Church, Oxford; conservative MP, Brightside division (1900-6), Central division (1908-29) of Sheffield; junior lord of Treasury, 1916-19; parliamentary and financial secretary to Ministry of Munitions, 1919-21; chairman of ways and means and deputy speaker, 1921-4, 1924-9; prominent Roman Catholic layman; PC, 1922; baron, 1932.

HOPE, JOHN ADRIAN LOUIS, seventh EARL OF HOPETOUN and first MARQUESS OF LINLITHGOW (1860-1908), first governor-general of the Commonwealth of Australia; succeeded to earldom, 1873; conservative lord-in-waiting to Queen Victoria, 1885-9; governor of Victoria and GCMG, 1889; paymaster-general, 1895-8; lord chamberlain, 1898; president of Institution of Naval Architects, 1895-1900; first governor-general of Commonwealth of Australia, 1900-2; KT and GCVO, 1900; resigned owing to insufficient salary, May 1902; created Marquess of Linlithgow, Oct. 1902; secretary of state for Scotland, 1905; sold Rosyth for naval base to government, 1903; died at Pau; a keen sportsman and huntsman.

HOPE, LAURENCE (pseudonym) (1865-1904), poetess. [See NICOLSON, ADELA FLORENCE.]

HOPE, VICTOR ALEXANDER JOHN, second MARQUESS OF LINLITHGOW (1887-1952), viceroy of India; succeeded father, first Marquess [q.v.], 1908; educated at Eton; civil lord of Admiralty, 1922-4; president, Navy League, 1924-31; chairman, royal commission on Indian agriculture, 1926-8; joint select committee on Indian constitutional reform, 1933-4; viceroy of India, 1936-43; introduced provincial autonomy, persuading Congress to take office, 1937; ensured stability of North-West Frontier by military operations, 1936-8; organized India for war despite Congress opposition and withdrawal of provincial ministries and country's rejection of Cripps's [q.v.] constitutional proposals, 1942; expanded his Council to 15 of whom 10 were Indians; established National Defence Council and department of supply and initiated Eastern Group Supply Council; chairman, Midland Bank, 1944-52; KT, 1928; GCIE, 1929; PC, 1935; GCSI, 1936; KG, 1943.

HOPE, SIR WILLIAM HENRY ST. JOHN (1854-1919), antiquary; BA, Peterhouse, Cambridge; carried out excavations at Dale Abbey, Repton Priory, Lewes Priory, and Alnwick Abbey; FSA, 1883; assistant secretary to Society of Antiquaries, 1885-1910; knighted, 1914; writings include numerous ecclesiological and heraldic works and over 200 papers.

HOPETOUN, seventh EARL OF (1860-1908), first governor-general of the Commonwealth of Australia. [See HOPE, JOHN ADRIAN LOUIS.]

HOPKINS, EDWARD JOHN (1818-1901), organist; organist at Temple Church, London, 1843-98; hon. Mus.Doc., Toronto, 1886; compiled *Temple Church Choral Service Book* (1867) and *Temple Psalter* (1883); wrote *The Organ, its History and Construction* (1855), a standard work.

HOPKINS, SIR FREDERICK GOWLAND (1861-1947), biochemist; educated at City of London School; articled to consulting analyst; studied chemistry at South Kensington and University College; assistant to (Sir) Thomas

Stevenson [q.v.]; entered Guy's Hospital, 1888; B.Sc., London, 1890; qualified, 1894; assistant in physiology department, 1894-8; lecturer on chemical physiology, Cambridge, 1898; reader, 1902; praelector in biochemistry, Trinity College, 1910-21; professor of biochemistry, 1914; Sir William Dunn professor, 1921-43; in *Journal of Physiology* (1912) published important paper giving precision to ideas about existence of vitamins and methods of exploring them; devoted himself to study of chemistry of intermediary metabolism and establishment of biochemistry as separate discipline; FRS, 1905; shared Nobel prize, 1929; president, Royal Society, 1930-5; British Association, 1933; knighted, 1925; OM, 1935.

HOPKINS, JANE ELLICE (1836-1904), social reformer; daughter of William Hopkins [q.v.]; worked among navvies at Cambridge; removed to Brighton; edited *Life and Letters* of James Hinton [q.v.] (1875); founded White Cross League, 1886; published *Active Service* (1872-4), *The Power of Womanhood* (1899), and *The Story of Life* (1903).

HOPKINS, Sir RICHARD VALENTINE NIND (1880-1955), civil servant; educated at King Edward's School, Birmingham, and Emmanuel College, Cambridge; first class, part i, classical tripos, 1901, and part ii, history tripos, 1902; entered Inland Revenue, 1902; member of Board, 1916; chairman, 1922; transferred to Treasury, controlling finance and supply services, 1927; second secretary, 1932; permanent secretary, 1942-5; withstood challenge to Treasury finance of J. M. (later Lord) Keynes [q.v.] before Macmillan committee, 1930; invited him to Treasury, 1940; CB, 1919; KCB, 1920; GCB, 1941; PC, 1945.

HOPKINSON, Sir ALFRED (1851-1939), lawyer, educationist, and politician; brother of John Hopkinson [q.v.]; educated at Owens College, Manchester, and Lincoln College, Oxford; called to bar (Lincoln's Inn), 1873; QC, 1892; professor of law, Owens College (1875-89), principal (1898-1904); vice-chancellor, Manchester University, 1900-13; unionist MP, Cricklade (1895-8), combined English universities (1926-9); knighted, 1910.

HOPKINSON, BERTRAM (1874-1918), engineer and physicist; son of John Hopkinson, FRS [q.v.]; BA, Trinity College, Cambridge; professor of mechanism and applied mechanics at Cambridge, 1903-18; FRS, 1910; joined Royal Engineers, 1914; carried out investigations for government on explosives and aircraft equipment; CMG, 1917; killed flying.

HOPWOOD, CHARLES HENRY (1829-1904), recorder of Liverpool, 1886-1904; called to bar, 1853; QC, 1874; treasurer of Middle Temple, 1895; edited *Registration Cases, 1863-72* (3 vols., 1868-79); liberal MP for Stockport, 1874-85, and for Middleton division of Lancashire, 1892-5; energetic supporter of principle of personal liberty, of trade-unionism, and adult suffrage; opposed severity of punishment and as recorder inflicted short sentences; founded Romilly Society, 1897; edited *Middle Temple Records* (1904).

HOPWOOD, FRANCIS JOHN STEPHENS, first BARON SOUTHBOROUGH (1860-1947), civil servant; educated at King Edward VI School, Louth; admitted solicitor, 1882; entered Board of Trade, 1885; permanent secretary, 1901-7; member, West Ridgeway committee on Transvaal and Orange River Colony, 1906; permanent under-secretary for colonies, 1907-10; accompanied Prince of Wales to Canada (1908) and Duke of Connaught to South Africa (1910); vice-chairman, Development Commission, 1910-12; civil lord of Admiralty, 1912-17; chairman, war trade committees, 1914-18; investigated Austrian peace proposals, 1917; secretary, Irish convention, 1917-18; chairman, Indian franchise committee (1918-19) and many other committees; KCB, 1901; PC, 1912; baron and GCVO, 1917; KCSI, 1920.

HORDER, PERCY (RICHARD) MORLEY (1870-1944), architect; educated at City of London School; articled to George Devey [q.v.]; FRIBA, 1904; works include many traditional country houses; at Cambridge Cheshunt College, Westcott House, and the National Institute of Agricultural Botany; at Oxford new buildings for Somerville College and the Institute for Research in Agricultural Economics; chemist shops for Lord Trent [q.v.] and Nottingham University College.

HORDER, THOMAS JEEVES, first BARON HORDER (1871-1955), physician; educated at Swindon high school; qualified from St. Bartholomew's Hospital, 1896; MD, 1899; FRCP, 1906; house-physician to Samuel Gee [q.v.]; medical registrar, 1904-11, assistant physician, 1912-21, senior physician, 1921-36, St. Bartholomew's; outstanding clinician of his time, combining bedside observation with laboratory investigation; a rationalist and individualist of organized common sense; patients included four successive British sovereigns; organized Fellowship for Freedom in Medicine, 1948; chairman, Empire Rheumatism Council, 1936-53; medical adviser, London Transport, 1940-55; knighted, 1918; baronet, 1923; baron,

1933; KCVO, 1925; GCVO, 1938; held number of honorary degrees.

HORE-BELISHA, (ISAAC) LESLIE, Baron Hore-Belisha (1893-1957), politician; educated at Clifton and St. John's College, Oxford; served in France and Salonika, 1914-18; president of Union, 1919; called to bar (Inner Temple), 1923; liberal MP, Devonport, 1923-45; organized liberal national party supporting national government, 1931; its chairman until 1940; parliamentary secretary, Board of Trade, 1931-2; financial secretary to Treasury, 1932-4; minister of transport, 1934-7; introduced 'Belisha beacons' and other well-publicized road safety measures; transferred trunk roads to care of state and sponsored plans for new arterial roads; secretary for war, 1937; introduced rapid and extensive reforms covering conditions of service, training, mechanization, tactical reorganization, etc.; his personality and methods aroused hostility; his justified anxiety about defences in France caused friction with Lord Gort [q.v.] and others and brought offer of transfer; resigned, Jan. 1940; minister of national insurance, caretaker government, 1945; thereafter joined conservatives but failed to enter Parliament; PC, 1935; baron, 1954.

HORE-RUTHVEN, ALEXANDER GORE ARKWRIGHT, first Earl of Gowrie (1872-1955), soldier and governor-general of Australia; educated at Eton; served in Nile expeditions, 1898-9; VC; military secretary to Lord Dudley [q.v.] in Ireland and Australia, 1905-10; served in France and Gallipoli, 1914-18; DSO; CB, and CMG; commanded Welsh Guards, 1920-4; 1st Infantry brigade, 1924-8; governor, South Australia, 1928-34, New South Wales, 1935-6; governor-general of Australia, 1936-44; of imperturbable goodwill and common sense; on excellent terms with John Curtin [q.v.]; deputy constable and lieutenant-governor, Windsor Castle, 1945-53; KCMG, 1928; GCMG and Baron Gowrie, 1935; PC, 1937; earl, 1945; president, MCC 1948.

HORNBY, CHARLES HARRY ST. JOHN (1867-1946), printer and connoisseur; educated at Harrow and New College, Oxford; rowed for Oxford and third class, *lit. hum.*, 1890; entered W. H. Smith & Son, 1892; partner, 1894; extended printing business and initiated bookbinding department; from his private Ashendene Press (1895-1935) published forty major works; collected medieval and renaissance manuscripts and printed books.

HORNBY, JAMES JOHN (1826-1909), provost of Eton; son of Admiral Sir Phipps Hornby [q.v.]; educated at Eton and Balliol College,

Oxford; first class classic and fellow of Brasenose, 1849; rowed in Oxford eight, 1849 and 1851; MA, 1851; DD, 1869; principal of Bishop Cosin's Hall, Durham, 1853-64; headmaster of Eton, 1868-84; provost, 1884-1909; pioneer of alpine climbing; hon. DCL, Durham, 1882.

HORNE, HENRY SINCLAIR, Baron Horne (1861-1929), general; joined Royal Artillery, 1880; served in South African war, 1899-1902; inspector, horse and field artillery, 1912; proceeded to France as brigadier-general commanding Royal Artillery, I Army Corps, 1914; took part in retreat from Mons, battles of Marne, Aisne, and first battle of Ypres; commanded second division of I Corps, 1915; accompanied Lord Kitchener [q.v.] to Dardanelles, 1915; commanded XV Corps and took part in battle of Somme, 1916; commanded First Army, 1916; assault on Vimy ridge successfully carried out by First Army, 1917; ridge saved during great German attack, 1918; advance of First Army during 1918 ended with occupation of Mons (Nov.); general, 1919; received thanks of parliament, £30,000, and created baron, 1919; commander-in-chief, Eastern command, 1919-23; retired, 1926.

HORNE, ROBERT STEVENSON, Viscount Horne of Slamannan (1871-1940), lawyer, politician, and man of business; educated at George Watson's College, Edinburgh and Glasgow University; advocate, 1896; KC, 1910; unionist MP, Hillhead division, Glasgow, 1918-37; third civil lord, Admiralty, 1918-19; minister of labour, 1919-20; president, Board of Trade, 1920-1; chancellor of the Exchequer, 1921-2; director, Suez Canal Company, Lloyds Bank, etc.; chairman, Burmah Corporation and (1934-40) Great Western Railway Company; KBE, 1918; PC, 1919; viscount, 1937.

HORNER, ARTHUR LEWIS (1894-1968), miners' leader; left school at eleven to become clerk in railway office; began course of theological training, but abandoned it for politics, 1911; joined independent labour party and became known as agitator; conscientious objector, absconded to Ireland, arrested and imprisoned on return; discharged from army for 'incorrigible conduct', 1919; joined British communist party, 1920; underwent further periods of imprisonment, 1921 and 1932; expelled from South Wales Miners' Federation, 1930; re-entered, 1934; elected president, 1936; political views prevented service on General Council of Trades Union Congress; played leading part in formation of National Union of Mineworkers, 1945; national secretary, 1946-59; received freedom of Merthyr Tydfil, 1959;

published autobiography, *Incorrigible Rebel* (1960).

HORNIMAN, ANNIE ELIZABETH FREDERICKA (1860-1937), pioneer of the modern theatre repertory movement; daughter of F. J. Horniman [q.v.]; studied at Slade School; secretary to W. B. Yeats [q.v.]; subsidized Abbey Theatre, Dublin, 1904-10; founded repertory theatre in Manchester (1907-21) with first-class actors performing wide range of plays, especially by new writers; CH, 1933.

HORNIMAN, FREDERICK JOHN (1835-1906), founder of the Horniman Museum; joined father's tea-packing business; travelled extensively, making natural history collections; built Horniman's Museum at Forest Hill, 1897, and presented it to London County Council, 1901; liberal MP for Falmouth and Penryn boroughs, 1895-1904.

HORRABIN, JAMES FRANCIS (1884-1962), artist, lecturer, cartoonist, and left-wing socialist; educated at Stamford grammar school and Sheffield School of Art; staff artist on *Sheffield Telegraph*, 1906; art editor, *Yorkshire Telegraph and Star*, 1909; worked on London *Daily News* (later, *News Chronicle*), and the *Star*, 1911-60; created characters of Japhet and Happy, and Dot and Carrie; served in Queen's Westminster Rifles, 1917-18; chosen to illustrate H. G. Wells's [q.v.] *Outline of History* (1920); also illustrated textbooks and atlases; an advocate of guild socialism, member of executive committee, National Guilds League, 1915; first editor of *Plebs*; labour MP, Peterborough, 1929-31; succeeded Arthur Creech Jones [q.v.] as chairman, Fabian Colonial Bureau, 1945-50; published various books about his cartoon characters, including *Japhet and Fido* (1922) and *The Japhet Book* (1925).

HORRIDGE, SIR THOMAS GARDNER (1857-1938), judge; solicitor, 1879; called to bar (Middle Temple), 1884; practised in Liverpool; KC, 1901; liberal MP, East Manchester, 1906-10; judge of King's Bench division, 1910-37; sound, dignified, and expeditious; knighted, 1910; PC, 1937.

HORSBRUGH, FLORENCE GERTRUDE, BARONESS HORSBRUGH (1889-1969), politician; educated at Lansdowne House, Edinburgh and St. Hilda's, Folkestone; conservative MP, Dundee, 1931-45; MBE, 1920; CBE, 1939; parliamentary secretary, Ministry of Health, 1939-45; parliamentary secretary, Ministry of Food, 1945; PC, 1945; MP, Moss Side division, Manchester, 1950; minister of education,

1951-4; first woman to hold cabinet post in a conservative government, 1953; GBE, 1954; life peer, 1959; delegate, Council of Europe and Western European Union, 1955-61; hon. LLD, Edinburgh, 1946; hon. fellow, Royal College of Surgeons of Edinburgh, 1946.

HORSLEY, JOHN CALLCOTT (1817-1903), painter; son of William Horsley [q.v.]; studied art at Royal Academy, where he exhibited from 1839; opposed to study of nude model; painted domestic scenes, e.g. 'Holy Communion', 'The Gaoler's Daughter', 'L'Allegro', and 'Il Penseroso'; also painted portraits; ARA, 1855; RA, 1856; treasurer, 1882-97; organized 'Old Masters' exhibitions at Academy, 1875-90; interested in music; friend of Mendelssohn and John Leech [q.v.]; published *Recollections of a Royal Academician* (1903).

HORSLEY, JOHN WILLIAM (1845-1921), philanthropist; chaplain to Clerkenwell prison, 1876-86; vicar of Holy Trinity, Woolwich, 1889; rector of St. Peter's, Walworth, 1894; mayor of Southwark, 1910; works on social questions include *Jottings from Jail* (1887) and *How Criminals are Made and Prevented* (1912).

HORSLEY, SIR VICTOR ALEXANDER HADEN (1857-1916), physiologist and surgeon; son of John Callcott Horsley, RA [q.v.]; educated at University College Hospital; professor-superintendent to Brown Institution (university of London), 1884-90; pursued there three main lines of study—action of thyroid gland, protective treatment against rabies, and localization of function in the brain; FRS, professor of pathology, University College, and surgeon to National Hospital for Paralysed and Epileptic, Queen Square, 1886; knighted, 1902; consultant to Mediterranean expeditionary force, 1915; died at Amarah; one of the foremost surgeons of his day; a prolific writer.

HORTON, SIR MAX KENNEDY (1883-1951), admiral; joined *Britannia*, 1898; entered submarine branch; commanding E.9 sank cruiser *Hela* and destroyer S.116 in first two months of war, 1914; DSO; served in Baltic, 1914-15, 1919-20; in home waters commanded new types of submarines; commander, 1914; captain, 1920; rear-admiral, 1932; commanded first cruiser squadron, Mediterranean, 1935-6; vice-admiral, 1936; commanded Reserve Fleet, bringing it to war readiness, 1937-9; Northern Patrol, 1939-40; flag officer, submarines, 1940-2; successfully deployed submarines in anticipation of German attack on Norway; admiral, 1941; commander-in-chief, Western Approaches, 1942-5; mounted sea-air offensive of very long-range and long-range aircraft and Support Groups of warships to drive U-boats from mid-Atlantic, 1943; a perfectionist criticized for ruthlessness of his training methods; deeply religious man famous for accuracy of his hunches; CB, 1934; KCB, 1939; GCB, 1945; Bath King of Arms, 1946; received freedom of Liverpool and state funeral there.

HORTON, PERCY FREDERICK (1897–1970), artist; educated at Brighton municipal school and Brighton College of Art; imprisoned as conscientious objector, 1914; studied at Central School of Arts and Crafts, London, 1918–20; teacher at Rugby School, 1920–2; won Royal exhibition to Royal College of Art, South Kensington; ARCA, 1924; art master, Bishop's Stortford College, 1924–9; draughtsman, Royal College of Art, 1929–48; accomplished portraitist; master, Ruskin School of Drawing and Fine Art, Oxford, 1949–64; drew and painted many academic portraits, and landscapes, mainly of Sussex; pictures to be seen in the Tate, the National Portrait Gallery, the Ashmolean, and other galleries.

HORTON, ROBERT FORMAN (1855–1934), Congregational divine and theological writer; educated at Tettenhall College, Shrewsbury School, and New College, Oxford; president of Union, 1877; first class, *lit. hum.*, 1878; fellow, 1879–86; ordained, 1884; pastor of Lyndhurst Road Congregational church, Hampstead, 1884–1930; president, National Free Church Council, 1905; great preacher and saintly personality; publications include *The Word of God* (1898) and many devotional pamphlets.

HOSE, CHARLES (1863–1929), civil servant in Sarawak, ethnologist, and naturalist; cadet in Sarawak civil service, 1884; rose to be divisional resident; member of supreme council and judge of supreme court of Sarawak, 1904; established friendly relations with and brought under control warlike natives; retired, 1907; a keen naturalist who enriched zoological and botanical records with many new species, etc., and benefited British national collections; successfully investigated principal cause of disease beri-beri; carried out important ethnographical research among Sarawak tribes; wrote several books on subject; made first reliable map of Sarawak.

HOSIE, SIR ALEXANDER (1853–1925), diplomatist and Chinese explorer; joined Chinese consular service, 1876; sent on special service to Chungking, Szechwan, 1882; undertook travels in the interior; served in Canton, Wenchow, Chefoo, Amoy, Tamsui, Wuhu;

twice in charge of consulate at Newchang; first consul-general at Chingtu, Szechwan, 1903; British delegate at Shanghai international opium commission, 1909; retired, having travelled in twenty-one out of twenty-two Chinese provinces, 1912; knighted, 1907; unrivalled in knowledge and presentation of possibilities of Chinese trade; author of many books on China.

HOSIER, ARTHUR JULIUS (1877–1963), pioneer farmer, engineer, and inventor; son, grandson, and great grandson of farmers; as a lad, worked on the farm before walking to school at Bradford on Avon; with his brother took tenancy of farm at nineteen; invented a side-rake and mechanical milk filter, 1904; worked as engineer, 1904–10; returned to farming with his brother and bought Wexcombe estate, 1910–20; experimented with 'Open-Air Dairying', the 'Hosier System', 1922; designed other equipment, including a portable poultry unit; OBE, 1949; hon. LLD Cambridge, 1951; Methodist lay preacher throughout his life.

HOSKINS, SIR ANTHONY HILEY (1828–1901), admiral; entered navy, 1842; served against Arab slavers in Mozambique; in Kaffir war, 1851–2; in China, 1857–8; commander, 1858; at reduction of Taku forts; commanded in North American, Channel, and Australian waters, 1869–78; CB, 1877; rear-admiral, 1879; lord commissioner of Admiralty, 1880; KCB, 1885; commander-in-chief in Mediterranean, 1889–91; admiral and senior naval lord of Admiralty, 1891; GCB, 1893.

HOSKYNS, SIR EDWYN CLEMENT, thirteenth baronet (1884–1937), divine; educated at Haileybury and Jesus College, Cambridge; ordained, 1908; dean of chapel (1919–37), librarian and president (1929–37), Corpus Christi College, Cambridge; pioneer of critical, evangelical Anglo-Catholicism; works include *The Riddle of the New Testament* (with F. N. Davey, 1931); succeeded father, 1925.

HOTINE, MARTIN (1898–1968), geodesist and photogrammetrist; educated at Southend high school and Royal Military Academy, Woolwich; commissioned in Royal Engineers, 1917; saw active service in Persia, Iraq, and India; studied at Magdalene College, Cambridge; research officer, Air Survey Committee, 1925; attached to geographical section, general staff, War Office, 1927–31; published *Surveying from Air Photographs* (1931); worked in Central Africa, 1931–4; served in Ordnance Survey and initiated retriangulation of Great Britain, 1934–9; deputy director of survey, British Expeditionary Force, 1939; director of survey,

East Africa; director of military survey, 1941-6; responsible for army maps and aeronautical charts for RAF; director, Directorate of Overseas Surveys, 1946-63; on research staff of United States Coast and Geodetic Survey, 1963-8; published *Mathematical Geodesy* (1969); CBE, 1945; CMG, 1949; awarded Founder's medal of Royal Geographical Society, 1947, and received other awards and honours.

HOUGHTON, WILLIAM STANLEY (1881-1913), dramatist; engaged in father's cloth business in Manchester, 1897-1912; from 1900 made play-making and acting his hobby; contributed theatrical notices to *Manchester Guardian*, 1905-13; performed dramatic works include *The Dear Departed* (1908), *Independent Means* (1909), *The Younger Generation* and *The Master of the House* (1910), *Fancy-Free* (1911), and *Hindle Wakes* (1912); influenced by Ibsen; portrayed Lancashire life; reached height of his art in *Hindle Wakes*.

HOULDSWORTH, SIR HUBERT STANLEY, first baronet (1889-1956), chairman of National Coal Board; educated at Heckmondwike grammar school and Leeds University; first class, physics, 1911; lecturer, 1916-26; called to bar (Lincoln's Inn), 1926; KC, 1937; chairman, Midlands coal-selling control committee, 1936-42; regional controller, South and West Yorkshire, 1942-4; controller-general, Ministry of Fuel and Power, 1944-5; chairman, East Midland division, National Coal Board, 1946-51; chairman, National Coal Board, 1951-6; knighted, 1944; baronet, 1956; pro-chancellor, Leeds, 1949-56; hon. LLD Leeds and Nottingham.

HOUSE, (ARTHUR) HUMPHRY (1908-1955), scholar; educated at Repton and Hertford College, Oxford; first class, *lit. hum.*, 1929; professor of English, Presidency College, Calcutta, 1936; lecturer, Calcutta University, 1937-8; William Noble fellow, Liverpool, 1940; director of English studies, Peterhouse, Cambridge, 1947-9; university lecturer in English literature, Oxford, 1948-55; senior research fellow, Wadham, 1950-5; publications include *Note-Books and Papers of Gerard Manley Hopkins* (1937), *The Dickens World* (1941), *Coleridge* (1953), and *Aristotle's Poetics* (1956).

HOUSMAN, ALFRED EDWARD (1859-1936), poet and classical scholar; educated at Bromsgrove School and St. John's College, Oxford; first class, classical moderations (1879); failed to obtain honours in *lit. hum.* through neglecting philosophy and history; worked in Patent Office, 1882-92; professor of Latin,

University College, London (1892-1911), Cambridge (1911-36); published at least one classical article or review yearly (1887-1936), restricting himself mainly to Latin after 1892; dealt with most of chief Latin poets from Lucilius to Juvenal; published four masterly papers on manuscripts of Propertius; edited Ovid's *Ibis* (1894), the five Books of Manilius (with a commentary designed to treat only of 'what Manilius wrote, and what he meant', 1903-30), Juvenal (1905), and Lucan (1926); an intellectually honest and patient classical scholar with a disciplined passion for truth, swift insight, and brilliant power; his lyrics (published in *A Shropshire Lad*, 1896, *Last Poems*, 1922, and *More Poems*, 1936), limited in theme but show great variety of metre and a Horatian felicity of expression; acutely sensitive and very reserved, but when at ease his conversation full of wit and charm; combined declared atheism with an hereditary attachment to high church party; refused almost all honours but accepted honorary fellowship of St. John's College, Oxford, 1911.

HOUSMAN, LAURENCE (1865-1959), writer; brother of A. E. Housman [q.v.] whose biography he wrote (1937); educated at Bromsgrove School; studied art in London; encouraged to write by Kegan Paul [q.v.]; art critic of force and wit, *Manchester Guardian*, 1895-1911; as playwright his unconventionality brought conflict with censor but *Victoria Regina* (Lyric, 1937) immediately successful; publications include *Green Arras* (poems, 1896), *An Englishwoman's Love-Letters* (anonymously, 1900), *Sheepfold* (novel partly based on life of Mrs Girling (q.v.), 1918), *The Little Plays of St. Francis* (1922), *Trimblerigg* (satirical novel, 1924), and autobiography, *The Unexpected Years* (1937); a pioneer feminist, pacifist, and socialist, although too impetuous, insensitive, and muddled to be altogether effective.

HOUSTON, DAME FANNY LUCY (1857-1936), philanthropist and eccentric; born RADMALL; married, secondly, the ninth Lord Byron (died 1917); thirdly, Sir Robert Paterson Houston who bequeathed her (1926) four-fifths of his fortune; contributed widely to charity; purchased *Saturday Review*; a strident, combative patriot; financed successful Schneider trophy team (1931) and flight over Everest (1933); DBE, 1917.

HOWARD, SIR EBENEZER (1850-1928), originator of the garden city movement and founder of Letchworth and Welwyn garden cities; clerk in City of London; stenographer in Chicago, 1872-7; influenced by outcome of teaching of Emerson, Lowell, and Whitman;

shorthand writer to official reporters to Houses of Parliament; carried on shorthand business of his own until 1920; influenced by Edward Bellamy's *Looking Backward*; published *Tomorrow* (1898); set himself to find remedy for unhealthy conditions produced by overcrowding in great cities and for depopulation of countryside; proposed to build new towns called 'garden cities', both residential and industrial, surrounded by rural belt; formed Garden City Association, 1899; estate of Letchworth, Hertfordshire, bought and development begun, 1903; principle that town should own land adopted in modified form; bought, and subsequently developed, Welwyn estate, Hertfordshire, 1919; president of International Garden Cities and Town Planning Association, 1909-28; exercised important influence on town planning; knighted, 1927.

HOWARD, EDMUND BERNARD FITZ-ALAN-, first VISCOUNT FITZALAN OF DERWENT (1855-1947); son of fourteenth and brother of fifteenth Duke of Norfolk [qq.v.]; uncle of Lord Rankeillour [q.v.]; known as Lord Edmund Talbot, 1876-1921; educated at Oratory School; officer in 11th Hussars; served in South Africa, 1899-1902; DSO; conservative MP, Chichester, 1894-1921; junior lord of Treasury, 1905; joint parliamentary secretary, 1915-21; chief conservative whip, 1913-21; viceroy of Ireland, 1921-2; unionist by conviction but recognized necessity of coming to terms with Sinn Fein; deputy earl marshal, 1917-29; leading Roman Catholic layman; PC, 1918; GCVO, 1919; viscount, 1921; KC, 1925.

HOWARD, ESME WILLIAM, first BARON HOWARD OF PENRITH (1863-1939), diplomatist; educated at Harrow; served in diplomatic service, 1885-92; rejoined, 1903; consul-general for Crete, 1903-6; minister to Switzerland (1911-13), Sweden (1913-19); on British delegation to peace conference; ambassador to Spain (1919-24), to United States (1924-30); KCMG, 1916; KCB and PC, 1919; GCMG, 1923; GCB, 1928; baron, 1930.

HOWARD, GEORGE JAMES, ninth EARL OF CARLISLE (1843-1911), amateur artist; grandson of sixth earl [q.v.]; educated at Eton and Trinity College, Cambridge; liberal MP for East Cumberland, 1879-80, 1881-5; joined liberal unionists, 1886; succeeded to earldom, 1889; closed public houses on his Yorkshire and Cumberland estates; skilled water-colour landscape painter; published *A Picture Songbook* (1910); trustee of National Gallery, to which was transferred his Mabuse's 'Adoration of the Magi', 1911; his son, CHARLES JAMES STANLEY HOWARD, tenth earl (1867-1912), was unionist MP for South

Birmingham, 1904-11, and parliamentary whip.

HOWARD, HENRY FITZALAN-, fifteenth DUKE OF NORFOLK (1847-1917); son of H. G. FitzAlan-Howard, fourteenth duke [q.v.]; succeeded father, 1860; educated at Oratory School, Edgbaston; as young man took up cause of his co-religionists, especially Irish Catholics; postmaster-general, 1895-1900; maintained close relations with Vatican; as earl marshal largely responsible for coronation ceremonies of Edward VII and George V; great builder of Gothic churches.

HOWARD, LESLIE (1893-1943), actor, producer, and film director; acted and produced mainly in America; films included *Berkeley Square*, *Scarlet Pimpernel*, *Pygmalion*, and *Gone with the Wind*; part-producer, British war films *Pimpernel Smith*, *The First of the Few*, *The Gentle Sex*, and *The Lamp Still Burns*; shot down in aircraft returning from Portugal.

HOWARD, ROSALIND FRANCES, COUNTESS OF CARLISLE (1845-1921), promoter of women's political rights and of temperance reform; daughter of Edward John, second Baron Stanley of Alderley [q.v.]; married George Howard, afterwards ninth Earl of Carlisle [q.v.], 1864; supported home rule; president of National British Women's Temperance Association (1903) and of Women's Liberal Federation (1891-1901, 1906-14); possessed remarkable business ability.

HOWARD DE WALDEN, eighth BARON, and fourth BARON SEAFORD (1880-1946), writer, sportsman, and patron and lover of the arts. [See SCOTT-ELLIS, THOMAS EVELYN.]

HOWE, CLARENCE DECATUR (1886-1960), Canadian minister; born in United States; graduated, civil engineering, Massachusetts Institute of Technology, 1906; taught at Dalhousie University, 1908-13; naturalized, 1913; chief engineer, Board of Grain Commissioners, 1913-16; consulting engineer in Port Arthur, 1916-35; liberal MP, Port Arthur, 1935-57; minister of transport, 1935-40; established National Harbours Board, 1936, Trans-Canada Airlines, 1937; minister of munitions, 1940-6, of reconstruction, 1944-6, of reconstruction and supply, 1946-8, of trade and commerce, 1948-57, and of defence production, 1951-7; chancellor of Dalhousie, 1957-60; PC, 1946.

HOWELL, DAVID (1831-1903), dean of St. David's; held various Welsh livings, 1861-97; dean of St. David's, 1897-1903; member of first

Cardiff school board, 1875; well versed in Welsh literature and hymnology; gifted orator of evangelical fervour; mediator between Church and Welsh non-conformity.

HOWELL, GEORGE (1833-1910), labour leader and writer; joined Chartists, 1847; prominent in 'nine hours' struggle, 1859; joined 'the Junta' which directed trade-union affairs, 1860; secretary to parliamentary committee of trades union congress, 1871-5; prominent in securing Trade Unions Acts of 1871 and 1876; liberal MP for Bethnal Green, 1885-95; received civil list pension, 1906; publications include *Trade Unionism New and Old* (1891), *Labour Legislation, Labour Movements and Labour Leaders* (1902); his economic library is in the Bishopsgate Institute.

HOWES, THOMAS GEORGE BOND (1853-1905), zoologist; assisted T. H. Huxley [q.v.] at Royal College of Science, South Kensington, till 1880; succeeded Huxley as professor of zoology, 1895; excellent teacher of biology and skilled draughtsman; published Atlases of *Elementary Biology* (1885) and *Zootomy* (1902); made researches into comparative anatomy of the vertebrata; FRS, 1897; LLD, St. Andrews, 1898.

HOWITT, ALFRED WILLIAM (1830-1908), Australian anthropologist; son of William and Mary Howitt [qq.v.]; left Nottingham for Australia, 1852; explored central Australia, 1859; brought back remains of R. O'Hara Burke [q.v.] and W. J. Wills [q.v.] to Melbourne, 1862; police magistrate of Gippsland, 1862-89; made study of Australian aboriginal; admitted a member of the Kurnai tribe; with Lorimer Fison [q.v.] published *Kamilaroi and Kurnai* (1880) and *The Kurnai Tribe* (1880); secretary of mines in Victoria, 1889, and commissioner of audit, 1896; chairman of royal commission on coalmining, 1905-6; chief work, *The Native Tribes of South-East Australia* (1904); CMG, 1906; died at Melbourne.

HOWITT, SIR HAROLD GIBSON (1886-1969), chartered accountant; educated at Uppingham; articled to chartered accountant, W. R. Hamilton, 1904; qualified, joined W. B. Peat & Co. (later Peat, Marwick, Mitchell & Co.), London, 1909; partner, 1911-61; served in Green Howards, 1914-18; DSO; MC; engaged in many public commissions and inquiries and valuation problems; member, Air Council, 1939-45; chairman and deputy chairman, BOAC, 1943-8; member of council, NAAFI, 1940-6; chairman, Building Materials Board; financial adviser to Ministry of Works, 1943-5; president, Institute of Chartered Accountants

in England and Wales, 1945-6; president, International Congress of Accountants, 1952; wrote *History of the Institute of Chartered Accountants in England and Wales 1880-1965* (1966); knighted, 1937; GBE, 1946; hon. DCL Oxford, 1953; hon. LLD, Nottingham, 1958; chairman, trustees of Uppingham School (1949-67).

HOWLAND, SIR WILLIAM PEARCE (1811-1907), Canadian statesman; born in New York; went to Canada, 1830; bought mills near Toronto, 1840; liberal MP for West York, Ontario, 1857; finance minister, 1862 and 1866; receiver-general, 1863; postmaster-general, 1864-6; at Canadian federation conference in London, 1866; minister of inland revenue in first confederation cabinet, 1867-8; lieutenant-governor of Ontario, 1868-73; KCMG, 1879; promoted building of Canadian Pacific Railway, 1880.

HUBBARD, LOUISA MARIA (1836-1906), social reformer; born in St. Petersburg; her *Work for Ladies in Elementary Schools* (1872), led to the establishment of training college for girls at Chichester, 1873; published *The Englishwoman's Year Book*, 1875-98; advocated nursing, massage, typewriting, and gardening as middle-class women's occupations; helped to form Working Ladies' Guild, 1876, and Teachers' Guild, 1884.

HUDDART, JAMES (1847-1901), Australian shipowner; joined uncle's shipping firm at Geelong, Australia, 1864; founded intercolonial steamship line, 1870; aimed at 'All Red Route' by starting Canadian-Australian Royal Mail Steamship line between Sydney and Vancouver (1893) and a line between Canada and England (1894); scheme failed through lack of English support.

HUDDLESTON, SIR HUBERT JERVOISE (1880-1950), soldier and administrator; educated at Bedford and Felsted; enlisted, 1898; commissioned, 1900; transferred to Egyptian army, 1910; commanded Camel Corps, 1914-16; brigade in Palestine, 1916-18; chief staff officer and adjutant-general under Sir Lee Stack [q.v.], 1923-4; first commandant, Sudan defence force, 1925-30; major-general, 1933; commanded districts in India, 1934-8; governor-general, Sudan, 1940-7; DSO, 1917; KCMG, 1940; GCMG, 1947.

HUDLESTON (formerly SIMPSON), WILFRED HUDLESTON (1828-1909), geologist; educated at Uppingham and St. John's College, Cambridge; made ornithological collections in Lapland, Algeria, Eastern Atlas, Greece, and Turkey, 1855-60; FGS, 1867; president,

1892-4; FRS, 1884; travelled in India, 1895; published sixty papers dealing with jurassic system and oolite gasteropods, 1887-96, and Indian, Syrian, and African geology.

HUDSON, CHARLES THOMAS (1828-1903), naturalist; BA, St. John's College, Cambridge (fifteenth wrangler), 1852; MA, 1855; LLD, 1866; headmaster of Bristol grammar school, 1855-60; conducted private school at Clifton, 1861-81; devoted leisure to microscopical research, and to study of Rotifera; president of Royal Microscopical Society, 1888-90; FRS, 1889; published *The Rotifera: or Wheel-Animalculae* (1886-7).

HUDSON, SIR ROBERT ARUNDELL (1864-1927), political organizer; obtained post in National Liberal Federation, Birmingham, 1882; removed with Federation to London, 1886; secretary of Federation and of Liberal Central Association, 1893; influential in guiding party through troublous period; successful chairman of joint finance committee of British Red Cross Society and Order of St. John of Jerusalem, 1914; knighted, 1906; GBE, 1918.

HUDSON, ROBERT GEORGE SPENCER (1895-1965), geologist, stratigrapher, and palaeontologist; educated at lower school of Lawrence Sheriffe, Rugby; student teacher, Elborrow boys' school, 1913; entered St. Paul's Training College, Cheltenham, 1914; joined Artists' Rifles, 1916; served in France as machine-gunner; entered University College, London, 1918; B.Sc., first class, geology, 1920; part-time demonstrator in geology, University College, 1920-2; M.Sc., 1922; assistant lecturer, Leeds University, 1922-7; lecturer, 1927-39; professor, 1939-40; edited *Transactions of the Leeds Geological Association*, 1927-40; consultant geologist, 1942; involved in petroleum geology; appointed to staff of Iraq Petroleum Company, 1946; hon. lecturer in geology, University College, London, 1947-58; lecturer in palaeontology, Trinity College, Dublin, 1959; professor of geology and mineralogy, 1961-5; important contributor to the palaeontology, stratigraphy, and palaeogeography of the Carboniferous of north of England and the Mesozoic of the Middle East; fellow, Geological Society of London, 1921; vice-president, 1955-7; president, Yorkshire Geological Society, 1940-2; FRS, 1961; fellow, University College, London; president, Irish Geological Association, 1964-5.

HUDSON, ROBERT SPEAR, first VISCOUNT HUDSON (1886-1957), politician; educated at Eton and Magdalen College, Oxford; in diplomatic service, 1911-23; conservative MP,

Whitehaven, 1924-9, Southport, 1931-52; parliamentary secretary, Ministry of Labour, 1931-5, of Health, 1936-7; minister of pensions, 1935-6; secretary, Department of Overseas Trade, 1937-40; minister of agriculture and fisheries, 1940-5; achieved agricultural revolution in production and attitudes, decentralizing powers and introducing guaranteed prices and markets; PC, 1938; CH, 1944; viscount, 1952.

HUDSON, WILLIAM HENRY (1841-1922), naturalist and writer; born near Buenos Aires; ran wild on farms and ranches of Rio de la Plata; watched bird life of great plains with care and passion; his career entirely changed by fever, which affected his heart, at age of fifteen; went to England, 1869; lived for some years in drab London boarding-house surroundings and in considerable poverty; published *The Purple Land that England Lost* (1885); *A Crystal Age* (1887); *Argentine Ornithology* (with Dr P. L. Sclater, 1888-9); *The Naturalist in La Plata* (1892) first book to bring him into prominence; followed it with *Birds in a Village* (1893); thenceforth his books had increasing sale; awarded civil list pension, 1901; his most interesting book *Far Away and Long Ago* (1918); his philosophy of life based upon observation of animal world; possessed an absolute freedom of spirit and detachment combined with great sensitiveness and receptivity and an almost mystical sense of natural beauty.

HUEFFER, FORD HERMANN (1873-1939). [See FORD, FORD MADOX.]

HÜGEL, FRIEDRICH VON, Baron of the Holy Roman Empire (1852-1925), theologian. [See VON HÜGEL.]

HUGGINS, SIR WILLIAM (1824-1910), astronomer; son of London silk mercer; educated at City of London School; after a few years in business, built observatory at Tulse Hill, 1856; applied to stars the methods of Kirchhoff's researches (1862) into chemical constitution of sun; with William Allen Miller [q.v.] devised the star spectroscope, and presented to Royal Society results of first investigations with it in 'Lines of the Spectra of Some of the Fixed Stars', 1863, showing that in structure the stars resemble the sun; FRS, 1865; awarded royal medal, 1866; president, 1900-6; president of the Royal Astronomical Society, 1876-8; made observations of nebula in Orion (1865) and of several comets; examined motions of Sirius (1868) and other stars; with refracting telescope determined (1870-5) the velocity of stars; photographed successfully the spectrum of Vega (1876), of larger stars, of the moon, and the planets; method applied to great nebula in

Orion, 1882; hon. LLD, Cambridge, 1870, DCL, Oxford, 1871; awarded medals from Royal and Royal Astronomical societies, and from foreign and American societies; awarded civil list pension, 1890; president of British Association, 1891; KCB, 1897; OM, 1902; observed the new star in Auriga, 1892; made valuable researches as to extent and presence of calcium in sun, 1897; published results of work in *An Atlas of Representative Stellar Spectra* (1900) and *The Scientific Papers of Sir William Huggins* (1909); helped much by his wife in his researches.

HUGHES, ARTHUR (1832-1915), painter; studied under Alfred Stevens [q.v.] at Somerset House school of design and at Royal Academy Schools; adopted pre-Raphaelite principles; exhibited at Royal Academy, 1856-1908; chief pictures 'The Knight of the Sun' and 'Home from Sea'; illustrated poems of Tennyson, Christina Rossetti, works of George Mac-Donald, etc.

HUGHES, EDWARD (1832-1908), portrait painter; taught by father and by John Pye [q.v.]; student at Royal Academy Schools; exhibited at Royal Academy from 1847, at first subject pictures, but later portraits; excelled in portraits of ladies, who included Queen Mary and Queen Alexandra; idealized sitters; work reproduced in *The Book of Beauty* (1896).

HUGHES, EDWARD DAVID (1906-1963), organic chemist; educated at Portmadoc grammar school and University College, Bangor; first class, chemistry, 1927; researched on prototropy, 1927-30; Ph.D., 1930; post doctoral fellow, University College, London, 1930; worked with Professor (Sir) C. K. Ingold [q.v.] on the classification of organic mechanisms; M.Sc., 1932; D.Sc. (London), 1936; appointed lecturer, 1937; professor of chemistry, University College, Bangor, 1943-8; professor of chemistry, University College, London, 1948; deputy head of department, 1957; head, 1961; FRS, 1949; fellow, UCL, 1954; vice-president, Chemical Society, 1956-9; published many papers on organic reaction mechanisms.

HUGHES, HUGH PRICE (1847-1902), Methodist divine; BA, London, 1869; MA, 1881; from 1884 leader in London of the Methodist 'forward movement'; started *Methodist Times*, 1885, and West London Mission, 1886; promoted Free Church Congress, 1892; first president, National Free Church Council, 1896, and president of Wesleyan Methodist Conference, 1898; magnetic evangelical preacher; a radical in politics, but an imperialist; supported South African war,

1899-1901; opposed conservative Education Act, 1902; published religious works.

HUGHES, JOHN (1842-1902), Wesleyan Methodist divine; entered ministry, 1867; an eloquent preacher and writer in Welsh of religious prose and verse under bardic name of 'Glanystwyth'.

HUGHES, SIR SAM (1853-1921), Canadian soldier and politician; born in Ontario; represented Victoria county in federal house as conservative, 1892-1921; served in South African war, 1899; minister of militia and defence, 1911-16; KCB, 1915; lieutenant-general, 1916.

HUGHES, WILLIAM MORRIS (1862-1952), Australian prime minister; born in London of Welsh parents; educated at Llandudno grammar school and St. Stephen's, Westminster; emigrated to Queensland, 1884; set up shop in dockside slum, Sydney, 1886; secretary, Sydney Wharf Labourers' Union, 1899-1915; founder and first president, Waterside Workers' Federation, 1902; called to bar, 1903; KC, 1919; labour member, New South Wales parliament, 1894-1901; member, federal House of Representatives, 1901-52; overcame handicaps of ill health and deafness to become superb orator and dominating figure nationally known as 'Billy'; minister for external affairs, 1904; attorney-general, 1908-9; favouring strong defence policy persuaded his party and government of Alfred Deakin [q.v.] to adopt policy of compulsory military training serving within Australia, 1909; attorney-general, 1910-13, 1914-21; party leader and prime minister, 1915-17; in England and France, 1916, vigorously advocating total war effort; at home his advocacy of conscription rejected; with liberals formed nationalist party, 1917, remaining prime minister until 1923; in England, 1918-19, and at Paris peace conference; favoured harsh terms; obtained separate recognition for Australia as signatory to treaty and in League of Nations; lost confidence of nationalist party as too socialistic; upon coalition with country party induced to cede leadership and premiership, 1923; voted with labour to defeat government, 1929; joined united Australia party, 1931; deputy leader, 1939-41, 1943-4; leader, 1941-3; minister, repatriation and health, 1934-5, 1936-7; external affairs, 1937-9; industry, 1939-40; navy, 1940-1; attorney-general, 1939-41; member, War Advisory Council, 1941-4; PC, 1916; CH, 1941.

HULME, FREDERICK EDWARD (1841-1909), botanist and artist; skilful sketcher of plants and flowers; his best work, *Familiar Wild Flowers* (9 vols., 1875-1909); also published works on ornament, heraldry, flower lore, and art students' textbooks; FSA, 1872.

HULTON, SIR EDWARD (1869-1925), newspaper proprietor; entered father's newspaper business in Manchester, 1885; produced *Daily Sketch*, 1909; retired, after selling for £6 million his fourteen periodicals to Allied Newspapers Limited, 1923; baronet, 1921; interests as newspaper proprietor commercial and free from political interference; interested in horse-racing.

HUME, ALLAN OCTAVIAN (1829-1912), Indian civil servant and ornithologist; son of Joseph Hume [q.v.]; joined Bengal civil service, 1849; CB for services in Indian Mutiny, 1860; retired, 1882; worked for Indian parliamentary system through Indian National Congress convoked under his guidance, 1885-94; collaborated in standard work on Indian game birds.

HUME, MARTIN ANDREW SHARP (1843-1910), author; visited relatives in Spain, 1860; travelled extensively in Central and South America; published *The Courtships of Queen Elizabeth* and *The Year after the Armada* (1896); editor of *Spanish State Papers* at Public Record Office (1898); published histories of Spain (1898) and modern Spain (1899), and other works on Spanish history and literature; later works include *Queens of Old Spain* (1907), *Queen Elizabeth and Her England* (1910); lectured on Spanish history at Cambridge and elsewhere; hon. MA, Cambridge, 1908.

HUME-ROTHERY, WILLIAM (1899-1968), first Isaac Wolfson professor of metallurgy in the university of Oxford; great grandson of Joseph Hume [q.v.]; educated at Cheltenham College and Magdalen College, Oxford; passed into Royal Military Academy, Woolwich, 1916, but contracted cerebrospinal meningitis and became completely deaf, 1917; first class, natural science, 1922; joined Royal School of Mines; worked under (Sir) H. C. Carpenter [q.v.] on intermetallic compounds; Ph.D. (London), 1925; research fellow, Oxford, working on constitution of alloys, 1925; D.Sc. (Oxford), 1935; FRS, 1937; fellow, Magdalen College, 1938; lecturer in metallurgical chemistry; reader, 1955; Isaac Wolfson professor of metallurgy, 1958-66; OBE, 1951; hon. degrees, Manchester and Sheffield, 1966; many other academic awards.

HUMPHREY, HERBERT ALFRED (1868-1951), engineer; educated at Cowper Street Middle Class School, Finsbury Technical Institute, and City and Guilds Central Institution;

with Brunner, Mond & Co., 1890-1901; set up as consulting engineer, 1901; invented gas pump; technical adviser on explosives, 1914-18; member, commission to Synthetic Ammonia Works, Oppau, 1919; consulting engineer and director, Synthetic Ammonia and Nitrates Ltd., Billingham, 1919-26; constructed 40,000 kW electric power-station there; consulting engineer to ICI, 1926-31; died in South Africa.

HUMPHREYS, SIR (RICHARD SOMERS) TRAVERS (CHRISTMAS) (1867-1956), judge; educated at Shrewsbury and Trinity Hall, Cambridge; called to bar (Inner Temple), 1889; bencher, 1922; junior crown counsel, Central Criminal Court, 1908; senior counsel, 1916-28; recorder of Chichester, 1921-6; of Cambridge, 1926-8; judge of King's Bench division, 1928-51; appeared in prosecution of H. H. Crippen, F. H. Seddon, G. J. Smith, Sir Roger Casement [q.v.], and Horatio Bottomley [q.v.]; presided at trials of Mrs Barney (1932), Mrs Rattenbury and George Stoner (1935), and J. G. Haig (1949); knighted, 1925; PC, 1946; published *Criminal Days* (1946), and *A Book of Trials* (1953).

HUNT, DAME AGNES GWENDOLINE (1866-1948), pioneer in work amongst cripples; herself crippled by osteomyelitis from age of ten; qualified as Queen's nurse and in midwifery; founded first open-air hospital for crippled children at Baschurch, 1900; with Sir Robert Jones [q.v.] revolutionized treatment; moved hospital to Oswestry as Robert Jones and Agnes Hunt Orthopaedic Hospital, 1921; built up system of after-care and preventive clinics; founded Derwen Cripples' Training College, 1927; DBE, 1926.

HUNT, ARTHUR SURRIDGE (1871-1934), papyrologist; educated at Cranbrook, Eastbourne College, and Queen's College, Oxford; joined B. P. Grenfell and D. G. Hogarth [qq.v.] in excavations for papyri in the Fayum; lecturer in papyrology (1908), professor (1913-34) at Oxford; many publications with Grenfell and sole editor after 1920 of Egypt Exploration Society's volumes of papyrus texts; FBA, 1913.

HUNT, GEORGE WILLIAM (1829?-1904), song writer. [See under MACDERMOTT, GILBERT HASTINGS.]

HUNT, WILLIAM (1842-1931), historian; educated at Harrow and Trinity College, Oxford; vicar of Congresbury, Somerset, 1867-82; sub-editor and contributor of nearly 600 articles to this Dictionary; joint-editor of *A History of the English Church* (8 vols., 1899-1910) and *The Political History of England* (12 vols., 1905-10), contributing one volume to each.

HUNT, WILLIAM HOLMAN (1827-1910), painter; born in Wood Street, Cheapside; son of a warehouseman; studied art at British Museum and National Gallery from 1843; at Academy Schools, 1844; made acquaintance of Millais and D. G. Rossetti [qq.v.], 1844; exhibited at Academy from 1847; introduced Rossetti to Millais; founded in 1848 with them the pre-Raphaelite Brotherhood, which was joined subsequently by Woolner, W. M. Rossetti, James Collinson, and F. G. Stephens [qq.v.]; consistently carried on the brotherhood's principles till death; his first pre-Raphaelite picture, 'Rienzi', hung in Royal Academy, 1849; visited France and Belgium with Rossetti, 1849; removed to Chelsea; found valuable patron in Thomas Combe [q.v.]; his 'Valentine rescuing Sylvia from Proteus' (1851), attacked by *The Times*, was powerfully defended by John Ruskin [q.v.] with whom he became intimate for life; 'The Hireling Shepherd' (1852) praised by Carlyle; exhibited at Academy 'Claudio and Isabella' and 'Strayed Sheep', 1853, and in 1854 'The Awakened Conscience' and 'The Light of the World' (now at Keble College, Oxford; later replica in St. Paul's Cathedral); travelled in Egypt and Palestine, 1854; began at Jerusalem 'The Finding of the Saviour in the Temple' (finished, 1860); painted on the Dead Sea 'The Scapegoat' (exhibited, 1856); settled at Pimlico, soon removing to Tor Villa, Campden Hill, 1856; refused associateship of the Royal Academy (1856), where he ceased to exhibit after 1874; helped to form Hogarth Club, 1858; designed furniture, setting fashion developed by William Morris [q.v.]; in Florence, where his first wife died, 1866; painted there 'Isabella and the Pot of Basil'; in Holy Land, 1869-71; painted at Jerusalem 'The Shadow of Death' (or 'The Shadow of the Cross'), 1870-1; after return to London (1871-5) was again in Holy Land, 1875-8; there painted 'Nazareth' and 'The Triumph of the Innocents'; on voyage out painted 'The Ship', 1875 (now in Tate Gallery); exhibited at the newly founded Grosvenor Gallery from 1877; sent there portraits of his sons, of Sir Richard Owen (1881), and of Dante Rossetti (1884), as well as 'Amaryllis' (1885); executed a water-colour, 'Christ among the Doctors' (1886), and 'May Morning on Magdalen Tower, Oxford' (1891); painted 'The Miracle of Sacred Fire', 1899; published *Pre-Raphaelitism and the Pre-Raphaelite Brotherhood* (2 vols., 1905); OM and hon. DCL, Oxford, 1905; buried in crypt of St. Paul's Cathedral.

HUNTER, SIR ARCHIBALD (1856-1936), general; educated at Glasgow Academy and Sandhurst; gazetted to 4th Foot, 1874; captain, 1882; served with Egyptian army, 1884-99; commanded Sudan frontier force, 1894-6; able commander under Lord Kitchener [q.v.], 1896-8; major-general, 1896; KCB, 1898; in South Africa (1899-1901) prominent in defence of Ladysmith and responsible for second great Boer surrender (30 July 1900); assisted Kitchener in reorganizing Indian army, 1903-8; general, 1905; governor of Gibraltar, 1910-13; held Aldershot command, training New Armies, 1914-17; conservative MP, Lancaster, 1918-22; GCB, 1911.

HUNTER, COLIN (1841-1904), sea painter; after four years in Glasgow shipping office turned to landscape painting, 1861; exhibited 97 pictures, mainly seascapes, at Royal Academy from 1868; his chief works include 'Trawlers, waiting for Darkness' (1873), 'Their Only Harvest' (1878, in Tate Gallery), and 'Signs of Herring' (1899); ARA, 1884; vigorous water-colour artist and etcher.

HUNTER, SIR ELLIS (1892-1961), industrialist; articled to firm of accountants after leaving Middlesborough high school; qualified, 1914; worked in steel department, Ministry of Munitions, 1914-18; local partner, W. B. Peat & Co., 1919; general partner, Peat, Marwick, Mitchell & Co., 1928; fellow, Institute of Chartered Accountants, 1927; deputy chairman (and later, managing director), Dorman Long, 1938; chairman, 1948-61; president, Iron and Steel Federation, 1945-53; took leading part in drawing up first post-war development plan; opposed nationalization; knighted, 1948; GBE, 1961.

HUNTER, SIR GEORGE BURTON (1845-1937), shipbuilder; principal partner of C. S. Swan, Wallsend, 1880; combined as Swan, Hunter & Wigham Richardson, 1903; chairman, 1895-1928; pioneer of system of building within glazed sheds; built the *Mauretania*; KBE, 1918.

HUNTER, PHILIP VASSAR (1883-1956), electrical engineer; educated at Wisbech grammar school and Faraday House; joined C. H. Merz [q.v.], 1904; collaborated in evolution of wartime ASDIC submarine detection; CBE, 1920; chief engineer and joint manager, Callender's Cable and Construction, 1919-46; worked on high-voltage power cables; initiated research laboratories, 1934; engineer-in-chief, British Insulated Callender's Cables, 1946; joint deputy chairman, 1947-52; president, British Ice Hockey Association, 1935.

HUNTER, SIR ROBERT (1844-1913), solicitor, and authority on commons and public rights; partner in firm of solicitors to Commons

Preservation Society, 1869; most notable case, Epping Forest, 1871-4; solicitor to General Post Office, 1882-1913; helped to found National Trust, 1895; knighted, 1894; KCB, 1911.

HUNTER, Sir WILLIAM GUYER (1827-1902), surgeon-general; FRCS, 1858; MD, 1867; FRCP, 1875; joined Bengal medical service, 1850; civil surgeon in Upper Sind during Mutiny, 1857; professor of medicine in Grant Medical College, Bombay, 1858; principal, 1876; surgeon-general, 1877; vice-chancellor of Bombay University, 1880; KCMG, 1884; hon. LLD, Aberdeen, 1894; member of London school board, 1886-7; conservative MP for Central Hackney, 1885-92.

HUNTER-WESTON, Sir AYLMER GOULD (1864-1940), lieutenant-general. [See Weston.]

HUNTINGTON, GEORGE (1825-1905), rector of Tenby from 1867; his ritualistic sympathies led to controversy with his bishop, 1877; publications include *Church's Work in our Large Towns* (1863), *Random Recollections* (1895), sermons, and addresses.

HURLSTONE, WILLIAM YEATES (1876-1906), musical composer and pianist; grandson of Frederick Yeates Hurlstone [q.v.]; composed from age of nine; published orchestral compositions, pianoforte concerto, and cantata.

HURST, Sir ARTHUR FREDERICK (1879-1944), physician; educated at Bradford and Manchester grammar schools and Magdalen College, Oxford; first class, physiology, 1901; BM (Oxon.) from Guy's Hospital, 1904; assistant physician, in charge of neurological department, 1906; full physician, 1918-39; pioneer in clinical science; built up team of colleagues at New Lodge Clinic, Windsor Forest; studied particularly the alimentary tract; edited *Guy's Hospital Reports*, 1921-39; changed name from Hertz, 1916; knighted, 1937.

HURST, Sir CECIL JAMES BARRINGTON (1870-1963), international lawyer; educated at Westminster School and Trinity College, Cambridge; first class, part ii, law tripos, 1891-2; called to bar (Middle Temple), 1893; assistant legal adviser, Foreign Office, 1902; member of arbitration commission to report on *Alsop* claims of United States and Chile, 1910; British agent, British-American Claims Arbitration Tribunal, 1912; KC, 1913; legal adviser to Foreign Office, 1918-29; chief editor, *British Year Book of International Law*, 1919-28; president, Grotius Society, 1940; judge, Permanent Court of International Justice, 1929-46; president, 1934-6; chairman, Home Office panel for appeals against detention under Regulation 18B; first president, War Crimes Commission, 1943-5; CB, 1907; KCB, 1920; KCMG, 1924; GCMG, 1926; treasurer, Middle Temple, 1940; president, Institute of International Law; hon. LLD Cambridge and Edinburgh.

HUSSEY, CHRISTOPHER EDWARD CLIVE (1899-1970), architectural historian and architectural contributor to *Country Life* for fifty years; educated at Eton and Christ Church, Oxford, 1919-21; joined editorial staff, *Country Life*, 1920; contributed nearly 1,400 signed articles in fifty years; editor, 1933-40; publications include *Eton College* (1922), *Petworth House* (1926), *The Picturesque* (1927), *English Country Houses: Early Georgian* (1955), *Mid-Georgian* (1956), and *Late Georgian* (1958), *English Gardens and Landscapes 1700-1750* (1967), and the *Life* of Sir Edwin Lutyens [q.v.] (1950); president, Society of Architectural Historians, 1964-6; hon. ARIBA (1935); FSA (1947); CBE, 1956; associate of the Institute of Landscape Artists.

HUTCHINSON, ARTHUR (1866-1937), mineralogist; educated at Clifton and Christ's College, Cambridge; first class, natural sciences tripos, parts i and ii, 1886-8; fellow 1892, assistant tutor 1901-26, master 1928-37, Pembroke College; university demonstrator in mineralogy, 1895, lecturer in crystallography, 1923, professor of mineralogy, 1926-31; discovered mineral stokesite, 1899; encouraged research in X-ray crystallography; FRS, 1922.

HUTCHINSON, FRANCIS ERNEST (1871-1947), scholar and canon of Worcester; educated at Lancing and Trinity College, Oxford; deacon, 1896; priest, 1897; chaplain, King's College, Cambridge, 1904-12; vicar of Leyland, 1912-20; secretary, delegacy for extramural studies, Oxford, 1920-34; chaplain (1928), fellow (1934), All Souls; canon of Worcester, 1934-43; sensitive student of Caroline poetry; FBA, 1944.

HUTCHINSON, HORATIO GORDON (HORACE) (1859-1932), golfer and author; captained first four Oxford golf teams, 1878-82; amateur champion, 1886, 1887; played for England against Scotland, 1902-7 (except 1905); dashing and characteristic style; published *Hints on the Game of Golf* (1886) and thereafter wrote prolifically on golf, shooting, fishing, etc.

HUTCHINSON, Sir JONATHAN (1828-1913), surgeon; educated at York medical school and St. Bartholomew's Hospital, London; assistant surgeon to London Hospital, 1859; full surgeon, 1863; FRCS, 1862; Hunterian professor, 1879-83; FRS, 1882; left staff of London Hospital as emeritus professor of surgery, 1883; president of Royal College of Surgeons, 1889; delivered Hunterian oration, 1891; knighted, 1908; specialist of great repute on ophthalmology, dermatology, and especially on syphilis; works include *Illustrations of Clinical Surgery* (1878-84) and *Archives of Surgery* (1889-1900); formed museum of specimens and drawings.

HUTCHINSON, RICHARD WALTER JOHN HELY-, sixth EARL OF DONOUGHMORE (1875-1948), chairman of committees of the House of Lords. [See HELY-HUTCHINSON.]

HUTCHISON, SIR ROBERT, first baronet, of Thurle (1871-1960), physician and paediatrician; educated at the Collegiate School and university, Edinburgh; qualified in medicine, 1893; MD, 1896; FRCP, 1903; on staff of Hospital for Sick Children, Great Ormond Street, and London Hospital, 1900-34; publications include *Clinical Methods* (with H. Rainy, 1897), *Food and the Principles of Dietetics* (1900), and *Lectures on Diseases of Children* (1904); baronet, 1939.

HUTCHISON, SIR WILLIAM OLIPHANT (1889-1970), landscape and portrait painter; educated at Kirkcaldy high school, Cargilfield, Rugby, and Edinburgh College of Art; served in Royal Garrison Artillery, 1914-18; practised in London as a portrait painter; exhibited at Royal Academy, 1921-9; director, Glasgow School of Art, 1932-43; moved to Edinburgh, resumed portrait and landscape painting, and became deeply involved with the Royal Scottish Academy, 1943; president, 1950-9; knighted, 1953; returned to London, 1959; president, Royal Society of Portrait Painters, 1965; among his portraits were the Queen, Prince Philip, the Queen Mother, and J. Ramsay MacDonald (now in the House of Commons).

HUTH, ALFRED HENRY (1850-1910), bibliophile; son of Henry Huth [q.v.]; travelled in East with Henry Thomas Buckle [q.v.], 1861-2; publications include *The Marriage of Near Kin* (1875) and life of Buckle (2 vols., 1880); helped to found Bibliographical Society, 1892; by will left fifty volumes to British Museum; sale of part of collection realized over £108,000 (1911-12).

HUTTON, ALFRED (1839-1910), swordsman; joined 79th Highlanders, 1859; captain, 1868; organized Cameron Fencing Club, for which he published *Swordsmanship* (1862); on

retiring from army (1873) practised modern fencing and revived older systems; published *Cold Steel* (1889), *Fixed Bayonets* (1890), *The Swordsman* (1891), *Old Sword Play* (1892), and *The Sword and the Centuries* (1901); FSA, 1894; founder and chairman of Central London Throat and Ear Hospital, 1874; bequeathed collection of fencing literature to Victoria and Albert Museum.

HUTTON, FREDERICK WOLLASTON (1836-1905), geologist; served in Indian mercantile marine; in army, 1855-65; in Crimea and Indian Mutiny; emigrated to New Zealand, 1866; professor of geology in university of New Zealand, 1890-3; curator of museum, 1893; interested in ornithology and ethnology; writings include *Elementary Geology* (1875), *The Lesson of Evolution* (1902), and works on fauna of New Zealand (1904); FRS, 1892.

HUTTON, GEORGE CLARK (1825-1908), Presbyterian divine; educated at Edinburgh University; hon. DD, 1906; United Presbyterian minister at Paisley from 1851 till death; advocated 'voluntary' movement in religion; strenuously advocated disestablishment of Church of Scotland in speeches and pamphlets; urged state secular education; moderator of synod, 1884; principal of United Presbyterian theological college, 1892; co-principal with G. C. M. Douglas [q.v.] of United Free Church College, Glasgow, till 1902; moderator of United Free Church general assembly, 1906; a trenchant controversialist.

HUTTON, WILLIAM HOLDEN (1860-1930), historian and dean of Winchester; BA, Magdalen College, Oxford, 1882; first class in modern history; fellow of St. John's College, Oxford, 1884-1923; deacon, 1885; priest, 1886; archdeacon of Northampton and canon residentiary of Peterborough Cathedral, 1911; university reader in Indian history, 1913-20; dean of Winchester, 1919-30; died at Freiburg im Breisgau; works include lives of Wellesley (1893), More (1895), Laud (1895); *The English Church . . . 1625-1714* (1903); and *Lives of the English Saints* (Bampton lectures, 1903).

HUXLEY, ALDOUS LEONARD (1894-1963), man of letters; son of Leonard Huxley, grandson of T. H. Huxley, great grandson of Dr Thomas Arnold, and nephew of Mrs Humphrey Ward [qq.v.]; educated at Eton and Balliol College, Oxford; first class, English literature, in spite of damaged eyesight, 1916; met other young writers and painters at the Garsington home of Philip Morrell and Lady Ottoline [q.v.]; first novel, *Crome Yellow* (1921), followed by *Antic Hay* (1923), *Those Barren Leaves*

(1925), and *Brave New World* (1932); published short stories, such as *Mortal Coils* (1922), and essays and travel books, including *Jesting Pilot* (1926); *Point Counter Point* (1928), best-seller in Britain and America; decided to settle in United States, 1938; eyesight improved; published *The Art of Seeing* (1942); published new series of books including *The Perennial Philosophy* (1945), *Science, Liberty and Peace* (1946), and *Brave New World Revisited* (1958) and further novels, including *After Many a Summer* (1939), which won the James Tait Black memorial prize, and *The Genius and the Goddess* (1955); interested in hypnosis and the use of psychedelic drugs, described in *The Doors of Perception* (1954) and *Heaven and Hell* (1956); Los Angeles home destroyed by fire, 1961.

HUXLEY, LEONARD (1860-1933), biographer, poet, and editor of the *Cornhill Magazine*; son of T. H. Huxley [q.v.]; first class, *lit. hum.*, Balliol College, Oxford, 1883; assistant master, Charterhouse, 1884-1901; assistant editor, *Cornhill Magazine* (1901), sole editor (1916-33); works include an outstanding biography of his father (1900) and *Anniversaries* (poems, 1920).

'HWFA MÔN' (1823-1905), archdruid of Wales. [See WILLIAMS, ROWLAND.]

HYDE, DOUGLAS (1860-1949), Gaelic revivalist, poet, and first president of Eire; graduated in modern literature, Trinity College, Dublin, 1884; LLD, 1888; learnt Irish in his native Roscommon; president, Gaelic League, 1893-1915; conducted successful campaign for revival of Irish language but repudiated any political aim; professor of modern Irish, University College, Dublin, 1908-32; first president of Eire, 1937-45; publications include collected plays (1905), and *Love Songs of Connacht* (1893), *Religious Songs of Connacht* (1906), *The Story of Early Gaelic Literature* (1895), *Literary History of Ireland* (1899), and *Medieval Tales from the Irish* (1899).

HYDE, SIR ROBERT ROBERTSON (1878-1967), founder of the Industrial Welfare Society; educated at Westbourne Park School and King's College, London; studied theology; ordained deacon, 1903; priest, 1904; curate, St. Saviour's, Hoxton; warden, Hoxton hostel settlement, founded in memory of F. D. Maurice [q.v.], 1907; combined this with the living of parish church of St. Mary, 1912-16; asked by Seebohm Rowntree [q.v.] to take charge of boys' welfare department in Ministry of Munitions, 1916; set up the Boys' Welfare Association, 1918; name changed to Industrial Welfare Society, 1919; co-operated with the Duke of York in establishing the Duke of York's camps for boys from industry and from public schools; MVO, 1932; KBE, 1949; retired as director of Industrial Welfare Society, 1950; published *The Boy in Industry and Leisure* (1921) and *Industry was my Parish* (1968).

HYLTON, JACK (1892-1965), dance-band leader, pianist, composer, and impresario; educated at higher grade school, Bolton; learned piano as a child; first professional engagement as pianist and singer with pierrot troupe, 1905; cinema organist at the Alexandra, Stoke Newington, 1913; worked as double act with Tommy Handley, 1920; became leader of Queen's Hall Roof resident band; made first records, 1921; opened London office and placed dance-bands in Piccadilly Hotel and the Kit-Cat Club, 1925; appeared with his band at London Hippodrome and Palladium, 1927-37; toured in Europe; disbanded his orchestra, 1940; became theatre impresario, 1935; presented number of west-end successes, 1941-6; composed many popular songs; appeared in six Royal Command Variety performances.

HYLTON-FOSTER, SIR HARRY BRAUSTYN HYLTON (1905-1965), Speaker of the House of Commons; educated at Eton and Magdalen College, Oxford; first class, jurisprudence, 1926; called to bar (Inner Temple), 1928; legal secretary to Viscount Finlay [q.v.] at Permanent Court of International Justice, 1928; served in RAF Volunteer Reserve and as deputy judge advocate in North Africa and Italy, 1939-45; recorder of Richmond, 1940-4, Huddersfield, 1944-50, and Kingston upon Hull, 1950-4; chancellor of dioceses of Ripon, 1947-54, and Durham, 1948-54; KC, 1947; conservative MP, York, 1950-5; MP, Cities of London and Westminster, 1959-65; solicitor-general; knighted, 1954; PC, 1957; Speaker of the House of Commons, 1959-65.

HYNDLEY, VISCOUNT (1883-1963), business man and administrator. [See HINDLEY, JOHN SCOTT.]

HYNDMAN, HENRY MAYERS (1842-1921), socialist leader; on staff of *Pall Mall Gazette*, 1871-80; took lead in forming (Social) Democratic Federation, 1881; agitated among unemployed; opposed South African war; left British socialist party and formed national socialist party, 1916; published *England for All* (1881), several books defending political Marxism, and two autobiographical works.

I

IBBETSON, SIR DENZIL CHARLES JELF (1847-1908), lieutenant-governor of the Punjab; passed into Indian civil service, 1868; assistant settlement officer at Karnal, 1871; published lucid report of district (1883) containing scholarly research into tribal and agricultural systems; reported on Punjab census of 1881 and based on it *Outlines of Punjab Ethnography* (1883); compiled *Punjab Gazetteer* (1883); head of department of public instruction, 1884; his report (1891) on working of Deccan Agriculturists' Relief Act of 1879 led to its amendment in interest of peasantry; CSI, 1896; chief commissioner of Central Provinces, 1898; joined viceroy's council, 1902; active in measures for prevention of famine; carried Co-operation Credit Act, 1904; KCSI, 1903; lieutenant-governor of the Punjab, 1907-8; repressed disorders in Lahore and Rawalpindi.

IBBETSON, SIR HENRY JOHN SELWIN-, seventh baronet, and BARON ROOKWOOD (1826-1902), politician. [See SELWIN-IBBETSON.]

IGNATIUS, FATHER (1837-1908), preacher. [See LYNE, JOSEPH LEYCESTER.]

ILBERT, SIR COURTENAY PEREGRINE (1841-1924), parliamentary draftsman; BA, Balliol College, Oxford; Hertford, Ireland, Craven, and Eldon law scholar; first class, *lit. hum.*; fellow of Balliol, 1864; called to bar (Lincoln's Inn), 1869; joined department for drafting parliamentary bills; law member, council of governor-general of India, 1882-6; assistant parliamentary counsel to Treasury, 1886; parliamentary counsel, 1899-1902; clerk of House of Commons, 1902-21; KCSI, 1895; KCB, 1908; GCB, 1911.

ILCHESTER, sixth EARL OF (1874-1959), landowner and historian. [See FOX-STRANGWAYS, GILES STEPHEN HOLLAND.]

ILIFFE, EDWARD MAUGER, first BARON ILIFFE (1877-1960), newspaper and periodical proprietor; joined father's periodical firm, 1894; with future Lords Camrose [q.v.] and Kemsley founded Allied Newspapers, Ltd., 1924, owning *Sunday Times* and acquiring (1928) *Daily Telegraph*; on dissolution of association, 1937, retained Kelly's Directories; acquired *Birmingham Post* and *Mail*, 1943; conservative MP, Tamworth, 1923-9; CBE, 1918; knighted, 1922; baron, 1933; GBE, 1946; president, International Lawn Tennis Club of Great Britain, 1945-59.

ILLING, VINCENT CHARLES (1890-1969), petroleum geologist; educated at King Edward VI grammar school, Nuneaton, and Sidney Sussex College, Cambridge; first class, parts i and ii, natural sciences tripos, 1911-12; Harkness scholar, 1914; demonstrator, applied geology, Imperial College of Science and Technology, South Kensington, 1913; developed course in petroleum geology; lecturer, 1915; assistant professor, 1921; professor of oil technology, 1935-55; made first of many visits to Trinidad, 1915; made geological studies in Poland, Romania, and France; published 'The Migration of Oil and Natural Gas' in *Journal of the Institution of Petroleum Technologists*, 1933; published in the United States 'The Role of Stratigraphy in Oil Discovery' (*Bulletin of American Association of Petroleum Geologists*, 1945); petroleum consultant, travelled widely, including making surveys in Venezuela; formed Petroleum Scientific Services Ltd., 1946; changed to V. C. Illing and Partners, 1950; advised Nigerian government on future development of oil resources; adviser to Gas Council on gas exploration in Britain, 1958; member of council, Geological Society, 1927-8; vice-president, Institute of Petroleum, 1942-5 and 1948; hon. member, American Association of Petroleum Geologists; FRS, 1945; fellow, Imperial College, 1958; emeritus professor of oil technology, London University, 1955.

IMAGE, SELWYN (1849-1930), artist; BA, New College, Oxford; disciple of John Ruskin [q.v.]; ordained, 1872; Slade professor of fine art at Oxford, 1910-16; best known by his designs for stained glass; also designed decorative title-pages, etc.; produced landscape drawings and water-colours.

IMMS, AUGUSTUS DANIEL (1880-1949), entomologist; educated at St. Edmund's College and Mason College, Birmingham, and Christ's College, Cambridge; BA, Cambridge, D.Sc., Birmingham, 1907; professor of biology, Allahabad, 1907-11; forest entomologist to Indian government, 1911-13; reader in agricultural entomology, Manchester, 1913-18; chief entomologist, Rothamsted, 1918-31; reader in entomology, Cambridge, 1931-45; publications include *General Textbook of Entomology* (1925) and *Insect Natural History* (1947); FRS, 1929.

INCE, SIR GODFREY HERBERT (1891-1960), civil servant; educated at Reigate grammar school and University College, London; first class, mathematics, 1913; joined Ministry of Labour, 1919; chief insurance officer, 1933; advised Australian government on unemployment insurance, 1936-7; in charge of military recruiting department, Ministry of Labour, organizing arrangements for call-up, 1939-41; director-general of manpower under Ernest Bevin [q.v.], responsible for all matters affecting call-up to forces and supply of civilian labour, 1941-4; under-secretary, 1940; permanent secretary, 1944-56; concerned with demobilization, resettlement, and Youth Employment Service; chairman, Cable and Wireless, 1956-60; CB, 1941; KBE, 1943; KCB, 1946; GCB, 1951; fellow, University College, London; hon. LLD, London.

INCE, WILLIAM (1825-1910), regius professor of divinity at Oxford; BA, Lincoln College, Oxford (first class, *lit. hum.*), 1846; MA, 1849; DD, 1878; fellow of Exeter College, 1847; tutor, 1850; sub-rector, 1857-78; honorary fellow, 1882; regius professor of divinity and canon of Christ Church, 1878-1910; active in administration of Christ Church; an evangelical and moderate Anglican; published *The Life and Times of St. Athanasius* (1896) and doctrinal pamphlets.

INCHCAPE, first EARL OF (1852-1932), shipowner. [See MACKAY, JAMES LYLE.]

INDERWICK, FREDERICK ANDREW (1836-1904), lawyer and antiquary; called to bar (Inner Temple), 1858; QC, 1874; leader in Probate and Divorce division from 1876; commissioner in lunacy, 1903; liberal MP for Rye, 1880-5; mayor of Winchelsea, 1892-3; FSA, 1894; published, besides legal and historical works, *Calendar of the Inner Temple Records, 1505-1714* (3 vols., 1896-1901).

INGE, WILLIAM RALPH (1860-1954), dean of St. Paul's; grandson of Edward Churton [q.v.]; scholar of Eton and King's College, Cambridge; first class, parts i and ii, classical tripos (1882-3); Bell, Porson, and Craven scholar, Chancellor's medallist and Hare prizeman; master at Eton, 1884-8; fellow and tutor, Hertford College, Oxford, and ordained deacon, 1888; priest, 1892; vicar, All Saints', Ennismore Gardens, 1905; Lady Margaret's professor of divinity, Cambridge, and fellow of Jesus, 1907; BD and DD, 1909; dean of St. Paul's, 1911-34; his preaching attracted congregations for the originality of his thought; became known as 'the gloomy dean' for criticism of popular illusions in weekly articles in *Evening Standard*, 1921-46; in *Outspoken Essays* (2 vols., 1919-22) expressed his views on theological and political problems; other publications include *Christian Mysticism* (1899), *Faith and its Psychology* (1909), *The*

Philosophy of Plotinus (Gifford lectures, 2 vols., 1918), *God and the Astronomers* (1933), CVO, 1918; FBA, 1921; KCVO, 1930; several honorary degrees.

INGHAM, ALBERT EDWARD (1900-1967), mathematician; educated at King Edward VI grammar school, Stafford, and Trinity College, Cambridge; first class, part ii, mathematics tripos, 1921; Smith's prizeman, 1923; fellow, Trinity College, 1922; research work, 1922-6; reader, Leeds University, 1926; fellow and director of studies, King's College, Cambridge; university lecturer, 1930; FRS, 1945; reader in mathematical analysis, 1953-7; outstanding member of the G. H. Hardy [q.v.] and J. E. Littlewood school of mathematical analysis; wrote *The Distribution of Prime Numbers* (1932); edited the papers in number-theory in Vol. II of G. H. Hardy's *Collected Papers* (1967).

INGLIS, SIR CHARLES EDWARD (1875-1952), professor of engineering; educated at Cheltenham College and King's College, Cambridge; 22nd wrangler, 1897; first class, part i, mechanical sciences, 1898; trained with Sir John Wolfe-Barry and (Sir) Alexander Gibb [qq.v.]; fellow of King's and assistant to (Sir) Alfred Ewing [q.v.], 1901; to Bertram Hopkinson [q.v.], 1903; lecturer in engineering, 1908; professor of mechanical sciences and head of engineering department at time of great expansion, 1919-43; researches included mechanical vibration and bridge stress; published *Applied Mechanics for Engineers* (1951); FRS, 1930; knighted, 1945; hon. LLD, Edinburgh.

INGLIS, ELSIE MAUD (1864-1917), physician and surgeon; born in India; studied medicine at Edinburgh, Glasgow, and Dublin; joint-surgeon to Edinburgh Bruntsfield Hospital and Dispensary and private practitioner in Edinburgh; inaugurated there maternity hospice staffed by women, 1901; founded Scottish Women's Suffrage Federation (1906) from which sprang Scottish women's hospitals committee (1914); joined Serbian unit, 1915; remained at post at Krushevatz during German and Austrian invasion, 1915-16; organized two units in aid of Serbian division in Russia, 1916; worked at Braila, Galatz, and Reni, remaining till withdrawal of Serbs (1917).

INGOLD, SIR CHRISTOPHER KELK (1893-1970), organic chemist; educated at Sandown grammar school and Hartley University College (later, Southampton University); B.Sc. (London), 1913; research worker at Imperial College, London, 1913-18; research chemist with Cassel Cyanide Co., Glasgow, 1918-20;

lecturer, organic chemistry, Imperial College, 1920; D.Sc. (London), 1921; FRS, 1924; professor, organic chemistry, Leeds University, 1924-30; professor, chemistry, University College, London, 1930-7; director of the laboratories, 1937-61; professor emeritus, 1961; a chemical genius, published 744 papers and a monumental book *Structure and Mechanism in Organic Chemistry* (1st edn., 1953; 2nd edn., 1969); contributions to chemistry recognized by many honours and awards and numerous honorary degrees; president, Chemical Society, 1952-4; knighted, 1958.

INGRAM, ARTHUR FOLEY WINNINGTON- (1858-1946), bishop of London. [See WINNINGTON-INGRAM.]

INGRAM, SIR BRUCE STIRLING (1877-1963), editor of the *Illustrated London News*; grandson of Herbert Ingram [q.v.], founder of the paper; educated at Winchester and Trinity College, Oxford; served apprenticeship with the paper, 1897-1900; editor, 1900-63; also editor of the *Sketch*, 1905-46; served in Royal Garrison Artillery during 1914-18 war; MC (1917); OBE (military, 1918); interested in archaeology and collecting pictures; published *Three Sea Journals of Stuart Times* (1936); presented 700 drawings of Dutch marine artists to National Maritime Museum, 1957; presented other paintings to museums and art galleries; hon. keeper of drawings, Fitzwilliam Museum, Cambridge; vice-president, Society for Nautical Research; vice-president, Navy Records Society; president, Illustrated Newspapers Ltd.; knighted, 1950; Legion of Honour, 1950; hon. D.Litt., Oxford, 1960.

INGRAM, JOHN KELLS (1823-1907), scholar, economist, and poet; brother of Thomas Dunbar Ingram [q.v.]; senior moderator, Trinity College, Dublin, 1842; fellow, 1844; senior fellow, 1884; senior lecturer, 1887; D.Litt., 1891; helped to found Dublin Philosophical Society (1842), and contributed to *Transactions* abstruse papers in pure geometry; contributed sonnets to *Dublin University Magazine*, 1840; composed 'Who fears to speak of Ninety-eight?' which, printed anonymously in *Nation* newspaper (1 Apr. 1843), brought him fame as a popular nationalist poet; member of Royal Irish Academy, 1847; studied mathematics, classics and metaphysics; visited Comte, 1855, and accepted his beliefs; professor of oratory at Trinity College, 1852-66; regius professor of Greek, 1866-77; librarian, 1879-87; vice-provost, 1898-9; started and edited *Hermathena*, 1874; trustee of the National Library of Ireland, 1881; hon. LLD, Glasgow, 1893; president of Royal Irish Academy,

1892-6; helped to found Dublin Statistical Society, 1847; president, 1878-80; prepared for the society 'Considerations on the State of Ireland', 1863; wrote papers on poor law, 1875-6; vindicated economics as integral branch of sociology; contributed articles to *Encyclopaedia Britannica* (9th edn.), of which those on political economy (1885) and slavery (1887) were separately published; his *History of Political Economy* (1888) was translated into most European languages and Japanese; he declared his positivist beliefs in *Outlines of the History of Religion* (1900); published other positivist works from 1901 to 1905, and issued *Sonnets and other Poems* (1900). He distrusted C. S. Parnell [q.v.] and his nationalist associates; opposed South African war, 1899.

INGRAM, THOMAS DUNBAR (1826-1901), Irish historical writer and lawyer; LLB, Queen's College, Belfast, 1853; called to bar, 1856; professor of Hindu law in Presidency College, Calcutta, 1866-77; author of *A Critical Examination of Irish History* (2 vols., posthumous, 1904) and other works.

INNES, JAMES JOHN McLEOD (1830-1907), lieutenant-general Royal (Bengal) Engineers; born in Bengal; joined Bengal Engineers, 1848; in Mutiny in charge of Machi Bhowan fort, and of mining operations at defence of Lucknow, which he described in *Lucknow and Oude in the Mutiny* (1895); awarded VC for gallantry at Sutanpur, 1858; accountant-general in Indian public works department, 1870-7; inspector-general of military works, 1882-6; lieutenant-general, 1886; CB, 1907; published *The Sepoy Revolt of 1857* (1907) and lives of Sir Henry Lawrence (1898) and Sir James Browne (1905).

INNES, SIR JAMES ROSE- (1855-1942), chief justice of South Africa. [See ROSE-INNES.]

INSKIP, THOMAS WALKER HOBART, first VISCOUNT CALDECOTE (1876-1947), lawyer and statesman; educated at Clifton and King's College, Cambridge; called to bar (Inner Temple), 1899; KC, 1914; served in naval intelligence, 1914-18; conservative MP, Central Bristol (1918-29), Fareham (1931-9); solicitor-general, 1922-4, 1924-8, 1931-2; attorney-general, 1928-9, 1932-6; strong evangelical; leading adversary of Prayer Book revision, 1927-8; minister for co-ordination of defence, 1936-9; performed thankless and secret task with patience and integrity; obtained transfer of naval aircraft from Air Force to Navy as Fleet Air Arm; secretary of state for dominions, 1939, 1940; lord chancellor, 1939-40; lord chief justice, 1940-6; knighted, 1922; PC, 1932; viscount, 1939.

INVERCHAPEL, BARON (1882-1951), diplomatist. [See CLARK KERR, ARCHIBALD JOHN KERR.]

INVERFORTH, first BARON (1865-1955), shipowner. [See WEIR, ANDREW.]

INVERNAIRN, BARON (1856-1936), ship-builder. [See BEARDMORE, WILLIAM.]

IQBAL, SIR MUHAMMAD (1876-1938), Indian thinker and poet; educated at Government College, Lahore, Trinity College, Cambridge, and Munich University; called to bar (Lincoln's Inn), 1908; thereafter practised in India; knighted, 1923; president, All-India Moslem League, 1930; delegate, first Round Table conference, 1930-1; his poems in Persian and Urdu on theme of development of individual personality within idealized Islamic community immensely influential.

IRBY, LEONARD HOWARD LOYD (1836-1905), lieutenant-colonel and ornithologist; son of Frederick Paul Irby [q.v.]; entered army, 1854; served at Sevastopol and at Lucknow; at Gibraltar (1864-72) devoted himself to ornithology; retired as lieutenant-colonel, 1874; published *Ornithology of the Straits of Gibraltar* (1875) and *Key List of British Birds* (1888); ardent lepidopterologist.

IRELAND, JOHN NICHOLSON (1879-1962), composer, organist, and pianist; son of Alexander and Annie Ireland [qq.v.]; educated at Leeds grammar school; studied at Royal College of Music, London; studied organ with Sir Walter Parratt [q.v.]; assistant organist and choirmaster, Holy Trinity church, Sloane Street, 1896; studied with (Sir) Charles Stanford [q.v.], 1897-1901; Mus.B. (Durham), 1908; organist and choirmaster, St. Luke's, Chelsea, 1904-26; compositions include Phantasy Trio in A minor (1909), First Violin Sonata in D minor, and Second Violin Sonata (1917); professor of composition, Royal College of Music, 1923-39; FRCM, hon. RAM, 1924; composed Cello Sonata, Piano Sonatina, Piano Concerto in E flat, *A London Overture* and *These Things Shall Be*, 1924-37; hon. doctorate, Durham, 1932; during war years composed *Sasnia* (1941) and the Epic March (1942); followed by *Satyricon* (1946) and music for the film *The Overlanders* (1947); John Ireland Society formed, 1959.

IRELAND, WILLIAM WOTHERSPOON (1832-1909), physician; MD, Edinburgh,

1855; joined East India Company's service, 1856; severely wounded at siege of Delhi, 1856; retired on pension, 1859; medical superintendent of Scottish National Institution for Imbecile Children, 1869; wrote with authority on idiocy and imbecility; applied medico-psychological knowledge to explain lives of celebrated men in *The Blot upon the Brain* (1885) and *Through the Ivory Gate* (1889); wrote also life of Sir Harry Vane, 1905.

IRONSIDE, ROBIN CUNLIFFE (1912-1965), painter and writer; educated at Bradfield and the Courtauld Institute; assistant keeper, Tate Gallery, 1937-46; assistant secretary, Contemporary Art Society, 1938-45; publications include *Wilson Steer* (1943), *Pre-Raphaelite Painters*, *British Painting since 1939*, *David Jones* (1948), and *Andrea del Sarto* (1965); contributed articles to *Horizon*; painted in oils and gouache esoteric subjects such as 'Rose being offered in a Coniferous Wood' and 'Musical Performance by Patients in a Condition of Hypomania'; exhibited at Redfern Gallery, London, 1944, and later, at other galleries in London and New York; made theatrical designs for *Der Rosenkavalier* (1948), *Silvia* (1952), *A Midsummer Night's Dream* (1954), and *La Sylphide* (1960); represented at Tate Gallery, Boston Museum, and Leicester Art Gallery.

IRONSIDE, WILLIAM EDMUND, first BARON IRONSIDE (1880-1959), field-marshal; educated at Tonbridge School and Royal Military Academy, Woolwich; commissioned in Royal Artillery, 1899; served in South African war, 1899-1902; in France, 1914-18; DSO, 1915; CMG, 1918; commanded allied forces in Archangel, 1918-19; major-general and KCB, 1919; commandant, Staff College, Camberley, 1922-6; commander, 2nd division, Aldershot, 1926-8; Meerut District, India, 1928-31; quartermaster-general, India, 1933-6; general, 1935; held Eastern Command, 1936-8; commander-in-chief designate, Middle East, and governor of Gibraltar, 1938-9; GCB, 1938; inspector-general, overseas forces, May 1939; chief of imperial general staff, Sept. 1939-May 1940; planned for armies to defeat Hitler on land; commander-in-chief, Home Forces, preparing against invasion, May-July 1940; field-marshal, 1940; baron, 1941; hon. LLD, Aberdeen; published *Archangel 1918-19* (1953).

IRVINE, SIR JAMES COLQUHOUN (1877-1952), chemist and educationist; educated at Allen Glen's School and Royal Technical College, Glasgow, and St. Andrews and Leipzig universities; junior lecturer, St. Andrews, 1901; D.Sc., 1903; professor of chemistry, 1909-21;

dean of science faculty, 1912-21; researched mainly on carbohydrates; principal, St. Andrews, 1921-52; improved and expanded university; chairman, committee on Indian Institute of Science, 1936; West Indies higher education committee, 1944; Inter-University Council for Higher Education in Colonies, 1946-51; served on Pilgrim Trust and Commonwealth Fund; FRS, 1918; CBE, 1920; knighted, 1925; KBE, 1948.

IRVINE, WILLIAM (1840-1911), Mogul historian; joined Indian civil service, 1863; magistrate and collector in North-West provinces, 1863-89; published *Rent Digest* (1868); wrote on Mogul history for *Journal of Asiatic Society of Bengal*, 1896-1908; published *The Army of the Indian Moghuls* (1903), and translated and edited N. Manucci's account of India (1658-1707), which was published by the Indian government, 1907.

IRVING, SIR HENRY (1838-1905), actor, whose original name was JOHN HENRY BRODRIBB; born at Keinton Mandeville, Somerset; came of yeoman stock; clerk in London firm of East India merchants, 1852-6; made first appearance as actor at Lyceum Theatre, Sunderland, as Gaston in *Richelieu*, 1856; in Edinburgh, 1857-9, where his mannerisms were criticized; first appeared in London, 1859; at Manchester under Charles Calvert [q.v.], 1860-5; first acted role of Hamlet (his first Shakespearian character), 1864; after much provincial work, he toured with J. L. Toole [q.v.], 1866; at Queen's Theatre, London, played Petrucchio to (Dame) Ellen Terry's Katherine, Dec. 1867, and Bill Sikes in Oxenford's *Oliver Twist*, 1869; made first notable success in London as Digby Grant in Albery's *Two Roses* at Vaudeville, 1870; joined company of H. L. Bateman [q.v.] at Lyceum Theatre, 1871; became famous by his acting in *The Bells*, Nov. 1871 to May 1872; took title parts of *Charles I* (1872), *Eugene Aram* (1873), and *Richelieu* (1873); showed originality in conception, but criticized for affectations, monotony, and weakness of voice; scored triumph as Hamlet, 1874; appeared as Macbeth (1875) and Othello (1876), and in *Richard III*, *The Lyons Mail* (1877), and *Louis XI* (1878); became lessee and manager of Lyceum (1878), and began association—lasting till 1902—with (Dame) Ellen Terry [q.v.]; played Hamlet to her Ophelia, 1878; produced *Merchant of Venice* (1880), *The Corsican Brothers* (1880), Tennyson's *The Cup* (1881), *Romeo and Juliet* (elaborately mounted), and *Much Ado About Nothing* (1882-3); made first of eight tours to America, 1883-4; on return produced *Twelfth Night*, May 1884; gave weird and striking

impersonation of Mephistopheles in Wills's *Faust*, 1885-7; revived *Macbeth*, 1888-9; with Ellen Terry he appeared before Queen Victoria at Sandringham, Apr. 1889; produced *Henry VIII* (with splendid staging), Jan. 1892, and *King Lear*, Nov. 1892; Tennyson's *Becket* (Feb. 1893) proved one of Irving's greatest personal and financial triumphs; made fourth and most successful American tour, 1893-4; produced *Waterloo* (1894), *King Arthur* (1895), *Cymbeline* (Sept. 1896), and *Richard III* (Dec.); suffered pecuniary losses; transferred control of Lyceum to a company, 1898; appeared in Sardou's *Robespierre* (1899), and in *Coriolanus* (1901); revived *Faust* and *Merchant of Venice* at Lyceum, 1902; last appearance there, 19 July; produced at Drury Lane Sardou's *Dante*, 1903; made eighth American tour, 1903-4; died in Bradford after acting in *Becket* earlier in the evening, 1905; his ashes buried in Westminster Abbey; contributed to the *Nineteenth Century*, 'An Actor's Notes', 1877-87; superintended the 'Henry Irving Shakespeare', 1888; received honorary degrees from Dublin, Cambridge, and Glasgow; first actor to be knighted, 1895; Rede lecturer at Cambridge, 1898; an intellectual actor, of magnetic personality, but of mannered elocution and gait, he attracted the intelligent playgoer back to the theatre and revived popular interest in Shakespeare; he neglected modern drama and depended much on sumptuous and elaborate mountings for which he secured assistance from leading artists and composers.

ISAACS, ALICK (1921-1967), medical scientist; grandson of Lithuanian immigrant; educated at Pollokshields secondary school and Glasgow University; MB, Ch.B., 1944; studied streptococci bacteria; Medical Research Council student, Sheffield University, 1947; worked on influenza; with Rockefeller fellowship worked on influenza viruses in Melbourne, 1948-50; returned to work at National Institute for Medical Research, Mill Hill, 1951; with Dr J. Lindenmann, discovered 'interferon'; of fundamental importance in biology, as well as defence against viruses; worked also with World Influenza Centre under World Health Organization; MD with honours, Glasgow, 1955; head of virus division, Mill Hill, 1961-7; FRS, 1966; hon. MD, Louvain, 1962.

ISAACS, SIR ISAAC ALFRED (1855-1948), chief justice and governor-general of Australia; graduated in law, Melbourne University, and called to Victorian bar, 1880; QC, 1899; member, Victorian legislative assembly, 1892-1901; solicitor-general, 1893; attorney-general, 1894-9, 1900-1; MHR, 1901-6; attorney-general, 1905-6; high court judge, 1906-30; chief justice, 1930-1; his judgments on Australian constitutional law especially important; held Commonwealth legislation to bind the states; first Australian-born governor-general, 1931–6; PC, 1921; KCMG, 1928; GCB, 1937.

ISAACS, RUFUS DANIEL, first Marquess of Reading (1860–1935), lord chief justice of England, ambassador to the United States, and viceroy of India; son of Jewish fruit merchant in Spitalfields; entered family business at fifteen; ship's boy on *Blair Athole*, 1876–7; jobber on stock exchange, 1880–4; called to bar (Middle Temple), 1887; established busy practice dealing especially with commercial and trade-union law; QC, 1898; in front rank of advocates; triumphs include the Taff Vale litigation (1902) and prosecution of Whitaker Wright (q.v., 1904); liberal MP, Reading, 1904–13; solicitor-general (Mar.), attorney-general (Oct.), 1910; excellent law officer; led in *Rex* v. *Mylius* (1911) and secured conviction of Seddon (1912); PC and KCVO, 1911; entered cabinet, 1912; implicated (1912–13) in controversy over Marconi Company for which his brother as joint managing director had obtained a Post Office contract; in repudiating the untrue charge that he had exercised influence in favour of his brother did not think it necessary to state that he himself held shares in the American company which had no interest in the contract; this error of judgment caused him to be fiercely attacked when the fact became known; acquitted by House of acting otherwise than in good faith; lord chief justice, 1913–21; baron, 1914; secured loan of 500 million dollars to be spent in United States, Oct. 1915; viscount, 1916; high commissioner for finance in United States and Canada, Sept.–Nov. 1917; earl, 1917; ambassador and high commissioner in United States, 1918–19; accelerated supplies of food and troops; viceroy of India 1921–6; watched over early application of Montagu–Chelmsford reforms; approached problems in liberal spirit and showed much patience before resorting to extreme measures; put forward government of India's request for revision of treaty of Sèvres with result that Lausanne treaty went far to relieve Moslem anxieties about treatment of Turkey; marquess, 1926; director, later president, Imperial Chemical Industries; took prominent part in Indian Round Table conference, 1930–1; foreign secretary, Aug.–Oct. 1931; lord warden of cinque ports, 1934–5.

ISHERWOOD, Sir JOSEPH WILLIAM, first baronet (1870–1937), ship designer; surveyor to Lloyd's Register, 1896–1907; practised as naval architect; invented 'longitudinal', 'combination', and 'bracketless' systems of ship construction and 'arcform' hull design; baronet, 1921.

ISITT, Dame ADELINE GENÉE (1878–1970), ballet dancer. [See Genée.]

ISLINGTON, Baron (1866–1936). [See Poynder, Sir John Poynder Dickson-.]

ISMAIL, Sir MIRZA MOHAMMAD (1883–1959), Indian administrator and statesman; born in Bangalore; educated with Maharaja of Mysore [q.v.]; BA, Madras, 1905; joined Maharaja's staff; private secretary, 1923; dewan, 1926–41; worked with Maharaja in perfect accord for development of state with some state socialism; created pleasure gardens and made state famous for its beauty; represented Mysore at Round Table conferences, 1930–2, and joint select committee meetings; prime minister, Jaipur, 1942–6; Hyderabad, 1946–7; CIE, 1924; knighted, 1930; KCIE, 1936; published *My Public Life* (1954).

ISMAY, HASTINGS LIONEL, Baron Ismay (1887–1965), general; educated at Charterhouse and the Royal Military College, Sandhurst; entered Indian Army, 1905; posted to 21st Prince Albert Victor's Own Cavalry, 1907; seconded to King's African Rifles, 1914; served in Somaliland with Camel Corps, 1919–20; DSO; passed through Staff College, Quetta, 1921; posted to RAF Staff College, Andover, 1924; assistant secretary, Committee of Imperial defence, under Sir Maurice (later, Lord) Hankey [q.v.], 1925–30; CB, 1931; military secretary to viceroy, Lord Willingdon [q.v.], 1931–3; War Office, 1933–6; deputy secretary, Committee of Imperial Defence, 1936–8; succeeded Hankey as secretary, 1938; Churchill's essential link with chiefs of staff; major-general, 1939; lieutenant-general, 1942; full general, 1944; CH, 1945; GCB, 1946; baron, 1947; retired from army, 1946; chief of viceroy's staff under Lord Mountbatten, 1946–8; chairman of council for Festival of Britain, 1948–51; secretary of state for Commonwealth Relations, 1951; first secretary-general of NATO, 1951–7; KG; PC, 1951; hon. degrees, Belfast, Bristol, and Cambridge; president, National Institute for the Blind, 1952; deputy lieutenant for Gloucestershire, 1950; published *Memoirs* (1960); known to all as 'Pug'.

ISMAY, JOSEPH BRUCE (1862–1937), shipowner; son of T. H. Ismay [q.v.], founder of White Star line; educated at Harrow; partner in Ismay, Imrie & Co. (1891), chairman, 1899; chairman (1904–12), International Mercantile Marine Company; directorships included London Midland & Scottish Railway; inaugurated cadet ship *Mersey*.

IVEAGH, COUNTESS OF (1881-1966), philanthropist. [See under GUINNESS, RUPERT EDWARD CECIL LEE.]

IVEAGH, first EARL OF (1847-1927), philanthropist. [See GUINNESS, EDWARD CECIL.]

IVEAGH, second EARL OF (1874-1967), philanthropist. [See GUINNESS, RUPERT EDWARD CECIL LEE.]

IWAN-MÜLLER, ERNEST BRUCE (1853-1910), journalist; BA, New College, Oxford (first class in *lit. hum.*), 1876; editor of the tory *Manchester Courier*, 1884-93; assistant editor of *Pall Mall Gazette*, 1893-6; leader-writer for *Daily Telegraph*, 1896-1910; visited South Africa during Boer war, Ireland (1907), and Paris (1908); well versed in foreign politics; published *Lord Milner in South Africa* (1902) and *Ireland To-day and To-morrow* (1907).

J

JACKS, LAWRENCE PEARSALL (1860-1955), Unitarian divine; educated at University School, Nottingham, and Manchester New College, London; MA, 1886; Hibbert scholar, Harvard, 1886; Unitarian minister, Renshaw Street chapel, Liverpool, 1888; church of the Messiah, Birmingham, 1894-1903; first editor, *Hibbert Journal*, 1902-47; lecturer in philosophy, Manchester College, Oxford, 1903; principal, 1915-31; publications include *The Education of the Whole Man* (1931); hon. degrees from several universities.

JACKS, WILLIAM (1841-1907), ironmaster and author; started ironworks at Glasgow, 1880; liberal MP for Leith Burghs, 1885-6, for Stirling, 1892-5; wrote lives of Bismarck (1899) and James Watt (1901); hon. LLD, Glasgow, 1899; bequeathed £20,000 to Glasgow University for chair of modern languages.

JACKSON, SIR BARRY VINCENT (1879-1961), theatre director; son of George Jackson, a wealthy provision merchant who founded the Maypole Dairies; while a boy, developed love for the theatre; founded the amateur theatrical Pilgrim Players with friends, including John Drinkwater [q.v.], 1907; the Players developed into the Birmingham Repertory Theatre, 1913; among productions were *The Immortal Hour* (1921), *Back to Methuselah*, and *Cymbeline* (1923); moved to London and presented *The Farmer's Wife* (1924) and *Hamlet* in modern dress (1925); knighted, 1925; planned Malvern summer festival, 1929; worked for three seasons at the Shakespeare Memorial Theatre, 1946-8; concentrated on Birmingham Repertory Theatre, 1948-61; wrote, translated, and adapted plays; hon. MA and D.Litt., Birmingham; LLD, St. Andrews; D.Litt., Manchester; director, Royal Opera House, 1949-55.

JACKSON, SIR CYRIL (1863-1924), educationist; BA, New College, Oxford; inspector-general of schools and permanent head of education department, Western Australia, 1896-1903; chief inspector, Board of Education, 1903-6; served on numerous commissions and committees on unemployment, boy labour, etc.; member of London County Council, Limehouse division, 1907-13; chairman of Council, 1915; KBE, 1917.

JACKSON, SIR (FRANCIS) STANLEY (1870-1947), cricketer and administrator; son of first Baron Allerton [q.v.]; educated at Harrow and Trinity College, Cambridge; captain of cricket, 1892-3; played for Yorkshire and England; captained England in test matches against Australia, 1905; raised and commanded 2nd/7th West Yorkshire regiment, 1914-17; conservative MP, Howdenshire division, 1915-27; parliamentary and financial secretary, War Office, 1922-3; chairman, conservative party organization, 1923-7; governor of Bengal, 1927-32; PC, 1926; GCIE, 1927; GCSI, 1932.

JACKSON, FREDERICK GEORGE (1860-1938), explorer, soldier, and big-game hunter; organized Jackson-Harmsworth polar expedition surveying Franz Josef Land, 1894-7; served with distinction in South African and 1914-18 wars; travelled through tropical Africa in search of sport, 1925-6; served on international commission of inquiry into slavery in Liberia (report, 1930).

JACKSON, SIR FREDERICK JOHN (1860-1929), explorer, naturalist, and administrator; joined service of Imperial British East Africa Company, 1888; first class administrative assistant, Uganda, 1894; vice-consul, 1895; deputy commissioner, 1896; helped to quell mutiny of Sudanese troops employed in Uganda, 1897; deputy commissioner, East Africa Protectorate, under Sir C. N. E. Eliot [q.v.], 1902; lieutenant-governor of Protectorate, 1907; governor of Uganda, 1911-17; formed collection of over 12,000 specimens of birds; CMG, 1902; KCMG, 1913.

JACKSON, FREDERICK JOHN FOAKES (1855-1941), divine; educated at Eton and Trinity College, Cambridge; first class, theology, and deacon, 1879; priest, 1880; chaplain and lecturer in divinity and Hebrew, Jesus College, 1882; fellow, 1886-1941; dean, 1895-1916; Briggs professor of Christian institutions, Union Theological Seminary, New York, 1916-34; honorary canon of Peterborough, 1901-26; publications include *History of the Christian Church* (1891), *A History of Church History* (1939), and first 3 vols. (with Kirsopp Lake, q.v.) of *The Beginnings of Christianity* (5 vols., 1920-33).

JACKSON, HENRY (1839-1921), regius professor of Greek at Cambridge; BA, Trinity College, Cambridge; fellow, 1864; praelector in ancient philosophy, 1875; vice-master, 1914; regius professor of Greek, 1906; OM, 1908; principal contribution to learning, his doctrine of Plato's 'later theory of Ideas'.

JACKSON, SIR HENRY BRADWARDINE (1855-1929), admiral of the fleet and pioneer of wireless telegraphy; entered navy, 1868; qualified as torpedo lieutenant; conceived idea of employing wireless waves to announce to capital ships approach of friendly torpedo boat, 1890; as commander of *Defiance* successfully experimented with coherers, 1895-6; met Guglielmo Marconi, 1896; contract placed with Marconi Company for supply of wireless installation to many ships of royal navy, 1900; communicated important paper on transmission of electric waves to Royal Society, 1902; assistant director of torpedoes, Admiralty, 1902; third sea lord and controller of navy, 1905-8; commanded third cruiser squadron, Mediterranean, 1908; chief of war staff, Admiralty, 1913; first sea lord, 1915; president, Royal Naval College, Greenwich, 1916-19; admiral of the fleet, 1919; chairman, radio research board, Department of Scientific and Industrial Research; FRS, 1901; KCVO, 1906.

JACKSON, SIR HERBERT (1863-1936), chemist; successively student, demonstrator, lecturer, professor of organic chemistry (1905), fellow (1907), and Daniell professor of chemistry (1914-18), King's College, London; his 'focus-tube' (1896) the prototype of later X-ray tubes; determined formulae for many kinds of glass, 1914-15; KBE and FRS, 1917; first director of research, British Scientific Instrument Research Association, 1918-33.

JACKSON, JOHN (1833-1901), professional cricketer; member of Nottinghamshire cricket eleven; leading bowler in England, 1857; played for Players *v.* Gentlemen, 1859-64; visited America, 1859, and Australia, 1863; often caricatured by John Leech [q.v.] in *Punch* as 'demon bowler'.

JACKSON, JOHN HUGHLINGS (1835-1911), physician; studied medicine at York and St. Bartholomew's Hospital; MRCS, 1856; MD, St. Andrews, 1860; physician at London Hospital (1863-94) and at National Hospital for Epileptics, Queen Square (1862-1906); FRCP, 1868; Gulstonian lecturer, 1869; Croonian lecturer, 1884; Lumleian lecturer, 1890; FRS, 1878; studied speech defect in brain disease, the occurrence of local epileptic discharges (known as Jacksonian epilepsy), and showed that certain regions of the brain were definitely related to certain limb movements; formulated theory of levels in the nervous system, showing that the highest and most recently developed functions go first in process of disease; also investigated 'uncinate' epilepsy; one of the first to use ophthalmoscope in England in diagnosing eye diseases; published many scientific papers.

JACKSON, MASON (1819-1903), wood-engraver; pupil of brother, John Jackson [1801-48, q.v.]; art editor to *Illustrated London News*, 1860; published *The Pictorial Press: its Origin and Progress* (1885).

JACKSON, SAMUEL PHILLIPS (1830-1904), water-colour artist; born at Bristol; son of Samuel Jackson [q.v.]; painted land- and sea-scapes; exhibited mainly at Royal Water Colour Society; member, 1876; work praised by John Ruskin [q.v.]; showed preference for Devon and Cornish coast scenes and Thames views; skilful interpreter of West Country atmosphere; an efficient photographer.

JACKSON, SIR THOMAS GRAHAM, first baronet (1835-1924), architect; BA, Wadham College; entered office of Sir G. G. Scott [q.v.], 1858; contributed greatly towards changing appearance of Oxford; his Oxford buildings include: Examination Schools (foundations laid 1876); new buildings at Brasenose College (1886-9, 1907-11); chapel of Hertford College (finished 1908); his outstanding achievement in buildings connected with schools, Giggleswick School chapel; his buildings especially notable for their ornament; RA, 1896; baronet, 1913.

JACKSON, WILLIAM LAWIES, first BARON ALLERTON (1840-1917), politician; made father's almost bankrupt tanning business one of the largest in England; conservative MP, North Leeds, 1880; financial secretary to Treasury, 1885-91; PC, 1890; FRS, 1891;

chairman of Jameson raid inquiry, 1896-7; baron, 1902; rendered great financial services to Leeds.

JACKSON, WILLIS, BARON JACKSON OF BURNLEY (1904-1970), electrical engineer and educationist; educated at Burnley grammar school and Manchester University; first class, B.Sc., 1925; M.Sc., 1926; lecturer in electrical engineering, Bradford Technical College, 1926-9; joined Metropolitan-Vickers Company, 1929; further experience at Manchester College of Technology, 1930-3; research at Oxford, 1933-6; D.Phil. (Oxford), D.Sc. (Manchester), 1936; returned to Metropolitan-Vickers; appointed professor, Electrotechnics Department, Manchester University, 1939; professor, electrical engineering, Imperial College of Science and Technology, London, 1946-53; FRS, 1953; director, research and education, Metropolitan-Vickers, 1954; returned to Imperial College, 1961; pro-rector, 1967; member of number of commissions and committees on education and research, including the University Grants Committee; chairman, Industrial Research Committee of FBI (1958-60); published many books, articles, and reports; chairman, Scientific Manpower Committee (1963-4); knighted, 1958; life baron, 1967; hon. degrees from many universities; president, Institution of Electrical Engineers (1959-60) and the British Association for the Advancement of Science (1967).

JACOB, SIR CLAUD WILLIAM (1863-1948), field-marshal; educated at Sherborne and Sandhurst; commissioned in Worcestershire regiment, 1882; transferred to Indian army, 1884; served mainly on frontier; accompanied Meerut division to France, 1914; commanded Dehra Dun brigade, 1915; Meerut division, 1915; 21st division, 1915-16; II Corps, 1916-19; KCB and lieutenant-general, 1917; chief of general staff, India, 1920-4; general, 1920; held Northern command, India, 1924-5; commander-in-chief, 1925; secretary, military department, India Office, 1926-30; field-marshal and GCB, 1926; GCSI, 1930.

JACOB, EDGAR (1844-1920), bishop of Newcastle and of St. Albans; grandson of John Jacob [q.v.], Guernsey topographer; educated at Winchester and New College, Oxford; domestic chaplain to Robert Milman [q.v.], bishop of Calcutta, 1872-6; vicar of Portsea, 1878-95; bishop of Newcastle, 1896-1903; of St. Albans, 1903-19; successfully worked for division of latter see.

JACOBS, WILLIAM WYMARK (1863-1943), writer; grew up in Wapping; educated at Birkbeck College; Civil Service clerk, 1879-99;

wrote short stories relating misadventures of sailormen ashore, ingenuities of artful dodger in country village, and tales of macabre; publications include *Many Cargoes* (1896), *The Skipper's Wooing* (1897), *Light Freights* (1901), *The Lady of the Barge* (1902, containing *The Monkey's Paw* dramatized by L. N. Parker, q.v.), and *Night Watches* (1914); collaborated with Parker in *Beauty and the Barge* (1904); wrote *At Sunwich Port* (1902) and several other novels.

JAGGER, CHARLES SARGEANT (1885-1934), sculptor; learnt silver-engraving with Mappin & Webb; studied at Sheffield School of Art and Royal College of Art; served in 1914-18 war and awarded MC; works include Artillery (Hyde Park Corner), GWR (Paddington), British Memorial to Belgium (Brussels) and many other war memorials, and stone groups at Imperial Chemical House, Millbank; ARA, 1926.

JAMES, ALEXANDER WILSON (1901-1953), footballer; joined Raith Rovers, 1922, Preston North End, 1925; transferred to Arsenal for £9,000, 1929, then the second-highest sum ever paid; an inside forward, his tactical flair made him the mainspring of Arsenal's success in winning the League four times and the Cup twice during his eight seasons; retired, 1937; received eight Scottish caps.

JAMES, ARTHUR LLOYD (1884-1943), phonetician; graduated in French, University College, Cardiff, 1905; in medieval and modern languages, Trinity College, Cambridge, 1910; taught at Islington Training College, 1910-14; lecturer in phonetics, University College, London, 1920-7; studied West African languages; head of phonetics department, School of Oriental Studies, 1927-41; reader, 1930; professor, 1933; linguistic adviser to BBC, 1929-41.

JAMES, HENRY, BARON JAMES OF HEREFORD (1828-1911), lawyer and statesman; interested in cricket; president of MCC, 1889; called to bar, 1852; an excellent criminal advocate; QC, 1869; treasurer of Middle Temple, 1888; liberal MP for Taunton, 1869-85; solicitor-general under Gladstone, Sept. 1873; attorney-general, Nov. 1873, and 1880-5; drafted and carried corrupt practices bill, 1883; PC, 1885; opposed Gladstone's Irish policy and joined liberal unionists; unionist MP for Bury, 1885-95; resumed private bar practice, appeared for *The Times*, before Parnell commission, 1888-9; summing up clients' case in a notable twelve days' speech; attorney-general of duchy of Cornwall, 1892-5; hon. LLD, Cambridge,

1892; baron, and joined unionist cabinet as chancellor of the duchy of Lancaster, 1895; resigned office, 1902; GCVO, 1902; opposed to A. J. Balfour's education policy and tariff reform proposals; served on judicial committee of Privy Council from 1896; able chairman of Coal Conciliation Board, 1898-1909; opposed rejection of budget by House of Lords, 1909.

JAMES, HENRY (1843-1916), novelist; born in New York; educated in New York, London, Paris, and Geneva; studied law at Harvard; settled in Europe, 1875; lived in London, 1876-98, and at Rye, 1898-1916; found material for novels in society life of London; naturalized, 1915; OM, 1916; work as novelist falls into three 'periods'; in first, chiefly occupied with 'international' subject, impact of American life upon European civilization; novels of this period, *Roderick Hudson* (1875), *The American* (1877), *Daisy Miller* (1879), and *The Portrait of a Lady* (1881); in second period dealt with subjects from English life, social, political, and artistic; novels of this period, *The Tragic Muse* (1890), *The Spoils of Poynton* (1897), *What Maisie Knew* (1897), and *The Awkward Age* (1899); began to develop extremely intricate style to match subtlety of perceptions and discriminations; in third period, which contains *The Wings of the Dove* (1902), *The Ambassadors* (1903), and *The Golden Bowl* (1904), returned to 'international' theme; explored with greater thoroughness than any previous novelist nature and possibilities of art of fiction; in later works subjects appear hardly equal to immense elaboration of treatment; also wrote short stories, plays, and reminiscences.

JAMES, JAMES (1832-1902), joint-composer with his father of 'Land of My Fathers', the Welsh national anthem, 1856; it attracted public favour at eisteddfodau at Pontypridd, 1857, and Llangollen, 1858, was included in *Gems of Welsh Melody* (1860), and was sung at Bangor national eisteddfod, 1874, serving thenceforth as Welsh national anthem.

JAMES, MONTAGUE RHODES (1862-1936), biblical scholar, antiquary, and palaeographer; scholar of Eton and King's College, Cambridge; first class, parts i and ii, classical tripos (1884-5); assistant, Fitzwilliam Museum (1886), director (1893-1908); fellow of King's (1887) and lecturer in divinity; dean (1889-1900), tutor (1900-2), provost (1905-18); vice-chancellor of Cambridge, 1913-15; provost of Eton, 1918-36; reconstructed painted windows of King's College chapel and revealed the fifteenth-century wall-paintings in Eton chapel; catalogued (1895-1932) the Western manuscripts at Cambridge, Eton, Lambeth, West-

minster Abbey, John Rylands, and Aberdeen University libraries, and others in private collections; his studies of Apocryphal literature made it a comprehensible documentation of human thought and life; also wrote extensively on medieval arts and literature, translated texts, and wrote ghost stories; FBA, 1903; OM, 1930.

JAMES, REGINALD WILLIAM (1891-1964), physicist; educated at the Polytechnic Day School, Regent Street, the City of London School and St. John's College, Cambridge; first class, parts i and ii, natural sciences tripos, 1911-12; worked in Cavendish Laboratory with (Sir) W. L. Bragg [q.v.], 1913-14; joined Shackleton's [q.v.] Antarctic Expedition as physicist, 1914-16; served in Royal Engineers, 1917-18; appointed physics lecturer, Manchester University, where Bragg was professor, 1919; senior lecturer, 1921; reader in experimental physics, 1934; worked with Bragg on experimental measurements of X-ray reflection; professor of physics, Cape Town, 1936-56; established research in crystallography; published *The Optical Principles of the Diffraction of X-rays* (1948); FRS, 1955; fellow, Cape Town University, 1949-56; president, Royal Society of South Africa, 1950-3; acting principal and vice-chancellor, Cape Town University, 1957.

JAMES, ROLFE ARNOLD SCOTT- (1878-1959), journalist, editor, and literary critic. [See SCOTT-JAMES.]

JAMESON, ANDREW, LORD ARDWALL (1845-1911), Scottish judge; MA, St. Andrews, 1865; hon. LLD, 1905; called to Scottish bar, 1870; sheriff of Roxburghshire, 1886, Ross, 1890, and Perthshire, 1891; raised to bench, 1905; frequent arbitrator in industrial disputes; interested in Scottish Free Church affairs.

JAMESON, SIR LEANDER STARR, baronet (1853-1917), South African statesman; born in Edinburgh; MD, London, 1877; went out as doctor to Kimberley, 1878; formed friendship with Cecil John Rhodes [q.v.]; undertook three missions to Matabele chief, Lobengula, 1889-90; in consequence, effective confirmation of mineral rights concession obtained by British South Africa Company; accompanied, as Rhodes's personal representative, expedition of A. R. Colquhoun, first administrator of Mashonaland, 1890; administrator of Mashonaland, 1891; accompanied force dispatched by company to punish Lobengula for raids by Matabele 'impis' on Mashonas under British protection; Bulawayo occupied, 1893; Matabeleland placed under Jameson's administration, 1894; CB, 1894; carried out famous 'raid' over Transvaal border in 'uitlander' interests,

Dec. 1895; surrendered to Boer commandant, P. A. Cronje, Jan. 1896; handed over to British authorities and sent to England for trial; imprisoned in Holloway jail, but quickly released; entered Cape parliament as member for Kimberley, 1900; succeeded Rhodes as leader of progressive party at Cape, 1902; prime minister of Cape Colony, 1904-8; attended imperial conference in London, and PC, 1907; with General Botha [q.v.] played chief part in South African national convention, 1908-9; entered first Union parliament as member for Harbour division of Cape Town, 1910; leader of opposition; baronet, 1911; retired, 1912; died in London.

JAMESON, Sir (WILLIAM) WILSON (1885-1962), professor of public health, medical officer, and medical adviser; educated at Aberdeen grammar school and King's College, Aberdeen; BA, 1905; transferred to Marischal College and qualified MB, Ch.B., with distinction, 1909; MD (Aberdeen), 1912; worked in London hospitals and in Eastbourne, 1913-14; MRCP (London), 1913; DPH, 1914; served in RAMC 1915-19; medical officer of health, Finchley, and deputy medical officer of health, Marylebone, 1920; published *Synopsis of Hygiene* (with F. T. Marchant, 1920); called to bar (Middle Temple), 1922; served in further posts as medical officer of health and lecturer, 1925-8; professor of public health, London School of Hygiene and Tropical Medicine, 1929; dean, 1931; part-time medical adviser, Colonial Office, 1940; chief medical officer, Ministry of Health and Board of Education, 1940-50; closely concerned with plans for National Health Service; medical adviser, King Edward's Hospital Fund for London, 1950-60; FRCP, 1930; knighted, 1939; KCB, 1943; GBE, 1949; received many hon. degrees; master of the Society of Apothecaries.

JAPP, ALEXANDER HAY (1837-1905), author and publisher; at first a tailor's bookkeeper in Edinburgh; attended classes in Edinburgh University, 1860-1; literary adviser in London to Isbister & Co., 1864; assisted in editing *Good Words*, *Sunday Magazine*, and *Contemporary Review*; LLD, Glasgow, 1879; FRS, Edinburgh, 1880; wrote much under various pseudonyms, the chief being 'H. A. Page' and 'A. F. Scot'; edited De Quincey's *Posthumous Works* (1891-3) and *De Quincey Memorials* (1891); his works include studies of Hawthorne (1872), of Thoreau (1878), of De Quincey (2 vols., 1877), and R. L. Stevenson (of whose *Treasure Island* he had negotiated the publication) (1905); he also published verse and studies in German literature, natural history, and anthropology.

JARDINE, DOUGLAS ROBERT (1900-1958), cricketer; born in Bombay; educated at Winchester and New College, Oxford; qualified as solicitor, 1926; played cricket and tennis for Oxford; became notable Surrey player, heading English averages, 1927, 1928; played 22 tests, 15 as captain, making 1,296 runs, averaging 48; played in all five test matches in Australia, 1928-9; captain, England against New Zealand, 1931; took team to Australia, 1932-3; involved in bitter controversy arising from 'body-line bowling'; took MCC team to India, 1933-4; published *In Quest of the Ashes* (1933) and *Cricket* (1936).

JARDINE, Sir ROBERT, first baronet (1825-1905), East India merchant and racehorse owner; partner in uncle's London firm of East India merchants, 1859; succeeded uncle as head of firm, and inherited his property in Perthshire and Dumfriesshire, 1881; liberal MP for Ashburton, 1865, for Dumfries Burghs, 1868-74 and 1880-92; baronet, 1885; keen agriculturist and breeder of stock; won Two Thousand Guineas and Derby with Pretender (1869), the Cesarewitch (1877), and the Waterloo (coursing) Cup (1873); art collector.

JARVIS, CLAUDE SCUDAMORE (1879-1953), soldier, administrator, and orientalist; served in South Africa, 1899-1902; in France, Egypt, and Palestine, 1914-18; joined Egyptian Frontiers Administration, 1918; governor of Sinai, 1922-36; settled tribal feuds and became legendary figure for knowledge of Arabic and Bedouin customs and law; publications include *Yesterday and Today in Sinai* (1931); joined staff of *Country Life*; CMG, 1936.

JARVIS, Sir JOHN LAYTON (JACK) (1887-1968), racehorse trainer; son and grandson of racehorse trainers; educated at Cranleigh School; apprenticed as jockey to his father; rode winner of Cambridgeshire at age of sixteen; set up as trainer, 1914; first major success (Ellangowan in the Two Thousand Guineas), 1921; won again in 1928 and 1939; also won One Thousand Guineas three times, the St. Leger once and the Derby twice (Blue Peter, 1939, and Ocean Swell, 1944); trained for the fifth Earl of Rosebery [q.v.], and his son, 1922-68; keen on shooting and hare-coursing; won the Waterloo Cup; knighted for services to racing, 1967; published autobiography, *They're Off* (1969).

JAYNE, FRANCIS JOHN (1845-1921), bishop of Chester; educated at Rugby and Wadham College, Oxford; first class in classics and law and modern history; fellow of Jesus College, Oxford, 1868; ordained, 1870; tutor of Keble College, Oxford, 1871; principal of St.

David's College, Lampeter, 1879-86; vicar of Leeds, 1886-9; bishop of Chester, 1889-1919; a moderate churchman with exceptional administrative talents.

JEAFFRESON, JOHN CORDY (1831-1901), author; BA, Pembroke College, Oxford, 1852; private tutor in London, 1852-8; published many novels, including *Live it Down* (1863); made some repute as an anecdotal anthologist in books 'about doctors' (1860), 'about lawyers' (1866), and 'about the clergy' (1870); called to bar, 1859, but did not practise; inspector of MSS for Royal Historical MSS Commission, 1874-87; author of *The Real Lord Byron* (1883), *The Real Shelley* (1885), *Lady Hamilton and Lord Nelson* (1888), and *The Queen of Naples and Lord Nelson* (1889).

JEANS, SIR JAMES HOPWOOD (1877-1946), mathematician, theoretical physicist, astronomer, and popular expositor of physical science and astronomy; educated at Merchant Taylors' and Trinity College, Cambridge; bracketed second wrangler, 1898; fellow, 1901; university lecturer in mathematics (1904-5), Stokes lecturer (1910-12), Cambridge; professor of applied mathematics, Princeton, 1905-9; FRS, 1906; an honorary secretary, Royal Society, 1919-29; president, British Association, 1934; professor of astronomy, Royal Institution, 1935-46; publications include *Dynamical Theory of Gases* (1904), *Theoretical Mechanics* (1906), *The Mathematical Theory of Electricity and Magnetism* (1908), and *Problems of Cosmogony and Stellar Dynamics* (1919); popular scientific works include *The Universe Around Us* (1929), *The Mysterious Universe* (1930), *The Stars in their Courses* (1931), and *Through Space and Time* (1934); knighted, 1928; OM, 1939.

JEBB, EGLANTYNE (1876-1928), philanthropist; niece of Sir R. C. Jebb [q.v.]; educated at Lady Margaret Hall, Oxford; engaged in social work and travel, 1900-14; initiated 'Save the Children Fund', 1919; movement spread into forty countries and sums contributed reached £5½ million; thousands of Greek, Bulgarian, Romanian, Armenian, Polish, and Russian children fed and provided for; work made permanent by League of Nations' Declaration of Geneva (children's charter), 1924; died at Geneva.

JEBB, SIR RICHARD CLAVERHOUSE (1841-1905), Greek scholar; born at Dundee; great-nephew of John Jebb [q.v.]; educated at Charterhouse and Trinity College, Cambridge; Porson scholar, 1859; Craven scholar, 1860; senior classic and first chancellor's medallist,

1862; fellow and classical lecturer, 1863; honorary fellow, 1888; public orator, 1869; helped to found Cambridge Philological Society, 1868; published *The Characters of Theophrastus* (1870), and translations into Greek and Latin verse (1873); professor of Greek at Glasgow, 1875-89; lectured upon modern as well as classical Greek; visited Greece, 1878; wrote *Attic Orators* (2 vols., 1876), *Primer of Greek Literature* (1877), *Modern Greece* (1880), monograph on Bentley (in 'Men of Letters' series, 1882), and *Homer* (1887); visited America, 1884; hon. LLD, Harvard; regius professor of Greek at Cambridge, 1889-1905; conservative MP for the university, 1891-1905; served on many education commissions; Rede lecturer, Cambridge, 1890; Romanes lecturer, Oxford, 1899; delivered lectures at Johns Hopkins University (1892) on *The Growth and Influence of Greek Poetry* (published 1893); edited Bacchylides, 1905; chief work was edition of Sophocles, 7 vols., 1883-96; an eighth volume containing the Fragments was unfinished; helped to found Society for Promotion of Hellenic Studies, 1879, and the British School of Archaeology at Athens, 1887; FBA, 1902; close friend of Tennyson; received honorary degrees from Edinburgh, Dublin, Cambridge, Oxford, and Bologna; professor of ancient history, Royal Academy, 1898; knighted, 1900; trustee of British Museum, 1903; OM, 1905; *Essays and Letters* and *Life and Letters* edited by widow (1907).

JEFFERSON, SIR GEOFFREY (1886-1961), neuro-surgeon; educated at Manchester grammar school and Manchester University; qualified, 1909; appointed demonstrator in anatomy under (Sir) Grafton Elliot Smith [q.v.], 1911; FRCS, 1911; MS (London), 1913; worked in Anglo-Russian Hospital, Petrograd, 1915-17; worked at Salford on effects of injury on the nervous system, localization of cerebral tumours and the problem of consciousness; helped to found the Society of British Neurological Surgeons; secretary, 1926-52; worked in Manchester and London; professor, neuro-surgery, Manchester, 1939-51; consultant adviser to Ministries of Health and Pensions, 1939; FRS, FRCP, 1947; CBE, 1943; knighted, 1950; member, Medical Research Council, 1948-52; chairman, Clinical Research Board, 1953-9.

JEFFERY, GEORGE BARKER (1891-1957), mathematician and educationist; educated at Strand School, Wilson's grammar school, Camberwell, and University College, London; B.Sc., 1912; assistant to L. N. G. Filon [q.v.], 1912-21; university reader in mathematics, 1921; professor of mathematics, King's College,

1922-4; Astor professor of pure mathematics, University College, 1924-45; published *Relativity for Physics Students* (1924); FRS, 1926; director, Institute of Education, 1945-57; recommended foundation of West African Examinations Council, 1950; dean, College of Handicraft, 1952-7.

JELF, GEORGE EDWARD (1834-1908), master of Charterhouse; son of Richard William Jelf [q.v.]; educated at Charterhouse and Christ Church, Oxford; BA, 1856; MA, 1859; DD, 1907; honorary canon of St. Albans, 1878; residentiary canon of Rochester, 1880-1907; master of Charterhouse, 1907; a moderate high churchman, he exercised considerable influence by his popular devotional publications.

JELLICOE, (JOHN) BASIL (LEE) (1899-1935), housing reformer; educated at Haileybury and Magdalen College, Oxford; deacon, 1922; priest, 1923; head of Magdalen College mission in Somers Town, 1921-7; largely responsible for formation (1924) of St. Pancras House Improvement Society which rehoused some hundreds of families in flats at low but economic rents.

JELLICOE, JOHN RUSHWORTH, first EARL JELLICOE (1859-1935), admiral of the fleet; son of captain in merchant service; passed second into *Britannia* and first out (1874); qualified as gunnery lieutenant and served on staff of *Excellent* gunnery school, 1884-5 (under (Lord) Fisher, q.v.) and 1886-9; with Fisher at Admiralty, 1889-92; commander, 1891; commanded *Victoria*, flagship of Sir George Tryon [q.v.], 1893; picked up when she was rammed by the *Camperdown* off Tripoli; commanded *Ramillies*, flagship of Sir Michael Culme-Seymour, Mediterranean, 1893-6; captain, 1897; member of ordnance committee, 1897-8; flag captain with Sir E. H. Seymour [q.v.] on China station, 1898-1901; converted Wei-hai-wei into naval base; chief of staff with international naval brigade advancing on Peking (1900) after Boxer rising; severely wounded in lung; naval assistant to the controller, Admiralty, 1902; commanded *Drake*, 1903-4; director of naval ordnance, 1905-7; stimulated accuracy of long-range gunnery by battle-practice; transferred responsibility for output of naval ordnance from War Office to Admiralty; rear-admiral, 1907; second-in-command, Atlantic fleet, and KCVO, 1907; controller and third sea lord, 1908-10; secured inclusion of eight battleships in 1909-10 programme; demanded improved armour-piercing shells; vice-admiral commanding Atlantic fleet, 1910-11; commanded second division of home fleet, 1911; second sea lord, 1912-14; commander-in-

chief, Grand Fleet, 4 Aug. 1914, with flag in *Iron Duke* based on Scapa Flow; admiral, 1915; kept up vigilant blockade; sent a detached force under (Lord) Beatty [q.v.] towards the Skagerrak to entice Germans northward (31 May 1916); on learning that contact had been established made all speed south; deployed into single line on his easternmost wing when German battle fleet reported to westward; crossed the enemy's **T,** compelling them to alter course and steer parallel to him; thus placed himself between them and Germany; opened fire at 6.23 p.m.; Germans fell back, and reappearing at 7.10 again retreated before devastating fire, launched a torpedo attack which Jellicoe evaded by manœuvre; under cover of night Germans finally succeeded in returning to base with heavy losses; the publication of their claim of victory before receipt of Jellicoe's dispatches resulted in mistaken tradition that Jutland, if not a British defeat, was at least a drawn battle; in fact Jellicoe was correct in thinking that the Germans would not again risk such an encounter; appointed OM, 1916; first sea lord, Dec. 1916; accelerated arming of merchantmen; with entry of United States into war was able to implement convoy system but was dismissed on Christmas Eve, 1917, Lloyd George having convinced himself that he was the embodiment of what he considered to be Admiralty negligence; viscount, 1918; admiral of fleet, 1919; enthusiastically received during tour of Empire (1919-20) to investigate naval defence; results included formation of naval base at Singapore and establishment of royal Indian navy and New Zealand naval division; governor-general, New Zealand, 1920-4; earl, 1925; president of British Legion, 1928-32; lies in St. Paul's Cathedral alongside Nelson and Collingwood; a selfless character of deep religious convictions; radiated friendliness and sympathy and inspired deepest devotion in his men who implicitly trusted his leadership; maintained imperturbable calm in battle and made his decisions with lightning speed.

JENKIN, CHARLES FREWEN (1865-1940), first professor of engineering science in the university of Oxford; son of H. C. F. Jenkin [q.v.]; educated at Edinburgh Academy and University and Trinity College, Cambridge; a practical engineer until 1908; professor at Oxford, 1908-29; reported (1920) for Royal Air Force on materials for aircraft construction; investigations included 'corrosion fatigue'; CBE, 1919; FRS, 1931.

JENKINS, DAVID LLEWELYN, BARON JENKINS (1899-1969), judge; educated at Charterhouse and Balliol College, Oxford; Craven scholar; served in Rifle Brigade in France, 1918;

called to bar (Lincoln's Inn) 1923; devilled for J. H. Stamp; KC, 1938; served with RASC, 1939-43; transferred to Political Warfare Executive, under Sir Robert Bruce Lockhart [q.v.], 1943-5; appointed attorney-general of Duchy of Lancaster, 1946; judge, Chancery Court, 1947; Appeal Court judge, 1949-59; lord of appeal in ordinary, 1959; chairman, lord chancellor's Law Reform Committee, and other committees concerned with law reform; bencher, Lincoln's Inn, 1945; hon. fellow, Balliol, 1950; governor, Sutton's Hospital in Charterhouse, 1953-65; chairman, Tancred studentship trustees; knighted, 1947; PC, 1949; life baron, 1959.

JENKINS, EBENEZER EVANS (1820-1905), Wesleyan minister and missionary; worked at Madras till 1865; superintendent of the Hackney circuit from 1865; president of the Wesleyan Conference, 1880.

JENKINS, JOHN EDWARD (1838-1910), politician and satirist; born at Bangalore; called to bar (Lincoln's Inn), 1864; retained by Aborigines Protection Society for British Guiana coolie commission, 1870; gained repute as author of *Ginx's Baby* (anonymous) (1870), a satire on sectarian education; originated Imperial Federation movement, 1871; first agent-general for Canada, 1874-6; radical MP for Dundee, 1874-80; wrote other political satires and novels.

JENKINS, SIR LAWRENCE HUGH (1857-1928), Indian judge; called to bar (Lincoln's Inn), 1883; judge, high court of Calcutta, 1896; chief justice, high court of judicature, Bombay, 1899-1908; member of council of India, 1908-9; took large part in drafting Morley-Minto reforms; chief justice of Bengal, 1909-15; dealt with complicated conspiracy cases; member of judicial committee of Privy Council, 1916; knighted, 1899; master of Indian law and custom.

JENKINSON, SIR (CHARLES) HILARY (1882-1961), deputy keeper of the public records; educated at Dulwich College and Pembroke College, Cambridge; first class, classical tripos, 1904; entered the Public Record Office, 1906; in charge of literary search room, 1912; carried out reorganization of this and repairing department and repository, 1912-38; secretary and principal assistant keeper, 1938; deputy keeper, 1947-54; brought in photographic service; served in Royal Garrison Artillery, 1916-18, and at War Office, 1918-20; Maitland memorial lecturer, Cambridge, 1911-35; lecturer, King's College, London, 1920-5, reader

in palaeography and archives, 1925-47; publications include *English Court Hand* (with Charles Johnson [q.v.], 1915), *Later Court Hands in England from the Fifteenth to the Seventeenth Century* (1927), and *Manual of Archive Administration* (1922); joint secretary (and later vice-president), British Records Association; chairman, the directorate of the National Register; fellow, Society of Antiquaries; president, Jewish Historical Society, and Society of Archivists; CBE, 1943; knighted, 1949; hon. fellow, University College, London; hon. LLD, Aberdeen.

JENKINSON, FRANCIS JOHN HENRY (1853-1923), librarian; BA, Trinity College, Cambridge; fellow of Trinity, 1878; university librarian, 1889-1923; specialized in study of *incunabula* and extended library's acquisitions particularly among fifteenth-century books; Sandars reader in bibliography at Cambridge, 1907-8; an authority on Lepidoptera; works include edition of *Hisperica Famina* (1908).

JENKS, EDWARD (1861-1939), writer on law and history; educated at Dulwich College and King's College, Cambridge; first class, law (1886) and history (1887); called to bar (Middle Temple), 1887; professor of law, Melbourne (1889-91), Liverpool (1892-6); reader in English law, Oxford, 1896-1903; principal and director of legal studies, Law Society, 1903-24; professor of English law, London, 1924-9; works include *The Government of Victoria (Australia)* (1891) and *Law and Politics in the Middle Ages* (1898); edited *A Digest of English Civil Law* (1905-17); FBA, 1930.

JENNER-FUST, HERBERT (1806-1904), cricketer; son of Sir Herbert Jenner [q.v.]; educated at Eton and Trinity Hall, Cambridge; LLB, 1829; LLD, 1835; called to bar, 1831; captain of first Cambridge cricket XI to meet Oxford, 1827; prominent in first-class cricket till 1836; especially as wicket-keeper; president of MCC, 1833.

JENNINGS, SIR (WILLIAM) IVOR (1903-1965), constitutional lawyer; educated at Bristol grammar school and St. Catharine's College, Cambridge; first class, part i, mathematical tripos, and parts i and ii, law tripos, 1923-5; Holt scholar, Gray's Inn, 1925; Barstow scholar, 1926; called to bar (Gray's Inn), 1928; lecturer in law, Leeds University, 1925-9; lecturer, London School of Economics and reader in English law, 1930-40; published eleven books including *Cabinet Government* (1936), *The Law and the Constitution* (1933), *Parliamentary Reform* (1934), *Parliament* (1939), and *A Federation for Western Europe* (1940);

principal, University College, Ceylon; first vice-chancellor, 1940-54; constitutional adviser and chief draughtsman to Pakistan, 1954-5; member, Malayan constitutional commission, 1956-7; master, Trinity Hall, Cambridge, 1954; vice-chancellor, 1961-3; Downing professor in the laws of England, 1962; other publications include (with C. M. Young) *Constitutional Laws of the British Empire* (1938), *The Approach to Self-Government* (1956), and *Party Politics* (3 vols., 1960-2); knighted, 1948; KC, 1949; KBE, 1955; FBA, 1955; bencher, Gray's Inn, 1958; hon. degrees in law from many British and overseas universities.

JEPHSON, ARTHUR JERMY MOUNTENEY (1858-1908), explorer; accompanied (Sir) H. M. Stanley [q.v.] through forests to Lake Albert for relief of Emin Pasha, 1888; imprisoned with Emin at Dufile, Aug. 1888; rejoined Stanley at Kavali (Feb. 1889), and subsequently rescued Emin; Queen's messenger, 1895; published *Emin Pasha and the Rebellion at the Equator* (1890), and native folk tales.

JEROME, JEROME KLAPKA (1859-1927), novelist and playwright; went on the stage and finally took to journalism; a founder of *The Idler* magazine, 1892; his best-known book, *Three Men in a Boat* (1889); his best-known play, *The Passing of the Third Floor Back* (1908).

JERRAM, SIR (THOMAS HENRY) MARTYN (1858-1933), admiral; entered navy, 1871; served mainly in training ships; commander, 1894; captain, 1899; rear-admiral, 1908; vice-admiral, 1913; commander-in-chief, China station, 1913-15; KCB, 1914; commanded second battle squadron in Grand Fleet (1915-16) and in *King George V* led battle line at Jutland; KCMG, 1916; admiral and retired, 1917; president of committee on pay scales (1918), of permanent welfare committee (1919), and naval prize tribunal (1925); GCMG, 1919.

JERROLD, DOUGLAS FRANCIS (1893-1964), author and publisher; educated at Westminster and New College, Oxford; secretary, Oxford Union; served in Royal Naval Division in Gallipoli and France during 1914-18 war; severely wounded; civil servant in Ministry of Food, 1918; and Treasury, 1920-3; became publisher, first with Benn Brothers, 1923-8, where he collaborated with (Sir) Victor Gollancz [q.v.]; later, director, 1929-45, and chairman, 1945-59, of Eyre and Spottiswoode; also edited the *English Review*, 1930-6, and the *New English Review*, 1945-50; held strongly right-wing views; chairman, Author's Club; prolific

writer; publications include a novel, *Storm over Europe* (1930), his autobiography, *Georgian Adventure* (1937), and *An Introduction to the History of England* (of which he only finished the first volume, 1949).

JERSEY, COUNTESS OF (1849-1945). [See VILLIERS, MARGARET ELIZABETH CHILD-.]

JERSEY, seventh EARL OF (1845-1915), colonial governor. [See VILLIERS, VICTOR ALBERT GEORGE CHILD-.]

JESSOP, GILBERT LAIRD (1874-1955), cricketer; educated at Cheltenham grammar school and Christ's College, Cambridge; captain of cricket, 1899; first played for Gloucestershire, 1894; captain, 1900-13; played in 18 test matches, first playing for England in 1899; nicknamed 'the Croucher' he was a fast scorer with a bewildering variety of strokes; played 5 innings of over 200, in 1903 scoring 286 in 175 minutes; made 53 centuries, 6 of them in less than 1 hour; scored over 1,000 runs in 14 seasons; brilliant fielder and fast bowler; published *A Cricketer's Log* (1922) and *Cricket and How to Play It* (1925).

JESSOPP, AUGUSTUS (1823-1914), schoolmaster and historical writer; BA, St. John's College, Cambridge; headmaster of King Edward VI School, Norwich, which he transformed into a modern public school, 1859-79; rector of Scarning, Norfolk, 1879-1911; works include *One Generation of a Norfolk House* (the Walpoles, 1878) and *The Coming of the Friars* (1889); edited *Visitations of Diocese of Norwich, 1492-1532* (1888), etc.

JEUNE, FRANCIS HENRY, BARON ST. HELIER (1843-1905), judge; eldest son of Francis Jeune [q.v.]; scholar of Harrow and Balliol College, Oxford; first class classic, 1865; obtained Stanhope (1863) and Arnold (1867) essay prizes; called to bar, 1868; original fellow of Hertford College, 1874; counsel for Tichborne plaintiff in action of ejectment, 1871-2; had large ecclesiastical practice; counsel for evangelicals in Mackonochie case; counsel for Louis Riel [q.v.] in application for leave to appeal; chancellor of several dioceses; QC, 1888; counsel for Edward King [q.v.] in ritual case, 1889-90; judge of Probate, Divorce, and Admiralty division, and knighted, 1891; president of Probate division, 1892; made court a model of efficiency and dispatch; judge-advocate-general without payment, 1892-1904; KCB, 1897; GCB, 1902; resigned presidency of Probate division through ill health and created peer, Jan. 1905.

JEX-BLAKE, SOPHIA LOUISA (1840-1912), physician; sister of T. W. Jex-Blake [q.v.]; medical student at Edinburgh, 1869-72; on being refused, with other women students, right to graduate, founded London School of Medicine for Women, 1874; gained legal right to practise in Great Britain, 1877; practised in Edinburgh, 1878-99.

JEX-BLAKE, THOMAS WILLIAM (1832-1915), schoolmaster and dean of Wells; educated at Rugby and University College, Oxford; fellow of Queen's College, Oxford, 1855; assistant master at Rugby, 1858-68; principal of Cheltenham College, 1868-74; headmaster of Rugby, which he restored to prosperity, 1874-87; dean of Wells, 1891-1910.

JINNAH, MAHOMED ALI (1876-1948), creator of Pakistan; son of Karachi merchant; called to bar (Lincoln's Inn), 1896; practised in Bombay; represented Bombay Moslems on imperial legislative council, 1909-19; joined Moslem League, 1913; resigned from Indian National Congress opposing non-co-operation policy, 1920; led independent party in legislative assembly from 1923; powerful in Moslem League and took lead in seeking Hindu-Moslem unity; attended Round Table conferences, 1930-1; remained in England; returned to India to devote himself to Moslem interests, 1934; inspired Moslem League resolution demanding Pakistan, 1940; took direct action leading to bitter communal fighting, 1946; first governor-general of Pakistan, 1947-8.

JOACHIM, HAROLD HENRY (1868-1938), philosopher; educated at Harrow and Balliol College, Oxford; first class, *lit. hum.*, 1890; fellow and tutor in philosophy, Merton, 1897-1919; Wykeham professor of logic, 1919-35; mainly inspired by Hegel; publications include *A Study of the Ethics of Spinoza* (1901), *The Nature of Truth* (1906, presenting the coherence theory), and translation, with commentary, of Aristotle's *De generatione et corruptione* (1922); FBA, 1922.

JOAD, CYRIL EDWIN MITCHINSON (1891-1953), writer and teacher; educated at Blundell's School, Tiverton, and Balliol College, Oxford; first class, *lit. hum.*, 1914; with labour exchanges department, Board of Trade (later Ministry of Labour), 1914-30; head, philosophy department, Birkbeck College, 1930-53; D.Litt., 1936; reader in philosophy, 1945-53; stimulating and lucid teacher and prolific writer; Plato and Aristotle his two great loves; lively member of first Brains Trust; moved from agnosticism to Christianity; published *The Recovery of Belief* (1952).

JOEL, JACK BARNATO (1862-1940), financier and sportsman; career resembled that of his younger brother S. B. Joel [q.v.] whom he succeeded as chairman of the Johannesburg Consolidated Investment Company in 1931; won Derby in 1911 and 1921 and headed list of winning owners in 1908, 1913, and 1914.

JOEL, SOLOMON BARNATO (1865-1931), financier and sportsman; nephew of Barnett Barnato and brother of J. B. Joel [qq.v.]; partner in Barnato Brothers; chairman (from 1898) of Johannesburg Consolidated Investment Company; director of De Beers Consolidated Mines, Standard Bank of South Africa, and many other enterprises; a lavish host; patron of the theatre and of cricket; racing successes included the Two Thousand Guineas, Derby, and St. Leger in 1915 with Pommern; headed list of winning owners, 1921.

JOHN, AUGUSTUS EDWIN (1878-1961), artist; attended art school at Tenby and Slade School of Art, London, 1894-8; injured skull in diving accident, 1898; studied under Henry Tonks [q.v.]; visited Amsterdam, 1898; first one-man exhibition, 1899; travelled abroad with other artists, including (Sir) William Orpen [q.v.]; art instructor, Liverpool University, 1901-2; began producing succession of masterly works, mainly portraits and landscapes, 1902 up to 1920, when his work deteriorated; paintings in many exhibitions including the retrospective at the Royal Academy, 1954, and at the National Portrait Gallery, 1975; contributed to the *Journal of The Gypsy Lore Society*; published *Chiaroscuro, Fragments of Autobiography* (1952), and *Finishing Touches* (1964); RA, 1928; OM, 1942.

JOHN, Sir WILLIAM GOSCOMBE (1860-1952), sculptor and medallist; trained at City and Guilds and Royal Academy schools; ARA, 1899; RA, 1909; academic sculptor compounding realism and romanticism; exhibited annually at Royal Academy for sixty-three years; executed public statues, portrait busts, and medals including Jubilee medal of George V; collection of works at National Museum of Wales; knighted, 1911.

JOHNS, CLAUDE HERMANN WALTER (1857-1920), Assyriologist; BA, Queens' College, Cambridge; rector of St. Botolph's, Cambridge, 1892-1909; lecturer in Assyriology at Queens', 1895, and in Assyrian at King's College, London, 1904; master of St. Catharine's College, Cambridge, and canon of Norwich, 1909-19; chief work *Assyrian Deeds and Documents* (1898-1923).

JOHNS, WILLIAM EARL (1893-1968), journalist, and creator of the children's popular fiction character, 'Biggles'; educated at Hertford grammar school; enlisted in Norfolk Yeomanry; served in Gallipoli and Salonika, 1916-17; transferred to Royal Flying Corps, 1917; shot down and became prisoner of war, 1918; remained in RAF, 1920-7; founder-editor, *Popular Flying*, 1932; first 'Biggles' stories in *The Camels are Coming* (1932); altogether, wrote ninety-six 'Biggles' books; wrote regularly for the *Modern Boy*, *Pearson's Magazine*, and *My Garden* and edited *Flying*; lectured to Air Training Corps, 1939; created 'Worrals of the W.A.A.F.', female counterpart of 'Biggles', and 'Gimlet', a commando; works translated into fourteen languages and issued in braille.

JOHNSON, ALFRED EDWARD WEBB-, BARON WEBB-JOHNSON (1880-1958), surgeon. [See WEBB-JOHNSON.]

JOHNSON, AMY, otherwise AMY MOLLISON (1903-1941), airwoman; educated at Boulevard secondary school, Hull; BA, Sheffield, 1925; obtained ground-engineer's licence and full navigation certificate; flew solo to Australia, 1930; made record flights to Tokyo, 1931, Cape Town, 1932, Karachi (with J. A. Mollison, whom she married, 1932), 1934, Cape and back, 1936; marriage dissolved, 1938; perished ferrying for Air Transport Auxiliary; CBE, 1930.

JOHNSON, CHARLES (1870-1961), assistant keeper of the public records and historian; educated at Giggleswick School and Trinity College, Oxford; first class, *lit. hum.*, 1892; entered Public Record Office, 1893; engaged in arrangement and reclassification of ancient miscellanea and files of the Chancery; collaborated with C. G. Crump [q.v.] and Arthur Hughes in producing *Dialogus de Scaccario* (1902); largely responsible for Domesday section of the *Victoria History of the County of Norfolk* (1906); with (Sir) C. H. Jenkinson [q.v.], produced *English Court Hand* (1915); also published *The Public Record Office* (1918), *The Care of Documents and Management of Archives* (1919), and *The Mechanical Processes of the Historian* (1922); principal initiator and guide in publication of *Revised Word-List* (1965); after retirement in 1930 edited some of the basic texts of English medieval history and made frequent contributions to reviews on historical topics; vice-president, Royal Historical Society and Society of Antiquaries, and member of other learned societies; FBA, 1934; CBE, 1951; took charge of records stored for safe-keeping in Culham College, 1939-46.

JOHNSON, HEWLETT (1874-1966), dean successively of Manchester and Canterbury; educated at Macclesfield grammar school and the Owens College, Manchester; B.Sc., 1894; associate member of Institute of Civil Engineers, 1898; entered Wadham College, Oxford, as theological student; ordained as deacon (1905), and priest (1906), at St. Margaret, Altrincham; honorary canon, Chester, 1919; rural dean, Bowden, 1922; edited *The Interpreter*; BD, Oxford, 1917; DD, 1924; dean of Manchester, 1924-31; dean of Canterbury, 1931-63; visited Russia, 1937, and became indefatigable speaker for Left Book Club in association with (Sir) Victor Gollancz and John Strachey [qq.v.]; published *The Socialist Sixth of the World* (1939); received Stalin peace prize, 1951; a controversial figure who came to be known as the 'Red Dean'; chairman, editorial board of the *Daily Worker*; published autobiography, *Searching for Light* (1968).

JOHNSON, JOHN DE MONINS (1882-1956), printer and scholar; educated at Magdalen College School and Exeter College, Oxford; in Egyptian civil service, 1905-7; senior demy, Magdalen College, Oxford, 1908-11; explored in Egypt, 1911, 1913-14; assistant secretary, delegates of Oxford University Press, 1915-25; printer to university, 1925-46; a leader of the renaissance of book printing; his unique collection of printed ephemera moved to Bodleian Library, 1968; published *Print and Privilege at Oxford to the Year 1700* (with S. Gibson, 1946); CBE, 1945.

JOHNSON, LIONEL PIGOT (1867-1902), critic and poet; scholar of Winchester College; edited *The Wykehamist*, 1884-6; BA, New College, Oxford (first class, *lit. hum.*), 1890; entered on literary career in London, 1890, contributing to magazines and reviews; joined Church of Rome, 1891; published his first volume of poems, 1895, in which he gave expression to a 'catholic puritanism'; in *Ireland and other Poems* (1897) he betrayed an intense love for Ireland; also wrote *The Art of Thomas Hardy* (1894) and *Post Liminium: Essays and Critical Papers* (posthumous, 1911).

JOHNSON, SIR NELSON KING (1892-1954), meteorologist; educated at Simon Langton School, Canterbury and Royal College of Science; B.Sc., 1913; pilot, Royal Flying Corps, 1915-19; joined Meteorological Office, 1919; worked mainly at Chemical Warfare Experimental Station, Porton; researches laid foundation of micro-meteorology; director, Meteorological Office, 1938-53, organizing wartime service, research, and international links; D.Sc. London, 1939; KCB, 1943.

JOHNSON, WILLIAM ERNEST (1858-1931), logician; educated at Perse School, Liverpool Royal Institution School and King's College, Cambridge; eleventh wrangler, 1882; first class, moral sciences tripos, 1883; Sidgwick university lecturer in moral science and fellow of King's, 1902-31; in treatise on *Logic* (3 parts, 1921-4) defined the subject-matter of logic as 'the analysis and criticism of thought' and attached great importance to the 'epistemic' aspect; his treatment of probability similar to that of (Lord) Keynes [q.v.]; FBA, 1923.

JOHNSON, WILLIAM PERCIVAL (1854-1928), archdeacon of Nyasa, 1896-1928; BA, University College, Oxford; joined Universities' Mission to Central Africa, under Bishop E. Steere [q.v.], 1876; deacon, 1876; priest, 1878; worked on and by Lake Nyasa, 1881-1928; made translations in several African languages.

JOHNSTON, CHRISTOPHER NICHOLSON, LORD SANDS (1857-1934), Scottish judge; educated at St. Andrews, Edinburgh, and Heidelberg universities; advocate, 1880; advocate-depute, 1892, 1895-9; KC, 1902; sheriff of Perth, 1905-16; conservative MP, Edinburgh and St. Andrews universities, 1916-17; knighted, 1917; senator of College of Justice, 1917-34, with judicial title of Lord Sands; procurator of the Church of Scotland, 1907-18; his memorandum the basis of the Church union of 1929.

JOHNSTON, EDWARD (1872-1944), calligrapher and designer of lettering; taught at Central School of Arts and Crafts and Royal College of Art; designed lettering for London Transport; published *Writing and Illuminating, and Lettering* (1906) and *Manuscript and Inscription Letters* (1909); CBE, 1939.

JOHNSTON, GEORGE LAWSON, first BARON LUKE (1873-1943), man of business and philanthropist; educated at Dulwich and Blair Lodge, Polmont; entered father's firm of Bovril, Ltd.; director, 1896; vice-chairman, 1900; chairman, 1916; managing director, 1931; other directorships included *Daily Express* (1900-17); did much work for hospitals; KBE, 1920; baron, 1929.

JOHNSTON, SIR HARRY HAMILTON (1858-1927), explorer and administrator; displayed precocious scientific, linguistic, and artistic interests; became interested in problem of partition of Africa, 1879; joined seventh Earl of Mayo on expedition through Southern Angola, 1882; penetrated alone into Congo basin, where he won friendship of (Sir) H. M. Stanley [q.v.], 1883; undertook scientific and

political mission to Mount Kilimanjaro and surroundings, 1884; vice-consul, Cameroon and Niger Delta, 1885; helped to open navigable mouths of Niger to legitimate trade; formed alliance with Cecil Rhodes [q.v.]; British consul, Portuguese East Africa, 1889; extended British protectorate over Shiré Highlands to include Nyasaland, etc.; British commissioner for South Central Africa, 1891-6; consul-general, Tunisia, 1897-9; special commissioner, Uganda, where he achieved administrative success and pursued geographical, ethnological, and naturalist studies, 1899-1901; occupied with Liberian affairs, 1904-6; KCB, 1896.

JOHNSTON, SIR REGINALD FLEMING (1874-1934), scholar, traveller, and administrator; educated at Edinburgh and Oxford universities; entered Hong Kong civil service, 1898; made series of journeys into the interior of China; seconded to Wei-hai-wei, 1904; district officer and magistrate, 1906-17; European tutor to last of the Ch'ing emperors, Hsüan T'ung, 1918-25; comptroller of the imperial household, 1924; contrived Emperor's flight to the legation quarter, 1924; commissioner, Wei-hai-wei, 1927-30; professor of Chinese, London, 1931-7; KCMG, 1930; to all intents and purposes a Confucianist; highly critical of Christian missionary activities in China; books include *From Peking to Mandalay* (1908), *Lion and Dragon in Northern China* (1910), *Buddhist China* (1913), *Twilight in the Forbidden City* (1934), and *Confucianism and Modern China* (1934).

JOHNSTON, THOMAS (1881-1965), politician and newspaper editor; educated at Lenzie Academy and Glasgow University; launched the socialist weekly, *Forward*, 1906; published *The History of the Working Classes in Scotland* (1920); labour MP, West Stirlingshire, 1922-4; MP, Dundee, 1925-9; West Stirlingshire, 1929-31 and 1935-45; under-secretary of state for Scotland, 1929; lord privy seal; regional commissioner for Scotland, 1939; secretary of state for Scotland, 1941-5; chairman, North of Scotland Hydro-Electric Board, 1945-59; chairman, Scottish Tourist Board; forestry commissioner; chairman, Broadcasting Council for Scotland; chancellor, Aberdeen University, 1951; president, Scottish History Society, 1950-2; CH, 1953; LLD Glasgow, 1945; published *Memories* (1952).

JOHNSTON, WILLIAM (1829-1902), Orangeman; BA, Trinity College, Dublin, 1852; MA, 1856; called to Irish bar, 1872; entered Orange order, 1848; proposed institution of triennial council of Orangemen, 1865; imprisoned for organizing and leading a demonstration against Party Processions Act,

1868; independent conservative MP for Belfast, 1868–78; inspector of Irish fisheries, 1878; dismissed for violent speeches against Land League and home rule, 1885; MP for South Belfast, 1885 till death; firm advocate of 'the three F's' (fair rent, free sale, fixity of tenure), and of temperance.

JOICEY, JAMES, first Baron Joicey (1846–1936), colliery proprietor; became chairman and managing director of family firm, trading from 1924 as the Lambton, Hetton, and Joicey Collieries, the largest concern in the northern coalfield; liberal MP, Chester-le-Street division of county Durham, 1885–1906; baronet, 1893; baron, 1906.

JOLOWICZ, HERBERT FELIX (1890–1954), academic lawyer; educated at St. Paul's School and Trinity College, Cambridge; first class, part i, classical tripos (1911), and part i, law tripos (1913); called to bar (Inner Temple), 1919; All Souls reader in Roman law, Oxford, 1920–31; lecturer, 1924; professor, 1931–48, Roman law, University College, London; regius professor of civil law, Oxford, 1948–54; edited *Journal* of Society of Public Teachers of Law, 1924–54; published *Historical Introduction to Roman Law* (1932) and *Roman Foundations of Modern Law* (1957).

JOLY, CHARLES JASPER (1864–1906), royal astronomer of Ireland; mathematical scholar of Trinity College, Dublin, 1882; obtained studentship, 1886; fellow, 1894–7; made researches in physics, and especially studied the properties of linear vector functions; royal astronomer of Ireland at Dunsink, 1897; edited and enlarged Hamilton's *Elements of Quaternions* (2 vols., 1899–1901); contributed numerous papers on quaternions and kindred subjects to Royal Irish Academy; published *A Manual of Quaternions* (1905); accompanied eclipse expedition to Spain, 1900; FRS, 1904; member of Alpine Club.

JOLY, JOHN (1857–1933), engineer, geologist, and physicist; educated at Rathmines School and Trinity College, Dublin; professor of geology, Dublin, 1887–1933; invented a meldometer, a hydrostatic balance, and a condensation method of calorimetry; reduced aluminium from topaz; propounded (with H. H. Dixon) the generally accepted cohesion theory of the ascent of sap in trees, 1893; pioneer in colour photography; as a keen yachtsman wrote on synchronous signalling, a collision predictor, floating breakwaters, etc.; paid much attention to radioactivity in geology, advancing in *The Surface History of the Earth* (1925) the radioactive explanation of thermal cycles; chiefly

responsible for the building of Trinity College science schools; originated Royal Dublin Society's Radium Institute; FRS, 1892.

JOLY DE LOTBINIÈRE, Sir HENRY GUSTAVE (1829–1908), Canadian politician; born at Épernay, France; called to Canadian bar, 1855; liberal member of Canadian house of assembly for Lotbinière, 1861; opposed federation movement; member of Quebec legislative assembly, 1867–74; QC, 1878; formed government, 1878–9; KCMG, 1895; minister of inland revenue and PC, 1897; lieutenant-governor of British Columbia, 1900–6; promoted interests of agriculture and forestry.

JONES, ADRIAN (1845–1938), sculptor; qualified as veterinary surgeon; retired from army as captain, 1890; perfectly understood horses, his chief models; monumental sculptures include Royal Marines Monument, St. James's Park (1903), Peace Quadriga, Constitution Hill (1912), and Cavalry War Memorial, Stanhope Gate (1924).

JONES, (ALFRED) ERNEST (1879–1958), physician and psycho-analyst; educated at Swansea grammar school, Llandovery College, University College, Cardiff; qualified from University College Hospital, London, 1900; MD, 1903; MRCP, 1904; FRCP, 1942; began practising psycho-analysis, 1906; director, psychiatric clinic, Toronto, 1908–12; established London practice, 1913; founded British Psycho-Analytical Society, 1919; and *International Journal of Psycho-Analysis*, 1920; editor until 1939; set up Institute of Psycho-Analysis, 1924, publishing International Psycho-Analytical Library; publications include *Sigmund Freud, Life and Work* (3 vols., 1953–7); hon. D.Sc., Wales.

JONES, Sir ALFRED LEWIS (1845–1909), man of business; born at Carmarthen; became partner of Messrs. Elder, Dempster's shipping firm at Liverpool, 1879; gained monopoly of West African shipping trade; founded Bank of British West Africa, 1897; revivified the Canaries, inaugurating there banana industry (1884) and tourist traffic, and establishing coaling station at Las Palmas; inaugurated new steamship service with Jamaica from Bristol (1901) for banana traffic; helped to found Liverpool School of Tropical Medicine, 1898; KCMG, 1901; honorary fellow of Jesus College, Oxford, 1905.

JONES, ARNOLD HUGH MARTIN (1904–1970), historian of Greece and Rome; educated at Cheltenham College and New College, Oxford; first class, hon. mods., and *lit.*

hum., 1924-6; Craven scholar, 1923; fellow, All Souls, 1926-46; reader in ancient history, Cairo University, 1929-34; lecturer in ancient history, Wadham College, Oxford, 1939-46; served in Ministry of Labour and War Office during 1939-45 war; professor of ancient history, University College, London, 1946; Cambridge, 1951; fellow, Jesus College; publications include *A History of Abyssinia* (with Elizabeth Mouroe, 1935), *The Herods of Judaea* (1938), *The Greek City from Alexander to Justinian* (1940), *Constantine and the Conversion of Europe* (1948), *Athenian Democracy* (1957), *Studies in Roman Government and Law* (1960), *Sparta* (1967), and *Augustus* (1970); FBA, 1947; president, Society for Promotion of Roman Studies, 1952-5; LLD Cambridge, 1965; DD Oxford, 1966; hon. fellow, New College.

JONES, ARTHUR CREECH (1891-1964), politician; educated at Whitehall boys' school Bristol; worked in solicitor's office, 1905; entered Civil Service as junior clerk; honorary secretary, Dulwich branch of ILP, 1913; organized opposition to conscription; imprisoned, 1916-19; secretary, National Union of Docks, Wharves, and Shipping Staffs, and editor, *Quayside and Office*, 1919-22; national secretary, administrative, clerical, and supervisory group, Transport and General Workers Union, 1922; interested in trade-union organization in colonies; published *Trade Unionism Today* (1928); organizing secretary, Workers' Travel Association, 1929; labour MP, Shipley division of Yorkshire, 1935-50; parliamentary secretary to Ernest Bevin [q.v.], minister of labour, 1940-4; founded, with Dr Rita Hinden, Fabian Colonial Bureau, 1940; parliamentary under-secretary of state, Colonial Office, 1945; secretary of state for colonies, 1946-50; contributed much to policy of development and preparing colonies for independence; governor, Ruskin College, 1923-56; governor, Queen Elizabeth House, Oxford, 1954; vice-president, Workers' Educational Association; MP, Wakefield, 1954-64; PC, 1946.

JONES, BERNARD MOUAT (1882-1953), chemist, principal of Manchester College of Technology, and vice-chancellor of Leeds University; educated at Dulwich College and Balliol College, Oxford; first class, chemistry, mineralogy, and crystallography, 1904; professor of chemistry, Government College, Lahore, 1906-13; Aberystwyth, 1919-21; DSO, 1917, for work at GHQ central laboratory identifying gases; principal, Manchester College of Technology, 1921-38; vice-chancellor, Leeds, 1938-48; in latter appointments obtained interest and co-operation of industry; hon. degrees, Durham, Leeds, and Wales.

JONES, (FREDERIC) WOOD (1879-1954), anatomist; MB, BS, London Hospital, 1904; professor of anatomy, London School of Medicine for Women, 1912-14; Adelaide University, 1919-27; Rockefeller professor of anthropology, Hawaii, 1927-30; professor of anatomy, Melbourne, 1930-7; Manchester, 1938-45; Royal College of Surgeons, 1945-52; in *Man's Place among the Mammals* (1929) refuted close relationship between man and apes; other publications include works on anatomy of the hand (1920) and the foot (1944); FRS, 1925; FRCS, 1930.

JONES, SIR (GEORGE) RODERICK (1877-1962), newsagency head; on father's death sent to live in South Africa; became Pretoria journalist in his teens; learned Afrikaans and gained confidence of Paul Kruger; interviewed (Sir) L. S. Jameson [q.v.] after his capture by Boers; covered South African war as correspondent; arrested by British as a spy; Reuters' senior cable correspondent in Cape Town; chief sub-editor, *Cape Times*; editor, South African desk, London, 1902; succeeded Baron Herbert de Reuter as head of company, 1915; chairman and managing director, 1919-41; travelled widely; received honours from France, Italy, Greece, and China; KBE, 1918; chief executive and director of propaganda, Ministry of Information, 1918; resigned from Reuters after dispute with other directors regarding agreement to accept government financial assistance; 1941; published *A Life in Reuters* (1951); member of council of Royal Institute of International Affairs; delegate to Congress of Europe at The Hague, 1948; chairman of governors of Roedean School, 1950-62; married Enid Bagnold, novelist and playwright, 1920.

JONES, SIR HAROLD SPENCER (1890-1960), astronomer; educated at Latymer Upper School, Hammersmith and Jesus College, Cambridge; first class, parts i and ii, mathematical tripos, 1909-11, and physics in part ii, natural sciences tripos, 1912; research fellow, 1913; appointed to Greenwich, 1913; astronomer, Royal Observatory, Cape of Good Hope, 1923-33; Sc.D., Cambridge, 1925; astronomer royal, Greenwich, 1933-55; organized removal to Herstmonceux Castle and obtained approval for 98-inch telescope; his demonstration of irregularities in rate of rotation of earth led to adoption of concept of Ephemeris Time, 1950; published, 1941, discussions of results of his world-wide programme of observations of Eros (1930-1) to determine value of 'solar parallax'; contributed to time measurement, horology, application of geo-magnetism to navigation, and organization of international science; FRS,

1930; knighted, 1943; KBE, 1955; hon. fellow, Jesus College, Cambridge, 1933.

JONES, SIR HENRY (1852-1922), philosopher; master of Ironworks School, Brynamman, 1873; MA, Glasgow University, 1878; influenced by Edward Caird [q.v.]; professor of philosophy and political economy, University College of North Wales, Bangor, 1884; professor of logic, rhetoric, and metaphysics, St. Andrews, 1891; professor of moral philosophy, Glasgow, 1894-1922; called himself a 'spiritual realist'; fullest statement of his metaphysic given in his last volume, *A Faith that Enquires* (1922); FBA, 1904; knighted, 1912.

JONES, HENRY ARTHUR (1851-1929), dramatist; commercial traveller in London, Bradford, and Exeter districts, 1869-79; thereafter relied for livelihood solely on writing of plays; these include *The Silver King* (1882), *Saints and Sinners* (1884), *Judah* (1890), *The Dancing Girl* (1891), *The Tempter* (1893), *The Case of Rebellious Susan* (1894), *Michael and his Lost Angel* (1896), *The Liars* (1897), *Mrs. Dane's Defence* (1900), *The Hypocrites* (1906), *The Divine Gift* (1913, never produced), *The Lie* (1914); played an important part in revival of English drama; excelled as a craftsman.

JONES, HENRY CADMAN (1818-1902), law reporter; educated at Trinity College, Cambridge; second wrangler, second Smith's prizeman, and fellow, 1841; called to bar, 1845; edited Chancery reports from 1857 till 1865.

JONES, SIR HENRY STUART- (1867-1939), classical scholar, Roman historian, and lexicographer; educated at Rossall School and Balliol College, Oxford; first class, *lit. hum.*, and fellow of Trinity, 1890; tutor, 1896; director, British School at Rome, 1903-5; engaged upon revision of Liddell and Scott's *Greek-English Lexicon*, 1911-39; organized remarkable body of specialists as voluntary collaborators and although incorporating much new material kept the edition within reasonable bounds; Camden professor of ancient history, Oxford, 1919-27; principal, University College of Wales, Aberystwyth, 1927-34; vice-chancellor, university of Wales, 1929-31; served on representative council of Welsh Church; publications include revision of Thucydides (1898-1900) for Oxford Classical Texts, *Companion to Roman History* (1912), *Fresh Light on Roman Bureaucracy* (1920), and chapters in vol. vii of *Cambridge Ancient History*; FBA, 1915; knighted, 1933.

JONES, (JAMES) SIDNEY (1861-1946), composer; wrote song 'Linger Longer Loo'

(1892), and operettas including *A Gaiety Girl* (1893), *An Artist's Model* (1895), *The Geisha* (1896), *A Greek Slave* (1898), *San Toy* (1899), and *A Persian Princess* (1909).

JONES, JOHN DANIEL (1865-1942), Congregational minister; son of J. D. Jones [q.v.]; educated at Towyn Academy, Chorley grammar school, and Owens College, Manchester; BA, 1886; BD, St. Andrews, and ordained to Newland church, Lincoln, 1889; minister, Richmond Hill church, Bournemouth, 1898-1937; chairman, Congregational Union, 1909-10, 1925-6; honorary secretary, 1919-42; moderator, Free Church Federal Council, 1921-3; of International Congregational Council, 1930-42; CH, 1927.

JONES, SIR JOHN EDWARD LENNARD- (1894-1954), scientist and administrator. [See LENNARD-JONES.]

JONES, SIR JOHN MORRIS- (1864-1929), Welsh poet and grammarian. [See MORRIS-JONES.]

JONES, JOHN VIRIAMU (1856-1901), physicist; son of Thomas Jones (1819-82, q.v.); educated at University College, London (B.Sc. and fellow), and Balliol College, Oxford; first class in mathematics, 1879, and natural science, 1880; first principal of University College of South Wales, Cardiff, 1883; first vice-chancellor of university of Wales, 1893; engaged in physical researches dealing especially with the determination of the ohm; FRS, 1894; member of Alpine Club; died at Geneva.

JONES, OWEN THOMAS (1878-1967), geologist; educated at Pencader grammar school, University College of Wales, Aberystwyth, and Trinity College, Cambridge; first class, parts i and ii, natural sciences tripos, 1902-3; joined Geological Survey of Great Britain, 1903; assisted in mapping part of south-Welsh coalfield; first professor of geology, Aberystwyth, 1910; D.Sc., Wales; professor of geology, Manchester University, 1919; professor, Cambridge, 1930-43; last paper written in Welsh dealt with distribution of the blue stones of Carn Meini of Pembrokeshire; president, Geological Society, 1936-8, and 1950-1; FRS, 1926; vice-president, Royal Society, 1940-1; hon. LLD, university of Wales; hon. fellow of geological societies abroad.

JONES, SIR ROBERT, first baronet (1857-1933), orthopaedic surgeon; graduated from Liverpool School of Medicine (1878) and specialized in orthopaedic surgery; met (Dame) Agnes Hunt [q.v.] and with her developed novel system of surgery, nursing, and after-care for crippled children resulting in establishment of a hospital at Baschurch and (1920) of Central Council for Care of Cripples; widely influential as inspector of military orthopaedics, 1916–19; knighted, 1917; baronet, 1926.

JONES, Sir ROBERT ARMSTRONG- (1857–1943), alienist. [See ARMSTRONG-JONES.]

JONES, THOMAS (1870–1955), civil servant, administrator, and author; educated at Lewis School, Pengam, University College of Wales, Aberystwyth, and Glasgow University; first class, economic science, 1901; joined independent labour party and Fabian Society, 1895; assistant in political economy, Glasgow, 1900–9; special investigator, Poor Law royal commission, 1906–9; secretary, Welsh campaign against tuberculosis, 1910; first secretary, National Health Insurance Commission (Wales), 1912; as deputy secretary of cabinet, 1916–30, exercised great influence behind the scenes; a negotiator for Irish treaty, 1921; first secretary, Pilgrim Trust, 1930–45; member, Unemployment Assistance Board, 1934–40; deputy chairman, Arts Council, 1939–42; founder-trustee, *Observer*, 1946; president, Aberystwyth, 1944–54; principal founder, Coleg Harlech, 1927; 1851 royal commissioner, 1921–55; his lifelong interest in social work enabled him to guide the philanthropy of the wealthy, especially in Wales; publications include biography of Lloyd George (1951) and *A Diary with Letters, 1931–1950* (1954); CH, 1929; hon. degrees from several universities.

JONES, THOMAS RUPERT (1819–1911), geologist and palaeontologist; educated at Ilminster, where he studied geology of district; studied medicine at Newbury; published *Monograph on the Cretaceous Entomostraca of England* (1849) and papers on the geology of Newbury (1854) and of the Kennet valley (1871); assistant secretary of Geological Society, 1851–62; professor of geology at Royal Military College, Sandhurst, 1862; interested in South African geology; FRS, 1872; president of Geologists' Association, 1879–81.

JONES, WILLIAM WEST (1838–1908), archbishop of Cape Town; BA, St. John's College, Oxford, 1860; MA, 1863; fellow, 1859–79; hon. fellow, 1893; hon. DD, 1874; vicar of Summertown, Oxford, 1864–74; bishop of Cape Town, 1874; see elevated to archbishopric, 1897; a strong high churchman, popular with English and Afrikaans alike.

JORDAN, (HEINRICH ERNST) KARL (1861–1959), entomologist; educated at Hildes-

heim high school and Göttingen University; *summa cum laude*, botany and zoology; entomologist, Tring zoological museum, created by future Lord Rothschild [q.v.], 1893–1939; director, 1930–3; naturalized, 1911; founded International Congress of Entomology, 1910; secretary until 1948; published large volume of papers on his researches including *Monograph of Charaxes* and *Revision of the Sphingidae*; worked especially on classification of fleas and beetles; FRS, 1932.

JORDAN, Sir JOHN NEWELL (1852–1925), diplomatist; BA, Queen's College, Belfast; joined Chinese consular service in Peking, 1876; appointed to legation in Peking, 1886; assistant Chinese secretary, 1889; full secretary, 1891; consul-general, Seoul, Korea, 1896; chargé d'affaires, 1898; minister-resident, 1901; envoy extraordinary and minister plenipotentiary to court of Peking, 1906; co-operated with Sir Alexander Hosie [q.v.] in successful attempt to stop export to China of Indian opium; retired, 1920; KCMG, 1904; KCB, 1909; GCMG, 1920.

JORDAN LLOYD, DOROTHY (1889–1946), biochemist. [See LLOYD.]

JOSEPH, HORACE WILLIAM BRINDLEY (1867–1943), philosopher; educated at Allhallows, Winchester, and New College, Oxford; first class, *lit. hum.*, 1890; fellow, 1891; senior philosophical tutor, 1895–1932; junior bursar, 1895–1919; publications include *An Introduction to Logic* (1906), *The Labour Theory of Value in Karl Marx* (1923), *Some Problems in Ethics* (1931), *Essays in Ancient and Modern Philosophy* (1935, containing the important lecture on 'The Concept of Evolution'), and *Knowledge and the Good in Plato's Republic* (1948); FBA, 1930.

JOUBERT DE LA FERTÉ, Sir PHILIP BENNET (1887–1965), air chief marshal; had French grandfather; educated at Harrow and the Royal Military Academy, Woolwich; commissioned in Field Gunners, 1907; learnt to fly, 1912; attached to Royal Flying Corps (Military Wing), 1913; in France with No. 3 Squadron, 1914; commanded No. 15 Squadron in France, 1915; commanded No. 5 Wing in Egypt, 1916–17; DSO; commanded No. 14 Wing in Italy; commanded RFC in Italy, 1918; permanently commissioned in Royal Air Force, 1918; attended Army Staff College, 1920; group captain, instructor RAF College, Andover, 1922; first RAF instructor, Imperial Defence College, 1926–9; commandant, RAF Staff College, 1930–3; air vice-marshal; commanded Fighting Area of Great Britain, 1934; air marshal, 1936; AOC-in-C Coastal Command, 1936; AOC India, 1937; recalled to England as air adviser on combined operations, 1939; assistant chief of Air Staff, special responsibility for application of radar in the RAF; air chief marshal and AOC-in-C Coastal Command, 1941-3; inspector-general of the RAF, 1943; deputy chief of staff, South East Asia Command, 1943-5; director, public relations, Air Ministry, 1946; published *The Third Service* (1955); CMG, 1919; CB, 1936; KCB, 1938.

JOURDAIN, FRANCIS CHARLES ROBERT (1865-1940), ornithologist; BA, Oxford, 1887; ordained, 1890; vicar of Clifton-by-Ashbourne, Derbyshire, 1894-1914; of Appleton, Berkshire, 1914-25; a leading authority on the breeding biology of birds of the palaearctic region; collaborated in *A Practical Handbook of British Birds* (2 vols., 1919-24) and other standard works.

JOWITT, WILLIAM ALLEN, EARL JOWITT (1885-1957), lord chancellor; educated at Marlborough and New College, Oxford; first class, jurisprudence, 1906; called to bar (Middle Temple), 1909; KC, 1922; liberal MP, the Hartlepools, 1922-4; Preston, 1929; joined labour government as attorney-general, knighted, and re-elected for Preston as labour member, 1929; remained in 'national' government, 1931; defeated as national labour candidate, Combined English Universities, Oct. 1931; resigned office, Jan. 1932; labour MP, Ashton-under-Lyne, 1939-45; solicitor-general, 1940-2; paymaster-general, 1942; minister without portfolio, 1943-4; of national insurance, 1944-5; lord chancellor, 1945-51; leader of opposition in House of Lords, 1952-5; baron, 1945; viscount, 1947; earl, 1951; pub-

lished *The Strange Case of Alger Hiss* (1953), and *Some Were Spies* (1954).

JOYCE, JAMES AUGUSTINE (1882-1941), poet, novelist, and playwright; educated at Belvedere College and University College, Dublin; specialized in languages; graduated, 1902; lived in Trieste (1904-14), Zürich (1914-18), Paris (1920-40); died in Zürich; financial independence assured by benefactors after 1918; published *Chamber Music* (poems, 1907), *Dubliners* (1914), *A Portrait of the Artist as a Young Man* (1916); his major work *Ulysses*, the record of a Dublin day, published in Paris, 1922; banned in United States (until 1934) and Great Britain (until 1936); highly influential in style and technique, notably in use of interior monologue; in *Finnegans Wake* (1939), again with central idea of recurrence, created own vocabulary; published also *Exiles* (play, 1918) and *Pomes Penyeach* (1927).

JOYCE, SIR MATTHEW INGLE (1839-1930), judge; called to bar (Lincoln's Inn), 1865; junior equity counsel to Treasury, 1886-1900; judge of High Court and knighted, 1900; PC on resignation, 1915.

JOYNSON-HICKS, WILLIAM, first VISCOUNT BRENTFORD (1865-1932), statesman. [See HICKS.]

JULIUS, SIR GEORGE ALFRED (1873-1946), consulting engineer; B.Sc., engineering, university of New Zealand, 1896; set up practice in Sydney, 1907; invented automatic racecourse totalizator; chairman, Commonwealth Council for Scientific and Industrial Research, 1926-45; of Standards Association of Australia, 1926-40; president, National Research Council, 1932-7; knighted, 1929.

K

KANE, ROBERT ROMNEY (1842-1902), writer on Irish land law; son of Sir Robert Kane [q.v.]; MA, Queen's College, Cork, 1862; hon. LLD, 1882; called to Irish bar, 1865; professor of jurisprudence at King's Inns, Dublin, 1873; legal assistant commissioner under Irish Land Act of 1881, 1881-92; county court judge, Ireland, 1892.

KARLOFF, BORIS (1887-1969), actor; born William Henry Pratt; brother of Sir John Thomas Pratt; educated at Merchant Taylors' School, Uppingham, and King's College, London; acted with repertory companies in

Canada; adopted name Boris Karloff, 1911; went to Hollywood and played villains in Douglas Fairbanks films; played the mesmerist in *The Bells* (1926), and the convict in *The Criminal Code* (1930); appeared in number of horror films, including *Frankenstein*, *The Mask of Fu Manchu*, *The Black Room*, and *Son of Frankenstein*; acted on Broadway stage in plays including *Arsenic and Old Lace* (1941), and *The Linden Tree* (1948); appeared in Hollywood colour films, including *The Secret Life of Walter Mitty* (1947); played in British TV series including *The Name of the Game* (1969).

KEARLEY, HUDSON EWBANKE, first VISCOUNT DEVONPORT (1856-1934), man of business; founded International Stores, 1880; liberal MP, Devonport, 1892-1910; parliamentary secretary, Board of Trade, 1905-9; conducted passage of Port of London bill; became first chairman of the Authority, 1909-25; first food controller, 1916-17; baronet, 1908; PC, 1909; baron, 1910; viscount, 1917.

KEAY, JOHN SEYMOUR (1839-1909), Anglo-Indian politician; went to India to manage branches of government Bank of Bengal, 1862; opened successful banking business and cotton mills at Hyderabad; sympathized with Indian nationalists; liberal MP for Elgin and Nairn, 1889-95.

KEBLE MARTIN, WILLIAM (1877-1969), botanist. [See MARTIN.]

KEEBLE, SIR FREDERICK WILLIAM (1870-1952), botanist, civil servant, and industrial adviser; educated at Alleyn's School, Dulwich, and Caius College, Cambridge; first class, part i, natural sciences tripos, 1891; lecturer, 1902, professor, 1907-14, botany, Reading; director, Royal Horticultural Society gardens, Wisley, 1914-19; joined Board of Agriculture, 1914; controller of horticulture, 1917-19; Sherardian professor of botany, Oxford, 1920-7; agricultural adviser to Imperial Chemical Industries, 1927-32; Fullerian professor, Royal Institution, 1938-41; editor, *Gardeners' Chronicle*, 1908-19; married Lillah McCarthy [q.v.], 1920; FRS, 1913; CBE, 1917; knighted, 1922.

KEEBLE, LILLAH, LADY (1875-1960), actress. [See MCCARTHY, LILLAH.]

KEETLEY, CHARLES ROBERT BELL (1848-1909), surgeon; entered St. Bartholomew's Hospital, 1871; FRCS, 1876; LRCP, 1873; on staff of West London Hospital from 1878 till death; introduced there antiseptic methods of surgery; advocated appendicotomy; wrote handbook on orthopaedic surgery (1900) and other medical works.

KEILIN, DAVID (1887-1963), biologist; born in Moscow; educated in Warsaw and Liège; studied entomology in Paris, 1905-15; research assistant to George H. F. Nuttall [q.v.], first Quick professor, Cambridge, 1915; Beit fellow, 1920-5; lecturer in parasitology, 1925-31; succeeded Nuttall as professor of biology and director, Molteno Institute, 1931-52; published thirty-nine papers between 1914 and 1923 on the reproduction of lice, the life-cycle of the horse bot-fly, the respiratory adaptations in fly

larvae, and other subjects; detected pigment in bot-fly which he named cytochrome, an important development in biochemistry; D.Sc. (Sorbonne), 1915; FRS, 1928; hon. fellow, Magdalene College, Cambridge, 1957; hon. degrees and other academic distinctions from number of universities abroad; published *The History of Cell Respiration and Cytochrome* (ed. J. Keilin, 1966).

KEITH, SIR ARTHUR (1886-1955), conservator of the Hunterian Museum, Royal College of Surgeons; educated at Gordon's College, Aberdeen; qualified from Marischal College, 1888; MD, Aberdeen, and FRCS, 1894; senior demonstrator, later lecturer in anatomy, London Hospital medical school, 1895-1908; conservator, Hunterian Museum, 1908-33; Fullerian professor of physiology, Royal Institution, 1918-23; president, British Association, 1927; rector, Aberdeen, 1930-3; studied human evolution, claiming higher antiquity for *Homo sapiens* than generally accepted, and involved in controversy over fraudulent Piltdown skull; publications include *Human Embryology and Morphology* (1902), *The Antiquity of Man* (1915), and *A New Theory of Human Evolution* (1948); FRS, 1913; knighted, 1921; hon. degrees from Aberdeen, Durham, Manchester, Birmingham, and Oxford.

KEITH, ARTHUR BERRIEDALE (1879-1944), Sanskrit scholar and constitutional lawyer; brother of Sir W. J. Keith [q.v.]; educated at Royal High School and University, Edinburgh, and Balliol College, Oxford; first class, classics, Edinburgh, 1897; oriental languages (1900) and *lit. hum.* (1901), Oxford; called to bar (Inner Temple), 1904; served in Colonial Office, 1901-14; regius professor of Sanskrit and comparative philology, Edinburgh, 1914-44; made important contributions to Vedic and classical Sanskrit studies; published *Responsible Government in the Dominions* (1909, revised, 1912, 1928) and many other books on same subject; *Constitutional History of the First British Empire* (1930) and of *India, 1600-1935* (1936); and works on British constitutional history; FBA, 1935.

KEITH, JAMES, BARON KEITH OF AVONHOLM (1886-1964), judge; educated at Hamilton Academy and Glasgow University; MA; first class, history, 1906; LLB, 1908; admitted to Faculty of Advocates, 1911; commissioned in Seaforth Highlanders, 1914; served in France and the Sudan; KC, 1926; dean, Faculty of Advocates, 1936; lord commissioner of judiciary and senator, College of Justice, with judicial title, Lord Keith, 1937; member, royal commission on marriage and divorce (the Morton commission), 1951; succeeded Lord Normand [q.v.] as lord of appeal in ordinary, 1953-61; life peer, 1953; PC; member of Judicial Committee of Privy Council; honorary bencher, Inner Temple, 1953; trustee, National Library of Scotland, 1925-37; chairman, Scottish Youth Advisory Committee, 1942-6, and of Scottish Probation Council, 1943-9.

KEITH, SIR WILLIAM JOHN (1873-1937), administrator in Burma; brother of A. B. Keith [q.v.]; a dominating influence in Burma secretariat (1896-1928) where he specialized in revenue and finance; leader of house in legislative council, and finance member and vice-president of executive council, 1923-8; knighted, 1925; KCSI, 1928.

KEKEWICH, SIR ARTHUR (1832-1907), judge; educated at Eton and Balliol College, Oxford; BA, 1854; fellow of Exeter, 1854-8; called to bar, 1858; QC, 1877; had large junior practice at Chancery bar, but was unsuccessful as leader; judge, 1886; knighted, 1887; expeditious judge, but his judgments were often reversed on appeal; a strong churchman and conservative.

KEKEWICH, ROBERT GEORGE (1854-1914), major-general; nephew of Sir A. Kekewich [q.v.]; entered army, 1874; lieutenant-colonel in Loyal North Lancashires, 1898; during South African war defended Kimberley from Boers, 15 Oct. 1899-15 Feb. 1900; CB, 1900; major-general, 1902.

KELLAWAY, CHARLES HALLILEY (1889-1952), scientist; educated at Church of England grammar school and the university, Melbourne; MB, BS, 1911; served in Europe in Australian Army Medical Corps, 1915-18; Foulerton research student, Royal Society, 1920-3; director, Walter and Eliza Hall Institute for Pathological Research, Melbourne, 1923-44; director of pathology, AAMC, 1939-42; scientific liaison officer to director-general, 1942-4; research director-in-chief, Wellcome Foundation, London, 1944-52; FRCP, 1929; FRS, 1940.

KELLY, SIR DAVID VICTOR (1891-1959), diplomatist; educated at St. Paul's School and Magdalen College, Oxford; first class, history, 1913; became Roman Catholic, 1914; served in France, 1915-19; entered diplomatic service, 1919; served in Argentina, Portugal, Belgium, and Sweden; in Foreign Office American department, counsellor, 1931-4; acting high commissioner, and chargé d'affaires, Egypt, 1934-8; head, Foreign Office Egyptian department, 1938-9; minister to Switzerland, 1940-2; am-

bassador to Argentina, 1942-6; to Turkey, 1946-9; to USSR, 1949-51; advocated co-ordination between foreign policy and financial and economic policy and cultivation of industrial, financial, and social leaders in countries of his appointment; commentator on Soviet affairs, mainly in *Sunday Times*, 1951-4; chairman, British Council, 1955-9; publications include autobiography, *The Ruling Few* (1952); CMG, 1935; KCMG, 1942; GCMG, 1950.

KELLY, FREDERICK SEPTIMUS (1881-1916), musician and oarsman; born at Sydney; educated at Eton and Balliol; rowed for Oxford; at Henley won Grand Challenge cup, Stewards' cup, Diamond skulls, Wingfield sculls; one of the greatest scullers of all time; studied pianoforte at Frankfurt am Main, 1903-8; left wide range of compositions; joined Royal Naval division, 1914; DSC for gallantry in Gallipoli, 1915; killed in France.

KELLY, JAMES FITZMAURICE- (1857-1923), historian of Spanish literature. [See FITZMAURICE-KELLY.]

KELLY, SIR JOHN DONALD (1871-1936), admiral of the fleet; entered navy, 1884; captain, 1911; rear-admiral, 1921; fourth sea lord, 1924-7; vice-admiral, 1926; KCB, 1929; admiral, 1930; commander-in-chief, home fleet, 1931-3, with task of restoring discipline after Invergordon mutiny; GCVO, 1932; commander-in-chief, Portsmouth, 1934-6; admiral of the fleet, 1936.

KELLY, MARY ANNE, 'EVA' (1826-1910), Irish poetess. [See O'DOHERTY, MARY ANNE.]

KELLY, WILLIAM (1821-1906), Plymouth brother and biblical critic; joined Plymouth Brethren, 1841; edited collected works of John Nelson Darby [q.v.], 34 vols., 1867-83; seceded from Darbyites on point of church discipline, 1879; a voluminous scriptural commentator and controversialist.

KELLY-KENNY, SIR THOMAS (1840-1914), general; joined army, 1858; took sixth division out to South African war, 1900; took part in engagements of Klip Kraal Drift, Paardeberg, Poplar Grove, and Driefontein; left in command in Free State, 1900; KCB, 1902; general, 1905.

KELTIE, SIR JOHN SCOTT (1840-1927), geographer; took up journalism; edited *Statesman's Year Book*, 1884-1927; inspector of geographical education for Royal Geographical Society, 1884; produced important report, 1886; librarian and subsequently secretary of

Society, publications of which he reorganized; inaugurated and edited *Geographical Journal*, 1893; editor and author of several geographical works; knighted, 1918.

KELVIN, first BARON (1824–1907), scientist and inventor. [See THOMSON, WILLIAM.]

KEMBALL, SIR ARNOLD BURROWES (1820–1908), general, colonel commandant Royal Artillery; born in Bombay; joined Bombay artillery, 1837; present at Ghazni and Kabul; in Persian war, 1856–7; distinguished in expedition against Ahwaz; CB, 1857; consul-general at Baghdad, 1859; KCSI, 1866; accompanied Shah of Persia to England, 1873; on international commission for delimiting Turco-Persian frontier, 1875; military commissioner with Turkish army in Serbo-Turkish and Russo-Turkish wars; lieutenant-general and KCB, 1878; interested in Persian and East African development; advocated construction of Uganda railway; general, 1880.

KEMBALL-COOK, SIR BASIL ALFRED (1876–1949), civil servant; scholar of Eton and King's College, Cambridge; second class, classics, 1898; entered transport department, Admiralty, 1900; director, naval sea transport, Ministry of Shipping, 1917–19; assistant British delegate, Reparation Commission, 1921–6; managing director, British Tanker Company, 1927–35; deputy divisional (later divisional) food officer, London, 1942–8; KCMG, 1925.

KEMBLE, HENRY (1848–1907), comedian; grandson of Charles Kemble and nephew of Fanny Kemble [qq.v.]; made professional début at Dublin, 1867; with company of (Sir) John Hare [q.v.], 1875; joined the Bancrofts (qq.v., 1876), with whom he played Dolly Spanker in *London Assurance* and Sir Oliver Surface in *The School for Scandal*; played Polonius to (Sir) Herbert Tree [q.v.] as Hamlet, 1891.

KEMP, STANLEY WELLS (1882–1945), zoologist and oceanographer; educated at St. Paul's School and Trinity College, Dublin; first senior moderator, natural science, 1903; assistant naturalist, fisheries, Irish Department of Agriculture, 1903–10; superintendent, zoological section, Indian Museum (from 1916 Zoological Survey of India), 1910–24; first director, Colonial Office *Discovery* investigations, 1924–36; led voyages, 1925–7, 1929–31; secretary, Marine Biological Association, and director, Plymouth laboratory, 1936–45; FRS, 1931.

KEMSLEY, first VISCOUNT (1883–1968), newspaper proprietor. [See BERRY (JAMES) GOMER.]

KENDAL, DAME MARGARET SHAFTO (1848–1935), better known as MADGE KENDAL, actress; born ROBERTSON of long line of actors; sister of T. W. Robertson [q.v.]; first appeared at age of five; by seventeen had played Ophelia and Desdemona with Walter Montgomery [q.v.] in London; married (1869) W. H. Kendal [q.v.] whose ostentatious cult of respectability she adopted, becoming increasingly censorious; her character possibly too firm and robust; an accomplished but not a great actress; especially fitted for elderly parts; retired, 1908; DBE, 1926; GBE, 1927.

KENDAL, WILLIAM HUNTER (1843–1917), actor-manager, whose original name was WILLIAM HUNTER GRIMSTON; went on professional stage, 1861; partner with Sir John Hare [q.v.], first at Court Theatre and later at St. James's Theatre; best parts those in *Peril*, *The Queen's Shilling*, *Diplomacy*, *A White Lie*, and *The Elder Miss Blossom*; overshadowed by his wife [see above].

KENNARD, SIR HOWARD WILLIAM (1878–1955), diplomatist; educated at Eton; entered diplomatic service, 1901; minister to Yugoslavia, 1925–9; to Sweden, 1929–31; to Switzerland, 1931–5; ambassador to Poland, 1935–9; objected to any Polish support of Hitler's aggression or racialism; British alliance with Poland obtained, 1939; remained with Polish government in exile, 1939–41; CMG, and CVO, 1923; KCMG, 1929; GCMG, 1938.

KENNAWAY, SIR ERNEST LAURENCE (1881–1958), experimental and chemical pathologist; scholar, New College, Oxford; first class, physiology, 1903; qualified from Middlesex Hospital, 1907; DM, Oxford, 1911; demonstrator in physiology, Guy's Hospital, 1909–14; chemical pathologist, Bland-Sutton Institute, Middlesex Hospital, 1914–21; joined Research Institute, Cancer Hospital, 1921; director and professor of experimental pathology, London, 1931–46; doyen of cancer research; over thirty years inspired research; FRS, 1934; knighted, 1947; hon. fellow, New College, Oxford, and many other academic honours.

KENNEDY, SIR ALEXANDER BLACKIE WILLIAM (1847–1928), engineer; nephew of John Stuart Blackie [q.v.]; gained first experience in marine engine construction; professor of engineering, University College, London, where he founded influential school, 1874–89; established pioneer engineering laboratory there, 1878; began practice, which soon became one of largest in country, as consulting engineer at Westminster, 1889; engineer to Westminster Electric Supply Corporation, of which he planned whole system and works, 1889; also engineer to other electric companies; closely connected with development of electric transport; undertook exploration of Petra, 1922; published *Petra; Its History and Monuments*, a valuable monograph (1925); FRS, 1887; knighted, 1905.

KENNEDY, (AUBREY) LEO (1885-1965), journalist; educated at Harrow and Magdalen College, Oxford; BA, 1906; joined *The Times* under G. E. Buckle [q.v.], 1910; worked in Paris and in Serbia, Romania, and Albania, 1911-12; served in army during 1914-18 war; MC; returned to *The Times* under Henry Wickham Steed [q.v.], 1919; worked in Europe, mainly concerned with power politics; published first book, *Old Diplomacy and New 1876-1922* (1922); founding member of Royal Institute of International Affairs, 1920; foreign leader-writer under G. Geoffrey Dawson [q.v.], 1923; published *Britain Faces Germany* (1937); wrote first draft of leading article in *The Times* suggesting that Prague might consider ceding the Sudeten lands, 7 Sept. 1938; later, wrote personal letter to *The Times* warning of dangers of further German expansion; retired after disagreement with new editor, R. M. Barrington-Ward [q.v.], 1941; joined European service of BBC, 1941-5; published a life of Lord Salisbury, 1953; and edited letters of fourth Earl of Clarendon [q.v.], 1956.

KENNEDY, HARRY ANGUS ALEXANDER (1866-1934), Scottish New Testament scholar; Free Church minister, Callander, 1893-1905; professor of New Testament language and literature, Knox College, Toronto (1905-9), and New College, Edinburgh (1909-25); published *The Sources of New Testament Greek* (1895), *St. Paul's Conceptions of the Last Things* (1904), *St. Paul and the Mystery Religions* (1913), *Philo's Contribution to Religion* (1919), and *The Theology of the Epistles* (1919).

KENNEDY, MARGARET MOORE (1896-1967), writer; educated at Cheltenham Ladies' College and Somerville College, Oxford; contemporary of Dorothy Sayers and Vera Brittain [qq.v.]; published first book, *A Century of Revolution* (1922); first novel, *The Ladies of Lyndon* (1923), followed by the best-seller, *The Constant Nymph* (1924); rewritten as play, and several film versions made; *Escape me Never*, produced at Apollo, with Elisabeth Bergner (1933); other novels include *The Feast* (1950), *Lucy Carmichael* (1951), and *Troy Chimneys* (James Tait Black memorial prize, 1953); also published biography of *Jane Austen* (1950) and a study of the art of fiction *The Outlaws on Parnassus* (1958); fellow, Royal Society of

Literature; married (Sir) David Davies, barrister, who became a County Court judge, and was knighted in 1952.

KENNEDY, Sir WILLIAM RANN (1846-1915), judge; grandson of Revd Rann Kennedy [q.v.]; BA, King's College, Cambridge (senior classic); called to bar (Lincoln's Inn), 1871; QC, 1885; judge in Queen's Bench division and knighted, 1892; lord justice of Court of Appeal and PC, 1907; FBA, 1909.

KENNET, first BARON (1879-1960), politician and writer. [See YOUNG, EDWARD HILTON.]

KENNET, (EDITH AGNES) KATHLEEN, LADY KENNET (1878-1947), sculptor exhibiting as KATHLEEN SCOTT; studied at Slade School and Académie Colarossi; influenced by her friend Rodin; married (1908) Robert Falcon Scott [q.v.]; granted rank of KCB's widow after his death in 1912; married secondly (1922) Edward Hilton Young (created Baron Kennet, 1935); executed Scott memorial, Waterloo Place, and many portrait busts and statues remarkably expressive of subject's personality; sitters included Asquith, Lloyd George, Neville Chamberlain, and W. B. Years.

KENNETT, ROBERT HATCH (1864-1932), biblical and Semitic scholar; first class, Semitic languages tripos, Queens' College, Cambridge, 1886; ordained, 1887; fellow, 1888; university lecturer in Aramaic, 1893-1903; regius professor of Hebrew and canon of Ely, 1903-32; bridged gulf between biblical and modern modes of thought; critical work independent and original.

KENNEY, ANNIE (1879-1953), suffragette; born in Yorkshire and became part-time mill worker at ten; met (Dame) Christabel and Emmeline Pankhurst [qq.v.], 1905, and became their loyal lieutenant; directed from France by Christabel in organizing extreme militancy, 1912-14; received several prison sentences; supported Pankhursts' war-work, 1914-18, and retired from public life on granting of women's suffrage, 1918; published *Memories of a Militant* (1924).

KENNINGTON, ERIC HENRI (1888-1960), artist; educated at St. Paul's School; studied at Lambeth School of Art and City and Guilds School; first exhibited at Royal Academy, 1908; a private in France and Flanders, 1914-15; official war artist, 1915-18, 1940-5; art editor, *Seven Pillars of Wisdom* (1926) by T. E. Lawrence [q.v.]; executed portraits in sculpture and drawing, especially in pastel; British war memorial at Soissons and

24th Division, Battersea Park; carved decorations on Shakespeare Memorial Theatre, Stratford; ARA, 1951; RA, 1959.

KENNY, COURTNEY STANHOPE (1846-1930), legal scholar; BA, Downing College, Cambridge; senior in law and history tripos, 1874; fellow of Downing, 1875; called to bar (Lincoln's Inn), 1881; liberal MP, Barnsley division of Yorkshire, 1885-8; university reader in English law, Cambridge, 1888; Downing professor of laws of England, Cambridge, 1907-18; FBA, 1909; an outstandingly successful law teacher; his most important work, *Outlines of Criminal Law* (1902, 14th edn., 1932).

KENNY, ELIZABETH (1880-1952), nurse; born in Australia; served in Australian Army Nursing Service, 1915-19; advocated non-rigid splinting and early movement for treatment of poliomyelitis paralysis; opened first clinic, 1933; her world-wide demonstrations and untiring fight for acceptance of her technique caused great controversy; although her method gradually superseded orthodox treatment its superiority never openly acknowledged by medical profession; received most acclaim in United States; hon. degrees from American universities; published *And They Shall Walk* (1951).

KENSIT, JOHN (1853-1902), protestant agitator; early joined anti-Romanist agitation; started protestant book depot in Paternoster Row, 1885; secretary of Protestant Truth Society, 1890; organized 'Wicliffite' itinerant preachers for the denunciation of ritualism, 1898; publicly objected to confirmation of bishops Mandell Creighton (1897), Winnington-Ingram and Gore (1901); fatally wounded in religious riot at Liverpool.

KENSWOOD, first BARON (1887-1963), professional violinist and economist. [See WHITFIELD, ERNEST ALBERT.]

KENT, DUCHESS OF (1906-1968). [See MARINA.]

KENT, DUKE OF (1902-1942). [See GEORGE EDWARD ALEXANDER EDMUND.]

KENT, ALBERT FRANK STANLEY (1863-1958), scientist; educated at Magdalen College School and Magdalen College, Oxford; demonstrator in physiology, Manchester (1887-9), Oxford (1889-91), St. Thomas's Hospital (1891-5); professor of physiology, Bristol, 1899-1918; D.Sc., 1915; director, department

of industrial administration, Manchester College of Technology, 1918-22; editor-in-chief, *Journal of Industrial Hygiene*; best known for his work on cardiac physiology.

KENT, (WILLIAM) CHARLES (MARK) (1823-1902), author; a strict Roman Catholic; editor and proprietor of the *Sun*, a London evening paper, 1853-71, and editor of the *Weekly Register*, a Roman Catholic periodical, 1874-81; called to bar, 1859, but did not practise; wrote for *Household Words* and *All the Year Round*, and grew intimate with Charles Dickens, who addressed to him his last letter (8 June 1870, now in British Museum); collected his *Poems* in 1870; prepared popular complete editions of works of great writers, including Charles Lamb, 1875; received civil list pension, 1887.

KENYON, SIR FREDERIC GEORGE (1863-1952), scholar and administrator; great-grandson of Baron Kenyon and grandson of Edward Hawkins [qq.v.]; scholar of Winchester and New College, Oxford; first class, *lit. hum.*, 1886; entered manuscripts department, British Museum, 1889; assistant keeper, 1898; director of British Museum, 1909-30; edited Aristotle's treatise on the Athenian constitution, works of Bacchylides and Hyperides, and Chester Beatty collection of biblical papyri; publications include *The Palaeography of Greek Papyri* (1889), *Greek Papyri in the British Museum* (3 vols., 1893-1907), and *Our Bible and the Ancient Manuscripts* (1895); FBA, 1903; president, 1917-21; secretary, 1930-49; CB, 1911; KCB, 1912; GBE, 1925; hon. fellow, Magdalen and New College, Oxford.

KENYON, GEORGE THOMAS (1840-1908), politician; educated at Harrow and Christ Church, Oxford; called to bar, 1869; conservative MP for Denbigh Boroughs, 1885-95, 1900-5; mainly instrumental in passing the Welsh Intermediate Education Act, 1889.

KENYON, JOSEPH (1885-1961), organic chemist; educated at Blackburn secondary school; laboratory assistant at age of fourteen, Municipal Technical School, Blackburn; John Mercer FRS scholar; B.Sc. (London), 1907; D.Sc., 1914; research assistant to (Sir) R. H. Pickard [q.v.]; made important contributions to stereochemistry; published over 160 papers on stereochemistry and its relation to reaction mechanisms; assistant lecturer, Blackburn Technical College, 1906; lecturer, 1907; worked with W. H. Perkin [q.v.] in Oxford on dyestuffs chemistry, 1916-20; head of department of chemistry, Battersea Polytechnic, 1920-50;

FRS, 1936; member, London University Board of Studies in Chemistry and the senate; vice-president, Chemical Society; fellow, Royal Institute of Chemistry.

KENYON-SLANEY, WILLIAM SLANEY (1847-1908), colonel and politician; born at Rajkot, India; educated at Eton and Christ Church, Oxford; joined army 1867; present at Tel-el-Kebir, 1882; retired with rank of colonel, 1892; conservative MP for Newport division of Shropshire, 1886-1908; ardent tariff reformer; author of the 'Kenyon-Slaney clause' in Education Act of 1902; PC, 1904.

KEOGH, Sir ALFRED (1857-1936), lieutenant-general; educated at Queen's College, Galway; qualified in medicine; entered army, 1880; deputy director-general, Army Medical Services, 1902-5; director-general, 1905-10 and 1914-18; reforms included establishment of Royal Army Medical College, Millbank, army school of hygiene, and central hospitals; brought teaching hospitals into Territorial scheme; rector, Imperial College of Science, 1910-22; KCB, 1906; GCVO and CH, 1918.

KEPPEL, Sir GEORGE OLOF ROOS- (1866-1921), soldier and Anglo-Indian administrator. [See Roos-Keppel.]

KEPPEL, Sir HENRY (1809-1904), admiral of the fleet; entered navy, 1822; commander, 1833; served in West and East Indies, and in China war, 1841-2; senior officer at Singapore, 1842; helped in suppressing Borneo piracy, 1843-4; served with distinction in Baltic campaign, 1854; CB, 1856; second-in-command on China station, 1856; KCB; groom-in-waiting to Queen Victoria, 1858-60; commander-in-chief on Cape and Brazilian stations, 1860; on China station, 1866; admiral, 1869; GCB, 1871; commander-in-chief at Devonport, 1872-5; admiral of the fleet, 1877; intimate friend of King Edward VII; published *A Sailor's Life under Four Sovereigns* (3 vols., 1899).

KER, WILLIAM PATON (1855-1923), scholar and author; educated at Glasgow University and Balliol College, Oxford; assistant to professor of humanity, Edinburgh, 1878-83; fellow of All Souls College, Oxford, 1879-1923; professor of English literature and history, University College of South Wales, Cardiff, 1883; Quain professor of English language and literature, University College, London, 1889-1922; director of department of Scandinavian studies, 1917-23; professor of poetry, Oxford, 1920; died in Italy; his works include *Epic and Romance* (1897); *The Dark Ages* (1904); *English Literature: Medieval* (1912); editions of Dryden's *Essays* (1900) and of Lord Berners's translation of Froissart's *Chronicles* (1901-3).

KERMACK, WILLIAM OGILVY (1898-1970), biochemist and mathematician; educated at Webster's Seminary, Kirriemuir and Aberdeen University; MA, first class, mathematics and natural philosophy; B.Sc., 1918; served with RAF; worked at Dyson Perrins Laboratory, Oxford, 1919-21, and at the Royal College of Physicians Laboratory, Edinburgh, 1921-49; permanently blinded in laboratory accident, 1924; continued research work; developed methods for synthesis of heterocyclic compounds with antimalarial activity; also returned to mathematics, and collaborated with A. G. McKendrick on mathematical theory of epidemics; first MacLeod-Smith professor of biological chemistry, Aberdeen, 1949-68; dean, Science Faculty, 1961-4; D.Sc., Aberdeen, 1925; hon. LLD, St. Andrews, 1937; FRS, 1944; fellow, Royal Society of Edinburgh, 1924; member of RSE Council, 1946-9.

KERR, ARCHIBALD JOHN KERR CLARK, Baron Inverchapel (1882-1951), diplomatist. [See Clark Kerr.]

KERR, JOHN (1824-1907), physicist; pupil of Lord Kelvin [q.v.] at Glasgow University; lecturer in mathematics to Glasgow Free Church training college for teachers, 1857-1901; made important discoveries in nature of light, 1875-6; author of *An Elementary Treatise on Rational Mechanics* (1867); hon. LLD, Glasgow, 1867; FRS, 1890; royal medallist, 1898; awarded civil list pension, 1902.

KERR, Sir JOHN GRAHAM (1869-1957), zoologist; educated at Royal High School and university, Edinburgh, and Christ's College, Cambridge; first class, parts i and ii, natural sciences tripos, 1894-6; fellow, Christ's College; demonstrator, animal morphology, Cambridge, 1897-1902; regius professor of zoology, Glasgow, 1902-35; devised system for camouflaging ships; MP, Scottish Universities, 1935-50; publications include *A Naturalist in Gran Chaco* (1950), *Zoology for Medical Students* (1921), *Introduction to Zoology* (1929); his collections from expeditions to Chaco region preserved at Glasgow; especially interested in marine biology and Scottish natural history; FRS, 1909; knighted, 1939.

KERR, (JOHN MARTIN) MUNRO (1868-1960), obstetrician and gynaecologist; educated at Glasgow Academy and University and

Berlin, Jena, and Dublin; MB, CM, 1890; MD, 1909; assistant to midwifery professor, Glasgow, 1894; Muirhead professor, Glasgow, 1911–27; regius professor, 1927–34; published many books on his specialities.

KERR, PHILIP HENRY, eleventh MARQUESS OF LOTHIAN (1882–1940), journalist and statesman; educated at Oratory School and New College, Oxford; first class, modern history, 1904; lost Roman Catholic faith and later became convinced Christian Scientist; youngest member of Milner's 'kindergarten'; secretary, Transvaal indigency commission, 1907; editor of *The State*, 1908–9; co-founder and first editor of the *Round Table*, 1910–16; private secretary to Lloyd George, 1916–21; played important part in dealing with the dominions, India, and United States; largely responsible for document on German nation which forms preface to treaty of Versailles; secretary to Rhodes trustees, 1925–39; succeeded to title, 1930; as a representative of the liberal party entered 'national' government as chancellor of the duchy of Lancaster, 1931; under-secretary of state for India, 1931–2; chairman, Indian franchise committee, 1932; ambassador to United States, 1939–40; revealed remarkable persuasive powers and statesmanlike qualities; CH, 1920; PC, 1939.

KERR, ROBERT (1823–1904), architect; cousin of Joseph Hume [q.v.]; did much to develop Royal Institute of British Architects; professor of arts of construction, King's College, London, 1861–90; designed many buildings in London and country; published many architectural works.

KERR, LORD WALTER TALBOT (1839–1927), admiral of the fleet; brother of ninth Marquess of Lothian [q.v.]; entered navy, 1853; served in naval brigade throughout Indian Mutiny, 1857–8; captain, 1872; flag captain to Sir Beauchamp Seymour (afterwards Lord Alcester), 1874–7, 1880–1; naval private secretary to Lord George Hamilton [q.v.], first lord of Admiralty, 1885; rear-admiral, 1889; fourth naval lord, 1892; second naval lord, 1893–5; vice-admiral commanding Channel squadron, 1895; first sea lord, 1899–1904; admiral, 1900; admiral of the fleet, 1904; a thorough seaman and wise administrator.

KETELBEY, ALBERT WILLIAM (1875–1959), composer; composition scholar, Trinity College, London, at thirteen; organist, St. John's church, Wimbledon, at sixteen; musical director, Vaudeville Theatre, at twenty-two; his romantic light music included 'In a Monastery Garden', 'In a Persian Market', and atmospheric music for silent films.

KETTLE, EDGAR HARTLEY (1882–1936), pathologist; student, pathologist, and ultimately director of Institute of Pathology and Medical Research, St. Mary's Hospital; professor of pathology, Welsh National School of Medicine, Cardiff (1924–7), St. Bartholomew's Hospital (1927–34), and British Postgraduate Medical School (1934–6); histopathology his special province; published *The Pathology of Tumours* (1916) and (with W. E. Gye) papers on silicosis and its association with pulmonary tuberculosis (1922–34); investigated whole problem of relation of dust to infection; FRS, 1936.

KETTON-CREMER, ROBERT WYNDHAM (1906–1969), biographer and historian; educated at Harrow and Balliol College, Oxford; health impaired by rheumatic fever; assisted his father in restoring Felbrigg (near Cromer); inherited the estate, 1933; first book, *The Early Life and Diaries of William Windham* (published 1930); other publications include *Horace Walpole* (1940), *Thomas Gray* (won James Tait Black memorial prize, 1955), *Felbrigg: The Story of a House* (1962), and *Norfolk in the Civil War* (1969); published also five volumes of collected essays on Norfolk personalities and events (1944–61); major in Home Guard, 1941–5; chairman, magistrates, 1948–66; high sheriff, 1951; trustee, National Portrait Gallery; FBA, 1968; governor, Gresham's School, Holt; hon. degree, East Anglia University, 1969; fellow-commoner, Christ's College, Cambridge; bequeathed Felbrigg to National Trust.

KEYES, ROGER JOHN BROWNLOW, first BARON KEYES (1872–1945), admiral of the fleet; entered navy, 1885; promoted commander for services in Boxer rising, 1900; naval attaché, Rome, 1905–8; captain, 1905; commanded *Venus*, 1908–10; inspecting captain, submarines, 1910; commodore in charge of submarine service, 1912–14; commanded submarines in North Sea, 1914–15; fought at Heligoland Bight; chief of staff to (Sir) John de Robeck [q.v.] at Dardanelles, 1915; prominent in planning naval operations and army landings; pressed unsuccessfully in October for second naval attempt to force Narrows; DSO; commanded *Centurion* in Grand Fleet, 1916–17; rear-admiral, 1917; director of plans, Admiralty, Oct. 1917–Jan. 1918; planned blockage of Ostend and Zeebrugge; vice-admiral, Dover patrol, Jan. 1918; implemented audacious operation of storming the German batteries and sinking blockships at Zeebrugge, 23 Apr. 1918; unsuccessful at Ostend; conducted unremitting offensive against enemy on Belgian coast; KCB, KCVO, 1918; baronet, 1919; commanded battle cruiser squadron, 1919-21; vice-admiral, 1921; deputy chief of naval staff, 1921-5; reached agreement with Royal Air Force on dual control of Naval Air Service; commander-in-chief, Mediterranean, 1925-8; Portsmouth, 1929-31; admiral, 1926; admiral of the fleet and GCB, 1930; conservative MP, North Portsmouth, 1934-43; liaison officer with King of Belgians, 1940; first director, Combined Operations Command, 1940-1; baron, 1943.

KEYNES, JOHN MAYNARD, Baron KEYNES (1883-1946), economist; scholar of Eton and King's College, Cambridge; twelfth wrangler and president of Union, 1905; member of 'the Apostles' and later of the 'Bloomsbury group'; in India Office, 1906-8; lecturer in economics, Cambridge, 1908-15; second (1919), first (1924-6) bursar, King's College, Cambridge; editor, *Economic Journal*, 1912-45; member, royal commission on Indian finance and currency, 1913-14; joined Treasury, 1915; principal representative at peace conference, 1919; vigorously disagreed with proposals on frontiers and reparations; resigned, June 1919; wrote *The Economic Consequences of the Peace* (1919) and became centre of controversy on European economics; closely associated with liberal party; wrote regularly in *Nation* and *Athenaeum*, 1923-31; collaborated with (Sir) Hubert Henderson in *Can Lloyd George do it?* (1929); member, committee on finance and industry, 1929-31; his conviction of possibility of curing unemployment and interest in nature of saving and investment and their relation to rising and falling prices resulted in *A Treatise on Money* (2 vols., 1930) and *General Theory of Employment, Interest and Money* (1936); maintained that economic system has no automatic tendency to full employment; expounded new theory of rate of interest and of forms of short-period equilibrium; for a time divided the economists but his general approach later widely accepted; possessed immense vitality, optimism, and conviction that problems were soluble; by investment and speculation amassed large fortune; collected books and paintings; through marriage (1925) with Lydia Lopokova interested himself in ballet; built and financed Arts Theatre, Cambridge, 1925; chairman, CEMA, later the Arts Council; wrote *Essays in Persuasion* (1931), *Essays in Biography* (1933), and *Two Memoirs* (1949); returned to Treasury, 1940; chiefly responsible for new concept of budgetary policy, 1941; played leading part at Bretton Woods conference (1944) from which emerged International Monetary Fund and International Bank; first British governor on both; engaged in negotiations with United States on lend-lease, 1944-5; obtained loan conditional on early establishment of convertibility, 1945; FBA, 1929; baron, 1942.

KHAN SAHIB (1883-1958), Indian politician; educated at Peshawar government high school and mission college; qualified in medicine; joined brother, Abdul Ghaffar Khan, organizing Red Shirt movement of Moslem Pathans and formed alliance with Congress party; imprisoned, then externed, 1931; chief minister, North-West Frontier Province, 1937; resigned, 1939, but supported war effort; chief minister, 1945; lost office and popularity when Pathans opted for Pakistan, 1947; minister of communications, Pakistan, 1954; chief minister, West Pakistan, 1955-7; assassinated.

KIDD, BENJAMIN (1858-1916), sociologist; clerk in Inland Revenue department, 1877-94; published *Social Evolution*, which had remarkable success due to its violent attack on socialism, 1894; its main theme is glorification of religion and attack on reason; other works include *Principles of Western Civilization* (1902) and *Science of Power* (posthumous, 1918).

KIGGELL, Sir LAUNCELOT EDWARD (1862-1954), lieutenant-general; born and educated in Ireland; from Sandhurst joined Royal Warwickshire Regiment, 1882; on staff in South Africa, 1899-1904; deputy-assistant-adjutant-general, Staff College, 1904-7; director, staff duties, War Office, 1909-13; commandant, Staff College, 1913-14; at War Office, 1914-15; chief of general staff to Haig [q.v.], Dec. 1915-Jan. 1918; failure of campaigns of 1916-17 due to some extent to his inability to abandon earlier orthodoxies in light of realities of modern warfare; lieutenant-general, 1917; lieutenant-governor, Guernsey, 1918-20; CB, 1908; KCB, 1916; KCMG, 1918.

KILBRACKEN, first Baron (1847-1932), civil servant. [See GODLEY, (JOHN) ARTHUR.]

KILLEARN, first Baron (1880-1964), diplomatist. [See LAMPSON, MILES WEDDERBURN.]

KILLEN, WILLIAM DOOL (1806-1902), ecclesiastical historian; after education at Belfast, entered Presbyterian ministry, 1829; minister of Raphoe, county Donegal, 1829-41; studied deeply church history; professor of church history at Presbyterian College, Belfast, 1841-89; president of the college, 1869; DD, Glasgow, 1845; LLD, 1901; his voluminous historical works include *The Ancient Church* (1859) and *The Ecclesiastical History of Ireland* (2 vols., 1875); he also published *Reminiscences of a Long Life* (1901).

KILMUIR, EARL OF (1900-1967), lord chancellor. [See FYFE, DAVID PATRICK MAXWELL.]

KIMBERLEY, first EARL OF (1826-1902), statesman. [See WODEHOUSE, JOHN.]

KIMMINS, DAME GRACE THYRZA (1870-1954), pioneer in work for crippled children; born Hannam; educated at Wilton House School, Reading; founded Guild of the Brave Poor Things, 1894; married C. W. Kimmins, 1897; founded home for cripples at Chailey, 1903, developed into Heritage Craft Schools; remained commandant until transfer to National Health Service, 1948; DBE, 1950.

KINAHAN, GEORGE HENRY (1829-1908), geologist; entered Irish Geological Survey, 1854; as district surveyor (1869) prepared geological maps; author of *Manual of the Geology of Ireland* (1878), *Economic Geology of Ireland* (1889), and other kindred works.

KINCAIRNEY, LORD (1828-1909), Scottish judge. [See GLOAG, WILLIAM ELLIS.]

KINDERSLEY, ROBERT MOLESWORTH, first BARON KINDERSLEY (1871-1954), banker and president of the National Savings Committee; educated at Repton; joined Lazard Brothers, 1906; chairman, 1919-53; member, Court of Bank of England, 1914-46; chairman, War Savings Committee, 1916; president, National Savings Committee, 1920-46; on Dawes reparations committee, 1924; KBE, 1917; GBE, 1920; baron, 1941.

KING, EARL JUDSON (1901-1962), clinical biochemist; born in Toronto; educated at Brandon College, McMaster University, Ontario, and Toronto University; graduated in chemistry and biology, 1921; Ph.D., 1926; research post, Banting Institute, Toronto, studied biochemistry of silicosis; worked at Lister Institute, London, 1928; and at Kaiser Wilhelm Institute, Munich; associate professor, Banting Institute, 1929; head of biochemical section, 1931; head of chemical pathology department, British Postgraduate Medical School, Hammersmith, 1934-62; reader in chemical pathology, London University, 1935; professor, 1945; continued work on silicosis; demonstrated that silica dissolved to produce toxic action on nearby pulmonary cells; extended research to effects of inhaled asbestos fibres; during 1939-45 war, developed methods for estimating effects of new anti-malarial drugs, and became consultant in medical biochemistry to Royal and Indian Army Medical Corps; chairman, Biochemical Society, 1957-9; editor, *Biochemical Journal*; first chairman, International Federation of Clinical Chemists, 1952-60; MD, Oslo

and Iceland; books and articles include *Microanalysis in Medical Biochemistry* (1946).

KING, EDWARD (1829-1910), bishop of Lincoln; BA, Oriel College, Oxford, 1851; influenced by tractarian movement and by Charles Marriott [q.v.]; curate of Wheatley, near Cuddesdon, 1854-8; chaplain of Cuddesdon theological college, 1858-63; appointed principal of college, and vicar of Cuddesdon, 1863; professor of pastoral theology at Oxford and canon of Christ Church, 1873; exerted much influence on religious life of Oxford; bishop of Lincoln, 1885; specially interested in youths and in confirmations; a high churchman, he taught real objective presence and practised confession; prosecuted for illegal ritualist practices by the Church Association, 1889; archbishop's judgment (1890) was substantially in King's favour, and was upheld on appeal, 1892; the trial enhanced his popularity in Lincolnshire; a staunch tory, active in opposition to education bills of liberal government; favoured franchise bill of 1884; had great faculty for sympathy, and perfect refinement of thought and bearing; published many devotional works, sermons, and pamphlets; two churches at Grimsby erected to his memory.

KING, SIR (FREDERIC) TRUBY (1858-1938), pioneer of 'mothercraft'; born in New Zealand; qualified in medicine, Edinburgh, 1886; founded Royal New Zealand Society for the Health of Women and Children (Plunket Society), 1907; established training centre, London, 1918; in many countries his mothercraft centres imitated and teaching followed with notable decrease in infant mortality; director of child welfare, New Zealand, 1921-7; knighted, 1925.

KING, SIR GEORGE (1840-1909), Indian botanist; studied medicine and botany at Aberdeen University; MB, 1865; hon. LLD, 1884; entered Indian medical service, 1865, Indian forest service, 1869; superintendent of Royal Botanic Garden, Calcutta, and of cinchona cultivation in Bengal, 1871; organized Botanical Survey of India; inaugurated economic method of separating quinine from cinchonas, 1887; CIE, 1890; KCIE, 1898; FRS, 1887; made study of flora of Malayan peninsula.

KING, HAROLD (1887-1956), organic chemist; educated at Friar's grammar school and University College, Bangor; first class, chemistry, 1908; worked for Wellcome laboratories, 1912-19; chemist, with special responsibility for study of drugs, Medical Research Council, working at National Institute for Chemical Research, 1919-50; chemotherapy research included arsenical drugs, anti-malarial drugs, diamidines, and work leading to discovery of methonium drugs; with Otto Rosenheim [q.v.] revised formulation of cholesterol; FRS, 1933; CBE, 1950.

KING, HAYNES (1831-1904), genre painter; born at Barbados; exhibited at Royal Academy, 1860-1904; influenced by Thomas Faed [q.v.]; works include 'Looking Out' (1860), 'The New Gown' (1892), 'Latest Intelligence' (1904).

KING, WILLIAM BERNARD ROBINSON (1889-1963), geologist; educated at Uppingham and Jesus College, Cambridge; first class, part ii, natural sciences tripos, 1912; Harkness scholar; appointed to Geological Survey of Great Britain, 1912; served in Royal Welch Fusiliers and Royal Engineers, 1914-18; OBE (military); demonstrator and assistant to Woodwardian professor of geology, Cambridge, 1920; Yates-Goldsmid professor of geology, University College, London, 1931; Woodwardian professor, 1943-55; fellow, Jesus College, Cambridge, 1920, and Magdalene College, 1922; published over fifty papers concerned mainly with the stratigraphy of Lower Palaeozoic and Quarternary, the Cambrian palaeontology of the Dead Sea and Persian Gulf, the floor of the English Channel, hydrogeology and military geology; MC, 1940; advised on geological conditions for invasion of Europe, 1941-3; president, Geological Society of London, 1953-5; FRS, 1949; many academic honours both in England and abroad.

KING, WILLIAM LYON MACKENZIE (1874-1950), Canadian statesman; born in Ontario; grandson of W. L. Mackenzie [q.v.]; graduated in political science, Toronto, 1895, LLB, 1896; AM, 1898, Ph.D., 1909, Harvard; deputy minister, Canadian labour department, 1900-8; MP, 1908-11; minister of labour, 1909-11; director of industrial research, Rockefeller Foundation, 1914-18; elected liberal party leader and re-entered parliament, 1919; prime minister, 1921-30, 1935-48; minister for external affairs, 1921-30, 1935-46; resigned when refused dissolution by Lord Byng [q.v.], 1926; brought down subsequent conservative government as unconstitutional; won ensuing election as champion of Canadian independence and British constitutional system; steadfastly developed Canadian autonomy; at imperial conferences would not agree that Commonwealth should act as single unit (1923) and adopted middle course on dominion status resulting in Balfour statement (1926); in opposition, 1930-5; obtained overwhelming majority, 1935; repudiated proposal of oil sanctions against Italy by Canadian representative at Geneva,

1935; told Hitler Canada would fight, 1937; increased military appropriations; declared war seven days after Great Britain, 1939, with virtually unanimous support of Commons; pledged not to introduce overseas conscription; returned with record majority, 1940; supplied vast quantities of war material under mutual aid; introduced conscription for home service and wages and prices freeze; made with Roosevelt Ogdensburg agreement on defence (1940) and Hyde Park declaration integrating two economies (1941); conciliated between Roosevelt and Churchill; required resignation of J. L. Ralston [q.v.] who pressed for conscription which King introduced three weeks later, 1944; won last election (1945) on programme of reconstruction; completed work for Canadian autonomy with Canadian Citizenship Act (1946); OM, 1947; retired, 1948.

KING-HALL, (WILLIAM) STEPHEN (RICHARD), BARON KING-HALL (1893-1966), writer, and broadcaster on politics and international affairs; educated at Lausanne, and at Osborne and Dartmouth; in action at Jutland in *Southampton*; served in Admiralty, 1919; passed through Royal Naval Staff College, 1920-1; posted to China Squadron, 1922-3; intelligence officer to Sir Roger (later Lord) Keyes [q.v.], Mediterranean Fleet, 1925-6; commander, working on Naval Staff, 1928-9; resigned to take research post in Royal Institute of International Affairs, 1929-35; produced in collaboration with Ian Hay (John Hay Beith, q.v.) *The Middle Watch*, 1929; made weekly broadcast on current affairs, 1930-7; produced *King-Hall News Letter*, 1936; national labour MP, Ormskirk, 1939-42; independent MP, 1942-5; founded Hansard Society, 1944; publications include *Western Civilization and the Far East* (1924), *Imperial Defence* (1926), and *Our Own Times* (1934-5); knighted, 1954; life peer, 1966.

KINGDON-WARD, FRANCIS (FRANK) (1885-1958), plant collector, explorer, and author; son of Harry Marshall Ward [q.v.]; educated at St. Paul's School and Christ's College, Cambridge; became professional plant collector undertaking some twenty-five expeditions in mountain regions of India, China, Tibet, and Burma; brought back Rhododendrons, Primulas, Gentians, Lilies, etc., and notably the blue poppy for cultivation and material for many books from *The Land of the Blue Poppy* (1913) to *Return to the Irrawaddy* (1956); contributed to study of plant geography; OBE, 1952.

KINGSBURGH, LORD (1836-1919), lord justice-clerk of Scotland. [See MACDONALD, JOHN HAY ATHOLE.]

KINGSCOTE, Sir ROBERT NIGEL FITZ-HARDINGE (1830-1908), agriculturist; in Scots Fusilier Guards, 1846-56; served in Crimea; CB, 1855; liberal MP for Western division of Gloucestershire, 1852-85; inherited estate at Kingscote, 1861; parliamentary groom-in-waiting to Queen Victoria, 1859-66; extra equerry to King Edward VII when Prince of Wales, 1867; commissioner of woods and forests, 1885-95; paymaster-general of the royal household, 1901; KCB, 1889; GCVO, 1902; president of Royal Agricultural Society, 1878; hon. LLD, Cambridge, 1894; member of royal commissions on agriculture, 1879 and 1893.

KINGSFORD, CHARLES LETHBRIDGE (1862-1926), historian and topographer; BA, St. John's College, Oxford; Board of Education official, 1890-1912; FBA, 1924; his works include *Henry V* (1901), editions of three unprinted *Chronicles of London* (1905) and of Stow's *Survey of London* (1908), *English Historical Literature in the Fifteenth Century* (1913), *Prejudice and Promise in Fifteenth Century England* (1925), *The Early History of Piccadilly* ... (1925).

KINGSTON, CHARLES CAMERON (1850-1908), Australian statesman; born at Adelaide; admitted to colonial bar, 1873; QC, 1889; radical member for West Adelaide in house of representatives of South Australia, 1881-1900; attorney-general, 1884-5 and 1887-9; chief secretary, 1892-3; premier and attorney-general, 1893-9; defeated bills for imposition of land taxes, and for employers' liability; helped to secure woman's suffrage, factory legislation, state banking, protective tariff, and payment of members; represented South Australia at Queen Victoria's diamond jubilee; hon. DCL, Oxford, and PC, 1897; took prominent part in securing enactment of Australian Commonwealth constitution bill, 1900; elected to legislative council of South Australia, 1900; minister of trade and customs in first Commonwealth administration, 1901-3.

KINNEAR, ALEXANDER SMITH, first BARON KINNEAR (1833-1917), judge; advocate at Scots bar, 1856; QC, 1881; lord ordinary, 1882-90; member of first division, 1890-1913; baron, 1897.

KINNEAR, Sir NORMAN BOYD (1882-1957), ornithologist; educated at Edinburgh Academy and Trinity College, Glenalmond; officer-in-charge, Bombay Natural History Society museum, 1907-19; organized survey of mammals of India, Burma, and Ceylon; joined zoology department, British Museum (Natural History), 1920; assistant keeper, 1928; deputy

keeper in charge of birds, 1936; keeper of zoology, 1945; director of museum, 1847-50; active in work for bird protection, nature reserves, National Trust, etc.; knighted, 1950.

KINNS, SAMUEL (1826-1903), writer on the Bible; Ph.D., Jena University, 1859; rector of Holy Trinity, Minories (1889-99), of which he wrote a history, 1890; published *Moses and Geology* (1882) and *Graven in the Rock* (1891).

KINROSS, first BARON (1837-1905), Scottish judge. [See BALFOUR, JOHN BLAIR.]

KIPLING, (JOSEPH) RUDYARD (1865-1936), author; born in Bombay; cousin of Stanley Baldwin; out of schooldays at United Services College, Westward Ho!, later wove *Stalky and Co.* (1899); joined staff of Lahore *Civil and Military Gazette*, 1882; wrote especially of the imperial race doing justice and upholding law; became known through stories and verse including *Departmental Ditties* (1886), *Plain Tales from the Hills*, *Soldiers Three*, and *Wee Willie Winkie* (1888); settled in London (1889) and made much of by the critics; travelled widely; eventually settled at Burwash (1902); published his novel *The Light That Failed* (1891), *Many Inventions* (1893), the two *Jungle Books* (1894-5), and *Captains Courageous* (1897); had now reached height of his fame; himself ranked *Just So Stories for Little Children* (1902) and 'Recessional' (1897) highest among his stories and poems respectively, thus choosing the two poles of family and Empire about which his genius turned; between *Barrack-Room Ballads* (1892) and *The Seven Seas* (1896) became exponent of an imperial ethic; the anti-imperialist reaction to South African war made him less universally popular; maintained his reputation with *Kim* (1901), *Puck of Pook's Hill* (1906), *Rewards and Fairies* (1910), and *A School History of England* (1911), publishing some of his most durable verse besides 'The Glory of the Garden' and 'If'; but his later style became over-mannerized and obscure and his authoritarian political faith unacceptable to many; awarded Nobel prize for literature, 1907; refused the laureateship (1895) and the OM thrice.

KIPPING, FREDERIC STANLEY (1863-1949), chemist; educated at grammar school and Owens College, Manchester, and Munich (Ph.D.); assistant to W. H. Perkin [q.v.], Heriot-Watt College, Edinburgh, 1887-90; collaborated in *Organic Chemistry* (1894-5); chief chemistry demonstrator, Central Technical College, 1890-7; professor of chemistry, University College, Nottingham, 1897-1936; published many papers on organic compounds of silicon and stereochemistry of nitrogen; FRS, 1897.

KIRK, SIR JOHN (1832-1922), naturalist and administrator; MD, Edinburgh University; physician and naturalist to David Livingstone [q.v.] on second Zambezi expedition, 1858-63; vice-consul, Zanzibar, 1866; assistant political agent, 1868; consul-general, 1873; political agent, 1880; persuaded Sultan to abolish slave-trade, 1873; checkmated German designs on Zanzibar; was instrumental in persuading Sultan to make great concessions of mainland territory to East African Association, 1887; FRS, 1887; KCMG, 1881.

KIRK, SIR JOHN (1847-1922), philanthropist; entered service of Ragged School Union, 1867; secretary, 1879; knighted and styled director, 1907; largely responsible for survival of Union after 1870 and for increase in activities and annual income.

KIRK, KENNETH ESCOTT (1886-1954), bishop of Oxford; educated at Royal grammar school, Sheffield, and St. John's College, Oxford; first class, *lit. hum.*, 1908; deacon, 1912; priest, 1913; chaplain in France and Flanders, 1914-19; tutor, Keble College, Oxford, 1914-22; fellow and chaplain, Trinity, 1922-33; BD, 1922; DD, 1926; reader in moral theology, 1927-33; regius professor of moral and pastoral theology and canon of Christ Church, 1933-7; bishop of Oxford, 1937-54; reorganized administration of diocese and maintained close relationship with university; a leading Anglo-Catholic; publications include *The Vision of God* (1931), *Commentary on the Epistle to the Romans* (1937); edited *The Study of Theology* (1939) and *The Apostolic Ministry* (1946); hon. fellow, St. John's and Trinity Colleges, Oxford.

KIRKPATRICK, SIR IVONE AUGUSTINE (1897-1964), diplomatist; educated at Downside; commissioned in Royal Inniskilling Fusiliers and served at Gallipoli and in Holland, 1914-18; entered foreign service, 1919; served for ten years in Western Department, 1920-30; in Rome, 1930-3; head of Chancery, Berlin, 1933-8; served under Sir Eric Phipps and Sir Nevile Henderson [qq.v.]; foreign adviser to BBC, 1941; controller European Services, 1941-4; identified Hess, after landing in Scotland, 1941; assistant under-secretary in charge of information work, Foreign Office, 1945-7; deputy under-secretary, 1947-9; permanent under-secretary of German Section, 1949-50; worked closely with Ernest Bevin [q.v.]; high commissioner, Germany, 1950-3; succeeded Sir William (later Lord) Strang as permanent under-secretary, 1953-6; chairman, Indepen-

dent Television Authority, 1957-62; CMG, 1939; KCMG, 1948; GCMG, 1953; KCB, 1951; GCB, 1956; published memoirs, *The Inner Circle* (1959), and *Mussolini, Study of a Demagogue* (1964).

KIRKWOOD, DAVID, first BARON KIRKWOOD (1872-1955), politician; trained as engineer on Clydeside; joined Amalgamated Society of Engineers at twenty; prominent in union affairs at Parkhead Forge; opposed dilution of labour, increases in rents, etc.; deported to Edinburgh as trouble-maker, 1916-17; labour MP, Dumbarton Burghs, 1922-50, East Dumbartonshire, 1950-1; the most vehement Clydesider in Parliament; twice suspended; in depression on Clydeside secured resumption of work on *Queen Mary*; PC, 1948; freedom of Clydebank and baron, 1951.

KITCHENER, HORATIO HERBERT, first EARL KITCHENER OF KHARTOUM AND OF BROOME (1850-1916), field-marshal; educated at Royal Military Academy, Woolwich; received commission in Royal Engineers, 1871; lent to Palestine Exploration Fund, 1874; sent to survey Cyprus, 1878; second-in-command of Egyptian cavalry, 1882; served in (Lord) Wolseley's expedition for relief of General Gordon [qq.v.], 1884-5; governor-general of Eastern Sudan, 1886; adjutant-general of Egyptian army, 1888; contributed to defeat of dervishes at Toski, 1889; CB, 1889; sirdar of Egyptian army, 1892; prepared army for conquest of Sudan, 1892-6; KCMG, 1894; major-general and KCB for services in River war, 1896; well-planned and well-executed campaign resulted in annihilation of Khalifa's army at Omdurman, with great loss of dervishes, and reoccupation of Khartoum, 1898; had interview at Fashoda on White Nile with Major Marchand, leader of small French expedition, resulting in its withdrawal, 1898; baron, 1898; governor-general of Sudan and completed its pacification, 1899; Lord Roberts's chief of staff in South Africa, 1899; frequently employed as second-in-command and representative of commander-in-chief in his absence; ordered attack and directed preliminary operations at Paardeberg, Feb. 1900; suppressed rebellion of Cape Boers round Priska and cleared southern portion of Orange Free State; as commander-in-chief organized tactics against guerrilla warfare of Boers, Nov. 1900-May 1902; viscount and OM, 1902; commander-in-chief in India, 1902-9; prevailed with (Lord) Morley [q.v.], secretary of state for India, to abolish system of dual military control; initiated numerous reforms, including improvement of central administration, redistribution of troops, modernization of system of training, and establishment of Staff College for India; field-marshal, 1909; British agent and consul-general in Egypt, 1911; kept Egypt quiet during period of unrest in Near East; earl, 1914; secretary of state for war, 1914; possessed first-hand knowledge of military resources of Empire but had little experience of organization of army at home or of working of War Office and cabinet; envisaged long war; increased British army from six regular and fourteen territorial divisions to seventy divisions, 3,000,000 men having voluntarily joined colours, 1914-16; KG, 1915; relations with colleagues in cabinet sometimes strained; went to Near East and advised abandonment of Dardanelles enterprise, 1915; went down with HMS *Hampshire* off Orkneys on way to Russia.

KITCHIN, GEORGE WILLIAM (1827-1912), dean of Winchester, 1883-94, and of Durham, 1894-1912; BA, Christ Church, Oxford; censor of Oxford non-collegiate students, 1868-83; wrote *History of France* (1873-7).

KITSON, JAMES, first BARON AIREDALE (1835-1911), ironmaster; placed (1854) in charge of father's Monkbridge ironworks, which became limited liability company, 1886; president of Iron and Steel Institute, 1889-91; first lord mayor of Leeds, 1896-7; benefactor to Leeds hospitals and art gallery; hon. D.Sc., 1904; honorary freeman, 1906; president of National Liberal Federation, 1883-90; MP for Colne Valley, 1892-1902; baronet, 1886; PC, 1906; baron, 1907.

KITTON, FREDERICK GEORGE (1856-1904), writer on Dickens; began life as wood-engraver and etcher; publications include *Dickensiana* (a bibliography, 1886), *Dickens and his Illustrators* (1899), and *The Dickens Country* (1905); his Dickens library presented to Guildhall Library, 1908.

KLEIN, MELANIE (1882-1960), psychoanalyst; born in Vienna of Jewish parentage; her marriage to A. S. Klein ended in divorce, 1923; began practise of analysis in Berlin, 1921; developed lay-technique for analysis of very young children; invited to London by Ernest Jones [q.v.], 1925, and settled there, 1926; naturalized, 1934; her application of Freudian techniques to child analysis aroused great controversy but influenced social attitudes to child care; publications include *The Psycho-Analysis of Children* (1932) and *Narrative of a Child Analysis* (1961).

KNIGHT, HAROLD (1874-1961), painter; educated at Nottingham high school and Nottingham School of Art; won British Institute travelling scholarship, 1895; studied in Paris;

exhibited at Royal Academy; married Laura Johnson, 1903, also a painter; careers closely connected. [See KNIGHT, DAME LAURA.]

KNIGHT, JOSEPH (1829-1907), dramatic critic; joined father's business of cloth merchant at Leeds at nineteen; devoted to literature through life; embarked on journalistic career in London, 1860; dramatic critic of *Athenaeum* from 1867 till death; chief contributor of lives of actors and actresses to this Dictionary; editor of *Notes and Queries* from 1883 to 1907; his numerous literary friends and associates included John Westland Marston [q.v.] and D. G. Rossetti [q.v.], whose life he wrote in 'Great Writers' series, 1887; FSA, 1893; a popular member of the Garrick Club from 1883.

KNIGHT, JOSEPH (1837-1909), landscape painter and engraver; lost right arm at seven; exhibited mainly Welsh subjects at Royal Academy and elsewhere; fellow of Society of Painter-Etchers, 1883.

KNIGHT, DAME LAURA (1877-1970), painter; educated at Brincliffe School, St. Quentin, and Nottingham School of Art; married Harold Knight [q.v.], 1903; 'A Cup of Tea' by Harold, and 'Dutch Interior' by Laura, accepted by RA, 1906; settled at Newlyn, Cornwall; 'Daughters of the Sun' exhibited at RA, 1909; moved to London, 1918; Harold now established as portrait painter; Laura worked on ballet scenes; ARA, 1927; introduced by (Sir) Alfred James Munnings to Bertram Mills [qq.v.]; produced studies of circus scenes; 'Charivari' exhibited at RA, 1929; Harold elected ARA, 1928; Laura, DBE, 1929; painted gipsies at Epsom and Ascot; Harold continued to paint fine portraits of celebrities; Laura elected RA, 1936 (first woman to become full member); Harold elected RA, 1937; Laura worked for War Artists' Advisory Committee during 1939-45 war; painted scenes at Nuremberg war criminals' trial, 1946; Harold died, 1961; Laura's work shown in retrospective exhibition, Royal Academy, 1965; Harold was ROI (Royal Institute of Oil Painters) and RP (Royal Society of Portrait Painters); his work represented in many public collections in England and abroad; Laura was honoured by many societies of artists, and her works exhibited in numerous galleries, including the Tait and National Portrait Gallery; hon. LLD St. Andrews, 1931, hon. D.Litt. Nottingham, 1951; publications include *Oil Paint and Grease Paint* (1936) and *The Magic of a Line* (1965).

KNOLLYS, EDWARD GEORGE WILLIAM TYRWHITT, second VISCOUNT KNOLLYS (1895-1966), business man and public servant; son of Francis Knollys [q.v.], first Viscount; educated at Harrow and New College, Oxford; served in Army and Royal Flying Corps, 1914-18; DFC, MBE, and Croix de Guerre; succeeded his father, 1924; joined Barclays Bank, and worked in Cape Town, 1929-32; director, Employers' Liability Assurance Corporation, 1932; managing director, 1933; first chairman of joint company when Employers' merged with Northern, 1960; governor and commander-in-chief, Bermuda; KCMG, 1941-3; first full-time chairman, British Overseas Airways Corporation, 1943-7; represented Britain at International Materials Conference, Washington, 1951; GCMG, 1952; chairman, Vickers Ltd., 1956-62; chairman, English Steel Corporation, 1959-65; FRSA, 1962, chairman, RAF Benevolent Fund; trustee, Churchill College, Cambridge.

KNOLLYS, FRANCIS, first VISCOUNT KNOLLYS (1837-1924), private secretary to King Edward VII; son of General Sir W. T. Knollys [q.v.]; educated at Royal Military College, Sandhurst; entered Civil Service; private secretary to Prince of Wales (afterwards King Edward VII), 1870-1910; joint private secretary to King George V, 1910-13; baron, 1902; viscount, 1911; a strong liberal; excelled in art of letter-writing.

KNOTT, RALPH (1878-1929), architect; entered office of Sir A. Webb [q.v.] as draughtsman; won competition for design of new London County Hall on south side of river at Westminster, 1908; two-thirds of building completed and opened, 1922; one of the most successful public buildings of the time; FRIBA, 1921.

KNOWLES, SIR JAMES THOMAS (1831-1908), founder and editor of the *Nineteenth Century*, and architect; joined his father's office as architect; practised his profession for thirty years; designed 'Thatched House Club', St. James's Street, 1865; laid out Leicester Square for Albert Grant [q.v.], 1874; published *The Story of King Arthur and his Knights* (1862) which attracted Tennyson and led to a close intimacy with the poet; founded Metaphysical Society (1869) which lasted till 1881, and included leaders of all schools of thought; with Gladstone, who joined the society, Knowles's relations were as close as those with Tennyson; editor of *Contemporary Review* from 1870 to 1877, when he founded the highly successful *Nineteenth Century* with himself as editor; 'signed writing' by eminent persons was Knowles's main editorial principle; KCVO, 1903.

KNOX, EDMUND ARBUTHNOTT (1847-1937), bishop of Manchester; brother of Sir G. E. Knox [q.v.]; first class, *lit. hum.* (1868), law and modern history (1869), Corpus Christi College, Oxford; fellow (1868), tutor (1875-84) of Merton College; ordained priest, 1872; suffragan bishop of Coventry, 1894-1903; bishop of Manchester, 1903-21; a leading evangelical and vigorous critic of Enabling Act (1919).

KNOX, SIR GEOFFREY GEORGE (1884-1958), diplomatist; born in New South Wales; educated at Malvern College; for the Levant service studied at Trinity College, Cambridge; served in Persia, Egypt, and Salonika; transferred to diplomatic service, 1920; served in Berlin, Constantinople, and Madrid; chairman, international Saar governing commission, 1932-5; minister to Hungary, 1935-9; ambassador to Brazil, 1939-42; CMG, 1929; KCMG, 1935.

KNOX, SIR GEORGE EDWARD (1845-1922), Indian civil servant; born at Madras; entered Indian civil service, 1864; posted to Meerut, 1867; judge of small causes court, Allahabad, 1877; legal remembrancer to local government of North-Western Provinces and Oudh, 1885; judge of high court of judicature, Allahabad, 1890; knighted, 1906; died at Naini Tal; his legal and linguistic equipment (Urdu, Sanskrit, Arabic, Persian) outstanding.

KNOX, ISA (1831-1903), poetical writer; born CRAIG; early contributed to *Scotsman* under name of 'Isa'; married cousin, John Knox, an iron merchant, 1866; won prize for Burns centenary poem at Crystal Palace, 1858; published verse and fiction.

KNOX, RONALD ARBUTHNOTT (1888-1957), Roman Catholic priest and translator of the Bible; son of Revd E. A. Knox [q.v.]; scholar of Eton and Balliol College, Oxford; Hertford, Ireland, and Craven scholar; president of Union, 1909; first class, *lit. hum.*, 1910; deacon, 1911; priest, 1912; fellow (1910), chaplain (1912), Trinity College, Oxford; became Roman Catholic, 1917; ordained, 1919; taught at Ware, 1918-26; Catholic chaplain, Oxford, 1926-39; thereafter undertook translation of Bible completed in 1955; maintained literary output ranging from detective stories to apologetics; publications include *Absolute and Abithofhell* (1913), *A Spiritual Aeneid* (1918), *Let Dons Delight* (1939), *Enthusiasm* (1950); appointed monsignor, 1936, protonotary apostolic, 1951, member of Pontifical Academy, 1956; honorary fellow of Trinity (1941), Balliol (1953); Romanes lecturer, 1957.

KNOX, WILFRED LAWRENCE (1886-1950), biblical scholar and divine; son of E. A. Knox [q.v.]; first class, *lit. hum.*, Trinity College, Oxford, 1909; ordained, 1913/14; warden, Oratory House, Cambridge, 1924-40; chaplain (1940), fellow (1946), Pembroke College; studied Hellenistic influence on New Testament writers; FBA, 1948.

KNOX-LITTLE, WILLIAM JOHN (1839-1918), divine and preacher; born in county Tyrone; BA, Trinity College, Cambridge; rector of St. Alban's, Cheetwood, Manchester, 1875-85; canon of Worcester, 1881; vicar of Hoar Cross, Staffordshire, 1885-1907; enjoyed great popularity as extempore preacher, especially at missions; high churchman; published sermons and other works.

KNUTSFORD, first VISCOUNT (1825-1914), politician. [See HOLLAND, SIR HENRY THURSTAN.]

KNUTSFORD, second VISCOUNT (1855-1931), hospital administrator and reformer. [See HOLLAND, SIR SYDNEY GEORGE.]

KOMISARJEVSKY, THEODORE (1882-1954), theatrical producer and designer; born in Venice; educated at military academy and Imperial Institute of Architecture, St. Petersburg; founded school of acting in Moscow, 1910; became theatrical producer; came to England, 1919; naturalized, 1932; produced widely for societies in the lead of theatrical taste in a manner unorthodox, provocative, sometimes brilliant, sometimes wayward; brilliant designer of own sets and costumes; required new depth of feeling and understanding from his actors; greatly influenced methods of direction, acting, setting, and lighting.

KORDA, SIR ALEXANDER (1893-1956), film producer; born in Hungary; became film director; moved to Vienna, 1919, Berlin, 1923, Hollywood, 1926, Paris, 1930; settled in London, 1931; formed London Film Productions, 1932; completed Denham studios, 1937; and later studios at Shepperton; films include *The Ghost Goes West* (1935), *The Scarlet Pimpernel* (1935), *Elephant Boy* (1936-7), *The Third Man* (1949), *The Wooden Horse* (1950); naturalized, 1936; knighted, 1942.

KOTZÉ, SIR JOHN GILBERT (1849-1940), South African judge; born at Cape Town; called to bar (Inner Temple), 1874; judge, High Court, Transvaal, 1877-81; chief justice, South African Republic, 1881-98; contested presidency, 1893; in 1897 caused constitutional crisis by reversing his decision (1884) that supreme power was vested in the Volksraad which was not subject to the jurisdiction of the Supreme Court; dismissed, 1898; attorney-general of Southern Rhodesia, 1900; KC, 1902; judge president, Cape eastern districts court, Grahamstown, 1904-13; judge, Cape provincial division, Supreme Court, Cape Town, 1913-20; judge president thereof, 1920-2; judge of appeal, 1922-7; knighted, 1917.

KRONBERGER, HANS (1920-1970), leader in the physics and engineering of nuclear reactors; born in Linz, Austria, of Jewish parents; escaped from Austria, and studied mechanical engineering at King's College, Newcastle; interned, 1940-2; returned to King's College; honours degree in physics, 1944; Ph.D. Birmingham, 1948; joined (Sir) F. E. Simon's [q.v.] team in 'Tube Alloys' project, 1946; worked with Heinz London [q.v.] at new atomic energy establishment, Harwell; research manager, components development laboratory at uranium diffusion plant, Capenhurst, 1951; head of development laboratories, 1953; chief physicist, research and development branch, Industrial Group of Atomic Energy Authority, 1956; director, 1958; scientist-in-chief, Reactor Group; member, Atomic Energy Authority, 1969; OBE, 1957; CBE, 1966; FRS, 1965.

KRUGER GRAY, GEORGE EDWARD (1880-1943), designer. [See GRAY.]

KUCZYNSKI, ROBERT RENE (1876-1947), demographer; born and educated in Germany; director, Statistical Office, Berlin-Schoenberg, 1906-21; research fellow (1933), reader (1938-41), in demography, London School of Economics; adviser to Colonial Office, 1944-7; naturalized, 1946; works include *Fertility and Reproduction* (1932), *Colonial Population* (1937), and *Demographic Survey of the British Colonial Empire* (3 vols., 1948-53).

KYLSANT, BARON (1863-1937), shipowner and financier. [See PHILIPPS, OWEN COSBY.]

KYNASTON (formerly SNOW), HERBERT (1835-1910), classical scholar; educated at Eton and St. John's College, Cambridge; first Porson scholar and senior classic, 1857; fellow, 1858; MA, 1860; DD, 1882; rowed in university boat, 1856 and 1857; member of Alpine Club; principal of Cheltenham College, 1874-88; canon of Durham and professor of Greek in the university, 1889; edited Theocritus (1869) and *Poetae Graeci* (1879).

L

LABOUCHERE, HENRIETTA (1841-1910), actress. [See HODSON.]

LABOUCHERE, HENRY DU PRÉ (1831-1912), journalist and politician; nephew of Henry Labouchere, first Baron Taunton [q.v.]; educated at Eton and Trinity College, Cambridge; in diplomatic service, 1854-64; wrote for *Daily News* and *World*, and established reputation as journalist; founded weekly journal *Truth*, notable for its exposure of fraudulent enterprises, 1876; liberal MP, Northampton, with Charles Bradlaugh [q.v.] as his colleague, 1880; held seat till 1906; became one of the most powerful radicals in Commons; attacked home and foreign policy of whigs and worked for reorganization of liberal party; designs frustrated by decision of Joseph Chamberlain [q.v.] to vote against first home rule bill, 1886; died near Florence.

LACEY, THOMAS ALEXANDER (1853-1931), ecclesiologist and controversialist; second class, *lit. hum.*, Balliol College, Oxford, 1875; ordained, 1876; an accomplished Latinist; with E. Denny composed *Dissertatio Apologetica de Hierarchia Anglicana* (1895); attended commission of inquiry into validity of Anglican orders (1896); on staff of *Church Times*; chaplain (1903), warden (1910-19), London diocesan penitentiary, Highgate; canon of Worcester, 1918-31.

LACHMANN, GUSTAV VICTOR (1896-1966), aeronautical engineer; born of Austrian parents at Dresden; educated at the Realgymnasium, Darmstadt; served in German army and air force during 1914-18 war; studied mechanical engineering and aerodynamics, Darmstadt Technical University, 1918-21; became doctor of engineering, Aachen Technical University, for thesis on the slotted wing, 1923; designer, Schneider aircraft works, Berlin, 1924; chief designer, Albatross aircraft works, Johannisthal, 1925-6; technical adviser, Ishikawajima aircraft works, Tokyo, 1926-9; engineer in charge of aerodynamics, Handley Page Ltd., England, 1929-32; chief designer, 1932-6; during 1939-45 war undertook non-military aircraft design studies; naturalized, 1949; head of Handley Page Research Department, 1953-65; fellow, Royal Aeronautical Society, 1938; publications include *Leichtflugzenbau* (Munich, 1925), and *Boundary Layer and Flow Control* (1961).

LAFONT, EUGÈNE (1837-1908), science teacher in India; born at Mons, Belgium; was admitted a Jesuit, 1854; inaugurated science teaching in Bengal at St. Xavier's College, Calcutta, 1865; rector of the college, 1873-1904; fellow of Calcutta University, 1877; hon. D.Sc., 1908; CIE, 1880; died at Darjeeling.

LAIDLAW, ANNA ROBENA (1819-1901), pianist; studied music in Edinburgh, Königsberg, and London; made successful appearances in Germany and Austria; praised by Schumann; pianist to Queen of Hanover until 1840; settled in London; retired on marriage to George Thomson, 1852.

LAIDLAW, JOHN (1832-1906), Presbyterian divine and theologian; student at Edinburgh University; hon. MA, 1854; hon. DD, 1880; studied theology in Edinburgh and Germany; minister at Perth, 1863-72, and Aberdeen, 1872-81; professor of systematic theology, New College, Edinburgh, 1881-1904; a conservative theologian; author of *The Biblical Doctrine of Man* (1879) and *The Miracles of Our Lord* (1890).

LAIDLAW, SIR PATRICK PLAYFAIR (1881-1940), physician; educated at the Leys School, St. John's College, Cambridge, and Guy's Hospital; worked at Wellcome Physiological Research Laboratories, 1909-13; lecturer in pathology, Guy's Hospital, 1913-22; investigated histamine shock; joined National Institute for Medical Research, 1922; deputy director, 1936-40; concentrated on bacteriological and virus research; developed two methods of inducing immunity to dog distemper; proved human epidemic influenza to be a virus infection; investigated parasitic amoebae and the treatment of amoebic dysentery by alkaloids of ipecacuanha; FRS, 1927; knighted, 1935.

LAIRD, JOHN (1887-1946), regius professor of moral philosophy, Aberdeen; first class, philosophy, Edinburgh, 1908; moral sciences tripos, Trinity College, Cambridge, 1910-11; professor of logic and metaphysics, Queen's University, Belfast, 1913-24; regius professor of moral philosophy, Aberdeen, 1924-46; publications include *Problems of the Self* (1917), *Study in Realism* (1920), *An Enquiry into Moral Notions* (1935), *Theism and Cosmology* (1940), and *Mind and Deity* (1941); FBA, 1933.

LAKE, KIRSOPP (1872-1946), biblical scholar; educated at St. Paul's School and Lincoln College, Oxford; second class, theology, 1895; priest, 1896; professor of early Christian literature, Leiden, 1904-14; Harvard, 1914-19;

of ecclesiastical history, Harvard, 1919; of history, 1932–8; edited 'Lake Group' of manuscripts, 1902; publications include *The Beginnings of Christianity* (with F. J. Foakes Jackson, q.v., 5 vols., 1920–33).

LAKE, Sir PERCY HENRY NOEL (1855–1940), lieutenant-general; educated at Uppingham; gazetted to 59th Foot, 1873; quartermaster-general (1893–8), chief of the general staff (1904–8), inspector-general (1908–10), Canadian Militia; lieutenant-general, 1911; chief of general staff, India, 1912–15; commander-in-chief, Mesopotamia, and unsuccessful in relieving Kut el Amara, 1916; KCMG, 1908; KCB, 1916.

LAMB, HENRY TAYLOR (1883–1960), painter; son of Sir Horace Lamb [q.v.]; educated at Manchester grammar school and medical school; studied painting in London and Paris; founder-member Camden Town and London groups; qualified at Guy's Hospital, 1916, and served as medical officer; MC, 1918; at first one-man exhibition, 1922, his portrait of Lytton Strachey [q.v.] (Tate Gallery) brought him public attention; painted other writers, including Evelyn Waugh [q.v.] and Lord David Cecil; ARA, 1940; RA, 1949; trustee, Tate Gallery (1944–51), and National Portrait Gallery (1942–60).

LAMB, Sir HORACE (1849–1934), mathematician; educated at Stockport grammar school, Owens College, Manchester, and Trinity College, Cambridge; second wrangler and second Smith's prizeman, 1872; fellow and lecturer, 1872–5; professor of mathematics, Adelaide, Australia, 1875–85; professor of pure (later also of applied) mathematics, Manchester, 1885–1920; honorary (Rayleigh) lecturer, Cambridge, 1920–34; lucid teacher and writer; his special subjects hydrodynamics, sound, elasticity, and mechanics; works include *Hydrodynamics* (1895) and *Infinitesimal Calculus* (1897); FRS, 1884; president of British Association, 1925; knighted, 1931.

LAMBART, FREDERICK RUDOLPH, tenth EARL OF CAVAN (1865–1946), field-marshal; educated at Eton and Sandhurst; gazetted to Grenadier Guards, 1885; succeeded father while serving in South Africa, 1900; retired, 1913; commanded 4th (Guards) brigade (Ypres and Festubert), 1914–15; Guards division (Loos), 1915; XIV Corps (Somme and third Ypres), 1916–17; took Corps to Italy, Nov. 1917; took over command of British troops in Italy, Mar. 1918; repulsed Austrian offensive; commanded small army for final offensive across Piave; elected representative Irish peer, 1915;

KP, 1916; KCB, 1918; GCMG, 1919; GOC-in-C, Aldershot command, 1920–2; headed War Office section, British delegation, Washington conference, 1921; chief of imperial general staff, 1922–6; chief of staff to Duke of York touring Australia and New Zealand, 1927; field-marshal, 1932; commanded troops at coronation, 1937; GCVO, 1922; GCB, 1926; GBE, 1927.

LAMBE, Sir CHARLES EDWARD (1900–1960), admiral of the fleet; entered navy, 1914; qualified in torpedo and at Naval Staff College; captain, 1937; successively assistant, deputy, and director of plans, joint planning staff, 1940–4; commanded aircraft carrier, *Illustrious*, 1944–5; acting rear-admiral, 1945; commander-in-chief, Far East, 1953–4, Mediterranean, 1957–9; second sea lord, 1955–7; first sea lord, 1959–60; admiral of the fleet, 1960; CVO, 1938; CB, 1944; KCB, 1953; GCB, 1957.

LAMBERT, BROOKE (1834–1901), social reformer; student at King's College, London, under F. D. Maurice [q.v.]; BA, Brasenose College, Oxford, 1858; MA, 1861; BCL, 1863; vicar of St. Mark's, Whitechapel, 1866–70; as vestryman and guardian he made thorough study of poor law and local government; in work on pauperism, 1871, anticipated scientific statistical researches of Charles Booth [q.v.]; resigned through ill health, 1870; went to West Indies to restore health; held living of Tamworth, 1872–8; helped to found London University Extension Society, 1879; vicar of Greenwich, 1880–1901, where he continued his activity in social and educational reform; a prominent freemason; published volumes of sermons.

LAMBERT, CONSTANT (1905–1951), musician; educated at Christ's Hospital and Royal College of Music; his music for ballet includes *Romeo and Juliet*, *Pomona*, *Horoscope*, *Tiresias*, and many arrangements for Sadler's Wells Ballet of which he was musical director until 1947; other works include *Music for Orchestra*, 'Elegiac Blues', *The Rio Grande*, Piano Sonata and Piano Concerto, *Aubade Héroïque*, and choral masque *Summer's Last Will and Testament*; frequently conducted at promenade concerts and Covent Garden.

LAMBERT, GEORGE (1842–1915), tennis player; went to Hampton Court Palace tennis court, 1866; head professional at Marylebone Cricket Club court at Lord's, 1869–89; champion, 1870–85.

LAMBERT, GEORGE, first VISCOUNT LAMBERT (1866–1958), yeoman farmer and Member of Parliament; son of Devon landowner; educated locally, and began farming 800 acres at nineteen; liberal MP, South Molton, 1891-1924, 1929-45; on royal commission on agriculture, 1893; civil lord of Admiralty, 1905-15; supported national government of 1931 and became national liberal; member, Devon county council, 1889-1952; foundation chairman of Seale-Hayne Agricultural College, Newton Abbot; PC, 1912; viscount, 1945.

LAMBERT, MAURICE (1901-1964), sculptor; brother of Constant Lambert [q.v.]; educated at Manor House School, Clapham; apprenticed to F. Derwent Wood [q.v.], 1918-23; first public exhibition, 1925; first one-man exhibition, 1929; alabaster carving accepted by Tate Gallery, 1932; first exhibited at Royal Academy, 1938; ARA, 1941; fellow, Royal Society of British Sculptors, 1938; master, Royal Academy Sculpture School, 1950-8; RA, 1952; works include bronze statue of Dame Margot Fonteyn, equestrian statue of George V, statue of Viscount Nuffield, and busts of Dame Edith Sitwell, J. B. Priestley, and Lord Devlin.

LAMBOURNE, first BARON (1847-1928), politician. [See LOCKWOOD, AMELIUS MARK RICHARD.]

LAMBURN, RICHMAL CROMPTON (1890-1969), author; known as Richmal Crompton; educated at St. Elphin's Clergy Daughters' School, Warrington, at Darley Dale, Derbyshire, and Royal Holloway College, London; BA London, 1914; taught at her old school, 1915-17; classics mistress, Bromley High School for Girls, 1917-24; gave up teaching after attack of poliomyelitis; concentrated on writing short stories; selection of stories about William Brown, schoolboy, published as *Just William* and *More William* (1922); between 1922 and 1969 'William' series ran to thirty-eight titles; translated into many foreign languages; thirty-nine other novels not so successful.

LAMINGTON, second BARON (1860-1940). [See BAILLIE, CHARLES WALLACE ALEXANDER NAPIER ROSS COCHRANE-.]

LAMPSON, MILES WEDDERBURN, first BARON KILLEARN (1880-1964), diplomatist; grandson of Sir Curtis Miranda Lampson, first baronet [q.v.]; educated at Eton; entered Foreign Office, 1903; served, as secretary to Prince Arthur of Connaught in Japan, and in Peking, 1906-18; acting, high commissioner, Siberia, 1919-20; at Washington conference, 1921-2; head, Central European Department, Foreign Office, 1922-6; minister, Peking, 1927-33; high commissioner, Egypt, 1933-6; first British ambassador, Egypt, 1936-46; special com-

missioner, South-East Asia, 1946-8; CB, 1926; KCMG, 1927; GCMG, 1937; PC, 1941; baron, 1943.

LANCHESTER, FREDERICK WILLIAM (1868-1946), engineer; educated at Hartley College, Southampton, South Kensington, and Finsbury Technical College; devised pendulum governor controlling speed of gas-engines and Lanchester gas-starter; produced his first experimental motor-car, 1895; second, 1897; formed Lanchester Engine Company, 1899; introduced first real motor-car, 1901; consulting engineer, Daimler Motor Company, 1910-30; laid foundations of aircraft design in *Aerial Flight* (2 vols., 1907-8); member, Advisory Committee on Aeronautics, 1909-20; FRS, 1922.

LANCHESTER, GEORGE HERBERT (1874-1970), automobile engineer and inventor; educated at Clapham high school, London; apprenticed at fourteen to his brother, Frederick William Lanchester [q.v.], works manager and designer, Forward Gas Engine Co., Birmingham; works manager, 1893-7; assisted in design of experimental motor cars; works manager, Lanchester Engine Co., 1899-1905; designer and chief engineer and technical director, Lanchester Motor Co., 1909-1936; designed Lanchester cars and armoured cars; joined Alvis Co., 1936; designed Silver Crest car; consultant, Stirling Armament Co., 1939; consultant, Russell Newbery Diesel Engine Co., 1945-52, and part time, 1952-61; fellow, Institution of Mechanical Engineers; president, Institution of Automobile Engineers, 1943-4; consultant editor, *Automobile Engineers' Reference Book.*

LANE, SIR ALLEN (LANE WILLIAMS) (1902-1970), publisher; born Allen Lane Williams; educated at Bristol grammar school; left school to work at Bodley Head, publishing house of John Lane [q.v.], a relative who insisted that he change his name by deed-poll, 1919; director, Bodley Head, 1925; chairman, 1930; published first experimental paperback reprints independently of Bodley Head, 1935; Penguin Books established, resigned from Bodley Head, 1936; first Pelicans issued, 1937; first Penguin Classic, *The Odyssey* (published, 1946); published unexpurgated text of D. H. Lawrence's *Lady Chatterley's Lover* and won *cause célèbre,* 1960; Penguin Books became public company, 1960; retired as managing director, 1967; knighted, 1952; CH, 1969; hon. degrees at Birmingham, Bristol, Manchester, Oxford, and Reading; hon. fellow, Royal College of Art.

LANE, SIR HUGH PERCY (1875-1915), art collector and critic; born in county Cork;

picture dealer in London, 1898; formed gallery of modern art in Dublin; knighted, 1909; director of Irish National Gallery, 1914; torpedoed on *Lusitania*; his will caused controversy between Dublin and London National galleries.

LANE, JOHN (1854–1925), publisher; clerk in Railway Clearing House, 1869–87; set up as bookseller and then as publisher in Vigo Street with Elkin Matthews, 1887–94; first book under imprint of 'Bodley Head' appeared, 1889; moved to the Albany, 1894; poetry published by firm includes works of Sir William Watson, Francis Thompson, and Richard Le Gallienne [qq.v.]; novels include those of W. J. Locke [q.v.]; produced quarterly *Yellow Book*, to which famous poets, essayists, dramatists, story-tellers, and artists contributed, 1894–7.

LANE, LUPINO (1892–1959), actor and theatre-manager; born into theatrical family of Lupino; established himself as a leading comedian; played cockney Bill Snibson in and also presented *Twenty to One* (London Coliseum, 1935) and *Me and My Girl* (Victoria Palace, 1937); in latter created 'The Lambeth Walk', the title used when play was filmed.

LANE, SIR (WILLIAM) ARBUTHNOT, first baronet (1856–1943), surgeon; educated in Scotland; entered Guy's Hospital, 1872; MRCS, 1877; FRCS and demonstrator of anatomy, 1882; appointed to staff, 1888; his wonderful manual dexterity brought him to front rank in abdominal surgery; his three main surgical procedures were operation for cleft palate at one day old, for treatment of simple fractures, and for removal of large gut which he believed to be a focus of sepsis; devised aseptic surgical excellence known as 'Lane technique'; at Hospital for Sick Children, Great Ormond Street, 1883–1916; consulting surgeon, Aldershot command, 1914–18; during war organized and opened Queen Mary's Hospital, Sidcup, for plastic surgery; retired, 1918; founded New Health Society (1925) as first organized body dealing with social medicine; resigned from medical register (1933) for greater freedom in this work; baronet, 1913; CB, 1917.

LANE POOLE, REGINALD (1857–1939), historian. [See POOLE.]

LANE-POOLE, STANLEY EDWARD (1854–1931), orientalist and historian. [See POOLE.]

LANG, (ALEXANDER) MATHESON (1877–1948), actor-manager and dramatist; cousin of Cosmo Gordon Lang [q.v.]; educated at Inverness College and St. Andrews; made

first stage appearance, 1897; notably successful as Mr Wu (1913), Matathias in *The Wandering Jew* (1920), and Count Pahlen in *Such Men are Dangerous* (1928); produced *Jew Süss* (1929); repeated many stage successes on the cinema screen.

LANG, ANDREW (1844–1912), scholar, folklorist, poet, and man of letters; educated at St. Andrews and Glasgow universities and Balliol College, Oxford; fellow of Merton, 1868; settled in London and devoted himself to journalism and letters, 1875; poetical works include *Ballads and Lyrics of Old France* (1872), *xxii Ballades in Blue China* (1880), and *Helen of Troy* (1882); as anthropologist showed that folklore is foundation of higher or literary mythology; works on this subject include *Custom and Myth* (1884), *Myth, Ritual, and Religion* (1887), and *The Making of Religion* (1898); classical works include prose translation of *Odyssey* (with S. H. Butcher, q.v., 1879), of *Iliad* (with Walter Leaf and Ernest Myers, qq.v., 1883), and three books on Homeric question, *Homer and the Epic* (1893), *Homer and his Age* (1896), and *The World of Homer* (1910); historical works include *Pickle the Spy* (1897), *The Companions of Pickle* (1898), *Prince Charles Edward* (1900), and *History of Scotland* (1900–7); also author of *Life and Letters of J. G. Lockhart* (1896), and of essays, novels, and children's books; a founder of Psychical Research Society.

LANG, JOHN MARSHALL (1834–1909), principal of Aberdeen University; educated at Glasgow University; hon. DD, 1873; hon. LLD, 1901; minister of East Parish of St. Nicholas, Aberdeen, 1856–65, and of Anderston church, Glasgow, 1865–8; there introduced improvements in ritual, including first organ used in Church of Scotland and psalms chanted in prose version; minister of Morningside, Edinburgh, 1868–73, and Barony of Glasgow, 1873–1901; served on school board, on commission for housing of poor, and kindred bodies in Glasgow; instituted Sunday evening services in Glasgow; visited Australia, 1897; convener of Assembly's commission of inquiry into religious condition of the people of Scotland, 1890–6; moderator of General Assembly, 1893; promoted Pan-Presbyterian Alliance for union of the Churches; principal of Aberdeen University, 1900–9; CVO, 1906; Baird lecturer at Glasgow, 1901; author of many devotional works.

LANG, (WILLIAM) COSMO GORDON, BARON LANG OF LAMBETH (1864–1945), archbishop of Canterbury; son of J. M. Lang [q.v.]; educated at Glasgow University and Balliol College, Oxford; president of Union, 1884; second class, *lit. hum.*, 1885; first, modern history, 1886; fellow of All Souls, 1888; deacon, 1890; priest, 1891; vicar of St. Mary the Virgin, Oxford, 1894–6; of Portsea, 1896–1901; began lifelong association with royal family as a chaplain to Queen Victoria; canon of St. Paul's and suffragan bishop of Stepney, 1901–9; archbishop of York, 1909–28; formed diocese of Sheffield, 1914; member of royal commission on divorce, 1909–12; signed minority report; criticized for ill-timed reference to 'sacred memory' of German Emperor, 1914; visited Grand Fleet, 1915; western front, 1917; United States and Canada, 1918; took leading part in National Mission of Repentance and Hope, 1916; chairman, commission on ecclesiastical courts and cathedrals commission; favoured 1549 communion service as alternative rite but accepted majority decision in Prayer Book discussions; worked closely with Randall Davidson [q.v.]; succeeded him as archbishop of Canterbury, 1928; obtained 'Appeal to All Christian People' of 1920 Lambeth conference; visited leading Orthodox ecclesiastics while cruising in Mediterranean; held joint theological commission at Lambeth, 1931; founded Church of England council on foreign relations, 1933; sent delegation to Romanian Church (1935) resulting in Romania joining Jerusalem, Constantinople, Cyprus, and Alexandria acknowledging Anglican orders by 'Economy'; served on Indian joint committee, 1933–4; agreed to total 'extinguishment' of tithe, 1936; broadcast on abdication widely criticized, 1936; GCVO, 1937; resigned and created baron, 1942.

LANG, WILLIAM HENRY (1874–1960), botanist; educated at Dennistoun school and the university, Glasgow; B.Sc., botany and zoology, 1894; qualified in medicine, 1895; D.Sc., 1900; taught by F. O. Bower [q.v.] and became authority on ferns; under D. H. Scott [q.v.] acquired interest in fossil botany; lecturer in botany, Glasgow, 1902; Barker professor of cryptogamic botany, Manchester, 1909–40; his investigation with Robert Kidston of the 'Rhynie fossils' made unique contribution to evolutionary theory; FRS, 1911; hon. LLD, Manchester and Glasgow.

LANGDON, STEPHEN HERBERT (1876–1937), Assyriologist; born in Michigan; educated in America and Europe; appointed reader in Assyriology, Oxford, as condition of Shillito foundation, 1908; professor, 1919–37; built up English school of Assyriologists; naturalized, 1913; FBA, 1931.

LANGDON-BROWN, SIR WALTER LANGDON (1870–1946), physician and regius professor of physic, Cambridge; educated at Bedford School and St. John's College, Cam-

bridge; first class, parts i and ii, natural sciences tripos, 1892–3; qualified at St. Bartholomew's Hospital and house-physician to S. J. Gee [q.v.], 1897; medical registrar, 1906; assistant physician, 1913; full physician, 1924–30; assistant physician (1900), full physician (1906–22), Metropolitan Hospital; FRCP, 1908; senior censor, 1934; regius professor of physic, Cambridge, 1932–5; first English physician to relate work of psychologists like Freud, Jung, and Adler to practice of clinical medicine; published *Physiological Principles in Treatment* (1908); knighted, 1935.

LANGEVIN, SIR HECTOR LOUIS (1826–1906), Canadian statesman; born at Quebec; called to bar of Lower Canada, 1850; mayor of Quebec, 1858–60; member of Canadian legislative assembly, 1857–67; QC, 1864; solicitor-general for Lower Canada, 1864–6; postmaster-general, 1866–7; helped to form dominion of Canada; member of dominion house of commons, 1867–96; secretary of state, 1867–9; minister of public works, 1869–73, 1879–91; postmaster-general, 1878–9; led French-Canadian conservative party from 1873; KCMG, 1881; died at Quebec.

LANGFORD, JOHN ALFRED (1823–1903), Birmingham antiquary and journalist; contributed to *Howitt's Journal*; joined Unitarians under George Dawson [q.v.]; carried on printing business at Birmingham 1852–5; closely associated with *Birmingham Daily Press*, 1855, and *Birmingham Daily Gazette*, 1862–8; an ardent liberal, he helped in party organization; joined Gladstonian section of party, 1886; author of *Century of Birmingham Life* (2 vols., 1868) and *Modern Birmingham* (2 vols., 1873–7), and poems and dramas.

LANGLEY, JOHN NEWPORT (1852–1925), physiologist; BA, St. John's College, Cambridge; fellow of Trinity College, Cambridge, 1877; carried out his first researches on new drug, pilocarpine, and then proceeded to study of secretory process; lecturer at Trinity College and university lecturer in physiology, 1883; professor of physiology, Cambridge University, 1903–25; owned and edited *Journal of Physiology*, 1894–1925; his acquisition of *Journal*, which he thoroughly reformed and made pattern in presentation of scientific work, decisive event for British physiology; climax of his achievement as investigator reached in his research into sympathetic nervous system, 1890–1906; subjected whole of sympathetic ganglionic system to exhaustive analysis; directed energies of workers in newly completed Cambridge physiological laboratory into channels of direct value in time of war, 1914–18; after 1914–18 war attracted even larger numbers to his school of physiology at Cambridge, which was remarkably productive of distinguished physiologists; FRS, 1883; Croonian lecturer, 1906.

LANGTON, SIR GEORGE PHILIP (1881-1942), judge; educated at Beaumont and New College, Oxford; second class, modern history, 1902; president, OUDS; called to bar (Inner Temple), 1905; specialized in maritime law; director, labour department, commissioner, labour disputes, Ministry of Munitions, 1916-18; controller, demobilization department, Ministry of Labour, 1918-19; secretary and adviser, British maritime law committee, 1922-30; KC, 1925; judge, Probate Divorce and Admiralty division, 1930-42; knighted and bencher, 1930.

LANKESTER, SIR EDWIN RAY (1847-1929), zoologist; son of Edwin Lankester [q.v.]; BA, Christ Church, Oxford; Radcliffe travelling fellow, 1870; studied marine zoology at Naples, 1871-2; fellow and tutor, Exeter College, Oxford, 1872; Jodrell professor of zoology, University College, London, 1874-91; Linacre professor of comparative anatomy, Oxford, 1891-8; director of natural history departments and keeper of zoology, British Museum, South Kensington, 1898-1907; Fullerian professor of physiology and comparative anatomy, Royal Institution, 1898-1900; FRS, 1875; KCB, 1907; soon recognized as leading British authority on zoology; most distinguished as morphologist; his pioneer researches on embryology of Mollusca have had lasting influence on science of embryology; his researches on protozoan parasites important for study of disease; edited *Quarterly Journal of Microscopical Science*, 1878-1920; founded Marine Biological Association, 1884.

LANSBURY, GEORGE (1859-1940), labour leader and politician; elected to Poplar board of guardians, 1892; known as the John Bull of Poplar; a non-smoker, teetotaller, and Anglican, whose socialism and uncompromising pacificism sprang from spiritual conviction; signed minority report as member (1905-9) of royal commission on poor laws; labour MP, Bow and Bromley division, 1910-12; 1922-40; supporter of women's suffrage and defender of conscientious objectors; a founder (1912) and editor (1919-23) of *Daily Herald*; first commissioner of works, 1929-31; established Hyde Park Lido; leader of labour party opposition, 1931-5; resigned over League of Nations sanctions; PC, 1929.

LANSDOWNE, fifth MARQUESS OF (1845-1927). [See PETTY-FITZMAURICE, HENRY CHARLES KEITH.]

LARKE, SIR WILLIAM JAMES (1875-1959), first director of the British Iron and Steel Federation; educated at Colfe's School, Lewisham; joined British Thomson-Houston Company, 1898; executive engineer, 1912-15; joined Ministry of Munitions, 1915; director-general of raw materials, 1919; director, National Federation of Iron and Steel Manufacturers, 1922-34, of British Iron and Steel Federation, 1934-46; OBE, 1917; CBE, 1920; KBE, 1921; hon. D.Sc., Durham.

LARMOR, SIR JOSEPH (1857-1942), physicist; graduated at Queen's College, Belfast; senior wrangler and fellow, St. John's College, Cambridge, 1880; professor of natural philosophy, Queen's College, Galway, 1880-5; university lecturer in mathematics, Cambridge, 1885-1903; Lucasian professor of mathematics, 1903-32; FRS, 1892; a secretary, 1901-12; knighted, 1909; unionist MP, Cambridge University, 1911-22; chiefly memorable for conception of matter as consisting entirely of electric particles, 'electrons', moving about in the aether according to electromagnetic laws (*Aether and Matter*, 1900); first to give the formula for rate of radiation of energy from an accelerated electron and to explain effect of a magnetic field in splitting lines of spectrum into multiple lines; edited works of Cavendish, Thomson, Stokes, and Kelvin [qq.v.], and own collected papers (2 vols., 1929).

LASCELLES, SIR FRANK CAVENDISH (1841-1920), diplomatist; entered diplomatic service, 1861; agent and consul-general, Bulgaria, 1879-87; British minister to Romania, 1887, to Persia, 1891; ambassador to Russia, 1894, to Berlin, 1896-1908; worked for Anglo-German amity till 1914; KCMG, 1886; PC, 1892.

LASCELLES, HENRY GEORGE CHARLES, sixth EARL OF HAREWOOD (1882-1947); educated at Eton and Sandhurst; served with Grenadier Guards, 1915-18; DSO; inherited fortune of Marquess of Clanricarde [q.v.], 1916; married Princess (Victoria Alexandra Alice) Mary, only daughter of King George V, 1922; KG, 1922; succeeded father, 1929; GCVO, 1934; chancellor, Sheffield University, 1944-7; interested in the arts, freemasonry, and horse-racing.

LASKI, HAROLD JOSEPH (1893-1950), political theorist and university teacher; educated at Manchester grammar school and New College, Oxford; first class, modern history, 1914; lecturer, McGill University, 1914-16; at Harvard (forming close friendship with Oliver Wendell Holmes), 1916-20; London School of Economics, 1920-6; professor of political science, 1926-50; after 1931 rejected pluralist theory of State, accepting Marxism; chairman of labour party and chief target of conservative electioneers, 1945; publications include *A Grammar of Politics* (1925), *Parliamentary Government in England* (1938), and *Reflections on the Constitution* (1951).

LAST, HUGH MACILWAIN (1894-1957), Roman historian and principal of Brasenose College, Oxford; grandson of George Macilwain [q.v.]; scholar of St. Paul's School and Lincoln College, Oxford; first class, *lit. hum.*, 1918; fellow, St. John's, 1919-36; university lecturer in Roman history, 1927; Camden professor and fellow of Brasenose, 1936-48; principal of Brasenose, 1948-56; president, Roman Society, 1934-7; contributed to *Cambridge Ancient History* but his publications otherwise mainly reviews of works of others; especially successful in supervising young graduates; his main interests Roman Republican constitution, Roman legal system, and early history of Christianity; emeritus fellow, Brasenose.

LÁSZLÓ DE LOMBOS, PHILIP ALEXIUS (1869-1937), painter; born in poor circumstances in Budapest; studied there and in Munich and Paris; painted Emperor Francis Joseph (1899), Pope Leo XIII (1900), and King Edward VII in 1907 when he moved to London and developed enormous practice among the famous; rapid painter of impeccable likenesses without deep psychological penetration; naturalized, 1914.

LASZOWSKA, (JANE) EMILY DE (1849-1905), novelist. [See GERARD.]

LATEY, JOHN (1842-1902), journalist; art and literary editor of *Penny Illustrated Paper*, 1861-1901; parliamentary reporter to *Illustrated London News* for fifteen years; co-editor (1881-2) with Mayne Reid [q.v.] of *The Boys' Illustrated News*; edited *Sketch*, 1899-1902; author of *The Showman's Panorama* (1880) and novels.

LATHAM, CHARLES, first BARON LATHAM (1888-1970), public servant; changed name from Lathan; left elementary school in Norwich at fourteen to work as clerk; after war service during 1914-18, qualified as member of London Association of Accountants; sole labour member, Hendon Urban District Council, 1926-31; alderman, London County Council, 1928-34 and 1946-7; chairman, finance committee, LCC, 1934; leader, LCC, 1940-7; baron, 1942; chairman, London Transport Executive, 1947-53; lord lieutenant, Middlesex, 1945-56; founder member, Administrative Staff College,

Henley, 1946-59, and Council of Europe, 1960-2.

LATHAM, HENRY (1821-1902), master of Trinity Hall, Cambridge; BA, Trinity College, Cambridge (eighteenth wrangler), 1845; appointed clerical fellow of Trinity Hall, 1847; senior tutor, 1855; broadened aims of the college by destroying its exclusively legal associations; attracted promising men from other colleges; resigned tutorship, 1885; succeeded Sir Henry Sumner Maine [q.v.] as master, 1888; rebuilt college and reconstructed lodge and hall; published *Pastor Pastorum* (1890) and other devotional works.

LATHAM, PETER WALKER (1865-1953), rackets and tennis champion; won world rackets championship, 1887; defended title against George Standing (1897) and Gilbert Browne (1902); resigned it 1902; gained British real tennis title, 1895; lost to C. Fairs, 1905; regained it and retired, 1907; for many years from 1888 head professional at Queen's Club.

LAUDER, SIR HARRY (1870-1950), comedian; born in Portobello; miner; became professional entertainer giving songs with interlude of patter; immediate success in London, 1900; performed notably as rollicking absurdly-kilted Highlander; usually wrote own words and music, drawing on traditional airs; songs included 'Roamin' in the Gloamin'', 'Stop your tickling, Jock'; frequently ended on serious note with song like 'The End of the Road'; indefatigable recruiter and entertainer in two wars; knighted, 1919.

LAUGHTON, SIR JOHN KNOX (1830-1915), naval historian; BA, Caius College, Cambridge; naval instructor in navy, 1853; transferred to Royal Naval College, Portsmouth, 1866; instructor at Greenwich Naval College, 1873-85; lectured there on naval history, 1876-89; professor of history, King's College, London, 1885-1914; founded Navy Records Society, 1893; first secretary, 1893-1912; knighted, 1907; edited *Memoirs relating to Lord Torrington* (1889), Lord Barham's papers (1907-10), etc.; wrote books on Nelson, etc.; contributor to this Dictionary.

LAURIE, JAMES STUART (1832-1904), inspector of schools, 1854-63; held educational posts in Ireland and Ceylon; called to bar, 1871; published school handbooks.

LAURIE, SIMON SOMERVILLE (1829-1909), educational reformer; elder brother of James Stuart Laurie [q.v.]; MA, Edinburgh

University, 1849; secretary to education committee of Church of Scotland at Edinburgh, 1855–1905; his reports as visitor and examiner for Dick Bequest Trust (1856–1907) gave masterly expositions of educational principles and practice; his report on the Edinburgh Merchant Company's 'hospitals' led to their reform by Act of Parliament, 1869; secretary to royal commission on endowed schools in Scotland, 1872; first Bell professor of education in Edinburgh University, 1876–1903; president of Teachers' Guild of Great Britain, 1891; hon. LLD of St. Andrews, Edinburgh, and Aberdeen universities; wrote on *Training of Teachers* (1882), *Institutes of Education* (1892), and *Educational Opinion from the Renaissance* (1903); his philosophical works include *Metaphysica, Nova et Vetusta* (1884) and *Ethica* (1885); Gifford lecturer in natural theology at Edinburgh, 1905–6; embodied these lectures in *Synthetica* (1906).

LAURIER, Sir WILFRID (1841–1919), Canadian statesman; French Canadian; born near Montreal; elected to legislature of Quebec, 1871; liberal member of Canadian parliament for Drummond–Arthabaska, 1874; entered cabinet of Alexander Mackenzie [q.v.], 1877; member for Quebec East, 1877–1919; in opposition, 1878–96; leader of liberal party in succession to Edward Blake [q.v.], 1888; prime minister of Canada, 1896–1911; GCMG, 1897; attacked by nationalist leader, Bourassa, for sending contingents to help of Great Britain in South African war (1900) and for policy of Canadian navy (1910); supported conservative government's war policy (1914–18), but refused to form coalition (1917).

LAUTERPACHT, Sir HERSCH (1897–1960), international lawyer; born in Eastern Galicia; obtained doctorates in law (1921), political science (1922), Vienna; an active Zionist; LLD, London School of Economics, 1925; assistant lecturer, 1927; naturalized, 1931; reader in public international law, London University, 1935–8; Whewell professor of international law, Cambridge, 1938–55; called to bar (Gray's Inn), 1936; bencher, 1955; KC, 1949; member, United Nations' International Law Commission, 1951–5; judge, International Court of Justice, 1954–60; revised *Manual of Military Law*, 1958; edited *British Year Book of International Law*, 1944–54, and *Annual Digest of Public International Cases*, 1929–56; FBA, 1948; knighted, 1956.

LAVERY, Sir JOHN (1856–1941), painter; born in Belfast; apprenticed to painter-photographer, Glasgow; studied at Glasgow School of Art, Heatherley's, and Académie

Julian; influenced for a time by J. A. M. Whistler [q.v.]; painted Queen Victoria's state visit to Glasgow Exhibition, 1888; thereafter uninterruptedly successful especially with portraits of women, including 'Miss Mary Burrell' (Glasgow) and 'Mrs. Lavery Sketching' (Dublin); presented collections of portraits of contemporary statesmen to Dublin and Belfast; also painted conversation pieces and scenes such as Casement trial; knighted, 1918; RA, 1921.

LAW, ANDREW BONAR (1858–1923), statesman; born in New Brunswick; brought up in Glasgow by his mother's relations from age of eleven; educated at Glasgow high school; entered his Kidston relations' firm of merchant bankers, 1874; junior partner in Glasgow firm of William Jacks & Co., iron-merchants, 1885; a director of Clydesdale Bank and chairman of Glasgow Iron Trade Association; unionist MP, Blackfriars and Hutchesontown division of Glasgow, 1900–6; parliamentary secretary to Board of Trade, 1902; supported Joseph Chamberlain's scheme of colonial preference and tariff reform, 1903; returned MP, Dulwich, at by-election, 1906; with Austen Chamberlain now recognized as most effective advocate of fiscal change in unionist party; denounced Lloyd George's budget of 1909 as socialism 'pure and unadulterated'; again returned for Dulwich, Jan. 1910; failed to capture North-West Manchester, Dec.; returned MP, Bootle division of Lancashire, at by-election, 1911; supported Lord Lansdowne and Balfour in decision to accept parliament bill, 1911; elected leader of opposition in House of Commons, 1911; with Sir Edward Carson shared leadership of faction which carried opposition to Irish home rule to brink of civil war; on outbreak of European war tendered Asquith support of unionist party in resisting German aggression, Aug. 1914; relegated to insignificant position of secretary for colonies in first coalition ministry, May 1915; led group in cabinet which pressed for evacuation of Dardanelles, autumn 1915; took charge of compulsory military service bill, Jan. 1916; invited by King George V to form administration, Dec.; on failure to secure co-operation first of Lloyd George and then of Asquith advised King to call on Lloyd George; chancellor of Exchequer and leader of House of Commons in Lloyd George's government; this 'most perfect partnership in political history' profoundly affected fortunes of war; issued series of War Loans on lower interest basis for long terms, one of the greatest achievements in history of British finance; his campaign for national war bonds (Oct. 1917) provided state with continuous flow of money until end of war; introduced war budgets of 1917 and 1918; the 1918 budget unparalleled in demands which it made on nation; throughout these anxious years couched his appeals for sacrifice in markedly sober speeches; supported continuance of coalition, 1918; MP, Central Glasgow, 1918; lord privy seal and leader of House of Commons, 1918; signatory of treaty of Versailles, 1919; resigned, owing to ill health, Mar. 1921; emerged from retirement in order to recommend Irish treaty, autumn 1921; prime minister, with purely conservative government, and leader of party, Oct. 1922; outlined programme of negation as best method of securing national recovery; presided at conference of allied prime ministers on subject of reparations in London (Dec. 1922) and Paris (Jan. 1923); conference broke down owing to failure of English and French to come to agreement over policy with regard to German finance; with utmost reluctance accepted settlement demanded by USA with regard to British debt, Jan. 1923; resigned office, May 1923.

LAW, DAVID (1831-1901), etcher and water-colour painter; student at Trustees' Academy, Edinburgh, 1845-50; engraver in ordnance survey office, Southampton, 1850-70; helped to found Royal Society of Painter-Etchers, 1880; his etchings after Turner and Corot were in great demand, 1875-90; best work done in water-colour.

LAW, SIR EDWARD FITZGERALD (1846-1908), expert in state finance; joined Royal Artillery, 1868; retired, 1872; started business agency in Russia; acting consul at St. Petersburg, 1880-1; commercial and financial attaché for Russia, Persia, and Asiatic Turkey, 1888; British delegate for negotiating commercial treaty with Turkey, 1890; reported on Greek finance, 1892-3, and on railway development in Asiatic Turkey, 1895; as commercial secretary at Vienna negotiated commercial treaty with Bulgaria, 1896-7; British delegate at Constantinople for determining Greece's war indemnity to Turkey, 1897; president of international commission on Greek finance, 1898; KCMG, 1898; finance member of government in India, 1900; completed currency reform and reduced arrears of land revenue, income and salt taxes; CSI, 1903; KCSI, 1906; an active champion of imperial preference and tariff reform; represented Great Britain on Cretan reform commission, and on committee to found bank of Morocco, 1906; died in Paris; buried at Athens.

LAW, THOMAS GRAVES (1836-1904), historian and bibliographer; grandson of Edward Law, first Earl of Ellenborough [q.v.], and brother of Augustus Henry Law [q.v.]; educated at Winchester, University College, London, and Stonyhurst; Roman Catholic priest, 1860-78; keeper to Signet Library,

Edinburgh, 1879-1904; helped to found Scottish History Society, 1886; hon. LLD, Edinburgh, 1898; wrote much on sixteenth-century religious history; *Collected Essays* were posthumously issued, 1904.

LAWES (afterwards LAWES-WITTEWRONGE), SIR CHARLES BENNET, second baronet (1843-1911), sculptor and athlete; son of Sir John Bennet Lawes, first baronet [q.v.], of Rothamsted; educated at Eton and Trinity College, Cambridge; BA, 1866; distinguished oarsman at Cambridge and Henley, runner and cyclist; exhibited sculpture at Royal Academy, 1872-1908; unsuccessful defendant in libel action brought by R. C. Belt, a sculptor, 1882; succeeded to baronetcy and Rothamsted property, 1900.

LAWES, WILLIAM GEORGE (1839-1907), missionary; worked at Niué in South Seas, 1861-72; completed translation of New Testament into Niué, 1886; settled at Port Moresby, New Guinea, 1874-94; with James Chalmers [q.v.] greatly helped British administration and development of British New Guinea; hon. DD, Glasgow, 1895; settled at Sydney, 1906, where he died.

LAWLEY, FRANCIS CHARLES (1825-1901), sportsman and journalist; BA, Balliol College, Oxford, 1848; fellow of All Souls, 1848-53; BCL, 1851; liberal MP for Beverley, 1852; private secretary to Gladstone; lost fortune in gambling and speculation; imputations of dishonesty in stock exchange dealings led to cancelling of his appointment as governor of South Australia, 1854; settled in America, 1854-65; sent vivid accounts of civil war to *The Times*; returned to London, 1865; wrote on sport in *Daily Telegraph* and *Baily's Magazine*.

LAWRENCE, ALFRED TRISTRAM, first BARON TREVETHIN (1843-1936), lord chief justice of England; educated at Mill Hill and Trinity Hall, Cambridge; first class, law, 1866; called to bar (Middle Temple), 1869; joined Oxford circuit; recorder of Windsor, 1885-1904; QC, 1897; judge, 1904-21; lord chief justice, 1921-2; president, War Compensation Court, 1920-2; baron and PC, 1921.

LAWRENCE, (ARABELLA) SUSAN (1871-1947), politician; studied mathematics at University College, London, and Newnham College, Cambridge; member of London County Council (municipal reform), 1910-12, (labour), 1913-28; elected to Poplar borough council, 1919; labour MP, East Ham North, 1923-4, 1926-31; parliamentary secretary to

Ministry of Health, 1929-31; chairman, labour party, 1929-30.

LAWRENCE, DAVID HERBERT (1885-1930), poet, novelist, and essayist; schoolmaster at Croydon until 1911; thenceforth devoted himself to literature; passionately sensitive to nature; published novels, *The White Peacock* (1911), *The Trespasser* (1912), and *Sons and Lovers* (1913), and *Love Poems and Others* (1913); by 1914 had won certain reputation as author; indifferent to material success and always possessed simple tastes; his desire to receive sympathetic response to his writings, which he believed to be of value to mankind, defeated by outbreak of 1914-18 war; felt keenly condemnation for indecency of his novel *The Rainbow* (published 1915); left England, largely because of hostile attitude towards himself and his writings, 1919; lived in Italy, Sicily, and Mexico, and then in Italy again; his later works include *Sea and Sardinia* (1921) and *Lady Chatterley's Lover* (1928, banned until 1960); both prosecutions caused great stir; died at Vence.

LAWRENCE, SIR (FREDERICK) GEOFFREY (1902-1967), judge; educated at City of London School and New College, Oxford; BA 1926; violinist; president, Oxford Musical Club; formed Magi String Quartet; called to bar (Middle Temple), 1930; KC, 1950; successfully defended Dr John Bodkin Adams on murder charge, 1957; recorder, Tenterden, 1948-51; recorder, Canterbury, 1952-62; chairman, West Sussex quarter sessions; chairman, General Council of the Bar, 1960-2; chairman, National Incomes Commission, 1962-4; knighted, 1963; High Court judge, 1965.

LAWRENCE, FREDERICK WILLIAM PETHICK-, first BARON (1871-1961), social worker and politician. [See PETHICK-LAWRENCE.]

LAWRENCE, GERTRUDE (1898-1952), actress; born Klasen of theatrical parentage; made first stage appearance in pantomime, 1910; became foremost of (Sir) Noël Coward's leading ladies especially popular in New York; played notably in *London Calling* (1923), *Private Lives* (1930), and *The King and I* (1952); a fine player of high vitality, keen wit, and style; published *A Star Danced* (1945); died in New York.

LAWRENCE, SIR HERBERT ALEXANDER (1861-1943), soldier and banker; son of Lord Lawrence [q.v.]; educated at Harrow and Sandhurst; gazetted to 17th Lancers, 1882; served in South Africa, 1899-1902; retired,

1903; committee member, Ottoman Bank, 1906; joined Glyn's Bank, 1907; a director, Midland Railway, 1913; commanded King Edward's Horse, 1904-9; commanded successively 127th (Manchester) brigade, 53rd and 52nd divisions, Dardanelles; in charge of Cape Helles evacuation; responsible for victory at Romani, Egypt, 1916; took 66th division to France and KCB, 1917; chief of general staff to (Lord) Haig [q.v.] from Jan. 1918; general, 1919; on Samuel commission on coal industry, 1925; GCB, 1926; director (1921), chairman (1926), Vickers; chairman of Glyn's, 1934-43.

LAWRENCE, SIR PAUL OGDEN (1861-1952), judge; educated at Malvern College; called to bar (Lincoln's Inn), 1882; QC, 1896; Chancery judge, 1918-26; lord justice of appeal, 1926-33; helped his sisters found Roedean School, 1885; knighted, 1919; PC, 1926.

LAWRENCE, THOMAS EDWARD (1888-1935), known as 'Lawrence of Arabia'; educated at Oxford high school and Jesus College; first class, modern history, 1910; assisted in excavations at Carchemish, 1911-14; served in Arab Bureau, 1914-16; adviser to Faisal (third son of Grand Sharif of Mecca), whose confidence he won by force of personality, 1916-18; brought Hejaz south of Aqaba, except Medina, under Arab-British control; given half-million pounds by (Lord) Allenby [q.v.] to raise Arab levies as mobile right wing; broke up Turkish Fourth Army and led Arab troops up to Damascus, 1 Oct. 1918; research fellow, All Souls, 1919; political adviser, Middle Eastern department, Colonial Office, 1921-2; obtained appointment of Faisal as King of Iraq and Abdullah as ruling Prince of Trans-Jordan; served in ranks of Royal Air Force, 1922-35, changing name to T. E. Shaw (1927); recorded Arabian exploits in *Seven Pillars of Wisdom* (1935; limited edition, 1926; abridged version, *The Revolt in the Desert*, 1927) and early Air Force days in *The Mint* (1955).

LAWRENCE, SIR WALTER ROPER, first baronet (1857-1940), Indian civil servant; settlement commissioner in Kashmir, 1889-95; private secretary to viceroy (Lord Curzon, q.v.), 1898-1903; chief of staff to Prince and Princess of Wales on Indian visit, 1905-6; works include *The Valley of Kashmir* (1895); KCIE, 1903; baronet, 1906; GCVO, 1918.

LAWS, ROBERT (1851-1934), pioneer missionary; studied theology and medicine at Edinburgh, Aberdeen, and Glasgow Universities; ordained in United Presbyterian Church of Scotland, 1875; helped to found (1875) Livingstonia mission station on Lake Nyasa; in full control, 1877-1927; developed Christian community of 60,000 including African pastors and over 700 schools.

LAWSON, EDWARD FREDERICK, fourth BARON BURNHAM (1890-1963), newspaper proprietor and soldier; educated at Eton and Balliol College, Oxford; reporter on *Daily Telegraph* in Paris and New York, 1913-14; commissioned in Royal Bucks. Hussars, 1914; DSO, MC; returned to *Daily Telegraph*, 1919; hon. colonel, 99th (Bucks. and Berks.) Field Brigade Royal Artillery, 1933; brigadier, Royal Artillery 48th (South Midland) Division, 1938; fought at Dunkirk, CB, 1940; major-general commanding Yorkshire Division, 1941; director, public relations, War Office and senior military representative, Ministry of Information, 1941-5; succeeded to title, 1943; retired from army, 1945; general manager, *Daily Telegraph*, 1927-39 and 1945-61; vice-chairman, Newspaper Proprietors (later Publishers) Association, 1934-9 and 1945-61; published *Peterborough Court, the Story of the Daily Telegraph* (1955).

LAWSON, EDWARD LEVY-, first BARON BURNHAM (1833-1916), newspaper proprietor. [See LEVY-LAWSON.]

LAWSON, GEORGE (1831-1903), ophthalmic surgeon; entered King's College Hospital, 1848; MRCS, 1852; FRCS, 1857; joined army as surgeon, 1854; served in Crimea; surgeon at Royal London Ophthalmic Hospital, Moorfields, from 1862; surgeon-oculist to Queen Victoria from 1886; published works on eye diseases.

LAWSON, GEORGE ANDERSON (1832-1904), sculptor; studied at Royal Scottish Academy Schools and in Rome; exhibited at Royal Academy from 1862; works include his popular 'Dominie Sampson' (1868), Burns memorial at Ayr, and the Wellington monument at Liverpool.

LAWSON, SIR HARRY LAWSON WEBSTER LEVY-, second baronet, second BARON, and VISCOUNT BURNHAM (1862-1933), newspaper proprietor; son of first Baron Burnham [q.v.]; educated at Eton and Balliol College, Oxford; first class, modern history, 1884; liberal MP, West St. Pancras (1885-92), Cirencester (1893-5); unionist MP, Mile End, 1905-6 and 1910-16; succeeded father, 1916; managing proprietor of *Daily Telegraph*, 1903-28; an ideal chairman, notably of the standing joint committee of teachers and local education authorities which formulated the 'Burnham scales'; and of three international labour conferences (1921,

1922, 1926); served on Indian statutory commission, 1927-30; CH, 1917; viscount, 1919.

LAWSON, SIR WILFRID, second baronet (1829-1906), politician and temperance advocate; son of Sir Wilfrid Lawson, an advanced liberal; keen sportsman, huntsman, angler, and agriculturist; liberal MP for Carlisle, 1859-65, 1868-85; supported motion for Sunday closing of public houses, 1863; introduced (1864) and frequently reintroduced (1869-74) his 'permissive', later known as 'local veto', bill; carried resolution for local option in 1880, in 1881, and in 1883; president of United Kingdom Alliance, 1879; succeeded to baronetcy and estates, 1867; supported motion of Sir Charles Dilke [q.v.] for inquiry into Queen Victoria's expenditure, 1872; advocated Sunday closing in Ireland, 1875-6 (measure carried, 1879); opposed parliamentary 'adjournment for the Derby', 1874; supported claim of Charles Bradlaugh [q.v.] for religious freedom, 1880; opposed liberal government's Egyptian policy, 1882-3; MP for Cockermouth, 1886-1900; supported Gladstone's home rule policy and opposed Balfour's coercion measures; MP for Camborne, 1903, and again for Cockermouth, 1906; passionately denounced South African war and defended free trade; of spontaneous humour, he seasoned his speeches with genial sarcasm and humorous quotation; easy writer of light verse; published *Cartoons in Rhyme and Line* (illustrated by Sir F. C. Gould, q.v.) (1905).

LAYCOCK, SIR ROBERT EDWARD (1907-1968), soldier; educated at Eton and the Royal Military College, Sandhurst; commissioned, Royal Horse Guards, 1927; GSO, Chemical Warfare in France, 1939; lieutenant-colonel in command of 'Layforce' commandos, 1941; fought in Crete and North Africa; commanded Special Service brigade, 1942; fought in Sicily and at Salerno, 1943; DSO; United States Legion of Merit; major-general, succeeding Lord Mountbatten as chief of combined operations, 1943-7; CB, 1945; governor and commander-in-chief, Malta, 1954-9; KCMG, 1954; high sheriff (1954-5) and lord lieutenant, Nottinghamshire, 1962; colonel commandant, Special Air Service and Sherwood Rangers Yeomanry, 1960.

LAYTON, SIR WALTER THOMAS, first BARON LAYTON (1884-1966), economist, editor, and newspaper proprietor; educated at King's College School, London, Westminster City School, University College, London, and Trinity College, Cambridge; first class, parts i and ii, economics tripos, 1906-7; lecturer in economics with J. M. (later Lord) Keynes [q.v.], 1908; fellow, Gonville and Caius College, 1909-14; Newmarch lecturer, University College, London, 1909-12; worked for Ministry of Munitions during 1914-18 war; member, Milner [q.v.] Mission to Russia, and Balfour Mission to USA, 1917; CBE, 1917; CH, 1919; published *An Introduction to the Study of Prices*, 1920; director, Economic and Financial Section, League of Nations; editor, *The Economist*, 1922-38; refashioned the paper; assisted in publication of Liberal Yellow Book, 1928; chairman, *News Chronicle*, 1930-50, and *Star* 1936-50; director, Reuters, 1945-53; director-general programmes, Ministry of Supply, 1940-2; chairman, Executive Committee, Ministry of Supply, 1941-2, chief adviser on programmes, Ministry of Production, 1941-3; head, Joint War Production Staff, 1942-3; post-1945, worked for Anglo-American understanding, European unity, and the United Nations; knighted, 1930; baron, 1947; hon. fellow, Gonville and Caius College, 1931; hon. degrees from Columbia and Melbourne universities.

LEACH, ARTHUR FRANCIS (1851-1915), historical writer; educated at Winchester and New College, Oxford; fellow of All Souls, 1874-81; assistant charity commissioner (endowed schools department), 1884; second charity commissioner, 1906-15; works include *English Schools at the Reformation (1546-1548)* (1896) and *Schools of Medieval England* (1915).

LEACOCK, STEPHEN BUTLER (1869-1944), professor and humorist; educated at Upper Canada College and Toronto University; graduated in modern languages, 1891; in philosophy, Chicago, 1903; lecturer in political science and history, McGill University, 1903; associate professor, 1905; William Dow professor and head of department of economics and political science, 1908-36; publications include many collections of humorous articles beginning with *Literary Lapses* (1910); also *Elements of Political Science* (1906) and *Our British Empire* (1940).

LEADER, BENJAMIN WILLIAMS (1831-1923), painter; born WILLIAMS; brother of Sir E. L. Williams [q.v.]; entered Royal Academy Schools, 1854; painter of landscapes in Worcestershire and Wales; his pictures include 'February Fill-Dyke' (1881) and 'In the Evening there shall be Light' (1882); RA, 1898.

LEADER, JOHN TEMPLE (1810-1903), politician and connoisseur; educated at Charterhouse, 1823 and at Christ Church, Oxford, 1828; knew from youth Lord Brougham [q.v.], his father's friend; liberal MP for Bridgwater, 1835; acted with Grote, Molesworth [qq.v.],

and the philosophical radicals; favoured the Chartists; unsuccessfully contested Westminster at a by-election (May 1837) against Sir Francis Burdett [q.v.]; MP for Westminster, Aug. 1837-47; prominent in London society; frequent traveller in Italy and France; saw much in London of Louis Napoleon, afterwards Napoleon III; in 1844 his career underwent sudden change, and he left England for permanent residence abroad; at Cannes joined Brougham, and bought property for building; chiefly spent his long exile at Florence, in and near which he bought and restored several old residences, including the gigantic castle of Vincigliata, where he was visited by many distinguished persons, including Queen Victoria (1888) and Gladstone; directed compilation of many archaeological treatises on Vincigliata and adjoining places; wrote with Giuseppe Marcotti lives of Sir John Hawkwood (1889) and Sir Robert Dudley, Duke of Northumberland (1895); left £7,000 for restoration of central bronze door of Duomo at Florence.

LEAF, WALTER (1852-1927), classical scholar and banker; educated at Harrow and Trinity College, Cambridge; bracketed senior classic, 1874; entered family business, silks and ribbons dealers, 1875; retired and devoted more attention to banking, 1892; a director of London and Westminster Bank, 1891; chairman, 1918; chairman, Institute of Bankers, 1919; president, International Chamber of Commerce, 1924; his reputation rests chiefly on his work as Homeric scholar; his publications include *Banking* (1926), an edition of the *Iliad* (1886-8, 2nd edn. 1900-2), *Troy . . .* (1912), *Homer and History* (1915), and *Strabo on the Troad* (1923).

LEAKE, GEORGE (1856-1902), premier of Western Australia; born at Perth, Western Australia; barrister of supreme court; crown solicitor, 1883-94; member of legislative assembly, 1890-1900; QC, 1898; twice premier of Western Australia, and attorney-general, May 1901-June 1902; died at Perth; a strong advocate of federation.

LEARMONTH, SIR JAMES RÖGNVALD (1895-1967), surgeon; educated at Girthon School, Gatehouse of Fleet, Kilmarnock Academy, and Glasgow University; MB, Ch.B., 1921; assistant to Professor Archibald Young, 1922-4; Rockefeller research fellow, Mayo Clinic, 1924-5; dispensary surgeon, Western Infirmary, Glasgow, 1925; Ch.M., 1927; FRCSE, 1928; worked in department of neuro-surgery, Mayo Clinic, 1928-32; regius professor of surgery, Aberdeen, 1932-9; specialized in disease of the arteries; professor of surgery, Edinburgh, 1939-56, and regius professor of clinical surgery, 1946-56; expert in peripheral vascular disease; surgeon to royal household in Scotland, 1934; surgeon to the King, 1949; CBE, 1945; KCVO, 1949; chevalier, Legion of Honour, 1950; hon. fellow, American College of Surgeons and Royal Medical Society.

LEATHERS, FREDERICK JAMES, first VISCOUNT LEATHERS (1883-1965), industrialist and public servant; left school at fifteen to work with Steamship Owners Coal Association (later merged with William Cory & Sons, Ltd.), 1898; managing director, 1916; concerned also with other companies dealing with coal or shipping services; adviser to Ministry of Shipping, 1914-18 and 1940-1; minister of war transport, 1941; baron, 1941; attended Casablanca, Washington, Quebec, and Cairo Conferences, 1943; negotiated lease-lend of American ships to Britain; CH, 1943; accompanied prime minister to Yalta and Potsdam; secretary of state for co-ordination of transport, fuel, and power, 1951-3; viscount, 1954; hon. LLD Leeds (1946) and Birmingham (1951); hon. member, Institution of Naval Architects; president, Institute of Chartered Shipbrokers.

LEATHES, SIR STANLEY MORDAUNT (1861-1938), historian and administrator; son of Stanley Leathes [q.v.]; scholar of Eton and Trinity College, Cambridge; first class, classics, 1882; fellow (1886), history lecturer (1892-1903); joint-editor, *Cambridge Modern History* (1901-12); secretary to the Civil Service Commission (1903), commissioner (1907), first commissioner (1910-27); KCB, 1919.

LE BAS, EDWARD (1904-1966), painter and collector; educated at Harrow and Pembroke College, Cambridge; studied at Royal College of Art, London; first exhibited at the Royal Academy, 1933; shared first one-man show with Dame Ethel Walker [q.v.] at Lefevre Galleries, 1936; specialized in still life, landscapes, and figures in interiors; influenced by W. R. Sickert [q.v.] and Vuillard; became wealthy collector after father's death, including work of young British painters; last one-man exhibition in England, 1939; exhibited annually at Royal Academy; ARA, 1943; RA, 1954; CBE, 1957; exhibited successfully in New York, 1956 and 1961; works to be seen in Tate Gallery, London, galleries in York and Leicester, New South Wales, and America.

LECKY, SQUIRE THORNTON STRATFORD (1838-1902), writer on navigation; served on merchant vessels from age of fourteen, becoming expert in navigation of Pacific; detected off Rio de Janeiro submerged 'Lecky Rock' 1865; showed many errors in South American charts and surveyed most of South American coast; served in Egyptian war of 1882; chief work on navigation was *Wrinkles in Practical Navigation* (1881, 15th edn. 1898); marine superintendent of Great Western Railway, 1884; younger brother of Trinity House; FRAS and FRGS; died at Las Palmas.

LECKY, WILLIAM EDWARD HARTPOLE (1838-1903), historian and essayist; born near Dublin; of Scottish descent; educated at Cheltenham, 1852-5, and at Trinity College, Dublin, studying desultorily history and philosophy; BA, 1859; MA, 1863; after a volume of poems (1859) he published anonymously *The Religious Tendencies of the Age* (1860) and *Leaders of Public Opinion in Ireland* (1862, revised edn. 1903), which met at first with little success; spent holidays abroad, especially in Spain and Italy; his essay on 'The Declining Sense of the Miraculous' (1863) subsequently formed first two chapters of his abstruse, discursive, but lucid *History of Rationalism* (2 vols., 1865), which brought him his first fame; a liberal in politics, he condemned Disraeli's reform bill of 1867; but supported Irish Church disestablishment and Irish Land Act of 1870; published *History of European Morals*, (2 vols., 1869); married in 1871 Elizabeth van Dedem, maid of honour to Queen Sophia of the Netherlands, and settled at 38 Onslow Gardens, Kensington; meanwhile collected material for his *History of England in the Eighteenth Century*, making extensive researches in Dublin (vols. 1 and 2 appeared in 1878; vols. 3 and 4 in 1882; vols. 5 and 6 in 1887; vols. 7 and 8 in 1890; cabinet edition, 12 vols., 1892; last 5 volumes devoted to History of Ireland); that work, which aimed at refuting Froude's calumnies of Irish people, was praised by Lord Acton [q.v.] and American critics; Lecky declined regius professorship of modern history of Oxford, 1892; hon. DCL, Oxford, 1888; Litt.D., Cambridge, 1891; LLD, Dublin (1879), St. Andrews (1885), Glasgow (1895); became a liberal unionist in 1886; wrote weighty letters to *The Times* and elsewhere (1886, 1892-3) in opposition to home rule; MP for Dublin University, 1895-1902; favoured establishment of a Roman Catholic university in Ireland; supported agricultural policy of Sir Horace Plunkett [q.v.] there; opposed old age pensions; favoured international arbitration; a fluent, rapid, but monotonous speaker; his later works were *Democracy and Liberty* (2 vols., 1896; a revised edition of 1899 gave an admirable estimate of Gladstone's work and character), *The Map of Life* (1899), and *Historical and Political Essays* (posthumous, 1908); FBA and OM, 1902; Lecky chair of history founded at Trinity College, Dublin, to which were left all his MSS.

LEDINGHAM, SIR JOHN CHARLES GRANT (1875–1944), bacteriologist and director of Lister Institute; educated at Banff Academy and Aberdeen University; MA, 1895; B.Sc., 1900; MB, Ch.B., 1902; studied pathology in Leipzig and London; assistant bacteriologist, Lister Institute, 1905; chief bacteriologist, 1909; director, 1931–43; professor of bacteriology, London University, 1920–42; secretary, National Collection of Type Cultures, 1920–30; with (Sir) J. A. Arkwright [q.v.] wrote *The Carrier Problem in Infectious Diseases* (1912); an associate editor of *System of Bacteriology* (1929–31); FRS, 1921; knighted, 1937.

LEDWARD, GILBERT (1888–1960), sculptor; son of Richard Arthur Ledward, sculptor [q.v.]; studied at Royal College of Art and Royal Academy School; professor of sculpture, Royal College of Art, 1926–9; war memorials include Guards Memorial, London; other works were direct stone carvings such as 'Monolith' (Tate Gallery), portrait busts, and Great Seal of the Realm (1953); ARA, 1932; RA, 1937; OBE, 1956.

LEDWIDGE, FRANCIS (1891–1917), poet; engaged in rural occupations in Slane district, county Meath; although a strong nationalist, joined army, 1914; killed in Belgium; wrote about the countryside; *Complete Poems* published posthumously, 1919.

LEE, SIR (ALBERT) GEORGE (1879–1967), engineer-in-chief, General Post Office; educated at Collegiate School, Llandudno; engineering assistant, Post Office Engineering Department, London, 1901; studied part-time at Northampton Institute, Finsbury Technical College, and King's College, London; B.Sc; increased experience on telephone transmission and promoted to sectional engineer, Bolton, 1908–12; served in Royal Engineers Signal Service during 1914–18; MC; British delegate to Inter-Allied Radio Conference, Paris, 1921; worked on trans-Atlantic telephony, 1923–6; engineer-in-chief, GPO, 1931–9; director, communications, research and development, Air Ministry, 1939; senior telecommunications officer, Ministry of Supply, 1944; vice-president, Institute of Radio Engineers of America, 1929; president, Institution of Electrical Engineers, 1937–8; OBE, 1927; knighted, 1937.

LEE, ARTHUR HAMILTON, VISCOUNT LEE OF FAREHAM (1868–1947), statesman, benefactor, and patron of the arts; educated at Cheltenham College and Woolwich; joined Royal Artillery, 1888; professor at Royal Military College, Kingston, Canada, 1893–8; military attaché, United States Army in Cuba, 1898;

Washington, 1899; retired, 1900; conservative MP, Fareham division, 1900–18; civil lord of Admiralty, 1903–5; parliamentary military secretary, Ministry of Munitions, 1915; personal military secretary to Lloyd George, 1916; director-general of food production, 1917–18; KCB, 1916; presented Chequers estate to nation for use of prime ministers, 1917; baron, 1918; PC, 1919; president of Board of Agriculture, 1919–21; first lord of Admiralty, 1921–2; second British delegate to Washington conference, 1921–2; viscount, 1922; with Samuel Courtauld [q.v.] founded Courtauld Institute and brought Warburg Institute to London.

LEE, FREDERICK GEORGE (1832–1902), theological writer; educated at St. Edmund Hall, Oxford; won Newdigate prize for English poem, 1854; vicar of All Saints', Lambeth, 1867–99; practised an advanced ritualism; vindicated the validity of Church of England orders, 1870; subsequently questioned their validity and founded Order of Corporate Reunion to restore to Church of England valid orders; consecrated by catholic prelates 'bishop of Dorchester', 1877; FSA, 1857; historical works, which are partisan and untrustworthy, include *Historical Sketches of the Reformation* (1879) and *The Church under Queen Elizabeth* (3rd edn. 1897); published also *History of Thame* (1886), verse, devotional and antiquarian works; joined Roman Catholic Church, 1901.

LEE, RAWDON BRIGGS (1845–1908), writer on dogs; succeeded father as editor of the *Kendal Mercury* till 1885; devoted much time to breeding of dogs; his English setter, Richmond, was sent to Australia to improve the breed; kennel editor of *Field*, 1883–1907; wrote accounts of fox terrier, 1889, and modern dogs of Great Britain, 3 vols., 1894 and 1897.

LEE, ROBERT WARDEN (1868–1958), lawyer; scholar of Rossall School and Balliol College, Oxford; first classes, classics, 1889–91; called to bar (Gray's Inn), 1896; professor, Roman-Dutch law, London University, 1906–14, Oxford (and fellow of All Souls), 1921–56; dean of law faculty, McGill University, Montreal, 1914–21; publications include *Introduction to Roman-Dutch Law* (1915) and *Elements of Roman Law* (1944); FBA, 1933.

LEE, SIR SIDNEY (1859–1926), Shakespearian scholar and editor of the *Dictionary of National Biography*; educated at City of London School, where interest in Elizabethan literature was stimulated by Dr E. A. Abbott [q.v.]; BA, Balliol College, Oxford; began his Shakespearian studies as undergraduate with two articles in *Gentleman's Magazine*, 1880; assistant editor to (Sir) Leslie Stephen [q.v.], on foundation of Dictionary of National Biography, 1883; his exact and scholarly methods well fitted for organizing editorial work; joint-editor, 1890; sole editor, 1891-1901 and 1910-12, retaining general oversight of Dictionary, 1901-16; preserved balance and uniformity of Dictionary; his greatest asset as editor was his personality; Dictionary completed in 63 volumes, 1900; first Supplement issued in 3 volumes, 1901; out of his Dictionary articles developed *Life of William Shakespeare* (1898) and *Queen Victoria* (1902); *Life of Shakespeare*, which passed through fourteen editions, a work of exegesis of first order; furnished reliable basis for sound aesthetic appreciation by his study of origin and formation of Shakespeare's text and influence of foreign literature on Shakespeare's subject-matter; subsequently developed former theme by publishing facsimiles of earliest editions of some of Shakespeare's works; developed latter theme by extensive examination, chiefly in articles and lectures, of foreign influence on Elizabethan literature in general and on Shakespeare in particular; his *Queen Victoria* first serious attempt to present queen's public and private life as whole; although information available, especially on latter part of reign, restricted, as pioneer piece of work remarkably successful; superintended summary of Dictionary which appeared as *Index and Epitome* (1903), volume of *Errata* (1904), and reissue of Dictionary and first Supplement (1909); Clark lecturer in English literature, Trinity College, Cambridge, 1901-2; toured universities and colleges of USA, 1903; edited second Supplement of Dictionary, 1910-12; neither first nor second Supplement preserved exactly standard of selection maintained in main work; first Supplement tended to restrict admission, while second Supplement was far more inclusive than main Dictionary; president (1890) of Elizabethan Literary Society at Toynbee Hall, which he and Frederick Rogers developed into centre of Elizabethan study; a founder of English Association, 1906; president, 1917; appointed to new chair of English language and literature at East London College, London University, 1913; retired 1924; his last work of importance, *Life of King Edward VII*, undertaken at request of King George V; first volume published, 1925; second, completed by (Sir) S. F. Markham, published, 1927; FBA, 1910; knighted, 1911.

LEE, VERNON (pseudonym) (1856-1935), author. [See PAGET, VIOLET.]

LEE-HAMILTON, EUGENE JACOB (1845-1907), poet and novelist; educated at Oriel College, Oxford; held minor diplomatic posts at Paris and Lisbon, 1871-3; disabled for twenty years through nervous disease; lived at Florence and became a centre of intellectual society; published *Imaginary Sonnets* (1888) and *The Sonnets of the Wingless Hours* (1894); was restored to health, 1897; published also a tragedy, two novels, and a metrical translation of Dante's *Inferno* (1898).

LEE-WARNER, SIR WILLIAM (1846-1914), Indian civil servant and author; BA, St. John's College, Cambridge; entered Indian civil service, 1867; served in India, 1869-95; secretary of political and secret department at India Office, 1895-1903; KCSI, 1898; GCSI, 1911; works include *Protected Princes of India* (1894), revised as *Native States of India* (1916).

LEES, GEORGE MARTIN (1898-1955), geologist; educated at St. Andrew's College, Dublin and Royal Military Academy, Woolwich; served with Royal Flying Corps; DSO; MC; joined Iraq civil service, 1919-21; from Royal School of Mines joined Anglo-Persian Oil Company, 1921; chief geologist, 1930-53; initiated search for oil in Britain, 1933, discovering East Midland oilfields, 1939; also successfully explored in Nigeria, Libya, and Abu Dhabi; president, Geological Society, 1951-3; FRS, 1948.

LEESON, SPENCER STOTTESBERY GWATKIN (1892-1956), schoolmaster and bishop; scholar of Winchester and New College, Oxford; first class, classical moderations, 1913; war degree, 1916; served in Flanders and naval intelligence, 1915-18; at Board of Education, 1919-24; called to bar (Inner Temple), 1922; assistant master, Winchester, 1924-7; headmaster, Merchant Taylors', 1927-35; moved school to outskirts of London; headmaster, Winchester, 1935-46; chairman, Headmasters' Conference, 1939-45; deacon, 1939; priest, 1940; rector of Southampton, 1946-9; bishop of Peterborough, 1949-56; eloquent speaker and immensely hard worker in educational causes; publications include *Study of the Gospel of Christ* (1941) and *Christian Education* (1947).

LE FANU, SIR MICHAEL (1913-1970), admiral of the fleet; educated at Bidford School and Royal Naval College, Dartmouth; specialized in gunnery, 1938; gunnery officer, *Aurora*, 1939; DSO, 1941; commander, liaison officer to American 3rd and 5th Fleets, 1945; Legion of Merit; captain, 1949; commanded 3rd Training Squadron, Londonderry, 1951; special duty, Admiralty, 1952-3; commanded HMS *Ganges*, 1954; commanded aircraft carrier, *Eagle*, 1957-8; rear-admiral, 1958; director-general, weapons, 1958-60; second-in-command, Far East station, 1960-1; controller of the navy,

1961-5; vice-admiral, 1961; admiral, 1965; commanded three Services in Middle East, Aden, 1965; first sea lord, 1968-9; admiral of the fleet, 1969; CB, 1960; KCB, 1963; GCB, 1968.

LEFROY, WILLIAM (1836-1909), dean of Norwich; BA, Trinity College, Dublin, 1863; BD, 1867; DD, 1889; obtained fame as evangelical preacher; incumbent of St. Andrew's chapel, Renshaw Street, Liverpool, 1866; archdeacon of Warrington, 1887; Donnellan lecturer at Dublin, 1887; member of Liverpool school board from 1876; dean of Norwich, 1889-1909; member of the Alpine Club; helped to build several English churches in Switzerland; died and was buried at Riffel Alp; published theological works.

LE GALLIENNE, RICHARD THOMAS (1866-1947), poet and essayist; educated at Liverpool College; published verse including *Volumes in Folio* (1889), literary criticisms, and romantic novel *The Quest of the Golden Girl* (1896); original member, Rhymers' Club; contributor to *Yellow Book*; reader for the Bodley Head; prominently associated with literary movement of nineties; moved to United States, 1901; later to France; published *The Romantic '90s* (1925).

LEGG, JOHN WICKHAM (1843-1921), physician and liturgiologist; MD, University College, London, 1868; joined staff of St. Bartholomew's Hospital, London, 1870; abandoned medicine, 1887; FSA, 1875; distinguished student of liturgies; works include editions of Quignon Breviary of 1535 (1888), *Second Recension of Quignon Breviary* (1908, 1912), *Westminster Missal* (1891-7), and *Sarum Missal* (1916), and *English Church Life* [1660-1833] (1914); high churchman.

LEGH, THOMAS WODEHOUSE, second BARON NEWTON (1857-1942), diplomatist and politician; educated at Eton and Christ Church, Oxford; entered Foreign Office, 1879; served in Paris, 1881-6; conservative MP, Newton division, 1886-98; succeeded father, 1898; paymaster-general and PC, 1915; controller, prisoner-of-war department, Foreign Office, 1916-19; obtained, exchange of thousands of prisoners; wrote biographies of Lord Lyons (1913) and fifth Marquess of Lansdowne (1929).

LEGROS, ALPHONSE (1837-1911), painter, sculptor, and etcher; born at Dijon; worked in Paris as scene-painter; exhibited at Salon, 1857; enrolled by Champfleury among the 'Realists'; exhibited 'Angelus' (1859), 'Ex Voto' (1861), and 'Le Lutrin' (1863); earned living by etchings and lithographs; encouraged by J. A. M.

Whistler [q.v.] to come to London, 1863; Slade professor of fine art at University College, London, 1875-92; designed fountains for gardens at Welbeck, 1897; exhibited paintings, etchings, and medals at Royal Academy, 1864-82; fellow of Society of Painter-Etchers, 1880, and honorary fellow of Royal Scottish Academy, 1911; works are in public galleries in Paris, Dijon, London, Manchester, and Liverpool, as well as in private collections.

LEHMANN, RUDOLF (1819-1905), painter; born near Hamburg; studied art at Paris, Munich, and Rome; first visited London, 1850; exhibited at Royal Academy from 1851, and at other English galleries, mainly subject pictures and portraits of prominent persons; lived in Italy, mostly at Rome, 1856-66; best-known portraits were those of Helen Faucit, Robert Browning, Viscount Goschen [qq.v.]; intimate friend of Browning; published *Reminiscences* (1894) and *Men and Women of the Century* (portrait sketches, 1896).

LEICESTER, second EARL OF (1822-1909), agriculturist. [See COKE, THOMAS WILLIAM.]

LEIGH, VIVIEN (1913-1967), actress; born Vivian Mary Hartley; educated at Convent of the Sacred Heart, Roehampton; at thirteen, travelled with parents abroad, 1926-31; studied at Royal Academy of Dramatic Art; adopted stage name, Vivien Leigh; appeared in film, *Things are Looking Up* (1934); played in *The Mask of Virtue* at Ambassadors Theatre, 1935; signed five-year contract with (Sir) Alexander Korda [q.v.]; played with Laurence (later Lord) Olivier in *Fire Over England* (1937); played Scarlett O'Hara in *Gone with the Wind* (1939); won an Oscar award; married Laurence Olivier, 1940; toured North Africa with Beatrice Lillie, 1943; toured Australia and New Zealand with Old Vic Company and Olivier, 1948, played Blanche du Bois on stage and in film in *A Streetcar Named Desire* (1949-50); awarded second Oscar; with Olivier at Stratford-upon-Avon in *Titus Andronicus* (1955); last appearance with Olivier, 1957; marriage dissolved, 1960; made new reputation in musical *Tovarich* in New York, 1963; knight's cross, Legion of Honour, 1957; died of tuberculosis, 1967; exterior lights of London's west-end theatres darkened for an hour.

LEIGH-MALLORY, SIR TRAFFORD LEIGH (1892-1944), air chief marshal; brother of G. L. Mallory [q.v.]; educated at Haileybury and Magdalene College, Cambridge; served with Royal Flying Corps, 1916-19; DSO and permanent commission, 1919; commanded No. 12 (fighter) group, 1937-40; No. 11 group,

1940-2; AOC-in-C, Fighter Command, 1942-3; air chief marshal and KCB, 1943; air commander-in-chief, Allied Expeditionary Air Force of some 9,000 aircraft, 1943-4; lost flying to take up appointment as allied air commander-in-chief, South East Asia.

LEIGHTON, STANLEY (1837-1901), politician and antiquary; educated at Harrow and Balliol College, Oxford; BA and MA, 1864; called to bar, 1861; conservative MP for North Shropshire, 1876-85, and Oswestry, 1885-1901; ardent supporter of the Church in parliament; FSA, 1880; founded Shropshire Parish Register Society, 1897; accomplished amateur artist; wrote and illustrated *Shropshire Houses, Past and Present* (1901).

LEININGEN, PRINCE ERNEST LEOPOLD VICTOR CHARLES AUGUSTE JOSEPH EMICH (1830-1904), admiral, reigning prince of Leiningen; born at Amorbach, Bavaria; entered British navy, 1849; served in Burmese war, 1851-2; served against Russians on the Danube, 1854; took part in Baltic campaign, 1856; commander and captain from 1858 of royal yacht *Alberta*, which (with Queen Victoria on board) accidentally ran down schooner yacht *Mistletoe* in Stokes bay, Aug. 1875; vice-admiral, 1881; commander-in-chief at the Nore, 1885-7; admiral, 1887; GCB, 1866; GCVO, 1898; died at Amorbach.

LEIPER, ROBERT THOMSON (1881-1969), professor of helminthology; educated at Warwick School, Leamington Technical College, Mason College, Birmingham, and Glasgow University; graduated in medicine, 1904; helminthologist, London School of Tropical Medicine, 1905; detected cause of 'Guinea worm' infection, Accra, 1905; and of Calabar swelling, West Africa, 1912; studied wide variety of helminths in the East, 1913; in charge of Bilharzia mission in Egypt, 1914-15; discovered Egyptian species of parasites causing human schistosomiasis; director, prosectorium, Zoological Society, London, 1919-21; professor of helminthology, London University, 1919; founded *Journal of Helminthology*, 1923; director of parasitology, London School of Hygiene and Tropical Medicine, 1924-45; set up Institute of Agricultural Parasitology, 1925; founder and director, Commonwealth Bureau of Helminthology, 1929-58; D.Sc. Glasgow, 1911; MD, 1917; FRS, 1923; FRCP, 1936; CMG, 1941.

LEISHMAN, THOMAS (1825-1904), Scottish divine and liturgiologist; MA, Glasgow University, 1843; DD, 1871; Presbyterian minister at Linton, Teviotdale, 1855-94; with

G. W. Sprott [q.v.] published an annotated edition of *The Book of Common Order*; advocated observance of the five great Christian festivals by the Church of Scotland, 1868; helped to found Scottish Church Society, 1892; thrice president; writings include *The Moulding of the Scottish Reformation* (1897); moderator of General Assembly, 1898.

LEISHMAN, SIR WILLIAM BOOG (1865-1926), bacteriologist; MB, CM, Glasgow; entered Army Medical Service, 1887; served in Waziristan campaign, 1894-5; assistant professor of pathology, Army Medical School (Netley), 1900; professor of pathology, Army Medical School (Millbank), 1903-13; War Office expert on tropical diseases, 1914; adviser in pathology to British expeditionary force in France, 1914-18; director of pathology, War Office, 1919-23; medical director-general, Army Medical Services, and lieutenant-general, 1923; knighted, 1909; FRS, 1910; famous for his work on anti-typhoid inoculation and kala-azar.

LEITH-ROSS, SIR FREDERICK WILLIAM (1887-1968), civil servant and authority on finance; educated at Merchant Taylors' School and Balliol College, Oxford; first class, *lit. hum.*, 1909; passed top into home Civil Service; posted to Treasury, 1909; private secretary to prime minister, H. H. Asquith (later first earl of Oxford and Asquith), 1911; returned to Treasury, 1913; British member, Finance Board, Reparation Commission, 1920-5; deputy controller, finance, Treasury, 1925; settled post-war claims, Egypt, 1929; attended Hague Conference with chancellor of the Exchequer, 1929; chief economic advisor, 1932; worked on currency reform, China, 1935; negotiated revised German payments agreement, 1938; director-general, Ministry of Economic Warfare, 1939-42; deputy director-general, European regional office, UNRRA; chairman, European committee, 1945-6; published *UNRRA in Europe* (1946); governor, National Bank of Egypt, 1946-51; director, National Provincial Bank, 1951; deputy chairman, 1952; director, Standard Bank; CB, 1925; KCB, 1933; KCMG, 1930; GCMG, 1937; published autobiography, *Money Talks* (1968).

LE JEUNE, HENRY (1819-1904), historical and genre painter; studied art at Royal Academy Schools; exhibited at Academy, 1840-94, and British Institution, 1842-63; curator of painting school of Royal Academy, 1848-64; ARA, 1863; painted subjects from Bible, Shakespeare, and Spenser; later devoted himself to painting children, as in 'Little Bo-Peep' and 'My Little Model'; musician and chess player.

LEMMENS-SHERRINGTON, HELEN (1834-1906), soprano vocalist; studied music at Brussels under Cornelis, 1852; first appeared at London concerts, 1856; married Nicolas Jacques Lemmens, 1857; leading English soprano from 1860, singing in English opera (1860-5) and in Italian opera (1866); showed great power in oratorio music, as in Mendelssohn's *Elijah* and Haydn's *Creation*; sang in first performance in England of Bach's High Mass, 1876; teacher of singing at Brussels conservatoire, 1881-91; occasionally revisited England; died at Brussels.

LEMON, SIR ERNEST JOHN HUTCHINGS (1884-1954), mechanical and railway engineer; trained at Glasgow Technical College and Heriot-Watt College, Edinburgh; chief wagon inspector, Midland Railway, 1911; works superintendent, Derby, 1917; divisional superintendent, LMS, 1923; carriage and wagon superintendent, 1927; chief mechanical engineer, 1931; operating and commercial vice-president, 1932-43; reorganized, modernized, and mechanized; director-general, aircraft production, 1938-40; chairman, post-war railway planning commission; knighted, 1941.

LEMPRIERE, CHARLES (1818-1901), writer and politician; son of John Lempriere, DD [q.v.]; BCL, St. John's College, Oxford, 1842; DCL, 1847; called to bar, 1844; agent of conservative party from 1850; sent on private mission to Mexico, to watch British interests, 1861; published *The American Crisis Considered* (1861), and *Notes on Mexico* (1862); colonial secretary of the Bahamas, 1867; compelled to resign owing to his tory opinions; wrote for the American *Tribune*; organized unsuccessful English colony at Buckhorn, West Virginia, 1872.

LENG, SIR JOHN (1828-1906), newspaper proprietor; sub-editor and reporter of *Hull Advertiser*, 1847-51; editor of bi-weekly *Dundee Advertiser*, 1851; raised paper to high rank; became part proprietor, 1852; issued *Advertiser* daily, 1861; established office in London, 1870; first to attempt illustrations in daily paper; founded first halfpenny daily in Scotland, 1859, the weekly *People's Journal*, 1858, and literary weekly, *People's Friend*, 1869; started *Evening Telegraph*, 1877; liberal MP for Dundee, 1889-1905; supported railway and factory labour legislation and home rule bill, 1893; knighted, 1893; hon. LLD, St. Andrews, 1904; thrice visited America and Canada, India (1896), and Near East; died at Delmonte, California.

LENG, SIR WILLIAM CHRISTOPHER (1825-1902), journalist; brother of Sir John Leng [q.v.]; contributed to *Dundee Advertiser*

from 1859; managing editor and owner (1864) of *Sheffield Daily Telegraph*, which became a powerful conservative organ; first to set up linotype machines; denounced trade-unionist terrorism at Sheffield, 1867, and obtained royal commission of inquiry into his allegations; knighted, 1887.

LENNARD-JONES, SIR JOHN EDWARD (1894-1954), scientist and administrator; educated at Leigh grammar school and Manchester University; first class, mathematics, 1915; researched and taught at Manchester, Cambridge, and Bristol; professor of theoretical physics, Bristol, 1927; Plummer professor of theoretical chemistry, Cambridge, 1932-53; principal of University College of North Staffordshire, 1953-4; engaged in armament research, 1939-45; known for work on theory of molecular orbitals and theory of liquids; FRS, 1933; KBE, 1946; Sc.D., Cambridge; hon. D.Sc., Oxford; president, Faraday Society, 1948-50.

LENNOX, CHARLES HENRY GORDON-, sixth DUKE OF RICHMOND and first DUKE OF GORDON (1818-1903), lord president of the Council. [See GORDON-LENNOX.]

LENO, DAN (1860-1904), music-hall singer and dancer, whose true name was GEORGE GALVIN; made first appearance in London as 'Little George' the contortionist, 1864; took to clog-dancing and singing, 1869; admired by Charles Dickens at Belfast, 1869; won clog-dancing championship of the world at Leeds, 1880; made first appearance as 'Dan Leno' in London, 1883; appeared in Drury Lane pantomime from 1888-9 annually till death; played at Sandringham before King Edward VII, Nov. 1901; most memorable songs a mixture of song and 'patter'; lavish in charity; wrote burlesque autobiography, *Dan Leno: his Book* (1901).

LENOX-CONYNGHAM, SIR GERALD PONSONBY (1866-1956), geodesist; educated at Edinburgh Academy and Royal Military Academy, Woolwich; commissioned in Royal Engineers; transferred to Survey of India, 1889; undertook important longitudinal and gravity measurements; superintendent of trigonometrical survey, 1912-20; colonel, 1914; reader in geodesy, Cambridge, 1922-47; developed department of geodesy and geophysics; FRS, 1918; knighted, 1919.

LESLIE, SIR BRADFORD (1831-1926), civil engineer; son of C. R. Leslie [q.v.], painter; apprenticed to civil engineer, I. K. Brunel [q.v.], 1847; in service of Eastern Bengal Railway Company, acting under W. Purdon as resident engineer in charge of large bridges and viaducts, 1858–62; re-entered service, 1865; chief resident engineer for extension of line in northern delta, 1867–71; this included his first great achievement in India, bridge over Gorai; constructed an original floating bridge over Hugli, 1873–4; municipal engineer of Calcutta, 1873–6; agent and chief engineer, East Indian Railway Company, 1876; superintended construction of bridge over Hugli at Naihati; KCIE, 1887.

LESTER, SEAN (JOHN ERNEST) (1888–1959), secretary-general of the League of Nations; educated at Methodist College, Belfast; became journalist; publicity officer, external affairs department, Irish Free State, 1922–9; Irish representative at League of Nations, 1929; League of Nations' high commissioner, Danzig, 1934–7; deputy secretary-general, 1937; opposed compromise with Nazis; secretary-general, 1940–7; received Woodrow Wilson award.

LE STRANGE, GUY (1854–1933), orientalist; son of H. L'E. S. le Strange [q.v.]; mainly interested in historical geography of Middle Eastern Moslem lands; works include *Palestine under the Moslems* (1890) and *The Lands of the Eastern Caliphate* (1905).

LETHABY, WILLIAM RICHARD (1857–1931), author and architect; worked under R. N. Shaw [q.v.]; began independent practice, 1891; closely associated with Philip Webb [q.v.]; an organizer and principal, LCC Central School of Arts and Crafts, 1894; appointed professor of design, Royal College of Art, 1900; surveyor, Westminster Abbey, 1906–28; publications include *Westminster Abbey and the King's Craftsmen* (1906) and *Mediaeval Art . . . 312–1350* (1904).

LETT, SIR HUGH, baronet (1876–1964), surgeon; educated at Marlborough College, as preclinical student, at Leeds, and the London Hospital; MB, B.Ch. (Victoria), Leeds, 1899, diploma, Royal Colleges, 1901; FRCS, 1902; surgical registrar, London Hospital, 1902; assistant surgeon, 1905; surgical tutor, 1909–12; in charge of urological department; full surgeon, 1915–34; attached to Anglo-American hospital, Wimereux, 1914–15; Belgian Field Hospital, Furnes, 1915; major, RAMC; CBE, 1920; consulting surgeon, London Hospital, 1934; president, Royal College of Surgeons, 1938–41; chairman, Hunterian Museum Trust, 1955–9; baronet, 1941; KCVO, 1947; president, British Medical Association, 1946–8; hon. fellow, Royal Society of Medicine and Hunterian Society; hon. DCL, Durham, hon. Sc.D., Cambridge.

LEVER, SIR (SAMUEL) HARDMAN, baronet, of Allerton (1869–1947), chartered accountant; educated at Merchant Taylors' School; partner in New York firm of accountants; entered Ministry of Munitions to advise on costing, 1915; financial secretary to Treasury, 1916; Treasury representative in United States, 1917–19; at Ministry of Transport, 1919–21; headed air mission to Canada (1938), Australia and New Zealand (1939); KCB, 1917; baronet, 1920.

LEVER, WILLIAM HESKETH, first VISCOUNT LEVERHULME (1851–1925), soap manufacturer; entered father's grocery business in Bolton, 1867; partner, 1872; with his brother began to trade on his own account, specializing in soap, for which he chose name 'Sunlight', 1884; began to manufacture it, 1885; inaugurated new town, Port Sunlight, on Mersey, near Bebington, Cheshire, as centre for his works and workpeople, 1888; Lever Brothers made limited company, 1890, public company, 1894; by purchase or by interchange of shares exercised wide control over soapmaking trade; liberal MP, Wirral division of Cheshire, 1906–9; brought and won libel action against Northcliffe press, 1907; established crushing mills in Nigeria for supply of palm-oil for his soap; assiduous in care for Port Sunlight, on which he lavished many gifts, including art gallery; an enthusiastic collector of pictures, etc.; presented Stafford House (which, as Lancaster House, housed London Museum) to nation; baronet, 1911; baron, 1917; viscount, 1922.

LEVERHULME, first VISCOUNT (1851–1925), soap manufacturer. [See LEVER, WILLIAM HESKETH.]

LEVESON-GOWER, (EDWARD) FREDERICK (1819–1907), politician; son of first Earl Granville [q.v.]; educated at Eton and Christ Church, Oxford; BA, 1840; called to bar, 1846; liberal MP for Derby, 1846–7, Stoke-on-Trent, 1852–7, and Bodmin, 1859–85; chairman of National School of Cookery, 1874–1903; visited India, 1850–1, and Russia, 1856; a conspicuous figure in society; edited his mother's *Letters* (1894) and published *Bygone Years* (1905).

LEVESON GOWER, SIR HENRY DUDLEY GRESHAM (1873–1954), cricketer. [See GOWER.]

LEVICK, GEORGE MURRAY (1876–1956), surgeon and explorer; educated at St. Paul's School; qualified from St. Bartholomew's hospital and commissioned as doctor in Royal Navy, 1902; chosen by Captain R. F. Scott [q.v.] for Antarctic expedition, 1910; spent two years with northern party exploring Victoria Land coast; studied Adélie penguin; surgeon-commander, 1915; retired, 1917; founded Public Schools Exploring Society, 1932, and honorary chief leader of first nine expeditions; recalled to Royal Navy to assist training of commandos, 1939-45.

LEVY-LAWSON, EDWARD, first BARON BURNHAM (1833-1916), newspaper proprietor; son of Joseph Moses Levy [q.v.]; assumed additional surname of Lawson, 1875; began career as dramatic critic to *Sunday Times*; became editor of *Daily Telegraph* shortly after its acquisition by his father, 1855; managing proprietor and sole controller, 1885; humanized his newspaper; paper's support transferred from Gladstone to Beaconsfield, 1879; after 1886 paper definitely unionist and imperialist; organized appeals to public for national and charitable efforts through *Daily Telegraph* funds; sponsored enterprises such as Assyrian expedition of George Smith [q.v.], 1873; baronet, 1892; created baron on retirement from active control of paper, 1903; KCVO, 1904.

LEVY-LAWSON, SIR HARRY LAWSON WEBSTER, VISCOUNT BURNHAM (1862-1933), newspaper proprietor. [See LAWSON.]

LEWIS, AGNES (1843-1926), discoverer of the 'Sinai Palimpsest'; born SMITH; married S. S. Lewis [q.v.], 1887; with her twin sister, Margaret Gibson, visited St. Catherine's convent, Mount Sinai, 1892; discovered among Syriac manuscripts palimpsest containing ancient text of Gospels; photographed and subsequently (1893) transcribed manuscript.

LEWIS, BUNNELL (1824-1908), classical archaeologist; BA, London, 1843; MA, 1849; fellow of University College, London, 1847; professor of Latin, Queen's College, Cork, 1849-1905; FSA, 1865; made researches into surviving Roman antiquities in Europe; contributed to *Archaeological Journal*, 1875-1907; bequeathed library and £1,000 for classical prize to University College.

LEWIS, CLIVE STAPLES (1898-1963), writer and scholar; educated at Malvern and University College, Oxford; first class *lit. hum.*, 1922, and English, 1923; fellow, Magdalen College, 1925; published narrative poem *Dymer* under pseudonym, Clive Hamilton, 1926; first allegorical work, *The Pilgrim's Progress* (1933); other publications, while he was lecturing at Oxford, include *The Allegory of Love* (1936), and *The Screwtape Letters* (1942); professor, English, medieval, and Renaissance literature,

Cambridge, 1954-63; fellow, Magdalene College, Cambridge; published *English Literature in the Sixteenth Century* (1954); wrote the Narnia series of children's tales, 1948-56; hon. fellow, Magdalen and University Colleges, Oxford, and Magdalene College, Cambridge; other publications include *The Problem of Pain* (1940), *The Four Loves* (1960), and *An Experiment in Criticism* (1961); FBA, 1955; hon. DD, St. Andrews, 1948; and D.Litt., Laval, Quebec, 1952.

LEWIS, EVAN (1818-1901), dean of Bangor; BA, Jesus College, Oxford, 1841; MA, 1863; held livings of Aberdare (1859-66) and Dolgelly (1866-84); chancellor (1872-6), canon residentiary (1877-84), and dean of Bangor (1884-1901); influenced by tractarian movement; his teaching as curate on the sacraments led to a long controversy, 1850-2; his best work was Welsh treatise on apostolic succession, 1851; his elder brother, DAVID LEWIS (1814-95), vice-principal of Jesus College, Oxford, 1845-6, joined Roman Church, 1846, and translated theological works from Latin and Spanish.

LEWIS, SIR GEORGE HENRY, first baronet (1833-1911), solicitor; joined father's firm of solicitors, 1851; established reputation in connection with Balham mystery, 1876 [see GULLY, JAMES MANBY]; obtained monopoly of 'society' cases; unrivalled in knowledge of criminals and in thoroughness of investigation; acted for incriminated nationalists before the Parnell commission, 1888-9; intimate of King Edward VII; CVO, 1905; knighted, 1892; baronet, 1902; advocate of Criminal Evidence Act of 1898, of Court of Criminal Appeal, 1907, and of Moneylenders Act, 1900.

LEWIS, JOHN SPEDAN (1885-1963), shopkeeper and industrial reformer; educated at Westminster School; at nineteen, joined father's business in Oxford Street shop, London; took charge of Peter Jones Ltd., acquired by his father, and began experiment of distributing preference shares to employees, 1916-20; became sole partner in John Lewis business, 1928; organized business on partnership lines for benefit of employees; formed public company, John Lewis Partnership, 1929; completed settlement of business held by trustees on behalf of employees, 1950; retired as chairman, 1955; president, Classical Association, 1956-7; published *Partnership for All* (1948) and *Fairer Shares* (1954).

LEWIS, JOHN TRAVERS (1825-1901), archbishop of Ontario; BA, Trinity College, Dublin, 1848; ordained, 1848; settled in Canada, 1849; first bishop of Ontario, 1861; metropolitan of

Canada, 1893; and archbishop of Ontario, 1894; his advocacy (1864) of national council for whole Anglican Church led to first Lambeth conference, 1867; hon. DD, Oxford, 1897; hon. LLD, Dublin.

LEWIS, PERCY WYNDHAM (1882–1957), writer and artist; son of American father and British mother; educated at Rugby and Slade School; lived mainly in Brittany and Paris, 1901–9; moving to London exhibited with Camden Town and London groups; director, Rebel Art Centre, 1914; published Vorticist review *Blast* (June 1914, July 1915); organized Vorticist exhibition, 1915; served in France as gunner and war artist, 1917–18; edited two issues of *Tyro*, 1921–2, and three of *Enemy*, 1927–9; his enthusiasm for Hitler, subsequently recanted, aroused hostility; his portrait of T. S. Eliot rejected by Royal Academy, 1938; in United States and Canada, 1939–45; art critic to *Listener*, 1946–51; awarded Civil List pension, 1951; hon. Litt.D., Leeds, 1952; *Childermass* presented by BBC, 1951; Tate Gallery exhibition 'Wyndham Lewis and Vorticism', 1956; became totally blind, 1954; publications include autobiographical and critical works and among his novels are *Tarr* (1918) and *Self Condemned* (1954); moved from abstract to representational art, notably portraits; a towering, undisciplined, and quarrelsome egotist, his greatest enemy was himself.

LEWIS, RICHARD (1821–1905), bishop of Llandaff; BA, Worcester College, Oxford, 1843; DD, 1883; travelled in Europe and Near East, 1843–4; ordained, 1844; rector of Lampeter Velfry, 1851–83; archdeacon of St. David's, 1875–83; bishop of Llandaff, 1883–1905; inaugurated Bishop of Llandaff's church extension fund, for erection of new churches and support of additional curates, thus greatly extending work of diocese; established annual diocesan conference, 1884; a broad and tolerant churchman, but uncompromising on question of church schools; took seat in House of Lords, 1885; unsympathetic with modern Welsh nationalism.

LEWIS, ROSA (1867–1952), hotel owner; went into service at twelve; in kitchens of Comte de Paris and Duc d'Orleans learned French cooking and the tastes of King Edward VII; cooked privately for fashionable hostesses; acquired Cavendish Hotel, 1902, and ran it with elegant distinction on Robin Hood tactics until her death.

LEWIS, SIR THOMAS (1881–1945), physician; B.Sc., University College, Cardiff, 1902; MB, BS, University College Hospital, London,

1905; lecturer in cardiac pathology, 1911; assistant physician, 1913; full physician, 1919–45; founded journal *Heart*, 1909; changed scope and title to *Clinical Science*, 1933; remained editor until 1944; changed name of department to that of clinical research and founded Medical Research Society, 1930; transferred control of journal to Society, 1939; investigations included the pulse and respiration; irregularities of heart's action especially auricular flutter and fibrillation; origin and course of excitation wave; 'soldier's heart' or 'effort syndrome' (for Medical Research Committee); vascular reactions of skin; and pain; FRS, 1918; knighted, 1921.

LEWIS, SIR WILFRID HUBERT POYER (1881–1950), judge; educated at Eton and University College, Oxford; third class, history, 1903; called to bar (Inner Temple), 1908; bencher, 1929; practised in Cardiff until 1914; served with Glamorgan Yeomanry, 1914–18; joined T. W. H. Inskip (later Viscount Caldecote, q.v.) in London; built up large and varied practice; junior (common law) counsel to Treasury, 1930; judge of King's Bench division, 1935–50; inherited and devoted leisure to a Pembrokeshire estate; gave much time to affairs of Welsh Church; knighted, 1935.

LEWIS, WILLIAM CUDMORE McCULLAGH (1885–1956), physical chemist; educated at Bangor grammar school, co. Down and Royal University of Ireland; first class, experimental science, 1905; researched at Liverpool and Heidelberg; demonstrator, later lecturer, University College, London, 1909–13; professor of physical chemistry, Liverpool, 1913–48; published *A System of Physical Chemistry* (2 vols., 1916); studied theory of chemical change, colloid science, and physico-chemical processes possibly underlying malignancy; FRS, 1926.

LEWIS, WILLIAM THOMAS, first BARON MERTHYR (1837–1914), engineer and coalowner; controller of Marquess of Bute's Welsh estates, 1881; main colliery interests latterly in Rhondda valley and Senghenydd districts; served on royal commissions dealing with coal industry and labour problems; promoted industrial peace; knighted, 1885; baronet, 1896; baron, 1911.

LEWIS, SIR WILLMOTT HARSANT (1877–1950), journalist; educated at Eastbourne, Heidelberg, and the Sorbonne; *New York Herald* correspondent during Boxer rising and Russo-Japanese war; edited *Manila Times*, 1911–17; worked in American information service in France, 1917–19; represented *New York Tribune* at peace conference, 1919; London *Times* correspondent, Washington, 1920–48;

great social figure; politically well informed, especially until advent of Roosevelt administration; KBE, 1931.

LEY, HENRY GEORGE (1887-1962), organist, pianist, and composer; joined choir of St. George's, Windsor, 1896; music scholar, Uppingham, 1903; exhibitioner, Royal College of Music, 1904; organ scholar, Keble College, Oxford, 1906; organist, Christ Church, Oxford, 1909; MA, 1913; D.Mus., 1919; university choragus, 1923-6; precentor, St. Peter's College, Radley, 1916-18; organ teacher, Royal College of Music, 1919-41; precentor, Eton College, 1926-45; president, Royal College of Organists, 1933-4; president, Incorporated Association of Organists, 1952-3; FRCM, 1928; hon. FRCO, 1920; and hon. RAM, 1942; hon. fellow, Keble College, Oxford.

LEYEL, HILDA WINIFRED IVY (Mrs C. F. LEYEL) (1880-1957), herbalist; born Wauton; married C. F. Leyel, 1900; studied Nicholas Culpeper [q.v.] and other herbalists; published *The Magic of Herbs* (1926) and many other similar works; opened Culpeper House (Baker Street, 1927) and others elsewhere; founded Society of Herbalists.

LIAQAT ALI KHAN (1895-1951), first prime minister of Pakistan; born in East Punjab; educated at Muhammad Anglo-Oriental College and Exeter College, Oxford; honours in jurisprudence, 1921; called to bar (Inner Temple), 1922; member, United Provinces legislative council, 1926; deputy president, 1931; general-secretary, All-India Moslem League and member, parliamentary board, 1936; became loyal associate of M. A. Jinnah [q.v.]; member, central Legislative Assembly, and deputy leader, Moslem League Party, 1940; finance minister, 1946; prime minister, Pakistan, 1947-51; assassinated.

LIBERTY, Sir ARTHUR LASENBY (1843-1917), fabric manufacturer; opened business in Regent Street, London (afterwards Liberty & Co.), 1875; dealt in oriental fabrics and produced British machine-made stuffs which equalled hand-made products of Asia; friend of pre-Raphaelite painters; knighted, 1913.

LIDDELL HART, Sir BASIL HENRY (1895-1970), military historian and strategist. [See HART.]

LIDDERDALE, WILLIAM (1832-1902), governor of the Bank of England; born at St. Petersburg; son of a Russian merchant; director (1870), deputy governor (1887), and governor (1889) of Bank of England; concerned in reduc-

tion of interest on national debt by G. J. (later Viscount) Goschen [q.v.], 1888; by his firm action liquidated Messrs. Baring's affairs and increased the City's confidence in the Bank, 1890; PC, 1891; concluded negotiations with government which took shape in Bank Act of 1892.

LIDGETT, JOHN SCOTT (1854-1953), theologian and educationist; educated at Blackheath Proprietary School and University College, London; MA, logic and philosophy, 1875; entered Wesleyan Methodist ministry, 1876; founder, 1891, and warden, 1891-1949, Bermondsey Settlement providing social and educational amenities; member, LCC education committee, 1905-28; leader of progressive party, 1918-28; member, London University senate, 1922-46; vice-chancellor, 1930-2; president, National Council of Evangelical Free Churches, 1906; of Wesleyan Methodist Conference, 1908; of united Church, 1932; moderator, Free Church General Council, 1923-5; founder member, British Council of Churches, 1942; sought unity of English Christendom; editor, *Methodist Times*, 1907-18; joint editor, *Contemporary Review*, 1911-53; publications include *The Spiritual Principle of the Atonement* (1897), *The Fatherhood of God* (1902), *The Christian Religion, its Meaning and Proof* (1907), *God in Christ Jesus* (1915); CH, 1933; hon. degrees from Aberdeen, Oxford, and London.

LIGHTWOOD, JOHN MASON (1852-1947), conveyancing counsel and legal writer; educated at Kingswood School, Bath, and Trinity Hall, Cambridge; bracketed eighth wrangler and fellow, 1874; called to bar (Lincoln's Inn), 1879; practised as conveyancer and draftsman; conveyancing counsel to the Court, 1932-47; wrote copiously on property law for legal journals; legal editor, *Law Journal*, 1925-39; publications include *Possession of Land* (1894).

LILLICRAP, Sir CHARLES SWIFT (1887-1966), naval constructor; educated at Stoke School, Devonport, Royal Naval Engineering College, Keyham, and Royal Naval College, Greenwich; first class professional certificate, 1910; joined Royal Corps of Naval Constructors, 1910; appointed to Naval Construction Department, Admiralty, 1914; constructor, responsible for cruiser design, 1917; lecturer in naval architecture, Royal Naval College, 1921; made special survey of welding, 1930; assistant director, naval construction, 1936; director, 1944-51; MBE, 1918; CB, 1944; KCB, 1947; officer of the Légion d'Honneur; president, Institute of Welding, 1956-8; vice-president, Royal Institution of Naval Architects,

1945; fellow, Imperial College of Science and Technology, 1964; hon. D.Sc. (Eng.), Bristol, 1951; president, Johnson Society, 1955-6.

LINCOLNSHIRE, MARQUESS OF (1843-1928), politician. [See WYNN-CARRINGTON, CHARLES ROBERT.]

LINDEMANN, FREDERICK ALEXANDER, VISCOUNT CHERWELL (1886-1957), scientist and politician; born in Baden Baden of naturalized British father of French Alsatian Catholic origin and American mother; educated at Blair Lodge, Polmont, and Darmstadt Hochschule; Ph.D., Physikalisch-Chemisches Institut, Berlin, 1910; worked on low temperature physics; contributed to theory of solids; devised a fibre electrometer; also became excellent pianist and tennis player; at Royal Aircraft Factory, Farnborough, 1915-19; learned to fly and solved problem of 'spin'; Dr Lee's professor of experimental philosophy, Oxford, and fellow of Wadham, 1919-56; obtained new Clarendon Laboratory and made it a leading physics department; elected student of Christ Church, 1921, where he lived from 1922; became close friend of Churchill who obtained his membership of Tizard committee on air defence, 1935-6; his unorthodox methods of obtaining his objectives caused committee's break up; personal assistant to Churchill and head of his statistics section, 1940-5; provided useful checks on departmental statistics; advised Churchill on many subjects, producing some 2,000 minutes for him; introduced 'bending' of wireless beams used by German night bombers; encouraged research into microwave radar; overestimated effectiveness of massive area bombing; sceptical of existence of German rocket bombs; nevertheless made immense contribution to war effort combining scientific expertise with clarity and brilliance of mind; paymaster-general, 1942-5 and 1951-3; a vegetarian, teetotaller, and non-smoker his background was wealthy, his contacts aristocratic, and his views extreme right wing; aggressive in the cause of science and unable to suffer fools he aroused friction but could be an amusing if cynical controversialist; FRS, 1920; baron, 1941; PC, 1943; CH, 1953; viscount, 1956.

LINDLEY, SIR FRANCIS OSWALD (1872-1950), diplomatist; son of Lord Lindley [q.v.]; educated at Winchester and Magdalen College, Oxford; third class, jurisprudence, 1893; entered diplomatic service, 1896; counsellor of embassy, Petrograd, 1915-18; consul-general for Russia, 1918-19; high commissioner, Vienna, 1919-20, minister, 1920-1; Athens, 1921-2; Oslo, 1923-9; ambassador, Lisbon,

1929-31, Tokyo, 1931-4; KCMG, 1926; PC, 1929; GCMG, 1931.

LINDLEY, NATHANIEL, BARON LINDLEY (1828-1921), lord of appeal; son of John Lindley, FRS [q.v.]; educated at University College School, London; called to bar (Middle Temple), 1850; published *A Treatise on the Law of Partnership* . . ., publicly noticed by judges, 1860; career assured by success as junior for Overend, Gurney & Co., in City financial crisis, 1866; reputation enhanced by case of *Knox* v. *Gye* (1871) and action concerning 'Frou-frou'; QC, 1872; judge of common pleas and knighted, 1875; lord justice of appeal, 1881; master of the rolls, 1897; FRS, 1897; lord of appeal in ordinary, 1900-5; life peer, 1900; remarkable for impartiality and versatility.

LINDRUM, WALTER ALBERT (1898-1960), billiards player; born in Western Australia; son and grandson of billiards champions; in a career lasting from fifteen to fifty broke all billiards records including one break of 4,137 (1932), six of 3,000-odd, and twenty-nine of 2,000-odd; twice won world championship (1933, 1934); made four tours of Britain, 1929-33; active in charitable work; OBE, 1958.

LINDSAY, ALEXANDER DUNLOP, first BARON LINDSAY OF BIRKER (1879-1952), educationist; son of T. M. Lindsay [q.v.]; educated at university of Glasgow, 1899; first class, *lit. hum.*, University College, Oxford, 1902; president of Union, 1902; Clark philosophy fellow, Glasgow, 1902-4; Shaw fellow, Edinburgh, 1904-9; fellow and classical tutor, Balliol College, Oxford, 1906-22; deputy controller of labour in France and lieutenant-colonel, 1917-19; CBE; professor, moral philosophy, Glasgow, 1922-4; master of Balliol, 1924-49; welcomed opening of Oxford to wider social classes; his democratic theories were outcome of his Christian beliefs and his moral fervour made him a national figure; adviser on education to labour party and Trades Union Congress; chairman, committee on work of Protestant colleges in India, 1930; unsuccessfully contested Oxford City on anti-Munich platform, 1938; vice-chancellor, Oxford University, 1935-8; sponsored appeal for funds; piloted schemes for expansion of science departments including new Clarendon Laboratory and absorption of Nuffield benefactions including Nuffield College; first principal, University College of North Staffordshire which he had worked to create 1949-52; publications include *The Essentials of Democracy* (1929) and *Religion, Science and Society in the Modern World* (1943); baron, 1945; hon. fellow, Balliol College; hon. LLD, Glasgow, St. Andrews, and Princeton.

LINDSAY, DAVID (1856-1922), explorer; born in South Australia; entered South Australia survey department, 1873; led expedition for scientific exploration of interior of Western Australia, financed by Sir Thomas Elder, 1891; results fell short of expectations, but auriferous area revealed.

LINDSAY, DAVID ALEXANDER EDWARD, twenty-seventh EARL OF CRAWFORD and tenth EARL OF BALCARRES (1871-1940), politician and art connoisseur; educated at Eton and Oxford; succeeded father, 1913; conservative MP, Chorley division, 1895-1913; party whip, 1903-13; lord privy seal, 1916-19; first commissioner of works, 1921-2; trustee, British Museum, National Gallery, etc.; chairman, Royal Fine Art Commission; chancellor of Manchester University from 1923; PC, 1916; KT, 1921; FRS, 1924.

LINDSAY, GEORGE MACKINTOSH (1880-1956), major-general; educated at Sandroyd and Radley; commissioned in Rifle Brigade, 1900; served in South African war, 1900-2; instructor, Musketry School, Hythe, 1913-15; in France, 1916-18, and moving spirit in creation of Machine-Gun Corps after 1915; chief inspector, Royal Tank Corps Centre, 1923-5; inspector, Royal Tank Corps, 1925-9; foremost advocate of mobile armoured warfare; brigadier general staff, Egypt, 1929-32; commanded 7th Infantry brigade (motorized), 1932-4; major-general, 1934; commander, Presidency and Assam District, India, 1935-9; colonel-commandant, Royal Tank Regiment, 1938-47; CMG, 1919; CB, 1936; CBE, 1946.

LINDSAY, JAMES GAVIN (1835-1903), colonel RE; joined Madras engineers, 1854; served in Indian Mutiny; as engineer-in-chief constructed northern Bengal railway, 1872; showed capacity in dealing with Bengal famine, 1873-4; colonel, 1882; built Sukkur-Sibi railway in second Afghan war, 1879-80; took part in relief of Kandahar; finished Southern Mahratta railway, 1891.

LINDSAY, JAMES LUDOVIC, twenty-sixth EARL OF CRAWFORD and ninth EARL OF BALCARRES (1847-1913), astronomer, collector, and bibliophile; son of twenty-fifth earl [q.v.]; erected Dunecht observatory, near Aberdeen, 1872; succeeded father, 1880; collector of French Revolution documents, etc.; published *Bibliotheca Lindesiana* (1883-1913); FRS, 1878; KT, 1896.

LINDSAY, (JOHN) SEYMOUR (1882-1966), designer and metalworker; educated at home; apprenticed to Leonard Ashford, designer and draughtsman, London, 1889; designer of electrical fittings for Higgins and Griffiths; served in London Rifle Brigade, DCM, 1914-16; returned to Higgins and Griffiths, 1919; started own ironwork business and worked for Sir Hubert Baker, Sir Edwin Lutyens, and (Sir) Albert Richardson [qq.v.]; work included ironwork in Bank of England, 1921-37, light fittings and ironwork, Government Buildings, New Delhi, and iron staircase and weathervane, Trinity House, Tower Hill; designed silver altar rails and plate for altar, Battle of Britain memorial chapel, Westminster Abbey, 1947; published articles in the *Architect*, *Iron and Brass Implements of the English House* (1927), and *An Anatomy of English Wrought Iron* (1964); fellow, Society of Antiquaries, 1942; fellow, Royal Society of Arts, 1949.

LINDSAY, (afterwards LOYD-LINDSAY), **ROBERT JAMES**, BARON WANTAGE (1832-1901), soldier and politician; joined Scots Guards, 1850; retired as lieutenant-colonel, 1859; served with distinction in Crimea; received VC, 1857; assumed name of Loyd-Lindsay on marriage, 1858; a pioneer of Volunteer movement, 1859; conservative MP for Berkshire, 1865-85; financial secretary to War Office, 1877-80; represented Red Cross Aid Society, which he helped to found (1870), in Franco-Prussian and Turko-Serbian wars; KCB, 1881; baron, 1885; a prominent freemason; leading agriculturist in Berkshire; discriminating art patron; helped to found Reading University College (afterwards Reading University).

LINDSAY, SIR RONALD CHARLES (1877-1945), diplomatist; son of twenty-sixth and brother of twenty-seventh Earl of Crawford [qq.v.]; educated at Winchester; entered diplomatic service, 1899; under-secretary, Egyptian ministry of finance, 1913-19; assistant under-secretary, Foreign Office, in charge Near Eastern affairs, 1921-4; 'representative', Constantinople, 1924; ambassador, 1925-6; negotiated treaty of Angora, 1926; ambassador, Berlin, 1926-8; permanent under-secretary, Foreign Office, 1928-30; ambassador, Washington, 1930-9; KCMG and PC, 1925; GCB, 1939.

LINDSAY, THOMAS MARTIN (1843-1914), historian; educated at Glasgow and Edinburgh universities; entered ministry of Free Church of Scotland, 1869; professor of church history at Free Church theological college, Glasgow, 1872; principal, 1902; interested in missions and social problems; chief works, *Luther and the German Reformation* (1900) and *A History of the Reformation in Europe* (1906-7).

LINDSAY, WALLACE MARTIN (1858-1937), classical scholar; brother of T. M. Lindsay [q.v.]; educated at Glasgow University and Balliol College, Oxford; first class, *lit. hum.*, 1881; fellow of Jesus; tutor, 1884-99; professor of humanity, St. Andrews, 1899-1937; established reputation with *The Latin Language* (1894); other works include editions of Martial (1903), Plautus (1904-5), and Terence (1926), and *Early Latin Verse* (1922) and *Glossaria Latina* (5 vols., 1926-32); FBA, 1905.

LINGEN, RALPH ROBERT WHEELER, Baron Lingen (1819-1905), civil servant; educated at Trinity College, Oxford; friend of Jowett, Froude, and Frederick Temple [qq.v.]; Ireland and Hertford scholar (1838-9); BA, 1840; won Latin essay (1843) and Eldon scholarship (1846); fellow of Balliol, 1841; hon. DCL, 1881; honorary fellow of Trinity, 1886; called to bar, 1847; secretary to Education Office, 1849-69; worked harmoniously under Lord Granville and Robert Lowe (Lord Sherbrooke) [qq.v.]; issued code advocating payment by results in accordance with report of Newcastle commission of inquiry, 1861; code severely criticized by Lord Robert Cecil (Lord Salisbury) and W. E. Forster [qq.v.]; CB, 1869; permanent secretary of the Treasury, 1869-85; vigilant guardian of public purse; KCB, 1878; baron, 1885; alderman of first London County Council, 1889-92.

LINLITHGOW, first Marquess of (1860-1908), first governor-general of the Commonwealth of Australia. [See Hope, John Adrian Louis.]

LINLITHGOW, second Marquess of (1887-1952), viceroy of India. [See Hope, Victor Alexander John.]

LINSTEAD, Sir (REGINALD) PATRICK (1902-1966), experimental organic chemist and university administrator; educated at City of London School and Imperial College; first class, chemistry, 1923; Ph.D., 1926; demonstrator, and later, lecturer, Imperial College, 1929; Firth professor of chemistry, Sheffield University, 1938; professor, organic chemistry, Harvard, 1939-45; deputy director of scientific research, Ministry of Supply, 1942; director, government Chemical Research Laboratory, Teddington, 1945-9; professor, organic chemistry, Imperial College, 1949; dean, Royal College of Science, 1953; rector, 1955; D.Sc. London, 1930; hon. MA Harvard, 1942; hon. D.Sc. Exeter, 1965; CBE, 1946; knighted, 1959; FRS, 1940; vice-president, Royal Society, 1959-65; trustee, National Gallery, 1962; member, Central Advisory Council for Education,

1956-60, and Committee on Higher Education, 1961-4; revised Cain and Thorpe's *The Synthetic Dyestuffs* (with Thorpe, 1933), and published *A Course in Modern Techniques of Organic Chemistry* (with J. A. Elridge and M. Whalley, 1955) and *A Guide to Qualitative Organic Chemical Analysis* (with B. C. L. Weedon, 1956).

LIPSON, EPHRAIM (1888-1960), economic historian; grievously deformed by childhood accident; scholar of Sheffield royal grammar school and Trinity College, Cambridge; first class, parts i and ii, historical tripos, 1909-10; migrated as private tutor to Oxford; reader in economic history and fellow of New College, 1922-31; left Oxford embittered by non-election as first Chichele professor of economic history; published vol. i, *Economic History of England* (The Middle Ages, 1915); vols. ii and iii (The Age of Mercantilism, 1931).

LIPTON, Sir THOMAS JOHNSTONE, baronet (1850-1931), grocer and yachtsman; born in Glasgow tenement; worked for five years in United States; opened in Glasgow (1871) first of series of grocery shops which by publicity backed by sound stock at fair prices made him a millionaire by 1880; lavish host and generous contributor to charity; spent a fortune on yacht-racing; made unsuccessful attempts to win the America's cup with successive *Shamrocks* (1899-1930); knighted, 1898; KCVO, 1901; baronet, 1902.

LISTER, ARTHUR (1830-1908), botanist; son of Joseph Jackson Lister [q.v.]; published *A Monograph of the Mycetozoa* (1894) and *Guide to the British Mycetozoa* (1895); FRS, 1898.

LISTER, JOSEPH, first Baron Lister (1827-1912), founder of antiseptic surgery; son of Joseph Jackson Lister, FRS [q.v.]; educated at Grove House School, Tottenham, and University College, London; especially influenced during his medical studies by Wharton Jones, professor of ophthalmic medicine and surgery, and William Sharpey [q.v.], professor of physiology; after taking MB (1852) carried out researches on physiological problems; went to Edinburgh to study method of James Syme [q.v.], celebrated surgeon, 1853; settled in Edinburgh; assistant surgeon to Royal Infirmary, 1856; professor of surgery, Glasgow University, and FRS, 1860; surgeon to Glasgow Infirmary, 1861; professor of clinical surgery, Edinburgh, 1869-77; at King's College, London, 1877-92; baronet, 1883; president of Royal Society, 1894-1900; baron, 1897; OM, 1902; devoted himself to prevention of mortality from injuries and wounds by studying inflammation and suppuration; influenced by researches of Louis Pasteur; employed carbolic acid to destroy germs and prevent septic infection; successfully applied treatment to cases of compound fracture; invented new operations and improved technique of old ones by introducing use of absorbable ligatures and drainage tubes; reduction of septic diseases caused practice of surgery to undergo complete revolution and enormously enlarged its field.

LISTER, SAMUEL CUNLIFFE, first BARON MASHAM (1815-1906), inventor; brother carried on worsted mill at Manningham till 1889; took out over 150 patents for inventions; evolved Lister-Cartwright (1845), the 'square motion' (1846), and 'square nip' (1850) woolcombing machines, which cheapened cloth, advanced Bradford's prosperity, and created Australian wool trade; successfully converted silk waste into silk velvets, poplins, and the like; invented (1848) a compressed air brake for railways; purchased Ackton colliery in Yorkshire, where the works were destroyed in coal strike of 1893; early advocated tariff reform; presented Lister Park to Bradford city; baron, 1891; hon. LLD, Leeds University; ardent art collector and sportsman; published account of his inventions, 1905.

LITHGOW, SIR JAMES, first baronet, of Ormsary (1883-1952), shipbuilder and industrialist; educated at Glasgow Academy and in Paris; apprenticed in family shipyard; partner, 1906; with younger brother assumed management on death of father, 1908; director, merchant shipbuilding, Admiralty, 1917-18; rationalized post-war Scottish shipbuilding; appointed chairman, 1930, National Shipbuilders' Security Ltd. to buy out uneconomic shipyards; chairman, Scottish National Development Council, 1931; rescued number of Scottish industries including Beardmores and Fairfields; president, Federation of British Industries, 1930-2; controller, merchant shipbuilding and member, Board of Admiralty, 1940-6; president, Iron and Steel Federation, 1943-5; baronet, 1925; GBE, 1945; CB, 1947.

LITTLE, ANDREW GEORGE (1863-1945), historian; educated at Clifton and Balliol College, Oxford; first class, modern history, 1886; at Dresden and Göttingen; lecturer in history, University College of South Wales, Cardiff, 1892; professor, 1898-1901; visiting lecturer (reader from 1920) in palaeography, Manchester, 1903-28; founded (1902) a society which became (1907-37) the British Society of Franciscan Studies; works include *The Grey Friars in Oxford* (1892), *Studies in English Franciscan History* (1917), and (with F. Pelster,

SJ) *Oxford Theology and Theologians c. A.D. 1282-1302* (1934); FBA, 1922.

LITTLE, SIR ERNEST GORDON GRAHAM GRAHAM- (1867-1950), physician, and member of parliament for London University. [See GRAHAM-LITTLE.]

LITTLE, WILLIAM JOHN KNOX- (1839-1918), divine and preacher. [See KNOX-LITTLE.]

LITTLER, SIR RALPH DANIEL MAKINSON (1835-1908), barrister; BA, London, 1854; called to bar, 1857; QC, 1873; treasurer of Middle Temple, 1900-1; CB, 1890; knighted, 1902; chairman of Middlesex sessions and county council from 1889 to death; was often criticized for long sentences on habitual criminals.

LITTLEWOOD, SIR SYDNEY CHARLES THOMAS (1895-1967), solicitor and principal architect of England's legal aid system in civil proceedings; worked as office boy in solicitor's office and as a police constable; served in army and RFC during 1914-18 war; prisoner of war; admitted as solicitor, 1922; partnership with Wilkinson Howlett & Co., 1928; member, Council of Law Society, 1940; member, Committee on Legal Aid, 1944; chairman, Legal Aid Committee; knighted, 1951; president, Law Society, 1959; president, London Rent Assessment Panel, 1965-7.

LIVEING, GEORGE DOWNING (1827-1924), chemist; BA, St. John's College, Cambridge; eleventh wrangler, 1850; senior in new natural sciences tripos, 1851; fellow of St. John's, 1853-60 and 1880-1924; professor of chemistry, Staff College and Royal Military College, Sandhurst, 1860; professor of chemistry, Cambridge, 1861-1908; carried out in collaboration with (Sir) James Dewar [q.v.] spectroscopic investigations, 1878-1900; subjects investigated include the reversal of the lines of metallic vapours, the spectrum of carbon, ultra-violet spectra, and sunspots; seventy-eight joint papers republished in a single volume, 1915; superintended erection of new university chemical laboratory from 1888; FRS, 1879; published *Chemical Equilibrium the Result of the Dissipation of Energy* (1885).

LIVENS, WILLIAM HOWARD (1889-1964), soldier and inventor; educated at Oundle and Christ's College, Cambridge; commissioned in Royal Engineers, 1914; joined the Special Brigade, formed to retaliate for German gas attacks; designed Livens gas flame projector and phosgene bomb; MC, DSO; worked in Ministry of Supply, 1939-45.

LIVESEY, Sir GEORGE THOMAS (1834–1908), promoter of labour co-partnership; joined South Metropolitan Gas Company, 1848; assistant manager, 1857; engineer and secretary, 1871; chairman of board of directors, 1885; adopted principle of sliding scale, 1876; admitted foremen (1886) and workmen (1889) to share in profits; capitalized bonus of workmen, who became shareholders (1894) and were admitted to board of directors (1898); sat on labour commission, 1891–4; knighted, 1902; erected Livesey library, Camberwell; Livesey professorship of coal gas and fuel industries was founded at Leeds University.

LIVINGSTONE, Sir RICHARD WINN (1880–1960), educationist; scholar of Winchester and New College, Oxford; first class, hon. mods. and *lit. hum.*, 1901–3; fellow, tutor, and librarian, Corpus Christi College until 1924; vice-chancellor, Queen's University, Belfast, 1924–33; president of Corpus, 1933–50; vice-chancellor, Oxford University, 1944–7; originator and general editor, Clarendon Series of Greek and Latin authors; helped to found Denman College; publications include *A Defence of Classical Education* (1916); knighted, 1931; hon. degrees from ten universities.

LLANDAFF, Viscount (1826–1913), lawyer and politician. [See Matthews, Henry.]

LLEWELLIN, JOHN JESTYN, Baron LLEWELLIN (1893–1957), politician and first governor-general of the Federation of the Rhodesias and Nyasaland; educated at Eton and University College, Oxford; called to bar (Inner Temple), 1921; conservative MP, Uxbridge, 1929–45; civil lord of Admiralty, 1937–9; parliamentary secretary, Ministry of Supply (1939–40), Aircraft Production (1940–1), Transport (1941–2); minister, Aircraft Production, May–Nov. 1942; minister resident for supply, Washington, 1942–3; minister of food, 1943–5; first governor-general, Federation of the Rhodesias and Nyasaland, 1953–7; PC, 1941; baron, 1945; CBE, 1939; GBE, 1953.

LLEWELLYN, Sir (SAMUEL HENRY) WILLIAM (1858–1941), artist and president of Royal Academy of Arts; studied at South Kensington and in Paris; began exhibiting at Royal Academy, 1884; ARA, 1912; RA, 1920; president, 1928–38; negotiated and supervised arrangements for exhibitions of Dutch, Italian, Persian, French, Chinese, and Scottish art; also exhibition of industrial art, 1935; his portraits sincere and graceful rather than character-searching presentations; sitters included Queen Mary, seventeenth Earl of Derby, and Archbishop Lang; KCVO, 1918; GCVO, 1931.

LLOYD, DOROTHY JORDAN (1889–1946), biochemist; first class, parts i and ii, natural sciences tripos, 1910–12, Newnham College, Cambridge; fellow, 1914–21; joined British Leather Manufacturers' Research Association, 1920; director, 1927–46; planned and contributed to *Progress in Leather Science, 1920–45* (3 vols., 1946–8).

LLOYD, EDWARD MAYOW HASTINGS (1889–1968), British and international civil servant, and world food expert; educated at Rugby and Corpus Christi College, Oxford; entered home Civil Service and posted to Department of Inland Revenue, 1913; transferred to War Office, 1914–17; Ministry of Food, 1917–19; seconded to League of Nations Secretariat, 1919–21; Ministry of Agriculture, 1921–6; assistant secretary, Empire Marketing Board, 1926; secretary, Market Supply Committee, 1933; assistant director, Food (Defence Plans) Department, 1936; principal assistant secretary, General Department, Ministry of Food, 1939; unhappy relationship with permanent secretary; transferred as economic advisor to Minister of State, Middle East, 1942; worked with Food and Agriculture Organization, 1946–7; under-secretary, Ministry of Food, 1947–53; president, Agricultural Economics Society, 1956; consultant to Political and Economic Planning, 1958–64; publications include *Stabilisation* (1923), *Experiments in State Control* (1924), and *Food and Inflation in the Middle East, 1940–1945* (1956); CMG, 1945; CB, 1952.

LLOYD, GEORGE AMBROSE, first Baron LLOYD (1879–1941), statesman; educated at Eton and Trinity College, Cambridge; coxed winning Cambridge boat, 1899, 1900; conservative MP, West Staffordshire, 1910–18; Eastbourne, 1924–5; attached to Arab Bureau, 1916–17; DSO, 1917; governor of Bombay, 1918–23; initiated Bombay development scheme and Lloyd barrage across Indus; GCIE, 1918; GCSI and PC, 1924; high commissioner in Egypt and baron, 1925; his views increasingly divergent from government; virtually compelled to resign by Arthur Henderson [q.v.], 1929; published *Egypt since Cromer* (2 vols., 1933–4); chairman, British Council, 1937–40; secretary of state for colonies, 1940–1; flew to Bordeaux in unsuccessful attempt to persuade French government to continue fighting, 1940.

LLOYD, Sir JOHN EDWARD (1861–1947), historian; educated at University College of Wales, Aberystwyth and Lincoln College, Oxford; first class, modern history, 1885; lecturer, Aberystwyth, 1885–92; lecturer in Welsh history (1892–9), Bangor; registrar (1892–1919); professor of history, 1899–1930; first chairman, board of Celtic studies, 1919-40; publications include *History of Wales to the Edwardian Conquest* (2 vols., 1911); FBA, 1930; knighted, 1934.

LLOYD, MARIE (pseudonym) (1870-1922), music-hall comedian. [See WOOD, MATILDA ALICE VICTORIA.]

LLOYD, SIR THOMAS INGRAM KYNAS-TON (1896-1968), civil servant; educated at Rossall School and Gonville and Caius College, Cambridge; served in Royal Engineers in 1914-18 war; entered home civil service and posted to Ministry of Health, 1920; transferred to Colonial Office, 1921; private secretary to permanent under-secretary of state, 1929; secretary, Palestine Commission, 1929; Colonial Service Department, 1930-8; secretary, royal commission on West Indies, 1938; assistant secretary, 1939; head of Defence Department, 1942; assistant under-secretary, West African and Eastern Departments, 1943; permanent under-secretary of state, 1947-56; CMG, 1943; KCMG, 1947; KCB, 1949; GCMG, 1951; director, Harrisons and Crosfield Ltd.; governor and member, Rossall School Council, 1956-68.

LLOYD GEORGE, DAVID, first EARL LLOYD-GEORGE OF DWYFOR (1863-1945), statesman; brought up at Llanystumdwy, Caernarvonshire, by widowed mother and her shoemaker brother; attendance at church school developed his hostility to English privilege; qualified as solicitor with honours, 1884; established reputation as fearless advocate and made mark as speaker on religious, temperance, and political subjects; elected liberal MP for Caernarvon Boroughs, 1890; retained seat until 1945; became leading Welsh political figure; his opposition to South African war made him notorious; escaped mob in Birmingham disguised as policeman, 1901; opposed rate-aid to voluntary schools, started revolt schools in Wales, and added to his reputation as champion of Welsh causes; became less single-minded after taking office; obtained a Welsh department in the Board of Education but his delays over Welsh Church disestablishment (not effected until 1920) irritated his followers; president of Board of Trade and PC, 1905; a patient negotiator in settling strikes; promoted Merchant Shipping Act, 1906, Port of London Act, 1908; chancellor of Exchequer, 1908; introduced first ('the People's') budget (1909) providing funds for social services and setting up road and development funds; most controversial measure the taxes on unearned increment in land values (abolished, 1920); rejection of budget by House of Lords led to general election, Jan. 1910, after

which it was passed; on appearance of *Panther* at Agadir, 1911, warned Germany that Britain would resist interference with her international interests; established contributory scheme of health and unemployment insurance, 1911; illjudged investment in American Marconi shares when government was concluding contract with British company (1912) led to investigation by select committee which cleared his honour; after German violation of Belgian frontier (1914) handled immediate crisis with courage and skill; introduced first war budget, Nov. 1914; doubled income-tax; increased tea and beer duties; in conduct of war favoured 'side-shows' and was never content with policy of maximum concentration in west; minister of munitions, May 1915-July 1916; secretary of state for war, July 1916; deeply agitated by progress of war; his resignation (5 Dec. 1916) precipitated crisis; succeeded Asquith as prime minister, 7 Dec.; obtained support of conservative, labour, and some liberal members; made immediate impact as most widely known, dynamic, and eloquent figure then in British politics; set up war cabinet of five; summoned dominion ministers to imperial war conference and war cabinet; persisted in urging convoy system on Admiralty; in land campaign fought for pooling of resources, unified conception of front from Flanders to Mesopotamia with attack at weakest points, and unity of command in France; his dislike of offensives in west increased by events of 1917; obtained establishment of Supreme War Council, Nov. 1917; appointment of Foch as commander-in-chief of allied armies in France, Apr. 1918; at home (1917) set up Irish convention, forecast responsible government for India, and enthusiastically supported Balfour declaration on national home for Jews; took extreme measures to reinforce army during German offensive, 1918; persuaded Commons that charges of issuing misleading statements on military matters were unfounded, May 1918; with Bonar Law appealed to country as coalition, 1918; returned with large unionist majority but only 133 liberal supporters; OM, 1919; spent five months in Paris negotiating peace treaty, 1919; sought pacification and economic survival of Europe at this and subsequent conferences culminating in Genoa conference, 1922; negotiated treaty with Ireland, 1921; averted war by firm handling of Chanak crisis, 1922, but gave conservatives opportunity to end coalition; resigned Oct. 1922; returned at head of fiftyfive national liberals; controlled party fund believed acquired by traffic in honours; increased it by journalism and investment in and skilful management of *Daily Chronicle* and other periodicals; used fund to investigate social problems on which he produced series of able reports; liberal party reunited by general election, 1923;

differences with Asquith ended in separation over general strike, 1926; after Asquith's resignation (1926) worked hard as chairman of parliamentary party to restore liberalism; advocated expansionist remedies for unemployment; despite lavish expenditure returned with only fifty-nine followers, 1929; opposed general election of 1931 and returned with family party of four; declined to stand again for party leadership; developed agricultural estate at Churt; settled there and wrote *The Truth About Reparations and War-Debts* (1932), *War Memoirs* (6 vols., 1933-6), and *The Truth About the Peace Treaties* (2 vols., 1938); launched unsuccessful programme for 'new deal' and Council of Action for Peace and Reconstruction, 1935; had become increasingly hostile to France and partial to Germany; visited Hitler who much impressed him, 1936; as he gradually recognized the German menace urged cooperation with Russia and attacked Chamberlain's policy of appeasement; owing to declining health refused invitation to enter (Sir) Winston Churchill's government and the ambassadorship in Washington, 1940; created earl, 1945; buried on bank of river Dwyfor.

LLOYD-GEORGE, GWILYM, first Viscount Tenby (1894-1967), politician; son of David Lloyd George, prime minister; educated at Eastbourne College and Jesus College, Cambridge; served in army during 1914-18 war; national liberal MP, Pembrokeshire, 1922-4; managing director, United Newspapers, 1925-6; MP, Pembrokeshire, 1929-50; parliamentary secretary, Board of Trade, 1939-41; parliamentary secretary, Ministry of Food, 1941-2; Minister of Fuel and Power, 1942-5; national liberal and conservative MP, Newcastle upon Tyne (North), 1951-7; Minister of Food, 1951-4; home secretary, 1954-7; viscount, 1957; president, Fleming Memorial Fund for Medical Research, 1961; hon. fellow, Jesus College, Cambridge, 1953.

LLOYD GEORGE, Lady MEGAN (1902-1966), politician; daughter of David Lloyd George, prime minister; educated at Garratts' Hall, Banstead, and in Paris; liberal MP Anglesey, 1929-51; served on war-time consultative committees for Ministries of Health and Labour; member of Speaker's conference on electoral reform, 1944; deputy leader, liberal parliamentary party, 1949; after 1951 much involved with Welsh affairs; joined labour party, 1955; labour MP, Carmarthen, 1957-66; CH, 1966; superb speaker in Welsh or English; member, BBC advisory council; bard of National Eisteddfod, 1935; first woman member of Welsh Church Commissioners, 1942.

LLOYD JAMES, ARTHUR (1884-1943), phonetician. [See James.]

LOATES, THOMAS (1867-1910), jockey; first rode a winner, 1883; won Derby, 1889; headed list of winning jockeys, 1889, 1890, and 1893; won on Isinglass the Two Thousand Guineas, Derby, and St. Leger, 1893, and on St. Frusquin the Two Thousand Guineas, 1896; a resourceful rider; amassed large fortune.

LOCH, Sir CHARLES STEWART (1849-1923), social worker; born in Bengal; BA, Balliol College, Oxford; influenced by T. H. Green and Arnold Toynbee [qq.v.]; secretary to council of Charity Organisation Society, 1875-1914; his enthusiastic idealism, combined with common sense and efficiency, infected paid staff and volunteers; made influence of Society felt in social legislation and in institution of hospital almoners; member of several royal commissions, 1893-1909; majority report on poor laws largely his work; Tooke professor of economic science and statistics, King's College, London, 1904-8; knighted, 1915.

LOCK, WALTER (1846-1933), warden of Keble College, Oxford, and professor of divinity; educated at Marlborough and Corpus Christi College, Oxford; first class, *lit. hum.*, 1869; priest, 1873; tutor (1870), sub-warden (1881), warden (1897-1920), Keble College; held Dean Ireland's (1895-1919) and Lady Margaret (1919-27) professorships.

LOCKE, WILLIAM JOHN (1863-1930), novelist; born in British Guiana; BA, St. John's College, Cambridge; schoolmaster; secretary, Royal Institute of British Architects, 1897-1907; died in Paris; his novels include *The Morals of Marcus Ordeyne* (1905), *The Beloved Vagabond* (1906), *The Glory of Clementina Wing* (1911), and *The Joyous Adventures of Aristide Pujol* (1912).

LOCKEY, CHARLES (1820-1901), tenor vocalist; sang in Rossini's *Stabat Mater*, 1842, and created tenor part in Mendelssohn's *Elijah* at Birmingham, 1846.

LOCKHART, Sir ROBERT HAMILTON BRUCE (1887-1970), diplomatist and writer; educated at Fettes College and in Berlin and Paris; rubber planter, Malaya; entered consular service; vice-consul, Moscow, 1912; acting consul-general, 1914-17; headed special mission to Russia, 1918; imprisoned and exchanged for Litvinov, 1918; commercial secretary, Prague, 1919; left foreign service, 1922; editor, Londoner's Diary, *Evening Standard*, 1928-37;

rejoined foreign office, 1939; British representative, provisional Czechoslovak Government in exile, 1940; deputy under-secretary of state, in charge of Political Warfare Executive, 1941-5; knighted, 1943; publications include *Memoirs of a British Agent* (1932), *Retreat from Glory* (1934), and *Comes the Reckoning* (1947).

LOCKWOOD, AMELIUS MARK RICHARD, first BARON LAMBOURNE (1847-1928), politician; joined army, 1866; conservative MP, Epping, 1892-1917; baron, 1917; exercised considerable independent influence in House of Commons; vice-president, Royal Society for Prevention of Cruelty to Animals; served on royal commission on vivisection, 1906-8; a noted horticulturist; PC, 1905; lord-lieutenant of Essex, 1919.

LOCKWOOD, SIR JOHN FRANCIS (1903-1965), university administrator; educated at Preston grammar school and Corpus Christi College, Oxford; assistant lecturer, Manchester University, 1927; senior assistant classics lecturer, University College, London, 1927; lecturer in Greek, 1930-40; tutor to arts students, University College, and London University reader in classics, 1940; professor of Latin, 1945; dean of faculty of arts, 1950-1; chairman, governing delegacy, Goldsmiths' College, 1951-8; master, Birkbeck College, 1951-65; public orator, London University, 1952-5; chairman, Collegiate Council, 1953-5; deputy vice-chancellor, 1954-5; vice-chancellor, 1955-8; member, United States-United Kingdom Educational Commission, 1956-61; chairman, Secondary School Examinations Council, 1958-64; chairman, Colonial Office working party on higher education in East Africa, 1958-9; chairman, West African Examinations Council, 1960-4; knighted, 1962; chairman, Voluntary Societies' Committee for Service Overseas, 1962.

LOCKYER, SIR (JOSEPH) NORMAN (1836-1920), astronomer; clerk in War Office, 1857; made pioneer observations of spectrum of sunspot, 1866, and of solar prominences, 1868; coined terms 'chromosphere', and 'helium', 1868; secretary to royal commission on scientific instruction and advancement of science, 1870; transferred to science and art department, South Kensington, 1875; director of Solar Physics Observatory and professor of astronomical physics, Royal College of Science, 1890-1913; FRS, 1869; Rumford medallist, 1874; CB, 1894; KCB, 1897; wrote numerous astronomical books.

LODGE, ELEANOR CONSTANCE (1869-1936), historian, and principal of Westfield College, London; sister of Sir Oliver and Sir Richard Lodge [qq.v.]; studied history at Lady Margaret Hall, Oxford; librarian, 1895; history tutor, 1899-1921; vice-principal, 1906-21; principal of Westfield College, 1921-31; works include *The English Rule in Gascony* (1926); CBE, 1932.

LODGE, SIR OLIVER JOSEPH (1851-1940), scientist and first principal of Birmingham University; brother of Sir Richard and Eleanor Constance Lodge [qq.v.]; studied at Royal College of Science and University College, London; D.Sc., 1877; professor of physics, Liverpool, 1881-1900; made fundamental contributions to wireless and experiments on relative motion of matter and ether; as first principal (1900-19) shaped development of Birmingham University; actively interested in, and especially influential as philosopher of, psychical research; concluded that mind survives death and expounded instrument theory of relation between body and mind; works include *The Ether of Space* (1909), *Advancing Science* (1931), *Raymond* (1916), and *My Philosophy* (1933); FRS, 1887; knighted, 1902; president, British Association, 1913.

LODGE, SIR RICHARD (1855-1936), historian and teacher; brother of Sir Oliver and Eleanor Constance Lodge [qq.v.]; educated at Christ's Hospital and Balliol College, Oxford; first class, modern history, 1877; fellow of Brasenose College, 1878; vice-principal, 1891; professor of modern history, Glasgow (1894-9), Edinburgh (1899-1925); dean of faculty of arts, 1911-25; works include *A History of Modern Europe* (1885) and *Richelieu* (1896); knighted, 1917.

LOFTIE, WILLIAM JOHN (1839-1911), antiquary; ordained, 1865; assistant chaplain, Chapel Royal, Savoy, 1871-95; FSA, 1872; frequent contributor to reviews; travelled much in Egypt; a keen Egyptologist; wrote much on British art and architecture; specially interested in London history and London buildings; chief works were *Memorials of the Savoy* (1878) and *A History of London* (2 vols., 1883-4).

LOFTUS, LORD AUGUSTUS WILLIAM FREDERICK SPENCER (1817-1904), diplomatist; son of second Marquess of Ely; attaché to British legation at Berlin, 1837-44; at Stuttgart and Baden Baden 1844-7; joined Sir Stratford Canning [q.v.] in special mission to European courts and witnessed revolutionary incidents in Germany and Austria, 1848; secretary of legation at Berlin, 1853; reported as to British consulates on German shores of Baltic after Crimean war; envoy extraordinary to

Emperor of Austria, 1858; warned Austrian government of England's friendship to Italy, 1859; transferred to Berlin (1860), where he favoured Denmark's claims in Schleswig-Holstein crisis; at Munich, 1863–6; returned to Berlin, 1866; GCB, 1866; PC, 1868; managed *solde de captivité* for French prisoners of war in Germany, 1870; at St. Petersburg, 1871–9; attended Tsar Alexander on visit to England, 1874; conferred with Prince Gortchakoff with a view to prevent Russo-Turkish war, 1876; suggested Anglo-Russian understanding, which was brought about by Lord Salisbury and the Russian ambassador in London; governor of New South Wales, 1879–85; published *Diplomatic Reminiscences* (4 vols., 1892–4).

LOGUE, MICHAEL (1840–1924), cardinal; entered Maynooth College, 1857; deacon, 1864; priest and professor of dogmatic theology, Irish College, Paris, 1866; curate of Glenswilly, county Donegal, 1874; dean at Maynooth, 1876; professor of theology, Maynooth, 1878; bishop of Raphoe, 1879; archbishop of Armagh, 1887; cardinal, with title of Santa Maria della Pace, 1893; exercised great influence in Irish politics, largely contributing to eventual deposition of C. S. Parnell [q.v.] from leadership of nationalist party; a great church builder; mediated between Irish people and British government, 1919–21.

LOHMANN, GEORGE ALFRED (1865–1901), Surrey cricketer; won great success as medium-pace bowler for Surrey from 1885 to 1890; thrice visited Australia; played for Players *v.* Gentlemen, 1886–96; a good hitting batsman and first-class fieldsman; raised Surrey cricket to leading position.

LOMBARD, ADRIAN ALBERT (1915–1967), aero-engineer; educated at John Gulson central advanced school, Coventry and Coventry Technical College; joined Rover Co., in drawing office, 1930; transferred to Morris Motors, 1935; returned to Rover Co., 1936; one of design team making Whittle W2B jet engine suitable for production, 1940; joined Rolls-Royce when they took over Whittle engine, 1943; supervised production of W2B engine, Derwent I, Nene, Derwent V, and Avon engines, 1944–5; chief designer (projects), Rolls-Royce Co.; chief designer (aero), 1952; chief engineer, 1954; supervised production of Conway and Spey engines; director, engineering (aero) and director, Rolls-Royce Co., 1958; served on Council, Royal Aeronautical Society, Air Registration Board, Aeronautical Research Council; CBE, 1967.

LONDON, HEINZ (1907–1970), physicist; born at Bonn, Germany; educated at Bonn,

Berlin, and Munich; studied low temperature physics at Breslau University; Ph.D. Breslau, 1933; moved to Oxford, 1933, and to Bristol, 1936; studied high frequency resistance problem; interned, 1939; released to work on atom bomb project at Bristol, Birmingham, Imperial College, ICI Witton and Winnington, and Ministry of Supply, Mold, 1940–4; naturalized, 1942; leader of Birmingham team, 1944; principal scientific officer, Harwell, 1945; senior principal scientific officer, 1950; deputy chief scientist, 1958; worked on isotope separation; invented 'dilution refrigerator'; awarded first Simon memorial prize, 1959; FRS, 1961.

LONDONDERRY, sixth MARQUESS OF (1852–1915), politician. [See VANE-TEMPEST-STEWART, CHARLES STEWART.]

LONDONDERRY, seventh MARQUESS OF (1878–1949), politician. [See VANE-TEMPEST-STEWART, CHARLES STEWART HENRY.]

LONG, WALTER HUME, first VISCOUNT LONG OF WRAXALL (1854–1924), statesman; educated at Harrow and Christ Church, Oxford; conservative MP, North Wiltshire, 1880–5; East Wiltshire, 1885–92; West Derby division of Liverpool, 1893–1900; South Bristol, 1900–6; South County Dublin, 1906–10; Strand division of Middlesex, 1910–18; St. George's Westminster, 1918–21; belonged to 'country party' on entry into parliament, 1880; parliamentary secretary to Local Government Board, 1886–92; took large part in framing and getting through House of Commons Local Government Act, which created county councils throughout Great Britain, 1888; president of Board of Agriculture, with seat in cabinet, 1895–1900; popularity of his appointment soon impaired by his vigorous and successful measures to stamp out rabies, which met with violent opposition; president of Local Government Board, 1900–5; secured passing of Metropolitan Water Act, in teeth of bitter opposition, 1902; successful chief secretary for Ireland, 1905; created Union Defence League, 1907; president of Local Government Board, 1915–16; secretary of state for colonies, 1916–18; first lord of Admiralty, 1919–21; FRS, 1902; created viscount, 1921.

LONGHURST, WILLIAM HENRY (1819–1904), organist and composer; chorister (1828), assistant organist (1836), and organist (1873–98) of Canterbury Cathedral; Mus.Doc., 1875; published church music.

LONGMORE, SIR ARTHUR MURRAY (1885–1970), air chief marshal; educated at Benges School, Hertford and Foster's Academy, Stubbington; entered *Britannia* as naval cadet, 1900; commissioned in Royal Navy, 1904; attended course of flying instruction at Eastchurch, 1910; awarded Royal Aero Club certificate No. 72; commanded Cromarty Air Station, and later, experimental seaplane station, Calshot; commanded No. 1 squadron, Royal Naval Air Service, 1914-16; lieutenant-commander, *Tiger*, present at battle of Jutland, 1916; wing-captain RNAS, Malta; lieutenant-colonel, Royal Air Force, 1918; commanded Adriatic Group at Taranto; DSO, 1919; group captain, Iraq, 1923; CB, 1925; commandant, RAF College, Cranwell; air vice-marshal, 1931; air marshal, 1933; commandant, Imperial Defence College, 1934; KCB, 1935; commander-in-chief, Training Command, 1939-40; initiated Empire air-training scheme; AOC-in-C, Middle East, 1940-1; inspector-general, RAF, 1941-2; GCB; member, Post-Hostilities Planning Committee, 1943-4; published autobiography *From Sea to Sky* (1946).

LONGSTAFF, TOM GEORGE (1875-1964), mountain explorer; educated at Eton and Christ Church, Oxford; MB, B.Ch., 1903; MD, 1906; qualified, St. Thomas's Hospital, 1903; climbed in Caucasus, 1903; explored eastern approaches to Nanda Devi, Himalayas, 1905; explored western approaches to Nanda Devi, climbed Trisul, and explored glaciers of Kamet group, 1907; explored Karakoram, continental watershed between Indus and Tarim Basin of Chinese Turkestan, 1909; climbed in Canadian Rockies, 1910-11; served in India during 1914-18 war; joined expeditions to Everest, Spitzbergen, and Greenland, 1922-34; hon. secretary, Royal Geographical Society, 1930-4; vice-president, 1934-7; president, Alpine Club, 1947; published autobiography, *This My Voyage* (1950).

LONSDALE, fifth EARL OF (1857-1944), sportsman. [See LOWTHER, HUGH CECIL.]

LONSDALE, FREDERICK (1881-1954), playwright; original name Lionel Frederick Leonard; son of seaman; educated locally in St. Helier; wrote musical comedies and plays with brilliant dialogue about the wealthy and well-bred, a subject which eventually became outdated; productions include *The King of Cadonia* and *The Early Worm* (1908), *The Balkan Princess* (1910), *The Maid of the Mountains* (a notable success, with José Collins [q.v.] as Teresa, 1927), *Aren't We All?* (1923), and *The Last of Mrs. Cheyney* (1925); moved to America (1938), then to France (1950).

LOPES, SIR LOPES MASSEY, third baronet (1818-1908), politician and agriculturist; educated at Winchester and Oriel College, Oxford;

BA, 1842; MA, 1845; conservative MP for Westbury, 1857-68, for South Devon, 1868-85; urged grievance of burden of local taxation; helped to carry Agricultural Ratings Act, 1879; civil lord of the Admiralty, 1874-80; PC, 1885; alderman of Devonshire county council, 1888-1904; a scientific farmer, he spent much money on improving his estates.

LORAINE, SIR PERCY LYHAM (1880-1961), twelfth baronet, diplomatist; educated at Eton and New College, Oxford, 1899; served in South African war; entered Foreign Office, 1904; served in Constantinople, 1904, Tehran, 1907, Rome, 1909, Peking, 1911, Paris, 1912, and Madrid, 1916; attached to peace conference delegation, 1918-19; minister, Tehran, 1921; Greece, 1926-9; high commissioner, Egypt and Sudan, 1929-33; Ankara, 1933-9; PC; ambassador, Italy, 1939-40; succeeded to baronetcy, 1917; CMG, 1921; KCMG, 1925; GCMG, 1937.

LORAINE, VIOLET MARY (1886-1956), actress; chorus girl in Drury Lane pantomime, 1902; principal boy there, 1911; played notably opposite (Sir) George Robey [q.v.] in *The Bing Boys are Here* (1916), with songs such as 'If you were the only girl in the world'; married and retired, 1921.

LORD, THOMAS (1808-1908), Congregational minister; held Midland pastorates (1834-79) till he settled at Horncastle, where he preached in his 101st year; original member of Peace Society; published devotional works.

LOREBURN, EARL (1846-1923), lord chancellor. [See REID, ROBERT THRESHIE.]

LORIMER, SIR ROBERT STODART (1864-1929), architect; son of Professor James Lorimer [q.v.]; apprenticed to Sir Rowand Anderson, architect, 1885; entered office of G. F. Bodley [q.v.], 1889; returning to Edinburgh, began long series of restorations, which were among his most pleasing works, 1892; designed many large Scottish country houses; his churches include St. Peter's (Roman Catholic), Morningside, Edinburgh (1906); designed chapel of Order of the Thistle, St. Giles' Cathedral, Edinburgh, 1909-11; recognized as leading architect of Scotland, 1914-18; after European war chiefly occupied on memorials, most important being Scottish National War Memorial, Edinburgh (1918-27); knighted, 1911; saviour of crafts in Scotland; restored to Scotland vital and characteristic architecture.

LOTBINIÈRE, SIR HENRY GUSTAVE JOLY DE (1829-1908), Canadian politician. [See JOLY DE LOTBINIÈRE.]

LOTHIAN, eleventh MARQUESS OF (1882-1940), journalist and statesman. [See KERR, PHILIP HENRY.]

LOUISE CAROLINE ALBERTA (1848-1939), princess of Great Britain and Ireland, Duchess of Argyll; sixth child of Queen Victoria; married (1871) the Marquess of Lorne (later ninth Duke of Argyll, q.v.); a gifted sculptress; made home an artists' rendezvous; wrote magazine articles as 'Myra Fontenoy'; first president of National Union for the Higher Education of Women.

LOUISE VICTORIA ALEXANDRA DAGMAR (1867-1931), princess royal of Great Britain and Ireland, Duchess of Fife; third child of Prince and Princess of Wales; married (1889) the sixth Earl of Fife, created duke on his marriage; declared princess royal, 1905; rescued from shipwreck off Cape Spartel, 1911.

LOVAT, fourteenth (sometimes reckoned sixteenth) BARON (1871-1933). [See FRASER, SIMON JOSEPH.]

LOVATT EVANS, SIR CHARLES ARTHUR (1884-1968), physiologist. [See EVANS.]

LOVE, AUGUSTUS EDWARD HOUGH (1863-1940), mathematician and geophysicist; second wrangler (1885), fellow (1886-99), St. John's College, Cambridge; FRS, 1894; Sedleian professor of natural philosophy, Oxford, 1898-1940; investigated theory of elasticity of solids in its mathematical setting and its application to problems of the earth's crust; discovered 'Love waves', 1911; formulated a theory of biharmonic analysis, 1929; his *Treatise on the Mathematical Theory of Elasticity* (1892-3) a standard work.

LOVELACE, second EARL OF (1839-1906), author of *Astarte*. [See MILBANKE, RALPH GORDON NOEL KING.]

LOVETT, RICHARD (1851-1904), author; spent boyhood (1858-67) in United States; BA London, 1873; MA, 1874; book editor of Religious Tract Society, 1882; secretary, 1899; wrote centenary history of London Missionary Society, 1899; wrote lives of James Gilmour (1892) and James Chalmers [q.v.] (1902); author of *The Printed English Bible* (1895).

LOW, ALEXANDER, LORD Low (1845-1910), Scottish judge; BA, St. John's College, Cambridge, 1867; passed to Scottish bar, 1870; raised to bench, 1890; his decision against minority's claim to 'Free Church' property (1900) was reversed by House of Lords, 1904.

LOW, SIR DAVID ALEXANDER CECIL (1891-1963), cartoonist and caricaturist; born in New Zealand; educated at Christchurch boys' high school; political cartoonist, *Spectator* and *Canterbury Times*; joined Sydney *Bulletin*, 1911; resident cartoonist, 1914; William M. Hughes [q.v.], his target for pictorial wit; cartoons published in *The Billy Book* (Sydney, 1918); arrived in London, 1919; political cartoonist, the *Star*, 1919-26; *Evening Standard*, 1926-49; derided Hitler and Mussolini; joined *Daily Herald*, 1950-3; *Manchester Guardian*, 1953; created 'Colonel Blimp'; selections published in book form included *Lloyd George & Co.* (1921), *Low's Political Parade* (1936), and *Low Visibility: A Cartoon History, 1945-53* (1953); other publications include *British Cartoonists, Caricaturists and Comic Artists* (1942) and *Low's Autobiography* (1956); knighted, 1962; hon. LLD, New Brunswick, 1958.

LOW, SIR ROBERT CUNLIFFE (1838-1911), general; son of Sir John Low [q.v.]; joined Indian army, 1854; served in Indian Mutiny at Delhi and Lucknow; lieutenant-colonel, 1878; director of the transport service on march from Kabul to Kandahar; CB, 1880; actively engaged in Upper Burma, 1886-7; KCB, 1887; commander-in-chief of Chitral relief expedition, 1895; lieutenant-general and GCB, 1896; commanded Bombay army, 1898-1909; general, 1900; keeper of crown jewels at Tower of London, 1909-11.

LOW, SIR SIDNEY JAMES MARK (1857-1932), author and journalist; educated at King's College School and Balliol College, Oxford; first class, modern history, 1879; brilliant editor of *St. James's Gazette*, 1888-97; ardent imperialist and friend of Rhodes, Cromer, Curzon, Milner [qq.v.], etc.; works include *The Governance of England* (1904) and *A Vision of India* (1906); knighted, 1918.

LOWE, SIR DRURY CURZON DRURY- (1830-1908), lieutenant-general. [See DRURY-LOWE.]

LOWE, EVELINE MARY (1869-1956), first woman chairman of the London County Council; born Farren; educated at Milton Mount College; trained as teacher at Homerton College; lecturer, 1893; vice-principal, 1894-1903, Homerton College; married G. C. Lowe, a Bermondsey veterinary surgeon, later a doctor, 1903; labour member, LCC, for West Bermondsey, 1922-46; member (1919-49), chairman

(1934-7), LCC education committee; deputy chairman, LCC, 1929-30; chairman, Council's jubilee year, 1939-40; hon. LLD, London.

LOWKE, WENMAN JOSEPH BASSETT- (1877-1953), model maker. [See BASSETT-LOWKE.]

LOWRY, CLARENCE MALCOLM (1909-1957), author; educated at Leys School and St. Catharine's College, Cambridge; established lifelong reputation as writer and drinker; married twice and lived variously in United States, Mexico, Canada (1940-54), and England; based *Ultramarine* (1933) on voyage 'before the mast' to China Seas, 1927; *Under the Volcano* (1947), claimed as work of genius, describes last day in life of drunken consul and wife in Mexico on Day of the Dead; *Lunar Caustic* and *Dark as the Grave* published posthumously (1968).

LOWRY, HENRY DAWSON (1869-1906), author; wrote Cornish stories for *National Observer* from 1891; on staff of *Pall Mall Gazette*, 1895, and *Morning Post*, 1897; published novels and *The Hundred Windows* (poems, 1904).

LOWRY, THOMAS MARTIN (1874-1936), chemist; educated at Kingswood School, Bath, and Central Technical College, South Kensington; assistant there to H. E. Armstrong [q.v.], 1896-1913; head of chemical department, Guy's Hospital medical school, 1913-20; professor of physical chemistry, Cambridge, 1920-36; mainly investigated optical rotatory power; FRS, 1914.

LOWTHER, HUGH CECIL, fifth EARL OF LONSDALE (1857-1944), sportsman; educated at Eton; succeeded brother, 1882; master in turn of Woodland Pytchley, Blankney, Quorn, and Cottesmore hounds; won St. Leger (1922) but normally unsuccessful as racehorse owner; notable boxer and yacht-racer; maintained splended establishments and endeared himself to populace as sporting grandee; friend of circus folk and London costermongers; with Eric Parker edited 'The Lonsdale Library of Sports, Games & Pastimes'; lord-lieutenant of Cumberland, 1917-44; GCVO, 1925; KG, 1928.

LOWTHER, JAMES (1840-1904), politician and sportsman; educated at Westminster and Trinity College, Cambridge; BA, 1863; MA, 1866; called to bar, 1864, but did not practise; conservative MP for York City, 1865-80, for North Lincs., 1881-5, for Isle of Thanet, 1888-1904; parliamentary secretary to the Poor Law Board under Disraeli, 1867-8; opposed Irish land bill, 1870; under-secretary for the colonies, 1874-8; chief secretary to lord-lieutenant of Ireland, 1878-80; PC, 1878; opposed establishment of county councils (1888); advocated protection; took part in Yorkshire local affairs; bred racehorses from 1873; senior steward of Jockey Club, 1889.

LOWTHER, JAMES WILLIAM, first VISCOUNT ULLSWATER (1855-1949); speaker of the House of Commons; educated at Eton and Trinity College, Cambridge; third class, law, 1878; called to bar (Inner Temple), 1879; conservative MP, Rutland, 1883; Penrith division, 1886-1921; charity commissioner, 1887; chairman of ways and means and deputy speaker, 1895; PC, 1898; speaker, 1905-21; handled series of difficult situations in House with tact, fairness, and humour; GCB and viscount, 1921; thereafter served on royal commissions and other public bodies.

LÖWY, ALBERT or ABRAHAM (1816-1908), Hebrew scholar; born in Moravia; studied at Vienna University; helped to found 'Die Einheit', a society for promoting welfare of Jews; came to London for support of scheme, 1840; with Jewish reformers in London founded West London synagogue, 1842, and became first minister, 1842-92; helped to form Anglo-Jewish Association, 1870; secretary, 1875-89; catalogued Lord Crawford's Samaritan literature, 1872, and Hebrew books of City of London, 1891; founded Society of Hebrew Literature, 1870; hon. LLD, St. Andrews, 1893.

LOYD-LINDSAY, ROBERT JAMES, BARON WANTAGE (1832-1901), soldier and politician. [See LINDSAY.]

LUARD, SIR WILLIAM GARNHAM (1820-1910), admiral; of Huguenot origin; joined navy, 1835; served in China war, 1841; commander, 1850; took part in capture of Rangoon and of Pegu (1852) and in operations in Japan in Straits of Shimonoseki, 1864; CB, 1864; captain superintendent of Sheerness dockyard, 1870-5; superintendent of Malta dockyard, 1878-9; vice-admiral, 1879; president of Royal Naval College, Greenwich, 1882-5; admiral, 1885; KCB, 1897.

LUBBOCK, SIR JOHN, fourth baronet, and first BARON AVEBURY (1834-1913), banker, scientist, and author; son of Sir J. W. Lubbock [q.v.], third baronet; educated at Eton, but early installed in his father's bank; succeeded father, 1865; liberal MP, Maidstone, 1870 and 1874, London University, 1880-1900; secured passage of Bank Holidays Act (1871), Act for Preservation of Ancient Monuments (1882), Early Closing Act (1904), etc.; PC, 1890; baron, 1900;

held leading position in banking world; his researches on ants his most valuable contribution to science; FRS, 1858; author of numerous scientific and ethical works.

LUBBOCK, PERCY (1879-1965), author; educated at Eton and King's College, Cambridge; first class, classical tripos, 1901; Pepys librarian, Magdalene College, Cambridge, 1906-8; contributed to *The Times Literary Supplement*, 1908-14; publications include *Elizabeth Barrett Browning in her Letters* (1906), *Samuel Pepys* (1909), *The Craft of Fiction* (1921), *Earlham* (1922), *Roman Pictures* (1923), *The Region Cloud* (1925), *Shades of Eton* (1929), and *Portrait of Edith Wharton* (1947); CBE, 1952.

LUBY, THOMAS CLARKE (1821-1901), Fenian; BA, Trinity College, Dublin, 1845; abandoned theological studies for nationalist propaganda; planned risings in Ireland, 1848-9; captured and imprisoned; went to Australia; on return started with James Stephens [q.v.] Fenian movement, 1853; founded Irish Republican Brotherhood, 1858; sent as envoy from Ireland to America to collect funds, 1863; on return revived waning enthusiasm, and launched *Irish People* newspaper as organ of party (Nov. 1863–Sept. 1865); sentenced to twenty years' penal servitude for treason-felony, 1865; set at liberty, 1871; settled in New York and engaged in journalism; distrusted home rule movement under C. S. Parnell [q.v.]; wrote *Lives of . . . Representative Irishmen* (1878).

LUCAS, eighth BARON (1876-1916), politician and airman. [See HERBERT, AUBERON THOMAS.]

LUCAS, SIR CHARLES PRESTWOOD (1853-1931), civil servant and historian; educated at Winchester and Balliol College, Oxford; first class, *lit. hum.*, 1876; headed Civil Service list, 1877; assistant under-secretary (1897), first head of dominions department (1907-11) of Colonial Office; KCMG, 1907; KCB, 1912; wrote many books on British Empire; fellow of All Souls, 1920-7.

LUCAS, EDWARD VERRALL (1868-1938), journalist, essayist, and critic; cultivated effortless manner of communicating his delight in art, travel, and letters; publications include a life of his idol Charles Lamb (1905), travel essays, short books on painters, collections of light essays, and anthologies; prolific contributor to and on staff of *Punch*; chairman of Methuen's, 1924-38; CH, 1932.

LUCAS, FRANK LAURENCE (1894-1967), author and scholar; educated at Colfe's grammar school, Lewisham, Rugby, and Trinity

College, Cambridge; first class, classical tripos; served on western front in 1914-18 war; fellow, classics, King's College, Cambridge; employed at Government Codes and Cyphers headquarters, Bletchley Park, 1939-45; reader, Cambridge University, 1947-62; CBE, 1946; publications include *Seneca and Elizabethan Tragedy* (1922), *Euripides and his Influence* (1924), *The River Flows* (1926), *Tragedy in Relation to Aristotle's Poetics* (1927), *Time and Memory* (1929), *Cécile* (1930), *Studies French and English* (1934), *The Decline and Fall of the Romantic Ideal* (1936), *Delights of Dictatorship* (1938), *Greek Poetry for Everyman* (1951), *Greek Drama for Everyman* (1954), *The Art of Living* (1959), and *The Drama of Chekhov, Synge, Yeats and Pirandello* (1963).

LUCAS, KEITH (1879-1916), physiologist; BA, Trinity College, Cambridge; fellow of Trinity, 1904; science lecturer, 1908; Croonian lecturer of Royal Society, 1919; FRS, 1913; researched on muscle and nerve problems; services enlisted for Royal Aircraft Factory, Farnborough, 1914; killed flying.

LUCKOCK, HERBERT MORTIMER (1833-1909), dean of Lichfield; educated at Shrewsbury and Jesus College, Cambridge; BA, 1858; MA, 1862; DD, 1879; won university theological prizes and scholarships; fellow, 1860; vicar of All Saints', Cambridge, 1862-3, 1865-75; principal of Ely theological college, 1876-87; residentiary canon of Ely, 1875-92; dean of Lichfield, 1892-1909; a high churchman, he exerted influence through his devotional writings, which included *After Death* (1879) and *The Intermediate State* (1890).

LUCY, SIR HENRY WILLIAM (1843-1924), journalist; employed by various newspapers and engaged in freelance journalism, 1864-72; engaged by *Daily News*, 1872; manager of its parliamentary staff and writer of parliamentary summary; as 'Toby, M.P.', wrote 'Essence of Parliament' for *Punch*, 1881-1916; knighted, 1909; his close personal relations with prominent politicians made his work first-hand.

LUDLOW, JOHN MALCOLM FORBES (1821-1911), social reformer; born at Nimach, India; educated in Paris; called to bar, 1843; practised as conveyancer, 1843-74; advocate of reforms in India and of abolition of slavery; member of anti-corn-law league; in Paris during revolution of 1848; friend of F. D. Maurice, Charles Kingsley, and Tom Hughes [qq.v.]; one of founders of Christian Socialist movement; promoted labour co-partnership, 1850; founded and edited weekly *Christian Socialist*, 1850; helped to found Working Men's College (1854), lecturing there on law and English and Indian history; wrote historical works, including *Popular Epics of the Middle Ages* (2 vols., 1865); secretary to royal commission on friendly societies, 1870-4; chief registrar of friendly societies, 1875-91; CB, 1887.

LUGARD, FREDERICK JOHN DEAL-TRY, BARON LUGARD (1858-1945), soldier, administrator, and author; educated at Rossall School and Sandhurst; commissioned in 9th Foot (Norfolk regiment); joined second battalion in India, 1878; skilled big-game hunter; seconded to military transport service, 1884; served in Sudan (1885) and Burma (1886); DSO, 1887; commanded force sent by African Lakes Company to defend Karongwa against slave raiders, 1888-9; sent to Uganda by Imperial British East Africa Company, 1890; secured treaty with Kabaka of Buganda and established some sort of order; the Company being unable to maintain its position, he returned to England (1892) to persuade government to undertake responsibility of administration; influential in securing dispatch of Sir Gerald Portal [q.v.], resulting in British protectorate, 1894; published *The Rise of Our East African Empire* (2 vols., 1893); sent by Royal Niger Company to Nikki; arrived ahead of the French and secured treaty with Borgu, 1894; CB, 1895; explored mineral concession in Ngamiland for British West Charterland Company, 1896-7; HM commissioner for Nigerian hinterland, 1897; raised and commanded West African frontier force, 1897-9; lieutenant-colonel, 1899; high commissioner, Northern Nigeria, 1900-6; KCMG, 1901; brought area under administrative control with minimum use of force and realistic and statesmanlike conception of relations between his administration and chiefs as dependent rulers to be guided and when necessary controlled; governor, Hong Kong, 1907-12; largely responsible for creation of Hong Kong University, 1911; GCMG, 1911; governor of North and South Nigeria, 1912-14; carried out amalgamation of two protectorates; governor-general of Nigeria, 1914-19; conspicuously developed system of indirect rule; regarded traditional native institutions as surest foundation upon which to build; published *The Dual Mandate in British Tropical Africa* (1922); acknowledged authority on colonial administration; member of permanent mandates commission of League of Nations (1922-36) and colonial advisory committee on education (1923-36); chairman of International Institute of African Languages and Cultures from 1926; PC, 1920; baron, 1928.

LUKE, first BARON (1873-1943), man of business and philanthropist. [See JOHNSTON, GEORGE LAWSON.]

LUKE, SIR HARRY CHARLES (1884-1969), colonial administrator; educated at Eton and Trinity College, Cambridge; appointed private secretary (1908) and aide-de-camp (1909) to the governor of Sierra Leone; second-lieutenant, London Yeomanry, 1909-11; private secretary to high commissioner, Cyprus, 1911; assistant secretary to government, 1912; political officer with Royal Navy in eastern Mediterranean, 1915-16; commissioner, Famagusta, Cyprus, 1918; British chief commissioner, Georgia, Armenia, and Azerbaijan, 1920; assistant governor, Jerusalem, 1920; colonial secretary, Sierra Leone, 1924-8; chief secretary, Palestine, 1928-30; lieutenant-governor, Malta, 1930-8; governor, Fiji, and high commissioner, Western Pacific, 1938-43; CMG, 1926; knighted, 1933; KCMG, 1939; D.Litt., Oxford, 1938; hon. LLD, Malta; publications include *The Fringe of the East* (1913) and *Cities and Men* (3 vols., 1953-6).

LUKE, JEMIMA (1813-1906), hymn writer; born THOMPSON; enthusiastic nonconformist; author of children's hymn, 'I think when I read that sweet story of old' (1840) and autobiography (1900).

LUKIN, SIR HENRY TIMSON (1860-1925), major-general; served in Zulu war, 1879; South African war, 1899-1902; commandant-general, Cape Colonial forces, 1904-12; inspector-general, Permanent Force, Union of South Africa, 1912; commanded 1st South African infantry brigade in Egypt, gaining victory at Agagiya, 1916; commanded brigade in France, Apr.; major-general commanding 9th division, 1916-18; KCB, 1918.

LUMLEY, LAWRENCE ROGER, eleventh EARL OF SCARBOROUGH (1896-1969), public servant; educated at Eton, Sandhurst, and Magdalen College, Oxford; served with 11th Hussars on western front, 1916-18; conservative MP, Hull East, 1922-9; York, 1931-7; governor of Bombay, 1937-43; succeeded uncle as eleventh earl, 1945; hon. colonel, Yorkshire Dragoons, 1956-62; hon. major-general, 1946; parliamentary under-secretary of state for India and Burma, 1945; chairman, commission on Oriental, Slavonic, East European, and African studies, 1945-6; president, Royal Asiatic Society, 1946-9; the East India Association, 1946-61; the Royal Asian Society, 1954-60; chairman, School of Oriental and African Studies, 1951-9; chairman, Commonwealth Scholarship Committee, 1960-3; lord-lieutenant, West Riding of Yorkshire and of the city of York, 1948; chancellor, Durham University, 1958; hon. DCL, Durham; LLD, Sheffield,

Leeds, and London; high steward, York Minster, 1967; GCIE, 1937; GCSI, 1943; KG, 1948; lord chamberlain, 1952–63; permanent lord in waiting, 1963; PC, 1952; GCVO, 1953; Royal Victorian Chair, 1963.

LUNN, Sir HENRY SIMPSON (1859–1939), founder of travel agency; Indian medical missionary (Methodist), 1887–8; founded *The Review of the Churches*, 1891; devoted himself to cause of reunion; founded his firm (1909) as result of arranging religious conferences; knighted, 1910.

LUPTON, JOSEPH HIRST (1836–1905), scholar and schoolmaster; BA, St. John's College, Cambridge (fifth classic), 1858; MA and fellow, 1861; ordained, 1859; DD, 1896; surmaster in St. Paul's School, 1864–99; published *Wakefield Worthies* (1864); published *Life of Dean Colet* (1887) and edited and translated many of Colet's works; Hulsean lecturer (1887) and Seatonian prizeman (1897) of Cambridge; other works were life of St. John of Damascus (1882) and an edition of More's *Utopia* (1895).

LUSH, Sir CHARLES MONTAGUE (1853–1930), judge; son of Sir Robert Lush [q.v.]; BA, Trinity College, Cambridge; called to bar (Gray's Inn), 1879; KC, 1902; judge, King's Bench division and knighted, 1910; retired, 1925; an eloquent advocate, but hesitant judge; wrote *Law of Husband and Wife* (1884).

LUSK, Sir ANDREW, baronet (1810–1909), lord mayor of London; started grocery business in Greenock, 1835; founded business in London, 1840; chairman of Imperial Bank from 1862; lord mayor of London, 1873; raised £150,000 for relief of Bengal famine; baronet, 1874; liberal MP for Finsbury, 1865–85; became liberal unionist, 1886.

LUTHULI, ALBERT JOHN (1898?–1967), president-general of the African National Congress; born in Southern Rhodesia; educated at Ohlange Institute and Methodist Institution, Edenvale; appointed as teacher, 1918; became lay preacher; awarded bursary to Adams College, Durban; became teacher trainer; elected chief of Umvoti Mission Reserve, 1936; served on advisory board to South African Sugar Association, delegate to International Missionary Conference in Madras, 1938; lecture tour on missions, United States, 1948; member, executive, African National Congress, 1945; president of Congress in Natal, 1951; led Defiance Campaign against apartheid, 1952; deposed from chieftainship by South African Government; president-general, ANC, 1952; banned from attending public gatherings,

1953–6; arrested on charge of treason, 1956; discharged, 1957; banned again, 1959; arrested again, 1960; prison sentence suspended on health grounds; awarded Nobel peace prize for 1960, 1961; under house arrest, 1964–7; killed by a freight train, 1967; awarded United Nations Human Rights prize, 1968.

LUTYENS, Sir EDWIN LANDSEER (1869–1944), architect; studied at South Kensington and under (Sir) Ernest George [q.v.]; set up London practice, 1888; built brilliant series of romantic country houses; architect for British pavilion, Paris exhibition, 1900; FRIBA, 1906; consulting architect, Hampstead Garden Suburb, 1908–9; designed British pavilion for international exhibition, Rome, Rand war memorial, and Johannesburg Art Gallery, 1909–11; joint-architect with Sir Herbert Baker [q.v.] for New Delhi, 1913–30; in the viceroy's house and ancillaries created one of the finest palaces in architectural history, characterized by extraordinary fertility of invention and aristocratic restraint; appointed to Imperial War Graves Commission, 1917; designed the Cenotaph (1919) and many war memorials including that to 'the Missing of the Somme' at Thiepval; his work romantic in inspiration, classic in discipline, yet abstract in design; buildings include Britannic House, Finsbury Circus (1920–2), Midland Bank, Poultry (1924–37), British Embassy, Washington (1926–9), Campion Hall, Oxford (1934); collaborated on many large blocks of flats including Grosvenor House, Park Lane; his magnificent designs for Liverpool Roman Catholic cathedral (1929–43) proved impracticable to complete in post-war conditions, ARA, 1913; RA, 1920; president, 1938–44; knighted, 1918; KCIE, 1930; OM, 1942.

LUTZ, (WILHELM) MEYER (1829–1903), musical composer; born in Bavaria; settled in England, 1848; toured provinces with Italian operatic singers; conductor of Gaiety Theatre, 1869–96, and organist of St. George's Cathedral, Southwark; composed, besides church music, operettas for the Gaiety Theatre.

LUXMOORE, Sir (ARTHUR) FAIRFAX (CHARLES CORYNDON) (1876–1944), judge; educated at King's School, Canterbury, and Jesus College, Cambridge; played Rugby for Cambridge and England; called to bar (Lincoln's Inn), 1899; KC, 1919; bencher, 1922; judge of Chancery division, 1929–38; lord justice of appeal, 1938–44; possessed unrivalled knowledge of trade-mark and patent law; chairman, committee on post-war agricultural education, 1941–3; active man of Kent and president of county cricket club; knighted, 1929; PC, 1938.

LYALL, SIR ALFRED COMYN (1835-1911), Indian civil servant and writer; son of Alfred Lyall [q.v.]; educated at Eton and Haileybury; joined Indian civil service, 1856; actively served in Mutiny, 1857-8; made commissioner of West Berar (1867), home secretary to Indian government (1873), and governor-general's agent in Rajputana (1874); foreign secretary to Indian government, 1878-81; took part in negotiations at Kabul and Kandahar, 1880, and advocated definite treaty with Russia in regard to Afghanistan, 1881; CB, 1879; KCB, 1881; lieutenant-governor of North-West Provinces and Oudh, 1882-7; founded new university of Allahabad; returned to England, 1887; member of India Council in London, 1887-1902; KCIE, 1887; GCIE, 1896; PC, 1902; filled distinguished place in English society; Rede lecturer at Cambridge, 1891; Ford's lecturer at Oxford, 1908; published *Verses Written in India* (1889), *Asiatic Studies* (2 series, 1882 and 1899), *Rise of British Dominion in India* (1893), and life of the Marquess of Dufferin (2 vols., 1905); hon. DCL, Oxford, 1889, LLD, Cambridge, 1891; FBA, 1903; trustee of British Museum, 1911; a liberal unionist, free trader, and opponent of women's suffrage.

LYALL, SIR CHARLES JAMES (1845-1920), Indian civil servant and orientalist; entered Indian civil service, 1865; held secretariat posts, etc., in India, 1867-94; chief commissioner of Central Provinces, 1895; KCSI, 1897; secretary of judicial and public department, India Office, 1898-1910; works include series on early Arabic literature.

LYALL, EDNA (pseudonym) (1857-1903), novelist. [See BAYLY, ADA ELLEN.]

LYGON, WILLIAM, seventh EARL BEAUCHAMP (1872-1938), politician; educated at Eton and Christ Church, Oxford; succeeded father, 1891; governor of New South Wales, 1899-1902; PC, 1906; lord president of the Council, 1910, 1914-15; first commissioner of works, 1910-14; leader of liberal party in the Lords, 1924-31; KG, 1914.

LYLE, CHARLES ERNEST LEONARD, first BARON LYLE OF WESTBOURNE (1882-1954), industrialist and politician; educated at Harrow, Trinity Hall, Cambridge, and Kahlsruhe University; international lawn tennis player; entered family sugar refining firm; chairman, Tate & Lyle, 1922-37; president, 1937-54; vigorous opponent of nationalization of his industry; conservative MP, Stratford, West Ham (1918-22), Epping (1923-4), Bournemouth (1940-5); knighted, 1923; baronet, 1932; baron, 1945.

LYNCH, ARTHUR ALFRED (1861-1934), author, soldier, and politician; born in Australia; Paris correspondent, *Daily Mail*, 1896-9; fought for Boers, 1900; sentenced to death for high treason, 1903; pardoned, 1907; qualified and practised medicine in North London, 1908-34; MP, Galway City (1901), West Clare (1909-18); works include *Principles of Psychology* (1923).

LYND, ROBERT WILSON (1879-1949), journalist and essayist; educated at Royal Academical Institution and Queen's College, Belfast; assistant literary editor (1908), literary editor (from 1912) of *Daily News* (after 1930 the *News Chronicle*); essayist as 'Y.Y.' in *New Statesman*, 1913-45; publications include *Dr. Johnson and Company* (1927).

LYNE, JOSEPH LEYCESTER (Father Ignatius) (1837-1908), preacher; educated at St. Paul's School and Trinity College, Glenalmond; ordained, 1860; developed advanced views; curate to George Rundle Prynne [q.v.] at Plymouth; studied Benedictine order at Bruges, 1861; on return to London replaced A. H. Mackonochie [q.v.] as curate of St. George's-in-the-East; resigned on assuming Benedictine habit; formed a monastic community at Claydon, near Ipswich, 1862; removed to Elm Hill, Norwich, 1863-6, where he frequently came into conflict with the bishop; built Llanthony Abbey, 1869; the community dwindled owing to quarrels; an eloquent preacher; made missionary tour through Canada and America, 1890-1.

LYNE, SIR WILLIAM JOHN (1844-1913), Australian politician; born in Tasmania; entered New South Wales legislative assembly, 1880; premier, 1899; minister of home affairs in first Commonwealth ministry, 1901; minister of trade and customs, 1903-5-7; treasurer, 1907-8; KCMG, 1900; protectionist with rather narrowly Australian outlook.

LYNSKEY, SIR GEORGE JUSTIN (1888-1957), judge; educated at St. Francis Xavier's College and the university, Liverpool; LLB, 1907; admitted solicitor, 1910; called to bar (Inner Temple), 1920; KC, 1930; judge, Salford Hundred Court of Record, 1937-44; of King's Bench division, 1944-57; chairman, Board of Trade inquiry, 1948; knighted, 1944.

LYON, CLAUDE GEORGE BOWES-, fourteenth EARL OF STRATHMORE AND KINGHORNE (1855-1944). [See BOWES-LYON.]

LYONS, SIR ALGERNON McLENNAN (1833-1908), admiral of the fleet; born at Bombay; entered navy, 1847; served on China and Mediterranean stations; distinguished himself in Crimean war; promoted commander, 1858; commodore in charge at Jamaica, 1875-8; rear-admiral, 1878; commander-in-chief in Pacific, 1881; in command of North America and West Indies station, 1886; admiral, 1888; commander-in-chief at Plymouth, 1893-6; admiral of the fleet, 1897; KCB, 1889; GCB, 1897.

LYONS, Sir HENRY GEORGE (1864-1944), geographer and scientist; educated at Wellington College and Woolwich; gazetted to Royal Engineers, 1884; posted to Cairo, 1890; transferred to organize Geological Survey of Egypt, 1896; director, combined geological and cadastral survey department, 1901-9; originated an observatory and meteorological office; entered Science Museum, South Kensington, 1912; director, 1920-33; created Royal Engineers' meteorological service and directed Meteorological Office, 1914-18; fellow (1906), foreign secretary (1928), treasurer (1929-39), Royal Society; wrote *The Royal Society, 1660-1940* (1944); knighted, 1926.

LYONS, JOSEPH ALOYSIUS (1879-1939), prime minister of Australia; born in Tasmania; a Roman Catholic; became schoolteacher; labour member of Tasmanian parliament, 1909-29; treasurer and minister for education and for railways, 1914-16; leader of opposition, 1916-23; premier, 1923-8; member of Commonwealth parliament, 1929-39; postmaster-general, 1929-31; acting-treasurer, 1930-1; opposed bringing Commonwealth Bank under Treasury control and expelled from labour party; joined with nationalists to form united Australia party and won elections, 1931; prime minister, 1932-9; treasurer, 1932-5; pledged to carry out the 'Premier's Plan' combining deflationary with some expansionist measures; his recovery policy successful; establishment of independent department of external affairs (1935) one of few positive achievements of later years; PC, 1932; CH, 1936.

LYTE, Sir HENRY CHURCHILL MAXWELL (1848-1940), deputy keeper of the public records and historian; educated at Eton and Oxford and wrote their histories; deputy keeper of the public records, 1886-1926; instituted series of *Calendars of the Chancery Rolls*; edited *Book of Fees* (2 parts, 1920-3); KCB, 1897; FBA, 1902.

LYTTELTON, ALFRED (1857-1913), lawyer and statesman; son of fourth Baron Lyttelton [q.v.]; educated at Eton and Trinity College, Cambridge; practised successfully at

bar, 1881-1903; liberal unionist MP, Leamington, 1895; chairman of commission to South Africa, 1900; colonial secretary, 1903-5; sanctioned, in face of opposition, introduction of Chinese coolies into Rand, 1904; prepared way for development of imperial conference; MP, St. George's, Hanover Square, 1906; first-class cricketer and for long amateur tennis champion.

LYTTELTON, ARTHUR TEMPLE (1852-1903), suffragan bishop of Southampton; son of fourth Baron Lyttelton [q.v.]; educated at Eton and Trinity College, Cambridge; BA, 1874; MA, 1877; DD, 1898; tutor of Keble College, Oxford, 1879-82; first master of Selwyn College, Cambridge, 1882-93; Hulsean lecturer, 1891; vicar of Eccles, Lancashire, 1893-8; suffragan bishop of Southampton, 1898-1903; a high churchman; published *Modern Poets of Faith, Doubt, and Unbelief* (1904).

LYTTELTON, EDWARD (1855-1942), schoolmaster, divine, and cricketer; son of fourth Baron Lyttelton [q.v.]; educated at Eton and Trinity College, Cambridge; captained Eton (1874) and Cambridge (1878) elevens; deacon, 1884; priest, 1886; master at Eton, 1882-90; headmaster of Haileybury, 1890-1905; of Eton, 1905-16; dean of Whitelands Training College, Chelsea, 1920-9; honorary canon of Norwich from 1931.

LYTTELTON, Sir NEVILLE GERALD (1845-1931), general; son of fourth Baron Lyttelton [q.v.]; educated at Eton; joined Rifle Brigade, 1865; commander-in-chief, South Africa, 1902-4; chief of the general staff, 1904-8; general, 1906; built up expeditionary force concentrating especially on training of staff and formation of Officers' Training Corps; commander-in-chief, Ireland, 1908-12; governor, Chelsea Hospital, 1912-31; KCB, 1902; GCVO, 1911.

LYTTON, second EARL OF (1876-1947). [See BULWER-LYTTON, VICTOR ALEXANDER GEORGE ROBERT.]

LYTTON, Sir HENRY ALFRED (1865-1936), actor, whose original name was HENRY ALFRED JONES; played with D'Oyly Carte Opera Company (1884, 1887-1903, 1908-34) and became its mainstay; played thirty characters including Jack Point in *The Yeomen of the Guard*, Reginald Bunthorne in *Patience*, and Ko-Ko in *The Mikado*; possessed a light, pleasant voice with crystal-clear diction, and excellent sense of comedy and timing; knighted, 1930.

M

MacALISTER, Sir DONALD, first baronet (1854-1934), physician, principal and vice-chancellor, and, later, chancellor of Glasgow University; educated at Liverpool Institute and St. John's College, Cambridge; senior wrangler and fellow, 1877; MD, 1884; president, General Medical Council, 1904-31; played large part in preparing *British Pharmacopœia* (1898, 1914); principal of Glasgow University, 1907-29; chancellor, 1929-34; activities in country's university business included chairmanship of vice-chancellors' committee; KCB, 1908; baronet, 1924.

MacALISTER, Sir (GEORGE) IAN (1878-1957), secretary of the Royal Institute of British Architects; scholar of St. Paul's School; exhibitioner, Merton College, Oxford; secretary, Royal Institute of British Architects, 1908-43; widened its influence in provinces and commonwealth; obtained Architects Registration Acts, 1931, 1938; organized new headquarters in Portland Place and knighted, 1934.

MACAN, Sir ARTHUR VERNON (1843-1908), gynaecologist, and obstetrician; BA, Trinity College, Dublin, 1864; MB and M.Ch., 1868; MAO, 1877; studied medicine at Berlin, 1869-72; served as volunteer in Prussian army, 1870; fellow of King and Queen's College of Physicians, Ireland; master of the Rotunda Hospital, Dublin, 1882; introduced newer obstetric methods despite opposition; applied Listerian principles in midwifery; president of Royal College of Physicians, Ireland, 1902-4; knighted, 1903; contributed to Dublin scientific journals.

MACARA, Sir CHARLES WRIGHT, first baronet (1845-1929), cotton spinner; managing partner of Henry Bannerman & Sons, cotton spinners and merchants, of Manchester, 1880; president, Federation of Master Cotton Spinners' Associations, 1894-1914; founded International Federation of Master Cotton Spinners' and Manufacturers' Associations, 1904; chairman, 1904-15; baronet, 1911.

McARTHUR, CHARLES (1844-1910), politician and writer on marine insurance; won repute by *The Policy of Marine Insurance Popularly Examined*, 1871; established own business as an average adjuster at Liverpool, 1874; president of Liverpool Chamber of Commerce, 1892-6; liberal unionist MP for Exchange division of Liverpool, 1897-1906, and for Kirkdale, 1907-10; championed shipping and pro-

testant church interests in parliament; wrote on *Evidences of Natural Religion* (1880).

MACARTHUR, MARY REID (1880-1921), women's labour organizer. [See ANDERSON.]

MacARTHUR, Sir WILLIAM PORTER (1884-1964), director-general of the Army Medical Services; educated at Queen's University, Belfast; MB, B.Ch. (RUI), 1908; DPH, Oxon, 1910; MD, Belfast, 1911; FRCP, Ireland, 1913; DTM & H, Cantab. 1920; FRCP, London, 1937; joined RAMC, 1909; served in Mauritius, 1911-14; in France, 1915-16; DSO; commanding officer and chief instructor, Army School of Hygiene, 1919-22; professor, tropical medicine, RAM College, London, 1922-9 and 1932-4; consulting physician to the army, 1929-34; director of studies and commandant, RAM College, 1935-8; deputy director-general, Army Medical Services, 1934-5; director-general, 1938-41; CB, 1938; KCB, 1939; hon. physician to the King, 1930-41; colonel commandant, RAMC, 1946-51; consultant in tropical diseases, Royal Masonic Hospital; additional member, faculty of medicine, Oxford; editor, *Transactions* of Royal Society of Tropical Medicine and Hygiene; lifelong hobby, the study of medical history; contributed to various journals and published numerous lectures.

MACARTNEY, Sir GEORGE (1867-1945), consul-general at Kashgar; son of Sir Samuel Halliday Macartney [q.v.] and his Chinese wife; educated at Dulwich College and in Paris; *B. ès. L.*, 1886; entered Indian foreign department, 1888; sent to Kashgar, 1890; position regularized as first British consul in Chinese Turkestan, 1908; consul-general, 1910-18; KCIE, 1913.

MACARTNEY, Sir SAMUEL HALLIDAY (1833-1906), official in the Chinese service; studied medicine at Edinburgh University, 1852-5; joined medical staff in Crimean war, 1855; MD, 1858; served in Indian Mutiny (1859) and in China (1860-2); entered Chinese service, 1862; commanded Chinese troops co-operating with C. G. Gordon [q.v.]; took Fung Ching and Seedong, 1863; mediated between Gordon and Li Hung Chang regarding murder of Taiping leaders at Soochow, 1864; in charge of arsenal at Nanking, 1865-75; secretary to the Chinese legation in London, 1877-1906; adviser of Chinese government; CMG, 1881; KCMG, 1885.

MACASSEY, Sir LYNDEN LIVINGSTON (1876-1963), industrial lawyer; educated at

Upper Sullivan School, Holywood, County Down, Bedford School, Trinity College, Dublin, and London University; trained as engineer; called to bar (Middle Temple), 1899; lecturer on economics and law, London School of Economics, 1901-9; KC, 1912; secretary to Royal Commission on London Traffic, 1903-6; Board of Trade arbitrator in shipbuilding and engineering disputes, 1914-16; head of dilution commission to the Clyde, 1916; director of shipyard labour, Admiralty, 1917-18; member, War Cabinet Committees on Labour and on Women in Industry, 1917-19; labour assessor for British government, Permanent Court of International Justice, 1920; KBE, 1917; bencher (1922) and treasurer (1935), Middle Temple; hon. fellow, chairman of governors, Queen Mary College, London University; president, Institute of Arbitrators; president, Scottish Amicable Life Assurance Society, Ltd.; published *Labour Policy—False and True* (1922).

MACAULAY, DAME (EMILIE) ROSE (1881-1958), author; educated at Oxford high school and Somerville College, Oxford; wrote twenty-three novels many exposing current absurdities with gentle irony; they include *Dangerous Ages* (Femina Vie Heureuse prize, 1921), *The Towers of Trebizond* (James Tait Black memorial prize, 1956), *Keeping up Appearances* (1928), and *Going Abroad* (1934); other writings were a biography of Milton, a study of *Some Religious Elements in English Literature* (1931), *Personal Pleasures* (essays, 1935), and travel books such as *Pleasure of Ruins* (1953); contributed prolifically to periodicals; DBE, 1958; hon. D.Litt., Cambridge, 1951.

MACAULAY, JAMES (1817-1902), author; MA and MD, Edinburgh, 1838; a strenuous opponent of vivisection; tutor in Italy and Spain, and Madeira; FRCS, Edinburgh, 1862; edited *Leisure Hour*, *Sunday at Home*, and the Religious Tract Society's periodicals; helped to found *Boy's Own Paper* and *Girl's Own Paper*, 1879; published accounts of travels in America, 1871, and Ireland, 1872; wrote narratives of adventure for boys and girls.

MACBAIN, ALEXANDER (1855-1907), Celtic scholar; graduate of King's College, Aberdeen, 1880; rector of Raining's school, Inverness, from 1881; published *Celtic Mythology and Religion* (1885) and a Gaelic dictionary (1896); edited many Celtic and Gaelic works; hon. LLD, Aberdeen, 1901; awarded civil list pension, 1905.

MACBETH, ROBERT WALKER (1848-1910), painter and etcher; son of Norman Macbeth [q.v.]; studied at schools of Royal Scottish Academy and of Royal Academy; frequent exhibitor at Academy; an able etcher of pictures by Velazquez and Titian; ARA, 1883; RA, 1903.

McBEY, JAMES (1883-1959), etcher and painter; educated at Newburgh village school; bank clerk in Aberdeen and Edinburgh, 1899-1910; taught himself etching; travelled and studied work of Rembrandt and Whistler; obtained London recognition as etcher and became much sought after during twenties; official artist, Egyptian Expeditionary Force, 1917-19; collections of his works are at Aberdeen Art Gallery, Boston Public Library (Mass.), and Washington National Gallery; became American citizen, 1942; died in Tangier.

MacBRYDE, ROBERT (1913-1966), painter of still life and figure subjects. [See COLQUHOUN, ROBERT.]

McCABE, JOSEPH MARTIN (1867-1955), rationalist; educated at Roman Catholic elementary school, Gorton, Manchester; solemnly professed as Franciscan at twenty-one; ordained priest, 1890; left order and Church, 1896; publications include *Twelve Years in a Monastery* and *Modern Rationalism* (1897); a founder of Rationalist Press Association, 1899; lecturer, journalist, and author in rationalist cause.

MacCALLUM, ANDREW (1821-1902), landscape painter; apprenticed to Nottingham hosiery business; studied art at Nottingham; art teacher in Manchester and Stourbridge, 1850-4; exhibited at Royal Academy, 1850-86; went to Italy, 1854; painted landscapes in Windsor Forest, in Switzerland, Germany, Italy, Paris, and Egypt; lectured on art; work represented in Tate Gallery, Victoria and Albert Museum, and in Nottingham Art Gallery.

McCALMONT, HARRY LESLIE BLUNDELL (1861-1902), sportsman; joined army, 1881; inherited some £4 million, 1894; won £57,455 with horse Isinglass, which won the Two Thousand Guineas, Derby, and St. Leger in 1893, and was sire of Cherry Lass and Glass Doll, winners of the Oaks, 1905 and 1907; conservative MP for Newmarket, 1895-1902; CB for services in South African war.

McCARDIE, SIR HENRY ALFRED (1869-1933), judge; called to bar (Middle Temple), 1894; joined Midland circuit; judge of King's Bench division, 1916-33; remarkable for the learning, careful phraseology, and prolixity of his judgments; convinced that a judge should consider social problems; criticized for his comments thereon; his removal sought after he recorded his opinion that action of R. E. H. Dyer [q.v.] in Amritsar was right; cases included the trials for murder (1922) of Henry Jacoby and Ronald True; committed suicide.

McCARRISON, SIR ROBERT (1878-1960), medical scientist; first class honours, medicine, Queen's College, Belfast, 1900; entered Indian medical service, 1901; served in Chitral, 1902-4, Gilgit, 1904-11; assigned to special study of goitre, 1913, and of deficiency diseases, 1918; on active service, 1914-18; first director, Nutrition Research Laboratories, Coonor, 1929-35, having persisted in research despite official discouragement; combined laboratory and field-work; retired as major-general, 1935, and settled in Oxford; deputy regional adviser, Emergency Medical Service, 1939-45; first director, postgraduate medical education, 1945-55; FRCP, 1914; CIE, 1923; knighted, 1933; hon. LLD, Belfast, 1919.

MacCARTHY, SIR (CHARLES OTTO) DESMOND (1877-1952), literary and dramatic critic; educated at Eton and Trinity College, Cambridge; an 'Apostle' with talent for criticism and conversation; became literary journalist; covered Vedrenne-Barker seasons at Royal Court Theatre for *Speaker* published as *The Court Theatre 1904-1907* (1907); edited *New Quarterly*, 1907-10; on staff of *New Statesman*, 1913-28, as dramatic critic, and literary editor (1920-7), reviewing as 'Affable Hawk'; senior literary critic, *Sunday Times*, 1928-52; editor, *Life and Letters*, 1928-33; seven volumes of collected writings published in lifetime including *Portraits* (1931) and *Shaw* (1951); reviewed from background of wide reading and knowledge; knighted, 1951; president, PEN in England, 1945; hon. LLD, Aberdeen, 1932.

McCARTHY, DAME (EMMA) MAUD (1858-1949), army matron-in-chief; born in Sydney; trained at London Hospital; served in South Africa, 1899-1902; in Queen Alexandra's Imperial Military Nursing Service, 1902-10; principal matron at War Office, 1910-14; matron-in-chief, British Armies in France, 1914-19, Territorial Army Nursing Service, 1920-5; GBE, 1918.

M'CARTHY, JUSTIN (1830-1912), Irish politician, historian, and novelist; leader-writer on *Daily News*, 1871; MP, county Longford, 1879, Derry City, 1886, North Longford, 1892-1900; chairman of anti-Parnellite nationalist party, 1890-6; wrote *History of Our Own Times* (1877), *Dear Lady Disdain* (1875), *Miss Misanthrope* (1878), etc.

McCARTHY, LILLAH (1875-1960), actress; studied elocution (with Hermann Vezin, q.v.) and voice production; acted with Wilson Barrett [q.v.], 1896-1904; at Court Theatre 1905-6 in plays by G. B. Shaw [q.v.] who 'blessed the day' when he found her; married Harley Granville-Barker [q.v.], 1906; played title-role in Masefield's *Nan* and Lady Sybil in Barrie's *What Every Woman Knows* (1908), and Margaret Knox in *Fanny's First Play* (1911); with husband made historic Shakespearian productions at Savoy, 1912-14; played Lavinia in *Androcles and the Lion* (1913); divorced husband, 1918; married (Sir) Frederick Keeble [q.v.], 1920; published autobiography, *Myself and My Friends* (1933).

McCLEAN, FRANK (1837-1904), civil engineer and amateur astronomer; BA, Trinity College, Cambridge (27th wrangler), 1859; partner in father's engineering firm, 1862-70; built private observatory near Tunbridge Wells; designed a star spectroscope; published results of systematic survey of spectra of stars brighter than magnitude $3\frac{1}{2}$, 1898; FRAS, 1877; hon. LLD, Glasgow, 1894; FRS, 1895; bequeathed large sums of money to Cambridge and Birmingham universities for physical research.

McCLINTOCK, SIR FRANCIS LEOPOLD (1819-1907), admiral; entered navy, 1831; served under Sir James Clark Ross [q.v.], 1848, Sir Erasmus Ommanney [q.v.], 1850, and Sir Edward Belcher [q.v.], 1852, in Arctic voyages; commanded expedition in search of Sir John Franklin [q.v.], 1857-9; published account of his voyage and fate of Franklin and his companions, 1859; knighted, 1860; commodore in charge at Jamaica, 1865-8; admiral superintendent of Portsmouth dockyard, 1872-7; vice-admiral, 1877; admiral, 1884; KCB, 1891.

McCLURE, SIR JOHN DAVID (1860-1922), schoolmaster; BA, London and Trinity College, Cambridge; professor of astronomy, Queen's College, London, 1889-94; headmaster, Mill Hill School, 1891-1922; raised it from being comparatively unknown nonconformist school of sixty boys to status of successful public school with over three hundred boys; knighted, 1913; contributed to educational, musical, and religious life of his time.

McCOAN, JAMES CARLILE (1829-1904), author and journalist; called to bar, 1856; practised in supreme consular court at Constantinople till 1864; published *Egypt as it is* (1877) and *Egypt under Ismail* (1889); wrote sympathetically of the Turks in *Turkey in Asia* (2 vols., 1879); protestant home ruler MP for Wicklow county, 1880-5.

MacCOLL, DUGALD SUTHERLAND (1859-1948), painter, critic, and art gallery, director; educated at Glasgow Academy, University College School, University College, London, and Lincoln College, Oxford; studied art under Frederick Brown [q.v.]; exhibited regularly at New English Art Club; art critic successively on *Spectator*, *Saturday Review*, and *Week-end Review;* editor, *Architectural Review*, 1901-5; keeper, Tate Gallery, 1906-11, Wallace Collection, 1911-24; energetic administrator and controversialist; publications include *Nineteenth Century Art* (1902) and biography of P. Wilson Steer [q.v.] (1945).

MacCOLL, MALCOLM (1831-1907), high-church divine and author; ordained, 1856; attracted notice of Gladstone, who presented him to City living of St George's, Botolph Lane, 1871-91, and canonry of Ripon, 1884; took frequent part in ecclesiastical and political controversies; supported Gladstone's Irish Church and home rule policies; visited Eastern Europe with Canon Liddon [q.v.], 1876, and denounced Bulgarian atrocities; hon. DD, Edinburgh, 1899.

MacCOLL, NORMAN (1843-1904), editor of the *Athenaeum* and Spanish scholar; BA, Downing College, Cambridge, 1866; MA, 1869; fellow, 1869; called to bar, 1875; editor of *Athenaeum*, 1871-1900; published *Select Plays of Calderon* (1888), and translations of Cervantes' *Exemplary Novels* (2 vols., 1902), and of his *Miscellaneous Poems* (posthumous, 1912); endowed MacColl lectureship in Spanish and Portuguese at Cambridge.

MacCORMAC, SIR WILLIAM, first baronet (1836-1901), surgeon; BA, Queen's University, Belfast, 1855; MA, 1858; MD, 1857; hon. M.Ch., 1879; D.Sc., 1882; hon. MD and M.Ch., Dublin, 1900; studied surgery in Belin; MRCS, 1857; FRCS Ireland, 1864; volunteered for surgical service in Franco-German war, 1870; lecturer on surgery at St. Thomas's Hospital, 1873-93; chief surgeon to National Aid Society in Turco-Serbian war; knighted for services as secretary to seventh international medical congress, 1881; baronet, 1897; KCVO, 1898; president of the Royal College of Surgeons, 1896-1900; government consulting surgeon to the field force in South African war, 1899-1900; KCB, 1901; publications include account of his work at Sedan (1870) and *Surgical Operations* (1885-9).

McCORMICK, WILLIAM PATRICK GLYN (1877-1940), vicar of St. Martin-in-the-Fields, London; educated at St. John's College, Cambridge; deacon, 1900; priest, 1901;

vicar in Cleveland (1903-10), Belgravia (1910-14), Johannesburg; of Croydon (1919-27), St. Martin-in-the-Fields (1927-40); army chaplain, 1914-19; DSO, 1917.

McCORMICK, SIR WILLIAM SYMINGTON (1859-1930), scholar and administrator; grandson of Revd William Symington [q.v.]; MA, Glasgow University; professor of English literature, University College, Dundee, 1890; first secretary of Carnegie Trust for Universities of Scotland, 1901; connected with every important government step to aid university education from 1906 onwards; chairman of Advisory Council for Scientific and Industrial Research, 1915; of Treasury University Grants Committee, 1919; knighted, 1911; FRS, 1928; died at sea; works include *The MSS. of Chaucer's Canterbury Tales* (posthumous, 1933).

McCREERY, SIR RICHARD LOUDON (1898-1967), general; educated at Eton and Sandhurst; commissioned in 12th Royal Lancers, 1915; served in France, 1915-18; MC; fine horseman; attended staff College, 1928; brigade-major, cavalry brigade; commanded 12th Lancers, 1935-8; colonel, 1938; commanded 2nd Armoured Brigade; DSO, 1940; major-general; commanded armoured division in England, 1940-2; chief of staff to General Alexander (later Earl Alexander of Tunis, q.v.), Cairo, 1942; lieutenant-general; commanded X corps, 1943; CB, KCB, 1943; commanded Eighth Army, 1944-5; commanded British Forces of Occupation, Austria, 1945-6; KBE, 1945; full general, 1946; commanded British Army of the Rhine, 1946-8; British Army representative, UN Military Staff Committee, 1948-9; GCB, 1949; colonel commandant, RAC, 1947-56.

McCUDDEN, JAMES THOMAS BYFORD (1895-1918), airman; joined Royal Flying Corps, 1913; went to France as air mechanic, 1914; learned to fly, 1916; became leading British fighting pilot; brought down fifty-four enemy aeroplanes; MC and DSO, 1917; VC, 1918; killed flying in France.

MacCUNN, HAMISH (JAMES) (1868-1916), musical composer; studied at Royal College of Music; orchestral and choral works include 'Land of the Mountain and Flood' (1887), 'Ship o' the Fiend', 'Dowie Dens o' Yarrow', 'Lord Ullin's Daughter', 'Lay of the Last Minstrel', and 'Bonny Kilmeny' (all 1888); composed operas.

MacDERMOT, HUGH HYACINTH O'RORKE, THE MACDERMOT (1834-1904), attorney-general for Ireland; called to bar

(King's Inns, Dublin), 1862; bencher, 1884; counsel in leading political cases in Ireland; succeeded father as titular 'Prince of Coolavin', 1873; liberal solicitor-general for Ireland, May-July 1885 and Feb.-Aug. 1886; attorney-general and PC, 1892-5.

MACDERMOTT, GILBERT HASTINGS (1845-1901), music-hall singer, whose real surname was FARRELL; made some position as an actor and writer of melodramas which included *Driven from Home*, 1871; leaped into fame (1878) by singing on music-hall stage the patriotic song by George William Hunt (1829-1904), 'We don't want to fight', which became popular watchword of war-party in England during Russo-Turkish war and gave the political terms 'jingo' and 'jingoism' to the English language; last 'lion comique' of English music-hall.

MacDERMOTT, MARTIN (1823-1905), Irish poet and architect; wrote occasional verse for *Nation* from 1840; delegate to Paris to obtain French republican support for Young Ireland movement, 1848; chief architect to Egyptian government in Alexandria from 1866; retired to London, 1878; prepared anthology of Irish poetry, 1894.

MACDONALD, SIR CLAUDE MAXWELL (1852-1915), soldier and diplomatist; joined army, 1872; minister at Peking, 1896; organized defence of legations during Boxer rising, 1900; minister at Tokyo, 1900; ambassador, 1905-12; promoted Anglo-Japanese friendship; KCB, 1898; military KCB, 1901; PC, 1906.

MACDONALD, SIR GEORGE (1862-1940), numismatist, classical scholar and archaeologist, and civil servant; first class, *lit. hum.*, Balliol College, Oxford, 1887; lecturer in Greek, Glasgow, 1892-1904; assistant secretary, Scottish Education Department (1904), secretary, 1922-8; catalogued Greek and Roman coins in Hunterian collection, Glasgow; authority on Romano-British history; FBA, 1913; KCB, 1927.

MacDONALD, GEORGE (1824-1905), poet and novelist; MA, King's College, Aberdeen, 1845; hon. LLD, 1868; settled in Manchester, 1853; published narrative poem, *Within and Without* (1855, admired by Tennyson and Lady Byron), and *Phantastes* (1858), a faerie romance in prose; thenceforth largely engaged in prose fiction—either of mystical character, as *David Elginbrod* (1863), or descriptive of Scottish humble life, as *Alec Forbes* (1865) and *Robert Falconer* (1868); long preached as a layman at Manchester; settled in London, 1860; friend of F. D. Maurice, Browning, Ruskin, Carlyle,

William Morris, and Tennyson [qq.v.]; lectured in London and in America, 1872; granted civil list pension, 1877; spent part of year at Bordighera, 1881-1902; his works include the children's stories *At the Back of the North Wind* (1871) and *The Princess and the Goblin* (1872), and *Unspoken Sermons* (3 vols., 1867-89) and *Letters from Hell* (1884); a collective edition, excluding novels, appeared in 1886 (10 vols.,), and his *Poetical Works* (2 vols.) in 1893.

MACDONALD, SIR HECTOR ARCHIBALD (1853-1903), major-general; joined army as private in Gordon Highlanders, 1870; served with distinction in second Afghan war, 1879-80; won sobriquet of 'Fighting Mac'; promoted second lieutenant, 1880; displayed heroism at battle of Majuba, 1881; shared in Nile expedition (1885), and in reorganization of Egyptian army; distinguished in Sudan campaign, 1888-91; commanded brigade of Egyptian infantry in expedition to Dongola (1896) and at Atbara (1898); displayed successful adroitness at Omdurman, Sept. 1898; CB, 1897; brigadier-general in Punjab, 1899-1900; major-general, Jan. 1900; in South African war prepared way for relief of Kimberley by seizure of Koodoesberg, Feb. 1900; engaged in actions which led to surrender of generals Cronje and Prinsloo, Feb.-May 1900; KCB, 1900; placed in command of Belgaum district, 1901, and of troops in Ceylon, 1902; owing to opprobrious accusation, he shot himself in Paris.

MACDONALD, HECTOR MUNRO (1865-1935), mathematical physicist; educated at Aberdeen University and Clare College, Cambridge; fourth wrangler, 1889; fellow, 1890-1908; made life study of radiation, transmission, and reflection of electric waves; publications include *Electric Waves* (1902) and *Electro-magnetism* (1934); FRS, 1901; professor of mathematics, Aberdeen, 1904-35; member of university court from 1907 especially interested in oversight of university lands and buildings.

MacDONALD, JAMES RAMSAY (1866-1937), labour leader and statesman; born illegitimate at Lossiemouth, Morayshire, of farming stock; educated locally and became pupil teacher; joined Social Democratic Federation (1885) and Fabian Society (1886); experienced some years of extreme poverty and intensive study in London; defeated as independent labour party candidate, Southampton, 1895; earned living by journalism and contributed to this Dictionary; obtained financial independence, opportunity for world-wide travel, and upper middle-class background through marriage (1896) to Margaret Ethel (*d.*

1911), daughter of John Hall Gladstone, a distinguished scientist and active social worker; secretary, labour representation committee (later the labour party), 1900-12; treasurer, 1912-24; MP, Leicester, 1906-18; early showed himself a natural parliamentarian; his books (among them *Socialism and Society*, 1905, and *Socialism*, 1907) marked him also as a theorist; chairman (1906-9) and for long a leading figure in independent labour party in which he implanted his own instinct for moderation; chairman, parliamentary labour group, 1911; unsupported in opposing a war credit and resigned, 5 Aug. 1914; his view that although war must be won Britain had been wrong to embark on it and must preserve a generous temper for sake of future peace widely misunderstood and misrepresented; became greatly mistrusted; heavily defeated at Leicester, 1918; his courage in face of bitter attack gained his party's respect; persuaded both independent labour party and labour party to reject communism, 1920; MP, Aberavon division, 1922-9; became chairman of parliamentary labour party and leader of opposition, 1922; PC, first labour prime minister, and foreign secretary, Jan. 1924; his diplomacy more successful than his domestic policy; defeated over Campbell case, Oct. 1924; sought to avert but acquiesced in general strike, 1926; MP, Seaham division, and prime minister for second time, 1929; chiefly interested in foreign affairs, brought about and presided over London naval conference, 1930; presided over Indian Round Table conference, 1930; tendered resignation of government, 23 Aug. 1931, after failure of cabinet to agree on reduction in payments to unemployed in face of financial crisis; formed all-party government in conjunction with conservative and liberal leaders; breach with his own party became permanent; on winning election, Oct. 1931, formed fourth administration; pressed on with programme of retrenchment and reform; continued to regard European situation key to domestic recovery and believed in personal diplomacy; after rise of Hitler realized necessity of rearmament and drafted white paper on national defence, 1935; resigned premiership and became lord president of the Council, June 1935; defeated at Seaham, 1935; MP for Scottish Universities, 1936-7.

MACDONALD, Sir JAMES RONALD LESLIE (1862-1927), major-general; gazetted to Royal Engineers, 1882; attached to military works department, India, 1885-91; chief engineer on preliminary survey for projected Uganda railway; acting commissioner, Uganda protectorate, 1893; largely instrumental in securing its safety from rebels, 1897-9; director of balloons (afterwards of railways) for China

expeditionary force, 1900; commanded military escort of political mission dispatched to Tibet under (Sir) Francis Younghusband [q.v.], 1903-4; major-general, 1908; general officer commanding in Mauritius, 1902-12; KCIE, 1904.

McDONALD, JOHN BLAKE (1829-1901), Scottish artist; painted, largely in chiaroscuro, dramatic episodes of Jacobite romance, and later landscape; ARSA, 1862; RSA, 1877.

MACDONALD, Sir JOHN DENIS (1826-1908), inspector-general of hospitals and fleets, 1880-6; joined navy as assistant surgeon, 1849; engaged in microscopic study and deep-sea investigations; FRS, 1859; professor of naval hygiene at Netley, 1872; KCB, 1902; published *Outlines of Naval Hygiene* (1881).

MACDONALD, JOHN HAY ATHOLE, LORD KINGSBURGH (1836-1919), lord justice-clerk of Scotland; called to Scottish bar, 1859; solicitor-general for Scotland, 1876-80; QC, 1880; lord advocate, 1885-6, 1886-8; lord justice-clerk and assumed judicial title, 1888; presided over second division of Court of Session, 1888-1915; specialized in criminal law; PC, 1885; KCB, 1900; FRS.

MacDONALD, Sir MURDOCH (1866-1957), engineer; educated at Dr Bell's Institution, Inverness; apprenticed with Highland Railway; joined Sir Benjamin Baker [q.v.] as assistant engineer, Aswan Dam, 1898; remained in Egyptian government service until 1921 responsible for heightening Aswan Dam, construction of Esna Barrage, and schemes for Sennar and Gebel Aulia dams; founded London consulting firm, 1921; liberal (later national liberal) MP for Inverness, 1922-50; CMG, 1910; KCMG, 1914; CB, 1917; president, Institution of Civil Engineers, 1932.

MACDONELL, ARTHUR ANTHONY (1854-1930), Sanskrit scholar; born in India; educated at Dresden and Göttingen; BA, Corpus Christi College, Oxford; deputy to Boden professor of Sanskrit at Oxford and keeper of Indian Institute, 1888; succeeded to these offices and to professorial fellowship at Balliol College, 1899; retired, 1926; raised funds for critical edition of Sanskrit epic, *Mahā-Bhārata*; visited India, 1907-8 and 1922-3; FBA, 1906; as Sanskrit scholar worked chiefly in Vedic field; works include editions of *Sarvānukramaṇī* (1886) and *Brhad-devtā* (1904), *Vedic Mythology* (1897), and *Vedic Grammar* (1910).

MacDONELL, Sir HUGH GUION (1832-1904), soldier and diplomatist; served in British

Kaffraria with Rifle Brigade, 1849–53; after holding several minor diplomatic posts (1858–85) was British envoy at Rio, 1885, Copenhagen, 1888, and Lisbon 1893–1902; tactfully dealt with Anglo-Portuguese difficulties in South African war; KCMG, 1892; GCMG, 1899; PC, 1902.

MACDONELL, Sir JOHN (1845–1921), jurist; brother of James Macdonell [q.v.]; MA, Aberdeen; called to bar (Middle Temple), 1873; master of Supreme Court, 1889; senior master and King's remembrancer, 1912–20; Quain professor of comparative law, University College, London, 1901–20; knighted, 1903; KCB, 1914; FBA, 1913; writer on legal subjects.

MACDONELL, Sir PHILIP JAMES (1873–1940), colonial judge; son of James Macdonell [q.v.]; educated at Clifton and Brasenose College, Oxford; first class, modern history, 1894; president of Union, 1895; called to bar (Gray's Inn), 1900; war correspondent of *The Times* in South Africa, 1900–2; secretary, Transvaal native commission, 1903; public prosecutor (1908–18), high court judge (1918–27), Northern Rhodesia; chief justice, Trinidad and Tobago (president, West Indian Court of Appeal), 1927–30; of Ceylon, 1930–6; knighted, 1925; PC, 1939.

MacDONNELL, ANTONY PATRICK, Baron MacDonnell (1844–1925), statesman; BA, Queen's College, Galway; entered Indian civil service, 1865; accountant-general and later revenue secretary to provincial government, Calcutta, 1881; largely responsible for Tenancy Act, 1885; home secretary to central government, 1886–9; chief commissioner, Central Provinces, 1890; lieutenant-governor, North-Western Provinces and Oudh, 1895; his period of office marked by severe visitations of plague and famine, which he combated with notable energy and ability; chairman of important famine commission, 1901; permanent under-secretary of state, Ireland, 1902–8; attempted to lift long-standing Irish quarrel above bitterness of party warfare, but his administration regarded with suspicion by both parties in Ireland; helped George Wyndham [q.v.], chief secretary for Ireland, to prepare and shape Irish land purchase bill, 1903; unjustly censured by cabinet for his assistance to Lord Dunraven [q.v.] and Irish Reform Association in preparing scheme of 'devolution' for Ireland; secured support of (Lord) Bryce [q.v.] when he became chief secretary for Ireland, 1905–7; his new scheme of devolution, which formed basis of Irish council bill (1907), approved by Bryce; unsupported by A. Birrell [q.v.] in his devolution scheme and in attempts to suppress disorder; after resignation

continued to take part in public life; created baron, 1908.

McDONNELL, Sir SCHOMBERG KERR (1861–1915), civil servant; principal private secretary to Marquess of Salisbury, 1888–92; 1895–9, 1900–2; secretary to Office of Works, 1902–12; KCB, 1902; died of wounds in Flanders.

McDOUGALL, WILLIAM (1871–1938), psychologist; educated at Owens College, Manchester, St. John's College, Cambridge (fellow, 1897), and St. Thomas's Hospital; Wilde reader in mental philosophy, Oxford, 1903; professor of psychology, Harvard (1920), Duke University, North Carolina 1927–38; works include his influential *Introduction to Social Psychology* (1908) and *Outlines of Psychology* (1923) and *Abnormal Psychology* (1926); powerful advocate of the idealistic outlook; investigated inheritance of acquired characteristics; FRS, 1912.

MACE, JAMES (Jem Mace) (1831–1910), pugilist; at first a showman and circus performer; became best boxer of his generation; middle-weight champion, 1860; beat Thomas King [q.v.] and Joe Goss for championship, 1862–6; last surviving representative of old prize ring.

McEVOY, ARTHUR AMBROSE (1878–1927), painter; entered Slade School of Fine Art, 1893; began as painter of poetical landscapes and restful interiors; later became popular as portrait painter; executed series of portraits in Imperial War Museum; capable of rendering masculine qualities, but his subjective treatment of women's portraits gave him unique place among contemporary English portrait painters.

McEWEN, Sir JOHN BLACKWOOD (1868–1948), principal of Royal Academy of Music; MA, Glasgow, 1888; entered Royal Academy of Music, 1893; professor of harmony and composition, 1898–1924; principal, 1924–36; pioneer of renascence of chamber music composition; knighted, 1931.

MACEWEN, Sir WILLIAM (1848–1924), surgeon; BM, CM, Glasgow University; assistant surgeon, Royal Infirmary, Glasgow, 1875; full surgeon, 1877; regius professor of surgery, Glasgow, 1892–1924; FRS, 1895; knighted, 1902; his most important contributions to surgery made in brain surgery and bone surgery; first surgeon deliberately to operate for brain disorders; laid basis of modern brain surgery, especially in mastoid disease; introduced method of implanting small grafts to replace missing parts of limb-bones and new method of rectifying knock-knee.

MACFADYEN, ALLAN (1860-1907), bacteriologist; MB, CM, Edinburgh University, 1883; MD, 1886; B.Sc. in hygiene, 1888; lecturer on bacteriology at College of State Medicine, subsequently amalgamated with Lister Institute of Preventive Medicine, from 1889; director, 1891; secretary, 1903; planned and organized Lister Institute; Fullerian professor of physiology at Royal Institution, 1901-4; lectures published as *The Cell as the Unit of Life* (posthumous, 1908).

MACFADYEN, Sir ERIC (1879-1966), rubber industry pioneer; educated at Lynams (the Dragon) School, Oxford, Clifton College and Wadham College, Oxford; president of the Union, 1902; served in South African war, 1901-2; entered Malayan civil service, 1902; resigned and became rubber planter, 1905; senior partner, Macfadyen Wilde & Company; chairman, Planters' Association; member, Federal Council, Federated Malay States, 1911-16 and 1919-20; joined board of Harrisons & Crosfield Ltd., East India merchants, 1919; director, 1919-55; liberal MP, Devizes, 1923-4; chairman, Ross Institute, 1946-58; chairman, Letchworth Garden City; knighted, 1943; life president, Imperial College of Agriculture, Trinidad.

McFADYEN, JOHN EDGAR (1870-1933), biblical scholar; educated at Glasgow University, Balliol College, Oxford, and the Free Church College, Glasgow; professor of Old Testament literature and exegesis, Knox College, Toronto, 1898-1910; of Old Testament language, literature, and theology, Trinity College, Glasgow, 1910-33.

MACFARLANE, Sir (FRANK) NOEL MASON- (1889-1953), lieutenant-general. [See MASON-MACFARLANE.]

MACFARREN, WALTER CECIL (1826-1905), pianist and composer; brother of Sir George A. Macfarren [q.v.]; studied at Royal Academy of Music; sub-professor of pianoforte there from 1846; composed pianoforte pieces in style of Mendelssohn, vocal works, church services, and overtures; published *Scale and Arpeggio Manual* (1882) and autobiographical *Memories* (1905).

MacGILLIVRAY, Sir DONALD CHARLES (1906-1966), colonial administrator; educated at Sherborne School and Trinity College, Oxford, 1920-9; entered Colonial administrative service, 1928; posted to Tanganyika,

1929; district officer, 1930-4; private secretary to governor, Sir Harold A. MacMichael [q.v.], 1935-8; posted to Palestine with Sir Harold, 1938-42; district officer, Galilee, and then, Samaria, 1942-4; under-secretary, Palestine Government; colonial secretary, Jamaica, 1947-52; deputy high commissioner, Malaya, 1952-4; civil high commissioner, Malaya, 1954-7; retired to Kenya; chairman, Council of Makerere College, Uganda, 1958-61; chairman, Council of East Africa University, 1961-4; director, UN Special Fund for East Africa Livestock Development Survey, 1964; MBE, 1936; CMG, 1949; KCMG, 1953; GCMG, 1957.

McGOWAN, HARRY DUNCAN, first BARON McGOWAN (1874-1961), businessman; educated at Huthesontown school and Allan Glen's School, Glasgow; at fifteen joined Nobel's Explosives Co. as office boy; became assistant to general manager; assisted in constructing Canadian Explosives Ltd., later Canadian Industries Ltd., largest chemical business in Canada, 1909-11; chairman and managing director, Explosives Trades Ltd., 1918; joined board, British Dyestuffs Corporation, 1919; agreed with Sir Alfred Mond (later Lord Melchett, q.v.) on merger of British chemical firms and formed ICI, 1926; chairman and sole managing director, ICI, 1930-50; KBE, 1918; baron, 1937.

McGRATH, Sir PATRICK THOMAS (1868-1929), statesman and journalist; born in Newfoundland; editor of Newfoundland *Evening Herald*, 1894-1907; assistant clerk, Newfoundland house of assembly, 1897-1900; chief clerk, 1900-11; president, legislative council, 1915-19; 1925-9; KBE, 1918.

MacGREGOR, Sir EVAN (1842-1926), civil servant; entered Admiralty, 1860; private secretary to successive senior naval lords, 1869-79; principal clerk in secretariat and head of military branch which dealt with fleet operations and political work, 1880; permanent secretary of Admiralty, 1884-1907; his period of office one of immense development both in fleet and in administration of navy, witnessing almost entire rebuilding of navy, construction of new harbours, barracks, and dockyards, reforms in naval education and training and in distribution and organization of fleet and construction of new great fleet; KCB, 1892.

MacGREGOR, JAMES (1832-1910), moderator of the General Assembly of the Church of Scotland, 1891; student of St. Andrews University, 1848-55; hon. DD, 1870; served churches at Paisley, Glasgow, and Edinburgh; first minister of St. Cuthbert's, Edinburgh; 1873-1910; a fervent and popular preacher; visited Canada, 1881, and Australia, 1889.

MacGREGOR, Sir WILLIAM (1846-1919), colonial governor; MD, Aberdeen, 1874; chief medical officer for Fiji, 1875; first administrator (styled lieutenant-governor, 1895) of British New Guinea, 1888; promoted policy of peaceful penetration and exploration; governor of Lagos, where he carried on campaign against malaria, 1899; governor of Newfoundland, 1904; organized and conducted important scientific expedition to Labrador; governor of Queensland, 1909-14; KCMG, 1889; GCMG, 1907; PC, 1914.

McGRIGOR, Sir RHODERICK ROBERT (1893-1959), admiral of the fleet; during 1914-18 war, served in destroyers in Dardanelles campaign, and in *Malaya* at the battle of Jutland; commanded *Renown*, 1940-1, in *Bismarck* action and bombardment of Genoa; rear-admiral, 1941; assistant chief of naval staff (weapons), 1941-3; Force commander for capture of Pantelleria and invasion of Sicily; naval commander, southern Italy, 1944; commanded first cruiser squadron, Scapa Flow, 1944-5; vice-chief of naval staff, 1945-7; admiral, 1948; commander-in-chief, Home Fleet, 1948-50, Plymouth, 1950-1; first sea lord, 1951-5; CB, 1944; KCB, 1945; GCB, 1951; admiral of the fleet, 1953.

MACHELL, JAMES OCTAVIUS (1837-1902), owner and manager of racehorses; joined army, 1854; retired as captain, 1863; won Derby with Hermit, 1867; superintended training of Isinglass [See McCALMONT, HARRY] and of many other winning horses; his own horses won 540 races, 1864-1902; a good athlete.

MACHRAY, ROBERT (1831-1904), archbishop of Rupert's Land; of Presbyterian parentage; MA, Aberdeen, 1851; BA, Sidney Sussex College, Cambridge, 1855; MA, 1858; vicar of Madingley, 1862-5; bishop of Rupert's Land, 1865; reorganized St. John's College, Winnipeg; chancellor of Manitoba University, 1877-1904; subdivided diocese into eight sees; metropolitan of Canada, 1875; archbishop of Rupert's Land and primate of all Canada, 1893.

McINDOE, Sir ARCHIBALD HECTOR (1900-1960), plastic surgeon; educated at Otago high school and university, New Zealand; MB, Ch.B., 1924; postgraduate, Mayo Clinic; MS (Rochester), 1930; joined cousin Sir Harold Gillies (q.v.) in London; FRCS (England), 1932; appointed consultant in plastic surgery to Royal Air Force, 1939; organized centre at

Queen Victoria Hospital, East Grinstead, for treatment and rehabilitation; on council, Royal College of Surgeons, 1948-60; a founder, British Association of Plastic Surgeons; skilful surgeon and powerful and independent personality; CBE, 1944; knighted, 1947.

McINTOSH, WILLIAM CARMICHAEL (1838-1931), zoologist; MD, Edinburgh, 1860; medical superintendent, Perth District Asylum, 1863-82; professor of zoology (1882-1917), St. Andrews, where he established first marine laboratory in United Kingdom; works include *Monograph of the British Marine Annelids* (4 vols., 1873-1923) and report on the polychaete worms obtained by *Challenger* expedition (1885); FRS, 1877.

MACINTYRE, DONALD (1831-1903), major-general, Bengal staff corps; won VC for gallantry in Lushai expedition, 1871-2; commanded 2nd Prince of Wales's Own Ghurkhas with Khyber column in Afghan war, 1878-9, and in Bazar valley expeditions; major-general, 1880; published account of travel and sport in Himalayas, 1889.

MacIVER, DAVID RANDALL- (1873-1945), archaeologist and anthropologist. [See RANDALL-MacIVER.]

MACKAIL, JOHN WILLIAM (1859-1945), classical scholar, literary critic, and poet; educated at Ayr Academy, Edinburgh University and Balliol College, Oxford; first class, *lit hum.*, 1881; fellow, 1882; entered Education Department (later Board of Education), 1884; assistant secretary, 1903-19; professor of poetry, Oxford, 1906-11 (lectures published in 3 vols., 1909-11); other publications include translation of the *Aeneid* (1885), of the *Eclogues* and *Georgics* (1889), *Select Epigrams from the Greek Anthology* (1890), *Latin Literature* (1895), *Life of William Morris* (1899), translation of the *Odyssey* (3 vols., 1903-10), *Virgil and his Meaning to the World of To-day* (1923), *Approach to Shakespeare* (1930), and an edition of the *Aeneid* (1930); fellow (1914), president (1932-6), British Academy; OM, 1935.

MACKAY, ÆNEAS JAMES GEORGE (1839-1911), legal and historical writer; BA, University College, Oxford, 1862; MA, 1865; admitted to Scottish bar, 1864; professor of constitutional law, Edinburgh University, 1874; advocate depute, 1881; sheriff principal of Fife and Kinross, 1886-1901; LLD, Edinburgh, 1882; works include *The Practice of the Court of Session* (2 vols., 1877-9), *William Dunbar* (1889), and *A History of Fife and Kinross* (1896).

MACKAY, ALEXANDER (1833-1902), promoter of education in Scotland; MA, St. Andrews; LLD, 1891; developed educational methods and organization as schoolmaster at Torryburn; editor of *Educational News*, 1878; president of Educational Institute of Scotland, 1881; member of Edinburgh school board, 1897; published educational works.

MACKAY, DONALD JAMES, eleventh BARON REAY (1839-1921), governor of Bombay (1885-90), and first president of British Academy (1902-7); born at The Hague; educated at Leiden; entered Dutch foreign office; settled in England, 1875; naturalized, 1877; created baron of United Kingdom, 1881; under-secretary of state for India, 1894-5; interested in international law and politics.

MACKAY, JAMES LYLE, first EARL OF INCHCAPE (1852-1932), shipowner; joined staff of Mackinnon, Mackenzie & Co., Calcutta, 1874; became senior partner; served on Bengal legislative council, 1891-3; returned to take charge of London office of British India Company, 1893; chairman, 1914; carried through fusion with P & O and other companies into a group with capital of £23 million and nearly 2 million tonnage; director and chairman of numerous shipping and banking concerns; thrice president of UK chamber of shipping; negotiated commercial treaty with China, 1901-2; served on council of India (1897-1911), Imperial Defence Committee (1917), and Geddes economy committee (1921-2); chairman, Indian retrenchment committee, 1922-3; disposed of wartime shipping for government for £35 millions at expense of £850; ardent free trader; crossed from liberal to conservative benches, 1926; KCIE, 1894; baron, 1911; viscount, 1924; earl, 1929.

MACKAY, MARY (1855-1924), novelist; known as MARIE CORELLI; daughter of Charles Mackay [q.v.]; showed precocious talent for piano-playing; published her first article, signed 'Marie Corelli', 1885; published her first novel, 1886; achieved popularity with *Barabbas* (1893) and *The Sorrows of Satan* (1895) which secured hysterical triumph and marked climax of her career as popular novelist in Great Britain, and reinforced her determination to flout critics; later works include *The Mighty Atom* (1896); voiced mass-sentiment of pre-war bourgeoisie; settled in Stratford-upon-Avon, to which she was a benefactress, 1901.

McKECHNIE, WILLIAM SHARP (1863-1930), constitutional historian; MA, Glasgow University; solicitor in Glasgow, 1890-1915; lecturer in constitutional law and history, Glasgow University, 1894; professor of conveyancing, Glasgow, 1916-27; his chief work, *Magna Carta* (1905), in which the Charter is represented as a feudal document only accidentally serving interests of others than the baronial class; other works include *The State and the Individual* (1896) and *The New Democracy and the Constitution* (1912).

McKENNA, REGINALD (1863-1943), statesman and banker; scholar of Trinity Hall, Cambridge; senior optime, 1885; rowed for Cambridge, 1887; called to bar (Inner Temple), 1887; liberal MP, North Monmouthshire, 1895-1918; financial secretary to the Treasury, 1905-7; PC, 1907; president of Board of Education, 1907-8; first lord of the Admiralty, 1908-11; convinced of German danger, obtained construction of eight *Dreadnoughts* in 1909 and five each in 1910-11; refused to give way to War Office on question of strategy and transferred to Home Office (1911-15); attempted to solve suffragette problem by 'Cat and Mouse Act', 1913; chancellor of Exchequer, May 1915-Dec. 1916; obtained $40 million securities from Prudential Assurance Company for new American contracts, 1915; by increasing income-tax and imposing import and other taxes met running cost of war in two budgets, Sept. 1915 and Apr. 1916; opposed conscription but remained in office until Asquith's resignation; refused chancellorship of Exchequer, 1922; director (1917), chairman (1919-43), Midland Bank; published *Post-War Banking Policy* (1928).

MACKENNAL, ALEXANDER (1835-1904), Congregational minister; educated at Glasgow University, 1851-4; hon. DD, 1887; BA, London, 1857; minister at Bowdon, Cheshire, 1877 to death; chairman of Congregational Union, 1887; frequently visited America from 1889; advocated co-operative union of Churches; secretary (1892-8) and president (1899) of National Free Church Council; published many theological works, and life of J. A. Macfadyen, DD, 1891.

MACKENNAL, SIR (EDGAR) BERTRAM (1863-1931), Australian sculptor; studied in Melbourne and Paris; works include obverse of new coinage (1910); King Edward VII memorial, St. George's chapel, Windsor (1921), and equestrian statue, Waterloo Place; war memorial, members of both Houses of Parliament, St. Stephen's Hall; brilliant all-round sculptor, particularly in marble; ARA, 1909; RA, 1922; KCVO, 1921.

MACKENZIE, SIR ALEXANDER (1842-1902), lieutenant-governor of Bengal; joined Indian civil service, 1862; under-secretary to local government, Bengal, 1866; home secretary to government of India, 1882; helped to shape Bengal Tenancy Act of 1885; CSI, 1886; chief commissioner of Central Provinces, 1887-90, and of Burma, 1890-5; KCSI, 1891; suppressed hill tribe raids and restored order, 1892; lieutenant-governor of Bengal, 1895-8; made sanitary survey of Calcutta; enlarged powers of Bengal municipalities; co-operated with Assam in Lushai expedition of 1895-6; expedited land settlement operations in Bihar and Orissa; dealt efficiently with the famine of 1896-7 and the plague; published *History of the Relations of Government with the Hill Tribes of the North-East Frontier of Bengal* (1884).

McKENZIE, ALEXANDER (1869-1951), professor of chemistry; educated at Dundee high school and St. Andrews; B.Sc., chemistry and natural philosophy, 1891; lecture assistant, 1891-3; university assistant, 1893-8; Ph.D., Berlin, 1901; assistant lecturer, chemistry, Birmingham, 1902; head, chemistry department, Birkbeck College, London, 1905-13; professor of chemistry, Dundee, 1914-38; researched in stereochemical field; FRS, 1916; hon. LLD, St. Andrews, 1939.

MACKENZIE, Sir ALEXANDER CAMPBELL (1847-1935), composer and principal of the Royal Academy of Music; studied there; violinist, organist, and conductor in Edinburgh, 1866-81; lived in Florence, 1881-8; principal, Royal Academy of Music, 1888-1924; broadened its musical education; with Royal College of Music founded an examining body in the Associated Board; conductor, Philharmonic Society's concerts, 1892-9; president, International Musical Society, 1908-12; compositions include operas *Colomba* and *The Cricket on the Hearth*; oratorio 'Rose of Sharon', 'The Cotter's Saturday Night' for chorus and orchestra; cantata 'The Bride'; Scottish rhapsodies; overture 'Cervantes'; knighted, 1895; KCVO, 1922.

MACKENZIE, Sir GEORGE SUTHERLAND (1844-1910), explorer and administrator; born at Bolarum, India; as representative of a London firm of East India merchants (1866) opened up trade route through Persian interior from Persian Gulf; pioneer explorer of Persian interior; managing director of Imperial British East Africa Company, 1888; developed East Africa, and explored interior as far as Uganda; CB, 1897; KCMG, 1902.

MACKENZIE, Sir JAMES (1853-1925), physician and clinical researcher; MA, Edinburgh University, 1878; practised at Burnley, 1879-1907; consulting practitioner, London,

1907; FRS, 1915; knighted, 1915; best known for his researches into the nature of irregularities of heart's rhythm; his works include *The Study of the Pulse* (1902) and *Diseases of the Heart* (1908).

M'KENZIE, Sir JOHN (1836-1901), minister of lands in New Zealand; emigrated from Ross-shire to New Zealand, 1860; minister of lands and immigration, 1881-1900; successfully purchased and divided large estates among small farmers; passed Lands for Settlement Act, 1894; member of legislative council and KCMG, 1901.

MACKENZIE, JOHN STUART (1860-1935), philosopher; educated at Glasgow high school and university and Trinity College, Cambridge; first class, moral sciences tripos, 1889; fellow, 1890-6; professor of logic and philosophy, University College, Cardiff, 1895-1915; works include *Manual of Ethics* (1893), *Elements of Constructive Philosophy* (1917), *Outlines of Social Philosophy* (1918), *Fundamental Problems of Life* (1928), and *Cosmic Problems* (1931); a neo-Hegelian idealist; FBA, 1934.

McKENZIE, (ROBERT) TAIT (1867-1938), Canadian sculptor and expert in physical culture; BA (1889), MD (1892), McGill University; practised as orthopaedic surgeon in Montreal; professor of physical education, university of Pennsylvania, 1904-31; exhibited sculpture giving direct plastic expression to his ruling passion for bodily fitness; other work included the Scottish-American (Princess Street, Edinburgh) and Cambridge war memorials.

MACKENZIE, Sir STEPHEN (1844-1909), physician; brother of Sir Morell Mackenzie [q.v.]; student at London Hospital, 1866; MRCS, 1869; MB, Aberdeen, 1873; MD, 1875; FRCP, 1879; physician at London Hospital, 1886, and at London Ophthalmic Hospital, 1884; made original researches into skin diseases and ophthalmology; knighted, 1903.

MACKENZIE, Sir WILLIAM (1849-1923), Canadian financier and railway builder; born in Upper Canada; became contractor, 1871; partner with (Sir) Donald Mann, 1886; their first line, Lake Manitoba Railway; Canadian Northern Railway Company incorporated, 1899; line completed from Port Arthur to Winnipeg, 1902; from Winnipeg to Edmonton, 1905; by 1915 trains were running from Montreal to Vancouver, financed all these operations; knighted, 1911; died at Toronto.

MACKENZIE, WILLIAM WARRENDER, first Baron Amulree (1860-1942), lawyer and industrial arbitrator; educated at Perth Academy and Edinburgh University, called to bar (Lincoln's Inn), 1886; KC, 1914; a chairman of committee on production, 1917-19; first president, Industrial Court, 1919-26; chairman, railway national wages board (1920-6), royal commissions on licensing (1929-31) and Newfoundland (1933), and many committees; secretary of state for air, 1930-1; KBE, 1918; baron, 1929; PC, 1930.

MACKENZIE KING, WILLIAM LYON (1874-1950), Canadian statesman. [See KING.]

McKERROW, RONALD BRUNLEES (1872-1940), scholar and bibliographer; educated at Harrow, King's College, London, and Trinity College, Cambridge; joint honorary secretary, Bibliographical Society, from 1912; founded and edited (1925-40) *Review of English Studies*; published *Introduction to Bibliography for Literary Students* (1927), an edition of Thomas Nashe (1904-10), and prepared substantial portion of the Oxford Shakespeare; FBA, 1932.

McKIE, DOUGLAS (1896-1967), historian of science; educated at Tredegar grammar school, Sandhurst, and University College, London; served in France, 1915-17; severely wounded, 1917, and left army, 1920; first class, chemistry, B.Sc., 1923; Ph.D., 1928; part-time assistant in department of history and philosophy of science, University College, 1925; lecturer, 1934; reader, 1946; professor, 1957; emeritus professor, 1964; publications include *The Discovery of Specific and Latent Heat* (with N. H. de V. Heathcote, 1935), *The Essays of Jean Ray* (1951); and books on Lavoisier (1935 and 1952), for which he was made chevalier of the Legion of Honour; also contributed to the *New Cambridge Modern History;* founded and edited the *Annals of Science,* 1936-67; FRS, Edinburgh, 1958; fellow, University College, London, Royal Institute of Chemistry, Royal Society of Arts, and Society of Antiquaries.

MACKINDER, SIR HALFORD JOHN (1861-1947), geographer and politician; educated at Epsom College and Christ Church, Oxford; first class, natural science and president of Union, 1883; second class, history, 1884; called to bar (Inner Temple), 1886; reader in geography, 1887-1905; principal, University College, Reading, 1892-1903; director, London School of Economics, 1903-8; taught economic geography, London University, 1895-1925; unionist MP, Camlachie division, Glasgow, 1910-22; British high commissioner, south Russia, 1919-20; chairman, imperial shipping

committee, 1920-45; of imperial economic committee, 1925-31; knighted, 1920; PC, 1926.

MacKINLAY, ANTOINETTE (1843-1904), contralto singer. [See STERLING.]

MacKINNON, SIR FRANK DOUGLAS (1871-1946), judge and author; educated at Highgate School and Trinity College, Oxford; second class, *lit. hum.*, 1894; called to bar (Inner Temple), 1897; pupil and close associate of Sir T. E. Scrutton [q.v.]; KC, 1914; bencher, 1923; treasurer, 1945; regular leader in commercial cases; judge of King's Bench division, 1924-37; sat frequently in Commercial Court and went regularly on circuit; lord justice of appeal, 1937-46; especially interested in eighteenth century; publications include edition of *Evelina* (1930), *The Murder in the Temple and other Holiday Tasks* (1935), *Grand Larceny* (1937), *On Circuit* (1940), and the lives of a number of lawyers in this Dictionary; knighted, 1924; PC, 1937.

MACKINNON, SIR WILLIAM HENRY (1852-1929), general; appointed to Grenadier Guards, 1870; colonel commandant, City of London imperial volunteers, Dec. 1899; took unit out to South Africa, Jan. 1900; its most outstanding experiences battles of Doornkop (29 May) and Diamond Hill (12 June), but all its achievements completely justified faith of its commandant, whose military reputation was enhanced by its exploits; director-general of newly formed Territorial Force, 1908; general officer commanding-in-chief, Western command, 1910; full general, 1913; KCB, 1908.

MACKINTOSH, SIR ALEXANDER (1858-1948), parliamentary correspondent; educated at Macduff and Aberdeen University; on *Aberdeen Free Press*, 1879-1922; joined London parliamentay staff, 1881; London editor, 1887; political correspondent, *Liverpool Daily Post*, 1923-38; knighted, 1932.

MACKINTOSH, CHARLES RENNIE (1868-1928), architect and painter; studied art and architecture in Glasgow; won competition for new building of Glasgow School of Art (1894) of great originality; Mackintosh exhibitions held in Venice, Munich, Dresden, Budapest, and Moscow; extremely influential on continent and founder of foreign school (*Jugendstil*); subsequently devoted himself with great success to water-colours.

MACKINTOSH, HAROLD VINCENT, first VISCOUNT MACKINTOSH OF HALIFAX (1891-1964), man of business and public servant; educated at Halifax New School and private grammar school; worked in family confectionary business in Germany, 1909-11; served in RNVR during 1914-18 war; chairman, family firm, 1920; president, Advertising Association, 1942-6; involved in negotiations for Halifax building societies amalgamation into 'The Halifax', 1928; chairman, National Savings Committee, 1943; president, National Sunday School Union, 1924-5; president, World Council of Christian Education and Sunday School Association, 1928-58; acquired A. J. Caley, chocolate business, Norwich; chairman, promotion committee, East Anglia University, 1959; president, Yorkshire Agricultural Society, 1928-9; president, Royal Norfolk Agricultural Society, 1960; published *Early English Figure Pottery* (1938); JP, 1925; deputy lieutenant, West Riding of Yorkshire, 1945; hon. LLD, Leeds, 1948; knighted, 1922; baronet, 1935; baron, 1948; viscount, 1957.

MACKINTOSH, HUGH ROSS (1870-1936), Scottish theologian; educated at George Watson's College and Edinburgh University; studied theology at New College, Edinburgh, and in Germany; ordained to Free Church ministry, 1897; professor of systematic theology, New College, Edinburgh, 1904-36; works include *The Doctrine of the Person of Jesus Christ* (1912) and *The Christian Experience of Forgiveness* (1927).

MACKINTOSH, JAMES MACALISTER (1891-1966), public health teacher and administrator; educated at Glasgow high school and Glasgow University; MA, 1912; served in France with 6th Cameron Highlanders, wounded, graduated MB, Ch.B., 1916; returned to France in RAMC, 1918; DPH, 1920; MD, 1923; called to bar (Gray's Inn) 1930; served as public health officer in Dorset, Burton-on-Trent and Leicestershire, 1920-30; county medical officer, Northampton, 1930-7; expert on rural housing; chief medical officer, Department of Health for Scotland, 1937-41; professor, public health, Glasgow, 1941-4; professor, public health, London, 1944-56, and dean, London School of Hygiene and Tropical Medicine, 1944-9; active in international field under World Health Organization; director WHO division of education and training, 1958-60; FRCP, London and Edinburgh, 1943; hon. LLD, Glasgow, 1950, and Birmingham, 1961; publications include *Housing and Family Life* (1952), *Teaching of Hygiene and Public Health in Europe* (with Professor Fred Grundy, 1957), and *Topics in Public Health* (1965).

MACKINTOSH, JOHN (1833-1907), Scottish historian; published *History of Civilization in Scotland*, 4 vols., 1878-88; LLD, Aberdeen, 1880.

MACKWORTH-YOUNG, GERARD (1884-1965), Indian civil servant and archaeologist; eldest son of (Sir) William Mackworth Young [q.v.]; educated at Eton and King's College, Cambridge; first class, part i, classical tripos, 1906; entered Indian civil service, 1907; journeyed in western Tibet, 1912; under-secretary, Punjab government, 1913; under-secretary, home department, government of India, 1916-19; deputy-commissioner, Delhi, 1921; deputy secretary, army department, government of India 1924; secretary, 1926; CIE, 1929; retired from ICS, 1932; published *The Epigrams of Callimachus* (1934), and *Archaic Marble Sculpture from the Acropolis* (with Humfry Payne, q.v., 1936); enrolled as student, British School of Archaeology, Athens, 1932; director, 1936-9; returned to India during 1939-45 war, joint secretary, war department; director, British School of Archaeology, Athens, 1945-6; published *What Happens in Singing* (1953).

McLACHLAN, ROBERT (1837-1904), entomologist; published *Catalogue of British Neuroptera* (1870) and *Synopsis of the Trichoptera of the European Fauna* (1874-84); FRS, 1877; president of Entomological Society, 1885-6, and editor from 1864 and proprietor (1902) of *Entomological Monthly Magazine*.

MACLAGAN, CHRISTIAN (1811-1901), Scottish archaeologist; devoted time and money to removal of slums in Stirling; made valuable researches into and published works on prehistoric remains in Scotland; prepared skilful rubbings from sculptured stones; lady associate of Scottish Society of Antiquaries, 1871.

MACLAGAN, Sir ERIC ROBERT DAL-RYMPLE (1879-1951), director of the Victoria and Albert Museum; son of W. D. Maclagan [q.v.], archbishop of York; educated at Winchester and Christ Church, Oxford; joined Victoria and Albert Museum in textiles department, 1905; transferred to architecture and sculpture, 1909; director of museum, 1924-45; popularized it and organized special exhibitions including English medieval art (1930) and William Morris centenary (1934); publications include *Guide to English Ecclesiastical Embroideries* (1907), catalogues of *Italian Plaquettes* (1924) and *Italian Sculpture* (with Margaret Longhurst, 1932), and *The Bayeux Tapestry* (King Penguin, 1943); CBE, 1919; knighted, 1933; KCVO, 1945; hon. LLD, Birmingham, 1944; hon. D.Litt., Oxford, 1945.

MACLAGAN, WILLIAM DALRYMPLE (1826-1910), successively bishop of Lichfield and archbishop of York; studied law at Edinburgh University; an officer in 51st regiment

Madras native infantry, 1847–9; BA, Peterhouse, Cambridge, 1857; ordained, 1856; served curacies in London until 1869; rector of Newington, 1869–75; of St. Mary Abbots, Kensington, 1875–8; bishop of Lichfield, 1878–91; interested in unity of Christendom; attended conference of Old Catholics at Bonn, 1887; archbishop of York, 1891–1908; established training college for clergy at York, 1892; discouraged advanced ritual; inaugurated Poor Benefices Fund; responsible with Archbishop Frederick Temple [q.v.] for reply to Pope Leo XIII's Bull denying validity of Anglican orders, 1896; crowned Queen Alexandra, 1902; resigned archbishopric, 1908; composed hymns (including 'The Saints of God') and hymn tunes.

MACLAREN, ALEXANDER (1826–1910), Baptist divine; student at Glasgow University; BA, London, 1845; successful minister at Southampton, 1846–58, and Union Chapel, Manchester, 1858–1903; rebuilt Union Chapel (1869) and added schools (1880); built new churches and missions in poor districts; exerted great influence as preacher of sermons showing both literary and exegetical skill; president of Baptist Union, 1875 and 1901, and of Baptist World Congress in London, 1905; hon. DD, Edinburgh, 1877; Glasgow, 1907; Litt.D., Manchester, 1902; visited Australia, 1883, and Italy, 1865 and 1903; published devotional works and scripture expositions; edited in *Expositor's Bible*, Colossians and Ephesians (1887) and the Psalms (3 vols., 1893–4).

MacLAREN, ARCHIBALD CAMPBELL (1871–1944), cricketer; captained Harrow, 1890; first played for Lancashire, 1890; his 424 against Somerset (1895) long remained record innings; played for England in Australia, 1894, 1897–8, and (captain) 1901–2; captained England in fourteen home matches against Australia (1899, 1902, 1909), winning only twice; captained MCC in Australia and New Zealand, 1922–3.

McLAREN, CHARLES BENJAMIN BRIGHT, first BARON ABERCONWAY (1850–1934), barrister and man of business; son of Duncan McLaren and half-brother of John, Lord McLaren [qq.v.]; educated at Edinburgh, Bonn, and Heidelberg universities; called (Lincoln's Inn, 1874) and practised at Chancery bar until 1897 when turned to direction of industrial concerns; chairman, John Brown & Co., Metropolitan Railway Co., and steel and colliery undertakings; liberal MP, Stafford (1880–6), Bosworth division (1892–1910); QC, 1897; baronet, 1902; PC, 1908; baron, 1911.

McLAREN, HENRY DUNCAN, second BARON ABERCONWAY (1879–1953), industrialist; son of first Baron Aberconway [q.v.]; succeeded father, 1934; educated at Eton and Balliol College, Oxford; liberal MP, West Staffordshire, 1906–10; Bosworth division, 1910–22; served in Ministry of Munitions during 1914–18 war; succeeded father as chairman of John Brown & Co.; formed and chairman, English Clays Lovering Pochin & Co., 1932; his garden at Bodnant given to National Trust, 1949; president, Royal Horticultural Society, 1931–53.

MACLAREN, IAN (pseudonym) (1850–1907), Presbyterian divine and author. [See WATSON, JOHN.]

McLAREN, JOHN, LORD McLAREN (1831–1910), Scottish judge; son of Duncan McLaren [q.v.]; passed to Scottish bar, 1856; helped to reorganize Scottish liberals and arrange Gladstone's 'Midlothian campaign', 1879–80; MP for Wigton, 1880; appointed lord advocate (Apr.), losing seat on seeking re-election; elected for Edinburgh, Jan. 1881; accepted Scottish judgeship under pressure from Gladstone and Sir William Harcourt [q.v.], 1881; an eminently successful judge; edited works on Scottish law; a student of astronomy and mathematics; hon. LLD, Edinburgh, 1882, Glasgow, 1883, and Aberdeen, 1906.

MACLAY, JOSEPH PATON, first BARON MACLAY (1857–1951), shipowner and shipping controller; educated in Glasgow; established trampship firm with T. W. McIntyre, 1885, to become one of largest shipping concerns on Clyde; active in Scottish evangelical and philanthropic life; appointed shipping controller and head of new Ministry of Shipping, 1916; requisitioned all British shipping and co-ordinated with Americans; this with convoys and control of imports overcame shipping shortage; baronet, 1914; PC, 1916; baron, 1922.

MACLEAN, SIR DONALD (1864–1932), politician; educated at Haverfordwest and Carmarthen grammar schools; admitted solicitor, 1887; practised in Cardiff and London; liberal MP, Bath (1906–10), Peebles (1910–22), North Cornwall (1929–32); deputy speaker, 1911–8; influential in formation (chairman, 1919–22) of liberal parliamentary party; president, Board of Education, 1931–2; PC, 1916; KBE, 1917.

MACLEAN, SIR HARRY AUBREY DE VERE (1848–1920), soldier; served in army, 1869–76; kaid of infantry and instructor to forces attached to court of sultans of Morocco, 1877–1909; CMG, for services rendered to British legation at Tangier, 1898; KCMG,

1901; kidnapped and held to ransom by rebel sherif, Raisuli, 1907; died at Tangier.

MACLEAN, JAMES MACKENZIE (1835–1906), journalist and politician; educated at Christ's Hospital; edited *Newcastle Chronicle*, 1855–8; a leader-writer for *Manchester Guardian*, 1859; edited *Bombay Gazette*, 1859–61; proprietor from 1864; an independent critic both of Indian aspirations and of Indian government; greatly influenced public opinion in Bombay; appointed a magistrate, 1862; helped in creation of semi-elective municipal corporations, 1872; returned home, 1879; had interest in and regularly contributed to *Western Mail*, Cardiff, from 1882; supporter of Lord Randolph Churchill [q.v.]; conservative MP for Oldham, 1885–92, and for Cardiff, 1895–1900; opposed South African war, 1899; supported free trade, 1903; broke with conservative party, and wrote for liberal journals; published *Recollections of Westminster and India* (1902).

McLEAN, NORMAN (1865–1947), orientalist; educated at Edinburgh high school and university and Christ's College, Cambridge; first class, classics (1888), Semitic languages (1890); fellow, 1893; tutor, 1911; master, 1927–36; university lecturer in Aramaic, 1903–31; with A. E. Brooke [q.v.] prepared larger Cambridge edition of Septuagint; FBA, 1934.

MACLEAR, GEORGE FREDERICK (1833–1902), theological writer; BA, Trinity College, Cambridge, 1855; first class in theological tripos, 1856; MA, 1860; DD, 1872; reader at the Temple, 1865–70; assistant at (1860–6) and headmaster of (1867–80) King's College School, London; warden of St. Augustine's Missionary College, Canterbury, 1880–1902; honorary canon of Canterbury Cathedral, 1885; works include lucid textbooks on Bible history (1862) and *The Conversion of the West* (4 vols., 1878).

MACLEAR, JOHN FIOT LEE PEARSE (1838–1907), admiral; son of Sir Thomas Maclear [q.v.]; born at Cape Town; entered navy, 1851; commander, 1868; went as commander of *Challenger* with Sir George Nares [q.v.] on a voyage of discovery round the world, 1872–6; surveyed Straits of Magellan in *Alert*, 1879–82; on surveying service, 1883; admiral, 1907; died at Niagara.

McLENNAN, SIR JOHN CUNNINGHAM (1867–1935), Canadian physicist; first class honours, physics, Toronto, 1892; assistant demonstrator in physics (1892), demonstrator (1899), associate professor (1902), director of physics laboratory (1904), professor (1907–32),

dean of graduate studies (1930–2); worked in England on anti-submarine measures (1917–19), on radium treatment of cancer (1932–5); authority on radioactivity and cosmic rays; succeeded in liquefying helium, 1923; reproduced 'auroral green line', 1925; FRS, 1915; KBE, 1935.

MACLEOD, FIONA (pseudonym) (1855–1905), romanticist. [See SHARP, WILLIAM.]

MACLEOD, HENRY DUNNING (1821–1902), economist; BA, Trinity College, Cambridge, 1843; MA, 1863; called to bar, 1849; published *The Theory and Practice of Banking* (1856); lectured on banking at Cambridge, London, Edinburgh, and Aberdeen, 1877–82; made valuable contributions to historical side of economic science; in *Elements of Political Economy* (1858) first applied term 'Gresham's Law' to principle that 'bad money drives out good'; awarded civil list pension, 1892; also wrote *The Theory of Credit* (2 vols., 1889–91) and *The History of Banking in Great Britain* (1896).

MACLEOD, IAIN NORMAN (1913–1970), politician; educated at Fettes College and Gonville and Caius College, Cambridge; president, Cambridge Bridge Club; fought in France, 1939–40 and 1943–4; joined conservative parliamentary secretariat, 1946; wrote weekly bridge column for *Sunday Times*; published *Bridge is an Easy Game* (1952); conservative MP, Enfield, 1950; minister of health, 1952–5; PC, 1952; minister of labour, 1955–9; secretary of state for colonies, 1959–61; accelerated movement towards independence of African colonies; chancellor of duchy of Lancaster and leader of House of Commons; chairman, conservative party organization, 1961–3; editor of *The Spectator*, 1963–5; chancellor of the Exchequer, 1970; published *Neville Chamberlain* (1961).

MACLEOD, JOHN JAMES RICKARD (1876–1935), physiologist and biochemist; educated at Aberdeen grammar school and Marischal College; MB, Ch.B., 1898; professor of physiology, Western Reserve University, Cleveland, Ohio (1903–18), Toronto (1918–28), Aberdeen (1928–35); from 1905 interested in diabetes; provided facilities, advice, and co-operation for experiments of (Sir) F. G. Banting [q.v.] and C. H. Best which resulted in discovery of insulin; shared Nobel prize with Banting, 1923; carried out research in many fields but most influential as a teacher and director of research; FRS, 1923.

McLINTOCK, SIR WILLIAM, first baronet (1873–1947), chartered accountant; educated at Dumfries academy and Glasgow high school;

qualified as chartered accountant; senior partner, Thomson McLintock & Co., Glasgow; opened London office, 1912; overhauled finances of royal household after war; a financial adviser at imperial wireless and cable conference, 1928; served on Economic Advisory Council, Industrial Arbitration Court, and many other bodies; CVO and KBE, 1922; GBE, 1929; baronet, 1934.

McLINTOCK, WILLIAM FRANCIS PORTER (1887-1960), geologist; educated at George Heriot's and university, Edinburgh; B.Sc., distinction in botany, 1907; assistant curator, Museum of Practical Geology, London, 1907-11; curator of geology, Royal Scottish Museum, Edinburgh, 1911-21; D.Sc., 1915; curator, Museum of Practical Geology, 1921-5; planned new Geological Museum, South Kensington, opened 1935, on lines of popular exposition of science; deputy director, 1937-45; director, 1945-50; reopened museum, 1947; reorganized Geological Survey to advise on water and fuel resources; CB, 1951.

MACLURE, EDWARD CRAIG (1833-1906), dean of Manchester; BA, Brasenose College, Oxford, 1856; MA, 1858; DD, 1890; vicar of Rochdale, 1877-90; dean of Manchester, 1890-1906; chairman of Manchester school board, 1891-1903; member of royal commission on secondary education, 1894; hon. LLD, Victoria University, Manchester, 1902.

MACLURE, SIR JOHN WILLIAM, first baronet (1835-1901), brother of E. C. Maclure [q.v.]; secretary of Lancashire cotton relief fund, 1862; conservative MP for Stretford, 1886-1901; baronet, 1898.

McMAHON, SIR (ARTHUR) HENRY (1862-1949), military political officer; son of C. A. McMahon [q.v.]; educated at Haileybury and Sandhurst; entered Indian political department, 1890; demarcated Baluchistan-Afghanistan boundary; arbitrator on Persian-Afghan boundary, 1903-5; agent in Baluchistan, 1905-11; foreign secretary to Indian government, 1911-14; negotiated treaty with China and Tibet, 1913-14; high commissioner, Egypt, 1914-16; conducted 'McMahon-Husain correspondence'; KCIE, 1906; GCVO, 1911; GCMG, 1916.

McMAHON, CHARLES ALEXANDER (1830-1904), general and geologist; joined Indian army, 1847; commanded troops in Sialkot district during Indian Mutiny; Punjab commissioner, 1872-85; lieutenant-general, 1892; edited *Records of Geological Survey of India*, vol. x, 1877; pioneer in study of petrology and in metamorphism and foliation of rocks; FGS, 1879; Lyell medallist, 1899; FRS, 1898.

MacMAHON, PERCY ALEXANDER (1854-1929), mathematician; born at Malta; educated at the Royal Military Academy, Woolwich; served with Royal Artillery in India; mathematical instructor, Royal Military Academy, 1882; professor of physics, Ordnance College, 1890-7; retired from army to devote himself to mathematical and scientific pursuits, 1898; deputy warden of standards, 1904-20; FRS, 1890; president, London Mathematical Society, 1894-6; chiefly studied theory of algebraic forms; his works include *Combinatory Analysis* (1915).

MacMICHAEL, SIR HAROLD ALFRED (1882-1969), colonial civil servant; educated at Bedford grammar school and Magdalene College, Cambridge; first class, part i, classical tripos, 1904; first class, Arabic examination for Sudan political service, 1905; entered Sudan political service, 1905; served in Kordofan province and Blue Nile province, 1905-15; senior inspector, Khartoum province, 1915-16; political officer to expedition to Darfur, 1916; DSO; assistant civil secretary, Khartoum; 1919-26; civil secretary, 1926; governor of Tanganyika, 1933-7; high commissioner and commander-in-chief, Palestine, 1938-44; Jewish terrorism increased; attempt to murder high commissioner and his wife; Lady Mac-Michael slightly wounded, 1944; worked on constitution for Malaya, 1945-6; and Malta, 1946; publications include *Sudan Notes and Records* (1918), *A History of the Arab in the Sudan* (2 vols., 1922), *The Anglo-Egyptian Sudan* (1934), and *The Sudan* (1954); CMG, 1927; KCMG, 1932; GCMG, 1941; hon. fellow, Magdalene College, Cambridge, 1939.

MACMILLAN, SIR FREDERICK OR-RIDGE (1851-1936), publisher; son of Daniel Macmillan [q.v.]; educated at Uppingham; entered family business and built up world-wide organization; took lead in establishing 'net book agreement', 1890; knighted for hospital work, 1909.

MACMILLAN, HUGH (1833-1903), religious writer; Free Church minister at Glasgow (1864-78) and Greenock (1878-1901); made wide fame through *Bible Teachings in Nature* (1867) and *The Ministry of Nature* (1871); hon. LLD, St. Andrews, 1871; hon. DD, Edinburgh, 1879, and Glasgow; FRS Edinburgh, and FSA Scotland; Cunningham (1894) and Gunning (1897) lecturer at Edinburgh; moderator of General Assembly of Free Church, 1897;

voluminous author of works mainly dealing with relations of religion and science.

MACMILLAN, HUGH PATTISON, Baron Macmillan (1873-1952), judge; son of Revd Hugh Macmillan [q.v.]; educated at Collegiate School, Greenock, Edinburgh, (first class, philosophy, 1893) and Glasgow (LLB, 1896) universities; advocate, 1897; KC, 1912; lord advocate in labour government, 1924; established London chambers; standing counsel, Canada (1928), Australia (1929); lord of appeal in ordinary, 1930-9, 1941-7; minister of information, 1939-40; chairmanships included Treasury finance and industry committee, 1929-31; royal commission on Canadian banking and currency, 1933; Pilgrim Trust, 1935-52; Political Honours committee, 1935-52; court of London University, 1929-43; BBC, advisory council, 1936-46; PC, 1924; life peer, 1930; GCVO, 1937; published autobiography *A Man of Law's Tale* (1952).

McMILLAN, MARGARET (1860-1931), educationist; educated at Inverness high school; joined independent labour party, Bradford, 1893; obtained first elementary school medical inspection, 1899; with Rachel McMillan, established children's clinic at Deptford (1910) and pioneer open-air nursery school (1917); built new Rachel McMillan training college, 1930; CBE, 1917; CH, 1930.

McMURRICH, JAMES PLAYFAIR (1859-1939), Canadian anatomist; BA, Toronto, 1879; professor of anatomy (1907-30), dean of graduate studies (1922-30); wrote extensively on morphology, embryology, and the history of anatomy; leading authority on sea-anemones; FRS Canada, 1909.

MACNAGHTEN, Sir EDWARD, fourth baronet, and Baron Macnaghten (1830-1913), judge; educated at Trinity College, Dublin, and Trinity College, Cambridge; bracketed senior classic, 1852; fellow of Trinity, Cambridge, 1853; called to bar (Lincoln's Inn), 1857; equity junior, 1857-80; QC, 1880; conservative MP, county Antrim, 1880, North Antrim, 1885; lord of appeal in ordinary and life peer, 1887; GCMG, 1903; GCB, and succeeded brother in baronetcy, 1911; took active part in both Houses in debates on Irish questions; as judge possessed remarkable power of combining learning, style, and humour.

McNAIR, JOHN FREDERICK ADOLPHUS (1828-1910), colonial official; joined Madras artillery, 1845; at Singapore efficient superintendent of convicts, 1858; controller of convicts (1867), colonial secretary (1868), member of the executive council of Straits Settlements (from 1869), and surveyor-general (from 1873); lieutenant-governor of Penang, 1881-4; chief commissioner in Perak, 1875-6; CMG, 1878; wrote accounts of Perak (1878) and Singapore convict prison (1899).

MacNALTY, Sir ARTHUR SALUSBURY (1880-1969), expert in public health; educated at Hartley College, Southampton, St. Catherines' Society, Oxford, and Corpus Christi College, Oxford; BM, 1907; DM, 1911; MRCP, 1925; DPH, 1927; FRCP, 1930; special interest in pulmonary tuberculosis; joined Local Government Board, as specialist to promote 'sanatorium' and 'dispensary' service, 1913; transferred to Ministry of Health; deputy chief medical officer, 1919; chief medical officer, 1935-40; editor-in-chief, official medical history of 1939-45 war, 1940-69; KCB 1936.

MACNAMARA, THOMAS JAMES (1861-1931), politician; liberal MP, North Camberwell, 1900-24; parliamentary and financial secretary to Admiralty, 1908-20; minister of labour, 1920-2; organizer and platform speaker for Lloyd George; PC, 1911.

McNAUGHTON, ANDREW GEORGE LATTA (1887-1966), Canadian soldier, scientist, and public servant; born at Moosomin, Saskatchewan; educated at Bishop's College School, Le noxville, Quebec, and McGill University; honours, electrical engineering, 1910; M.Sc., 1912; lecturer at McGill; served in Canadian army in France and Belgium, 1914-18; DSO, 1917; CMG, 1919; hon. LLD McGill University, 1920; director of military training, acting brigadier-general; attended Staff College, 1921; deputy chief, Canadian general staff, 1923; attended Imperial Defence College, 1927; major-general, chief, Canadian general staff, 1929-35; president, Canadian National Research Council; commanded Canadian troops in Britain, 1939-43; retired from army with rank of full general, 1944; Canadian minister of defence 1944-5; chairman, Canadian section, Canada-United States Permanent Joint Board on Defence, 1945-59; president, Atomic Energy Control Board of Canada, 1946-8; Canadian representative, UN Atomic Energy Commission, 1946-50; permanent Canadian delegate to UN, 1948-9; chairman, Canadian section, International Joint Commission, 1950-2; PC (Canada), 1944; CB, 1935; CH, 1946.

MACNEICE, (FREDERICK) LOUIS (1907-1963), writer; educated at Marlborough and Merton College, Oxford; first class, *lit. hum.*, 1930; edited *Oxford Poetry* with Stephen Spender; published first book of poems, *Blind*

Fireworks (1930); lecturer in classics, Birmingham University, 1930-6; lecturer in Greek, Bedford College, London, 1936-40; lecturer at Cornell University, USA, 1940; joined BBC Features Department, 1941; director, British Institute, Athens, 1950; left BBC, 1961; hon. doctorate, Queen's University, Belfast, 1957; CBE, 1958; publications include *Poems* (1935), *Letters from Iceland* (with W. H. Auden, 1937), *The Earth Compels* (1938), *Autumn Journal* (1939), *The Last Ditch* (1940), *Springboard* (1944), *The Poetry of W. B. Yeats* (1941), *Collected Poems 1925-1948* (1949), *Autumn Sequel* (1954), *Visitations* (1957), and *The Strings are False* (1965).

McNEIL, HECTOR (1907-1955), journalist and politician; educated at Glasgow secondary schools and university; joined *Scottish Daily Express* becoming night news editor and finally leader-writer; labour MP, Greenock, 1941-55; parliamentary private secretary to Philip Noel-Baker, 1942-5; minister of state, foreign affairs, 1946-50; forceful speaker at United Nations general assemblies; Scottish secretary, 1950-1; PC, 1946; died in New York.

McNEILE, (HERMAN) CYRIL (1888-1937), soldier, and novelist under the pseudonym of SAPPER; educated at Cheltenham and Woolwich; officer in Royal Engineers, 1907-19; wrote series of thrillers beginning with *Bull-dog Drummond* (1920).

McNEILL, JAMES (1869-1938), Indian and Irish civil servant; an Ulster Catholic; educated at Belvedere, Dublin, and Emmanuel College, Cambridge; served in Bombay presidency, 1890-1915; joined Mr de Valera's political movement, 1916; helped to draft Irish Free State constitution, 1922; high commissioner in London, 1923-8; governor-general, Irish Free State, 1928-32; challenged efforts to reduce his position to obscurity and forced executive council to obtain his removal.

McNEILL, SIR JAMES McFADYEN (1892-1964), shipbuilder; educated at Clydebank high school and Allan Glen's School, Glasgow; apprenticed with John Brown & Co. Ltd., 1908; Lloyd's scholar in naval architecture, Glasgow University; B.Sc., 1915; served in Royal Field Artillery in 1914-18 war; MC, 1918; assistant naval architect, Clydebank, 1922-8; principal naval architect and technical manager, 1928; collaborated with Cunard Company in producing *Queen Mary* and *Queen Elizabeth*; delivered classic paper on launching to Institution of Naval Architects, 1935; managing director, Clydebank yard, 1948; CBE, 1950; deputy chairman, 1953-62; on completion of Royal

Yacht *Britannia*, KCVO, 1954; retired from executive duties, 1959; director of number of other firms; hon. LLD, Glasgow, 1939; FRS, 1948; vice-president, Institution of Naval Architects; president, Institution of Engineers and Shipbuilders in Scotland, 1947-9; president, Shipbuilding Conference, 1956-8.

McNEILL, JOHN (otherwise EOIN) (1867-1945), Irish scholar and politician; brother of James McNeill [q.v.]; BA, Royal University of Ireland, 1888; entered civil service; a founder (1893) and first secretary, Gaelic League; first professor of early Irish history, University College, Dublin, 1908-45; leader of extreme Irish nationalists, 1914; countermanded mobilization on learning of plans for Easter rising, 1916; released from life imprisonment, 1917; MP, National University, 1918-22; member of the Dail, 1921-7; speaker, 1921-2; minister of education, 1922-5; Free State representative, Ulster boundary commission, 1924; resigned after newspaper forecast of contents of award, 1925; publications include *Phases of Irish History* (1919) and *Celtic Ireland* (1921).

McNEILL, SIR JOHN CARSTAIRS (1831-1904), major-general; joined Bengal native infantry, 1850; served with distinction at Lucknow, 1857-8; won VC in Maori war, 1864; on staff of Red River expedition, 1870; chief of staff in Ashanti war, 1873-4; CB, 1874; KCMG, 1880; major-general and KCB, 1882; commanded second infantry brigade in Sudan campaign, 1885; criticized for lack of caution; retired, 1890; GCVO, 1901.

MacNEILL, JOHN GORDON SWIFT (1849-1926), Irish politician and jurist; closely connected with protestant Ireland; BA, Christ Church, Oxford; called to Irish bar, 1875; professor of constitutional and criminal law, King's Inns, Dublin, 1882-8; joined Home Government Association founded by Isaac Butt [q.v.], 1870; nationalist MP, South Donegal, 1887-1918; had real veneration for parliament; professor of constitutional law, National University of Ireland, 1909-26.

McNEILL, RONALD JOHN, BARON CUSHENDUN (1861-1934), politician; educated at Harrow and Christ Church, Oxford; editor, *St. James's Gazette*, 1900-4; assistant editor, eleventh edition *Encyclopaedia Britannica*, 1906-11; MP, East Kent, 1911-27; parliamentary under-secretary, foreign affairs, 1922-4-5; financial secretary to Treasury, 1925-7; chancellor, duchy of Lancaster, 1927-9; diehard conservative of impulsive temper; PC, 1924; baron, 1927.

MACPHAIL, Sir (JOHN) ANDREW (1864-1938), pathologist and author; graduated in medicine at McGill University, 1891; professor of medical history, 1907-37; edited *Canadian Medical Journal* and *University Magazine*; translated *Marie Chapdelaine* (1921); other publications include *The Master's Wife* (1939); knighted, 1918.

MACPHERSON, (JAMES) IAN, first Baron Strathcarron (1880-1937); educated at George Watson's College and Edinburgh University; called to bar (Middle Temple), 1906; liberal MP, Ross and Cromarty, 1911-35; under-secretary for war, 1916-18; vice-president, Army Council, 1918-19; chief secretary for Ireland, 1919-20; minister of pensions, 1920-2; recorder of Southend, 1931-7; PC, 1918; KC, 1919; baronet, 1933; baron, 1936.

MACPHERSON, Sir JOHN MOLES-WORTH (1853-1914), Anglo-Indian legislative draftsman; son of John Macpherson, MD [q.v.]; born in Calcutta; called to English bar, 1876; deputy secretary to Indian government in legislative department, 1877; permanent secretary, 1896-1911; knighted, 1911.

McQUEEN, Sir JOHN WITHERS (1836-1909), major-general; born in Calcutta; joined 27th Bengal native infantry, 1854; recommended for VC for bravery at the Secundarabagh, 1857; commanded 5th Punjab rifles in Jowaki expedition, 1877-8; of great service in Afghan war, 1878-80; CB, 1879; brigadier-general in command of Hyderabad contingent, 1885; commanded expedition against Black Mountain tribes, 1888; KCB, 1890; major-general, 1891; GCB, 1907.

MACQUEEN-POPE, WALTER JAMES (1888-1960), theatre manager, publicist, and historian; educated at Tollington School; business-manager for Sir Alfred Butt at the Queen's, St. James's, Lyric, and other theatres; manager, Alexandra Palace (1922-5); manager number of other west-end theatres; public relations officer for ENSA during 1939-45 war; wrote books and lectured on the history of particular theatres.

MACREADY, Sir (CECIL FREDERICK) NEVIL, first baronet (1862-1946), general; son of W. C. Macready [q.v.]; educated at Marlborough, Cheltenham, and Sandhurst; joined Gordon Highlanders, 1881; served in Egyptian campaign (1882) and South Africa (1899-1902); director of personal services, War Office, 1910-14; major-general, 1910; reported on administration in Dublin and Belfast, 1913; GOC, Belfast, 1914; adjutant-general, British expe-

ditionary force, 1914-16; adjutant-general to the forces, 1916-18; commissioner, Metropolitan Police, 1918-20; GOC-in-C, Ireland, 1920-3; lieutenant-general, 1916; general, 1918; KCB, 1912; KCMG, 1915; GCMG, 1918; baronet, 1923.

MACRORIE, WILLIAM KENNETH (1831-1905), bishop of Maritzburg; BA, Brasenose College, Oxford, 1852; MA, 1855; consecrated, by Bishop Gray [q.v.] of Cape Town; bishop of Maritzburg in Natal in opposition to Bishop Colenso [q.v.], 1869; an uncompromising high churchman; resigned bishopric, 1891; canon of Ely and assistant bishop, 1892.

M'TAGGART, JOHN M'TAGGART ELLIS (1866-1925), philosopher; BA, Trinity College, Cambridge; first class in moral sciences tripos, 1888; prize fellow of Trinity, 1891; college lecturer in moral sciences, 1897-1923; FBA, 1906; an atheist who believed in human immortality; his earlier work devoted to expounding and defending method and some results of Hegel's *Logic* in *Studies in the Hegelian Dialectic* (1896), *Studies in Hegelian Cosmology* (1901), *Commentary on Hegel's Logic* (1910); other works include *Some Dogmas of Religion* (1906) and *The Nature of Existence* (2 vols., 1921-7).

McTAGGART, WILLIAM (1835-1910), artist; fellow student at Trustees' Academy, Edinburgh, with (Sir) W. Q. Orchardson, Tom Graham, and John MacWhirter [qq.v.]; ARSA, 1859; RSA, 1870; exhibited at Royal Academy, 1866-75; vice-president of the Royal Scottish Water Colour Society, 1878; early engaged in portraiture, genre pictures, and land- and seascapes; later confined his work to landscape and the sea.

MacWHIRTER, JOHN (1839-1911), landscape painter; fellow student with William McTaggart [q.v.] at Trustees' Academy, Edinburgh; made direct study of nature; annually visited continent; ARSA, 1867; removed to London, 1869; ARA, 1879; hon. RSA, 1892; RA, 1893; published *Landscape Painting in Water Colours* (1901); painted popular landscapes and studies of trees; his best-known work, 'June in the Austrian Tyrol', in Tate Gallery.

MADDEN, Sir CHARLES EDWARD, first baronet (1862-1935), admiral of the fleet; entered navy, 1875; specialized in torpedo; staff officer in the *Vernon*, torpedo school, 1893-6, 1899-1901; naval assistant to (Sir) H. B. Jackson [q.v.], 1905, and to (Lord) Fisher [q.v.],

1906-7, at Admiralty at time of great reforms; captain of the *Dreadnought* and chief of staff, home fleet, 1907-8; fourth sea lord, 1910-11; rear-admiral, 1911; chief of staff to (Lord) Jellicoe [q.v.], 1914-16; commander, first battle squadron, and second-in-command to (Lord) Beatty [q.v.], 1916-19; commander-in-chief, Atlantic fleet, 1919-22; admiral, 1919; admiral of the fleet, 1924; first sea lord, 1927-30; assented under protest to reduced figure of fifty cruisers agreed at London naval conference, 1930; KCB and KCMG, 1916; baronet, 1919; GCVO, 1920; OM, 1931.

MADDEN, FREDERIC WILLIAM (1839-1904), numismatist; assistant in coin department of British Museum, 1859-68; chief librarian at Brighton public library, 1888-1902; published *The Coins of the Jews* (1881) and a manual of Roman coins (1861).

MADDEN, KATHERINE CECIL (1875-1911), novelist. [See Thurston.]

MADDEN, THOMAS MORE (1844-1902), Irish gynaecologist; MRCS, 1862; raised Irish ambulance corps in Franco-Prussian war, 1870; master of the National Lying-in Hospital, Dublin, 1878; FRCS, Edinburgh, 1882; recognized as a foremost gynaecologist; voluminous writings include *Uterine Tumours* (1887), *Clinical Gynaecology* (1893), and accounts of his family.

MAFFEY, JOHN LOADER, first Baron Rugby (1877-1969), public servant; educated at Rugby and Christ Church, Oxford; entered Indian civil service, 1899; transferred to political department, 1905; served with Mohmand field force, 1908; political agent, Kyber, 1909-12; deputy commissioner, Peshawar, 1914-15; chief political officer with forces in Afghanistan, 1919; deputy secretary, foreign and political department, government of India, 1915-16; private secretary to the viceroy, 1916-20; chief secretary to Duke of Connaught [q.v.], 1920-1; chief commissioner, North-West Frontier Province, 1921-4; organized rescue of Mollie Ellis from tribesmen; disagreed with British government frontier policy, and resigned from ICS, 1924; governor-general, Sudan, 1926-33; permanent under-secretary of state, Colonial Office, 1933-7; first British representative in Eire, 1939-49; director, Rio Tinto Company and Imperial Airways; CIE, 1916; CSI, 1920; KCVO, 1921; KCMG, 1931; KCB, 1934; GCMG, 1935; baron, 1947.

MAGRATH, JOHN RICHARD (1839-1930), provost of Queen's College, Oxford; born in Guernsey; BA, Oriel College, Oxford; fellow of Queen's College, 1860; ordained, 1863; tutor of Queen's, 1864; dean, 1864-77; chaplain, 1867-78; bursar, 1874-8; pro-provost, 1877; provost, 1878-1930; member of hebdomadal council, 1878-99; curator of university chest, 1885-1908; delegate of University Press, 1894-1920; vice-chancellor, 1894-8; alderman, 1889-95; supported movement for higher education of women; greatly interested in northern schools linked with Queen's, particularly St. Bees; identified himself with reforms of statutory commission of 1877; a keen sportsman; his works include *The Queen's College* (1921).

MAGUIRE, JAMES ROCHFORT (1855-1925), president of British South Africa Company; BA, Merton College, Oxford; friend of Cecil Rhodes [q.v.]; Parnellite MP, North Donegal, 1890; MP, West Clare, 1892-5; one of the emissaries sent by Rhodes to obtain mineral concessions from Matabele chief, Lobengula, at Bulawayo, 1883; vice-president, British South Africa Company, 1906; president, 1923.

MAHAFFY, Sir JOHN PENTLAND (1839-1919), provost of Trinity College, Dublin; born in Switzerland; BA, Trinity College, Dublin (first senior moderator in classics and logics), 1859; fellow, and ordained, 1864; first professor of ancient history at Dublin, 1869; provost of Trinity, 1914-19; GBE, 1918; reputation chiefly rests on works dealing with life, literature, and history of ancient Greeks; these include *Prolegomena to Ancient History* (1871), *Greek Social Life from Homer to Menander* (1874), *History of Classical Greek Literature* (1880), *Story of Alexander's Empire* (1887), *Greek Life and Thought from Alexander to the Roman Conquest* (1887), *The Greek World under Roman Sway* (1890), and *Problems in Greek History* (1892); turned his attention to papyri, 1890; produced *Flinders Petrie Papyri* (1891, etc.) and *The Empire of the Ptolemies* (1895); a remarkably versatile writer of great shrewdness and sagacity.

MAHON, Sir BRYAN THOMAS (1862-1930), general; gazetted to 8th Hussars, 1883; served in Egypt, playing active part in operations in Dongola (1896) and Nile valley (1897) which led to final destruction of dervish power, 1893-1900; left Egypt for South Africa on special service, Jan. 1900; commanded column which achieved relief of Mafeking, on which his reputation chiefly rests, 17 May; governor of Kordofan, 1901-4; commanded district of Belgaum, India, 1904-9; Lucknow division, 1909-13; appointed to command 10th (Irish) division of new armies, 1914; took division to Gallipoli, 1915; commander-in-chief, Salonika army, 1915-16; commander-in-chief, Ireland, 1916-

18; military commander, Lille, 1918-19; KCB, 1922.

MAIR, WILLIAM (1830-1920), Scottish divine; served ministries in established church, 1861-1903; moderator of General Assembly, 1897; pioneer in cause of Church reunion and authority on Scottish ecclesiastical law.

MAITLAND, AGNES CATHERINE (1850-1906), principal of Somerville College, Oxford; published *The Rudiments of Cookery* (35th thousand, 1910) and other cookery books and novels; principal of Somerville College, Oxford, 1889-1906; largely increased numbers and extended buildings; developed tutorial system and had college library erected.

MAITLAND, SIR ARTHUR HERBERT DRUMMOND RAMSAY-STEEL-, first baronet (1876-1935), politician and economist. [See STEEL-MAITLAND.]

MAITLAND, FREDERIC WILLIAM (1850-1906), Downing professor of the laws of England at Cambridge; son of John Gorham Maitland [q.v.]; educated at Eton and Trinity College, Cambridge; senior in moral sciences tripos, 1872; Whewell international law scholar, 1873; president of Union; obtained blue for running; BA, 1873; MA, 1876; hon. LLD, 1891; called to bar, 1876; reader in English law at Cambridge, 1884; Downing professor, 1888-1906; founded Selden Society for encouraging study of history of English law, editing several volumes, 1887; literary director, 1895; published *Bracton's Note-Book* (3 vols., 1887) and *History of English Law before the time of Edward I* (with Sir Frederick Pollock [q.v.], 2 vols., 1895); traced Roman influence in English law in thirteenth century in *Bracton and Azo* (Selden Soc., 1895) and in *Roman Canon Law in the Church of England* (1898); made researches into legal effect of the Reformation; edited and translated MSS of Year Books in old legal Anglo-French, temp. Edward II (4 vols., 1903-7); advocated simplification and codification of English law; ardent alpinist; Ford's lecturer, Oxford, 1897; Rede lecturer, Cambridge, 1901; hon. DCL, Oxford, 1899, LLD, Glasgow, Cracow, and Moscow; original FBA, 1902; hon. fellow, Trinity College, Cambridge; died at Las Palmas; other publications include *Life and Letters of Leslie Stephen* (1906) and posthumously *The Constitutional History of England* (1908) and *Collected Works* (1911).

MAITLAND, JOHN ALEXANDER FULLER- (1856-1936), musical critic and connoisseur; educated at Trinity College, Cambridge; musical critic, *Pall Mall Gazette*, 1882-4,

Guardian, 1884-9, *The Times*, 1889-1911; contributed vol. iv (*The Age of Bach and Handel*, 1902) to *Oxford History of Music* and edited second edition of *Grove's Dictionary of Music and Musicians* (1904-10).

MAJOR, HENRY DEWSBURY ALVES (1871-1961), theologian; educated at home and St. John's College, Auckland, New Zealand; BA, 1895; ordained as deacon, 1895; priest, 1896; curate, St. Mark's, Remuera, 1895-9; MA, first class in natural sciences, 1896; acting vicar, Waitotara, 1899; vicar, St. Peter's, Hamilton, 1900-2; entered Exeter College, Oxford, 1903; first class, theology, 1905; chaplain, Ripon Clergy College, 1906; vice-principal, 1906; joined Churchman's Union (later the Modern Churchmen's Union); became fanatical advocate of modernism, 1910; founded and edited *Modern Churchman*, 1911-56; rector, Copgrove, 1915-18; principal, Ripon Hall, Oxford, 1919-46; vicar, Merton, 1929-61; DD, Oxford 1924; publications include *Life and Letters of W. B. Carpenter* (1925), *The Gospel of Freedom* (1912), *A Resurrection of Relics* (1922), *English Modernism; its Origin, Methods, Aims* (1927), *Reminiscences of Jesus by an Eye-Witness* (1925), and *The Roman Church and the Modern Man* (1934).

MALAN, DANIEL FRANÇOIS (1874-1959), South African prime minister; obtained MA in philosophy, Stellenbosch, and doctorate in divinity, Utrecht; Dutch Reformed Church minister, 1906-15; editor, *Die Burger*, 1915-24; leader, Cape nationalist party, 1915-53; MP, Calvinia, 1919-38; Piketberg, 1938-54; minister, interior public health and education, 1924-33; made Afrikaans second official language; reformed Senate and civil service; gave South Africa her own flag; broke with J. B. M. Hertzog [q.v.] and became leader of 'purified' nationalists, 1934, which party gradually increased in parliamentary strength; further break with Hertzog and N. C. Havenga brought him nationalist leadership, 1940; made election agreement with Havenga and became prime minister, 1948; formed government exclusively Afrikaner and republican; abolished dual citizenship and appeal to Privy Council; but wished to remain within Commonwealth; introduced apartheid despite world opinion; resigned as Cape leader, 1953, and as prime minister, 1954.

MALAN, FRANÇOIS STEPHANUS (1871-1941), South African statesman; BA, Victoria College, Stellenbosch, 1892; LLB, Cambridge, 1894; called to Cape bar; editor, *Ons Land*, 1895-1908; minister for agriculture and education, 1908-10; member, national convention, 1908-9; MP, Malmesbury, 1910-24; minister for education (1910-21), agriculture

(1919–21), mines and industries (1912–24); acting minister, native affairs, 1915–21; elected to senate, 1927; president, 1940–1; PC, 1920.

MALCOLM, Sɪʀ DOUGAL ORME (1877–1955), scholar and imperialist; educated at Eton and New College, Oxford; first class, *lit. hum.*, 1899; fellow, All Souls, 1899–1955; Lord Mallard, 1928–55; entered Colonial Office, 1900; joined Milner's 'kindergarten' as private secretary to Lord Selborne [q.v.] in South Africa, 1905–10; assisted with memorandum on South African Union, foundation of *Round Table*, and *The Problem of the Commonwealth* (1916) by Lionel Curtis [q.v.]; private secretary to Lord Grey [q.v.] in Canada, 1910–11; transferred to Treasury, 1912; retired, 1912, on nomination as director of British South Africa Company in charge of its affairs in Rhodesia; president, 1937–55; ceded political functions to colonial governments, 1923; sold railways, 1947; negotiated new mining rights, 1950; opposed federation with Nyasaland; KCMG, 1938.

MALET, Sɪʀ EDWARD BALDWIN, fourth baronet (1837–1908), diplomatist; born at The Hague; son of Sir Alexander Malet, second baronet [q.v.]; attaché to his father at Frankfort, 1854; served under Lord Lyons [q.v.], at Washington, 1862–5, and Paris, 1867–71; negotiated meeting of Jules Favre and Bismarck at Ferrières, 1870; in charge of embassy at Paris, Mar.–June 1871; CB, 1871; secretary of legation at Peking, 1871, Rome, 1875, and Constantinople, 1878; British agent and consul-general in Egypt, 1879–83; KCB, 1881; helped to restore financial stability and soothe native unrest in Egypt; reconstituted government machinery, and developed scheme of reorganization, 1882–3; British envoy at Brussels, 1883, and Berlin, 1884–95, where he settled rival British and German claims in the Congo and in Samoa; PC and GCMG, 1885; GCB, 1886; a British member of the international court of arbitration at The Hague, 1899; succeeded brother, 1904; author of *Shifting Scenes* (1901) and an unfinished memoir of his service in Egypt (posthumous, 1909).

MALET, LUCAS (pseudonym) (1852–1931), novelist. [See HARRISON, MARY Sᴛ. LEGER.]

MALLESON, (WILLIAM) MILES (1888–1969), actor, dramatist, and stage director; educated at Brighton College, Emmanuel College, Cambridge, and the (Royal) Academy of Dramatic Art; acted at the Court Theatre under J. B. Fagan [q.v.], and at the Lyric Hammersmith under (Sir) Nigel Playfair [q.v.]; a most resourceful stage clown; wrote serious plays, including *The Fanatics* (1927) and *Six Men of*

Dorset (1938); first chairman, Screen Writers' Association, 1937; played with Old Vic Company, 1945–6; and 1949–50; played the main part in Molière's *The Miser* (his own adaptation); translated series of Molière plays.

MALLOCK, WILLIAM HURRELL (1849–1923), author; nephew of R. H., W., and J. A. Froude [qq.v.]; BA, Balliol College, Oxford; his satires include *The New Republic* (1877), in which he seeks to demonstrate impossible position of undogmatic belief; his novels include *The Old Order Changes* (1886); his political writings include *Social Equality* (1882), in which he refutes erroneous ideas about distribution of wealth, especially ownership of land.

MALLON, JAMES JOSEPH (1875–1961), warden of Toynbee Hall; educated at the Owens College, Manchester; did social work in Ancoats Settlement, Manchester; secretary, National League to Establish a Minimum Wage, 1906; hon. secretary Trade Boards Advisory Council; warden, Toynbee Hall, 1919–54; helped to found Toynbee Hall Theatre and the Children's Theatre; governor, BBC, 1937–9 and 1941–6; member, executive committee, British Empire Exhibition, 1924; member, executive committee and hon. treasurer, Worker's Educational Association; wrote articles on social and economic subjects for *Observer*, *Daily News*, and *Manchester Guardian*; published (with E. C. T. Lascelles), *Poverty Yesterday and Today* (1930); hon. doctorate, Liverpool University, 1944; CH, 1939; awarded Margaret McMillan medal, 1955.

MALLORY, GEORGE LEIGH (1886–1924), mountaineer; educated at Winchester, where R. L. G. Irving introduced him to climbing; BA, Magdalene College, Cambridge; assistant master, Charterhouse, 1910–15; returned there after serving in European war, 1919; lecturer and assistant secretary, board of extra-mural studies, Cambridge, 1923; took part in Mount Everest expeditions, 1921, 1922, and 1924; perished in attempt on summit, 1924.

MALLORY, Sɪʀ TRAFFORD LEIGH LEIGH- (1892–1944), air chief marshal. [See LEIGH-MALLORY.]

MALONE, SYLVESTER (1822–1906), Irish ecclesiastical historian; vicar-general of Kilrush, 1872–1906; canon and archdeacon; made valuable researches in his *Church History of Ireland, 1169–1532* (1867); promoter of preservation of Irish language.

MANECKJI BYRAMJI DADABHOY, Sɪʀ (1865–1953), Indian lawyer, industrialist, and parliamentarian. [See DADABHOY.]

MANLEY, NORMAN WASHINGTON (1893-1969), Jamaican statesman and lawyer; born in Jamaica, his father part Negro, part English, his mother, part Irish descent; educated at Jamaica College and Jesus College, Oxford; Rhodes scholar; enlisted in Royal Field Artillery, 1915; MM; returned to Oxford, 1919; BA, 1921; BCL, called to bar (Gray's Inn), 1921; practised in Jamaica, 1922; KC, 1932; founded Jamaica's first political party, the people's national party, 1938; MP, 1949; chief minister, 1955-62; retired from politics, 1969; hon. LLD, Howard University, 1946, and university of the West Indies (posthumously), 1970).

MANLEY, WILLIAM GEORGE NICHOLAS (1831-1901), surgeon-general; MRCS, 1851; joined army medical staff, 1855; present at Sevastopol; won VC in New Zealand war, 1863-6; in charge of divison of British ambulance corps in Franco-Prussian war, 1870; at siege of Paris, 1870; in Afghan war, 1878-9; principal medical officer in Egyptian war, 1882; surgeon-general, 1884; CB, 1894.

MANN, ARTHUR HENRY (1850-1929), organist; chorister at Norwich Cathedral; B.Mus., New College, Oxford, 1874; organist of King's College, Cambridge, 1876-1929; with his active co-operation lay clerks replaced by choral scholars and a residential school for choirboys established; his Festival Choir, which continued under his name until 1912, established, 1887; it presented a fine series of works by Elgar, Beethoven, Brahms, etc.; rearranged Handel MSS at Fitzwilliam Museum, 1889-92; fellow of King's, 1922.

MANN, CATHLEEN SABINE (1896-1959), painter; daughter of portrait painter Harrington Mann; studied with him, at Slade School, and with (Dame) Ethel Walker [q.v.]; exhibited at Royal Academy and with Royal Society of Portrait Painters; official war artist (mainly portraitist), 1939-45; her marriage (1926) to tenth Marquess of Queensberry dissolved, 1946; took own life.

MANN, SIR JAMES GOW (1897-1962), master of the Armouries of the Tower of London and director of the Wallace Collection; educated at Winchester and New College, Oxford; BA, 1920; B.Litt., 1922; assistant keeper of fine arts, Ashmolean Museum, Oxford, 1922-3; assistant to keeper of Wallace Collection, 1924-32; deputy director, Courtauld Institute, London University, 1932-6; keeper, Wallace Collection, 1936-46; director, 1946-62; master of the Armouries, Tower of London, 1939; director, Society of Antiquaries of London,

1944; president, 1949-54; surveyor of Royal Works of Art, 1946-62; trustee, British Museum and the College of Arms; knighted, 1948; KCVO, 1957; FBA, 1952; publications include *The Wallace Collection Catalogue of Sculpture* (1931) and *Catalogue of the European Arms and Armour* of the Wallace Collection (1962).

MANN, THOMAS (1856-1941), trade-unionist and communist; known as TOM MANN; engineer in Birmingham and London; joined Amalgamated Society of Engineers, 1881, Social Democratic Federation, 1885; demanded shorter working hours; helped in London dock strike, 1889; first president, Dockers' Union, 1889-93; member of royal commission on labour, 1891-4; signed minority report; secretary, independent labour party, 1894-7; a founder and first president of International Federation of Ship, Dock, and River Workers, 1896; largely responsible for launching Workers' Union, 1898; labour organizer in Australia, 1902-10; edited *Industrial Syndicalist*, 1910-11, advocating direct action; imprisoned for inciting to mutiny, 1912; joined British socialist party, 1916; founder-member, British communist party, 1920; chairman, National Minority Movement, 1924-32; paid four visits to Russia; general secretary, Amalgamated Society of Engineers, 1919-20, Amalgamated Engineering Union, 1920-1.

MANNERS, (LORD) JOHN JAMES ROBERT, seventh DUKE OF RUTLAND (1818-1906), politician; brother of sixth duke [q.v.]; MA, Trinity College, Cambridge, 1839; published verse account of foreign travel in 1839-40, *England's Trust and other Poems* (1841, containing a couplet on English nobility which obtained permanent currency), *English Ballads and other Poems* (1850), and notes of Irish and Scotch tours (1848-9); conservative MP for Newark, 1841-7; under Disraeli's influence joined 'Young England Party'; freely criticized Peel's administration, 1843-4; advocated public holidays, 1843, factory reform, 1844, and a general system of allotments; with Disraeli and George Smythe (later Viscount Strangford, q.v.) toured through industrial districts of Lancashire, 1844; figures in *Coningsby* (1844), *Sybil* (1845), and *Endymion* (1880); advocated disestablishment of the Irish Church and supported proposed grant to Maynooth, 1845; differences of opinion on religious and free trade questions led to dissolution of 'Young England Party', 1847; unsuccessfully contested as protectionist Liverpool, 1847, and City of London, 1849; conservative member for Colchester, 1850-7, North Leicestershire, 1857-85, and Melton, 1885-8; PC, 1852; first commissioner of works, and cabinet minister, Feb.-Dec. 1852,

1858-9, and 1866-8; accepted Disraeli's reform bill of 1867; postmaster-general, 1874-80 and 1885-6; reduced minimum telegram charge to sixpence, Oct. 1885; opposed extension of franchise without redistribution, 1884-5; chancellor of duchy of Lancaster, 1886-92; succeeded to dukedom, 1888; KG, 1891; made Baron Roos of Belvoir, 1896; hon. LLD, Cambridge, 1862, DCL, Oxford, 1876; GCB, 1880.

MANNING, BERNARD LORD (1892-1941), scholar; first class, parts i and ii, history tripos (1914-15), Jesus College, Cambridge; fellow, 1919; bursar, 1920-33; senior tutor, 1933-41; university lecturer in medieval history, 1930-41; publications include *The Making of Modern English Religion* (1929) and chapters in vol. vii of *Cambridge Medieval History*.

MANNING, JOHN EDMONDSON (1848-1910), Unitarian divine; BA, London, 1872; MA, 1876; minister at Upper Chapel, Sheffield, 1889-1902, of which he wrote a history, 1900; tutor at Unitarian home missionary college, Manchester.

MANNS, Sir AUGUST (1825-1907), conductor of the Crystal Palace concerts from 1855 to 1901; born at Stolzenburg, Pomerania; bandmaster in von Roon's regiment at Königsberg, 1851; came to England, 1854; transformed wind band at Crystal Palace into full orchestra; conducted Saturday concerts there for forty years; introduced works by Schumann, Schubert, and Brahms; frequently (from 1861) devoted programme to living English composers; conducted Handel triennial festivals, 1883-1900.

MANSBRIDGE, ALBERT (1876-1952), founder of Workers' Educational Association; left Battersea grammar school at fourteen; obtained clerical work; cashier, Co-operative Permanent Building Society, 1901-5; a director, 1910; student and teacher at evening classes; founded Workers' Educational Association, 1903; general secretary, 1905-14; founded branches throughout country and obtained co-operation of universities in providing tutorial classes for study at university level; founded Central Library for Students (later National Central Library), 1916; World Association for Adult Education, 1918; Seafarers' Educational Service, 1919; British Institute of Adult Education, 1921; member, royal commission on Oxford and Cambridge, 1919-22; appointed to Oxford statutory commission, 1923; publications include *The Kingdom of the Mind* (1944); hon. MA, Oxford, 1912; hon. LLD, Manchester, 1922, Cambridge, 1923; CH, 1931.

MANSEL-PLEYDELL, JOHN CLAVELL (1817-1902), Dorset antiquary; BA, St. John's College, Cambridge, 1839; built Milborne Reformatory, 1856; founded Dorset Natural History Club, 1875; made valuable geological finds in Dorsetshire; published the *Flora* (1874), *Birds* (1888), and *Mollusca* (1898) of Dorsetshire.

MANSERGH, JAMES (1834-1905), civil engineer; worked in Brazil, 1855-9; practised at Westminster from 1866 till death; specialized in water and sewage works; constructed reservoirs and aqueduct from valleys of Elan and Cherwen rivers to Birmingham, 1894-1904; carried out sewage disposal for several midland towns; prepared schemes for sewerage of Lower Thames valley; on royal commission on metropolitan water-supply, 1892-3; FRS, 1901; president of Institution of Civil Engineers, 1900-1.

MANSFIELD, Sir JOHN MAURICE (1893-1949), vice-admiral; entered navy, 1906; chief of staff to commander-in-chief, Western Approaches, 1941-3; rear-admiral, 1943; commanded cruiser squadron, Mediterranean, 1943-5; supported landings in Italy and southern France; commanded British naval forces liberating Greece; DSO, 1945; flag officer, Ceylon (1945-6), submarines (1946-8); vice-admiral, 1946; KCB, 1948.

MANSFIELD, KATHERINE (pseudonym) (1888-1923), writer. [See MURRY, KATHLEEN.]

MANSFIELD, ROBERT BLACHFORD (1824-1908), author and oarsman; brother of Charles Blachford Mansfield [q.v.]; educated at Eton and University College, Oxford; BA, 1846; called to bar, 1849; pioneer of English golf, 1857; rowed in university boat race, 1843; pioneer of English rowing in Germany; recorded his German experiences in *The Log of the Water-Lily* (1851) and *The Water-Lily on the Danube* (1852).

MANSON, JAMES BOLIVAR (1879-1945), painter and director of Tate Gallery; educated at Alleyn's School, Dulwich; studied at Heatherley's and Académie Julian, Paris; admirer of Impressionists; most successful in landscape and still life; first secretary, Camden Town (1911) and London (1913) groups; assistant (1912), assistant keeper (1917), director (1930-7), Tate Gallery.

MANSON, Sir PATRICK (1844-1922), physician and parasitologist; MB, CM, Aberdeen, 1865; medical officer for Formosa to Chinese imperial maritime customs, 1866; at

Amoy, 1871–82; while working on elephantoid diseases discovered developmental phase in life of *filaria* worms in tissues of the blood-sucking insect, mosquito; settled in Hong Kong, where he instituted school of medicine which developed into university and medical school of Hong Kong, 1883–9; physician to Seamen's Hospital, London, 1892; propounded his mosquito-malaria theory, which was confirmed by other researchers; physician and adviser to Colonial Office, 1897; instrumental in foundation of London School of Tropical Medicine, 1899; FRS, 1900; KCMG, 1903; the 'father of tropical medicine'; published works on tropical diseases.

MANSON, THOMAS WALTER (1893–1958), biblical scholar; educated at Tynemouth high school and Glasgow University; MA, 1917; entered Westminster and Christ's colleges, Cambridge; first class, part ii, oriental languages tripos (Hebrew and Aramaic), 1923; ordained in Presbyterian Church of England, 1925; minister, Falstone, Northumberland, 1926–32; Yates professor of New Testament Greek, Mansfield College, Oxford, 1932–6; Rylands professor of biblical criticism, Manchester, 1936–58; served on New Testament and Apocrypha panels for New English Bible; publications include *The Church's Ministry* (1948) and *The Servant-Messiah* (1953); FBA, 1945.

MAPLE, Sir JOHN BLUNDELL, baronet (1845–1903), merchant and sportsman; joined (1862) father's furnishing firm in Tottenham Court Road, which was converted into limited liability company, with Maple as chairman, 1891; conservative MP for Dulwich, 1887–1903; knighted, 1892; baronet, 1897; racehorse breeder and owner; won 544 races; headed list of winning owners, 1891; won Cesarewitch, 1894, and Two Thousand Guineas, 1895; rebuilt University College Hospital, 1897–1906.

MAPLESON, JAMES HENRY (1830–1901), operatic manager; studied violin and pianoforte at Royal Academy of Music, 1884, and singing in Italy; managed Italian opera season from 1858 variously at Drury Lane, Lyceum Theatre, and Covent Garden; produced Gounod's *Faust* (1863) and Bizet's *Carmen* (1878); from 1878 took touring companies to America in winter; engaged Adelina Patti, 1881–5; his repertory of Italian opera lost vogue from 1887; published *Memoirs* (2 vols., 1888).

MAPOTHER, EDWARD DILLON (1835–1908), surgeon; MD, Queen's University, Dublin, 1857; FRCS Ireland, 1862; surgeon of St. Vincent's Hospital, Dublin, 1859; professor of hygiene in Royal College of Surgeons, 1864; professor of anatomy and physiology, 1867;

president, 1879; first medical officer of health for Dublin; wrote much on diseases of the skin and public health.

MAPPIN, Sir FREDERICK THORPE, first baronet (1821–1910), benefactor to Sheffield; head of father's cutlery business, 1841–59; senior partner in firm of Turton & Sons, steel manufacturers, 1859–85; mayor of Sheffield, 1877–8; a founder and benefactor of Sheffield Technical School; largely endowed Sheffield University; first senior pro-chancellor, 1905; director of Midland Railway, 1869–1900; enthusiastic Volunteer; whig MP for East Retford, 1880–5, and for Hallamshire, 1885–1906; baronet, 1886; added eighty pictures to Mappin Art Gallery, founded by uncle; first honorary freeman of Sheffield, 1900.

MAPSON, LESLIE WILLIAM (1907–1970), biochemist; educated at Cambridge and County high school, and Fitzwilliam House (later College), Cambridge, 1928–30; began research in biochemistry under (Sir) F. G. Hopkins and J. B. S. Haldane [qq.v.]; studied nutrition and enzyme catalysis; Ph.D., 1934; lecturer in biochemistry, Portsmouth College of Technology, 1934–8; scientific officer, Food Investigation Board, DSIR, 1938; seconded to Dunn Nutritional Laboratory, MRC, Cambridge, 1939; principal scientific officer, DSIR, 1950; senior principal scientific officer, 1956–67; FRS, 1969; hon. professor, East Anglia University.

MARETT, ROBERT RANULPH (1866–1943), anthropologist; son of Sir R. P. Marett [q.v.]; educated at Victoria College, Jersey and Balliol College, Oxford; first class, *lit. hum.*, 1888; studied philosophy, Berlin University; fellow, Exeter College, Oxford, 1891; tutor in philosophy, 1893; sub-rector, 1893–8; rector, 1928–43; secretary, committee for anthropology, 1907–27; reader in social anthropology, 1910–36; developed the conception of 'Präanimismus'; publications include *Faith, Hope and Charity in Primitive Religion* (1932) and *Sacraments of Simple Folk* (1933); FBA, 1931.

MARGESSON, (HENRY) DAVID (REGINALD), first VISCOUNT MARGESSON (1890–1965), politician; educated at Harrow and Magdalene College, Cambridge; served with 11th Hussars during 1914–18 war; MC; conservative MP, Upton division of West Ham, 1922–3; MP Rugby, 1924–42; junior lord of the Treasury, 1926; parliamentary secretary to the Treasury, 1931; PC, 1933; Government chief whip, 1931–40; secretary of state for war, 1940–2; displaced by his permanent under-secretary Sir (P.) James Grigg [q.v.]; viscount, 1942.

MARGOLIOUTH, DAVID SAMUEL (1858-1940), classical scholar and orientalist; scholar of Winchester and New College, Oxford; first class, *lit hum.*, 1880; won Hertford, Ireland, Craven, Derby, Boden Sanskrit, Pusey and Ellerton Hebrew and Kennicott Hebrew scholarships; fellow, 1881; Laudian professor of Arabic, 1889-1937; ordained, 1899; works include an edition of *The Poetics of Aristotle* (1911), *Arabic Papyri of the Bodleian Library* (1893), *A Compendious Syriac Dictionary* (with his wife, 1896-1903), *Lines of Defence of the Biblical Revelation* (1900), *The Synoptic Gospels as Independent Witnesses* (1903), *Mohammed and the Rise of Islam* (1905), *Mohammedanism* (1911), and catalogue of Arabic papyri in the John Rylands library, 1933; his editions of Arabic texts include Yāqūt's *Dictionary of Learned Men* (1907-27); his translations, *The Eclipse of the Abbasid Caliphate* (7 vols., 1920-1); FBA, 1915.

MARIE LOUISE, Princess, whose full names were Franziska Josepha Louise Augusta Marie Christiana Helena (1872-1956), daughter of Prince and Princess Christian and granddaughter of Queen Victoria; married Prince Aribert of Anhalt, 1891; marriage annulled, 1900; returned to England and devoted life to charitable, social, and artistic causes; published *My Memories of Six Reigns* (1956); GBE, 1919; GCVO, 1953.

MARILLIER, HENRY CURRIE (1865-1951), journalist and expert on tapestries; born in South Africa; educated at Christ's Hospital and Peterhouse, Cambridge; joined *Pall Mall Gazette*, 1893; Swan Electric Engraving Co., 1896; Merton Abbey Tapestry Works of William Morris (q.v.), 1905; developed craft of tapestry repair; wound up company, 1940; published history of its tapestries, 1927; compiled subject-index and illustrated catalogue of European tapestries given to Victoria and Albert Museum, 1945; publications include *The Liverpool School of Painters 1810-67* (1904).

MARINA, Duchess of Kent (1906-1968), youngest daughter of Prince and Princess Nicholas of Greece; educated in Paris; married Prince George [q.v.], fourth son of King George V and Queen Mary, 1934; eldest son, Prince Edward, born, 1935; Princess Alexandra born, 1936; Prince Michael born, 1942; Prince George killed in flying accident, 1942; commandant (later chief commandant), WRNS, 1940; president, Royal National Life-boat Institution; president, All England Lawn Tennis Club; chancellor, Kent University; patron, National Association for Mental Health; colonel-in-chief, Queen's Own Royal West Kent Regiment and the Corps of Royal Electrical and Mechanical Enginners; CI and GBE, 1937; GCVO, 1948.

MARJORIBANKS, EDWARD, second Baron Tweedmouth (1849-1909), politician; educated at Harrow and Christ Church, Oxford; a keen sportsman through life; toured round world, 1872-3; abandoned law for politics; liberal MP for North Berwickshire, 1880-94; comptroller of Queen Victoria's household and PC, 1886; parliamentary secretary to the Treasury and chief liberal whip under Gladstone, 1892; succeeded to peerage, 1894; joined Lord Rosebery's cabinet as lord privy seal and chancellor of the duchy of Lancaster, 1894-5; married (1873) Lady Fanny (*d.* 1904), sister of Lord Randolph Churchill [q.v.], a successful society and political hostess; suffered financial losses, 1904; first lord of the Admiralty in Campbell-Bannernan's ministry, 1905-8; maintained policy of England's naval supremacy; incurred public censure for alleged premature disclosure of naval estimates to German Emperor, 1908; KT, 1908; lord president of Council under Asquith, 1908; died of cerebral malady.

MARKHAM, Sir ALBERT HASTINGS (1841-1918), admiral and Arctic explorer; entered navy, 1856; commanded *Alert* in Arctic expedition of Sir G. S. Nares [q.v.], 1875-6; record of latitude reached without dogs lasted until 1895; rear-admiral, 1891; KCB, 1903.

MARKHAM, Sir CLEMENTS ROBERT (1830-1916), geographer and historical writer; cousin of Sir A. H. Markham [q.v.]; served in navy, 1844-51; entered civil service, 1853; in charge of geographical work of India Office, 1867-77; CB, 1871; FRS, 1873; accompanied Arctic expedition of Sir G. S. Nares [q.v.], 1875; president of Hakluyt Society, 1889-1909, and of Royal Geographical Society, 1893-1905; KCB, 1896; promoted Antarctic exploration.

MARKHAM, VIOLET ROSA (1872-1959), public servant; granddaughter of Sir Joseph Paxton [q.v.]; daughter of Chesterfield colliery owner; made home in London meeting place for notable people in politics, the arts, and social service; joined liberal party; served from 1914 on, and for many years chairman of, Central Committee on Women's Training and Employment; member, executive committee, National Relief Fund; member, Industrial Court, 1919-46; joined Assistance Board, 1934; deputy chairman, 1937-46; member, appeals tribunal on internment, 1939-45; chairman, investigation committee on welfare of Service women, 1942; mayor of Chesterfield, 1927; married (1915)

Lt.-Col. James Carruthers (died 1936); published autobiography *Return Passage* (1953); CH, 1917; hon. Litt.D, Sheffield, 1936; hon. LL.D, Edinburgh, 1938.

MARKS, DAVID WOOLF (1811-1909), Goldsmid professor of Hebrew at University College, London, from 1844 to 1898; secretary to the Hebrew congregation at Liverpool, 1833; first minister with Albert Löwy [q.v.] of newly established West London congregation, 1842-95; prepared reformed prayer book, and obtained licence for marriages, 1857; hon. DD, Cincinnati; published lectures on Mosaic law.

MARKS, SIMON, first BARON MARKS OF BROUGHTON (1888-1964), retailer and business innovator; son of Jewish immigrant from Poland who founded Marks and Spencer Ltd.; educated at Manchester grammar school; studied languages and business methods in Europe; joined father's firm, 1907; director, 1911; chairman, 1916-64; joined by his friend and brother-in-law, Israel (later Lord) Sieff, 1915; joined Royal Artillery during, 1914-18 war; seconded to Chaim Weizmann [q.v.], to set up Zionist headquarters in London, 1917; Marks and Spencer incorporated as public company, 1926; registered 'St. Michael', the firm's brand name, 1928; deputy chairman, London and South Eastern regional production board and adviser to petroleum warfare department during 1939-45 war; one of first directors, British Overseas Airways; benefactor to Royal College of Surgeons, University College, London, Manchester grammar school, British Heart Foundation, and the cause of Israel; knighted, 1944; baron, 1961; vice-president Zionist Federation; hon. D.Sc. (Economics), London; LLD, Manchester; Ph.D., Hebrew University, Jerusalem; hon. fellow, Royal College of Surgeons, Weizmann Institute of Science, and University College, London.

MARLOWE, THOMAS (1868-1935), journalist; educated at Queen's College, Galway; joined the *Star* (1888), *Evening News* (1894); managing editor, *Daily Mail*, 1899-1926.

MARQUIS, FREDERICK JAMES, first EARL OF WOOLTON (1883-1964), politician and business man; educated at Manchester grammar school and Manchester University; B.Sc., 1906; schoolteacher, 1906-10; research fellow, Manchester University, 1910; MA, 1912; warden, David Lewis Hotel and Club Association, Liverpool, and warden, university settlement, 1912; secretary, Leather Control Board, 1914; secretary, federation of the boot industry; joined board of Lewis's, Liverpool; joint managing director, 1928; chairman, 1936-51;

member, advisory councils to Overseas Development Committee, 1928-31, Board of Trade, 1930-4, and Post Office, 1933-47; president, Retail Distributors Association, 1930-3; chairman, 1934; director-general, Ministry of Supply, 1939; minister of food, 1940; minister of reconstruction, 1943-5; joined conservative party, 1945; party chairman, 1946-55; lord president of the Council, 1951-2; chancellor of the duchy of Lancaster, 1952-5; minister of materials, 1953-4; knighted, 1935; baron, 1939; PC, 1940; CH, 1942; viscount, 1953; earl, 1956; chairman, executive committee, British Red Cross Society, 1943-63; chancellor, Manchester University, 1944; hon. degrees, Manchester, Liverpool, Cambridge, McGill, and Hamilton College.

MARR, JOHN EDWARD (1857-1933), geologist; educated at Lancaster grammar school and St. John's College, Cambridge; first class, natural sciences tripos (geology), 1878; fellow, 1881-1933; university lecturer in geology, 1886-1917; professor, 1917-30; authority on the palaeozoic strata; influential teacher; FRS, 1891.

MARRIOTT, SIR JOHN ARTHUR RANSOME (1859-1945), historian, educationist, and politician; educated at Repton and New College, Oxford; lecturer, modern history, Worcester College, 1885-1920; secretary, Oxford university extension delegacy, 1895-1920; conservative MP, Oxford City (1917-22), York (1923-9); successful popularizer and voluminous writer on history, biography, and politics; knighted, 1924.

MARRIOTT, SIR WILLIAM THACKERAY (1834-1903), judge-advocate-general; BA, St. John's College, Cambridge, 1858; curate of St. George's, Hulme, 1858; renounced orders, 1861; called to bar, 1864; acquired lucrative practice in railway and compensation cases; QC, 1877; liberal MP for Brighton, 1880-4; re-elected for Brighton as conservative, 1884-93; PC, 1885; judge-advocate-general, 1885-92; knighted, 1888; chancellor of Primrose League, 1892; counsel for ex-Khedive Ismail Pasha in claims against Egyptian government, 1887-8; made unsuccessful financial speculations; died at Aix-la-Chapelle.

MARRIS, SIR WILLIAM SINCLAIR (1873-1945), Indian civil servant; educated at Wanganui, Canterbury College, New Zealand, and Christ Church, Oxford; posted to North-West Provinces and Oudh, 1896; under-secretary, Indian home department, 1901; deputy secretary, 1904-6; lent to Transvaal government, 1906-8; intimately associated with Milner's 'kindergarten' and member of *Round*

Table group; joint-secretary, home department, India, 1917; drafted Montagu-Chelmsford report, Nov. 1917–Apr. 1918; reforms commissioner, 1919–21; governor of Assam, 1921–2; of United Provinces, 1922–7; principal, Armstrong College, Newcastle upon Tyne, 1929–37; advocated union of two Newcastle colleges; translated Horace (1912), Catullus (1924), and Homer (*Odyssey* 1925, *Iliad* 1934) into English verse; KCIE, 1919; KCSI, 1921.

MARSDEN, ALEXANDER EDWIN (1832–1902), surgeon; son of William Marsden [q.v.]; MRCS, 1854; MD, St. Andrews, 1862; FRCS Edinburgh, 1868; served as surgeon at Scutari and Sevastopol; surgeon to Royal Free Hospital, 1853–84; and to Brompton Cancer Hospital; published works on cancer.

MARSDEN, Sir ERNEST (1889–1970), atomic physicist and science administrator; educated at Queen Elizabeth's grammar school, Blackburn, and Manchester University; worked with (Lord) Rutherford [q.v.] on radioactivity, 1907–9; John Harling fellow, Manchester University, 1911; lecturer and research assistant, 1912–13; professor of physics, Victoria University College, Wellington, New Zealand, 1914; served in France, 1914–18; MC, 1919; returned as teacher and administrator to Victoria College, 1919; assistant director of education, 1922–6; permanent secretary, New Zealand Department of Scientific and Industrial Research, 1926–47; trained in England in radar production, 1939; scientific adviser to New Zealand fighting services (lieutenant-colonel); attended Empire Scientific Congress, 1946; New Zealand scientific liaison officer, London, 1947–54; CBE, 1935; FRS, CMG, 1946; knighted, 1958; hon. degrees, Oxford (1946), Manchester (1961), and Victoria University, Wellington (1965).

MARSH, Sir EDWARD HOWARD (1872–1953), civil servant, scholar, and patron of the arts; great grandson of Spencer Perceval [q.v.]; educated at Westminster and Trinity College, Cambridge; first class, parts i and ii, classical tripos, 1893–5; an 'Apostle'; entered Colonial Office, 1896; worked for (Sir) Winston Churchill at Colonial Office (1906–8), Board of Trade (1908–10), Home Office (1910–11), Admiralty (1911–15), Munitions (1917), War Office (1919–21), Colonial Office (1921–2), and Treasury (1924–9); in Dominions Office, 1930–7; retired, 1937; corrected proofs of Churchill's literary writings from *Marlborough* (4 vols., 1933–8) onwards; also sixteen works by Somerset Maugham; began collecting pictures, 1896; acquired Horne collection, 1904; became patron of contemporary British painting and literature; edited five volumes of *Georgian*

Poetry (1912–22); literary executor of Rupert Brooke [q.v.] whose collected poems he published, 1918; translations include La Fontaine's Fables (2 vols., 1931) and *Odes of Horace* (1941); published reminiscences, *A Number of People* (1939); trustee of Tate Gallery; governor of Old Vic; chairman, Contemporary Art Society, 1936–52; KCVO, 1937.

MARSHALL, ALFRED (1842–1924), economist; educated at Merchant Taylors' School and St. John's College, Cambridge; second wrangler, 1865; fellow of St. John's, 1865–77, 1885–1908; came into intellectual circle of which Henry Sidgwick [q.v.] was chief, 1867; passed from study of metaphysics to that of political economy; lecturer in moral science, St. John's, 1868; first principal of University College, Bristol, 1877–81; professor of political economy, Bristol, 1877–83; fellow and lecturer in political economy, Balliol College, Oxford, 1883–5; professor of political economy, Cambridge, 1884–1908; served on royal commissions on labour, 1891–4; the aged poor, 1893, and local taxation, 1899; on Indian currency committee, 1899; his works include *The Economics of Industry* (1879), *Principles of Economics* (1890), *Industry and Trade* (1919), and *Money, Credit, and Commerce* (1923).

MARSHALL, GEORGE WILLIAM (1839–1905), genealogist; LLB, Peterhouse, Cambridge, 1861; LLM, 1864; LLD, 1874; called to bar, 1865; founded *The Genealogist*, 1877; works include *The Genealogist's Guide* (1879) and *Handbook to the Ancient Courts of Probate* (1889); FSA, 1872; helped to found Parish Register Society, 1896; Rouge Croix pursuivant of arms, 1887; York herald, 1904; made for College of Arms unique collection of parish registers.

MARSHALL, Sir GUY ANSTRUTHER KNOX (1871–1959), entomologist; born in Amritsar; grandson of Sir Jonathan Frederick Pollock [q.v.]; educated at Charterhouse; followed various employments in Natal and Rhodesia; encouraged as amateur entomologist by Sir Edward Poulton [q.v.]; scientific secretary, Entomological Research Committee (Tropical Africa), 1909–13; director, Imperial Bureau (later the Commonwealth Institute) of Entomology, 1913–42; made it centre of information on insect pests; specialized on beetles of the family Curculionidae; FRS, 1923; CMG, 1920; knighted, 1930; KCMG, 1942; hon. D.Sc., Oxford.

MARSHALL, Sir JOHN HUBERT (1870–1958), archaeologist; educated at Dulwich College and King's College, Cambridge; first class, parts i and ii, classical tripos, 1898-1900; at British School at Athens, 1898-1901; director-general of archaeology, India, 1902-28; on special duties, 1928-34; organized Archaeological Survey of India; prepared Antiquities Law; gave first attention to conserving upstanding structures and recreated original beauty of setting by restoring gardens; notable explorations included ancient Taxila, near Rawalpindi (3 vols., 1951) and Indus Valley Civilization (3 vols., 1931); restored Buddhist site of Sanchi, central India (3 vols., with A. Foucher, 1940); CIE, 1910; knighted, 1914; FBA, 1936; hon. fellow, King's College, Cambridge, 1936.

MARSHALL, JULIAN (1836-1903), art collector and author; in family flax-spinning business at Leeds, 1855-61; formed collection of engravings embracing leading works of ancient and modern schools, of musical autographs, and of book-plates; contributed to *Grove's Dictionary of Music*; wrote *Annals of Tennis* (1878) and other kindred works.

MARSHALL, SIR WILLIAM RAINE (1865-1939), lieutenant-general; educated at Repton and Sandhurst; commissioned in the Sherwood Foresters, 1885; captain, 1893; served with distinction at Bothaville, 1900; commanded a brigade at Gallipoli landings, 1915; promoted major-general commanding a division; commanded III (Indian) Corps, Mesopotamia, 1916-17; commander-in-chief, Mesopotamia, 1917-19; enforced surrender of Turks on upper Tigris, Oct. 1918; lieutenant-general, 1919; held Southern command, India, 1919-23; KCB, 1917; KCSI, 1918; GCMG, 1919.

MARSHALL HALL, SIR EDWARD (1858-1929), lawyer. [See HALL.]

MARTEL, SIR GIFFARD LE QUESNE (1889-1958), lieutenant-general; educated at Wellington College and Royal Military Academy, Woolwich; commissioned in Royal Engineers, 1909; army and combined Services welterweight champion, 1912-1913; army champion, 1920, imperial Services champion, 1921-1922; served in France, 1914-16; at tanks headquarters in France, 1916-18; major, 1917; DSO, MC; returned to Royal Engineers experimental establishment; devised a box girder bridge; built a tankette; commanded first mechanized RE field company, 1926-9; devised a 'stepping-stone' bridge and a 'mat bridge'; recognized in Germany as pioneer of new type of warfare; instructor, Quetta Staff College, 1930-3; assistant director, mechanization, War Office, 1936, deputy director, 1938-9; commanded 50th Northumbrian division, 1939-40; improvised counter-attack at Arras, May 1940;

commander, Royal Armoured Corps, 1940-2; lieutenant-general, 1942; head of military mission, Moscow, 1943-4; CB, 1940; KBE, 1943; KCB, 1944.

MARTEN, SIR (CLARENCE) HENRY (KENNETT) (1872-1948), provost of Eton; educated at Eton and Balliol College, Oxford; first class, modern history, 1895; returned to Eton as history master; vice-provost, 1929; provost, 1945-8; entrusted with Princess Elizabeth's historical education, 1938; KCVO, 1945.

MARTIN, ALEXANDER (1857-1946), Presbyterian theologian and church leader; son of Hugh Martin [q.v.]; educated at George Watson's College, the University, and New College, Edinburgh; first class, philosophy, 1880; minister, Morningside Free Church, 1884-97; professor, apologetics and practical theology, New College, 1897-1927; principal, 1918-35.

MARTIN, (BASIL) KINGSLEY (1897-1969), editor; educated at Hereford cathedral school, Mill Hill and Magdalene College, Cambridge; first class, historical tripos, parts i and ii, 1920-1, after serving in France with Friends' Ambulance Unit, 1917-18; bye-fellow, Magdalene, 1920; assistant lecturer in politics, London School of Economics, 1924; leader-writer, *Manchester Guardian*, 1927-30; editor, *New Statesman* (later *New Statesman and Nation*), 1931-60; co-founder (with W. A. Robson) and joint-editor, *Political Quarterly*; publications include *The Triumph of Lord Palmerston* (1924), *French Liberal Thought in the Eighteenth Century* (1929), *Father Figures* (1966), and *Editor* (1968).

MARTIN, SIR CHARLES JAMES (1866-1955), physiologist and pathologist; educated at Birkbeck and King's colleges and St. Thomas's Hospital, London; B.Sc., 1886; MRCS, LSA, 1889; MB, 1890; demonstrator in physiology, Sydney, 1891-7; Melbourne, 1897-1901; professor, 1901-3; director, Lister Institute of Preventive Medicine, 1903-30; professor of experimental pathology, London, 1912-30; investigations included bubonic plague in India, vitamins and deficiency diseases; pathologist, Third Australian General Hospital, Lemnos, 1915-18; identified cause of enteric fever; devised vitamin 'soup cube', creating nutrition division at Institute; director, nutrition division and professor of biochemistry and general physiology, Adelaide, 1931-3; at Cambridge continued investigations including myxomatosis and wartime food problems; FRS, 1901; CMG, 1919; knighted, 1927; FRCP,

1913; fellow, King's College, London; several honorary degrees.

MARTIN, HERBERT HENRY (1881-1954), secretary of the Lord's Day Observance Society; educated at Alderman Norman's Endowed School, Norwich; a 'Wycliffe preacher' of Protestant Truth Society, attacking Roman Catholicism, 1898-1923; worked for Imperial Alliance for Defence of Sunday, 1923-5; secretary, Lord's Day Observance Society, 1925-51; opposed Sunday opening of cinemas (unsuccessfully, 1931) and theatres (1941).

MARTIN, HUGH (1890-1964), ecumenical student leader and publisher; educated at Glasgow Academy, Royal Technical College, Glasgow, and Glasgow University; MA, 1913; studied theology at Baptist Theological College of Scotland, 1909-14; assistant secretary, Student Christian Movement, 1914; treasurer, World Students Christian Federation, 1928-35; chairman, preparatory committee for 1924 Conference on Politics, Economics, and Citizenship; organized 'SCM Press', 1929; set up Religious Book Club, 1937; director, Religious Division, Ministry of Information, 1939-43; managing director, SCM Press, 1943; Free Church leader, British Council of Churches; vice-president, 1950; moderator, Free Church Federal Council, 1952-3; edited *Baptist Hymn Book Companion* (1962); hon. DD, Glasgow University, 1943; CH, 1955.

MARTIN, SIR THEODORE (1816-1909), man of letters; educated at Edinburgh high school and university, 1830-3; hon. LLD, 1875; practised as solicitor in Edinburgh until 1846, when he migrated to London to become parliamentary agent; his parliamentary work the main occupation of his life; contributed before leaving Edinburgh humorous prose and verse to *Tait's* and *Fraser's* magazines under pseudonym of Bon Gaultier; soon collaborated with William Edmondstone Aytoun [q.v.]; together they published *Bon Gaultier Ballads* (1845, 16th edn., 1903), a notable collection of witty parodies; devoted to the drama, was fascinated by the acting of Helen Faucit [q.v.], for whom he adapted from the Danish *King René's Daughter*, 1849; married Miss Faucit at Brighton, 1851, and settled for life at 31 Onslow Square, Kensington; acquired in 1861 country residence Bryntysilio; wrote on dramatic themes in *Fraser's Magazine*, 1858-65, the *Quarterly Review*, and *Blackwood*; translated Oehlenschläger's German romantic dramas *Aladdin* (1854) and *Correggio* (1857), the works of Horace, (1860, 1882) and of Catullus (1861), Dante's *Vita Nuova* (1862), Goethe's *Faust* (pt. i, 1865; pt. ii, 1866) poems and ballads of

Heine (1878), and the *Aeneid*, i-vi (1896); on recommendation of his friend Sir Arthur Helps [q.v.], Martin prepared for Queen Victoria a life of the Prince Consort (5 vols., 1875-80); CB, 1878; KCB, 1880; KCVO, 1896; wrote life of Lord Lyndhurst, 1883; lord rector of St. Andrews University, 1881; privately circulated *Queen Victoria as I knew her*, 1901 (published, 1908); a trustee of Shakespeare's birthplace from 1889 till death, and an active member of the Royal Literary Fund from 1868.

MARTIN, SIR THOMAS ACQUIN (1850-1906), industrial pioneer in India and agent-general for Afghanistan; founded engineering firm in Calcutta, which took over Bengal Iron and Steel Company (1889) and worked iron deposits at Manharpur; pioneer of light railways in India; built many jute mills and controlled large collieries and engineered water supplies in Bengal; appointed agent to Amir of Afghanistan (1887), for whom he built an arsenal, a mint, and various factories; accompanied Amir's son to England on diplomatic mission, and knighted, 1895.

MARTIN, VIOLET FLORENCE (1862-1915), novelist under the pseudonym of MARTIN Ross; member of ancient Galway family; wrote books in collaboration with her cousin, E. A. Œ. Somerville [q.v.], describing Anglo-Irish life; best-known novels *The Real Charlotte* (1894) and *The Irish R.M.* series (begun 1899).

MARTIN, WILLIAM KEBLE (1877-1969), botanist; educated at Marlborough, Christ Church, Oxford, and Cuddesdon Theological College; ordained, 1902; curate, Beeston, Ashbourne, and Lancaster, 1902-9; vicar, Wath-upon-Dearne, near Rotherham, 1909-17; temporary chaplain to the forces, 1917; rector, Haccombe and Coffinswell, near Torquay, 1921; Great Torrington, North Devon, 1934-43; Combe-in-Teignhead with Milber, 1943-9; a keen botanist and flower artist; edited (with Gordon T. Fraser), *The Flora of Devon* (1939); published *A History of the Ancient Parish of Wath-upon-Dearne* (1920), and *Concise British Flora in Colour* (1965); fellow, Linnean Society, 1928; hon. D.Sc., Exeter, 1966; four stamps issued with his designs, 1967.

MARTINDALE, CYRIL CHARLIE (1879-1963), priest and scholar; educated at Harrow and Campion Hall, Oxford; entered Jesuit novitiate, 1897; first class, *lit. hum.*, 1905; ordained, 1911; taught at Stoneyhurst, 1913-16; classics lecturer, Oxford, 1916-27; joined staff of Farm Street church, Mayfair, 1927; noted preacher and broadcaster; unable to leave Denmark, 1940-5; returned to Farm Strett, 1945-53; published over eighty books and hundreds of pamphlets, including biographies of R. H. Benson and C. D. Plater, *Waters of Twilight* (1914), and *The Goddess of Ghosts* (1915).

MARTINDALE, HILDA (1875-1952), civil servant; educated in Germany, at Brighton high school, Royal Holloway College, and Bedford College; studied hygiene and sanitary sciences; appointed temporary factory inspector, 1901; senior lady inspector, 1908; served in Ireland, Birmingham, and London; superintending inspector, 1921; deputy chief inspector, 1925; director, women establishments, Treasury, 1933-7; CBE, 1935; publications include *Women Servants of the State* (1938).

MARTIN-HARVEY, SIR JOHN MARTIN (1863-1944), actor-manager; with Sir Henry Irving [q.v.] at Lyceum, 1882-96; outstandingly successful with *The Only Way* (adaptation of *Tale of Two Cities*) first produced at Lyceum, 1899; produced *Hamlet* (1904), *Richard III* (1910), and *The Taming of the Shrew* (1913); rakish vigour of his 'The Rat' in *The Breed of the Treshams* (1903) contrasted with sensitive study as Count Skariatine in *A Cigarette Maker's Romance* (1901); his *Œdipus Rex* (1912) profoundly impressive; knighted, 1921.

MARWICK, SIR JAMES DAVID (1826-1908), legal and historical writer; founded legal firm of Watt & Marwick in Edinburgh, 1855; town clerk of Edinburgh, 1860-73, and of Glasgow, 1873-1903; extended city of Glasgow by annexing fourteen suburban burghs, 1881-91; FRS Edinburgh, 1884; hon. LLD, Glasgow, 1878; knighted, 1888; helped to found Scottish Burgh Records Society, Edinburgh, editing its publications from 1868 to 1897.

MARY, (VICTORIA MARY AUGUSTA LOUISE OLGA PAULINE CLAUDINE AGNES) (1867-1953), queen consort of King George V; born in Kensington Palace; only daughter of Duke and Duchess of Teck; devoted sister of Earl of Athlone [q.v.]; known until marriage as Princess May; extended her education beyond drawing-room accomplishments with help of governess but mainly by own determination with aid of excellent memory; proficient in French, German, European history, and knowledge of art; engaged to Duke of Clarence, 1891; after his death to Duke of York, 1893; married in St. James's Palace chapel 6 July 1893; had five sons and one daughter; with husband set new pattern of family life providing simple and sensible upbringing for children whom she protected from occasional over-harsh

discipline from their father; travelled with Duke in *Ophir* for first opening of Australian federal Parliament, 1901, and extended tour of Empire; became Princess of Wales, 1901; made arduous tour of India with husband, 1905-6; with his succession to throne became known as Queen Mary, 1910; with King visited India (1911-12), Berlin (1913), Paris (1914); toured industrial areas of Great Britain with genuine interest and sympathetic understanding; indefatigable during war of 1914-18; thereafter supported the King through difficult post-war years in execution of duties for which he was physically unfitted especially after his illness in 1928-9 during which she was a tower of strength; after his death (1936) moved to Marlborough House; sustained abdication of eldest son with calm dignity; attended coronation of King George VI, 1937; pursued her interest in art collection and cultural and industrial projects; her eyesight injured in car accident, May 1939; removed to Badminton during war of 1939-45; attended funeral of her fourth and favourite son, Duke of Kent [q.v.], 1942; returned to London, 1945, and resumed public engagements until after death of King George VI in 1952; died at Marlborough House 24 Mar. 1953; buried beside her husband in St. George's chapel, Windsor; possessed of great physical strength, self-discipline, and mental vigour; formidable in her rigidity for conduct she yet exercised great practical sympathy; a great queen consort, selflessly loyal to monarchy, she won respect and affection of nation by her devotion to duty and refusal to be deflected from it by personal griefs.

MASEFIELD, JOHN EDWARD (1878-1967), poet laureate; educated at King's School, Warwick, and in the *Conway*, where he learnt seamanship; travelled and worked in the United States; returned to England and contributed to the *Outlook*, the *Academy* and the *Speaker*, 1897-1906; published *Salt Water Ballads* (1902); worked on *Manchester Guardian*, 1907; published first major poem, *The Everlasting Mercy* (1911); followed by *Reynard the Fox* (1919); wrote naval histories, and criticism, including *William Shakespeare* (1911); also novels and plays, including *Lost Endeavour* (1910), and *Good Friday* (1916); and children's books such as *The Box of Delights* (1935); appointed poet laureate, 1930; OM, 1935; wrote *The Nine Days Wonder* on Dunkirk (1940); hon. degrees from Oxford, Liverpool, and St. Andrews; C.Litt., 1961; president, Society of Authors, 1937; and National Book League, 1944-9.

MASHAM, first BARON (1815-1906), inventor. [See LISTER, SAMUEL CUNLIFFE.]

MASKELYNE, MERVYN HERBERT NEVIL STORY- (1823–1911), mineralogist. [See STORY-MASKELYNE.]

MASON, ALFRED EDWARD WOODLEY (1865–1948), novelist; educated at Dulwich and Trinity College, Oxford; provincial actor, 1888–94; liberal MP, Coventry, 1906–10; secret service agent, 1914–18; wrote *The Four Feathers* (1902) and other adventure stories; historical novels including *Musk and Amber* (1942); Inspector Hanaud detective series; and several plays.

MASON, ARTHUR JAMES (1851–1928), theological scholar and preacher; BA, Trinity College, Cambridge; fellow of Trinity, 1873; assistant tutor, 1874; honorary canon and diocesan missioner, Truro, 1878; vicar of All Hallows, Barking, 1884–95; honorary canon of Canterbury, 1893; Lady Margaret's professor of divinity and fellow of Jesus College, Cambridge, 1895; master of Pembroke College, Cambridge, 1903; withdrew to Canterbury, 1912; the trusted adviser and helper of successive archbishops in current ecclesiastical affairs; belonged to older school of high churchmen; wrote theological and historical works.

MASON-MACFARLANE, SIR (FRANK) NOEL (1889–1953), lieutenant-general; educated at Rugby and Royal Military Academy, Woolwich; gazetted to Royal Artillery, 1909; served in 1914–18 and Afghan (1919) wars; military attaché, Berlin, 1937–9; believed war inevitable and advocated choosing time unfavourable to Hitler; major-general and director of military intelligence, BEF in France, 1939–40; improvised 'MacForce', 1940; DSO; deputy governor, Gibraltar, 1940; head, British military mission, Moscow, 1941–2; governor, Gibraltar, 1942–4; chief commissioner, Allied Control Commission, Italy, 1944; labour MP, North Paddington, 1945–6; CB, 1939; KCB, 1943.

MASSEY, (CHARLES) VINCENT (1887–1967), diplomatist, patron of education and arts, and first Canadian-born governor-general of Canada; born in Toronto; educated at St. Andrews College, Toronto, university of Toronto, and Balliol College, Oxford; lecturer in modern history, Toronto University, 1913; dean of residence, Victoria College, 1913–15; served in Canadian army (lieutenant-colonel), 1914–18; president of family business, Massey-Harris, 1921–5; set up Massey Foundation to encourage the arts; first diplomatic representative to United States of ministerial rank, 1926; president, National Liberal Federation, 1932–5; high commissioner, London, 1935–46; trustee, National Gallery, London, 1941–6; chairman,

1943–6; trustee, Tate Gallery, 1942–6; chairman, board of trustees, National Gallery of Canada, 1948–52; governor-general of Canada, 1952–9; PC, 1941; CH, 1946; hon. DCL Oxford and hon. degrees from other universities; hon. fellow, Royal Society of Canada; publications included *The Sword of Lionheart and Other Wartime Speeches* (1943) and *What's Past is Prologue, Memoirs* (1963).

MASSEY, GERALD (1828–1907), poet; after scanty education was put to work at Tring at age of eight; studied for himself; published at Tring *Poems and Chansons*, 1848; joined Chartists; helped to edit *The Spirit of Freedom* (1849); soon turned to Christian socialism; wrote for the *Christian Socialist*; brought out *Voices of Freedom and Lyrics of Love* (1850), and *The Ballad of Babe Christabel and other Poems* (1854); his lyrical impulse was widely acknowledged; other volumes of verse followed; complete poetical works, Boston, 1857, London, 1861; a selection, *My Lyrical Life*, appeared in 1899; his career suggested that of Felix Holt to George Eliot [q.v.]; he became a journalist and popular lecturer, living for a time at Edinburgh, and from 1862 to 1877 near Little Gaddesden in a farmhouse provided by Lord Brownlow; wrote on Shakespeare's sonnets, 1866 (reissued, 1888); thrice lectured in America; developed faith in spiritualism and finally took to writing on old Egyptian civilization.

MASSEY, WILLIAM FERGUSON (1856–1925), prime minister of New Zealand; born in county Derry; emigrated to New Zealand and took up farming, 1870; conservative MHR, 1894–1925; chief opposition whip, 1895–1903; leader of conservative opposition, 1903–12; prime minister, as leader of 'reform party', 1912–25; formed national government, 1915; led New Zealand very ably through European war.

MASSINGBERD, SIR ARCHIBALD ARMAR MONTGOMERY- (1871–1947), field-marshal. [See MONTGOMERY-MASSINGBERD.]

MASSINGHAM, HAROLD JOHN (1888–1952), author and journalist; son of H. W. Massingham [q.v.]; educated at Westminster and Queen's College, Oxford; contributed on literary and natural history topics to *Nation* and *Athenaeum*, 1916–24; *Field*, 1938–51; books on English countryside include *English Downland* (1936) and *Cotswold Country* (1937); other publications include *Downland Man* (1926), *The English Countryman* (1942), and edition of *The Writings of Gilbert White of Selborne* (1938).

MASSINGHAM, HENRY WILLIAM (1860-1924), journalist; joined staff of *Eastern Daily Press*, 1877; went to London, 1883; editor of *Star*, 1890; of *Labour World*, 1891; occupied various positions on staff of *Daily Chronicle*, 1892-5; editor, 1895; resigned because of his opposition to South African war, 1899; editor of *The Nation*, 1907-23; formed this liberal weekly journal into powerful organ of advanced but independent opinion; joined labour party and transferred his 'Wayfarer's Diary' to *New Statesman*, 1923; showed passionate energy for human welfare.

MASSON, DAVID (1822-1907), biographer and editor; MA, Aberdeen University, 1839; studied divinity at Edinburgh, 1839-42; but abandoned thoughts of entering ministry; visited London, 1843; introduced by Thomas Carlyle to editor of *Fraser's Magazine*, 1844; wrote for W. & R. Chambers textbooks on Roman, ancient, medieval, and modern history, 1847-56; removed to London, 1847; intimate with the Carlyles, Thackeray, Douglas Jerrold, and Mark Lemon [qq.v.]; professor of English literature at University College, London, 1853-65; published *Essays, chiefly on English Poets* (1859); started (1859) *Life of Milton* (6 vols., 1859-80), the standard authority; started, 1859, and edited till 1867, *Macmillan's Magazine*; professor of rhetoric and English literature at Edinburgh University, 1865-95; a popular teacher; supporter of women's higher education; edited *Privy Council Register of Scotland* (1880-99); Rhind lecturer, 1886; historiographer royal for Scotland, 1896; hon. RSA, 1896; hon. LLD, Aberdeen, Litt.D., Dublin; voluminous writings include editions of Goldsmith (1869), Milton (3 vols., 1874), and De Quincey (14 vols., 1889-90); biographies of Drummond of Hawthornden (1873) and De Quincey (1878); *Edinburgh Sketches and Memories* (1892), and (posthumously) *Memories of London in the Forties* (1908) and *Memories of Two Cities* (1911).

MASSON, SIR DAVID ORME (1858-1937), chemist; son of David Masson [q.v.]; educated at Edinburgh Academy and University; research fellow in chemistry, 1882-6; professor of chemistry, Melbourne, 1886-1923; president, Australasian Association for Advancement of Science, 1911-13; influential in founding and served on Commonwealth Council for Scientific and Industrial Research; initiated Australian Chemical Institute, the Australian National Research Council (with Sir T. W. E. David, q.v.), and other bodies; FRS, 1903; KBE, 1923.

MASSY, WILLIAM GODFREY DUNHAM (1838-1906), lieutenant-general; BA, Trinity College, Dublin, 1859; LLD, 1873; entered army, 1854; served at the Redan, where his gallantry earned him the sobriquet of 'Redan' Massy; commanded Royal Irish Lancers, 1871-9; prominent in battle of Charasiab (Oct. 1879), in Afghan war; cut off by Afghans at Killa Kazi, Dec. 1879; removed from command for rash advance, 1880; CB, 1887; in command of troops in Ceylon, 1888-93; lieutenant-general, 1893.

MASTERMAN, CHARLES FREDERICK GURNEY (1874-1927), politician, author, and journalist; BA, Christ's College, Cambridge; liberal MP, West Ham (North), 1906-11; South-West Bethnal Green, 1911-14; under-secretary, Local Government Board, 1908; under-secretary of state, home department, 1909; financial secretary to Treasury, 1912; chancellor, duchy of Lancaster, with seat in cabinet, 1914-15; director, Wellington House (propaganda department), 1914-18; MP, Rusholme division of Manchester, 1923-4; works include *The Condition of England* (1909).

MASTERS, MAXWELL TYLDEN (1833-1907), botanist; MRCS, 1856; MD, St. Andrews, 1862; lecturer on botany at St. George's Hospital medical school, 1855-68; published valuable researches in his *Vegetable Teratology* (1869); principal editor of *Gardener's Chronicle*, 1865; active supporter of Royal Horticultural Society; wrote much on passion flowers and conifers in botanical works and journals; revised Henfrey's *Elementary Course of Botany* (1870) and wrote *Botany for Beginners* (1872) and *Plant Life* (1883); FRS, 1870.

MATHESON, GEORGE (1842-1906), theologian and hymn writer; known as 'the blind preacher'; blind from boyhood; BA, Glasgow University, 1861; MA, 1862; minister of Innellan church, 1868-86, and of St. Bernard's parish church, Edinburgh, 1886-99; issued popular theological and devotional works which had a wide vogue; they include *The Growth of the Spirit of Christianity* (2 vols., 1877), *The Psalmist and the Scientist* (1887), *Sacred Songs* (1890, 3rd edn. 1904), and studies of representative men and women of the Bible (4 ser., 1902-7); DD, Edinburgh, 1879; LLD, 1902; FRS Edinburgh, 1902.

MATHEW, SIR JAMES CHARLES (1830-1908), judge; senior moderator, Trinity College, Dublin, 1850; called to bar (Lincoln's Inn), 1851; had vast City practice as a junior; among Treasury counsel on prosecution of the Tichborne claimant, 1873; made judge of Queen's Bench division and knighted, 1881; obtained institution of Commercial Court, 1895; as first judge, gave concise and terse judgments; as chairman of royal commission of inquiry into evictions in Ireland, created precedent of refusing to allow cross-examination by counsel, which led to resignation of many members of commission, 1892; judge of Court of Appeal, 1901-5; ready, facile, and humorous speaker; and ardent radical and devout Roman Catholic.

MATHEW, THEOBALD (1866-1939), lawyer and wit; son of Sir J. C. Mathew and brother-in-law of John Dillon [qq.v.]; educated at the Oratory School and Trinity College, Oxford; called to bar (Lincoln's Inn), 1890; bencher, 1916; practised as junior on common law side; recorder of Margate (1913-27), Maidstone (1927-36); joint-editor of *Commercial Cases* (1896); contributed 'Forensic Fables' to *Law Journal* (collected in 4 vols., 1926-32).

MATHEW, Sir THEOBALD (1898-1964), director of public prosecutions; educated at the Oratory School, London; served with Irish Guards during 1914-18 war; MC, 1918; called to bar (Lincoln's Inn), 1921; admitted solicitor and partner in Charles Russell & Co., 1928; joined Home Office, 1941; head of criminal division, 1942-4; director, public prosecutions, 1944-64; KBE, 1946.

MATHEWS, BASIL JOSEPH (1879-1951), writer and teacher on the missionary and ecumenical movement; educated at Oxford high school and university; editor for London Missionary Society, 1910-19; head, Press Bureau of Conference of British Missionary Societies, 1920-4; literary secretary, World's Committee, YMCA, Geneva, 1924-9; visiting lecturer, later professor, Christian world-relations, Boston, 1932-44; Union College, university of British Columbia, 1944-9; wrote over forty books; hon. LLD, British Columbia, 1949.

MATHEWS, CHARLES EDWARD (1834-1905), alpine climber and writer; solicitor at Birmingham from 1856; clerk of the peace, 1891-1905; founded Children's Hospital there, 1864; lifelong friend of Joseph Chamberlain [q.v.]; helped to found Alpine Club, 1857; president, 1878-80; prominent in conquest of Alps; wrote critical and exhaustive *Annals of Mont Blanc*, (1898).

MATHEWS, Sir CHARLES WILLIE, baronet (1850-1920), lawyer; called to bar (Middle Temple), 1872; director of public prosecutions, 1908-20; appeared in criminal trials and notorious civil cases; knighted, 1907; baronet, 1917.

MATHEWS, Sir LLOYD WILLIAM (1850-1901), general and prime minister of Zanzibar;

entered navy, 1863; served in Ashanti campaign, 1873-4; retired with rank of lieutenant, 1881; in command of Zanzibar army, with rank of brigadier-general, 1877; obtained abolition of slavery in Zanzibar, 1890; Zanzibar declared British protectorate, 1890; prime minister and treasurer of reconstituted government, 1891; introduced modern agricultural methods; CMG, 1880; KCMG, 1894; died at Zanzibar.

MATHEWS, Dame VERA (ELVIRA SIBYL MARIA) LAUGHTON (1888-1959), director of the Women's Royal Naval Service; daughter of Sir John Knox Laughton [q.v.]; educated at convents and King's College, London; served with Women's Royal Naval Service, 1917-19; first editor, *Time and Tide*; married Gordon Dewar Mathews (died 1943), 1924; director, Women's Royal Naval Service, 1939-46; CBE, 1942; DBE, 1945.

MATHIESON, WILLIAM LAW (1868-1938), historian; educated at Edinburgh Academy and University; historian of *Politics and Religion . . .* [in Scotland] (1902), of *English Church Reform . . .* (1923) and *British Slavery and its Abolition . . .* (1926).

MATTHEWS, ALFRED EDWARD (1869-1960), actor; son of original Christy Minstrel; became touring then west-end actor; in demand for plays by Pinero, Galsworthy, and Barrie [qq.v.]; skilled in technique of light comedy acting; became star as Lord Lister in *The Chiltern Hundreds* (1947) and its sequel (1954); OBE, 1951.

MATTHEWS, HENRY, Viscount Llandaff (1826-1913), lawyer and politician; son of Henry Matthews [q.v.]; born in Ceylon; educated at universities of Paris and London; called to bar (Lincoln's Inn), 1850; QC, 1868; conservative, MP, Dungarvan, 1868-74, East Birmingham, 1886-95; home secretary, 1886-92; baron, 1895; a founder of Westminster Cathedral.

MATTHEWS, Sir WILLIAM (1844-1922), civil engineer; pupil in office of Sir John Coode [q.v.], harbour engineer; partner in Coode's firm, consulting engineers to crown agents for colonies, 1892; frequently employed by Admiralty on works at naval bases; KCMG, 1906.

MATURIN, BASIL WILLIAM (1847-1915), Catholic preacher and writer; BA, Trinity College, Dublin; ordained, 1870; member of Society of St. John the Evangelist, Cowley, 1873-97; received into Roman Church, 1897; ordained, 1898; torpedoed on *Lusitania* returning from third visit to America.

MAUD CHARLOTTE MARY VICTORIA
(1869-1938), princess of Great Britain and
Ireland, Queen of Norway; fifth child of Prince
and Princess of Wales; married (1896) Prince
Christian Frederick Charles George Valdemar
Axel (second son of Crown Prince of Denmark),
elected to throne of Norway as King Haakon
VII, 1905.

MAUDE, AYLMER (1858-1938), translator
and expounder of Tolstoy's works; educated at
Christ's Hospital and in Moscow where he lived
until 1897; close friend of Tolstoy, whom he
translated jointly with his wife (21 vols., 1928-
37); wrote *Life of Tolstoy* (2 vols., 1908-10).

MAUDE, SIR (FREDERICK) STANLEY
(1864-1917), lieutenant-general; born at
Gibraltar; educated at Eton and Sandhurst;
joined 2nd Coldstream Guards, 1884; entered
Staff College, Camberley, 1895; served in South
African war, 1899-1901; DSO, 1901; military
secretary to Earl of Minto, governor-general of
Canada, 1901-5; CMG, 1905; held staff appointments in England and Ireland, 1906-14;
served in France, Aug. to Nov. 1914; CB, 1915;
major-general and appointed to command 13th
division at Dardanelles, 1915; took prominent
part in evacuation of Suvla and Helles, 1915-16;
took division to Mesopotamia, 1916; assumed
command of army in Mesopotamia and created
KCB, 1916; recovered Kut and captured Baghdad, 1917; died of cholera at Baghdad.

MAUGHAM, FREDERICK HERBERT,
first VISCOUNT MAUGHAM (1866-1958), lord
chancellor; born in Paris; grandson of Robert
Maugham [q.v.]; educated at Dover College
and Trinity Hall, Cambridge; senior optime,
1888; rowed for Cambridge, 1888-9; president
of Union, 1889; called to bar (Lincoln's Inn),
1890; KC, 1913; Chancery judge (knighted),
1928-34; lord justice of appeal (PC), 1934; lord
of appeal in ordinary (life peer), 1935-8, 1939-
41; lord chancellor, 1938-9; made solid contribution to English law especially in patents
and trademarks; publications include *The Tichborne Case* (1936), *The Truth About the Munich
Crisis* (1944), *U.N.O. and War Crimes* (1951),
and *At the End of the Day* (1954); viscount,
1939; hon. fellow, Trinity Hall, 1928.

MAUGHAM, WILLIAM SOMERSET
(1874-1965), writer; educated at King's School,
Canterbury and Heidelberg University; medical student, St. Thomas's Hospital London,
1892-5; MRCS, LRCP, 1897; published
first novel, *Liza of Lambeth* (1897); first stage
play success, *Lady Frederick* (1907); served in
Intelligence Department during 1914-18 war;
published *Of Human Bondage* (1915), *The Moon

and Sixpence* (1919), *The Painted Veil* (1925),
and *Cakes and Ale* (1930); settled in south of
France, 1928-40; returned there in 1945; pub-
lished *The Summing Up* (1938) and *The Razor's
Edge* (1944); founded the Somerset Maugham
Award for young writers, 1947; CH, 1954;
C.Lit., 1961; fellow, Royal Society of Literature;
commander of the Legion of Honour; hon.
D.Litt., Oxford and Toulouse.

MAURICE, SIR FREDERICK BARTON
(1871-1951), major-general; son of (Major-
General Sir) John Frederick Maurice [q.v.];
educated at St. Paul's School and Sandhurst;
married Helen Margaret Marsh, sister of (Sir)
Edward Marsh [q.v.], 1899; commissioned in
Derbyshire Regiment, 1892; served in Tirah
campaign, 1897-8; South African war, 1899-
1902; instructor to Staff College under Sir
William Robertson [q.v.], 1913; served in
France, 1914-15; director, military operations,
War Office, under Robertson, Dec. 1915-18;
resigned Apr. 1918, and wrote to newspapers
accusing Lloyd George's government of deceiving parliament about strength of British
Army in France; Lloyd George defeated censure motion, quoting War Office figures that he
knew to be inaccurate; Army Council retired
Maurice from army and refused inquiry; principal, Working Men's College, 1922-33; professor of military studies, London University,
1927; principal, East London College, 1933-44;
member of London University senate, 1946;
president, British Legion, 1932-47; published
works include military biographies, *Forty Days
in 1914* (1919), *Governments and War* (1926), and
a *History of the Scots Guards* (2 vols., 1934);
KCMG, 1918; hon. LLD, Cambridge, 1926.

MAURICE, SIR JOHN FREDERICK (1841-
1912), major-general; son of F. D. Maurice
[q.v.]; joined Royal Artillery, 1862; served in
War Office and in Ashanti, South Africa, Egypt,
and Sudan, 1873-85; professor of military art
and history at Staff College, 1885; commanded
artillery, Woolwich district, 1895; major-
general, 1895; KCB, 1900; works include prize
essay which greatly influenced army reform;
edited first two volumes of official *History of the
War in South Africa, 1899-1902* (1906-7).

MAVOR, OSBORNE HENRY (1888-1951),
better known as playwright JAMES BRIDIE, educated at Glasgow Academy and Glasgow University; fellow student of Walter Elliot [q.v.];
qualified in medicine, 1913; served in Royal
Army Medical Corps, 1914-18 war; consulting
physician, Victoria Infirmary; professor of
medicine, Anderson College, Glasgow; wrote
some forty plays, including *Tobias and the Angel*
(1930), *The Anatomist* (1931), *Mr Bolfry*,

(1943), *Daphne Laureola* (1949); established Citizens' Theatre, Glasgow, 1943; hon. LLD, Glasgow, 1939; CBE, 1946.

MAWDSLEY, JAMES (1848-1902), trade-union leader; secretary of Amalgamated Society of Cotton Spinners, 1878; developed trade-union policy in Lancashire; negotiated conciliation scheme—the Brooklands agreement—referring trade disputes to arbitration, 1893; member of royal commission on labour questions, 1891-4.

MAWER, SIR ALLEN (1879-1942), scholar; graduated in English, London (1897) and Cambridge (1904); lecturer in English, Sheffield, 1905-8; professor, Armstrong College, Newcastle, 1908-21; Liverpool, 1921-9; provost, University College, London, 1930-42; published *The Vikings* (1913), *Place-Names of Northumberland and Durham* (1920), and chiefly responsible for English Place-Name Society publications, 1924-43; FBA, 1930; knighted, 1937.

MAWSON, SIR DOUGLAS (1882-1958), scientist and explorer; born at Shipley, Yorkshire; educated at Fort Street school and university of Sydney, Australia; BE, mining, 1902; B.Sc., 1905; lecturer in mineralogy and petrology, Adelaide, 1905; D.Sc., Adelaide, 1909; professor of geology and mineralogy, 1920-52; physicist on Antarctic expedition of (Sir) Ernest Shackleton [q.v.], 1907; climbed Mount Erebus, 1908; reached south magnetic pole, 1909; led Australian Antarctic expedition, 1914-18; organized Banzare expedition, 1929-31, which led to annexation for Australia of two and a half million square miles between the Ross Dependency and Enderby Land; made important mineral discoveries in South Australia, including uranium; knighted, 1914; OBE, 1920; FRS, 1923.

MAXIM, SIR HIRAM STEVENS (1840-1916), engineer and inventor; born in Maine, USA; chief engineer to United States Electric Lighting Company, 1878; came to England and opened workshop in London, *c.*1882; naturalized, 1900; knighted, 1901; chief inventions electrical pressure regulator (1881), rapid-firing gun with completely automatic action (adopted by British army, 1889, and navy, 1892), steam-driven flying-machine (1889-94), and maxi-mite, a smokeless powder (1889).

MAXSE, SIR (FREDERICK) IVOR (1862-1958), general; son of Admiral Frederick Augustus Maxse [q.v.]; educated at Rugby and Sandhurst; commissioned, 1882; captain, Coldstream Guards, 1891; seconded to Egyptian Army, 1897-9; fought at Atbara and Khartoum;

DSO, 1898; in South African war on staff of Lord Roberts [q.v.], 1899-1902; brigadier-general, 1st Guards brigade, 1910; major-general with brigade in France, 1914; commanded 18th division in France, 1915-17; lieutenant-general in command XVIII Corps, 1917-18; inspector-general of training in France, Apr. 1918; held Northern Command in Britain, 1919-23; general, 1923-6; took up commercial fruit growing after retirement from army; CB, 1900; CVO, 1907; KCB, 1917.

MAXSE, LEOPOLD JAMES (1864-1932), journalist and political writer; son of F. A. Maxse [q.v.]; educated at Harrow and King's College, Cambridge; owned and edited *National Review*, 1893-1932; convinced imperialist and democrat; lover of France; foresaw both German wars; considered League of Nations useless and advocated armaments and plain speaking with Germany.

MAXTON, JAMES (1885-1946), politician; educated at Hutcheson's grammar school and the university, Glasgow; elementary school-teacher and spare-time expositor of socialism; persuasive platform orator of burning eloquence; called for general strike on Clyde and imprisoned for sedition, 1916; organizer, independent labour party, 1919; chairman, 1926-31, 1934-9; MP, Bridgeton division of Glasgow, 1922-46; published *Lenin* (1932) and *If I were Dictator* (1935).

MAXWELL, SIR ALEXANDER (1880-1963), civil servant; educated at Plymouth College and Christ Church, Oxford; first class, *lit. hum.*, 1903; entered Home Office, 1904; acting chief inspector, reformatory and industrial schools, 1917; assistant secretary, 1924; chairman, Prison Commission, 1928; deputy under-secretary of state, Home Office, 1932; permanent under-secretary, 1938-48; CB, 1924; KBE, 1936; KCB, 1939; GCB, 1945; member, royal commission on capital punishment, 1949; governor, Bedford College, 1948-50.

MAXWELL, GAVIN (1914-1969), writer and conservationist; educated at Stowe and Hertford College, Oxford; strong interest in natural history; served with Scots Guards, 1939-44; bought island of Soay off Skye; worked in London as portrait painter, 1949-52; visited Sicily, 1953; travelled in southern Iraq, 1956; president, British Junior Exploration Society; fellow, Royal Society of Literature, Royal Geographical Society, and Royal Zoological Society (Scotland); publications included *Harpoon at a Venture* (1952), *God Protect me from my Friends* (1956), a study of Sicilian life, *A*

Reed Skaken by the Wind (1957), a study of marsh Arabs, *Ring of Bright Water* (1962), his first study of relationship between man and otter, *The House of Elrig* (1965), an autobiography, and *Lords of the Atlas*, a history of the Moroccan house of Glacus.

MAXWELL, SIR HERBERT EUSTACE, seventh baronet, of Monreith (1845-1937), country gentleman, politician, and writer; succeeded father, 1877; conservative MP, Wigtownshire, 1880-1906; lord-lieutenant, 1903-35; works include *Memories of the Months* (1897-1922), *Life of Wellington* (1899); edited the *Creevey Papers*, 1903; PC, 1897; FRS, 1898; KT, 1933.

MAXWELL, SIR JOHN GRENFELL (1859-1929), general; grandson of Vice-Admiral J. P. Grenfell [q.v.] and cousin of first Baron Grenfell [q.v.]; gazetted to 42nd Foot, 1879; served in Egypt, 1882-1900; present at battle of Tel-el-Kebir, 1882; accompanied Lord Wolseley [q.v.] on his fruitless attempt to relieve General Gordon [q.v.], 1884-5; took part in Sudan frontier operations, 1885-9; served under Lord Kitchener [q.v.] in reconquest of Sudan, 1896-8; proceeded to Cape, Feb. 1900; appointed military governor of Pretoria, June; chief staff officer to Duke of Connaught in Ireland, London, and Malta, 1902-8; commanded British troops in Egypt, 1908-12; resumed command, 1914-16, an important and exacting position; commander-in-chief in Ireland, 1916; commander-in-chief, Northern command, 1916-19; full general, 1919; KCB, 1900; died at Cape Town.

MAXWELL, MARY ELIZABETH (1837-1915), better known as MISS BRADDON, novelist; sister of Sir Edward N. C. Braddon [q.v.]; her best-known novel, *Lady Audley's Secret* (1862), had an immediate and very great success and made her fortune; published about eighty novels between 1862 and 1911; also wrote many plays, edited several magazines, and contributed to periodicals; plots of novels concerned with crime, but not a mere sensationalist; married John Maxwell, publisher, 1874.

MAXWELL FYFE, DAVID PATRICK, EARL OF KILMUIR (1900-1967), lord chancellor. [See FYFE.]

MAXWELL LYTE, SIR HENRY CHURCHILL (1848-1940), deputy keeper of the public records and historian. [See LYTE.]

MAY, GEORGE ERNEST, first BARON MAY (1871-1946), financial expert; educated at Cran-leigh; entered Prudential Assurance Company, 1887; secretary, 1915-31; made its American investments available to government, 1915; manager, American dollar securities committee, 1915-19; deputy quartermaster-general in charge of canteen administration, 1916-19; chairman of committee on national expenditure which recommended economies mainly in social services amounting to over £96½ million and brought about downfall of labour government, 1931; chairman, and specially responsible for iron and steel industry, Import Duties Advisory Committee, 1932-41; KBE, 1918; baronet, 1931; baron, 1935.

MAY, PHILIP WILLIAM [PHIL MAY] (1864-1903), humorous draughtsman; a caricature by him of Irving, Bancroft, and Toole (1883) attracted notice of Lionel Brough [q.v.], who introduced him to editor of *Society*, for which, he executed drawings; in Australia, 1885-8, contributing to the *Sydney Bulletin*; studied art in Paris; made sketches for *St. Stephen's Review*; from 1892 issued *Phil May's Winter Annual*; made reputation as comic artist of low life in *Daily Graphic* and other illustrated papers; published *Phil May's Sketch Book: Fifty Cartoons* (1895) and his vivid *Guttersnipes: Fifty Original Sketches* (1896); member of the *Punch* table from 1896 to death; a rapid cartoonist of vigour and vivacity; sociable and generous to a fault; his *Picture Book* edited with life by G. R. Halkett, 1903.

MAY, SIR WILLIAM HENRY (1849-1930), admiral of the fleet; entered navy, 1863; navigating officer of the *Alert* in Arctic expedition led by Sir G. S. Nares [q.v.], 1875-6; joined torpedo-school ship *Vernon*, 1877-80; captain, 1887; naval attaché to European states, 1891-3; rear-admiral, 1901; third sea lord and controller of navy, 1901-5; during his controllership *Dreadnought* policy initiated; commander-in-chief, Atlantic fleet, 1905-7; second sea lord, 1907-9; successfully protested against reduction in naval expenditure, 1907; commander of home fleet, 1909-11; devoted his attention to reforms in naval tactics; commander-in-chief, Devonport, 1911-13; admiral of the fleet, 1913; member of Dardanelles commission, 1916-17; KCVO, 1904.

MAYBURY, SIR HENRY PERCY (1864-1943), civil engineer; educated at Upton Magna School; county engineer, Kent, 1904-10; joined Road Board, 1910; director of roads in France, 1917-19; director-general of roads, Ministry of Transport, 1919-28; consulting engineer to minister, 1928-32; chairman, London traffic advisory committee, 1924-33; KCMG, 1919; GBE, 1928.

MAYOR, JOHN EYTON BICKERSTETH (1825–1910), classical scholar and divine; revelled in classical literature from age of six; educated at Christ's Hospital, Shrewsbury, and St. John's College, Cambridge; third classic, 1848; fellow, 1849; master at Malborough, 1849–53; prepared edition of Juvenal, (1853, 3rd edn. 1881); classical tutor at St. John's from 1853; an accomplished linguist; published lives of Nicholas Ferrar (1855), Matthew Robinson (1856), Ambrose Bonwicke (1870), and William Bedell(1870), and edited Ascham's *Scholemaster* (1863); edited transcript of admissions to St. John's College; published Baker's *History of St. John's College* (2 vols., 1869); university librarian, 1864–7; completed catalogue of MSS; university professor of Latin, 1872 to death; edited (with J. R. Lumby, q.v.) Bede's *Ecclesiastical History*, bks. iii and iv (1878); visited Leiden and Rome, 1875; advocate of vegetarian diet from middle life; president of St. John's, 1902; hon. DCL, Oxford, 1895, LLD, Aberdeen, 1892, and St. Andrews, 1906, DD, Glasgow, 1901; original FBA, 1902; had power of accumulating knowledge but small faculty of construction; projected uncompleted commentary on Seneca and a Latin dictionary; edited works by Cicero, Pliny, Homer, Quintilian, and volumes for the Rolls and Early English Text societies; published also *A Bibliography of Latin Literature* (1875), a *First Greek Reader* (1868), and *First German Reader* (1910).

MEADE, RICHARD JAMES, fourth EARL OF CLANWILLIAM in the Irish peerage, and second BARON CLANWILLIAM in the peerage of the United Kingdom (1832–1907), admiral of the fleet; entered navy, 1845; served in Russian war, 1852; wounded at storming of Canton, 1857; junior sea lord, 1874–80; CB, 1877; succeeded to earldom, 1879; commanded flying squadron, 1880–2; KCMG, 1882; commander-in-chief on North America station, 1885–6, and at Portsmouth, 1891–4; admiral, 1886; KCB, 1887; admiral of the fleet and GCB, 1895.

MEADE-FETHERSTONHAUGH, SIR HERBERT (1875–1964), admiral; entered *Britannia*, 1889; lieutenant, served in *Iphigenia* in China, 1897; severely injured in cable accident in *Venerable*, 1904; in Admiralty yacht, *Enchantress*, 1906, commander, 1908; divisional leader in destroyer, *Goshawk*, 1912; in Heligoland Bight action, 1914; DSO; captain, 1914; in Dogger Bank action, 1915; in *Caroline* at battle of Jutland, 1916; chief of staff to commander-in-chief, Rosyth, 1919; commanded *Renown* on Prince of Wales's visit to India, China, and Japan, 1921; CVO, 1922; captain, Royal Naval College, Dartmouth, 1923; rear-admiral, CB, 1925; commanded Mediterranean destroyer flotillas, 1926–8; KCVO, 1929; vice-admiral, 1930; admiral, 1934; GCVO, extra equerry to the King, 1934; sergeant at arms, House of Lords, 1939–46.

MEAKIN, JAMES EDWARD BUDGETT (1866–1906), historian of the Moors; edited *Times of Morocco*, founded (1884) by his father; visited Mohammedan settlements in Asia and Africa, 1893; settled down to journalism in England, 1897; helped to found British Institute of Social Service, 1905; published *The Moorish Empire* (1899), *The Land of the Moors* (1901), and *The Moors* (1902).

MEATH, twelfth EARL OF (1841–1929), clerk in Foreign Office. [See BRABAZON, REGINALD.)

MEDD, PETER GOLDSMITH (1829–1908), theologian; BA, University College, Oxford, 1852; MA, 1855; fellow, 1852–77; rector of North Cerney, Cirencester, 1876–1908; hon. canon of St. Albans, 1877; Bampton lecturer, Oxford, 1882; edited Andrewes's *Greek Devotions* (1892); wrote *The One Mediator* (1884) and other theological works.

MEDLICOTT, HENRY BENEDICT (1829–1905), geologist; BA in civil engineering, Trinity College, Dublin, 1850; MA, 1870; joined Indian geological survey, 1854; professor of geology at Rurki from 1862; volunteer in Indian Mutiny; made study of structure of Himalayas; superintendent of Indian survey, 1876; FRS, 1877; president of Asiatic Society of Bengal, 1879–81; wrote on geology of Punjab (1874) and of India (2 vols., 1879, with W. T. Blanford, q.v.).

MEE, ARTHUR HENRY (1875–1943), journalist; on staff of Lord Northcliffe [q.v.]; productions included a *Self-Educator, History of the World, Natural History,* and *Popular Science* (between 1905 and 1913), and *I See All* (1928–30) and *Arthur Mee's Thousand Heroes* (1933–4); first issued *Children's Encyclopedia*, 1908; edited monthly *My Magazine*, 1908–33, and weekly *Children's Newspaper*, 1919–43.

MEEK, CHARLES KINGSLEY (1885–1965), anthropologist and colonial administrator; educated at Rothesay Academy, Bedford School, and Brasenose College, Oxford; BA, 1910; entered Colonial administrative service; posted to Nigeria, 1912; commissioner, decennial census, 1921; published *The Northern Tribes of Nigeria* (1925); government anthropologist, Nigeria, 1924–9; resident, Southern Nigeria, 1929–33; resigned due to ill health; Heath Clark lecturer, London University,

1938-9; senior research fellow, Brasenose College, 1943; advised Colonial Office on 'Devonshire' courses, training programme, 1947; tutor and supernumerary fellow, Brasenose, 1950; publications include *A Sudanese Kingdom* (1931), *Tribal Studies in Northern Nigeria* (1931), *Law and Authority in a Nigerian Tribe: a study in indirect rule* (1937), and *Land Tenure and Land Administration in Nigeria and the Cameroons* (1957).

MEGHNAD SAHA (1893-1956), scientist; educated at Government Collegiate School, Dacca, Dacca College and Presidency College, Calcutta; B.Sc., 1913; M.Sc., 1915; lecturer in mathematics, University College of Science, Calcutta, 1916; studied astrophysics, 1918-25; published paper on physical theory of stellar spectra, 1921; worked in London with Alfred Fowler [q.v.], and in Berlin with W. Nernst; professor, Allahabad University, 1923-38; concerned with spectroscopic and ionospheric studies, 1925-38; Palit professor, Calcutta, 1938-55; concerned mainly with nuclear physics; established Institute of Nuclear Physics, Calcutta, 1948; member of National Planning Commission, 1939-41, and of Indian Education Commission, 1948; published numerous scientific papers in Indian and foreign journals; FRS, 1927.

MEIGHEN, ARTHUR (1874-1960), Canadian statesman; educated at St. Mary's Collegiate Institute and Toronto University; first class, mathematics, 1896; admitted to Manitoba bar, 1903; conservative MP, 1908-26; solicitor-general in ministry of (Sir) Robert Borden [q.v.], 1913-17; secretary of state, 1917; minister of interior, 1917-20; his concern with conscription and public ownership of railways in wartime government made him a leading figure; prime minister, 1920-1 and 1926; member of Senate, 1932-41; minister without portfolio in government of R. B. (later Viscount) Bennett [q.v.], 1932-5; resumed leadership conservative party, 1941; retired from public life, 1942.

MEIKLEJOHN, JOHN MILLER DOW (1836-1902), writer of school books; MA, Edinburgh, 1858; war correspondent in Danish-German war, 1864; first professor of education in St. Andrews University, 1876; raised standard of school books in his *English Language* (1886), *The British Empire* (1891), *The Art of Writing English* (1899), and *English Literature* (1904).

MELBA, DAME NELLIE (1861-1931), prima donna; born HELEN PORTER MITCHELL; born, educated, and trained at Melbourne, Australia;

came to London 1886; studied in Paris under Mme Mathilde Marchesi and adopted name 'Melba'; her appearance in *Rigoletto* in Brussels (1887) extraordinary triumph; sang regularly at Covent Garden and in the capitals of the world, 1888-1926; helped in her acting by Sarah Bernhardt; studied *Faust* and *Roméo et Juliette* with Gounod, *Otello* with Verdi, *La Bohème* with Puccini; introduced Nedda in *Pagliacci* at Covent Garden (1893) at request of Leoncavallo; first sang *Hélène*, composed for her by Saint-Saëns, at Monte Carlo, 1904; her voice extraordinarily fresh and beautiful, with power of expansion, and perfectly even through $2\frac{1}{2}$ octaves; DBE, 1918; GBE, 1927.

MELCHETT, first BARON (1868-1930), industrialist, financier, and politician. [See MOND, ALFRED MORITZ.]

MELDRUM, CHARLES (1821-1901), meteorologist; MA, Marischal College, Aberdeen, 1844; professor of mathematics, Royal College of Mauritius, 1848; founded Mauritius Meteorological Society, 1851; in charge of Mauritius government observatory, 1862-96; studied laws of cyclones in Indian ocean; FRS, 1876; CMG, 1886.

MELLANBY, SIR EDWARD (1884-1955), medical scientist and administrator; brother of John Mellanby [q.v.]; educated at Barnard Castle School and Emmanuel College, Cambridge; first class, part ii, natural sciences, 1905; research student under (Sir) Frederick Gowland Hopkins [q.v.], 1905-7; qualified in medicine at St. Thomas's Hospital, 1909; lecturer and then professor of physiology, King's College for Women, 1913-20; MD, Cambridge, 1915; professor of pharmacology, Sheffield, and honorary physician, Royal Infirmary, 1920-33; secretary, Medical Research Council, 1933-49; Fullerian professor, Royal Institution, 1936-7; his research established main cause of rickets as deficiency of vitamin D, 1919; expert in biochemistry and physiology; chairman of international conferences on vitamins and nutrition; concerned with schemes for wartime diet, 1939-45; KCB, 1937; GBE, 1948; FRS, 1925; FRCP, 1928; hon. FRCS Ed., 1946.

MELLANBY, JOHN (1878-1939), physiologist; educated at Emmanuel College, Cambridge; MD, 1907; lecturer (from 1920 professor in London University) in charge of physiological department, St. Thomas's Hospital medical school, 1909-36; Waynflete professor of physiology, Oxford, 1936-9; obtained valuable results in work on the proteins of the blood, coagulation, and the secretion of the pancreas; FRS, 1929.

MELLON, SARAH JANE (1824-1909), actress; born WOOLGAR; made début at Plymouth, 1836; long associated with Adelphi from 1843; original Lemuel there in Buckstone's *Flowers of the Forest* (1847); at Lyceum appeared as Florizel in burlesque of *Perdita* (1856) and as Ophelia (1857); married Alfred Mellon [q.v.], 1858; played Catherine Duval at Adelphi in *The Dead Heart* (1859), Mrs Cratchit in *The Christmas Carol* (1860), and Anne Chute in *The Colleen Bawn* (1860); subsequently lost vogue; reappeared as Mrs O'Kelly in *The Shaughraun*, and created Miss Sniffe in *A Bridal Tour* (1880); retired, 1883; versatile actress in tragedy, comedy, burlesque, or farce.

MELVILLE, ARTHUR (1855-1904), artist; studied at Royal Scottish Academy and in Paris, 1878; travelled in Egypt, Persia, and Turkey, 1881-3; moulded the Glasgow artistic movement; exhibited several oil portraits at Edinburgh, including 'The Flower Girl' (1883) and 'Portrait of a Lady', before settling in London, 1888; visited Spain, 1889-92 and 1904, and Venice, 1894; ARSA, 1886; member of Royal Water Colour Society, 1900; works include 'A Moorish Procession', 'Christmas Eve', 'The Capture of a Spy', and 'The Little Bull Fight'.

MENDELSOHN, ERIC (1887-1953), architect; born in East Prussia of German-Jewish parents; educated at the Gymnasium, Allenstein; graduated in architecture at Munich, 1912; served in German Army, 1914-18; designed buildings in steel and concrete, including Einstein Observatory, Potsdam, 1920, and numerous factories, stores, houses, and flats in Berlin and other German cities; after advent of Hitler, moved to London, 1933; in partnership with Serge Chermayeff, designed De La Warr Pavilion, Bexhill, 1934; naturalized, 1938; FRIBA, 1939; left England for Palestine, 1939; moved to United States, 1941; set up practice in San Francisco, 1945; built many hospitals, synagogues, and community centres; one of first architects to realize potentialities of steel, concrete, and glass.

MENDL, SIR CHARLES FERDINAND (1871-1958), press attaché; educated at Harrow; in business in Paris until outbreak of war, 1914; served as interpreter with 25th Infantry brigade; invalided out, 1915; worked in intelligence in Paris for the Admiralty, 1918; Paris representative of Foreign Office news department, 1920; knighted, 1924; press attaché, Paris embassy, 1926-40; served five ambassadors, including Lord Tyrrell and Sir Eric Phipps [qq.v.]; cultivated wide contacts with the social and political world, enjoyed close friendship with Pertinax of the *Echo de Paris*.

MENON, VAPAL PANGUNNI (1894-1966), Indian public servant and author; educated at Ottopalam high school; employed by Imperial Tobacco Company, and as contractor in Kolar gold fields; clerk in Home Department, government of India, 1914; transferred to Reforms Office, 1930; on secretarial staff, Round Table Conference, London, 1931; assistant secretary, Reforms office, 1933; under-secretary, 1934; deputy secretary, 1936; CIE, 1941; reforms commissioner, 1942; CSI, 1946; KCSI, 1948; secretary, States Department; governor, Orissa, 1951; member, Finance Commission; retired, 1952; published *The Story of the Integration of the Indian States* (1956), *The Transfer of Power in India* (1957), and *An Outline of Indian Constitutional History* (1965).

MENZIES, SIR FREDERICK NORTON KAY (1875-1949), medical officer of health; educated at Llandovery College; MB, Edinburgh, 1899; MD, 1903; public health diploma, 1905; lecturer in public health, University College, London, and deputy medical officer of health, Stoke Newington, 1907; a medical officer, London County Council, 1909-24; county medical officer, 1926-39; welded hospitals of Metropolitan Asylums Board and Boards of Guardians into integrated service on their transfer to the Council; KBE, 1932.

MENZIES, SIR STEWART GRAHAM (1890-1968), head of the secret intelligence service; educated at Eton; joined Grenadier Guards, 1909; transferred to Life Guards, 1910; served in France, 1914; assigned to Intelligence, 1915; DSO, MC; military liaison officer with MI6, secret intelligence service, 1919; colonel, 1932; retired from Life Guards, 1939; appointed 'C', head of MI6, 1939; responsible also for supervision of Government Code and Cypher School, which broke the German 'Enigma' code; in charge of SIS throughout 1939-45 war and with the cold war period up to 1951; CB, 1942; KCMG, 1943; KCB, 1951.

MERCER, CECIL WILLIAM (1885-1960), novelist under name of DORNFORD YATES; first cousin of 'Saki' (H. H. Munro, q.v.); educated at Harrow and University College, Oxford; president, OUDS, 1906-7; called to bar (Inner Temple), 1909; assisted (Sir) Travers Humphreys [q.v.] in Crippen case; commissioned in 3rd County of London Yeomanry in 1914-18 war; settled at Pau in south of France, 1919; wrote many romantic comedies and thrillers including *The Brother of Daphne* (1914), *Berry and Co.* (1921), *Jonah and Co.* (1922), and *She Fell Among Thieves* (1925); after 1939-45 war lived at Umtali, Southern Rhodesia; published

As Berry and I Were Saying (1952) and *B— Berry and I Look Back* (1958).

MERCER, JAMES (1883-1932), mathematician; bracketed senior wrangler, Cambridge, 1905; mathematical lecturer, Christ's College, 1912-26; FRS, 1922; contributed striking theorems to theories of integral equations and orthogonal series and to modern theory of divergent series.

MEREDITH,   GEORGE   (1828-1909), novelist and poet; grandson of Melchizedek Meredith (*d.* 1804), a prosperous tailor and naval outfitter of Portsmouth [see the novel *Evan Harrington*]; was privately educated at Portsmouth and Southsea and at the Moravian school at Neuwied, where he became adept at German, 1843-4; articled to a London solicitor of Bohemian tastes, 1845; soon turning to journalism, he contributed poems to *Household Words* and to *Chambers's Journal*; making acquaintance of a son of Thomas Love Peacock [q.v.] he married Peacock's daughter, 9 Aug. 1849; boarding at Weybridge he came to know there Sir Alexander and Lady Duff Gordon, and afterwards settled in a cottage at Lower Halliford. In 1858 he was deserted by his wife (*d.* 1861), who had borne him a son, Arthur Gryffydh (1853-90); meanwhile, besides writing for *Fraser's Magazine*, he published *Poems* (with dedication to Peacock), 1851; *The Shaving of Shagpat: an Arabian Entertainment* (1855), and *Farina, a Legend of Cologne* (1857); *The Ordeal of Richard Feverel*, the first of his great novels, came out in 1859; the book introduced him to Swinburne and the pre-Raphaelite group, while through the Duff Gordons his acquaintance with other notable people quickly grew, but few copies of the book were sold, and his means were long scanty and precarious; he worked regularly for the *Ipswich Journal*, 1859-75; contributed to *Once a Week* six poems, 1859, and a new novel, *Evan Harrington*, serially through 1860; made his residence at Copsham near Esher, 1860, where his intimate circle soon included Frederick Augustus Maxse [q.v.], J. A. Cotter Morison [q.v.], and others; found his chief recreation in long walks in Surrey and occasionally in France and Switzerland; lodged for a time with Rossetti and Swinburne at the Queen's House, Chelsea, 1861-2; published *Modern Love and Poems of the English Roadside* (1862); became contributor to the *Morning Post*, 1862; was reader to Chapman & Hall from 1862 to 1894; brought out in 1864 *Emilia in England* (later renamed *Sandra Belloni*); married his second wife Sept. 1864, and after some years at Norbiton finally settled for life at Flint Cottage facing Boxhill, 1867; there his second wife died, 17 Sept. 1889,

leaving a son and daughter; he published *Rhoda Fleming* (1865); contributed serially to *Fortnightly Review* in 1866 *Vittoria*, a sequel to *Emilia*, which was expanded in separate issue; went to Italy as special correspondent for *Morning Post* during war with Austria, 1866; contributed serially to the *Cornhill*, *The Adventures of Harry Richmond* (separately issued in 1871); published *Odes in Contribution to the Song of French History* (1871), *Beauchamp's Career* (1876, after serial issue in condensed form in the *Fortnightly*, 1875), and the *Egoist* (1879, after serial issue in *Glasgow Weekly Herald*); delivered (1 Feb. 1877) at London Institution a characteristic lecture on 'The Idea of Comedy and the Uses of the Comic Spirit' (printed in *New Quarterly Magazine*, 1877, and separately, 1897); published *The Tragic Comedians*, embodying the love story of Ferdinand Lassalle, the German socialist, in Dec. 1880 (after serial issue in *Fortnightly*), and *Poems and Lyrics of the Joy of Earth* (1883); though both his novels and poetry won growing appreciation from critical circles, the public showed small interest until publication of *Diana of the Crossways* (1885) (after serial issue in the *Fortnightly*); there followed *Ballads and Poems of Tragic Life* (1887), *A Reading of Earth* (1888), two of his most characteristic volumes of verse; and the last three novels, *One of Our Conquerors* (1891), *Lord Ormont and his Aminta* (serially issued in *Pall Mall Magazine*, 1894), and *The Amazing Marriage* (begun in 1879 and serially issued in *Scribner's Magazine* through 1895); in his last years he published many expressions of opinion on public questions, but was from 1893 disabled from active exercise by paraplegia; he received in old age many marks of public regard; addresses were presented by his admirers on both his seventieth and eightieth birthdays; he was president of the Society of Authors from 1892 to death, and was admitted to Order of Merit, 1905; he died at Flint Cottage, 18 May, and was buried in Dorking cemetery, 23 May 1909; a memorial service was held in Westminster Abbey on day of funeral; there appeared posthumously *Celt and Saxon*, an unfinished story (*Fortnightly*, 1910), and *The Sentimentalists*, a conversational comedy, was produced at Duke of York's Theatre, Mar. 1910; *Last Poems* came out in 1910; two collections were made of his work: the edition de luxe (36 vols., 1896-1911) and the memorial edition (27 vols., 1909-11); a collection of his letters appeared in 1912.

MEREDITH, Sir **WILLIAM RALPH** (1840-1923), Canadian politician; born in Upper Canada; called to bar of Upper Canada; represented London as conservative in legislative assembly of Ontario, 1872; leader of opposition,

1878; retired from politics and appointed chief justice, common pleas division of Ontario high court of justice, 1894; chief justice of Ontario, 1912; took important part in codifying laws of province and in affairs of Toronto University.

MERIVALE, HERMAN CHARLES (1839–1906), playwright and novelist; son of Herman Merivale [q.v.]; educated at Harrow and Balliol College, Oxford; BA, 1861; called to bar, 1864; edited *Annual Register*, 1870–80; collaborated in *All for Her* (1875) and *Forget Me Not* (1879), plays which attained great success; his *The White Pilgrim* (1883) shows high qualities of poetic drama; skilful adapter of foreign dramas; wrote excellent farces and burlesques, including *The Butler* (1886), and *The Don* (1888), written for J. L. Toole [q.v.]; published novel, *Faucit of Balliol* (3 vols., 1882), and a children's fairytale, *Binko's Blues* (1884); lost fortune through default of trustee and awarded civil list pension, 1900.

MERRIMAN, FRANK BOYD, BARON MERRIMAN (1880–1962), judge; educated at Winchester; called to bar (Inner Temple), 1904; served with Manchester Regiment, 1914–18; OBE, 1918; KC, 1919; recorder of Wigan, 1920–8; conservative MP, Rusholme division, Manchester, 1924–33; solicitor-general, 1928–9 and 1932–3; president, Probate, Divorce, and Admiralty Division of the High Court, 1933; PC, 1933; bencher, Inner Temple, 1927; chairman, Bishop of London's Commission on City churches, 1941–6; hon. LLD, McGill University; knighted, 1928; baron, 1941; GCVO, 1950.

MERRIMAN, HENRY SETON (pseudonym) (1862–1903), novelist. [See SCOTT, HUGH STOWELL.]

MERRIMAN, JOHN XAVIER (1841–1926), South African statesman; son of N. J. Merriman [q.v.]; taken to South Africa, 1848; educated in England; returned to Cape, where he practised as land surveyor, 1861; elected to Cape house of assembly, 1869; joined cabinet of (Sir) J. C. Molteno [q.v.] as commissioner of crown lands and public works, 1875–7; secretary of war during Kaffir war, 1877–8; associated with J. W. Sauer, 1881–1913; commissioner of public works, Scanlen ministry, 1881–4; treasurer-general, Rhodes ministry, 1890–3; Schreiner ministry, 1898–1900; prime minister, 1908–10; concerned with hastening South African union and restoring finances of Cape Colony.

MERRIVALE, first BARON (1855–1939), judge and politician. [See DUKE, HENRY EDWARD.]

MERRY, WILLIAM WALTER (1835–1918), classical scholar; BA, Balliol College, Oxford;

fellow and classical tutor, Lincoln College, 1859–84; rector, 1884–1918; ordained, 1860; public orator, 1880–1910; completed large edition of *Odyssey* begun by James Riddell [q.v.]; edited plays of Aristophanes, etc.

MERRY DEL VAL, RAFAEL (1865–1930), cardinal; studied for priesthood at Ushaw College, Durham (1883–5), and in Rome (1885–91); ordained priest, 1888; secretary to papal commission on Anglican orders, 1896; apostolic delegate to Canada, 1897; consistorial secretary, 1903; pontifical secretary of state, 1903–14; cardinal priest with titular church of Santa Prassede, 1903; archpriest of St. Peter's, 1914; secretary of Holy Office, 1914–30; died in Rome.

MERSEY, first VISCOUNT (1840–1929), judge. [See BIGHAM, JOHN CHARLES.]

MERTHYR, first BARON (1837–1914), engineer and coal-owner. [See LEWIS, WILLIAM THOMAS.]

MERTON, SIR THOMAS RALPH (1888–1969), scientist; educated at Farnborough School, Eton, and Balliol College, Oxford; B.Sc., 1910; lieutenant, RNVR, in the secret service, 1916; D.Sc., Oxford, 1916; lecturer in spectroscopy, King's College, London; research fellow, Balliol, reader in spectroscopy, 1919; professor, 1920; FRS, 1920; retired to his Herefordshire estate, 1923; member, Air Defence Committee; experiments with cathode rays contributed to radar; other research included paint to reduce light reflected from bombers by searchlights; post-war research in methods of ruling diffraction gratings led to production of cheap infra-red spectrometers; treasurer, Royal Society, 1939–56; became collector of renaissance paintings; member, scientific advisory board, National Gallery; chairman, 1957–65; trustee, National Gallery and National Portrait Gallery, 1955–62; knighted, 1944; KBE, 1956.

MERZ, CHARLES HESTERMAN (1874–1940), electrical engineer; educated at Bootham School, York, and Armstrong College, Newcastle; established a consultative firm with William McLellan, 1902; designed Neptune Bank power station, Wallsend (1902), first station in Britain to generate 3-phase current at voltage of 5,500; electrified railway between Newcastle and Tynemouth (1902) and Melbourne suburban railways (1912); organizer and first director of Admiralty department of experiment and research; chairman of committee whose report resulted in the 'grid' system; superb expert witness.

MESTON, JAMES SCORGIE, first BARON MESTON (1865-1943), Indian civil servant and man of affairs; educated at Aberdeen grammar school and university and Balliol College, Oxford; posted to North-Western Provinces and Oudh, 1885; financial secretary, 1899-1903; financial secretary to Indian government, 1906-12; lieutenant-governor, United Provinces, 1912-18; influential in preparing way for political advance; finance member, viceroy's executive council, 1918-19; chairman, committee on financial adjustments between provinces and centre, 1920; with Lionel Curtis main designer of Royal Institute of International Affairs; first chairman of governors, 1920-6; chancellor, Aberdeen University, 1928-43; president, liberal party organization, 1936-43; KCSI, 1911; baron, 1919.

METCALFE, Sir CHARLES HERBERT THEOPHILUS, sixth baronet (1853-1928), civil engineer; born at Simla; great-nephew of Baron Metcalfe [q.v.]; BA, University College, Oxford; succeeded father, 1883; apprenticed to engineering firm of Fox & Sons; his most important work accomplished in South Africa, where he lived more and more frequently, 1882-1914; friend of Cecil Rhodes [q.v.], whose dreams of northward expansion in Africa he realized by his railway work; together with firm of Fox, consulting engineer for various lines constituting Rhodesia railway system, Benguela Railway through Portuguese West Africa, etc.

METHUEN, Sir ALGERNON METHUEN MARSHALL, baronet (1856-1924), publisher, whose original name was ALGERNON STEDMAN; BA, Wadham College, Oxford; private schoolmaster, 1880-95; opened publishing office in London as Methuen & Co., 1889; firm's range catholic, but specialized in educational and topographical works; firm's authors included Kipling, Corelli, Belloc, J. B. Bury, Chesterton, Conrad, 'Anthony Hope', E. V. Lucas; baronet, 1916.

METHUEN, PAUL SANFORD, third BARON METHUEN (1845-1932), field-marshal; educated at Eton; joined Scots Fusilier Guards, 1864; commanded Methuen's Horse in Bechuanaland, 1884-5; succeeded father, 1891; commanded Home district, 1892-7; lieutenant-general, 1898; commanded 1st division in South Africa, 1899-1902; unjustly criticized for miscarriage of his attack at Magersfontein, 11 Dec. 1899; obtained many minor successes; wounded and captured at Tweebosch, 7 Mar. 1902; general, 1904; held Eastern command, 1905-8; GOC-in-C, South Africa, 1908-12; field-marshal, 1911; governor and commander-in-chief, Malta, 1915-19; constable of the Tower,

1919-32; KCVO, 1897; KCB, 1900; GCB, 1902; GCVO, 1910; GCMG, 1919.

MEUX (formerly LAMBTON), **SIR HEDWORTH** (1856-1929), admiral; entered navy, 1870; captain, 1889; naval private secretary to first lord of Admiralty, 1894-7; landed with naval brigade in time to join garrison in Ladysmith, Oct. 1899; rear-admiral, 1902; commanded cruiser division of Mediterranean fleet, 1904-6; vice-admiral, 1908; commander-in-chief, China, 1908-10; changed his name to Meux on coming into large fortune under will of Lady Meux, 1911; admiral, 1911; commander-in-chief, Portsmouth, 1912-16; admiral of the fleet, 1915; secured safe passage of transports conveying British expeditionary force to France; conservative MP, Portsmouth, 1916-18; KCVO, 1906; KCB, 1908.

MEW, CHARLOTTE MARY (1869-1928), poet; lived nearly all her life in Bloomsbury; published poems, stories, essays, studies in periodicals, and two volumes of verse (1915 and, posthumous publication, 1929).

MEYER, FREDERICK BROTHERTON (1847-1929), Baptist divine; BA, London University; served Baptist ministries in York and Leicester; Melbourne Hall, Leicester, built for his pastorate after 1878; later, minister at Regent's Park chapel and Christ Church, Westminster Bridge Road, London; a prolific writer; travelled widely for religious purposes.

MEYER, Sir WILLIAM STEVENSON (1860-1922), Indian civil servant; born in Moldavia; joined Indian civil service in Madras, 1881; deputy secretary, finance department, government of India, 1898; Indian editor, *Imperial Gazetteer of India*, 1902-5; financial secretary, government of India, 1905; secretary of military finance, 1906; finance member, government of India, 1913-18; his administration of military finance unjustly criticized for parsimony; first high commissioner of India, 1920-2; largely instrumental in procuring open market for government requirements not available in India; KCIE, 1909.

MEYNELL, ALICE CHRISTIANA GERTRUDE (1847-1922), poet, essayist, and journalist; born THOMPSON; joined Roman Church, c.1872; married Wilfrid Meynell, 1877; her first volume of poems, *Preludes* (published, 1875); of essays, *The Rhythm of Life* (1893); championship of Coventry Patmore and George Meredith, and poems addressed to her by Francis Thompson [qq.v.], did much to secure early prestige for her writings; a marked difference noticeable between her earlier and later poems; published several further volumes of essays and edited various anthologies; admirer of seventeenth-century poetry.

MEYRICK, EDWARD (1854-1938), entomologist; educated at Marlborough and Trinity College, Cambridge; first class, classics, 1877; a classical master at Marlborough, 1887-1914; reclassified whole order Lepidoptera and described some 20,000 new species; published *A Handbook of British Lepidoptera* (1895); FRS, 1904.

MEYRICK, FREDERICK (1827-1906), divine; BA, Trinity College, Oxford, 1847; MA, 1850; fellow, 1847; travelled in Europe; founder and secretary of Anglo-Continental Society, 1853; rector of Blickling, Norfolk, 1868-1906; non-residentiary canon of Lincoln, 1869; helped to organize Bonn conferences on reunion, 1874-5; an ardent evangelical controversialist; wrote several anti-Roman pamphlets, as well as *The Church in Spain* (1892) and *Memories of Life at Oxford* (1905).

MICHELL, ANTHONY GEORGE MALDON (1870-1959), engineer; born in London; educated at Perse School, Cambridge, and Melbourne University; BCE, 1895; MCE, 1899; set up practice, centred on hydraulic engineering, 1903; invented Michell thrust-block, 1905, and 'crankless engine', 1922; published *Lubrication: its principles and practice* (1950); FRS, 1934; Kernot memorial medal of Melbourne University, 1938; James Watt international medal of Institution of Mechanical Engineers, London, 1942.

MICHELL, SIR LEWIS LOYD (1842-1928), South African banker and politician; joined London and South African Bank, 1863; sent to Port Elizabeth, Cape Colony, 1864; manager, Standard Bank of South Africa and Port Elizabeth, *c.*1872; transferred to Cape Town, 1885; sole general manager of Standard Bank, South Africa, 1895-1902; executor and trustee of Cecil Rhodes [q.v.]; directed financing of British army's requirements during South African war, 1899-1902; chairman, De Beers Consolidated Mines and director, British South Africa Company, 1902; member, Cape house of assembly, 1902; minister without portfolio, in Jameson's cabinet, 1903-5; knighted, 1902; wrote life of Rhodes (1910).

MICHIE, ALEXANDER (1833-1902), writer on China; prominent in Chinese commerce at Hong Kong and Shanghai from 1853; helped in negotiations with Taiping rebels, 1861; explored Yangtze valley and Szechuan, 1869; special

correspondent to *The Times* in Chino-Japanese war, 1895; wrote *The Englishman in China* (2 vols., 1900) and *China and Christianity* (1900).

MICKLETHWAITE, JOHN THOMAS (1843-1906), architect; pupil of George Gilbert Scott [q.v.], 1862; partner with fellow pupil, Somers Clarke, 1876-92; churches designed by him include St. Hilda, Leeds, St. Peter, Bocking, and St. Bartholomew, East Ham; executed much internal decoration, chancels, and screens; surveyor to dean and chapter of Westminster Abbey, 1898; restored south transept and west front; architect to St. George's chapel, Windsor, 1900; FSA, 1870; vice-president, 1902; helped to found Alcuin Club and Henry Bradshaw Society; published *Ornaments of the Rubric*, (1897).

MIDDLETON, JAMES SMITH (1878-1962) secretary of the labour party; educated at elementary schools; worked in his father's printing establishment, Workington; secretary, Workington trades council and local branch of ILP; moved to London, 1902; assistant secretary, labour party, 1904; close friendship with Mr and Mrs Ramsay MacDonald; secretary, War Emergency Workers' National Committee, 1914-18; secretary, labour party, 1934-44.

MIDLANE, ALBERT (1825-1909), hymn writer; tinsmith and ironmonger; joined Plymouth brethren; wrote over 800 hymns, the best known, 'There's a Friend for Little Children', being composed in 1859; hymns marked by religious emotion and love of children; published several volumes of verse.

MIDLETON, ninth VISCOUNT and first EARL OF (1856-1942), statesman. [See BRODRICK (WILLIAM) ST JOHN (FREMANTLE).]

MIERS, SIR HENRY ALEXANDER (1858-1942), mineralogist, administrator, and scholar; grandson of John Miers [q.v.]; scholar of Eton and Trinity College, Oxford; second class, mathematics, 1881; studied crystallography and mineralogy in Strasbourg; first class assistant, mineralogy department, British Museum, South Kensington, 1882-95; instructor in crystallography, Central Technical College, 1886-95; Waynflete professor of mineralogy, Oxford, 1895-1908; fellow of Magdalen, 1908-42; principal, London University, 1908-15; vice-chancellor and professor of crystallography, Manchester, 1915-26; member, royal and standing commissions on museums; published *Mineralogy: an Introduction to the Scientific Study of Minerals* (1902); mineral miersite named after him; FRS, 1896; knighted, 1912.

MILBANKE, RALPH GORDON NOEL KING, second EARL OF LOVELACE (1839-1906), author of *Astarte*; grandson of poet Byron; spent a year in Iceland, 1861; a bold alpinist and accomplished linguist; his privately printed *Astarte* (1905), vindicating his grandmother, Lady Byron, from aspersions cast on her, and incriminating Lord Byron, provoked replies from (Sir) John Murray [q.v.] and Richard Edgcumbe.

MILDMAY, ANTHONY BINGHAM, second BARON MILDMAY OF FLETE (1909-1950), gentleman rider; educated at Eton and Trinity College, Cambridge; with Baring Brothers, 1930-3; thereafter devoted himself to becoming successful amateur steeplechase jockey; rode 21 winners in 1937-8 and 32 in 1946-7; took third and fourth places respectively in Grand National in 1948 and 1949; beloved by racecourse crowds for his courage and skill; member, National Hunt Committee, 1942-50; succeeded father, 1947; drowned while bathing.

MILFORD, SIR HUMPHREY SUMNER (1877-1952), publisher; grandson of Charles Richard Sumner, bishop of Winchester [q.v.]; scholar of Winchester and New College, Oxford; first class, *lit. hum.*, 1900; assistant to Charles Cannan [q.v.], secretary to the delegates of Oxford University Press, 1900; transferred to London office, 1906; manager of London business and publisher to the university of Oxford, 1913-45; under his management, the Oxford University Press became one of the three or four largest publishers in the country; originator of the *Oxford Dictionary of Quotations*; editor of the *Oxford Book of English Verse of the Romantic Period*; hon. D.Litt., Oxford, 1928; knighted, 1936.

MILFORD HAVEN, first MARQUESS OF (1854-1921), admiral of the fleet. [See MOUNT-BATTEN, LOUIS ALEXANDER.]

MILL, HUGH ROBERT (1861-1950), geographer and meteorologist; educated privately owing to tubercular illnesses; B.Sc., Edinburgh, 1883; fellow, Scottish Marine Station, 1884; FRSE, 1885; D.Sc., 1886; lecturer in geography and physiography, Heriot-Watt College, Edinburgh, 1887-92; librarian, Royal Geographical Society, 1892-1900; university extension lecturer, 1887-1900; director, British Rainfall Organization, 1901-19; established it on strictly scientific basis; rainfall expert to Metropolitan Water Board, 1903-19; confidant and inspirer of many polar explorers; edited geographical material for eleventh edition of *Encyclopaedia Britannica*; publications include *The Siege of the South Pole*

(1905), biography of Sir Ernest Shackleton [q.v.] (1923), and *Record of the Royal Geographical Society, 1830-1930* (1930).

MILLAR, GERTIE (1879-1952), actress; first success at fourteen years of age in Manchester pantomime; played in London theatres in *The Toreador* (1901), and other musical comedies, including *Our Miss Gibbs* (1909), *The Quaker Girl* (1910), and *A Country Girl* (1914); married in 1902 Lionel Monckton, who composed the music for her shows; regarded as the Gaiety Girl *par excellence*; after Monckton's death, married second Earl of Dudley [q.v.], 1924.

MILLER, SIR JAMES PERCY, second baronet (1864-1906), sportsman; joined army, 1888; served in South Africa, 1900-1; won Derby with Sainfoin, 1890; purchased mare Roquebrune (1894), who, mated with Sainfoin, produced Rock Sand, winner of Two Thousand Guineas, Derby, and St. Leger in 1903; headed list of winning owners, 1903 and 1904.

MILLER, WILLIAM (1864-1945), historian and journalist; educated at Rugby and Hertford College, Oxford; first class, *lit. hum.*, 1887; *Morning Post* correspondent for Italy and Balkans, 1903-37; publications include *Mediaeval Rome* (1901), *The Latins in the Levant* (1908), *The Ottoman Empire, 1801-1913* (1913), *Essays on the Latin Orient* (1921), and *Greece* (1928); FBA, 1932.

MILLIGAN, GEORGE (1860-1934), Scottish divine and biblical scholar; son of William and brother of Sir William Milligan [qq.v.]; studied at Aberdeen, Edinburgh, Göttingen, and Bonn; ordained, 1887; minister, Caputh, Perthshire, 1894-1910; regius professor of biblical criticism, Glasgow, 1910-32; clerk to the senate, 1911-30; works include *The Vocabulary of the Greek Testament* (1914-29); moderator of General Assembly, 1923.

MILLIGAN, SIR WILLIAM (1864-1929), laryngologist and otologist; son of William Milligan [q.v.]; grandson of D. M. Moir [q.v.]; MD, Aberdeen University; successful aural surgeon and laryngologist in Manchester; advocate of radium; knighted, 1914.

MILLS, BERTRAM WAGSTAFF (1873-1938), circus proprietor; worked until 1914 exhibiting carriages built by his father; put on circus at Olympia every Christmas, 1920-37; started a touring circus, 1929; member of the London County Council, 1928-38.

MILLS, PERCY HERBERT, first VISCOUNT MILLS (1890-1968), politician and industrialist; educated at North Eastern county school,

Barnard Castle; left school at fifteen and articled to chartered accountants in London, 1905; entered W. & T. Avery Ltd. (later Averys Ltd.), 1919; general manager, 1924; managing director, 1933-55; deputy director, ordnance factories, 1939; controller-general, machine tools, 1940-4; head of production division, Ministry of Production, 1943-4; knighted, 1942; head of economic sub-commission of British element of control commission, Germany, 1945; ignored instructions and blew up submarine yards of Bloehm & Voss; KBE, 1946; president, Birmingham Chamber of Commerce, 1947; chairman, National Research Development Corporation, 1949-55; hon. adviser on housing to Harold Macmillan; baronet, 1953; baron, 1957; conservative minister of power, 1957-9; paymaster general, 1959-61; minister without portfolio, and deputy leader, House of Lords, 1961-2; viscount, 1962; chairman, electronic subsidiary, Electric and Musical Industries, 1962-8.

MILLS, Sir WILLIAM (1856-1932), engineer; trained as marine engineer; invented a boat-disengaging gear; established first British aluminium foundry, Sunderland, 1885; introduced (1915) and manufactured (Birmingham) hand grenades known by his name; knighted, 1922.

MILLS, WILLIAM HOBSON (1873-1959), organic chemist; educated at Uppingham and Jesus College, Cambridge; first class, part i, natural sciences tripos, 1896, and part ii, chemistry, 1897; fellow of Jesus College, 1899; head of chemical department, Northern Polytechnic Institute, London, 1902-12; demonstrator to Jacksonian professor of natural philosophy, Cambridge, Sir James Dewar [q.v.], 1912-19; university lecturer in organic chemistry, 1919; reader in stereochemistry, 1931-8; president, Jesus College, 1940-8; worked mainly on stereochemistry and cyanine dyes; after retirement, studied sub-species of British bramble and donated collection to university botany department; FRS, 1923.

MILNE, ALAN ALEXANDER (1882-1956), author; educated at Westminster and Trinity College, Cambridge; editor, the *Granta*; assistant editor, *Punch*, under (Sir) Owen Seaman [q.v.], 1906; during 1914-18 war served as signalling officer in Royal Warwickshire Regiment; left *Punch* to work on stage comedies, 1919; *Mr Pim Passes By* (1920), *The Truth About Blayds* (1921), *The Dover Road* (1922), *To Have the Honour* (1924), *The Fourth Wall* (1928), *Toad of Toad Hall*, a dramatization of *The Wind in the Willows* by Kenneth Grahame [q.v.] (1929); wrote verses for children, *When*

We Were Very Young (1924), *Winnie-the-Pooh* (1926), and *The House at Pooh Corner* (1928).

MILNE, Sir (ARCHIBALD) BERKELEY, second baronet (1855-1938), admiral; son of Sir Alexander Milne [q.v.]; entered navy, 1869; commander, 1884; captain, 1891; succeeded father, 1896; spent eight years between 1882 and 1900 in royal yachts; in command, HM yachts, 1903-5; rear-admiral, 1904; second-in-command, Atlantic fleet (1905-6), Channel fleet (1908-9); admiral, 1911; commander-in-chief, Mediterranean, 1912-14; on outbreak of war obtained Admiralty authority to concentrate his force at Malta; criticized for allowing German battle cruiser *Goeben* and cruiser *Breslau* to escape eastwards to Dardanelles, but his conduct and dispositions subsequently received public approval of Admiralty; retired, 1919; KCVO, 1904; KCB, 1909; GCVO, 1912.

MILNE, EDWARD ARTHUR (1896-1950), mathematician and natural philosopher; educated at Hymers College, Hull, and Trinity College, Cambridge; served in munitions inventions department, Ministry of Munitions, 1916-19, and on Ordnance Board, 1939-44; assistant director, Solar Physics Observatory, Cambridge, 1920-4; devoted himself to theory of stellar atmospheres; university lecturer in astrophysics, 1922-5; Beyer professor of applied mathematics, Manchester, 1925-8; Rouse Ball professor and fellow of Wadham College, Oxford, 1929-50; worked on theory of stellar structure and development of kinematic relativity; FRS, 1926.

MILNE, GEORGE FRANCIS, first Baron Milne (1866-1948), field-marshal; educated at the Gymnasium, Old Aberdeen; gazetted from Woolwich to Royal Artillery, 1885; with Lord Kitchener [q.v.] on Nile (1898) and in South Africa (1899-1902); DSO, 1902; served in France, 1914-15; assumed command British forces in Salonika, 1916 (in final offensive the first Allies to enter Bulgaria); commanded in Constantinople until Nov. 1920; GOC-in-C, Eastern command, 1923-6; chief of imperial general staff, 1926-33; colonel commandant, Royal Artillery, 1918-48; master gunner, St. James's Park, 1929-46; general, 1920; field-marshal, 1928; KCB, 1918; baron, 1933; founder and head of Old Contemptibles and Salonika Reunion Association.

MILNE, JOHN (1850-1913), mining engineer and seismologist; educated at King's College, London, and Royal School of Mines; professor of geology and mining, imperial college, Tokyo, 1875; first professor of seismology, imperial university, Tokyo; established seismic survey of

Japan; secretary to seismological committtee of British Association, 1895-1913; devised seismograph; travelled widely.

MILNE-WATSON, SIR DAVID MILNE, first baronet (1869-1945), man of business; educated at Merchiston Castle, Edinburgh, and Oxford, and Marburg universities; joined Gas Light and Coke Company, 1897; general manager, 1903; managing director, 1916; governor and managing director, 1918-45; made it world's biggest gas company; encouraged new developments, established research laboratories, and maintained good labour relations; president, National Gas Council, 1919-43; chairman, Joint Industrial Council for Gas Industry, 1919-44; knighted, 1927; baronet, 1937.

MILNER, ALFRED, VISCOUNT MILNER (1854-1925), statesman; born in Hesse-Darmstadt; educated at Tübingen, King's College, London, and Balliol College, Oxford; first class, classical mods. (1874) and *lit. hum.* (1876); won Hertford (1874), Craven (1877), Eldon (1878), and Derby (1878) scholarships; president of Union, 1875; in society of his Oxford friends developed passion for public work, political and social; fellow, New College, Oxford, 1876; called to bar (Inner Temple), 1881; joined staff of *Pall Mall Gazette*, 1882-5; in London maintained association with his Oxford friends, notably Arnold Toynbee [q.v.], by whom he was profoundly influenced; took part in University Extension Society founded by S. A. Barnett [q.v.]; co-operated in foundation of Toynbee Hall, Whitechapel, 1884; private secretary to G. J. (afterwards Viscount) Goschen [q.v.], whose ideas on social reform and foreign and imperial policy coincided with his own, 1884; actively co-operated with Goschen in forming Liberal Unionist Association, 1886; secretary to Goschen as chancellor of Exchequer, 1887-9; director-general of accounts, Egypt, 1889; under-secretary for finance, Egypt, 1890-2; rendered great services to Great Britain's task in Egypt, especially by his book, *England in Egypt* (1892); chairman, Board of Inland Revenue, 1892-7; had large share in introducing new form of death duties devised by Sir William Harcourt [q.v.]; high commissioner for South Africa, 1897-1905; went out with open mind as to rights in dispute between Boers and Britons and resolved to form judgement on spot; learned Dutch; attempted to effect friendly and informal relations with President Kruger; came to understanding with Cecil Rhodes [q.v.]; concluded that there was no solution of political troubles of South Africa except reform in Transvaal or war, 1898; warned Cape Dutch against disloyalty; although

his objects and those of Joseph Chamberlain [q.v.], secretary of state for colonies, were identical in South Africa, namely to secure justice and reasonable liberty for uitlanders and to ensure Great Britain's right to be alone responsible for whole of South Africa's foreign relations, his forward policy was somewhat distrusted by Chamberlain; forwarded to Chamberlain petition of uitlanders recounting their grievances, Mar. 1899; himself set out grievances in famous cable pronouncing case for intervention by British government to be overwhelming, May; met Kruger at abortive conference at Bloemfontein, 31 May; over-hasty in breaking off conference; on outbreak of war (Oct.) British colonies, in spite of his urgent representations, almost defenceless; Mafeking and Kimberley enabled to hold out largely through his insistence; his months of anxiety relieved by Lord Roberts's capture of Pretoria and resignation of ministry of W. P. Schreiner [q.v.], June 1900; administrator of Orange River Colony and Transvaal, 1900; went to England, where he was received with extraordinary honour, May 1901; while there made preparations for starting his schemes of reorganization and social reform in new colonies; with Lord Kitchener [q.v.] signed treaty of Vereeniging on behalf of British government, May 1902; differed from Kitchener in his greater rigidity; immediately set about task of repatriating Boers on their farms; organized permanent system of education; succeeded, in spite of Boer opposition, in establishing English as medium of instruction; greatly encouraged improved methods and results in farming; stimulated land settlement by British farmers in order to introduce English ideas into country districts; set before him ultimate aim of union of all South African colonies and early secured co-operative measures; aroused controversy by sympathy with idea of suspending Cape constitution; visited by Chamberlain, now in general accord with his South African policy, 1902; obtained consent of Alfred Lyttelton [q.v.], secretary of state for colonies, to import Chinese labour into Rand, 1904; policy aroused violent opposition, especially from liberal party; acquiesced in grant of representative institutions to Transvaal, 1905; left South Africa, 1905; in spite of his very real success in repairing ravages of South African war, failed to touch hearts or win confidence of Boers; took little part in politics for some time after return, but carried on remunerative work in City and occupied himself with Rhodes Trust; supported movements for national service and tariff reform; opposed Lloyd George's budget (1909), which he advised Lords to reject, parliament bill (1911), and home rule; during European war successfully presided over committee to increase food production of country; member of Lloyd George's small war cabinet, 1916; accompanied him to allied conference in Rome, Jan. 1917; sent to Amiens front in order to report on serious state of affairs owing to great German attack and breakdown of co-operation between two allied commands, Mar. 1918; took upon himself responsibility for momentous decision of enforcing unity of command; owing to his initiative, General Foch placed in supreme command of allied armies on western front, thereby procuring turning-point of war; secretary of state for war, 1918; his great reform in administration, inauguration of army education branch; his sane utterances with regard to peace absurdly denounced as pro-German; secretary of state for colonies, 1918–21; visitied Egypt in order to report on her future relations with Great Britain, 1919–20; chancellor-elect of Oxford University, 1925; baron, 1901; viscount, 1902; KG, 1921; a great public servant, whose chief contribution to contemporary political thought was conviction of need for imperial unity.

MILNER, JAMES, first BARON MILNER OF LEEDS (1889–1967), politician and lawyer; educated at Easingwold grammar school, Leeds modern school, and Leeds University; LLB, 1911; served in the army in France during 1914–18 war; MC with clasp; returned to family firm, J. H. Milner & Son, Leeds; joined labour party; served on city council, 1923–9; deputy lord mayor and president, Leeds labour party, 1928–9; labour MP South East Leeds, 1929–51; parliamentary private secretary to Christopher (later Viscount) Addison [q.v.]; member, select committee on capital punishment, 1931; member, Indian franchise committtee, 1932; chairman, committee of ways and means, deputy speaker, 1943; PC, 1945; chairman, British group, Inter-Parliamentary Union; baron, deputy speaker, House of Lords, 1951; hon. LLD, Leeds; deputy-lieutenant, West Riding of Yorkshire; vice-president, Association of Municipal Corporations and Building Societies Association.

MILNER, VIOLET GEORGINA, VISCOUNTESS MILNER (1872–1958), editor of the *National Review*; daughter of Admiral Frederick Augustus Maxse [q.v.], and sister of (Sir) Ivor and Leo Maxse [qq.v.]; married, first, Lord Edward Cecil [q.v.], 1894 (died, 1918); secondly, Sir Alfred (later Viscount) Milner [q.v.], 1921; edited *National Review*, 1932–48; brilliant raconteuse; published autobiography, *My Picture Gallery 1886–1901* (1951).

MILNER HOLLAND, SIR EDWARD (1902–1969), lawyer. [See HOLLAND.]

MILNES, ROBERT OFFLEY ASHBURTON CREWE-, second BARON HOUGHTON, and MARQUESS OF CREWE (1858–1945), statesman. [See CREWE-MILNES.]

MINETT, FRANCIS COLIN (1890–1953), veterinary pathologist; educated at King Edward's School, Bath, and the Royal Veterinary College, London; MRCVS, 1911; B.Sc. (veterinary science), London, 1912; Royal Army Veterinary Corps, 1914–24; research officer, Ministry of Agriculture laboratory, Weybridge, working on foot-and-mouth disease, 1924–7; director of research institute in animal pathology, Royal Veterinary College, 1927–39; D.Sc., London, 1927; combined duties as director with those of professor, pathology, 1933–9; director, Imperial Veterinary Research Institute, Mukteswar, India, 1939–47; animal husbandry commissioner, government of Pakistan, 1947–50; director, farm livestock research station, Animal Health Trust, 1950–3; CIE, 1945.

MINTO, fourth EARL OF (1845–1914), governor-general of Canada and viceroy of India. [See ELLIOT, GILBERT JOHN MURRAY KYNYNMOND.]

MINTON, FRANCIS JOHN (1917–1957), artist; educated at Reading School; studied at St. John's Wood Art School; collaborated with Michael Ayrton on costumes and decor for (Sir) John Gielgud's production of *Macbeth* (1942); leading figure among post-war romantic painters; taught at Camberwell Art School, the Central School of Arts and Crafts, and the Royal College of Art; produced large number of paintings reflecting his travels in Spain, the West Indies, and Morocco; exhibited regularly at Royal Academy from 1949; elected member of the London Group, 1949; from 1950 felt himself to be out of contact with international fashion; died from overdose of drugs.

MIRZA MOHAMMAD ISMAIL, SIR (1883–1959), Indian administrator and statesman. [See ISMAIL.]

MITCHELL, SIR ARTHUR (1826–1909), Scottish commissioner in lunacy (1870–95) and antiquary; MA, Aberdeen, 1845; MD, 1850; hon. LLD, 1875; member of English commission on criminal lunacy, 1880; FSA Scotland, 1867; made study of superstition in Scottish Highlands; first Rhind lecturer in archaeology, Edinburgh; CB, 1886; KCB, 1887; hon. FRCP Ireland, 1891; published *The Past in the Present* (1880) and edited *Macfarlane's Topographical Collections* (3 vols., 1906–8).

MITCHELL, JOHN MURRAY (1815-1904), Presbyterian missionary and orientalist; MA, Marischal College, Aberdeen, 1833; hon. LLD, 1858; missionary in Bombay, 1838; made many converts among Marathis; founded flourishing Free Church mission at Poona, 1843; in Bengal, 1867-73; formed 'Union Church' at Simla; minister of Scottish church at Nice, 1888-98; Duff missionary lecturer at Edinburgh, 1903; published *Hinduism Past and Present* (1885), *The Great Religions of India* (posthumous, 1905), and metrical translations from Indian poets.

MITCHELL, SIR PETER CHALMERS (1864-1945), zoologist; MA, Aberdeen, 1884; first class, natural science, Christ Church, Oxford, 1888; university demonstrator in comparative anatomy, 1888-91; lecturer in biology, Charing Cross Hospital, 1892, London Hospital, 1894; secretary, Zoological Society of London, 1903-35; made it the leading institution of its kind; mainly responsible for creation of Whipsnade Zoological Park; biological editor, eleventh edition *Encyclopaedia Britannica*; scientific correspondent of *The Times*; publications include biography of T. H. Huxley (1900); FRS, 1906; knighted, 1929.

MITCHELL, SIR PHILIP EUEN (1890-1964), colonial administrator; educated at St. Paul's School and Trinity College, Oxford; joined colonial service, 1912; assistant resident Nyasaland; served in King's African Rifles, 1914-18; MC; district officer, Tanganyika, 1919-26; assistant secretary, native affairs, 1926; secretary, 1928; chief secretary, 1934; governor, Uganda, 1935-40; deputy chairman, Governors' Conference, 1940; on staff of Sir A. P. (later Earl) Wavell [q.v.], major-general, 1941; governor, Fiji, 1942; governor, Kenya, 1944-52; CMG, 1933; KCMG, 1937; GCMG, 1947.

MITCHELL, REGINALD JOSEPH (1895-1937), aircraft designer; trained as an engineer; joined Supermarine Aviation works, Southampton, 1916; chief engineer and designer, 1919-37; designed high-speed float-seaplanes for Schneider trophy races (1922-31) and the Spitfire fighter aircraft used in the 1939-45 war; CBE, 1931.

MITCHELL, SIR WILLIAM GORE SUTHERLAND (1888-1944), air chief marshal; born in Australia; educated at Wellington College; with Royal Flying Corps, 1914-18; wing commander, RAF, 1919; director of training, 1929-33; commandant, Cranwell, 1933-4; AOC, Iraq, 1934-7; air member for personnel, 1937-9; AOC-in-C, Middle East, 1939-40; inspector-general of RAF, 1940-1; air chief marshal, 1939; KCB, 1938.

MITFORD, ALGERNON BERTRAM FREEMAN-, first BARON REDESDALE in the second creation (1837-1916), diplomatist and author; great-grandson of William Mitford [q.v.], the historian; educated at Eton and Christ Church, Oxford; entered Foreign Office, 1858; attaché in Japan, 1866-70; resigned from diplomatic service, 1873; secretary to Board of Works, 1874-86; as heir to cousin assumed additional name and arms of Mitford and went to live at Batsford Park, Gloucestershire, 1886; conservative MP, Stratford-upon-Avon division of Warwickshire, 1892-5; baron, 1902; works include *Tales of Old Japan* (1871) and his autobiography, *Memories* (1915).

MOBERLY, ROBERT CAMPBELL (1845-1903), theologian; son of George Moberly [q.v.]; scholar of Winchester and New College, Oxford; BA, 1867; won Newdigate prize, 1867; MA, 1870; DD, 1892; a senior student of Christ Church, 1867-80; principal of diocesan theological college, Salisbury, 1878; honorary canon of Chester, 1890; regius professor of pastoral theology at Oxford and canon of Christ Church, 1892-1903; contributed 'The Incarnation as the Basis of Dogma' to *Lux Mundi* (1889); chief work was *Atonement and Personality* (1901).

MOCATTA, FREDERIC DAVID (1828-1905), Jewish philanthropist; in father's bullion broker's business, 1843-74; promoter of Charity Organisation Society, 1869; interested in housing of working classes and liberal benefactor of London hospitals; organized Board of Guardians of the Jewish Poor (founded 1859); generous supporter of Jewish charities; encouraged Jewish literature and research; FSA, 1889; published *The Jews and the Inquisition* (1877).

MÖENS, WILLIAM JOHN CHARLES (1833-1904), Huguenot antiquary; settled in Hampshire, devoting himself to yachting and antiquarian research; held captive in South Italy by brigands for four months, 1865; published experiences in *English Travellers and Italian Brigands* (1866); helped to found Huguenot Society of London, 1885; president, 1899-1902; edited for society registers of Walloons at Norwich (1887-8), of French church, Threadneedle Street (1896), and of Dutch church, Colchester (1905); FSA, 1886.

MOERAN, ERNEST JOHN (1894-1950), composer; educated at Uppingham and Royal College of Music; studied under Dr John

Ireland; compositions include many songs, folksong arrangements and choral pieces; chamber music; G minor Symphony (1937), 'Sinfonietta' (1945), violin Concerto (1942), 'cello Concerto (1945), and Sonata (1947).

MOFFATT, JAMES (1870-1944), divine; educated at Glasgow Academy, University, and Free Church of Scotland College; ordained, 1896; professor of Greek and New Testament exegesis, Mansfield College, Oxford, 1911-15; of church history, Glasgow Free Church College (1915-27) and Union Theological Seminary, New York (1927-39); publications include *The Historical New Testament* (1901), *Introduction to the Literature of the New Testament* (1911), and translations of New (1913) and Old (1924) Testaments; edited 'Moffatt New Testament Commentary'.

MOIR, FRANK LEWIS (1852-1904), song composer; composed ballads, church music, and organ voluntaries; best-known songs were 'Only Once More', 1883, and 'Down the Vale', 1885.

MOLLISON, AMY (1903-1941), airwoman. [See JOHNSON.]

MOLLISON, JAMES ALLAN (1905-1959), airman; educated at Glasgow and Edinburgh academies; held short service commission in RAF, 1923-8; became airline pilot in Australia; assisted by Lord Wakefield [q.v.] with aeroplanes for record-breaking flights; solo flight from Australia to England, 1931; England to the Cape, 1932; east-west north Atlantic, 1932; married Amy Johnson [q.v.], 1932; flew south Atlantic east to west, 1933; awarded Britannia Trophy, 1933; made further record flights with his wife, 1934; MBE, for work with Air Transport Auxiliary, 1946; published *Death Cometh Soon or Late* (1932) and *Playboy of the Air* (1937).

MOLLOY, GERALD (1834-1906), rector of the Catholic University of Dublin from 1883 till death; professor of theology at Maynooth, 1857; professor of natural philosophy, Catholic University, Dublin, 1874; rector, 1883; D.Sc., Royal University of Ireland, 1879; commissioner of inquiry into educational endowments in Ireland, 1885-94; vice-chancellor of Royal University, 1903; published *Geology and Revelation* (1870) and *Gleanings in Science* (1888).

MOLLOY, JAMES LYNAM (1837-1909), composer; MA, Catholic University, Dublin, 1858; called to English bar, 1863; composed songs (of which 'Darby and Joan', 'The Kerry Dance', 'Love's Old Sweet Song' had wide

vogue) and operettas; wrote *Our Autumn Holiday on French Rivers* (1874).

MOLLOY, JOSEPH FITZGERALD (1858-1908), miscellaneous writer; published *Songs of Passion and Pain* (1881), *London under the Four Georges* (4 vols., 1882-3), and *London under Charles II* (2 vols., 1885), lives of Peg Woffington (2 vols., 1884) and Edmund Kean (2 vols., 1888), *Romance of the Irish State* (2 vols., 1897), and many novels.

MOLONY, SIR THOMAS FRANCIS, first baronet (1865-1949), lord chief justice of Ireland; thrice law prizeman, and senior moderator (1886), history and political science, Trinity College, Dublin; called to Irish (1887) and English (1900) bar; QC, 1899; second serjeant-at-law, 1911; solicitor-general for Ireland, 1912; attorney-general, 1913; judge, 1913-15; lord justice of appeal, 1915-18; lord chief justice of Ireland, 1918-24; baronet, 1925; vice-chancellor, Dublin University, 1931-49.

MOLYNEUX, SIR ROBERT HENRY MORE- (1838-1904), admiral. [See MORE-MOLYNEUX.]

MONASH, SIR JOHN (1865-1931), Australian general; educated at Scotch College and Melbourne University; practised as a civil engineer specializing after 1900 in reinforced concrete construction; commissioned in Australian Citizen Forces, 1887; colonel, 1913; commanded 4th Infantry brigade at Gallipoli, 1915; 3rd Australian division (1916-18) at Messines, Passchendaele, Ypres, and Amiens; Australian Army Corps, 1918; lieutenant-general; withstood German offensive; launched allied offensive, Aug. 1918; brilliant planner; director-general, Australian repatriation and demobilization; general and retired, 1930; chairman, Victorian Government State Electricity Commission, 1920-31; vice-chancellor, Melbourne University, 1923-31; president, Australasian Association for Advancement of Science, 1924-6; KCB, 1918; GCMG, 1919.

MONCKTON, WALTER TURNER, first VISCOUNT MONCKTON OF BRENCHLEY (1891-1965), lawyer and politician; educated at Harrow and Balliol College, Oxford; president of the Union, 1913; served in France during 1914-18 war; MC, 1919; called to bar (Inner Temple), 1919; KC, 1930; recorder, Hythe, 1930-7; chancellor, diocese of Southwell, 1930-6; attorney-general to Prince of Wales, 1932-6; constitutional adviser to Nizam of Hyderabad and Nawab of Bhopal [qq.v.], 1933-6; attorney-general, Duchy of Cornwall, 1936-51; close confidant of Edward VIII during abdication crisis, 1936; KCVO, knighted, 1937; chairman, aliens advisory committee, 1939; director-general, Press and Censorship Bureau; deputy-director, and then director-general, Ministry of Information, 1940; director-general, propaganda and information services, Cairo, 1941; acting minister of state, 1942; solicitor-general, 1945; visited Hyderabad and negotiated with government of India on behalf of Nizam, 1946-8; conservative MP, Bristol West, 1951-7; minister of labour and national service, 1951-5; minister of defence, 1955-6; paymaster-general, 1956-7; viscount; president, MCC, 1957; chairman, Midland Bank, 1957-64; chairman, Iraq Petroleum Company, 1958-65; governor, Harrow; standing counsel, Oxford University (1938-51); hon. degrees, Oxford, Bristol, and Sussex; hon. fellow and visitor, Balliol, 1957; first chancellor, Sussex University, 1963; KCMG, 1945; PC, 1951; GCVO, 1964.

MONCREIFF, HENRY JAMES, second BARON MONCREIFF (1840-1909), Scottish judge; son of first baron [q.v.]; BA and LLB, Trinity College, Cambridge, 1861; passed to Scottish bar, 1863; whig advocate depute, 1865-6; reappointed under Gladstone, 1868 and 1880; joined liberal unionists, 1886; sheriff of Renfrew and Bute, 1881; Scottish judge, 1888-1905.

MONCRIEFF, SIR ALEXANDER (1829-1906), colonel and engineer; joined army, 1855; served in Crimea; colonel, 1878; invented Moncrieff system of raising and lowering guns, and devised means of laying and sighting them when thus mounted; designed hydropneumatic carriage for similar purposes, 1869; FRS, 1871; CB, 1880; KCB, 1890; a keen sportsman, amateur artist, and golfer.

MOND, ALFRED MORITZ, first BARON MELCHETT (1868-1930), industrialist, financier, and politician; son of Ludwig Mond [q.v.]; educated at Cheltenham, Cambridge, and Edinburgh; called to bar (Inner Temple), 1894; managing director of father's chemical business; advocated and worked for principles of co-ordination and co-operation as bases of industrial enterprise; became prominent for number and importance of enterprises with which he was connected and for scale of amalgamations incorporated in firm of Imperial Chemical Industries, Ltd., 1926; instituted Mond-Turner conference to discuss problems arising between employers and employed, 1927; liberal MP, Chester, 1906; Swansea, 1910-23; Carmarthen, 1924-8; first commissioner of works, 1916-21; minister of health, 1921-2; became protectionist after European war; convert to conservatism, 1926; an enthusiastic Zionist;

baronet, 1910; baron, 1928; PC, 1913; FRS, 1928.

MOND, LUDWIG (1839-1909), chemical technologist, manufacturer, and art collector; born and educated at Cassel; employed in soda works near Cassel, 1859; in England (1862) took out patent for recovery of sulphur from Leblanc alkali waste; joined firm at Widnes to push process, 1867; bought English patent of ammonia-soda process, and with (Sir) J. T. Brunner started alkali works at Winnington near Northwich, 1873; firm became Brunner, Mond & Co. (1881), with Mond as managing director (in 1909 employing 4,000 workmen); invented Mond producer-gas plant (patented 1883) for the production of ammonia and cheap producer-gas; his efforts to recover chlorine wasted in ammonia-soda process, and his use of nickel compounds to purify producer-gas, led to discovery of nickel carbonyl, and of a method for extracting metallic nickel from its ores; Mond formed Mond Nickel Company, with mines in Canada and works near Swansea, 1888; active in founding Society of Chemical Industry and its *Journal*, 1881; FRS, 1891; received hon. doctorates from Padua, Heidelberg, Manchester, and Oxford; left fortune of over £1 million; took out forty-nine patents and published many scientific papers; founded Davy-Faraday laboratory for chemical research at Royal Institution, 1896; left large sums to Royal Society and Heidelberg for research; benefactor to town of Cassel and to Jewish charities; bequeathed (contingently on wife's death) art collection, mainly early Italian pictures, to National Gallery.

MOND, SIR ROBERT LUDWIG (1867-1938), chemist, industrialist, and archaeologist; son of Ludwig Mond and brother of first Lord Melchett [qq.v.]; carried out and encouraged others in research in father's firm; hon. life secretary, Davy-Faraday laboratory, Royal Institution; financed, organized, and published results of archaeological expeditions in Egypt, Palestine, etc.; knighted, 1932; FRS, 1938.

MONKHOUSE, WILLIAM COSMO (1840-1901), poet and critic; clerk in Board of Trade, 1856; assistant secretary to finance department at death; published poems, *A Dream of Idleness* (1865) and *Corn and Poppies* (1890); a novel (1868); *Masterpieces of English Art* (1869); lives of Turner (1879) and Tenniel (1901), and *Earlier English Water Colour Painters* (1890).

MONRO, SIR CHARLES CARMICHAEL, baronet (1860-1929), general; grandson of Alexander Monro, tertius (1773-1859, q.v.); gazetted to 2nd Foot, 1879; served in South

African war, 1899-1900; chief instructor, Hythe School of Musketry, 1901-3; commandant, 1903-7; virtually responsible for evolution of new system of infantry fire-tactics; commanded 13th infantry brigade, Ireland, 1907-12; major-general, 1910; commanded 2nd division, Aldershot, 1914; proceeded to France, Aug.; received command of I Corps, Dec.; commanded Third Army, with rank of general, July 1915; placed in command of Mediterranean expeditionary force at Gallipoli, Oct.; successfully advocated and effected (Jan. 1916) complete withdrawal from peninsula; commanded First Army in France, 1916; commander-in-chief, India, 1916-20; successfully developed Indian military power, 1917-18; ably handled unrest in India following war; governor of Gibraltar, 1923-8; KCB, 1915; baronet and Bath King of Arms, 1921.

MONRO, CHARLES HENRY (1835-1908), author; educated at Harrow; BA, Caius College, Cambridge (first class classic), and fellow, 1857; law lecturer, 1872-96; planned but left unfinished a complete translation of Justinian's *Digest* (2 vols., 1904-9); memorial fellowship and Celtic lectureship founded at Caius College.

MONRO, DAVID BINNING (1836-1905), classical scholar; educated at Glasgow University and Balliol College, Oxford; BA (first class classic), 1858; Ireland scholar, 1858; fellow of Oriel, 1858; vice-provost, 1874; provost, 1882; vice-chancellor of university, 1901-4; published a Homeric grammar, 1882 (an authoritative work), a school edition of *Iliad* (2 vols., 1884-9), and *Odyssey* xiii-xxiv (1901); edited complete text, 1896; sought in philology the solution of Homeric problems; founded Oxford Philological Society, 1870; original FBA, 1902; hon. DCL, Oxford, 1904, LLD, Glasgow, 1883, Litt.D., Dublin, 1892; died in Switzerland.

MONRO, HAROLD EDWARD (1879-1932), poet, editor, and bookseller; educated at Radley and Gonville and Caius College, Cambridge; published several volumes of verse and founded *Poetry Review* (1912), Poetry Bookshop, Bloomsbury, 1913-32, *Poetry and Drama* (1913-14), and *(Monthly) Chapbook* (1919-25.)

MONRO, SIR HORACE CECIL (1861-1949), civil servant; educated at Repton and Clare College, Cambridge; second class, classics, 1883; entered Local Government Board, 1884; private secretary to successive presidents; an assistant secretary to the Board, 1897; permanent secretary, 1910-19; KCB, 1911.

MONSON, SIR EDMUND JOHN, first baronet (1834-1909), diplomatist; educated at

Eton and Balliol College, Oxford; BA (first class in history), 1855; MA and fellow of All Souls, 1858; held various minor posts in diplomatic service, 1856-65; consul in Azores, 1869-71; consul-general at Budapest, 1871; British representative at Cetinje during war of Serbia and Montenegro with Turkey, 1876-7; CB, 1878; employed in Uruguay (1879-84), Buenos Aires (1884), Copenhagen (1884-8), Athens (1888), and Brussels (1892); KCMG, 1886; GCMG, 1892; ambassador at Vienna and PC, 1893, and Paris, 1896-1904; GCB, 1896; tactfully settled disputes with French in Newfoundland, New Hebrides, and East and West Africa; hon. DCL, Oxford, 1898, LLD, Cambridge, 1905; GCVO, 1903; baronet, 1905; received grand cross of Legion of Honour.

MONTAGU OF BEAULIEU, second BARON (1866-1929), pioneer of motoring. [See DOUGLAS-SCOTT-MONTAGU, JOHN WALTER EDWARD.]

MONTAGU, EDWIN SAMUEL (1879-1924), statesman; son of first Baron Swaythling [q.v.]; BA, Trinity College, Cambridge; president of Union, 1902; liberal MP, Chesterton division, Cambridgeshire, 1906-22; private secretary to Asquith, 1906-10; parliamentary under-secretary of state for India, 1910-14; financial secretary to Treasury, 1914-16; chancellor of duchy of Lancaster with seat in cabinet, 1915; minister of munitions, 1916; resigned, Dec. 1916; secretary of state for India, June 1917-Mar. 1922; his first task, declaration of goal of British policy regarding constitutional change in India, namely 'progressive realization of responsible government' in India, Aug. 1917; toured round provinces of India with small delegation, Nov. 1917-May 1918; outcome of this delegation, drafting of *Report on Indian Constitutional Reforms*, 1918; handled his problems with elasticity and resilience, and by pertinacity, drive, and determination rallied bulk of opinion to his scheme, which passed into law as Government of India Act, 1919; central feature, extension of self-government operating through 'dyarchy'; forced to resign owing to divergences with his colleagues over Turkish policy.

MONTAGUE, LORD ROBERT (1825-1902), politician and controversialist; MA, Trinity College, Cambridge, 1849; conservative MP for Huntingdonshire, 1859-74; champion of church rates and trade unions; vice-president of the Committee of Council on Education, charity commissioner, and PC, 1867; criticized education bill of 1870; conservative home rule MP for Westminster, 1874-80; condemned conservative policy in Afghan war; joined Church of Rome (1870), but rejoined English Church on ethical and political grounds (1882) and attacked Romanists in published volumes.

MONTAGU, SAMUEL, first BARON SWAYTHLING (1832-1911), foreign exchange banker and Jewish philanthropist; founded with brother and brother-in-law the foreign exchange and banking firm of Samuel Montagu & Co., 1853; acquired large exchange business and helped to make London the clearing-house of the international money market; engaged in large transactions in silver; liberal MP for Whitechapel, 1885-1900; chief author of Weights and Measures Act, 1897; ardent supporter of bimetallism; member of select committee on alien immigration, 1888; took leading part in Jewish religious, social, and charitable work; founded Jewish Working Men's Club, 1870; formed federation of smaller East End synagogues, 1887; gave to London County Council £10,000 for Tottenham housing scheme, 1903; travelled abroad in interests of oppressed Jews; visited Russia to investigate condition of Jews (1886), but was expelled; president of Russo-Jewish committee, 1896-1909; collector of works of art and of old English silver; FSA, 1897; baronet, 1894; baron, 1907.

MONTAGU-DOUGLAS-SCOTT, LORD CHARLES THOMAS (1839-1911), admiral [See SCOTT.]

MONTAGU-DOUGLAS-SCOTT, LORD FRANCIS GEORGE (1879-1952), soldier, Kenya farmer, and political leader. [See SCOTT.]

MONTAGUE, CHARLES EDWARD (1867-1928), man of letters and journalist; BA, Balliol College, Oxford; on staff of *Manchester Guardian*, 1890-1914, 1919-25; second-in-command, 1898; a brilliant, many-sided journalist; made his mark as dramatic critic; served in European war, 1914-19; works include *A Hind Let Loose* (1910), *Disenchantment* (1922), *Fiery Particles* (1923), *The Right Place* (1924), *Rough Justice* (1926), *Action* (1928).

MONTAGUE, FRANCIS CHARLES (1858-1935), historian; brother of C. E. Montague [q.v.]; first class, *lit. hum.*, Balliol College, Oxford, 1879; professor of history, University College, London, and lecturer at Oriel College, Oxford, 1893-1927.

MONTEATH, SIR JAMES (1847-1929), Indian civil servant; passed Indian civil service examination, 1868; appointed to Bombay presidency, 1870; chief secretary to government of Bombay, 1896; successfully dealt with Bombay famines, 1896-7 and 1899-1902; introduced land revenue code amendment bill, which greatly benefited Bombay agriculturists, 1901; KCSI, 1903.

MONTEFIORE, CLAUDE JOSEPH GOLDSMID- (1858-1938), Jewish biblical scholar and philanthropist; first class, *lit. hum.*, Balliol College, Oxford, 1881; joint-editor, *Jewish Quarterly Review*, 1888-1908; joint-founder, Jewish Religious Union for the Advancement of Liberal Judaism and Liberal Jewish synagogue, London; president, Anglo-Jewish Association (1896-1921) and University College, Southampton (1915-34); works include *The Synoptic Gospels* (2 vols., 1909) and *Rabbinic Literature and Gospel Teachings* (1930).

MONTGOMERIE, ROBERT ARCHIBALD JAMES (1855-1908), rear-admiral; entered navy, 1869; served in Egyptian war, 1882; in charge of naval transport in Nile expedition, 1885-6; commanded field battery in Vitu expedition, 1890; CB, 1892; conducted bombardment of Puerto Cabello, 1903; inspecting captain of boys' training ships, 1904; CMG, 1904; rear-admiral, 1905; CVO, 1907; champion heavyweight boxer of navy.

MONTGOMERY-MASSINGBERD, SIR ARCHIBALD ARMAR (1871-1947), field-marshal; educated at Charterhouse and Woolwich; commissioned in Royal Artillery, 1891; served in South Africa, 1899-1902; chief of staff to (Lord) Rawlinson [q.v.] in France, 1914-18; GOC-in-C, Southern command, 1928-31; general, 1930; adjutant-general to the forces, 1931-3; chief of the imperial general staff, 1933-6; field-marshal, 1935; KCMG, 1919; KCB, 1925; GCB, 1934.

MONTMORENCY, JAMES EDWARD GEOFFREY DE (1866-1934), legal scholar. [See DE MONTMORENCY.]

MONTMORENCY, RAYMOND HARVEY DE, third VISCOUNT FRANKFORT DE MONTMORENCY (1835-1902), major-general. [See DE MONTMORENCY.]

MONEYPENNY, WILLIAM FLAVELLE (1866-1912), journalist, and biographer of Disraeli; joined editorial staff of *The Times*, 1893; chosen to undertake authoritative biography of Lord Beaconsfield; published first volume, 1910, second, 1912; work completed by G. E. Buckle [q.v.].

MOODY, HAROLD ARUNDEL (1882-1947), medical practitioner and founder (1931) of the League of Coloured Peoples; negro, born in Jamaica; MB, BS, King's College, London,

1912; practised in Peckham; sought to improve status of coloured people; prominent Congregationalist.

MOOR, Sir FREDERICK ROBERT (1853-1927), South African statesman; born in Natal; Kimberley diamond digger, 1872-9; afterwards farmer in Natal; represented Weenen county in Natal legislative assembly, 1886; on grant of responsible government (1893) held portfolio of native affairs until 1897, and again 1899-1903; prime minister, 1906-10; worked wholeheartedly for unification of South Africa; held portfolio of commerce and customs in federal ministry of Louis Botha [q.v.], 1910; senator, 1910-20; knighted, 1911.

MOOR, Sir RALPH DENHAM RAYMENT (1860-1909), first high commissioner of Southern Nigeria, 1900-3; commandant of constabulary in Oil Rivers protectorate, 1891; vice-consul, 1892; consul, 1896; commissioner and consul-general of newly formed Niger Coast protectorate, 1896-1900.

MOORE, ARTHUR WILLIAM (1853-1909), Manx antiquary; educated at Rugby and Trinity College, Cambridge; second class historical tripos, 1875; MA, 1879; blue for Rugby football; speaker of House of Keys, 1898-1909; championed house in disputes with governor; wrote on meteorology of the island; CVO, 1902; founded Manx Language Society, 1899; wrote on Manx folklore (1891), music (1896), and records (1905); his *History of the Isle of Man* (1900) is an authoritative work.

MOORE, EDWARD (1835-1916), principal of St. Edmund Hall, Oxford, and Dante scholar; BA, Pembroke College, Oxford; fellow of Queen's, 1858; principal of St. Edmund Hall, 1864-1913; preserved its independence; canon of Canterbury, 1903; works include *Contributions to the Textual Criticism of the 'Divina Commedia'* (1889) and *Oxford Dante* (1894).

MOORE, GEORGE AUGUSTUS (1852-1933), novelist; son of G. H. Moore [q.v.]; born and brought up in Ireland; educated at Oscott College, Birmingham; lived in London (1880-1901), in Dublin (1901-11), and at 121 Ebury Street, London (1911-33); works include *A Modern Lover* (1883), *A Mummer's Wife* (1885), *Confessions of a Young Man* (1888), *Esther Waters* (1894), *Hail and Farewell*, candid and intimate autobiography (3 vols., 1911-14), *The Brook Kerith* (1916), *A Story-Teller's Holiday* (1918), *Avowals* (1919), *Héloïse and Abélard* (1921), *Conversations in Ebury Street* (1924); sought lucidity and 'the melodic line', treating epic themes in prose beautiful and dignified yet

preserving illusion of a story melodiously spoken.

MOORE, GEORGE EDWARD (1873-1958), philosopher; grandson of George Moore [1803-80, q.v.]; educated at Dulwich College and Trinity College, Cambridge; first class, part i, classical tripos, 1894; Craven scholarship, 1895; first class, part ii, moral sciences tripos, 1896; fellow of Trinity, 1898; university lecturer in moral science, 1911-25; professor of philosophy, 1925-39; editor of *Mind*, 1921-47; close friend at Cambridge of Bertrand (later Earl) Russell [q.v.]; published *Principia Ethica* in same year as Russell's *Principles of Mathematics* (1903); leading figure in twentieth-century revolution in philosophy, insisting that philosophy should adhere to commonsense; outstanding teacher and lecturer; published *Ethics* (1912), *Philosophical Studies* (1922), and *Some Main Problems of Philosophy* (1953); FBA, 1918; OM, 1951.

MOORE, MARY (1861-1931), actress and theatre manager. [See WYNDHAM, MARY, LADY.]

MOORE, STUART ARCHIBALD (1842-1907), legal antiquary; FSA, 1869; called to bar, 1884; published volumes (1888 and 1903) on the history and law relating to foreshore and fishery rights; keen yachtsman; edited antiquarian works for Camden Society and Roxburghe Club.

MOORE, TEMPLE LUSHINGTON (1856-1920), architect; pupil (1875-8) of George Gilbert Scott, junior, with whom he maintained close professional association, 1878-90; employed pure Gothic style; designs include seventeen important new churches (1885-1917), nave of Hexham Abbey (1902-8), Anglican cathedral, Nairobi (1914), chapels of Pusey House, Oxford, and Bishop's Hostel, Lincoln, and several houses.

MOORE-BRABAZON, JOHN THEODORE CUTHBERT, first BARON BRABAZON OF TARA (1884-1964), aviator and politician. [See BRABAZON, JOHN THEODORE CUTHBERT MOORE-.]

MOORHOUSE, JAMES (1826-1915), bishop of Melbourne and afterwards of Manchester; BA, St. John's College, Cambridge; prebendary of St. Paul's, 1874; bishop of Melbourne, 1876-86; presided over synod at Sydney which framed constitution of Church in Australia; bishop of Manchester, 1886-1903.

MORAN, PATRICK FRANCIS (1830-1911), cardinal archbishop of Sydney; nephew of Cardinal Cullen [q.v.]; educated at Rome; priest, 1953; vice-principal of St. Agatha's

College, Rome, 1856–66; private secretary to Cardinal Cullen, 1866–72; archbishop of Sydney, 1884; cardinal, 1885; militant churchman and a keen controversialist; built many Roman Catholic institutions in New South Wales; prominent in Australian politics; advocated home rule, and supported Australian federation; publications include *The Catholic Archbishops of Dublin* (1864), and *The Catholics of Ireland . . . in the 18th Century* (1899).

MORANT, GEOFFREY MILES (1899–1964), anthropologist and statistician; educated at Battersea secondary school and University College, London; B.Sc. (applied statistics), 1920; M.Sc., 1922; D.Sc., 1926; joined staff of the Department of Applied Statistics; leader of biometric school of physical anthropologists; papers on 'Tibetan skulls' (1923), and 'Racial History of Egypt' (1925); published *The Races of Central Europe*, with preface by J. B. S. Haldane [q.v.] (1939); joined Ministry of Information, 1939; assisted Medical Research Council's Army Personnel Research Committee, 1942; worked at Physiological laboratory (later RAF Institute of Aviation Medicine), 1944–59; OBE, 1952.

MORANT, Sir ROBERT LAURIE (1863–1920), civil servant; BA, New College, Oxford; assistant director of special inquiries and reports, Education Department, 1895; assistant private secretary to eighth Duke of Devonshire [q.v.], 1902; passing of Education Act (1902) largely due to him; permanent secretary, Board of Education, which he entirely remodelled, 1903–11; chairman, National Health Insurance Commission, which he organized, 1911–19; first secretary, Ministry of Health, which he constructed, 1919–20; CB, 1902; KCB, 1907.

MORE-MOLYNEUX, Sir ROBERT HENRY (1838–1904), admiral; joined navy, 1852; served in Crimea, 1854; on West Coast of Africa, 1859; commander, 1865; on North America and West Indies station, 1867; commanded *Ruby* in Russian war, 1877–8; at bombardment of Alexandria; CB; protected Suakin till arrival of Sir Gerald Graham [q.v.]; KCB, 1885; captain superintendent of Sheerness dockyard, 1885–8; admiral, 1899; president of Royal Naval College, Greenwich, 1900–3; GCB, 1902; died at Cairo.

MORESBY, JOHN (1830–1922), admiral and explorer; son of Sir Fairfax Moresby [q.v.]; entered navy, 1842; took part in suppression of Taiping rebellion, 1861; commanded paddle sloop *Basilisk* on Australia station, 1871; with *Basilisk* carried out important explorations in Torres straits and on New Guinea coasts, 1872–

4; his surveys covered 1,200 miles of unknown coast-line and about 100 islands; senior naval officer, Bermuda dockyard, 1878–81; rear-admiral and assessor to Board of Trade, 1881; vice-admiral, 1888; admiral, 1893.

MORFILL, WILLIAM RICHARD (1834–1909), Slavonic scholar; BA, Oriel College, Oxford, 1857; MA, 1860; travelled much in Slavonic countries, studying their history and literature; university reader in Russian at Oxford, 1889; Ph.D., Prague; FBA, 1903; wrote grammars of Polish (1884), Serbian (1887), Russian (1889), Czech (1889), and Bulgarian (1897), and histories of Russia (1885 and 1902), Poland (1893), and of Slavonic literature (1883); he also knew Welsh, old Irish, and Turkish; bequeathed Slavonic library to Queen's College, Oxford.

MORGAN, CHARLES LANGBRIDGE (1894–1958), novelist, critic, and playwright; served in Royal Navy, 1907–13; entered at Brasenose College, Oxford, 1913, rejoined navy on outbreak of war, 1914; took part in Antwerp expedition, interred in Holland, 1914–17; published *The Gunroom* (1919); returned to Oxford University, 1919–21; president OUDS; joined editorial staff, *The Times*, 1921; succeeded A. B. Walkley [q.v.], as dramatic critic, 1926–39; published *My Name is Legion* (1925), *Portrait in a Mirror* (1929) (Femina Vie Heureuse prize, 1930), *The Fountain* (1932) (Hawthornden prize, 1933), *Epitaph on George Moore* (1935), and *Sparkenbroke* (1936); successful play, *The Flashing Stream* (produced in London, 1938); served with the Admiralty during 1939–45 war; published further novels, *The Voyage* (1940) (James Tait Black memorial prize), *The Empty Room* (1941), *The Judge's Story* (1947), *The River Line* (1949), *A Breeze of Morning* (1951), and *Challenge to Venus*, (1957); last play, *The Burning Glass* (produced, 1953); his books translated into nineteen languages; member of Institute of France, 1949; president, International PEN, 1953–6.

MORGAN, CONWY LLOYD (1852–1936), comparative psychologist and philosopher; studied at School of Mines and Royal College of Science, London; professor of geology and zoology (from 1910 of psychology and ethics), Bristol, 1884–1919; principal, 1887–1909; first vice-chancellor, Bristol University, 1909–10; a pioneer in study of animal psychology; works include *Animal Life and Intelligence* (1890–1) and *Instinct and Experience* (1912); FRS, 1899.

MORGAN, EDWARD DELMAR (1840–1909), linguist and traveller; lived in St. Petersburg; travelled in Persia, 1872, Little Russia, and Lower Congo, 1883; honorary secretary of Hakluyt Society, 1886–92, editing Anthony Jenkinson's travels, 1886; translated works by the Central Asian explorer Przhevalsky, 1876–9.

MORGAN, Sir FREDERICK EDGWORTH (1894–1967), lieutenant-general; educated at Clifton College and Royal Military Academy, Woolwich; commissioned in Royal Artillery, 1913; served in India and France during 1914–18 war; served in India, 1919–35; Staff College, Quetta, 1927–8; commanded support group, 1st Armoured division in France, 1940; commanded I Corps, 1943; chief of staff to supreme allied commander, 1943–4; prepared plan for invasion of Europe; criticized by General Sir B. L. Montgomery (later Viscount Montgomery of Alamein); deputy chief of staff to General Eisenhower, 1944–5; chief of operations to UNNRA in Germany, 1945–6; retired from army, 1946; controller, atomic energy, 1951; controller atomic weapons, Atomic Energy Authority, 1954–6; CB, 1943; KCB, 1944; commander US Legion of Merit and French Legion of Honour; colonel commandant, Royal Artillery, 1948–58; publications include *Overture to Overlord* (1950), *Memoirs* (1958), and *Peace and War, a Soldier's Life* (1961).

MORGAN, Sir GILBERT THOMAS (1872–1940), chemist; educated at Finsbury Technical College and Royal College of Science; professor of chemistry, Royal College of Science, Dublin (1912–16), at Finsbury (1916–19), at Birmingham (1919–25); director, Chemical Research Laboratory, 1925–37; mainly interested in chemical reactions under high pressures and synthetic resins; FRS, 1915; knighted, 1936.

MORGAN, JOHN HARTMAN (1876–1955), lawyer; educated at Caterham School, University College of South Wales, and Balliol College, Oxford; MA, London, 1896; on literary staff, *Daily Chronicle*, and postgraduate student at London School of Economics, 1901–3; leader-writer for *Manchester Guardian*, 1904–5; Home Office representative with British Expeditionary Force to inquire into conduct of Germans in the field, 1914; called to bar (Inner Temple), 1915; attended peace conference as assistant adjutant-general, 1919; member of Inter-Allied Council of Central Commission for disarmament of Germany; retired from army with rank of brigadier-general, 1923; convinced that Germany had no intention of disarming; counsel for defence of Sir Roger Casement [q.v.], 1916; professor of constitutional law, University College, London, 1923–41; KC, 1926; reader in constitutional law, Inns of Court, 1926–36; legal adviser to American War Crimes Commission,

1947–9; published *The House of Lords and the Constitution* (1910), *War, its Conduct and Legal Results* (with T. Baty) (1915), *Gentlemen at Arms* (1918), *Viscount Morley, an Appreciation* (1924), *Assize of Arms* (1945), and *The Great Assize* (1948).

MORIARTY, HENRY AUGUSTUS (1815–1906), captain in the navy; prepared vessels for bombardment of Sveaborg, 1855; navigated *Great Eastern* when employed for laying Atlantic cables, 1865–6; CB, 1866; Queen's harbourmaster, Portsmouth, 1869–74, captain, 1874; nautical expert before parliamentary committees; published volumes of sailing directions, 1887–93.

MORISON, STANLEY ARTHUR (1889–1967), typographer; educated at Owen's School, Islington; left school at fourteen; clerk with London City Mission, 1905–12; assistant, the *Imprint*, 1913–14; joined staff of Burns & Oates, Roman Catholic publishers; conscientious objector, imprisoned, 1916; typographer, Cloister Press, 1921–2; freelance consultant; part-time consultant, Monotype Corporation, 1922–54; part-time typographical adviser, Cambridge University Press, 1923–59; typographical advisor to *The Times*, 1930–60; edited *The History of The Times* (1935–52), edited *The Times Literary Supplement* (1945–47); member, editorial board, *Encyclopaedia Britannica*, 1961; James F. R. Lyell reader in bibliography, Oxford, 1956–7; Litt.D., Cambridge, 1950; Litt.D., Marquette (Wisconson, USA) and Birmingham; FBA, 1954; Royal Designer for Industry, 1960; published *Block-letter Text* (1942); his collection of books in 'Morison Room' at Cambridge University Library.

MORISON, Sir THEODORE (1863–1936), educationist and writer; son of J. A. C. Morison [q.v.]; educated at Westminster and Trinity College, Cambridge; professor, Mohammedan Anglo-Oriental College, Aligarh, 1889; principal, 1899–1905; member, council of India, 1906–16; principal, Armstrong College, Newcastle, 1919–29; director, British Institute, Paris, 1933–6; KCIE, 1910; KCSI, 1917.

MORLAND, Sir THOMAS LETHBRIDGE NAPIER (1865–1925), general; born at Montreal, Canada; gazetted into King's Royal Rifle Corps, 1884; transferred to West African frontier force, 1898; took part in six minor campaigns in Nigeria, 1898–1903; inspector-general, West African frontier force, 1905–9; major-general, 1913; commanded 5th division, 1914; X Army Corps, 1915–18; XIII Army Corps, 1918–19; commander-in-chief, army of occupation, Cologne, 1920–2; at Aldershot,

1922-3; general, 1922; KCB, 1915; died at Montreux.

MORLEY, third EARL OF (1843-1905), chairman of committees of the House of Lords. [See PARKER, ALBERT EDMUND.]

MORLEY, JOHN, VISCOUNT MORLEY OF BLACKBURN (1838-1923), statesman and man of letters; educated at Cheltenham College and Lincoln College, Oxford; freelance journalist in London, 1860-3; joined staff of *Saturday Review*, 1863; became acquainted with George Meredith and John Stuart Mill [qq.v.], to both of whom he owed much; editor of *Fortnightly Review*, which he made influential organ of liberal opinion, 1867-82; in close sympathy with Frederic Harrison [q.v.], who assisted him with *Fortnightly*, and leading positivists; published his first study of Burke, 1867; attempted unsuccessfully to enter parliament, 1868-9; admirer of T. H. Huxley [q.v.]; his studies of Frenchmen of the Revolution and their precursors including *Voltaire* (1872) and *Rousseau* (1873) conveyed a message of rationalism and progress to his generation; through his radical opinions and views on national education brought into contact with Joseph Chamberlain [q.v.]; worked with Chamberlain and Sir Charles Dilke [q.v.] at programme of disestablishment, secular education, land reform, and progressive taxation; adopted pacific outlook on foreign and imperial policy; published *Burke* ('English Men of Letters' series), 1879; editor of *Pall Mall Gazette*, which he changed from conservatism and imperialism to radicalism and Cobdenism, 1880; again attempted unsuccessfully to enter parliament, 1880; published *Life of Cobden*, one of his best writings and a classic among English political biographies, 1881; MP, Newcastle upon Tyne, 1883-95; although only moderately successful as speaker his moral leadership gave him position of independence and influence; broke with Chamberlain over Irish question; chief secretary for Ireland, 1886; took leading part in round table conference, the object of which was to bring about concordat with Chamberlain and Sir G. O. Trevelyan [q.v.] on Irish question, 1887; during years of opposition his energies largely absorbed in denouncing Lord Salisbury's policy of coercion; one of the most popular orators on liberal platforms; edited 'English Men of Letters' series; published *Walpole* in 'Twelve English Statesmen' series, 1889; chief secretary for Ireland, 1892-5; achieved his main task of helping Gladstone to prepare and carry through House of Commons second home rule bill; MP, Montrose Burghs, 1896-1908; with Sir William Harcourt [q.v.] resisted policy which found expression in South African war; published *Oliver Cromwell*, 1900;

published *Life of Gladstone*, his most important work (3 vols., 1903); secretary of state for India, 1905-10; planned series of reforms which aimed at gradually associating people of India with civil administration and government; created viscount, 1908; helped to conduct parliament bill limiting Lords' veto through Upper House, 1911; with John Burns [q.v.] resigned from cabinet on government's decision to intervene in 1914-18 war; in retirement occupied himself with affairs of Manchester University, of which he had been elected chancellor (1908); OM, 1902; remained to end of his life agnostic, liberal, and individualist.

MORLEY HORDER, PERCY (RICHARD) (1870-1944), architect. [See HORDER].

MORRELL, LADY OTTOLINE VIOLET ANNE (1873-1938), half-sister of sixth Duke of Portland; married Philip Edward Morrell, 1902; centre and patroness of distinguished literary and artistic circle in London, at Garsington Manor, Oxfordshire (1913-24), and after 1924 at 10 Gower Street, London.

MORRIS, EDWARD PATRICK, first BARON MORRIS (1859-1935), premier of Newfoundland; born at St. John's, Newfoundland; educated at Ottawa University; MP, St. John's West, 1885-1919; entered cabinet, 1889; leader of independent liberal party, 1898-1900; attorney-general and minister of justice under Sir Robert Bond [q.v.], 1902-7; leader of 'people's' party, 1908; premier, 1909-18; brought Newfoundland prominently into councils of the Empire; knighted, 1904; PC, 1911; KCMG, 1913; baron, 1918.

MORRIS, SIR HAROLD SPENCER (1876-1967), president of the Industrial Court; educated at Westminster, Clifton, and Magdalen College, Oxford; called to bar (Inner Temple), 1899; member, South-Eastern circuit; commissioned in Coldstream Guards, 1916; transferred to Flying Corps as deputy assistant adjutant-general, 1918; MBE, 1919; KC, 1921; recorder of Folkestone, 1921-6; national liberal MP, East Bristol, 1922-3; chairman, court of investigation into dispute in the woollen industry, 1925; chairman, Railways National Wages Board, 1925; president, Industrial Court, 1926-45; knighted, 1927; chairman, Coal Wages Board, 1930-5; published *The Barrister* (1930) and *Back View* (1960).

MORRIS, SIR LEWIS (1833-1907), poet and Welsh educationist; born at Carmarthen; BA (first class, *lit. hum.*), Jesus College, Oxford, 1856; MA, 1858; chancellor's prize for English essay, 1858; fellow, 1877; called to bar, 1861;

practised as conveyancer in London till 1880; published anonymously *Songs of Two Worlds*, sonorous and optimistic verse (3 series, 1871, 1874, 1875; republished in one vol., 1878); imitated Tennyson's *Tithonus* in a series of blank verse monologues, collected as *The Epic of Hades* (1877), which was popular with the middle classes and reached a 45th edition; there followed *Gwen: a Drama in Monologue* (1879), *The Ode of Life* (descriptive poems, 1880), *Songs Unsung* (1883, the first volume issued under author's name), *Gycia: a tragedy* (1886), and *Songs of Britain* (1887); collected editions appeared in 1882 (3 vols.) and 1890; *A Vision of Saints* in 1890, and subsequently other collections of lyrics; his work was ridiculed in *Saturday Review*; published a volume of essays, *The New Rambler* (1905); honorary secretary (1878), honorary treasurer (1889), and vice-president (1896) of University College of Wales, Aberystwyth; helped in establishing university of Wales, 1893; hon. D.Litt., 1906; knighted, 1895; an advanced liberal; failed in attempts to enter parliament.

MORRIS, MICHAEL, Baron Morris and Killanin (1826-1901), lord chief justice of Ireland; senior moderator in ethics and logic, Trinity College, Dublin, 1846; hon. LLD., 1887; called to Irish bar, 1849; recorder of Galway, 1857-65; QC, 1863; independent conservative MP for Galway, 1865; solicitor-general for Ireland, July 1866; attorney-general, Nov.; Irish PC; puisne judge of court of common pleas, 1867; lord chief justice of Ireland, 1887; baronet, 1885; member of judicial committee of English Privy Council, receiving life peerage, 1889; bencher of Lincoln's Inn, 1890; commissioner of Irish national education; vice-chancellor of Royal University, Ireland, 1899; made hereditary baron of Killanin, 1900; an opponent of home rule, but a caustic critic of English rule in Ireland.

MORRIS, PHILIP RICHARD (1836-1902), painter; won travelling studentship at Royal Academy Schools, 1858; exhibited at Academy, 1858-1901; ARA, 1877; early painted sea pictures, later religious subjects; best-known works were 'Sons of the Brave' and 'The First Communion'.

MORRIS, TOM (1821-1908), golfer; apprenticed to Allan Robertson, golfer of St. Andrews and golf ball maker; won open golf championship, 1861-2-4-6; green keeper to St. Andrews Golf Club, 1863-1903.

MORRIS, WILLIAM O'CONNOR (1824-1904), Irish county court judge and historian; BA, Oriel College, Oxford, 1848; called to Irish

bar, 1854; professor of common and criminal law in King's Inns, Dublin, 1862; contributed to *Edinburgh Review*, and wrote articles on land tenure in *The Times*, 1870; county court judge for Louth, 1872, and Kerry, 1878; disapproved of Land Act of 1881; transferred as judge for Sligo and Roscommon, 1886; published superficial but independent studies of Hannibal (1890), Napoleon (1890), Moltke (1893), Nelson (1898), and Wellington (1904); *Ireland from 1494 to 1868* (1894) and *Ireland from '98 to '98* (1898).

MORRIS, WILLIAM RICHARD, Viscount Nuffield (1877-1963), industrialist and philanthropist; educated at village school, Cowley; started making bicycles at sixteen with capital of £4; designed motor-cycle, 1902; produced Morris-Oxford car at Motor Show, 1912; acquired property in Cowley for expansion; Cowley factory produced mainly munitions during 1914-18 war; OBE, 1917; Morris Motors Ltd. incorporated, 1919; produced 50,000 cars a year, 1926; established Pressed Steel Company, Cowley, and Morris Motors (1926) Ltd.; produced Morris-Minor, 1931; acquired Wolseley, MG, and SU Carburettor companies; endowed chair of Spanish studies; Oxford, 1926; endowed medical school at Oxford, 1936; aided financially the Royal College of Surgeons, 1948, Oxford hospitals, Guy's, St. Thomas's, Great Ormond Street, and hospitals in Birmingham, Coventry, and Worcester; provided Nuffield Orthopaedic Centre, Oxford, Nuffield Fund for Cripples (1935-7), and Fund for orthopaedic services in Australia, New Zealand, and South Africa (1935-45); financed Nuffield Provincial Hospitals Trust, 1939; founded Nuffield College, 1937, and Nuffield Foundation to promote research, 1943; manufactured tanks and aircraft, 1937-8; merged Morris Motors with Austin Motor Company to form British Motor Corporation, 1952; baronet, 1929; baron, 1934; viscount, 1938; GBE, 1941; CH, 1958; hon. DCL, Oxford (1931); MA, 1937; FRS, 1939; hon. FRCS, 1948; hon. fellow, St. Peter's, Pembroke, Worcester, and Nuffield colleges, Oxford.

MORRIS AND KILLANIN, Baron (1826-1901), lord chief justice of Ireland. [See Morris, Michael.]

MORRIS-JONES, Sir JOHN (1864-1929), Welsh poet and grammarian; educated at Christ College, Brecon; BA, Jesus College, Oxford; turned from mathematics to devote himself wholly to study of Welsh; a founder of 'Dafydd ap Gwilym' Society at Oxford; specialized in philology and Welsh verse; lecturer in Welsh,

Bangor University College, 1889; professor of Welsh, 1895–1929; by his series of adjudications at national eisteddfod, Llandudno, raised standard of poetic diction; his chief works, *Cerdd Dafod* (1925) and *Welsh Grammar* (1913); knighted, 1918.

MORRISON, HERBERT STANLEY, BARON MORRISON OF LAMBETH (1888–1965), labour cabinet minister; left school at fourteen to become an errand boy; worked as shop assistant and switchboard operator; circulation manager for first official labour paper, the *Daily Citizen*, 1912–15; part-time secretary, London labour party, 1915; mayor of Hackney, 1919; member London County Council, 1922–45; leader 1934–40; labour MP, Hackney South, 1923–4, and 1929–31; minister of transport, 1929; created London Passenger Transport Board; lost parliamentary seat, 1931; re-elected for Hackney South, 1935–45; Clement Attlee preferred for labour leadership; minister of supply in national government, 1940; home secretary and minister of home security, 1940; created National Fire Service; played notable part in preparing for labour victory, 1945; MP, Lewisham East, 1945–51, and South Lewisham, 1951–9; lord president of the Council, 1945–7; leader, House of Commons, 1947–51; CH, 1951; foreign secretary, 1951; not so successful as in previous ministries; over-ridden by Attlee in proposal for direct action against Mossadeq in Persia; deputy prime minister, 1945–51; strong claims to succeed Attlee as leader but Hugh Gaitskell [q.v.] elected, 1955; visiting fellow, Nuffield College, Oxford, 1947; publications include *Socialization and Transport* (1933), *Government and Parliament, a Survey from the Inside* (1954), and *An Autobiography* (1960); life peer, 1959; president, British Board of Film Censors, 1960; great leader of the London labour party and LCC, and a fine parliamentarian.

MORISON, WALTER (1836–1921), man of business and philanthropist; educated at Eton and Balliol; liberal MP, Plymouth, 1861–74; liberal unionist MP, Skipton division of Yorkshire, 1886–92, 1895–1900; inherited large fortune which he increased; entertained many eminent friends at Malham Tarn, Yorkshire; benefactions include gifts to northern universities, Giggleswick School, King Edward's Hospital Fund (annual contribution for many years £10,000), Palestine Exploration Fund, Oxford University (£30,000), and Bodleian Library (£50,000).

MORRISON, WILLIAM SHEPHERD, first VISCOUNT DUNROSSIL (1893–1961), Speaker of the House of Commons; educated at George Watson's College and Edinburgh University;

MA, 1920; served in Royal Field Artillery, MC, during 1914–18 war; called to bar (Inner Temple), 1923; conservative MP, Cirencester and Tewkesbury, 1929–59; chairman, 1922 Committee, 1932–6; KC, 1934; recorder, Walsall, 1935; financial secretary to the Treasury, 1935; minister of agriculture and fisheries, 1936; PC, 1936; chancellor, duchy of Lancaster and minister of food, 1939–40; postmaster-general, 1940–2; minister, town and country planning, 1943–5; Speaker of the House of Commons, 1951–9; governor-general, Australia, 1960–1; viscount, GCMG, 1959.

MORSHEAD, SIR LESLIE JAMES (1889–1959), lieutenant-general; born in Australia; educated at Mount Pleasant state school, Ballarat and the Teachers' Training College, Melbourne; schoolmaster up to 1914; commissioned in First Australian Imperial Force, 1914; landed at Anzac, 1915; invalided home with enteric fever, and, on recovery, commanded 33rd battalion (3rd division) in France, 1916–18; DSO, CMG; on demobilization became sheep farmer, and then joined the Orient Steam Navigation Company; on outbreak of 1939–45 war commanded 18th Australian brigade; promoted major-general and commanded 9th Australian division, 1941; besieged in Tobruk; lieutenant-general in command Australian Imperial Force, Middle East, 1942; at El Alamein battle; commanded 1st Australian Corps in recapture of Borneo; KBE and KCB, 1942; general manager in Australia, Orient Steam Navigation Company, 1948.

MOSELEY, HENRY GWYN JEFFREYS (1887–1915), experimental physicist; son of Professor Henry Nottidge Moseley [q.v.]; BA, Trinity College, Oxford; physics lecturer, Manchester University, 1910–14; carried out important researches on X-ray spectra of elements; killed in action at Gallipoli.

MOTT, SIR BASIL, first baronet (1859–1938), civil engineer; trained at Royal School of Mines; constructed first (Monument–Stockwell, 1890) and second (Bank–Shepherd's Bush, 1900, with Sir Benjamin Baker, q.v.) deep-level tubes; responsible with his partners for many tube extensions, bridge schemes, and the Mersey tunnel (opened 1934); introduced escalators into Great Britain; chairman, St. Paul's preservation works committee, 1925; baronet, 1930; FRS, 1932.

MOTT, SIR FREDERICK WALKER (1853–1926), neuro-pathologist; MD, University College, London; held various posts at Charing Cross Hospital school, from 1884; pathologist to London County Council asylums,

1895-1923; best known for his determination of association between syphilitic infection and other bodily changes and mental disorders; did notable work on shell-shock during 1914-18 war; FRS, 1896; KBE, 1919; his numerous honours include Croonian lectureship (1900), Lettsomian lectureship (1916), and Harveian oratorship (1925) of Royal College of Physicians, London; his works include *Archives of Neurology and Psychiatry*.

MOTTISTONE, first BARON (1868-1947), politician and soldier. [See SEELY, JOHN EDWARD BERNARD.]

MOULE, GEORGE EVANS (1828-1912), missionary bishop in mid-China; second son of Revd Henry Moule [q.v.]; BA, Corpus Christi College, Cambridge; went as missionary to China, 1857; first bishop of mid-China, 1880-1906.

MOULE, HANDLEY CARR GLYN (1841-1920), bishop of Durham; eighth son of Revd Henry Moule [q.v.]; BA, Trinity College, Cambridge; principal of Ridley Hall, Cambridge, 1880-99; Norrisian professor of divinity, 1899-1901; bishop of Durham, 1901-20; an evangelical; wrote theological and devotional works.

MOULLIN, ERIC BALLIOL (1893-1960), professor of electrical engineering; educated privately and at Downing College, Cambridge; first class, mechanical sciences tripos, 1916; lecturer, Royal Naval College, Dartmouth, 1917-19; teacher and research worker, Cambridge Engineering Laboratory, assistant director of studies, King's College, 1919-29; Donald Pollock reader in engineering, Oxford University, 1929-39; fellow, Magdalen College, 1931; joined Admiralty Signals Establishment, Portsmouth, 1939; transferred to Metropolitan Vickers Electrical Co. Ltd., Manchester, 1942; professor, electrical engineering, Cambridge, 1945-60; professorial fellow, King's College, 1946; Sc.D., Cambridge, 1939; hon. LLD, Glasgow, 1958; president, Institution of Electrical Engineers, 1949-50; member, Radio Research Board, Department of Scientific and Industrial Research, 1934-42; publications include *The Theory and Practice of Radio Frequency Measurements* (1926), *The Principles of Electromagnetism* (1932), *Spontaneous Fluctuations of Voltage* (1938), *Radio Aerials* (1949), and *Electromagnetic Principles of the Dynamo* (1955).

MOULTON, JAMES HOPE (1863-1917), classical and Iranian scholar and student of Zoroastrianism; son of W. F. Moulton [q.v.];

BA, King's College, Cambridge; Greenwood professor of Hellenistic Greek and Indo-European philology, Manchester, 1908; works include *Prolegomena* to unfinished *Grammar of New Testament Greek* (1906) and *Early Zoroastrianism* (1913); torpedoed in Mediterranean returning from India.

MOULTON, JOHN FLETCHER, BARON MOULTON (1844-1921), lord of appeal in ordinary; brother of W. F. Moulton [q.v.]; BA, St. John's College, Cambridge (senior wrangler 1868; fellow of Christ's College Cambridge, 1868-75; called to bar (Middle Temple), 1874; QC, 1885; specialized in patent actions; liberal MP, Clapham division of Battersea, 1885-6, South Hackney, 1894-5, Launceston division of Cornwall, 1898-1906; lord justice of appeal, knight, and PC, 1906; lord of appeal in ordinary and life peer, 1912; created KCB, 1915, GBE, 1917, for brilliant organization of explosives supply department, 1914-18.

MOUNT STEPHEN, first BARON (1829-1921), financier and philanthropist. [See STEPHEN, GEORGE.]

MOUNT TEMPLE, BARON (1867-1938), politician. [See ASHLEY, WILFRID WILLIAM.]

MOUNTBATTEN, EDWINA CYNTHIA ANNETTE, COUNTESS MOUNTBATTEN OF BURMA (1901-1960), daughter of Colonel W. W. Ashley, PC, MP, later Baron Mount Temple [q.v.], granddaughter of Sir Ernest Cassel [q.v.], and great-granddaughter of the seventh Earl of Shaftesbury [q.v.]; at age of nineteen inherited large fortune from her grandfather; married Lieutenant Lord Louis Mountbatten, RN, 1922; undertook numerous charitable activities; on outbreak of 1939-45 war, served with Order of St. John; superintendent-in-chief, St. John Ambulance Brigade, 1942; inaugurated welfare services for allied prisoners of war and internees, 1943-5, when her husband was supreme allied commander, South-East Asia; accompanied Lord Mountbatten to India when he was appointed last viceroy and first governor-general of independent India, 1947-8; chairman, St. John and Red Cross Services Hospitals, 1948; superintendent-in-chief, St. John Ambulance Brigade Overseas, 1950; CI, 1947; GBE, 1947; DCVO, 1946; GCSt.J, 1945.

MOUNTBATTEN, LOUIS ALEXANDER, first MARQUESS OF MILFORD HAVEN, formerly styled PRINCE LOUIS ALEXANDER OF BATTENBERG (1854-1921), admiral of the fleet; son of Prince Alexander of Hesse; born at Graz, Austria; naturalized and entered British navy, 1868; director of naval intelligence, 1902-5; rear-admiral, 1904; commander-in-chief Atlantic fleet, 1908-10; vice-admiral, 1910; first sea lord, 1912; resigned, Oct. 1914; relinquished German titles and created marquess, 1917; admiral, 1919.

MOUNTEVANS, first BARON (1880-1957), admiral. [See EVANS, EDWARD RATCLIFFE GARTH RUSSELL.]

MOUNTFORD, EDWARD WILLIAM (1855-1908), architect; won open competition for Sheffield town hall, 1890; designed several London buildings, including the Central Criminal Court, Old Bailey; style developed from Renaissance to classic method.

MOWAT, SIR OLIVER (1820-1903), Canadian statesman; born at Kingston, Ontario; called to bar of Upper Canada, 1841; leader of chancery bar at Toronto; QC, 1856; on commission to consolidate statutes of Upper Canada; radical member of legislative assembly, 1857; advocated federation of Upper and Lower Canada, and representation by population, 1859; as postmaster-general carried out considerable economies, 1863; took part in conference for federation, 1864; vice-chancellor of Ontario, 1864-72; premier, 1872-96; responsible for Ballot Act, 1874, Manhood Suffrage Act, 1888, and Acts to simplify and cheapen judicial procedure; championed provincial rights in matters of legal appointments, regulation of companies, and control of liquor traffic; gained victory for Ontario over Manitoba in delimitation of boundaries, 1884; KCMG, 1892; GCMG, 1893; as dominion minister of justice, 1896-7, suggested compromise with Roman Catholics in Manitoba school question; lieutenant-governor of Ontario, 1897; advocated reciprocity with America.

MOWATT, SIR FRANCIS (1837-1919), civil servant; clerk in Treasury, 1856; assistant secretary, 1888; permanent secretary, 1894-1903; raised reputation of department for promptness and efficiency; owing to misunderstanding offered to resign, 1900; served on numerous royal commissions and committees; CB, 1884; KCB, 1893; GCB, 1901; PC, 1906.

MOYNE, first BARON (1880-1944), statesman and traveller. [See GUINNESS, WALTER EDWARD.]

MOYNIHAN, BERKELEY GEORGE ANDREW, first BARON MOYNIHAN (1865-1936), surgeon; qualified MB (London, 1887) from Leeds medical school; FRCS, 1890; MS, 1893; professor of clinical surgery, Leeds,

1909-26; assistant surgeon, Leeds General Infirmary, 1896; surgeon, 1906-26; considered the most accomplished surgeon in England; advocated gentleness in manipulation; studied medical care before and after operation; made advances in asepsis and introduced use of rubber gloves; set out his surgical doctrine in *Abdominal Operations* (1905); wrote also on surgical treatment of diseases of the stomach, pancreas, gastric and duodenal ulcers, and gallstones; founded 'Chirurgical Club' for visiting surgical clinics, 1909; instigated *British Journal of Surgery*, 1913; member, army medical advisory board, 1917-36; president, Royal College of Surgeons, 1926-32; knighted, 1912; KCMG, 1918; baronet, 1922; baron, 1929.

MOZLEY, JOHN KENNETH (1883-1946), divine; scholar of Pembroke College, Cambridge; first class, classical tripos (1905), and theological tripos (1906); ordained, 1909-10; fellow of Pembroke, 1907-19; dean, 1909-19; principal, Leeds clergy school, 1920-5; lecturer, Leeds parish church, 1920-30; canon (1930-41), chancellor (1931-41), St. Paul's Cathedral; publications include *The Doctrine of the Atonement* (1915) and *The Impassibility of God* (1926).

MUDDIMAN, SIR ALEXANDER PHILLIPS (1875-1928), Indian civil servant; joined Indian civil service, 1899; secretary to government and nominated official member of central legislature, 1915; president of Council of State, 1921; ordinary member of governor-general's council in charge of home department, 1924; governor of United Provinces, 1928; knighted, 1922; KCSI, 1926; died in India.

MUIR, EDWIN (1887-1959), writer; educated at Kirkwall grammar school; office boy and clerk in Glasgow and Greenock; contributed verses to *New Age* (1913); published *We Moderns* (1918); assistant to A. R. Orage [q.v.] on *New Age*, 1919; published voluminous work as critic and novelist, 1924-42, including a life of John Knox (1929); came to poetry late in life; worked for British Council, 1942-50; warden of Newbattle Abbey, 1950-5; Charles Eliot Norton professor, Harvard, 1955-6; published *An Autobiography* (1954); collected poems (published, 1960); CBE, 1953; hon. degrees from Prague, 1947, Edinburgh, 1947, Rennes, 1949, Leeds, 1955, and Cambridge, 1958.

MUIR, (JOHN) RAMSAY (BRYCE) (1872-1941), historian and politician; educated at University College, Liverpool, and Balliol College, Oxford; first class, *lit. hum.*, 1897, and modern history, 1898; lecturer (1899), professor of modern history, Liverpool, 1906-13; at Manchester, 1914-21; director, liberal summer school, from 1921; liberal MP, Rochdale, 1923–4; chairman (1931–3), president (1933–6), National Liberal Federation; chairman, education and propaganda committee, and vice-president, liberal party organization, 1936–41; publications include *Short History of the British Commonwealth* (2 vols., 1920–2).

MUIR, SIR ROBERT (1864–1959); pathologist; educated at Teviot Grove Academy and Edinburgh University; MA, 1884; MB, CM, first class honours, 1888; assistant pathologist, Edinburgh, 1892–8; lecturer on pathological bacteriology, 1894–8; wrote *Manual of Bacteriology* with James Ritchie (1897); professor of pathology, St. Andrews, 1898; professor of pathology, Glasgow, 1899–1936; published *Textbook of Pathology* (1924); served on Medical Research Council, 1928–32; FRS, 1911; knighted, 1934; did outstanding work on diseases of the blood cells, immunology, and cancer.

MUIR, SIR WILLIAM (1819–1905), Indian administrator and principal of Edinburgh University; brother of John Muir [q.v.]; joined East India Company's service, 1847; head of intelligence department at Agra during Mutiny, 1857; related experiences in *Agra in the Mutiny* (1896); secretary to Lord Canning's government at Allahabad, 1858; foreign secretary under Lord Lawrence and KCSI, 1867; lieutenant-governor of North-West Provinces, 1868–74; founded Muir College and University at Allahabad; financial member of Lord Northbrook's council, 1874–6; member of council of India in London, 1876–85; principal of Edinburgh University, 1885–1905; president of Royal Asiatic Society, 1884; hon. DCL, Oxford, 1882, LLD, Edinburgh and Glasgow, Ph.D., Bologna, 1888; published standard *Life of Mahomet* (4 vols., 1858–61), *Annals of the Early Caliphate* (1883), and *Mameluke Dynasty of Egypt* (1896); Rede lecturer at Cambridge, 1881; with his brother endowed Shaw professorship of Sanskrit at Edinburgh, 1862.

MUIRHEAD, JOHN HENRY (1855–1940), philosopher; educated at Glasgow Academy and University, Balliol College, Oxford, and Manchester New College, London; lecturer in mental and moral science, Royal Holloway College, 1888–96; professor of philosophy, Birmingham, 1896–1922; editor, 'Library of Philosophy', 1888–1940; a founder (1891) of Ethical Society; works include *The Elements of Ethics* (1892) and *Coleridge as Philosopher* (1930); prominent representative of British school of idealists; FBA, 1931.

MÜLLER, ERNEST BRUCE IWAN- (1853–1910), journalist. [See IWAN-MÜLLER.]

MULLINS, EDWIN ROSCOE (1848–1907), sculptor; studied art at Lambeth, Royal Academy, and Munich; executed works such as 'Cain', 'Innocence', 'Rest', and several portraits, including Dr Martineau (bust), Gladstone (statuette), Queen Victoria, William Barnes (statues); decorated many London buildings; published *A Primer of Sculpture* (1892).

MUNBY, ARTHUR JOSEPH (1828–1910), poet and civil servant; BA, Trinity College, Cambridge, 1851; MA, 1856; called to bar, 1855; in ecclesiastical commissioners' office, 1858–88; published poetic idylls, *Benoni* (1852), *Verses New and Old* (1865), and *Dorothy* (1880); work praised by Browning for craftsmanship; later works include *Poems, chiefly Lyric and Elegiac* (1901) and *Relicta* (1909); FSA; supporter of Working Men's College.

MUNNINGS, SIR ALFRED JAMES (1878–1959), painter; educated at Redenhall grammar school and Framlingham College; apprenticed to lithographers at Norwich and studied at Norwich School of Art; blinded in right eye, 1898; designed posters for Caley's chocolates and crackers; specialized in painting and drawing horses; joined Newlyn group around Stanhope Forbes [q.v.], 1911; official war artist with Canadian Cavalry Brigade, 1917; produced many equestrian portraits, 1919–59, including Lord Harewood [q.v.], with the Princess Royal, and Lord Birkenhead [q.v.]; immortalized many famous racehorses, including Humorist, Hyperion, and Brown Jack; ARA, 1919; RA, 1925; president, RA, 1944–9; knighted, 1944; KCVO, 1947.

MUNRO, HECTOR HUGH (1870–1916), writer of fiction; born in Burma; took up journalism in London; works include *The Westminster Alice* (political sketches, 1902), *Reginald* (1904), *Reginald in Russia* (1910), *Chronicles of Clovis* (1911), *Beasts and Super-Beasts* (1914), all short stories, and *The Unbearable Bassington* (1912), a novel; employed pseudonym 'Saki'; killed in action in France.

MUNRO, JAMES (1832–1908) premier of Victoria, Australia; emigrated to Victoria, 1858, working as printer till 1865; founded Federal Banking Company, 1882, and Real Estate Bank, 1887; liberal minister of public instruction, Aug.–Oct. 1875; treasurer and premier, 1890–1902; agent-general in London, 1902; temperance advocate.

MUNRO-FERGUSON, RONALD CRAUFORD, VISCOUNT NOVAR (1860–1934), politician. [See FERGUSON.]

MURDOCH, WILLIAM LLOYD (1855-1911), Australian cricketer; born in Victoria, Australia; practised as solicitor; played cricket for New South Wales, 1875-84; known as 'W. G. Grace of Australia'; acquired fame first as wicket-keeper and finally as batsman; member (1878) and captain (1880-2, 1884, 1890) of Australian teams to England; made record Australian individual score (211) against England, 1884; visited America, 1878, and South Africa, 1891-2; settled in England, 1891; captained Sussex team, 1893-9; died at Melbourne; published handbook of cricket, 1893.

MURISON, ALEXANDER FALCONER (1847-1934), jurist and author; first class, classics, Aberdeen; called to bar (Middle Temple), 1881; professor, Roman law (1883), jurisprudence (1901), deputy professor, Roman-Dutch law (1913), concurrently until 1925, University College, London; dean of law faculty, 1912-24; deputy professor, civil law, Oxford, 1916-19; KC, 1924; left unfinished manuscript collation of codices of Justinian's *Institutes*; editor, *Educational Times*, 1902-12; translated into verse Horace, Pindar, and some of Virgil and Homer.

MURRAY, ALEXANDER STUART (1841-1904), keeper of Greek and Roman antiquities in the British Museum from 1886; brother of G. R. M. Murray [q.v.]; MA, Edinburgh University; assistant at British Museum, 1867; as keeper reorganized galleries of Greek and Roman antiquities; frequently visited classical sites; LLD, Glasgow, 1891; FSA, 1889; FBA, 1903; independent works include *A History of Greek Sculpture* (2 vols., 1880-3), *Handbook of Archaeology* (1892), *Greek Bronzes* (1898), and *The Sculptures of the Parthenon* (1903).

MURRAY, ANDREW GRAHAM, VISCOUNT DUNEDIN (1849-1942), judge; educated at Harrow, Trinity College, Cambridge, and Edinburgh University; admitted advocate, 1874; senior advocate depute, 1888-90; sheriff of Perthshire, 1890-1; solicitor-general for Scotland, 1891-2, 1895-6; lord advocate, 1896-1903; QC, 1891; PC, 1896; conservative MP, Buteshire, 1891-1905; secretary for Scotland with seat in cabinet, 1903-5; lord justice-general of Scotland and lord president of the Court of Session, 1905-13; infused animation and efficiency into every branch of the court's activities; lord of appeal in ordinary, 1913-32; admirable exponent of legal doctrine; especially interested in patent law and feudal law; keeper of the great seal of the principality of Scotland, 1900-36; KCVO, 1908; GCVO, 1923; baron, 1905; viscount, 1926.

MURRAY, SIR ARCHIBALD JAMES (1860-1945), general; educated at Cheltenham and Sandhurst; gazetted to 27th regiment (Royal Inniskilling Fusiliers), 1879; served with distinction in South Africa, 1899-1902; DSO; director of military training, 1907-12; inspector of infantry, 1912-14; chief of the general staff, British expeditionary force, Aug. 1914-Jan. 1915; deputy chief of the imperial general staff, Feb.-Sept. 1915; chief of the imperial general staff, Sept.-Dec. 1915; commanded forces in Egypt, Jan. 1916-June 1917; advanced to Palestine frontier; suffered serious defeat in second attempt to capture Gaza with insufficient forces and superseded; GOC-in-C, Aldershot, 1917-19; general, 1919; KCB, 1911; GCB, 1928.

MURRAY, CHARLES ADOLPHUS, seventh EARL OF DUNMORE (1841-1907), explorer; joined army, 1860; lord-in-waiting to Queen Victoria, 1874-80; explored Kashmir and Tibet, 1892; published account in *The Pamirs* (1893); wrote *Ormisdale*, a novel (1893); joined Christian Scientists late in life.

MURRAY, DAVID CHRISTIE (1847-1907), novelist and journalist; reporter to *Birmingham Morning News* till 1865; parliamentary reporter for *Daily News*, 1871; contributed articles to *Referee*, collected as *Guesses at Truth* (1908); travelled much on continent and in colonies; represented *The Times* in Russo-Turkish war, 1877-8; a prolific and vigorous novelist; novels include *Rainbow Gold* (1885) and *Aunt Rachel* (1886); he published also *A Novelist's Notebook* (1887) and *Recollections* (posthumous, 1908).

MURRAY, SIR (GEORGE) EVELYN (PEMBERTON) (1880-1947), civil servant; son of Sir G. H. Murray [q.v.]; educated at Eton and Christ Church, Oxford; entered civil service, 1903; secretary to the Post Office, 1914-34; chairman, Board of Customs and Excise, 1934-40; KCB, 1919.

MURRAY, GEORGE GILBERT AIMÉ (1866-1957), classical scholar and internationalist; born in Sydney, son of (Sir) Terence Murray [q.v.]; educated at Merchant Taylors' School and St. John's College, Oxford; first class, *lit. hum.*; fellowship, New College, 1888; described by Sir Richard Jebb [q.v.] as 'the most accomplished Greek scholar of the day'; professor of Greek, Glasgow University, 1889-99; returned to New College, Oxford, 1905; regius professor of Greek, Oxford, 1908-36; FBA, 1910; founder of the League of Nations Union and chairman of the executive council, 1923-38; publications includes *Rise of the Greek Epic* (1907), *Four Stages of Greek Religion*

(1912), extended in 1925 to *Five Stages, Euripides and his Age* (1913), and *Foreign Policy of Sir Edward Grey* (1915); OM, 1941.

MURRAY, SIR GEORGE HERBERT (1849-1936), civil servant; educated at Harrow and Christ Church, Oxford; entered Foreign Office, 1873; transferred to Treasury, 1880; chairman, Board of Inland Revenue, 1897-9; secretary, Post Office, 1899-1903; permanent secretary, Treasury, 1903-11; KCB, 1899; GCB, 1908; PC, 1910; GCVO, 1920.

MURRAY, GEORGE REDMAYNE (1865-1939), physician; educated at Eton, Trinity College, Cambridge, and University College Hospital, London; MB (Camb.), 1889; discovered cure for myxoedema by hypodermic injection of animal thyroid, 1891; professor of comparative pathology, Durham, 1893-1908; of medicine (and physician to Royal Infirmary), Manchester, 1908-25; physician, Royal Victoria Infirmary, Newcastle, 1898-1908.

MURRAY, GEORGE ROBERT MILNE (1858-1911), botanist; assistant at British Museum, 1876; keeper of botanical department, 1895-1905; director of scientific staff of Antarctic expedition of Captain R. F. Scott [q.v.], 1901; FRS, 1897; made special study of marine algae; wrote works on *Cryptogamic Botany* (1889) and on seaweeds (1895); edited *Phycological Memoirs* (1892-5).

MURRAY, SIR JAMES AUGUSTUS HENRY (1837-1915), lexicographer; educated at Cavers and Minto schools; headmaster of Hawick Subscription Academy, 1857; took up study of languages; on staff of Mill Hill School, 1870-85; BA, London University, 1873; editor of *New English Dictionary on Historical Principles* projected by Clarendon Press, 1879; removed to Oxford in order to devote himself to this work, 1885; conceived plan and settled scope of greatest lexicographical achievement of present age, although his editorial responsibility only covers half of it (A-D, H-K, O, P, T); also edited Sir David Lyndesay's *Works*, part v, *The Complaynte of Scotlande*, and *The Romance and Prophecies of Thomas of Erceldoune*; wrote *Dialect of the Southern Counties of Scotland* (1873); FBA, 1902; knighted, 1908.

MURRAY, SIR JAMES WOLFE (1853-1919), lieutenant-general; entered army from Woolwich, 1872; engaged for many years in staff and extra-regimental service; KCB, for services in South African war, 1901; quartermaster-general in India and major-general, 1903; lieutenant-general, 1909; chief of imperial general staff, 1914-15.

MURRAY, SIR JOHN (1841-1914), marine naturalist and oceanographer; born at Cobourg, Ontario; came to Scotland to complete his education, 1858; naturalist in charge of collections on *Challenger* expedition directed by (Sir) C. W. Thomson [q.v.], 1872-6; chief assistant in 'Challenger Office', Edinburgh, 1876; director of office and editor of *Report on the Scientific Results of the Voyage of H.M.S. Challenger* (1880-95), 1882-95; explored Faroe Channel, 1880-2; carried out bathymetrical survey of Scottish freshwater lochs, 1897-1909; leased, explored, and financed scientific expeditions to Christmas Island; explored North Atlantic, 1910; FRS, 1896; KCB, 1898; works include *On the Origin and Structure of Coral Reefs and Islands* (1880), *Deep Sea Deposits* (1891), and *Depths of the Sea* (1912), the two latter in collaboration.

MURRAY, SIR JOHN (1851-1928), publisher; son of John Murray (1808-92, q.v.); entered father's publishing house, Albemarle Street, London; head of firm, 1892; edited *Quarterly Review*, 1922-8; KCVO, 1926; his works include editions of Gibbon's *Autobiography* (1897) and Byron's *Correspondence* (1922).

MURRAY, JOHN (1879-1964), educationist and politician; educated at Robert Gordon's College, Aberdeen, Aberdeen University, and Christ Church, Oxford; first class, lit. hum., 1905; prize fellow, Merton College, 1905; student and tutor, Christ Church, 1908-15; joined Labour Department, Ministry of Munitions, 1915; assistant commissioner, Labour Adviser's Department; coalition liberal MP, Leeds West, 1918-23; principal, University College of the South West, Exeter, 1926-51; hon. LLD, Aberdeen, 1930; Litt.D., Columbia, NY, 1939; D.Litt., Exeter, 1956; member, de la Warr Committee on higher education in East Africa, 1936-7; contributed to *Hibbert Journal* and *Contemporary Review*.

MURRAY, SIR (JOHN) HUBERT (PLUNKETT) (1861-1940), Australian administrator of Papua; born in Sydney; son of Sir T. A. Murray [q.v.]; first class, lit. hum., Magdalen College, Oxford, 1885; called to bar (Inner Temple), 1886; practised as a barrister in Australia; commanded New South Wales Irish Rifles in South African war; chief judicial officer, British New Guinea, 1904-40; lieutenant-governor, 1908-40; built up strong and enduring administration with native welfare as object, based upon an understanding of native mentality; insisted upon exclusion of Asiatic immigration; by land ordinance restricted purchase of land from natives to government; prescribed areas for the competing missions and placed education in their hands with government subsidy; undertook extensive patrols throughout country and insisted upon justice and mutual understanding; his government recognized as model administration of a native community and fully appreciated by Papuans; KCMG, 1925.

MURRAY, MARGARET ALICE (1863–1963), Egyptologist; born in Calcutta; educated privately in England and Germany; entered Calcutta general hospital as 'lady probationer', 1883; returned to England, 1886; entered University College, London, to study, Egyptology, under (Sir) Flinders Petrie [q.v.], 1894; junior lecturer, 1899; assistant; 1909; lecturer, 1921; senior lecturer and fellow, University College, 1922; assistant professor, 1924–35; D.Litt., 1931; assisted Petrie in excavation at Abydos, 1902; excavated in Malta, 1920–3, Minorca, 1930–1, and Petra, 1937; joined Petrie's expedition to ancient Gaza and concentrated on Hyksos cities; published over eighty books and articles on ancient Egypt, including *Ancient Egyptian Legends* (1904), *Elementary Egyptian Grammar* (1905), *Index of Names and Titles in the Old Kingdom* (1908), *Egyptian Temples* (1931), *The Splendour that was Egypt* (1949), and *The Genesis of Religion* (1963); interested in folklore and witchcraft; president, Folklore Society, 1953–5; fellow, Royal Anthropological Institute, 1926; published *The Witch-cult in Western Europe* (1921); and autobiography, *My First Hundred Years* (1963).

MURRAY, Sir OSWYN ALEXANDER RUTHVEN (1873–1936), civil servant; son of Sir J. A. H. Murray [q.v.]; educated at high school and Exeter College, Oxford; first class, *lit. hum.* (1895), jurisprudence (1896); entered Admiralty, 1897; as director of victualling and clothing (1905–11) made drastic reforms; assistant secretary, 1911–17; permanent secretary, 1917–36; influential in preserving adequate navy; KCB, 1917; GCB, 1931.

MURRY, JOHN MIDDLETON (1889–1957), author; educated at Christ's Hospital and Brasenose College, Oxford; wrote for *Westminster Gazette*, 1912–13; and then for *The Times Literary Supplement*; worked in political intelligence department of War Office, 1916; chief censor, 1919; OBE, 1920; married Katherine Mansfield [q.v.], 1918; editor of *Athenaeum*, 1919–21; founded the *Adelphi*; controlled it 1923–48; published works include studies of Shakespeare, Keats, Blake, D. H. Lawrence, and *Jonathan Swift* (1954), *Unprofessional Essays* (1956), and *Love, Freedom and Society* (1957).

MURRY, KATHLEEN (1888–1923), writer under the pseudonym of KATHERINE MANSFIELD; born in New Zealand; educated in England; married John Middleton Murry, the critic, 1918; died in France; her works include collections of short stories: *In a German Pension* (1911), *Prelude* (1918), *Je ne parle pas français* (1919), *Bliss* (1920), *The Garden Party* (1922), and posthumous volumes; *Poems* (1923), *Journal* (1927), *Letters* (1928).

MUSGRAVE, Sir JAMES, baronet (1826–1904), benefactor of Belfast; established firm of ironfounders in Belfast; chairman of Belfast harbour commission, 1887–1903; greatly improved harbour and constructed 'Musgrave Channel' and docks; founded Musgrave chair of pathology in Queen's College, Belfast, 1901; baronet, 1897.

MUYBRIDGE, EADWEARD (1830–1904), investigator of animal locomotion, whose original name was EDWARD JAMES MUGGERIDGE; born at Kingston-on-Thames; emigrated to America; as government director of photographic surveys, he proved that the trotting horse has at times all feet off the ground, 1872; published researches in *The Horse in Motion* (1878); invented the 'zoopraxiscope', 1881, which projected animated pictures on a screen; published further researches in *Animal Locomotion* (profusely illustrated, 1887, abridged as *Animals in Motion*, 1899) and other works; led way to invention of cinematograph; bequeathed £3,000, with his zoopraxiscope and lantern slides, to Kingston public library.

MYERS, CHARLES SAMUEL (1873–1946), psychologist; first class, natural sciences tripos, parts i and ii, Gonville and Caius College, Cambridge, 1893–5; MB, 1898; successively demonstrator, lecturer, and (1921–2) reader in experimental psychology, Cambridge; professor, King's College, London, 1906–9; consultant psychologist, British armies in France, 1916–18; with H. J. Welch founded National Institute of Industrial Psychology, 1921; editor, *British Journal of Psychology*, 1911–24; publications include *Text Book of Experimental Psychology* (1909) and *Mind and Work* (1920); FRS, 1915.

MYERS, ERNEST JAMES (1844–1921), poet and translator; son of Revd Frederic Myers and brother of F. W. H. Myers [qq.v.]; BA, Balliol College, Oxford; works include four volumes of verse (1877–1904), and prose translations of Pindar's *Odes* (1874) and of last eight books of *Iliad* (in collaboration, 1882); enthusiast for Greece.

MYERS, FREDERIC WILLIAM HENRY (1843–1901), poet and essayist; son of Revd

Frederic Myers [q.v.]; educated at Cheltenham College and Trinity College, Cambridge; BA, and fellow, 1864; classical lecturer, 1865; inspector of schools, 1872–1900; published several volumes of poems, including 'St. Paul', and 'The Renewal of Youth' (1867–1882); contributed monograph on Wordsworth to 'English Men of Letters' series (1881); also wrote on Shelley, and published *Essays Classical and Modern* (1883); became interested in spiritualism and helped to found the Society for Psychical Research, 1882; joint author of *Phantasms of the Living* (2 vols., 1886), and contributor to the Society's Proceedings on the 'Subliminal Self'.

MYERS, LEOPOLD HAMILTON (1881–1944), novelist; son of F. W. H. Myers [q.v.]; educated at Eton and Trinity College, Cambridge; published *The Orissers* (1923), *The Clio* (1925), and four philosophical novels set in India (1929–40) republished collectively as *The Near and the Far* (1943).

MYRDDIN-EVANS, Sir GUILDHAUME (1894–1964), civil servant; educated at Llandovery and Christ Church, Oxford; first class mathematical mods., 1914; served in South Wales Borderers, 1914–17; invalided from army, joined prime minister's secretariat, 1917; assistant secretary to the cabinet, 1919; assistant principal, Treasury, 1919–29; transferred to Ministry of Labour, 1929; head of International Labour Division, 1938–59; led most of the British delegations to international labour con-

ferences; representative of British government on governing body of ILO, 1945; chairman for three periods of office; first civil servant to be elected president of International Labour Conference, 1949; CB, 1945; KCMG, 1947; published (with (Sir) Thomas Chegwidden) *The Employment Exchange Service of Great Britain* (1934).

MYRES, Sir JOHN LINTON (1869–1954), archaeologist and historian; scholar of Winchester and New College, Oxford; first class, *lit. hum.*, 1892; fellow of Magdalen College, Oxford, 1892–5; Craven fellow, 1892; excavated in Crete and Cyprus; lecturer in classical archaeology, Oxford, 1895–1907; professor of Greek, Liverpool University, 1907–10; Wykeham professor in ancient history, Oxford, 1910–39; FBA, 1923; general secretary British Association, 1919–32; knighted, 1943; publications include *History of Rome* (1902), *Dawn of History* (1911), *Who were the Greeks?* (1930), *Herodotus, Father of History* (1953), and *Geographical History in Greek Lands* (1953).

MYSORE, Sir Shri KRISHNARAJA WADIYAR BAHADUR, Maharaja of (1884–1940); succeeded father as head of the state, 1895; invested with full administrative powers, 1902; introduced a system of popular government (1922) and later a system approaching constitutional monarchy; welcomed prospect of federal India; a strictly orthodox Hindu; GCSI, 1907; GBE, 1917.

N

NAIR, Sir CHETTUR SANKARAN (1857–1934), Indian jurist, etc. [See SANKARAN NAIR.]

NAIRNE, ALEXANDER (1863–1936), scholar, theologian, and mystic; educated at Haileybury and Jesus College, Cambridge; first class, classics and theology; fellow, 1887–93, 1917–32; deacon, 1887; priest, 1888; rector of Tewin, Hertfordshire, 1894–1912; canon of Chester Cathedral (1914–22), of St. George's chapel, Windsor (1921–36); professor of Hebrew and Old Testament exegesis, King's College, London, 1900–17; regius professor of divinity, Cambridge, 1922–32; publications include *The Epistle of Priesthood* (1913).

NAMIER, Sir LEWIS BERNSTEIN (1888–1960), historian; born in Poland of Jewish parents; educated at Lwow and Lausanne Universities, London School of Economics, and

Balliol College, Oxford; first class, modern history, 1911; naturalized, 1913; private in Royal Fusiliers, 1914; working in Foreign Office, 1915–20; tutor at Balliol, 1920–1; working at historical research, 1924–9; published *The Structure of Politics at the Accession of George III* (1929) and *England in the Age of the American Revolution* (1930); political secretary to the Jewish Agency for Palestine, 1929–31; professor of modern history, Manchester University, 1931–53; adviser to Chaim Weizmann [q.v.], 1939; political work with Jewish Agency, 1940–5; appointed to editorial board for a history of parliament, 1951–60; this work published in 3 vols, 1964; FBA, 1944; hon. fellow, Balliol, 1948; knighted, 1952.

NARBETH, JOHN HARPER (1863–1944), naval architect; trained at Royal Naval College, Greenwich; entered Royal Corps of Naval

Constructors, 1885; appointed to Admiralty, 1887; chief constructor, 1912; assistant director of construction, 1919-23; designed battleships of *King Edward VII*, *Lord Nelson*, and *Dreadnought* classes, minor war vessels, and aircraft carriers; CB, 1923.

NARES, SIR GEORGE STRONG (1831-1915), admiral and Arctic explorer; great-grandson of Sir George Nares [q.v.]; entered navy, 1845; captain of HMS *Challenger* dispatched by government to explore Southern oceans, 1872-4; led government Arctic expedition in *Alert* and *Discovery*, 1875-6; FRS, 1875; KCB, 1876; rear-admiral, 1887; vice-admiral, 1892.

NASH, PAUL (1889-1946), artist; educated at St. Paul's School and Slade School of Fine Art; advised by Sir William Richmond [q.v.] to 'go in for Nature'; water-colourist of great individuality, regarding surrealism as licence to paint as he chose within terms of modern practice; his paintings peopled by a quality of light and imagination which later developed into a fantastic tangible personality expressed by monoliths and monster trees; founded 'Unit One' group of imaginative painters, 1933; official artist in both world wars; illustrated books; designed scenery, fabrics, posters, etc.; paintings include 'The Menin Road' and others in Imperial War Museum, 'Totes Meer' (Tate Gallery), and Dymchurch beach series begun in 1921.

NASH, SIR WALTER (1882-1968), prime minister of New Zealand; born at Kidderminster; left school at eleven; worked in variety of jobs; emigrated to Wellington, New Zealand, 1909; worked as commercial traveller; joined New Zealand labour party, 1916; represented the party at Second Socialist International, Geneva, 1920; elected secretary of his party, 1922-32; elected to parliament, 1929; minister of finance and customs, 1935-40; minister of finance and deputy prime minister, 1940-9; minister to Washington, 1941; member of Pacific War Council and war cabinet in London; engaged on post-war planning; attended many international conferences; leader of opposition to government of (Sir) Sidney Holland [q.v.], 1950; prime minister, 1957-60; less impressive as prime minister than as minister of finance; continued as labour leader until 1963, and as MP until he died; PC, 1946; CH, 1959; GCMG, 1965; hon. doctorates from universities in England and New Zealand.

NATHAN, HARRY LOUIS, first BARON NATHAN (1889-1963), lawyer and public servant; educated at St. Paul's School, London; admitted as solicitor, 1913; served in France

with Royal Fusiliers and badly wounded, 1916; returned as junior partner to Herbert Oppenheimer; through Sir Alfred Mond (later, Lord Melchett, q.v.) became legal adviser to Zionist Organization and to Economic Board of Palestine, and companies in Palestine; liberal MP, North East Bethnal Green, 1929; supported 'national government', 1931; became independent liberal, 1933; joined labour party, 1934; labour MP, Central Wandsworth, 1937; baron, 1940; under-secretary of state, War Office, 1945; minister of civil aviation; PC, 1946; resigned 1948; chairman Charitable Trusts Committee, 1950; published (with A. R. Barrowclough) *Medical Negligence* (1957); chairman, Wolfson Foundation; president, Royal Geographical Society; chairman, Royal Society of Arts; chairman, governors of Westminster Hospital Group; chairman, executive, British Empire Cancer Campaign.

NATHAN, SIR MATTHEW (1862-1939), soldier and civil servant; gazetted to Royal Engineers, 1880; major, 1898; governor, Gold Coast (1900-3), Hong Kong (1903-7), Natal (1907-9); secretary, Post Office, 1909-11; chairman, Board of Inland Revenue, 1911-14; under-secretary for Ireland, 1914-16; secretary, Ministry of Pensions, 1916-19; governor of Queensland, 1920-5; KCMG, 1902; GCMG, 1908.

NAWANAGAR, MAHARAJA SHRI RANJITSINHJI VIBHAJI, MAHARAJA JAM SAHEB OF (1872-1933), Indian ruler and cricketer; educated at Rajkumar College, Rajkot, and Trinity College, Cambridge; played cricket for Cambridge University (1893), Sussex (1895-1904, captain, 1899-1903), England (1896-1904, except 1898); popularized India with British public which knew him as perhaps the greatest cricketer of his generation; installed as Jam Saheb, 1907; gained much revenue from the development of his ports; constructed irrigation works, rebuilt Jamnagar, and improved communications; threw resources of his state into the war of 1914-18 and served on the staff in France; active in formation of the Chamber of Princes (1921) becoming chancellor, 1932-3; KCSI, 1917; GBE, 1919; GCSI, 1923.

NEHRU, JAWAHARLAL (1889-1964), national leader and first prime minister of India; son of Pandit Motilal Nehru [q.v.]; educated at Harrow and Trinity College, Cambridge; read for the bar (Inner Temple), 1910-12; joined M. K. Gandhi's [q.v.] non-co-operation campaign, 1920; imprisoned with his father, Motilal, 1921; between Dec. 1921 and June 1945 spent almost nine years in prison; secretary, All-India Congress; left India with wife Kamala to seek medical treatment for her, 1926; came in touch with European socialists; succeeded Motilal as president of Congress, 1929; argued for complete independence for India; in gaol much of period of civil disobedience, 1930-5; attended discussions with Sir Stafford Cripps and Maulana Azad [qq.v.], 1942; after failure of talks joined 'Quit India' movement and returned to prison; on release in 1945 found M. A. Jinnah's [q.v.] movement for Pakistan a powerful force; vice-president of viceroy's council in Lord Wavell's [q.v.] provisional government, 1946; resigned to partition; became prime minister of independent India amid massacre and violence; appalled by assassination of Gandhi, 1948; brief war with Pakistan over Kashmir; India, a republic within the Commonwealth, 1950; Nehru used his influence to nurture parliamentary democracy and to achieve social reform; chairman, Planning Commission, 1950; introduced three Five Year Plans; advocated non-alignment in foreign policy; approved invasion of Portuguese Goa, 1961; during years 1959 to 1964 faced increasing difficulties both internally and externally; published *Glimpses of World History* (1934-5), *Autobiography* (1936), and *The Discovery of India* (1946).

NEHRU, PANDIT MOTILAL (1861-1931), Indian lawyer and Congress leader; educated at Muir College, Allahabad; advocate of High Court, Allahabad, 1895; president, Congress annual session, Amritsar, 1919; suspended practice to campaign for civil disobedience; entered Indian legislative assembly, 1923; president, Swaraj party, 1925; merged it (1926) with and became leader of Congress party; chairman, 'All Parties Conference' and its committee which outlined plan of dominion status, 1928.

NEIL, ROBERT ALEXANDER (1852-1901), classical and oriental scholar; graduate of Aberdeen University, 1870; hon. LLD, 1891; scholar of Peterhouse, Cambridge; Craven university scholar, 1875; second classic, 1876; fellow of Pembroke College and successful classical lecturer there from 1876 to death; read Sanskrit with E. B. Cowell [q.v.] and with him edited the *Divyāvadāna* (1886) and joined in translating the *Jātaka* (6 vols., 1897-1907); university lecturer on Sanskrit, 1883; edited Aristophanes' *Knights*, 1901; interested in ancient and medieval architecture.

NEIL, SAMUEL (1825-1901), author; rector of Moffat Academy, 1859-73; edited *Moffat Register* (1857) and *British Controversialist* (40 vols., 1850-73); published works on philosophy, rhetoric, and elocution as well as *Shakespeare: a Critical Biography* (1861) and *Home of Shakespeare described* (1871); prominent in educational

and philanthropic affairs in Edinburgh from 1873; helped to found Educational Institute of Scotland; edited *The Home Teacher* (6 vols., 1886).

NEILSON, GEORGE (1858-1923), historian and antiquary; procurator fiscal of police, Glasgow, 1891; stipendiary police magistrate of Glasgow, 1910; charter scholar and expert palaeographer; correspondent of F. W. Maitland [q.v.]; his works include *Trial by Combat* (1890), *Peel, its Meaning and Derivation* (1894), *Annals of the Solway* (1899), *John Barbour* (1900), and *Huchown of the Awle Ryale* (1902); co-editor of *Acta Dominorum Concilii, 1496-1501* (1918).

NEILSON, JULIA EMILIE (1868-1957), actress; educated in Wiesbaden and at the Royal Academy of Music; first stage appearance at Lyceum Theatre, 1888; married Fred Terry [q.v.], 1891; acted in plays with Lewis Waller, Rutland Barrington, (Sir) Herbert Beerbohm Tree, (Sir) John Hare, and (Sir) George Alexander [qq.v.], 1888-1900; in management and acting with husband, 1900-30, in series of successful productions, including *Sweet Nell of Old Drury*, *The Scarlet Pimpernel*, and *Henry of Navarre*; last stage appearance in *The Widow of Forty* (1944).

NELSON, ELIZA (1827-1908), dramatist and actor. [See under CRAVEN, HENRY THORNTON.]

NELSON, SIR FRANK (1883-1966), organizer of Special Operations Executive; educated at Bedford grammar school and Heidelberg; worked in mercantile firm in Bombay, and rose to be senior partner; served in Bombay Light Horse during 1914-18 war; chairman, Bombay Chamber of Commerce; its representative in Bombay legislative council, 1922-4; president, Associated Indian Chamber of Commerce, 1923; knighted, 1924; conservative MP, Stroud, 1924-31; employed on intelligence work in Basle, 1939; appointed by Hugh (later, Lord) Dalton [q.v.] to be head of SO2, one section of SOE, 1940; responsible for all SOE activities, 1941; built up espionage organization in Europe in spite of serious difficulties; established SOE and gained the confidence of the chiefs of staff; resigned due to failing health, 1942; KCMG; subsequently held appointments as air commodore in Washington and Germany.

NELSON, GEORGE HORATIO, first BARON NELSON OF STAFFORD (1887-1962), chairman of the English Electric Company Ltd.; educated under Silvanus Thompson [q.v.] at City and Guilds Technical College, London; Brush student; joined British Westinghouse Company,

Manchester; chief outside engineer, 1911; chief electrical superintendent, 1914; manager, Sheffield works of Metropolitan Vickers Electrical Co., 1920; managing director, English Electric, 1930; chairman and managing director, 1933; successful with production of bombers during 1939-45 war; served on many government committees and advisory councils; president, Federation of British Industries, 1943-4; served on number of academic bodies; president Institution of Electrical Engineers, 1955; and Institution of Mechanical Engineers, 1957-8; hon. fellow, Imperial College, London; hon. LLD, Manchester; knighted, 1943; baronet, 1955; baron, 1960.

NELSON, SIR HUGH MUIR (1835-1906), premier of Queensland; emigrated with father from Kilmarnock to Queensland, 1853; minister for railways, 1888-90, and treasurer, Oct. 1893-8; restored confidence after bank crisis of 1893; KCMG, 1896; hon. DCL, Oxford, and PC, 1897; president of legislative council, 1898; lieutenant-governor of Queensland; opposed Australian federation; died near Toowoomba.

NERUDA, WILMA MARIA FRANCISCA (1839-1911), violinist. [See HALLÉ, LADY.]

NESBIT, EDITH (1858-1924), writer of children's books, poet, and novelist. [See BLAND.]

NETTLESHIP, EDWARD (1845-1913), ophthalmic surgeon; brother of Henry, John Trivett, and Richard Lewis Nettleship [qq.v.]; student at Moorfields Eye Hospital; FRCS, 1870; ophthalmic surgeon to St. Thomas's Hospital and surgeon to Royal London Ophthalmic Hospital; FRS, 1912.

NETTLESHIP, JOHN TRIVETT (1841-1902), animal painter and author; brother of Edward, Henry, and Richard Lewis Nettleship [qq.v.]; studied art at Heatherley's; executed black-and-white sketches of biblical scenes; exhibited studies of lions, tigers, etc., at Royal Academy, 1874-1901; painted in India a cheetah hunt for Gaekwar of Baroda, 1880; early admirer of Browning; published *Essays on Robert Browning's Poetry* (1868, new edn. 1895) and a volume on George Morland (1898).

NEUBAUER, ADOLF (1832-1907), orientalist; born at Kottesó, Hungary; studied for rabbinate at Prague and Munich; studied medieval Jewish MSS at Paris, 1857-68; published *La Géographie du Talmud*, 1868 (crowned by the French Academy); catalogued Hebrew MSS in the Bodleian, Oxford, 1868-86; sub-librarian of Bodleian, 1873-99; obtained for library many items from the 'Genizah' (i.e. synagogue depository) at Cairo; edited from Bodleian and Rouen MSS Arabic text of Hebrew dictionary of Abu-'l-Walid (eleventh century), 1875; two volumes on French rabbis (1877, 1893) embodied much recondite research; reader in rabbinic Hebrew at Oxford University, 1884-1900; other works include *Medieval Jewish Chronicles* (2 vols., 1887-95); MA, Oxford, 1873; hon. fellow of Exeter College, 1890; hon. Ph.D., Heidelberg.

NEVILLE, HENRY (1837-1910), actor, whose full name was THOMAS HENRY GARTSIDE NEVILLE; first appeared in London at Lyceum, 1860; made reputation at Olympic Theatre, 1862-6; original Bob Brierley in *The Ticket of Leave Man* (1863), Job Armroyd in *Lost in London* (1867), and Jean Valjean in *The Yellow Passport*, his own version of *Les Misérables* (1868); successful lessee of Olympic, 1873-9; original George Kingsmith in H. A. Jones's *Saints and Sinners* (1884); thenceforth acted the romantic hero in melodrama; toured in America, 1890-1; taught acting and wrote about the stage.

NEVINSON, CHRISTOPHER RICHARD WYNNE (1889-1946), artist; son of H. W. Nevinson [q.v.]; educated at Uppingham; studied at St. John's Wood and Slade schools of art; official artist in both wars, painting views from aircraft and scenes of devastation; other works include town and river scenes from windows, landscapes, and portrait heads; technique mainly naturalistic, sometimes stiffened with semi-Cubist angles; ARA, 1939.

NEVINSON, HENRY WOODD (1856-1941), essayist, philanthropist, and journalist; educated at Shrewsbury and Christ Church, Oxford; second class, *lit. hum.*, 1879; studied literature and the army in Germany; *Daily Chronicle* correspondent in Greco-Turkish and South African wars; correspondent on western front and at Dardanelles, 1914-18; published *The Dardanelles Campaign* (1918); organized relief for Macedonians (1903), Albanians (1911); investigated slavery in Portuguese Angola and published *A Modern Slavery* (1906); crusaded also for women's suffrage and against 'Black and Tan' outrages; his 'middle' articles in the *Nation* collected in several volumes; three autobiographical volumes abridged as *Fire of Life* (1935).

NEWALL, DAME BERTHA SURTEES (1877-1932), better known as DAME BERTHA PHILLPOTTS, educationist and Scandinavian scholar; first class, medieval and modern languages tripos, Girton College, Cambridge, 1901; studied Scandinavian culture, 1901-13;

first Lady Carlisle fellow, Somerville College, Oxford, 1913; principal, Westfield College, 1919–21; mistress of Girton, 1922–5; university lecturer, 1926–32; married (1931) H. F. Newall [q.v.]; works include *Edda and Saga* (1931); DBE, 1929.

NEWALL, CYRIL LOUIS NORTON, first BARON NEWALL (1886–1963), marshal of the Royal Air Force; educated at Bedford School and Sandhurst; commissioned in Royal Warwickshire Regiment, 1905; transferred to King Edward's Own Gurkha Rifles, 1909; served on North-West frontier of India; learned to fly, 1911; joined Royal Flying Corps, 1914; served throughout 1914–18 war; awarded Albert medal, CMG, and CBE; joined Royal Air Force, 1919; commanded School of Technical Training, Halton, 1922–5; in charge of operations and intelligence, 1926–31; member of Air Council; commanded RAF in Middle East, 1931–4; air member of Air Council for supply and organization, 1935; chief of air staff, 1937–40; marshal of the Royal Air Force, 1940; governor-general, New Zealand, 1941–6; GCB, 1938; GCMG, 1940; OM, 1940; baron, 1946.

NEWALL, HUGH FRANK (1857–1944), astrophysicist; son of R. S. Newall [q.v.] constructor of 'Newall telescope'; educated at Rugby and Trinity College, Cambridge; junior optime, 1880; fellow, 1909; assistant master, Wellington College, 1881–5; demonstrator in experimental physics, Cambridge, 1886; offered funds for transfer of Newall telescope to Cambridge and own services as observer, 1889; professor of astrophysics, 1909–28; first director, Solar Physics Observatory at Cambridge, 1913–28; made frequent eclipse expeditions; FRS, 1902; married (1931) Dame Bertha Surtees Phillpotts [see above].

NEWBERRY, PERCY EDWARD (1869–1949), Egyptologist; educated at King's College School; excavated for Egypt Exploration Fund, 1890–5; surveyed Theban Necropolis, 1895–1901; Brunner professor of Egyptology, Liverpool, 1906–19; of ancient Egyptian history and archaeology, Cairo, 1929–33; notebooks and papers presented to Griffith Institute, Oxford.

NEWBOLD, SIR DOUGLAS (1894–1945), civil secretary to the Sudan Government; scholar of Uppingham and Oriel College, Oxford; entered Sudan political service, 1920; made three expeditions into Libyan desert; governor of Kordofan, 1932–9; civil secretary to Sudan government, 1939–45; policy of development towards self-government delayed through exigencies of war; prime mover in foundation of Khartoum University College; KBE, 1944.

NEWBOLT, SIR HENRY JOHN (1862–1938), poet and man of letters; educated at Clifton and Corpus Christi College, Oxford; called to bar (Lincoln's Inn) and practised, 1887–99; works include spirited poems of the sea such as 'Drake's Drum' (1896) and others classical and romantic in form such as 'Moonset' and 'Nightjar'; also fiction, criticism, and vols. iv and v of official naval history of 1914–18 war; expressed his faith in Christianity and English tradition; undertook much public service; knighted, 1915; CH, 1922.

NEWBOLT, WILLIAM CHARLES EDMUND (1844–1930), divine and preacher; BA, Pembroke College, Oxford; ordained deacon for parish of Wantage, 1868; worked under W. J. Butler [q.v.]; vicar of Dymock, Gloucestershire, 1870–77; of Malvern Link, 1877–88; principal of Ely theological college, 1887–90; canon of St. Paul's Cathedral, London, 1890–1930; an unswerving champion of tractarian orthodoxy; chiefly to be remembered for his work for clergy; conducted retreats and heard confessions of priests; a leading speaker in convocation.

NEWMAN, ERNEST (1868–1959), musical critic; real name, William Roberts; educated at Liverpool College and University College, Liverpool; clerk in Bank of Liverpool, 1889–1903; self-educated in music; published *Gluck and the Opera* (1895); on staff of Midland Institute school of music, 1903–5; music critic, *Manchester Guardian*, 1905–6, and *Birmingham Daily Post*, 1906–19; writing for the *Observer*, 1919; music critic, *Sunday Times*, 1920–58; published works on Wagner, Strauss, Elgar, Hugo Wolf, and Liszt, including *The Life of Richard Wagner* (4 vols., 1933–47); D.Litt., Exeter, 1959.

NEWMAN, SIR GEORGE (1870–1948), pioneer in public and child health; educated at Bootham School, King's College, London, and Edinburgh University; qualified, 1892; MD, and DPH, Cambridge, 1895; county medical officer, Bedfordshire, 1897–1900; medical officer, Finsbury, 1900–7; chief medical officer, Board of Education, 1907–19; of Ministry of Health, 1919–35; drafted scheme of school medical inspection and correlated it with other health services; promoted medical education; active also in Quaker interests; knighted, 1911; KCB, 1918; GBE, 1935.

NEWMAN, WILLIAM LAMBERT (1834–1923), scholar and philosopher; BA, Balliol College, Oxford; Hertford scholar, 1853; Ireland scholar, 1854; fellow of Balliol, 1854; through his lectures exercised a unique influence on teaching of history and political economy at Oxford; retired to Cheltenham, owing to ill health, 1870; published an edition of Aristotle (4 vols., 1887, 1902).

NEWMARCH, CHARLES HENRY (1824-1903), divine and author; wrote *Recollections of Rugby* (1848) and *Remains of Roman Art in Cirencester* (1850); BA, Corpus Christi College, Cambridge, 1855; rector of Belton, Rutland, 1856-93.

NEWNES, Sir GEORGE, first baronet (1851-1910), newspaper and magazine proprietor; at sixteen joined London firm of dealers in fancy goods; conceived and produced at Manchester successful weekly periodical, *Tit-Bits*, 1881; transferred publication to London, 1884; instituted prize competitions and insurance policies in connection with the paper; brought out *Strand Magazine*, 1891; started the *Westminster Gazette* as liberal organ, 1893; his firm became a limited company, 1891, and was reconstructed, 1897; liberal MP for Newmarket, 1885-95, and for Swansea, 1900-10; baronet, 1895; presented library buildings to Putney, 1897; fitted out Borchgrevinck's Norwegian South Polar expedition, 1898.

NEWSAM, Sir FRANK AUBREY (1893-1964), civil servant; educated at Harrison College, Barbados, and St. John's College, Oxford; commissioned in Royal Irish Regiment; served in Ireland and India during 1914-18 war; MC; joined home civil service and posted to Home Office, 1920; private secretary to Sir John Anderson (later Viscount Waverley, q.v.), 1925-8; principal private secretary to Sir William Joynson-Hicks (later Viscount Brentford), J. R. Clynes, Sir Herbert (later Viscount) Samuel, and Sir John Gilmour [qq.v.], 1928-33; assistant-secretary; CVO, 1933; responsible for preparation of Betting and Lotteries Act, 1934, and Public Order Act, 1936; assistant under-secretary in charge of security, 1940; deputy under-secretary, 1941; permanent secretary, Home Office, 1948-57; KBE, 1943; KCB, 1950; GCB, 1957.

NEWSHOLME, Sir ARTHUR (1857-1943), expert in public health; qualified from St. Thomas's Hospital, 1880; medical officer of health, Brighton, 1888-1908; made statistical study of diseases; medical officer, Local Government Board, 1908-19; laid foundations of national health services, dealing with tuberculosis, maternity and child welfare, and venereal diseases; KCB, 1917.

NEWTON, second BARON (1857-1942), diplomatist and politician. [See LEGH, THOMAS WODEHOUSE.]

NEWTON, ALFRED (1829-1907), zoologist; born at Geneva; BA, Magdalene College, Cambridge, 1853; made ornithological researches in Lapland (with John Wolley, ornithologist, 1853 and 1858), in West Indies (1857), and Spitsbergen (1864); first professor of zoology at Cambridge, 1866-1907; pioneer of zoological study there; FRS, 1870; wrote *Ornithology of Iceland* (1863); *Ootheca Wolleyana* (1863-1902); *Dictionary of Birds* (1893-6); edited *Ibis* from 1865 to 1870.

NEWTON, ERNEST (1856-1922), architect; entered office of R. N. Shaw [q.v.], 1873; began to work independently, 1879; mainly engaged on small suburban houses, 1879-89; his earliest important work, House of Retreat, Lloyd Square, Clerkenwell (1880); Bullers Wood, Chislehurst (1889), a marked advance on stylistic tendency of good work of time; designed many country houses; president, Royal Institute of British Architects, 1914; RA, 1919; CBE, 1920; serenity of his touch influenced contemporary English domestic architecture.

NICHOL SMITH, DAVID (1875-1962), scholar. [See SMITH.]

NICHOLS, ROBERT MALISE BOWYER (1893-1944), poet; educated at Winchester and Trinity College, Oxford; served in France, 1915; became soldier-poet with *Invocation* (1915) and *Ardours and Endurances* (1917); professor of English, Tokyo, 1921-4; published further poems in *Aurelia* (1920) and selected poems in *Such was my Singing* (1942); prose fiction *Fantastica* (1923) and *Under the Yew* (1928); dramas *Guilty Souls* (1922), *Twenty Below* (with J. Tully, 1927), and *Wings Over Europe* (with M. Browne, 1932).

NICHOLSON, Sir CHARLES, first baronet (1808-1903), chancellor of the university of Sydney, New South Wales; MD, Edinburgh University, 1833; settled in Sydney, 1834; speaker of New South Wales, 1847-56; president of legislative council of new colony, Queensland, 1860; helped to found Sydney University; obtained royal charter, 1858; chancellor, 1854-62; knighted, 1852; baronet, 1859; returned permanently to England, 1862; hon. DCL, Oxford, 1857, LLD, Cambridge, 1857, Edinburgh, 1886.

NICHOLSON, Sir CHARLES ARCHIBALD, second baronet, of Luddenham (1867-1949), architect; son of Sir Charles Nicholson [q.v.]; succeeded him, 1903; educated at Rugby and New College, Oxford; pupil of J. D. Sedding [q.v.]; set up practice, 1893; FRIBA, 1905; mainly employed in church architecture using English Gothic style with complete freedom.

NICHOLSON, CHARLES ERNEST (1868-1954), yacht designer; educated at Mill Hill School; joined family firm, Camper and Nicholsons, Ltd., 1886; chief designer, 1889-1954; chairman and managing-director; designed all types of sailing yachts, including four America's Cup challengers, notably *Shamrock IV* (1914) and *Endeavour* (1934); first designer to see possibilities of Bermuda rig; built wooden hulls for flying boats, and wooden cargo vessels, 1914-18; OBE, 1949; rare combination of artist, technical genius, and business man.

NICHOLSON, EDWARD WILLIAMS BYRON (1849-1912), scholar and librarian; BA, Trinity College, Oxford; librarian of London Institution, 1873-82; Bodley's librarian, Oxford, 1882-1912; thoroughly reorganized and greatly extended Bodleian library; interests included biblical criticism, Celtic antiquities, comparative philology, folklore, music, palaeography, numismatics, athletics; wrote on many of these subjects.

NICHOLSON, GEORGE (1847-1908), botanist; assistant at Kew, 1876; curator, 1886-1901; helped Sir Joseph Hooker [q.v.] to extend the arboretum at Kew, for which his own collection was purchased, 1889; his herbarium of British plants was presented to Aberdeen University; chief work was *The Dictionary of Gardening* (4 vols., 1885-9, 2 suppl. vols., 1900-1).

NICHOLSON, JOSEPH SHIELD (1850-1927), economist; BA, King's College, London, and Trinity College, Cambridge; professor of political economy and mercantile law, Edinburgh University, 1880-1925; maintained highest Scottish traditions; obtained position of exceptional authority on questions of imperial economic policy, of currency and banking; did much to bring academic life into touch with that of community; pioneer of economic history in Scotland; his main work, *Principles of Political Economy* (1893-1901); his numerous minor writings constitute guide to economic controversies of half-century ending 1925.

NICHOLSON, REYNOLD ALLEYNE (1868-1945), orientalist; son of H. A. Nicholson [q.v.]; educated at Aberdeen University and Trinity College, Cambridge; first class, classical tripos (part i) and Indian languages tripos; fellow, 1893; university lecturer in Persian, 1902-26; professor of Arabic, 1926-33; FBA, 1922; edited, translated, and interpreted the literature of Súfism, notably the *Mathnawí* of Rúmí (8 vols., 1925-40).

NICHOLSON, SIR SYDNEY HUGO (1875-1947), organist, church musician, and founder of the Royal School of Church Music; brother of Sir C. A. Nicholson [q.v.]; educated at Rugby, New College, Oxford, and Royal College of Music; B.Mus., Oxford, 1902; organist, Lower Chapel, Eton, 1903; acting organist, Carlisle Cathedral, 1904-8; organist, Manchester Cathedral, 1908-18, Westminster Abbey, 1918-28; founded School of English Church Music as affiliation of church choirs, 1928 (royal charter, 1945); and College of St. Nicolas as teaching institution of church music, 1929; knighted, 1938.

NICHOLSON, WILLIAM GUSTAVUS, BARON NICHOLSON (1845-1918), field-marshal; joined Royal Engineers from Woolwich, 1865; served almost continuously in India, 1871-99; KCB and adjutant-general in India, 1898; director of transport in South African war and engaged in operations in Orange Free State and Transvaal, 1900; director of military operations, War Office, 1901; chief British military attaché with Japanese army, 1904; quartermaster-general, 1905; general, 1906; chief of imperial general staff, 1908-12; GCB, 1908; field-marshal, 1911; baron, 1912.

NICHOLSON, SIR WILLIAM NEWZAM PRIOR (1872-1949), artist; educated at Magnus grammar school, Newark; studied under (Sir) Hubert von Herkomer [q.v.] and at Académie Julian; collaborated with brother-in-law James Pryde [q.v.] as 'Beggarstaff Brothers' in series of posters of revolutionary simplicity, 1894-6; executed notable woodcuts for William Heinemann [q.v.]; designed costumes for *Peter Pan* (1904) and other stage productions; later became known as painter of still life, landscape, and portraits; work revealed uncanny power to extract elegant beauty from the commonplace with exquisite refinement of subdued tone, pale, clear colour, and confident handling of paint; knighted, 1936.

NICKALLS, GUY (1866-1935), oarsman; rowed for Eton and (1887-91) for Oxford; out of 81 races at Henley won 67, including 5 Diamond sculls, 6 Goblets, 7 Stewards', 4 Grand Challenge cups, one Ladies' plate, and one Olympic eights (at age of forty-one); member of the stock exchange, 1891-1922.

NICOL, ERSKINE (1825-1904), painter; studied art at Trustees' Academy, Edinburgh; taught art at Dublin, 1846-50; settled in Edinburgh, 1850; ARSA, 1851; RSA, 1859; went to London, 1862; exhibited at Royal Academy, 1851-79; ARA, 1866; painter in oils and watercolours; depicted humours of Irish peasant life;

chief works include 'Irish Merry Making' (1856), 'Donnybrook Fair' (1859), 'Unwillingly to School' (1877).

NICOLL, Sir WILLIAM ROBERTSON (1851-1923), journalist and man of letters; MA, Aberdeen University, 1870; trained for Free Church ministry; ordained to first charge, Dufftown, Banffshire, 1874; moved to Kelso, 1877; resigned owing to ill health, 1885; moved to London; editor of *The Expositor*, 1885-1923; of *The British Weekly*, which he made very influential, 1886-1923; founder and editor of *The Bookman*, 1891-1923; knighted for services to liberal party, 1909.

NICOLSON, ADELA FLORENCE (1865-1904), poetess under the pseudonym of LAURENCE HOPE; settled in India; married Malcolm Hassels Nicolson [q.v.], 1889; chief work was *The Garden of Kama* (1901), lyrics showing oriental passion; committed suicide.

NICOLSON, Sir ARTHUR, eleventh baronet, and first BARON CARNOCK (1849-1928), diplomatist; entered Foreign Office, 1870; assistant private secretary to Lord Granville [q.v.], 1872; third secretary, Berlin, 1874; second secretary, Constantinople, 1879; secretary of legation, Tehran, 1885-8; consul-general, Budapest, 1888-93; secretary of embassy, Constantinople, 1893; minister, Tangier, 1895; ambassador, Madrid, 1904; British delegate, conference on Morocco, Algeciras, 1906; ambassador, St. Petersburg, 1906; under-secretary of state for foreign affairs, 1910-16; advocated policy of keeping Russia in *entente*; baron, 1916.

NICOLSON, Sir HAROLD GEORGE (1886-1968), diplomatist and author; son of (Sir) Arthur Nicolson (later Lord Carnock, q.v.); educated at Wellington College and Balliol College, Oxford; passed into diplomatic service, 1909; served in Madrid and Constantinople; married Vita Sackville-West [q.v.], 1913; served in London during 1914-18 war; with British delegation to Paris peace conference, 1919; CMG, 1920; posted to Tehran, 1925; critical of policy in Middle East; recalled to London, 1927; posted to Berlin; resigned from foreign service, 1929; contributed to 'Londoner's Diary' in *Evening Standard*, 1930-2; took over Sissinghurst Castle, Kent, where he and his wife created a celebrated garden, 1930; national labour MP, West Leicester, 1935-45; parliamentary secretary under Alfred Duff Cooper (later Viscount Norwich, q.v.), 1940-1; governor, BBC, 1941-6; bitterly opposed to Munich agreement, 1938; frequent broadcaster; contributed weekly book reviews to the *Observer*, 1949-53; KCVO, 1953; hon. fellow, Balliol

College; chairman or vice-chairman of several academic bodies; published many books including literary biographies of *Paul Verlaine* and others, 1921-6; *Some People* (1927), *Lord Carnock* (1930), *Public Faces* (1932); *Peacemaking 1919* (1933), *Curzon, the Last Phase* (1934), *Dwight Morrow* (1935), *Diplomacy* (1939), *King George V: His Life and Reign* (1952), *Good Behaviour* (1955), and *The Age of Reason* (1960); his diaries kept from 1930 to 1962 were edited by his son, *Diaries and Letters, 1930-62* (3 vols., 1966-8).

NICOLSON, MALCOLM HASSELS (1843-1904), general; entered Indian army, 1859; served in Abyssinian campaign, 1867-8, and in Afghan war, 1878-80; distinguished in Zhob valley campaign, 1890; C.B., 1891; lieutenant-general, 1899.

NICOLSON, VICTORIA MARY, LADY (1892-1962), writer and gardener. [See SACKVILLE-WEST.]

NIGHTINGALE, FLORENCE (1820-1910), reformer of hospital nursing; born in Florence; early engaged in cottage visiting and in nursing; visited hospitals in London and abroad from 1844 to 1855; toured in Egypt, 1849-50; inspected Roman Catholic schools and hospital at Alexandria; on way home she visited Kaiserswerth Institute for deaconesses and nurses, 1850; trained as sick nurse there, 1851; superintendent of Hospital for invalid Gentlewomen, Chandos Street, 1853; accepted invitation of Sidney Herbert [q.v.], secretary for war, to take out nurses to Crimea soon after outbreak of Crimean war; reached Scutari, 4 Nov. 1854; known there as 'The Lady-in-chief'; by powers of organization and discipline she revolutionized the conditions at the barrack hospital; established kitchen and laundry, and looked after soldiers' wives and children; christened by wounded patients 'The Lady with the Lamp'; by sanitary reforms she lessened the cases of cholera, typhus, and dysentery; visited Balaclava hospitals, where she was attacked with fever, May 1855; returned to Scutari, and provided reading and recreation rooms for the soldiers; on return to England she visited Queen Victoria at Balmoral, Sept. 1856; Nightingale School and Home for Nurses established as memorial of her services at St. Thomas's Hospital, 1860; published *Notes on ... Hospital Administration of the British Army* (1857); was consulted by foreign governments in American civil war, 1862-4, and Franco-German war, 1870-1; prominent in establishing nursing home in Liverpool Infirmary, 1862, the East London Nursing Society, 1868, the Workhouse Nursing Association, 1874, and the Queen's Jubilee Nursing Institute,

1890; familiar, through correspondence, with all phases of Indian social life, she powerfully advocated native education and village sanitation; her *Notes on Nursing* (1860) went through many editions; received Order of Merit, 1907, freedom of City of London, 1908, and honours from Germany, France, and Norway; raised nursing to an honoured vocation in England; a memorial tablet set up on her birthplace at Florence.

NIXON, Sir JOHN ECCLES (1857-1921), general; entered army from Sandhurst, 1875; served with 18th Bengal cavalry, 1878-1903; served on staff in Chitral relief force, 1895; in Tochi field force, 1897-8; served in South African war, 1901-2; CB, 1902; returned to India on conclusion of war; second-class district commander, 1903; inspector-general of cavalry, 1906; divisional commander, 1908; lieutenant-general, 1909; KCB, 1911; general-officer-commanding Southern army of India, 1912; general, 1914; transferred to command of Northern army, 1915; after entry of Turkey into 1914-18 war appointed to command Mesopotamian expedition, 1915; secured control of Basra vilayet; Kut occupied (Sept.); being satisfied of sufficiency of his forces, received permission from home government to advance on Baghdad (Oct.); failed in attempt to relieve (Sir) Charles V. F. Townshend [q.v.], who, after being defeated at Ctesiphon (Nov.), was besieged in Kut (Dec. 1915-Apr. 1916); relinquished command, Jan. 1916; summoned to appear before Mesopotamia commission of inquiry in England (Aug.); received largest share of blame for failure of expedition on account of his undue optimism, but his explanation accepted as satisfactory by Army Council, 1918; GCMG, 1919.

NOBLE, Sir ANDREW, first baronet (1831-1915), physicist and artillerist; served with Royal Artillery, 1849-60; secretary to committees formed to investigate new rifled breech-loading field gun advocated by (Lord) Armstrong [q.v.], 1858; partner in Armstrong's Elswick Ordnance Company, 1860; chairman of Armstrong, Whitworth & Co., 1900-15; KCB, 1893; baronet, 1902; carried out important experiments in gunnery and explosives to which were due exact science of ballistics and revolution in composition of gunpowder and design of guns.

NOBLE, MONTAGU ALFRED (1873-1940), Australian cricketer; in 39 matches against England (1897-1909) scored 1,905 runs and took 115 wickets; in a match on his first visit to England (1899) batted for 8½ hours; captained Australia on his fourth visit, 1909.

NOBLE, Sir PERCY LOCKHART HARNAM (1880-1955), admiral; born in India; entered *Britannia*, 1894; commanded destroyer *Ribble*, 1907-8; qualified as signal specialist, Portsmouth, 1908; captain, 1918; director, operations division, naval staff, 1928-9; rear-admiral, 1929; fourth sea lord and vice-admiral, 1935-7; in command, China station, 1937-40; CB, 1932; KCB, 1936; admiral, 1939; commander-in-chief, Western Approaches, 1941-2; laid firm and lasting foundation for success in anti-submarine war; head, British Admiralty delegation, Washington, 1942-4; GBE, 1944.

NODAL, JOHN HOWARD (1831-1909), journalist and writer on dialect; of Quaker parentage; sub-editor of *Manchester Courier*, 1864, and of *Manchester City News*, 1871-1904; on staff of *Saturday Review*, 1875-85; compiled *Glossary of the Lancashire Dialect* (with George Milner, 2 parts, 1875-82) and *Bibliography of Ackworth School* (1889).

NOEL-BUXTON, NOEL EDWARD, first BARON NOEL-BUXTON (1869-1948), politician and philanthropist; son of Sir T. F. Buxton [q.v.]; educated at Harrow and Trinity College, Cambridge; liberal MP, Whitby (1905-6), North Norfolk (1910-18); labour MP, North Norfolk (1922-30); advocated peace terms based on justice throughout both wars; minister of agriculture, 1924, 1929-30; president, Save the Children Fund, 1930-48; endowed Noel Buxton Trust for public and charitable purposes; baron, 1930.

NORFOLK, fifteenth DUKE OF (1847-1917). [See HOWARD, HENRY FITZALAN-.]

NORGATE, KATE (1853-1935), historian; much influenced by J. R. Green [q.v.]; works include *England under the Angevin Kings* (2 vols., 1887), *John Lackland* (1902), *The Minority of Henry the Third* (1912), and *Richard the Lion Heart* (1924).

NORMAN, CONOLLY (1853-1908), alienist; FRCS Ireland, 1876; FRCP Ireland, 1890; made special study of insanity; medical superintendent of the Richmond Asylum, Dublin, 1886-1908; introduced humane methods and advocated system of boarding out; hon. LLD, Dublin University, 1907.

NORMAN, Sir FRANCIS BOOTH (1830-1901), lieutenant-general; joined Bengal army, 1848; served in Indian Mutiny, in Bhutan and Hazara campaigns, 1868, and in Afghan war, 1878-80; present at battle of Kandahar, 1880; CB; commanded brigade in Burmese war, 1885-6; KCB, 1886; major-general, 1889.

NORMAN, SIR HENRY WYLIE (1826–1904), field-marshal and administrator; went out from London to his father, who was in business in Calcutta, 1842; joined Bengal army, 1844; took active part in Sikh war, 1848–9; served in Kohat pass expedition, 1850; conspicuous in actions round Delhi, and at Aligarh and Agra, 1857; proceeded with Sir Colin Campbell (later Baron Clyde, q.v.) to relief of Lucknow; present at battle of Cawnpore, Dec. 1857, and at capture of Lucknow, Mar. 1858; in campaign in Oudh, 1858–9; CB, 1859; ADC to Queen Victoria, 1863–9; major-general, 1869; first secretary to Indian government in military department, 1862–70; member of governor-general's council, 1870–7; opposed to forward policy of Lord Lytton [q.v.] and resigned, 1877; KCB, 1873; lieutenant-general, 1877; member of council of India, 1878–83; general, 1882; governor of Jamaica, 1883–9; tactfully settled constitutional crisis; GCMG and GCB, 1887; governor of Queensland, 1889–95; declined governor-generalship of India, 1893; president of royal commission on conditions of West India sugar-growing colonies, 1896; governor of Chelsea Hospital, 1901–4; field-marshal, 1902.

NORMAN, MONTAGU COLLET, BARON NORMAN (1871–1950), governor of the Bank of England; grandson of G. W. Norman [q.v.]; educated at Eton, King's College, Cambridge, in Germany and Switzerland; with Brown, Shipley & Co., merchant bankers, 1894–1915; served in South Africa, 1900–1; DSO; elected to court of Bank of England, 1907; entered service of the Bank, 1915; deputy governor, 1918; governor, 1920–44; with Stanley Baldwin negotiated settlement of American debt, 1923; advised return to gold standard, 1925; assisted reorganization of industry after 1931 crisis; completed transition of Bank from commercial to central banking, reorganized it, established its authority, and widened the conception of its responsibilities; strenuously engaged in war and post-war finance until overtaken by illness, 1944; PC, 1923; baron, 1944.

NORMAN-NERUDA, WILMA MARIA FRANCISCA (1839–1911), violinist. [See HALLÉ, LADY.]

NORMANBROOK, BARON (1902–1967), secretary of the cabinet. [See BROOK, NORMAN CRAVEN.]

NORMAND, WILFRED GUILD, BARON NORMAND (1884–1962), judge; educated at Fettes College, Oriel College, Oxford, and Edinburgh University; first class, *lit. hum.*, 1906; LLB, Edinburgh, 1910; admitted to Faculty of Advocates, 1910; served in Royal Engineers,

1915–18; KC, 1925; edited *Juridical Review*; solicitor general for Scotland, 1929 and 1931–3; unionist MP West Edinburgh, 1931; lord advocate, PC, 1933; succeeded Lord Clyde [q.v.] as lord justice general and lord president of the Court of Session, 1935–47; lord of appeal in ordinary, 1947–53; life peer, 1947; hon. bencher, Middle Temple, 1934; trustee, National Library of Scotland, 1925–62; trustee, British Museum, 1950–3; president of number of academic and legal bodies in Scotland; hon. LLD, Edinburgh; hon. fellow, Oriel College, Oxford, and University College, London.

NORTH, SIR DUDLEY BURTON NAPIER (1881–1961), admiral; entered *Britannia*, 1896; sub-lieutenant, 1902; lieutenant, 1903; took part in battle of Heligoland in battle-cruiser, *New Zealand*, 1914; and, as commander, in battles of Dogger Bank, 1915, and Jutland, 1916; captain, CMG, 1919; naval equerry to members of Royal family; MVO, 1919; CVO, 1920; CSI, 1922; commanded 'C' class cruisers, 1922–4; Royal tour with Prince of Wales, 1925; flag captain, 1926–7; director of operations, Admiralty, 1930; rear-admiral, 1932; chief of staff to Admiral Sir John Kelly [q.v.]; commander-in-chief, Home Fleet; commanded Royal yachts at Portsmouth, 1934; CB, 1935; vice-admiral, 1936; KCVO, 1937; flag officer commanding North Atlantic station and admiral superintendent, Gibraltar, 1939; admiral, 1940; reprimanded for criticizing bombardment of Oran, 1940; failed to intercept French squadron passing Gibraltar for Dakar, 1940; relieved in spite of protests; joined Home Guard; appointed flag officer, Yarmouth, 1942; appointed by George VI admiral commanding Royal yachts, 1945; GCVO, 1947; attempts to secure inquiry into dismissal failed, but Harold Macmillan, when prime minister, vindicated North's honour.

NORTH, JOHN DUDLEY (1893–1968), aircraft designer; educated at Bedford School; apprenticed to marine engineering; advised by Charles Grey Grey [q.v.] to transfer to Aeronautical Syndicate; joined Grahame-White company; chief engineer, 1912; designed and supervised construction of number of aeroplanes, 1912–15; superintendent, aeroplane division, Austin Motor Company, 1915; responsible for constructing RE7 and RE8 aircraft for RFC; joined Boulton & Paul Ltd., Norwich, 1917; designed successful fighter and bomber aircraft; also concerned with design of R101 airship; designed gun turrets for bombers; contributed number of scientific papers; member of council, Society of British Aircraft Constructors, 1931–62; served on council of Air Registration Board; hon. fellow, Royal Aeronautical Society; hon. D.Sc., Birmingham; CBE, 1962.

NORTHBROOK, first EARL OF (1826-1904), statesman. [See BARING, THOMAS GEORGE.]

NORTHCLIFFE, VISCOUNT (1865-1922), journalist and newspaper proprietor. [See HARMSWORTH, ALFRED CHARLES WILLIAM.]

NORTHCOTE, HENRY STAFFORD, BARON NORTHCOTE (1846-1911), governor-general of the Australian Commonwealth; son of first Earl of Iddesleigh [q.v.]; educated at Eton and Merton College, Oxford; BA, 1869; MA, 1873; clerk in Foreign Office, 1868; secretary to British members of the claims commission at Washington relating to the *Alabama*, 1871-3; private secretary to his father when chancellor of Exchequer, 1877-80; conservative MP for Exeter, 1880-99; financial secretary to War Office, 1885-6; surveyor-general of ordnance, 1886-7; CB, 1880; baronet, 1887; governor of Bombay, 1899-1903; baron and GCIE, 1900; efficiently dealt with plague and famine, and preserved the fine Gujarat breed of cattle; passed useful measures of land revenue reform, and District Municipalities Act; governor-general of Australian Commonwealth, 1903-7; GCMG, 1904; travelled widely through Commonwealth and encouraged immigration; PC, 1909.

NORTHCOTE, JAMES SPENCER (1821-1907), president of Oscott College and archaeologist; BA, Corpus Christi College, Oxford (first class classic), 1841; MA, 1844; ordained, 1844; joined Roman communion, 1846; in Italy studied Christian archaeology, 1847-50; editor of *Rambler*, 1852-4; ordained priest, 1855; canon of St. Chad's cathedral church, Birmingham, 1860; provost of cathedral chapter, 1884; president of Oscott College, 1861-77; remodelled studies on public school lines; chief work was *Roma Sotteranea* (1869), the standard English treatise on the catacombs.

NORTHUMBERLAND, eighth DUKE OF (1880-1930). [See PERCY, ALAN IAN.]

NORTON, first BARON (1814-1905), statesman. [See ADDERLEY, CHARLES BOWYER.]

NORTON, EDWARD FELIX (1884-1954), lieutenant-general; born in Argentina; educated at Charterhouse and Royal Military Academy, Woolwich; commissioned, 1902; in France with Royal Horse Artillery, 1914-18; DSO, MC; senior instructor, Staff College, Quetta, 1929-32; CB, 1939; acting governor and commander-in-chief, Hong Kong, 1940-1; retired with honorary rank of lieutenant-general, 1942; on second Mount Everest expedition, 1922; led 1924 expedition in which Mallory [q.v.] and

Irvine were killed; Founder's medal, Royal Geographical Society, 1926.

NORTON, JOHN (1823-1904), architect; pupil of Benjamin Ferrey [q.v.]; honorary secretary of Arundel Society, 1848-98; had large practice in domestic and ecclesiastical buildings; worked in Gothic style.

NORTON-GRIFFITHS, SIR JOHN, first baronet (1871-1930), engineer; engineer in South Africa; served in South African war, 1899-1902; in England founded firm of Griffiths & Co., public works contractors; conservative MP, Wednesbury (1910-18), Wandsworth Central (1918-24); organized and equipped second regiment of King Edward's Horse, Aug. 1914; organized, and served with, coal-miners and other underground workers for military purposes in 1914-18 war; sent on special mission to Romania, 1916; KCB, 1917; baronet, 1922; died near Alexandria.

NORWAY, NEVIL SHUTE (1899-1960), novelist and aeronautical engineer; educated at Shrewsbury and Balliol College, Oxford; chief-calculator, Airship Guarantee Company, 1924; worked on construction of R100; deputy chief engineer under (Sir) Barnes Wallis, 1929-30; founded and managed Airspeed, Ltd., 1931-8; served in Royal Naval Volunteer Reserve, 1939-43; war correspondent, 1944-5; settled in Australia, 1945; wrote many successful novels under name of Nevil Shute, including *No Highway* (1948), *A Town like Alice* (1950), and *On the Beach* (1957).

NORWICH, first VISCOUNT (1890-1954), politician, diplomatist, and author. [See COOPER, ALFRED DUFF.]

NORWOOD, SIR CYRIL (1875-1956), educationist; educated at Merchant Taylors' School and St. John's College, Oxford; first class, *lit. hum.*, 1898; headed civil service entry and posted to Admiralty, 1899; sixth-form master, Leeds grammar school, 1901-6; headmaster, Bristol grammar school, 1906-16; master, Marlborough College, 1916-26; headmaster, Harrow School, 1926-34; published *The English Tradition of Education* (1929); president, St. John's College, Oxford, 1934-46; chairman, committee on curriculum and examinations in secondary schools; Norwood report published 1943; many recommendations included in Education Act of 1944; knighted, 1938.

NOVAR, VISCOUNT (1860-1934), politician. [See FERGUSON, RONALD CRAUFORD MUNRO-.]

NOVELLO, CLARA ANASTASIA (1818–1908), oratorio and operatic prima donna; daughter of Vincent Novello [q.v.]; studied music in Paris; first appeared in public at Windsor, 1832; admired by Charles Lamb; praised by Mendelssohn and Schumann; able interpreter of Handel's solos; sang much in Germany and Italy, where she was immensely popular; original soprano soloist in Rossini's *Stabat Mater*; appeared in English opera, 1843; married Count Gigliucci, 1843; specially distinguished for rendering of 'I know that my Redeemer liveth' in Handel's *Messiah*; retired, 1860; lived in Rome and Fermo; published, *Reminiscences*, 1910.

NOVELLO, IVOR (1893–1951), actor-manager, dramatist, and composer; born, David Ivor Davies; changed to Ivor Novello by deed poll, 1927; educated at Magdalen College School, Oxford; published his first songs in his teens; achieved fame with 'Keep the Home Fires Burning', 1914; joined the Royal Naval Air Service and worked at the Air Ministry during 1914–18 war; acted in films and on the stage; became an actor-manager, 1924; wrote and produced numerous comedies, 1928–51, including *Glamorous Night* (1935), *Careless Rapture* (1936), *The Dancing Years* (1939), *Perchance to Dream* (1944), and *King's Rhapsody* (1949).

NOYCE, (CUTHBERT) WILFRID (FRANCIS) (1917–1962), mountaineer and writer; educated at Charterhouse and King's College, Cambridge; first class, part i, classical tripos, and in preliminary examination for part ii, modern languages tripos, 1939–40; commissioned in King's Royal Rifle Corps, 1941; served in India in intelligence and as chief instructor, Aircrew Mountain Centre, Kashmir, 1942–6; assistant master, Malvern College, 1946–50; master at Charterhouse, 1950–61; fine rock climber, climbed in English lake district, the Alps, and the Himalayas; member of Everest expedition, 1953; took part in expeditions in the Karakoram and Pamirs, 1960–2; died in climbing accident; published *Mountains and Men* (1947), an autobiography, *Michael Angelo, a Poem* (1953), *Poems* (1960), *South Col* (1954), *Climbing the Fish's Tail* (1958), and *To the Unknown Mountain* (1962); other books include a novel, *The Gods are Angry* (1957) and *They Survived* (1962).

NOYES, ALFRED (1880–1958), poet; educated in Aberystwyth and Exeter College, Oxford; published *The Loom of Years* (1902), *Drake* (2 vols., 1906–8), and *Forty Singing Seamen* (1907); professor, modern English literature, Princeton, 1914–23; CBE, 1918; published *The Torch-Bearers* (3 vols., 1922–30), *The Unknown God* (1934), and *Voltaire* (1936); gradually became blind, 1945–58; published autobiography, *Two Worlds for Memory* (1953); final collection of poems published, 1963; hon. degrees from Yale, Glasgow, Syracuse, and Berkeley, California.

NUFFIELD, VISCOUNT (1877–1963), industrialist and philanthropist. [See MORRIS, WILLIAM RICHARD.]

NUNBURNHOLME, first BARON (1833–1907), shipowner and politician. [See WILSON, CHARLES HENRY.]

NUNN, JOSHUA ARTHUR (1853–1908), colonel, army veterinary service; MRCVS, 1877; FRCVS, 1886; served as veterinary surgeon in Afghan war, 1879–80; reported on diseases of cattle for Punjab government, 1880–5; principal of Lahore veterinary school, 1890–6; CIE, 1895; principal veterinary officer in England (Eastern command), 1904–5, in South Africa, 1905–6, and India, 1906–7; CB, 1906; works include *Stable Management in India* (1896) and *Veterinary Toxicology* (1907).

NUNN, SIR (THOMAS) PERCY (1870–1944), educationist; B.Sc., London, 1890; BA, 1895; vice-principal, London Day Training College, 1905; director, 1922; transformed it into Institute of Education of London University, 1932; professor of education from 1913; retired, 1936; knighted, 1930.

NUTT, ALFRED TRÜBNER (1856–1910), publisher, folklorist, and Celtic scholar; became head of father's foreign bookselling and publishing firm in London, 1878; publisher of 'Tudor Library', 'Tudor Translations', and works by W. E. Henley [q.v.] and fairy-tale collections; founded the *Folklore Journal*; president of *Folklore Society*, 1897–8; helped to found Irish Texts Society, 1898; published *Studies on the Legend of the Holy Grail* (1888) and many papers on Celtic folklore.

NUTTALL, ENOS (1842–1916), bishop of Jamaica, primate and first archbishop of the West Indies; went to Jamaica as lay Wesleyan missionary, 1862; took Anglican orders, 1866; contributed important share to reorganization of disestablished Church of England in Jamaica; bishop of Jamaica, 1880; primate of West Indies, 1893; styled archbishop, 1897.

NUTTALL, GEORGE HENRY FALKINER (1862–1937), bacteriologist; born at San Francisco; educated in Europe; MD, California, 1884; studied in Germany; university lecturer in bacteriology and preventive medicine, Cambridge, 1900; Quick professor of biology, 1906-31; founded and edited the *Journal of Hygiene* (1901-37) and *Parasitology* (1908-33); founded study of humoral immunity; studied the micro-organism *Clostridium Welchii* and the part played by arthropods and ticks in spread of disease; FRS, 1904.

NYE, SIR ARCHIBALD EDWARD (1895-1967), lieutenant-general; educated at Duke of York's Royal Military School, Dover; enlisted in the ranks and served in France as non-commissioned officer, 1914; commissioned in Leinster Regiment, 1915; twice wounded; MC;

granted regular commission, Royal Warwickshire Regiment, 1922; qualified as barrister (Inner Temple), 1932; commanded 2nd battalion, Royal Warwickshires, 1937; major-general, director of staff duties, War Office, 1939; lieutenant-general, vice-chief, imperial general staff, under Sir Alan Brooke (later Viscount Alanbrooke, q.v.), 1941-5; governor, Madras, 1946-8; United Kingdom high commissioner, Delhi, 1948-52; CB, 1942; KBE, 1944; KCB, 1946; GCIE, 1946; GCSI, 1947; high commissioner to Canada, 1952-6; GCMG, 1951; chairman, committee to consider reorganisation of War Office, 1962—the Nye Committee.

O

OAKELEY, SIR HERBERT STANLEY (1830-1903), musical composer; son of Sir Herbert Oakeley, third baronet [q.v.]; BA, Christ Church, Oxford, 1853; MA, 1856; Reid professor of music in Edinburgh University, 1865-91; knighted, 1876; prolific composer of vocal and instrumental music; wrote hymn tunes 'Edina', 1862, and 'Abends', 1871.

OAKLEY, SIR JOHN HUBERT (1867-1946), surveyor; educated at Uppingham and Royal Agricultural College, Cirencester; entered family firm D. Smith, Son and Oakley, surveyors; partner, 1892; senior partner, 1898-1946; served frequently as an arbitrator, also on advisory committee on crown lands (1933-9) and other public bodies; knighted, 1919; GBE, 1928.

OATES, LAWRENCE EDWARD GRACE (1880-1912), Antarctic explorer; entered army, 1900; captain, 1906; joined Antarctic expedition led by Captain R. F. Scott [q.v.], 1910; member of sledging party which reached South Pole, Jan. 1912; on the return journey 'walked willingly to his death in a blizzard to try and save his comrades beset by hardships'.

O'BRIEN, CHARLOTTE GRACE (1845-1909), Irish author and social reformer; lived in Brussels with father, William Smith O'Brien [q.v.], 1854-6; wrote much on Irish politics; urged improved steamship accommodation for Irish emigrant girls; works include a novel, *Light and Shade* (1878), and *Lyrics* (1880), and articles on flora of Shannon district; *Selections* appeared in 1909.

O'BRIEN, CORNELIUS (1843-1906), Roman Catholic archbishop of Halifax, Nova

Scotia, from 1882; born in Prince Edward Island; studied for priesthood in Rome, 1864-71; rector of cathedral of Charlottetown, 1871-4; founded collegiate school in Halifax, the germ of the future university; president of Royal Society of Canada, 1896; published theological works and novels.

O'BRIEN, IGNATIUS JOHN, BARON SHANDON (1857-1930), lord chancellor of Ireland; called to Irish bar, 1881; his success ensured by defence of Canon Keller, parish priest of Youghal, 1887; QC, 1899; sympathized with liberal government; serjeant-at-law, 1910; solicitor-general and PC for Ireland, 1912; lord chancellor, 1913; baronet, 1916; ousted from chancellorship by conservative pressure, 1918; baron, 1918; moved to England; called to English bar (Middle Temple), 1923.

O'BRIEN, JAMES FRANCIS XAVIER (1828-1905), Irish politician; studied medicine in Paris; at New Orleans enlisted in William Walker's expedition to Nicaragua, 1856; met James Stephens [q.v.] at New Orleans and joined local Fenian organization, 1858; assistant surgeon in American civil war, 1861; arrested in Fenian rising near Cork, Mar. 1867; death sentence for treason commuted to penal servitude; released, Mar. 1869; nationalist MP for South Mayo, 1885-95; and for Cork City from 1895 till death; seceded from C. S. Parnell [q.v.], 1891; was general secretary of United Irish League of Great Britain.

O'BRIEN, PETER, BARON O'BRIEN (1842-1914), lord chief justice of Ireland; called to Irish bar, 1865; QC, 1880; serjeant, 1884; solicitor-general, 1887; attorney-general, 1888;

lord chief justice, 1889-1913; baronet, 1891; baron, 1900.

O'BRIEN, WILLIAM (1852-1928), Irish nationalist leader; editor of *United Ireland*, organ of Land League movement, 1881; imprisoned and journal suppressed, 1881-2; nationalist MP, Mallow, 1883; with John Dillon [q.v.] started 'plan of campaign', 1886; again imprisoned, 1887; MP, North-East Cork, 1887-92; arrested, but escaped to U.S.A., 1890; refused to succeed C. S. Parnell [q.v.] as leader, 1890; MP, Cork City, 1892; attempted to solve Irish problem by conciliation and union of all classes, creeds, and political parties; convened land conference, which led to Land Purchase Act (1903), 1902; left Irish parliamentary party; MP, Cork City, 1910; founded 'All for Ireland' League, supported by independent parliamentary party; retired with followers at general election, 1918.

O'CALLAGHAN, SIR FRANCIS LANGFORD (1839-1909), civil engineer; entered public works department of India, 1862; CIE, 1883; constructed railway through Bolan pass to Quetta, and made CSI, 1887; chief engineer and consulting engineer for state railways, 1889; secretary to public works department, 1892-4; KCMG, 1902; MICE, 1872.

O'CASEY, SEAN (1880-1964), Irish dramatist and author; born John Casey, of Protestant parents; suffered from eye disease, precluding formal education; worked as casual labourer until thirty years of age; involved in Irish politics; started serious play writing, 1916; encouraged by Lady Gregory and W. B. Yeats [qq.v.]; *The Shadow of a Gunman* produced at Abbey Theatre, Dublin, 1923; followed by *Juno and the Paycock* (1924), and *The Plough and the Stars* (1926); awarded Hawthornden prize for *Juno*, 1926; emigrated to London; anti-war play *The Silver Tassie* produced at the Apollo, 1929; *Within the Gates* at the Royalty, 1934; not understood by London audiences; *Red Roses for Me* produced at the Embassy, 1946, favourably received by the critics; *Drums of Father Ned* provided for Dublin theatre festival, 1958; but withdrawn by the author; autobiography (6 vols.) published 1939-54.

O'CONNOR, CHARLES YELVERTON (1843-1902), civil engineer, emigrated to New Zealand, 1865; under-secretary for public works, 1883-90; marine engineer for colony, 1890-1; engineer-in-chief to Western Australia, 1891-1902; constructed Fremantle harbour, 1892-1902, and Coolgardie water supply, 1898-1903; CMG, 1897; wrote reports on his chief undertakings.

O'CONNOR, JAMES (1836-1910), Irish journalist and politician; an early Fenian; imprisoned for sedition, 1865-70; sub-editor of *United Ireland*, 1881; imprisoned in Kilmainham, Dec. 1881; edited anti-Parnellite *Weekly Nationalist Press*, 1887; nationalist MP for West Wicklow, 1892-1910; published *Recollections of Richard Pigott* (1889).

O'CONNOR, THOMAS POWER (1848-1929), journalist and politician; BA, Queen's College, Galway; took up journalism; employed in London office of *New York Herald*; Parnellite MP, Galway, 1880; MP, Scotland division of Liverpool, 1885-1929; founded *T.P.'s Weekly*, 1902; as journalist, not publicist but observer and chronicler of life; as parliamentarian, subordinate member of Irish party which he served by ready pen; first president, Board of Film Censors, 1917.

O'CONOR, CHARLES OWEN, styled THE O'CONOR DON (1838-1906), Irish Roman Catholic politician; liberal MP for Roscommon County, 1860-80; frequently spoke on Irish education and land tenure; hon. LLD, Royal University of Ireland, 1892; mainly responsible for Irish Sunday Closing Act, 1879; active member of Bessborough land commission, 1880; chairman of royal commission on financial relations between Great Britain and Ireland, 1896; Irish PC, 1881; president of Royal Irish Academy and Irish Language Society.

O'CONOR, SIR NICHOLAS RODERICK (1843-1908), diplomatist; entered diplomatic service, 1866; served under Lord Lyons [q.v.] in Paris, 1877-83; secretary of legation at Peking, 1885; successfully negotiated difficulties with China in regard to temporary British occupation of Port Hamilton, 1885, and annexation of Upper Burma, 1886; CMG and CB, 1886; consul-general in Bulgaria, 1887; concluded commercial treaty between Great Britain and Bulgaria, 1889; envoy to Emperor of China at Peking, 1892; active in negotiations with powers after Chino-Japanese war, 1894-5; KCB, 1895; ambassador at St. Petersburg, 1895; attended coronation of Tsar Nicholas II, 1896; GCMG and PC, 1896; had disagreement with Russian foreign minister in regard to Russian lease from China of Port Arthur, 1897; GCB, 1897; transferred to Constantinople, 1898-1908; settled Turco-Egyptian boundary in Sinaitic peninsula, and British frontier in hinterland of Aden; died at Constantinople.

O'DOHERTY, KEVIN IZOD (1823-1905), Irish and Australian politician; contributed to *Nation*, 1848; transported for treason to Australia, 1849; pardoned, 1856; took medical degrees in Ireland, 1857 and 1859; settled in Brisbane; member of Queensland legislative assembly and of legislative council, 1877-85; nationalist MP for North Meath, 1885-8; returned to Brisbane, 1888, where he died in poverty.

O'DOHERTY, MARY ANNE (1826-1910), Irish poetess; born KELLY; wife of K. I. O'Doherty [q.v.]; as 'Eva' wrote Irish patriotic verse for *Nation* and other Irish papers; *Poems* published at San Francisco, 1877, *Selections*, 1908; died at Brisbane.

O'DONNELL, PATRICK (1856-1927), cardinal; educated at Catholic University, Maynooth; ordained priest, 1880; professor of theology, Maynooth, 1880; bishop of Raphoe, 1888-1922; archbishop of Armagh, 1924; cardinal, 1925; presided at national convention, 1900; supported policy of John Redmond [q.v.] during European war, 1914-18.

O'DWYER, SIR MICHAEL FRANCIS (1864-1940), Indian administrator; born in Ireland; posted to the Punjab in Indian civil service, 1885; revenue commissioner, North-West Frontier Province, 1901-8; lieutenant-governor, Punjab, 1913-19; averted trouble in early years of war by firm but sympathetic administration; approved action of General R. E. H. Dyer [q.v.] at Amritsar, 1919; believed in necessity of British control in India; shot by an Indian at meeting of Royal Central Asian Society in London; KCSI, 1913; GCIE, 1917.

OGDEN, CHARLES KAY (1889-1957), linguistic psychologist and originator of Basic English; educated at Rossall School and Magdalene College, Cambridge; first class, part i, classical tripos, 1910; founded *Cambridge Magazine*, 1912-22; became editor *Psyche*, 1922, and planned and edited *The History of Civilisation* and *The International Library of Psychology, Philosophy and Scientific Method* (1922-32); introduced Ludwig Wittgenstein [q.v.] to English readers in *Tractatus Logico-Philosophicus* (1922); published *Basic English* (1930), *The Basic* Vocabulary (1930), *Debabelization* (1931), and *The Basic* Words (1932); assigned copyright in Basic English to the crown and received compensation of £23,000, 1946; Basic English Foundation established, 1947.

OGILVIE, SIR FREDERICK WOLFF (1893-1949), economist and university vice-chancellor; educated at Clifton and Balliol College, Oxford; served in army, 1914-19; lecturer in economics, Trinity College, Oxford, 1919; fellow, 1920; professor of political economy, Edinburgh, 1926-34; published *The Tourist Movement* (1933); president and vice-chancellor, Queen's University, Belfast, 1934-8; director-general, British Broadcasting Corporation, 1938-42; undertook special war duties for British Council, 1943-5; principal, Jesus College, Oxford, 1944-9; knighted, 1942.

OGLE, JOHN WILLIAM (1824-1905), physician; BA, Trinity College, Oxford, 1847; MA and BM, 1851; DM, 1857; FRCP, 1855; vice-president, 1884; physician to St. George's Hospital, 1866-76; made special study of nervous diseases; founded *St. George's Hospital Reports*, 1866; contributed largely to medical journals.

O'HANLON, JOHN (1821-1905), Irish hagiographer and historical writer; emigrated to Quebec, 1842; ordained Roman Catholic missionary priest at Missouri, 1847; returned to Ireland, 1853; parish priest of St. Mary's, Irishtown, 1880-1905; canon, 1886; wrote *The Lives of the Irish Saints* (10 vols., 1875-1905), *Life and Scenery in Missouri* (1890), and poems.

O'HIGGINS, KEVIN CHRISTOPHER (1892-1927), Irish statesman; Sinn Fein MP, Queen's County, 1918; assistant minister for local government in revolutionary ministry, 1919; minister for economic affairs in provisional government of Arthur Griffith [q.v.], 1922; vice-president of executive council and minister for justice, 1922; established Civic Guard, 1922, new judiciary, 1924; aimed at free and undivided Ireland within British Commonwealth; assassinated near Dublin.

O'KELLY, SEAN THOMAS (1882-1966), president of Ireland; educated at the Christian Brothers' School and O'Connell Schools, Dublin; worked as boy-messenger at National Library, 1898-1902; full-time journalist, 1903; wrote for Arthur Griffith's [q.v.] newspaper, the *United Irishman*, and became manager of Gaelic weekly, *An Claidheamh Solais*, 1903; elected as Sinn Féin member, Dublin City Council, 1906-32; founder-member, Sinn Féin volunteers, 1913; took part in Easter rebellion, 1916; imprisoned with Eamon de Valera; Sinn Féin director of organization, 1918; successful candidate for College Green, Dublin, 1918; returned as deputy at every general election between 1918 and 1944; elected speaker of the Dáil; with De Valera opposed ratification of Anglo-Irish treaty, 1921; founder member of Fianna Fáil, 1926; deputy leader (with De Valera as leader); vice-president, Executive Council, minister for local government and public health, 1932-9; minister of finance, 1939-45; president, 1945-59; made state visits abroad; addressed joint session of Congress in USA, 1953; succeeded as president by De Valera, 1959; broadcast political reminiscences, 1959 (afterwards published by *Irish Press*); received many hon. degrees and honours both in Ireland and abroad.

OLDHAM, HENRY (1815-1902), obstetric physician; studied medicine at Guy's Hospital; MRCS, 1837; FRCP, 1857; MD, St. Andrews, 1858; physician-accoucheur and lecturer on midwifery at Guy's Hospital, 1849-69; president of Obstetrical Society of London, 1863-5. His nephew, CHARLES JAMES OLDHAM (1843-1907), ophthalmic surgeon at Brighton, was an educational benefactor.

OLDHAM, JOSEPH HOULDSWORTH (1874-1969), missionary; educated at Edinburgh Academy and Trinity College, Oxford; went to India to work for Scottish YMCA, 1897; invalided home, 1900; studied theology and missionary work at New College, Edinburgh and Halle University, Germany, 1901; ministerial assistant, Free St. George's, Edinburgh; organizing secretary, World Missionary Conference, 1910; Secretary, 'Edinburgh Continuation Committee'; editor, *International Review of Missions*, 1912-27; published *The World and the Gospel* (1916); in co-operation with Lord Archbishop Randall Davidson [q.v.], secured in peace treaty clause guaranteeing freedom for missions in former German colonies; secretary, International Missionary Council, 1921; with Sir Frederick (later Lord) Lugard [q.v.], drafted *Education Policy in British Tropical Africa* (Cmd. 2374, 1925); administrative director, International Institute of African Languages and Cultures, 1931-8; chairman, research committee for Universal Christian Council for Life and Work, 1934; edited the *Christian News Letter*, 1939-48; hon. DD Edinburgh and Oxford; CBE, 1951; publications include *The World and the Gospel* (1916), *Christianity and the Race Problem* (1924), *The New Christian Adventure* (1929), *Life is Commitment* (1953), and *New Hope in Africa* (1955).

O'LEARY, JOHN (1830-1907), Fenian leader; while law student of Trinity College, Dublin, he was imprisoned for share in attack on police near Clonmel, 1848; studied medicine at Queen's College, Galway, 1851-3; contributed to *Nation*; joined advanced Fenian movement, and went to America on its behalf, 1859; edited in Dublin Fenian weekly journal, *Irish People*, 1863; paper seized by government, and O'Leary with others was arrested, 1865; he spent nine years in prison; retired to Paris, 1874; returned to Dublin, 1885, where he was prominent in literary society; published *Recollections of Fenians and Fenianism* (1896).

OLIVER, DAVID THOMAS (1863-1947), lawyer; educated at Caterham School; called to bar (Middle Temple), 1888; LLB, Ireland (1888), London (1898), Cambridge (1902); law lecturer, Trinity Hall, Cambridge; fellow, 1920-47; lecturer, first for Gonville and Caius College, subsequently for university, in Roman-Dutch law, 1920-8.

OLIVER, FRANCIS WALL (1864-1951), palaeobotanist and ecologist; educated at Bootham School, York, and Trinity College, Cambridge; first class, parts i and ii, natural sciences tripos, 1885-6; lecturer, University College, London, 1888; Quain professor of botany, 1890-1929; studied fossil seeds, and, later, dynamic aspects of ecology, in particular, shingle beaches and salt-marsh vegetation; professor, Cairo University, 1929-35; studied desert vegetation.

OLIVER, FREDERICK SCOTT (1864-1934), man of business and publicist; educated at George Watson's College, Edinburgh University, and Trinity College, Cambridge; entered firm of Debenham & Freebody (1892) and made it a leading concern; books include *Alexander Hamilton* (1906), *Federalism and Home Rule* (1910), *Ordeal by Battle* (1915), and *The Endless Adventure* (3 vols., 1930-5).

OLIVER, SIR HENRY FRANCIS (1865-1965), admiral of the fleet; entered *Britannia*, 1878; qualified as navigator; served in cruisers; captain, 1903; established navigation training school; MVO, 1905; naval assistant to Sir John (later Lord) Fisher, first sea lord [q.v.], 1907; director, naval intelligence; rear-admiral, 1913; chief of Admiralty war staff; KCB, 1916; deputy chief of naval staff to Sir J. R. (later first Earl) Jellicoe [q.v.], 1917-18; KCMG, 1918; rear-admiral; vice-admiral and commander-in-chief Home Fleet, 1919; second sea lord, 1920; admiral, 1923; commanded Atlantic Fleet, 1924-7; admiral of the fleet, 1928-33; GCB, 1928; hon. LLD, Edinburgh, 1920.

OLIVER, SAMUEL PASFIELD (1838-1907), geographer and antiquary; joined Royal Artillery, 1859; served in China; toured in Japan, 1861; explored Madagascar, closely studying its history and ethnology, 1862-3; joined exploring expedition to Central America, 1867; made archaeological explorations in Brittany (1868), on the Mediterranean coasts (1872), and Cornwall (1873); resigned commission, 1878; FRGS, 1866; FSA, 1874; published *On and Off Duty* (1881), *Madagascar and the Malagasy* (1866), and *Madagascar* (2 vols., 1886); compiled abridged official accounts of

The Second Afghan War, 1878-80 (1908) and edited journals of early travel.

OLIVER, Sir THOMAS (1853-1942), physician and authority on industrial hygiene; educated at Ayr Academy and Glasgow University; MB, CM, 1874; practised in Newcastle from 1879; lecturer in physiology, Newcastle medical school, 1880; professor, 1889-1911; professor of medicine, 1911-27; investigated industrial diseases; president, Durham College of Medicine, 1926-34; vice-chancellor, Durham University, 1928-30; knighted, 1908.

OLIVIER, SYDNEY HALDANE, Baron OLIVIER (1859-1943), civil servant and statesman; educated at Tonbridge and Corpus Christi College, Oxford; entered Colonial Office, 1882; colonial secretary, Jamaica, 1900-4; governor, 1907-13; permanent secretary, Board of Agriculture, 1913-17; assistant comptroller and auditor of Exchequer, 1917-20; secretary for India, 1924; a socialist; secretary, Fabian Society, 1886-9; publications include *The Anatomy of African Misery* (1927) and *White Capital and Coloured Labour* (2nd edn. 1929); KCMG, 1907; PC and baron, 1924.

OLPHERTS, Sir WILLIAM (1822-1902), general; joined Bengal artillery, 1839; distinguished himself at Jhirna Ghaut, 1842, and Gwalior campaign, 1843; raised battery of horse artillery and marched across India to join Sir Charles Napier [q.v.] in Sind; commanded battery at Peshawar in expedition against frontier tribes, 1851; served in Crimean war, 1854, and throughout Indian Mutiny, 1857-9; distinguished for great bravery during relief of Lucknow; lieutenant-colonel, VC, and CB, 1858; in expedition against Waziris gained sobriquet of 'Hell-fire Jack', 1859-60; held commands at Peshawar, Gwalior, Ambala, and Lucknow, 1861-75; major-general, 1877; general, 1883; colonel commandant of Royal Artillery, 1888; KCB, 1886; GCB, 1900.

OLSSON, JULIUS (1864-1942), painter; self-trained; settled at St. Ives, painting chiefly moonlit seascapes and Cornish coast in stormy weather; ARA, 1914; RA, 1920.

OMAN, Sir CHARLES WILLIAM CHADWICK (1860-1946), historian; scholar of Winchester and New College, Oxford; first class, *lit. hum.* (1882), modern history (1883); fellow of All Souls, 1883; taught history at New College; Chichele professor of modern history, 1905-46; MP, Oxford University, 1919-35; publications include *A History of the Art of War in the Middle Ages* (1898, and 2 vols., 1924) and *History of the*

Peninsular War (7 vols., 1902-30); FBA, 1905; KBE, 1920.

OMAN, JOHN WOOD (1860-1939), divine; educated at Edinburgh University (first class honours in philosophy, 1882, and the Gray and Rhind scholarships) and the United Presbyterian Church College, Edinburgh; minister, Clayport Street church, Alnwick, 1889-1907; professor of systematic theology and apologetics, Westminster College, Cambridge, 1907-35; principal, 1922-35; moderator of the General Assembly of Presbyterian Church of England, 1931; FBA, 1938; works include *Vision and Authority* (1902), *Grace and Personality* (1917), and *The Natural and the Supernatural* (1931).

OMMANNEY, Sir ERASMUS (1814-1904), admiral; entered navy, 1826; took part in battle of Navarino, 1827; served under Captain James Clark Ross [q.v.] in Baffin's Bay, 1835; commander, 1840; captain, 1846; second-in-command of Franklin search expedition, 1850; organized system of sledge journeys; discovered first traces of Franklin's fate, Aug. 1850; received Arctic medal, 1851; during Crimean war blockaded Archangel, 1854; did blockade work in Gulf of Riga, 1855; in West Indies, 1857, and in Mediterranean, 1859; CB, 1867; FRS, 1868; vice-admiral, 1871; knighted, 1877; admiral, 1877; KCB, 1902.

OMMANNEY, GEORGE DRUCE WYNNE (1819-1902), theologian; brother of Sir E. Ommanney [q.v.]; BA, Trinity College, Cambridge, 1842; MA, 1845; vicar of Draycot, 1875-88; prebendary of Wells, 1884; voluminous and lucid writer on Athanasian creed.

ONIONS, CHARLES TALBUT (1873-1965), lexicographer and grammarian; educated at board school, Camp Hill branch of King Edward VI's foundation, Birmingham, and Mason College, Birmingham; BA (London), 1892; MA, 1895: invited to join staff of English Dictionary at Oxford by (Sir) J. A. H. Murray [q.v.], 1895; prepared portions of the work under supervision, 1906-13; independent editorial work, 1913; revised and completed work of William Little on *Shorter Oxford English Dictionary* (1933); completed editing of *Shakespeare's England* (1916); in naval intelligence division, Admiralty, 1918; Oxford university lecturer in English, 1920; reader in English philology, 1927-49; revised the *Anglo-Saxon Reader* (1922); fellow, Magdalen College, 1923; librarian, Magdalen, 1940-55; president, Philological Society, 1929-33; FBA, 1938; hon. degrees from Oxford, Leeds, and Birmingham; CBE, 1934; hon. director, Early English Text Society, 1945; editor, *Medium Aevum*, 1932-56;

concerned with *Oxford Dictionary of English Etymology*, 1945–65 (published 1966); contributed chapter on English language to *The Character of England* (1947), edited by Sir Ernest Barker [q.v.].

ONSLOW, WILLIAM HILLIER, fourth EARL OF ONSLOW (1853–1911), governor of New Zealand; educated at Eton and Exeter College, Oxford; conservative under-secretary of state for colonies, 1887; parliamentary secretary to Board of Trade, 1888; delegate to sugar bounties conference, 1887–8; KCMG, 1887; governor of New Zealand, 1889–92; GCMG, 1889; under-secretary of state for India, 1895–1900, and for colonies, 1900–3; joined cabinet as president of Board of Agriculture, and PC, 1903; chairman of committees in House of Lords, 1905–11; president of Royal Statistical Society, 1905–6; alderman of London County Council, 1896–9.

OPPÉ, ADOLPH PAUL (1878–1957), art historian and collector; educated at Charterhouse, St. Andrews, and New College, Oxford; first class, *lit. hum.*, 1901; lecturer, St. Andrews University, 1902: lecturer in ancient history, Edinburgh University, 1904; at Board of Education, 1905–38; deputy director, Victoria and Albert Museum, 1910–13; discovered drawings of Francis Towne [q.v.], 1910; published work on artists, including *Rowlandson* (1923), *Cotman* (1923), *Turner, Cox and De Wint* (1925), *Hogarth* (1948), and *Alexander and John Robert Cozens* (1952); CB, 1937; FBA, 1952; hon. LLD, Glasgow, 1953.

OPPENHEIM, EDWARD PHILLIPS (1866–1946), novelist; educated at Wyggeston grammar school, Leicester; produced steady output of romances of intrigue in high places frequently set in Monte Carlo; among his best are *The Kingdom of the Blind* (1917) and *The Great Impersonation* (1920).

OPPENHEIM, LASSA FRANCIS LAWRENCE (1858–1919), jurist; born near Frankfort on Main; settled in London, 1895; naturalized, 1900; Whewell professor of international law, Cambridge, 1908–19; chief work, *International Law: A Treatise* (1905–6).

OPPENHEIMER, SIR ERNEST (1880–1957), South African financier; born in Germany; junior clerk in London firm of Dunkelsbuhler & Co., 1896; naturalized, 1901; organized gold-mining enterprize, the Anglo-American Corporation of South Africa, 1917; extended his interests to diamond mining and became chairman of De Beers Company, 1929; established Diamond Corporation, 1930; went in to copper mining in Northern Rhodesia and set up Rhodesian Anglo-American, Ltd., 1928; opened up the Orange Free State goldfield and created Orange Free State Investment Trust, 1944; knighted, 1921; represented Kimberley in Union Parliament as supporter of J. C. Smuts [q.v.], 1924–38; hon. DCL, Oxford, 1952; generous benefactor to Queen Elizabeth House, Oxford, to medical research in South Africa, and to scientific research at Leeds University.

ORAGE, ALFRED RICHARD (1873–1934), editor, and exponent of social credit; keen Fabian; edited the *New Age* (1907–22) and raised it to high prestige; in 1923–30 raised funds in America for institute at Fontainebleau of Russian mystic Gurdjieff; founded the *New English Weekly* (1932) as social credit organ.

ORAM, SIR HENRY JOHN (1858–1939), engineer vice-admiral; entered navy, 1879; assistant engineer, Admiralty, 1884; chief engineer, 1889; staff-engineer, 1893; fleet-engineer, 1897; deputy engineer-in-chief, 1902; engineer-in-chief, 1907–17; KCB, 1910; FRS, 1912.

ORCHARDSON, SIR WILLIAM QUILLER (1832–1910), artist; fellow student with Tom Graham and John Pettie [qq.v.] at Trustees' Academy, Edinburgh, under Scott Lauder [q.v.]; exhibited at Royal Scottish Academy, from 1848; painted historical pictures; his 'Challenged' (1865) first brought him fame; exhibited Shakespearian subjects, 1865–74; ARA, 1868; visited Venice, 1870, painting Venetian subjects, 1870–4; exhibited at Academy 'The Queen of the Swords', 1877; RA, 1877; thenceforth his main subjects were either dramatic episodes of social life or incidents from careers of the great; they comprised 'The Social Eddy' (1878), 'Hard Hit' (1879), and 'Napoleon on Board the *Bellerophon*' (1880), the 'Marriage de Convenance' series (1884 and 1886), 'The First Cloud' (1887); he also won success as portrait painter, showing appreciation of character and subtle draughtsmanship; hon. RSA, 1871; hon. DCL, Oxford, 1890; knighted, 1907; memorial exhibition of his work at Royal Academy winter exhibition, 1911.

ORCZY, EMMA MAGDALENA ROSALIA MARIE JOSEPHA BARBARA, BARONESS ORCZY (1865–1947), novelist; born in Hungary; educated in Brussels and Paris; studied art in London; married (1894) Montague Barstow, illustrator; prolific writer of historical romances, notably *The Scarlet Pimpernel* (1905).

ORD, BERNHARD (BORIS) (1897–1961), musician; educated at Clifton College, 1907–14;

ARCO; won scholarship to Royal College of Music; studied organ under Sir Walter Parratt [q.v.]; war service in Artists' Rifles and RFC, 1916-18; organ scholar, Corpus Christi College, Cambridge, 1919; founded University Madrigal Society, 1920; Mus.B., 1922; fellow, King's College, 1923; freelance musician as player on piano or harpsichord, or as conductor; worked at Cologne opera house, 1927; succeeded A. H. Mann [q.v.] as organist of King's College and the university, 1929; university lecturer in music, 1936; conductor, Cambridge University Musical Society, 1938-54; served in RAF, 1941-6; returned to Cambridge, and conducted concerts until disseminated sclerosis forced retirement, 1954; resigned university lectureship, 1957-8; CBE, 1958; hon. Mus.D., Durham (1955) and Cambridge (1960).

ORD, WILLIAM MILLER (1834-1902), physician; studied at St. Thomas's Hospital medical school, 1852; MB, London, 1857; MD, 1877; FRCP, 1875; dean of medical school, 1876-87; physician, 1877-98; elucidated disease now known as myxoedema; gave Bradshaw lecture on subject at Royal College of Physicians, 1898; president of Medical Society of London, 1885; contributed much to medical literature; a foremost clinical teacher.

ORDE, CUTHBERT JULIAN (1888-1968), painter; educated at Framlingham College; served in RFC and RAF, 1914-19; exhibited at the Royal Academy, Royal Society of Portrait Painters, and Royal Institute of Oil Painters; painted many portraits including those of Sir Winston Churchill and air chief marshal, Sir John Salmond [q.v.]; undertook war paintings, 1939-45; book of drawings of pilots of Fighter Command published, 1942; portraits carried out for Royal Aeronautical Society, Senior United Service Club, and other military and flying organizations.

O'RELL, MAX (pseudonym) (1848-1903), humorous writer. [See BLOUET, LÉON PAUL.]

ORMEROD, ELEANOR ANNE (1828-1901), economic entomologist; daughter of George Ormerod [q.v.]; helped Royal Horticultural Society in forming collection of economic entomology, 1868; first woman fellow of Meteorological Society, 1878; published twenty-four *Annual Reports of Observations of Injurious Insects*, 1877-1900; led useful crusade against insect pests; consulting entomologist to Royal Agricultural Society, 1882-92; lecturer at Royal Agricultural College, Cirencester, 1881-4; hon. LLD, Edinburgh, 1900; her publications include *Guide to the Methods of Insect Life* (1884),

A Text Book of Agricultural Entomology (1892), and *Handbook of Insects Injurious to Orchard and Bush Fruits* (1898); benefactor to Edinburgh University; *Autobiography* published, 1904.

ORMSBY-GORE, WILLIAM GEORGE ARTHUR, fourth BARON HARLECH (1885-1964), politician and banker; educated at Eton and New College, Oxford; unionist MP, Denbigh, 1910-18, and Stafford, 1918-38; commissioned in Shropshire Yeomany, 1908; on active service in Egypt, 1914-16; intelligence officer, Arab Bureau, 1916-17; parliamentary private secretary to Lord Milner [q.v.], 1917; assistant secretary to the cabinet, assisting Sir Mark Sykes [q.v.]; friend of Chaim Weizmann [q.v.]; British liaison officer with Zionist Mission to Palestine, 1918; member, British delegation to Paris peace conference, 1918; member, Permanent Mandates Commission, 1921-2; accompanied E. F. L. Wood (later Earl Halifax, q.v.) to West Indies, 1921-2; parliamentary under-secretary, Colonial Office, 1922; PC, 1927; postmaster-general, 1931; first commissioner of works, 1931-6; colonial secretary, 1936; resigned when he succeeded to peerage, 1938, being outspoken critic of Nazi Germany; commissioner for Civil Defence, north-east region, 1939-40; British high commissioner, South Africa, 1941-4; director, Midland Bank, 1944; chairman, 1952-7; GCMG, 1938; KG, 1948; lord lieutenant, Merionethshire, 1938-57; trustee, National Gallery, Tate, and British Museum; president, National Museum of Wales and National Library of Wales; hon. fellow, New College, 1936; pro-chancellor, university of Wales, 1945-57; published *Florentine Sculptors of the Fifteenth Century* (1930), and wrote four volumes in the series, *Guide to the Ancient Monuments of England* (1935, 1936, and 1948).

ORPEN, SIR WILLIAM NEWENHAM MONTAGUE (1878-1931), painter; born in county Dublin; studied at the Metropolitan School of Art, Dublin and the Slade School, London; painted remarkable series of interiors with small figures; also half-length figures in the open air and fanciful self-portraits; official war artist, 1917-19; his groups include 'Hommage à Manet' (Manchester) and paintings of the peace conference, 1919; a most fashionable portrait painter of high professional skill; ARA, 1910; RA, 1919; KBE, 1918.

ORR, ALEXANDRA SUTHERLAND (1828-1903), biographer of Robert Browning [q.v.]; sister of Lord Leighton [q.v.]; married Sutherland George Gordon Orr, 1857; in India during Mutiny; settled in London, 1869; became intimate with Browning, whom she first met in Paris in 1855; generously supported Browning Society, 1881; published *Handbook to the Works of Robert Browning*, 1885 (6th edn. 1892) and his *Life and Letters*, 1891 (revised edn. 1908).

ORR, WILLIAM McFADDEN (1866-1934), mathematician; born in county Down; educated at Queen's College, Belfast, and St. John's College, Cambridge; senior wrangler, 1888; professor of applied mathematics, Royal College of Science, Dublin, 1891-1926, and University College, Dublin, 1926-33; best known for his work on the stability of the steady motions of a liquid; FRS, 1909.

ORTON, CHARLES WILLIAM PREVITÉ- (1877-1947), historian and editor. [See PREVITÉ-ORTON.]

ORTON, JOHN KINGSLEY (JOE) (1933-67), playwright; educated at Marriots Road school, Leicester; took secretarial course, Clark's College, Leicester; won scholarship to RADA, 1953; unsuccessful as actor; influenced by Kenneth Leith Halliwell; they lived together, and together wrote series of novels, including *The Silver Bucket* (1958) and *The Boy Hairdresser* (1956); both sent to prison for six months for defacing public library books, 1962; Orton's radio play *The Ruffian on the Stair* broadcast, 1964; wrote stage plays *Entertaining Mr. Sloane* (1963, produced in 1964), *Loot* (1964-5, produced in 1965), and *What the Butler Saw* (1967, produced in 1969); also wrote one-act plays and film scripts; films made of *Sloane* (1969) and *Loot* (1970); murdered by Halliwell, who then committed suicide, 1967.

ORWELL, GEORGE (pseudonym) (1903-1950), author. [See BLAIR, ERIC ARTHUR.]

ORWIN, CHARLES STEWART (1876-1955), agricultural economist; educated at Dulwich College and the South Eastern Agricultural College, Wye; worked as land agent until 1903 when he became lecturer at Wye College; land agent in Lincolnshire, 1906; first director of Oxford Institute for Research in Agricultural Economics, 1913-45; estates bursar, Balliol College, 1926-46; hon. fellow, 1946; first D.Litt. in Oxford School of social studies, 1939; editor, *Journal* of the Royal Agricultural Society, 1912-27; member, first Agricultural Wages Board, 1917-21; published many books on agricultural subjects, including *The Future of Farming* (1930) and *A History of English Farming* (1949).

OSBORNE, WALTER FREDERICK (1859-1903), painter; studied art at Royal

Hibernian Academy Schools and Antwerp; painted field and street life in England, Ireland, and Brittany; regularly exhibited at Royal Hibernian and Royal academies; made sketches in Spain, 1895, and Holland, 1896; RHA, 1886; works include 'A Cottage Garden' (in National Gallery of Ireland), 'Life in the Streets', 1902 (Tate Gallery), and many good portraits.

O'SHEA, JOHN AUGUSTUS (1839-1905), Irish journalist; educated at Catholic University, Dublin; Paris correspondent of *Irishman*, conducted by Richard Pigott [q.v.]; special correspondent to *Standard*, 1869-94; reported siege of Paris, Carlist war, and Bengal famine; published experiences in *An Iron-bound City* (2 vols., 1886), *Roundabout Recollections* (2 vols., 1892), and other works.

O'SHEA, WILLIAM HENRY (1840-1905), Irish politician; married Katharine Page Wood, 1867; made acquaintance of C. S. Parnell [q.v.], 1880; home rule MP for county Clare, 1880-5; attempted compromise between Parnell and liberal leaders, 1882-4; was returned through Parnell's influence MP for Galway, 1886; obtained (Nov. 1890) divorce from wife, who married Parnell, the co-respondent, June 1891.

OSLER, ABRAHAM FOLLETT (1808-1903), meteorologist; managed from 1831 father's Birmingham glass business; made first self-recording pressure plate anemometer and rain gauge, 1835; invention adopted at Greenwich (1841) and elsewhere; applied self-recording methods to cup anemometer for measuring horizontal motion of the air; examined horary and diurnal variations of the wind; showed relation of atmospheric disturbances to the great trade winds; set up in Birmingham standard clock with transit instrument and astronomical clock, 1842; devised instrument for craniometry; FRS, 1855; benefactor to Birmingham University.

OSLER, SIR WILLIAM, baronet (1849-1919), regius professor of medicine at Oxford; nephew of Edward Osler [q.v.]; born at Bond Heath, Ontario; studied medicine at universities of Toronto and McGill, Montreal; professor of institutes of medicine, McGill, 1874-84; professor of medicine, university of Pennsylvania (Philadelphia), 1884-9; professor of medicine, Johns Hopkins University, Baltimore, 1889-1904; built up there first organized clinical unit in any Anglo-Saxon country; combined best features of English and German systems; FRCP, 1884; FRS, 1898; regius professor of medicine at Oxford, 1904-19; baronet, 1911; principal work, *The Principles and Practice of Medicine*

(1891, 9th edn. 1920); a great teacher and man of wide sympathies.

O'SULLIVAN, CORNELIUS (1841-1907), brewers' chemist; studied at Royal School of Mines; student assistant to Prof. A. W. von Hofmann, chemist; became head of analytical staff of Messrs. Bass & Co., Burton-on-Trent; voluminous writings on technology of brewing include 'On Maltose' (1876) and 'Presence of Raffinose in Barley' (1886); elucidated distinct character of maltose; FRS, 1885.

OTTÉ, ELISE (1818-1903), scholar and historian; born at Copenhagen; stepdaughter of Benjamin Thorpe [q.v.], who taught her Anglo-Saxon and Icelandic; published *History of Scandinavia* (1874) and Danish and Swedish grammars and translations.

OTTLEY, SIR CHARLES LANGDALE (1858-1932), rear-admiral; entered navy, 1871; qualified as a torpedo officer; devised the automatic mooring gear for submarine mines which bears his name; captain, 1899; served as naval attaché, 1899-1904; director of naval intelligence, 1905-7; principal naval delegate at Hague peace conference, 1907; secretary, Committee of Imperial Defence, 1907-12; KCMG, 1907.

OUIDA (pseudonym) (1839-1908), novelist. [See DE LA RAMÉE, MARIE LOUISE.]

OULESS, WALTER WILLIAM (1848-1933), painter; educated at Victoria College, Jersey; studied at the Royal Academy Schools; exhibited regularly at the Academy, 1869-1928; quickly recognized as one of the most trustworthy portrait painters of the day; sitters included Charles Darwin (1875), John Bright (1879), Cardinal Newman (1881), King Edward VII (1900) and King George V (1905), as Prince of Wales, and Thomas Hardy (1922); ARA, 1877; RA, 1881.

OVERTON, JOHN HENRY (1835-1903), canon of Peterborough and church historian; educated at Rugby and Lincoln College, Oxford; BA, 1858; MA, 1860; prebendary of Lincoln Cathedral, 1879; rector of Epworth, 1883-98, and of Gumley from 1898 till death; hon. DD, Edinburgh, 1889; canon of Peterborough, 1903; voluminous writings include *The English Church in the 18th Century* (with C. J. Abbey, 2 vols., 1878), *The Evangelical Revival in the 18th Century* (1886), and biographies of William Law (1881), John Wesley (1891), and *The Nonjurors* (1902).

OVERTOUN, BARON (1843-1908), Scottish churchman and philanthropist. [See WHITE, JOHN CAMPBELL.]

OWEN, SIR (ARTHUR) DAVID (KEMP) (1904-1970), international civil servant; educated at Leeds grammar school and Leeds University; assistant lecturer in economics, Huddersfield Technical College; director, social survey committee, Sheffield, 1929-33; secretary, Civic Research Division, Political and Economic Planning (PEP), 1933-6; co-director, Pilgrim Trust Unemployment Enquiry, 1936-7; Stevenson lecturer in citizenship, Glasgow University, 1937-40; general secretary, PEP, 1940-1; personal assistant to Sir R. Stafford Cripps [q.v.], 1941-4; in charge of League of Nations affairs, Foreign Office, 1944-5; member, UK delegation to UN conference, San Francisco, 1945; deputy executive secretary, preparatory commission, United Nations, 1945-6; assistant secretary-general, in charge of economic affairs, UN, 1946; executive chairman, Technical Assistance Board, 1951-66; co-administrator, UN Development Programme, 1966-9; secretary-general, International Planned Parenthood Federation, 1969-70; KCMG, 1970; hon. doctorates from Leeds (1954) and Wales (1969).

OWEN, JOHN (1854-1926), bishop of St. David's; BA, Jesus College, Oxford; educated as Calvinistic Methodist, but joined English Church, 1879; professor of Welsh and lecturer in classics, St. David's College, Lampeter, 1879; ordained deacon, 1879; priest, 1880; warden and headmaster, Llandovery College, 1885-9; dean of St. Asaph, 1889; took part in 'tithe war', 1890; principal of St. David's College, Lampeter, 1892; active in establishing university of Wales and in drafting its charter; secured new charter for Lampeter; bishop of St. David's, 1897-1926; greatly improved conditions in large, poor, and difficult diocese; largely instigated passing of parliamentary default bill, which resulted in county councils accepting Education Act of 1902; laboured incessantly in opposition to disestablishment of Welsh Church, 1906-14; on accomplishment of disestablishment accepted situation courageously and helped to refashion Church's constitution, 1915-19; active in procuring passing of Welsh Church Temporalities Act, which greatly mitigated severity of original Act, 1919; conducted proceedings in parish church of Llandrindod when bishop of St. Asaph was elected first archbishop and metropolitan of Wales; devoted himself to building up Church in Wales, 1920-6; reluctantly consented to division of his diocese by creation of diocese of Swansea and Brecon, 1923; endowed with exceptional energy and burning enthusiasm.

OWEN, ROBERT (1820–1902), theologian; BA, Jesus College, Oxford, 1842; MA, 1845; BD, 1852; fellow, 1845–64; accepted tractarian view of catholic Church of England; advocated its disestablishment and disendowment; published *An Introduction to the Study of Dogmatic Theology* (1858), *Institutes of Canon Law* (1884), and *The Kymry* (1891).

OXFORD AND ASQUITH, COUNTESS OF (1864–1945). [See ASQUITH, EMMA ALICE MARGARET.]

OXFORD AND ASQUITH, first EARL OF (1852–1928), statesman. [See ASQUITH, HERBERT HENRY.]

P

PAGE, SIR ARCHIBALD (1875–1949), engineer and electricity supply administrator; trained in Edinburgh and Glasgow; entered Glasgow corporation electricity department, 1899; general manager, Clyde Valley Electric Power Company, 1917–20; electricity commissioner, 1920–5; general manager, County of London Electric Supply Corporation, 1925–7; Central Electricity Board, 1927–35; chairman, 1935–44; directed 'grid' construction; knighted, 1930.

PAGE, SIR FREDERICK HANDLEY (1885–1962), aircraft designer and manufacturer; educated at Cheltenham grammar school and Finsbury Technical College (under Professor Silvanus Thompson, q.v.); chief electrical designer, Johnson and Phillips Ltd., 1906; joined the (later Royal) Aeronautical Society, 1907; adopted an aeronautical career, constructing aeroplanes and gliders; registered Handley Page Ltd., 1909; lecturer in aeronautics, Northampton Polytechnic; installed a wind tunnel; his first passenger-carrying monoplane flew across London, 1911; produced twin-engined bomber for Royal Naval Air Service and later four-engined machine; first aeroplanes to fly from England to India, 1918; CBE, 1918; promoted airlines in Europe, India, South Africa, and Brazil, 1919; produced airliners and bombers for RAF, 1920–39; knighted, 1942; transport aircraft assisted in Berlin air-lift, 1948–9, and equipped African routes of BOAC; fellow, City and Guilds of London Institute, 1939; helped to set up Cranfield College of Aeronautics, 1946; chairman, Society of British Aircraft Constructors; vice-chairman, Air Registration Board; president, Institute of Transport and of Royal Aeronautical Society; Handley Page memorial lecture given annually.

PAGE, H. A. (pseudonym) (1837–1905), author and publisher. [See JAPP, ALEXANDER HAY.]

PAGE, SIR LEO FRANCIS (1890–1951), magistrate; born at Hobart, Tasmania; educated at Beaumont College, Royal Military Academy, Woolwich, and University College, Oxford; Royal Flying Corps, 1914–16; called to bar (Inner Temple), 1918; justice of the peace for Berkshire, 1925–51; secretary, commissions of the peace, 1940–5; essentially a conservative reformer; published *Justice of the Peace* (1936), *Crime and the Community* (1937), and *The Sentence of the Court* (1948); knighted, 1948.

PAGE, THOMAS ETHELBERT (1850–1936), classical scholar, teacher, editor, and political critic; educated at Shrewsbury and St. John's College, Cambridge; second classic, 1873; sixth-form master, Charterhouse, 1873–1910; produced standard school editions of Horace's *Odes* and the *Aeneid*; editor-in-chief, 'Loeb Classical Library', 1912–36; served on Surrey county council and education committee, and governing bodies of Charterhouse, Cranleigh, and Shrewsbury; exponent of liberalism as political philosophy; CH, 1934.

PAGE, WILLIAM (1861–1934), historian and antiquary; partner with W. J. Hardy in firm of record agents and legal antiquaries; joint-editor (1902), sole editor (1904–34), *Victoria County History*; became sole proprietor (1928) and made over all material to London University, 1931.

PAGET, SIR BERNARD CHARLES TOLVER (1887–1961), general; educated at Shrewsbury and Sandhurst; commissioned into Oxfordshire and Buckinghamshire Light Infantry, 1907; served in India, 1908–14; adjutant, 5th (service) battalion; in France, 1915; brigade-major, 42nd Infantry brigade of 14th Light division, 1915–17; MC; on GHQ staff of Sir Douglas (later Earl) Haig [q.v.], 1917; DSO, 1918; severely wounded; graduated, Staff College, 1920; major, 1924; instructor, Staff College, 1926; student, Imperial Defence College, 1929; commander, regimental depot, Cowley, Oxford, 1930; chief instructor, Quetta Staff College, 1932–5; commanded 4th Quetta

Infantry brigade, 1936-7; major-general, 1937; commandant, Staff College, Camberley, 1938-9; commanded 18th division, 1939; took part in ill-fated Norwegian campaign, 1940; GOC-in-C, South Eastern Command; lieutenant-general, 1941; commander-in-chief, Home Forces, 1941-3; general; prepared plans and training for Normandy invasion; GOC-in-C Middle East Forces, 1944-6; retired, 1946; colonel of his regiment; principal, Ashridge College; Governor, Royal Hospital, Chelsea; CB, 1940; KCB, 1942; GCB, 1946.

PAGET, FRANCIS (1851-1911), bishop of Oxford; son of Sir James Paget, first baronet [q.v.]; educated at Shrewsbury and Christ Church, Oxford; Hertford scholar and chancellor's Latin verse prizeman; BA, 1873; MA, 1876; DD, 1885; tutor, 1876; regius professor of pastoral theology and canon of Christ Church, 1885-92; contributed an essay on 'Sacraments' to *Lux Mundi*, 1889; dean of Christ Church, 1892-1901; bishop of Oxford, 1901-11; declined see of Winchester, 1903; member of commission on ecclesiastical discipline, 1904-6; published introduction to Hooker's *Ecclesiastical Polity*, Bk. v (1899), *Faculties and Difficulties for Belief and Disbelief* (1887), *The Spirit of Discipline* (1891), and other sermons.

PAGET, DAME (MARY) ROSALIND (1855-1948), social reformer, nurse, and midwife; daughter of John Paget [q.v.]; cousin of Eleanor Rathbone [q.v.]; nurse at London Hospital; with others founded Midwives' Institute (Royal College of Midwives), 1881; obtained registration of midwives, 1902; DBE, 1935.

PAGET, LADY MURIEL EVELYN VERNON (1876-1938), philanthropist; daughter of twelfth Earl of Winchelsea; married (Sir) Richard Arthur Surtees Paget, second baronet, [q.v.] 1897; founder and honorary secretary (1905-38) of Invalid Kitchens of London; organized the Anglo-Russian Hospital in Russia, 1915-17; inaugurated and administered hospital and child-welfare work in Czechoslovakia, Latvia, Estonia, Lithuania, and Romania (1919-22) and relief for British subjects in Russia (1924-38); OBE, 1918; CBE, 1938.

PAGET, SIR RICHARD ARTHUR SURTEES, second baronet, of Cranmore (1869-1955), barrister and physicist; educated at Eton and Magdalen College, Oxford; called to bar (Inner Temple), 1895; succeeded his father, 1908; investigated problems of language and speech and made new approach in communication with deaf and dumb by sign language; published *Human Speech* (1930).

PAGET, SIDNEY EDWARD (1860-1908), painter and illustrator; studied art at schools of Royal Academy, where he exhibited, 1879-1905; works include 'Lancelot and Elaine' and portraits of Lord Winterstoke and Dr Weymouth; illustrated in black and white in English and American magazines.

PAGET, STEPHEN (1855-1926), biographer and essayist; son of Sir James Paget and brother of Francis Paget [qq.v.]; BA, Christ Church, Oxford; FRCS, 1885; practised surgery until 1897; won confidence as secretary to association of physicians and surgeons formed to watch working of 1877 Act concerning experiments on animals, 1888; published *Experiments on Animals* (1900); founded Research Defence Society, 1908; literary works include *Memoirs and Letters of Sir James Paget* (1899), *Francis Paget* (in collaboration, 1912), *Henry Scott Holland* (1921), and several volumes of essays.

PAGET, VIOLET (1856-1935), author under the pseudonym VERNON LEE; half-sister of Eugene Lee-Hamilton [q.v.]; works include *Studies of the Eighteenth Century in Italy* (1880), *Euphorion: being Studies of the Antique and the Mediaeval in the Renaissance* (1884), *Renaissance Fancies and Studies* (1895), and *Ariadne in Mantua* (a play, 1903).

PAIN, BARRY ERIC ODELL (1864-1928), humorist; BA, Corpus Christi College, Cambridge; became well known as a novelist and writer of short stories, mainly of a humorous character; his works include *Eliza* (the first of a series, 1900), *Mrs. Murphy* (1913), and *Edwards* (1915).

PAINE, CHARLES HUBERT SCOTT- (1891-1954), pioneer of aviation and of high-speed motor-boats. [See SCOTT-PAINE.]

PAKENHAM, SIR FRANCIS JOHN (1832-1905), diplomatist; secretary of legation at Buenos Aires, 1864; served at Rio de Janeiro, Stockholm, Brussels, Washington, and Copenhagen, 1865-78; consul-general at Santiago, 1878-85; British commissioner for claims after war between Chile and Bolivia and Peru, 1883; envoy at Buenos Aires, 1885-96, and Stockholm, 1896-1902; KCMG, 1898; died at Alameda.

PAKENHAM, SIR WILLIAM CHRISTOPHER (1861-1933), admiral; entered navy, 1874; captain, 1903; as naval attaché in Japan (1904-6) wrote brilliant reports of Russo-Japanese war; fourth sea lord, 1911-13; rear-admiral, 1913; as rear-admiral commanding the Australian Fleet (1915-16) gave able support to

(Lord) Beatty [q.v.] at Jutland; succeeded him in command of battle-cruiser force, 1917-19; vice-admiral, 1918; president, Royal Naval College, Greenwich, 1919-20; commander-in-chief, North America and West Indies station, 1920-3; admiral, 1922; KCB, 1916; KCVO, 1917; KCMG, 1919; GCB, 1925.

PALAIRET, SIR **(CHARLES) MICHAEL** (1882-1956), diplomatist; great-grandson of Thomas Allen, mineralogist [q.v.]; educated at Eton; nominated for diplomatic service, 1905; third secretary, Rome, 1907; on staff of British delegation to the peace conference, 1918-19; counsellor, Tokyo, 1922-5; counsellor, Rome, 1928; minister, Bucharest, 1929; Stockholm, 1935; Vienna, 1937; recalled and legation closed on Nazi invasion of Austria, 1938; minister, Athens, 1939; accompanied Greek king to Crete and subsequently to London, 1941; ambassador accredited to Greek monarch, 1942; retired, 1943, but returned to Foreign Office as temporary assistant under-secretary of state until 1945; KCMG, 1938.

PALGRAVE, SIR **REGINALD FRANCIS DOUCE** (1829-1904), clerk of the House of Commons from 1886 to 1900; son of Sir Francis Palgrave [q.v.]; admitted solicitor, 1851; obtained clerkship in House of Commons, 1853; CB, 1887; KCB, 1892; edited *Rules . . . of Procedure of the House of Commons* (8th-11th eds.), 1886-96; proficient water-colour artist; published *The House of Commons* (1869) and *The Chairman's Handbook* (1877).

PALLES, CHRISTOPHER (1831-1920), lord chief baron of the Exchequer in Ireland; BA, Trinity College, Dublin; called to Irish bar, 1853; QC 1865; solicitor-general for Ireland, 1872; attorney-general, 1872; chief baron of Irish court of Exchequer, 1874-1916; PC (Ireland), 1872, (England), 1892; a prominent educationist.

PALMER, SIR **ARTHUR POWER** (1840-1904), general; entered Indian army, 1857; served in Mutiny, 1857, in North-West Frontier campaign, 1863-4, and Abyssinian expedition, 1868; quartermaster-general to Kuram field force, 1878-80; assistant adjutant-general in Bengal, 1880-5; active in expedition to Suakin and raid on Thakul; CB, 1885; commanded force in Northern Chin Hills in Burma expedition, 1892-3; KCB, 1894; major-general, 1893; lieutenant-general, 1897; in Tirah campaign, 1897-8; commanded Punjab frontier force, 1898-1900; general, 1899; commander-in-chief in India, 1900-2; GCIE, 1901; GCB, 1903.

PALMER, SIR **CHARLES MARK,** first baronet (1822-1907), shipowner and iron-master; joined in founding at South Shields firm for manufacture of coke, 1839; founded shipping service for carrying coals from north of England to London; established (1851) shipyard near Jarrow, which developed from a small village into a large town; first mayor, 1875; constructed battleships; first to manufacture rolled armour plate; liberal MP for North Durham, 1874-85, and for Jarrow, 1885-1907; baronet, 1886.

PALMER, SIR **ELWIN MITFORD** (1852-1906), finance officer in India and Egypt; assistant comptroller-general to Indian government, 1871-85; director-general of accounts in Egypt, 1885; CMG, 1888; financial adviser to khedive, 1889-98; first governor of National Bank of Egypt at Cairo, 1898; KCMG, 1892; KCB, 1897.

PALMER, GEORGE **HERBERT** (1846-1926), musician; ordained, 1869; priest-organist, St. Barnabas's church, Pimlico, 1876-83; his life-work, rediscovery of true plain chant tradition and adaptation of ancient melodies to English texts; helped to found Plainsong and Medieval Music Society, 1888; issued many volumes of liturgical music.

PALMER, GEORGE **WILLIAM** (1851-1913), biscuit manufacturer; son of George Palmer [q.v.], founder of Huntley & Palmer's biscuit factory at Reading; associated with the business from youth; liberal MP, Reading, 1892-5, 1898-1904; PC, 1906; generous benefactor to University College, afterwards Reading University.

PALMER, WILLIAM **WALDEGRAVE,** second EARL OF SELBORNE (1859-1942), statesman; son of first Earl of Selborne [q.v.]; succeeded father, 1895; educated at Winchester and University College, Oxford; first class, modern history, 1881; liberal, 1885; liberal unionist, 1886-92, MP, Petersfield division; West Edinburgh, 1892-5; undersecretary of state for colonies, 1895-1900; first lord of Admiralty, 1900-5; brought (Lord) Fisher [q.v.] to Admiralty, 1902; established Osborne, Dartmouth, and Royal Naval War College; introduced common entry system, fleet reserves, modern equipment, and approved *Dreadnought* design; PC, 1900; GCMG, 1905; high commissioner for South Africa, 1905-10; issued memorandum recommending central national government, 1907; achieved South African Union, 1909, largely by his personal statesmanship; KG, 1909; prominent as 'diehard' and advocate of House of Lords reform; president, Board of Agriculture, 1915–16; unsuccessfully advocated guaranteed minimum prices; chairman, agricultural policy sub-committee, 1917; of joint parliamentary committee on government of India bill, 1919; leading ecclesiastical statesman and chairman of House of Laity, 1924–42.

PANETH, FRIEDRICH ADOLF (1887–1958), scientist; born in Vienna; educated at Schotten Gymnasium and university of Vienna; Ph.D., Vienna, 1910; assistant in Radium Research Institute attached to Vienna Academy of Science, 1912; assistant professor, university of Hamburg, 1919; head of inorganic department of chemical institute, Berlin University, 1922; head of chemical institute, Königsberg University, 1929; reader in atomic chemistry, Imperial College of Science and Technology, London, 1938; naturalized, 1939; professor of chemistry, Durham, 1939; head of chemistry division of joint British-Canadian atomic energy team in Montreal, 1943–5; returned to Durham and established Londonderry Laboratory for radio-chemistry until retirement, 1953; greatest authority of his time on volatile hydrides; made important contributions to the study of the stratosphere; FRS, 1947.

PANKHURST, DAME CHRISTABEL HARRIETTE (1880–1958), suffragette; daughter of Emmeline Pankhurst [q.v.]; educated at Manchester University; first class, LLB, 1906; assisted her mother to form the Women's Social and Political Union, 1903; helped to disrupt liberal meeting in Manchester, arrested and imprisoned, 1905; became organizing secretary of WSPU, 1907; imprisoned again for militant suffragette activities, 1908; fled to Paris and edited *The Suffragette*, 1912–14; returned to England on outbreak of war and spoke at many recruiting meetings; stood, un-successfully, as coalition candidate for Smeth-wick, 1918; after the 1914–18 war, settled in the United States; DBE, 1936.

PANKHURST, EMMELINE (1858–1928), leader of the militant movement for women's suffrage; born GOULDEN; married R. M. Pank-hurst, a radical Manchester barrister, 1879; took up suffrage work; widowed, 1898; registrar of births and deaths, Rusholme, until 1907; with her eldest daughter, Christabel, founded new women's suffrage society, Women's Social and Political Union, 1903; began to adopt sensational methods of propaganda, 1905; imprisoned in Holloway jail, 1908; militant suffragette campaign intensified, 1909; visited United States, 1909, 1911, 1913; constantly arrested, released as result of hunger-striking, and re-arrested, 1912–14; maintained truce to militancy

during 1914–18 war; took little part in agitation leading to first instalment of women's suffrage, 1918.

PANTIN, CARL FREDERICK ABEL (1899–1967), zoologist; educated at Tonbridge School and Christ's College, Cambridge; first class, parts i and ii, natural sciences tripos, 1921–2; joined staff of Marine Biological Laboratory, Plymouth, 1922; fellow, Trinity College, Cambridge, 1929; university lecturer in zoology; Sc.D., 1933; published papers on the nervous system of the sea anemone; FRS, 1937; reader in invertebrate zoology, 1937; professor, 1959–66; president, Linnaean Society, 1958–60; president, Marine Biological Association, 1960–6; chairman, Board of Trustees of British Museum, Natural History; hon. fellow, Christ's College; hon. doctorates, São Paulo and Durham.

PARES, SIR BERNARD (1867–1949), his-torian of Russia; educated at Harrow and Trinity College, Cambridge; first visited Russia, 1898; reader in Russian history, Liver-pool, 1906–8; professor, Russian history, lan-guage, and literature, Liverpool, 1908–17, Lon-don, 1919–36; director, School of Slavonic and East European Studies, 1922–39; pioneer of Russian studies and zealous interpreter of Russia; attached to Russian army, 1914–17; publications include *History of Russia* (1926), *The Fall of the Russian Monarchy* (1939), and *Russia* (1940); KBE, 1919.

PARIS, SIR ARCHIBALD (1861–1937), major-general; joined Royal Marine Artillery, 1879; commanded 'Kimberley column' in South Africa under Lord Methuen [q.v.]; forced to surrender, 1902; commanded Royal Naval division at Antwerp, throughout Galli-poli campaign, and in France, 1914–17; major-general, 1915; KCB, 1916.

PARISH, WILLIAM DOUGLAS (1833–1904), writer on dialect; son of Sir Woodbine Parish [q.v.]; BCL, Trinity College, Oxford, 1858; vicar of Selmeston, Sussex, from 1863 to death; published dictionary of Sussex dialect and provincialisms, 1875.

PARKER, ALBERT EDMUND, third EARL OF MORLEY (1843–1905), chairman of commit-tees of the House of Lords, 1889–1905; edu-cated at Eton and Balliol College, Oxford; BA, 1865; lord-in-waiting to Queen Victoria under Gladstone, 1868–74; under-secretary for war, 1880–5; first commissioner of public works, Feb.–Apr. 1886; resigned on home rule ques-tion, 1886; active in Devonshire affairs.

PARKER, CHARLES STUART (1829-1910), politician and author; educated at Eton and University College, Oxford; with G. J. (later Viscount) Goschen and G. C. Brodrick [qq.v.] formed Oxford Essay Club, 1852; BA and MA, 1855; fellow, 1854-68; honorary fellow, 1899; organized university volunteer corps; ardent alpinist; friend of Gladstone; urged university reform and national system of elementary education; private secretary to Edward Cardwell [q.v.], 1864-6; liberal MP for Perthshire, 1868-74, and for Perth, 1878-92; member of commissions on public schools (1868-74), military education (1869), and Scotch educational endowments (1872); published lives of Sir Robert Peel (3 vols., 1891-9) and Sir James Graham (2 vols., 1907); hon. LLD, Glasgow, hon. DCL, Oxford, 1908; PC, 1907.

PARKER, ERIC (FREDERICK MOORE SEARLE) (1870-1955), author and journalist; educated at Eton (King's scholar) and Merton College, Oxford; junior assistant editor, *St. James's Gazette*, 1900; edited *County Gentleman*, 1902-7; published *The Sinner and the Problem* (1901); edited the *Gamekeeper*, 1908-10; appointed shooting editor, the *Field*, 1911-32; published *Promise of Arden* (1912), *A West Surrey Sketch Book* (with William Hyde, 1913), and *Eton in the Eighties* (1914); served as officer in Queen's Royal West Surrey Regiment, 1914-18; published numerous books on sport and wild life, 1918-54; edited 'The Lonsdale Library of Sports, Games and Pastimes', 1928; edited the *Field*, 1929-37.

PARKER, Sir (HORATIO) GILBERT (GEORGE), baronet (1862-1932), author and politician; born in Ontario; wrote thirty-six books, many of them novels about Canada, including *The Trail of the Sword* (1894) and *The Seats of the Mighty* (1896); although sensational, unconvincing, inaccurate, and turgidly written, they appealed to uncritical readers; conservative MP, Gravesend, 1900-18; promoter of imperial unity and small ownership; knighted, 1902; baronet, 1915; PC, 1916.

PARKER, JOHN (1875-1952), founder and editor of *Who's Who in the Theatre*; born Jacob Solomons, in New York, of a Welsh mother and Polish father; educated at Whitechapel Foundation School, London; on advice of his mother adopted name of John Parker, subsequently legalized in 1917; contributed articles to the *Free lance*, owned by Clement Scott [q.v.]; London manager and critic of the *New York Dramatic News*, 1903-20; published *Who's Who in the Theatre*, and compiled eleven editions

almost singlehanded, 1912-52; hon. secretary, the Critics' Circle, 1924-52.

PARKER, JOSEPH (1830-1902), Congregationalist divine; studied for Congregational ministry in London, 1852; ordained minister at Banbury, 1853-8; minister of Cavendish chapel, Manchester, 1858-69; DD, Chicago, 1862; minister of Poultry chapel, London, 1869; instituted Thursday noonday service; removed congregation to City Temple, 1874; chairman of Congregational Union, 1884 and 1901; president of National Free Church Council, 1902; original and fervid preacher; visited America five times; chief publication was *The People's Bible* (25 vols., 1885-95); voluminous writings include *Ecce Deus* (a reply to *Ecce Homo*, 1867), *The City Temple Pulpit* (1899), *A Preacher's Life* (autobiography, 1899), and some fiction.

PARKER, LOUIS NAPOLEON (1852-1944), musician, playwright, and inventor of pageants; son of American father and English mother; naturalized, 1914; born in Normandy and educated in Europe; studied at Royal Academy of Music; music teacher, Sherborne School, 1873; director of music, 1877-92; resigned through deafness and became dramatist; successes include *The Cardinal* (1903), *Beauty and the Barge* (with W. W. Jacobs, q.v., 1904), *Pomander Walk* (1910), *Disraeli* (written for George Arliss, q.v., 1911), and *Drake* (1912); invented and organized pageants, first at Sherborne (1905), thereafter at Warwick, Bury St. Edmunds, Colchester, York, and Dover.

PARKER, ROBERT JOHN, BARON PARKER OF WADDINGTON (1857-1918), judge; educated at Westminster, Eton, and King's College, Cambridge; called to bar (Lincoln's Inn), 1883; junior equity counsel to Treasury, 1900; chancery judge, 1906; specialized in patent cases; lord of appeal, life peer, and PC, 1913; in House of Lords gained authoritative position as judge of final appeal; dealt in masterly fashion with prize appeals heard during 1914-18 war; most striking characteristic of his judgments was their intellectual compactness; brought before House of Lords detailed scheme for formation of League of Nations, 1918.

PARKIN, Sir GEORGE ROBERT (1846-1922), educationist and imperialist; born in Canada; BA, New Brunswick University, Fredericton; headmaster, collegiate school, Fredericton, 1871-89; devoted himself to cause of Imperial Federation League, 1889; principal, Upper Canada College, Toronto, 1895-1902; first organizing secretary of Rhodes Scholarships and settled permanently in England, 1902;

KCMG, 1920; his works include *Imperial Federation* (1892).

PARKINSON, Sir (ARTHUR CHARLES) COSMO (1884-1967), civil servant; educated at Epsom College and Magdalen College, Oxford; first class, classical hon. mods. and *lit. hum.*, 1905-7; entered Civil Service and posted to Admiralty, 1908; transferred to Colonial Office, 1909; assistant private secretary to secretary of state, 1914; served with King's African Rifles, 1917-19; rejoined Colonial Office, 1920; assistant secretary, Dominions Office, 1925; returned to Colonial Office as head of East African department, 1928; assistant under-secretary of state, 1931; permanent under-secretary of state, 1937; permanent under-secretary, Dominions Office, 1940; returned to Colonial Office, 1940-2; seconded for special duties, visiting colonies, 1942-5; published *The Colonial Office from Within* (1947); served on number of public bodies; hon. LLD, St. Andrews; OBE, 1919; CMG, 1931; KCMG, 1935; KCB, 1938; GCMG, 1942.

PARMOOR, first Baron (1852-1941), lawyer and politician. [See CRIPPS, CHARLES ALFRED.]

PARR, LOUISA (*d.* 1903), novelist; born TAYLOR; married George Parr, 1869; novels include *How it All Happened* (1868), *Dorothy Fox* (1871), and *Adam and Eve* (1880); wrote life of Mrs Craik (1897).

PARRATT, Sir WALTER (1841-1924), organist and composer; as a boy filled various posts as organist in and near Huddersfield, his native town; organist of Great Witley church, Worcestershire, 1864; of Wigan parish church, 1868; of Magdalen College, Oxford, 1872-82; of St. George's chapel, Windsor, 1882-1924; chief professor of organ, Royal College of Music, 1882; founded 'school of organ playing'; private organist to Queen Victoria, 1892; master of queen's music, 1893; professor of music, Oxford, 1908-18; composer chiefly of church music; knighted, 1892.

PARRY, Sir CHARLES HUBERT HASTINGS, baronet (1848-1918), composer, musical historian, and director of the Royal College of Music; son of Thomas Gambier Parry [q.v.]; educated at Eton and Exeter College, Oxford; studied piano under Edward Dannreuther; influenced by Wagner; professor at Royal College of Music, 1883; director, 1895; choragus at Oxford, 1884; professor of music, 1899-1908; knighted, 1898; baronet, 1902; formative period as composer closed, 1880; new period opened with 'Scenes from Shelley's Prometheus Unbound', for solo voices, chorus,

and orchestra (performed, 1880), and closed with most famous of his works, setting to Milton's ode 'At a solemn Music' (performed, 1887); other settings include music to Aristophanes' *The Birds* (1884), *The Frogs* (1891), *The Clouds* (1905), and *The Acharnians* (1914); fresh stage in his career marked by performance of his first oratorio, *Judith* (1888), followed by *Job* (1892) and *King Saul* (1894); in last period, beginning with symphonic ode 'War and Peace' (1903), wrote series of works dealing with his thoughts on problems of human race; assistant editor of *Grove's Dictionary of Music and Musicians*; his own works include *Art of Music* (1893), *Oxford History of Music*, vol. iii (1902), and *Johann Sebastian Bach* (1909).

PARRY, JOSEPH (1841-1903), musical composer; went with family to America, 1854; won prizes for composition at Danville (Pennsylvania) eisteddfod (1860) and at national eisteddfod at Swansea (1863), Llandudno (1864), and Chester (1866); removed to London, 1868; Mus.Bac., Cambridge, 1871; Mus. Doc., 1878; professor of music at University College of Wales, Aberystwyth, 1873-9, and at University College, Cardiff, 1888-1903; composed operas *Blodwen* (1878), *Virginia* (1882), *Sylvia* (1889), oratorios, cantatas, and hymn tunes (including 'Aberystwyth') and anthems; edited *Cambrian Minstrelsie* (6 vols., 1893) and *Dr. Parry's Book of Songs*; his son, JOSEPH HAYDN PARRY (1864-94), was professor at Guildhall School of Music, 1890, and composed operas.

PARSONS, ALFRED WILLIAM (1847-1920), painter and illustrator; RA, 1911; president, Royal Society of Painters in Water Colours, 1914-20; painted landscapes; chief picture, 'When Nature Painted All Things Gay' (Tate Gallery, 1887).

PARSONS, Sir CHARLES ALGERNON (1854-1931), engineer and scientist; son of the third and brother of the fourth Earl of Rosse [qq.v.]; educated at Trinity College, Dublin, and St. John's College, Cambridge; eleventh wrangler, 1877; trained as an engineer; constructed his first turbo-dynamo, 1884; founded his own firm, 1889; installed turbo-alternators in many power-stations; raised the voltage generated to 11,000 in 1905 and 36,000 in 1928; founded separate firm for marine steam turbines; the experimental vessel *Turbinia* created a sensation by its speed at naval review of 1897; the turbine adopted by the Admiralty and the Cunard Company, 1905; devised geared turbines for slower vessels; also, later, high pressure geared turbines; interested also in optics; manufactured searchlight reflectors, some hundred different kinds of glass for optical purposes, and 36- and 74-inch reflecting telescopes; FRS, 1898; president, British Association, 1919; considered the most original British engineer since James Watt; KCB, 1911; OM, 1927.

PARSONS, SIR JOHN HERBERT (1868–1957), ophthalmologist and physiologist; educated at Bristol grammar school, University College, Bristol, and University College, London; B.Sc., physiology, 1890; MB, St. Bartholomew's Hospital, 1892; clinical assistant, Moorfields Eye Hospital; FRCS, 1900; consultant surgeon, Moorfields Hospital and University College Hospital, 1904–39; published *Elementary Ophthalmic Optics* (1901), *The Ocular Circulation* (1903), *The Pathology of the Eye* (4 vols., 1904–8), *Diseases of the Eye* (1907), *An Introduction to the Study of Colour Vision* (1915), and *An Introduction to the Theory of Perception* (1927); president, Royal Society of Medicine, 1936–8; chairman, editorial committee, the *British Journal of Ophthalmology*, 1917–48; CBE, 1919; FRS, 1921; knighted, 1922; hon. D.Sc., Bristol, 1925; LLD, Edinburgh, 1927.

PARSONS, LAURENCE, fourth EARL OF ROSSE (1840–1908), astronomer; son of third earl [q.v.]; BA, Trinity College, Dublin, 1864; representative Irish peer, 1868; KP, 1890; lord-lieutenant of King's County, 1892–1908; made researches into astrophysics at Birr Castle, Parsonstown; published 'Account of Observations of Great Nebula in Orion from 1848 to 1867', 1867; made from 1868 prolonged investigations into radiation of heat from the moon; chancellor of Dublin University, 1885; hon. DCL, Oxford, 1870, LLD, Dublin, 1879, and Cambridge, 1890; FRS and FRAS, 1867; president of Royal Irish Academy, 1896–1901.

PARSONS, SIR LEONARD GREGORY (1879–1950), paediatrician; educated at King Edward VI grammar school and Mason College, Birmingham; MB, BS (London), 1905; appointed to Birmingham Children's Hospital, 1910; first professor of child health, Birmingham University, 1928–46; obtained new buildings for Children's Hospital and establishment of Institute of Child Health, 1945; insisted upon scientific investigation of children's diseases and the importance of ante-natal factors; knighted, 1946; FRS, 1948.

PARSONS, RICHARD GODFREY (1882–1948), bishop; educated at Durham School and Magdalen College, Oxford; first class, *lit. hum.* (1905), and theology (1906); ordained, 1907; fellow, praelector, and chaplain, University College, Oxford, 1907–11; principal, Wells theological college, 1911–16; vicar of Poynton, 1916–19; rector of Birch-in-Rusholme, 1919–31; bishop of Middleton, 1927–32; canon and sub-dean of Manchester, 1931–2; bishop of Southwark, 1932–41; of Hereford, 1941–8.

PARTINGTON, JAMES RIDDICK (1886–1965), chemist and historian of science; educated at Southport Science and Art School and Manchester University; first class, chemistry, 1906; undertook research in physical organic chemistry; Beyer fellow, 1910–11; M.Sc., 1911; D.Sc., 1918; worked on specific heats of gases in Berlin, 1911–13; assistant lecturer and demonstrator in chemistry, Manchester University, 1913–19; professor of chemistry, East London (later Queen Mary) College, 1919–51; served in army and Ministry of Munitions, 1914–18; MBE (military), 1918; publications include *An Advanced Treatise on Physical Chemistry* (5 vols., 1949–54), *Origins and Development of Applied Chemistry* (1935), and *A History of Chemistry* (4 vols., 1961–70); first chairman, Society for the Study of Alchemy and Early Chemistry (1937–8); president, British Society for the History of Science (1949–51); fellow, Queen Mary College, 1959.

PARTRIDGE, SIR BERNARD (1861–1945), *Punch* artist and cartoonist; son of Richard Partridge [q.v.]; educated at Stonyhurst; junior (1891), second (1901), principal (1909–45) *Punch* cartoonist; knighted, 1925.

PASSFIELD, BARON (1859–1947), social reformer and historian. [See WEBB, SIDNEY JAMES.]

PATEL, VALLABHBHAI JAVERABHAI (1875–1950), deputy prime minister of India; brother of Vithalbai Patel [q.v.]; born in Gujerat; called to bar (Middle Temple), 1913; practised at Ahmedabad; joined civil disobedience campaign of M. K. Gandhi [q.v.], 1917; led several campaigns and became known as 'Sardar' (leader); president of Congress, 1931; formed parliamentary subcommittee which later guided Congress ministries, 1935; imprisoned, 1930, 1932–4, 1940–1, 1942–5; played a leading part in negotiations, 1945–7; home member, 1946; deputy prime minister with portfolios of home, information and broadcasting, and States, 1947–50; maintained discipline of Congress party and integrated the States into a unitary system.

PATEL, VITHALBAI JHAVABHAI (1870–1933), first Indian president of the Indian legislative assembly; called to bar (Lincoln's Inn), 1908; practised in Bombay; entered

Bombay legislative council (1914), Indian legislative assembly (1923); president, 1925-30; stormy career of agitation for independence; second only to M. K. Gandhi [q.v.] in influence.

PATERSON, SIR ALEXANDER HENRY (1884-1947), prison reformer; educated at Bowdon College and University College, Oxford; third class, *lit. hum.*, 1906; worked amongst poor in Bermondsey; published *Across the Bridges* (1911); undertook supervision of boys released from Borstal institutions, 1908; assistant director, Central Association for the Aid of Discharged Prisoners, 1911; served with 22nd County of London Battalion (Bermondsey) in France, 1914-18; became active member of Toc H; in Ministry of Labour, 1919-22; prison commissioner, 1922-47; sought prisoners' rehabilitation by abolishing practices destructive of self-respect, by providing recreational and educational facilities, and by introduction of open prisons; in Borstal institutions replaced enforced obedience by educational method based on house system designed to invite co-operation and inculcate new standards; advised on penal administration in Burma and Ceylon (1925), West Indies (1937), East Africa (1939), West Africa (1943), Malta and Gibraltar (1944); visited penal institutions in United States (1931) and French Guiana (1937); vice-president, International Penal and Penitentiary Commission, 1938; acting president, 1943; visited internment camps in Canada, 1940-1; director, Czech Refugee Trust Fund, 1941-5; knighted, 1947.

PATERSON, SIR WILLIAM (1874-1956), mechanical engineer; educated at Heriot-Watt College, Edinburgh; filed patents covering processes and equipment for water purification, 1898-1902; formed company dealing with purification of water for industrial purposes and drinking supplies, 1902; designed and installed plant for Cheltenham drinking supply, using chlorine for sterilization; advised Indian Army medical authorities on purification of water supplies; during 1939-45 war designed mobile filtration and sterilizing unit for water supplied to troops in the field; designed domestic (Anderson) shelter for protection against blast from bombing, named after John Anderson (later Viscount Waverley) [q.v.], a lifelong friend; knighted, 1944; hon. member, Institution of Water Engineers, 1948.

PATERSON, WILLIAM PATERSON (1860-1939), Scottish divine; educated at the Royal High School and University, Edinburgh, and in Germany; ordained minister of Church of Scotland, 1887; minister in Crieff, 1887-94; professor of systematic theology, Aberdeen,

1894-1903; of divinity, Edinburgh, 1903-34; dean of faculty of divinity, 1912-28; exerted far-reaching influence in theological education and church leadership; moderator, General Assembly, Church of Scotland, 1919; prominent in negotiations leading to reunion, 1929; works include *The Apostles' Teaching* (1903), *Rule of Faith* (1912), *The Nature of Religion* (1925), and *Conversion* (1939).

PATIALA, SIR BHUPINDRA SINGH, MAHARAJA OF (1891-1938); succeeded father, 1900; invested with full powers, 1910; attended imperial war conference (1918), League of Nations Assembly (1925), and Round Table conference (1930); chancellor, Chamber of Princes, 1926-30, 1933-5, 1937; captained Indian visiting cricket team, 1911; GCIE, 1911; GCSI, 1921; GCVO, 1922.

PATON, DIARMID NOËL (1859-1928), physiologist; son of Sir J. N. Paton [q.v.]; MB, CM, Edinburgh; superintendent of research laboratory, Royal College of Physicians, Edinburgh, 1889; regius professor of physiology, Glasgow, 1906-28; a pioneer in study of metabolism and nutrition; FRS, 1914; his chief work, *Poverty, Nutrition, and Growth* (in collaboration, 1926).

PATON, JOHN BROWN (1830-1911), nonconformist divine and philanthropist; BA, London, 1849; MA, 1854; Congregational minister in Sheffield, 1854-63; first principal of Congregational Institute, Nottingham, 1863-98; DD, Glasgow, 1882; took part in many movements for social regeneration; founded National Home Reading Union, 1889, Bible Reading and Prayer Union, 1892, English Land Colonization Society, 1892, and Boys' and Girls' League of Honour, 1906; edited with R. W. Dale [q.v.] *The Eclectic Review*, 1858-61; consulting editor of *Contemporary Review*, 1882-8; publications include *Criticisms and Essays* (2 vols., 1892-7) and *Christ and Civilization* (1910).

PATON, JOHN GIBSON (1824-1907), missionary to the New Hebrides; city mission-ary in Glasgow; ordained by reformed Pres-byterian Church of Scotland, 1858, and went with wife to establish mission station in island of Tanna in New Hebrides, where he stayed four years; fled from native persecution to New South Wales, where he pleaded for supply of missionaries; settled on island of Aniwa, 1866-81; from 1881 made headquarters at Melbourne; died there; his *Autobiography* edited by his brother James (2 pts., 1889) and *Later Years and Farewell* by his son (1910).

PATON, JOHN LEWIS (ALEXANDER) (1863–1946), schoolmaster; son of J. B. Paton [q.v.]; educated at Shrewsbury and St. John's College, Cambridge; first class, classical tripos, parts i and ii, 1886–7; fellow, 1887; at Leys School, 1887–8; Lower Bench master, Rugby, 1888–98; headmaster, University College School, 1898–1903; high master, Manchester grammar school, 1903–24; first president, Memorial University College, St. John's, Newfoundland, 1925–33.

PATON, SIR JOSEPH NOËL (1821–1901), artist; brother of Waller Hugh Paton [q.v.]; influenced by William Blake's designs; ARSA, 1847; RSA, 1850; exhibited at Royal Academy, 1856–69; painted mythological or historical scenes, as 'Hesperus' (1858), 'Luther at Erfurt' (1861), and 'The Fairy Raid' (1867); Queen Victoria's limner for Scotland, 1865; knighted, 1867; painted religious pictures, including 'Mors Janua Vitae' (1866), 'Vigilate et Orate' (1885), and 'Beati Mundo Corde' (1890), which were sent on tour through the country; correct draughtsman but lacked appreciation of colour; published verse and collected armour; hon. LLD, Edinburgh, 1876.

PATON, WILLIAM (1886–1943), Presbyterian minister and secretary of International Missionary Council; educated at Whitgift School, Pembroke College, Oxford, and Westminster College, Cambridge; on staff, Student Christian Movement, 1911–21; ordained, 1917; secretary, National Christian Council of India, Burma, and Ceylon, 1922–7; of International Missionary Council and editor *International Review of Missions*, 1927–43.

PATTISON, ANDREW SETH PRINGLE- (1856–1931), philosopher; originally ANDREW SETH; educated at the Royal High School and University, Edinburgh (first class, classics and philosophy, 1878), and in Germany; professor of logic and philosophy, University College, Cardiff (1883–7), of logic, rhetoric, and metaphysics, St. Andrews (1887–91), of logic and metaphysics, Edinburgh (1891–1919); works include *Scottish Philosophy* (1885), *Hegelianism and Personality* (1887), *The Idea of God in the Light of Recent Philosophy* (1917), and *Studies in the Philosophy of Religion* (1930); FBA, 1904.

PAUL, CHARLES KEGAN (1828–1902), author and publisher; educated at Eton and Exeter College, Oxford; BA, 1849; chaplain at Eton, 1853; became vegetarian and positivist, 1854; joined Free Christian Union (Unitarian), 1870; edited *New Quarterly Magazine*, 1873; resigned Church of England living, 1874; published account of William Godwin (2 vols.,

1876); took over publishing firm of H. S. King, 1877; published *The Nineteenth Century*, works by Tennyson, Thomas Hardy, Meredith, and Stevenson, the 'International Scientific' series, and *Last Journals of General Gordon*; joined Church of Rome, 1890; published *Biographical Sketches* (1883), *Memories* (1899), verse, and translations.

PAUL, HERBERT WOODFIELD (1853–1935), author and politican, educated at Eton and Corpus Christi College, Oxford; president of Union and first class, *lit. hum.*, 1875; liberal MP, South Edinburgh (1892–5), Northampton (1906–9); second civil service commissioner, 1909–18; works include lives of Gladstone (1901), Matthew Arnold (1902), Froude (1905), and *History of Modern England* (5 vols., 1904–6).

PAUL, WILLIAM (1822–1905), horticulturist; founded nursery business at Waltham Cross; published *The Rose Garden* (1848, 10th edn. 1903) and *American Plants* (1858); issued *The Rose Annual*, 1858–81, and *The Florist and Pomologist*, 1868–74; a fluent lecturer; collected rich library of old gardening books.

PAUNCEFOTE, JULIAN, BARON PAUNCEFOTE (1828–1902), diplomatist; born at Munich; called to bar, 1852; attorney-general of Hong Kong, 1865–72; knighted, 1872; chief justice of Leeward Islands, 1874; legal under-secretary at Foreign Office, 1876; KCMG and CB, 1880; permanent under-secretary, 1882; commissioner at Paris concerning free navigation of Suez canal, 1885; GCMG, 1885; KCB, 1888; envoy extraordinary to United States, 1889; raised to rank of ambassador, 1893; dealt tactfully with questions of Canadian seal fishing in Bering Sea (1892), the boundary between Venezuela and British Guiana (1895–9), and Spanish–American war (1898); senior British delegate at Hague peace conference, 1899; baron, 1899; signed 'Hay–Pauncefote treaty' (1901), granting all nations equal passage through Panama canal; died at Washington; hon. LLD, Harvard and Columbia, 1900.

PAVY, FREDERICK WILLIAM (1829–1911), physician; student at Guy's Hospital; MB, 1852; studied diabetes under Claude Bernard in Paris; lecturer at Guy's Hospital from 1854; physician, 1871; FRCP, 1860; Harveian orator, 1886; president of various medical societies; first chairman of permanent committee of international congress of medicine, 1909; FRS, 1863; hon. LLD, Glasgow, 1888; pioneer among modern chemical pathologists; founded modern theory of diabetes; initiated 'Pavy's test' for sugar; writings include treatises on diabetes (1862), food and dietetics (1873), and carbohydrate metabolism (1906).

PAYNE, EDWARD JOHN (1844-1904), historian; BA, Charsley's Hall, Oxford, 1871; fellow of University College, 1872-99; called to bar, 1877; honorary recorder of Wycombe, 1883-1904; published *History of European Colonies* (1875), *Colonies and Colonial Federation* (1904), and his unfinished *History of the New World called America* (2 vols., 1892-9); wrote much on music; accomplished violinist.

PAYNE, HUMFRY GILBERT GARTH (1902-1936), archaeologist; son of E. J. Payne [q.v.]; educated at Westminster and Christ Church, Oxford; first class, *lit. hum.*, 1924; director, British School of Archaeology at Athens, 1929-36; his *Necrocorinthia* (1931) placed him in front rank of classical archaeologists.

PAYNE, JOHN WESLEY (JACK) (1899-1969), bandleader; interested in dance music during service with RFC, 1918; formed band employed by Hotel Cecil, London, 1925; director of dance music, BBC, 1928-32; first bandleader to use signature tune; band appeared in films, *Say It with Music* (1932) and *Sunshine Ahead* (1935); entertained the services, 1939-45; also composed dance tunes; recordings highly successful; published *This is Jack Payne* (1932) and *Signature Tune* (1947); made important contribution to British dance music.

PAYNE, JOSEPH FRANK (1840-1910), physician; son of Joseph Payne [q.v.]; BA, Magdalen College, Oxford, 1862; fellow 1865-83; honorary fellow, 1906; B.Sc., London, 1865; MB, Oxford, 1867; MD, 1880; FRCP, 1873; assistant physician at St. Thomas's Hospital, 1871; physician, 1887; chief medical witness in Staunton poisoning case, 1877; published *Manual of General Pathology* (1888) and work on skin diseases (1889); president of Epidemiological (1892) and Pathological (1897) societies; reported on plague at Vetlanka, 1879; wrote much on history of medicine; works include lives of Linacre (1881), Thomas Sydenham (1900), *English Medicine in Anglo-Saxon Times* (1904); Harveian orator (1896) and Fitz-Patrick lecturer (1903-4) at College of Physicians; Harveian librarian, 1899; member of royal commission on tuberculosis, 1890; collected fine medical library.

PEACOCK, Sir EDWARD ROBERT (1871-1962), merchant banker; born in Ontario; educated at Almonte high school and Queen's University, Kingston; English teacher, Upper Canada College, under headmaster (Sir) George Parkin and with close friend Stephen Leacock [qq.v.], 1897-1902; joined Dominion Securities Corporation, Toronto financial house, 1902; European representative in London, 1907; administered group of electric power and traction companies in Spain, Mexico, and Brazil, 1915-24; director, Bank of England, with Montagu (later Lord) Norman [q.v.] as governor, 1921-4 and 1929-46; joined Baring Brothers, 1924-54; appointed to boards of other companies, including Canadian Pacific Railway and Hudson's Bay Company; chairman, overseas committee, Canadian National War Services Funds Advisory Board, 1939-45; receiver-general, Duchy of Cornwall, 1929; GCVO, 1934; chairman, finance committee, King Edward's Hospital Fund for London, 1929-54; member, Rhodes Trust, 1925, and of other commissions and trusts; hon. degrees from Oxford, Edinburgh, and Queen's University, Kingston; helped in formation of Commonwealth Development Finance Corporation.

PEACOCKE, JOSEPH FERGUSON (1835-1916), archbishop of Dublin; BA, Trinity College, Dublin; rector of St. George's, Dublin, 1873, of Monkstown, county Dublin, 1878; bishop of Meath, 1894; archbishop of Dublin, 1897-1915.

PEAKE, ARTHUR SAMUEL (1865-1929), theologian and biblical scholar; BA, St. John's College, Oxford; theological fellow, Merton College, Oxford, 1890-7; tutor in charge of curriculum, Hartley Primitive Methodist College, Manchester, 1892; first Rylands professor of biblical criticism and exegesis, Manchester University, 1904; possessed vast knowledge of current literature, native and foreign, in field of biblical studies; his original works diffused methods and results of sound biblical learning.

PEAKE, Sir CHARLES BRINSLEY PEMBERTON (1897-1958), diplomatist; educated at Wyggeston School, Leicester, and Magdalen College, Oxford; served as officer in the Leicester Regiment, 1914-18; MC; entered diplomatic service, 1922; appointed head of news department of Foreign Office, 1939; personal assistant to Lord Halifax [q.v.], ambassador, Washington, 1941; British representative to French National Committee, 1942-3; ambassador, Belgrade, 1946-51; ambassador, Athens, 1951-7; CMG, 1941; KCMG, 1948; GCMG, 1956.

PEAKE, FREDERICK GERARD (1886-1970), pasha, founder of the Arab Legion; educated at Stubbington House, Fareham, and

Sandhurst; commissioned in Duke of Wellington's Regiment, and posted to India, 1906; seconded to Egyptian Army, 1914; transferred to Camel Corps, 1916; dislocated his neck, but recovered; commanded company of Egyptian Camel Corps, 1918; assisted T. E. Lawrence's [q.v.] campaign against Turks; returned to HQ in Cairo; sent to administer Aqaba and environs; returned to Egyptian Camel Corps in Palestine, 1919; on disbandment of Camel Corps, joined Palestine Police, 1920; under authority of Sir Herbert (later Viscount) Samuel [q.v.], high commissioner, raised small force of officers and men which became the Arab Legion, 1920; attended Cairo conference, 1921; when Sharif Abdullah accepted Emirate of Trans-Jordan under British protection, Arab Legion increased, and the Emirate defended from tribal insurgents, 1924; law and order established throughout the area; succeeded as Legion's commander by Glubb Pasha, 1930; ranks in Arab Legion, brigadier (1920-2), major-general (1922-6), general or pasha (1926-30); OBE, 1923; CBE, 1926; CMG, 1939; retired from army with rank of lieutenant-colonel, 1939; published *A History of Jordan and its Tribes* (1958).

PEAKE, HAROLD JOHN EDWARD (1867-1946), archaeologist; honorary curator, Newbury Museum; superintended local archaeological investigations; president, Royal Anthropological Institute, 1926-8; publications include *The English Village* (1922) and, with H. J. Fleure, *The Corridors of Time* (10 vols., 1927-56).

PEAKE, MERVYN LAURENCE (1911-1968), artist and author; born in China; educated at Tientsin grammar school, Eltham College, and the Royal Academy Schools; taught at Westminster School of Art, 1935-9; served in Royal Artillery and Royal Engineers, 1939-43; worked in Ministry of Information, 1943-5; part-time teacher, Central School of Art, Holborn, 1949-60; his stage play, *The Wit to Woo* failed, 1957; published novels, *Titus Groan* (1946), *Gormenghast* (1950), and *Titus Alone* (1959); illustrated books, including Lewis Carroll's *The Hunting of the Snark* (1941) and the *Alice* books (1946 and 1954), Coleridge's *The Rime of the Ancient Mariner* (1943), R. L. Stevenson's *Dr. Jekyll and Mr. Hyde* (1948) and *Treasure Island* (1949); illustrated his own children's books and published collections of sketches including *The Craft of the Lead Pencil* (1946) and *The Drawings of Mervyn Peake* (1949); published collections of poems, including *Shapes and Sounds* (1941), *The Glassblowers* (1950), and *A Reverie of Bone* (1967); hon. fellow, Royal Society of Literature.

PEARCE, ERNEST HAROLD (1865-1930), bishop of Worcester; educated at Christ's Hospital and Peterhouse, Cambridge; ordained, 1889; metropolitan district secretary, British and Foreign Bible Society, 1892; vicar of Christ Church, Newgate Street, London, 1895-1912; established close connection with Mansion House and City life; ecclesiastical correspondent to *The Times*, 1899; canon of Westminster Abbey, 1911; assistant chaplain-general, War Office, 1915-19; bishop of Worcester, 1919-30; maintained lifelong connection with Christ's Hospital; his works include *The Annals of Christ's Hospital* (1901), *The Monks of Westminster* (1916), *Thomas de Cobham, Bishop of Worcester* (1923), *The Register of Thomas de Cobham, 1317-1327* (1930).

PEARCE, SIR GEORGE FOSTER (1870-1952), Australian statesman; son of a blacksmith; at the age of eleven became a farm labourer and at fifteen an apprentice carpenter; organized trade unions, 1892-1901; elected to first federal Senate, 1901; minister for defence, 1908-9, 1910-13, 1914-21, and 1931-4; minister for home and territories, 1921-6; vice-president, executive council, 1926-9; minister for external affairs, 1934-7; responsible for transportation and provisions of Australian troops, 1914-18; represented Australia at Washington conference, 1921-2; led Australian delegation, League of Nations Assembly, Geneva, 1927; expelled from federal labour party when he supported conscription, 1916; lost parliamentary seat, 1937; chairman, Defence Board of Business Administration, 1940-7; PC, 1921; KCVO, 1927.

PEARCE, SIR (STANDEN) LEONARD (1873-1947), electrical engineer; studied at Finsbury Technical College and served apprenticeship; deputy chief electrical engineer (1901-4), chief electrical engineer (1904-25), Manchester corporation; erected Barton power-station; electricity commissioner, 1925; engineer-in-chief, London Power Company, 1926-47; collaborated with Sir Giles Gilbert Scott in designing Battersea power-station; knighted, 1935.

PEARCE, STEPHEN (1819-1904), portrait and equestrian painter; pupil of Sir Martin Archer Shee [q.v.], 1841; exhibited at Royal Academy, 1849-85; painted portraits of Arctic explorers (now in National Portrait Gallery) and equestrian portraits, the chief being 'Coursing at Ashdown Park', 1869; wrote *Memories of the Past* (1903).

PEARCE, SIR WILLIAM GEORGE, second baronet, of Carde (1861-1907), shipbuilder; son of Sir William Pearce [q.v.]; BA and LLB, Trinity College, Cambridge, 1884; MA, 1888; chairman of Fairfield Shipbuilding Company, Glasgow, 1888; conservative MP for Plymouth, 1892-5; left property estimated at £150,000 to Trinity College, Cambridge.

PEARS, Sir EDWIN (1835-1919), barrister-at-law, publicist, and historical writer; called to bar (Middle Temple), 1870; practised at consular bar, Constantinople, 1873-1914; president, 1881; knighted, 1909; died at Malta; wrote *Life of Abdul Hamid* (1917) and two monographs on Byzantine history.

PEARSALL, WILLIAM HAROLD (1891-1964), ecologist and professor of botany; educated at Ulverston grammar school and Victoria University, Manchester; first class, botany, 1913; M.Sc., 1915; on active service in France with RGA, 1916-19; lecturer, Leeds University, and D.Sc., Manchester, 1920; reader in botany, Leeds, 1922-38; professor of botany, Sheffield, 1938; FRS, 1940; Quain professor, University College, London, 1944-57; published studies of lakes and their vegetation (1917-21); and of bogs and soils, 1937; publications include *Mountains and Moorlands* (1950) and *Report on an Ecological Survey of Serengeti National Park, Tanganyika* (1956); helped to establish the Freshwater Biological Association, 1929; president, British Ecological Society, 1936; edited *Journal of Ecology* and *Annals of Botany*; hon. D.Sc. Durham and Birmingham; president, Institute of Biology, 1957-8; chairman, scientific policy committee, Nature Conservancy, 1953-63.

PEARSALL SMITH, (LLOYD) LOGAN (1865-1946), writer. [See SMITH.]

PEARSON, ALFRED CHILTON (1861-1935), classical scholar; first class, classics, Christ's College, Cambridge, 1881-3; taught at Dulwich College, 1893-1900; professor of Greek, Liverpool (1919-21), Cambridge (1921-8); works include school editions of Euripides, an edition of Sophocles' *Fragments* (3 vols., 1917), and edition of Sophocles for 'Oxford Classical Texts' (1924); FBA, 1924.

PEARSON, CHARLES JOHN, Lord PEARSON (1843-1910), Scottish judge; BA, Corpus Christi College, Oxford (first class, *lit. hum.*), 1865; called to English and Scottish bars, 1870; knighted, 1887; sheriff of Renfrew and Bute, 1888, and Perthshire, 1889; conservative MP for Edinburgh and St. Andrews universities, 1890-6; QC, 1890; solicitor-general for Scotland, 1890-2, 1895-6; lord advocate and PC, 1891; raised to Scottish bench, 1896.

PEARSON, Sir CYRIL ARTHUR, first baronet (1866-1921), newspaper proprietor; brought out *Pearson's Weekly*, 1890, and *Daily Express*, 1900; purchased *Standard*, 1904; sold these papers on going blind, 1910-12; founded St. Dunstan's, Regent's Park, for blinded soldiers and sailors, 1915; baronet, 1916; GBE, 1917.

PEARSON, KARL (1857-1936), mathematician and biologist; educated at University College School, King's College, Cambridge (third wrangler, 1879), and in Germany; studied a wide variety of subjects; called to bar (Inner Temple), 1882; Goldsmid professor of applied mathematics and mechanics (1884-1911), Galton professor of eugenics (1911-33), University College, London; Gresham professor of geometry, 1891; applied the calculus of probabilities to biological data; his enthusiasm for Darwinian theory of evolution by natural selection never flagged and statistical verifications of it were his principal interest; his statistical methods applicable in many fields of scientific research, widely studied, and responsible for establishment of many statistical laboratories; founder (1901) and editor of *Biometrika*; his *Grammar of Science* (1892) widely read; among the most influential university teachers of his time; admired and feared rather than loved; FRS, 1896.

PEARSON, WEETMAN DICKINSON, first VISCOUNT COWDRAY (1856-1927), contractor; partner in grandfather's firm of contractors, Bradford, afterwards London, 1875; went to Mexico, 1889; acquired extensive tracts of land, rich in oil; firm's contracts included completion of Blackwall tunnel under Thames, extension of Dover harbour, and building of dam across Blue Nile above Khartoum; liberal MP, Colchester, 1895-1910; baronet, 1894; baron, 1910; viscount, 1917; a generous public benefactor.

PEASE, Sir ARTHUR FRANCIS, first baronet, of Hummersknott (1866-1927), coal-owner and industrialist; grandson of Joseph Pease [q.v.]; chairman and managing director of Pease & Partners, Darlington, 1906; became known as prominent negotiator when organized demand for minimum wage arose among miners; second civil lord of Admiralty, 1918-19; baronet, 1920.

PEASE, EDWARD REYNOLDS (1857-1955), founder member and secretary, Fabian Society; educated at home; met Frank Podmore [q.v.], biographer of Robert Owen [q.v.]; together they founded the Fabian Society, 1884; trained and worked as a cabinet-maker under the influence of William Morris [q.v.], 1866-9;

paid secretary, Fabian Society, 1890–1913; hon. secretary, 1913–39; published *History of the Fabian Society* (1916).

PEASE, JOSEPH ALBERT, first BARON GAINFORD (1860–1943), politician and man of business; son of Sir J. W. Pease [q.v.]; educated as a Quaker and at Trinity College, Cambridge; notable sportsman; entered family iron and coal business; liberal MP, Tyneside (1892–1900), Saffron Walden (1901–10), Rotherham (1910–17); chief whip and PC, 1908; chancellor, duchy of Lancaster, 1910–11; president, Board of Education, 1911–15; postmaster-general, 1916; baron, 1917; chairman (1922–6), vice-chairman (1926–32), British Broadcasting Corporation.

PEASE, SIR JOSEPH WHITWELL, first baronet, of Hutton Lowcross and Pinchinthorpe (1828–1903), director of mercantile enterprise; entered Pease banking and other firms at Darlington, 1845; chairman of North Eastern Railway, 1894; liberal MP for South Durham, 1865–85, and Barnard Castle, 1885–1903; president of Peace Society; first Quaker baronet, 1882.

PEAT, STANLEY (1902–1969), professor of chemistry in University College of North Wales, Bangor; educated at Rutherford College and Armstrong College, Newcastle; first class, chemistry, 1924; research under Professor (Sir) W. N. Haworth at Newcastle, and from 1925, at Birmingham; Ph.D., 1928; lecturer in biochemistry, Birmingham, 1928–34; returned to chemistry department, 1934; worked with Haworth on chemistry of sugars and polysaccharides; results published in *Journal of the Chemical Society*; D.Sc., and reader, 1944; during 1939–45 war worked for government departments and became interested in synthesis of starch; FRS, 1948; professor of chemistry, Bangor, 1948–69; published results of research on starch and other plant polysaccharides; dean of faculty of science, Bangor; consultant to number of research associations.

PEEK, SIR CUTHBERT EDGAR, second baronet (1855–1901), amateur astronomer and meteorologist; BA, Pembroke College, Cambridge, 1880; MA, 1884; visited Iceland, 1881; set up observatories at Wimbledon, 1882, and Rousdon, 1884; observed transit of Venus in Queensland, 1882; at Rousdon made systematic astronomical and meteorological observations, of which he published annual reports; FRAS, 1884; FSA, 1890.

PEEL, ARTHUR WELLESLEY, first VISCOUNT PEEL (1829–1912), speaker of the House of Commons; youngest son of Sir Robert Peel, second baronet, prime minister; educated at Eton and Balliol College, Oxford; liberal MP, Warwick, 1865–85, Warwick and Leamington, 1885–95; speaker, 1884–95; first to use power of applying closure for combating Irish 'obstruction', 1885; viscount, 1895.

PEEL, SIR FREDERICK (1823–1906), railway commissioner; second son of Sir Robert Peel, second baronet; educated at Harrow and Trinity College, Cambridge; BA (sixth classic), 1845; MA, 1849; called to bar, 1849; liberal MP for Leominster, 1842–52, and for Bury, 1852–7, 1859–65; under-secretary for colonies, 1851–5; introduced clergy reserves bill, 1853; under-secretary for war, 1855–7; severely censured for failures of Crimean war; PC, 1857; financial secretary to the Treasury, 1859–65; KCMG, 1869; member of railway and canal commission from 1873 to death.

PEEL, JAMES (1811–1906), landscape painter; exhibited landscapes at Royal Academy, 1843–88, and Royal Society of British Artists from 1845; work shows sincere feeling for nature and excellent technique.

PEEL, WILLIAM ROBERT WELLESLEY, first EARL and second VISCOUNT PEEL (1867–1937), statesman; educated at Harrow and Balliol College, Oxford; called to bar (Inner Temple), 1893; member of London County Council (1900–19), chairman (1914–16); unionist MP, Southern division, Manchester (1900–6), Taunton (1909–12); succeeded father [q.v.], 1912; under-secretary for war, 1919–21; chancellor of duchy of Lancaster, 1921–2; first commissioner of works, 1924–8; secretary of state for India, 1922–4, 1928–9; member of India (1930–1), chairman of Burma (1931–2), Round Table conferences; member of Indian joint select committee, 1933; chairman, royal commissions on dispatch of business at common law (1934–6) and Palestine (1936–7); PC, 1919; earl, 1929.

PEERS, SIR CHARLES REED (1868–1952), antiquary; educated at Charterhouse and King's College, Cambridge; trained and worked as architect, 1893–1902; appointed architectural editor, *Victoria County Histories of England*, 1903; inspector (later, chief inspector) of ancient monuments, Office of Works, 1910–33; laid down principles governing architectural conservation in Britain; surveyor, Westminster Abbey, 1935; consulting architect to York Minster and Durham Cathedral; hon. editor, *Archaeological Journal*, 1900–3; secretary, Society of Antiquaries, 1908–21; CBE, 1924;

knighted, 1931; trustee, British Museum, 1929; trustee, London Museum, 1930.

PEERS, EDGAR ALLISON (1891-1952), Hispanic scholar and educationist; educated at Dartford grammar school, London University, and Christ's College, Cambridge; BA, London, 1910; first class, medieval and modern languages tripos, Cambridge, 1912; modern languages master, Mill Hill, Felsted, and Wellington, 1913-19; appointed to Gilmour chair of Spanish, Liverpool, 1920; published numerous text-books and study aids, notably *Spain, A Companion to Spanish Studies* (1929), *A Handbook to the Study and Teaching of Spanish* (1938), and *A Critical Anthology of Spanish Verse* (1948); edited quarterly *Bulletin of Spanish* (from 1949 *Hispanic*) *Studies*, 1923-52; founded the Institute of Hispanic Studies, Liverpool, 1934; published *Studies of the Spanish Mystics* (2 vols., 1927-30), *History of the Romantic Movement in Spain* (2 vols., 1940), and numerous travel books on Spain; hon. LLD, Glasgow, 1947.

PEET, THOMAS ERIC (1882-1934), Egyptologist educated at Merchant Taylors' School and Queen's College, Oxford; excavated for Egypt Exploration Society; lecturer in Egyptology, Manchester, 1913-28; Brunner professor, Liverpool, 1920-33; Laycock student, Worcester College, Oxford, 1923; university reader, 1933-4; edited *Journal of Egyptian Archaeology*, 1923-34; excelled in interpretation of Egyptian history and mathematics.

PEILE, SIR JAMES BRAITHWAITE (1833-1906), Indian administrator; son of Thomas Williamson Peile [q.v.]; BA, Oriel College, Oxford (first class classic), 1855; joined Indian civil service, 1856; settled claims of ruler of Bhavnagar against British government, 1859; under-secretary to Bombay government, 1860; passed enactments in favour of depressed land-owners in Bombay; registrar-general of assurances, 1867; director of public instruction, Bombay, 1869-73; as political agent of Kathiawar (1874-8) pacified discontent of native chiefs; built bridges, roads, and railways, 1877; averted threatened famine, 1877; member of famine commission, 1878-80; acting chief secretary to Bombay government, 1879-82; CSI, 1879; member of council at Bombay, 1882-6; carried through Local Boards and District Municipalities Acts, 1884, and amendment to Bombay land revenue code, 1886; vice-chancellor of Bombay University, 1884; member of supreme council, 1886-7; member of India Council in London, 1887-1902; urged increased powers for provincial councils;

member of royal commission on administration of Indian expenditure, 1895-1900; KCSI, 1888.

PEILE, JOHN (1837-1910), master of Christ's College, Cambridge, and philologist; educated at Repton and Christ's College, Cambridge; won Craven scholarship, 1859, and chancellor's medal, 1860; senior classic and fellow, 1860; tutor, 1870-84; master, 1887-1910; MA, 1863; Litt.D., 1884; teacher of Sanskrit at Cambridge till 1867; published volume on Greek and Latin philology, 1869, and a primer of philology, 1877; university reader in comparative philology, 1884-91; vice-chancellor, 1891-3; advocated degrees for women; president of council of Newnham College, 1895; hon. Litt.D., Trinity College, Dublin, 1892; FBA, 1904; compiled biographical register (vol. i, 1900) and history (1900) of Christ's College.

PELHAM, HENRY FRANCIS (1846-1907), Camden professor of ancient history, Oxford, from 1889; son of John Thomas Pelham, bishop of Norwich [q.v.]; educated at Harrow and Trinity College, Oxford; BA (first class, *lit. hum.*) and fellow, 1869; won chancellor's English essay prize, 1870; tutor, 1870-89; fellow of Brasenose, 1889; president of Trinity, 1897; original FBA, 1902; hon. LLD, Aberdeen, 1906; FSA, 1890; prominent in founding British Schools at Rome and Athens; published *Outlines of Roman History* (1893) and *Essays on Roman History* (posthumous, 1911).

PÉLISSIER, HARRY GABRIEL (1874-1913), comedian; directed troupe of entertainers named 'The Follies', famous for their 'potted plays' and 'potted opera', 1896-1912; opened on his own account in London, 1907; accomplished composer, producer, and comedian.

PELL, ALBERT (1820-1907), agriculturist; hon. MA, Trinity College, Cambridge, 1842; settled on farm at Hazelbeach, 1846; founder and first chairman (1866) of central chamber of agriculture; conservative MP for South Leicestershire, 1868-85; as guardian in Leicestershire and London, became an authority on poor law; advocated abolition of outdoor relief; chairman of central poor law conference, 1877-98; studied agricultural questions in Canada and America, 1879; sat on royal commissions on city guilds and aged poor; member of council of Royal Agricultural Society, 1886; pioneer of teaching of agriculture at Cambridge; hon. LLD, 1894; his *Reminiscences* edited by Thomas Mackay, 1908.

PEMBER, EDWARD HENRY (1833-1911), lawyer; educated at Harrow and Christ Church, Oxford; BA, 1854; called to bar (Lincoln's Inn),

1858; QC, 1874; treasurer, 1906-7; had large practice at parliamentary bar; conducted bill for creating Manchester Ship Canal, 1885; counsel for Cecil Rhodes [q.v.] before parliamentary committee to inquire into Jameson raid, 1897; prominent in London literary society; brilliant talker and accomplished musician; privately printed much 'vers de société', adaptations from Greek and Latin, and classical plays in English; contributed to *Grove's Dictionary of Music*.

PEMBERTON, THOMAS EDGAR (1849-1905), biographer of the stage; published novels, including *A Very Old Question* (1877), and comediettas; collaborated in plays *Sue* and *Held Up* with Bret Harte, whose life he wrote, 1902; dramatic critic of *Birmingham Daily Post*, 1882-1900; wrote memoirs of E. A. Sothern (1889), the Kendals (1891), T. W. Robertson (1892), and works on Dickens.

PEMBREY, MARCUS SEYMOUR (1868-1934), physiologist; educated at Christ Church, Oxford, and University College Hospital; lecturer in physiology, Charing Cross Hospital (1895-1900), Guy's Hospital (1900-20); professor, 1920-33; researched mainly on respiratory processes; FRS, 1922.

PENLEY, WILLIAM SYDNEY (1852-1912), actor-manager; grandson of Aaron Penley [q.v.]; went on stage, 1871; most successful in humorous parts such as Lay Brother Pelican in Chassaigne's *Falka* (1883), the title-role of *The Private Secretary* (1884-6), and, most notably, Lord Fancourt Babberley in *Charley's Aunt*, of which he gave 1,466 consecutive performances, 1892-7; managed several theatres.

PENNANT, GEORGE SHOLTO GORDON DOUGLAS-, second BARON PENRHYN (1836-1907), landowner. [See DOUGLAS-PENNANT.]

PENRHYN, second BARON (1836-1907), landowner. [See DOUGLAS-PENNANT, GEORGE SHOLTO GORDON.]

PENROSE, DAME EMILY (1858-1942), college principal; daughter of F. C. Penrose [q.v.]; studied languages and archaeology in Europe; first class, *lit. hum.*, Somerville College, Oxford, 1892; principal, Bedford College, 1893-8; professor of ancient history, 1894-8; principal, Royal Holloway College, 1898-1907; Somerville College, 1907-26; DBE, 1927.

PENROSE, FRANCIS CRANMER (1817-1903), archaeologist and astronomer; son of John Penrose [q.v.] and Elizabeth Penrose, 'Mrs Markham' [q.v.]; educated at Winchester and Magdalene College, Cambridge; BA, 1842; honorary fellow, 1884; rowed in university boat race, 1840-2; studied art and architecture in Europe, 1842-5; at Rome criticized pitch of pediment of Pantheon; at Athens (1845-7) made careful measurements of Greek classical buildings; published *Principles of Athenian Architecture* (1851, enlarged edn. 1888); surveyor of St. Paul's Cathedral from 1852, where he did much restoration and decoration; interested in astronomical study; published *The Prediction and Reduction of Occultations and Eclipses* (1869); observed eclipses of sun at Jerez (1870) and at Denver, USA (1878); made researches into orientation of temples; FRAS, 1867; FRS, 1894; built entrance gate at Magdalene and wing at St. John's, Cambridge; designed British School at Athens, 1882-6; president of Royal Institute of British Architects, 1894-6; FSA, 1898; hon. Litt.D., Cambridge, and DCL, Oxford, 1898.

PENSON, DAME LILLIAN MARGERY (1896-1963), historian; educated at Birkbeck and University Colleges, London; first class, history, 1917; Ph.D., 1921; junior administrative officer, Ministry of National Service, 1917-18; worked in War Trade Intelligence Department, 1918-19; lecturer, Birkbeck College, 1921-30; also part time at East London (later Queen Mary) College, 1923-5; professor of modern history, Bedford College, 1930; published *The Colonial Agents of the British West Indies* (1924), and assisted G. P. Gooch and H. W. V. Temperley [qq.v.] with *British Documents on the Origins of the War 1898-1914* (11 vols., 1926-38); produced, with Temperley, *Foundations of British Foreign Policy* (1938) and *A Century of Diplomatic Blue Books* (1938); dean of the faculty of arts (1938-44); chairman of the academic council, 1945; vice-chancellor, 1948-51; the first woman to hold such an office; acting chairman, United States Educational Commission in the United Kingdom, 1953-4; assisted in work for higher education in the colonies; DBE, 1951; hon. LLD, Cambridge, 1949; hon. DCL, Oxford, 1956; other hon. degrees; hon. fellow, Royal College of Surgeons; hon. vice-president, Royal Historical Society, 1959-63.

PENTLAND, first BARON (1860-1925), politician. [See SINCLAIR, JOHN.]

PEPLER, SIR GEORGE LIONEL (1882-1959), town planner; educated at Bootham School, York and the Leys School, Cambridge; trained as surveyor; specialized in town planning, 1905-14; chief technical planning officer, Local Government Board, 1914-46; helped to formulate Town and Country Planning Act, 1947; president, Town Planning Institute,

1919–20 and 1949–50; chairman, Institution of Professional Civil Servants, 1937–42; CB, 1944; knighted, 1948.

PEPPIATT, Sir LESLIE ERNEST (1891–1968), solicitor; educated at Bancroft's School, Woodford; articled to solicitor, 1907; admitted as solicitor, 1913; served in London Regiment, 1914–18; MC, and bar, 1918; legal adviser to British-American Tobacco Company, London, 1921–35; partner in Freshfields, Leese & Munns, solicitors to the Bank of England, 1935–62; principal legal adviser to the Bank; member of council of Law Society, 1940; president, 1958; knighted, 1959; concerned with management of important charities, including King George VI Memorial Fund; chairman, Discipline Committee, Architects Registration Council, 1949– 55; chairman, Departmental Committee of Betting on Horse-racing, 1960.

PERCIVAL, JOHN (1834–1918), school-master and bishop; BA, Queen's College, Oxford (double first class); first headmaster of Clifton College, for which he won recognized place among great public schools, 1862–79; president of Trinity College, Oxford, where he was less successful, 1879–87; canon of Bristol, 1882–7; headmaster of Rugby, 1887–95; bishop of Hereford, 1895–1917.

PERCY, ALAN IAN, eighth DUKE OF NORTH-UMBERLAND (1880–1930); brother of Earl Percy [q.v.]; joined Grenadier Guards, 1900; served in South African war, 1901–2, and with Egyptian army, 1907–10; official 'eye-witness' in France, 1914–16; subsequently served in War Office intelligence department; succeeded father, 1918; took prominent part in public life; extreme right-wing conservative; author of political and military essays and articles.

PERCY, EUSTACE SUTHERLAND CAMPBELL, BARON PERCY OF NEWCASTLE (1887–1958), politician and educationist; educated at Eton and Christ Church, Oxford; first class, modern history, 1907; entered diplomatic service, 1909; served in Washington embassy, 1910–14; Foreign Office, 1914–18; attended peace conference as assistant to Lord Robert Cecil (later Viscount Cecil of Chelwood, q.v.), 1919; resigned from diplomatic service, 1919; conservative MP, Hastings, 1921–37; PC, and president, Board of Education, 1924–9; published *Democracy on Trial* (1931), *Government in Transition* (1934), *The Study of History* (1935), and *John Knox* (1937); rector, King's College, Newcastle upon Tyne, 1937–52; published *The Heresy of Democracy* (1954) and *Some Memories* (1958); created Baron Percy of Newcastle, 1953.

PERCY, HENRY ALGERNON GEORGE, EARL PERCY (1871–1909), politician and travel-ler; educated at Eton and Christ Church, Oxford; first class, *lit. hum.*, 1893, Newdigate prize for English verse, 1892; conservative MP for South Kensington, 1895–1909; parliamen-tary under-secretary for India, 1902–3, and for foreign affairs, 1903–5; published accounts of travel in *Notes of a Diary in Asiatic Turkey* (1898) and *The Highlands of Asiatic Turkey* (1901); trustee, National Portrait Gallery, 1901; hon. DCL, Durham, 1907.

PEREIRA, GEORGE EDWARD (1865–1923), soldier and traveller; joined army, 1884; served in China, South Africa, Manchuria; military attaché, Peking, 1905–10; served in 1914–18 war; brigadier-general, 1919; left Peking for Lhasa, Jan. 1921; arrived Oct. 1922, after travelling nearly 7,000 miles; died at Kanze, Szechwan, *en route* to explore Tibetan-Szechwan border; travelled nearly 45,000 miles on foot in China; CMG, 1905; CB, 1917.

PERKIN, ARTHUR GEORGE (1861–1937), organic chemist; son of Sir W. H. and brother of W. H. Perkin [qq.v.]; educated at City of London School, Royal College of Chemistry, and at Leeds; lecturer and research chemist, Leeds, 1892; professor of colour chemistry and dyeing, 1916–26; an international authority on natural colouring matters; FRS, 1903.

PERKIN, Sir WILLIAM HENRY (1838–1907), chemist; pupil and assistant of Hofmann at Royal College of Chemistry; discovered 'azodinaphthyldiamine', 1855, and the first aniline dye, 'aniline purple' or 'mauve', 1856; opened chemical factory at Greenford Green for its manufacture; improved methods of silk dyeing; developed coal-tar industry by cheapen-ing process of manufacturing artificial alizarin; published description of 'Perkin synthesis' of unsaturated organic acids, 1867; devoted main energies to investigation of constitution of chemical molecules; president of Chemical Society, 1883–5, and Society of Chemical Industry, 1884–5; FRS, 1866; royal medallist, 1879; master of Leathersellers' Company, 1896; knighted on jubilee of aniline discovery (1906), and received hon. degrees from English and foreign universities; 'Perkin memorial fund' and 'Perkin medal' founded for researches in chemical industry.

PERKIN, WILLIAM HENRY (1860–1929), organic chemist; son of Sir W. H. Perkin [q.v.]; educated at City of London School; entered Royal College of Science, South Kensington, 1877; studied chemistry at Würzburg and at Munich (under Adolf von Baeyer); professor of chemistry, Heriot-Watt College, Edinburgh, 1887; professor of organic chemistry, Owens College, Manchester, 1892; organized school of organic chemistry, prosecuting analytic and synthetic investigations, directing and encouraging work of others, and providing necessary facilities; Waynflete professor of chemistry, Oxford, 1912; produced change in attitude of Oxford to chemistry; established strong school of original research; new laboratory completed through munificence of C. W. Dyson Perrins, 1915; fellow of Magdalen College; FRS, 1890; his researches include work on formation of rings of carbon atoms and berberine; a lucid teacher; a skilled amateur musician and gardener; collaborated with F. S. Kipping [q.v.] in writing three chemical works.

PERKINS, SIR ÆNEAS (1834-1901), general; colonel commandant Royal Engineers (late Bengal) from 1895; joined Bengal engineers, 1851; served in relief of Delhi, 1856; constructed 'Perkins' battery' at Delhi; executive engineer at Morshedabad and Darjeeling, 1865-78; superintending engineer in North-West Provinces, 1870; as commanding royal engineer of Kuram field force (1878) facilitated advance on Kabul; CB, 1879; present at victory of Charasiab and entry into Kabul; commanding royal engineer in march to Kandahar, 1880; inspector-general of military works, 1881; chief engineer of Central Provinces, 1883-6, and of Punjab, 1886-9; major-general, 1887; commanded Oudh division, 1890-1; lieutenant-general, 1890; KCB, 1897.

PERKINS, ROBERT CYRIL LAYTON (1866-1955), entomologist; educated at Merchant Taylors' School and Jesus College, Oxford; while still at school, met Edward Saunders [q.v.] who encouraged his interest in aculeate hymenoptera; spent most of ten years, 1891-1900, collecting in the Hawaiian Islands; worked for Hawaiian board of agriculture, 1902-4; director, entomology division, Hawaiian sugar planters experimental station, 1904-12; settled in Devon and worked mainly on taxonomy; D.Sc., Oxford, 1906; gold medallist, Linnean Society, 1912; FRS, 1920.

PERKS, SIR ROBERT WILLIAM, first baronet (1849-1934), Methodist, industrialist, and politician; educated at Kingswood School, Bath, and King's College, London; railway lawyer associated with Metropolitan District Railway, harbour works at Barry, Rio de Janeiro, and Buenos Aires, the Andes and Severn tunnels; liberal MP, Louth division, 1892-1910; originator of influential nonconformist parliamentary committee; organized

Methodist million guinea fund and worked successfully for Methodist union (1932).

PEROWNE, EDWARD HENRY (1826-1906), master of Corpus Christi College, Cambridge; BA, Corpus Christi College, Cambridge (senior classic), 1850; MA, 1853; DD, 1873; fellow, 1850; master, 1879-1906; vice-chancellor of university, 1879-81; a strong evangelical; published theological works.

PEROWNE, JOHN JAMES STEWART (1823-1904), bishop of Worcester; brother of E. H. Perowne [q.v.]; BA, Corpus Christi College, Cambridge, 1845; MA, 1848; DD, 1873; fellow, 1849; vice-principal of St. David's College, Lampeter, 1862-72; canon of Llandaff, 1869-78; member of Old Testament revision company, 1870; fellow of Trinity College, Cambridge, 1873-5; Hulsean professor of divinity, 1875-8; dean of Peterborough, 1878-91; member of ecclesiastical county commission, 1881; bishop of Worcester, 1891-1901; obtained suffragan bishopric and facilitated division of diocese; published translation and commentary on the Psalms (1864), and edited *Cambridge Bible* and *Greek Testament for Schools*, and the *Letters of Bishop Thirlwall* (1881).

PERRING, WILLIAM GEORGE ARTHUR (1898-1951), director of the Royal Aircraft Establishment; apprenticed at Royal Naval Dockyard, Chatham, 1913; won scholarship to Royal Naval College, Greenwich, 1919-22; first class professional certificate, naval architecture; research scholar, National Physical Laboratory, Teddington, 1923-5; joined Royal Aircraft Establishment, Farnborough, 1925; worked on wind tunnels; promoted to take charge of design and construction of new high-speed tunnel, 1937; superintendent, scientific research, 1940; deputy director, Royal Aircraft Establishment, 1941; director, 1946-51; CB, 1949.

PERRINS, CHARLES WILLIAM DYSON (1864-1958), collector and benefactor; educated at Charterhouse and the Queen's College, Oxford; served in 4th battalion, Highland Light Infantry, 1888-92; joined family business, Lea and Perrins, makers of Worcester sauce; mayor of Worcester, 1897; high sheriff, Worcestershire, 1899; life governor, Birmingham University; member of Council, Malvern College; gave Malvern hospital and public library in conjunction with Carnegie Trust; benefactor to Oxford University; hon. DCL, Oxford, 1919; acquired manuscripts and printed books of finest quality, 1900-20; publications about his library include the *Descriptive Catalogue* of his illuminated manuscripts by Sir George Warner [q.v.]; he was a generous benefactor to the Victoria and

Albert Museum, the National Gallery, the Ashmolean, and the British Museum; his library of printed books sold to save Royal Worcester Porcelain Factory from closure, 1946-7.

PERRY, SIR (EDWIN) COOPER (1856-1938), physician and administrator; scholar of Eton and King's College, Cambridge; senior classic, 1880; MRCS (London Hospital), 1885; dean of medical school (1888), superintendent (1892-1920), Guy's Hospital; vice-chancellor (1917-19), principal (1920-6), London University; prominent in foundation of London School of Hygiene and Tropical Medicine, 1924; knighted (1903) for services in founding Royal Army Medical College, Millbank; GCVO (1935) for services to King Edward's Hospital Fund.

PERRY, WALTER COPLAND (1814-1911), schoolmaster and archaeologist; educated at Manchester College, York, and Göttingen; Ph.D., 1837; classical tutor at Manchester College, 1837-8; Congregational minister at George's Meeting, Exeter, 1838-44; called to bar, 1851; schoolmaster at Bonn, 1844-75; secured the formation (1878) of large collection of casts at British Museum, of which he published catalogue, 1884; voluminous writings include *German University Education* (1845), *The Francks* (1857), *Greek and Roman Sculpture* (1882), and *Sicily in Fable, History, Art, and Song* (1908).

PERTH, sixteenth EARL OF (1876-1951), first secretary-general of the League of Nations. [See DRUMMOND, JAMES ERIC.]

PETAVEL, SIR JOSEPH ERNEST (1873-1936), engineer and physicist; educated at Lausanne and University College, London; designed the 'Petavel gauge' for measuring pressures set up in exploding gaseous mixtures; professor of engineering and director of Whitworth laboratories, Manchester, 1908-19; authority on aeronautics; director, National Physical Laboratory, 1919-36; FRS, 1907; KBE, 1920.

PETERSON, SIR MAURICE DRUMMOND (1889-1952), diplomatist; son of (Sir) William Peterson [q.v.]; educated at Rugby and Magdalen College, Oxford; first class, modern history, 1911; entered foreign service, 1913; foreign trade department (later Ministry of Blockade), 1916-18; eastern department, Foreign Office, under Lord Curzon [q.v.], 1918-20; second secretary, Washington, 1920; first secretary, Prague, 1923, and Tokyo, 1924; counsellor, Madrid, 1929-31; minister, Bul-

garia, 1936; ambassador, Baghdad, 1938, and to Franco Spain, 1939-40; ambassador, Turkey, 1944-6; Moscow, 1946-8; CMG, 1933; KCMG, 1938; GCMG, 1947; published memoirs, *Both Sides of the Curtain* (1950).

PETERSON, SIR WILLIAM (1856-1921), classical scholar and educationist; educated at Edinburgh University and Corpus Christi College, Oxford; first principal of University College, Dundee, 1882-95; principal of McGill University, Montreal, 1895-1919; KCMG, 1915.

PETHICK-LAWRENCE, FREDERICK WILLIAM, BARON PETHICK-LAWRENCE (1871-1961), social worker and politician; educated at Eton and Trinity College, Cambridge; first class, part i, mathematics tripos, and part i, natural sciences tripos, 1894-5; president of the Union, 1896; fellow, Trinity College, 1897-1903; called to bar (Inner Temple), 1899; social worker in East End of London together with his wife, Emmeline Pethick, 1901; owner and editor, the *Echo*, 1902-5; took up cause of women's suffrage; joint editor with his wife of *Votes for Women* (1907); both sentenced to imprisonment, 1912; disagreed with Emmeline Pankhurst [q.v.]; conscientious objector during 1914-18 war; labour MP, West Leicester, 1923-31; financial secretary to the Treasury, 1929-31; re-elected to parliament, 1935; PC, 1937; published autobiography, *Fate Has Been Kind* (1942); secretary of state for India and Burma, 1945; baron; led cabinet mission to India, 1946; no agreement possible between Congress and the Muslim League; resigned, 1947; continued to support government in legislation granting independence to India and Burma.

PETIT, SIR DINSHAW MANOCKJEE, first baronet (1823-1901), Parsi merchant and philanthropist; born at Bombay; partner in father's broking business at Bombay, 1852-64; started Manockjee Petit cotton mill at Bombay, 1860; acquired chief interest in other mills, and converted Bombay into great industrial centre; sheriff of Bombay, 1886-7; first Parsi member of legislative council of governor-general, 1866-8; knighted, 1887; baronet, 1890; founded at Bombay hospitals for women and children, lepers, and animals.

PETRE, SIR GEORGE GLYNN (1822-1905), diplomatist; entered diplomatic service, 1846; held minor posts at several European capitals till 1868; secretary of embassy at Berlin, 1868-72; British envoy at Buenos Aires, 1881-4, and at Lisbon, 1884-93; conciliated Portuguese ministers in long dispute concerning free transit of English vessels and missions over waterways of Zambezi, Shiré, and Pungwe rivers, 1890-1; CB, 1886; KCMG, 1890.

PETRIE, Sir DAVID (1879-1961), public servant; educated at Aberdeen University, 1900; entered Indian police and posted to Punjab, 1900; assistant to deputy inspector-general, Punjab CID, 1909; assistant director, Central CID India, 1911; on special duty investigating subversion and terrorism, 1914-18; on staff of the Prince of Wales during Indian tour, 1922; director, Intelligence Bureau, Home Department, government of India, 1924-31; member, Public Services Commission, India, 1931-2; chairman, 1932-6; King's Police Medal, 1914; CIE, 1915; OBE, 1918; CBE, 1919; CVO, 1922; knighted, 1929; retired, 1936; commissioned in Intelligence Corps and posted to Cairo, 1940; recalled to London to report on organization of MI5, 1940; director-general, MI5, 1941-6; KCMG, 1945.

PETRIE, WILLIAM (1821-1908), electrician; studied magnetism and electricity at Frankfort-on-Main, 1840; made study of electric lighting problems, 1846-53; invented self-regulating arc lamp, 1847-8; efforts to promote electric illumination financially disastrous; subsequently devoted himself to improving electro-chemical processes.

PETRIE, Sir (WILLIAM MATTHEW) FLINDERS (1853-1942), Egyptologist; son of William Petrie and grandson of Matthew Flinders [qq.v.]; educated at home; studied the Pyramids and published *The Pyramids and Temples of Gizeh* (1883); for Egypt Exploration Fund excavated Naucratis with F. L. Griffith and E. A. Gardner [qq.v.], 1884-5; thereafter worked independently, financed by popular subscriptions to his 'British School of Archaeology in Egypt'; discoveries included early royal tombs at Abydos, Pre-Dynastic cultures, the Hyksos, and the Sinaitic monuments; revolutionized methods of excavators by careful noting and comparison of details; introduced sequence-dating for Egyptian pottery; excavated in Palestine after 1926; professor of Egyptology, London, 1892-1933; FRS, 1902; FBA, 1904; knighted, 1923.

PETTER, (WILLIAM) EDWARD (WILLOUGHBY) (1908-1968), aircraft designer; son of (Sir) Ernest Willoughby Petter; educated at Marlborough College and Gonville and Caius College, Cambridge; first class, mechanical sciences tripos, 1929; apprentice, Westland Aircraft Works, 1929; personal assistant to managing director, 1932; technical director, with his father as chairman, 1935; concerned

with production of Lysander monoplane; designed the Whirlwind fighter plane; chief engineer, English Electric Company, 1944; concerned with design of the Canberra; designed the supersonic Lightning; managing director and chief engineer, Folland Aircraft Ltd., 1950; designed Gnat and Midge, 1954; resigned when Follands taken over by Hawker Siddeley Group, 1959; fellow, Royal Aeronautical Society, 1944; CBE, 1951.

PETTIGREW, JAMES BELL (1834-1908), anatomist; studied arts at Glasgow University and medicine at Edinburgh; MD, 1861; Croonian lecturer at Royal Society, 1860; FRS, 1869; curator of museum of Royal College of Surgeons, Edinburgh, 1869; FRS, Edinburgh, 1872; FRCP, 1873; Chandos professor of medicine and anatomy, St. Andrews, 1875; hon. LLD, Glasgow, 1883; studied flight of insects and birds, and experimented in artificial flight; published *Animal Locomotion* (in 'International Scientific' series, 1873), and *Design in Nature* (3 vols., 1908).

PETTY-FITZMAURICE, EDMOND GEORGE, Baron Fitzmaurice (1846-1935), better known as Lord Edmond Fitzmaurice, statesman and historian; son of the fourth and brother of the fifth Marquess of Lansdowne [qq.v.]; educated at Eton and Trinity College, Cambridge; president of Union, 1866; first class, classics, 1868; liberal MP, Calne (1868-85), Cricklade division (1898-1905); commissioner at Constantinople under treaty of Berlin, 1880-1; under-secretary of state for foreign affairs, 1883-5, 1905-8; chancellor of duchy of Lancaster, 1908-9; chairman, Wiltshire county council (1896-1906), court of quarter-sessions (1899-1905); wrote biographies of William, Earl of Shelburne (3 vols., 1875-6) and Lord Granville (2 vols., 1905); baron, 1906; PC, 1908; FBA, 1914.

PETTY-FITZMAURICE, HENRY CHARLES KEITH, fifth Marquess of Lansdowne (1845-1927); son of fourth Marquess of Lansdowne [q.v.]; educated at Eton and Balliol College, Oxford; influenced by Benjamin Jowett [q.v.]; succeeded father, 1866; inherited liberal traditions; junior lord of Treasury, 1869; under-secretary for war, 1872-4; under-secretary of state for India, 1880; resigned, owing to his lack of sympathy as Irish landlord with Gladstone's Irish policy, 1880; governor-general of Canada, 1883-8; his period of office saw completion of Canadian Pacific Railway (1886), dispute with USA over delimitation of Newfoundland fisheries, and second rebellion of Louis Riel (q.v., 1885); viceroy of India, 1888-94; his term of office marked by no sensational incidents, although troublesome situation arose in connection with Indian currency, with which he dealt successfully; secretary of state for war, 1895-1900; supported new commander-in-chief, Lord Wolseley [q.v.], and carried through reforms which transferred much administrative power to secretary of state; greatly but unjustly criticized for initial British military failures in South African war, 1899; secretary of state for foreign affairs, 1900-5; excellently equipped for his office; responsible for negotiation of the two great alliances with Japan (1902) and France (1904); made unsuccessful approaches for alliance with Germany; came into conflict with USA over blockade of Venezuela (1903) and question of Alaskan boundary (1903); leader of conservatives in House of Lords, 1903; on accession of liberals to power (1905) had to decide how far revisionary powers of Upper House should be exercised in constant collisions of opinion between two Houses; introduced bill for internal reform of House of Lords, 1911; abstained from voting on parliament bill, but failed to carry with him whole of his followers, 1911; held meeting with Bonar Law and a few other colleagues at Lansdowne House which pledged unionist party to support France and had effect of deciding cabinet to embark on war, 2 Aug. 1914; joined first coalition administration as member of inner committee responsible for conduct of war; at invitation of prime minister, Asquith, set out in private memorandum his views as to possibility of peace 'of accommodation', Nov. 1916; this memorandum afterwards held to be prime cause of break-up of first coalition government; published his famous letter to *Daily Telegraph* on same lines as memorandum of 1916, 29 Nov. 1917; although written with approval of his colleagues, the letter, which appeared at inopportune moment, was repudiated by government and press, and brought on him reproaches for disloyalty to allied cause and 'excommunication' from conservative party; during last years suffered from ill health and lived in retirement; felt very keenly destruction of his Irish home, Derreen, county Kerry, by 'Irregulars', 1922; KG, 1894.

PEULEVÉ, HENRI LEONARD THOMAS (HARRY) (1916-1963), British agent in enemy-occupied France; educated in number of schools in England, France, and Algiers; trained in telegraphy and wireless in London; technical assistant, Baird Television Company; cameraman, BBC; commissioned in RAOC, 1939; seconded to SOE, 1942; parachuted into France, 1942; injured, but escaped back to England; landed in France, 1943; trained resistance fighters and led them in sabotage operations; arrested as black marketeer, but undiscovered as a British agent, 1944; wounded,

attempting to escape; sent to Buchenwald; escaped, 1945; DSO; chevalier of Legion of Honour; MC, and croix de guerre; worked for Shell Oil Company after the 1939-45 war.

PHEAR, SIR JOHN BUDD (1825-1905), judge in India and author; BA, Pembroke College, Cambridge (sixth wrangler), 1847; MA, 1850; fellow of Clare, 1847-65; called to bar, 1854; judge of high court of Bengal, 1864-76; knighted, 1877; chief justice of Ceylon, 1877-9; revised civil and criminal code; chairman of Devonshire quarter-sessions, 1881-95; published *The Aryan village in India and Ceylon* (1880) and mathematical and legal works.

PHELPS, LANCELOT RIDLEY (1853-1936), provost of Oriel College, Oxford, and authority on poor law administration; educated at Charterhouse; at Oriel College (1872-1936) successively scholar, fellow, provost (1914-29), and honorary fellow; influential tutor, 1893-1914; deacon, 1879; priest, 1896; member, royal commission on the poor laws, 1905-9; chairman, departmental committee on relief of casual poor, 1929-30.

PHILBY, HARRY ST. JOHN BRIDGER (1885-1960), explorer and orientalist; born in Ceylon (Sri Lanka); educated at Westminster School and Trinity College, Cambridge; first class, modern languages tripos, 1907; entered Indian civil service, 1907; distinguished himself as linguist; assistant political officer, Mesopotamian Expeditionary Force, 1915-17; CIE, 1917; attached to mission to Abdul Aziz ibn Saud, 1917; made first east-west crossing of Arabia; disagreed with British government policy in Arabia and resigned from the Indian civil service, 1925; became unofficial counsellor to Ibn Saud and embraced Moslem faith, 1930; devoted much time to Arabian exploration; imprisoned under Section 18B, Defence Regulations, 1940; his criticism of Saudi inefficiency led to exile from Saudi Arabia, 1955; settled at Beirut; among his published works were *Heart of Arabia* (2 vols., 1922), *Sheba's Daughters* (1939), and *Arabian Highlands* (1952); father of Harold (Kim) Philby, intelligence agent of USSR.

PHILIP, SIR ROBERT WILLIAM (1857-1939), physician and founder of tuberculosis dispensaries; educated at Royal High School and University, Edinburgh; qualified, 1882; opened tuberculosis dispensary, Edinburgh, 1887; established 'co-ordinated Edinburgh system' of open-air school, sanatorium, tuberculosis colony, and hospital; his conception of uniform tuberculosis service adopted by British government, 1912; knighted, 1913; professor of tuberculosis, Edinburgh, 1917–39.

PHILIPPS, Sir IVOR (1861–1940), major-general and man of business; brother of Viscount St. Davids and Lord Kylsant [qq.v.]; served in Indian army, 1883–1903; DSO, 1900; major, 1901; liberal MP, Southampton, 1906–22; raised and served in France with 38th Welsh division, 1915–16; honorary major-general, 1916; governor of Pembroke Castle; director of City companies; KCB, 1917.

PHILIPPS, Sir JOHN WYNFORD, thirteenth baronet, and first Viscount St. Davids (1860–1938), financier; brother of Sir Ivor Philipps and Lord Kylsant [qq.v.]; educated at Felsted and Keble College, Oxford; liberal MP, Mid-Lanarkshire (1888–94), Pembrokeshire (1898–1908); chairman of twelve trust companies and of South American railway companies; baron, 1908; succeeded father in baronetcy, 1912; PC, 1914; viscount, 1918.

PHILIPPS, OWEN COSBY, Baron Kylsant (1863–1937), shipowner and financier; brother of Viscount St. Davids and Sir Ivor Philipps [qq.v.]; founded the King Line on the Clyde, 1888; director (1902), later chairman, Royal Mail Steam Packet Company; acquired the Pacific Steam Navigation Company, the Union-Castle Mail Steamship Company (1912), Harland & Wolff, Belfast (1924), the White Star Line (1927); sentenced to a year's imprisonment (1931) for publishing and circulating a false prospectus in 1928; liberal MP, Pembroke and Haverfordwest district (1906–10); conservative MP, Chester (1916–22); KCMG, 1909; GCMG, 1918; baron, 1923.

PHILLIMORE, JOHN SWINNERTON (1873–1926), classical scholar and poet; cousin of first Baron Phillimore [q.v.]; educated at Westminster and Christ Church, Oxford; first class, *lit. hum.*, 1895; Hertford, Craven, and Ireland scholar; president of Union, 1895; professor of Greek, Glasgow University, 1899; of humanity, 1906; a notable teacher; joined Roman Church, 1906; his works include text of Propertius (1901), *Index Verborum* of Propertius (1906), translation of Philostratus's *Apollonius of Tyana* (1912), and two volumes of poems (1902 and 1918).

PHILLIMORE, Sir RICHARD FORTESCUE (1864–1940), admiral; entered navy, 1878; captain, 1904; commanded *Inflexible* at battle of the Falkland Islands (1914) and at Dardanelles (1915); principal beach master, Gallipoli landings, 1915; rear-admiral, 1915; head of British naval mission in Russia, 1915–

16; commanded first battle cruiser squadron, Grand Fleet (1916–18), aircraft carriers, 1918; vice-admiral, 1920; commander-in-chief, Plymouth, 1923–6; admiral, 1924; KCMG, 1918; GCB, 1929.

PHILLIMORE, Sir WALTER GEORGE FRANK, second baronet, and first Baron Phillimore (1845–1929), judge, ecclesiastical lawyer, and international jurist; son of Sir R. J. Phillimore, first baronet [q.v.]; grandson of Joseph Phillimore [q.v.]; cousin of J. S. Phillimore [q.v.]; educated at Westminster and Christ Church, Oxford; first class, *lit. hum.* and law and modern history; fellow of All Souls; called to bar (Middle Temple), 1868; joined Western circuit; counsel in many famous ecclesiastical cases; for many years leader of Admiralty court; succeeded father, 1885; judge of Queen's Bench division, 1897–1913; not one of the greatest common law judges; lord justice of appeal and PC, 1913–16; distinguished himself in court of appeal; chairman, committee on a League of Nations, 1917–18; baron, 1918; acquired European reputation as international jurist; a devout churchman.

PHILLIPS, Sir CLAUDE (1846–1924), art critic; BA, London University; called to bar (Inner Temple), 1883; travelled abroad for artistic study; art critic to *Daily Telegraph*, 1897; keeper of Wallace Collection, 1897–1911; knighted, 1911; wrote five valuable numbers of 'Portfolio' monographs.

PHILLIPS, MORGAN WALTER (1902–1963), labour party organizer; left school at twelve to work in coal-mine; became active trade unionist; chairman, Bargoed Steam Coal Lodge, 1924–6; secretary, Bargoed labour party, 1923–5; awarded South Wales Miners' Federation Scholarship to Central Labour College, London, 1926–8; secretary and agent, West Fulham labour party, 1928–30; agent for Whitechapel, 1934; member Fulham Borough Council, 1934–7; joined headquarters staff at Transport House, 1937; served in Ministry of Information, 1939; party organizer for Eastern Counties, 1940–1; national secretary, executive committee, labour party, 1944–61; member of 1945 campaign committee; responsible for 1955 election campaign; supported Hugh Gaitskell [q.v.]; played leading part in re-creation of Socialist International; chairman, 1948–57; general secretary, labour party, 1960; published *Labour in the Sixties* (1960).

PHILLIPS, STEPHEN (1864–1915), poet and dramatist; educated at Oundle; member of theatrical company of (Sir) Frank Benson [q.v.], c.1885–92; adopted letters as profession, *c*.1898; poetical works include *Orestes and Other Poems* (1884), *Primavera* (1890, containing, with other poems, 'To a Lost Love' and 'A Dream'), 'Eremus' (1894), 'Christ in Hades' and 'The Apparition' (1897), 'Marpessa' and 'The Wife' (1898, the last four contained in *Poems*, 1898), *New Poems* (1908), and *The New Inferno* (1911); at his best in his shorter lyrics; dramatic works include *Paolo and Francesca* (1900), *Herod* (1901), *Ulysses* (1902), *The Sin of David* (1904), and *Nero* (1906); gained great reputation as dramatic poet, but his genius, although intense, of limited range; plays spoiled by over theatricality in production; sank into obscurity in latter years.

PHILLIPS, Sir TOM SPENCER VAUGHAN (1888–1941), admiral; entered navy, 1903; assistant director, plans division, 1930–2; director, 1935–8; rear-admiral, 1939; vice-chief of naval staff, May 1939–Oct. 1941; vice-admiral, 1940; admiral, 1941; commander-in-chief, Eastern fleet, 1941; on Japanese attack on Malaya, 8 Dec., set sail in *Prince of Wales* accompanied by *Repulse* for scene of invasion; attacked by Japanese aircraft, 10 Dec., with loss of both ships and own life; KCB, 1941.

PHILLIPS, WALTER ALISON (1864–1950), historian; educated at Merchant Taylors' and Merton College, Oxford; first class, modern history, 1885; Lecky professor, Trinity College, Dublin, 1914–39; publications include *Modern Europe, 1815–1899* (1901), *The Confederation of Europe* (1914), and *The Revolution in Ireland, 1906–1923* (1923).

PHILLIPS, WILLIAM, (1822–1905), botanist and antiquary; enthusiastic Volunteer in youth; engaged in botanical research, 1861; founded Shropshire Archaeological and Natural History Society, 1878; published *Guide to the Botany of Shrewsbury* (1878) and *Manual of the British Discomycetes* (1887); did much archaeological research in Shropshire; edited *Shropshire Notes and Queries*, and *The Ottley Papers* (1893–8), *Quarter Sessions Rolls of 1652–9*, and Shropshire parish registers.

PHILLPOTTS, Dame BERTHA SURTEES (1877–1932), educationist and Scandinavian scholar. [See NEWALL.]

PHILLPOTTS, EDEN (1862–1960), writer; born in India; educated at Mannamead school (later incorporated with Plymouth College); employed as clerk in Sun Fire office, 1879–89; assistant editor, *Black and White* magazine; published numerous books, including *Lying Prophets* (1897) and *Children of the Mist* (1898); collaborated with Jerome K. Jerome [q.v.] in

The Prude's Progress, his first play (1895); followed by *The Farmer's Wife* (produced in 1924) and *Yellow Sands* (1926); poems included *Wild Fruit* (1910); essays, *My Devon Year* (1904); humorous stories, *The Human Boy* (1899) and *The Human Boy Again* (1908).

PHILPOT, GLYN WARREN (1884–1937), painter; studied at Lambeth Art School, in Rouen and Paris; used underpainting and glaze in manner of Titian and Goya; works include 'Manuelito' (1909), 'Zarzarrosa' (1910), 'The Marble Worker' (1912), and 'Under the Sea' (1919); RA, 1923; much in demand as a portrait painter; after 1931 changed his style, endeavouring by direct painting with brighter colours, by simplification of form and more sensitive line to reveal essence of his vision.

PHIPPS, Sir ERIC CLARE EDMUND (1875–1945), diplomatist; educated on continent; entered diplomatic service, 1899; counsellor, 1919; minister plenipotentiary, 1922; minister in Vienna, 1928–33; ambassador in Berlin, 1933–7; in Paris, 1937–9; gave warning of Hitler's ambitions and danger of French demoralization; KCMG, 1927; PC, 1933; GCMG, 1934; GCVO, 1939; GCB, 1941.

PIATTI, ALFREDO CARLO (1822–1901), violoncellist and composer; born at Bergamo; studied at Milan, 1832–7; played for Liszt at Munich; first appeared in London, 1844; led violoncellos at leading concerts from 1851; played at Monday Popular Concerts from 1859 to 1896; died near Bergamo; unsurpassed in execution, tone, and expression; composed sonatas, concertos, and caprices.

PICK, FRANK (1878–1941), vice-chairman, London Passenger Transport Board; educated at St. Peter's School, York; qualified as solicitor and entered North Eastern Railway Company, 1902; joined London Underground group, 1906; complementary to Lord Ashfield [q.v.] in developing London transport; managing director, 1928; vice-chairman, London Passenger Transport Board, 1933–40; encouraged modern art in posters, lettering, and station design.

PICKARD, BENJAMIN (1842–1904), trade-union leader; worked in mines from age of twelve; secretary of West Yorkshire Miners' Association, 1876, and of Amalgamated Yorkshire Miners' Association, 1881; played leading part in miners' dispute of 1893; liberal MP for Normanton, 1885–1904.

PICKARD, Sir ROBERT HOWSON (1874–1949), chemist; studied at Mason College, Birmingham; B.Sc., London, 1895; Ph.D., Munich, 1898; head of chemistry department,

Blackburn Technical School, 1899-1920; principal, 1908-20; principal, Battersea Polytechnic, 1920-7; director, British Leather Manufacturers' Research Association, 1920-7, British Cotton Industry Research Association, 1927-43; vice-chancellor, London University, 1937-9; FRS, 1917; knighted, 1937.

PICKFORD, WILLIAM, Baron Sterndale (1848-1923), judge; BA, Exeter College, Oxford; called to bar (Inner Temple), 1874; joined Northern circuit, 1875; a 'local' at Liverpool; moved to London, 1892; QC, 1893; continued for a time on Northern circuit, of which he became *de facto* leader; did much work in Commercial Court, 1895-1907; recorder of Oldham, 1901; of Liverpool, 1904; leading counsel for Great Britain at inquiry in Paris on Dogger Bank incident, 1905; additional judge of High Court, King's Bench division, and knighted, 1907; proved ideal judge, the substance of his decisions being almost invariably right; active in promoting movements for unification of maritime law; lord justice of appeal and PC, 1914; member of Dardanelles commission, 1916; chairman, 1917; president of Probate, Divorce, and Admiralty division and created baron, 1918; master of the rolls, 1919.

PICKLES, WILLIAM NORMAN (1885-1969), general practitioner and epidemiologist; educated at Leeds grammar school; studied medicine at Yorkshire College and Leeds General Infirmary; qualified, 1909; MB, BS London, 1910; MD, 1918; locum tenens for Dr Hime of Aysgarth, 1912; purchased practice in partnership with Dean Dunbar, 1913; remained there until 1964, except for service in Royal Naval Volunteer Reserve as surgeon-lieutenant during 1914-18 war; traced source of epidemic of catarrhal jaundice, 1929; published in *Lancet* records of an outbreak of Sonne dysentery, 1932, and recorded in *British Medical Journal* epidemic of myalgia, or Bornholm disease, 1933; published *Epidemiology in Country Practice* (1939)—a medical classic; Milroy lecturer, Royal College of Physicians, London, 1942; Cutter lecturer at Harvard, 1948; MRCP, 1939; FRCP, 1963; hon. fellow, Royal College of Physicians, Edinburgh, 1955; hon. D.Sc. Leeds, 1950; first president, College of General Practitioners, 1953-6; CBE, 1957; hon. fellow, Royal Society of Medicine, 1965; hon. vice-president, British Medical Association.

PICTON, JAMES ALLANSON (1832-1910), politician and author; son of Sir James Allanson Picton [q.v.]; MA, London, 1855; Congregational minister at Manchester (1856-62), Leicester (1862-9), and Hackney (1869-79); his secular lectures on Sunday afternoons antici-

pated Pleasant Sunday Afternoon movement; member of London school board, 1870-9; radical MP for Leicester, 1884-94; works include *Oliver Cromwell* (1882), a life of his father (1891), *The Bible in School* (1901), *Pantheism* (1905), and *Spinoza* (1907).

PIERCY, WILLIAM, first Baron Piercy (1886-1966), economist and banker; left school at twelve; worked for Pharaoh Gane, timber brokers, 1898; studied at London School of Economics, 1910-13; first class, B.Sc. (Econ.); lecturer in history and public administration, LSE, 1914; worked as civil servant, 1914-18; CBE, 1919; joint managing director, Pharaoh Gane; played leading part in organizing first unit trusts; member, London Stock Exchange, 1934-42; headed British Petroleum Mission, Washington, during 1939-45 war; principal assistant secretary, Ministry of Supply and Ministry of Aircraft Production; personal assistant to deputy prime minister, C. R. (later Earl) Attlee; baron, 1945; first chairman, Industrial and Commerical Finance Corporation, Ltd., 1945; chairman, Estate Duties Investment Trust, 1952-66; appointed to court of Bank of England, 1946, 1950, and 1956; governor, LSE; member of senate and court of London University; president, Royal Statistical Society, 1954-5; chairman of Wellcome Trust, 1960-5; chairman of other similar bodies.

PIGOU, ARTHUR CECIL (1877-1959), economist; educated at Harrow and King's College, Cambridge; first class, history tripos, 1899; first class, part ii, moral sciences tripos, 1900; president, Cambridge Union; fellow of King's, 1902; Girdler's lecturer in economics, 1904; succeeded Alfred Marshall [q.v.] as professor of political economy, 1908; held chair until 1943; published *Principles and Methods of Industrial Peace* (1905), *Wealth and Welfare* (1912), renamed *The Economics of Welfare* in later editions; served in the Friends' Ambulance Unit during 1914-18 war; served on Cunliffe committee, 1918-19, and Chamberlain committee, 1924-5; his views attacked by J. M. (later Lord) Keynes [q.v.]; published further books, including, *A Study in Public Finance* (1928) and *The Theory of Unemployment*, 1933; engaged in controversy with Keynes, but later recognized importance of Keynes's theories.

PINERO, Sir ARTHUR WING (1855-1934), playwright; acted under Sir Henry Irving and Sir Squire Bancroft [qq.v.]; left stage, 1884; in fifty-five years wrote fifty-four plays of great variety and superb technical skill; at first excelled in farce but later, influenced by Ibsen, wrote serious, adult plays; works include *The Magistrate* (1885), *The Profligate* (1889), *The Second*

Mrs. Tanqueray (1893), *The Gay Lord Quex* (1899), *Letty* (1903), *His House in Order* (1906), *Mid-Channel* (1909), *The Enchanted Cottage* (1922); first English dramatist to cast plays to type; knighted, 1909.

PINSENT, Dame ELLEN FRANCES (1866–1949), pioneer worker in the mental health services; sister of Lord Parker [q.v.]; married (1888) H. C. Pinsent, solicitor; on royal commission on feeble-minded, 1904–8; honorary commissioner, Board of Control, 1913; commissioner, 1921–32; DBE, 1937.

PIPPARD, (ALFRED JOHN) SUTTON (1891–1969), engineer; educated at Yeovil School and Merchant Venturers' College, Bristol; first class, civil engineering, 1911; M.Sc., 1914; joined technical section, Air Department, Admiralty, 1914; contributed to improvement and safety of aircraft; MBE, 1918; published (with J. L. Pritchard) *Aeroplane Structures* (1919); D.Sc., Bristol, 1920; worked as consulting aeronautical engineer; lectured at Imperial College, London, 1919–22; professor of engineering, University College, Cardiff, 1922–8; professor, civil engineering, Bristol, 1928–33; professor, civil engineering, Imperial College, 1933–56; helped to design Dokan dam in Iraq, 1952–4; chairman, investigating committee on pollution of tidal Thames, 1951–61; publications include (with J. F. Baker), *The Analysis of Engineering Structures* (1936); FRS, 1954; president, Institution of Civil Engineers, 1958–9; pro-rector, Imperial College, 1955–6; hon. degrees from Bristol, Birmingham, and Brunel.

PIRBRIGHT, Baron (1840–1903), politician. [See DE WORMS, HENRY.]

PIROW, OSWALD (1890–1959), South African lawyer and politician; born of German parents; educated in the Transvaal, Germany, and England; called to bar (Middle Temple), 1913; practised at Pretoria bar, 1913; took silk, 1925; entered parliament as supporter of J. B. M. Hertzog [q.v.] and national party, 1924; minister of justice, 1929; minister of defence and of railways and harbours, 1934; toured western Europe and met Hitler, Mussolini, Salazar, and Franco, 1938; launched New Order movement in South Africa, based on Nazi ideology, 1940; his supporters defeated, he declined to stand for re-election, 1943; published *J. B. M. Hertzog* (1958).

PIRRIE, WILLIAM JAMES, Viscount PIRRIE (1847–1924), shipbuilder; born at Quebec; entered shipbuilding firm of Harland & Wolff, Belfast, 1862; partner, 1874; later chairman of board; identified with all important contemporary developments in naval architecture and marine engineering; many improvements due to him or his firm; his firm sole builders for White Star and Bibby lines; comptroller-general of merchant shipbuilding, 1918; lord mayor of Belfast, 1896–7; baron, 1906; viscount, 1921.

PISSARRO, LUCIEN (1863–1944), painter, wood-engraver, and printer of books; born in Paris, son of Camille Jacob Pissarro, the Impressionist painter, with whom he studied; established in England the Eragny Press (1894–1914), using 'Vale type' of C. de S. Ricketts [q.v.] and his own 'Brook type' and woodcuts; a landscape painter closely associated with W. R. Sickert and P. Wilson Steer [qq.v.] and exhibiting with New English Art Club; his painting in main stream through which Impressionism reached England; his technique (based on modified form of Seurat's *pointillisme*) of considerable influence; instrumental in founding Monarro Group; naturalized, 1916.

PITMAN, Sir HENRY ALFRED (1808–1908), centenarian; BA, Trinity College, Cambridge, 1832; MB, 1835; MD, 1841; FRCP, 1845; registrar of College of Physicians, 1858–89; assistant physician (1846) and physician (1857) at St. George's Hospital; active in organization of College of Physicians; knighted, 1883; largely responsible for first edition of *Nomenclature of Diseases* (1869).

PLASKETT, JOHN STANLEY (1865–1941), astronomer; born in Canada; worked as engineer; graduated in mathematics and physics, Toronto University, 1899; joined Ottawa Observatory, 1903; created astrophysical department, designed and installed several new instruments, and worked on solar rotation and measurement of stellar radial velocities; obtained establishment of Dominion Astrophysical Observatory at Victoria, British Columbia, with 72-inch reflector, 1918; retired, 1935; FRS, 1923.

PLATER, CHARLES DOMINIC (1875–1921), Catholic divine and social worker; educated at Stonyhurst; entered Jesuit novitiate, 1894; rector of Jesuit hall, Oxford, 1916; obtained for it, under title of Campion Hall, status of permanent private hall of university, 1918; founded Catholic Social Guild, 1909.

PLATTS, JOHN THOMPSON (1830–1904), Persian scholar; born at Calcutta; held educational posts in North-West and Central provinces of India, 1858–72; teacher of Persian in Oxford University, 1880; hon. MA, 1881; compiled Hindustani and Persian grammars and dictionaries, and translated Persian and Hindustani texts.

PLAYFAIR, SIR NIGEL ROSS (1874-1934), actor-manager; son of W. S. Playfair [q.v.]; educated at Harrow and University College, Oxford; exhibited his gift for dry but good-humoured comedy on the professional stage, 1902-18; among his scholarly and finished productions at the Lyric Theatre, Hammersmith (1920-34), of eighteenth-century comedy and twentieth-century satire was the famous revival of *The Beggar's Opera*, 1920; knighted, 1928.

PLAYFAIR, WILLIAM SMOULT (1835-1903), obstetric physician; brother of first Baron Playfair [q.v.]; MD, Edinburgh, 1856; professor of surgery at Calcutta Medical College, 1859-60; MRCP, 1863; FRCP, 1870; professor of obstetric medicine in King's College and obstetric physician to King's College Hospital, 1872-98; hon. LLD, St. Andrews, 1885, and Edinburgh, 1898; author of *Science and Practice of Midwifery* (1876) and *System of Gynaecology* (with Sir T. C. Allbutt, q.v.) (1896).

PLENDER, WILLIAM, BARON PLENDER (1861-1946), accountant; educated at Royal Grammar School, Newcastle upon Tyne; qualified as chartered accountant and entered Deloitte & Co., London; partner, 1897; senior partner, 1904-46; served on many government committees and commissions; Treasury controller, German, Austrian and Turkish banks, 1914-18; first independent chairman, national board for coal industry, 1921-5; knighted, 1911; GBE, 1918; baronet, 1923; baron, 1931.

PLEYDELL, JOHN CLAVELL MANSEL-(1817-1902), Dorset antiquary. [See MANSEL-PLEYDELL.]

PLIMMER, ROBERT HENRY ADERS (1877-1955), biochemist; son of Alfred Aders, who died in 1887; Mrs Aders married Henry George Plimmer, FRS, 1887; Robert adopted his stepfather's surname; educated at Dulwich College and University College, London; B.Sc., 1899; studied biochemistry at Geneva and Berlin; Ph.D. (Berlin) and D.Sc. (London), 1902; assistant in department of physiology, University College, London, 1904; studied chemistry of proteins; fellow, University College, 1906; assistant professor, physiological chemistry, 1907; university reader, 1912; first secretary, the Biochemical Club (later Society), 1911-19; worked in directorate of hygiene, War Office, 1914-18; biochemist, Rowett Institute for Research in Animal Nutrition, Aberdeen,

1919-22; professor of chemistry, St. Thomas's Hospital medical school, 1922-42; worked at Postgraduate Medical School, Hammersmith, 1943-55; published *Organic and Bio-Chemistry* (1915), *Analyses and Energy Values of Foods* (1921), and *History of the Biochemical Society* (1949).

PLUCKNETT, THEODORE FRANK THOMAS (1897-1965), legal historian; educated at schools in Leicester and Newchurch, Lancashire, and University College, London, 1915; research student, Emmanuel College, Cambridge, 1918; LLB, 1920; studied at Harvard Law School, 1921; instructor there, 1923-6; assistant professor, 1926-31; professor of legal history, London School of Economics, 1931-63; joint literary director, Selden Society, 1937; director, 1946; president, Royal Historical Society, 1948-52; president, Society of Public Teachers of Law, 1953-4; FBA, 1946; fellow, University College, London, 1950; hon. fellow, Emmanuel College, Cambridge, 1952; hon. LLD, Glasgow and Birmingham; hon. Litt.D., Cambridge; publications include *Concise History of the Common Law* (1929), *Legislation of Edward I* (1949), *Early English Legal Literature* (1958), *The Mediaeval Bailiff* (1954), and *Edward I and Criminal Law* (1960).

PLUMER, HERBERT CHARLES ONSLOW, first VISCOUNT PLUMER (1857-1932), field-marshal; educated at Eton; joined the 65th Foot (The York and Lancaster Regt.), 1876; captain, 1882; served in Sudan campaign, 1884; entered Staff College, 1885; led Matabele relief force and by prompt action saved the settlers (brevet lieutenant-colonel), 1896; raised the Rhodesian Horse at Bulawayo, 1899; commanded local forces north of Kimberley; commanded northernmost of three columns advancing on Pretoria; pursued De Wet for 800 miles; major-general, 1902; quartermaster-general, 1903-5; lieutenant-general, 1908; commander-in-chief, York, 1911-14; commanded the II Corps in France (1914-15), Second Army (1915-18); held the Ypres salient, 1915-17; captured Messines, Wytschaete, and the Oostaverne line, June 1917; fought series of operations at Passchendaele, July-Nov. 1917; in Italy (Nov. 1917-Mar. 1918) regrouped allied forces and saved the situation after Caporetto; back in France exchanged twelve good divisions for tired ones; lost Messines, Wytschaete, and Passchendaele but held the Ypres salient; commander in occupied territory, 1918-19; secured immediate dispatch of food but ruthless in putting down internal trouble; baron and field-marshal, 1919; as governor of Malta (1919-24) and high commissioner in Palestine (1925-8) combined firmness with sympathy;

KCB, 1906; GCMG, 1916; GCVO, 1917; GCB, 1918; GBE, 1924; viscount, 1929.

PLUMMER, HENRY CROZIER KEATING (1875-1946), astronomer and mathematician; educated at St. Edward's School and Hertford College, Oxford; first class, mathematics, 1897; second class, natural science, 1898; second assistant, Oxford University Observatory, 1901-12; professor of astronomy, Dublin, and royal astronomer for Ireland, 1912-21; of mathematics, Artillery College, Woolwich, 1921-40; publications include *Introduction Treatise on Dynamical Astronomy* (1918), *Principles of Mechanics* (1929), *Probability and Frequency* (1940), and many original papers.

PLUNKETT, EDWARD JOHN MORETON DRAX, eighteenth BARON OF DUNSANY (1878-1957), writer; educated at Eton and Sandhurst; succeeded to Dunsany title, 1899; served in Coldstream Guards during South African war; published stories, *The Gods of Pegana* (1905); first play, *The Glittering Gate* produced at Abbey Theatre, Dublin, 1909; *The Gods of the Mountain* produced in London, 1911; served as captain in Royal Inniskilling Fusiliers, 1914-18; lectured in United States, 1919-20; published novels, *The Chronicles of Rodriguez* (1922), *The Blessing of Pan* (1927), *The Curse of the Wise Woman* (1933), and *Rory and Bran* (1936); published three autobiographical works, *Patches of Sunlight* (1938), *While the Sirens Slept* (1944), and *The Sirens Wake* (1945); published a narrative poem, *A Journey* (1943); hon. Litt.D., Dublin, 1940.

PLUNKETT, SIR FRANCIS RICHARD (1835-1907), diplomatist; British envoy at Tokyo, 1883-8; KCMG, 1886; senior British delegate at conferences on revising treaties between Japan and European powers; minister at Stockholm, 1888-93, and Brussels, 1893-1900; British commissioner at sugar conference at Brussels, 1898-9; ambassador at Vienna, 1900-5; GCMG, 1894; PC, 1900; GCB, 1901; GCVO, 1903; died in Paris.

PLUNKETT, SIR HORACE CURZON (1854-1932), Irish statesman; educated at Eton and University College, Oxford; cattle-rancher in Wyoming (1879-89) and visited America annually thereafter; organized co-operative creameries in Southern Ireland, 1889; member of congested districts board, 1891-1918; unionist MP, South County Dublin, 1892-1900; president, Irish Agricultural Organization Society, 1894-9; vice-president, Department of Agriculture and Technical Instruction for Ireland, 1899-1907; spread at home and over-seas his philosophy of self-reliance and co-operation in rural life; established foundation in London to advance agricultural co-operation, 1919; chairman, conference on Empire agricultural co-operation, 1924; in politics converted to home rule but opposed partition; chairman, Irish convention, 1917-18; founded Irish Dominion League, 1919; senator, Irish Free State, 1922-3; Irish PC, 1897; KCVO, 1903; FRS, 1902.

PLUNKETT-ERNLE-ERLE-DRAX, SIR REGINALD AYLMER RANFURLY (1880-1967), admiral; brother of Baron of Dunsany [q.v.]; educated at Cheam School; entered *Britannia*, 1896; lieutenant, 1901; torpedo specialist; attended Military Staff College, Camberley, 1912; commander; war staff officer to Admiral (later Earl) Beatty [q.v.]; served in *Lion* at Heligoland Bight, Dogger Bank, and Jutland; captain, 1916; DSO; first director, Naval Staff College, Greenwich, 1919-22; president, allied naval control commission, Berlin, 1922-4; commanded *Marlborough*, 1926-7; rear-admiral, 1928; commanded first battle squadron in Mediterranean, 1929; director of manning, Admiralty, 1930-2; commander-in-chief, America and West Indies, 1932-4; vice-admiral, 1932; admiral, 1936; commander-in-chief, Plymouth, 1935-8; led British section, Anglo-French military mission to Russia, 1939; commander-in-chief, the Nore, 1939-41; retired, but volunteered as commodore of convoy, 1943-5; privately printed the *Art of War* (1943), a criticism of British defence policy; founder-member of the Naval Society; CB, 1928; KCB, 1934.

PODE, SIR (EDWARD) JULIAN (1902-1968), steel executive; joined training ship *Conway* and went to sea at fifteen; discharged, through impaired eyesight, 1919; joined William Wing & Son, chartered accountants, Sheffield, 1919; qualified, 1924; district accountant, Guest, Keen & Nettlefolds Ltd., 1926; secretary, Guest, Keen Baldwins Iron & Steel Co. Ltd., 1930; joint commercial manager, 1938; assistant managing director, 1943; joint managing director, 1945; managing director, Steel Company of Wales Ltd., 1947; deputy chairman, 1961; chairman, 1962-7; served on number of advisory councils and on boards of other companies, including Lloyds Bank; high sheriff, Glamorgan, 1948; JP, 1951-66; knighted, 1959; president, Iron and Steel Federation, 1962-4.

PODMORE, FRANK (1855-1910), writer on psychical research; BA, Pembroke College, Oxford, 1877; clerk in Post Office, 1879-1907; member of council of Society for Psychical Research, 1882-1909; argued for psychological, as opposed to spiritualist, causality; helped to found, and originated title of, Fabian Society, 1884; writings include *Studies in Psychical Research* (1897), *Modern Spiritualism* (1902), *Life of Robert Owen* (1906), and *Telepathic Hallucinations* (1910).

POEL, WILLIAM (1852-1934), actor, stage-director, and author; son of William Pole [q.v.]; revolutionized stage production by presenting without scenery seventeen of Shakespeare's plays beginning with the first quarto *Hamlet*, himself playing the title-role (1881); other productions include Marlowe's *Dr. Faustus*, Milton's *Samson Agonistes*, and *Everyman*; founded Elizabethan Stage Society, 1895; wrote several plays and *Shakespeare in the Theatre* (1913).

POLAND, Sir HARRY BODKIN (1829-1928), criminal lawyer; called to bar (Inner Temple), 1851; appeared for crown in success-ful prosecution of Sir J. D. Paul [q.v.], 1855; counsel to Treasury at Central Criminal Court and adviser to Home Office in criminal matters, 1865-88; cases in which he appeared for crown include those of *Lennie* mutineers (1876), Wain-wrights (1876), *Franconia* (1876), and Stauntons (1877); defended Governor E. J. Eyre [q.v.] in both civil and criminal proceed-ings, 1868; recorder of Dover, 1874-1901; QC, 1888; became leading counsel for defence in criminal cases; retired and knighted, 1895.

POLE, Sir FELIX JOHN CLEWETT (1877-1956), railway general manager and industrialist; educated at village school; at fourteen, joined Great Western Railway as telegraph clerk; rejected for military service in 1914-18 war because of defective eyesight; assistant general manager, 1919; general manager, 1921-9; reorganized the Great Western Railway, en-larged under Railways Act of 1921; chairman, Associated Electrical Industries, 1929-45; knighted, 1924.

POLLARD, ALBERT FREDERICK (1869-1948), historian; educated at Felsted School and Jesus College, Oxford; first class, modern history, 1891; fellow of All Souls, 1908; assistant editor, and contributor of some 500 lives, *Dictionary of National Biography*, 1893-1901; professor, constitutional history, University College, London, 1903-31; founded Historical Association, 1906; president, 1912-15; edited *History*, 1916-22; obtained foundation of Insti-tute of Historical Research, 1920; director, 1920-31; honorary director, 1931-9; leading authority on Tudor period; his point of view protestant, English, and liberal; publications include *England under Protector Somerset*

(1900), *Henry VIII* (1902), and studies of Edward VI, Mary, and Elizabeth (1910) in *Political History of England*, *The Evolution of Parliament* (1920), *Wolsey* (1929), and frequent contributions to *The Times* and *The Times Literary Supplement*; FBA, 1920.

POLLARD, ALFRED WILLIAM (1859-1944), librarian, bibliographer, and English scholar; educated at King's College School and St. John's College, Oxford; first class, *lit. hum.*, 1881; entered printed books department, British Museum, 1883; keeper, 1919-24; co-editor, the *Library*, 1904-20; sole editor, 1920-34; honorary secretary Bibliographical Society, 1893-34; honorary professor of bibliography, London, 1919-32; co-editor 'Globe Chaucer', 1898; of *Towneley Plays* (1897), and *Macro Plays* (1904); his *Shakespeare Folios and Quartos: a Study in the Bibliography of Shake-speare's Plays, 1594-1685* (1909) a milestone in Shakespearian criticism; FBA and CB, 1922.

POLLEN, JOHN HUNGERFORD (1820-1902), artist and author; educated at Eton and Christ Church, Oxford; BA, 1842; MA, 1844; fellow of Merton, 1842-52; pro-vicar of St. Saviour's, Leeds, 1847-52; designed ceiling in Merton College chapel, 1850; joined Church of Rome, 1852; professor of fine arts at Catholic University, Dublin, 1855-7; joined William Morris, Burne-Jones [qq.v.], and others in fresco decoration of Oxford Union Society, 1858; friend of Turner, Millais, and Ruskin [qq.v.]; decorated many private houses in Eng-land, and built church of St. Mary, Rhyl, 1863; editor of art and industrial departments of South Kensington Museum, 1863-76; private secretary to Marquess of Ripon [q.v.] from 1876; visited India, 1884; did much to reform taste in domestic furniture and decoration; published *Universal Catalogue of Books on Art* (3 vols., 1870-7) and many official catalogues for South Kensington.

POLLITT, GEORGE PATON (1878-1964), chemical industrialist and farmer; educated in Bruges, the Owens College, Manchester, and Basle University; Ph.D. (Zurich Polytechnic), 1902; joined staff of Brunner Mond & Co., manufacturers of heavy chemicals, 1907; served in army during 1914-18 war; DSO with two bars; director, Brunner Mond, 1919; managing director, Synthetic Ammonia & Nitrates Ltd.; director, Imperial Chemical Industries Ltd., a merger of all major chemical manufacturers in Britain, 1926; resigned as executive director, 1934, but remained a director till 1945; took up farming, 1932; published booklet, *Britain Can Feed Herself* (1942); zone commander, Home Guard, 1939-45; high sheriff, Shropshire,

1945-6; lived and farmed in Southern Rhodesia, 1947-52.

POLLITT, HARRY (1890-1960), general secretary and chairman, British communist party; educated at elementary school; worked in weaving mill from age of twelve to fifteen; then apprenticed to Gorton Tank, locomotive-building plant; joined independent labour party, 1909; member of Boilermakers' Society, 1912; political speaker at socialist meetings, 1911-14; opposed British participation in 1914-18 war; secretary, London district, Boilermakers' Society, 1919; foundation member, British communist party, 1920; secretary, National Minority Movement, aiming to bring trade unions under communist control, 1924; sentenced to twelve months' imprisonment for seditious libel, 1925; general secretary, British communist party, 1929; chairman, 1956-60.

POLLOCK, BERTRAM (1863-1943), headmaster, and bishop of Norwich; brother of Viscount Hanworth [q.v.]; scholar of Charterhouse and Trinity College, Cambridge; priest, 1891; assistant master, Marlborough, 1886-93; master, Wellington College, 1893-1910; bishop of Norwich, 1910-42; chaplain-in-ordinary to the King, 1904-10; KCVO, 1921.

POLLOCK, ERNEST MURRAY, first VISCOUNT HANWORTH (1861-1936), judge; brother of Bertram Pollock [q.v.]; educated at Charterhouse and Trinity College, Cambridge; called to bar (Inner Temple), 1885; bencher, 1914; KC, 1905; conservative MP, Warwick and Leamington, 1910-23; KBE, 1917; solicitor-general, 1919-22; attorney-general, PC, and baronet, 1922; master of the rolls, 1923-35; baron, 1926; viscount, 1936; brought much industry and enthusiasm to his duty as custodian of records.

POLLOCK, SIR FREDERICK, third baronet (1845-1937), jurist; son of Sir W. F. Pollock [q.v.]; succeeded father, 1888; scholar of Eton and Trinity College, Cambridge; second classic, 1867; fellow, 1868; called to bar (Lincoln's Inn), 1871; bencher, 1906; KC, 1920; Corpus professor of jurisprudence, Oxford, 1883-1903; chairman, royal commission on public records, 1910; works include *Principles of Contract at Law and in Equity* (1876), a life of Spinoza (1880), *The Land Laws* (1883), *The Law of Torts* (1887), *First Book of Jurisprudence for Students of the Common Law* (1896), *The Genius of the Common Law* (1912), and *Essays in the Law* (1922); collaborated with F. W. Maitland [q.v.] in *History of English Law* (2 vols., 1895); edited the *Law Quarterly Review*, 1885-1919; editor-in-chief of the *Law Reports*, 1895-1935; *The*

Holmes-Pollock Letters . . . 1874-1932 (1942) commemorates his friendship with Oliver Wendell Holmes; deeply interested in philosophy which inspired his ambition to define nature of legal concepts; advanced study and prestige of English law; cosmopolitan in sympathy and experience; FBA, 1902; PC, 1911.

POLLOCK, HUGH McDOWELL (1852-1937), Irish politician; represented Belfast chamber of commerce at Irish convention, 1917-18; member of Northern Ireland parliament, finance minister and deputy prime minister, 1921-37; secured acceptance of principle that equal taxation with Great Britain implied equal social benefits; CH, 1936.

POLLOCK, SIR (JOHN) DONALD, baronet, of Edinburgh (1868-1962), physician, industrialist and philanthropist; studied science at Glasgow University, and medicine at Edinburgh; MB, CM, 1892; MD, 1895; practised in London, 1895-1908; private medical adviser and companion to Duke of Leinster, an epileptic, 1908-14; served at naval hospital, Granton during 1914-18 war; OBE, 1919; honorary surgeon-commander, RNVR, 1922; helped to form shipbreaking firm, Metal Industries Ltd., chairman, 1922-51; salvaged some of the German warships sunk at Scapa Flow, 1933-9; produced oxygen by new method; amalgamated with British Oxygen Company; chairman, British Oxygen, 1932-7; member, Scottish Milk Marketing Board, Carnegie Trust, and other trusts; contributed to charities and missionary work; lord rector, Edinburgh University, 1939-45; hon. LLD, Edinburgh; hon. D.Sc., Oxford; fellow, Royal Society of Edinburgh; vice-president, British Association, 1951; baronet, 1939.

PONSONBY, ARTHUR AUGUSTUS WILLIAM HARRY, first BARON PONSONBY OF SHULBREDE (1871-1946), politician and author; son of Sir H. F. Ponsonby [q.v.] whose life he published in 1942; educated at Eton and Balliol College, Oxford; in diplomatic service, 1894-1902; principal private secretary to Campbell-Bannerman, 1906-8; liberal MP, Stirling Burghs, 1908-18; a lifelong pacifist; urged negotiated peace throughout 1914-18; labour MP, Brightside division of Sheffield, 1922-30; under-secretary, Foreign Office, 1924; negotiated commercial and general treaties with Russia; under-secretary, dominion affairs, 1929; parliamentary secretary, Ministry of Transport, 1929-31; baron, 1930; chancellor, duchy of Lancaster, 1931; leader, labour party opposition in Lords, 1931-5; vigorously denounced rearmament; authority on English diaries.

PONSONBY, VERE BRABAZON, EARL OF BESSBOROUGH (1880-1956), governor-general of Canada; educated at Harrow and Trinity College, Cambridge; called to the bar (Inner Temple), 1903; represented Marylebone East on London County Council, 1907-10; MP, Cheltenham, 1910; MP, Dover, 1913-20; served in Gallipoli and France during 1914-18 war; chairman, San Paulo Railway and Margarine Union, 1919-29; joint chairman, Unilever, with (Sir) D'Arcy Cooper [q.v.], 1929; governor-general of Canada, 1931-5; chairman of Rio Tinto Company; worked in Foreign Office on welfare of French in Britain, 1940-5; published *Lady Bessborough and her Family* (1940); CMG, 1919; GCMG, PC, 1931; Irish earldom raised to earldom of United Kingdom, 1937.

POOLE, REGINALD LANE (1857-1939), historian; son of E. S. Poole and brother of S. E. Lane-Poole [qq.v.]; educated at Balliol and Wadham colleges, Oxford; lecturer, Jesus College, 1886-1910; research fellow, Magdalen, 1898-1933; university lecturer in diplomatic history, 1896-1927; keeper of university archives, 1909-27; assistant editor (1886), editor (1901-20), *English Historical Review*; works include *Sebastian Bach* (1882), *Illustrations of the History of Medieval Thought* (1884), *The Exchequer in the Twelfth Century* (1912), and *Lectures on the History of the Papal Chancery* (1915); FBA, 1904.

POOLE, STANLEY EDWARD LANE- (1854-1931), orientalist and historian; brother of R. L. Poole [q.v.]; educated at Corpus Christi College, Oxford; prepared *Catalogue of Oriental and Indian Coins in the British Museum* (14 vols., 1875-92); professor of Arabic, Trinity College, Dublin, 1898-1904; prolific writer.

POOLEY, SIR ERNEST HENRY, baronet (1876-1966), public servant; educated at Winchester and Pembroke College, Cambridge; called to bar (Lincoln's Inn), 1901; legal assistant, Board of Education, 1903; assistant clerk to Drapers' Company, 1905; clerk, 1908-44; master, 1944-5; warden, 1952-3 and 1962-3; published *The Guilds of the City of London* (1945); served in RNVR and RGA during 1914-18 war; concerned with King Edward's Hospital Fund for London, 1928-66; member of senate and court of London University, 1929-48; hon. LLD, 1948; hon. fellow, Pembroke College, Cambridge, and Queen Mary College, London; member of several educational committees; succeeded Lord Keynes [q.v.] as chairman, Arts Council, 1946; knighted, 1932; KCVO, 1944; GCVO, 1957; baronet, 1953.

POORE, GEORGE VIVIAN (1843-1904), physician and authority on sanitation; studied medicine at University College Hospital; MRCS, 1866; MB and BS, London, 1868; MD, 1871; FRCP, 1877; professor of medical jurisprudence at University College Hospital, 1876-83; physician, 1883-1903; published *Text Book of Electricity in Medicine and Surgery* (1876), *A Treatise on Medical Jurisprudence* (1901), *Essays on Rural Hygiene* (1893), and *The Dwelling House* (2nd edn. 1898).

POPE, GEORGE UGLOW (1820-1908), missionary and Tamil scholar; born in Nova Scotia; brother of William Burt Pope [q.v.]; went to Madras for missionary work, 1839; ordained, 1843; worked in Tinnevelly and Tanjore till 1859; founder and headmaster, of Ootacamund grammar school, 1859; principal of Bangalore College, 1870; hon. DD, Lambeth, 1864; teacher of Tamil and Telugu in Oxford University, 1884; hon. MA, 1886; published in Tamil *First Catechism of Tamil Grammar* (1842, reissued with translation, 1895) and other Tamil handbooks; edited classical Tamil works, *Kurral* (1886), *Nāladiyār* (1893), and *Tiruvaçagam* (1900); began catalogue of Tamil books at British Museum, 1890; friend of Robert Browning [q.v.], whose *Death in the Desert* he edited (1897).

POPE, SAMUEL (1826-1901), barrister; called to bar (Middle Temple), 1858; treasurer, 1888-9; recorder of Bolton and QC, 1869; leader of the parliamentary bar; temperance advocate and freemason.

POPE, WALTER JAMES MACQUEEN- (1888-1960), theatre manager, publicist and historian. [See MACQUEEN-POPE.]

POPE, WILLIAM BURT (1822-1903), Wesleyan divine; born in Nova Scotia; served Wesleyan ministries from 1842 to 1867; tutor of systematic theology at Didsbury, 1867-86; hon. DD, Edinburgh, 1877; president of Wesleyan conference, 1877; publications include *Compendium of Christian Theology* (3 vols., 1875); edited *London Quarterly Review*, 1860-86.

POPE, SIR WILLIAM JACKSON (1870-1939), chemist; educated at Finsbury Technical College and the Central Institution; professor of chemistry, Manchester (1901-8), Cambridge (1908-39); made great advances in stereochemistry by applying the principles of symmetry to molecular structure; first to demonstrate basic principles of the enantiomorphism and the optical activity of chemical compounds; FRS, 1902; KBE, 1919.

POPHAM, ARTHUR EWART (1889-1970), keeper of prints and drawings at the British Museum; educated at Dulwich College, University College, London and King's College, Cambridge; assistant, Department of Prints and Drawings, British Museum, 1912; served during 1914-18 in the Artists' Rifles, Royal Naval Air Service, and RFC, croix de guerre; returned to British Museum, 1918-54; with (Sir) K. T. Parker, launched *Old Master Drawings*, 1926; museum publications for which he was responsible include *A Handbook of the Drawings and Watercolours* in the department (1939), and *Italian Drawings of the Fourteenth and Fifteenth Centuries* (1950, with Philip Pouncy); he also published *Drawings of the Early Flemish School* (1926), *Drawings of Leonardo da Vinci* (1946), *Corregio's Drawings* (1957), and the *Drawings of Parmigianino* (3 vols., 1971); FBA, 1949; CB, 1954; hon. fellow, King's College, Cambridge, 1955; after retirement as keeper acted as advisor to National Gallery of Canada, Ottawa.

POPHAM, Sir (HENRY) ROBERT (MOORE) BROOKE- (1878-1953), air chief marshal. [See BROOKE-POPHAM.]

PORTAL, MELVILLE (1819-1904), politician; educated at Eton and Christ Church, Oxford; BA, 1842; MA, 1844; president of Union, 1842; called to bar, 1845; conservative MP for Northern division of Hampshire, 1849-57; chairman of Hampshire quarter-sessions, 1879-1903; reformed treatment of prisoners; published *The Great Hall of Winchester Castle* (1899).

PORTAL, Sir WYNDHAM RAYMOND, third baronet, and VISCOUNT PORTAL (1885-1949), industrialist and public servant; educated at Eton and Christ Church, Oxford; managing director, Portals, Ltd., makers of banknote paper, 1919; succeeded father as chairman and third baronet, 1931; chairman, Bacon Development Board, 1935; adviser on Welsh distressed areas from 1934; additional parliamentary secretary, Ministry of Supply, 1940-2; minister of works and buildings, 1942-4; initiated prefabricated houses; presided over XIV Olympiad, 1948; baron, 1935; PC, 1942; viscount, 1945; GCMG, 1949.

PORTER, Sir ANDREW MARSHALL, first baronet (1837-1919), judge; son of Revd J. S. Porter [q.v.]; called to Irish bar, 1860; QC, 1872; liberal MP, county Londonderry, and Irish solicitor-general, 1881; attorney-general and PC (Ireland), 1883; master of the rolls in Ireland, 1883-1906; baronet, 1902.

PORTER, SAMUEL LOWRY, BARON PORTER (1877-1956), judge; educated at the Perse School and Emmanuel College, Cambridge; called to bar (Inner Temple), 1905; served as captain in 1914-18 war; KC, 1925; recorder, Newcastle-under-Lyme, 1928-32, and Walsall, 1932-4; judge, King's Bench, knighted, 1934; lord of appeal in ordinary, life peerage, PC, 1938; chairman, tribunal appointed to inquire into budget leakage of 1936, resulting in resignation of J. H. Thomas [q.v.]; chairman, national reference tribunal of the coal-mining industry; hon. fellow, Emmanuel College, 1937; hon. LLD, Birmingham, 1940, and Cambridge, 1947; GBE, 1951.

POSTAN, EILEEN EDNA LE POER (1889-1940), better known as EILEEN POWER, historian and teacher; educated at Bournemouth and Oxford high schools and Girton College, Cambridge; first class, historical tripos, parts i and ii, 1909-10; director of studies in history, Girton, 1913-21; lecturer (1921), reader (1924) and professor (1931-40) of economic history, London School of Economics; medievalist interested especially in peasant life, economic position of women, and the wool trade; works include *Medieval English Nunneries, c.1275-1535* (1922), *Some Medieval People* (1924), and *The Wool Trade in English Medieval History* (1941); with R. H. Tawney edited *Tudor Economic Documents* (3 vols., 1924); married Michael Moissey Postan, 1937.

POSTGATE, JOHN PERCIVAL (1853-1926), classical scholar; BA, Trinity College, Cambridge; fellow, 1878; classical lecturer, 1884-1909; professor of comparative philology, University College, London, 1880-1910; in charge of *Classical Review*, 1899-1907; of *Classical Quarterly*, 1907-11; largely instrumental in founding Classical Association, 1903; professor of Latin, Liverpool, 1909-20; FBA, 1907; his emendations less convincing than his collation and criticism of manuscript evidence; his works include *Select Elegies of Propertius* (1881), editions of Books VII and VIII of Lucan (1896, 1917), text and translation of Tibullus (1913), *New Latin Primer* (1888), and edition of *Corpus Poetarum Latinorum* (1893-4, 1904-5).

POTT, ALFRED (1822-1908), principal of Cuddesdon College; educated at Eton and Magdalen College, Oxford; BA, 1844; MA, 1847; BD, 1854; fellow, 1853; first principal of Cuddesdon theological college, 1854-8; held various livings, 1867-99; archdeacon of Berkshire, 1869-1903; learned in ecclesiastical law.

POTTER, (HELEN) BEATRIX, afterwards MRS HEELIS (1866-1943), writer and illustrator of children's books; published *Peter Rabbit* (1900), *The Tailor of Gloucester* (1902), followed by tales of Jemima Puddle-Duck, Mrs Tiggy-Winkle, Squirrel Nutkin, and many others soon established as nursery classics; married (1913) William Heelis, solicitor.

POTTER, STEPHEN MEREDITH (1900–1969), writer and radio producer; educated at Westminster School; commissioned in Coldstream Guards, 1918–19; graduated at Merton College, Oxford, 1922; lecturer in English, Birkbeck College, London, 1926–38; writer-producer for British Broadcasting Corporation, 1938; wrote 'How' programmes with Joyce Grenfell, such as 'How to give a party'; theatre critic for the *New Statesman and Nation*, 1945; book critic for *News Chronicle*, 1946; editor of weekly *Leader*, 1949–51; publications include *The Young Man* (1929), an autobiographical novel, *Coleridge and S.T.C.* (1935), *The Muse in Chains: a Study in Education* (1937), *The Theory and Practice of Gamesmanship; or the Art of Winning Games Without Actually Cheating* (1947), followed by *Some Notes on Lifemanship* (1950), *One Upmanship* (1952), and *Supermanship* (1958); he also published *Steps to Immaturity* (1959), an autobiography.

POULTON, SIR EDWARD BAGNALL (1856–1943), zoologist; scholar of Jesus College, Oxford; first class, natural science (zoology), 1876; Burdett-Coutts scholar in geology, 1878; president of Union, 1879; lecturer in natural science at Jesus and Keble colleges, 1880; Hope professor of zoology, 1893–1933; foremost exponent of protective colouring in all its forms—concealing, warning, and mimetic; published *The Colours of Animals* (1890); consistently urged Darwinian principle of evolution through small variations and the importance of observation and experiment in the field; FRS, 1889; president, British Association, 1937; knighted, 1935.

POUND, SIR (ALFRED) DUDLEY (PICKMAN ROGERS) (1877–1943), admiral of the fleet; entered navy, 1891; formed nucleus of plans division at Admiralty, 1917; director of operations (home), 1917–20; director, plans division, 1922–5; Admiralty representative at Lausanne conference, 1922–3; chief of staff to Sir Roger (later Lord) Keyes [q.v.], Mediterranean, 1925–7; rear-admiral, 1926; assistant chief of naval staff, 1927–9; commanded battle cruiser squadron, 1929–31; second sea lord, 1932–5; admiral, 1935; chief of staff to Sir William Fisher [q.v.], Mediterranean, 1935–6; commander-in-chief, Mediterranean, 1936–9; first sea lord, 1939–43; admiral of the fleet, 1939;

organized naval strategy resulting in victory of Cape Matapan, evacuation British forces from Greece and Crete, retention of Malta, North African landings, recovery of control of Mediterranean, landings in Sicily; scuttling of *Graf Spee* and sinking of *Bismark*; creation of Western Approaches command and establishment of naval and air headquarters at Liverpool; criticized for lack of naval action at Trondheim, loss of *Prince of Wales* and *Repulse* in Far East, and escape of *Scharnhorst* and *Gneisenau* from Brest; KCB, 1933; GCVO., 1937; GCB, 1939; OM, 1943.

POWELL, CECIL FRANK (1903–1969), physicist; educated at Sir A. Judd's commercial school, Tonbridge, and Sidney Sussex College, Cambridge; first class, parts i and ii, natural sciences tripos, 1924–5; research worker in Cavendish Laboratory under C. T. R. Wilson [q.v.]; research assistant to Professor A. M. Tyndall [q.v.] at Bristol University, 1928; lecturer, 1931; reader, 1946; Melville Wills professor of physics, 1948–63; director of H. H. Wills Physics Laboratory, 1964–9; pro-vice-chancellor, Bristol, 1964–7; FRS, 1949; awarded Nobel prize for physics, 1950; president, Association of Scientific Workers, 1954, and World Federation of Scientific Workers, 1957; carried out research in nuclear physics and seismology, played leading role in establishing CERN (the European Centre for Nuclear Research at Geneva); chairman, Scientific Policy Committee of CERN, 1961–3; involved in setting up Pugwash Conferences on Science and World Affairs, 1957; chairman, 1967; hon. fellow, Sidney Sussex College, Cambridge; many other academic honours and distinctions; chairman, Nuclear Physics Board of British Science Research Council, 1965–8; published many scientific papers, including *Nuclear Physics in Photographs* (with G. P. S. Occhialini, 1947) and *The Study of Elementary Particles by the Photographic Method* (with P. H. Fowler and D. H. Perkins, 1959).

POWELL, FREDERICK YORK (1850–1904), regius professor of modern history at Oxford; after two years at Rugby visited Sweden, and studied German and Icelandic; BA, Christ Church, Oxford, 1872; called to bar, 1874; lecturer in law at Christ Church, 1874–94; student of Christ Church, 1884–94; taught Old English, Old French, and Old German, in addition to law; published *Early England* for 'Epochs of English History' series (1876); compiled jointly with Gudbrandr Vigfusson [q.v.] an *Iceland Prose Reader* (1879) and *Corpus Poeticum Boreale* (2 vols., 1881), collection of ancient northern poetry with biographies of the poets, as well as *Origines Islandicae* (posthumous, 1905); helped to found *English Historical Review*, 1886; wrote history of *England to Death of Henry VII* (1885), and edited *English History from Contemporary Writers* (1885); regius professor of modern history at Oxford, 1894–1904; translated *Faereyinga Saga* (1896) and quatrains from *Omar Khayyám* (1901); friend of Verlaine, Verhaeren, and Rodin; president of Irish Text Society; active in foundation of Ruskin College, Oxford, 1899; 'socialist and jingo'; hon. LLD, Glasgow, 1901; *Life and Letters* edited by O. Elton [q.v.] (2 vols., 1906).

POWELL, SIR (GEORGE) ALLAN (1876–1948), public servant; educated at Bancroft's School and King's College, London; joined Metropolitan Asylums Board, 1894; clerk to Board, 1922–30; organized public assistance department, London County Council, 1930–2; vice-chairman, Food Council, 1925–9; chairman, 1929–32; member, Import Duties Advisory Committee, 1932–9; chairman, British Broadcasting Corporation, 1939–46; knighted, 1927; GBE, 1944.

POWELL, SIR RICHARD DOUGLAS, first baronet (1842–1925), physician; studied medicine at University College and Hospital, London; owed much to Sir William Jenner [q.v.]; assistant physician, Charing Cross Hospital, 1871; to Middlesex Hospital, 1878; full physician, 1880; consulting physician, 1900; also attached to Brompton Hospital; connected with the court for thirty-eight years; very active in life of London medical societies; baronet, 1897; his works include *Pulmonary Tuberculosis* (6 editions, 1872–1921).

POWELL, ROBERT STEPHENSON SMYTH BADEN-, first BARON BADEN-POWELL (1857–1941), lieutenant-general, and founder of the Boy Scouts and Girl Guides. [See BADEN-POWELL.]

POWER, SIR ARTHUR JOHN (1889–1960), admiral of the fleet; entered *Britannia* and won King's gold medal for best cadet, 1904; lieutenant, 1910; appointed to *Excellent* to specialize in gunnery, 1913; gunnery officer in 1914–18 war; commander, 1922; instructor, Naval Staff College, 1927–9; Imperial Defence College, 1933; commanded naval gunnery school, *Excellent*, 1935–7; commander, *Ark Royal*, 1938–40; assistant chief of naval staff (Home), rear-admiral, 1940–2; flag officer commanding fifteenth cruiser squadron, 1942; flag officer, Malta, vice-admiral, 1943; second-in-command, Eastern Fleet, 1944; lord commissioner of the Admiralty, second sea lord, 1946; admiral, 1946; commanded Mediterranean Fleet, 1948;

commander-in-chief, Portsmouth, 1950–2; admiral of the fleet, 1952; CB, 1941; KCB, 1944; GBE, 1946; GCB, 1950.

POWER, SIR D'ARCY (1855–1941), surgeon and historian; qualified from St. Bartholomew's Hospital, 1882; assistant surgeon, 1898; surgeon, 1904–20; consulting surgeon and hon. archivist; hon. librarian, Royal College of Surgeons, 1929–41; works included *William Harvey* (1897), many lives of surgeons for this Dictionary, and editions of medical texts; KBE, 1919.

POWER, EILEEN EDNA LE POER (1889–1940), historian and teacher. [See POSTAN.]

POWER, SIR JOHN CECIL, first baronet (1870–1950), company director and public benefactor; entered family export business and subsequently dealt in real estate; associated with development of Kingsway; conservative MP, Wimbledon, 1924–45; gave £20,000 towards foundation of Institute of Historical Research, 1920; honorary treasurer and generous benefactor, Royal Institute of International Affairs, 1921–43; of British Council, 1934–50; baronet, 1924.

POWER, SIR WILLIAM HENRY (1842–1916), expert in public health; medical inspector, Local Government Board, 1871–87; carried out original work in connection with infectious diseases; principal medical officer to Board, 1900–8; FRS, 1895; KCB, 1908.

POWICKE, SIR (FREDERICK) MAURICE (1879–1963), historian; educated at Stockport grammar school, the Owens College, Manchester, and Balliol College, Oxford; first class, history, 1903; Langton research fellow, Manchester University, 1902–5; assistant lecturer, Liverpool University, 1905–6; assistant lecturer, Manchester, under T. F. Tout [q.v.], 1906–8; fellow, Merton College, Oxford, 1908; published series of articles in *English Historical Review*, 1906–9; professor of modern history, Queen's University, Belfast, 1909–19; professor, medieval history, Manchester, 1919–28; regius professor of modern history, Oxford, 1928–47; publications include *The Loss of Normandy* (1913), *Stephen Langton* (1928), *King Henry III and the Lord Edward* (1947), and *The Thirteenth Century, 1216–1307* the fourth volume of the Oxford History of England; also produced, in collaboration with A. B. Emden, a new edition in three volumes of *Universities of Europe in the Middle Ages* (1895) by Hastings Rashdall [q.v.], and *The Medieval Books of Merton College* (1931); president, Royal Historical Society, 1935–7; FBA, 1927; knighted,

1946; hon. fellow, Merton, Balliol, and Oriel Colleges, Oxford; hon. doctorates from number of foreign universities.

POWNALL, SIR HENRY ROYDS (1887-1961), lieutenant-general; educated at Rugby and Royal Military Academy, Woolwich; commissioned into Royal Field Artillery, 1906; served in France, 1914-19; MC, DSO; on staff of Staff College, Camberley, 1926-9; bar to DSO on North-West Frontier of India, 1931; served under Sir Maurice (later Lord) Hankey [q.v.] on secretariat of Committee of Imperial Defence, 1933-6; commandant, School of Artillery, Larkhill, 1936-8; director, military operations and intelligence, War Office, 1938; chief of staff to Lord Gort [q.v.], commander, BEF, 1939-40; KBE; commander-in-chief, British Forces in Northern Ireland, 1940-1; vice-chief, Imperial General Staff, 1941; commander-in-chief, Far East; chief of staff to Sir A. P. (later Earl) Wavell [q.v.], South-West Pacific Command, 1942; lieutenant-general, Ceylon Command, 1942-3; succeeded Sir Maitland (later Lord) Wilson [q.v.] as commander-in-chief, Persia-Iraq, 1943; chief of staff to Admiral Mountbatten (later Earl Mountbatten of Burma), South-East Asia Command, 1943-5; KCB; colonel commandant, Royal Artillery, 1942-52; chancellor, Order of St. John, 1951; assisted Churchill as military consultant on *History of the Second World War*.

POWYS, JOHN COWPER (1872-1963), novelist and miscellaneous writer; educated at Sherborne and Corpus Christi College, Cambridge; lecturer for Oxford University Extension Delegacy, 1898-1909; made first lecture tour in United States, 1905; published first novel, *Wood and Stone*, in New York, 1915; other published novels include *Wolf Solent* (1929), *A Glastonbury Romance* (New York, 1932), and *Maiden Castle* (New York, 1936); historical novels include *Porius* (1951), *Atlantis* (1954), *The Brazen Head* (1956), and related prose retelling of the *Iliad*, *Homer and the Aether* (1959); philosophical essays include *The Meaning of Culture* (New York, 1929) and *The Art of Happiness* (1935); published also studies of great writers such as *Dostoievsky* (1947) and *Rabelais* (1948), his *Autobiography* (1934), and *Letters to Louis Wilkinson* (1958); hon. D.Litt., Wales, 1962.

POYNDER, SIR JOHN POYNDER DICKSON-, sixth baronet, and BARON ISLINGTON (1866-1936), politician and administrator; succeeded uncle, 1884; educated at Harrow and Christ Church, Oxford; MP, Chippenham division (conservative, 1892-1905, liberal, 1905-10); member of London County Council,

1898-1904; governor of New Zealand, 1910-12; chairman, royal commission on Indian public services, 1912-14; under-secretary of state for the colonies (1914-15), for India (1915-18); chairman, National Savings Committee, 1920-6; baron, 1910; PC, 1911.

POYNTER, SIR EDWARD JOHN, first baronet (1836-1919), painter and president of the Royal Academy; son of Ambrose Poynter [q.v.]; born in Paris; studied art in London and Paris; exhibited at Royal Academy, 1861-1919; ARA, 1868; RA, 1877; first Slade professor of fine art at University College, London, 1871-5; director for art and principal of National Art Training School, South Kensington, 1875-81; director of National Gallery, 1894-1904; president of Royal Academy, 1896-1918; knighted, 1896; baronet, 1902; gained reputation as painter of subject-pictures in oil- and watercolour; reached highest level in four large oil-pictures for Earl of Wharncliffe, 1872-9; executed decorative designs, including frescoes; implanted French principles in English art teaching.

POYNTING, JOHN HENRY (1852-1914), physicist; BA, Trinity College, Cambridge; fellow of Trinity, 1878; professor of physics, Mason College, Birmingham (afterwards Birmingham University), 1880-1914; FRS, 1888; most important contributions to physics two Royal Society papers which revolutionized ideas about motion of energy in the electric field; advanced knowledge of pressure of light; *Collected Scientific Papers* (published, 1920).

PRAIN, SIR DAVID (1857-1944), botanist and administrator; MA, Aberdeen, 1878; MB, CM, 1883; entered Indian medical service; curator of herbarium, Royal Botanic Garden, Calcutta, 1887; superintendent of Garden, director of Botanical Survey of India, and government quinologist, 1898-1905; professor of botany, Calcutta Medical College, 1895-1905; published *Bengal Plants* (1903); FRS, 1905; director, Royal Botanic Gardens, Kew, 1905-22; editor, *Botanical Magazine*, 1907-20; chairman of council, John Innes Horticultural Institution, 1909-44; knighted, 1912.

PRATT, HODGSON (1824-1907), peace advocate; joined East India Company's service at Calcutta, 1847; helped to found Vernacular Literature Society, 1851; settled in England, 1861; active in industrial cooperative movement; president of Working Men's Club Union, 1885-1902; helped to found International Arbitration and Peace Association, 1880; founded and edited its *Journal*, 1884; died at Le Pecq; memorial lecture and scholarship established, 1911.

PRATT, JOSEPH BISHOP (1854-1910), engraver; pupil of David Lucas; from 1896 engraved almost exclusively subjects after Georgian painters, including Raeburn's 'Mrs. Gregory' and Lawrence's 'Mrs. Cuthbert'; also engraved state portraits by Sir Luke Fildes [q.v.] of King Edward VII, 1902, and Queen Alexandra, 1906.

PREECE, Sir WILLIAM HENRY (1834-1913), electrical engineer; in service of Post Office, 1870-1904; engineer-in-chief, 1892-9; FRS, 1881; KCB, 1899; studied telegraphy and telephony; early pioneer of wireless telegraphy; introduced improvements in railway signalling.

PRENDERGAST, Sir HARRY NORTH DALRYMPLE (1834-1913), general; son of Thomas Prendergast [q.v.]; born in India; entered army, 1854; VC for gallantry during Indian Mutiny, 1857; in command of British Burma division occupied Mandalay, 1885; KCB, 1885; general, 1887; GCB, 1902.

PRESTAGE, EDGAR (1869-1951), historian and professor of Portuguese; educated at Radley and Balliol College, Oxford; practised as solicitor, 1896-1907; married and settled in Portugal, 1907; press officer, British legation, Lisbon, 1917-18; Camoens professor of Portuguese, King's College, London, 1923; published numerous articles, books, and papers on Portuguese history including *Diplomatic Relations of Portugal with France, England and Holland from 1640 to 1668* (1925) and *The Portuguese Pioneers* (1933); FBA, 1940.

PREVITÉ-ORTON, CHARLES WILLIAM (1877-1947), historian and editor; first class, historical tripos, parts i and ii, St. John's College, Cambridge, 1907-8; fellow, 1911; professor of medieval history, 1937-42; edited *English Historical Review*, 1925-37; joint-editor, *Cambridge Medieval History*; prepared shorter version (1952); FBA, 1929.

PRICE, FREDERICK GEORGE HILTON (1842-1909), antiquary; succeeded father as chief acting partner of Child's Bank, Temple Bar, of which he published an account, 1875; compiled *Handbook of London Bankers* (1877); ardent Egyptologist, numismatist, and geologist; formed various antiquarian collections, of which that illustrating prehistoric and Roman London became nucleus of London Museum, 1911; the others realized on his death some £15,000, 1909-11; director of Society of Antiquaries, 1894-1909; died at Cannes; also wrote *The Signs of Old Lombard Street* (1887).

PRICE, THOMAS (1852-1909), premier of South Australia; born near Wrexham; emigrated to Adelaide, 1883; labour member of house of assembly, 1893; secretary (1900) and parliamentary leader (1901) of labour party; first labour premier, commissioner of public works, and minister of education, 1905-9.

PRICHARD, HAROLD ARTHUR (1871-1947), philosopher; educated at Clifton and New College, Oxford; first class, *lit. hum.*, 1894; fellow of Hertford, 1895-8; of Trinity, 1898-1924; White's professor of moral philosophy, 1928-37; published *Kant's Theory of Knowledge* (1909), and influential paper 'Does Moral Philosophy Rest on a Mistake?' (*Mind*, 1912), developing his views further in 'Duty and Ignorance of Fact' (British Academy lecture, 1932), and 'Moral Obligation' (1949); FBA, 1932.

PRIMROSE, ARCHIBALD PHILIP, fifth EARL OF ROSEBERY (1847-1929), statesman and author; grandson of fourth Earl of Rosebery [q.v.]; nephew of fifth Earl Stanhope [q.v.]; educated at Eton and Christ Church, Oxford; succeeded grandfather, 1868; announced his adhesion to liberal cause, 1869; by 1871 outstanding figure among his contemporaries; elected lord rector of Aberdeen University, 1878; of Edinburgh University, 1880; active in pressing Scottish claims; under-secretary of Home Office with special charge of Scottish business in House of Lords, 1881; visited New Zealand and Australia, where in course of a series of public speeches he developed his view of imperial relations, based on conception of 'commonwealth of nations', 1883-4; urged appointment of select committee to consider reform of House of Lords, 1884; first commissioner of works with seat in cabinet as lord privy seal, 1885; secretary for foreign affairs, 1886; in his foreign policy adopted decided attitude on several occasions; in distrusting Russia, maintaining an attitude of friendship towards France, and in laying stress on common interests of Germany and Great Britain, secured continuity with policy of Lord Salisbury; declared his adhesion to Gladstonian liberalism in speeches up and down country, 1887; did more than any other contemporary statesman to dissociate idea of empire from aggrandizement and to reconcile liberal opinion to new conception of imperial relations; brought forward further proposals for reform of House of Lords, 1888; chairman of new London County Council, 1889-90; secretary of state for foreign affairs, 1892; KG, 1892; came into conflict with Sir William Harcourt [q.v.] over Uganda and Egypt; intervened successfully in great coal strike, 1893; prime minister, 1894; differences within cabinet soon apparent, friction between prime minister and Harcourt being especially marked; again pressed for reform of House of Lords, 1894; resigned, 1895; as result of Gladstone's attitude towards Armenian massacres, which threatened to involve Great Britain in war, resigned leadership of liberal party, Oct. 1896; formation of Liberal League, of which he was president, marked definite split between two wings, insular and imperial, of liberal party, Feb. 1902; criticized Anglo-French agreement, 1905; dissociated himself from home rule; severed himself from liberal party, 1905; denounced Lloyd George's budget, 1909; twice brought forward resolutions for reform of House of Lords, 1910; denounced parliament bill, but voted for it as lesser evil than creation of peers to ensure passing of bill; his works include *William Pitt* (1891) and *Sir Robert Peel* (1899) in 'Twelve English Statesmen' series, *Oliver Cromwell* (1899), *Napoleon: the Last Phase* (1900), *Lord Randolph Churchill* (1906), *Chatham: his Early Life and Connections* (1910), and *William Windham* (1913); had very successful career as owner of racehorses, winning Derby three times; a notable orator; drawn into politics largely by force of circumstances, and unsuited for public life as he found it; his name not associated with any notable parliamentary measure; proved unequal to reuniting party shattered by home rule; but will be remembered as missionary of imperial ideas and advocate of reform of House of Lords as means to creation of strong second chamber.

PRIMROSE, Sir HENRY WILLIAM (1846-1923), civil servant; BA, Balliol College, Oxford; entered Treasury, 1869; secretary to HM Office of Works, 1887-95; chairman of Board of Inland Revenue, 1899-1907; member of royal commissions on civil service (1912) and railways (1913); KCB, 1899; PC, 1912.

PRINGLE, WILLIAM MATHER RUTHERFORD (1874-1928), politician; MA, Glasgow University; called to bar (Middle Temple), 1904; liberal MP, North-West Lanarkshire, 1910-18; for Penistone division of Yorkshire, 1923-4; devoted adherent of Asquith; worked unremittingly to try to heal breach in liberal party; had great knowledge of parliamentary procedure.

PRINGLE-PATTISON, ANDREW SETH (1856-1931), philosopher. [See PATTISON.]

PRINSEP, VALENTINE CAMERON [VAL PRINSEP] (1838-1904), artist; born at Calcutta; son of Henry Thoby Prinsep [q.v.]; intimate from youth with G. F. Watts [q.v.]; pupil of Gleyre in Paris; influenced by Lord Leighton

[q.v.]; exhibited at Royal Academy from 1862 to death; chief works were 'Miriam watching the Infant Moses', 'Bacchus and Ariadne', 'The Gleaners', and 'A Minuet'; painted Lord Lytton's Indian durbar, 1876; ARA, 1878; RA, 1894.

PRIOR, MELTON (1845-1910), war artist; draughtsman for *Illustrated London News* from 1868; war correspondent for the paper from 1873, depicting all the great wars between Ashanti campaign, 1873, and Russo-Japanese war, 1904; accompanied Edward, Prince of Wales, to Athens, 1875; and George, Prince of Wales, to Canada, 1901, and to Delhi durbar, 1903; his *Campaigns of a War Correspondent* edited by S. L. Bensusan, 1912.

PRITCHARD, Sir CHARLES BRADLEY (1837-1903), Indian civil servant; son of Charles Pritchard (1808-93, q.v.); joined Indian civil service, 1858; at first collector of Bombay salt revenue; suppressed smuggling; reformed system of manufacture and sale of opium and native spirits from 1877; commissioner of customs, 1881, and of salt, 1882; CSI, 1886; commissioner of Sind, 1887-9; revenue member of government of Bombay, 1890; KCIE, 1891; member for public works department on viceroy's council, 1892.

PRITCHETT, ROBERT TAYLOR (1828-1907), gunmaker and draughtsman; invented with W. E. Metford [q.v.] 'Pritchett bullet', 1853, and three-grooved rifle, 1854; exhibited at Royal Academy from 1851; joined staff of *Punch*; painted many water-colour drawings of royal functions; published accounts of visits to Holland, 1871, and Norway, 1878; artist to Lord [q.v.] and Lady Brassey on tours in *Sunbeam*, 1883 and 1885; illustrated books and magazines; collected curios and pipes; wrote and illustrated *Smokiana* (1890).

PROBERT, LEWIS (1841-1908), Welsh divine; Congregational minister in Rhondda valley and Portmadoc, 1867-96; principal of Congregational College, Bangor, 1896-1908; chairman of Welsh Congregational Union, 1901; published Welsh theological works.

PROCTER, FRANCIS (1812-1905), divine; BA, St. Catharine's College, Cambridge, 1835; fellow and tutor, 1842-7; vicar of Witton, Norfolk, 1847-1905; author of *A History of the Book of Common Prayer* (1855), often reprinted; edited *Sarum Breviary* (3 vols., 1857-86).

PROCTOR, ROBERT GEORGE COLLIER (1868-1903), bibliographer; grandnephew of John Payne Collier [q.v.]; BA, Corpus Christi College, Oxford, 1890; catalogued 3,000 *incunabula* at the Bodleian library, 1891-3; assistant in printed books department, British Museum, from 1893 to death; completed in 1898 *Index of Early Printed Books to the year MD*; devised a new Greek type; interested himself in Icelandic literature; disappeared while on solitary walking tour in Tyrol; *Bibliographical Essays* collected in 1905.

PROPERT, JOHN LUMSDEN (1834-1902), physician and art critic; MB, London, 1857; good etcher; revived taste for miniatures; published *A History of Miniature Art* (1887).

PROTHERO, SIR GEORGE WALTER (1848-1922), historian; BA, King's College, Cambridge; fellow, 1872; university lecturer in history, 1884; professor of modern history, Edinburgh, 1894-9; editor of *Quarterly Review*, 1899-1922; co-editor of *Cambridge Modern History*, 1901-12; general editor of Foreign Office peace handbooks, 1917-19; FBA, 1903; KBE, 1920; his works include *Simon de Montfort* (1877), and *Select Statutes . . . of the Reigns of Elizabeth and James I* (1894).

PROTHERO, ROWLAND EDMUND, BARON ERNLE (1851-1937), administrator, author, and minister of agriculture; brother of Sir George Prothero [q.v.]; educated at Marlborough and Balliol College, Oxford; first class, modern history, 1875; published *The Pioneers and Progress of English Farming* (1888; republished as *English Farming, Past and Present*, 1912); edited *Quarterly Review*, 1893-9; chief agent for eleventh Duke of Bedford [q.v.], 1898-1918; conservative MP, Oxford University, 1914-19; president, Board of Agriculture, 1916-19; introduced guaranteed price for wheat; PC, 1916; baron, 1919; works include *Life and Correspondence of Arthur Penrhyn Stanley* (1893) and *Letters and Journals of Lord Byron* (6 vols., 1898-1901).

PROUT, EBENEZER (1835-1909), musical composer; BA, London, 1854; organist in London chapels from 1861; professor of pianoforte at Crystal Palace School of Art, 1861-85; editor of *Monthly Musical Record*, 1871-5; introduced Wagner's operas to English public; composed organ concerto in E minor, 1871, cantatas, orchestral symphonies, and overtures; publications include primers on Instrumentation (1877), Harmony (1889), Counterpoint (1890), Fugue (1891), and *The Orchestra* (2 vols., 1897); professor at Royal Academy of Music, 1879, and Guildhall School of Music, 1884, and at Dublin University, 1894; hon. Mus.D., Dublin.

PRYDE, JAMES FERRIER (1866-1941), artist; educated at George Watson's College and

Royal Scottish Academy School; with brother-in-law, (Sir) William Nicholson [q.v.], as 'Beggarstaff Brothers' produced posters employing new technique using cut-out paper to build striking silhouette, 1894-6; published gouache of Irving as Dubosc in *The Lyons Mail* (1906); painted romantic imaginative landscapes, usually architectural in subject and often melancholy in feeling; exhibited regularly at International Society; of considerable influence on other artists and designers.

PRYNNE, GEORGE RUNDLE (1818-1903), hymn writer; BA, Catharine Hall, Cambridge, 1840; MA, 1861; incumbent of St. Peter's, Plymouth, from 1848 to death; his championship of the views of Dr Pusey [q.v.] on Anglican catholicism and extreme ritualism involved him in much controversy and litigation, 1850-2; vice-president of English Church Union, 1901; chief work was *The Eucharistic Manual* (1865, 10th edn. 1895); author of 'Jesu, meek and gentle', 1856, and other hymns.

PUDDICOMBE, ANNE ADALISA (1836-1908), novelist writing under the pseudonym of ALLEN RAINE; daughter of Benjamin Evans, solicitor; married Beynon Puddicombe, 1872; wrote simple love stories; found difficulty in finding a publisher for her first work, *A Welsh Singer* (1897); her novels, which include *Torn Sails* (1898), *By Berwen Banks* (1899), and *On the Wings of the Wind* (1903), had a wide circulation.

PUGH, SIR ARTHUR (1870-1955), trade-union official; educated at an elementary school; apprenticed to farmer and butcher and later worked in steel works in Wales and Lincolnshire; assistant secretary, British Steel Smelters' Association, 1906; general secretary, Iron and Steel Trades Confederation, 1917; member, Parliamentary Committee (later, General Council), Trades Union Congress, 1920-36; chairman, TUC, 1925; involved in the negotiations leading up to the general strike, 1926, and in its settlement; CBE, 1930; knighted, 1935.

PULLEN, HENRY WILLIAM (1836-1903), pamphleteer; BA, Clare College, Cambridge, 1859; MA, 1862; vicar choral of York Minster, 1862, and Salisbury Cathedral, 1863-75; obtained fame by allegorical pamphlet *The Fight at Dame Europa's School* (1870, 193rd thousand, 1874), accusing England of cowardice in observing neutrality in Franco-Prussian war; chaplain in *Alert* in Arctic expedition of Sir George Nares [q.v.], 1875-6; edited John Murray's 'Handbooks' of travel from 1876; criticized educational system in stories of school life.

PUMPHREY, RICHARD JULIUS (1906-1967), professor of zoology in the university of Liverpool; educated at Marlborough and Trinity Hall, Cambridge; first class, part ii, natural sciences tripos (1929, zoology and comparative anatomy); Ph.D., 1932; Rockefeller fellow, Pennsylvania University, and Beit fellow, Cambridge, 1936-9; demonstrator in experimental biology, 1930; carried out research into the physiology of hearing and vision in insects and birds; Sc.D., Cambridge, 1949; FRS, 1950; did research on radar, 1939-45; Derby professor of zoology, Liverpool University, 1949.

PURCELL, ALBERT ARTHUR WILLIAM (1872-1935), agitator; delegate to Salford trades council, 1907-22; on Trades Union Congress general council, 1919-28 (president, 1924); a leading member of general strike committee, 1926; president, International Federation of Trade Unions, 1924-7; labour MP, Coventry (1923-4), Forest of Dean (1925-9); believed in well-organized industrial action combined with political pressure.

PURSE, BENJAMIN ORMOND (1874-1950), blind social worker and expert on blind welfare; educated at Henshaw's Institution for the Blind, Manchester; on government advisory committee for blind welfare, 1917-42; superintended administration of relief, training, and general employment, National Institute for the Blind, 1920-43.

PURSER, LOUIS CLAUDE (1854-1932), classical scholar; senior moderator, classics, Trinity College, Dublin, 1875; fellow, 1881-1927; professor of Latin, 1898-1904; vice-provost, 1924-7; with R. Y. Tyrrell [q.v.] edited *The Correspondence of Cicero* (1886-99); com-pleted a revised edition by 1932; also edited Cicero's Letters for 'Oxford Classical Texts' (1901-3), and Apuleius' *Story of Cupid and Psyche* (1910); FBA, 1923.

PURVIS, ARTHUR BLAIKIE (1890-1941), industrialist and buyer of war supplies; educated at Tottenham grammar school; joined Nobel Explosives, 1910; responsible for purchase of explosives materials in America, 1914-18; president, Canadian Explosives, Ltd., 1925; re-organized it into Canadian Industries, Ltd.; director-general, British purchasing in United States, 1939; chairman, Anglo-French Purchasing Board, 1940, British Supply Council in North America, Dec. 1940; obtained co-ordination of British purchasing; took over French commitments, June 1940; killed in air crash at Prestwick.

PYE, SIR DAVID RANDALL (1886-1960), engineer and administrator; educated at Tonbridge and Trinity College, Cambridge; first class, mechanical sciences tripos, 1908; joined Oxford engineering school, 1909; fellow, New College, 1911; master at Winchester, 1915-16; worked on design and testing, Royal Flying Corps, 1916-19; lecturer and fellow, Trinity College, Cambridge, 1919; deputy director, scientific research, Air Ministry under H. E. Wimperis [q.v.], 1925; director, 1937; closely associated with development of jet aircraft engines; provost, University College London, 1943-51; president, Institution of Mechanical Engineers, 1952; published *The Internal Combustion Engine* (2 vols., 1931-4); CB, FRS, 1937; knighted, 1952; enthusiastic climber, vice-president, Alpine Club, 1956.

PYNE, LOUISA FANNY BODDA (1832-1904), soprano vocalist. [See BODDA PYNE.]

Q

QUARRIER, WILLIAM (1829-1903), philanthropist; entered a pin factory at Glasgow at seven; started boot business at twenty; founded shoeblack, news, and parcels brigades in Glasgow, 1864; opened orphan home, 1871; purchased farm of forty acres at Bridge of Weir (1876), where he opened the 'Orphan Homes of Scotland' (1878), which subsequently consisted of fifty villas, with church, school, poultry farm, consumptive sanatoria, and homes for epileptics.

QUICK, SIR JOHN (1852-1932), lawyer, politician, and judge of the Commonwealth Arbi-tration Court; LLB, Melbourne, 1877; admitted to Victorian bar (1878) and practised in Bendigo; member of Victorian parliament, 1880-9; worked tirelessly for federation and delegate to the federal convention, 1897; member of Commonwealth parliament, 1901-13; chairman, royal commission on Commonwealth tariffs, 1904-7; postmaster-general, 1909-10; deputy president, Commonwealth Arbitration Court, 1922-30; knighted, 1901.

QUICK, OLIVER CHASE (1885-1944), divine; son of R. H. Quick [q.v.]; educated at Harrow and Corpus Christi College, Oxford;

deacon, 1911; priest, 1912; canon of Newcastle (1920–3), Carlisle (1923–30), St. Paul's (1930–4), Durham (1934–9, professor of divinity), and Christ Church, Oxford (1939–43, regius professor of divinity).

QUICKSWOOD, BARON (1869–1956), politician and provost of Eton. [See CECIL, HUGH RICHARD HEATHCOTE GASCOYNE-.]

QUILLER-COUCH, SIR ARTHUR THOMAS ('Q') (1863–1944), Cornishman, man of letters, and professor of English literature; educated at Newton Abbot, Clifton and Trinity College, Oxford; second class, *lit. hum.*, 1886; first used pseudonym 'Q' in *Oxford Magazine*; wrote for the *Speaker*, 1890–9; published *Astonishing History of Troy Town* (1888), one of many novels set in Fowey where he settled (1892); romantic, versatile, and prolific writer; conspicuous as an anthologist and stylist; edited notably *The Oxford Book of English Verse* (1900); King Edward VII professor of English literature, Cambridge, 1912–44; knighted, 1910.

QUILTER, HARRY (1851–1907), art critic; BA, Trinity College, Cambridge, 1874; MA, 1877; called to bar, 1878; art critic and journalist mainly for *Spectator* and *The Times*, 1876–87; evoked anger and attack from J. A. M. Whistler [q.v.] by his frank criticisms; exhibited oil-paintings in London, 1884–93; edited ambitious *Universal Review*, 1888–90; conducted 'rational' boarding-schools at Mitcham and Liverpool, 1894–6; published *What's What*, 1902; art collections realized over £14,000, 1906; published verse, prose, essays, and works on art.

QUILTER, ROGER CUTHBERT (1877–1953), composer; son of (Sir) Cuthbert Quilter, the first baronet [q.v.]; educated at Eton, and studied music at Frankfurt; composed numerous songs, including *Three Shakespeare Songs* (1905); worked for the theatre, 1911–36; composed music for *Where the Rainbow Ends*, *A Children's Overture*, and *Julia*; leading singers of his day, singing his songs, included Plunket Greene and Gervase Elwes [qq.v.].

QUILTER, SIR WILLIAM CUTHBERT, first baronet (1841–1911), art collector and politician; brother of H. Quilter [q.v.]; a founder and director of National Telephone Co., 1881; developed Bawdsey estate near Felixstowe, 1883; agriculturist, cattle breeder, and yachtsman; liberal MP (1885) and unionist MP (1886–1905) for Sudbury; baronet, 1897; formed great art collection which realized £87,780 in 1909.

QUIN, WINDHAM THOMAS WYND-HAM-, fourth EARL OF DUNRAVEN AND MOUNT-EARL in the peerage of Ireland and second BARON KENRY of the United Kingdom (1841–1926), Irish politician; son of third earl [q.v.]; entered army, 1862; as war correspondent for *Daily Telegraph* present at capture of Magdala (1868) and at Versailles (1871); succeeded father, 1871; settled at Adare, county Limerick; under-secretary for colonies, 1885–6, 1886–7; chairman, House of Lords committee on sweating, 1888–90; of Irish land conference, report of which became basis for Wyndham Land Act (1903), 1902–3; active on Irish Reform Association, which advocated policy of 'devolution'; believed in solution of Irish difficulty on federal lines; nominated to serve on first senate of Irish Free State, 1921; a noted sportsman, especially famous for horse-breeding and yachting.

R

RACKHAM, ARTHUR (1867–1939), illustrator; educated at City of London School and Lambeth Art School; drew for *Cassell's Magazine* and other periodicals; his book-illustrations include Grimms' *Fairy Tales* (1900) and Washington Irving's *Rip van Winkle* (1905); style angular, imaginative, and detailed, in the Gothic manner.

RACKHAM, BERNARD (1876–1964), museum curator and authority on ceramics; brother of Arthur Rackham [q.v.], the illustrator; educated at City of London School and Pembroke College, Cambridge; first class, classical tripos, 1898; assistant keeper, South Kensington Museum (later the Victoria and Albert) in ceramics department; registered ceramic collection acquired from the Museum of Practical Geology; prepared the *Catalogue* of Lady Charlotte Schreiber Collection of English pottery, porcelain, and enamels (3 vols., 1924–30); other publications include *English Pottery* (with (Sir) Herbert Read, q.v.), *Catalogue* (2 vols., 1935) of the Glaisher Collection at the Fitzwilliam Museum Cambridge, *Catalogue of the Le Blond Collection of Corean Pottery* (1918), and a translation from Danish of Emil Hannover's *Keramisk Haandbag* in *Pottery and Porcelain* (3 vols, 1925); his interest in maiolica led to catalogues and guides including *Early Netherlands Maiolica* (1926), *Guide to Italian Maiolica* (1933), and *Catalogue of Italian Maiolica* (2 vols., 1940); interest in stained glass produced a *Guide to the Collections of Stained Glass* in the museum (1936), and *Ancient Glass of Canterbury Cathedral* (1949); after retirement as keeper, ceramics department (1914-38), continued to publish works on ceramics such as *A Key to Pottery and Glass* (1940) and *Early Staffordshire Pottery* (1951); fellow, Society of Antiquaries; CB, 1937; president, English Ceramic Circle.

RADCLIFFE-CROCKER, HENRY (1845-1909), dermatologist; studied medicine at University College Hospital, London; MRCS, 1873; LRCP, 1874; BS, 1874; MD, 1875; FRCP, 1887; physician and dermatologist at University College Hospital from 1879; member of English and foreign dermatological societies; prominent in affairs of British Medical Association; died at Engelberg, Switzerland; distinguished leprologist; published *Diseases of the Skin* (1888, 3rd edn., 2 vols., 1903) and *The Atlas of Diseases of the Skin* (2 vols., 1896); his library, with £1,500 to found scholarship in dermatology, was presented by his widow to University College Hospital.

RAE, WILLIAM FRASER (1835-1905), author; called to bar, 1861; abandoned law for journalism; special correspondent to *Daily News* in Canada and America; newspaper articles were collected in volumes on Canada, and America, Austria and Egypt; translated Taine's *Notes on England*, 1873; wrote much on 'Junius' problem and on eighteenth-century political history; chief work was *Sheridan, a Biography* (2 vols., 1896); edited from original MSS Sheridan's plays, 1902; also wrote fiction.

RAGGI, MARIO (1821-1907), sculptor; born at Carrara; came to London, 1850; executed memorials to Beaconsfield (in Parliament Square) and Gladstone (at Manchester); exhibited busts at Royal Academy, 1878-95.

RAIKES, HUMPHREY RIVAZ (1891-1955), chemist, and principal and vice-chancellor of the university of Witwatersrand, Johannesburg; educated at Tonbridge, Dulwich, and Balliol College, Oxford; Abbott scholar, 1911; first class, chemistry, 1914; served with the Buffs in France, 1914; transferred to Royal Flying Corps, 1915; chief experimental officer, RFC, 1918; tutorial fellow, Exeter College, Oxford, 1924; chief instructor, Oxford University Air

Squadron (wing commander, RAF), 1925; principal, university of Witwatersrand, 1927; vice-chancellor, 1948; encouraged development of postgraduate studies in pure and applied sciences; approved inclusion of all races in university; AFC, 1918; hon. degrees from universities of Bristol, Cambridge, Cape Town, Toronto, and Witwatersrand.

RAILTON, HERBERT (1858-1910), black-and-white draughtsman; abandoned architecture for book illustration; depicted old buildings with rare delicacy.

RAINE, ALLEN (pseudonym) (1836-1908), novelist. [See PUDDICOMBE, ANNE ADALISA.]

RAINES, SIR JULIUS AUGUSTUS ROBERT (1827-1909), general; entered army, 1842; served in Crimean war, 1854-5; commanded 95th regiment throughout Indian Mutiny, 1857-9; CB, 1859; commanded expedition into Arabia, 1865-6; lieutenant-general, 1877; general, 1881; KCB, 1893; GCB, 1906; published *The 95th (Derbyshire) Regiment in Central India* (1900).

RAINY, ROBERT (1826-1906), Scottish divine; MA, Glasgow University, 1844; DD, 1863; studied for Free Church ministry at Free Church New College, Edinburgh, 1844; minister at Free High Church, Edinburgh, 1854-62; professor of church history in Free Church College, 1862; principal, 1874-1901; moderator of assembly, 1887, 1900, 1905; advocated 'voluntary' policy; achieved union of reformed Presbyterian synod with Free Church, 1876; as convener of 'highland committee' from 1881 raised much money for church endowment; helped to pass the Declaratory Act, 1892; secured union with United Presbyterian Church, 1900; first moderator; writings include *Three Lectures on the Church of Scotland* (1872, a reply to Dean Stanley), *The Bible and Criticism* (1878); his son, ADAM ROLLAND RAINY (1862-1911), surgeon-oculist, was liberal MP for Kilmarnock Burghs, 1906-11.

RAIT, SIR ROBERT SANGSTER (1874-1936), historian and principal of Glasgow University; educated at Aberdeen and New College, Oxford; first class, modern history and fellow, 1899; professor, Scottish history and literature, Glasgow, 1913-29; principal and vice-chancellor, 1929-36; historiographer-royal for Scotland, 1919-29; works include *The Parliaments of Scotland* (1924); knighted, 1933.

RALEIGH, SIR WALTER ALEXANDER (1861-1922), critic and essayist; son of Alexander Raleigh [q.v.]; BA, University College,

London, and King's College, Cambridge; professor of modern literature, University College, Liverpool, 1889; professor of English language and literature, Glasgow University, 1900; first holder of new chair of English literature at Oxford and fellow of Magdalen College, 1904; Merton professor of English literature and fellow of Merton, 1914; contributed greatly to development of school of English language and literature, his lectures arousing great enthusiasm; his thoughts after 1914 occupied by European war; Clark lecturer in English literature, Trinity College, Cambridge, 1899 and 1911; Leslie Stephen lecturer, 1907; knighted, 1911; his works include *Style* (1897), *Milton* (1900), *Wordsworth* (1903), *The English Voyages of the Sixteenth Century* (1906), *Shakespeare* (1907), *Six Essays on Johnson* (1910), an edition of the *Complete Works of George Savile, first Marquess of Halifax* (1912), *The War in the Air* (1922, vol. i of the official history of the Royal Air Force), and many war pamphlets.

RALSTON, JAMES LAYTON (1881–1948), statesman, lawyer, and soldier; born in Nova Scotia; educated at Amherst Academy and Dalhousie Law School; KC, 1914; practised successively in Amherst, Halifax, and Montreal; federal liberal member, Nova Scotia legislature, 1911–20; served in France, 1917–18; DSO; lieutenant-colonel, 1918; colonel, 1924; minister of national defence, Canada, 1926–30; of finance, 1939–40; of defence, 1940–4; his demand (Oct. 1944) for conscription for overseas service resulted in government crisis and his resignation.

RAM, Sir (LUCIUS ABEL JOHN) GRANVILLE (1885–1952), parliamentary draftsman; educated at Eton and Exeter College, Oxford; called to bar (Inner Temple), 1910; served with Hertfordshire Yeomanry in Egypt, Gallipoli, and France, 1914–18; solicitor to Ministry of Labour, 1923; third parliamentary counsel to the Treasury, 1925; second parliamentary counsel, 1929; first parliamentary counsel, 1937–47; head of consolidation branch, parliamentary counsels' office, 1947–52; CB, 1931; KCB, 1938; KC, 1943.

RAMAN, Sir (CHANDRASEKHARA) VENKATA (1888–1970), physicist; educated at college in Vizagapatam and Presidency College, Madras; BA, 1904; MA, 1908; joined Government Finance Department, Calcutta, 1907–17; did scientific experimental work in spare time at the Indian Association for the Cultivation of Science; Palit professor of physics, Calcutta University, 1917; published *Molecular Diffraction of Light* (1922); gave name to the Raman effect, one of the main methods for determining molecular vibrational frequencies and for the specific identification of features of molecular structure; studied the sounds of musical instruments; awarded Nobel prize, 1930; director of Indian Institute of Science, Bangalore, 1933–48; first president, Indian Academy of Sciences, 1934–70; director, Raman Research Institute of Bangalore, 1949–70; published *The Physiology of Vision* (1968); first national professor appointed by Indian government; travelled much in Europe and North America; a legendary figure of Indian science; FRS, 1924; knighted, 1929; hon. doctorates and other academic distinctions from universities and societies in India, Britain, Europe, and America.

RAMÉE, MARIE LOUISE DE LA, 'OUIDA' (1839–1908), novelist. [See De La Ramée.]

RAMSAY, ALEXANDER (1822–1909), Scottish journalist; editor of *Banffshire Journal* from 1847 to death; edited *Aberdeen-Angus Herd Book*, 1872–1905; hon. LLD, Aberdeen, 1895; wrote *History of Agricultural Society of Scotland* (1879).

RAMSAY, Sir BERTRAM HOME (1883–1945), admiral; entered navy, 1898; saw active service in Somaliland expedition, 1903–4; qualified as signal officer, 1909; as war staff officer, 1913; lieutenant-commander, 1914; served in Dover Patrol, 1915–18; commanded *Broke* in Ostend operations, 1918; naval officer, Imperial Defence College, 1931–3; commanded *Royal Sovereign*, 1933–5; rear-admiral, 1935; asked to be relieved as chief of staff to Sir Roger Backhouse [q.v.] owing to incompatibility and went on half-pay, 1935; retired, 1938; vice-admiral, 1939; flag officer in charge, Dover, Aug. 1939–Apr. 1942; organized Dunkirk evacuation, 26 May–4 June 1940; KCB; deputy naval commander, North African landings, 1942; naval commander, Eastern Task Force, invasion of Sicily, 1943; KBE; allied naval commander-in-chief for Normandy invasion; restored to active list and promoted admiral, 1944; killed when aircraft crashed taking off from Toussus-le-Noble.

RAMSAY, Sir JAMES HENRY, tenth baronet, of Bamff (1832–1925), historian; son of Sir George Ramsay, ninth baronet [q.v.], and nephew of William Ramsay [q.v.]; born at Versailles; BA, Christ Church, Oxford; student, 1854–61; succeeded father, 1871; FBA, 1915; his lifework, writing of the political history of England down to end of Middle Ages: *Lancaster and York . . . 1399-1485* (1892), *The Foundations of England* (1898), *The Angevin Empire, 1154-1216* (1903), *The Dawn of the Constitution, 1216-1307* (1908), and *The Genesis of Lancaster,*

1307-1399 (1913); a chronicler rather than an historian, who suffered from comparative isolation, but the narrative part of whose work is invaluable.

RAMSAY, SIR WILLIAM (1852-1916), chemical discoverer; educated at Glasgow University; Ph.D., Tübingen; taught at Glasgow University and carried out investigations in organic chemistry, 1872-80; professor of chemistry, University College, Bristol, 1880-7; principal, 1881; professor of general chemistry, University College, London, 1887-1913; FRS, 1888; KCB, 1902; Nobel prizeman, 1904; his scientific work in London of three periods; first line of investigation physico-chemical, including determination of molecular complexity of pure liquids; second, discovery of chemically inert elementary gases, argon (in collaboration with J. W. Strutt, third Baron Rayleigh, q.v.), helium, neon, krypton, and xenon; third, proof that emanation of radium produces helium during its atomic disintegration; greatest chemical discoverer of his time.

RAMSAY, SIR WILLIAM MITCHELL (1851-1939), classical scholar and archaeologist and foremost authority of his day on the topography, antiquities, and history of Asia Minor in ancient times; educated at Aberdeen and St. John's College, Oxford; first class, *lit. hum.*, 1876; explored Anatolia, 1881-91, 1899-1914; regius professor of humanity, Aberdeen, 1886-1911; works include *Historical Geography of Asia Minor* (1890) and *The Church in the Roman Empire before A.D. 170* (1893); knighted, 1906.

RAMSAY-STEEL-MAITLAND, SIR ARTHUR HERBERT DRUMMOND, first baronet (1876-1935), politician and economist. [See STEEL-MAITLAND.]

RAMSDEN, OMAR (1873-1939), goldsmith; educated in Illinois, Sheffield, and at the Royal College of Art, South Kensington; set up independent practice in London, 1904; executed plate for Coventry, Bermuda, Colombo, and Westminster cathedrals; showed great feeling for material, finish in detail, and opulence in design.

RANDALL, RICHARD WILLIAM (1824-1906), dean of Chichester; BA, Christ Church, Oxford, 1846; MA, 1849; DD, 1892; vicar of All Saints', Clifton, 1868-92; hon. canon of Gloucester, 1891; dean of Chichester, 1892-1902; strong ritualist; wrote devotional works.

RANDALL-MacIVER, DAVID (1873-1945), archaeologist and anthropologist; edu-cated at Radley and Queen's College, Oxford; first class, *lit. hum.*, 1896; Laycock student of Egyptology, Worcester College, 1900-6; directed Nubian expedition for University Museum, Philadelphia, 1907-11; librarian, American Geographical Society, 1911-14; served in 1914-18 war; settled in Rome; publications include *Mediaeval Rhodesia* (1906), *Villanovans and Early Etruscans* (1924), and other works on Italian archaeology; FBA, 1938.

RANDEGGER, ALBERTO (1832-1911), musician; born at Trieste; settled in London, 1854; organist at St. Paul's, Regent's Park, 1859-70; joined staff of Royal Academy of Music, 1868, and Royal College of Music, 1896; composed grand opera *Bianca Capello*, operettas, and cantatas; wrote primer on *Singing*; conducted Italian opera for Sir Augustus Harris [q.v.], 1887-98, and Norwich triennial festivals, 1881-1905.

RANDLES, MARSHALL (1826-1904), Wesleyan divine; member of legal conference, 1882; tutor of systematic theology at Didsbury, 1886; president of conference, 1896; advocated total abstinence in *Britain's Bane and Antidote* (1864); theological works include *For Ever, an Essay on Everlasting Punishment* (1871).

RANDOLPH, FRANCIS CHARLES HINGESTON- (1833-1910), antiquary. [See HINGESTON-RANDOLPH.]

RANDOLPH, SIR GEORGE GRANVILLE (1818-1907), admiral; entered navy, 1830; served off Spain, in the Mediterranean, and in East Indies; in Black Sea and Crimean operations, 1854-5; commodore at Cape of Good Hope, 1867-9; CB, 1869; vice-admiral, 1877; admiral, 1884; KCB, 1897; published *Problems in Naval Tactics* (1879); FRGS.

RANJITSINHJI, MAHARAJA JAM SAHEB OF NAWANAGAR (1872-1933), Indian ruler and cricketer. [See NAWANAGAR.]

RANKEILLOUR, first BARON (1870-1949), parliamentarian. [See HOPE, JAMES FITZALAN.]

RANKIN, SIR GEORGE CLAUS (1877-1946), judge; first class, philosophy, Edinburgh, 1897; moral sciences tripos, Trinity College, Cambridge, 1899-1900; president of Union, 1901; called to bar (Lincoln's Inn), 1904; bencher, 1935; judge, Calcutta high court, 1918-26; chairman, civil justice committee, 1924; chief justice of Bengal, 1926-34; member, judicial committee of Privy Council, 1935-44; leading authority on Indian jurisprudence; knighted, 1925.

RANSOM, WILLIAM HENRY (1824-1907), physician and embryologist; fellow pupil with T. H. Huxley [q.v.] at University College, London; fellow, 1896; MD, London, 1850; FRCP, 1869; physician to Nottingham General Hospital, 1854-90; FRS, 1870; published original researches into ovology; devised gas-heating disinfecting stove, 1870; helped to establish University College, Nottingham; published *The Inflammation Idea in General Pathology*, (1906).

RANSOME, ARTHUR MICHELL (1884-1967), journalist and author; educated at Rugby and Yorkshire College, Leeds; worked for Grant Richards, publishers, at seventeen; freelance writer of stories and articles, 1901-13; won libel suit over book on Oscar Wilde [q.v.], 1912-13; went to Russia, 1913; taught himself Russian; correspondent for *Daily News*, 1914; published *Old Peter's Russian Tales* (1916); stayed in Russia till 1919; published *Six Weeks in Russia in 1919* (1919) and *The Crisis in Russia* (1921); wrote for *Manchester Guardian* in association with C. P. Scott [q.v.], 1924; *Guardian* correspondent in Egypt and China, 1925-6; published *Swallows and Amazons* (1930) about the Walker children, followed by *Swallowdale* (1931) and *Peter Duck* (1932); also wrote books about bird-watching and fishing, such as *Rod and Line* (1929) and *Mainly about Fishing* (1959); hon. D. Litt., Leeds, 1952; CBE, 1953.

RAPER, ROBERT WILLIAM (1842-1915), classical scholar; BA, Trinity College, Oxford; first class *lit. hum.*; life fellow of Trinity, 1871; tutor, and subsequently lecturer and bursar; vice-president, 1894; declined presidency three times; founded Oxford University Appointments Committee, 1894.

RAPSON, EDWARD JAMES (1861-1937), Sanskrit scholar; educated at Hereford Cathedral School and St. John's College, Cambridge; fellow, 1887-93; assistant keeper, coins and medals, British Museum, 1887-1906; professor of Sanskrit, University College, London (1903-6), Cambridge (1906-36); edited Kharosthi texts discovered by Sir Aurel Stein [q.v.] (3 parts, 1920-9) and vol. i *Cambridge History of India* (1922); FBA, 1931.

RASHDALL, HASTINGS (1858-1924), moral philosopher, theologian, and historian of universities; educated at Harrow and New College, Oxford; ordained, 1884; fellow of Hertford College, Oxford, 1888; of New College, 1895-1917; taught philosophy for the school of *literae humaniores*; canon of Hereford, 1909; dean of Carlisle, 1917-24; an active member of the Christian Social Union; FBA,

1909; a figure of importance in English religious thought as a leader of liberal school of Anglicanism; had little sympathy with mystical element in religion, and was above all a moralist; his works include *The Universities of Europe in the Middle Ages* (expanded from his chancellor's English essay prize [1883], 1895), *The Theory of Good and Evil* (which embodies his teaching on moral philosophy, 1907), and *The Idea of Atonement in Christian Theology* (his Bampton lectures [1915], 1919).

RASSAM, HORMUZD (1826-1910), Assyrian explorer; born in Asiatic Turkey; son of Nestorian Christian; converted in youth to protestantism; assisted Sir A. H. Layard [q.v.] in excavations at Nimroud, 1845, and at Kouyunjik (i.e. Nineveh), 1847; for short period at Magdalen College, Oxford, 1848; continued with Layard excavations at Nineveh and Nimroud, 1849, and in Babylon, 1850-1; discovered palace of Aššurbani-âpli at Nineveh; political interpreter at Aden, 1854-61; magistrate and political resident; represented British interests in Zanzibar, 1861; sent with two others commissioned by British government to protest against imprisonment at Magdala, by Theodore, king of Abyssinia, of C. D. Cameron [q.v.] and H. A. Stern [q.v.], 1864; was imprisoned in chains at Magdala, 1866; released on arrival of Sir Robert Napier (later Lord Napier of Magdala, q.v.), 1868; narrated his experiences in *British Mission to Theodore, King of Abyssinia* (2 vols., 1869); again employed in Asiatic Turkey to inquire into condition of Christian communities, 1877; from 1876 to 1882 made many important finds in Assyria and Babylonia; after 1882 resided at Brighton, writing on Assyriology and, as a strong evangelical, engaging in religious controversy.

RATHBONE, ELEANOR FLORENCE (1872-1946), social reformer; daughter of William Rathbone (1819-1902, q.v.); educated at Kensington high school and Somerville College, Oxford; second class, *lit. hum.*, 1896; independent MP, combined English universities, 1929-46; leading advocate of family allowances; prominent in women's suffrage movement; fought prolonged battle to raise conditions of Indian women; violently opposed appeasement and continuously in conflict with government, 1931-9; worked untiringly for refugees; publications include *The Disinherited Family* (1924), *The Case for Family Allowances* (1940), and *War Can Be Averted* (1937).

RATHBONE, WILLIAM (1819-1902), philanthropist; after much travel became partner in father's Liverpool mercantile firm, 1841; established Liverpool Training School for

Nurses, 1862; helped in founding in London National Association for providing trained nurses, 1874; supported Liverpool cotton famine relief fund, 1862-3; liberal MP for Liverpool, 1868-80, for Carnarvonshire, 1880-5, for North Carnarvonshire, 1885-95; president of University College, Liverpool, from 1892, and of University College of North Wales from 1891; LLD, Victoria University, 1895; published works on social questions.

RATTIGAN, SIR WILLIAM HENRY (1842-1904), Anglo-Indian jurist; born at Delhi; called to bar (Lincoln's Inn), 1873; DL, Göttingen, 1885; formed extensive practice as a barrister at Lahore; vice-chancellor of Punjab University, 1887-95; DL, 1896; LLD, Glasgow, 1901; served on viceroy's legislative council, 1892-3; wrote *A Digest of Civil and Customary Law of the Punjab* (1880, 7th edn. 1909); knighted, 1895; QC, 1897; settled in England, 1901; liberal unionist MP for North-East Lanark, 1901-4.

RAU, SIR BENEGAL NARSING (1887-1953), Indian judge and diplomatist; educated at Madras University and Trinity College, Cambridge; entered Indian civil service, 1909; served in Bengal and Assam, 1910-25; secretary to legislative department and legal adviser, Assam, 1925-35; draftsman, law department, government of India, 1935-8; high court judge, Calcutta, 1938-44; retired from civil service, prime minister, Kashmir, 1944-5; reforms office, government of India, 1945; constitutional adviser to Constituent Assembly, 1946; member, Indian delegation to United Nations General Assembly, 1948; Indian representative, Security Council, 1950-1; president, Security Council, 1950; judge, International Court of Justice at The Hague, 1951-3; CIE, 1934; knighted, 1938.

RAVEN, CHARLES EARLE (1885-1964), professor of divinity; educated at Uppingham and Gonville and Caius College, Cambridge; first class, part i, classical tripos, and part ii, theological tripos, 1907-8; worked for Liverpool City Council, 1908-9; lecturer in divinity, fellow and dean, Emmanuel College, Cambridge, 1910; assistant master, Tonbridge, 1915-17; army chaplain in France, 1917-18; rector, Bletchingley, Surrey, 1919-23; residentiary canon, Liverpool Cathedral, 1924-32; chaplain to the King, 1920; published *Christian Socialism 1848-1854* (1920) and *Apollinarianism* (1923); DD, Cambridge; regius professor of divinity, Cambridge, 1932-50; canon, Ely Cathedral; fellow, Christ's College; master, 1939-50; other publications include *In Praise of Birds* (1925), *John Ray, Naturalist* (1942), and *English Naturalists from*

Neckam to Ray (1947), which won a James Tait Black memorial prize; pacifist; one of the sponsors of the Peace Pledge Union; vice-chancellor, Cambridge, 1947-50; warden, Madingley Hall; travelled widely preaching and lecturing; religious publications include *Jesus and the Gospel of Love* (1931), *Evolution and the Christian Concept of God* (1936), and *Good News of God* (1943); trustee, British Museum, 1950; FBA, 1948; fellow, Linnean Society; hon. degrees from number of universities.

RAVEN, JOHN JAMES (1833-1906), archaeologist and campanologist; BA, Emmanuel College, Cambridge, 1857; MA, 1860; DD, 1872; headmaster of Bungay (1859-66) and Yarmouth (1866-85) grammar schools; hon. canon of Norwich, 1888; FSA, 1891; published accounts of the church bells of Cambridgeshire (1869), Suffolk (1890), and England (1906), works on Suffolk, and sermons.

RAVEN-HILL, LEONARD (1867-1942), artist, illustrator, and cartoonist; studied at City and Guilds Art School and in Paris; joined *Punch* round table, 1901; a political cartoonist, 1910-35; illustrations included *Stalky and Co.* for Rudyard Kipling and *Kipps* for H. G. Wells [qq.v.].

RAVERAT, GWENDOLEN MARY (1885-1957), artist; daughter of (Sir) George Howard Darwin [q.v.] and granddaughter of Charles Darwin [q.v.]; studied at Slade School, 1908; married Jacques Pierre Raverat, 1911; he died, 1925; her work included painting, drawing, and theatre design; wood-engraving, her outstanding achievement; founder-member of the Society of Wood Engravers; illustrated *Spring Morning*, poems by her cousin, Frances Cornford [q.v.] (1915); did engravings for *The Cambridge Book of Poetry for Children*, selected by Kenneth Grahame [q.v.] (1932); published *Period Piece*, an account of her childhood (1952).

RAVERTY, HENRY GEORGE (1825-1906), soldier and oriental scholar; served in Indian army, 1843-64; present at Multan and Gujarat, 1848-9; published *Thesaurus of English Hindustani Technical Terms* (1859), Pushtu or Afghan grammar (1855), dictionary (1860), and anthology (1860), as well as *Notes on Afghanistan and Baluchistan* (4 parts, 1881-8); translated from Persian Minhāj's *Ṭabaḳāt i Nāṣirī*, and made many contributions to *Journal* of Asiatic Society, Bengal; left unpublished MSS on Afghanistan history.

RAVILIOUS, ERIC WILLIAM (1903-1942), artist; studied at Eastbourne Art School and

Royal College of Art; instructor in design, 1929-38; executed wood-engravings for Golden Cockerel, Nonesuch, Curwen, and Cresset presses and commercial concerns; appliqué decorations for Wedgwood pottery; and designs for engraving on glass; painted water-colour landscapes of delicate and precise design; official war artist, 1940-2; lost flying from Iceland.

RAWLING, CECIL GODFREY (1870-1917), soldier and explorer; entered army, 1891, explored unsurveyed Western Tibet and Rudok, 1903, upper Tsangpo (Brahmaputra), 1904-5; Dutch New Guinea, 1909; DSO, 1917; killed in action in Flanders.

RAWLINSON, GEORGE (1812-1902), canon of Canterbury; BA, Trinity College, Oxford, 1838; MA, 1841; president of Union, 1840; fellow of Exeter College, 1840; Bampton lecturer, 1859; Camden professor of ancient history, 1861-89; canon of Canterbury, 1872; rector of All Hallows, Lombard Street, 1888; edited with brother, Sir Henry C. Rawlinson [q.v.], *The History of Herodotus* (4 vols., 1858-60; abridged edn. 2 vols., 1897); published *The Five Great Monarchies* [viz. Chaldaea, Assyria, Babylonia, Media, and Persia] *of the Ancient Eastern World* (4 vols., 1862-7), *The Sixth Great Monarchy* [i.e. Parthia] (1873), *The Seventh Great Monarchy* [i.e. Sassanian or new Persian] (1876), and histories of Ancient Egypt (2 vols., 1881) and Phoenicia (1889), and many kindred works; wrote life of his brother, 1898.

RAWLINSON, SIR HENRY SEYMOUR, second baronet, and BARON RAWLINSON (1864-1925), general; son of Sir H. C. Rawlinson, first baronet [q.v.]; gazetted to King's Royal Rifles, 1884; served in India and Burma, 1884-9; transferred to Coldstream Guards, 1891; passed into Staff College, Camberley, 1893; succeeded father, 1895; served at battles of Atbara and Omdurman, 1898; served in South African war, 1899-1902; brigadier-general and commandant of Staff College, 1903; major-general, 1909; commanded third division, 1910-14; commanded IV Corps, 1914-15; led it at Neuve Chapelle, Aubers Ridge, Festubert, and Loos; in temporary command of First Army, 1915; lieutenant-general in command of Fourth Army, 1916; engaged in battle of Somme, July to Nov.; general, 1917; British military representative on Supreme War Council, 1918; commanded Fifth Army, which he reconstituted as Fourth, 1918; attacking astride Somme, fought and won four great battles and eighteen actions, Aug. to Nov.; carried out evacuation of northern Russia by allied forces, 1919; commander-in-chief, India, 1920-5; carried out reforms in

army, introduced process of Indianization, and opened up North-West Frontier; baron, 1919; died at Delhi.

RAWLINSON, WILLIAM GEORGE (1840-1928), art collector and writer on J. M. W. Turner [q.v.]; collected and studied drawings and engraved work of Turner; produced catalogues of *Liber Studiorum* (1878) and of *The Engraved Work of J. M. W. Turner, R. A.* (1908, 1913); sold his collections.

RAWSON, SIR HARRY HOLDSWORTH (1843-1910), admiral; entered navy, 1857; served in China war, 1858-61; reported on capabilities of defence of Suez canal, 1878; principal transport officer in Egyptian campaign, 1882; CB; commander-in-chief at Cape of Good Hope and West coast of Africa station, 1895-8; in command of capture of M'Weli and bombardment of Zanzibar, 1895-6; commanded Benin expedition, 1897; KCB, 1897; vice-admiral, 1898; commanded Channel squadron, 1898-1901; governor of New South Wales, 1902-9; admiral, 1903; GCB, 1906; GCMG, 1909.

RAYLEIGH, third BARON (1842-1919), mathematician and physicist. [See STRUTT, JOHN WILLIAM.]

RAYLEIGH, fourth BARON (1875-1947), experimental physicist. [See STRUTT, ROBERT JOHN.]

READ, SIR CHARLES HERCULES (1857-1929), antiquary and art connoisseur; assistant, department of British and medieval antiquities and ethnography, British Museum, 1880; keeper, 1896; successively secretary and president, Society of Antiquaries; knighted, 1912; FBA, 1913; his works include official museum publications and contributions to publications of learned societies; died at Rapallo.

READ, CLARE SEWELL (1826-1905), agriculturist; farmed family estates in Norfolk from 1854 to 1896; conservative MP for East Norfolk, 1865-7, for South Norfolk, 1868-80, and West Norfolk, 1884-6; served on cattle plague commission, 1865, and on several parliamentary agricultural committees; secretary to Local Government Board, 1874-6; assistant commissioner to inquire into agricultural conditions in Canada and America, 1879; wrote many agricultural papers; chairman of Farmers' Club, 1868 and 1892.

READ, GRANTLY DICK- (1890-1959), obstetrician. [See DICK-READ.]

READ, SIR HERBERT EDWARD (1893-1968), writer on art, critic, and poet; educated at an endowed school for orphans in Halifax and Leeds University; commissioned into the Green Howards, 1914; MC; DSO; worked at the Treasury, 1919-22; assistant keeper, Victoria and Albert Museum, 1922-31; Watson Gordon professor of fine art, Edinburgh University, 1931-3; editor, the *Burlington Magazine*, 1933-9; founded, with Roland Penrose, the Institute of Contemporary Arts, 1947; knighted, 1953; director, Routledge & Kegan Paul; publications include *The Innocent Eye* (1933), a fragment of autobiography, *Art Now* (1933), *Art and Industry* (1934), *Art and Society* (1937), a novel, *The Green Child* (1935), *In Defence of Shelley* (1936), and *Education through Art* (1943); he also wrote *Moon's Farm*, a poem for radio (1951), and Concise Histories of *Modern Painting* (1959) and of *Modern Sculpture* (1964).

READ, HERBERT HAROLD (1889-1970), geologist; educated at Simon Langton School, Canterbury and Royal College of Science, London; first class, geology, 1911; M.Sc.; joined Geological Survey of Great Britain, 1914; served in Royal Fusilier's, 1914-17; rejoined Geological Survey, 1917; concerned mainly with mapping in Northern Scotland; George Herdman professor, Liverpool University, 1931-9; professor, Imperial College, London, 1939-55; dean, Royal School of Mines, 1943-5; pro-rector, Imperial College, 1952-5; acting head of Imperial College, 1954-5; hon. D.Sc., Columbia; hon. Sc.D., Dublin; hon. MRIA; FRS, 1939; published eight revisions of geological classic, Rutley's *Elements of Mineralogy*, and *The Granite Controversy* (1957).

READ, SIR HERBERT JAMES (1863-1949), civil servant; educated at Allhallows School, Honiton and Brasenose College, Oxford; first class, mathematics, 1884; entered Colonial Office, 1889; assistant under-secretary supervising East and West African departments, 1916-24; governor of Mauritius, 1924-30; recognizing importance of scientific research was largely responsible for inception of Commonwealth Institutes of Entomology and Mycology; KCMG, 1918; GCMG, 1935.

READ, JOHN (1884-1963), organic chemist; educated at Sexey's School, Bruton, Finsbury Technical College, London, and Zurich University; Ph.D., 1907; assistant to (Sir) W. J. Pope [q.v.] in Manchester and Cambridge; studied at Emmanuel College, Cambridge; hon. MA, 1912; carried out research in organic stereochemistry; professor, organic chemistry, Sydney, Australia, 1916-23; Purdie professor of chemistry, St. Andrews, 1923-63; FRS, 1935;

Sc.D., Cambridge, 1935; last professor of chemistry to hold a life appointment in a British university; produced many original papers and textbooks; published *Humour and Humanism in Chemistry* (1947) and *A Direct Entry to Organic Chemistry* (1948).

READ, WALTER WILLIAM (1855-1907), Surrey cricketer; by prolific scores raised county to leading position; played for Gentlemen *v.* Players, 1877-95; for England *v.* Australia, 1884-93; visited Australia, 1882-3, and 1887-8, and South Africa, 1891-2; compiled *Annals of Cricket* (1896).

READE, THOMAS MELLARD (1832-1909), geologist; architect and civil engineer at Liverpool from 1860; laid out Blundell-sands estate, 1868; wrote much on glacial and postglacial geology of Lancashire, on mineral structure of slaty rocks, and on geomorphology; published *Origin of Mountain Ranges* (1886) and *Evolution of Earth Structure* (1903); FGS, 1872; Murchison medallist, 1896.

READING, first MARQUESS OF (1860-1935), lord chief justice of England, etc. [See ISAACS, RUFUS DANIEL.]

REAY, eleventh BARON (1839-1921), governor of Bombay and first president of the British Academy. [See MACKAY, DONALD JAMES.]

REDESDALE, first BARON (1837-1916), diplomatist and author. [See MITFORD, ALGERNON BERTRAM FREEMAN-.]

REDMAYNE, SIR RICHARD AUGUSTINE STUDDERT (1865-1955), mining engineer; educated privately and at the College of Physical Science, Newcastle upon Tyne; trained in colliery work; worked in Natal, 1891-3; manager, Seaton Delaval Collieries, Northumberland, 1894-1902; professor of mining, Birmingham University, 1902-8; pioneered higher educational methods for training mining engineers; first chief inspector of mines, Home Office, 1908-19; concerned with framing Coal Mines Act, 1911; chairman, Imperial Mineral Resources Bureau (amalgamated in 1925 with Imperial Institute), 1918-35; first president of the Institution of Professional Civil Servants, 1922-55; president, Institution of Civil Engineers, 1934-5; CB, 1912; KCB, 1914; companion of the Order of St. John of Jerusalem and chevalier of the Legion of Honour.

REDMOND, JOHN EDWARD (1856-1918), Irish political leader; born at Ballytrent, county Wexford; educated at Clongowes and Trinity College, Dublin; clerk in House of Commons, 1880; MP, New Ross, county Wexford, 1881;

sent on political mission to Irish of Australia and America, 1882-4; chief supporter of C. S. Parnell [q.v.] on split in Irish party, 1890; leader of Parnellite group on Parnell's death, 1891; resigned North Wexford, held since Redistribution Act (1885), 1891; stood for Cork, but heavily defeated; returned for Waterford, sole seat captured by Parnellites; a leading debater in House; adopted statesmanlike policy and conciliatory attitude towards government and anti-Parnellites; chairman of reunited Irish party, 1900; supported land purchase policy of George Wyndham [q.v.]; incurred temporary loss of prestige for himself and party in Ireland through attitude over liberal government's 'devolution' scheme, 1906; recovered ground by 1910; gave tardy support to Irish volunteer movement, 1914; on outbreak of 1914-18 war did utmost to promote recruiting in Ireland, but opposed her inclusion in first national service bill, 1916; surprised and horrified by Irish rebellion, 1916; lost the confidence of Ireland, which passed into control of extreme nationalists under Mr de Valera; three main points gained by Irish during his leadership, control of local government, ownership of land, and statutory establishment of Irish parliament with executive responsible to it; nationalist, devoid of hostility to British Empire, who aimed at free Ireland within the Empire.

REDMOND, WILLIAM HOEY KEARNEY (1861-1917), Irish nationalist; brother of J. E. Redmond [q.v.]; imprisoned as nationalist 'suspect', 1881; joined brother on political mission to Australia and America, 1883; MP, Wexford, 1883, North Fermanagh, 1885; East Clare, 1891-1917; re-imprisoned, 1888; violent Parnellite; served in 1914-18 war; killed in Flanders.

REDPATH, ANNE (1895-1965), painter; educated at Hawick high school, Edinburgh College of Art, and Moray House College of Education; qualified as art teacher, 1917; won travelling scholarship, enabling her to travel in Europe; exhibited regularly at Royal Scottish Academy and held exhibitions annually; up to 1949 painted still life and domestic interiors; later, travelled in Europe again and painted landscapes in new ranges of colour; lost use of right arm for a time, 1959; learnt to paint with left hand; many of her finest paintings, townscapes and continental church interiors; OBE, 1955; hon. LLD, Edinburgh, 1955; associate Royal Scottish Academy, 1947; academician, 1952; ARA, 1960; ARWS, 1962; president, Scottish Society of Women Artists, 1944-7.

REDPATH, HENRY ADENEY (1848-1908), biblical scholar; BA, Queen's College, Oxford,

1871; MA, 1874; D.Litt., 1901; completed *A Concordance to the Septuagint* begun by Edwin Hatch [q.v.] (3 vols., 1892-1906); Grinfield lecturer on Septuagint at Oxford, 1901-5.

REED, AUSTIN LEONARD (1873-1954), men's outfitter; educated at Reading School; joined his father's business of hosier and hatter, 1888; studied business methods in United States; opened his own business in City of London, 1900; expanded business into the provinces, 1913; Austin Reed Ltd. offered shares to public, 1920; Regent Street shop opened, 1926; chairman, Regent Street Association, 1927; president, City of London Trade Association and master of the Glovers' Company.

REED, SIR EDWARD JAMES (1830-1906), naval architect; edited *Mechanic's Magazine*, 1853; secretary to newly founded Institution of Naval Architects, 1860; employed by navy to convert wooden men-of-war into armourclads, 1862; chief constructor of the navy, 1863-70; designed and launched *Bellerophon*, on which he introduced longitudinal and bracket frame system, 1865; joined ordnance works of Sir J. Whitworth [q.v.] at Manchester, 1870; designed ships for foreign navies; consulting naval engineer to Indian government and crown colonies; FRS, 1876; CB, 1868; KCB, 1880; liberal MP for Pembroke Boroughs, 1874-80, and for Cardiff, 1880-95, 1900-5; lord of the Treasury, 1886; chairman of load line (1884) and manning of ships (1894) committees; published history of Japan (2 vols., 1880), *Treatise on the Stability of Ships* (1884), papers on naval subjects, and verse.

REED, EDWARD TENNYSON (1860-1933), caricaturist; son of Sir E. J. Reed [q.v.]; educated at Harrow; on staff of *Punch*, 1890-1912; palaeontology, archaeology, and heraldry the basis of his humour.

REED, SIR (HERBERT) STANLEY (1872-1969), newspaper editor and politician; educated privately; joined staff of *Times of India* in Bombay, 1897; travelled widely in India for his newspaper and became its special correspondent to the *Daily Chronicle*, 1900; accompanied Prince and Princess of Wales on tour of India, 1905; editor, *Times of India*, 1907; director of publicity to government of India, 1914-18; vice-president, Central Publicity Board, 1918; knighted, 1916; KBE, 1919; retired from India, 1923; conservative MP, Aylesbury, 1938-50; supported Indian independence; published *The India I knew, 1897-1947* (1952) and, with (Sir) Patrick Cadell, *India: The New Phase* (1928); founded the *Indian Year Book*, 1922.

REEVES, SIR WILLIAM CONRAD (1821-1902), chief justice of Barbados; son of negro mother; called to bar, 1863; practised at Barbados; member of local house of assembly, Barbados; 1874; solicitor-general; championed ancient Barbados constitution against schemes of crown government from 1876; attorney-general, 1882; KC, 1883; chief justice, 1886; knighted, 1889.

REGAN, CHARLES TATE (1878-1943), zoologist and director of the British Museum (Natural History); educated at Derby School and Queens' College, Cambridge; second class, natural sciences tripos, 1901; assistant, British Museum (Natural History), 1901; keeper, department of zoology, 1921; director, 1927-38; defined and surveyed some forty orders of fishes producing comprehensive classification; FRS, 1917.

REICH, EMIL (1854-1910), historian; born in Hungary; settled in England, 1893, as writer and lecturer at Oxford, Cambridge, and London; voluminous works include *Graeco-Roman Institutions* (1890), *Hungarian Literature* (1897), lectures on Plato (1906), and *General History of Western Nations—Antiquity* (2 vols., 1908-9); he laid stress as historian on geographical and economic conditions.

REID, ARCHIBALD DAVID (1844-1908), painter; studied at Trustees' Academy and Royal Scottish Academy, Edinburgh; visited Holland, Italy, and Spain; pupil of Julian at Paris; ARSA, 1892; painted portraits, but mainly sea pieces and landscapes, both in watercolour and in oil; works include 'A Lone Shore' (1875) and 'Harvest Scene' (1878).

REID, FORREST (1875-1947), novelist and critic; educated at Royal Academical Institution, Belfast, and Christ's College, Cambridge; reviewed regularly for *Spectator* and other papers; novels, set in Ulster and mainly concerned with boyhood and adolescence seen in retrospect, include *Uncle Stephen* (1931), *The Retreat* (1936), *Young Tom* (1944), and *Brian Westby* (1934).

REID, SIR GEORGE HOUSTOUN (1845-1918), colonial politician; born in Scotland; taken to Australia, 1852; admitted to colonial bar, 1879; represented East Sydney in legislative assembly, 1880-4, 1885-1901; premier of New South Wales, 1894-9; free trader, but half-hearted federalist; premier of Australian Commonwealth through coalition with labour party, 1904-5; first Commonwealth high commissioner in London, 1910-15; PC, 1897; KCMG, 1909; GCMG, 1911; GCB, 1916.

REID, JAMES SMITH (1846-1926), classical scholar and ancient historian; BA, Christ's College, Cambridge; bracketed senior classic, 1868; fellow of Christ's, 1869-72; of Gonville and Caius College, 1878; first professor of ancient history at Cambridge, 1899-1925; FBA, 1917; a distinguished Ciceronian scholar; edited works of Cicero; as professor, chiefly concerned with questions of Roman constitutional history.

REID, SIR JOHN WATT (1823-1909), medical director-general of the navy; LRCS Edinburgh, 1844; MD, Aberdeen, 1856; entered navy as assistant surgeon, 1845; served as naval surgeon in Black Sea, 1854; in China war, 1857-9; in Ashanti war, 1873-4; deputy inspector-general, 1874-80; inspector-general, 1880-8; KCB, 1882.

REID, SIR ROBERT GILLESPIE (1842-1908), Canadian contractor and financier; emigrated from Perthshire to Australia, 1865; subsequently settled at Montreal; carried out many important building and engineering contracts; built bridges across Niagara river at Buffalo, and across the Rio Grande between Texas and Mexico, and railways in Newfoundland; president of Reid-Newfoundland Company, which had control of Newfoundland railways, the St. John's dry dock, and immense tracts of land; knighted, 1907.

REID, ROBERT THRESHIE, EARL LOREBURN (1846-1923), lord chancellor; born at Corfu; educated at Cheltenham and Balliol College, Oxford; first class, *lit. hum.*; Ireland scholar; a noted athlete; called to bar (Inner Temple), 1871; joined Oxford circuit; quickly made his mark, particularly in commercial suits; liberal MP, Hereford, 1880; QC, 1882; MP, Dumfries, as supporter of Gladstone's Irish policy, 1886-1905; solicitor-general and knighted, May 1894; attorney-general, Oct. 1894; arbitrated successfully in boundary dispute between Venezuela and British Guiana, 1899; standing counsel for Oxford University, 1899-1906; supported Boer cause during South African war, 1899-1902; lord chancellor, with title of Lord Loreburn, 1905; principal legal achievement as lord chancellor establishment of Court of Criminal Appeal, 1907; advocated extension of jurisdiction of county courts; as liberal placed in difficult position of presiding over almost entirely hostile House of Lords; wholeheartedly supported parliament bill, 1911; earl, 1911; resigned, 1912.

REID, SIR THOMAS WEMYSS (1842-1905), journalist and biographer; early took up journalism; chief reporter on *Newcastle Journal*, 1861;

editor of *Preston Guardian*, 1864; head of reporting staff of *Leeds Mercury*, 1866; obtained admission of provincial reporters to press gallery of House of Commons, 1881; editor of *Leeds Mercury*, 1870-87; of moderate liberal views; supported Gladstone's home rule policy and Forster's education bill; manager of Cassell's publishing firm from 1887 to death; founded (1890) and edited till 1897 the *Speaker*, a weekly liberal organ; friend and supporter of Lord Rosebery; knighted, 1894; hon. LLD, St. Andrews, 1893; wrote lives of Charlotte Brontë (1877), W. E. Forster (2 vols., 1888), Lord Houghton (2 vols., 1890), Lord Playfair (1899), and William Black (1902); a successful novelist; memoirs edited by S. J. Reid, 1905.

REID DICK, Sir WILLIAM (1878-1961), sculptor. [See DICK.]

REILLY, Sir CHARLES HERBERT (1874-1948), professor of architecture; educated at Merchant Taylors' and Queens' College, Cambridge; first class, mechanical sciences, 1896; articled to John Belcher [q.v.]; set up in partnership; Roscoe professor of architecture, Liverpool, 1904-33; champion of university training for architects; obtained foundation Lever chair in civic design; regular contributor to *Manchester Guardian* and other papers; publications include *The Theory and Practice of Architecture* (1932) and *Outline Plan for Birkenhead* (1947) in which he suggested grouping small houses round urban ('Reilly') greens; FRIBA, 1912; knighted, 1944.

REITZ, DENEYS (1882-1944), South African soldier, author, and politician; recorded his adventures with Boer forces (1899-1902) in *Commando* (1929); practised as solicitor; served in South-West and East Africa and in France, 1914-18; minister of lands, 1920-4, 1933-5; of agriculture, 1935-8; of mines, 1938-9; of native affairs and deputy prime minister, 1939-43; high commissioner in London, 1943-4.

RELF, ERNEST FREDERICK (1888-1970), scientist and authority on aerodynamics; entered Portsmouth Dockyard as apprentice shipwright, 1903; attended at Royal Dockyard School; won scholarship to Royal College of Science, 1909; ARCS, first class honours, 1912; appointed assistant in aeronautics section, Engineering Department, National Physical Laboratory, Teddington, under (Sir) Leonard Bairstow [q.v.], 1912; succeeded (Sir) R. V. Southwell [q.v.] as superintendent, Aerodynamics Department, 1925-45; FRS, 1936; CBE, 1944; FRAe.S, 1926; FIAS (United States), 1933; published some forty scientific papers on wide variety of aeronautical subjects;

principal, College of Aeronautics, Cranfield, 1945-50; member, Aeronautical Research Council, 1918-68; talented musician, playing on piano and organ and composing chamber music and songs.

RENDALL, MONTAGUE JOHN (1862-1950), headmaster of Winchester College; scholar of Harrow and Trinity College, Cambridge; first class, parts i and ii, classics, 1884-5; played Association football for Cambridge; joined staff at Winchester College, 1887; second master, 1899; headmaster, 1911-24; visited principal schools in dominions for Rhodes trustees; chairman, public schools empire tours committee; a governor of British Broadcasting Corporation, 1927-32; CMG, 1931.

RENDEL, Sir ALEXANDER MEADOWS (1829-1918), civil engineer; son of James Meadows Rendel, and brother of George Wightwick Rendel [qq.v.]; head of father's practice, 1856; designed docks and harbours; consulting engineer to Indian State Railways, 1872; KCIE, 1887.

RENDEL, GEORGE WIGHTWICK (1833-1902), civil engineer; son of James Meadows Rendel [q.v.]; assistant to father; partner in ordnance works of Sir William (later Lord) Armstrong [q.v.], at Elswick, 1858; responsible for hydraulic method of mounting and working heavy guns and introduction of cruiser, intermediate between armour-clad and unprotected war vessel; extra professional civil lord of Admiralty, 1882-5; settled near Naples as director of Armstrong Pozzuoli company, 1887; friend of Empress Frederick of Germany and Lord Rosebery.

RENDEL, HARRY STUART GOODHART-(1887-1959), architect. [See GOODHART-RENDEL.]

RENDLE, ALFRED BARTON (1865-1938), botanist; first class, natural sciences tripos (botany), St. John's College, Cambridge, 1887; assistant, department of botany, British Museum (Natural History), 1888; keeper, 1906-30; works include *Classification of Flowering Plants* (2 vols., 1904-25) and *Flora of Jamaica* (with W. Fawcett, 5 vols.); edited *Journal of Botany*, 1924-37; FRS, 1909.

RENNELL, first BARON (1858-1941), diplomatist and scholar. [See RODD, JAMES RENNELL.]

REPINGTON, CHARLES À COURT (1858-1925), soldier and military writer; joined Rifle

Brigade, 1878; entered Staff College, Camberley, 1887; military attaché, Brussels and The Hague, 1898; served in South African war, 1899-1900; his military career closed owing to an indiscretion, 1902; military correspondent first of *Morning Post*, then of *The Times*, 1904-18; returned to *Morning Post* and later joined *Daily Telegraph*; publications include *The First World War* (1920) and *After the War* (1922).

REYNOLDS, JAMES EMERSON (1844-1920), chemist; born in county Dublin; studied medicine in Edinburgh; abandoned medicine for chemistry; analyst to Royal Dublin Society, 1868-75; discovered thiocarbamide, 1868; professor of chemistry, Royal College of Surgeons, Dublin, 1870-5, at Trinity College, Dublin, 1875-1903; FRS, 1880.

REYNOLDS, OSBORNE (1842-1912), engineer and physicist; BA, Queens' College, Cambridge; professor of engineering, Owens College, Manchester, 1868-1905; conducted long series of investigations into mechanical questions and physical phenomena such as lubrication, laws of flow of water in pipes, etc.; FRS, 1877; gold medallist of Royal Society, 1888; collected scientific papers published, 1900-3.

RHODES, CECIL JOHN (1853-1902), imperialist and benefactor; fifth son of Francis William Rhodes, vicar of Bishop's Stortford, Hertfordshire; educated at Bishop's Stortford grammar school, 1861-9; owing to failure of health, went to South Africa to join his eldest brother, who was growing cotton in Natal, 1870; removed to Orange Free State on discovery of diamonds there, 1871; worked with his brother a moderately prosperous claim; meanwhile matriculated at Oriel College, Oxford, 13 Oct. 1873; revisited Oxford at frequent intervals until he succeeded in graduating as a passman BA and MA, 17 Dec. 1881; during this period Rhodes with partners increased his holdings in Kimberley diamond fields, obtaining a large interest in the De Beers mines there; in 1880 he helped to establish De Beers mining company; during 1875 he made solitary journey through Bechuanaland and Transvaal; formed the aspiration to work with Dutch settlers and to federate South Africa under British rule with Cape Dutch assent; was elected to Cape legislature for newly formed constituency of Barkly West (1880) and retained seat for life; in the South African parliament he from the first sought to maintain wide powers of local self-government while extending British settlement and influence; soon concentrated political activities on northern expansion of the British

dominion; aided in securing great part of Bechuanaland for the Cape government in spite of rivalry of Transvaal Republic, 1882; when the whole of Bechuanaland was formally annexed, Rhodes was made deputy commissioner, 1884; and he negotiated in person with President Kruger the withdrawal of Boer claims to portions of the territory; after he had made terms with Lobengula, king of Matabeleland, the British South Africa Company was incorporated by royal charter to administer territory north of Bechuanaland, 1889; this territory was named Rhodesia after the projector of the scheme; in 1887 and 1888 Rhodes succeeded in amalgamating the diamond mines about Kimberley under the style of the De Beers Consolidated Mines, a corporation of which he became chairman and ruler; by skilful negotiation this new company acquired the interest in the Kimberley mine which had been controlled by Rhodes's chief rival in the diamond fields, Barnett Isaacs Barnato [q.v.]; at the same time Rhodes acquired important share in the newly discovered gold mines on Witwatersrand in the Transvaal and helped to form the corporation known as the Consolidated Goldfields of South Africa; in the organization of Rhodesia by the chartered company, of which (Sir) L. S. Jameson [q.v.] became administrator in 1890, Rhodes played an energetic part; he directed the war with the Matabeles, 1893-4, whereby he greatly extended the company's territory; from July 1890 to Jan. 1896 he was also prime minister of the Cape; his ministry carried through reforms in local administration and native policy, and sought to unite English and Dutch interests; in pursuit of his ideal of imperial federation he subscribed in 1888 £10,000 to the home rule party in England; in 1893 he reconstructed his ministry owing to internal differences; at the end of 1895 Rhodes secretly encouraged the uitlander population on the Rand and in the Transvaal to look to an armed insurrection for redress of their grievances against the Transvaal government; the failure of Jameson's precipitate raid in support of this movement (27 Dec. 1895) seriously involved Rhodes; inquiries by both the Cape parliament and the British House of Commons pronounced Rhodes guilty of grave breaches of duty; resigning office of premier (6 Jan. 1896) and for a time the directorship of the Chartered Company, which he resumed in 1898, he devoted himself to the development of Rhodesia; through his personal influence with the chiefs he effected a permanent peace with the Matabeles after a new outbreak, 1896; and he greatly extended railway and telegraph communication; he was made hon. DCL, Oxford, 1899, in spite of some protests; on outbreak of South African war he moved to Kimberley and was besieged there (15 Oct. 1899-16 Feb. 1900);

next year he spent much time in Europe, returning to the Cape in Feb. 1902; he died after long suffering from heart disease on 26 Mar. at the village of Muizenburg, being buried among the Matoppo Hills; by his will he endowed some 170 scholarships at Oxford for students from the colonies, the United States, and Germany; £100,000 was left to his old college, Oriel.

RHODES, FRANCIS WILLIAM (1851–1905), colonel; brother of C. J. Rhodes [q.v.]; joined army, 1873; served in Sudan and Nile expedition, 1884–5; colonel, 1889; military secretary to Lord Harris [q.v.] at Bombay, 1890–3; military member of council in Matabeleland, 1894; imprisoned and fined for complicity in Jameson raid, 1895–6; war correspondent to *The Times* at Omdurman, 1898; distinguished himself in South African war, 1899–1900; CB, 1900; died at Cape Town.

RHONDDA, VISCOUNT (1856–1918), statesman, colliery proprietor, and financier. [See THOMAS, DAVID ALFRED.]

RHONDDA, VISCOUNTESS (1883–1958), founder and editor of *Time and Tide*. [See THOMAS, MARGARET HAIG.]

RHYS, ERNEST PERCIVAL (1859–1946), author and editor; educated at Bishop's Stortford and in Newcastle; qualified as mining engineer; commenced writing in London, 1886, as reviewer, essayist, and poet; contributed to Rhymers' Club collections of poems, 1892, 1894; edited 'Everyman's Library' (1906–46) in 983 volumes.

RHŶS, SIR JOHN (1840–1915), Celtic scholar; BA, Jesus College, Oxford; fellow of Merton College, 1869; first Jesus professor of Celtic at Oxford, 1877; fellow and bursar of Jesus, 1881–95; principal, 1895–1915; knighted, 1907; PC, 1911; served on numerous commissions dealing with education and social advancement especially in Wales, 1881–1915; prolific writer on Celtic philology, phonology, inscriptions, history, religion, ethnology, and folklore.

RICHARDS, FRANCIS JOHN (1901–1965), agriculturalist; educated at Burton grammar school and Birmingham University; B.Sc., first class, botany, 1924; M.Sc., 1926; joined Research Institute of Plant Physiology, Imperial College, London, under Professor V. H. Blackman and F. G. Gregory [qq.v.], 1926; plant physiologist, Rothamsted Experimental Station, 1927, by normal promotion, rose to be deputy chief scientific officer, 1963; responsible

for nutritional aspects of Institute's research; director, research unit, Wye College, Kent, 1961; published, with F. G. Gregory, in the *Annals of Botany*, first of series of papers on mineral nutrition of barley, 1929; interested in problem of patterns of leaf arrangements in plants—phyllotaxis; FRS, 1954; D.Sc., Birmingham.

RICHARDS, FRANK (pseudonym) (1876–1961), writer. [See HAMILTON, CHARLES HAROLD ST. JOHN.]

RICHARDS, SIR FREDERICK WILLIAM (1833–1912), admiral; entered navy, 1848; captain, 1866; commanded *Devastation*, first steam turret battleship designed without any sail power, 1873–7; CB for services at battle of Gingihlovo and relief of Echowe during Zulu war, 1879; KCB for services at Laing's Nek, 1881; junior naval lord at Admiralty, 1882; largely responsible for able report on manœuvres of 1888, a determining cause of Naval Defence Act, 1889; vice-admiral, 1888; second naval lord, 1892; admiral, 1893; first naval lord, 1893–9; promoted shipbuilding and construction of harbours and dockyards; GCB, 1895; admiral of fleet, 1898; a leading administrator in naval history.

RICHARDSON, SIR ALBERT EDWARD (1880–1964), architect; educated at Boys British School, Highgate; articled to architect, 1895; served with other architects, including Frank T. Verity, 1898–1906; as Verity's leading assistant, designed 'mansion flats' in Cleveland Row, St. James's, and Bayswater Road, and middle section of 169–201 Regent St., 1906–9; also designed façade of Regent Street Polytechnic; designed other buildings, including, in association with Horace Farquharson, Moorgate Hall, Finsbury Pavement, EC2, 1913–17; published, with C. Lovett Gill, *London Houses from 1660 to 1820* (1911), followed by *Monumental Classic Architecture in Great Britain and Ireland in the Eighteenth and Nineteenth Centuries* (1914); served in RFC, 1916–18; professor of architecture, Bartlett School of Architecture, University College, London, 1919–46; professor emeritus, 1946; published, with H. O. Corfiato, *The Art of Architecture* (1938); ARA, 1936; RA, 1944; PRA, 1954; member, Royal Fine Art Commission, 1939–56; KCVO, 1956; FSA; hon. MA (Cantab); hon. Litt.D. (Dublin); hon. RWS; hon. fellow, St. Catherine's College, Cambridge; after 1939–45 war formed new partnership with E. A. S. Houfe, and designed new buildings in the City and carried out restorations of war-damaged buildings, including St. James's, Piccadilly.

RICHARDSON, ETHEL FLORENCE LINDESAY (1870-1946), novelist under name of HENRY HANDEL RICHARDSON; educated at Presbyterian Ladies' College, Melbourne; studied pianoforte at Leipzig Conservatorium; married (1895) J. G. Robertson, professor of German literature, London, 1903-33; published *Maurice Guest* (1908), *The Getting of Wisdom* (autobiographical novel, 1910), trilogy of early colonial society *The Fortunes of Richard Mahony* (1930), *The Young Cosima* (1939), and *Myself When Young* (1948).

RICHARDSON, HENRY HANDEL (1870-1946), novelist. [See RICHARDSON, ETHEL FLORENCE LINDESAY.]

RICHARDSON, LEWIS FRY (1881-1953), physicist and meteorologist; educated at Bootham School, York, Durham College, and King's College, Cambridge; first class, part i, natural sciences tripos, 1903; superintendent, Eskdalemuir Observatory, 1913; served with Friends' Ambulance Unit, 1916-19; head, physics department, Westminster Training College, 1920; principal, Paisley Technical College and School of Art, 1929-40; researched in problems of meteorology and mathematics; published *Weather Prediction by Numerical Process* (1922); carried out mathematical investigation of causes of war; published *Arms and Insecurity* and *Statistics of Deadly Quarrels* (1960); FRS, 1926.

RICHARDSON, SIR OWEN WILLANS (1879-1959), physicist; educated at Batley grammar school and Trinity College, Cambridge; first class, parts i and ii, natural sciences tripos, 1899-1900; fellow, Trinity College, 1902; Clerk Maxwell scholar, 1904; D.Sc., London, 1904; carried out research in thermionic emission at Cavendish Laboratory with (Sir) J. J. Thomson [q.v.]; professor of physics, Princeton, 1906-14; continued research in electron physics; FRS, 1913; Wheatone professor of physics, King's College, London, 1914; Yarrow research professor, 1924-44; published *The Electron Theory of Matter* (1914), *The Emission of Electricity from Hot Bodies* (1916), and *Molecular Hydrogen and its Spectrum* (1934); awarded Nobel prize in physics, 1928; hon. degrees, Leeds, St. Andrews, and London; fellow, Kings College London, 1925; hon. fellow, Trinity College, Cambridge, 1941; president, Physical Society, 1926-8; knighted, 1939.

RICHEY, JAMES ERNEST (1886-1968), geologist; educated at St. Columba's College and Trinity College, Dublin; BAI (engineering), 1909; joined Geological Survey, 1911; com-

pleted mapping of broad road dyke crossing Loch Ba, Scotland; helped elucidate history of Tertiary volcanoes; served with Royal Engineers in France, 1914-19; MC; surveyed Ardnamurchan peninsula, 1919-24; district geologist, 1925; on retirement in 1946, became consulting geologist, and lectured at Queen's College, Dundee; FRS, 1938; fellow, Royal Society of Edinburgh; vice-president, 1956-9; hon. fellow of number of geological societies and president or chairman of many others.

RICHMOND, SIR BRUCE LYTTELTON (1871-1964), editor; grandson of Henry Austin Bruce, first baron Aberdare [q.v.]; educated at Winchester and New College, Oxford; called to bar (Inner Temple), 1897; assistant editor, *The Times*, 1899; took over from (Sir) James Thursfield [q.v.] as editor, *The Times Literary Supplement*, 1902-37; gave first opportunities as critics to Virginia Woolf and T. S. Eliot [qq.v.]; published *The Pattern of Freedom* (1940), prose and verse anthology; member, executive committee, Royal College of Music; hon. doctorates, Oxford and Leeds; knighted, 1935.

RICHMOND, SIR HERBERT WILLIAM (1871-1946), admiral, and master of Downing College, Cambridge; son of Sir William Blake Richmond [q.v.]; entered navy, 1885; assistant director, operations division, 1913-15; commanded successively battleships *Commonwealth*, *Conqueror*, and *Erin*; rear-admiral, 1920; president, Naval War College, 1920-3; commander-in-chief, East Indies, 1923-5; commandant, Imperial Defence College, 1926-8; vice-admiral, 1925; admiral, 1929; convinced of importance of British strategy based on sea power; reprimanded for articles contributed to *The Times* (1929) on naval reduction and refused further employment; retired, 1931; Vere Harmsworth professor of imperial and naval history, Cambridge, 1934-6; master of Downing, 1936-46; publications include *The Navy in the War of 1739-48* (1920), *The Navy in India, 1763-83* (1931), *Economy and Naval Security* (1931), *Sea Power in the Modern World* (1934), *Statesmen and Sea Power* (1946), and *The Navy as an Instrument of Policy, 1558-1727* (1953); FBA, 1937; KCB, 1926.

RICHMOND, SIR IAN ARCHIBALD (1902-1965), archaeologist; educated at Ruthin School and Corpus Christi College, Oxford; studied at British School in Rome, 1924-6; first major work, *The City Wall of Imperial Rome* (1930); followed by *Roman Britain* (Britain in Pictures, 1947, and Pelican History of England, 1955); edited three editions of John Collingwood Bruce's *Handbook to the Roman Wall* (1947, 1957, and 1966); prepared revised edition of *Archaeology of Roman Britain* by R. G. Collingwood [q.v.], published in 1969; contributed to number of archaeological journals; authority on Roman military architecture; lecturer in classical archaeology and ancient history, The Queen's University, Belfast, 1926–30; director, British School, Rome, 1930–2; completed and published Thomas Ashbys' [q.v.] *The Aqueducts of Ancient Rome* (1935); lecturer in Roman-British history and archaeology, Durham University, 1935; reader, 1943; professor, 1950–6; dean of faculty of art; public orator; directed series of excavations on Hadrian's Wall and elsewhere; professor of archaeology of the Roman Empire, Oxford, 1956; fellow of All Souls; gave many public lectures; published *Archaeology and the After-life in Pagan and Christian Imagery* (1950); served on Royal Commission on Historical Monuments, 1944; FSA, 1931; president, Society of Antiquaries, 1964; CBE, 1958; knighted, 1964; FBA, 1947; hon. doctorates from number of universities, and member of many learned societies.

RICHMOND, SIR WILLIAM BLAKE (1842–1921), artist; son of George Richmond, RA [q.v.]; entered Royal Academy Schools, 1858; studied in Italy, 1865–9; Slade professor of fine art, Oxford, 1879–83; ARA, 1888; RA, 1895; KCB, 1897; painted subject pictures from classical mythology; successful portrait painter; best-known work his mosaic decorations for St. Paul's Cathedral.

RICHMOND AND GORDON, sixth DUKE OF (1818–1903), lord president of the Council. [See GORDON-LENNOX, CHARLES HENRY.]

RICKETTS, CHARLES DE SOUSY (1866–1931), painter, printer, stage-designer, writer, and collector; apprenticed to wood-engraving; became lifelong friend of C. H. Shannon [q.v.] with whom he owned and edited *The Dial* (1889–97) and produced *Daphnis and Chloe* (1893); founded, and designed founts, bindings, and illustrations for, the Vale Press (83 vols, 1896–1904); thereafter devoted himself to painting, his mood usually melancholy and romantic; revolutionized stage design by using one prevailing colour and large pattern; his décors include *Saint Joan* (1924) and *Henry VIII* (1925); with Shannon made valuable art collection bequeathed to museums and galleries; writings include *The Prado and its Masterpieces* (1903), *Titian* (1910), *Pages on Art* (1913); RA, 1928.

RIDDELL, CHARLES JAMES BUCHANAN (1817–1903), major-general RA, meteorologist; joined Royal Artillery, 1834; super-intendent of meteorological observatory, Toronto, 1839; assistant superintendent of ordnance magnetic observatories of Royal Military Repository, Woolwich, 1840–4; FRS, 1842; served in Crimea and Indian Mutiny; CB, 1858; major-general, 1866.

RIDDELL, CHARLOTTE ELIZA LAWSON (1832–1906), novelist, known as MRS J. H. RIDDELL; wrote thirty novels, including *George Geith of Fen Court* (1864); made commerce a frequent theme; co-proprietor and editor of *St. James's Magazine* from 1861.

RIDDELL, GEORGE ALLARDICE, BARON RIDDELL (1865–1934), newspaper proprietor; admitted solicitor, 1888; chairman, *News of the World*, 1903–34; very friendly with Lloyd George; liaison officer between British delegation and the press at Paris peace and other conferences; published his diaries, 1908–23 (3 vols., 1933–4); knighted, 1909; baronet, 1918; baron, 1920.

RIDDING, GEORGE (1828–1904), head-master of Winchester and first bishop of Southwell; educated at Winchester and Balliol College, Oxford; Craven scholar and BA, 1851; MA, 1853; DD, 1869; fellow of Exeter College, 1851; tutor, 1853–63; hon. fellow, 1890; headmaster of Winchester, 1866–84; 'second founder of Winchester', extending buildings, playing-fields, and staff; modernized curriculum; founded school mission at Landport in Portsmouth, 1882; first bishop of Southwell, 1884–1904; created a corporate spirit in diocese; published *The Revel and the Battle* (sermons) (1897).

RIDGEWAY, SIR JOSEPH WEST (1844–1930), soldier and administrator; entered Bengal Infantry, 1860; political secretary to (Lord) Roberts [q.v.], 1879; distinguished himself on English commission sent to determine with Russian commission ill-defined northern boundary of Afghanistan, 1884–6; sent to St. Petersburg to resume negotiations, a final and satisfactory agreement being reached, 1887; under-secretary for Ireland, 1887–92; governor of Isle of Man, 1893–5; of Ceylon, 1896–1903; chairman of committee sent to Africa which reported in favour of granting responsible government to Transvaal and Orange River Colony, 1906; president, British North Borneo Company, 1910; KCSI, 1886.

RIDGEWAY, SIR WILLIAM (1853–1926), classical scholar; educated at Trinity College, Dublin, and Gonville and Caius College, Cambridge; fellow of his college, 1880; professor of Greek, University College, Cork, 1883; Disney professor of archaeology, Cambridge, 1892–1926; Brereton reader in classics, 1907–26; opposed granting degrees to women and abolition of compulsory Greek; published *The Origin of Metallic Currency and Weight Standards*, which threw flood of light on early life of Mediterranean lands (1892); published first volume of *Early Age of Greece* (1901); second volume posthumously published (1931); this, his chief contribution to history, aroused bitter controversy; other works include *The Origin and Influence of the Thoroughbred Horse* (1905) and *The Origin of Tragedy* (1910); FBA, 1910; knighted, 1919.

RIDLEY, HENRY NICHOLAS (1855–1956), plant geographer and economic botanist; educated at Haileybury and Exeter College, Oxford; Burdett-Coutts scholar in geology, 1880; worked in botanical department, British Museum, 1880; visited Brazil on expedition sponsored by Royal Society, 1887; director of botanical gardens, Singapore, 1888–1911; travelled extensively in South-East Asia, 1890–1915; actively engaged in establishing rubber plantation industry in Malaya; awarded gold medal of Rubber Planters' Association, 1914; published *Flora of the Malay Peninsula* (5 vols., 1922–5), *Spices* (1912), and *The Dispersal of Plants Throughout the World* (1930); FRS, 1907; CMG, 1911; gold medal of Linnean Society, 1950; died less than two months before 101st birthday.

RIDLEY, Sir MATTHEW WHITE, fifth baronet, and first VISCOUNT RIDLEY (1842–1904), home secretary; educated at Harrow and Balliol College, Oxford; BA (first class classic), 1865; MA, 1867; fellow of All Souls, 1865–74; conservative MP for North Northumberland, 1868–85; under-secretary to Home Office, 1878; financial secretary to Treasury, 1885; MP for Blackpool, 1886–1900; PC, 1892; home secretary, 1895–1900; viscount, 1900; chairman of North Eastern Railway from 1902; developed town of Blyth; as chairman of board of commissioners he transformed harbour and dock there; president of Royal Agricultural Society, 1888.

RIEU, CHARLES PIERRE HENRI (1820–1902), orientalist; born at Geneva; studied Arabic and Sanskrit at Bonn; Ph.D., 1843; assistant in department of oriental MSS at British Museum, 1847; keeper, 1867–95; compiled catalogues of oriental, Persian, and Turkish MSS there; Adams professor of Arabic at Cambridge, 1894–1902.

RIGBY, Sir JOHN (1834–1903), judge; BA, Trinity College, Cambridge (second wrangler and second Smith's prizeman), and fellow,

1856; MA, 1859; called to bar, 1860; QC, 1880; secured large practice in Court of Appeal and House of Lords; liberal MP for Wisbech, 1885–6, for Forfarshire, 1892–4; solicitor-general and knighted, 1892; attorney-general, 1894; judge of Court of Appeal and PC, 1894–1901; a master of the science of equity.

RIGG, JAMES HARRISON (1821–1909), Wesleyan divine; ordained, 1849; served in successive circuits till 1868; contributor from 1853 and sole editor (1886–98) of *London Quarterly Review*; explained theological position in *Principles of Wesleyan Methodism* (1850), and *Modern Anglican Theology* (1857); principal of Westminster (Wesleyan) training college for day-school teachers, 1868–1903; favoured denominational schools, 1870; member of London school board, 1870–6; member of royal commission on elementary education, 1886–8; president of Wesleyan conference, 1878 and 1892; carried 'Sandwich compromise' (1890), giving larger share to laity in work of conference; voluminous writings include *The Living Wesley* (1875), *Oxford High Anglicanism* (1895), and *Reminiscences Sixty Years Ago* (1904).

RIGG, JAMES McMULLEN (1855–1926), biographer, historian, and translator; son of J. H. Rigg [q.v.]; BA, St. John's College, Oxford; contributed over 600 articles to this Dictionary; editorial work includes *A Calendar of State Papers.... Rome* (2 vols. [1558–78], 1916, 1926); wrote *St. Anselm of Canterbury* (1896).

RINGER, SYDNEY (1835–1910), physician; studied medicine at University College, London; MB, 1860; MA, 1863; MRCP, 1863; FRCP, 1870; physician to University College Hospital, 1865–1900; Holme professor of clinical medicine, 1887–1900; wrote *A Handbook of Therapeutics* (1869, 13th edn. 1897); made many physiological researches; FRS, 1885.

RIPON, first MARQUESS OF (1827–1909), statesman. [See ROBINSON, GEORGE FREDERICK SAMUEL.]

RISLEY, Sir HERBERT HOPE (1851–1911), Indian civil servant and anthropologist; educated at Winchester and New College, Oxford; BA, 1872; joined Indian civil service, 1873; under-secretary (1879) and secretary (1902) to Indian government in home department; employed in compiling statistics of castes and occupations of people of Bengal, 1885; CIE, 1892; census commissioner, 1899; director of ethnography for India, 1901; CSI, 1904; KCIE, 1907; secretary to public and judicial department at India Office, 1910; chief anthropological works were *Anthropometric Data* (2 vols.,

1891), *Tribes and Castes of Bengal* (1891–2), *Ethnographical Glossary* (2 vols., 1892), and *The People of India* (1908).

RITCHIE, ANNE ISABELLA, LADY RITCHIE (1837–1919), novelist and woman of letters; daughter of William Makepeace Thackeray [q.v.]; wife of Sir Richmond T. W. Ritchie [q.v.]; works include *The Village on the Cliff* (1867) and *Old Kensington* (1873).

RITCHIE, CHARLES THOMSON, first BARON RITCHIE OF DUNDEE (1838–1906), statesman; after education at City of London School, he joined his father's firm, William Ritchie & Son, East India merchants, jute spinners, and manufacturers, of London and Dundee; conservative MP for the Tower Hamlets, 1874–85, for St. George's in the East, 1885–92, and for Croydon, 1895–1905; became chairman of a select committee appointed on his own motion (22 Apr. 1879) to consider measures to protect West Indian cane sugar from the injurious effects of the bounty system whereby European countries stimulated the competing production of beet sugar; financial secretary to the Admiralty, 1885–6; took steps to accelerate process of shipbuilding and to decrease the cost; became president of Local Government Board, July 1886; was admitted to cabinet, 1887; carried revolutionary local government bill for England and Wales, 1888, which created 'county councils' for counties and large towns, and superseded in London the metropolitan board of works; at the same time he devised complementary scheme of district and parish councils, which he was compelled to drop, but which was made law by his liberal successor in 1894; was responsible for other important administrative acts, including the Public Health Acts of 1891; president of Board of Trade, 1895–1900; carried Conciliation Act for settlement of labour disputes (1896), the Light Railways Act (1896), the Merchant Shipping Act (1898), the Railway Employment (Prevention of Accidents) Act (1900), and Companies Act (1900); established an intelligence branch of commercial, labour, and statistical departments of Board of Trade, Oct. 1899; became home secretary Nov. 1900; passed Factory and Workshop Act (1900), Youthful Offenders Act (1901), and Licensing Act (1902); was appointed chancellor of the Exchequer, when A. J. Balfour became prime minister, Aug. 1902; dropped in his budget of 1903, in spite of dissent within the cabinet, the shilling a quarter duty on corn which his predecessor, Sir Michael Hicks Beach [q.v.], had imposed the year before; resisted in the cabinet Joseph Chamberlain's proposals of tariff reform; resigned owing to consequent breach in the cabinet, 14 Sept. 1903; explained to his constituents that he was opposed to any fiscal arrangement with the colonies which would compel a tax on the food of the people; baron, 17 Dec. 1905; died at Biarritz, 19 Jan. 1906; noted in political circles for grasp of complicated detail and shrewd commonsense.

RITCHIE, DAVID GEORGE (1853–1903), philosopher; MA, Edinburgh University, 1875; hon. LLD, 1898; BA, Balliol College, Oxford, 1878; fellow of Jesus College, 1878; tutor of Balliol, 1882–6; professor of logic and metaphysics at St. Andrews, 1894–1903; president of Aristotelian Society, 1898–9; works, influenced by Hegel, include *Darwinism and Politics* (1889), *Principles of State Interference* (1891), *Natural Rights* (1895), and *Plato* (1902); *Philosophical Studies* edited with memoir by Prof. R. Latta, 1905.

RITCHIE, SIR RICHMOND THACKERAY WILLOUGHBY (1854–1912), civil servant; born at Calcutta; BA, Trinity College, Cambridge; entered India Office, 1877; private secretary to secretary of state for India, 1895–1902; secretary in political and secret department, India Office, 1902–9; permanent undersecretary of state, 1909; KCB, 1907.

RIVAZ, SIR CHARLES MONTGOMERY (1845–1926), Indian civil servant; entered Indian civil service, 1864; superintendent of Kapurthala State, 1876–85; commissioner of Lahore, 1887–92; financial commissioner of Punjab, 1892–7; appointed to executive council of governor-general, as member in charge of home department and revenue and agricultural departments, 1897; promoted Punjab Land Alienation Act, 1900; lieutenant-governor of Punjab, 1902–7; KCSI, 1901; a firm and progressive administrator, who ensured for the agricultural population a just and efficient official rule.

RIVERDALE, first BARON (1873–1957), industrialist. [See BALFOUR, ARTHUR.]

RIVIERE, BRITON (1840–1920), painter; son of William Riviere [q.v.]; came of artistic family; educated at Cheltenham and St. Mary Hall, Oxford; exhibited in oil and water-colour at Royal Academy from 1858; ARA, 1877; RA, 1880; much influenced by artists of new Scottish school, such as Orchardson and Pettie [qq.v.]; best known for paintings of lions and other animals, although he himself cared least for these.

ROBECK, SIR JOHN MICHAEL DE, baronet (1862–1928), admiral of the fleet. [See DE ROBECK.]

ROBERTS, ALEXANDER (1826-1901), classical and biblical scholar; MA, King's College, Old Aberdeen, 1847; DD, Edinburgh, 1864; member of New Testament revision company, 1870-84; professor of humanity at St. Andrews, 1872-99; voluminous works include *Greek the Language of Christ and His Apostles* (1888).

ROBERTS, FREDERICK SLEIGH, first EARL ROBERTS (1832-1914), field-marshal; son of General Sir Abraham Roberts [q.v.]; born at Cawnpore; educated at Sandhurst and Addiscombe; joined Bengal artillery, 1851; served in Indian Mutiny, 1857-8; won VC, 1858; accompanied Sir Robert Napier (later Lord Napier of Magdala, q.v.) on Abyssinian expedition as assistant quatermaster-general, 1868; CB, 1871; quartermaster-general of army in India, 1875; advocate of 'forward' policy in India, i.e. prevention of control by Russia of passes of the Himalaya; commander of Punjab frontier force, 1878; major-general and KCB for victory over Afghans at Peiwar Kotal, 1878; defeated Afghans at Charasia and occupied Kabul, 1879; conducted celebrated march from Kabul to Kandahar, resulting in pacification of Afghanistan, 1880; GCB, baronet, and commander-in-chief of Madras army, 1880; commander-in-chief in India, 1885-93; baron, 1892; left India for good, 1893; field-marshal, 1895; commander-in-chief in Ireland, 1895-9; appointed to supreme command in South Africa, end of 1899; at once increased number of mounted troops, and with help of Lord Kitchener [q.v.] remodelled transport; decided to invade Free State; received unconditional surrender of General Cronje at Paardeberg, Feb. 1900; occupied Bloemfontein (Mar.), Johannesburg (May), Pretoria (June); defeated General Louis Botha [q.v.] at Diamond Hill (June); captured Machadodorp, temporary seat of President Kruger's government (July); returned to England after annexation of Transvaal in Oct.; earl and KG for services in saving situation in South African war, 1900; commander-in-chief, 1900-5; devoted remainder of his life to cause of national service, becoming president of National Service League, 1905; colonel-in-chief of Indian expeditionary force dispatched to France, 1914; died at St. Omer on way to visit it; wrote *Forty-one Years in India* (1897).

ROBERTS, GEORGE HENRY (1869-1928), politician and labour leader; educated at a Norwich national school; apprenticed to printing trade; became active in Norwich local politics as a leader of independent labour party; labour MP, Norwich, 1906; left the party, 1914; lord commissioner of Treasury, 1915; parliamentary secretary of Board of Trade, 1916; minister of labour with seat in cabinet, 1917; broke entirely with labour party, 1918; food controller, 1919-20; defeated, standing as unionist, 1923.

ROBERTS, ISAAC (1829-1904), amateur astronomer; master builder in Liverpool, 1859-88; made many geological researches; FGS, 1870; experimented in stellar photography at Liverpool from 1883; his photograph of the Pleiades revealed new knowledge of structure of the group, 1886; successfully photographed great nebula in Andromeda (1888) and in Orion (1889); settled at Crowborough, Sussex, 1890; published selections of stellar photographs, 1893 and 1899; FRS, 1890; hon. D.Sc., Dublin, 1892; gold medallist, Royal Astronomical Society, 1895; left residue of estate for founding scholarships at Liverpool, Bangor, and Cardiff.

ROBERTS, ROBERT DAVIES (1851-1911), educational administrator; B.Sc., London, 1870; D.Sc., 1878; BA, Clare College, Cambridge (second in natural sciences tripos), 1875; MA, 1878; fellow, 1884-90; university lecturer in geology, Cambridge, 1884; published *Earth's History: an Introduction to Modern Geology* (1893); secretary to London Society for Extension of University Teaching, 1885-94; in charge of Cambridge syndicate for university extension, 1894-1902; registrar of London University extension board, 1902; chairman of executive committee of university of Wales, 1910-11.

ROBERTS-AUSTEN, SIR WILLIAM CHANDLER (1843-1902), metallurgist; associate of Royal School of Mines, South Kensington, 1865; chemist of Royal Mint, 1870-82; chemist and assayer, 1882-1902; professor of metallurgy at Royal School of Mines, 1880-1902; member of War Office explosives committee, 1899; initiated alloys research work at Institution of Mechanical Engineers, 1889; invented automatic recording pyrometer, 1891; FRS, 1875; hon. general secretary of British Association, 1897-1902; president of Iron and Steel Institute, 1899-1901; hon. DCL, Durham, 1897, D.Sc., Manchester, 1901; CB, 1890; KCB, 1899; published *An Introduction to Metallurgy* (1891, 6th edn. 1910).

ROBERTSON, ALEXANDER (1896-1970), organic chemist and farmer; educated at Auchterless school, Turiff higher grade school, and Aberdeen University; commissioned into 2nd Seaforth Highlanders, 1917-18; MA, 1919; studied chemistry at Glasgow University, B.Sc., 1922; Carnegie research fellow under Professor G. G. Henderson [q.v.], 1922-4; Ph.D., 1924; as Rockefeller fellow worked with

Professor (Sir) Robert Robinson at Manchester University, 1924-6; assistant lecturer in chemistry, 1926-8; reader in chemistry, East London (later Queen Mary) College, London, 1928-30; reader in biochemistry, London School of Hygiene and Tropical Medicine, 1930-3; succeeded (Sir) Ian M. Heilbron [q.v.] as Heath Harrison professor of organic chemistry, Liverpool, 1933-57; pro-vice-chancellor, 1949-54; FRS, 1941; hon. LLD, Aberdeen; served on number of scientific committees; after retirement from Liverpool, became full-time farmer; adviser and consultant to research institutes; member, Agricultural Research Council, 1960-5.

ROBERTSON, ARCHIBALD (1853-1931), bishop of Exeter; educated at Bradfield and Trinity College, Oxford; first class, *lit. hum.*, 1876; fellow, 1876-86; dean, 1879-83; deacon, 1878; priest, 1882; principal, Bishop Hatfield's Hall, Durham, 1883-97; of King's College, London, 1897-1903; vice-chancellor, London University, 1902-3; bishop of Exeter, 1903-16; works include *Select Writings and Letters of Athanasius* (1892), *Roman Claims to Supremacy* (1896), and *Regnum Dei* (1901).

ROBERTSON, Sir CHARLES GRANT (1869-1948), historian and academic administrator; educated at Highgate School and Hertford College, Oxford; first class, *lit. hum.*, 1892, and modern history, 1893; fellow of All Souls, 1893-1948; domestic bursar, 1897-1920; tutor in modern history, Exeter, 1895-9, Magdalen, 1905-20; principal, Birmingham University, 1920-38; vice-chancellor, 1927-38; wrote mainly on political and constitutional history of eighteenth and nineteenth centuries; knighted, 1928.

ROBERTSON, Sir DENNIS HOLME (1890-1963), economist; educated at Eton and Trinity College, Cambridge; first class, part i, classical tripos, 1910; and part ii, economics tripos, 1912; fellow, Trinity College, 1914; served in 11th battalion, London Regiment, 1914-19; MC; published *A Study of Industrial Fluctuation* (1915); university lecturer in economics, 1924-8; Girdler's lecturer, 1928-30; reader, 1930-8; monetary specialist; published *Money* (1922) and *Banking Policy and the Price Level* (1926); reacted strongly against J. M. (later Lord) Keynes's [q.v.] *General Theory of Employment, Interest and Money* (1936); Sir Ernest Cassel professor at London School of Economics, 1938-44; economic adviser at the Treasury during 1939-45 war; succeeded A. C. Pigou [q.v.] as professor of political economy, Cambridge, 1944-57; continued to oppose Keynesianism; published *Lectures on Economic*

Principles (1957-9); member of Royal Commission on Equal Pay, 1944-6, and other committees; received many hon. degrees at home and abroad; CMG, 1944; knighted, 1953; FBA, 1932; fellow of Eton, 1948-57; president, Royal Economic Society, 1948-50.

ROBERTSON, DONALD STRUAN (1885-1961), classical scholar; brother of Agnes Arber, botanist; educated at Westminster School and Trinity College, Cambridge; first class, parts i and ii, classical tripos, 1906-8; fellow, Trinity College, 1909; assistant lecturer, with Ernest Harrison and Francis Cornford [qq.v.], Trinity College, 1911; commissioned in Army Service Corps, 1914-19; returned to Trinity, 1919; succeeded A. C. Pearson [q.v.] as regius professor of Greek, 1928; vice-master, Trinity College, 1947-51; publications include *A Handbook of Greek and Roman Architecture* (1929), text of the *Metamorphosis* of Apuleius (Paris, 3 vols, 1940-5), and *The Early Age of Greece* (2nd vol., from the notes of Sir William Ridgeway [q.v.] in collaboration with Andrew Gow, 1931); FBA, 1940; hon. degrees from Durham, Glasgow, and Athens.

ROBERTSON, DOUGLAS MORAY COOPER LAMB ARGYLL (1837-1909), ophthalmic surgeon; MD, St. Andrews, 1857; FRCS Edinburgh, 1862; president, 1886-7; published researches proving value of Calabar bean in ophthalmology, 1862; ophthalmic surgeon to Royal Infirmary, Edinburgh, 1867-97; described 'the Argyll Robertson pupil' in *Edinburgh Medical Journal*, 1869-70; hon. LLD, Edinburgh, 1896; died in India on a third visit.

ROBERTSON, GEORGE MATTHEW (1864-1932), alienist; MB, Ch.B., Edinburgh, 1885; studied at Bethlem Royal and National hospitals, London; physician-superintendent, Perth (1892-9), Stirling (1899-1910) district asylums; at Royal Edinburgh Asylum, Morningside, 1908-32; university lecturer on mental diseases, 1908; first professor of psychiatry, 1920-32; instituted nursing homes and hospital for uncertified cases and residential clinic for problem children.

ROBERTSON, Sir GEORGE SCOTT (1852-1916), Anglo-Indian administrator; entered Indian medical service, 1878; surgeon to British political agent in Gilgit, 1889; British agent in Gilgit, 1894; held out, till relieved, in Chitral with small, ill-provisioned force against combined Pathan and rebel Chitral armies, 4 Mar.-20 Apr., 1895; KCSI, 1895; retired, 1899; liberal MP, Central Bradford, 1906.

ROBERTSON, SIR HOWARD MORLEY (1888-1963), architect; educated at Malvern College, Architectural Association School of Architecture, and École Supérieure des Beaux Arts, Paris, 1908-12; French Architectural Diploma, 1913; served in army during 1914-18 war; MC; Legion of Honour; joined with John Murray Easton in private practice, 1919; and joined teaching staff of the AA School of Architecture; principal, 1926; director of education, 1929-35; private practice merged with Stanley Hall, 1931; FRIBA, 1925; hon. AIA; designed with Easton, exhibition hall of Royal Horticultural Society; awarded RIBA architecture medal; designed exhibition pavilions in Paris, Brussels, and New York and interiors for hotels and liners; designed printing works for the Bank of England and reconstructed university buildings post-war; last major work, the Shell Centre on south bank of Thames; D.Litt., Reading, 1957; president, Architectural Association, 1947; president, RIBA council, 1952-4; ARA, 1949; RA, 1958; knighted, 1954; publications include articles 'Obbligato to Architecture' in the *Builder* 1962, *The Principles of Architectural Composition* (1924), and *Modern Architectural Design* (1932).

ROBERTSON, JAMES PATRICK BANNERMAN, BARON ROBERTSON (1845-1909), lord president of the Court of Session in Scotland; MA, Edinburgh University, 1864; hon. LLD, 1890; lord rector, 1893; passed to Scottish bar, 1867; QC, 1885; solicitor-general for Scotland, 1885 and 1886; conservative MP for Buteshire, 1885-6, 1886-91; lord advocate and PC, 1889; carried Local Government (Scotland) Act, 1889; lord president of the Court of Session, 1891; life peer, and member of judicial committee of Privy Council, 1899; chairman of Irish university commission (report published, 1904); hostile to Joseph Chamberlain's tariff policy.

ROBERTSON, JOHN MACKINNON (1856-1933), writer and politician; worked with Charles Bradlaugh [q.v.] on *National Reformer*, 1884-91; editor, 1891-3; liberal MP, Tyneside division, 1906-18; parliamentary secretary to Board of Trade, 1911-15; wrote two histories of free thought (1899 and 1929); Shakespearian scholar, literary critic, and social scientist; PC, 1915.

ROBERTSON, SIR JOHNSTON FORBES- (1853-1937), actor; educated at Charterhouse and the Royal Academy schools; acted under Samuel Phelps [q.v.]; speedily made a name; went into management, 1895; his choice of plays frequently in advance of public taste; achieved greatest artistic success as Hamlet (1897); other productions included *The Devil's Disciple* (1900), *Mice and Men* (1902), *The Light That Failed* (1903); in *The Passing of the Third Floor Back* (1908) revealed the sweetness and goodness of his character; retired from English stage and knighted, 1913.

ROBERTSON, SIR ROBERT (1869-1949), explosives expert and government chemist; graduated in arts and science, St. Andrews; entered Royal Gunpowder Factory, Waltham Abbey, 1892; in charge of laboratory, 1900; superintending chemist, research department, Royal Arsenal, Woolwich, 1907-21; developed manufacture and use of TNT and 'amatol'; government chemist, 1921-36; director, Salters' Institute of Industrial Chemistry, 1936-49; FRS, 1917; KBE, 1918.

ROBERTSON, SIR WILLIAM ROBERT, first baronet (1860-1933), field-marshal; enlisted in 16th Lancers, 1877; troop-sergeant-major, 1885; gazetted second lieutenant in 3rd Dragoon Guards in India, 1888; appointed to intelligence department, 1892; entered Staff College, 1896; head of foreign section, intelligence department, War Office, 1900-7; colonel, 1903; brigadier-general, general staff, Aldershot, 1907-10; major-general, 1910; commandant, Staff College, 1910-13; introduced realistic war training; director of military training, 1913-14; quartermaster-general, GHQ, France, Aug. 1914-Jan. 1915; chief of general staff, Jan.-Dec. 1915; a convinced Westerner opposed to Gallipoli campaign; urged an organization for control of allied strategy; accepted appointment as chief of imperial general staff (Dec. 1915) on condition of a small war council co-ordinating strategy in all war theatres with himself as its military adviser in direct touch with his commanders without intervention of Army Council; asked also for reorganization of general staff, War Office; obtained evacuation of Helles; created reserves in Egypt supplying France and Mesopotamia; general, 1916; found it increasingly difficult to work with Lloyd George who sought greater political control of strategy and, without consulting Robertson, at an allied conference in Jan. 1917 proposed an attack on Austria, and, in Feb., that Nivelle should control British army in France; opposed Lloyd George's plan for an offensive in Palestine (Feb. 1918) and his suggestion of permanent military representation giving technical advice to Supreme War Council and controlling an allied general reserve; refused as wrong in principle the appointment of military representative at Versailles; transferred to Eastern command at home (Feb. 1918); commander-in-chief, Home Forces (June 1918), Army on the Rhine (1919-20); KCVO, 1913; KCB, 1915; GCB, 1917;

GCMG and baronet, 1919; field-marshal, 1920; GCVO, 1931.

ROBERTSON SCOTT, JOHN WILLIAM (1866-1962), journalist, author, and founder-editor of the *Countryman*; educated at Quaker and grammar schools; contributed articles to *Manchester Guardian* under C. P. Scott [q.v.] and other national papers; joined W. T. Stead [q.v.] on *Pall Mall Gazette*, 1887; moved to *Westminster Gazette*, 1893; and to *Daily Chronicle*, 1899; invited by J. St. Loe Strachey [q.v.] to contribute farming articles to the *Country Gentleman*, 1902; wrote hundreds of articles for journals such as the *Field*; during 1914-18, lived in Japan; edited *New East*, and published *Japan, Great Britain and the World* (1916) and *The Foundations of Japan* (1922); returned to England, 1921, and published articles contributed to the *Nation* as *England's Green and Pleasant Land* (1925); founded the quarterly *The Countryman*, 1927; editor, 1927-47; active in local government; JP; CH, 1947; hon. MA, Oxford, 1951; published autobiography *The Day Before Yesterday* (1951).

ROBEY, SIR GEORGE EDWARD (1869-1954), comedian; originally surnamed Wade; educated at school in Dresden and Leipzig University; employed as clerk in Birmingham; appeared at concerts as amateur comedian; changed his name to Robey and made first professional music-hall appearance at the Oxford in London, 1891; rapidly established his name as 'The Prime Minister of Mirth'; appeared in number of revues, including *The Bing Boys are Here* with Alfred Lester and Violet Loraine [qq.v.] during the 1914-18 war; successfully played Falstaff in *Henry IV, Part I* (1935); CBE, 1919; knighted, 1954.

ROBINS, THOMAS ELLIS, BARON ROBINS (1884-1962), business man; born in Philadelphia, USA; educated at Pennsylvania University and Christ Church, Oxford; Rhodes scholar, 1904-7; assistant editor, *Everybody's Magazine*, New York, 1907-9; returned to England; private secretary to sixth Earl Winterton MP [q.v.], 1909-12; became naturalized British subject, and joined City of London Yeomanry, 1912; served in Middle East; DSO, 1914-21; secretary, Conservative Club, London, 1921-8; general manager, Salisbury, Southern Rhodesia, of British South Africa Company (Chartered), 1928; resident director, 1934; director of number of South African companies, including the Anglo-American Corporation; commanded 1st battalion, Rhodesia Regiment, 1940-3; served on General Staff, India, 1943; knighted, 1946; chairman, Central Africa Rhodes Centenary

Exhibition, Bulawayo, 1953; trustee, Rhodes-Livingstone Institute and Rhodes National Gallery; president, Royal African Society, London; returned to London as president, Chartered Company, 1957; baron, 1958, knight of St. John of Jerusalem.

ROBINSON, (ESMÉ STUART) LENNOX (1886-1958), Irish dramatist and theatre director; educated at home and at Bandon grammar school; his one-act play *The Clancy Name* staged at the Abbey Theatre, Dublin, 1908; manager, Abbey Theatre, 1910-14 and 1919-23; director, 1924; his plays produced at the Abbey Theatre included *Patriots* (1912), *The Big House* (1926), and *Birds Nest* (1938); notable productions while he was director were Sean O'Casey's *Juno and the Paycock* (produced 1924) and *The Silver Tassie* (1929); published many essays, short stories, and a novel, together with *Curtain Up* (1942), *Lady Gregory's Journals* (1946), and *Ireland's Abbey Theatre* (1951); hon. D.Litt., Trinity College, Dublin, 1948.

ROBINSON, FREDERICK WILLIAM (1830-1901), novelist; published *The House of Elmore* (1855) and some fifty succeeding novels dealing mainly with semi-religious themes or low life; the chief were *Grandmother's Money* (1860), *High Church* (1860), *Christie's Faith* (1867) and *Anne Judge, Spinster* (1867); edited *Home Chimes*, 1884-93; friend of Swinburne, Theodore Watts-Dunton, and Sir Henry Irving [qq.v.].

ROBINSON, GEORGE FREDERICK SAMUEL, first MARQUESS OF RIPON (1827-1909), statesman; born at 10 Downing Street, while his father, first Viscount Goderich, later first Earl of Ripon was prime minister; was known as Viscount Goderich, 1833-59; attaché to Sir Henry Ellis [q.v.] on abortive mission to Brussels to negotiate peace between Austria and Piedmont, 1849; joined Christian Socialist movement of F. D. Maurice [q.v.] and formed intimacy with Thomas Hughes [q.v.], 1849; pleaded for democracy in *The Duty of the Age*, a Christian socialist tract which Maurice suppressed as being too radical, 1852; actively engaged in Volunteer movement, 1859; liberal MP for Huddersfield, 1853-7, and for West Riding of Yorkshire, 1857-8; succeeded his father as Earl of Ripon and his uncle as Earl de Grey, 1859; became, under Palmerston, under-secretary at the War Office, June 1859; at the India Office, Jan.-July 1861, and again at the War Office, July 1861-Apr. 1863; secretary for war and PC with seat in the cabinet, 1863-6; secretary for India, 1866; lord president of Council under Gladstone, 1868-73; KG, 1869; hon. DCL, Oxford, 1870; chairman of joint high commission for the settlement of American claims against Great Britain, 1871; made marquess, 23 June 1871; resigned cabinet office, Aug. 1873; was received into Roman Catholic communion, 7 Sept. 1874; appointed governor-general of India, 28 Apr. 1880; settled critical difficulties in Afghanistan, 1880; repealed restrictions on the vernacular press, and encouraged development of self-government in India, 1882; provoked controversy by proposals to subject Europeans and Americans to trial by native Indian magistrates or judges ('the Ilbert bill'); accepted a satisfactory compromise; developed system of provincial settlements; retired from India, Dec. 1884; became first lord of the Admiralty in Gladstone's first home rule cabinet, 1886; received freedom of city of Dublin, 1898; colonial secretary, 1892–5; lord privy seal with leadership of liberal party in House of Lords, 1905–8.

ROBINSON, (GEORGE) GEOFFREY (1874–1944), twice editor of *The Times*. [See DAWSON.]

ROBINSON, HENRY WHEELER (1872–1945), Baptist divine and Old Testament scholar; educated at Regent's Park College, Edinburgh University, and Mansfield College, Oxford; second class, oriental studies, 1898; minister, Pitlochry, 1900–3; Coventry, 1903–6; tutor, Rawdon Baptist College, 1906–20; principal, Regent's Park College, 1920–42; transferred college to Oxford; reader in biblical criticism, Oxford, 1934–41; chairman, board of faculty of theology, 1937–9; Speaker's lecturer, 1942–5; author of many works and co-editor 'Library of Constructive Theology'.

ROBINSON, SIR JOHN (1839–1903), first prime minister of Natal; left Hull with parents for Natal, 1850; succeeded father as manager of *Natal Mercury*, 1860; member of council for Durban, 1863–82, and from 1884; represented Durban at colonial conference in London, 1887; KCMG, 1889; advocated from 1882 self-government for colony; first prime minister, colonial secretary, and minister of education, 1893–7; published *A Lifetime in South Africa* (1900).

ROBINSON, SIR JOHN CHARLES (1824–1913), art connoisseur and collector; first superintendent of South Kensington Museum art collections, 1852–69; surveyor of Queen's pictures, 1882–1901; adviser to private collectors; knighted, 1887; CB, 1901.

ROBINSON, SIR JOHN RICHARD (1828–1903), journalist; sub-editor of Unitarian

journal the *Inquirer*, 1848; editor of evening *Express*, 1855; manager of *Daily News*, 1868–1901; obtained descriptive accounts of Franco-Prussian war, 1870, and of Turkish atrocities in Bulgaria, 1876; knighted, 1893; showed sympathy with Boers in South African war, 1899–1902; conspicuous member of Reform Club.

ROBINSON, JOSEPH ARMITAGE (1858–1933), dean of Westminster Abbey and of Wells Cathedral; educated at Liverpool College and Christ's College, Cambridge; fourth classic, 1881; fellow, 1881–99; deacon, 1881; priest, 1882; edited new series of Cambridge 'Texts and Studies' on Christian apologetics; Norrisian professor of divinity, 1893–9; rector of St. Margaret's and canon of Westminster, 1899–1902; dean of Westminster (1902–11), of Wells (1911–33); theological and historical writer; FBA, 1903; KCVO, 1932.

ROBINSON, SIR JOSEPH BENJAMIN-first baronet (1840–1929), South African mineowner; born in Cape Colony; secured valuable diamond claims at Hebron and in area afterwards famous as Kimberley, 1868–71; reached Rand goldfield, 1886; acquired Randfontein block, 1889; baronet, 1908; a bold, far-sighted, energetic, but unlovable industrial pioneer.

ROBINSON, PHILIP STEWART [PHIL ROBINSON] (1847–1902), naturalist and author; born at Chunar, India; assisted father in editing *Pioneer*, 1869; editor of *Revenue Archives* of Benares province, 1872; pioneer of Anglo-Indian literature, descriptive of natural history; chief works were *In my Indian Garden* (1878), *Tigers at Large* (1884), and *The Valley of Teetotum Trees* (1886).

ROBINSON, ROY LISTER, BARON ROBINSON (1883–1952), forester; born in Australia; educated at St. Peter's College and Adelaide University; B.Sc., 1903; Rhodes Scholar, Magdalen College, Oxford, 1905–9; first class, natural science (geology), 1907; Burdett-Coutts scholar; diploma in forestry; represented university at lacrosse (1906–9), athletics (1907–9), and cricket (1908–9); assistant inspector, Board of Agriculture and Fisheries, 1909; secretary to the forestry sub-committee of the Reconstruction Committee, 1916; technical commissioner, Forestry Commission, 1919; chairman, Forestry Commission, 1932–52; took leading part in report on 'Post-War Forest Policy' (Cmd. 6447, 1943); attended six Commonwealth Forestry Conferences between 1918 and 1952; OBE, 1918; knighted, 1931; raised to peerage, 1947; hon. LLD, Aberdeen, 1951.

ROBINSON, VINCENT JOSEPH (1829–1910), connoisseur of oriental art; published *Eastern Carpets*, 1882; director of Indian section at Paris exhibition, 1889; FSA, 1889; CIE, 1891; described his own collections in *Ancient Furniture and other Objects of Art* (1902); helped to revive European taste for oriental art.

ROBINSON, Sir (WILLIAM) ARTHUR (1874–1950), civil servant; educated at Appleby grammar school and Queen's College, Oxford; entered Colonial Office, 1897; assistant secretary, Office of Works, 1912–18; permanent secretary, Air Ministry, 1918–20; Ministry of Health, 1920–35; chairman, Supply Board, 1935–9; permanent secretary, Ministry of Supply, 1939–40; KCB, 1919; GCB, 1929.

ROBINSON, WILLIAM HEATH (1872–1944), cartoonist and book-illustrator; studied at Islington Art School and Royal Academy Schools; his caricatures of machinery as complicated jerry-built apparatus designed for absurd purposes gave rise to adjective 'Heath Robinson'.

ROBINSON, WILLIAM LEEFE (1895–1918), airman; joined Royal Flying Corps, 1915; VC for shooting down at Cuffley, Hertfordshire, first enemy airship brought down on British soil, 1916; prisoner of war in Germany, 1917–18, succumbing to his bad treatment there, after his return to England.

ROBISON, ROBERT (1883–1941), biochemist; studied at University College, Nottingham; B.Sc., London, 1906; D.Phil., Leipzig; worked with F. S. Kipping [q.v.]; joined Lister Institute of Preventive Medicine, 1913; head of biochemical department, Lister Institute, and professor of biochemistry, London University, 1931–41; published *The Significance of Phosphoric Esters in Metabolism* (1932); FRS, 1930.

ROBSON, WILLIAM SNOWDON, Baron Robson (1852–1918), lawyer and politician; BA, Caius College, Cambridge; called to bar (Inner Temple), 1880; QC, 1892; MP, Bow and Bromley, 1885; South Shields, 1895–1910; solicitor-general and knighted, 1905; attorney-general, 1908; piloted budget of 1909; presented British case at Atlantic fisheries arbitration at The Hague, 1910; lord of appeal in ordinary and PC, 1910.

ROBY, HENRY JOHN (1830–1915), educational reformer and classical scholar; BA, St. John's College, Cambridge (senior classic); fellow of St. John's, 1854–61; second master, Dulwich College, 1861–5; secretary to schools inquiry commission, 1864; to endowed schools commission, 1869; largely responsible for reforms introduced by commissions; partner in Manchester sewing-cotton manufacturing firm, 1874–94; liberal MP, Eccles division of Manchester, 1890–5; works include *Grammar of the Latin Language from Plautus to Suetonius* (1871–4, 7th edn. 1904), distinguished for its wealth of illustration, *Introduction to Justinian's Digest* (1884), and *Roman Private Law* (1902).

ROCHE, ALEXANDER ADAIR, Baron Roche (1871–1956), judge; educated at Ipswich grammar school and Wadham College, Oxford; first class, hon. mods, 1892; first class, *lit. hum.*, 1894; hon. fellow, Wadham, 1917; called to bar (Inner Temple), 1896; bencher Inner Temple, 1917; treasurer, 1939; practised on North-Eastern circuit, specializing in maritime cases; extensive practice in Commercial and Admiralty Courts; took silk, 1912; appointed judge, King's Bench and knighted, 1917; lord justice, PC, 1934; lord of appeal in ordinary, life peerage, 1935; retired, 1938; chairman, Agricultural Wages Board, 1940–3.

RODD, JAMES RENNELL, first Baron Rennell (1858–1941), diplomatist and scholar; educated at Haileybury and Balliol College, Oxford; second class, *lit. hum.*, 1880; entered diplomatic service, 1883; minister, Stockholm, 1904–8; ambassador, Rome, 1908–19; refrained from exerting direct pressure upon Italy to join Allies, 1915; on Lord Milner's commission on status of Egypt, 1919–20; British representative, General Assembly, League of Nations, 1921, 1923; president, court of conciliation between Austria and Switzerland, 1925; conservative MP, St. Marylebone, 1928–32; published collections of poems, reminiscences, and classical and medieval studies including *Homer's Ithaca* (1927) and *Rome of the Renaissance and Today* (1932); KCMG, 1899; GCVO, 1905; PC, 1908; GCMG, 1915; GCB, 1920; baron, 1933.

ROE, Sir (EDWIN) ALLIOTT VERDON VERDON- (1877–1958), aircraft designer and constructor. [See VERDON-ROE.]

ROGERS, ANNIE MARY ANNE HENLEY (1856–1937), educationist; daughter of J. E. T. and sister of L. J. Rogers [qq.v.]; honorary secretary of committee of the Association for the Higher Education of Women in Oxford (1894–1920) and of the governing body of the Society of Oxford Home-Students (1893–1930); tutor and member of council (1894–1936) of St. Hugh's College.

ROGERS, BENJAMIN BICKLEY (1828–1919), barrister and translator of Aristophanes; BA, Wadham College, Oxford; fellow of

Wadham, 1852-61; called to bar (Lincoln's Inn), 1856; works include annotated verse translations of Aristophanes' *Clouds* (1852), *Peace* (1867), *Wasps* (1875), *Lysistrata* (1878), and *Thesmophoriazusae* (1904).

ROGERS, EDMUND DAWSON (1823-1910), journalist and spiritualist; manager of *Norfolk News*, 1845; started *Eastern Daily Press*, first daily paper in eastern counties, 1870; established National Press Agency, 1873; helped to found British National Association of Spiritualists, 1873; founded (1881) and edited from 1894 weekly spiritualist journal, *Light*; his autobiographical *Life and Experiences* appeared in 1911.

ROGERS, JAMES GUINNESS (1822-1911), Congregational divine; BA, Trinity College, Dublin, 1843; minister at Newcastle upon Tyne, 1846-51, Ashton under Lyne, 1851-65, and Clapham, 1865-1900; chairman of the Congregational unions of Lancashire, 1865, Surrey, 1868, and England and Wales, 1874; hon. DD, Edinburgh, 1895; friend of Gladstone; publications include *The Church Systems of England in the Nineteenth Century* (1881), *The Unchanging Faith* (1907), and an *Autobiography* (1903); edited *Congregationalist*, 1879-86, and *Congregational Review*, 1887-91.

ROGERS, SIR LEONARD (1868-1962), pioneer of tropical medicine; grandson of John Rogers [1778-1856, q.v.]; educated at Tavistock grammar school, Devon county school, and Plymouth College; entered St. Mary's Hospital medical school, 1886; FRCS, MB, and BS, London, 1892; entered Indian medical service, 1893; MD, London, 1897; MRCP, 1898; carried out veterinary research on rinderpest, 1898-1900; transferred to Bengal civil medical department, 1900; acted for professor of pathology, Calcutta; confirmed as professor, 1906; made many discoveries regarding the nature and treatment of tropical diseases, including use of emetine in amoebic dysentery; carried out important research with cholera and leprosy; organized construction of Calcutta School of Tropical Medicine and founded British Empire Leprosy Relief Association, 1923; on retirement from IMS, lecturer, London School of Tropical Medicine, 1921; member, India Office medical board, 1922; president, 1928; CIE, 1911; knighted, 1914; KCSI, 1932; FRS, 1916; president, Indian Science Congress, 1919; president, Royal Society of Tropical Medicine and Hygiene, 1933-5; published autobiography, *Happy Toil* (1950).

ROGERS, LEONARD JAMES (1862-1933), mathematician; son of J. E. T. and brother of

Annie Rogers [qq.v.]; educated at Balliol College, Oxford; first class, mathematics, and B.Mus., 1884; professor of mathematics, Leeds, 1888-1919; worked on theory of reciprocants, and in fields of elliptic functions, theta functions, and basic hypergeometric series; FRS, 1924; skilled musician, linguist, mimic, skater, and knitter.

ROLLESTON, SIR HUMPHRY DAVY, baronet (1862-1944), physician; son of George Rolleston [q.v.]; educated at Marlborough, St. John's College, Cambridge, and St. Bartholomew's Hospital; successively assistant physician, and consulting physician to St. George's Hospital and Victoria Hospital for Children; consulting physician, Royal Navy, 1914-18; regius professor of physic, Cambridge, 1925-32; eminent clinician and pathologist and leading authority on presentation of medical articles; joint-editor, second edition Allbutt's *System of Medicine* (1905-11); published *Diseases of the Liver, Gall-bladder, and Bile-ducts* (1905) and many other works; KCB, 1918; baronet, 1924; GCVO, 1929.

ROLLS, CHARLES STEWART (1877-1910), engineer and aviator; son of first Baron Llangattock; BA, Trinity College, Cambridge, 1898; MA, 1902; studied practical engineering; pioneer of motor cars in England; formed motor car business of C. S. Rolls & Co., and established Rolls-Royce, Ltd., 1904; ardent aeronaut and expert aviator; crossed and recrossed English Channel in aeroplane without stopping, June 1910; killed at Bournemouth in aeroplane accident; first English victim of aviation; wrote on motors and on his experiences as motorist and aviator.

ROMER, MARK LEMON, BARON ROMER (1866-1944), judge; son of Sir Robert Romer and grandson of Mark Lemon [qq.v.]; educated at Rugby and Trinity Hall, Cambridge; junior optime, 1888; called to bar (Lincoln's Inn), 1890; built up Chancery practice; KC, 1906; bencher, 1910; judge of Chancery division, 1922-9; lord justice of appeal, 1929-38; lord of appeal in ordinary, 1938-44; chairman of committee on draft bills on real property enacted in 1925; member, lord chancellor's Law Revision Committee; knighted, 1922; PC, 1929; life peer, 1938.

ROMER, SIR ROBERT (1840-1918), judge; BA, Trinity Hall, Cambridge (senior wrangler); called to bar (Lincoln's Inn), 1867; QC, 1881; practised in court of Sir J. W. Chitty [q.v.]; judge of Chancery division and knighted, 1890; lord justice of appeal, 1899-1906; PC and FRS, 1899; GCB, 1901.

RONALD, Sir LANDON (1873-1938), musician; son of Henry Russell [q.v.]; studied at Royal College of Music; for some years accompanist to Melba [q.v.]; conductor, Royal Albert Hall Orchestra, from 1909; principal, Guildhall School of Music, 1910-38; compositions mainly ballads; knighted, 1922.

RONAN, STEPHEN (1848-1925), lord justice of appeal in Ireland; BA, Queen's College, Cork; called to Irish bar, 1870; junior crown prosecutor for county of Kerry; 1873; counsel to attorney-general, 1883; called to English bar (Inner Temple), 1888; employed on Parnell commission as a junior counsel for *The Times*, 1888; QC (Ireland), 1889; senior crown prosecutor for Cork city and county, 1891; queen's advocate-general for Ireland, 1892; KC (England), 1909; lord justice of appeal and PC (Ireland), 1915-24; a great advocate.

ROOKWOOD, Baron (1826-1902), politician. [See Selwin-Ibbetson, Sir Henry John.]

ROOPER, THOMAS GODOLPHIN (1847-1903), writer on education; educated at Harrow and Balliol College, Oxford; BA, 1870; private tutor to eleventh Duke of Bedford [q.v.], 1871-7; inspector of schools from 1877; improved teaching of geography and methods of teaching of infants; stimulated manual education, and encouraged school gardens; helped to found Hartley University College, Southampton; writings include *School and Home Life* (1896), *Educational Studies and Addresses* (1902), *School Gardens in Germany* (1902); *Selected Writings* edited with memoir by R. G. Tatton, 1907.

ROOS-KEPPEL, Sir GEORGE OLOF (1866-1921), soldier and administrator; joined army, 1886; political agent in Khyber, 1899-1908; CIE, 1900; chief comissioner, North-West Frontier Province, and agent to governor-general, 1908-19; KCIE, 1908; GCIE, 1917.

ROOSE, EDWARD CHARLES ROBSON (1848-1905), physician; studied at Guy's Hospital and in Paris; MRCP Edinburgh, 1875; FRCP, 1877; MD, Brussels, 1877; had large fashionable London practice; wrote many popular medical works.

ROOTES, WILLIAM EDWARD, first Baron Rootes (1894-1964), motor car manufacturer; educated at Cranbrook school; apprenticed to Singer Cars, Coventry, 1913; joined RNVR, 1915; moved to Royal Naval Air Service, 1917; started motor business with brother Reginald (Claud), 1919; business moved to Devonshire House, Piccadilly, 1926; acquired Hillman and Humber factories in Coventry, 1929; planned Humber Snipe and Hillman Wizard; launched Hillman Minx, 1932; as business prospered took over other firms, including Singer Cars and Sunbeam Motor Car Co. Ltd., 1930-7; served on many national and official bodies; chairman, Joint Aero Engine Committee, 1940-1; chairman, Motor Vehicle Maintenance Advisory Committee, 1941; chairman, supply council, Ministry of Supply, 1941-2; KBE, 1942; chairman, Coventry Industrial Reconstruction and Co-ordinating Committee; engaged in post-war expansion; chairman, Dollar Export Board, 1951-64; new plant opened at Linwood, 1963; produced Hillman Imp; obtained new capital from American firm of Chrysler, 1964; GBE, 1955; baron, 1959; chairman, promotion committee, new university of Warwick, 1964.

ROPES, ARTHUR REED (1859-1933), lyric writer and librettist under pseudonym of Adrian Ross; eleventh wrangler (1882), first class, history (1883), fellow (1884-90), King's College, Cambridge; wrote lyrics for many musical plays including *The Circus Girl* (1896), *San Toy* (1899), *A Country Girl* (1902), *The Quaker Girl* (1910), *Monsieur Beaucaire* (1919), and English versions of *The Merry Widow* (1907) and *Lilac Time* (1922).

ROSCOE, Sir HENRY ENFIELD (1833-1915), chemist; son of Henry Roscoe [q.v.]; BA, University College, London; Ph.D., Heidelberg; researched with R. W. von Bunsen on measurement of chemical action of light; professor of chemistry, Owens College, Manchester, 1857-85; knighted, 1884; liberal MP, South Manchester, 1885; PC, 1909; his most important contribution to chemistry was preparation of pure vanadium.

ROSCOE, KENNETH HARRY (1914-1970), professor of engineering; educated at the high school, Newcastle under Lyme, and Emmanuel College, Cambridge; first class, mechanical science tripos, 1938; trainee technician, Metropolitan Cammell Co. Ltd.; assistant works manager, 1939; served in Royal Engineers, 1939; prisoner of war 1940-5; escaped; MC; research student, Cambridge, 1945; demonstrator, 1946; lecturer; fellow, Emmanuel College, 1948; active in university OTC; tutor and director of studies, 1952-65; domestic bursar, Emmanuel, 1955; SSA, 1953; produced over thirty papers on soil tests, and organized soil mechanics group; reader, 1965; professor, 1968.

ROSE, JOHN HOLLAND (1855-1942), historian; educated at Bedford modern school,

Owens College, Manchester and Christ's College, Cambridge; reader in modern history, 1911-19; Vere Harmsworth professor of naval history, 1919-33; joint-editor, *Cambridge History of the British Empire*; publications include *Life of Napoleon I* (1902) and two volumes on the younger Pitt; FBA, 1933.

ROSE-INNES, SIR JAMES (1855-1942), chief justice of South Africa; BA (1874), LLB (1877), Cape university; practised at bar, Cape Town, 1878-1902; QC, 1890; member, Cape legislative assembly, 1884-1902; attorney-general, 1890-3, 1900-2; KCMG, 1901; chief justice, Transvaal, 1902-10; judge of appeal, South African Supreme Court, 1910-14; chief justice of South Africa, 1914-27; zealous champion of native rights; PC, 1915.

ROSEBERY, fifth EARL OF (1847-1929), statesman and author. [See PRIMROSE, ARCHIBALD PHILIP.]

ROSENHAIN, WALTER (1875-1934), metallurgist; born in Berlin of Jewish stock; naturalized in Australia; educated at Melbourne University and St. John's College, Cambridge; superintendent, metallurgical division, National Physical Laboratory, 1906-31; systematically investigated the constitution of metallic alloys, especially of iron; FRS, 1913; his *Introduction to the Study of Physical Metallurgy* (1914) widely influential.

ROSENHEIM, (SIGMUND) OTTO (1871-1955), organic chemist and biochemist; born at Würzburg, Germany; educated at Würzburg and Bonn University; Ph.D.; research student, Manchester University, 1894-5; adopted British nationality, 1900; research student of pharmacological chemistry, King's College, London, 1901; lecturer in chemical physiology, King's College, 1904; reader in biochemistry, 1915; resigned from teaching duties, 1920; worked at National Institute for Medical Research, 1923-42; his research made major contributions to the chemistry of natural products; FRS, 1927; fellow of the Linnean Society, member of the Medical Research Council.

ROSS, ADRIAN (pseudonym) (1859-1933), lyric writer and librettist. [See ROPES, ARTHUR REED.]

ROSS, SIR ALEXANDER GEORGE (1840-1910), lieutenant-general; born at Meerut; joined Indian army, 1857; served in Mutiny; present at capture of Magdala, 1868; in Jowaki expedition, 1877; commanded 1st Sikh infantry in Afghan war, 1878-9, in campaign against Mahsud Waziris, 1881, and in Zhob valley

expedition, 1890; lieutenant-general, 1897; CB, 1887; KCB, 1905.

ROSS, SIR (EDWARD) DENISON (1871-1940), orientalist; educated at Marlborough and University College, London; studied oriental languages on the continent; Ph.D., Strasbourg; professor of Persian, University College, London, 1896-1901; principal, Calcutta Madrasah, 1901-11; found and edited (3 vols., 1910-28) an unfinished Arab manuscript history of Gujarat; in charge of collections of Sir Aurel Stein [q.v.] at British Museum, 1914; director, School of Oriental Studies and professor of Persian, London University, 1916-37; counsellor, British Embassy, Istanbul, 1939-40; knighted, 1918.

ROSS, SIR FREDERICK WILLIAM LEITH- (1887-1968), civil servant and authority on finance. [See LEITH-ROSS.]

ROSS, SIR IAN CLUNIES (1899-1959), veterinary scientist and scientific administrator; born at Bathurst, Australia; educated at Newington College, Sydney, and Sydney University; B.V.Sc., 1921; research fellowship for postgraduate studies, London School of Tropical Medicine and Molteno Institute, Cambridge, 1922-4; lecturer in veterinary parasitology, Sydney, 1925; D.V.Sc., Sydney, 1928; research in Japan, 1929-30; officer-in-charge, McMaster Laboratory, Sydney, for research in animal health, 1931; published textbook on internal parasites of sheep in collaboration with H. M. Gordon (Sydney, 1936); chairman, International Wool Secretariat, London, 1937; executive officer, Council for Scientific and Industrial Research, 1946; first chairman, Commonwealth Scientific and Industrial Research Organisation, 1949-59; CMG, and knighted, 1954; hon. degrees, Melbourne, New England, and Adelaide universities.

ROSS, SIR JOHN (1829-1905), general; son of Sir Hew Dalrymple Ross [q.v.]; entered army, 1846; served in Canada, 1847-52, in the Crimea, 1854-5, in Indian Mutiny, 1857-9; helped to raise at Lucknow (1858) camel corps, which he commanded at Gowlowlie, Calpi, and Jugdespore; CB, 1861; commanded Laruf field force in Malay peninsula, 1875-6; major-general, 1877; in command of second division of Kabul field force, 1878-80; KCB, 1881; commanded Poona division, 1881-6; commander-in-chief in Canada, 1888; GCB, 1891.

ROSS, SIR JOHN, first baronet, of Dunmoyle (1853-1935), last lord chancellor of Ireland; educated at Trinity College, Dublin; called to Irish bar, 1879; QC, 1891; conservative MP,

Londonderry City, 1892-5; land judge, 1896-1921; lord chancellor of Ireland, 1921-2; Irish PC, 1902; baronet, 1919.

ROSS, JOSEPH THORBURN (1849-1903), artist; art student at Royal Scottish Academy, 1877-80; ARSA, 1896; treated portraiture, land- and sea-scape, and incident; frequently exhibited abroad; his 'The Bass Rock' in National Gallery of Scotland.

ROSS, MARTIN (pseudonym) (1862-1915), novelist. [See MARTIN, VIOLET FLORENCE.]

ROSS, SIR RONALD (1857-1932), discoverer of the mosquito cycle in malaria; qualified at St. Bartholomew's Hospital; entered Indian medical service, 1881; began association with (Sir) Patrick Manson [q.v.], 1894; found the characteristic pigmented oocysts of the malaria parasite in the anopheles mosquito, 1897; demonstrated cycle of development for bird malaria, 1898; pioneer in the further study of malaria; lecturer (1899), professor (1902-12) of tropical medicine, Liverpool School of Tropical Medicine; of tropical sanitation, 1912-17; physician for tropical diseases, King's College, Hospital, London, 1912; FRS, 1901; Nobel prize for medicine, 1902; KCB, 1911; first director-in-chief, Ross Institute and Hospital for Tropical Diseases, Putney Heath, 1926-32; editor, *Science Progress*, 1913-32.

ROSS, WILLIAM STEWART, 'SALADIN' (1844-1906), secularist; educated at Glasgow University; early wrote verse and fiction; became under style of Stewart & Co. educational publisher in London, 1872; joint-editor from 1880, and sole editor and proprietor from 1884, of *Agnostic Journal and Secular Review*; works include *Lays of Romance and Chivalry* (1881) and *God and His Book* (1887).

ROSSE, fourth EARL OF (1840-1908), astronomer. [See PARSONS, LAURENCE.]

ROSSETTI, WILLIAM MICHAEL (1829-1919), man of letters and art critic; brother of Christina Rossetti and D. G. Rossetti [qq.v.]; served in Excise Office (later Inland Revenue Board), 1845-94; friend of A. C. Swinburne [q.v.]; works include four editions of D. G. Rossetti's collected works and *Memoir* (1895), and editions of Christina Rossetti's *New* and *Collected Poems* (latter containing memoir, 1904).

ROTHENSTEIN, SIR WILLIAM (1872-1945), painter, and principal of Royal College of Art; educated at Bradford grammar school; studied at Slade School and Académie Julian,

Paris; executed remarkable series of portrait drawings and lithographs; professor of civil art, Sheffield, 1917-26; principal, Royal College of Art, 1920-35; trustee, Tate Gallery, 1927-33; member, Royal Fine Art Commission, 1931-8; liberal-minded protagonist of the artist in society; published memoirs in three volumes; knighted, 1931.

ROTHERMERE, first VISCOUNT (1868-1940), newspaper proprietor. [See HARMSWORTH, HAROLD SIDNEY.]

ROTHERY, WILLIAM HUME- (1899-1968), first Isaac Wolfson professor of metallurgy in the university of Oxford. [See HUME-ROTHERY.]

ROTHSCHILD, SIR LIONEL WALTER, third baronet and second BARON ROTHSCHILD (1868-1937); son of first Baron Rothschild [q.v.]; succeeded father, 1915; educated at Magdalene College, Cambridge; worked in family bank, 1889-1908; formed at Tring a zoological collection, opened as public museum (1892) and bequeathed to British Museum; FRS, 1911; conservative MP, Aylesbury division, 1899-1910.

ROTHSCHILD, SIR NATHAN MEYER, second baronet, and first BARON ROTHSCHILD (1840-1915), banker and philanthropist; son of Lionel Nathan de Rothschild [q.v.]; liberal MP, Aylesbury, 1865-85; succeeded to baronetcy, 1876; head of firm, 1879; created baron, 1885; first professing Jew to enter House of Lords; PC and GCVO, 1902; his munificent benefactions include gifts to Jews throughout the world, whose acknowledged leader he became.

ROTTER, GODFREY (1879-1969), government scientist and explosives and munitions specialist; educated at City of London School and University College of North Wales, Bangor; completed honours course in chemistry, 1902; research student; joined explosives experimental establishment, Woolwich Arsenal, under O. J. Silberrad [q.v.]; from 1907 worked under (Sir) Robert Robertson [q.v.], assisting in manufacture of Tetryl and 'Amatol'; invented the Rotter Impact Testing Machine and Fuse No. 106; D.Sc., Wales, 1919; succeeded Robertson as director of explosives research, 1921; during inter-war years, closely concerned with important developments in production of more powerful explosives; after retirement in 1942 continued to act as consultant to Ministry of Supply; awarded George Medal for evacuating explosive stores after serious accidents, 1943-4; OBE, 1918; CBE, 1925; CB, 1936; fellow, Royal Institute of Chemistry and Institute of Physics.

ROUND, HENRY JOSEPH (1881-1966), wireless pioneer; educated at Cheltenham grammar school and Royal College of Science, London; ARCS (mechanics), 1901; joined Marconi Company, 1902; worked in United States and South America, 1902-12; returned to England and invented important improved valve and transmission system, 1913-14; seconded to military intelligence; supervised establishment of valved direction-finding stations, instrumental in detecting intentions of German navy, leading to battle of Jutland; MC, 1918; invented valves for transmitter which made first European telephonic crossing of Atlantic, 1919; first wireless telephony world news service inaugurated, 1920; designed transmitter for 2LO, first station to be taken over by BBC; chief of Marconi Research Group, 1921; designed number of improved transmitters and receivers; set up private practice as research consultant, 1931; worked on ASDIC and echo sounding, 1941-62; most prolific inventor for electronics industry; published *The Shielded Four-Electrode Valve* (1927).

ROUND, JOHN HORACE (1854-1928), historian; grandson of Horatio (Horace) Smith [q.v.]; BA, Balliol College, Oxford; first class in modern history; influenced by William Stubbs [q.v.]; devoted himself to historical and genealogical studies; first showed real promise in three essays contributed to *Domesday Studies* (1888, 1891); contributed notes to Pipe Roll Society's volume of *Ancient Charters, Royal and Private, Prior to A.D. 1200* (1888); his *Geoffrey de Mandeville, a Study of the Anarchy* (1892), by opening a fresh line of approach to Anglo-Norman history, placed him among leading historians of his day; his *Feudal England* (1895) includes a convincing description of methods of assessing Danegeld in the eleventh century and demonstrates that tenure by knight service was the creation of William the Conqueror; the chief importance of his *Commune of London* (1899) lies in a study of the origin of the exchequer and in two contributions to early London history; other works include studies in *Peerage and Family History* (1901), *Peerage and Pedigree* (1910), and *The King's Serjeants and Officers of State* (1911); hon. historical adviser to crown in peerage cases, 1914-22; founded modern study of Domesday Book; gave new value to genealogical studies; first modern historian to base a narrative on charters; a fierce controversialist, who attacked E. A. Freeman [q.v.] and other scholars.

ROUSBY, WILLIAM WYBERT (1835-1907), actor and theatrical manager; after provincial engagements from 1849 joined Samuel Phelps [q.v.] at Sadler's Wells Theatre, 1853; played Romeo at Manchester, 1864; with wife

[see ROUSBY, CLARA MARION JESSIE] appeared at Queen's Theatre, London, in leading parts in *The Fool's Revenge, 'Twixt Axe and Crown, As You Like It,* and *King Lear,* 1868-74; proprietor of Theatre Royal, Jersey, from 1879.

ROUSE, WILLIAM HENRY DENHAM (1863-1950), schoolmaster and classical scholar; educated at Haverfordwest grammar school, Doveton College, Calcutta, and Christ's College, Cambridge; first class, parts i and ii, classical tripos, 1885-6; fellow, 1888-94; university teacher of Sanskrit, 1903-39; headmaster, Perse School, 1902-28; taught classics by direct method of speaking only Greek or Latin; joint-editor, *Classical Review* (1907-20) and (from its foundation) Loeb Classical Library; publications include translations from Homer, *Greek Votive Offerings* (1902), *A Greek Boy at Home* (1909), and *Chanties in Greek and Latin* (1922).

ROUTH, EDWARD JOHN (1831-1907), mathematician; born at Quebec; son of Sir Randolph Routh [q.v.]; BA, London, 1849; MA, 1853; joined Peterhouse, Cambridge, 1850; senior wrangler and Smith's prizeman, 1854; fellow, 1855; hon. fellow, 1883; lecturer in mathematics, 1855-1904; had unprecedented success as private coach; produced among his pupils 28 senior wranglers and 43 Smith's prizemen; FRAS, 1866; FRS, 1872; published *Rigid Dynamics* (1860, 7th edn. 2 vols., 1905), *Statics* (2 vols., 1891), and *Dynamics of a Particle* (1898); won Adams prize with 'Treatise on the Stability of a given State of Motion, particularly Steady Motion', which greatly advanced knowledge of dynamics, 1877; hon. LLD, Glasgow, 1878; hon. Sc.D., Cambridge, 1883, and Dublin, 1892.

ROWE, JOSHUA BROOKING (1837-1908), antiquary and naturalist; admitted solicitor, 1860; made study of Devonshire natural history and archaeology; FSA, 1875; part-founder of Devon and Cornwall Record Society; writings include *The Ecclesiastical History of Plymouth* (4 parts, 1873-6), and *The History of Plympton Erle* (1906).

ROWLANDS, SIR ARCHIBALD (1892-1953), civil servant; educated at Penarth county school, University College of Wales (first class, modern languages, 1914), and Jesus College, Oxford; civil servant, War Office, 1920; principal, 1923; assistant secretary, 1936; defence finance adviser, government of India, 1937; deputy under-secretary, Air Ministry, 1939; permanent secretary, Ministry of Aircraft Production, 1940; adviser to viceroy of India, on war administration, 1943; finance member, viceroy's Executive Council, 1945; permanent secretary, Ministry of Supply, 1946-53; KCB, 1941; GCB, 1947.

ROWLANDS, DAVID, 'DEWI MON' (1836-1907), Welsh scholar and poet; BA, London, 1860; held congregational pastorates in Wales, 1861-72; tutor of Brecon Congregational College, 1872-97; principal, 1897-1907; wrote much Welsh and English verse; literary editor of *Cambrian Minstrelsie*, 1893, and of Welsh Congregational Hymnal, 1895; chairman of Congregational Union of Wales, 1902.

ROWLATT, SIR SIDNEY ARTHUR TAYLOR (1862-1945), judge; educated at Fettes and King's College, Cambridge; first class, parts i and ii, classical tripos, 1882-4; fellow, 1886; called to bar (Inner Temple), 1886; bencher, 1908; junior counsel, Board of Inland Revenue, 1900-5; Treasury, 1905-12; judge of King's Bench division, 1912-32; expert in revenue cases; chairman, Indian sedition committee, 1917-18, royal commission on lotteries and betting, 1932-3, general claims tribunal, 1939-45; knighted, 1912; KCSI, 1918; PC, 1932.

ROWLEY, HAROLD HENRY (1890-1969), Old Testament scholar; educated at the Wyggeston school, Leicester, Bristol Baptist College and Bristol University; BD (London), 1913; BA (theology), Bristol, 1913; joined St. Catherine's Society, and studied at Mansfield College, Oxford; B.Litt., 1929; served with YMCA in Egypt, 1916; Baptist missionary teaching Old Testament in Shantung Christian College, China; assistant lecturer in Semitic languages, Cardiff, 1930; professor, Bangor, 1935; professor of Semitic languages and literatures, Manchester, 1945; editor, the *Book List* of the Society for Old Testament Study; joint editor, *Journal of Semitic Studies*, 1956-60; published numerous works including *The Aramaic of the Old Testament* (1929), *Darius the Mede and the Four World Empires* (1935), *From Joseph to Joshua* (1950), *Prophecy and Religion in Ancient China and Israel* (1956), *From Moses to Qumran* (1963), and *The Zadokite Fragments and the Dead Sea Scrolls* (1952); FBA, 1947; president, Society for Old Testament Study, 1950; president, Baptist Union, 1957-8; hon. degrees from many universities and other academic distinctions.

ROWNTREE, BENJAMIN SEEBOHM (1871-1954), sociologist; son of Joseph Rowntree [q.v.]; educated at Bootham School, York, and the Owens College, Manchester; joined family firm of H. I. Rowntree & Co., 1889; director, 1897; chairman, 1923-41; leader in field of scientific management and industrial welfare; director, welfare department, Ministry of Munitions, 1915-18; published *Human Needs of Labour* (1918), *Human Factor in Business* (1921); investigated state of poor in York, 1897-8; published *Poverty, a Study of Town Life* (1901); carried out second York survey, 1936; published *Poverty and Progress* (1941); collaborated with Lord Astor [q.v.] in studies of British agriculture; published *The Agricultural Dilemma* (1935); hon. LLD, Manchester, 1942; CH, 1931.

ROWNTREE, JOSEPH (1836-1925), cocoa manufacturer and philanthropist; son of Joseph Rowntree (1801-59, q.v.); partner with his brother, cocoa manufacturer at York, 1869; sole owner of H. I. Rowntree & Co., 1883; chairman of limited company, 1897-1923; built up welfare organizations for his employees.

ROWTON, BARON (1838-1903), politician and philanthropist. [See CORRY, MONTAGU WILLIAM LOWRY.]

ROXBURGH, JOHN FERGUSSON (1888-1954), first headmaster of Stowe School; educated at Charterhouse and Trinity College, Cambridge; first class, part i, classical tripos, 1910; L. ès L., Paris University; master at Lancing, 1911-22; during 1914-18 war served with Corps of Signals; headmaster, Stowe, 1922-49; published, *The Poetic Procession* (1921) and *Eleutheros* (1930).

ROY, CAMILLE JOSEPH (1870-1943), French-Canadian man of letters; educated at Quebec seminary, Laval University, the Catholic Institute, and Sorbonne, Paris; ordained, 1894; taught in Quebec seminary from 1892; professor of French literature, Laval, 1918; of Canadian literature, 1927-43; four times rector; protonotary apostolic, 1925; by lecturing and writing brought French-Canadian literature to notice of world; fellow (1904), president (1928-9), Royal Society of Canada.

ROYCE, SIR (FREDERICK) HENRY, baronet (1863-1933), engineer; chief electrical engineer for pioneer scheme of street lighting in Liverpool, 1882; founded own electrical firm at Manchester, 1884; produced his first car, 1904; combined with selling agent, C. S. Rolls [q.v.], 1906; concentrated on the 40-50 h.p. 'Silver Ghost' (1906-25), followed by 'Phantoms' and 'Wraiths'; produced 'Eagle' (1915) and other aero engines; company won Schneider trophy, 1929, 1931; made prototype designs of 'Merlin' engine used in 1939-45 war; baronet, 1930.

ROYDEN, (AGNES) MAUDE (1876-1956), preacher; sister of Thomas (later Lord) Royden [q.v.]; educated at Cheltenham Ladies College and Lady Margaret Hall, Oxford; worked at Victoria Women's Settlement, Liverpool, 1900-3; lectured for women's suffrage, 1908-14; edited the *Common Cause*, 1913-14; assistant preacher, City Temple, 1917-20; preached at 'Fellowship Services' at Kensington town hall and later at the Guildhouse, Eccleston Square, 1920-36; undertook preaching tours to the United States, Australia, New Zealand, India, and China; CH, 1930; hon. DD, Glasgow, 1931; hon. LLD, Liverpool, 1935.

ROYDEN, Sir THOMAS, second baronet, and BARON ROYDEN (1871-1950), shipowner; educated at Winchester and Magdalen College, Oxford; entered family shipping firm, 1895; succeeded father as chairman and second baronet, 1917; director, Cunard Steam-Ship Company, 1905-50; deputy chairman, 1909-22; chairman, 1922-30; formulated two-liner project for Atlantic crossing; director, Cunard White-Star, Ltd. (1934-50), and many other companies; conservative MP, Bootle, 1918-22; CH, 1919; baron, 1944.

RUDOLF, EDWARD DE MONTJOIE (1852-1933), philanthropist; served in Office of Works, 1871-90; founded Church of England Waifs and Strays (later Children's) Society, 1881, to take in homeless children and where possible board them out; secretary, 1890-1919; helped Benjamin Waugh [q.v.] to found National Society for the Prevention of Cruelty to Children, 1884; deacon, 1898; priest, 1907; prebendary of St. Paul's Cathedral, 1911-33.

RUFFSIDE, VISCOUNT (1879-1958), Speaker of the House of Commons. [See BROWN, DOUGLAS CLIFTON.]

RUGBY, first BARON (1877-1969), public servant. [See MAFFEY, JOHN LOADER.]

RUGGLES-BRISE, Sir EVELYN JOHN (1857-1935), prison reformer; educated at Eton and Balliol College, Oxford; first class, *lit. hum.*, 1880; entered Home Office, 1881; prison commissioner, 1892; chairman of Prison Commission, 1895-1921; brought into operation reforms in conditions of labour, classification of prisoners, treatment of mentally defective, medical care, development of prisoners' aid societies, and training of prison staffs; devised special treatment for young prisoners which developed into the Borstal system; chairman, International Prison Commission, 1910; KCB, 1902.

RUGGLES GATES, REGINALD (1882-1962), botanist, geneticist, and anthropologist. [See GATES.]

RUMBOLD, Sir HORACE, eighth baronet (1829-1913), diplomatist; born in Calcutta; entered diplomatic service, 1849; secretary to legation in China, 1859; held minor diplomatic posts, 1862-96; ambassador at Vienna, 1896-1900; succeeded brother, 1877; PC, 1896; GCB, 1897.

RUMBOLD, Sir HORACE GEORGE MONTAGU, ninth baronet (1869-1941), diplomatist; son of Sir Horace Rumbold [q.v.]; succeeded father, 1913; educated at Eton; minister at Berne, 1916-19; in Poland, 1919-20; high commissioner and ambassador, Constantinople, 1920-4; deputy to Lord Curzon [q.v.], first Lausanne conference, 1922-3; chief delegate, second conference; signed peace treaty with Turkey, 1923; ambassador, Madrid, 1924-8, Berlin, 1928-33; vice-chairman, Palestine royal commission, 1936; KCMG, 1917; PC, 1921; GCB, 1934.

RUNCIMAN, WALTER, first BARON RUNCIMAN (1847-1937), shipowner; brother of James Runciman [q.v.]; ran away to sea, 1859; obtained master mariner's certificate, 1871; began business as shipowner at South Shields, 1885; built first new steamer (1889), thus founding the Moor Line; acquired controlling interest in Anchor Line, 1935; president, Shipping Federation, 1932-7; liberal MP, Hartlepools division, 1914-18; baronet, 1906; baron, 1933.

RUNCIMAN, WALTER, first VISCOUNT RUNCIMAN OF DOXFORD (1870-1949), statesman; son of Lord Runciman [q.v.]; educated at South Shields high school and Trinity College, Cambridge; entered father's shipping business; liberal MP, Oldham (1899-1900), Dewsbury (1902-18), Swansea West (1924-9), St. Ives (1929-37, from 1931 as liberal national); parliamentary secretary, Local Government Board, 1905-7; financial secretary to Treasury, 1907-8; president, Board of Education, 1908-11; of Agriculture, 1911-14; of Board of Trade, 1914-16, 1931-7; unsuccessfully attempted mediation between Czech government and Sudeten Germans, 1938; lord president of Council, 1938-9; PC, 1908; viscount, 1937.

RUNDALL, FRANCIS HORNBLOW (1823-1908), inspector-general of Indian irrigation; born in Madras; joined Madras engineers, 1841; superintending engineer of the northern circle, 1859; chief engineer in construction of

Orissa canals from 1861; chief irrigation engineer and joint-secretary to the Bengal government, 1867; commenced the Son irrigation canals; inspector-general of irrigation and deputy secretary to Indian government, 1872-4; CSI, 1875; colonel commandant RE, 1876; general, 1885; examined Nile delta, 1876-7; lectured and wrote on Indian irrigation.

RUNDLE, Sir (HENRY MACLEOD) LESLIE (1856-1934), general; gazetted to Royal Artillery, 1876; in Egyptian army (1883-98) served on Sudan frontier and with Lord Kitchener [q.v.] in Nile, Dongola, and Khartoum expeditions; DSO, 1887; adjutant-general, 1892-6; major-general, 1896; KCB, 1898; commanded 8th division, South African Field Force, 1900-2; commanded at battles of Biddulphsberg and Wittebergen; KCMG, 1900; lieutenant-general, 1905; held Northern command, 1905-7; general, 1909; governor of Malta, 1909-15; commander-in-chief, Central Force, 1915-16.

RUSDEN, GEORGE WILLIAM (1819-1903), historian of Australia and New Zealand; emigrated with father from England to Maitland, New South Wales, 1834; clerk to executive council, Victoria, 1852; clerk of parliaments, 1856; retired on pension, 1882, and wrote histories of Australia (3 vols., 1883) and New Zealand (3 vols., 1883).

RUSHBROOKE, JAMES HENRY (1870-1947), Baptist divine; studied at Midland Baptist College and in Europe; minister successively in Derby, Highgate, and Hampstead Garden Suburb; editor, *Goodwill*, 1915-20; Baptist commissioner for Europe, 1920-5; eastern secretary, Baptist World Alliance, 1925-8; general secretary, 1928-39; president, 1939-47.

RUSHBURY, Sir HENRY GEORGE (1889-1968), painter, draughtsman, and engraver; studied at Birmingham School of Art, 1903-9; worked as assistant to Henry Payne, designing stained glass, 1909-12; official war artist during 1914-18 war; first exhibit at Royal Academy, 1913, followed by sequence of over 200 works; illustrated Sidney Dark's *Paris* (1926), Sir James Rennell Rodd's (later Lord Rennell, q.v.) *Rome of the Renaissance and Today* (1932), and Lady Iris Wedgwood's *Fenland Rivers* (1936); did series of mural decorations for Chelmsford Town Hall, 1937; official war artist, 1939-45; keeper of the Royal Academy, 1949-64; ARA, 1927; RA, 1936; hon. associate, Royal Institute of British Architects, 1948; CVO, 1955; CBE, 1960; KCVO, 1964.

RUSHCLIFFE, BARON (1872-1949), politician.[See BETTERTON, HENRY BUCKNALL.]

RUSSELL, ARTHUR OLIVER VILLIERS, second BARON AMPTHILL (1869-1935); succeeded father, first Baron Ampthill [q.v.], 1884; educated at Eton and New College, Oxford; president of university Boat Club and Union, 1891; governor of Madras, 1900-6; acting viceroy, 1904; opposed India bills of 1919 and 1935; pro-grand master of English freemasonry, 1908-35.

RUSSELL, BERTRAND ARTHUR WILLIAM, third EARL RUSSELL (1872-1970), philosopher and social reformer; grandson of first Earl Russell [q.v.], and baron Stanley of Alderley [q.v.]; brought up by his grandmother, widow of first Earl Russell; entered Trinity College, Cambridge, 1890; joined the Apostles; seventh wrangler, part i, mathematical tripos, 1893; first class, moral sciences tripos, 1894; fellow, Trinity College, 1895; published first philosophical book, *An Essay on the Foundations of Geometry* (1897); visited Germany and the United States, lecturing; lectured at Cambridge for J. E. M'Taggart [q.v.], 1899; published *The Principles of Mathematics* (1903); FRS, 1908; lecturer, Trinity College, 1910; love affair with Lady Ottoline Morrell [q.v.], first of many such affairs; published *Principia Mathematica* (3 vols., 1910, 1912, and 1913); president, Aristotelian Society, 1911; close contacts with Ludwig Wittgenstein [q.v.]; published *The Problems of Philosophy* (1912); lectured at Harvard, and met T. S. Eliot [q.v.], 1914; joined No-Conscription Fellowship and removed from Trinity lecturership, 1915; imprisoned for seditious article, 1918, and wrote *Introduction to Mathematical Philosophy* (1919); lectured at London School of Economics (lectures published as *The Analysis of Mind*, 1921); reinstated by Trinity College, 1919; travel in Russia and China led to *The Practice and Theory of Bolshevism* (1920) and *The Problem of China* (1922); resigned lecturership at Trinity College, 1921; earned a living by writing, including *The Analysis of Matter* (1927), and a number of lighter books such as *Marriage and Morals* (1929) and *Religion and Science* (1935); lectured in United States, 1924-31; inherited title from brother, Frank, 1931; held visiting lecturerships in Chicago, Los Angeles, and New York, 1938-42; rejoined Trinity College, Cambridge, 1944; took part in BBC's popular Brains Trust programmes; gave first Reith lectures on radio, 1949, published as *Authority and the Individual*; OM, 1949; Nobel prize for literature, 1950; campaigned against nuclear weapons, 1954; first president of CND, 1958; published *Portraits from Memory* (1956), *My Philosophical Develop-*

ment (1959), and *Autobiography* (3 vols., 1967–9).

RUSSELL, SIR CHARLES, first baronet (1863–1928), solicitor; son of Lord Russell of Killowen [q.v.]; admitted solicitor, 1888; employed by British agent in Bering Sea arbitration (1893) and by Marquess of Queensbury in Oscar Wilde action (1895); solicitor to Canadian government, 1896; baronet, 1916.

RUSSELL, SIR (EDWARD) JOHN (1872–1965), agriculturalist; educated at Technical school, Birmingham, University College, Aberystwyth, and Owens College, Manchester; B.Sc., first class honours, chemistry, 1896; D.Sc. (London), 1901; studied agricultural co-operatives, Copenhagen, 1900; lecturer, Wye Agricultural College, 1901; moved to Rothamsted Experimental Station, 1907; director, 1912–43; built up finances of Rothamsted, 1910–34; initiated Imperial (later Commonwealth) Agricultural Bureaux; after retirement, continued to publish reports, lectures, and other papers; publications include *Soil Conditions and Plant Growth* (1st edn. 1912, 7th edn. 1937), *World Population and Food Supplies* (1954), *A History of Agricultural Science in Great Britain 1620–1954* (1966), and *The Land Called Me* (1956), an autobiography; FRS, 1917; OBE, 1918; knighted, 1922.

RUSSELL, EDWARD STUART (1887–1954), biologist; educated at Greenock Academy and Glasgow University; MA, 1907; carried out research in animal behaviour, 1907–9; appointed to Board of Agriculture and Fisheries, 1909; inspector of fisheries during 1914–18 war; director of fishing investigations for England and Wales, 1921–45; fisheries scientific adviser, 1945–7; chairman, consultative committee, International Council for Exploration of the Sea, 1938–46; published *Form and Function* (1916), *Study of Living Things* (1924), *The Interpretation of Development and Heredity* (1930), *The Behaviour of Animals* (1934), *The Directiveness of Organic Activities* (1945), *The Diversity of Animals* (1962), and *The Overfishing Problem* (1942); president, Linnean Society, 1940–2; OBE, 1930.

RUSSELL, FRANCIS XAVIER JOSEPH (FRANK), BARON RUSSELL OF KILLOWEN (1867–1946), judge; son of Lord Russell of Killowen [q.v.]; educated at Beaumont and Oriel College, Oxford; first class, jurisprudence, 1890; called to bar (Lincoln's Inn), 1893; KC, 1908; judge of Chancery division, 1919–28; lord justice of appeal, 1928–9; lord of appeal in ordinary, 1929–46; PC, 1928; life peer, 1929.

RUSSELL, GEORGE WILLIAM (1867–1935), Irish poet, painter, economist, and journalist, better known by pseudonym 'AE' (AEon); educated at Rathmines School, Dublin; became friend of W. B. Yeats [q.v.] who introduced him to theosophy; abandoned natural vocation of painting; joined Irish Agricultural Organization Society, 1897; edited *Irish Homestead*, 1906–23, and *Irish Statesman*, 1923–30; became a leading figure in Irish literary movement with *Homeward: Songs by the Way* (1894); other works include *Co-operation and Nationality* (1912), *The Candle of Vision* (1918; expounding his religious philosophy), and *The Interpreters* (1922) and *The Avatars* (1933), fictional fantasies expounding his political idealism.

RUSSELL, HENRY CHAMBERLAINE (1836–1907), astronomer; born at West Maitland, New South Wales; BA, Sydney University, 1858; assistant at Sydney observatory, 1859; government astronomer, 1870–1905; established meteorological stations throughout colony; organized Australian observations of transit of Venus, 1874 (account published, 1892); vice-chancellor of Sydney University, 1891; FRS, 1886; CMG, 1890.

RUSSELL, HERBRAND ARTHUR, eleventh DUKE OF BEDFORD (1858–1940); son of ninth Duke of Bedford [q.v.]; educated at Balliol College, Oxford; joined Grenadier Guards, 1879; succeeded brother, 1893; devoted himself to management of his estates; established research stations in forestry and arboriculture; developed his collection of animals at Woburn Abbey; fellow (1872), president (1899–1936), Zoological Society; with Sir P. C. Mitchell [q.v.] created Whipsnade; chairman, Bedfordshire county council, 1895–1928; president, Imperial Cancer Research Fund, 1910–36; supported his wife's hospital work; KG, 1902; FRS, 1908.

RUSSELL, MARY ANNETTE, COUNTESS RUSSELL (1866–1941), writer; known as ELIZABETH; born in Sydney; married (1891) Count Henning August von Arnim (died 1910); his Pomeranian estate the background to *Elizabeth and her German Garden* (1898) and other stories; married (1916) the second Earl Russell.

RUSSELL, MARY DU CAURROY, DUCHESS OF BEDFORD (1865–1937); born TRIBE; married Lord Herbrand Russell [q.v.], 1888; directed model hospital at Woburn, 1903–37; became surgeon's assistant (1917), radiographer, and radiologist; made record flights to India (1929) and South Africa (1930).

RUSSELL, THOMAS O'NEILL (1828–1908), a founder of the Gaelic movement in Ireland; urged in *Irishman* revival of ancient Irish tongue from 1858; commercial traveller in America for nearly thirty years; returned to Ireland, 1895; largely helped to form Gaelic League, 1893, and Feis Ceoil, 1897; published novels and dramas and edited much Gaelic poetry.

RUSSELL, PASHA, SIR THOMAS WENTWORTH (1879–1954), Egyptian civil servant; educated at Cheam, Haileybury and Trinity College, Cambridge; entered Egyptian civil service, 1902; served in every province, directing police activities, 1903–11; assistant-commandant of police, Alexandria, and then Cairo, 1911–17; commandant, Cairo city police with title of Pasha, 1917; after Egyptian independence in 1922, served under twenty-nine different ministers of interior; director, Egyptian Central Narcotics Intelligence Bureau, 1929; vice-president, League of Nations advisory committee on opium traffic, 1939; retired, the last British officer in Egyptian service, 1946; published *Egyptian Service* (1949); OBE, 1920; CMG, 1926; KBE, 1938.

RUSSELL, SIR WALTER WESTLEY (1867–1949), painter and teacher; studied under Frederick Brown [q.v.]; painted portraits, interiors, and landscapes; taught at Slade School, 1895–1927; ARA, 1920; RA, 1926; keeper, 1927–42; instrumental in replacing visitor system by permanent teaching staff; close friend of Henry Tonks and P. Wilson Steer [qq.v.]; link between Academy and opposing forces; knighted, 1935.

RUSSELL, WILLIAM CLARK (1844–1911), novelist; son of Henry Russell [q.v.]; in British merchant service, 1858–66; after engaging in journalism he produced some sixty nautical tales of adventure, chief being *John Holdsworth, Chief Mate* (1875) and *The Wreck of the Grosvenor* (1877); some of his contributions to *Daily Telegraph* on sea topics (1882–9) were republished in *My Watch Below* (1882) and *Round the Galley Fire* (1883); his writings led to improved conditions in merchant service; he also wrote lives of Dampier (1889), Nelson (1890), and Collingwood (1891), and naval ballads.

RUSSELL, SIR WILLIAM HOWARD (1820–1907), war correspondent; educated at Trinity College, Dublin; reported for *The Times* the Irish general election, 1841, and episodes in repeal agitation in Ireland, 1843; called to bar (Middle Temple), 1850; war correspondent for *The Times* in Crimea, 1854; applied the phrase 'the thin red line' to the British infantry at Balaclava; called attention to the sufferings of English army there through the winter of 1854–5, and inspired the work of Florence Nightingale [q.v.]; hon. LLD, Trinity College, Dublin, 1855; acted as *The Times* correspondent in Indian Mutiny, 1858, and in American civil war, 1861–2; wrote a frank account of battle of Bull Run (July 1861), which made him unpopular in America; founded *Army and Navy Gazette*, 1860; only occasionally worked for *The Times* after 1863; saw battle of Königgrätz in Prusso-Austrian war, 1866; also served in Franco-Prussian war, 1870, and (for *Daily Telegraph*) in Zulu war, 1879; accompanied Edward, Prince of Wales, through Near East (1869) and India (1875–6), published accounts of both tours; knighted, 1895; CVO, 1902; published, besides experiences as correspondent, accounts of travels in Canada (3 vols., 1863–5) and the United States (2 vols., 1882).

RUSSELL, WILLIAM JAMES (1830–1909), chemist; Ph.D., Heidelberg, 1855; lecturer in chemistry at St. Mary's Hospital, 1868–70, and at St. Bartholomew's, 1870–97; professor of natural philosophy, Bedford College, 1860–70; FRS, 1872; president of Chemical Society, 1889–91, and of Institute of Chemistry, 1894–7.

RUSSELL FLINT, SIR WILLIAM (1880–1969), artist. [See FLINT.]

RUTHERFORD, ERNEST, BARON RUTHERFORD OF NELSON (1871–1937), physicist; born at Nelson, New Zealand; first class honours, mathematics and physics, Canterbury College, New Zealand, 1893; with an 1851 Exhibition scholarship went to Trinity College, Cambridge, to work under (Sir) J. J. Thomson [q.v.], 1895; investigated conduction of electricity through gases; devised methods for determining velocity and rate of recombination of gaseous ions; proved uranium radiation different in nature from X-rays and consists of at least two types (α and β) of radiation; Macdonald research professor of physics, McGill University, Montreal, 1898; published paper 'Thorium and Uranium Radiation', 1899; became leader of team, gaining as well as giving intellectual stimulus; with Frederick Soddy carried out chemical investigation of radioactive material; in 1902 first put forward the disintegration theory—that radioactivity is a phenomenon accompanying spontaneous transformation of the atoms of radioactive elements into different kinds of matter; published *Radioactivity* (1904); his theory at first received with extreme scepticism but by 1907 was generally accepted; Langworthy professor of physics, Manchester, 1907; embarked on final stages of his research on the 'alpha particle'; by counting the number of α-particles produced in the disintegration of radium deduced the electric charge on each particle; at the same time determined the number of molecules in a unit volume of gas and the elementary charge of the electron; this agreed with the figure deduced by Max Planck in developing his quantum theory; finally proved the α-particle to be helium, 1908; between 1906 and 1914 developed the nuclear theory of the atom; engaged on anti-submarine work, 1915-17; published evidence for the first artificial transmutation of matter, 1919; Cavendish professor of experimental physics, Cambridge, 1919-37; proved that long-range particles from nitrogen are hydrogen nuclei; developed methods of producing fast charged particles by the application of high voltages to vacuum tubes; unsurpassed in influence upon contemporaries he directly inspired many of the spectacular discoveries of the years after 1933; chairman, Advisory Council of Department of Scientific and Industrial Research, 1930-7; FRS, 1903; president, 1925-30; Nobel prize, 1908; knighted, 1914; OM, 1925; baron, 1931; buried in Westminster Abbey.

RUTHERFORD, MARK (pseudonym) (1831-1913), novelist, philosophical writer, literary critic, and civil servant. [See WHITE, WILLIAM HALE.]

RUTHERFORD, WILLIAM GUNION (1853-1907), classical scholar; educated at St. Andrews University and Balliol College, Oxford; BA, 1876; classical master at St. Paul's School, 1876-83; published *First Greek Grammar* (1878), *The New Phrynichus* (1881), and edition of *Babrius* (1883); fellow and tutor of University College, Oxford, 1883; headmaster of Westminster, 1883-1901; hon. LLD, St. Andrews, 1884; edited *Thucydides, Bk. IV* (1889), *Mimiambi* of Herondas (1892), and scholia to Aristophanes (3 vols., 1896-1905); translated St. Paul's Epistle to Romans (1900).

RUTLAND, seventh DUKE OF (1818-1906), politician. [See MANNERS, (LORD) JOHN JAMES ROBERT.]

RUTTLEDGE, HUGH (1884-1961), mountaineer; educated at Cheltenham and Pembroke College, Cambridge; entered Indian civil service, 1908; appointed to United Provinces; deputy commissioner, Himalayan district of Almora, 1925; explored in the Himalayas; retired from ICS, 1932; led expedition to Everest, 1933; doubts expressed about his leadership, but appointed as leader again, 1936; expedition defeated by weather; published

Everest (1933) and *Everest: the Unfinished Adventure* (1937).

RYDE, JOHN WALTER (1898-1961), physicist; educated at St. Paul's School and City and Guilds Technical College, Finsbury; studied under Professor Silvanus P. Thompson [q.v.], 1914; served in Royal Engineers, 1916-18; appointed to staff of Research Laboratories, General Electric Co. Ltd.; chief scientist, 1953-61; developed first successful high-pressure mercury vapour lamp; devised Ryde Night Illumination Diagrams for use by Hydrographic Department of Admiralty; FRS, 1948; fellow, Royal Astronomical Society, and Institute of Physics; chairman, Davy-Faraday Committee, Royal Institution.

RYDER, CHARLES HENRY DUDLEY (1868-1945), geographer; assistant superintendent, Survey of India, 1891; superintendent 1916; surveyor-general, 1919-24; with (Sir) Francis Younghusband [q.v.] in Tibet, 1903-4; returned via Gartok with C. G. Rawling [q.v.], 1904-5; resurveyed North-West Frontier from Baluchistan to Chitral, 1906-13; on Turco-Persian boundary commission, 1913-14.

RYE, MARIA SUSAN (1829-1903), social reformer; sister of Edward Caldwell Rye [q.v.]; founded law-stationer's business for employment of girls, 1859; founded 'Female Middle Class Emigration Society', 1861; from 1868 devoted herself to emigration of 'gutter children'; opened house at Peckham for waifs and strays, 1869; drafted children after training in house at Niagara, 'Our Western Home' (opened Dec. 1869); pioneer of pauper emigration.

RYE, WILLIAM BRENCHLEY (1818-1901), keeper of printed books in the British Museum; assistant in British Museum library, 1839; arranged Grenville library, 1846, and reference library, 1857; assistant keeper of printed books, 1857-69; keeper, 1869-75; chief work was *England as seen by Foreigners* (1865); a skilful etcher.

RYLE, HERBERT EDWARD (1856-1925), successively bishop of Exeter and of Winchester and afterwards dean of Westminster; son of Bishop J. C. Ryle [q.v.]; BA, King's College, Cambridge; fellow, 1881; ordained, 1882; principal of St. David's College, Lampeter, 1886-8; Hulsean professor of divinity, Cambridge, 1887; president of Queens' College, Cambridge, 1896; bishop of Exeter, 1901; of Winchester, 1903; dean of Westminster, 1911-25; his main interest directed to study of Old Testament; an excellent teacher, who strove to commend the methods and more assured results of higher criticism; his works include *The Early Narratives of Genesis* (1892), *The Canon of the Old Testament* (1892), and *Philo and Holy Scripture* (1895).

RYLE, JOHN ALFRED (1889–1950), physician; grandson of J. C. Ryle [q.v.]; educated at Brighton College and Guy's Hospital; medical registrar, 1919; assistant physician, 1920; FRCP, 1924; mainly interested in gastro-intestinal illness; standardized, and invented tube for, fractional test meal; regius professor of physic, Cambridge, 1935–43; professor of social

medicine, and director of Institute of Social Medicine, Oxford, 1943–50.

RYRIE, SIR GRANVILLE DE LAUNE (1865–1937), major-general and high commissioner for Australia; born and educated in New South Wales; manager, Micalago sheep-station; served in South Africa, 1900–1; major, 1900; commanded 2nd Light Horse brigade in Gallipoli, Sinai, and Palestine, 1915–18; Australian division of Light Horse in Syria, 1918; Australian troops in Egypt, 1919; major-general, 1919; MP, North Sydney (1911–22), Warringah (1922–7); assistant minister for defence, 1920–3; high commissioner in London, 1927–32.

S

SACKVILLE-WEST, EDWARD CHARLES, fifth BARON SACKVILLE (1901–1965), man of letters; cousin of Victoria Sackville-West [q.v.]; educated at Eton and Christ Church, Oxford; first novel, *Piano Quintet* (1925), followed by *The Ruin* (1926) with reminiscences of Knole, the family seat, *Simpson* (1931), and *Sun in Capricorn* (1934); a work on De Quincey, *A Flame in Sunlight* (1936), awarded James Tait Black memorial prize; joined Features and Drama department of BBC, 1939; translated Paul Claudel's *L'Otage*, and wrote *The Rescue*, a play, with music by Benjamin Britten; also published literary essays, *Inclinations* (1949) and *Record Guide* (with Desmond Shawe-Taylor, 1951); inherited title, 1965, but Knole had been transferred to the National Trust in 1946.

SACKVILLE-WEST, LIONEL SACKVILLE, second BARON SACKVILLE (1827–1908), diplomatist; entered diplomatic service, 1847; served under Lord Lyons [q.v.] at Paris through Franco-German war; British envoy at Buenos Aires, 1872–8, Madrid, 1878–81, and Washington, 1881–8; life threatened by American Fenians, 1882; KCMG, 1885; helped to settle differences between America and Canada regarding fishing rights, 1887–8; was recalled from Washington owing to alleged intervention in American presidential election, Oct. 1888; succeeded brother in barony, 1888; GCMG, 1889; thenceforth lived in retirement at Knole, Sevenoaks.

SACKVILLE-WEST, VICTORIA MARY (1892–1962), writer and gardener; brought up at Knole, the family seat; educated by governesses; travelled widely in Europe; married (Sir) Harold

Nicolson [q.v.], 1913; prolific writer; published *Knole and the Sackvilles* (1922), *The Land* (a long poem which won the Hawthornden prize, 1927), *The Edwardians* (1930), a novel based on Knole, and essays and travel books, including *Twelve Days* (1928); had passionate love affair with Violet Trefusis, 1918–21; later had close relationship with Virginia Woolf [q.v.], 1925–41; on retirement of Harold Nicolson from foreign service (1929), they bought Sissinghurst Castle, a ruined Elizabethan mansion, restored it, and created a beautiful garden; published other poems, including *Sissinghurst* (1931), novels, and biographies, 1930–9; joined Women's Land Army during 1939–45 war; continued to publish biographies and a last novel *No Signposts in the Sea* (1961); wrote gardening column for the *Observer*, 1946–61; CH, 1948.

SADLEIR, MICHAEL THOMAS HARVEY (1888–1957), writer and publisher; son of (Sir) Michael Ernest Sadler [q.v.]; adopted Sadleir as a *nom de plume*; educated at Rugby and Balliol College, Oxford; joined publishing firm of Constable, 1912; director, 1920; chairman, 1954; served in war trade intelligence department, 1915–18; accomplished book-collector; published *Excursions in Victorian Bibliography* (1922), *Trollope: a Commentary* (1927), and *Trollope: a Bibliography* (1928); published other bibliographical and biographical works including *Bulwer and His Wife, 1803–1836*; published a number of novels, including *Fanny by Gaslight* (1940).

SADLER, SIR MICHAEL ERNEST (1861–1943), educational pioneer and art patron; scholar of Rugby and Trinity College, Oxford; first class, *lit. hum.*, 1884; president of Union,

1882; secretary, extension lectures subcommittee (an independent delegacy from 1892), 1885-95; steward (1886-95), student (1890-95), Christ Church; obtained conference on secondary education, 1893; member, royal commission on secondary education, 1894-5; director of special inquiries and reports, Education Department, 1895-1903; published eleven volumes of information and made office a great research bureau; resigned on ceasing to be consulted on educational policy; professor of education, Manchester, 1903-11; investigated secondary education for nine local authorities and worked also on continuation and technical education; vice-chancellor, Leeds, 1911-23; president, Calcutta University commission (1917-19) whose report set pattern for secondary as well as university education throughout India; master, University College, Oxford, 1923-34; collected old masters and modern paintings, encouraging unknown artists; KCSI, 1919.

SAHA, MEGHNAD (1893-1956), scientist. [See MEGHNAD SAHA.]

ST. ALDWYN, first EARL (1837-1916), statesman. [See HICKS BEACH, SIR MICHAEL EDWARD.]

ST. DAVIDS, first VISCOUNT (1860-1938), financier. [See PHILIPPS, SIR JOHN WYNFORD.]

ST. HELIER, BARON (1843-1905), judge. [See JEUNE, FRANCIS HENRY.]

ST. JOHN, SIR SPENSER BUCKINGHAM (1825-1910), diplomatist and author; son of James Augustus St. John [q.v.]; private secretary to Sir James Brooke [q.v.] in Labuan, 1848-56; temporary commissioner of Labuan, 1851-5; British consul-general at Brunei, 1856; explored interior; chargé d'affaires in Haiti, 1863, and in Dominican republic, 1871; resident minister in Haiti, 1872-4; minister residentiary in Peru, 1874-83; KCMG, 1881; envoy extraordinary to Mexico, 1884-93; minister to Sweden, 1893-6; GCMG, 1894; wrote *Life of Sir James Brooke* (1879), an account of Haiti (1884), and other works.

ST. JOHN, VANE IRETON SHAFTESBURY (1839-1911), author and journalist; brother of Sir S. B. St. John [q.v.]; pioneer of boys' journals.

ST. JUST, first BARON (1870-1941), banker and politician. [See GRENFELL, EDWARD CHARLES.]

SAINTSBURY, GEORGE EDWARD BATEMAN (1845-1933), literary critic and historian; educated at King's College School,

London, and Merton College, Oxford; schoolmaster, 1868-76; assistant editor, *Saturday Review*, 1883-94; regius professor of rhetoric and English literature, Edinburgh, 1895-1915; works include *Short History of French Literature* (1882), *Essays in English Literature* (1890 and 1895), *Miscellaneous Essays* (1892), *Short History of English Literature* (1898), *History of Criticism and Literary Taste in Europe* (3 vols., 1900-4), *History of English Prosody* (3 vols., 1906-10), *History of English Prose Rhythm* (1912), a *History of the French Novel* (2 vols., 1917-19), and editions of many English classics; considered the absorption of the reader the true test of literary greatness; covered with much gusto a wider range than any other English critic; FBA, 1911.

SAKLATVALA, SHAPURJI (1874-1936), politician; son of a Bombay Parsee merchant; worked for firm of his uncle, J. N. Tata [q.v.], in London, 1905-25; assisted in formation of Communist Party of Great Britain; labour (1922-3), communist (1924-9) MP, North Battersea; active agitator in England and India; visited Russia, 1934.

SALADIN (pseudonym) (1844-1906), secularist. [See ROSS, WILLIAM STEWART.]

SALAMAN, CHARLES KENSINGTON (1814-1901), musical composer; studied pianoforte in London and Paris; gave annual orchestral concerts in London from 1833; instituted 'Concerti da Camera', 1835; in Rome, 1846-8; gave musically illustrated lectures in London and provinces from 1855; founded Musical Association, 1876; composed songs (the most famous of which was his setting of Shelley's 'I arise from dreams of thee', 1836), orchestral and pianoforte pieces; published *Jews as they are* (1882).

SALAMAN, JULIA (1812-1906), portrait painter. [See GOODMAN.]

SALAMAN, REDCLIFFE NATHAN (1874-1955), authority on the potato; educated at St. Paul's School and Trinity Hall, Cambridge; first class, part i, natural sciences tripos, 1896; MD, London, 1904; gave up medical work because of tuberculosis and took up study of genetics with particular reference to the potato, 1905; published *Potato Varieties* (1926); director, Potato Virus Research Institute, Cambridge, 1926-39; published *The History and Social Influence of the Potato* (1949); FRS, 1935; hon. fellow, Trinity Hall, 1955.

SALISBURY, fourth MARQUESS OF (1861-1947). [See CECIL, JAMES EDWARD HUBERT GASCOYNE-.]

SALISBURY, third MARQUESS OF (1830-1903), prime minister. [See CECIL, ROBERT ARTHUR TALBOT GASCOYNE-.]

SALISBURY, FRANCIS OWEN (FRANK) (1874-1962), painter; apprenticed at fifteen to his brother's stained-glass works at St. Albans; won scholarship to Royal Academy Schools; Landseer scholarship enabled him to go to Italy, 1896; first exhibited at Royal Academy, 1899; heraldry and pageantry among his interests; painted murals for House of Lords, Royal Exchange, and elsewhere; recorded historical events, including 'The Burial of the Unknown Warrior', and the coronation of George VI; painted number of portraits of celebrities including presidents of the United States and British prime ministers; also painted religious subjects and worked in stained glass; important exhibition held at Royal Institute Galleries, 1953; CVO, 1938; LLD, St. Andrews, 1935; RP, 1917; RI, 1936; master, Worshipful Company of Glaziers; published memoirs, *Portrait and Pageant* (1944).

SALMON, SIR ERIC CECIL HEYGATE (1896-1946), administrator; scholar of Malvern College and Corpus Christi College, Oxford; served in Ministry of Health, 1919-34; dealt chiefly with local finance and legislation and acted as secretary to series of departmental committees; deputy clerk, London County Council, 1934-9; clerk, 1939-46; responsible for co-ordination of civil defence services, working closely with London regional commissioners; knighted, 1943.

SALMON, GEORGE (1819-1904), mathematician and divine; B.A., Trinity College, Dublin (first mathematical moderator), 1838; MA, 1844; BD and DD, 1859; fellow, 1841; divinity lecturer, 1845; Donegal lecturer in mathematics, 1848-66; published *Conic Sections* (1847), *Higher Plane Curves* (1852), *Modern Higher Algebra* (1859), and *Geometry of Three Dimensions* (1862, 5th edn. 1912); regius professor of divinity, 1866-88; a strong protestant and liberal evangelical; acute theological critic; published *Non-Miraculous Christianity* (1881), *Introduction to the New Testament* (1885), *The Infallibility of the Church* (1889), and *The Human Element in the Gospels* (posthumous, 1907), and sermons; provost of Trinity College, 1888-1902; member of Royal Irish Academy, 1843, and of many foreign academies; received many hon. degrees; FRS, 1863; royal medallist, 1868; original FBA, 1902; chancellor of St. Patrick's Cathedral, 1871; he endowed fund and exhibitions for divinity students at Dublin.

SALMOND, SIR JOHN MAITLAND (1881-1968), marshal of the Royal Air Force; educated at Wellington College and Royal Military Academy, Sandhurst; gazetted to King's Own Royal Lancaster Regiment, 1901; seconded to West African Frontier Force, 1903; served in Nigeria, 1903-6; on return to England, learned to fly, 1912; instructor, Central Flying School, Upavon; commanded No. 3 Squadron, RFC, in France, 1914; DSO, 1915; brigadier-general, 1916; director-general of military aeronautics, War Office, 1917; took over command of RFC in France from H. M. Trenchard (later first viscount, q.v.), 1918; commanded Inland Area defences; air officer commanding in Iraq, 1922; in charge of air defence of Great Britain, appointed to Air Council, 1929; succeeded Trenchard as chief of air staff, 1930-3; marshal of the Royal Air Force; on retirement, director, Imperial Airways; chairman, Air Defence Cadet Corps, 1938; director, armament production, Ministry of Aircraft Production, 1939-41; co-ordinated Air/Sea Rescue and Flying Control organizations, 1941-3; president, Royal Air Force Club, 1946; CMG, 1917; CVO, 1918; KCB, 1919; GCB, 1931.

SALMOND, SIR (WILLIAM) GEOFFREY (HANSON) (1878-1933), air chief marshal; commissioned in Royal Artillery, 1898; obtained Royal Aero Club certificate, 1912; graduated at Staff College and major, 1914; commanded 5th Wing, Royal Flying Corps, Egypt, 1915-16; Middle East Brigade, 1916-17; Middle East command, 1918-21; air vice-marshal and KCMG, 1919; air member for supply and research, Air Council, 1922-7; AOC, India, 1927-31; AOC-in-C, Air Defence of Great Britain, 1931-3; chief of Air Staff, 1933; air marshal, 1929; air chief marshal, 1933.

SALOMONS, SIR JULIAN EMANUEL (1835-1909), Australian lawyer and politician; emigrated from Birmingham to Sydney; called to bar at Gray's Inn and of New South Wales, 1861; member of New South Wales legislative assembly, 1869-71, 1887-99; solicitor-general, 1869-70; vice-president of executive council and representative of government in legislative council, 1887-9, 1891-3; agent-general for colony in London, 1899-1900; knighted, 1891.

SALTER, SIR ARTHUR CLAVELL (1859-1928), judge; BA, King's College, London; called to bar (Middle Temple), 1885; joined Western circuit, 1886; KC, 1904; recorder of Poole, 1904-17; conservative MP, Basingstoke division of Hampshire, 1906-17; judge of King's Bench division and knighted, 1917; chairman of railway and canal commission, 1928; his most famous criminal cases those of

Colonel Rutherford (1919) and Horatio Bottomley [q.v.] (1922).

SALTER, HERBERT EDWARD (1863-1951), Oxford historian; educated at Winchester and New College, Oxford; first class, *lit. hum.*, 1886; first class, theology, 1887; ordained, 1888; vicar, Shirburn, South Oxfordshire, 1899-1909; leading authority on Oxford history since Anthony Wood [q.v.]; edited the *Eynsham Cartulary* (1904), *Oseney Cartulary* (1929, 1931, 1934-6), *Oxford Balliol Deeds* (1913), *Oxford City Properties* (1926), *Oriel College Records* (1926), and *Medieval Archives of the University of Oxford* (1917, 1919); published *Medieval Oxford* (1936) and *Early History of St. John's College* (1939); research fellow, Magdalen College, 1919-39; FBA, 1930; hon. freeman, city of Oxford, 1930; hon. D.Litt., Oxford, 1933.

SALTING, GEORGE (1835-1909), art collector and benefactor; born at Sydney, New South Wales; educated at Eton and Sydney University; BA, 1857; settled in England, 1857; developed taste for art on a visit to Rome, 1858; for forty years made collections of art treasures, especially of Chinese porcelain, which he loaned to Victoria and Albert Museum; also collected Italian and Spanish majolica, small sculptures, statuettes, and English miniature portraits from Tudor times onward; spent £40,000 at Spitzer sale in Paris, 1893; left fortune of £1,287,900; bequeathed his collections to National Gallery, British Museum, and Victoria and Albert Museum.

SALVIDGE, SIR ARCHIBALD TUTTON JAMES (1863-1928), political organizer; in service of Brent's Brewery Company, Liverpool; built up Liverpool Working-men's Conservative Association; principal figure in Liverpool conservatism; chairman of National Union of Conservative and Unionist Associations, 1913; knighted, 1916; KBE, 1920; PC, 1922.

SALVIN, FRANCIS HENRY (1817-1904), writer on falconry and cormorant fishing; served in army, 1839-64; went on hawking tour through north of England, 1843; made successful flights with goshawks; revived cormorant fishing in England; collaborated in *Falconry in the British Isles* (1855), and in *Falconry, its Claims, History, and Practice* (1859).

SAMBOURNE, EDWIN LINLEY (1844-1910), artist in black and white; apprenticed as marine engineer at Greenwich; first contributed to *Punch*, 1867; member of staff, 1871; illustrated 'Essence of Parliament'; cartoonist-in-chief, 1900-10; illustrated books, including Kingsley's *Water Babies*, 1885; combined artistic grace and dignity with firmness and delicacy of line.

SAMPSON, GEORGE (1873-1950), scholar; elementary schoolteacher, headmaster, and LCC schools inspector; sought improved teaching of English; published *English for the English* (1921); edited works of Berkeley, Burke, Newman, George Herbert, Emerson, Keats, Bagehot, Coleridge, etc.; prepared one-volume epitome of *Cambridge History of English Literature* (1941).

SAMPSON, JOHN (1862-1931), Romani scholar; brother of R. A. Sampson [q.v.]; first librarian, University College (later the university of), Liverpool, 1892-1928; studied the Welsh Romani dialect, roaming amongst the gipsies and training himself in phonetics, Sanskrit, and comparative philology; works include *The Dialect of the Gypsies of Wales* (1926) and *Welsh Gypsy Tales* (1907-32); produced definitive text of Blake's *Poetical Works* (1905).

SAMPSON, RALPH ALLEN (1866-1939), astronomer; brother of John Sampson [q.v.]; third wrangler, St. John's College, Cambridge, 1888; professor of mathematics, Durham, 1896-1910; astronomer royal for Scotland and professor of astronomy, Edinburgh, 1910-37; chiefly occupied with his *Tables of the Four Great Satellites of Jupiter* (1910) and his 'Theory' of these satellites; FRS, 1903.

SAMSON, CHARLES RUMNEY (1883-1931), air commodore; entered navy, 1898; qualified as a pilot, 1911; made first flight from a ship's deck, 1912; commandant, naval air station, Eastchurch, 1912-14; greatly advanced naval flying; fought with dash and ingenuity in France in aircraft and armoured cars, 1914-15; served in Dardanelles campaign; commanded a seaplane carrier off coasts of Palestine, Syria, and Arabia; commanded anti-submarine and Zeppelin group, Great Yarmouth, 1917-18; group captain, Royal Air Force, 1919; air officer commanding, Mediterranean, 1921-2; air commodore, 1922; chief staff officer, Middle East command, 1926-7; organized and led pioneer bomber formation flight from Cairo to the Cape.

SAMUEL, HERBERT LOUIS, first VISCOUNT SAMUEL (1870-1963), liberal politician, administrator, and philosopher; of Jewish birth; nephew of Montagu Samuel (later Lord Swaythling, q.v.); educated at University College School, London, and Balliol College, Oxford; first class, history, 1893; assisted in social work in East End; liberal MP Cleveland division of Yorkshire, 1902-18; under-secretary of state, Home Office, 1905; PC, 1908; chancellor of Duchy of Lancaster, 1909; postmaster-general, 1910; inadvertently involved in 'Marconi Scandal'; president, Local Government Board, 1914; postmaster-general, 1915; chancellor of Duchy of Lancaster, 1915; home secretary, 1916; in charge of Irish affairs at time of Easter rebellion and arrest of Sir Roger Casement [q.v.]; resigned as home secretary when Lloyd-George supplanted Asquith, 1916; defeated in 'coupon' election, 1918; first high commissioner, Palestine, 1920–5; pursued policy of creating multi-national commonwealth; chairman, Royal Commission on coal-mining industry, 1925–6; negotiated with Trades Union Congress for settlement of general strike, 1926; KGCB, 1926; liberal MP Darwin, 1929–35; as acting liberal leader, advocated 'national' government, 1931; home secretary, 1931–2; viscount, 1937; leader of liberals in House of Lords; approved Munich agreement, 1938; retired from politics and concentrated on philosophical studies, 1955; travelled widely and took part in BBC Brains Trust programmes; hon. degrees from number of universities; president of many learned societies; visitor of Balliol College, Oxford, 1949; OM, 1958.

SAMUEL, MARCUS, first Viscount Bearsted (1853–1927), joint-founder of the Shell Transport and Trading Company; took up business of shipping oil from Russia to Far East; Shell Transport and Trading Company formed, 1897; British and Dutch oil interests in Borneo amalgamated, 1907; greatly assisted Great Britain during 1914–18 war; knighted, 1898; lord mayor of London, 1902–3; baronet, 1903; baron, 1921; viscount, 1925.

SAMUELSON, Sir BERNHARD, first baronet (1820–1905), ironmaster and promoter of technical education; manager of Manchester firm of engineers, 1842–6; established railway works at Tours, 1846; purchased (1848) and greatly developed factory of agricultural implements at Banbury; erected blast-furnaces near Middlesbrough for ironworking, 1853, which he transferred to Newport, 1863; both firms were turned into limited liability companies, 1887; built Britannia ironworks at Middlesbrough, which produced gigantic output of iron, tar, and by-products; unsuccessful in attempts to make steel from Cleveland ore; liberal MP for Banbury, 1859 and 1865–85, and for North Oxfordshire, 1885–95; PC, 1895; pioneer of tariff reform movement, 1901; served on royal commissions on scientific instruction (1870), on technical instruction (chairman, 1881), and elementary education (1887); chairman of parliamentary committees on patent

laws (1871–2) and railways (1873); FRS, 1881; part-founder and president (1883–5) of Iron and Steel Institute; presented technical institute to Banbury, 1884; baronet, 1884.

SANDAY, WILLIAM (1843–1920), theological scholar; BA, Corpus Christi College, Oxford; Dean Ireland's professor of exegesis of Holy Scripture, Oxford, 1882; fellow and tutor of Exeter, 1883; Lady Margaret professor of divinity and canon of Christ Church, 1895–1919; FBA, 1903; devoted himself to scientific study of New Testament, especially the Gospels; works include *Portions of the Gospels according to St. Mark and St. Matthew from the Bobbio MS.* (1886), and preliminary studies for a Life of Christ (never written), most important being *The Life of Christ in Recent Research* (1907).

SANDBERG, SAMUEL LOUIS GRAHAM (1851–1905), Tibetan scholar; son of Indian missionary; BA, Dublin University, 1870; called to bar, 1874; ordained, 1879; chaplain in Bengal, 1886–1904; published works and magazine articles on Tibetan dialects, literature, and topography; chief were *Manual of Colloquial Tibetan* (1894), *A Tibetan-English Dictionary* (1902), and *The Exploration of Tibet* (1904).

SANDERSON, Baron (1868–1939), educationist. [See Furniss, Henry Sanderson.]

SANDERSON, EDGAR (1838–1907), historical writer; BA, Clare College, Cambridge, 1860; MA, 1865; headmaster of grammar schools at Stockwell, 1871–3, Macclesfield, 1873–7, and Huntingdon, 1877–81; works include *History of the World* (1898) and *The British Empire in the Nineteenth Century* (6 vols., 1898–9).

SANDERSON, FREDERICK WILLIAM (1857–1922), schoolmaster; BA, Durham University and Christ's College, Cambridge; assistant master, Dulwich College, 1885; successful in forming engineering side of school; headmaster of Oundle School, 1892–1922; appointed with definite object of reorganizing teaching, introducing fresh subjects of study, and raising numbers and status of school; succeeded in all these objects; established science and engineering sides; new laboratories and workshops built and co-operative method introduced for engineering and other subjects; publicly advocated his views on education.

SANDERSON, Sir JOHN SCOTT BURDON-, baronet (1828–1905), regius professor of medicine at Oxford. [See Burdon-Sanderson.]

SANDERSON, THOMAS HENRY, Baron Sanderson (1841-1923), civil servant; junior clerk in Foreign Office, 1859; private secretary to Lord Derby, 1866-8, 1874-8, and to Lord Granville [qq.v.], 1880-5; senior clerk, 1885; assistant under-secretary, 1889; permanent under-secretary, 1894-1906; KCMG, 1887; KCB, 1893; GCB, 1900; baron, 1905; an efficient administrator of the old school and gifted composer of official dispatches.

SANDERSON, THOMAS JAMES COBDEN- (1840-1922), bookbinder and printer. [See COBDEN-SANDERSON.]

SANDHAM, HENRY (1842-1910), painter and illustrator; born in Montreal; travelled in Europe and settled in Boston, 1880; member of Royal Canadian Academy, 1880; painted battle and historical scenes and portraits; chief works were 'The March of Time' and 'The Dawn of Liberty'; a good water-colour artist.

SANDS, Lord (1857-1934), Scottish judge. [See JOHNSTON, CHRISTOPHER NICHOLSON.]

SANDYS, (ANTHONY) FREDERICK (AUGUSTUS) (1829-1904), pre-Raphaelite painter; son of drawing-master at Norwich; published anonymously 'A Nightmare', a lithographic caricature of Millais's 'Sir Isumbras at the Ford', 1857; formed friendship with D. G. Rossetti [q.v.] (1857) and pre-Raphaelite group, including J. A. M. Whistler, and later with George Meredith [qq.v.]; devoted much time to wood block designs, which were praised by Millais and Rossetti; exhibited at Royal Academy subject pictures, including 'Oriana', 'La Belle Ysonde', 'Morgan le Fay', 'Cassandra' (praised by Swinburne), 'Medea', and portraits; gained repute for crayon heads; of bohemian temperament and habits; work mainly in English private collections or in America.

SANDYS, Sir JOHN EDWIN (1844-1922), classical scholar; BA (senior classic), St. John's College, Cambridge, 1867; fellow; 1867; lecturer, 1867-1907; public orator, 1876-1919; presented nearly 700 distinguished men for honorary degrees with speeches which, of their kind, were almost perfect; FBA, 1909; knighted, 1911; the foremost editor of Greek orators; a great humanist, whose spiritual home was Greece; his chief work, *History of Classical Scholarship* (1903-8).

SANDFORD, GEORGE EDWARD LANGHAM SOMERSET (1840-1901), lieutenant-general; joined Royal Engineers, 1856; served in China with General Gordon [q.v.], 1862; executive engineer in Indian public works department, 1873; served in Afridi expedition, 1878, and Afghan war, 1878-9; quatermaster-general in India, 1882-3; commanding royal engineer in Burmese expedition, 1885-6; CB, 1886; director-general of military works in India, 1886-93; CSI, 1890; commanded Meerut district, 1893-8; lieutenant-general, 1898.

SANGER, GEORGE, known as 'Lord George Sanger' (1825-1911), circus proprietor and showman; brother of John Sanger [q.v.]; started with brother independent show at Stepney Fair, 1848; inaugurated travelling show, 1853; established 'world's fair' at Plymouth, 1860; purchased Astley's amphitheatre and exhibited spectacles there (1871-93) and also at Agricultural Hall and at Margate; frequently toured the continent; in later life was hampered by American competition; was shot by an employee at Finchley; published autobiography *Seventy Years a Showman* (1910).

SANKARAN NAIR, Sir CHETTUR (1857-1934), Indian jurist, administrator, and politician; practised at Madras high court bar; judge, 1907-15; member of viceroy's executive council (1915-19), of India Council (1919-21); chairman, Indian central committee to co-operate with Simon statutory commission, 1928; unable to reconcile its discordant elements; knighted, 1912.

SANKEY, JOHN, Viscount Sankey (1866-1948), lord chancellor; educated at Lancing and Jesus College, Oxford; BCL, 1891; called to bar (Middle Temple), 1892; practised in South Wales; KC, 1909; judge of King's Bench division, 1914-28; chairman, enemy aliens advisory committee, 1915; of coal industry commission which recommended nationalization, 1919; lord justice of appeal, 1928-9; lord chancellor, 1929-35; chairman, inter-imperial relations committee of imperial conference, and of federal structure committee of Indian Round Table conference; appointed Law Revision Committee; knighted, 1914; GBE, 1917; PC, 1928; baron, 1929; viscount, 1932.

SANKEY, Sir RICHARD HIERAM (1829-1908), lieutenant-general Royal (Madras) Engineers; joined Madras engineers, 1846; under-secretary of public works department at Calcutta, 1857; conspicuous in field-work in Indian Mutiny at Allahabad, Cawnpore, and Lucknow; field engineer to the Gurkha force; chief engineer, Mysore, 1864-77; originated irrigation department and built new roads and government offices; commanding royal engineer of Kandahar field force, 1878; CB, 1879; secretary in public works department, Madras,

1879–83; major-general, 1883; lieutenant-general, 1884; KCB, 1892.

SANSOM, SIR GEORGE BAILEY (1883–1965), diplomatist and scholar; educated at a lycée at Caen, France, and in Germany; entered British consular service, 1903; posted to Japan; spent greater part of official career in Tokyo; head of commercial department of embassy, 1923–40; acquired profound knowledge of Japanese history and culture; published *An Historical Grammar of Japanese* (1928) and *Japan: a Short Cultural History* (1931); adviser to economic mission in Singapore, 1941; minister in Washington, 1942–7; professor, Japanese studies, Columbia University, 1947–9; director, East Asian Institite, 1949–55; hon. professor, Stanford University, California, 1955–65; published *The Western World and Japan* (1950) and *History of Japan* (3 vols., 1958–61–64); hon. fellow, Japanese Academy, 1934; hon. degrees, Columbia and Leeds; CMG, 1926; KCMG, 1935; GBE, 1947.

SANTLEY, SIR CHARLES (1834–1922), singer; employed in commerce in Liverpool, devoting spare time to music; voice early developed into fine baritone; professional singer, 1855; studied at Milan; made début in opera at Pavia, 1857; returned to England, 1857; sang in English and Italian opera in London, Manchester, and elsewhere in England, Dublin, Milan, Barcelona, and America; one of his most notable appearances was Valentine in first performance of *Faust* in England, 1863; prominent figure at Birmingham festivals from 1861; joined Roman Church, 1880; knighted, 1907.

SAPPER (pseudonym) (1888–1937). [See McNEILE, (HERMAN) CYRIL.]

SARGANT, SIR CHARLES HENRY (1856–1942), judge; educated at Rugby and New College, Oxford; first class, *lit. hum.*, 1879; called to bar (Lincoln's Inn), 1882; bencher, 1908; junior counsel to Treasury (equity), 1908–13; judge of Chancery division, 1913–23; chairman, royal commission on awards to inventors, 1919–23; lord justice of appeal, 1923–8; knighted, 1913; PC, 1923.

SARGEAUNT, JOHN (1857–1922), teacher and scholar; BA, University College, Oxford; master of classical sixth form, Felsted School, 1885; of classical sixth form (second highest division), Westminster School, 1890–1918; his exceptional capacity as a teacher founded upon the versatility of his learning and breadth of his tastes; his works include *Annals of Westminster School* (1898).

SARGENT, SIR (HAROLD) ORME GARTON (1884–1962), diplomatist; educated at Radley and in Switzerland; entered diplomatic service, 1906; served in Berne, 1917; and Paris, 1919–25; CMG, 1925; transferred to Foreign Office; counsellor, 1926; head of Central Department; assistant under-secretary, 1933; deputy under-secretary, 1939; permanent under-secretary, 1946–9; CB, 1936; KCMG, 1937; KCB, 1947; GCMG, 1948; opposed appeasement of Nazi Germany, but was ineffective in making his views acceptable to politicians; 'a philosopher strayed into Whitehall'.

SARGENT, SIR (HENRY) MALCOLM (WATTS) (1895–1967), conductor, composer, pianist, and organist; educated at Stamford School; articled to Haydn Keeton, organist of Peterborough Cathedral, 1912; organist, Melton Mowbray, 1914; pianist, composer, and conductor at Leicester concerts; invited by Sir Henry J. Wood [q.v.] to conduct in 1921 Promenade Concert season; joined staff of Royal College of Music, 1923; conducted for a number of organizations including the Diaghilev Ballet (1927), the Royal Choral Society (1928), and the D'Oyly Carte Opera Company (1930); conducted first performances of works by Vaughan Williams [q.v.] and (Sir) William Walton; succeeded Sir Adrian Boult as conductor, BBC Symphony Orchestra, 1950; appeared in Brains Trust programmes; gifts of personality and showmanship demonstrated in Promenade Concerts; awarded many musical honours, including hon. D.Mus. (Oxford), 1942; knighted, 1947.

SARGENT, JOHN SINGER (1856–1925), painter; born in Florence of American parentage; began his serious artistic training on entering the studio of Carolus Duran in Paris, 1874; sent his first contribution to Salon, 1877; established for himself recognized position among younger artists of Paris as painter of portraits and subject pictures; his portrait of Madame Gautreau, exhibited at Salon of 1884, achieved *succès de scandale*; settled in London (Tite Street, Chelsea), 1885; his first considerable public success in England was his study of childhood, 'Carnation, Lily, Lily, Rose' (1885–6); his work as a portrait painter eventually gained for him an international prestige almost unparalleled; RA, 1897; his outstanding successes of period 1888–1914 include Miss Ellen Terry as Lady Macbeth (1889), Lady Agnew of Lochaw (1893), Coventry Patmore (1894), W. Graham Robertson (1894), (General Sir) Ian Hamilton (1898), Lord Ribblesdale (1902), Henry James (1913); co-operated with other artists in decorating special libraries floor of Boston (USA) public library, 1890–1916; his later pictures include 'Gassed' (1918) and 'Some General Officers of the Great War' (1922); always retained his American citizenship.

SARKAR, Sir JADUNATH (1870-1958), Indian historian; educated at Rajshahi and Calcutta collegiate schools and Presidency College, Calcutta; first class, English, Calcutta University, MA, 1892; entered provincial education service, 1898; Indian educational service, 1918; vice-chancellor, Calcutta University, 1926-8; nominated member, Bengal legislative council, 1929-32; founder-member of Indian Historical Records Commission, 1919; published *History of Anrangzib* (5 vols., 1912-24), *Fall of the Mughal Empire* (4 vols., 1932-50), *Shivaji and his Times* (1919), and *House of Shivaji* (1940); CIE, 1926; knighted, 1929.

SASSOON, Sir PHILIP ALBERT GUSTAVE DAVID, third baronet, of Kensington Gore (1888-1939), politician and connoisseur; grandson of Sir A. A. D. Sassoon [q.v.]; educated at Eton and Christ Church, Oxford; succeeded father, 1912; unionist MP, Hythe, 1912-39; private secretary to (Lord) Haig [q.v.], 1915-18; under-secretary for air, 1924-9, 1931-7; aroused public interest in and promoted welfare of Royal Air Force; first commissioner of works, 1937-9; his houses renowned as centres of art and entertainment; PC, 1929.

SASSOON, SIEGFRIED LORAINE (1886-1967), poet and prose-writer; educated at Marlborough College and Clare College, Cambridge; lived as country gentleman, hunting and writing poems, 1906-12; commissioned in Royal Welch Fusiliers, 1915; MC; unsuccessfully recommended for VC; wounded, 1917; critical of conduct of the war; returned to France and wounded again, 1918; published war poems *The Old Huntsman* (1917) and *Counter-Attack* (1918); literary editor, the *Daily Herald,* 1919; brought out further volumes of poetry, *Selected Poems* (1925), *Satirical Poems* (1926), and *The Heart's Journey* (1927); published *Memoirs of a Fox-Hunting Man* (1928) (awarded Hawthornden and James Tait Black memorial prizes), *Memoirs of an Infantry Officer* (1930), and *Sherston's Progress* (1936); wrote autobiographical, *The Old Century and Seven More Years* (1938), *The Weald of Youth* (1942), and *Siegfried's Journey* (1945); CBE, 1951; hon. D.Litt., Oxford, 1965.

SASTRI, VALANGIMAN SANKARANARAYANA SRINIVASA (1869-1946), Indian social worker and politician; joined Servants of India Society, 1907; president,

1915-27; elected to viceroy's legislative council, 1916; member, Council of State, 1921; a representative of India at imperial conference, 1921; obtained admission in principle (South Africa dissenting) of Indian claims to equality of citizenship within Empire; visited Australia, New Zealand, and Canada, 1922; obtained agreement with South Africa, 1927; first 'agent' of Indian government in South Africa, 1927-9; visited Malaya and East Africa, 1929; attended Round Table conferences, 1930-1; vice-chancellor, Annamalai University, 1935-40; PC, 1921; CH, 1930.

SATOW, SIR ERNEST MASON (1843-1929), diplomatist and historian; educated at Mill Hill School; BA, University College, London; entered Japanese consular service as student interpreter, 1861; Japanese secretary to British legation, 1868; consul-general, Bangkok, 1884; minister, 1885; minister, Montevideo, 1888; envoy extraordinary and minister plenipotentiary, Morocco, 1893; minister plenipotentiary, Tokyo, 1895-1900; minister, Peking, 1900-6; credit of agreement between China and Powers signed at Peking in 1901 due to him; British member of court of arbitration at The Hague, 1906-12; second British delegate to second Hague peace conference, 1907; KCMG, 1895; GCMG, 1902; PC, 1906; his writings on international law and history include *A Guide to Diplomatic Practice* (1917); accomplished in both theory and practice of diplomacy and conversant with languages and cultures of China and Japan.

SAUMAREZ, THOMAS (1827-1903), admiral: entered navy, 1841; served in South America and West Africa; commander, 1854; led in attack on Taku forts, 1858; CB, 1873; admiral, 1886.

SAUNDERS, SIR ALEXANDER MORRIS CARR- (1886-1966), biologist, sociologist, and academic administrator. [See CARR-SAUNDERS.]

SAUNDERS, EDWARD (1848-1910), entomologist; published *The Hemiptera Heteroptera* (1892), *The Hymenoptera Aculeata* (1896) of the British Isles, and *Wild Bees, Wasps and Ants* (1907); FRS, 1902.

SAUNDERS, SIR EDWIN (1814-1901), dentist; dental surgeon and lecturer in dental surgery at St. Thomas's Hospital, 1839-54; FRCS, 1855; dentist to royal family from 1846; obtained foundation of diploma in dental surgery, 1859; founder (1857) and president (1864 and 1879) of Odontological Society; knighted, 1883; president of British Dental

Association, 1886; author of *Advice on the Care of the Teeth* (1837).

SAUNDERS, HOWARD (1835-1907), ornithologist and traveller; went to South America, 1855; crossed Andes and explored the Amazon river, 1860-2; wrote papers on South American, Spanish, and Swiss ornithology; joint-editor of *Ibis*, 1883-8 and 1894-1900; published *An Illustrated Manual of British Birds* (1889) and edited *Yarrell's British Birds* (1882-5) and other works.

SAUNDERSON, EDWARD JAMES (1837-1906), Irish politician, well known as Colonel Saunderson (militia batt., Royal Irish Fusiliers); lived at Nice during boyhood; returned to Ireland, 1858; whig MP for county Cavan, 1865-74; opposed Irish disestablishment, 1869; opposed nationalist movement from 1882; conservative MP for North Armagh from 1885 till death; active opponent of home rule bill of 1893; grand master of Orange lodges at Belfast, 1901-2; PC, 1898; lord-lieutenant of Cavan, 1900; supported South African war; capable artist and boat-builder; ardent protestant.

SAVAGE, ETHEL MARY (1881-1939), better known as Miss ETHEL M. DELL, novelist; married Lieutenant-Colonel G. T. Savage, 1922; enormously successful with *The Way of an Eagle* (1912) and 34 subsequent novels which provided an uncritical public with wholesome well-sustained romantic narratives certain of a happy ending.

SAVAGE-ARMSTRONG, GEORGE FRANCIS (1845-1906), poet; brother of Edmund John Armstrong [q.v.]; BA, Trinity College, Dublin, 1869; hon. MA, 1872; published much verse, including *Poems Lyrical and Dramatic* (1869) and *Stories of Wicklow* (1886); professor of history and English literature, Queen's College, Cork, 1870-1905; hon. D.Litt., Queen's University, 1891.

SAVILL, THOMAS DIXON (1855-1910), physician; studied medicine at St. Thomas's and St. Mary's hospitals and in Europe; MB, 1881; MD, 1882; FRCP, 1882; medical superintendent of Paddington infirmary, 1885-92; physician of West End Hospital for Diseases of the Nervous System, 1893; made original researches into nervous diseases and hysteria; published *A System of Clinical Medicine* (2 vols., 1903-5); died at Algiers.

SAXE-WEIMAR, PRINCE EDWARD OF (1823-1902), field-marshal. [See EDWARD OF SAXE-WEIMAR.]

SAXL, FRIEDRICH ('FRITZ') (1890-1948), historian of art and, with Aby M. Warburg, founder of Warburg Institute; born in Vienna; studied at Vienna and Berlin universities and Austrian Historical Institute, Rome; compiled *Catalogue of Astrological and Mythological Illuminated Manuscripts of the Latin Middle Ages* (3 vols., 1915-53); developed Warburg's library into Institute for research on classical heritage; director, 1929-48; moved it to London, 1933; naturalized, 1940; FBA, 1944.

SAYCE, ARCHIBALD HENRY (1845-1933), orientalist and comparative philologist; educated at Grosvenor College, Bath, and Queen's College, Oxford; first class, *lit hum.*, 1868; fellow, 1869; tutor and ordained, 1870; professor of Assyriology, 1891-1915; travelled widely; instrumental in securing for the British Museum Aristotle's *Constitution of Athens* and Herondas' *Mimes*; laid foundations of Sumerian grammar with 'An Accadian Seal' (1870); translated very difficult astronomical tablets from Nineveh, 1874; deciphered cuneiform inscriptions of Van, 1882; also active in Hittite, Elamite, and especially Assyrian studies; published *Assyrian Grammar* (1872), *Principles of Comparative Philology* (1874-5), and *Early History of the Hebrews* (1897); showed importance of oriental archaeology for understanding the Bible.

SAYERS, DOROTHY LEIGH (1893-1957), writer; educated at Godolphin School, Salisbury, and Somerville College, Oxford; first class, modern languages, 1915; advertiser's copywriter with S. H. Benson, Ltd., 1916-31; introduced private detective Lord Peter Wimsey in *Whose Body?* (1923); followed by many other mystery stories from *Clouds of Witness* (1926) and *Strong Poison* (1930) to *The Nine Tailors* (1934) and *Gaudy Night* (1935); also wrote *The Zeal of Thy House* (1937) and *The Devil to Pay* (1939), plays for the Canterbury Festival; and the radio drama, *The Man Born to be King* (broadcast between December, 1941, and October, 1942); published translations of Dante's *Inferno* (1949) and *Purgatorio* (1955); hon. D.Litt., Durham, 1950.

SCARBOROUGH, eleventh EARL OF (1896-1969), public servant. [See LUMLEY, LAWRENCE ROGER.]

SCHAFER, SIR EDWARD ALBERT SHARPEY- (1850-1935), physiologist; educated at University College School and University College, London; qualified in medicine; first Sharpey scholar, 1871; assistant professor of physiology, 1874-83; Jodrell professor, 1883-99; professor of physiology, Edinburgh, 1899-

1933; researches included histology, cerebral localization, endocrinology, pulmonary circulation, and nerve function; his prone-pressure method of artificial respiration adopted by Royal Life Saving Society; founder (1908) and until 1933 editor of *Quarterly Journal of Experimental Physiology*; works include *The Essentials of Histology* (1885), *Experimental Physiology* (1912), and *The Endocrine Organs* (1916); FRS, 1878; president, British Association, 1912; knighted, 1913; added 'Sharpey' to his name (1918) to mark his indebtedness to William Sharpey [q.v.].

SCHARLIEB, DAME MARY ANN DACOMB (1845–1930), gynaecological surgeon; born BIRD; married W. M. Scharlieb, barrister, 1865; accompanied him to Madras, 1866; qualified, Madras medical school, 1877; entered London School of Medicine for Women, 1878; MB, 1882; instrumental in establishing Royal Victoria Hospital for Caste and Gosha Women, Madras, *c.*1884–6; returned to London, 1887; MD, London University, 1888; lecturer on diseases of women, School of Medicine for Women, 1889; senior surgeon, New Hospital for Women, 1892–1903; gynaecologist, Royal Free Hospital, 1902; DBE, 1926.

SCHILLER, FERDINAND CANNING SCOTT (1864–1937), philosopher; educated at Rugby and Balliol College, Oxford; first class, *lit. hum.*, 1886; instructor in philosophy, Cornell University, 1893; tutorial fellow, Corpus Christi College, Oxford, 1897–1926; professor, university of Southern California, 1929–36; pragmatist, insisting upon importance of action which he considered to be the supreme test of truth; FBA, 1926.

SCHLICH, SIR WILLIAM (1840–1925), forester; born in Germany; entered Indian forest department, 1866; conservator of Bengal forests, 1872–9; acting inspector of forests to government of India, 1881; inspector, 1883–5; professor of forestry, Royal Indian Engineering College, Cooper's Hill, 1885; transferred with forestry branch to Oxford, 1905; built up school of forestry in university; reader, with status of professor, 1911; naturalized, 1886; FRS, 1901; KCIE, 1909; his works include *Manual of Forestry* (1889–96).

SCHOLES, PERCY ALFRED (1877–1958), musical writer and encyclopedist; ill health limited his schooling, but he studied music assiduously; gained B.Mus., Oxford, from St. Edmund Hall, 1908; editor, *The Music Student* (later *The Music Teacher*), 1908–21; music critic, the *Evening Standard*, 1913–20; music critic, the *Observer*, 1920–5; musical editor,

Radio Times, 1926–8; published a number of books, including a biography of Dr Charles Burney (2 vols., 1948, James Tait Black memorial prize), *Concise Oxford Dictionary of Music* (1952), and *Oxford Junior Companion to Music* (1954); degrees from Oxford included D.Mus. (1943), MA (1944), and D.Litt. (1950); hon. fellow, St. Edmund Hall; FSA, 1938; OBE, 1957.

SCHREINER, OLIVE EMILIE ALBERTINA (1855–1920), authoress; sister of W. P. Schreiner [q.v.]; married S. C. Cronwright, 1894; best-known work *The Story of an African Farm* (1883).

SCHREINER, WILLIAM PHILIP (1857–1919), South African lawyer and statesman; born in Cape Colony; BA, Downing College, Cambridge (senior jurist); called to bar (Inner Temple) and Cape bar, 1882; QC, 1892; prime minister of Cape Colony, 1898–1900; high commissioner for Union of South Africa in London, 1914–19.

SCHUNCK, HENRY EDWARD (1820–1903), chemist; studied chemistry at Berlin and Giessen University; Ph.D.; made exhaustive original researches into colouring matters of vegetable substances, including indigo and chlorophyll, as well as the madder plant; published papers 'On Rubian and its Products of Decomposition' from 1851 to 1855; first showed chemical nature of alizarin [see PERKIN, SIR WILLIAM HENRY]; FRS, 1850; Davy gold medallist, 1889; president of Society of Chemical Industry, 1896–7; hon. D.Sc., Manchester, 1899; presented in 1895 £20,000 to Owens College, Manchester, for chemical research.

SCHUSTER, SIR ARTHUR (1851–1934), mathematical physicist; brother of Sir F. O. Schuster [q.v.]; educated at Frankfort Gymnasium, Geneva, and Owens College, Manchester; Ph.D., Heidelberg, 1873; naturalized, 1875; worked at Cavendish laboratory, Cambridge, 1876–81; professor of applied mathematics, Manchester, 1881–8; of physics, 1888–1907; worked mainly in field of spectroscopy (obtaining first photograph of spectrum of solar corona, 1882), electricity in gases, terrestrial magnetism, optics, and the mathematical theory of periodicity; fellow (1879), secretary (1912–19), foreign secretary (1920–4), vice-president (1919–24), Royal Society; president, British Association, 1915; secretary, International Research Council, 1919–28; knighted, 1920.

SCHUSTER, CLAUD, BARON SCHUSTER (1869–1956), civil servant; educated at Winchester and New College, Oxford; called to bar

(Inner Temple), 1895; secretary, London Government Act Commission, 1899-1902; worked with (Sir) Robert Morant [q.v.] at Board of Education, 1903-11, and in national insurance, 1911-15; clerk of the Crown in Chancery and permanent secretary, lord chancellor's office, 1915-44; served under ten lord chancellors; head of legal branch, Allied Control Commission (British zone) in Austria, 1944-6; president, the Ski Club of Great Britain, 1932-4, and the Alpine Club, 1938-40; published *Peaks and Pleasant Pastures* (1911) and *Postscript to Adventure* (1950); knighted, 1913; CVO, 1918; KC, 1919; KCB, 1920; GCB, 1927; raised to peerage, 1944; hon. fellow, St. Catharine's College, Cambridge.

SCHUSTER, SIR FELIX OTTO, first baronet (1854-1936), banker; brother of Sir Arthur Schuster [q.v.]; educated at Frankfort Gymnasium, Geneva, and Owens College, Manchester; naturalized, 1875; partner in family merchant banking business, 1879; director, Union Bank of London, 1887; governor, 1895-1918; convinced freetrader; baronet, 1906; carried through amalgamations with Smith, Payne, and Smiths, London and Yorkshire, Prescott's, and in 1919 with National Provincial Bank of England; finance member, council of India, 1906-16.

SCHWABE, RANDOLPH (1885-1948), etcher, draughtsman, and teacher; studied at Slade School, 1900-6; principal, 1930-48; his drawings and prints beautifully precise and reasonable statements of fact.

SCOTT, ARCHIBALD (1837-1909), Scottish divine and leader of the General Assembly of the Church of Scotland; BA, Glasgow University, 1856; hon. DD, 1876; held ministries in Glasgow, Linlithgow, and Edinburgh; chairman of Edinburgh school board, 1878-82; incumbent of St. George's, Edinburgh, from 1890 to death; leader of General Assembly from 1887; moderator, 1896; advocated reunion of Scottish Presbyterians; published *Endowed Territorial Work* (1873), *Buddhism and Christianity* (Croall lecture, 1890), and *Sacrifice* (Baird lecture, 1894).

SCOTT, CHARLES PRESTWICH (1846-1932), journalist; first class, *lit. hum.*, Corpus Christi College, Oxford, 1869; editor (1872-1929) of *Manchester Guardian* which he purchased in 1905 on death of his cousin J. E. Taylor [q.v.]; gave paper serious standing as critic of arts and letters and raised it to a leading place as a moral force in world politics; supported Gladstone's home rule policy; opposed South African war; prepared opinion for reforms

of liberal administrations; presented reasoned criticism of Grey's foreign policy; supported Lloyd George during the war; later urged liberal co-operation with labour governments and sought to promote liberal reunion; MP, Leigh division, 1895-1905.

SCOTT, LORD CHARLES THOMAS MONTAGU-DOUGLAS- (1839-1911), admiral; son of fifth Duke of Buccleuch [q.v.]; entered navy, 1853; served in Baltic, Black Sea, and China campaigns; distinguished in Indian Mutiny; on China station, 1868-71; commanded *Bacchante*, with royal princes on board, in Mediterranean and West Indies, 1879-82; CB, 1882; commander-in-chief on Australia station, 1889-92; vice-admiral, 1894; KCB, 1898; admiral, 1899; commander-in-chief at Plymouth, 1900-3; GCB, 1902.

SCOTT, CLEMENT WILLIAM (1841-1904), dramatic critic; son of William Scott (1813-72, q.v.); junior clerk in War Office, 1860-79; dramatic critic for *Sunday Times*, 1863-5, for *London Figaro*, 1870, and for *Daily Telegraph*, 1871-98; edited *The Theatre*, 1880-9; pioneer of picturesque dramatic criticism, some of which he issued in volume form; adapted many French dramas, chiefly by Sardou; contributed to *Punch* from 1880 sentimental verse, collected in *Lays of a Londoner* (1882) and *Lays and Lyrics* (1888).

SCOTT, CYRIL MEIR (1879-1970), composer and writer; began to compose at age of seven; studied at Hoch Conservatorium, Frankfurt, under Lazarro Uzielli and Engelbert Humperdinck, 1891-3; studied piano in Liverpool with Steudner-Welsing, and returned to Frankfurt to study with Iwan Knorr; returned to Liverpool, 1898; published translations of Baudalaire and Stefan George (1909-10); also three volumes of poetry; Hans Richter played his 'Heroic Suite' at Manchester and Liverpool, 1900; published piano pieces and Violin Sonata, 1909, and Piano Quintet, 1925; overture *Princess Maleine* performed in Vienna, 1913; opera, *The Alchemist* performed at Essen, 1925; publications include *The Philosophy of Modernism* (1917), *The Initiate* (1920), and *Music—its Secret Influence through the Ages* (1933).

SCOTT, DUNKINFIELD HENRY (1854-1934), palaeobotanist; son of Sir G. G. Scott [q.v.]; read classics at Christ Church, Oxford, 1872-6; trained as engineer, 1876-9; studied botany at Würzburg, 1880-2; in charge, under T. H. Huxley [q.v.], of botanical work at Normal School of Science, 1885-92; honorary director, Jodrell laboratory, Kew, 1892-1906; acquired interest in fossil plants from and collaborated with W. C. Williamson [q.v.]; works include *Introduction to Structural Botany* (1894), *Studies in Fossil Botany* (1900), and *Extinct Plants and Problems of Evolution* (1924); his collection in British Museum (Natural History) includes more than 3,000 slides of carboniferous plants; FRS, 1894.

SCOTT, Lord FRANCIS GEORGE MONTAGU-DOUGLAS- (1879-1952), soldier, Kenya farmer, and political leader; educated at Eton and Christ Church, Oxford; served in Grenadier Guards in South African War; aide-de-camp to Earl of Minto, viceroy of India, 1905-10; served in 1914-18 war; DSO; emigrated to Kenya, 1920; elected to legislature, 1925; leader, European elected members, 1931; member, governor's executive council, 1932-6 and 1937-44; KCMG, 1937; military secretary to commander-in-chief, East African Forces, 1943-6.

SCOTT, GEORGE HERBERT (1888-1930), airship commander; joined Royal Naval Air Service, 1914; during 1914-18 war developed system of mooring airships at head of mast or tower; commanded 'rigid' airship R34 on her Atlantic flight—the first Atlantic airship flight—July 1919; joined technical staff, Royal Airship Works, Cardington, Bedfordshire, 1920; appointed to post in airship directorate, 1924; assistant director, 1930; killed in airship R101 disaster near Beauvais, France.

SCOTT, Sir GILES GILBERT (1880-1960), architect; grandson of Sir George Gilbert Scott [q.v.]; educated at Beaumont College; pupil of the architect, Temple Moore [q.v.], 1898; won competition for new Liverpool Anglican cathedral; first contract placed, 1903; other church works include chapels for Charterhouse School and Ampleforth Abbey, the Roman Catholic cathedral, Oban, and the Carmelite church, Kensington; his university buildings include the University Library, Cambridge, the extension to the Bodleian, Oxford, and an addition to Clare College, Cambridge; appointed architect for new Waterloo Bridge, London, 1932; and rebuilding of House of Commons, 1948-50; ARA, 1922; knighted, 1924; OM, 1944; hon. DCL, Oxford, 1933, LLD, Liverpool, 1925, and Cambridge, 1955.

SCOTT, HUGH STOWELL (1862-1903), novelist, writing under the pseudonym of HENRY SETON MERRIMAN; abandoned underwriter's office for foreign travel and novelwriting; published first novel, *Young Mistley* (2 vols.), anonymously in 1888; his works, which embody much study of foreign nationalities, include *The Slave of the Lamp* (1892), *From one*

Generation to another (1892), *With Edged Tools* (1894), *The Sowers* (1896), *In Kedar's Tents* (1897), *The Isle of Unrest* (1900), *Barlasch of the Guard* (1902); memorial edition of fourteen novels appeared in 14 vols. in 1909-10.

SCOTT, Sir (JAMES) GEORGE (1851-1935), administrator in Burma and author; educated at Stuttgart, King's College School, London, Edinburgh and Oxford universities; special correspondent for *Standard* in Malaya (1875-6) and Tongking (1884); headmaster, St. John's College, Rangoon, 1879-81; introduced association football there; entered Burma civil service, 1885; resident, Northern (1891), Southern (1902-10) Shan States; works include *The Burman* (2 vols., 1882) and *Gazetteer of Upper Burma and the Shan States* (5 vols., 1900-1); KCIE, 1901.

SCOTT, JOHN (1830-1903), shipbuilder and engineer; joined father's shipbuilding firm on the Clyde, becoming its head, 1868; developed marine steam-engine; introduced water-tube boilers into corvettes for French and English navies; associated with Samson Fox [q.v.] in developing corrugated flues; ardent bibliophile, yachtsman, and Volunteer; CB, 1887; vice-president of Institution of Naval Architects, 1903; FRS Edinburgh, and FSA Scotland.

SCOTT, Sir JOHN (1841-1904), judicial adviser to the Khedive; BA, Pembroke College, Oxford, 1864; MA, 1869; honorary fellow, 1898; obtained cricket blue, 1863; called to bar, 1865; published *Bills of Exchange* (1869); English judge of court of international appeal at Alexandria, 1874-82; vice-president, 1881; puisne judge of high court at Bombay, 1882-90; judicial adviser to the Khedive, 1891-8; recreated Egyptian judicial system and simplified procedure; KCMG, 1894; hon. DCL, Oxford, 1898; deputy judge-advocate-general of the army, 1898.

SCOTT, JOHN WILLIAMSON ROBERTSON (1866-1962), journalist, author, and founder-editor of *The Countryman*. [See ROBERTSON SCOTT.]

SCOTT, KATHLEEN (1878-1947), sculptor. [See KENNET, (EDITH AGNES) KATHLEEN, LADY KENNET.]

SCOTT, LEADER (pseudonym) (1837-1902), writer on art. [See BAXTER, LUCY.]

SCOTT, Sir LESLIE FREDERIC (1869-1950), judge, politician, and chairman of committees; son of Sir John Scott [q.v.]; educated at

Rugby and New College, Oxford; called to bar (Inner Temple), 1894; KC, 1909; conservative MP, Exchange division, Liverpool, 1910-29; solicitor-general, 1922; lord justice of appeal, 1935-48; chairman, ministers' powers (1931-2), land utilization (1941-2) and other committees; knighted, 1922; PC, 1927.

SCOTT, Sir PERCY MORETON, first baronet (1853-1924), admiral; entered navy, 1866; specialized in gunnery; active in converting gunnery school at Whale Island into model naval barracks and training establishment; developed and introduced number of valuable inventions in signal apparatus and gunnery appliances; as captain of *Terrible* rendered valuable service in South African war by devising land mountings and carriages for heavy naval guns destined for relief of Ladysmith, 1899; rendered similar services during Boxer rising in China, 1900; rear-admiral, 1905; inspector of target practice, 1905-7; commanded second cruiser squadron of Channel fleet, 1907; vice-admiral, 1908; admiral, 1913; appointed to Admiralty for special service during 1914-18 war; created Anti-Aircraft Corps; baronet, 1913; his gunnery inventions of great importance.

SCOTT, ROBERT FALCON (1868-1912), naval officer and Antarctic explorer; entered navy, 1880; led national Antarctic expedition in *Discovery*, 1901-4; expedition surveyed coast of South Victoria Land, interior of Antarctic continent, made southern record, discovered King Edward VII Land, sounded Ross Sea, and investigated nature of ice-barrier; captain and CVO, 1904; commanded new Antarctic expedition in *Terra Nova*, 1910; reached Pole, 18 Jan. 1912, shortly after Norwegian expedition under Roald Amundsen; perished with remainder of party not far from One Ton depot owing to lack of food and bad weather conditions on return jouney (*c*.29 Mar.); bodies, together with Scott's diaries, specimens, etc., discovered by search party eight months later.

SCOTT-ELLIS, THOMAS EVELYN, eighth Baron Howard de Walden and fourth Baron Seaford (1880-1946), writer, sportsman, and patron and lover of the arts; educated at Eton and Sandhurst; succeeded father, 1899; inherited great estates in London and elsewhere; acquired property in Kenya and Wales; supporter of turf, lover of sailing, skilled fencer, exponent of falconry and hawking; keen medievalist; and editor and benefactor of *The Complete Peerage*; wrote several plays; produced operatic trilogy (as T. E. Ellis) with music by Josef Holbrooke; collected modern paintings; benefactor of many branches of art and charity.

SCOTT-JAMES, ROLFE ARNOLD (1878-1959), journalist, editor, and literary critic; educated at Mill Hill School and Brasenose College, Oxford; literary editor, *Daily News*, 1905-11; editor, the *New Weekly*, 1914; served in France in 1914-18 war; MC, 1918; leader-writer, *Daily Chronicle*, 1919-30; assistant-editor, *Spectator*, 1933-5 and 1939-45; editor, *London Mercury*, 1934; published *Modernism and Romance* (1908), *Personality in Literature* (1913), *The Influence of the Press* (1913), *The Making of Literature* (1928), *Fifty Years of English Literature 1900-1950* (1951), and short studies of Thomas Hardy (1951) and Lytton Strachey (1955); OBE, 1954.

SCOTT-PAINE, CHARLES HUBERT (1891-1954), pioneer of aviation and of high-speed motor-boats; from early youth felt enthusiasm for ships and aircraft; joined Pemberton Billing Ltd. (later Supermarine), manufacturing seaplanes, 1913; general manager, 1914; managing director, 1916; Supermarine won Schneider Cup outright, 1931; resigned from Supermarine, 1923; founded British Power Boat Co., Ltd., 1927; most famous of his boats, 'Miss Britain III', first successful all-metal motor-boat, from which were developed motor gunboats and rescue launches giving notable service in 1939-45 war; settled in Greenwich, Connecticut, and supplied similar craft for the US Navy, 1939.

SCRUTTON, Sir THOMAS EDWARD (1856-1934), judge; educated at Mill Hill School, London University, and Trinity College, Cambridge; first class, moral sciences (1879), law (1880); senior Whewell scholar (1879) and four times Yorke prizeman; president of Union, 1880; called to bar (Middle Temple), 1882; KC, 1901; bencher, 1908; established practice in commercial law and copyright business; published *The Contract of Affreightment as expressed in Charterparties and Bills of Lading* (1886; 14th edn. 1939) and *The Laws of Copyright* (1883); among the busiest practitioners in the Commercial Court, 1895-1910; judge of King's Bench division, 1910-16; efficient but unpopular by reason of petulant rudeness to counsel; lord justice of appeal, 1916-34; developed remarkable mastery of legal principles; mellowed with age; interests included poetry, travel, music, and church architecture; knighted, 1910; PC, 1916.

SEALE-HAYNE, CHARLES HAYNE (1833-1903), liberal politician; called to bar, 1857; liberal MP for Ashburton, Devonshire, 1885-1903; paymaster-general, 1892-5; PC, 1892; by will endowed industrial college near Newton Abbot.

SEAMAN, SIR OWEN, baronet (1861-1936), poet, satirist, and parodist; educated at Shrewsbury and Clare College, Cambridge; first class, classical tripos, 1883; professor of literature, Newcastle upon Tyne, 1890-1903; gifted composer of light verse; joined *Punch*, 1897; assistant editor, 1902; editor, 1906-32; wrote brilliant parodies, political satire, and memorial verses of great dignity; assiduous in encouragement of talent; knighted, 1914; baronet, 1933.

SECCOMBE, THOMAS (1866-1923), critic and biographer; BA, Balliol College, Oxford; assistant editor, *Dictionary of National Biography*, 1891-1900; professor of English, Royal Military College, Sandhurst, 1912-19; of English literature, Queen's University, Kingston, Ontario, 1921-3.

SEDDON, RICHARD JOHN (1845-1906), premier of New Zealand; born at St. Helens, Lancashire; left iron foundry at Liverpool for Victoria; 1863; settled as store-keeper in goldmine diggings at Waimea Creek, New Zealand, 1866; removed to goldfields at Kumara, 1874; first mayor; Member of Parliament for Hokitika, 1879; and for Kumara (renamed Westland, 1890), 1881-1906; joined Young New Zealand reform party; supported great shipping strike (1890), and advocated state ownership and state socialism generally; minister for mines, public works, and defence, 1891-6; abolished subletting of government contracts; minister of marine, 1892; premier from May 1893 till death; also minister of native affairs, Sept. 1893-9; carried out predecessor's policy of woman's suffrage, Sept. 1893; consolidated criminal code and introduced local option for control of liquor traffic; held, with the premiership till death, the offices of minister for labour (Jan. 1896), colonial treasurer (June), and minister of defence (1899); was for a time in addition commissioner of customs (1899) and trade (1899-1900); attended Queen Victoria's diamond jubilee, 1897; PC and hon. LLD, Cambridge, 1897; passed old age pensions bill, 1898; arranged for universal penny postage (1901), and nationalized coal-mines and fire insurance; visited South Africa on his way to attend King Edward VII's coronation, 1902; minister of immigration and of education, 1902; passed Preferential Trade Act favouring British imports; condemned Chinese labour in South Africa, 1904; died at sea on voyage home from Australia.

SEDGWICK, ADAM (1854-1913), zoologist; great-nephew of Adam Sedgwick (1785-1873, q.v.); BA, Trinity College, Cambridge; reader in animal morphology, Cambridge, 1890; FRS, 1886; fellow of Trinity, 1880; professor of zoology, Cambridge, 1907; at Imperial College of Science and Technology, South Kensington, 1909.

SEE, SIR JOHN (1844-1907), premier of New South Wales; accompanied parents from England to New South Wales, 1853; built up large produce and shipping business at Sydney; member of New South Wales legislative assembly, 1880-1904; postmaster-general, Oct. to Dec. 1885; treasurer, 1891-4; introduced protectionist tariff; chief secretary and minister for defence, 1899-1901; premier, 1901-4; member of legislative council, 1901-4; KCMG, 1902.

SEEBOHM, FREDERIC (1833-1912), historian; educated at Bootham School, York; called to bar, 1856; settled at Hitchin as partner in bank, 1857; strong Quaker; works include *The Oxford Reformers* (1867), *The English Village Community* (1883), *The Tribal System in Wales* (1895), and *Tribal Custom in Anglo-Saxon Law* (1902); last three books embody his principal contribution to historical studies; joined widespread revolt against romantic Germanistic interpretation of medieval history; traced back English system of communal farming to so-called manorial system (which, in his view, already existed in Roman 'villa'), and Celtic tribal community to authority of patriarch.

SEELEY, HARRY GOVIER (1839-1909), geologist and palaeontologist; cousin of Sir J. R. Seeley [q.v.]; assisted Adam Sedgwick (1785-1873, q.v.) at Woodwardian museum, Cambridge, 1859-71; published *Index to the Fossil Remains of Aves, Ornithosauria and Reptilia* (1869); professor of geography at King's College, London, 1876; lecturer and professor of geology and mineralogy at Cooper's Hill, 1890-1905, and at King's College, London, 1896-1905; chief work was *Researches on the Fossil Reptilia* (10 parts, 1888-96, in Philos. Trans. of Royal Society); FRS, 1879; Lyell medallist, Geological Society, 1885.

SEELY, JOHN EDWARD BERNARD, first BARON MOTTISTONE (1868-1947), politician and soldier; educated at Harrow and Trinity College, Cambridge; served in South Africa, 1900-1; DSO, 1900; MP, Isle of Wight, 1900-6; resigned from conservative party, 1904; liberal MP, Abercromby division, Liverpool, 1906-10; Ilkeston division, Derbyshire, 1910-22; Isle of Wight, 1923-4; under-secretary, Colonial Office, 1908-11, War Office, 1911-12; secretary of state for war, 1912-14; ordered troop movements in Ireland giving false impression of immediate action against Ulster in which all but Ulster-domiciled officers must participate or be dismissed, Mar. 1914; (Sir) Hubert Gough and other officers at the Curragh preferred dismissal; to a cabinet statement prepared for Gough, Seely added an assurance that troops would not be used to enforce home rule; this led to his resignation as having privately bargained with troublesome officers; commanded Canadian Cavalry brigade in France, 1914-18; under-secretary for air, 1919; chairman, national savings committee, 1926-43; PC, 1909; baron, 1933.

SELBIE, WILLIAM BOOTHBY (1862-1944), Congregational divine; educated at Manchester grammar school, Brasenose and Mansfield colleges, Oxford; Congregational minister, Highgate, 1890-1902, Emmanuel church, Cambridge, 1902-9; principal, Mansfield College, 1909-32; Wilde lecturer, 1921-4; inspiring preacher.

SELBORNE, second EARL OF (1859-1942), statesman. [See PALMER, WILLIAM WALDEGRAVE.]

SELBY, first VISCOUNT (1835-1909), speaker of the House of Commons. [See GULLY, WILLIAM COURT.]

SELBY, THOMAS GUNN (1846-1910), Wesleyan missionary in China; served at Fatshan, 1868-76; started North River Mission at Shiu Chau Foo, 1878; circuit pastor in England, 1822; works include *Life of Christ* in Chinese (*c.*1890) and *Chinamen at Home* (1900).

SELFRIDGE, HARRY GORDON (1858-1947), man of business; born at Ripon, Wisconsin, USA; bank clerk at fourteen; joined mail-order firm Field, Leiter & Co., Chicago, as clerk, 1879; junior partner, 1890; retired, 1904; came to London, 1906; opened Oxford Street store with 130 departments and £900,000 capital, 1909; resigned with honorary title of president, 1939; naturalized, 1937.

SELIGMAN, CHARLES GABRIEL (1873-1940), ethnologist; qualified at St. Thomas's Hospital, 1896; recorded his ethnological fieldwork in *The Melanesians of British New Guinea* (1910), *The Veddas* (1911), and *The Pagan Tribes of the Nilotic Sudan* (1932); professor of ethnology, London, 1913-34; FRS, 1919.

SELINCOURT, ERNEST DE (1870-1943), scholar and literary critic; educated at Dulwich and University College, Oxford; second class, *lit. hum.*, 1894; university lecturer in English, 1899; professor, Birmingham, 1908-35; vice-principal, 1931-5; edited works of Keats, Spenser, and William and Dorothy Wordsworth;

professor of poetry, Oxford, 1928-33; FBA, 1927.

SELOUS, FREDERICK COURTENEY (1851-1917), hunter and explorer; hunter and ivory trader in South Africa, 1871-81; entered service of British South Africa Company and acted as intermediary between Cecil Rhodes [q.v.] and Matabele chief, Lobengula, 1890; guide and chief of pioneers who secured Mashonaland for British crown; DSO, 1916; killed near Kissaki in (German) East Africa.

SELWIN-IBBETSON, SIR HENRY JOHN, seventh baronet, and BARON ROOKWOOD (1826-1902), politician; BA, St. John's College, Cambridge, 1849; MA, 1852; conservative MP for South Essex, 1865-8, for Western division of Essex, 1868-84, and for Epping division, 1884-92; baron, 1892; under-secretary to Home Office, 1874-8; parliamentary secretary to the Treasury, 1878; piloted bill for opening Epping Forest to public (1878).

SELWYN, ALFRED RICHARD CECIL (1824-1902), geologist; assistant geologist on Geological Survey of Great Britain, 1845; made special study of geology of North Wales; director of Geological Survey of Victoria, Australia, 1852-69; and of Canada, 1869-94; paid special attention to goldfields, mineral areas, and water-supply; published official reports; FRS, 1874; LLD, McGill University, 1881; CMG, 1886; president of Royal Society of Canada, 1896; died at Vancouver.

SEMON, SIR FELIX (1849-1921), laryngologist; born at Danzig; MD, Berlin; came to London, 1874; physician in charge of throat department, St. Thomas's Hospital, 1882-97; laryngologist to National Hospital for Paralysed and Epileptic, 1887-1909; FRCP, 1885; knighted, 1897; naturalized, 1901; KCVO, 1905; especially skilled in diagnosis of cancer of larynx.

SEMPILL, nineteenth BARON (1893-1965), airman. [See FORBES-SEMPILL, WILLIAM FRANCIS.]

SENANAYAKE, DON STEPHEN (1884-1952), first prime minister of Ceylon (Sri Lanka); educated at St. Thomas's College, Colombo; founder-member of the Ceylon National Congress, 1919; member legislative council, 1924; minister of agriculture, 1931; leader of State Council, 1942; prime minister of independent Ceylon, 1948, PC, 1950.

SENDALL, SIR WALTER JOSEPH (1832-1904), colonial governor; BA, Christ's College, Cambridge (first class classic), 1858; MA, 1867;

inspector of schools, 1860-70, and director of education, 1870-2, in Ceylon; poor law inspector in England, 1873-8; assistant secretary of Local Government Board, 1878-85; governor-in-chief of Windward Islands, 1885-9, and of Barbados, 1889-92; high commissioner of Cyprus, 1892-8; and of British Guiana, 1898-1901; CMG, 1887; KCMG, 1889; GCMG, 1899; hon. LLD, Edinburgh; edited *Literary Remains of C. S. Calverley* (1885).

SEQUEIRA, JAMES HARRY (1865-1948), dermatologist; educated at King's College School and London Hospital; MB, 1890; dermatologist to London Hospital, 1902-27; made lifelong study of venereal disease; pioneer working in radiology; published *Diseases of the Skin* (1911).

SERGEANT, (EMILY FRANCES) ADELINE (1851-1904), novelist; published *Poems*, 1866; educated at Queen's College, London; private governess, 1870-80; engaged in much social work in London, 1887-1901; joined Roman communion, 1899; described religious development in *Roads to Rome* (1901); her novels (over ninety), include *Jacobi's Wife* (1882), *Esther Denison* (1889), and *The Story of Phil Enderby* (1898).

SERGEANT, LEWIS (1841-1902), journalist and author; BA, St. Catharine's College, Cambridge, 1865; for long period leader-writer to *Daily Chronicle*; editor of *Educational Times*, 1895-1902; wrote much on modern Greece; other works include *The Franks* (1898) and verse and fiction.

SERVICE, ROBERT WILLIAM (1874-1958), versifier; educated at Hillhead high school, Glasgow; emigrated to British Columbia, 1895; clerk in Canadian Bank of Commerce; in British Columbia and the Yukon, 1904-12; made a fortune as verse writer with 'The Shooting of Dan McGrew' and 'The Cremation of Sam McGee'; published *Songs of a Sourdough* (1907), *Ballads of a Cheechako* (1909), and *Rhymes of a Rolling Stone* (1912); served with American ambulance unit and Canadian army intelligence in 1914-18 war; published *Rhymes of a Red-Cross Man* (1916) and a number of novels; settled in France, 1945, and published more books of verse, including *Rhymes of a Roughneck* (1950), and *Rhymes for My Rags* (1956).

SETH, ANDREW (1856-1931), philosopher. [See PATTISON, ANDREW SETH PRINGLE-.]

SETON, GEORGE (1822-1908), Scottish genealogist; heraldic and legal writer; BA,

Exeter College, Oxford, 1845; MA, 1848; called to Scottish bar, 1846; superintendent of civil service examinations in Scotland, 1862-89; FRS Edinburgh and FSA Scotland; voluminous writings include *The Law and Practice of Heraldry in Scotland* (1863) and a history of the Seton family (2 vols., 1896).

SETON-WATSON, ROBERT WILLIAM (1879-1951), historian; educated at Winchester and New College, Oxford; first class, modern history, 1902, studied at Berlin University and the Sorbonne; visited Vienna, intending to write history of Austria since Maria Theresa; published *The Future of Austria-Hungary* (1907), and other books on racial problems in the Hapsburg empire; appointed Masaryk professor of Central European history, King's College, London, 1922; with Sir Bernard Pares [q.v.], founded and edited the *Slavonic Review*; published other works, including *A History of the Roumanians* (1934), *Disraeli, Gladstone, and the Eastern Question* (1935), and *Britain in Europe 1789-1914* (1937); hon. degrees, Prague, Zagreb, Bratislava, Belgrade, Cluj, and Birmingham; FBA, 1932; first professor of Czechoslovak studies, Oxford University, 1945-9; hon. fellow, New College and Brasenose College, Oxford.

SEVERN, WALTER (1830-1904), watercolour artist; born near Rome; son of Joseph Severn [q.v.]; held post in Education Department, 1852-85; made art furniture; revived art needlework and embroidery; skilful landscape painter in water-colours; founded Dudley Gallery Art Society, 1865; president, 1883-1904.

SEWARD, SIR ALBERT CHARLES (1863-1941), botanist and geologist; educated at Lancaster grammar school and St. John's College, Cambridge; first class, parts i and ii, natural sciences tripos, 1885-6; fellow and tutor, Emmanuel College, 1899-1906; university lecturer in botany, 1890; professor, 1906-36; master of Downing College, 1915-36; vice-chancellor, Cambridge, 1924-6; FRS, 1898; foreign secretary, 1934-41; president, Geological Society of London, 1922-4, International Botanical Congress, 1930, International Union of Biological Sciences, 1931, British Association, 1939; publications include *Fossil Plants as Tests of Climate* (1892), *Fossil Plants for Students of Botany and Geology* (4 vols., 1898-1919), and *Plant Life through the Ages* (1931); edited *Darwin and Modern Science* (1909), *Science and the Nation* (1917); and general editor, Cambridge Botanical Handbooks; knighted, 1936.

SEWELL, ELIZABETH MISSING (1815-1906), author; sister of James Edwards Sewell [q.v.]; influenced by Oxford Movement; published *Amy Herbert* (1844) and *Laneton Parsonage* (3 parts, 1846-8), novels embodying her Anglican views; published *The Experience of Life* (1852), *Ursula* (1858), and other novels; took private pupils at Bonchurch from 1852 to 1891; founded at Ventnor St. Boniface School for middle-class girls, 1866; wrote also educational and devotional works; *Autobiography* edited by E. L. Sewell, 1907.

SEWELL, JAMES EDWARDS (1810-1903), warden of New College, Oxford, from 1860 to 1903; educated at Winchester and New College, Oxford; BA, 1832; MA, 1835; DD, 1860; fellow, 1829; vice-chancellor of Oxford University, 1874-8.

SEWELL, ROBERT BERESFORD SEYMOUR (1880-1964), zoologist; educated at Weymouth College and Christ's College, Cambridge; first class, parts i and ii, natural sciences tripos, 1902-3; trained at St. Bartholomew's Hospital, 1905-7; qualified as doctor; joined Indian medical service, 1908; joined marine survey as surgeon-naturalist, Calcutta, 1910; became authority on the *Copepoda*; seconded as professor of biology, Calcutta Medical College, 1911-13; port health officer, Aden, and service in Sinai and Palestine, 1914-18; superintendent, zoological survey, India, 1919; director, 1925-31; leader, John Murray expedition to Indian Ocean, 1933-4; in retirement, edited number of volumes of *Fauna of British India*; CIE, 1931; FRS, 1934; president, Linnean Society, 1952-5, and Ray Society, 1950-3; many distinctions from learned societies in India.

SEXTON, SIR JAMES (1856-1938), politician and labour leader; worked as labourer; joined National Union of Dock Labourers, 1889; secretary, 1893-1922; founder-member, independent labour party, 1893; served many years on general council of Trades Union Congress; president, 1905; MP, St. Helens, 1918-31; first labour correspondent of *Labour Gazette*; knighted, 1931.

SEXTON, THOMAS (1848-1932), Irish politician; leader-writer on the *Nation*, 1867; joined Home Rule League; Parnellite MP, Sligo (1880-5), South Sligo (1885-6), West Belfast (1886-92); anti-Parnellite MP, North Kerry, 1892-6; eloquent speaker and skilful obstructionist; managed *Freeman's Journal*, 1892-1912.

SEYMOUR, SIR EDWARD HOBART (1840-1929), admiral of the fleet; grandson of Rear-Admiral Sir Michael Seymour [q.v.], first baronet, and nephew of Admiral Sir Michael

Seymour [q.v.]; entered navy, 1852; served in Crimean war, 1854-6; served in China, being present at capture of Canton, 1857-8; took part in attack on Taku forts, 1860; served against Taiping rebels, 1862; served on west coast of Africa, 1869-70; captain of *Iris* during Egyptian war, 1882; promoted to flag rank, 1889; second-in-command of Channel squadron, 1892-4; admiral-superintendent of naval reserves, 1894-7; commander-in-chief, China station, 1897-1901; commanded international naval brigade during Boxer rising, 1900; admiral, 1901; commanded at Devonport, 1903-5; admiral of the fleet, 1905; retired, 1910; GCB, 1901; OM, 1902; a good linguist.

SHACKLETON, SIR DAVID JAMES (1863-1938), labour leader, politician, and civil servant; worked as a cotton operative; secretary, Darwen Weavers' Association, 1894-1907; labour MP, Clitheroe division, 1902-10; chairman, national labour party, 1905; president, Trades Union Congress, 1908-9; senior labour adviser, Home Office, 1910; national health insurance commissioner, 1911-16; first permanent secretary, Ministry of Labour, 1916-21; chief labour adviser, 1921-5; KCB, 1917.

SHACKLETON, SIR ERNEST HENRY (1874-1922), explorer; apprentice in mercantile marine, 1890; served in White Star, Shire, and Union Castle lines; junior officer on national Antarctic expedition under Commander R. F. Scott [q.v.] in *Discovery*, 1901; left sea, 1903; began to mature plans for reaching South Pole; sailed in *Nimrod*, 1907; reached lat. 88° 23′ S. on Antarctic plateau; sent parties to South Magnetic Pole and summit of Mount Erebus; commanded trans-Antarctic expedition in *Endurance*, 1914; when ship was crushed in ice, made voyage of 800 miles in small boat through stormy seas to South Georgia in order to bring help to his party, 1916; organized winter equipment of North Russian expeditionary force, 1918-19; sailed in *Quest*, 1921; knighted, 1909; died in South Georgia.

SHADWELL, CHARLES LANCELOT (1840-1919), college archivist and translator of Dante; son of Lancelot Shadwell [q.v.]; BA, Christ Church, Oxford; fellow of Oriel, 1864-98; provost, 1905-14; published *Registrum Orielense, 1500-1900* (1893, 1902), and verse translation of Dante's *Purgatorio*.

SHAND (afterwards BURNS), ALEXANDER, BARON SHAND (1828-1904), judge; studied law at Edinburgh University, 1848-52; passed to Scottish bar, 1853; sheriff of Kincardine, 1862, and of Haddington and Berwick, 1869; raised to bench, 1872; settled in London, 1890; PC and member of the judicial committee of Privy Council, 1890; baron, 1892; lord of appeal in House of Lords, 1892-1904.

SHAND, ALEXANDER INNES (1832-1907), journalist and critic; MA, Aberdeen University, 1852; admitted to Scottish bar, 1865; prolific contributor to *The Times*, *Blackwood's*, and *Saturday Review*; British commissioner at Paris exhibition, 1893; fine rider, shot, and angler; works include *Shooting the Rapids* (novel, 1872), life of Sir Edward Hamley (1895), *The War in the Peninsula* (1898), *Shooting* (1899), *Dogs* (1903), *Old World Travel* (1903), and *Days of the Past* (1905).

SHANDON, BARON (1857-1930), lord chancellor of Ireland. [See O'Brien, IGNATIUS JOHN.]

SHANNON, CHARLES HASLEWOOD (1863-1937), lithographer and painter; apprenticed to wood-engraving; studied at Lambeth Art School; met Charles Ricketts [q.v.], his lifelong companion; with him produced *The Dial* (1889-97) and made outstanding art collection; executed lithographs and large oil canvases at first of pervasive silvery quality, later of opulent contours and rich darks; ARA, 1911; RA, 1921.

SHANNON, SIR JAMES JEBUSA (1862-1923), painter; born in New York state; studied art in England; gained widespread clientele for his portraits, which constitute the bulk of his paintings; RA, 1909; knighted, 1922.

SHARP, CECIL JAMES (1859-1924), musician, author, and collector and arranger of English folk-songs and dances; BA, Clare College, Cambridge; held posts in Adelaide, Australia, 1882-92; music-master, Ludgrove School, 1893-1910; work on his *Book of British Song* (1902) probably led him to realize essential importance of traditional art; began researches into survival of traditional music in England, 1903; as result, published *Folk-Songs from Somerset* (1904), which aroused much interest; elected to committee of Folk-Song Society, 1904; carried on active propaganda on behalf of claims of folk-song; published *English Folk-Songs; Some Conclusions* (1907); visited Appalachian mountains in search of songs, 1916-18; the other main subject of his activities was the folk-dance, in which he was a pioneer; researched into morris-, sword-, and country-dances; founded English Folk-Dance Society, 1911; director, 1912-24.

SHARP, WILLIAM, 'FIONA MACLEOD' (1855-1905), romanticist; educated at Glasgow Academy and University; in lawyer's office, 1874-6; travelled in Australia for health, 1876-8; clerk in London, 1878-81; introduced in 1881 to D. G. Rossetti [q.v.], whose life he wrote, 1882; published volumes of verse—including *Romantic Ballads and Poems of Phantasy* (1888); editor of 'Canterbury Poets', 1884; wrote lives of Shelley (1887), Heine (1888), and Browning (1890), and *The Children of To-morrow* (1889), a romantic tale; visited Canada and America (1889), Scotland, Germany, and Rome (1890); published *Sospiri di Roma* (1891), *Life of Joseph Severn* (1892), and dramatic *Vistas* (1894); began to write mystical prose and verse under pseudonym of 'Fiona Macleod', 1893; such works include *Pharais: a Romance of the Isles* (1894), *The Mountain Lovers* (1895), *The Sin Eater* (Celtic tales, 1895), and plays including *The House of Usna* (1900) and *The Immortal Hour* (1900); uniform edition of 'Fiona's' works appeared in 1910; Sharp kept his identity with 'Fiona' a secret till death; under his own name continued to write stories, including *The Gypsy Christ* (1895) and *Wives in Exile* (1896), as well as *Literary Geography* (1904); died in Sicily, near Mount Etna.

SHARPE, RICHARD BOWDLER (1847-1909), ornithologist; first librarian to Zoological Society, 1866-72; senior assistant of zoological department, British Museum, 1872-95; assistant keeper of vertebrates, 1895; hon. LLD, Aberdeen, 1891; greatly increased ornithological collection; prepared British Museum Catalogue of Birds (27 vols., 1874-98), and *Handlist of the Genera and Species of Birds* (5 vols., 1899-1909); edited Allen's 'Naturalist's Library' (16 vols.), contributing *The Birds of Great Britain* (4 vols., 1894-7); wrote also monographs on kingfishers (1868-71), swallows (1885-94), and birds of paradise (1891-8); founded British Ornithologists' Club, 1892.

SHARPEY-SCHAFER, SIR EDWARD ALBERT (1850-1935), physiologist. [See SCHAFER.]

SHATTOCK, SAMUEL GEORGE (1852-1924), pathologist; born BETTY; qualified in medicine, University College, London; curator of pathological museum, Royal College of Surgeons, 1897-1924; FRS, 1917; for forty years played important part in revolution of outlook on pathology.

SHAUGHNESSY, THOMAS GEORGE, first BARON SHAUGHNESSY (1853-1923), Canadian railway administrator; born at Milwaukee, USA; employed in purchasing department, Milwaukee and St. Paul Railway, 1869; general store-keeper, 1879; general purchasing agent to

Canadian Pacific Railway Company, at Montreal, 1882; assistant general manager, 1885; vice-president, 1891; president, 1899-1918; chairman of board of directors, 1918; knighted, 1901; KCVO, 1907; baron, 1916; progress of company largely due to him; created its Atlantic fleet.

SHAW, ALFRED (1842-1907), cricketer; played regularly for Nottinghamshire and for Players *v.* Gentlemen, 1865-87; played for England *v.* Australia in first test match in England, Sept. 1880; visited America twice (1868 and 1879), and Australia five times, thrice as captain of English team; was privately engaged by Lord Sheffield [q.v.] in Sussex, 1883-94; played for Sussex, 1894-5; called 'The Emperor of Bowlers'.

SHAW, SIR EYRE MASSEY (1830-1908), head of the London Metropolitan Fire Brigade, 1861-91; BA, Trinity College, Dublin, 1848; MA, 1854; in army, 1854-60; chief constable of Belfast, 1859-61; reorganized Belfast fire service; perfected organization of metropolitan system; wrote many treatises on fire brigade subjects; CB, 1879; KCB, 1891; freeman of City of London, 1892.

SHAW, GEORGE BERNARD (1856-1950), playwright; born in Dublin into a loveless and genteelly poor household over-shadowed by his father's tippling, but filled with music and musicians by his mother; educated at Wesley Connexional School, in the National Gallery of Ireland, and by wide reading; entered estate agent's office, 1871; cashier, 1872-6; joined mother who had moved to London as a music teacher, 1876; in 1878-83 wrote but failed to publish five novels; *Immaturity*, *The Irrational Knot*, *Love Among the Artists*, *Cashel Byron's Profession*, and *An Unsocial Socialist*; made remarkable intellectual progress, conquest of shyness, and many important friendships; converted to vegetarian diet, 1881; book-reviewer, *Pall Mall Gazette*, 1885-8; art critic, *World*, 1886-9; music critic (as Corno di Bassetto), *Star*, 1888-90; on *World*, 1890-5; joined Fabian Society, 1884; edited *Fabian Essays in Socialism* (1889); published *The Quintessence of Ibsenism* (1891); wrote *Widowers' Houses* for performance by Independent Theatre (1892); followed it by *The Philanderer* (1893), *Mrs. Warren's Profession* (banned by lord chamberlain until 1925), *Arms and the Man*, *Candida*, and *You Never Can Tell*; dramatic critic, *Saturday Review*, 1895-8; obtained first successful production of a play with *The Devil's Disciple* (New York, 1897); corresponded at this time with (Dame) Ellen Terry [q.v.] for whom he wrote *Captain Brassbound's Conversion*; vestryman, St. Pancras,

London, 1897-1903; published *The Common Sense of Municipal Trading* (1904); married (1898) Charlotte Frances Payne-Townshend, a wealthy Anglo-Irish-woman; completed *Caesar and Cleopatra*, produced by Mrs Patrick Campbell [q.v.] (1899); wrote *Man and Superman* (1901-3), *John Bull's Other Island* (1904), *Major Barbara* (1905), and *The Doctor's Dilemma* (1906); these and others of his plays produced by Harley Granville-Barker [q.v.] during the Vedrenne-Barker seasons at the Royal Court Theatre, 1904-7; established as leading dramatist; continued with productions of *Getting Married* (1908), *The Shewing-Up of Blanco Posnet* (banned in Great Britain but produced at Abbey Theatre, Dublin, 1909), *Misalliance* (1910), *Fanny's First Play* (written for Lillah McCarthy, 1911), *Androcles and the Lion* (1913), *Pygmalion* (with Sir Herbert Tree [q.v.] and Mrs Patrick Campbell, 1914), *Heartbreak House* (New York, 1920, London, 1921), *Back to Methuselah* (New York, 1922, Birmingham Repertory, 1923), *Saint Joan* (New York, 1923, London, with (Dame) Sybil Thorndike, 1924); Nobel prize, 1925; established Anglo-Swedish Literary Foundation for translation of Swedish literature into English; published *The Intelligent Woman's Guide to Socialism and Capitalism* (1928), *The Adventures of the Black Girl in Her Search for God* (1932), *Everybody's Political What's What?* (1944), collected plays (1931), and collected prefaces (1934); his later plays produced at the Malvern Festivals included *The Apple Cart* (1929), *Too True to be Good* (1932), *Geneva* (1938), and *In Good King Charles's Golden Days* (1939); obtained new fame and fortune through filming of his plays, notably *Pygmalion* and *Major Barbara*; left residue of estate to institute a British alphabet of at least forty letters; his house at Ayot St. Lawrence became the property of the National Trust.

SHAW, HENRY SELBY HELE-(1854-1941), engineer. [See HELE-SHAW.]

SHAW, JAMES JOHNSTON (1845-1910), county court judge; professor of metaphysics in Magee College, Londonderry, 1869-78; called to Irish bar, 1878; Whately professor of political economy, Trinity College, Dublin, 1876-91; commissioner of national education in Ireland, 1891; county court judge of Kerry, 1891, and of Antrim, 1909; framed statutes for Queen's University, Belfast, 1908; pro-chancellor, 1909; recorder of Belfast, 1909; his *Occasional Papers* collected in 1910.

SHAW, JOHN BYAM LISTER (1872-1919), painter and illustrator; ARWS, 1913; has been called 'a kind of belated pre-Raphaelite'.

SHAW, RICHARD NORMAN (1831-1912), architect; pupil of William Burn [q.v.]; travelled abroad, 1854-5; published *Architectural Sketches from the Continent* (1858); chief assistant to George Edmund Street [q.v.], 1859; started practice in London with William Eden Nesfield [q.v.], 1862; ARA, 1872; RA, 1877; trained as 'Gothic' architect; his town houses include many studios for artist clients in so-called 'Queen Anne' style; chief piece of city architecture New Scotland Yard (1891); designed a number of country houses, including Chesters, Northumberland, and Bryanston, Dorset (1890-4), both on the grand scale; pre-eminent among Victorian architects for sincerity of his art.

SHAW, THOMAS, BARON SHAW, later first BARON CRAIGMYLE (1850-1937), lawyer and politician; educated at Dunfermline high school and Edinburgh University; admitted advocate, 1875; liberal MP, Hawick District, 1892-1909; solicitor-general for Scotland, 1894-5; QC, 1894; lord advocate, 1905-9; PC, 1906; lord of appeal, with life peerage (Lord Shaw), 1909-29; baron (Lord Craigmyle), 1929.

SHAW, THOMAS (TOM) (1872-1938), labour leader and politician; a cotton operative; secretary, International Federation of Textile Workers, 1911-29, 1931-8; director of national service, West Midland region, 1914-18; labour MP, Preston, 1918-31; minister of labour and PC, 1924; secretary of state for war, 1929-31; joint-secretary, Labour and Socialist International, 1923-5.

SHAW, WILLIAM ARTHUR (1865-1943), archivist and historian; educated at Owens College, Manchester; BA, 1883; for Public Record Office edited *Treasury Books and Papers* and *Treasury Books* series; wrote many articles and publications chiefly on economics and seventeenth-century history; FBA, 1940.

SHAW, SIR (WILLIAM) NAPIER (1854-1945), meteorologist; educated at King Edward VI's School, Birmingham, and Emmanuel College, Cambridge; sixteenth wrangler and first class, natural sciences tripos, 1876; fellow, 1877-1906; assistant director, Cavendish laboratory, 1898-1900; secretary, Meteorological Council, 1900-5; director, Meteorological Office, 1905-20; professor, meteorology, Royal College of Science, 1920-4; published *Forecasting Weather* (1911) and *Manual of Meteorology* (4 vols., 1926-31); FRS, 1891; knighted, 1915.

SHAW-LEFEVRE, GEORGE JOHN, BARON EVERSLEY (1831-1928), statesman; son of Sir J. G. Shaw-Lefevre and nephew of Charles Shaw-Lefevre, Viscount Eversley [qq.v.]; BA, Trinity College, Cambridge; liberal MP, Reading, 1863-85; instrumental (1866) in forming Commons Preservation Society, of which he was almost continuously chairman until death; secretary to Board of Trade, 1868; under-secretary to Home Office, 1871; secretary to Admiralty, 1871-4; first commissioner of works, 1880-3, postmaster-general, with seat in cabinet, 1883; MP, Central Bradford, 1886-95; first commissioner of works, with seat in cabinet, 1892-4; president, Local Government Board, 1894-5; found scope on London County Council for devotion to public administrative work; baron, 1906; a statesman and administrator of extraordinary industry and public spirit.

SHEARMAN, SIR MONTAGUE (1857-1930), judge; BA, St. John's College, Oxford; distinguished as both scholar and athlete; called to bar (Inner Temple), 1881; KC, 1903; judge of King's Bench division and knighted, 1914; presided at some notable trials, including those of Harold Greenwood (1920), Frederick Bywaters and Mrs Edith Thompson (1922), and murderers of Sir Henry Wilson (q.v., 1922).

SHEEPSHANKS, SIR THOMAS HERBERT (1895-1964), civil servant; educated at Winchester and Trinity College, Oxford; served in army in France, 1914-18; entered Ministry of Health, 1919; assistant secretary, 1936; seconded to air raids precautions department, Home Office, 1937; principal assistant secretary, Ministry of Home Security, 1939; deputy secretary, 1942; deputy secretary, Ministry of National Insurance, 1944; under-secretary, Treasury, 1945-6; permanent secretary, Ministry of Town and Country Planning, 1946; permanent secretary, Ministry of Local Government and Planning, 1951-5; CB, 1941; KBE, 1944; KCB, 1948.

SHEFFIELD, third EARL OF (1832-1909), patron of cricket. [See HOLROYD, HENRY NORTH.]

SHEFFIELD, fourth BARON (1839-1925). [See STANLEY, EDWARD LYULPH.]

SHELFORD, SIR WILLIAM (1834-1905), civil engineer; educated at Marlborough; apprenticed as mechanical engineer, 1852; assistant to Sir John Fowler [q.v.], 1856-60; engaged in construction of metropolitan railway; resident engineer on London, Chatham, and Dover Railway, 1860-5; practised on his own account from 1865; engineer of the Hull and Barnsley Railway, 1881-5; visited Canada (1885), Italy (1889), and the Argentine (1890) to report on railway schemes; main work was construction of railways in West Africa from 1893 till 1904; studied and wrote on engineering works of rivers and estuaries; reported on river Tiber and its floods, 1879 and 1885; CMG, 1901; KCMG, 1904.

SHENSTONE, WILLIAM ASHWELL (1850-1908), writer on chemistry; studied chemistry at school of Pharmaceutical Society of Great Britain; demonstrator of practical chemistry there; science master at Clifton College from 1880 to death; collaborated with (Sir) W. A. Tilden and others in much chemical research, especially in ozone; FRS, 1898; published works on chemistry.

SHEPHERD, GEORGE ROBERT, first BARON SHEPHERD (1881-1954), labour party national agent; educated at board-school; shoe shop assistant; organizer for Midland division of independent labour party, 1908; assistant national agent, ILP, 1924; national agent, ILP, 1929-45; baron, 1946; chief whip, 1949; PC, 1951.

SHEPPARD, HUGH RICHARD LAWRIE (1880-1937), dean of Canterbury; educated at Marlborough and Trinity Hall, Cambridge; deacon, 1907; priest, 1908; vicar, St. Martin-in-the-Fields, 1914-27; kept church open day and night and made it famous through development of religious broadcasting; with William Temple [q.v.] inaugurated Life and Liberty Movement, 1917; dean of Canterbury, 1929-31; canon of St. Paul's, 1934-7; CH, 1927.

SHERBORN, CHARLES WILLIAM (1831-1912), engraver; apprenticed to London silver-plate engraver; engraved for London jewellers, 1856-72; his finest achievement series of over 350 book-plates produced between 1881 and 1912; regular exhibitor at Royal Academy.

SHEREK, (JULES) HENRY (1900-1967), theatre impresario; educated in Germany and Switzerland; enlisted in British army at fifteen, and served in Near East, 1915-18; worked in United States, 1923-5; returned to London and joined his father's theatrical agency, 1925; produced Robert Sherwood's *Idiot's Delight* and *The Petrified Forest* (1938-42); served in army, 1941-4; from 1945, produced over 110 plays in London, New York, and Paris; had great success with *Edward, My Son* (1947); presented T. S. Eliot's [q.v.] *The Cocktail Party* (1949), *The Confidential Clerk* (1953), and *The Elder Statesman* (1958); also produced *Boys in Brown* (1947), *Under Milk Wood* (1956), and *The Affair* (1961); published *Not in front of the Children* (1959).

SHERIDAN, CLARE CONSUELO (1885-1970), artist and sculptor; daughter of Moreton Frewen, and cousin of (Sir) Winston Churchill; educated in France and Germany; married Wilfred Sheridan, 1910 (killed 1915); after death of daughter, took up modelling; studied under (Sir) William Reid Dick [q.v.]; modelled heads of Asquith and F. E. Smith (later Lord Birkenhead, q.v.), and Churchill; visited Moscow, 1920, and made busts of Lenin, Kamenev, and other Soviet leaders; travelled in America and became correspondent for New York *World*, 1922; obtained many scoops, including interviews with Attaturk and Mussolini; toured South Russia on a motor cycle called 'Satanella', 1924; lived in Algeria, 1925-37; on death of her son, took up wood carving, of which one notable example, an oak madonna, stands in Brede church, where she was buried; her many publications include *Mayfair to Moscow* and *Russian Portraits* (1921), *Across Europe with Satanella* (1925), *A Turkish Kaleidoscope* (1926), *Arab Interlude* (1936), and *To the Four Winds* (1957).

SHERRIFF, GEORGE (1898-1967), explorer and plant collector; educated at Sedburgh and the Royal Military Academy, Woolwich; commissioned in Royal Garrison Artillery and fought in France, 1918; posted to India and served on North-West Frontier, 1919; British vice-consul, Kashgar, Chinese Turkestan, 1927; acting consul-general till resignation in 1931; travelled widely, including journey into Tibet, with Frank Ludlow, collecting plants, including 69 species of *Rhododendron* and 59 *Primula*, 1933-4; discovered *Meconopsis sherriffii*; in Central Bhutan, collecting, 1937; with Ludlow and (Sir) George Taylor of the Natural History Museum, explored the drainage of the Tsangpo, 1938; resumed military duties in Assam, 1939; in charge of British Mission, Lhasa, 1943; continued to carry out exploration and plant collection in the Tsangpo Gorge, 1946; provided the British Museum with the finest collection of Himalayan plants in the world; on return from India, carried out number of public duties in Scotland; OBE, 1947; Victoria medal of honour from Royal Horticultural Society, 1948.

SHERRINGTON, SIR CHARLES SCOTT (1857-1952), physiologist; educated at Ipswich grammar school and Gonville and Caius College, Cambridge; first class, parts i and ii, natural sciences tripos, 1881-3; qualified at St. Thomas's Hospital, London; MRCS, 1884; LRCP, 1886; Cambridge MB, 1885; MD, 1892; Sc.D., 1904; fellow, Caius College, 1887; professor-superintendent, Brown Animal Sanatory Institution, London, 1891; Holt professor of physiology,

Liverpool, 1895; Waynflete professor of physiology, Oxford, 1913-35; published *The Integrative Action of the Nervous System* (1904), *Man on his Nature* (1940), and *The Endeavour of Jean Fernel* (1946); FRS, 1893; president, RS, 1920-5; president, British Association, 1922; GBE, 1922; OM, 1924; Nobel prize for medicine, 1932; his researches opened up an entirely new chapter in the physiology of the central nervous system.

SHERRINGTON, HELEN LEMMENS- (1834-1906), soprano vocalist. [See LEMMENS-SHERRINGTON.]

SHIELDS, FREDERIC JAMES (1833-1911), painter and decorative artist; served with mercantile lithographers in London and Manchester, 1847-56; influenced by pre-Raphaelite works at Manchester exhibition, 1857; successful water-colour artist and book-illustrator; became intimate with D. G. Rossetti [q.v.], whose memorial window in Birchington church he designed; settled in London, 1876; later devoted himself to decorative design and oil-painting; most important work was decoration of walls of chapel of Ascension, Bayswater Road, from 1889 onwards; bequeathed fortune to foreign missionary societies.

SHIELS, SIR (THOMAS) DRUMMOND (1881-1953), physician and politician; educated at elementary school and night school in Glasgow; employed by firm of photographers; served with 9th (Scottish) division in 1914-18 war; MC; studied medicine at Edinburgh University; MB, B.Ch., 1924; labour MP, Edinburgh (East), 1924-31; under-secretary of state for India, 1929; under-secretary of state, Colonial Office, 1929-31; medical secretary, British Social Hygiene Council, 1931; public relations officer, Post Office, 1946-9; secretary, Inter-Parliamentary Union (British Group), 1950-3; knighted, 1939.

SHIPLEY, SIR ARTHUR EVERETT (1861-1927), zoologist; BA (first class, parts i and ii, natural sciences tripos, 1882 and 1884), Christ's College, Cambridge; devoted himself to study of zoology under influence of F. M. Balfour [q.v.]; took up subject of parasitic worms; university demonstrator in comparative anatomy, 1886; lecturer in advanced morphology of Invertebrata, 1894; reader in zoology, 1908-20; fellow of Christ's College, 1887; played leading part in administration of affairs of college and university; tutor in natural sciences, Christ's College, 1892; member of council of senate of university, 1896-1908; master of Christ's College, 1910; noted for his hospitality; vice-chancellor, 1917-19; maintained close con-

nection with Colonial Office; chairman, Tropical Agriculture Committee, 1919; served on various royal commissions; GBE, 1920; his numerous works include *The Zoology of the Invertebrata* (1893) and *Text-book of Zoology* (with E. W. MacBride, 1901); edited, with (Sir) S. F. Harmer, *Cambridge Natural History* (1895-1909).

SHIPPARD, SIR SIDNEY GODOLPHIN ALEXANDER (1837-1902), colonial official; born at Brussels; BA, Hertford College, Oxford, 1863; BCL and MA, 1864; DCL, 1878; called to bar, 1867; migrated to South Africa, 1867; attorney-general of Griqualand West, 1875-8; puisne judge of supreme court of Cape Colony, 1880-5; chief magistrate of British Bechuanaland, 1885-95; CMG, 1886; KCMG, 1887; an able Roman-Dutch lawyer.

SHIRLEY, FREDERICK JOSEPH JOHN (1890-1967), headmaster of Worksop College and King's School, Canterbury; educated at Oxford high school and St. Edmund Hall; served in RNVR, 1914-18; honours in law (London University), 1920; called to bar (Lincoln's Inn); on teaching staff of Framlingham College, 1919; ordained deacon, 1920; priest, 1921; part-time rector, Sternfield, 1923-5; headmaster, St. Cuthbert's College, Worksop, 1925, school founded by Nathaniel Woodard [q.v.]; moved to King's School, Canterbury, 1935-62; canon, Canterbury cathedral; treasurer and librarian; published number of historical studies, including *Richard Hooker and Contemporary Political Ideas* (1949); DD, Oxford, 1949; Ph.D., London, 1931; hon. fellow, St. Edmund Hall, Oxford; fellow, Royal Historical Society and Society of Antiquaries.

SHIRREFF, MARIA GEORGINA (1816-1906), promoter of women's education. [See GREY.]

SHOENBERG, SIR ISAAC (1880-1963), electronic engineer; born in Russia of Jewish parents; educated at Kiev Polytechnic Institute; joined S. M. Aisenstein at St. Petersburg, 1905, in what became Russian Marconi Company in 1907; chief engineer; came to London and studied at Royal College of Science, 1914; joined Marconi Wireless and Telegraph Company; head of patents department; general manager; became British subject, 1919; general manager, Columbia Gramophone Company, 1928; merged with Gramophone Company (HMV) to form Electric and Musical Industries (EMI), 1931; director of research and head of patents; concerned with invention of electronic television picture-generating tube, 1932; recommended 405-line standard adopted by BBC,

1935; director, EMI, 1955; knighted, 1962; Shoenburg memorial lecture instituted by Royal Television Society.

SHORE, THOMAS WILLIAM (1840-1905), geologist and antiquary; founded Hampshire Archaeological Society, and wrote much in its *Transactions* on Hampshire geology; FGS, 1878; settled in London, 1896, and worked on London archaeology; publications include *A History of Hampshire* (1892); a memorial volume of geological papers was collected in 1908.

SHORT, SIR FRANCIS (FRANK) JOB (1857-1945), etcher and engraver; trained as engineer; studied at Stourbridge Art School and South Kensington; revived art of mezzotint and aquatint; reproduced paintings by Turner, Reynolds, Constable, De Wint, Watts, and others; own work includes 'The Lifting Cloud' (1901), 'A Silver Tide' (1912), etc.; taught engraving at South Kensington, 1891-1924; president, Royal Society of Painter-Etchers, 1910-38; ARA, 1906; RA, 1911; treasurer, 1919-32; knighted, 1911.

SHORT, (HUGH) OSWALD (1883-1969), aeronautical engineer; trained as engineer with help from brothers; founded showman-aeronaut business with brother, Eustace, 1898; manufactured observation balloons; official aeronauts and engineers to (Royal) Aero Club, 1907-8; with second brother, Horace, formed Short Brothers, 1908; suppliers to Royal Flying Corps (Naval Wing), 1911; responsible for Shorts' designs for flying boats of stressed-skin construction; chairman and managing director, Short Brothers Ltd.; obtained contracts for Empire flying boats for Imperial Airways, and Sunderlands and Stirlings for RAF; resigned as chairman and became honorary life president, 1943; president, Society of Aviation Artists; freeman of City of London; FRAS; FZS.

SHORTER, CLEMENT KING (1857-1926), journalist and author; clerk in exchequer and audit department, Somerset House, 1877-90; editor of *English Illustrated Magazine*, 1890; of *Illustrated London News*, 1891; founded and edited for *Illustrated London News* Company *Sketch*, 1893; parted from *Illustrated London News* and founded *Sphere*, 1900; remained editor until his death; founded *Tatler*, 1901; his works include *Charlotte Brontë and her Circle* (1896), *Napoleon and his Fellow-Travellers* (1908), *Napoleon in his own Defence* (1910), and *George Borrow and his Circle* (1913).

SHORTHOUSE, JOSEPH HENRY (1834-1903), novelist; of Quaker parentage; entered father's chemical works at Birmingham, 1850; influenced by John Ruskin [q.v.] and pre-Raphaelitism; joined Anglican communion, 1861; his psychological romance, *John Inglesant*, begun in 1866 and finished in 1876, was privately printed in 1880 and published in 1881; work attracted Gladstone, Huxley, Manning, and had wide vogue; other novels include *Sir Percival* (1886), *A Teacher of the Violin* (1888); also wrote on *The Platonism of Wordsworth* (1882).

SHORTT, EDWARD (1862-1935), politician; educated at Durham School and University; called to bar (Middle Temple), 1890; KC, 1910; recorder of Sunderland, 1907-18; liberal MP, Newcastle upon Tyne, 1910-22; chief secretary for Ireland, 1918-19; home secretary, 1919-22; president, Board of Film Censors, 1929-35; PC, 1918.

SHREWSBURY, ARTHUR (1856-1903), Nottinghamshire cricketer; played regularly for Nottinghamshire from 1875 to 1902; four times visited Australia; leading English batsman from 1882 to 1893; especially successful in 1887; scored sixty centuries in first-class cricket; showed strong defence and unwearying patience; established athletic outfitter's business in Nottingham, 1880; shot himself at Gedling, Nottinghamshire.

SHUCKBURGH, EVELYN SHIRLEY (1843-1906), classical scholar; educated at Ipswich grammar school and Emmanuel College, Cambridge; thirteenth classic, 1866; fellow, 1866-74; president of Union, 1865; assistant master at Eton, 1874-84; edited many elementary school classics; translated Polybius (1889), and *Cicero's Letters* (1889-1900); edited Suetonius's *Life of Augustus* (1896); Litt.D., Cambridge, 1902; published *Life of Augustus* (1903); also wrote histories of Rome (1894) and Greece (to 146 BC, 1901; AD 14, 1905), and a history of Emmanuel College, 1904.

SHUCKBURGH, SIR JOHN EVELYN (1877-1953), civil servant; son of Evelyn Shirley Shuckburgh [q.v.]; scholar of Eton and King's College, Cambridge; first class, classical tripos, 1899; India Office, 1900-21; assistant under-secretary of state, Middle East department, Colonial Office, 1921-31; deputy under-secretary, 1931-42; Cabinet Office, historical section, 1942-8; published *An Ideal Voyage* (1946); CB, 1918; KCMG, 1922.

SHUTE, NEVIL (pseudonym) (1899-1960), novelist. [See NORWAY, NEVIL SHUTE.]

SIBLY, SIR (THOMAS) FRANKLIN (1883-1948), geologist and university administrator; educated at Wycliffe and University College,

Bristol; B.Sc., London, 1903; researched on carboniferous limestones; chairman, Geological Survey Board, 1930-43; lecturer in geology, King's College, London, 1908-13; professor, University College, Cardiff, 1913-18; Armstrong College, Newcastle, 1918-20; principal, University College, Swansea, 1920-6; London University, 1926-9; vice-chancellor, Reading, 1929-46; chairman, Committee of Vice-Chancellors and Principals, 1938-43; knighted, 1938; KBE, 1943.

SICKERT, WALTER RICHARD (1860-1942), painter; studied under J. A. M. Whistler [q.v.]; drew frequently at music-halls; visited and later settled at Dieppe; met leading French Impressionists and writers; returned to London, 1905; associated with P. Wilson Steer, Frederick Brown [qq.v.] as leaders of London Impressionists; wrote with wit and common sense in defence of new ideals in art; changed technique under influence of Pissarro, painting mainly interiors with figures; sponsored Allied Artists' Association (1908) holding annual no-jury exhibition; associated with Spencer Frederick Gore and others as Camden Town Group (1911) and London Group (1913); turned to painting of pure landscape; later style looser, brighter, and more decorative; ARA, 1924; RA, 1934-5; painter of robust virility and great craftsmanship.

SIDEBOTHAM, HERBERT (1872-1940), journalist; educated at Manchester grammar school and Balliol College, Oxford; first class, *lit. hum.*, 1895; leader-writer and military critic, *Manchester Guardian*, 1895-1918; military critic and 'Student of Politics', *The Times*, 1918-21; after 1923 wrote as 'Scrutator' in *Sunday Times*, 'Candidus' in *Daily Sketch*, and 'Student of Politics' in *Daily Telegraph*.

SIDGREAVES, Sir ARTHUR FREDERICK (1882-1948), man of business; educated at Downside; with D. Napier & Son, Ltd., 1902-20; export manager, Rolls-Royce, 1920; general sales manager, 1926; managing director, 1929-46; undertook intensive production of 'Merlin' engine in war of 1939-45; knighted, 1945.

SIDGWICK, ELEANOR MILDRED (1845-1936), principal of Newnham College, Cambridge; sister of A. J. Balfour [q.v.]; married (1876) Henry Sidgwick [q.v.]; treasurer (1880-1919), principal (1892-1910), and generous benefactor of Newnham College; after 1916 with brother G. W. Balfour [q.v.] studied psychical phenomena and evidence of continuing personality after death; honorary secretary (1907-32), president (1908), Society for Psychical Research.

SIDGWICK, NEVIL VINCENT (1873-1952), chemist; nephew of Henry Sidgwick [q.v.]; educated at Rugby and Christ Church, Oxford; first class, chemistry, 1895; first class, *lit. hum.*, 1897; in Tübingen, working under von Pechmann, 1899-1901; Sc.D., 1901; tutorial fellow, Lincoln College, Oxford, 1901-48; published *Organic Chemistry of Nitrogen* (1910), *The Electronic Theory of Valency* (1927), *Some Physical Properties of the Covalent Link in Chemistry* (1933), and *The Chemical Elements and their Compounds* (2 vols., 1950); reader in chemistry, Oxford University, 1924-45; professor, 1935; delegate of the University Press; FRS, 1922; CBE, 1935.

SIEPMANN, OTTO (1861-1947), teacher of modern languages; born in Rhineland village; studied at Strasbourg University; modern languages master, Inverness College, 1888; Clifton College, 1890-1921; pioneer and leading authority on modern language teaching with special attention to pronunciation; produced French and German primers, annotated classics, and 'Word-and-Phrase' books; naturalized, 1905.

SIEVEKING, Sir EDWARD HENRY (1816-1904), physician; of a Hamburg family; studied at Berlin and Bonn universities; MD, Edinburgh, 1841; hon. LLD, 1884; practised at Hamburg, 1843-7; FRCP, 1852; vice-president, 1888; assistant physician to St. Mary's Hospital, London, 1851-66; physician, 1866-87; helped to found Epsom College, 1855; president of Harveian Society, 1861; knighted, 1886; invented an aesthesiometer, 1858; collaborated in *Manual of Pathological Anatomy* 1854, and wrote many papers on nervous diseases, climatology, and nursing.

SIFTON, Sir CLIFFORD (1861-1929), Canadian statesman; born in Ontario; called to bar, 1882; liberal member for North Brandon in Manitoba legislature, 1888; attorney-general and minister of education, 1891; minister of the interior, 1896; contributed greatly to development of Western Canada; resigned, owing to disagreement with prime minister, Sir Wilfrid Laurier [q.v.], 1905; prime agent in defeat of liberal government, 1911; organized commission for conservation of natural resources, 1909; strongly advocated Canadian constitutional equality with Great Britain; KCMG, 1915.

SILBERRAD, OSWALD JOHN (1878-1960), scientist; educated at Dean Close School, Cheltenham, City and Guilds Technical College, Finsbury, and Würzburg University; head, experimental establishment at Woolwich of the War Office explosives committee of which Lord

Rayleigh [q.v.] was chairman, 1902; developed explosives, including Tetryl and undertook chemical research; consulting research chemist and director, Silberrad Research Laboratories, 1906; produced erosion resistant bronze for ships' propellers, new method of dyestuffs' manufacture, and other discoveries, including a new chlorinating agent and a new method of blasting petroleum wells.

SILLITOE, Sir PERCY JOSEPH (1888–1962), policeman, and head of MI5; educated at St. Paul's choir school; worked for Anglo-American Oil Company, 1905–7; trooper in British South Africa police, 1908; transferred to Northern Rhodesia police, 1911; political officer, Tanganyika, 1916–20; district officer in Colonial service, 1920–2; chief constable of Chesterfield, 1923; chief constable, Sheffield, 1926; Glasgow, 1931; Kent joint force, 1943; director-general, MI5, in succession to Sir David Petrie [q.v.], 1946–53; in these years Allan Nunn May and Klaus Fuchs were tried; Pontecorvo, Burgess, and MacLean defected; promoted organization for security in the old Commonwealth countries; on retirement from MI5 head of De Beers International Diamond Security Organization; chairman, Security Express Ltd.; CBE, 1936; knighted, 1942; KBE, 1950; published *Cloak without Dagger* (1955).

SILVERMAN, (SAMUEL) SYDNEY (1895–1968), lawyer, member of Parliament, and penal reformer; of Jewish parentage; educated at Liverpool Institute and Liverpool University; imprisoned as conscientious objector, 1916–18; taught at Helsinki University, 1921–5; took law degree at Liverpool, 1927; worked successfully as solicitor in Liverpool; city councillor, 1932; labour MP, Nelson and Colne, 1935–68; wrote for the *Gazette*, labour party weekly of Nelson area; supported war against Nazi Germany, but opposed 'unconditional surrender' policy; chairman, British section of World Jewish Congress; active member of Campaign for Nuclear Disarmament; expelled from parliamentary party; led advocates of abolition of death penalty; succeeded in getting bill passed in parliament for abolition, 1964; published (with Reginald Paget) *Hanged, and Innocent?* (1953).

SIMMONS, Sir JOHN LINTORN ARABIN (1821–1903), field-marshal and colonel commandant, Royal Engineers; joined Royal Engineers, 1837; served in Canada, 1839–45; employed under railway commissioners, 1846–54; British commissioner with Omar Pasha's Turkish army on the Danube, 1854; helped in defence of Silistria; commanded 20,000 men at battle of Giurgevo (July); repulsed Russians at Eupatoria; took part with Turkish army in siege of Sevastopol; CB, Oct. 1855; accompanied Omar Pasha in attempt to relieve Kars (Nov.); made major-general in Turkish army; British commissioner for delimitation of new boundary between Turkey and Russia in Asia, 1857; British consul at Warsaw, 1858–60; commanding royal engineer at Aldershot, 1860–5; director of Royal Engineers' Establishment at Chatham, 1865–8; lieutenant-governor of Royal Military Academy at Woolwich and KCB, 1869; governor, 1870–5; lieutenant-general and colonel commandant Royal Engineers, 1872; member of royal commission on railway accidents, 1874–5; inspector-general of fortifications at the War Office, 1875–80; military delegate at Berlin congress, 1878; general, 1877; GCB, 1878; member of royal commission on defence of British possessions, 1880–2; governor of Malta, 1884–8; GCMG, 1887; field-marshal, 1890.

SIMON, ERNEST EMIL DARWIN, first BARON SIMON OF WYTHENSHAWE (1879–1960), industrialist and public servant; educated at Rugby and Pembroke College, Cambridge; first class, part i, mechanical sciences tripos, 1901; became head of family engineering business on death of father, 1899; active in liberal politics; lord mayor, Manchester, 1921; liberal MP, 1923–4 and 1929–31; joined labour party, 1946; knighted, 1932; baron, 1947; hon. LLD, Manchester University, 1944; chairman, council, Manchester University, 1941–57; chairman, British Broadcasting Corporation, 1947–52; with Sidney and Beatrice Webb [qq.v.], founded the *New Statesman*; published *A City Council from Within* (1926) and *Rebuilding Britain, a Twenty Year Plan* (1945); endowed Simon Population Trust.

SIMON, Sir FRANCIS (FRANZ) EUGEN (1803–1956), physicist; born in Berlin of Jewish parentage; educated in Berlin, Göttingen, and Munich; compulsory service in German army during 1914–18 war; D.Phil., Berlin, 1921; worked in Physikalisch-Chemisches Institut of Berlin University on specific heats at low temperatures, 1921–31; developed desorption method for liquefaction of helium; professor, physical chemistry, Technische Hoehschule, Breslau, 1931–3; visiting professor, California University, 1931; first person to liquefy helium in United States; invited by F. A. Lindemann (later Viscount Cherwell, q.v.) to Clarendon Laboratory, Oxford, 1933; reader in thermodynamics, 1936; naturalized, 1938; professor and student of Christ Church, 1945; worked on separation of uranium isotopes and other aspects of atomic energy, 1940–56; FRS, 1941; CBE, 1946; knighted, 1954; Dr Lee's professor, experimental philosophy, head of Clarendon Laboratory, 1956, one month before his death.

SIMON, SIR JOHN (1816–1904), sanitary reformer and pathologist; educated privately at Greenwich and in Germany; apprenticed to Joseph Henry Green [q.v.] at St. Thomas's Hospital, 1833–40; MRCS, 1838; hon. FRCS, 1844; president, 1878–9; senior assistant surgeon at King's College Hospital, 1840–7, and lecturer on pathology, 1847; subsequently surgeon at St. Thomas's Hospital, with which he was associated for life; FRS, 1945; first medical officer of health for the City of London, 1848; medical officer of general board of health, 1855–8, and of Privy Council under Public Health Act, 1858–71; published valuable annual reports, reprinted in *Public Health Reports* (2 vols., 1887); chief medical officer of newly formed Local Government Board, 1871–6; CB, 1876; crown member of Medical Council, 1876–95; KCB, 1887; first Harben medallist of the Royal Institute of Public Health (1896) and first Buchanan medallist of the Royal Society (1897) for services to sanitary science; received hon. degrees from Oxford, Cambridge, Edinburgh, Dublin, and Munich; intimate friend of John Ruskin [q.v.]; published *English Sanitary Institutions* (1890) and his *Personal Recollections* (1898, revised edn. 1903).

SIMON, JOHN ALLSEBROOK, first VISCOUNT SIMON (1873–1954), statesman and lord chancellor; educated at Bath grammar school, Fettes, and Wadham College, Oxford; first class, *lit. hum.*, and president of Union, 1896; fellow, All Souls, 1897; called to bar (Inner Temple), 1899; KC, 1908; liberal MP, Walthamstow, 1906–18, and Spen Valley, 1922–40; solicitor-general and knighted, 1910; PC, 1912; attorney-general, 1913; refused lord chancellorship, 1915; home secretary, 1915–16; resigned on conscription issue; chairman, Indian statutory commission, 1927–30; formed liberal national party, 1931; foreign secretary, 1931–5; home secretary, 1935–7; chancellor of Exchequer, 1937–40; lord chancellor and viscount, 1940–5; KCVO, 1911; GCSI, 1930; GCVO, 1937; high steward, Oxford University, 1948.

SIMON, OLIVER JOSEPH (1895–1956), printer; nephew of (Sir) William Rothenstein [q.v.]; educated at Charterhouse and Jena; served in Gallipoli, Egypt, and Palestine in 1914–18 war; joined Curwen Press, 1920; chairman, 1949; with Hubert Foss [q.v.], founded Double Crown Club, 1924; published *Signature*, a periodical covering design, printing, and calligraphy, 1935–54; recognized internationally as typographical authority; published *Introduc-*

tion to Typography (1945) and *Printer and Playground* (1956); OBE, 1953.

SIMONDS, JAMES BEART (1810–1904), veterinary surgeon; consulting veterinary surgeon to Royal Agricultural Society, 1842–1904; president of Royal College of Veterinary Surgeons, 1862–3; reported on continental and London cattle plague, 1857 and 1865; chief inspector of veterinary department of Privy Council office, 1865–71; principal of Royal Veterinary College, 1871–81; *Autobiography* privately printed, 1894.

SIMONSEN, SIR JOHN LIONEL (1884–1957), organic chemist; educated at Manchester grammar school and Manchester University; first class, chemistry, 1904; D.Sc., 1909; assistant lecturer, 1907; professor of chemistry, Presidency College, Madras, 1910; helped to found Indian Science Congress Association, 1914; general secretary, 1914–26; chief chemist, Forest Research Institute and College, Dehra Dun, 1919–25; professor of organic chemistry, Indian Institute of Science, Bangalore, 1925–8; professor of chemistry, University College of North Wales, Bangor, 1930–42; worked on chemistry of terpenes and sesquiterpenes; director of research, Colonial Products Research Council, 1943–52; FRS, 1932; hon. degrees from universities of Birmingham, Malaya, and St. Andrews; knighted, 1949.

SIMPSON, SIR GEORGE CLARKE (1878–1965), meteorologist; educated at Diocesan School, Derby, and the Owens College, Manchester; first class, physics, 1900; exhibitioner doing research at Geophysikalisches Institüt, Göttingen, 1902; lecturer in meteorology, Manchester University, 1905; joined Indian Meteorological Office under (Sir) Gilbert T. Walker [q.v.], 1905; made study of thunderstorm electricity; took part in Captain R. F. Scott's [q.v.] Antarctic expedition, 1909–12; worked with Indian Munitions Board during 1914–18 war; director, British Meteorological Office, succeeding Sir (William) Napier Shaw [q.v.], 1920–38; in charge of Kew Observatory, 1939–47; continued work on electrical structure of thunderstorms; FRS, 1915; CBE, 1919; CB, 1926; KCB, 1935; D.Sc. (Sydney, 1914, and Manchester, 1906); hon. LLD, Aberdeen; hon. FRSE, 1947; member of council of British Association, 1927–35.

SIMPSON, SIR JOHN WILLIAM (1858–1933), architect; associate (1882), fellow (1900), president (1919–21), Royal Institute of British Architects; specialized in design of public buildings, including the National Hospital, Queen Square, and Crown Agents' Offices, Millbank;

responsible in collaboration for general layout, stadium, palaces of industry and engineering, Wembley Exhibition, 1924; KBE, 1924.

SIMPSON, MAXWELL (1815-1902), chemist; BA, Trinity College, Dublin, 1837; studied chemistry at University College, London; MB, 1847; obtained synthetically succinic and other di- and tri-basic acids; professor of chemistry in Queen's College, Cork, 1872-91; FRS, 1862; fellow of Royal University of Ireland, 1882-91; hon. D.Sc., 1882; hon. MD, Dublin, 1864, LLD, 1878.

SIMPSON, PERCY (1865-1962), scholar; educated at Denstone College and Selwyn College, Cambridge; taught at Denstone, 1887-95; and St. Olave's grammar school, Southwark, 1899; began work on annotating plays of Ben Jonson; worked for Clarendon Press, Oxford, 1913; first librarian of new English faculty library, 1914; fellow, Oriel College, 1921; university reader in English textual criticism, 1927; Goldsmith's reader in English literature, 1930-5; hon. fellow, Oriel College, 1943; published *Shakespearian Punctuation* (1911) and *Proof-Reading in the Sixteenth, Seventeenth and Eighteenth Centuries* (1935); with help from his wife, completed work begun by C. H. Herford [q.v.] on editing works of Ben Jonson; hon. degrees from Cambridge and Glasgow; hon. fellow, Selwyn College, Cambridge.

SIMPSON, WILFRED HUDLESTON (1828-1909), geologist. [See HUDLESTON.]

SIMPSON, SIR WILLIAM JOHN RITCHIE (1855-1931), physician and pioneer in tropical medicine; MB, CM, Aberdeen, 1876; MD and medical officer of health, Aberdeen, 1880; first medical officer of health, Calcutta, 1886-98; professor of hygiene, King's College, London, and lecturer on tropical hygiene, London School of Tropical Medicine, 1898-1923; director of tropical hygiene and physician, Ross Institute and Hospital for Tropical Diseases, 1926-31; investigated plague and health conditions in many colonies; knighted, 1923.

SIMS, CHARLES (1873-1928), painter; studied art in London and Paris, 1890-2; exhibited at Royal Academy, 1896; took to outdoor figure-painting, 1900; began series of mother-and-child pictures, 1901; held successful first 'one-man' show at Leicester Galleries (1906) which established his reputation; 'The Fountain' (1908) and 'The Wood beyond the World' (1913) purchased by Chantrey trustees for Tate Gallery; RA, 1915; an official artist in France, 1918; keeper of Royal Academy Schools, 1920-6; an assiduous and fertile technical

experimenter; delighted in painting open-air lyrics.

SINCLAIR, SIR ARCHIBALD HENRY MACDONALD, first baronet, of Ulbster, Caithness, and first VISCOUNT THURSO (1890-1970), politician; educated at Eton and Sandhurst; commissioned into the 2nd Life Guards, 1910; succeeded to baronetcy, 1912; served in France, 1914-18; second in command to his friend, (Sir) Winston Churchill, 1916; personal military secretary to Churchill at War Office, 1919-21; private secretary at Colonial Office, 1921-2; liberal MP Caithness and Sutherland, 1922-45; liberal chief whip, 1930; joined 'national' government as secretary of state for Scotland, 1931; resigned with Sir Herbert Samuel (later first Viscount Samuel, q.v.) in protest against Ottawa agreements, 1932; succeeded Samuel as chairman parliamentary liberal party, 1935; with Churchill, condemned Munich agreement; declined office under Chamberlain, 1939; supported Churchill, 1940; secretary of state for air, 1940-5; advocated strategic bombing; defeated in 1945 election; accepted peerage, 1950; created viscount, 1952; leader of liberals in House of Lords; lord-lieutenant, Caithness, 1919-64; CMG, 1922; knighted, 1941; PC, 1931.

SINCLAIR, SIR EDWYN SINCLAIR ALEXANDER- (1865-1945), admiral. [See ALEXANDER-SINCLAIR.]

SINCLAIR, JOHN, first BARON PENTLAND (1860-1925), politician; followed military career, 1879-87; liberal MP, Dumbartonshire, 1892-5; assistant private secretary to (Sir) Henry Campbell-Bannerman; MP, Forfarshire, 1897-1909; secretary for Scotland, 1905-12; carried through National Galleries (Scotland) Act (1906), Scottish Education Act (1908), and small landholders (Scotland) bill (1911); baron, 1909; governor of Madras, 1912-19.

SINGER, CHARLES JOSEPH (1876-1960), historian of medicine and science; educated at City of London School and Magdalen College, Oxford; MRCS, LRCP, St. Mary's Hospital, Paddington, 1903; BM, Oxford, 1906; MRCP, London, 1909; registrar, Cancer Hospital and physician, Dreadnought Hospital, 1909-14; DM, Oxford, 1911; FRCP, 1917; captain, RAMC, 1914-18; lecturer in history of medicine, University College, London, 1920; professor, 1931-4; publications include *Studies in the History and Methods of Science* (2 vols., 1917-21), *The Evolution of Anatomy* (1925), a *Short History of Medicine* (1928), *of Biology* (1931), and *of Science* (1941), and edited a *History of Technology* (5 vols., 1954-8); D.Litt.,

Oxford, 1922; hon. D.Sc., Oxford, 1936; hon. fellow, Magdalen College, Oxford, and of the Royal Society of Medicine.

SINGLETON, Sir JOHN EDWARD (1885–1957), judge; educated at Lancaster grammar school and Pembroke College, Cambridge; called to bar (Inner Temple), 1906; served in Royal Field Artillery during 1914–18 war; KC, 1922; conservative MP, Lancaster, 1922–3; judge of appeal, Isle of Man, 1928–33; recorder of Preston, 1928–34; judge of King's Bench and knighted, 1934; lord justice of appeal and PC, 1948–57; bencher, Inner Temple, 1929; treasurer, 1952; hon. fellow, Pembroke College, Cambridge, 1938; hon. LLD, Liverpool, 1949; published *Conduct at the Bar* (1933); British chairman, Anglo-American Palestine commission, 1945.

SINGLETON, MARY (1843–1905), authoress under the pseudonym of VIOLET FANE. [See CURRIE, MARY MONTGOMERIE, LADY CURRIE.]

SINHA, SATYENDRA PRASANNO, first BARON SINHA (1864–1928), Indian statesman; born in Bengal; came to England, 1881; called to bar (Lincoln's Inn), 1881; standing counsel to Bengal government, 1903; officiated as advocate-general, Bengal, 1905; confirmed, 1908; legal member, governor-general's council, 1909; first Indian member of Indian government; president, Indian National Congress, 1915; member, imperial war cabinet and conference in London, 1917; parliamentary under-secretary of state for India, 1919; governor of Bihar and Orissa, 1920–1; member of judicial committee of Privy Council, 1926; knighted, 1914; KC, 1918; baron, 1919; PC, 1919; died in Bengal.

SITWELL, DAME EDITH LOUISA (1887–1964), poet and critic; daughter of Sir George Reresby Sitwell, fourth baronet [q.v.], and sister of Osbert [q.v.]; first volume of poetry, *The Mother* (published, 1915); collaborated with Osbert on *Twentieth-Century Harlequinade* (1916); successful performance of *Façade*, concert piece, with music by (Sir) William Walton (1923); published more poems, including *Bucolic Comedies* (1923), *The Sleeping Beauty* (1924), *Troy Park* (1925), and *Gold Coast Customs* (1929); lived in Paris until her friend Helen Rootham died in 1938; further published poems include *Street Songs* (1942), *Green Song* (1944), *The Shadow of Cain* (1947), *Gardeners and Astronomers* (1953), and *The Outcasts* (1962); published prose works include *The English Eccentrics* (1933), *Fanfare for Elizabeth* (1946), and *The Queens and the Hive* (1962); her anthologies include *A Poet's Notebook* (1943) and *The Atlantic Book of British and American*

Poetry (1959); DBE, 1954; hon. doctorates, Oxford, Leeds, Durham, and Sheffield.

SITWELL, SIR (FRANCIS) OSBERT SACHEVERELL (1892–1969), fifth baronet, writer; educated at Eton, 1906–9; commissioned in Sherwood Rangers, 1911; transferred to Grenadier Guards (Special Reserve), 1912; first poem, *Babel* (published, 1916); also *Twentieth-Century Harlequinade, and Other Poems* (in collaboration with his sister (Dame) Edith [q.v.]); left army, 1919; published novels, *Before the Bombardment* (1926), *The Man who Lost Himself* (1929), and *Miracle on Sinai* (1933); travel writings, *Winters of Discontent* (1932) and *Escape with Me* (1939); short stories, *Collected Stories* (1953); and poems *England Reclaimed* (1927) and *On the Continent* (1958); finest achievement, his five-volume autobiography, *Left Hand, Right Hand* (1944–50); succeeded to baronetcy, 1943; CBE, 1956; CH, 1958; hon. LLD, St. Andrews, 1946; hon. D.Litt., Sheffield, 1951; trustee, Tate Gallery, 1951–8.

SITWELL, SIR GEORGE RERESBY, fourth baronet (1860–1943), antiquarian; succeeded father, 1862; educated at Eton and Christ Church, Oxford; conservative MP, Scarborough, 1885–6, 1892–5; wide range of interests included genealogy, heraldry, and garden planning; travelled extensively in Italy; retired there to Castello di Montegufoni, 1925; publications include *The Barons of Pulford . . .* (1889), *The First Whig* (1894), *The Letters of the Sitwells and Sacheverells* (1900), and *On the Making of Gardens* (1909); erratic and difficult parent of Osbert, Sacheverell, and Edith Sitwell.

SKEAT, WALTER WILLIAM (1835–1912), philologist; BA, Christ's College, Cambridge; first Elrington and Bosworth professor of Anglo-Saxon, Cambridge, 1878–1912; works include *Anglo-Saxon Gospels* (1871–87), Ælfric's *Lives of Saints* (1881–1900), great edition of *Piers Plowman* (completed 1886), seven-volume edition of Chaucer (1894–7), and *Etymological Dictionary* (1879–82).

SKELTON, RALEIGH ASHLIN (1906–1970), cartographical historian; educated at Aldenham School and Pembroke College, Cambridge; first class, French and German, part i, modern languages tripos, 1927; assistant keeper, department of printed books, British Museum, 1931; served in Royal Artillery, 1939–45; worked in BM map room, 1945; deputy keeper, 1953; FSA, 1951; chairman, IGU commission on early maps, 1961; involved in organization of *Monumenta Cartographica vetustioris aevi A.D. 1200–1500*; served on committies of number of learned societies; secretary, Hakluyt Society,

1946-66; corresponding editor, *Imago Mundi*, 1950; secretary, 1959-66; general editor, 1966-70; revised and translated Leo Bagrow's *History of Cartography* (1964); lectured and advised on cartography at Harvard and Chicago and in Newfoundland; collaborated with T. E. Marston and G. D. Painter on publication of *The Vinland Map and the Tartar Relation* (1965); contributed to English map making; publications include *County Atlases of the British Isles, 1579-1703* and *Explorers' Maps* (1958); after retirement from British Museum in 1967 continued up to his death to work on compilations of historical maps.

SKIPSEY, JOSEPH (1832-1903), the collier poet; worked in Northumberland coalpits from the age of seven; taught himself to read and write; varied work in mines with other employment until 1882; published *Poems* (1859), *Poems, Songs and Ballads* (1862), *The Collier Lad, and other Lyrics* (1864), *Carols from the Coalfields* (1886); edited the 'Canterbury Poets', 1884-5; custodian of Shakespeare's birthplace, Stratford-upon-Avon, 1889-91; was commended by D. G. Rossetti and Burne-Jones [qq.v.]; awarded civil list pension, 1880.

SLANEY, WILLIAM SLANEY KENYON- (1847-1908), colonel and politician. [See KENYON-SLANEY.]

SLATER, SIR WILLIAM KERSHAW (1893-1970), scientist and agricultural administrator; educated at Hulme grammar school, Oldham, and Manchester University; B.Sc., first class, chemistry, 1914; served in Explosives Directorate, 1914-18; assistant lecturer in chemistry, Manchester University, 1919-21; research in biochemistry, Manchester, 1922; medical research fellowship, University College, London, 1923; joined Leonard and Dorothy Elmhirst at Dartington, 1929; trustee, Dartington Hall; joined Agricultural Improvement Council staff, 1942; secretary, AIC, 1944; secretary, Agricultural Research Council, 1949-60; KBE, 1951; FRS, 1957; D.Sc., Manchester; hon. D.Sc., Queen's University, Belfast; president, Royal Institute of Chemistry, 1961-3; general secretary, British Association for Advancement of Science, 1962-5; member, Colonial Research Council; published *Man Must Eat* (1964).

SLIM, WILLIAM JOSEPH, first VISCOUNT SLIM (1891-1970), field-marshal; educated at King Edward's School, Birmingham; joined Birmingham University OTC; commissioned in Royal Warwickshire Regiment, 1914; served in Middle East, 1915-16; MC; transferred to Indian Army, 1919; posted to 1/6th Gurkha Rifles, 1920; entered Staff College, Quetta,

1926-8; at army headquarters, India, 1928-34; Indian Army instructor, Staff College, Camberley, 1934; in command 2/7th Gurkhas, 1937; commanded 10th Indian infantry brigade in Eritrea, 1940, and 10th India division in Iraq and Syria, 1941; as lieutenant-general, evacuated 'Burcorps' from Burma, 1942; commanded XV Corps; in Arakan campaign, 1942-3; took command of 14th Army, 1943; conducted offensive in Arakan and in drive to Chindwin river, 1944; disapproved of Chindit operations of Major-General Orde Wingate; carried out successful advance into Burma, taking Meiktila and Mandalay, 1945; recaptured Rangoon; intervention by Lord Alanbrooke [q.v.], CIGS, cancelled decision of Sir Oliver Leese to replace him in command of 14th Army; promoted to full general, and superseded Leese as commander-in-chief, Allied Land Forces, South-East Asia; commandant, Imperial Defence College, 1946-8; chief of imperial general staff; field-marshal, 1948; governor-general, Australia, 1953-60; honoured by universities; freeman of City of London; colonel of three regiments; constable and governor, Windsor Castle, 1964; published *Defeat into Victory* (1956), *Courage and other Broadcasts* (1957), and *Unofficial History* (1959); CBE, 1942; DSO, 1943; CB and KCB, 1944; GBE, 1946; GCB, 1950; GCMG, 1952; GCVO, 1954; KG, 1959; viscount, 1960.

SMART, SIR MORTON WARRACH (1877-1956), manipulative surgeon; educated at George Watson's College, Edinburgh, and Edinburgh University; qualified at Royal College of Surgeons; MB, Ch.B., 1902; MD, 1914; served in Royal Naval Volunteer Reserve, 1914-18; DSO, 1917; founded London Clinic for Injuries; pioneer of manipulative surgery and authority on rehabilitation; consultant in physical medicine, RAF, 1939-45; CVO, 1932; KCVO, 1933; GCVO, 1949; published *The Principles of Treatment of Muscles and Joints by Graduated Muscular Contractions* (1933).

SMART, WILLIAM GEORGE (BILLY) (1893-1966), circus proprietor; at age of fifteen worked on his father's fairground at Slough; after one or two failures, succeeded in setting up his own fairground business; expanded into travelling circus, 1946; first circus to be screened by BBC, 1957; grew to be one of largest in the world; permanent quarters, with zoo, at Winkfield; extensive works for charity; regular television appearances; a showman on the grand scale.

SMARTT, SIR THOMAS WILLIAM (1858-1929), South African politician; LRCS Ireland; in medical practice, Cape Colony, 1880;

member for Wodehouse, Cape house of assembly, 1893; colonial secretary under Sir J. G. Sprigg [q.v.], 1896-8; intimately associated with Cecil Rhodes [q.v.]; commissioner of crown lands and public works under (Sir) L. S. Jameson [q.v.], 1904-8; supported General Botha [q.v.] in work for South African Union; with Jameson organized unionist party, 1910; leader, 1912; secretary of agriculture under General Smuts [q.v.], 1921-4; KCMG, 1911; died at Cape Town.

SMEATON, DONALD MACKENZIE (1846-1910), Indian civil servant; MA, St. Andrews University; entered Indian civil service, 1865; secretary of land revenue department and director of agriculture, Burma, 1882-6; chief secretary to chief commissioner, 1887; published account of hill races in *Loyal Karens of Burma* (1887); financial commissioner of Burma, 1891-1902; represented Burma on supreme legislative council, 1898-1902; retired in 1902 to England; liberal MP for Stirlingshire, 1906-10.

SMILES, SAMUEL (1812-1904), author and social reformer; educated at Haddington grammar school; apprenticed to local medical practitioners, 1826; studied at Edinburgh University, 1829-32; engaged in practice at Haddington, 1832-8; published *Physical Education* (1837, new edns. 1868 and 1905); travelled in Holland and first visited London, 1838; editor of *Leeds Times*, an advanced radical organ, 1838-42; assistant secretary of Leeds and Thirsk railway, 1845; secretary of South Eastern Railway, 1854-66; president of the National Provident Institution, 1866-71; devoted leisure to advocacy of political and social reform on lines of Manchester school, and to biography of industrial leaders or of humble self-taught students; published *Life of George Stephenson* (1857), *Lives of the Engineers* (3 vols., 1861-2), and many similar works; achieved great popular success with *Self-help* (1859), *Character* (1871), *Thrift* (1875), *Duty* (1880), and *Life and Labour* (1887); wrote also on *The Huguenots . . . in England and Ireland* (1867) and *The Huguenots in France* (1874); hon. LLD, Edinburgh, 1878; other biographies treated of George Moore the philanthropist (1878), and of John Murray the publisher (1891); autobiography posthumously issued, 1905.

SMILLIE, ROBERT (1857-1940), labour leader and politician; worked in Lanarkshire mines; president, Scottish Miners' Federation, 1894-1918, 1921-40; active in founding Miners' Federation of Great Britain; president, 1912-21; displayed force and skill in giving evidence before royal commission on coal industry, 1919;

close friend of Keir Hardie [q.v.]; founder-member of independent labour party; a pacifist; MP, Morpeth, 1923-9.

SMITH, SIR ARCHIBALD LEVIN (1836-1901), judge; BA, Trinity College, Cambridge, 1858; rowing blue (1857-9); good cricketer; president of MCC, 1899; called to bar (Inner Temple), 1860; standing junior counsel to Treasury, 1879; judge of Queen's Bench division and knighted, 1883; special commissioner to inquire into *The Times* allegations against C. S. Parnell [q.v.], 1888; promoted to Court of Appeal, 1892; master of the rolls, 1900.

SMITH, ARTHUR HAMILTON (1860-1941), archaeologist; son of Archibald Smith [q.v.]; educated at Winchester and Trinity College, Cambridge; entered department of Greek and Roman antiquities, British Museum, 1886; assistant keeper, 1904; keeper, 1909-25; catalogued the sculpture and engraved gems; invented cyclograph for photographing curved surfaces; FBA, 1924.

SMITH, ARTHUR LIONEL (1850-1924), historian; educated at Christ's Hospital; BA, Balliol College, Oxford; Lothian prize essayist, 1874; fellow and lecturer, Trinity College, Oxford, 1874-9; modern history tutor, Balliol, 1879; fellow, 1882; helped to build up position of history school in university; a devoted and successful college tutor; keenly interested in college athletics; dean, 1907; master, 1916; an ardent member of standing joint committee of university extension delegacy, 1908; lectured to working-class audiences all over England during 1914-18 war; served on Archbishops' Committee on Church and State (1914) and on Christianity and Industrial Problems, 1916-17; his works include *Church and State in the Middle Ages* (Ford's lectures, 1905, 1913).

SMITH, SIR CECIL HARCOURT- (1859-1944), archaeologist and director of the Victoria and Albert Museum. [See HARCOURT-SMITH.]

SMITH, SIR CHARLES BEAN EUAN- (1842-1910), soldier and diplomatist. [See EUAN-SMITH.]

SMITH, SIR CHARLES EDWARD KINGS-FORD (1897-1935), air-pilot; born at Hamilton, Brisbane; educated at Sydney Cathedral School and Technical College; commissioned in Royal Air Force, 1917-19; founded Interstate Services as Australian taxi firm, 1926; flew round Australia in ten days, 1927; in *Southern Cross* made first trans-Pacific flight (1928), flew from Australia to England (1929), and completed flight round world (1930); made solo flight from

England to Australia in seven days five hours, 1933; left England Nov. 1935 and last seen over Calcutta; knighted, 1932.

SMITH, DAVID NICHOL (1875-1962), scholar; educated at George Watson's College, Edinburgh, and Edinburgh University; first class, English, 1895; revised the index of George Saintsbury's [q.v.] *History of Nineteenth Century Literature*; studied at Sorbonne; produced edition of Boileau's *L'Art Poetique* (1898); assistant to (Sir) Walter Raleigh [q.v.] at Glasgow University, 1902; professor of English, Armstrong College, Newcastle upon Tyne, 1904; Goldsmith's reader, Oxford, 1908; fellow, Merton College, 1921; Merton professor of English Literature, 1929-46; worked at Huntington Library, 1936-7; at Smith College and Chicago University, 1946-7; Nuffield lecturer, New Zealand, 1950-1; publications include *Eighteenth Century Essays on Shakespeare* (1903), *Some Observations on Eighteenth Century Poetry* (1937), and *John Dryden* (1950); edited Swift's *Tale of a Tub* (with A. C. Guthkelch, 1920), *The Letters of Jonathan Swift to Charles Ford* (1935), and *The Poems of Samuel Johnson* (with E. L. McAdam, 1941); FBA, 1932; hon. degrees from universities at home and abroad.

SMITH, DONALD ALEXANDER, first BARON STRATHCONA AND MOUNT ROYAL (1820-1914), Canadian financier; born in Scotland; went to Canada as clerk in Hudson's Bay Company, 1838; a 'chief factor', 1862; head of company's Montreal department, 1868; sent by Canadian government to negotiate with Louis Riel [q.v.]; conservative member of federal parliament, 1871-9; with group of colleagues, which included a cousin, George Stephen (afterwards Baron Mount Stephen, q.v.), completed greater part of Great Northern Railway, 1879, and Canadian Pacific Railway, 1885; knighted, 1886; re-entered parliament for Montreal constituency, 1887; governor of Hudson's Bay Company, 1889; high commissioner for Canada, 1896; baron, 1897; raised 'Strathcona's Horse' during South African war; lived latter years in England and died there.

SMITH, SIR ERNEST WOODHOUSE (1884-1960), fuel technologist; educated at Arnold School, Blackpool, and Manchester University; D.Sc., 1918; research chemist, Institution of Gas Engineers, Leeds University, 1908; Birmingham City gas department, 1910-20; technical director, Woodall-Duckham Companies, 1920-44; director-general gas supply, Ministry of Fuel and Power, 1942-3; a founder of Gas Research Board; president, Institute of Fuel, 1943-5; CBE, 1930; knighted, 1947.

SMITH, SIR FRANCIS (VILLENEUVE-) (1819-1909), chief justice of Tasmania; early accompanied father to Van Diemen's Land; educated at University College, London; BA, 1840; called to bar (Middle Temple), 1842, and of Van Diemen's Land, 1844; crown solicitor of colony, 1849; attorney-general, 1854-6; attorney-general in first responsible ministry, 1856-7; attorney-general and premier, 1857-60; puisne judge of supreme court, 1860-70; knighted, 1862; chief justice, 1870-84.

SMITH, SIR FRANK EDWARD (1876-1970), industrial scientist; educated at Smethwick central school and Royal College of Science; RCS in physics (first class), 1899; assistant to (Sir) Richard Glazebrook [q.v.], National Physical Laboratory, Kew, 1900; research on accurate electrical standards and methods of measurement; constructed first 'Lorenz machine'; proposed International Congress on Electrical Units and Standards; held in London, 1908; established standard units; war work, 1914-18; invented first magnetic mine; FRS, 1918; director, Scientific Research and Experimental Department, Admiralty, 1920; supervised construction of Admiralty Research Laboratory, Teddington; secretary, Department of Scientific and Industrial Research, 1929-39; secretary, Royal Society, 1929-38; closely concerned with administration of many other scientific bodies; adviser to Anglo-Iranian Oil (later British Petroleum), 1939-56; director, Birmingham Small Arms Co. Ltd., 1947-57; chairman, Technical Defence Committee, MI5, 1940-6; OBE, 1918; CBE, 1922; GBE, 1939; CB, 1926; KCB, 1931; GCB, 1942; many academic honours.

SMITH, SIR FREDERICK (1857-1929), major-general, Army Veterinary Service; educated at Royal Veterinary College, London; commissioned to artillery; investigated cause of sore backs among horses belonging to 12th Lancers in India, 1880; professor, Army Veterinary School, Aldershot, 1886-91; served in South Africa, 1899-1905; director-general, Army Veterinary Service, and honorary major-general, 1907-10; KCMG, 1918; a prolific writer on veterinary subjects.

SMITH, FREDERICK EDWIN, first EARL OF BIRKENHEAD (1872-1930), lord chancellor; BA (first class in jurisprudence, 1895), Wadham College, Oxford; president of Union, 1893; Vinerian scholar and fellow of Merton College, Oxford, 1896; called to bar (Gray's Inn), 1899; began to practise in Liverpool; engaged in several important cases, 1902-6; conservative MP, Walton division of Liverpool, 1906-18; moved to London; achieved parliamentary reputation by his maiden speech; unfortunately his extravagance and fondness for gaiety soon gave impression that he was a reckless partisan, fighting for his own hand, and not a responsible or serious person either as lawyer or as parliamentarian; made many enemies; attacked mass of attempted liberal legislation with great bitterness, but believed that vital issues could only be dealt with by government comprising men of all constitutional parties; opposed Lloyd George's budget (1909) in Commons, but urged friends in Lords to allow it to pass; opposed parliament bill, Welsh disestablishment, and home rule in parliament and country, but when party leaders entered into conference (June 1910) advocated party truce and formation of national government; adhered to extreme right wing of conservative party which advocated resistance by Lords to parliament bill; at first absorbed in preparations for resistance of Ulster to home rule, but later advocated settlement based on exclusion of Ulster; during these years carried on ever-increasing practice at bar; KC, 1908; after outbreak of 1914-18 war constantly engaged on constructive work for government; head of Press Bureau, 1914; solicitor-general in first coalition government, May 1915; attorney-general, Nov. 1915; greatly concerned with law of prize; most famous cases in which he was engaged in court, SS *Zamora* and SS *Ophelia* (1916), trial of Sir Roger Casement (1916), trial of Wheeldons (1917), and Rhodesian land case (1918); member of cabinet until end of war; MP, West Derby division of Liverpool, 1918; lord chancellor, 1919-22; acquired great authority among his legal colleagues; strove earnestly to do justice; heard many famous cases; among other law reforms accomplished amendment of law relating to transfer of land, 1922; supported Irish treaty, 1921; greatly lamented fall of coalition government, 1922; secretary of state for India, 1924-8; baronet, 1918; baron, 1919; viscount, 1921; earl, 1922.

SMITH, GEORGE (1824-1901), publisher, and founder and proprietor of the *Dictionary of National Biography*; apprenticed to Smith and Elder & Co., publishers, in which his father was a partner, 1838; took charge of some of the firm's publishing work, 1843; became head of the firm on his father's death, 1846; published works by Ruskin, Charlotte Brontë, and Thackeray [qq.v.], including *Modern Painters* (1848), *Jane Eyre* (1848), and *Esmond* (1851); founded *The Cornhill Magazine* (1859) and the *Pall Mall Gazette* (1865); extended publishing activities to include medical books and new authors among whom were Robert Browning, Matthew Arnold, and (Sir) Leslie Stephen [qq.v.]; founded the *Dictionary of National*

Biography, with Sir Leslie Stephen as first editor, 1882.

SMITH, Sir GEORGE ADAM (1856-1942), Old Testament scholar and theologian; educated at Royal High School, University, and New College, Edinburgh, and in Germany; minister, Queen's Cross Free Church, Aberdeen, 1882-92; professor, Old Testament language, literature, and theology, Free Church College, Glasgow, 1892-1909; principal, Aberdeen University, 1909-35; publications include *Book of Isaiah* (2 vols., 1888-90), *Historical Geography of the Holy Land* (1894), *Modern Criticism* (1901), and *Jerusalem* (2 vols., 1908); knighted and FBA, 1916.

SMITH, GEORGE BARNETT (1841-1909), author and journalist; on editorial staff of *Globe*, 1865-8, and *Echo*, 1868-76; compiled lives of Shelley (1877), Gladstone (1879), John Bright (1881), Victor Hugo (1885), and *History of the English Parliament* (2 vols., 1892); a skilful etcher; awarded civil list pension, 1891.

SMITH, GEORGE VANCE (1816?-1902), Unitarian biblical scholar; divinity student at Manchester College, York (removed to Manchester in 1840 and to London in 1853), where he held many offices, 1839-50, finally serving as principal, 1850-3, and as professor of critical and exegetical theology, 1853-7; MA and Ph.D., Tübingen, 1857; minister of St. Saviourgate chapel, York, 1858-75; joined New Testament revision company, 1870; DD, Jena, 1873; principal of Presbyterian College, Carmarthen, 1876-88; voluminous writings include *The Bible and Popular Theology* (1871) and *The Prophets and their Interpreters* (1878).

SMITH, GOLDWIN (1823-1910), controversialist; at Eton, 1836-41; matriculated from Christ Church, Oxford, 1841; elected demy of Magdalen College, 1842; showed unusual classical aptitude; made a record as winner of classical prizes; formed in youth lasting liberal opinions on religious and political questions; BA, 1845; MA, 1848; Stowell law fellow at University College, 1846-50; ordinary fellow, 1850-67; tutor, 1850-4; attacked clerical ascendancy in university; joint-secretary of Oxford University commission (1850) and of executive commission which framed new regulations under Oxford University Reform Act, 1855-7; an effective writer in *Saturday Review*, 1855-8; member of royal commission on national education, 1858; regius professor of modern history at Oxford, 1858-66; stimulated thought in Oxford by his professorial lectures; engaged in political agitation and pamphleteering; defined his distrust of imperialism in *The*

Empire (1863) (letters reprinted from the *Daily News*); was denounced as a mischievous propagandist by Disraeli; diagnosed Irish difficulties in *Irish History and Irish Character* (1862); influenced by John Bright [q.v.], he actively advocated cause of North in American civil war; visited America, 1864; joined in attack on Governor Eyre [q.v.] for alleged ill treatment of negroes in Jamaica, 1867; published lectures on *Three English Statesmen* (Pym, Cromwell, and Pitt, 1867), to provide funds for Jamaica committee; suffered anxiety from illness (1866) of father, who committed suicide next year; resolved on prolonged residence in America; accepted offer of professorship of history at newly founded Cornell University, Ithaca, NY, 1868; bade farewell to Oxford in a pamphlet on the *Reorganization of the University* (1868); reached Ithaca, Nov. 1868; devoted himself zealously to professorial duties there; resented American hostility to England, 1869–70, and Disraeli's attack on him as a 'social parasite' in *Lothair* (1870); made a tour in Canada, 1870; settled at Toronto, 1871; married there (1875) widow of Henry Boulton, of 'The Grange', which became his permanent residence for life; at Toronto Goldwin Smith was very active in journalism, contributing literary and political essays to daily, weekly, and monthly periodicals of Canada, America, and England; he strongly encouraged sentiment of national independence in Canada from 1871, and a commercial union with the United States from 1888, recommending the amalgamation of the English-speaking race on American continent; on frequent visits to England he took much part in public life there, but declined invitations to stand for parliament or to become master of University College, Oxford, 1881; hon. DCL, Oxford, 1882; strenuously opposed Gladstone's home rule policy, 1886; always remained faithful to unionism (cf. *Irish History and the Irish Question*, 1906); opposed South African war, 1899; explained his pacifist position in *In the Court of History* (1902); and warned America against imperialism in *Commonwealth and Empire* (1902); other scattered political writings were collected in *Lectures and Essays* (1881), and *Essays on Questions of the Day* (1893); continued his historical and literary studies in a monograph on Cowper (1880), *Specimens of Greek Tragedy* (1893), *The United States: an Outline of Political History, 1492–1871* (1893), and *The United Kingdom: a Political History* (2 vols., 1899); also issued many speculative treatises deprecating orthodox religion, e.g. *Guesses at the Riddle of Existence* (1897); despite the unpopularity of his political and religious views in Canada he won much affectionate respect there by his enlightened activity in educational matters, by his advocacy of purity in public life,

and by his philanthropy and private charity; president of American Historical Association, 1904, he laid cornerstone of 'Goldwin Smith Hall' at Cornell, 1904; dying at 'The Grange', Toronto, he was buried in St. James's cemetary there; bequeathed residue of large fortune to Cornell for promotion of liberal studies; *Reminiscences* (1911) and *Correspondence* (1913), both edited by Arnold Haultain.

SMITH, Sir GRAFTON ELLIOT (1871–1937), anatomist and anthropologist; born in New South Wales; MB, Ch.M., Sydney, 1892; studied cerebral morphology at Sydney and Cambridge; first professor of anatomy, Government Medical School, Cairo, 1900–9; became critic of methods of anthropologists, classical scholars, and Egyptologists; worked on Archaeological Survey of Nubia; professor of anatomy, Manchester (1909–19), University College, London (1919–36); works include *The Migrations of Early Culture* (1915) and *Human History* (1930); FRS, 1907; knighted, 1934.

SMITH, Sir HENRY BABINGTON (1863–1923), civil servant and financier; son of Archibald Smith [q.v.]; educated at Eton and Trinity College, Cambridge; entered Education Office, 1887; private secretary to G. J. (later Viscount) Goschen [q.v.], 1891; transferred to Treasury, 1892; secretary to Post Office, 1903–9; administrateur directeur-général, National Bank of Turkey, 1909; served on numerous financial commissions and committees; KCB, 1908.

SMITH, HENRY SPENCER (1812–1901), surgeon; entered St. Bartholomew's Hospital, 1832; MRCS, 1837; FRCS, 1843; senior assistant surgeon at newly founded St. Mary's Hospital, 1851; dean of medical school, 1854–60; stimulated in England microscopic study of tissues by his translation of German works.

SMITH, HERBERT (1862–1938), Yorkshire miners' leader; born in workhouse; went into pit at age of ten; active in local government; president, Yorkshire Miners' Association (1906–38), Miners' Federation of Great Britain (1922–9), International Miners' Federation (1921–9); prominent in miners' strike, 1926; completely fearless and foremost in rescue work.

SMITH, Sir HUBERT LLEWELLYN (1864–1945), civil servant and social investigator; first class, mathematics, Corpus Christi College, Oxford, 1886; first commissioner of labour, Board of Trade, 1893; permanent secretary, 1907–19; planned new unemployment insurance scheme; organized Ministry of Munitions, 1915; chief government economic adviser,

1919–27; director, *New Survey of London Life and Labour*, 1928–35; wrote history of Board of Trade (1928) and East London (1939); KCB, 1908; GCB, 1919.

SMITH, JAMES HAMBLIN (1829–1901), mathematician; BA, Caius College, Cambridge, 1850; MA, 1853; private coach at Cambridge till death; lecturer in classics at Peterhouse, 1868–72; published handbooks on Statics (1868), Hydrostatics (1868), Trigonometry (1868), Algebra (1869), Geometry (1872), Arithmetic (1872), Heat (1877), and Geometrical Conic Sections (1887).

SMITH, JOHN ALEXANDER (1863–1939), philosopher and classical scholar; educated at Edinburgh University and Balliol College, Oxford; first class, *lit. hum.*, 1887; fellow, 1891; Waynflete professor of moral and metaphysical philosophy, 1910–36; fine Aristotelian scholar; joint-editor, Oxford translation of Aristotle, 1908–12; translated *De Anima* (1931); maintained idealist tradition of T. H. Green and Edward Caird [qq.v.]; much influenced by Benedetto Croce and Giovanni Gentile.

SMITH, (LLOYD) LOGAN PEARSALL (1865–1946), writer; born in New Jersey; educated at Quaker Penn Charter School, Haverford College, Harvard, and Balliol College, Oxford; naturalized, 1913; publications include *Trivia* (enlarged and revised, 1918), *More Trivia* (1922), and *Afterthoughts* (1931), published collectively as *All Trivia* (1933); *The Life and Letters of Sir Henry Wotton* (1907), *The English Language* (1912), *Words and Idioms* (1925), and *Milton and his Modern Critics* (1940).

SMITH, LUCY TOULMIN (1838–1911), scholar; born at Boston, Massachusetts; daughter of Joshua Toulmin Smith (1816–69, q.v.); resided at Highgate, London, from 1842 to 1894; edited works for Early English Text, Camden, and New Shakespeare societies from 1870, as well as *Leland's Itinerary* (5 vols., 1906–10); librarian of Manchester College, Oxford, 1894–1911.

SMITH, Sir MATTHEW ARNOLD BRACY (1879–1959), painter; educated at Halifax grammar school and Giggleswick School; studied at Manchester School of Art and the Slade; went to Pont Aven, Brittany, 1908; exhibited at Salon des Indépendents in Paris, 1911–12; returned to London, 1914; exhibited with London Group, 1916; served in Artists Rifles and Labour Corps, 1916–19; in France again producing long series of painting of female nudes, and landscapes, 1923–40;

evacuated from France, 1940; continued to work at still-lifes and large decorative subjects despite ill health and defective eyesight; retrospective exhibition, Tate Gallery, 1953; CBE, 1949; knighted, 1954; hon. D.Lit., London, 1956.

SMITH, REGINALD BOSWORTH (1839–1908), schoolmaster and author; educated at Marlborough and Corpus Christi College, Oxford; BA (first class classic), 1862; MA, 1865; president of Union, 1862; fellow and tutor of Trinity College, Oxford, 1863; classical master at Harrow, 1864–1901; published lectures on *Mohammed and Mohammedanism* (1874), *Carthage and the Carthaginians* (1878), and *Life of Lord Lawrence* (4 edns. 2 vols., 1883); wrote on current political, religious, and educational matters in the reviews; defended Turkish character, urging permanent English occupation of Sudan; opposed disestablishment of Church of England; settled in Dorset, 1901; ardent ornithologist; published *Bird Life and Bird Lore* (1905).

SMITH, REGINALD JOHN (1857–1916), barrister and publisher; BA, King's College, Cambridge; called to bar (Inner Temple); joined publishing firm of Smith, Elder, & Co., 1894; assumed sole control, 1899; editor of *Cornhill Magazine*, 1898; publications include centenary editions of Thackeray and Browning, Brontë and Gaskell definitive editions, and thin-paper edition of *Dictionary of National Biography* (1908).

SMITH, RODNEY (1860–1947), evangelist; known as 'Gipsy Smith'; worked under William Booth [q.v.], 1877–82; made first of many visits to United States, 1889; joined Manchester Methodist Mission; toured round world, 1894; first missioner, National Free Church Council, 1897–1912; subsequently preached for Methodist Home Mission Committee.

SMITH, Sir ROSS MACPHERSON (1892–1922), airman; born at Adelaide; served at Gallipoli, 1915; transferred to Royal Flying Corps, 1916; after war, began series of long-distance flights, including first flight from Cairo to Calcutta (1918) and first flight from England to Australia (Nov.–Dec. 1919); killed during trial flight at Brooklands; KBE, 1919.

SMITH, SAMUEL (1836–1906), politician and philanthropist; educated at Edinburgh University; apprenticed to Liverpool cotton-broker, 1853; after visit to North American cotton-growing districts (1860) started cotton-broking business at Liverpool; visited India (1862–3) to test cotton-growing possibilities;

president of Liverpool chamber of commerce, 1876; liberal MP for Liverpool, 1882–5, and for Flintshire, 1886–1905; friend of Gladstone; promoted Criminal Law Amendment Act, 1885; attacked opium traffic between China and India; advocate of bimetallism; published account of a second visit to India in *India Revisited* (1886); champion of native races; urged church disestablishment; PC, 1905; revisited India, 1904–5 and 1906; died at Calcutta; bequeathed £50,000 to Liverpool institutions; published works on Indian cotton trade and Indian problems, and autobiographical *My Life Work* (1902).

SMITH, SARAH (1832–1911), authoress under the pseudonym of HESBA STRETTON; first short story, 'The Lucky Leg', accepted by Dickens for *Household Words*, 1859; contributed to *All the Year Round* Christmas numbers till 1866; her *Jessica's First Prayer* (1866) won lasting popularity and has been translated into every European language and most African and Asiatic tongues; published some fifty other volumes; helped to found London Society for Prevention of Cruelty to Children, 1884.

SMITH, SIR SYDNEY ALFRED (1883–1969), professor of forensic medicine; born in New Zealand; trained as pharmacist; part-time medical student, Victoria College, Wellington; studied at Edinburgh University; MB, Ch.B., first class; research scholar, 1912; DPH, 1913 MD, 1914; assistant, department of forensic medicine; returned to New Zealand; medical officer for health, Dunedin, 1914; principal medico-legal expert to Egyptian government, 1917–28; responsible for investigation of murder of Sir Lee Stack [q.v.], 1924; regius professor of forensic medicine, Edinburgh, 1928–53; dean of medical faculty, 1931; expert witness in many cases; member, General Medical Council, 1931–56; consultant to World Health Organization, 1953; rector, Edinburgh University, 1954–7; hon. LLD, 1955; CBE, 1944; knighted, 1949; founded British Association in Forensic Medicine; publications include *Forensic Medicine* (1925) and *Mostly Murder* (1959); edited four editions of *Principles and Practice of Medical Jurisprudence*.

SMITH, THOMAS (1817–1906), missionary and mathematician; graduated in mathematics at Edinburgh University; hon. MA, 1858, DD, 1867, LLD, 1900; ordained to Scottish mission at Calcutta, 1839; on disruption (1843) joined Free Church; inaugurated first Zenana missions to India, 1854; conducted home mission in Edinburgh, 1859–79; professor of evangelistic theology, New College, Edinburgh, 1880–93; moderator of Free Church General Assembly,

1891; works include *Plane Geometry* (1857), *The Life of Euclid* (1902), and *Medieval Missions* (1880).

SMITH, SIR THOMAS, first baronet (1833–1909), surgeon; educated at Tonbridge and St. Bartholomew's Hospital; MRCS, 1854; coached pupils for examinations from 1854, publishing *Manual of Operative Surgery on the Dead Body* (1859); FRCS, 1858; demonstrator of operative surgery at St. Bartholomew's, 1859; surgeon, 1873–98; surgeon to Children's Hospital, Great Ormond Street, 1868–83; baronet, 1897; KCVO, 1901; a dexterous operator and sure guide in diagnosis.

SMITH, THOMAS (1883–1969), superintendent of the Light Division, National Physical Laboratory; educated at Leamington School, Warwick School, and Queens' College, Cambridge, 1904–6; master, Oundle School; joined National Physical Laboratory, 1907; head of Optics Division, 1909; superintendent, Light Division, 1940–8; president, Optical Society, 1925–7; president, Physical Society, 1936–8; FRS, 1932; first president, International Commission for Optics, 1947–9; hon. member, Optical Society of America, 1957; author of some 170 scientific papers concerning the design and testing of optical systems and colorimetry; in association with J. Guild, proposed a 'System of Colorimetry', adopted by the International Commission on Illumination.

SMITH, THOMAS ROGER (1830–1903), architect; entered office of Philip Hardwick [q.v.]; designed several buildings in Bombay, 1864; work in England included many board schools, churches, and hospitals; president of Architectural Association, 1860–1, and 1863–4; district surveyor for Southwark and North Lambeth (1874–82) and for West Wandsworth from 1882; master of Carpenters' Company, 1901; professor of architecture at University College, London, 1880–1903; published manual of *Acoustics* (1861) and works on architecture.

SMITH, VINCENT ARTHUR (1848–1920), Indian historian and antiquary; son of Aquilla Smith [q.v.]; entered Indian civil service, 1871; served in North-West Provinces and Oudh till 1900; CIE, 1919; works include *Early History of India* (1904) and *Oxford History of India* (1918).

SMITH, VIVIAN HUGH, first BARON BICESTER (1867–1956), banker; educated at Eton and Trinity Hall, Cambridge; joined firm of Hay's Wharf, of which his father was chairman; then joined Morgan Grenfell & Co.; director of many companies, and governor of the most important, the Royal Exchange Assurance;

baron, 1938; keen interest in racing, successful with his horses in many major steeplechases; lord-lieutenant of Oxfordshire, 1934-54.

SMITH, WALTER CHALMERS (1824-1908), poet and preacher; MA, Marischal College, Aberdeen, 1841; Free Church minister at Glasgow and Edinburgh, 1862-94; moderator of General Assembly, 1893; DD, Glasgow, 1869; LLD, Aberdeen, 1876, and Edinburgh, 1893; published many volumes of verse; a complete edition appeared in 1902.

SMITH, WILLIAM FREDERICK DAN-VERS, second VISCOUNT HAMBLEDEN (1868-1928), philanthropist; son of W. H. Smith [q.v.]; BA, New College, Oxford; head of newspaper-distributing business of W. H. Smith & Son, 1891; opened bookshop branch, 1905; conservative MP, Strand division of Westminster, 1891-1910; succeeded to title, 1913; did much work for voluntary hospitals.

SMITH, WILLIAM SAUMAREZ (1836-1909), archbishop of Sydney; born at St. Helier, Jersey; BA, Trinity College, Cambridge (first class classic), 1858; MA, 1862; Tyrwhitt Hebrew scholar and fellow, 1860; principal of St. Aidan's, Birkenhead, 1869-89; bishop of Sydney, 1890-7; and first archbishop, 1897 till death; hon. DD, Cambridge, 1890, Oxford, 1897; published theological works and verse.

SMITH-DORRIEN, SIR HORACE LOCK-WOOD (1858-1930), general; gazetted to 95th Foot, 1877; served in Zulu war, 1879; Egyptian war, 1882; present at battle of Ginnis, 1885; passed through Staff College; served in India, 1889-98; present at battle of Omdurman, 1898; served in South African war, 1899-1902; present at battle of Paardeberg and on advance to Pretoria, 1900; major-general, 1901; adjutant-general, India, 1901-3; commanded 4th division, Quetta, 1903-7; commander-in-chief, Aldershot, 1907; transferred to Southern command, 1912; advocated army expansion on Territorial basis; given command of II Corps, Aug. 1914; with II Corps bore brunt of fighting at Mons; decided on his own responsibility to stand fast and await German attack; this decision, which led to battle of Le Cateau (26 Aug.), averted dangerous situation; while general headquarters was anxious for vigorous attack and placed little faith in Kitchener's new armies, he put high value on them and was content to wait; commanded Second Army, 1915; after first German gas attack (Apr.) recommended withdrawal to line nearer Ypres; resigned, May; commanded First Army for home defence, 1915; governor of Gibraltar, 1918-23; KCB, 1907.

SMITHELLS, ARTHUR (1860-1939), chemist; educated at Glasgow University and Owens College Manchester; professor of chemistry, Leeds, 1885-1923; active in establishment of the university; his researches on flame structure brought him contacts with gas industry; president, Society of British Gas Industries, 1911; director, Salters' Institute of Industrial Chemistry, London, 1923-37; president, Institute of Chemistry, 1927-30; FRS, 1901.

SMUTS, JAN CHRISTIAN (1870-1950), statesman; born in Cape Colony of Dutch origin; graduated in science and literature, Stellenbosch, 1891; first class, both parts law tripos, Christ's College, Cambridge, 1894; admitted to Cape bar, 1895; disillusioned by Jameson raid went to Transvaal and admitted, 1896; state attorney, 1898; worked unsuccessfully for peace, 1899; wrote *A Century of Wrong*; commanded raid into Cape Colony to stir up rebellion, 1901-2; on commission sent by Boer delegates to negotiate in Pretoria, 1902; persuaded Boers to accept peace; sent to England to seek responsible government, 1906; colonial secretary and minister of education under Louis Botha [q.v.] in Transvaal, 1907-10; his plan for Union the basis of discussion at national convention, 1908-9; minister of defence, Union government, 1910-19, as member of South African Party under Botha; formed South African Defence Force and used it to suppress strike on Rand, 1914; after outbreak of war crushed rebellion quickly and leniently, 1914; with Botha conducted successful campaign in South-West Africa, 1915; commissioned as lieutenant-general in British Army to command imperial forces in East Africa, 1916; inflicted decisive defeat but at great cost in casualties; represented Union at imperial conference and war cabinet, 1917; remained as member of British war cabinet; mainly responsible for independent organization of Royal Air Force; chairman, war priorities committee; a chief sponsor of League of Nations; published *The League of Nations: A Practical Suggestion* (1918); inventor of mandates system; fought in vain for magnanimous peace; mediated between British government and Southern Irish leaders, 1921; prime minister, South Africa, 1919-24; used forcible measures in face of industrial unrest in South Africa, culminating in suppression of Rand rebellion, 1922; defeated at election of 1924; during long period in opposition wrote *Holism and Evolution* (1926) his philosophy of the whole; entered coalition as deputy prime minister under J. B. M. Hertzog [q.v.], 1933-9; fused their respective parties as United South African National Party, 1934; opposed Hertzog's proposed neutrality and became prime minister,

1939–48; field-marshal, 1941; paid many visits to Europe and North Africa and frequently consulted by (Sir) Winston Churchill; rector of St. Andrews, 1931–4; chancellor, universities of Cape Town (1936–50) and Cambridge (1948–50); KC, 1906; PC and CH, 1917; FRS, 1930; president, British Association, 1931; OM, 1947.

SMYLY, SIR PHILIP CRAMPTON (1838–1904), surgeon and laryngologist; BA, Trinity College, Dublin, 1859; MB, 1860; MD, 1863; FRCS Ireland, 1863; president, 1878–9; surgeon to Meath Hospital, 1861–1904; president of Laryngological Association, 1889, and of Irish Medical Association, 1900; introduced laryngoscope to Ireland, 1860; knighted, 1892; a musician and freemason.

SMYTH, DAME ETHEL MARY (1858–1944), composer, author, and feminist; studied at Leipzig Conservatorium; composed Serenade in D and overture 'Antony and Cleopatra', 1890; Mass in D, 1893; Concerto for violin and horn, 1927; and oratorio 'The Prison', 1931; obtained performance of operas *Fantasio* (Germany, 1898), *Der Wald* (1902), *The Wreckers* (Germany, 1906, England, 1909), *The Boatswain's Mate* (after story by W. W. Jacobs, q.v., 1916), dance dream *Fête Galante* (1923), and farcical comedy *Entente Cordiale* (1926); first women to compose in largest forms of opera, oratorio, and concerto; fought for her own career as musician, for recognition as composer, and for women's suffrage; her series of autobiographical works reveals her high spirits, serious purpose, and English eccentricity; DBE, 1922.

SMYTH, SIR HENRY AUGUSTUS (1825–1906), general and colonel commandant, Royal Artillery, from 1894; joined Royal Artillery, 1843; served in Bermuda, 1847–51, in Nova Scotia, 1851–4; commanded a field battery at Sevastopol, 1855; in Canada, 1861–5; in India, 1867–74; commanded artillery at Sheerness, 1876, and in southern district, 1877–80; on ordnance committee at Woolwich, 1881–3; major-general, 1882; commandant of Woolwich garrison, 1882–6; lieutenant-general, 1886; crushed rising in Zululand, 1887; acting governor of Cape Colony, 1889–90; CMG, 1889; KCMG, 1890; governor of Malta, 1890–3; general, 1891.

SMYTHE, FRANCIS SYDNEY (1900–1949), mountaineer; educated at Berkhamsted School and Faraday House Electrical Engineering College; climbed new routes up Brenva face of Mont Blanc, 1927; 1928; led successful Kamet expedition, 1931; took part in Everest expeditions, 1933–6–8; interpreted mountain adven-

ture to non-climbers in books, articles, and lectures; excellent mountain photographer; publications include *Climbs and Ski Runs* (1929), *Kamet Conquered* (1932), and *Camp Six* (1937).

SNEDDEN, SIR RICHARD (1900–1970), chief executive of the Shipping Federation; educated at George Watson's Boys' College and Edinburgh University; MA, 1921; LLB, 1922; joined National Confederation of Employers' Organizations (later CBI), 1923; called to bar (Middle Temple), 1925; member of British delegations to all ILO conferences from 1923 to 1969; member, governing body, 1952–60; vice-president, International Organization of Employers, 1955–6; president, 1957–8; assistant secretary, Shipping Federation, 1929; secretary, 1933; chief executive, 1936–62; CBE, 1942; hon. captain RNR; knighted, 1951; CVO, 1967; member of Queen's Body Guard for Scotland; director of number of companies, including Monotype and Consolidated Gold Fields; hon. LLD, Edinburgh, 1953.

SNELL, HENRY, BARON SNELL (1865–1944), politician and secularist; self-educated; worked as farm servant and in public houses; joined secularist movement and independent labour party; a lecturer for and secretary of Union of Ethical Societies; secretary, Secular Education League, 1907–31; on London County Council, 1919–25; chairman, 1934–8; labour MP, East Woolwich, 1922–31; baron, 1931; deputy leader, House of Lords, 1940–4; PC, 1937; CH, 1943.

SNELL, SIR JOHN FRANCIS CLEVERTON (1869–1938), electrical engineer; educated at Plymouth grammar school and King's College, London; borough electrical engineer, Sunderland, 1896; consulting engineer in London, 1906–18; chairman, electricity commission, 1919–38; took leading part in introducing the 'grid' system; knighted, 1914.

SNELUS, GEORGE JAMES (1837–1906), metallurgist; student at Royal School of Mines, 1864–7; patented process of eliminating phosphorus from molten pig-iron by oxidation in basic lined enclosure, 1872; process was commercially developed by Sidney Gilchrist Thomas [q.v.] in 1879; Snelus was awarded gold medal at Paris exhibition and Bessemer gold medal of Iron and Steel Institute, 1878; FRS, 1887; general manager of West Cumberland Iron and Steel Company from 1872 to 1900; an enthusiastic Volunteer and rifle shot.

SNOW, HERBERT (1835–1910), classical scholar. [See KYNASTON.]

SNOW, SIR THOMAS D'OYLY (1858–1940), lieutenant-general; educated at Eton;

commissioned in 13th Foot, 1879; major-general, 1910; commanded 4th division (1911-14) at Le Cateau and the Marne; the 27th division (1914-15) at first gas attack, Ypres; VII Corps (1915-17) on the Somme, at Arras, and Cambrai; transferred to Western command at home (1917-19) on account of injury; lieutenant-general, 1917; KCB, 1915.

SNOWDEN, PHILIP, VISCOUNT SNOWDEN (1864-1937), statesman; son of a weaver; junior exciseman, 1886-93; became chronically crippled with inflammation of spinal cord, 1891; propagandist for independent labour party, 1895-1905; national chairman, 1903-6, 1917-20; MP, Blackburn (1906-18), Colne Valley division (1922-31); became authority on national finance; pacifist and champion of conscientious objectors, 1914-18; chancellor of the Exchequer, 1924, 1929-31; obtained large changes in favour of Britain in Young plan, 1929; after onset of world depression disclosed gravity of country's situation (Feb. 1931); balanced stopgap budget (Apr.) pending report of May economy committee; remained in 'national' government (Aug.) without enthusiasm but from strong sense of duty; with estimated deficit reaching £170 millions balanced budget (10 Sept.) half by cuts, half by new taxation; in this speech and in announcing suspension of gold standard (21 Sept.) reached his highest level; viscount and lord privy seal (Nov.); resigned over proposal of preferential tariffs, Sept. 1932; published autobiography, 1934; PC, 1924.

SODDY, FREDERICK (1877-1956), chemist; educated at Eastbourne College, University College, Aberystwyth, and Merton College, Oxford; first class, chemistry, 1898; demonstrator, McGill University, Montreal; collaborated with Ernest (later Lord) Rutherford [q.v.] in formulating the theory of atomic disintegration; returned to London, 1903; lecturer in physical chemistry and radioactivity, Glasgow University, 1904-14; research led to conception of isotopes; professor, chemistry, Aberdeen University, 1914-19; Dr Lee's professor, chemistry, Oxford, 1919-36; published *The Interpretation of Radium* (1909) and *The Interpretation of the Atom* (1932); FRS, 1910; Nobel prize for chemistry, 1921.

SOISSONS, LOUIS EMMANUEL JEAN GUY DE SAVOIE-CARIGNAN DE, VISCOUNT D'OSTEL, BARON LONGROY (1890-1962), architect and town planner. [See DE SOISSONS.]

SOLLAS, WILLIAM JOHNSON (1849-1936), geologist; first class, natural sciences tripos (geology), St. John's College, Cambridge, 1873; professor of geology, Bristol (1880), Trinity College, Dublin (1883), Oxford (1897-1936); a leading authority on palaeolithic man; wrote on almost every branch of geology and published *Ancient Hunters and their Modern Representatives* (1911); FRS, 1889.

SOLOMON, SIR RICHARD (1850-1913), South African statesman; born in Cape Town; attorney-general of Cape Colony, 1898, of Transvaal, 1902; high commissioner for Union of South Africa in London, 1910; KCMG, 1902; KCB, 1905.

SOLOMON, SIMEON (1840-1905), painter and draughtsman; brother of Abraham Solomon [q.v.]; exhibited at Royal Academy from 1858, mainly scriptural subjects, showing sincerity and beauty of colour; became acquainted with Burne-Jones, Rossetti, Swinburne, and Lord Houghton [qq.v.]; from 1865 exhibited classical subjects; his 'Bacchus' (1867) praised by Walter Pater [q.v.]; visited Florence (1866) and Rome (1869); published *A Vision of Love revealed in Sleep* (1871), which was praised by Swinburne; also made a reputation by water-colour and pencil studies; ceased to exhibit after 1872 when vicious indulgences ruined his career; thenceforth he endured great poverty, earning a precarious livelihood by sentimental sketches; his 'A Greek Acolyte' is in Birmingham Art Gallery.

SOLOMON, SOLOMON JOSEPH (1860-1927), painter; studied at Royal Academy Schools; settled in London where he gradually made a name as painter of portraits and classical and biblical scenes; RA, 1906; his works include 'Samson' (1887), 'Niobe' (1888), 'Israel Zangwill' (1894), and 'Echo and Narcissus' (1895); suggested and helped to organize 'camouflage' during European war.

SOMERS-COCKS, ARTHUR HERBERT TENNYSON, sixth BARON SOMERS (1887-1944), chief scout for Great Britain and the British Commonwealth and Empire, and governor of Victoria, Australia; succeeded great-uncle, 1899; educated at Charterhouse and New College, Oxford; all-round athlete; played cricket for Worcestershire; served in 1st Life Guards, 1906-11, 1914-18; DSO; retired as lieutenant-colonel, 1922; active in Boy Scout movement from 1920; deputy chief scout, 1936; chief scout, 1941-4; governor of Victoria, 1926-31; KCMG, 1926.

SOMERSET, LADY ISABELLA CAROLINE, LADY HENRY SOMERSET (1851-1921); daughter of last Earl Somers; married Lord

Henry Somerset, 1872; devoted herself to temperance work; president of British Women's Temperance Association, 1890-1903; of World's Women's Christian Temperance Union, 1898-1906; founded Duxhurst, farm colony for inebriate women near Reigate, 1895.

SOMERVELL, DONALD BRADLEY, BARON SOMERVELL OF HARROW (1889-1960), politician and judge; educated at Harrow and Magdalen College, Oxford; first class, chemistry, 1911; fellow of All Souls, 1912; called to bar (Inner Temple), 1916; served in India and Mesopotami, staff captain 53rd Infantry brigade, 1914-19; KC, 1929; conservative MP, Crewe, 1931-45; solicitor-general, and knighted, 1933; attorney-general, 1936; PC, 1938; home secretary, 1945; recorder of Kingston upon Thames, 1940-6; lord justice of appeal, 1946-54; lord of appeal in ordinary and life peer, 1954-60; treasurer, Inner Temple, 1957; hon. DCL, Oxford, 1959; chairman of governors, Harrow, 1947-53.

SOMERVILLE, EDITH ANNA ŒNONE (1858-1949), writer; cousin of Violet Florence Martin (Martin Ross) [q.v.] with whom she wrote fourteen books including *The Real Charlotte* (1894), *Some Experiences of an Irish R.M.* (1899), and *In Mr. Knox's Country* (1915); continued writing after her cousin's death (1915) still giving her name as co-author; master, West Carbery foxhounds; organist, Castlehaven parish church, 1875-1949; painter; feminist.

SOMERVILLE, SIR JAMES FOWNES (1882-1949), admiral of the fleet; entered navy, 1897; wireless specialist; director of signals, 1925-7; on staff of Imperial Defence College, 1929-31; commodore, Royal Naval Barracks, Portsmouth, 1932-4; rear-admiral, 1933; director of personal services, 1934-6; commanded Mediterranean destroyer flotillas, 1936-8; vice-admiral, 1937; commander-in-chief, East Indies, 1938-9; retired with pulmonary tuberculosis but recalled on outbreak of war; served under (Sir) Bertram Ramsay [q.v.] during Dunkirk evacuation; commanded Force H at Gibraltar, 1940-2; commander-in-chief, Eastern fleet, 1942-4; admiral, 1942; head of Admiralty delegation, Washington, 1944-5; admiral of the fleet, 1945; KCB, 1939.

SOMERVILLE, MARY (1897-1963), educationist and broadcasting executive; educated at Abbey School, Melrose, Selkirk high school, and Somerville College, Oxford; with assistance of John (later Lord) Reith, appointed schools assistant to J. C. Stobart, director of education, BBC, 1925; responsible for broadcasting to schools and secretary to Central Council of School Broadcasting, 1929; assistant controller, talks division, 1947; controller, 1950-5; first woman controller; made most valuable contribution to educational broadcasting; OBE, 1935; award from Institute of Education by Radio-Television in America, 1955.

SOMERVILLE, SIR WILLIAM (1860-1932), agriculturist; educated at Royal High School, Edinburgh; graduated in agriculture, Edinburgh (1887), forestry, Munich (1889); lecturer in forestry, Edinburgh, 1889-91; professor, agriculture and forestry, Newcastle, 1891-9; of agriculture, Cambridge, 1899-1902; assistant secretary, Board of Agriculture, 1902-6; Sibthorpian professor of rural economy, Oxford, 1906-25; developed the simple field experiment; publications include *Text-Book of the Diseases of Trees* (1894); KBE, 1926.

SONNENSCHEIN, EDWARD ADOLF (1851-1929), classical scholar and writer on comparative grammar and metre; BA, University College, Oxford; professor of Greek and Latin at newly founded Mason College, Birmingham (afterwards Birmingham University), 1883-1918; Plautine scholar; published editions of *Captivi* (1879), *Mostellaria* (1884), and *Rudens* (1891); took up reform of grammar teaching and published 'Parallel Grammar' series, which was warmly welcomed in Great Britain and on the continent; joined J. P. Postgate [q.v.] in forming Classical Association, 1903; much of his grammatical research summed up in *The Unity of the Latin Subjunctive* (1910) and *The Soul of Grammar* (1927); insisted upon the humanities taking their proper place in the modern university; took up question of war-guilt during European war; a very exact scholar.

SORABJI, CORNELIA (1866-1954), Indian barrister and social reformer; born in Bombay presidency; educated at Deccan College, Poona; first class degree, 1886; studied at Somerville Hall, Oxford; BCL, 1892; undertook educational and social work with women in India, 1894; adviser in court of wards, Bengal, Bihar, and Orissa, 1904-22; called to bar (Lincoln's Inn), 1923; practised as barrister in India, 1924-9; settled in London, 1929; published *India Calling* (1934) and *India Recalled* (1936); helped to edit *Queen Mary's Book for India* (1943).

SORBY, HENRY CLIFTON (1826-1908), geologist; of old Sheffield family of cutlers; versatile scientist; made researches in spectroscope and microscope, meteorology, archaeology, and architecture; interested in Egyptian hieroglyphics; pioneer of microscopic petrology; published papers on *Calcareous Grit of Scar-*

borough (1851), on *Slaty Cleavage* (1853), and *On the Microscopic Structure of Crystals* (1858); studied microscopic structure of irons and steels; FGS, 1850; president, 1878-80; FRS, 1857; royal medallist, 1874; hon. LLD, Cambridge, 1879; helped to found Firth College and bequeathed to Sheffield University his collections, with money to found a chair in geology.

SORLEY, WILLIAM RITCHIE (1855-1935), philosopher; educated at Edinburgh and Cambridge universities; professor of logic and philosophy, Cardiff, 1888-94; of moral philosophy, Aberdeen (1894-1900), Cambridge (1900-33); based his theistic argument on the moral values; works include *Moral Values and the Idea of God* (1918) and *A History of English Philosophy* (1920); FBA, 1905.

SOTHEBY, SIR EDWARD SOUTHWELL (1813-1902), admiral; joined navy, 1828; served in Syria, 1840; commander, 1841; commanded on East Indies and China station, 1855-8; distinguished in Oudh during Indian Mutiny; CB, 1858; extra ADC to Queen Victoria, 1858-67; commanded Portland coast-guard division, 1863; KCB, 1875; admiral, 1879.

SOUTAR, ELLEN (1848-1904), actress. [See FARREN.]

SOUTHBOROUGH, first BARON (1860-1947), civil servant. [See HOPWOOD, FRANCIS JOHN STEPHENS.]

SOUTHESK, ninth EARL OF (1827-1905), poet and antiquary. [See CARNEGIE, JAMES.]

SOUTHEY, SIR RICHARD (1808-1901), Cape of Good Hope official; went to South Africa, 1820; engaged in farming; served in Kaffir war, 1834-5; resident agent with Kaffir tribes, 1835-6; farmer at Graaffreinet, 1836-46; civil commissioner and resident magistrate of Swellendam, 1849-52; acting secretary to Cape government, 1852-4; auditor-general, 1859; acting colonial secretary, 1860-1; treasurer and accountant-general and member of executive council, 1861; colonial secretary, 1864-72; opposed responsible government for Cape; lieutenant-governor of province of Griqualand West, 1873-5; member of house of assembly, 1876-8; CMG, 1872; KCMG, 1891.

SOUTHWARD, JOHN (1840-1902), writer on typography; son of Liverpool printer; conducted *Liverpool Observer* from 1857 to 1865; printer's reader in London, 1867-8; travelled in Spain, 1868; edited *Printers' Register*, 1886-90; authoritative works include *Dictionary of Typography* (1872), *Practical Printing* (1882), and

Modern Printing (1898); was largely responsible for Bigmore and Wyman's *Bibliography of Printing* (3 vols., 1880-6).

SOUTHWELL, SIR RICHARD VYNNE (1888-1970), professor of engineering science; educated at King Edward VI School, Norwich, and Trinity College, Cambridge; first class, parts i and ii, mechanical sciences tripos, 1909-10; fellow, Trinity College, 1912; during 1914-18 war worked at Royal Aircraft Establishment, Farnborough; superintendent, Aeronautics Department, National Physical Laboratory, 1920-5; carried out research on space frames of rigid airships; returned to Trinity, 1925; structural consultant to Colonel V. C. Richmond, designer of R101; professor of engineering science, Oxford, succeeding C. F. Jenkin [q.v.], 1929-42; developed relaxation method of analysis; published *Introduction to the Theory of Elasticity of Engineers and Physicists* (1936); succeeded Sir H. T. Tizard [q.v.] as rector, Imperial College, London, 1942-8; FRS, 1925; knighted, 1948; hon. fellow, Brasenose College, Oxford, Trinity College, Cambridge, and Imperial College, London; president, International Congress of Applied Mechanics, 1948-52; general secretary, British Association, 1948-56.

SOUTHWELL, THOMAS (1831-1909), naturalist; in service of Gurney's bank at Lynn, Fakenham, and Norwich from 1846 to 1896; completed Henry Stevenson's *Birds of Norfolk* (3rd vol., 1890); published *The Seals and Whales of the British Seas* (1881); edited *Notes on Natural History of Norfolk* (1902) from Sir Thomas Browne's MSS.

SOUTHWOOD, VISCOUNT (1873-1946), newspaper proprietor. [See ELIAS, JULIUS SALTER.]

SOUTTAR, SIR HENRY SESSIONS (1875-1964), surgeon; educated at Oxford high school and Queen's College, Oxford; double first, mathematics, 1895-8; qualified at the London Hospital; MRCS; LRCP; BM, Oxford, 1906; FRCS, 1909; held number of resident hospital appointments; assistant surgeon, London Hospital, 1915; designed many of his instruments and became expert in use of radium; carried out pioneer cardiac operation; published *The Art of Surgery* (1929) with his own illustrations; vice-president, Royal College of Surgeons of England, 1943-4; helped to found the faculties of dental surgery and anaesthetists; president, British Medical Association, 1945-6; retired as consulting surgeon, 1947; hon. MD, Dublin, 1933; CBE for work during 1914-18 war; knighted, 1949.

SPARE, AUSTIN OSMAN (1886-1956), artist; studied at Lambeth School of Art and Royal College of Art; exhibited at Royal Academy at age of sixteen; collaborated with Clifford Bax in *Golden Hind* quarterly, 1922-4.

SPEARMAN, CHARLES EDWARD (1863-1945), psychologist; educated at Leamington College; studied psychology in Germany; reader in experimental psychology, University College, London, 1907; Grote professor, philosophy of mind and logic, 1911; professor of psychology, 1928-31; formulated general laws in *The Nature of 'Intelligence' and the Principles of Cognition* (1923); the nature and interrelations of human abilities in *The Abilities of Man* (1927); wrote *Psychology Down the Ages* (2 vols., 1937); FRS, 1924.

SPENCE, Sir JAMES CALVERT (1892-1954), paediatrician; educated at Elmfield College, York, and Durham College of Medicine, Newcastle upon Tyne; qualified, 1914; served in RAMC, 1914-19; MC and bar, 1917-18; medical registrar and chemical pathologist, Royal Victoria Infirmary, Newcastle, 1922-6; assistant physician, 1928-42; research in child health and aspects of social paediatrics; professor of child health, Newcastle, 1942-54; member, Medical Research Council, 1944 and 1952; knighted, 1950.

SPENCER, Sir HENRY FRANCIS (1892-1964), industrialist; left school at thirteen to work as sandboy in Cannon Iron Foundry, Bilston; started small iron and steel merchants business at Coseley, 1914; served with Royal Engineers, 1916-18; elected to Coseley council; business prospered; acquired connection with Ebbw Vale Company and John Lysaght & Co., manufacturer of flat-rolled steel products; opened new stockholding centre at Wolverhampton; joined Richard Thomas organization as sheet sales controller, 1935; managing director, Richard Thomas & Baldwins, 1952; member of executive committee, British Iron and Steel Federation; planned the Llanwern Works, South Wales; opened as Spencer Works, 1962; chairman, British Institute of Management; knighted, 1963.

SPENCER, HERBERT (1820-1903), philosopher; educated at day school at Derby, his native place, and from 1833 to 1836 mainly at Hinton Charterhouse near Bath under his uncle Thomas Spencer [q.v.]; made little progress in the classics, but was much interested in natural science; after acting as assistant schoolmaster at Derby (1837) he was employed as engineer on Birmingham and Gloucester Railway, 1837-41; wrote letters to the *Nonconformist* urging limita-

tion of the functions of the state, 1842, which he republished as *The Proper Sphere of Government*, 1843; sub-editor of the *Pilot*, the organ of the 'Complete Suffrage Movement', published at Birmingham, 1844; occupied himself anew with engineering. 1844-6; experimented with mechanical inventions, 1846-7, subsequently inventing an invalid bed, 1866; sub-editor of the *Economist* in London, 1848-53; visited house of John Chapman [q.v.], the advanced publisher, 1849; there came to know George Eliot [q.v.]; made acquaintance of Huxley and Tyndall [qq.v.]; published *Social Statics* (1851), advocating an extreme individualism; contributed articles to the *Leader*, *Westminster Review*, and other periodicals, collecting many of these in *Essays* (1st ser. 1857, 2nd ser. 1863, 3rd ser. 1864); published *Principles of Psychology* (1855, recast in two vols. 1870-2); during the writing of this book his health gave way, and was never fully restored; during 1858 he planned a system of synthetic philosophy, covering metaphysics, biology, psychology, sociology, and ethics, for publication of which he in 1860 invited subscriptions in both England and America, but this plan, which was for a time successful, broke down about 1865; the deaths of his uncle William and his father in 1866 improved his financial position; in 1861 he issued *Education*, a treatise aiming at a natural development of the child's intelligence, which became a leading textbook; he published *First Principles* of his synthetic philosophy, 1862, and *Principles of Biology*, vol. i, 1864, vol. ii, 1867; joined Jamaica committee for prosecution of Governor Eyre [q.v.], 1866; was elected to Athenaeum Club by the committee, 1868; was offered in 1871 lord rectorship of St. Andrews University, but declined this, and all later offers of honours; in order to deal with the principles of sociology he employed assistants to collect systematically large masses of facts, of which eight volumes under general title of *Descriptive Sociology* were issued by 1881, while additional volumes appeared after Spencer's death; contributed his widely read *Study of Sociology* (1873) to 'International Scientific' series, edited by his American friend, Prof. E. L. Youmans; issued *Principles of Sociology*, vol. i, 1876, vol. ii, 1882, vol. iii, 1896; meanwhile wrote *Data of Ethics* (1879), forming part i of the *Principles of Ethics* (vol. i, 1892, vol. ii, 1893); formed with Frederic Harrison and John Morley [qq.v.] and others an anti-aggression league, 1882; visited America, 1882; collected four articles from *Contemporary Review* to form *The Man versus the State* (1884) in which he urged anew the limitation of state functions; attacked the positivist religion in *The Nature and Reality of Religion*, 1885 (suppressed by Spencer, but reissued without his knowledge as *The Insuppressible Book*); completed his

Synthetic Philosophy in vol. iii of the *Principles of Sociology*, 1896; allowed his portrait to be painted by Sir Hubert von Herkomer [q.v.] for presentation to the Scottish National Portrait Gallery, Edinburgh, 1896; issued *Various Fragments* (1897) and *Facts and Comments* (1902); died, 8 Dec. 1903, after much suffering from shattered nervous system, at Brighton, where he had lived since 1898; he was cremated at Golders Green, his ashes being buried in Highgate cemetery; left bulk of property in trust for carrying on his *Descriptive Sociology*; the central doctrines of his philosophy, which exerted an immense influence on his era throughout Europe and America, India and Japan, were on its social side individualism and opposition to war, and on its scientific side evolution and explanation of phenomena from the materialistic standpoint; *Autobiography* published, 1904; *Life and Letters* by D. Duncan, 1908; an epitome of the *Synthetic Philosophy* by F. Howard Collins, 5th edn. 1901.

SPENCER, JOHN POYNTZ fifth EARL SPENCER (1835-1910), statesman and viceroy of Ireland; educated at Harrow, 1848-54, and Trinity College, Cambridge, 1854-7; MA, 1857; hon. LLD, 1864; liberal MP for South Northamptonshire, 6 Apr. 1857; succeeded father in earldom, 27 Dec.; devoted to sport and shooting; helped to form National Rifle Association, 1859; chairman, 1867-8; groom of the stole to Prince Consort, 1859-61, and to Prince of Wales, afterwards King Edward VII, 1862-7; made KG by Lord Palmerston, 1865; lord-lieutenant of Ireland under Gladstone without seat in cabinet, 1868-74; firm in maintenance of the law; president of the Council, with seat in cabinet, 1880-2; reappointed lord-lieutenant of Ireland on retirement of Lord Cowper [q.v.], in order to pursue a conciliatory policy; retained seat in cabinet; went to Ireland, 5 May 1882, with Lord Frederick Cavendish [q.v.] as Irish secretary; on assassination of Lord Frederick next day, Spencer was compelled to invoke repressive measures and was consequently vilified by mass of the Irish people; reduced Irish disorder; deprecated local government proposals of his cabinet colleagues, Sir Charles Dilke and Joseph Chamberlain [qq.v.], just before fall of government, 1885; resented an implied censure on the part of the conservatives in office on his use of 'coercion', 1885; supported Gladstone's home rule scheme, 1886; again president of the Council, 1886; while in opposition he urged the cause of home rule, 1886-92; first lord of the Admiralty, 1892-5; insistent on requirements of national security; formulated 'Spencer programme' of shipbuilding, which involved increased expenditure, despite objections of Gladstone while prime

minister; succeeded Lord Kimberley [q.v.] as liberal leader in the House of Lords, 1902; amid dissensions in liberal party he was often suggested as best fitted for its leadership; incapacitated by illness from 1905; lord-lieutenant of Northamptonshire, 1872-1908; chairman of Northamptonshire county council, 1889; president of Royal Agricultural Society, 1898; chancellor of Victoria University, Manchester, 1892-1907; sold great part of Althorp library to form John Rylands library at Manchester, 1892.

SPENCER, LEONARD JAMES (1870-1959), mineralogist and geologist; educated at Bradford Technical College, Royal College of Science, Dublin, and Sidney Sussex College, Cambridge; first class, parts i and ii, natural sciences tripos, 1892-3; joined mineralogy department, British Museum, 1894; keeper of minerals, 1927-35; published *The World's Minerals* (1911) and *A Key to Precious Stones* (1936); edited the *Mineralogical Magazine*, 1900-55; and *Mineralogical Abstracts*, 1920-55; Sc.D., Cambridge, 1921; FRS, 1925; CBE, 1934.

SPENCER, SIR STANLEY (1891-1959), artist; born at Cookham-on-Thames; studied at Slade School 1908-12; served in RAMC and the army, 1915-18; inspired by Cookham village scenes related to biblical stories; his most productive period, 1919-32; made drawings for Burghclere Memorial, 1922-3; later, his imaginative work deteriorated; ARA, 1932; resigned, 1935; RA, 1950; CBE, 1950; knighted, 1959; retrospective exhibition, Tate Gallery, 1955; Tate Gallery has his 'Apple Gatherers' (1912-13), 'The Robing of Christ' and 'The Disrobing of Christ' (1922), 'Resurrection, Cookham' (1923-7), and 'Resurrection: Port Glasgow' (1947-50).

SPENCER, SIR WALTER BALDWIN (1860-1929), biologist and ethnographer; educated at Owens College, Manchester, and Exeter College, Oxford; BA (first class in natural science), 1884; professor of biology, Melbourne University, 1887-1919; FRS, 1900; KCMG, 1916; in collaboration with F. J. Gillen, 1894-1912, and afterwards alone, brought to light unknown world of authentic stone-age folk of Central Australia; his works (with F. J. Gillen) include *Native Tribes of Central Australia* (1889), *Northern Tribes of Central Australia* (1904), and *The Arunta* (1927).

SPENDER, JOHN ALFRED (1862-1942), journalist and author; son of Lily Spender [q.v.]; educated at Bath College and Balliol

College, Oxford; editor; *Eastern Morning News*, Hull, 1886-91; moved to Toynbee Hall; wrote *The State and Pensions in Old Age* (1892); assistant editor, *Westminster Gazette*, 1893; editor, 1896-1922; wrote daily front-page leader preaching robust and reasoned liberalism; exerted greater influence per copy than any other newspaper; on terms of intimacy with half the cabinet after 1906; supported Asquith against Lloyd George; biographer of Campbell-Bannerman, Asquith, Cowdray, and Sir R. A. Hudson [qq.v.]; other works include *Fifty years of Europe* (1933) and *The Government of Mankind* (1938); CH, 1937.

SPENS, Sir WILLIAM (WILL) (1882-1962), educational administrator; educated at Rugby and King's College, Cambridge; first class, part i, natural sciences tripos, 1903; studied theology; fellow, Corpus Christi College, Cambridge, 1907; tutor, 1911; worked in Foreign Office, 1915-18; CBE, 1918; chevalier of Legion of Honour; member of council of Senate, 1920; master, Corpus Christi College, 1927-52; vice-chancellor, 1931-3; chairman, Cambridge University Appointments Board, 1930-9; knighted, 1939; regional controller for civil defence in East Anglia, 1939-45; high steward, Great Yarmouth, 1948; chairman, governing body, Rugby School, 1944-58; chairman, Joint Committee of Governors, Headmasters, Headmistresses and Bursars, 1948-59; chairman, National Insurance Advisory Committee, 1947-56; published *Belief and Practice* (1915); member of Archbishops' Commission on Doctrine, 1922-38; hon. LLD, Columbia (1933) and St. Andrews (1939).

SPEYER, Sir EDGAR, baronet (1862-1932), financier, philanthropist, and patron of music; born in New York; director of Speyer Brothers, London, 1887; naturalized, 1892; financed Metropolitan District Railway Company; chairman and baronet, 1906; chairman, Underground Electric Railways Co., 1906-15; active liberal and friend of Asquith; PC, 1909; financed Queen's Hall promenade concerts and encouraged many cultural and charitable enterprises; accused of pro-German activities he moved to New York, 1915; naturalization and privy counsellorship revoked, 1921.

SPIERS, RICHARD PHENÉ (1838-1916), architect; assistant to Sir Matthew Digby Wyatt [q.v.], 1861; silver and gold medallist, Royal Academy, 1863; travelled abroad and studied water-colour painting, 1865-6; master of Royal Academy architectural school, 1870-1906; his executed works include additions to houses, restorations of churches, and a few new buildings; some of his most valuable work in architec-

tural research contained in *Architecture, East and West* (collected papers, 1905).

SPILSBURY, Sir BERNARD HENRY (1877-1947), pathologist; second class, natural science, Magdalen College, Oxford, 1899; qualified from St. Mary's Hospital, Paddington, 1905; of methodical habits, phenomenal memory, and passion for detail; conducted some 25,000 post-mortems; honorary pathologist to Home Office; became leading 'detective-pathologist' whose evidence was invariably accepted by the jury; gave evidence at nearly every murder trial in south of England including those of Crippen (1910), Seddon (1912), 'Brides in Bath' case (1915), Mahon (1924), Thorne (1925), Fox (1930), Rouse (1931), Barney (1932), Chaplin (1938), Cummins (1942), and Loughans (1944); knighted, 1923; took own life.

SPOFFORTH, FREDERICK ROBERT (1853-1926), Australian cricketer; born at Balmain; played in second test match between England and Australia, Melbourne, 1877; performed remarkable bowling feats for Australian team at Lord's (May 1878) and at Oval (Aug. 1882); styled 'demon' bowler.

SPOONER, WILLIAM ARCHIBALD (1844-1930), warden of New College, Oxford; BA, New College; fellow; 1867; lecturer, 1868; tutor, 1869; dean, 1876-89; lectured on ancient history and philosophy; ordained, 1872; warden, 1903-24; devoted to interests of the college; made reconstructed lodgings centre of generous hospitality; did valuable service as 'Greats' examiner; supported promotion of natural science studies in university; poor law guardian and active member of Oxford Charity Organisation Society; suffered from weak eyesight due to albinoism and lapse of speech known as 'Spoonerism'.

SPRENGEL, HERMANN JOHANN PHILIPP (1834-1906), chemist; born near Hanover; Ph.D., Heidelberg, 1858; engaged in chemical research in Oxford and London from 1859; patented many safety explosives, liquid and solid; invented mercurial air pump for production of high tenuity (1865), which facilitated many later inventions; FRS, 1878; published researches in *Origin of Melinite and Lyddite* (1890) and *The Discovery of Picric Acid* (1902), in which he complained of infringement of rights to priority in his own discoveries.

SPRIGG, Sir JOHN GORDON (1830-1913), South African statesman; born at Ipswich; settled in South Africa, 1861; member for East London, 1869-1904; prime minister of Cape Colony, 1878-81, 1886-90, 1896-8; attended premiers' conference in London and created PC, 1897; prime minister, 1900–4; federalist member for East London, 1908; KCMG, 1886.

SPRIGGE, SIR (SAMUEL) SQUIRE (1860–1937), medical editor and author; educated at Uppingham and Gonville and Caius College, Cambridge; qualified at St. George's Hospital, 1887; assistant editor, *Lancet*, 1893; editor, 1909–37; works include *Life and Times of Thomas Wakley* (1897) and *Physic and Fiction* (1921); knighted and FRCS, 1921; FRCP, 1927.

SPRING, (ROBERT) HOWARD (1889–1965), journalist and novelist; left school at twelve to work as errand boy and office boy in Cardiff; worked in newspaper offices; learnt shorthand and studied languages at evening classes; reporter on *South Wales News*; moved to *Yorkshire Observer*; then to the *Manchester Guardian* under C. P. Scott [q.v.], 1915; came to notice of Lord Beaverbrook [q.v.], and joined *Evening Standard* as book reviewer, following Arnold Bennett [q.v.] and J. B. Priestly, 1931–9; first book *Darkie & Co.* published in 1932, followed by novels, *Shabby Tiger* (1934), *Rachel Rosing* (1935), *O Absalom!* (1938, retitled *My Son, My Son*), and *Fame is the Spur* (1940); reviewed for *Country Life*, 1941; published autobiographical works, collected in one volume as *The Autobiography of Howard Spring* (1972); after the 1939–45 wrote a number of other novels including *There is No Armour* (1948), *I met a Lady* (1961), and *Winds of the Day* (1964).

SPRING-RICE, SIR CECIL ARTHUR (1859–1918), diplomatist; grandson of first Baron Monteagle [q.v.]; educated at Eton and Balliol College, Oxford; entered Foreign Office, 1882; at Berlin, 1895–8; British commissioner on *Caisse de la Dette Publique*, Cairo, 1901–3; secretary of embassy, St. Petersburg, 1903–6; KCMG, 1906; British minister to Persia, 1906–8, to Sweden, 1908–13; ambassador at Washington, 1913–18; acted as conciliatory influence between Great Britain and United States during 1914–18 war; his tact contributed to America joining Allies; author of 'I vow to thee, my country'; died at Ottawa.

SPROTT, GEORGE WASHINGTON (1829–1909), Scottish divine and liturgical scholar; born in Nova Scotia; BA, Glasgow, 1849; hon. DD, 1880; ordained in Church of Scotland, 1852; chaplain to Scottish troops in Kandy, 1857–65; prominent in formation of Church Service Society, 1865; published critical edition of *Book of Common Order* (1868) and *Scottish Liturgies of James VI* (1871); moderator of synod, 1873; minister of North Berwick, 1873–1903; visited Presbyterian churches in

Canada, 1879; published lectures (1879) in *Worship and Offices of the Church of Scotland* (1882); advocate of church reunion; retired in 1903 to Edinburgh and engaged in literary work; edited liturgiological works and wrote a life of his father, 1906.

SPRY, CONSTANCE (1886–1960), artist in flower arrangement; born Fletcher; educated at Alexandra School and College, Dublin; head of women's staff, department of aircraft production, Ministry of Munitions, 1914–18; principal, LCC day continuation school, Homerton; as Constance Spry, opened shop in London, supplying exquisite flower arrangements to royal and society functions; adviser to minister of works on floral decorations at coronation of Queen Elizabeth II; OBE, 1953; wrote a number of books on floral arrangement and design; work for Royal Gardeners' Orphan Fund raised large sums for children in need.

SPY (pseudonym) (1851–1922), cartoonist. [See WARD, SIR LESLIE.]

SQUIRE, SIR JOHN COLLINGS (1884–1958), poet and man of letters; educated at Plymouth grammar school, Blundell's, and St. John's College, Cambridge; literary editor, *New Statesman*, 1913; founded and edited the *London Mercury*, 1919–34; founded The Invalids cricket club; chairman, the Architecture Club, 1922–8; FSA, hon. ARIBA; published works include *A Book of Women's Verse* (1921) and the *Cambridge Book of Lesser Poets* (1927); his *Collected Poems* published in 1959; knighted, 1933.

SQUIRE, WILLIAM BARCLAY (1855–1927), musical antiquary; BA, Pembroke College, Cambridge; assistant, department of printed books, British Museum, 1885; assistant keeper, with special charge of printed music, 1912–20; honorary curator of King's music library, 1924; music critic to various periodicals; edited English madrigals; FSA, 1888.

STABLES, WILLIAM (GORDON) (1840–1910), writer for boys; MD and CM, Aberdeen University, 1862; made a first voyage to Arctic, 1859; assistant surgeon in navy, 1863–71; for two years in merchant service, visiting India, Africa, and South Seas; from 1875 a prolific writer of boys' books of adventure and of historical novels; early pioneer of caravanning, 1886.

STACK, SIR LEE OLIVER FITZMAURICE (1868–1924), soldier and administrator; born at Darjeeling; joined army, 1888; Egyptian army, 1899; commanded Shambé field force, 1902;

Sudan agent and director of military intelligence, Cairo, 1908; civil secretary to Sudan government, 1914; acting sirdar of Egyptian army and governor-general of Sudan, 1917-24; KBE, 1918; shot in Cairo street.

STACPOOLE, FREDERICK (1813-1907), engraver in mixed mezzotint of works by contemporary artists; most successful engravings were Lady Butler's 'Roll Call', 1874, and Holman Hunt's 'Shadow of Death', 1877; exhibited at Royal Academy, 1842-99; ARA, 1880; also painted in oils.

STACPOOLE, HENRY DE VERE (1863-1951), novelist; educated at Malvern; studied medicine, St. George's and St. Mary's Hospitals, London; qualified, 1891; worked as ship's doctor and wrote some fifty novels, including *The Blue Lagoon* (1908) and *The Street of the Flute-Player* (1912).

STAFFORD, SIR EDWARD WILLIAM (1819-1901), premier of New Zealand; educated at Trinity College, Dublin; emigrated to Nelson, New Zealand, 1843; elected to house of representatives, 1855; premier and colonial secretary, 1856-61; created three new provinces; transferred land revenue to provincial councils; established itinerant courts of justice and native juries; visited England, 1859; again premier of New Zealand, 1865-9, holding also offices of colonial secretary, 1865-6, colonial treasurer, 1865-6, and postmaster-general, 1865-6 and 1869; premier for a third time, Sept.-Oct. 1872; lived in England from 1874; KCMG, 1879; GCMG, 1887.

STAINER, SIR JOHN (1840-1901), organist and composer; chorister of St. Paul's Cathedral, 1849; graduated B.Mus. from Christ Church, Oxford, 1859; organist of Magdalen College, Oxford, 1860, and of Oxford University, 1861; BA, St. Edmund Hall, 1864; MA, 1866; D.Mus., 1865; founded Oxford Philharmonic Society, 1866; organist of St. Paul's Cathedral, 1872-88; organist to Royal Choral Society, 1873-88; professor of organ (1876) and principal (1881) of National Training College of Music; knighted, 1888; professor of music in Oxford University, 1889-99; master of Musicians' Company, 1900; hon. fellow of Magdalen College, Oxford; hon. Mus.D., Durham, 1858, and hon. DCL, 1895; compositions include oratorios, 'The Daughter of Jairus' (1878), 'Crucifixion' (1887), forty anthems, church services, and over 150 hymn tunes, organ compositions, madrigals, and part songs; edited *Christmas Carols, New and Old*, 1884; publications include *Music of the Bible* (1879), and *Early Bodleian Music* (2 vols., 1902); first editor of Novello's 'Music Primers', to

which he contributed volumes on *The Organ* and *Harmony*; formed unique collection of old song books.

STALBRIDGE, first BARON (1837-1912), railway administrator and politician. [See GROSVENOR, RICHARD DE AQUILA.]

STALLYBRASS, WILLIAM TEULON SWAN (1883-1948), principal of Brasenose College and vice-chancellor of Oxford; changed name from Sonnenschein, 1917; scholar of Westminster and Christ Church, Oxford; second class, *lit. hum.*, 1906; called to bar (Inner Temple), 1909; fellow and tutor in jurisprudence, Brasenose, 1911; vice-principal, 1914; principal, 1936-48; vice-chancellor, 1947-8; university lecturer in criminal law and law of evidence, 1924; reader, 1927-39; honorary treasurer, University Cricket Club, 1914-46.

STAMER, SIR LOVELACE TOMLINSON, third baronet (1829-1908), bishop-suffragan of Shrewsbury; educated at Rugby and Trinity College, Cambridge; BA, 1853; MA, 1856; DD, 1888; rector of Stoke-upon-Trent, 1858-92; succeeded father, 1860; built at Stoke three churches, five mission churches, and five schools; started night schools there, 1863; chairman of Stoke school board, 1871-88; formed North Staffordshire Coal and Ironstone Workers' Permanent Relief Society, 1870; chairman, 1870-1908; founded Staffordshire Institution for Nurses, 1872; archdeacon of Stoke-upon-Trent, 1877; bishop-suffragan of Shrewsbury, 1888-1905; vicar of St. Chad's, Shrewsbury, 1892-6; rector of Edgmond, 1896-1905; published sermons and charges.

STAMFORDHAM, BARON (1849-1931), private secretary to King George V. [See BIGGE, ARTHUR JOHN.]

STAMP, JOSIAH CHARLES, first BARON STAMP (1880-1941), statistician and administrator; educated at private boarding-school; entered Inland Revenue Department as boy clerk, 1896; B.Sc., London, 1911; encouraged by Graham Wallas [q.v.]; assistant secretary, Board of Inland Revenue, 1916-19; secretary and director, Nobel Industries, 1919-26; president, London, Midland, and Scottish Railway, 1926-41; director, Bank of England, 1928-41; member of committee on national debt and taxation, 1924-7; of Economic Advisory Council, 1930-41; British representative on Dawes (1924) and Young (1929) committees on German reparations; treasurer (1928-35), president (1936), British Association; joint-secretary and editor (1920-30), president (1930-2), Royal Statistical Society; vice-chairman (1925-35), chairman (1935-41), London School of Economics; devoted to Wesleyan connection and to pursuit of academic scholarship; especially gifted in handling men; publications include *British Incomes and Property* (1916), *Fundamental Principals of Taxation in the Light of Modern Developments* (1921), and *Wealth and Taxable Capacity* (1922); FBA, 1926; KBE, 1920; GBE, 1924; GCB, 1935; baron, 1938; killed in air raid.

STAMP, SIR (LAURENCE) DUDLEY (1898-1966), geographer; brother of Josiah (later Lord) Stamp [q.v.]; educated at University School, Rochester, and King's College, London; first class honours, B.Sc., 1917; served in army, 1917-19; demonstrator in geology, King's College; BA, first class, geography, 1921; D.Sc., 1921; professor of geology and geography, Rangoon, 1923-6; Sir Ernest Cassel reader in economic geography, London School of Economics, 1926; professor, 1945; professor, social geography, 1948-58; formed Land Utilization Survey of Britain; published report *The Land of Britain: its use and misuse* (1948); awarded Founder's medal of Royal Geographical Society, 1949; chief adviser on rural land utilization, Ministry of Agriculture, 1942-55; member of Nature Conservancy and other similar bodies; wrote extensively on problems of population growth; published lectures in *Land for Tomorrow, the Under-developed World* (1952); compiled textbooks, including *The World* (1929); president, Royal Geographical Society, 1963-6; active in many other learned bodies; president, International Geographical Union, 1952-6; CBE, 1946; knighted, 1965; awarded number of hon. degrees and other academic distinctions; remembered in Dudley Stamp Memorial Trust.

STANFORD, SIR CHARLES VILLIERS (1852-1924), composer, conductor, and teacher of music; educated in Dublin; organist and BA, Trinity College, Cambridge; studied music in Germany; attracted attention by music to Tennyson's *Queen Mary*, 1876; professor of composition and orchestral playing, Royal College of Music, London, 1883-1924; professor of music, Cambridge, 1887-1924; knighted, 1902; most versatile British composer of latter half of nineteenth century; made great but unsuccessful efforts in cause of English opera; his large-scale choral works, notably 'Requiem' (1897) and 'Stabat Mater' (1907), esteemed in their day and influenced later writers; composed seven symphonies for orchestra alone, best known being 'The Irish' (1887) and 'L'Allegro ed il Penseroso' (1895); his six Irish rhapsodies more representative of his genius; a prolific writer of chamber music; his church music

firmly established in English cathedrals and churches; his reputation most securely established as writer of solo songs; a noted conductor, especially of Bach Choir, London (1885-1902).

STANIER, SIR WILLIAM ARTHUR (1876-1965), railway engine designer; son of railwayman; educated at Wycliffe College, Stonehouse; joined Great Western Railway as office boy at fifteen; apprenticed at Swindon Works, 1892; posted to drawing office, 1897; inspector of materials, 1900; technical inspector to divisional locomotive carriage and wagon superintendent, 1902; promoted to superintendent, 1906; assistant locomotive works manager, 1912; manager, 1920; works assistant to chief mechanical engineer, 1922; took locomotive *King George V* to United States, 1927; chief mechanical engineer, London, Midland, and Scottish Railway, 1932; responsible for new standard locomotives, the 'Pacifics', 'Jubilees', and other engines; with Sir Ralph Wedgwood [q.v.] advised government of India on their state-owned railways, 1936; president, Institution of Locomotive Engineers, 1936-7 and 1938-9; scientific adviser to Ministry of Production, 1942; knighted, 1943; chairman, Power Jets Ltd.; FRS, 1944; awarded many honours and medals.

STANLEY, ALBERT HENRY, BARON ASHFIELD (1874-1948), chairman of London Passenger Transport Board; born in Derby; educated in Detroit; general superintendent, Detroit Street Railway Co.; general manager, street railway department, New Jersey Public Service Corporation, 1904-7, London Underground companies, 1907-10; managing director, 1910-19; chairman and managing director, 1919-33; with objective of single authority, acquired other transport firms; negotiated with London County Council for common management for Underground companies and Council's tramway system; chairman, London Passenger Transport Board, 1933-47; member, British Transport Commission, 1947-8; president, Board of Trade, 1916-19; MP, Ashton-under-Lyne, 1916-20; knighted, 1914; PC, 1916; baron, 1920.

STANLEY, SIR ARTHUR (1869-1947), philanthropist; son of sixteenth Earl of Derby [q.v.]; educated at Wellington; in diplomatic service, 1891-8; conservative MP, Ormskirk, 1898-1918; chairman, Royal Automobile Club, 1905-7, 1912-36; chairman, executive committee, British Red Cross Society, 1914-43; of joint war committee, Red Cross and Order of St. John, 1914-18; treasurer, St. Thomas's Hospital, 1917-43; president, British Hospitals Association, 1920-47; GBE, 1917; GCVO, 1944.

STANLEY, EDWARD GEORGE VIL-LIERS, seventeenth EARL OF DERBY (1865-1948), son of sixteenth Earl of Derby [q.v.]; educated at Wellington College; gazetted to Grenadier Guards, 1885; aide-de-camp to father in Canada, 1889-91; private secretary to Lord Roberts [q.v.] in South Africa, 1900; conservative MP, West Houghton division of Lancashire, 1892-1906; junior lord of Treasury, 1895-1900; financial secretary, 1900-3; post-master-general, 1903-5; succeeded father, 1908; chairman West Lancashire Territorial Association, 1908-28; president, 1928-48; raised five battalions of King's Regiment, 1914; director of recruiting, 1915-16; his scheme by which men were invited voluntarily to attest willingness to serve a first step towards conscription; under-secretary (1916), secretary of state for war, Dec. 1916-Apr. 1918; sought to hold the balance between civilian colleagues and soldiers; several times offered resignation in support of Sir William Robertson [q.v.] but finally withdrew having satisfied himself that (Lord) Haig [q.v.] would work to war cabinet policy, 1918; ambassador in Paris, 1918-20; inaugurated Franco-British Society, 1920; visited Dublin incognito for unsuccesssful conversation with Mr de Valera, Apr. 1921; secretary of state for war, 1922-4; president, Liverpool Chamber of Commerce, 1910-43; chancellor, Liverpool University, 1908-48; lord-lieutenant of Lancashire, 1928-48; president, Pilgrims Society, 1945-8; intimate friend of King George V; possessed what Englishmen admire: geniality, generosity, public spirit, great wealth, and successful racehorses; won St. Leger six times and Derby twice; PC, 1903; KCVO, 1905; GCVO, 1908; KG, 1915; GCB, 1920.

STANLEY, EDWARD LYULPH, fourth BARON SHEFFIELD, of Roscommon, in the peerage of Ireland, and fourth BARON STANLEY OF ALDERLEY, in the peerage of the United Kingdom (1839-1925); son of second Baron Stanley of Alderley [q.v.]; BA, Balliol College, Oxford; liberal MP, Oldham, 1880-5; member of London school board, 1876-85, 1888-1904; vice-chairman, 1897-1904; took active part in controversies over organization and administration of public education; successfully resisted proposals to strengthen influence of voluntary or denominational bodies; failed to promote scheme of higher education based on authority of school board; a laborious administrator and educational pioneer; succeeded brother as Baron Stanley, 1903; succeeded kinsman as Baron Sheffield, 1909.

STANLEY, FREDERICK ARTHUR, sixteenth EARL OF DERBY (1841-1908), governor-general of Canada; second son of fourteenth earl

[q.v.]; in Grenadier Guards, 1858-65; conservative MP for Preston, 1865-8; civil lord of the Admiralty, 1868; MP for North Lancashire, 1868-85, and for Blackpool, 1885-6; joined cabinet as secretary of state for war, and PC, 1878-80; GCB, 1880; colonial secretary, 1885-6; created Baron Stanley, 1886; president of the Board of Trade, 1886-8; governor-general of Canada, 1888-93; stimulated imperial sentiment in dominion; succeeded to earldom, 1893; prominent in Liverpool affairs; first lord mayor, 1895-6; first chancellor of university from 1903; successful racehorse owner; won Oaks in 1893 and 1906; KG, 1897; GCVO, 1905.

STANLEY, HENRY EDWARD JOHN, third BARON STANLEY OF ALDERLEY (1827-1903), diplomatist and orientalist; son of second baron [q.v.]; pensioner of Trinity College, Cambridge, 1846-7; entered diplomatic service; secretary of legation at Athens, 1854-9; began extensive travels in the East; adopted Moslem religion; succeeded to peerage and estates in Cheshire and Anglesey, 1869; champion of Church of England in Wales; supported Indian National Congress movement; keen sportsman and strict total abstainer; published *Essays on East and West*, 1865; translated for Hakluyt Society many Spanish and Portuguese works of travel.

STANLEY, SIR HENRY MORTON (1841-1904), explorer, administrator, author, and journalist, whose real name was JOHN ROW-LANDS; born at Denbigh, 29 June 1841; owing to father's early death and mother's neglect, he was brought up in St. Asaph workhouse, 1847-56; after some humble desultory employment he shipped as cabin boy on American packet for New Orleans, 1859; was adopted at New Orleans by cotton-broker, Henry Stanley (*d.* 1861), whose name he assumed; became a naturalized American; served as volunteer with confederate army (1861-2), but being taken prisoner he enlisted in United States artillery, from which he was discharged owing to ill health; suffered much distress; served in USA navy, 1864-5; made some progress as newspaper correspondent; sought adventure in Asia Minor, 1866; described General Hancock's expedition against Indians in *Missouri Democrat*, 1867; subsequently republished his articles in *My Early Travels and Adventures in America and Asia Minor* (1885); showed great resource as special correspondent of *New York Herald* during Abyssinian war, 1868-9; reported for *Herald* disturbances in Spain, 1869; was ordered (16 Oct. 1869) by proprietor, Gordon Bennett, to 'find Livingstone', who was lost in interior of Africa; before proceeding on this mission he attended opening of Suez canal, and visited Egypt and Palestine and Persia; started from

Bagamoyo in Zanzibar for Lake Tanganyika, where Livingstone was believed to be, 21 Mar. 1871; met Livingstone at Ujiji, 10 Nov. 1871; travelled about the district till 18 Feb. 1872 with Livingstone, who declined to accompany Stanley home; described his journey in *How I found Livingstone* (1872); met in England with much scorn from men of science, but was warmly welcomed by the general English public; described Ashanti war (1873-4) for *New York Herald*; embodied experiences in *Coomassie and Magdala* (1874); on learning of Livingstone's death Stanley undertook expedition to equatorial Africa under joint commission from *New York Herald* and London *Daily Telegraph*; left Zanzibar, 11 Nov. 1874; initiated conversion of kingdom of Uganda to Christianity; first traced course of the Congo; laid foundation of Anglo-Egyptian dominions of Upper Nile; emerged upon the shores of the Atlantic (9 Aug. 1877), having opened up the heart of the continent for the first time; published *Through the Dark Continent* (1878); was refused help in England for developing the Congo district; proposed the scheme to King Leopold II of Belgium, who accepted it, Aug. 1878; helped to organize Congo region for the Belgians (1879-84), and thus inaugurated the Congo State; issued *The Congo and the founding of its Free State* (1885); lectured in Germany, England, and America on the commercial possibilities of Central Africa, 1884-6; undertook leadership of his final expedition to equatorial Africa in order to rescue Emin Pasha, who had been abandoned in command of Egyptian army when the Mahdi overran the Sudan; started from the mouth of the Congo, Feb. 1887; divided his forces at Yambuya, June; leaving strong rearguard there, he pushed on through dense tropical forest to the Albert Nyanza, where he arrived, 13 Dec.; sent message to Emin, who visited him there with reluctance; returned to find his rearguard in tragic plight; brought remnant to Lake Albert, Jan. 1889; was joined there by Emin and his party; while marching back to Bagamoyo (4 Dec. 1889), he discovered much new country which ultimately went to form the British East African Protectorate; on arriving in England, although he was attacked for failure to 'rescue' Emin and for sacrificing his rearguard, he was received with enthusiasm and accorded many honours; described his journey in *In Darkest Africa* (1890), which had an immense sale; married at Westminster Abbey, 12 July 1890, Dorothy Tennant; lectured in America and Australasia; was re-naturalized as British subject; unionist MP for North Lambeth, 1895-1900; toured through South Africa, 1897-8; described experiences in *Through South Africa* (1898); GCB, 1899; stricken by paralysis, 15 Apr. 1903; died in London, 10 May 1904;

buried in Pirbright churchyard; *Autobiography* edited by his widow, 1909.

STANLEY, Sir **HERBERT JAMES** (1872-1955), colonial administrator; educated at Eton and Balliol College, Oxford; BA, 1897; private secretary in several posts, including secretary to Lord Gladstone [q.v.], governor-general, South Africa, 1897-1913, official secretary to governor-general, 1913; resident commissioner, Rhodesia, 1915; imperial secretary to governor-general, 1918; governor, Northern Rhodesia, 1924; governor, Ceylon, 1927; British high commissioner, South Africa, 1931; governor, Southern Rhodesia, 1935-41; CMG, 1913; KCMG, 1924; GCMG, 1930.

STANLEY, OLIVER FREDERICK GEORGE (1896-1950), politician; son of seventeenth Earl of Derby [q.v.]; educated at Eton; served in France, 1915-18; conservative MP, Westmorland (1924-45), Bristol West (1945-50); parliamentary private secretary to president of Board of Education, 1924-9; parliamentary under-secretary, Home Office, 1931-3; minister of transport, 1933-4; largely responsible for Road Traffic Act, 1934; minister of labour, 1934-5; president, Board of Education, 1935-7, Board of Trade, 1937-40; completed reciprocal trade agreement with United States; secretary of state for war, 1940; for colonies, 1942-5; initiated policy of colonial development; chancellor, Liverpool University, 1948-50; PC, 1934.

STANLEY, WILLIAM FORD ROBINSON (1829-1909), scientific-instrument maker and author; started in London a drawing instrument business, 1854; invented cheap 'Panoptic Stereoscope', 1855; patented application of aluminium to manufacture of mathematical instruments, 1861; greatly improved theodolite and other surveying instruments; published standard work on *Mathematical Drawing Instruments* (1866); made many scientific inventions; FGS, 1884; FRAS, 1894; accomplished musician, artist, photographer, and architect; opened at Norwood, Stanley public hall (1903) and technical school (1907); also published *Experimental Researches into . . . Fluids* (1881) and *Surveying and Levelling Instruments* (1890).

STANMORE, first **BARON** (1829-1912), colonial governor. [See GORDON, ARTHUR CHARLES HAMILTON-.]

STANNARD, HENRIETTA ELIZA VAUGHAN (1856-1911), novelist under the pseudonym of JOHN STRANGE WINTER; born PALMER; contributed to *Family Herald* from 1874 to 1884 short stories and serials dealing with military life; married Arthur Stannard, 1884; published *Bootles' Baby* (1885), a work which was admired by John Ruskin [q.v.] and was universally popular; edited *Winter's Weekly*, 1892-5; president of Society of Women Journalists, 1901-3.

STANNUS, HUGH HUTTON (1840-1908), architect, author, and lecturer; studied art foundry work at Sheffield; assistant of Alfred Stevens [q.v.], with whom he worked on the Wellington monument in St. Paul's Cathedral; studied architecture at Royal Academy Schools, 1872; FRIBA, 1887; did much private structural and decorative work; taught modelling at Royal Academy, 1881-1900; director of architectural studies at Manchester School of Art, 1900-2; published works on decoration, form-design, and architecture, and *Alfred Stevens and his Work* (1891).

STANSFIELD, MARGARET (1860-1951), pioneer in physical training for women; educated at a day-school in Bloomsbury; teacher in physical education, Bedford high school, 1887; founded and organized Bedford Physical Training College, 1903-1945; founder-member, Ling Association of Teachers of Physical Education; vice-president or president continuously, 1901-20; OBE, 1939.

STANSGATE, first VISCOUNT (1877-1960), parliamentarian. [See BENN, WILLIAM WEDGWOOD.]

STANTON, ARTHUR HENRY (1839-1913), divine; BA, Trinity College, Oxford; curate of St. Alban, Holborn, 1862-1913; under Alexander Heriot Mackonochie [q.v.], 1862-82; adopted advanced ritualistic views, and with Mackonochie subjected to series of ritual prosecutions, 1867-82; inhibited from preaching; was attacked again, 1906; exercised great spiritual influence within and beyond his parish; an eloquent preacher.

STAPLEDON, SIR (REGINALD) GEORGE (1882-1960), pioneer of grassland science; educated at United Services College, Westward Ho! and Emmanuel College, Cambridge; joined Royal Agricultural College, Cirencester, 1910; advisory officer, agricultural botany, University College of Wales, 1912; director, Welsh Plant Breeding Station, Aberystwyth, 1919-42; authority on grassland development; established research station, Drayton, Stratford-upon-Avon, 1942-6; first president, British Grassland Society, 1945; FRS, and knighted, 1939; published number of books, including *The Land: Now and Tomorrow* (1935) and *The Way of the Land* (1943).

STARK, ARTHUR JAMES (1831-1902), painter; son of James Stark [q.v.], landscape painter; studied animal painting under Edmund Bristow [q.v.]; exhibited at Royal Academy from 1848 to 1887, and elsewhere; perfected painting of horses at a studio at Tattersall's; depicted homely English scenes and landscapes; one of the last artists of Norwich school; works include 'Interior of a Stable' (1853), 'A Farmyard' (1875), and 'Dartmoor Drift' (1877).

STARLING, ERNEST HENRY (1866-1927), physiologist; educated at Guy's Hospital, London; became head of physiology department there; Jodrell professor of physiology, University College, London, 1899-1923; FRS, 1899; first Foulerton research professor of Royal Society, 1922; formed lifelong intellectual alliance with Sir W. M. Bayliss [q.v.]; his most important original investigations cover the secretion of lymph and other body fluids, the discovery of secretin (1902), and laws which govern activity of the heart; successfully advocated more efficient use of respirators by Italian army, 1917; his works include *Principles of Human Physiology* (1912).

STEAD, WILLIAM THOMAS (1849-1912), journalist and author; editor of *Northern Echo*, Darlington liberal daily paper, 1871-80; assistant editor of *Pall Mall Gazette*, 1880-3; editor, 1883-90; inaugurated 'new journalism'; directly responsible for dispatch of General Gordon [q.v.] to Khartoum (1884) and for Criminal Law Amendment Act (1885); started *Review of Reviews*, 1890; took up psychical research; drowned when *Titanic* sank.

STEBBING, (LIZZIE) SUSAN (1885-1943), philosopher; educated at Girton College, Cambridge; MA, London, 1912; lecturer in philosophy, King's College, London, 1913-15, Bedford College, 1915; reader, 1927; professor, 1933-43; made notable contributions to philosophical analysis.

STEED, HENRY WICKHAM (1871-1956), editor of *The Times*; educated at Sudbury grammar school and in Berlin and Paris; *The Times* correspondent, Berlin, 1896; Rome, 1897-1902; Vienna, 1902-13; head of foreign department, *The Times*, 1914-19; editor, 1919-22; bought and edited *Review of Reviews*, 1923-30; lecturer on Central European history, King's College, London, 1925-38; broadcaster on world affairs in overseas service, BBC, 1937-47; published *The Hapsburg Monarchy* (1913) and *Through Thirty Years* (2 vols., 1924).

STEEL, ALLAN GIBSON (1858-1914), cricketer; BA, Trinity Hall, Cambridge; in

Cambridge eleven, 1878–81; captain, 1880; admirable slow bowler; played in Gentlemen's eleven v. Players and in early Australian test matches, in which he proved powerful batsman; called to bar (Inner Temple), 1883; KC, 1901.

STEEL, FLORA ANNIE (1847–1929), novelist; born WEBSTER; married H. W. Steel, Indian civil service, 1867; lived in India, 1868–89; threw herself wholeheartedly into lives of Indians; became familiar with life of countryside and customs of people; penetrated behind purdah; exercised remarkable influence over Indian women; advocated their education; first government inspectress of girls' schools, 1884; her works include *The Potter's Thumb* (1894), *On the Face of the Waters* (1896), *In the Permanent Way* (1897), and *The Hosts of the Lord* (1900).

STEEL-MAITLAND, SIR ARTHUR HERBERT DRUMMOND RAMSAY-, first baronet (1876–1935), politician and economist; educated at Rugby and Balliol College, Oxford; first class, *lit. hum.*, president of Union, and rowed for Oxford, 1899; first class, jurisprudence, Eldon scholar, and fellow of All Souls, 1900; special commissioner to royal commission on the poor laws, 1906–7; conservative MP, East Birmingham (1910–29), Tamworth (1929–35); chairman, unionist party, 1911; parliamentary under-secretary for the colonies, 1915–17; joint parliamentary under-secretary for the Foreign Office, 1917–19; minister of labour, 1924–9; responsible for Unemployment Insurance Act, 1927; baronet, 1917; PC, 1924.

STEER, PHILIP WILSON (1860–1942), painter; son of portrait and landscape painter; educated at Hereford Cathedral School and privately; studied at Gloucester Art School and under Bouguereau and Cabanel in Paris; settled in Chelsea, 1885; influenced by J. A. M. Whistler [q.v.] and Manet; painted notable seascapes at Walberswick, exhibiting clean, decisive, and exquisite sense of tone; strove continually for greater exactitude and clearer definition; increasingly interested in watercolour; among his finest portraits is 'Mrs. Raynes' (Tate Gallery); supporter and exhibitor, New English Art Club; taught at Slade School, 1895–1930; OM, 1931.

STEGGALL, CHARLES (1826–1905), organist and composer; professor of organ at Royal Academy of Music, 1851–1903; Mus.Bac. and Mus.Doc., Cambridge, 1852; organist of Lincoln's Inn chapel, 1864–1905; composed church music and wrote *Instruction Book for the Organ* (1875).

STEIN, SIR EDWARD SINAUER DE (1887–1965), merchant banker. [See DE STEIN.]

STEIN, SIR (MARK) AUREL (1862–1943), scholar, explorer, archaeologist, and geographer; born in Budapest; studied at Vienna, Leipzig, Tübingen, and Oxford universities, and at British Museum; principal, Oriental College, Lahore, and registrar, Punjab University, 1888–99; worked for Indian education service, 1899–1910; for Archaeological Survey of India, 1910–29; published Sanskrit edition of *Râjataringini* of Kalhana (1892); translation (1900); engaged in: (*a*) four Central Asian expeditions (1900–1, 1906–8, 1913–16, and 1930); made valuable discoveries and collections; published scientific records in *Ancient Khotan* (2 vols., 1907), *Serindia* (5 vols., 1921), *The Thousand Buddhas* (1921), and *Innermost Asia* (4 vols., 1928); (*b*) reconnaissances in Baluchistan and Persia, connecting the earliest cultures of the Indus and Euphrates, 1927–36; published results in *Archaeological Reconnaissances in North-Western India and South-Eastern Iran* (1937) and *Old Routes of Western Iran* (1940); (*c*) elucidation of Alexander's campaign from battle of Arbela to return to Babylon; (*d*) examination of Roman frontier with Parthia; naturalized, 1904; KCIE, 1912; FBA, 1921; died in Kabul.

STENTON, SIR FRANK MERRY (1880–1967), historian; educated at Minster grammar school, Southwell, University Extension College, Reading, and Keble College, Oxford; first class, history, 1902; history master, Llandovery College, 1904; research fellow in local history, Reading University College, 1908; professor of modern history, Reading, 1912–46; vice-chancellor, 1946–50; published essays and lectures include *Norman London* (1915), *The Danes in England* (1926), and *The Road System of Medieval England* (1937); made extensive contributions to the *Victoria History of the Counties of England*; one of founders of study of place-names in an early work *The Place-Names of Berkshire* (1911); also published *The First Century of English Feudalism 1066–1166* (1932) and *Anglo-Saxon England* (1943); chairman of editorial committee of the *History of Parliament* (1951–65); president, Royal Historical Society, 1937–45, and other learned societies; trustee, National Portrait Gallery, 1948–65; closely concerned with purchase of Whiteknights Park for Reading University, 1946–7; FBA, 1926; hon. fellow, Keble College, 1947; knighted, 1948; hon. doctorates and other academic honours.

STEPHEN, SIR ALEXANDER CONDIE (1850–1908), diplomatist; entered diplomatic service, 1876; at Philippopolis, 1880–1; CMG, 1881; in Khorassan, north-east province of

Persia, 1882-5; CB, 1884; present at Penjdeh when affray between troops of Afghanistan and Russia threatened war between England and Russia, 1885; at Sofia, 1886; at Vienna and Paris, 1887; chargé d'affaires at Coburg, 1893-7; minister resident to Saxony and Coburg, 1897-1901; KCMG, 1894; KCVO, 1900; groom-in-waiting to King Edward VII, 1901-8.

STEPHEN, CAROLINE EMELIA (1834-1909), philanthropist; sister of Sir Leslie Stephen [q.v.]; joined Society of Friends, 1879; wrote *Quaker Strongholds* (1891), *Light Arising* (1898), and *A Vision of Faith* (posthumous, 1911).

STEPHEN, GEORGE, first BARON MOUNT STEPHEN (1829-1921), financier and philanthropist; born in Scotland; entered business in Montreal, 1850; successful cloth manufacturer; with group of colleagues completed St. Paul and Pacific and Canadian Pacific railways; president of Canadian Pacific Railway Company, 1880-8; settled in England, 1893; created baronet, 1886; baron, 1891.

STEPHEN, SIR LESLIE (1832-1904), first editor of this Dictionary, man of letters, and philosopher; grandson of James Stephen [q.v.]; third son of Sir James Stephen [q.v.], and younger brother of Sir James FitzJames Stephen [q.v.]; educated at Eton and King's College, London; entered Trinity Hall, Cambridge, 1850; twentieth wrangler, 1854; fellow, 1854-67; tutor, 1856; ordained 1855; interested himself in social life of college and in its athletic prestige; won mile race at university athletic sports, 1860; made his first Alpine ascent of Col du Géant, 1857; joined Alpine Club, 1858; first climbed the Schreckhorn, 1861; acquired high reputation as mountaineer, 1862-7; president of Alpine Club, 1865-8; published *Playground of Europe*, a collection of mountaineering sketches (1871); meanwhile abandoned his old religious convictions, resigned college tutorship, 1862, and relinquished holy orders, 1875; he aided his friend Henry Fawcett [q.v.] in electoral contests, 1863-4; visited America, 1863, as an enthusiastic supporter of slave emancipation; began literary career in London, 1864; revisiting America in 1868, formed close friendships with James Russell Lowell, Charles Eliot Norton, and others; wrote much for *Saturday Review* and for *Pall Mall Gazette*; editor from 1871 to 1882 of *Cornhill Magazine*, where first appeared his *Hours in a Library* (critical essays: 1st ser. 1874, 2nd ser. 1876, 3rd ser. 1879); wrote on religious and philosophical speculation in *Fraser's Magazine* and *Fortnightly Review*, collecting contributions in *Essays on Free*

Thinking and Plain Speaking (1873); published his chief work, *History of English Thought in the Eighteenth Century* (1876); inaugurated in 1878 with *Johnson* the 'English Men of Letters' series, to which he also contributed *Pope* (1880), *Swift* (1882), *George Eliot* (1902), and *Hobbes* (1904); and produced *Science of Ethics* (1882); first editor of this Dictionary, 1882-91; suffered in health from 1889, but on retirement from editorship in 1891 remained a chief contributor; first Clark lecturer at Trinity College, Cambridge, 1883; published in later life biography of Henry Fawcett (1885), *An Agnostic's Apology* (1893), life of his brother Sir James FitzJames Stephen (1895), *Social Rights and Duties* (2 vols., 1896), *The English Utilitarians* (3 vols., 1900), *Studies of a Biographer* (2 ser. 1899-1902), and *English Literature and Society in the Eighteenth Century—Ford lectures at Oxford, 1903* (1904); president of London Library from 1892 to death; hon. degrees from Edinburgh, Oxford, Cambridge, and Harvard; KCB, 1902; *Collected Essays*, 10 vols., with introds. by James Bryce and Herbert Paul [qq.v.], 1907; *Life and Letters* by F. W. Maitland [q.v.], 1906.

STEPHENS, FREDERIC GEORGE (1828-1907), art critic; joined pre-Raphaelite Brotherhood, 1848; exhibited portraits of mother (1852) and father (1854) at Royal Academy; contributed to *Germ*; art critic to *Athenaeum*, 1861-1901; wrote there valuable series of articles on 'The Private Collections of England'; publications include unfinished *Catalogue of Prints and Drawings (Satire) in the British Museum* (4 vols., 1870-83); championed claims of D. G. Rossetti [q.v.] to initiation of pre-Raphaelite movement in his life of Rossetti, 1894; was model for head of Christ in 'Christ Washing Peter's Feet' by Madox Brown [q.v.].

STEPHENS, JAMES (1825-1901), organizer of the Fenian conspiracy; at first a civil engineer; assisted William Smith O'Brien [q.v.] in Ballingary affray, 1848; planned unsuccessful plot of kidnapping prime minister, Lord John Russell, Sept. 1848; escaped to Paris, teaching English there for some years; inaugurated Irish Republican Brotherhood on military basis, 1858; visited America (1858-9) to collect funds to provide arms, and stimulated movement there; published (1862) a scheme for the future government of Ireland in the event of success attending conspiracy; founded the *Irish People* as organ of his party, 1863; on second visit to America (1864) promised a rising in Ireland in Sept. 1865; *Irish People* offices raided, Sept. 1865; Stephens arrested and imprisoned, 11 Nov.; escaped, 24 Nov.; went to Paris (Mar. 1866) and revisited New York, where in Dec. he was denounced as an impostor; returned to

Paris, 1867; wrongly suspected of share in American dynamite plots and expelled from France, 1885; lived thenceforth in Ireland.

STEPHENS, JAMES (1880?-1950), writer; born in Dublin; became attorney's clerk; obtained early recognition of genius in poetry and prose; published *Insurrections* (verse, 1909), *The Charwoman's Daughter* and *The Crock of Gold* (1912), *The Demi-Gods* (1914), and *Reincarnations* (1918); registrar, National Portrait Gallery, Dublin, 1915-24; lived in Paris, 1912-14, in Paris and London, 1924-50; broadcaster of verse and stories; awarded civil list pension, 1942.

STEPHENS, JAMES BRUNTON (1835-1902), Queensland poet; educated at Edinburgh University, 1852-4; private tutor on continent of Europe, 1854-7; six years schoolmaster at Greenock; emigrated to Queensland (1866) and engaged there in tutorial work; published poem *Convict Once* (1871); clerk to colonial secretary, 1883-1902; published fiction and verse; collected edition of poems appeared in 1902.

STEPHENS, WILLIAM RICHARD WOOD (1839-1902), dean of Winchester; BA, Balliol College, Oxford, 1862; MA, 1865; DD, 1901; held Sussex livings, 1870-94; dean of Winchester, 1894-1902; FSA, 1894; voluminous works include lives of his father-in-law, Dean Hook (2 vols., 1878), and E. A. Freeman (2 vols., 1895) and *History of English Church from Norman Conquest to Edward I* (1901); revised Dean Hook's *Church Dictionary*, 1887.

STEPHENSON, SIR **FREDERICK CHARLES ARTHUR** (1821-1911), general; joined Scots Guards, 1837; served throughout Crimean war; in China war, 1858 and 1860; CB, 1858; lieutenant-general, 1878; commanded army of occupation in Egypt, 1883; KCB, 1884; commanded frontier field force and defeated Mahdists at Ginnis, Dec. 1885; GCB, 1886; colonel of Coldstream Guards, 1892; constable of Tower of London, 1898-1911.

STEPHENSON, GEORGE ROBERT (1819-1905), civil engineer; employed by uncle, George Stephenson [q.v.], in drawing office of Manchester and Leeds Railway, 1837-43; resident engineer on new lines of South Eastern Railway; superintended construction of many branch lines in Kent; also constructed lines in Schleswig-Holstein, Jutland, and New Zealand; joint engineer-in-chief for East London Railway, 1864; built many bridges, fixed and swinging, at home and abroad; proprietor of locomotive works at Newcastle upon Tyne, 1859-86; president of Institution of Civil Engineers, 1875-7.

STEPHENSON, (JOHN) CECIL (1889-1965), artist; educated at Bishop Auckland, Darlington Technical College, Leeds School of Art, Royal College of Art, and the Slade School of Art; head of art, architectural department, Northern Polytechnic, London, 1922-55; joined avant-garde group of artists—Ben Nicholson, (Dame) Barbara Hepworth, and Henry Moore in early 1930s; painting influenced by interest in machinery—'The Pump' (1932), 'The Lathe' (1933), and 'Mechanism' (1933); exhibited in London, 1934 and 1937; painted pictures of the blitz during 1939-45 war; 'Painting 1937' purchased by Tate Gallery, 1963; memorial exhibition at Drian Galleries, London, 1966.

STEPHENSON, MARJORY (1885-1948), biochemist; educated at Berkhamsted high school and Newnham College, Cambridge; worked in R. H. A. Plimmer's laboratory, University College, London, 1911-14; in biochemical department, Cambridge, under Sir F. G. Hopkins [q.v.] from 1919; studied biochemical activities of bacteria; reader in chemical microbiology, 1947-8; published *Bacterial Metabolism* (1930); FRS, 1945.

STEPHENSON, THOMAS ALAN (1898-1961), marine biologist and artist; educated at Kingswood School, Bath, and University College of Wales; M.Sc. and D.Sc., 1920 and 1923; research work at Aberystwyth, 1920-3; lecturer in zoology, University College, London, 1923; produced *British Sea Anemones* (2 vols., 1928 and 1935, with his own illustrations); member, Great Barrier Reef Expedition, 1928; professor of zoology, Cape Town University, 1930; carried out extensive study of South African shore ecology; professor of zoology, Aberystwyth, 1940; with his wife, made study of Atlantic and Pacific coasts of North America; highly skilled miniature artist; published *Seashore Life and Pattern* (1944); *Life between Tidemarks on Rocky Shores* completed by his wife and published in 1972; FRS, 1951.

STERLING, ANTOINETTE (1843-1904), contralto singer; born at Sterlingville, New York State; taught singing in Mississippi State from 1868; studied singing in Germany and London; first appeared in London, 1873; successful in oratorio music and German *Lieder*; married John MacKinlay, 1875; from 1877 attained much popularity by her rendering of 'The Lost Chord', 'Caller Herrin'', and other ballads; toured in Australia, 1893; ardent Christian Scientist.

STERN, SIR ALBERT GERALD (1878-1966), banker and administrator in the production of the first tanks; educated at Eton and Christ Church, Oxford; entered family merchant banking business, Stern Brothers; partner, 1904; commissioned in Royal Naval Volunteer Reserve, 1914; secretary, Admiralty Landships Committee under (Sir) Eustace Tennyson-d'Eyncourt [q.v.], 1915; took over project for the production of armoured fighting vehicles; in co-operation with (Sir) Ernest Swinton [q.v.] organized production of first tank, 1916; chairman, Tank Supply Committee, Ministry of Munitions; transferred from navy to army, and concentrated on improvement of design and quantity of tanks, 1916-18; commissioner for mechanical warfare (overseas and Allies), lieutenant-colonel, Tank Corps, 1917; CMG, 1917; KBE, 1918; published *Tanks 1914-1918: the Log-Book of a Pioneer* (1919); returned to banking, 1919; director, Midland Bank and other banks; head of Stern Brothers; chairman, Special Vehicle Development Committee, Ministry of Supply, 1939-45; master, Drapers Company, 1946-7; high-sheriff of Kent, 1945-6; chairman, board of governors, Queen Mary College, London.

STERNDALE, BARON (1848-1923), judge. [See PICKFORD, WILLIAM.]

STERRY, CHARLOTTE (1870-1966), lawn tennis champion; daughter of Henry Cooper of Caversham; married Alfred Sterry, a solicitor, 1901; won the ladies singles title at Wimbledon five times, 1895, 1896, 1898, 1901, and 1908; the only British player to defeat Mrs Lambert Chambers [q.v.] between 1903 and 1919; learnt her tennis at Ealing Lawn Tennis Club; won mixed doubles championship with H. S. Mahony five times, 1894-8, with H. L. Doherty, 1900, and with X. E. Casdagli, 1908; also won many other championships; one of the most popular players of her day; notable for her supreme steadiness, equable temperament, and great tactical ability.

STEVENS, MARSHALL (1852-1936), one of the founders and first general manager of the Manchester Ship Canal Company; entered father's shipping business, Plymouth; moved to the Mersey; took lead in obtaining consent to Manchester Ship Canal project; general manager, 1885-96; first managing director, Trafford Park Estates; coalition unionist MP, Eccles, 1918-22.

STEVENSON, SIR DANIEL MACAULAY, baronet (1851-1944), merchant, civic administrator, and philanthropist; coal-exporter; member, Glasgow corporation, 1892-1914; lord provost, 1911-14; chancellor, Glasgow University, 1934-44; founded lectureship in citizenship and chairs of Italian and Spanish, Glasgow; of international history, London; baronet, 1914.

STEVENSON, DAVID WATSON (1842-1904), Scottish sculptor; studied in Rome (1876) and Paris; ARSA, 1877; RSA, 1886; executed groups for Prince Consort memorial, Edinburgh, and monuments of Wallace and Burns.

STEVENSON, JAMES, BARON STEVENSON (1873-1926), administrator; managing director of John Walker & Sons, distillers, Kilmarnock; did valuable work for munitions 1915-18; surveyor-general of supply, War Office, 1919-21; as chairman of Exhibition Board, in control of Wembley British Empire Exhibition, 1924 and 1925; baronet, 1917; baron, 1924.

STEVENSON, JOHN JAMES (1831-1908), architect; MA, Glasgow University; pupil of Sir George Gilbert Scott [q.v.] in London, 1858-60; evolved a simple type of brick design; built several board schools and churches; built 'The Red House', Bayswater Hill; restored many colleges at Oxford and Cambridge; first architect to design interior decorations of sea vessels; published *House Architecture* (2 vols., 1880); FSA, 1884; FRIBA, 1879.

STEVENSON, SIR THOMAS (1838-1908), scientific analyst and toxicologist; MB, London, 1863; MD, 1864; FRCP, 1871; lecturer at Guy's Hospital on chemistry, 1870-98, and forensic medicine, 1878-1908; analyst to Home Office, 1872-81; senior scientific analyst, 1881-1908; expert witness in leading poisoning cases from 1881; knighted, 1904; edited medical works.

STEVENSON, WILLIAM HENRY (1858-1924), historian and philologist; edited Nottingham borough records (4 vols., 1882-9); engaged to calendar muniments of Merton College, Oxford, 1888; research fellow of Exeter College, 1895-1904; produced eleven volumes of Calendars of Close Rolls, 1892-1908; published edition of *Asser's Life of King Alfred* (1904); fellow and librarian of St. John's College, Oxford, 1904; authority on place-names.

STEWART, CHARLES (1840-1907), comparative anatomist; MRCS, 1862; conservator of Hunterian Museum at Royal College of Surgeons from 1884; Hunterian professor of comparative anatomy and physiology, 1886-1902; Fullerian professor at Royal Institution, 1894-7; president of Linnean society, 1890-4; helped to found Anatomical Society, 1887; FRS, 1896; hon. LLD, Aberdeen, 1899.

STEWART, SIR HALLEY (1838-1937), founder of trust bearing his name; Congregational pastor, 1863-77; founder (1877) and first editor, *Hastings and St. Leonards Times*; vice-chairman, London Brick Company; advanced liberal; MP, Spalding division (1887-95), Greenock (1906-10); established trust (1924) for 'research towards the Christian ideal in all social life'; knighted, 1932.

STEWART, ISLA (1855-1910), hospital matron; nurse at St. Thomas's Hospital, 1879-85; matron and superintendent of nursing at St. Bartholomew's Hospital, 1887-1910; founded League of St. Bartholomew's Hospital Nurses, and Matrons' Council for Great Britain and Ireland, 1894; collaborated in *Practical Nursing* (2 vols., 1899-1903).

STEWART, JAMES (1831-1905), African missionary and explorer; educated at Edinburgh and St. Andrews universities; studied theology at New College, Edinburgh, 1855-9, and medicine at the university, 1859; influenced in 1857 by David Livingstone [q.v.], he resolved on establishing mission in Central Africa; reached Livingstone's headquarters at Shupanga, 1862; explored Shiré and Zambezi districts, 1862-3; hon. FRGS, 1866; MB and CM, Glasgow, 1866; returned to Africa, reaching Lovedale, Jan. 1867; principal of Lovedale Missionary Institute, which he consolidated and developed, 1870-90; urged foundation of Livingstonia in memory of Livingstone, 1875; organized new settlement at Bandawe, 1876; explored Lake Nyasa, 1877; was consulted by colonial administrators on native questions; established new East African mission, 1891; returned to Scotland, 1892; hon. DD, Glasgow, 1892; moderator of Free Church General Assembly, 1899; delivered (1902) Duff missionary lectures in Edinburgh on *Dawn in the Dark Continent* (published 1903); returned to Lovedale, 1904, and died there; author of Kaffir phrase book (1898) and grammar (1902).

STEWART, JOHN ALEXANDER (1846-1933), philosopher; first class, *lit. hum.*, Lincoln College, Oxford, 1870; classical lecturer, Christ Church, 1870-82; tutor, 1882-97; White's professor of moral philosophy, 1897-1927; published *Notes on the Nicomachean Ethics of Aristotle* (2 vols., 1892) and *The Myths of Plato* (1905); distrusted philosophic dogmatism and adhered to no school.

STEWART, SIR (PERCY) MALCOLM, first baronet, of Stewartby, county Bedford (1872-1951), industrialist; son of (Sir) Halley Stewart [q.v.]; educated at University School, Hastings, King's School, Rochester, the Royal High School, Edinburgh, and in Germany; managing director, B. J. Forder & Son Ltd., subsequently London Brick Company, 1900-50; managing director, British Portland Cement Manufacturers, 1912-45; chairman and president, Cement Makers' Federation, 1918-51; commissioner for special areas, 1934-6; benefactor to numerous charities; OBE, 1918; baronet, 1937; hon. LLD, Manchester, 1937.

STEWART, SIR (SAMUEL) FINDLATER (1879-1960), civil servant; educated at Edinburgh University; MA, 1899; entered India Office, 1903; assistant under-secretary of state and clerk of the Council of India, 1924; secretary, statutory commission on India, headed by Sir John (later Viscount) Simon [q.v.], 1927; permanent under-secretary of state, India and (later) Burma Office, 1930-40; chairman, Home Defence Executive, 1940; chairman, Anglo-American co-ordinating committee, 1943-5; chairman, British and French Bank, 1945; CIE, 1919; CSI, 1924; KCIE, 1930; KCB, 1932; GCIE, 1935; GCB, 1939; hon. LLD, Edinburgh and Aberdeen.

STEWART, WILLIAM DOWNIE (1878-1949), New Zealand politician; educated at Otago boys' high school and university; LLB and admitted solicitor, 1900; MP, Dunedin, 1914-35; contracted rheumatoid arthritis in France, 1916; minister of internal affairs, 1921-3; customs, 1921-8; industry and commerce, 1923-6; finance, 1926-8; leading advocate of coalition, 1931; minister of finance, customs and attorney-general, 1931; stood for orthodox financial methods; resigned on raising of exchange rate, 1933; encyclopaedic knowledge of New Zealand politics.

STEWART, SIR WILLIAM HOUSTON (1822-1901), admiral; son of Sir Houston Stewart [q.v.]; entered navy, 1835; served in Carlist war, 1836-7; and Syrian war, 1840; commander, 1848; in Pacific, 1851-3; retook revolted Chilian colony of Punta Arenas; distinguished in Crimean war; captain superintendent of Chatham dockyard, 1863-8; controller of Portsmouth dockyard, 1872-81; KCB, 1877; admiral, 1881; GCB, 1887.

STEWART-MURRAY, KATHARINE MARJORY, DUCHESS OF ATHOLL (1874-1960), public servant; daughter of Sir James Ramsay [q.v.]; educated at Wimbledon high school and Royal College of Music; married (1899) Marquess of Tullibardine (later, Duke of Atholl); active in Scottish social service and local government; DBE, 1918; conservative MP, Kinross and West Perthshire, 1923-38; parliamentary secretary, Board of Education,

1924-9; resigned seat in parliament in opposition to Chamberlain policy of appeasement of Hitler; chairman, British League for European Freedom, 1944-60; published *Women and Politics* (1931), *Searchlight on Spain* (1938), and *Working Partnership* (1958).

STILES, SIR HAROLD JALLAND (1863-1946), surgeon; first class honours, Edinburgh, 1885; surgeon to Royal Hospital for Sick Children and Chalmers Hospital from 1898; with Sir Robert Jones [q.v.] raised standard of orthopaedic surgery, 1914-18; regius professor of clinical surgery, Edinburgh Royal Infirmary, 1919-25; introduced sterilization by high-pressure steam; made notable cancer researches; knighted, 1918; KBE, 1919.

STILES, WALTER (1886-1966), plant physiologist; educated at Latymer upper school, Hammersmith, and Emmanuel College, Cambridge, first class, parts i and ii, natural sciences tripos, 1907-9; assistant lecturer in botany, Leeds University, 1910; research in plant physiology; during 1914-18 war worked on food preservation research; professor of botany, University College, Reading, 1919; Mason professor, Birmingham University, 1929-51; prolific author both in journals and in books; publications include *Carbon Assimilation* (with I. Jørgensen, 1917), *Permeability* (1924), *Respiration in Plants* (with W. Leach, 1931), *An Introduction to the Principles of Plant Physiology* (1936), and *Trace Elements in Plants and Animals* (1946); Sc.D., Cambridge, 1922; FRS, 1928; fellow, Linnean Society and other learned bodies; hon. D.Sc., Nottingham, 1963; professor emeritus, 1952, and life governor, Birmingham University, 1954.

STILL, SIR (GEORGE) FREDERIC (1868-1941), paediatrician; educated at Merchant Taylors' School and Gonville and Caius College, Cambridge; first class, classical tripos, 1888; qualified from Guy's Hospital, 1893; physician, Hospital for Sick Children, Great Ormond Street; physician for diseases of children, King's College Hospital, 1899-1935; first professor of diseases of children, King's College, 1906-35; established entity of a chronic rheumatoid arthritis peculiar to childhood; works include *Common Disorders and Diseases of Childhood* (1909) and *History of Paediatrics* (1931); KCVO, 1937.

STIRLING, SIR JAMES (1836-1916), judge; BA, Trinity College, Cambridge (senior wrangler), 1860; called to bar (Lincoln's Inn), 1862; attorney-general's 'devil', 1881-6; Chancery judge and knighted, 1886; lord justice of appeal, 1900-6; PC, 1900.

STIRLING, JAMES HUTCHISON (1820-1909), Scottish philosopher; educated at Glasgow University; MRCS Edinburgh, 1842; FRCS, 1860; abandoned medicine for philosophy; published *The Secret of Hegel* (2 vols., 1865) and *Analysis of Sir William Hamilton's Philosophy* (1865); sought to refute Huxley's theory of protoplasm, 1869; issued *Text Book to Kant* (1881) and other works; hon. LLD, Edinburgh, 1867, Glasgow, 1901; received civil list pension, 1889.

STIRLING, WALTER FRANCIS (1880-1958), lieutenant-colonel; educated at Kelly College, Tavistock, and Sandhurst; joined Royal Dublin Fusiliers, 1899; served in South African war; DSO; seconded to Egyptian Army, 1906-12; joined Royal Flying Corps, 1914; rejoined British army and served at Gallipoli and in Palestine; chief staff officer to T. E. Lawrence [q.v.]; MC, and bar to DSO; lieutenant-colonel, 1920; governor, Jaffa district, Palestine, 1920-3; advisor, Albanian government, 1923-31; intelligence work in Balkans, 1940; military commander, East Syria, 1943; *The Times* correspondent, Damascus, 1945-9; published *Safety Last* (1953).

STOCKDALE, SIR FRANK ARTHUR (1883-1949), tropical agriculturist; educated at Wisbech grammar school and Magdalene College, Cambridge; BA, 1904; mycologist and lecturer in agricultural science, West Indian imperial department of agriculture, 1905-9; assistant director of agriculture, British Guiana, 1909-12; director, Mauritius, 1912-16, Ceylon, 1916-29; agricultural adviser, Colonial Office, 1929-40; comptroller, development and welfare, West Indies, 1940-5; co-chairman, Anglo-American Caribbean Commission, 1942-5; adviser on development planning, Colonial Office, 1945-8; vice-chairman, Colonial Development Corporation, 1948-9; KCMG, 1937.

STOCKS, JOHN LEOFRIC (1882-1937), philosopher; educated at Rugby and Corpus Christi College, Oxford; first class, *lit. hum.*, 1905, and captain of hockey; fellow and tutor, St. John's College, 1906-24; DSO, 1916; professor of philosophy, Manchester, 1924-36; vice-chancellor, Liverpool University, 1936-7.

STODDART, ANDREW ERNEST (1863-1915), cricketer; chosen to play cricket for Middlesex, 1885; scored 485, then highest individual innings recorded, 1886; played regularly for Middlesex, a strong county team, till 1898; visited Australia with teams successful except on last occasion, 1887, 1891, 1894, 1897; in front rank of English amateurs as high scorer and all-round player.

STOKES, ADRIAN (1887-1927), pathologist; born at Lausanne; MB, Trinity College, Dublin; as officer in Royal Army Medical Corps during 1914-18 war did invaluable work on tetanus, typhoid, jaundice, etc.; professor of bacteriology and preventive medicine, Trinity College, Dublin, 1919; Sir William Dunn professor of pathology, London University, 1922; died at Lagos on Rockefeller yellow fever commission investigation.

STOKES, SIR FREDERICK WILFRID SCOTT (1860-1927), civil engineer; articled to a civil engineer, 1878; assistant to R. C. Rapier, of Messrs. Ransomes & Rapier, Ipswich, 1885; engineer and managing director of London office, 1896; managing director, 1897; chairman, 1907; his inventions include improved rotary kilns for cement-making, dam and canal sluices, and a trench mortar; KBE, 1917.

STOKES, SIR GEORGE GABRIEL, first baronet (1819-1903), mathematician and physicist; educated at schools at Dublin and Bristol and at Pembroke College, Cambridge; senior wrangler, first Smith's prizeman, and fellow, 1841; master, 1902-3; formed at Cambridge close friendship with Lord Kelvin [q.v.]; contributed to *Cambridge Mathematical Journal*; Lucasian professor of mathematics from 1849 till death; early developed Lagrange's theory of motion of fluids; created modern theory of motion of viscous fluids; made valuable researches into optics; published *The Dynamical Theory of Diffraction* (1849); pioneer in discovery of and development of spectrum analysis; discovered nature of fluorescence (1852) and explored great range of invisible ultraviolet spectrum; virtual founder of modern science of geodesy, 1849; fellow (1851), secretary (1854-85), president (1885-90), and Copley medalist (1893) of the Royal Society; prominent in foundation of observatory for solar physics at South Kensington, 1878; Burnett lecturer at Aberdeen, 1883-5; published lectures on *Light* (3 vols., 1884-7); Gifford lecturer at Edinburgh, 1891-3; president of British Association, 1869; conservative MP for Cambridge University, 1887-91; baronet, 1889; received many foreign honours and hon. degrees from Oxford, Cambridge, Edinburgh, Dublin, Glasgow, and Aberdeen; his writings were collected in five volumes of *Mathematical and Physical Papers* (1880-1905); his *Scientific Correspondence* was edited by (Sir) Joseph Larmor [q.v.] (2 vols., 1907).

STOKES, SIR JOHN (1825-1902), lieutenant-general RE; joined Royal Engineers, 1843; in Cape Colony, 1846-51; served in Kaffir wars, 1846-7, 1850-1; instructor in surveying at Royal Military Academy, Woolwich, 1851; served in Crimea; British member of European commission of the Danube, 1856-71; improved mouths and navigation of lower Danube and regulated pilotage; urged on Great Britain measures which led to 'Danube Loan Act'; drafted articles respecting the Danube in the treaty of London, 1871; CB, 1871; commanding royal engineer of South Wales military district, 1872-5, of Chatham district, 1875; British commissioner on international commission at Constantinople concerning Suez canal dues, 1873; reported on Khedive's financial difficulties and negotiated for the representation of British government on Suez Canal Company's board, 1875-6; became director, 1876; vice-president, 1887; KCB, 1877; as member of Channel tunnel committee, opposed its construction, 1882; deputy adjutant-general for Royal Engineers at War Office, 1881-6; lieutenant-general, 1887.

STOKES, WHITLEY (1830-1909), Celtic scholar; son of William Stokes [q.v.]; BA, Trinity College, Dublin, 1851; studied Irish philology from early age; called to bar (Inner Temple), 1855; practised in London till 1862, when he went to India; secretary to legislative department at Calcutta, 1865; law member of council of the governor-general, 1877-82; president of Indian law commission, 1879; published works on Indian law, 1865-91; CSI, 1877; CIE, 1879; returned to England, 1882; collaborated in series of *Irische Texte* at Leipzig, 1884-1909, in *Urkeltischer Sprachschatz*, 1894, and with John Strachan [q.v.] in *Thesaurus Palaeohibernicus*, 1901 and 1903; published many Irish texts with translations and glossaries, as well as Cornish and Breton works; original FBA, 1902; hon. fellow, of Jesus College, Oxford; his library of Celtic printed books at University College, London.

STOLL, SIR OSWALD (1866-1942), theatrical impresario; stepson of owner of Parthenon music-hall, Liverpool; acquired or opened number of music-halls including London Coliseum and Alhambra; provided inexpensive, wholesome entertainment; chairman, Stoll Picture Theatre (Kingsway), Ltd., 1919-42; initiated War Seal (Sir Oswald Stoll) Foundation for disabled officers; knighted, 1919.

STONE, DARWELL (1859-1941), Anglo-Catholic theologian; educated at Owens College, Manchester, and Merton College, Oxford; deacon, 1883; priest, 1885; vice-principal, Dorchester Missionary College, 1885; principal, 1888-1903; contributed extensively to *Church Quarterly Review*; librarian, Pusey House, Oxford, 1903-9; principal, 1909-34; leader of extreme Anglo-Catholic group; editor, Lexicon of Patristic Greek, 1915-41; works include *History of the Doctrine of the Holy Eucharist* (2 vols., 1909).

STONER, EDMUND CLIFTON (1899-1968), physicist; educated at Bolton grammar school and Emmanuel College, Cambridge; first class, parts i and ii, natural sciences tripos, 1920-1; carried out research at Cavendish Laboratory under Sir E. Rutherford (later Lord Rutherford, q.v.); lecturer in physics, Leeds University, 1924; reader, 1927; professor of theoretical physics, 1939; succeeded R. Widdington [q.v.] as Cavendish professor of physics, Leeds, 1951-63; research on X-ray absorption, 1924; published papers on theory of magnetic properties of metallic ferromagnets, such as nickel, 1938-9; also on theory of permanent magnets, magnetic recording tapes, and magnetic thin films, 1948; FRS, 1937; Sc.D., Cambridge, 1938.

STONEY, BINDON BLOOD (1828-1909), civil engineer; BA, Trinity College, Dublin, 1850; MA and MAI, 1870; hon. LLD, 1881; chief engineer (1862-98) to Dublin port authority; improved channel between Dublin bay and city, rebuilt quay walls and Grattan and O'Connell bridges; FRS, 1881; president of Institution of Civil Engineers of Ireland, 1871-2; member of Royal Irish Academy; published *The Theory of Strains in Girders and Similar Structures* (2 vols., 1866).

STONEY, GEORGE GERALD (1863-1942), engineer; son of G. J. Stoney [q.v.]; second senior moderator, Trinity College, Dublin, 1886; BA, engineering, 1887; joined (Sir) Charles Parsons [q.v.], 1888; assisted him in establishing reaction steam turbine; supervised design of *Turbinia*, 1894; chief designer, steam-turbine department, 1895; works technical manager, 1910; resigned, 1912; professor of mechanical engineering, College of Technology, Manchester, 1917-26; returned to Parsons as director of research, 1926-30; FRS, 1911.

STONEY, GEORGE JOHNSTONE (1826-1911), mathematical physicist; elder brother of Bindon Blood Stoney [q.v.]; BA, Trinity College, Dublin, 1848; MA and Madden prizeman, 1852; assistant at Parsonstown observatory, 1848-52; professor of natural philosophy, Queen's College, Galway, 1852-7; secretary of Queen's University, Ireland, 1857-82; superintendent of civil service examinations in Ireland till 1893; made original contributions to study of physical optics, of molecular physics, of the kinetic theory of gases, and of conditions limiting planetary atmospheres; introduced word 'electron' into scientific vocabulary; wrote

also on abstract physics, music, and musical echoes; for twenty years secretary of Royal Dublin Society; first Boyle medallist, 1899; hon. D.Sc., Queen's University, Ireland, 1879; hon. Sc.D., Trinity College, Dublin, 1902; FRS, 1861; settled in London, 1893; advocate of women's higher education.

STOOP, ADRIAN DURA (1883-1957), rugby footballer; educated at Dover College, Rugby, and University College, Oxford; joined Harlequins club while a schoolboy, 1900; played for Oxford University, 1902-4, captain, 1904; vice-captain and secretary, Harlequins, 1905; captain and secretary, 1906-14; fifteen caps for England, 1905-12; served in Mesopotamia in 1914-18 war; MC; president, Harlequins, 1920-49; selector for England; president, Rugby Union, 1932-3.

STOPES, MARIE CHARLOTTE CARMICHAEL (1880-1958), scientist and sex reformer; educated at St. George's, Edinburgh, North London Collegiate School, and University College, London; first class, botany, 1902; Ph.D., Munich, 1904; D.Sc., London, 1905; lecturer in palaeobotany, Manchester and London, 1909-20; fellow, University College, London, 1910; published *Ancient Plants* (1910) and *Catalogue of Cretaceous Flora* (2 vols., 1913-15); married Humphrey Verdon-Roe, 1918; with him, founded Mothers' Clinic for Birth Control, London, 1921; devoted herself to sex education and family planning; published *Married Love* (1918), *Wise Parenthood* (1918), *Radiant Motherhood* (1920), and *Enduring Passion* (1928).

STOPFORD, Sir FREDERICK WILLIAM (1854-1929), general; commissioned to Grenadier Guards, 1871; served with Egyptian expeditionary force, 1882; with Suakin expeditionary force, 1885; Ashanti expedition, 1895; South African war, 1899-1902; chief of staff of I Army Corps, 1902; major-general and director of military training, War Office, 1904; commanded London district, 1906; lieutenant of Tower of London, 1912; commanded First Home Defence Army, 1914; commanded IX Corps of New Army troops at Suvla Bay landing, Dardanelles operations, 1915; KCMG, 1900.

STOPFORD, JOHN SEBASTIAN BACH, Baron Stopford of Fallowfield (1888-1961), anatomist and vice-chancellor; educated at Liverpool College, Manchester grammar school, and Manchester University medical school; MB, Ch.B., 1911; demonstrator of anatomy under (Sir) Grafton Elliot Smith [q.v.], 1912; lecturer, MD, 1915; professor,

1919-37; worked in military orthopaedic centre during 1914-18 war; dean of the medical school, 1923-7; pro-vice-chancellor, 1928-30; FRS, 1927; published *Sensation and the Sensory Pathway* (1930); head, personal chair of experimental neurology, 1937-56; vice-chancellor, 1934-56; chairman, Universities Bureau of the British Empire, and other academic bodies; vice-chairman, trustees of Nuffield Foundation, 1943; continued to work as medical scientist; first chairman, Manchester Regional Hospital Board, 1948-53; MBE, 1920; knighted, 1941; KBE, 1955; FRCP, 1942; hon. FRCS, 1955; life peer, 1958; freeman of the City of Manchester, and hon. doctorates from five universities.

STORRS, Sir RONALD HENRY AMHERST (1881-1955), Near Eastern expert and governor; educated at Charterhouse and Pembroke College, Cambridge; first class, classical tripos, 1903; entered Egyptian civil service, 1904; oriental secretary, British Agency, Cairo, under Sir Eldon Gorst, Lord Kitchener, and Sir Henry McMahon [qq.v.], 1909-17; military governor, Jerusalem, 1917-20; civil governor, Jerusalem and Judaea, 1920-6; governor, Cyprus, 1926-32; governor, Northern Rhodesia, 1932-4; published *Orientations* (1937); CMG, 1916; CBE, 1919; KCMG, 1929; hon. LLD, Aberdeen and Dublin.

STORY, ROBERT HERBERT (1835-1907), principal of Glasgow University; son of Robert Story (1790-1859, q.v.); educated at Edinburgh and St. Andrews universities; succeeded father as parish minister of Rosneath in Church of Scotland, 1860-86; helped to found Church Service Society, 1865; moderator of General Assembly, 1894; senior clerk, 1895-1907; chaplain in ordinary to Queen Victoria and King Edward VII from 1886; professor of church history in Glasgow University, 1886-98; principal, 1898-1907; largely extended university buildings; hon. DD, Edinburgh, 1874; hon. LLD, St. Andrews, 1900; prominent freemason; works include lives of Robert Lee [q.v.] (1870) and William Carstares (1874), and *The Apostolic Ministry of the Scottish Church* (Baird lecture, 1897) and sermons.

STORY-MASKELYNE, MERVYN HERBERT NEVIL (1823-1911), mineralogist; grandson of Nevil Maskelyne [q.v.]; BA, Wadham College, Oxford, 1845; MA, 1849; hon. fellow, 1873; pioneer teacher of mineralogy and chemistry in Oxford University from 1851; professor of mineralogy, 1856-95; keeper of the minerals at the British Museum, 1857-80; rearranged and greatly extended the collections, publishing a catalogue (1853) and a guide (1868); made important researches into

meteorites and the diamond; based on his lectures *The Morphology of Crystals* (1895); FRS, 1870; vice-president, 1897-9; FGS, 1854; Wollaston medallist, 1893; hon. D.Sc., Oxford, 1903; MP for Cricklade as liberal, 1880-6, and as liberal unionist, 1886-92; member of Wiltshire county council, 1889-1904.

STOUT, GEORGE FREDERICK (1860-1944), philosopher; first class, classical and moral sciences triposes, St. John's College, Cambridge; fellow, 1884; university lecturer in moral sciences, 1894-6; Anderson lecturer in comparative psychology, Aberdeen, 1896-8; Wilde reader in mental philosophy, Oxford, 1898-1903; professor of logic and metaphysics, St. Andrews, 1903-36; editor, *Mind*, 1891-1920; works include *Analytic Psychology* (1896), *Manual of Psychology* (2 vols., 1898-9), *Studies in Philosophy and Psychology* (1930), *Mind and Matter* (1931), and *God and Nature* (1932); FBA, 1903.

STOUT, Sir ROBERT (1844-1930), prime minister and chief justice of New Zealand; emigrated to New Zealand, 1863; solicitor and banker; attorney-general and minister for lands and immigration, 1878; premier, attorney-general, and minister of education in Stout-Vogel ministry, 1884-7; chief justice of New Zealand, 1899-1926; a keen educationist and great advocate; KCMG, 1886.

STRACHAN, DOUGLAS (1875-1950), artist in stained glass; executed windows for Palace of Peace, The Hague; Scottish National War Memorial, Edinburgh Castle; Glasgow Cathedral and University chapel; St. Giles' Cathedral and St. Margaret's chapel, Edinburgh; church of St. Thomas, Winchelsea; of St. John, Perth.

STRACHAN, JOHN (1862-1907), classical and Celtic scholar; educated at Aberdeen University and Pembroke College, Cambridge; Porson university scholar, 1883; second chancellor's medallist, 1885; after studying Sanskrit and Celtic at Jena, graduated BA at Cambridge, 1885; professor of Greek at Owens College, Manchester, where he also taught Celtic and Sanskrit, 1885-1907; collaborated with Whitley Stokes [q.v.] in *Thesaurus Palaeohibernicus* (2 vols., 1901-3); wrote memoirs on Irish philology and *An Introduction to Early Welsh* (posthumous, 1909); hon. LLD, Aberdeen, 1900.

STRACHAN-DAVIDSON, JAMES LEIGH (1843-1916), classical scholar; BA, Balliol College, Oxford; fellow of Balliol, 1866; senior dean, 1875-1907; master, 1907-16; devoted his life to service of his college and university; lectured on Roman history; works include

Cicero and the Fall of the Roman Republic (1894) and *Problems of the Roman Criminal Law* (1912).

STRACHEY, SIR ARTHUR (1858-1901), judge; son of Sir John Strachey [q.v.]; LLB, Trinity Hall, Cambridge, 1880; called to bar, 1883; judge of high court of Bombay, 1895-9; chief justice of high court, Allahabad and knighted, 1899; died at Simla.

STRACHEY, SIR EDWARD, third baronet (1812-1901), author; brother of Sir John and Sir Richard Strachey [qq.v.]; friend of F. D. Maurice [q.v.]; succeeded to uncle's baronetcy and estates in Somerset, 1858; took prominent part in local affairs; made study of oriental languages and of biblical criticism; works include *Jewish History and Politics* (1874), *Miracles and Science* (1854), *Talk at a Country House*, largely autobiographical (1895); edited Globe edition of Malory's *Morte d'Arthur* (1868).

STRACHEY, SIR EDWARD, fourth baronet, and first BARON STRACHIE (1858-1936), politician and landowner; son of Sir Edward Strachey, third baronet, and brother of J. St. L. Strachey [qq.v.]; succeeded father, 1901; liberal MP, South Somerset, 1892-1911; baron, 1911; parliamentary secretary, Board of Agriculture, 1909-11; paymaster-general, 1912-15; PC, 1912; active in Somerset local administration and landowners' organizations.

STRACHEY, (EVELYN) JOHN ST. LOE (1901-1963), politician and writer; son of John St. Loe Strachey [q.v.], editor of the *Spectator*; educated at Eton and Magdalen College, Oxford; editor, with Robert (later Lord) Boothby, of the *Oxford Fortnightly Review*; joined staff of *Spectator*; joined the labour party, 1923; editor, the *Socialist Review* and the *Miner*; labour MP, Aston, 1929-31; parliamentary private secretary to (Sir) Oswald Mosley; supported the communist party; assisted (Sir) Victor Gollancz [q.v.] in founding the Left Book Club, 1936; wrote for *Left News*; broke with communists, 1940; joined RAF; transferred to Air Ministry; made reputation as air commentator for BBC; labour MP, Dundee, 1945-50; under-secretary of state for air, 1945; minister of food, 1946; involved in the abortive Tanganyika ground-nut scheme, 1949; labour MP, West Dundee, 1950-63; secretary of state for war, 1950-1; supported Hugh Gaitskell [q.v.] as successor to Attlee, 1955; publications include *Revolution by Reason* (1925), *The Coming Struggle for Power* (1932), *Why You Should be a Socialist* (1938), *A Programme for Progress* (1940), *The End of Empire* (1959), *On the Prevention of War* (1962), and *The Strangled Cry* (1962).

STRACHEY, (GILES) LYTTON (1880-1932), critic and biographer; son of Sir Richard Strachey [q.v.]; educated at Liverpool and Trinity College, Cambridge; prominent member of Bloomsbury Group; works include *Eminent Victorians* (1918), *Queen Victoria* (1921), and *Elizabeth and Essex* (1928); inaugurated new type of biography fusing fact and reflection in a brief, brilliant creative work of art.

STRACHEY, SIR JOHN (1823-1907), Anglo-Indian administrator; entered Bengal civil service, 1842; served at first in North-West Provinces; 1862; president of permanent sanitary commission, 1864; chief commissioner of Oudh, 1866; member of governor-general's council, 1868; lieutenant-governor of North-West Provinces, 1874-6; created agriculture department, extended survey, and constructed railways; restored historic Mogul buildings at Agra, 1876; finance member of governor-general's council, 1876; retired owing to his serious under-estimate of cost of war in Afghanistan, 1880; knighted, 1872; GCSI, 1878; settled in Florence, 1880-3; published lectures on *India* (1888), and *Hastings and the Rohilla War* (1892); member of council of India, 1885-95; hon. DCL, Oxford, 1907.

STRACHEY, JOHN ST. LOE (1860-1927), journalist; son of Sir Edward Strachey, third baronet [q.v.]; BA, Balliol College, Oxford; called to bar (Inner Temple), 1885; took up journalism; editor of *Cornhill Magazine*, 1896; editor and proprietor of *Spectator*, 1898-1925; made it for years most influential unionist weekly paper; active and influential unionist free trader; his important registration scheme of 'Surrey Veterans', etc., widely copied before 1914; published numerous books.

STRACHEY, SIR RICHARD (1817-1908), lieutenant-general; joined Bombay engineers, 1836; executive engineer on Ganges canal, 1843; served in Sutlej campaign at Badiwal, Aliwal, and Sobraon, 1846; made scientific explorations in Kumaon, Himalayas, and Tibet, 1847-8; collected many new botanical specimens and added much to geological knowledge; stayed in England arranging his collections, 1850-5; FRS, 1854; royal medallist, 1897; secretary of newly formed Central Provinces, 1856; rebuilt railway station at Allahabad; consulting engineer to government at Calcutta, 1858; reorganized public works department and initiated adequate forest service; secretary and head of public works department, 1862-5; CSI, 1866; inspector-general of irrigation, 1866; major-general on retirement, 1871; lieutenant-general and member of council of India, 1875; arranged for purchase of East Indian Railway, 1877; chairman, 1889-1906; British commissioner at prime meridian conference at Washington, 1884; president of Royal Geographical Society, 1887-9; represented India at international monetary conference at Brussels, 1892; hon. LLD, Cambridge, 1892; founded scientific study of Indian meteorology; chairman of the meteorological council which controlled the meteorological office in London, 1893-5; invented mechanical instruments in connection with meteorological problems; CSI, 1866; GCSI, 1897; collaborated with brother John [q.v.] in *The Finances and Public Works of India* (1882).

STRACHIE, first BARON (1858-1936), politician and landowner. [See STRACHEY, SIR EDWARD.]

STRADLING, SIR REGINALD EDWARD (1891-1952), civil engineer; educated at Bristol grammar school and Bristol University; B.Sc., 1912; served with Royal Engineers, 1914-17; MC; lecturer in civil engineering, Birmingham University, 1918; Ph.D., Birmingham, 1922; D.Sc., Bristol, 1925; head of civil engineering and building, Bradford Technical College, 1922; director, building research, Department of Scientific and Industrial Research, 1924-39; chief adviser, Ministry of Home Security, 1939; chief scientific adviser, Ministry of Works, 1944-9; adviser on civil defence, Home Office, 1945-8; dean, Military College of Science, Shrivenham, 1949-52; CB, 1934; knighted, 1945; FRS, 1942.

STRAKOSCH, SIR HENRY (1871-1943), financier; born in Austria; educated in Vienna and England; joined Union Corporation, Ltd.; managing director, 1902; chairman, 1924-43; financial adviser to and representative of South African and Indian governments; member, financial committee, League of Nations, 1920-37; chairman, Economist Newspaper, Ltd., 1929-43; naturalized, 1907; knighted, 1921.

STRANG, WILLIAM (1859-1921), painter and etcher; pupil of Alphonse Legros [q.v.]; his 747 etchings include imaginative compositions (chiefly book illustrations) and portraits; executed portrait drawings and paintings; RA, 1921.

STRANGWAYS, ARTHUR HENRY FOX (1859-1948), schoolmaster, music critic, and founder-editor of *Music and Letters*; educated at Wellington and Balliol College, Oxford; studied music at Berlin Hochschule; schoolmaster at Dulwich (1884-6), Wellington (1887-1910); music critic, *The Times*, 1911-25; the *Observer*, 1925-39; founder-editor, *Music and Letters*,

1920-36; translated songs of Schubert, Schumann, Brahms, etc.

STRANGWAYS, GILES STEPHEN HOLLAND FOX-, sixth EARL OF ILCHESTER (1874-1959), landowner and historian. [See FOX-STRANGWAYS.]

STRATHALMOND, first BARON (1888-1970), industrialist. [See FRASER, WILLIAM.]

STRATHCARRON, first BARON (1880-1937). [See MACPHERSON, (JAMES) IAN.]

STRATHCLYDE, BARON (1853-1928), lawyer and politician. [See URE, ALEXANDER.]

STRATHCONA AND MOUNT ROYAL, first BARON (1820-1914), Canadian financier. [See SMITH, DONALD ALEXANDER.]

STRATHMORE AND KINGHORNE, fourteenth EARL OF (1855-1944). [See BOWES-LYON, CLAUDE, GEORGE.]

STRATTON, FREDERICK JOHN MARRIAN (1881-1960), astrophysicist; educated at King Edward's grammar school, Mason University College, Birmingham, and Gonville and Caius College, Cambridge, third wrangler, mathematics tripos, 1904; fellow, Caius College, 1906; mathematics lecturer, 1906-14; served in Royal Engineers during 1914-18 war; DSO; tutor (later, senior tutor), Caius College, 1919-28; professor, astrophysics, and director, Solar Physics Observatory, 1928-47; general secretary, British Association, 1930-5; president, Royal Astronomical Society, 1933-5; president, Caius College, 1946-8; FRS, 1947.

STREET, ARTHUR GEORGE (1892-1966), farmer, author, and broadcaster; left Dauntsey's School, Devizes, at fifteen to work on his father's farm; worked on farm in Canada; rejected for army service, 1914; took over his father's farm, 1917; wrote first farming article for *Daily Mail*, 1929; published *Farmer's Glory* (1932), and over thirty other books, including *Land Everlasting* (1934) and *Country Calendar* (1935); made first broadcast, 1933; lectured on farming in Canada and America, 1937; contributed regular column to *Farmers Weekly*; a notable interpreter of country life and farming techniques, and a television celebrity.

STREET, SIR ARTHUR WILLIAM (1892-1951), civil servant; educated at county school, Sandown; entered civil service as a boy clerk; served abroad in 1914-18 war; MC; clerk at Board of Agriculture and Fisheries, principal,

1922; second secretary, 1936–8; permanent under-secretary, Air Ministry, 1939–45; deputy chairman, National Coal Board, 1946–51; CIE, 1924; CMG, 1933; CB, 1935; KBE, 1938; KCB 1941; GCB, 1946.

STREETER, BURNETT HILLMAN (1874–1937), divine; educated at King's College School, London, and Queen's College, Oxford; first class, *lit. hum.* (1897), theology (1898); fellow and dean, 1905; chaplain, 1928; provost, 1933–7; ordained, 1899; canon of Hereford, 1915–34; distinguished New Testament scholar; works include *Studies in the Synoptic Problem* (1911), *The Four Gospels: A Study of Origins* (1924), and *Reality: A New Correlation of Science and Religion* (1926); FBA, 1925.

STRETTON, HESBA (pseudonym) (1832–1911), authoress. [See SMITH, SARAH.]

STRICKLAND, GERALD, BARON STRICK-LAND (1861–1940), colonial administrator and politician; born in Malta; succeeded as sixth Count della Catena, 1875; educated at Oscott and Trinity College, Cambridge; graduated in law, called to bar (Inner Temple), and president of Union, 1887; helped to frame Maltese constitution, 1887; chief secretary, Malta, 1889–1902; governor, Leeward Islands (1902), Tasmania (1904), Western Australia (1909), New South Wales (1913–17); conservative MP, Lancaster division, 1924–8; helped to draft Maltese constitution, 1921; member, Maltese legislative assembly, 1921–30; formed Anglo–Maltese party; prime minister and minister of justice of Maltese coalition government with labour, 1927–30; leader of elected members, council of government, 1939–40; unceasing in opposing Italian influence; founded *Times of Malta* and other papers; KCMG, 1897; baron, 1928.

STRIJDOM, JOHANNES GERHARDUS (1893–1958), South African prime minister; BA, Victoria College, Stellenbosch, 1912; LLB, admitted to bar, 1918; MP, Waterberg, 1929–58; keen nationalist and republican; minister of lands under D. F. Malan [q.v.], 1948; succeeded Malan as prime minister, 1954–8; secured removal of Cape Coloured voters from common roll.

STRONG, EUGÉNIE (1860–1943), classical archaeologist and historian of art; born SELLERS; educated in Europe and at Girton College, Cambridge; life fellow; married (1897) S. A. Strong [q.v.]; assistant director, British School at Rome, 1909–25; publications include *Roman Sculpture from Augustus to Constantine* (1907) and *Art in Ancient Rome from the Earliest Times to Justinian* (2 vols., 1929).

STRONG, LEONARD ALFRED GEORGE (1896–1958), writer; educated at Brighton College and Wadham College, Oxford; assistant master, Summer Fields School, Oxford, 1917–19 and 1920–30; versatile writer of novels, plays, children's books, poems, and film scripts; publications include *The Brothers* (1932), *The Open Sky* (1939), *The Sacred River* (1949), and *Personal Remarks* (1953).

STRONG, SIR SAMUEL HENRY (1825–1909), chief justice of Canada; accompanied father to Canada, 1836; called to bar at Toronto, 1849; QC, 1963; vice-chancellor of Ontario, 1869; puisne judge of supreme court of Canada, 1875; chief justice, 1892–1902; knighted, 1893; member of judicial committee of Privy Council, 1897.

STRONG, SANDFORD ARTHUR (1863–1904), orientalist and historian of art; BA, St. John's College, Cambridge, 1884; MA, 1890; studied Sanskrit at Cambridge; librarian of Indian Institute, Oxford, 1885; professor of Arabic at University College, London, 1895–1904; librarian to Duke of Devonshire [q.v.] at Chatsworth, 1895–1904; librarian to House of Lords, 1897–1904; published Pali and Arabic texts; works include reproductions of drawings at Wilton House (1900) and Chatsworth (1902); his *Critical Studies and Fragments* appeared posthumously in 1905; memorial 'Arthur Strong Oriental Library' at University College, London.

STRONG, THOMAS BANKS (1861–1944), bishop successively of Ripon and Oxford; brother of S. A. Strong [q.v.]; educated at Westminster and Christ Church, Oxford; second class, *lit. hum.*, 1883; lecturer, 1884; student, 1888; censor, 1892; dean, 1901–20; vice-chancellor of Oxford, 1913–17; GBE, 1918; deacon, 1885; priest, 1886; bishop of Ripon, 1920–5; of Oxford, 1925–37; publications include *Manual of Theology* (1892), *Christian Ethics* (1896), and *The Doctrine of the Real Presence* (1899).

STRUTHERS, SIR JOHN (1857–1925), educationist and civil servant; educated at Glasgow University and Worcester College, Oxford; inspector of schools in Scotland, 1886; senior examiner, Scottish education department, 1898; assistant secretary, 1900; secretary, 1904–22; KCB, 1910; contributed greatly to reform of Scottish education.

STRUTT, EDWARD GERALD (1854–1930), agriculturist; brother of third Baron Rayleigh [q.v.]; undertook management of Rayleigh estates in Essex, 1876; founded London firm of Strutt & Parker, land agents and surveyors, primarily to act as receiver for Lincoln and Essex estates of Guy's Hospital, 1877; successfully developed large-scale arable dairy farming on Rayleigh estates; founded Lord Rayleigh's Dairies, Ltd., retail selling organization in London; advocated, for maintaining agricultural population, increase of arable land, more small holdings, agricultural education and research, light railways, government assistance in building rural cottages, no adverse legislation against capital invested in soil; gave valuable advice to government in framing agricultural policy, 1914-18.

STRUTT, JOHN WILLIAM, third BARON RAYLEIGH (1842-1919), mathematician and physicist; educated privately; BA, Trinity College, Cambridge (senior wrangler), 1865; first Smith's prizeman, 1865; fellow of Trinity, 1866-71; published first of his 446 scientific papers, 1869; FRS, 1873; succeeded father, 1873; early investigations concerned with psychical research; published *Treatise on the Theory of Sound*, 1877; Cavendish professor of experimental physics at Cambridge, 1879-84; directed research on redetermination of electrical units in absolute measure; published series of papers on subject, 1881-3; retired to Terling Place, Essex, to pursue research in private laboratory, 1885; secretary of Royal Society, 1885-96; discovered argon (in collaboration with (Sir) William Ramsay, q.v.), 1894; carried out important investigations in physical optics; an original recipient of OM, 1902; Nobel prizeman, 1904; PC, 1905; chancellor of Cambridge University, 1908; president of Royal Society, 1905; Rumford medallist, 1914; his researches covered fields of physics, chemistry, and mathematics, but only striking discovery that of argon; his scientific supremacy due to aptitude for arranging existing knowledge.

STRUTT, ROBERT JOHN, fourth BARON RAYLEIGH (1875-1947), experimental physicist; son of third Baron Rayleigh [q.v.]; succeeded father, 1919; educated at Eton and Trinity College, Cambridge; first class, parts i and ii, natural sciences tripos, 1897-8; fellow, 1900-6; professor of physics, Imperial College of Science and Technology, 1908-19; carried out research in private laboratory, notably investigating the age of minerals and rocks by measurement of their radioactivity and helium content; publications include *The Becquerel Rays and the Properties of Radium* (1904); FRS, 1905; foreign secretary, 1929-34; president, British Association, 1938; of Royal Institution, 1945-7.

STUART, SIR JOHN THEODOSIUS BURNETT- (1875-1958), general. [See BURNETT-STUART.]

STUART-JONES, SIR HENRY (1867-1939), classical scholar. [See JONES.]

STUBBS, SIR REGINALD EDWARD (1876-1947), colonial governor; son of William Stubbs [q.v.]; educated at Radley and Corpus Christi College, Oxford; first class, *lit. hum.*, 1899; entered Colonial Office, 1900; colonial secretary, Ceylon, 1913-19; governor, Hong Kong, 1919-25, Jamaica, 1926-32, Cyprus, 1932-3, Ceylon, 1933-7; vice-chairman, West India royal commission, 1938-9; KCMG, 1919; GCMG, 1928.

STUBBS, WILLIAM (1825-1901), historian and bishop successively of Chester and Oxford; went to Ripon grammar school; servitor of Christ Church, Oxford, 1844; BA (with first class in *lit. hum.* and third in mathematics), 1848; studied historical documents in the Christ Church library; fellow of Trinity College, 1844-50; ordained deacon, 1848, and priest, 1850; vicar of Navestock, Essex, 1850-66; issued *Registrum Sacrum Anglicanum* (1858, new edn. 1897); took private pupils, including H. P. Liddon and A. C. Swinburne [qq.v.]; appointed librarian at Lambeth, 1862; edited *Chronicle Memorials of Richard I* for Rolls Series, 1864-5; regius professor of history at Oxford, 1866-84; and ex-officio fellow of Oriel College; published many of his public professorial lectures in *Seventeen Lectures on the Study of Medieval and Modern History* (1886); other of his lectures which appeared posthumously treated of *European History* (1904), *Early English History* (1906), *Germany, 476-1500* (2 vols., 1908); disabled by examination and tutorial system at Oxford from carrying out his ideal of a thorough 'historical school', he devoted himself to private research; continued his contributions to Rolls Series, admirably editing the text of many medieval historical authorities (19 vols., 1864-89); his introductory essays to these volumes were collected posthumously in 1902; in 1870 he issued *Select Charters and other Illustrations of English Constitutional History from the Earliest Times to the Reign of Edward I*; made his chief fame by his *Constitutional History of England down to 1485* (vol. i, 1873, vol. ii, 1875, vol. iii, 1878), a massive work of historic synthesis, which gave a new direction to study of medieval English history; supporting the high church party at Oxford, Stubbs was rector of the Oriel living of Cholderton, Wiltshire, 1875-9, and canon of St. Paul's, 1879-84, working during his periods of residence at St. Paul's on the muniments and impressing his hearers by weighty sermons; he took a leading part as member of the royal commission on ecclesiastical courts, 1881-3; bishop of Chester, 1884-7, abandoning historical work with his edition of William of Malmesbury's works (Rolls Series, 2 vols., 1887-9); made energetic attempt to build new churches in his diocese; interested himself in education and archaeology, and was active at Lambeth conference of 1888; translated to see of Oxford, 1888, he remained there till death; finding routine episcopal work irksome, he obtained appointment of suffragan bishop of Reading, 1889; but he proved his episcopal efficiency in his *Ordination Addresses* (1901) and *Visitation Charges* (1904), both published posthumously; took part while bishop of Oxford in academic affairs; acted as assessor on trial of Edward King [q.v.], bishop of Lincoln, for ritualistic practices, 1889-90, and approved judgment of Archbishop Benson [q.v.]; preached with difficulty owing to failing health in St. George's chapel, Windsor, the day after Queen Victoria's funeral, 3 Feb. 1901; received Prussian order 'pour le mérite', 1897, and numerous academic honours; *Letters* edited by W. H. Hutton [q.v.], 1904.

STUDD, SIR (JOHN EDWARD) KYNASTON, first baronet (1858-1944), philanthropist; educated at Eton and Trinity College, Cambridge; played cricket for Eton and Cambridge; captain, 1884; joined Quintin Hogg [q.v.] at Regent Street Polytechnic as honorary secretary, 1885; president, 1903-44; senior sheriff of London, 1922; lord mayor, 1928; knighted, 1923; baronet, 1929.

STURDEE, SIR FREDERICK CHARLES DOVETON, first baronet (1859-1925), admiral of the fleet; entered navy, 1871; became torpedo expert; commanded first battle squadron, 1910; vice-admiral, 1913; chief of war staff under Prince Louis of Battenberg [q.v.], 1914; won decisive victory over Admiral von Spee at battle of Falkland Islands, Dec. 1914; commanded fourth battle squadron, 1915-18; took part in battle of Jutland, 1916; baronet, 1916; admiral, 1917; commander-in-chief at the Nore, 1918-21; admiral of the fleet, 1921.

STURGIS, JULIAN RUSSELL (1848-1904), novelist; born at Boston, Massachusetts; educated at Eton and Balliol College, Oxford; BA, 1872; MA, 1875; called to bar, 1876; works include novels, *John-a-Dreams* (1878), *My Friends and I* (1884), and *Stephen Calinari* (1901), as well as *Little Comedies* (1879) and *Comedies New and Old* (1882).

STURT, GEORGE (1863-1927), author under the pseudonym of GEORGE BOURNE; entered firm of wheelwrights at Farnham in which his ancestors had worked since 1706, 1885; his works include *The Bettesworth Book*

(1901), *The Wheelwright's Shop* (1923), and *A Small Boy in the Sixties* (1927).

STURT, HENRY GERARD, first BARON ALINGTON (1825-1904), sportsman; educated at Eton and Christ Church, Oxford; BA, 1845; MA, 1848; conservative MP for Dorchester, 1847-56, and for Dorset county, 1856-76; baron, 1876; in racing partnership with Sir Frederic Johnstone from 1868; won Derby with St. Blaise (1883) and Common (1891), besides other important races.

SUETER, SIR MURRAY FRAZER (1872-1960), rear-admiral; entered Royal Navy, 1886; qualified as torpedo specialist, 1896; captain, 1909; inspecting captain, airships, 1910; director, air department, Admiralty, 1912; influential in formation of Royal Naval Air Service, 1914; encouraged development of armoured cars and tanks, 1915; superintendent, aircraft construction, 1915-17; involved in difference of opinion with Board of Admiralty, 1917; placed on retired list, rear-admiral, 1920; conservative MP, Hertford, 1921-45; knighted, 1934; publications include *The Evolution of the Submarine Boat, Mine and Torpedo* (1907) and *The Evolution of the Tank* (1937).

SULLIVAN, ALEXANDER MARTIN (1871-1959), barrister; son of A. M. Sullivan [q.v.]; educated at Trinity College, Dublin; called to Irish bar, 1892; called to English bar (Middle Temple), 1899; KC, Ireland, 1908; KC, England, 1919; first King's Serjeant in Ireland, 1920; defended Sir Roger Casement [q.v.], 1916; after establishment of Irish Free State, moved to England, 1922; bencher, Middle Temple, 1925; treasurer, 1944; represented Halliday Sutherland [q.v.] in libel action brought by Marie Stopes [q.v.]; published *Old Ireland* (1927) and *The Last Serjeant* (1952).

SUMNER, VISCOUNT (1859-1934), judge. [See HAMILTON, JOHN ANDREW.]

SUMNER, BENEDICT HUMPHREY (1893-1951), historian; educated at Winchester and Balliol College, Oxford; served in King's Royal Rifle Corps during 1914-18 war; fellow of All Souls, 1919; fellow of Balliol, 1925; tutor in modern history, 1922-44; warden, All Souls, 1945-51; closely concerned with establishment of (Royal) Institute of International Affairs (Chatham House); publications include *Russia and the Balkans* (1937) and *Peter the Great and the Ottoman Empire* (1949).

SUTHERLAND, ALEXANDER (1852-1902), Australian journalist; emigrated with family from Glasgow to Sydney, 1864; BA,

Melbourne University, 1874; MA, 1876; London representative of *South Australian Register*, 1898; registrar of Melbourne University, 1901; works include *Thirty Short Poems* (1890), *History of Australia* (1897), and *Origin and Growth of the Moral Instinct* (1898).

SUTHERLAND, HALLIDAY GIBSON (1882-1960), physician, author, and controversialist; educated at Glasgow high school and Merchiston Castle School, Edinburgh; MB, Ch.B., Edinburgh, 1906; MD, 1908; opened tuberculosis dispensary, St. Marylebone, London, with open-air school, 1911; served in Royal Navy and Royal Air Force during 1914-18 war; physician, St. Marylebone Hospital, 1919; joined London County Council medical service, 1925; contested views of Marie Stopes [q.v.] on birth-control; won action brought by Dr Stopes for libel, 1924; deputy medical officer of health, Coventry, 1941; directed Mass Radiography Centre, Birmingham, 1943-51; publications include *Arches of the Years* (1933), *In My Path* (1936), and *Irish Journey* (1956).

SUTHERLAND, SIR THOMAS (1834-1922), chairman of Peninsular and Oriental Steamship Company, 1881-1914; clerk in company's London office, 1853; superintendent of company's Japan and China agencies, 1860; inspector and then assistant manager of company after return to London, 1867; managing director, 1873; chairman of board, 1881; rendered invaluable services to British shipowners by promoting 'Programme de Londres' relating to Suez canal 1883; liberal (subsequently liberal unionist) MP, Greenock, 1884-1900; KCMG, 1891.

SUTRO, ALFRED (1863-1933), playwright and translator of Maurice Maeterlinck; educated at City of London School and in Brussels; wholesale merchant, 1883-94; became friend of Maeterlinck and translated almost all his works; a leading English dramatist after 1904; plays include *The Walls of Jericho* (1904), *John Glayde's Honour* (1907), *The Choice* (1919), and *Living Together* (1929).

SUTTON, SIR BERTINE ENTWISLE (1886-1946), air marshal; educated at Eton and University College, Oxford; served in Royal Flying Corps, 1916-18; DSO, 1917; commissioned in Royal Air Force; instructor at Imperial Defence College, 1929-32; served in India, 1932-6; air vice-marshal, 1937; commanded 22 Group, 1936-9, 21 Group, 1939-40; commandant, RAF Staff College, 1941; air member for personnel, 1942-5; air marshal and KBE, 1942.

SUTTON, HENRY SEPTIMUS (1825-1901), author; born at Nottingham; from an early age a friend of Philip James Bailey and Coventry Patmore [qq.v.]; betrayed the influence of Emerson in his mystic prose work, *Evangel of Love* (1847); chief of *Manchester Examiner and Times* reporting staff, 1853; published mystical poems, *Clifton Grove Garland* (1848) and *Quinquenergia* (1854), which were praised by James Martineau and criticized by Carlyle [qq.v.]; edited *Alliance News*, 1854-98; joined Swedenborgians in 1857, and expounded Swedenborg's writings in various volumes.

SUTTON, SIR JOHN BLAND-, baronet (1855-1936), surgeon; qualified at Middlesex Hospital; assistant surgeon, 1886; full surgeon, 1905-20; presented the hospital with Institute of Pathology, 1913; surgeon, Chelsea Hospital for Women, 1896-1911; dexterous and self-reliant; the leading exponent of gynaecological surgery and inspiring teacher; works include *Ligaments, their Nature and Morphology* (1887) and *Tumours Innocent and Malignant* (1893); knighted, 1912; baronet, 1925.

SUTTON, MARTIN JOHN (1850-1913), scientific agriculturist; senior partner in seed firm of Sutton & Sons, Reading, 1887; wrote standard book, *Permanent and Temporary Pastures* (6th edn. 1902); strong churchman and philanthropist.

SWAFFER, HANNEN (1879-1962), journalist; educated at Stroud Green grammar school, Kent; reporter on Folkestone paper; joined *Daily Mail*, under the future Lord Northcliffe [q.v.], 1902; editor, *Weekly Dispatch*; helped transform *Daily Mirror* into mass-circulation paper; invented 'Mr. Gossip' for *Daily Sketch*, 1913; and 'Mr London' for *Daily Graphic*; editor, the *People*; drama critic, *Daily Express*, 1926; joined *Daily Herald*, 1931; espoused socialism, and became a spiritualist; said to write nearly a million words a year; resigned from labour party, 1957; publications include *Northcliffe's Return* (1925), *Really Behind the Scenes* (1929), *Hannen Swaffer's Who's Who* (1929), *Adventures with Inspiration* (1929), *When Men Talk Truth* (1934), *My Greatest Story* (1945), and *What Would Nelson Do?* (1946).

SWAIN, JOSEPH (1820-1909), wood-engraver; manager of engraving department of *Punch*, 1843-4; executed on his own account all engraving of *Punch* from 1844 to 1900, including *Punch* cartoons by Sir J. Tenniel [q.v.]; reproduced works by leading artists.

SWAN, JOHN MACALLAN (1847-1910), painter and sculptor; studied art in London and Paris; exhibited at Royal Academy from 1878;

made special study of animals; best-known pictures are 'The Prodigal Son' (in Tate Gallery), 'Maternity' (in Amsterdam), and 'A lioness defending her cubs'; sculptured work includes bronze bust of Cecil Rhodes [q.v.] and eight colossal lions for Rhodes's monument in Cape Town; ARA, 1894; RA, 1905; hon. LLD, Aberdeen; many of his studies were acquired for the nation.

SWAN, SIR JOSEPH WILSON (1828-1914), chemist and electrical inventor; partner in firm of chemists, Newcastle upon Tyne; studied improvements in collodion process invented by Frederick Scott Archer [q.v.], and produced preparation of collodion for photographic use; patented 'carbon process' for printing permanent photographs (afterwards known as 'autotype'); manufactured celebrated gelatine-silver bromide plates; carried out developments in incandescent electric lighting; FRS, 1894; knighted, 1904.

SWANN, SIR OLIVER (1878-1948), air vice-marshal; originally Schwann; entered navy; deputy to Sir Murray Sueter in creating Royal Naval Air Service, 1912-14; commanded sea-plane carrier *Campania*, 1914-18; transferred to Royal Air Force, 1918; air member for personnel, 1922-3; AOC, Middle East, 1923-6; air vice-marshal, 1922; KCB, 1924.

SWAYNE, JOSEPH GRIFFITHS (1819-1903), obstetric physician; studied at Bristol medical school, Guy's Hospital, and Paris; MRCS, 1841; MB, London, 1842; MD, 1845; succeeded father as lecturer on midwifery at Bristol medical school, 1850-95; physician accoucheur to Bristol general hospital, 1853-75; published *Obstetric Aphorisms for the Use of Students* (1856, 10th edn. 1893).

SWAYTHLING, first BARON (1832-1911), foreign exchange banker and Jewish philanthropist. [See MONTAGU, SAMUEL.]

SWEET, HENRY (1845-1912), phonetician, comparative philologist, and anglicist; BA, Balliol College, Oxford; reader in phonetics at Oxford, only official position which he obtained, 1901; greatest British philologist and chief founder of modern phonetics; works include *History of English Sounds from the Earliest Period* (1874, enlarged form, 1888), *Anglo-Saxon Reader* (1876), *Handbook of Phonetics* (1877), *The Oldest English Texts* (1885), *Elementarbuch des gesprochenen Englisch* (1885, English edn. 1890), *A New English Grammar* (1892, 1898), *The History of Language* (1900), and *The Sounds of English* (1908).

SWETE, HENRY BARCLAY (1835-1917), regius professor of divinity at Cambridge; BA, Caius College, Cambridge; fellow, 1858; professor of pastoral theology, King's College, London, 1882-90; professor of divinity, Cambridge, 1890-1915; his works deal with Greek version of Old Testament, exegesis of New Testament, Christian doctrine, patristic studies, history and interpretation of Apostles' Creed, and Christian worship.

SWETTENHAM, SIR FRANK ATHELSTAN(E) (1850-1946), colonial administrator; entered Straits Settlements service, 1871; adviser to Sultan of Selangor, 1874-5; deputy commissioner, Perak expedition, 1875-6; resident, Selangor, 1882-4, 1887-9; of Perak, 1884-5, 1889-95; initiated federation of Perak, Selangor, Negri Sembilan, and Pahang; resident-general, federated states, 1895-1901; high commissioner, Malay States and governor, Straits Settlements, 1901-4; chairman, royal commission on finances of Mauritius, 1909; joint-director, Press Bureau, 1915-18; publications include *Malay Sketches* (1895), *Unaddressed Letters* (1898), and *The Real Malay* (1899); KCMG, 1897; GCMG, 1909; CH, 1917.

SWIFT, SIR RIGBY PHILIP WATSON (1874-1937), judge; LLB, London; called to bar (Lincoln's Inn), 1895; bencher, 1916; joined Northern circuit; KC, 1912; conservative MP, St. Helens, 1910-18; recorder of Wigan, 1915-20; judge of King's Bench division, 1920-37; presided over many trials of general interest; his great power over juries due to ability to marshal facts and present them attractively; his direction of jury in *R. v. Woolmington* (1935) held to be wrong and the conviction quashed by House of Lords.

SWINBURNE, ALGERNON CHARLES (1837-1909), poet; spent childhood in Isle of Wight; early developed love for climbing, riding, and swimming; entered Eton, Easter 1849; won second Prince Consort's prizes for French and Italian (1852) and first prizes (1853); showed proficiency in Greek verse; left school, July 1853; matriculated from Balliol College, 24 Jan. 1856; soon dropped youthful high church proclivities for nihilism in religion and republicanism in politics; contributed to *Undergraduate Papers*, 1857-8; attracted at Balliol kindly notice of Benjamin Jowett [q.v.], who proved a lifelong friend; won a second in classical moderations and Taylorian scholarship for French and Italian, Easter 1858; finally left Oxford, Nov. 1859; took lodgings in London, 1860; published *The Queen Mother and Rosamond. Two Plays* (1861); in London he lived on terms of close intimacy with D. G. Rossetti [q.v.]; made reputation in cultivated society by brilliance of his conversation; at house of Monckton Milnes formed friendship with Sir Richard Burton [q.v.]; came to know George Meredith [q.v.], 1862; acquired enthusiastic admiration for the work of Victor Hugo, whom he met in Paris, 1882; contributed prose and verse to *Spectator*; joined Rossetti for a time (1862–3) at 16 Cheyne Walk, Chelsea; on visit to Paris (Mar. 1863) made acquaintance of J. A. M. Whistler [q.v.]; suffered from epileptic fits; issued, in Apr. 1865; *Atalanta in Calydon*, which was enthusiastically received; published *Chastelard* (Dec.) and, in Apr. 1866, *Poems and Ballads*, which excited storm of hostility and was withdrawn by publisher, Moxon, from circulation; replied to his critics in *Notes on Poems and Reviews*, Oct.; issued *A Song of Italy* (Feb. 1867), and *An Appeal to England*, a political pamphlet in verse (Nov.); formed intimacy with Adah Isaacs Menken [q.v.], 1867; published in prose *William Blake* and *Notes on the Royal Academy* (1868); spent much time with Mazzini and his circle; injured his health by imprudences; published *Ode on the Proclamation of the French Republic* (Sept. 1870), and his republican *Songs before Sunrise* (1871); denounced Robert Williams Buchanan [q.v.] in his pungent *Under the Microscope* (1872); published *Bothwell* (1874) and *George Chapman: a Critical Essay* (Dec. 1874); issued *Essays and Studies* (prose) and *Songs of Two Nations* (1875), *Erechtheus: a Tragedy*, and *A Note on the Muscovite Crusade* (1876), *A Note on Charlotte Brontë* (1877), and *Poems and Ballads* (2nd ser. 1878); in Sept. 1879 he removed, owing to broken health, to house of Theodore Watts-Dunton [q.v.], The Pines, Putney, where he spent the rest of his life in much retirement; grew increasingly deaf, but his general health was rapidly restored; published, among other later works, *Studies in Song*, and *A Study of Shakespeare* (1880), *Mary Stuart, a Tragedy* (1881), *Tristram of Lyonesse* (1882), *A Century of Roundels* (1883), *A Midsummer Holiday* (1884), *Marino Faliero, a Tragedy* (1885), *A Study of Victor Hugo* and *Miscellanies* (1886), *Locrine* (1887), *A Study of Ben Jonson* and *Poems and Ballads* (3rd ser. 1889), *The Sisters* (1892), *Studies in Prose and Poetry* and *Astrophel* (1894), *The Tale of Balen* (1896), *Rosamund, Queen of the Lombards* (1899), and *A Channel Passage* (1904).

SWINBURNE, SIR JAMES, ninth baronet (1858–1958), pioneer of electrical engineering and of plastics; educated at Clifton College; apprenticed to locomotive works, and later, to engineering firm; worked for (Sir) Joseph Swan [q.v.], 1881–5, and for R. E. B. Crompton [q.v.], 1885–8; invented watt-hour metre and hedgehog transformer; consultant in London, 1894;

recognized as one of leading authorities of the electrical industry; FRS, 1906; interested in development of artificial silk; established Damard Lacquer Co., Birmingham, 1910; chairman, Bakelite Ltd., 1926–48; president, Plastics Institute, 1937–8; versatile inventor who made great contribution to patent jurisprudence; succeeded to baronetcy, 1934.

SWINFEN, first BARON (1851–1919), judge. [See EADY, CHARLES SWINFEN.]

SWINTON, ALAN ARCHIBALD CAMPBELL (1863–1930), electrical engineer; apprentice at Armstrong engineering works, Elswick-on-Tyne, 1882; set up independent practice as electrical contractor and consulting engineer in London, 1887; published first photograph produced by X-rays in England, 1896; one of the first radiographers employed by medical profession; intimately connected with early development of steam turbine; FRS, 1915.

SWINTON, SIR ERNEST DUNLOP (1868–1951), major-general; educated at University College School, Rugby, Cheltenham, Blackheath proprietary school, and Royal Military Academy, Woolwich; commissioned in Royal Engineers, 1888; served in South African war; DSO, 1900; chief instructor in fortification and geometrical drawing, RMA, Woolwich, 1907–10; historical section, Committee of Imperial Defence, 1910; assistant secretary to Committee of Imperial Defence, 1913; official war correspondent, 1914–15; advocated use of caterpillar tractors for trench-crossing; developed idea of tanks, supported by (Sir) Murray Suetor [q.v.]; commanded tank training force, 1916; accompanied Lord Reading [q.v.] to United States, 1917 and 1918; temporary rank of major-general; controller of information in civil aviation department, Air Ministry, 1919–21; Chichele professor of military history, Oxford, 1925–39; CB, 1917; KBE, 1923; publications include *The Defence of Duffer's Drift* (1904) and *Eyewitness* (1932).

SYDENHAM OF COMBE, BARON (1848–1933), administrator. [See CLARKE, GEORGE SYDENHAM.]

SYKES, SIR FREDERICK HUGH (1877–1954), chief of air staff and governor of Bombay; somewhat chequered education; served in South African war with Imperial Yeomanry Scouts; joined regular army, 15th Hussars, 1901; obtained air pilot's certificate, 1911; commander, military wing, Royal Flying Corps, 1912; chief of staff to Sir David Henderson [q.v.], 1914; appointed to command RFC, 1914, aroused hostility of Hugh (later Marshal of the

RAF Viscount) Trenchard; commanded Royal Naval Air Service at Gallipoli; CMG, 1915; assistant adjutant-general, War Office, 1916; promoted major-general, and succeeded Trenchard as chief of air staff, 1918-19; replaced by Trenchard and retired from service, 1919; unionist MP, Hallam division, Sheffield, 1922-8; governor of Bombay, 1928-33; MP, Nottingham, 1940-5; published *From Many Angles* (1942); KCB and GBE, 1919; GCIE, 1928; GCSI, 1934; PC, 1928.

SYKES, SIR MARK, sixth baronet (1879-1919), traveller, soldier, and politician; brought up as Roman Catholic and had roving education abroad; as a young man travelled in Syria, Mesopotamia, and Southern Kurdistan; honorary attaché to British embassy, Constantinople, 1905-7; conservative MP, Central Hull, 1911; succeeded father, 1913; largely responsible for 'Sykes-Picot Agreement', 1916; by which definite spheres of interest in the Near East were assigned to Russia, France, and Great Britain; thenceforth attached to Foreign Office and employed as chief adviser on Near Eastern policy; sent out to Egypt, 1916 and 1917; to Palestine, 1918; champion of Arab independence and pro-Zionist; died in Paris; wrote books of travel.

SYKES, SIR PERCY MOLESWORTH (1867-1945), soldier and administrator; educated at Rugby and Sandhurst; served in India, South African war, and Indian army (1902-20); lieutenant-colonel, 1914; first British consul, Kerman and Persian Baluchistan, 1894; founded consulate, Seistan and Kain, 1898; consul-general, Khorasan, 1906-13; raised and commanded South Persia Rifles, 1916-18; wrote history of Persia (2 vols., 1915) and Afghanistan (2 vols., 1940); KCIE, 1915.

SYLLAS, STELIOS MESSINESOS (LEO) DE (1917-1964), architect. [See DE SYLLAS.]

SYME, DAVID (1827-1908), Australian newspaper proprietor and economist; studied theology in Scotland and philosophy in Germany; sailed for San Francisco, 1851; settled in Melbourne, 1852; bought *The Age* newspaper, 1856; championed protection in working-class interest and attacked capitalism; first in Australia to propose protective duties on imports, and the opening of land to small farmers; obtained disestablishment, payment of members, and free compulsory secular education; paper boycotted by landowners and government, 1862; wielded great influence in appointments of Victorian ministries; through paper obtained anti-sweating and factory Acts and the levy of an income tax; supported Australian federation

and conscription and formation of Australian navy; left £50,000 to Victorian charities; published *Outlines of an Industrial Science* (1877), *Representative Government in England* (1882), and philosophical works.

SYMES, SIR (GEORGE) STEWART (1882-1962), soldier and administrator; educated at Malvern College and Sandhurst; commissioned into Hampshire Regiment, 1900; served in India, South Africa, Aden, and the Somali Coast, 1900-4; DSO, 1904; transferred to Egyptian Army, 1905; aide-de-camp to Sir Reginald Wingate [q.v.], sirdar and governor-general, Sudan; assistant director of intelligence, Khartoum, 1909-12; private secretary to the governor-general, 1913-16; went with Wingate to Egypt, 1917; concerned with Arab attacks under T. E. Lawrence [q.v.], in support of Sir Edmund (later Viscount) Allenby [q.v.]; governor, Northern district, Palestine, 1920-5; chief secretary, 1925-8; resident and commander-in-chief, Aden, 1928-31; succeeded Sir Donald Cameron [q.v.] as governor of Tanganyika, 1931-4; governor-general, Sudan, 1934-40; published *Tour of Duty* (1946); KBE, 1928; GBE, 1939; CMG, 1917; KCMG, 1932.

SYMES-THOMPSON, EDMUND (1837-1906), physician; educated at King's College and King's College Hospital; MB, 1859; MD, 1860; FRCP, 1868; physician (1869) and consulting physician (1889) to Brompton hospital; professor of physic at Gresham College from 1867; president of Harveian Society, 1883; and of British Balneological Society, 1903; published *Winter Health Resorts in the Alps* (1888); helped to establish sanatoria at Davos; other works were *Lectures on Pulmonary Tuberculosis* (1863) and *On Influenza* (1890).

SYMONDS, SIR CHARTERS JAMES (1852-1932), surgeon; born in New Brunswick; qualified at Guy's Hospital; assistant surgeon, 1882; in charge of throat department, 1886-1902; full surgeon, 1902-12; one of the greatest bedside teachers; skilled operator; excelled in diagnosis and decision before and care after operation; KBE, 1919.

SYMONS, ARTHUR WILLIAM (1865-1945), poet, translator, critic, and editor; educated in Devonshire; friend of many artists and writers of the nineties in England and France; published several volumes of verse and criticism, notably *Symbolist Movement in Literature* (1899); contributed regularly to *Athenaeum* and *Saturday* and *Fortnightly* reviews; wrote plays, edited anthologies and other works, and translated from six languages; never recovered former self after mental breakdown in 1908.

SYMONS, WILLIAM CHRISTIAN (1845-1911), decorative designer, painter in oil and water-colours; educated at Lambeth Art School and Royal Academy; exhibited at Academy from 1869 to death; joined Church of Rome, 1870; designed stained-glass windows; began mosaic decorations in Westminster Cathedral, 1899; paintings include 'The Convalescent Connoisseur' (at Dublin) and 'In Hora Mortis' (at Sheffield); excellent in flower pieces.

SYNGE, JOHN MILLINGTON (1871-1909), Irish dramatist; BA, Trinity College, Dublin, 1892; visited Germany, 1894, and Italy, 1896; settled in Paris, 1895; first visited Aran Islands, 1898; in Paris in 1899 met W. B. Yeats [q.v.], who persuaded him to describe primitive life in Aran and elsewhere; revisited Aran Islands annually from 1899 to 1902; soon wrote the plays *The Shadow of the Glen*, performed 1903, and *Riders to the Sea*, performed 1904; there followed *The Well of the Saints* (1905), *The Playboy of the Western World* (1907), and *The Tinker's Wedding* (1907); lived from 1903 in Ireland, where he found material for his descriptive essays 'In Wicklow' and 'In West Kerry', which he contributed to *Manchester Guardian*; literary adviser to Abbey Theatre, Dublin, 1904-9; produced his last and unfinished play, *Deirdre of the Sorrows*, published in 1910; his works collected in 4 vols., 1910.

T

TADEMA, Sir LAWRENCE ALMA- (1836-1912), painter. [See ALMA-TADEMA.]

TAFAWA BALEWA, ALHAJI, Sir ABU BAKAR (1912-1966), prime minister of the Federation of Nigeria; born of peasant stock in Northern Nigeria; educated at the middle school, Bauchi, and Katsina Training College; appointed as teacher, 1932; headmaster, Bauchi middle school, 1944; scholarship to Institute of Education, London University, 1945; appointed as education officer and inspector of schools; member, Northern House of Assembly; Northern member of Central House of Representatives, 1952; minister of works; member of Council of Ministers in Federal government, 1954; as minister of transport visited United States, 1955; federal prime minister, 1957-66; faced with increasing difficulties after independence in 1960; after period of disorders and disturbances, murdered in course of a *coup d'état* by young army officers, 1966; undertook pilgrimage to Mecca, 1957; OBE, 1952; CBE, 1955; KBE, 1960; PC, 1961.

TAGORE, Sir RABINDRANATH (1861-1941), Indian writer; born in Calcutta into intellectual Hindu family; remarkable lyrical poet inspired especially by Bengali landscape; wrote also essays, reviews, dramas, short stories, and novels; showed combined influences of ancient India, modern Europe, and popular Bengal; in philosophical writings interpreted main Indian tradition for the West; sought synthesis of Eastern and Western cultures by founding at Santiniketan, Bolpur, a school (1901), agricultural school (1914), and international university (1921); obtained world-wide reputation by translation of his works and extensive travel in service of world unity; Nobel prize, 1913; knighted, 1915.

TAIT, JAMES (1863-1944), historian; educated at Owens College, Manchester, and Balliol College, Oxford; first class, modern history, 1887; assistant lecturer, English history and literature, Manchester, 1887; lecturer, ancient history, 1896; professor of ancient and medieval history, 1902-19; chairman, Manchester University Press, 1925-35; contributed to this Dictionary, *English Historical Review*, and Lancashire volumes of *Victoria County History*; publications include *Mediaeval Manchester and the Beginnings of Lancashire* (1904) and *Medieval English Borough* (1936); first president, English Place-Name Society, 1923-32; FBA, 1921-43.

TAIT, PETER GUTHRIE (1831-1901), mathematician and physicist; educated at Edinburgh Academy and University, and at Peterhouse, Cambridge; senior wrangler, first Smith's prizeman, and fellow, 1852; hon. fellow, 1885; wrote with William John Steele *Dynamics of a Particle* (1856, 7th edn. 1900); professor of mathematics in Queen's College, Belfast, 1854-60, where he made acquaintance of Sir William R. Hamilton [q.v.]; published *Elementary Treatise on Quaternions* (1867); professor of natural philosophy at Edinburgh, 1860-1901; FRS Edinburgh, 1860; became acquainted in 1861 with Lord Kelvin [q.v.], who collaborated with him in *Natural Philosophy*, vol. i, 1867; embodied further researches in works on *Heat* (1884), *Light* (1884), and *Properties of Matter* (1885); published *Thermodynamics* (1868) and *Recent Advances in Physical Science* (1876); collaborated with Balfour

Stewart [q.v.] in *The Unseen Universe* (1875); issued treatises on *Dynamics* (1895) and *Newton's Laws of Motion* (1899); he investigated properties of ozone, and verified Lord Kelvin's discovery of the 'latent heat of electricity'; Rede lecturer (on thermo-electricity) at Cambridge, 1873; made prolonged researches into knots, the foundations of the kinetic theory of gases, and the flight of a golf ball; royal medallist of Royal Society, 1886; hon. Sc.D., University of Ireland, 1875; hon. LLD, Glasgow, 1901; scientific papers collected in 2 vols., 1898-1900; memorial research scholarships and professorship founded at Edinburgh University; his third son, FREDERICK GUTHRIE TAIT (1870-1900), who was killed in South African war, was champion amateur golfer in 1896 and 1898.

TAIT, SIR (WILLIAM ERIC) CAMPBELL (1886-1946), admiral; entered navy, 1902; commodore (1937), rear-admiral (1938-40), Royal Naval Barracks, Portsmouth; vice-admiral 1941; director of personal services, Admiralty, 1940-1; commander-in-chief, South Atlantic, 1942-4; established combined headquarters at Cape Town where Royal Navy and South African army and air force worked in complete co-operation; governor and commander-in-chief, Southern Rhodesia, 1944-6; admiral, 1945; KCB, 1943.

TALBOT, EDWARD STUART (1844-1934), bishop successively of Rochester, Southwark, and Winchester; educated at Charterhouse and Christ Church, Oxford; first class, *lit. hum.*, 1865, law and modern history, 1866; senior student, 1866-9; first warden of Keble College, 1869-88; deacon, 1869; priest, 1870; vicar of Leeds, 1889-95; bishop of Rochester, 1895; divided diocese and became bishop of Southwark, 1905; of Winchester, 1911-23; a high churchman able to work with evangelicals and liberals.

TALBOT, SIR GEORGE JOHN (1861-1938), judge; nephew of E. S. Talbot [q.v.]; educated at Winchester and Christ Church, Oxford; first class, *lit. hum.*, 1884; fellow of All Souls, 1886; called to bar (Inner Temple), 1887; KC, 1906; bencher, 1914; developed busy parliamentary and ecclesiastical practice; judge of King's Bench division, 1923-37; PC, 1937.

TALLACK, WILLIAM (1831-1908), prison reformer; of Quaker family; secretary to Society for Abolition of Capital Punishment (1863-6) and to Howard Association (1866-1901); visited prisons on continent, in colonies, and in Egypt and United States; author of *Penological and Preventive Principles* (1888) and autobiographical *Howard Letters and Memories* (1905).

TALLENTS, SIR STEPHEN GEORGE (1884-1958), civil servant; educated at Harrow and Balliol College, Oxford; entered civil service, 1909; served in Board of Trade; commissioned in Irish Guards and severely wounded, 1915; worked in Ministry of Munitions, 1915; transferred to Ministry of Food, 1916; chairman, Milk Control Board, 1918; carried out relief work in Poland, and British Commissioner, Baltic provinces, 1919-21; imperial secretary, Northern Ireland, 1922-6; secretary, Empire Marketing Board, 1926-33; public relations officer, Post Office, 1933-5; controller, public relations, British Broadcasting Corporation, 1935-40; controller, Overseas Service, 1940-1; principal assistant secretary, Ministry of Town and Country Planning, 1943-6; publications include *Man and Boy* (1943) and *Green Thoughts* (1952); CB, 1918; CBE, 1920; CMG, 1929; KCMG, 1932.

TANGYE, SIR RICHARD (1833-1906), engineer; of Cornish Quaker family; engaged in teaching till 1851; joined a Birmingham engineering firm, 1852; set up with brothers tool and machinery manufactory at Birmingham, 1855; supplied hydraulic jacks to launch the *Great Eastern*; growth of business led to building of 'Cornwall Works' near Birmingham, 1862, and works in Belgium, 1863, and London, 1868; instituted Saturday half-holiday for employees, 1873; active in religious, municipal, and political life of Birmingham; staunch liberal and free trader; founded Birmingham *Daily Argus*, 1891; knighted, 1894; generous benefactor to Birmingham municipal art gallery and school of art; ardent collector of Cromwellian literature and relics; published *The Two Protectors* (1899), and autobiographical *One and All* (1890) republished as *The Rise of a Great Industry* (1905).

TANNER, JOSEPH ROBSON (1860-1931), historian; educated at Mill Hill School and St. John's College, Cambridge; first class, historical tripos, 1882; president of Union, 1883; lecturer, 1883-1921; fellow, 1886-1931; tutor, 1900-12; tutorial bursar, 1900-21; active in university administration; chiefly interested in English seventeenth-century naval history (particularly Samuel Pepys) and modern constitutional history.

TANSLEY, SIR ARTHUR GEORGE (1871-1955), plant ecologist; educated at University College, London, and Trinity College, Cambridge; first class, parts i and ii, natural sciences tripos, 1893-4; joined F. W. Oliver [q.v.], professor of botany, University College; launched botanical journal *The New Phytologist*, 1902; edited *Types of British Vegetation* (1911);

first president, British Ecological Society, 1913; editor, *Journal of Ecology*, 1917-38; lecturer in botany, Cambridge, 1907-23; studied psychology under Freud in Vienna, 1923-4; Sherardian professor of botany, Oxford, 1927-37; published *The British Islands and their Vegetation* (1939); first chairman, Nature Conservancy, 1949-53; FRS, 1915; knighted, 1950; hon. fellow, Trinity College, Cambridge, 1944.

TARN, SIR WILLIAM WOODTHORPE (1869-1957), ancient historian; educated at Eton and Trinity College, Cambridge; first class, classical tripos, 1891; called to bar (Inner Temple), 1894; ill health caused retirement from bar and concentration on study of ancient history, 1905; wrote first nine chapters in vols. vi and vii, *Cambridge Ancient History*, and one chapter in vol. ix; other publications include *Hellenistic Civilisation* (1927) and *The Greeks in Bactria and India* (1938); FBA, 1928; hon. LLD, Edinburgh, 1933; hon. fellow, Trinity College, Cambridge, 1939; knighted, 1952.

TARTE, JOSEPH ISRAEL (1848-1907), Canadian statesman and journalist; born in province of Quebec; conducted *L'Événement* of Quebec for over twenty years; removed to Montreal (1891) and published the weekly *Le Cultivateur*; was elected as conservative to the federal parliament at Ottawa, 1891; denounced conservative irregularities in public administration in Quebec, 1891; joined liberal opposition; minister of public works in Laurier administration, 1896-1902; editor of *La Patrie*, 1902-7.

TASCHEREAU, SIR HENRI ELZÉAR (1836-1911), chief justice of Canada; born in Beauce county, Quebec; called to Quebec bar, 1857; QC, 1867; conservative member of Canadian legislative assembly, 1861-7; puisne judge of superior court, 1871-8; judge of supreme court, 1878-1902; chief justice, 1902-6; knighted, 1902; member of judicial committee of Privy Council, 1904; wrote on Canadian jurisprudence.

TASCHEREAU, SIR HENRI THOMAS (1841-1909), Canadian judge; first cousin of Sir Henri Elzéar Taschereau [q.v.]; BL, Laval University, 1861; BCL, 1862; hon. LLD, 1890; called to Quebec bar, 1863; sat as liberal in dominion parliament, 1872-8; puisne judge of superior court of Quebec, 1878-1907; chief justice of the King's Bench, 1907-9; knighted, 1908.

TATA, SIR DORABJI JAMSETJI (1859-1932), Indian industrial magnate and phil-anthropist; son of J. N. Tata [q.v.]; educated at Cambridge and Bombay universities; partner in Tata & Sons, 1887; succeeded father as head of firm (1904); made it largest industrial concern in India; established iron and steel works at Jamshedpur, 1911; endowed Indian Institute of Science, Bangalore, 1911; patron of learning and sport; created Tata Trust (£2 millions) for charitable endowments without distinction of caste or creed, 1932; knighted, 1910.

TATA, JAMSETJI NASARWANJI (1839-1904), pioneer of Indian industries; born in Gujerat; entered father's business, 1858; in China, 1859-63; visited England to study conditions of Lancashire cotton mills, 1872; opened mills at Nagpur (1877), and subsequently at Coorla near Bombay; produced cotton of finer texture and acclimatized Egyptian cotton; obtained reduction of heavy freight charges between Bombay and Far East; opposed imposition of excise duty on products of Indian mills; greatly extended Indian iron-ore industry from 1901 and utilized heavy monsoon rainfall of Western Ghauts for electric power in Bombay factories; endowed scholarships for Indians for study in Europe; Indian Institute of Science at Bangalore established according to his plans; wrote economic pamphlets.

TATLOW, TISSINGTON (1876-1957), general secretary of the Student Christian Movement; educated at St. Columba's College, Rathfarnham, and Trinity College, Dublin; ordained deacon, 1902; priest, 1904; general secretary, Student Christian Movement, 1903-29; rector, All Hallows, Lombard Street, London, 1926; rector, St. Edmund the King, 1937; hon. canon, Canterbury Cathedral, 1926; founded Anglican Fellowship, 1912; founded Institute of Christian Education, 1936; European treasurer, Faith and Order Movement, later, a constituent part of World Council of Churches; hon. DD, Edinburgh, 1925; one of the chief architects of ecumenical movement.

TATTERSFIELD, FREDERICK (1881-1959), chemist; educated at Wheelwright grammar school, Dewsbury, and Leeds University; first class, chemistry, London (external student), 1908; D.Sc., 1927; research work with International Paint and Antifouling Co., Ltd., 1908-14; Friends Ambulance Unit, 1914-17; engaged in research at Rothamstead, 1918-47; founded department of insecticides and fungicides; originator of modern research on insecticides; research on pyrethrum led to founding of Kenya pyrethrum industry; OBE, 1947.

TAUNTON, ETHELRED LUKE (1857–1907), ecclesiastical historian; ordained Roman Catholic priest, 1883; works include *History of the Growth of Church Music* (1887), *The English Black Monks of St. Benedict* (2 vols., 1898), *The History of the Jesuits* (1901); composed church music.

TAWNEY, RICHARD HENRY (1880–1962), historian; educated at Rugby and Balliol College, Oxford; fellow, Balliol, 1918–21; hon. fellow, 1938; member, executive committee, Workers' Educational Association, 1905; taught political economy, Glasgow University, 1906–8; teacher, WEA tutorial classes in Rochdale and Manchester, 1908–14; published *The Agrarian Problem in the Sixteenth Century* (1912); joined Fabian Society, 1906; member of executive, 1921–33; joined independent labour party, 1909; enlisted as private, 1915; wounded, 1916; stood unsuccessfully as labour candidate in elections, 1918, 1922, and 1924; member, Coal Industry Commission, 1919; published *The Acquisitive Society* (1921); wrote on education as well as industry; urged abolition of fees in secondary schools; member, consultative committee of the Board of Education, 1912–31; lecturer in economic history, London School of Economics, 1917 and 1920–49; reader, 1923; professor, 1931; hon. fellowships and doctorates, British and foreign universities; FBA, 1935; co-edited the *Economic History Review*, 1927–34; published *Religion and the Rise of Capitalism* (1926) and *Equality* (1931); member, University Grants Committee, 1943–8; vice-president, WEA, 1944–8; published *Business and Politics under James I: Lionel Cranfield as Merchant and Minister* (1958); thought and writings outlined distinctive version of socialism, influencing the labour party of his time.

TAYLOR, ALFRED EDWARD (1869–1945), philosopher; educated at Kingswood School, Bath, and New College, Oxford; first class, *lit. hum.*, 1891; fellow of Merton, 1891–8; lecturer, Greek and philosophy, Owens College, Manchester, 1896–1903; professor of philosophy, McGill University, Montreal, 1903–8; of moral philosophy, St. Andrews, 1908–24, Edinburgh, 1924–41; international authority on Plato; strong Anglo-Catholic; publications include *The Faith of a Moralist* (1930); FBA, 1911.

TAYLOR, CHARLES (1840–1908), master of St. John's College, Cambridge; educated at King's College School, London, and St. John's College, Cambridge; ninth wrangler and second class classic, 1862; obtained first class in theology, 1863; Crosse and Tyrwhitt scholar,

1864; helped to found and edit (1862–84) *The Oxford, Cambridge and Dublin Messenger of Mathematics*; president of Mathematical Association, 1892; ordained, 1866; college lecturer in theology, 1873; edited *Sayings of the Jewish Fathers in Hebrew and English* (1877); member of Alpine Club; appointed master of St. John's College and hon. DD, 1881; vice-chancellor of the university, 1887–9; presented Taylor-Schechter collection of Hebrew MSS to university library; published volumes on geometrical conics and theological works, including *The Teaching of the Twelve Apostles* (1886) and *The Oxyrhynchus Logia* (1899).

TAYLOR, CHARLES BELL (1829–1909), ophthalmic surgeon; MRCS, 1852; MD, Edinburgh, 1854; FRCS, Edinburgh, 1867; settled in Nottingham, 1859; gained European renown as operator for cataract; supported anti-vivisection and anti-vaccination societies, to which he left most of his property.

TAYLOR, EVA GERMAINE RIMINGTON (1879–1966), geographer and historian of science; educated at Camden school for girls, North London collegiate school for girls, and Royal Holloway College; London B.Sc., first class, chemistry, 1903; studied at Oxford University; diploma of geography; research assistant, School of Geography, 1908–10; wrote geography textbooks, 1910–16; lecturer, Clapham Training College and Froebel Institute, 1916–18; London D.Sc., 1929; professor of geography, Birkbeck College, 1930–44; chairman, committee of Royal Geographical Society, concerned with distribution of industrial population, 1938; served on other planning committees; hon. fellow, Royal Geographical Society, 1965; vice-president, Hakluyt Society; hon. LLD, Aberdeen, 1949; fellow, Birkbeck College, 1960; publications include *Tudor Geography, 1485–1583* (1930), *Late Tudor and Early Stuart Geography, 1583–1650* (1934), *the Mathematical Practitioners of Tudor and Stuart England* (1954), *The Haven-Finding Art: a history of navigation from Odysseus to Captain Cook* (1956), and *The Mathematical Practitioners of Hanoverian England* (1966).

TAYLOR, FRANK SHERWOOD (1897–1956), chemist, historian of science, and director of the Science Museum, South Kensington; educated at Sherborne School and Lincoln College, Oxford; graduated in chemistry, 1921, after serving in Honourable Artillery Company in 1914–18 war; teaching in public schools up to 1933; assistant lecturer, inorganic chemistry, Queen Mary College, London, 1933–8; curator, Science Museum,

Oxford, 1940-50; director, Science Museum, South Kensington, 1950-6; Ph.D., London, 1931; published numerous books on scientific subjects, including *The World of Science* (1936) and *An Illustrated History of Science* (1955).

TAYLOR, SIR GORDON GORDON- (1878-1960), surgeon. [See GORDON-TAYLOR.]

TAYLOR, HELEN (1831-1907), advocate of women's rights; stepdaughter of John Stuart Mill [q.v.], with whom she lived at Avignon and co-operated in *The Subjection of Women* (1869); edited Buckle's works (1872), Mill's autobiography (1873), and essays (1874); settled in London on Mill's death (1873), and engaged in social and political life; radical member of London school board, 1876-84; her agitation caused drastic reforms in London industrial schools; she actively opposed Irish coercion policy of liberal government, 1880-5; advocate of land nationalization and taxation of land values; friend of Henry George; helped to found Democratic Federation, 1881; advocated female suffrage; carried on parliamentary election campaign in North Camberwell, although her nomination was subsequently refused, 1885; retired to Avignon, 1885; returned to England, 1904, and died at Torquay; an able public speaker.

TAYLOR, HENRY MARTYN (1842-1927), mathematician; BA, Trinity College, Cambridge; fellow, 1866; on mathematical staff, 1869-94; mainly interested in geometry in its intuitive aspect; went blind soon after retirement; devoted himself to provision of Braille books on mathematics and natural science, inventing suitable symbols and contrivances; his Embossed Scientific Book Fund taken over by Royal Society; FRS, 1898.

TAYLOR, ISAAC (1829-1901), archaeologist and philologist; eldest son of Isaac Taylor (1787-1865, q.v.); BA, Trinity College, Cambridge (nineteenth wrangler), 1853; MA, 1857; held livings in London (1865-75) and at Settrington, Yorkshire (1875-1901); canon of York, 1885; wrote sympathetically of Islam in *Leaves from an Egyptian Note Book* (1888); philological works include *Words and Places* (1864, enlarged and revised as *Names and Their Histories*, 1896), and *The Alphabet* (2 vols., 1883); other works include *Memorials of the Taylor Family of Ongar* (2 vols., 1867) and *The Origin of the Aryans* (1889); hon. LLD, Edinburgh, 1879, D.Litt., Cambridge, 1885; original member of Alpine Club, 1858.

TAYLOR, JAMES HAWARD (1909-1968), geologist; educated at Clifton College and King's College, London; B.Sc., first class, geology, 1931; undertook research in petrology; awarded Henry fellowship for research in the United States; AM, 1934; demonstrator, King's College; joined staff of Geological Survey of Great Britain, 1935; London Ph.D., 1936; succeeded Professor W. T. Gordon as professor of geology, King's College, 1949-68; fellow, Geological Society, 1938; during 1939-45 war, worked with team formed to aid ironstone mining industry; continued this work after returning to King's College; investigated lead-ore deposits in Northern Rhodesia; geological adviser to Iron and Steel Board, 1954; chairman, British Ceramic Association, 1957; president, Mineralogical Society, 1963-5; FRS, 1960; fellow, King's College, London, 1962; chairman, British National Committee for Geology, 1964; drowned while observing underwater limestone formation off the Seychelles Islands, 1968.

TAYLOR, SIR JOHN (1833-1912), architect; surveyor of royal palaces, public buildings, and royal parks, 1866-98; consulting architect, 1898-1908; KCB, 1897; his designs include Bow Street police court and station and bankruptcy buildings.

TAYLOR, JOHN EDWARD (1830-1905), art collector and newspaper proprietor; son of John Edward Taylor [q.v.]; called to bar, 1853; sole proprietor from 1848 of *Manchester Guardian*, which he converted into a daily in 1855; helped in formation of Press Association, 1868; acquired *Manchester Evening News*, 1868; early supporter and benefactor of Owens College, Manchester; trustee of Manchester College from 1854 to death; the sale of his art collections (1911) realized some £358,500.

TAYLOR, JOHN HENRY (1871-1963), golfer; left school at eleven to become a caddie at Westward Ho! golf club; groundsman, 1888; greenkeeper-cum-professional, Burnham Club, 1891; professional, Royal Mid-Surrey Club, 1899-1946; first president, Professional Golfers' Association, 1901; played in Open championship at Prestwick, 1893; first English professional to win the Open (at Sandwich), 1894; won again at St. Andrews, 1895 and 1900; won for fourth and fifth time at Deal, 1909, and Hoylake, 1913; won also in France and Germany, and was second in United States Open, 1900; one of the great 'triumvirate' with James Braid [q.v.] and Harry Vardon; hon. member, Royal and Ancient Golf Club, 1950; president, Royal North Devon (formerly Westward Ho!, where he started as caddie), 1957; published *Golf: My Life's Work* (1943).

TAYLOR, LOUISA (*d.* 1903), novelist. [See PARR.]

TAYLOR, SIR THOMAS MURRAY (1897-1962), lawyer and educationist; educated at Keith grammar school and Aberdeen University; first class, classics, 1919; studied law; LLB, 1922; called to bar in Edinburgh, 1925; KC, 1945; junior advocate depute in Crown office, 1929-34; home advocate depute, 1934; professor of law, Aberdeen University, 1935-48; principal, 1948; member, Committee of Vice-Chancellors and Principals; sheriff of Argyll, 1945-8; sheriff of Renfrew, 1946-8; CBE, 1944; knighted, 1954; hon. DD, Edinburgh, 1952; hon. LLD, St. Andrews, 1950, and Glasgow, 1960; chairman, Crofting Commission, 1951-4; ordained an elder in the Church of Scotland, 1936; member, executive committee, World Council of Churches, 1948-54; chairman, commission on prevention of war, 1961; published *The Discipline of Virtue* (1954) and *Where One Man Stands* (1960).

TAYLOR, SIR THOMAS WESTON JOHNS (1895-1953), scientist and academic administrator; educated at City of London School and Brasenose College, Oxford; served in Essex Regiment, 1914-18; first class, chemistry, fellow, Brasenose, 1920; university lecturer, inorganic chemistry, 1927; Rhodes travelling fellow, 1931; secretary, (and later, director) British Central Scientific Office, Washington, 1943; head of operational research division, South-East Asia Command, 1944; CBE, 1946; principal, University College (later, the university) of the West Indies, 1946-52; knighted, 1952; principal, University College of the South West (later, Exeter University), 1952-3.

TAYLOR, WALTER ROSS (1838-1907), Scottish ecclesiastic; educated at Edinburgh University; minister of Kelvinside Free church, Glasgow, from 1868 to death, as moderator he constituted first general assembly of the United Free Church, 1900; active in Free Church crisis of 1904; hon. DD, Glasgow, 1891; published *Scottish Church Life in the Nineteenth Century* (1900).

TAYLOR, WILLIAM (1865-1937), designer of scientific instruments; studied at Finsbury Technical College; founded own firm with his brother, 1886; designed (and mass produced) high quality camera lenses; also the 'dimple' golf ball, and a widely used engraving machine; FRS, 1934.

TEALE, THOMAS PRIDGIN (1831-1923), surgeon; BA, Brasenose College, Oxford; began practice in Leeds, 1856; surgeon, Leeds General Infirmary, 1864-84; with (Sir) T. C. Allbutt [q.v.], pioneer in 'team work' in medicine and surgical treatment of scrofulous neck; favoured operations of lithotrity and ovariotomy; made advances in ophthalmic surgery; pioneer in advocating and practising anaesthesis and antiseptic surgery; stressed importance of sanitation and domestic hygiene; FRS, 1888.

TEALL, SIR JETHRO JUSTINIAN HARRIS (1849-1924), geologist; BA, St. John's College, Cambridge; fellow, 1875-9; published in parts illustrated monograph on *British Petrography* (1886-8); joined Geological Survey of Great Britain, 1888; director, 1901-14; fellow of Geological Society, 1873; president, 1900-2; FRS, 1890; knighted, 1916; his publications include numerous petrological papers contributed to geological magazines and official memoirs of Geological Survey.

TEARLE, (GEORGE) OSMOND (1852-1901), actor; made début at Liverpool, 1869; appeared first in 1871 as Hamlet, which he played some 800 times; made first tour in America, 1880; subsequently toured there as Wilfred Denver in *The Silver King*.

TEARLE, SIR GODFREY SEYMOUR (1884-1953), actor; elder son of Osmond Tearle [q.v.]; toured with his father's Shakespearian company, 1899-1901; played Othello at Royal Court theatre, invited by J. B. Fagan [q.v.], 1921; played in many other stage productions including *The Garden of Allah* by Robert Hichens [q.v.] (1920), *Hamlet* (1930), *Anthony and Cleopatra* (1946), *Othello* (1948), and *The Heiress* (1950); produced *The Fake* by Frederick Lonsdale [q.v.] (1924), and *The Flashing Stream* by Charles Morgan [q.v.] (1938); first president, Equity, 1932; knighted, 1951.

TEDDER, ARTHUR WILLIAM, first BARON TEDDER (1890-1967), marshal of the Royal Air Force; educated at Whitgift grammar school, Croydon, and Magdalene College, Cambridge; joined colonial service, 1914; resigned to do war service; commissioned in Royal Flying Corps, 1916; in command, No. 70 Fighter Squadron, 1917; squadron leader, Royal Air Force, 1919; instructor and assistant commandant, RAF Staff College, 1929-31; group captain in command, Air Armament School, 1931-3; director of training, Air Ministry, 1934-6; air officer commanding, Far East, 1936-8; air vice-marshal, 1937; director-general, research and development, Air Ministry, 1938; came into conflict with Lord Beaverbrook [q.v.], 1940; deputy commander, Middle East, 1940; commander-in-chief, 1941; air marshal, 1941; air chief marshal, 1942;

commander-in-chief, Mediterranean Air Command, under General Eisenhower, 1943; deputy supreme commander, allied air forces in Europe, 1943–5; marshal of the Royal Air Force, 1945; chief of air staff, 1946–9; baron, 1946; published *Air Power in War* (1948); chairman, British Joint Services Mission and United Kingdom representative on military committee of North Atlantic Treaty Alliance, 1949–51; hon. fellowships and doctorates; CB, 1937; KCB and GCB, 1942; chancellor, Cambridge University, 1950; vice-chairman, board of governors, BBC, 1950; chairman, Standard Motor Co., 1954–60; published memoirs *With Prejudice* (1966).

TEGART, SIR CHARLES AUGUSTUS (1881–1946), Indian police officer; educated at Portora Royal School, Enniskillen, and Trinity College, Dublin; entered Indian police and assigned to Bengal, 1901; acting deputy commissioner, Calcutta, 1906; deputy commissioner, 1913–17; commissioner, 1923–31; legendary figure in dealing with terrorism; on council of India, 1931–6; adviser in Palestine, 1937; knighted, 1926; KCIE, 1937.

TEICHMAN, SIR ERIC (1884–1944), diplomatist and traveller; educated at Charterhouse and Gonville and Caius College, Cambridge; entered Chinese consular service, 1907; travelled in Central Asia; authority on Tibetan border and loess highlands of north-west; published valuable accounts of travels; Chinese secretary, Peking, 1924–35; adviser, Chungking, 1942–3; KCMG, 1933.

TEMPERLEY, HAROLD WILLIAM VAZEILLE (1879–1939), historian; educated at Sherborne and King's College, Cambridge; fellow of Peterhouse, 1905; master, 1938–9; university reader in modern history, 1919; professor, 1931–9; edited *History of the Peace Conference of Paris* (6 vols., 1920–4) and (with G. P. Gooch) *British Documents on the Origins of the War, 1898–1914* (13 vols., 1926–38); publications include *The Foreign Policy of Canning, 1822–1827* (1925) and *The Crimea* (1936); FBA, 1927.

TEMPEST, DAME MARIE (1864–1942), actress; born MARY SUSAN ETHERINGTON; educated at Ursuline convent, Thildonck, in Paris, and at Royal Academy of Music under Manuel Garcia [q.v.]; made début in *Boccaccio*, 1885; sealed popularity in *Dorothy*, 1887; appeared successfully in New York, 1890–5; in musical comedies at Daly's, 1895–1900; entered straight comedy in *Becky Sharp*, 1901; accepted as front-rank comedienne and appeared thereafter in works by leading playwrights; possessed

superb figure, intelligent and winsome face, fine voice, elegance, and industry; thrice married; DBE, 1937.

TEMPLE, FREDERICK (1821–1902), archbishop of Canterbury; born at Santa Maura; educated at Blundell's School and Balliol College, Oxford; first class, classics and mathematics; friend and contemporary of A. H. Clough, A. P. Stanley, Matthew Arnold, and (Lord) Lingen [qq.v.]; lecturer and fellow of Balliol; junior dean, 1845; ordained, 1846; examiner to education department, 1848–9; principal of Kneller Hall, Twickenham, 1849–55; inspector of men's training colleges, 1855–7; wrote on 'National Education' in *Oxford Essays* (1856); headmaster of Rugby School, 1857–69; greatly enlarged the school and modernized the curriculum; member of commission on secondary school education, 1864–8; chairman of governors of Rugby, 1892–1902; provoked unmerited suspicions of his orthodoxy by his essay on 'The Education of the World' in *Essays and Reviews* (1860); consecrated, despite strong opposition, bishop of Exeter, 1869; zealous advocate of educational, social, and temperance reform; instrumental in founding secondary schools in Exeter and Plymouth; helped in creation of diocese of Truro, 1876; expressed liberal views in speeches in House of Lords on university tests bill (1870) and on bill for opening churchyards to nonconformists (1880); Bampton lecturer at Oxford, 1884; bishop of London, 1885–96; mainly instrumental in building Church House, Westminster, 1887; carried through parliament pluralities act amendment bill, 1885; member of royal commission on education, 1888; promoted conciliation scheme in dockers' strike, 1889; archbishop of Canterbury, 1896–1902; made two visitations of his diocese, 1898 and 1902; deprecated use of incense and processional lights and pronounced against the reservation of the sacrament, 1899; hon. LLD, Cambridge, 1897; hon. freeman, Exeter (1897) and Tiverton (1900); officiated at Queen Victoria's funeral and King Edward VII's coronation, 1901–2; supported A. J. Balfour's education bill, 1902; rugged and simple in character and speech; memorial speech room at Rugby opened, 1909; chief works were *Sermons preached in Rugby School Chapel* (3rd ser. 1861–71) and *The Relations between Religion and Science* (Bampton lectures, 1884).

TEMPLE, SIR RICHARD, first baronet (1826–1902), Anglo-Indian administrator; educated at Rugby and Haileybury College; joined East India Company's service, 1847; assistant to John Lawrence [q.v.] in Punjab from 1851; helped James Wilson [q.v.] in inaugurating new financial system, 1856-60; chief commissioner of Central Provinces, 1862-7; organized educational department, long-term settlements of land, and freehold tenancies; established municipality at Nagpur, 1864; CSI, 1866; KCSI, 1867; financial member of council, 1868-74; conducted famine campaign in Bihar, 1874; lieutenant-governor of Bengal, 1874-7; baronet, 1876; special commissioner for famine relief measures in Southern India; governor of Bombay, 1877-80; GCSI and CIE, 1878; active in dispatch of native troops to Malta (1878) and to Afghanistan (1878-80); improved port of Bombay and extended forest area; returned to England (1880) and devoted himself to literary ·work and politics; works include *India in 1880* (1880), *Men and Events of My Time* (1882), lives of John, Lord Lawrence (1889), and James Thomason (1893); hon. DCL, Oxford, 1880, LLD, Cambridge, 1883; member of London school board, 1884-94; conservative MP for Evesham, 1885-92, and for Kingston, Surrey, 1892-5; PC and FRS, 1896; published *The Story of My Life* (1896).

TEMPLE, SIR RICHARD CARNAC, second baronet (1850-1931), soldier and oriental scholar; son of Sir Richard Temple [q.v.]; succeeded father, 1902; educated at Harrow and Trinity Hall, Cambridge; served in Indian army, 1877-1904; chief commissioner, Andaman and Nicobar Islands, 1894-1904; editor, *Indian Antiquary*, 1892-1931; works include *Legends of the Panjâb* (3 vols., 1883-90) and editions of several seventeenth-century travel documents; FBA, 1925.

TEMPLE, WILLIAM (1881-1944), archbishop of Canterbury; son of Frederick Temple [q.v.]; educated at Rugby and Balliol College, Oxford; first class, *lit. hum.*, and president of Union, 1904; fellow, Queen's College, 1904-10; deacon, 1908; priest, 1909; headmaster, Repton, 1910-14; rector of St. James's, Piccadilly, 1914-18; resigned to devote himself to Life and Liberty Movement; canon of Westminster, 1919-21; bishop of Manchester, 1921-9; chairman of 'Copec', 1924; sought to mediate in coal strike, 1926; concurred with Prayer Book policy of Randall Davidson [q.v.]; archbishop of York, 1929-42; of Canterbury, 1942-4; inspired throughout by three particular desires: (i) for social and national righteousness: president, Workers' Educational Association, 1908-24; member, labour party, 1918-25; headed committee investigating unemployment which reported in *Men Without Work* (1938); in talks on wireless preached Christian social gospel; published *Christianity and Social Order* (1942); his honesty and sense of justice contributed to

settlement for Education Act of 1944; (ii) for Church union: chairman, Lambeth Conference committee on unity, 1930; drafted report on proposed Church of South India; played leading part in Ecumenical Movement; chairman, Edinburgh conference, 1937; inaugurated British Council of Churches, 1942; (iii) for reasoned exposition of Christian faith: maintained close contact with Student Christian Movement; undertook missions to universities; publications include *Nature, Man, and God* (1934).

TEMPLEWOOD, VISCOUNT (1880-1959), statesman. [See HOARE, SIR SAMUEL JOHN GURNEY, second baronet.]

TENBY, first VISCOUNT (1894-1967). [See LLOYD-GEORGE, GWILYM.]

TENNANT, SIR CHARLES, first baronet (1823-1906), merchant and art patron; grandson of Charles Tennant (1768-1838, q.v.), founder of chemical works at St. Rollox, Glasgow; partner in firm, 1846; works absorbed in 1900 in United Alkali Co., of which Tennant became chairman; chairman of Union Bank of Scotland; he acquired portraits by Reynolds, Gainsborough, and Romney; became trustee of National Gallery, 1894; his private art collection housed in Queen Anne's Gate, London; liberal MP for Glasgow, 1879-80, and for Peebles and Selkirk, 1880-6; baronet, 1885; member of Joseph Chamberlain's tariff reform commission, 1904.

TENNANT, SIR DAVID (1829-1905), speaker of the house of assembly of the Cape of Good Hope; born at Cape Town; admitted attorney-at-law of supreme court there, 1849; member of house of assembly for Piquetberg, 1866-96; speaker, 1874-96; knighted, 1877; KCMG, 1892; agent-general for colony in London, 1896-1901.

TENNANT, MARGARET MARY EDITH (MAY) (1869-1946), pioneer in public social work; born ABRAHAM; first woman factory inspector in England, 1893-6; married (1896) Harold John Tennant, liberal MP; member, Central Committee on Women's Employment, 1914-39; CH, 1917.

TENNIEL, SIR JOHN (1820-1914), artist and cartoonist; studied at Academy Schools and Clipstone Street Life Academy; joined staff of *Punch* as second cartoonist, 1850; first cartoonist, 1864-1901; drew over 2,000 cartoons, several volumes of which have been published; his cartoons noted for dignity, humour, and fairness; excelled in drawings of beasts and allegorical figures; illustrated Lewis Carroll's *Alice's Adventures in Wonderland* (1865) and *Alice Through the Looking-Glass* (1872); knighted, 1893.

TENNYSON-D'EYNCOURT, Sir EUSTACE HENRY WILLIAM, first baronet (1868-1951), naval architect; his father, cousin of Alfred Tennyson [q.v.]; educated at Charterhouse, and apprenticed to Armstrong, Whitworth & Co.; trained as naval architect; employed in design office and appointed to take charge of it, 1902-12; succeeded (Sir) Philip Watts [q.v.] as director, naval construction, Admiralty, 1912-24; designs included *Royal Sovereign* class battleships, the bulge form of anti-submarine design, aircraft carriers, the battle-cruiser, *Hood*, and battleships, *Nelson* and *Rodney*; headed 'landships' committee, 1915, which led to production of tanks; director, Armstrong, Whitworth & Co., 1924-8; on board of Parsons Marine Steam Turbine Co., 1928-48; KCB, 1917; FRS, 1921; baronet, 1930; hon. degrees from Durham and Cambridge.

TERRY, DAME (ALICE) ELLEN (1847-1928), actress; made first appearance on stage, 1856; joined Theatre Royal, Bristol, 1862, Haymarket Theatre, 1863; married G. F. Watts [q.v.], 1864; separated, 1865; returned to stage; acted for first time with (Sir) Henry Irving [q.v.], 1867; lived in retirement with E. W. Godwin [q.v.], 1868-74; returned to stage, 1874; hitherto not happy or successful on stage, but reached turning-point of her career as Portia in *Merchant of Venice* staged by Sir Squire Bancroft [q.v.] at old Prince of Wales's Theatre (1875); divorced by Watts and married C. C. Wardell, 1877; received second great opportunity as Olivia in W. G. Wills's adaptation of *Vicar of Wakefield* (1878); engaged by Henry Irving to play Ophelia at Lyceum Theatre, 1878; her association with Irving lasted up to 1902; played leading female parts in all Irving's productions up to 1896; her most perfect part Beatrice in *Much Ado About Nothing*; her 'jubilee' celebrated with great enthusiasm, 1906; married J. Usselmann, 1907; latterly played parts of old women; made last appearance on stage, 1925; lectured on Shakespearian subjects; GBE, 1925.

TERRY, CHARLES SANFORD (1864-1936), historian and musician; BA, Clare College, Cambridge, 1886; lecturer in history, Newcastle (1890-8), Aberdeen (1898-1903); professor, 1903-30; edited the *Albemarle Papers, 1746-1748* (2 vols., 1902); wrote chiefly on seventeenth-century Scottish history and on J. S. Bach.

TERRY, FRED (1863-1933), actor; brother of Dame Ellen Terry [q.v.]; played Sebastian to her Viola in *Twelfth Night*, 1884; appeared frequently in London; went into management (1900) with his wife, Julia Neilson, presenting romantic-historical plays including *Sweet Nell of Old Drury* (1900), *The Scarlet Pimpernel* (1905), and *Henry of Navarre* (1909); retired, 1927.

TERRY, Sir RICHARD RUNCIMAN (1865-1938), musician and musical antiquary; director of music, Downside (1896-1901), Westminster Cathedral (1901-24); supremely talented in choir training; from early manuscripts presented the work of the Tudor polyphonic school; knighted, 1922.

THANKERTON, BARON (1873-1948), judge. [See WATSON, WILLIAM.]

THESIGER, FREDERIC AUGUSTUS, second BARON CHELMSFORD (1827-1905), general; eldest son of first baron [q.v.]; joined army, 1844; served in Crimea and in Indian Mutiny; deputy adjutant-general in Abyssinian expedition, 1868; CB and ADC to Queen Victoria, 1868; adjutant-general in East Indies, 1869-74; commanded troops at Shorncliffe, 1874-6; major-general, 1877; commanded troops in South Africa in Kaffir war, 1878, and in Zulu war, 1879; after disaster at Isandhlwana (22 Jan.) relieved Col. Pearson's force at Etshowe; marched on to Umvolosi and defeated Zulu army under Cetywayo at Ulundi (July); meanwhile (June) superseded in command by Sir Garnet (afterwards Viscount) Wolseley [q.v.]; KCB, 1878; GCB, 1879; lieutenant-general, 1882; general, 1888; lieutenant of Tower of London, 1884-9; GCVO, 1902.

THESIGER, FREDERIC JOHN NAPIER, third BARON and first VISCOUNT CHELMSFORD (1868-1933), viceroy of India; son of second Baron Chelmsford [q.v.]; succeeded father, 1905; educated at Winchester and Magdalen College, Oxford; first class, jurisprudence, captain of cricket, and fellow of All Souls, 1892; called to bar (Inner Temple), 1893; governor of Queensland (1905-9), of New South Wales (1909-13); viceroy of India, 1916-21; signed Montagu-Chelmsford report (1918) which set India on path towards responsible government; first lord of Admiralty, 1924; re-elected fellow of All Souls, 1929; warden, 1932-3; PC, 1916; viscount, 1921.

THIRKELL, ANGELA MARGARET (1890-1961), novelist; daughter of John William Mackail [q.v.] and grand-daughter of (Sir) Edward Burne-Jones [q.v.]; educated at

St. Paul's school for girls, London, and at 'finishing school' in Paris; married and emigrated to Australia, 1918; began writing articles and short stories; returned to England, 1930; *Three Houses* (1931), followed by over thirty successful novels, including *Ankle Deep* and *High Rising* (1930), *Wild Strawberries* (1934), *Pomfret Towers* (1938), *The Brandons* (1939), and *Cheerfulness Breaks In* (1940); last novel, *Three Score and Ten* (1961, completed by A. Lejeune.)

THISELTON-DYER, Sir WILLIAM TURNER (1843-1928), botanist; BA, Christ Church, Oxford; professor of natural history, Royal Agricultural College, Cirencester, 1868; professor of botany, Royal College of Science, Dublin, 1870-2; at Royal Horticultural Society, South Kensington and Chiswick, 1872-5; assistant director, Royal Botanic Gardens, Kew, 1875; director of Kew Gardens, which he developed in every way, 1885-1905; founded *Kew Bulletin* for exchange of information among colonial institutions associated with Kew, 1887; made Jodrell laboratory 'the best botanical laboratory in Europe'; FRS, 1880; KCMG, 1899; his editorial work includes *Flora Capensis* (1896-1925), *Flora of Tropical Africa* (1897-1913), *Icones Plantarum* (1896-1906).

THODAY, DAVID (1883-1964), botanist; educated at Tottenham grammar school and Trinity College, Cambridge; senior scholar; first class, parts i and ii, natural sciences tripos, 1905-6; research student; university demonstrator, 1909-11; lecturer in physiological botany, Manchester, 1911-18; professor of botany, Cape Town University, 1918-23; professor of botany, University College of North Wales, 1923-49; Sc.D., Cambridge, 1933; wrote many papers including series on the succulent *Kleinia articulata*; spent two years in Egypt as professor of plant physiology, 1949-51; returned to Bangor to research on Loranthaceae (mistletoes), 1952-62; designed Thoday potometer and Thoday respirometer; published *Botany: a Text-book for Senior Students* (1915); FRS, 1942; hon. D.Sc. (university of Wales), 1960.

THOMAS, BERTRAM SIDNEY (1892-1950), explorer; employed in Post Office, 1908-14; with army in Belgium (1914-16), Mesopotamia (1916-18); in political department, Mesopotamia, 1918-22; British representative's assistant, Trans-Jordan, 1922-4; financial adviser (later prime minister), Muscat, 1924-31; crossed Great Southern Desert (Dhufar-Dauha) by camel, 1930-1; director, Middle East Centre of Arabic Studies, 1944-8.

THOMAS, DAVID ALFRED, Viscount RHONDDA (1856-1918), statesman, colliery proprietor, and financier; became associated with Cambrian collieries in Rhondda Valley, *c*.1882; Gladstonian liberal MP, Merthyr Tydfil, 1888; held seat till 1910; went to United States on business of Ministry of Munitions, 1915; created baron for his services, 1916; promoted viscount, 1918; president of Local Government Board, 1916; as food minister carried out successful rationing during 1914-18 war (1917-18.)

THOMAS, DYLAN MARLAIS (1914-1953), poet; educated at Swansea grammar school; worked as reporter on *South Wales Daily Post*, 1931-2; first poem published, 1933; published *18 Poems* (1934), *Twenty-five Poems* (1936), *The World I Breathe* (1939); worked with BBC as actor and reader of poetry; *Deaths and Entrances* (1946) sealed his promise as a poet; followed by *Collected Poems* (1952); *Under Milk Wood*, 'play for voices', first performed in New York, 1953; died of alcoholic poisoning.

THOMAS, FOREST FREDERIC EDWARD YEO- (1902-1964), French resistance organizer. [See YEO-THOMAS.]

THOMAS, FREDERICK WILLIAM (1867-1956), orientalist; educated at King Edward VI's high school, Birmingham, and Trinity College, Cambridge; first class, parts i and ii, classical tripos, 1887-9; first class, Indian languages tripos, 1890; fellow, Trinity College, 1892; assistant librarian, India Office, 1898; librarian, 1903-27; Boden professor of Sanskrit, Oxford, 1927-37; FBA, 1927; CIE, 1928; hon. degrees from Munich, Allahabad, and Birmingham; pioneer in Asian philology; authority on Jainism.

THOMAS, FREEMAN FREEMAN-, first MARQUESS OF WILLINGDON (1866-1941), governor-general of Canada and viceroy of India. [See FREEMAN-THOMAS.]

THOMAS, GEORGE HOLT (1869-1929), pioneer in aircraft manufacture; newspaper manager up till 1906; thereafter devoted money, energy, and imagination to stimulating public interest in aviation; organized air display at Brooklands, 1909; regular aerodrome laid out there as result; criticized authorities for backward state of British military aviation; national air service instituted, 1911; took up rights of French aircraft pioneer, Henri Farman, and founded Aircraft Manufacturing Company; acquired English rights to manufacture Gnome and Le Rhone engines; produced 'DH' series of aeroplanes, invaluable in 1914-18 war; later took up question of air transport; died at Cimiez, France.

THOMAS, SIR **HENRY** (1878-1952), Hispanologist and bibliographer; educated at King Edward VI grammar school, Aston, and Mason College (later Birmingham University); assistant, department of printed books, British Museum, 1903; Spanish scholar; publications include *Spanish and Portuguese Romances of Chivalry* (1920) and *Spanish Sixteenth-Century Printing* (1926); responsible for *Short-Title Catalogue of Books printed in Spain and of Spanish Books printed elsewhere in Europe before 1601 now in the British Museum* (1921); president Anglo-Spanish Society, 1931-47; FBA, 1936; knighted, 1946; D.Litt. and hon. LLD, Birmingham, D.Litt., London.

THOMAS, HERBERT HENRY (1876-1935), geologist; educated at Sidney Sussex College, Cambridge, and Balliol College, Oxford; geologist (1901-11), petrographer (1911-35) to Geological Survey of Great Britain; identified 'blue stones' of Stonehenge as from Prescelly mountains in Pembrokeshire, 1923; FRS, 1927.

THOMAS, SIR HUGH EVAN- (1862-1928), admiral. [See EVAN-THOMAS.]

THOMAS, HUGH HAMSHAW (1885-1962), palaeobotanist; educated at Wrexham and Downing College, Cambridge; first class, part i, natural sciences tripos, 1906; carried out research on fossil plants; collaborated with (Sir) A. C. Seward [q.v.]; studied Jurassic plants of Yorkshire, preserved compressed on shales; wrote paper on *Williamsoniella* and another (with Nellie Bancroft) on cycad leaf cuticles; served in Royal Field Artillery and Royal Flying Corps during 1914-18 war; MBE (military); on return to Cambridge played important part in foundation of University Air Squadron; dean and steward, Downing College, 1920; university lecturer in botany, 1923; published paper on Caytoniales, 1925, and made other outstanding contributions to palaeobotany; served with RAF Volunteer Reserve on photographic interpretation, 1939-43; FRS, 1934; awarded Linnean Society's Gold Medal, 1960; in later years active in promoting international cooperation in botany through membership of number of societies.

THOMAS, JAMES HENRY (1874-1949), trade-union leader and politician; educated at Newport national schools; engine-cleaner, Great Western Railway, 1889; fireman, later engine-turner, 1894-1906; president, Amalgamated Society of Railway Servants, 1905-6; organizing secretary, 1906-10; assistant secretary, 1910-13; a leader in nation-wide railway strike, 1911; assistant secretary, National Union of Railwaymen, 1913; general secretary, 1917; parliamentary general secretary, 1919-31; president, Trades Union Congress, 1920; MP, Derby (labour), 1910-31, (national labour), 1931-6; visited North America with Balfour mission and PC, 1917; directed railway strike and negotiated settlement, 1919; leader in refusing support to extremist miners' leaders thereby causing collapse of 'Triple Alliance', 1921; unable to prevent general strike, 1926; colonial secretary, 1924; lord privy seal with special mandate to deal with unemployment, 1929-30; dominions secretary, 1930-5; deprived of union membership and ostracized by labour party for continuing in 'national' government after 1931; colonial secretary, 1935-6; found to have made 'unauthorized disclosure' of budget proposals, 1936; popular figure, of engaging common sense and humour.

THOMAS, JAMES PURDON LEWES, VISCOUNT CILCENNIN (1903-1960), politician; educated at Rugby and Oriel College, Oxford; conservative MP, Hereford, 1931-55; parliamentary private secretary to J. H. Thomas [q.v.], 1932-6; p.p.s. to Anthony Eden (later the Earl of Avon), 1937-8; government whip, 1940-3; financial secretary to Admiralty, 1943-5; deputy chairman, conservative party, 1945-51; first lord, Admiralty, 1951-6; PC, 1951; viscount, 1955.

THOMAS, MARGARET HAIG, VISCOUNTESS RHONDDA (1883-1958), founder and editor, *Time and Tide*; daughter of Viscount Rhondda [q.v.]; educated at Notting Hill high school and Somerville College, Oxford; survived torpedoing of *Lusitania*, 1915; succeeded father, 1918; founded *Time and Tide*, 1920; editor, 1926-58; spent some quarter million pounds on *Time and Tide*.

THOMAS, PHILIP EDWARD (1878-1917), critic and poet; BA, Lincoln College, Oxford; works include *Richard Jefferies, His Life and Work* (1909), imaginative idylls *Rest and Unrest* (1910), *Light and Twilight* (1911), studies of *Swinburne* (1912), and *Walter Pater* (1913); *Collected Poems* published, 1920; killed in action at Arras.

THOMAS, SIR (THOMAS) SHENTON (WHITELEGGE) (1879-1962), colonial governor; educated at St. John's School, Leatherhead, and Queens' College, Cambridge; assistant district commissioner, East African Protectorate, 1909; transferred to Uganda, 1918; and then to Nigeria, 1921; colonial secretary, Gold Coast, 1927; governor, Nyasaland, 1929; governor, Gold Coast, 1932; governor and commander-in-chief, Straits Settlements, high commissioner, Malay States, and British agent, North Borneo and Sarawak, 1934; stayed on when 1939-45 war broke out; warned Colonial Office of vulnerability of Malaya to Japanese attack; position weakened after sinking of *Prince of Wales* and *Repulse* by appointment of A. Duff Cooper (later Viscount Norwich, q.v.) as resident minister, 1941; on fall of Singapore interned by Japanese, 1942-5; retired, 1946; chairman, Overseas League, 1946-9, and active on other committees; tried without success to correct official history of the war in Malaya, 1956; OBE, 1919; CMG, 1929; KCMG, 1931; GCMG, 1937; hon. fellow, Queens' College, Cambridge, 1935.

THOMAS, SIR WILLIAM BEACH (1868-1957), journalist and author; educated at Shrewsbury and Christ Church, Oxford; notable athlete; president, Athletics Club, 1890-1; taught at Bradfield, 1891-6, and Dulwich, 1897-8; invited by J. L. Garvin [q.v.] to write for *Outlook*; on staff of *Saturday Review*; writer on country life and, later, war correspondent for *Daily Mail*; contributor to *Spectator*; publications include *The English Year* (3 vols., 1913-14, in collaboration with A. K. Collett), *With the British on the Somme* (1917), *The Story of the Spectator* (1928), *The English Landscape* (1938), and *The Poems of a Countryman* (1945); KBE, 1920.

THOMAS, WILLIAM MOY (1828-1910), novelist and journalist; private secretary to Sir Charles Wentworth Dilke [q.v.]; contributed to *Household Words*, 1851-8; re-edited Lord Wharncliffe's *Letters and Works of Lady Mary Wortley Montagu* (2 vols., 1861); dramatic critic and contributor to *Daily News*, 1868-1901; first editor of *Cassell's Magazine*; published novels and tales.

THOMPSON, ALEXANDER HAMILTON (1873-1952), historian; educated at Clifton College and St. John's College, Cambridge; extra-mural teacher for Cambridge University, 1897; lecturer in English, Armstrong College, Newcastle upon Tyne, 1919; reader in medieval history, Leeds University, 1922, professor, 1924; head of department, 1927-39; numerous publications, include *History of English Literature* (1901), *Military Architecture in England during the Middle Ages* (1912), and *The English Clergy and their Organization in the later Middle Ages* (1947); FBA, 1928; hon. ARIBA, hon. fellow, St. John's College, Cambridge, 1938; CBE, 1938; hon. doctorates, Durham, Leeds, and Oxford.

THOMPSON, D'ARCY WENTWORTH (1829-1902), Greek scholar; educated at Christ's Hospital, London, and Pembroke College, Cambridge; BA (sixth classic), 1852; classical master in Edinburgh Academy (1852-63), where R. L. Stevenson [q.v.] was one of his pupils, 1861-2; professor of Greek in Queen's College, Galway, 1863-1902; gave Lowell lectures at Boston, published in his *Wayside Thoughts* (1867); his fame rested on *Day Dreams of a Schoolmaster* (1864).

THOMPSON, SIR D'ARCY WENTWORTH (1860-1948), zoologist and classical scholar; son of D'Arcy Wentworth Thompson [q.v.]; educated at Edinburgh Academy and University and Trinity College, Cambridge; first class, natural sciences tripos, parts i and ii 1882-3; professor of biology, University College, Dundee, 1884-1917; natural history, United College, St. Andrews, 1917-48; member, Fishery Board for Scotland, 1898-1939; British representative, International Council for Exploration of Sea, from 1902; publications include *On Growth and Form* (1917), translation of Aristotle's *Historia Animalium* (1910), and Glossary of *Greek Birds* (1895) and of *Greek Fishes* (1945); FRS, 1916; knighted, 1937.

THOMPSON, EDMUND SYMES- (1837-1906), physician. [See SYMES-THOMPSON.]

THOMPSON, EDWARD JOHN (1886-1946), writer; educated at Kingswood School, Bath; Wesleyan minister, 1909-23; chaplain, Mesopotamia and Palestine, 1916-18; taught at Wesleyan College, Bankura, Bengal, 1910-23; studied Sir Rabindranath Tagore [q.v.] and other Bengali poets; lecturer in Bengali, Oxford, 1923-33; research fellow, Indian history, Oriel College, 1936-46; supported Indian self-government; published poetry (notably of Indian scenery), novels, and historical works.

THOMPSON, SIR EDWARD MAUNDE (1840-1929), palaeographer and director of the British Museum; born in Jamaica; entered British Museum, 1861; engaged on preparation of 'Class Catalogue' of manuscripts; assistant keeper, department of manuscripts, 1871; engaged on cataloguing of recent manuscript accessions; joined (Sir) E. A. Bond [q.v.] in founding Palaeographical Society, 1873; keeper, 1878; principal librarian (to which title of director prefixed, 1898), 1888; a masterful administrator who promoted cataloguing, developed educational and popular side of museum, and encouraged excavations abroad; retired, 1909; took active part in founding British Academy and was an original fellow (1902); KCB, 1895; his publications include

Handbook of Greek and Latin Palaeography (1893, enlarged 1912); and editions of chronicles and other documents, notably *Chronicon Galfridi Le Baker de Swynebroke, 1303-1356* (1889).

THOMPSON, FRANCIS (1859-1907), poet and prose writer; educated at Ushaw College; studied medicine without success at Owens College, Manchester, 1876-82; made helpless efforts to earn a livelihood, 1883-8; settled in London, 1885; fell a prey to opium and became a street waif; contributed poems and prose essays to *Merry England* from 1888; published first volume of *Poems* (1893) which included 'The Hound of Heaven'; *Sister Songs* followed in 1895, and *New Poems* in 1897; contributed literary criticism to *Academy* and *Athenaeum*; prose work includes *Health and Holiness* (1905) and *Essay on Shelley* (1909); died of consumption.

THOMPSON, SIR HENRY, first baronet (1820-1904), surgeon; studied medicine at University College, London; MB, 1851; FRCS, 1853; Hunterian professor of surgery and pathology, 1883; surgeon (1863), professor of clinical surgery (1866), and emeritus professor (1874) at University College Hospital; made successful removals of stone in the bladder by lithotrity; operated on King Leopold I, 1863, and Napoleon III, 1873; helped to found the first cremation society in London, 1874, and to form company which erected Golder's Green crematorium, 1902; keen astronomer; presented large photographic telescope to Greenwich observatory; exhibited paintings at Royal Academy from 1865 to 1885; collected old white and blue Nanking china; frequently entertained men famous in all branches of society; knighted, 1867; baronet, 1899; works include *Stricture of the Urethra* (1854), *Practical Lithotomy and Lithotrity* (1863), *Cremation* (1874), *Food and Feeding* (1880), and novels; a collected edition of his surgical works appeared at Paris in 1880.

THOMPSON, SIR (HENRY FRANCIS) HERBERT, second baronet (1859-1944), Egyptologist; son of Sir Henry Thompson [q.v.]; succeeded father, 1904; educated at Marlborough and Trinity College, Cambridge; called to bar (Inner Temple), 1882; studied Egyptology and became friend of F. Ll. Griffith and W. E. Crum [qq.v.]; specialized in demotic and Coptic; published *The Demotic Magical Papyrus of London and Leiden* (1904), *The Coptic (Sahidic) Version of Certain Books of the Old Testament* (1908), demotic and Coptic texts in *Theban Ostraca* (1913), and *A Family Archive*

from Siût (2 vols., 1934); edited papyrus manuscript of St. John's Gospel, 1924; compiled handlist of demotic papyri in British Museum; FBA, 1933.

THOMPSON, HENRY YATES (1838-1928), book-collector; BA, Trinity College, Cambridge; travelled widely in Europe, Asia, etc., 1862-75; formed famous collection of one hundred illuminated manuscripts, partly dispersed at three sales, 1919, 1920, and 1921; his beneficiaries included Liverpool, Harrow, and Newnham College, Cambridge.

THOMPSON, JAMES MATTHEW (1878-1956), scholar; educated at the Dragon School, Oxford, Winchester, and Christ Church, Oxford; first class, *lit. hum.*, 1901; ordained deacon, 1903; priest and fellow, Magdalen College, 1904; tutorial fellow, modern history, 1920; university lecturer, French history, 1931-8; editor, *Oxford Magazine*, 1945-7; publications include biographies of Napoleon I (1951), Napoleon III (1954), and Robespierre (2 vols., 1935); FBA, 1947.

THOMPSON, LYDIA (1836-1908), actress; made début in ballet at Her Majesty's Theatre, London, 1852; acted in burlesque in provinces, 1864-5; appeared at Drury Lane, 1865, and Prince of Wales's Theatre, 1866; toured in America, Australia, and India, 1868-74; at New York began association with Willie Edouin [q.v.], 1870-1; reappeared in London in Farnie's burlesque of *Blue Beard* (1874); thrice revisited New York; opened the Strand Theatre, London, Sept. 1886, with *The Sultan of Mocha*.

THOMPSON, REGINALD CAMPBELL (1876-1941), Assyriologist; educated at St. Paul's School and Gonville and Caius College, Cambridge; first class, oriental languages, 1898; in Egyptian and Assyrian department, British Museum, 1899-1905; published version of rock-inscription of Darius at Bisitun, 1907; excavated at Carchemish (1911), Wadi Sargah (1913), and Nineveh (1927-32); fellow, Merton College, Oxford, 1923-41; Shillito reader, 1937-41; edited *The Epic of Gilgamish* (1930); published dictionaries of Assyrian chemistry (1936), geology (1936), and botany (1949); FBA, 1934.

THOMPSON, SILVANUS PHILLIPS (1851-1916), physicist; BA, London University; principal and professor of applied physics and electrical engineering of City and Guilds Technical College, Finsbury, 1885-1916; FRS, 1891; published works on electricity and magnetism, and lives of famous scientists.

THOMPSON, WILLIAM MARCUS (1857-1907), journalist; joined staff of London *Standard*, 1877; parliamentary reporter, 1884-90; developed radical and Irish nationalist principles; called to bar, 1880; defended professionally trade unions and political offenders, including John Burns [q.v.] for share in Trafalgar Square riots, 1886-8; editor of *Reynolds's Newspaper*, 1894-1907; radical member of London County Council for West Newington, 1895-8.

THOMSON, ARTHUR (1858-1935), anatomist; MB, Edinburgh, 1880; university lecturer in human anatomy, Oxford, 1885; extraordinary professor, 1893; Dr Lee's professor, 1919-33; took leading share in development of Oxford medical school; professor of anatomy, Royal Academy, 1900-34.

THOMSON, SIR BASIL HOME (1861-1939), colonial governor and assistant commissioner of the Metropolitan Police; son of William Thomson [q.v.]; served in colonial service; became prison governor, 1896; secretary, prison commission, 1908; assistant commissioner, Metropolitan Police, 1913; director of intelligence, Scotland Yard, 1919-21; KCB, 1919.

THOMSON, CHRISTOPHER BIRD-WOOD, BARON THOMSON (1875-1930), soldier and politician; born at Nasik, India; nephew of Sir G. C. M. Birdwood and H. M. Birdwood [qq.v.]; entered Royal Engineers, 1894; served in South African war, 1899-1902; served under Sir H. H. Wilson [q.v.], director of military operations, War Office, 1911; military attaché, Bucharest, 1915-17; served in Palestine, 1917-19; left army, 1919; unsuccessfully attempted to enter parliament as socialist, 1919-23; secretary of state for air in first labour government and created baron, 1924; largely responsible for government's decision on three years' scheme of air development, including construction of two airships, R100 and R101; returned to Air Ministry on formation of second labour government, 1929; perished in R101 disaster near Beauvais, France.

THOMSON, SIR GEORGE PIRIE (1887-1965), chief press censor, Ministry of Information, during 1939-45 war; grandson of William R. Pirie [q.v.]; educated at George Watson's College, Edinburgh; joined Royal Navy at fifteen; spent twenty-five years in navy; rear-admiral, 1939; appointed to press censorship, 1939, at a time of disorganization of relations with Fleet Street; succeeded Admiral C. V. Usborne as chief press censor, 1940-5; won confidence of press and restored good relations;

secretary, Services, Press and Broadcasting Committee, 1945; public relations officer, Latin American Centre; OBE, 1919; CBE, 1939; CB, 1946; knighted, 1963; published *Blue Pencil Admiral* (1947).

THOMSON, GEORGE REID (1893-1962), lord justice clerk; educated at South African College, Cape Town, and Corpus Christi College, Oxford; Rhodes scholar; served in army during 1914-18 war; studied law at Edinburgh University, 1919-22; admitted advocate, 1922; built up substantial practice; KC, 1936; advocate depute, 1940-5; lord advocate; PC, 1945; labour MP, East Edinburgh, succeeding (Baron) F. W. Pethick-Lawrence [q.v.], 1945; lord justice clerk, 1947-62; hon. fellow, Corpus Christi College, and hon. LLD, Edinburgh, 1957.

THOMSON, HUGH (1860-1920), illustrator and pen-and-ink draughtsman; executed illustrations for *English Illustrated Magazine*, Macmillan's 'Highways and Byways' county series, works by Goldsmith, Miss Mitford, Mrs Gaskell, Jane Austen, Fanny Burney, Thackeray, George Eliot, Austin Dobson, Charles Reade, Shakespeare, Sheridan, Barrie, Thomas Hughes, Hawthorne, etc.

THOMSON, JOCELYN HOME (1859-1908), chief inspector of explosives; son of William Thomson (1819-90, q.v.); observed transit of Venus in Barbados, 1882; secretary to War Office explosives committee, 1888; suggested name of 'cordite' for smokeless powder; chief inspector of explosives, 1899-1908; published works on petroleum lamps.

THOMSON, JOHN (1856-1926), physician and writer on children's diseases; MD, Edinburgh University; physician, New Town Dispensary, Edinburgh; established clinic for diseases of children; extra physician, Royal Hospital for Sick Children, Edinburgh, 1889-1918.

THOMSON, SIR JOSEPH JOHN (1856-1940), physicist; educated at Owens College, Manchester, and Trinity College, Cambridge; second wrangler, second Smith's prizeman, and fellow, 1880; Adams prizeman and university lecturer in mathematics, 1883; Cavendish professor of experimental physics, 1894-1919; master of Trinity, 1918-40; professor of natural philosophy, Royal Institution, 1905-20; proved that an electric charge must possess inertia and showed how to calculate its mass, 1881; published *Notes on Recent Researches in Electricity and Magnetism* (1893) which included account of the discharge of electricity through gases; after discovery of X-rays (1895) proved that electric currents were carried by positive and negative ions generated in the gas by X-rays; ascertained and measured their physical properties and found them to be such as would be expected if they were generated by the disruption of the molecules of the gas; proved that cathode rays were rapidly moving material particles projected from the negative pole, the mass of each particle being a minute fraction of that of the atom of hydrogen; announced this revolutionary discovery of the electron, Apr. 1897; henceforth devoted most of his personal experimental work to improving technique and applying it to ascertain the masses, energies, and electric charges of the other particles occurring in electric currents through gases, especially the positive rays; calculated amount of scattering of X-rays by the electrons in atoms; fellow (1884), president (1915-20), Royal Society; president, British Association, 1909; Nobel prize, 1906; knighted, 1908; OM, 1912; active in formation and served (1919-27) on Advisory Council of Department of Scientific and Industrial Research; vital personality of enormous mental energy; even-tempered and accessible; made Cavendish laboratory the greatest research school in experimental physics; buried in Westminster Abbey.

THOMSON, WILLIAM, BARON KELVIN (1824-1907), scientist and inventor; born at Belfast; second son of James Thomson (1786-1849, q.v.), professor of mathematics in the Royal Institution there; removed with family to Glasgow in 1832, when his father became professor of mathematics there; matriculated at Glasgow University, 1834; early made his mark in mathematics and physical science; studied writings of Lagrange, Laplace, and Fourier; entered Peterhouse, Cambridge, 1841; won Colquhoun silver sculls, 1844; helped to found Cambridge University Musical Society; second wrangler and first Smith's prizeman, Jan. 1845; visited Faraday's laboratory at Royal Institution, London, and Regnault's laboratory at Paris University; fellow of Peterhouse, 1846-52 and 1872-1907; professor of natural philosophy in Glasgow, 1846-99; gathered round him at Glasgow band of enthusiastic students of mathematical physics; devoted himself to developing the new doctrine propounded by Sadie Carnot in 1824 and by James Prescott Joule [q.v.] in 1847 that work and heat were convertible; formulated between 1851 and 1854 in communication made to Royal Society of Edinburgh the two great laws of thermodynamics—of equivalence and of transformation; subsequently rounded off his thermodynamic work by enunciating the doctrine of available energy; FRS, 1851; married in 1852 his second cousin Margaret

(d. 1870), daughter of Walter Crum, FRS; made while staying at Creuznach in 1855 the acquaintance of Helmholtz, who became a close friend; throughout life sought to utilize science for practical ends; by means of mathematical analysis evolved theory of electric oscillations, which forms the basis of wireless telegraphy, 1853; experimented on electric telegraph cables, 1854, propounding the 'law of squares'; became director of the Atlantic telegraph company, 1856; invented the mirror galvanometer; served as electrician on the *Agamemnon* which laid the cable across the Atlantic, 1858; the experiment was ruined by a colleague's neglect of Thomson's counsels; Thomson triumphantly redeemed the defeat by superintending the laying of a new cable, 1866; knighted for these services, 1866; president of Society of Telegraph Engineers, 1874; president of mathematical and physical section of British Association at Glasgow, 1876; meanwhile studied atmospheric electricity, and improved the system of electrical measurement and the adoption of rational units; suggested the formation of commission of electrical standards of British Association; advocated the metric system; worked at mathematical theory of magnetism; joined with Peter Guthrie Tait [q.v.] in a *Treatise on Natural Philosophy* (vol. i, part i, 1867, part ii, 1874); made important contributions to the theory of elasticity in a paper on 'Vortex Atoms', Edinburgh, 1867; undertook important researches into gyrostatic problems; married secondly, Frances Anna Blandy, 1874; built mansion at Netherhall near Largs; navigated sailing yacht the *Lalla Rookh* on the Clyde; reformed the mariner's compass, 1873-8; devised apparatus for taking flying soundings, 1872; invented a tide-predicting machine; studied the propagation of wave motion; president of British Association at Edinburgh, 1871; president of physical and mathematical section at York, 1881, when he showed the possibility of utilizing the power of Niagara in generating electricity; interested himself in electric lighting; was founder of firm Kelvin & White, Ltd., Glasgow, for manufacture of his inventions; always fond of illustrating recondite scientific notions by ingenious models; Copley medallist of Royal Society, 1883; delivered at Baltimore in 1884 twenty lectures on *Molecular Dynamics and the Wave Theory of Light* (published 1904); president of the Royal Society, 1890-4; baron, 1892; celebrated jubilee of Glasgow professorship, June 1896; received OM and made PC, 1902; chancellor of Glasgow University, 1904; buried in Westminster Abbey, 23 Dec. 1907; ardent unionist in politics; cherished through life a strong religious faith; mathematical and physical papers edited by Sir Joseph Larmor [q.v.] (5 vols., 1882-1911).

THOMSON, Sir WILLIAM (1843–1909), surgeon; BA, Queen's College, Galway, 1867; MD and M.Ch., Queen's University, Ireland, 1872; hon. MA, 1881; FRCS Ireland, 1874; president, 1896–8; visiting surgeon to Richmond Hospital, Dublin, from 1873 to death; first general secretary of Royal Academy of Medicine in Ireland, 1882; knighted, 1897; established field hospital in Pretoria in South African war, 1900; CB, 1901; published report on poor law medical service of Ireland, 1891, and surgical works.

THORNE, WILLIAM (WILL) JAMES (1857–1946), labour leader; began work at age of six; employed in Birmingham brickyards and gas-works; moved to Beckton gas-works, London; joined Social Democratic Federation, 1884; chairman, 1930; established (1889) National Union of Gasworkers and General Labourers, later the National Union of General and Municipal Workers; general secretary, 1889–1934; obtained eight-hour day for gasworkers, 1889; main speaker at inaugural meeting of London dock strike, 1889; member, parliamentary committee, Trades Union Congress, 1894–1933; president, annual congress, 1912; on West Ham town council, 1891–1946; mayor, 1917–18; labour MP, West Ham (South), 1906–45; labour delegate to Russia, 1917; PC, 1945.

THORNTON, ALFRED HENRY ROBINSON (1863–1939), painter; educated at Harrow and Trinity College, Cambridge; studied under Frederick Brown [q.v.] at Westminster and Slade schools of art; member of New English Art Club and London Group; landscapes somewhat influenced by French Impressionists but fundamentally English.

THORNTON, Sir EDWARD (1817–1906), diplomatist; son of Sir Edward Thornton [q.v.]; BA, Pembroke College, Cambridge, 1840; MA, 1877; held diplomatic posts in Mexico, 1845–54, at Montevideo, 1854–9, and Buenos Aires, 1859–63; CB, 1863; British envoy at Rio de Janeiro, 1865–7; at Washington, 1867–81; helped to settle questions of boundary and fishing disputes between Canada and United States; KCB, 1870; PC, 1871; arbitrator in disputes of United States with Brazil, 1870, and Mexico, 1873–6; British ambassador at St. Petersburg, 1881–4; arranged for delimitation of northern frontier of Afghanistan by joint commission of Russians and British, 1884; GCB, 1883; meanwhile appointed ambassador to Constantinople, 1884, but the post was temporarily filled by Sir William White (1824–91, q.v.), 1884–6; Thornton resigned, and retired to England, 1886; hon. DCL, Oxford, LLD, Harvard.

THORNYCROFT, Sir JOHN ISAAC (1843–1928), naval architect; born in Rome; son of Thomas Thornycroft and brother of Sir W. H. Thornycroft [qq.v.]; studied engineering at Glasgow University; assisted father to establish shipyard at Chiswick, 1866; built first recorded small high-speed boat, 1871; constructed first torpedo boat, for Norwegian government, 1873; built first torpedo boat for English navy, 1877; constructed many torpedo boats and later torpedo-boat destroyers; introduced improvements in hull form, propeller design, and boilers; during 1914–18 war constructed coastal motor-boats known as 'scooters', which proved invaluable, especially in attacks on Zeebrugge and Ostend; FRS, 1893; knighted, 1902; a pioneer in naval architecture.

THORNYCROFT, Sir (WILLIAM) HAMO (1850–1925), sculptor; son of Thomas Thornycroft and brother of Sir J. I. Thornycroft [qq.v.]; studied sculpture under father and at Royal Academy Schools; RA, 1888; knighted, 1917; his best-known works include 'Artemis' (1880), 'Teucer' (1881), 'The Mower' (1884), 'A Sower' (1886), statues of General Gordon (1888) and Oliver Cromwell (1899) in London, and of King Alfred (1901) at Winchester.

THORPE, Sir THOMAS EDWARD (1845–1925), chemist; educated at Owens College, Manchester, and Heidelberg; professor of chemistry, Andersonian College, Glasgow, 1870; Yorkshire College of Science, Leeds, 1874; Royal College of Science, South Kensington, 1885–94, 1909–12; government chemist, 1894–1909; chiefly investigated in field of inorganic chemistry; FRS, 1876; knighted, 1909; his works include *Dictionary of Applied Chemistry* (1893).

THRELFALL, Sir RICHARD (1861–1932), physicist and chemical engineer; educated at Clifton and Gonville and Caius College, Cambridge; professor of physics, Sydney, 1886–99; established good laboratory; joined Albright & Wilson, chemical manufacturers, Birmingham, 1899; original member, Advisory Council on Scientific and Industrial Research, 1916–26; chairman, Chemical and Fuel research boards; first director, Chemical Research boards; first director, Chemical Research Laboratory, Teddington; noted for ingenuity, accuracy, and manipulative skill; published *Laboratory Arts* (1898); FRS, 1899; KBE, 1917.

THRING, GODFREY (1823–1903), hymnologist; BA, Balliol College, Oxford, 1845; rector of Alford, 1858–93; prebendary of Wells, 1876; published volumes of hymns, which included 'The radiant morn' and 'Fierce raged the tempest'.

THRING, HENRY, Baron Thring (1818–1907), parliamentary draftsman; brother of Godfrey Thring [q.v.]; educated at Shrewsbury and Magdalene College, Cambridge; third classic, 1841; fellow; called to bar, 1845; made study of statute law; framed colonial bill for Sir William Molesworth [q.v.], 1850, which drew attention to its draftsman; drafted Gladstone's Succession Act of 1853; recast merchant shipping law in drawing the Merchant Shipping Act of 1854; drafted series of bills, culminating in Companies Act of 1862; Home Office counsel, 1860; drafted for Lord Derby's government the 'ten minutes' bill, which became law as Representation of the People Act, 1867; was made parliamentary counsel to the Treasury in 1869 on the creation of the office by Robert Lowe (later Viscount Sherbrooke, q.v.) for the drafting of government bills; among measures which he prepared were Irish Church Act, 1869, Irish Land Act, 1871, Army Act, 1871, and home rule bill, 1886; he greatly improved style of drafting; explained methods in his *Practical Legislation*, 1902; member (1868) and subsequently chairman of statute law committee, which indexed, expurgated, republished, and consolidated existing statutes; initiated publication of state trials from 1820 to 1858; KCB, 1873; baron, 1886; superintended compilation of first edition of War Office *Manual of Military Law*.

THRUPP, GEORGE ATHELSTANE (1822–1905), author of *History of the Art of Coach-building* (1877); entered father's coach business; head of firm, 1866; had high reputation in England and on continent; master of Coachmakers' Company, 1883.

THUILLIER, Sir HENRY EDWARD LANDOR (1813–1906), surveyor-general of India; joined Bengal artillery, 1832; deputy surveyor-general and superintendent of revenue surveys from 1847; prepared first postage stamps used in India, 1854; joint-author of *The Manual of Surveying in India* (1851); surveyor-general, 1861–78; transferred preparation of atlas of India from England to Calcutta; FRS, 1869; CSI, 1870; knighted, 1879; general, 1881; colonel commandant of Royal Artillery, 1883.

THURSFIELD, Sir JAMES (1840–1923), naval historian and journalist; BA, Corpus Christi College, Oxford; fellow of Jesus College, 1864–81; *The Times* leader-writer, 1881; devoted himself to study of naval policy and history; knighted, 1920; his works include *Peel* (1891) and *Nelson and other Naval Studies* (1909).

THURSO, first Viscount (1890–1970), politician. [See Sinclair, Sir Archibald Henry Macdonald.]

THURSTON, KATHERINE CECIL (1875–1911), novelist; acquired fame by her *John Chilcote, M.P.* (1904); other works were *The Gambler* (1906) and *Max* (1908).

TILLETT, BENJAMIN (BEN) (1860–1943), labour leader; founded Dockers' Union, 1887; secretary until its amalgamation (1922) in Transport and General Workers' Union; secretary to political and international department, 1922–31; leader in London dock strike, 1889; and in formation of National Transport Workers' Federation, 1910; member, general council, Trades Union Congress, 1921–31; president, annual congress, 1929; MP, North Salford, 1917–24, 1929–31.

TILLEY, VESTA (1864–1952) male impersonator; born Matilda Alice Powles; her father, manager of variety hall in Gloucester, helped her to make stage début at age of three; appeared on stage in boy's clothing at age of five; on London stage as 'The Great Little Tilley', 1878–1920; celebrated principal boy in pantomime; appeared in *Sindbad* (1882) and *Beauty and the Beast* (1890); popularized numerous songs including 'Burlington Bertie' and 'Jolly Good Luck to the Girl who Loves a Soldier'; married (Sir) Walter de Freece (died 1935), 1890.

TINSLEY, WILLIAM (1831–1902), publisher; joined brother Edward in publishing business in London, 1854; published novels by G. A. Sala and Miss Braddon, 1861–3; started *Tinsley's Magazine*, 1868; for a time chief producer of works by leading novelists; failed in business, 1878; wrote *Random Recollections* (2 vols., 1900).

TINWORTH, GEORGE (1843–1913), modeller; member of Messrs. Doulton's pottery works, Lambeth, 1867–1913; executed reliefs (chiefly of scenes from biblical history), including part of York Minster reredos, and statues.

TITCHMARSH, EDWARD CHARLES (1899–1963), mathematician; educated at King Edward VII School, Sheffield, and Balliol College, Oxford; served in army, 1918–19; studied mathematics at Oxford under G. H. Hardy [q.v.], 1921; first class, 1922; BA, 1922; MA, 1924; senior lecturer, University College, London, 1923; reader, 1925; fellow, Magdalen College, Oxford, 1924–30; professor of pure mathematics, Liverpool, 1929; succeeded

Hardy as Savilian professor of geometry, Oxford, 1931-63; fellow, New College, Oxford; curator, Mathematical Institute; prodigious mathematical output; numerous papers; publications include *The Theory of Functions* (1932), the *Zeta-Function of Riemann* (1930), *Introduction to the Theory of the Fourier Integrals* (1937), *Mathematics for the General Reader* (1948), and *Eigenfunction Expansions Associated with Second-order Differential Equations* (part i, 1946, and part ii, 1958); president, London Mathematical Society, 1945-7; FRS, 1931; hon. D.Sc., Sheffield.

TIWANA, NAWAB SIR (MUHAMMAD) UMAR HAYAT (1874-1944), statesman; commissioned in 18th Tiwana Lancers, 1901; served in Somaliland (1903), Tibet (1904), France and Mesopotamia (1914-18), and Afghan war (1919); member, Council of State, 1921-8, council of India, 1929-34; KCIE, 1916; GBE, 1934.

TIZARD, SIR HENRY THOMAS (1885-1959), scientist and administrator; son of Thomas Henry Tizard [q.v.]; educated at Westminster School and Magdalen College, Oxford; first class, chemistry, 1908; tutorial fellow, Magdalen, 1911; served in Royal Flying Corps, testing aircraft, 1915-17; Ministry of Munitions, 1917-18; carried out research on aviation fuels; reader in chemical thermodynamics, Oxford, 1920; assistant-secretary, Department of Scientific and Industrial Research, 1920; permanent secretary, DSIR, 1927-9; rector, Imperial College, London, 1929-42; chairman Aeronautical Research Committee, 1933-43; chairman, Air Defence Committee, 1935-40, the committee responsible for development of radar; resigned in consequence of disagreements with Lindemann (Viscount Cherwell) [q.v.]; adviser, Ministry of Aircraft Production, 1940-2; queried policy of bombing built-up areas of Germany; president, Magdalen College, Oxford, 1942-6; chairman, Defence Research Policy Committee and Advisory Council on Scientific Policy, 1947-52; CB, 1927; KCB, 1937; GCB, 1949; FRS, 1926; president, British Association, 1948; hon. fellow, Oriel and Magdalen, Oxford, and Imperial College and University College, London.

TIZARD, THOMAS HENRY (1839-1924), oceanographer, hydrographic surveyor, and navigator; entered navy, 1854; served on surveying vessel *Rifleman* on China station, 1860-7; appointed to *Challenger*, 1872-6; as navigating officer, in close collaboration with (Sir) G. S. Nares [q.v.] and scientific staff of expedition; in charge of home survey, 1880-9;

assistant hydrographer of navy and FRS, 1891; CB, 1899.

TODD, SIR CHARLES (1826-1910), government astronomer and postmaster-general of South Australia; assistant at Greenwich (1841-7) and Cambridge University (1848-54) observatories; pioneer in astronomical photography; director of colonial observatory at Adelaide, 1855-1905; organized extensive meteorological service; inaugurated intercolonial telegraphic system between Adelaide and Melbourne, 1858; system extended to Sydney, 1858, and Brisbane, 1861; amalgamated telegraph and postal systems of South Australia, 1869; postmaster-general, 1870-1905; supervised construction of internal telegraph line across continent, and established communication between Adelaide and England via Port Darwin and Singapore, 1872; CMG, 1872; KCMG, 1893; FRS, 1889; hon. MA, Cambridge, 1886; died at Adelaide.

TOMLIN, THOMAS JAMES CHESSHYRE, BARON TOMLIN (1867-1935), judge; educated at Harrow and New College, Oxford; first class, jurisprudence, 1889; called to bar (Middle Temple, 1891; Lincoln's Inn, 1892; bencher of latter, 1918); junior equity counsel to Board of Inland Revenue and other departments; KC, 1913; 'went special', 1919; judge of Chancery division, 1923; lord of appeal in ordinary, 1929-35; life peer and PC, 1929; chairman, royal commissions on awards to inventors (1923-33) and civil service (1929-31).

TOMLINSON, GEORGE (1890-1952), politician; educated at elementary school and evening classes; began work in Lancashire cotton mill at twelve years of age; urban district councillor, 1914; president, Rishton District Weavers' Association; elected to Lancashire County Council, 1931; labour MP, Farnworth, 1938-52; joint parliamentary secretary, Ministry of Labour and National Service, under Ernest Bevin [q.v.], 1941; minister of works and PC, 1945; minister of education, 1947; hon. LLD, Liverpool University, 1947.

TOMLINSON, HENRY MAJOR (1873-1958), writer; left school at thirteen to work in shipping office; joined staff of *Morning Leader*, 1904; leader-writer, *Daily News*, 1912; literary editor under H. W. Massingham [q.v.] of the *Nation*, 1917-23; publications include *The Sea and the Jungle* (1912), *Gallions Reach* (1927), *All Our Yesterdays* (1930), and *The Trumpet Shall Sound* (1957); hon. LLD, Aberdeen, 1949.

TOMSON, ARTHUR (1859-1905), landscape painter; studied art at Düsseldorf; painted poetic landscapes of Sussex and Dorset; exhibited at Royal Academy, 1883-92; wrote *Jean-François Millet and the Barbizon School* (1903).

TONKS, HENRY (1862-1937), painter and teacher of art; educated at Clifton; qualified in medicine at London Hospital; senior medical officer, Royal Free Hospital; studied art under Frederick Brown [q.v.]; became his assistant at Slade School, 1894; Slade professor, 1917-30; outstanding teacher and vigorous defender of traditional spirit, especially in draughtsmanship; exhibited regularly with New English Art Club; works include 'Strolling Players' (Manchester) and 'The Birdcage' (Ashmolean).

TOOLE, JOHN LAWRENCE (1830-1906), actor and theatrical manager; son of James Toole, City toastmaster commemorated by Dickens and Thackeray; at first a clerk in London; made professional début as comedian in Dublin, 1852, where he played the role of Paul Pry for first time; appeared at Edinburgh as the 'Artful Dodger' in *Oliver Twist* (1854); first acted in London at St. James's Theatre, 1854; played there Weazle in *My Friend the Major* (1854); presented Bottom in *Midsummer Night's Dream* at Edinburgh (1855); toured annually in the provinces from 1857; met (Sir) Henry Irving [q.v.] at Edinburgh, 1857; appeared as Tom Cranky in *The Birthplace of Podgers* at Lyceum (1858); played at New Adelphi from 1858 to 1867; the original Spriggins in *Ici on parle Français*; other popular roles were Bob Cratchit in *The Christmas Carol* (1859), Caleb Plummer—a histrionic masterpiece—in Boucicault's *Dot* (1862), and Mr Tetterby in *The Haunted Man* (1863); he was praised by Dickens for his rendering of *Stephen Digges* (1864); toured with Irving, 1866; created Michael Garner in *Dearer than Life* at Liverpool (1867); began long association with the Gaity under John Hollingshead [q.v.], 1869; successful as Buzfuz in *Bardell* v. *Pickwick* (1871), as Thespis in Gilbert and Sullivan's first extravaganza (1871), as Ali Baba (1872), and as Bob Acres (1874); paid his only visit to America, 1874, where he met with lukewarm reception; created Charles Liquorpond in H. J. Byron's *A Fool and his Money* (1878); leased (1879-95) the Folly Theatre (called Toole's, 1882), producing farcical comedies and burlesques by a permanent stock company; staged several travesties by Burnand of popular plays, caricaturing Wilson Barrett [q.v.] and Irving; obtained success in Merivale's *The Butler* (1886) and *The Don* (1888); was warmly welcomed in Australia, 1890-1; appeared as Ibsen in J. M. Barrie's *Ibsen's Ghost* (1891), and as Jasper Phipps in Barrie's *Walker, London,* (1892); his *Remin-*

iscences, compiled by Joseph Hatton [q.v.], appeared in 1889; successful in eccentric drollery, with propensity to gagging, Toole was the last low comedian of the old school.

TOPLEY, WILLIAM WHITEMAN CARLTON (1886-1944), bacteriologist; educated at City of London School and St. John's College, Cambridge; first class, natural sciences, 1907; qualified from St. Thomas's Hospital, 1909; director, clinical pathological department, Charing Cross Hospital, 1911-22; professor of bacteriology, Manchester, 1922-7, London, 1927-41; secretary, Agricultural Research Council, 1941-4; published *The Principles of Bacteriology and Immunity* (with G. S. Wilson, 2 vols., 1929) and *Outline of Immunity* (1933); FRS, 1930.

TORRANCE, GEORGE WILLIAM (1835-1907), musician and divine; chorister and organist at Dublin, 1847-54; composed oratorios 'Abraham' (1854) and 'The Captivity' (1864) and secular music; BA, Trinity College, Dublin, 1864; MA, 1867; hon. Mus.D., 1879; ordained, 1865; served curacies and incumbencies in Australia, 1870-97; returned to Ireland, 1897; canon of St. Canice's Cathedral, Kilkenny, 1900.

TOUT, THOMAS FREDERICK (1855-1929), historian and teacher; BA, Balliol College, Oxford; pupil of William Stubbs [q.v.]; professor of modern history, St. David's College, Lampeter, 1881-90; of medieval and modern history, Victoria University of Manchester, 1890-1925; took prominent part in securing severance (1903) of Manchester University from older federal body, in building up faculty of arts, establishing undenominational faculty of theology, and founding Manchester University settlement and press; developed history school, in his last years devoting himself especially to organization of post-graduate school; his works include numerous successful textbooks, vol. iii (1216-1377) of Longman's *Political History of England* (1905), *State Trials of the Reign of Edward I, 1289-1293* (with (Professor) Hilda Johnstone, 1906); after 1908 devoted himself to study of administrative history, his work in which constitutes his peculiar contribution to medieval history; this work embodied in two books, *The Place of the Reign of Edward II in English History* (1914) and *Chapters in the Administrative History of Medieval England* (6 vols., 1920-31); FBA, 1911.

TOVEY, SIR DONALD FRANCIS (1875-1940), musician; his education and musical presentation undertaken by Miss Sophie

Weisse; studied under (Sir) Walter Parratt and (Sir) C. H. Parry [qq.v.]; BA, Balliol College, Oxford, 1898; in front rank as pianist playing classical chamber music with Lady Hallé [q.v.], Joachim, Hausmann, Casals, and other eminent artists; Reid professor of music, Edinburgh, 1914-40; founded Reid Orchestra, 1917; conducted, played, and wrote programme commentaries reprinted as *Essays in Musical Analysis* (6 vols., 1935-9); originated new system of university training in music; wrote articles on music for eleventh edition of *Encyclopaedia Britannica*; edited and finished (1931) and wrote commentary on Bach's 'Kunst der Fuge'; compositions include pianoforte concerto (1903), 'cello concerto (for Casals, 1933), and opera, *The Bride of Dionysus* (1929); knighted, 1935; accounted by Joachim the most learned man in music who had ever lived.

TOWNSEND SIR JOHN SEALY EDWARD (1868-1957), pioneer in physics of ionized gases; educated at Corrig School and Trinity College, Dublin; first class, mathematics, 1890; fellow, Trinity College, Cambridge, and assistant demonstrator, Cavendish Laboratory, 1899; Wykeham professor, experimental physics and fellow, New College, Oxford, 1900-41; made electrical laboratory centre of research on ionized gases; publications include *Electricity in Gases* (1915), *Electrons in Gases* (1947), and *Electromagnetic Waves* (1951); FRS, 1903; knighted, 1941.

TOWNSEND, MEREDITH WHITE (1831-1911), editor of the *Spectator*; educated at Ipswich grammar school; went to India (1848) to assist John Clark Marshman [q.v.] in editing the *Friend of India* at Serampore; editor, 1852; proprietor, 1853; exerted great influence on Indian policy; gave valuable support to Lords Dalhousie and Canning [qq.v.]; returned to England through ill health, 1859; bought *Spectator*, 1860, and took into partnership Richard Holt Hutton [q.v.]; supported the North in the American civil war; wrote chiefly on foreign politics and on India; resigned editorship to J. St. Loe Strachey [q.v.], 1898; contributed also for many years political article in the *Economist*; collaborated with J. L. Sanford [q.v.] in *The Great Governing Families of England* (2 vols., 1865); republished articles from reviews in *Asia and Europe* (1901).

TOWNSHEND, SIR CHARLES VERE FERRERS (1861-1924), major-general; gazetted to Royal Marine light infantry, 1881; served in Sudan expedition, 1884-5; transferred to Indian army, 1886; served in Hunza-Nagar expedition, 1891-2; commanded garrison of Chitral fort with great skill and judgement,

1895; transferred to Egyptian army, 1896; present at battles of Atbara and Khartoum, 1898; served in South African war, 1900; his abilities being regarded as unbalanced, passed through lean years, including periods of service in India and South Africa; major-general, 1911; commanded Rawalpindi brigade, 1913; commanded sixth (Indian) division, one of two divisions operating in Mesopotamia under command of Sir J. E. Nixon [q.v.], 1915; drove Turks from Kurna on Tigris and pursued them to Amara, where bulk of forces surrendered; defeated Turks and captured Kut al Amara, Sept.; after failing to capture Baghdad, retreated to and finally surrendered Kut, 1916; interned near Constantinople, 1916-18; KCB, 1917; independent conservative MP, Wrekin division of Shropshire, 1920-2; died in Paris.

TOWSE, SIR (ERNEST) BEACHCROFT (BECKWITH) (1864-1948), soldier and pioneer of blind welfare; educated at Wellington College; served with Gordon Highlanders in India and South Africa; wounded with loss of sight, 1900; VC; vice-chairman, National Institute for Blind, 1901-23; chairman, 1923-44; welfare officer in France, 1914-18; inaugurated Special Fund for Blind Ex-Servicemen, 1923; vice-president, British Legion, 1927-48; member, Honourable Corps of Gentlemen-at-Arms, 1903-39; KCVO, 1927.

TOYNBEE, PAGET JACKSON (1855-1932), Dante scholar; son of Joseph and brother of Arnold Toynbee [qq.v.]; educated at Haileybury and Balliol College, Oxford; works include the *Dante Dictionary* (1898; concise edn. 1914), *Dante in English Literature from Chaucer to Cary* (2 vols., 1909), revision of fourth edition of *Oxford Dante* (1924), and volumes of the letters of Horace Walpole; FBA, 1919.

TRACEY, SIR RICHARD EDWARD (1837-1907), admiral; entered navy, 1852; served in Baltic campaign, 1854, and on East Indies and China station, 1862; on active service in Japan, 1863-4; organized Japanese naval school at Tsukiji, 1867-8; did similar service for Chinese navy; second-in-command of Channel squadron, 1889-92; admiral superintendent at Malta, 1892; president of Royal Naval College at Greenwich, 1897-1900; KCB and admiral 1898.

TRAFFORD, F. G. (pseudonym) (1832-1906), novelist. [See RIDDELL, CHARLOTTE ELIZA LAWSON.]

TRAILL, ANTHONY (1838-1914), provost of Trinity College, Dublin; BA, Trinity College, Dublin; fellow, 1865-1904; provost,

1904-14; successfully resisted scheme of (Viscount) Bryce [q.v.] to amalgamate Trinity with colleges of Royal University of Ireland, 1906; took principal share in effecting internal constitutional reform of college; keen unionist of Ulster type; gave lifelong devotion to Irish Church.

TRAILL-BURROUGHS, SIR FREDERICK WILLIAM (1831-1905), lieutenant-general. [See BURROUGHS.]

TRAVERS, MORRIS WILLIAM (1872-1961), chemist; educated at Blundell's School, Tiverton, and University College, London; B.Sc., 1893; published paper on preparation of acetylene in the *Proceedings* of the Chemical Society, 1893; demonstrator, University College; worked with (Sir) William Ramsay [q.v.] on helium; with Ramsay discovered 'krypton', 'neon', and 'xenon'; assistant professor, 1898; professor of chemistry, University College, Bristol, 1904; director, Indian Institute of Science, Bangalore, 1906-14; worked during 1914-18 war with Baird and Tatlock producing scientific glassware; founded firm of Travers and Clark Ltd., specializing in construction of plant for gasification of coal, 1920-6; hon. professor and research fellow, Bristol, 1927; consultant on explosives, Ministry of Supply, 1939; retired as professor emeritus, Bristol, 1949; FRS, 1904; president, Faraday Society, 1936-8; president, Society of Glass Technology, 1921-2; publications include *The Experimental Study of Gases* (1901), *The Discovery of Rare Gases* (1928), *William Ramsay and University College, London, 1852-1952* (1952), and *A Life of Sir William Ramsay, K.C.B., F.R.S.* (1956).

TREE, SIR HERBERT BEERBOHM (1852-1917), actor-manager, whose original name was HERBERT BEERBOHM; began professional acting career, 1879; appeared in over fifty plays between 1880 and 1887, most conspicuous successes being Prince Borowsky in Sydney Grundy's *Glass of Fashion* (1883) and Revd. Robert Spalding in *The Private Secretary* (1884); lessee and manager of Haymarket Theatre, London, 1887-97; produced and acted in over thirty plays, including plays by Ibsen, Wilde, Maeterlinck, and Shakespeare; parts taken include Iago, Hamlet, and Falstaff; opened Her Majesty's Theatre, 1897; devoted himself increasingly to over-elaborate productions of Shakespeare; gave celebrated performance of Mark Antony in forum scene of *Julius Caesar*; staged plays by Stephen Phillips [q.v.], appearing as Herod, Ulysses, and Nero; extraordinarily versatile actor; knighted, 1909.

TRELOAR, SIR WILLIAM PURDIE, baronet (1843-1923), carpet manufacturer and philanthropist; entered father's mat manufactory, 1858; eventually became manager; alderman of ward of Farringdon Without, 1892; sheriff, 1899-1900; lord mayor of London, 1906; as result of appeal for cripples' home, Lord Mayor Treloar Cripples' Hospital and College, Alton, Hampshire, founded, 1908; knighted, 1900; baronet, 1907.

TRENCH, FREDERIC HERBERT (1865-1923), poet and playwright; BA, Keble College, Oxford; in service of Board of Education, 1891-1909; 'artistic director', Haymarket Theatre, 1909-11; his works include *Deirdre Wed and other Poems* (1901), *New Poems* (containing 'Apollo and the Seaman', 1907), *Ode from Italy in time of War* (1915), *Napoleon* (a four-act play, 1919); died at Boulogne-sur-Mer.

TRENCHARD, HUGH MONTAGUE, first VISCOUNT TRENCHARD (1873-1956), marshal of the Royal Air Force; little education; failed entry to Dartmouth; entered army as militia candidate after twice failing examination, 1893; served in India, in South African war, and in Nigeria, 1903-10; obtained air pilot's certificate and seconded to Royal Flying Corps, 1912; commandant, Military Wing, 1914; commanded RFC in France, 1915-18; chief of air staff, Jan.-Apr. 1918; unable to work with Lord Rothermere [q.v.], air minister; returned to post under (Sir) Winston Churchill, 1919-29; first marshal of Royal Air Force 1927; 'Father of the Royal Air Force'; commissioner, Metropolitan Police, 1931-5; chairman, United Africa Company, 1936-53; DSO, 1906; KCB, 1918; baronet, 1919; GCB, 1924; baron, 1930; GCVO, 1935; viscount, 1936; OM, 1951; hon. LLD, Cambridge, and DCL, Oxford.

TRENT, first BARON (1850-1931), man of business and philanthropist. [See BOOT, JESSE.]

TREVELYAN, SIR CHARLES PHILIPS, third baronet, of Wallington (1870-1958), politician; eldest son of (Sir) George Otto Trevelyan [q.v.]; educated at Harrow and Trinity College, Cambridge; liberal MP, Elland division of Yorkshire, 1899-1918; parliamentary secretary, Board of Education, 1908-14; advocated 'peace by negotiation' throughout 1914-18 war; labour MP, Central division, Newcastle, 1922-31; president, Board of Education, 1924 and 1929-31; succeeded his father, 1928.

TREVELYAN, GEORGE MACAULEY (1876-1962), historian; son of (Sir) George Otto Trevelyan [q.v.]; great nephew of Lord

Macaulay [q.v.], the historian; educated at Harrow and Trinity College, Cambridge; first class, historical tripos, 1896; fellow, Trinity College, 1898; published *England in the Age of Wycliffe* (1899) and *England under the Stuarts* (1904); lecturer, 1898-1903; wrote trilogy on Garibaldi (1907, 1909, 1911); raised and commanded British Red Cross ambulance unit on Italian front, 1914-18; Italian decorations and CBE, 1920; published *Life of John Bright* (1913), *Lord Grey of the Reform Bill* (1920), *British History in the Nineteenth Century* (1922), and *History of England* (1926); regius professor of modern history, Cambridge, 1927; published *England under Queen Anne* (3 vols., 1930, 1932, 1934); master, Trinity College, Cambridge, 1940-51; president of the English Association, 1951; trustee, British Museum and National Portrait Gallery; chancellor, Durham University, 1950-8; high steward, city of Cambridge, 1946; FBA, 1925; FRS, 1950; OM, 1930; hon. fellow, Oriel College, Oxford; many hon. doctorates; other publications include *Clio, a Muse* (1913), *Grey of Fallodon* (1937), *The Poetry and Philosophy of George Meredith* (1906), anthologies from Meredith (1955) and *Carlyle* (1953), and *A Layman's Love of Letters* (1954).

TREVELYAN, SIR GEORGE OTTO, second baronet (1838-1928), historian, man of letters, and statesman; son of Sir C. E. Trevelyan, first baronet, and nephew of Lord Macaulay [qq.v.]; educated at Harrow and Trinity College, Cambridge; went out to India as private secretary to his father; financial member of governor-general's council, 1863; influenced by new liberal movement of 1860s, and enthusiastic for cause of Italian freedom; MP, Tynemouth, 1865-8; supporter of army reform, particularly of abolition of purchase system; MP, Border (Hawick) Burghs, 1868-86; civil lord of Admiralty, 1868-70; in opposition, advocated extension of working-class franchise to county divisions; parliamentary secretary to Admiralty, 1881; chief secretary for Ireland, 1882-4; entered cabinet as chancellor of duchy of Lancaster, 1884; secretary for Scotland, 1886; introduced crofters' bill, 1886; resigned on home rule question, 1886; accepted home rule and MP, Bridgeton division of Glasgow, 1887; secretary for Scotland, 1892-5; retired from politics, 1897; OM, 1911; his works include *Life and Letters of Lord Macaulay* (1876), *The Early History of Charles James Fox* (1880), and a history of the American Revolution in six volumes (1899-1914).

TREVELYAN, HILDA (1877-1959), stage name of Hilda Blow, actress; first stage appearance at age of twelve, 1889; toured in *The*

Little Minister by (Sir) J. M. Barrie [q.v.], 1899; first played Wendy in *Peter Pan* (1904); followed by many revivals; in fifty years, appeared in ten other Barrie parts up to retirement (1939); married Sydney Blow, actor and playwright, 1910.

TREVES, SIR FREDERICK, baronet (1853-1923), surgeon; studied medicine at London Hospital, 1871-5; assistant surgeon, London Hospital, 1879; full surgeon, 1884; 'demonstrator' of anatomy in medical school attached to London Hospital; built up reputation as a demonstrator and a leading surgeon; Hunterian professor of anatomy, Royal College of Surgeons, 1885; applied himself to new field of abdominal surgery; advocated operative treatment of appendicitis; surgeon-extraordinary to Queen Victoria, 1900; acquired world fame by his operation for appendicitis on King Edward VII, 1902; sergeant-surgeon to King Edward, 1902; to King George V, 1910; KCVO, 1901; baronet, 1902; his works include surgical textbooks, *The Pathology, Diagnosis and Treatment of Obstruction of the Intestine* (1884), and books of travel and reminiscence; died at Vevey.

TREVETHIN, first BARON (1843-1936), lord chief justice of England. [See LAWRENCE, ALFRED TRISTRAM.]

TREVOR, WILLIAM SPOTTISWOODE (1831-1907), major-general Royal (Bengal) Engineers; born in India; joined Bengal engineers, 1849; served in Burmese war, 1852-3, and in Mutiny, 1857; garrison engineer at Fort William, Calcutta, 1861; completed the Ganges and Darjeeling road, 1862-3; field engineer of Bhutan field force, 1865; awarded VC for gallantry at Dewan-Giri, Feb. 1865; chief engineer of British Burma, 1875-80; secretary to Indian government in public works department, 1882-7; retired as major-general, 1887.

TRISTRAM, ERNEST WILLIAM (1882-1952), painter and art historian; educated at Carmarthen grammar school and the Royal College of Art, South Kensington; main interest, British medieval wall paintings; on staff of Royal College of Art, 1906; professor of design 1925-48; published *English Medieval Painting* (1927, with Tancred Borenius) and *English Medieval Wall Painting* (3 vols., 1944-55); collections of sketches and records in Victoria and Albert Museum and Buckfast Abbey, Devon; hon. D.Litt., Oxford (1931) and Birmingham (1946); hon. ARIBA, 1935.

TRISTRAM, HENRY BAKER (1822-1906), divine and naturalist; BA, Lincoln College,

Oxford, 1844; MA, 1846; frequently visited Palestine and Egypt; hon. LLD, Edinburgh, and FRS, 1868; canon residentiary of Durham from 1874 to death; travelled in Japan, China, and North-West America, 1891; works include *The Natural History of the Bible* (1867), *The Topography of the Holy Land* (1872), and *The Fauna and Flora of Palestine* (1884); ornithological collection passed to Liverpool public museum.

TRITTON, SIR WILLIAM ASHBEE (1875-1946), engineer; educated at Christ's College, Finchley, and King's College, London; served engineering apprenticeship; general manager, William Foster & Co., Ltd., Lincoln, 1905-11; managing director, 1911-39; chairman, 1939-46; with Walter Gordon Wilson designed and demonstrated prototype of the tank, produced in accordance with requirements laid down by joint naval and military committee, 1916; director of tank construction, 1917-18; knighted, 1917.

TROTTER, WILFRED BATTEN LEWIS (1872-1939), surgeon, physiologist, and philosopher; qualified at University College Hospital, 1896; surgeon, 1906-39; professor of surgery, 1935-8; honorary surgeon (1928), sergeant-surgeon (1932) to the King; made first major exposition of herd psychology (1908-9), amplified as *Instincts of the Herd in Peace and War* (1916); investigated the physiology of cutaneous sensation in man; made advances in surgery of brain and spinal cord and of malignant disease; almost legendary for his diagnostic acumen and refined surgical technique; his addresses published as *Collected Papers* (1941); FRS, 1931.

TROUBRIDGE, SIR ERNEST CHARLES THOMAS (1862-1926), admiral; son of Sir T. St. V. H. C. Troubridge, third baronet [q.v.]; entered navy, 1875; as naval attaché at Tokyo, witnessed battle of Chemulpo (Russo-Japanese war, 1904) and furnished British Admiralty with valuable reports; private secretary to first lord of Admiralty, 1910; rear-admiral, 1911; chief of naval war staff, 1912; commanded cruiser squadron, Mediterranean fleet, 1913; exonerated before court martial for failure to intercept German cruisers *Goeben* and *Breslau* in Mediterranean, 1914; organized evacuation of Serbian army and refugees, 1915-16; served Serbian government, 1916-19; vice-admiral, 1916; admiral, 1919; KCMG, 1919; died at Biarritz.

TROUBRIDGE, SIR THOMAS HOPE (1895-1949), vice-admiral; son of Sir E. C. T. Troubridge [q.v.]; entered navy, 1908; naval attaché, Berlin, 1936-9; commanded aircraft carrier *Furious*, home fleet, 1940-1; battleship *Nelson*, Force H, Gibraltar, 1941; aircraft carrier *Indomitable* in Eastern fleet (1941-2), convoying to Malta (1942); DSO; commanded Central Task Force for capture of Oran, Nov. 1942; rear-admiral, 1943; commanded naval force under Sir Bertram Ramsay [q.v.], invasion of Sicily, July 1943; landed 1st British division north of Anzio, Jan. 1944; commanded carrier force covering invasion of southern France, Aug. 1944; fifth sea lord, 1945-6; KCB, 1945; commanded naval air stations in Britain, 1946-8; vice-admiral, 1947.

TROUP, ROBERT SCOTT (1874-1939), forestry expert; educated at Aberdeen University and Cooper's Hill College; entered Indian forest service, 1897; worked at Forest Research Institute, Dehra Dun, 1906-15; leading authority on Indian silviculture; assistant inspector-general of forests, 1915-19; professor of forestry, Oxford, 1920-39; first director (1924-35) and largely responsible for foundation of Imperial Forestry Institute, Oxford; works include *Silvicultural Systems* (1928); FRS, 1926; CMG, 1934.

TRUEMAN, SIR ARTHUR ELIJAH (1894-1956), geologist and administrator; educated at High Pavement School and University College, Nottingham; first class, geology, London, 1914; D.Sc., London, 1918; assistant lecturer, University College, Cardiff, 1917-20; lecturer, Swansea, 1920-30; professor, geology, 1930-3; professor, Bristol, 1933-7; professor, Glasgow, 1937-46; deputy chairman, University Grants Committee, 1946-9; chairman, UGC, 1949-53; carried out research on stratigraphical palaeontology of coal measures of England and Wales; edited and contributed to authoritative work on coalfields of Britain, 1954; FRS, Edinburgh and London; hon. LLD, Rhodes, Glasgow, Wales, and Leeds; KBE, 1951.

TRUMAN, EDWIN THOMAS (1818-1905), dentist and inventor; educated at King's College Hospital; dentist to royal household from 1855 till death; MRCS, 1859; invented improved method of preparing guttapercha to protect the Atlantic cable, 1860; established guttapercha factory at Vauxhall, 1860; invented guttapercha stoppings for dental work; made large collection of George Cruikshank's work; wrote on dentistry.

TRUSCOT, BRUCE (pseudonym) (1891-1952). [See PEERS, EDGAR ALLISON.]

TSHEKEDI KHAMA (1905-1959), African leader; son of Khama, chief of the Bamangwato;

educated at Serowe and South African Native College, Fort Hare; regent of the Bamangwato for his nephew, Seretse Khama, 1925–33; deposed but reinstated, 1933; opposed to Seretse's marriage to an Englishwoman while at Oxford; both Tshekedi and Seretse banished, 1950–6; after reconciliation, both returned as private persons (Seretse subsequently first president of Botswana Republic, 1966); Tshekedi died in London, 1959.

TUCKER, ALFRED ROBERT (1849–1914), missionary bishop of Uganda; BA, Christ Church, Oxford; ordained, 1882; bishop of Eastern Equatorial Africa, 1890; on division of diocese (1899) became bishop of Uganda; organized church built up by Church Missionary Society; retired, 1911; canon of Durham, 1911–14.

TUCKER, SIR CHARLES (1838–1935), general; educated at Marlborough; commissioned in 22nd regiment, 1855; transferred to 80th Staffordshire Volunteers (later 2nd South Staffordshire regiment), 1860; served in Bhutan (1865) and Perak (1875) expeditions and Zulu war (1878–9); major-general, 1893; commanded 7th division in South Africa, 1899–1901; lieutenant-general, 1902; held Scottish command, 1903–5; noted for forcefulness of expression; KCB, 1901; GCVO, 1905; GCB, 1912.

TUCKER, HENRY WILLIAM (1830–1902), secretary of the Society for the Propagation of the Gospel; BA, Magdalen Hall, Oxford, 1854; MA, 1859; assistant secretary to SPG, 1865–79; secretary, 1879–1901; prebendary of St. Paul's, 1881; promoted colonial and missionary work of the society; died at Florence.

TUCKWELL, GERTRUDE MARY (1861–1951), philanthropic worker; educated at home by her father; teacher in London, 1885–93; secretary to her aunt, Mrs Mark Pattison (Lady Dilke) [q.v.], 1893; on Lady Dilke's death, president, Women's Trade Union League; authority on industrial matters; justice of the peace; CH, 1930; published *The State and Its Children* (1894) and a biography of Sir Charles Dilke (with Stephen Gwynn [q.v.], 2 vols., 1917).

TUKE, HENRY SCOTT (1858–1929), painter; son of D. H. Tuke [q.v.]; entered Slade School of Art, 1875; studied painting in Italy and France; settled in Cornwall; chiefly painted studies of nude youths in sunlit atmosphere against background of sea or shore; one of the earliest and finest of these, 'August Blue' (1894), now in Tate Gallery; RA, 1914; a highly individual painter.

TUKER, SIR FRANCIS IVAN SIMMS (1894–1967), lieutenant-general; educated at Brighton College and Sandhurst; served in 2nd King Edward's Own Gurkha Rifles, 1914–20; entered Staff College, Camberley, 1925; commanded 1st battalion, Gurkha Rifles, on North-West Frontier, India, 1937; OBE; published articles on future pattern of war and policing of North-West Frontier; director, military training, GHQ India, 1939; major-general in command 34th Indian division, 1941; commanded 4th Indian division in Western Desert, 1942–4; CB; DSO; recalled to India, 1944; GOC Ceylon, 1945; commanded IV Indian Corps in Burma, 1945; lieutenant-general; commander-in-chief Eastern Command, 1946–7; KCIE, 1946; colonel, 2nd King Edward's Own Gurkha Rifles, 1946; published *The Pattern of War* (1948), *While Memory Serves* (1950), *Approach to Battle* (1963), and *Gurkha* (1957).

TULLOCH, WILLIAM JOHN (1887–1966), professor of bacteriology; educated at Dundee high school and St. Andrews University; MB, Ch.B., 1909; assistant to Professor L. R. Sutherland, Department of Pathology, Dundee; assistant in comparative pathology, Durham College of Medicine, 1911; lecturer in bacteriology, Dundee; MD, 1914; served in Royal Army Medical College, Millbank, 1916–19; worked on wound infections and cerebrospinal fever; major, RAMC; OBE (Military Division), 1919; professor of bacteriology, Dundee, 1921; also bacteriologist to Dundee Royal Infirmary; continued research activities; member, General Medical Council, 1949–62, and General Dental Council, 1956–9; dean of faculty of medicine, Dundee; 1945–56; D.Sc. on retirement, 1962.

TUPPER, SIR CHARLES, first baronet (1821–1915), Canadian statesman; born in Nova Scotia; medical practitioner; entered Nova Scotia legislative assembly as conservative member for Cumberland county, 1855; re-animated his party; premier, 1864; procured free education for Nova Scotia; strong advocate of Canadian federation; formed close alliance with Sir John A. Macdonald [q.v.], 1864; entered federal cabinet, 1870; committed conservatives to 'national policy' of protection while in opposition, 1873–8; first minister of railways and canals, 1879–84; largely responsible for completion of Canadian Pacific Railway; KCMG, 1879; Canadian high commissioner in London, 1884–96; baronet, 1888; prime minister of Canada, Apr.–June 1896; PC, 1908.

TUPPER, SIR CHARLES LEWIS (1848–1910), Indian civil servant; educated at Harrow and Corpus Christi College, Oxford; BA, 1870; joined Indian civil service, 1871; secretary to Punjab government, 1882-90; chief secretary from 1890; financial commissioner, 1899; CSI, 1897; KCIE, 1903; helped to create Punjab University, 1882; vice-chancellor, 1900-1; compiled *The Customary Law of the Punjab* (3 vols., 1880) and *Our Indian Protectorate* (1893).

TURING, ALAN MATHISON (1912-1954), mathematician; educated at Sherborne School and King's College, Cambridge; first class, part ii, mathematics tripos, 1934; fellow, King's, 1935; at Princeton, 1936-8; worked in Foreign Office during 1939-45 war; OBE, 1946; senior principal scientific officer, National Physical Laboratory, Teddington, 1945-8; developed computing machine; reader at Manchester, 1948-54; assistant director, Manchester Automatic Digital Machine; pioneer of computer problems; FRS, 1951.

TURNBULL, HUBERT MAITLAND (1875-1955), pathologist; grandson of Adam Black [q.v.]; educated at Charterhouse and Magdalen College, Oxford; medically qualified at London Hospital, 1902; director, Institute of Pathology, London Hospital, 1906-46; professor, morbid anatomy, London University, 1919; DM, Oxford, 1906; hon. D.Sc., 1945; FRCP, 1929; FRS, 1939; first to identify post-vaccinal encephalomyelitis, 1922-3.

TURNER, SIR BEN (1863-1942), labour pioneer; officer, General Union of Weavers and Textile Workers, 1889; president, 1902-22; of National Union of Textile Workers, 1922-33; president, annual Trades Union Congress, and conducted negotiations with Sir Alfred Mond (later Lord Melchett, q.v.), 1928; MP, Batley and Morley, 1922-4, 1929-31; parliamentary secretary for mines, 1929-30; knighted, 1931.

TURNER, CHARLES EDWARD (1831-1903), Russian scholar; educated at St. Paul's School and Lincoln College, Oxford; went to Russia, 1859; lector of the English language in St. Petersburg University, 1864-1903; works include *Studies in Russian Literature* (1882), *The Modern Novelists of Russia* (1890), and English handbooks for Russian students.

TURNER, CUTHBERT HAMILTON (1860-1930), ecclesiastical historian and New Testament scholar; educated at Winchester and New College, Oxford; theological lecturer, St. John's College, 1885; assistant lecturer to William Bright [q.v.], regius professor of ecclesiastical history, Oxford, 1888; fellow of Magdalen College, 1889-1930; Dean Ireland's professor of exegesis, 1920-30; concentrated

primarily upon study of textual material of early Western canon law; his main vocation as a scholar consisted in editing successive *fasciculi* of his great work *Ecclesiae Occidentalis Monumenta Juris Antiquissima* (1899-1930); his minor works include an important article on 'Chronology of the New Testament' in vol. i of Hastings's *Dictionary of the Bible* (1898), *Studies in Early Church History* (1912), *Early Worcester MSS.* (1916), an edition of the *Novum Testamentum S. Irenaei* (1923), *Early Printed Editions of the Greek New Testament* (1924), an article on St. Mark's Gospel in the *New Commentary on Holy Scripture* (1928), and 'Notes on Marcan Usage' in *Journal of Theological Studies*, vols. xxv-xxix; travelled extensively in pursuit of learning, and came in contact with many eminent scholars with whom he maintained a regular correspondence.

TURNER, EUSTACE EBENEZER (1893-1966), organic chemist; educated at Coopers' Company School and East London (later Queen Mary) College, London; B.Sc., first class, chemistry, 1913; D.Sc., 1920; research on biaryls with G. M. Bennett [q.v.]; lecturer, Goldsmith's College, 1914-15; working on organo-arsenic compounds with (Sir) W. J. Pope [q.v.], at Sidney Sussex College, Cambridge, 1916-19; lecturer in organic chemistry, Sydney University, 1920-1; senior lecturer, Queen Mary College, London, 1923-8; reader, Bedford College, 1928; professor, 1944; head of chemistry department, 1946-60; professor emeritus, 1960; research director, Biorex Laboratories; contributed to knowledge of stereochemistry and biaryls; published (with Margaret M. Harris) *Organic Chemistry* (1952); FRS, 1939; assistant editor, *British Chemical Abstracts*; fellow, Royal Institute of Chemistry and Queen Mary College; freeman of City of London.

TURNER, GEORGE CHARLEWOOD (1891-1967), headmaster of Marlborough College and university principal of Makerere College, Uganda; educated at Marlborough College and Magdalen College, Oxford; served in army, 1914-19; MC, 1918; assistant master, Marlborough under (Sir) Cyril Norwood [q.v.], 1919; succeeded Norwood as master, 1925; principal Makerere University College, Kampala, 1939; hon. fellow, Makerere, 1944; CMG, 1945; headmaster, Charterhouse School, 1947-52; chairman, Headmasters' Conference; governor, Gordonstoun School.

TURNER, GEORGE GREY (1877-1951), surgeon; educated at Durham University; MB, BS, first class, 1898; MRCS, 1899; MS, 1901; FRCS, 1903; joined staff, Royal Victoria

Infirmary, Newcastle; served in Royal Army Medical Corps, 1914-18; returned to Royal Infirmary, Newcastle; professor of surgery, 1927-34; director of surgery, British Postgraduate Medical School, Hammersmith, 1934-46; served on council of Royal College of Surgeons, 1926-50; hon. LLD, Glasgow, 1939; D.Ch., Durham, 1935.

TURNER, HAROLD (1909-1962), ballet dancer; studied dancing with Alfred Haines in Manchester; stage début, 1927; worked with Léonide Massine and (Dame) Marie Rambert; partnered Tamara Karsavina in *Les Sylphides* and *Le Spectre de la Rose* (1930); appeared in Camargo Society performances and with Vic-Wells Ballet (later the Royal Ballet); permanent member, 1935, taking lead in classical ballets and in *Checkmate, Les Patineurs*, and *The Rake's Progress*; served in RAF during 1939-45 war; on retirement taught at Royal Ballet School, 1955; director, Covent Garden Opera Ballet; first outstanding male dancer to win acclaim under his own English name; returned to play with Lydia Sokolova in *The Good-humoured Ladies* (1962); died after a rehearsal at Covent Garden.

TURNER, HERBERT HALL (1861-1930), astronomer; BA, Trinity College, Cambridge; chief assistant, Royal Observatory, Greenwich, 1884-93; Savilian professor of astronomy, Oxford, and fellow of New College, 1893-1930; FRS, 1896; convinced advocate of use of photography in astronomy; at Greenwich and Oxford threw himself into international scheme for an astrographic chart and catalogue; took up study of seismology and variable stars; died at Stockholm.

TURNER, JAMES SMITH (1832-1904), dentist; MRCS and LDS, 1863; dental surgeon and lecturer at Middlesex Hospital from 1864; lecturer on dental surgery mechanics at Royal Dental Hospital, 1871-80; instrumental in the passing of the Dentists Act, 1878; helped to found British Dental Association, 1879; president of Odontological Society, 1884.

TURNER, WALTER JAMES REDFERN (1889-1946), writer; educated at Scotch College, Melbourne; music critic, *New Statesman*, 1916-40; dramatic critic, *London Mercury*, 1919-23; literary editor, *Daily Herald*, 1920-3, *Spectator*, 1942-6; publications include *The Man who Ate the Popomack* (satirical comedy, 1922), autobiographical *Blow for Balloons* (1935) and *Henry Airbubble* (1936), and volumes of poetry.

TURNER, SIR WILLIAM (1832-1916), anatomist, teacher, and academic administrator;

studied at St. Bartholomew's Hospital, London, under Sir James Paget [q.v.]; MB, London University; senior demonstrator in department of anatomy, Edinburgh University, as private assistant to John Goodsir [q.v.], professor of anatomy, 1854; contended successfully against difficulties due to unofficial character of position and his English birth; succeeded Goodsir, 1867; established reputation as eminent teacher and held chair till 1903; took active part in administrative and legislative work of university; member of royal commission appointed to consider question of improvement of conditions to be imposed as qualification for practice of medical profession, 1881; opposed proposal to create central examining board with exclusive power of conferring licence to practise and advocated university's privilege of conferring licence through its degree; his views, expressed in minority report, embodied in Act of 1886; knighted, 1886; principal of Edinburgh University, 1903-16; first Englishman and first occupant of medical chair to become head of Scottish university; his period of office saw notable expansion in university activities.

TURNER, WILLIAM ERNEST STEPHEN (1881-1963), chemist and first professor of glass technology in the university of Sheffield; educated at King Edward VI grammar school, Five Ways, Birmingham, and Mason (University) College; first class, chemistry, B.Sc. (external) London University, 1902; M.Sc., Birmingham, 1904; D.Sc. (external), London, 1911; demonstrator and lecturer, University College of Sheffield, 1904; wrote articles for *Sheffield Daily Telegraph* on employment of scientists in industry, 1909; secretary, university scientific advisory committee, 1914; advocated courses in 'glass technology', 1915; new department for glass research set up, 1916; professor, 1920-45; founder and secretary, Society of Glass Technology, 1916; editor of its journal; professor emeritus, Sheffield; 1945; played major role in formation of International Commission on Glass; OBE, 1918; FRS, 1938; hon. D.Sc. Tech., Sheffield, 1954; FSA, 1958; active in many organizations concerned with glass and ceramics.

TURNOR, CHRISTOPHER HATTON (1873-1940), agricultural and social reformer; educated at Royal Agricultural College, Cirencester, and Christ Church, Oxford; inherited Lincolnshire family estates, 1903; improved conditions and quality of farming; used his mansion at Stoke Rochford to further educational reform by summer schools for WEA and teachers, and social and religious conferences.

TURNOUR, EDWARD, sixth EARL WINTER-TON and BARON TURNOUR (1883-1962), politician; grandson of first Duke of Abercorn [q.v.]; educated at Eton and New College, Oxford; conservative MP, Horsham, 1904-51; succeeded father as Irish peer, 1907; parliamentary private secretary to financial secretary to Admiralty; active opponent of Parliament Bill of 1911 and Home Rule for Ireland; served during 1914-18 war in Gallipoli and with T. E. Lawrence [q.v.]; parliamentary under-secretary for India, 1922-9; PC, 1924; delegate to third India Round Table Conference, 1932; declined United Kingdom peerage, 1935; chancellor, Duchy of Lancaster, 1937; deputy to secretary of state for air, Lord Swinton [q.v.], 1938; supported Neville Chamberlain; paymaster-general, 1939; displaced, 1940-5; closely associated with Emanuel (later Lord) Shinwell; 'Father' of the House of Commons, 1945; accepted United Kingdom peerage, 1952; master of the Chiddingfold Hunt; a 'character'; published *Pre-War* (1932), *Orders of the Day* (1953), and *Fifty Tumultuous Years* (1955).

TURPIN, EDMUND HART (1835-1907), organist and composer; organist in London from 1860 to death; hon. secretary of Royal College of Organists from 1875; hon. Mus.Doc., Lambeth, 1889; warden of Trinity College of Music, 1892; edited *Musical Standard*, 1880-6, 1889-90; composed oratorios, church music, pianoforte and organ music.

TURRILL, WILLIAM BERTRAM (1890-1961), botanist; educated at Oxford high school; junior assistant, department of botany, Oxford, 1906; temporary assistant, Royal Botanic Gardens, Kew, 1909; permanent assistant, 1914; B.Sc. (London), first class, botany, 1915; M.Sc., 1922; D.Sc., 1928; served with RAMC, 1916-18; returned to Kew, 1919; conducted evening classes at Chelsea Polytechnic on plant taxonomy, plant ecology, and plant genetics, 1918-39; in charge of large portion of Kew herbarium and library, transferred to Oxford during 1939-45 war; keeper of herbarium and library, Kew, 1946-57; published *The Plant Life of the Balkan Peninsula* (1929) and *The Royal Botanic Gardens, Kew. Past and Present* (1959); editor, the *Botanical Magazine*, 1948-61; OBE, 1953; vice-president, Linnean Society, 1949-50; FRS, 1958.

TUTTON, ALFRED EDWIN HOWARD (1864-1938), crystallographer and alpinist; educated at Owens College, Manchester and Royal College of Science; made crystallography his life work, alpine photography his recreation; inspector of technical schools under Board of Education, 1895-1924; works include *Crystallography and Practical Crystal Measurement* (1911); FRS, 1899.

TWEED, JOHN (1869-1933), sculptor; studied in Glasgow, London and Paris; friend of Rodin; made name as portrait and memorial sculptor; executed memorials to Duke of Wellington (St. Paul's, 1912), Joseph Chamberlain (Westminster Abbey, 1916), Kitchener (Horse Guards Parade, 1925), Rifle Brigade (Grosvenor Place, 1925), and Peers' war memorial (House of Lords, 1932).

TWEEDMOUTH, second BARON (1849-1909), politician. [See MARJORIBANKS, EDWARD.]

TWEEDSMUIR, first BARON (1875-1940), author and governor-general of Canada. [See BUCHAN, JOHN.]

TWINING, EDWARD FRANCIS, BARON TWINING (1899-1967), colonial governor; educated at Lancing and Sandhurst; commissioned into Worcestershire Regiment, 1918; served in Ireland; MBE, 1921; seconded to King's African Rifles in Uganda, 1923; transferred to colonial administrative service, 1928; assistant district commissioner, Uganda, 1929; director of labour, Mauritius, 1939-43; CMG; administrator, St. Lucia, 1944; governor, North Borneo, 1946-9; KCMG; governor, Tanganyika, 1949-58; GCMG, 1953; last two years as governor over-shadowed by rise of Julius Nyerere; life peer, 1958; published *History of the Crown Jewels of Europe* (1960) and *European Regalia* (1967).

TWYMAN, FRANK (1876-1959), designer of optical instruments; educated at Simon Langton School, Canterbury, Finsbury Technical College, and Central Technical College, South Kensington; assisted W. E. Ayrton [q.v.]; assistant to Otto Hilger, optical instrument maker, 1898; succeeded Hilger as head of firm, 1902; managing director, Adam Hilger Ltd., 1904-46; director, Hilger and Watts, 1948-52; designed spectrometer and spectrograph instruments; published *Prism and Lens Making* (1943) and *Metal Spectroscopy* (1951); authority on design of optical instruments.

TYABJI, BADRUDDIN (1844-1906), Indian judge and reformer; born at Bombay; educated at Bombay and in England; first Indian called to English bar (Middle Temple), 1867; member of Bombay legislative council, 1882-4; president of third Indian National Congress, 1887; ardent advocate of higher education of

Indian women; first Indian Moslem judge of Bombay high court, 1895; acting chief justice, 1903; died in London and buried at Bombay.

TYLER, THOMAS (1826–1902), Shakespearian scholar; BA, London, 1859; MA, 1871; published treatise on *Ecclesiastes* (1874) (commended by Ewald); studied Hittite antiquities; wrote also *The Philosophy of 'Hamlet'* (1874) and *Shakespeare's Sonnets* (1880) in which he identified Mary Fitton [q.v.] with the 'dark lady' of the sonnets.

TYLOR, Sir EDWARD BURNETT (1832–1917), anthropologist; brother of Alfred Tylor, the geologist [q.v.]; made acquaintance of Henry Christy, the ethnologist [q.v.], 1856; accompanied him to Mexico; keeper of university museum, Oxford, 1883; reader in anthropology, 1884; first professor of anthropology, 1896–1909; knighted, 1912; helped to secure place for anthropology among acknowledged sciences; chief work, *Primitive Culture* (1871).

TYLOR, JOSEPH JOHN (1851–1901), engineer and Egyptologist; son of Alfred Tylor [q.v.]; partner in father's brass-founding firm in London, 1872–91; AMICE, 1877; from 1891 studied Egyptology and published monographs on *Wall Drawings and Monuments of El Kab* (1895–1900).

TYNAN, KATHARINE (1861–1931), poet and novelist. [See HINKSON.]

TYNDALE-BISCOE, CECIL EARLE (1863–1949), missionary and educationist in Kashmir; educated at Bradfield and Jesus College, Cambridge; coxed winning Cambridge boat, 1884; deacon, 1887; priest, 1890; unconventional but remarkably successful headmaster, CMS school, Srinagar, 1890–1947.

TYNDALL, ARTHUR MANNERING (1881–1961), physicist; educated at private school in Bristol and University College of Bristol; B.Sc. (London), 1903; assistant college lecturer; promoted to lecturer, 1907; university lecturer, 1909; acting head, physics department, 1910; Henry Overton Wills professor of physics, Bristol, 1919–48; directed research work on cosmic rays and mobility of ions in gases; published *Mobility of Positive Ions in Gases* (1938); pro-vice-chancellor, 1940–5 and 1946–7; acting vice-chancellor, 1944–5; FRS, 1933; vice-president, Royal Society, 1942; president, Institute of Physics, 1946–8; CBE, 1950; hon. LLD, Bristol, 1958; member editorial board of the *Philosophical Magazine*.

TYRRELL, GEORGE (1861–1909), modernist; influenced by Father Robert Dolling [q.v.]; joined Roman Church, 1879; postulant at College of Society of Jesus, Malta, 1879; teacher there, 1885–7; ordained priest, 1891; lectured on philosophy at St. Mary's Hall, Stonyhurst, 1894–6; was transferred to literary staff at Farm Street, London, 1896; produced three orthodox works, 1897–9; an article on Hell in *Weekly Register*, Dec. 1899, compelled his retirement to Richmond, Yorkshire; there he completed *Oil and Wine* (1902) and *Lex Orandi* (1903), which show influence of pragmatism; was dismissed from Society of Jesus (1906) for unorthodox 'Letter to a Professor of Anthropology'; replied to Vatican decrees against modernism, Sept.–Oct. 1907; expounded his religious development in *Through Scylla and Charybdis* (1907); answered Cardinal Mercier's attack on modernism in *Medievalism* (1908); his *Christianity at the Cross-Roads* followed in 1909; he regarded doctrinal system of the Church as a 'pseudo-science' but admitted some measure of doctrinal development, and maintained essential identity of modern catholic church with the church of the apostles; collected contributions to periodical literature in *The Faith of the Millions* (2 vols., 1901–2); autobiography and life published by Maud D. Petre, 1912.

TYRRELL, ROBERT YELVERTON (1844–1914), classical scholar; BA, Trinity College, Dublin; fellow, 1868; professor of Latin, 1871; regius professor of Greek, 1880; professor of ancient history, 1900; senior fellow, 1904; FBA, 1902; joint commentator on *Correspondence of Cicero* (1879–1900).

TYRRELL, WILLIAM GEORGE, BARON TYRRELL (1866–1947), diplomatist; educated in Germany and at Balliol College, Oxford; entered Foreign Office, 1889; private secretary to permanent under-secretary, 1896–1903; secretary, Imperial Defence Committee, 1903–4; précis writer to Sir Edward Grey (later Viscount Grey of Fallodon, q.v.), 1905–7; his principal private secretary, 1907–15; head of political intelligence department, 1918; assistant under-secretary, 1918–25; attended peace conference and showed himself fierce critic of Lloyd George, 1919; lifelong believer in friendship with France; accompanied Lord Grey to Washington, 1919; principal adviser to Lord Curzon [q.v.] at Lausanne conference, 1922–3; permanent under-secretary, 1925–8; culminated shifting of policy towards co-operation with France in Locarno treaties, 1925; ambassador in Paris, 1928–34; president, British Board of Film Censors, 1935–47; KCMG, 1913; KCVO, 1919; GCMG, 1925; KCB, 1927; GCB, 1934; PC, 1928; baron, 1929.

TYRWHITT, Sir REGINALD YORKE, first baronet (1870-1951), admiral of the fleet; son of the Revd Richard St. John Tyrwhitt [q.v.]; entered *Britannia*, 1883; commanded *Hart* and began long career in destroyers; promoted captain, 1908; commodore in command of Harwich Force, 1914-18; rear-admiral, 1918; engaged in actions at Heligoland Bight (1914), Cuxhaven raid (1914), and Dogger Bank (1915); accepted surrender of German U-boats, 1918; CB, 1914; DSO, 1916; KCB, 1917; created baronet, 1919; vice-admiral, 1925; commanded China Station, 1927-9; admiral and GCB, 1929; commanded at the Nore, 1930-3; admiral of the fleet, 1934; hon. DCL, Oxford, 1919.

TYRWHITT-WILSON, Sir GERALD HUGH, fifth baronet, and fourteenth BARON BERNERS (1883-1950), musician, artist, and author; succeeded uncle, 1918; educated at Eton; honorary attaché, embassies at Constantinople (1909-11), Rome (1911-19); wrote five humorous novels and two volumes of autobiography; supplied music for ballet including three scores for Sadler's Wells Ballet Company; landscape painter in oil.

U

ULLSWATER, first VISCOUNT (1855-1949), speaker of the House of Commons. [See LOWTHER, JAMES WILLIAM.]

UNDERHILL, EDWARD BEAN (1813-1901), missionary advocate; grocer in Oxford, 1828-43; founded Hanserd Knollys Society, 1845; joint-secretary from 1849, and sole secretary 1869-76, of Baptist Missionary Society; visited India and West Indies, and denounced cruelty to natives in Jamaica, 1865; went to Cameroons, 1869; president of Baptist Union, 1873; wrote lives of Baptist missionaries, and edited early Baptist writings for Hanserd Knollys Society.

UNDERHILL, EVELYN (1875-1941), religious writer; married (1907) Hubert Stuart Moore; converted to Christianity, 1907; to Anglicanism, 1921; profoundly influenced by Friedrich von Hügel [q.v.]; devoted herself to giving spiritual direction, retreats, etc.; wrote for *Spectator* and *Time and Tide*; publications include *Mysticism* (1911), *The Life of the Spirit and the Life of To-day* (1922), and *Worship* (1936); a Christian pacifist, 1939.

UNWIN, SIR RAYMOND (1863-1940), architect; educated at Magdalen School, Oxford; with R. B. Parker designed Letchworth Garden City and Hampstead Garden Suburb; lecturer in town planning, Birmingham, 1911-14; in charge of town planning successively, Local Government Board, Ministry of Munitions, Ministry of Health, 1914-28; visiting professor, Columbia University, 1936-40; knighted, 1932.

UNWIN, SIR STANLEY (1884-1968), publisher; educated at School for the Sons of Missionaries, Blackheath, and Abbotsholme School, Derbyshire, 1897-9; gained experience in German book trade; joined publishing firm of his father's stepbrother, T. Fisher Unwin, 1904; bought bankrupt firm of George Allen & Co., 1912; established himself as spokesman on affairs of British book trade; his list of authors included Bertrand (Lord) Russell [q.v.], Benedetto Croce, August Strindberg, Albert Sorel, James Elroy Flecker [q.v.], Jules Romain, and Sir J. C. Squire [q.v.]; during 1914-18 war published books by conscientious objectors; post-war list included Arthur Waley [q.v.], Sidney J. Webb, G. D. H. Cole, R. H. Tawney, G. Lowes Dickinson, G. Gilbert Murray, H. W. Nevinson [qq.v.], and Sigmund Freud; strong links with Fabian Society; published best-seller, *Mathematics for the Million* (1936), and *Science for the Citizen* (1938), both by Lancelot Hogben; also published J. R. R. Tolkien's *The Hobbit* (1937) and *The Lord of the Rings* (1954-8), and *The Kon-Tiki Expedition*, by Thor Heyerdahl (1950); travelled widely on book-trade business; actively interested in Publishers' Association and British Council; president, Publishers' Association, 1933; of strong non-conformist faith; a keen tennis player; knighted, 1946; KCMG, 1966; LLD, Aberdeen; published *The Truth about Publishing* (1926) and *The Truth about a Publisher* (1960).

UNWIN, WILLIAM CAWTHORNE (1838-1933), engineer; B.Sc. (London), 1861; scientific assistant to (Sir) William Fairbairn [q.v.], Manchester, 1856-62; managed engineering works, 1862-7; instructor in marine engineering, Royal School of Naval Architecture and Marine Engineering, 1869-72; professor of hydraulics and mechanical engineering, Royal Indian Engineering College,

1872-84; of civil and mechanical engineering, Central Institution of the City and Guilds of London (in London University from 1900), 1884-1904; researched in strength of materials, hydraulics, and water- and steam-power; prominent in development of Niagara Falls hydro-electric scheme, introduction and application of internal combustion engine, and study of stability of masonry dams; FRS, 1886.

URE, ALEXANDER, BARON STRATHCLYDE (1853-1928), lawyer and politician; MA, Glasgow University; admitted member of Faculty of Advocates, 1878; liberal MP, Linlithgowshire, 1895-1913; QC for Scotland, 1897; solicitor-general for Scotland, 1905-9; lord advocate and PC, 1909; an enthusiastic supporter of Lloyd George's 1909-10 budget; lord justice-general for Scotland and lord president of Court of Session, 1913-20; baron, 1914; better as advocate than judge; excelled in cross-examination.

URWICK, WILLIAM (1826-1905), nonconformist divine and chronicler; BA, Trinity College, Dublin, 1848; MA, 1851; Congregational minister of Hatherlow, Cheshire, 1851-74; and St. Albans, 1880-95; professor of Hebrew at New College, London, 1874-7; wrote accounts of non-conformity in Cheshire (1864), Hertfordshire (1884), and Worcester (1897).

UTHWATT, AUGUSTUS ANDREWES, BARON UTHWATT (1879-1949), judge; educated at Ballarat College, Melbourne University, and Balliol College, Oxford; BCL, 1903; called to bar (Gray's Inn), 1904; bencher, 1927; legal adviser, Ministry of Food, 1916-19; junior counsel (Chancery) to Treasury and Board of Trade, and to attorney-general in charity matters, 1934; Chancery judge (knighted), 1941; lord of appeal in ordinary, 1946-9; life peer and PC, 1946; chairman, committee on compensation and betterment, 1941.

UVAROV, SIR BORIS PETROVITCH (1889-1970), entomologist; born in Russia; educated at Uralsk and St. Petersburg University; first class, biology, 1910; director, Tiflis Bureau of Plant Protection, 1915; lecturer, State University, Tiflis, and keeper of entomology and zoology, State Museum, Georgia, 1919; appointed to Imperial Bureau (later Commonwealth Institute) of Entomology under (Sir) Guy A. K. Marshall [q.v.], 1920; made extensive study of locusts and grasshoppers; initiated theory of locust phases; made outstanding contribution to control of locust plagues; published *Locusts and Grasshoppers* (1928), *Insect Nutrition and Metabolism* (1928), and *Insects and Climate* (1931); supervised unit investigating outbreaks of locust plagues and organized international co-operation; initiated international plan for control, 1938; in charge of Anti-Locust Research Centre, London, 1945, foremost laboratory in the world for research on locusts; became naturalized British subject, 1943; retired as director, but continued as consultant, 1959; also made major contribution to taxonomy; CMG, 1943; KCMG, 1961; FRS, 1950; hon. D.Sc., Madrid University, 1935; president, Royal Entomological Society of London, 1959-61.

V

VACHELL, HORACE ANNESLEY (1861-1955), novelist; educated at Harrow and Sandhurst; served in Rifle Brigade; wrote more than fifty novels and fourteen plays, including *Brothers* (1904), *The Hill* (1905), *The Fourth Dimension* (1920), and *The Fifth Commandment* (1932); his most famous play, *Quinneys* (1915).

VALLANCE, GERALD AYLMER (1892-1955), journalist; educated at Fettes and Balliol College, Oxford; first class, hon. mods., 1913; served in army, 1914-18; became brigade-major, 2nd Indian division; general secretary, National Maritime Board, 1919-28; joined *Economist*, 1929; editor, *News Chronicle*, 1933-6; assistant editor, *New Statesman*, 1937-55; published *The Centre of the World* (1935), *Hire Purchase* (1939), *Very Private Enterprise* (1955), and *The Summer King* (1956).

VALLANCE, WILLIAM FLEMING (1827-1904), marine painter; studied art at Trustees' Academy, Edinburgh; early work was portraiture and genre; after 1870 depicted Irish life and character; finally painted sea and shipping; painted also in water-colours; ARSA, 1875; RSA, 1881.

VANBRUGH, DAME IRENE (1872-1949), actress; born BARNES; sister of Violet Vanbrugh [q.v.]; educated at Exeter high school and in London; trained under Sarah Thorne; joined J. L. Toole [q.v.], 1889-92; successively with (Sir) Herbert Tree, (Sir) George Alexander, and Arthur Bourchier [qq.v.], 1893-8; made notable creations in plays by Pinero including *The Gay Lord Quex* (1899), *His House in Order* (1906), and *Mid-Channel* (1909); in Barrie's *Walker, London* (1892), *The Admirable Crichton* (1902), and *Rosalind* (1912); in Somerset Maugham's *Grace* (1910) and *The Land of Promise* (1914); and in A. A. Milne's *Mr. Pim Passes By* (1920); also appeared in films; married (1901) Dion Boucicault, the younger [q.v.], who became her manager (1915) and with whom she acted until his death (1929); DBE, 1941.

VANBRUGH, VIOLET (1867-1942), actress; born BARNES; sister of Dame Irene Vanbrugh [q.v.]; in Sarah Thorne's repertory company, 1886-8; accompanied W. H. and (Dame) Madge Kendal [qq.v.] to America, 1889-91; played Ann Boleyn in Irving's *King Henry VIII* (1892); married (1894) Arthur Bourchier [q.v.] whom she divorced, 1917; played modern leading parts, often in Bourchier's productions; notable as Queen Katherine in *King Henry VIII* (1910) and as Mistress Ford in *The Merry Wives of Windsor* (1911); appeared occasionally in films, including *Pygmalion*.

VANDAM, ALBERT DRESDEN (1843-1903), publicist and journalist; educated in Paris, 1858; engaged there as journalist till 1871, when he settled in London; correspondent for *Globe* in Paris, 1882-7; works include *An Englishman in Paris* (2 vols., 1892) and translation of Bartholomew Sastrow's autobiography (1902).

VANE-TEMPEST-STEWART, CHARLES STEWART, sixth MARQUESS OF LONDONDERRY (1852-1915), politician; conservative MP, county Down, 1878-84; succeeded father, 1884; assumed additional name of Stewart, 1885; viceroy of Ireland, 1886-9; first president, Board of Education, 1902-5; signed Ulster covenant, 1912; PC, 1886; KG, 1888.

VANE-TEMPEST-STEWART, CHARLES STEWART HENRY, seventh MARQUESS OF LONDONDERRY (1878-1949), politician; son of sixth Marquess of Londonderry [q.v.]; educated at Eton and Sandhurst; served in Royal Horse Guards, 1897-1906, 1914-17; conservative MP, Maidstone, 1906-15; succeeded father, 1915; on Irish convention, 1917-18; finance member, Air Council, and KG, 1919; under-secretary for air, 1920-1; leader of Senate and minister of education, Northern Ireland, 1921-6; PC, 1925; returned to England, his extensive property including coal-mines, 1926; first commissioner of works, 1928-9, 1931; secretary of state for air, 1931-5; encouraged development of Hurricane and Spitfire but was criticized for having done too little in face of German rearmament.

VAN HORNE, SIR WILLIAM COR-NELIUS (1843-1915), Canadian railway builder and financier; born in Illinois; general manager, Canadian Pacific Railway, 1881; president of company, 1888-99; naturalized Canadian, 1888; honorary KCMG, 1894.

VANSITTART, EDWARD WESTBY (1818-1904), vice-admiral; entered navy, 1831; served in East Indies, and on China station; at reduction of Karachi, 1839, and capture of Woosung batteries, 1842; suppressed piracy in China, 1852-5; accompanied Prince of Wales to Canada, 1860; CB, 1867; vice-admiral, 1879.

VANSITTART, ROBERT GILBERT, BARON VANSITTART (1881-1957), diplomatist; educated at Eton; entered diplomatic service as attaché, Paris, 1903; served at Tehran and Cairo, 1907-11; transferred to Foreign Office, 1911-20; private secretary to secretary of state, 1920-4; private secretary to prime minister, 1928-30; permanent under-secretary, 1930-8; opposed appeasement of Nazi Germany; 'kicked upstairs' as diplomatic adviser, 1938-41; published *The Mist Procession* (1958); MVO, 1906; CB, 1927; KCB, 1929; GCB, 1938; CMG, 1920; GCMG, 1931; PC, 1940; baron, 1941.

VAUGHAN, BERNARD JOHN (1847-1922), Jesuit priest; brother of Cardinal Vaughan and R. W. B. Vaughan [qq.v.]; educated at Stonyhurst; Jesuit novice, 1866; ordained priest, 1880; sent to church of Holy Name, Manchester, 1883; superior, 1888; attracted attention by his preaching, organization, and devoted work amongst poorest classes; transferred to Jesuit house, Farm Street, London, 1899; increased his reputation by series of sermons on 'sins of society', 1906; an ardent social reformer.

VAUGHAN, DAVID JAMES (1825-1905), honorary canon of Peterborough and social reformer; younger brother of Charles John Vaughan [q.v.]; educated at Rugby and Trinity College, Cambridge; Bell university scholar, 1845; fifth classic, 1848; MA, 1851; fellow, 1850-8; issued (with John Llewelyn Davies, q.v.) translation of Plato's *Republic*, 1852 (often reprinted); vicar of St. Martin's, Leicester, 1860-93, and master of Wyggeston's Hospital, 1860-1905; hon. canon of Peterborough, 1872; hon. DD, Durham, 1894; influenced by F. D. Maurice [q.v.]; started at Leicester in 1862 a working men's college with night classes, provident society, and book club, which was transferred to new building erected at Leicester in his memory, 1908; chairman of first Leicester school board, 1871; published theological works and sermons.

VAUGHAN, DAME HELEN CHARLOTTE ISABELLA GWYNNE- (1879-1967), botanist and leader of women's services in both world wars. [See GWYNNE-VAUGHAN.]

VAUGHAN, HERBERT ALFRED (1832-1903), cardinal; elder brother of R. W. B. Vaughan and B. J. Vaughan [qq.v.]; educated at Stonyhurst, in Belgium, at Downside, and in Rome; ordained, 1854; vice-president of St. Edmund's, Ware, 1855; joined congregation of Oblates, 1857; resolving to found missionary college in England, sailed for Caribbean Sea to collect funds, 1863; did much work in Panama despite opposition of civil authorities during smallpox epidemic, 1864; after two years' wandering in South America he was recalled to England by Cardinal Manning [q.v.], 1865; built St. Joseph's College for missionary students at Mill Hill, 1866; sent out first missionaries to United States, 1871; bought and edited *The Tablet*, 1868-71; bishop of Salford, 1872-92, opening pastoral seminary there, 1875, as well as St. Bede's College, 1880; improved the diocesan organization and developed the work of the deaneries; founded the Rescue and Protection Society for building homes and poor law schools for catholics; started Voluntary Schools Association, 1884; took leading part in Manchester social problems; archbishop of Westminster from 1892 till death; cardinal, 1893; made St. Mary's College, Oscott, the common seminary for southern and midland group of dioceses; urged withdrawal of papal admonition against catholic attendance at Oxford and Cambridge, 1895; built Westminster Cathedral (foundation-stone laid, 1895); played official part in controversy over validity of Anglican orders, 1894-7; did much to secure the Education Act of 1902, which obtained state aid for catholic schools; of direct, impulsive, and candid character.

VAUGHAN, KATE (1852?-1903), actress and dancer, whose real name was CATHERINE CANDELON; made début as dancer, 1870; took prominent part in burlesques at Gaiety Theatre, London, from 1876 to 1883; organized Vaughan-Conway comedy company, 1886-7; acted Lydia Languish and Lady Teazle, 1887; visited Australia (1896) and South Africa (1902) for health; died at Johannesburg; inaugurated modern school of skirt-dancing.

VAUGHAN, WILLIAM WYAMAR (1865-1938), schoolmaster; son of H. H. Vaughan [q.v.]; educated at Rugby, New College, Oxford, and Paris University; assistant master, Clifton, 1890-1904; headmaster, Giggleswick, 1904-10; Wellington College, 1910-21; Rugby, 1921-31; served on many educational bodies.

VAUGHAN WILLIAMS, RALPH (1872-1958), composer; grandson of Sir Edward Vaughan Williams [q.v.]; educated at Charterhouse and Trinity College, Cambridge; studied at Royal College of Music; Mus.Bac., 1894; Mus.D., 1901; influenced by English folk-songs and hymnology, including plainsong; edited *The English Hymnal* (1906); served during 1914-18 war in Royal Army Medical Corps and Royal Garrison Artillery; early music consisted of songs, including 'Linden Lea' (1902) and 'Silent Noon' (1908); collection of folk-songs led to three orchestral 'Norfolk Rhapsodies' (1906-7) and *Fantasia on Christmas Carols* (1912); composed nine symphonies, including *Sea Symphony* (1910), *Pastoral* (1922), and *Sinfonia Antartica* (1952); choral works include *Sancta Civitas* (1926); dramatic music includes *Hugh the Drover* (1926), and *The Pilgrim's Progress* (1951); chamber music includes 'On Wenlock Edge' (1909); conducted the Bach Choir, 1921-8; taught composition at Royal College of Music for twenty years; hon. Mus.D., Oxford, 1919; hon. fellow, Trinity College, Cambridge, 1935; OM, 1935.

VEITCH, SIR HARRY JAMES (1840-1924), horticulturist; uncle of J. H. Veitch [q.v.]; joined father's nursery garden business at Chelsea, 1868; keenly interested in employment of travellers who introduced new plants to cultivation; knighted, 1912; published *The Manual of Coniferae* (1881) and *The Manual of Orchidaceous Plants* (1887-94).

VEITCH, JAMES HERBERT (1868-1907), horticulturist; entered father's nursery at Chelsea, 1885; FLS, 1889; made tour round world (1891-3) and introduced large winter cherry into England; managing director of James Veitch & Sons, which became a limited company, 1898; compiled elaborate history of the firm in *Hortus Veitchii* (1906).

VENN, JOHN (1834-1923), logician and man of letters; son of Henry Venn (1796-1873, q.v.); BA, Gonville and Caius College, Cambridge; fellow, 1857: ordained, 1858; studied and taught logic for thirty years from 1862; availed himself of provisions of Clerical Disabilities Act, 1883; from about 1890 onwards devoted himself to university history; his works include *The Logic of Chance* (1866), *Symbolic Logic* (1881), *The Principles of Empirical Logic* (1889), *Biographical History of Gonville and Caius College* (1897), *Alumni Cantabrigienses* (vols. i and ii, 1922); FRS, 1883.

VENTRIS, MICHAEL GEORGE FRANCIS (1922-1956), architect and archaeologist; educated at Stowe and Architectural Association school, London; served in Royal Air Force during 1939-45 war; keenly interested in languages and scripts; deciphered Linear B Minoan Script and published results in *Journal* *of Hellenic Studies*, 1953; published *Documents in Mycenaean Greek* (1956); OBE, 1955; killed in car accident, 1956.

VERDON-ROE SIR (EDWIN) ALLIOTT VERDON (1877-1958), aircraft designer and constructor; educated at St. Paul's School and King's College, London; studied marine engineering; engineer, British and South African Royal Mail Company, 1899-1902; designed and built aeroplane, 1907; with his brother, H. V. Roe, founded Avro Company and designed other successful aeroplanes, including the Avro 504, 1910-28; sold company to Armstrong Siddeley Motor Co. and formed Saunders-Roe Co., building flying boats, 1928-58; knighted, 1929.

VEREKER, JOHN STANDISH SURTEES PRENDERGAST, in peerage of Ireland sixth and in peerage of United Kingdom first VISCOUNT GORT (1886-1946), field-marshal; succeeded father, 1902; educated at Harrow and Sandhurst; gazetted to Grenadier Guards, 1905; GSO 2, operations branch, GHQ, France, 1916-17; commanded 4th battalion, Grenadier Guards, 1917-18; 1st battalion, 1918; fought at Pilckem ridge, Cambrai, Arras, and Flesquières; MC, DSO with two bars; VC; director of military training, India, 1932-6; commandant, Staff College, Camberley, 1936-7; chief of imperial general staff, 1937-9; general, 1937; KCB, 1938; took expeditionary force to France, Sept. 1939; stationed on Belgian frontier; on German invasion (May 1940) outflanked on right by German breakthrough and in last phase on left by collapse of Belgian resistance; unable to break through southwards towards French and doubtful of French ability to thrust northward; moved two divisions to fill gap between British and Belgian armies, thereby saving the BEF; withdrew towards Dunkirk for evacuation; GCB; inspector-general, 1940-1; governor and commander-in-chief, Gibraltar, 1941-2; obtained construction of invaluable airstrip and improved local Anglo-Spanish relations; governor and commander-in-chief, Malta, 1942-4; successfully organized defence against air attack; field-marshal, 1943; high commissioner, Palestine and Trans-Jordan, 1944-5; viscount of United Kingdom, 1945.

VERNEY, ERNEST BASIL (1894-1967), physiologist and pharmacologist; educated at Tonbridge School and Downing College, Cambridge; first class, part i, natural sciences tripos, 1916; trained at St. Bartholomew's Hospital; MRCS and LRCP, 1918; joined Royal Army Medical Corps; house physician, St. Bartholomew's and East London Hospital for

Children, 1919-20; MB; B.Ch. (Cantab), 1921; MRCP (London); assistant in department of physiology, University College, London, 1921; worked on renal physiology; assistant in medical unit, University College Hospital, 1924; professor of pharmacology, University College, 1926; FRCP, London, 1928; moved to Cambridge, 1934; first Sheild professor of pharmacology, 1946-61; retired as emeritus professor; held personal chair, Melbourne, Australia, 1961-4; noted for his work on the kidney and on water balance; FRS, 1936; hon. fellow, Downing College, 1961; number of other academic awards.

VERNEY, MARGARET MARIA, LADY VERNEY (1844-1930), historical writer; born WILLIAMS; married Captain E. H. Verney (afterwards third baronet), 1868; interested in village life and education in Anglesey and Buckinghamshire; completed for press vols. i and ii of *Memoirs of the Verney Family* (1892); herself produced vol. iii (1894), vol. iv (1899), and two supplementary eighteenth-century volumes (1930).

VERNON-HARCOURT, LEVESON FRANCIS (1839-1907), civil engineer; educated at Harrow and Balliol College, Oxford; first class, mathematics and natural science; BA, 1862; resident engineer at East and West India Docks, 1866-70, and on Rosslare harbour works, 1872-4; surveyed Upper Thames valley, 1877; professor of civil engineering at University College, London, 1882-1905; wrote *Rivers and Canals* (2 vols., 1882) and *Harbours and Docks* (1885); served on many commissions and made many engineering reports on rivers and harbours in Great Britain and India; MICE, 1871.

VERRALL, ARTHUR WOOLLGAR (1851-1912), classical scholar; BA, Trinity College, Cambridge (second classic); fellow of Trinity, 1874; lectured there, 1877-1911; first King Edward VII professor of English literature at Cambridge, 1911; works include editions of plays of Aeschylus, *Euripides the Rationalist* (1895), *Essays on Four Plays of Euripides* (1905), and an edition of the *Bacchae* (1910).

VESTEY, WILLIAM, first BARON VESTEY (1859-1940), director of the Union Cold Storage Company; founded the company with his brother Sir Edmund Vestey; built freezing works in Argentina and formed Blue Star line of refrigerated steamers; baronet, 1913; baron, 1922.

VEZIN, HERMANN (1829-1910), actor; born at Philadelphia, USA; BA, Pennsylvania

University, 1847, and MA, 1848; made first appearance as actor in England, in minor Shakespearian parts, 1850; played with Charles Kean [q.v.] in London, 1852-3; toured in provinces, 1853-7; in America, 1857-9; achieved his reputation at Surrey Theatre in chief Shakespearian characters, 1859; married Mrs Charles Young [see VEZIN, JANE ELIZABETH], 1863; obtained first rank as James Harebell in *The Man o' Airlie*, 1867; supported S. Phelps [q.v.] in revivals of old comedies, 1874; scored triumph as Jacques in *As You Like It* (1875, 1880, and 1885); created Buckthorpe in Gilbert's *Randall's Thumb* (1871) and Edgar in Tennyson's *Promise of May* (1882); acted with (Sir) Henry Irving [q.v.], 1888-9; last appeared in London as Rowley in *The School for Scandal* (Apr. 1909); the most scholarly and intellectual actor of his generation.

VEZIN, JANE ELIZABETH (1827-1902), actress; formerly MRS CHARLES YOUNG; accompanied parents to Australia in childhood; appeared as singer and dancer at eight; married Charles Frederick Young, comedian, 1846; took part in G. V. Brooke's Australian tour of 1855; first appeared in London, 1857; played leading Shakespearian parts with S. Phelps [q.v.], 1857-8; acted Rosalind to Hermann Vezin's Orlando at Sadler's Wells, 1860; after divorce from Young (1862), she married Hermann Vezin [q.v.], 1863; played Desdemona to Phelps's Othello at Drury Lane, 1864; other roles were Lady Teazle (1867), Peg Woffington in *Masks and Faces* (1867), and Lady Macbeth (1876); she played with the Bancrofts [qq.v.], 1879, and with Edwin Booth, 1880-1.

VIAN, SIR PHILIP LOUIS (1894-1968), admiral of the fleet; educated at Hillside School and the Royal Naval Colleges, Osborne and Dartmouth; served in destroyers 1914-18; was present at battle of Jutland; specialized in gunnery and continued to serve in destroyers, 1919-36; commanded *Arethusa*, 1937-9; on outbreak of 1939-45 war, transferred to convoy escort duty; in *Cossack* led force which rescued 300 prisoners from *Altmark* in Josing Fiord, Norway, 1940; in destroyer flotilla attack on the German battleship *Bismarck*, 1941; rear-admiral, in command of Force K escorting Russian convoys, 1941; commanded fifteenth cruiser squadron at Alexandria; escorted convoys to Malta; flagship *Naiad* sunk, 1942; took part in invasion of Sicily, 1943; commanded Eastern Task Force in invasion of Normandy, 1944; commanded aircraft carrier squadron working with Americans in the Pacific, 1944-5; vice-admiral, 1945; fifth sea lord, in charge of naval aviation, 1946-8; admiral, 1948; commander-in-chief, Home

Fleet, 1950-2; admiral of the fleet, 1952; DSO, 1940, with bars, 1940 and 1941; KBE, 1942; CB and KCB, 1944; GCB, 1952; published memoirs, *Action This Day* (1960).

VICKERS, KENNETH HOTHAM (1881-1958), historian, and principal, University College, Southampton; educated at Oundle and Exeter College, Oxford; lecturer in history, University College, Bristol, 1905-8; lecturer, London County Council, 1907-9; tutor to London University tutorial classes, 1908-13; fellow, Royal Historical Society, 1909; professor of modern history, Durham University, 1913-1922; principal, University College, Southampton, 1922-46; publications include *England in the later Middle Ages* (1913) and *A Short History of London* (1914); hon. LLD, Southampton, 1953.

VICKY cartoonist, (1913-1966), [See WEISZ, VICTOR.]

VICTORIA (1819-1901), QUEEN OF THE UNITED KINGDOM OF GREAT BRITAIN AND IRELAND, AND EMPRESS OF INDIA; born on 24 May 1819 at Kensington Palace, the only child of King George III's fourth son, Edward Augustus, duke of Kent and Mary Louisa Victoria, daughter of the duke of Saxe-Coburg and Saalfield and sister of Prince Leopold of Saxe-Coburg, husband of Princess Charlotte, the only child of George IV, who died in childbirth in 1817 [qq.v.]. Victoria was the first member of the royal family to be vaccinated against smallpox, 1819. In 1820, her father died, leaving her mother, the duchess, as her sole guardian; her first eighteen years were spent in comparative seclusion at Kensington Palace, being taught by a governess, Louise (later baroness) Lehzen. Victoria became heir-presumptive to the throne on the death of George III in 1830; she met her cousin, Prince Albert of Saxe-Coburg, for the first time, when he visited England with his elder brother, Ernest, in 1836.

On the death of William IV in June 1837, she ascended the throne; her accession was greeted with enthusiasm by the British public, mindful of the contrast she presented with her two uncles who had preceeded her. Lord Melbourne, whig prime minister, became her private secretary and political mentor; her uncle Leopold, by this time king of the Belgians, sent his friend and confidant, Baron Stockmar, to direct her education in the broader fields of diplomacy. The young queen demonstrated her new independence by excluding her mother, the Duchess of Kent, from all public business, and taking up residence in Buckingham palace.

In the 1837 general election the tory leader,

Sir Robert Peel, seized the opportunity to criticize the influence which Melbourne was appearing to assert over the queen; the tories lost the election, and the whigs remained in office. Victoria was crowned on 28 June 1838, but the approbation with which she had been greeted by her people was soon tempered by censure; much unfavourable comment was aroused by the illness and death of Lady Flora Hastings [q.v.], lady-in-waiting to the Duchess of Kent, after unfounded allegations of immoral conduct had been made against her by some of the queen's attendants. In the spring of 1839, the whigs lost the support of the House of Commons, and Melbourne had to resign. The queen was deeply distressed, and tried to avoid calling upon Sir Robert Peel to form a government by asking the Duke of Wellington to come to her aid. The duke declined to take office and the queen had no alternative but to summon Sir Robert. This change of government brought about the first political crisis of her reign. Peel wished to replace some of the ladies in the higher posts of the royal household who undoubtedly had whig sympathies, but the queen demurred, and Peel refused to accept the responsibility of forming a government. To the queen's relief Lord Melbourne returned to office.

Victoria became engaged to marry Prince Albert on 15 Oct. 1839; the marriage took place on 10 Feb. 1840. Her choice was not universally popular; but a new era began in the life of the queen. Baroness Lehzen went into retirement, and little by little, Albert assumed responsibility for the tutelage of his passionate, imperious, obstinate but shrewd young wife. Their first child Victoria Adelaide Mary Louisa was born on 21 Nov. 1840; a son and heir to the throne, Albert Edward (later Edward VII) was born on 9 Nov. 1841. During the next sixteen years the queen bore her husband seven more children (Alice, 1843; Alfred, 1844; Helena, 1846; Louise, 1848; Arthur, 1850; Leopold, 1853; and Beatrice, 1857).

In the two years between 1840 and 1842 three attempts were made on the life of the queen. On 10 June 1840, while she was driving in the Green Park, two shots were fired at her by a young man armed with a pistol; she was unhurt and, apparently unperturbed, went on her way. The culprit was found to be insane. In 1842 two further attempts were made, both fortunately unsuccessful. Meanwhile, in 1841, the whigs were defeated in a general election. Peel became prime minister, and the queen, having learnt by experience, accepted his political guidance. During the early forties Queen Victoria and Albert made and received a number of state visits. They visited Louis-Philippe in France and Leopold in Belgium, and amid considerable

pomp and ceremony, Louis-Philippe made a return visit to England, and the Tsar, Nicholas I, was also entertained in London by the queen. The queen's relations with Peel steadily improved, and she found herself to be in complete accord with Lord Aberdeen, the foreign secretary. The tory government had come to office pledged to maintain the corn laws, but the pressure of events, in particular, the failure of the potato crop in Ireland and poor harvests in England, constrained Sir Robert to reconsider the wisdom of this policy. Victoria herself felt that the removal of restrictions on the import of food was essential to relieve the hardships of the poor, and firmly supported her prime minister when he decided that the corn laws must be repealed. Unwilling to introduce the necessary legislation himself Peel resigned in Dec. 1845. The queen sent for Lord John Russell, since Melbourne was now in failing health; she and Prince Albert besought him not to give the Foreign Office to Lord Palmerston, fearing that his intransigence would endanger peace with France. But Russell could not sustain a government without Palmerston, and Palmerston would accept no post other than that of foreign secretary. Lord Grey refused to serve in the government if Palmerston took the Foreign Office, and Lord Russell informed the queen that he could not form a new ministry. Gratefully, the queen turned again to Peel, who returned to office, and at the cost of splitting his party into warring groups, guided the bill for the repeal of the corn laws through parliament. On 26 June, the night that the bill passed its third reading in the House of Lords, the protectionist rebels in Peel's party and the whigs joined forces to defeat the government on the second reading of a coercion bill for Ireland.

Peel, perforce, resigned, and the queen called upon Lord John Russell, agreeing with much misgiving, to Palmerston's appointment as foreign secretary. Her misgivings were not unfounded. During the next five years Palmerston persistently flouted the views of the queen and her husband. On one occasion he reinserted a paragraph in the draft of a dispatch which Prince Albert had deleted. His sympathy with the revolutionary movements on the continent of Europe caused the royal couple great anxiety. At last, the recalcitrant foreign secretary went too far. Louis-Philippe was dethroned in 1848 and sought asylum in England. In the closing weeks of 1851 Louis Napoleon in a *coup d'état* made himself absolute head of the French government. Palmerston, without consulting the queen or his colleagues, committed the British government to a friendly recognition of the new regime. This was too much, even for Lord John Russell, and to the delight of the queen, Palmerston was peremptorily instructed to resign, and his place was taken by Lord Granville [q.v.].

Victoria and Albert, in the meantime, were enthusiastically engaged on a project which they hoped would be a demonstration of peace and good will among all nations. The design for a great exhibition was Prince Albert's, and he pursued it doggedly despite considerable opposition. However, when the queen carried out the inaugural ceremony in the Crystal Palace in Hyde Park on 1 May 1851 the exhibition proved to be a brilliant success.

The eclipse of Palmerston lasted a very short time. Within two months of his dismissal by Lord John Russell, he brought about the defeat of the government. Lord Derby formed a conservative administration with Disraeli as chancellor of the Exchequer. A new generation of statesmen was rising to power. Melbourne had died in 1848, and Peel in 1850. In 1852, the Duke of Wellington died. The queen and Prince Albert were very concerned at the weakness of her governments at this time, primarily the result of the dissensions caused by the repeal of the corn laws. When, in Dec. 1852 Disraeli's budget was rejected by the House of Commons, the queen called upon Lord Aberdeen to form a coalition government which she hoped would have strength and durability. Palmerston agreed to take the Home Office, and after a brief period under Lord John Russell, the Foreign Office went to Lord Clarendon [q.v.]. For the first time during Victoria's reign the country was to be involved in war. In the autumn of 1853 Russia made unaceptable demands upon Turkey, and war was declared. The queen hoped that Britain could avoid involvement, and Aberdeen was anxious to keep the peace. Palmerston resigned, however, when Aberdeen showed his reluctance to intervene, in alliance with France, to aid Turkey. Popular chauvinism throughout the country denounced the queen and Albert as subservient to Russian aggression, and although these slanders on the queen and her husband were repudiated in both Houses of Parliament, Albert continued to be regarded as the scapegoat for the vacillation of the Aberdeen ministry. Negotiations with Russia eventually broke down and war was declared on 28 Feb. 1854. Once her country was involved in war the queen showed intense interest in its progress, and did all that she could to mitigate the hardships of her troops in the Crimea. The gross mismanagement of the war, however, led to the downfall of Aberdeen, and the queen had to call upon Palmerston to head a new administration in 1855. Having taken this step, she extended her full confidence to her new prime minister, and whatever her private sentiments, subordinated everything to the paramount national interest.

When the emperor, Napoleon III, expressed a wish to discuss personally with the queen the possibility of his taking command of the French army in the Crimea, the queen invited him to Windsor with the empress. The visit passed off well, and Napoleon returned to Paris deeply impressed with his reception in England, having become a knight of the garter, and abandoned his proposal to head the French forces in the war theatre. Later in the year, Victoria and Albert returned this visit, and while they were in Paris, met for the first time Count (later Prince) Bismarck, then Prussian minister at Frankfurt. Sebastopol fell after a siege of almost a year and peace with Russia appeared to be in sight. The Princess Royal became engaged to marry the eldest son of the prince of Prussia (afterwards Emperor Frederick I), and this betrothal was received without enthusiasm in England, *The Times* going so far as to describe it as an act of truckling 'to a paltry German dynasty'. The marriage took place in 1858.

A peace treaty was signed with Russia on 30 Mar. 1856 amid great rejoicing, but the queen continued to take a constant interest in the welfare of the army, and instituted the Victoria Cross for conspicuous valour in war. In the following year, she conferred upon Albert the title of prince consort. In 1851 work had been completed on the palace at Osborne in the Isle of Wight, and in 1854 work had similarly been finished on Balmoral Castle. For the rest of the queen's life a part of every spring and autumn was spent at one or other of these homes. She was at Balmoral when the Indian mutiny was at its height, and was beset with anxiety at the news of the atrocities already committed and the threat of more to come, her concern exacerbated by the difficulties of communication between India and Britain. In Feb. 1858 Palmerston's government was defeated. Difficulties had arisen in consequence of a bomb attack upon the emperor and empress of the French. It was discovered that the perpetrator of this outrage, in which ten people were killed, was an Italian refugee who had planned the attack in England where his bomb was manufactured. The French demanded that the right of asylum in Britain, freely accorded to political malcontents, should be rigorously curtailed. Palmerston acquiesced to the limited extent of introducing legislation to make conspiracy to murder a felony rather than a misdemeanour. The queen approved, but the Commons would have no truckling with Palmerston's old friend Napoleon. The bill was rejected, and Palmerston promptly resigned. The queen sought to keep the ministry in office, not for Palmerston's sake, but because she was reluctant to lose Lord Clarendon. But Palmerston persisted in resigning, and she at once summoned Lord Derby.

Friendly relations were soon re-established with France, and the queen and the prince consort paid a state visit to Cherbourg, and then went on to visit their daughter in Germany. While they were abroad the government at home was busy with settling affairs in India. The mutiny was suppressed, and the administration was reorganized. The powers and territory of the East India Company were transferred to the crown. Competitive examinations for appointment in the new Indian civil service were substituted for nomination, and the Indian army was put under the authority of the council of state. The act giving effect to these changes received the royal assent on 2 Aug. 1858.

Early in the following year, the queen's first grandchild was born. A son and heir was born at Berlin to the Princess Royal. This child became in due course Kaiser Wilhelm II. The queen was thirty-nine years old. Her peace of mind at this time was greatly disturbed by the part being played in Europe by Napoleon III. War between Austria and France could threaten the security of Prussia. The queen's anxieties were not allayed when the Derby government was defeated, and Palmerston returned to power, not with Lord Clarendon at the foreign office, but Lord John Russell. Her differences with her ministers, however, were settled without any serious consequences. The queen opened parliament in 1861, and, for the last time, made the speech from the throne in person, announcing the forthcoming marriage of her daughter, Princess Alice to Prince Louis of Hess.

1861 was a year of tragedy for the queen. In March her mother died, and later, the health of the prince consort caused grave anxiety. Early in December, he persuaded the queen's ministers to modify their draft note to the United States regarding the *Trent* affair in which the captain of a federal ship-of-war had boarded a British ship and taken prisoner two confederate envoys. Palmerston would have sent a peremptory note of protest, but Albert succeeded in getting a more conciliatory message accepted by the prime minister; the matter was settled between London and Washington, and the possibility of war averted. A few weeks later, the prince consort died. The prince's untimely death at the age of forty-two left the queen distraught and desolate. No heavier blow could have befallen her. For years she had been utterly dependent on his advice and guidance. From this time on she would have to face her many difficulties alone. For the next two years she remained in strict seclusion, but although she never ceased to mourn bitterly for her husband, she gradually took up again her duties in the business of the state, and tried to carry on the policies of which the prince would have approved. She found

some comfort in her widowhood from the devotion of her Scottish servants. In particular, she took solace from the constant attendance of her personal retainer, John Brown, who had been a servant at Balmoral since 1849, and whose rugged manners and outspoken common sense she accepted as the manifestation of loyalty and service. In the spring of 1863 the Prince of Wales married Princess Alexandra, daughter of Prince Christian, the heir to the throne of Denmark. The queen took no part in the ceremony. She did, however, visit Germany, staying *en route* with her uncle Leopold, visiting Prince Albert's birthplace in Coburg, and meeting her daughter, the crown princess, and the crown prince of Prussia.

At this time the queen was watching anxiously the situation arising from the conflicting claims of Prussia and Denmark to the duchies of Schleswig-Holstein. Her sympathies were with Prussia, but those of her ministers were with the Danes. Early in 1864 war broke out between Denmark on the one hand and Austria and Prussia on the other. The queen was able to persuade her government to remain neutral. The Danes were defeated and the Prussians and Austrians occupied the disputed territory. Queen Victoria was deeply distressed by the death of Palmerston in October and her uncle Leopold in Dec. 1865. Lord John Russell took over as prime minister, but his government was short-lived. In the following year he retired from public life, and Derby again became prime minister, with Disraeli once more at the Treasury. During 1866 Prussia and Austria could not agree on the occupation of the duchies they had wrested from Denmark. War broke out again. Prussia was victorious; Hanover became a Prussian province, and Prussia became the undisputed head of all north Germany.

In 1867 the queen published an account of the early years of the prince consort and her *Leaves from a Journal of our Life in the Highlands*. She laid the foundation-stone of the Albert Hall to be erected in memory of her late husband. She also received visits from the Khedive of Egypt and the Sultan of Turkey. Disraeli's reform bill passed through parliament.

In February 1868 Disraeli became prime minister, but before the year was out the government was defeated, and Mr Gladstone took over as liberal prime minister for the first time. His first measure was the disestablishment of the Irish church, and although the queen disapproved, she did not oppose the legislation involved. France and Prussia were at war in 1870; the queen had little sympathy with the policies of Napoleon III but, when he and the Empress Eugénie fled, defeated, from France, she gave them both asylum at Chislehurst.

Edward (later Viscount) Cardwell [q.v.], Gladstone's secretary of state for war, introduced reforms in the army, abolishing the purchase of commissions and making the commander-in-chief responsible to the War Office; the latter proposal the queen regarded as an encroachment upon her prerogative, but she felt powerless to resist the determination of her ministers. Her fourth daughter, Princess Louise married the Marquis of Lorne, eldest son of the Duke of Argyll [qq.v.], the first time for some three and a half centuries since an English sovereign had sanctioned the union of a princess with a man who was not a member of a reigning house. At this time the queen's reluctance to appear regularly in public aroused considerable popular criticism, and anti-royalist feeling was rife until the queen decided that she must modify the habits of seclusion she had followed since Albert's death. In Jan. 1874, the queen's second son, Prince Alfred, married the Grand Duchess Marie Alexandrovna, the only daughter of Czar Alexander II. Gladstone's liberal government fell and Disraeli returned to power. Victoria and her new prime minister were in entire agreement and sympathy, and Disraeli's influence with her was such that he was able to stem the tide of anti-royalist sentiment by persuading her to take a more overt part in public life. The Prince of Wales made a highly successful tour of India in 1875, and in the following year, Disraeli passed through parliament a bill proclaiming the queen empress of India; in token of the appreciation of his devotion the queen conferred a peerage upon him and he became Earl of Beaconsfield. She staunchly supported him and used all her influence with her relatives in Europe to restrain the ambitions of Russia in the Balkans; when Beaconsfield returned from the congress of Berlin in triumph having, he claimed, secured 'peace with honour', she invested him with the order of the garter. In that year, 1878, the queen lost her second daughter, Princess Alice, who had been her constant companion since the death of Prince Albert; this was the first death of one of her children, and the queen deeply felt this further loss.

As the decade of the seventies drew to a close the queen was beset by new and grave anxieties; the prince imperial, the only child of the ex-empress Eugénie, was killed in the war the British were fighting against the Zulus in South Africa; Britain was also involved in war in Afghanistan; and Gladstone incurred the resentment of the queen by his Midlothian campaign of passionate denunciation of the imperialist policies of Beaconsfield's government. The queen was in despair when the liberals triumphed in a general election in 1880, and much against her will, she was forced once

more to accept Gladstone as prime minister. She urged the new government to make a sustained effort to bring the war in Afghanistan, and later, the war in South Africa, to a successful conclusion; she also attempted to induce the government to take prompt and effective action against the insurgent followers of Arabi Pasha in Egypt. The death of the Earl of Beaconsfield in 1881 was for the queen another personal bereavement.

Not only was the queen assailed by anxieties regarding problems arising for her government overseas. For the fifth time, her life was threatened in 1882 by a madman. A shot was fired at her, happily without hitting her, while she was on Windsor railway station. Lord Frederick Cavendish [q.v.], chief secretary for Ireland, and Thomas Henry Burke, the under-secretary were less fortunate. They were savagely murdered by extremists in Dublin. The queen had little sympathy for Irish nationalists and approved of every effort made to enforce law and order throughout the country. In the following year, 1883, there were further troubles in Egypt. The people of the Soudan rose in revolt, and the British government decided to abandon the territory after relieving garrisons in places threatened by the rebels. General Gordon [q.v.] was sent out with orders to put this policy into effect. He was besieged in Khartoum by the Mahdi's forces, and after holding out for over ten months, was killed. The queen blamed the government for having been dilatory in its efforts to relieve Khartoum, and she regarded the death of Gordon as an unmitigated disaster. The vacillation of Gladstone's government in dealing with the Mahdi was not only unpopular with the queen. In mid-1885 the government was defeated and, for a few months, Lord Salisbury succeeded Gladstone. The queen was further heartened by the marriage of her youngest daughter Princess Beatrice to Prince Henry of Battenberg.

Early in 1886 a general election (held in the previous autumn) brought Gladstone back to office, determined, however much the queen was opposed to his policy, to grant home rule to Ireland. The new session was more stormy than any the queen had watched since Peel had abolished the Corn Laws. To her relief, Gladstone could not carry all his supporters with him in this measure and was forced to resign. Lord Salisbury, who had the queen's confidence as the heir of Beaconsfield, was her prime minister when the country acclaimed her jubilee in the fiftieth year of her reign as she made one of her rare appearances in her capital, London.

At this time, the queen, although she was no longer young, made frequent visits abroad to

France and Germany and Italy and Spain. Since the death of John Brown, her faithful Scottish retainer, she had included Indians among her personal attendants, and one of them, the munshi, Abdul Karim, instructed her in Hindustani.

During the jubilee festivities the queen's peace of mind was clouded by anxiety for the welfare of her eldest daughter, the crown princess of Prussia, whose husband was suffering from a mysterious illness which eventually proved to be cancer of the throat. When in 1888 he became emperor, he was already a dying man, and his death occurred in June that year. Thus the high hopes of his rule in Germany to which the queen had looked forward for nearly thirty years were brought to nought. The visit to England two years later of her grandson, Kaiser Wilhelm II, although it passed off happily enough, did not succeed in allaying the anxiety of the queen and her government at the growing power and ambitions of the German emperor. In 1892 Gladstone briefly returned to office, and once more brought in a bill for home rule for Ireland. The bill passed the Commons, but to the queen's delight, was rejected by the Lords. On 2 Mar. 1894 Gladstone went to Windsor to relinquish the seals of office. The queen received him with coldness, and did not meet him again before he died in 1898. After a brief interlude, when the Earl of Rosebery became prime minister, Lord Salisbury returned to office, and remained there up to the time of the queen's death.

On 23 Sept. 1896 the queen achieved the distinction of having reigned longer than any other British sovereign. In the following year her diamond jubilee was celebrated with great splendour; millions of the queen's subjects acclaimed her progress through London with a fervour that brought her to tears. But her last years were clouded with worry. British and Egyptian troops at last crushed the long-drawn-out insurrection in the Soudan when Lord Kitchener [q.v.] routed the Mahdi at Omdurman in 1898; this victory, however, was overshadowed by the disasters which befell British forces in the early stages of the war in South Africa which broke out in 1899. The queen was indefatigable in her efforts to relieve the sufferings of her troops and their families. Throughout her long life she had enjoyed good health, but now her eyesight was failing, and her physical condition gradually deteriorated under the pressure of the worry and sorrow brought about by the South African war. On Tuesday, 22 Jan. 1901 the queen died. She was nearly 82 years old. During her last years she had inspired a passion of loyalty in her people; the coolness caused by her long period of seclusion after the bereavement of Albert's death was forgotten,

and the queen in her person had become the symbol of the British people's unity and strength.

VICTORIA ADELAIDE MARY LOUISE (1840–1901), PRINCESS ROYAL OF GREAT BRITAIN and GERMAN EMPRESS; born at Buckingham Palace, 21 Nov. 1840, was eldest child of Queen Victoria and Prince Albert; showed in youth much interest in her studies; was devotedly attached to her eldest brother, the Prince of Wales, afterwards King Edward VII; accompanied her parents and brother to France on a visit to Napoleon III, Aug. 1855; was affianced to Prince Frederick William, afterwards crown prince of Germany, 29 Sept. 1855; received from parliament dowry of £40,000 with annuity of £4,000, 19 May 1857; was married at St. James's Palace, 25 Jan. 1858; lived much in retirement at Berlin, where her eldest son William was born, 27 Jan. 1859; chafed against strict etiquette of Prussian court and its conservative sentiment; openly avowed her liberal convictions; acquired strong influence over her husband; became crown princess, Jan. 1861; resented the rise of Bismarck to supreme power, Mar. 1862; paid long visit to English court, Sept.–Dec. 1863; opposed Bismarck's policy in Schleswig-Holstein, 1864; organized hospitals and care for the wounded during Austro-Prussian conflict of 1866; went with her husband to Paris for opening of international exhibition, May 1867; although she was generally credited in Germany with French sympathies she was active in works of beneficence for the German soldiers during Franco-Prussian war, 1870–1; sought converse with men of culture and artists; tried to introduce into Germany English methods of promoting the industrial arts; helped to establish Berlin Industrial Art museum, Nov. 1881; urged more scientific training for sick-nurses; aided in forming Victoria House and Nursing School, Berlin, 1881; encouraged schemes for ameliorating social condition of working classes; was a pioneer in developing women's higher education in Germany; continued to suffer much annoyance owing to Bismarck's exclusion of her husband from political affairs; nursed her husband during his illness from cancer, 1887; was with him at San Remo when his father Emperor William I died, 9 Mar. 1888, and he became Emperior Frederick; was present at her husband's death at Potsdam, 15 June 1888; during her husband's short reign she quarrelled anew with Bismarck, owing to her obstinate encouragement of a match between her second daughter, Princess Victoria, and Alexander of Battenberg, Prince of Bulgaria; in her widowhood she was still exposed to Bismarck's harshness until the minister's dismissal by the new emperor, her son, Mar. 1890; at her son's request, she spent a week in Paris to test French sentiment, Feb. 1891, but was driven away by threats of a hostile demonstration; settled for rest of life at Cronberg, where she built a palatial residence, Friedrichshof; still followed the current course of politics, literature, and art, and continued her varied philanthropic labours; frequently revisited England; was attacked by cancer in autumn of 1898; died at Friedrichshof, 5 Aug. 1901; was buried at Potsdam beside her husband; was survived by two (of her four) sons and by her four daughters.

VICTORIA ALEXANDRA ALICE MARY (1897–1965), PRINCESS ROYAL OF GREAT BRITAIN; daughter of future King George V and Queen Mary; known as Princess Mary; educated privately; expert horse woman; interested in needlework; during 1914–18 war took part in nursing with VAD and active in the Girl Guide movement; entered Hospital for Sick Children as VAD probationer, 1918; married Viscount Lascelles (later sixth Earl of Harewood) [q.v.], 1922; created Princess Royal, 1932; shared with her husband, interest in horse-racing and hunting; colonel-in-chief, Royal Scots (The Royal Regiment), and other regiments; commandant-in-chief, British Red Cross Detachments, 1926; chief controller, ATS (later WRAC), 1940; controller commandant, 1941; both her sons on active service; eldest son, George, wounded and taken prisoner in Italy; husband died, 1947; chancellor, Leeds University, 1951; counsellor of state on numerous occasions between 1939 and 1957; undertook number of tours abroad; CI, 1919; GBE, 1927; GCVO, 1937; hon. DCL, Oxford; hon. LLD, Cambridge and other universities; hon. FRCS, 1927; conscientiously devoted to her royal duties; particularly attached to Yorkshire.

VICTORIA ALEXANDRA OLGA MARY (1868–1935), princess of Great Britain and Ireland; second daughter and indispensable helpmeet of King Edward VII and Queen Alexandra; closely united to her brother King George V.

VICTORIA EUGÉNIE JULIA ENA (1887–1969), queen consort of King Alfonso XIII of Spain; daughter of Prince and Princess Henry of Battenberg [qq.v.], and granddaughter of Queen Victoria; became a Roman Catholic and married King Alfonso XIII of Spain, 1906; on wedding day both escaped unharmed from bomb attack; encouraged provision of improved hospitals and assisted reorganization of Spanish Red Cross; also encouraged education; had four sons and two daughters; two sons afflicted with haemophilia, third, a deaf-mute; went into exile

in Italy, 1931; separated from her husband, but was with him when he died, 1941; lived in Switzerland, and made frequent visits to England, 1945–60; Dame Grand Cross of Sovereign Order of Malta.

VILLIERS, GEORGE HERBERT HYDE, sixth EARL OF CLARENDON (1877–1955), public servant; educated at Eton; lamed for life in accident; deputy lieutenant and J.P., Hertfordshire, 1909; farmed in Canada, 1911–14; succeeded to earldom and returned to England, 1914; lord-in-waiting to George V, 1921; chief conservative whip in Lords, 1922–5; parliamentary under-secretary, dominion affairs, 1925–7; chairman, British Broadcasting Corporation, 1927–30; governor-general, South Africa, and PC, 1931–7; KG, 1937; lord chamberlain, 1938–52; GCMG, 1930; GCVO, 1939; chancellor, Order of St. Michael and St. George, 1942–5.

VILLIERS, JOHN HENRY DE, first BARON DE VILLIERS (1842–1914), South African judge. [See DE VILLIERS.]

VILLIERS, MARGARENT ELIZABETH CHILD-, COUNTESS OF JERSEY (1849–1945); daughter of second Baron Leigh; married (1872) seventh Earl of Jersey [q.v.]; entertained widely at Middleton Park, Bicester, and Osterley Park, Isleworth; a founder and president (1901–27), Victoria League; actively interested in children's welfare; DBE, 1927.

VILLIERS, VICTOR ALBERT GEORGE CHILD-, seventh EARL OF JERSEY and tenth VISCOUNT GRANDISON (1845–1915), colonial governor; educated at Eton and Balliol College, Oxford; succeeded father, 1859; governor and commander-in-chief, New South Wales, 1891–3; dealt successfully with labour problem and financial crisis.

VINCENT, SIR (CHARLES EDWARD) HOWARD (1849–1908), politician; joined Royal Welsh Fusiliers, 1868; went to Russia to study military organization; published *Elementary Military Geography* (1872); left army, 1873; called to bar, 1876; represented *Daily Telegraph* in Russo-Turkish war of 1876; lieutenant-colonel of Central London Rangers, 1875–8; examined continental police systems, 1877–8; first director of criminal investigation at Scotland Yard, 1878–84; published *A Police Code and Manual of Criminal Law* (1882); conservative MP for Central Sheffield, 1885–1908; member of London County Council, 1889–96; to his persistence were mainly due Acts concerning alien immigration, 1905, and appointment of a public trustee, 1906; advocated colonial preference from 1889; CB, 1880; knighted, 1896; KCMG, 1898.

VINCENT, Sir EDGAR, sixteenth baronet, Viscount D'Abernon (1857-1941), financier and diplomatist; brother of Sir C. E. H. Vincent [q.v.]; educated at Eton; commissioned in Coldstream Guards, 1877-82; financial adviser to Egyptian government, 1883-9; KCMG, 1887; governor, Imperial Ottoman Bank, 1889-97; conservative MP, Exeter, 1899-1906; chairman, royal commission on imperial trade, 1912; baron, 1914; chairman, Central Control Board (Liquor Traffic), 1915-20; GCMG, 1917; PC, 1920; ambassador, Berlin, 1920-6; close friend of Stresemann; instigated Anglo-German commercial treaty (1924) and pact of mutual guarantee embodied in treaty of Locarno (1925); viscount and GCB, 1926; subsequent public services included economic mission to South America (1929), and work for art galleries, museums, medical research, industrial psychology, etc.; FRS, 1934; succeeded brother in baronetcy, 1936.

VINCENT, JAMES EDMUND (1857-1909), journalist and author; educated at Winchester and Christ Church, Oxford; BA, 1880; called to bar, 1884; chancellor of diocese of Bangor, 1890; edited *National Observer*, 1894-7, and *Country Life*, 1897-1901; works include lives of Duke of Clarence (1893), Edward VII (1902 and 1910), *Highways and Byways in Berkshire* (1906).

VINES, SYDNEY HOWARD (1849-1934), botanist; B.Sc., London, 1873; first in natural sciences tripos, Christ's College, Cambridge, 1875; fellow and lecturer, 1876; by his enthusiasm laid foundations of new school of botany; university reader, 1883; Sherardian professor, Oxford, 1888-1919; edited *Annals of Botany*, 1887-99; FRS, 1885.

VINOGRADOFF, Sir PAUL GAVRILO-VITCH (1854-1925), jurist and historian; born at Kostroma, Russia; graduated in history and philosophy, Moscow University; studied in Berlin under Mommsen and Brunner; came to England in order to research into feudal land system, 1883; formed friendship with F. W. Maitland [q.v.]; published *Villainage in England* (Russian, 1887, English, 1892); extraordinary professor of history, Moscow, 1884-7; full professor, 1887; as a liberal but not revolutionary, worked for progress of general education in Russia; resigned, 1901; settled in England; Corpus Christi professor of jurisprudence, Oxford, 1903-25; introduced seminar teaching; valuable series of 'Oxford Studies in Social and Legal History' produced under his aegis; himself published, among other

works, *The Growth of the Manor* (1905), *English Society in the Eleventh Century* (1908), *Roman Law in Mediaeval Europe* (1909), and *Outlines of Historical Jurisprudence* (1920-2); during 1914-18 war worked untiringly for mutual understanding of Russia and England; knighted, 1917; renounced Russian nationality after triumph of Bolshevism, 1918; died in Paris.

VOIGT, FREDERICK AUGUSTUS (1892-1957), journalist; parents born in Germany; educated at Haberdashers' Aske's School and Birkbeck College; first class, German, 1915; served in Royal Garrison Artillery, 1916-18; foreign correspondent, *Manchester Guardian*, in Germany, 1920-33; drew attention to nature of National Socialism; diplomatic correspondent, London, 1933-9; edited *The Nineteenth Century and After*, 1938-46; publications include *Unto Caesar* (1938) and *Pax Britannica* (1949).

VON HÜGEL, FRIEDRICH, Baron of the Holy Roman Empire (1852-1925), theologian; born at Florence; educated as Roman Catholic in Florence, Brussels, and England; made his home in England after 1873; naturalized, 1914; studied natural science, philosophy, and religious history and literature, adopting 'critical' views of Old Testament; brought by biblical studies into contact with Duchesne, Loisy, Mignot, and Semeria; intimate with George Tyrrell [q.v.]; centre of 'modernist' group; founded London Society for Study of Religion, 1905; repelled by 'immanentism' of many modernists as tending to undermine adoration; his works include *Mystical Element of Religion* (1908), *Eternal Life* (1912), *The Reality of God* (posthumous, 1931); his influence on contemporary religious thought due to his profoundly Christian personality, passionate desire for communion with God, and sympathy for all genuine religion.

VOYSEY, CHARLES (1828-1912), theistic preacher; BA, St. Edmund Hall, Oxford; ordained, 1851; vicar of Healaugh, near Tadcaster, 1864; deprived of living for unorthodox preaching and writing, 1871; began movement in London which ultimately developed into 'Theistic Church'; an attractive preacher.

VOYSEY, CHARLES FRANCIS ANNESLEY (1857-1941), architect; son of Charles Voysey [q.v.]; between 1890 and 1914 enjoyed extensive domestic practice; built long, low houses, with roughcast walls, roofs of steep pitch with gables, low leaded windows, and specially designed furniture; FRIBA, 1929.

W

WACE, HENRY (1836-1924), dean of Canterbury; BA, Brasenose College, Oxford; ordained, 1861; curate, St. James's church, Piccadilly, 1863-70; began to write regularly for *The Times*, 1863; maintained his association with the paper for over twenty years, largely as writer of leading articles; chaplain of Lincoln's Inn, 1872-96; professor of ecclesiastical history, King's College, London, 1875; principal, 1883-97; rector of St. Michael's church, Cornhill, 1896; dean of Canterbury, 1903-24; a churchman of wide learning and author of many works; a controversialist of decided views, notably in Convocation; and a stout champion of the Reformation settlement.

WADDELL, HELEN JANE (1889-1965), author and translator; educated at Victoria College and Queen's University, Belfast; BA, first class honours in English, 1911; MA, 1912; published *Lyrics from the Chinese* (1913) and a play *The Spoiled Buddha* (1919); went up to Somerville College, Oxford, 1920; interested in medieval Latin lyric and humanism; lectured for St. Hilda's Hall; taught at Bedford College, London, 1922-3; studied in Paris, 1923-5; lectured at Lady Margaret Hall on 'The Wandering Scholars', 1926; published numerous books, including *The Wandering Scholars* (1927), *Medieval Latin Lyrics* (1929), *Peter Abelard* (1933), and *The Desert Fathers* (1936); edited *A Book of Medieval Latin for Schools* (1931) and *Cole's Paris Journal* (1931); translated *Manon Lescaut* (1931), and a collection of stories, *Beasts and Saints* (1934); assistant editor (under F. A. Voigt, q.v.) *The Nineteenth Century*; last book *Poetry in the Dark Ages* (1948); hon. degrees from number of universities.

WADDELL, LAWRENCE AUGUSTINE (later AUSTINE) (1854-1938), medical officer in Indian government service, traveller, and orientalist; MB, M.Ch., Glasgow, 1878; entered Indian medical service, 1880; medical officer, Darjeeling district, 1888-95; professor of chemistry and pathology, Calcutta Medical College, 1896-1902; studied Buddhist sites and doctrine; chief medical officer, Tibetan expedition, 1904; professor of Tibetan, University College, London, 1906-8.

WADE, SIR WILLOUGHBY FRANCIS (1827-1906), physician; educated at Rugby and Trinity College, Dublin; BA, 1849; MB, 1851; hon. MD, 1896; MRCS England, 1851; FRCP, 1871; physician to Birmingham general hospital, 1865-92; knighted, 1896; first called attention to presence of albuminuria in diphtheria; wrote on diphtheria and gout.

WADSWORTH, ALFRED POWELL (1891-1956), journalist and economic historian; educated at Rochdale central school; at fourteen, employed by *Rochdale Observer* under W. W. Hadley [q.v.]; joined *Manchester Guardian*, 1917; labour correspondent under C. P. Scott [q.v.]; assistant editor, 1940; editor, 1944-56; made *Guardian* a national and international influence; published *The Cotton Trade and Industrial Lancashire 1600-1790* (1931, with Julia Mann) and *The Strutts and the Arkwrights 1758-1830* (1958, with R. S. Fitton); hon. MA (1931), and LLD (1955), Manchester; visiting fellow, Nuffield College, Oxford.

WADSWORTH, EDWARD ALEXANDER (1889-1949), painter; educated at Fettes; studied engineering and art in Munich; at Bradford and Slade art schools; worked in tempera in early Italian tradition; combined engineering skill and creative talent in highly individual style using symbolism of the sea; ARA, 1943.

WAGER, LAWRENCE RICKARD (1904-1965), geologist, explorer, and mountaineer; educated at Leeds grammar school and Pembroke College, Cambridge; first class, parts i and ii, natural sciences tripos (geology), 1925-6; research, 1926-9; lecturer, Reading University, 1929; carried out extensive exploration in East Greenland, 1929-39; studied the Skaergaard layered igneous intrusion; took part in 1933 Everest expedition; with (Sir) P. Wyn Harris, reached 28,000 feet; squadron leader, RAF, during 1939-45 war; succeeded Arthur Holmes [q.v.] as professor of geology, Durham, 1944; FRS, 1946; professor of geology, Oxford, 1950-61; fellow, University College; work recognized by numerous awards.

WAGGETT, PHILIP NAPIER (1862-1939), divine and preacher; educated at Charterhouse and Christ Church, Oxford; deacon, 1885; priest, 1886; head of Charterhouse mission, Southwark, 1889-92; joined Society of St. John the Evangelist, Cowley, 1892; became famous as preacher and missioner; head of St. Anselm's House, Cambridge, 1911; army chaplain, 1914-18; vicar of St. Mary's the Great, Cambridge, 1927-30.

WAIN, LOUIS WILLIAM (1860-1939), artist; his drawings of cats as human beings in comic situations enjoyed great popularity until the 1914-18 war; sank into poverty and became insane.

WAKE-WALKER, SIR WILLIAM FREDERIC (1888-1945), admiral; entered navy, 1903; qualified in torpedo, 1912; director, torpedoes and mining, Admiralty, 1935-8; rear-admiral, 1939; successfully co-ordinated technical measures dealing with German magnetic mine, Nov. 1939-May 1940; in control of vessels off Dunkirk, May-June 1940; commanded first cruiser squadron, 1941-2; sighted and took part in final attack on *Bismarck*, May 1941; vice-admiral, 1942; third sea lord and controller of navy, 1942-5; responsible for landing-craft development; KCB, 1943; admiral, 1945.

WAKEFIELD, CHARLES CHEERS, VISCOUNT WAKEFIELD (1859-1941), man of business and philanthropist; educated at Liverpool Institute; founded in City of London C. C. Wakefield & Co., dealing in lubricating oils and appliances, 1899; used trade name 'Castrol'; generous benefactor in City; sheriff, 1907-8; knighted, 1908; lord mayor, 1915-16; baronet, 1917; supported Imperial Cadet movement; gave £25,000 to Imperial Institute, 1932; financed flights by Sir Alan Cobham and Amy Johnson [q.v.] and Sir Henry Segrave's motor speed trials; owned *Miss England* speed-boats; chairman, RAF Benevolent Fund; benefactor of hospitals, of British Academy, British Museum, etc., baron, 1930; viscount, 1934; GCVO, 1936.

WAKLEY, THOMAS (1851-1909), only son of Thomas Henry Wakley [q.v.]; LRCP, 1883; succeeded father as editor of *Lancet*.

WAKLEY, THOMAS HENRY (1821-1907), surgeon and journalist; eldest son of Thomas Wakley (1795-1862, q.v.); studied medicine at University College, London, and Paris; MRCS, 1845; assistant surgeon to Royal Free Hospital, 1848; FRCS, 1849; part proprietor of *Lancet*, 1857; co-editor with son Thomas, 1886; founded *Lancet* relief fund for medical practitioners, 1889.

WALCOT, WILLIAM (1874-1943), architect and graphic artist; born near Odessa; studied architecture in St. Petersburg and Paris; settled in London and turned to architectural watercolours and etchings, notably of buildings of antiquity; FRIBA, 1922.

WALEY, ARTHUR DAVID (1889-1966), orientalist; grandson of Jacob Waley [q.v.] and brother of Sir Sigismund David Waley [q.v.]; educated at Rugby and King's College, Cambridge; first class, part i, classical tripos, 1910;

joined sub-department of oriental prints and drawings, British Museum, 1912; taught himself Chinese and Japanese; translated Chinese poems and published in *New Statesman*, the *Bulletin* of the School of Oriental Studies, and the *Little Review*; published *A Hundred and Seventy Chinese Poems* (1918), followed by *An Introduction to the Study of Chinese Painting* (1923), and his translation of the *Genji Monogatari*, by Murasaki Shikibu (6 vols., 1925-33); other Japanese translations were selections from the *Uta* (1919) and Nō plays (1921); retired from British Museum as assistant keeper, 1929, but continued his work, and published *The Opium War through Chinese Eyes* (1958); hon. fellow of King's College, 1945; FBA, 1945; awarded Queen's medal for poetry, 1953; CBE, 1952; CH, 1956; hon. doctorates, Aberdeen and Oxford; anthology of his writings edited by Ivan Morris and published as *Madly Singing in the Mountains* (1970).

WALEY, SIR (SIGISMUND) DAVID (1887-1962), public servant; grandson of Jacob Waley [q.v.], and brother of Arthur David Waley [q.v.]; educated at Rugby and Balliol College, Oxford; first class, hon. mods, and *lit. hum.*, 1910; passed top of examination list into home civil Service; posted to the Treasury, 1910; private secretary to Edwin Montagu [q.v.]; served in the army, 1916-18; MC, 1918; returned to Treasury to work on international finance with (Sir) Frederick Leith-Ross [q.v.], 1919; present at Paris peace conference, 1919, and Lausanne conference, 1922-3; assistant secretary, 1924; principal assistant secretary, 1931; worked on structure of exchange control during 1939-45 war; visited Athens to advise on economic reconstruction of Greece, 1944; dealt with reparations and inter-Allied debts at Potsdam and Moscow; third secretary, 1946-7; assisted in creating post-war international financial structure; after retirement from Treasury, joined boards of City companies, and became treasurer, British Epilepsy Association, and chairman, Furniture Development Council (1949-57), Sadler's Wells Trust, and Mercury Theatre Trust; CB, 1933; KCMG, 1943.

WALKDEN, ALEXANDER GEORGE, BARON WALKDEN (1873-1951), railway trade unionist; educated at Merchant Taylors' School; employed as clerk by Great Northern Railway, 1889; helped to organize Railway Clerks' Association, 1897; full-time union secretary, 1906-36; member, General Council, TUC, 1921-36; president, 1932-3; labour MP, Bristol, 1929-31 and 1935-45; baron, 1945.

WALKER, SIR BYRON EDMUND (1848-1924), Canadian banker; born in Canada West;

entered Canadian Bank of Commerce, Hamilton, 1868; held variety of posts in bank; general manager, 1886; director, 1906; president, 1907; Canadian banking system owes much to his skill; an authority on Canadian finance; greatest tribute to his financial ability lies in stability of Canadian finance, 1914-18; gave much time and thought to Toronto University, of which he was chancellor, 1923; promoted cause of the arts; knighted, 1910; died at Toronto.

WALKER, SIR EMERY (1851-1933), process-engraver and typographical expert; worked for Typographic Etching Co., the pioneer firm of process-engravers, 1873-83; founded own firm, 1886; a friend of William Morris [q.v.] with whom he co-operated in work of Kelmscott Press and in founding Arts and Crafts Exhibition Society, 1888; with T. J. Cobden-Sanderson [q.v.] founded Doves Press, 1900; most influential in improving book production; Sandars reader, Cambridge, 1924-5; knighted, 1930.

WALKER, ERNEST (1870-1949), musician; second class, *lit. hum.*, Balliol College, Oxford, 1891; B.Mus., 1893; D.Mus., 1898; assistant organist, Balliol, 1891-1901; organist, 1901-13; director of music, 1901-25; active in university musical affairs; his vocal and instrumental compositions scholarly, restrained, and sensitive; wrote *A History of Music in England* (1907).

WALKER, DAME ETHEL (1861-1951), painter and sculptress; studied at Putney, Westminster and the Slade Art Schools; influenced by Frederick Brown, W. R. Sickert [qq.v.], and the New English Art Club; painted mainly figures in interiors and portraits; also visionary decorations inspired by theosophical speculation; ARA, 1940; CBE, 1938; DBE, 1943; work well represented in the Tate Gallery.

WALKER, FREDERIC JOHN (1896-1944), captain, Royal Navy; entered navy, 1909; anti-submarine specialist; senior officer, thirty-sixth escort group, 1941-2; destroyed five submarines; captain, 1942; senior officer, second escort group, 1943-4; destroyed sixteen submarines; DSO, three bars.

WALKER, FREDERICK WILLIAM (1830-1910), schoolmaster; educated at Rugby and Corpus Christi College, Oxford; first class, *lit. hum.*, 1853; Boden (Sanskrit), Vinerian and Tancred law scholar, 1854; MA, 1856; fellow, 1859; hon. fellow, 1894; called to bar, 1858; high master of Manchester grammar school, 1859-76; reorganized school and finances and constitution; high master (1876-1905) of St. Paul's School, which was removed from City of

London to Hammersmith (1884); greatly increased the school's reputation for scholarship; gained victory in long struggle—from 1890 to 1899—with charity commissioners regarding the school's constitution; sat on commission for education of officers in the army, 1900; hon. Litt.D., Manchester, 1899; raised standard of public school education throughout country; friend of the positivist leaders and of Benjamin Jowett [q.v.].

WALKER, SIR FREDERICK WILLIAM EDWARD FORESTIER FORESTIER- (1844-1910), general. [See FORESTIER-WALKER.]

WALKER, SIR GILBERT THOMAS (1868-1958), applied mathematician and meteorologist; educated at St. Paul's School, London, and Trinity College, Cambridge; senior wrangler, 1889; first class, part ii, mathematics tripos, 1890; fellow, 1891; director-general, observatories, India, 1904-24; tried unsuccessfully to solve problem of variability of monsoon rainfall; professor of meteorology, Imperial College of Science and Technology, 1924-34; publications include *The Theory of Electromagnetism* (1910); FRS, 1904; CSI, 1911; knighted, 1924.

WALKER, SIR JAMES (1863-1935), chemist; educated at Dundee high school and Edinburgh University; B.Sc., 1885; Ph.D., Leipzig, 1889; professor of chemistry, University College, Dundee, 1894-1908; Edinburgh, 1908-28; established at Edinburgh fine chemical laboratory and research school; his *Introduction to Physical Chemistry* (1899) influential in furthering its cause; FRS, 1900; knighted, 1921.

WALKER, JOHN (1900-1964), numismatist; educated at John Street and Whitehill schools and university of Glasgow; first class, Hebrew and Arabic, 1922; teacher, St. Andrew's Boys' School, Alexandria, 1927; assistant lecturer in Arabic, Glasgow, 1928; worked for Egyptian Ministry of Education, 1928-30; assistant keeper in department of coins and medals, British Museum, 1931; produced catalogue of Arab-Sassanian coins, 1939-41; D.Litt., Glasgow; served in Air Staff Intelligence during 1939-45 war; secretary, Royal Numismatic Society, 1948; deputy keeper, 1949; keeper, 1952-64; editor, *Numismatic Chronicle*; lecturer at School of Oriental and African Studies, 1937-47; produced catalogue of Arab-Byzantine coins, 1956; produced many original and scholarly numismatic publications, including a contribution to history of Cyprus in fourth century BC by his reading of a new Phoenician inscription on a coin found in the island; his most important work, the *Catalogue of Muham-*

madan Coins in the British Museum (2 vols.); awarded medals and other honours; FBA, 1958; CBE, 1963.

WALKER, SIR MARK (1827-1902), general; brother of Sir Samuel Walker [q.v.]; joined army, 1846; served in Crimea; present at Balaclava and Inkerman where his gallantry won for him the VC; went to India, 1859; served through the China campaign, 1859-61; present at capture of Chusan, the assault on Taku forts, and surrender of Peking; was quartered in England, 1862-70; served in India, 1871-9; CB, 1875; major-general, 1878; commanded 1st brigade at Aldershot, 1883; commanded infantry at Gibraltar, 1884-8; lieutenant-general, 1888; general and KCB, 1893.

WALKER, SIR NORMAN PURVIS (1862-1942), dermatologist and president, General Medical Council; educated at Edinburgh Academy and University; MB, CM, 1884; studied dermatology in Vienna and Prague; assistant physician, skin department, Edinburgh Royal Infirmary, 1892-1906; full physician, 1906-24; consulting physician, 1925-42; direct representative for Scotland, General Medical Council, 1906-41; president, 1931-9; knighted, 1923.

WALKER, SIR SAMUEL, first baronet (1832-1911), lord chancellor of Ireland; BA, Trinity College, Dublin (first senior classical moderator), 1854; called to Irish bar, 1855; QC, 1872; attained large practice in equity and common law; ably defended C. S. Parnell [q.v.], 1881; bencher of King's Inns, 1881; a liberal in politics; solicitor-general for Ireland under Gladstone, 1883-5; liberal MP for Londonderry county, 1884-5; acting Irish secretary during 1884; attorney-general for Ireland, May-June 1885; Irish PC, 1885; again attorney-general for Ireland, Feb.-Aug. 1886; lord chancellor of Ireland and president of Irish court of appeal, Aug. 1892-July 1895, and Dec. 1905-11; chairman of commission on Irish fisheries, 1897; baronet, 1906; enthusiastic shot and angler.

WALKER, VYELL EDWARD (1837-1906), cricketer; a member of distinguished cricketing family of Southgate, Middlesex; educated at Harrow; first played for Gentlemen *v.* Players (1856), subsequently captaining the team ten times; helped to found Middlesex County Cricket Club, 1864; secretary, 1864-70; captain, 1864-72; president, 1898; president of MCC, 1891; orthodox batsman, powerful hitter, and deceptive slow 'lob' bowler; on retiring from cricket about 1874, he took part in family

brewing firm; succeeded to the family estate at Southgate, 1889; presented land for public recreation ground there, 1890.

WALKER, SIR WILLIAM FREDERIC WAKE- (1888-1945), admiral. [See WAKE-WALKER.]

WALKLEY, ARTHUR BINGHAM (1855-1926), dramatic and literary critic; BA, Corpus Christi College, Oxford; clerk in secretary's office, General Post Office, 1877-1911; assistant secretary in charge of telegraph branch, 1911-19; dramatic critic to *Star*, 1888-1900; to *Speaker*, 1890-9; to *The Times*, 1900-26; professed himself an 'impressionist'; gave warm welcome to Ibsen's plays; a master of the short miscellaneous essay; his works include *Dramatic Criticism* (1903) and *Drama and Life* (1907).

WALLACE, ALFRED RUSSEL (1823-1913), naturalist; educated at Hertford grammar school; master at Leicester collegiate school, 1844-6; made acquaintance of Henry Walter Bates [q.v.], the naturalist, with whom he went on collecting trip to the Amazon, 1848; returned to England, 1852; visited Malay Archipelago, 1854-62; while in Moluccas (1858) independently discovered principle of natural selection as key to method of evolution, his most important achievement on theoretical side; his theory of origin of species identical with that of Charles Darwin [q.v.], to whom he sent his views; modern theory of evolution published in joint paper, 1858; spent remainder of his life after return to England (1862) in writing and lecturing; material which he amassed in Malay Archipelago very valuable to biology; published great work, *The Malay Archipelago* (1869); other works include *Contributions to the Theory of Natural Selection* (1870) and *The Geographical Distribution of Animals* (1876); his most solid work done on geographical distribution; FRS, 1893; first Darwin medallist of the Royal Society, 1890; OM, 1910.

WALLACE, SIR CUTHBERT SIDNEY, baronet (1867-1944), surgeon; educated at Haileybury; qualified from St. Thomas's Hospital, 1891; assistant surgeon, 1900-13; full surgeon, 1913-30; dean of medical school, 1907-9, 1918-28; consultant surgeon in France, 1914-18; consultant adviser in surgery, Emergency Medical Service, 1939-45; president, Royal College of Surgeons, 1935-8; KCMG, 1919; baronet, 1937.

WALLACE, SIR DONALD MACKENZIE (1841-1919), newspaper correspondent, editor, and author; studied at various universities at home and abroad; visited Russia, 1870-5; foreign correspondent of *The Times* at Constantinople, 1878-84; private secretary to Lord Dufferin [q.v.], viceroy of India, 1884-8; KCIE, 1887; director of *The Times* foreign department, 1891-9; principal work, *Russia* (1877).

WALLACE, (RICHARD HORATIO) EDGAR (1875-1932), novelist, playwright, and journalist; educated at London board school; enlisted and drafted (1896) to South Africa; war correspondent, 1899-1902; first success with *The Four Just Men* (1905) and West African series including *Sanders of the River* (1911); produced over 170 novels of which the best known were thrillers; wrote with simplicity, vigour, and pace, with great variety and originality of plot; his successful plays include *The Ringer* (1926) and *On the Spot* (1931).

WALLACE, THOMAS (1891-1965), professor of horticultural chemistry; educated at Rutherford College, Newcastle upon Tyne, and Armstrong College, Durham University; B.Sc., with distinction in chemistry, 1913; served in France and Gallipoli during 1914-18 war; MC; joined staff of Long Ashton Agricultural and Horticultural Research Station, Bristol University, 1919; agricultural research chemist and advisory officer in agricultural chemistry, 1919-23; deputy director, 1924-43; proved that leaf scorch in plums and apples was curable by application of potash; with his team, published *The Diagnosis of Mineral Deficiencies in Plants by Visual Symptoms* (1943); director and professor of horticultural chemistry, Bristol, 1943-57; served on numerous committees of Ministry of Agriculture and other advisory bodies; governor, Royal Agricultural College, Cirencester, 1944; edited *Journal of Horticultural Science*, 1943-58; travelled widely abroad on advisory visits; D.Sc., Durham, 1931; CBE, 1947; FRIC, 1946; FRS, 1953; in 1939-45 war organized Long Ashton Local Defence Volunteers.

WALLACE, WILLIAM ARTHUR JAMES (1842-1902), colonel, Royal Engineers; joined Royal Engineers, 1860; entered railway branch of public works department of India, 1864; executive engineer, 1871; officiating consulting engineer to Indian government at Lucknow, 1877; field engineer to Kuram valley column in Afghan campaign, 1879; engineer-in-chief of northern Bengal railway at Saidpur, 1880; director of railway corps in Egyptian campaign, 1882; present at Tel-el-Kebir; chief engineer for guaranteed railways at Lahore, 1884-92; CIE, 1890.

WALLAS, GRAHAM (1858-1932), political psychologist; educated at Shrewsbury and Corpus Christi College, Oxford; member of Fabian Society, 1886-1904; lecturer, London School of Economics, 1895; professor of political science, 1914-23; attempted to improve mental processes of public men and promote cause of social and political invention; his influential works include *Human Nature in Politics* (1908), *The Great Society* (1914), *Our Social Heritage* (1921), and *The Art of Thought* (1926).

WALLER, CHARLES HENRY (1840-1910), theologian; BA, University College, Oxford, 1863; MA, 1867; DD, 1891; theological tutor at St. John's Hall, Highbury, 1865; professor of biblical exegesis, 1882; principal, 1884-98; works include *A Grammar of the Words in the Greek Testament* (2 parts, 1877-8) and commentaries on Deuteronomy and Joshua (1882).

WALLER, LEWIS (1860-1915), actor-manager, whose real name was WILLIAM WALLER LEWIS; born in Spain; acted with various companies in London and provinces, 1883-95; successively managed, alone or jointly, Haymarket, Lyceum, Imperial, and Lyric theatres; his finest impersonation that of the king in *Henry V* (1900); other brilliant parts, Hotspur in *Henry IV, Part I* (1896), D'Artagnan in *The Three Musketeers* (1898), and title-role of *Monsieur Beaucaire* (1902-3).

WALLER, SAMUEL EDMUND (1850-1903), painter of genre pictures; studied architecture under father at Gloucester; exhibited at Royal Academy, 1871-1902; chief works were 'The Empty Saddle' (1879), 'Sweethearts and Wives' (1882), and 'The Morning of Agincourt' (1888).

WALLS, TOM KIRBY (1883-1949), actor-producer; educated at Northampton county school; from 1912 established himself in west end with portrayals of eccentric old gentlemen; in 1922-31 produced and acted with Mr Ralph Lynn in *Tons of Money, It Pays to Advertise*, and series of plays by Mr Ben Travers including *A Cuckoo in the Nest, Rookery Nook, Plunder*, and *Turkey Time* which were subsequently filmed; maintained racing stable; won Derby, 1932.

WALPOLE, Sir HUGH SEYMOUR (1884-1941), novelist and man of letters; born in Auckland, New Zealand; educated at King's School, Canterbury, Durham School, and Emmanuel College, Cambridge; in 1909-14 published six novels including *Mr. Perrin and Mr. Traill* (1911); served in Russia, 1914-17, witnessing first revolution; used experiences in *The Dark Forest* (1916) and *The Secret City* (1919); established as successful writer and prominent figure in literary and artistic circles; popular lecturer in United States; of his forty-two novels his personal preference was *John Cornelius* (1937); probably best remembered for *The Cathedral* (1922); *Harmer John* (1926); the series *Rogue Herries* (1930), *Judith Paris* (1931), *The Fortress* (1932), and *Vanessa* (1933); and notably the series *Jeremy* (1919), *Jeremy and Hamlet* (1923), and *Jeremy at Crale* (1927); collected books, manuscripts, and paintings; gave generous help to writers; knighted, 1937.

WALPOLE, Sir SPENCER (1839-1907), historian and civil servant; elder son of Spencer Horatio Walpole [q.v.]; educated at Eton; entered War Office as clerk, 1857; private secretary to father at Home Office, 1858-9, and at War Office, 1866-7; inspector of fisheries for England and Wales, 1867; published life of grandfather, Spencer Perceval [q.v.] (1874); published first two volumes of *History of England from 1815 to 1856*, 1878 (vols. iii-vi, 1890); governor of the Isle of Man, 1882-93; secretary to Post Office, 1893-9; KCB, 1898; hon. D.Litt., Oxford, and FBA, 1904; continued his history in *A History of Twenty-five Years* (*1856-80*) (vols. i and ii, 1904, vols. iii and iv (left incomplete), 1908); other works include *Life of Lord John Russell* (2 vols., 1889), and *Essays Political and Biographical* (posthumous, 1908).

WALSH, STEPHEN (1859-1929), trade-union leader and politician; miner in Lancashire, 1872; vice-president, Miners' Federation of Great Britain, 1922; labour MP, Ince division of Lancashire, 1906-29; parliamentary secretary, Ministry of National Service, 1917; to Local Government Board, 1917-18; secretary of state for war, 1924.

WALSH, WILLIAM PAKENHAM (1820-1902), bishop of Ossory, Ferns, and Leighton from 1878 to death; BA, Trinity College, Dublin, 1841; MA, 1853; Donnellan lecturer, 1860; DD, 1873; chaplain of Sandford Church, Ranelagh, Dublin, 1858-73; dean of Cashel, 1873-8; published *Ancient Monuments and Holy Writ* (1878).

WALSHAM, Sir JOHN, second baronet (1830-1905), diplomatist; BA, Trinity College, Cambridge, 1854; MA, 1857; attaché to British legation at Mexico, 1860-6; second secretary at Madrid, 1866-70, The Hague, 1870-3; secretary of legation at Madrid, 1875-8, at Berlin, 1878-83; minister plenipotentiary at Paris, 1883-5; British envoy at Peking, 1885-92, and at Bucharest, 1892-4; KCMG, 1895.

WALSHAM, WILLIAM JOHNSON (1847-1903), surgeon; studied medicine at St. Bartholomew's Hospital, 1867; MB and CM, Aberdeen, 1871; MRCS England, 1871; FRCS, 1875; served as lecturer and surgeon at St. Bartholomew's from 1871; full surgeon, 1897; excelled in surgery of harelip and orthopaedy; chief work was *Surgery: its Theory and Practice* (1887, 8th edn. 1903).

WALTER, Sir EDWARD (1823-1904), founder of the Corps of Commissionaires; third son of John Walter (1776-1847, q.v.); served in army, 1843-53; founded self-supporting Corps of Commissionaires for employment of discharged soldiers and sailors, 1859; the Corps numbered 4,500 men in 1960; knighted, 1885; KCB, 1887.

WALTER, JOHN (1873-1968), a proprietor of *The Times*, founded by his great-great-grandfather; grandson of John Walter the third [q.v.]; educated at Eton and Christ Church, Oxford; served nominally as assistant to C. F. Moberly Bell [q.v.], manager of *The Times*; travelled widely abroad as representative of the paper; publicity attaché, British Embassy, Madrid, 1916-18; after negotiations with Lord Northcliffe [q.v.] regarding partnership (1908), succeeded his father, Arthur Walter, as chairman, 1910; became joint chief proprietor with John Aster (later Lord Astor of Hever), 1922; remained proprietor until *The Times* and *The Sunday Times* merged under Lord Thomson of Fleet, 1967, when his family's long connection with *The Times* ended.

WALTON, ARTHUR (1897-1959), physiologist; educated at Daniel Stewart's College and Edinburgh University; B.Sc. (Agric.), 1923; trained at Animal Breeding Research Department, Edinburgh; joined School of Agriculture and Animal Research Station, Cambridge; Ph.D., 1927; deputy director; contributed to knowledge of sperm physiology leading to artificial insemination; outstanding research in agriculture.

WALTON, FREDERICK PARKER (1858-1948), academic jurist; second class, *lit. hum.*, Lincoln College, Oxford, 1883; LLB, Edinburgh, and admitted advocate, 1886; dean of law faculty, McGill University, 1897-1915; director, Khedivial School of Law, Cairo, 1915-23; pioneer in study of comparative law; publications include *Handbook of Husband and Wife, according to the Law of Scotland* (1893), *Historical Introduction to the Roman Law* (1903), and *The Egyptian Law of Obligations* (1920); collaborated with Sir Maurice Amos [q.v.] in *Introduction to French Law* (1935).

WALTON, SIR JOHN LAWSON (1852–1908), lawyer; called to bar, 1877; QC, 1890; gained reputation by victory in action against William Smoult Playfair [q.v.], 1896; liberal MP for South Leeds from 1892 till death; attorney-general under Campbell-Bannerman, 1905–8; knighted, 1905; introduced trades disputes bill, 1906; offended labour party by defending clause making executives of trade unions responsible for members' breaches of law.

WALTON, SIR JOSEPH (1845–1910), judge; educated at Stonyhurst; BA, London, 1865; called to bar, 1868; practised at Liverpool in commercial and shipping cases; QC, 1892; recorder of Wigan, 1895–1901; chairman of General Council of the Bar, 1899; judge of King's Bench division, 1901–10; knighted, 1901; president of Medico-Legal Society, 1905.

WANKLYN, JAMES ALFRED (1834–1906), analytical chemist; MRCS, 1856; studied chemistry at Heidelberg under Bunsen; demonstrator of chemistry at Edinburgh, 1859–63; professor at London Institution, 1863–70; lecturer at St. George's Hospital, 1877–80; public analyst in various towns; made many valuable chemical researches independently and in collaboration; gave much attention to milk analysis and water analysis in London; works include *Milk Analysis* (1873), *Water Analysis* (with E. T Chapman, 1868; 11th edn. 1907); collaborated in *Bread Analysis* (1881), *Air Analysis* (1890), and *Sewage Analysis* (1899).

WANTAGE, BARON (1832–1901), soldier and politician. [See LINDSAY (afterwards LOYD-LINDSAY), ROBERT JAMES.]

WARBURG, EDMUND FREDERIC (1908–1966), botanist; grandson of Sir Edmund W. Byrne [q.v.]; educated at Marlborough and Trinity College, Cambridge; first class, part ii, natural sciences tripos (botany), 1930; introduced to cultivation *Daboecia azorica*; with his father, (Sir) Oscar Emanuel Warburg, wrote account of the genus *Cistus*; Ph.D., 1937; research fellow, Trinity College, 1933; assistant lecturer, Bedford College, London, 1938; served in RAF, 1941–5; university lecturer, Oxford, and curator of Druce [q.v.] Herbarium, 1948; reader in plant taxonomy; fellow of New College, 1964; FLS, 1934; member of council of Botanical Society of the British Isles; assisted with *The Atlas of the British Flora*; editor of *Watsonia*, 1949–60; president, Botanical Society, 1965–6; edited third edition of *A Census Catalogue of British Mosses* (1963); joint author (with A. R. Clapham and T. G. Tutin) of *Flora of the British Isles* (1952, 1962); vice-

president, Berkshire, Buckinghamshire, and Oxfordshire Naturalists' Trust; remembered by the Warburg Reserve and by a moss, *Anoectangium warburgii* which he found in the Outer Hebrides, 1946.

WARBURTON, ADRIAN (1918–1944), photographic reconnaissance pilot; educated at St. Edward's School, Oxford; commissioned in Royal Air Force, 1939; began photographic reconnaissance flights from Malta, Sept. 1940; legendary for brilliant flying, accurate photography, resourcefulness, and aggressiveness; photographed Italian fleet, Taranto, from fifty feet; made low-level reconnaissance of Pantelleria; wing commander, 1942; DSO and bar; lost flying to Italy from England.

WARD, SIR ADOLPHUS WILLIAM (1837–1924), historian; son of John Ward, diplomatist [q.v.]; educated in Germany, 1841–54; BA, Peterhouse, Cambridge; fellow, 1862; professor of history and English language and literature, Owens College, Manchester, 1866; active in founding history school and raising academic status of college; vice-chancellor, Victoria University, 1886–90, 1894–6; principal, 1889–97; master of Peterhouse, 1900–24; vice-chancellor, 1901; editor-in-chief, *Cambridge Modern History* (1901–12); co-editor, *Cambridge History of English Literature* (1907–16); co-editor, *Cambridge History of British Foreign Policy, 1793–1919* (1922–3); FBA, 1902; knighted, 1913; possessed unrivalled knowledge of political evolution of Europe since Middle Ages; his works include *History of English Dramatic Literature to the Death of Queen Anne* (1875), *Great Britain and Hanover* (1899), *The Electress Sophia and the Hanoverian Succession* (1903), *Germany, 1815–1890* (1916–18).

WARD, SIR EDWARD WILLIS DUNCAN, first baronet (1853–1928), soldier and military administrator; entered commissariat branch of Control Department, 1874; commissioned to newly instituted Army Service Corps, 1888; served in South African war, 1899–1900; responsible for feeding troops and inhabitants during siege of Ladysmith; permanent under-secretary of state for war, 1901–14; chairman of commission which designed Officers' Training Corps; KCB, 1900; baronet, 1914; died in Paris.

WARD, FRANCIS (FRANK) KINGDON- (1885–1958), plant collector, explorer, and author. [See KINGDON-WARD.]

WARD, HARRY LEIGH DOUGLAS (1825–1906), writer on medieval romances; educated at Winchester and University College, Oxford; BA, 1847; assistant in department of manuscripts at British Museum, 1849–93; catalogued Icelandic MSS there; published elaborate *Catalogue of Romances in the British Museum* (3 vols., 1883–1910), a standard textbook.

WARD, HARRY MARSHALL (1854–1906), botanist; studied science at Owens College, Manchester and Christ's College, Cambridge; BA, 1880; fellow, 1883; MA, 1883; Sc.D., 1892; hon. D.Sc., Victoria, 1902; investigated coffee leaf disease in Ceylon, 1880–2; Berkeley research fellow at Owens College, 1882; assistant lecturer, 1883–6; professor of botany at Cooper's Hill, 1885–95, and at Cambridge, 1895–1906; FLS, 1886; FRS, 1888; royal medallist, 1893; helped to found *Annals of Botany*, 1887, contributing many original papers; showed destructive effect of light on bacteria, 1894; publications include *Timber and some of its Diseases* (1889), *Diseases of Plants* (1889), *The Oak* (1892), and *Grasses* (1901).

WARD, HENRY SNOWDEN (1865–1911), photographer and author; joined printing firm of Percy Lund & Co., Bradford, 1885, for which he founded and edited the *Practical Photographer*, 1890; settled in London, 1891, and edited many photographic periodicals; published *Practical Radiography* (1896); with his wife wrote and illustrated topographical works relating to Shakespeare and Dickens; died in New York.

WARD, IDA CAROLINE (1880–1949), phonetician and West African language scholar; B.Litt., Durham, 1902; joined phonetics department, University College, London, 1919; School of Oriental and African Studies, 1932; head of African department, 1937; professor, West African languages, 1944–8; published works on Efik (1933), Ibo (1936), and Yoruba (1952).

WARD, JAMES (1843–1925), philosopher and psychologist; educated at Spring Hill College, Birmingham, with intention of entering Congregationalist ministry, 1863–9; studied in Germany; abandoned his religious beliefs and never entered ministry; BA, Trinity College, Cambridge; first class, moral sciences tripos, 1874; fellow, 1875; began by research in natural science, thence turning to domain of mind; after about 1894 turned from psychology to philosophy; professor of mental philosophy and logic, Cambridge, 1897–1925; FBA, 1902; his works include *Naturalism and Agnosticism* (Gifford lectures delivered at Aberdeen, 1895–8, 1899) and *The Realm of Ends, or Pluralism and Theism* (Gifford lectures delivered at St. Andrews, 1907–10, 1911), in both of which his philosophy is worked out; in the

first he gave final blow to competency of nature and its alliance with agnosticism as promulgated by Herbert Spencer [q.v.] and other contemporary thinkers; in the second he elaborated theory of 'spiritualistic monadism'; *Psychological Principles* (1918), *A Study of Kant* (1922).

WARD, JOHN (1866–1934), politician and soldier; worked as navvy; joined Social Democratic Federation, 1886; founded Navvies' Union, 1889; member, management committee, General Federation of Trade Unions, 1901–29; liberal MP, Stoke-upon-Trent, 1906–29; lieutenant-colonel, 21st Middlesex Regiment, serving in France and Far East, 1914–18; associated with anti-Bolshevik rising, Vladivostok, 1918.

WARD, Sir JOSEPH GEORGE, first baronet (1856–1930), prime minister of New Zealand; born near Melbourne; liberal member, house of representatives, for Awarua, 1887; postmaster-general, 1891, colonial treasurer, 1893; minister of industries and commerce, 1894–6; colonial secretary, postmaster-general, and minister of industries and commerce, 1899; prime minister, 1906–12; tariff reformer; minister of finance and postmaster-general, national government, 1915–19; prime minister, 1928–30; KCMG, 1901; baronet, 1911.

WARD, Sir LANCELOT EDWARD BARRINGTON- (1884–1953), surgeon. [See BARRINGTON-WARD.]

WARD, Sir LESLIE (1851–1922), cartoonist; son of E. M. Ward [q.v.]; contributed regularly to *Vanity Fair* under pseudonym of 'Spy' from 1873; for thirty-six years produced coloured character-portraits of large number of well-known people; knighted, 1922.

WARD, MARY AUGUSTA (1851–1920), better known as MRS HUMPHRY WARD, novelist and social worker; daughter of Thomas Arnold (1823–1900, q.v.); born in Tasmania; brought to England, 1856; married T. H. Ward, 1872; best-known novel, *Robert Elsmere* (1888); others are *David Grieve* (1892), *Marcella* (1894), *Lady Rose's Daughter* (1903), and *The Marriage of William Ashe* (1905); founded (1890) settlement in London which developed into Passmore Edwards Settlement (1897); anti-suffragist; carried on pro-Allied propaganda in America during 1914–18 war.

WARD, ROBERT McGOWAN BARRINGTON- (1891–1948), journalist. [See BARRINGTON-WARD.]

WARD, WILFRID PHILIP (1856-1916), biographer and Catholic apologist; son of W. G. Ward [q.v.]; educated at Ushaw College, Durham, and Gregorian University, Rome; editor of *Dublin Review*, 1906; works include lives of his father (1889-93), Cardinal Wiseman (1897), and Cardinal Newman (1912).

WARD, WILLIAM HUMBLE, second EARL OF DUDLEY (1867-1932), lord-lieutenant of Ireland and governor-general of Australia; educated at Eton; succeeded father, 1885; parliamentary secretary to Board of Trade, 1895-1902; lord-lieutenant of Ireland, 1902-5; governor-general of Australia, 1908-11; commanded Worcestershire Yeomanry, 1914-18.

WARDLAW, WILLIAM (1892-1958), chemist and university teacher; educated at Rutherford College and Armstrong (later King's) College, Durham; B.Sc., 1913; chemist, Ministry of Munitions, 1914; assistant lecturer and demonstrator in chemistry, Birmingham University, 1915; lecturer, 1921; senior lecturer, 1929; professor of physical chemistry, London University, 1937-57; joint-secretary, Scientific Advisory Committee of War Cabinet, 1941-5; CBE, 1949; president, Chemical Society, 1954-6, and Royal Institute of Chemistry, 1957-8.

WARDLE, SIR THOMAS (1831-1909), promotor of the silk industry; on leaving school joined father's silk-dyeing business at Leek, Staffordshire; founded silk and cotton-printing business at Leek on father's death, 1882; commercially utilized Indian wild silk from 1872; investigated for India Office the conditions of sericulture in India, 1885-6; stimulated by his report Bengal silk industry; revived Kashmir silk industry, 1897; president of Silk Association of Great Britain, 1887-1909; knighted, 1897; ardent geologist and palaeontologist; published many works on silk and its manufacture.

WARE, SIR FABIAN ARTHUR GOULSTONE (1869-1949), editor, and originator of Imperial War Graves Commission; educated at London and Paris universities; assistant director (1901), director (1903-5), of education, Transvaal; editor, *Morning Post*, 1905-11; served with Red Cross in France, 1914-16; created graves registration service; director-general, graves registration and inquiries, 1916-19, 1939-44; vice-chairman, Imperial War Graves Commission, 1917-48; made it sole imperial executive organization; obtained co-operation of foreign governments; gave British war cemeteries the familiar beauty of an English garden; KBE, 1920; KCVO, 1922.

WARING, ANNA LETITIA (1823-1910), hymn writer; published volumes of hymns, which included 'Father, I know that all my life' (written in 1846); influenced James Martineau [q.v.].

WARING, SIR HOLBURT JACOB, first baronet (1866-1953), surgeon; educated at the Owens College, Manchester; qualified in medicine at St. Bartholomew's Hospital, London; MRCS, LRCP, 1890; FRCS, 1891; B.Sc., London, 1888; MB, 1890; BS, 1891; MS, 1893; assistant surgeon, St. Bartholomew's, 1902; surgeon, 1909; consultant surgeon, St. Barts, Metropolitan Hospital, Royal Dental Hospital, and Ministry of Pensions; president, Royal College of Surgeons, 1932-5; dean of faculty of medicine, London University, 1920; vice-chancellor, 1922-4; governor, Imperial College of Science, 1930-47; publications include *A Manual of Operative Surgery* (1898) and *Surgical Treatment of Malignant Disease* (1928); colonel, RAMC, 1914-18; CBE, 1919; knighted, 1925; baronet, 1935.

WARINGTON, ROBERT (1838-1907), agricultural chemist; son of Robert Warington [q.v.]; assistant at Rothamsted laboratory, 1859; at Royal Agricultural College, Cirencester, 1862-7; made researches into effect of ferric oxide and alumina on soluble phosphates and other salts; chemist at J. B. Lawes's manure and citric acid works at Barking and Millwall, 1867-76; was associated with Rothamsted investigations, 1876-91; pursued original researches into nitrification of the soil; Lawes lecturer in America, 1891; Sibthorpian professor of agriculture at Oxford, 1894-7; contributed many articles to scientific publications; chief work was *Chemistry of the Farm* (1881), which reached nineteen editions; FRS, 1886.

WARLOCK, PETER (pseudonym) (1894-1930), musical composer and writer on music. [See HESELTINE, PHILIP ARNOLD.]

WARNE, FREDERICK (1825-1901), publisher; assistant to George Routledge [q.v.], 1839; partner in Routledge's publishing firm, 1851; began independent business, 1865; established New York branch, 1881; inaugurated 'Chandos Classics', 1868; published coloured picture books for children, including 'Aunt Louisa Toy Books', 1871-80; retired, 1895.

WARNEFORD, REGINALD ALEXANDER JOHN (1891-1915), airman; born in India; entered Royal Naval Air Service, 1915; VC for destruction of Zeppelin airship in Flanders, 1915; killed flying near Paris.

WARNER, CHARLES (1846-1909), actor, whose real name was CHARLES JOHN LICKFOLD; made first appearance on stage at command performance of *Richelieu* at Windsor Castle, Jan. 1861; first appeared in London, 1864; supported S. Phelps [q.v.], at Drury Lane in minor Shakespearian and other parts, 1866-8; made first pronounced success as Charley Burridge in H. J. Byron's *Daisy Farm* (1871); successful roles in London (1873-8) included Charles Middlewick in Byron's *Our Boys*, and Young Mirabel in Farquhar's *The Inconstant*; made reputation in melodrama as Tom Robinson in *It's Never too Late to Mend* and as Coupeau in *Drink* (1878-9); went on prosperous tours in Australia, 1887-90, and in America, 1904 and 1907; committed suicide in New York, 1909; highly strung actor in melodrama; sympathetic interpreter of old comedy.

WARNER, SIR GEORGE FREDERIC (1845-1936), palaeographer and scholar; educated at Christ's Hospital and Pembroke College, Cambridge; entered department of manuscripts, British Museum, 1871; assistant keeper, 1888; keeper and Egerton librarian, 1904-11; especially interested in palaeography and illuminated manuscripts; edited the travels of Sir John Mandeville (1889) and *Libelle of Englyshe Polycye* (1926); FBA, 1906; knighted, 1911.

WARNER, SIR PELHAM FRANCIS (1873-1963), cricketer and cricket writer; born in Trinidad; educated at Harrison College, Barbados, Rugby, and Oriel College, Oxford; cricket blue, 1895; played for Middlesex against W. G. Grace [q.v.] and Gloucestershire, 1894; called to bar (Inner Temple), 1900; made first tour abroad to West Indies with Lord Hawke [q.v.], 1897; selected, not without controversy, to captain England test team in Australia, 1903-4; recovered the Ashes; selected again as captain in 1911-12; owing to illness played in only one game; England won the series; served in Inns of Court Officers' Training Corps during 1914-18 war; captained Middlesex, 1908-20; Middlesex won county championship, 1920; excelled as captain rather than as batsman, but, although handicapped by poor health, made nearly 30,000 runs, including 60 centuries; began writing for the *Sportsman*, 1897; invited by J. A. Spender [q.v.] to write for *Westminster Gazette*, 1903; cricket correspondent, *Morning Post*, 1921-33; editor, the *Cricketer*, 1921; publications include *My Cricketing Life* (1921), *Cricket Between Two Wars* (1942), *Lords, 1787-1945* (1946), and *Long Innings* (1951); chairman, Test Match Selection Committee, 1926, 1931-2, and 1935-8; joint manager, touring side to Australia, 1932-3,

involved in 'body line' controversy; president, MCC, 1950; MBE, 1919; knighted, 1937; stand at Lords named after him, 1958.

WARNER ALLEN, (HERBERT) (1881-1968), journalist and author [See ALLEN.]

WARR, CHARLES LAING (1892-1969), dean of the Chapel Royal in Scotland; educated at Glasgow Academy and the universities of Edinburgh and Glasgow; MA, Edinburgh, 1914; commissioned into 9th Argyll and Sutherland Highlanders, 1914; dangerously wounded at Ypres, 1915; decided to enter the church; assistant minister, Glasgow Cathedral, 1917-18; ordained; minister, St. Paul's, Greenock, 1918-26; minister, St. Giles' Cathedral, Edinburgh, 1926-62; dean of Chapel Royal, 1926-69; dean of Order of the Thistle, 1926-69; chaplain to King George V, 1936; chaplain also to Edward VIII, George VI, and Queen Elizabeth II; chaplain of number of other organizations in Scotland; director of hospitals and Convenor of church committees; trustee, National Library of Scotland and Iona Cathedral; publications include *The Unseen Host* (1916), *Echoes of Flanders* (1916), *Scottish Sermons and Addresses* (1930), the *Presbyterian Tradition* (1933), and *The Glimmering Landscape* (autobiography, 1960); CVO, 1937; KCVO, 1950; GCVO, 1967; many hon. degrees.

WARRE, EDMOND (1837-1920), headmaster of Eton; educated at Eton and Balliol College, Oxford; fellow of All Souls, 1859-61; assistant master at Eton, 1860; opened school boarding house, to which he quickly attracted distinguished pupils, 1861; headmaster of Eton, 1884-1905; increased reputation and numbers of school by raising moral tone and intellectual standard; opened new buildings, including drill hall, lower chapel, Queen's schools and, Warre schools; invented school office; started Eton mission at Hackney Wick; provost of Eton, 1909-18; decline of latter years, due to ill health, alone prevented him from being regarded as greatest of Eton headmasters.

WARRE-CORNISH, FRANCIS WARRE (1839-1916), teacher, author, and bibliophile; educated at Eton and King's College, Cambridge; assistant master at Eton, 1861; vice-provost and librarian, 1893-1916.

WARREN, SIR CHARLES (1840-1927), general and archaeologist; son of Major-General Sir Charles Warren [q.v.]; joined Royal Engineers, 1857; for Palestine Exploration Fund made reconnaissance of Philistia, Jordan valley, and Gilead, and excavated extensively in

Jerusalem, 1867-70; employed in Griqualand West, 1876-7, 1879-80; commanded Diamond Fields Horse in Kaffir war, 1877-8; chief surveying instructor, Chatham, 1880-4; special officer attached to Admiralty, Egyptian campaign, 1882; successfully commanded Bechuanaland expedition, 1884-5; chief commissioner, London Metropolitan Police, 1886-8; responsible for police arrangements for Queen Victoria's jubilee, 1887; commanded at Singapore, 1889-94; commanded Thames district, 1895; lieutenant-general, 1897; commanded 5th division, South African war, 1899-1900; general, 1904; colonel commandant, Royal Engineers, 1905; KCMG, 1882; FRS, 1884; works include *The Recovery of Jerusalem* (in collaboration, 1871), *Underground Jerusalem* (1874), and *The Temple or the Tomb* (1880).

WARREN, SIR THOMAS HERBERT (1853-1930), president of Magdalen College, Oxford, scholar, and man of letters; BA, Balliol College, Oxford; first class, classical mods. (1873) and *lit. hum.* (1876); Hertford scholar (1873) and Craven scholar (1878); prize fellow of Magdalen, 1877; president, 1885-1928; gathered round him distinguished staff of teachers; increased number of undergraduates; raised standard of college both intellectually and in other ways; did much to promote studies of natural science and modern languages in Oxford; vice-chancellor, 1906-10; professor of poetry, 1911-16; KCVO, 1914; his works include two volumes of verse: *By Severn Sea* (1897) and *The Death of Virgil* (1907).

WARRENDER, SIR GEORGE JOHN SCOTT, seventh baronet (1860-1917), admiral; entered navy, 1873; captain, 1899; largely responsible for naval operations during Boxer rising, 1900; succeeded father, 1901; reached flag rank, 1908; commanded second battle squadron of Grand Fleet, 1914; commanded force dispatched to intercept German raiders on East coast (Dec.); commander-in-chief, Plymouth, 1915-16.

WARRINGTON, PERCY EWART (1889-1961), founder of public schools; educated at Stapenhill School and Hatfield College, Durham; ordained deacon, 1914; priest, 1915; served at St. Matthew's Church, Rugby, and later at St. Peter's, Congleton; accepted benefice of Monkton Combe, 1918-61; secretary, Church Trust Society; Wrekin College acquired, 1921; Stowe and Canford founded, 1923; Westonbirt, 1928; Felixstowe, for girls, 1929; his vicarage, administrative centre of thirteen public schools; also assisted in funding a high school in Kenya, St. Peter's Hall (now

College), Oxford, and Clifton Theological College; a visionary who made enemies; forced to resign, 1933; continued to be vicar of Monkton Combe; after 1939-45 war, involved in work for the aged; founded Claremont House at Corsham, and Waterhouse at Monkton Combe.

WARRINGTON, THOMAS ROLLS, BARON WARRINGTON OF CLYFFE (1851-1937), judge; educated at Rugby and Trinity College, Cambridge; called to bar (Lincoln's Inn), 1875; QC, 1895; bencher, 1897; judge of the Chancery division, 1904-15; lord justice of appeal, 1915-26; PC, 1915; baron, 1926.

WARWICK, COUNTESS OF (1861-1938). [See GREVILLE, FRANCES EVELYN.]

WATERHOUSE, ALFRED (1830-1905), architect; after architectural study in Manchester, France, Italy, and Germany, he practised in Manchester, 1853-65; won competition for Manchester assize courts (1859) and for town hall (1877), which shows individual type of Gothic; also designed Owens College, Salford jail at Manchester, University College and Royal Infirmary at Liverpool, and Yorkshire College at Leeds, 1878, the Natural History Museum at South Kensington, 1868-80 (which exhibits unwonted exuberance of detail), the Prudential Assurance office, Holborn, and St. Paul's School, Hammersmith; employed terracotta; also designed National Liberal Club, 1884; did ecclesiastical restoration and decoration; at Cambridge he made additions to Caius, Jesus, and Pembroke colleges, and designed the Union building, 1866; at Oxford he was responsible for Balliol College hall and the new Union debating hall; did much domestic reconstruction, including Yattendon Court, his own residence, 1877; took his son Paul [q.v.] into partnership, 1891; in great demand as assessor in competitions; exhibited watercolours at Royal Academy; president of Royal Institute of British Architects, 1888-91; ARA, 1878; RA, 1885; treasurer, 1898; hon. LLD, Manchester, 1895; received many foreign diplomas.

WATERHOUSE, PAUL (1861-1924), architect; son of Alfred Waterhouse [q.v.]; B.A., Balliol College, Oxford; partner in father's office, 1891; carried on practice alone, 1905-19; with son, 1919-24; completed many of father's works; his own include chemical laboratories, Oxford University (1913); three university buildings, St. Andrews; numerous hospital buildings, offices, and banks, some private houses, and a few churches; FRIBA, 1895; president, 1921-2; authority on town planning.

WATERLOW, Sir ERNEST ALBERT (1850-1919), painter; ARA, 1890; RA, 1903; president of Royal Society of Painters in Water Colours, 1897-1914; knighted, 1902; painted chiefly landscapes.

WATERLOW, Sir SYDNEY HEDLEY, first baronet (1822-1906), lord mayor of London and philanthropist; was apprenticed to uncle, Thomas Harrison, the government printer, 1836-43; joined brothers in adding printing branch to father's stationery business, 1844; supplied printing and stationery to railways; his firm became limited company, 1876, with Waterlow as managing director, 1877-95; alderman of the City of London, 1863-83; sheriff, 1866-7; lord mayor of London, 1872; knighted, 1867; master of Stationers' Company, 1872-3; baronet, 1873; originated Industrial Dwellings Company, Limited, 1863; liberal MP for Dumfriesshire, 1868-9, for Maidstone, 1874-80, and for Gravesend, 1880-5; presented Lauderdale House, Highgate, and twenty-nine acres of land (now known as Waterlow Park) to London County Council, 1889; KCVO, 1902.

WATKIN, Sir EDWARD WILLIAM, first baronet (1819-1901), railway promoter; joined father's cotton firm at Manchester; started Saturday half-holiday movement in Manchester, 1844; helped to raise money for opening three public parks there, 1845; founded *Manchester Examiner*, 1845; secretary to Trent Valley Railway (1845), which was soon transferred to London and North Western Railway Company; general manager of Manchester, Sheffield, and Lincolnshire Railway, 1853-61; chairman of directors from 1864 to 1894; opened independent line, afterwards Great Central Railway, from Sheffield to London, 1899; chairman of South Eastern Railway, 1866-94, and of Metropolitan Railway, 1872-94; advocated channel tunnel railway between Dover and Calais from 1869; liberal MP for Stockport, 1864-8, and for Hythe, 1874-85; liberal unionist MP for Hythe, 1886-95; high sheriff of Cheshire, 1874; knighted, 1868; baronet, 1880.

WATKINS, HENRY GEORGE ('GINO') (1907-1932), Arctic explorer; educated at Lancing and Trinity College, Cambridge; skilled climber and flying enthusiast; led expedition to Edge Island, Spitsbergen (1927), survey in Labrador (1928-9), British Arctic air route expedition (1930-1); drowned in Greenland.

WATSON, ALBERT (1828-1904), principal of Brasenose College, Oxford, and classical scholar; educated at Rugby and Wadham College, Oxford; BA, 1851; MA, 1853; fellow of Brasenose, 1852; tutor, 1854-67; lecturer, 1868-73; principal, 1886-9; edited *Selected Letters of Cicero* (1870).

WATSON, ARTHUR ERNEST (1880-1969), managing editor of the *Daily Telegraph*; son of journalist; educated at Rutherford College, Newcastle upon Tyne, Alleyn's School, Dulwich, and Armstrong College, Newcastle; worked on *Newcastle Daily Leader*, 1898-1902, joined parliamentary staff of *Daily Telegraph*, 1902; sub-editor; served in the army, 1914-18; night editor; news editor; assistant editor, 1923; managing editor, 1924; worked in harmony with Sir William Berry (later Viscount Camrose, q.v.), principal proprietor, 1928; opposed Munich and supported (Sir) Winston Churchill, 1938-40; remained in full control of the paper till retirement in 1950.

WATSON, Sir CHARLES MOORE (1844-1916), soldier and administrator; entered army, 1866; served in Egypt, 1882-6; deputy inspector of fortifications, 1896-1902; colonel, 1906; KCMG, 1905.

WATSON, Sir DAVID MILNE MILNE-, first baronet (1869-1945), man of business. [See MILNE-WATSON.]

WATSON, FOSTER (1860-1929), historian of education; MA, Owens College, Manchester; professor of education (afterwards emeritus professor), University College of Wales, Aberystwyth, 1895-1929; authority on history of education 1500-1660, publishing *The English Grammar Schools to 1660* (1908), and on J. L. Vives [q.v.]; contributor to dictionaries and encyclopaedias.

WATSON, GEORGE LENNOX (1851-1904), naval architect; entered shipbuilding firm at Govan, 1867; started own business in Glasgow as naval architect, 1872; built for Edward, Prince of Wales, the yacht *Britannia*, which won prizes valued at £9,973 between 1893 and 1897; designed British challengers' yachts for contest with American yachts from 1887 to 1901; also designed cargo and passenger steamers and steam yachts; published lectures on yachting, 1881.

WATSON, (GEORGE) NEVILLE (1886-1965), mathematician; educated at St. Paul's School, London, and Trinity College, Cambridge; senior wrangler, 1907; first class, mathematical tripos, 1908; Smith's prizeman, 1909; fellow, Trinity College, 1910; assistant lecturer, University College, London, 1914; explained why wireless waves could be propagated over great distances; collaborated with (Sir) Edmund T. Whittaker [q.v.] on second edition of Whittaker's *A Course of Modern Analysis* (1915); Mason professor of pure mathematics, Birmingham University, 1918–51; wrote treatise on Bessel functions, 1922; wrote twenty-five papers elucidating the work of S. Ramanujan, an Indian mathematician; edited *Proceedings* of the London Mathematical Society; president, 1932–3; FRS, 1919; hon. FRSE; hon. degrees from Edinburgh and Dublin.

WATSON, HENRY WILLIAM (1827–1903), mathematician; BA, Trinity College, Cambridge; second wrangler and Smith's prizeman, 1850; fellow, 1851; tutor, 1851–3; member of 'Apostles'; mathematical master at Harrow, 1857–65; one of founders of Alpine Club, 1857; FRS, 1881; Sc.D., Cambridge, 1884; author of mathematical works, including *Treatise on the Kinetic Theory of Gases* (1876).

WATSON, SIR (JAMES) ANGUS (1874–1961), business man and philanthropist; educated privately; began work as junior clerk; successfully associated in business with W. H. Lever (later Viscount Leverhulme, q.v.); founded Angus Watson & Co. Ltd., fish canners, chiefly known for Skipper sardines; also founded Imperial Canneries, Norway; active congregationalist; chairman, Congregational Union of England and Wales, 1935–6; served as Northern Divisional food officer, Ministry of Food, 1939–45; JP; partner, with (Sir) Evelyn Wrench [q.v.], of the *Spectator*; helped to found publishing firm, Ivor Nicholson & Watson; published *My Life* (1937) and *The Angus Clan, Years 1588–1950* (1955); hon. DCL, Durham; knighted, 1945; generously supported many religious, social and educational causes.

WATSON, JOHN (1850–1907), Presbyterian divine and author, and writer under the pseudonym of IAN MACLAREN; of highland stock; MA, Edinburgh University, 1870; Free Church minister at Logiealmond ('Drumtochty'), Perthshire, 1875–7; in charge of Presbyterian Church of England at Sefton Park, Liverpool, 1880–1905; a powerful and cultured preacher; influenced civic life; helped in creation of Liverpool University; made reputation in England and America by publication, under pseudonym of 'Ian Maclaren', of *Beside the Bonnie Brier Bush* (1894), which was translated into many European languages; other works of fiction were *The Days of Auld Lang Syne* (1895), *Kate Carnegie and those Ministers* (1897), and *His Majesty Baby* (1902); he made lecture tours in America, 1896 and 1899; published Lyman Beecher lectures on preaching

delivered at Yale University as *The Cure of Souls* (1896); hon. DD, 1896; theological works include *The Mind of the Master* (1896); moderator of Presbyterian synod, 1900; president of National Free Church Council, 1907; died on third American lecture tour at Mount Pleasant, Iowa; buried at Liverpool.

WATSON, JOHN CHRISTIAN (1867–1941), first labour prime minister of Australia; born in Chile; educated in New Zealand; compositor on Sydney *Star*; president, Sydney trades and labour council, 1893; labour leader, New South Wales legislature, 1894–1901; MHR, 1901–9; labour party leader, 1901–7; perfected discipline and made it constitutional party; withdrew support of Alfred Deakin [q.v.] over arbitration bill and became prime minister, Apr.–Aug. 1904; defeated over same bill; resigned party leadership through ill health, 1907.

WATSON, SIR (JOHN) WILLIAM (1858–1935), poet; Yorkshireman; attracted notice with *Wordsworth's Grave and Other Poems* (1890); work greatly admired by Gladstone; constantly inspired by current events; deplored neglect of poetry and the weakening of the British Empire; works include *Odes and Other Poems* (1894), *Ode on the Day of the Coronation of King Edward VII* (1902), and *The Muse in Exile* (1913); knighted, 1917.

WATSON, SIR MALCOLM (1873–1955), malariologist; educated at high school and university, Glasgow; qualified, 1895; MD, 1903; served in Malayan medical service, 1901–8; inspired by work of (Sir) Ronald Ross [q.v.], specialized in mosquito control; general and consultant practice with rubber planters and others, 1908–28; at Ross's request, joined Ross Institute of Tropical Hygiene, 1928; director, 1933–42; knighted, hon. LLD, Glasgow, 1924; publications include *The Prevention of Malaria in the Federated Malay States* (1911, 2nd edn. 1921) and *Rural Sanitation in the Tropics* (1915).

WATSON, SIR PATRICK HERON (1832–1907), surgeon; MD, Edinburgh University, 1853; LRCS Edinburgh, 1853; FRCS, 1855; president, 1878 and 1905; served as surgeon in Crimea, 1855–6; lecturer on systematic and clinical surgery at Royal College of Surgeons, Edinburgh; surgeon to Royal Infirmary, Edinburgh, 1863–83; performed ovariotomy and excision of joints before introduction of Listerian methods; hon. LLD, Edinburgh, 1884; hon. FRCS Ireland, 1887; knighted, 1903.

WATSON, ROBERT SPENCE (1837–1911), political, social, and educational reformer;

articled to father as solicitor at Newcastle upon Tyne, 1860; honorary secretary of Newcastle Literary and Philosophical Institution, 1862–93; president, 1900; helped to found Durham College of Science (now Armstrong College) 1871; first president, 1910; hon. DCL, 1906; member of first Newcastle school board, 1871–94; pioneer of university extension in north; member of Alpine Club, 1862; visited Wazan, the sacred city of Morocco, 1879, and wrote an account of visit, 1880; founded Newcastle Liberal Association, 1874; president of National Liberal Federation, 1890–1902; PC, 1907; president of Peace Society; advocate of Indian nationalist movement and of free institutions in Russia; umpire from 1864 in over 100 trade disputes; hon. LLD, St. Andrews, 1881; works include *Life of Joseph Skipsey* (1909).

WATSON, ROBERT WILLIAM SETON- (1879–1951), historian. [See SETON-WATSON.]

WATSON, WILLIAM, BARON THANKERTON (1873–1948), judge; son of Baron Watson [q.v.]; educated at Winchester and Jesus College, Cambridge; third class, law, 1894–5; admitted advocate, 1899; KC, 1914; procurator of Church of Scotland, 1918–22; advocate depute, 1919; unionist MP, South Lanarkshire (1913–18), Carlisle (1924–9); solicitor-general for Scotland, 1922; lord advocate, 1922–4–9; lord of appeal in ordinary and life peer, 1929–48; PC, 1922; hon. bencher, Gray's Inn, 1928.

WATT, GEORGE FIDDES (1873–1960), portrait painter; left school at fourteen, apprenticed to Aberdeen lithographer; studied at Gray's School of Art and Royal Scottish Academy; made reputation as portrait painter of men, including Lord Haldane, Asquith, and Balfour; associate Scottish Academy, 1910; full member, 1924; hon. LLD, Aberdeen, 1955.

WATT, MARGARET ROSE (1868–1948), introduced Women's Institutes into England and Wales; born Robertson in Canada; married (1893) A. T. Watt, MOH, British Columbia; joined Women's Institute movement, 1909; obtained first foundations in England and Wales, 1915; instigated formation of Associated Country Women of the World, 1930.

WATTS, GEORGE FREDERIC (1817–1904), painter and sculptor; born in Queen Street, Bryanston Square, London, son of George Watts, a piano maker and tuner in straitened circumstances; read much as a boy, though he had no regular schooling; early enjoyed the run of the studio of William Behnes [q.v.], sculptor; took some lessons in oil-

painting; began to earn a livelihood at sixteen by portraits in pencil and chalk; entered Royal Academy Schools, 1835; hired studio in Clipstone Street, 1837; exhibited at Academy 'A Wounded Heron' and two portraits of ladies, 1837; came to know Nicholas Wanostrocht or 'Felix' [q.v.], schoolmaster at Blackheath and professional cricketer, for whom he drew and lithographed seven illustrations of cricket with portraits; befriended by Constantine Ionides, 1839, for whom and whose family he painted twenty portraits during his career; won a premium of £300 with a cartoon of Caractacus for the decoration of Westminster Palace, 1842; travelled through France to Italy; was introduced at Florence to Lord and Lady Holland; remained as their guest at Borgo San Frediano, 1843–7; painted portraits of his hosts and friends, and began many vast canvases inspired by Italian history and legend; also practised modelling; returned to London, Apr. 1847; won a second premium of £500 for Westminster Palace cartoon of King Alfred, 1848; painted decorations for Holland House, 1848; rented a studio at 30 Charles Street, Berkeley Square, with a view to monumental emblematic paintings, 1848–53; joined a distinguished social circle, including (Sir) Robert Morier [q.v.], and others, who soon formed the Cosmopolitan Club; projected a series of portraits of the distinguished men of his time, beginning with Lord John Russell; settled with Mr and Mrs Henry Thoby Prinsep at Little Holland House, Kensington, 1850–75; painted 'Triumph of the Red Cross Knight' for a wall in a cramped corridor at Westminster Palace; visited Venice with R. Spencer Stanhope, 1853; painted in fresco 'Justice—a Hemicycle of Lawgivers' for Lincoln's Inn Hall, 1853–9; painted portraits of Thiers, Prince Jerome Buonaparte, and others in Paris, 1855–6; joined Sir Charles Newton [q.v.], who was excavating the site of Halicarnassus, 1857; painted Lord Stratford de Redcliffe at Constantinople, 1857; on returning to England painted Tennyson, 1857, and Gladstone, 1859; executed frescoes for Lord Lansdowne at Bowood; designed figures of St. Matthew and St. John for mosaics in St. Paul's Cathedral; married in 1864 (Dame) Ellen Terry [q.v.], but they parted June 1865, and Watts obtained a divorce in 1877; worked on symbolic pictures 'Love and Death' and 'The Court of Death', on classical subjects, including 'Orpheus and Eurydice', and on his finest female portraits, including Lady Margaret Beaumont, 1860–70; executed monument of Lord Lothian at Blickling church and the statue of Lord Holland in Holland Park (with (Sir) Edgar Boehm, q.v.); ARA and RA 1867; built for himself and occupied from 1876 to 1903 the new Little Holland House; exhibited at

Grosvenor Gallery from its opening in 1877; devoted much time to a colossal equestrian statue called 'Physical Energy', one cast of which formed part of the memorial to Cecil Rhodes at Cape Town, 1905, and another is in Kensington Gardens; 56 of Watts's pictures collected by Charles Richards were exhibited at Manchester Institution, 1880; 200 pictures at the Grosvenor Gallery in the winter of 1881-2; there were also exhibitions at Paris, 1883, and New York, 1885, and one on a great scale at New Gallery, 1897; hon. DCL, Oxford, and LLD, Cambridge, 1882; married his second wife Mary Fraser Tytler, 1886; built country house called Limnerslease near Guildford, 1894; added a gallery which was opened to the public, Apr. 1904; twice declined a baronetcy, 1885 and 1894; presented to National Portrait Gallery 15 portraits and 2 drawings of distinguished contemporaries, 1895, and to National Gallery of British Art, 20 symbolic paintings, 1897; executed monumental statue of Tennyson at Lincoln, 1898; received OM, 1902; died at Limnerslease and was buried at Compton.

WATTS, HENRY EDWARD (1826-1904), author; born at Calcutta; after education at Exeter grammar school returned to Calcutta, 1846; went to Australia; edited *Melbourne Argus*, 1859; on staff of *Standard* in London from 1868; translated *Don Quixote* (1888), with a life of Cervantes (issued separately, 1895), and wrote *Spain* for 'Story of the Nations' series (1893).

WATTS, JOHN (1861-1902), jockey; associated with Richard Marsh, trainer to Prince of Wales (afterwards King Edward VII) from 1879; won Oaks and Derby four times—the last time on Prince of Wales's Persimmon, on which he also won the St. Leger and the Ascot Cup, 1896; won the Two Thousand Guineas twice, and the St. Leger five times; won in all 1,414 races; trained horses at Newmarket from 1900.

WATTS, SIR PHILIP (1846-1926), naval architect; came of shipbuilding family; apprenticed as shipwright in royal dockyard, Portsmouth, 1860; studied at Royal School of Naval Architecture, South Kensington, 1866-70; appointed to assist chief constructor's staff at Admiralty in making calculations with regard to designs of new ships, 1870; draughtsman of third class on constructor's staff, 1872; constructor, 1883; appointed to staff of Chatham dockyard, 1884; naval designer and general manager to Armstrong & Co., at warship-yard, Elswick-on-Tyne, 1885-1902; brought yard to foremost position by designing and constructing warships for Japan, Argentine, Brazil, Chile, Norway, Italy, Portugal, Romania, and Turkey,

and by building several British warships designed by Admiralty; director of naval construction, Admiralty, 1902-12; constructed from own designs new and more powerful type of battleship, 1904-5; appointment of (Lord) Fisher [q.v.] as first sea lord (1904) gave opportunity for realization of Watts's desire for powerful ships; aspirations resulted in construction of remarkable series of ship-types, of which *Dreadnought* battleship and *Indomitable* battle-cruiser were chief examples; main armament of *Dreadnought* type set fashion to whole world; all effective battle cruisers in royal navy at outbreak of 1914-18 war of his design; largely responsible for England's superiority at sea; a director of Armstrong, Whitworth & Co., 1916; FRS, 1900; KCB, 1905.

WATTS-DUNTON, WALTER THEODORE (1832-1914), critic, novelist, and poet; took up literary criticism, 1874; wrote for *Athenaeum*; intimate with the Rossetti group; controlled life and affairs of A. C. Swinburne [q.v.], 1879-1909; works include *The Coming of Love*, a narrative poem (1897), and *Aylwin*, a novel (1898); learned in gipsy lore.

WAUCHOPE, SIR ARTHUR GRENFELL (1874-1947), soldier and administrator; educated at Repton; gazetted, 2nd Black Watch, 1896; severely wounded at Magersfontein, 1899; DSO, 1900; served in India, 1903-14; in France, 1914-15; commanded his battalion at Loos; and in Mesopotamia, 1916; wounded at Shaikh Sa'ad; commanded Highland battalion at Sannaiyat, fall of Baghdad, and at Mushahida, 1917; major-general, 1923; chief of British section, Berlin control commission, 1924-7; GOC, 44th Home Counties division, 1927-9; in Northern Ireland, 1929-31; high commissioner, Palestine and Trans-Jordan, 1931-8; general, 1936; eagerly interested in and financial supporter of many Palestine projects and institutions; thought to have shown insufficient strictness in handling early stages of Arab rebellion, 1936; KCB, 1931; GCMG, 1933; GCB, 1938.

WAUGH, BENJAMIN (1839-1908), philanthropist; Congregational minister from 1865 to 1887; instituted at Greenwich, Wastepaper and Blacking Brigade for neglected children; member of London school board for Greenwich, 1870-6; edited *Sunday Magazine*, 1874-96; assisted Hesba Stretton [q.v.] in establishing London Society for the Prevention of Cruelty to Children, 1884, which was converted into the National Society, 1888, and was incorporated by royal charter, 1895; mainly responsible for Act for protection of children, 1889, and more stringent Acts of 1894, 1904, and 1908.

WAUGH, EVELYN ARTHUR ST. JOHN (1903-1966), novelist; educated at Lancing and Hertford College, Oxford; assistant master in several schools, 1924-7; wrote essay on *Pre-Raphaelite Brotherhood*, printed privately, 1926; commissioned by Duckworth's to write book on Rossetti (1928); published first novel, *Decline and Fall* (1928), followed by *Vile Bodies* (1930), *Black Mischief* (1932), and *A Handful of Dust* (1934); travelled in Africa as correspondent for the *Graphic* and *The Times*, and as war correspondent for the *Daily Mail* in Abyssinia; published *Remote People* (1931), *Waugh in Abyssinia* (1936), and *Scoop* (1938); won Hawthornden prize with biography of Edmund Campion, 1935; served in army during 1939-45 war; wrote further books, *Put Out More Flags* (1942), *Brideshead Revisited* (1945), and *The Loved One* (1948); served with mission to Yugoslavia, 1944; produced historical novel about Saint Helena, mother of Constantine the Great (1950); published trilogy on wartime experiences, *Men at Arms* (1952), which was awarded the James Tait Black memorial prize, *Officers and Gentlemen* (1955), and *Unconditional Surrender* (1961); trilogy issued in one volume as *Sword of Honour* (1962); also published *Love Among the Ruins* (1953), *The Ordeal of Gilbert Pinfold* (1957), a biography of his friend Monsignor Ronald Knox [q.v.] (1957), *A Tourist in Africa* (1960), and the first volume of his autobiography, *A Little Learning* (1964).

WAUGH, JAMES (1831-1905), trainer of racehorses; succeeded Matthew Dawson [q.v.] at Gullane (1859) and Russley (1866); trained chief winners in Austria-Hungary and Germany, 1872-80; settled for life at Newmarket, 1880; trained St. Gatien, winner of the Derby and Cesarewitch, 1884, and of the Ascot Cup, 1885.

WAVELL, ARCHIBALD PERCIVAL, first EARL WAVELL (1883-1950), field-marshal; educated at Winchester and Sandhurst; gazetted, Black Watch; served in India, 1903-10; with Russian army, 1911-12; in France, 1914-16; liaison officer, Grand Duke Nicholas's army in Turkey, 1916; in Palestine, 1917-20; commanded 2nd division, Aldershot, 1935-7, forces in Palestine, 1937-8, Southern command, 1938-9; formed Middle East command, July 1939; defeated Italians at Sidi Barrani, 9 and 10 Dec. 1940; by mid-Feb. 1941 held whole of Cyrenaica; was urged, and finally agreed, to send forces to Greece; by mid-Apr. Greece and Cyrenaica lost and Tobruk invested; Crete lost, May 1941; urged, against his judgement and with insufficient forces, to operate against Iraq, Syria, and Rommel in desert; the last a costly

failure; commander-in-chief, India, July 1941; supreme commander, South-West Pacific, Dec. 1941-3; fought Burma war with minimum help from home; viceroy of India, 1943-7; released Congress leaders for discussion, 1945; showed limitless patience in seeking settlement; KCB, 1939; GCB, 1941; GCSI, GCIE, PC, viscount, and field-marshal, 1943; earl, 1947; publications include *The Palestine Campaigns* (1928).

WAVELL, ARTHUR JOHN BYNG (1882-1916), soldier and explorer; educated at Winchester; served in army, 1900-6; made Mecca pilgrimage successfully in assumed character, 1908; attempted to explore south central Arabia, 1910; raised force known as 'Wavell's Own' in East Africa during 1914-18 war; MC, 1914; killed in ambush; wrote *A Modern Pilgrim in Mecca* (1912).

WAVERLEY, first VISCOUNT (1882-1958), administrator and statesman. [See ANDERSON, JOHN.]

WEAVER, SIR LAWRENCE (1876-1930), architectural critic; architectural editor of *Country Life*, 1910-16; controller of supplies division of food production department, 1917; commercial secretary, land department, Ministry of Agriculture, 1919; director-general, 1920-2; dealt with development of small holdings and building of cottages and farms; founded National Institute of Agricultural Botany, Cambridge, 1919; KBE, 1920: his works include '*Country Life*' *Book of Cottages* (1913).

WEBB, ALFRED JOHN (1834-1908), Irish biographer; one of earliest advocates of home rule; supported C. S. Parnell [q.v.] till 1887; anti-Parnellite M.P. for West Waterford, 1890-5; frequently visited India; president of Indian National Congress, 1898; published *A Compendium of Irish Biography* (1877).

WEBB, ALLAN BECHER (1839-1907), dean of Salisbury and bishop in South Africa; educated at Rugby and Corpus Christi College, Oxford; BA, 1862; MA, 1864; DD, 1871; fellow of University College, 1863-7; vice-principal of Cuddesdon College, Oxford, 1864-5; bishop of Bloemfontein, Orange Free State, 1870-83; encouraged extension of sisterhood work there; bishop of Grahamstown, 1883-98; did much to heal schism caused by Colenso controversy; dean of Salisbury, 1901-7.

WEBB, SIR ASTON (1849-1930), architect; apprenticed as architect, 1867; set up independent practice, 1873; his works include principal block of Victoria and Albert Museum (1891) and Royal College of Science (1900-6); in partnership with E. I. Bell designed Birmingham law courts (1886-91) and new Christ's Hospital, Horsham (1894-1904); University of Birmingham (1906-9); president, RIBA, 1902-4; PRA, 1919-24; knighted, 1904.

WEBB, CLEMENT CHARLES JULIAN (1865-1954), theologian, philosopher, and historian; son of Benjamin Webb [q.v.]; educated at Westminster and Christ Church, Oxford; first class, *lit. hum.*, 1888; fellow, Magdalen College, 1889-1922; tutor in philosophy, Magdalen College, 1890; senior proctor, 1905-6; first Oriel professor of philosophy of Christian religion, 1920; fellow, Oriel College, 1922-30; publications include Gifford lectures, *God and Personality* (1918), and *Divine Personality and Human Life* (1920), *Religious Thought in the Oxford Movement* (1928), and *A Study of Religious Thought in England from 1850* (1933); FBA, 1927; D.Litt., Oxford, 1930; hon. D.Litt., St. Andrews, 1921; DD, Glasgow, 1937.

WEBB, FRANCIS WILLIAM (1836-1906), civil engineer; at fifteen pupil of Francis Trevithick, then locomotive superintendent of London and North Western Railway, with which Webb was thenceforth associated till death; chief mechanical engineer and locomotive superintendent, 1871; carried out improvements in locomotive construction; introduced compound locomotives (1882) with three and subsequently with four cylinders; developed town of Crewe; initiated cottage hospital; twice mayor, 1886 and 1887; MICE, 1872; vice-president, 1900; left money to found nursing institution at Crewe and orphanage for children of railway employees.

WEBB, GEOFFREY FAIRBANK (1898-1970), art historian; educated at Birkenhead School and Magdalene College, Cambridge; served as able seaman in RNVR, 1917-19; friend of Roger Fry [q.v.]; wrote articles on architecture and sculpture in the *Burlington Magazine*; edited, with Bonamy Dobrée, *The Complete Works of Sir John Vanbrugh* [q.v.] (4 vols., 1927, Nonesuch edition); university demonstrator, Cambridge, 1933; lecturer in fine arts, 1938; published letters and drawings of Nicholas Hawksmoor [q.v.], 1931, and a short life of Sir Christopher Wren [q.v.], 1937; lecturer, Courtauld Institute of Fine Art, 1934-7; Slade professor of fine art, Cambridge, 1938-49; served on Intelligence Staff, Admiralty, 1939-43; adviser on monuments, fine arts, and archives, SHAEF, 1944; croix de guerre, 1945; Legion of Honour, 1946; bronze medal of freedom (USA), 1947; secretary, Royal

Commission on Historical Monuments (England), 1948-62; member, Royal Fine Arts Commission, 1943-62; CBE, 1953; FBA, 1957; FSA, 1945; hon. ARIBA, 1934; published *Architecture in Britain: the Middle Ages* (1956).

WEBB, (MARTHA) BEATRICE (1858-1943). [See under WEBB, SIDNEY JAMES.]

WEBB, MARY GLADYS (1881-1927), novelist, essayist, and poet; born MEREDITH; brought up in Shropshire; married H. B. L. Webb, schoolmaster, 1912; her chief novels are *The Golden Arrow* (1916), *Gone to Earth* (1917), *The House in Dormer Forest* (1920), *Seven for a Secret* (1922), *Precious Bane* (1924); also published *The Spring of Joy* (essays, 1917) and *Poems* (posthumous, 1929).

WEBB, PHILIP (SPEAKMAN) (1831-1915), architect; entered office of George Edmund Street [q.v.], at Oxford; began practice in London, about 1856; produced decorative designs, among others for firm of William Morris [q.v.]; built one church (Brampton, Cumberland), and fifty or sixty houses, including 1 Palace Green, Kensington; West House, Chelsea; 19 Lincoln's Inn Fields; Joldwynds, near Dorking; Rounton Grange, Yorkshire; Clouds, Wiltshire; Standen, East Grinstead; and Red House, Upton (for Morris); copied no particular style; co-founder with Morris of Society for Protection of Ancient Buildings, 1877.

WEBB, SIDNEY JAMES, BARON PASSFIELD (1859-1947), social reformer and historian; educated on continent, at Birkbeck Institute and City of London College; civil servant, 1878-91; called to bar (Gray's Inn), 1885; LLB, London, 1886; joined Fabian Society, 1885; with G. B. Shaw, Sydney (later Lord) Olivier, and Graham Wallas [qq.v.] dominated its counsels for many years; progressive member, LCC, for Deptford, 1892-1910; especially influential as chairman of the Council's technical education board; married (1892) (MARTHA) BEATRICE POTTER (1858-1943), daughter of wealthy industrialist and granddaughter of Richard Potter [q.v.], whom he met in course of her co-operative studies; they shared collectivist ideals and as partnership henceforth devoted themselves to public service; published together *The History of Trade Unionism* (1894), *Industrial Democracy* (2 vols., 1897), and *English Local Government* (9 vols., 1906-29); expected social progress to come from influence of an élite; subjected key persons to permeation of their ideas; with R. B. (later Viscount) Haldane [q.v.] obtained teaching university for London; launched London School of Economics and

Political Science, 1895; drafted minority report of royal commission on poor laws on which Beatrice served (1905-9); organized nationwide but unsuccessful campaign in its support; founded *New Statesman*, 1913; turned towards labour party, on executive of which Sidney served, 1915-25; drafted its policy statement *Labour and the New Social Order* (1918); organized exhaustive campaign for Sidney's election as labour MP, Seaham division of Durham, 1922-9; (he was president of Board of Trade and PC, 1924; baron, 1929; secretary for dominions and colonies, 1929-30; for colonies, 1930-1; OM, 1944); visited Russia, 1932, publishing their impressions in *Soviet Communism: a New Civilisation?* (2 vols., 1935); Beatrice Webb's diaries, the first two volumes woven into narrative by herself, give candid presentation of two great public servants; their ashes buried in Westminster Abbey.

WEBB, THOMAS EBENEZER (1821-1903), lawyer; classical scholar of Trinity College, Dublin; LLD, 1857; professor of moral philosophy, 1857-67; fellow, 1863-71; called to Irish bar, 1861; QC, 1874; regius professor of laws at Trinity College, 1867-87; public orator, 1879-87; county court judge for Donegal, 1887-1903; attacked Gladstone's scheme of home rule; translated Goethe's *Faust* into verse, 1880; published philosophic works and *The Mystery of William Shakespeare* (1902).

WEBB-JOHNSON, ALFRED EDWARD, BARON WEBB-JOHNSON (1880-1958), surgeon; educated at Newcastle under Lyme high school and the Owens College, Manchester; MB, Ch.B., 1903; FRCS, 1906; resident medical officer, Middlesex Hospital, London, 1908; assistant surgeon, 1911; served in Royal Army Medical Corps, 1914-18; DSO, 1916; CBE, 1919; dean, Middlesex Hospital medical school, 1919-25; chairman, plans committee for new hospital, completed in 1935; consultant surgeon, Queen Alexandra Military Hospital; chairman, Army Medical Advisory Board, 1946-57; surgeon to Queen Mary, 1936-53; knighted, 1936; KCVO, 1942; GCVO, 1954; baronet, 1945; baron, 1948; president, Royal College of Surgeons, 1941-9; hon. LLD, Liverpool and Toronto.

WEBBER, CHARLES EDMUND (1838-1904), major-general, Royal Engineers; joined Royal Engineers, 1855; served in Indian Mutiny, taking part in battle of Jhansi and capture of Gwalior; assistant instructor in military surveying at Woolwich, 1861-6; attached to Prussian army in seven weeks' war, 1866; commanded company which assisted in constructing and organizing telegraph service of Post Office, 1869-71; founded Society of Tele-

graph Engineers, 1871; president, 1882; assistant adjutant in Zulu war, 1879; in charge of telegraphs in Egyptian campaign, 1882; present at Tel-el-Kebir; CB, 1882; served in Nile expedition, 1884-5; on retirement as major-general (1885) was engaged in electrical pursuits.

WEBSTER, BENJAMIN (1864-1947), actor; grandson of B. N. Webster [q.v.]; educated at King's College School; called to bar, 1885; became actor (1887) in order to be near and married (1892) May Whitty: **WEBSTER,** DAME MARY LOUISE (MAY) (1865-1948), actress; granddaughter of M. J. Whitty [q.v.]; trained by (Dame) Madge Kendal [q.v.]; capable organizer as chairman of women's organizations during war; DBE, 1918; thereafter appeared notably in *The Last of Mrs. Cheyney*, *There's Always Juliet*, and *Night Must Fall*; played in film version of last, 1937, and remained in Hollywood; other films include *The Lady Vanishes* and *Mrs. Miniver*.

WEBSTER, SIR CHARLES KINGSLEY (1886-1961), historian; educated at Merchant Taylors' School, Crosby, and King's College, Cambridge; first class, part ii, historical tripos, 1907; fellow, King's College, 1910; contributed to *English Historical Review*; professor of modern history, Liverpool University, 1914; served in intelligence section, War Office, under H. W. V. Temperley [q.v.], 1917; seconded to Foreign Office, 1918; published *The Congress of Vienna* (1919); secretary to military section of British delegation to peace conference; returned to Liverpool chair, 1919-22; professor of international relations, Aberystwyth, 1922-31; travelled widely abroad; published *The Foreign Policy of Castlereagh 1815-22* (2 vols., 1925 and 1931); professor of international relations, London School of Economics, 1931; during 1939-45 war worked for Royal Institute of International Affairs, lectured in America, and worked in Foreign Office; KCMG, 1946; concerned with preparatory arrangements for the United Nations; FBA, 1930; president, 1950; many academic honours; published *The Foreign Policy of Palmerston 1830-41* (2 vols, 1951); collaborated with Noble Frankland on official history of the Anglo-American bombing offensive in 1939-45 war.

WEBSTER, (GILBERT) TOM (1886-1962), sporting cartoonist and caricaturist; educated at Royal Wolverhampton School; at fourteen, employed as clerk by Great Western Railway; won prizes offered by newspapers for humorous drawings; joined staff of Birmingham *Sports Argus*; served in army during 1914-18 war; supplied sporting cartoons to London *Evening*

News, 1919; noticed by Lord Northcliffe [q.v.], and given staff post on *Daily Mail*, 1919; highest-paid cartoonist in the world, 1924; originated 'Tishy' the racehorse with twisted forelegs; commissioned to do sporting panorama in colour for the *Queen Mary*, 1936; retired from *Daily Mail*, 1940; resumed work with Kemsley Newspapers, 1944; joined *News Chronicle*, 1953-6; published each year *Tom Webster's Annual*.

WEBSTER, RICHARD EVERARD, VIS-COUNT ALVERSTONE (1842-1915), judge; son of Thomas Webster, QC [q.v.]; called to bar (Lincoln's Inn), 1868; QC, 1878; conservative MP, Launceston, 1885, Isle of Wight, 1886; attorney-general, 1885-6-92, 1895-1900; baronet, master of the rolls, PC, and baron, 1900; lord chief justice, 1900-13; viscount, 1913.

WEBSTER, WENTWORTH (1829-1907), Basque scholar and folklorist; BA, Lincoln College, Oxford, 1852; MA, 1855; ordained, 1854; Anglican chaplain at St. Jean-de-Luz, 1869-81; settled at Sare, 1881, where he died; received civil list pension, 1894; studied and wrote on Basque language and literature; published *Basque Legends* (1878) and *Spain* (1882).

WEDGWOOD, JOSIAH CLEMENT, first BARON WEDGWOOD (1872-1943), politician; direct descendant of Josiah Wedgwood [q.v.]; educated at Clifton; naval architect; served in South African war; resident magistrate, Ermelo, Transvaal, 1902-4; MP, Newcastle under Lyme, 1906-42; left liberal for labour party, 1919; served in Belgium, 1914-15; Gallipoli, 1915; DSO; on royal commission on Mesopotamia, 1916; vice-chairman, parliamentary labour party, 1921-4; chancellor, duchy of Lancaster and PC, 1924; chairman, committee on House of Commons records, 1929-42, and largely responsible for initiation of *History of Parliament*; devoted to causes: taxation of land values, Indian independence, and Jewish national home; baron, 1942.

WEDGWOOD, SIR RALPH LEWIS, first baronet (1874-1956), railway administrator, grandson of James Meadows Rendel [q.v.] and brother of Lord Wedgwood [q.v.]; educated at Clifton and Trinity College, Cambridge; first class, parts i and ii, moral sciences tripos, 1895-6; joined North Eastern Railway, 1896; company secretary, 1904; goods and passenger manager, 1914; served with Royal Engineers, 1914; director of docks in France, 1916; general manager, North Eastern Railway, 1922; chief general manager, London and North Eastern

Railway, 1923-39; CMG, 1917; CB, 1918; knighted, 1924; baronet, 1942.

WEEKS, RONALD MORCE, BARON WEEKS (1890-1960), industrialist and soldier; educated at Charterhouse and Caius College, Cambridge; captain, university association football, 1911; joined Pilkington Brothers, Ltd., 1912; on active service, 1915-18; captain, Rifle Brigade, 1917; MC (bar), 1917-18; DSO, 1918; rejoined Pilkingtons, 1919; director, 1928; chairman, 1939; director-general, army equipment, 1941; deputy chief, imperial general staff, lieutenant-general, 1942; joined Vickers, 1945; chairman, 1949-56; chairman, National Advisory Council for Education in Industry and Commerce, 1948-56; CBE, 1939; KCB, 1943; baron, 1956; hon. fellow, Caius College, 1945; hon. doctorates, Liverpool, 1946, Sheffield, 1951, and Leeds, 1957.

WEIR, ANDREW, first BARON INVERFORTH (1865-1955), shipowner; educated at high school, Kirkcaldy; entered Commercial Bank of Scotland, but left bank for shipping office in Glasgow and bought sailing ship, 1885; built up large fleet of sailing ships; moved to London and steamships, 1896; surveyor-general of supply, 1917; minister of munitions, 1919-21; PC, 1919; baron, 1919; entered field of communications and became president, Radio Communications and other companies; founder and chairman, United Baltic Corporation, 1919.

WEIR, SIR CECIL McALPINE (1890-1960), industrialist and public servant; educated at Morrison's Academy, Crieff, and in Switzerland and Germany; joined family firm, Schrader, Mitchell and Weir; partner, 1910-56; in 1914-18 war served in the Cameronians; MC; KBE, 1938; director, Union Bank of Scotland, 1939-47; at Board of Trade, 1940-6; director-general, equipment and stores, Ministry of Supply, 1942-6; chairman, Dollar Exports Board, 1949-51; KCMG, 1952; headed United Kingdom delegation to the High Authority, European Coal and Steel Community, 1952-5.

WEIR, HARRISON WILLIAM (1824-1906), animal painter and author; apprentice to George Baxter, the colour printer, 1837-44; employed as wood-engraver on *Illustrated London News* from its foundation, 1842; made special study of birds and animals; occasionally exhibited at British Institution and Royal Academy; member of New Water Colour Society, 1849; illustrated for J. G. Wood [q.v.] his *Illustrated Natural History* (1853); compiled *The Poetry of Nature* (1867); wrote and illustrated *Every Day in the Country* (1883) and *Our Cats and all about them* (1889); an experienced poultry breeder, practical horticulturist, and designer; received civil list pension, 1891.

WEIR, WILLIAM DOUGLAS, first VISCOUNT WEIR (1877-1959), industrialist and public servant; educated at Glasgow high school; joined family engineering firm, G. and J. Weir; manager, 1902; chairman, 1912; director of munitions, Scotland, 1915; controller, aeronautical supplies, Ministry of Munitions, 1917; director-general, aircraft production, 1917; air minister, 1918; chairman, advisory committee on civil aviation, 1919; adviser to air minister, 1935-8; director-general, explosives, Ministry of Supply, 1939-41; knighted, 1917; PC, 1918; baron, 1918; GCB, 1934; viscount, 1938; hon. LLD, Glasgow, 1919.

WEISZ, VICTOR (1913-1966), the cartoonist VICKY; born in Germany of Hungarian Jewish parents; attended Berlin Art School; worked on *12 Uhr Blatt*; published anti-Hitler cartoon at fifteen years old; harassed by Gestapo; escaped to England; failed to obtain post with *Daily Herald*, 1938; studied English language and British humour, encouraged by (Sir) Gerald Barry [q.v.], editor of *News Chronicle*, 1939; staff cartoonist, *News Chronicle*, 1941-55; granted British nationality, 1946; succeeded (Sir) David Low [q.v.] as best cartoonist in Britain; moved to *Daily Mirror*, and then to *Evening Standard*, 1958; also drew weekly cartoon for *New Statesman*, 1954; and portraits for *New Statesman* 'profiles'; prodigious output; no public personality escaped his ridicule; died of overdose of sleeping pills.

WEIZMANN, CHAIM (1874-1952), Zionist leader and first president of Israel; born in Jewish Pale of Settlement, Russia; educated at high school, Pinsk, and Freiburg University, D.Sc., 1899; lecturer, Manchester University, 1904; reader in biochemistry, 1913; naturalized British subject, 1910; during 1914-18 war pressed Zionist claims on British politicians whilst working for Ministry of Munitions and Admiralty; headed Zionist Commission to Palestine, 1918; presented Zionist case at peace conference, 1919; president, World Zionist Organization, 1920; created Jewish Agency for Palestine, 1929; urged claims of Zionists on League of Nations, 1947; first president, Israel, 1948; helped to found Hebrew University, Jerusalem, and Weizmann Institute of Science, Rehoroth.

WELBY, REGINALD ERLE, first BARON WELBY (1832-1915) civil servant; BA, Trinity College, Cambridge; entered Treasury, 1856; assistant financial secretary of Treasury, 1880;

permanent secretary, 1885-94; baron, 1894; chairman of London County Council, 1900; enthusiastic free trader and exponent of rigid economy in public service.

WELCH, ADAM CLEGHORN (1864-1943), Scottish divine; educated at George Watson's College and Edinburgh University; United Presbyterian minister, 1887-1913; professor of Hebrew and Old Testament exegesis, New College, Edinburgh, 1913-34; challenged existing position in Old Testament criticism in a lecture (1921) and subsequent publications.

WELDON, WALTER FRANK RAPHAEL (1860-1906), zoologist; son of Walter Weldon [q.v.]; abandoned medical studies for zoology; BA, St. John's College, Cambridge (first class, natural sciences), 1881; fellow, 1884; demonstrator in zoology, 1882-4; university lecturer in invertebrate morphology, 1884-91; made statistical study of variation of common shrimp, from 1889 to 1891, thus founding study of 'biometrics'; FRS, 1890; Jodrell professor of zoology at University College, London, 1891-9; secretary to committee of Royal Society for statistical inquiries into measurements of plants and animals; prepared for Royal Society report on death-rate due to selective destruction of shore crabs, 1894; Linacre professor of comparative anatomy, Oxford, 1899-1906; co-editor of *Biometrika*, 1901-6; published *A Note on the Offspring of Thoroughbred Chestnut Mares* (1906); Weldon memorial prize for biometric research founded at Oxford, 1907.

WELLCOME, SIR HENRY SOLOMON (1853-1936), manufacturing chemist and patron of science; born in United States; qualified at Philadelphia College of Pharmacy; set up as manufacturing chemist with S. M. Burroughs in England, 1880; sole proprietor, 1895; introduced 'tabloid' drugs; founded laboratories and museums incorporated (1924) as Wellcome Research Institution of Wellcome Foundation; promotor of missionary enterprise and geographical and archaeological exploration; naturalized, 1910; FRS and knighted, 1932.

WELLDON, JAMES EDWARD COWELL (1854-1937), successively headmaster of Dulwich and Harrow, bishop of Calcutta, and dean of Manchester and of Durham; scholar of Eton and King's College, Cambridge; senior classic, 1877; fellow, 1878; master, Dulwich College, and ordained, 1883-5; headmaster, Harrow, 1885-98; bishop of Calcutta and metropolitan of the Indian Church, 1898-1902; canon of Westminster, 1902-6; dean of Manchester, 1906-18; of Durham, 1918-33;

works include translations of Aristotle (1883–92), edition of *De Civitate Dei* (1924), *The Hope of Immortality* (1898), and *The English Church* (1926).

WELLESLEY, DOROTHY VIOLET, DUCHESS OF WELLINGTON (1889–1956), poet; married Lord Gerald Wellesley, 1914; husband became Duke of Wellington, 1943; published collected poems, *Early Light* (1955); autobiography, *Far Have I Travelled* (1952).

WELLESLEY, SIR **GEORGE GREVILLE** (1814–1901), admiral; entered navy, 1828; took part in operations on coast of Syria, 1840; commanded in Pacific, 1849–53; commanded squadron at Sveaborg, 1855; CB, 1856; in charge of Indian navy, 1857–62; admiral superintendent, Portsmouth, 1865–9; commander-in-chief, North America station, 1869–70, and 1873–5; admiral, 1875; first sea lord, 1877–9; KCB, 1880; GCB, 1887.

WELLESLEY, SIR **VICTOR ALEXANDER AUGUSTUS HENRY** (1876–1954), diplomatist; educated at Wiesbaden and Heidelburg; entered diplomatic service, 1899; second secretary, Rome, 1905–6; commercial attaché, Spain, 1908–12; controller, commercial and consular affairs, Foreign Office, 1916–19; counsellor in charge, Far Eastern department, 1920–4; assistant under-secretary of state, 1924–5; deputy under-secretary, 1925–36; CB, 1919; KCMG, 1926; published *Diplomacy in Fetters* (1944) and *Recollections of a Soldier-Diplomat* (1947).

WELLINGTON, DUCHESS OF (1889–1956), poet. [See WELLESLEY, DOROTHY VIOLET.]

WELLINGTON, HUBERT LINDSAY (1879–1967), painter, draughtsman, and art teacher; educated at the Crypt grammar school, Gloucester, Gloucester School of Art, and the Slade School of Fine Art, London; teacher at Stafford School of Art, 1904–16; inclined towards French culture; subscribed to *Nouvelle Revue Française*; served in France during 1914–18 war; lectured in London, 1919–23; published monographs on (Sir) William Rothenstein [q.v.], 1923, and (Sir) Jacob Epstein [q.v.], 1924; registrar and lecturer, Royal College of Art, 1923–32; principal, Edinburgh College of Art, 1932–43; professor of history of fine art, Slade School, 1946–9; exhibited at New English Art Club, Allied Artists Association, and the London Group; comprehensive exhibition of paintings and drawings, Messrs Agnews, 1963; examples of work in Tate Gallery, Ashmolean Museum, and Gloucester and Southampton art galleries.

WELLS, HENRY TANWORTH (1828–1903), portrait painter in oils and miniature; exhibited at Royal Academy miniatures, 1846–60, and large oil portraits and groups from 1861; ARA, 1866; RA, 1870; deputy-president, 1895; best-known group is 'The Queen at the Royal Courts of Justice', 1887; portraits include those of Duke of Devonshire (1872), Lord Selborne (1874), Sir Redvers Buller (1889), Sir Lowthian Bell (1895); most popular work was 'Victoria Regina', painting of Queen Victoria, as princess, receiving news of her accession, 1880 (now in Tate Gallery); limner to Grillion's Club from 1870.

WELLS, HERBERT GEORGE (1866–1946), author; son of unsuccessful tradesman; his penurious childhood profoundly influenced his novels and social attitudes; educated at a commercial school and by wide reading; apprenticed successively to pharmacy and drapery; student assistant, Midhurst grammar school, 1883–4; obtained studentship at Normal School of Science, South Kensington; studied under T. H. Huxley [q.v.]; failed third year examinations; taught at Holt Academy, Wrexham, and Henley House School, Kilburn; B.Sc., London, 1890; tutor, University Tutorial College, 1891; found ready market for short stories such as *The Stolen Bacillus* (1895); with *The Time Machine* (1895) began long series of fantastic and imaginative ('Wellsian') romances, usually based on scientific fact; ranging from *The Invisible Man* (1897) to *The Shape of Things to Come* (1933) they earned him reputation of prophet; foresaw war in air, tanks, and atomic bomb; wrote also 'real' novels drawing on his own experiences including *Kipps* (1905), *Tono-Bungay* (1909), and *The History of Mr. Polly* (1910); pursued topic of freer love in *Ann Veronica* (1909), *The New Machiavelli* (1911), and *Marriage* (1912); increasingly preoccupied with ideal of 'World State'; believed in perfectibility of mankind but continually frustrated by inadequacy of material from which he sought to shape his higher species; flounced in and out of party socialism and other movements, notably the Fabian Society; saw *The War That Will End War* (1914) and later the League of Nations as prelude to his ideal but was disillusioned; began assault on education; proved himself a great public teacher in *The Outline of History* (1920), *The Science of Life* (1931), and *The Work, Wealth and Happiness of Mankind* (1932); showed increasing pessimism in *The Fate of Homo Sapiens* (1939); in his frustration underestimated the influence on thinking of three decades excited by his genius to stimulate imagination; his *Experiment in Autobiography* (2 vols., 1934) a remarkable contribution to social history.

WEMYSS, ROSSLYN ERSKINE, BARON WESTER WEMYSS (1864-1933), admiral of the fleet; entered *Britannia*, 1877; served in royal yacht *Osborne*, 1887-9; commander, 1898; second-in-command, *Ophir*, on Duke of York's tour of the dominions, 1901; captain, 1901; first captain, *Osborne*, 1903-5; commanded *Balmoral Castle* taking Duke of Connaught [q.v.] to South Africa, 1910; rear-admiral, 1911; commanded twelfth cruiser squadron (Aug. 1914-Feb. 1915) in charge of western patrol in English Channel, co-operating with French; governor of Lemnos, 1915; organized and equipped naval base at Mudros for Dardanelles campaign; in charge of landing operations at Helles, Apr. 1915; KCB and vice-admiral, 1916; commander-in-chief, East Indies and Egypt station, 1916-17; supported Murray's advance to Sinai; attempted to send food ship to Kut; supported Allenby's advance into Palestine; deputy sea lord, 1917; replaced Jellicoe as first sea lord, Dec. 1917; co-ordinated war staff into team and inspired them with his own enthusiasm; maintained naval interests at armistice and peace conference; admiral, 1919; resigned (Nov. 1919), promoted admiral of the fleet and created baron; an attractive, courageous personality universally known as 'Rosy'; published *The Navy in the Dardanelles Campaign* (1924).

WEMYSS - CHARTERIS - DOUGLAS, FRANCIS, tenth EARL OF WEMYSS AND MARCH (1818-1914), politician; BA, Christ Church, Oxford; conservative MP, East Gloucestershire, 1841-6, Haddingtonshire, 1847-83; with Edward Horsman [q.v.], Robert Lowe (afterwards Viscount Sherbrooke, q.v.), and others formed 'cave of Adullam' against reform bill of 1866; opposed to socialism; enthusiastic supporter of Volunteer movement; instrumental in inaugurating National Rifle Association, 1859; succeeded father, 1883.

WERNHER, SIR JULIUS CHARLES, first baronet (1850-1912), financier and philanthropist; born at Darmstadt; educated at Frankfort; engaged by Porges, diamond merchant of Paris and London, to assist partner in diamond buying in South Africa, 1870; settled at camp which afterwards became Kimberley, 1871; partner in firm and its sole representative at fields, 1873; through skill and honesty made firm (known as French Company) one of the most important diamond producing companies; settled in London, 1884; directed from London firm's gold mining in Rand from 1887; assisted Alfred Beit [q.v.] and Cecil Rhodes [q.v.] to amalgamate chief diamond mines of Kimberley as De Beers Consolidated Mines, 1888; firm of Wernher, Beit & Co. created, 1890; naturalized, 1898; baronet, 1905; gave munificently to charities and education.

WEST, SIR ALGERNON EDWARD (1832-1921), chairman of the Board of Inland Revenue, 1881-92; served in Admiralty and India Office, 1852-66; private secretary to W. E. Gladstone, 1868-72; commissioner of inland revenue, 1872-81; KCB, 1886; PC, 1894.

WEST, EDWARD CHARLES SACKVILLE-, fifth BARON SACKVILLE (1901-1965), man of letters. [See SACKVILLE-WEST.]

WEST, EDWARD WILLIAM (1824-1905), oriental scholar; studied engineering at King's College, London, 1839-42; superintended cotton presses in Bombay, 1844-50; made special study of Parsi religion; published copies of Buddhist records of Kanheri and other caves, 1861; won reputation by Iranian researches; at Munich (1867-73) published translation of the Pahlavi texts of Zoroastrianism; hon. Ph.D., Munich, 1871; revisited India (1874-6) to procure Pahlavi manuscripts; soon settled in England and translated Pahlavi texts for Max Müller [q.v.] for his *Sacred Books of the East*; member of Royal Asiatic Society, 1884-1901; gold medallist, 1901; collaborated with Prof. Martin Haug in *Book of Ardā-Vīrāf, Pahlavi and English* (1872), and wrote on *Pahlavi Literature* (1897).

WEST, LIONEL SACKVILLE-, second BARON SACKVILLE (1827-1908), diplomatist. [See SACKVILLE-WEST.]

WEST, SIR RAYMOND (1832-1912), Indian civil servant, judge, and jurist; entered East India Company's service, 1855; joined judicial department, 1860; registrar of high court, 1863; judge of Bombay high court, 1873-86; member of executive council of governor of Bombay, 1886-92; KCIE, 1888; published annotated edition of Bombay code (1867-8) and, in collaboration, important *Digest of Hindu Law* (1867-9).

WEST, VICTORIA MARY SACKVILLE- (1892-1962), writer and gardener. [See SACKVILLE-WEST.]

WESTALL, WILLIAM (BURY) (1834-1903), novelist and journalist; engaged in journalism mainly on the continent from 1870; at Geneva met Prince Kropotkin and Stepniak, and translated Stepniak's *Russia under the Czars*; his many novels include *Sons of Belial* (1895) and *Her Two Millions* (1897).

WESTCOTT, BROOKE FOSS (1825-1901), bishop of Durham; educated at King Edward VI's School, Birmingham, 1837-44, and Trinity College, Cambridge, 1844-9; won Battie university scholarship, 1846, and several university prizes for classics; BA (twenty-fourth wrangler), 1848; bracketed senior classic, 1849; fellow, 1849; hon. fellow, 1890; had as pupils at Cambridge J. B. Lightfoot, E. W. Benson, and F. J. A. Hort [qq.v.]; obtained Norrisian prize for theological essay, 1850; ordained, 1851; master at Harrow, 1852-70; while there published *The Canon of the New Testament* (1855), *Characteristics of the Gospel Miracles* (1859), *Introduction to the Study of the Gospels* (1860), *The Bible in the Church* (1864), and *The Gospel of the Resurrection* (1866); residentiary canon of Peterborough, 1869-83; regius professor of divinity at Cambridge, 1870-90; after abolition of tests in 1871, he published *Religious Office of the Universities* (sermons and papers, 1873); improved academic arrangements for the theological lectures; framed fresh regulations for BD and DD degrees, 1871; an influential teacher; lectured on early church history and on Christian doctrine; founded Cambridge clergy training school (now called Westcott House); advocated university extension; canon of Westminster, 1883-90; member of New Testament revision company, 1870-81; published after twenty-eight years' labour with F. J. A. Hort, his epoch-making critical text of the New Testament in Greek (2 vols., 1881); published valuable commentaries on St. John's Gospel (1882), St. John's Epistles (1883), and Epistle to the Hebrews (1889); wrote much on Origen from 1877 to 1889; collected his sermons in *Christus Consummator* (1886) and *Social Aspects of Christianity* (1887); fellow of King's College, Cambridge, 1882; hon. fellow, 1890; hon. DCL, Oxford, 1881; hon. DD, Edinburgh, 1884, and Dublin, 1888; bishop of Durham, 1890-1901; hon. DD, Durham, 1890; frequently brought together employers of labour and trade-union officials; took influential part in settling Durham coal strike, 1892; helped to secure industrial boards of conciliation in the county; first church dignitary to address Northumberland Miners' Gala, 1894; fostered intimate relations with the younger clergy of diocese; cherished a lifelong interest in foreign missions; first president of Christian Social Union, 1889; advocate of peace and international arbitration; published, during his episcopate, *Religious Thought in the West* (1891) (early essays reprinted), *The Gospel of Life* (1892) (Cambridge lectures); collected sermons and addresses in *The Incarnation and Common Life* (1893), *Christian Aspects of Life* (1897), *Lessons from Work* (1901); a commentary on Ephesians was published posthumously; his theological position, at which he arrived independently, strongly resembled that of F. D. Maurice [q.v.]; *Life and Letters* by son, Arthur Westcott (2 vols. with bibliography, 1903).

WESTER WEMYSS, BARON (1864-1933), admiral of the fleet. [See WEMYSS, ROSSLYN ERSKINE.]

WESTLAKE, JOHN (1828-1913), jurist; BA, Trinity College, Cambridge; called to bar (Lincoln's Inn), 1854; QC, 1874; Whewell professor of international law, Cambridge, 1888-1908; author of works on private and public international law.

WESTLAND, SIR JAMES (1842-1903), Anglo-Indian financier; joined Indian civil service, 1861; assistant magistrate in Bengal districts, 1862-9; under-secretary of financial department of government of India, 1870; officiating accountant-general of Bengal, 1873, and of Central Provinces, 1873-6; accountant and comptroller-general, 1878; head of Egyptian accounts department, Mar.-June 1885; secretary of Indian finance department from 1886; temporary finance member of government, Aug. 1887-Nov. 1888; CSI, 1888; KCSI, 1895; chief commissioner of Assam, July-Oct. 1889; finance member of viceroy's council, 1893-9; established gold standard and raised rupee to fixed value of 1s. 4d.; member of India Council in London, 1899.

WESTON, DAME AGNES ELIZABETH (1840-1918), organizer of 'Sailors' Rests'; brought up at Bath; opened correspondence with soldiers and later (1868) with sailors; issued printed monthly letter for distribution to ships' companies from 1871; began work in Devonport for Royal Naval Temperance Society, 1873; opened first 'Sailors' Rest' there, 1876; later opened one at Portsmouth; GBE, 1918.

WESTON, SIR AYLMER GOULD HUNTER- (1864-1940), lieutenant-general; commissioned in Royal Engineers, 1884; served in South Africa, 1899-1901; cut railway near Bloemfontein; DSO, 1900; assistant director, military training, 1911-14; commanded 11th Infantry brigade (1914-15) in France, 29th division at Helles landing (1915), VIII Corps (1915-18) in Gallipoli and France; lieutenant-general, 1919; unionist MP, North Ayrshire, 1916-35; KCB, 1915.

WESTON, FRANK (1871-1924), bishop of Zanzibar; BA, Trinity College, Oxford; ordained deacon, 1894; priest, 1895; joined Universities' Mission to Central Africa, 1898; principal of St. Andrew's Training College,

Kiungani, Zanzibar, 1901-8; protested against Kikuyu concordat and certain modernist teaching in England, 1913; during 1914-18 war successfully organized and led corps of carriers for Zanzibar government; president of second Anglo-Catholic congress, London, 1923; died at Hegongo; a great orator; had real understanding of Africans.

WET, CHRISTIAAN RUDOLPH DE (1854-1922), Boer general and politician. [See DE WET.]

WEYMAN, STANLEY JOHN (1855-1928), novelist; BA, Christ Church, Oxford; his historical romances include *The House of the Wolf* (1890), *A Gentleman of France* (1893), *Under the Red Robe* (1896), *The Castle Inn* (1898), *The Long Night* (1903); his developed historical imagination and skill apparent in *Chippinge* (1906) and *Ovington's Bank* (1922).

WEYMOUTH, RICHARD FRANCIS (1822-1902), philologist; educated at University College, London; fellow, 1869; BA, London, 1846; MA, 1849; first Litt.D. of London University, 1868; headmaster of Mill Hill School, 1869-86; joined Philological Society, 1851; published *Early English Pronunciation* (1874), *The Resultant Greek Testament* (1886), and *The New Testament in Modern Speech* (posthumous, 1903); awarded civil list pension, 1891.

WHARTON, SIR WILLIAM JAMES LLOYD (1843-1905), hydrographer of the navy; entered navy, 1857; served in surveying vessel on North America station, 1865-9; made surveys in Mediterranean, examining undercurrents in Bosphorus, 1872-6; published *Hydrographical Surveying*, 1882; hydrographer to the navy, 1884-1904; rear-admiral, 1895; CB, 1895; KCB, 1897; president of geographical section of British Association in Cape Town, 1905; FRS, FRAS, and FRGS; died of fever at Cape Town.

WHEATLEY, JOHN (1869-1930), labour politician; coal-miner, 1880-93; afterwards employed in various kinds of business; set up successful publishing business in Glasgow, 1912; joined independent labour party, 1908; studied local government problems; elected to Lanarkshire county council, 1909; labour MP, Shettleston division of Glasgow, 1922-30; minister of health, 1924; responsible for Housing (Financial Provisions) Act; became identified with revolutionary socialist views.

WHEELER, SIR WILLIAM IRELAND DE COURCY (1879-1943), surgeon; graduated in natural science, Trinity College, Dublin, 1899; MD, 1902; joined staff of Mercer's (1904) and other hospitals; served in France, 1914-18; knighted, 1919; migrated to England, 1932; a leading surgeon well known in United States as well as in Dublin and Great Britain.

WHEELHOUSE, CLAUDIUS GALEN (1826-1909), surgeon; educated at Leeds school of medicine; MRCS, 1849; LSA, 1850; FRCS, 1864; surgeon to public dispensary, Leeds, from 1851; lecturer at medical school there; hon. D.Sc., Leeds, 1904; president of council of British Medical Association, 1881-4; advocated 'Wheelhouse' external urethrotomy for impermeable strictures, 1876.

WHETHAM, WILLIAM CECIL DAMPIER (1867-1952), scientist and agriculturist. [See DAMPIER.]

WHIBLEY, CHARLES (1859-1930), scholar, critic, and journalist; BA, Jesus College, Cambridge; adopted profession of letters; assisted W. E. Henley [q.v.] in conduct of *Scots Observer* (afterwards called *National Observer*) and in producing series of 'Tudor Translations', to which he contributed several introductions; allied himself with H. J. C. Cust [q.v.], editor of *Pall Mall Gazette*, 1893; Paris correspondent for the paper, 1894; formed association with *Blackwood's Magazine*, c.1897; for more than twenty-five years contributed monthly article entitled 'Musings without Method'; his works include several volumes of literary and historical essays and much editing; died at Hyères.

WHIBLEY, LEONARD (1863-1941), classical scholar; educated at Bristol grammar school and Pembroke College, Cambridge; first class, parts i and ii, classical tripos, 1885-6; fellow, 1889-1941; university lecturer in ancient history, 1899-1910; edited *Companion to Greek Studies* (1905) and, with P. J. Toynbee [q.v.], Thomas Gray's letters (3 vols., 1935).

WHIDDINGTON, RICHARD (1885-1970), physicist; educated at William Ellis School, Highgate and St. John's College, Cambridge; first class, parts i and ii, natural sciences tripos, 1907-8; London (external) B.Sc.; fellow, St. John's College, 1911; D.Sc., London; served in RFC establishment, Farnborough, during 1914-18 war; university lecturer in experimental physics, Cambridge, 1919; Cavendish professor of physics, Leeds, 1919-51; FRS, 1925; served in various government posts during 1939-45 war; CBE, 1946; published, with J. G. Crowther, *Science at War* (1947); pro-vice-chancellor, Leeds, 1949-51; president, Physical Society, 1952-4; scientific adviser to

Central Treaty Organization, 1959–63; published sixteen papers on the properties of X-rays and electrons, and thirty on electron studies seeking correlation with optical rather than with X-radiation.

WHIPPLE, ROBERT STEWART (1871–1953), manufacturer and collector of scientific instruments and books; son of George Mathews Whipple [q.v.]; educated at King's College School, Wimbledon; assistant, Kew Observatory, 1888–1896; private assistant to (Sir) Horace Darwin [q.v.]; manager and secretary, Cambridge Scientific Instrument Co., 1898; joint managing director, 1909–35; chairman, 1939–49; presented collection of scientific instruments and books to Cambridge University, 1944.

WHISTLER, JAMES ABBOT McNEILL (1834–1903), painter; born at Lowell, Massachusetts; educated at Military Academy, West Point, 1851–4; engaged on United States geodetic survey; studied art at Paris under Gleyre, 1855–9; published etchings *The French Set* (1858); his 'At the Piano', rejected by Salon, 1859, was accepted by Royal Academy, 1860; he came to London (1859) and worked at etching with his brother-in-law, (Sir) F. S. Haden [q.v.]; for first year was chiefly occupied in etchings of Thames scenery; visited Brittany, 1861; settled at Chelsea, 1863; influenced by Japanese art in his 'Lange Leizen', and more especially in 'The Little White Girl', 1863; visited Germany, 1865, and Chile, 1866; chief pictures between 1866 and 1872, when he ceased to exhibit paintings at the Academy, include 'Portrait of my Mother' (Paris), his 'Nocturnes', and portraits of Carlyle (at Glasgow) and Miss Alexander (Tate Galley); he sent to first exhibition of Grosvenor Gallery eight pictures, including 'The Falling Rocket, a Nocturne in Black and Gold', 1877; brought libel action against John Ruskin [q.v.] for condemning this picture, 1878; was awarded a farthing damages and became bankrupt, 1879; executed profitable etchings and pastels in Venice, 1879–80; settled at 13 Tite Street, Chelsea, 1881, and painted there the best pictures of his later years, including 'The Blue Girl' and the 'Yellow Buskin'; toured in Belgium and Holland, 1885, and in Touraine and in Holland, 1888–9; published *The Gentle Art of Making Enemies* (1890); moved to Paris, 1892; defendant in action at Paris brought by Sir William Eden, 1895; published report of the litigation in *The Baronet and the Butterfly* (1899); president of Society of British Artists, 1886–8, and of International Society of Sculptors, Painters, and Engravers from 1898 till death; recipient of many foreign honours; hon.

LLD, Glasgow; exerted immense influence on contemporary art.

WHISTLER, REGINALD JOHN, (REX) (1905–1944), artist; educated at Haileybury; studied at Slade School; executed murals for Tate Gallery refreshment room, for Marquess of Anglesey at Plâs Newydd, and for Lady Mountbatten at Brook House, London; book-illustrations include series for A. E. W. Mason's *Königsmark* (Tate Gallery); designed scenery and costumes for ballets and plays; book-jackets, textiles, china, carpets, bookplates, etc.; used rococo style with freshness and wit; killed serving with Guards armoured division in Normandy.

WHITAKER, SIR (FREDERICK) ARTHUR (1893–1968), maritime civil engineer; educated at Liverpool Institute high school and Liverpool University; first class, engineering, 1914; worked on construction of naval dockyard at Rosyth, 1915–19; M.Eng., Liverpool, 1917; naval agent and member, Jamaica Marine Board, 1920; worked on oil installations, jetties, and dredging at Devonport and Portsmouth and in Malta, 1920–30; superintending civil engineer in charge of construction of Singapore naval base, 1933; deputy civil engineer-in-chief, Admiralty, 1934; succeeded Sir Athol Anderson as CE-in-C, 1940–54; responsible for construction of the Orkney causeways, 1940–2; member, Dover Harbour Board and other organizations; CB, 1941; KCB, 1945; commander of Légion d'Honneur, 1947; partner in Livesey & Henderson, Consulting Engineers, 1954–62; president, Institution of Civil Engineers, 1957–8; hon. D.Eng., Liverpool, 1960.

WHITBY, SIR LIONEL ERNEST HOWARD (1895–1956), medical scientist and regius professor, physic, Cambridge University; educated at King's College, Taunton, Bromsgrove, and Downing College, Cambridge; served in Royal West Kent regiment in 1914–18 war; MC, leg amputated, 1918; MB, B.Ch., Cambridge, 1923; MD, Cambridge, and MRCP, 1927; assistant pathologist, Bland Sutton Institute, 1923; his research established sulpha-pyridene (M & B 693); in charge of army blood transfusion service, 1939–45; knighted, 1945; regius professor, physic, Cambridge, 1945; master, Downing College, 1947; vice-chancellor, Cambridge, 1951–3; FRCP, 1933; hon. fellow, Lincoln College, Oxford; hon. degrees, Glasgow, Toronto, and Louvain.

WHITE, CLAUDE GRAHAME- (1879–1959), pioneer aviator and aircraft manufacturer. [See GRAHAME-WHITE.]

WHITE, Sir **(CYRIL) BRUDENELL (BINGHAM)** (1876-1940), general; joined Queensland permanent artillery, 1899; entered Staff College, Camberley, 1906; director, military operations, Australian Military Forces, 1912-14; organized Australian mobilization, first landings at Gallipoli, evacuation, and subsequent expansion of Australian forces in Egypt; DSO, 1915; chief of staff to (field-marshal, Lord) Birdwood [q.v.] in France, 1915-19; chief of general staff, Australia, 1919-23, 1939-40; chairman, Public Service Board, 1923-8; major-general, 1917; general, 1940; KCMG, 1919.

WHITE, Sir **GEORGE STUART** (1835-1912), field-marshal; born in county Antrim; entered army from Sandhurst, 1853; served in Indian Mutiny; VC for gallantry in Afghan war (1879), 1880; CB, 1880; took leading part in ending Burmese war, 1885-7; KCB, 1886; major-general, 1889; in command at Quetta and leader of Zhob valley expedition, 1889; GCIE for pacification of Baluchistan, 1893; commander-in-chief, India, 1893-7; greatly developed 'forward' policy which produced succession of frontier campaigns; GCB, 1897; quartermaster-general at War Office, 1897; in command in Natal during South African war, 1899-1901; defended Ladysmith, 2 Nov. 1899-28 Feb. 1900; governor of Gibraltar, 1900-5; field-marshal, 1903; OM, 1905.

WHITE, HENRY JULIAN (1859-1934), Latin biblical scholar and dean of Christ Church, Oxford; first class, theology (Christ Church), 1883; deacon, 1885; priest, 1886; assisted John Wordsworth [q.v.] in editing Vulgate New Testament, 1884-1911; thereafter sole editor; contributed to 'Old Latin Biblical Texts'; lecturer and chaplain, Merton College, 1895-1905; professor, New Testament exegesis, King's College, London, 1905-20; dean of Christ Church, 1920-34; FBA, 1932.

WHITE, JOHN CAMPBELL, Baron Overtoun (1843-1908), Scottish churchman and philanthropist; MA, Glasgow University, 1864; joined father's chemical firm at Shawfield, 1867; generous benefactor to United Free Church after the crisis of 1904, and to United Free Church mission in Livingstonia; supported liberal party in Scotland; baron, 1893.

WHITE, LEONARD CHARLES (1897-1955), general secretary, Civil Service Clerical Association; educated at Paston grammar school; entered Post Office, 1914; served in navy, 1916-17; clerical officer, Admiralty naval ordnance, 1920; clerical assistant, Civil Service Clerical Association, 1925; assistant secretary,

1928; assistant general secretary, 1936; general secretary, 1942-55; considerable contributor to strengthening national staff side of Civil Service Whitley Council.

WHITE, WILLIAM HALE (1831-1913), novelist (under the pseudonym of MARK RUTHERFORD], philosophical writer, literary critic, and civil servant; entered civil service, 1854; assistant director of contracts at Admiralty, 1879-91; works include *The Autobiography of Mark Rutherford* (1881) and *Mark Rutherford's Deliverance* (1885), his best-known novels, three series of *Pages from a Journal* (1900, 1910, 1915), translations of Spinoza, and books on Wordsworth and Bunyan.

WHITE, Sir **WILLIAM HALE-** (1857-1949), physician. [See HALE-WHITE.]

WHITE, Sir **WILLIAM HENRY** (1845-1913), naval architect; trained at Royal School of Naval Architecture, South Kensington; joined Admiralty staff as professional secretary to (Sir) E. J. Reed [q.v.], chief constructor of navy, 1867; secretary of council of naval construction, 1872; joined firm of Armstrong & Co., 1883-5; director of naval construction at Admiralty, 1885-1902; designs for battleships include those of *Royal Sovereign* class (1889) and *King Edward VII* class (1902); designs for cruisers include *Cressy* and *Drake* classes and twenty 'protected' cruisers; assistant controller of navy, 1885; KCB, 1895; author of *Manual of Naval Architecture* (1877) and many papers on the same subject.

WHITEHEAD, ALFRED NORTH (1861-1947), mathematician and philosopher; educated at Sherborne School and Trinity College, Cambridge; fourth wrangler, 1883; fellow, 1884-1947; lecturer, 1884-1910; published *A Treatise on Universal Algebra, with Applications* (1898); FRS, 1903; with Bertrand (Earl) Russell published *Principia Mathematica* (3 vols., 1910-13), greatest single contribution to logic since Aristotle; professor of applied mathematics, Imperial College of Science and Technology, 1914-24; of philosophy, Harvard, 1924-37; developed his philosophy of nature in *An Enquiry Concerning the Principles of Natural Knowledge* (1919) and *The Concept of Nature* (1920) putting forward doctrine that ultimate components of reality are events; defined his mature philosophy in *Process and Reality, an Essay in Cosmology* (1929); called it 'Philosophy of Organism'; FBA, 1931; OM, 1945.

WHITEHEAD, JOHN HENRY CONSTANTINE (1904-1960), mathematician; educated at Eton and Balliol College, Oxford;

first class, mathematics, 1926; studied at Princeton, 1929; published *The Foundations of Differential Geometry* (1932); lecturer in mathematics, 1932, and fellow and tutor, Balliol College, 1933; served in Admiralty and Foreign Office during 1939-45; Waynflete professor, pure mathematics, Magdalen College, Oxford, 1947; FRS, 1944.

WHITEHEAD, ROBERT (1823-1905), inventor; apprenticed as engineer at Manchester, 1837-44; joined uncle in Marseilles, 1844; set up in business for himself at Milan, 1847; employed at Trieste, 1848-56; started engineering works at Fiume, 1856; designed engines for several Austrian warships; invented Whitehead torpedo, 1866; solely constructed torpedoes and their accessories from 1872; established branch at Portland, 1890; invented 'servo-motor' to regulate steering, and continually increased the speed and improved the precision of torpedoes; received many foreign decorations.

WHITEING, RICHARD (1840-1928), journalist and novelist; apprenticed to Benjamin Wyon [q.v.], medallist and engraver of seals; set up for himself as engraver; made first essay in journalism and joined staff of *Morning Star*, 1866; Paris correspondent for London *World* and New York *World*, 1867; left Paris, 1886; on staff of *Daily News*, 1886-99; his works include *The Island* (1888) and *No. 5 John Street*, his most successful work (1899).

WHITELEY, WILLIAM (1831-1907), 'universal provider'; draper's assistant at Wakefield, 1848-52, and in London, 1852-63; opened small draper's shop in Westbourne Grove, 1863; by 1876 he had fifteen shops and 2,000 employees, and provided all kinds of goods; business premises six times destroyed by fire, 1882-7; he converted firm into limited company, 1899; was shot dead by Horace George Rayner, 24 Jan. 1907; left a million pounds for Whiteley homes for the aged poor, constructed at Burr Hill, Surrey.

WHITELEY, WILLIAM (1881-1955), politician; miner's son; left Brandon Colliery school at age of twelve; worked in coal-pit; joined labour party, 1906; played football for Sunderland; miners' agent, 1912; MP, Blaydon, 1922-31 and 1935-55; labour whip, 1926; junior lord of the Treasury, 1929-31; controller of the household, 1940; chief whip, 1942-55; PC, 1943; parliamentary secretary to Treasury, 1942-51; CH, 1948.

WHITEWAY, SIR WILLIAM VALLANCE (1828-1908), premier of Newfoundland; left Totnes for Newfoundland, 1843; called to

Newfoundland bar, 1852; QC, 1862; speaker of house of assembly, 1865-9; solicitor-general, 1873-8; premier and attorney-general, 1878-85, and again, 1889-94; counsel for Newfoundland at Halifax fisheries commission, 1877; KCMG, 1880; unseated for corrupt practices, 1894; again premier, 1895-7; attended diamond jubilee and colonial conference in London, and was made PC and DCL, Oxford, 1897; carried first railway bill for Newfoundland, 1880; negotiated earlier contracts with Robert G. Reid [q.v.].

WHITFIELD, ERNEST ALBERT, first BARON KENSWOOD (1887-1963), professional violinist and economist; educated at Archbishop Tenyson's and University College Schools, London, and Vienna and London Universities; took commercial appointment in Vienna, 1907, and rose to be departmental manager; forced by failing vision to seek new profession; made début as solo violinist, 1913; returned to London, 1914; introduced to braille by Sir (Cyril) Arthur Pearson [q.v.]; appointed leader of orchestra at Wyndham's Theatre, 1917; first appearance as soloist in London at Queen's Hall Promenade Concerts, 1918; formed Guild of Singers and Players, 1920; continued to perform on concert platform, 1921-3, but also studied economics; B.Sc., London, 1926; Ph.D., 1928; member, executive council, National Institute for the Blind, 1928; damaged hand forced abandonment of musical career, 1935; worked for blind in United States and Canada during 1939-45 war; governor, BBC, 1946-50; created baron, 1951; president, National Federation of the Blind, 1951-5; served on number of welfare committees and contributed to various periodicals on blind welfare.

WHITLA, SIR WILLIAM (1851-1933), physician; qualified at Queen's College, Belfast, 1873; MD, 1877; professor of materia medica and therapeutics, 1890-1918; pro-chancellor, 1924-33; MP, Queen's University, 1918-23; author of *Elements of Pharmacy . . .* (1882), *Dictionary of Treatment* (1891), and *Practice and Theory of Medicine* (1908); knighted, 1902.

WHITLEY, JOHN HENRY (1866-1935), speaker of the House of Commons; educated at Clifton and London University; entered family cotton-spinning business; liberal MP, Halifax, 1900-28; whip, 1907-10; chairman of ways and means, deputy speaker, and PC, 1911; speaker, 1921-8; maintained unruffled demeanour in disorder; chairman (1917-18) of committee on relations between employers and employed which recommended joint consultative machinery known by his name; chairman, royal commission on labour in India (1929-31), of governors of British Broadcasting Corporation, 1930-5.

WHITLEY, WILLIAM THOMAS (1858-1942), historian of art; wrote regularly in *Morning Post* and art periodicals; works include *Thomas Gainsborough* (1915) and three books on art in England, 1700-1837 (1928-30).

WHITMAN, ALFRED CHARLES (1860-1910), writer on engravings; attendant at prints department of British Museum, 1885-1903; departmental clerk, 1903-10; chief works on mezzotint engraving were monographs on Samuel Cousins (1904) and Charles Turner (1907).

WHITMORE, Sir GEORGE STODDART (1830-1903), major-general; born in Malta; joined Cape mounted rifles, 1847; distinguished in Kaffir wars, 1847 and 1851-3, and in Crimea; in command of Hawke's Bay militia, New Zealand, in last Maori war, 1866-9, of which he published an account, 1902; member of legislative council in New Zealand from 1863; colonial secretary and defence minister, 1877-9; commandant of the colonial forces and major-general, 1874; CMG, 1869; KCMG, 1882.

WHITNEY, JAMES POUNDER (1857-1939), ecclesiastical historian; educated at Owens College, Manchester, and King's College, Cambridge; first class, mathematics and history, 1881; deacon, 1883; priest, 1885; principal, Bishop's College, Lennoxville, Canada, 1900-5; chaplain, St. Edward's church, Cambridge, 1906-9; professor of ecclesiastical history, King's College, London (1908-18), at Cambridge (1919-39); joint-editor, *Cambridge Medieval History* (1907-22).

WHITTAKER, Sir EDMUND TAYLOR (1873-1956), mathematician, astronomer, and philosopher; educated at Manchester grammar school and Trinity College, Cambridge; bracketed second wrangler, 1895; first class, part ii, mathematical tripos, and fellow of Trinity, 1896; professor, astronomy, Dublin, 1906; professor, mathematics, Edinburgh, 1912-46; instituted mathematical laboratory, 1914; made numerous, important contributions to mathematics and theoretical physics; published *A Course of Modern Analysis* (1902); *Treatise on the Analytical Dynamics of Particles and Rigid Bodies* (1904); *The Calculus of Observations* (with G. Robertson) (1924); and *A History of the Theories of Aether and Electricity* (1910); knighted, 1945; FRS, 1905; hon. fellow, Trinity College, 1949.

WHITTARD, WALTER FREDERICK (1902-1966), Chaning Wills professor of geology in the university of Bristol; educated at county-secondary school, Battersea, Chelsea Polytechnic, and Imperial College, London; first class, geology and zoology; ARCS; B.Sc., London (external), 1924; research on the Valentian rocks of Shropshire; Ph.D., London; entered Sidney Sussex College, Cambridge, 1926; Ph.D., Cambridge; chief geologist on East Greenland expedition under (Sir) J. M. Wordie [q.v.], 1929; assistant lecturer, Imperial College, 1931; lecturer, 1935; professor of geology, Bristol, 1937-66; continued to publish papers on geology of Shropshire; during war of 1939-45, geological adviser to South-west region of England; involved in mapping geology of English Channel, 1955-6; FRS, 1957.

WHITTEN BROWN, Sir ARTHUR (1886-1948), air navigator and engineer. [See Brown.]

WHITTY, Dame MARY LOUISE (MAY) (1865-1948), actress. [See under Webster, Benjamin.]

WHITWORTH, GEOFFREY ARUNDEL (1883-1951), founder of the British Drama League; educated privately and at New College, Oxford; worked for Chatto and Windus, publishers, 1907-28; founder and secretary, British Drama League 1919-48; hon. secretary, Shakespeare Memorial Theatre Committee, 1930-51, CBE, 1947.

WHITWORTH, WILLIAM ALLEN (1840-1905), mathematical and religious writer; BA, St. John's College, Cambridge (sixteenth wrangler), 1862; MA, 1865; fellow, 1867-84; co-edited *The Messenger of Mathematics* till 1880; held livings in Liverpool and London; Hulsean lecturer at Cambridge, 1903-4; prebendary of St. Paul's, 1900; published doctrinal works and sermons; mathematical writings include *Trilinear Co-ordinates* (1866) and *Choice and Chance* (1867).

WHYMPER, EDWARD (1840-1911), wood-engraver and alpinist; second son of Josiah Wood Whymper [q.v.]; joined father's wood-engraving business; illustrated many books; prepared water-colour sketches for *Peaks, Passes and Glaciers* (2nd ser. 1862); made the first ascent of Pointe des Écrins, Dauphiné, of several peaks of Mont Blanc chain, 1864, and Aiguille Verte, 1865; his first successful ascent of the Matterhorn was followed by a disastrous descent, four of his party being killed, July 1865; he related experiences in *Scrambles in the Alps* (1871); visited interior of Greenland, 1867 and 1872; vice-president of Alpine Club,

1872-4; studied in the Andes of Ecuador the effect of atmosphere of high altitudes on human beings, 1888; there made geological and geographical researches of mountain systems, which he embodied in *Travels among the Great Andes of the Equator* (1892); royal medallist of Royal Geographical Society, 1892; devised 'Whymper' tent for mountaineering expeditions; compiled handbooks to Chamonix (1896) and Zermatt (1897); died at Chamonix.

WHYMPER, JOSIAH WOOD (1813-1903), wood-engraver; taught himself wood-engraving, and obtained first success by etching of New London Bridge, 1831; did much work for John Murray [q.v.], the SPCK, and RTS; had as pupils Frederick Walker and Charles Keene [qq.v.]; friend of Sir John Gilbert [q.v.]; member of New Water Colour Society, 1854.

WHYTE, ALEXANDER (1836-1921), divine; educated at King's College, Aberdeen, and New College, Edinburgh; minister of St. George's Free church, Edinburgh, 1870-1916; moderator of General Assembly, 1898; principal of New College, Edinburgh, 1909-18.

WIART, Sir ADRIAN CARTON DE (1880-1963), lieutenant-general. [See CARTON DE WIART.]

WICKHAM, EDWARD CHARLES (1834-1910), dean of Lincoln; educated at Winchester; BA, New College, Oxford, 1857; MA, 1859; DD, 1894; tutor, 1858-73; initiated system of intercollegiate lectures; headmaster of Wellington College, 1873-93; edited Horace (vol. i, 1874, vol. ii, 1893); dean of Lincoln, 1894-1910; married Agnes, daughter of W. E. Gladstone, 1873; died at Sierre, Switzerland; published *The Prayer-Book* (1895) and *Horace for English Readers* (1903).

WIGGINS, JOSEPH (1832-1905), explorer of the sea route to Siberia; apprenticed to uncle, a shipowner at Sunderland, 1846; Board of Trade examiner in navigation there, 1868; attempted to establish trade route by sea between Western Europe and Asiatic Russia; sailed from Dundee and rounded White Island, Aug. 1874; was accompanied by Henry Seebohm [q.v.] to Siberia, 1876; successfully brought back from the Ob a cargo of wheat, 1878; reached Yeniseisk, 1887-8; in larger vessel *Orestes* he reached mouth of Yenisei, delivering there a cargo of rails, 1893; was shipwrecked near Yugor Strait, 1894; awarded Murchison medal of Royal Geographical Society, 1894.

WIGHAM, JOHN RICHARDSON (1829-1906), inventor; apprentice in, and sub-sequently owner of, brother-in-law's hardware business 'Joshua Edmundson & Co.', Dublin; made experiments in gas-lighting, and designed several private and public gasworks; president of Dublin chamber of commerce, 1894-6; invented important applications of gas to lighthouse illumination; invented 'composite' burner, 1868, and first group-flashing arrangement in lighthouses, 1871; installed at Galley Head a quadriform light, 1878; his system was supported by John Tyndall [q.v.] in controversy with Joseph Chamberlain and Sir James N. Douglass [qq.v.]; as member of Society of Friends he refused knighthood, 1887.

WIGRAM, CLIVE, first BARON WIGRAM (1873-1960), private secretary to King George V; educated at Winchester; commissioned in Royal Artillery, 1893; served in India and South Africa, 1893-1908; military secretary to commander-in-chief, Aldershot, 1908-10; assistant private secretary to George V, 1910-31; private secretary, 1931-6; KCVO, 1928; PC, 1932; GCB, 1933; baron, 1935.

WIGRAM, WOOLMORE (1831-1907), campanologist; BA, Trinity College, Cambridge, 1854; MA, 1858; hon. canon of St. Albans, 1886; published *Change-ringing Disentangled* (1880); member of Alpine Club, 1858-68; made first successful ascent of La Dent Blanche, July 1862.

WILBERFORCE, ERNEST ROLAND (1840-1907), bishop successively of Newcastle and Chichester; son of Samuel Wilberforce [q.v.]; educated at Harrow and Exeter College, Oxford; BA, 1864; MA, 1867; DD, 1882; vicar of Seaforth, Liverpool, 1873-8; canon of Winchester, 1878-82; first bishop of Newcastle, 1882-95; inaugurated fund for diocesan purposes; bishop of Chichester, 1895-1907; dealt successfully with ritualistic practices in his diocese; ardent temperance advocate; chairman of Church of England Temperance Society, 1896; visited South Africa, 1904.

WILBRAHAM, Sir PHILIP WILBRAHAM BAKER, sixth baronet (1875-1957), ecclesiastical lawyer and administrator; educated at Harrow and Balliol College, Oxford; fellow, All Souls, 1899-1906; called to bar (Lincoln's Inn), 1901; bencher, 1942; specialized in ecclesiastical matters; succeeded to baronetcy, 1912; chancellor, diocese of Chester, 1913; chancellor, York, 1915; Truro, 1923; Chelmsford, 1928; and Durham, 1929; dean of the Arches, Canterbury, and auditor, chancery court, York, 1934-55; first secretary, Church Assembly, 1920-39; first Church estates commissioner, 1939-54; KBE, 1954; DCL, 1936.

WILD, (JOHN ROBERT) FRANCIS (1873–1939), Antarctic explorer; became naval rating, 1900; member of Antarctic expeditions of R. F. Scott (1901), Sir E. H. Shackleton [qq.v.] (1907, and as second-in-command, 1914 and 1921–2), and (Sir) Douglas Mawson (1911); energetic, resourceful, and an incomparable sledger.

WILDE, JOHANNES (JÁNOS) (1891–1970), art historian; born at Budapest; educated at State Gymnasium and university of Budapest; studied also at Vienna University, 1915–17; Ph.D., 1918; published translation into Hungarian of Adolf Hildebrand's *Das Problem der Form in der bildenden Kunst* (1910); assistant keeper, Prints and Drawings Department, Budapest Museum of Fine Arts, 1918–22; prepared, in collaboration with Karl Swoboda, edition of collected work of Max Dvořák (1921); assistant keeper, Kunsthistorisches Museum, Vienna, 1923; keeper; became Austrian citizen, 1928; left Austria with his Jewish wife, 1939; learnt English and worked at Aberystwyth University Library; formed friendship with A. E. Popham [q.v.], and invited by British Museum to compile catalogue of Michelangelo drawings, 1940; interned as enemy alien in Canada, 1940; released, 1941; resumed work on Michelangelo; became British subject, 1947; reader in history of art, London University, and deputy director, Courtauld Institute of Art, 1948–58; given title of professor, 1950; FBA, 1951; CBE, 1955; one of most important achievements systematic use of X-rays in study of pictures; greatest achievement, his catalogue of Michelangelo drawings in British Museum, published in 1953.

WILDE, WILLIAM JAMES (JIMMY) (1892–1969), professional boxer; son of Welsh coal-miner; left school at thirteen to work in the pits; engaged in boxing at fairground booths; took up boxing professionally, 1910; fought 138 fights including seven British or world-title contests, 1911–23; flyweight champion of the world, 1916–23; won outright Lord Lonsdale challenge belt; known as 'the Mighty Atom'; beaten only four times; after retirement from the ring worked as manager and promotor.

WILDING, ANTHONY FREDERICK (1883–1915), lawn-tennis player; born in New Zealand; BA, Trinity College, Cambridge; called to bar, 1906; All England singles champion at Wimbledon, 1910–14; four times winner of championship doubles; killed in Artois during 1914–18 war.

WILKIE, SIR DAVID PERCIVAL DALBRECK (1882–1938), surgeon; educated at Edinburgh Academy and University; MB,

1904; MD, 1908; assistant surgeon, Royal Infirmary, Edinburgh, 1912; acknowledged master of abdominal surgery; professor of systematic surgery, Edinburgh, 1924–38; particularly interested in development of research department; director, surgical unit of the municipal hospitals, 1932–8; took notable part in administrative and social affairs of the university; knighted, 1936.

WILKINS, AUGUSTUS SAMUEL (1843–1905), classical scholar; BA, St. John's College, Cambridge (fifth classic), and MA, London, 1868; Litt.D., 1885; president of Union, 1868; won university prize for English essay; disqualified as nonconformist for fellowship; Latin lecturer, 1868, and professor, 1869–1903, at Owens College, Manchester; edited Cicero's rhetorical works, including *De Oratore*, lib. i–iii (1879–92), and Horace's *Epistles* (1885); wrote primers of *Roman Antiquities* (1877) and *Roman Literature* (1890); received hon. degrees from St. Andrews and Dublin.

WILKINS, SIR (GEORGE) HUBERT (1888–1958), polar explorer, climatologist, and naturalist; born, South Australia; educated at Adelaide School of Mines, 1903–8; appointed photographer, Canadian Arctic Expeditions, 1913; served in Australian Flying Corps as photographer in France, 1916–18; MC, with two bars; second-in-command, J. L. Copes' expedition to Antarctic, 1920–1; naturalist with (Sir) Ernest Shackleton [q.v.] in *Quest* expedition; on expedition for British Museum in Northern Australia, 1923–5; published *Undiscovered Australia* (1928); flew from Barrow, Alaska, to Spitsbergen, 1928; published *Flying the Arctic* (1928); organized submarine expedition to Arctic, 1931; published *Under the North Pole* (1931); four Antarctic expeditions with Lincoln Ellsworth; geographer, climatologist, and Arctic adviser with United States forces, 1942–5; served in United States Navy Office and Weather Bureau, 1946–53; knighted, 1928; hon. D.Sc., Alaska, 1955.

WILKINS, WILLIAM HENRY (1860–1905), biographer; BA, Clare College, Cambridge, 1887; MA, 1899; published *The Alien Invasion* (1892), and four novels, 1892–4; discovered at Lund, in Sweden, 1897, and issued the unpublished correspondence of Queen Sophia Dorothea, consort of George I, and her lover Count Philip Königsmarck (2 vols., 1900); wrote life of Isabel Lady Burton (1897), Queen Caroline (1901), Mrs Fitzherbert and George IV (1905), and other biographical works.

WILKINSON, ELLEN CICELY (1891–1947), trade-unionist and politician; second class, history, Manchester, 1913; national woman organizer, Union of Shop, Distributive and Allied Workers from 1915; represented union on four trade boards, 1919-25; joined independent labour party, 1912; communist party, 1920-4; labour MP, Middlesbrough East (1924-31), Jarrow (1935-47); parliamentary private secretary to Susan Lawrence [q.v.], 1929-31; headed India League delegation to India, 1932; led Jarrow march to London, 1936; passionately opposed non-intervention in Spain, 1937; carried hire-purchase bill through parliament, 1938; parliamentary secretary, Ministry of Home Security, 1940-5; minister of education, 1945-7; implemented Act of 1944; PC, 1945; vivid personality and hard fighter.

WILKINSON, GEORGE HOWARD (1833-1907), successively bishop of Truro and of St. Andrews; BA, Oriel College, Oxford, 1854; MA, 1859; DD, 1883; vicar of Bishop Auckland, 1863-7; incumbent of St. Peter's, Eaton Square, 1870-83; bishop of Truro, 1883-91; after visit to South Africa for health was elected bishop of St. Andrews, 1893; primus of Scottish Episcopal Church, 1904; published devotional works.

WILKINSON, (HENRY) SPENSER (1853-1937), military historian and journalist; educated at Owens College, Manchester, and Merton College, Oxford; leader-writer, *Manchester Guardian* (1882-92), *Morning Post* (1895-1914); Chichele professor of military history, Oxford, 1909-23; works include the immensely influential *The Brain of an Army* (1890) and *The Brain of the Navy* (1895); moving spirit in foundation of Navy League, 1894.

WILKINSON, Sir NEVILE RODWELL (1869-1940), soldier, herald, and artist; educated at Harrow and Sandhurst; held commission in Coldstream Guards, 1890-1907; studied at National Art Training School; Ulster King of Arms, 1908-40; his sixteen-roomed model house ('Titania's Palace'), complete in every detail, finished in 1923 and widely exhibited; knighted, 1920; KCVO, 1921.

WILKINSON, NORMAN (1882-1934), stage designer; studied at Birmingham School of Art; lacked application to co-ordinate his diverse talents; for Harley Granville-Barker [q.v.] designed sets for *A Winter's Tale* and *Twelfth Night* (1912), *A Midsummer Night's Dream* and *The Dynasts* (1914); after 1918 designed for Sir Nigel Playfair [q.v.], the Stage Society, and Stratford Festival.

WILKS, Sir SAMUEL, baronet (1824-1911), physician; student at Guy's Hospital, 1842;

MRCS, 1847; MB, 1848; MD, 1850; FRCP, 1856; president, 1896-9; Harveian orator, 1879; physician of Guy's Hospital, 1866-85; lecturer on pathology and medicine there; published *Lectures on Pathological Anatomy* (1859); edited *Guy's Hospital Reports* (1854-65); also published *Lectures on Diseases of the Chest* (1874) and of the nervous system (1878); baronet, 1897.

WILL, JOHN SHIRESS (1840-1910), legal writer; called to bar, 1864; QC, 1883; liberal MP for Montrose Burghs, 1885-96; county court judge at Liverpool, 1906; wrote authoritative works on law relating to gas and electric lighting.

WILLCOCKS, Sir JAMES (1857-1926), general; born in India; gazetted to 100th Foot in Punjab, 1878; appointed to newly formed army transport department in India, 1884; adjutant of 100th Foot, 1887-9; in command of Ashanti field force for relief of Kumasi, which he reached in spite of appalling difficulties, July 1900; commanded northern army in India, 1910; commanded Indian army corps which proceeded to France, 1914; resigned, owing to differences with Sir Douglas (afterwards Earl) Haig [q.v.], 1915; governor of Bermuda, 1917-22; KCMG, 1900; died at Bharatpur.

WILLCOX, Sir WILLIAM HENRY (1870-1941), physician and toxicologist; B.Sc., London, 1892; MB and DPH, St. Mary's Hospital, 1900; lecturer on chemical pathology, 1900-30; on forensic medicine, 1906-35; outpatients' physician, 1907-13; physician, 1913-35; junior analyst, Home Office, 1904-8; senior analyst, 1908-19; medical adviser, 1919-41; work on Crippen (1910) and Seddon (1912) cases especially notable; consulting physician in Gallipoli and Mesopotamia, 1914-18; KCIE, 1921; specially interested in rheumatic diseases and industrial toxicology.

WILLES, Sir GEORGE OMMANNEY (1823-1901), admiral; joined navy, 1838; present at bombardments of Odessa and Sevastopol, 1854; commanded rocket boats in attack on Peiho forts, 1860; CB, 1861; chief of staff at Admiralty, 1870-3; ADC to Queen Victoria, 1870-4; admiral superintendent at Devonport, 1876-9; commander-in-chief in China, 1881-4; KCB, 1884; admiral, 1885; commander-in-chief at Portsmouth, 1885-8; GCB, 1892.

WILLETT, WILLIAM (1856-1915), builder, and the originator of 'daylight saving'; with father created reputation for Willett-built houses; advocated introduction of first daylight saving bill, 1908; bill passed, 1916; Summer Time Act passed, 1925.

WILLIAMS, ALFRED (1832-1905), alpine painter; drew on wood for book-illustrations, 1849-56; engaged in maltster's business at Salisbury, 1861-86; member of Alpine Club, 1878; exhibited paintings of alpine scenery at Alpine Club.

WILLIAMS, ALWYN TERRELL PETRE (1888-1968), headmaster and bishop; educated at Rossall School and Jesus College, Oxford; first class, *lit. hum.*, 1910, and modern history, 1911; fellow, All Souls, 1911; ordained deacon, 1913; priest, 1914; history master at Winchester, 1915-24; succeeded M. J. Rendall [q.v.] as headmaster, 1924-34; dean of Christ Church, Oxford, 1934-9; succeeded H. Hensley Henson [q.v.] as bishop of Durham, 1939-52; bishop of Winchester, 1952-61; concerned with production of New English Bible; chairman, Joint Committee of Churches and Bible Societies, 1950; prelate to the Most Noble Order of the Garter; installed Sir Winston Churchill; hon. fellow, Jesus and All Souls; DD, Oxford; hon. degrees from Durham, St. Andrews, Glasgow, and Southampton; published *The Anglican Tradition in the Life of England* (1947).

WILLIAMS, (ARTHUR FREDERIC) BASIL (1867-1950), historian; educated at Marlborough and New College, Oxford; professor of history, McGill, 1921-5, Edinburgh, 1925-37; publications include lives of William Pitt, Earl of Chatham (1913), Cecil Rhodes (1921), and Stanhope (1932), and *The Whig Supremacy 1714-60* (1939); FBA, 1935.

WILLIAMS, CHARLES (1838-1904), war correspondent; first editor of *Evening Standard*, and of *Evening News*, 1881-4; correspondent for *Standard* in Franco-Prussian war (1870), in Armenian campaign (1877), at Berlin congress (1878), in Afghanistan (1879-80), and in Nile expedition (1884); war correspondent for *Daily Chronicle* in Eastern Europe, 1885 and 1897, and at Omdurman and Khartoum, 1898; published independent life of Sir Evelyn Wood (1892); founder and president (1896-7) of the Press Club.

WILLIAMS, CHARLES HANSON GREVILLE (1829-1910), chemist; consulting and analytical chemist in London, 1852-3; assistant to (Sir) William H. Perkin [q.v.], 1863-8; partner in Star Chemical Works, Brentford, 1868-77, and chemist to Gas Light and Coke Co., London, 1877-1901; discovered cyanine or quinoline-blue; wrote much on chemistry of coal gas, and published *Chemical Analysis* (1858); FRS, 1862.

WILLIAMS, CHARLES WALTER STANSBY (1886-1945), author and scholar; educated at St. Albans School and University College, London; publisher's reader, Oxford University Press, 1908-45; published first book of verse *The Silver Stair* (1912) followed by over thirty volumes of poetry, plays, literary criticism, fiction, biography, and theological argument; increasingly devoted himself in his writings to propagation and elaboration of doctrines of romantic love and the co-inherence of all human creatures; devoted member of Church of England.

WILLIAMS, EDWARD FRANCIS, BARON FRANCIS-WILLIAMS (1903-1970), author, journalist, and publicist; educated at Queen Elizabeth's grammar school, Middleton; while at school had two poems published by *Country Life* and *Weekly Westminster Gazette*; joined the *Bootle Times*, 1910; joined independent labour party; moved to Liverpool *Daily Courier*; freelance journalist in London, 1922; came to notice of Lord Beaverbrook [q.v.], and appointed financial reporter, *Sunday Express*, 1926; city editor, *Daily Herald*, 1929; editor, 1936; resigned after disagreement with Lord Southwood [q.v.], chairman of Odhams, 1940; controller, press and censorship, Ministry of Information, 1941; CBE, 1945; public relations adviser to Clement Attlee, 1945-7; correspondent of *Observer* in Washington, and columnist for *News Chronicle* and *New Statesman*; governor of BBC, 1951-2; became television personality; Regent's professor, California University, 1961; Kemper Knapp professor, Wisconsin University, 1967; life peer, 1962; publications include *War by Revolution* (1940), *Democracy's Last Battle* (1941), *Press, Parliament and People* (1946), *Fifty Years' March: the Rise of the Labour Party* (1949), *Ernest Bevin: Portrait of a Great Englishman* (1952), *A Prime Minister Remembers* (with Earl Attlee, 1961), *The American Invasion* (1962), *A Pattern of Rulers* (1965), and his autobiography *Nothing So Strange* (1970); he also wrote a number of novels.

WILLIAMS, SIR EDWARD LEADER (1828-1910), engineer; apprenticed to father, who was chief engineer to Severn navigation commission, 1846; resident engineer on Shoreham harbour works, 1849-52; engineer on Admiralty pier, Dover, 1852-5, and to River Weaver Trust, 1856; established through traffic from river Weaver to Trent and Mersey canal by means of hydraulic lift for transferring canal boats; became (1882) chief engineer for projected Manchester ship canal, which, 35½ miles long, was opened by Queen Victoria in 1894; knighted, 1894; MICE, 1860; vice-president, 1906-7.

WILLIAMS, SIR GEORGE (1821-1905), founder of the Young Men's Christian Association; apprenticed to draper at Bridgwater, 1836; entered employ of Messrs. Hitchcock & Rogers, drapers of St. Paul's Churchyard, 1841, subsequently becoming partner; started among employees Mutual Improvement Society, 1842, and established Young Men's Christian Association, 1844; work spread to Europe and colonies; first international conference at Paris, 1855; leased Exeter Hall as headquarters, 1880-1907; new buildings opened in Tottenham Court Road, 1912; president, 1886; knighted, 1894; present at jubilee of World's Alliance of YMCAs at Paris, 1905.

WILLIAMS, SIR HAROLD HERBERT (1880-1964), critic and scholar; educated at Liverpool College and Christ's College, Cambridge; first class, part i, theological tripos, 1903; ordained deacon and priest, 1904; lecturer and chaplain, Ripon Theological College, 1905; curate, Crediton, 1906-9; relinquished orders; university extension lecturer in English literature, Cambridge and London; published *Two Centuries of the English Novel* (1911), *Modern English Writers* (1918), and *Outlines of Modern English Literature 1890-1914* (1920); served in army during 1914-18 war; called to bar (Inner Temple), 1920; JP; chairman, Hertford County Council, 1947-52; chairman, Local History Council, 1950-63; Sandars reader in bibliography, Cambridge, 1950; worked on bibliography of Swifts' works; published edition of Swifts' poems (1937) and the *Journal to Stella* (1948); authority on *Gullivers Travels*; published five volumes of Swifts' *Letters* (1963-5); president, Bibliographical Society, 1938-44; FBA, 1944; FSA, 1948; knighted, 1951; hon. D.Litt., Durham, 1954.

WILLIAMS, HUGH (1843-1911), ecclesiastical historian; student from 1864 and tutor, 1867-9, at Calvinistic Methodist College, Bala; BA, London, 1870; MA, 1871; professor of Greek and mathematics at Bala, 1873-91; professor of church history, 1891-1911; hon. DD, Glasgow, 1904; edited *Gildas* with English translation and notes, 2 parts, 1899-1901; his *Christianity in Early Britain*, a careful piece of research, appeared in 1912; published biblical commentaries in Welsh.

WILLIAMS, IVY (1877-1966), the first woman to be called to the English bar; educated privately and travelled in Europe; joined the Society of Oxford Home-Students (later St. Anne's College); taught by Edward Jenks and (Professor Sir) William S. Holdsworth [qq.v.]; BCL, 1902; LLD, London, 1903; called to bar (Inner Temple), 1922; tutor and lecturer in law, Society of Oxford Home-Students, 1920-45; published *The Sources of Law in the Swiss Civil Code* (1923) and *The Swiss Civil Code: English Version, with Notes and Vocabulary* (1925); DCL, Oxford, the first woman to be awarded this degree; served as delegate to Hague Conference for the Codification of International Law, under Sir Maurice L. Gwyer [q.v.], 1930; hon. fellow, St. Anne's College, 1956; taught herself to read braille and published booklet on braille for the National Institute for the Blind, 1948; generous benefactor; endowed two law scholarships at Oxford, one for women only.

WILLIAMS, JOHN CARVELL (1821-1907), nonconformist politician; secretary to Society for Liberation of Religion from State Control, 1847-77; vice-president, 1903; founded the *Liberator*, a monthly periodical, 1853; chairman of Congregational Union, 1900; liberal MP for South Nottinghamshire, 1885-6, and for Mansfield, 1892-1900; promoted Burials Act of 1880, and Marriage Acts of 1886 and 1898.

WILLIAMS, SIR JOHN COLDBROOK HANBURY- (1892-1965), industrialist. [See HANBURY-WILLIAMS.]

WILLIAMS, SIR JOHN FISCHER (1870-1947), international lawyer; educated at Harrow and New College, Oxford; first class, *lit. hum.*, and fellow, 1892; called to bar (Lincoln's Inn), 1894; KC, 1921; British legal adviser, Reparation Commission, 1920-30; chairman, royal commission resulting in Tithe Act, 1936; British member, Permanent Court of Arbitration, The Hague, from 1936; knighted, 1923.

WILLIAMS, NORMAN POWELL (1883-1943), divine; scholar of Christ Church, Oxford; first class, *lit. hum.*, 1906; deacon, 1908; priest, 1909; chaplain fellow, Exeter College, 1909-27; Lady Margaret professor of divinity and canon of Christ Church, 1927-43; subdean, 1939-43; works include *The Ideas of the Fall and of Original Sin* (1927); a liberal catholic.

WILLIAMS, RALPH VAUGHAN (1872-1958), composer. [See VAUGHAN WILLIAMS.]

WILLIAMS, SIR ROLAND BOWDLER VAUGHAN (1838-1916), judge; son of Sir Edward Vaughan Williams [q.v.]; BA, Christ Church, Oxford; called to bar (Lincoln's Inn), 1864; QC, 1889; judge of Queen's Bench division, 1890; assigned bankruptcy jurisdiction of high court, 1891; lord justice of appeal, 1897-1914; PC, 1897; published *The Law and Practice of Bankruptcy* (1870, 13th edn. 1925).

WILLIAMS, ROWLAND, 'Hwfa Môn' (1823-1905), archdruid of Wales; Congregational minister in London and Wales; distinguished himself at national eisteddfod at Caernarvon, 1862, Mold, 1873, and at Birkenhead, 1878; chief bardic adjudicator from 1875 to 1892; archdruid of the bardic Gorsedd from 1894; published collected works, 1883 and 1903.

WILLIAMS, THOMAS, BARON WILLIAMS OF BARNBURGH (1888-1967), politician; left school at eleven to work at Thrybergh Hall Colliery; at Wath Main Colliery, became interested in trade-union affairs; dismissed from work for criticizing working arrangements; steward Working Men's Club, Wath upon Dearne; worked again in pit at Bamburgh Main Colliery, 1919; member, local district council; backed by Yorkshire Miners' Association, labour MP, Don Valley, 1922-59; parliamentary private secretary to N. E. (later lord) Noel-Buxton [q.v.], minister of agriculture, 1924; parliamentary private secretary to Margaret G. Bondfield [q.v.], minister of labour, 1929; joint parliamentary secretary to R. S. (later viscount) Hudson [q.v.], minister of agriculture, 1940; labour minister of agriculture, 1945-51; PC, 1941; hon. LLD, Cambridge University (1951), and Nottingham (1955); life peer, 1961; published autobiography, *Digging for Britain* (1965).

WILLIAMS, WATKIN HEZEKIAH, 'Watcyn Wyn' (1844-1905), Welsh poet; conducted Hope Academy at Ammanford, 1880-1905; won bardic chair at Aberdare, 1885, and crown at World's Fair eisteddfod at Chicago, 1893; published in Welsh lyrical and humorous poems and a survey of Welsh literature, 1900; autobiography edited by J. Jenkins, 1907.

WILLIAMS-FREEMAN, JOHN PEERE (1858-1943), archaeologist; educated at Haileybury and Woolwich; MB (1885), MD (1888), Durham; practised at Weyhill, Hampshire, 1889-1928; published *Field Archaeology as illustrated by Hampshire* (1915); realized value of air photography to archaeology.

WILLIAMSON, ALEXANDER WILLIAM (1824-1904), chemist; educated at Dijon and Heidelberg; studied chemistry at Giessen under Liebig (Ph.D., 1846), and mathematics in Paris with Comte, 1846-9; professor of practical chemistry at University College, London, 1849-87; trained at the college many Japanese

youths from 1863 onwards; from 1849 he made researches into theory of etherification, which resulted in the theory of the constitution of salts and the doctrine of valency; his papers on the subject were reissued as an Alembic Club reprint, 1902; he published other papers mainly on gas analysis; president of Chemical Society, 1863-5, 1869-71; president of British Association, 1873; general treasurer, 1874-91; FRS, 1855; royal medallist, 1862; FRS Edinburgh, 1883; hon. member of Royal Irish Academy, 1885; received hon. degrees from Dublin, Edinburgh, and Durham, and many foreign academic honours; prominent in introducing science degrees at London University; chief gas examiner to Board of Trade, 1876-1901; published *Chemistry for Students* (1865).

WILLIAMSON, JOHN THOBURN (1907-1958), geologist and diamond merchant; born in Canada; educated at Macdonald high school and McGill University; BA, honours in geology, 1928; MA, 1930; Ph.D., 1933; geologist, Northern Rhodesia, 1934; Tanganyika, 1935; diamond prospecting, 1937-40; registered claim to mine at Mwadui, 1940; rapid development from 1948 onwards; many valuable diamonds discovered; Mwadui made a garden city; generous benefactions to Makerere College, Uganda, for science and to assist mining development in East Africa; hon. D.Sc., McGill, 1956.

WILLINGDON, first MARQUESS OF (1866-1941), governor-general of Canada and viceroy of India. [See FREEMAN-THOMAS, FREEMAN.]

WILLIS, HENRY (1821-1901), organbuilder; articled to John Gray, 1835-42; organist at Islington chapel-of-ease, 1860-91; capable player on double-bass; started organbuilding in London in 1845; rebuilt Gloucester Cathedral organ, 1847; built organs at Great Exhibition, 1851 (later used in Winchester Cathedral), at St. George's Hall, Liverpool, 1855, at Great Exhibition of 1862, at Royal Albert Hall, 1871, and at St. Paul's Cathedral, 1872; devised important inventions in organbuilding; extended range of pedal board from G to C; keen yachtsman.

WILLIS, WILLIAM (1835-1911), lawyer; BA, London, 1859; LLD, 1865; called to bar (Inner Temple), 1861; QC, 1877; fervid and voluble advocate; president of Baptist conference, 1903; liberal MP for Colchester, 1880-5; carried motion for exclusion of bishops from House of Lords, 1884; county court judge, 1897; had wide knowledge of English literature;

published several lectures, including *The Shakespeare-Bacon Controversy* (1902) and kindred works.

WILLOCK, HENRY DAVIS (1830–1903), Indian civil servant; born in Persia; nephew of John Francis Davis [q.v.]; joined Indian civil service, 1852; commanded company of volunteers in Mutiny; only civilian to receive the medal with three clasps; after serving various districts as magistrate and collector, and as judge of Benares, he was judge of Azimgarh, 1876–84.

WILLOUGHBY, DIGBY (1845–1901), soldier adventurer; raised Willoughby's Horse, and served in Basuto war, 1880; general commander of Madagascar forces, 1884–5; after first Matabele war helped in administration of Rhodesia, 1893; one of council of defence at Bulawayo in second Matabele war, 1896.

WILLS, SIR GEORGE ALFRED, first baronet, of Blagdon (1854–1928), president of the Imperial Tobacco Company, philanthropist; entered Bristol tobacco firm of W. D. & H. O. Wills, 1874; with his cousin W. H. Wills (afterwards Baron Winterstoke, q.v.) conducted negotiations which led to foundation of Imperial Tobacco Company of Great Britain and Ireland, 1901; chairman, 1911; president, 1924; benefactor to Bristol General Hospital; presented Leigh Woods to National Trust, 1909; with his brother presented magnificent buildings and playing fields to Bristol University; chairman of council, 1913; baronet, 1923.

WILLS, WILLIAM HENRY, BARON WINTERSTOKE (1830–1911), benefactor to Bristol; after education at Mill Hill School, joined family tobacco business at Bristol; partner, 1858; chairman of directors on its conversion into limited company; chairman of Imperial Tobacco Company, 1901–11; liberal MP for Coventry, 1880–5, and for East Bristol, 1895–1900; baronet, 1893; at Bristol he erected art gallery, branch library, and statue of Burke; made large donations to Bristol University; benefactor to Mansfield College, Oxford, and Mill Hill School; baron, 1906; high sheriff of Somerset, 1905–6.

WILMOT, JOHN, BARON WILMOT OF SELMESTON (1895–1964), politician; educated at Hither Green central school; left at fifteen to work as office boy; studied in night classes at Chelsea Polytechnic and King's College, London; joined Westminster Bank; rose rapidly to general manager and then to be secretary, Anglo-Russian Bank; served as aircraftsman in Royal Naval Air Service, 1914–18; helped to found East Lewisham labour party, 1919; MP, East Fulham, 1933–5; MP, Kennington, 1939–45; parliamentary private secretary to E. H. J. N. (later Lord) Dalton [q.v.], at Ministry of Economic Warfare, 1940; and at Board of Trade, 1942; joint parliamentary secretary, Ministry of Supply, 1944; MP, Deptford, 1945–7; minister of supply and aircraft production, 1945–7; recommended nationalization of iron and steel; opposed by H. S. Morrison (later Lord Morrison of Lambeth, q.v.) and R. Stafford Cripps [q.v.]; dismissed by C. R. Attlee, 1947; accepted peerage, 1950; returned to his business interests; on boards of number of companies; chairman of six.

WILMOT, SIR SAINTHILL EARDLEY- (1852–1929), forester; born in Tasmania; joined Indian forest service, 1873; assistant conservator of forests, North-West Provinces and Oudh, 1874–90; conservator in Oudh, 1890–8; in Burma, 1900–3; inspector-general of Indian forests, 1903–9; inaugurated Imperial Forest Research Institute, Dehra Dun, 1906; member of Development Commission, 1910–15; forestry adviser to commissioners, 1915–19; KCIE, 1911.

WILSHAW, SIR EDWARD (1879–1968), president, Cable and Wireless (Holding) Ltd., and financier; left school at fourteen and joined Eastern Telegraph Co. Ltd. as clerk; secretary, Eastern Telegraph Cable Co., 1922; Eastern cable and Marconi wireless telegraph interests merged, 1929; general manager and secretary, 1929; chief general manager, 1933; joint managing director, Cable and Wireless Ltd., 1935; chairman and sole managing director, 1936; when Cable and Wireless became state-owned public company, founded Electra House Group, investment trust; first governor, 1947–64; president, Chartered Institute of Secretaries, 1930–1; LLD, London University; KCMG, 1939.

WILSON, SIR ARNOLD TALBOT (1884–1940), soldier, explorer, civil administrator, author, and politician; son of J. M. Wilson [q.v.]; educated at Clifton and Sandhurst; posted to 32nd Sikh pioneers, 1904; transferred to Indian political department, 1909; served in Persia (1907–13), on Turco-Persian boundary commission, 1913–14; deputy chief political officer, Indian expeditionary force 'D' at Basra, 1915; DSO, 1916; deputy civil commissioner (1916–18), acting civil commissioner (1918–20), Persian Gulf; KCIE, 1920; with Anglo-Persian Oil Company, 1921–32; national conservative MP, Hitchin division, 1933–40; pressed for rearmament and compulsory military service after 1933; firm believer in British Empire as civilizing influence; made close study of and wrote on industrial assurance and workmen's compensation; recorded his opinions and experiences in a number of books; shot down serving as air gunner in Royal Air Force.

WILSON, ARTHUR (1836-1909), shipowner; generous benefactor to Hull; host of King Edward VII (while Prince of Wales) at Tranby Croft when allegations were made against a guest of cheating at baccarat, Sept. 1890.

WILSON, SIR ARTHUR KNYVET, third baronet (1842-1921), admiral of the fleet; entered navy, 1855; VC for gallantry at second battle of El Teb, 1884; commanded torpedo training ship, *Vernon*, at Portsmouth, 1889; rear-admiral, 1895; controller of navy and third sea lord, 1897; vice-admiral and commander of Channel squadron, 1901; elaborated successful tactical system; admiral of fleet, 1907; first sea lord of Admiralty, 1910-12; OM, 1912; succeeded brother, 1919.

WILSON, CHARLES HENRY, first BARON NUNBURNHOLME (1833-1907), shipowner and politician; joined with brother Arthur [q.v.] father's shipowning firm at Hull; joint-manager, 1867; extended Baltic service, and inaugurated new services from 1870; subsequently absorbed other shipping lines; liberal MP for Hull, 1874-85, and for West Hull, 1885-1905; ardent free trader; baron, 1905.

WILSON, SIR CHARLES RIVERS (1831-1916), civil servant and financier; BA, Balliol College, Oxford; entered Treasury, 1856; comptroller-general of National Debt Office, 1874-94; served on council of Suez Canal Company, 1875-96; vice-president of Anglo-French commission on revenue and expenditure of Egyptian khedive, Ismail, 1878; finance minister in Nubar Pasha's ministry under Khedive Ismail, 1878; dismissed, 1879; KCMG, 1880; president of Grand Trunk Railway of Canada, 1895-1909.

WILSON, CHARLES ROBERT (1863-1904), historian of British India; BA, Wadham College, Oxford, 1887; D.Litt., 1902; officer in charge of Indian government records, 1900; published *The Early Annals of the English in Bengal* (3 parts, 1895-1911).

WILSON, CHARLES THOMSON REES (1869-1959), physicist; educated at the Owens College, Manchester and Sidney Sussex College, Cambridge; B.Sc., 1887; first class, parts i and ii, natural sciences tripos, 1890-2; fellow,

Sidney Sussex, 1900; lecturer, experimental physics, 1901-19; reader, electrical meteorology, 1919-25; Jacksonian professor of natural philosophy, 1925-34; outstanding achievements in field of physical optics; FRS, 1900; Nobel prize for physics, 1927; CH, 1937; hon. Sc.D., Cambridge, 1947.

WILSON, SIR CHARLES WILLIAM (1836-1905), major-general R.E.; educated at Cheltenham and Bonn University; joined Royal Engineers, 1855; made survey of Jerusalem, 1864-5, which led to formation of Palestine Exploration Fund, 1865; chairman, 1901; with E. H. Palmer [q.v.] surveyed Sinaitic peninsula, 1868-9; first director of topographical department of War Office, 1870; FRS, 1874; CB, 1877; head of ordnance survey in Ireland between 1876 and 1886; British military consul-general in Anatolia, Asia Minor, 1879-82; KCMG, 1881; military attaché to British agency in Egypt, 1882-3; hon. DCL, Oxford, 1883; chief of intelligence department in Nile expedition, Sept. 1884; KCB, 1885; hon. LLD, Edinburgh, 1886; director-general of the ordnance survey of the United Kingdom, 1886-93; director-general of military education at War Office, 1892-8; revisited Palestine, 1899 and 1903; wrote *Life of Clive* (1890), and handbooks for Asia Minor (1892) and Constantinople (1895).

WILSON, EDWARD ADRIAN (1872-1912), naturalist and Antarctic explorer; M.B., Cambridge; junior surgeon to national Antarctic expedition under Commander R. F. Scott [q.v.], 1901-4; joined new Antarctic expedition under Scott as chief of scientific staff, 1910; member of party which reached South Pole, Jan. 1912; perished with rest on return journey.

WILSON, FRANK PERCY (1889-1963), scholar and bibliographer; educated at Camp Hill and King Edward's grammar schools, Birmingham, and Birmingham University; first class, English, 1911; MA, 1912; studied at Lincoln College, Oxford; B.Litt., 1913; served in army and severely wounded, 1914-16; university lecturer, Oxford, 1921; reader, 1927; published an edition of Thomas Dekker's [q.v.] *Foure Birds of Noahs Arke* (1924) and the *Plague Pamphlets of T. Dekker* (1925); also *The Plague in Shakespeare's London* (1927); professor of English, Leeds, 1929-36; visiting lecturer to number of American universities; joint general editor (with Bonamy Dobrée), the *Oxford History of English Literature*; Hildred Carlile professor of English literature, Bedford College, London, 1936-47; succeeded (Sir) W. W. Greg as general editor, the Malone Society, 1948-60; succeeded David Nichol Smith [q.v.] as Merton professor at Oxford, 1947-57; senior research fellow, Merton College, 1957-60; published lectures and papers include *Elizabethan and Jacobean* (1945), *Marlowe and the Early Shakespeare* (1953), *The Proverbial Wisdom of Shakespeare* (1961), and in collaboration with his wife, the *Oxford Dictionary of English Proverbs* (1970); FBA, 1943; hon. LLD, Birmingham, 1947; hon. fellow, Lincoln College, Oxford, 1948; president, Bibliographical Society, 1950-2; Malone Society, 1960-3; and Modern Humanities Research Association, 1961.

WILSON, GEORGE FERGUSSON (1822-1902), inventor; entered father's candle-making business, E. Price & Son, 1840; patented process of utilizing cheap malodorous fats for candle-making, 1842; sold business for £250,000, 1847; formed Price's Patent Candle Company, 1847; introduced 'New Patent Night Lights', 1853; discovered process of manufacturing pure glycerine, 1854; FRS, 1855; in later life an enthusiastic gardener at Wisley, Surrey; FLS, 1875.

WILSON, SIR GERALD HUGH TYR-WHITT-, fifth baronet, and fourteenth BARON BERNERS (1883-1950), musician, artist, and author. [See TYRWHITT-WILSON.]

WILSON, SIR HENRY HUGHES, baronet (1864-1922), field-marshal; obtained commission, Longford militia, 1882; gazetted to Rifle Brigade, 1884; served in South African war, 1899-1901; assistant-adjutant-general, military training branch of War Office, 1903; commandant of Staff College, Camberley, with rank of brigadier-general, 1907; formed friendship with (Marshal) Foch and became leading exponent of policy of close co-operation with France in event of continental war; director of military operations, War Office, 1910; major-general, 1913; supported Ulster cause during Curragh incident, 1914; sub-chief of general staff on landing of British expeditionary force in France, Aug. 1914; largely responsible for work of general headquarters; chief liaison officer with French headquarters, 1915; commanded IV Army Corps, 1915-16; head of mission sent to Russia to discuss supply of war material, 1916-17; lieutenant-general, 1917; returned to England, 1917; appointed to Eastern command, 1917; came into close contact with prime minister, Lloyd George, whom he accompanied to conference of allied powers at Rapallo, Nov.; chosen British military representative on Supreme War Council created at Rapallo; superseded Sir William Robertson [q.v.] as chief of imperial general staff, 1918; held post until 1922; present at allied conference at Doullens when Foch received control of com-

bined allied armies, 1918; field-marshal, 1919; baronet, 1919; dissatisfied with terms of peace of Versailles and began to drift away from Lloyd George; main objects of his deepest antipathy Bolshevik governors of Russia and Sinn Fein leaders in Ireland; conservative MP, North Down (Ireland), 1922; assassinated by Sinn Feiners in London; a politically minded soldier.

WILSON, HENRY MAITLAND, first BARON WILSON (1881-1964), field-marshal; educated at Eton and Sandhurst; commissioned into the Rifle Brigade, 1900; served in South African war; and in India, 1907; adjutant to Oxford University OTC, 1911; served as staff officer, 1914-18; DSO, 1917; at Staff College, 1919; returned to his regiment and served on North-West Frontier of India; instructor at Staff College; promoted to major-general, 1935; GOC-in-C, British troops in Egypt and lieutenant-general, 1939, with General A. P. (later Earl) Wavell [q.v.] as C-in-C, Middle East; GOC, British Forces in Palestine and Transjordan, 1941; GOC-in-C, 9th Army; C-in-C, Middle East, 1943; won confidence of Churchill; supreme commander in Mediterranean; field-marshal; head of Joint Staff Mission in Washington, 1945-7; present at Yalta and Potsdam conferences; CB, 1937; KCB, 1940; GCB, 1944; GBE, 1941; baron, 1946; active in affairs of service organizations; colonel commandant, Rifle Brigade, 1939-51; constable of Tower of London, 1935-60; president, Old Etonian Association, 1948-9.

WILSON, HENRY SCHÜTZ (1824-1902), author; followed a commercial career till 1870; member of Alpine Club, 1871-98; works include *Alpine Ascents and Adventures* (1878) and *Studies in History, Legend, and Literature* (1884).

WILSON, HERBERT WRIGLEY (1866-1940), writer on naval matters and journalist; scholar of Durham School and Trinity College, Oxford; author of *Ironclads in Action* (1896; republished as *Battleships in Action*, 1926) and other naval works; a chief inspirer of agitation (1908-9) for more battleships; assistant editor, *Daily Mail*, 1898-1938.

WILSON, SIR JACOB (1836-1905), agriculturist; studied at Royal Agricultural College, Cirencester, 1854-6; land agent to Earl of Tankerville at Chillingham, 1866; joined Royal Agricultural Society, 1860; hon. director of shows, 1875-92; knighted at Society's fiftieth show at Windsor, 1889; actively urged passing of Animals Acts of 1878 and 1884; chairman of departmental committee of inquiry into pleuro-pneumonia, 1888; agricultural adviser to the Board of Agriculture, 1892-1902; KCVO, 1905.

WILSON, JAMES MAURICE (1836-1931), schoolmaster, divine, and antiquary; father of Sir A. T. Wilson [q.v.]; educated at Sedbergh and St. John's College, Cambridge; senior wrangler, 1859; fellow, 1860; mathematical master in charge of new science department, Rugby, 1859-79; ordained, 1879; headmaster of Clifton, 1879-90; a born teacher and preacher; became increasingly interested in biblical criticism and social work; first headmaster to introduce a summer camp for working boys run by the school; vicar of Rochdale and archdeacon of Manchester, 1890-1905; canon of Worcester, 1905-26; as librarian studied and catalogued many manuscripts; active in civic work.

WILSON, Sir (JAMES) STEUART (1889-1966), singer, and musical administrator and scholar; son of Revd James Maurice Wilson [q.v.], and brother of (Sir) Arnold Wilson [q.v.]; educated at Winchester and King's College, Cambridge, 1906-9; first public appearance as singer, New Theatre, Cambridge, 1909, followed by other performances of music by Ralph Vaughan Williams [q.v.]; served in army during 1914-18 war; severely wounded; helped to found the English Singers, 1920; music teacher, Bedales School, 1921-3; made translations of German *Lieder*, and produced new texts for Hayden's *Creation*, Brahm's *German Requiem*, and the vocal finale of Mahler's *Fourth Symphony*; worked in United States, 1939-42; overseas music director, BBC, 1943; music director, Arts Council, 1945-8; music director, BBC, 1948; knighted; deputy to (Sir) David Webster, general administrator, Royal Opera House, 1950-5; principal, Birmingham School of Music, 1957-60.

WILSON, JOHN COOK (1849-1915), philosopher; educated at Derby grammar school and Balliol College, Oxford; won Chancellor's Latin essay prize, 1873; fellow of Oriel College, Oxford, 1874; Wykeham professor of logic at Oxford, 1889-1915; owing to various causes published little; main energies devoted to teaching, whereby he became strongest philosophical influence in Oxford of his generation; first principle of his philosophy may be described as the principle that there is no first principle.

WILSON, JOHN DOVE (1833-1908), Scottish legal writer; called to Scottish bar, 1857; sheriff substitute of Aberdeen, 1870-90; issued *The Practice of the Sheriff Courts of Scotland in Civil Causes* (1875); hon. LLD, Aberdeen, 1884; professor of law at Aberdeen, 1891-1901; died at San Remo.

WILSON, JOHN DOVER (1881-1969), Shakespeare scholar; educated at Lancing and Gonville and Caius College, Cambridge; awarded Harness prize for essay on *John Lyly* (1905); assigned two chapters of *Cambridge History of English Literature* by (Sir) A. W. Ward [q.v.]; English lektor, Helsingfors (Helsinki), 1906-9; lecturer in English literature, Goldsmith's College, London, 1909; published *Life in Shakespeare's England* (1911); inspector for Board of Education, 1912; professor of education, King's College, London, 1924; began *Journal of Adult Education*, 1926; published articles on texts of Shakespeare's plays; joined Sir Arthur Quiller-Couch [q.v.], as editor, *New Cambridge Shakespeare* (1919); produced commentaries on *Hamlet, the Essential Shakespeare* (1932) and *The Fortunes of Falstaff* (1943); regius professor of rhetoric and English literature, Edinburgh, 1935-45; published *Shakespeare's Happy Comedies* (1962); FBA, 1931; CH, 1936; hon. fellow, Gonville and Caius College, 1936; Litt.D., Cambridge, 1926; hon. doctorates from universities at home and abroad; trustee, Shakespeare Birthplace Trust and National Library of Scotland.

WILSON, JOHN GIDEON (1876-1963), bookseller; went to board school in Glasgow till thirteenth year, then apprenticed to John Smith & Son, Glasgow, booksellers; worked at Constable & Co., London, 1908; then at Jones and Evans, 1910; served in army in France, 1916-19; manager, J. & E. Bumpus Ltd., 1923; acquired the business from Messrs. Debenhams, and became chairman and managing director, 1935; most famous English bookseller of his time; retired in eighty-third year, 1959; chairman, Educational Board of the Booksellers Association; designed training syllabus for young booksellers, *The Business of Bookselling* (1930); edited *The Odd Volume*, published to support Book Trade Provident Society; CBE, 1948.

WILSON, (JOHN) LEONARD (1897-1970), bishop of Singapore and Birmingham; educated at Newcastle grammar school and St. John's School, Leatherhead; served in the army in France, 1916-19; studied at Knutsford Training School and the Queen's College, Oxford; BA, 1922; completed training at Wycliffe Hall, Oxford; ordained, 1924; joined Church Missionary Society, 1927; worked in Egypt; curate, St. Margaret's, Durham, 1929; vicar, Eighton Banks, 1930-5; Roker, 1935-8; archdeacon, Hong Kong, 1938; bishop of Singapore, 1941; interned in Changi Gaol, 1943; tortured; dean of Manchester, 1949; succeeded Ernest William Barnes [q.v.] as bishop of Birmingham, 1953-69; CMG, 1946; KCMG, 1968; prelate of the Order, 1963; Lambeth DD, 1953; hon. D.Sc., university of Aston, Birmingham, 1969; hon.

fellow, Queen's College, Oxford; published *Marriage, Sex, and the Family* (1959).

WILSON, JOSEPH HAVELOCK (1858-1929), founder of the National Union of Seamen; at sea, 1870-85; founded National Union, 1887; liberal MP, Middlesbrough, 1892-1900, 1906-10; coalition liberal MP, South Shields, 1918-22; secured reforms for merchant seamen and workmen; advocated peaceful negotiation.

WILSON, SAMUEL ALEXANDER KINNIER (1874-1937), neurologist; born in New Jersey; educated at Edinburgh (MB, 1902), Paris, and Leipzig; worked at National Hospital for the Paralysed and Epileptic, Queen Square, London, 1904-37; junior neurologist, King's College Hospital, 1919; senior neurologist, 1928-37; founded *Journal of Neurology and Psychopathology*, 1920; detected relationship between liver disease and putaminal destruction, 1912.

WILSON, Sir SAMUEL HERBERT (1873-1950), colonial administrator; entered Royal Engineers from Woolwich, 1893; assistant secretary, Imperial Defence Committee, and secretary, Oversea Defence Committee, 1911-14, 1918-21; governor, Trinidad and Tobago, 1921-4, Jamaica, 1924-5; permanent under-secretary, Colonial Office, 1925-33; unified colonial service and established personnel division, 1930; reorganized Office on subject rather than territorial basis; built up system of expert advisory services; KBE, 1921; KCMG, 1923; GCMG, 1929; KCB, 1927.

WILSON, WALTER GORDON (1874-1957), engineer; educated as naval cadet and at King's College, Cambridge; first class, part i, mechanical sciences tripos, 1897; designed Wilson-Pilcher and Armstrong Whitworth motor-cars; worked with J. & E. Hall, Dartford, 1908-14; designed Hallford lorry; rejoined Royal Navy, 1914; with (Sir) William Tretton [q.v.], designed the tank, 1915; served in the Tank Corps; CMG, 1917; invented Wilson self-changing gearbox.

WILSON, WILLIAM (1875-1965), physicist; educated at village school and Agricultural College, Aspatria, West Cumberland; studied at Royal College of Science, London, 1893-6; left without a degree; teacher at Towcester school; mathematics master, Beccles College, and then Craven College, Highgate, 1898; taught in schools in Germany; studied mathematics at Leipzig University, 1902-6; Ph.D.; assistant lecturer, Wheatstone Laboratory, King's College, London, 1906-17; studied Hamiltonian mechanics, and applied formulae published in

Philosophical Magazine, 1915-16, to the elliptic orbits of the hydrogen atom; further researches led to quantum theory of the emission and absorption of radiation; D.Sc., London, 1917; reader, physics department, King's College, London, 1920; Hildred Carlile professor, Bedford College, London, 1921-44; FRS, 1923; published *Theoretical Physics* (3 vols., 1931-40).

WILSON, WILLIAM EDWARD (1851-1908), astronomer and physicist; set up private observatory at Daramona, 1870; began in 1886 pioneer researches into temperature of the sun, which he published in 1900; FRS, 1896; hon. D.Sc., Dublin, 1901.

WIMBORNE, first VISCOUNT (1873-1939), politician. [See GUEST, SIR IVOR CHURCHILL.]

WIMPERIS, HARRY EGERTON (1876-1960), scientist; educated at Royal College of Science and Gonville and Caius College, Cambridge; graduated in mechanical sciences, 1899; worked with Armstrong Whitworths; served during 1914-18 war as experimental scientist with Royal Naval Air Service and Royal Air Force; superintendent, Air Ministry scientific laboratory, 1918; director, 1925-37; set up the committee on air defences, with (Sir) Henry Tizard [q.v.] as chairman, which promoted radar, 1934; advised Australian government on aeronautical research, 1937; CBE, 1928; CB, 1935.

WIMSHURST, JAMES (1832-1903), engineer; apprenticed at Thames Ironworks till 1853; surveyor of Lloyd's, 1853; chief of Liverpool Underwriters' Registry, 1865-74; chief shipwright surveyor to Board of Trade, 1874-99; built several electrical influence machines; invented 'Wimshurst duplex machine'; his machines were used for exciting 'Röntgen rays' and for treatment of cancer; FRS, 1898; published works on engineering.

WINDUS, WILLIAM LINDSAY (1822-1907), artist; studied art at Liverpool Academy; associate, 1847; full member, 1848; visited London, 1850; accepted Pre-Raphaelite principles; exhibited at Royal Academy in 1856 'Burd Helen', which was admired by Rossetti and Ruskin [qq.v.]; 'Too Late' (1858, now in Tate Gallery) was condemned by Ruskin; never exhibited after death of wife in 1862; poetic and imaginative painter; forty-five of his works exhibited at Walker Art Gallery, Liverpool, 1908.

WINFIELD, Sir PERCY HENRY (1878-1953), lawyer and legal scholar; educated at King's Lynn grammar school and St. John's

College, Cambridge; first class, parts i and ii, law tripos, 1898-9; called to bar (Inner Temple), 1903; lecturer, Cambridge, 1911; fellow, St. John's College, 1921; Rouse Ball professor of English law, 1928-43; hon. bencher, Inner Temple, 1938; KC, 1943; FBA, 1934; hon. LLD, Harvard (1929), Leeds (1944), and London (1949); knighted, 1949; publications include *Chief Sources of English Legal History* (1925), *Province of the Law of Tort* (1931), and *Textbook of the Law of Tort* (1937); edited the *Cambridge Law Journal*, 1927-47.

WINGATE, Sir (FRANCIS) REGINALD, first baronet (1861-1953), soldier and governor-general of the Sudan; educated at St. James's Collegiate School and Royal Military Academy, Woolwich; served with Royal Artillery in India, Aden, and Egypt; aide-de-camp to Sir Evelyn Wood [q.v.], sirdar of the Egyptian army, 1884; director, military intelligence, 1889; governor-general, Sudan, 1899-1916; high commissioner, Egypt, 1917-19; superseded by Lord Allenby [q.v.]; published *Mahdiism and the Egyptian Sudan* (1891); KCMG, 1898; KCB, 1900; GCVO, 1912; GCB, 1914; GBE, 1918; baronet, 1920; DSO, 1889.

WINGATE, ORDE CHARLES (1903-1944), major-general; born in India of Plymouth Brethren parentage; educated at Charterhouse and Woolwich; gazetted to Royal Artillery, 1923; in Sudan Defence Force, 1928-33; in England, 1933-6; intelligence staff, Palestine, 1936-9; organized Jewish night squads; DSO; summoned to Middle East to organize assistance to rebels in Abyssinia, 1940; crossed frontier with Emperor Haile Selassie, Jan. 1941; entered Addis Ababa, May 1941; sent to India; proposed formation 'long range penetration group' to operate behind Japanese lines for reconquest of Burma; trained mixed force; crossed Chindwin river, Feb. 1943, and fought some six weeks behind enemy front; major-general, 1943; trained larger force, winter 1943-4; led new operation landing greater part of force from air, Mar. 1944; crashed over north Assam, 24 Mar. buried in Arlington cemetery, Virginia.

WINNINGTON-INGRAM, ARTHUR FOLEY (1858-1946), bishop of London; educated at Marlborough and Keble College, Oxford; second class, *lit. hum.*, 1881; ordained, 1884; domestic chaplain to bishop of Lichfield, 1885-8; head of Oxford House, Bethnal Green, 1888-97; rector of Bethnal Green, 1895-7; made Church real influence in east end; suffragan bishop of Stepney and canon of St. Paul's, 1897-1901; bishop of London, 1901-39; gifted mission preacher; inspired many voca-tions; his best work personal and pastoral; paid number of visits to Canada and United States; and to France and Salonika in 1914-18; KCVO, 1915.

WINSTANLEY, DENYS ARTHUR (1877-1947), historian; educated at Merchant Taylors' and Trinity College, Cambridge; first class, parts i and ii, historical tripos, 1899-1900; fellow and lecturer, 1906; tutor, 1919; senior tutor, 1925; vice-master, 1935-47; chief publications *Unreformed Cambridge* (1935), *Early Victorian Cambridge* (1940), and *Later Victorian Cambridge* (1947).

WINSTEDT, Sir RICHARD OLOF (1878-1966), Malayan civil servant and Malay scholar; son of naturalized Swede; educated at Magdalen College School and New College, Oxford; appointed to Federated Malay States civil service, 1902; posted to Perak; wrote four government-published books on Malay life and customs, 1907-11, and a collection of Malay folk stories; published *Malay Grammar* (1913), *Colloquial Malay: a simple grammar with conversations* (1916), and *An English-Malay Dictionary* (3 vols., 1914-17); transferred to education department, Singapore, 1916; published more books and articles on Malay language and literature; D.Litt., Oxford, 1920; first president of Raffles College, Singapore, 1921-31; director of education and member of Legislative Council, 1924-31; CMG, 1926; general adviser, Unfederated State of Johore, 1931-5; published a history of Johore and a history of Malaya, 1932-5; KBE, 1935; lecturer in Malay, School of Oriental Studies, London; reader, 1937-46; FBA, 1945; further publications include *A History of Malay Literature* (1939) and *The Malays—a Cultural History* (1947); president, Royal Asiatic Society; hon. LLD, university of Malaya, 1951.

WINSTER, Baron (1885-1961), politician. [See FLETCHER, REGINALD THOMAS HERBERT.]

WINTER, Sir JAMES SPEARMAN (1845-1911), premier of Newfoundland; called to Newfoundland bar, 1867; QC, 1880; speaker of house of assembly, 1877-8; solicitor-general, 1882-5; attorney-general, 1885-9; judge of supreme court, 1893-6; premier, 1897-1900; represented Newfoundland at fisheries conference at Washington, 1887-8; KCMG, 1888; died at Toronto.

WINTER, JOHN STRANGE (pseudonym), (1856-1911) novelist. [See STANNARD, HENRIETTA ELIZA VAUGHAN.]

WINTERSTOKE, Baron (1830-1911), benefactor to Bristol. [See WILLS, WILLIAM HENRY.]

WINTERTON, sixth EARL, BARON TURNOUR (1883–1962), politician. [See TURNOUR, EDWARD.]

WINTON, SIR FRANCIS WALTER DE (1835–1901), major-general and South African administrator. [See DE WINTON.]

WISE, THOMAS JAMES (1859–1937), book-collector, bibliographer, editor, and forger; clerk and later partner in London oil firm; collected work of English poets and formed the Ashley Library (sold to British Museum after his death); made it in the seventeenth and eighteenth centuries of very great, and in the nineteenth of unapproached, completeness; paid great attention to bibliographical detail and issued full and sectional catalogues; also sold books and in 1934 was proved to have printed and sold forgeries to collectors whom he advised.

WITHERS, HARTLEY (1867–1950), financial editor and author; educated at Westminster and Christ Church, Oxford; joined *The Times* City office, 1894; City editor, 1905–10; editor *The Economist*, 1916–21; published series of readable and intelligible works on economics.

WITT, SIR ROBERT CLERMONT (1872–1952), art collector; educated at Clifton and New College, Oxford; qualified as solicitor; senior partner, Stephenson, Harwood and Tatham; published *How to Look at Pictures* (1902); a founder of National Art-Collections Fund, 1903; first hon. secretary, 1903–20; chairman, 1920–45; trustee, National Gallery and Tate; a founder of Courtauld Institute of Art, 1932; presented library of photographs and reproductions to Courtauld Institute, 1944; CBE, 1918; knighted, 1922.

WITTEWRONGE, SIR CHARLES BENNET LAWES- (1843–1911), sculptor and athlete. [See LAWES (afterwards LAWES-WITTEWRONGE).]

WITTGENSTEIN, LUDWIG JOSEF JOHANN (1889–1951), philosopher; born in Vienna; educated at Linz and Berlin-Charlottenburg; engineering research student, Manchester University, 1908–11; studied philosophy at Trinity College, Cambridge, 1912–13; served in Austrian army in 1914–18 war, and as architect in Vienna, 1926–8; returned to Cambridge, 1929; Ph.D.; fellow, Trinity College, 1930–6; naturalized, 1938; professor of philosophy, 1939–47; published in England, *Tractatus Logico-Philosophicus* (1922), *Philosophical Investigations* (1953), and *Remarks on the Foundations of Mathematics* (1956); a philosopher of rare genius and originality.

WODEHOUSE, JOHN, first EARL OF KIMBERLEY (1826–1902), statesman; educated at Eton and Christ Church, Oxford; BA, 1847; succeeded grandfather as Baron Wodehouse, 1846; under-secretary of state for foreign affairs under Lords Aberdeen and Palmerston, 1852–6 and 1859–61; British minister at St. Petersburg, 1856–8; was sent on special mission to settle Schleswig-Holstein dispute, 1863; lord-lieutenant of Ireland, 1864–6; dealt resolutely with Fenian movement; created Earl of Kimberley, Norfolk, 1866; joined Gladstone's first cabinet as lord privy seal, 1868–70; colonial secretary, July 1870–4; during his administration Griqualand West was annexed, 27 Oct. 1871, Kimberley was named after him, and responsible government was granted to Cape Colony, 1872; he authorized expedition against Ashantis, 1873; formed Rupert's Land into province of Manitoba, and brought British Columbia into the dominion; introduced licensing bill into the Lords, 1872; again colonial secretary, 1880–2; showed irresolution in dealing with revolt of Boers, 1881; held India Office, 1882–5, and Feb.–Aug. 1886; had charge of franchise bill of 1884, and redistribution bill of 1885 in the Lords; KG, 1885; succeeded Lord Granville [q.v.] as leader of liberal party in House of Lords, 1891; secretary for India and lord president of the Council, 1892–4; foreign secretary under Lord Rosebery, 1894–5; made agreement, resented by Germany, with Congo Free State, May 1894; defended military operations in South African war, 1899; hon. DCL, Oxford, 1894; chancellor of London University, 1899–1902.

WOLFE, HUMBERT (1886–1940), poet and civil servant; born UMBERTO WOLFF in Milan of Jewish parentage; educated at Bradford grammar school and Wadham College, Oxford; entered civil service, 1908; worked on organization of labour exchanges and unemployment insurance; head of employment and training, Ministry of Labour, 1934; deputy secretary, 1938–40; successful in recruiting manpower, 1939; wrote on Herrick, Tennyson, Shelley, etc.; published volumes of verse including *London Sonnets* (1920), *Requiem* (1927), *Kensington Gardens in Wartime* (1940); an expert translator.

WOLFE-BARRY, SIR JOHN WOLFE (1836–1918), civil engineer; son of Sir Charles Barry [q.v.]; pupil of Sir John Hawkshaw [q.v.]; works include extension to London District and Underground railways, Tower bridge (in collaboration, 1894), numerous docks, including Barry docks and railway, South Wales; from 1867 consulting engineer to many railways and public undertakings; served on many royal commissions; KCB, 1897.

WOLFF, Sir HENRY DRUMMOND CHARLES (1830-1908), politician and diplomatist; born at Malta; only child of Joseph Wolff [q.v.]; educated at Rugby; entered Foreign Office, 1846; attached to British legation at Florence, 1852-3; private secretary to colonial secretary, Sir Edward Bulwer Lytton (later Baron Lytton, q.v.), 1858; CMG, 1859; secretary to high commissioner of Ionian islands, 1859-64; KCMG, 1862; arranged transfer of islands to Greece, 1864; from 1864 to 1870 promoted various financial undertakings; visited Franco-Prussian battlefields, 1870, narrating experiences in *Morning Post* (reprinted 1892); conservative MP for Christchurch, 1874-80; with G. J. (later Viscount) Goschen [q.v.] went to Egypt on commission of inquiry into Egyptian finance, 1876; GCMG, 1878; British commissioner for organization of Eastern Roumelia, 1878-80; KCB, 1879; MP for Portsmouth, 1880-5; formed in the House of Commons 'fourth party' with Lord Randolph Churchill, (Sir) John Gorst, and Arthur Balfour [qq.v.], 1880-5; suggested to Churchill the formation of the 'Primrose League', 1883; PC, 1885; was sent on special mission to Constantinople to negotiate with Turkey on the future of Egypt, Aug. 1885; British commissioner to reorganize Egyptian administration, 1885-6; negotiated second convention with Turkey (May 1887) stipulating evacuation of Egypt by British after three years, but the convention, owing to opposition of France and Russia, was not ratified by the Sultan; served as British envoy in Persia, Dec. 1887-91; helped to establish Imperial Bank of Persia; GCB, 1889; was transferred to Bucharest, July 1891, and to Madrid, 1892-1900; published *Rambling Recollections* (2 vols., 1908); his daughter, writing as 'Lucas Cleeve', was a prolific novelist.

WOLFF, MARTIN (1872-1953), academic lawyer; born in Berlin; studied and taught law in Berlin and Bonn, 1900-35; ousted by Nazis, 1935; invited to All Souls College, Oxford, 1938; naturalized, 1945; published *Private International Law* (1945), *Traité de droit comparé* (3 vols, with Arminjon and Nolde, 1950-2); hon. DCL, Oxford, 1952.

WOLFIT, Sir DONALD (1902-1968), actor-manager; educated at Magnus School, Newark; joined Charles Doran's touring company, 1920; played small part in *The Scarlet Pimpernel* with Fred Terry [q.v.] (1923); appeared in London in *The Wandering Jew* with Matheson Lang [q.v.] (1924); played at the Old Vic, 1929-30; played in west end and on tour in Canada, 1930-5; played leading Shakespearian parts at Stratford, 1936-7; successful tour of *Othello* and other Shakespearian plays, 1938; combined Home Guard duties with producing lunch-hour Shakespeare, 1939-40; acted in *Richard III* and *King Lear* (1941-4); toured for ENSA; toured again, including visit to Canada and United States, 1945-7; had great success in (Sir) Tyrone Guthrie's production of Marlowe's *Tamburlaine the Great* (1951); presented series of classical plays, and made last appearance as Mr Barrett in *Robert and Elizabeth* (1966-7); also played in films, including *Room at the Top, Becket*, and *Decline and Fall*; CBE, 1950; knighted, 1957.

WOLLASTON, ALEXANDER FREDERICK RICHMOND (1875-1930), naturalist and explorer; BA, King's College, Cambridge; studied medicine at London Hospital; took part, as doctor and collector of plants and insects, in British Museum expedition to Ruwenzori mountains, Central Africa, 1906; joined British Ornithologists' Union expedition to Dutch New Guinea, 1909; organized expedition to New Guinea on his own account, reaching glaciers and studying hitherto unknown stone-age natives, 1912-13; fellow of King's College, 1920; took part, as doctor and naturalist, in first Mount Everest expedition, 1921; his works include *Pygmies and Papuans* (1912).

WOLMARK, ALFRED AARON (1877-1961), painter and decorative artist; born of Jewish parents in Warsaw; naturalized, 1894; studied at Royal Academy Schools, 1895-8; influenced by Rembrandt; visited Poland, 1903, and painted works based on Jewish historical subjects; later, under influence of modern French painting, executed series of studies of Breton fisherfolk and habour life, 1911; turned to interior decoration, the theatre, and stained glass, 1911-15; executed window for St. Mary's Church, Slough, 1915; painted series of cityscapes in New York, 1919-20; held many exhibitions in London, New York, and Paris, including portrait drawings at Lefèvre Galleries, 1928; one of prime movers in setting up the Ben Uri Art Society, 1915.

WOLSELEY, GARNET JOSEPH, Viscount WOLSELEY (1833-1913), field-marshal; born in county Dublin; joined army, 1852; served with distinction in second Burma war, 1852-3, Crimean war, 1854-6, Indian Mutiny, 1857-9, and China war, 1860; assistant quartermaster-general in Canada, 1861; colonel, 1865; deputy quartermaster-general, 1867; commanded Red River expedition which put down rebellion of Louis Riel [q.v.], 1870; KCMG, and CB, 1870; assistant adjutant-general at War Office, 1871; ardent supporter of military reforms of (Viscount) Cardwell [q.v.]; commanded expedition sent against King Koffee of Ashanti,

1873; defeated him at Amoaful and occupied Kumasi, 1874; major-general, GCMG, and KCB, 1874; administrator and general commanding in Natal, 1875; first administrator of Cyprus, 1878; sent to retrieve situation in Zulu war, 1879; captured King Cetywayo and defeated native chief, Sekukuni; established administration in Zululand and granted Transvaal constitution of crown colony; quartermaster-general at War Office, 1880-2; made preparation for war first principle of policy of army reform, and, in spite of violent opposition of commander-in-chief, second Duke of Cambridge [q.v.], carried each item in his programme after severe struggle; adjutant-general, 1882; crushed rebellion of Egyptian army under Arabi Pasha at Tel-el-Kebir and occupied Cairo, 1882; promoted general and created baron, 1882; conducted Nile campaign for relief of General Gordon [q.v.], 1884-5; created viscount and KP, 1885; commander-in-chief in Ireland, 1890-5; commander-in-chief of British army, re-creation of which constitutes his chief title to fame, 1895-9; works include *The Soldier's Pocket Book* (1869) and *Life of Marlborough* (1894); died at Mentone.

WOLVERHAMPTON, first VISCOUNT (1830-1911), statesman. [See FOWLER, HENRY HARTLEY.]

WOOD, CHARLES (1866-1926), composer; educated at Cathedral School, Armagh; studied at Royal College of Music, London, under (Sir) C. V. Stanford, and (Sir) Frederick Bridge [qq.v.], 1883-8; teacher of harmony and counterpoint at College, 1889-1926; organist of Gonville and Caius College, Cambridge, 1891-1926; settled in Cambridge; fellow of Caius, 1894; Mus.D., Cambridge, 1895; professor of music, Cambridge, 1924; had more beneficent and far-reaching effect on contemporary musical production than any other teacher; his compositions include 'Ode to Music' (1890) and a setting of the 'Passion according to St. Mark'.

WOOD, SIR CHARLES LINDLEY, fourth baronet, and second VISCOUNT HALIFAX (1839-1934); son of first Viscount Halifax [q.v.]; succeeded father, 1885; educated at Eton and Christ Church, Oxford; groom of the bedchamber to Prince of Wales, 1862-77; president, English Church Union, 1868-1919, 1927-34; led Anglo-Catholics in face of the ritual prosecutions accompanying the Act for the Regulation of Public Worship (1874); his attempt to initiate conversations between Canterbury and Rome resulted in the latter condemning Anglican orders, 1896; his approach to Cardinal Mercier, archbishop of Malines, led to series of conversations (1921-6) unfruitful but

significant as a precedent; his authority among Anglo-Catholics unrivalled.

WOOD, EDWARD FREDERICK LINDLEY, first EARL OF HALIFAX (1881-1959), statesman; fourth son of Charles Lindley Wood, Viscount Halifax [q.v.]; educated at Eton and Christ Church, Oxford; first class, history, 1903; fellow, All Souls; conservative MP, Ripon, 1910-25; assistant secretary, Ministry of National Service, 1917-19; undersecretary, Colonial office, 1921; president, Board of Education, 1922-4; PC; viceroy of India, 1925-31; baron Irwin, 1925; GCSI, GCIE, 1926; KG, 1931; president, Board of Education, 1932; chancellor, Oxford University, 1933; succeeded father as Viscount Halifax, 1934; secretary of state for war, 1935; lord privy seal and leader of House of Lords, 1935-7; lord president of the Council, 1937; foreign secretary, 1938-41; Chamberlain took responsibility for negotiations with Hitler, 1938; ambassador, Washington, 1941-6; earl, 1944; OM, 1946; chairman, general advisory council, BBC, 1947; published *Fullness of Days* (1957).

WOOD, FRANCIS DERWENT (1871-1926), sculptor; studied sculpture under Professor Lantéri at National Art Training School, South Kensington; assistant to Alphonse Legros [q.v.] at Slade School of Art, 1890-2; studied at Royal Academy Schools, 1894-7; professor of sculpture, Royal College of Art, 1918-23; RA, 1920; employed neo-baroque style; his works include Machine Gun Corps memorial Hyde Park Corner (1925), portrait busts, and statues.

WOOD, SIR (HENRY) EVELYN (1838-1919), field-marshal; entered navy, 1854; transferred to army during Crimean war; won VC in Indian Mutiny, 1859; passed through Staff College, 1862-4; accompanied (Viscount) Wolseley [q.v.] to Ashanti, as special service officer, 1873; CB, 1874; accompanied Lieutenant-General Thesiger (afterwards second Baron Chelmsford, q.v.) to South Africa, 1878; KCB for services in Zulu war, 1879; sent to Natal, 1881; royal commissioner for settlement of Transvaal, 1881; major-general and GCMG, 1882; accompanied Wolseley to Egypt, 1882; first British sirdar of Egyptian army, 1882; held Eastern command, 1886, Aldershot command, 1889; lieutenant-general, 1891; quartermaster-general at War Office, 1893; reorganized system of transporting troops; full general, 1895; adjutant-general to forces, 1897; field-marshal, 1903.

WOOD, SIR HENRY JOSEPH (1869-1944), musical conductor; studied at St. John's Wood and Slade art schools and at Royal Academy of Music; conducted for D'Oyly Carte, Carl Rosa, and other opera companies; musical director, series of promenade concerts at Queen's Hall under Robert Newman's management, 1895; developed concerts into annual event sponsored between 1915-26 by Chappells; after 1927 (first broadcast) by British Broadcasting Corporation; BBC Symphony Orchestra formed, 1930; after destruction of Queen's Hall (1941) transferred to Royal Albert Hall; known from 1944 as 'The Henry Wood Promenade Concerts'; first British conductor to found stable and permanent orchestra fully trained at rehearsal; reflected most important trends in contemporary music giving constant encouragement to British composers and inviting foreign composers to conduct their own works; trained and conducted many provincial choral societies and orchestras; took orchestral classes at Royal Academy of Music to which he presented his library of orchestral scores and sets of parts; knighted 1911; CH, 1944.

WOOD, Sir (HOWARD) KINGSLEY (1881-1943), politician; educated at Central Foundation Boys' School, London; admitted solicitor, 1903; set up City practice; municipal reform member, London County Council, Woolwich, 1911-19; conservative MP, West Woolwich, 1918-43; parliamentary private secretary, minister of health, 1919-22; parliamentary secretary, Ministry of Health, 1924-9, Board of Education, 1931; postmaster-general, 1931-5; carried out internal reorganization; obtained new financial relationship with Treasury; made full use of commercial publicity; minister of health, 1935-8; secretary of state for air, 1938-40; worked indefatigably to increase fighting strength; lord privy seal, 1940; chancellor of Exchequer, 1940-3; sought advice from J. M. (later Lord) Keynes [q.v.] whose influence was discernible in budget introduction of compulsory saving and control of cost of living by subsidies, 1941; continued new concept of budget as instrument of economic policy, 1942-3; introduced PAYE proposals; knighted, 1918; PC, 1928.

WOOD, MATILDA ALICE VICTORIA (1870-1922), music-hall comedian, professionally known as 'Marie Lloyd'; first appeared on music-hall stage, 1884; her songs all written and composed for her; notable for cheery vitality, knowledge of vulgar, especially cockney, English, and swift, significant expression.

WOOD, Sir ROBERT STANFORD (1886-1963), educationist; educated at City of London School and Jesus College, Cambridge; first

class, part i, classical tripos, 1908; after two years as teacher, became inspector of schools, 1911; principal private secretary to president of Board of Education, 1926-8; director of establishments, 1928-36; principal assistant secretary for technical education, 1936-40; seconded for service on home security with Sir John Anderson (later Viscount Waverley, q.v.), 1938; deputy secretary, Board of Education, 1940-6; largely responsible for section on further education in White paper of 1943; principal, University College of Southampton, succeeding K. H. Vickers [q.v.], 1946; led college to full university status; first vice-chancellor, Southampton University, 1952; CB, 1939; KBE, 1941; hon. fellow, Jesus College, 1952; hon. member, Goldsmiths' Company of London.

WOOD, THOMAS (1892-1950), composer; educated at Christ Church and Exeter College, Oxford, and Royal College of Music; D.Mus., 1920; director of music, Tonbridge, 1919-24; lecturer, Exeter College, 1924-7; compositions include 'Forty Singing Seamen' (1925), 'A Seaman's Overture' (1927), and 'Chanticleer' (1947); devoted to sea, foreign travel, and British Empire; published *Cobbers* (1934).

WOOD, THOMAS McKINNON (1855-1927), politician; represented Central Hackney on London County Council as progressive, 1892-1907; liberal MP, St. Rollox division of Glasgow, 1906-18; parliamentary secretary, Board of Education, 1908; under-secretary of state, Foreign Office, 1908; financial secretary, Treasury, 1911; secretary for Scotland with seat in cabinet, 1912; financial secretary, Treasury, and chancellor, duchy of Lancaster, 1916.

WOODALL, WILLIAM (1832-1901), politician; carried on china business at Burslem; chairman of Burslem school board, 1870-80; sat on royal commission on technical education, 1881-4; liberal MP for Stoke-upon-Trent, 1880-5, and for Hanley, 1885-1900; introduced woman suffrage bills in the house, 1884, 1889, and 1891; financial secretary to the War Office, 1892-5; generous benefactor to Burslem.

WOODGATE, WALTER BRADFORD (1840-1920), oarsman; brother of Sir Edward Woodgate [q.v.]; BA, Brasenose College, Oxford; rowed for Oxford, 1862 and 1863; at Henley won Grand Challenge cup, Stewards' cup, Diamond sculls, Goblets, and Wingfield sculls; writer on miscellaneous subjects.

WOODS, Sir ALBERT WILLIAM (1816-1904), Garter King of Arms; entered College of Arms, 1837; Lancaster herald, 1841-69; Garter principal King of Arms, 1869-1904; knighted,

1869; CB, 1887; KCMG, 1890; KCB, 1897; GCVO, 1902; grand director of ceremonies in freemasonry, 1860-1904; past grand warden, 1875; FSA, 1847.

WOODS, DONALD DEVEREUX (1912-1964), chemical microbiologist; educated at Northgate School, Ipswich, and Trinity Hall, Cambridge; first class, parts i and ii, natural sciences tripos, 1932-3; Beit memorial fellow; Ph.D., 1937; worked on metabolic processes of bacteria with Dr Marjory Stephenson [q.v.], 1933-9; Halley-Stewart research fellow, Middlesex Hospital, London, 1939; 'war work' at Chemical Defence Research Establishment, Porton, 1940-6; reader in chemical microbiology, Oxford, 1946; fellow, Trinity College, Oxford, 1951; FRS, 1952; first Iveagh professor of chemical microbiology, 1955; noted for his research on the nature of the anti-sulphanilamide factor in yeast extracts and on the nature and functions of folic acid.

WOODS, EDWARD (1814-1903), civil engineer; assistant engineer, 1834, and chief engineer, 1836-52, of Liverpool and Manchester Railway; made early investigations into workings of railways and of locomotives; reported to British Association on resistance of railway trains, 1837; greatly extended railways in South America from 1853; president of Institution of Civil Engineers, 1886-7.

WOODS, HENRY (1868-1952), palaeontologist; educated at higher grade school in Cambridge and St. John's College, Cambridge; first class, parts i and ii, natural sciences tripos, 1889-90; joined department of geology, 1892; lecturer in palaeontology, 1899-1934; published *Catalogue of Type Fossils in the Woodwardian Museum, Cambridge* (1891); organized exhibits in new geological museum (the Sedgwick), 1904; published *Monograph of the Cretaceous Lamellibranchia of England*, (2 vols., 1899-1913) and *Monograph of the Fossil Macrurous Crustacea of England* (1925-31); FRS, 1916.

WOODS, SIR JOHN HAROLD EDMUND (1895-1962), civil servant; educated at Christ's Hospital; served in army and severely wounded, 1914-18; studied at Balliol College, Oxford; entered civil service and posted to Treasury, 1920; seconded to Imperial Defence College, 1928; assistant private secretary to Prince of Wales, 1929; MVO, 1930; returned to Treasury; private secretary to Philip (later Viscount) Snowden [q.v.], chancellor of the Exchequer, 1931; principal private secretary to Neville Chamberlain, 1936, and to Sir John (later, Viscount) Simon [q.v.], 1937-40; responsible for control of expediture on defence material,

1940; permanent secretary, Ministry of Production, 1943; at Board of Trade, 1945, under Sir Stafford Cripps [q.v.] and Harold Wilson; member, Economic Planning Board, 1947; chairman, Treasury organization committee, 1950; retired, and joined board of English Electric Company, 1951; member of number of important committees related to industry and commerce; director, Sadlers Wells Trust Ltd.; visitor, Nuffield College, Oxford; governor of Administrative Staff College; treasurer, British School at Rome; CB, 1943; KCB, 1945; GCB, 1949.

WOODWARD, SIR ARTHUR SMITH (1864-1944), palaeontologist; educated at Macclesfield grammar school and Owens College, Manchester; entered geology department, British Museum, South Kensington, 1882; assistant keeper, 1892; keeper, 1901-24; catalogued the fossil fishes (4 vols., 1889-1901) and established himself as greatest palaeoichthyologist of his time; after 1912 devoted considerable time and energy to Piltdown skull discovered by Charles Dawson but revealed (1953) to be a fraud; FRS, 1901; knighted, 1924.

WOODWARD, HERBERT HALL (1847-1909), musical composer; Mus.B., Corpus Christi College, Oxford, 1866; BA, 1867; minor canon, 1881, and precentor, 1890, of Worcester Cathedral; composed church music, including anthem 'The Radiant Morn', 1881.

WOOLAVINGTON, BARON (1849-1935), philanthropist and racehorse owner. [See BUCHANAN, JAMES.]

WOOLDRIDGE, HARRY ELLIS (1845-1917), painter, musician, and critic; exhibitor at Royal Academy; Slade professor of fine art at Oxford, 1895-1904; re-edited Chappell's *Popular Music of the Olden Time* as *Old English Popular Music* (1893); edited with Robert Bridges [q.v.] *The Yattendon Hymnal* (1895-9); contributed first two volumes to *Oxford History of Music* (1901, 1905).

WOOLDRIDGE, SIDNEY WILLIAM (1900-1963), geomorphologist; educated at Glendale county school, Wood Green, and King's College, London; first class, geology, 1921; joined staff of department of geology and geography, King's College; M.Sc., 1923; D.Sc., 1927; worked on glaciation of London Basin, 1928-30; lecturer, King's College, 1927; studied structure and erosional history of the Weald; published (with D. L. Linton) *Structure, Surface and Drainage in South-east England* (1939); criticized Greater London Plan, 1944; member of Waters Committee on sand and gravel resources, 1946; CBE, 1954; reader, London University, 1942; served in Observer Corps; professor of geography, Birkbeck College, 1944; professor, King's College, 1947; chairman, Council for Field Studies, and committee of management, Institute of Archaeology; hon. member, Geologists' Association; fellow, King's College, London, 1956; FRS, 1959; president, Institute of British Geographers, 1949-50; publications include *Physical Basis of Geography, an Outline of Geomorphology*, (with R. S. Morgan 1937).

WOOLF, (ADELINE) VIRGINIA (1882-1941), novelist and critic; daughter of and educated by Sir Leslie Stephen [q.v.]; a central figure of Bloomsbury circle, one of whom (Leonard Sidney Woolf) she married, 1912; possessed brilliant, imaginative creative intelligence conveyed to reader in novels of extraordinary sensibility to the beautiful in nature and art and in critical essays of just and penetrating judgment; works include *Orlando* (1928), *The Waves* (1931), *The Years* (1937); drowned herself in final collapse of a nervous system of extreme fragility.

WOOLF, LEONARD SIDNEY (1880-1969), author, publisher, and political worker; born of Jewish parents; educated at St. Paul's and Trinity College, Cambridge; first class, part i, classical tripos, 1902; entered colonial service, and posted to Ceylon, 1904; left colonial service and married Virginia, daughter of Sir Leslie Stephen [qq.v.], 1912; published *The Village in the Jungle* (1913) and *The Wise Virgin* (1914); joined the Fabian Society; interested in the 'Co-operative Movement'; studied international relations; published *International Government* (1916) and *Empire and Commerce in Africa* (1921); edited *International Review*, 1919; also international section of the *Contemporary Review*, 1920-1; interested in the Hogarth Press, 1917; joint-editor, *Political Quarterly* (1931-59), and literary editor (1959-62); literary editor, the *Nation* (1923-30); publications include *Imperialism and Civilization* (1928), *Barbarians at the Gate* (1939), *The War for Peace* (1940), and *Principia Politica* (1953); also five autobiographical volumes, *Sowing* (1960), *Growing* (1961), *Beginning Again* (1964), *Downhill all the Way* (1967), and *The Journey not the Arrival Matters* (1969); hon. doctorate, Sussex University, 1964.

WOOLGAR, SARAH JANE (1824-1909), actress. [See MELLON.]

WOOLLARD, FRANK GEORGE (1883-1957), pioneer of mass production in the motor car industry; educated at City of London School and the Goldsmiths' and Birkbeck colleges; apprenticed to railway engineering; joined E. G. Wrigley & Co., Birmingham; designed axles and gearboxes of first Morris-Cowley motor-cars; general manager, engines' branch, Morris Motors, Ltd., 1923; director, Morris Motors, 1926-32; managing director, Rudge Whitworth Ltd., 1932-6; director, Midland Motor Cylinder Co., 1936-53; published *Principles of Mass and Flow Production* (1954).

WOOLLEY, SIR (CHARLES) LEONARD (1880-1960), archaeologist; educated at St. John's, Leatherhead, and New College, Oxford; first class, *lit. hum.*, 1903; assistant to (Sir) Arthur Evans [q.v.], keeper, Ashmolean Museum; excavation work in Nubia, 1907; leader, British Museum expedition to Carchemish, 1912-14 and 1919; research into Sumerian civilization at Ur, 1922-35; digging at Atchana, 1937-9 and 1946-9; publications include *The Development of Sumerian Art* (1935), *Excavations at Ur, A Record of Twelve Years Work* (1954), and *Alalakh, excavations at Tell Atchana* (1955); knighted, 1935; hon. fellow, New College, Oxford; hon. ARIBA, 1926.

WOOLTON, first EARL OF (1883-1964), politician and business man. [See MARQUIS, FREDERICK JAMES.]

WORBOYS, SIR WALTER JOHN (1900-1969), industrialist and promotor of good design in industry; born in Australia; educated at Scotch College and university of Western Australia; Rhodes scholar at Lincoln College, Oxford; Ph.D., 1925; research chemist, Synthetic Ammonia & Nitrates Ltd., subsidiary of Brunner, Mond & Co.; chairman, Imperial Chemical Industries, plastics division, 1942-8; director, ICI, 1948-59; chairman, Council of Industrial Design, 1953-60; concerned with establishment of Design Centre; fellow, Royal Society of Arts, 1949; chairman, 1967-9; chairman, BTR Industries, 1960, and British Printing Corporation, 1965; served on many educational bodies; hon. fellow, Lincoln College, Oxford, 1957; knighted, 1958; hon. doctorate, Brunel University, 1967.

WORDIE, SIR JAMES MANN (1889-1962), polar explorer and scholar; cousin of Sir James Mann [q.v.]; educated at Glasgow Academy and Glasgow University; B.Sc., distinction in geology, 1910; read geology at St. John's College, Cambridge; university demonstrator in petrology, 1914-17 and 1919-23; brought into touch with Frank Debenham and (Sir) Raymond Priestley [qq.v.]; joined Antarctic expedition of Sir Ernest Shackleton [q.v.], 1914;

returned with important geological specimens; served with Royal Artillery in France, 1917–18; went to Spitzbergen with W. S. Bruce [q.v.], 1919 and 1920; fellow, St. John's College, Cambridge, 1921; went on other Arctic and Antarctic expeditions between 1921 and 1937; served on *Discovery* committee of Colonial Office, 1923–49; chairman, Scott Polar Research Institute, Cambridge, 1937–55; adviser to (Sir) Vivian Fuchs in planning of Trans-Antarctic Expedition; chairman, British national committee, International Geophysical Year (1954–8); first chairman, Mount Everest Foundation; chairman, British Mountaineering Council (1953–6); hon. secretary, Royal Geographical Society, 1934–48; president, 1951–4; tutor, St. John's College, 1923; director of studies in geography, 1921–52; senior tutor, 1933; president, 1950; master, 1952–9; CBE, 1947; knighted, 1957; awarded many medals and hon. degrees.

WORDSWORTH, Dame ELIZABETH (1840–1932), principal of Lady Margaret Hall, Oxford; daughter of Christopher and sister of John Wordsworth [qq.v.]; first principal, Lady Margaret Hall, 1878–1909; opened St. Hugh's Hall (later College), 1886; of wide sympathies and common sense; encouraged opening (1897) of Lady Margaret Hall Settlement in Lambeth; DBE, 1928.

WORDSWORTH, JOHN (1843–1911), bishop of Salisbury; elder son of Christopher Wordsworth (1807–85, q.v.); educated at Winchester and New College, Oxford; BA, 1865; MA, 1868; hon. DD, 1885; Craven scholar, 1867; fellow of Brasenose, 1867; prebendary of Lincoln, 1870; published *Fragments and Specimens of Early Latin* (1874); from 1878 worked at critical edition of Vulgate text of New Testament, of which Matthew to Acts was published between 1889 and 1905; a minor edition of the whole Vulgate New Testament appeared, posthumously, in 1912; deputy regius professor of divinity, Oxford, 1877–9; Bampton lecturer, 1881; first Oriel professor of the interpretation of Scripture and canon of Rochester, 1883–5; bishop of Salisbury, 1885–1911; an efficient ecclesiastical lawyer; an assessor in bishop of Lincoln case, 1889–90; advocate of reunion of Christendom; delivered at Chicago Hale lectures on the national church of Sweden, 1910 (lectures published, 1911); wrote in Latin on validity of Anglican orders; published history of episcopate of Charles Wordsworth [q.v.], 1899, and scholarly theological and doctrinal works, including *Ministry of Grace* (1901); hon. LLD, Dublin, 1890, and Cambridge, 1908; hon. DD, Berne, 1892; FBA, 1905.

WORKMAN, HERBERT BROOK (1862–1951), Methodist divine and educationist; educated at Kingswood School and the Owens College, Manchester; BA (London), 1884; MA, 1885; trained for ministry at Didsbury College; circuit minister, 1888–1903; principal, Westminster Training College, 1903–30; secretary, Methodist Education Committee, 1919–40; publications include *The Church of the West in the Middle Ages* (2 vols., 1898–1900), *Persecution in the Early Church* (1906), and *John Wyclif* (2 vols., 1926); DD, Aberdeen, 1914.

WORMALL, ARTHUR (1900–1964), biochemist; educated at boys' modern school, Leeds, and Leeds University; B.Sc., chemistry; joined RAF, 1918; demonstrator in biochemistry, Leeds, 1919; lecturer, 1926; senior lecturer, 1933; D.Sc., 1930; as Rockefeller medical research fellow worked in United States, 1928; visited Uganda to investigate sleeping sickness; noted for investigations into immunochemistry; first professor of biochemistry, St. Bartholomew's Hospital Medical College, London, 1936–64; pioneer in use of isotopic tracers; with G. E. Francis, organized first course in the United Kingdom on the use of stable isotopes in biological investigations, 1949; organized similar course at São Paulo, Brazil, 1952; awarded hon. doctorate; FRS, 1956.

WORMS, HENRY DE, Baron PIRBRIGHT (1840–1903), politician. [See DE WORMS.]

WORRELL, Sir FRANK MORTIMER MAGLINNE (1924–1967), cricketer; born at Bridgetown, Barbados; educated at Combermere School; played cricket for Barbados at seventeen; scored first century, 1942–3; began test match career against England, 1947–8; scored 294 in three tests and first of nine test centuries; played for Radcliffe in Central Lancashire League, 1948; BA (Admin.), Manchester, 1959; played with first West Indies side to win a test series in England, 1950; scored six centuries and headed test averages with 539 at 89·83; scored 261 at Nottingham; had successful series as a bowler in Australia, 1951–2; took 6 for 38 at Adelaide in the only test the West Indies won; headed tour averages when West Indies came to England in 1957; captained West Indies in last four matches of the tour; captained side in Australia, 1960–1; Worrell Trophy instituted; led West Indies to victory in his last tour in England, 1963; in 51 tests, made 3,860 runs at 49·48 and took 69 wickets at 38·73; did much to raise the status and self-respect of West Indian cricketers; knighted, 1964; dean of students, Trinidad, university of the West Indies; member of Jamaican Senate, 1962–4.

WORTHINGTON, Sir HUBERT (1886-1963), architect; educated at Sedbergh and Manchester University; joined office of (Sir) Edwin Lutyens [q.v.], 1912-13; served in France during 1914-18 war; lectured post-war at RIBA and Liverpool and Manchester Universities; professor of architecture, Royal College of Art, 1923-8; Slade lecturer in architecture, Oxford, 1929; designed new wing of Radcliffe Science Library, garden buildings at Merton College, library for New College, remodelled Old Bodleian Library, and designed building for St. Catherine's Society; also designed school buildings and private houses; restored war damage in Manchester Cathedral, the Inner Temple, London, and other buildings including Westminster School, the Merchant Taylors' Hall, and the Brewers' Hall; designed post-war buildings at Oxford including the remodelled Radcliffe Camera; principal architect for North Africa and Egypt, for Commonwealth Graves Commission; worked on designs at Imperial College, London, Manchester University, and Roedean School; OBE, 1929; knighted, 1949; ARA, 1945; RA, 1955; member Royal Fine Art Commission, 1945-50; vice-president, Royal Institute of British Architects.

WORTHINGTON, Sir PERCY SCOTT (1864-1939), architect; educated at Clifton and Corpus Christi College, Oxford; took over father's practice in Manchester; work included buildings for Manchester University, Royal Infirmary, Masonic Temple, grammar school; pioneer in hospital building; knighted, 1935.

WORTHINGTON-EVANS, Sir (WORTHINGTON) LAMING, first baronet (1868-1931), politician. [See EVANS.]

WRENBURY, first BARON (1845-1935), judge. [See BUCKLEY, HENRY BURTON.]

WRENCH, Sir (JOHN) EVELYN (LESLIE) (1882-1966), founder of the Royal Over-Seas League and the English-Speaking Union; educated at Summer Fields, Oxford, and Eton; left school at sixteen, travelled on continent and set up a firm producing picture postcards, 1900-4; editor, *Overseas Daily Mail*, 1904-12; also manager, export department, Amalgamated Press, 1907, and sales manager, 1909; formed Over-Seas Club, 1910 (which became Over-Seas League, 1923); joined Royal Flying Corps, 1917; principal private secretary to Lord Rothermere [q.v.], air minister; deputy controller for dominions and United States, Ministry of Information; founded the English-Speaking Union, 1918; secretary, Over-Seas League; editor, *Over-Seas*; chairman, English-Speaking Union; editor, the *Spectator*, 1925-32;

American relations officer to government of India, 1942-4; publications include *Uphill, the First Stage in a Strenuous Life* (1934), *Struggle, 1914-20* (1935), *Immortal Years: 1937-44, as viewed from five continents* (1945), and *Transatlantic London: three centuries of association between England and America* (1949); also published biographies, *Francis Yeats-Brown* (1948). *Geoffrey Dawson and our Times* (1955), and *Alfred Lord Milner: the Man of No Illusions* (1958); concerned with foundation of Anglo-Kin Society, 1958; president, Dickens Fellowship, 1961-4; senior trustee, Cecil Rhodes Memorial Foundation; CMG, 1917; knighted, 1932; KCMG, 1960; hon. degrees from Bristol and St. Andrews.

WRIGHT, Sir ALMROTH EDWARD (1861-1947), bacteriologist; son of C. H. H. and brother of Sir C. T. H. Wright [qq.v.]; graduated in modern literature (1882) and medicine (1883), Trinity College, Dublin; professor of pathology, Army Medical school, Netley, 1892-1902, St. Mary's Hospital, London, 1902-46; introduced anti-typhoid inoculation; developed new school of 'therapeutic immunization' by vaccines for treatment of microbic infections; researched on wound infections, 1914-18; continually devised new and ingenious techniques; knighted and FRS, 1906; KBE, 1919.

WRIGHT, CHARLES HENRY HAMILTON (1836-1909), Hebraist and theologian; younger brother of Edward Perceval Wright [q.v.]; BA, Trinity College, Dublin, 1857; MA, 1859; DD, 1879; incorporated MA, Exeter College, Oxford, 1862; Ph.D., Leipzig, 1875; won at Dublin prizes for Hebrew and Arabic; English chaplain at Dresden, 1863-8, and Boulogne-sur-Mer, 1868-74; served benefices in Ireland, 1874-91, and at Liverpool, 1891-8; Bampton lecturer at Oxford, 1878; Donnellan lecturer at Dublin, 1880-1; Grinfield lecturer on the Septuagint, Oxford, 1893-7; clerical superintendent of Protestant Reformation Society, 1898-1907; published critical and exegetical works on the Old Testament.

WRIGHT, Sir CHARLES THEODORE HAGBERG (1862-1940), librarian; son of C. H. H. and brother of Sir A. E. Wright [qq.v.]; educated on continent and at Trinity College, Dublin; secretary and librarian, London Library, 1893-1940; responsible for its rebuilding, extension, and classification; student of Russian literature; knighted, 1934.

WRIGHT, EDWARD PERCEVAL (1834-1910), naturalist; founded and edited (1854-66) quarterly *Natural History Review*; BA, Trinity

College, Dublin, 1857; lecturer in zoology at Trinity College, 1858-68; MA, 1859; MD, 1862; made study of botany and geology in Ireland (1865), the Seychelles (1867), Sicily and Portugal (1868); professor of botany, 1869-1904, and keeper of the herbarium, 1869-1910, at Trinity College; secretary of Royal Irish Academy, 1874-7 and 1883-99; awarded Cunningham medal, 1883; president of Royal Society of Antiquaries, Ireland, 1900-2.

WRIGHT, JOSEPH (1855-1930), philologist; began work, aged six, as donkey-boy, near Bradford; later 'doffer', then wool-sorter; taught himself to read and write; attended classes at Mechanics' Institute, Bradford; schoolmaster, 1876-82; studied philology at Heidelberg, 1882-5; Ph.D., 1885; lecturer, Association for Higher Education of Women in Oxford, and deputy lecturer in German, Taylor Institution, 1888; special lecturer in Teutonic philology, 1890; deputy to professor of comparative philology, Oxford, 1891; his greatest achievement, compilation of *The English Dialect Dictionary*, originated by English Dialect Society, undertaken 1891; responsible for actual work and business side of *Dictionary*, 1896; published it himself (6 vols., 1896-1905); Corpus Christi professor of comparative philology, Oxford, 1901-24; active in development of modern language studies at Oxford; FBA, 1904; his works include *The English Dialect Grammar* (1905) and many other grammars and primers.

WRIGHT, SIR NORMAN CHARLES (1900-1970), agricultural and nutritional scientist; educated at Christ Church choir school, Oxford, University College, Reading, Christ Church, Oxford, and Gonville and Caius College, Cambridge; Ph.D., 1925; carried out research in England and the United States, 1924-6; physiologist, Hannah Dairy Research Institute, Ayr, 1928; director, 1930; special adviser to Imperial Council of Agricultural Research in India, 1936-7; D.Sc., Oxford, 1937; member, Anglo-American Scientific Mission to Middle East Supply Centre, 1944-5; British member, first FAO mission to Greece, 1946; succeeded Sir Jack C. Drummond [q.v.] as chief Scientific adviser, Ministry of Food, 1947; and later, Ministry of Agriculture, Fisheries and Food; deputy director-general, UN Food and Agriculture Organization, 1959-63; member of many councils and committees concerned with agricultural research; chairman, FAO Programme Committee in Rome, 1953-9; secretary, British Association for the Advancement of Science, 1963-8; first hon. president, British Dietetic Association, 1963-9; member of other advisory committees concerned with nutrition;

CB, 1955; knighted, 1963; hon. LLD, Leeds, 1967.

WRIGHT, ROBERT ALDERSON, BARON WRIGHT (1869-1964), lord of appeal and jurist; educated at Trinity College, Cambridge; first class, parts i and ii, classical tripos, 1895-6, and part ii, moral sciences tripos, 1897; fellow of Trinity, 1899; called to bar (Inner Temple), 1900; specialized in commercial work in chambers of future Lord Justice Scrutton [q.v.]; KC, 1917; judge of King's Bench; knighted, 1925; presided at trial of Lord Kylsant [q.v.], 1931; PC and life peer, 1932; lord of appeal in ordinary, 1932-47 (except for 1935-7 when he was master of the Rolls); wrote articles and lectured on legal subjects; FBA, 1940; chairman, Law Revision Committee, 1935; chairman, United Nations War Crimes Commission, 1945; KCMG, 1948; published 'War Crimes under International Law' in *Law Quarterley Review* (vol. lxii, 1946); hon. fellow, Trinity College, Cambridge; deputy high steward, Cambridge University.

WRIGHT, SIR ROBERT SAMUEL (1839-1904), judge; BA, Balliol College, Oxford (first class classic), 1861; BCL, 1863; MA, 1864; won many English and classical university prizes, 1859-62; Craven scholar, 1861; fellow of Oriel, 1861-80; hon. fellow, 1882; published *Golden Treasury of Ancient Greek Poetry* (1866); called to bar, 1865; collaborated with Sir Frederick Pollock [q.v.] in *An Essay on Possession in the Common Law* (1888); junior counsel to the Treasury, 1883; judge of Queen's Bench division, 1890-1904; knighted, 1891; frequently sat as extra Chancery judge, or judge in bankruptcy.

WRIGHT, WHITAKER (1845-1904), company promoter; went from England as assayer to United States, 1866; pioneer in mining boom at Leadville, 1879; settled in Philadelphia, but lost his fortune; returned to England (1889) and engaged in company promoting; floated West Australian Exploring and Finance Corporation, 1894, and London and Globe Finance Corporation, 1895; amalgamated the companies in 1897 with Marquess of Dufferin [q.v.] as chairman and himself as managing director; floated many subsidiary companies between 1896 and 1898, including the Lake View Consols, 1896, the depreciation of whose shares led in 1900 to liquidation of the London and Globe Company, and involved the ruin of many members of the stock exchange and small investors; Wright was charged with misrepresentation and falsification of accounts, 1903; escaped to New York, but was extradited; tried and found guilty at Old Bailey, Jan. 1904; committed suicide after sentence.

WRIGHT, WILLIAM ALDIS (1831-1914), Shakespearian and biblical scholar; BA, Trinity College, Cambridge; librarian of Trinity, 1863-70; senior bursar, 1870-95; fellow, 1878; vice-master, 1888-1914; secretary to Old Testament revision company, 1870-85; joint-editor with William George Clark [q.v.] of 'Cambridge Shakespeare' (1863-6); edited with Clark 'Globe Shakespeare' (1864) and single plays in Clarendon Press series (1868-72); continued series alone (1874-97); other works include edition of Bacon's *Essays* (1862) and *Advancement of Learning* (1869), Edward FitzGerald's *Letters and Literary Remains* (final edition, 1902-3), and abridgement of Smith's *Dictionary of the Bible* (1865); edited *Journal of Philology*, 1863-1913; his editions noted for accuracy of texts and concise learning of notes.

WRIGHT, Sir (WILLIAM) CHARLES, second baronet (1876-1950), ironmaster and steelmaker; succeeded father, 1926; joined father's steelmaking business, 1893; merged into Baldwins, Ltd., 1902; director, 1903; chairman and managing director from 1925; controller, iron and steel production, Ministry of Munitions, 1917-18; president, Iron and Steel Federation, 1937-8; deputy controller, iron and steel, Ministry of Supply, 1939; controller, 1940-2; KBE, 1920; GBE, 1943.

WRONG, GEORGE MACKINNON (1860-1948), Canadian historian; first class, mental and moral philosophy, Toronto, and ordained, 1883; lecturer in history and ethnology, 1892; professor, 1894-1927; founded *Review of Historical Publications relating to Canada*, 1897; gave editorial assistance to publisher of *Chronicles of Canada* (32 vols., 1914-16); wrote two volumes for *Chronicles of America*; leader in founding Champlain Society, 1905; publications include *The Rise and Fall of New France* (2 vols., 1928) and *Canada and the American Revolution* (1935).

WROTH, WARWICK WILLIAM (1858-1911), numismatist; assistant in medal room at British Museum, 1878-1911; compiled six catalogues of Greek coins (1886-1903), of *Imperial Byzantine Coins* (2 vols., 1908), and coins of *Vandals, Ostrogoths and Lombards* (1911); published with brother *The London Pleasure Gardens of the Eighteenth Century* (1896); FSA, 1889.

WROTTESLEY, Sir FREDERIC JOHN (1880-1948), judge; educated at Tonbridge and Lincoln College, Oxford; called to bar (Inner Temple), 1907; practised successfully at parliamentary bar; KC, 1926; recorder of Wolverhampton, 1930-7; chairman, gas legislation

committee, 1931; conducted inquiry into marketing of sugar-beet, 1936; judge of King's Bench division, 1937-47; studied criminology and the prison system and took his place among foremost criminal judges; lord justice of appeal, 1947-8; knighted, 1937; PC, 1947.

WROTTESLEY, GEORGE (1827-1909), soldier and antiquary; joined Royal Engineers, 1845; served in Crimea; ADC to Sir John Fox Burgoyne [q.v.], 1855-68; secretary to several War Office committees; president of army signalling committee, 1863; commanding officer, Royal Engineers at Woolwich, 1875; retired as major-general, 1881; published *Life of Sir John Burgoyne*, (2 vols., 1873), and many works on genealogy; helped to found William Salt Society, 1879; his genealogical researches were embodied in the Society's *Staffordshire Collections*, 34 vols.

WYLD, HENRY CECIL KENNEDY (1870-1945), English philologist and lexicographer; B.Litt., Corpus Christi College, Oxford, 1899; lecturer, English language, Liverpool, 1899; Baines professor, English language and philology, 1904-20; Merton professor, English language and literature, Oxford, 1920-45; publications include *A Short History of English* (1914), *A History of Modern Colloquial English* (1920), and *Universal Dictionary of the English Language* (1932).

WYLIE, CHARLES HOTHAM MONTAGU DOUGHTY- (1868-1915), soldier and consul. [See DOUGHTY-WYLIE.]

WYLIE, Sir FRANCIS JAMES (1865-1952), first warden of Rhodes House, Oxford; educated at St. Edward's School, Oxford, Glasgow University, and Balliol College, Oxford; first class, *lit. hum.*, 1888, lecturer, Balliol, 1891; fellow, Brasenose, 1892; superviser Rhodes Trust scheme at Oxford, and first warden, Rhodes House, 1903-31; knighted, 1929; hon. fellow, Brasenose, 1931.

WYLLIE, Sir WILLIAM HUTT CURZON (1848-1909), lieutenant-colonel in the Indian army; son of Sir William Wyllie [q.v.]; joined Indian staff corps, 1869; served in Afghan campaign, 1878-80; held various posts as commissioner and political agent, mainly in Rajputana, from 1881 to 1898; agent to governor-general in Central India, 1898-1900, and in Rajputana, 1900-1; CIE, 1881; lieutenant-colonel, 1892; organized famine relief measures in Rajputana, 1899-1900; political ADC to secretary of state for India in London, 1901-9; KCIE, 1902; CVO, 1907; assassinated at Imperial Institute, London, on 1 July 1909 by an Indian student.

WYLLIE, WILLIAM LIONEL (1851-1931), painter; studied at Royal Academy Schools; spent much time at sea: did work for White Star shipping line and the navy; works include 'L'Entente Cordiale' (1905), 'Blocking of Zeebrugge Waterway, St. George's Day, 1918' and etchings of Port of London; ARA, 1889; RA, 1907.

WYNDHAM, SIR CHARLES (1837-1919), actor-manager, whose original name was CHARLES CULVERWELL; went on London stage, 1862; manager of Criterion Theatre, London, 1876-99; appeared in title-role of T. W. Robertson's *David Garrick*, his best-known part, 1886; opened Wyndham's Theatre, London, 1899; knighted, 1902; excelled in comedy.

WYNDHAM, GEORGE (1863-1913), statesman and man of letters; educated at Eton and Sandhurst; joined Coldstream Guards, 1883; private secretary to A. J. Balfour, 1887; conservative MP, Dover, 1889-1913; parliamentary under-secretary to War Office, 1898; chief secretary for Ireland, 1900-5; his Irish administration, last attempt (successful while it lasted) to maintain Union and carry out policy of economic development, his main practical achievement; his Land Act passed, 1903; miscellaneous writer.

WYNDHAM, JOHN (1903-1969), writer. [See HARRIS, J. W. P. L. B.]

WYNDHAM, LADY MARY (1861-1931), better known as MARY MOORE, actress and theatre-manager; married James Albery (died 1889, q.v.), 1878; joined company of (Sir) Charles Wyndham [q.v.], 1885; became his leading lady (1886), partner (1896), wife (1916); particularly talented in playing silly, helpless but attractive women; founded Wyndham Theatres, Ltd., 1924.

WYNDHAM-QUIN,   WINDHAM THOMAS, fourth EARL OF DUNRAVEN (1841-1926), Irish politician. [See QUIN.]

WYNN-CARRINGTON,   CHARLES ROBERT, third BARON CARRINGTON and MARQUESS OF LINCOLNSHIRE (1843-1928), politician; liberal MP, High Wycombe, 1865; succeeded father, 1868; governor, New South Wales, 1885-90; president, Board of Agriculture and Fisheries, with seat in cabinet, 1905; lord privy seal, 1911-12; Viscount Wendover and Earl Carrington, 1895; Marquess of Lincolnshire, 1912; KG, 1906.

WYNYARD, DIANA (1906-1964), actress; born Dorothy Isobel Cox; educated at Woodford School, Croydon; after voice training secured parts on stage, mainly in light comedy, 1925; joined Liverpool Repertory Company under William Armstrong [q.v.], 1927; made London début at St. Martin's theatre in *Sorry You've Been Troubled* (1929); established as leading lady in the west end; played Charlotte Brontë in *Wild Decembers* (1933), followed by *Sweet Aloes* (1934), Eliza Doolittle in *Pygmalion* at the Old Vic (1937), *Design for Living* (1939), *No Time for Comedy* (1941), and *Watch on the Rhine* (1942); toured for ENSA in *Gaslight* (1943) and *Love from a Stranger* (1944); joined Shakespeare company at Stratford, 1948; took leading parts in many of Shakespeare's plays, 1948-9; toured Australia, 1949-50; CBE, 1953; appeared in films including *Cavalcade*, *The Prime Minister*, and *An Ideal Husband*.

WYON, ALLAN (1843-1907), medallist and seal engraver; son of Benjamin Wyon [q.v.]; carried on family business from 1884 till death; engraver of the royal seals, 1884-1901; FSA, 1889; compiled *The Great Seals of England* (1887).

Y

YAPP, SIR ARTHUR KEYSALL (1869-1936), national secretary of the Young Men's Christian Association; general secretary, Derby YMCA, 1892-8; secretary for Lancashire, 1898-1907; introduced YMCAs in Volunteer camps (1901) and later with Territorial Force; general secretary, Manchester, 1907-12; secretary, National Council of YMCAs, Incorporated, London, 1912-29; deputy president, 1929-36; inaugurated war emergency service, 1914; introduced red triangle sign; established over 10,000 centres on fighting fronts; hon. director, food economy campaign, 1917; KBE, 1917.

YARROW, SIR ALFRED FERNANDEZ, first baronet (1842-1932), marine engineer and shipbuilder; apprenticed in London; established own firm at Poplar, 1866; produced improved steam launches; made first torpedo-boat, 1876; later constructed vessels for most large navies; carried out systematic experiments and speed trials; the Yarrow straight-tube boiler adopted by the navy; transferred work to the

Clyde, 1907; built twenty-nine destroyers and a fleet of shallow-draught gunboats for Mesopotamia, 1914-18; his designs characterized by detailed ingenuity and foresight; gave generous endowments for research and founded Yarrow children's convalescent home, Broadstairs; baronet, 1916; FRS, 1922.

YATE, SIR CHARLES EDWARD, baronet (1849-1940), Indian administrator and British politician; gazetted to 49th Royal Berkshire regiment, 1867; served in Indian political service (1871-1906), mainly in north-west; agent to governor-general and chief commissioner of Baluchistan, 1900-4; colonel, 1901; conservative MP, Melton division, 1910-24; upheld old standards of administration and severely criticized Indian policy of E. S. Montagu [q.v.]; baronet, 1921.

YATES, DORNFORD (pseudonym) (1885-1960), novelist. [See MERCER, CECIL WILLIAM.]

YEATS, JACK BUTLER (1871-1957), painter; brother of W. B. Yeats [q.v.]; studied at Westminster School of Art, 1888; illustrator for London magazines; exhibited drawings and water-colours, 1890-1900; settled in Ireland, 1900, and produced oil paintings exhibited in London and Dublin; retrospective exhibition, National Gallery, London, 1942; national loan exhibition, Dublin, 1945.

YEATS, WILLIAM BUTLER (1865-1939), Irish poet and playwright; born in Dublin of predominantly Irish Protestant origin; educated in London, and at the high school and Metropolitan School of Art, Dublin; settled in London, moving in literary, aesthetic, theosophical, and spiritualistic circles; established himself as an imaginative writer with *The Celtic Twilight* (1893) and *Poems* (1895), followed by the more sophisticated and stylized *The Secret Rose* (1897) and *Wind Among the Reeds* (1899); stimulated by friendship with John O'Leary [q.v.] and Maud Gonne (the subject of his love poetry) organized literary societies among Fenians in Ireland; with help of Lady Gregory [q.v.] produced *The Countess Cathleen* in Dublin (1899) and subsequently established the Abbey Theatre as a national institution; a vigorous defender of J. M. Synge [q.v.]; his own most popular play *Cathleen ni Houlihan* (1902); in *Responsibilities* (1914) and *The Wild Swans at Coole* (1917) moved away from the wavering moods and rhythm of his earlier verse, and kept close to particulars; in *A Vision* (1925) recorded his astrologico-historical speculations revealing his preoccupation with the impact of eternity in time which dominated his later verse; reached peak of poetic achievement in *The*

Tower (1928) and *The Winding Stair* (1929); Nobel prize for literature, 1923; senator, Irish Free State, 1922-8; with G. B. Shaw and G. W. Russell [qq.v.] founded Irish Academy of Letters, 1932.

YEO, GERALD FRANCIS (1845-1909), physiologist; MB and M.Ch., Trinity College, Dublin, 1867; MD, 1871; MRCP and MRCS Ireland; professor of physiology, King's College, London, 1875-90; FRCS England, 1878; made researches on cerebral localization in monkeys; inaugurated international physiological congresses (triennial), 1891; FRS, 1889; published *Manual of Physiology* (1884).

YEO-THOMAS, FOREST FREDERIC EDWARD (1902-1964), French resistance organizer; educated at Dieppe Naval College and the Lycée Condorcet, Paris; dispatch rider with United States forces, 1918; fought with Poles, captured, and escaped from Bolsheviks, 1920; employed in various jobs in France, 1921-32; secretary, fashion house of Molyneux, 1932-9; joined RAF, 1939; intelligence officer with 308 Polish Squadron, 1941; joined Special Operations Executive, 1942; parachuted into France to organize resistance groups, 1943; obtained ear of (Sir) Winston Churchill and secured increased supplies and weapons for French resistance; returned to France, arrested, imprisoned, and tortured, 1944; organized escape, 1945; one of most outstanding workers behind German lines; awarded George Cross; MC and bar; Polish Cross of Merit; croix de guerre; commander, Legion of Honour; helped to bring to trial several Nazi war criminals; returned to Molyneux, 1946-8; Paris representative of Federation of British Industries, 1950.

YERBURY, FRANCIS ROWLAND (FRANK) (1885-1970), secretary of the Architectural Association and a founder and first director of the London Building Centre; at sixteen became office boy at the Architectural Association; secretary, 1911-37; influential publicist of modern European building; publications include *Modern European Buildings* (1928), *Modern Dutch Buildings* (1931), and *Small Modern English Houses* (1929); with Vincent Vincent created the London Building Centre, 1931; resigned from AA to direct the Centre, 1937; made it a model for other cities and countries; vice-president, British Design and Industries Association; hon. ARIBA, 1928; OBE, 1952.

YONGE, CHARLOTTE MARY (1823-1901), novelist and story-teller for children; born, and educated by her father, at Otterbourne, near Winchester; was influenced from

1835 by John Keble [q.v.], vicar of neighbouring parish of Hursley, who urged her to expound her religious views in fiction; *The Heir of Redclyffe* (1853) first brought her popular success; *Heartsease* (1854) and *The Daisy Chain* (1856) followed; her early historical romances included *The Lances of Lynwood* (1855), *The Prince and the Page* (1865), *The Dove in the Eagle's Nest* (1866), and *The Caged Lion* (1870); she edited *Monthly Packet*, 1851-98, to which she contributed 'Cameos from English History' (collected in 8 series between 1868 and 1896); published full life of Bishop Patteson [q.v.], 1873; issued in all 160 books, including *The Book of Golden Deeds* (1864), *History of France* (in Freeman's 'Historical Course', 1879), and *Life of Hannah More* (1888); *Life and Letters* by Christabel Coleridge, 1903.

YORKE, ALBERT EDWARD PHILIP HENRY, sixth EARL OF HARDWICKE (1867-1904), politician; godson of Prince of Wales, afterwards King Edward VII; hon. attaché at Vienna, 1886-91; succeeded to earldom, 1897; moderate member of LCC for Marylebone, 1898; under-secretary to India Office, 1900-2 and 1903-4; and to War Office, 1902-3.

YORKE, FRANCIS REGINALD STEVENS (1906-1962), architect; educated at Chipping Camden School and what later became the Birmingham School of Architecture; qualified as architect, 1930; designed number of modern style houses; travelled widely; published *The Modern House* (1934) and, in association with (Sir) Frederick Gibberd, *The Modern Flat* (1937); editor, *Specification*, 1935-62; founder member of Modern Architectural Research Groups; in partnership with Eugene Rosenberg and Cyril Mardall, post-war, designed number of buildings, including Barclay School at Stevenage, the Ladyshot Estate at Harlow, and Gatwick Airport; fellow, RIBA 1943; CBE, 1962.

YORKE, WARRINGTON (1883-1943), physician; MB, Ch.B, Liverpool, 1905; joined School of Tropical Medicine, 1907; directed Runcorn research laboratory, 1909-14; investigated blackwater fever in Nyasaland (1907-9) and sleeping sickness in Rhodesia (1911-12); professor of parasitology, 1914-29; of tropical medicine, 1929-43; assistant physician, tropical diseases, Liverpool Royal Infirmary, 1920-9; physician, 1929-43; FRS, 1932.

YOUL, SIR JAMES ARNDELL (1811-1904), Tasmanian colonist; accompanied parents to Van Diemen's Land, 1819; settled at Symmons Plains, 1827-54, as successful agriculturist;

agent for Tasmania in London, 1861-3; induced English government to establish mail service to Australia; acting agent-general, 1888; he first introduced salmon and trout into Tasmania, 1864, and New Zealand, 1868; CMG, 1874; KCMG, 1891.

YOUNG, SIR ALLEN WILLIAM (1827-1915), sailor and polar explorer; joined merchant service, 1842; took part in Franklin search expedition, 1857-9, and North Atlantic Telegraph expedition, 1860; co-operated in government Arctic expedition, 1875; led successful search for explorer Benjamin Leigh Smith, 1882; knighted, 1877; CB, 1881.

YOUNG, MRS CHARLES (1827-1902), actress. [See VEZIN, JANE ELIZABETH.]

YOUNG, EDWARD HILTON, first BARON KENNET (1879-1960), politician and writer; third son of Sir George Young [q.v.]; educated at Eton and Trinity College, Cambridge; president of the Union, editor, *Cambridge Review*, and first class, natural sciences tripos, 1900; called to bar (Inner Temple), 1904; assistant editor, *The Economist*, 1908-10; city editor, *Morning Post*, 1910-14; joined Royal Naval Volunteer Reserve, 1914; lost an arm at Zeebrugge, 1918; DSC, with bar; fought in Russia; DSO, 1920; liberal MP, Norwich, 1915, 1918-23, 1924-9; financial secretary, Treasury, 1921; chief whip, Lloyd George liberals, PC, 1922; joined conservative party, 1926; GBE, 1927; MP Sevenoaks, 1929-35; minister of health, 1931-5; baron, 1935; chairman of committee administering exemption from military service, 1939-45; chairman, Capital Issues Committee, 1939-59; publications include *By Sea and Land* (1920) and *A Bird in the Bush* (1936); married Kathleen Scott [q.v.], 1922.

YOUNG, FRANCIS BRETT (1884-1954), novelist; educated at Epsom College and Birmingham University; MB, Ch.B., 1907; practised in Brixham, and served in Royal Army Medical Corps in 1914-18 war; published novels include *The Dark Tower* (1915), *The Young Physician*, (1919), *Portrait of Clare* (1927), and *The House under the Water* (1932); also verse drama *The Island* (1944); hon. D.Litt., Birmingham, 1950.

YOUNG, GEOFFREY WINTHROP (1876-1958), mountaineer; second son of Sir George Young [q.v.]; brother of Edward Hilton Young (Lord Kennet, q.v.); educated at Marlborough and Trinity College, Cambridge; BA, 1898; assistant master, Eton, 1900-5; inspector, secondary schools, 1905-13; consultant for Europe to Rockefeller Foundation, 1925-33;

reader in comparative education, London University, 1932-41; commanded Ambulance Units in Belgium and Italy, 1914-19; leg amputated; one of greatest British mountaineers; publications include *Mountain Craft* (1920), *On High Hills* (1927), and *Collected Poems* (1936); president, the Alpine Club, 1941-4.

YOUNG, GEORGE, LORD YOUNG (1819-1907), Scottish judge; LLD, Edinburgh University, 1871; passed to Scottish bar, 1840; advocate depute, 1849; sheriff of Inverness, 1853, and of Haddington and Berwick, 1860; solicitor-general for Scotland, 1862-5, 1865-7, and 1868-9; liberal MP for Wigtown, 1865-74; lord advocate, 1869-74; called to bar (Middle Temple), 1869; PC, 1872; author of Public Health Act for Scotland, 1871, the Scottish Education Act, 1872, and Law Agents Act, 1873; judge of the Court of Session with title of Lord Young, 1874-1905.

YOUNG, SIR GEORGE, third baronet, of Formosa Place (1837-1930), administrator and author; brother of Sir William Mackworth Young [q.v.] and nephew of W. M. Praed [q.v.]; succeeded father, 1848; educated at Eton and Trinity College, Cambridge; fellow of Trinity, 1862; secretary to various commissions, 1870-80; responsible for drawing up report on Irish Land Acts; charity commissioner under Endowed Schools Act, 1882; chief charity commissioner for England and Wales, 1903-6; keenly interested in education; took active part in abolition of tests and promoting university education of women at Cambridge; translator of poetry; an intrepid alpine climber.

YOUNG, GEORGE MALCOLM (1882-1959), scholar; educated at St. Paul's School, London, and Balliol College, Oxford; fellow, All Souls, 1905; joined Board of Education, 1908; trustee, National Portrait Gallery, 1937; trustee, British Museum, 1947-57; publications include *Gibbon* (1932), *Charles I and Cromwell* (1935), *Victorian England, Portrait of an Age* (1936), and *Stanley Baldwin* (1952); hon. fellow, Balliol College, 1953.

YOUNG, GERALD MACKWORTH- (1884-1965), Indian civil servant and archaeologist. [See MACKWORTH-YOUNG.]

YOUNG, SIR HUBERT WINTHROP (1885-1950), soldier and administrator; son of Sir William Mackworth Young [q.v.]; educated at Eton and Woolwich; transferred to Indian army, 1908; organized transport and supplies, Hejaz operations of T. E. Lawrence [q.v.], 1918; DSO; in Eastern department, Foreign Office,

1919-21; assistant secretary, Middle East department, Colonial Office, 1921-6; colonial secretary, Gibraltar, 1926-9; counsellor to high commissioner, Iraq, 1929-32; governor of Nyasaland, 1932-4, Northern Rhodesia, 1934-8, Trinidad and Tobago, 1938-42; knighted, 1932; KCMG, 1934.

YOUNG, SIR ROBERT ARTHUR (1871-1959), physician; educated at Westminster City School and King's College, London; qualified at Middlesex Hospital; first class, physiology, B.Sc., 1891; MB, 1894; MD, 1895; MRCP, 1897; FRCP, 1905; consulting physician, Middlesex Hospital; chest specialist; CBE, 1920; knighted, 1947.

YOUNG, SYDNEY (1857-1937), chemist; studied at Owens College, Manchester, and Strasbourg; lecturer (1882), professor (1887-1904) in chemistry, University College, Bristol; in Dublin, 1904-28; with (Sir) William Ramsay [q.v.] investigated vapour pressures of solids and liquids, 1882-7; studied behaviour of mixed liquids when distilled; works include *Fractional Distillation* (1903) and *Stoichiometry* (1908); FRS, 1893.

YOUNG, WILLIAM HENRY (1863-1942), mathematician; educated at City of London School and Peterhouse, Cambridge; fourth wrangler, 1884; fellow, 1886-92; successful tripos coach; married (1896) Grace Chisholm, a brilliant mathematician with whom he collaborated; settled successively in Göttingen, Geneva, and Lausanne; developed theory of integration; worked on theory of Fourier series; and on the differential calculus of functions of more than one variable; part-time professor, Liverpool, 1913-19; professor, Aberystwyth, 1919-23.

YOUNG, SIR WILLIAM MACKWORTH (1840-1924), Indian civil servant; brother of Sir George Young [q.v.]; nephew of W. M. Praed [q.v.]; educated at Eton and King's College, Cambridge; entered Indian civil service, 1862; posted to Punjab, 1863; secretary to Punjab government, 1880-7; lieutenant-governor of Punjab (and KCSI), 1897-1902; his relations with viceroy, Lord Curzon [q.v.], strained over dismemberment of Punjab; active supporter of missionary work.

YOUNGER, GEORGE, first VISCOUNT YOUNGER OF LECKIE (1851-1929), politician; took charge of family brewing business, 1868; convener, county council of Clackmannan, 1895-1906; conservative MP, Ayr Burghs, 1906-22; a talented party manager; chairman, unionist party organization, 1917; responsible for success of coalition in 'coupon' election, 1918; baronet, 1911; viscount, 1923.

YOUNGER, ROBERT, BARON BLANESBURGH (1861-1946), judge; brother of Viscount Younger of Leckie [q.v.]; educated at Edinburgh Academy and Balliol College, Oxford; second class, jurisprudence, 1883; visitor, Balliol, 1934-46; called to bar (Inner Temple), 1884; QC, 1900; bencher, Lincoln's Inn, 1907; Chancery judge, 1915-19; lord justice of appeal, 1919-23; lord of appeal in ordinary, 1923-37; served on number of committees arising out of war of 1914-18; principal British representative, Reparation Commission, 1925-30; knighted, 1915; GBE, 1917; PC, 1919; life peer, 1923.

YOUNGHUSBAND, SIR FRANCIS EDWARD (1863-1942), soldier, diplomatist, explorer, geographer, and mystic; born in India; educated at Clifton and Sandhurst; joined 1st King's Dragoon Guards in India, 1882; explored the Karakoram and Pamirs; travelled from Peking to Rawalpindi via Chinese Turkestan (1887) and from India to Kashgar and back (1890-1); transferred to foreign department, 1889; political officer, Hunza, 1892; Chitral, 1893-4, Haraoti and Tonk, 1898-1902; resident, Indore, 1902-3; headed mission to Lhasa and obtained treaty, 1903-4; resident, Kashmir, 1906-9; enthusiastic spirit behind first three Everest expeditions; founder, World Congress of Faiths, 1936; KCIE, 1904; KCSI, 1917.

YOXALL, SIR JAMES HENRY (1857-1925), educationist; educated at Westminster training college for teachers; schoolmaster, 1878-92; elected to executive council of National Union of Teachers, 1889; general secretary of Union, 1892-1924; increased its membership and improved its organization and financial position; edited its magazine, *The Schoolmaster*, from 1909; invaluable to Union in establishing good relations with Whitehall and in conciliating new local authorities; after 1914-18 war successfully engaged in series of negotiations over teachers' pay; liberal MP, Nottingham West, 1895-1918; knighted, 1909.

YPRES, first EARL OF (1852-1925), fieldmarshal. [See FRENCH, JOHN DENTON PINKSTONE.]

YULE, GEORGE UDNY (1871-1951), statistician; son of Sir George Udny Yule [q.v.]; educated at Winchester and University College, London; demonstrator under Karl Pearson [q.v.] at University College, 1893; assistant professor of applied mathematics, 1896; Newmarch lecturer in statistics, University College, 1902-9; university lecturer in statistics, Cambridge, 1912-31; fellow, St. John's College, 1922; director of studies, natural sciences, 1923-35; CBE, 1918; FRS, 1921; published *Introduction to the Theory of Statistics* (1911) and *The Statistical Study of Literary Vocabulary* (1944).

Z

ZANGWILL, ISRAEL (1864-1926), author and philanthropist; BA, London University; teacher, then journalist; discarded his Jewish religion, but clung passionately to his race; as result of invitation of newly founded Jewish Publication Society of America to write Jewish novel, produced *Children of the Ghetto* (1892); this book, which handled the Jew with knowledge, affection, and justice, solidly established his reputation; it was followed by *Ghetto Tragedies* (1893) and *Dreamers of the Ghetto* (1898); also wrote other novels, comedies, and tragedies; from 1896 onwards threw himself into Zionist cause under influence of Dr Theodor Herzl; founded the unsuccessful Jewish Territorial Organization.

ZETLAND, second MARQUESS OF (1876-1961), public servant and author. [See DUNDAS, LAWRENCE JOHN LUMLEY.]

ZIMMERN, SIR ALFRED ECKHARD (1879-1957), scholar and authority on international institutions; educated at Winchester and New College, Oxford; first class *lit. hum.*, 1902; lecturer in ancient history, fellow and tutor, New College, 1904-9; inspector, Board of Education, 1912-15; Wilson professor of international politics, Aberystwyth, 1919-21; director, School of International Studies, Geneva, 1925-39; Montague Burton professor of international relations, Oxford, 1930-44; knighted, 1936; publications include *The Greek Commonwealth* (1911), *Nationality and Government* (1918), *The League of Nations and the Rule of Law* (1936), and *The American Road to World Peace* (1953).

ZULUETA, FRANCIS DE (FRANCISCO MARIA JOSÉ) (1878-1958), academic lawyer; Spanish by birth; grandson of Sir Justin Sheil [q.v.]; educated at Beaumont, the Oratory School, and New College, Oxford; first class *lit. hum.*, 1901, and jurisprudence, 1902; fellow, Merton College, 1902; Vinerian law scholar, 1903; called to the bar (Lincoln's Inn), 1904; tutorial fellow, New College, 1907; All Souls reader in Roman law, 1912–17; naturalized, 1914; regius professor of civil law, Oxford, 1919–48; published *Digest* titles on Ownership and Possession (1922), *Roman Law of Sale* (1945), and *Institutes of Gaius* (1946–53); hon. fellow, Merton; hon. doctorates, Paris and Aberdeen; fellow, Accademia dei Lincei.